Developments in Environmental Modelling, 5

Analysis of Ecological Systems: State-of-the-Art in Ecological Modelling

Developments in Environmental Modelling

Series Editor: S.E. Jørgensen
Langkaer Vaenge 9,
3500 Vaerløse,
Copenhagen,
Denmark

1. ENERGY AND ECOLOGICAL MODELLING
edited by W.J. Mitsch, R.W. Bosserman and J.M. Klopatek
1981 839 pp.

2. WATER MANAGEMENT MODELS IN PRACTICE:
A CASE STUDY OF THE ASWAN HIGH DAM
by D. Whittington and G. Guariso
1983 xxii + 246 pp.

3. NUMERICAL ECOLOGY
by L. Legendre and P. Legendre
1983 xvi + 419 pp.

4A. APPLICATION OF ECOLOGICAL MODELLING IN
ENVIRONMENTAL MANAGEMENT, PART A
edited by S.E. Jørgensen
1983 viii + 735 pp.

5 ANALYSIS OF ECOLOGICAL SYSTEMS: STATE-OF-THE-ART
IN ECOLOGICAL MODELLING
edited by W.K. Lauenroth, G.V. Skogerboe and M. Flug
1983

Developments in Environmental Modelling, 5

Analysis of Ecological Systems: State-of-the-Art in Ecological Modelling

Edited by

WILLIAM K. LAUENROTH *(Natural Resource Ecology Laboratory)*,
GAYLORD V. SKOGERBOE *(Agricultural and Chemical Engineering Department)* and
MARSHALL FLUG *(National Park Service, Water Resources Laboratory)*
Colorado State University, Fort Collins, CO 80523, U.S.A.

Proceedings of a Symposium held from 24 to 28 May 1982 at Colorado State University, Fort Collins, Colorado, U.S.A.

Sponsored by the International Society for Ecological Modelling (ISEM)

Hosted by the Natural Resource Ecology Laboratory, Colorado State University

ELSEVIER SCIENTIFIC PUBLISHING COMPANY
Amsterdam – Oxford – New York 1983

Published jointly by
INTERNATIONAL SOCIETY FOR ECOLOGICAL MODELLING (ISEM)
Langkaer Vaenge 9, 3500 Vaerløse, Copenhagen, Denmark

and

ELSEVIER SCIENTIFIC PUBLISHING COMPANY
Molenwerf 1,
P.O. Box 330, 1000 AH Amsterdam, The Netherlands

Distributors for the United States and Canada:

ELSEVIER SCIENCE PUBLISHING COMPANY INC.
52, Vanderbilt Avenue
New York, N.Y. 10017

Library of Congress Cataloging in Publication Data
Main entry under title:

Analysis of ecological systems.

(Developments in environmental modelling ; 5)
Includes bibliographies.
1. Ecology--Simulation methods--Congresses.
I. Lauenroth, William K. II. Skogerboe, Gaylord V.
III. Flug, Marshall. IV. International Society for
Ecological Modelling. V. International Conference on
State-of-the-Art in Ecological Modelling (3rd : 1982 :
Colorado State University) VI. Series.
QH541.15.S5A5 1983 574.5'0724 83-1564
ISBN 0-444-42179-3

ISBN 0-444-42179-3 (Vol. 5)
ISBN 0-444-41948-9 (Series)

Printed in Denmark by Fair-Print AS, Roskilde

PREFACE

The papers in this volume were presented at the 3rd International Conference on State-of-the-Art in Ecological Modelling held in Fort Collins, Colorado on May 24-28, 1982. The Conference was sponsored by the International Society for Ecological Modelling (ISEM) in cooperation with the National Park Service Water Resources Laboratory. The host for the Conference was the Natural Resource Ecology Laboratory at Colorado State University.

The International Society for Ecological Modelling sponsors conferences, workshops and training courses to advance the development of ecological and environmental modelling. The 1st International Conference on State-of-the-Art in Ecological Modelling was held in Copenhagen, Denmark, August 28-September 2, 1978. The 2nd International Conference was held at Liege, Belgium, April 18-24, 1980. The theme for this the 3rd International Conference was Application of Ecological Modelling to Environmental Management.

The Conference consisted of an opening session with three invited papers addressing the topic of Trends in Ecological Modelling. The contributed papers were divided into five general sessions. These were: Theory and Methods of Ecological Modelling; Application of Ecological Models to Animals; Application of Ecological Models to Land Resources; Application of Ecological Models to Water Resources; and Application of Ecological Models to Energy Development. The final session of the Conference was devoted to summaries and syntheses of the topics covered under the various sessions. Working groups comprised of session chairpersons were charged with producing statements about the state-of-the-art in ecological modelling for their various topics as represented by the contributions to the conference. These statements are included here as the final section of the proceedings.

The conference and these proceedings are the result of a tremendous amount of hard work on the parts of many persons. The Department of Conferences and Institutes at Colorado State University in particular, Craig Summer did an excellent job with the local arrangements. Betsy Zakely of the Department of Agricultural and Chemical Engineering took care of many of the early details in planning the conference and implementing the plan for announcements and request for manuscripts. Dr. Jim Gibson, Director of the Natural Resource Ecology Laboratory at Colorado State University provided both moral and material support and Jan Hill and the Publications section of the Natural Resource Ecology Laboratory provided technical assistance and program and manuscript preparation of the highest quality. Dr. Ray Herrmann of the National Park Service was generous in his support of the conference. Dr. Evan Vlachos of the Sociology Department at Colorado State University was indispensable in lending his inimicable style to the summary and synthesis session. Numerous people helped in the reviewing and editing process of this volume including Charlie Bicak, Dan Milchunas, Bob Dougherty. Arlene Boaman in the Natural Resource Ecology Laboratory handled many of the details of arranging for receipt of the manuscripts. Sven E. Jorgensen of the International Society of Ecological Modelling gave the support of the Society to the Symposium and to this publication.

William K. Lauenroth
Gaylord V. Skogerboe
Marshall Flug

October 1982 Fort Collins, Colorado

LIST OF CONTRIBUTORS

D. A. Adams. North Carolina State University, Box 5488, Raleigh, North Carolina 27650.

V. Alavian. University of Illinois, Department of Civil Engineering, Urbana, Illinois 61801.

J. M. Alcamo. University of California, Department of Civil Engineering, Davis, California 95616.

D. W. Anderson. University of Saskatchewan, Department of Soil Science, Saskatoon, Canada.

G. T. Auble. Western Energy and Land Use Team, U.S. Fish and Wildlife Service, Fort Collins, Colorado.

M. Averner. NASA Ames Research Center, Mailstop 239-10, Moffet Field, California 94035.

D. Bachelet. Natural Resource Ecology Laboratory, Colorado State University, Fort Collins, Colorado 80523.

L. M. Bahr. Louisiana State University, Center for Wetland Resources, Baton Rouge, Louisiana 70803.

D. N. Baker. USDA-ARS Crop Simulation Research Unit, P.O. Box 5367, Mississippi State, Mississippi 39762.

M. C. Barber. University of Georgia, Institute of Ecology, Athens, Georgia 30602.

S. M. Bartell, Oak Ridge National Laboratory, P.O. Box S, Building 1505, MS 2, Oak Ridge, Tennessee 37830.

G. Baskerville. Forest Resources, New Brunswick Department of Natural Resources, Frederick, New Brunswick, Canada.

A. Bauer. USDA-ARS, Mandan, North Dakota 58554.

D. F. Behrend. SUNY College of Environmental Science and Forestry, Syracuse, New York 13210

K. B. Benson. University of Louisville, Systems Science Institute, Louisville, Kentucky 40292.

F. Beonio-Brocchieri. Commission of the European Communities, Joint Research Centre, I-21020 Ispra, Varese, Italy.

A. A. Berryman. Washington State University, Department of Entomology, Pullman, Washington 99164.

S. H. Berwick. HDR Sciences, 804 Anacape Street, Santa Barbara, California 93101.

S. K. Bhagat. Washington State University, Department of Civil Engineering, Pullman, Washington 99163.

B. Bietz, International Convironmental Consultants, Ltd., Suite 345, 4400 McCloud Trail SW, Calgary, Alberta, Canada T2G 4Y7.

G. E. Bingham. Lawrence Livermore Environmental Laboratory, Livermore, California 94550.

A. L. Black. USDA-ARS, Mandan, North Dakota 58554.

A. Blackwell, The Center for Dynamic Systems Studies, The University of Texas at Arlington, Arlington, Texas 76019.

C. C. Blackwell, P.O. Box 19023, The University of Texas at Arlington, Department of Mechanical Engineering, Arlington, Texas 76019.

W. B. Bowden. University of South Carolina, Columbia, South Carolina 29208.

S. P. Burggraf. Florida A&M University, 301 Lee Hall, Tallahassee, Florida 32307.

J. L. Capinera. Colorado State University, Department of Zoology and Entomology, Fort Collins, Colorado 80523.

C. Caupp. Utah State University, UMC 41 Utah State University, Logan, Utah 84322.

M. A. Chartock. University of Oklahoma, Department of Zoology, Science & Public Policy, Norman, Oklahoma 73019.

C. E. Clapp. USDA-ARS, University of Minnesota, 313 Soil Science Building, St. Paul, Minnesota 55108.

K. Clark. Wildlife Branch, Ontario Ministry of Natural Resources, Queen's Park, Toronto, Ontario, Canada M7A 1W3.

T. Clark. Wildlife Branch, Ontario Ministry of Natural Resources, Queen's Park, Toronto, Ontario, Canada M7A 1W3.

W. R. Clark. Iowa State University, 124 Science II, Department of Animal Ecology, Ames, Iowa 50011.

C. J. Cleveland. Louisiana State University, Marine Science Department, Baton Rouge, Louisiana 70803.

J. Colbert. USDA, Forest Service, 809 NE Sixth Avenue, Portland, Oregon 97232.

C. V. Cole. USDA-ARS, Natural Resource Ecology Laboratory, Colorado State University, Fort Collins, Colorado 80523.

R. A. Cole. New Mexico State University, Las Cruces, New Mexico 88003.

W. H. Conner. Louisiana State University, Center for Wetland Resources, Baton Rouge, Louisiana 70803.

A. Cooperrider. U.S. Bureau of Land Management, Denver Service Center, Bldg. 50, Denver Federal Center, Denver, Colorado 80225.

R. Costanza. Louisiana State University, Center for Wetland Resources, Baton Rouge, Louisiana 70803.

M. B. Coughenour. Syracuse University, Biological Research Lab, 130 College Place, Syracuse, New York 13210.

G. M. Cressman. North Carolina State University, Department of Forestry, Raleigh, North Carolina 27650.

N. L. Crookston. University of Idaho, Moscow, Idaho 83843

W. Cuff. Maritime Forest Research Centre, Environment Canada, Forestry Service, Box 4000, Fredericton, New Brunswick, Canada E3B 5P7.

J. W. Day. Louisiana State University, Center for Wetland Resources, Baton Rouge, Louisiana 70803.

L. A. Deegan. Louisiana State University, Marine Science Department, Baton Rouge, Louisiana 70803.

N. M. de Rooy. Delft Hydraulics Laboratory, P.O. Box 177, Delft, The Netherlands 2600MH.

J. K. Detling. Natural Resource Ecology Lab, Colorado State University, Fort Collins, Colorado 80523.

R. L. Dwyer. Martin Marietta Corporation, Environmental Center, 1450 South Rolling Road, Baltimore, Maryland 21227.

P. T. Dyke. USDA-ARS, P.O. Box 748, Temple, Texas 76501.

E. T. Elliott. Natural Resource Ecology Laboratory, Colorado State University, Fort Collins, Colorado 80523.

M. J. Enderle. Electricite de France, Direction des Etudes et Recherches, 1 Avenue du Genera-de-Gaulle, 92141 clamart Cedex, France.

M. Erlandsen. National Agency of Environmental Protection, The Freshwater Laboratory, Lysbrogade 52, DK-8600 Silkeborg, Denmark

D. Euler. Wildlife Branch, Ontario Ministry Natural Resources, Queen's Park, Toronto, Ontario, Canada M7A 1W3.

T. J. Fahey. Cornell University, School of Natural Resources, Ithaca, New York 14853.

L. Fahrig. Carleton University, Biology Department, Ottawa, Ontario, Canada K1S 5B6.

K. Fedra. International Institue for Applied Systems Analysis, A-2361, Laxenburg, Austria.

J. T. Finn. University of Massachusetts, Department of Forestry & Wildlife, Amherst, Massachusetts 01001.

R. A. Fisher. Canadian Forestry Service, Box 4000, Fredericton, New Brunswick, Canada E3B 5P7.

P. A. Flebbe. NASA Langley Research Center, Hampton, Virginia 23665.

M. Flug. ERC, Colorado State University, Fort Collins, Colorado 80523.

D. E. Ford. Ford, Thornton, Norton and Associates, Ltd., P.O. Box 927, Vicksburg, Mississippi 34180.

R. C. Francis. NMFS, Resource Ecology and Fisheries Management Division, 2727 Montlake Blvd. E., Seattle, Washington 98112.

R. B. Frederick. Iowa State University, 124 Science II, Department of Animal Ecology, Ames, Iowa 50011.

J. Fruci. Louisiana State University, Center for Wetland Resources, Baton Rouge, Louisiana 70803.

D. P. Galya. Stone & Webster Engineering Corporation, 245 Summer Street, Boston, Massachusetts 02107.

R. H. Gardner. Oak Ridge National Laboratory, P.O. Box 5, Building 1505, MS 2, Oak Ridge, Tennessee 37830.

W. M. Getz. University of California, Department of Entomological Science, Berkeley, California 94720.

M. W. Gilliland. University of Nebraska, Department of Civil Engineering, Omaha, Nebraska 68132.

J. E. Glasser. University of Georgia, Institute of Ecology, Athens, Georgia 30602.

F. W. Glover. University of Colorado, Boulder, Colorado 80302.

T. C. Gopalkrishnan. KISR, Hydraulic and Coastal Engineering Group, Kuwait.

M. L. Granstrom. Rutgers University, Department of Civil and Environmental Engineering, Piscataway, New Jersey 08853.

W. E. Grant. Texas A&M University, Nagle Hall, College Station, Texas 77843.

W. J. Grenney. Utah State University, UMC 41, Logan, Utah 84322.

S. C. Gupta. USDA-ARS, University of Minnesota, 313 Soil Science Building, St. Paul, Minnesota 55108.

R. Haar. Center for Quantitative Science, University of Washington, Seattle, Washington 98115.

P. W. Hadley. Aerojet General Corporation, 1304 Fig Place, Davis, California 95616.

E. Halfon. Canada Centre for Inland Waters, Burlington, Ontario, Canada L7R 4A6.

J. D. Hanson. USDA-ARS, 6828 Willshire, Cheyenne, Wyoming 82009.

T. B. Hardy. University of Nevada at Las Vegas, Las Vegas, Nevada 89109.

D. R. F. Harleman. Massachusetts Institute of Technology, Department of Civil Engineering, Cambridge, Massachusetts 02139.

M. A. Harwell. Ecosystems Research Center, Cornell University, E-221 Biological Sciences Building, Ithaca, New York 14853.

J. B. Haufler. Michigan State University, Department of Fisheries & Wildlife, East Lansing, Michigan 48824.

P. T. Haug. 1005 Lake Park Drive, Tumwater, Washington 98502.

R. L. Hays. USDI Fish and Wildlife Service, Western Energy and Land Use Team, Fort Collins, Colorado 80523.

J. E. Heasley. Colorado State University, Forest and Wood Science, Fort Collins, Colorado 80523.

E. Herricks. University of Illinois, Environmental Biology, 208 North Romine, Urbana, Illinois 61801.

D. W. Hilbert. Department of Zoology-Entomolgy, Colorado State University, Fort Collins, Colorado 80523.

J. Ho. Utah State University, Logan, Utah 84322.

J. E. Hobbie. Ecosystems Center, Marine Biological Laboratory, Woods Hole, Massachussetts 02543.

T. J. Horst. Stone & Webster Engineering Corporation, 245 Summer Street, Boston, Massachusetts 02107.

S. Hsueh. New Jersey Department of Environmental Protection, Trenton, New Jersey 08625.

H. W. Hunt. Natural Resource Ecology Laboratory, Colorado State University, Fort Collins, Colorado 80523.

D. A. Jameson. Range Science Department, 240 Natural Resource, Colorado State University, Fort Collins, Colorado 80523.

T. Jenq. New Jersey Department of Environmental Protection, Trenton, New Jersey 08625.

T. G. Jenkins. Roman L. Hruska U.S. Meat Animal Research Center, USDA, ARS, Clay Center, Nebraska 68933.

C. A. Jones. USDA-ARS, P.O. Box 748, Temple, Texas 76501.

G. B. Jones. Burgess & Niple, Ltd., 5085 Reed Road, Columbus, Ohio 43220.

L. A. Jorgensen. Copenhagen Teknikum, Pr. Charlottegade, 2200 Copenhagen N., Denmark.

S. E. Jorgensen. Langkaer vaenge 9, Vaerlose, Denmark 3500.

L. Kamp-Nielsen. University of Copenhagen, Freshwater Biological Lab, Helsingorsgade 51, DK 3400 Hillerod, Denmark.

P. Kangas. Eastern Michigan University, Biology Department, Ypsilanti, Michigan 48197.

R. Kaufmann. Complex Systems Research Center, University of New Hampshire, O'Kane House, Durham, New Hampshire 03824.

H. M. Kennedy. Louisiana Department of Natural Resources, Baton Rouge, Louisiana.

J. R. Kercher. Lawrence Livermore National Laboratory, P.O. Box 808, L-524, Livermore, California 94550.

J. R. Kerr. Burgess & Niple, Ltd., 5085 Reed Road, Columbus, Ohio 43220.

D. J. Kiell. Newfoundland and Labrador Hydro, St. John's, Newfoundland, Canada.

D. A. King. University of California at Davis, Lawrence Livermore National Lab., Livermore, California 94550.

T. B. Kirchner. Natural Resource Ecology Laboratory, Colorado State University, Fort Collins, Colorado 80523.

E. E. Klaas. Iowa State University, 124 Science II, Department of Animal Ecology, Ames, Iowa 50011.

R. Klomp. Delft Hydraulics Laboratory, P.O. Box 177, Delft, The Netherlands 2600MH.

D. H. Knight. University of Wyoming, Department of Botany, Laramie, Wyoming 82071.

L. J. Koong. Roman L. Hruska U.S. Meat Animal Research Center, USDA, ARS, Clay Center, Nebraska 68933.

J. N. Kremer. University of Southern California, Department of Biological Sciences, Los Angeles, California 90007.

W. E. Larson. University of Minnesota, Department of Soil Science, St. Paul, Minnesota 55108.

W. K. Lauenroth. Natural Resource Ecology Laboratory, Colorado State University, Fort Collins, Colorado 80523.

L. P. Lefkovitch. Engineering and Statistical Research Institute, Agriculture Canada, Ottawa, Canada, C. number I-405.

S. G. Leibowitz. Louisiana State University, Center for Wetland Resources, Baton Rouge, Louisiana 70803.

R. Levins. Harvard School of Public Health, 655 Huntington Avenue, Boston, Massachusetts 02115.

C. Lind. University of Louisville, Systems Science Institute, Kentucky 40292.

D. R. Linden. USDA-ARS, University of Minnesota, 313 Soil Science Building, St. Paul, Minnesota 55108.

J. A. Logan. Department of Zoology and Entomology, Colorado State University, Fort Collins, Colorado 80523.

F. J. Los. Delft Hydraulics Laboratory, P.O. Box 177, Delft, The Netherlands 2600MH.

M. D. MacNeil. University of Nebraska, Lincoln, Nebraska 68523.

D. S. Mahamah. Washington State University, Department of Civil Engineering, Pullman, Washington 99163.

R. F. Malone. Louisiana State University, Civil Engineering Department, Baton Rouge, Louisiana 70803.

F. K. Martinson. U.S. Bureau of Land Management, Denver Service Center, Building 50, Denver Federal Center, Denver, Colorado 80225.

J. H. Matis. Texas A&M University, Department of Statistics, College Station, Texas 77843.

B. A. Maurer. Statistical Support Unit, College of Agriculture, University of Arizona, Tucson, Arizona 85721.

D. A. Mauriello. U.S. Environmental Protection Agency, Office of Toxic Substances, TS-796, 401 M Street, SW, Washington, DC 20460.

M. J. McCormick. NOAA D.O.C., Great Lakes Research Lab, 2300 Washtenaw Avenue, Ann Arbor, Michigan 48104.

J. McDonnell. Wildlife Branch, Ontario Ministry of Natural Resources, Queen's Park, Toronto, Ontario, Canada M7A 1W3.

J. M. McKinion. USDA-ARS Crop Simulation Research Unit, P.O. Box 5367, Mississippi State, Mississippi 39762.

H. F. Mejer. Copenhagen Teknikum, Pr. Charlottegade, 2200 Copenhagen N, Denmark.

P. J. Mello. Michigan State University, Department of Fisheries and Wildlife, East Lansing, Michigan 48824.

C. E. Mericas. Louisiana State University, Department of Civil Engineering, Baton Rouge, Louisiana 70803.

A. Merriam. Florida State University, Institute of Science and Public Affairs, 361 Bellamy Building, Tallahasse, Florida 32306.

H. G. Merriam. Carelton University, Biology Department, Ottawa, Ontario, Canada K1S 5B6.

J. J. Messer. Utah State University, Utah Water Research Lab, UMC82, Logan, Utah 84322.

W. Mitsch. University of Louisville, Systems Science Management, Louisville, Kentucky 40292.

J. A. E. Molina. University of Minnesota, Department of Soil Science, St. Paul, Minnesota 55108.

B. Moore. Complex Systems Research Center, University of New Hampshire, Durham, New Hampshire 03824.

J. A. Morgan. Research Agronomist, USDA-ARS, A.E.R.C., Colorado State University, Fort Collins, Colorado 80523.

J. Morris. University of South Carolina, Columbia, South Carolina.

C. Neill. Louisiana State University, Center for Wetland Resources, Baton Rouge, Louisiana 70803.

R. A. Nisbet. HDR Sciences, 804 Anacape Street, Santa Barbara, California 93101.

J. L. Norton. Ford, Thornton, Norton & Associates, Ltd., P.O. Box 927, Vicksburg, Mississippi 39180.

R. V. O'Neill. Oak Ridge National Laboratory, P.O. Box S, Building 1505, MS 2, Oak Ridge, Tennessee 37830.

G. T. Orlob. University of California, Davis, California 95616.

M. D. Otis. Colorado State University, Department of Radiology and Radiation Biology, Fort Collins, Colorado 80521.

M. F. Overton. North Carolina State University, Department of Civil Engineering, Raleigh, North Carolina 27650.

W. J. Parton. Natural Resource Ecology Laboratory, Colorado State University, Fort Collins, Colorado 80523.

B. C. Patten. University of Georgia, Department of Zoology, Athens, Georgia 30602.

J. B. Pederson. University of Massachusetts, Harbor Campus, Environmental Science, Boston, Massachusetts 02125.

J. Persson. Swedish University of Agricultural Science, Department of Soil Science, Uppsala, Sweden.

B. J. Peterson. Ecosystems Center, Marine Biological Laboratory, Woods Hole, Massachussetts 02543.

C. G. Prewitt. Colorado State University, Fort Collins, Colorado 80523.

C. J. Puccia. Harvard School of Public Health, 665 Huntington Avenue, Boston, Massachusetts 02115.

G. Racey. Wildlife Branch, Ontario Ministry of Natural Resources, Queens Park, Toronto, Ontario, Canada M7A 1W3.

K. H. Reckhow. Duke University, 307 Biological Sciences Building, Durham, North Carolina 27712.

K. L. Reed. HDR Sciences, 804 Anacape Street, Santa Barbara, California 93101.

R. W. Rice. University of Arizona, Tucson, Arizona 85712.

J. E. Roelle. USDI Fish and Wildlife Service, Western Energy and Land Use Team, Fort Collins Colorado.

K. Rose. University of Washington, School of Fisheries, Seattle, Washington 98195.

G. Rossi. Commission of the European Communities, Joint Research Centre, I-21020 Ispra, Varese, Italy.

S. W. Running. University of Montana, School of Forestry, Missoula, Montana 59812.

G. V. Sabol. New Mexico State University, Las Cruces, New Mexico 88003.

J. Sangregorio. Commission of the European Communities, Joint Research Center, I-21020 Ispra, Varese, Italy.

W. P. Saunders. University of Massachusetts, Massachussetts Cooperative Fishery Research Unit, 204 Holdsworth Hall, Amherst, Massachusetts 01003.

D. Scavia. NOAA D.O.C., Great Lakes Environmental Research Laboratory, 2300 Washtenaw Ave., Ann Arbor, Michigan 48104.

L. J. Schmitt. University of Illinois, 3216 Newmark, Civil Engineering Lab, 208 Romine, Urbana, Illinois 61801.

M. Schum. San Diego State University, San Diego, California 92182.

M. J. Shaffer. USDA-ARS, University of Minnesota, 313 Soil Science Building, St. Paul, Minnesota 55108.

P. Shanahan. Environmental Research and Technology, 696 Virginia Road, Concord, Massachusetts 01742.

K. Sheehan. University of Idaho, Moscow, Idaho 83843.

J. F. Sisler. Colorado State University, Range Science Department, Fort Collins, Colorado 80523.

J. W. Skiles. Colorado State University, Department of Range Science, Fort Collins, Colorado 80523.

F. H. Sklar. Louisiana State University, Center for Wetland Resources, Baton Rouge, Louisiana 70803.

D. E. Smika. USDA-ARS, Akron, Colorado 80720.

J. G. C. Smits. Delft Hydraulics Laboratory, P.O. Box 177, Delft, The Netherlands 2600MH.

L. Somlyódy. International Institute for Applied Systems Analysis, Laxenburg, Austria.

C. B. Stalnaker. USDI Fish and Wildlife Service, Instream Flow and Aquatic Systems Group, Fort Collins, Colorado 80523.

J. C. Stans. Delft Hydraulics Laboratory, P.O. Box 177, Delft, The Netherlands 2600MH.

G. M. Sullivan. Auburn University, Auburn, Alabama 36830.

G. L. Swartzman. University of Washington, Center for Quantitative Science HR-20, Seattle, Washington 98115.

J. R. Taylor. University of Louisville, Department of Biology, Louisville, Kentucky 40292.

K. W. Thornton. Ford, Thornton, Norton & Associates, Ltd., P.O. Box 927, Vicksburg, Mississippi 39180.

N. Thyssen. National Agency of Environmental Protection, The Freshwater Laboratory, Lysbrogade 52, DK-8600 Silkeborg, Denmark

M. Turner. Martin Marietta Environmental Center, 1450 South Rolling Road, Baltimore, Maryland 21227.

P. R. Turner. New Mexico State University, Las Cruces, New Mexico 88003.

C. Uchrin. Cook College, Rutgers University, Department of Environmental Science, P.O. Box 231, New Brunswick, New Jersey 08903.

J. A. van Pagee. Delft Hydraulics Laboratory, P.O. Box 177, Delft, The Netherlands 2600MH.

J. Varnell. San Diego State University, San Diego, California 92182.

J. H. G. Verhagen. Delft Hydraulics Laboratory, P.O. Box 177, 2600 MH Delft, The Netherlands.

J. M. Vevea. Colorado State University, Department of Forest and Wood Science, Fort Collins, Colorado 80523.

R. Villasenor. HDR Sciences, 804 Anacape Street, Santa Barbara, California 93101.

K. A. Voos. USDI Fish and Wildlife Service, Instream Flow and Aquatic Systems Group, Fort Collins, Colorado 80523.

C. J. Vorosmarty. Complex Systems Research Center, University of New Hampshire, O'Kane House, Durham, New Hampshire 03824.

B. G. Walkenshaw. Burgess & Niple, Ltd., 5085 Reed Road, Columbus, Ohio 43220.

M. Wang. 9435 Mirror Pond Drive, Fairfax, Virginia 22032.

F. A. Ward. New Mexico State University, Las Cruces, New Mexico 88003.

T. J. Ward. New Mexico State University, Box 3CE, Las Cruces, New Mexico 88003.

V. J. Watson. University of Wisconsin, Laboratory of Limnology, Institue for Environmental Studies, 524 WARF, University of Wisconsin, Madison, Wisconsin 53706.

K. R. Wegener. Louisiana State University, Department of Civil Engineering, Baton Rouge, Louisiana 70823.

J. S. Wei. North Carolina State University, Department of Marine, Earth, and Atmospheric Sciences, Raleigh, North Carolina 27650.

D. A. Weinstein. Ecosystems Research Center, Cornell University, E-141 Biological Sciences Center, Ithaca, New York 14853.

W. H. West. Bureau of Land Management, 3019 Sumac, Fort Collins, Colorado 80526.

B. L. Wharram. NOAA D.O.C., Great Lakes Environmental Research Laboratory, 2300 Washtenaw Avenue, Ann Arbor, Michigan 48104.

F. W. Whicker. Colorado State University, Department of Radiology and Radiation Biology, Fort Collins, Colorado 80523.

R. G. Wiegert. Institute of Ecology, University of Georgia, Athens, Georgia 30602.

J. R. Williams. USDA-ARS, P.O. Box 748, Temple, Texas 76501.

W. O. Willis. USDA-ARS, Fort Collins, Colorado 80526.

R. G. Woodmansee. Range Science Department, Colorado State University, Fort Collins, Colorado 80523.

T. M. Zaret. University of Washington, Department of Zoology, Seattle, Washington 98115.

G. Zimmerman. Colorado State University, Department of Range Science, Fort Collins, Colorado 80523.

CONTENTS

SECTION VII: SUMMARY AND SYNTHESIS

SECTION I:

Trends in Ecological Modelling

TRENDS IN ECOLOGICAL MODELLING[1]

S. E. Jørgensen[2] and H. F. Mejer[3]

Abstract. A general theory of ecology is missing but seems to be approached from three directions: theory of ecosystem evolution, cybernetics of ecosystem and holistic approaches in ecological modelling. A unification of these three approaches might lead to the desired general theory in ecology. A unification seems possible by Pattens theory on indirect effect and the environs as well as by the authors' theory on ecosystem development towards higher exergy and buffer capacity. This last theory is reviewed and additional support to the theory presented. It is shown how a sudden change in the impact on an ecosystem (a lake is studied) will lead to a decrease (or increase) in exergy but that the ecosystem reacts to the changes by development towards higher exergy level under these new circumstances.

INTRODUCTION - THREE APPROACHES

The basic theory of physics and chemistry was constructed during the first decades in the century. Many signs indicate that a more general theory for ecology and systems ecology is "just around the corner". Such a general ecological theory seems to be approached from three different directions, which all will contribute substantially to the final results, see figure 1.

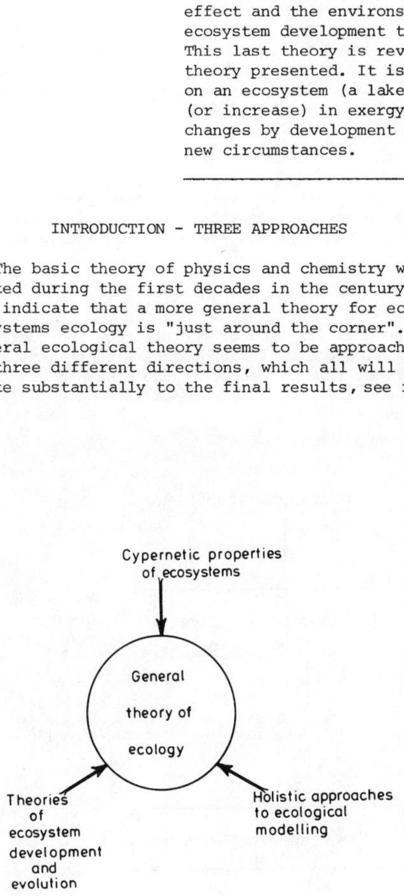

Figure 1. Three approaches toward an ecological theory.

[1]Paper presented at ISEM's 3rd Internat'l Conf., Colorado State University May 24-28, 1982

[2]Royal Danish School of Pharmacy, 2 University Park, 2100 Copenhagen Ø, Denmark

[3]Copenhagen Teknikum, Pr. Charlottegade, 2200 Copenhagen N, Denmark.

It is not new to consider evolution on the ecosystem level (Slobodkin, 1964; Dunbar, 1972; Darnell, 1970; Blackburn, 1973; Whittaker & Woodwell, 1972). Dunbar (1960) envisions selection applying "at the level of the ecosystem as well as at the level of the individual" and states that "ecosystems can compete". Margalef (1968) distinguishes between highly evolved or "mature" ecosystems and more primitive or "immature" ones.

E. Odum (1969) lists 24 trends in the development of ecosystems. Some of these are properties of populations and have received considerable attention (niche specialization, size, life cycles, growth form). Population ecologists have contributed much information of the adaptive value of different strategies in this respect. Phenomena of community energetics (such as the biomass/energy flow ratio, net community production, relative importance of grazing and detritus food chain) or homeostatic mechanisms (such as symbiosis and nutrient conservation) are of great importance in understanding ecosystems.

The ecosystem evolution theory is mainly concerned with the explanation of trends in the development of ecosystems, while the second approach, which we could call the cybernetic approach, is concerned with the function of the ecosystem.

Straskraba (1980) operates with system dynamics of four orders

1) Dynamics of the first order.

Ecosystem models are commonly expressed as system of ordinary time-variant equations with fixed parameters. In matrix notation:

$$S = f(S_1 S_o, \vec{p}, \vec{z}, t) \qquad (1)$$

where

$$S_i = \varepsilon\ S$$
$$x_{i,j} = \varepsilon\ S_i$$
$$x_{i',j} = \Sigma\ r_{i,j,k}$$
$$r_{i,j,k} = f(S, p, \vec{z}, t)$$

x_i = trophic levels

x_{ij} = species

$r_{i,j,k}$ = subprocesses

p = parameters

z = driving variables

The dynamics represented by equation (1) is characterized by a system, which is nonadaptive and which has fixed parameters and fixed structure. The trajectory of the system is solely determined by the system state, given its state in time t = o,S_o, vector of fixed parameters $\vec{p}$ and time t. This is covered by far the most simulating ecosystem models.

2) Dynamics of the second order.

In this case the systems parameter vector is dependent on a control vector, $\vec{u}$, which might represent either external variables, state variables or other control variables. In matrix notation:

$$S = f(S_1 S_o, \vec{p}(\vec{u}), \vec{u}, \vec{z}, t) \qquad (2)$$

The structure of the system described by equation (2) remains constant, but the function changes in accordance with adaptation of parameters. It opens the possibilities for inclusion of self-adaptation on three different levels of organization:

a) the level of processes,
b) the levels of individuals or population of species
c) the level of compartments.

In all instances the general numerical realization will be via adaptive parameters, although the nature of the adaptation and its description will, of course, be different.

An example is the light adaptation of photosynthesis due to variation in the chlorophyll content or shifting of temperature optima and maxima due to changes in the environmental temperature pattern. (Groden 1977 and Fedra 1979). It is noticeable that such types of adaptation are accompanied with an increase in the internal organization of the system, (Jørgensen and Mejer 1979), measured by use of exergy or negative entropy, which are related by the following equation:

$$Ex = T(S^{eq} - S) \qquad (3)$$

where S^{eq} is the entropy at thermodynamic equilibrium with the environment. It is, in other words, possible to describe this self-adaptation by use of exergy as goal function, as it has been attempted by Jørgensen and Mejer (1981).

3) Dynamics of the third order.

In this case the structure of the system is becoming dependent on its state:

$$S = f(S(\vec{u})\ S_o\ \vec{p}\ \vec{u}\ \vec{z}\ t) \qquad (4)$$

This description implies that some relations between elements are ruled out, while others are entering according to the control information signals. Correspondingly some elements vanish and, others, not introduced before become essential. Such a system will be denoted a self-organizing system or a variable structure system.

In ecosystems such changes in structure are observed as the presence of different species at the same trophic level in different situations. The numerical realization is possible by two methods:

a) by use of adaptation of parameters
b) by use of models with variable structure.

The latter type of models is represented by use of the catastrophy theory in ecosystem modelling. (Dubois 1979). In this case a mathematical analysis of the model reveals that for some values of the control variables two states are possible. The switch from one state to another is dependent on previous states, which implies that the model shows a hysteresis effect. This is only a pure mathematical consequence of the equation system.

The first mentioned method using adaptation of parameters has been used by Radtke and Straskraba (1980), and is also illustrated in Jørgensen and Mejer (1981) and Jørgensen (1982).

4) Dynamics of fourth order.

In this case the control of the system will become dependent on the system state. We could call this dynamics of the systems evolution. In matrix notation it is:

$$S = f(S(\vec{u})\ S_o\ \vec{p}(\vec{u})\ \vec{u}(S)\ \vec{z}\ t) \qquad (5)$$

The goal function or the strategy and control performation change in accordance with the state of the system. This is caused in the ecosystem by changes in the genetic pool. These ideas are conceptualized in figure 2.

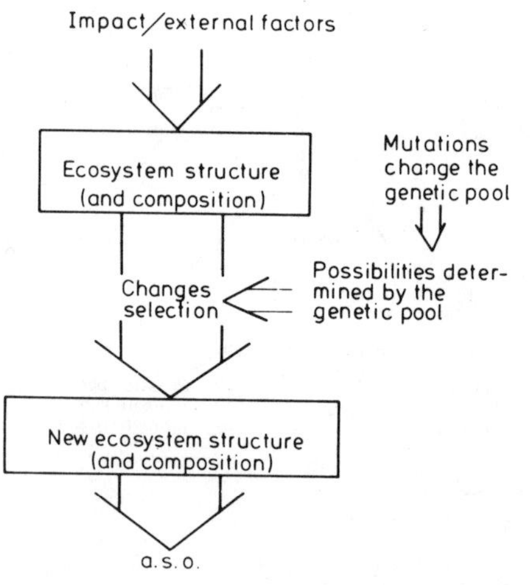

Figure 2. Concepts of the presented theory.

The impact on the ecosystem determines the structure, which is currently changed as the impact is changed. However, the possibility of changing the structure is dependent on the genetic pool, which is slowly changed. Models describing change in the genetic pool have not yet been developed, it could be possible to change the limits of crucial parameters stochastically in accordance with known mutation frequencies. As mutations occur stochastically it seems necessary to introduce a stochastic element in this part of the description of the model. It is foreseen that such fourth order models will appear during the coming decade.

The third direction of approach is more pragmatic, and it is based on the experience of ecological modelling. Not only from a computer point of view is it impossible to cope with the ecosystem complexity in a direct way but also from the biological point of view. To describe in detail all the individual subunits and their behaviour under all possible circumstances and to know all the parameters involved in such a detailed description exceeds man's possibilities. It implies that other methods which we could name holistic methods have to be found. Table 1 contains some hypothesized general principles of system development, which have been used in self-designing models (Fontaine 1981). This approach leads to the same result as description of third order dynamics by use of adaptation of parameters, mentioned above.

Table 1. Hypothesized general principles of system development

Proposed for	Principle (Objective function)	Reference
All systems at all temporal spatial, and organizational levels of resolution	Maximum energy flow	Lotka (1922); Odum & Pinkerton (1955)
	Minimum entropy	Glansdorff & Prigogine (1970)
	Maximum exergy	Mejer & Jørgensen (1979)
Ecological systems at organizational level of resolution	Optimum ascendency	Ulanowicz (1980)
greater than populations	Maximum persistent organic matter	Whittaker & Woodwell (1971) O'Neill et al. (1975)
Ecological systems; applicable scale unclear	Maximum biomass; maintenance metabolism ratio	Margalef (1968)
Econimic systems	Maximum profit (specific growth rate)	Various authors

UNIFICATION OF THE VARIOUS APPROACHES

The obvious question is: could the above mentioned different approaches moving toward a general theory be unified? We believe that it is worth while to attempt a unification, because that might lead to the basic ecological theory needed.

Patten (1978 and 1981) has stated that indirect effects probably vastly outweigh direct ones in ecosystems. This is a very interesting observation, as these indirect effects are related to structure or order of the system and the extremely well developed cybernetic nature of the system can be explained as the results of all the indirect effects. Evolution proceeds, however, by natural selection of heritable variations of individual organisms based on direct influences of environment. The ecological niche represents the point of direct contact between organisms and their environment. To encompass indirect influences, (1978 and 1981) has extended niches to new structures, named environs, which are units of organisms-environment coevolution. The motive force for coevolution is closure of outputs back upon inputs of the organisms members of ecosystems. Closure is achieved by biogeochemical cycling and mutualistic interactions, direct and indirect, between organisms. To the extent that closure does not occur there is no imperative for organisms-environment coevolution, which at the system level based on indirect effects is compatible with the normal evolution at the individual organism level based on direct effects. The organism is the unit of the latter, while environs are the units of coevolution.

Patten's theory is, in other words, able to unify the cybernetic nature of ecosystems with a theory of evolution working hand in hand with the coevolution of of the system.

The same is the case with the theory set up by Jørgensen and Meyer (1977), (1979) and (1981). As mentioned above we are using exergy as goal function for the evolution of ecosystems. Exergy is related to entropy, as seen from equation (3), but has some advantages compared with entropy:

1) It contains a level, temperature and an (indirect) "flow", namely the negative entropy, compared with the environment, which is consistent with all other descriptions of energy concepts.

2) The exergy is conserved by transformation of entropy-free energy. If, e.g. two reservoirs are exchanging entropy-free energy it is obvious that the resulting thermodynamic information, I = negative entropy, is not generally zero, since:

$$I_1 + I_2 = \frac{\Delta Ex_1}{T_1} + \frac{\Delta Ex_2}{T_2} \qquad (6)$$

However, the resulting exergy $\Delta(Ex_1 + Ex_2) = 0$.

3) One bit of information corresponds to kT ln2 in exergy. In other words the temperature is of importance for the level of information. Information from a system having a high temperature can cause more constructive changes in the environment than information from a low temperature system. The exergy is opposite the thermodynamic information directly measuring the amount of order the system is able to induce in other systems.

4) Exergy is not a state variable, but is dependent on the state of the environment, which makes this function well suited to describe the level of order or organization that the ecosystem has, compared with a selected reference state.

The following example illustrates exergy as a measure of organization: The order of bricks in a cathedral has more exergy than if the bricks formed a cube. This level of exergy, however, is higher than the same number of bricks without any order, while the bricks as such have higher exergy than a complete mixture of the same molecules.

The theory of maximum exergy claims that an ecosystem attempts to meet changes in external factors by developing a new structure with higher exergy under the conditions determined by the external factors. An ecosystem is in other words not stable in the sense that it maintains a certain structure. All ecosystems on earth, that is, the entire ecosphere, has since life appeared about 4 billion years ago been subject to changes in the external factors which have determined the selection in the Darwin sense (1859). The evolution as a continuous process, starting 4 billion years ago and up to the present day, has resulted in a steady growth in the exergy of the ecosphere due to an increase in the genetic pool available for delivering biological elements to the ecological structure of the ecosystems. The wide spectrum of species on earth today is fitted to almost all natural conditions, which implies that there will be species in all ecological niches contributing to the total exergy of the ecosphere. The thermodynamics of evolution states that an ecosystem evolves towards maximum biomass. It can easily be shown that this principle is included in the hypothesis on development and evolution towards maximum of exergy. (The difference between development and evolution is that the latter involves a change in the genetic pool).

This implies that the response of ecosystems to new external conditions is linked to the development and evolution of ecosystems.

The theory states further that it is a necessity for an ecosystem to meet perturbations by changing the structure in such a way that the ecological buffer capacity, β, defined as

$$\beta = \frac{d(\text{a forcing function})}{d(\text{a state variable})}$$

is increased.

There are an infinite number of ecological buffer capacities corresponding to all possible combinations of forcing functions and state variables. They express the ability of the system to meet perturbations and exergy measures all buffer capacities.

A constant change in the external factors will of course also change the selection pressure on the species. The selection is however not only serving the survival of the best fitted species but also the prevailing structure of the ecosystem. The relatively quick changes in ecological structures are caused by alterations in the external factors, which again are constantly modifying the selection pressure. However, the pool of genetic material is simultaneously slowly growing and being modified, which opens for a new combination of selection pressure and possibilities to meet this pressure. Many mutations will not be better fitted to the steadily changing external factors, but there will always be a probability that some mutations, better fitted to a set of external factors valid at a given time, will occur.

As everything is linked to everything in an ecosystem, the evolution of species must work hand in hand with the selection of ecological structure. As seen from this discussion the principle of development or evolution towards higher exergy, is able to explain how ecosystems react to perturbation as well as evolution in the Darwin sense. The principle of the theory is illustrated in fig.2.

An ecosystem is a soft system, which implies that its structure is continuously changing in accordance with the external factors, on which the ecosystem has no influence. The usually applied stability concept, where resistance to displacement is equivalent to stability cannot be used in ecology. It is characteristic for ecosystems that they are easily displaced from one state to another, because the ecosystem is soft and flexible, which makes it resistant to large and catastrophic changes.

As seen from this presentation the principle on maximazation of exergy can explain how evolution of the species works hand in hand with the coevolution of the system.

We might even claim that the theory eliminates the paradox og Darwins theory: How can the species survive when they are all struggling against each other? The answer is that the expression "survival of the fittest" must be understood in a wider sense. "fittest" means fittest not only to the environment in the broadest sense but also fittest to the entire ecosystem, including all the other species and the structure they are forming. This is consistent with Patten and Legovic (1982), as they have demonstrated that the out weighing indirect effects are changing a negative direct effect between species (e.g. prey-predator) to a positive effect. The species, which are struggling for themselves only, and thereby reduce the order (exergy) of the system as such, will not survive, as they are also dependent on the proper function of the entire system. It would be going too far in this context to go into detail about the philosophical aspects of the theory, but it should just be mentioned that the theory seems to give a socially acceptable interpretation of Darwinism.

The theory has been supported by use of models, which have been developed for well examined ecosystems. The models contain parameter values taken from the literature (Jørgensen et al. 1979) and the structure is in accordance with observations. Often already calibrated and validated models were used. A model is never able to give a complete picture of an ecosystem, but is seems right to conclude, that a model is able to describe the response of the ecosystem to the change in certain external factors, provided that the model is based upon a sound knowledge of the ecosystem and the problem in focus, see. e.g. the discussion in (Jørgensen 1980). All the models follow the principles of development towards maximum exergy and thereby buffer capacities. But the strength in this support is due to a development of the model qualitatively parallel to previous generally observed trends in ecosystem succession. One example taken from papers previously published on this theory (Jørgensen and Mejer (1979) and (1981)) should be mentioned here:

Marshes and hypereutrophic lakes have relatively simple structures and low species diversity, because a high and varying nutrient concentration gives the ecosystem with the relatively low complexity the highest exergy (and buffer capacity for nutrient). Correspondingly, the theory can explain why oligotrophic lakes have high complexity and give favour to the rapid growing phytoplankton species (see figure 3).

Finally, it should be mentioned that the principle of exergy maximization also opens the possibilities for our models to include the holistic and cybernetic features of the ecosystem. As indicated in the introduction this is possible by a current adjustment of the parameters by use of the exergy principle. Experience with this type of model is presented in Jørgensen and Mejer (1981) and from this it can be consluded that the results of the model follows the actual development observed in lakes, when the impact on the system is changed, for instance that smaller phytoplankton species dominate when the nutrient loading is reduced (see figure 3).

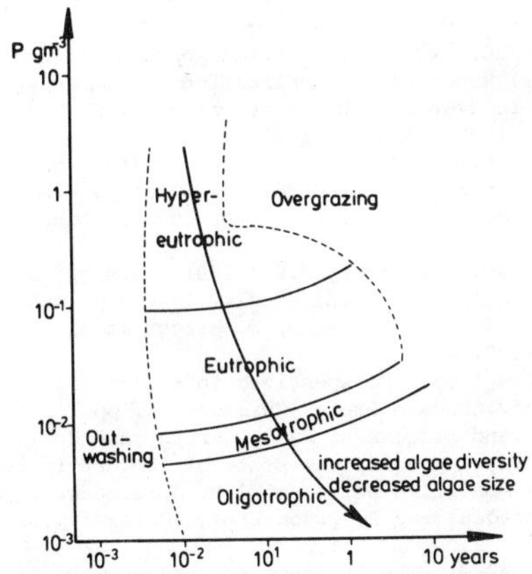

Figure 3. Trends in lake development.

From this discussion it seems possible to conclude, that the theory of exergy seems able to unify the three approaches mentioned in the introduction and can explain some of the trends in the development of ecosystems. As pointed out above the theory is also consistent with Patten's theories which unify the cybernetic nature of ecosystems, the evolution of species and coevolution of ecosystems.

FURTHER SUPPORT TO THE THEORY

In addition to the previous results obtained with the theory some recently obtained results should be mentioned in this context as further support.

The lake model presented in figure 4 has been applied and the exergy and d(exergy)/dt have been computed in accordance with the equations set up in Mejer and Jørgensen (1979).

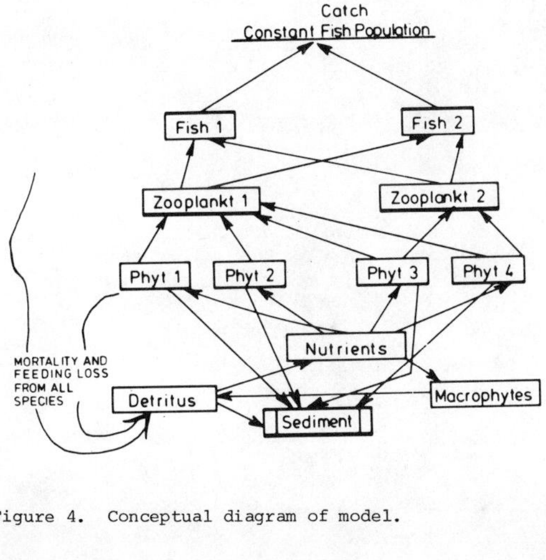

Figure 4. Conceptual diagram of model.

Again the models follow the exergy principle and show simultaneously a development which is in accordance with general observations on the development of the ecosystem.

The results obtained can be summarized as follows:

1) A more equal allocation of the biomass of the classes in each trophic level is observed when the nutrient loading is reduced (higher Shannon index in oligotrophic lakes than in hypereutrophic lakes).

2) Regular fluctuation in temperature or light favour positive d(exergy)/dt - it gives higher buffer capacities to changes in temperature and light.

3) If sudden changes in temperature or nutrient concentration are introduced the exergy level is changed correspondingly, but afterwards the exergy is always increased (d(exergy)/dt > 0). In other words the sudden change creates a new situation, but the system develops afterwards always toward a higher exergy level (sse figure 5).

4) Sudden changes in oligotrophic systems, but not in eutrophic systems, favour macrophyres, which might explain why macrophytes often can be found in oligotrophic systems (Ott 1982).

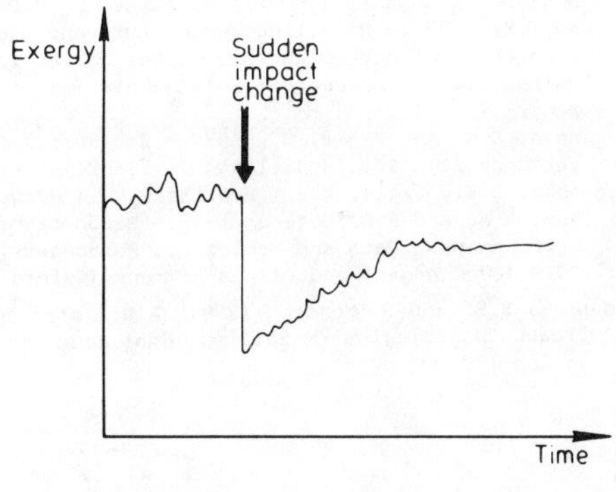

Figure 5. Ex - f (time) by sudden inpact change.

CONCLUSIONS

An ecological principle, which attempts to unify the three approaches toward a general ecological theory, has been presented. The theory is supported here and in previous papers on the same topic by application of well examined ecological models applicable to certain ecosystems. The models follow the principle and show simultaneously the same trends in ecosystem development as generally found (see e.g. figure 3) and referred to in the literature.

Further support is, however, required before the exergy principle can be used as a core in ecosystem theory. Especially it might be questioned whether ecological buffer capacities and exergy can always be considered as parallel concepts.

To get further good support it seems appropriate to test the principles on extreme ecosystems and question whether the principle can explain actually observed trends of such ecosystems. Such tests have been planned by applying observations and models of hot springs (see Wiegert and Mitchell, 1973), and results of microcosmos models with extreme low and high nutrient concentrations.

LITERATURE

Blackburn, T.R. 1973. Information and the Ecology of Scholars. Science, 181: 1141-1146.

Darnell, R.M. 1970. Evolution and the Ecosystem. Amer. Zool., 10: 9-15.

Darwin, C. 1859. The Origin of Species by Means of Natural Selection. Murray, London. Reprinted 1958, Mentor, New York.

Dubois, D. 1979. Catastrophe Theory Applied to Water Quality Regulation of Rivers. State of the Art in Ecological Modelling, 7: 751-758.

Dunbar, M.J. 1969. The Evolution of Stability in Marine Environments: Natural Selection at the Level of the Ecosystem. Amer.Natur, 94: 129-136.

-- , 1972. The Ecosystem as a Unit of Natural Selection. Trans.Conn.Acad.Arts Sci., 44: 111-130.

Fedra, K. 1979. Modelling Biological Processes in the Aquatic Environment. IIASA Working Paper WP-79-20: 56 pp.

Fontaine, Thomas D. 1981. A Self-Designing Model for Testing Hypotheses of Ecosystem Development. In Progress in Ecological Engineering and Management by Mathematical Modelling edited by D.Dubois.

Groden, T.W. 1977. Modelling Temperature and Light Adaptation of Phytoplankton. Center for Ecological Modelling, Rensselaer Polytechnic Institute, Report 2: 1-17.

Jørgensen, S.E. and Mejer, H. 1977. Ecological Buffer Capacity. Ecol.Modelling, 3: 39-62.

Jørgensen, S.E., Friis, M.B., Henriksen, J., Jørgensen, L.A. and H.F. Mejer. 1979. Handbook of Environmental Data and Ecological Parameters. ISEM (Copenhagen) and Pergamon Press (Oxford).

Jørgensen, S.E. and H. Mejer. 1979. A Holistic Approach to Ecological Modelling. Ecol.Modelling, 7: 169-189.

Jørgensen, S.E. 1980. Lake Management. Pergamon Press, New York, London and Oxford.

Jørgensen, S.E. 1981. A Holistic Approach to Ecological Modelling by Application of Thermodynamics. In Systems and Energy edited by W. Mitsch et al. (1982) Ann Arbor.

Jørgensen, S.E. and H. Mejer. 1981. Application of Exergy in Ecological Models. In Progress in Ecological Modelling, edited by D. Dubois, Liege, p.311-47.

Jørgensen, S.E. and Mejer, H.F. 1981. Exergy as Key Function in Ecological Models. In Energy and Ecol. Modelling edited by W. Mitsch et al. p.587-590.

Margalef, R. 1968. Perspectives in Ecological Theory. Univ.Chacago Press, Chicago, 112 pp.

Mejer, H.F. and Jørgensen. 1979. Exergy and Ecological Buffer Capacity. In State of the Art in Ecological Modelling edited by S.E. Jørgensen. ISEM (Copenhagen) and Pergamon Press (Oxford) p.829-846.

Odum, E.P. 1969. The Strategy of Ecosystem Development. Science, 164: 262-270.

Ott, Jørgen A. 1982. Adaptive Strategies at the Ecosystem Level. Examples from two Benthic Marine Systems. Hydrobiology, in press.

Patten, B.C. 1978. Systems Approach to the Concept of Environment. Ohio J. Sci. 78: 206-222.

-- , 1981. Environs: Realivistic Elementary Particles for Ecology. Am.Nat., in press.

Patten and Legovic. 1982. Examination of Indirect Effects. Science, in press.

Radtke, E. and M. Straskraba. 1980. Selv-Optimization in a Phytoplankton Model. Ecol.Modelling, 9: 247-268.

Slobodkin, L.B. 1964. The Strategy of Evolution. Amer.Sci., 52: 342-357.

Straskarba, Milan. 1980. Cybernetic-Categories of Ecosystem Dynamics. ISEM-Journal 2: 81-96.

Whittaker, R.H. and G.M. Woodwell. 1972. Evolution of Natural Communities. In Ecosystem Structure and Function edited by J.A. Wiens. Oregon State University Press, Corvallis: 137-159.

Wiegert, R.G. and Mitchell, R. 1973. Ecology of Yellowstone thermal effluent systems. Hydrobiologia 41, p. 251-271.

ON THE QUANTITATIVE DOMINANCE OF INDIRECT EFFECTS IN ECOSYSTEMS[1]

Bernhard C. Patten[2]

Abstract.--System structure and function analyses of a small scale compartment model of energy flow in an oyster reef ecosystem indicate that, between any two components, indirect effects are quantitatively dominant over direct causal interactions. These dominant indirect influences are expressed over paths of great lengths, underscoring the significance of long term, rather than short term, phenomena in ecosystems.

INTRODUCTION

It would not be unreasonable to date the origin of systems ecology with Möbius' (1877) study of oyster beds in the North Sea. Möbius coined the word biocoenosis (Gr. bíosis, way of life; Gr. koinos, common) to represent the collection of organisms he saw as mutually interdependent in forming a life system based on the substrate and interstices provided by the oyster. Soon after, Forbes (1887) captured a similar sense of system in his famous paper, "The lake as a microcosm." He earlier (1883, p. 33) saw superorganismal connotations in the organization of animal assemblages: "A group or association of animals is like an organism." The subsequent perception of orderly change in plant (Cowles 1988, 1901, 1911) and animal (Shelford 1911) communities led to concepts of succession and climax. "The concept of the climax as a complex organism inseparably connected with its climate and often continental in extent was introduced by Clements (1916)" (Clements 1936, p. 253). Lacking any system theory language to draw on, and with the organism model from Darwin everywhere the sine qua non of biological organization, it was natural for the emerging systems doctrine of ecology to first be cast in an idiom like superorganism. von Uexküll (1926) in fact converted the organism literally into a superorganism with his invention of "function-circles" which circumscribed chains of cause and effect as unique personal worlds attached to each living thing. Soon thereafter, Tansley (1935) culminated the early formulation of holistic thought in ecology by introducing the term ecosystem (Gr. oikos, house; Gr. systema, place together). Systems ecology had reached a secure benchmark, from which Lindeman's (1942) "The trophic-dynamic aspect of ecology" would subsequently initiate its modern development.

[1]Paper presented at the Third International Conference on State-of-the-Art in Ecological Modelling, Colorado State University, May 24-28, 1982.

[2]Bernard C. Patten is Professor of Zoology and member of the Institute of Ecology, University of Georgia, Athens, Georgia U.S.A.

In recent years, in this development, evidence has begun to appear in the empirical and modeling literature implicating indirect effects as potentially important in the ecosystem. For example, Zaret and Paine (1973) described system-wide consequences of introducing a piscivorous predator, Cichla ocellaris, into tropical Lake Gatun, Panama. This fish not only reduced its secondary consumer prey populations, but also generated higher order indirect effects throughout other trophic levels. The entire lacustrine ecosystem went through a successional redesign in response to the Cichla invasion. Patten et al. (1975, p. 369) simulated this type of perturbation with a large scale model of a cove in Lake Texoma, Texas-Oklahoma, and observed a similar realignment of many system compartments (plant and animal populations, organic matter, nutrients) and energy balance characteristics (gross production, respiration).

Lane and Levins (1977) indicated that laboratory experiments on the effects of nutrient enrichment of algal cultures could seldom be accurately extrapolated to field conditions. In the field, species are enmeshed in a network of interacting variables which change the conditions from those established under laboratory control. These authors used loop analysis (Levins 1973, 1975a, b) to show that unexpected, often counterintuitive, indirect effects develop from network interactions, and that therefore one-link direct effects cannot be extended to actual ecological situations. Their principal conclusions were: that the effect of nutrient change on any variable in a system is a function of the whole network of variables; a given effect in situ may be opposite that of a known physiological process observed in the laboratory; and a nutrient input can affect a system component even where there is no direct physiological link between the two. Similar conclusions were demonstrated in nutrient manipulation trials with the abovementioned Lake Texoma model (Patten et al. 1975, p. 365, 366).

Indirect effects are becoming increasingly recognized and invoked in explanations of field and laboratory observations. Lynch (1978) studied seasonal replacement of Daphnia by Ceriodaphnia in a Minnesota pond. He observed complex interactions not easily explainable by direct effects. For example, coexploitation of resources led to enhancement, rather than inhibition, of certain age classes. Such facilitation could only result from chains of events in a complex system, and emphasized "the importance of studying species in the context of their natural environment" (Lynch 1978, p. 563). Paine (1980) reviewed results of adding or removing species to or from communities. Whole "cascades" of effects, i.e. indirect effects, are typically observed. The affected groups of species he termed "modules," resource sets "and their specific consumers which, under conditions favoring maximal coevolution, will behave as a functional unit" (Paine 1980, p. 668-669). Modules correspond to the more formal systems constructs known as "output environs" (Patten 1978, 1982).

Hurlbert and Mulla (1981) investigated consequences of mosquitofish (Gambusia) predation on plankton communities in twelve experimental ponds in California. They predicted an increase in the ratio of calanoid to cyclopoid copepods, but observed the opposite. Their explanation (p. 146) involved a complex causal chain of indirect relationships. Elliott et al. (1982) examined food chain control of productivity and nutrient dynamics in an experimental plankton community. These authors observed effects of fish predation on lower trophic levels, nutrients, zooplankton population structure, fish production and plankton productivity, and related these changes to oddness or eveness of food chain length. Schoenberg and Carlson (1982) studied the direct and indirect effects on phytoplankton of introducing Daphnia and Bosmina zooplankton into in situ column enclosures in a hypereutrophic lake in Ohio. They had to analyze both proximate and ultimate mechanisms to explain their results.

Population-theoretic ecologists, who generally are nonsystems oriented (e.g., Simberloff 1980), find indirect effects expressed if populations are allowed to interact via a community matrix (May 1973). Thus, for example, Holt (1977) demonstrated with standard population mathematics similar conclusions to those above: "Some ecologists have suggested that all (-,-) relations between species should be labeled 'competition,' all (+,-) relations 'predation,' all (+,+) relations 'mutualism,' and so forth (e.g., Odum 1959). This classificatory scheme seems inappropriate when the relationship is imposed, as it were, by the structure of the food web. As was noted in the introduction, the elements $a_{ij} \equiv \partial F_i / \partial R_j$ making up the community matrix embody all the direct pairwise interactions occurring in a community at equilibrium. In multispecies community models, however, the sign of a_{ij} may bear little relation to the net effect of species j upon species i at equilibrium; the latter reflects the structure of the entire community" (Holt 1977, p. 217-218). This author went on to term the real relations between species, expressed through indirect network effects, as "apparent competition," "apparent predation," etc. when the corresponding direct interactions, a_{ij}, were not, respectively, competition or predation. He concluded (p. 219), "An important task for future work is to investigate the relation between models of direct interactions and apparent interactions arising in multispecies communities."

This paper concerns exactly this subject. It describes a single theoretically-derived result of some generality, namely, in ecosystem models, indirect effects are quantitatively very dominant over direct ones. This result has previously been demonstrated for two other models, one a carbon model

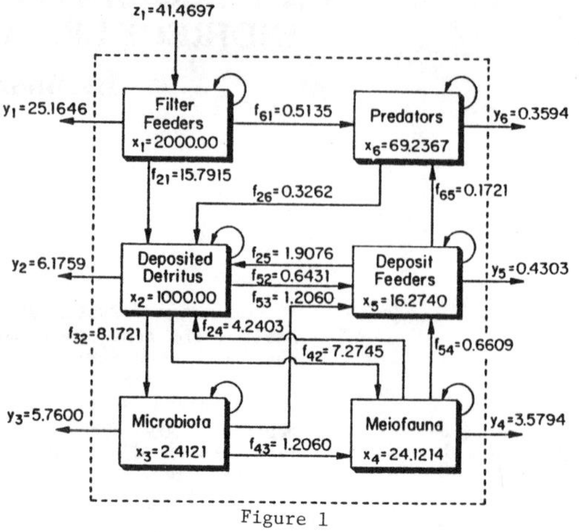

Figure 1

of the nekton subsystem of the Gulf of Mexico continental shelf ecosystem (Patten 1982), and the other a static biomass model based on the Lake Texoma simulation model (Patten, et al. 1982). Here, a simpler model is employed. Returning to the system that contributed the biocoenosis to ecology, marking the beginnings of ecological holism, a bioenergetics model for an oyster reef ecosystem will be examined.

OYSTER REEF MODEL

Figure 1 depicts a compartment model for energy flow (kcal $m^{-2}d^{-1}$) and storage (kcal m^{-2}) in an intertidal oyster reef community. The quantities are representative of such systems in the coastal marshes and estuaries of the southeastern United States. This model has been described previously (Dame and Patten 1981), and a more detailed analysis of its structure and function is given elsewhere (Patten, Dame and Barber 1983). Only a minimal description will be provided here.

The filter feeding compartment (1) consists of the American oyster (Crassostrea virginica) and, secondarily, a mussel (Brachidontes exustus). The shells of these bivalve mollusks form the frame and inner spaces which create habitat, and in additon the animals filter particles suspended in the ambient water around them. These particles form the trophic base for the community. Deposited detritus (2) comprises the feces and pseudofeces produced by the feeding and excreting mechanisms of the shellfish. Microbiota (3) consists of bacteria, yeasts and fungi associated with detritus. The meiofauna (4) are defined as benthic animals which pass through a 1 mm seive, but are retained by a 0.063 mm seive. Deposit feeders (5) consist of macrofauna which feed in the sediments. Predators (6) are the animals directly benefitting from this mode (predator-prey) of interaction with other forms. The energy flow processes that couple these compartments internally, and also those which provide input and output linkage to environment, are detailed in Dame and Patten (1981) together with documentation of the numerical data. The single system input (z_1) consists of phytoplankton and suspended detrital particles acquired by filter feeding. Output processes include respiration (y_k, k=1,3,4,5,6), mortality (y_1, y_6) and resuspension

(y_2). The intrasystem flows are realized through feeding and egestion, variously, by the different compartments.

An energy model like this, or any single commodity compartment model, is a very contracted version of reality. In ecosystems, energy, matter and information pass in diverse forms between components, giving rise not only to "brute" energy-matter exchanges (Engelberg and Boyarsky 1979), but also to intricate mechanisms of signal flow which form the basis for a view of the ecosystem as a cybernetic system (Patten and Odum 1981). The figure 1 model is thus a poor replica of any living community, but that notwithstanding, its energy flows can be taken to represent causality and influence in principle, just as though all relevant interactive details were present. That is, whatever the mechanisms or quantities interchanged in an interaction, they ultimately would have to be expressed in some quantitative form comparable to the flows in figure 1. Therefore, in investigating this model, it is justified to interpret it as a causal, influence-propagating system, and such language rather than the conceptually restrictive vocabulary of ecological energetics will be used in the discussion below.

SYSTEM STRUCTURE

MacArthur (1955) initiated the diversity-stability controversy in ecology by suggesting that diversity of components (species) induces more diverse interactive networks which serve to buffer disturbances. The issue moved away from its original network basis, largely dissipating in the technicalities of such questions as "What is diversity?", "What is stability?" The reason for this is probably that network analysis procedures were as yet unknown in ecology. Such methods are available now, permitting the structure and function of compartment-type models to be examined.

Path Analysis

Let $A=(a_{ij})$ be the adjacency matrix (table 1a) which represents the connectivity of the figure 1 system. This matrix has $a_{ij}=1$ if a direct causal interaction (energy flow) exists from compartment j (column) to compartment i (row), and $a_{ij}=0$ otherwise. Each $a_{ij}=1$ denotes a directed path of length 1 (j→i) in the model. If A is multiplied by itself, the product A^2 indicates the number of indirect paths of length 2 (j→k→i) from compartment j to compartment i. In general, a product matrix A^ℓ represents the number of length ℓ paths from j to i. For $\ell=0$, $A^\ell=I$, the identity matrix. In a system, the total number of paths of all lengths through ℓ is the sum, $\Sigma_{k=0}^{\ell} A^k$. The total number of incoming paths of length ℓ to compartment i is the sum of the i'th row of A^ℓ, $\Sigma_{j=1}^{n} a_{ij}^{\ell}$, where n is the number of compartments in the system. Similarly, the number of outgoing paths of length ℓ from compartment j is the sum of the j'th column of A^ℓ, $\Sigma_{i=1}^{n} a_{ij}^{\ell}$. In the following description of properties of the oyster reef model, typically only relationships between compartments 1 to 5 and 6, and 2 to 5, will be mentioned. For generality, however, the reader should examine each of the tables more closely.

Table 1.--Oyster reef model first order matrices (a) A for paths, and (b) P for causality. Example entry in P: 9.948-1 = 9.948×10^{-1}.

Compartments

(a)

to \ from	1	2	3	4	5	6	Row Sum
1	1	0	0	0	0	0	1
2	1	1	0	1	1	1	5
3	0	1	1	0	0	0	2
4	0	1	1	1	0	0	3
5	0	1	1	1	1	0	4
6	1	0	0	0	1	1	3
Column Sum	3	4	3	3	3	2	18

(b)

to \ from	1	2	3	4	5	6	Row Sum
1	9.948-1	0	0	0	0	0	9.948-1
2	1.974-3	9.944-1	0	4.395-2	2.930-2	1.178-3	1.071
3	0	2.043-3	1.530-1	0	0	0	1.551-1
4	0	1.818-3	1.250-1	9.121-1	0	0	1.039
5	0	1.608-4	1.250-1	6.850-3	9.614-1	0	1.093
6	6.419-5	0	0	0	2.644-3	8.975-1	1.000
Column Sum	9.969-1	9.985-1	4.030-1	9.629-1	9.934-1	9.987-1	5.353

Length 1 Paths

Direct paths ($\ell=1$) appear in table 1a and figure 1. There is none from compartment 1 to 5, whereas there are from compartments 1 to 6 (1→6) and 2 to 5 (2→5). Compartment 1 has one incoming direct path (1→1), given by the row 1 sum, and three outgoing ones (1→1, 1→2 and 1→6), identified by the sum of column 1. Note that the paths enumerated are intrasystem, not inputs or outputs. Compartment 2 has five incoming (1→2, 2→2, 4→2, 5→2, 6→2) and four outgoing (2→2, 2→3, 2→4 and 2→5) paths of length 1. And so forth for the remaining compartments. There are 18 direct paths in the model.

Length 2 Paths

Table 2a lists the second order matrix, A^2, for the oyster system, enumerating paths of length 2 ($\ell=2$). There is one such path from compartment 1 to 5 (1→2→5), and two from 1 to 6 (1→1→6and 1→6→6). Four length 2 paths exist from compartment 2 to 5: 2→2→5, 2→3→5, 2→4→5 and 2→5→5. It is left to the reader to examine the cumulative number of paths, $\Sigma_{k=1}^{\ell} A^k$, here and subsequently. Compartment 1 has only one afferent length 2 path (1→1→1) but nine efferent ones: 1→1→1, 1→1→2, 1→1→6, 1→2→2, 1→2→3, 1→2→4, 1→2→5, 1→6→2 and 1→6→6. Compartment 2 has 16 such paths incoming and 13 outgoing. Compartments 5 and 6 have 14 and 8 afferent paths of length 2, vs. 9 and 6 efferent ones, respectively. The reader is again invited to identify some of these paths as an exercise that will reveal much redundancy, but at the same time establish that each path is unique. It is this latter characteristic, and its later consequences, that are usually difficult to appreciate. The model contains 56 paths of length two.

Length 3 Paths

The third order structure of the oyster reef model, A^3, elaborating length 3 paths, appears in table 3a. There are indicated to be six such paths from compartment 1 to 5: 1→1→2→5, 1→2→2→5, 1→2→3→5, 1→2→4→5, 1→2→5→5 and 1→6→2→5. There are four from compartment 1 to 6: 1→1→1→6, 1→1→6→6, 1→2→5→6 and 1→6→6→6. From compartment 2 to 5 there are twelve: 2→2→2→5, 2→2→3→5, 2→2→4→5, 2→2→5→5, 2→3→3→5, 2→3→4→5, 2→3→5→5, 2→4→2→5, 2→4→4→5, 2→4→5→5, 2→5→2→5 and 2→5→5→5. The total number of incoming (row sums) and outgoing (column sums) paths of length 3 for the compartments is beginning now to become a challenge to identify. For example, there are 49 incoming and 41 outgoing paths of length 3 associated with compartment 2. A total of 176 length 3 paths exists in the system. For brevity, a skip will now be made to paths of length 10, and then, finally, length 50.

Length 10 Paths

There are, incredibly, 23695 paths of length 10 from compartment 1 to 5, 11032 from 1 to 6, and 34729 from 2 to 5 (table 4a). These seem to be unbelieveable numbers at first, but they make sense after some thought. The reason is that the length of a cyclic path is infinite. Energy, matter, information or, in general, influence, may pass around such a path until it either dissipates or exits the cycle. Thus, the series $\Sigma_{k=1}^{\ell} A^k$ is bound to be divergent as $\ell \to \infty$. Any systems with feedback in their structure therefore are possessors of infinite path lengths, and this includes even small scale systems such as the oyster reef compartment model. Compartment 2 in this system has 149187 incoming paths (row 2 sum) and 125491 outgoing paths. There are over a half million length 10 paths in the model.

Table 2.--Oyster reef model second order matrices (a) A^2 for paths, and (b) P^2 for influences.

Compartments

(a)

to \ from	1	2	3	4	5	6	Row Sum
1	1	0	0	0	0	0	1
2	3	3	2	3	3	2	16
3	1	2	1	1	1	1	7
4	1	3	2	2	1	1	10
5	1	4	3	3	2	1	14
6	2	1	1	1	2	1	8
Column Sum	9	13	9	10	9	6	56

(b)

to \ from	1	2	3	4	5	6	Row Sum
1	9.897-1	0	0	0	0	0	9.897-1
2	3.927-3	9.890-1	9.156-3	8.399-2	5.732-2	2.346-3	1.146
3	4.033-6	2.344-3	2.342-2	8.979-5	5.987-5	2.406-6	2.592-2
4	3.590-6	3.723-3	1.331-1	8.320-1	5.329-5	2.142-6	9.689-1
5	3.174-7	5.823-4	1.402-1	1.284-2	9.244-1	1.894-7	1.078
6	1.279-4	4.251-7	3.305-4	1.811-5	5.179-3	9.951-1	1.001
Column Sum	9.937-1	9.956-1	3.062-1	9.290-1	9.870-1	9.974-1	5.209

Table 3.--Oyster reef model third order matrices (a) A^3 for paths, and (b) P^3 for influences. Smaller values than corresponding non-diagonal entries in P^2 (Table 2b) are underlined.

Compartments

(a)

to \ from	1	2	3	4	5	6	Row Sum
1	1	0	0	0	0	0	1
2	8	11	8	9	8	5	49
3	4	5	3	4	4	3	23
4	5	8	5	6	5	4	33
5	6	12	8	9	7	5	47
6	4	5	4	4	4	2	23
Column Sum	28	41	28	32	28	19	176

(b)

to \ from	1	2	3	4	5	6	Row Sum
1	9.845-1	0	0	0	0	0	9.845-1
2	5.859-3	9.837-1	1.906-2	1.205	8.410-2	3.505-3	1.217
3	8.640-6	2.379-3	3.602-3	1.853-4	1.263-4	5.162-6	6.306-3
4	1.092-5	5.487-3	$\underline{1.244-1}$	7.590-1	1.603-4	6.521-6	8.891-1
5	1.465-6	1.037-3	$\underline{1.386-1}$	1.807-2	8.887-1	8.747-7	1.046
6	1.911-4	1.963-6	7.002-4	5.201-5	7.610-3	9.926-1	1.001
Column Sum	9.906-1	9.926-1	2.863-1	8.978-1	9.807-1	9.961-1	5.144

Length 50 Paths

Pausing still further along the way to infinity, table 5a denotes the 50'th order structure of the model. For effect if not comprehension, there are 1.975×10^{24} length 50 paths from compartment 1 to 5, 9.194×10^{23} from compartment 1 to 6, and 2.894×10^{24} from compartment 2 to 5.

A length 50 path is very long in such a small model, and the question is what, if any, is its significance or that of any long path in a system? Intuition says that most of the causality in an interaction is direct, with some indirect remnants perhaps once or twice removed. To assess this objectively, however, requires an influence analysis.

SYSTEM FUNCTION

MacArthur (1955) did not know how to count paths in a system as has just been done above, but his intuition gave him the path as the central element of his stability measure. Letting p_k be the probability of food passing along the k'th path, where p_k is the product of the chain of probabilities p_{ij} associated with each link, j→i, in the chain, then stability is $-\Sigma_{k=1}^{m} p_k \log p_k$, where m is the number of paths (of all lengths) in the system. To measure influence, the same probability approach MacArthur used can be employed.

Influence Analysis

Let the matrix $P=(p_{ij})$, $0 \leqq p_{ij} \leqq 1$, represent the direct effects of compartment j on compartment i in a compartmental system, expressed as the probability that substance in j will be transferred to i in some interval of time. P is a one-step transition matrix in Markov chain theory, and can be computed readily from storage and flow information, such as that which appears in figure 1 (Matis and Patten 1981, see also Patten, Dame and Barber 1983 for specifics pertaining to the oyster reef model). For the oyster model a six hour period was used, and the resulting matrix, P, appears in table 1b. Note the contracted exponential notation. P has the same form as the table 1a structure matrix because p_{ij} represents the direct causality associated with direct path a_{ij} of length 1 connecting j to i. Powers of P, P^{ℓ}, denote the influence of j on i over the paths defined by A^{ℓ}. When $\ell>1$, this influence is indirect, when $\ell=1$ it is direct, and when $\ell=0$, $P^0=I$, the identity matrix, by definition. Since energy, matter or information and the influence they represent are all dissipative in systems, unlike the divergent series, $\Sigma_{k=0}^{\infty} A^k$, for paths, the influence series $\Sigma_{k=0}^{\infty} P^k$ converges. Its finite portion $\Sigma_{k=0}^{\ell} P^k$ represents the total influence propagated over all paths of all lengths through ℓ from compartment j to i. The matrix to which the infinite sum converges is termed a transitive closure matrix (Patten et al. 1976, p. 523), and it has an easily computable form: $\lim_{\ell\to\infty} \Sigma_{k=0}^{\ell} P^k = (I-P)^{-1}$. Thus, to obtain the total influence between compartments in a system it is only necessary to convert P to $(I-P)^{-1}$. Once this is done, the ratio of partial to total influence can be easily calculated as the quotient of corresponding elements in the matrices $\Sigma_{k=0}^{\ell} P^k$ and $(I-P)^{-1}$. Total influence propagated to compartment i over paths of length ℓ from all compartments is the sum of the i'th row of P^{ℓ}, $\Sigma_{j=1}^{n} p_{ij}^{\ell}$, and that propagated away from compartment j over length ℓ paths to all

Table 4.--Oyster reef model tenth order matrices (a) A^{10} for paths, and (b) P^{10} for influences. Smaller values than corresponding nondiagonal entries in P^3 (Table 3b) are underlined.

Compartments

(a)

to \ from	1	2	3	4	5	6	Row Sum
1	1	0	0	0	0	0	1
2	23696	34729	23697	27201	23696	16168	149187
3	11033	16168	11032	12664	11033	7528	69458
4	16169	23696	16168	18560	16169	11033	101795
5	23695	34729	23696	27201	23696	16169	149186
6	11032	16169	11033	12664	11032	7527	69457
Column Sum	85626	125491	85626	98290	85626	58425	539084

(b)

to \ from	1	2	3	4	5	6	Row Sum
1	9.494-1	0	0	0	0	0	9.494-1
2	1.883-2	9.494-1	7.290-2	2.988-1	2.410-1	1.137-2	1.592
3	4.029-5	2.303-3	1.581-4	6.662-4	5.254-4	2.430-5	3.718-3
4	1.416-4	1.396-2	6.616-2	4.011-1	1.915-3	8.512-5	4.833-1
5	3.353-5	4.009-3	1.098-1	3.899-2	6.753-1	2.014-5	8.282-1
6	6.203-4	4.520-5	3.003-2	5.831-4	2.203-2	9.755-1	1.002
Column Sum	9.690-1	9.697-1	2.521-1	7.401-1	9.408-1	9.870-1	4.859

Table 5.--Oyster reef model fiftieth order matrices (a) A^{50} for paths, (b) P^{50} for influences, and (c) $\Sigma_{k=0}^{50} P^k$ for cumulative influences over all paths through length 50. Smaller values than nondiagonal entries in P^{10} (Table 4b) are underlined.

Compartments

(a)

to \ from	1	2	3	4	5	6	Row Sum
1	1	0	0	0	0	0	1
2	1.975+24	2.894+24	1.975+24	2.267+24	1.975+24	1.347+24	1.243+25
3	9.194+23	1.347+24	9.194+23	1.055+24	9.194+23	6.273+23	5.788+24
4	1.347+24	1.975+24	1.347+24	1.547+24	1.347+24	9.194+23	8.483+24
5	1.975+24	2.894+24	1.975+24	2.267+24	1.975+24	1.347+24	1.243+25
6	9.194+23	1.347+24	9.194+23	1.055+24	9.194+23	6.273+23	5.788+24
Column Sum	7.136+24	1.046+25	7.136+24	8.191+24	7.136+24	4.869+24	4.493+25

(b)

to \ from	1	2	3	4	5	6	Row Sum
1	7.712-1	0	0	0	0	0	7.712-1
2	7.752-2	7.996-1	1.504-1	4.544-1	5.660-1	4.941-2	2.097
3	1.836-4	1.938-3	3.622-4	1.098-3	1.360-3	1.168-4	5.059-3
4	1.504-3	2.000-2	5.035-3	2.074-2	1.238-2	9.472-4	6.061-2
5	6.732-4	1.176-2	2.610-2	2.361-2	1.460-1	4.207-4	2.086-1
6	2.693-3	9.444-4	8.643-2	4.149-3	5.480-2	8.835-1	9.547-1
Column Sum	8.537-1	8.343-1	1.905-1	5.040-1	7.806-1	9.344-1	4.097

(c)

to \ from	1	2	3	4	5	6	Row Sum
1	4.492+1	0	0	0	0	0	4.492+1
2	2.141	4.545+1	5.581	1.912+1	1.991+1	1.334	9.354+1
3	4.944-3	1.073-1	1.194	4.482-2	4.641-2	3.077-3	1.401
4	3.386-2	8.662-1	1.757	1.161+1	3.358-1	2.096-2	1.459+1
5	1.300-2	3.849-1	3.552	1.690	2.258+1	8.017-3	2.823+1
6	7.273-2	1.799-2	2.823-1	1.073-1	1.875	4.797+1	5.026+1
Column Sum	4.719+1	4.683+1	1.237+1	3.257+1	4.475+1	4.934+1	2.330+2

compartments is the sum of the j'th column, $\Sigma_{i=1}^{n} p_{ij}^{\ell}$. With these various measures, influence can be assessed just as paths were previously to determine, for example, the significance of direct vs. indirect effects, and of short vs. long paths.

Direct Causality

Table 1b exhibits the direct causes transmitted from compartment j to compartment i in the oyster reef model, expressed as fractions of the energy stored in the column compartments transferred each six hours to the row compartments over the length 1 paths identified in table 1a. The effect of compartment 1 on compartment 5 is zero because no such path provides this linkage. The direct effect of compartment 1 on 6 is 6.419×10^{-5}, and that for 2 on 5 is 1.608×10^{-4}. Total incoming causality (row sums) is, for example, 0.995 and 1.071 for compartments 1 and 2, respectively, and corresponding outgoing values (column sums) are 0.997 and 0.998. Comparisons with the other compartments can be made directly from the table. The total direct causality propagated within the system during each six hours is 5.353.

Table 6.--Oyster reef model matrices: (a) transitive closure, $(I-P)^{-1}$, representing total influences; (b) ratios of direct (Table 1b) to total (Table 6a) influences; and (c) ratios of cumulative influences through 50th order (Table 5c) to total influences (Table 6a).

Compartments

(a)

to \ from	1	2	3	4	5	6	Row Sum
1	1.929+2	0	0	0	0	0	1.929+2
2	9.646+1	2.494+2	4.986+1	1.401+2	1.977+2	1.187+2	8.523+2
3	2.327-1	6.017-1	1.301	3.380-1	4.769-1	2.863-1	3.236
4	2.327	6.017	2.882	1.476+1	4.769	2.863	3.361+1
5	1.570	4.059	4.937	4.302	2.915+1	1.932	4.595+1
6	6.678	4.335	5.273	4.594	3.113+1	4.060+2	4.580+2
Column Sum	3.002+2	2.645+2	6.425+1	1.641+2	2.632+2	5.298+2	1.586+3

(b)

to \ from	1	2	3	4	5	6
1	5.157-3	0	0	0	0	0
2	2.046-5	3.987-3	0	3.136-4	1.482-4	9.924-6
3	0	3.396-3	1.176-1	0	0	0
4	0	3.023-4	4.338-2	6.181-2	0	0
5	0	3.961-5	2.532-2	1.592-3	3.298-2	0
6	9.611-6	0	0	0	8.492-5	2.457-3

(c)

to \ from	1	2	3	4	5	6
1	2.328-1	0	0	0	0	0
2	2.220-2	1.822-1	1.119-1	1.364-1	1.007-1	1.124-2
3	2.125-2	1.784-1	9.176-1	1.326-1	9.731-2	1.075-2
4	1.455-2	1.440-1	6.098-1	7.867-1	7.041-2	7.321-3
5	8.281-3	9.483-2	7.194-1	3.928-1	7.745-1	4.151-3
6	1.089-2	4.151-3	3.354-2	2.336-2	6.022-2	1.181-1

Total Influence

Table 6a gives the transitive closure matrix for the model, expressing total influence propagated within the system. Compared to the direct effects (Table 1b), the total effects are enormous. The six hour influence of compartment 1 on 5 is 1.570 vs. no direct effect. The total influence of 1 on 6 is 6.678, which is 104035 times greater than the direct effect of 6.419×10^{-5}. The total influence of compartment 2 on 5 is 4.059, which is 25243 times larger than the direct value, 1.608×10^{-4}. The grand total of all influence propagated each six hours within the system is 1586. Table 6b lists the direct/total effects ratios, the quotients of corresponding entries in tables 1b and 6a. It is obvious that: <u>indirect influences are quantitatively dominant over direct causes in the oyster reef model</u>. With this result, now it is of interest to investigate if these dominant indirect influences are expressed mainly over short paths or longer ones. In other words, what is the relationship of influence to path length, and its concomitant, number of paths?

Order 2 Influence

Table 2b presents the indirect influences propagated over the length 2 paths (table 2a). From compartment 1 to 5 the six hour value is 3.174×10^{-7}, from 1 to 6 it is 1.279×10^{-4}, and from 2 to 5 it is 5.823×10^{-4}. The shellfish (compartment 1) are observed to have a measureable effect on detritus feeders (5), even though there is no direct energy flow between these two compartments (figure 1, table 1a). The second order influence of compartment 1 on 6 is two times greater than the corresponding direct effect, and that of compartment 2 on 5 is almost four times more. Comparing other corresponding entries in tables 2b and 1b, in all nondiagonal cases the second order effects are greater than direct causes. For diagonal entries the higher order self influences are smaller. Incoming and outgoing influences over length 2 paths appear, respectively, as row and column sums in table 2b. For example, the values for compartment 1 are 0.989 and 0.994, and for compartment 2 1.146 and 0.996. Thus, filter feeders propagate slightly more second order influence than they receive from the system, and detritus is a net recipient of length 2 path effects. It is evident that while second order influences generally exceed direct effects in the model, there is no suggestion that major fractions of the total indirect effects observed above (table 6a) will be accounted for early in the sequence of path length influences. The total second order influence propagated within the system during six hours is 5.209, less than the direct effects (5.353) due to smaller diagonal values.

Order 3 Influence

Third order effects are listed in table 3b. For the influence of compartment 1 on 5 the value is 1.465×10^{-6}, almost five times greater than the corresponding second order effect (table 4b). The third order effect of 1 on 6 is 1.911×10^{-4}, one and a half times greater than the path length 2 effect. And the value for the compartment 2 to 5 influence is 1.037×10^{-3}, almost twice the corresponding second order effect. In only two cases (underlined) are third order results less than corresponding second order ones. Afferent third order effects on , e.g., compartments 1 and 2 are 0.985 and 1.217, and the corresponding outgoing values are 0.991 and 0.993. Shellfish (1) are net receivers and detritus (2) net producers of third order intrasystem influence. While third order effects exceed direct effects (compare tables 3b and 1b), again the increment toward the final total influence (table 6a) is small. Also, total third order influence (5.144) is less than second order (5.209) due to continued reduction in self effects represented by principal diagonal elements.

Order 10 Influence

Influences over the length 10 paths are listed in table 4b. For the energy flows from compartments 1 to 5 and 6 the respective values are 3.353×10^{-5} and 6.203×10^{-4}; the effect of 2 on 5 is 4.009×10^{-3}. For only four (underlined) of the nondiagonal entries in table 4b are the length 3 values greater. Thus, indirect effects are generally still tending to grow at the 10'th order level. The reason is the enormous number of paths. Influence per path decreases with increasing path length. As an example, for the effect of compartment 1 on 6, influence/path number ratios are $6.419 \times 10^{-5}/1$ for first order, $1.279 \times 10^{-4}/2 = 6.395 \times 10^{-5}$ for second, $1.911 \times 10^{-4}/4 = 4.778 \times 10^{-5}$ for third and $6.203 \times 10^{-4}/11032 = 5.623 \times 10^{-8}$ for order ten. However, as shown in the previous section, number of paths increases geometrically with path length. In terms of total incoming vs. outgoing influence of order 10, compartments 1 and 2 have 0.949 and 1.592 incoming, and 0.969 and 0.970 outgoing. At this level, compartment 1 is a slightly stronger net donor of influence than previously, and compartment 2 is a considerably stronger net receiver. Total tenth order influence (4.859) is down from that previously observed (5.144).

Order 50 Influence

The indirect influences associated with the enormous numbers of length 50 paths in the model are shown in table 5b. Again, for such a small model, intuition leads to the expectation that significant decay of incremental effects would have set in by now. The data run contrary to this expectation, however. The 50'th order effects of compartment 1 on 5 and 6 are 6.732×10^{-4} and 2.693×10^{-3}, and for the effect of 2 on 5 the value is 1.176×10^{-2}. Four nondiagonal elements in table 5b (underlined) are smaller than corresponding values in the previously displayed matrix (table 4b), but this is still a small number. At the level of length 50 paths in this small model, the disappearance of influence with path length, which must occur for the series $\Sigma_{k=0}^{\infty} P^k$ to converge, has not begun to happen yet in earnest. Note, in table 5b, that the 50'th order effect of compartment 4 on 2 (4.544×10^{-1}) is greater than the tenth order effect (table 4b), but this latter value was less than the previous third order value (table 3b). Growth or diminution of influence is not necessarily monotonic. Influence per path is smaller than previously. For the selected example of the effect of compartment 1 on 6, the order 50 value is $2.693 \times 10^{-3}/9.194 \times 10^{23} = 2.929 \times 10^{-27}$. However, again, the presence of so many paths accounts for significant influence. Total 50'th order afferent effects to compartments 1 and 2 are 0.771 and 2.097, and corresponding efferent values 0.854 and 0.834. At this level, compartment 1 (shellfish) has become a considerably stronger net donor of influence than before, and detritus (compartment 2) a very strong net receiver. Total 50'th order influence within the system (4.097) is less than at previously observed levels.

Table 5c lists the cumulative influence in the model associated with all paths of all lengths through 50. Large as these values are in comparison to the direct causes given in table 1b, they still do not, in general, constitute large fractions of the total influence. Table 6c gives these fractions (quotients of corresponding entries in table 5c and 6a). For the compartment interactions 1 to 5 and 6, and 2 to 5, these ratios are only 8.281×10^{-3}, 1.089×10^{-2} and 9.483×10^{-2}. However, some entries in Table 6c do indicate that significant portions of total influence have been accounted for by paths through length 50. The largest of these, for nondiagonal elements, are 0.719 (3 to 5), 0.610 (3 to 4) and 0.393 (4 to 5), but most of the remaining values are well below 10%. In general, then, little of the ultimate total influence within this system is accounted for by paths through length 50.

Summary Points

The principal conclusions to be derived from the foregoing results are as follows: (1) Indirect effects in realistically structured compartment models, and thus by implication in real ecosystems as well, are quantitatively dominant over direct causal interactions. (2) These dominant indirect influences are expressed over paths of very great lengths, not of short lengths, in the structural networks of ecosystems. (3) Because long times are needed to traverse long paths, the quantitatively most significant effects within ecosystems are long term, not short term, in character.

DISCUSSION

There is nothing exceptional about the figure 1 oyster reef model. It is a typical static, compartmental representation of an ecosystem, and thus may be treated generically in arriving at some general understandings about indirect vs. direct effects. Real ecosystems are, of course, much larger, and there the described properties from a small model would be greatly amplified. In that sense, the conclusions and interpretations below are conservative.

The human mind seems to have a blind spot for systems. It readily perceives direct causes and immediate effects, but the reckoning of things held together by invisible, ephemeral or momentary interactions several or more times removed is left to instinct, intuition and, some would say, imagination. Thus, it is no real surprise that many ecologists, in 1982, would still deny the existence of an entity like ecosystem. Figure 1 is a typical means of portraying a system; the direct causes are evident, but the indirect ones elaborated in the preceding section are quite hidden. There is little to suggest the dominant role of the latter in binding parts into wholes. Adjacency matrices (table 1a) and community matrices (derivable from table 1b; Matis and Patten 1981) are other examples of system representations in wide use. All of them convey the impression of a system as a set of foci (the compartments) for direct causes, not as a diffuse net of indirect effects everywhere dispersed. Yet that is the image evoked by the foregoing path and influence analyses.

Path length is a key feature. Figure 2 illustrates several general characteristics of the oyster reef model associated with path length. Number of paths increases geometrically as a function of path length (figure 2a) because of the divergence characteristics of the power series of the adjacency matrix, $\Sigma_{k=1}^{\infty} A^k$. Influence from a source to a receiving compartment increases, mostly monotonically, to a maximum at some relatively large path length, then decreases, again mostly monotonically, and asymptotically, to zero (figure 2b). The latter must occur in order for the influence series $\Sigma_{k=0}^{\infty} P^k$ to converge. Monotonicity is not universal, as revealed in a former study (Patten et al. 1982), and also here by the order 3, 10 and 50 influences from compartment 4 to 2 (tables 3b, 4b and 5b). Nonmonotonic patterns of influence development and decay with path length may be an expression of cyclic network structure; repeating numbers in the higher order path matrices (tables 4a, 5a) reflect such structure in the present model.

Incidentally, returning to the oyster model as an energy system for a moment, the above analysis of energy propagation should leave little doubt in anyone's mind that energy cycles profusely in ecosystems, contrary to textbook statements and conventional wisdom. The longest possible noncyclic path in the figure 1 model is length 5, and at this short path length energy movement around this system has hardly begun to happen.

Returning to figure 2b, this also shows the pattern of cumulative influence from one compartment to another over all paths up to a particular length. This tends to be hyperbolic, but again may not be monotonic. Figure 2c indicates the total influences propagated between all compartments at different path lengths. The exponential decay exhibited reflects dissipation with passage through the network. Figure 2d shows influence per path. This is greatest for direct paths, then decreases geometrically with path length.

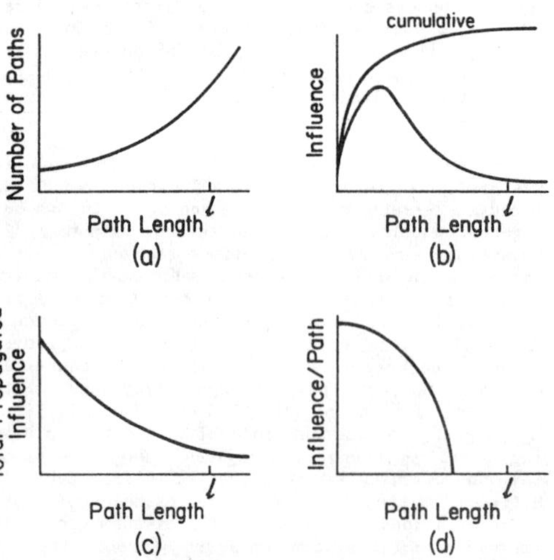

Figure 2

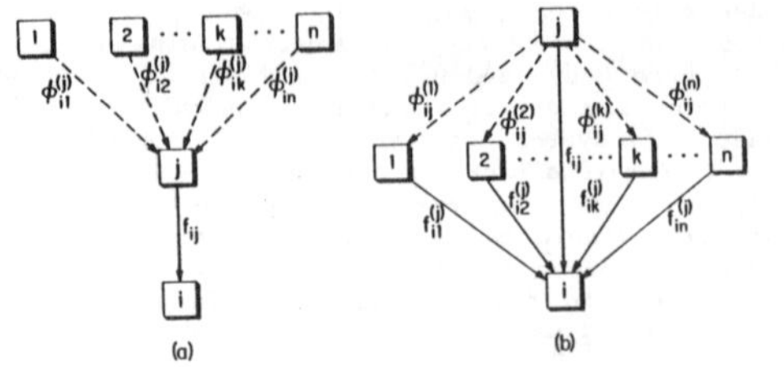

Figure 3

How should indirect effects be interpreted and evaluated? Whatever their significance, it must be acknowledged that ultimately this has to be through direct interactions. That is, indirect influences arrive as direct ones; they are implicit in the latter. Why, then, can't they be simply ignored? The reason is because of the kind of cause-and-effect questions which science usually asks. Consider figure 3a. The direct effect (solid arrow) of system component j on i is preceded by complex chains of causality leading indirectly (broken arrows) from other system components, k=1,...,n, to j. Let f_{ij} be the direct effect of j on i, and $\phi_{ik}^{(j)}$ be the indirect influences propagated over all paths of all lengths from k=1,...,n to j. Once these influences have arrived at j, their origins can be forgotten. They become implicit in f_{ij}, and to know j's effect on i, it is sufficient to know only the direct cause, f_{ij}.

The flaw in this reasoning derives from failure to distinguish between <u>input environment</u> and <u>output environment</u> (Patten 1978, 1982; Patten and Auble 1981), concepts which correspond to von Uexküll's (1926) "world-as-sensed" and "world-of-action," respectively, in his superorganismic conception of the organism-environment relationship. Figure 3a depicts the input environment of component j, traced back to each other component k=1,...,n which directly or indirectly affects j. But it erroneously represents the portion of the output environment of j that leads to i. This is shown in figure 3b. Here, in addition to the direct effect f_{ij}, there are also indirect influences, $\phi_{ij}^{(k)}$, of j on i transmitted through each system component, k=1,...,n, which enters into a direct relationship, f_{ik}, with i. Each such direct effect implicitly carries indirect influences, $f_{ik}=\Sigma_{j=1}^{n} f_{ik}^{(j)}$, to i, where $f_{ik}^{(j)}$ is the embodied constituent of the direct cause f_{ik} contributed by j. These constituents left j at different times in the past, and now they arrive contemporaneously with the direct interaction f_{ij}. Furthermore, this study has shown the latter to be quantitatively insignificant compared to the indirect effects, i.e.,

$$\sum_{\substack{k=1 \\ k\neq j}}^{n} f_{ik}^{(j)} \gg f_{ij}$$

Consequently, to know j's effect on i it is insufficient to have knowledge only of f_{ij}. This direct effect is small compared to the totality of indirect effects from j that arrive at i as f_{ik}, k=1,...,n, in the same interval of time, coded as implicit $f_{ik}^{(j)}$. These latter indirect effects serve to carry the historical influence of j within the system, integrated over relatively long periods of time, forward to the interval in question, where they finally impact i, not as j, but as k=1,...,n. Therefore, i's relationship to j cannot be based on f_{ij} alone. It must be based on all of its interactions f_{ik}, k=1,...,n, that past activity of j has influenced. That is, what i really receives from j in a given interval of time is a small quantity of uncoded direct causality by which it may develop some weak knowledge of j, plus a huge quantity of coded indirect influence which relates it strongly, although implicitely, to j's historical role in the system. This happens concurrently with respect to all components j that can reach (influence) all components k that directly interact with i.

The receipt by i of so much implicit information about the system ensures its conformity in the selective process with the structural and functional organization which the system represents. That is, natural selection of i based on i's experience of all k=1,...,n variables with which it directly interacts would appear geared to adapt i not to any particular component, j, but rather to the entire system as the selective unit. Thus, the property of dominant indirect effects in systems appears to provide a mechanism by which the individually selected organism becomes automatically integrated into the whole ecosystem in evolution. It is this integration that seems to have eluded evolutionary theory in the past, causing it to become a nonsystems science during the same period when Möbius and his successors were forging systems ecology. Further clarification of the significance of indirect effects should provide a much needed link between evolution and the ecosystem.

ADDENDUM

Critique and review comments provided by R.V. O'Neill (Oak Ridge National Laboratory), R.G. Wiegert (University of Georgia) and R.E. Ulanowicz (Chesapeake Biological Laboratory) have indicated a need to clarify the distinction between "direct" and "indirect" paths and influences. This is not so straightforward as represented, and requires reconciling the mathematical methods of computation to the ecological ends intended.

In the System Structure section above, a number of length 2 and length 3 paths were identified for the oyster reef model. Computed influences P^2 and P^3 associated with these paths were interpreted as indirect. In general, here and previously (Patten 1982, Patten et al. 1982), matrices P^k were considered to quantify direct effects when k=1, and indirect effects when $k \geq 2$. However, note that many k'th order paths from compartment j to i do not involve a third compartment, and hence are not indirect in the sense intended. For example, length 2 and 3 paths like 1→1→2, 1→1→6, 1→2→2, 1→6→6, 2→2→5, 2→5→5, and 1→1→1→6, 1→1→6→6, 1→6→6→6, 2→2→2→5, 2→2→5→5, 2→5→5→5 and 2→5→2→5 are of this type. They are indirect mathematically in being of order k>1, but they are not indirect in the ecologically meaningful sense of involving a third intermediary compartment.

Thus, the generating matrix P (table 1b) of higher order effects represents only minimum values for direct effects, not total direct effects. To obtain the latter, influences associated with k'th order paths, k>1, that do not contain a third compartment must be added to the first order direct effects. It is thus appropriate, consistent with the mathematical analysis, to recognize both first order (instantaneous) and higher order (time delayed) direct effects, and to consider matrices P^k of k'th order influences as partitionable into direct and indirect components, P_δ^k and P_ι^k, respectively. The total direct plus indirect effects in a system, $(I-P)^{-1}$, are then

$$(I-P)^{-1} - I = P + \sum_{k=2}^{\infty} P^k$$

$$= [P + \sum_{k=2}^{\infty} P_\delta^k] + [\sum_{k=2}^{\infty} P_\iota^k],$$

where the first bracketed terms represent total direct and the second total indirect influence generated between the compartments. The hypothesis of this study can thus be expressed as

$$\sum_{k=2}^{\infty} P_\iota^k >> P + \sum_{k=2}^{\infty} P_\delta^k.$$

Suppose the diagonal elements of the adjacency matrix A (table 1a) are zeroed, A_o. This eliminates the self loops, j→j, shown in figure 1. All of the length k paths enumerated by A^k which possessed one or more such self loops would be eliminated in the accounting of paths provided by A_o^k. In the oyster model, as an example, all the length 2 and 3 paths listed two paragraphs above, except the last one, would be so eliminated. Since the eliminated paths represent those over which higher order direct effects are propagated, a means to calculate these effects is needed. Constructing a generating matrix P_o by zeroing the diagonal entries of P is an approach, but one that is not adequate because it has the effect of shunting energy normally held by the compartments (note the generally large diagonal elements in table 1b) out of the system where it cannot contribute at all to subsequent influence propagation. Another approach is needed.

P is a probability matrix whose elements p_{ij} represent flows f_{ij} which have been normalized to storages x_j. That is,

$$P = I + hA'',$$

where the entries in A" are $a''_{ij} = f_{ij}/x_j$ for j≠i, and $a''_{ij} = -\Sigma_{k=0}^{n} f_{kj}/x_j$ for j=i, x_j the energy content of compartment j; h is a scalar whose units (time) cancel those (time^{-1}) of A", and whose magnitude is such that I + hA" is nonnegative, with $0 \leq p_{ij} \leq 1$ (Matis and Patten 1981). Note that p_{ij} = prob[f_{ij}] is also interpretable as a joint probability, p_{ij} = prob[$f_{ij} \wedge \Sigma_{k=0}^{n} f_{kj}$], since prob [$\Sigma_{k=0}^{n} f_{kj}$] = 1. M.C. Barber (University of Georgia) has suggested, as an alternative approach, return to input-output flow analysis to construct a matrix P" of conditional probabilities p_{ij}'' whose elements represent flows f_{ij} normalized to outflows $\Sigma_{k=0}^{n} f_{kj}$ rather than to storages. With $p_{jj} \varepsilon P$ the probability of energy remaining in compartment j:

$$p_{ij}''=\text{prob}[f_{ij} \mid \sum_{k=0}^{n} f_{kj}] = \text{prob}[f_{ij} \wedge \sum_{k=0}^{n} f_{kj}]/\text{prob}[\sum_{k=0}^{n} f_{kj}]$$

$$=p_{ij}/1-p_{jj}$$

$$=ha_{ij}''/1-(1+ha_{jj}'')$$

$$=ha_{ij}''/-h\ a_{jj}''$$

$$=a_{ij}''/|a_{jj}''|\ ,$$

and to display the normalization with respect to outflows,

$$p_{ij}'' = \frac{f_{ij}}{x_j} / \sum_{k=0}^{n} \frac{f_{kj}}{x_j} = f_{ij}/\sum_{k=0}^{n} f_{kj} \quad .$$

Note that $p_{jj}'' = a_{jj}''/|a_{jj}''| = -1$. The relationship of the elements of P'' to those of P is then

$$p_{ij}'' = p_{ij}/h|a_{jj}''|,\ j \neq i$$
$$= (p_{jj}-1)/h|a_{jj}''|,\ j=i \quad ;$$

the entries in each column j of p_{ij} are normalized to the turnover rate $|a_{jj}''|$ of compartment j, and in addition they are scaled by the factor h.

The diagonal elements $p_{jj}'' = -1$ may now be zeroed without losing any energy from the system. The resulting matrix P_o'' represents direct, first order effects in terms of the new scaling. If the series $\Sigma_{k=0}^{\infty} (P_o'')^k = (I-P_o)^{-1}$ is generated, the convergent matrix will denote the total (direct plus indirect) influence propagated in the system over all paths of all lengths which contain no self loops. That is,

$$(I-P_o'')^{-1} - I = P_o'' + \sum_{k=2}^{\infty} (P_o'')^k$$

$$= [P_o'' + \sum_{k=2}^{\infty} (P_o'')_{\delta}^{k}] + [\sum_{k=2}^{\infty} (P_o'')_{\iota}^{k}].$$

In this case, however, since no paths with self loops are involved, the only paths which can contribute higher order direct effects are those of the form $\ldots j \to i \to j \to i \ldots$, which may exist for odd values of $k \geq 3$ in cases where compartments j and i are mutually linked by a direct connection (k even terms cannot contribute because each path of length k must begin with j and end with i). There can, however, be at most only one such path for each odd $k \geq 3$. Therefore, higher order direct effects will tend to be very small, $\Sigma_{k=2}^{\infty} (P_o'')_{\delta}^{k} \approx 0$, with most direct effects accounted for by P_o'', and the remainder of $(I-P_o'')^{-1}-I$, i.e., $\Sigma_{k=2}^{\infty} (P_o'')_{\iota}^{k}$, representing indirect effects. These are minimal indirect effects, however, for all paths containing one or more intermediary compartments other than j and i which possessed even one self loop would be eliminated from the accounting of total effects.

Table 7.--Oyster reef model matrices: (a) P_o'', representing first order direct effects; and (b) $(I-P_o'')^{-1}-(I+P_o'')$, representing a virtual lower bound on indirect effects, relative to the direct effects measured by P_o''.

Compartments

(a)

to \ from	1	2	3	4	5	6
1	0	0	0	0	0	0
2	3.808-1	0	0	5.000-1	7.600-1	4.758-1
3	0	3.670-1	0	0	0	0
4	0	3.267-1	1.476-1	0	0	0
5	0	2.888-2	1.476-1	7.793-2	0	0
6	1.238-2	0	0	0	6.857-2	0

(b)

to \ from	1	2	3	4	5	6
1	0	0	0	0	0	0
2	1.561-1*	3.885-1	2.775-1	2.800-1*	3.405-1*	1.848-1*
3	1.971-1	1.426-1*	1.019-1	2.863-1	4.039-1	2.425-1
4	2.045-1	2.021-1*	1.057-1*	2.971-1	4.192-1	2.516-1
5	6.053-1	1.276-1	4.279-2*	8.793-2	1.241-1	7.447-2
6	4.150-3*	1.073-2	1.305-2	1.137-2	8.507-3*	5.106-3

*Values smaller than those for corresponding direct effects.

Tables 7a and b show, respectively, the minimal direct and indirect effects matrices for the oyster reef model. Any further contribution to the direct effects (higher order terms) will be small, but that to indirect effects (by higher order paths with intermediary compartments and self loops) will tend to be large. Therefore, it seems reasonable to conclude that, when the definitive algorithm for sorting paths and influences properly into direct and indirect categories becomes available, the principal conclusion of this paper will be upheld. Indirect effects will still considerably exceed direct ones in small models, and for large models and real ecosystems they will probably be quantitatively dominant--perhaps more believeably than in terms of the incredibly large numbers that our first attempts at developing this interesting and unexpected principle of systems have yielded.

ACKNOWLEDGMENTS

I am grateful to my two coauthors on other cited versions of this work. M. Craig Barber wrote the software and executed it for the path and influence analyses. Richard F. Dame authored the oyster reef model. Support from the National Science Foundation, grants DEB-8110639 and DEB-8114823 is acknowledged. University of Georgia, Contributions in Systems Ecology, No. 58.

LITERATURE CITED

Clements, F.E. 1916. Plant succession: An analysis of the development of vegetation. Carnegie Inst. Wash. Publ. 290:1-388.

Clements, F.E. 1936. Nature and structure of the climax. J. Ecol. 24: 252-284.

Cowles, H.C. 1899. The ecological relations of the vegetation on the sand dunes of Lake Michigan. Bot. Gaz. 27:95-117, 167-202, 281-308, 361-391.

Cowles, H.C. 1901. The physiographic ecology of Chicago and vicinity. Bot. Gaz. 31:145-182.

Cowles, H.C. 1911. The causes of vegetative cycles. Bot. Gaz. 51: 161-183.

Dame, R.F., and B.C. Patten. 1981. Analysis of energy flows in an intertidal oyster reef. Mar. Ecol. Prog. Ser. 5:115-124.

Elliott, E.T., L.G. Castañares, D. Perlmutter, and K.G. Porter. 1982. Trophic-level control of production and nutrient dynamics in an experimental plankton community. Oikos, submitted.

Engelberg, J., and L.L. Boyarsky. 1979. The noncybernetic nature of ecosystems. Am. Nat. 114:317-324.

Forbes, S.A. 1883. The food relations of the Carabidae and Coccinellidae. Bull. Ill. Mus. Nat. Hist.

Forbes, S.A. 1887. The lake as a microcosm. Bull. Ill. Nat. Hist. Survey 15:537-550.

Holt, R.D. 1977. Predation, apparent competition, and the structure of prey communities. Theor. Pop. Biol. 12:197-229.

Hurlbert, S.H., and M.S. Mulla. 1981. Impacts of mosquitofish (Gambusia affinis) predation on plankton communities. Hydrobiologia 83:125-151.

Lane, P., and R. Levins. 1977. The dynamics of aquatic systems. 2. The effects of nutrient enrichment on model plankton communities. Limnol. Oceanogr. 22:454-471.

Levins, R. 1973. The qualitative analysis of partially specified systems. Ann. N.Y. Acad. Sci. 231:123-138.

Levins, R. 1975a. Problems of signed digraphs in ecological theory. Pages 264-277 in Ecosystem analysis and prediction, S.A. Levin, ed. Soc. Indust. App. Math., Philadelphia.

Levins, R. 1975b. Evolution in communities near equilibrium. Pages 16-50 in Ecology and evolution of communities, M. Cody and J. Diamond, eds. Belknap Press, Cambridge, Massachusetts.

Lindeman, R.L. 1942. The trophic-dynamic aspect of ecology. Ecol. 23: 399-418.

Lynch, M. 1978. Complex interactions between natural coexploiters--Daphnia and Ceriodaphnia. Ecol. 59:552-564.

MacArthur, R.H. 1955. Fluctuations of animal populations, and a measure of community stability. Ecol. 36:533-536.

Matis, J.H., and B.C. Patten. 1981. Environ analysis of linear compartmental systems: the static, time invariant case. Proc. 42nd Session, Int. Stat. Inst., Manila, Phillipines, Dec. 4-14, 1979, in press.

May, R.M. 1973. Stability and complexity in model ecosystems. Princeton Univ. Press, Princeton, New Jersey, 235 pp.

Möbius, K. 1877. Die Auster und die Austernwirtschaft. Wiegundt, Hempel und Parey, Berlin. Transl. 1880, The oyster and oyster culture. Rep. U.S. Fish. Comm. 1880:683-751.

Odum, E.P. 1959. Fundamentals of ecology, 2nd ed. Saunders, Philadelphia.

Paine, R.T. 1980. Food webs: linkage, interaction strength and community infrastructure. J. An. Ecol. 49:667-685.

Patten, B.C. 1978. Systems approach to the concept of environment. Ohio J. Sci. 78:206-222.

Patten, B.C. 1982. Environs: relativistic elementary particles for ecology. Am. Nat. 119:179-219.

Patten, B.C., and G.T. Auble. 1981. System theory of the ecological niche. Am. Nat. 118: 345-369.

Patten, B.C., M.C. Barber, and T.H. Richardson. 1982. Path analysis of a reservoir ecosystem model. Proc. Int. Symp. on Reservoir Ecology and Management, Quebec, Canada, June 2-3, 1981, in press.

Patten, B.C., R.W. Bosserman, J.T. Finn, and W.G. Cale. 1976. Propagation of cause in ecosystems. Pages 457-579 in Systems analysis and simulation in ecology, vol. 4, B.C. Patten, ed. Academic Press, New York, 593 pp.

Patten, B.C., R.F. Dame, and M.C. Barber. 1983. Energy environs of an intertidal oyster reef community. Ecol. Mon., submitted.

Patten, B.C., D.A. Egloff, T.H. Richardson, and 38 coauthors. 1975. Total ecosystem model for a cove in Lake Texoma. Pages 205-421 in Systems analysis and simulation in ecology, vol. 3, B.C. Patten, ed. Academic Press, New York, 601 pp.

Patten, B.C. and E.P. Odum. 1981. The cybernetic nature of ecosystems. Am. Nat. 118:886-895.

Schoenberg, S.A., and R.E. Carlson. 1983. Direct and indirect effects of zooplankton introduction on phytoplankton in a hypereutrophic lake. Limnol. Oceanogr., submitted.

Shelford, V.E. 1911. Ecological succession. II. Pond fishes. Biol. Bull. 21:127-151.

Simberloff, D. 1980. A succession of paradigms in ecology: essentialism to materialism and probabilism. Synthese 43:3-39.

Tansley, A.G. 1935. The use and abuse of vegetational concepts and terms. Ecol. 16:284-307.

von Uexküll, J. 1926. Theoretical biology. Kegan, Paul, Trench, Tubner, London.

Zaret, T.M., and R.T. Paine. 1973. Species introduction in a tropical lake. Science 182: 449-455.

FIGURE CAPTIONS

Figure 1. Intertidal oyster reef model. Numbers within the rectangles indicate steady state standing crop energy storages (x_i, $i=1,\ldots,6$, in kcal m^{-2}), and those associated with arrows denote energy flows (input z_1, output y_j, and internal f_{ij}, $i,j=1,\ldots,6$, in kcal m^{-2} d^{-1}).

Figure 2. Generalized functions of path length. (a) Number of paths from one compartment to another. (b) Influence and cumulative influence propagated from one compartment to another. (c) Total influence propagated between all compartments in a network. (d) Influence per path propagated from one to another compartment.

Figure 3. Input and output environments of system component j with respect to i in a network. (a) Direct effect, f_{ij}, of j on i, embodying indirect influences $\phi_{ik}^{(j)}$ in j's input environment propagated over all paths of all lengths from $k=1,\ldots,n$ to j. (b) Direct, f_{ij}, and indirect, $\phi_{ij}^{(k)}$, effects from j to i through components $k=1,\ldots,n$ in the output environment of j leading to i; $f_{ik}^{(j)}$ is the portion of the direct effect f_{ik} from k to i due to j. Solid arrows denote direct effects; broken arrows indicate antecedent indirect or direct effects.

MODELING REALITY: FACT, FANCY, AND FICTION[1]

R. G. Woodmansee and W. K. Lauenroth[2]

Abstract.--Mathematical modeling of ecosystem processes and behavior has developed as a valuable methodology for use in comprehending nature. Appropriate use of models requires recognition of various levels within ecological hierarchies for application of acceptable representation of system structure and function. This paper addresses ecological hierarchies and components that can be used to represent and are acceptable within these hierarchies. The levels discussed are landscapes, catenas, and soil-plant associations.

INTRODUCTION

Mathematical modeling is an invaluable tool for evaluating our understanding of ecological systems. The development of ecological modeling has had much publicity, many criticisms, and many successes. During this development many of the growing pains common to all developing disciplines have been experienced, i.e., elaborate models of unreal systems, overly simplified models of complex systems that have no contact with reality, grievous misunderstandings by modelers of the systems they were modeling, and often considerable overselling of modeling techniques and their potential. We believe the growing process has now yielded a mature discipline that now constitutes an important component of the suite of tools used by ecologists to evaluate ecological systems. However, many steps in the modeling process must be adhered to if mistakes of the past are not to be repeated in the future. We will describe in this paper a set of precautions that are of importance to modelers to increase the probability that their models will be accepted by the ecological community.

Our specific objectives are:

1. To discuss some perceptions of reality as they relate to ecological systems.

2. To discuss some representations of reality or models of ecological systems and,

3. To argue that modelers should strive to represent reality whether or not they use fact, fancy or fiction to build their models.

---

[1]Paper presented at the Third International Conference on State-of-the-Art in Ecological Modeling, Colorado State University, May 24-28, 1982.

[2]R. G. Woodmansee is currently Program Director, Ecosystem Studies Program, National Science Foundation, Washington, D.C., and formerly Associate Professor, Range Science Department, Colorado State University; W. K. Lauenroth is Associate Professor, Range Science Department, and Research Ecologist, Natural Resource Ecology Laboratory, Fort Collins, Colorado 80523, U.S.A.

To achieve these objectives we must first define several terms. The first term is reality, which is defined as a real event, entity, or state of affairs. A fact is (1) something that has actual existence or an actual occurrence or (2) a piece of information presented as having objective reality. Fancy is (1) an image or representation of something formed in the mind or (2) the free play of creative imagination. Fiction is (1) an assumption of a possibility as a fact irrespective of the question of its truth or (2) something invented by the imagination.

REALITY

Natural systems are real. However, the description and understanding of those ecological systems is limited by human abilities to comprehend complex systems. For humans to understand extremely complex systems like ecosystems, we generally must order and simplify reality. One means of ordering and simplifying nature is to develop a hierarchical structure in which we can go up the hierarchy to more general but abstract types or down the hierarchical structure to more specific and easily comprehended units. The goal of many ecologists is to understand ecological systems sufficiently well so that they can be described, comprehended, and often manipulated either for aesthetics, productivity, or possibly maintenance and preservation. The use of mathematical models to pursue this goal has been a revolution for ecologists. However, the use of models has many inherent problems, not the least of which is making certain that the developed models represent real systems. An understanding of some basic principles of systems is required to understand matter and energy flow or population dynamics in real ecosystems. The primary understanding that is required here is the recognition that ecological systems exist in a hierarchy of structure. Woodmansee and Adamsen (1983) have discussed one form of an ecological hierarchy that deals primarily with a set of systems that can be viewed in the context of matter and energy flow (fig. 1). In this scheme the global ecosystem is composed of many regional ecosystems. Regional ecosystems are in turn composed of land systems, land systems are composed of landscapes, and landscapes are composed of catenas, catenas are composed of soil plant associations, soil plant associations are composed of organisms,

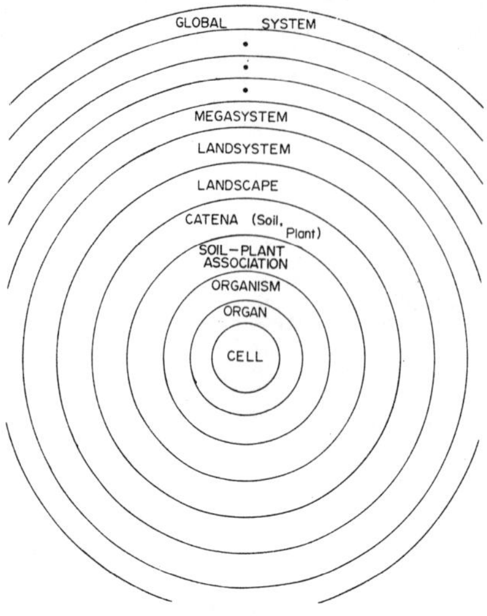

Figure 1.--Diagram indicating the integration of concepts from biological and pedological hierarchies.

associated soils, and the atmospheric column above the land surface, organisms are in turn composed of organs, which are made up of cells. In this paper we will discuss those parts of the hierarchy that pertain to organisms, soil-plant associations, catenas and landscapes. Progressing from the bottom to the top of this hierarchy is in the direction of increasingly abstract definitions of reality.

An organism is relatively easy to define in real terms. A cow, for example, is well bounded by its skin. It can be seen, touched and smelled. It can be weighed and it occupies a discrete volume. In a like manner, many plants are easily recognized. They also have a distinct structure which, at least aboveground, is easily visible. They also can be touched, felt, and weighed. They occupy a discrete volume and there is little difficulty defining the aerial portions of a plant. Belowground portions of the plant present somewhat different and more complex problems. Some roots are, in fact, part of the plant if they are alive. Other roots may not be considered part of the plant if they are dead, however, some of the dead roots may be attached to the plant and thus the definition of the real living organism becomes more complex.

The next level in the hierarchy, the soil-plant association, is a unit that can be considered to be a homogeneously distributed association of plants, i.e., a distinctly defineable community which exists in a relatively homogeneous soil, a particular soil type (see fig. 2). The term homogeneous must be viewed with some discretion in that there is certainly variability that exists within plant communities as well as in soils but we assume that the variability can be described statistically. Many models have been developed to represent whole systems and have dealt with the concept of the soil-plant association (Innis 1978, Botkin et al. 1972, and many others). Those models when formulated assume absolute homogeneity.

The catena, the next higher level of organization, is a physically related sequence of soil-plant associations (fig. 2). The relationship among soil-plant associations in a catena is expressed through the transport of matter from one soil-plant association to another, typically along a downhill gradient with the transport medium being water.

Landscapes are even more abstract in terms of their definition. The landscape level of organization is composed of various catenas. The landscape can be thought of as a small watershed, an agricultural field in which one or several different crops may be planted, a pasture in grazing lands that is bounded by fences or geographic barriers.

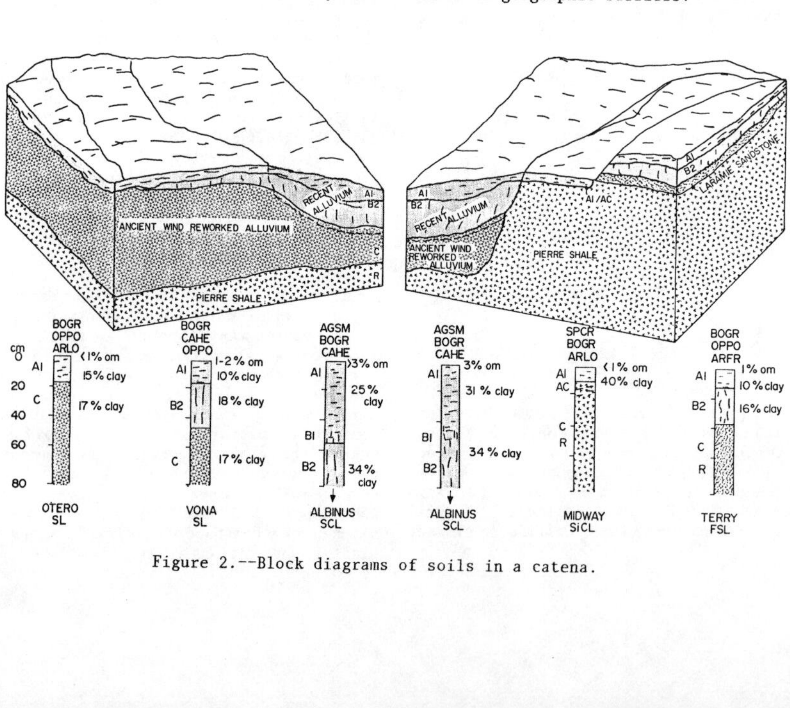

Figure 2.--Block diagrams of soils in a catena.

Even though these levels within the ecological hierarchy get more abstract and thus more complicated to define as one goes up the hierarchy, the levels exist. They are the units that we study in ecological studies, they are the units that we manage as a result of whatever management strategy we use. Thus, it is important to recognize that different levels exist and to attempt to represent those systems in as realistically a manner as possible. Further, it is important to recognize that many processes are unique to particular levels of organization. For example, photosynthesis is an organism process, not a soil/plant association process. Plant competition is a soil/plant association process, not a landscape process. Biomass production can be defined as a soil/plant association response, but crop yield is a landscape response.

REPRESENTATION OF REALITY

Factual Representation

Many factual representations of aspects of the ecological system exist in the literature. Parton (1978) described waterflow in grassland soils and plants in a relatively realistic manner. His formulations were based largely on the physical mechanisms of the waterflow through soil and evaporation/transpiration processes in the soils and plants. His model results compare very favorably with data and thus we might consider his model to be a relatively factual representation of waterflow and soil-plant associations.

A difficulty with discussing reality in regards to a waterflow model like Parton's is that even though the model may be a good representation of reality, the data which is used to develop the model and which is used to evaluate model results may be less factual. For example, many soil-water calculations are based on either neutron activation techniques or on coring techniques that extract columns of soil from the soil body and then the water content of those soils is determined. Estimates of availability of water for plants or evaporation are based on assumed relationships between soil water potential and soil water content. Development of such a data base requires many assumptions and has many opportunities for error in determining the actual soil water relationships in soils. One source of error is simply sampling variability. Many people with experience in determining soil water content by either neutron activation techniques or soil coring techniques know that location of the sampling device is a subjective placement. The coring device, for example may be on the order of 2-5 cm in diameter. The sampling design may allow for 4-6 cores per treatment. The area of the treatment may be several ha or larger in size and may include more than one soil/plant association. The soils may have different water-holding capacities (see fig. 2).

In the example above what represents reality? An average of six cores taken randomly? A stratified sample with six cores? Should samples be taken from the inner spaces between plants? A sample taken from within the plant canopy? Should the samples be a combination based on the percentage of their area versus area covered by plant canopy? These questions must generally be answered in the context of a small sample size that is in turn dependent on financial resources to conduct the actual sampling. The result of such restricted sampling usually yields data with a high variability even within the presumed homogeneously distributed soil-plant association. Another serious problem is in the development of soil water retention curves which are the supposed actual relationship between soil water content and soil water potential. The determination of soil water retention curves is a very laborious, very assumption rich procedure which very often yields information that is not reconcilable with field experience yet the use of soil water retention curves is imparative for conceptual understanding of water relations of soil-plant systems. Very often these two sources of error mask the routine sampling error and existing error that is a common part of our statistical analysis. Thus, which is correct, model results or field data?

Another class of representations of ecological systems that we will categorize as being factual are the many temperature-organism response relationships developed either in growth chambers, greenhouses, or in some cases field conditions. We consider many of these relationships to be "factual" because ecologists and physiologists have had sufficient experience to recognize that many of these relationships can be repeated over and over again in various environments and consequently we have a great deal of confidence in their representing the system accurately. An example of such a relationship is shown in Fig. 3 from Hilbert and Logan (1982).

Fanciful Representations

There are many examples in the literature of model representations based on creative imagination. We will use as examples those representations of ecological systems that are developed for soil-plant relationships, particularly nutrient cycling.

Some of our best understanding of biogeochemical cycles is based largely on integration techniques afforded by simulation models. These models, however, must represent many components and processes within the system that occur within the soil. Even though we have made tremendous advances in the past decade understanding soil systems and nutrient cycling within soil/plant associations, our actual knowledge based on data is limited. For example, we cannot separate live and dead roots with confidence. We can separate bacteria from fungi with less confidence. We cannot separate the effects of microbial grazing from the decomposition by bacteria and fungi. However, the information available has been integrated, and soil systems have been represented mathematically. The resulting simulation models have yielded intuitively reasonable results (McGill et al. 1981, Stewart et al.

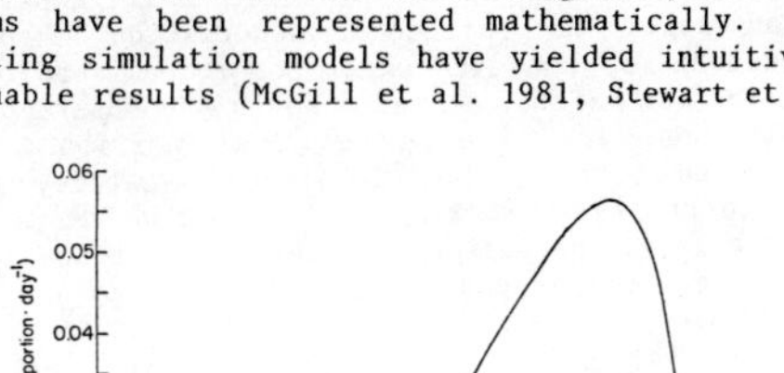

Figure 3.--Nymphal developmental rate as a function of temperature.

1983, Hunt et al. 1983). Many of the representations inherent in these models are purely the product of creative imagination and intuition. The conceptual basis of these relationships is solidly based on principles of microbial ecology and soil chemistry. However, the direct measures are wanting. Thus we have a category of model representations that are fancy, often not verifiable, but immensely useful.

Another category of fanciful representations are those depicted by Parton, Singh and Coleman (1978) in which roots of grassland soil-plant associations were modeled. The model representations of components and processes are based largely on what the authors felt must exist in order for the system to operate. They assumed various age and maturity categories of roots, and they assumed that several processes such as root growth, suberization, and death must follow an organized pattern. Data were not available to verify the results of the model, however, the model produced intuitively correct results. We consider this another example of a fanciful representation that has proven to be extremely useful.

Fictitious Representations

The category of fictitious representations include such entities as highly complex models of systems that simply don't exist (are not real) or very simple models of systems that are inherently complex and cannot be adequately represented simplistically. An example of the complex model of an unreal system would be ELM, a simulation model of grassland ecosystems (Innis 1978). This model has been extremely valuable, particularly to those individuals who were involved its development and analysis. It represented interactions of important components and processes within ecosystems. It is a model that represents the average square meter of a soil-plant association. However, the data that were used to develop the model were taken from various sources; averaged for large units that may be of several soil-plant associations and then combined to form a product as an interacting whole. An example of a fictitious representation within ELM is the relationship between the plant community which was a collection of functional groups of plants that included warm season grasses, cool season grasses, forbs, shrubs and cactus and a supposed important grazer, cattle. The plant community represented was the mean of the plants in a large pasture, a landscape. The particular association of plants represented may not even exist under real conditions. If it in fact exists, it would be in the upper portion of the landscape, i.e., the upper portion of the catenas on rather sandy locations. Cattle, under light to moderate stocking rates (as modeled in ELM) do not graze those positions within the landscape during the growing season except occasionally or under rare conditions (Senft et al. 1982). Thus the interaction of the grazing herbivores and the plant communities as modeled do not exist. The model is, according to our argument, declared a fictitious representation. However, we emphasize that the model has proven to be extremely useful in focusing attention on the interactions of the various components of soil-plant associations.

Other examples of fictitious representations of ecological phenomena are the abundant examples of regression equations that are used to predict responses of systems. In many cases, these regression relationships are simple linear representations. These are simply best fit relationships. They often can be very useful in showing correlations of various factors within systems but they cannot be used to explain biological, chemical and physical phenomena. They are often useful but not in an explanatory sense.

In summary, we contend that nature can be described in various hierarchical designs, one of which is discussed herein. Each level within the hierarchy is a system made up of appropriate components, processes and interactions. The choice of a specific level depends upon the objectives of the study. As scientists we must find ways to describe systems of interest if we are to represent those systems with mathematical models. Additionally, we must choose acceptable conceptual formulations that are consistent with the objectives of our inquiry. Various combinations of fact, fancy, or fiction can be successfully used to represent systems, but we must strive to recognize the difference between the three formulations and take care to inform our readers of which is being used.

ACKNOWLEDGMENTS

This research supported in part by Colorado State Agricultural Experiment Station Grant No. 0079361 and Long-Term Ecological Research (LTER), National Science Foundation, Grant No. DEB-8114822.

LITERATURE CITED

Botkin, D.B., J.F. Janak, and J.R. Wallis. 1972. Some ecological consequences of a computer model of forest growth. Journal of Ecology 60:849-872.

Hilbert, D.W., and J.A. Logan. 1982. A simulation model of the migratory grasshopper Melanoplus sanguinipes. (This volume)

Hunt, H.W., J.W.B. Stewart, and C.V. Cole. 1983. A conceptual model for the basis of interactions among carbon, nitrogen, sulfur and phosphorus transformations. In Proceedings of SCOPE Symposium, Stockholm, Sweden (in press).

Innis, G.S. (editor). 1978. Grassland simulation model. 298 p. Springer-Verlag, New York.

McGill, W.B., H.W. Hunt, R.G. Woodmansee, and J.O. Reuss. 1981. PHOENIX, a model of the dynamics of carbon and nitrogen in grassland soils. In Terrestrial nitrogen cycles, processes, ecosystem and management impacts (F.E. Clark and T. Rosswall, editors). Ecological Bulletin (Stockholm) 33:49-116.

Parton, W.J., J.S. Singh, and D.C. Coleman. 1978. A model of production and turnover of roots in shortgrass prairie. Journal of Applied Ecology 47:515-542.

Senft, R., L.R. Rittenhouse, and R.G. Woodmansee. 1980. Predicting patterns of cattle behavior on shortgrass prairie. Proceedings of the Western Section, American Society of Animal Science 31:276-279.

Stewart, J.W.B., C.V. Cole, and D.G. Maynard. 1983. Interactions of biogeochemical cycles in grassland ecosystems. In Proceedings of SCOPE Symposium, Stockholm, Sweden (in press).

Woodmansee, R.G., and F.J. Adamsen. Biogeochemical cycles and ecological hierarchies. In Nutrient cycling in agricultural ecosystems (R.L. Todd, editor). Ann Arbor Sciences Publishers, Inc., Ann Arbor, Michigan (in press).

SECTION II:

Theory and Methods of Ecological Modelling

LIVING SYSTEMS THEORY: CONCEPTUAL BASIS FOR ECOSYSTEM MODELLING[1]

Peter T. Haug[2]

Abstract -- This paper summarizes a recent general theory of living systems proposed by J. G. Miller, reviews some relevant modern ecosystem theory, and proposes an interim conceptual ecosystem paradigm, based on synthesis of the two. The paradigm offers a broad conceptual basis that is currently absent from ecosystem theory. Such orientation should be useful for modelling ecosystems within a logical, consistent theoretical framework that applies not only to the "natural" ecosystem, but also which incorporates man as both component and manipulator of living systems in general.

INTRODUCTION

Ecosystem simulation models tend to be *ad hoc* with little or no broad conceptual basis in ecosystem theory. Instead, each model tends to have a relatively narrow purpose, usually to simulate matter-energy flows among a subset of selected ecosystem components. A model of a particular ecosystem constructed for a very specific purpose lacks generality, transferability to other ecosystems, and often broad applicability even within the ecosystem for which it was designed. Although such models recognize implicitly the existence of cybernetic mechanisms in ecosystems, none, to my knowledge, applies the general constructs of living systems theory as postulated by James Grier Miller (1978). Matter-energy and information are explicitly separated in Miller's theory (table 1), and information is recognized as the controlling mechanism in the system.

Ecological theory is in a state of flux. There is no widely accepted ecosystem paradigm (Woodwell 1976, Johnson 1977, Regier and Rapport 1978, Woodwell 1978, Robertson 1979, Haug 1981, Suter 1981), and ecologists have not yet agreed on unifying concepts (van Dobben and Lowe-McConnell 1975). Even the concept of ecosystem as cybernetic system (Patten 1959), though defensible, is still controversial (Engelberg and Boyarsky 1979, McNaughton and Coughenour 1981, Knight and Swaney 1981, Patten and Odum 1981). Attempts to develop an ecosystem model based on general systems theory have been few (e.g., Overton 1972, White and Overton 1977), and large-scale simulation models ofecosystems and society have been severely criticized (Lee 1973, Holcomb Research Institute 1976).

Despite these problems, we all operate on the basis of models. Forrester (1968:3-1 and 3-2) observes: "Any set of rules and relationships that describe something is a model of that thing. In this sense, all of our thinking depends on models.. . . The human mind is well adapted to building models that relate objects in space. Also, the mind is excellent at manipulating models that associate words and ideas. But the unaided human mind, when confronted with modern social and technological systems, is not adequate for constructing and interpreting dynamic models that represent changes through time in complex systems."

Models can range from vague concepts, which obscure a multitude of conscious and unconscious assumptions and biases, across a broad spectrum, terminating in computer-based simulation models that explicitly document each assumption impeccably with bare-faced, unabashed candor. (Don't we wish!)

Where on this spectrum can we operate effectively? To answer this question, ecosystem analysts need to work toward developing a smooth continuum from theory to applications. Ecosystem modellers have been operating near the latter end, using empirical, *ad hoc* models. They have been restricted to this end of the spectrum primarily for two reasons: (1) There is no single, widely accepted operational theory of ecosystems, no guiding paradigm of ecology within which to embed a theoretically viable approach to ecosytem modelling; and (2) There probably are not enough reliable data to support a model developed from such a

[1]Paper presented at the Third International Conference on State-of-the-Art in Ecological Modelling, Colorado State University, Fort Collins, May 24-28, 1982.

[2]Peter T. Haug is Systems Ecologist, Bureau of Land Management, Washington, D. C., U. S. A.

Table 1. System Levels and Critical Subsystems. These seven levels of living systems, each with 19 subsystems critical to the survival and reproduction of that level, are identified in J. G. Miller's living systems theory (Miller 1978).

LEVELS OF LIVING SYSTEMS

Cell
Organ
Organism
Group
Organization
Society
Supranational System

CRITICAL SUBSYSTEMS

Subsystems That Process Both Matter-Energy and Information

Reproducer
Boundary

Subsystems That Process Matter-Energy	Subsystems That Process Information
Ingestor	Input Transducer
	Internal Transducer
Distributor	Channel and Net
Converter	Decoder
Producer	Associator
Matter-energy Storage	Memory
	Decider
	Encoder
Extruder } Motor }	Output Transducer
Supporter	

paradigm, should one exist. There are other, more practical considerations, not the least of which is funding, but the underlying reasons are as stated.

Credibility and acceptance of models waxes and wanes according to the vicissitudes of modellers themselves (McLeod 1981), and of politics and other environmental influences. Efforts to construct a unifying theory for ecosystem modelling could help bridge this credibility gap. Perhaps it is a quest for the Holy Grail, but if order really does exist in the universe, such a quest is inevitable sooner or later. Why not begin now?

In an attempt to suggest a conceptual framework for such a quest, this paper reviews explicit assumptions and definitions from living systems theory as they relate to ecosystems, explores the cybernetic nature of ecosystems within this context, and suggests an operational ecosystem paradigm for simulation modelling, derived from living systems theory as proposed by Miller (1978).

An ecosystem is assumed here to be a living system. As such, an ecosystem should be analyzable and simulatable within the context of living systems theory as postulated by Miller (1978). This assumption has been both challenged (Boulding 1979, Kuhn 1979) and defended (Miller 1979). The arguments are reviewed below. However, even if the ecosystem is shown not to conform strictly to Miller's definition of a living system, most of the concepts and ideas that follow are still worth considering when building simulation models of ecosystems.

The paradigm proposed here is simplified from Miller's living systems theory, yet is capable of being expanded back to the more complex structure of the original, when our state of knowledge about ecosystems advances sufficiently. The simplified theory provides an unconventional way of conceiving an ecosystem in order to model it. Because it is not system-specific, this simplification, and eventually the expanded theory, has potentially broad application for analyzing virtually any type of ecosystem based on general principles of living systems. All references to living systems theory are to Miller (1978), unless otherwise specified.

Overview

Miller's theory identifies precisely seven levels of living systems, each level comprising 19 process-oriented subsystems that are critical to the survival of each level of system (table 1). Subsystems process matter-energy and/or information. The interim paradigm proposed here views ecosystems in terms of living systems theory, aggregating Miller's 19 critical subsystems into eight for operational practicality (table 2).

Definitions

Because misunderstandings and other problems in interdisciplinary work often arise from semantic differences, a glossary of key definitions is provided (Appendix A).

LIVING SYSTEMS THEORY: AN ECOSYSTEM PARADIGM

An ecosystem can be viewed as a set of interacting processes carried out by components, some of which are living. Components process matter-energy and/or information in space and time. Dynamics of the system can be described and sometimes measured in terms of changes in components and processes. The paradigm proposed here was developed by integrating concepts from ecological thinking within the organizational framework provided by living systems theory. Validity of this paradigm has yet to be determined.

This paradigm recognizes in concept that a dichotomy between matter-energy and information exists. Although it sometimes may be possible to separate the processing of matter-energy from the processing of information (e.g., in behavioral studies), I know of no instance where this has been done in ecosystem simulation. Simulation models to date have focused only on how matter-energy is processed, with little or no explicit attempt to understand how information that controls the matter-energy flows is processed. This is understandable in view of the great complexity of living systems.

For example, the relative complexity, or degree of organization, of organisms defines their relative quality: Scientists have long chosen to equate increasing complexity in life forms with "higher quality" life forms, man representing the epitome as the most complex known. A complex organism contains a higher information content than a less complex one.

(Information is bound in all system structures, not just perceptual organs. For example, an enzyme is a complex structure designed to respond to one message.) Quantitatively this means the former embodies more bits of information than the latter. Odum and Odum (1981) call this "high embodied energy," in contrast to the lower embodied energy of the less complex organism, because more energy was required to organize the matter into the greater degree of complexity. One could conclude it is therefore the value man places on the quantity and arrangement of stored matter-energy that determines the quality of any information, rather than any intrinsic value it might contain.

An example of the way man does this is found in the diverse taxonomic systems we use to classify biological entities, from subcellular entities to species to biomes (Levin 1979, Mayr 1981). Such classifications do not exist in nature; rather they are figments of man's fertile imagination. That they work so well is testimony to man's insight, logic, and creativity in processing the information borne on the matter-energy he observes.

We classify things according to how we perceive them, in order to understand them better (Rigler 1975:16-17): "Any natural system is too complex for us to predict the behavior of every component. In fact, we could neither describe, nor recognize every component. Ecosystems are no exception. The potential array of component parts is too vast to be encompassed by our thoughts or theories. To make the system comprehensible, we simplify it by grouping its components into classes. However, the complexity of all systems, and the variety of the components are such that there are many equally valid methods of classifying the components. No one of these can be said to be correct. Yet the classification we adopt is all important. It usually reflects the way in which we were conditioned to see the world. In turn, it will influence the world view of those who use it. Past theory is subtly interwoven in the fabric of the system, as is the probability that it will inspire a new one. . . Classification is an essential, arbitrary process whereby we simplify nature, fossilize our current world view, and affect the probability of changing our theories. Because it embodies our world view, an accepted classification scheme is not easily discarded."

This paper offers a different perspective on ecosystems, and, in particular, a new way of classifying ecosystem components in order to model the system efficiently. As Wiegert (1975:314) notes: "Simulation models of ecosystems need not be all inclusive but they must be all encompassing, i.e. they must cover all kinds of interactions present in the system without including all the interactions." Living systems theory offers a perspective, a viewpoint, of systems that may permit us to simplify ecosystem modelling without loss of those critical interactions.

Introduction to the Paradigm

The ecosystem is conspicuously absent from Miller's seven-level hierarchy of living systems (table 1). He appears to consider it above the supranational system (1978: 25): "The universe contains a hierarchy of systems, each more advanced or 'higher' level made up of systems of lower levels. . . Cells are the lowest level of living systems. Cells are composed of atoms, molecules, and multimolecular organelles; . . . supranational systems, of societies and organizations. Higher levels of systems may be of mixed composition, living and nonliving. They include ecological systems, planets, solar systems, galaxies, and so forth."

Table 2. Aggregation of Miller's (1978) 19 Critical Subsystems into Eight Operational Equivalents. Verbal nouns are used to emphasize the functional, process-oriented role of the aggregated subsystems.

Aggregated Subsystem	Miller's Subsystems and What They Process[1/]	
Controller	Associator	I
	Decider	I
	Converter	M-E
	Producer	M-E
Transformer	Internal transducer	I
	Decoder	I
	Encoder	I
Integrator	Boundary	M-E/I
	Supporter	M-E
Retainer	Matter-energy storage	M-E
	Memory	I
Reproducer	Reproducer	M-E/I
Importer	Ingestor	M-E
	Input transducer	I
	Distributor	M-E
Transporter	Motor	M-E
	Channel and net	I
Exporter	Extruder	M-E
	Output transducer	I

[1/] I = Information; M-E = Matter-Energy

Miller specifically excludes these latter systems from his theory. However, in response to questions about the ecosystem's being a living system (Boulding 1979, Kuhn 1979), he seems to state that an ecosystem is a living system under his definition (Miller 1979:713): "All 19 subsystems are present in the exquisitely controlled 'balance of nature' seen in ecosystems."

Miller thus views an ecosystem as a macrocosm at the level of the biosphere (ecosphere) or biome, rather than the smaller scale of ecosystem that most ecologists study. Nonetheless, (1) if his theory is applicable at a macro scale (which he suggests), and (2) if ecologists recognize a hierarchy of ecosystems (which they generally do), then the ecosystem should be analyzable and simulatable as a living system within Miller's theory.

Living systems theory can be summarized as follows (Haug 1982): The theory postulates eight critical subsystems that process matter-energy, nine that process information, and two that process both matter-energy and information. It states that all information is borne on markers of matter-energy.

Table 3. A Generic Classification of Subsystems Based on Distribution of Processes Among System Components (Miller 1978:31-32).

Subsystem Taxon	Distribution
Local subsystem	Boundary of subsystem congruent with a component boundary; Process limited to one component.
Combined subsystem	Subsystem located in smaller region than component; Shares component with one or more other subsystems to perform a process.
Laterally dispersed subsystem	Located in more than one component.
Joint subsystem	Simultaneously a part of more than one local system.
Upwardly dispersed	Boundary not congruent with a subsystem component boundary; Process carried out by a system at a higher level in the hierarchy of systems.
Downwardly dispersed	Boundary not congruent with a subsystem component boundary; Process carried out by a subsystem at a lower level.
Outwardly dispersed	Boundary not congruent with a component boundary; Process carried out by another system, which sometimes is not a living system.

Markers can be altered in two ways: (a) The matter-energy can be transformed into forms that are more easily transmitted. (This is done by the _input_, _internal_, and _output transducers_.); or (b) The matter-energy code for the information being carried can itself be translated into another code more easily understood within the system. (This is done by the _encoder_ and _decoder_ in a living system). Information-processing subsystems that do not alter the markers are the _associator_, _decider_, _channel and net_, and _memory_. Subsystems that process matter-energy only are the _ingestor_, _distributor_, _converter_, _producer_, _matter-energy storage_, _extruder_, _motor_, and _supporter_. Subsystems that process both matter-energy and information are the _boundary_ and _reproducer_.

There are practical limitations in trying to identify all 19 of Miller's critical subsystems when analyzing and modelling ecosystems. However, many parallels exist between the way some subsystems process matter-energy and the way their analogs process information. An ecosystem processes matter-energy and information by means of its interacting critical subsystems, as suggested in table 2.

Critical Subsystems as "Adhocracies"

The term _adhocracy_ was coined by Toffler (1970:119) in reference to mission-oriented organizational systems in government and industry: "Teams are assembled to solve specific short-term problems. Then . . . they are disassembled and their human components reassigned. Sometimes these teams are thrown together to serve only for a few days. Sometimes they are intended to last a few years. . . The project or task force is temporary by design. . ."

Miller's critical subsystems are adhocracies of sorts: Each is task-oriented in the way it processes matter-energy and/or information, to the ultimate goal of maintaining stability and permanence of the system. Each subsystem may also be "temporary by design," in that it can disassemble and re-assemble as necessary to perform its _ad hoc_ function(s) periodically.

In Miller's words (1978:30), "In every system it is possible to identify one sort of unit, each of which carries out a distinct and separate process, and another sort of unit, each of which is a discrete, separate structure. The totality of all structures in a system which carry out a particular process is a _subsystem_. A subsystem, thus, is identified by the process it carries out. It exists in one or more identifiable structural units of the system. These specific, local, distinguishable structural units are called _components_ or _members_ or _parts_.. . . There is no one-to-one relationship between process and structure. One or more processes may be carried out by two or more components."

Subsystems thus can be classified in terms of how processes are distributed among components. This distribution differs in different systems. Processes can be centralized or dispersed widely (table 3). A process may be "(a) localized in a single component, (b) combined with others in a single component, (c) laterally dispersed to other components in the system, (d) upwardly dispersed to the suprasystem or above, (e) downwardly dispersed to subsystems or below, or (f) outwardly dispersed to other systems external to the hierarchy it is in. Which allocation pattern is employed is a fundamental aspect of any given system" (Miller 1978:31).

To complicate matters, the process that a component performs at any instant in time may change later as the component itself undergoes change (e.g., the varying behavior of insects and amphibians during different stages of their life cycles). Such changes in process have implications for the periodic reorganization of systems, as in the case of ecosystem succession. Although this topic is not discussed further here, it should be kept in mind as a practical problem in applying this paradigm.

To simplify the interim paradigm still more, processes carried out by each subsystem can be classified in one of three categories: inputs, internal processes, and outputs. Input and output processes provide linkages with other subsystems and, through the importer and exporter, with the environment (table 4). Internal processes filter, metabolize, transform, transport, support, store, retrieve, or otherwise operate on matter-energy and/or information within a particular subsystem.

The Controller in Ecosystems

At the very heart of the question whether an ecosystem is a living system by Miller's definition is the concept of control. Kuhn (1979:693) identifies "two closely related questions -- the meaning of decision and the nature of ecological systems." He states that ecosystems are uncontrolled systems: "Each organism . . . is a controlled system. The whole is . . . an uncontrolled suprasystem of controlled subsystems, which is my definition of an ecological system."

Boulding (1979:690) argues similarly for two basic models of the world: "One is the physiological model, best represented by the organism, the human body, and all the phenotypes of the biological world, which has its closest parallel in the social system in the organization like the firm, the church, or the planned economy. Besides this, however, there is also the model of the ecosystem, and the evolutionary model of continual interaction and change of species (selection) under continually changing parameters (mutation). This is a very different way of thinking about physiology.. . . The forest and the pond are not really even 'communities' as the biologists like to call them.. . . They are not organisms. They do not even resemble organisms very much. It would certainly be impossible to find Miller's 19 subsystems in an ecosystem."

The fundamental objection in these two arguments centers around the concept of control: A physiological system has a teleological controlling mechanism that maintains the system's equilibrium and increases the probability of its survival under changing environmental conditions. This is Kuhn's "controlled" system, Boulding's "physiological model." Systems above the level of the organism, including ecosystems, are "uncontrolled" in Kuhn's view. Patten and Odum (1981) concur, differentiating between "nonteleological systems, including ecosystems" and "homeostatic or goal-seeking organismal systems." In Boulding's view, ecosystems are "evolutionary," and, one presumes, subject to random controlling mechanisms, which is to say they are not centrally controlled.

Miller, however, explains the control mechanisms in living systems in terms of decision processes performed by the decider subsystem. He differentiates between uncontrolled and decentralized systems (1979:713): "An uncontrolled system in which outcomes come about by strictly random processes is an impossibility in my view. It would not be a system. A decentralized system, in which decision-making is done in many parts of the system and coordinated to produce an outcome is another matter." Thus, "At each level systems differ markedly in the degree of centralization or decentralization of their decision processes" (Miller 1979:712).

Miller's view of teleology in decision processes is as follows: "When I use the term 'decider' I do not necessarily imply any assumption of voluntarism and free will." (1978:67). "This use of the terms 'decider' or 'deciding' is different from the usage of many lay persons and some social scientists. They assume a degree of voluntarism and often of subjectivity. 'Decider,' in my conceptual system, includes any process which cuts down the number of options between input and output, or the number of statistical degrees of freedom. This definition not only covers the choices among alternatives made consciously by human organisms but the motivated, goal-directed behavior that is determined outside the person's awareness. It is used as well for the processes of information-reduction that take place in the nuclei of the central nervous systems and at synapses. 'Deciding,' under this definition applies also to predator-prey processes in ecosystems by which the numbers of animals of various species are determined. Decision-making is determining outcome. The essence of the deciding process is the elimination of alternatives and selection of a pathway, or option" (Miller 1979:713).

Miller concludes by stating (1979:713-714): "I cannot agree that ecological systems are uncontrolled. On the contrary, all 19 subsystems are present in the exquisitely controlled 'balance of nature' seen in ecosystems. The matter-energy processing subsystems present no problem. Flows of energy from the sun, and of water and minerals from the earth, to plants start the food chain. The matter-energy ingested in this way is distributed, converted, reeassembled in numerous ways, and eventually returned to earth, where the endless chain begins again. Information flows also through this chain . . . a channel and net stretches through the system, and information is conveyed from plants to animals and among animal species over it, controlling their behavior. The decider of such systems is decentralized, not lacking."

An example of how this works in an ecosystem is provided by Gutierrez and Fey (1980:199): "Secondary succession is the action of an ecosystem that returns the system variables to a climax state. Any disturbance that removes the system from that climax state activates the restoration process. While humanlike volition cannot be attributed to a grassland, the restoration forces operate like a control system or servomechanism designed to approach and maintain the climax equilibrium. The form of the ecosystem control system is interesting because it differs from the form of control system usually designed by humans. Generally speaking, negative-feedback controls are used in human and engineering systems to maintain or achieve goals. Natural ecological control seems to take the form of positive loops tha drive populations and related variables toward environmentally determined limits and hold them there. The difference is important because negative-loop control can result in a slow, oscillatory, and/or energy inefficient response unless the control system is very carefully and competently designed. Limited positive-loop control, on the other hand, tends to be rapid, efficient, and nonoscillatory within broad ranges of parameter values and relationship formats."

The Cybernetic Nature of Ecosystems

Although questions have been raised concerning the cybernetic nature of ecosystems (Engelberg and Boyarsky 1979), it is generally accepted among ecologists that ecosystems are cybernetic (e.g., Patten 1959, McNaughton and Coughenour 1981, Knight and Swaney 1981, Patten and Odum 1981, Odum and Odum 1981). Despite their other differences, Miller and his three reviewers also agree that living systems are cybernetic systems (Miller 1978 and 1979, Boulding 1979, Kuhn 1979, Parsons 1979). Miller, however, views the general cybernetic control processes differently from Gutierrez and Fey (1980). He states (1978:1030):

"Variables are maintained in steady-state ranges by negative feedbacks. The governance or control of living systems, as matter-energy and information processes go on, with flows from input to output,

Table 4. Aggregated Subsystem Processes.

Inputs	Internal Processes	Outputs
	AGGREGATED SUBSYSTEM (Miller's critical subsystems)	
	INTEGRATOR (Boundary, Supporter)	
- Matter-energy and information from environment or suprasystem and from internal components or subsystems of the system.	- Bars indiscriminate flows of matter-energy and information into and out of system through selective screening (filtering); - Maintains steady-state differential between interior of system and its environment; - Works with importer and exporter to transfer matter-energy and information between the system and its environment; - Supports or holds internal components in positions where they can interact positively within the system.	- Matter-energy and information to the environment and suprasystem and to internal components or subsystems of the system (via transporter).
	IMPORTER (Ingestor, Input Transducer)	
- Matter-energy (including markers bearing information) from the environment.	- Transfers matter-energy and information from environment across integrator into system, under control of integrator; - Transforms (as necessary) markers to other matter-energy forms (new markers) suitable for transmission within system.	- Matter-energy (including markers) into the system (via transporter).
	EXPORTER (Extruder, Output Transducer)	
- Matter-energy (including markers bearing information) from within the system (from transporter subsystem).	- Transfers matter-energy and information from system across integrator into environment under control of integrator; - Transforms (as necessary) markers to other matter-energy forms (new markers) suitable for transmission over channels in the system's environment (the suprasystem).	- Matter-energy (including markers) out of the system (sometimes via transporter).
	TRANSFORMER (Converter, Producer, Internal Transducer, Decoder, Encoder)	
- Matter-energy from integrator or other subsystem; - Matter-energy markers bearing information about significant alternatives in the system's subsystems or components (i.e., those subsystems or components that transmitted the markers), and coded in "public" or "private" code.	- Transforms matter-energy from one form to forms more useful (often intermediate or emphemeral) to the system; - Synthesizes matter-enrgy into relatively enduring, stable associations; - Creates higher levels of matter-energy complexity (negentropy); - Changes matter-energy markers into other matter-energy forms that can be transmitted within the system; - Transforms public code into private code that can be used internally by the system (as in pattern recognition -- comparing signals from importer with stored or learned information); - Transforms private code into public code that can be interpreted by other systems in the environment.	- Matter-energy in forms more useful to system; - Matter-energy in the form of stable associations - Matter-energy markers suitably altered and coded for use within system or outside it.

Table 4. (Continued)

	TRANSPORTER (Distributor, Channel and Net, Motor)	
- Matter-energy from outside system (via integrator) or from within system (from exporters of other subsystems); - Matter-energy markers bearing information, including addresses and signatures.	- Transports matter-energy to relevant components and subsystems within system; - Transports information (on matter-energy markers) to relevant components and subsystems within system; - Transports the entire system, or parts of the system, in space and time.	- Matter-energy to the environment (via integrator), or to other relevant subsystems within the system (via their importers); - Matter-energy markers bearing information to relevant components and subsystems; - Matter-energy motion in time-space.
	RETAINER (Matter-energy Storage, Memory)	
- Matter-energy in various forms; - Matter-energy markers bearing information.	- Accepts into storage matter-energy in various forms (and in different degrees of organization), including information markers; - Retains and maintains matter-energy to prevent entropic degradation of matter-energy, of matter-energy markers bearing information, or of code in which information is coded; and to prevent loss or distortion of matter-energy or information; - Retrieves selectively, from storage, matter-energy in various forms and information borne on markers.	- Stored matter-energy in various forms; - Stored information on markers.
	REPRODUCER (Reproducer)	
- Matter-energy and information containing a template and instructions for replication.	- Creates a new generation of system that resembles the original system in essential ways.	- Matter-energy and information constituting a new system that replicates the parent system in essence.
	CONTROLLER (Associator, Decider)	
- Information markers from all components and subsystems.	- Synthesizes a set of "association bonds" or relationships among bits of information (called knowledge or wisdom), so that bit A elicits B in the future with a probability greater than zero (i.e., inputting A will elicit items B,...,N as outputs, each with its own probability); - Organizes information into meaningful patterns (i.e., establishes meaning, including thresholds); - Establishes purposes or goals for the system: comparison signals for the feedback adjustment process; - Reviews inputs and interprets patterned information about the state of the system in terms of deviation from the comparison signal, in terms of what adjustment processes are available to the system to decrease this deviation, and in terms of alternative resolutions to the adjustment processes; - Reviews patterned information tht represents alternative choices; - Selects or synthesizes "optimal" solution to the deviation problem by logically decreasing number of available options, based on information analyzed. (This information is always incomplete); - Implements decision by transmitting command signals to other subsystems to coordinate and carry out processes leading to "optimal" solution.	- Information markers to selected relevant subsystems to adjust their processes so as to effect overall control of the system in response to its condition, i.e., the state perceived to exist at some time previous to the decision.

maintains related variables within steady-state ranges. Negative feedbacks damp down either increases or decreases in a particular variable to maintain its stability. . . The energetics of the flow of markers is the same as that of any other matter-energy flow. That is one reason why matter-energy and information flows have similarities, although there are differences. . . For instance, there are no matter-energy processes that correspond to encoding and decoding of information. Information flows also convey meaning, which matter-energy flows do not."

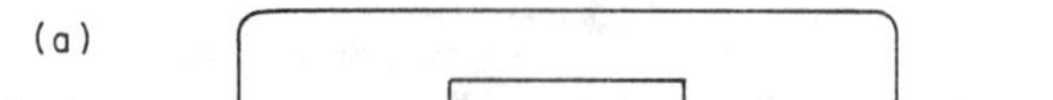

Figure 1. -- (a) Input-output system with feedback; (b) Feedback control system. (Modified from Patten and Odum 1981.) Model a is appropriate for nonteleological systems, including ecosystems, and model b, for man-made automatic control systems and homeostatic or goal-seeking organismal systems. In a, control mechanisms are diffuse, whereas in b, they are concentrated in specific structures. Both are cybernetic, according to these authors.

Patten and Odum (1981) describe these mechanisms specifically for ecosystems. Their discussion seems to provide explicit answers to Kuhn's and Boulding's remarks about the uncontrolled behavior of ecosystems. This view of Patten and Odum seems to support Miller's perspective of an ecosystem controlled by decentralized mechanisms that could be conceived as constituting a "decider" subsystem, yet these authors do not accept the teleological concept of goals in the ecosystem (fig. 1). They write (1981:894):

"The ecosystem is the level of organization concerned with the orderly, not chaotic, processing of energy-matter in the biosphere. We do not believe it is the goal of ecosystems to do this. They are not superorganisms. Rather, it is a constraint of existence that if living systems are not orderly, antientropic, then they will not persist. The balance of nature calls for a conjugate action-reaction kind of organization that creates order where there should be chaos as a matter of implicit design that simply evolved over geologic time. That design, we believe, has the basic form represented in [fig. 1] in which a diffuse, decentralized, and anatomically distinct information system regulates the primary energy-matter subsystems to achieve order. There is feedback in the movement of energy-matter in the biosphere, and there are informational processes operating which modulate this movement. The closed-loop structure appears to mandate ecosystem level coevolution, which offers potential for the introduction of systems perspectives into the evolutionary process. The systems side of evolution is a real issue to which ecosystems may contribute new dimensions. If ecosystems are not cybernetic, then by what other means could the perceived harmony of the biosphere have evolved?"

Although Patten and Odum assert that an ecosystem is nonteleological (fig. 1), the above quotation suggests that the "constraint of existence" which produces orderly living systems derives from some kind of design embedded in the genetic code (for homeostasis in organisms) and in other coevolved templates (for stereotyped interactions and optimization strategies among system components in living systems above the organism level). These designs program living systems to behave as they do, i.e., to seek the goals built into the designs.

Whether these designs derive from coevolutionary processes or from other causes is irrelevant from an operational standpoint: The fact is that evidence supporting the existence of such designs in living systems is very strong. The trend of systems to evolve toward orderly, increased complexity, improbable or not, is clearly established. The ultimate goal of any living system is to persist, the exception being the case of organisms that are programmed to yield their lives during the reproductive cycle in order that the species may persist.

If we accept the premise that such design exists, irrespective of its ultimate source, then we can view any living system as represented by model b in figure 1. This is the model I believe Miller postulates for all levels of living systems. The goal-seeking mechanisms are found in Miller's decider subsystem or the controller subsystem proposed here. Although perhaps widely dispersed, this subsystem establishes purposes or goals for the system, based on instructions coded on various templates during evolutionary development of system components, hence during coevolution of the system itself. Those purposes or goals are the comparison signals for the feedback adjustment processes regulating the homeostatic mechanisms that provide system stability.

Evidence that man recognizes such goals can be found in the literature (e.g., Maynard Smith 1978, Walters 1978, Somero 1978). The meaning that man attributes to various ecological phenomena is simply a reflection of his understanding of how the system evolved and how it currently functions. Haug (1982) observes: "This perspective views evolution as an optimization game in which the reproductive success of a species is maximized. The successful reproduction and maintenance of multiple-species interactions with each other, and with their environment, constitutes coevolution and thus evolution of the ecosystem, *per se*. Man's ignorance of the specific underlying mechanisms, however, severely limits his ability to evaluate the role of information in the coevolved ecosystem. His ability to place meaningful values on information is thus limited also. Nonetheless, he can and does speculate on these values when modelling the system.

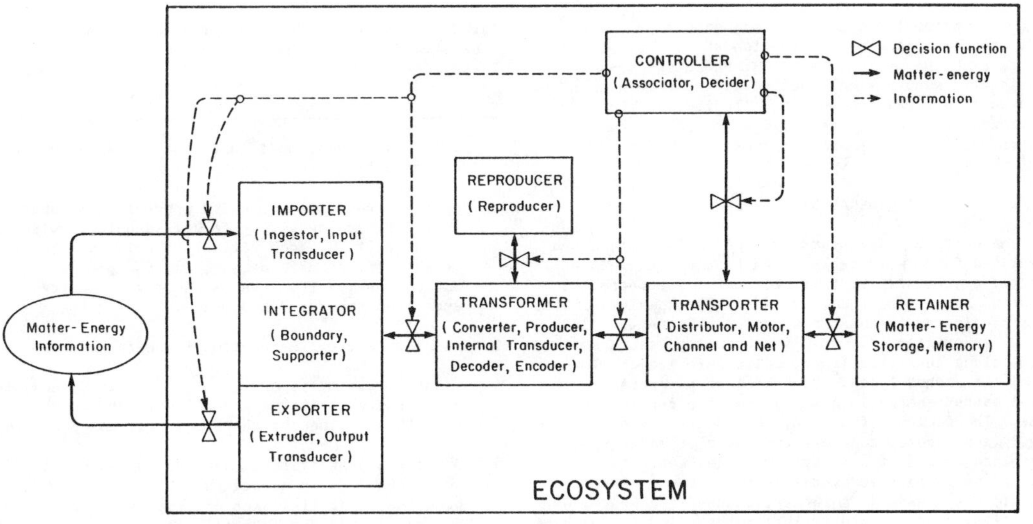

Figure 2. -- Conceptual model of an ecosystem, based on Miller's living systems theory. Critical subsystems aggregated from Miller's 19 subsystems are shown in capital letters. Miller's subsystems are shown in parentheses. Solid lines represent the transfer of matter-energy between subsystems. Flows are regulated by decision functions. Although actual flows of information are along matter-energy paths, conceptual flows of information occur along dashed lines. Information is thus shown to control decision functions.

"Evolution, as we understand it, guarantees that organisms and subsystems value information, at least implicitly: Optimization implies goals. From the existence of optimization goals one can infer negative feedback loops that control a system and help it organize its behavior toward attaining those goals. Behavior can be conscious (as in man), instinctual (as in other animals), physiological (as in cells and organs), or combinations of these. Homeostasis of any type mandates goal-oriented behavior."

THE ROLE OF INFORMATION: SYSTEM CONTROL

If information plays a critical role in ecosystem functioning, then the role of information needs to be examined closely. Haug (1982) explored this concept. Briefly, information can be viewed as an abstraction of reality borne on matter-energy markers. Although it is the matter-energy that living systems process, the information borne on the markers is what conceptually controls ecosytem processes. Information quantity can, in theory, be measured objectively in terms of bits. Information quality is not amenable to measurement or quantification, except by subjective criteria.

How matter-energy markers are processed depends on the subjective interpretation of receivers rather than on the transmitters. Information as a control mechanism for ecosystem processes has three aspects: (a) facts about the existence or occurrence of a phenomenon; (b) the perception of that phenomenon by one or more components within the ecosystem; and (c) the meaning of that phenomenon to those ecosystem components, as they perceive it.

How a subsystem perceives information it receives determines how that information is processed (Haug 1982): "If a phenomenon is perceived by a subsystem as furthering that subsystem's survival or reducing stresses, the subsystem will respond in one way. If the phenomenon is perceived as jeopardizing survival or increasing stress, the subsystem will respond quite differently. As the subsystem processes information about a phenomenon, its perception of that phenomenon is what really controls the subsystem's response. This is precisely the way we model system behavior, based on our own perceptions of a control mechanism." It is because we really don't know very much about how an ecosystem processes information that we use ad hoc models.

Pathology and Stability

Miller (1978) points out that the essential role of information in coordinating a system's activities is to maintain a system's "health." When coordinated action ceases, pathology sets in: "Any state of a system is pathological in which one or more variables remain for a significant period beyond their ranges of stability, or in which the costs of adjustment processes required to keep them within their ranges of stability are significantly increased" (1978:81).

In a living system, Miller says, "if one component cannot handle a stress, more and more others are recruited to help. Eventually the entire capacity of the system may be involved in coping with the situation" (1978:36). However, "If there are duplicate components capable of performing the same process, the system is less vulnerable to stress and therefore more likely to survive longer, because if one component is inactivated, the other can carry out the process" (1978:31).

This view of pathology has interesting implications when applied to such ecosystem concepts as stability, resilience, thresholds, ecological niche, ecological succession, and the relationship of diversity to these. Thus, the system is maintained through the interaction of both continual and periodic adjustment processes that are controlled by information about the condition or state of the system. As B. L. Bandurski observed (personal communication), health is truly more than the absence of diseased components. It is also the assurance of intact process pathways within the living system and between it and its environs.

A GENERAL SYSTEM DIAGRAM

The critical subsystems aggregated in table 2 and described further in table 4 can be arranged into a general system diagram (fig. 2) that illustrates the relationships among them. Haug (1982) explains: "This type of diagram is sometimes called a Forrester diagram after its originator.. . . Conceptually, matter-energy flows along the solid lines, while information flows along the dashed lines. Each rectangle represents the total matter-energy in a subsystem at a particular time. The decision function, represented by an hour-glass figure, controls the rates of matter-energy flow into or out of a subsystem. The level of matter-energy in a subsystem controls the decision functions by means of information about the state, or condition, of that level. This information is fed back through the controller subsystem.

"An important departure from Forrester's conventions . . . is the idea of a controller subsystem that coordinates matter-energy flows among other subsystems (fig. 2). Information flowing to the controller is not included in the conceptual diagram except via the matter-energy flows. This diagrammatic convention acknowledges the fact that information is borne on matt-energy markers, both into and out of the controller subsystem, but that decisions reached by the controller regulate and coordinate the matter-energy flows over the information network, at least in concept."

This diagram is also compatible with figure 1b. Comparison of figures 1a, 1b, and 2 will quickly show why the question of teleology in the ecosystem is important. If one accepts the premise that ecosystems are nonteleological, then control in a system is fragmented and, according to Miller (1978), there is no system. If a teleological perspective is embraced, then the problem is to discover the underlying goals and purposes toward which the system is attempting to optimize. Either way, the problem is not trivial, but I believe the latter perspective holds more promise for producing a unified theory of ecosystems.

A General Equation

Haug (1982) discusses a difference equation and a procedure (table 5) that might be used to develop general a simulation model of an ecosystem using this paradigm. It is basically a modification of Forrester's (1961, 1968) modelling approach applied within the framework of fig. 2.

SUMMARY AND CONCLUSIONS

The recent publication of a general theory of living systems by James Grier Miller presents a challenge to ecologists and ecosystem modellers to test the theory for its applicability to ecosystem analysis. This paper proposes one approach that might be used in developing a unified theory of ecology, one which includes man as both component and manipulator of the ecosystem.

Table 5. A General Procedure for Developing an Ecosystem Simulation Model Based on Living Systems Theory (Haug 1982).

1. Identify conventional indicator variables and other key variables of interest.
2. Wherever possible, identify conventional input, internal, and output processes associated with each variable (i.e., each variable's role in ecosystem metabolism, transformation, andtransport of matter-energy) from the perspective of current ecosystem theoryand knowledge.
3. Identify control mechanisms for each process.
4. Identify submodels of the eight critical subsystems by grouping variables according to the roles they play in carrying out the eight critical processes.
5. Develop a quantitative submodel for each critical subsystem. Build equations for the input/output processes that link each critical subsystem with others and for the internal processes.
6. Eliminate (or at least hold in abeyance) all unnecessary and unused variables from earlier steps.

ACKNOWLEDGEMENTS

I would like to thank four reviewers, each of whom contributed valuable critical ideas and suggestions, not only in reviewing this paper, but also in much informal interaction during the past several months: Cliff Hawkes, Linda Joyce, Craig Loehle, and John Mitchell. Jolie Haug, Stan Ponce, and Tom Hoekstra also supplied valuable critical comments on the paper and Norm French on the concepts. In particular, I would like to thank Bruce L. Bandurski for encouraging me during the past three years to undertake this project and for his valuable insights and comments on living systems theory. The Bureau of Land Management is also thanked for allowing me the latitude to attempt this project.

LITERATURE CITED

Boulding, K. E. 1979. Universal physiology. Contemporary Sociology 8:687-691.

Engelberg, J. and L. L. Boyarsky. 1979. The noncybernetic nature of ecosystems. The American Naturalist 114:317-324.

Forrester, J. W. 1961. Industrial dynamics. 464 p. The M.I.T. Press, Cambridge, MA.

Forrester, J. W. 1968. Principles of systems. Pages numbered variously. Wright-Allen Press, Cambridge, MA.

Gutierrez, L. T. and W. R. Fey. 1980. Ecosystem succession: a general hypothesis and a test model of a grassland. 231 p. The MIT Press, Cambridge, MA.

Haug, P. T. 1981. A paradigmatic modeling approach to environmental decisionmaking, pp. 195-209 in R. K. Ragade and J. A. Dillon Jr. (eds.), Paradigms in Changing Times, Proceedings of the Tenth Annual Conference of the Southeastern Region of the Society for General Systems Research.

Haug, P. T. 1982. Applying living systems theory to ecosystem simulation: the role of information. Proceedings of the Summer Computer Simulation Conference, Denver, CO, 19-21 July 1982. In press.

Holcomb Research Institute. 1976. Environmental modeling and decision making: the United States experience. 152 p. Praeger Publishers, New York.

Johnson, P. L. (ed.). 1977. An ecosystem paradigm for ecology. 19 p. Oak Ridge Associated Universities, Oak Ridge, TN.

Knight, R.L. and D. P. Swaney. 1981. In defense of ecosystems. The American Naturalist 117:991-992.

Kuhn, A. 1979. Differences vs. similarities in living systems. Contemporary Sociology 8:691-696.

Lee, D. B., Jr. 1973. Requiem for large-scale models. Journal of the American Institute of Planners 39(3):163-178.

Levin, D. A. 1979. The nature of plant species. Science 204: 381-384.

Maynard Smith, J. 1978. Optimization theory in evolution. Annual Review of Ecology and Systematics 9:31-56.

Mayr, E. 1981. Biological classification: toward a synthesis of opposing methodologies. Science 214:510-516.

McLeod, J. 1981. Project MAC (methodology, acceptance, credibility). Simulation 37(4):ix-xii.

McNaughton, S. J. and M. B. Coughenour. 1981. The cybernetic nature of ecosystems. The American Naturalist 117:985-990.

Miller, J. G. 1978. Living systems. 1102 p. McGraw-Hill, New York.

Miller, J. G. 1979. Response to reviewers of Living Systems. Contemporary Sociology 8:705-715.

Odum, H. T. and E. C. Odum. 1981. Energy basis for man and nature. 337 p. McGraw-Hill, New York.

Overton, W. S. 1972. Toward a general model structure for a forest ecosystem. pp. 37-47 in Proceedings -- Research on Coniferous Forest Ecosystems -- A Symposium, Bellingham, WA, 23-24 March 1972.

Parsons, T. 1979. Concrete Systems and 'Abstracted' Systems. Contemporary Sociology 8:696-705.

Patten, B. C. 1959. An introduction to the cybernetics of the ecosystem: the trophic-dynamic aspect. Ecology 40:221-231.

Patten, B. C. and E. P. Odum. 1981. The cybernetic nature of ecosystems. The American Naturalist 118:886-895.

Regier, H. J. and D. J. Rapport. 1978. Ecological paradigms, once again. Bulletin of the Ecological Society of America 59:2-6.

Rigler, F. H. 1975. The concept of energy flow and nutrient flow between trophic levels. pp. 15-26 in W. H. van Dobben and R. H. Lowe-McConnell (eds.) Uniufying Concepts in Ecology. Dr W. Junk B. V. Publishers, The Hague.

Robertson, J. 1979. On paradigms and ecology. Bulletin of the Ecological Society of America 60:182-186.

Somero, G. N. 1978. Temperature adaptation of enzymes: biological optimization through structure-function compromises. Annual Review of Ecology and Systematics 9:1029.

Suter, G. W., II. 1981. Ecosystem theory and NEPA assessment. Bulletin of the Ecological Society of America 62:186-192.

Toffler, A. 1970. Future shock. 505 p. Random House, New York.

van Dobben, W. H. and R. H. Lowe-McConnell (eds.). 1975. Unifying concepts in ecology. 302 p. Dr W. Junk B. V. Publishers, The Hague.

Walters, C. J. 1978. Ecological optimization and adaptive management. Annual Review of Ecology and Systematics 9:157-188.

White, C. and W. S. Overton. 1977. Users manual for the FLEX2 and FLEX3 model processors for the FLEX modelling paradigm. 119 p. + appendices. Forest Research Laboratory, Oregon State University, Corvallis.

Wiegert, R. G. 1975. Simulation models of ecosystems. Annual Review of Ecology and Systematics 6:311-338.

Woodwell, G. W. 1976. A confusion of paradigms (musings of a president-elect). Bulletin of the Ecological Society of America 57(4):8-10.

Woodwell, G. W. 1978. Paradigms lost. Bulletin of the Ecological Society of America 59:136-140.

APPENDIX A

Glossary

Adjustment process - any process of a subsystem that helps maintain a steady state in a system by keeping variables within their ranges of stability despite stresses (Miller 1978:35).

Indicator - a structure, process, or mathematical derivative of these that provides an objective, reliable measure of specific states of a system variable. Indicators "can be used to evaluate the current condition of a system, to measure the amount of departure of variables from established norms, or by extrapolation to forecast changes in system states in the immediate or more distant future.. . .
Indicators are preferably in centimeter-gram-second units, in monetary units, or in some other units that can be quantified. Indicators may, however, concern qualitative aspects of systems as well as quantitative ones. It is important that an indicator validly

measure the variable with which it is supposed to vary" (Miller 1978:1038).

Information - "the degrees of freedom that exist in a given situation to choose among signals, symbols, messages, or patterns to be transmitted" (Miller 1978:11). "All information flows are sequences of patterns that pass over a channel in space-time, from a transmitter to a receiver. Each pattern is conveyed on a marker, a bundle of matter-energy. The pattern cannot be transmitted unless it is borne on a marker" (Miller 1978:1027).

Markers - "observable bundles, units, or changes of matter-energy whose patterning bears or conveys the informational symbols ..." (Miller 1978:12).

Meaning (of information) - "the significance of information to a system which possesses it: it constitutes a change in that system's processes elicited by the information, often resulting from associations made to it on previous experience with it" (Miller 1978:11).

Pathology (of a system) - "state of a system . . . in which one or more variables remain for a significant period beyond their ranges of stability, or in which the costs of adjustment processes required to keep them within their ranges of stability are significantly increased" (Miller 1978:81).

Process - "all change over time of matter-energy or information in a system ... If the equation describing a process is the same no matter whether the temporal variable is positive or negative, it is a reversible process; otherwise it is irreversible, or better, less readily reversible" (Miller 1978:23).

Structure - "the arrangement of [a system's] subsystems and components in three-dimensional space at a given moment of time" (Miller 1978:22).

Subsystem - "the totality of all the structures in a system which carry out a particular process . . . . A subsystem, thus, is identified by the process it carries out. It exists in one or more identifiable structural units of the system" (Miller 1978:30).

Transducer - a device actuated by power from one system and supplying power, usually in another form, to a second system; a converter of energy or information into another form.

Variable - "any property of a unit or relationship within a system which can be recognized by an observer who chooses to attend to it, which can potentially change over time, and whose change can potentially be measured by specific operations" (Miller 1978:17).

DEFINING THE RESILIENCE THRESHOLDS OF ECOSYSTEMS[1]

Alan A. Berryman[2]

Abstract.--Resilience thresholds define the unstable boundaries separating distinct behavioral domains of ecological systems. Because they represent transient system states, however, they are difficult to define empirically. Models are utilized to define the resilience properties of a bark beetle-conifer system, to identify the unstable threshold, and to develop risk assessment methods.

INTRODUCTION

The concept of resilience was first introduced into the ecological literature by Holling (1973) to describe the qualitative stability boundaries of ecosystems with more than one stable state. Resilient systems, therefore, are those which return towards their original equilibrium states following large disturbances while non-resilient, or fragile, ecosystems tend toward different equilibrium states following relatively minor perturbations. Implicit in this concept is the assumption that equilibrium domains are separated by unstable boundaries or thresholds. When the system is displaced across one of these boundaries it comes under the influence of a different attractor and exhibits radically different or divergent dynamic behavior.

A number of examples of ecological systems which seem to exhibit divergent behavior can be found in the literature, including spruce budworms (Ludwig et al. 1978), bark beetles (Berryman 1978a,b), pacific salmon (Peterman 1977), Dungeness crabs (Wickham and Botsford, unpubl. ms.), pathogens (May 1977), and grazing systems (Noy-Meir 1965). All these systems seem to posess at least two behavioral domains; a domain of low-density stable equilibrium behavior and another of high-density stable or unstable equilibirum behavior.

From the viewpoint of the ecosystem manager, it is important to define the boundaries of these behavioral domains for it is the boundaries which specify the resilience of the system to management perturbations; e.g., harvesting, pest control, and so forth. The problem of defining these resilience thresholds, however, is no simple matter, for thresholds represent transient system states which are rarely observed in nature. Hence, it is usually impossible to determine resilience thresholds empirically, at least by non-destructive methods. Destructive testing, which is the normal procedure for defining resilience thresholds of materials such as wood, metals, etc., is usually an impractical or dangerous approach in ecological settings. Because of these problems, modeling becomes one of the main tools available for defining the resilience properties of ecological systems.

The purpose of the present paper is to show how analysis of a specific ecological system, lodgepole pine forests (*Pinus contorta* var. *latifolia*) infested by the mountain pine beetle (*Dendroctonus ponderosae*), enables us to describe the resilience properties of the system and to identify the threshold separating low-density (endemic) beetle behavior from high-density (epidemic) behavior. Before proceeding to this example, however, we need to develop a general mathematical framework for analysis.

THEORETICAL CONSIDERATIONS

The general notions of resilience have been discussed by Holling (1973), Long (1974), and, with particular reference to insect epidemics, by Berryman (1978a, 1982). None of these discussions, however, develops a theory of resilience within the overall concepts of population ecology. As we are specifically interested in the resilience of population systems we will attempt to do this here.

Starting with a general expression for population growth in a consistent stable environment, we can write

$$\frac{1}{N}\cdot\frac{dN}{dt} = f(N) \qquad (1)$$

where f is an unknown function expressing the specific growth rate of the population in respect to population density, N. Expanding by Taylor's theorem we obtain

$$\frac{1}{N}\cdot\frac{dN}{dt} = r_0 + aN + bN^2 + cN^3 \quad . \; . \; . \; . \qquad (2)$$

[1]Paper to be presented at the International Conference on State-of-the-Art in Ecological Modeling, Colorado State University, May 24-28, 1982. This research was conducted under Projects 0102 and 0052 of the College of Agriculture Research Center, Washington State University, and was supported by the National Science Foundation (GB-30752 and GB-34718) and the Competitive Research Grants Office of the U.S. Department of Agriculture (Grant No. 7800861).

[2]Alan A. Berryman is professor of Entomology and Forestry and Range Management, Washington State University, Pullman, Washington, U.S.A., 99164-6432.

where r_0 is the intercept of f at N(0) and a, b, c, etc. are population growth parameters. The simplest expression for constrained population growth is the truncated Taylor series

$$\frac{1}{N}\cdot\frac{dN}{dt} = r_0 + aN \qquad (3)$$

which gives us the well-known "logistic" equation (Lotka 1956) when the parameter a is negative. The signs of the population growth coefficients represent the dominance of two kinds of interactions between members of the population. When competition dominates the growth parameter is negative while when cooperation dominates the sign of the parameter is positive.

Berryman (1981) showed that the effect of cooperation at low population densities can be introduced into the logistic equation by adding a third term; i.e.,

$$\frac{1}{N}\cdot\frac{dN}{dt} = r_0 + aN + bN^2 \qquad (4)$$

Unlike Eq. 3, which has but a single equilibrium, Eq. 4 may give rise to two equilibirum points when r_0 and b are negative. The first low-density equilibrium point is unstable, while the second high-density equilibrium is stable. As a general result, unstable equilibria are always created when cooperative (+) interactions dominate, while stable equilibiria are created by the dominance of competitive (-) interactions (Berryman 1981). In the case of Eq. 4, the unstable point is a resilience threshold which separates extinction behavior from growth towards a stable equilibirum. This boundary, therefore, is sometimes called the extinction threshold.

There seems to be no inherent reason why additional terms of the Taylor series cannot be included in the population growth equation. For instance,

$$\frac{1}{N}\cdot\frac{dN}{dt} = r_0 + aN + bN^2 + cN^3 + dN^4 \qquad (5)$$

When r_0, b and d are negative in Eq. 5, we can have an extinction threshold, a low-density equilibirum, a second unstable threshold and a final stable equilibirum. Thus, we can have three behavioral domains: extinction behavior, low-density equilibirum behavior and high-density equilibrium behavior.

The important points which emerge from this theoretical discourse is that unstable equilibria, resilience thresholds or boundaries, only occur where cooperative effects dominate over competitive interactions. That is, where increases in population density result in increased reproduction and survival of the average individual. Hence, we might expect to see unstable thresholds in those population systems where aggregations, swarms, packs, schools, etc., have a beneficial effect on the reproduction and survival of the species. For a more exhaustive treatment of these general notions see Berryman (1981, 1982).

BEETLES AND PINE FORESTS

Bark beetle populations normally persist in the forest ecosystem at very low densities, infesting individual trees or small groups of trees which have been severely weakened by root diseases (Stark and Cobb 1969), lightning strike (Hodges and Pickard 1971), and other environmental stresses. They are prevented from infesting healthy individuals because these trees have effective defensive mechanisms which produce large quantities of resin and "pitch out" the attacking beetles (Berryman 1972). Occasionally, however, beetle populations seem to build up and spread though relatively healthy forests, usually as a result of environmental disturbances which stress the whole forest (e.g., drought) or which increase the resident beetle population (e.g., windthrown trees). This transition from low-density endemic equilibirum to epidemic behavior suggests that unstable thresholds may be involved in the dynamics of beetle-conifer interactions.

The hypothesis that structural stability of bark beetle populations is characterized by an unstable outbreak threshold is further supported by the fact that beetles cooperate during the colonization of their host. When a single "pioneer" beetle bores into a tree it produces chemical pheromones which attract others of the same species to the attacked tree. The subsequent "mass attack" exhausts the tree's defenses and the invading beetles can reproduce successfully in the dying host. It is interesting that the beetles utilize the defensive chemicals of the host (monoterpenes) as the base material for pheromone synthesis (hydroxylated monoterpenes). Thus, pheromone synthesis has been found to be directly related to the amount of resin produced in the host's defensive responses (Hughes 1973a,b). On the other hand, trees which produce very large quantities of resin seem to inhibit pheromone synthesis or emission (Raffa 1980). This inhibition, however, decreases as the density of "pioneer" attacks increases, probably because the resin flow is rationed amongst a greater number of beetles (ibid).

Host Colonization Models

Based on the above biological considerations, Raffa (1980) constructed a detailed simulation model of the colonization of coniferous hosts by bark beetle populations. The structural stability of this individual tree colonization model was analyzed by Stenseth, Berryman and Raffa (in Prep.). The model was shown to have an unstable equilibirum, separating two qualitative outcomes; (1) the tree wins and the beetle is "pitched out", (2) the beetle wins and the tree dies. This threshold was shown to be very sensitive to the number of flying beetles (i.e., beetle population size), and to the vigor or health of the host. The model predicted that the critical beetle attack density required to kill a tree increases in direct relationship to the vigor of the tree. This prediction is supported by independent data (fig. 1a).

Population Models

Raffa (1980) extended his simulation model to describe the interaction between beetle populations and stands of trees. Numerical analysis executed under the assumption that all the trees in the forest have identical vigor demonstrated that an unstable outbreak threshold separates endemic from epidemic beetle behavior (fig. 1b). This analysis also demonstrated that the outcome of beetle invasion of the stand is strongly influenced by the initial density of "pioneer" attacks on individual trees. As initial "pioneer" attacks occur at random (Hynum and Berryman 1980), then their density is also determined by the size of the flying beetle population. Hence, in a stand of uniform vigor, the initiation of an outbreak is completely determined by the size of the beetle population. The system, therefore, can be described by a generalized Taylor series; e.g.,

$$\frac{1}{N}\cdot\frac{dN}{dt} = r_0 + f_1(N,W) + f_2(N,V) + f_3(N,T) \quad (6)$$

where r_0 is the maximum specific reproductive rate of the beetle, f_1 is a monotronic decreasing function expressing competition for W weakened

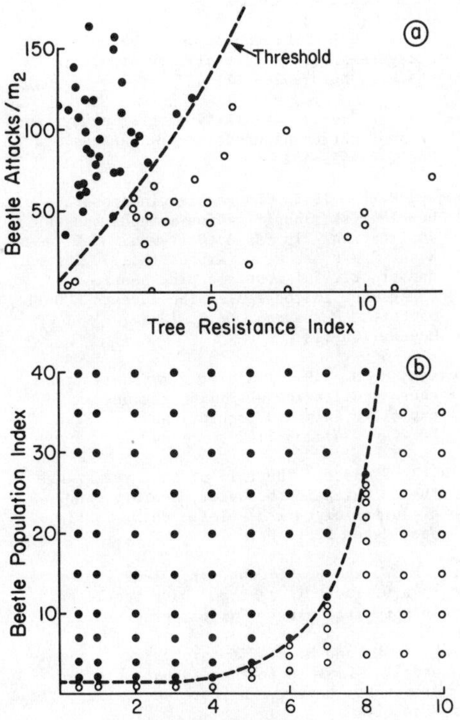

Fig. 1(a). The threshold mountain pine beetle attack density required to kill lodgepole pines of different vigor (resistance); • = dead trees, o = live trees (after Waring and Pitman 1980);
(b). Results of simulation experiments in hypothetical lodgpole pine stands of varying vigor (resistance) and flying mountain pine beetle populations; • = epidemic behavior, o = endemic behavior (after Raffa 1980).

hosts, f_2 is a monotonic increasing function describing the cooperative overwhelming of host resistance, V, and f_3 is a monotonic decreasing function representing competition for a total of T hosts.

Application

Detailed simulation and analytical models suggest that the resilience threshold of bark beetle-conifer systems can be described by an exponential function (fig. 1b); i.e.,

$$N^* = e^{\alpha V} \quad (7)$$

where N* is the critical beetle population size required to successfully invade a stand of vigor V; i.e., N* represents the unstable beetle equilibrium density in a stand of given vigor. If the size of the beetle population can be obtained by sampling, or estimated from environmental variables; e.g.,

$$N = g(X_1, X_2 \cdot \cdot \cdot \cdot) \quad (8)$$

and if V can be estimated as a function of stand variables or measured directly from a sample of trees, e.g.;

$$V = h(Y_1, Y_2 \cdot \cdot \cdot \cdot) \quad (9)$$

then the epidemic threshold is specified by

$$N^* = e^{\alpha h(Y_1, Y_2, \cdot \cdot \cdot \cdot)} \quad (10)$$

The threshold function can be approximated by estimating the independent variables N and V in a number of stands, some of which are suffering beetle epidemics and others with low density beetle populations (fig. 2; in this figure the threshold is the line segregating outbreak from low-density beetle populations.). Once the threshold function has been empirically defined, it can be utilized to identify the risk of epidemics ocurring in particular stands, the risk being inversely proportional to the shortest distance between the stand's current status and the outbreak threshold. This is because stands which are close to the threshold are most likely to be pushed into the outbreak zone by minor perturbations of V or N. In figure 2 the area to the right of the threshold has been divided into three risk zones. The width of the high risk zone should, ideally, reflect the risk aversion of the forest manager and the variability of the physical environment.

Finally, we should note that the concept of resilience is the inverse of risk. In this sense coniferous stands which are remote from the outbreak threshold are the most resilient because the beetle population will return to its low-density equilibirum following major environmental or management perturbations.

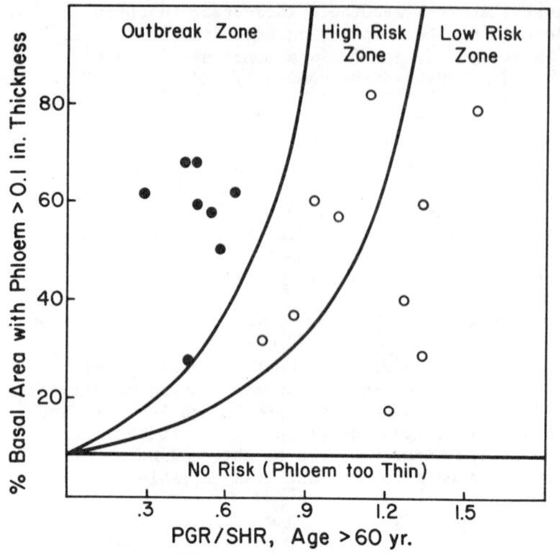

Fig. 2. Approximate threshold function and risk decision lines for mountain pine beetle interacting with lodgepole pine stands, where population density is assumed proportional to the pholem thickness distribution, and vigor is assumed proportional to periodic growth rate (PGR), stand density/species composition (SHR) and age. The threshold function was estimated from a series of endemic (o) and epidemic (•) stands (after Berryman 1978b and Mahoney 1978).

LITERATURE CITED

Berryman, A. A. 1972. Resistance of conifers to invasion by bark beetle-fungus associations. BioSci 22:598-602.

Berryman, A. A. 1978a. Towards a theory of insect epidemiology. Res. Popul. Ecol. 19:181-196.

Berryman, A. A. 1978b. A synoptic model of the lodgepole pine/mountain pine beetle interaction and its potential application in forest management. In A. A. Berryman, G. D. Amman, R. W. Stark and D. L. Kibbee (Eds.). Theory and practice of mountain pine beetle management in lodgepole pine forests. College of Forest Resources, Univ. Idaho, Moscow:98-105.

Berryman, A. A. 1981. Population systems: A general introduction. Plenum Press, N.Y.

Berryman, A. A. 1982. Biological control, thresholds, and pests outbreaks. Environ. Ent. (in press).

Hodges, J. D. and L. S. Pickard. 1971. Lightning in the ecology of the southern pine beetle, Dendroctonus frontalis (Coleoptera:Scolytidae). Can. Entomol. 103:44-51.

Holling, C. S. 1973. Resilience and stability of ecological systems. Ann. Rev. Ecol. Syst. 4:1-23.

Hughes, P. R. 1973a. Dendroctonus: Production of pheromones and related compounds in response to host monoterpenes. Z. angew. Entomol. 73:294-312.

Hughes, P. R. 1973b. Effect of α-pinene exposure on trans-verbenol synthesis in Dendroctonus ponderosae Hopk. Die. Naturwissenchaften 60:261-262.

Hynum, B. G., and A. A. Berryman. 1980. Dendroctonus ponderosae (Coleoptera: Scolytidae): Pre-aggregation landing and gallery initiation on lodgepole pine. Can. Entomol. 112:185-191.

Long, G. E. 1974. Model stability, resilience, and management of an aquatic community. Oecologia 17:65-85.

Lotka, A. J. 1956. Elements of Mathematical biology. Dover, N.Y.

Ludwig, D., D. D. Jones and C. S. Holling. 1978. Qualitative analysis of insects outbreak systems: The spruce budworm and forest. J. Anim. Ecol. 47:315-332.

May, R. M. 1977. Thresholds and breakpoints in ecosystems with a multiplicity of stable states. Nature 269:471-477.

Noy-Meir, I. 1975. Stability of grazing systems: An application of predator-prey graphs. J. Ecol. 63:459-481.

Mahoney, R. L. 1978. Lodgepole pine/mountain pine beetle risk classification methods and their application. In A. A. Berryman, G. D. Amman, R. W. Stark and D. L. Kibbee (Eds.), Theory and practice of mountain pine beetle management in lodgepole pine forests. College of Forest Resources, Univ. Idaho, Moscow:106-113.

Peterman, R. M. 1977. A simple mechanism that causes collapsing stability regions in exploited salmonid populations. J. Fish. Res. Bd. Can. 34:1130-1142.

Raffa, K. F. 1980. The role of host resistance in the colonization behavior, ecology, and evolution of bark beetles. Ph.D. Thesis, Washington State Univ.

Stark, R. W. and F. W. Cobb, Jr. 1969. Smog injury, root diseases and bark beetle damage in ponderosa pine. Calif. Agric. 23:13-15.

Waring, R. H. and G. B. Pitman. 1980. A simple model of host resistance to bark beetles. Oregon St. Univ., Sch. For., Corvallis, Res. Note 65.

POPULATION STABILITY IN A PATCHY ENVIRONMENT

L. Fahrig[1], L. P. Lefkovitch[2], and H. G. Merriam[1]

Abstract.--To see how habitat patchiness affects the stability of resident populations, a weighted graph-theoretic model has been constructed. The weighted vertices are populations in habitat patches, and the weighted edges are connections between the patches. The model is represented by a transition matrix which operates on a vector of age class abundances of a species in the patches. Changes in population sizes over time are simulated stochastically. The model is applied to populations of Peromyscus leucopus in woodlots in farmland which are interconnected by fencerows. Nature reserve planners can use this type of model to determine spatial arrangements of patches which permit longterm stable populations.

INTRODUCTION

One of the goals of the applied ecologist is to discover a set of principles that can be used to predict the effects of human activity on the environment. It has been suggested that the theory of island biogeography may be a model of the effects of human activity on species diversity in man-dominated areas (Forman et al. 1976, Davis and Glick 1978, Matthiae and Stearns 1981, Levenson 1981). In particular, it was noted that, in areas where large areas of forest have been cleared, the small remaining pieces of forest (woodlots) have a lower species diversity than did the original forest (Galli et al. 1976, Moore and Hooper 1975, Gottfried 1979, Whitcomb et al. 1981). It has been supposed that these patches of forest behave as "habitat islands", and are subject to the species-area relationship predicted by island biogeographic theory. MacArthur and Wilson (1967) suggested that island species diversity is a function of the immigration rate from the mainland and the extinction rate on the islands; species diversity increases with an increasing rate of immigration (decreasing distance from the mainland) and decreasing rate of extinction. Both rates are affected by island size, so that greater species diversity is associated with large islands (Simberloff 1976, Goeden 1979). This reasoning has led to a debate between those who feel that land use planners should preserve the largest possible forested areas (Terborgh 1974, Whitcomb et al. 1981), and those who feel that, when a particular quantity of land is available for preservation, several small reserves are to be preferred since, when combined, they will contain more species (Higgs and Usher 1980).

There are two main reasons for the failure of island biogeographic theory to produce clear-cut rules to be used in the design of nature reserves. First, (Gilpin and Diamond 1980), for any particular set of taxa, species diversity will increase with increasing number (and decreasing size) of patches, but there is a point where this relationship breaks down and, as patches become smaller and smaller, the extinction rate rises and diversity declines. It is clear, then, that there is no general solution to the patch size problem, since each species will have a different optimal balance between patch size and number associated with the lowest probability of local extinction. An increase in stability is here considered to be equivalent to an increase in the probability of survival.

The second problem is that the island theory presupposes the presence of a large "mainland" source area which can supply immigrants to the habitat islands. However, such a source area need not exist (Pickett and Thompson 1978). There is often no large forested area sufficiently close to the woodlots of an agricultural area to provide them with immigrants, so that the survival of a species within the whole area depends on the population dynamics within the system of woodlots and the rates and directions of movements of the organisms among them. Since each species responds differently to the size and spatial arrangements of the woodlots, it is preferable to make planning decisions based on studies that i) consider those species which seem most likely to be affected by patch isolation and size reduction and ii) take into consideration the rates and directions of movement of these populations among the patches.

Before presenting the model itself, some intuitive predictions concerning the effect of spatial configurations of patches on the stability of their resident populations are useful. Suppose each of four woodlots may or may not be "connected" to each of the other three. A connection could imply either that the woodlots are sufficiently close that migration can occur between them, or that there is some physical connection (eg: fencerow) that serves as a "corridor" along which the organisms can move. It is reasonable to believe that the stability of the overall population in the four woodlots (ie: its probability of survival) increases with an increasing number of connections among the woodlots, since this would increase the probability that colonists can replenish woodlots in which populations have become extinct. This assumption leads to the predictions in figure 1.

---

[1]Department of Biology, Carleton University, Ottawa, Canada.
[2]Engineering and Statistical Research Institute, Agriculture Canada, Ottawa, Canada, C. number I-405.

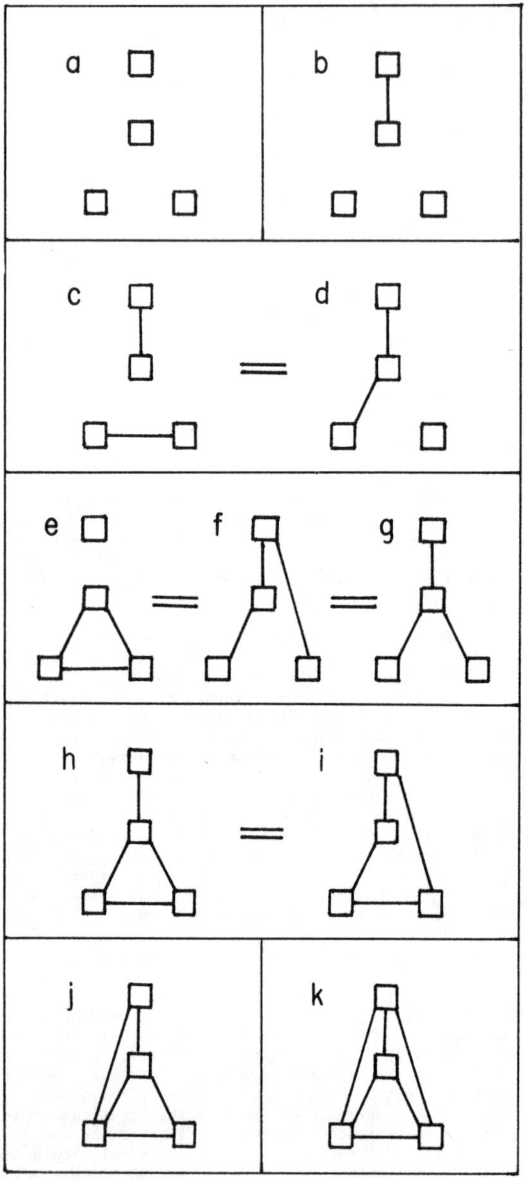

Figure 1. Groups of four woodlots arranged in suggested order of increasing overall population stability. An "=" indicates that the two groups are predicted to be equally stable.

The 11 possible spatial arrangements of the four woodlots are listed in their proposed order of overall increasing stability. Diamond (1975) made similar predictions, stating that "if there are several disjunctive reserves, these should ideally be grouped equidistant from each other rather than grouped linearly. An equidistant grouping means that populations from each reserve can readily recolonise, or be recolonised from, another reserve. In a linear arrangement, the terminal reserves are relatively remote from each other, reducing exchange of colonists". This agrees with the prediction that the arrangement in fig. 1e is likely to be more stable than that in fig. 1d.

A further prediction is that, within a particular arrangement of woodlots, the one having the highest number of connections (ie: the most "central" woodlot) should be the most stable, since it has the highest number of possible sources for population replenishment in the case of extinction.

This paper considers the possible effect of the spatial arrangement of habitat patches on the stability of the populations in each, and on the patches combined. The model which is formulated can be applied to any single species, provided that the necessary population statistics are available, can include the effects of patch size on population stability, and can be used to analyse any spatial arrangement of any number of patches.

THE MODEL

Description

The model is structured to follow the population sizes and age structures in a series of interconnected patches through time. There are two components to the population dynamics. They are i) within-patch dynamics and ii) between-patch dynamics.

Within a patch, individuals can be added to or lost from each age class. Additions into the youngest age class occur through births, and into the other age classes through ageing, so that individuals "graduate" into the next age class. Losses occur through deaths.

Between-patch dynamics consist of additions, due to immigration of organisms from adjacent patches, and losses due to emigration of organisms to connected patches.

The basic operator of the model is composed of a series of transition matrices, whose elements are parameters representing population processes appropriate for a certain time interval, t. The population sizes of the various age classes in the different patches at time t + 1, is then obtained as the product of the transition matrix for time t with the vector of population sizes at time t.

Construction

The model will be described by means of an example, namely the group of interconnected patches represented in figure 2, which can be represented by the graph in figure 3.

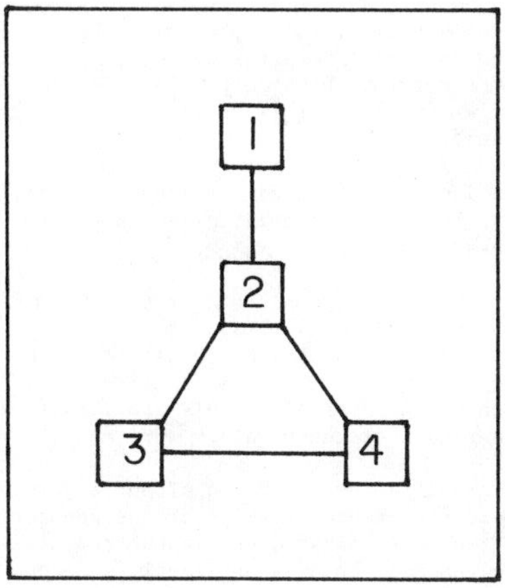

Figure 2. A suggested arrangement of four woodlots.

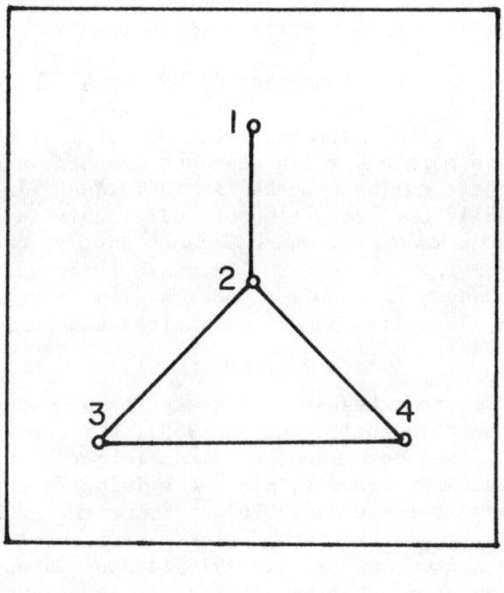

Figure 3. Graph corresponding to the group of woodlots in figure 2.

The adjacency matrix of the graph in figure 3 is

$$A = \begin{bmatrix} 0 & 1 & 0 & 0 \\ 1 & 0 & 1 & 1 \\ 0 & 1 & 0 & 1 \\ 0 & 1 & 1 & 0 \end{bmatrix}$$

where a "1" in position $A_{i,j}$ means that there is a connection between patches i and j and a "0" means that there is no connection. If this adjacency matrix were multiplied by a vector of population sizes in the different woodlots, the "1's" would effect the movement of all individuals in j to i. Since it is unlikely that all of the individuals in j would move to i, the elements would actually be numbers less than unity, each representing the probability of organisms moving from j to i. A becomes, in consequence, a "weighted-edge" graph. The probability that organisms do not move out of a woodlot will be represented by elements on the principal diagonal of the matrix. The transition matrix now has a typical off-diagonal element denoted by $P_{i,j}$, defined as the probability of moving from j to i; and diagonal element

$$1 - \sum_{\substack{i=1 \\ j \neq i}}^{m} P_{j,i},$$

where,

m is the number of woodlots.

To include the effects of births and deaths on the woodlot populations, the main diagonal values in the matrix must be further modified.

Let a) the death rate in woodlot i over one time interval = d_i and
b) the birth rate in woodlot i over one time interval = b_i.

The elements of the transition matrix differ from those just described in that the ith diagonal element is

$$1 - \sum_{\substack{i=1 \\ j \neq i}}^{m} P_{j,i} + b_i - d_i;$$

this matrix is denoted by X.

A further complication, which does not introduce any new principles, arises if there are several relevant age classes in the population, say z, since each of the elements in X is replaced by a matrix Z, of dimensions z x z, in which the diagonal blocks are essentially Leslie matrices modified by the migration rates, while the off-diagonal elements represent just the age-specific migration probabilities. Since the values of p, d, and b may be functions of time, there are several transition matrices required, one for each time interval; the whole series may be repeated over a longer time period, for example, for each complete breeding season. If the transition matrix for a particular time period is X_t and the vector of the number of individuals in the patches at time t is n_t, then the whole model, for one breeding season, is

$$n_1 = X_o\, n_o,$$

$$n_2 = X_1\, n_1,$$

$$n_3 = X_2\, n_2,$$

$$\vdots$$

$$n_r = X_{r-1}\, n_{r-1} = X_{r-1} \cdots X_1 X_o n_o$$

$$= M n_o$$

where,

r is the total number of time intervals in one season.

In order to determine the effect of a particular arrangement of patches on the stability of one resident species, the appropriate X-matrices need to be estimated and applied repetitively to $\underset{\sim}{n}$ until near stability of the populations in the patches and of the overall population for all patches combines is achieved.

Although the model has been described as if it were deterministic, its actual implementation is stochastic; in each simulation, the probability at any step is used to draw a sample from an appropriate distribution (binomial, Poisson, Normal etc.) and these values substituted for the deterministic estimates. In consequence, the eigenvectors of M, although approximately giving information on the mean of the process, are not as informative as the simulations.

There are many advantages to this model. First, it can be applied to any species of interest, as long as the appropriate p, b, and d values are available for the different ages classes (or some method of determining them is available). Second, any type of patch and any number of them can be considered, and patch size can be included in the model as a variable in the functions determining one or all of p, b, and d. Third, although the time unit in the model is arbitrary, so implying that all the information required to predict the population at a particular time is contained within the population at the previous time, this too can easily be generalized to refer to specific events and states in the past history.

APPLICATION OF THE MODEL

An application of the model to Peromyscus leucopus populations in southern Ontario will serve two functions, first, to demonstrate that the model can be applied to a natural system, and secondly, to test the hypotheses outlined in the introduction (see fig. 1). The P. leucopus populations considered live in woodlots surrounded by cultivated, grazed, or fallow fields. The preferred habitat of P. leucopus is woodland (preferably dry sites) and they almost never venture into non-wooded areas (McCarley 1963, Whitaker 1967, Smith and Speller 1970), although they will move long distances (Dueser et al. 1981). There is some evidence that they will travel along fencerows (for protective cover) during migrations in an agricultural setting (Wegner and Merriam 1979). Gottfried (1979) found that, in 10 woodlots surrounded by cornfields, densities of P. leucopus decreased with the size of the woodlot and with increasing isolation of the woodlot. Also, densities in the woodlots were significantly lower than those in nearby large forests. All of this information suggests that the model is appropriate for P. leucopus.

Determination of p-values

Since P. leucopus populations consist of three "trappable" age classes, namely juveniles, subadults, and adults, and since Madison (1977) found that subadults migrate significantly more than do juveniles or adults, it is necessary to use the more complicated version of the model.

The dispersal rates of P. leucopus appear to be independent of density (Hansen and Batzli 1978, Nadeau et al. 1981), possibly because in northern areas the population density may not reach a sufficiently high level (Middleton and Merriam 1981).

Another factor which might affect the migration tendency of an animal is its sex. Although Harland et al. (1979) found that male P. leucopus migrate more than do females, Nadeau et al. (1981) and Taylor (1978) found no significant differences between male and female migration tendencies. Since Harland et al. (1979) also found that sex ratios of the different populations were consistently 1:1, coupled with the contradictory evidence, sex will be ignored as a factor. It is assumed, therefore, that the p-values for P. leucopus do not differ from week to week (ie: for different X-matrices).

Madison (1977) found that the proportion of transient individuals, calculated over a period of about two weeks, were:

juveniles	0.154
subadults	0.571
adults	0.111.

Since the time interval in this application is one week, one half of these values were used here.

A final factor which can affect the rate of dispersal is the characteristics of the connection. Longer connections may decrease the probability that a mouse reaches the adjacent woodlot, and so decrease the effective dispersal rate. Other connection characteristics such as the corridor width, the vegetation of the corridor etc. may also affect the p-values. Here it is assumed that the corridors are all identical, but there are no new problems if this is false, so this assumption does not lose generality. In a field situation where corridors are not the same, some quantitative estimate of the effects of their various characteristics on dispersal is necessary.

Determination of b-values

The b-values determine the number of individuals which are born and which graduate from one age class to the next during a week. An individual element, b_i, is actually the proportion of individuals in an age class k in woodlot i which becomes part of the age class k + 1 in one week. There are three such age class changes to consider; adults give "birth" to juveniles, juveniles become subadults, and subadults become adults.

Since it takes about 3 weeks for a newborn mouse to become a juvenile (Sheppe 1965), the number of juveniles added to a woodlot is equal to the number born three weeks previously, minus the infant mortality over those three weeks. To determine the probability of a female giving birth during a particular week, the data of Hansen and Batzli (1978) and of Sheppe (1965) have been used. Hansen and Batzli give proportions of females pregnant at different times in the breeding period, but since these values were obtained in Illinois, where the breeding period is longer, and since northern populations exhibit a more distinct bimodal breeding cycle than do southern populations (Sheppe 1965), their data have been modified to conform with Sheppes' (1965) data of the proportions of juveniles in the population at any one time during the summer. Millar et al. (1979) found the average P. leucopus litter size to be 4.94. By combining the results of Harland et al. (1979), Millar and Gyug (1981) and Hansen and Batzli (1978), the following values for nestling survival rates have been obtained:

mid-Mar - mid-June	70%
mid-June - mid-Aug	40%
mid-Aug - mid-Oct	20%

The b_i values for juveniles over the summer are then calculated using the equation

$$b_i = \frac{\frac{\text{proportion of females}}{\text{pregnant at time t-3}}}{3} \times 4.94 \times \text{nestling survival}$$

The resulting b_i values are given in table 1.

Table 1.--Probability that a female present in woodlot i at time 6-3 gives "birth" to a juvenile at time t.

Time	
March-week 3	0.03
March-week 4	0.03
April-week 1	0.03
April-week 2	0.03
April-week 3	0.03
April-week 4	0.05
May-week 1	0.15
May-week 2	0.15
May-week 3	0.25
May-week 4	0.35
June-week 1	0.45
June-week 2	0.45
June-week 3	0.45
June-week 4	0.45
July-week 1	0.45
July-week 2	0.45
July-week 3	0.35
July-week 4	0.35
August-week 1	0.20
August-week 2	0.20
August-week 3	0.10
August-week 4	0.10
September-week 1	0.10
September-week 2	0.25
September-week 3	0.25
September-week 4	0.35
October-week 1	0.45
October-week 2	0.25

Since it takes about 3 weeks for a juvenile to become a subadult (Madison 1977), the addition of subadults through ageing of juveniles depends on the number of juveniles present three weeks previously and the probability of survival from the juvenile to the subadult age classes (as calculated from the d-values).

It takes about 2 weeks for a subadult to become an adult (Harland et al. 1978, Millar et al. 1979), so that the addition of adults through ageing of subadults, therefore, depends on the number present two weeks previously and the probability of survival from the subadult to the adult age classes (as calculated from the d-values).

Since the additions due to births and graduations into the different age classes depend on population sizes more than one week before t, calculations separate from those of the matrix multiplication are required.

Determination of d-values

The probability of *P. leucopus* dying during a particular week appears to be independent of the age class of the mouse and also the population density. Millar and Getz (1977) found no significant differences among death rates of juveniles, subadults and adults, and no significant relationship between death rates and population density. They found that the average death rate was 55.3%/60 days, which is equivalent to a weekly death rate of 6.45%. This mortality rate is also used to calculate the probability of survival of juveniles to subadulthood and subadults to adulthood in the calculations of b-values for subadults and adults.

The Simulation Experiments

Eleven experiments were performed, corresponding to the eleven hypothetical arrangements presented in the introduction (fig. 1). A computer program was written in order to perform the necessary calculations.

In each experiment, the four woodlots begin the first season with 9 adult mice each. Population sizes are calculated for each week, for each age class in each woodlot from the third week of March until the second week in October. Each simulation was run for 100 "years", and it was assumed that the adult population in each woodlot in the spring was 15% of the population in the previous fall (Middleton 1979 , Taylor 1978). The mean yearly final sizes (in the second week of October) of the total populations in the four woodlots over 100 years, and the corresponding variances for those woodlot populations which survived throughout the entire 100 year simulation were calculated.

Results

The means, variances, and survival data are presented in table 2. Other analyses will be postponed until field data are collected. However, a number of general results are of interest.

i) The *P. leucopus* populations in completely isolated woodlots became extinct within 3 to 4 years.
ii) If two woodlots were connected to each other, but to no other woodlots, their populations became extinct within the 100 years. However, the length of time required for this extinction was quite variable. A linkage of two woodlots was quite unstable, although less so than a completely isolated one.
iii) In almost every arrangement containing three or more interconnected woodlots, the *P. leucopus* population survived for 100 years (although the populations in individual woodlots sometimes became extinct). There was one exception to this result; the entire population in "h" died out. The reasons for this are not yet apparent, but perhaps can be explained by regarding the triangle as forming one unit of a pair, and so "h" approximates "b".
iv) The prediction of Diamond (1975) (see introduction) that woodlots arranged in a triangle or circle have more stable populations than those arranged linearly was not supported in this experiment. This is evident from two comparisons. First, the three woodlots in the triangle in "e" had similar average final populations and variances to those of the three linear woodlots in "d". Second, the four woodlots in the circle "i" had similar average final populations and variances to those arranged linearly in "f".
v) The suggestion that the woodlot having the highest number of connections is the most stable was not supported. Central woodlots had approximately equivalent average final population sizes and variances to non-central woodlots in the same interconnected group (see results for arrangements d, f, g and j).
vi) The population means of those populations that survived 100 years were all very similar (between 30 and 35 mice).
vii) Slightly smaller variances are associated with an increase in the number of connections. However, this was not conclusive.

Table 2. Average yearly finals (A.Y.F.), variances of the A.Y.F.'s (for populations that survived 100 years) and survival in years. In each case, the numbering of woodlots is the same as that in fig. 3.

	Woodlot	A.Y.F.	Var A.Y.F.	Survival (yrs.)
a 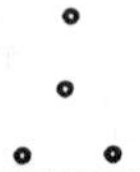	1	0.03	-	3
	2	0.06	-	3
	3	0.06	-	4
	4	0.05	-	3
b 	1	30.26	-	87
	2	29.38	-	88
	3	0.07	-	3
	4	0.31	-	4
c 	1	8.48	-	22
	2	7.85	-	23
	3	30.64	-	96
	4	30.15	-	96
d 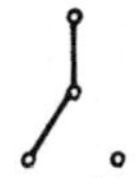	1	30.64	172.4	100
	2	31.67	159.6	100
	3	32.28	148.2	100
	4	0.08	-	3
e 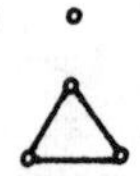	1	0.25	-	4
	2	34.19	132.4	100
	3	32.86	81.5	100
	4	34.98	148.0	100
f 	1	31.94	158.1	100
	2	32.31	165.2	100
	3	32.42	119.9	100
	4	31.93	108.6	100
g 	1	32.03	143.4	100
	2	33.13	144.4	100
	3	31.99	130.8	100
	4	32.84	140.4	100
h 	1	4.34	-	35
	2	12.22	-	37
	3	8.38	-	38
	4	8.66	-	37
i 	1	34.02	139.8	100
	2	33.33	133.3	100
	3	35.14	176.3	100
	4	33.98	182.8	100
j 	1	30.51	93.4	100
	2	32.05	103.6	100
	3	31.39	144.4	100
	4	31.96	130.9	100
k 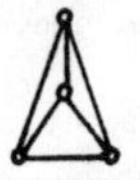	1	30.58	114.2	100
	2	32.62	80.0	100
	3	31.46	83.9	100
	4	31.60	109.6	100

DISCUSSION

The results of the simulation experiments reported here indicate that the spatial arrangement of habitat patches is important for several reasons. First, it is evident that links among habitat patches can increase the probability of survival of resident populations. Second, it appears that, for a particular species which inhabits a certain type of patch, there is a minimum number of patches which need to be connected in order to ensure the survival of the species in that area. It does not seem to make much difference how these patches are interconnected, as long as the minimum number of patches is involved. Although increasing the number of connections may slightly decrease the year-to-year variance in population size, this effect appears to be less important than the number of interconnected patches.

If in some species patch size is an important factor affecting movement among patches, or population growth within a patch, it should be expected that there will be an equilibrium population, depending on a balance between size and number of interconnected patches (for a particular total habitat area). Our model can be extended to determine these equilibria for real or hypothetical species for which the relationships between patch size and movement and population growth rate are known or can be postulated. This problem will be addressed in future research.

From this study it is apparent that the spatial arrangement of interconnected habitat patches associated with a stable and abundant population depends on the species under consideration and, in particular, on the interplay between the intrinsic growth and movement rates. For example, a species having a higher growth rate/movement rate ratio than does _P. leucopus_ would probably require a smaller minimum number of interconnected patches for population stability than the minimum number required by P. leucopus (ie: 3), since movement would have less effect on the populations than reproduction. If the growth rate/movement rate were lower than that for _P. leucopus_, the minimum number of interconnected patches required would need to be higher than that for _P. leucopus_. Further studies using the model will test this prediction.

The results of these simulations indicate that nature reserve planners may need to consider the question of the spatial arrangement of reserves in terms of one or a few critical species, especially those considered most likely to be deleteriously affected by decreased patch size and increased patch isolation. The model could then be used to provide guidance on the most suitable spatial arrangement of interconnected patches, both in terms of their number and also their sizes, to ensure the longterm stability of the populations of concern.

LITERATURE CITED

Davis, A.M. and T.F. Glick. 1978. Urban ecosystems and island biogeography. Environ. Conserv. 5: 299-304.

Diamond, J.M. 1975. The island dilemma: Lessons of modern biogeographic studies for the design of nature reserves. Biol. Conserv. 7: 129-146.

Dueser, R.D., M.L. Wilson and R.K. Rose. 1981. Attributes of dispersing meadow voles in open-grid populations. Acta Theorologica 26: 139-162.

Forman, R.T.T., A.E. Galli and C.F. Leck. 1976. Forest size and avian diversity in New Jersey woodlots with some land use implications. Oecologia 26: 1-8.

Galli, A.E., C.F. Leck and R.T.T. Forman. 1976. Avian distribution patterns within different sized forest islands in central New Jersey. Auk 93: 356-364.

Gilpin, M.E. and J.M. Diamond. 1980. Subdivision of nature reserves and the maintenance of species diversity. Nature 285: 567-568.

Goeden, G.B. 1979. Biogeographic theory as a management tool. Environ. Conserv. 6: 27-32.

Gottfried, B.M. 1979. Small mammal populations in woodlot islands. Am. Midl. Nat. 102: 105-112.

Hansen, L. and G.O. Batzli. 1978. The influence of food availability on the white-footed mouse: populations in isolated woodlots. Can. J. Zool. 56: 2530-2541.

Harland, R.M., P.J. Blancher and J.S. Millar. 1979. Demography of a population of Peromyscus leucopus. Can. J. Zool. 57: 323-328.

Higgs, A.J. and M.B. Usher. 1980. Should nature reserves be large or small? Nature 285: 568-569.

Levenson, J. 1981. Woodlots as biogeographic islands in southeastern Wisconsin. In Burgess, R.L. and D.M. Sharpe (eds.). Forest island dynamics in man-dominated landscapes. Springer-Verlag, New York.

MacArthur, R.H. and E.O. Wilson. 1967. The theory of island biogeography. Princeton University Press, Princeton, N.J.

Madison, D.M. 1977. Movements and habitat use among interacting Peromyscus leucopus as revealed by radiotelemetry. Can. Field Natur. 91: 273.

Matthiae, P.E. and F. Stearns. 1981. Mammals in forested islands in southeastern Wisconsin. In Burgess, R.L. and D.M. Sharpe (eds.). Forest island dynamics in man-dominated landscapes. Springer-Verlag, New York.

McCarley, H. 1963. Distributional relationships of sympatric populations of Peromyscus leucopus and P. gossypinus. Ecology 44: 784-788.

Middleton, J.D. 1979. Insular biogeography in a rural mosaic: the evidence of Peromyscus leucopus. M.Sc. thesis. Carleton University.

Middleton, J.D. and H.G. Merriam. 1981. Woodland mice in a farmland mosaic. J. Appl. Ecol. 18: 703-710.

Millar, J.S. and L.W. Gyug. 1981. Initiation of breeding by northern Peromyscus in relation to temperature. Can. J. Zool. 59: 1094-1098.

Millar, J.S., F.B. Wille and S.L. Iverson. 1979. Breeding by Peromyscus in seasonal environments. Can. J. Zool. 57: 719-727.

Millar, D.H. and L.L. Getz. 1977. Comparisons of population dynamics of Peromyscus and Clethrionomys in New England. J. Mammal. 58: 1-16.

Moore, N.W. and M.D. Hooper. 1975. On the number of bird species in British woods. Biol. Conserv. 8: 239-250.

Nadeau, J.H., R.T. Lombardi and R.H. Tamarin. 1981. Population structure and dispersal of Peromyscus leucopus on Muskeget Island. Can. J. Zool. 59: 793-799.

Pickett, S.T.A. and J.N. Thompson. 1978. Patch dynamics and the design of nature reserves. Biol. Conserv. 13: 27-37.

Sheppe, W.A. 1965. Island populations and gene flow in the deer mouse, Peromyscus leucopus. Evolution 19: 480-495.

Simberloff, D.S. 1976. Experimental zoogeography of islands: effects of island size. Ecology 57: 629-648.

Smith, D.A. and S.W. Speller. 1970. The distribution of Peromyscus maniculatus gracilis and Peromyscus leucopus noveboracensis (Rodentia: Cricidae) in a southeastern Ontario woodlot. Can. J. Zool. 48: 1187-1199.

Taylor, D.G. 1978. The population biology of white-footed mice in an isolated and a non-isolated woodlot in southeastern Ontario. M.Sc. Thesis. Carleton University.

Terborgh, J. 1974. Preservation of natural diversity: The problem of extinction prone species. Bioscience 24: 715-722.

Wegner, J.F. and H.G. Merriam. 1979. Movements by birds and small mammals between a wood and adjoining farmland habitats. J. Appl. Ecol. 16: 349-358.

Whitaker, J.O. 1967. Habitat relationships of four species of mice in Vigo County, Indiana. Ecology 48: 867-872.

Whitcomb, F.R., C.S. Robbins, J.F. Lynch, B.L. Whitcomb, M.K. Klimkiewicz and D. Bystrak. 1981. Effects of forest fragmentation on avifauna of the eastern deciduous forest. In Burgess, R.L. and D.M. Sharpe (eds.). Forest island dynamics in man-dominated landscapes. Springer-Verlag, New York.

SOME NEW RESULTS IN RESOURCE MANAGEMENT ORIENTED NATURAL PROCESS DYNAMICS MODELLING[1]

Charles C. Blackwell[2]

Abstract.--Results are presented which demonstrate the positive effects derived from including a simple model of the translocation process in a two-state model of the Isle Royale vegetation-moose system. Comparison with results of previous work and with best estimate observations of moose population is included. The study is directed to the time period 1915-1950.

INTRODUCTION

Difficulties in modelling the predation process have contributed to less than satisfactory results from efforts to model ecosystems at all levels of detail and complexity. May (1980), who has based much theoretical ecology work on extensions of the classical predation model due to Lotka (1923) and Volterra (1926), has indicated the desirability of more comprehensive models of this process. Several researchers (Ivlev 1945, Monod 1949, Watt 1959, Holling 1965, Wiegert 1977, and DeAngelis et al. 1975), in treating situations involving quite different organisms and environments, have generated a variety of functional forms which proved to be more effective for their immediate application, but still were deficient of satisfactory general applicability and accuracy throughout the ranges of variables which are desired and needed.

In an effort to obtain the properties which these investigators indicated were needed, Blackwell et al. (1977) offered a perspective which might provide the mathematical model(s) which was(were) needed. Two synthesized model forms were offered which were compatible with the suggested perspective, and results of their use in the classical Lotka-Volterra equations were presented.

Since that time, Blackwell (1980) has offered the suggestion that from a macroscopic perspective, the predation process and the grazing process are members of a category of processes which could be called the resource consumption process and that a model of this process might describe these and other previously independently treated processes. If so, then specific "consumers" and "resources" and their characteristic search and "evasion" strategies would be reflected by a unique set of constants in the same model equation form.

[1] Paper presented at the Third International Conference on State-of-the-Art in Ecological Modelling, Colorado State University, May 24-28, 1982.

[2] Charles C. Blackwell is Professor of Mechanical Engineering, The University of Texas at Arlington, Arlington, Texas U. S. A.

PURPOSE AND RANGE OF APPLICABILITY OF THE MODEL

The ultimate purpose of the models being sought is to allow the prediction of the behavior of a system of interest in a scenario of interest with at least sufficient accuracy to support management decisions and all that is implied thereby (such as state and parameter estimation and optimal control (management) implementation). The resource consumption process, having profound impact on any system including organisms, must be commensurately accurately modelled.

A system model must maintain its integrity over the required ranges of the system variables during scenarios of interest. In the case of the presently presented example, the resource magnitude varies by a hundredfold, and the consumer magnitude varies by twentyfold.

TESTING AND EVALUATION OF A MODEL OF THE RESOURCE CONSUMPTION PROCESS

One procedure of test and evaluation which is being pursued is to use the proposed resource consumption model in a model of an ecosystem for which quantitative observations have been taken, and to compare predictions and observations. This should be done for as many systems as possible. Results cannot be simply interpreted, however, inasmuch as there are other process models in the system model, and one cannot expect to fix the cause of failure on any one process model without additional independent information.

Criterion for Model Success and Restrictions on the Selection of Parameter Values

In order to be defended as a successful process model and not a simple "curve fit" (e.g. econometric "models"), a candidate model must satisfy at least three criteria: it must describe the process behavior asymptotically in the extremes and it must accommodate the intermediate conditions quantitatively, it must be compatible with scientific principles and laws pertinent to the process, and the parameters in the model must be compatible with validly acquired field and laboratory data related to the process. It is hypothesized that an ecosystem model validly constructed with such process models will predict behavior which is compatible with valid observations of the actual system.

We will describe a model as successful (not proved valid, but merely shown to be successful) if it predicts observed system behavior within the limits of uncertainty of those observations and if all the criteria of the previous paragraph are satisfied.

DEMONSTRATION OF THE PERFORMANCE OF THE MODEL

Choice of System for the Demonstration

Because it has been referred to as possibly the most thoroughly studied ecosystem (Caughley 1980), and because it was already the subject of four papers on modelling efforts, (Dixon and Cornwell 1970, Rykiel and Kuenzel 1971, Gazis et al. 1973, and Beyer et al. 1979), the Isle Royale vegetation-moose ecosystem was chosen as the subject of a modelling effort.

Unfortunately, only Beyer et al. were concerned with the 1915-1950 period (when moose and no wolves were on the island), the others being concerned with the moose-wolf interaction predominantly. The Beyer et al. model is a Monte Carlo type structure, and cannot be compared to a deterministic model of the present type, but results could be compared.

Rykiel and Kuenzel (1971) develop the model of the vegetation-moose-wolf interaction from a systems ecology perspective, and the paper was a very useful source of interpretation of the published literature into characteristic process constants. The first test of the present model was done by taking constants directly from Rykiel and Kuenzel when they could be found there, and inferring them from there as a first source, if possible. Details are presented in Blackwell (1980).

Vegetation-Moose Model Functional Form for the First Evaluation

The rationale for selecting the process detail of the model for the first evaluation is presented in Blackwell (1980), but the facts pertinent to the present demonstration are presented next. It was concluded (compatibly with Rykiel and Kuenzel) that the most simple acceptable model of the ecosystem must account of the following processes:

1. Respiration and mortality of the vegetation available for browse, and all feeding on this vegetation by other organisms than moose. We assign the symbol T_{11} to this contribution to the rate of change of the browse state. The aggregation or lumping of these processes together was done by Rykiel and Kuenzel and was commonly done by the previously cited authors.
2. Browsing of vegetation by moose (T_{12}). This is the resource consumption process in the present example.
3. Gross primary production (T_{13}).
4. Respiration and mortality of the moose and predation on the moose by parasites, etc. (wolves were not on the island during this period) (T_{21}).
5. Assimilation of the browse by the moose (T_{22}). This term will be taken to be a constant fraction of T_{12}, the fraction being the assimilation efficiency of the moose.

The first three processes determine the dynamics of the vegetation state, and the last two determine the dynamics of the moose state:

$$dv/dt = -T_{11} - T_{12} + T_{13} \tag{1}$$

and

$$dm/dt = -T_{21} + T_{22} \tag{2}$$

where,

dv/dt = the time rate of change of the vegetation state: magnitude units per unit of time

dm/dt = the time rate of change of the moose state: magnitude units per unit of time

T_{ij} are defined immediately above.

Units of the Variables

The available browse state was chosen to be dry mass per unit area of the island (kg/sqm), and the moose state was chosen to be live mass per unit area of the island (kg/sqm). The unit of time which is used is the year (yr). Energy units are an attractive alternative to biomass units and were used by Rykiel and Kunzel (1971).

The Functional Forms of the Models of the Processes

Since the object of the effort was to test and evaluate a model of the resource consumption process, every effort was made to use only frequently published process model forms for the other processes. That intention was abandoned of apparent necessity, but it was adhered to as much as possible.

Available Browse Respiration, Mortality, and Miscellaneous Feeding

A simple linear model of this process was used by Rykiel and Kuenzel (1971) and has ample precedent in the literature:

$$T_{11} = -a_1 v \tag{3}$$

where,

v = the available browse state

the units of a_1 are yr^{-1} and its value was taken from Rykiel and Kuenzel (1971).

Browsing of Vegetation by Moose

This process is the "resource consumption process" for which the proposed model is to be tested. From Blackwell (1980):

$$T_{12} = -a_2 m(\exp[a_3 m/v] + a_4 v\exp[v/a_5 m]) \tag{4}$$

where,

a_2 = the ad lib browse rate of the moose on the available browse

m = the moose biomass

a_3 = an index of the ease with which available browse can be accessed by the moose

a_4 = an index of the likelihood of accessing available biomass under conditions of scarcity.

This expression is what was judged to be the portion

of the model described in Blackwell (1980) which was applicable to the expected variable ranges. The coefficients were inferred from Rykiel and Kuenzel (1971) ad hoc as described in Blackwell (1980).

Gross Primary Production

Rykiel and Kuenzel (1971) use a constant value for the model of this process, which is at variance with the needs of a scenario characterized by large fluctuations in magnitude of the primary producer. A review of Montieth (1971) lead to the selection of a Monod (1949) form which agreed with Rykiel and Kuenzel (1971) in the saturation range of values of the available browse biomass:

$$T_{13} = +a_6 v/(a_7 + v). \quad (5)$$

Respiration, Mortality, and Predation: Moose State

As is the case for the available browse state, a linear model is taken:

$$T_{21} = -b_1 m \quad (6)$$

where,

b_1 = the effective respiration coefficient, and was taken from Rykiel and Kuenzel (1971).

Assimilation of Browse by Moose

This term is the reflection of the resource consumption process into the consumer state dynamics. It is assumed to be a constant fraction of the browse rate, the fraction being the assimilation efficiency:

$$T_{22} = b_2(T_{12}/(-a_2)) \quad (7)$$

where,

b_2 = the browse assimilation constant (yr^{-1}).

The value of b_2 was inferred from Rykiel and Kuenzel (1971)

Values of the Process Constants

The process constants, as taken from Rykiel and Kuenzel (1971) directly or inferred are presented in table 1. The values of each of the states at 1915 (the "initial conditions") are also parameters to the model prediction.

Table 1.--Process constants taken from Rykiel and Kuenzel (1971) or inferred from Rykiel and Kuenzel (1971) and other sources

Constant	Value	Constant	Value
a_1	.6250	a_7	.03
a_2	12.22	b_1	45.12
a_3	29.21	b_2	45.21
a_4	.3680	v[1915]	3.000
a_5	29.16	m[1915]	.0001401
a_6	2.063		

Results of the First Ecosystem Model Prediction

As indicated in Blackwell (1980), The model predictions of moose population drastically differed from observations as reported by Mech (1966). Figure 1 displays model prediction of moose population history, and reported best estimate observations. Estimated maximum uncertainty bars are also displayed, where indicated by Mech.

Parameter Adjustment Results

No uncertainty bounds for the parameters were available from Rykiel and Kuenzel, but parameter adjustments were made using a parameter estimation system (Blackwell et al. 1980) with minimum squared fractional residual between observation and prediction as the adjustment criterion. Table 2 presents the resulting values of the parameters, and the resulting predicted moose population history is shown in figure 1.

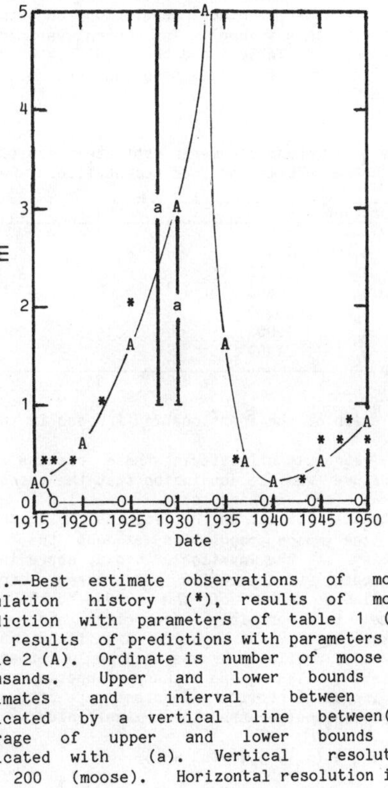

Figure 1.--Best estimate observations of moose population history (*), results of model prediction with parameters of table 1 (O), and results of predictions with parameters of table 2 (A). Ordinate is number of moose in thousands. Upper and lower bounds of estimates and interval between are indicated by a vertical line between(-). Average of upper and lower bounds is indicated with (a). Vertical resolution is 200 (moose). Horizontal resolution is 1 (year).

As can be seen, the values of the parameters after adjustment differ from the original values by orders of magnitude in some cases, and thus the model was not successful by the adopted criterion. However the quality of fit obtained was encouraging, and a literature search was initiated to verify the original estimates of the parameter values.

Improved Estimates of the Parameter Values

It was found (Blackwell 1981) that information

Table 2.--Process constant values obtained from adjustment to minimize the sum squared normalized residuals.

Constant	Value	Constant	Value
a_1	.2082	a_7	.05813
a_3	4.695	b_2	21.69
a_4	.2760	v[1915]	.5000
a_5	101.3	m[1915]	.0001333
a_6	.4500		

in the literature which became available after 1971 indicated that some parameter estimates should be revised, and new estimates were made from those sources.

The model was found to be extremely sensitive to the values of some of the parameters, with slight adjustments changing the results from a total die-off to a relatively minor population adjustment (see Blackwell, this volume). The nominal parameters are presented in Table 3 and the resulting dynamics are shown in figure 2. It is seen that the model is not successful.

Table 3.--Process parameter estimates resulting from using information from recent literature.

Constant	Value	Constant	Value
a_1	1.905	a_7	.0350
a_2	5.220	b_1	13.03
a_3	.0880	b_2	13.27
a_4	40.00	v[1915]	.0800
a_5	1.000	m[1915]	.0001337
a_6	.1359		

Inclusion of the Translocation Process in the Model

The result of reviewing these results and the literature was the conclusion that the translocation process (and perhaps seed sprouting) makes a substantial impact on the dynamics during the period when the moose population exceeds the carrying capacity of the available browse appreciably. In this case, translocation of energy from the unavailable sector of the browse to the available browse is certainly occurring and (it was hypothesized) might provide significantly greater amounts of available browse than would otherwise be available. The gross primary production term, T_{13}, was altered to provide a constant rate of growth when v was much smaller than the nominal value:

$$T_{13} = (a_8 + a_6 v)/(a_7 + v) \quad (8)$$

where,

a_8 = the constant added to provide the translocation effect.

a_6 and a_7 are defined earlier.

Actually, the ratio a_8/a_7 defines the magnitude of the modelled translocation input, and the ratio a_6/a_7 defines the maximum per unit gross primary production rate. The translocation rate decreases with the magnitude of v, which it should. In fact no way was found to provide a sufficient rate of decrease with v with this function form.

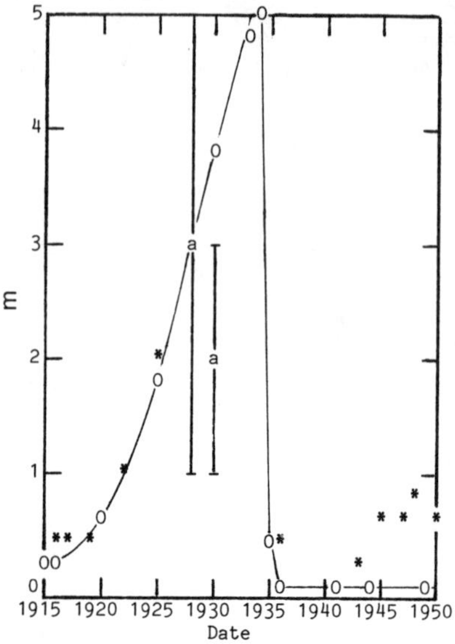

Figure 2.--Results of the model prediction using the revised parameter estimates (from table 3). The model predicts that the moose will not survive the die-off.

Selecting a value of a_8 (2.4×10^{-5}) to provide a translocation rate of 1% of the equilibrium gross primary production (6.9×10^{-4}), and changing no other parameters from those yielding figure 2 results in predicted moose population dynamics shown by figure 3. It is seen that the population decrease reported to occur in 1933 is accommodated, but the ensuing dynamics do not match well.

Final Parameter Adjustment

The results presented in figure 3 provide an excellent basis for a review of the parameters which were derived from the literature and of the implications of their use in the functional forms of the process models. Two particularly significant considerations emerge from such a review, and the first is that the estimates of the initial (1915) available browse and the gross primary production rate do not match. The other is that the values of the saturation gross primary production rate and the half saturation constant result in a predicted available browse maximum growth rate which is an order of magnitude greater than that which one would conclude to be reasonable (Marks 1974).

Since a large uncertainty in the estimation of the respiration and mortality coefficient was expected, it was decided to adjust that parameter to provide agreement between estimated standing crop and estimated gross primary production. The Monod form is not capable of allowing the description of the three observable properties; saturation level, maximum growth rate, and equilibrium gross primary production rate independently. It was decided to sacrifice agreement of saturation since it was believed that inaccuracy in that quality would be less crucial to the accuracy of the model than would inaccuracy in the other two properties. The resulting parameter values are presented in Table 4, and the model prediction is presented in figure 4.

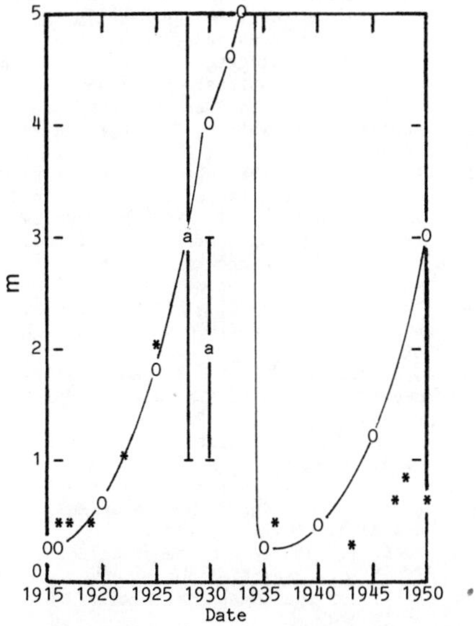

Figure 3.--Results of the model prediction using the revised parameter estimates (from table 3) and adding the model of translocation. The model predicts that the moose survive the die-off, but the high growth rate from 1947 to 1950 does not agree with the observations.

Table 4.--Process parameter estimates resulting from final review of consistency of the parameters.

Constant	Value	Constant	Value
a_1	1.148	a_7	.3419
a_2	5.220	a_8	.00016
a_3	.0880	b_1	13.03
a_4	40.0	b_2	13.27
a_5	1.000	v[1915]	.8000
a_6	.4883	m[1915]	.0001337

DISCUSSION

The agreement of the system model prediction with the best estimate observations would appear to be quite acceptable in light of the apparent uncertainty of the best estimate observations. All model parameters which can be compared to experimental or field values are in essential agreement therewith with the exception of the gross primary production rate term, as has already been explained. In this case, saturation gross primary production rate and half saturation constant value accuracy was sacrificed to obtain accuracy of two properties more important to the system dynamics, the equilibrium gross primary production rate and the maximum per unit gross primary production rate (which occurs at low values of primary producer magnitude, relative to equilibrium).

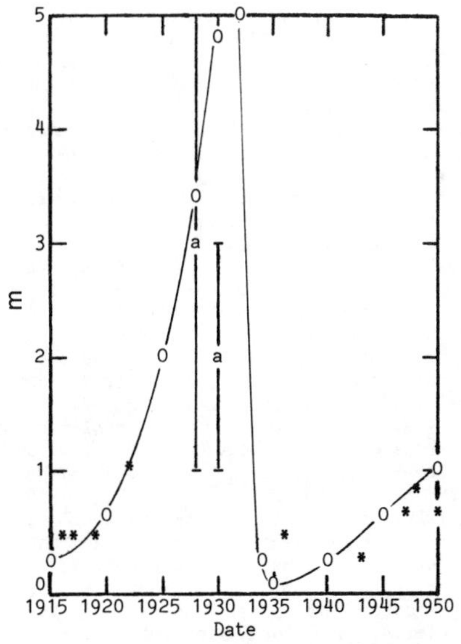

Figure 4.--Model predictions resulting from parameter values presented in Table 4.

The Monod model form cannot realistically describe the gross primary production process because it cannot accommodate the fundamental individuality of primary producers in the three areas: initial production rate, nominal conditions production rate, and efficiency of solar conversion. At least three parameters are needed. The resource consumption rate process model which is used herein to describe the browsing process can accommodate this need.

An evaluation of the effectiveness of this model in the description of the gross primary production process is planned, as is the application of the system model to other observed systems.

Other Systems Which May Be Suitable for Model Evaluation

As pointed out by Caughley (1980), the George deer herd (McCullough 1979) and the caribou of St. Matthew's Island (Klein 1959) are possible candidates for model test and evaluation. In addition, the snowshoe hare population dynamics phenomena and bacteria culture dynamics (Hansen and Hubbell 1980) may be suitable vehicles for model evaluation.

LITERATURE CITED

Beyer, W. A., D. R. Harris, and R. J. Ryan. 1979. A stochastic model of the Isle Royale biome. Rocky Mtn. J. of Math. 9:3-18.

Blackwell, A. L. 1981. Extremes in the resource consumption process: a mathematical model of the irruption of moose population on Isle Royale. Tech. rpt. no. 81-001. 32 p. Department of Mechanical Engineering, The University of Texas at Arlington, Arlington, Texas.

Blackwell, C. C., A. L. Blackwell, C. W. Hutchinson, and D. G. Harrison. 1977. More realistic alternatives to the Lotka-Volterra predation function. Abst. Bull. of the Ecological Soc. America 58(2):51.

Caughley, G. 1980. A population of mammals. Book review: D. R. McCullough. 1979. The George reserve deer herd. Population ecology of a K-selected species. University of Michigan Press, Ann Arbor. 272p. Science 207(4437):1338-1339.

DeAngelis, D. L., R. A. Goldstein, and R. V. O'Neill. 1975. A model for trophic interaction. Ecology 56(4):881-892.

Dixon, K. R. and G. W. Cornwell. 1970. A mathematical model for predator and prey populations. Res. Popul. Ecol. XII:127-136.

Gazis, D. C., E. W. Montroll, and J. E. Ryniker. 1973. Age-specific, deterministic model of predator-prey populations: application to Isle Royale. IBM J. Res. Develop. 17:47-53.

Hansen, S. R., and S. P. Hubbell. 1980. Single nutrient microbial competition: qualitative agreement between experimental and theoretically forecast outcomes. Science 207(4432):1491-1493.

Holling, C. S. 1965. The functional response of predators to prey density and its role in mimicry and population regulation. Mem. Ento. Soc. Canad. 45:62p.

Ivlev, V. S. 1945. The biological productivity of waters. Uspekhi Sovrem. Biol. 19:98-120.

Klein, D. R. 1959. Saint Matthew Island reindeer-range study. 48 p. Special Scientific Report-Wildlife No. 3. U. S. Fish and Wildlife Service, Washington, D. C.

Lotka, A. J. 1923. Contribution to quantitative parasitology. J. Washington Acad. Sci. 13:152-158.

McCullough, D. R. 1979. The George Reserve deer herd. 271 p. The University of Michigan Press, Ann Arbor, Mich.

Marks, P. L. 1974. The role of pin cherry in the maintenance of stability in northern hardwood ecosystems. Ecol. Monogr. 44:73-88.

May, R. M. 1980. Harvests from the sea: some consequences of interactions within and between exploited populations. p. 47-48. In Abstracts of Papers of the 146th National Meeting of the American Association for the Advancement of Science.

Mech, L. D. 1966. The wolves of Isle Royale. Fauna of the National Parks of the United States. Fauna Series 7. 210 p. United States Government Printing Office, Washinton, D. C.

Monod, J. 1949. The growth of bacterial cultures. Annu. Rev. Microbiol. 3:371-394.

Monteith, J. L. 1965. Light distribution and photosynthesis in field crops. Ann. Bot., N. S. 29(113):17-37.

Rykiel, E. J., and N. T. Kuenzel. 1971. Analog computer models of "The Wolves of Isle Royale." pp 513-541. B. C. Patten, (ed.) In Systems analysis and simulation in ecology. Vol I. 607 p. Academic Press, New York.

Volterra, V. 1926. Variazioni e fluttuazioni del numero d'individui in specie animali conviventi. Men. Acad. Lindie 2:31-113.

Watt, K. E. F. 1959. A mathematical model for the effect of densities of attacked and attacking species on the number attacked. Canad. Ento. 91:129-144.

Wiegert, R. G. 1977. A model of a thermal spring food chain. p. 289-315. C. A. S. Hall and J. W. Day (eds.), In Ecosystem modeling in theory and practice: an introduction with case histories. 684 p. John Wiley and Sons, New York.

A METHODOLOGY FOR DETERMINING THE PREDICTIVE ACCURACY OF MATHEMATICAL MODELS FROM A LIMITED DATA BASE[1]

Kevin R. Wegener[2] and Ronald F. Malone[3]

Abstract.--A split-sample testing scheme has been combined with a Monte Carlo analysis to develop a methodology for defining the accuracy of mathematical models when used in a predictive mode. The data pool available to the modeler is subdivided into calibration and verification subsets in an iterative and random manner. This calibration/verification procedure is repeated a large number of times. The methodology was found to have significant advantages when compared to more traditional approaches to defining model reliability.

INTRODUCTION

This paper presents a methodology which can be used to determine the predictive accuracy of mathematical models. Predictive accuracy is a measure of how well a mathematical model can simulate events not included in their calibration process (Craig, 1970). In recent years a large number of water quality models have been utilized to evaluate the impacts of alternative courses of management upon this nation's lakes and streams. It is essential that uncertainties associated with the use of mathematical models to represent complex water quality phenomena be recognized. If these uncertainties are not recognized, the engineer or scientist utilizing the model runs the risk of misinforming those making management decisions with incorrect or misleading modeling results.

A clear distinction must be made between the ability of a model to fit calibration data and its ability to predict unseen or future events. The question of "how good is this model?" should be addressed by evaluating how realistically the model simulates the process in question (Craig, 1970) rather than how well the model curve fits observed calibration data points. In practical applications, goodness of fit criteria utilized in calibration such as the standard error of estimate or the correlation coefficient do not necessarily reflect a model's ability to simulate in the predictive mode. The methodology presented here evaluates a model in its predictive operations. This is accomplished by devising a Monte Carlo experiment which iteratively calibrates and verifies the model while utilizing a split-sample testing scheme. This technique permits the modeler to obtain an evaluation of how well the model performs in its most important predictive mode.

CASE STUDY

The total phosphorus loading associated with runoff from a 44 acre (18 hectare) urban drainage basin was investigated in conjunction with a restoration project of the University Lake in Baton Rouge, Louisiana (Knaus, 1981). The lake receives excessive nutrient loadings and has a mean concentration of 0.428 mg/l total phosphorus. A statistical (or parametric) model relating total phosphorus loading and runoff volume through a power function was suggested for use (Diniz, 1979).

$$L = a\,Q^{b} \tag{1}$$

where,

L = total phosphorus load for the storm event in grams
Q = total runoff for the storm event in cubic meters
a,b = regression coefficients

A log-transformation of the variables makes the model amenable to linear regression.

$$\ln(L) = \ln(a) + b\,\ln(Q) \tag{2}$$

The form of this model acknowledges the occurrence of the well known washoff phenomenon reported in the literature (Huber, 1975).

Runoff volumes and total phosphorus loadings were measured for thirteen springtime storm events on the drainage basin of the University #7 storm-drain. The sampling was conducted from January through May 1980. Total phosphorus loading and runoff values for the storm events are depicted graphically on figure 1 with a least squares linear fit of the log-transformed data. Regression coefficients were computed for ln(a) to be 0.1907 g and for b to be 0.8614 g/m^3. A value of r^2 for the log-transformed model was computed to be 0.911. Comparison of r^2 only provides an indication of goodness of fit of a model. It does not indicate

---

[1]Paper presented at the Third International Conference on State of the Art in Ecological Modeling, Colorado State University, May 24-28, 1982.

[2]Kevin R. Wegener is a Graduate Research Assistant, Louisiana State University, Baton Rouge, Louisiana U.S.A.

[3]Ronald F. Malone is Assistant Professor of Civil Engineering, Louisiana State University, Baton Rouge, Louisiana U.S.A.

the ability to predict events unseen by the model, although this is frequently implied.

ITERATIVE PREDICTIVE ACCURACY DETERMINATION

In the general sense, an iterative predictive accuracy determination examines all possible calibration and verification outcomes available to the modeler from a limited data set. This is accomplished by superimposing a Monte Carlo experiment upon whatever model and calibration procedure is selected by the modeler. Each Monte Carlo iteration randomly divides the data set into calibration and verification points. Model performance is then evaluated for both the calibration and predictive mode of operation. The calibration and verification procedures are then repeated until all possible combinations of data are used. This procedure is analagous to that followed by a modeler in the field who must collect calibration data from a number of observed events in order to predict a number of future or unseen events.

A FORTRAN program was used to determine the predictive accuracy of the runoff loading model. The options to be selected for use during an execution of the program are described in the RESULTS section below. A run consists of the execution of this program with a unique set of options. The flowchart of the predictive accuracy routine is presented in figure 2. The same iterative technique was used to assess the predictive accuracy of a dissolved oxygen (D.O.) model on a small southern stream (Crane, 1981). The function of each component of the predictive accuracy determination routine is addressed in the following sections.

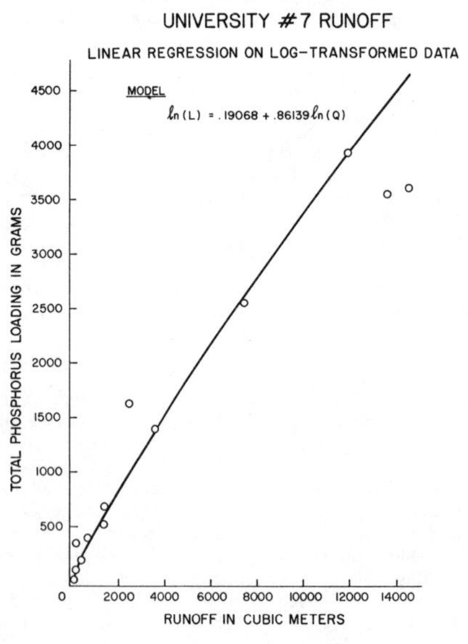

Figure 1.--Runoff and TP loadings from complete storm events.

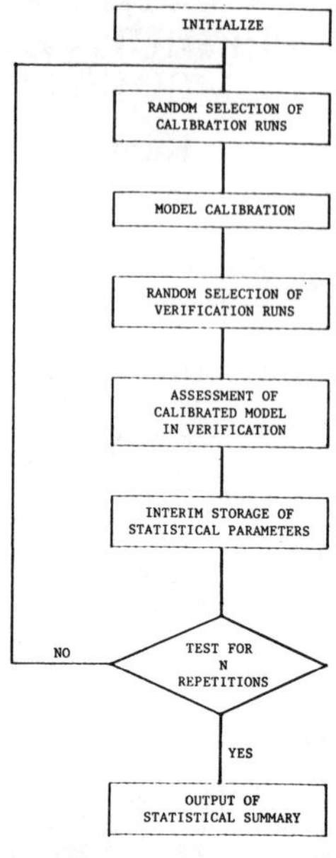

Figure 2.--Flow chart for iterative determination of predictive accuracy.

Random Selection of Calibration Data Subset

A subset of calibration data is randomly selected from the total pool of data for calibration. As each entry is selected, it is compared to the preceding entries so that no duplicate entries are made. The selection/comparison sequence is repeated until the required number of calibration entries has been selected. In the previously described case study, a psuedo-random number generator, GGUBS (IMSL, 1980), was used to select entries from the input data array.

Model Calibration

In this step of the procedure the model under examination can be calibrated by any selected technique. Most commonly, calibration is conducted by a least squares analysis although other criteria may be used. Selection of the calibration technique is dependent on the nature of the model in question. In the case study, a least squares linear regression was performed on the log-transformed data using the FORTRAN library routine RLONE (IMSL, 1980). Since log-transformed data were used, the error term is multiplicative for the original model. Although a true "best-fit" calibration was not accomplished for the original power function, the bias on the estimate of the residual sum squares is expected to be small for models with r^2 near unity (Freund and Minton, 1979). However, different calibration procedures such as a non-linear least squares technique may produce different results.

Random Selection of Verification Data Subset

The psuedo-random number generator is used to select entries to the verification subset in a fashion similar to the selection of calibration data. The selected entries are compared to each preceding verification entry as well as to the entries of the concurrent calibration subset. Duplicate points are rejected to prevent the same point from being used in calibration and verification concurrently or to prevent the verification subset from containing the same point more than once simultaneously. This ensures that the model verification is independent of the calibration.

Comparison of Model Predictions to Verification Data

The calibrated model is used to predict values for each of the entries of the verification data subset. In the case study, the regression coefficients developed are used in the model to predict values of the dependent variable, total phosphorus loading. The predictive accuracy of the model is determined by evaluating the difference between observed and predicted loadings of total phosphorus. The residual sum of squares is calculated separately for the verification subset and the calibration subset.

Interim Storage of Statistical Values

The statistical parameters and regression coefficients resulting from the model calibration and verification from each iteration is stored for further use. The procedure consisting of the above five steps is then repeated for m iterations. Selection of calibration and verification subsets is conducted independently from all previous iterations. A running mean of statistical parameters and coefficients is maintained throughout the

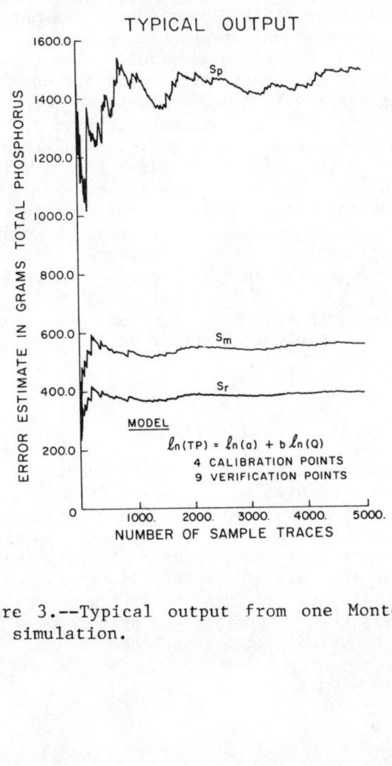

Figure 3.--Typical output from one Monte Carlo simulation.

simulation. These values vary because of the random selection of calibration and verification data. Making use of the running means of the residual sum of squares calculated over successive iterations, the standard error of estimate for the calibration (S_m) and prediction (S_p) modes is calculated. The standard error of estimate is approximated as

$$S = \left(\sum_{j=1}^{m} \sum_{i=1}^{n} [(L_i - \hat{L}_i)^2/df]/m\right)^{\frac{1}{2}} \qquad (3)$$

where,

- m = number of iterations
- n = number of calibration or verification points used
- $\hat{L}_i$ = model-predicted value for dependent variable used for calibration or verification
- L_i = observed value for dependent variable
- df = number of degrees of freedom

Calculation of the standard error of estimate for calibration and verification differs only in the data subset selected for comparison and in the degrees of freedom which apply. In the case of the power function model (eq. 2) the degrees of freedom are "n-2" and "n" for calibration and verification, respectively. A third statistical parameter, S_r or the standard deviation of the residual errors of calibration can be calculated from the equation by assigning n degrees of freedom for the calibration calculation. This parameter provides a baseline measure of residual errors with no correction of the degrees of freedom for model bias.

RESULTS

The predictive accuracy of the model of equation 2 was evaluated using the methodology described above. An example of the output generated by one run utilizing 5,000 iterations with four calibration and nine verification points is presented in figure 3. With an increasing number of iterations, the values of S for calibration (S_m) and for prediction (S_p) converge to values of 459 g and 511 g, respectively. The run has an unsupervised calibration routine, that is, selection of entries to the calibration subset is completely random with no selection criteria supervising the selection.

The error estimates represent square roots of the running means of squared residual errors. The graphical depiction of S_m and S_r demonstrates that the two are proportional and respond similarly with the random draw of the calibration data subsets. They have both become very stable beyond the 4000th iteration. S_p, on the other hand, demonstrates considerable instability and bears little resemblance to the two measures of goodness-of-fit. The sharp peaks on the graph represent single iterations for which the calibrated model was an exceptionally poor predictor of the verification subset. With an increasing number of iterations, the effect of these occurrences is dampened. At 5000 iterations, an approximate value of 1490 grams TP has been reached. Using four calibration points, therefore, the resulting error in prediction is over double the error in calibration. But it must be recognized that the generated calibration and prediction error estimates are unique to the particular selection criteria employed (totally random selection) and the number of calibration points used for each iteration. Varying the number of verification points has little effect on the estimated values of S_p.

Number of Calibration Points

A sufficient number of calibration points must be selected to insure that a reasonable predictive accuracy is obtained. The response of a model to increasing the number of calibration points can be analyzed by running a series of predictive accuracy determinations while varying the calibration subset size. A plot of the final error estimates versus the number of calibration points in the subset can then be prepared. The power function loading model responds in a fashion typical of the parametric runoff models (fig. 4). As the number of calibration points increase, both S_p and S_m converge to S_r. Substantial improvements in the predictive accuracy are realized initially. As the number of calibration points becomes larger, the incremental increase in predictive accuracy (or decrease in S_p) becomes smaller. Hypothetically, when S_p and S_r converge, the calibration subset size is adequate to completely represent the entire data pool and presumably the population of events from which the data pool was drawn. Recognizing this, the ratio (S_p/S_r) can be calculated to determine the amount of improvement in S_p that may be obtained by increasing the calibration data subset size. For this particular case, it is seen that the S_p/S_r ratio for the power function model calibrated with twelve calibration points is 1.22. It is concluded that increasing the data pool could still contribute significantly to the model's predictive accuracy. Each model and calibration procedure responds differently to increases in the number of calibration points; therefore, generalizations are difficult.

Bracketing Subsets

In many instances, it is not realistic for the practicing engineer or scientist to calibrate a model with a strictly random selection. For the runoff model, it would be poor statistical practice to calibrate the model on a subset consisting of flows in either the low range or the high range and then extrapolate the model to verify on the remaining data. A calibration subset selection criterion could be used to ensure that an equal number of data from the low, middle and upper

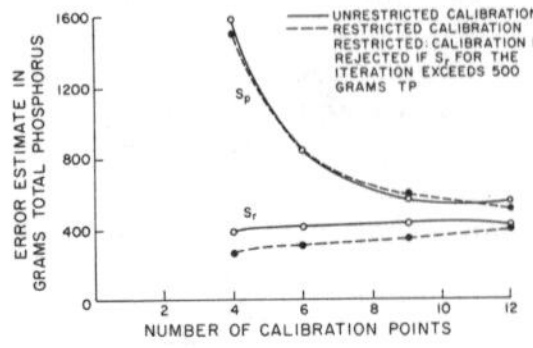

Figure 5.--Effect of calibration acceptance criteria on unsupervised log-transformed model.

ranges of the independent variable be selected for the model calibration. By following this procedure with the nine calibration storm events, values of S_m and S_p were estimated as 480 g and 461 g, respectively. This corresponds to a reduction of 2 percent and 24 percent from the unsupervised case. When using twelve calibration points and one verification point, the data bracketing technique results in the same estimate of S_p as determined by the unsupervised data selection technique. The methodology underscores the pitfall of using a model to extrapolate predictive values beyond the limits of the calibration data. Other types of supervision criteria could also be used.

Calibration Acceptance Criteria

The predictive accuracy determination procedure readily adapts to a wide variety of methods to enhance calibration. For instance, the authors hypothesized that a significant portion of the error in the predictive accuracy determination may be attributed to random draws of the calibration subset which result in a poor fit of the model. To test this hypothesis, an acceptance criterion was established as a maximum allowable value for S_r. Each calibration subset which failed to meet this criterion was rejected and a new subset drawn. Establishment of an acceptance criterion of 1000 grams TP had no effect on the simulation. A criterion of 500 grams TP resulted in the values presented in figure 5. S_r was reduced by an average of approximately 20 percent over the various calibration set sizes tested. S_p changed very little with use of the acceptance criterion with increases ranging from 5 percent using four calibration points to 3 percent using twelve calibration points. In essence, use of a rejection criterion had minimal effect on predictive accuracy.

Model Comparisons

A comparison of models can be readily undertaken with a predictive accuracy determination. In contrast to recent works which use calibration criteria for a standard of comparison, the predictive accuracy determination more directly evaluates the models in their mode of application. For purposes of illustration, the predictive accuracy routine was used to compare the performances of the model of equation 2 against a linear model:

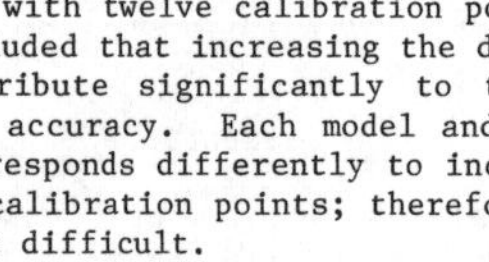

Figure 4.--Comparison of error estimates for various sizes of calibration subsets.

$$L = a + bQ \tag{4}$$

Estimates of S_p are depicted graphically on figure 6 for equations 2 and 4. As can be seen from this comparison, the linear model (eq. 4) displays greater predictive accuracy than the power function relationship. The linear model also displays a smaller S_r, suggesting that as S_p/S_r approaches unity with increasing data set sizes, the predictive accuracy would remain greater than that of the power function model. This is attributable at least in part to the log-transformation of the data prior to application of the least squares technique. A more complete model comparison in this case would have to include a nonlinear regression technique to calibrate the model, since it is difficult to separate the effects of model structure from the inadequacies of the calibration procedure.

When used with an unsupervised calibration procedure, the predictive accuracy determination encompasses both the sample reuse method (Geisser, 1975) and the half-sample cross validation technique (McCarthy, 1976). The predictive accuracy technique allows for a broader evaluation of the model and associated calibration procedures. This method also appears to enhance the split-sampling technique discussed by Schroder, McCuen and Rawls (1980) through its consideration of all possible combinations for calibration and verification subsets.

DISCUSSION

It is apparent that as mathematical models are used increasingly as a basis for making management decisions, more attention must be given to model selection and calibration techniques. The predictive accuracy determination provides a format to evaluate model performance as well as to enhance the modeler's understanding of the modeling process.

The utility of the method is underscored by the variety of options that can be evaluated. The methodology is easily adapted to a wide variety of models. Additionally, since the method is analagous to the calibration and verification procedures typically undertaken by modelers, it is easily explained and readily accepted.

This methodology appears to have potential for defining statistical terms required for Bayesian analyses (Davies, 1972) and for Kalman Filter techniques. Some refinement in the parameters employed here (S_p, S_m, and S_r) may be required prior to specific application. For example, variances here are averaged across models. The implications that these values have concerning the establishment of confidence bounds for individual predictions is not clear. It must be recognized that the predictive error estimate, S_p, reflects a mean modeling response and individual model reliability may vary widely.

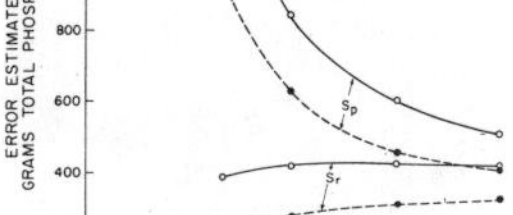

Figure 6.--Comparison of two models using predictive accuracy determination.

The method also lends itself to incorporation into calibration procedures. The method eliminates both conscious and unconscious bias that sometimes affects the selection of both calibration and verification subsets and subsequent model refinement. The impact of subtleties of the calibration procedure is clearly more evident.

CONCLUSIONS

1. A significant enhancement of the model used in the case study was realized by supervising the selection of calibration points and controlling the size of the calibration data subset.
2. Predictive error was found to be substantially greater than calibration error when a small number of calibration points was employed. As the number of calibration points was increased, predictive error and calibration error appeared to converge.
3. A data pool subset size of 12 appeared to be sufficient to calibrate both the power function and linear models examined in this paper.
4. No enhancement of predictive accuracy was realized by placing a maximum calibration error criterion on the calibration procedure.
5. The iterative predictive accuracy determination provides a useful estimate of the ability of a model to predict unseen events.

LIST OF REFERENCES

Craig, George D. 1970. Predictive accuracy of quarterly and annual aggregate saving functions. p. 1131. Journal of the American Statistical Association, vol. 65, no. 331.

Crane, Stephen M. 1981. Reliability analysis of BOD kinetics in a small, southern stream governed by the discharge of an oxidation pond. M.S. Thesis, 124 p. Louisiana State University, Baton Rouge, LA.

Davis, Donald R., Chester C. Kisiel, and Lucien Duckstein. 1972. Bayesian Decision Theory Applied to Design in Hydrology. p. 33-40. Water Resources Research, vol. 8, no. 1.

Diniz, E. V. 1979. Water quality prediction for urban runoff--an alternative approach. p. 112. Proceedings, SWMM Users Group Meeting. U.S. Environmental Protection Agency.

Freund, Rudolf J., and Paul D. Minton. 1979. Regression methods: a tool for data analysis. 261 p. Marcel Dekker, Inc., New York, NY.

Geisser, Seymour. 1975. The predictive sample reuse method with applications. p. 320-328. Journal of the American Statistical Association, vol. 70, no. 350.

Huber, W. C., et al. 1975. Storm Water Management Model, User's Manual, Version II. 367 p. Environmental Protection Technology Series. USEPA, Cincinnati, OH.

International Mathematical and Statistical Libraries, Inc. 1980. Regression analysis. In IMSL Library Reference Manual. Eighth edition. International Mathematical and Statistical Libraries, Inc., Houston, TX.

Knaus, Ronald M., et al. 1981. Second annual report: lakes restoration project. 208 p. U.S. EPA Grant Number CS805836-10. Institute for Environmental Studies, Louisiana State University, Baton Rouge, LA.

McCarthy, Philip J. 1976. The use of balanced half-sample replication in cross-validation studies. p. 569-604. Journal of the American Statistical Association, vol. 71, no. 355.

Shroder, Michael L., Richard H. McCuen, and Walter J. Rawls. 1980. The effect of data independence in model calibration and model testing. p. 49-55. Water Resources Bulletin, vol. 16, no. 1.

GLOBAL STABILITY CONCEPTS FOR COMPLEX ECOSYSTEM MODELS[1]

David A. Mauriello[2]

Abstract.---A simulation approach to the global stability properties of complex ecosystem models is introduced. A Convergence Coefficient is developed which quantifies the deviation between perturbed and unperturbed system trajectories. The amplitude and duration of the deviation from the undisturbed system trajectory provides a measure of system resilience and persistence. This approach to the global stability properties of complex ecological systems is illustrated with an aquatic ecosystem model. This model exhibits globally stable behavior for the range of perturbations tested.

INTRODUCTION

The question of the stability of ecological systems has received a great deal of attention from both mathematicians and ecologists. The ecological literature contains numerous analyses of the stability properties of ecological models of varying complexity. In the majority of these investigations, the concept of stability utilized is the classical Lagrangian definition of a stable system as one which returns to a state of dynamic equilibrium following a perturbation. Stability characteristics are determined from analytic solutions to linearized equations which approximate system behavior in the immediate vicinity of equilibrium points (May 1973).

However, conclusions derived from such local stability analyses yield only limited insights into the dynamics of real ecosystems because of a number of conceptual and methodological difficulties:

(a) Stability properties are valid only for small deviations from equilibrium states. Except in the case of linear systems, no conclusions can be made concerning system responses to large perturbations.

(b) Real ecosystems and complex ecosystem models may possess numerous equilibria and local stability domains (Austin and Cook 1974, Holling 1973, Sutherland 1974). Local stability properties must be determined in the vicinity of each of these equilibrium points in order to describe the global stability properties of such a model. In addition, local stability analyses would determine neither the number or the extent of such local stability domains.

(c) Stability criteria predicated on equilibrium concepts are inadequate tools for understanding systems which normally fluctuate over wide ranges or are subjected to time-varying environmental forcing functions (Botkin and Sobel 1975).

The deficiencies of traditional approaches to analyzing ecological stability have been widely recognized. Holling (1973) proposed a global view of ecosystem stability based on the dual concepts of "stability" and "resilience". These concepts described the ability of ecosystems to absorb changes in forcing functions, state variables and parameters. Botkin and Sobel (1975) emphasized "persistence" and "recurrence" as useful global properties, while Beddington et al. (1976) suggested that characteristic recovery times be used as a measure of system resilience following a perturbation. Other studies have examined the transient behavior of ecological models and real ecosystems following perturbations of state variables and parameters (Beddington et al. 1976, Dudzik et al. 1975, Harwell et al. 1981, Hurd and Wolf 1974, O'Neill 1976).

{1} Paper presented at the Third International Conference on the State-of-the-Art in Ecological Modelling, Colorado State University, May 24-28, 1982.

{2} David A. Mauriello is an ecologist with the U.S. Environmental Protection Agency, Office of Toxic Substances, TS-796, 401 M Street, SW, Washington, DC 20460.

These studies represent efforts to identify global properties relevant to ecologists and depart from those criteria based primarily on mathematical convenience. In this paper we will attempt to define appropriate global stability criteria and introduce a simulation technique for examining the global behavior of complex ecological models.

MODELS AND ECOSYSTEMS

Any methodological approach to analyzing the global stability characteristics of complex ecosystem models must provide useful insights into the dynamics of real ecosystems. Observations of the behavior of real ecosystems yield several general conclusions which are especially relevant to development of a global perspective on ecological stability.

(a) Real ecosystems tend to be persistent. Observed fluctuations in density or population size do not normally result in extinction.

(b) Large scale perturbations are the rule rather than the exception in natural systems. The assumption that deviations from equilibrium states are arbitrarily small is often violated.

(c) Real ecosystems are seldom in a state of dynamic equilibrium. Environmental heterogeneity, in both time and space, may prevent natural ecosystems from reaching or remaining in equilibrium states. This heterogeneity may serve to enhance diversity and prevent the competitive exclusion of species from the system (Hutchinson 1961, Mauriello 1976).

Efforts to incorporate realistic descriptions of ecological processes have resulted in increasingly complex nonlinear ecosystem models. Such models are not easily analyzed by the classical methods employed to determine stability. Analytical methods for systems of nonlinear differential equations are not well defined and are difficult to apply to any but the simplest of models (Diamond 1976, Goh 1976, Jeffries 1976, Lin and Kahn 1977).

The dynamic behavior of complex ecosystem models is extremely diverse. Models may possess one or more stable equilibrium points, each of which is the focal point of a "zone of attraction" or local stability region. Where no asymptotically stable equilibria exist, state variable trajectories may converge on stable, bounded limit cycles. In addition, convergence of system trajectories towards equilibrium points or stable limit cycles may be masked by fluctuations resulting from environmental driving functions such as solar radiation or temperature. Finally, structural properties of complex system models, the number and size of local stability domains, may be altered by changes in parameter values, resulting in discontinuities in dynamic system behavior (Jones 1977, Peterman 1977, DeAngelis et al. 1975).

A GLOBAL STABILITY SYNTHESIS

The concepts of persistence and resilience can provide ecologists with a unifying global stability perspective on the dynamic behavior of both models and real ecosystems. Persistence, the maintenance of all system components through time, can be related to the existence of one or more local stability domains. The concept of resilience relates directly to the topography of the local stability regions and the transient behavior of state variable trajectories following a perturbation.

These general concepts are incorporated into the following series of questions which can be employed to define the global stability characteristics of both models and real ecosystems.

1. Does the system persist? Are all ecosystem components preserved following a perturbation?

2. How resilient are the components of the system? What is the duration required before perturbed state variable trajectories converge on unperturbed trajectories? Are alternative state variable trajectories possible?

3. What is the range of perturbations for which the system will continue to persist?

4. What are the boundaries of the local stability domains?

5. How are global stability properties modified by changes in parameter values?

In the next section we will introduce a technique which can be readily applied to complex ecosystem models to yield insights into the global stability properties discussed above.

THE CONVERGENCE COEFFICIENT

An intuitive definition of a stable system is one which has the capability to recover from a perturbation. In the mathematical sense, system state variables approach stable equilibrium points or converge on limit cycle trajectories. As a further generalization, perturbed state variable trajectories converge on unperturbed trajectories. These concepts can be formalized into the following definition of stability.

Let X(t) represent the state variable trajectories of an unperturbed system and Y(t) represent the state variable trajectories of the system following a perturbation.

$$\text{If} \quad [Y(t0) - X(t0)] < N(k)$$

$$\text{and} \qquad (1)$$

$$[Y(t) - X(t)] < k \quad \text{for } t > t0$$

then the dynamic behavior of the system is stable in the sense of Lyapunov (Hochstadt 1964). Given a bounded perturbation, deviations between perturbed and unperturbed state variable trajectories will be bounded in a stable system. The deviations between perturbed and unperturbed state variable trajectories become arbitrarily small with the passage of time. This is an operational definition of stability which is valid for both large and small perturbations as well as for time-varying systems.

From this definition we can develop a Convergence Coefficient to quantify the deviation between perturbed and unperturbed dynamic state variable trajectories.

The Convergence Coefficient, C(t), can be defined as follows:

$$C(t) = ABS\{[\ Y(t) - X(t)\ ]/\ X(t)\} \qquad (2)$$

where,

X(t) = Vector of unperturbed state variable trajectories

Y(t) = Vector of perturbed state variable trajectories

C(t) is a normalized, time-varying vector of deviation trajectories of system state variables.

If C(t) is calculated for all state variables in a system model, a comprehensive picture of system response to perturbations can be obtained. If perturbed state variable trajectories converge on unperturbed or reference trajectories, the Convergence Coefficient approaches zero, satisfying the Lyapunov stability criterion. The perturbed trajectories thus lie within the same local stability domain as the reference trajectories. Within a local stability domain, the rate at which C(t) approaches zero provides a relative measure of system resilience.

If the Convergence Coefficient, C(t), does not approach zero, then the perturbed state variable trajectories lie outside the local stability domain of the reference state variable trajectories. If the perturbed trajectory lies in another local stability domain, the state variable will persist in the system, although with a different pattern of dynamic behavior. If not, the population represented by that state variable will not persist in the system following the perturbation. By varying the amplitude of perturbations in a series of simulation runs, the boundaries of local stability domains may be determined. If the amplitude of initial value perturbations exceeds the dimensions of the local stability region, then C(t) for one or more of the state variables will not approach zero. The perturbed state varaible trajectories may persist in another local stability domain or the system components they represent will disappear from the system.

The Convergence Coefficient may also be utilized to examine the structural stability properties of complex ecosystem models. For a given set of parameter values, local stability domains can be mapped and dynamic behavioral response to perturbations determined. A global analysis for each new set of parameter values will demonstrate how global stability properties are affected by parameter changes.

The Convergence Coefficient, C(t), can provide meaningful insight into the global stability criteria discussed previously. It can be used with ecosystem models of varying complexity, avoiding the inherent limitations of traditional methods of stability analysis. In the next section we will utilize the Convergence Coefficient to examine the global stability properties of a model of plankton dynamics in a marine upwelling area.

THE MODEL

The model we will use to illustrate the Convergence Coefficient is a three compartment model of phytoplankton and zooplankton dynamics in a marine upwelling area. The model is similar to several others described by Riley (1965), and Steele (1974), with the addition of time-varying functions for solar radiation, vertical mixing, and water temperature. The water column is divided into two layers. The upper layer, where interactions between phytoplankton and zooplankton occur, is well lighted, homogeneously mixed, and lies above the seasonal thermocline. Below this lies an unlighted zone which mixes with the upper layer only when the thermocline is absent. Particulate organic matter sinks into this zone from above and eventually decomposes. This material is later recycled to the photic zone during periods of turbulent mixing. Thus, the lower layer serves as a reservoir of nutrients for the upper. Nutrient concentration in the lower layer is considered constant in this model. In addition to nutrients supplied to phytoplankton from exchanges with the lower layer, zooplankton excretion provides a pathway for rapidly recycling nutrients to phytoplankton in the photic zone.

The state equations for the three state variables, Phytoplankton (P), Zooplankton (Z), and Nitrate (N), are as follows:

Phytoplankton

dP/dt = P(GROW-RESP-SINK-AMIX) - Z(GRAZ)

where, (3)

P = phytoplankton biomass, mg/cu-m
Z = zooplankton biomass, mg/cu-m
GROW = phytoplankton growth rate, 1/day
RESP = respiration loss rate, 1/day
SINK = sinking loss rate, 1/day
AMIX = mixing loss rate, 1/day
GRAZ = zooplankton grazing rate, 1/day

Zooplankton

dZ/dt = Z(ASIM-ZRESP-ZMOR) (4)

where,

ASIM = zooplankton growth rate, 1/day
ZRESP = zooplankton respiration loss, 1/day
ZMOR = zooplankton mortality rate, 1/day

Nitrate

dN/dt = UPWEL+REGEN-UPTK (5)

where,

UPWEL = rate of nutrient supply from lower zone, mg/cu-m/day
REGEN = nutrient regeneration from zooplankton excretion, mg/cu-m/day
UPTK = nutrient uptake by phytoplankton, mg/cu-m/day

We now examine the functional process descriptions incorporated into each of the state equations.

Phytoplankton Functional Relations

Growth

Phytoplankton growth rate is modelled as the product of a maximum growth rate and normalized functions relating growth to levels of limiting variables for solar radiation, nitrate, and water temperature. The functions for photosynthesis, nutrient uptake, and temperature dependence are shown in figure 1. Photosynthesis is an increasing function of light intensity until a saturation level is exceeded (Steele 1962). Likewise, growth rates increase with temperature but decline rapidly beyond the optimum temperature (Lassiter and Kearns 1973). The limitation of phytoplankton growth by nutrient availability is described by Michaelis-Menten kinetics (MacIsaac and Dugdale 1969). The rate at which phytoplankton biomass increases is given by the following relationship:

GROW = UMAX(PHOT)(ANUT)(ATEMP) (6)

where,

UMAX = maximum growth rate, 1/day
PHOT = photosynthesis function
ANUT = nutrient uptake function
ATEMP = temperature dependence function

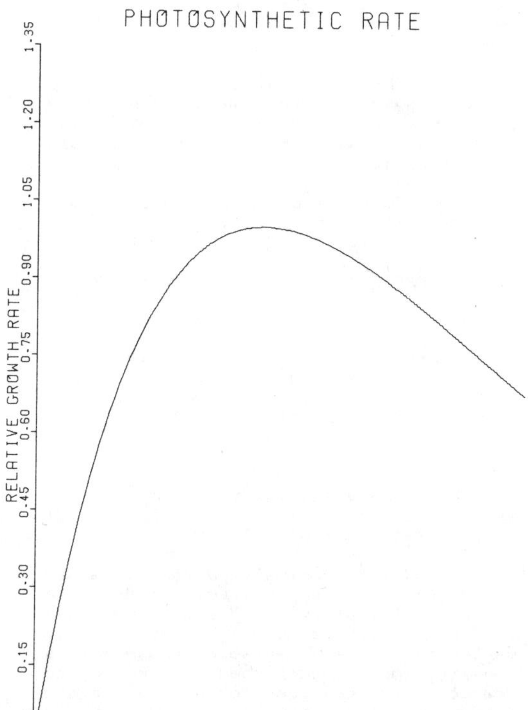

Figure 1(a).-- Normalized phytoplankton growth rate as a function of light intensity.

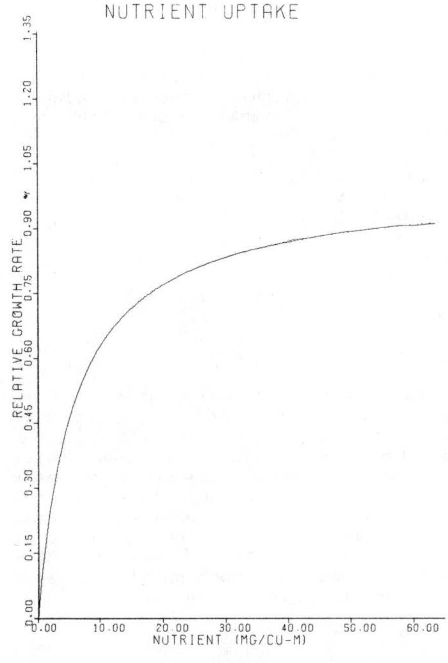

Figure 1(b).-- Normalized phytoplankton growth rate as a function of nutrient concentration.

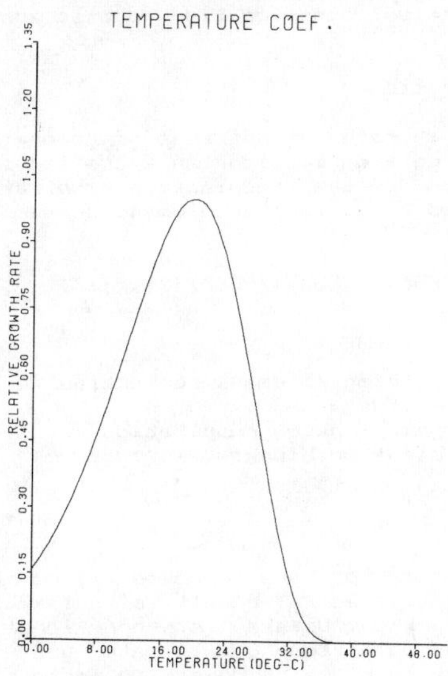

Figure 1(c).-- Normalized phytoplankton growth rate as a function of ambient water temperature.

Respiration Losses

Phytoplankton respiration losses are directly proportional to water temperature and the metabolic Q10 value:

RESP = RO{EXP[TCOEF(TEMP)]} (7)

where,

RO = 0 degree respiration rate
TCOEF = Q10 coefficient
TEMP = water temperature, degree C

Sinking Losses

Phytoplankton biomass is lost as cells sink from the photic zone into the aphotic zone. The loss rate is proportional to an intrinsic sinking rate and the depth of the photic zone. The depth of the photic zone is variable and dependent on the magnitude of the light extinction coefficient, which is an empirically derived function of phytoplankton density:

SINK = VS/DEP (8)

where,

VS = phytoplankton sinking rate, m/day
DEP = depth of photic zone, m

Mixing Losses

Phytoplankton biomass is also lost when cells are carried from the photic zone to the aphotic zone by turbulent mixing processes. The rate of mixing, AMIX, varies seasonally, depending on water temperature (figure 3).

Grazing Losses

Phytoplankton losses to zooplankton grazing are modelled as the product of a temperature-dependent maximum grazing rate and a grazing coefficient which incorporates phytoplankton availability and a grazing threshold into a Michaelis-Menten type function (figure 2):

GRAZ = GRMAX(GRCOEF) (9)

where,

GRMAX = maximum grazing rate, 1/day
GRCOEF = grazing coefficient (figure 2)

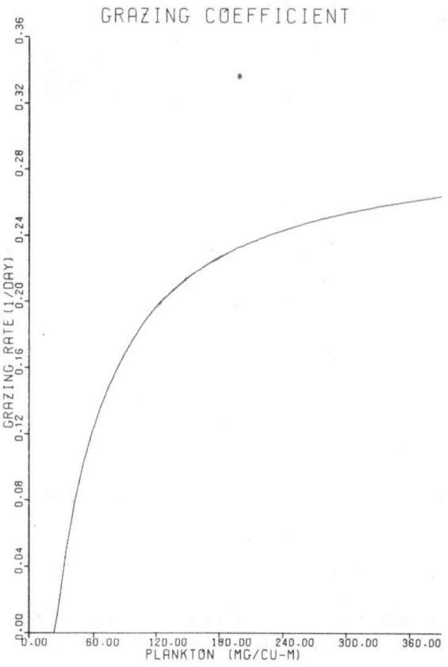

Figure 2.-- Zooplankton grazing rate as a function of phytoplankton biomass.

Zooplankton Functional Relations

Growth

The growth of zooplankton biomass results from the ingestion and assimilation of phytoplankton through the grazing process. The ingestion rate is given by the grazing coefficient, GRCOEF. A portion, ASIM, of the phytoplankton ingested is assimilated as new zooplankton biomass while the remainder is egested:

$$ASIM = EFF(GRAZ) \quad (10)$$

where,

EFF = assimilation efficiency
GRAZ = zooplankton ingestion rate, 1/day

Respiration

Zooplankton respiratory losses are a function of a temperature dependent respiration factor, similar to phytoplankton, and the rate of phytoplankton ingestion. As more food is ingested, the zooplanton respiration rate is increased:

$$ZRESP = ZRO\{EXP[TCOEF(TEMP)]\}+ZRM(GRAZ) \quad (11)$$

where,

ZRO = 0 degree respiration rate
TCOEF = Q10 coefficient
TEMP = water temperature, degrees C
ZRM = respiration multiplier
GRAZ = zooplankton ingestion rate, 1/day

Mortality

Zooplankton mortality results primarily through the predatory activities of planktivorous fish and invertebrates. Since these populations are not included in the model, we will consider the rate of zooplankton losses to be a small constant fraction of zooplankton biomass, ZMOR.

Nutrient Functional Relations

Upwelling

Nutrients, principally nitrates, are supplied to phytoplankton in the photic zone as the result of turbulent mixing from the lower layer and zooplankton excretion. The amount of nutrient made available through the upwelling process is a function of the seasonal mixing rate and is proportional to the gradient of nutrient concentrations between the photic zone and that portion of the water column which lies below the seasonal thermocline:

$$UPWEL = AMIX(NO-N) \quad (12)$$

where,

AMIX = rate of turbulent mixing
NO = nutient concentration, lower layer
N = nutrient concentration, photic zone

Regeneration

The rate of nutrient regeneration resulting from zooplankton excretion is a function of the zooplankton respiration rate and a nutrient to biomass conversion factor:

$$REGEN = CONV(ZRESP)(Z) \quad (13)$$

where,

CONV = nutrient/biomass conversion factor
ZRESP = zooplankton respiration rate
Z = zooplankton biomass, mg/cu-m

Uptake

Nutrients are removed from the photic zone as a result of uptake by phytoplankton. Uptake is proportional to the phytoplankton growth rate and the nutrient to biomass conversion factor:

$$UPTK = CONV(GROW)(P) \quad (14)$$

where,

CONV = nutrient/biomass conversion factor
GROW = phytoplankton growth rate, 1/day
P = phytoplankton biomass, mg/cu-m

Environmental Functions

Time-varying environmental forcing functions for incident solar radiation, water temperature, and turbulent mixing are included in the model, as shown in figure 3.

Solar Radiation

The seasonal variation of incident solar radiation is approximated in the model by a sinusoidal function which reaches a maximum during the summer months of June and July:

$$RAD = RMIN+[(RMAX-RMIN)/2][1-COS(ARG)] \tag{15}$$

where,

RAD = average daily radiation, ly/day
RMIN = minimum daily radiation, ly/day
RMAX = maximum daily radiation, ly/day
ARG = (2/365)(TIME)
TIME = time, days

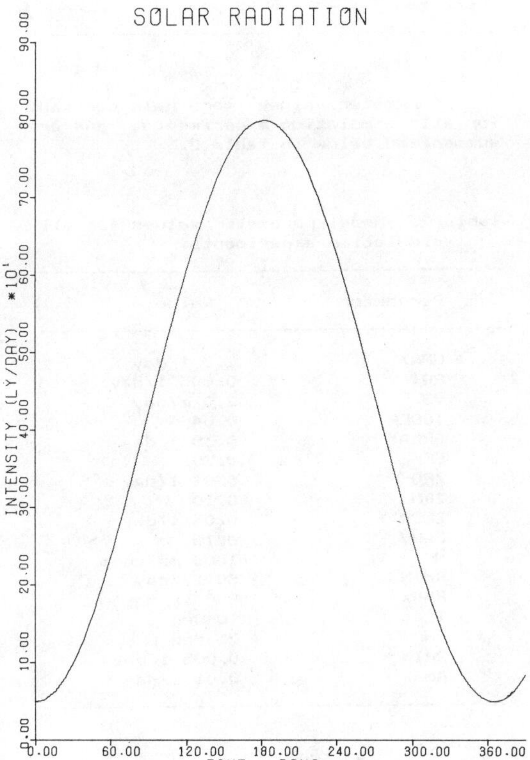

Figure 3(a).-- Seasonal variation of incident solar radiation.

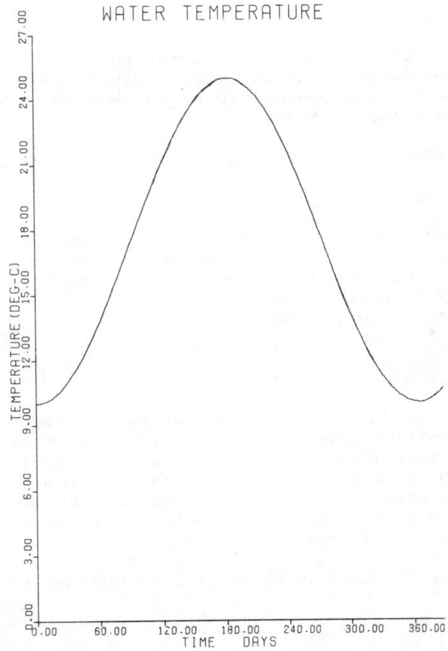

Figure 3(b).-- Seasonal variation of ambient water temperature.

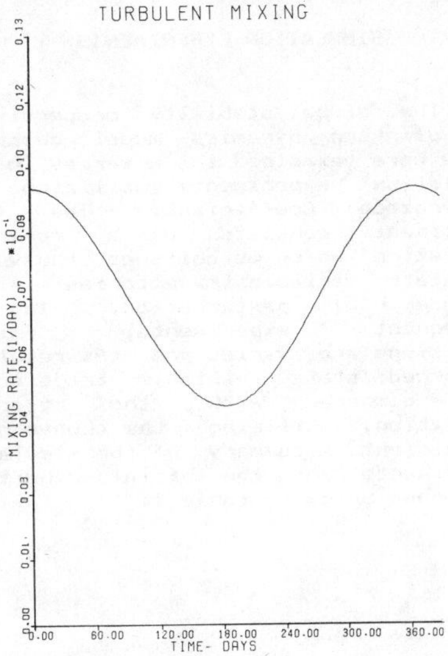

Figure 3(c).-- Seasonal variation in the rate of turbulent mixing in the water column. The mixing rate represents the daily fraction of the upper layer of the water column exchanged with the lower layer.

Water Temperature

Like solar radiation, water temperature varies from minimum levels during the winter to a maximum during the summer months:

$$TEMP = TL + [(TH-TL)/2][1-COS(ARG)] \quad (16)$$

where,

TEMP = daily water temperature, deg C
TL = minimum daily temperature, deg C
TH = maximum daily temperature, deg C
ARG = (2/365)(TIME)
TIME = time, days

Mixing

In the hypothetical marine upwelling area considered in this model, turbulent mixing reaches a peak during the winter months when the thermocline is absent. Conversely, mixing between layers is minimized during the summer when the thermocline is established:

$$AMIX = AMIN + [(AMAX-AMIN)/2][1-COS(ARG1)] \quad (17)$$

where,

AMIX = daily mixing rate
AMIN = minimum daily mixing rate
AMAX = maximum daily mixing rate
ARG1 = (2/365)(TIME-182.5)
TIME = time, days

SIMULATION EXPERIMENTS

The global stability properties of the plankton dynamics model described above were examined in a series of six simulation experiments utilizing the Convergence Coefficient. The first experiment consists of a reference simulation which establishes the vector of state variable trajectories in the absence of perturbation. In the subsequent experiments, initial conditions are varied and the resulting perturbed state variable trajectories are compared with the reference simulation, utilizing the Convergence Coefficient. A summary of the simulation experiments and the initial conditions are shown below in table 1.

Table 1.--Summary of simulation experiments. Initial conditions for state variables during reference and perturbation simulation runs.

	Concentrations (mg/cu-m)		
Experiment	P	Z	N
Reference Simulation	50	50	350
Phytoplankton Increased	500	50	350
Phytoplankton Decreased	5	50	350
Zooplankton Increased	50	500	350
Zooplankton Decreased	50	5	350
Zooplankton Decreased/ Nitrate Increased	50	5	3500

Parameter values were held constant for all simulation experiments, and are summarized below in table 2.

Table 2.--Model parameter values for all simulation experiments.

Parameter	Value
UMAX	2.0 1/day
RO	0.005 1/day
VS	2.0 m/day
TCOEF	0.04
GRMAX	0.30 1/day
EFF	0.70
ZRO	0.01 1/day
ZRM	0.10
ZMOR	0.05 1/day
CONV	0.15
NO	1000 mg/cu-m
RMIN	50 ly/day
RMAX	800 ly/day
TL	10 deg C
TH	25 deg C
AMIN	0.005 1/day
AMAX	0.01 1/day

The results of the reference simulation run are shown in figure 4. A two year simulation period was used in order to reduce the influence of initial conditions on the reference vector of state variable trajectories. Nitrate concentrations in the upper layer reach a maximum during the winter months as a result of high rates of turbulent mixing and low phytoplankton populations. Phytoplankton populations reach a peak in early spring only to be reduced by zooplankton grazing and nutrient limitation during the summer stratification period. A lesser phytoplankton peak in late summer is rapidly grazed by a growing zooplankton population, even though the supply of nutrients is increasing.

In the second experiment, the initial phytoplankton population is increased by a factor of ten as compared with the reference simulation. The trajectories of the Convergence Coefficients for each state variable are shown in figure 5. C(t) for the phytoplankton population approaches zero almost immediately, indicating rapid convergence between perturbed and unperturbed trajectories. The Convergence Coefficient trajectories for zooplankton and nitrate also approach zero, but more slowly than that for phytoplankton. Zooplankton populations exceed their reference levels in response to the higher initial phytoplankton population. In contrast, nitrate concentrations are reduced at first because of greater uptake by phytoplankton.

When the initial concentration of phytoplankton is reduced to one tenth of the reference level, a similar pattern of convergence is exhibited (figure 6). The dynamic behavior of the state variables is almost a mirror image of the second experiment. The zooplankton population declines in response to lower phytoplankton concentrations, while the nitrate concentration is initially elevated because of reduced uptake by phytoplankton. The trajectories of the Convergence Coefficients all approach zero by the end of the first year of the simulation.

In the fourth and fifth experiments, initial zooplankton populations are manipulated while the other state variables are held at their reference levels. When initial zooplankton levels are increased by a factor of ten, the Convergence Coefficients for all three state variables approach zero by the end of the first year of the simulation run (figure 7). Grazing by zooplankton reduces phytoplankton populations below their reference levels, resulting in a corresponding increase in nitrate

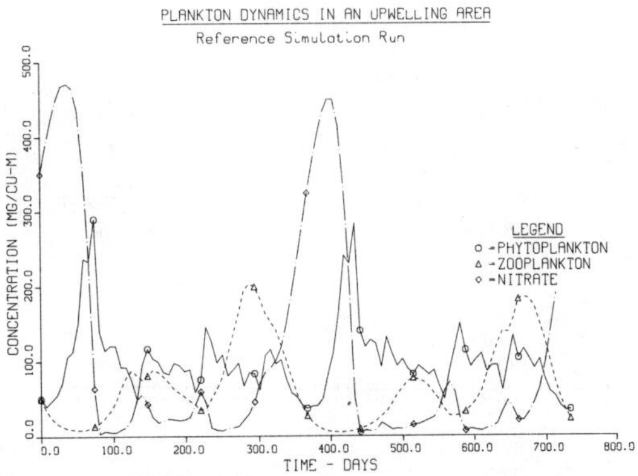

Figure 4.-- Reference simulation run. Dynamics of plankton populations in an upwelling area.

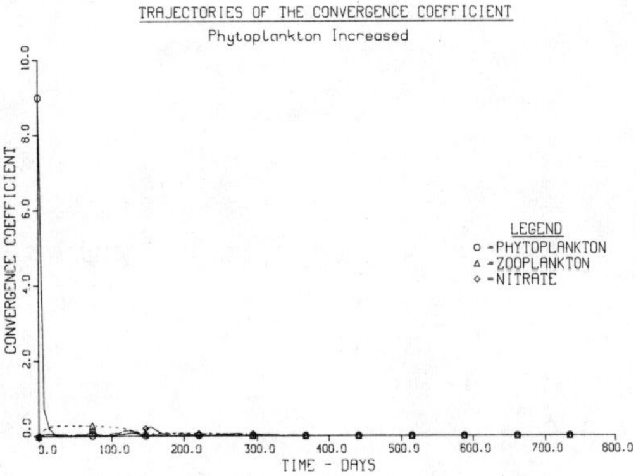

Figure 5.-- Trajectories of the Convergence Coefficient when initial phytoplankton biomass is increased by a factor of ten.

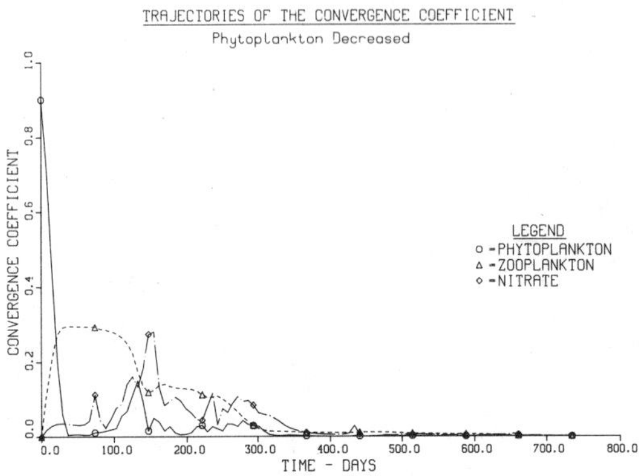

Figure 6.-- Trajectories of the Convergence Coefficient when initial phytoplankton biomass is reduced to one tenth the reference simulation level.

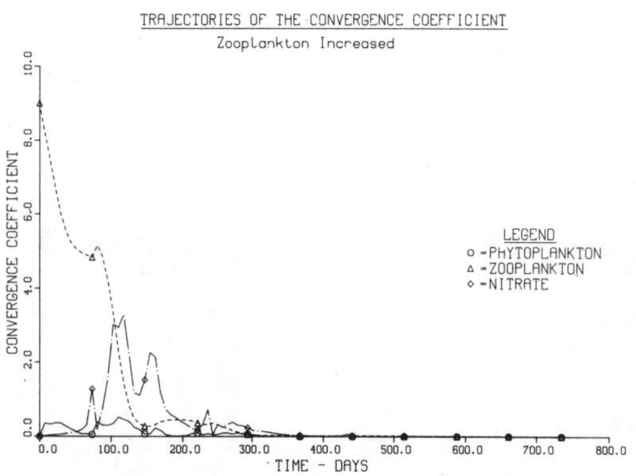

Figure 7.-- Trajectories of the Convergence Coefficient when the initial zooplankton biomass is increased by a factor of ten over the reference level.

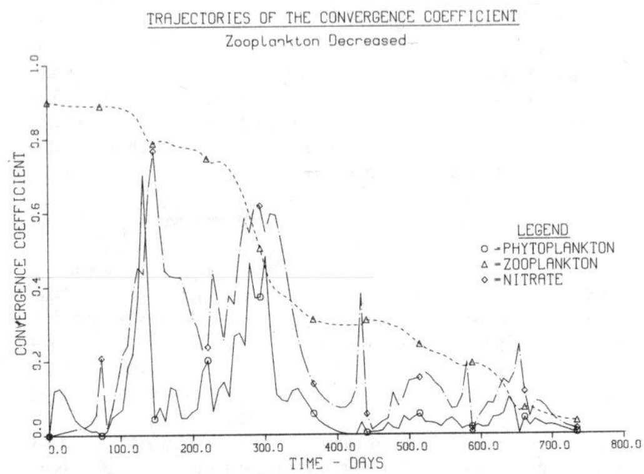

Figure 8.-- Trajectories of the Convergence Coefficient when initial zooplankton biomass is reduced by a factor of ten.

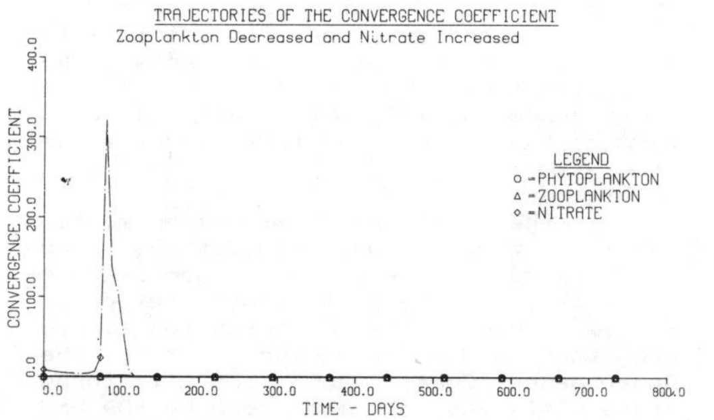

Figure 9.-- Trajectories of the Convergence Coefficient. Initial levels of zooplankton and nitrate are decreased and increased respectively by a factor of ten as compared with reference run.

concentration. The dynamic behavior of the system is somewhat different when the zooplankton population is initially reduced. The reduction in the zooplankton population frees the phytoplankton from grazing pressure for much of the first year of the simulation run. Phytoplankton dynamics are thus regulated primarily by nutrient limitation. Nitrate levels remain below those of the reference trajectory. The Convergence Coefficients for phytoplankton and nitrate exhibit a series of damped oscillations which approach zero only late in the second year of the simulation, as the zooplankton population slowly converges towards the reference trajectory (figure 8).

In the final simulation experiment, the nitrate concentration is increased at the same time that the initial zooplankton population is reduced. Released from grazing pressure, the phytoplankton population grows rapidly because of the nutrient enrichment. This rapid growth phase ends abruptly as the nitrates in the photic zone are exhausted. Thus nutrient limitation controls the phytoplankton dynamics to an even greater extent than when only the zooplankton were reduced. This is confirmed by examining the trajectories of the Convergence Coefficients in figure 9. The large spike in C(t) for nitrate results from the rapid depletion of the nutrient supply from uptake by the exploding phytoplankton population. The subsequent collapse of the phytoplankton population, due to nutrient limitation, results in the rapid convergence of the state variable trajectories.

CONCLUSIONS

The results of the simulation experiments indicate that the model of the marine plankton upwelling ecosystem is stable for the range of perturbations tested. This system appears to be resilient to large scale perturbations of the component populations and resource levels. These results are consistent with the observations of O'Neill (1976), Steele (1974), and Sjoberg (1977) that such pelagic ecosystems are inherently resilient and persistent. Sjoberg (1977) points out that the internal recycling pathways and the nature of the herbivore functional response in large part determine the extent to which such systems are resistant to perturbations. This conclusion is given credence by the results of the fifth experiment, where the initial zooplankton level was reduced. In the absence of grazing pressure, convergence of the state variable trajectories was not as rapid as in the other experiments.

The introduction of the Convergence Coefficient has provided a convenient methodology by which to explore the global properties of complex, time-varying ecosystem models. It is precisely these global stability characteristics which are of prime concern to ecologists. It is hoped that the adoption of a global perspective by ecologists will result in a greater degree of understanding of the dynamics of real ecosystems.

LITERATURE CITED

Austin, M.P. and B.G. Cook. 1974. Ecosystem stability: a result from an abstract simulation. Journal of Theoretical Biology 45:435-458.

Beddington, J.R., Free, C.A., and J.H. Lawton. 1976. Concepts of stability and resilience in predator-prey models. Journal of Animal Ecology 45:791-816.

Botkin, Daniel B., and Matthew J. Sobel. 1975. Stability in time-varying ecosystems. The American Naturalist 109:625-646.

DeAngelis, D.L., Goldstein, R.A., and R.V. O'Neill. 1975. A model for trophic interaction. Ecology 56:881-892.

Diamond, Phil. 1976. Domains of stability and resilience for biological populations obeying difference equations. Journal of Theoretical Biology 61:287-306.

Dudzik, M., Harte, J., Levy, D., and J. Sandusky. 1975. Stability indicators for nutrient cycles in ecosystems. Report No. 3264. Lawrence Berkeley Laboratory. 59p.

Goh, B.S.. 1976. Global stabiltiy in two species interactions. Journal of Mathematical Biology 3:313-318.

Harwell, Mark A., Cropper, Wendell P., and Harvey L. Ragsdale. 1981. Transient characteristics of a nutrient cycling model. Ecological Modelling 12:105-131.

Hochstadt, Harry. 1975. Differential equations: a modern approach. 294 p. Dover Publications. New York.

Holling, C.S.. 1973. Resilience and

stability of ecological systems. Annual Review of Ecology and Systematics 4:1-23.

Hurd, Lawrence E., and Larry L. Wolf. 1974. Stability in relation to nutrient enrichment in arthropod consumers of old-field successional ecosystems. Ecological Monographs 44:465-482.

Hutchinson, G.E.. 1961. The paradox of the plankton. The American Naturalist 95:137-145.

Jeffries, Clark. 1976. Stability of predation ecosystem models. Ecology 57-1321-1325.

Jones, Dixon D.. 1977. Catastrophe theory applied to ecological systems. Simulation 58:1-15.

Lassiter, R.R., and D.K. Kearns. 1973. Phytoplankton population changes and nutrient fluctuations in a simple aquatic ecosystem model. In: Middlebrooks, J.E. (ed.) Modeling the Eutrophication Process. Utah Water Research Laboratory. pp. 131-138.

Lin, Juan, and Peter B. Kahn. 1977. Qualitative behavior of predator-prey communities. Journal of Theoretical Biology 65:101-132.

MacIsaac, J.J., and R.C. Dugdale. 1969. The kinetics of nitrate and ammonia uptake by natural populations of marine phytoplankton. Deep Sea Research 16:45-57.

Mauriello, David A.. 1976. Disequilibrium and the paradox of the plankton. Ph.D. Thesis. 183 p. Rutgers University. New Brunswick, New Jersey.

May, Robert M.. 1973. Stability and complexity in model ecosystems. 235 p.. Princeton University Press. Princeton, New Jersey.

O'Neill, R.V.. 1976. Ecosystem persistence and heterotrophic regulation. Ecology 57:1244-1253.

Peterman, Randall. 1977. A simple mechanism which causes collapsing stability regions in exploited salmonid populations. Journal of the Fisheries Research Board of Canada 34:1130-1142.

Riley, G.A.. 1965. A mathematical model of regional variations in plankton. Limnology and Oceanography 10(Supp.):R202-215.

Sjoberg, Stig. 1977. Are pelagic systems inherently unstable? A model study. Ecological Modelling 3:17-37.

Steele, John H.. 1959. The quantitative ecology of marine phytoplankton. Biological Reviews 34:129-158.

Steele, John H.. 1962. Environmental control of photosynthesis in the sea. Limnology and Oceanography 7:137-150.

Steele, John H.. 1974. The structure of marine ecosystems. 114 p.. Harvard University Press. Cambridge, Massachusetts.

Sutherland, John P.. 1974. Multiple stable points in natural communities. The American Naturalist 108:859-873.

ECOLOGICAL MODELLING AND MANAGEMENT OF SPRUCE BUDWORM INFESTED FIR-SPRUCE FORESTS OF NEW BRUNSWICK, CANADA[1]

Wilf Cuff[2] and Gordon Baskerville[3]

Abstract.--Four of 13 existing models have been applied to forest protection/management strategy evaluation in New Brunswick, Canada. Each successful model was built in association with the user. It may be possible to apply models built independently of a user but it will require a search for an appropriate niche.

INTRODUCTION

The eastern spruce budworm, *Choristoneura fumiferana* (Clem.) is a pest of the fir-spruce forests of eastern North America. Although epidemics are known to have occurred periodically since the first decade of the 18th century (Blais 1965) it was only with the development of a pulp and paper industry in the present century that man became a serious competitor with the budworm for the softwood forests (Baskerville 1976).

By 1943 entomologists were forecasting another budworm outbreak in New Brunswick (Miller & Kettela 1975). This prompted intensive research on the biology and population dynamics of this insect (Morris 1963), and on the potential of forest management practices to mitigate the problem. The threat became a reality in 1950 when the forecasts were confirmed by the appearance of areas of defoliation in the central highlands of northern New Brunswick. Aerial spraying of insecticide began in 1952 to prevent mortality on some 78,000 ha of forest. By 1957 the area sprayed had peaked at 2 million ha, about a third of the productive forest land of New Brunswick. The infested area dropped off in the 1960's but increased again in the 1970's; forest protection (spraying) operations have continued to the present, annually adjusting to budworm dynamics which alter both the location and intensity of the problem.

About 13 ecological models relating to New Brunswick have been published during the period of forest protection. In this paper we briefly describe these models, with the aim of finding out why only a fraction of them has been used in forest management in New Brunswick.

[1]Paper presented at the Third International Conference on State-of-the-Art in Ecological Modelling, Colorado State University, May 24-28, 1982.

[2]Wilfred R. Cuff is Research Scientist, Maritimes Forest Research Centre, Environment Canada Forestry Service, Fredericton, New Brunswick, Canada.

[3]Gordon L. Baskerville is Assistant Deputy Minister, Forest Resources, New Brunswick Department of Natural Resources, Fredericton, New Brunswick, Canada.

HISTORICAL PERSPECTIVE

The first model recorded in the literature (Webb et al. 1956) has the double distinction of being among the simplest to date and the only one that has been used (and extensively so) in forest protection operations. The model is applied annually to delineate those *hazard* areas of New Brunswick where further budworm attack could cause tree mortality; this annual hazard map is the basis for the design of the protective spray operations in the coming year.

The hazard-rating model defines the risk to tree populations across the province as a function of the degree of defoliation in past years, the degree of recovery of the foliage destroyed by budworm attacks in past years, the degree of defoliation in the current year, and the prospective intensity of defoliation in the coming year (as indicated by egg-mass counts in the current year). Each of these variables is classified into various quantitative or qualitative categories and assigned ratings. For example: current defoliation has ratings of 0 through 4, designating no defoliation, up to 25%, 26-65%, 66-95%, and all shoots missing; recovery from past defoliation has ratings of -3, -2, -1, and 0, designating current foliage crop apparently normal and trees evidently capable of rapid recovery (-3), shoot production moderate but obviously affected by declining tree vigor (-2), and so on. The overall hazard rating of an area of forest is estimated by summing the component ratings.

A full-province map of hazard rating is generated in terms of four categories: low, moderate, severe, and extreme. Initially, spraying was recommended for severe and extreme hazard ratings but over the years tactics have varied somewhat from this norm (Miller & Kettela 1975). One of the changes listed by these authors suggested a lapse of faith in the model, while another suggested that the model may not be appropriate for endemic budworm conditions. In fact, an analysis of this model using the MFRC-UBC model (see below) suggested that strict adherence to the norm would have produced disastrous results (Baskerville 1976). Commonsense, based on observations and experience, seems to have been a valuable adjunct to the hazard-rating model. Unfortunately, no one thought to keep a log of these annual adjustments to the model output. Thus the observations and experience behind these adjustments have not been captured for others to use.

Other models followed, in two "outbreaks" of effort. The first bout of modelling effort culminated two long-term field research programs conducted in northern New Brunswick: the Green River Project (1945-1959) and the Aerial Spraying Project (1952-1959). Three models were published in a multi-authored volume edited by R.F. Morris (1963) and one separately by Watt (1964).

The field programs compiled 168 life tables (Morris & Miller 1954) covering the epidemic phase of one spruce budworm cycle. These measurements were used in the application of Watt's (1961) method of deductive-inductive modelling to the life tables from each Project. Each of the resulting models was based on the following expression (p. 31 of Morris 1963) describing population change from one year to the next

$$N_{E_{n+1}}/N_{E_n} = S_E\ S_S\ S_L\ S_P\ S_A\ P_{♀}\ P_F\ F \pm N_D/N_{E_n} \qquad (1)$$

where,

$N_{E_{n+1}}$ = spruce budworm density (eggs) in year n + 1

S_E = survival of budworm eggs to eclosion

S_S = survival of larval instars 1 and 2

S_L = survival of larval instars 3-6 inclusive

S_P = survival of pupae

S_A = survival of adults to time of oviposition

$P_{♀}$ = proportion of adults that are female

P_F = proportion of F that can be achieved by the females of a given population

F = maximum fecundity (per female) of a budworm *population*

N_D = change in egg density due to moth dispersal

This expression was reorganized slightly (equation 7.1 of Morris 1963) to describe generation survival, and thereby conforming more closely to the life table conventions developed by the Projects (section 2.2 of Morris 1963).

The objective of the modelling exercise was to relate the dependent variables (being the S and P terms of equation 1) to a set of independent variables which had been collected for or along with the data for the life tables (p. 27 of Morris 1963). The modelling exercise used data from various study plots in the two large areas: no aerial insecticide spraying (Green River Project) and varying amounts of spraying as determined by the annual hazard calculations (Aerial Spraying Project). The sub-models, or regression equations, for the dependent variables in the unsprayed area are given by Mott (table 15.1 of Morris 1963) and in the sprayed area by Macdonald (table 27.1 of Morris (1963) for unsprayed check treatment; table 27.2 for the sprayed generation; table 27.3 for the generation sprayed one year earlier; and table 27.4 for the generation sprayed two years earlier).

While these Projects influenced the thinking of many participants in the protection program, there was no systematic application of the models to forest protection practice. Nor have the models been used as tools for management decision making.

A third model, based on Morris' (1959) method of key-factor analysis, was also published in Morris (1963). The regression models described above were based on epidemic population levels from 1950 on (p. 116 & 124 of Morris 1963) and incorporated a number of independent variables. The model reproduced below added the pre-epidemic larval data collected from 1945 in the Green River area, and incorporated only the "key" factors, to produce

$$\log N_{n+1} = 0.98 + 0.76 \log N_n + 0.18\ (T_{max} - 66.5) \qquad (2)$$

where,

N_{n+1} = large-larval (instars 3-6 inclusive) density in year n + 1

T_{max} = mean maximum daily temperature during the period of large-larval development (°F)

Although based on a more extensive data set, this model also has had no application.

A year later, Watt (1964) published the first comprehensive *strategy evaluation simulation* model. Although Watt intended that the model relate to a hypothetical insect population, much of the forest and pest information relates to New Brunswick and its spruce budworm population.

The model describes the effects of weather, past pest densities, disease, parasites, and insecticides on reproduction, dispersal, and mortality in a pest population; it also describes the effects of the pest on tree growth loss and mortality; and, lastly, it estimates the pest control costs for 15 control strategies and the financial losses from tree damage (growth loss and mortality).

The effect of weather and past pest densities on the abundance of the pest follows equation 2 above, but introduces a constant factor to represent genetic and post-spray generation reactions to insecticide application. Disease (viral) incidence is a linear function of its incidence in the previous year. Parasite attack uses a model developed earlier by Watt with parameter values estimated by Miller (p. 231 of Morris 1963). Insecticide effects are assumed constant for each dosage of DDT.

The forest is assumed to consist only of balsam fir 20 cm in diameter at breast height, spaced evenly with a stand density of 647 trees per ha and a radial increment at breast height of 0.24 cm per year. Tree mortality is modelled as a function of defoliation in such a way that mortality lags behind defoliation, that defoliation has a cumulative effect on tree mortality, and that there is a threshold in tree mortality relative to cumulative defoliation. Growth loss is modelled as a linear function of cumulative percentage defoliation 4,5, and 6 years previously.

Three control agents are included in the model: DDT, parasites, and viruses. Spraying costs decline exponentially above a minimum area sprayed. Viral costs are estimated by multiplying insecticide spraying costs by a constant factor. Parasite costs are fixed, per unit, costs. Lastly, financial losses from tree damage follow from estimates of merchantable timber losses and the market value of timber.

These processes are assumed to operate within each of 625 regions of 41 km^2 each. Migrations of pests, parasites, and incidence of disease are modelled as proportions of the gradient from one area to the next.

Simulations with this model were conducted and various conclusions drawn but the model has, surprisingly, had no impact on management strategy evaluation.

Following this paper there exists a lull of about a decade in published modelling activity relating to any aspect of the spruce budworm and its effects. Modelling activity began again in October, 1972, with a 5-day workshop held in Fredericton, New Brunswick, and lead by staff from the Institute of Animal Resource Ecology at the University of British Columbia (UBC) in Vancouver. The UBC team provided the modelling expertise while the information on budworm and forest dynamics came from the local participants. The result of the workshop was a simulation model which described the spatial and temporal dynamics of the spruce budworm and forest on about two-thirds of New Brunswick. The model evolved through 1973 and in January, 1974, the International Institute for Applied Systems Analysis (IIASA) in Austria became interested in using the budworm-forest management problem as a case study of the application of applied systems analysis to regulation of a renewable resource. Further modelling studies then followed, based in various countries.

The original simulation model is structured along the lines of that produced by Watt (1964). But, whereas Watt (1964) stayed closely to an empirical basis in the Green River and Aerial Spraying Projects, the *MFRC-UBC* model did not, for reasons explained in the following paragraph.

The *site model* component of the MFRC-UBC model deals with the spruce budworm population in terms of eggs, small larvae (instars 1 & 2), large larvae, pupae, and adults. In some cases the functions for each of these variables can be traced back to results in Morris (1963); in other cases the origin of the functions is obscure, even to the point of contradicting results in Morris (1963). This arose because, during the workshops, it was argued that data from the Green River area are not characteristic of most of New Brunswick and that the observational experience of local scientists is preferable to inapplicable results from the Green River region.

The MFRC-UBC model addresses a forest, not a stand (as Watt (1964) had done earlier). The model used provincial forest inventory data to establish forest conditions for each geographical unit. Relevant tree species are lumped and grouped into 75 age classes, and described in terms of the surface area of branches and the quantity of foliage on the branches. Branch surface area increases with tree age and only those trees thought to be old enough to be susceptible to budworm attack are included in the measure of surface area. Foliage is categorized in terms of new (this year's) and old (past year's) foliage. Foliage matures from new to old, is eaten by spruce budworm, is produced (by a photosynthetic assimilation rate proportional to total foliage), old foliage dies, and new foliage is brought into the variable by tree ageing.

Budworm Control Policy and Forest Management Policy subroutines also exist (fig. 1 of Jones 1979) but they have not yet been documented.

The basic budworm dynamics seem to be an extension of an idea of R.F. Morris for a possible form of Reproduction Curve (N_{n+1} vs. N_n; *see* equation 2 above), with the addition of a vertical scaling of the curve to stand age. (Compare fig. 20 of Jones (1979) to line C in fig. 18.3 of Morris (1963).)

It is relevant that the original Reproduction Curve was put forward as an hypothesis as it clearly shows the nature of the MFRC-UBC model: a consensus of opinion of the workshop participants, opinions based on many years of observation but not necessarily on scientific data. It would not be possible to build such a model on the basis of scientific fact alone because neither a sufficiently comprehensive data base for New Brunswick nor a mature theoretical base of proven relevance to New Brunswick exist. Nevertheless, the model-developers did contribute to pest control and related forest management strategy evaluation in New Brunswick. A Task Force set up by the New Brunswick government to evaluate budworm control alternatives (Baskerville 1976) used the model - about 1000 exploratory simulations - to illustrate possible developments of a number of indicators well into the next century.

Various other models were developed through the participation of IIASA, related to the development of policy alternatives. Notably, Winkler and Dantzig (*see* references to unpublished reports given on p. 30 of Clark et al. 1979) used dynamic programming to estimate the conditions of tree age, foliage, and budworm (egg) density for which insecticide spraying or tree harvesting are optimal (i.e. maximize long-term logging profits). The optimal spraying rules are easily described (fig. 14 of Clark et al. 1979) and are of the order of complexity of the hazard-rating model. A simplification of the Winkler-Dantzig model was used to search for policies in the Task Force. This model was also used to allow various decision makers to compare "policies" based on their respective priorities and objective functions. This exercise proved to be an important step in sensitizing decision makers both to the existence of viable alternative policies and to the different ways that they used to evaluate the effectiveness of policies.

Another offshoot from the MFRC-UBC model is a simple, three-equation analytical model of the spruce budworm and the forest (Ludwig et al. 1978). The three variables are budworm density, average tree size (a measure of total surface area of branches in a stand), and "energy reserve" (a measure of the condition of the foliage and health of the trees). The model was used to analyze the qualitative consequences of what is essentially the Reproduction Curve outlined above and given in Jones (1979). It has not been used either in forest protection practices or in the evaluation of forest management strategies.

The year 1978 also saw the publication of another simulation model, a *WO*od *S*upply and *FO*rest *P*roductivity (WOSFOP) model developed by T.H. Hall for the Forest Management Branch of the New Brunswick Department of Natural Resources. WOSFOP is a development of the forest submodel of the MFRC-UBC model and is the only model to date to be developed within the New Brunswick management agency. The model projects the state of a forest through time using a user-specified age distribution function and development curves (stand volume and tree size vs. age) for each stand, assuming various growing regimes: existing growing stock; natural regeneration from untreated cutover, depending on type of stand cut; plantations; and response to spacing carried out in 0-20 yr old regeneration stands. These projections are modified by human use, based on stated conditions for stand operability.

To quote Hall (1978): "WOSFOP accepts a user's (manager's) dynamic description, calculates and sorts stocks according to age, stand volume and tree size for each Forest Unit, removes these stocks for a specified annual cut according to one of a variety of harvesting rules, updates the forest and calculates the values of a variety of parameters, in terms of which forest development is expressed, over time."

Even though WOSFOP is a relatively new and exceedingly simple model, and one which places heavy demands on the user (must specify both functions and parameter values for each simulation), the model has been extensively used in the New Brunswick Department of Natural Resources for both management planning and the design of timber licence areas. Further, various versions of WOSFOP have spread to at least three other provinces and to more than 20 forest companies.

WOSFOP allows the manager to quickly see, and compare, the possible outcomes of various possible management policies for fairly large land areas. Subsequent development has lead to models for testing the probable outcomes of local cutting patterns, for forecasting yield in plantations, and for forecasting initial response on areas that have been cutover. These, and other, models are under active use and development in the Department of Natural Resources and several pulp companies in New Brunswick.

In 1981 Ker & Van Raalte published tree biomass equations for balsam fir and white spruce, relating ovendry weights of various biomass components to diameter at breast height, tree height, crown diameter, crown length, and various derived variables (see also Baskerville 1965). In the same year Baskerville & Kleinschmidt (1981) published a dynamic model of growth in defoliated fir stands, and Royama (1981) looked at spruce budworm population dynamics from the viewpoint of stochastic difference equations. In 1982 MacLean published a vulnerability-rating model, of the same order of complexity as the hazard-rating model (Webb et al. 1956). None of these models has yet been applied.

A summary of the models discussed in this section is given in table 1.

WHAT DETERMINES SUCCESS?

Our review of the ecological models relating to New Brunswick and their role in the management of budworm infested fir-spruce forests has shown that there has been only limited application of the existing models. One (hazard-rating) model has found extensive use in forest protection practices. Three (MFRC-UBC, Winkler-Dantzig, and WOSFOP) models have found use in the evaluation of management strategies.

The MFRC-UBC and Winkler-Dantzig models have been used mainly in the Task Force evaluating budworm control alternatives (Baskerville 1976). The hazard-rating model has been used for almost 30 years to guide the insecticide spraying program. WOSFOP is too new to allow comment on its durability but its rate of spread (to three provinces and more than 20 forest companies) is suggestive of a long working life. One might then ask: what characteristics of models lead to their application and, once used one time, to their durability?

A common characteristic of the models that have been applied is that they were built in close association with the intended user. The hazard-rating model and WOSFOP were built virtually within the user's environment. The MFRC-UBC model included representatives of the user organization in the workshops set up to construct the model. The Winkler-Dantzig model was not constructed in the user's environment but had close enough ties with the MFRC-UBC model to be considered as part of that modelling effort.

A close association between model-developer and model-user is probably beneficial in two ways. Firstly, it ensures that the model, once built, will have a niche in contemporary forest management practice. Secondly, it ensures that the model will relate in scale, structure, and output to contemporary forest management practice.

A close association between model-developer and model-user is most desirable when the model can be finished in a relatively short time. The MFRC-UBC model-developers used informed opinion where scientific information/theory was lacking. WOSFOP passed the difficult decisions regarding the functional forms and parameter values on to the user. However, where long-term experimental programs are set up in an attempt to relate the advancement of basic science to models being built for the purposes of contemporary forest management practice, the consequences can be undesirable for forest management and research alike. The major experimental conclusions will not be produced rapidly enough to be useful to management; as the experimental program sacrifices rigorous science to the demands of contemporary application, the scientific validity of the results will be limited when they do become available. What would seem most useful in the long run is sound research working with mathematical models to enhance our knowledge of systems behavior. This approach will lead to a solid scientific base, but will not contribute much to effective resource management now. There is a need for research into the science of using science in resource management.

But what to do with those models that have been (and will be) developed independently of the user? Perhaps it would be useful to turn the management problem upside down; to realize that it is possible *not only* to set management priorities and look for useful information *but also* to study the available scientific information and look for questions of management interest to which our current scientific information can contribute. Clearly this would require some scientist to become very knowledgeable with respect to management needs and interests. The approach is not goal-oriented and hence bears some risk, but it does offer another avenue of advancing the use of science in resource management.

For example: the deductive-inductive models from the Green River Project have been said not to be usable because they relate only to a forest cover type found in northwestern New Brunswick. But looked at from the viewpoint of a limited set of scientific tools, one can say that the existing models are useful to management in northwestern New Brunswick. And if the models were tested in a scientifically rigorous manner in other regions of the province one may find that they have a systematic bias that can be compensated for.

Table 1.--Those variables which the authors have highlighted in describing their models, in terms of floral, faunal, and forest protection components.

Model (Major Publication)	Components Floral	Faunal	Forest Protection
Webb et al. (1956)	- Risk of tree mortality		
Morris (1963) (p. 105; 172-173)		- Survival of budworm eggs, small larvae, large larvae, & pupae - % adults that are ♀ - % maximum fecundity	
Morris (1963) (p. 120)		- Large-larval budworm density	
Watt (1964)	- Tree growth loss - Tree mortality	- Large-larval budworm density - Virus density - Parasite density	- Insecticide spraying - Pest control costs - Financial losses from tree damage
Jones (1979) (MFRC-UBC)	- Tree age classes - Branch surface area - Foliage/branch	- Density of budworm eggs, small larvae, large larvae, pupae, & adults	- Unspecified financial costs & losses
Clark et al. (1979) (Winkler-Dantzig)	- Tree age - Foliage density	- Density of budworm eggs	- long-term logging profits
Ludwig et al. (1978)	- Average tree size - Tree energy reserve	- Density of budworm	
Hall (1978) (WOSFOP)	- Tree age distribution - Tree size - Stand volume		
Ker & Van Raalte (1981)	- Dry weight of tree biomass components		
Baskerville & Kleinschmidt (1981)	- Foliage weight - Number of stems - Stand volume increment		
Royama (1981)		- Density of budworm	
MacLean (1982)	- Tree vulnerability		

To apply the existing unused models it will initially be necessary to discover if there are niches for them, either in forest management or in forest protection. This search itself will be useful in linking science and management since it can only be carried out from a knowledge both of the existing models and of forest management/protection practices. Once niches are found, the models (or the contemporary forest management practice) will likely have to be modified somewhat for effective use; it will thus be necessary for the research organization or the management/protection organization to provide the staff (or money for contract research) to do (get done) the required modifications.

We now turn to the second question posed at the beginning of this section: what characteristics of models lead to their durability in management organizations? Of course it helps if the model relates to an ongoing problem. The hazard-rating model has lasted not just because it works but also because the budworm continues to defoliate trees and kill them. Even here, however, the life of the model may be limited as attention turns to growth loss rather than tree mortality. But the main factor involved seems to be model simplicity. The hazard-rating model can be readily understood by any potential user. WOSFOP is based on concepts well understood in forestry and its computer implementation provides input/output in variables used by decision makers in a user-friendly, interactive manner. The MFRC-UBC and the Winkler-Dantzig models are relatively complex models and suffer from never having been described in full in readily available publications. Their main use, then, has occurred as one of the model-developers took up a position with the New Brunswick government. These two models may be of continuing use, but it will take someone to fully explain the models to potential users and to implement the models in a user-friendly, interactive manner.

On the other hand, it may also be that these models have served their purpose in illustrating how models can help managers organize information for decision making. Having sensitized managers to the potential, perhaps these models will be discarded in favour of new ones designed with the lessons in mind.

Throughout the paper we have concentrated on the application of the models per se to forest management. This is not to demean the value of the results of model use, as reported in the scientific literature. Many such ideas have in time become conventional wisdom.

ACKNOWLEDGMENTS

We thank Dr. Bob Fisher for useful discussions; Ms. Nancy Gogan and Mrs. Ellen VanTassel for typing the manuscript.

LITERATURE CITED

Baskerville, G.L. 1965. Dry matter production in immature balsam fir stands. Forest Science Monograph 9, 42p.

Baskerville, G. 1976. Report of the Task Force for evaluation of budworm control alternatives. Department of Natural Resources, Province of New Brunswick, 210p. Fredericton, Canada.

Baskerville, G., and S. Kleinschmidt. 1981. A dynamic model of growth in defoliated fir stands. Canadian Journal of Forest Research 11: 206-214.

Blais, J.R. 1965. Spruce budworm outbreaks in the past three centuries in the Laurentide Park, Quebec. Forest Science 11: 130-138.

Clark, William C., Dixon D. Jones, and C.S. Holling. 1979. Lessons for ecological policy design: a case study of ecosystem management. Ecological Modelling 7: 1-53.

Hall, T.H. 1978. Toward a framework for forest management decision-making in New Brunswick. Department of Natural Resources, Province of New Brunswick, TR1-78, 83p. Fredericton, Canada.

Jones, Dixon D. 1979. The budworm site model. p. 91-155. *In* Norton, G.A., and C.S. Holling, editors. Pest management: Proceedings of an international conference, October 25-29, 1976. 349p. Pergamon Press, Oxford.

Ker, M.F., and G.D. Van Raalte. 1981. Tree biomass equations for *Abies balsamea* and *Picea glauca* in northwestern New Brunswick. Canadian Journal of Forest Research 11: 13-17.

Ludwig, D., D.D. Jones, and C.S. Holling. 1978. Qualitative analysis of insect outbreak systems: the spruce budworm and forest. Journal of Animal Ecology 47: 315-332.

MacLean, D.A. 1982. Vulnerability rating of forests in New Brunswick and Nova Scotia to budworm attack. Canadian Forestry Service Information Report M-X-132, 21p. Maritimes Forest Research Centre, Fredericton, New Brunswick.

Miller, C.A., and E.G. Kettela. 1975. Aerial control operations against the spruce budworm in New Brunswick, 1952-1973. p. 94-112. *In* Prebble, M.L., editor. Aerial control of forest insects in Canada. 330p. Department of the Environment, Ottawa, Canada.

Morris, R.F. 1959. Single-factor analysis in population dynamics. Ecology 40: 580-588.

Morris, R.F., editor. 1963. The dynamics of epidemic spruce budworm populations. Memoirs of the Entomological Society of Canada 31, 332p.

Morris, R.F., and C.A. Miller. 1954. The development of life tables for the spruce budworm. Canadian Journal of Zoology 32: 283-301.

Royama, T. 1981. Fundamental concepts and methodology for the analysis of animal population dynamics, with particular reference to univoltine species. Ecological Monographs 51: 473-493.

Watt, Kenneth E.F. 1961. Mathematical models for use in insect pest control. The Canadian Entomologist Supplement 19, 62p.

Watt, K.E.F. 1964. The use of mathematics and computers to determine optimal strategy and tactics for a given insect pest control problem. The Canadian Entomologist 96: 202-220.

Webb, F.E., D.G. Cameron, and D.R. Macdonald. 1956. Studies of aerial spraying against the spruce budworm in New Brunswick V. Techniques for large-scale egg and defoliation ground surveys 1953-55. Canada Department of Agriculture, Forest Biology Division, Interim Report 1955-8, 24p. + illustrations. Forest Biology Laboratory, Fredericton, New Brunswick.

SUPPORTING DECISIONS ON WESTERN SPRUCE BUDWORM IN FOREST MANAGEMENT USING SIMULATION MODELS[1]

J. J. Colbert[2], Katharine Sheehan[3], and Nicholas L. Crookston[4]

Abstract.--A system of computer models is being developed to assist forest managers and planners in making decisions about western spruce budworm, an important defoliator of western forests. Models of budworm population dynamics and stand growth and yield are described, and use of this decision-support system is reviewed.

INTRODUCTION

The western spruce budworm (Choristoneura occidentalis Freeman) has been defoliating western North American forests for at least as long as entomologists have been recording insect damage. The Canada/United States Spruce Budworms (CANUSA) Program is a 6-year, accelerated research, development, and applications program charged with focusing resources and expertise from the two nations to find new and improve current techniques for managing Choristoneura species in North America. The Program began in late 1977 and will end in the fall of 1983.

CANUSA-East coordinates efforts on C. fumiferana (Clemens) in the northeastern United States and Canada. Most of the infested acreage in the East is in Canada, but budworm is also a problem in the United States from the Lake States to Maine.

CANUSA-West has sponsored between twenty and thirty studies on western spruce budworm in North America each year, about half within the Forest Service and half at eleven universities in the United States and Canada. The annual budget of about 1.6 million dollars has accelerated research on western spruce budworm to provide forest managers with new and improved management methods.

Although budworms have been the subject of many models, none of them suited CANUSA-West's objectives. Models of spruce budworm in the East are centered on the budworm itself; host foliage serves only as a substrate or source of nutrition for the budworms (Morris 1963, Jones 1976, Gage and Sawyer 1979). These models do not allow pest-management decisions or options on budworm to be included as part of overall forest management and planning; for example, the effects of defoliation by budworm on merchantable yield are not readily evaluated. CANUSA-West was fortunate to have a model of stand growth and yield called the PROGNOSIS model (Stage 1973) already in use in part of the range of western spruce budworm. Thomson (1979) had described several functional relationships specific to population dynamics of western spruce budworm, but did not have enough information to complete a model. Integration of budworm management into forest management thus required a population model to predict interactions among stand conditions, budworm population dynamics, and stand growth. The Douglas-fir tussock moth model and its linkage to PROGNOSIS provided the initial paradigm for CANUSA-West's modeling efforts (Chapter 11, Brookes et al. 1978, Colbert et al. 1979).

The combined budworm/stand-growth models, together with other existing models that predict the economic consequences of changes in stand yields or optimize production of desired resources, form the nucleus of the decision-support system for budworm-related forest management and planning. These models can be used by forest managers to investigate a variety of silvicultural options, as well as the effect of direct suppression with insecticides. Planners will be able to incorporate budworm impacts and management into the long-term plans for Federal, State, or private forests.

From the beginning, CANUSA-West studies were closely integrated to form the basis for an emerging set of computer models. Individual researchers (or small teams) have taken lead responsibility for specific studies. They submit proposals for technical review, collect and analyze data, develop models if appropriate, and publish their results. These "working groups," designated by program management to focus research efforts on certain topics, consisted of 5 to 10 scientists working on related studies--such as effects of budworm defoliation on tree growth, budworm population dynamics, or sampling and surveys of budworm and hosts. Annual Working Group meetings of all CANUSA-West supported researchers (and other interested scientists) with resource managers have promoted exchange of information and review of proposed models.

---

[1]Presented at the Third International Conference on State-of-the-Art in Ecological Modeling, Colorado State University, May 24-28, 1982.

[2]J. J. Colbert is Research Coordinator for CANUSA-West, USDA Forest Service, Portland, Oregon U.S.A.

[3]Katharine Sheehan is Research Associate/ Insect Modeler, University of Idaho, working at Pacific Northwest Forest and Range Experiment Station, Portland, Oregon U.S.A.

[4]Nicholas L. Crookston is Research Associate, University of Idaho, working at Forestry Sciences Laboratory, Moscow, Idaho U.S.A.

As the program progresses, these working groups are being replaced with more specifically focused groups, which meet as needed to evaluate results, identify areas requiring further research, and formulate models. For example, a small group of scientists working on phenology of budworm and hosts will develop model components to allow accurate simulation of budworm feeding and defoliation.

CANUSA-West has supported five representatives in Forest Pest Management and Timber Management Units of the USDA Forest Service. These people have been responsible for identifying the needs of users and for explaining how the models will be used to meet these needs. A formal, 3-day workshop was held in March 1981 so planners and forest and pest managers could describe their current operations and desires for management and planning tools. The design of the decision-support system was elaborated as a result of this workshop.

Integration of research results and models of individual components is coordinated by a team of modelers, carefully selected to include complementary backgrounds and training. Nilima Srivastava, program statistician who received an M.S. in statistics at Oregon State University, has been working with one of the fastest growing data sets--population dynamics of western spruce budworm. Nicholas L. Crookston, who received his M.S. in forest science at the University of Idaho, helped develop the mountain pine beetle extension of PROGNOSIS and assisted in linking the Douglas-fir tussock moth model to PROGNOSIS; he is working on the extensive modifications to PROGNOSIS that are necessary to link it with the budworm models, and is responsibile for linking the two models. Peter J. McNamee, a student of C. S. Holling at the University of British Columbia, served as insect modeler from 1979 through 1980; he worked on the first two versions of the budworm model. Since October 1981, the insect modeler has been Katharine Sheehan, who studied forest entomology (with emphasis on computer programing and statistical analysis) at the Division of Biological Control, University of California at Berkeley. William P. Kemp, a University of Idaho graduate student assisting in development of the budworm models, had extensive experience with biology and modeling of eastern spruce budworm as a study coinvestigator and in earlier graduate studies at the University of Maine and Michigan State University. Additional mathematical and modeling expertise is provided by J. J. Colbert (Ph. D. in mathematics, Washington State University), who was an investigator in the Douglas-fir tussock moth program and assisted in linking the tussock moth model to PROGNOSIS.

WESTERN SPRUCE BUDWORM AND ITS HOSTS

Western spruce budworm is one of several *Choristoneura* species found in western forests; although these insects have been studied for decades, their taxonomy is just now being settled (Powell 1980). Western spruce budworm has been described as the most destructive defoliator in western North America. Details on budworm biology, damage to host trees, and history of outbreaks are reviewed in McKnight (1968), Johnson and Denton (1975), Furniss and Carolin (1977), and Dolph (1980), and are summarized below.

The budworm feeds on a variety of hosts, primarily Douglas-fir (*Pseudotsuga menziesii* (Mirb.) Franco), grand fir (*Abies grandis* (Dougl. ex D. Don) Lindl.), subalpine fir (*A. lasiocarpa* (Hook.) Nutt.), white fir (*A. concolor* (Gord. & Glend.) Lindl. ex Hildebr.), Engelmann spruce (*Picea engelmannii* Parry ex Engelm.), and western larch (*Larix occidentalis* Nutt.).

The insect can be found throughout most of the hosts' ranges from British Columbia to New Mexico and Arizona; most of the infested acreage is in the United States. Only once have budworms been found at high densities west of the Cascade divide in Oregon, and they have never been a problem in the highly productive area of western Washington (Dolph 1980). East of the Cascade divide throughout the inland Northwest to the northern Rocky Mountains, however, budworms have infested between 4 and 6 1/2 million acres (1.6 - 2.6 million ha) each year over the past several decades. Budworm defoliation has been reported by at least one USDA Forest Service Ranger District every year since 1922 (Johnson and Denton 1975). Outbreak episodes do not follow a particular pattern: some last only 3 or 4 years, others may continue for over 20 (Furniss and Carolin 1977).

Each budworm egg mass is laid on an individual leaf or needle in late summer. Within a few days, the eggs hatch and the young larvae (caterpillars) crawl to hiding places on the tree, where they spend the winter. They emerge in the spring when new foliage is just beginning to flush, and they feed on the succulent, developing needles. By midsummer when they pupate, extremely dense populations may have destroyed almost all of the foliage from the current year. Adults emerge about 10 days after pupation. Little is known about adult flight; large-scale redistribution, as is known to occur regularly with spruce budworm in the East, has not been documented in the West. Adults are capable of flight sufficient to redistribute them between neighboring stands.

The major impact of defoliation is a decrease in volume growth. This decrease usually lags defoliation by 1 or 2 years, and growth may remain depressed for several years after an outbreak. Feeding on developing buds and cones can destroy a significant proportion of the seed crop. Feeding can also kill regeneration of hosts in the understory, but the major damage comes through reduced growth and top-kill of large trees. Rarely does extreme defoliation kill large trees outright, although a tree stressed by defoliation may succumb to secondary attack by another pest.

Control efforts have generally been by aerial application of chemical insecticides; often hundreds of thousands of acres have been sprayed in a single year. Bacterial and viral pesticides have also been tested; *Bacillus thuringiensis* (Berliner) has recently been registered for use against budworm.

WESTERN SPRUCE BUDWORM MODEL

The initial version of the budworm population-dynamics model was based on a CANUSA workshop held in early 1980. Thirty researchers and potential users brainstormed to produce a model based, usually, on educated guesses. Peter J. McNamee (ESSA Ltd., Vancouver, B.C.) and a modeling team from LGL Ecological Research Associates constructed that first version (McNamee et al. 1980). A second version linking the budworm model to PROGNOSIS was developed by McNamee and Crookston.

The framework for the completely revised, third version of the budworm model has been described in detail by Colbert et al. (1981). Researchers and potential users have reviewed this report, and they will continue to work closely with the modeling team to develop and test specific functions for the model. This model will be available in a stand-alone version or as a submodel linked to PROGNOSIS.

The budworm life cycle has been split into two parts, which are modeled separately. Figure 1 illustrates the two components: BWMOD (egg hatch through adult emergence) and BWFLY (adult dispersal and egg laying).

BWFLY

BWFLY evaluates the number of mated females in each stand and has those females lay a portion of their eggs in that stand. Some adult mortality occurs here. Next, adults are redistributed across stands. Additional mortality occurs during this dispersal. Finally, the remaining eggs are distributed within each stand according to the number of adults that wind up in that stand. Figure 2 shows a flowchart for this subroutine.

Much of the biological information necessary to model these steps will come from a study being conducted by Robert W. Campbell and Torolf R. Torgersen (Pacific Northwest Forest and Range Experiment Station, Corvallis, Oregon) on Douglas-fir and grand fir in eastern Oregon, central Idaho, and western Montana. They are examining the ratio of observed eggs (number of egg masses per unit of foliage area multiplied by number of eggs per mass) to expected eggs (number of females emerged per unit of foliage multiplied by number of eggs laid per female). A ratio greater than one would show that the net result of adult dispersal was an influx of moths into the stand; a ratio less than one would indicate net dispersal out of the stand. This study will also produce information on within-tree distribution of egg masses, which will complement work done in other geographic locations and on other hosts by Roy C. Beckwith (Pacific Northwest Forest and Range Experiment Station, La Grande, Oregon), working on Douglas-fir in northcentral Washington; John M. Schmid (Rocky Mountain Forest and Range Experiment Station, Flagstaff, Arizona), working on white fir in southern Colorado and northern New Mexico; and William E. Waters and W. Jan A. Volney (University of California, Berkeley), working on Douglas-fir and grand fir in southern Oregon and on white fir and Douglas-fir in New Mexico. Additional information on adult dispersal may come from a study by Roy F. Shepherd (Canadian Forestry Service, Pacific Forest Research Centre, Victoria, B.C.), using pheromone traps in British Columbia. Beckwith, working with Douglas-fir and grand fir, and Michael R. Wagner (Northern Arizona University, Flagstaff), working with white fir and Douglas-fir, are also studying the egg-laying potential of females reared on different hosts.

BWMOD

BWMOD is summarized by the flowchart in figure 3. Rates of egg mortality (including the effects of parasites, nonviable eggs, and unknown causes) are being recorded as part of life-table studies conducted by Beckwith, Campbell/Torgersen, and Waters/Volney. Beckwith and Donald G. Burnell (University of Idaho, Moscow) are examining dispersal of larvae in fall and spring; Thomas H. Egan (Oregon State University, Corvallis) is measuring overwintering mortality. Dale O. Everson, Donald G. Burnell, and William P. Kemp (University

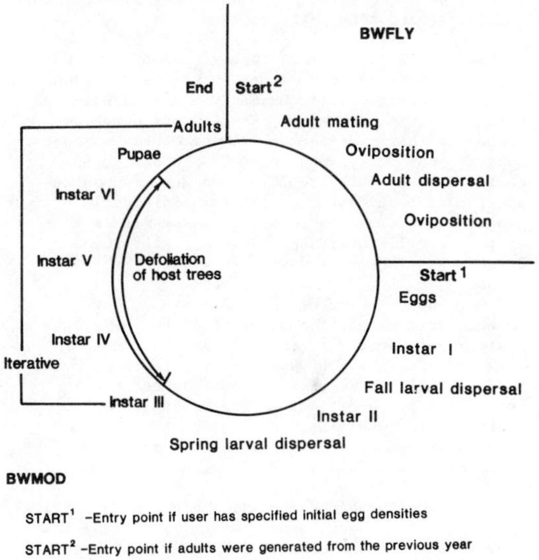

Figure 1.--Life cycle of the western spruce budworm showing events modeled by BWFLY and BWMOD.

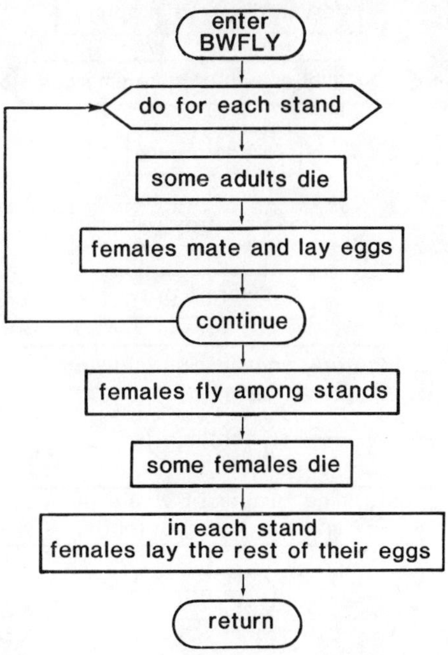

Figure 2.--BWFLY simulates budworm life cycle from adult emergence through ovipositon.

of Idaho, Moscow) are studying effects of weather on both budworm and hosts. They are developing stochastic weather predictions (based on parameters that may be derived from local weather records) to drive development of both budworm and host foliage.

Beginning with larval feeding and continuing until adult emergence, the following steps are repeated. First, additional foliage growth is simulated, using predicted weather. Budworm larvae then feed on available foliage and may disperse if not enough foliage is available. Budworms themselves age (again, dependent on predicted weather) and suffer mortality, and the cycle begins again. Beckwith (Douglas-fir and grand fir in the Pacific Northwest) and Wagner (white fir and Douglas-fir in the Southwest) are studying rates of budworm development and feeding. Effects of foliage biochemical composition on budworm survival and development are being studied by Rex G. Cates (University of New Mexico); J. Wayne Brewer, John L. Campinera, and Jesse A. Logan (Colorado State University, Fort Collins) have examined the effects

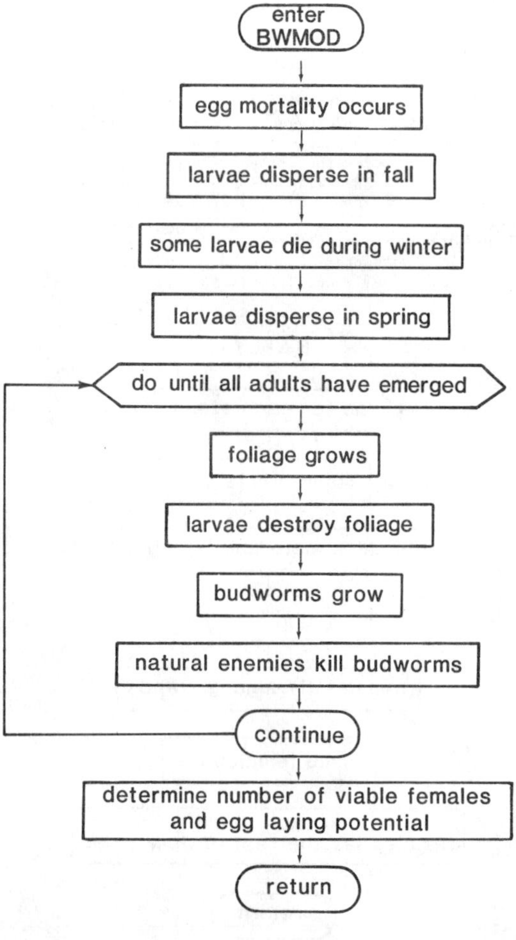

Figure 3.--BWMOD simulates budworm life cycle from egg hatch to adult emergence.

of nitrogen, phosphorus, and potassium (through fertilization) on budworm growth, survival, and fecudity. Beckwith, Campbell/Torgersen (concentrating on birds and ants), and Waters/Volney (concentrating on parasites) are working on distribution and effects of natural enemies, as part of the life-table studies. The budworm cohort structure will be modeled explicitly so that effects of age-specific feeding and mortality rates can be included. Finally, the number of emerging adults, the proportion of females, and the egg-laying potential of those females (determined from the life-table studies) are calculated.

THE STAND-PROGNOSIS MODEL

PROGNOSIS was first described by Stage (1973); it has since been revised and expanded by a team of modelers led by Stage (Wykoff et al. 1982). This model simulates growth of individual trees in a distance-independent manner; information routinely measured during forest inventories serves as inputs to this model. Diameter increment, height increment, change in crown length, and mortality rates are predicted by using a sample-based inventory of individual trees. Budworm population density, previous and current defoliation measurements, or both, may be entered into the budworm component of the combined model.

A variety of silvicultural practices can be simulated, such as harvesting, thinning, fertilizing, and various site-preparation techniques. Information about the stand related to wildlife habitats and watershed protection is provided. The actions of two other forest insects besides the budworm--Douglas-fir tussock moth (Orgyia pseudotsugata (McD.)) and mountain pine beetle (Dendroctonus ponderosae Hopkins)--may also be simulated (Colbert 1978, Crookston 1978, Monserud 1978). Foresters can thus use PROGNOSIS to predict growth and mortality in a stand for scenarios in which they specify initial conditions, silvicultural practices, and some pest problems.

PROGNOSIS is currently being used by the USDA Forest Service, Bureau of Land Management, state forestry departments in Washington and Idaho, the British Columbia Forestry Service, and private companies in the United States and Canada. Separate geographical extensions of PROGNOSIS have been calibrated for northern Idaho and western Montana and for eastern Montana. Several additional geographical extensions, either under development or proposed, will cover Utah, Nevada, California, Oregon, Washington, and western Wyoming.

This description of PROGNOSIS has so far concentrated on how the model predicts the growth of a single, independent stand. Stands often do not grow independently, however. The effects of fire, insects, and diseases on a stand may be greatly influenced by conditions in surrounding stands. In addition, for many resources--such as water quality or wildlife--interactions among stands must be considered by managers and planners.

To simulate moth dispersal among stands requires simulation of several stands at once. CANUSA has therefore supported development by Crookston and Stage (in preparation) of a "parallel-processing extension" (PPE), which allows this parallel simulation. The PPE is diagramed in figure 4.

Another important use of the PPE is to allow stand-management alternatives to be simulated efficiently. Weighted-average yield tables can be computed for all of the stands in the run and decision trees can be created automatically. A decision tree is a diagram for composing alternative sequences of events affecting a stand (Stage 1975). The choice of one branch rather than another may depend upon random events or management decisions. Either way, the PPE displays the outcome of following the various sequences of events, and the average or expected value when the choice among alternatives is a chance event.

To portray a stand accurately through the stochastic procedures in PROGNOSIS, a large list of individual trees is required. Budworm activity is simulated using a much smaller list of trees. This requires special aggregation-disaggregation routines

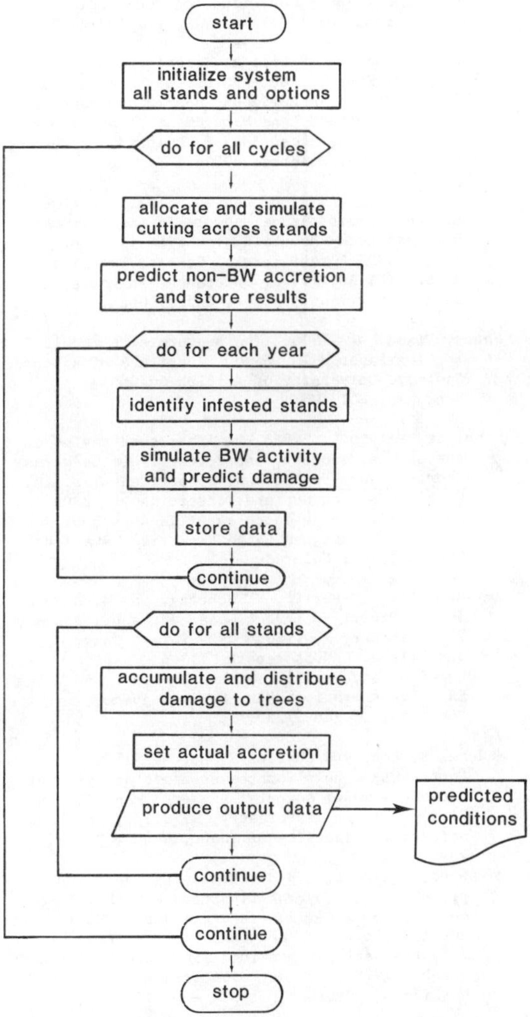

Figure 4.--Simplified flowchart for PROGNOSIS with parallel processing and budworm models.

for linking these tree lists. Once aggregation is complete, foliage biomass for host trees must be predicted. Tree-budworm interactions operate through foliage biomass. The distribution and amount of foliage for various host species and the effects of defoliation history on foliage biomass is estimated, using results of a study by Thomas M. Hinckley (University of Washington, Seattle).

The effects of budworm on conifer regeneration can be simulated using models developed by Dennis Ferguson (Intermountain Forest and Range Experiment Station, Moscow, Idaho) (Stage and Ferguson in press). Studies of how budworm affects establishment and growth of young trees are being conducted by Ferguson and Clinton E. Carlson (Intermountain Forest and Range Experiment Station, Missoula, Montana). After a simulated budworm outbreak, the effects of defoliation are distributed to the detailed list of trees in the stand, according to the predicted amount of--and variation in--defoliation. Burnell and David R. M. Scott (University of Washington, Seattle) are developing damage models for the northwestern United States, following procedures of G. Alan Van Sickle and Rene Alfaro (Canadian Forestry Service, Pacific Forest Research Centre, Victoria, B.C.)

USING THE MODELS FOR DECISION SUPPORT

Several characteristics separate these models from other broad-scale models. Because information collected by current survey or inventory methods is the input for the models, previously measured stands may be simulated; also, field crews need not be retrained to measure additional stands. PROGNOSIS and models for resources planning are already used in much of the range of the budworm, again reducing the amount of training needed to use this system. Indeed, Boise Cascade Corporation used PROGNOSIS to estimate budworm effects on yields to evaluate a proposed suppression project. Careful consideration has been given to identifying and modeling variables that influence managers' decisions--such as effects of budworm defoliation on merchantable yields.

The system of models has already been used to direct CANUSA-West funding; previous versions have helped identify areas requiring further research. As the models are completed and calibration proceeds, we expect sensitivity analyses to show where additional research is needed in the final year of the program and to provide direction for postprogram research.

The models are already being used to explain integration of individual studies to resource managers. Models allow them to see how specific research results can be used to answer their questions. Sometimes the connections between their questions or expressed needs and research results require the model for interpretation.

Researchers developing new programs for budworm management will compare their results and evaluate how well the models mimic nature. They will use the models to extrapolate beyond their data, and the models may provide insight for further studies and demonstration projects. Aberrant behavior of the model will lead to additional analyses and define new studies to provide further biological and silvicultural insight for improved forest management. Models resulting from the Douglas-fir tussock moth program provided similar leads.

The combined PROGNOSIS/Budworm models will provide yield tables that include impacts of budworm defoliation on forest productivity. These yield tables are used by forest planners and economists with models such as FORPLAN (Johnson et al. 1980), for forest resources planning, which optimizes land-use allocations, harvest schedules, and so on. Changes in forest productivity may greatly affect the outcome of these forest-planning models, making accurate predictions of budworm impacts on forest growth essential.

Forest managers will use the models in a "gaming" sense, to assist them in developing silvicultural plans to reduce the likelihood of budworm outbreaks, minimize losses should an outbreak occur, and reduce losses during infestations. A computer program has been developed for making economic assessments of the various alternatives being considered (Medema and Hatch 1979).

The model will also predict results of direct suppression with chemical or microbial insecticides; costs of direct and indirect treatments can be accurately and easily compared to potential losses. Treatment sequences through stand rotations (100-200 years) can be examined to develop more efficent control strategies.

Our current plan calls for completing all research and development of the models by fall of 1983. Documentation and training will follow during the subsequent 2 years.

LITERATURE CITED

Brookes, Martha H., R. W. Stark, and Robert W. Campbell (Eds.) 1978. The Douglas-fir tussock moth: A synthesis. USDA Technical Bulletin 1585, 331 p. USDA, Washington, D.C.

Colbert, J. J. 1978. Douglas-fir tussock moth population outbreak model. p. 253-255. In Proceedings, Society of American Foresters and Canadian Institute of Forestry national meeting. [St. Louis, Missouri, October 1978] Society of American Foresters, Washington, D.C.

Colbert, J. J., W. Scott Overton, and Curtis White. 1979. Documentation of the Douglas-fir tussock moth outbreak-population model. USDA Forest Service General Technical Report PNW-89, 85 p. Pacific Northwest Forest and Range Experiment Station, Portland, Oregon.

Colbert, J. J., Nicholas L. Crookston, William P. Kemp, and Nilima Srivastava. 1981. Description of the combined PROGNOSIS/western spruce budworm model-Version 3.0. USDA Forest Service, unpublished CANUSA report, 26 p. On file, CANUSA-West, Pacific Northwest Forest and Range Experiment Station, 809 NE Sixth Avenue, Portland, Oregon 97232.

Crookston, Nicholas L. 1978. Predicting the outcome of management activities in mountain pine beetle susceptible lodgepole pine stands. p.276-280. In Proceedings, Society of American Foresters and Canadian Institute of Forestry national meeting. [St. Louis, Missouri, October 1978] Society of American Foresters, Washington, D.C.

Crookston, Nicholas L., and Albert R. Stage. In preparation. Simulataneous processing of forest stand prognoses to include interstand relations and decision tree analysis. (Contact USDA Forest Service, Forestry Sciences Laboratory, 1221 South Main Street, Moscow, Idaho 83843.)

Dolph, R. E., Jr. 1980. Budworm activity in Oregon and Washington 1947-1979. USDA Forest Service, Pacific Northwest Region, FIDM-033-1980, 54 p. Portland, Oregon.

Furniss, R. L., and V. M. Carolin. 1977. Western forest insects. USDA Forest Service Miscellaneous Publication 1339, 654 p. Washington, D.C.

Gage, Stuart H., and Alan J. Sawyer. 1979. A simulation model for eastern spruce budworm populations in a balsam fir stand. Proc. Tenth Annual Pittsburg Conference on Modeling and Simulation, April 25-27, 1979, p. 1103-1115.

Johnson, K. Norman, Daniel B. Jones, and Brian M. Kent. 1980. Forest planning model (FORPLAN) user's guide and operations manual. Draft. (Contact Land Management Planning Staff, USDA Forest Service, 3825 E. Mulberry Street, Fort Collins, Colorado 80524.)

Johnson, P. C., and R. E. Denton. 1975. Outbreaks of the western spruce budworm in the American northern Rocky Mountain area from 1922 through 1971. USDA Forest Service General Technical Report INT-20, 144 p. Intermountain Forest and Range Experiment Station, Ogden, Utah.

Jones, Dixon, D. 1976. The budworm site model. Working Paper W-16, 66 p. Institute of Resource Ecology, University of British Columbia, Vancouver V6T 1W5.

McKnight, Melvin E. 1968. A literature review of the spruce, western, and 2-year cycle budworms Choristoneura fumiferana, C. occidentalis, and C. biennis (Lepidoptera:Tortricidae). USDA Forest Service Research Paper RM-44, 35 p. Rocky Mountain Forest and Range Experiment Station, Fort Collins, Colorado.

McNamee, P., R. Everitt, N. Sonntag, and M. Stanley. 1980. Report -- Insect model workshop, January 28 - February 1, 1980. USDA Forest Service, unpublished CANUSA report, 113 p. On file, CANUSA-West, Pacific Northwest Forest and Range Experiment Station, 809 NE Sixth Avenue, Portland, Oregon 97232.

Medema, E. Lee, and Charles R. Hatch. 1979. Computerized help for the economic analysis of prognosis-model outputs: A user's manual. 71 p. College of Forestry, Wildlife, and Range Sciences, University of Idaho, Moscow 83843.

Monserud, Robert A. 1978. Combining the stand prognosis and Douglas-fir tussock moth outbreak models. p. 268-272. In Proceedings, Society of American Foresters and Canadian Institute of Forestry national meeting. [St. Louis, Missouri, October 1978] Society of American Foresters, Washington, D.C.

Morris, R. F. (Ed.) 1963. The dynamics of epidemic spruce budworm populations. Memoirs of the Entomological Society of Canada No. 31, 332 p.

Powell, Jerry A. 1980. Nomenclature of nearctic conifer-feeding *Choristoneura* (Lepidoptera: Tortricidae): Historical review and present status. USDA Forest Service General Technical Report PNW-100, 18 p. Pacific Northwest Forest and Range Experiment Station, Portland, Oregon.

Stage, Albert R. 1975. Forest stand prognosis in the presence of pests: Developing the expectations. p. 233-245 *In* David M. Baumgartner (Ed.) Management of lodgepole pine ecosystems: Proceedings of the symposium. [Washington State University, Pullman, October 9-11, 1973] Washington State University Cooperative Extension Service, Pullman, Washington.

Stage, Albert R., and Dennis E. Ferguson. In press. Regeneration modeling as a component of forest succession simulation. *In* Joseph E. Means (Ed.) Forest succession and stand development research in the Northwest: Proceedings of the symposium. [Oregon State University, Corvallis, March 26, 1981] School of Forestry, Oregon State University, Corvallis Oregon 97331.

Stage, A. R. 1973. Prognosis model for stand development. USDA Forest Service Research Paper INT-137, 32 p. Intermountain Forest and Range Experiment Station, Ogden, Utah.

Thomson, A. J. 1979. Evaluation of key biological relationships of western budworm and its host trees. BC-X-186, 19 p. Pacific Forest Research Centre, Victoria, B.C.

Wykoff, William R., Nicholas L. Crookston, and Albert R. Stage. 1982. User's guide to the stand prognosis model. USDA Forest Service General Technical Report INT-133, Intermountain Forest and Range Experiment Station, Ogden, Utah.

TWO DECADES OF BUDWORM SIMULACRA: A RETROSPECTIVE VIEW OF COMPREHENSIVE SIMULATION MODELS[1]

R. A. Fisher[2]

Abstract.--Three comprehensive simulation models including budworm, forest, and management actions have been developed based on distinct hypotheses concerning budworm outbreak mechanisms. Sampling problems preclude determination of the relative importance of proposed mechanisms. Nevertheless a control policy exists which in all the models reduces spray intervention without reducing harvested volumes.

INTRODUCTION

Models related to the spruce budworm (Choristoneura fumiferana (Clem.)) have taken many forms (Cuff and Baskerville, this volume). Morris (1963) and Clark et al. (1979) include representative reasons for modelling the system.

This paper is restricted to relating three simulation models of the budworm-forest system to each other and to the experimental program on which these models are based (Morris 1963). These models represent three different views of budworm population dynamics. They are comprehensive in that each considers the budworm and factors in nature affecting its survival, the host forest, and management policies applied to extensive areas of forest. Each was intended to help in the evaluation of different pest control management policies. Watt (1964), Baskerville (1976), and Stedinger (1978) describe the results of applying the chosen models to the study of forest management policy.

Characteristic differences among the models will be outlined. The review of recommendations to forest managers will demonstrate that conflicts between models with different biological structures need not extend to their policy implications.

THE BUDWORM/FOREST ECOSYSTEM

The spruce budworm was named and first identified in Virginia; its range extends from Minnesota and Ontario, eastward to Newfoundland. The budworm and the forest have coexisted for many generations of the host species (Morris 1963). Periodically, high insect densities occur, causing tree growth loss and mortality. In New Brunswick damage has been documented beginning about 1770, 1806, 1878, 1912, and 1949 (Morris 1963).

Budworm adults disperse actively, creating the potential to attack new stands over 450 km from their source (Greenbank et al. 1980). Maps of budworm density show a patchy distribution at both large and small scales (Webb et al. 1961).

The host forest, extending throughout the budworm's domain, varies in type from extensive regions of pure balsam fir to heterogeneous patches of mixed fir, spruce, and birch lying between hardwood ridges and black spruce bogs. In addition, the presence of hardwood forests, non-host conifers, and agricultural lands creates discontinuities in the host forest.

This paper will only consider phenomena in part of the budworm's range, namely Maine and New Brunswick. Even within the limited size of New Brunswick there is still a large range of forest conditions (Anon. 1981). Despite this variety the bulk of the host fir in northern New Brunswick in the 1950's (as described by Morris 1963) appears to have started active growth either before 1900 or shortly after 1912, when budworm-caused mortality was reliably observed (Morris 1963; p.24). The pre-1900 growth also resulted from natural forces, presumably budworm or windthrow subsequent to budworm mortality so that large regions of fir began growth at about the same time (Baskerville 1975).

Clark (1979) includes an analysis of spatial variation in the outbreak pattern for budworm. Equally interesting would be a comparison of variation in forest condition to variation in budworm population level at various spatial scales. Examination of maps generated for identifying regions of high budworm hazard in New Brunswick, combined with Baskerville's (1975) observations of tree age distribution, the maps of Webb et al. (1961), and the analyses of Clark (1979) are revealing. Both the forest and insect show great local variation. Still, at larger scales, there is a regional pattern demonstrating that certain regions or times show higher densities than others. This would imply that both stand conditions and budworm densities are not solely determined by local factors, but remain at least partially synchronized with neighbouring areas, even in regions larger than those considered here.

DATA SOURCES

Budworm research presently represents an extensive program (Grimble 1981). One of the earlier major programs of budworm - related research began

[1]Paper presented at the Third International Conference on State-of-the-Art in Ecological Modelling, Colorado State University, May 24-28, 1982.

[2]Robert A. Fisher is Biometrician, Maritimes Forest Research Centre, Environment Canada Forestry Service, Fredericton, New Brunswick, Canada.

with the establishment of the Green River Project in 1944 by R.F. Morris. Twelve authors, mainly entomologists, contributed to the articles in the summary monograph (Morris 1963), reporting the results of studies in the intervening period.

Underlying the reported studies were population counts obtained at the egg, larval, pupal, and adult stages together with possible sources of mortality and measures of stand condition. Results for two areas are reported. In the first, which was never sprayed, some eighty life tables were assembled. These summarized data from sample plots chosen in a variety of mature and immature forest types, with plots resampled each year. Life tables were also created for the second area which was subject to spray protection beginning in 1952. The plots selected were deemed representative of the variety of forest conditions in the region but do not constitute a designed sample.

Also, with the advent of spray protection in 1952, a second major sampling program began which was designed to identify tree hazard conditions in budworm-infested areas of the province. Typically the data consist of some thousand sample points (Prebble 1975). Stand conditions and budworm egg mass density are determined for each point. In addition other insect stages are sampled as needed to provide additional resolution for definition of areas to be sprayed.

A third source of data stems from inventories of New Brunswick's forests (Anon. 1981). Approximately every ten years the whole forest is inventoried within a five year period. The last inventory (1974-77) categorized forest stands into softwood, hardwood, hardwood-softwood, and softwood-hardwood in each of three diameter classes. The inventory now underway will provide more information on species composition, and identify and organize stands into "development classes" (Clifford, W. pers. comm. 1981). The new inventory structure is intended to allow better projection of forest growth (Baskerville 1982). This implies that the forest is now viewed as a dynamic entity rather than a fibre warehouse.

Modellers now have a large selection of data to draw on, even though none of these were collected with comprehensive ecosystem models in mind. Reducing these data to a manageable form is a continuing problem. As early as in the preparation of the monograph, Morris (1963) used models as a means to summarize data.

THE WATT MODEL

The first of three models to be considered was developed by K.E.F. Watt. Prior to implementing the model he worked on the Green River Project, writing several sections of the monograph (Morris 1963). He describes a hypothetical insect dispersing through a hypothetical forest (Watt 1964).

The forest consisted of 625 blocks of land on a square grid. All blocks were identical; with the characteristics of a typical fir stand at Green River circa 1950. The progress of an initial infestation in the central blocks was simulated by calculating larval densities in all blocks using a formula presented in the monograph (Morris 1963).

$$\log N_{n+1} = .98 + .76 \log N_n + .18\ (TMAX - 66.5)$$

where: N_n = the larval density in year n
TMAX = the mean maximum daily temperature during the period of larval development.

The identical temperature pattern as observed in the Green River unsprayed area during the development of the outbreak was used to fit the equation and again to generate population levels in the simulation model. Thus the model, as applied, always resulted in the development of outbreak pest densities.

The insects dispersed to adjacent blocks by a diffusive process, allowing the outbreak to spread. Spraying, parasite release and disease in various combinations were simulated as control mechanisms by modifying N_{n+1} calculated from the formula. Different diffusion rates were investigated. Total costs of control measures over the period 1925-1960 were calculated for a variety of control scenarios. Tree mortality was estimated from defoliation estimated in turn from insect density. Mortality was affected by both duration and intensity of pest infestation.

The model when run was judged to behave reasonably. Insecticides prevented tree mortality, while predators and parasites slowed down the population growth but failed to eliminate tree damage. When implemented the model used all the memory resources of the University of Toronto IBM 7090. I doubt that a significantly more complex budworm model could have been implemented in 1963.

I will use reproduction curves as Morris (1963) did to contrast the different population dynamics implicit in the models considered. In Figure 1 the axes represent log population density in two succeeding generations. The diagonal (labelled "A") represents stable population levels. Because of the formula used generation survival in Watt's model is always

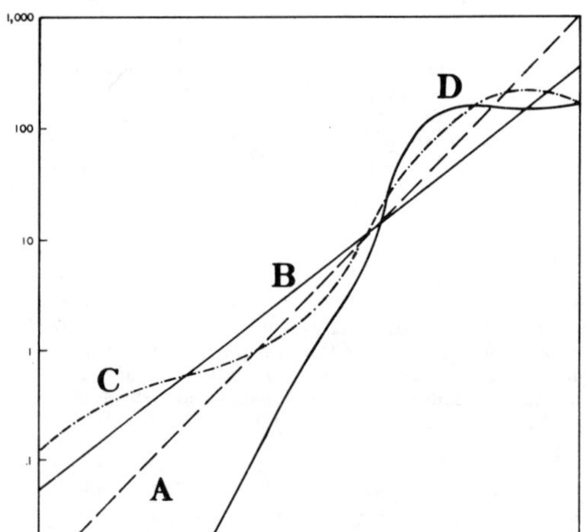

Figure 1. Typical forms for reproduction curves. "A": stable population, "B": Watt model, "C": MFRC-UBC models, "D": Stedinger model.

represented by a straight line (labelled "B" in fig. 1) with less slope than that for a stable population. Changes in temperature, predators or other factors in his model shift the position of Watt's survival curve but not its form. For any set of conditions there is always one point where the lines intersect, representing a single stable equilibrium.

The major conclusion described by Watt (1964), arising from his simulations was that, "For all kinds of control, control is vastly more effective if applied 10 years or more before peak densities". Nine years intervened before the next formulation of a comprehensive model began.

THE MFRC-UBC MODELS

The second model (Baskerville 1976) to be discussed evolved over a period of time with contributions from a large number of participants. Initial modelling activity occured during an October 1972 workshop held at the Maritimes Forest Research Centre (MFRC). Participants included staff of the Institute of Animal Resource Ecology (IARE) at the University of British Columbia (UBC) along with biologists at MFRC. C.S. Holling from IARE had formerly been involved in the analysis of predation data from the Green River Project. G.L. Baskerville (program manager MFRC) had earlier undertaken forest biomass studies in the same area.

A computer simulation model was implemented at the workshop and was subsequently extensively modified. Descriptions of the initial model exist only in the form of draft documentation, (Stander 1973), memos, and computer listings. A second model with much different budworm population dynamics (Jones 1979) was presented at a workshop in 1975 (which I attended). Several modifications to the budworm dynamics were tested at the workshop, but no structural changes were retained in subsequent work.

The budworm population dynamics in the 1975 model (Jones 1979) described mortality losses at each of the egg, small larval, large larval, pupal, and adult stages. Sources of mortality included larval dispersal, predation, parasitism, starvation, and insecticide application. The functional forms included in this model for the first time relate budworm survival to host forest conditions.

The forest was described in terms of its age-class distribution (through the corresponding branch surface area), together with new and old foliage. Each block of land (site) was related to a map sheet from the New Brunswick inventory (Anon. 1981), and given the appropriate host forest area and age class structure. During the simulations as each site aged its ability to support budworm changed, and new forest regenerated as portions of the old forest died from budworm effects, old age, or harvesting.

Clark et al. (1979) used a model with 264 sites comprising 75 age classes, whereas Baskerville (1976) used 450 sites and 120 age classes as a more complete representation of the New Brunswick forest.

The Baskerville (1976) version of the model was used extensively to study the implications of different harvesting and spray rules in the modelled forest. The simulations indicated that less pesticide was applied to prevent forest mortality when the control policy used resulted in spraying areas before budworm density became very high. This is very similar to Watt's policy of spraying early in the outbreak.

The qualitative form of the reproduction curve in the MFRC-UBC models contrast strongly with Watt's. Its characteristics for certain stand conditions are similar to the sinuous curve hypothesized by Morris (1963; p.126) (labelled "C" in fig. 1). With all other conditions held constant there are two stable equilibria separated by an unstable equilibrium point. The lower equilibrium is a consequence of the avian predators submodel which traces its origins back to earlier predation studies of Holling (e.g. Holling and Buckingham 1976) rather than budworm studies. Morris (1963; p.127) speculated that such a mechanism might operate to maintain low level populations between outbreaks. The upper equilibrium results since, in the model, foliage depletion caused by budworm feeding reduces budworm survival.

Unlike Watt's model, the shape of this curve is not constant over time, since it changes with stand conditions. In young stands the upper equilibrium may not be present and in mature stands the lower equilibrium will disappear. In effect the curve rises and falls with stand age. Harvesting and budworm-induced mortality can therefore affect generation survival for many years by modifying the stand age structure. Defoliation will also modify the upper equilibrium over a shorter time span. The onset of an outbreak in this model is no longer dependant on a period of years with favourable weather, but on a significant portion of the forest being in mature age classes.

THE STEDINGER MODEL

A third model (Stedinger 1978) proposed another alternative form to describe budworm population dynamics. Stedinger's Ph. D. thesis (1978) includes a model developed to study budworm control policies in the state of Maine, and acknowledges the assistance of D.G. Mott, who earlier wrote several portions of the monograph summarizing the Green River Project (Morris 1963).

In deriving his model, Stedinger undertook an extensive comparison of functional relations contained in the MFRC-UBC models with available data reported in the literature. Virtually all of the relations in his model of budworm population dynamics are completely changed from those in the MFRC-UBC models. The thesis (Stedinger 1978) includes a detailed comparison of components in each model. The avian predator submodel was discussed at length and excluded from the Maine model. The forest component in his models consisted of either a line of ten sites or an 8 by 10 block of sites, each consisting of all aged stands of 55 to 65% host species. These models allowed adult dispersal in either one or two dimensions respectively.

The major conclusion regarding policy from this study was that Watt's spray recommendations were again supported by this new model. (Stedinger 1978; p. 5-28). He also discussed how his simulation results differed from those of the MFRC-UBC models in detail.

The difference in underlying structure is apparent in the typical reproduction curves resulting from his budworm model (labelled "D" in fig. 1). In Stedinger's model the distance between the unstable and stable equilibrium points is less than for the MFRC-UBC models. More dramatically the lower equilibrium no longer exists, since no functional form equivalent to the avian predator submodel appears in the Stedinger model. In this model outbreaks only occur when sufficient adult invasion occurs to raise the local population density above the current position of the unstable equilibrium point. Populations below that density rapidly become extinct; as in the MFRC-UBC model the shape and position of the curve depend on the current age class structure and defoliation.

REPRODUCTION CURVES

The typical reproduction curves described for each model show a variation in form that reflect particular hypotheses of how budworm outbreaks develop. The regression-based linear model in Watt (1964) supposes that in a mature stand budworm densities sufficient to cause damage will always appear, with the actual timing strongly affected by weather. The MFRC-UBC models assume that budworm generation survival is directly related to stand condition, and also that predators have a limited ability to regulate budworm populations. Stedinger (1978) assumes that stands will not suffer major budworm attack unless external invasion creates pre-outbreak densities. I believe all the authors would accept that weather, stand condition, predators, and invasion would affect budworm survival, certainly in extreme conditions. The difficulty is in the weight to be applied to each of the factors.

The models have been used to evaluate spray policies to be applied to a forest in which a budworm outbreak can be anticipated. In this context the shape and position of the reproduction curves are not important to forest management, since all the models provide the same prescription: apply control early in outbreak development.

On the other hand, it seems that if our policy decisions become sensitive to choosing between these potential causes of outbreak, then our forests will be impossible to manage. The problem is that the different hypotheses are not only tenable given presently available data, but also will probably remain tenable given any foreseeable amount of biological sampling effort. Morris (1955) describes both sample size and manpower requirements to sample larval densities. Sampling at or below 10 larvae per 10 square feet, approximately the density at which the Watt model has its equilibrium, requires obtaining and examining branches from 25 or more trees. Each such sample would require over 15 man-days of effort. Some density samples needed to establish the form of the budworm reproduction curve would require examination of several hundred trees. All such samples must be taken over a period of a few weeks. Relating the densities found to stand parameters, predators, or weather would require extensive replication. I do not forsee such data becoming available.

At high densities, where the two later models have upper equilibria the sampling problems are still acute but have quite a different nature. Here major defoliation of the trees occurs, and while mortality is not immediate, changes that precede death will potentially introduce changes in foliage quality and the microclimate surrounding the budworm larvae. At these densities, it is our inability to establish the current stand condition during such rapid change, which limits our attempts to relate budworm survival to local conditions.

APPLYING THE MODELS

In the face of this sampling uncertainty there seems to be little value in attempting to refine our policy models to find the most appropriate functional forms and parameters.

If we cannot establish the suitable form for the reproduction curve, then the modeller's original goal of developing optimal control strategies cannot be met. Rather than choosing an optimal policy under a given set of assumptions, we must seek a policy which is acceptable in conjunction with all the contending models. Fortunately all the models suggest spraying early in the outbreak development as a better strategy than allowing it to build. The value of this approach is reinforced by Baskerville and MacLean (1979) who have shown that loss of growth through defoliation has long term effects on stand volume even if mortality is prevented.

Only the Task Force Report for Evaluation of Budworm Control Alternatives (Baskerville 1976) addressed making the harvest part of the control policy. It seems apparent that harvesting high risk stands would be a better policy under the other two models as well, if only by reducing the land area currently capable of supporting and exporting large numbers of budworm.

Integrating harvest and early intervention spray policies into forest management requires rules for implementation. The models suggest the direction of change but are not precise enough to establish the amount. Hall (1978) describes a program to study wood-supply based on expected harvest requirements and yields on a stand by stand basis. Effects of budworm damage can be incorporated by using data on actual yields from stands presently under pest control management. MacLean (1982) describes a method of assessing the forest's vulnerability to budworm attack.

Since the present wood-supply model does not use information on stand location or access costs, these must be included before the added costs of preferentially harvesting more vulnerable stands can be assessed. The ability to implement policy change by steps, and to improve it through monitoring undertaken in normal operations, makes such changes feasible. The policies proposed as best by the Baskerville (1976) Task Force Report were not sensitive to small errors in policy implementation when tested in the MFRC-UBC model (unpublished data). Except when the local insect populations are extreme, the forest changes slowly, allowing management to compensate for recent errors in policy application.

The strategy which now seems most attractive is to concentrate our efforts on attainable data, by developing the wood supply models to be responsive to moderate levels of budworm attack. Budworm-induced tree mortality will remain a problem for such models, since mortality resulting from outbreaks is variable in some common stand types (MacLean 1980).

Hypotheses to explain budworm population dynamics have certainly not been exhausted. One such, formulating the dynamics as a second-order density-dependant process (Royama 1981) implies that budworm populations can fall to low levels in time, even though the pest control programs limit defoliation and tree mortality. Such behaviour can not occur in our present models, but was observed in certain stands by Morris (1963; p.225) and by MacLean (1980). The implications of such proposed budworm dynamics should be tested in a responsive forest model.

The course suggested then is to build the best possible wood supply model for the forest, determine how the forest reacts to budworm, and test policies in models with all currently proposed forms of budworm population dynamics. So long as the different versions of budworm dynamics do not affect the relative values of the proposed alternate policies, there is little benefit to the forest manager in determining which version is most appropriate.

CONCLUSION

The models presented here had a common goal but represent conflicting views of the ecosystem. This has not limited their use as a means of studying management policy, but only reinforced the caveats given throughout by the authors, for example Baskerville (1976); "What they (the models) provide is an expression of how the world might look over a period of time for the given explicit assumptions and rules (policy implementation)". Models which encompass quite different views of the reproduction curve all point in the same direction. If you must take action do it early; if you plan to harvest and control damage, harvest where its most beneficial to your control program.

In response to hearing many statements of the form, "We should do this because the model says such-and-such", I must make a final comment. It is not suggested that either a research plan or a management policy should be implemented because they have major effects when used in our models. The inherent uncertainty in any of our comprehensive ecological models limits them to suggesting directions for policy change. The decision to change should not be made on the basis of the model's performance, but on a comprehensive reexamination of the manager's problem in light of possibilities suggested by the model.

ACKNOWLEDGEMENTS

I thank Dr. Wilf Cuff for useful discussion and assistance in the manuscript preparation; Ms. Nancy Gogan for typing the manuscript; and Dr. David MacLean for reviewing the manuscript.

LITERATURE CITED

Anonymous. 1981. The New Brunswick Forest Inventory. Department of Natural Resources, Fredericton 141 p.

Baskerville, Gordon L. 1975. Spruce budworm: Super Silviculturist. For. Chron. 51: 138-140.

Baskerville, Gordon L. 1976. Report of the Task Forces for Evaluation of Budworm Control Alternatives. Department of Natural Resources, Fredericton, N.B. 210 p.

Baskerville, Gordon L. 1982. The spruce/fir wood supply in New Brunswick. Department of Natural Resources, Fredericton, 15 p.

Baskerville, Gordon, L., and David A. MacLean 1979. Budworm-caused mortality and 20-year recovery in immature balsam fir stands. Maritime Forest Research Centre Information Report M-X-102, Fredericton, 23 p.

Clark, William C. 1979. Spatial Structure and Population Dynamics in an Insect Epidemic Ecosystem. Thesis, University of British Columbia, Vancouver, 505 p. mimeographed.

Clark, William C., Dixon D. Jones and C.S. Holling. 1979. Lessons for ecological policy design: A case study of ecosystem management. Ecol. Modelling 7: 1-53.

Cuff, Wilf and Gordon L. Baskerville. 1982. Ecological modelling and management of spruce budworm infested fir-spruce forests of New Brunswick, Canada. Proceedings of the Third International Conference on State-of-the-Art in Ecological Modelling, Colorado State University, Fort Collins.

Greenbank, D.O., G.W. Schaefer, and R.C. Rainey. 1980. Spruce Budworm (Lepidoptera: Fortricidae) Moth Flight and Dispersal: New Understanding From Canopy Observations, Radar, and Aircraft. Mem. Entomol. Soc. Can. No. 110 49 p.

Grimble, David A. 1981. CANUSA Mid-program Report. Northeast Forest Experimental Station, Broomall, 14 p.

Hall, T.H. 1978. Toward a framework for forest management decision-making in New Brunswick. Department of Natural Resources, Fredericton, Report TR1-78, 83 p.

Holling, C.S. and S. Buckingham. 1976. A behavioral model of predator-prey functional responses. Behav. Sci. 21: 183-195.

Jones, Dixon D. 1979. The Budworm Site Model. In: Pest Management: Proceedings of an International Conference, October 25-29, 1976. Pergamon 349 p.

MacLean, David A. 1980. Vulnerability of fir-spruce stands during uncontrolled spruce budworm outbreaks: A review and discussion. For. Chron. 56: 213-221.

MacLean, David A. 1982. Vulnerability rating of forests in New Brunswick and Nova Scotia to budworm attack, Maritimes Forest Research Centre Information Report M-X-132 Fredericton, 21 p.

Morris, R.F. 1955. The development of sampling techniques for forest insect defoliators, with particular reference to the spruce budworm. Can. J. Zool. 33: 225-294.

Morris, R.F. (Ed.). 1963. The Dynamics of Epidemic Spruce Budworm Populations. Mem. Entomol. Soc. Can. No. 31. 332 p.

Prebble, M.L. (Ed.). 1975. Aerial Control of Forest Insects in Canada. Environment Canada, Ottawa 328 p.

Royama, T. 1981. Fundamental concepts and methodology for the analysis of animal population dynamics, with particular reference to univoltine species. Ecological Monographs 51: 473-493.

Stander, Jeffrey M. 1973. A simulation model of the spruce budworm and the forest in New Brunswick. Unpublished manuscript, University of British Columbia, Vancouver.

Stedinger, Jery. 1978. Spruce Budworm Management Models. Thesis, Harvard University, Harvard, Mass. 351p., mimeographed.

Watt, K.E.F. 1964. The use of mathematics and computers to determine optimal strategy and tactics for a given insect pest control problem. Can. Ent. 96: 202-220.

Webb, F.E., J.R. Blais and R.N. Nash. 1961. A cartographic history of spruce budworm outbreaks and aerial forest spraying in the Atlantic region of North America 1948-59. Can. Ent. 93: 360-379.

MODELING THE EFFECTS OF TEMPERATURE ON ARTHROPOD POPULATION SYSTEMS[1]

Jesse A. Logan and David W. Hilbert[2]

Abstract.--Application of model descriptions of temperature effects on critical life history parameters has led to insightful and sometimes novel interpretation of field observations. We discuss in detail how the effects of temperature on individual population parameters dramatically influence community stability in an acarine prey-predator system. Evidence is presented which indicates that a similar mechanism may be responsible for observed temperature mediated instabilities in several acarine prey-predator associations. Potential physiological differences associated with trophic structure are hypothesized to explain the apparent general nature of temperature effect on arthropod community structure. We further discuss how temperature-mediated community dynamics logically lead to a pulse-stable model of adaptive change. Such a model is used to help explain the frequently "sudden" appearance of pesticide-resistant arthropod populations.

INTRODUCTION

Literally thousands of papers have been published relating effects of temperature to various insect life history parameters. There was, in fact, enough information in print by 1901 for Bachmetjew to publish a 160-page paper reviewing the literature to that date. During the ensuing 81 years, quantifying the effect of temperature on arthropod life history systems has been a recurring theme in the literature on insect ecology and applied entomology. The reason for this sustained interest is twofold and is related to the nature of entomology as a science. Entomology, at least in this country, is a hybrid--one-half science, one-half profession. From the basic science point of view, temperature is of interest as a principal physiological driver for poikilothermic organisms. From the view of a profession, the pragmatic value of understanding temperature-related pest phenology was recognized early in the history of economic entomology. The applied importance of understanding temperature effects on arthropod life history parameters is primarily responsible for development of models predicting temperature effects. From an economic point of view it was not enough to empirically describe the effect temperature had, but it was also necessary to predict the occurrence of temperature-mediated events.

The earliest model for describing temperature-dependent biological rate processes was that of Oettingen (1879), which proposed a linear relationship between the rate of the process and temperature (fig. 1). This model was first applied to insects by Simpson in 1903 for description of developmental rates in codling moth larvae. Simpson was among the first to use the concept of "thermal units of development," which has become commonly known as the day-degree model. Sanderson and Peairs (1913) further expanded the concept of linearity to predict a developmental zero. The linear, day-degree model was quickly accepted by entomologists as a useful tool and was applied to a wide variety of insect species. General acceptance of this model is indicated by Chapman's (1931) strong advocacy in his influential text *Animal Ecology with Especial Reference to the Insects*.

There were several good reasons for the general acceptance of the day-degree model. First, it seemed to work well for a wide variety of insects. The model was also easily modified; for example, by Glenn (1922,

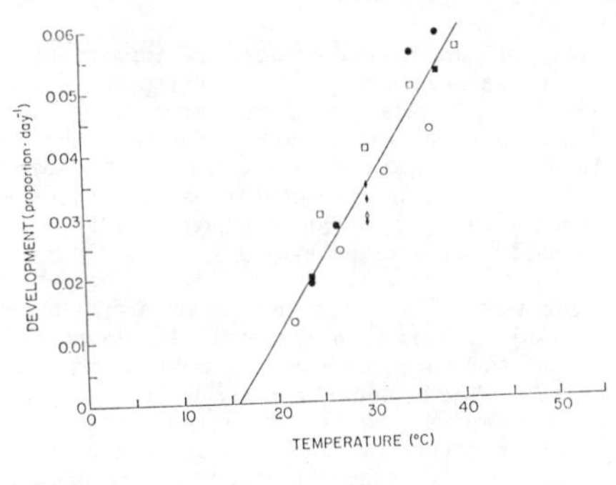

Figure 1.--Linear (day degree) model of developmental rate as a function of temperature (from Hilbert and Logan, in press).

---

[1]Paper presented at the Third International Conference on State-of-the-Art in Ecological Modeling, Colorado State University, May 24-28, 1982. Work reported herein was supported in part by Colorado State University Agricultural Experiment Station Project 53.

[2]Jesse A. Logan is a Population Ecologist and David W. Hilbert is a Graduate Research Assistant at the Natural Resource Ecology Laboratory, Colorado State University, Fort Collins, Colorado, U.S.A.

1931), who used the concept of linearity to develop an hour-degree model and also used a piecewise linear model to account for supra-optimum, high temperature decline in developmental rate. Finally, and perhaps most important, the assumption of linearity provided a model that was easily fit to data.

As useful a concept as the day-degree model was, several problems in applying the linear concept were recognized from the earliest critical work dealing with description of temperature-dependent rate processes. Verworn (1899), for example, noted that the general shape of the rate function was highly nonlinear and asymmetric. In 1924, Crozier clearly depicted departure from linearity at lower temperatures. Shelford discusses these and other problems in his 1929 ecology text. The assumption of linearity may also have led to the confusion in interpretation of empirical results, particularly those relating to prediction of events under fluctuating temperature regimes. Hagstrum and Hagstrum (1970; table 1) summarize the literature comparing fluctuating to constant temperatures for 33 insect species; percent change in developmental rates listed range from -35 to +82%. There are at least two explanations for such discrepancies: There may, in fact, be some curious physiological effect of alternating temperatures; or, alternatively, as shown by Tanigoshi and Logan (1979) and Hilbert and Logan (in press), the observed discrepancies may result from the assumption of linearity when, in fact, the true underlying relationship is highly nonlinear and asymmetric.

Primarily because of problems in application of linear models, several alternate models were proposed. Among the earliest was Janisch (1925), who analogized the integral of the rate function as a hanging-cable problem. Janisch's asymmetric catenary model recognized and accounted for nonlinearity over the entire range of biologically active temperatures. Andrewartha and Birch, in their classic 1954 book, strongly recommended use of the logistic equation to describe rate/temperature developmental curves in insects and suggested that it was superior to all previous descriptions. They also noted that parameters were easily estimated from empirical data and finally concluded, "From now on it should be used in preference to all the older expressions."

Until quite recently, however, the vast majority of models relating insect life history parameters to temperature have been linear, day-degree models. Some of the reasons for reluctance to abandon the day-degree model are conceptual. As stated by Gilbert et al. in their 1976 book:

> Over the intermediate range of temperatures which the insect normally experiences in the field, the rate of development increases linearly with temperature. Why this should be so is a mystery, since rates of enzyme action (which are presumably basic to development) usually increase exponentially, not linearly, with temperature.

We concur that if in fact the relationship were linear, it would indeed be a mystery. In essentially all cases when the rate response to temperature has been precisely measured, departures from linearity are obvious. For many insects, departures from linearity occur at temperatures well within those experienced in the field. In fact adaptation of an organism to its environment may critically depend on a nonlinear response to temperature (Hilbert et al., submitted). This is not only true for periods of environmental stress (i.e., diapause, estivation, etc.) but may also assume particular importance in establishing community structure. We will further discuss such temperature-mediated structural stability in prey-predator communities.

The principal reason for lack of widespread use of nonlinear models is difficulty in parameter estimation. Even though the methods for nonlinear parameter estimation are conceptually simple, they are computationally cumbersome. However, the increased accessibility of computers which occurred during the 1960s did much to mitigate these computational problems. Combined with the availability of computing power was the motivation to use this power in an entomological context. Concurrent with the development of an integrated pest management (IPM) philosophy of pest control was the realization that computer modeling and systems analysis could play a central role in development of IPM research. Several new nonlinear models were developed during the 1970s. The next section of this paper provides a critique of these models.

NONLINEAR RATE/TEMPERATURE DEVELOPMENTAL MODELS

The first of these "new generation" models was published in a paper by Stinner et al. in 1974. The basic problem that motivated this work involved the simulation of *Heliothis zea* (Bodie) populations in North Carolina. Because of an interesting relationship between cannibalism and *apparent* phenology in the field (Stinner et al. 1977), it was necessary to accurately model temperature-driven developmental rates. Stinner's model (fig. 2) is important for several reasons. It was the first serious improvement over Janisch's work, which was published in the 1920s. The model provided for a realistic decrease in developmental rates at supra-optimum temperatures. Finally, and perhaps most important, use of the temperature relationship in a full simulation model resulted in an improved understanding of the basic biology of the modeled system.

As important as Stinner's work was, some difficulties were noted in application of his model to other arthropod systems. The first is a minor technical difficulty of a discontinuous first derivative at the optimum temperature (T_o). This is only a potential problem in some numerical analysis schemes (i.e., models). A second and perhaps more serious deficiency is the symmetry about T_o. Apparently, in the system for which the model was developed, the symmetry about T_o was a reasonably good approximation. For other

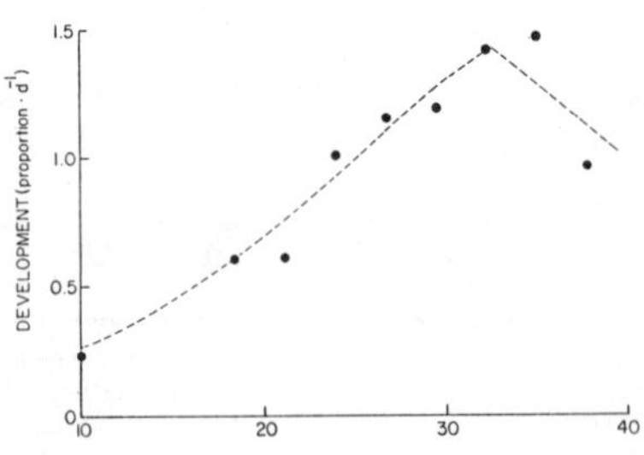

Figure 2.--Stinner's (Stinner et al. 1974) nonlinear model (from Logan et al. 1976).

species in other habitats, the response is noticeably asymmetric about T_o.

Motivated primarily by observed lack of symmetry about T_o, Logan et al. (1976) developed a uniformly valid temperature-response curve by applying the method and matched asymptotic expansions. The resulting functional relationship is continuous with continuous derivatives throughout the range of biological activity and is capable of describing the marked asymmetry that initially motivated the work (fig. 3). Furthermore, this model is mathematically simple and allows a descriptively straightforward interpretation of parameters. Although the model provided a good fit to a wide range of rate/temperature data sets, it has been criticized as being too empirical and lacking a mechanistic basis.

In an attempt to describe the relationship between temperature and developmental rate at a biochemical level, Sharpe and De Michele (1977) published a model based on enzyme kinetics (fig. 4). Although mathematically somewhat complex (six parameters requiring at least seven independent temperatures), the model has been successfully fit to a large number of data sets. The biochemical basis for the model is ontologically satisfying; however, it should be noted that the temperature response, particulary during high temperature decline, may not be caused by enzyme denaturization. In small organisms with a relatively large surface-to-volume ratio, such as most terrestrial arthropods, breakdown of the cuticular wax layer and subsequent desiccation (Beament 1958) most likely has an overriding influence.

A final model we discuss was developed by Hilbert and Logan (in press) primarily to provide a low-temperature threshold modification of the Logan et al. (1976) model (fig. 5). Similar to the Logan et al. model, this model was empirically derived by applying matched asymptotoic expansions and is a relatively simple mathematical relationship. The model also provided a good fit to the data set that motivated model development.

A comparison of the four recently developed rate/temperature models indicates that for a specific application one may be more desirable than another. However, it seems to us that their commonalties are perhaps more important than their differences. In our work, which is centered at the population or community level of organization, there are several characteris-

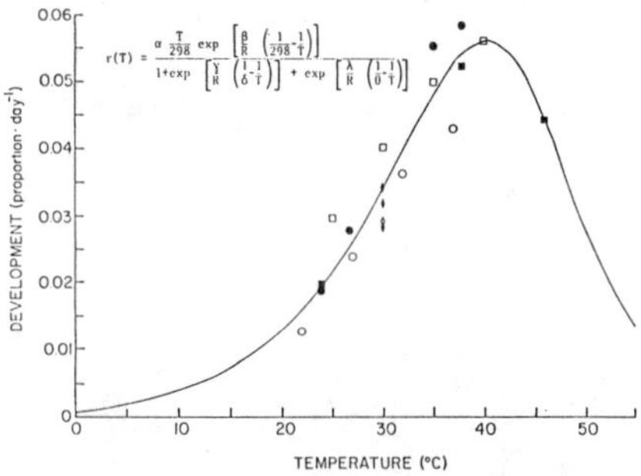

Figure 4.--Sharpe and DeMichele (1977) enzyme kinetics model (see original publication for parameter definitions).

tics of rate/temperature models that are of recurring importance. First, in all applications we have found nonlinearity essential for adequate biological descriptions and inference. Asymmetry about T_o is often important and will play a critical role in the example we discuss in a later section of this paper. Likewise, it has often been necessary to explicitly include thermal maximum and/or thermal minimum. In our opinion, these three characteristics of thermal limits, nonlinearity, and asymmetry provide sufficient conditions for adequate description of biological rate/temperature phenomena. In the next section we discuss how these models can be used to describe dynamic population behavior.

APPLICATION OF RATE/TEMPERATURE MODELS

The temperature models we have discussed to this point describe rate of a process (e.g., developmental rate) as a function of temperature. Field applications, however, require description of rate as a function of time; in fact, the models we have discussed to this point may be thought of as physiological time filters. One approach to development of such a time filter is to consider temperature, and therefore the rate function, constant for some small interval of time. Mathematical statement of this process (diagramatically shown in figure 6) is given by

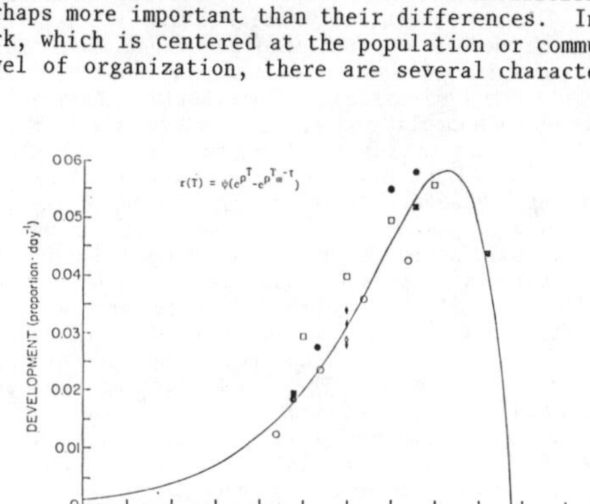

Figure 3.--Logan's (Logan et al. 1976) nonlinear model. Functional form given (see original publication for parameter definition).

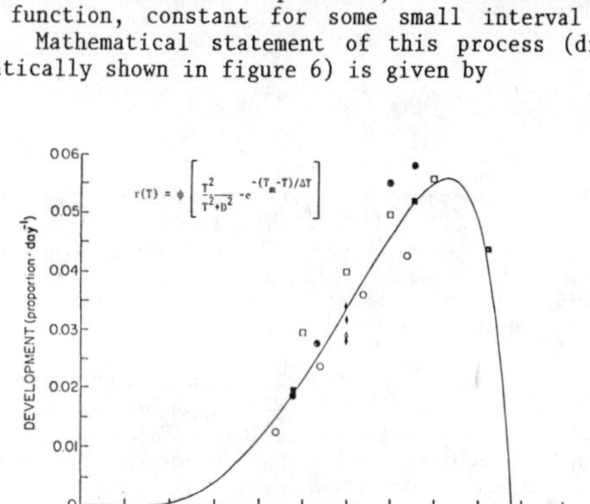

Figure 5.--Hilbert and Logan (in press) model (see original publication for parameter description).

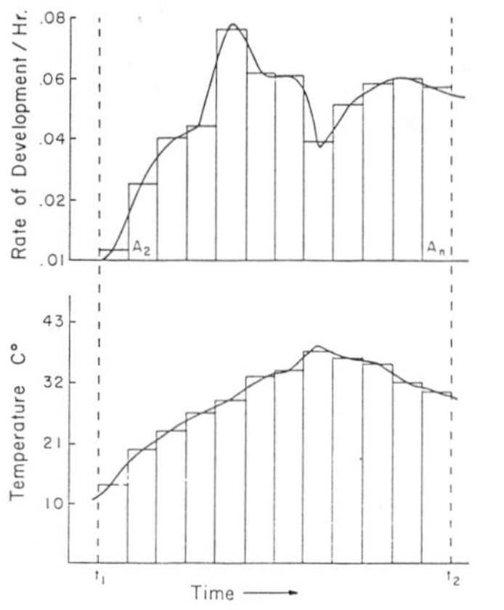

Figure 6.--Discrete time approximation to the continuous temperature and continuous rate function. For each interval on the temperature curve there is a corresponding developmental rate. When the area of the rectangles representing the rate curve sum to unity, the process is completed.

$$D = \sum_{i=1}^{n} d(T) \cdot \Delta t_i = 1 \quad , \qquad (1)$$

where,

- d = temperature (T) dependent rate function
- D = time required for the modeled process to be completed
- Δt_i = width of the small time increment
- n = number of increment.

Equation (1) indicates that when the area represented by the rectangles of figure 6 sum to unity, the modeled process is completed. It is easy to imagine that as Δt becomes smaller and smaller, the rectangles become an increasingly improved approximation to the continuous temperature curve. This fact is expressed mathematically as

$$D = \int_{t_1}^{t_2} d[T(t)]dt = 1 \quad , \qquad (2)$$

where the interpretation of variables are the same as in equation (1). If suitable functional forms of d and T are chosen (Logan et al. 1976, Tanigoshi and Logan 1979), equation (2) can easily be integrated by numerical schemes, resulting in an efficient algorithm for computing critical events.

Three implicit assumptions are associated with equation (2). The first is that under variable temperatures, the organism adjusts "instaneously" to changes in ambient conditions. Under natural temperature regimes, where temperature changes tend to be gradual and continuous, this generally appears to be a valid assumption. Equation (2) also assumes that the temperature of the microhabitat of the organism can be accurately measured, at least in a correlative fashion (Logan et al. 1979). Interesting advances are being made in models that relate easily measured biometeorological parameters to microhibitat conditions (Parton and Logan 1981). The final assumption is that the adaptive behavior of the organism under test conditions mimics that in the natural environment. This final assumption is perhaps the most tenuous of the three, and validity needs to be judged on an individual basis.

Data for determination of the functional form of d in equations (1) and (2) are most often obtained from laboratory studies. For example, developmental times for a particular instar would be measured over a range of various constant temperatures. The empirical value of $d(T_i)$ at a particular temperature would then be computed as $1/D(T_i)$. Some optimization procedure (e.g., Marquardt's nonlinear least squares algorithm) is then used to fit model parameters to the empirical data. This process has resulted in many useful models; however, it is not always possible or desirable to perform these types of studies. For many insects it is difficult, if not impossible, to recreate critical habitat in a laboratory setting. Techniques for parameter estimation from field data, such as the abundant data base in the literature, is a germane area for research.

To this point we have discussed modeling temperature effects on individual physiological processes. This area of ecology in the past has been referred to as autecology, or, more topically, as physiological ecology. However, the title of this paper implies modeling temperature effects at the population or higher levels of organization. The question is then: How do you go from the individual to the population level? One way this link has been established is classical life tables analysis. The advent of mathematical modeling, in particular that of computer simulation modeling, has introduced a new dynamic dimension to the analysis of population phenomena. A computer simulation model may, in fact, include the same information as a life table analysis but would also attempt to describe the time dependent densities of involved populations. The essence of mathematical modeling is to capture the temporal (and perhaps spatial) dynamics of interacting components.

Incorporation of temperature effects on individual physiology into simulation models of arthropod population systems has been extensively documented (e.g., Logan, submitted; Reusink 1976; Welch 1979). Simulation models of arthropod pest systems in which temperature is a principal driving variable have made and are making significant contributions in analysis of arthropod population systems. Simulation models provide for easy inclusion of a great deal of biological realism. For this reason they are ontologically satisfying. However, there are inherent weaknesses in this approach. Because of the ease of including complexity, simulation models may become unwieldly. Large simulation models are also expensive in computer space and time. It may therefore be prohibitively costly to directly use simulation models for analysis of temperature effects at the population or community level of organization. These difficulties are often mitigated by use of a simulation model to generate data for estimating parameters in more conservative, mathematically concise models. These models, in turn, are more amenable to analysis and inference than the original simulation model. The concordant approach of using computer simulation and analytic models for analysis of population systems is shown in figure 7.

Although figure 7 may be inadequate because it fails to indicate potential information feedback, it illustrates that specific types of models should be

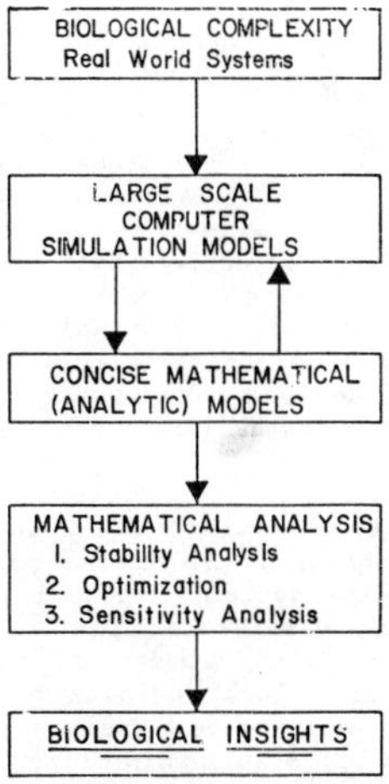

Figure 7.--Relationship between simulation models, analytic models, and biological reality.

used in a manner that exploits their potential strengths and avoids ill-suited applications. Such a concordant modeling approach applied to analysis of temperature effects on structural stability of an arthropod prey-predator community is discussed in the next section.

MODELING TEMPERATURE EFFECTS ON COMMUNITY STABILITY: A CASE HISTORY

In the late 1960s, S. C. Hoyt initiated a biological control program for spider mite pests on apples in Washington State. This integrated program was based on control of the spider mite pest *Tetranychus mcdanieli* (McGregor) by a predaceous phytoseiid mite, *Metaseiulus occidentalis* (Nesbitt). While this program has been extremely successful, problems occurred during initial phases of the program if rust mite populations were not present for early spring build-up of *M. occidentalis* populations. The potential for

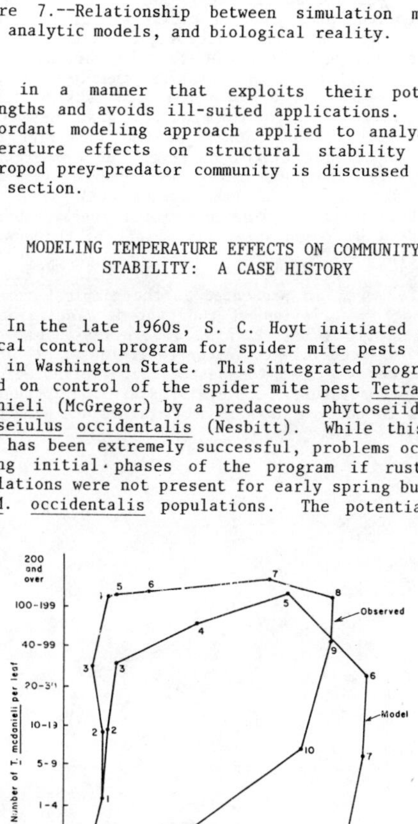

Figure 8.--Simulation model results compared with field observations for *Tetranychus mcdanieli* and *Metaseiulus occidentalis*.

failure of biological control under these circumstances increased during periods of prolonged high temperatures. For this reason, particular emphasis was placed on the role that temperature played in the interactive population dynamics of the two species.

Tanigoshi et al. (1975a, 1975b) conducted research detailing life history information for both species over a range of temperatures similar to that expected under summer field conditions. Logan (1977) then used this information to develop a detailed, discrete time computer simulation model which explicitly included temperature dependence of developmental rates, oviposition, and prey consumption. Although this simulation model provided a valuable research tool, and performed relatively well under field conditions (fig. 8); it was typical of simulation models in being analytically intractable and expensive to run.

As an alternative to the simulation model, we felt it would be heuristically valuable to develop the simplest, most concise model of the system that seemed biologically reasonable. One such model, considered in detail by R. M. May (e.g., 1977) is given by

$$\dot{N}_1 = r_1 N_1 (1 - N_1/K) - k N_1 N_2/(N_1 + D)$$
$$\dot{N}_2 = r_2 N_2 (1 - m N_2/N_1) \quad (3)$$

where

$\dot{N}_i$ = time derivatives for the ith population
K = prey-carrying capacity
k = maximum consumption rate per predator
D = prey density that results in consumption equal to 0.5 k
m = maintenance feeding level for the predator
r_1 and r_2 = maximum population growth rates for the prey and predator, respectively.

Equation (3) seems to represent the essential characteristics of the *T. mcdanieli*/*M. occidentalis* system. Prey density is limited by a finite carrying capacity. Effect of the producer trophic level is therefore implicitly included through modification of the realized population growth rate of the prey. The functional response of the predator to prey density is represented by a Type II (Holling 1959) response. Phytoseid predators typically exhibit a Type II response (Logan, submitted). Finally, carrying capacity for the predator is proportionate to prey population density. Realized predator population growth rate ($\dot{N}_2$) is thus a function of prey availability.

Descriptions such as equation (3) have frequently been criticized as being overly simplistic (Wangersky 1978). Two considerations help mitigate this concern. First is the generally high biotic potential of mite populations. As a result, stable age distributions are reached quite rapidly. Since a stable age distribution is a basic assumption of equation (3), mites are among the few terrestrial arthropod populations for which an instantaneous representation is a good approximation (Taylor 1979). Secondly, the use of simulation models for parameter estimation adds biological credibility to model equation (3). For example, the maximum population growth rate possible at a given temperature (r_i) was estimated from the simulation model (Logan 1977) by computing the population growth rate at each time step and allowing enough time

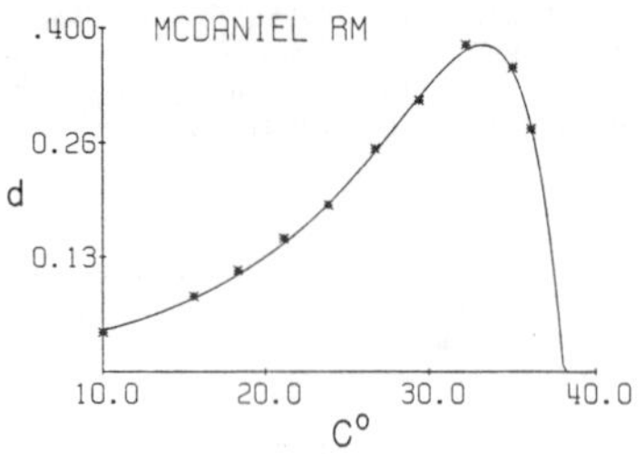

Figure 9.--Functional relationship (Logan et al. 1976) for r as a function of temperature fit to "data" generated from simulation model of T. mcdanieli (Logan 1977).

for convergence of the time step estimation to a constant. This procedure results in expressions of complex temperature dependent biology (e.g., developmental rates, fecundity, etc.) by the simple parameter r. The temperature dependence of r is then described by using the simulation data for a least squares fit of a continuous function (fig. 9). A similar procedure was followed for the remaining parameters of equation (3) (Logan 1977; Wollkind and Logan 1978; Logan, unpublished).

A comparative plot of temperature dependent r for T. mcdanieli and M. occidentalis is given in figure 10. This plot provides for interesting speculation. At low to moderate temperatures the population growth rate of the predator is greater than that of the prey. At temperatures above 23°C the relationship reverses, with the prey growth rate becoming much greater than that of the predator. In other words, temperature differentially affects various system components. This differential effect of temperature on system parameters has dramatic consequences for system behavior.

As Huffaker and Flaherty (1966) noted, the potential efficacy of a biological control agent cannot be judged by the relative magnitude of population growth rates alone. When we appraise the effect of temperature on parameters of the entire system (a process called structural stability analysis), the results are even more interesting. The structural stability criterion of model equation (3) as a function of temperature is shown in figure 11. This criterion is obtained by first performing a linear stability analysis of the model (May 1977), which results in relationships between model parameters and model stability. Relationships are then derived that describe parameter values as a function of temperature (fig. 10, Logan 1977, Tanigoshi and Logan 1979, Wollkind et al. 1980). Therefore, this process finally results in a stability criterion as a function of temperature. As seen from figure 11, model instability occurs at approximately 32°C.

A greater appreciation for the structural stability of model equation (3) may be gained from phase space plots of the system. One such plot is shown in figure 12. Prey density is given on the horizontal (x) axis and predator density on the vertical (y) axis. The broken lines indicate predator and prey isoclines, respectively; and their point of intersection is a potential stable point (i.e., steady-state). Phase space plots at different temperatures beginning with identical prey-predator ratios are given in figure 13. Note that the density scale and basic system behavior are quite different for the various plots. At low temperatures the populations immediately cycle back and remain at a low population density steady state. Even at high temperatures just below the threshold of destabilization, population cycles damp quickly to steady state. However, at temperatures only slightly above the threshold, behavior is dramatically different. The populations cycle back, but immediately rebound. This behavior is perhaps better illustrated by comparative time plots of the populations (fig. 14).

This model is presented as the simplest reasonable model description of the system. Inclusion of more realistic terms, such as a more reasonable description of predator starvation (Wollkind et al. 1980) have further verified the destabilizing effect of increasing temperature. Implications are that potential problems in control at high temperature are predicted

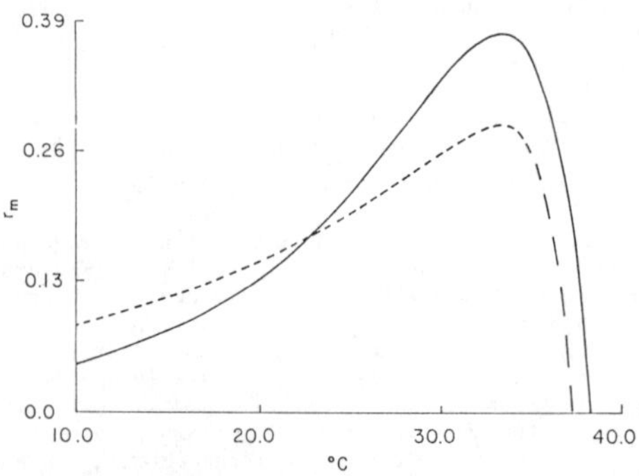

Figure 10.--Comparative r values for T. mcdanieli (solid line) and M. occidentalis (broken line) as a function of temperature. Functional forms were fit to "data" generated from the simulation model (see figs 7 and 9). Point of crossover is approximately 23°C.

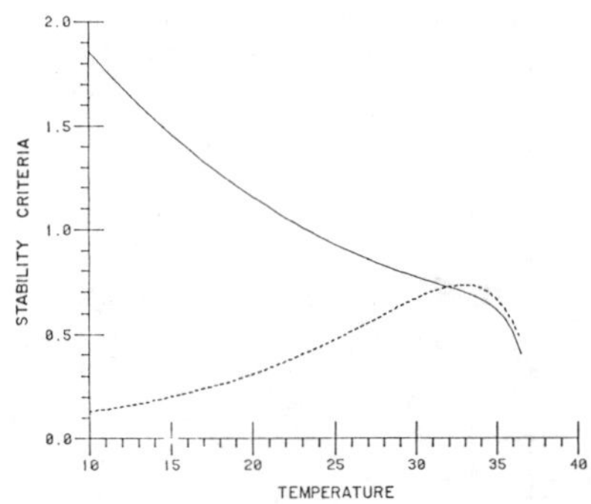

Figure 11.--Structural stability for equation (2) as a function of temperature (see text). As long as the solid line (the ratio of predator r to prey r) remains above the broken line (a composit function of all parameters in the model other than prey r), the model is stable. Conversely, when the broken line crosses the solid line, instability results. The point of instability occurs at approximately 32°C.

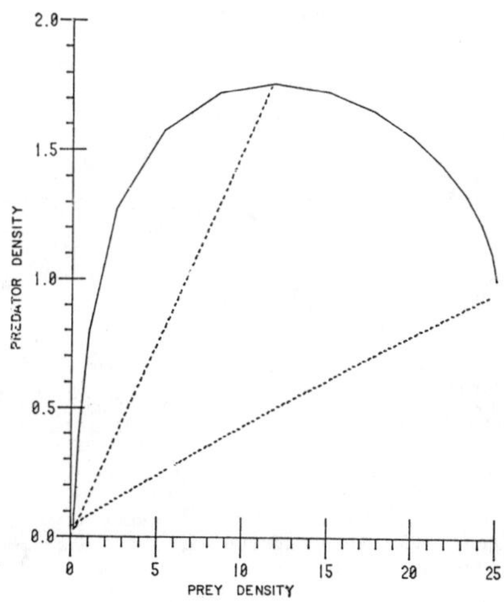

Figure 12.--Phase space plot of model equation (3). Predator and prey isoclines are the prey-predator densities that result in zero growth for the predator and prey populations, respectively. The intersection of the isoclines is therefore a steady-state solution to the model. For this example, the trajectory from an initial prey-predator ratio of 25:1 cycles quickly to steady state, indicating a stable interaction.

from the population dynamics of the two species alone, irrespective of the effect of temperature on host plant or other components of the system. This prediction is compatible with, and perhaps provides insights into, field observations (Tanigoshi and Logan 1979).

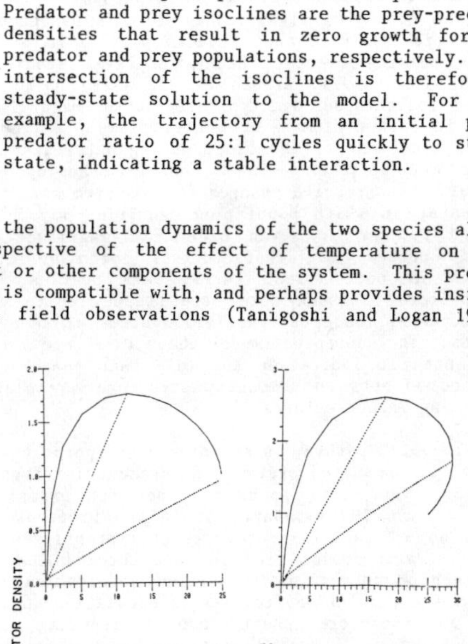

Figure 13.--Phase plane trajectories starting with identical prey-predator ratios (25:1) and run under various constant temperatures: (a) 21°C, (b) 30°C, (c) 31.5°C, (d) 32.5°C.

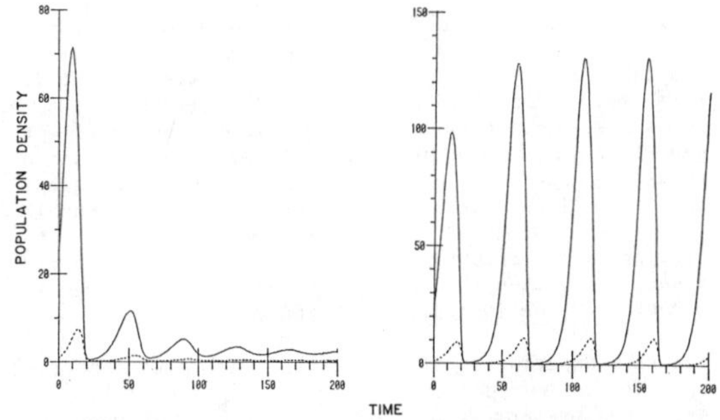

Figure 14.--Population time traces for the phase space plots given in figures 13c and 13d, respectively. Solid line is prey density; broken line is predator density. Implications for biological control are self-apparent. For the situation depicted in 14a, pest population would be maintained at a low density. Population behavior in 14a is contrasted by that in 14b, where pest populations rebound, indicating a cycle of spider mite outbreaks.

ADAPTIVE SIGNIFICANCE

Results from the previous section can be interpreted as: at low temperatures the predator wins, and at high temperatures the prey wins. For one specific example, this outcome may be due to chance alone. However, accumulating evidence indicates that the _T. mcdanieli/M. occidentalis_ relationship is not an isolated example but is perhaps characteristic of temperature effects on arthropod prey-predator interactions. Conventional wisdom has long correlated phytophagous insect outbreaks with atypical temperature events, usually hot/dry conditions. Destabilizing effects of high temperature on tetranychid/phytoseiid interactions have been documented for _Oligonychus pratensis/Amblyseius fallacis_ (Logan, unpublished); _T. kanzawai/Phytoseiulus persimilis_ (Shrukaji et al. 1978); _T. urticae/P. persimilis_ (Shaw 1982); and _T. mcdanieli/M. occidentalis_ (see previous section). For mite prey-predator interactions, at least, it appears that in general high temperatures lead to destabilization and subsequent outbreaks of phytophagous species.

The apparent general nature of high temperature prey escape leads to the question: why should this be the case? The answer may simply lie in the fact that for small terrestrial organisms (such as insects and mites) with a high surface/volume ratio, dehydration and subsequent dessication is a serious problem faced at high temperatures. Water intake is a simple problem for a phytophagous species; all they have to do is reach down and take a bite of plant material. It is a much different matter for a predator. In order to get a drink a predaceous insect must actively pursue and overcome its prey, an increasingly unlikely event as the predator becomes more water-stressed. The prey species may also gain an advantage through the indirect effect of temperature on host plant response. Apparently, for many phytophagous species the host plant becomes a higher quality food resource as temperature increases (White 1969). Therefore, both prey and predator experience a general increase in physiological rates with increasing temperature, but the prey species experiences an additional boost due to an improved food quality.

Regardless of the exact mechanistic pathways, the disruptive nature of high temperatures in arthropod prey-predator associations is apparent. Temperature generally acts as an accelerator to the system, but it is a complex accelerator which disproportionately affects system components. This disproportionate effect generally favors the prey over its predator and often results in prey escape, or outbreaks of phytophagous species. If our hypothesized temperature relationship is correct, then a logical conclusion is: if the prey/predator relationship is a closely co-evolved association, then the observed relationship between T. mcdanieli/M. occidentalis is to be expected.

The expectation of prey escape with high temperature leads to a pulse-stable model of co-evolution in arthropod prey-predator systems. This model requires several assumptions:

1. Most of the time populations are comprised of relatively genetically isolated demes. This is true for endemic spider mite populations characterized by stable low densities. In general, arthropod populations may be more genetically discontinuous than is apparent from geographic distribution (Ehrlich and Raven 1969).

2. For many arthropod communities, assumption (1) is true because populations are maintained at stable low densities due to community structure. The potential for stable community structure maintained by predators and parasites is well documented in applied entomology. Historically, community stability has often been disrupted through agronomic practices which have deleterious effects on natural enemies. In many examples stability has been regained through reintroduction and/or conservation of natural enemies (Huffaker 1974, Debach 1974).

3. Unusual weather events, typically involving high temperature, are capable of destabilizing community structure (see previous section).

4. Subsequent density-induced dispersal of phytophagous species (and predators) leads to genetic interchange between previously isolated demes. High density-induced dispersal is well documented for many insects and mites. Dispersal at high densities may be induced by overexploitation of the food resource (e.g., spider mites) or for other more complex reasons (e.g., tent caterpillars and grasshoppers).

5. Genetic interchange between previously isolated demes results in new genotypic combinations, leading to the potential for rapid adaptation to major selection pressures such as predation. This adaptation leads to a co-evolutionary surge and excape from predation by the prey. Community stability characteristics have therefore shifted from stable, low population densities to limit cycle or outbreak conditions.

6. Co-evolution of the predator will result in response to prey escape. The community structure should therefore revert to the previous stable, low density configuration. Time required for reversion will depend on how closely coupled, temporally and spatially, the predator is to its prey.

Assumptions, and the resulting general scenario of the previous paragraph, lead to the conceptual outbreak model shown in figure 15. The scenario just

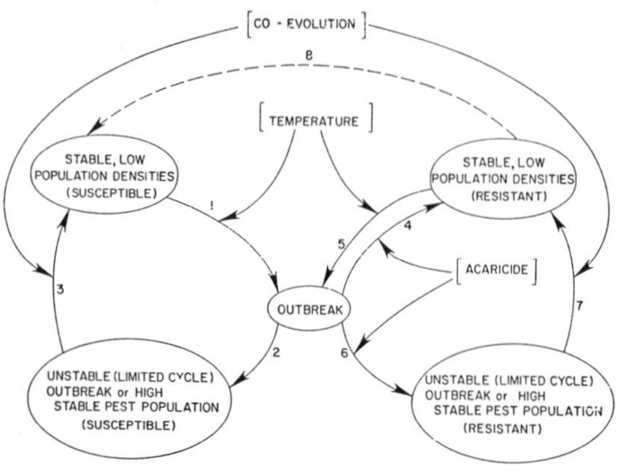

Figure 15.--Diagram of potential community pathways after temperature-induced perturbations (see text for explanation).

described is indicated by the 1-2-3 pathway. Ecological (and economic) significance of this hypothesized pathway is the role that insects play as major ecosystem pulsing events. Ecosystem level studies in general (Chew 1974, French et al. 1979) have found that macroarthropods are typically not of major importance in either biomass or energy flows. While this may be true most of the time, during outbreaks arthropods may be the dominant consumer causing catastrophic retrogressive changes in ecosystem structure. Outbreaks, in which population densities may be increased by six or seven orders of magnitude over endemic phases, are characteristic of many insect species in both forest (e.g., bark beetles, tent caterpillars, tussock moth) and steppe (white grub, grasshoppers, range caterpillar) systems of temperate regions. The conceptual model shown in figure 15 is a first step in indicating the role that temperature-mediated effects on community structure may play in initiating insect outbreaks.

The 1-2-3 pathway is envisioned as operating in a natural or unmanaged system. Management intervention can circumvent this pathway. One such example is shown in the 1-4 pathway. As in the previous scenario, an unusual temperature event destabilizes community structure resulting in an outbreak. However, management intervention (pesticide application to a susceptable population) results in population suppression, and therefore density-induced dispersal, etc., does not occur. Repeated pesticide applications will result in selection for a stable community resistant to the pesticide. An unusual temperature event would then lead to an outbreak and subsequent pesticide application; however, the resistant populations would not be suppressed. In such cases, the "sudden" appearance of resistant populations may actually be a gradual development of resistance due to chronic pesticide applications, followed by prey escape due to temperature-mediated community destabilization. Community trajectory should then follow the 5-6-7 pathway, which is in every way analogous to the 1-2-3 pathway for nonresistant populations. An interesting correlary to the 1-4-5-6-7 pathway is that eventual re-establishment of biological control should be expected in pesticide-temperature perturbed systems. Temporal resurgence of pesticide-resistant natural enemies has in fact been observed in several agroecosystems (e.g., Hoyt and Caltagirone 1974).

Finally, in absence of pesticides one would expect a slow (genetic) drift of the community to the nonresistant state. In practice, loss of resistance is extremely slow, if it occurs at all. One possible explanation for this is close genetic linkage between loci responsible for detoxification of naturally occurring and artificial toxins. Natural selection for detoxification of naturally occurring plant defense compounds, therefore, maintains a cross resistance to synthetic pesticides.

SUMMARY

We have attempted to show the pervasive nature of temperature to organization and function of arthropod systems. This influence is exhibited at biochemical, individual, population, interdeme, community, and ecosystem levels of organization. Understanding the ecological roles played by arthropods necessitates understanding temperature effects at all these levels of organization. Improved model descriptions of temperature effects has indeed facilitated such understanding!

LITERATURE CITED

Andrewartha, H. G., and L. C. Birch. 1954. The distribution and abundance of animals. University of Chicago Press, Chicago.

Bachmetjew, P. 1901. Experimentelle entomologische Studien. I. Temperaturverhältnise bei Insekten. Leipzig, 160 p. [cited in Chapman 1931].

Beament, J. W. L. 1958. The effect of temperature on the waterproofing mechanism of an insect. Journal of Experimental Biology 35:494-519.

Chapman, R. N. 1931. Animal ecology with especial reference to insects. 464 p. McGraw-Hill, New York.

Chew, R. M. 1974. Consumers as regulators of ecosystems: An alternative to energetics. Ohio Journal of Science 74:359-370.

Crozier, W. J. 1924. On the critical thermal increment for the locomotion of a diplopod. Journal of General Physiology 6:647-652 [cited in Shelford 1929].

Debach, P. 1974. Biological control by natural enemies. 323 p. Cambridge University Press, London.

Erlich, P. R., and P. H. Raven. 1969. Differentiation of populations. Science 165:1228-1232.

French, N. R., R. K. Steinhorst, and D. M. Swift. 1979. Grassland biomass trophic pyramids. p. 59-88. In Perspectives in grassland ecology (N. R. French, ed). Springer-Verlag, New York.

Gilbert, N., A. D. Gutierrez, B. D. Frazer, and R. E. Jones. 1976. Ecological relationships. 157 p. W. H. Freeman, San Francisco.

Glenn, P. A. 1922. Relation of temperature to development of the codling moth. Journal of Economic Entomology 15:193-198.

Glenn, P. A. 1931. Use of temperature accumlations as an index to the time of appearance of certain insect pests during the season. Transactions of the Illinois Academy of Sciences 24:167-180.

Hagstrum, D. W., and W. R. Hagstrum. 1970. A simple divice for producing fluctuating temperatures, with an evaluation of the ecological significance of fluctuating temperatures. Annals of the Entomological Society of America 63:1385-1389.

Hilbert, D. W., and J. A. Logan. An empirical model of nymphal development for the migratory grasshopper (*Melanoplus sanguinipes*) (Orthoptera: Acrididae). Environmental Entomology (in press).

Hilbert, D. W., J. A. Logan, and D. M. Swift. A unifying hypothesis of egg development and diapause of the migratory grasshopper, *Melanoplus sanguinipes* (Orthoptera: Acrididae). Environmental Entomology (submitted).

Holling, C. S. 1959. The components of predation as revealed by a study of small-mammal predation of the European pine sawfly. Canadian Entomologist 91:293-320.

Hoyt, S. C., and L. E. Caltagirone. 1974. The developing programs of integrated control of pests of apples in Washington and peaches in California. p. 395-419. In Biological control (C. B. Huffaker, ed.). Plenum, New York.

Huffaker, C. B., ed. 1974. Biological control. 511 p. Plenum, New York.

Huffaker, C. B., and D. L. Flaherty. 1966. Potential of biological control of twospotted spider mites on strawberries in California. Journal of Economic Entomology 59:786-792.

Janisch, E. 1925. Über die temperaturabhangigkeit biologischer vorgänge und ihre kervenmässige analyse. Pfluger's Archiv fuer die Gesamte Physiologie des Menschen und der Tiere 209:414-436 [cited in Shelford 1929].

Logan, J. A. 1977. Population model of the association of *Tetranychus mcdanieli* (Acarina: Tetranchidae) with *Metaseiulus occidentalis* (Acarina: Phytoseiidae) in the apple ecosystem. Ph.D. dissertation. 137 p. Washington State University, Pullman.

Logan, J. A. Recent advances and new directions in phytoseiid population models. University of California Experiment Station Bulletin (submitted).

Logan, J. A., R. E. Stinner, R. L. Rabb, and J. S. Bacheler. 1979. A descriptive model for predicting spring emergence of *Heliothis zea* populations in North Carolina. Environmental Entomology 8:141-146.

Logan, J. A., D. J. Wollkind, S. C. Hoyt, and L. K. Tanigoshi. 1976. An analytic model for description of temperature dependent rate phenomena in Arthropoda. Environmental Entomology 5:1133-1140.

May, R. M. 1977. Models for two interacting populations. p. 49-70. In Theoretical ecology (R. M. May, ed.). 317 p. Saunders, Philadelphia.

Oettingen, A. von. 1879. Archivfür naturo, doplater lignosen. Phenologie von der Liv. Esth. und Kurlands 8:1-112 [cited in Chapman 1931].

Parton, W. J., and J. A. Logan. 1981. A model for diurnal variations in soil and air temperature. Agricultural Meteorology 23:205-216.

Ruesink, W. G. 1976. Status of the systems approach to pest management. Annual Review of Entomology 21:27-44.

Sanderson, E. D., and L. M. Peairs. 1913. The relationship of temperature to insect life. New Hampshire College, Agricultural Experiment Station Technical Bulletin 7, 125 pp. [cited in Shelford 1929].

Simpson, C. B. 1903. The codling moth. USDA Bureau of Entomology Bulletin No. 41, 105 p. [cited in Shelford 1929].

Sharpe, P. J. H., and D. DeMichel. 1977. Reaction kinetics of poikilotherm development. Journal of Theoretical Biology 64:649-670.

Shaw, P. B. 1982. Analysis of an acarine predator-prey system. Ph.D. dissertation. University of California, Davis (in press) [dissertation summary provided by P. B. Shaw].

Shelford, V. E. 1929. Laboratory and field ecology. 608 p. Williams & Wilkins, Baltimore.

Shinkaji, N., T. Hamamura, and W. Ashihara. 1978. High critical temperature for the biological control of the Kanzawa spider mite by Phytoseiulus persimilis Athias. Henrist (Acarina: Phytoseiidae). Bulletin of the Fruit Tree Research Station, E 2:71-82.

Stinner, R. E., A. D. Gutierrez, and G. D. Butler. 1974. An algorithm for temperature-dependent growth rate simulation. Canadian Entomologist 106:519-524.

Stinner, R. L., R. L. Rabb, and J. R. Bradley, Jr. 1977. Natural factors operating in the population dynamics of Heliothis zea in North Carolina. Proceedings of the 15th International Congress of Entomology, p 632-642. Entomological Society of America, College Park, Maryland.

Tanigoshi, L. K., and J. A. Logan. 1979. Tetranychid development under variable temperature regimes. p. 165-175. In Recent advances in acrology, volume I (J. G. Rodriguez, ed.). Academic Press, New York.

Tanigoshi, L. K., S. C. Hoyt, R. W. Browne, and J. A. Logan. 1975a. Influence of temperature on population increase of Metaseinlus occidentalis (Acarina: Phytoseiidae). Annals of the Entomological Society of America 68:979-986.

Tanigoshi, L. K., S. C. Hoyt, R. W. Browne, and J. A. Logan. 1975b. Influence of temperature on population increase of Tetranychus mcdanieli (Acarina: Tetranychidae). Annals of the Entomolgical Society of America 68:972-978.

Taylor, F. 1979. Convergence to the stable age distribution in populations of insects. American Naturalist 113:511-530.

Verworn, Max. 1899. General physiology. An outline of the science of life [translated by F. S. Lee]. 615 p. London [cited in Shelford 1929].

Wangersky, P. J. 1978. Lotka-Volterra population models. Annual Review of Ecology and Systematics 7:189-218.

Welch, S. M. 1979. The application of simulation models to mite pest management. p. 31-40. In Recent advances in Acarology, volume I (J. G. Rodriguez, ed.). Academic Press, New York.

White, T. R. C. 1969. An index to measure weather-induced stress of trees associated with outbreaks of psyllids in Australia. Ecology 50:905-909.

Wollkind, D. J., and J. A. Logan. 1978. Temperature-dependent predator-prey mite ecosystem on apple tree foliage. Journal of Mathematical Biology 6:265-283.

Wollkind, D. J., A. Hastings, and J. A. Logan. 1980. Models involving differential and integral equations appropriate for describing a temperature dependent predator-prey ecosystem on apples. p. 266-277. In Modeling and differential equations in biology (T. A. Burton, ed.). Marcel Dekker, New York.

AQUATIC ECOSYSTEM MODELS FOR RISK ASSESSMENT[1]

Steven M. Bartell, Robert V. O'Neill, and Robert H. Gardner[2]

Abstract.--A stochastic modelling procedure was developed for use in evaluation of risks posed to aquatic ecosystems by toxic chemicals. Two models (FOAM and SWACOM) were used in the procedure to estimate probable fate and effects of potential toxicants and calculate uncertainties associated with these estimates. Results of simulations illustrate that structure-activity relationships can be used in FOAM to estimate transport, degradation, and accumulation of anthracene in lentic systems. SWACOM can extrapolate bioassay data to approximate risk of indirect effects on higher trophic levels caused by toxicants in aquatic systems.

INTRODUCTION

Assessment of risks to ecosystems posed by toxicants requires quantification of transport, estimation of dose, and translation of dose to effects on organisms that inhabit the system (Crawford and Leggett 1980). The large number of potential toxic chemicals precludes experimental approaches to routine screening and risk assessment. Simulation models, because of their flexibility in design, relative ease of implementation, and economy have been offered as surrogates for experimental protocols in evaluation of chemical fate and effects. Models have proved useful for predictions of fate of several organic toxicants (Bartell et al. 1981, Baughman and Lassiter 1978, Park et al. 1980).

Previous use of models emphasized deterministic solutions of sets of differential equations representing transport, degradation, and accumulation. Risk assessment requires that probabilites be assigned to model output. Given the difficulties inherent in dealing with toxicants in ecosystems, models must include uncertainties generated by incomplete knowledge of system function, of fate-determining processes, and of effects on complex food webs.

Our purpose here is to present one approach to incorporation of uncertainites into models. We present results from two models: the first describes the fate of anthracene spilled onto the surface of a pond, and the second estimates risks of algal blooms and reductions in game fish in a standard water column exposed to phenol and cadmium.

ECOSYSTEM RISK ASSESSMENT

Figure 1 summarizes our approach to translation of fates and effects data to probabilistic estimates of risk. In this approach uncertainties associated with laboratory data are carried through calculations of individual fate or effects models to produce estimates of probable fate or probable effects. The fate of anthracene is estimated by the Fates of Aromatics Model (FOAM) (Bartell et al. 1981). Transport parameters are estimated from regressions of chemical characteristics of polycyclic aromatic hydrocarbons (PAH) against rates of transport, degradation, and accumulation (Bartell et al. 1981). These quantitative structure-activity relationships (QSAR) also provide estimates of uncertainty or variability associated with each transport parameter. Residual variance in each regression was used to calculate a coefficient of variation for the parameter value. To apply these uncertainties, we performed 200 simulations with FOAM using transport parameters chosen randomly from distributions defined by the QSAR regressions. Statistical analysis of the predicted time-varying concentrations was used to summarize the expected system concentrations of anthracene and their uncertainty.

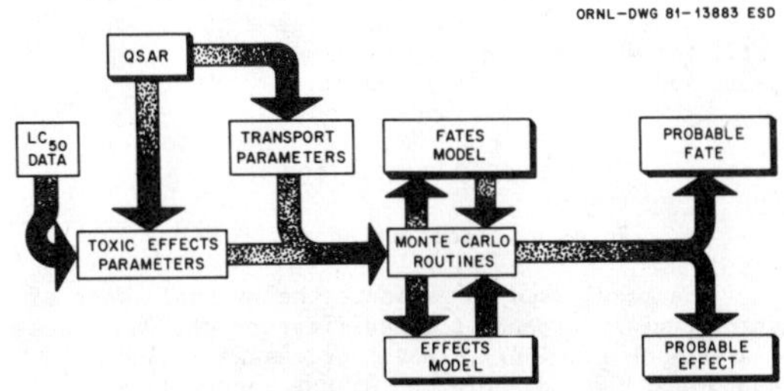

Figure 1.--Scheme for extrapolating laboratory data concerning fate and effects of toxicants to predictions of probable fate and probable effects in aquatic systems.

Similarly, repeated simulations with the Standard Water Column Model (SWACOM) (O'Neill and Giddings 1979) were used to translate laboratory bioassay data to probable effects of phenol and cadmium on an aquatic food web. Available toxicity data were primarily in the form of mortalities. Therefore, assumptions concerning the mode of action of the toxicant were required to determine appropriate changes in model parameters. We assumed that organisms respond to all chemicals according to a General Stress Syndrome (GSS). That is, organisms increase respiration rates, decrease photosynthetic and grazing rates, and become more susceptible to

---

[1]Paper presented at the Third International Conference on State-of-the-Art in Ecological Modelling, Colorado State University, May 24-28, 1982.

[2]Steven M. Bartell, Research Associate; Robert V. O'Neill, Senior Ecologist; Robert H. Gardner, Research Staff; Environmental Sciences Division, Oak Ridge National Laboratory, Oak Ridge, Tennessee, USA.

predation with increasing concentrations of toxicants. The GSS assumption permits us to define toxic effects parameters that modify rates of the above growth processes in order to simulate mortalities with SWACOM similar to those measured in the laboratory. Structure-activity relationships have also been correlated with toxic effects for some chemical classes (Cohen et al. 1974, Giddings, 1979), and these relationships can augment bioassay data used to estimate toxic effects parameters in the overall modelling approach (fig. 1), although these relationships were not used in the present examples.

The expected effect and uncertainties associated with extrapolating laboratory data for single species are preserved by assigning a probability distribution to each effects parameter. To calculate probable effects, 500 Monte Carlo simulations were performed with independent sets of parameters chosen from the distributions. Risk associated with an undesirable effect, such as 50% reduction in game fish, was estimated as the frequency of simulations that produced or exceeded this effect.

MODEL DESCRIPTION

Fates of Aromatics Model (FOAM)

The model used to estimate the probable fate of anthracene in a pond is a modification of FOAM (Bartell et al. 1981). Model processes include evaporation and dissolution of anthracene from a surface slick of synthetic oil, volatilization and photolytic degradation of dissolved anthracene, sorption to suspended particulate matter and sediments, and accumulation by rooted aquatic vegetation, zooplankton, and benthic invertebrates.

FOAM was developed to evaluate the hypothesis that fate of polycyclic aromatic hydrocarbons could be predicted from basic chemical properties of PAH. This focus distinguishes FOAM from transport models that require measurements of rates of transport as model input (e.g., Baughman and Lassiter 1978, Park et al. 1980). FOAM uses regression equations to estimate parameters that influence processes of dissolution, photolysis, volatilization, and desorption. For a specific PAH, anthracene in this case, FOAM parameters were estimated from the molecular weight, melting point, octanol:water partition coefficient, and light absorbance spectrum (300-500 nm). These molecular descriptors are easily obtained for most PAH. FOAM predicts time-varying concentration of the PAH in system components listed above, keeps track of PAH flux along all system pathways, and simulates seasonal changes in biomass of all food-web components.

Standard Water Column Model (SWACOM)

SWACOM (O'Neill and Giddings 1979, O'Neill et al. in press) is an adaptation of an earlier model, CLEAN (Park et al. 1974). SWACOM is a spatially homogeneous ecosystem model that comprises one gamefish, 10 phytoplankton, 5 zooplankton, and 3 forage-fish populations. The model simulates the annual cycle of a northern temperate-zone lake and incorporates seasonal light, temperature, and nutrient effects. Changes in biomass of food web components are modeled as functions of nutrient uptake, photosynthesis, respiration, grazing, mortality, and predation. This structure makes SWACOM amenable for incorporating toxic effects, lethal and chronic, as modifiers of the basic physiological process rates in the equations that determine growth. SWACOM estimates risk associated with both direct and indirect effects of toxicants. It considers impacts on all trophic levels rather than emphasizing a single taxonomic group. Therefore, the model can incorporate data for a variety of bioassay organisms and predict synergistic effects. For example, a toxicant with a direct effect on fish can also affect components of the food web that decrease the fish's food supply, and thereby amplify the direct toxic effect.

RESULTS

Fate Assessment with FOAM

Figure 2(a-d) shows the predicted concentrations of anthracene in four system components that result from 200 Monte Carlo simulations. Different degrees of uncertainty are evident for each model component, although the mean solution is similar to the deterministic solution (all parameters fixed at their mean values). These similarities indicate minimal tendency for FOAM to systematically over or underestimate anthracene concentrations when transport parameters are chosen at random from distributions defined by the structure-activity relationships.

The pattern of anthracene distribution agrees qualitatively with measurements made on experimental streams (Bartell et al. 1981). During the early part of the simulations, the predicted concentration of dissolved anthracene remains constant, bounded by variability associated with the maximum water solubility of anthracene. Up to day 35, high rates of dissolution produce inputs in excess of anthracene solubility and result in rapid accumulation of precipitated anthracene in the sediments (fig. 2-d). Following day 35, depletion of anthracene from the surface slick slowed the rate of dissolution, and the dissolved concentration subsequently decreases because of accumulation by other system components.

Rooted vegetation accumulated more anthracene than other food-web components (fig. 2-b). By day 50, anthracene content had increased to 30 $\mu g/m^2$ (0.35 μg/g dry wt). After day 50, metabolic degradation exceeded uptake and the anthracene concentration in plants decreased. Maximum and minimum values of the 200 simulations indicate that uncertainty in prediction for this component approached 35% of the mean values.

Results of the simulations show that predictions of anthracene concentration in benthic invertebrates were more precise than those of rooted plants (fig. 2-c). Maximum accumulation (4 $\mu g/m^2$) was nearly one order of magnitude less than that of the vegetation. The benthic invertebrates continued to accumulate anthracene throughout the simulation by ingesting sediments and detritus produced by other food-web components.

The highest concentrations of anthracene should accumulate in sediments (fig. 2-d). However, maximum and minimum values indicate that the estimates varied up to 200% of the mean concentrations. Uncertainty associated with the rate of desorption, as well as metabolic transformation and inputs to sediment from benthic invertebrates contributed to the relative imprecision.

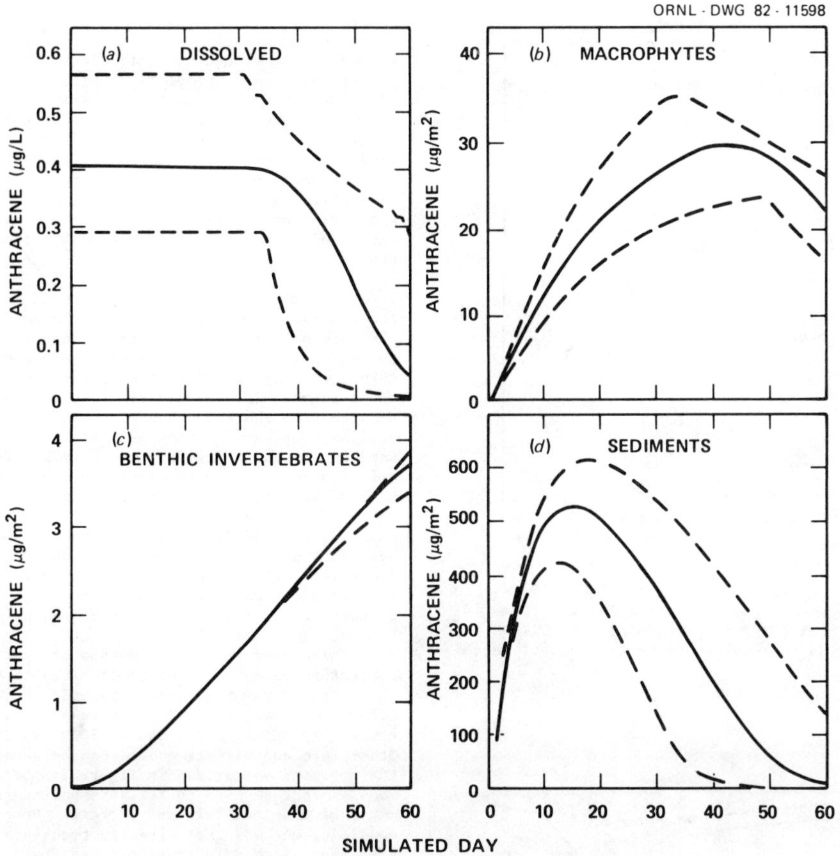

Figure 2.--Predicted concentrations of anthracene in water, macrophytes, benthic invertebrates, and sediments. Results from 200 Monte Carlo simulations with FOAM. Daily mean concentrations (———); maximum and minimum concentrations (— — —). Note different y-axis scales.

Effects Assessment with SWACOM

Figures 3 and 4 summarize the results of risk assessment of phenol and cadmium using SWACOM and the approach illustrated in figure 1. Two ecosystem endpoints were considered: probability of increase in noxious blue-green algae and probability of decrease in game fish. Results are presented for a range of environmental concentrations that might occur in a large river (Barnthouse et al. 1982). Note that the lines in figures 3 and 4 do not pass through the origin. That is, there is some risk of increasing algal biomass or decreasing game fish biomass even though exposure concentrations of phenol approach zero. This reflects residual uncertainty in simulating ecosystem effects. There is always some finite probability of a small decrease in game fish due to natural variability in the system.

The results for phenol show that risk increases monotonically with exposure concentration (fig. 3). The predicted relationship between increasing concentration of phenol and probability of game-fish reduction is not always simple. For example, at concentrations betwen 0.006 and 0.038 mg/L, the risk of a 25% reduction in game fish changes very little. This is due to the complex interactions in SWACOM that simultaneously change all trophic levels in the ecosystem. The increasing risk of an algal bloom is also counterintuitive. These are examples of the indirect effects that our approach to ecosystem risk assessment is capable of showing. Even though phenol is toxic to all algae, a reduction in grazing organisms more than compensates for the direct effect. As a result, noxious algae tend to increase as phenol concentrations increase.

The risks associated with cadmium (fig. 4) are also significant. At the concentrations tested, there is an increasing risk of reducing game-fish biomass. The risk of an increase in blue-green algae shows an unexpected decrease at higher concentrations. This effect results from the pattern of sensitivities of the food-web components in the system. In addition, these cadmium levels are high enough to cause direct toxic effects on the algae.

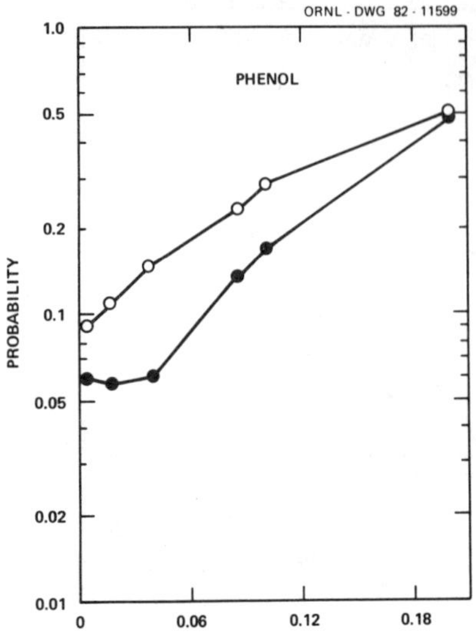

Figure 3.--Probabilities of a fourfold increase in blue-green algae (○) or a 25% reduction in game fish (●) as extrapolated by SWACOM in relation to exposure concentrations of phenol.

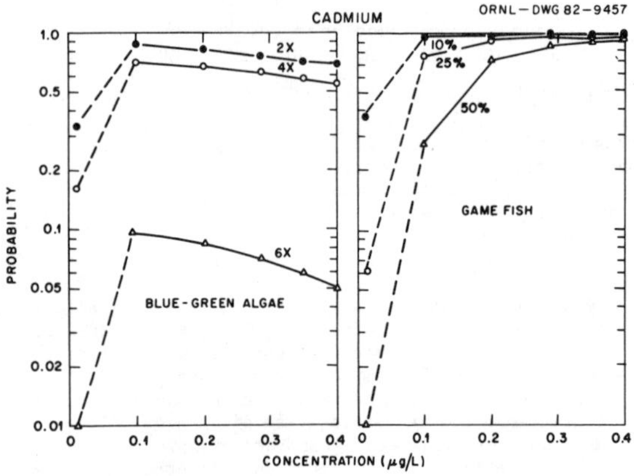

Figure 4.--Probabilities of 2X, 4X, and 6X increase in blue-green algal biomass or 10, 25, and 50% reduction in game fish biomass extrapolated by SWACOM in relation to exposure concentrations of cadmium.

DISCUSSION

The applications discussed above serve as examples of ecosystem risk assessment. For predictions of fate of organic toxicants, the Monte Carlo methods translate basic PAH chemistry into predictions of concentrations in individual system components. Statistical analysis of the 200 simulations provides estimates of the probability of exceeding predefined concentrations. Uncertainties that result from incomplete knowledge of the chemistry of the toxicant are carried directly through the calculations. The structure-activity regressions used to estimate model parameters can be modified as more data become available. As new data are included in the regressions, new estimates of variablity can be assigned to transport parameters and additional Monte Carlo simulations can be performed to re-evaluate the probable fate of the chemical.

The results in figures 3 and 4 illustrate that the relationships between toxicant concentration and risk are not simply linear or exponential. The curves show unexpected changes in slope, and the curves for the two chemicals do not simply parallel one another. The complexity of these responses results from the interconnected nonlinear interactions considered in the analysis. Of particular interest is the capability to apply statistical analyses to the results of risk assessment of SWACOM, as with fate predictions of FOAM. This makes possible a quantification of the relative contributions of direct and indirect effects to overall risk. It is now possible to determine the decrease in risk associated with increased information on additional organisms or more precise data concerning the modes of action of the chemicals.

This paper presents a method for extrapolating laboratory data to estimates of risks to ecosystems that result from transport, accumulation, and toxic effects of chemicals. These extrapolations are, of course, specific to the use of FOAM and SWACOM. Other fate and effects models can be substituted in the approach summarized in figure 1. Deterministic models might be used to assist experimental measurements of chemical fate and effects. However, use of modelling methods that directly translate uncertainties associated with transport and toxicological data to statements of probable fate or probable effects appears as a more realistic approach to risk assessment for ecosystems.

ACKNOWLEDGMENTS

We express our gratitude to C. W. Gehrs and S. G. Hildebrand for support and encouragement of this research. Comments by L. W. Barnthouse, D. L. DeAngelis, C. D. Powers, G. W. Suter, G. W. Thompson, D. S. Vaughan, and W. Van Winkle are much appreciated. Research supported in part by the Environmental Protection Agency under Interagency Agreement 40-740-78, and in part by the National Science Foundation's Ecosystem Studies Program under Interagency Agreement DEB-77-25781 with the U.S Department of Energy under contract W-7504-eng-26 with Union Carbide Corporation. Publication No. 2006, Environmental Sciences Division, ORNL.

LITERATURE CITED

Barnthouse, L. W., S. M. Bartell, D. L. DeAngelis, R. H. Gardner, R. V. O'Neill, C. D. Powers, G. W. Suter II, G. P. Thompson, and D. S. Vaughan. Preliminary environmental risk analysis for indirect coal liquefaction. ORNL/TM-8373. Oak Ridge National Laboratory. Oak Ridge, Tennessee (in press).

Bartell, S. M., P. F. Landrum, J. P. Giesy, and G. J. Leversee. 1981. Simulated transport of polycyclic aromatic hydrocarbons in artificial streams. p. 133-143. In W. J. Mitsch, R. W. Bosserman, and J. M. Klopatek (eds.), Energy and ecological modelling. Elsevier, N.Y.

Baughman, G. L., and R. R. Lassiter. 1978. Prediction of environmental pollution concentration. p. 35-54. In J. Cairns, K. L. Dickson, and A. W. Maki (eds.), Estimating the hazard of chemical substances to aquatic life. ASTM ASTP, American Society for Testing Materials, Philadelphia, Penn.

Cohen, J. L., W. Lee, and E. J. Lien. 1974. Dependence of toxicity on molecular structure: Group theory analysis. J. Pharm. Sci. 63:1068-1072.

Crawford, D. J., and R. W. Leggett. 1980. Assessing the risk of exposure to radioactivity. Amer. Sci. 68:524-536.

Giddings, J. M. 1979. Acute toxicity to Selenastrum capricornutum of aromatic compounds from coal conversion. Bull. Environ. Contam. Toxicol. 23:360-364.

O'Neill, R. V., R. H. Gardner, L. W. Barnthouse, G. W. Suter, S. G. Hildebrand, and C. W. Gehrs. Ecosystem risk analysis. Environ. Toxicol. Chem. (in press).

O'Neill, R. V., and J. M. Giddings. 1979. Population interactions and ecosystem function. p. 103. In G. S. Innis and R. V. O'Neill (eds.), Systems analysis of ecosystems. International Cooperative Publishing House, Fairland, Md.

Park, R. A., and 24 coauthors. 1974. A generalized model for simulating lake ecosystems. Simulation 23:33-50.

Park, R. A., C. I. Connolly, J. A. Albanese, L. S. Clesceri, G. W. Heitzman, H. H. Herbrandson, B. H. Indyke, J. R. Loehe, S. Ross, D. D. Sharma, and W. W Shuster. 1980. Modeling transport and behavior of pesticides and other organic materials in aquatic environments. Report No. 7. Center for Ecological Modeling, Rensselaer Polytechnic Institute, Troy, N.Y, 163 p.

A SIMULATION COMPARISON AND EVALUATION OF PARAMETER SENSITIVITY METHODS APPLICABLE TO LARGE MODELS[1]

Kenneth A. Rose[2]

Abstract. New problems which arise with the parameter sensitivity analysis of large complex models are detailed. Methods proposed in the literature which have the potential to address these problems are summarized and all applied to an example model. Comparisons among the sensitivity conclusions provided by the different methods enables the determination that the method of fractional factorial design with ANOVA is best suited for use on large models.

INTRODUCTION

Simulation models are being constructed at an increasing rate and with increasing complexity. Concordant with this activity has been the development of new sensitivity analysis methods. Sensitivity analysis provides information about which of the model components are important in determining the behavior of a model output variable. In this general context, sensitivity analysis has been performed on state variables (e.g., Blake and Gentil 1979), parameters (e.g., Mohn 1979), environmental variables (Kremer and Nixon 1978), initial conditions (e.g., Thornton et al. 1979), and aspects of the numerical integration techniques used (e.g., Shahane and Maloy 1979).

In this paper, limiting ourselves to the classical sensitivity analysis of parameters, methods are evaluated for their applicability for use on large models. First the problems associated with the sensitivity analysis of large models are detailed. This is followed by a discussion of methods reviewed which are unsuccessful in dealing with these problems. Candidate methods, those having the potential to address the problems, are then summarized as two step procedures; the generation or sampling of parameter values and the statistical analysis of the resulting output. Methods are categorized based on whether parameter values are sampled systematically (fractional factorial design) or randomly (Monte Carlo simulation). General descriptions of fractional factorial designs and two types of random sampling, simple and a form of stratified sampling termed Latin Hypercube sampling (LHS), are provided. Each description is followed by a section that summarizes the different statistical quantities that have been proposed as sensitivity measures. These include analysis of variance (ANOVA) and linear regression for fractional factorial designs and correlation (C), partial correlation (PC), partial rank correlation (PRC), and Kolmogorov-Smirnov (K-S) statistics for random sampling. Finally, the candidate methods are all applied to a 27 parameter example model and sensitivity conclusions compared. Comparisons are made to determine which of the methods, if any, are well suited for use with large models. For ease of comparison, a measure of similarity between the sensitivity conclusions from two different methods is defined based on the top 10 ranked parameters according to each method.

PROBLEMS ASSOCIATED WITH SENSITIVITY ANALYSIS OF LARGE MODELS

Parameter sensitivity analysis of large, complex models must deal with new problems which many commonly used methods have traditionally ignored. These problems are:

(1) The computational requirements of the analysis. The computer costs of repeated runs of a large model can very quickly become excessive.

(2) The possibility of important parameter interaction effects (Gardner et al. 1981). Interaction effects occur when the response of the model to varying values of a parameter depends on the values chosen for other parameters. Methodologies incapable of incorporating interaction effects can produce sensitivity conclusions for a given parameter that depend on the values assumed for the other parameters. Figure 1 illustrates this using a simple two parameter situation. The value of a model output variable is observed for two values of P1 with P2=K and P2=K'. Based on the response of the output variable to the changed values of P1 for P2=K, P1 would be concluded as "unimportant" whereas if P1 was changed with P2=K', P1 would be inferred as "important".

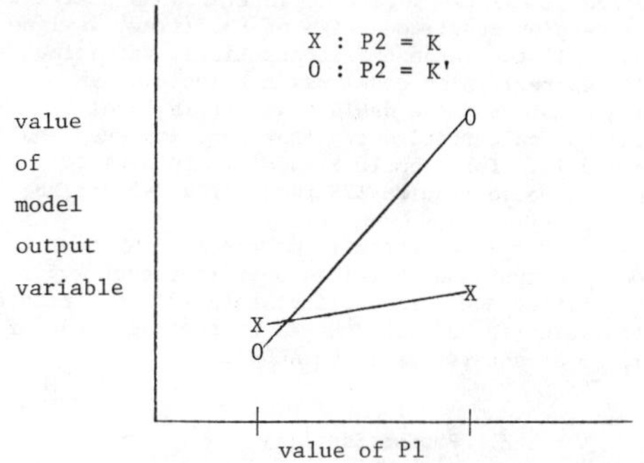

Figure 1. Hypothetical model responses depicting an interaction effect between parameters P1 and P2.

[1]Paper presented at the Third International Conference on State-of-the-Art in Ecological Modeling, Colorado State University, May 24-28, 1982.

[2]Kenneth A. Rose is Pre-Doctoral Research Assistant, Center for Quantitative Science, School of Fisheries, University of Washington, Seattle, WA. 98195 U.S.A.

involves the generation of perturbed parameter values from probability distributions. Random sampling can be further divided into two types, simple and Latin hypercube sampling (LHS).

Once output is generated via one of the sampling schemes, the methods within a category primarily differ by the use of different, though related, statistical measures. For brevity and ease of comparison, only the basic idea or concept of each method as it was proposed in the literature will be presented. The reader should be aware that for some methods variations and/or generalizations have also been proposed.

Fractional Factorial Designs

In factorial designs, runs are made such that all 'levels' of a parameter are combined with all 'levels' of every other parameter. The simplest case is having two levels (a nominal value and plus or minus a constant percentage) which, for d parameters, implies 2^d runs. A fractional factorial design involves running only a fraction (1/2 to some power c) of the total possible runs (denoted 2^{d-c}) (Box et al. 1978). The information lost appears as the estimated effects of various parameter combinations on the output being aliased or inseparable from the available runs. The best that can be done is an estimate of the sum of the effects of the aliased parameter combinations. Proper selection of which runs to make enables individual parameter effects to be aliased only with high order interaction effects and low order interaction effects with interactions of moderate order. By assuming that higher order interaction effects are negligible, individual parameter effects may be estimated. Low order interaction effects may not all be completely separable in some designs but usually their relative importance as a group can be determined. Box et al. (1978) define the resolution R of a fractional factorial design as one in which no K parameter interaction is aliased with any other combinations containing R-K parameters.

Analysis of Output From Fractional Factorial Designs

Two statistical analyses of the output have been proposed. Steinhorst et al. (1978) apply ANOVA to the values of the output variable from a fractional factorial design and use the F-values to rank parameters. Kohberger et al. (1978) advocate the use of a linear regression model containing linear, quadratic, and all two parameter interaction terms over ANOVA. Inclusion of all two parameter interactions restricts the regression model to design of $R \geq 5$ (i.e., designs in which all two parameter interactions are estimable). The linear regression model was not included as a candidate method since designs with this level of resolution would require too many runs for even small sized models. For example a model containing 10 parameters would require 128 runs for a R=5 design.

Note that since there is no random component to the model output the F-values only represent empirical quantities not probability statements. Parameters can equivalently be ranked based on the magnitude of their sum of squares from the ANOVA.

Random Sampling

By far the most common random sampling procedures assume the parameters are uncorrelated (see O'Neill et al. 1980 for a correlated example). A probability distribution is assumed for each parameter and for each run every parameter is sampled from its distribution. To gain as much information as possible from the fewest runs parameter values which sufficiently cover the parameter space must be selected in an efficient manner. For a limited number of runs, simple random sampling can result in a clustering of parameter values leaving gaps in the coverage of the parameter space and a redundancy in the information contained in some runs. This redundance leads to the use of some form of stratified sampling.

Conceptually, stratified sampling ensures good coverage with little redundance by dividing the parameter space into sections and sampling from each section with certain probabilities. McKay et al. (1979) compare simple, stratified, and LHS on an example model and conclude that of the three, provided certain monotonicity conditions hold, LHS provides the most precise estimates of the mean, standard deviation, and empirical distribution function of the output variable.

Until this point, no mention of a specific probability distribution has been made. This is because the choice of the distributions to assume for parameters is an unsolved practical problem. In most situations only partial information, if any, is available on the true variability and distribution of a parameter. In light of this incomplete information various justifications for the use of specific distributions have been made. Tiwari and Hobbie (1976) use information theory to derive the least biased density functions for several commonly encountered situations. Majkowski et al. (1981) justify the use of lognormal distributions when the variability associated with parameters is "large". The question of how to select a probability distribution and whether the choice influences the sensitivity conclusions obtained still needs further investigation.

Analysis of the Output from Random Sampling

Four statistical quantities have been proposed for use as sensitivity measures. Spear and Hornberger (1980) categorize each run as exhibiting acceptable behavior or not based on criteria derived from observed field data (e.g., in their system, for a run to be acceptable the peak biomass value could not exceed the initial value by more than a factor of ten). For each parameter, a two sample Kolmogorov-Smirnov test (K-S) is applied comparing the empirical distribution functions of the parameter values from runs that exhibited acceptable behavior with those of runs of unacceptable behavior. The K-S statistic is defined as the maximum vertical distance between two empirical distribution functions (Conover 1971). It measures the degree to which the values of a parameter differ between the two types of behavior. The premise is that the greater the difference between the values of a parameter corresponding to acceptable and unacceptable runs, the more important that parameter is in determining model behavior.

The other three statistical measures used to rank parameters by importance are closely related. Under the name of 'error analysis', O'Neill et al. (1980) use the correlation (C) and Gardner et al. (1980a) the partial correlation (PC) between each parameter and the output variable. Gardner et al. (1981) compare the sensitivity conclusions provided by the two measures and offer intuitive reasons for the use of each. McKay et al. (1976) use partial rank correlation (PRC), the partial correlation between the rank transformed values of the parameters and output variable.

(3) The size and shape of the parameter space for which sensitivity conclusions are valid. Many sensitivity methods only explore a small region in the parameter space (size) and unrealistically assume all parameters have equal relative variabilities (shape). For example, a commonly used method, which I call the IPP method, consists of ranking parameters based on the magnitude of change of the output variable between the standard run (all parameters at their nominal values) and runs in which individual parameters are perturbed by plus or minus 10%. Complex models can respond in very different and unpredictable (non-linear) ways to variations in parameters dependent on the size and shape of the parameter space explored. Continuing the IPP example, different sensitivity conclusions may result if parameters were perturbed by plus or minus 50%, or by different percentages related to the variability associated with parameter estimates.

(4) The complication of interpreting sensitivity conclusions caused by large models offering many potential output variables on which to base the analysis. Additionally, not only will sensitivity conclusions vary from output to output variable, but the conclusions for a given output variable will vary with time in the simulation.

LITERATURE REVIEW

Parameter sensitivity methods were evaluated for their potential to address the problems associated with large models, specifically problems (1) - (3) (see Rose and Swartzman 1981 for details). Methods were not evaluated against problem (4), multiple time dependent outputs complicating sensitivity conclusions, since no satisfactory solution to this problem was found. A discussion of this problem and potential solutions is found at the end of this section.

The historical measure of sensitivity, the partial derivative (Tomovic 1963), has seen widespread use on both linear and small, non-linear models (e.g., Patten 1969; McElwee and Yukler 1978). Steinhorst et al. (1978) demonstrated the impracticality of using the partial derivative measure on large models. Among the more recently proposed methods, those which failed to address any of the problems (1) - (3) were deemed unsatisfactory and not included as candidate methods which are summarized in more detail in the next section. The drawbacks of these unsuccessful methods can best be summarized by discussing them in three general groups based on how they dealt with the problems.

Several methods attempted to establish that interaction effects (problem (2)) and non-linear model responses (problem (3)) were unimportant for the particular model under investigation. This was done either at the expense of additional model runs (Miller 1974; Mohn 1979) or via the linking between how parameters were perturbed and the functional form chosen for the output variable (Majkowski et al. 1981). With this approach if interactions or non-linear responses are found to be important much time may have been wasted and an alternative method must be used.

Other methods recognized problems (2) and (3) and attempted to incorporate interaction effects and non-linear responses (Cukier et al. 1978; Wong 1980), but these had excessive computational requirements.

A third group of methods solely addressed the computational cost problems (Blanning 1975; Hosni and Doering 1979). These methods reduce the number of model runs required by basing the sensitivity analysis on a simplified model which the investigator derives from the original. Neither method specifies an objective way to reduce the size of the original model and furthermore it is questionable whether sensitivity conclusions based on the simplified version can be extrapolated to the original model.

The remaining problem of multiple outputs complicating sensitivity conclusions, problem (4), is one for which I am not presently aware of a completely satisfactory solution. The general approach has been to form relative measures on which to base sensitivity conclusions. These relative measures vary from the simple standardization of the partial derivative (Patten 1969; Smith 1970; Miller 1974) to basing the sensitivity analysis on an aggregated measure of the relative changes in all state variables over time (Miller 1974; Kohberger et al. 1978).

For a sensitivity analysis to be computationally cost effective, as much information as possible should be gleaned from the model runs. The use of measures which aggregate over model variables and/or time prevents one from obtaining information concerning the importance of a parameter within its own process related dynamics or during different times in the simulation. Possible alternatives are to analyze each state variable at selected times as well as having several aggregated measures of model behavior or to apply multivariate statistics to the output. Among the possibilities in the candidate methods presented below are canonical correlation applied to all state variables at a fixed time for random sampling (Gardner et al. 1980b) or multivariate ANOVA for fractional factorial designs (Steinhorst 1979).

CANDIDATE METHODS

This section summarizes methods which have the potential to address problems (1) - (3). To highlight the similarities among these methods they are presented as two step procedures. The first step is the generation or sampling of parameter values to be used in model runs and the second step is the statistical analysis of the resulting output which provides quantities on which parmeters can be ranked. Figure 2 summarizes the sampling schemes and statistics used.

Parameter sampling schemes are grouped into two categories, fractional factorial design and random sampling. Fractional factorial design involves the systematic variation of parameters. Random sampling

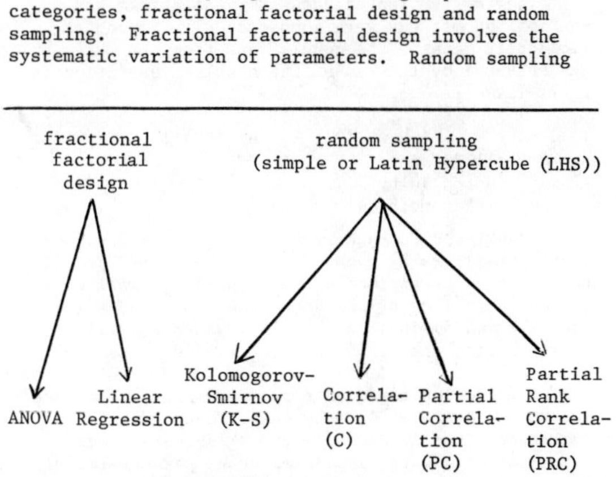

Figure 2. Summary of the candidate methods in terms of the parameter sampling procedures and statistical analyses used.

BASIS OF COMPARISON

Based on their descriptions all of the candidate methods are potentially applicable for use with large models. To determine which of these methods is best suited for such an analysis all the methods were applied to an example model and their sensitivity conclusions compared. This section describes: 1) the example model chosen, 2) the basis of comparison of the sensitivity conclusions, 3) the implementation of the candidate methods to the example model, and 4) the selection of the output variables and times.

Example Model

The model used will not be discussed in detail since its exact form is not the focus of this paper. The model is a modified version of the phytoplankton dynamics described by MS. CLEANER (Desormeau 1978; Collins 1980) adapted to a microcosm experiment. It describes the growth of a single species of phytoplankton and nutrient changes in the microcosm. The model consists of seven state variables (biomass, available internal phosphorus and nitrogen, unavailable internal phosphorus and nitrogen, and environmental phosphorus and nitrogen) and 27 parameters. The highly non-linear nature of the process equations is typical of many current models. The equations and parameter definitions, final tuned values, and coefficients of variation (standard deviation (SD)/mean, denoted CV) are presented in Appendices 1 and 2. Initial parameter values and CV's were estimated from a literature review by Smith (1980). In the absence of available literature estimates a CV=.25 was assumed. The CV values are considered as representative of the relative differences in individual parameter variabilities. The initial values of the state variables, environmental conditions (light and temperature), and the mean and standard deviation of phytoplankton biomass from replicate microcosm trials at 9 times during the experiment were extracted from Kindig (1979). The model was tuned to the biomass time trace terminating at 31 days and the final tuned values (Appendix 2) were used as the nominal or mean values for application of the sensitivity analysis methods.

Sensitivity Conclusions

The top ten ranked parameters as determined by each of the methods were compared. The degree of similarity between the conclusions from two methods was measured by the sum of the absolute deviation in ranks (SADR):

$$\mathrm{SADR} = \sum_{\substack{\text{(all Pj ranked}\\ \text{in top 10 by}\\ \text{either method)}}} \left| \mathrm{rank}_1(\mathrm{Pj}) - \mathrm{rank}_2(\mathrm{Pj}) \right|$$

where rank_1 and rank_2 correspond to the rank values given parameters by each of the methods. To provide the reader with an idea of the degree of similarity implied bv values of the SADR two comparisons (see Tables 1 and 5) include the SADR values as well as corresponding vectors of parameters and ranks.

A criterion based on a ranking of all 27 parameters was not used since for many comparisons large SADR values resulted even when methods indicated nearly identical highly ranked or important parameters. These large SADR values were caused by the shifting of ranks among the unimportant parameters. In almost all situations the most detail required by an investigator would be the ranks of the important parameters. Rarely would the exact ranks of the unimportant parameters be used.

The IPP method was one of methods deemed unsatisfactory by Rose and Swartzman (1981) since by singly perturbing parameters while holding the remaining parameters at their nomial values interaction effects are completely ignored. Throughout this paper, the degree of agreement between the conclusions from a particular method and IPP is used as evidence of whether the method incorporates interaction effects or not.

Implementation of the Candidate Methods

To compare the candidate methods the three sampling schemes were implemented. A 2^{27-21} fractional factorial design of resolution R=4 requiring 64 runs was applied. The standard $\pm$ 10% perturbation was used with the direction of perturbation (plus or minus) randomly assigned to each parameter.

Random sampling was implemented assuming normal distributions with standard deviations determined from the CV and tuned values presented in Appendix 2. Normal random variables were generated from a mixed congruential generator using the Box-Mueller transformation (Kennedy and Gentle 1980). The LHS method was implemented using an inverse normal program to obtain intervals of equal area for each parameter. The criterion for a run being categorized as exhibiting acceptable behavior for the K-S test was that the phytoplankton biomass is within the mean plus or minus two standard deviations of the microcosm observed biomass at nine different times in the simulation.

Output Variables and Times Monitored

The values of internal phosphorus at time 10 days (INTPHOS(10)) and biomass at time 30 days (BIOMASS(30)) were used to compare sensitivity conclusions since they represented variables and times for which interaction effects were, and were not important, respectively. The relative importance of interactions was determined by the agreement (unimportant) and disagreement (important) between the conclusions from fractional factorial design with ANOVA, which explicitly accounts for interactions, and IPP with $\pm$ 10% perturbations in the same directions (Table 1). Further evidence of the relative importance of interactions was the ranking by ANOVA of the largest aliased group of two parameter interactions 16th for BIOMASS(30) and 1st for INTPHOS (10).

COMPARISONS AND RESULTS

A series of four types of comparisons both among the random sampling based methods and between the random sampling and fractional factorial design method were made and are presented below. The first comparison was to establish the number of runs needed for each of the random sampling methods to generate consistent conclusions. Consistency is used here to mean that if the prescribed number of runs were made for two independent sets of runs the sensitivity conclusions would agree between the two sets. The next comparison consisted of contrasting the sensitivity conclusions among the different random sampling based methods, given enough runs were made for consistency. The final two comparisons were between the best method among the random sampling based methods as determined from the first two comparisons and fractional factorial design with ANOVA.

Table 1. SADR values and corresponding vectors of parameters and ranks for ANOVA and IPP with ±10% perturbations in the same directions. BIOMASS (30) and INTPHOS(10) were used as output variables.

BIOMASS(30) rank	IPP (±10%)	ANOVA (±10%)	INTPHOS(10) rank	IPP (±10%)	ANOVA (±10%)
1	P5	P1	1	P27	P27
2	P1	P5	2	P1	P19
3	P8	P8	3	P26	P27
4	P26	P16	4	P16	P4
5	P16	P26	5	P8	P22
6	P9	P9	6	P17	P1
7	P2	P2	7	P2	P6
8	P21	P21	8	P7	P23
9	P7	P27	9	P9	P24
10	P4	P4	10	P4	P10
11		P7	13		P26
12.5	P27		14		P8
			15	P10	P9
			16		P2
			17	P19	P17
			19		P16
			20	P6	
			21	P14	
			23		P7
			25.5	P22,P23, P24	
	(SADR = 9.5)			(SADR = 194.5)	

Consistency and Random Sampling

The first step in the comparison of the random sampling based methods is to determine the number of model runs that each statistic requires to provide consistent sensitivity conclusions. To do this, two independent sets totaling 1000 runs each (denoted seed1 and seed2) were made. The four statistical measures (C, PC, PRC, and K-S) were applied to each set consisting of the first 200,300,...,1000 runs using BIOMASS(30) as the output variable. For each of the statistics and number of model runs the value of the SADR was computed comparing the conclusions from the seed1 and seed2 sets (Table 2). K-S provides vastly different sensitivity conclusions between the two sets even when each consisted of 1000 runs. For the PC and PRC measures the SADR generally decreases with the increasing number of runs. Beyond 400 runs for PRC and 500 runs for PC consistent sensitivity conclusions are obtained. This is not true for the C measure (i.e. SADR=24 at 200 runs and SADR=29 at 1000 runs). Also apparent in Table 2 is a general ordering of the SADR values across the four statistics. For a given number of runs the magnitude of SADR is larger for K-S and generally decreases from C to PC to PRC.

The differences in the number of runs required for each statistic to product consistent results are due to what the statistics are attempting to measure. C measures the linear association between the parameter and state variable. PC measures the linear association after accounting for the variability of the state variable that is explained by the remaining parameters. PRC is analogous to PC except it measures the monotonic association. If the parameters are statistically uncorrelated then for a given state variable and parameter, the PC coefficient must be a greater than or equal to the C coefficient (Kleinbaum and Klupper 1978 p. 166). For all of the results the inequality relation always held for the relatively

Table 2. SADR values comparing the consistency of sensitivity conclusions of C, PC, PRC, and K-S for increasing numbers of model runs. Comparisons are based on BIOMASS(30) as the output variable and are between the two vectors of ranked parameters (seed1 and seed2) provided by each of the statistics.

number of runs in seed1 and seed2	K-S	C	PC	PRC
200	132.5	24	36	18
300	160.5	23	16	14
400	173	52	13	4
500	124	31	2	6
600	129	15.5	7	8
700	129	10	4	2
800	102	18	4	2
900	113.5	26	2	2
1000	105	29	2	2

important parameters (i.e., top 15), although not always for lower ranked parameters. This was probably due to some spurious correlations among these relatively unimportant parameters. The smaller the upper value a C coefficient can assume relative to the PC coefficient, the more parameters that must subsequently be ranked based on a narrower range of possible values. Slight changes in a C coefficient can cause larger changes in rank than the same change in a PC coefficient Table 3 demonstrates this by showing the magnitudes of the C and PC coefficients between BIOMASS(30) and the top 3 ranked parameters (P5,P26,P9) for 800 runs in the seed1 and seed2 sets. The remaining 23 parameters must be ranked on C coefficients that can range from 0 to .2875 whereas the PC coefficient can range from 0 to .6936 (seed2).

Table 3. Values of the C and PC coefficients between the top 3 ranked parameters and BIOMASS(30) for 800 runs in seed1 and seed2.

	seed1 C	seed1 PC	seed2 C	seed2 PC
P5	.7751	.9266	.7218	.9312
P26	.4170	.7791	.4309	.8110
P9	.2525	.6893	.2875	.6936

The difference in the number of runs required for consistency are also dependent upon the number of parameters that are monitored. A comparison of the top 5 ranked parameters according to C produced SADR values of 4 or less beginning at 200 runs. Similarly, when the top 20 were compared PC yielded an SADR=60 for 1000 runs in seed1 and seed2. Nevertheless for any number of parameters monitored PC will also provide more consistent conclusions for less model runs.

The explanation for PRC requiring less runs than PC is that the rank transformation lessens the influence that extreme values of the output or in the parameter space can have on the PC coefficient. Parametric statistics such as PC can be greatly affected by outliers (Draper and John 1981). Two sets of runs, one with an outlier and one without, that exhibit the same montonic relationships among parameters and output would yield different PC coefficients but identical PRC coefficients.

The inconsistencies in the K-S measure are probably rooted in the method's discrete classification of runs into two groups (acceptable and unacceptable behavior).

Comparison of Random Sampling Based Methods

Table 4 presents the SADR values comparing sensitivity conclusions based on K-S (1000 runs, seed1 and seed2), C (1000 runs seed1 and seed2), PC (500 runs seed1), and PRC (400 runs, seed1) for BIOMASS(30) as the output variable. The three correlation based measures yield similar sensitivity conclusions and all three differ from the K-S conclusions. Similar results were obtained for when INTPHOS(10) was used as the output variable.

Table 4. SADR values comparing the sensitivity conclusions based on K-S (1000 runs, seed1 and seed2), C (1000 runs, seed1 and seed2), PC (500 runs, seed1), and PRC (400 runs, seed1). BIOMASS (30) was used as the output variable.

	K-S (seed1)	K-S (seed1)	C (seed1)	C (seed2)	PC (seed1)	PRC (seed1)
K-S (seed1)	0	105	95	103.5	108	112
K-S (seed2)	---	0	107.5	120	123	128
C (seed1)	---	---	0	29	6	8
C (seed2)	---	---	---	0	24	14
PC (seed1)	---	---	---	---	0	2
PRC (seed1)	---	---	---	---	---	0

At this stage in the comparison, we can conclude that PRC provides sensitivity conclusions that agree with PC and C but can be obtained with fewer model runs. The next step is to determine if LHS can yield the same conclusions at even less computational cost.

The SADR comparing PRC based on simple random sampling (from Table 4) and LHS of sizes 64 runs (the same as the number required by fractional factorial design) and 200 runs (recommended by Iman and Beckman 1980) are presented in Table 5. The LHS of size 200 runs provides satisfactorily close conclusions to those of PRC with simple random sampling of 400 runs. Hence LHS with PRC is the most efficient of the random sampling based methods.

Table 5. SADR values and corresponding vectors of parameters and ranks comparing PRC based on simple random sampling (from Table 4) with LHS of sizes 64 and 200 runs. BIOMASS(30) was used as the output variable.

rank	PRC (400 runs)	PRC-LHS (64 runs)	PRC (400 runs)	PRC-LHS (200 runs)
1	P5	P5	P5	P5
2	P26	P26	P26	P26
3	P9	P9	P9	P9
4	P2	P2	P2	P8
5	P16	P24	P16	P16
6	P8	P10	P8	P21
7	P21	P7	P21	P2
8	P4	P13	P4	P4
9	P19	P8	P19	P7
10	P7	P21	P7	P10
11		P4		P19
12		P16		
13	P10		P10	
15	P24		P24	
23		P19		

Comparison of Fractional Factorial Design With ANOVA and LHS with PRC

Table 6 compares IPP, PRC (LHS of size 200 runs from Table 5), and ANOVA. SADR values are presented for both BIOMASS(30) and INTPHOS(10) as output variables. As discussed above in the context of the relative difference in the importance of interactions between the two output variables, IPP and ANOVA yield almost identical conclusions for BIOMASS(30) and very different conclusions for INTPHOS(10). Interestingly, unlike IPP and ANOVA, the degree of agreement between IPP and PRC is unaffected by the importance of interactions. I interpret this as evidence that LHS with PRC does not adequately incorporate interaction effects. If LHS did account for interactions we would expect the opposite result of what was observed. Since ANOVA explicitly accounts for interactions and IPP ignores interactions we would expect the agreement between PRC and ANOVA to be unaffected by the choice of output variables and that between PRC and IPP to be affected.

Table 6. SADR values comparing the sensitivity conclusions from IPP (±10%), ANOVA (±10%), and PRC (LHS-200 runs) for BIOMASS(30) and INTPHOS(10) as output variables.

BIOMASS(30)

	IPP (±10%)	PRC-LHS (200 runs)	ANOVA (±10%)
IPP (±10%)	0	31.5	9.5
PRC-LHS (200 runs)	---	0	42
ANOVA (±10%)	---	---	0

INTPHOS(10)

	IPP (±10%)	PRC-LHS (200 runs)	ANOVA (±10%)
IPP (±10%)	0	55	194.5
PRC-LHS (200 runs)	---	0	116
ANOVA (±10%)	---	---	0

Before LHS with PRC can be dismissed as not adequately incorporating interaction effects the effect on sensitivity conclusions of IPP and ANOVA covering different size and shape regions in the parameter space must be examined.

Shape and Size Effects on Sensitivity Conclusions

The size and shape of the region in the parameter space covered by the fractional factorial design can be easily characterized as ±10% since all parameters were perturbed in this manner. The size and shape covered by LHS is more difficult to characterize since parameter values are generated from normal probability distributions with some parameters having different variabilities (i.e., different CV). To determine the parameter space covered by LHS two approximations were made:

1) Since 16 of the 27 parameters have CV=.25 and only one parameter had a larger CV, this value was assumed to represent the typical CV used in the normal distribution

2) The spread of parameter values generated from a normal distribution with a CV=.25 can be represented by a range that would contain 80% of the generated values.

With these approximations the size and shape of the parameter space covered by LHS can be characterized as ±32% (i.e., a normal distribution with a CV=.25 would generate 80% of its values in the range tuned value ±.32 X tuned value). The shape covered by LHS is determined by the CV values other than .25 associated with some parameters.

To separate the effects that size and shape have on the sensitivity conclusions from fractional factorial design and LHS, three additional fractional factorial designs (±32%, ±.312XSD, and ±1.28XSD) were applied with the same directions of perturbation as the ±10% design. The interpretation of these designs and the original ±10% design is presented in Table 7. A ±32% design corresponds to the same size parameter space as LHS but with the shape associated with the ±10% fractional factorial design. The other two designs cover the same shape as LHS since they incorporate individual parameter variabilities by perturbing parameters in different amounts dependent on their CV values. These designs differ in that the ±.312XSD design covers the size of the parameter space corresponding to the ±10% design whereas the ±1.28XSD design corresponds to the size covered by LHS (i.e., ±32%). This can be seen by noting that for the typical CV=.25, a normal distribution specified by a mean (tuned value) and a standard deviation equal to ±.312XSD would generate 80% of its values in the range tuned value ±.1Xtuned value (i.e., ±10%). Similarly for a CV=.25, a normal distribution with a standard deviation equal to 1.28XSD would generate 80% of its values in the range tuned value ±.32Xtuned value (i.e., ±32%).

Table 7. The size and shape of the region in the parameter space covered by the four fractional factorial designs (±10%, ±32%, ±.312XSD, and ±1.28XSD) in relation to LHS.

		SHAPE	
		all parameters have equal variability	some parameters have different variabilities (LHS)
SIZE	±10%	±10%	±312XSD
	±32% (LHS)	±32%	±1.28XSD

Table 8 shows the SADR comparing LHS with PRC and the four fractional factorial designs with ANOVA for both BIOMASS(30) and INTPHOS(10). When interactions are unimportant (BIOMASS(30)), the sensitivity conclusions from LHS agree with those from the fractional factorial designs which cover the same shape parameter space regardless of the size covered (±.312XSD and ±1.28XSD). However when interactions are important (INTPHOS(10)) there is no agreement between LHS and any of the fractional factorial designs. Thus when interactions are unimportant the elimination of differences in the size and shape of the parameter space covered by LHS and fractional factorial designs result in both generating similar sensitivity conclusions. Yet when interactions are important even LHS and fractional factorial designs which covere similar regions in the paramater space produce different conclusions. This information,

Table 8. SADR values comparing PRC (LHS-200 runs) and the four fractional factorial designs (±10%, ±32%, ±.312XSD, ±1.28XSD) with ANOVA for BIOMASS (30) and INTPHOS(10) as output variables.

BIOMASS(30)	ANOVA (±10%)	ANOVA (±32%)	ANOVA (±.312XSD)	ANOVA (±1.28XSD)	PRC-LHS (200 runs)
ANOVA (±10%)	0	43.5	36	41	42
ANOVA (±32%)	---	0	68	79	80
ANOVA (±.312XSD)	---	---	0	10	7
ANOVA (±1.28XSD)	---	---	---	0	10
PRC-LHS (200 runs)	---	---	---	---	0

INTPHOS(10)	ANOVA (±10%)	ANOVA (±32%)	ANOVA (±.312XSD)	ANOVA (±1.28XSD)	PRC-LHS (200 runs)
ANOVA (±10%)	0	159	192.5	168	116
ANOVA (+32%)	---	0	124.5	182	99.5
ANOVA (±.312XSD)	---	---	0	107	88
ANOVA (±1.28XSD)	---	---	---	0	155
PRC-LHS (200 runs)	---	---	---	---	0

coupled with the relatively constant agreement between LHS and IPP regardless of the importance of interactions (see Table 6), implies that LHS with PRC fails to adequately account for parameter interaction effects. This is a general result which would also hold if C and PC were used instead of PRC and simple random sampling instead of LHS. Neither C, PC or simple random sampling have capabilities that differ from their respective counterparts, PRC and LHS.

Although the relative changes in SADR values among PRC, ANOVA, and IPP indicates that PRC does incorporate interactions, the SADR value of 31.5 for BIOMASS(30) in Table 6 is bothersome since it implies that even when interactions are unimportant PRC and IPP yield similar but not identical conclusions. Applying the same argument as was used to modify the fractional factorial design to cover the same shape parameter space as LHS, two additional IPP designs, ±.31XSD and ±1.28XSD, were applied. With BIOMASS(30) the larger of the two SADR values was 15 implying both IPP designs yield similar conclusions to PRC.

CONCLUSIONS

The fractional factorial design with ANOVA is the best method of those reviewed for parameter sensitivity analysis of large models. It is a generally applicable method which can easily incorporate information about the size and shape of the parameter space that should be covered. This model not only required the least number of model runs but also was the only method to explicitly account for interaction effects among parameters.

A potentially redeeming feature of LHS is that, unlike fractional factorial designs, the number of model runs required does not increase with the number of parameters in the model. In a situation where minimizing computational costs was especially important and the model consisted of fairly simple equations so that interactions were most likely unimportant LHS may seem perferable to fractional factorial design. Yet in such a situation one could use the more economical IPP method and obtain similar conclusions to those from LHS.

LITERATURE CITED

Blake, G. and S. Gentil. 1979. A discrete lake model and its sensitivity study. pp. 527-544. In (S.E. Jorgensen, ed.) State-of-the-Art in Ecological Modelling. Pergamon Press, New York.

Blanning, R.W. 1975. The construction and implementation of metamodels. Simulation 24(6):177-184.

Box, G.E.P., W.G. Hunter, and J.S. Hunter. 1978. Statistics for experimenters: An introduction to design, data analysis, and model building. John Wiley & Sons, New York.

Collins, C.D. 1980. Formulation and validation of a mathematical model of phytoplankton growth. Ecology 6(13):639-649.

Conover, W.J. 1971. Practical nonparametric statistics. John Wiley and Sons, New York.

Cukier, R.I., R.H. Levine, and K.E. Shuler. 1978. Nonlinear sensitivity analysis of multiparameter model systems. J. Compu. Physics 26:1-42.

Desormeau, C.J. 1978. Mathematical modeling of phytoplankton kinetics with application to two apline lakes. Report No. 4, Center for Ecological Modeling, Rensselaer Polytechnic Institute.

Draper, N.R., and J.A. John. 1981. Influential observations and outliers in regression. Technometrics 23(1):21-26.

Gardner, R.H., D.D. Huff, R.V. O'Neill, J.B. Mankin, J. Carney, and J. Jones. 1980a. Application of error analysis to a marsh hydrology model. Water Resources Res. 16(4):659-664.

Gardner, R.H., J.B. Mankin, and W.R. Emanuel. 1980b. A comparison of three carbon models. Ecol. Model. 8:313-333.

Gardner, R.H., R.V. O'Neill, J.B. Mankin, and J.H. Carney. 1981. A comparison of sensitivity analysis and error analysis based on a stream ecosystem model. Ecol. Model. 12:173-190.

Hosni, Y.A., and R.D. Doering. 1979. Sensitivity analysis for multi-variable models: Methodology and application for the economics of energy management systems. Comput. and Indus. Engng. 3:235-241.

Iman, R.L., and W.J. Beckman. 1980. Small sensitivity analysis techniques for computer models, with an application to risk assessment. Commun. Statist.-Theor. Meth. A9(17):1749-1842.

Kennedy, W.J., Jr. and J.E. Gentle. 1980. Statistical computing. Marcel Deeker, New York.

Kindig, A.C. 1979. Experimental report ME25: Investigations for strptomycin-induced algal competitive dominance reversals. Food and Drug Administration Contract No. 223-76-8348. College of Fisheries, University of Washington.

Kleinbaum, D.G., and L.L. Kupper. 1978. Applied regression analysis and other multivariate methods. Duxbury Press, Massachusetts.

Kohberger, R.C., D. Scavia, and J.W. Wilkinson. 1978. A method for parameter sensitivity analysis in differential equation models. Water Resources Res. 14(1):25-28.

Kremer, J.N., and S.W. Nixon. 1978. A coastal marine ecosystem: Simulation and analysis. Ecological Studies 24. Springer-Verlag, New York.

Majkowski, J., J.M. Ridgeway, and D.R. Miller. 1981. Multiplicative sensitivity analysis and its role in development of simulation models. Ecol. Model. 12:191-208.

McElwee, C.D., and M.A. Yukler. 1978. Sensitivity of groundwater models with respect to variations in transmissivity and storage. Water Resources Res. 14(3):451-459.

McKay, M.D., W.J. Conover, and D.E. Whiteman. 1976. Report on the application of statistical techniques to the analysis of computer codes. Informal Report LA-NUREG-G526-MS. Los Almos Scientific Laboratory.

McKay, M.D., R.D. Bechman, and W.J. Conover. 1979. A comparison of three methods for selecting values of input variables in the analysis of output from computer code. Technometrics 21(2):239-245.

Miller, D.R. 1974. Sensitivity analysis and validation of simulation models. J. Theor. Bio. 48:345-360.

Mohn, R.K. 1979. Sensitivity analysis of two harp seal (*Pagophilus groenlanditus*) population models. J. Fish. Res. Bd. Can. 36:404-410.

O'Neill, R.V., R.H. Gardner, and J.B. Mankin. 1980. Analysis of parameter error in a non-linear model. Ecol. Model. 8:297-311.

Patten, B.C. 1969. Ecological systems analysis and fisheries science. Trans. of the Amer. Fish. Soc. 98(3):570-581.

Rose, K.A., and G.L. Swartzman. 1981. A review of parameter sensitivity methods applicable to ecosystem models. NUREG CR-2016. Center for Quantitative Science, University of Washington.

Shahane, A.N., and J.R. Maloy. 1979. The water quality planning model. pp. 451-485. In (S.E. Jorgensen, ed) State-of-the-Art in Ecological Modelling. Pergamon Press, New York.

Smith, E.P. 1980. Parameter estimation for microcosm models. Methods Report No. 3. Food and Drug Administration Contract No. 223-76-8348. College of Fisheries, University of Washington.

Smith, F.E. 1970. Analysis of ecosystems. pp. 7-18 In (D.E. Recihle, ed) Analysis of Temperate Forest Ecosystems. Springer-Verlag, New York.

Spear, R.C., and G.M. Hornberger. 1980. Eutrophication in Peel Inlet-II. Identification of critical uncertainties via generalized sensitivity analysis. Water Res. 14:43-49.

Steinhorst, R.K. 1979. Parameter identifiability, validation, and sensitivity analysis of large models. In (G.S. Innis and R.V. O'Neill, eds.) Systems Analysis of Ecosystems. Co-Operative Publishing House, Maryland.

Steinhorst, R.K., H.W. Hunt, G.S. Innis, and K.P. Haydock. 1978. Sensitivity analysis of the ELM model. pp. 231-255. In (G.S. Innis, ed) Grassland Simulation Model. Ecological Studies 26. Springer-Verlag, New York.

Thornton, K.W., A.S. Lessem, D.E. Ford, and C.A. Stirgus. 1979. Improving ecological simulation through sensitivity analysis. Simulation 32(5): 155-166.

Tiwari, J.L., and J.E. Hobbie. 1976. Random differential equations as models of ecosystems-II. Initial condition and parameter specifications in terms of maximum entrophy distributions. Math. Biosci. 31:37-53.

Tomovic, R. 1963. Sensitivity analysis of dynamic systems. McGraw-Hill Book Co. Inc., New York.

Wong, C.S.Y. 1980. Criteria sensitivity analysis: A new approach to parameter sensitivity. Appl. Math. Modelling 4:7-15.

APPENDIX 1.

Single species phytoplankton model used to compare sensitivity conclusions from the candidate methods.

ENVIRONMENTAL VARIABLES

depth of water column (Z) (.1 m)
water temperature (WT) (20°C)

STATE VARIABLES AND MODEL EQUATIONS

phytoplankton biomass (BIOMASS) (g/m^3)

$$\frac{d}{dt}(\text{BIOMASS}) = (\text{PROD-RESP-MORT-EXCB})*\text{BIOMASS}$$

environmental nitrogen (ENVNIT) (g/m^3)

$$\frac{d}{dt}(\text{ENVNIT}) = (\text{EXCNIT} + \text{MORT}*(\frac{\text{INTNIT}}{\text{BIOMASS}})-\text{UPNIT})*\text{BIOMASS}$$

environmental phosphorus (ENVPHOS) (g/m^3)

$$\frac{d}{dt}(\text{ENVPHOS}) = (\text{EXCPHOS} + \text{MORT}*(\frac{\text{INTPHOS}}{\text{BIOMASS}}) - \text{UPPHOS})*\text{BIOMASS}$$

available internal nitrogen (INTNIT) (g/m^3)

$$\frac{d}{dt}(\text{INTNIT}) = (\text{UPNIT} - \text{EXCNIT} - \text{MORT}*(\frac{\text{INTNIT}}{\text{BIOMASS}}) - \text{ASSNIT})*\text{BIOMASS}$$

available internal phosphorus (INTPHOS) (g/m^3)

$$\frac{d}{dt}(\text{INTPHOS}) = (\text{UPHOS} - \text{EXCOHOS} - \text{MORT} * (\frac{\text{INTPHOS}}{\text{BIOMASS}}))*\text{BIOMASS}$$

unavailable or assimilated internal nitrogen (UNINIT) (g/m^3)

$$\frac{d}{dt}(\text{UNINIT}) = (\text{ASSNIT} - \text{MORT} * (\frac{\text{UNINIT}}{\text{BIOMASS}}))*\text{BIOMASS}$$

unavailable or assimilated internal phosphorus (UNIPHOS) (g/m^3)

$$\frac{d}{dt}(\text{UNIPHOS}) = (\text{ASSPHOS} - \text{MORT}* (\frac{\text{UNIPHOS}}{\text{BIOMASS}}))*\text{BIOMASS}$$

PROCESS EQUATIONS

production (PROD)

$$PROD = P1*LIGHT*NUTR \quad (1/day)$$

assimilation of nitrogen (ASSNIT) (g NIT/g BIOMASS/day)

$$ASSNIT = P2 * LIGHT * \left(\frac{INTNIT}{INTNIT + (P4 * BIOMASS)}\right)$$

assimilation of phosphorus (ASSPHOS) (g PHOS/g BIOMASS/day)

$$ASSPHOS = P3*LIGHT*\left(\frac{INTPHOS}{INTPHOS + (P6*BIOMASS)}\right)$$

respiration (RESP) (1/day)

$$RESP = P9* \exp\,[\,alog\,(P8)* WT /10] + P7 * PROD$$

mortality (MORT) (1/day)

$$MORT = P21* [\exp\{P22*(P23-LIGHT)\}]; \text{ if } NUTR < P23$$

$$MORT = P21; \text{ if } NUTR \geq P23$$

excretion of biomass (EXCB) (1/day)

$$EXCB = P18*[PROD*(1-LIGHT)]$$

excretion of nitrogen (EXCNIT) (g NIT/g BIOMASS/day)

$$EXCNIT = P19 * [PROD*(1-LIGHT)]$$

excretion of phosphorus (EXCPHOS) (g PHOS/g BIOMASS/day)

$$EXCPHOS = P20 * [PROD* (1-LIGHT)]$$

uptake of nitrogen (UPNIT) (g NIT/g BIOMASS/day)

$$UPNIT = P16 * LIGHT/ \left[\left(1 + \frac{P14}{ENVNIT}\right)*\left(1 + \frac{P12* INTNIT}{P10 * BIOMASS}\right) \right]$$

uptake of phosphorus (UPPHOS) (g PHOS/g BIOMASS/day)

$$UPPHOS = P17 * LIGHT/ \left[\left(1+ \frac{P15}{ENVPHOS}\right)*\left(1 + \frac{P13*INTPHOS}{P11*BIOMASS} \right) \right]$$

light effect (LIGHT) (unitless)

$$LIGHT = 2.718 * (W1-W2) / (EXPL * Z)$$

where:

$$W1 = \exp\,[(-WT /P27)* \exp\,(-EXPL * Z)]$$

$$W2 = \exp\,(-WT /P27)$$

$$EXPL = .1 + P5 * BIOMASS$$

nutrient effect (NUTR) (unitless)

$$NUTR = \frac{INTNIT}{INTNIT + (P26 * BIOMASS)} \; ; \text{ if } W3 \leq P24$$

$$NUTR = \frac{INTPHOS}{INTPHOS + (P25 * BIOMASS)} \; ; \text{ if } W3 > P24$$

where:

$$W3 = \frac{WTNIT * 14}{INTPHOS * 32}$$

APPENDIX 2.

Parameter definitions, tuned values, and coefficients of variation (CV) estimated from Smith (1980).

PARAMETER	DEFINITION	TUNED VALUE	CV
P1	maximum production rate (1/day)	1.7	0.0185
P2	maximum assimilation rate of nitrogen (g NIT/g BIOMASS/day)	0.015	0.25
P3	maximum assimilation rate of phosphorus (g PHOS/g BIOMASS/day)	0.002	0.25
P4	assimilation half saturation of nitrogen (g NIT/g BIOMASS)	0.024	0.25
P5	light extinction coefficient due to phytoplankton (m^2/g BIOMASS)	1.0	0.25
P6	assimilation half saturation of phosphorus (g PHOS/g BIOMASS)	0.01	0.18
P7	fraction of production used in respiration (unitless)	0.1	0.25
P8	Q_{10} for respiration (unitless)	2.0	0.0725
P9	respiration constant (1/day)	0.0125	0.25
P10	inhibitor half saturation of nitrogen (g NIT/g BIOMASS)	0.005	0.25
P11	inhibitor half saturation of phosphorus (g PHOS/g BIOMASS)	0.0001	0.245
P12	inhibitor constant of nitrogen (g NIT/g BIOMASS)	0.052	0.25
P13	inhibitor constant of phosphorus (g PHOS/g BIOMASS)	0.01	0.25
P14	uptake half saturation of nitrogen (g NIT/m^3)	0.1	0.17
P15	uptake half saturation of phosphorus (g PHOS/m^3)	0.02	0.215
P16	maximum uptake rate of nitrogen (g NIT/g BIOMASS/day)	0.02	0.118
P17	maximum uptake rate of phosphorus (g PHOS/g BIOMASS/day)	0.01	0.315
P18	maximum excretion rate of BIOMASS (1/day)	0.095	0.163
P19	maximum excretion rate of nitrogen (g NIT/g BIOMASS/day)	0.005	0.25
P20	maximum excretion rate of phosphorus (g PHOS/g BIOMASS/day)	0.001	0.25
P21	intrinsic mortality rate (1/day)	0.02	0.25

PARAMETER	DEFINITION	TUNED VALUE	CV
P22	nutrient availability effect on mortality (unitless)	0.1	0.25
P23	value of NUTR below which nutrient availability affects mortality (unitless)	0.1	0.25
P24	value of ratio of nitrogen to phosphorus which determines which nutrient is limiting (unitless)	30.0	0.083
P25	half saturation of nitrogen for nutrient limitation (g NIT/g BIOMASS)	0.001	0.25
P26	half saturation of phosphorus for nutrient limitation (g PHOS/g BIOMASS)	0.015	0.025
P27	optimal light intensity for photosynthesis (ly/min)	0.05	0.08

STATISTICAL PATTERNS OF NON-EQUILIBRIUM SYSTEMS[1]

Charles J. Puccia and Richard Levins[2]

Abstract.--Most applied models assume the system under study is in equilibrium or moving-equilibrium. Recent theoretical work has considered chaotic behavior or sustained-bounded-motion (SBM) of systems. We show how to determine the necessary conditions for SBM.

We establish, in some cases, the patterns of covariation or the bounds on variance. It is possible to use the data to identify sources of external perturbation to a SBM system. Alternative measures of stability, resiliency and rates of change of systems are provided. Theoretical models help explain variable time-series data collected for environmental assessment or monitoring.

INTRODUCTION

There are three classes of models based on the assumptions about the stability near equilibrium of the systems they represent.

First, there are models which assume local equilibrium. The observed variation of variables is the return from a perturbation to the equilibrium. The behavior of the system as it returns to equilibrium is predicted by calculating the eigenvalues. The problems of interest are conditions for local stability, oscillation and rate of return. One method of doing this is to turn Routh-Hurwitz criteria into a sign digraph formulation (Levins 1974):

The first part of the criteria is that feedback at level i (F_i) is less than zero, for all levels; each level corresponds to a subset of the model with i variables. The second part is a series of determinants of the feedbacks, the first of which is $F_1F_2+F_3>0$. The next set for a system of 5 variables would be $(F_1F_2+F_3)F_3+F_1(F_1F_4-F_5)>0$. The general rule is that negative feedback is greater than positive feedback at all levels to exclude monotonic instability. While long-length negative feedback cannot be too great compared to short-length negative feedback to exclude oscillatory instability.

An alternative representation of the Routh-Hurwitz criteria is to directly examine the characteristic polynomial.

$$-F_0\lambda^n - F_1\lambda^{n-1} - F_2\lambda^{n-2} - \ldots - F_n = 0 \quad (1)$$

Divide by $-(1/\lambda^n)F_n$ and take $F_0=1$. (IF $F_0\neq 1$, divide both sides by F_0 so new $F_0=1$.)

$$\left(\frac{1}{\lambda}\right)^n + \frac{F_{n-1}}{F_n}\left(\frac{1}{\lambda}\right)^{n-1} + \frac{F_{n-2}}{F_n}\left(\frac{1}{\lambda}\right)^{n-2} + \ldots = 0 \quad (2)$$

From the Routh-Hurwitz criteria,

$$\frac{-F_{n-1}}{nF_n} = \sum \frac{1}{\lambda} \quad (3)$$

or,

$$\frac{F_{n-1}}{nF_n} = \frac{1}{n}\sum \frac{1}{\lambda} = \overline{\left(\frac{1}{\lambda}\right)} \quad (4)$$

From (1),

$$\frac{F_1}{n} = \frac{1}{n}\sum \lambda = \bar{\lambda} \quad (5)$$

And, since the arithmetic mean is greater than the harmonic mean,

$$\bar{\lambda} > \frac{1}{\overline{\left(\frac{1}{\lambda}\right)}},$$

then

$$\frac{F_1}{n} < \frac{-nF_n}{F_{n-1}} \quad (6)$$

or,

$$F_1F_{n-1} < -n^2F_n \quad (7)$$

for all real roots of the same sign.

More generally, if

$$F_kF_{n-k} > -\binom{n}{k}^2 F_n \quad (8)$$

---

[1]Paper presented at the Third International Conference on State-of-the-Art in Ecological Modelling, Colorado State University, May 24-28, 1982.

[2]Charles J. Puccia is a Research Associate in Marine Ecology and Complex Systems and Richard Levins is John Rock Professor of Population Sciences, at Harvard University, School of Public Health, Boston, Massachusetts, U.S.A.

implies complex roots have a frequency fast enough compared to the rate of return to equilibrium from the real roots, so the oscillation is readily visible. The return time of the system is approximately $\ln(\frac{1}{\lambda})$.

Finally, the local equilibrium models calculate the eigenvectors of a system. The dominant eigenvector is the linear measure of least variability. A calculation of eigenvectors -- at least the sign patterns -- can also be obtained via signed digraphs. The eigenvector describe relatively conserved linear properties. For example, in a description of two competitors (fig. 1), the eigenvectors will be perpendicular to the equilibrium point, but the dominant eigenvector is oriented across the first and third quadrant whereas the minor eigenvector is oriented across the second and fourth. Consequently, the cluster of data points after a perturbation will be around the minor eigenvector, hence be closer to the equilibrium point, although with some spread.

The second class of models expand to include moving equilbirum. Parameters of the system change, but slow enough so the variables can track the equilibrium associated with new parameter values. The magnitude of change in the variables depends on the location of the parameters and the structure of the network of a signed digraph. The calculation has been given (Levins 1974) as:

$$\frac{\partial X^*_i}{\partial C} = \sum_{\substack{j=1 \\ k=1}}^{n} \frac{\partial f_j}{\partial C} P_{ij}^{(k)} \frac{F_{(n-k)}^{(-)}}{F_n} \qquad (9)$$

where, X^*_i denotes the change in equilibrium value of the 1th variable, $\partial f_j/\partial C$ denotes the changes in growth rate of the jth variable due to a change in parameter C; $P_{ij}^{(k)}$ is the path from j to i passing through k variables, $F_{(n-k)}^{(-)}$ is the feedback of the complementary subsystem, i.e., those variables not included along the path j to i; and, F_n is the feedback of the overall system.

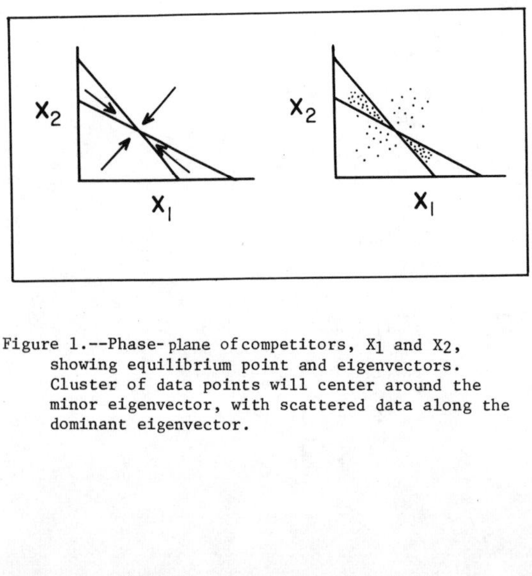

Figure 1.--Phase-plane of competitors, X_1 and X_2, showing equilibrium point and eigenvectors. Cluster of data points will center around the minor eigenvector, with scattered data along the dominant eigenvector.

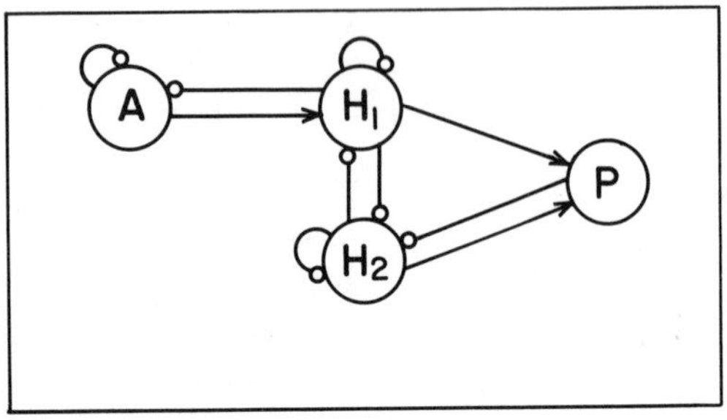

Figure 2.--Hypothetical aquatic system: Algae (A), Herbivore (H_1 and H_2), and Predator (P). ——o = negative effect; ——> = positive effect; (self-loop) = self-damping.

A sample hypothetical model (fig. 2) of an aquatic system has an algae (A) consumed by a herbivore (H_1) which is in competition with another herbivore (H_2) for nutrients as well as the herbivore secreting vitamins used by a predator (P) on the second herbivore. These observations can be made (Levins 1974):

Application of an algalcide to the algae can actually cause an algal bloom if the Herbivore-Predator subsystem is unstable ($F_3 > 0$). A greater supply of nutrients to the second herbivore, H_2, does not alter either herbivore population size, since the predator is left as an isolated, zero feedback complement (c.f. eq. (9)). There is no correlation between the two herbivore populations because of the input to H_2. However, the algal population will increase. Lastly, a greater supply of nutrients to H_1 increases the first herbivore population and decreases the second: an inverse correlation between herbivores.

These points can be made:

1. the biggest impact can be farthest from the point of entry;

2. the magnitude of change in a variable may neither be a measure of nor related to physiological sensitivity;

3. the correlation of two variables with negative loops between them depends on where the parameter change is effecting the growth rate of the variables of the system (the input point).

The uses of the qualitative signed digraphs are:

1. it provides a method to calculate a table of predictions to validate a graph;

2. it can locate sources of input;

3. it can predict covariance pattern of variables to test adequacy of graph;

4. it can identify which experiments and measurements are critical to further understanding and which just supply ancillary data.

The final set of models and the ones which we will discuss in more detail recognize that if the environment or human factors cause parameter changes which are fast enough, then the variables cannot track the environment and the equilibrium. The state of the system depends both on instantaneous equilibrium and on the pattern of environmental change. Thus, we can no longer talk about equilibrium. Instead, the focus of attention is on mean values. Moreover, we would like to know what can be carried over from the signed digraph methods to the present, since they are powerful methods and being able to use them would assist analysis.

TIME-AVERAGING

The basic operation which has been put forward (Levins 1979) is to define the expectation operator, analagous to the statistical operator, as

$$E([\cdot]) \equiv \lim_{t\to\infty} \frac{1}{t}\int_0^t [\cdot]\, dt \tag{10}$$

It is not necessary for our systems to be stochastic to apply these methods since the terminology is for convenience and ease of interpretation.

The fundamental theorem is (Levins 1979) that for a bounded variable,

$$E\left(\frac{dX}{dt}\right) = \lim_{t\to\infty} \frac{1}{t}\int_0^t \left(\frac{dX}{dt}\right) dt = 0 \tag{11}$$

In words, on average the rate of increase of a bounded variable equals its rate of decrease over all time. We note in passing that we can also carry out the analysis for a period of time in which $E(\frac{dX}{dt}) = R$, a finite number. It just complicates the arithmetic so it will be omitted in what follows.

Next, we consider a variable in which the growth rate depends solely on its own value multiplied by a function of itself with the other variables. Thus,

$$\frac{dX}{dt} = f(x)\left[g(x,y,\ldots)\right] \tag{12}$$

If $f(x) \neq 0$, then

$$\frac{1}{f(x)}\frac{dx}{dt} = g(x,y,\ldots) \tag{13}$$

The left-hand side of (13) is the derivative of X as a function of X, i.e.,

$$\frac{df_a(x)}{dt}.$$

Thus, by (11), is

$$E\left(\frac{df_a(x)}{dt}\right) = 0 = g(x,y,\ldots) \tag{14}$$

For example, the simple logistic with a variable resource, a(t), becomes

$$\frac{dx}{dt} = x(a(t)-x) \tag{15}$$

Then,

$$E\left(\frac{dx}{dt}\right) = 0 = \bar{x}(\bar{a}-\bar{x}) + \mathrm{cov}(a,x) - \mathrm{var}\, x \tag{16}$$

Now divide (15) by x, so

$$\frac{1}{x}\frac{dX}{dt} = \frac{d}{dt}(\ln X) = a(t) - x \tag{17}$$

and

$$E\left(\frac{d}{dt}\ln X\right) = 0 = \bar{a} - \bar{x} \tag{18}$$

Substitute (18) into (16) yields,

$$\mathrm{cov}(a,x) = \mathrm{var}\, x \tag{19}$$

Since the variance is always positive, then

$$\mathrm{cov}(a,x) > 0 \tag{20}$$

The environment and variable are positively correlated. Moreover, $\mathrm{cov}(a,x) = \sigma_a\sigma_x\rho$, hence

$$\sigma_a\sigma_x\rho = \sigma_x^2 \tag{21}$$

Square (21),

$$\sigma_a^2\sigma_x^2\rho = \sigma_x^4 \tag{22}$$

Divide by σ_x^2

$$\sigma_a^2\rho^2 = \sigma_x^2 \tag{23}$$

And, given $0 \leq \rho^2 \leq 1$

then

$$\sigma_a^2 \geq \sigma_x^2 \tag{24}$$

The environment is always more variable then the species.

From an ecological modelling point of view, we have to ask why does covariance matter? First, it is an indicator of a source of change. In the simple predator-prey model,

$$\frac{dx}{dt} = x(a-y) \tag{25}$$

$$\frac{dx}{dt} = y(x-d) \tag{26}$$

A variance in the growth of the prey, a=a(t), with a constant death rate for the predator d, yields,

$$E\left(\frac{dX}{dt}\right) = 0 = \bar{a}\,\bar{x} + \mathrm{cov}(a,x) - \bar{x}\,\bar{y} - \mathrm{cov}(x,y) \tag{27}$$

$$E\left(\frac{d}{dt}\ln X\right) = 0 = \bar{a} - \bar{y} \tag{28}$$

$$E\left(\frac{dy}{dx}\right) = 0 = \bar{y}\,\bar{x} + \mathrm{cov}(x,y) - d\,\bar{y} \tag{29}$$

$$E\left(\frac{d}{dt}\ln y\right) = 0 = \bar{x} - d \tag{30}$$

Solving (27) through (30) yields:

$$\text{Cov }(x,y) = 0$$

$$\text{Cov }(a,x) = 0 \qquad (31)$$

Thus, the prey is uncorrelated with its resource (via a(t)) and the predator, y, is not correlated with its resource (x, the prey).

On the other hand, if the growth rate of the prey is constant, while the death rate of the predator has a source of variation, so d = d(t), then

$$E\left(\frac{dx}{dt}\right) = 0 = a\,\bar{x} - \sigma_x^2 - \bar{x}\,\bar{y} - \text{cov}(x,y) \qquad (32)$$

$$E\left(\frac{d}{dt}\ln x\right) = 0 = a - \bar{x} - \bar{y} \qquad (33)$$

$$E\left(\frac{dy}{dt}\right) = 0 = \bar{x}\,\bar{y} + \text{cov}(x,y) - \bar{d}\,\bar{y} - \text{cov}(d,y) \qquad (34)$$

$$E\left(\frac{d}{dt}\ln y\right) = 0 = \bar{x} - \bar{d} \qquad (35)$$

Solve (32) through 35) to obtain

$$\text{cov }(x,y) = \text{cov }(d,y) \qquad (36)$$

and,

$$\bar{x}^2 - \sigma_x^2 = \text{cov}(x,y) \qquad (37)$$

or,

$$\text{cov }(x,y) = \overline{x^2} \qquad (38)$$

Thus,

$$\text{cov }(x,y) > 0 \qquad (39)$$

The predator is positively correlated with its prey.

Modifying equation (25), so that the growth rate of the prey is independent of its abundance -- as in marine benthic organisms which have local abundances dependent on pelagic larval settlement -- then,

$$\frac{dx}{dt} = a - x(y + c) \qquad (40)$$

$$\frac{dy}{dt} = y\,(x - d) \qquad (41)$$

Again, allowing for a time-varying death rate of the predator, d=d(t), then the expected value operation on (40) and (41) yields:

$$0 = \bar{a} - \bar{x}\,\bar{y} - \text{cov}(x,\ y) - c\,\bar{x} \qquad (42)$$

$$0 = \bar{x}\,\bar{y} + \text{cov}(x,\ y) - \bar{d}\,\bar{y} - \text{cov}(d,\ y) \qquad (43)$$

Divide (41) by y and take the expectation to obtain

$$\bar{x} = \bar{d} \qquad (44)$$

Likewise, divide (40) by x to obtain

$$\overline{\left(\frac{a}{x}\right)} = \bar{y} + c \qquad (45)$$

If we divide (42) by $\bar{x}$, we get

$$\frac{a}{\bar{x}} - (\bar{y} + c) = \frac{\text{cov}(x,\ y)}{\bar{x}} \qquad (46)$$

Substituting (45),

$$\frac{a}{\bar{x}} - a\overline{\left(\frac{1}{x}\right)} = \frac{\text{cov}(x,y)}{\bar{x}} \qquad (47)$$

or

$$a\left(\frac{1}{\bar{x}} - \overline{\left(\frac{1}{x}\right)}\right) = \frac{\text{cov}(x,\ y)}{\bar{x}} \qquad (48)$$

The arithmetic mean is greater than the harmonic mean, so the inverse of each yields $(\overline{1/x}) > (1/\bar{x})$ and

$$\text{cov}(x,\ y) < 0 \qquad (49)$$

In this case, the predator is negatively correlated with the prey.

Returning to the question of why covariances matter, we see that when a system is being driven by variation through the resource (the prey or primary production), then the resource will covary with the environment but not the higher trophic levels. On the other hand, for variation entering through the top predator, it causes a percolation down to lower trophic levels, each correlated with its resource; the predator covaries with the prey.

Another point is that we learn something about the state of the organisms or variables of the system based on the sign of the covariation. Thus, in the system represented by (25) and (26) with d(t), the cov(x, y) is positive; predator and prey abundance increase and decrease together. The predators' food supply, x, is scarce when its own population is small, the animals are hungry and often have low reproduction. The population is vulnerable to a stressed environment. In the system of (40) and (41), the cov(x, y) is negative. The predator population is minimal when food is abundant (x high). The reproduction of the predator is high and animals are healthy. The population is less vulnerable to a stressed environment.

Lastly, there is the general rule that cov(x, dx/dt) = 0. Thus, the correlation of a variable with its own rate of change is zero. Hence, any observations over a short, finite time which yield a positve correlation are not proof of density independent population regulation; nor, does a negative correlation demonstrate density-dependent regulation. It is simply an artefact of short-term observation.

We shift our attention to the variance. Starting with a single species model

$$\frac{dx}{dt} = x(a(t) - x^2) \qquad (50)$$

Divide by x, then take expected value of the equation to get,

$$\bar{a} = \overline{x^2} + \sigma_x^2 \qquad (51)$$

The variance of x, σ_x^2 , has to be less than $\bar{a}$. The mean of the environmental variability limits the variability of the population. The sensitivity of the mean of the population to environmental variation is obtained as,

$$2\bar{x}\,\frac{\partial \bar{x}}{\partial \sigma_a{}^2} + \frac{\partial \sigma_x^2}{\partial \sigma_a} = 0 \qquad (52)$$

$$\frac{\partial \bar{x}}{\partial \sigma_a{}^2} = -\left(\frac{\partial \sigma_x^2}{\partial \sigma_a^2}\right)/\ 2\bar{x} \qquad (53)$$

Thus, an increase in environmental variability through industrial activity which is slow to cause x to track the change will be a reduction in $\bar{x}$. In fact, $\bar{x}$ will go to zero as σ_x^2 moves towards $\bar{a}$, and becomes trapped there. The system bifurcates at x = 0.

In non-equilibrium systems, the sensitivity of the system to changing parameters is a change in the mean value of the variables. For the system,

$$\frac{dx_j}{dt} = x_j f_j(x_1,x_2,x_3,\ldots;\ a,b,c) \qquad (54)$$

where $x_1,x_2,\ldots x_k,\ldots$ are variables and a,b,c parameters, we expand in a Taylor series around the mean value:

$$\frac{dx_j}{dt} = x_j\left[f_j(\bar{x}_1,\bar{x}_2,\ldots\bar{x}_n) + \frac{\partial f_j}{\partial x_i}(x_i-\bar{x}_j) + \frac{1}{2}\frac{\partial^2 f_j}{\partial x_i^2}(x_i-\bar{x}_j)^2 + \frac{\partial^2 f_j}{\partial x_i x_k}(x_i-\bar{x}_j)(x_k-\bar{x}_j) + \ldots\right. \qquad (55)$$

Then,

$$E\left(\frac{d\ln x_j}{dt}\right) = f_j(\bar{x}_1,\bar{x}_2,\ldots\bar{x}_n) + \frac{1}{2}\frac{\partial^2 f_j}{\partial \bar{x}_i^2}\sigma_{x_i}^2 + \frac{\partial^2 f_j}{\partial \bar{x}_i\,\partial \bar{x}_k}\mathrm{cov}(x_i,x_k) + \ldots \qquad (56)$$

Analagous to the results of the moving equilibrium, e.g. (9), we obtain for a change in parameter, say, c, that:

$$0 = \sum\left[\left(\frac{\partial f_j}{\partial \bar{x}_i}\right)\left(\frac{\partial \bar{x}_i}{\partial c}\right) + \frac{1}{2}\left(\frac{\partial^3 f_j}{\partial \bar{x}_i^3}\sigma_x^2\frac{\partial \bar{x}_i}{\partial c} + \frac{\partial^2 f_i}{\partial \bar{x}_i^2}\frac{\partial \sigma x_i^2}{\partial c}\right) + \left(\frac{\partial^2 f_j}{\partial \bar{x}_i^2\,\partial \bar{x}_k}\sigma_{x_i x_k}\right)\frac{\partial \bar{x}_i}{\partial c} + \left(\frac{\partial^2 f_j}{\partial \bar{x}_j^2\,\partial \bar{x}_i}\right)\frac{\partial \sigma_{x_i x_j}}{\partial c} + \ldots + \frac{\partial f_j}{\partial c} + \frac{\partial^3 f_j}{\partial c\,\partial x^2} + \frac{\partial}{\partial c}\left(\frac{\partial^2 f_j}{\bar{x}_i\,\bar{x}_k} + \ldots\right)\right] \qquad (57)$$

The first term on the right-hand side of (57) is the same as that which leads to the derivation of (9), only applied to mean values instead of equilibrium values. There are new terms compared to the moving equilibrium: σ^2, cov, and higher moments; and the $\partial \bar{x}_i/\partial c$ now has the coefficient

$$\frac{\partial f_j}{\bar{x}_i} + \frac{1}{2}\left(\frac{\partial^3 f_i}{\partial \bar{x}_i^3}\sigma_x^2 + \frac{\partial^2 f_j}{\partial \bar{x}_j{}^2\,\partial \bar{x}_k}\,\underline{\sigma_{x_i x_j}}\right.$$

$$+ \ldots$$

Therefore, in constructing the solution of $\partial \bar{x}/\partial c$, we have a matrix with coefficients corresponding to those of the moving equilibrium analysis and sensitivities of higher moments entering as inputs.

As illustration, consider a nutrient (N) which is used by two algae (A_1 and A_2), with the second alga being consumed by a herbivore (H). The signed digraph of the moving equilibrium model (Fig. 3) also applies to the non-linear time-averaging model with the following stipulation:

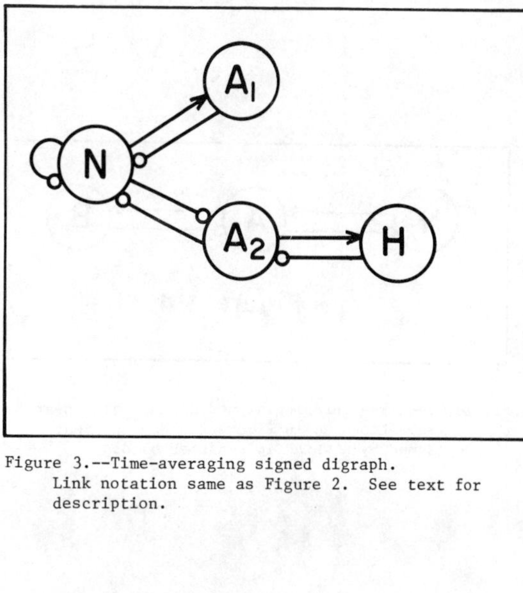

Figure 3.--Time-averaging signed digraph. Link notation same as Figure 2. See text for description.

If the nonlinearity occurs in the second alga, then any change in parameters effecting only the growth rate of A_2 will have no effect on the mean values of $\bar{A}_1$, $\bar{N}$, or $\bar{A}_2$ but will effect $\bar{H}$. There is an effective input to $\bar{H}$ as

$$\frac{\partial f_H}{\partial c} + \frac{1}{2}\frac{\partial^2 f_H}{\partial \bar{H}^2}\frac{\partial \sigma_A^2}{\partial c} + \ldots$$

Consequently, some nonlinearities are more important than others insofar as the effect the average of the rest of the system.

A detailed examination of $\bar{F}_n$, the feedback of the signed digraph of the average model, follows. (To distinguish it from the overall feedback of the signed digraph for the equilibrium model, a bar is placed over the symbol.)

In the N, A_1, A_2, H model, $\bar{F}_n < 0$. Then for a large $|\bar{F}_n|$ there will be little change in the variances or covariances of the system. The terms in brackets of eq. (57) will be small and the usual signed digraph analysis can be applied. For small $|\bar{F}_n|$ the means will change greatly as the statistics change:

$$\frac{\partial \bar{x}}{\partial c} = \frac{\frac{\partial \sigma_x^2}{\partial c}}{\bar{F}_n}$$

becomes very large as $|\bar{F}_n|$ gets very small.

The variances and covariances are incorporated into the links of the model (the α_{ij}'s) and therefore are part of the calculation for the feedback of the system, $\bar{F}_n$. Consequently, changes in parameters that cause increased or decreased variances and covariances can lead to increased or decreased $|\bar{F}_n|$, depending on the actual equation form. It is possible for $\bar{F}_n \longrightarrow 0$, and lead to a birfurcation.

Not every link can effect the $\bar{F}_n$, hence, lead to a bifurcation. There are at least three known possibilities for the effect of variance. First, the simple linear chain hierarchy model (Fig. 4):

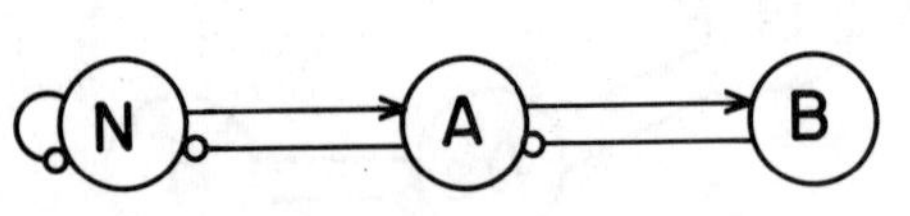

Figure 4.--Time-averaging signed digraph of linear chain. Links as in Figure 2. N = nutrient consumed by A which is consumed by B.

A change in σ_A^2 that effects A ——o N link will not effect $\bar{F}_3$ since it is not part of the $\bar{F}_3$ calculation.

Second, variance that enchances positive feedback terms may overwhelm the negative and lead to instability bifurcation.

Third, if all loops are initially negative then an increased variance can reverse the sign of the link provided $(\partial^3 f/\partial x^3)\sigma^2$ has the opposite sign from $\partial f/\partial x$. For the Michaelis-Menton equation,

$$\frac{\partial f}{\partial x} = \frac{vK}{(x+K)^2} \tag{58}$$

$$\frac{\partial^2 f}{\partial x^2} = \frac{-2vK}{(x+K)^3} \tag{59}$$

$$\frac{\partial^3 f}{\partial x^3} = \frac{6vK}{(x+K)^4} \tag{60}$$

Thus, $\partial^3 f/\partial x^3$ and $\partial f/\partial x$ are of the same sign. A change in variance cannot bring on a bifurcation.

For a sigmoidal function

$$\frac{\partial f}{\partial x} = e^{-rx^2} \tag{61}$$

$$\frac{\partial^2 f}{\partial x^2} = -2rxe^{-rx^2} \tag{62}$$

$$\frac{\partial^3 f}{\partial x^3} = (4r^2x^2 - 2r)e^{-rx^2} \tag{63}$$

At small values of x, $\partial^3 f/\partial x^3$ is of opposite sign to $\partial f/\partial x$. Thus, a variance which changes with environment can bring on a bifurcation through changing the signs of links.

The value of $\bar{F}_n$ is

$$\bar{F}_n = \sum\left(\frac{\partial f_j}{\partial x_i} + \frac{\partial^3 f_j}{\partial x_i}\sigma_x^2 + \frac{\partial^2 f_j}{\partial x_i^2 \partial x_j}\sigma_{x_i x_j} + \ldots\right) \tag{64}$$

If the covariance and higher order terms are zero, then

$$\bar{F}_n = e^{-rx^2}\left(1 + (4r^2x^2 - 2r)\sigma_x^2\right) \tag{65}$$

or,

$$\bar{F}_n = e^{-rx^2}\left(1 + (2rx^2 - 1)2r\sigma_x^2\right) \tag{66}$$

With $2rx^2 < 1$ and σ^2 large compared to x^2 near zero, say, then

$$\bar{F}_n \cong \left(1 - 2r\sigma_x^2\right) \tag{67}$$

$\bar{F}_n$ reverses sign and can make $\bar{F}_n \longrightarrow 0$ to bring on a bifurcation as $2r\sigma_x^2$ becomes larger than 1.

CONCLUSIONS

We have tried to illustrate the use of the time-averaging method applied to non-equilibrium systems and reach these conclusions.

1. The signed digraph method of analysis for moving equilibrium can be used with two changes. The sensitivity of higher moments appear as inputs and higher moments may enter into the α_{ij} (the links). The analysis applies to average values not equilibrium values.

2. Nonlinearities will only effect some means and not others.

3. The higher moments of environmental variation with $\partial f/\partial\sigma_a^3 = 0$ still give inputs as $\partial\sigma_x^2/\partial\sigma_a^2$.

4. If $\bar{F}_n$ is large, the change in $\bar{x}$'s will be small compared to changes in variance, covariance and higher moments. Thus, the statistics of the environment effect mostly higher moments of the variables. When $\bar{F}_n$ is small, the means of the x's will respond to the statistical pattern of the environment.

5. As $\bar{F}_n \longrightarrow 0$, the system bifurcates.

6. For $\bar{F}_n \longrightarrow 0$, necessary conditions are:

(a) there exist positive and negative parts to $\bar{F}_n$; nonlinear terms must strengthen the positive feedback paths.

(b) when all loops are negative, then $\partial^3 f/\partial x^3$ must have an opposite sign from $\partial f/\partial x$. This will allow reversal of a loop. For Michaelis-Menton type relations, this will not be true; for sigmoidal relations with low values of x a sign reversal is possible, provided the constant of the sigmoidal curve is big enough.

LITERATURE CITED

Levins, R. 1974. The qualitative analysis of partially specified systems. Annals of the New York Academy of Sciences 231: 123-138.

Levins, R. 1979. Coexistence in a variable environment. American Naturalist 114: 765-783.

ACKNOWLEDGMENT

This project has been financed with Federal funds from the EPA under grant number R807713. The contents do not necessarily reflect the views and policies of the Environmental Protection Agency.

STOCHASTIC COMPARTMENT MODELS OF ECOLOGICAL SUCCESSION: EVALUATION OF THEIR USEFULNESS TO NATURAL RESOURCE MANAGERS[1]

W. E. Grant[2] and J. H. Matis[3]

Abstract -- The application of linear compartment model theory to natural resource management problems involving ecological succession is discussed. The mathematical development of compartment models from the one-compartment deterministic case to the multi-compartment stochastic case is presented, using a parallel development to emphasize natural relationships between the various models, and important ecological interpretations are noted. The possibilities of defining problems involving ecological succession in terms of linear compartment model theory is particularly appealing because parameterization of such models does not require an extensive data base. The simple design of linear compartment models also increases the likelihood of their use by natural resource managers.

INTRODUCTION

The use of systems analysis and simulation in ecology and natural resource management is increasing. Sophisticated methodologies with theoretical foundations in mathematics and statistics have been applied to model a variety of ecological systems. The challenge is to effectively couple these powerful analytical procedures with available data bases to address relevant questions in natural resource management. Today many of the most urgent management questions involve the quantitative description of ecological succession in response to human activities. Resource managers frequently do not require detailed information from highly mechanistic models as input to the decision-making process. Often the most relevant information can be represented as probabilities and rates of successional change between very general community types in response to specific management alternatives.

Compartment models are used widely in ecology and natural resource management to simulate and analyze the storage and flow dynamics of a variety of substances, ranging from radionuclides (Patten and Witkamp 1967) to biomass (Patten 1975), in a variety of ecological systems. Use of compartment models to represent the process of ecological succession is a logical extension of the concept, although the "substance" transfered between compartments becomes abstract with flows representing the transition from one seral stage to another. Many applications of compartment models to succession have employed simulation techniques exclusively (Bledsoe and Van Dyne 1971, Bledsoe 1968). However, if compartment models are formulated with constant transfer rates between compartments, analytical treatment of the resulting matrix of transfer rates can yield estimates of the probability that a "unit" will reach a given compartment and the mean time of transfer of a "unit" from one compartment to another (Finn 1976; Barber 1978a, 1978b, 1979; Carney, et al. 1981, Matis and Patten 1981). These quantities have obvious applicability to the analysis of successional systems, where the probability of reaching a given seral stage within a given period of time is of primary concern to the resource manager.

In this paper we demonstrate the application of linear compartment model theory to natural resource management problems involving ecological succession. More specifically, we (1) present the mathematical development of compartment models from the one-compartment deterministic case to the multi-compartment stochastic case, using a parallel development to emphasize natural relationships between the various models; (2) note important ecological interpretations of the models; and (3) apply the multi-compartment stochastic model to analyze an old-field successional system that has been simulated previously (Bledsoe and Van Dyne 1971).

MODEL DEVELOPMENT AND ECOLOGICAL INTERPRETATION

One-Compartment System: Deterministic Model

A one-compartment successional system can be represented diagrammatically as in Figure 1. Since two successional stages are represented, some might choose to classify this as a two-compartment closed system, but we choose to classify it as a one-compartment "open" system since there is only one transient state.

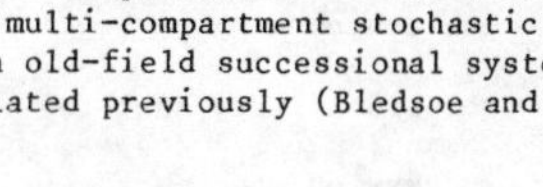

Figure 1. Diagram of the general one-compartment (open) model of a successional system. The transfer rate from compartment 1 out of the system is represented by k.

[1] Paper presented at the Third International Conference on State-of-the-Art in Ecological Modelling, Colorado State University, May 24-28, 1982.

[2] W. E. Grant is Associate Professor of Wildlife and Fisheries Sciences, Texas A&M University, College Station, Texas, U.S.A.

[3] J. H. Matis is Professor of Statistics, Texas A&M University, College Station, Texas, U.S.A.

The equation representing the deterministic one-compartment model is

$$\dot{P}1(t) = -k\ P1(t) \qquad (1)$$

where $P1(t)$ = the proportion of communities in stage 1 of succession at time t

k = the proportion of communities in stage 1 that transfer to stage 2 per unit time, a constant flow rate.

$\dot{P}1(t)$ = time derivative, $dP1(t)/dt$

The solution to equation 1 with initial condition $P1(0)=1$ is

$$P1(t) = \exp(-kt) \qquad (2)$$

and it follows that

$$P2(t) = 1-\exp(-kt) \qquad (3)$$

where $P2(t)$ = the proportion of the communities in stage 2 of succession at time t.

With the initialization of the system with all of the communities in stage 1, $P1(0)=1$, $P2(0)=0$, the model predicts the succession of communities from stage 1 to stage 2 over time. That is, at any point in time the model predicts the proportion of communities in each successional stage. Further, if we define

$Y1(t)$ = the number of an original n communities, all of which initially were in stage 1 of succession, that still are in stage 1 at time t,

then

$$Y1(t) = Y1(0)\ P1(t). \qquad (4)$$

One-Compartment System: Stochastic Model

For the stochastic one-compartment model, we redefine $P1(t)$, $P2(t)$, and k as follows.

$P1(t)$ = the probability that a given community is in stage 1 of succession at time t

$P2(t)$ = the probability that the given community is in stage 2 of succession at time t

k = an instantaneous rate defined such that the probability that a community in stage 1 of succession at time t will transfer to stage 2 by time $t + \Delta t$ is equal to $k\ \Delta t$, where Δt is indefinitely small

Note that these definitions provide a broader theoretical foundation for the problem in that model predictions now refer to likelihoods of a community being in a given successional stage rather than to an exact proportion of communities in a given successional stage.

If we define f1 as the expected lifetime of a given community in stage 1, it can be shown that f1 is a random variable with an exponential distribution;

$$f1(t) = k\ \exp(-kt). \qquad (5)$$

The probability density function described by equation 5 can be used to predict the probability that the expected lifetime of a community in stage 1 of succession falls between two specified time limits.

The integral of the probability density function is the cumulative distribution function;

$$F1(t) = \int_0^t f1(t)\ dt = 1-\exp(-kt). \qquad (6)$$

Equation 6 can be used to predict the probability that the expected lifetime of a community in stage 1 is less than t time units. Note that the probability that the expected lifetime of a community in stage 1 is less than zero is zero, $F1(0)=0$, and the probability that the expected lifetime of a community in stage 1 is less than infinity is one, $F1(\infty)=1$.

A slight variation of the cumulative density function is the survivorship function;

$$S1(t)=1-F1(t)=\exp(-kt). \qquad (7)$$

Equation 7 can be used to predict the probability that the lifetime of a community in stage 1 is greater than t time units; that is, that the community still is in stage 1 at time t. Note that this is equivalent to $P1(t)$ with initial conditions $P1(0)=1.0$. Thus, equation 7 is the stochastic analog of the deterministic solution presented in equation 2, but $S1(t)$ represents a probability whereas $P1(t)$, as defined for the deterministic model, represents a proportion.

An additional concept of the stochastic model that has a particularly useful ecological interpretation is that of the mean residence time in a given compartment, or the mean length of time that the community will be in a given stage of succession. The mean residence time for the one-compartment model is calculated as

$$u = 1/k. \qquad (8)$$

Strictly speaking, the deterministic model has no such concept. The half-life of the deterministic model is the analog of the median lifetime of the stochastic model.

Finally, the concept of a survivorship function presented in equation 7, which can be used to predict the probability that the community still is in stage 1 of succession at time t, can be extended to a distribution of n communities all initially in stage 1. If we define

$Y1(t)$ = the number of the original n communities, all of which initially were in stage 1 of succession, that still are in stage 1 at time t,

then $Y1(t)$ is a binomially distributed random variable, that is

$$P\ \{Y1(t)=y\} = \binom{n}{y}\ [P1(t)]^y\ [1-P1(t)]^{n-y}. \qquad (9)$$

Therefore, we can calculate the mean and variance of Y1 from a binomial distribution and construct a prediction interval for the number of communities out of n original communities all initially in stage 1 of succession that still are in stage 1 at time t. The mean of $Y1(t)$ is $E[Y1(t)] = Y1(0)\ P1(t)$ which is analogous to the deterministic number in (4), however, the variance $V[Y1(t)] = Y1(0)\, P1(t)\, [1-P1(t)]$ and the resulting prediction intervals are available only through stochastic modeling.

Multi-Compartment System: Deterministic Model

For simplicity, we describe a two-compartment (open) model (Fig. 2) to represent model development for the multi-compartment successional system. In theory, what follows is equally applicable to n compartments.

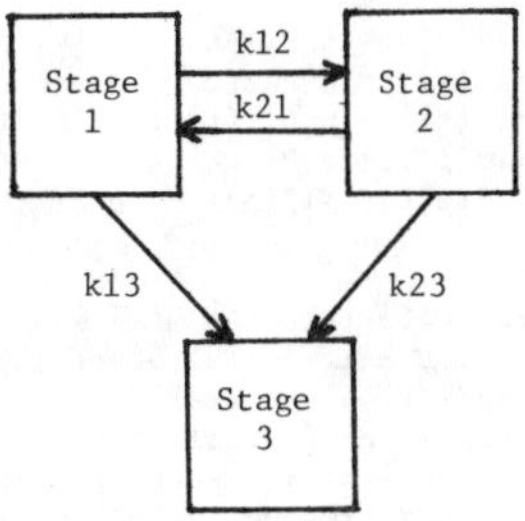

Figure 2. Diagram of the general two-compartment (open) model of a successional system. The transfer rate from compartment i to compartment j is represented by kij.

The set of equations representing the general deterministic model is

$$\dot{P}i1(t) = -(k12 + k13)\, Pi1(t) + k21\, Pi2(t) \quad (10)$$
$$\dot{P}i2(t) = k12\, Pi1(t) - (k21 + k23)\, Pi2(t) \quad (11)$$
$$\dot{P}i3(t) = k13\, Pi1(t) + k23\, Pi2(t) \quad (12)$$

where Pij(t) = the proportion of communities initially in stage i which are in stage j of succession at time t

kij = the proportion of communities in stage i that transfer to stage j per unit time, a constant flow rate.

$\dot{P}ij(t)$ = time-derivative of Pij(t), dPij(t)/dt.

If we define the matrix

$$P(t) = \begin{bmatrix} P11(t) & P21(t) \\ P12(t) & P22(t) \end{bmatrix} \quad (13)$$

then

$$\dot{P}(t) = -AP(t)$$

$$\text{where } A = \begin{bmatrix} k12 + k13 & -k21 \\ -k12 & k21 + k23 \end{bmatrix}.$$

Note that the coefficient matrix A is the analog of the constant flow rate k of the one-compartment model. Also note that we may solve explicitly for Pi1(t) and Pi2(t) and then calculate Pi3(t) by subtraction. The solution to equation 13 is

$$P(t) = \exp(-At) = (I - At + (At)^2/2! + \ldots) \quad (14)$$

Equation 14 can be solved in closed form by finding eigenvalue-eigenvector matrices of A (Matis and Patten 1981). As with the one-compartment deterministic model, the multi-compartment model predicts the proportion of communities in each successional stage at any point in time.

To illustrate the calculations, consider a simple two-compartment model which is the same as the general model presented in Figure 2, except that transfers represented by k21 and k13 do not exist. The coefficient matrix is

$$A = \begin{bmatrix} k12 & 0 \\ -k12 & k23 \end{bmatrix}.$$

The solution is

$$P(t) = \begin{bmatrix} \exp(-k12t) & 0 \\ k12(\exp(-k12t)- \exp(-k23t))/(k23-k12) & \exp(-k23t) \end{bmatrix}$$

Multi-Compartment System: Stochastic Model

For the stochastic multi-compartment model, we redefine Pij(t) and kij as follows.

Pij(t) = the probability that a given community initially in stage i is in stage j of succession at time t

kij = an instantaneous rate defined such that the probability that any particular community in stage i of succession at time t will transfer to stage j by time $t + \Delta t$ is equal to $kij\Delta t$, where Δt is indefinitely small

The total rate, denoted ki., at which a unit leaves a specified compartment, denoted compartment i, is the sum

$$ki. = \sum_{\substack{j=1 \\ j \neq i}}^{n} kij \quad (15)$$

Note that the first passage time out of a specified compartment in a multi-compartment system is equivalent to the particle lifetime in a single compartment system. It follows that the first passage time is an exponential random variable

$$Fi(t) = ki.\, \exp(-ki.t) \quad (16)$$

Now let Sij(t) denote the probability that a unit in stage i at time 0 is in stage j at time t; and let the survivorship matrix be denoted

$$S(t) = \begin{bmatrix} S11(t) \cdots Sn1(t) \\ S1n(t) \cdots Snn(t) \end{bmatrix}.$$

The coefficient matrix A is constructed as in the multi-compartment deterministic model, but the elements (kij) now have probabilistic interpretations. Thus, for the general two-compartment stochastic model (Fig. 2)

$$A = \begin{bmatrix} k12 + k13 & -k21 \\ -k12 & k21 + k23 \end{bmatrix},$$

and for the simple 2-compartment model

$$A = \begin{bmatrix} k12 & 0 \\ -k12 & k23 \end{bmatrix}.$$

If we define the matrix of probabilities as

$$P(t) = \begin{bmatrix} P11(t) & P21(t) \\ P12(t) & P22(t) \end{bmatrix}$$

then, as in the multi-compartment deterministic model,

$$\dot{P}(t) = -AP(t) \quad (17)$$

and

$$P(t) = \exp(-At) \quad (18)$$

Again, equation 18 can be solved in closed form by finding eigenvalue-eigenvector matrices of A.

Note that the matrix of survivorship functions is also

$$S(t) = \exp(-At) = P(t), \quad (19)$$

which for the simple two-compartment model is

$$S(t) = \begin{bmatrix} \exp(-k12t) & 0 \\ \frac{k12}{k12-k23}(\exp(-k23t) - \exp(-k12t)) & \exp(-k23t) \end{bmatrix} \quad (20)$$

Equation 19 can be used to predict the probability that a community in stage j of succession at time zero is in stage i at time t. For example, for the simple two-compartment model the matrix of survivorship functions (equation 20) indicates that the probability that a community in stage 1 at time zero will still be in stage 1 at time t is exp(-k12t). Likewise, the probability that a community in stage 1 at time zero will be in stage 2 at time t is $\frac{k12}{k12-k23}$ (exp(-k23t) - exp(-k12t)) and the probability that a community in stage 2 at time zero will be in stage 2 at time t is exp(-k23t). There is no chance that a community in stage 2 at time zero will be in stage 1 at time t.

Let Mij denote the mean residence time in stage j of a unit that started in stage i and let the matrix of mean residence times be

$$M = \begin{bmatrix} M11 & \cdots & Mn1 \\ \vdots & & \vdots \\ M1n & \cdots & Mnn \end{bmatrix}$$

It can be shown that

$$M = A^{-1}, \quad (21)$$

which for the simple two-compartment model is

$$M = A^{-1} = \begin{bmatrix} 1/k12 & 0 \\ 1/k23 & 1/k23 \end{bmatrix}. \quad (22)$$

Equation 21 can be used to predict the mean length of time (residence time) that a community in stage j of succession at time zero spends in stage i before first transfering out of the system, that is, to the last compartment or "climax" community. For example, for the simple two-compartment model the matrix of mean residence times (equation 22) indicates that the mean residence time for a community in stage 1, given that it was in stage 1 at time zero, is 1/k12. Likewise the mean residence time for a community in stage 2, given that it was in stage 1 at time zero, is 1/k23 and the mean residence time for a community in stage 2, given that it was in stage 2 at time zero, is 1/k23. A community in stage 2 at time zero spends no time in stage 1.

Finally, the concept of mean residence times can be extended to a distribution of n communities. If we define

Y11(t) = the number of the original n communities, all of which initially were in stage 1 of succession, that are in stage 1 at time t

Y12(t) = the number of the original n communities, all of which initially were in stage 1 of succession, that are in stage 2 at time t,

then the vector

$$\begin{bmatrix} Y11(t) \\ Y12(t) \end{bmatrix}$$

has a multinomial distribution. Therefore,

$$E\,(Y11(t)) = n\,P11(t) \quad (23)$$
$$E\,(Y12(t)) = n\,P12(t) \quad (24)$$
$$V\,(Y11(t)) = n\,P11(t)\,(1 - P11(t)) \quad (25)$$
$$V\,(Y12(t)) = n\,P12(t)\,(1 - P12(t)) \quad (26)$$
$$Cov\,(Y11(t),\,Y12(t)) = -n\,P11(t)\,P12(t). \quad (27)$$

That is, we can calculate the mean and variance of Y11 and Y12 from a multinomial distribution and construct a prediction interval for the number of communities, out of n original communities all initially in stage 1 of succession, that are in stage 1 and stage 2, respectively, at time t. Similarly, the vector

$$\begin{bmatrix} Y21(t) \\ Y22(t) \end{bmatrix}$$

also has a multinomial distribution with parameters P21(t) and P22(t).

EXAMPLE APPLICATION OF THE MULTI-COMPARTMENT STOCHASTIC MODEL

To demonstrate use of the multi-compartment stochastic model within an ecological framework, we have chosen to reexamine a compartment model used by Bledsoe and Van Dyne (1971) to simulate old-field succession on abandoned farmland in Oklahoma, U.S.A. This successional system was studied at 30 different sites by Smith (1940) in the 1930's, at which time he characterized various successional stages ranging from two to four years since abandonment to 20 to 30 years since abandonment. Successional stages were identified by the relative abundance of various plant species at each site. Abundance, in turn, was estimated subjectively as the relative percent cover of each plant species. By studying the data tabulated by Smith, Bledsoe and Van Dyne (1971) constructed a five-compartment model (Fig. 3) and were successful in simulating the general sequence of successional changes described by Smith.

The general form of the Bledsoe and Van Dyne model was that of a linear constant-coefficient compartment model:

$$\dot{x}i(t) = \sum_{\substack{j=1 \\ j \neq i}}^{n} bij\; xj(t) \quad (28)$$

where xi(t) = the content of the ith compartment at time t

bij = a constant coefficient representing the transfer rate from compartment i to compartment j.

Or, using matrix notation for the five-compartment case,

$$\dot{x} = Bx \tag{29}$$

where

$$x = \begin{bmatrix} x1(t) \\ x2(t) \\ x3(t) \\ x4(t) \\ x5(t) \end{bmatrix}, \quad \dot{x} = \begin{bmatrix} \dot{x}1(t) \\ \dot{x}2(t) \\ \dot{x}3(t) \\ \dot{x}4(t) \\ \dot{x}5(t) \end{bmatrix},$$

and, for the model represented in Figure 3,

$$B = \begin{bmatrix} b11 & 0 & 0 & 0 & 0 \\ b12 & b22 & 0 & 0 & 0 \\ b13 & b23 & b33 & 0 & 0 \\ 0 & 0 & b34 & b44 & 0 \\ 0 & 0 & 0 & b45 & 0 \end{bmatrix}.$$

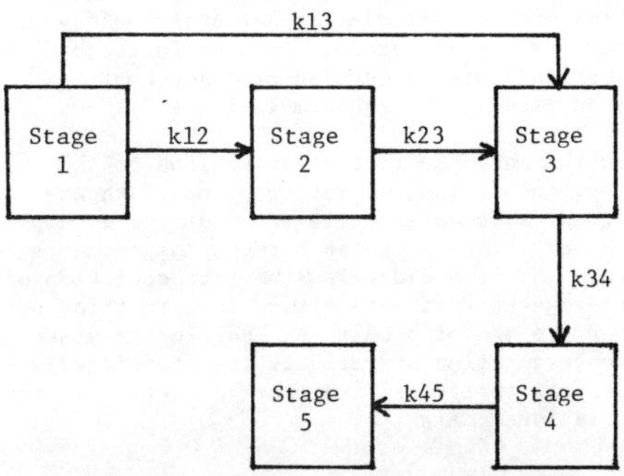

Figure 3. Diagram of the five-compartment (four-compartment open) model of old-field succession in Oklahoma, U.S.A. described by Bledsoe and Van Dyne (1971, p. 489). The transfer rate from compartment i to compartment j is represented by kij.

Through simulation trials on an analog computer, the matrix of transfer coefficients was estimated as

$$B = \begin{bmatrix} -2.0 & 0 & 0 & 0 & 0 \\ 1.6 & -1.0 & 0 & 0 & 0 \\ 0.4 & 1.0 & -0.7 & 0 & 0 \\ 0 & 0 & 0.7 & -0.8 & 0 \\ 0 & 0 & 0 & 0.8 & 0 \end{bmatrix}. \tag{30}$$

That is, elements of the B matrix were adjusted such that the level in a given compartment, which represents the abundance of organic material of plants that characterize that stage of succession, varied over time in a manner consistent with Smith's qualitative description of succession. Results of the final simulation are presented in Figure 4.

To analyze this successional system using linear compartment model theory, we represent the system as a four-compartment (open) stochastic model (Fig. 3). The matrix of transfer coefficients is

$$A = \begin{bmatrix} (k12 + k13) & 0 & 0 & 0 \\ -k12 & k23 & 0 & 0 \\ -k13 & -k23 & k34 & 0 \\ 0 & 0 & -k34 & k45 \end{bmatrix} \tag{31}$$

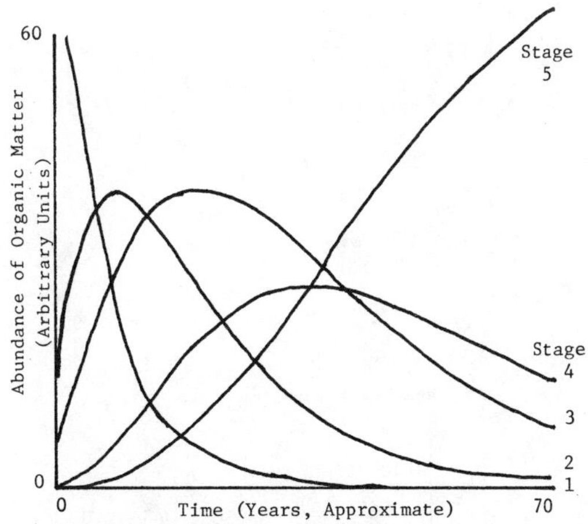

Figure 4. Results of the final simulation of old-field succession in Oklahoma, U.S.A. described by Bledsoe and Van Dyne (1971, p. 492).

or,

$$A = \begin{bmatrix} 2.0 & 0 & 0 & 0 \\ -1.6 & 1.0 & 0 & 0 \\ -0.4 & -1.0 & 0.7 & 0 \\ 0 & 0 & -0.7 & 0.8 \end{bmatrix}. \tag{32}$$

The matrix of survivorship functions is (33)

$$S11(t) = \exp(-(k12 + k13)t)$$

$$S12(t) = C1(\exp(-k23t) - \exp(-(k12 + k13)t))$$

$$S13(t) = (C2 + C3)\exp(-k34t) - C2\exp(-k23t) - C3\exp(-(k12 + k13)t)$$

$$S14(t) = (C4 + C5 + C6)\exp(-k45t) - C4\exp(-k34t) - C5\exp(-k23t) - C6\exp(-(k12 + k13)t)$$

$$S22(t) = \exp(-k23t)$$

$$S23(t) = C7\,(\exp(-k34t) - \exp(-k23t))$$

$$S24(t) = (C8 + C9)\exp(-k45t) - C8\exp(-k34t) - C9\exp(-k23t)$$

$$S33(t) = \exp(-k34t)$$

$$S34(t) = \frac{k34}{k34-k45}(\exp(-k45t) - \exp(-k34t))$$

$$S44(t) = \exp(-k45t)$$

where $C1 = \dfrac{k12}{k12+k13-k23}$

$$C2 = \frac{k23\ C1}{k23-k34}$$

$$C3 = \frac{k13-k23\ C1}{k12+k13-k34}$$

$$C4 = \frac{k34\ (C2+C3)}{k34-k45}$$

$$C5 = \frac{-k34\ C2}{k23-k45}$$

$$C6 = \frac{-k34\ C3}{k12+k13-k45}$$

$$C7 = \frac{k23}{k23-k34}$$

$$C8 = \frac{k34\ C7}{k34-k45}$$

$$C9 = \frac{-k34\ C7}{k23-k45}$$

The matrix of mean residence times is

$$M = A^{-1} = \begin{bmatrix} \frac{1}{(k12+k13)} & 0 & 0 & 0 \\ \frac{k12}{k23(k12+k13)} & \frac{1}{k23} & 0 & 0 \\ \frac{1}{k34} & \frac{1}{k34} & \frac{1}{k34} & 0 \\ \frac{1}{k45} & \frac{1}{k45} & \frac{1}{k45} & \frac{1}{k45} \end{bmatrix} \quad (34)$$

or,

$$M = A^{-1} = \begin{bmatrix} 0.5 & 0 & 0 & 0 \\ 0.8 & 1.0 & 0 & 0 \\ 1.4286 & 1.4286 & 1.4286 & 0 \\ 1.25 & 1.25 & 1.25 & 1.25 \end{bmatrix} \quad (35)$$

Also, note that the vector of expected times before first leaving the system, that is, before first reaching stage 5 of succession, for a community initially in stage i is given by

$$M_{\xi} = \begin{bmatrix} \frac{k12}{k23(k12+k13)} + \frac{1}{k12+k13} + \frac{1}{k34} + \frac{1}{k45} \\ \frac{1}{k23} + \frac{1}{k34} + \frac{1}{k45} \\ \frac{1}{k34} + \frac{1}{k45} \\ \frac{1}{k45} \end{bmatrix} \quad (36)$$

or,

$$M_{\xi} = \begin{bmatrix} 3.9786 \\ 3.6786 \\ 2.6786 \\ 1.25 \end{bmatrix} \cdot \quad (37)$$

DISCUSSION

Stochastic compartment models would appear to have considerable utility to natural resource managers concerned with problems involving ecological succession. The possibilities of defining such problems in terms of linear compartment model theory are particularly appealing for several reasons. For example, managers often are concerned with natural systems for which relatively little information is available. They may need to rely solely on expert opinion to obtain estimates of the average time that it takes for succession to proceed from one general community type to another. Or, as was the case in the example application above, available information may be highly qualitative. Yet such information is the only input needed to parameterize the transfer coefficient matrix of a compartment model. Even if managers are concerned with systems for which a large data base exists, the statistical problems of parameter estimation for complex mechanistic models of succession may be overwhelming. Indeed, the very complexity of such models may decrease the likelihood of their utilization by natural resource managers. On the other hand, compartment models are simple in basic design, and although they are not mechanistic in the sense of representing complex ecological processes and relationships explicitly, the transfer coefficients are ecologically interpretable, even if highly aggregated, parameters.

The predictions concerning system behavior that can be obtained through analytical treatment of the transfer coefficient matrix likewise have appealing ecological interpretations. Concepts such as the survivorship function and the mean residence time relate directly to the most fundamental questions of the natural resource manager; namely, what is the probability that a community in a given stage of succession will remain in that stage for a specified period of time and what is the average length of time that it will remain in that stage before changing to another stage. As seen in the example application above, such predictions cannot be obtained readily through simulation.

With regard to further extensions of the application of linear compartment model theory to ecological succession, relaxation of the assumption of constant transfer rates between compartments is possible. This would allow the representation of transfer rates that vary either as a function of time or the age of a unit in a particular state. The representation of variable transfer rates within the general analytical framework presented in this paper is forthcoming.

LITERATURE CITED

Barber, M. C. 1978a. A retrospective Markovian model for ecosystem resource flow. Ecol. Modelling 5:125-135.

Barber, M. C. 1978b. A Markovian model for ecosystem flow analysis. Ecol. Modelling 5:193-206.

Barber, M. C. 1979. A note concerning time parameterization of Markovian models of ecosystem flow analysis. Ecol. Modelling 6:323-328.

Bledsoe, L. J. 1968. Compartment models and their use in the simulation of secondary succession. M.S. Thesis, Colorado State University, Fort Collins, Colorado, U.S.A.

Bledsoe, L. J., and G. M. Van Dyne. 1971. A compartment model simulation of secondary succession. pp. 479-511. In (B. C. Patten, ed.) Systems Analysis and Simulation in Ecology. Academic Press, New York, New York, U.S.A.

Carney, J. H., D. L. DeAngelis, R. H. Gardner, J. B. Mankin, and W. M. Past. 1981. Calculation of probabilities of transfer, recurrence intervals, and positional indices for linear compartment models. Oak Ridge National Laboratory, Environmental Sci. Div. Publ. No. 1544. Oak Ridge, Tennessee, U.S.A. 70 pp.

Finn, J. T. 1976. Measures of ecosystem structure and function derived from analysis of flows. J. of Theoretical Biol. 56:363-380.

Matis, J. H., and B. C. Patten. 1981. Environ analysis of linear compartmental systems: The static, time invariant case. Int. Stat. Inst. Bulletin, Vol. 48 (in press).

Patten, B. C., and M. Witkamp. 1967. Systems analysis of 134-Cesium kinetics in terrestrial microcosms. Ecology 48:813-824.

Patten, B. C. 1975. A reservoir cove ecosystem model. Trans. Amer. Fish. Soc. 104:596-619.

Smith, C. C. 1940. Biotic and physiographic succession of abandoned eroded farmland. Ecol. Monogr. 10:421-484.

AQUATIC PREDATOR FEEDING IN PATCHY ENVIRONMENTS: ERROR INTRODUCED INTO MODELS BY ASSUMING PREY SPATIAL HOMOGENEITY[1]

R. T. Haar[2], G. L. Swartzman[3]

Abstract. Assuming spatial homogeneity of prey in aquatic predator-prey models leads to over estimates of predator ration. A Monte Carlo simulation was done to study the magnitude of this error assuming a patchy prey. While the error was generally found to be less than 10%, examination of several test cases based on real data showed that a predator which fed in a random manner would starve.

INTRODUCTION

The assumption of a spatially homogeneous model is almost ubiquitous in ecological simulation modeling. However, most organisms in ecosystems are patchily distributed, have onshore or offshore preferences, may prefer pelagic or demersal environments, and may migrate between habitats dirunally or seasonally.

The assumptions of spatial homogeneity introduce bias or error into the computation of the rates of interaction between compartments of an ecosystem model, i.e., nutrient uptake by phytoplankton, phytoplankton being grazed by zooplankton, zooplankton being grazed by fish. These rates form the basis for determining growth and mortality rates in most model compartments, and as such are the rates for controlling model dynamics. For example, Spigarelli and Thommes (1979) in a field study of fish moving in and out of a thermal plume showed that the migration pattern of the fish resulted in a far different energetics balance than would result if the fish were in an averaged temperature environment.

The functional response of organisms to changes in prey density are often measured in laboratory experiments and a variety of curves have been fitted to data from such experiments (Ivlev 1961; Holling 1965).

No aquatic ecosystems have uniform patterns of organisms, but at best are well mixed. In recent years a spate of literature has appeared on spatial heterogeneity and patchiness. Early work in this area appeared as theoretical discussions of 'contagious' or 'aggregated' statistical distributions (Feller 1943; Skellam 1952). Good application of these methods was made by Lloyd (1967) to soil anthropods, by Cassie and by Morisita (1959a, b, and 1962) to a variety of organisms.

The question we are addressing in this study is what effect prey patchiness has on the computation of predator feeding rates in aquatic ecosystem simulation models. Typical functions used to represent the prey density effect on ration (Andersen and Ursin 1973; Ivlev 1961), have the consumer becoming saturated with increasing ration, that is, the increase in ration per unit increase in ration per unit increase in prey density decreases with increasing prey density. Such functions (i.e. saturating functions) have the mathematical property of concavity, which allows us to utilize Jensen's inequality (Hewitt and Stromberg 1969) to show that the homogeneous model will always overestimate ration. This is true for any prey pattern where there is a stochastic element to local prey density, e.g., the predator sees different densities at different times even though the over-all mean density is constant (i.e., there is no gradient or trend). As we shall see below the magnitude of this error increases with increasing patchiness.

MODEL DESCRIPTION

Model Structure

A spatially homogeneous aquatic model assumes that every predator of a given type encounters the same prey density and that every predator of the same size has the same ration and growth rate. Results from such models purport to represent average conditions. However, the average of the responses of the predators in the system is not the response of a predator to the system average. In an attempt to understand how the above two responses differ, we have done a Monte Carlo simulation study calle MCFEED on the effect of a patchy prey density on the feed ing rate of a uniformily distributed predator. While we do not suppose, in reality, that if the prey are distributed at random, the predators will remain so, we use this assumption in the following, as a first step toward rxamining the biases in spatially homogeneous models. The effects of this assumption are discussed below.

All the predators in MCFEED are the same size (think of them as a cohort from a mixed model), and have the same feeding and growth parameters. Each predator is in a BOX of volume THETA, the interpredation of which is discussed below. There are X(i) prey, all of unit size, in the ith box, where X(i) is a random variable. The prey concentration in the ith box is denoted C(i) and is equal to X(i)/THETA. The effect of prey density on predator ration and growth rate can be functionally separate from the effects of predator body size, environmental temperature, etc., by assuming that thes are all constant over the simulation period (1 day). Then the ration and growth rate of the predator are linear functions of some f(C(i)), which we may call the prey density effect on ration in the ith box. Our conclusion that the homogeneous model always overestimates the ration arises from the overestimate of the prey density effect on ration by the homogeneous model. This in turn is a consequence of the concavity of f(C). For our simulation we used a common form of f(C), namely the Michaelis-Menton or Holling disk equation:

$$f(C) = C / [C + K3] \tag{1}$$

---

[1]Paper presented at the Third International Conference on State-of-the-Art in Ecological Modeling, Colorado State University, May 24-28, 1982.

[2]Robert L. Haar is Statistician, Center for Quantitative Science, School of Fisheries, University of Washington, Seattle, WA U.S.A.

[3]Gordon L. Swartzman is Res. Assoc. Professor, Center for Quantitative Science, School of Fisheries, University of Washington, Seattle, WA U.S.A.

In the box, define

$$Y(i) = f(C(i)) = C(i)/[C(i) + K3]$$
$$= X(i)/[X(i) + K3*THETA]. \quad (2)$$

Also, let us define E[Y] as the expectation of Y(i). Similarly, in the homogeneous model, we define

$$TY = f(MU) = MU/[MU + K3], \quad (3)$$

where MU is the mean prey concentration. For each i we can compute a bias variable

$$BY(i) = Y(i) - TY, \quad (4)$$

which, by Jensen's inequality, has an expected value, E(BY), less than zero.

Prey Distribution

Prey are distributed into NB boxes according to a negative bionomial distribution with parameters M (= MU*THETA) and K. MU is the mean prey concentration, as noted above, and THETA is the sample volume, thus the mean sample count in a sample of volume THETA becomes the product of these. The negative bionomial distribution has been found to provide an adequate statistical description for a variety of biological data (Feller 1943; Lloyd 1967; Cassie 1959b). With this parameterization the number of prey in the ith box, X(i), has a probability density function

$$Prob(X=x) = C(K+x-1,x) * [(M/(M+K))**x] \quad (5)$$
$$* [K/(M+K))**K]$$

where C(m,n) is the nth bionomial coefficient in the expansion of (x+y)**m. Under this form the expectation of X, E(X), is M, and the 1/K can be looked at as an indicator of the patchiness of the pattern that the sample is drawn from. Using an expression for variance from Lloyd (1967), we can show that as K gets very large our prey distribution approaches the Poisson distribution.

A table look-up method was used to generate the prey numbers, X(i), in each box. This involved generating the cumulative distribution for the variate, and comparing random numbers drawn from a uniform (0,1) distribution to the cumulative distribution of a negative bionomial variate to determine an integer value which then becomes the negative bionomial variate. Because the negative bionomial distribution takes on all positive integers for values, which is impossible in practice, the cumulative distribution was truncated at N*, where N* is the smaller of the N's such that the cumulative distribution function at N is within 0.1% of 1.0 or prey density effect at N was within 0.1% of saturation.

The uniform (0,1) random variables are generated using a mixed congruential random number generator (PRIME FORTRAN) that we have tested and found to be satisfactory (A. Bindman, personal communication).

PRELIMINARY RESULTS AND EXPERIMENTAL STRATEGY

A number of preliminary simulations were done to determine how large NB had to be to stabilize the values E(BY). MU was fixed at 10 cubic meters. Four sets of simulations were done with values of TY equal to either 0.20 (THETA = 1.0, K3 = 40) or 0.90 (THETA = 10.0, K3 = 1.111), and with K set to either 0.05 or 10.0. E(BY) was calculated for every 100 predators drawn at random and plotted against NB. The run with TY = 0.9 and K = 0.05 is shown in Figure 1 and is typical. It indicates that a sample of 8000 is sufficient for stability of E(BY) and, in fact, a much smaller number (2000) would be sufficient for accuracy within 5%.

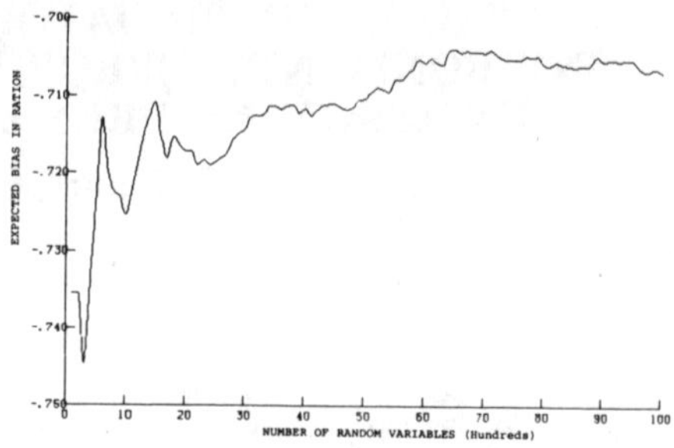

Figure 1. Value of expected bias in ration with increasing numbers of replicates.

The basic object of our study is to determine the effect of the patchiness parameter, K, on E(BY). A number of other parameters in the model also affect this expectation: MU, THETA, and K3. Quite a few simulations were done to explore how E(BY) varied in different regions of this four dimensional parameter space (MU, K, THETA, and K3). A series of simulations to determine the effect of fixing TY and K while varying MU, K3, and THETA, showed that E(BY) was invariant along hyperbolas of the form MU * THETA = M in the MU-THETA space. Analysis of the situation led to the conclusion that these were indeed invariant surfaces, thus allowing the use of three parameters, M, K and TY.

While this result aided our explorations of the behavior of E(BY) by reducing the dimension of the space, the reparameterization was not so useful for determining the biological significance of the simulation results. Towards this end we established a set of scenarios in terms of specific predator-prey interactions and interpreted the results of the parameter space explorations in terms of the interactions.

RESULTS

Parameter Space Explorations

The results of the parameter space exploration are presented in a series of tables, 1a through 1d, for prey concentrations set at different percentages of saturation. We note that as M approaches 0.0, the bias in ration, E(BY), approaches -TY, which follows from the fact that as M approaches 0.0, X takes on the value zero. This implies that Y is zero and we have the above result. Also, E(BY) goes to zero as M get large, since the larger the volume a predator searches the more it smoothes out the effect of patchiness.

Another result is that the patchiness parameter, K, has less of an effect on E(BY) than does either the mean, M, or TY, the fraction of saturation.

Estimating Model Parameters

We may view TY as a scale parameter. At prey densities high relative to the half saturation level, K3, TY is high. The mean, M, is the mean number of prey taken in a day (or a given feeding period) from a particular volume THETA at the particular value of TY used. Since M = MU*THETA, the mean number of prey in a sample of volume THETA, we can view THETA as the volume swept clear by a filter feeder or the volume

Table 1a. Absolute value of the expected negative bias in ration for average prey density at 20% of saturation.

patchiness	M (the average number of prey in a box)					
K	100.00	10.00	2.00	1.00	0.50	0.10
0.01	17.39	17.24	17.29	17.35	17.45	18.03
0.05	12.92	12.85	13.04	13.25	13.71	15.77
0.10	10.15	10.17	10.50	10.85	11.55	14.80
0.50	4.165	4.335	5.054	5.877	7.278	13.61
1.00	2.477	2.70	3.60	4.59	6.34	13.4
5.00	0.606	0.86	1.99	3.20	5.40	13.2
10.00	0.322	0.58	1.70	3.00	5.20	13.2
15.00	0.226	0.50	1.62	2.90	5.16	13.2

Table 1b. Absolute value of the expected negative bias in ration for average prey density at 50% of saturation.

patchiness	M (the average number of prey in a box)					
K	100.00	10.00	2.00	1.00	0.50	0.10
0.01	46.02	46.01	46.19	46.38	46.68	47.76
0.05	37.85	38.00	38.75	39.50	40.86	44.94
0.10	31.85	32.15	33.42	34.78	36.96	43.65
0.50	15.47	16.24	19.42	22.85	27.95	41.99
1.00	9.643	10.60	14.60	19.05	25.60	41.65
5.00	2.463	3.60	8.80	14.40	23.00	41.40
10.00	1.323	2.40	7.80	13.70	22.60	41.40
15.00	0.930	2.07	7.40	13.50	22.40	41.40

Table 1c. Absolute value of the expected negative bias in ration for average prey density at 80% of saturation.

patchiness	M (the average number of prey in a box)					
K	100.00	10.00	2.00	1.00	0.50	0.10
0.01	74.67	74.80	75.34	75.76	76.29	77.67
0.05	62.37	63.06	65.29	67.00	69.37	74.66
0.10	52.64	53.75	57.46	60.61	64.46	73.26
0.50	24.00	26.29	35.10	42.98	52.46	71.45
1.00	13.77	16.10	26.70	36.90	49.14	71.06
5.00	2.86	4.60	15.98	29.10	45.40	70.80
10.00	1.473	2.98	14.10	27.90	44.70	80.80
15.00	1.023	2.50	13.30	27.40	44.50	70.80

Table 1d. Absolute value of the expected negative bias in ration for average prey density at 90% of saturation

patchiness	M (the average number of prey in a box					
K	100.00	10.00	2.00	1.00	0.50	0.10
0.01	83.92	84.21	85.06	85.59	86.20	87.65
0.05	69.20	70.59	74.09	76.28	79.01	84.61
0.10	57.46	59.62	65.37	69.41	73.84	83.19
0.50	23.09	26.94	40.00	50.04	61.07	81.34
1.00	11.83	15.20	30.20	43.20	57.50	80.95
5.00	1.968	3.44	17.80	34.40	53.40	80.70
10.00	0.983	2.12	15.50	33.00	52.70	80.70
15.00	0.679	1.70	14.70	32.50	52.50	80.65

searched by a raptorial predator, where MU is the mean overall prey concentration. The parameter K can be estimated from multiple samples of a population, but as Morisita (1962) pointed out, the estimate depends on the number and size of the samples, as well as the actual spatial pattern of the prey.

The assumption implicit in the equation, M = MU * THETA, is that the volume THETA is taken as a random sample. This assumption is violated in natural ecosystems, as is evident from the so-called "food paradox", which is that given average prey densities the organism would not have sufficient food to meet its metabolic needs (Lasker and Zweifel 1978). Feeding at rates measured in the laboratory and on mean plankton concentrations found in the sea, some organisms would starve to death (e.g., larval anchovy, Lasker and Zweifel 1978). Stomach contents of fish larvae reflect prey densities encountered by those fish that survive. As Beyer (1976) showed, this is a biased estimate of average larval gut contents, since those larvae that died of starvation undoubtedly encountered lower prey densities. This result is a consequence of the non-uniformity of prey distribution. The assumption of random sampling is often unrealistic for adult fish, since they may search for the high density prey schools in a situation where school density and hence mean over-all ration would be significantly reduced over what they would get if the prey were spatially homogeneous. By biasing their search towards higher than average prey concentrations they would greatly increase their ration over what would be expected if they searched at random.

In the following section we will discuss reasonable ranges of parameters for MCFEED in three different predator-prey situations where the assumption of random searching is probably not unreliable. These include (1) zooplankton feeding on phytoplankton, (2) sockeye salmon parr feeding on zooplankton, and (3) larval anchovy feeding on dinoglagellates.

Zooplankton-phytoplankton Interaction

Cassie (1959b) estimates K for a number of marine plankters. While most of these were for zooplankton, he obtained three estimates of K for Coscinodiscus, a large diatom. The values obtained ranged from 10 to 14.3. At the same time he obtained estimates of MU which ranged from 20,000 to 161,000 cells/cubic meter, with a mean of 88,000 cells/cubic meter.

Values of volume swept clear for a calanoid copepod ranged from 1-4 ml/hr (Lam and Frost 1976). Using these values for reference, we parameterized MCFEED as follows:

THETA = 0.00003 cubic meters/day
K3 = 100,000 cells/cubic meter.

Two series of simulations were done; one with K = 10 and one with K = 15. In each series, the bias in ration was estimated at values of MU ranging from 10,000 to 200,000 cells/cubic meter. The results were similar for both series and are shown in Figure 2 for K = 10. At mean phytoplankton densities (88,000 cells/cubic meter) a bias of just over 5% is indicated.

Sockeye Parr Feeding in Lake Washington

Based on data from Eggers (1975) for sockeye salmon parr feeding on Cyclops and Diaptomus in October in Lake Washington, THETA was estimated as 1000 liters per day and MU as 8.5 individuals per liter. Since the volume searched, THETA, was not directly available, a rough

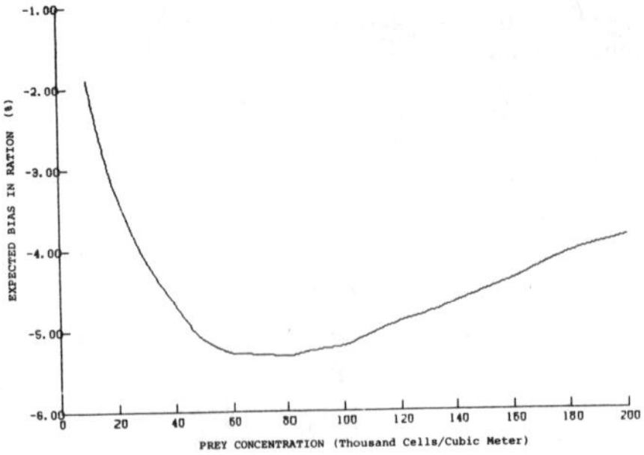

Figure 2. Bias in ration as a function of prey density for k = 15.0.

estimate was made by dividing stomach contents, M, by MU. The K value was taken from Cassie's (1959b) values for marine crustacea which ranged from 1.0 to 8.0. The half saturation level, K3, for sockeye parr is not known and thus values were chosen at three levels giving prey density effects TY = 0.2, 0.5, and 0.8.

Only at the smallest value of K (greatest patchiness) is the bias significant. The ration biases are shown in Table 2 for different values of K and TY.

Table 2. Expected bias in ration for sockeye parr feeding on zooplankton. Results are in percentage of absolute error (* - extrapolated from runs with smaller M's).

TY	K		
	1.0	5.0	10.0
0.2	<2.0	<0.5	<0.2*
0.5	<8.0	<2.0	<1.0
0.8	<10.0	<2.0	<1.0

Larval Anchovy Feeding

Parameters for anchovy and their prey were estimated based on data from Hunter's (1972) laboratory and feeding experiments. Anchovy were fed on Gymnodinium and rotifers at a concentration, MU, of 200 cells per milliter. A one centimeter fish searched a volume of 12 liters in a 12 hour feeding period (extrapolated from one liter per hour). This gave a value of M of 2.4 million cells per day -- a very large quantity. Due to the impracticality of running MCFEED with such large values of M we did not do this run. However, with values of K from 5 to 15 it is unlikely that the bias in this case ever exceeds 2%.

Vlymen (1977) constructed a model for larval anchovy feeding in the above case when a particular micro-patch pattern is assumed. He derived values corresponding to the "K" parameter of the negative bionomial distribution which vary between 0.001 and 0.15. However, it is unlikely that they are realistic. Vlymen did not assume random search by anchovy larvae but rather that the probability of staying in a patch is related to the patch density - a non-random searching property. We think that Vlymen's model relates to patch feeding at a different grain than we are considering here when dealing with anchovy feeding on a daily basis. Our THETA is 12 liters. In MCFEED we generated a prey density seen by a single larvae and replicated it many times to represent different prey densities encountered by many larvae over a large area. Vlymen confined all feeding to a single 12 liter volume (of mean overall density) with patches conceived of as embedded within those 12 liters. Thus Vlymen's model deals with patch feeding within the patches we are generating -- sort of micro-grain patchiness. Patchiness of the sort Vlymen is dealing with is difficult to detect and has not been demonstrated from field samples.

CONCLUSIONS

Results of applying MCFEED to real predator-prey interactions demonstrate that, although high bias from ration computation can arise, in most cases where random feeding is a plausible assumption, the biases are low, always less than 10% and often less than 2%. This bias is inconsequential when it is considered that the quantitative effect of prey density on feeding rate is not accurately known. Parameters for prey density effect on ration are usually estimated by model calibration, and even where estimates from data exist, they are extremely variable (Kremer and Nixon 1978). One source of this variability may be the patchiness of the prey, however the main source of variability is most likely due to the various searching behaviors of predators, which probably are not random. As discussed earlier with the feeding paradox, most predators find ways of putting themselves into higher than average prey densities while feeding. The method by which this is accomplished depends on the resources of the predator, the nature of the prey pattern, and the causal mechanisms which control prey pattern.

General statistical distributions used to describe prey pattern (like that considered in MCFEED) are inadequate for use in models which have predation dependent upon prey configuration or pattern, since many configurations can lead to the same statistical description (Lloyd 1967). This type of statistical model also assumes random searching by the predator, an assumption which would be hard to support for many important predators. On the other hand, those few models which have predators searching for prey (Eggers 1975; Kerr 1971), generally assume that the prey are distributed at random. While this is better than the homogeneity (uniformity) assumption, prey patchiness can invalidate conclusions drawn from this type of model.

Vlymen's (1977) model of larval anchovy feeding considered both non-random sampling by the predator and non-random distribution of prey. Vlymen synthesized a large amount of detailed information about larval anchovy searching and feeding behavior from Hunter's (1972) experimental work, but the model of prey distribution that he used was purely speculatibe and was based on a statistical description of prey pattern which was inadequate with regard to yielding information about pattern structure. Vlymen's work illustrated the need for very careful and detailed observation of both prey pattern and predator behavior before an adequate synthesis can be made.

While Hunter's (1972) work on anchovy feeding behavior is an excellent example of the degree of process elucidation required, detailed work on pattern is not well developed. Figure 3 shows a classification of processes leading to patch formation and stabilization. Work in this area has thus far been the province of the biological oceanographers, who concentrate on the interactions between spatially passive biological processes and physical processes operating at a space scale of kilometers (Kierstead and Slobodkin 1953; Wroblewski, et al. 1975; Dubois 1975). Models of patchiness incorporating these processes at this scale portray predator-prey interactions as being dependent

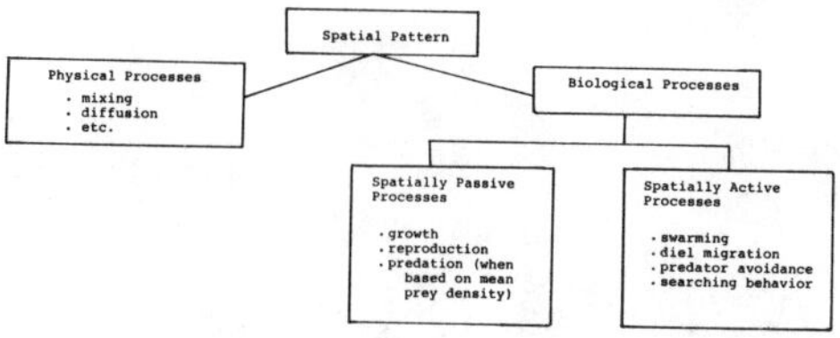

Figure 3. Hierarchial classification of some of the causal mechanisms contributing to the formation and stabilization of spatial pattern in aquatic systems.

on the assumption that the predator sees a uniform distribution of prey but that the mean concentration changes over space. This sort of patchiness model is useful for understanding spatial variability of oceanographic observations but sheds little light on the effects of patchiness on predation processes. There is a need for model approaches which consider the role of spatially active processes in predation since these processes control the consumption of the predator as well as the mortality of the prey.

LITERATURE CITED

Andersen and Ursin. 1973. A multi-species extension to the Beverton and Holt theory of fishing, with accounts of phosphorus circulation and primary production. ICES 1973/H:20. 49 pp.

Beyer, J.E. 1976. FISH 1 - Survival of fish larvae - a single server queue approach. Inst. Math. Stat, Oper, Res., Rech. Inst. of Denmark.

Cassie, R.M. 1959a. An experimental study of factors inducing aggregation in marine plankton. N.Z.J. Sci. 2:339-365.

Cassie, R.M. 1959b. Micro-distribution of plankton samples. N.Z.J. Sci. 2:436-448.

Cassie, R.M. 1959c. Some correlations in replicate plankton samples. N.Z.J. Sci. 2:436-448.

Cassie, R.M. 1962. Frequency distribution models in the ecology of plankton and other organisms. J. Anim. Ecol. 31:65-92.

Dubois, D.M. 1975. Simulation of the spatial structuration of a patch of prey-predator plankton populations in the southern bight of the North Sea. Mem. Soc. Roy. Sci. Liege VII(6):75-82.

Eggers, D.M. 1975. A synthesis of the feeding behavior and growth of juvenile sockeye salmon in the limnetic environment. Pd.D. Dissertation, Univ. of Wash., Seattle.

Feller, W. 1943. On a general class of "contagious" distributions. Ann. of Math. Stat. 14:389-400.

Holling, C.S. 1965. The functional response of predator to prey density and its role in mimicry and population regulation. Mem. Ent. Soc. Canada #45. 60 pp.

Hewitt, E., and K. Stromberg. 1969. Real and Abstract Analysis. Springer-Verlag.

Hunter, J.R. 1972. Swimming and feeding behavior of larval anchovy Engraulis mordax. Fishery Bulletin 70(3):821-838.

Ivlev, V.S. 1961. Experimental ecology of the feeding of fishes. New Haven, Conn., Yale Univ. Press.

Kerr, S.R. 1971. Prediction of fish growth efficiency in nature. J. Fish. Res. Bd. Canada 28(6):809-814.

Kierstead, H., and L.B. Slobodkin. 1953. The size of water masses containing plankton blooms. J. Mar. Res. 12(1):141-147.

Kremer, J.N., and S.W. Nixon. 1978. A Coastal Marine Ecosystem. Springer-Verlag.

Lam, R.K., and B.W. Frost. 1976. Model of copepod filtering response to changes in size and concentration of food. Limn, Oceano. 21:490-500.

Lasker, R., and J.R. Zweifel. 1975. Growth and survival of first-feeding northern anchovy larvae (Engraulis mordax) in patches containing different proportions of large and small prey. In (J. Steele, ed.) Spatial Pattern in Plankton Communities. Plenum Press.

Lloyd, M. 1967. "Mean Crowding". J. Anim. Ecol. 36:1-30.

Morisita, M. 1959a. Measuring of the dispersion of individuals and analysis of the distributional patterns. Mem. Fac. Sci. Kyushu Univ., Ser. E (Biol.) 2(4):217-235.

Morisita, M. 1959b. Measuring of interspecific association and similarity between communities. Mem. Fact. Sci. Kyushu Univ., Ser. E (Biol.) 3(1):65-80.

Morisita, M. 1962. I_δ - Index, a measure of dispersion of individuals. Res. Popul. Ecol. IV:1-7.

Skellam, J.G. 1952. Studies in statistical ecology. I. Spatial Pattern, Biometrika 39:346-362.

Spigarelli, S.A., and M.M. Thommes. 1979. Temperature selection and estimated thermal acclimation by rainbow trout (Salmo gairdneri) in a thermal plume. J. Fish. Res. Bd. Canada 36:366-376.

Vlymen, W.J. 1977. A mathematical model of the relationship between larval anchovy (Engraulis mordax) growth, prey microdistribution, and larval behavior. Env. Biol. Fish. 2(3):211-233.

Wroblewski, J.S., J.J. O'Brien, and T. Platt. 1975. On the physical and biological scales of phytoplankton patchiness in the ocean. Mem. Soc. Roy. Sci. Liege VII(6):43-57.

OPTIMUM ALLOCATION OF SAMPLING EFFORT: MINIMIZING THE UNCERTAINTY IN A CALCULATED RESULT[1]

Robert L. Hays[2]

Abstract.--The problem addressed is how to optimally allocate fixed total effort for inventory so as to minimize the uncertainty in the output of any deterministic model used to analyze the inventory results. It is assumed that the independent variables are statistically independent. The propagation of error equation was used to estimate the uncertainty in the dependent variable. It was modified so that the sample size for each independent variable was a simple function of the effort allocated to that variable. In addition, a numerical approximation was developed for use on variables whose effect on the dependent variable was highly nonlinear. The approach was initially tested using a simple polynomial equation, and also using a habitat suitability model for the pronghorn, with one cover type and four habitat variables. The shape of the objective function's surface in hyperspace was found to be a broad, flat-bottomed bowl. It is argued that this is likely to be a general shape. Plans for continued work and potential extensions are discussed. It is concluded that the approach shows promise.

INTRODUCTION

It is widely recognized that today's decisionmakers are faced with alternatives that will have large impacts on society and natural resources and bear high risks. To minimize these risks, enormous amounts of time and money are being spent by organizations of all types on the collection and analysis of data. However, it is a common complaint that the information provided to the decisionmakers is incomplete, irrelevant, or of low or dubious reliability. This implies that much room for improvement exists in decision support activities.

For certain decision processes, lead times and budgets are sufficient to permit the collection of new data. If we set a goal of providing the best possible information to the decisionmaker(s), we have the problem of identifying the optimum allocation of the available funds. Organizations strive to maximize their "gains" minus "costs", however defined. It is obvious that there must be an optimum allocation of an organization's resources to the catagory of information collection and processing to support decisionmaking versus all other competing uses of these resources. There is a curious dearth of solid guidance on how this optimum can be determined. An attractive explanation for this lack is that the components of "gains" and/or "costs" are hard to measure or to relate to the level of investment in information collection and processing.

In any case, managers do budget fixed resources for decision support. How should these resources be used? Conventional wisdom recommends that an information needs analysis be conducted. This is supposed to identify precisely what items of information are needed by the decisionmaker to weigh alternatives. Information needs analyses also have widely recognized limitations, in that it is often impossible to anticipate what alternatives must be evaluated and what criteria should be used (for example, Braybrooke and Lindblom 1963). In spite of the limitations, an organization must have defined targets of specific information needs in order for a decision support system to be effective. Of course, it is possible to plan optimum allocation only if such targets exist.

Given a list of needed information items, the problem becomes that of selecting analytical models and developing the values of the independent variables used as input to these models. Often there are alternative models that estimate the same dependent variable, but require different inputs. Typically, also, data exist on which to base estimates of some or all of the input variables. These data are often old or were collected in a similar but different area. At this point, the manager is normally advised to exercise good judgement in selecting models and allocating effort for sampling among the input variables. He is also advised to pay particular attention to the reliability of existing data, the cost of providing the specific data required by the alternative models, and the importance of the information item produced as output by each model. That is, to subjectively solve a mind-boggling optimum allocation problem that is not even precisely stated.

The "minimum cost plus loss" approach has been developed to determine the optimum sample size for estimating the value of a single variable (Yates 1960;

[1]Paper presented at the Third International Conference on State-of-the-Art in Ecological Modelling, Colorado State University, May 24-28, 1982.

[2]Robert L. Hays is Plant Ecologist, Western Energy and Land Use Team, USDI Fish and Wildlife Service, Fort Collins, Colorado U.S.A.

Cochran 1977). It balances the costs of data collection against the expected losses predicted if the estimate of the variable is in error. A few simple cost and loss functions are suggested. Hamilton (1979) discussed five common forms of loss functions. He also stressed that optimum sampling design is sensitive to changes in the cost and loss functions used.

A landmark paper by Burkhart et al. (1978) extended this approach to deal with multiple variables, inventories, and decisions. These authors were able to analyze a rather complex optimum allocation problem to plan the inventory of a sizable portion of a National Forest. They were able to deal with many variables and analytically determine some solutions that addressed a real management problem. However, for all its strengths, their approach suffers from several weaknesses which limit the breadth of its applicability. The technique requires deterministic quantitative models of the expected loss for any given degree of error in the estimation of each variable. Not only are few data available on which to base such models, but there is also a lack of methods or theory to guide development of them on a routine basis. Secondly, the optimum allocation solution found is not constrained by budget limitations. If adequate funds are not available to implement the chosen scheme, it is not clear how to optimally reduce the sampling effort. Thirdly, the technique assumes that the only source of uncertainty is sampling error. Thus, an "ocular estimate" of a variable, such as the site index for a forest stand, is treated as if it were just as reliable as an estimate based on extensive quantitative measurements. In addition to differing in precision and accuracy, these sampling methods also differ greatly in unit cost. Fourthly, each "variable" is treated as a single measurement even though some of these "measurements" (e.g., site index) are actually the results of analyses of raw data using simple models. Methods to optimize sampling of the variables used as input for any such simple models used (e.g., timber volume as a function of stand density, D.B.H, or tree height) are not addressed. Fifthly, a certain degree of roughness in the method still exists. Multiple ways of sampling are evaluated, but the most flexible one is found to be less effective than one with an arbitrary constraint on sampling. This suggests that the method contains an unrecognized defect. Also, the use of statistics based on infinite populations must be forced to fit the finite populations actually dealt with in the example. Lastly, the assumption that many variables can be assigned fixed values (e.g., the variance in measurements of site index is assumed to be constant across all stands) is weak.

An approach which completely ignores losses was developed by statisticians years ago. These techniques optimize sampling for one variable (e.g., sophisticated stratification). Because these tools have been widely known, it is an appealing approach to arbitrarily reduce the complexity of the problem to that level. Hence, managers are asked to state how precise an estimate is needed for each input variable, or how many samples of each variable they can afford. Given these inputs, the statisticians can invoke sophisticated, quantitative computer programs to provide impressive and precise sampling designs. Hamilton (1979) holds that such precision or sample size targets are usually determined subjectively. Unfortunately, because there is no reason to believe that these subjectively determined targets are optimum, one cannot assume that sampling designs based upon them are optimum in terms of the overall information need.

Another approach for optimizing the allocation of sampling effort is to rely on experience from similar studies done in the past. Reviewing the results of one study can help in planning for future studies if one can identify important information items that were previously ignored. However, there exist strong biases that work against developing the optimum allocation of effort by trial and error. These include the fact that uncertainty estimates (e.g., confidence limits) are often calculated for the input variables, but not for the final calculated results (i.e., output of the analytical models). A variable with a high uncertainty estimate appears as weak link and tends subsequently to receive greater share of the budget, even though the final calculated results may not be very sensitive to changes in this variable. Secondly, in large organizations, there is often inadequate feedback from the ultimate information user to the technical people who allocate effort at the data collection level. These individual technical specialists have little opportunity to grasp the overall problem. In addition, they are usually discipline-oriented and are naturally biased in favor of enlarging the budgets supporting their own discipline's activities. There may also be a bias toward favoring their discipline if by so doing there are potentially greater personal rewards.

The above factors seem to me to be major contributors to the deficiencies in decision support systems. There is a major need for an objective and quantitative method for estimating the optimum allocation of a fixed budget for data collection among the variables that need to be measured. It should be possible to apply this method without explicit loss models, but the method should provide the estimates of uncertainty needed to apply loss models if they are available. Such a method is needed to support the design of data collection for research purposes, as well as for applying models in decisionmaking. This paper presents the status of an effort whose objective is to develop such a method for single output deterministic models.

The balance of this paper is organized as follows: The problem is stated mathematically, and a theoretical solution is proposed. The limitations of this solution are identified, and a numerical method is identified for circumventing a common limitation. Next the results of test applications are described. Finally, there is a discussion of the nature of the general solution, work in progress or planned, and potential extensions of this method and its implementing software.

THEORY

Assume that there is a need to estimate the true value of a variable Υ for a particular area at a particular time. (Symbols are given in table 1.) Let this estimated value be $\tilde{Y}$. Also, let there be certain other variables in that area at that time that can be described by their true values μ_i (i.e., their population means). Let the estimate of μ_i a posteriori (i.e., after the inventory) be the mean of the inventory sample ($\bar{x}_i$). Further, let there exist a deterministic model, the "analysis function", that relates $\tilde{Y}$ to m of the $\bar{x}_i$:

$$\tilde{Y} = f(\bar{x}_1, \bar{x}_2, \bar{x}_3 \dots \bar{x}_m) \quad . \tag{1}$$

Table 1.--Symbols.

	A priori estimate	A posteriori estimate	"True" population parameter
Information desired	$\hat{Y}$	$\tilde{Y}$	Υ
Average value of an independent variable	$\hat{x}_i$	$\bar{x}_i$	μ_i
Uncertainty in the estimated dependent variable at probability p	$\pm\hat{\delta}_{\hat{Y}}$	$\pm\hat{\delta}_{\tilde{Y}}$	
Uncertainty in the estimated independent variable at probabiliy p	$\pm\hat{\delta}_{\hat{x}_i}$	$\pm\hat{\delta}_{\bar{x}_i}$	
Standard error of the mean of an independent variable	$\hat{S}_{\bar{x}_i}$	$S_{\bar{x}_i}$	
Standard deviation of sample values of an independent variable	$\hat{S}_{x_i}$	S_{x_i}	σ_i
Value of a single individual measurement (or single sample) of an independent variable		x_i	

Let equation (1) be assumed valid for the decision-making process at hand, (i.e., $\Upsilon = \tilde{Y}$). Also, let there be an estimate of μ_i:$\hat{x}_i$ (which could come either from a small preliminary sample, or by extrapolation of data from another place or time). Next, for each $\hat{x}_i$, let there be a scaler estimate of the uncertainty $\hat{\delta}_{\hat{x}_i}$ (same units as $\hat{x}_i$) at a stated probability p:

$$\hat{x}_i \pm \hat{\delta}_{\hat{x}_i} \text{ at p} \quad . \tag{2}$$

For example, 103.6 m ± 1.3 m at 95%, which is a calculated sample mean ± 1.96 times the standard error. The value of $\hat{\delta}_{\hat{x}_i}$ is a function of both the probability density distribution of $\hat{x}_i$ and the value of p. Since the former characteristic is of interest here, let p be a constant across all variables. Furthermore, let the amount of money to be spent during the inventory on the collection of data to calculate the $\bar{x}_i$ be C_i, with total costs (T) constant.

The problem is, minimize the estimated uncertainty in $\tilde{Y}$ ($\hat{\delta}_{\hat{Y}}$ at p) subject to $C_i \geqq 0$, and total costs = T. Note that the description of uncertainty $\hat{\delta}_q$ is used here as a kind of precision limit in the sense that if individual estimates of q were made (e.g., means from many separate sampling replications), each estimate would have a probability of p of falling within $\pm\hat{\delta}_q$ of the value found from one sample (Ku 1966).

What remains is to relate C_i to n_i, n_i to $\hat{\delta}_{\hat{x}}$ and $\hat{\delta}_{\hat{x}_i}$ to $\hat{\delta}_{\hat{Y}}$ and to identify an algorithm to find the minimum.

Relating C_i to n_i

Several models exist to predict the cost of sampling n_i values of x_i. The simplest is $C_i = n_i c_i$, where c_i is the unit cost of measuring x_i (in units of money or work time). Cochran (1977) discusses these and more complex and even non-linear cost functions. The choice of a cost function should be determined by the conditions at hand. However, for simultaneous sampling of multiple variables, it should often be adequate to use equations of the form $C_i = n_i c_i$, with:

$$T = \sum_{i=1}^{m} C_i \quad .$$

If there is a fixed cost associated with the total sampling effort, this can be easily accommodated by subtracting the fixed cost from the budget, leaving T as the amount remaining for allocation among variables to be inventoried.

Solving for n_i gives:

$$n_i = C_i/c_i \quad . \tag{3}$$

I will restrict further discussion here to the cost function of equation (3), although any other function could be readily substituted if it can be solved in terms of n_i.

Relating n_i to $\hat{\delta}_{\hat{x}}$

It is standard statistical practice to use the mean of the sample ($\bar{x}_i$) to estimate the population mean (μ_i). It is also standard to use the 95% confidence limits on the standard error of the mean ($\pm t S_{\bar{x}_i}$) as the estimate of the uncertainty in x_i ($\pm\hat{\delta}_{\bar{x}_i}$ at 95%). The standard error of the mean is calculated as $S_{\bar{x}_i} = S_{x_i}/\sqrt{n_i}$, where S_{x_i} is the standard deviation of the sample and n_i is the number of estimates in the sample. One can use these same relationships to predict the uncertainty in the $\bar{x}_i$ which will be calculated. Let $\hat{S}_{\bar{x}_i}$ be the estimate of $S_{\bar{x}_i}$. Also, let the estimate of $\hat{S}_{\bar{x}}$ be $\hat{S}_{x_i}/\sqrt{n_i}$ with $\hat{S}_{x_i}$ and n_i being estimates of S_{x_i} and n_i. This gives:

$$\hat{\delta}_{\hat{x}} = t\ \hat{S}_{x_i}/\sqrt{n_i} \quad . \tag{4}$$

Relating $\hat{\delta}_{\hat{x}_i}$ to $\hat{\delta}_{\hat{Y}}$

A posteriori we can use the classical propagation of error equation:

$$\hat{\delta}_{\tilde{Y}} = \left[\sum_{i=1}^{m} \left(\frac{\partial \tilde{Y}}{\partial \bar{x}_i} \right)^2 \left(\hat{\delta}_{\bar{x}_i} \right)^2 \right]^{\frac{1}{2}} \quad . \tag{5}$$

where,

$\hat{\delta}_{\tilde{Y}}$ = estimated uncertainty in $\tilde{Y}$ at a probability of p (with the partial derivatives evaluated at the $\hat{x}_i$)

$\hat{\delta}_{\bar{x}_i}$ = estimated uncertainty in $\bar{x}_i$ (Ku 1966, after Cramer 1946).

Ku (1966) concluded that these relationships are adequate for practical purposes, as long as the following conditions are met:

1. Each standard deviation is small in comparison with its respective mean;
2. "Second and higher order partial derivatives evaluated at their means do not give rise to abnormally large measures" (p. 272);
3. The n_i are large. [For small n, Ku (1966) concluded that the relationship holds but that the distribution of $\tilde{Y}$ is better approximated by a t distribution with an effective degrees of freedom calculated after Welsh 1947];
4. The errors in measurement of the individual $\bar{x}_i$ are statistically independent [Ku (1966) also gives an equation for explicitly considering statistical dependence]; and
5. The $\bar{x}_i$ are normally distributed.

Box et al. (1978:566) pointed out that it is preferable to estimate the partial derivatives numerically even when they can be found analytically because this permits a simple check for nonlinearity [and, hence, a violation of Ku's (1966) second condition].

Moffat (1978) discussed in detail the use of this approach to estimate uncertainty in experimental results. He and Box et al. (1978) also pointed out the utility of equation (5) in optimizing the design of single sample experiments. The problem of how to accommodate functions that violate Ku's second condition was addressed (see below).

We can apply equation (5) by substituting into it estimates and equations (3) and (4), giving us the "objective function" for optimization:

$$\hat{\delta}_Y = \left[\sum_{i=1}^{m} \left(\frac{\partial \hat{Y}}{\partial \hat{x}_i} \right)^2 \left(\frac{\sqrt{c_i}\ t\ \hat{S}_{x_i}}{\sqrt{C_i}} \right)^2 \right]^{\frac{1}{2}} \quad , \tag{6}$$

where,

$\partial\hat{Y}/\partial\hat{x}_i$ = the partial derivative of equation (1), using $\hat{Y}$ for Y and $\hat{x}_i$ for $\bar{x}_i$, and evaluated at $\hat{x}_i$.

If it is an acceptable assumption that t can be considered constant for planning purposes, it can be shown (Appendix 1) that the minimum of $\hat{\delta}_{\hat{Y}}$ occurs at:

$$C_i = \frac{T\ \frac{\partial \hat{Y}}{\partial \hat{x}_i}\ \sqrt{c_i}\ \hat{S}_{x_i}}{\sum_{j=1}^{m} \frac{\partial \hat{Y}}{\partial \hat{x}_j}\ \sqrt{c_j}\ \hat{S}_{x_j}} \quad , \tag{7}$$

Of course, initial estimates of $\hat{x}_i$, c_i, and $\hat{S}_{x_i}$ must be available. The partial derivatives can be evaluated numerically.

Departures from Ku's Conditions

Often, in practical application, Ku's (1966) second condition (i.e., a reasonably linear function) is not met. This is true because the function often is highly nonlinear (or discontinuous or stepwise) over a range of $\bar{x}_i \pm 3S_{\bar{x}_i}$.

Box et al. (1978) suggested the following test for adequate linearity. Evaluate the function at $\bar{x}_i$, and then at $\bar{x}_i \pm 3S_{\bar{x}_i}$. Inspect these results for obvious nonlinearity. It is easy to use a quantitative criterion based on fitting a line to the function evaluated at $\bar{x}_i$ and $\bar{x}_i - 3S_{\bar{x}_i}$. If the difference between the expected value from that line at $\bar{x}_i + 3S_{\bar{x}_i}$ and the function evaluated at $\bar{x}_i + 3S_{\bar{x}_i}$ is less than a given percent of the function's value at $\bar{x}_i$, adequate linearity can be assumed. This test could fail sometimes if the function strongly fluctuates up and down over this interval. In this situation, it might appear to be linear if only three points are examined.

Inspection of equation (6) shows that $\hat{\delta}_{\hat{Y}}$ is made up of contributions from each $\bar{x}_i$. The contribution is a measure of the uncertainty in $\bar{x}_i$, times the sensitivity of $\hat{Y}$ to changes in $\bar{x}_i$ (using the partial derivative as a linear approximation). A numerical approximation (the "propagated distribution approximation") was developed for estimating a scaler contribution for an x_i analogous to that in equation (6) for use when the partial derivative is not a good approximation (fig. 1).

Assume x is a normally distributed population of means of samples of size $n = C_i/c_i$, with mean $\bar{x} = \hat{x}_i$ and standard deviation $S_{\bar{x}} = \hat{S}_{\bar{x}_i}$. Consider that portion of the distribution in the interval $\bar{x} + wS_x$. Divide the interval $\bar{x} \pm wS_x$ into K equal intervals symmetrical around $\bar{x}$. Let the expected relative frequency of x in interval j (F_{xj}) be approximated by the width of the interval times the value of the normal curve at the midpoint of the interval:

$$F_{xj} = \left(\frac{2wS_x}{K}\right) \frac{1}{S_x\sqrt{(2\pi)}} \text{EXP}\left[-1/2\left(\frac{x_j - x}{S_x}\right)^2\right] \quad (8)$$

Next, the value of x_j is transformed into the equivalent value of Y (Y_j) by substituting in x_j for the appropriate $\bar{x}_i$ in equation 1 (while using $\hat{x}_i$ for all other $\bar{x}_i$). Then, by convention, the mean of the resulting distribution of Y values is (approximately):

$$\bar{Y} = \sum_{j=1}^{K} Y_j F_{xj} \ , \quad (9)$$

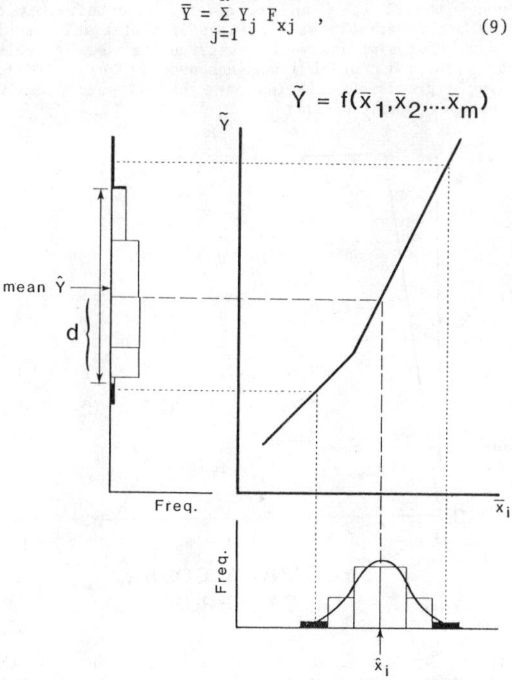

Figure 1.--Propagated distribution approximation (see text).

Next, the K individual segments of the Y distribution are ordered in terms of decreasing distance from $\bar{Y}$. Then, summation of small increments of area are calculated until the area exceeds 1-p. Linear interpolation is used to estimate the precise value of d, such that $\bar{Y} \pm d$ excludes 1-p. The value of d^2 can be substituted for the ith term in equation (6), as the contribution of x_i to the summation.

To determine appropriate values for the parameters (w and K), nine functions were evaluated. (In this evaluation, I dealt with t = 1 for simplicity. Hence $\delta_Y = S_Y$, and $\delta_x = S_x$.) For the simple linear function $Y = f(x) = x$, δ_Y was known to be equal to δ_x. A criterion of $\delta_Y = \delta_x \pm 5\%$ was used, and it was found that w = ±3Sx and K = 14 were adequate.

The other eight functions included the sigmoid curve, "plateau" and "mountain" shaped stepwise linear functions, and even a step function. The standard was taken to be the value of d calculated using w = ±5 Sx and K = 300. It was concluded that the propagated distribution approximation was within ±5% with w = ±3Sx and K = 14, except in the following cases:

1. Y is essentially constant over x, except that the tip of one tail extends beyond a "cliff" [Y=f(x) has a very large slope in that region];
2. Y = f(x) is a step function; or
3. Y = f(x) is a sawtooth pattern with a "wave length" approximating $(wS_x)/K$.

Of course, the necessary degree of precision for these unusual cases can be produced by increasing w and K.

This approximation can be used in conjunction with equation (7) as follows: The contribution to $\hat{\delta}_Y^2$ in equation (6) due to the ith term in the summation (d_i^2) is given by:

$$d_i^2 = \left(\frac{\partial\hat{Y}}{\partial\hat{x}_i} \frac{\sqrt{c_i}\ t\ \hat{S}_{x_i}}{\sqrt{C_i}}\right)^2 = \left(\frac{\partial\hat{Y}}{\partial\hat{x}_i}\hat{S}_{x_i}\right)^2\left(\frac{\sqrt{c_i}}{\sqrt{C_i}}\right)^2$$

Let

$$r_i = \frac{\partial\hat{Y}}{\partial\hat{x}_i}\hat{S}_{x_i}$$

Substituting gives:

$$d_i^2 = r_i^2\left(\frac{\sqrt{c_i}}{\sqrt{C_i}}\ t\right)^2$$

Taking the square roots and rearranging gives:

$$r_i = \left(\frac{\sqrt{c_i}\ t}{\sqrt{C_i}}\right)^{-1} d_i = \frac{d_i\ \sqrt{C_i}}{t\ \sqrt{c_i}} \ . \quad (10)$$

If an initial set of C_i (C_i^0) are used, the propagated distribution approximation can be used to calculate d^2. All other terms on the right side of equation (10) are known for this set of C_i^0, so r_i can be calculated. This value of r_i can be substituted into equation (7) for the analogous values of $(\partial\hat{Y}/\partial\hat{x}_i)\,\hat{S}_{x_i}$ and $(\partial\hat{Y}/\partial\hat{x}_j)\,\hat{S}_{x_j}$:

$$C_i = \frac{T\sqrt{c_i}\left(\dfrac{d_i\sqrt{C_i^0}}{t\sqrt{c_i}}\right)}{\sum\limits_{j=1}^{m}\left[\sqrt{c_j}\left(\dfrac{d_j\sqrt{C_j^0}}{t\sqrt{c_j}}\right)\right]}$$

Assuming t is constant, and cancelling terms gives:

$$C_i = \frac{T\, d_i\sqrt{C_i^0}}{\sum\limits_{j=1}^{m}(d_j\sqrt{C_j^0})} \quad . \qquad (11)$$

However, there is no reason to believe that this approximation will hold over a broad range of C_i. Hence equation (11) should be used iteratively to get increasingly better estimates of the optimum C_i.

Because equation (11) is an approximation, it may sometimes produce overshooting of the optimal C_i. To prevent this, less extreme adjustments in the C_i at each iteration can be used:

$$C_i^{0+1} = \frac{(C_i + uC_i)}{(1 + u)} \quad . \qquad (12)$$

where,

C_i^{0+1} = the value used

C_i = the estimate produced by equation (11)

u = an arbitrary weighting coefficient. As u increases the change in the C_i becomes more conservative.

TEST APPLICATIONS

Equation (7) was applied to the following function:

$$Y = x_1^3 + 2x_2^2 + 3x_3$$

The following parameters were used: $x_1 = 0.5$, $x_2 = 1$, $x_3 = 0.25$, $S_{x^1} = S_{x^2} = S_{x^3} = 1$, $c_1 = c_2 = c_3 = 1$, and $T = 3$. Initial values of $C_i = 1$ were used. Partial derivatives were estimated numerically using a 5% increment. One iteration was adequate to calculate the optimum allocation correct to at least 4 decimal places. Next, equation (12) was applied with u = 1. The correct allocation was found to ±2% with four iterations. Applying the propagated distribution approximation with w = 3 and K = 14 gave very similar results.

A second test was conducted that was more realistic and quite demanding. In this case, the objective was to optimize sampling of four variables in one cover type used in a deterministic analytical model for the habitat quality for the pronghorn. The model (Allen, in prep.) was simplified slightly to exclude variables that were irrelevant for lands of the Black Butte Coal Company, Point of Rocks, Wyoming. Actual field data on the vegetation and landform of that site were used (Allen, in prep.). The standard deviations of sample values were about one-half to slightly greater than the sample means.

The pronghorn model is sufficiently complex that its output cannot be estimated intuitively. The model includes three levels of hierarchical equations that involve finding the minimum of weighted geometric means of intermediate function values. At the lowest level in the hierarchy, stepwise linear functions of single habitat variables are used (e.g., fig. 2). There is no obvious simple way to anticipate the optimum allocation of sampling effort to minimize the uncertainty in the final calculated habitat suitability index.

Simple and very approximate cost models were used, with the unit cost of three variables taken as equal to 2/3 of a unit of effort. The unit cost of the other variable was taken as 1/10 of a unit of effort. For simplicity, t was given the constant value of 1.96. The problem was encoded in Fortran IV for single precision solution on a 64 bit word size CDC computer (machine epsilon $\cong 10^{-14}$).

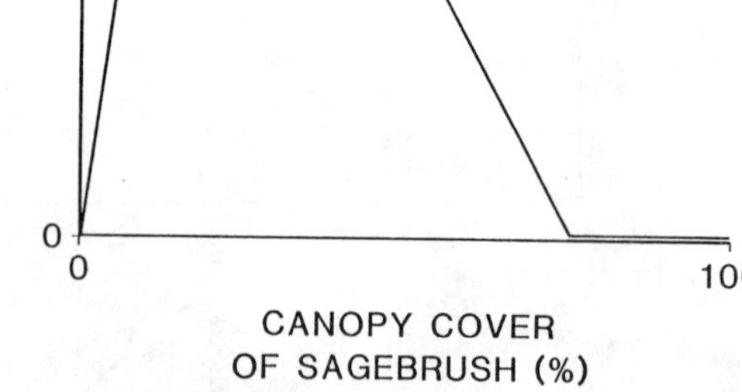

Figure 2.--A typical intermediate function from the lowest level in the hierarchy of the pronghorn habitat suitability model. $\tilde{Y}$ = f[g(canopy cover of sagebrush, h(canopy cover of grass), ...]

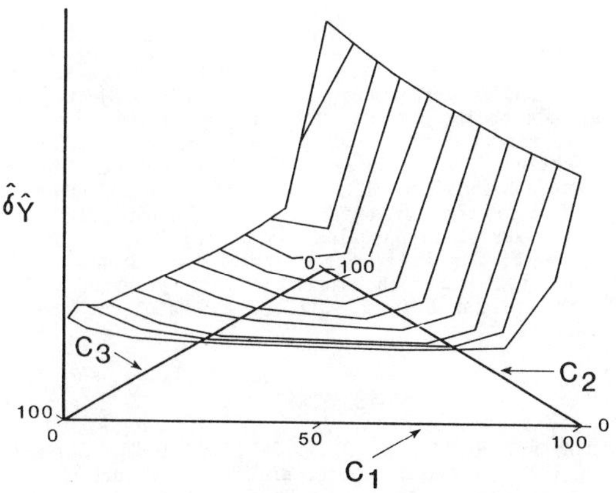

Figure 3.--A priori estimated uncertainty in the habitat suitability of pronghorn as a function of variable allocations of effort in sampling independent variables. (Calculated estimate points are connected by straight lines.)

Preliminary evaluation of the objective function [eq. (6)] for these data and arbitrary C_i values, using the propagated distribution approximation throughout, showed that $\hat{\delta}_{\hat{Y}}$ received a negligible contribution from one variable. Subsequently, that variable was removed from consideration in optimization (C was set = 0). Using the remaining three variables, the shape of the hypersurface of the objective function in three-dimensional space was mapped (fig. 3). The result was a broad bowl shape, with a nearly flat bottom.

At the approximation solution, the optimum allocation of effort to the three independent variables was quite unequal, with a factor of about 2 1/2 between the smallest and the largest allocation. At this allocation, $\hat{\delta}_{\hat{Y}}$ was 17% less than what it was for an equal allocation of effort among the original four variables and 10% less than that for an equal allocation among the retained three variables.

The algorithm of equation (11), with or without applying equation (12), resulted in convergence on C_i that were within ±4 units of effort to the optimum solution found by mapping. It appears that this discrepancy is a small error due to unknown causes. However, the "solution" found produces a value of $\hat{\delta}_{\hat{Y}}$ within 4% of the minimum. Further investigation of this anomaly is in progress.

DISCUSSION

The shape of the surface of the objective function found in the test application intuitively seems to be quite general. If the allocation to a single C_i is increased and all others held constant, n_i should increase no matter what the cost function is. This should reduce the width of the distribution of the $\bar{x}_i$ values in proportion to $\hat{S}_{x_i}/\sqrt{n_i}$. Hence, the contribution of $\bar{x}_i$ to $\hat{\delta}_{\hat{Y}}$ should monotonically and smoothly decrease (e.g., fig. 4a).

The result of looking at jointly varying two C_i's with the total effort fixed should usually be similar to that depicted in figure 4b. Generalizing to more

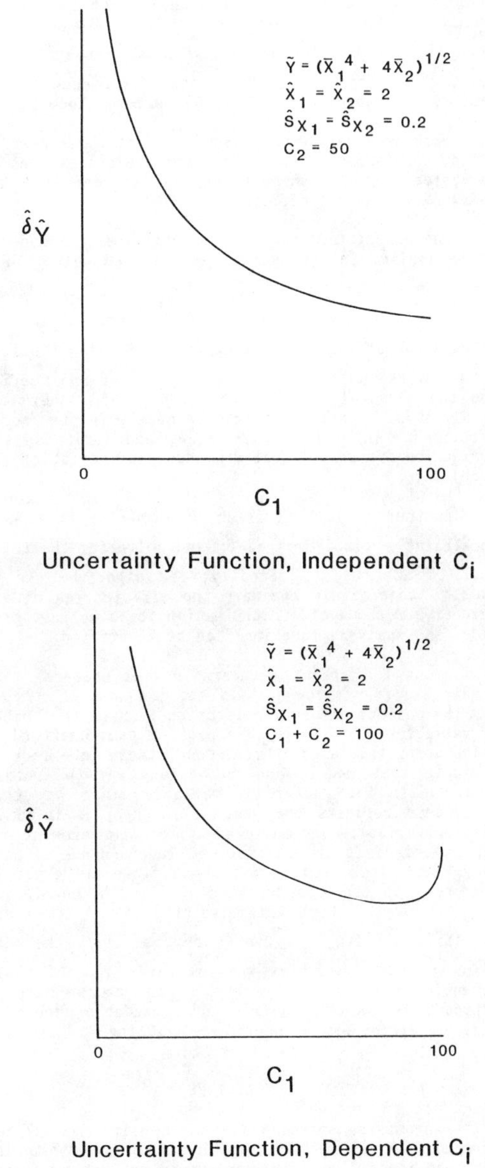

Figure 4.--A priori estimated uncertainty in a function as affected by variable sample sizes (C_i = 1).

variables is expected to produce the type of bowl-shaped surface found in the test application (fig. 3).

If the shape of the hypersurface of the objective function turns out to be a flat-bottomed bowl in general, it should be both feasible and affordable to locate an allocation of effort among a large number of variables that gives an expected uncertainty that is within a few percent of the global minimum. Such allocations should be robust with respect to errors, such as in initially estimated values of the x_i and Sx_i. It should also be robust with respect to at least small degrees of statistical dependence of inventory variables [Ku's (1966) second condition].

Work is underway or planned to test these predictions, to develop software to eliminate allocations to variables that are unimportant, and to fine tune the software to improve efficiency.

The propagated distribution approximation can also be applied in estimating the expected value for $\tilde{Y}$ (which need <u>not</u> be $\tilde{Y} = f(\bar{x}_1, \bar{x}_2, \ldots \bar{x}_m)$ if the function is highly nonlinear). Of course, it can also be used to estimate $\hat{\delta}_{\tilde{Y}}$. It currently is used routinely for that purpose in certain parts of engineering and the physical sciences (e.g., Ku, 1969, Abernathy et al. 1973). An analysis can be made more precise by letting t vary with sample size, and including the correction for statistical dependence of variables (Ku 1966).

Another useful extension would be to take as a constraint a fixed desired $\hat{\delta}_{\hat{Y}}$ and solve for the total cost to achieve it. Finally, the algorithm can be used to objectively evaluate the efficiencies of alternative methods of stratification for sampling, provided the analysis functions can be identified.

Among the areas of uncertainty or weakness that remain are the following: First, the cost of computing the optimum allocation is a function of the number of variables. How large of a problem can we afford to solve using this algorithm? Second, there is a lack of reliable cost models and no obvious way to combine cost models for individual variables in a way that accurately reflects the cost of inventories in which several variables are measured by the same crew at one time. Third, is the statistical independence of sampling error for real variables sufficiently great to permit us to assume independence? Fourth, how can we best develop initial estimates for the $\bar{x}_i$ ($\hat{x}_i$) and Sx_i ($S\hat{x}_i$)? Fifth, how can we best stratify the sampling for all variables when multiple variables are inventoried at once? Finally, sixth, can we make an approach as complex as this understandable enough so that managers and planners are willing to use it?

CONCLUSIONS

A promising approach for another incremental improvement in our ability to plan efficient inventories has been identified. However, it is not yet completely operational, and has not been tested on a real-world problem of the magnitude of the one dealt with by Burkhart et al. (1978). In addition, even at full development of this approach, we will still not have all the tools needed for a fully satisfactory planning system.

ACKNOWLEDGMENTS

Mr. A. Farmer has provided technical assistance, advice and encouragement throughout this project. Also contributing important technical advice or assistance were: Drs. R. Schnabel, P. Boggs, J. Skiles, F. Fontane, D. Zachmann, Ms. J. Donaldson, Mr. J. Koontz, K. Burnham, H. Ku, R. Moffat, and J. Hunter. Ms. J. Shoemaker drew the figures. Word processing and manuscript preparation was provided by C. Gulzow, D. Ibarra, Jan Hill, Susan Taylor, and Donald Hill. Useful comments on the manuscript were provided by Ms. C. Short, and Drs. J. Skiles, S. Matulich, and M. Schamberger.

LITERATURE CITED

Abernathy, R. B., B. D. Powell, and D. L. Colbert. 1973. Handbook, uncertainty in gas turbine measurements. 182 p. U.S. Dept. of Defense, U.S. Air Force, Arnold Engineering Development Center, AEDC-TR-73-5.

Allen, Arthur W. Habitat suitability index models: pronghorn. U.S. Department of Interior, Fish and Wildlife Service, Office of Biological Services (in prep.).

Box, George E. P., William G. Hunter, and J. Stuart Hunter. 1978. Statistics for experimenters. 653 p. John Wiley and Sons, New York.

Braybrooke, D., and C. E. Lindblom. 1963. A strategy of decision. 268 p. The Free Press, New York.

Burkhart, Harold E., R. Dean Stuck, William A. Leuschner, and Marion R. Reynolds. 1978. Allocating inventory resources for multiple-use planning. Canadian Journal of Forest Research 8:100-110.

Cochran, W. G. 1977. Sampling techniques. Third edition. 428 p. John Wiley and Sons, New York.

Cramer, Harold. 1946. Mathematical methods of statistics. 575 p. Princeton University Press, Princeton, New Jersey.

Hamilton, David A., Jr. 1979. Setting precision for resource inventories: the manager and the mensurationist. Journal of Forestry 77(10):667-670.

Hersey, Mayo D. 1965. A development of the theory of errors with reference to economy of time. Journal of Research National Bureau of Standards 69B:139-146.

Ku, H. H. 1966. Notes on the use of propagation of error formulas. Journal of Research National Bureau of Standards 69B:139-146.

Ku, Harry H., ed. 1969. Precision measurements and calibration. Statistical concepts and procedures. N.B.S. Special Publication 300, Volume 1. 436 p. U.S. Department of Commerce, National Bureau of Standards, Washington, D.C.

Moffat, Robert J. 1978. Planning experimental programs. 131 p. Dept. of Mechanical Engineering Stanford University, Stanford, Calif.

Welsh, B. L. 1947. The generaliziation of "Student's" problem when several different population variances are involved. Biometrika 34:28-35.

Yates, F. 1960. Sampling methods for censuses and surveys. Third edition. 440 p. Griffin and Co., London.

APPENDIX 1. DERIVATION OF EQUATION (7)

Equation (6) is:

$$\hat{\delta}_{\hat{Y}} = \left[\sum_{i=1}^{m} \left(\frac{\partial \hat{Y}}{\partial \hat{x}_i}\right)^2 \left(\frac{\sqrt{C_i}\ t\ S_{x_i}}{\sqrt{C_i}}\right)^2 \right]^{\frac{1}{2}} \tag{6}$$

Squaring and rearranging yields:

$$\hat{\delta}_{\hat{Y}}^2 = \sum_{i=1}^{m} \left(\frac{\partial \hat{Y}}{\partial \hat{x}_i}\ \sqrt{C_i}\ t\ S_{x_i}\right)^2 \frac{1}{C_i} \tag{A1}$$

Let:

$$\beta_i = \left(\frac{\partial \hat{Y}}{\partial \hat{x}_i}\ \sqrt{C_i}\ t\ \hat{S}_{x_i}\right)^2 \tag{A2}$$

Substituting (A2) into (A1) gives:

$$\hat{\delta}_{\hat{Y}}^2 = \sum_{i=1}^{m} \frac{\beta_i}{C_i} = \sum_{i=1}^{m-1} \frac{\beta_i}{C_i} + \frac{\beta_m}{C_m} \tag{A3}$$

By definition,

$$T = \sum_{i=1}^{m} C_i = \sum_{i=1}^{m-1} C_i + C_m \tag{A4}$$

For $j \neq i$, and $i \neq m$

$$\sum_{k=1}^{m-1} C_k = \sum_{j=1}^{m-1} C_j + C_i \ , \quad i \neq m,\ j \neq i \tag{A5}$$

Substituting (A5) into (A4) gives:

$$T = \sum_{j=1}^{m-1} C_j + C_i + C_m \ , \quad j \neq i \tag{A6}$$

Solving for C_m gives:

$$C_m = T - \sum_{j=1}^{m-1} C_j - C_i \ , \quad j \neq i \tag{A7}$$

Let

$$\varepsilon_i = T - \sum_{j=1}^{m-1} C_j \ , \quad j \neq i \tag{A8}$$

Substituting (A8) into (A7) yields:

$$C_m = \varepsilon_i - C_i \tag{A9}$$

Substituting (A9) into (A3) gives:

$$\hat{\delta}_{\hat{Y}}^2 = \sum_{i=1}^{m-1} \frac{\beta_i}{C_i} + \frac{\beta_m}{\varepsilon_i - C_i} \tag{A10}$$

Taking the partial derivatives of (A10) (assuming the t values (A2) are constant):

$$\frac{\partial\ \hat{\delta}_{\hat{Y}}^2}{\partial\ C_i} = -\frac{\beta_i}{C_i^2} + \frac{\beta_m}{(\varepsilon_i - C_i)^2} \tag{A11}$$

for any $i \neq m$, and

$$\frac{\partial\ \hat{\delta}_{Y}^2}{\partial\ C_m} = 0$$

At the minimum of $\hat{\delta}_{\hat{Y}}^2$, the partial derivatives equal zero. From (A11), it is clear that the mth partial is always zero. Setting the partials equal to zero for $i \neq m$ [from (A11)] and solving for C_i proceeds as follows:

$$0 = \frac{\beta_i}{C_i^2} + \frac{\beta_m}{(\varepsilon_i - C_i)^2} \tag{A12}$$

Eliminating ε_i by resubstituting (A8) gives:

$$0 = \frac{\beta_i}{C_i^2} + \frac{\beta_m}{\left(T - \sum_{j=1}^{m-1} C_j - C_i\right)^2} \ , \quad j \neq i \tag{A13}$$

The denominator of the last term equals C_m [eq. (A7)], so:

$$0 = -\frac{\beta_i}{C_i^2} + \frac{\beta_m}{C_m^2} \tag{A14}$$

Since β_m/C_m^2 is a constant, let it equal α:

$$0 = -\frac{\beta_i}{C_i^2} + \alpha \tag{A15}$$

Rearranging:

$$\beta_i = \alpha\ C_i^2 \tag{A16}$$

Taking square roots gives:

$$\beta_i^{\frac{1}{2}} = \alpha^{\frac{1}{2}} C_i \tag{A17}$$

Summing across all i gives:

$$\sum_{j=1}^{m} \beta_j^{\frac{1}{2}} = \sum_{i=1}^{m} \alpha^{\frac{1}{2}} C_i = \alpha^{\frac{1}{2}} \sum_{i=1}^{m} C_i = \alpha^{\frac{1}{2}} T \tag{A18}$$

Rearranging:

$$\alpha^{\frac{1}{2}} = \frac{\sum_{j=1}^{m} \beta_j^{\frac{1}{2}}}{T} \tag{A19}$$

Solving (A17) for $\alpha^{\frac{1}{2}}$ and substituting into (A19) gives:

$$\frac{\beta_i^{\frac{1}{2}}}{C_i} = \frac{\sum_{j=1}^{m} \beta_j^{\frac{1}{2}}}{T} \tag{A20}$$

Solving for C_i gives:

$$C_i = \frac{T\,\beta_i^{\frac{1}{2}}}{\sum_{j-1}^{m} \beta_j^{\frac{1}{2}}} \tag{A21}$$

Eliminating β in (A21) by substituting (A2) gives:

$$C_i = \frac{T\,\frac{\partial \hat{Y}}{\partial \hat{x}_i} \sqrt{c_i}\, t\, \hat{S}_{x_i}}{\sum_{j=1}^{m} \frac{\partial \hat{Y}}{\partial \hat{x}_j} \sqrt{c_j}\, t\, \hat{S}_{x_j}} = \frac{t\ T\,\frac{\partial \hat{Y}}{\partial \hat{x}_i} \sqrt{c_i}\, \hat{S}_{x_i}}{t \sum_{j=1}^{m} \frac{\partial \hat{Y}}{\partial \hat{x}_j} \sqrt{c_j}\, \hat{S}_{x_j}} \tag{A22}$$

Cancelling the t's gives:

$$C_i = \frac{T\,\frac{\partial \hat{Y}}{\partial \hat{X}_i} \sqrt{c_i}\, \hat{S}_{x_i}}{\sum_{j=1}^{m} \frac{\partial \hat{Y}}{\partial \hat{X}_j} \sqrt{c_j}\, \hat{S}_{x_j}} \tag{A23}$$

Q.E.D.

[Note: Hersey (1965) reached essentially the same conclusion about 1913, apparently on the basis of a geometrical formulation of the problem.)

PREMOD AND MODAID: SOFTWARE TOOLS FOR WRITING SIMULATION MODELS[1]

T. B. Kirchner and J. M. Vevea[2]

Abstract.--MODAID is a library of software "tools" that help simplify the process of writing simulation models in FORTRAN. These tools are programs, functions, and subroutines that accomplish tasks that are common to most ecological models. PREMOD is a precompiler used to help build simulations models. PREMOD is designed to be used in an interactive (on-line) mode, and can be used by persons having limited knowledge of programming.

INTRODUCTION

MODAID and PREMOD are software tools that help simplify the process of writing simulation models in FORTRAN. MODAID includes a library of FORTRAN subprograms that can be classified into six categories based upon their function: (1) routines that facilitate the generation of output, (2) routines that facilitate making changes in values of parameters used in the model, (3) routines that facilitate the updating of state variables during a simulation, (4) routines that facilitate the analysis of simulation models, (5) mathematical functions, and (6) pseudorandom number generators. These subprograms are designed to be referenced directly from within simulation models. MODAID also contains three auxiliary programs designed to be executed independently of the simulation model. One auxiliary program is used to create a table required by some of the MODAID subprograms. The other auxiliary programs can be used to help assemble or modify FORTRAN programs.

PREMOD is a precompiler used to help build simulation models. A precompiler is a program that uses information supplied to it to write computer code. PREMOD allows users who have a minimum amount of knowledge of FORTRAN to build simulation models. Nevertheless, PREMOD can be used to construct large, complex simulation models. PREMOD operates by inserting code into a source file. The source file is a skeleton of code common to all simulation models generated by PREMOD. The code that is inserted into the source file is either supplied by the user or generated by PREMOD. The user enters FORTRAN statements that define the mathematical relationships of the system to be simulated. The FORTRAN statements created by PREMOD include several common block definitions, dimension statements for some arrays required by MODAID routines, calls to several MODAID routines, calls to routines that solve the rate equations of the model, and code for several optional subroutines.

[1]Paper presented at the Third International Conference on State-of-the-Art in Ecological Modeling, Colorado State University, May 24-28, 1982.

[2]T. B. Kirchner is Research Ecologist, Natural Resource Ecology Laboratory, and J. M. Vevea is a Ph.D. candidate, Department of Forest and Wood Science, Colorado State University, Fort Collins, Colorado U.S.A.

The program that is constructed by PREMOD is designed to be used primarily as an interactive (on-line) simulation model. However, simulations may be run in a batch mode by supplying the program with an appropriate input file. The basic structure of the model permits the user to run a simulation, to plot or print the results of that simulation, to change the values of parameters and the initial values of state variables, and then to execute another simulation.

PREMOD builds models in which the rates of change of state variables are defined either by first order ordinary differential equations or by difference equations. PREMOD allows the user to construct models of different levels of complexity, including models that contain routines that will do sensitivity and error analyses. The user specifies the type of rate equations and the complexity of the code to be entered by choosing among five options. If none of the options are specified, the model constructed by PREMOD will solve a system of differential equations using a Runge-Kutta algorithm of order four.

The purpose of this paper is to describe the general features of MODAID, PREMOD, and models built using PREMOD.

MODAID

MODAID routines do many of the tasks commonly encountered in simulation models, such as updating state variables, generating random numbers, or producing graphical output. These routines are designed to be both easy to use and portable between computer systems. The descriptions of specific MODAID routines that follow are grouped according to their general function.

Output Routines

MODAID contains four routines directly associated with producing output from a simulation. These routines are PLTSET, PLOTSV, MODCTL, and VOUT. The purpose of PLOTSV is to write an unformatted record of the values of all variables the user wants to save for plotting or printing. One record is written into the plot file each time PLOTSV is called. Those variables to be saved for plotting or printing using PLOTSV must be stored in a common block labeled PLOTS1. PLTSET is used to write on the plot file a header record that gives the number of variables in each record written by PLOTSV.

MODCTL reads the plot file and allows the user to produce line printer plots, formatted printed output, plots on graphics terminals, and plots on pen plotters. A library of Tektronix plotting routines is used for interactive graphics. However, MODAID is designed such that conversion to other graphics libraries would not be difficult. The printing and plotting of variables is controlled by commands from the user (Table 1). These commands enable the user to specify the kind of output (plotted or printed) and the names of variables to be plotted or printed. The user can also specify that some functions of variables be printed or plotted. The functions that can be specified are the sum, difference, mean, product, and quotient of two or more variables and the natural logarithm and integral of single variables. In this context, functions of variables are considered variables, permitting functions to be nested. For example, a command could be given to MODCTL that would cause the natural logarithm of the sum of ADULTS and LARVAE to be plotted against TIME, where TIME, ADULTS, and LARVAE are variables in the PLOTS1 common block. Plots and tables are normally labeled with the names of the variables or functions specified in the command. However, optional commands permit the user to specify labels to replace the names of variables that are printed on the output, to specify an interval of simulation time from which data are to be plotted or printed, to scale the axes of plots, and to add titles and legends to plots.

The technique used to save the data for plotting --i.e., the generation of the plot file--allows MODCTL to be used in two ways; MODCTL can be called from the model program after a simulation is completed, or the plot file can be saved and output generated from it at a later time using another program to call MODCTL. The first method is useful when the model is run interactively because it allows the user to view the results of a simulation, make any desired changes in parameter values, and then run another simulation. The second method may be useful when running large models because the model can be run as a batch job. After the job has been executed, the user can view the results of the simulation using an interactive program. Thus, the user does not have to wait at a terminal while the simulation is executing.

The remaining output routine, VOUT, is designed to be used in association with PARSEL, one of the MODAID routines that help facilitate error and sensitivity analyses. VOUT can be used to save the results of the numerous simulations required of models using Monte Carlo methods for error and sensitivity analyses. VOUT writes a file of the values of certain selected variables used in the model. These variables are a subset of those that occur in the PLOTS1 common block and in a common block labeled PARAMS, and they are specified by the user in a file read by PARSEL. The values of these variables are written each time VOUT is called, which is normally once after each simulation. VOUT can be directed to write either formatted or unformatted records.

Change Routine

The MODAID routine CHANGE allows the user to change the values of variables used in the model. CHANGE can be called directly from the user's program or can be called from within MODCTL by entering the command C(HANGE) (table 1). The names of variables to be changed must appear in the common block PARAMS. These variables are usually parameters of the rate equations of the model and other variables that control the simulation, such as variables that define the length of the simulation, the time step over which the rate equations are to be solved, and the initial values of the state variables. In the following discussion, any variable that occurs in the PARAMS common block will be referred to as a parameter.

Changes in parameter values are controlled by commands from the user (table 2). In addition to specifying a change in the value of a parameter, the user can enter commands that (1) cause the values of specified variables to be listed, (2) save the current set of parameter values by writing them to a specific file, (3) re-initialize the parameters from a file saved previously, (4) return control to the calling routine, or (5) terminate execution. If CHANGE is called after a simulation, the user can enter a command that will allow the simulation to continue from a specified time in the previous simulation. In this case, variables in the PLOTS1 common block are initialized from the values on the plot file at the specified time, and the plot file is positioned so that records from the next simulation are added, starting at the specified time. The portion of the plot file before that time remains unchanged. Thus, the user can run the simulation for an interval of time, look at the results, and then restart the simulation from any time in the previous simulation that occurs in the plot file.

Analytical Tools

There are three main categories of analysis to which simulation models are often subjected: (1) validation, or comparison of model values with experimental data, (2) sensitivity analysis, or determination of the sensitivity of state variables to changes in the values of parameters, and (3) error analysis, or determination of the probability distribution of the values of state variables given the probability distributions of the parameters. There are MODAID routines that facilitate each kind of analysis.

MODCTL provides a means of validating models by allowing the user to simultaneously plot the results of a simulation and experimental data. The experimental data used in these plots must be stored in arrays in the PARAMS common block along with an INTEGER variable giving the length of the arrays. The model results are plotted as lines, and the experimental data are plotted as discrete points.

A common technique for doing sensitivity and error analyses requires that numerous Monte Carlo simulations be done, with the values of parameters changed before each simulation. The subroutine PARSEL can be used to select the values of parameters for these analyses. PARSEL requires that the names of the parameters to be randomly sampled during the analyses be provided in a formatted data file. A statistical distribution and the parameters for that distribution must be defined for every parameter in the list. The names of any variables to be used as "output" variables are also listed in this file. Output variables are treated as the dependent variables in sensitivity and error analyses. An output variable can be any variable that occurs in the PLOTS1 common block.

A call to PARSEL before a simulation causes it to select a value for each of the specified parameters

Table 1.--The following 16 commands can be used with MODCTL. The part of the command within the parentheses is optional. Those commands marked with * are "on-" and "off-"type commands. If these commands are used with no argument, with a + sign or with the word on, they are activated. If these commands are used with any other argument, they are deactivated.

Command	Description
A(NNOTATE)*	indicates that the user wants to override the default titles used by the L(IST), P(RINTERPLOT), and T(EKPLOT) commands.
C(HANGE)	deactivates the output mode of MODCTL and activates a set of commands (Table 2) that allow parameter values to be examined, changed, and/or stored.
F(ILTER)*	allows restricting the L(IST), P(RINTERPLOT), and T(EKPLOT) commands to data within a specified range of TIME. The command F T1, T2 defines the interval as being from T1 to T2, where T1 and T2 are numerical values.
G(RAPHCOORD)	will turn on a graphics cursor to allow the user to determine the corrdinates of a point on a graph. The cursor (either a set of crosshairs or an arrow) is positioned on the point and a letter other than S is pressed twice. The coordinates will then be printed on the screen. Pressing S twice turns off the graphics cursor and returns control to MODCTL. This command should be used only with a Tektronix terminal equipped with a graphics cursor (the 4010 and 4050 series terminals).
H(ARDCOPY)*	causes MODCTL to request a hard copy after every T(EKPLOT) command has produced a complete graph.
I(NKPLOT)*	causes MODCTL to produce a pen plot on a Tektronix 4662 plotter every time that a T(EKPLOT) is requested.
K(IND)*	is used to change the parameters specifying the type of Tektronix terminal that you are using.
L(IST)	causes MODCTL to produce a list of values of specified state variables for each time at which they were stored. The names of the variables to be listed are entered in response to a second prompt. This list can be written on a file other than OUTPUT by including an integer argument with the list command. The argument must be separated from the L(IST) command by a space and indicates the unit number of the file on which the list is to be written. L(IST) is subject to modification by the A(NNOTATE) and F(ILTER) commands.
M(ODIFY)*	causes MODCTL to list the actual range of values to be plotted by the P(RINTERPLOT) or T(EKPLOT) commands and the extreme values it has selected for the axes. The user is then given the opportunity to replace these extreme values with a set of his own choosing.
N(ICTTL)*	causes MODCTL to allow the user to specify a y title and the full x title for the plots. The N(ICTTL) command also activates the A(NNOTATE) command.
P(RINTERPLOT)	produces a printer plot of specified variables. These variables are entered after a second prompt is given. The first variable specified is used as the x axis (independent) variable. The remaining variables, up to five in number, are used as y axis (dependent) variables. An integer argument, such as that of the L(IST) command, may be specified as an alternate unit number of a file onto which the plot is to be written.
Q(UIT)	causes the execution of the model to stop.
R(ERUN)	causes a return to the calling routine. In models generated by PREMOD, another simulation will then occur.
T(EKPLOT)	is like P(RINTERPLOT) but produces a Tektronix plot. These plots are normally generated with the y variables plotted using lines. However, if any argument follows the command, the y variables will be plotted using symbols rather than lines.
V(ERIFY)	causes a normal Tektronix plot to be produced. However, the user is also given the opportunity to enter the names of arrays of variables to be plotted. These arrays must occur in the PARAMS common block--i.e., they must be declared as parameters in PREMOD. The data must be stored with the x values in one array, the y values in another, and the number of points stored in a variable. These data are then plotted using symbols. Up to five y variables may be plotted at one time, and these can be any combination of lines (state variables) and points (data) so long as at least one line is plotted. The purpose of V(ERIFY) is to allow the user to visually compare simulated values and data.
W(HAT)	causes MODCTL to list those state variables, parameters, and plotting functions that can be used. W(HAT) can be used with any combination of the arguments V(ARIABLES), P(ARAMETERS), and/or F(UNCTIONS) to limit the list to the type or types specified. With no arguments, W(HAT) assumes that all three are desired.

Table 2.--The following 11 commands can be used with CHANGE to examine, store, and change parameter values.

Command	Description
C(HANGE)	allows the user to change the value of a parameter. Two arguments are required: (1) the name of the variable, and (2) the new value to be assigned to that variable. The arguments are separated by commas or spaces. When one of the arguments to the C(HANGE) command is an array name for which no subscript is given, it is assumed that the entire array will be changed. C(HANGE) will then prompt the user to enter a value for each element of the array.
E(XTEND)	allows the user to specify a value of TIME at which a simulation is to be restarted. The variables of the PLOTS1 common block--i.e., the state variables-- are initialized with the values that occurred at that time. The values of the state variables before that time are retained on the "plot file" (TAPE1). Thus, the user can run a simulation for an interval of time, look at the results, and then continue the simulation starting from the end of the previous simulation or from any point of time within the previous simulation.
F(LAG)	allows the user to enter a list of variable names to an "active" list. This list will be assumed to be the argument list whenever the user gives a C(HANGE) or L(IST) command without any agruments.
L(IST)	allows the user to see the current values of variables specified in the argument list of the command or those in the active list [see F(LAG)]. If one of the arguments to the L(IST) command is an array name for which no subscript is given, the entire array will be listed. However, all succeeding arguments are ignored.
M(ODCTL)	causes a return to MODCTL.
N(EW)	allows the user to replace a previous "active" list of variable names with a new list of names [see F(LAG)].
O(LD)	causes a previously saved set of parameter values to be read. The unit number of the file from which the data are to be read must be given as as argument, and the data must have been written on the file previously using the S(TORE) command.
Q(UIT)	causes the execution of the model to stop.
R(ERUN)	causes a return to the routine that called MODCTL. In PREMOD-generated models, another simulation will then occur.
S(TORE)	causes the current values of parameters to be written on a file. The unit number of the file must be specified as an argument.
W(HAT)	causes a list of the names of the parameters to be displayed.

using an appropriate random number generator. PARSEL can be directed to sample the parameters using either simple random sampling (SRS) or Latin hypercube sampling (LHS) (McKay, Beckman, and Conover 1979). Both of these sampling designs assume that the parameters are independently distributed. In SRS a value for each parameter is selected from that parameter's statistical distribution. However, LHS is similar to a highly fractionalized factorial design in which the distributions of the parameters to be sampled are divided into 10 quantiles of equal probability. For $\underline{n}$ parameters, a matrix of size $\underline{n} \times 10$ is used to define the quantiles of the distribution of each parameter to be sampled in each of the next 10 simulations. This matrix is constructed so that (1) each quantile of every parameter is sampled only once, and (2) in each of the 10 runs the number of quantiles having the same range of probabilities is minimized. A new matrix is constructed before the start of a new set of 10 simulations.

Simple random sampling can be used for either error or sensitivity analysis. However, for models in which the output variables are monotonic functions of the parameters, Latin hypercube sampling can determine, in fewer simulations, the response surface relating an output variable to the model parameters (McKay, Beckman, and Conover 1979; Iman, Helton, and Cambell 1978). Therefore, LHS may be used to help reduce the cost of sensitivity analysis. However, LHS may bias the distributions of the output variables. Thus, LHS is usually an inappropriate design for error analyses.

The sensitivity of an output variable to a parameter can be described by the degree of correlation between that pair of variables. However, simple correlations between such pairs are confounded by the effects that the remaining parameters have on the output variables because each of the specified parameters is sampled before every simulation. The use of partial correlation coefficients helps solve this problem. The partial correlation coefficient is a measure of the degree to which the output variable is correlated with the parameter after removing the linear effects of all the other variables. The MODAID routine VASSES (Variable ASSESsment) calculates for each output variable the partial correlation coefficient with each of the selected parameters. VASSES calculates and prints the correlation coefficients after every 10 simulations (table 3). The parameters are listed in descending order of the absolute values of the partial correlation coefficients.

The subroutine VDIST (Variables' DISTributions) can be used to help perform an error analysis on the model. VDIST calls the subroutine VOUT to save the values of the output variables for each simulation. After the last simulation, VDIST produces for each output variable a frequency table and histogram of the distribution of the output variable and calculates the minimum, maximum, mean, standard deviation, standard error, variance, skewness, and kurtosis of the distribution (table 3). If the distribution is significantly ($\underline{p} < 0.05$) positively skewed, the statistics and histogram for the log-transformed data are calculated.

The maximum number of stochastic simulations to be run is specified by the user. However, if partial correlation coefficients are calculated by VASSES, the user can specify an additional set of criteria that will terminate a run before the maximum number of simulations is reached. These criteria are that, for each output variable, (1) $\underline{n}$ parameters have partial correlation coefficients that are significant at a probability level of $\underline{p}$ and (2) that the order of these $\underline{n}$ variables remains the same for three consecutive sets of 10 simulations. The user specifies the $\underline{n}$ and $\underline{p}$ to be used in these criteria. The purpose of these additional criteria is to provide a means of terminating a run if it appears that more simulations will not significantly affect the results of the sensitivity analysis.

Flow Routines

FLOW, FLOWL, and FLOWUP are three MODAID subroutines designed to let the user schedule and update changes in the values of state variables. FLOW and FLOWL are used to define the quantity to pass, or flow, from one state variable to another and to schedule the time at which the flow is to take place. Arguments in the calls to FLOW and FLOWL include the amount to be transferred, the time at which the transfer is to occur, and the names of the variables from which and to which the flow is to be made. FLOWUP is used to update the state variables--i.e., to subtract the appropriate quantity from one state variable and add it to another--at the appropriate times. FLOW permits any quantity to be flowed from a state variable, whereas FLOWL limits the maximum quantity to be flowed from a state variable to the quantity in the state variable at the time of the call to FLOWL. Thus, FLOWL is designed to be used only in cases where non-negative values of state variables are assumed to exist.

Mathematical Functions

MODAID contains a library of mathematical functions frequently used in simulation models. These functions include all of those described by Parton and Innis (1972) plus a subroutine for interpolating piecewise linear functions. Table 4 lists and describes these functions.

Pseudorandom Number Generators

MODAID has a library of nine pseudorandom number generators. These routines will generate pseudorandom numbers for normal (ANORM), lognormal (ALGNRM), exponential (EXPONT), chi-square (CHISQ), Poisson (IPOISS), triangular (TRIANG), uniform (UNIF), and multivariate normal (MVNORM) distributions. MODAID includes its own 0 to 1 uniform random number generator, RANDU, which is used by the other random number functions. RANDU provides a more nearly uniform distribution than that obtained by using a standard linear-congruential generator. Also, RANDU returns a count of the number of times it has been called.

Auxiliary Programs

MODAID contains three auxiliary programs, MAKTAB, MACRO, and DEMAC. MAKTAB creates the tables required by MODCTL, CHANGE, and several of the analytical routines. These tables are the means by which variable names can be associated with locations in common blocks. MAKTAB requires as input a file listing the FORTRAN definitions of the PLOTS1 and PARAMS common blocks plus all REAL, INTEGER, LOGICAL, and DIMENSION statements that affect variables in those blocks.

MACRO is a macro processor--i.e., a program that will insert blocks or lines of code in place of keywords (macros) in a source file. MACRO typically is

Table 3.--(a) A typical table of correlation coefficients produced by VASSES for use in a sensitivity analysis. The model used for the analysis represents competition between two species using Volterra's equations. For species i, POP(i) is the size of the population after a fixed time interval, K(i) is the carrying capacity, R(i) is the instantaneous rate of growth, and ALPHA and BETA are the competition coefficients. Simple random sampling was used to choose values for the six parameters of the model in each of 100 simulations. (b) Output produced by VDIST, showing the distribution of the values of POP(1) for the 100 simulations. The distribution is neither significantly skewed nor kurtotic.

Parameter		Simple correlation coefficient	Partial correlation coefficient	Significance probability for partial correlation
(a)		Correlations for output variable POP (1) after 100 runs (93 degrees of freedom)		
K	(1)	0.91187	0.99584	p is less than 0.001
K	(2)	-0.23230	-0.96117	p is less than 0.001
ALPHA		-0.35059	-0.95649	p is less than 0.001
BETA		0.05944	0.56779	p is less than 0.001
R	(2)	-0.00332	-0.41893	p is less than 0.001
R	(1)	0.02145	0.16891	p is less than 0.999
		Correlations for output variable POP (2) after 100 runs (93 degrees of freedom)		
K	(2)	0.96840	0.99842	p is less than 0.001
K	(1)	-0.13917	-0.96080	p is less than 0.001
BETA		-0.18934	-0.92883	p is less than 0.001
ALPHA		0.02450	0.71064	p is less than 0.001
R	(2)	0.08731	0.31986	p is less than 0.010
R	(1)	-0.07950	-0.02057	p is less than 0.999

(b) The frequency distribution for POP (1)

Interval	Number of observations	Frequency	
0.7220E+01-0.1729E+02	4	0.0400	****
0.1729E+02-0.2736E+02	8	0.0800	********
0.2736E+02-0.3743E+02	7	0.0700	*******
0.3743E+02-0.4749E+02	5	0.0500	*****
0.4749E+02-0.5756E+02	14	0.1400	**************
0.5756E+02-0.6763E+02	12	0.1200	************
0.6763E+02-0.7770E+02	15	0.1500	***************
0.7770E+02-0.8777E+02	13	0.1300	*************
0.8777E+02-0.9784E+02	6	0.0600	******
0.9784E+02-0.1079E+03	6	0.0600	******
0.1079E+03-0.1180E+03	7	0.0700	*******
0.1180E+03-0.1280E+03	3	0.0300	***

Mean = 0.66969+02
Standard deviation = 0.28671E+02
Variance = 0.82200E+03
Skewness = -0.0019
Kurtosis = 2.3575

Standard error = 0.28671E+01
Minimum = 0.72198E+01
Maximum = 0.12804E+03
Number of cases = 100

used to insure that the definitions of common blocks are consistent throughout a program by copying the same definition of each common block into specified routines. MACRO will also replace literals, which are keywords within a line of code that are to be replaced by strings of characters. The replacement of literals by MACRO is often used to change the dimensions of variables in a program. The definitions of the macros and literals are provided to MACRO as a file.

DEMAC is designed to deprocess a program that has been processed previously by MACRO. DEMAC will recreate the original source file processed by MACRO. Macro processing such as that provided by MACRO and DEMAC is already available in many computers. MACRO and DEMAC are designed primarily for users at installations where such capabilities do not already exist.

MODEL STRUCTURE

At the start of model construction, PREMOD allows the user to select up to four options. There are five options from which to choose, but the first two of these are mutually exclusive. The structure of the model assembled by PREMOD depends upon the options selected by the user (fig. 1). If no options are chosen (the default case), the program of the simulation

Table 4.--The mathematical functions of the MODAID library, taken from Parton and Innis (1972).

MODAID function	Function name	Equation
AF(x,a,b)	Allometric	ax^b
ATANF(x,a,b,c,d)	Arctangent	$b + \frac{c}{\pi}$ ARCTAN$[\pi d(x - a)]$
CF(x,a,b,c)	Cosine	$a + \cos[\frac{2\pi}{b}(x - c)]$
CLI(x,a,b,x1,x2)	Linear interpolation	$\frac{(a - b)}{(x1 - x2)} (X - x2) + b$; $x1 \neq x2$ $(a + b)/2$; $x1 = x2$
EF(x,a,b)	Exponential	ae^{bx}
EMF(x,a,b)	Modified exponential	ab^x
FL(x,a,b,c)	Logistic	$\frac{a}{1 + be^{-cx}}$
GFN(x,a,b)	Natural growth	$a(1 - e^{-bx})$
GGEF(x,a,b,c,d)	Generalized Gompertz	$ab^{-cb^{-dx}}$
GPDF	Generalized Poisson density	$\frac{b - x}{b - a}^{c}$ $\underline{\exp}$ $\frac{c}{d}[1 - (\frac{b - x}{b - a})^{d}]$
QEF(x,a,b,c)	Quadratic	$a + bx + cx^2$
RFI(x,a,b,c,d)	National function I	$\frac{a}{(1 + b/x)} + \frac{c}{(1 + d/x)}$
SF(x,a,b,c)	Sine	$a + \sin[\frac{2\pi}{b}(x - c)]$
TABLE(TBL,x,x1,x2,d)	Tabular interpolation	TBL(1) ; $x \leq x1$ $\frac{TBL(J) - TBL(I)}{d} (x - x1 - Kd) + TBL(I)$; $K = (x - x1)/d$; $x1 < x < x2$, $I = K + 1$, $J = K + 2$ $TBL(\frac{x2 - x1}{d} + 1)$; $x2 \leq x$
TRILF(x,a,b)	Piecewise linear	a ; $x < x1$ $(\frac{b - a}{x2 - x1}) (x - x1) + a$; $x1 \leq x \leq x2$ b ; $x > x2$
ZFL(x,a,b)	Linear	$a + bx$

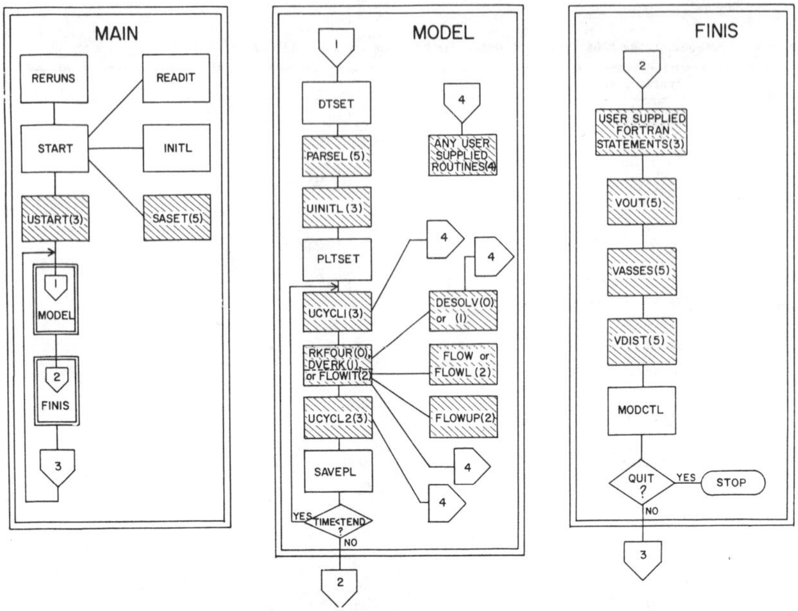

Figure 1.--A simulation model constructed by PREMOD is divided into three overlays, denoted by double-sided rectangles. The MAIN overlay calls the MODEL overlay to execute a simulation, followed by the FINIS overlay to analyze or view the results of a simulation. Subprograms are represented as single-sided rectangles. Those subprograms that are optional in a model are cross-hatched, and the numbers of the options that allow their inclusion are shown in parentheses. Option 0 is default, and options 1 to 5 are as described in the text.

model will contain a minimum number of subprograms, and it will be designed to solve a system of differential equations using RKFOUR, a Runge-Kutta algorithm of order four.

Options 1 and 2 specify alternative algorithms for solving the rate equations of the model but change the structure of the model very little compared with the default case. Option 1 causes PREMOD to replace the call to RKFOUR by a call to the International Mathematical and Statistical Library (IMSL) subroutine DVERK. DVERK attempts to keep the global error of the numerical solution of the differential equations proportional to a tolerance level specified by the user. DVERK uses Runge-Kutta algorithms of orders five and six and tends to increase the time required to do a simulation compared with using RKFOUR. Option 2 specifies that difference equations will be used to define the changes in the state variables. The difference equations are defined and solved using the MODAID subroutines FLOW, FLOWL, and FLOWUP.

Option 3 allows the user to enter FORTRAN statements into five subprograms within the simulation model: USTART, UINITL, UCYCL1, UCYCL2, and FINIS. USTART is called only once per run before the first simulation. UINITL is called before every simulation in a run. UCYCL1 and UCYCL2 are called each time step during the simulation, just before and just after solving the rate equations for the time step, respectively. FINIS is called once after each simulation in a run, and it contains a call to MODCTL. Except for FINIS, only those routines to which the user adds code are included in the simulation model. PREMOD inserts the PLOTS1 and PARAMS common blocks in each of these routines. Thus, the user does not have to be concerned with the communication of the values of state variables and parameters between routines.

Option 4 allows the user to enter the code for complete subprograms. PREMOD will add the PLOTS1 and PARAMS common blocks to any routine that contains the character strings *PLOTS1 and *PARAMS, respectively. Each string must appear in columns 1 through 7 of a record (card image). PREMOD will insert the appropriate common block at the point in the code where the string appears.

If option 5 is selected, PREMOD inserts into the simulation model the code necessary for doing sensitivity and/or error analyses using MODAID routines. PREMOD will also help the user build the data file required by PARSEL for such analyses. The simulation model constructed under option 5 will execute in either a deterministic mode--i.e., without performing sensitivity or error analysis--or in a stochastic mode. Variables read from a data file control the mode in which the model executes and determine the number of simulations performed when sensitivity or error analyses are done.

The code of the simulation model is divided into three overlays to reduce the total amount of memory required by the program during its execution. Overlays are sections of larger programs that can occupy the same storage locations in memory at different times. On the Colorado State University computer system (CDC Cybers), this reduction in memory requirements greatly improves the interactive response time and reduces the cost of an execution. The program generated by PREMOD consists of a main overlay, MAIN, and two secondary overlays, MODEL and FINIS (fig. 1). The secondary overlays are loaded into memory only when called from the main overlay and are unloaded after control is returned to the main overlay. Subprograms within the main overlay can be called from either secondary overlay, but subprograms in the secondary overlays can be called only from within the overlay in which they occur. On computer systems where the use of overlays is not advantageous, the overlays can be converted easily into subroutines.

Subprograms in the main overlay serve two purposes: (1) to carry out initializations of variables that need to be done only once per run, and (2) to contain common blocks of variables whose values must be preserved between successive simulations. RERUNS is called from MAIN to initialize some variables required by MODAID routines. MAIN calls START to initialize variables used in the model. START determines whether the model is to run in a deterministic or a stochastic mode. In a stochastic run, START calls SASET to initialize some variables used by the MODAID sensitivity and error analysis software. START calls READIT, which initializes the variables in the PLOTS1 and PARAMS common blocks by reading their values from a data file. START then calls INITL to transfer the initial values of the state variables into an array called ZINIT. ZINIT is placed in the PARAMS common block by PREMOD, and it is used to re-initialize state variables to their original values between successive simulations. Because ZINIT is in the PARAMS common block, the user can change the initial values of the state variables using the MODAID routine CHANGE.

The only subroutine in the main overlay to which the user can add code is USTART. USTART is called after START and can be used for either of the two purposes defined above for subprograms of the main overlay. If the user has not added code to USTART, it is not called and is omitted from the program.

The last two calls in MAIN are to the two secondary overlays, MODEL and FINIS. The call to MODEL causes a simulation to execute. FINIS is then called to allow the user to view or to analyze the results of that simulation. The calls to MODEL and FINIS are in a loop, so that another simulation will be executed each time control is returned to MAIN from FINIS.

The MODEL overlay contains two kinds of subprograms: those called once before solving the rate equations and those called each time step of a simulation. The subroutines called only once are used to initialize variables and files before the start of each simulation. There are from two to four such initialization routines, depending upon the options selected when the simulation model was constructed. These routines are UINITL, PARSEL, DTSET, and PLTSET. From two to four subroutines are also called each time step of a simulation. These routines are UCYCL1, UCYCL2, SAVEPL, and either RKFOUR, DVERK, or FLOWIT.

UINITL is included in the program only if the user has added code to it. PARSEL is included in those models built under option 5 of PREMOD. MODEL always calls DTSET and PLTSET at the start of a simulation. DTSET re-initializes the state variables using the values stored in ZINIT and ensures that the time interval used to save data for plotting and printing, DTPL, is defined. PLTSET re-initializes the file of data saved for plotting.

If no options were chosen during construction of the model, only RKFOUR and SAVEPL are called every time step of the simulation. RKFOUR solves the differential equations of the model for the time step. If option 1 was chosen, DVERK replaces RKFOUR as the solution algorithm. The differential equations are defined in subroutine DESOLV or in subprograms called from DESOLV, and DESOLV is called by RKFOUR or DVERK. If difference equations are used in the model (option 2), the subroutine FLOWIT replaces RKFOUR. The difference equations of the model occur in FLOWIT or in subprograms called from FLOWIT. FLOWIT calls FLOWUP and increments TIME by DT before returning control to MODEL. SAVEPL calls PLOTSV to save data for plotting or printing and increments TPLT, the time at which the next set of values is to be saved. SAVEPL is called only when the value of simulation time, TIME, is greater than or equal to TPLT. UCYCL1 and UCYCL2 are included in the simulation model if the user has added code to them. UCYCL1 and UCYCL2 are called just before and just after solving the rate equations for the time step, respectively.

The MODEL overlay is unloaded and the FINIS overlay is loaded after a simulation is terminated. Termination occurs when TIME attains a value greater than or equal to TEND. FINIS is used primarily to analyze or to generate output of the results of the previous simulation. FINIS contains a call to MODCTL to allow the user to plot or print the results and to make changes in parameter values before another simulation is run. If option 5 was selected during construction of the model, FINIS may also contain calls to VOUT, VDIST, and VASSES. These routines are used with PARSEL in doing sensitivity and error analyses. MODCTL is normally called only during deterministic simulations, whereas VOUT, VDIST, and VASSES are called only during stochastic simulations. Any code entered into FINIS by the user ordinarily is executed before any of the MODAID routines are called.

In stochastic runs, PARSEL controls the execution of additional simulations. PARSEL terminates the run if the stopping criteria specified by the user have been met. Otherwise, PARSEL selects a new set of values for parameters and another simulation is executed. In deterministic runs, on the other hand, the user controls the execution of simulations from within MODCTL. The user can terminate a run by entering the command Q(UIT) or execute another simulation by entering the command R(ERUN). If R(ERUN) is entered, the FINIS overlay is unloaded, the MODEL overlay is reloaded, and another simulation proceeds.

USING PREMOD

PREMOD requires the user to enter three kinds of information to define the system to be simulated: (1) names of state variables and parameters used in the model, (2) FORTRAN statements that define the mathematical relationships of the system, and (3) initial values for the state variables and parameters. In addition, the user must enter information that determines the structural complexity of the program that PREMOD constructs.

All information is entered in response to a series of prompts that PREMOD displays. The prompts describe the form and content of the information to be entered. We have attempted to include enough detail in the prompts to make clear what is being expected by PREMOD. However, if the user requires additional information to help in making an entry, he can use the HELP utility of PREMOD.

The user activates the HELP utility by entering the word HELP in response to a prompt from PREMOD. PREMOD will then display additional textual material describing the information required at the point where HELP was requested. The user will be given an opportunity to return to the model construction or to select from a menu another topic of interest (table 5). The information under these topics constitutes an abbreviated reference manual that can be accessed at any time to aid in construction of the model. Once the user has viewed the topic that he requested, he can choose another topic or return to model construction at the point where he requested HELP.

PREMOD requires the user to enter the name of any variable to be placed in either the PLOTS1 or PARAMS common blocks. These variables include state variables, any other variables to be printed or plotted using MODCTL, any variables used in sensitivity or error analyses, and parameters used in the model. Each variable name entered is checked to ensure that it is consistent with the requirements of American National Standards Institute (ANSI) FORTRAN, that the name has not been entered previously, and that the name is not in a list of names reserved by PREMOD for use in the simulation model. PREMOD describes any errors that it detects and then allows the user to make another entry.

Table 5.--After entering the HELP command in response to a prompt from PREMOD, the user can request information about any of the following topics.

1. Introduction to PREMOD
2. Conventions Used in PREMOD
    - 2.1 Use of brackets [<>]
    - 2.2 Keys to press
    - 2.3 Length of variable names
    - 2.4 Dimensions of variables
    - 2.5 Typing of variables
    - 2.6 Prompts
    - 2.7 Clearing the screen
    - 2.8 Reserved names
    - 2.9 HELP requests
3. Additional Features
    - 3.1 Options
    - 3.2 COPY utility
    - 3.3 LIST* utility
    - 3.4 Override of the return and end statements
    - 3.5 Flows
    - 3.6 Keystroke file
    - 3.7 Data file
    - 3.8 MODAID tables
    - 3.9 Model file
4. General Structure of the Model
    - 4.1 Subprograms and overlays
    - 4.2 Input/output files
5. FORTRAN Conventions
    - 5.1 Arithmetic operations
    - 5.2 Implicit typing
    - 5.3 Explicit typing
    - 5.4 Comments
    - 5.5 Executable statements
    - 5.6 Mapping arrays

The user must enter FORTRAN statements to define the rate equations of the model. If option 3 or 4 was chosen, the user is also given the opportunity to add additional FORTRAN statements to the simulation model. Whenever FORTRAN statements are to be entered, PREMOD gives a prompt that includes a display of the columns in which statement numbers, a continuation character, and the FORTRAN statements may be placed legally. However, PREMOD does not attempt to check for errors in the FORTRAN statements entered by the user.

The user normally enters any FORTRAN statements directly as input from the interactive terminal. However, the user may also use a COPY utility to specify that the statements should be copied from an alternate input file. The COPY utility is activated by entering either COPY FROM ALTFILE or COPY FROM _label_ in response to a prompt from PREMOD. The _label_ is an arbitrary label of up to 10 characters specified by the user. If the first form of the command is used, PREMOD copies statements starting with the current line position in the alternate input file. If the second form of the COPY command is entered, PREMOD searches the alternate input file for an occurrence of the label in the first 10 columns of each record in the file and then copies statements from the file starting with the first record after the label. Copying is terminated when the word END is encountered in the first three columns of a record or when an end of file is reached. At that point, PREMOD again displays the prompt requesting the input of FORTRAN statements, and control is returned to the user. Thus, the COPY utility enables the user to write code for a model and place it on a file before using PREMOD and then to copy that code into the program constructed by PREMOD. The user can select particular blocks of code to be copied by using the _label_ feature of the COPY command.

There are two patterns for the entry of the names of state variables for models that utilize differential equations. In the default mode of entry, variable names and differential equations are entered state variable by state variable--i.e., the user is prompted to enter the name of a state variable, to enter a list of names of any parameters associated with the differential equation of that state variable, and then to enter one or more FORTRAN statements that define the derivative. This pattern of entry is repeated until the names of all of the state variables have been entered.

The user activates the alternative pattern of entry by entering the command LIST* in response to the prompt for entering the name of a state variable. In the _list_ mode, the user can enter a list of names of state variables. Then the user enters a list of names of parameters, followed by FORTRAN statements that define derivatives for each state variable in the list. After the user completes the entry of the differential equations, the _list_ mode is deactivated and the default mode of entry is reactivated. The _list_ mode can be especially useful where the definition of the derivatives of the state variables are to be copied from the alternate input file.

PREMOD requires similar kinds of entries for both differential and difference equation models. However, the list mode is automatically assumed for the entry of names of state variables for models that use difference equations because a single difference equation usually defines a flow between two state variables. The FORTRAN statements that define the difference equations usually include calls to the MODAID subroutines FLOW and FLOWL. The syntax of the calls to FLOW and FLOWL is given in the prompt from PREMOD.

The requests for information made by PREMOD after all of the state variables have been entered and defined depend upon the options that the user has selected. If either option 3 or 4 was chosen, the user is given the opportunity to enter the names of additional variables to be saved for printing or plotting. If option 5 was chosen as well, the user is also given the opportunity to enter names of additional variables to be used as output variables in sensitivity or error analyses. These additional variables are placed in the PLOTS1 common block and are expected to be defined in the FORTRAN statements that the user can enter when option 3 or 4 is selected.

Under options 3 and 4, the user is given the opportunity to add FORTRAN statements to the model in response to a series of prompts from PREMOD. Each prompt either displays the name of the routine to which the user can add or indicates that complete subprograms are to be entered. The user is not required to enter code in response to any of the prompts.

If option 5 is chosen, PREMOD requests the user to select which of the analytical routines of MODAID are to be included in the model. PREMOD then prompts the user through the entry of information required by those routines, including the data needed for PREMOD to create the file of variable names and statistical parameters that is read by PARSEL.

The last data that PREMOD requires the user to enter are the initial values for the variables of the PLOTS1 and PARAMS common blocks. These data are written into a file that is read by the simulation model. PREMOD prompts the user through the entry of values for every variable by listing the name of each variable and, if it is an array, by indicating how many values must be entered. In addition to values for the variables defined by the user, PREMOD requests the user to enter values for TEND, DT, and DTPL.

After all of the requested information has been entered, PREMOD generates the data files and MODAID tables required to execute the model and assembles the code for the simulation model by merging the code from the source file, code entered by the user, and code generated by PREMOD. PREMOD also creates a keystroke file, which is a file that lists all of the entries made by the user in building the simulation model. PREMOD can be directed to reconstruct a model directly from such a keystroke file. Thus, the user can make corrections or additions to the keystroke file and then give it to PREMOD to construct a new simulation model. This feature is especially helpful when new variables are added to the model because PREMOD will take care of any necessary changes in the common blocks, in dimensions of arrays required by the MODAID routines, and in the MODAID tables.

SUMMARY

MODAID is a library of software "tools" that help simplify the process of writing simulation models. These tools are programs, functions, and subroutines that accomplish tasks that are common to most ecological models. Examples of the kinds of tasks handled by MODAID routines are scheduling and updating discrete flows between compartments, solving a wide variety of mathematical functions, supplying pseudorandom numbers from many different distributions, printing output in tabular and graphical forms, and supplying high-quality interactive graphics. MODAID also contains a macro processor and a set of routines that help simplify error and sensitivity analyses.

PREMOD is a precompiler used to help build simulation models. A precompiler is a program that uses information supplied to it to write computer code. PREMOD is designed to be used in an interactive (on-line) mode. The user is prompted through the entry of the information needed to build the code for the model. In general, this information consists of the names given to the state variables to be simulated, the names of any parameters to be used in defining the rate equations associated with the state variables, the rate equations for the system, and initial values for the state variables and parameters. PREMOD uses this information to generate a FORTRAN program for the simulation model. This program makes use of MODAID routines and is designed to be used primarily in an interactive mode. In addition, PREMOD generates a data file to be read by the simulation model and several other files required by the MODAID routines. PREMOD is designed to be used by persons having limited knowledge of programming. However, PREMOD allows the user to choose options that will allow more experienced modelers to construct relatively complicated models, including models that contain the code necessary for automatically carrying out sensitivity and error analyses.

ACKNOWLEDGMENTS

We thank Vicki Kirchner and Charles Van Baker for their valuable contributions to this effort. The development of PREMOD was supported by contract DE-AS08-79NV10057 between Colorado State University and the U.S. Department of Energy.

MODAID was developed at Colorado State University under the support of contract DE-AS08-79NV10057 from the U.S. Department of Energy, contract 16-917-CA from the Northern Forest Fire Laboratory of the U.S. Department of Agriculture, contract RP1635 from the Electric Power Research Institute, grant R806016-01 from the U.S. Environmental Protection Agency, and grants DEB-76-82821, DEB-7906009, and DEB-8105281 from the National Science Foundation.

LITERATURE CITED

Iman, R. L., J. C. Helton, and J. E. Cambell. 1978. Risk methodology for geologic disposal of radioactive waste: Sensitivity analysis techniques. Technical Report 78-0912, 111 p. Sandia Laboratories, Albuquerque, New Mexico.

McKay, M. D., R. J. Beckman, and W. J. Conover. 1979. A comparison of three methodologies for selecting values of input variables in the analysis of output from a computer code. Technometrics 21:239-245.

Parton, W. J., and G. Innis. 1972. Some graphs and their functional forms. US/IBP Grassland Biome Technical Report No. 153, 41 p. Colorado State University, Fort Collins.

CURVE FITTING FOR EXOGENOUS VARIABLES: PIECEWISE POLYNOMIALS AND THE L1 NORM[1]

Gregory Zimmerman and J. D. Hanson[2]

Abstract. A mathematical approximation curve fitting procedure was used to fit functions to ecological data. Advantages of this method include the ability to achieve excellent approximations to complicated underlying functions while smoothing through noisy data. This method should be useful if representations of exogenous processes are needed.

INTRODUCTION

While conducting a modeling project, it is often necessary to functionally represent processes external to the actual model. For example, field observations or output of a model of some other system component can be summarized as a function of time. The objective is to obtain an equation which adequately represents the information in the data, while not tracking the noise. Standard regression techniques are advantageous when both good data and some theoretical basis for a given functional form are available. In low information situations, however, data may be erratic and the form of the underlying function may not be known or it may change over different regions of the domain. In these instances, standard regression techniques are inadequate.

Recent advances in approximation theory can provide more appropriate techniques for statistical curve fitting. These advances include piecewise polynomials (often termed "splines") which are useful for interpolation and approximation. Fitting functions to experimental data is an approximation and not an interpolation problem because points from the experiment are not exact and the function should not be forced through each point.

METHODOLOGY

Complicated functions can be approximated by fitting polynomials to disjunct regions of the function. In order to obtain a certain degree of smoothnness, derivatives of the polynomials at the knots (the shared endpoint of two polynomials) are forced to match. Piecewise continuous polynomials have three distinct advantages over analytic functions. First, excellent approximations to complicated functions can be obtained without requiring a large number of parameters. Second, they can be applied with a variety of fitting criteria and side conditions to obtain approximating curves with specific properties. Finally, they have various optimality properties of interest (Chambers 1977). Also piecewise continuous polynomials exhibit decreased dependence on the behavior of the fit across different segments of the data or function, especially as requirements for smoothness are relaxed (Brannigan 1981). The latter point is particularly important in the presence of erratic noise.

Piecewise polynomial fitting techniques must perform two tasks. First, the fitting technique must solve the overdetermined equations defined by the approximation in each interval. The technique must secondly determine the number and position of the knots. Various algorithms have been devised for both tasks.

The most widely used criterion for solving overdetermined equations is least squares. In a vector geometry setting, this is defined as the L2 norm. However, the least absolute deviations criterion (the L1 norm) is currently receiving wide coverage in the literature (Gentle 1977). Simple examples of L1 and L2 estimators are the median and the mean, respectively. This can be represented mathematically as:

$$\text{L1 operator: } E|X-\theta|, \Rightarrow \theta = \text{Median} \quad (1)$$
$$\text{L2 operator: } E(X-\theta)^2 \Rightarrow \theta = \text{Mean} \quad (2).$$

The L1 approach actually predates the least squares solutions, but only since the advent of modern computers and the simplex linear programming algorithm, have L1 solutions been readily obtainable.

Under certain conditions, the L2 estimates of analytic functions provide estimators with known statistical distributions, allowing statistical inferences to be made with known risks. Yet in the presence of highly kurtotic noise this advantage is nullified. Also, the behavior of the entire fit may be adversely affected by a single erratic point. Ecological data, in particular, are notorious for such erratic points.

Piecewise polynomials are not unduly affected by erratic points (Brannigan 1981) and L1 fits have the ability to filter out data inconsistencies (Gentle 1977). L2 fits are best when consistent errors are present. Eventhough the sampling distributions of piecewise polynomials and L1 estimators are not known, they may be better in some senses because they may more adequately represent the data. The question subsequently becomes the relative merit of estimates having known analytic properties versus those which more accurately represent the center. Thus, consideration must be given on a case by case basis to determine whether to use L1 and L2 piecewise polynomials or traditional analytical techniques

[1]Paper presented at the Third International Conference on the State-of the Art in Ecological Modelling, Colorado State University, 24-28 May 1982.
[2]The authors are Research Associate, Department of Range Science, Colorado State University, Fort, Collins, Colorado 80523, USA and Research Scientist, U.S Department of Agriculture, Agricultural Research Service, 8408 Hildreth Road, Cheyenne, Wyoming 82009, USA.

(least squares regression on analytic functions) for fitting curves to data.

The second, and more exacting, problem in piecewise polynomial fitting is to decide on the number and position of the knots. Examples of algorithms for this are given by Brannigan (1981) and Avila and Taylor (1979). Brannigan requires assigning a probability density function to the error terms, then an information criterion is used. Avila and Taylor require that the analyst decide on an upper bound for the error between the observed and predicted data at any one point within an interval of approximation. The more straightforward Avila-Taylor algorithm was investigated extensively on both actual and artifical data. Optional input required by the algorithm includes the order of the polynomial to be fit, the degree of smoothness desired (this is controlled by specifying the number of matching derivatives at the knots) and the error tolerance. An optimal fit can be accomplished by trying various combinations of these inputs.

Use of Algorithm on Artifical Data

In order to determine the suitability of the algorithm for fitting piecewise polynomials to ecological data, an artifical data set was concocted. The series began with an area of rapid increase and had a small peak followed by a larger peak. Subsequently, the series decreased and ended with a very short peak. Four different types of polynomials were fit to this data set. The first approximation was a fifth order polynomial derived using the L1 operator and had smoothness of C1 (first derivatives equal at the knots). The error tolerance was 0.5. The second fit was an L1 polynomial of order 5 and smoothness C2 (second derivatives equal at the knots); with error tolerance of 3.0. The final approximations were fifth order polynomials using the L2 norm with error tolerances of 0.2 and 1.0, respectively, and C1 smoothness. These trials were conducted to display possible types and efficiencies of Avila-Taylor approximations.

Use of Algorithm on Actual Data

Motivation for the use of piecewise polynomial fits in our work was provided by two separate problems. First, functions describing herbivore diets over a given season or year were needed to produce input for a forage allocation model. Data from esophogeal fistulated animals were available but these data were erratic. The processes controlling selection of the various forage plants by grazing herbivores certainly changes over time, thus one low order function could not be expected to adequately describe the data. Also, sparsity of the data curtailed the usefulness of other techniques, such as Fourier series analysis. Piecewise polynomials seemed a reasonable choice and the L1 norm was thought to be most appropriate. Another problem was to summarize plant production data into a function of time. Once again, the changing nature of controlling processes, especially precipitation, indicated that piecewise polynomials would be useful. The following examples show the types of information obtainable from the technique.

Diet Composition Data

The percentage composition in esophageal fistual collections of cattle diets were determined by Samuel and Howard (1982). Graminious species analyzed included western wheatgrass, blue grama, dryland sedge, and needle-and-thread. Three to five samples were collected at weekly intervals. Individual data for each sampling date were summarized into means and medians. Variability for means and medians of dietary composition was estimated using variances and mean absolute deviations (MAD), respectively. In each case, the performance of the algorithm was improved by adding 100 points to the data set by linear interpolation. The L1 norm, with tolerance set to MAD was used to fit functions to the medians, while the L2 norm, with tolerance set to the standard error of the means, was used for the means. All fits were required to have C1 smoothness; that is only the first derivative and the value of the polynomials involved were forced to be equal at common knots. Fifth order polynomials were used over each segment.

Production Data

Double sampling weight-estimates were made for plant production for three graminoid species in the High Plains of southeastern Wyoming (Hanson, unpublished). Measurements were taken biweekly for 20 weeks at sites representing the tops and bottoms of catenae. Data included estimates of the mean and variance of standing green phytomass at each site. The primary factor influencing phytomass along these catenae is probably moisture availability. For these trials, piecewise polynomials were fit for blue grama and western wheatgrass on the crest and swale of the catenae. The L2 norm was used to fit the double sampling means and tolerance was set equal to the standard error of the mean; third order polynomials were used with C1 smoothness.

RESULTS AND DISCUSSION

Artifical Data

The Avila-Taylor algorithm was used on an artifical data set to show the versitility and smoothing properties of the techniques (Table 1). The technique smoothed through more variation in the data as the error tolerance was increased (Fig. 1). In this artifical data, all areas of the curve were smoothed at a tolerance of 3.0. However, by reducing the error tolerance 0.5 the fit followed the data very closely. This shows how the error tolerance can be adjusted to reflect the degree of noise (or inversely the confidence) in the data. Also, as the

Table 1. Summary of inputs and results of four piecewise polynomial fits of an artifical data set.

	Fit #			
	1	2	3	4
Operator	L1	L1	L2	L2
Order	5	3	5	5
Smoothness	C1	C2	C1	C1
Error Tolerance	0.5	3.0	0.2	1.0
# of Intervals	4	1	7	3
Largest Error	0.4952	2.5364	0.1999	0.9881

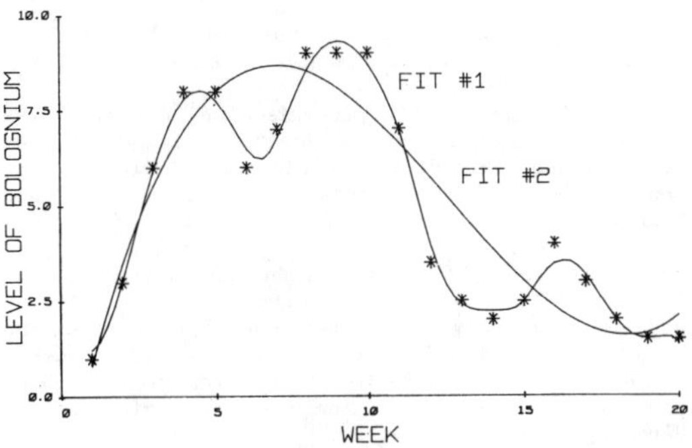

Figure 1. Resulting polynomial curves for Fits 1 and 2 of Table 1.

error tolerance is decreased, the number of intervals necessary to describe the data increases. In the case of this data set seven intervals were required in order to have an error tolerance of 0.2. By selecting high degrees of smoothness or lower order order polynomials, a similar increase in the number of intervals required would result.

Monte-Carlo simulations are being done in order to investigate the behavior of the algorithm under various noise structures. Different density functions will be used in the noise generators, and the utility of L1 versus L2 operators will be shown for particular types of noise. In addition, the sampling distributions of the piecewise polynomial estimates will be investigated (cf. Rosenberg and Carlson 1973).

Actual Data

Diet Compostion Data

Figure 2 shows median percentages of needle-and-thread and blue grama for cattle diets over a sixteen week period in 1976. Two fifth order polynomials were fit to the data for each species. For needle-and-thread, the error tolerance was set to 4.6%; two intervals were required to fit the data. The maximum error, defined as

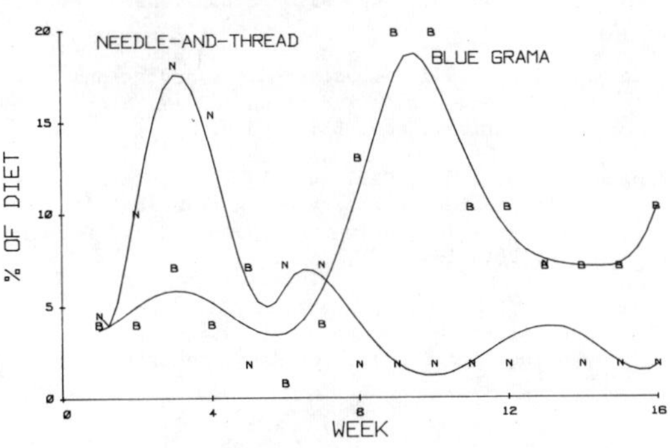

Figure 2. Piecewise polynomial fits for two grass species in cattle diets.

max(abs(observed-predicted)) (3)

across all intervals, was 4.3%. Spikes around weeks 5 and 13 were filtered and the constant phase around week 10 was smoothed. The algorithm required 1.9 sec to solve the problem. Using an error tolerance of 6.3 for blue grama resulted in a two segment fit with a maximum error of 3.3% (Fig.2). The algorithm smoothed out the dietary variation early in the season. About 2 sec were required to attain this solution.

Quite similar results were obtained for fits to median and mean for 1977 blue grama dietary data (Fig. 3). The error tolerances of the two fits were 5.68% and 5.35% for the medians and means, respectively. The associated maximum errors for the fit were 5.5% and 5.3%. The approximation using medians peaked 0.5 weeks earlier and was 4.92% lower than the actual data. Similarly, when the means were used, the predicted peak was 0.25 weeks earlier and 5.19% lower than actual. A dietary increase of blue grama around week 12 was detected by the L1 norm but not by the L2 norm. The primary differences between these two fits were (1) the L1 norm required only two segments while the L2 norm required three segments, and (2) the median approximation required 6.1 sec while the mean approximation required only 3.5 sec to solve the problem.

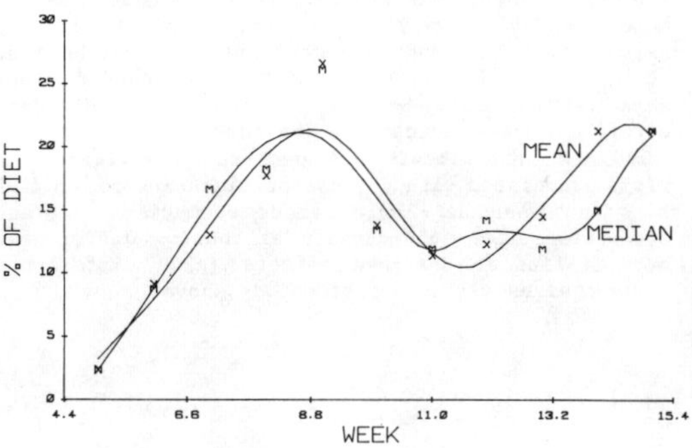

Figure 3. Median (M) and mean (X) fits of blue grama in cattle diets.

Some biological data sets contain peaks that could be considered noise; dryland sedges in cattle diets are such a case (Fig. 4). In this example most of the peaks are smoothed through. In the observed versus the fitted data the ranges of dietary sedges were 13.2 and 8.4%, respectively. This represents a 36% decrease in the range. For the L2 norm there was a 23% decrease in the range of values from the observed to the predicted diet. The error tolerances for the L1 and L2 approximations were 3.5% and 3.77%, respectively; the maximum errors for both fits were the same (3.2%). The fits for this data set were somewhat different than those for blue grama as previously discussed. In the present case, three polynomial segments were required for the medians and two for the means. Thus, a generalization concerning whether the L1 norm requires more or less segments can not be made. However, the L1 algorithm is a numerical problem thus requires more computing time than the analytical L2 solution. For the dryland sedge example, 4.6 sec and 2.0 sec were required for the L1 and L2 norms, respectively.

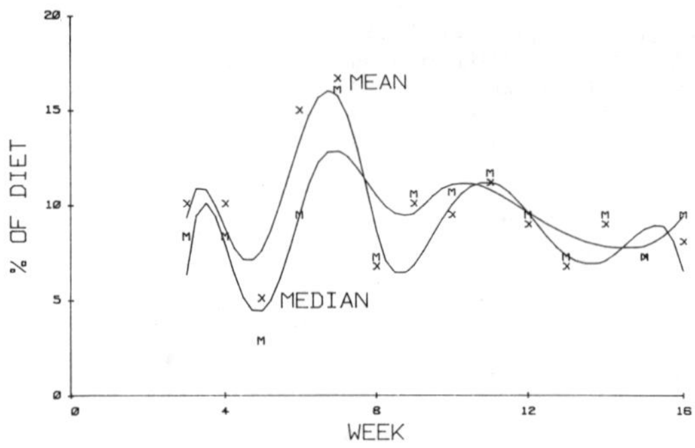

Figure 4. Median (M) and mean (X) fits of needle-and-thread in cattle diets.

Production Data

Standing green phytomass data for western wheatgrass and blue grama growing at two locations along a catena are shown in Figures 5 and 6. Note the oscillation in standing crop of both species on the upper slope; this probably reflects the lower and more temporally varying soil moisture at the top versus the higher and more constant soil moisture at the bottom. The fitted line follows the original data very closely because of the small standard errors for the species. We are currently undertaking simulation experiments designed to investigate the risks associated with the use of piecewise polynomial fits in producing single number summaries of plant production, and subsequently attibuting differences between fits to designed effects (e.g. differences between sites within a pasture, as above).

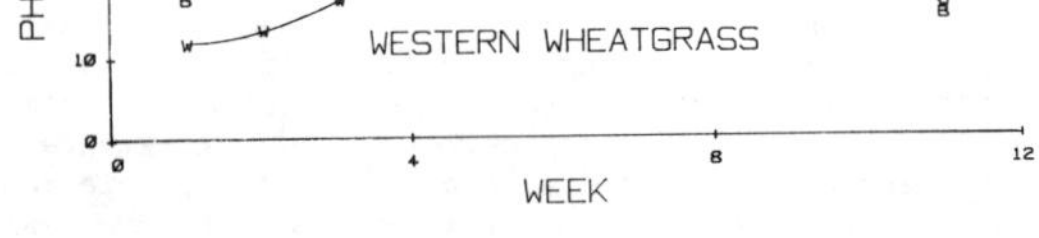

Figure 5. Two grass species phytomass at the lowland position of a catena.

CONCLUSION

When using the technique described above certain principles must be considered. First, whether to use medians or means will depend largely on the data set involved. Many data sets can best be described by the median, therefore, the median should be used. Second, more computing time is needed for the L1 fit, thus giving preference to the L2 operator for cases where L2 fits are adequate. And finally, by reducing the order of the polynomials more intervals will be required, thus increasing execution time.

The efficacy of piecewise polynomial curve fitting for analyzing ecological data has been demonstrated, thus providing the analyst with another tool. Like all techniques, it must not be used blindly. Yet, by examining the results of various choices provided by the technique, good representations of data can be accomplished. We recommend that when faced with a curve fitting problem, the analyst consider several approaches and choose the one which provides the best fit and meets the desired criteria based on the ecological problem at hand. This might include the criterion of "looking right."

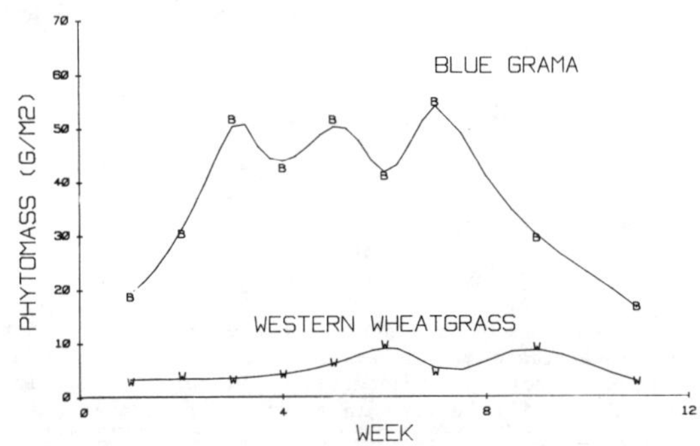

Figure 6. Two grass species phytomass at the upland position of a catena.

LITERATURE CITED

Avila, P. G. and G. D. Taylor. 1979. Adaptive L1-L2 curve fitting. Int. J. Num. Meth. Eng. 14:871-903.

Brannigan, M. 1981. An adaptive piecewise polynomial curve fitting proceedure curve fitting proceedure for data analysis. Comm. Stat. -Theory and Meth. A10:1823-1848.

Chambers, J. M. 1977. Computational methods for data analysis. John Wiley & Sons, New York. 268 pp.

Gentle, J. E. 1977. Least absolute deviations estimation: An introduction. Comm. Stat. B-Simulation. Comp. B6:313-328.

Rosenberg, B. and D. Carlson. 1977. A simple approximation of the sampling distribution of LAV regression estimates. Comm. Stat. B-Simul. Comp. B6:421-437.

Samuel, M. J. and G. S. Howard. 1982. Botanical composition of summer cattle diets on the Wyoming High Plains. J. Range Manage. 35:305-308.

MINIMUM SENSITIVITY PARAMETER ESTIMATION FOR DYNAMIC ECOSYSTEM MODELS[1]

Ann Lowes Blackwell and Charles C. Blackwell, Jr.[2]

Abstract.--In lieu of a full data set for model calibration, all the investigator's intuitive knowledge of the properties of the ecosystem being modelled must be used to find a set of parameter estimates. Some ecosystems have been described as robust, in the sense of minimum sensitivity to parameters. With parameter values known only between preassigned limits, some rough estimates about the system performance in the past, and fewer data points than unknown parameters, the property of system robustness is used to find a unique set of values of the uncertain parameters.

INTRODUCTION

Plans to manage, regulate and/or alter the environment are all too often formulated without the use of comprehensive tools, such as dynamic environmental system models, for evaluating the effects of such management, regulation, or alteration. In many instances suitable models may not exist. In other circumstances suitable models may exist but are not fully calibrated for the particular system of concern.

The dimensionality of a mathematical model of an ecological system is dictated by the geographic extent of the system to be modeled, the heterogeneity of the subsystems contained within that geographic region, and the number of significant processes which must be modeled within each subsystem in order to achieve the degree of accuracy required of the output of the model. The boundaries of the geographic region to be modeled may be quite extensive and may encompass several heterogeneous subsystems which must be analyzed concurrently because of significant interactions which occur among the subsystems on similar time scales. The reflection of system geographic extent and subsystem hetereogeneity can be observed in mathematical models of, for example, the biodynamics of lakes (MacLeod et al. in press), estuaries (Summers et al. 1980), rivers (Hydrologic Engineering Center 1975), grazing lands (Cole 1976), forest lands (Holling 1978), and agroecosystems (Shoemaker 1977).

The large dimensionality of models of environmental systems incurs a large number of parameters. Consequently, the data needed to calibrate extant models are time consuming and expensive to acquire. With notable exceptions, the availability of devices for the continuous automatic monitoring of the many variables required for the calibration of models of natural systems is rare. When all abiotic and biotic quantities which need to be measured are considered, the availability of automatic monitoring devices becomes the exception rather than the rule. When continuous automatic monitoring is attainable, a variety of analytical tools can be utilized. The method of instrumental variables (Young and Whitehead 1975, Beck and Young 1976, and Whitehead 1980) and the extended Kalman filter (McCarthy and Weaver 1977) are powerful techniques to apply to systems for which continuous or continuously sampled data are available.

When development plans are imminent it is seldom feasible to issue long moratoria until a complete set of calibration data can be obtained. In lieu of having a full data set available for model calibration, anecdotal information and all the investigator's intuitive knowledge of the system must be used to find a set of parameter estimates. These may include compartment aggregation (Cale and Odell 1979 and O'Neill and Rust 1979), state elimination (Schaffer 1981), consideration of known parameter relationships (Davidson et al. 1975), interpolation of output data (DiCola et al. 1978 and Lewis and Nir 1978), or application of other known system characteristics, such as population distribution characteristics (Aalto and Newsome 1980).

SOME PROPERTIES OF ECOSYSTEMS MODELS AND DATA

System Robustness

Clark (1976) stated that "to be a reasonable model of a biological . . . phenomenon, any dynamical system proposed should be structurally stable." If "the behavior of a system of solution trajectories" does "not change drastically if the model is subject to small perturbations," Clark defined the model as "structurally stable." This definition is similar to that given by May (1973) and includes the effect of parameter perturbations. Clark's definition of structural stability is equivalent to what some

---

[1]Paper presented at the Third International Conference on State-of-the-Art in Ecological Modelling, Colorado State University, May 24-28, 1982.

[2]A. L. Blackwell is Research Engineer/Scientist Associate, C. C. Blackwell is Professor, both of The Center for Dynamic Systems Studies, Department of Mechanical Engineering, The University of Texas at Arlington, Arlington, Texas, U. S. A.

investigators have termed "robustness." The property of model robustness in the sense of parameter sensitivity has been the subject of investigation by Miller (1974) and Miller et al. (1976). According to Miller, any variation in the behavior of the model output due to variation in the parameters of the model (within an admissible range from the nominal values) should be at least a reasonable representation of the system behavior. If robustness (defined as minimum sensitivity of the system performance measure to parameter variation) is a property of the system under consideration, that attribute can be used as an aid in the parameter estimation process.

Parameter Ranges

It is advantageous to formulate natural system models so that the parameters are functionally related to measurable biological, ecological, physical, etc. properties of the elements of the system. Often the expected ranges of these parameters (or their probability distributions) have already been determined from other studies. It is not uncommon to find in the literature an admissible range of values of the parameters (Gunnison 1978). This range of values may be due to genetic variation within a population, ecotypic variations among populations, uncontrolled environmental variations between experimental conditions not accounted for in the model, experimental error, etc. If the range of uncertainty in the parameters is due to experimental error, the actual value of each of the parameters may be unrelated to each other within that range. If the range of values represents ecotypic, population, or environmental variations, the actual values of the parameters may be related within that range.

Output Tolerances

The values of the system states or of system measurements (which are represented by algebraic functions of the states of the dynamic model) may be reported within a broad estimated range at certain points in time. For example, field data collected by observers on Isle Royale, Michigan, report that the moose population on the island was estimated between 1000 and 5000 animals in 1928 (Mech 1966). In later investigations, the moose counts on the island were given with much narrower confidence limits; Peterson (1977) reported a count of 1150 $\pm$ 260 animals in 1969. Since in this more recent data the range is comparatively small, the measurement may be considered to characterize the actual behavior of the system at that point in time.

MINIMUM SENSITIVITY PARAMETER ESTIMATION

A Resource Consumption Model

A model of the resource consumption process in which the destiny of the consumer is influenced by the availability of the resource has been proposed by Blackwell (1980). The model consists of two first order, nonlinear, homogeneous differential equations. The resource is represented by x_1 and the consumer is represented by x_2.

$$\frac{dx_1}{dt} = - a_1 x_1 - a_2 x_2[\exp(-\frac{a_3 x_2}{x_1}) + a_4 x_1 \exp(-\frac{x_1}{a_5 x_2})] + \frac{a_6 x_1}{a_7 + x_1} \tag{1a}$$

$$\frac{dx_2}{dt} = - b_1 x_2 + b_2 x_2[\exp(-\frac{a_3 x_2}{x_1}) + a_4 x_1 \exp(-\frac{x_1}{a_5 x_2})] \tag{1b}$$

$$x_1(0) = x_{10} \tag{1c}$$

$$x_2(0) = x_{20} \tag{1d}$$

The model (1) was applied to the moose-browse interaction on Isle Royale, Michigan, from the time of the establishment of the moose population in the early 1900's until the time of arrival of wolves on the island about the mid-twentieth century. The moose-browse system during that time is characterized by very large variations in consumer (moose) numbers and significant variations in the density of resource (available browse). In addition, resource to consumer ratios range from abundantly high to critically low. The biomass density of the browse dry weight available to the moose, which includes vegetation which is both within reach of and palatable to moose, is represented by state x_1. The biomass density of moose live weight is represented by state x_2.

A description of the development of the mathematical model, including the simplifying assumptions, processes modeled, state variables selected, and rationale for the mathematical form of the equations, is contained in Blackwell (1981). A brief history is given of the field observations which were made of the moose-browse interactions on Isle Royale during this period. Also included is a discussion of the means used to derive the feasible values of parameters and initial conditions for this model.

There is considerable uncertainty in some of the parameter values. The variation in the behavior of the model for the feasible ranges of the parameters and initial conditions is large. Values of the parameters were selected which were within the feasible ranges and which produced model behavior similar to that of the moose population numbers and available browse, as these variables were described by researchers on Isle Royale. The parameters selected were

$x_1(1915) = 0.08345$ kg browse m^{-2}

$x_2(1915) = 0.00016677$ kg moose m^{-2} (250 animals)

$a_1 = 1.905$ kg browse yr^{-1} per kg browse

$a_2 = 5.220$ kg browse yr^{-1} per kg moose

$a_3 = 0.08820$ kg browse per kg moose

$a_4 = 4.000$ per kg browse m^{-2}

$a_5 = 1.000$ kg browse per kg moose

$a_6 = 0.1359$ kg browse m^{-2} yr^{-1}

$a_7 = 0.0350$ kg browse m^{-2}

$b_1 = 13.03$ kg moose yr^{-1} per kg moose

$b_2 = 13.24$ kg moose yr^{-1} per kg moose.

From the simulation of $x_2(t)$ using these paramters and initial conditions, two points were selected to represent known data. The data and parameter limits were selected as

$t_0 = 1915$

$t_1 = 1941$

$xt_1 = .0013894$ kg m^{-2} (2083 animals)

$t_2 = 1950$

$xt_2 = 0.00090169$ kg m^{-2} (1352 animals)

$1.805 \leq a_1 \leq 2.005$

$0.1349 \leq a_6 \leq 0.1369$

$13.02 \leq b_1 \leq 13.04.$

It was assumed that the following observations were available concerning the moose population:

(a) in 1934 the moose population was estimated to be between 3000 and 5000 animals

(b) in 1946 the moose population was estimated to be less than 850 animals

The Estimation Procedure

In the above example a unique set of values for the parameters a_1, a_6, and b_1 must be found when

(a) the values of these parameters are known to lie within specific limits

(b) the value of one of the states (x_2) is known to lie between specific limits at two points in time and

(c) two data points are known

With only this information a_1, a_6, and b_1 cannot be determined uniquely in general.

If the model (1) is known to be a valid description of the system and it is known that the state x_2 is minimally sensitive to the parameters at the times at which data was taken, that property of robustness can be used to determine a set of unique values for a_1, a_6, and b_1. The task becomes to find the solution to a constrained nonlinear programming problem in which the objective is to find the values of the parameters a_1, a_6, and b_1 which minimize the sensitivities of the state x_2, at the times at which x_2 is known (1941 and 1950), to variations in these parameters.

Objective Function

Six sensitivities are involved

$$s_{x_2a_1}(1941),\ s_{x_2a_1}(1950),\ s_{x_2a_6}(1941),$$

$$s_{x_2a_6}(1950),\ s_{x_2b_1}(1941),\ s_{x_2b_1}(1950),$$

where

$$s_{x_2p_i}(t_j)$$

is the sensitivity of state x_2 to parameter p_i at time t_j. The requirement for the simultaneous minimization of several functions leads to a multiple objective optimization problem. Several methods which have been used to handle the problem of multiobjective decision making are described by Montgomery and Bettencourt (1977), Giesy (1978), Lin (1979), Corley (1980), and Mukai (1980). In lieu of any specific criteria by which to weight the various sensitivities, I elected to formulate the objective function as the sum of the squares of all six sensitivites.

Parameter Constraints

In order to transform the constrained nonlinear programming problem into an unconstrained problem, a combination of the method of transformation of variables (Box 1966) and the exterior point penalty function method (Zangwill 1967) was used. The inverse error function (IMSL 1979) was employed to transform the constrained parameters, a_1, a_6, and b_1 into the unconstained parameters $\tilde{a}_1$, $\tilde{a}_6$, and $\tilde{b}_1$.

$$\tilde{p} = \text{inv erf}\ [2.0\ \frac{p - p_{min}}{p_{max} - p_{min}} - 1.0], \tag{2}$$

where p is the original parameter constrained to lie between p_{min} and p_{max}, and $\tilde{p}$ is the transformed unconstrained parameter.

Output Constraints

The requirements that the estimated values of the state at the times of measurement, $\hat{x}_2(1941)$ and $\hat{x}_2(1950)$, should differ from the measured values of the data, 2083 and 1352 animals, respectively, by not more than a preassigned positive value, ϵ, become constraints h_1 and h_2.

$$h_1 = \epsilon - |\ \hat{x}_2(1941) - 2083\ | > 0 \tag{3a}$$

$$h_2 = \epsilon - |\ \hat{x}_2(1950) - 1352\ | > 0 \tag{3b}$$

ϵ was selected as 1% of the value of the data.

The requirements that the estimated value of the state $\hat{x}_2$ was between 3000 and 5000 in 1941 and was less than 850 in 1946 were formulated as

$$h_3 = 5000 - \hat{x}_2(1934) > 0 \tag{3c}$$

$$h_4 = \hat{x}_2(1934) - 3000 > 0 \tag{3d}$$

$$h_5 = 850 - \hat{x}_2(1946) > 0 \tag{3e}$$

where $\hat{x}_2(1934)$ is the estimated moose population (numbers) in 1934 and $\hat{x}_2(1946)$ is the estimated moose population (numbers) in 1946. In order to provide equal weighting, the constraints h_1 through h_5 were normalized.

Optimization

The optimization problem becomes

$$\min_{\tilde{a}_1,\tilde{a}_6,\tilde{b}_1} J \tag{4}$$

where

$$J = [s_{x_2a_1}|_{1941}]^2 + [s_{x_2a_6}|_{1941}]^2 + [s_{x_2b_1}|_{1941}]^2$$

$$+ [s_{x_2a_1}|_{1950}]^2 + [s_{x_2a_6}|_{1950}]^2 + [s_{x_2b_1}|_{1950}]^2$$

$$+ K_1\{(W_1)[(\hat{x}_2(1941) - 2083)/2083]\}^2$$

$$+ K_2\{(W_2)[(\hat{x}_2(1950) - 1352)/1352]\}^2$$

$$+ K_3\{(W_3)[(5000 - \hat{x}_2(1934))/5000]\}^2$$

$$+ K_4\{(W_4)[(\hat{x}_2(1934) - 3000)/3000]\}^2$$

$$+ K_5\{(W_5)[(850 - \hat{x}_2(1946))/850]\}^2$$

$$K_1 = \begin{cases} 1 \text{ if constraint } h_1 \text{ is violated} \\ 0 \text{ otherwise} \end{cases}$$

$$K_2 = \begin{cases} 1 \text{ if constraint } h_2 \text{ is violated} \\ 0 \text{ otherwise} \end{cases}$$

$$K_3 = \begin{cases} 1 \text{ if constraint } h_3 \text{ is violated} \\ 0 \text{ otherwise} \end{cases}$$

$$K_4 = \begin{cases} 1 \text{ if constraint } h_4 \text{ is violated} \\ 0 \text{ otherwise} \end{cases}$$

$$K_5 = \begin{cases} 1 \text{ if constraint } h_5 \text{ is violated} \\ 0 \text{ otherwise.} \end{cases}$$

and

$$\tilde{a}_1 = \text{inv erf } [2.0 \frac{a_1 - 1.805}{2.005 - 1.805} - 1.0]$$

$$\tilde{a}_6 = \text{inv erf } [2.0 \frac{a_6 - .1349}{.1369 - .1349} - 1.0]$$

$$\tilde{b}_1 = \text{inv erf } [2.0 \frac{b_1 - 13.02}{13.04 - 13.02} - 1.0]$$

where $(\ )|t_j$ indicates that the function is evaluated at time t_j.

The sensitivities were defined as

$$s_{x_2 p_i}(t_j) = (\partial x_2/\partial p_i)|_{t_j} \qquad (5)$$

according to Frank (1978).

The exterior point penalty method requires that a series of monotonically increasing values of W_1 through W_5 be utilized to perform a series of unconstrained minimizations upon J. This yields a series of solutions that more and more closely satisfy the constraints h_1 through h_5. When h_1 through h_5 are all satisfied within the prescribed tolerance the procedure is terminated.

The integration algorithm used to solve (1) and the sensitivity equations was a fifth and sixth order Runge-Kutta routine available in IMSL (1979).

The use of the quadratic penalty function $(W_j P_j)^2$ (j = 1, 2, 3, 4, 5) allows the use of the Marquardt-Levenberg minimization algorithm (Marquardt 1963) to find the solution of (4). The Marquardt-Levenberg algorithm is an unconstrained nonlinear programming algorithm that uses a modified Gauss-Newton method. In the neighborhood of the solution, the Marquardt-Levenberg algorithm requires that the Jacobian of the (unsquared) terms of the objective function be maximum rank. A sufficient condition for the system (1), subject to the constraints on the parameters and state estimates, to be identifiable is that the Jacobian be maximum rank at the minimum (Cobelli and DiStephano 1980). Consequently, convergence of the algorithm to a global minimum assures that conditions are met for uniquely identifiable parameters which satisfy all constraints. The Marquardt-Levenberg algorithm used was from IMSL (1979).

The time history of variable $x_2(t)$ from the system (1) using the parameters which generated the data from which the two sample points were drawn, the data points, and the time history of $\hat{x}_2(t)$ using the minimum sensitivity parameter estimates are illustrated in figure 1. Plots of the time history of $x_2(t)$ using the parameters a_1, a_6, and b_1 at two combinations of their extreme admissible values are shown in figure 2. Although the admissible tolerance in the parameters is small in this example, the time history of the state variables is diverse within that range. Using the minimum sensitivity parameters the time history of the variable $x_2(t)$ varies very little from that of the original model.

DISCUSSION

The feasibility of obtaining a unique set of parameter estimates with an incomplete data set for a mathematical model of an ecological system presumed to be robust has been demonstrated. Minimum parameter sensitivity may be desirable for a variety of considerations. It is important to note that the method which was developed to determine the parameter set is independent of the question of system robustness itself. The validity of the technique depends only upon the sufficiency conditions for identifiability (maximum rank of the Jacobian) and the mutual compatiblity of the parameter and output constraints. (The feasible parameter space should not be empty.)

The practicality of the application of the technique to the spectrum of models being utilized to describe the dynamics of ecological systems is a significant consideration. In order to develop all the equations required for finding the minimum sensitivity parameters, it is necessary to compute the partial derivatives of each of the state and ancillary variables with respect to each of the uncertain parameters. Since many natural system models (e. g. MacLeod et al. In press) contain a large number of state and ancillary equations and parameters, this can involve a considerable amount of mathematical computation and additional programming. If the computational capacity of the computer being used is not exceeded, the derivation, coding, and solution of these additional equations may be intimidating but not necessarily infeasible.

RECOMMENDATIONS FOR FURTHER RESEARCH

The models to which this concept has been applied have been limited to those described by differential and algebraic equations consisting of analytic functions. In simulation models, functions are often fit with straight line approximations to the function between values of the independent variable (e. g. Cole 1973). At the intersection between the two linear segments (where the slopes change) the derivatives are undefined; hence the functions are not analytic. The question of the applicability of the technique to models which are realized by nonanalytic functions is an important area for future research as is the applicability to models which are described by discrete time equations.

The number of unknown parameters which can practically be estimated is a common concern in the parameter estimation process and this concern is certainly reiterated with respect to the technique developed here. In addition, the effect of the range of uncertainty of the unknown parameters and how the magnitude of that range affects the difficulty of finding the global minimum of a potentially

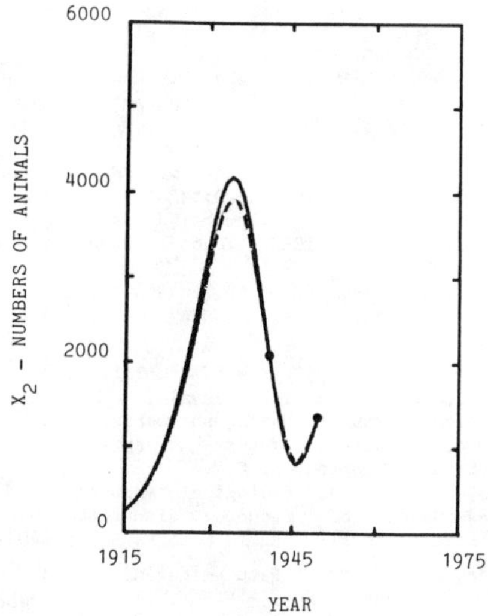

Figure 1.--Number of moose on Isle Royale fitted to equation (1b): minimum sensitivity parameters (— — —), original parameters (———), data (●).

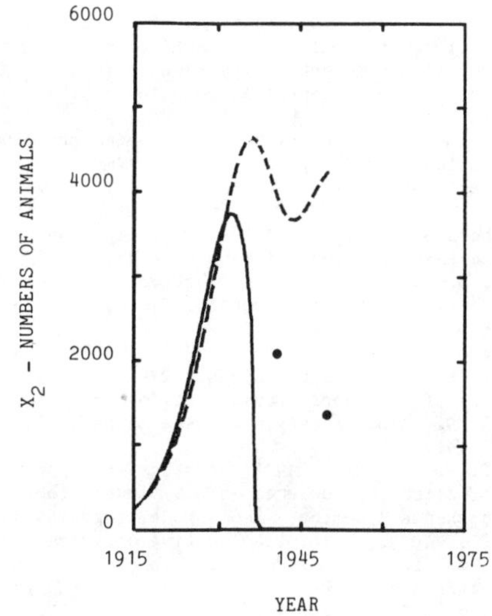

Figure 2.--Number of moose on Isle Royale fitted to equation (1b): a_1 = 2.005, a_6 = .1349, b_1 = 13.02, (———), a_1 = 1.805, a_6 = .1369, b_1 = 13.04, (— — —). Data are indicated by (●).

multimodal objective function is also an important consideration for practical applications.

The manner in which the vector valued objective function is formulated into a scalar objective function affects the resulting minimum sensitivity parameters. It would be interesting to compare various Pareto optimal solutions to the multiple objective problem.

LITERATURE CITED

Aalto, S., and G. E. Newsome. 1980. Some methods of estimating the parameters of the Leslie matrix using incomplete population data. Can. J. Fish. Aquat. Sci. 37:1140-1148.

Beck, M. B., and P. Young. 1976. Systematic identification of DO-BOD model structures. J. Environ. Eng. Div., Am. Soc. Civ. Eng. 102(No. EE5, Proc. Paper 12461, October):909-927.

Blackwell, A. L. 1981. Extremes in the resource consumption process: a mathematical model of the irruption of moose population on Isle Royale. Tech. rpt. no. 81-001. 32 p. Department of Mechanical Engineering, The University of Texas at Arlington, Arlington, Texas.

Blackwell, C. C. 1980. The resource consumption process: a macroscopic perspective. Tech. rpt. no. 80-001. 47 p. Department of Mechanical Engineering, The University of Texas at Arlington, Arlington, Texas.

Box, M. J. 1966. A comparison of several current optimization methods, and the use of transformations in constrained optimization. Comput. J. 9:67-77.

Cale, W. G., Jr., and P. L. Odell. 1979. Concerning aggregation in ecosystem modeling. p. 55-77. In E. Halfon [ed.] Theoretical systems ecology. 516 p. Academic Press, New York.

Clark, C. W. 1976. Mathematical bioeconomics: the optimal management of renewable resources. 350 p. John Wiley and Sons, New York.

Cobelli, C., and J. J. DiStephano, III. 1980. Parameter and structural identifiability concepts and ambiguities: a critical review and analysis. Am. J. Physiol. 239 (Regulatory Integrative Comp. Physiol. 8):R7-R24.

Cole, G. W. [ed.] 1976. ELM: Version 2.0. Range Science Dept. Science Series No. 20. 663 p. Colorado State University, Ft. Collins, Colorado.

Corley, H. W. 1980. A new scalar equivalence for Pareto optimization. IEEE Trans. Automat. Contr. AC-25:829-830.

Davidson, B., Y. S. Shieh, and S. M. Yih. 1975. Transient analysis of water quality in the Delaware estuary system via one-dimensional, intra-tidal, time-varying models with application to the identification of the tidal dispersion coefficient using discrete-space continuous-time ocean salinity data. p. 399-415. In G. C. Vansteenkiste [ed.] IFIP working conference on modelling and simulation of water resources, 1st. Computer simulation of water resources systems. Proceedings. 686 p. Elsevier-North Holland Publishing Company, New York.

Di Cola, G., L. Guerri, and N. Verheyden. 1978. Parameter estimation in a compartmental aquatic ecosystem. p. 823-831. In N. S. Rajbman [ed.] Identification and system parameter estimation, Vol. 2. Elsevier-North Holland Publishing Company, New York.

Frank, P. M. 1978. Introduction to system sensitivity theory. 386 p. Academic Press, New York.

Giesy, D. P. 1978. Calculation of Pareto-optimal solutions to multiple-objective problems using threshold-of-acceptability constraints. IEEE Trans. Automat. Contr. AC-23:1114-1115.

Gunnison, D. 1978. Mineral cycling in salt marsh-estuarine ecosystems: ecosystem structure, function, and general compartmental model describing mineral cycles. Tech. rpt. no. D-78-3. 93 p. Environmental Effects Laboratory, U. S. Army Engineer Waterways Experiment Station, Vicksburg, Miss.

Holling, C. S. [ed.] 1978. Adaptive environmental assessment and management. 377 p. John Wiley and Sons, New York.

Hydrologic Engineering Center. 1975. Water quality for river-reservoir systems. 210 p. U. S. Army Corps of Engineers, Davis, California.

IMSL. 1979. IMSL library reference manual, Edition 7. Vol. 3.

Lewis, S., and A. Nir. 1978. A study of parameter estimation procedures of a model for lake phosphorus dynamics. Ecol. Modelling 4:99-117.

Lin, J. G. 1979. Multiple-objective optimization by a multiplier method of proper equality constraints - Part I: Theory. IEEE Trans. Automat. Contr. AC-24:567-573.

McCarthy, D. W., and R. E. C. Weaver. 1977. Simulation and estimation of agricultural systems through Kalman filtering. p. 703-705. In W. G. Vogt and M. H. Mickle [eds.] Modeling and simulation. Vol. 8. Proc. of the eighth annual Pittsburg conf. 1155 p. Instrument Society of America, Pittsburg, Penn.

MacLeod, B. B., R. A. Park, C. D. Collins, J. B. Albanese, and D. Merchant. In press. "Documentation of the aquatic ecosystem model MINI.CLEANER." Draft. Environmental Research Laboratory, Office of Research and Development, U. S. Environmental Protection Agency, Athens, Georgia.

Marquardt, D. W. 1963. An algorithm for least squares estimation of nonlinear parameters. J. Soc. Indust. Appl. Math. 11:431-441

May, R. M. 1973. Stability and complexity in model ecosystems. 235 p. Princeton Univ. Press, Princeton, New Jersey.

Mech, L. D. 1966. The wolves of Isle Royale. Fauna of the National Parks of the United States. Fauna Series 7. 210 p. United States Government Printing Office, Washington, D. C.

Miller, D. R. 1974. Sensitivity analysis and validation of simulation models. J. Theor. Biol. 48:345-360.

Miller, D. R., G. Butler, and L. Bramall. 1976. Validation of ecological system models. J. Environ. Mgmt. 4:383-401.

Montgomery, D. C., and V. M. Bettencourt, Jr. 1977. Multiple response surface methods in computer simulation. Simulation 29:113-121.

Mukai, H. 1980. Algorithms for multicriterion optimization. IEEE Trans. Automat. Contr. AC-25:177-186.

O'Neill, R. V., and B. Rust. 1979. Aggregation error in ecological models. Ecol. Modelling 7:91-105.

Peterson, R. O. 1977. Wolf ecology and prey relationships on Isle Royale. National Park Service Scientific Monograph Series. No. 11. 210 p. United States Government Printing Office, Washington, D. C.

Schaffer, W. M. 1981. Ecological abstraction: the consequence of reduced dimensionality in ecological models. Ecol. Monogr. 51:383-401.

Shoemaker, C. A. 1977. Pest management models of crop ecosystems. p. 545-574. In C. A. S. Hall and J. W. Day, Jr. [eds.] Ecosystem modeling in theory and practice: an introduction with case histories. 684 p. John Wiley and Sons, New York.

Summers, J. K., W. M. Kitchens, H. N. McKellar, Jr., and R. F. Dame. 1980. A simulation model of estuarine subsystem coupling and carbon exchange with the sea. II. North Inlet estuary model structure, output and validation. Ecol. Modelling 11:101-140.

Whitehead, P. G. 1980. An instrumental variable method of estimating differential-equation models of dispersion and water quality in non-tidal rivers. Ecol. Modelling 9:1-14.

Young, P., and P. Whitehead. 1975. A recursive approach to time-series analysis for multivariable systems. p. 39-52. In G. C. Vansteenkiste [ed.] Modeling and simulation of water resources systems. Elsevier-North Holland Publishing Company, New York.

Zangwill, W. I. 1967. Non-linear programming via penalty functions. Management Science 13:344-358.

A MODULAR APPROACH TO COMPREHENSIVE SYSTEM SIMULATION: A CASE STUDY OF LAKES AND WATERSHEDS[1]

K. Fedra[2]

Abstract.--Within the framework of a comparative case study of Austrian lake ecosystems, a modular approach to the simulation of complex environmental systems was developed. The purpose of this simulation system, which directly links ecological models with a management and policy oriented framework, is to provide a formal basis for research, planning, and management.

INTRODUCTION

The Conceptual Framework

Environmental problems, resulting from the continuing development of society and its technology, have escalated into proportions which confound the intellectual potential of any individual who seeks to understand and resolve them. The size and complexity of the systems, the amount and variety of information to be handled and processed, the considerable gaps and uncertainties in our knowledge and basic understanding of systems behavior and finally the diversity of opinions and value systems, make it imperative to search for new approaches to cope with environmental issues in a broader framework.

The analysis of environmental systems and the policies that govern their management has to be cast into the appropriate socio-economic, political, and institutional framework. It is within this framework that any measure will ultimately have to be adopted, implemented, and managed. Consequently, the analysis will usually be complex, as complex as the environmental system and its economic and political framework itself. Feasible solutions usually will also be complex. Rarely will a singule answer, one project, one incentive, one regulation, improve a problem situation when the emphasis is on an integrated view of the system, on the overall "net" environmental effects together with their social and economic impacts. Real systems usually involve a variety of actors--government levels, organizations, industries, interest groups--with conflicting perceptions, contradictory value systems and diverging goals and objectives.

Simple solutions, which might make things much easier and much more straightforward, tend to be radical and are usually somewhat utopian when it comes to their practical implementation. If one accepts a given political and socio-economic system as largely given--which is a political rather than a scientific issue, but is usually realistic within the timespans of short- and medium-term environmental problem solving--then complexity is inevitable.

For the analysis of the lakes and watersheds discussed in this paper, this means extending the "classical" approach of load-response modeling of lakes, which requires the loading to be specified as an input (e.g., Chen and Orlob, 1975; Park et al., 1974; Vollenweider, 1975, Imboden and Gächter, 1978) towards a more comprehensive examination of lakes as integrated elements within their physical as well as their socio-economic watersheds (fig. 1). Here the pollution affecting a lake--usually excess nutrient loading, causing accelerated eutrophication--is treated explicitly; this allows study of alternative watershed management policies in their effect on nutrient loading, and subsequently, via the load response model, on lake water quality. But it also permits for an attempt at representation of any feedback that changes in water quality will have on the watershed.

Such a comprehensive view requires a spatial and functional decomposition of the systems at the level of the relevant impacts, i.e., the use of the lake, and at the level of the implementation of any management action--which is the individual community, industrial enterprise, or treatment plant. At the same time, decisions and resulting management actions are also taken at higher hierarchical levels--treatment associations, the provincial government, or the federal level, thus defining a corresponding hierarchy of systems and subsystems to be considered. Also, physical structures (e.g., watersheds) and political structures (e.g., communities or provinces) will not always fully overlap. This again results in at least two levels of structural or geographical conceptualization of any larger environmental system.

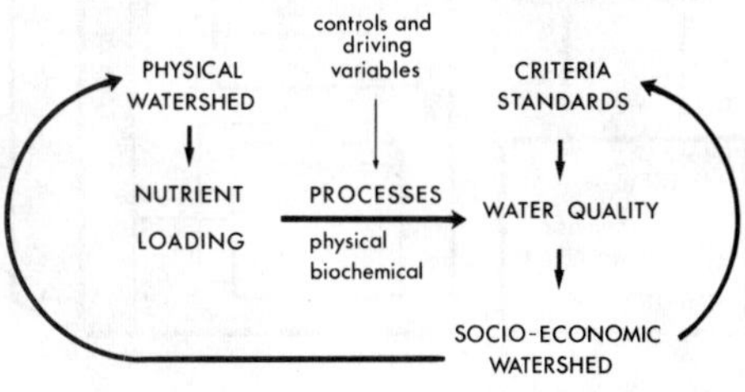

Figure 1.--The conceptual framework.

[1]Paper presented at the Third International Conference on State-of-the-Art in Ecological Modelling, Colorado State University, May 24-28, 1982.

[2]Kurt Fedra is a Research Scholar at the International Institute for Applied Systems Analysis, Laxenburg, Austria.

The Approach

Within the framework of a comparative case study of Austrian lake ecosystems, a modular approach to the simulation of complex environmental systems was developed. The explicit purpose of this system simulation is to provide a reliable and easily accessible basis for planning, management and decision making, and ultimately, integrated decision support systems.

The approach is based on a flexible system of coupled functional modules. These modules describe, at a variable and problem-specific level of detail and complexity, individual elements or processes in a comprehensive environmental system. Such elements or processes are (in the example of the lake systems) rain-runoff, erosion, pollutant release resulting from various forms of land use, tourism, and industrial activities, wastewater treatment, loading response of waterbodies, etc. Each of these modules not only contains a representation of the basic physical and biological processes, but, if appropriate, is also supplemented by routines describing socio-economic, and operational management attributes of the system (fig. 2).

With the emphasis on modularity and flexibility in design, the models can easily be adapted to different systems, and different aspects in the study of one system. Starting with a basic framework defining the elements of the systems and their interactions, as well as the indicators and performance criteria to be used, a skeleton system of dummy modules is constructed. In the simplest case, this may only consist of single number, e.g., an average export coefficient for a certain type of land use, or an average efficiency for a certain type of wastewater treatment. In this prototype system, the individual modules can now selectively be upgraded and extended into more complex and detailed ones, if necessary; for example, the spatial resolution could be increased, or institutional constraints can be added to the management options. At any point in the

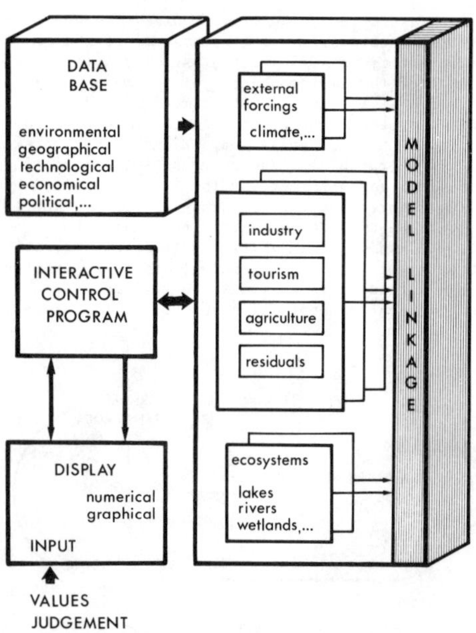

Figure 2.--A generalized flow chart of the approach.

development, however, the overall system is functional, since only one module will be exchanged at a time. As the individual modules are relatively small and easy to modify, extensive tests of their performance, sensitivity to uncertainty (Fedra et al., 1981), and structural modifications (Fedra, 1982b), can be made without excessive computational burden.

Parallel with any upgrading of the system's complexity, aggregating routines are introduced at the interface of the individual modules, which optionally preserve the basic original lumped structure of the system--and thus allow the production of comprehensible output from the simulations. Important elements of these simulation systems are the on-line linking with appropriate data bases, and auxiliary programs to generate a hierarchy of output formats of different levels of detail.

The approach is demonstrated in two contrasting examples of physically quite different systems. In the first example, an off-line simulation system of a chain of four deep lakes in the Austrian Salzkammergut is presented, where the effects of various wastewater diversion and treatment alternatives are examined. In the second example, an interactive planning and management simulation system, containing numerous stochastic elements, is described. The system simulated is an extremely shallow lake, subject to conflicting interests of industrial and agricultural development in the watershed, and the dominating tourist industry.

THE SYSTEMS STUDIED

Salzkammergut

The Environmental System

In the Upper Austrian Salzkammergut, a system of four coupled lakes, namely Attersee, Mondsee, Fuschlsee and Irrsee (fig. 3) provide a good example of the necessity of looking at catchments as a whole when dealing with the water quality problems of a given waterbody. Not unlike the situation in the Great Lakes basin (in miniature, packed into 469 km^2 of catchment area) the phosphorous loading of the downstream Attersee stems largely (>50%) from the outflow of the upstream Mondsee, which in turn receives the outflows from Fuschlsee and Irrsee (compare, e.g., Loehr et al., 1980). These deep, stratified lakes span almost the whole range of trophic conditions from the yet oligotrophic Attersee to the eutrophic Mondsee or Fuschlsee. The area is of mountainous character (up to 1700 meters above sea level), with woodlands, pastures, and uncultivated areas--frequently bare rock--dominating the landcover (table 1).

Table 1. Morphometric data of the Salzkammergut lakes.

Lake	Surface	Depth	Volume	Retention	Catchment
	km^2	m	mill m^3	years	km^2
Attersee	46.	171	3934	7.	469
Mondsee	14.	65	510	1.8	247
Fuschlsee	2.7	67	100	2.6	30
Irrsee	3.5	32	53	1.3	28

The Problem Situation

The lakes of the Salzkammergut are not only of considerable importance for tourism and consequently for

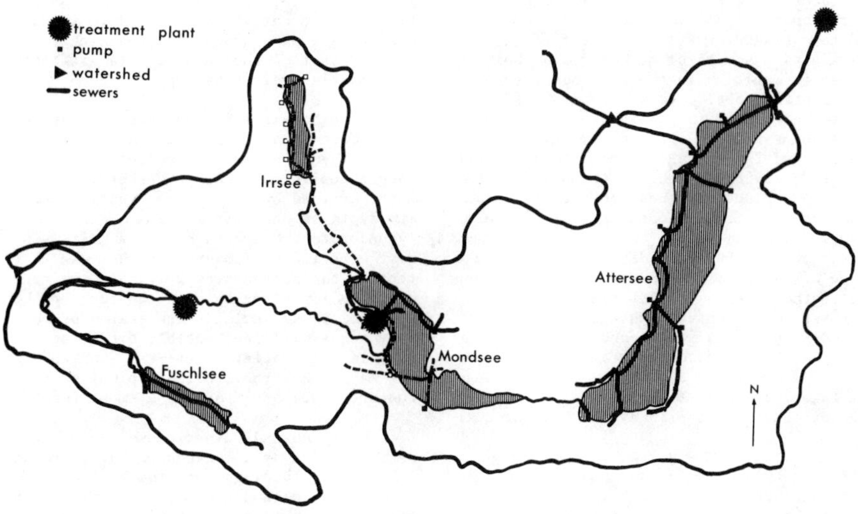

Figure 3.--Attersee: catchment, lakes, and treatment system.

regional economy, they also constitute relatively large reservoirs of high quality water; in the Attersee, the cool water of the hypolimnion would still be of potable quality, and about twenty years ago, even the now eutrophic lake Fuschl would have met drinking water standards. Considerable increase in tourism together with changes in average lifestyle, resulted not only in quantitative increases of wastewater, but also in changes in its composition. As a result, a conspicuous deterioration in water quality--decreasing visibility, algae blooms, and anoxic hypolimnic waters--had to be observed over the last two decades. This development (fig. 4) has led to the initiation of sanitation measures, namely sewage diversion and treatment (compare fig. 3). Due to the specific situation of the lakes, forming chains of coupled water bodies, any sanitation has to consider the whole system of lakes and watersheds.

The questions addressed are "classical" questions of lake water quality management: what are the relative impacts of different sources of pollution, specifically for coupled lakes? Are the current control measures, concentrating on point sources, sufficient to reverse the trend towards quality deterioration? If not, which additional or alternative control measures are necessary?

The Model Representation

The model representation of the system is designed around a lake water quality model used for trophic state prediction in terms of phosphorous concentrations, algae biomass and production, and oxygen. The model used is based on Imboden and Gächter's (1978), describing time-variable stratification and vertical mixing. Extensive numerical experiments with this model were carried out, studying calibration under uncertainty and prediction accuracy (Fedra, 1981a; Fedra et al., 1981). The model is extended such that it can handle an arbitrary number of lakes when called with the appropriate identification from the main program.

The main program handles the data base, and the communication of the individual modules, including the lake model (fig. 2). Using a daily timestep, the program generates (or reads in the case of descriptive simulation) a daily weather record for up to nine stations in the catchment. This weather record is then used to drive a rain-runoff model (Fedra, 1982), which is run for the day, for each of up to 26 subwatersheds

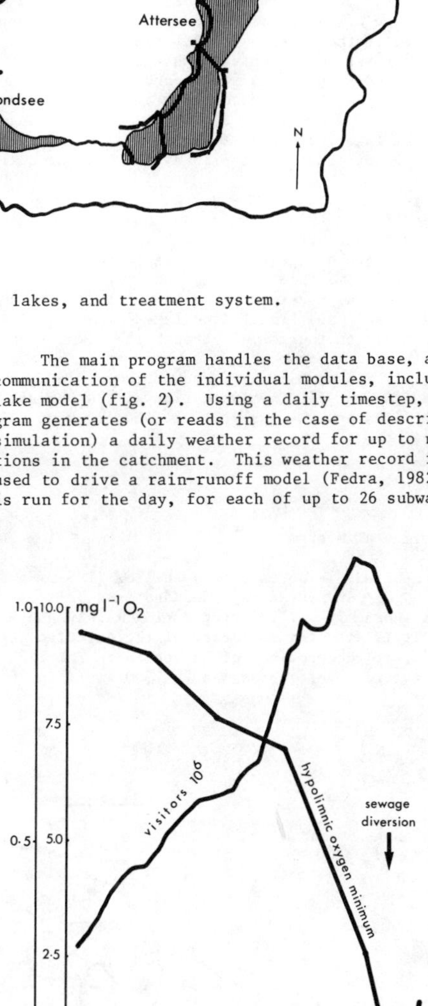

Figure 4.--Attersee region: development of tourism and water quality (Fuschlsee).

of the system. This rain-runoff model generates the inflow to the lakes (from their direct catchment), as well as a storm-runoff volume for each subcatchment, which is used to estimate erosion related diffuse sources of nutrients. According to the geomorphological and land use information in the data base, each subcatchment will contribute a certain amount of diffuse nutrient load to the inflow to the lakes. Subsequently, the next module scans through all the communities, and sums the domestic and industrial point-sources. These are then added to the diffuse sources if unsewered, or transferred to the appropriate treatment or diversion system. After treatment, the remaining effluents are added to the waterways (as in the case of the treatment plant downstream of lake Fuschl), directly into a lake (as in the case of the Mondsee), or pumped out of the watershed (as in the case of the Attersee diversion system). To these loads, an atmospheric background loading is added. Inflows of water and nutrients, which, in the case of the two downstream lakes Mondsee and Attersee also include the outflow from the respective upstream lakes, are then fed into the lake models, which are run for one time-step to evaluate water quality.

In addition to this dynamic mass budgeting of water, nutrients, and oxygen in the lakes, the program system keeps track of the number of visitors and the corresponding income from tourism, the costs of wastewater diversion and treatment, including sludge production and disposal, and finally the lakes' water quality and its feedback on tourism. Although the few indicators used to describe the system's behavior in no way cover the full range of impacts and interdependencies, they are sufficient for the evaluation of the current control policies and the qualitative comparison of alternatives.

Neusiedlersee

The Environmental System

Lake Neusiedl is an extremely shallow (1.5 m) lake of about 150 km^2 surface area, embedded in a belt of dense reed (Phragmites), covering approximately another 150 km^2. It is situated southeast of the Austrian capital Vienna, in the province of Burgenland. The lake's catchment extends over approximately 1300 km^2 (fig. 5) in the easternmost part of the country, in a low plain (120 m.a.s.l.), opening towards the Small Hungarian Plain in the east, and bordered in the northwest and west by groups of mountains reaching 748 m. The southernmost part of the lake and its artificial outflow are on Hungarian territory.

Lake Neusiedl is the largest Austrian lake in terms of surface area. A detailed description of the limnology of the lake has only recently been published in a comprehensive monograph (Löffler 1979). The lake is characterized by its extreme shallowness (with a maximum depth of only 1.8 m), and a peculiar and only partly understood hydrography (Gattinger 1979). One of the most conspicuous features of the lake is the large extent of the surrounding wetlands, the reed belt. It plays an important role in the nutrient budget of the system. These wetlands are reknowned for their rich and unique wildlife, notably birds, and together with a series of smaller lakes--the "Lacken"--southeast of the main lake, they, being typical of the landscape of the area, are an important element for tourism.

Lake Neusiedl experienced obvious water quality deterioration as a consequence of a dramatic increase in nutrient loadings over the last decades, attributed to changes in land use patterns (strong increase in intensive agriculture, particularly wine growing), the development of food processing industries, feedlots and chicken farms in its catchment, and a continuing increase of recreational activities around the lake's shoreline. Besides one major point source, the river Wulka, and a series of smaller, but well identified point sources such as the discharge from local sewage treatment works (numbering 26 in the catchment, including 4 under construction), a considerable influence of diffuse sources, including atmospheric sources (largely related to wind erosion of fertilized soils), is suspected.

The catchment is dominated by agricultural land use, and particularly wine growing around the lake. Agricultural production also includes grain, sugar beets, vegetables, and fruits, processed in the local sugar and canning industries. The dominant economic sector in the region, however, is summer tourism, most of it concentrated around the lake (fig. 5), and much of it depending on the direct use of the lake for bathing, boating, sailing, and surfing.

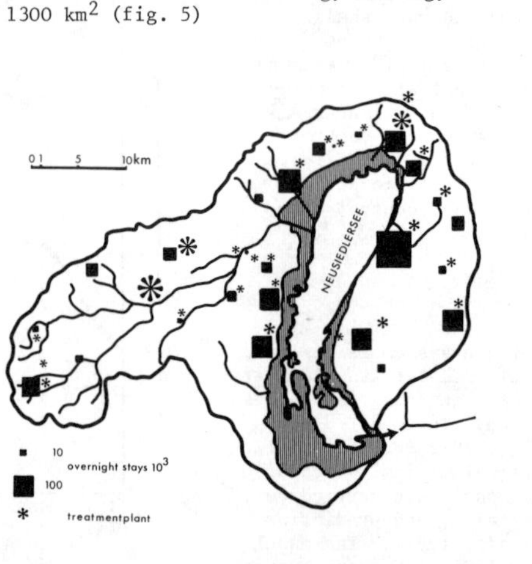

Figure 5.--Neusiedlersee: catchment, communities, and treatment plants (major settlements scaled according to overnight stays in 1979).

The Problem Situation

Since the early seventies, a conspicuous deterioration of the lake's water quality could be observed. A rapid increase in nutrient concentrations (nitrogen, phosphorus) in the lake was reported (fig. 6), paralleled by changes in plankton composition and an increase in plankton biomass (e.g., Löffler, 1979). This development, resulting in at least local algae blooms, is certainly most undesirable, as it quite obviously endangers the lake's attractiveness for recreation.

Recreational use of the lake however, or the income from tourism, is one of the most important elements in the economy of the region. Only recently, apparent water quality problems and several controversial development projects around the lake have triggered a heated political debate. The specific problems in the management of the lake system have to be understood as resulting from three major conflicting objectives in the development of the region, namely:

(a) development of tourism, leading to construction activities, usually adversely affecting landscape and biota, especially in the shore zones of the lake; it requires road building, and the construction of channels, marinas, and bathing beaches to provide direct access to the open lake; it also involves traffic congestion and crowding, and increased domestic waste and sewage;

(b) intensification of industrial and agricultural production; this involves obvious and direct forms of pollution such as emissions of air pollutants from industrial plants, wastewater and solid waste production, or the production of excess manure from animal husbandry and feedlots; it also includes less obvious forms of pollution, resulting from fertilizer leaching, wind erosion of fertilized soils, or the enhancement of erosion resulting from the conversion of rangeland into vineyards, from land fills, and irrigation; finally, the landscape and its diversity may be adversely affected by a shift towards increasingly larger monocultural land use;

(c) the preservation of environmental quality (as an essential resource for tourism), especially lake water quality, but also wildlife and the scenic beauty of the area, ranging from the conservation of historical rural architecture to the preservation of the typical pannonian landscape.

These objectives, which are clearly conflicting, have to be seen in the context of the region's overall development within Austria: situated at the "closed" border towards Hungary, the industrial and infrastructural development of Burgenland had a late and slow start in comparison with the overall Austrian development after World War II. A large number of commuters, attracted by labor opportunities mainly in the nearby capital, Vienna (pop. 1.7 mill.), is characteristic of the region.

The rapid development of tourism (fig. 6) together with the installation of food processing industries using locally grown fruits, vegetables, and sugar beets, as well as a remarkable development in agricultural activities, especially wine growing (relative increase of about 50% in the cultivated area from 1966 to 1975, and 48% from 1970 to 1978), are therefore important elements of the region's economic development. This development, however, is obviously paralleled by a decrease in environmental quality, and in particular, an increase in nutrient loading and the observed deterioration of water quality.

The Model Representation

In the model representation, the catchment is disaggregated into 46 communities, grouped into three districts within Austria, and one in Hungary. Together with these communities, 26 treatment plants and three major industries (food processing and canning, and sugar production) are treated explicitly. In the core of the simulation system is a phosphorous-based model of the lake and its surrounding reed belt (fig. 7). The model uses available and particulate (detritus) phosphorus in the open water, the water in the reed system, and in the sediments of the reed system, to describe

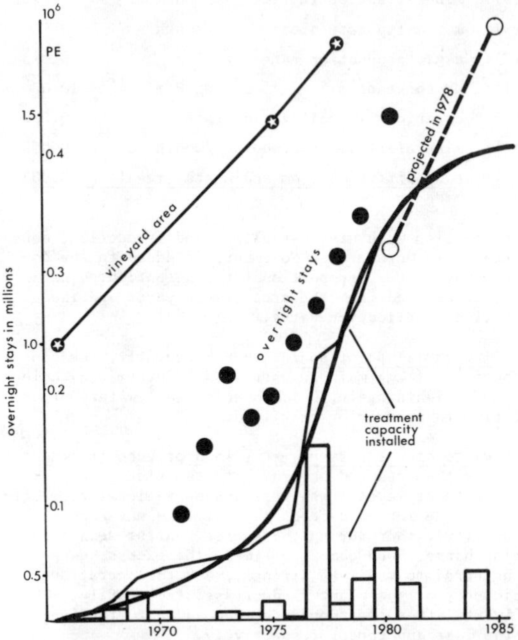

Figure 6.--Neusiedlersee: development of tourism, wine growing, and treatment capacity.

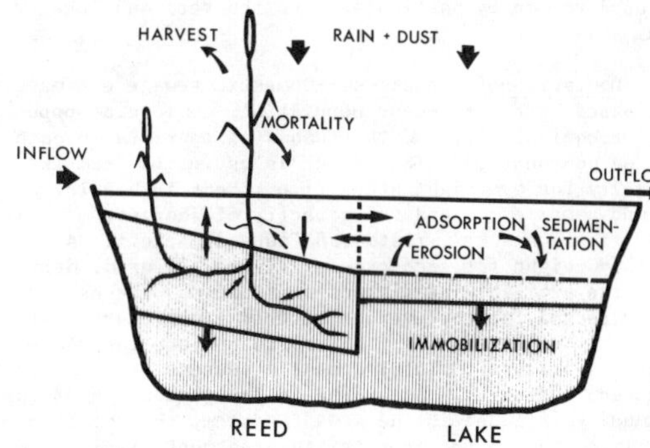

Figure 7.--Neusiedlersee: a diagram of the lake and reed model.

the dynamics of phosphorus, algae, and reed biomass. The latter can be exported from the system by harvesting, which not only reduces reed biomass, but also affects (depending on the harvest technology selected) the diffusion between the sediments and the water of the reed system (by pore squeezing due to heavy machinery), the mixing between reed and lake, and reed mortality. In addition to nutrient uptake and growth, mortality, respiration, and mineralization, various forms of mixing and transport are included. Nutrients are transported through the reed zones to the open lake to balance the outflow from the lake, and also by eddy diffusion along the border between reeds and open water. Since the latter process is considered to be largely wind driven, it is made a function of average wind velocities, and under ice, the mixing coefficient is correspondingly set to values of molecular diffusion. In addition to the settling of detritus (largely in the calm reed zone), adsorption and coprecipitation of phosphates in the open lake are included as a governing mechanism in the model. A detailed description of the model together with examples of its application is available (Fedra, 1982b). Some basic constants, initial values of the state variables, and parameters are listed in Table 2.

The model system is operated on a monthly time-step: after intialization and going through optional interactive parameter editing, the programs, for each month, generate a climatic record (temperature, winds, runoff in the main river, and outflow from the lake), which is either read from tables for historical periods or generated by random perturbations of the historic records, preserving their statistical properties. The program then calls a series of subroutines, which, for each of the community areas, estimates different sources of pollution, again taking phosphorus as a proxy for pollution affecting water quality:

Diffuse waterborne pollution.--This is made a function of runoff, each community's drainage situation (described by one lumped parameter describing slopes, drainage density, and geology), and land use. The latter includes also fertilizer application and irrigation. Also, a specific monthly weight, depending on the distribution of groundcover and farming activities throughout the year is applied.

Diffuse windborne pollution.--Wind driven erosion of fertilized soils is again made a function of the above landcover information, as well as on each community's position (direction and distance) towards the lake. For each month, an erosion and transport-relevant wind rose is generated, using eight main directions, which is used to convey particulates to the reed and lake system.

Domestic point sources.--Domestic sewage estimates are based on the resident population, commercial population equivalents, and the number of tourists in each of the communities. The latter is estimated from the capacity for overnight stays (i.e., beds in hotels, or second homes etc.) and the capacity of short-term visitors (i.e., beaches, restaurant capacity etc.). A monthly weight for each type of visitor is used, derived from the available statistics on tourism. The estimate is corrected for the average monthly temperature, water quality, or in case the lake is frozen (and therefore attracts skaters). Total domestic sources are summed for each community. According to the percentage of the community included in the sewage system, the total is partly routed to the appropriate treatment plant; from the unsewered part, a certain fraction, depending on the drainage situation, is routed to the lake system. As a specific feature of the area, wastewater produced

Table 2.--Neusiedlersee water quality model: constants, states, and parameters.

Neusiedlersee: input summary: lake and reed system			
area reed		in km^2	100.000
area lake		in km^2	150.000
water volume reed		mill. m^3	50.000
water volume lake		mill. m^3	150.000
initial conditions (January):			
reed biomass		mg P/m^2	15000.000
detritus reed system		mg P/m^3	100.000
available p reed system		mg P/m^3	100.000
detritus p reed (sediments)		mg P/m^2	25000.000
available p reed (sediments)		mg P/m^2	20000.000
detritus lake system		mg P/m^3	100.000
available p lake system		mg P/m^3	30.000
algae biomass		mg P/m^3	2.000
total phosphorous content in tons:			6029.800
parameter values:			
p (1):	sedimentation in reed	/month	0.300
p (2):	mineralization in reed	/month	0.020
p (3):	reed production rate		0.010
p (4):	reed carrying capacity	mg P/m^2	18000.000
p (5):	reed mortality	/month	0.010
p (6):	ratio of reed nutrients from water		0.250
p (7):	sedimentation in lake	/month	0.015
p (8):	mineralization in lake	/month ^oC	0.025
p (9):	mortality rate algae	/month	0.400
p(10):	algae production rate		0.075
p(11):	MM constant p	mg P/m^3	10.000
p(12):	organic p immobilization rate		0.004
p(13):	mineralization sediments	/month ^oC	0.003
p(14):	eddy diffusion from sediments (reed)		0.003

in connection with wine harvesting and production, concentrated in October and November, is added to the domestic wastewater, proportional to the estimated harvest, determined from the area of vineyards and the fertilizer application rates.

Industrial point sources.--Subsequently, the contributions of the main industries are determined, using monthly weights again, which specify wastewater output relative to average or peak load.

Wastewater treatment.--The load of each treatment plant is then reduced according to the plants' capacity, the treatment level installed, and operational reliability. If the inflow exceeds capacity, as may happen especially either during the summer tourist season or during harvest and campaign times, the excess sewage is transferred to the lake directly, and the operational efficiency of the plant is decreased temporarily. The effluents after treatment are then added to the input to the lake and reeds, respectively.

Water quality evaluation.--After summing the nutrient inputs reaching the reed belt and the lake directly for a month, the lake program is called to update the water quality status of the lake.

In addition to this dynamic nutrient budgeting, the program keeps track of revenues from tourism, the costs of wastewater treatment and reed harvesting (which can be specified monthly). Most of the estimates used in these models also include some stochastic perturbation in an attempt to account for uncertainties and natural fluctuations in these processes and relationships.

SYSTEM SIMULATION

In the simulation of these two systems, two different strategies were employed and compared. In the case of the Attersee, a traditional "off-line" approach was used. The required input file was prepared, and the programs subsequently run for a specified period, producing output to be read on a screen terminal, printed, or plotted. As it turned out, frequent changes in the output formats had to be made, since the information produced in one single run was by far too much to be all printed or stored. On the other hand, it turned out to be difficult to foresee which aspects of the systems' behavior would be of specific interest in each run. Although certain summary information was always found to be useful--such as aggregated regional statistics comparable to the ones usually compiled by government agencies--the preparation of an adaptive output format was found to be difficult. Such a format would, automatically, switch the level of detail and resolution; for example, if a water quality standard is violated, it would help to trace back the immediate causes. This, however, requires subjective evaluation, which is difficult to build into any program.

In summary, the major problem of the off-line simulation of the relatively large, comprehensive system, was found in the intelligent reduction and selective handling of model-generated information.

In the case of the Neusiedlersee, an interactive approach was chosen in contrast. Also, parallel video graphics (fig. 8) were added to the simulation programs for the display of geographically organized information. This not only allowed more information to be displayed simultaneously, it also allowed the recognition of spatial patterns, which are otherwise difficult to discern.

In the interactive version, the level of detail of the output can be selected adaptively. At any month, the system will only display the aggregated regional statistics (visitors, and relative contributions of various sources (i.e., diffuse, according to land use classes, atmospheric, domestic, industrial), the totals, the distribution of loading between lake and reed, and some residual statistics. Lake water quality status is indicated on a verbal scale.

Water quality is meant to represent the visitor's perception of water quality, i.e., it is based on algae biomass, detritus, and temperature, which are primarily responsible for color, taste, and odor. A weighted average of these variables is translated into a ten grade scale of quality classes, ranging from "excellent" via good, fair, moderate, to bad, disgusting, and finally dangerous. Since these qualifications represent a visitor's perception, water quality is always excellent when the lake is frozen.

For each month however, each of the individual communities' contributions, the loading and performance

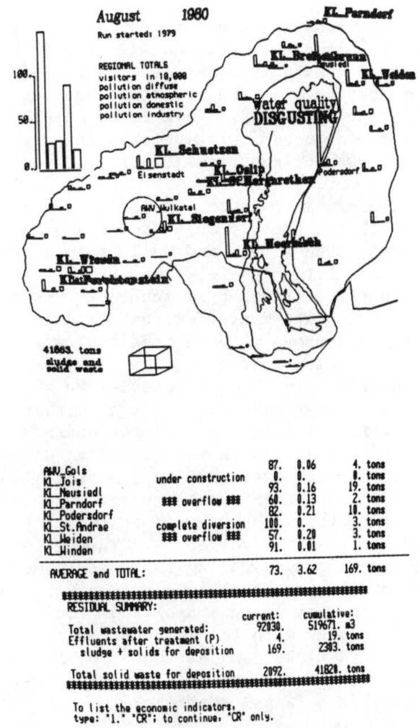

Figure 8. Display units of the interactive simulation system.

of all treatment plants, and economic indicators (revenues from tourism, and costs of reed management and wastewater treatment) can be displayed. Also, a detailed listing of the lake and reed systems' status (including all state variables and the average flows during the last month) can be displayed under input control, using interactive menu techniques with input feasibility checks.

At the end of each year, after the display of aggregated summaries of the above information, the system offers another round of editing of the control parameters: hotels can be built, or treatment plants upgraded, etc., guided by the interactive protocol of the system. These decisions (eventually with a certain time lag in case of major structural measures) will then be effective in the next year of simulation. At the end of a simulation run, an optional hardcopy version of all outputs can be printed on the line printer.

DISCUSSION

Detail, Scale, and Aggregation

To make the simulation systems described above operational, they have to be built on numerous gross simplifications. Each of these might be anathema to the specialist in the respective field--but only ruthless and sometimes naive abstraction and simplification will make such comprehensive systems work. The degree of resolution and the level of sophistication of the individual modules within the whole system has to be balanced, with special emphasis on the specific problem areas addressed in the study, but also, more pragmatically, as a consequence of data and resource limitations. The level of detail required certainly depends on the questions asked--and in complex and usually badly defined problem situations, as are normally encountered in environmental studies, these questions will have to be reformulated repeatedly during the course of the study. As a consequence, the structure and resolution of the simulation system will have to be adopted. This requires a large degree of flexibility, and a program design that allows for easy modifications.

Scale is a critical factor in these systems. Many phenomena, which are utterly unpredictable on a micro-scale, turn out to be simple and well-behaved, and thus predictable, on a larger scale. The error, for example, in the rain-runoff simulation is inversely proportional to the size of the catchments, or the forecasts of numbers of visitors improve with the regional aggregation. Similarly, in the case of the shallow Neusiedlersee, attempts to simulate the lake's behavior on a day-to-day basis or with spatial disaggregation within the lake failed--the description of average monthly properties however turned out to be satisfactory (within the limits imposed by data availability), especially when interpreting the resulting monthly values as mean values of probability distributions, based on the in-lake distribution of the observed variables. This, however, is rarely surprising considering the underlying randomness in many of the processes involved--given the model representation is more or less unbiased. Relative precision also increases with the application of successively larger yardsticks, and although the absolute error will also increase, the result might still serve its purpose within a larger framework.

Aggregation not only improves the relative precision of the simulation. It is essential in making the output from the models comprehensible. The aggregation used in this approach, however, is of the most trivial kind: whereas the underlying models operate with full resolution, it is only the output that is aggregated geographically. The real challenge would be a functional aggregation together with the geographical one, and running the models on the aggregated level until certain critical conditions, requiring further resolution, occur. Then the model would have to switch to a functionally disaggregated level. To obtain the initial conditions for these newly started subelements however, a formal disaggregating scheme is necessary. Assigning initial conditions to the subunits, when only the sum totals of these values are known, requires some additional knowledge of the relationships of the subunits to each other. This information could be obtained from a few full resolution trial runs. Although new sources of errors are introduced with such procedures, it might allow a considerable increase in the numbers and levels of units treated simultaneously.

To simulate large systems, and at the same time span the various hierarchical levels of organization usually operating within such systems, some kind of aggregation is essential. Also, much of the information available, specifically on the socio-economic attributes of environmental systems, will usually be available only in aggregated form. This aggregation follows the hierarchical structure of organization of the political and administrative systems, but since these data are an important part of the ultimate testing ground of any model generated predictions, the model output should be structured and formulated accordingly. These well-known and familiar structures are also most appropriate for communications--whereas information organized in terms of physiographical units, say for example, watersheds, would be difficult to translate into familiar terms.

Testability, Reliability, and Credibility

To be acceptable to any responsible planner, manager, or politician, a model generated output has to be credible. To be credible, it has to be testable. To be testable, it has to be fully transparent and understandable, or at least able to demonstrate reasonable performance in critical test situations. Although formal computer based methods are, in principle, open and transparent, it would be unrealistic to expect anybody to really carefully study someone else's several thousand lines of code and huge data bases. In the demonstrations of the program systems described above, confidence--with reservations--in the output was established when the models could reproduce known historical behavior of the systems. In general, the limitations of this kind of approach were well understood and accepted by scientists, planners, managers, and politicians.

Significantly, most of the test persons asked for trial runs under extreme and more or less unrealistic conditions: for example, diverting all wastewater from the lakes, or totally banning fertilizers and phosphorous detergents for the whole basin. Although unrealistic in the sense of possible implication, such runs helped to improve the systems' credibility, even if they did not produce some intuitively expected behavior. Another interesting observation in comparing people's reactions was in the obvious preference given to the graphical display and the single work description of the water quality standard (although arbitrarily derived from the simulated variables) over the detailed numerical and seemingly "quantitative" output on the screen terminal. Although the information in the graphical display is qualitative and comparative only, it was found more attractive and meaningful than all the tables of double precision numbers. The numerical output was only occasionally referred to, when something dramatic happened on the screen--overflowing treatment plants, or a lake water quality indicator flashing "DISGUSTING."

These attitudes however, should at the same time be taken as a warning that attractive display may well be distracting from the inevitable weaknesses of the approach, and pretend a degree of reality that can be very misleading and eventually dangerous. To overcome or at least reduce this problem, it is worthwhile attempting to attach a measure of reliability to any forecast. Another possibility is to translate seemingly precise numbers either into coarse classes (as in the case of the water quality indicators), or to present them in terms of distributions (Fedra, 1981a). The latter should be familiar information to people who are used to working with the results of statistical surveys or polls.

Modularity and Learning by Modeling

The basic purpose of these comprehensive systems is to serve as a formal basis for communication and learning. Attractive output formats such as the video display of familiar designs such as statistical information drawn into maps, can help to bridge an important communication gap between modelers and people without technical training. In contrast, it is almost hopeless to get somebody involved in studying large quantities of computer printouts. Some easy to understand summary information, preferrably graphical, is always required.

Another important element is in the immediate feedback from the interactive simulations: the possibility to play with the system, to explore in an experimental fashion its reactions to external changes, seem to be attractive and powerful in helping people of different backgrounds to communicate over one common problem. This, however, requires a user oriented protocol which makes changes easy, and keeps frustrations from delays or unstable programs to a minimum. It is essential to be able to make modifications to the system more or less immediately--for example, adding digestion facilities for sludge and then either burning it or recycling it to agriculture is an option one might want to add to the programs of the Nersiedlersee area, and explore the consequences in, say, one morning.

Adding additional features--or removing them, if found redundant for a given question to streamline the system--requires a flexible, that is modular, program design. The approach presented is understood as a kind of kit: a kit of tools or toys, to be combined and recombined in an attempt to build a model of certain aspects of environmental systems. The approach is essentially pluralistic. It is well understood that there is no single model or set of models, that would meet all the requirements of a diverse group of people using it. Attempts to design "the supermodel" are bound to fail, since just the problem formulation will most probably change before even half of the code has been generated and debugged.

Instead of asking "how to improve models?"--a question traditionally answered by including more and more detail and more and more sophisticated process descriptions, and at the same time asking for more data, one could ask: why and for what should a model be used, and what can we expect to learn from the model? Within which framework is it going to be used? When trying to answer such questions, problems of model complexity and detail may soon become redundant. When putting models of ecosystems into a sufficiently larger environmental context, it may well turn out that more or less qualitative answers are sufficient. Since, for example, the coupling between algae biomass and the benefits from the use of the lake, including a visitor's willingness to pay for his stay at the lakeshore, are in no way understood quantitatively, what good would it be for the planning of the lake area's development to try and predict biomass levels with the utmost precision anyway? However, one certainly would want to know whether a certain policy will result in "improving" the water quality, or if the lake is likely to turn disgusting.

At the same time, however, it is most desirable to utilize all the available information to develop and test models. In case of ecosystems, much of this information is on the micro-scale of samples and laboratory experiments. Consequently, in order to gain confidence in qualitative predictions of very coarse and simple models, one might want to compare them with a much more detailed, high resolution "state-of-the-art" model. In the modular simulation framework discussed above, this can be done very easily: plugging out one model and plugging in another one instead. Depending on the resolution required, the data available, and the emphasis of the study, a comparison of various models can be made, checking the robustness of the simulation system under uncertainty (in model selection).

It is important to note that whatever the degree of sophistication may be in such a system, its result will always require human judgement and interpretation, which should, as much as possible, be external to the program. At best, such systems can serve as an open, reliable (in the sense of repeatable) basis for communication and explorative learning through simulation, thus providing some background and aid for decisions, which, as it involves human values, will always be political in nature.

LITERATURE CITED

Chen, C.W., and G.T. Orlob. 1975. Ecological Simulation for Aquatic Environments. p. 475-588. In Systems Analysis and Simulation in Ecology, Vol. III. B.C. Patten (ed.). Academic Press, New York.

Fedra, K. 1980a. Austrian Lake Ecosystems Case Study: achievements, problems, and outlook after the first year of research. CP-80-41. International Institute for Applied Systems Analysis. 148 p. Laxenburg, Austria.

Fedra, K. 1980b. Mathematical modelling - a management tool for aquatic ecosystems? Helgoländer Meeresuntersuchungen 34: 221-235.

Fedra, K. 1981a. Estimating model prediction accuracy: a stochastic approach to ecosystems modeling. In Progress in Ecological Engineering and Management by Mathematical Modelling: Proceedings of the Second International Conference on the State-of-the-Art in Ecological Modelling, Liege, Belgium. R. Dubois (ed.). p. 87-105.

Fedra, K. 1981b. Pelagic foodweb analysis: hypothesis testing by simulation. Kieler Meeresforschung, Sonderheft 5: 240-258.

Fedra, K. 1982a. A simple watershed-model for daily runoff estimation: a module for a regional water quality simulation system. Working Paper. International Institute for Applied Systems Analysis. Laxenburg, Austria (forthcoming).

Fedra, K. 1982b. Water quality, tourism, and agriculture: a simulation study of conflicting objectives (Lake Neusiedl, Austria). Working Paper. Laxenburg, Austria (forthcoming).

Fedra, K., G. van Straten, and M.B. Beck. 1981. Uncertainty and Arbitrariness in Ecosystems Modelling: A Lake Modelling Example. Ecological Modelling 13: 87-110.

Imboden, D., and R. Gächter. 1978. A dynamic lake model for trophic state prediction. Ecological Modelling 4: 77-98.

Loehr, R.C., C.S. Martin, and W. Rast. 1980. Phosphorus Management Strategies for Lakes. Ann Arbor Science, Ann Arbor, Michigan.

Löffler, H. (ed.). 1979. Neusiedlersee: limnology of a shallow lake in Central Europe. Monographiae Biologicae 37, 543 p. Dr. W. Junk, The Hague.

Park, R.A., R.V. O'Neill, J.A. Bloomfield, H.H. Shugart, R.S. Booth, R.A. Goldstein, J.B. Mankin, J.F. Koonce, D. Scavia, M.S. Adams, L.S. Clesceri, E.M. Colon, E.H. Dettmann, J. Hoopes, D.D. Huff, S. Katz, J.F. Kitchell, R.C. Kohberger, E.J. LaRow, D.C. McNaught, J. Peterson, J. Titus, P.R. Weiler, J.W. Wilkinson, and C.S. Zahorcak. 1974. A Generalized Model for Simulating Lake Ecosystems. Simulation 23(2): 33-50.

Vollenweider, R.A. 1975. Input-Output Models, with special reference to the phosphorus loading concept in limnology. Schweiz Z. Hydrol. 37: 53-84.

A MODEL FOR DETERMINING 'APPROPRIATENESS' IN USING ENVIRONMENTAL SYSTEMS IN APPROPRIATE TECHNOLOGY APPLICATIONS[1]

Ane D. Merriam[2], Shirley P. Burggraf[3]

Abstract.--This paper explores the economic and energetic measurement methodologies of the relative values of natural vs. mechanical systems for solving environmental problems.

I. INTRODUCTION

A. Man's Original Reliance on Natural Systems

Man's curiosity about, attraction to, fascination, study and management of natural systems are as old as civilization (Bryan, 1973). The Roman aqueducts channeled water from the productive watersheds to their people. The Chinese culture coevolved with its intensive agriculture accelerated by irrigation efforts. The Seminole Indians were noted for living closely associated with the swamplands and meandering coastlines of the Southeast. In fact, modern cities across the landscape have arisen in direct relationship to available natural resources including water for transportation and consumption, forests for timber, rich soils for farming, and productive estuaries for fishing harvests.

Much of the collective accumulated wisdom presumably gleaned through historical cohabitation with the earth's diverse natural systems, however, seems not to be considered when modern society confronts daily decisions concerning difficult resource management problems. Building in floodplains and along hurricane threatened coastlines, intensive farming by ecologically irrational methods, rebuilding in fire-prone and mudslide areas, draining wetland water recharge areas to encourage more water demanding growth, and paving over of water purifying areas for growth that generates more runoff are but a few examples of how modern man rejects the sustenance of his natural roots (Pimentel, 1980).

B. Urbanization, Economic Growth and Mechanical Systems

The growth of what might be termed ecologically dysfunctional behavior has corresponded, of course, with the occurrence of population concentration and economic growth trends that accelerated with the industrial revolution. This process reached its apex in the United States in the 1950s and 1960s, a period which also correlated with unprecedented consumption of cheap fossil fuels. Economic growth and development created a wedge which seems progressively to have separated man from nature (Odum, 1971).

The sudden rise in cost of fossil fuels which occurred in the 70's, however, caused both a slowdown in economic growth and a realization that supplies of fuels and resources are limited. Beginning with legislative initiatives that were undertaken in the early 70's when the economy was still relatively strong, sewage treatment facilities and advanced waste treatment plants were built across the nation with public funds. With the slowdown of economic growth which began in the mid-70's, however, society experienced increasing difficulty in paying for such expensive public installations. Public money allocated to resource management has subsequently become considerably less available, yet the problems remain (Tschinkel, 1982).

C. Old Solutions for New Problems

If Florida is representative, there is now an increased interested in relying more on natural systems for solving resource management problems; i.e., letting nature do what nature has always done in the areas of air and water purification, soil conservation, and climate control. "Save-our-Rivers" (F.S.(a) 1982) and "Save-our-Coasts" (F.S.(b), 1982) can be viewed as in some sense a second-generation treatment of the 1970s "appropriate technology" thinking which responded to the realization that energy supplies are finite. The extent to which such approaches are rational, however, requires considerably more analysis and evaluation to determine which lands and waters should be saved, flooded, used for recycling sewage, or saved from paving. The public will simply not, nor should it, support an open-ended commitment to natural systems vs. mechanical ones.

The "Save" groups implicitly assume that the value of saving natural systems is undervalued by the economy in some way. It is the explicit hypothesis of this paper that this assumption is true at least in many important cases and that a methodology can be developed for evaluating the specific values. In passing, some of the reasons why the "fair" values of natural systems may have been implicitly underestimated by society in the past probably include at least the following:

(1) The availability of cheap fossil fuels and cheap capital for building and operating mechanical systems for resource management.

(2) The existence of "externalities" in decision making; i.e., a separation of the effects of decisions from those who make the decisions (e.g., the developer who decides to cut trees in order to build more units on a specific property and therefore make more profit may be totally unaffected by the runoff problems and long-run deterioration of air, water, and soil that result from his decision.)

---

[1]Presented at the Third International Conference on State-of-the-Art in Ecological Modeling, Colorado State University, Fort Collins, Colorado, May 24-28, 1982.

[2]Director, Water Resources, Institute of Science and Public Affairs, Florida State University, 361 Bellamy Building, Tallahassee, Florida 32306.

[3]Professor of Economics, Florida A&M University, 301 Lee Hall, Tallahassee, Florida 32307.

(3) Lack of appropriate concepts and data for making decisions about natural systems vs. mechanical ones in that many of nature's "services" are poorly understood and because they are free do not suggest measurement (i.e., we don't pay currency for their service).

D. An Analytical Approach to Nature's Services

Now that the energy and capital required for mechanical systems of resource management have become prohibitively expensive, there is a need to reevaluate the value of services which nature can perform as an alternative to expensive and frequently destructive mechanical systems.

Prairie settlers who surrounded their dwellings with cottonwoods and junipers understood that trees provided protection from extremes of wind and temperature. Trees are also known to absorb air pollution, noise and water pollution. One study found that a 500-meter wide greenbelt surrounding a factory reduced nitric oxide concentrations in the air by 27 percent, sulfur dioxide by 26 percent, and hydrogen sulfide by 1 percent (Bach, 1971). Trees can also absorb urban water runoffs and serve as an advanced treatment plant for sewage, as well as prevent soil erosion. These "services" have a very definite if difficult-to-measure economic value, which, given the increasing expensiveness of mechanical systems and decreasing availability of funds for building them, needs to be evaluated as carefully as possible.

II. METHODOLOGY OF PAPER

We are interested in approaching the concept of "appropriate technology" from several angles. A technology might be deemed appropriate by an energy engineer if it performs the desired task by expending the least possible amount of energy. To the economist, on the other hand, "appropriateness" means economic efficiency as measured in market costs with possible consideration of "externalities" added in. To local decision makers, "appropriate" undoubtedly means the cheapest available system for doing the job. (This may not be the same as economic efficiency; the key may rest in what is politically 'available' or feasible.) We propose to measure the services of natural systems in two very distinct ways--an economic analysis and an energy analysis--and interpret their relationship for policy purposes. Both approaches will be applied to two specific cases: (1) An urban greenbelt in the center of Sacramento, California; and (2) a swamp near Gainesville, Florida, which has been incorporated into the city's water treatment system as a substitute for a tertiary treatment plant.

The economic measurement of the value of the greenbelt and the swamp will involve estimating the cost of replacing the services performed by nature with mechanical systems. The concept is very straightforward in the case of the swamp in that it only involves putting a cost figure on a tertiary treatment plant. In the case of the greenblet, it involves estimating the costs of replacing the beneficient effects that trees have on air, water, soil, and climate, a much more complicated estimation.

The alternative analysis of the two natural systems (the greenbelt and the swamp) is done from an energy perspective. We attempt to measure the amount of work done by the natural system; i.e., the amounts of sunlight, water and carbon dioxide absorbed or alternatively the amounts of oxygen produced, using a methodology developed by Odum (1973).

The two measurements are then compared for what they show about the relative "appropriateness" of specific natural systems for performing certain environmental tasks.

III. METHODOLOGY APPLIED TO TWO CASE STUDIES

A. Method

For two case studies, a terrestrial urban ecosystem in Sacramento, California and another in a cypress strand wetland environment near Gainesville, Florida, calculations are made to determine:

- The dollar cost of providing the pollution control via the technology available and the fuel costs associated with that approach;
- The energetic costs of using a natural system to provide the same pollution control service and a conversion to dollars.

The economic calculations are straightfoward, using standard construction, operation, and maintenance figures for a technology and specific performance standards. The energetic calculations include another component--that of energy quality. While it is intuitive to most engineers that a Btu of coal compared with a Btu of electricity is different, it is equally intuitive to biologists that a Btu of sunlight is different from a Btu of biomass. Odum (1975) developed this notion to provide the basis for energy quality conversions to be incorporated in the energy analysis calculations. Merriam and Gilliland (1981) discuss the concept further and apply it in a paper which serves as a precursor to this one.

In addition to the energy quality scales, other conversion factors used in the energy analysis convert energy values to dollar values. As is pointed out in Figure 1, the U.S. economy is a function of dollar flows (as measured by the Gross National Product or GNP) plus the subsidies reflected in the energy flows of natural systems providing recycling and purifying services to society. In 1975, the total annual economic production is represented in a GNP of \$1,437 billion. Yet, the GNP only measures the value of the goods and services produced, or is a reflection of the economic value of the work produced as a result of energy flowing through and being processed by machines. However, there are many instances of passive subsidies provided by natural systems, where dollars are not exchanged, but energy is. So, a better estimate is to measure the amount of fuel actively consumed by society in generating the GNP, plus the solar fuels consumed by nature to provide pollution control subsidies to the GNP-economic system. In 1975 this is the sum of 73 quadrillion Btu's of fuel (Calculation 1, Appendix) and 26.6 quadrillion Btu's of solar fuels (Calculation 2, Appendix), for a total of 99.6×10^{15} (quadrillion) Btu's. Therefore, in 1975, technology using fuels contributed two-thirds of the power behind the GNP and environmental systems using sunlight contributed one-third. With this calculation, the ratio of dollars to energy units may be determined. For 1975 it is measured as 99.6×10^{15} Btu's/\$$1,437 \times 10^9$ or about 70,000 Btu's/\$1.00.

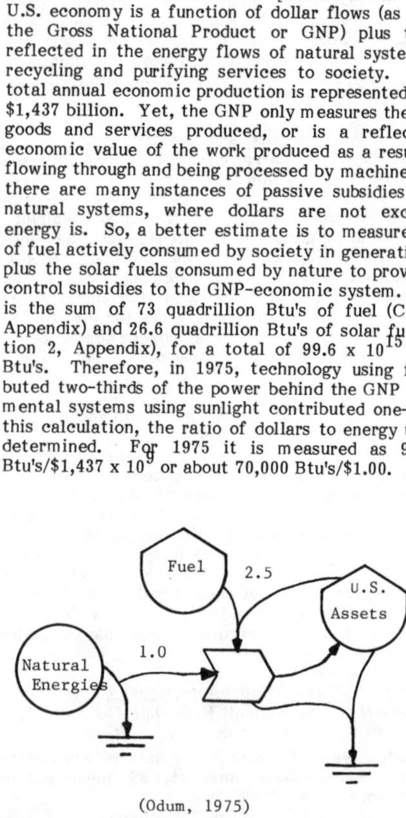

(Odum, 1975)

B. Results

Two separate cases where urban processes were coupled with environmental systems clean-ups were evaluated. The first is the Capitol Park complex in Sacramento, California. It is comprised of 40 acres, five acres of which house the Governor's Office and legislative chambers. The remaining 35 acres constitute an open botanical garden of 350 different species of mature trees.

The second is a 66-acre complex located in Waldo, Florida comprised of 40 acres of cypress strand used for wastewater recycling and 26 acres for buffering. The natural system treats to tertiary standards approximately 100,000 gallons per day of sewage effluent which, before being discharged into the strand, receives between primary and secondary treatment.

These analyses focused on:

- The capital asset value of the structure in the vegetation and the annual value of the services provided by the ecosystem using energy and dollar units.
- Traditional economic analysis of the cost of providing the pollution control services via technology--considering the dollar and energy expenditures of the replacement technologies.
- Comparison of the results of each analysis in the following ratios:

(a) $\frac{\text{Energy costs of technology}}{\text{Energy costs of natural system}}$

(b) $\frac{\text{Dollar costs of technology}}{\text{Dollar costs of natural system}}$

Table 1(a) presents the energy costs (and converted dollar costs) of the terrestrial urban ecosystem in California performing the pollution control services of air, water, and soil purification and reduction in heating and cooling of the Capitol complex buildings. Table 1(b) reflects the replacement technology costs for those same pollution control measures, with associated calculated and observed energy consumption of the technology envoked.

Table 1(a).--Energy and dollar values for Capitol Park predicted using energy analysis method. (Merriam and Gilliland, 1981).

	Energy (Btu's)	Dollars (1975)
Capital Asset Value	30-140 billion	0.4-2.0 million
Annual Value of Services	0.9-4.7 billion per year	0.014-0.070 million per year

Table 1(b).--Energy and dollar values for Capitol Park calculated using economic methods as fossil fuel and dollars saved. (Merriam and Gilliland, 1981).

	Fossil fuel saved in million Btu's per year	Dollars saved in $/year
Heating/cooling	2000-6000	4100-12,300
Water treatment	160-270	3000-5000
Air purification	160-5900	2000-7000
Soil conservation	130-6500	1000-50,000
Total range of annual services	2660-19,270	10,100-137,300

Table 2(a) presents the energy cost (and converted dollar costs) of using a wetlands ecosystem for wastewater purification and recycling using the energy analysis technique. Table 2(b) reflects the dollar costs and energy consumption of the replacement technology (i.e., tertiary or advanced waste treatment).

Table 2(a).--Energy and dollar values for Waldo, Florida cypress strand--sewage treatment using energy analysis methods. (Ordway, 1976 and Wharton and Odum et al., 1977).

	Energy (Btu's)	Dollars (1975)
Capital Asset Value	25.2 billion	0.36 million
Annual Value of Sewage Treatment Services	1.26 billion	0.02 million

Table 2(b).--Energy and dollar values for Waldo, Florida cypress strand--sewage treatment using economic analysis. (Ordway, 1976 and Huffnagle, 1982).

	Energy (Btu's)	Dollars (1975)
Capital Asset Value	99 billion	1.85 million
Annual Value of Services	5.4 billion	0.10 million

Table 3(a) compares the energy(i) and economic(ii) costs of the natural approach versus the technology approach for the Sacramento case. Table 3(b) compares the energy(i) and economic(ii) costs of the Waldo-Gainesville case.

Table 3(a).--Ratio of natural systems solution verus mechanical solution for Sacramento, California case.

(i) Energy Costs:

$$\frac{\text{Using technology}}{\text{Using trees}} = \frac{10.9 \text{ billion Btu's/yr*}}{2.8 \text{ billion Btu's/yr**}} = 3.89$$

(ii) Dollar Costs:

$$\frac{\text{Using technology}}{\text{Using trees}} = \frac{\$73{,}500\text{/yr***}}{\$42{,}000\text{/yr****}} = 1.75$$

*Calculation 3, Appendix
**Calculation 4, Appendix
***Calculation 5, Appendix
****Calculation 6, Appendix

Table 3(b).--Ratio of natural systems solution versus mechanical solution for Waldo-Gainesville, Florida case.

(i) Energy Costs:

$$\frac{\text{Using advanced waste treatment}}{\text{Using cypress strand}} = \frac{104.4 \text{ billion Btu's*}}{26.46 \text{ billion Btu's**}} = 3.95$$

(ii) Dollar Costs:

$$\frac{\text{Using advanced waste treatment}}{\text{Using cypress strand}} = \frac{\$1.95 \text{ million***}}{\$0.38 \text{ million****}} = 5.13$$

*Calculation 7, Appendix
**Calculation 8, Appendix
***Calculation 9, Appendix
****Calculation 10, Appendix

IV. SUMMARY AND CONCLUSIONS

The conclusions concerning both the Gainesville swamp and the Sacramento greenbelt are that the economic values exceed the energy values by substantial multiples.

It is now time to consider what is really being compared. The economic replacement cost is in some sense also an energy measure of the alternative mechanical systems in that the replacement cost of a natural system is really the market value placed on the energy, labor, capital, and materials necessary to build and operate the mechanical system. Since labor is human energy, and capital and materials can be viewed as embodied energy, the "economic" value is also a form of energy value associated with the alternative mechanical system.

Additionally, the calculations in Tables 3(a) and 3(b) point to an interesting observation. In the case of Waldo, Florida the decision to use the wetlands for sewage treatment was an "active decision", which may be reflected in a higher economic ratio value (5.13) than does the Sacramento case, which is an example of a "passive coupling" (1.75 economic ratio value). Therefore, the higher economic ratio values may indicate greater economic efficiencies for man for those instances where nature is already adapted to the end use prescribed. (In this case, the swamp is pre-adapted to absorb nutrients and recycle waste water, whereas the greenbelt example is more of a fortuitous coupling of nature and man.)

It can therefore be concluded that when the ratio of the economic value of service replacement by a mechanical system to the energy value of the natural system exceeds 1, the natural system is the appropriate system in the energy sense as well as in the economic sense. The same physical inappropriateness of splitting atoms for the purpose of boiling water which would offend an energy engineer should also offend the economist for reasons of economic inefficiency. "Appropriate technology," then, has very real meaning in at least two contexts. There is no particular conflict between energy appropriateness and economic appropriateness of a system as long as care is taken to measure the services of nature correctly, services which have historically been neglected by economists and virtually all decision makers. Because nature gives its services freely rather than selling them, their values are easily overlooked in our economic system.

A policy implication of this form of analysis which needs to be stated clearly is that the choice between mechanical and natural systems is hardly a matter of "either . . . or." It is rather a matter of looking for potentialities in nature for doing jobs more efficiently ("appropriately") in both the environmental and economic senses. Where they exist, they can provide both economic and environmental savings.

Beyond the initial problem of reckoning with energy flows and monetary values, another problem may be developing for decision makers. While the economy is slowing, energy prices are fluctuating and economic theory is being tested at the highest political levels. The principles of economic efficiency and optimal responses are being shifted, shuffled and perhaps ruffled. It has been argued that just as intermediation is natural to an expanding economy, disintermediation is the natural and appropriate activity in a steady or contracting one (Hawken, 1981). So, perhaps the economic system itself is asserting the more direct couplings of human and natural productivities by discouraging further proliferation of the middleman technologies of pollution control. If so, where better should the process of disintermediation come to a head but in the cohabitation and coevolution of man and nature--the ultimate balance between concentration and dilution, containment and dispersal. Meanwhile, the economists and ecologists continue to be dyslexic in their pursuit of valuing natural systems' contributions to society.

The conflicts and benefits of employing both the economic and energetic approaches are reflected in a recent similar modelling struggle between population ecologists and geneticists in the development of population biology models (Levins, 1966). Perhaps, like the ecologists and geneticists before, current ecologists and economists merely need to fine tune their respective peceptional lenses. A suggestion made in another context, but perhaps applicable here, is:

> . . . it is important to see distant things as if they are close and to take a distanced view of close things. (Musashi, 1974).

LITERATURE CITED

Bach, Wilfred. 1971. Steps to better living on the urban heat island. Landscape Architecture. Vol. 61, p. 138, January.

Bryan, W. 1973. The effects of growth on the urban environment. In Land use, energy flow, and decision making in human society. Vol. 2, Interdisciplinary Systems Group. Univ. California, Davis, CA.

Florida Statutes. 1982(a). Save-Our-Rivers. Formerly SB620, Chapter 8133, Laws of Florida.

Florida Statutes. 1982(b). Save-Our-Coasts. HB593 (not passed); HB16 (passed); HB1015 (passed/vetoed); HB1162 (not passed); HB1008 (not passed); HB594, SB255 (passed); HB 700, SB 595 (passed); HB 1176 (not passed).

Hawken, Paul. 1981. Disintermediation: An economics buzzword that neatly explains a lot of the good that is going on. Co-Evolution Quarterly, No. 29, Spring. POINT, Sausalito, Ca.

Huffnagle, V. 1982. Personal communication. Department of Environmental Regulation, Tallahassee, Florida.

Levins, Richard. 1966. The strategy of model building in population biology. American Scientist 54: 421-431.

Merriam, A.D. and Gilliland, M.W. 1981. The solar energy machine is green. In Energy and ecological modelling. Developments in Environmental Modelling 1. W.J. Mitsch, R.W. Bosserman, and J.M. Klopatek (eds.). Elsevier Scientific Pubs. Comp., Amsterdam.

Musashi, Miyamoto. 1974. A book of five rings. The Overlook Press. Woodstock, NY.

Odum, H.T. 1971. Environment, power and society. Wiley-Interscience, NY. 317 pp.

Odum, H.T. 1975. Energy quality principles for estimating the environmental carrying capacity for man. Given at the Environmental Health Conference at Pahlavi University, Tehran, Iran.

Ordway, J. 1976. General costs for cypress wetlands recycling; Projected costs of cypress wetland disposal for Waldo, Florida. In Cypress wetlands for water management, recycling and conservation. (Odum, H.T. et al.) Center for Wetlands, University of Florida, Gainesville, Florida.

Pimentel, David, E. Garnick, A. Berkowitz, S. Jacobson, S. Napolitano, P. Black, S. Valdes-Cogliano, B. Vinzant, E. Hudes, and S. Littman. 1980. Environmental quality and natural biota. Bioscience. November, 30(11): 750-755.

Tschinkel, Victoria. 1982. Testimony on behalf of the National Governor's Association and the State of Florida. U.S. Senate Committee on Environment and Public Works, February 26, Washington, D.C.

Wharton, C.H., H.T. Odum, K. Ewel, M. Duever, A. Lugo, R. Boyt, J. Bartholomew, E. DeBellevue, S. Brown, M. Brown, and L. Duever. 1977. Forested wetlands of Florida--their management and use. Division of State Planning, Tallahasssee, Florida.

APPENDIX

Calculations:

1. Fossil fuel component (coal, oil, natural gas) is 94 percent.

2. Solar energy component calculated as solar isolation on the U.S. divided by 2000 to convert sunlight to its equivalent fossil fuel energy form. Assumption is that the solar energy is coupled to the U.S. system by providing air, water and soil purification; climate control; recreation, etc.

3. Midrange of the sum of four services calculated in Merriam and Gilliland, 1981:

   (a) Climate control: 2000-6000 x 10^6 Btu's/yr
   (b) Water treatment: 160-270 x 10^6 Btu's/yr
   (c) Air purification: 170-5900 x 10^6 Btu's/yr
   (d) Soil conservation: 130-6500 x 10^6 Btu's/yr

4. Midrange of annual value of services calculated as gross primary production at 16,800 kcal/m^2/year or 2.38 x 10^{19} kcal/year for the 35-acre park; 3.968 Btu's/kcal yields 9.44 billion Btu's; divided by 2 to convert to its equivalent in organic matter, yields 4.7 billion Btu's.

5. Dollars calculated in Merriam and Gilliland 1981 for four services, midrange of the sum:

   (a) Climate control: $4100-12,300/yr
   (b) Water treatment: $3000-5000/yr
   (c) Air purification: $2000-7000/yr
   (d) Soil conservation: $1000-50,000/yr

6. Conversion from energy costs (calculation 4) to dollar conversion at 70,000 Btu/$ yields $67,143. All values divided by 5 using the conservative conversion to fossil fuel units and the low end of the range.

7. Dollar values supplied by the Department of Environmental Regulation and crosschecked by J. Ordway's projections in Odum et al. 1977. Conversion to energy values per Merriam and Gilliland 1981 using an estimate of 54,000 Btu's for each dollar of materials purchased in the construction and manufacturing of the mechanical system (tertiary treatment plant).

8. Total energy value of cypress strand calculated using gross primary productivity values reported in Odum, et al 1976 as 180 billion Btu's in standing stock for the 66 acres of cypress strand and 9 billion Btu's for annual gross primary productivity. Yet, that value represents all of the services the strand could provide; therefore, each value is multiplied by 14%--which is estimated by Wharton, Odum, et al. 1977, as the average percent of work attributable to water purification by swamps, yielding values of 25.2 billion Btu's capital asset and 1.26 billion Btu's operation and maintenance, when totaled = 26.46 billion Btu's.

9. Actual costs incurred at the Waldo site: $1.85 million capital asset and $0.10 million operation and maintenance.

10. Conversion of 70,000 Btu's/$.

SIMULATION OF SIMPLE FOOD CHAINS[1]

E. T. Elliot, R. G. Wiegert, and H. W. Hunt[2]

Abstract.--We have developed a food chain model with one resource and three biotic components which, using the appropriate parameters, demonstrates oscillatory behavior attributable to the interaction of the top two biotic components. The model includes feedback controls for resources (renewable) and space (non-renewable). The influences of refuge level, potential predator growth rate, predator space limitation, and enrichment all influence the oscillatory nature and state of the model. Changes in seasonal inputs (forcing functions) in natural systems when superimposed upon an intrinsic oscillatory property under appropriate parameter values could give extremely complex behavior. Changes in the parameters affecting the top predator cause alterations which are propagated throughout the entire system.

INTRODUCTION

Despite the considerable effort spent developing models of predator-prey relationships and coupled predator-prey models (food chains), there is little congruity in the factors used for controlling these relationships in ecological models. The array of past experimental and theoretical efforts give testament to this. The effects of predators on prey have been studied, as well as the effects of prey on predators; and consideration has been given to the simultaneous interaction of the two trophic levels. Food limitation from the perspective of one level is predator control from the perspective of the other. Primary producer-herbivore interactions may differ from herbivore-carnivore relationships, and the understanding of top carnivores and omnivory may considerably influence the function of particular ecosystems. Spatial heterogeneity and refugia should be considered as controlling factors. Models, both mechanistic and analytical, have been developed to consider many aspects of organismal interactions, either as isolated subsystems or within the context of a whole-ecosystem model driven by seasonal abiotic factors. Since ecological-model construction is our attempt to order or explain organismal systems as we observe them--and there is an incredible array of these systems in existence--it is no wonder that so many models have been developed to describe feeding relationships in ecosystems.

[1]Paper presented at the Third International Conference on State-of-the-Art in Ecological Modeling, Colorado State University, May 24-28, 1982.

[2]E. T. Elliott is a Research Ecologist with the Natural Resource Ecology Laboratory, Colorado State University, Fort Collins; R. G. Wiegert is Professor of Zoology, University of Georgia, Athens, Georgia; and H. W. Hunt is Senior Research Ecologist, Natural Resource Ecology Laboratory, Colorado State University, Fort Collins, Colorado, U.S.A.

One method for examining the behavior of a food-chain model is to sequentially increase the amount of resources flowing into the system (enrichment). Smith (1969) developed such a model containing either four or five levels (the resource pool is counted as a level). He observed that the addition of the fourth trophic level (biotic component) produced different results from that of a system containing three biotic levels, independent of the parameters used. However, at low-resource densities, the top level was not supported and the five-level system reverted to four levels. In model runs that contained three biological levels, the standing stock of plants was high and increased with enrichment, whereas the standing stock of resources was low and actually decreased with enrichment. These trends were "reversed completely" with the addition of the top carnivore. He found that increasing the complexity of the model by creating a food web had only "minor effects" on the general response of the model, and that the model behaved most closely to that predicted from observing the structure of the dominant path of the model. If there was not a dominant path, the model responded in a manner intermediate between the two trophic levels. The same was true when the amount of resources entering the system was enough to support the third biotic level but only a portion of the fourth. When the top trophic level of the model was self-limiting because of space limitation or self-predation, intermediate model response was again observed; but if the top level was completely self-limiting, it had little, if any, influence on the general response of the model (Smith 1967). Variation of inputs in time and time delays in the model had destabilizing effects, while spatial heterogeneity almost invariably stabilized the model (Smith 1972).

The model of Smith (1969), using Michaelis-Menton-type transfer functions, demonstrated increases or decreases in the biomass of all components as enrichment proceeded, as subsequently hypothesized by Fretwell (1977). However, this behavior has in general not been observed in models of other workers (Rosenzweig 1973, Wollkind 1976, Oksanen et al. 1981), who used different transfer functions than Smith (1969).

Based on the assumption that maximum gross primary production is the "ultimate regulator" controlling the length of food chains, Oksanen et al. (1981) reanalyzed previous work on exploitation ecosystems by including a secondary carnivore and applying a three-dimensional graphical analysis model. However, no explicit consideration was given to the dynamics of the resources that plants use, and self-limitation (e.g., territoriality) did not limit any component. They predicted that below some threshold limit, population dynamics should be noncyclic but that, as productivity increased, cyclic fluctuations occur. Rosenzweig (1973) observed that when a carnivore level was added to his model, the herbivore population density was stabilized. Wollkind (1976), using a model similar to that of Rosenzweig (1973) but including intraspecies (intratrophic-level) interactions, found that these interactions stabilized the population responses. More than ten years earlier Rosenzweig and MacArthur (1963) recognized that "stability is always conferred when the predator is severely limited at its equilibrium density by one of its resources other than its supply of prey." They also observed that predator-prey interactions may be stabilized by the presence of a prey hiding place, a low-predation pressure at moderate predator and high prey densities or "another predator-limiting resource." Even so, the latter set of considerations has rarely been applied to foodchain or so-called exploitative ecosystem models.

Kremer (1979) analyzed the oscillatory behavior of the Naragansett Bay model (Kremer and Nixon 1978) using analyses usually employed only upon simple models, which are usually unforced, rather than the mechanistic type such as the Narragansett Bay model. Dwyer and Kremer (submitted) continued this investigation by observing the frequency-domain sensitivity response of the model under unforced conditions during an oscillatory summer phase (June 30) and a non-oscillatory winter phase (January 1). One of the two dominant frequencies observed in both forced and unforced modes of the model could not be accounted for by "any single mechanism within the phytoplankton or zooplankton formulation of the model" and led to the conclusion that "these periodicities can only be the result of interactions between model state variables." But they could not determine which interactions were responsible.

Wiegert (1979) developed a generalized model that includes separate terms for resource and for space limitation for each biotic component. There is no Allee effect contained in the structure of the model. Wiegert (1979) investigated the influence of enrichment on the steady state achieved by a system containing a resource, producer, and herbivore. With increasing inputs to the resource component of the model, the system moved from a situation where the predator could not survive at low enrichment to where the predator controlled the prey population but was itself limited by both the amount of prey available and space, thus allowing resources to accumulate and culminating in a "food chain block." Wiegert was careful to point out that in an ecosystem this accumulated resource is likely to be used by an alternate primary producer. However, when the predator growth rate was reduced, the model proceeded to a state where space was the limitation of both prey and predator. Interestingly, at the reduced predation rate and at the highest enrichment levels, the system had two possible steady states, depending upon the initial densities of the two biotic components. In addition to the steady state described above, there was a situation where the prey was limited only by predation but the predator was both resource and space controlled. Thus, the structure of the model allowed multiple steady states without extinction that were dependent on initial conditions.

MODEL STRUCTURE

The structure of the generalized model presented in this paper is that of the predator-prey model reported by Wiegert (1979), except for three changes: (1) The feedback terms for space and resources are multiplicative, rather than additive, as he suggested, is more appropriate in the addendum of his paper; (2) an additional trophic level (top predator) was added in some cases; and (3) some parameters were changed to those more representative of a planktonic system.

The resource feedback term contains a threshold saturation density below which ingestion decreases and a refuge level below which the food is unavailable for consumption. The space feedback term is a self-limiting control with a threshold response density, where individuals of a population first begin to interfere with one another's growth (usually a value greater than zero), and a maximum asymptotic density, where there is no net increase in the size of the population. The feedback functions are linear between the lower and upper thresholds of each feedback term. Also, the time step for Euler integration is variable and is set so that the largest relative change in any component over that time period is no greater than 5%. Changing this percentage (1 to 20%) may slightly change the shape of the oscillatory curves (when present), but it does not affect the general outcome of the result. A constant time step gives identical results but considerably increases the cost of some of the model runs. The parameters (table 1) are not "tuned" to the results of Elliott et al. (1983), and the units are simply mass units and (unit time)$^{-1}$.

A flow diagram of the model is shown in figure 1. The feedback functions FF_{12}, FF_{23}, and FF_{34} are the resource controls for primary producers (X_2), herbivores (X_3), and carnivores (X_4), respectively; the feedback functions FF_{22}, FF_{33}, and FF_{44} are the self-limiting space controls for the same three components. These are each functions ranging from 0 to 1. The only input to the system is into the resource component (X_1), which varies with enrichment. Losses move out of each component as some constant function of the state variables at each time step, and there is no recycling within the system. Flow from one component to the next is a multiplicative function of the rate coefficient, recipient standing stock, and the two feedback terms.

RESULTS AND DISCUSSION

The results obtained by Wiegert (1979) for the predator-prey model using an additive expression for the feedback terms were reproduced approximately using the multiplicative rather than the additive convention of our model, but some of the parameter values had to be adjusted to obtain these results. A multiple steady state was obtained at high enrichment values, and the effect of progressive enrichment was qualitatively different when the predator growth rate was low and high (as in Weigert 1979). He demonstrated that when growth potential of the predator was low, the steady state of resources decreased temporarily and

Table 1.--Baseline parameters for the model run with outputs shown in figure 2.

Component	Input-Output (unit time)$^{-1}$			Resource Feedback (mass units)		Space Feedback (mass units)	
	Gross Specific Growth Potential	Specific Loss Rate	Maximum Specific Net Growth Potential	Threshold Saturation Density	Refuge Level	Threshold Response Density	Maximum Asymptotic Density
Resource (X1)	*	0.10	0.10	-	-	-	-
Producer (X2)	1.15	0.46	0.69	20.0	5.0	10.0	30.0
Herbivore (X3)	0.74	0.37	0.37	15.0	2.0	5.0	20.0
Predator (X4)	0.27	0.20	0.07	5.0	0.5	1.0	20.0

*Enrichment rate = 20 mass units · (unit time)$^{-1}$

Initial values of state variables: X1 = 10 mass units
X2 = 2 mass units
X3 = 5 mass units
X4 = 1 mass units

then increased again with increasing enrichment. In the same sequence of enrichment the feedback controls moved from a situation where the predator was controlled by space and resources to where both prey and predator were controlled only by space (see Wiegert 1979:262, fig. 7). We observed a similar response with multiplicative control, but the magnitude of increases and decreases in the resource pool size at steady state was not so great. Additive feedback function control and multiplicative control differ in magnitude of response and are not qualitatively different, at least for this model.

As an extension of previous work on planktonic foodchain dynamics (Elliott et al., in press), we added a top predator to the predator-prey model of Wiegert (1979) and adjusted the model parameters to those more representative of a planktonic ecosystem. Table 1 contains the parameters used for the baseline run for the following discussion, the results of which are presented in figure 2. With these parameters, the model demonstrated a limit cycle behavior. The cause of the model oscillations appear basically to be similar to those of classical stable limit cycles usually demonstrated by simpler models--namely, that it is "a combination of relatively ineffective prey self-limitation and a prey population that grows fast compared with the predator population" (May 1973). This is true of the herbivore-predator interaction; and, as is more generally true, the influence of the top predator is propagated through the rest of the food chain, thus creating oscillations in all components.

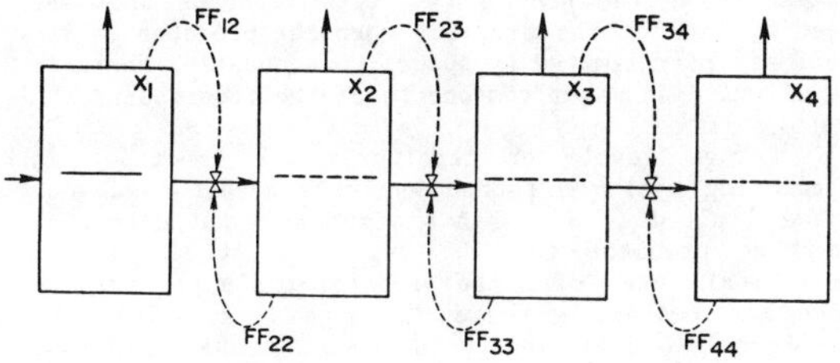

Figure 1.--Flow diagram of simple foodchain with resource pool (X_1), producers (X_2), herbivores (X_3) and predators (X_4). Flows between components are a multiplicative function of the gross specific growth potential, state variable of recipient, resource feedback function, and space feedback function. Loss rates are a constant function of the state variables. There is no recycling of material. Lines within boxes designate components in subsequent figures.

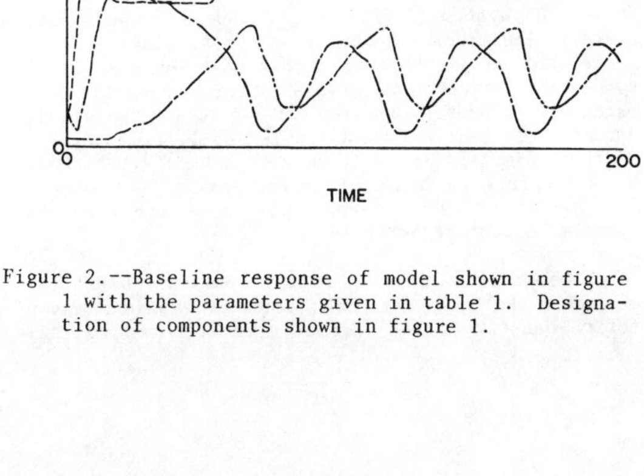

Figure 2.--Baseline response of model shown in figure 1 with the parameters given in table 1. Designation of components shown in figure 1.

As can be seen in figure 2, the slope of the increasing phase of the oscillation for the herbivore was greater than that for the predator. This is not surprising, since the maximum specific net growth potential for the herbivore was much greater than that for the carnivore (0.37 vs. 0.07 per unit time). At the high enrichment level (20 mass units), the herbivore was not limited by food, and the effect of crowding when the herbivore was at the peak of the oscillation was still 0.7 on a scale of 1.0, where 1.0 in a multiplicative expression means no limitation.

When the top predator was removed from the system (using the abovestated parameters, table 1), the oscillations ceased, demonstrating that the herbivore-carnivore interaction was responsible for these oscillations. However, when the carnivore was excluded and the parameters for the herbivore-carnivore interaction were used to replace the producer-herbivore parameters (all else remaining the same), the system again demonstrated limit cycle oscillations. Wiegert (1979) did not observe limit cycles in his model without explicit time delays, but his parameters did not meet the criteria for limit cycle oscillations. The limit cycles observed in our model are actually a form of time delay where the response of the predator and prey are delayed (out of phase), thus causing a stable limit cycle. These oscillations are intrinsic to the system and do not depend on the timestep used to solve the difference equations.

For the results shown in figure 1, space limits the predator to 20 mass units (table 1). When this value was reduced to 12, there were no oscillations (figure 3d), and values of anywhere from 13 to 30 gave the same response as a value of 20 except that the steady state amount of resources changed. Thus, space limitation conferred a certain "stability" to the model. The structure of the model used here does not allow extinction to occur, strictly speaking, since state variables may not be zero but may approach zero asymptotically. Assuming natural populations periodically falling to small numbers are more likely to become extinct, an oscillating system would be less stable than a nonoscillatory system with state variables equal to the average values of those in the oscillating system. Hence, more stability (as defined here) is conferred to a food chain system where the top predator is space limited.

The following discussion deals with the interacting effects of enrichment, predator growth potential, and herbivore refuge level, either with or without predator space limitations. The parameters given in table 1 will serve as the "baseline" upon which comparisons will be made (figure 2). Enrichment levels used were 7, 10, 15, and 20 mass units entering the resource component per unit time. At an enrichment of 7 the top predator is just barely supported. With an enrichment level of 6, the top predator asymptotically declines to zero. As enrichment increased in the "baseline" system (figs. 6,n,j,f and b, respectively) a stable equilibrium occurred at the two lower enrichments, but an oscillating steady state existed at the two higher enrichment levels. This demonstrates a principle conveyed by May (1974) that "population dynamics may switch from a stable equilibrium point to a stable limit cycle as 'life gets better'". Although he meant this in terms of increasing "effective r or K", enrichment also seems applicable since it increases the actual net r in this case.

The model's response at the four enrichment levels when the space feedback term for maximum asymptotic density of the predator is reduced to 12 is

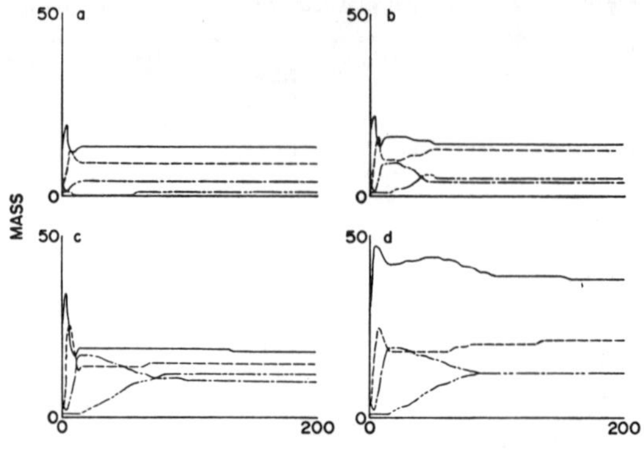

Figure 3.--Response of model to increasing enrichment [7, 10, 15, and 20 mass units · (unit time)$^{-1}$ for 3a,b,c and d, respectively] when the top predator has reduced available space (maximum asymptotic density of X_4 = 12). Designation of components shown in figure 1.

shown in figure 3. The system does not undergo oscillations but maintains a stable steady state. As enrichment proceeds from 7 to 10, the mass of predators increases significantly, while the producer and herbivore masses decrease. The resource level remains about the same. With an increase in enrichment from 10 to 15, all components increase; the herbivore and predator components increase the most. As the input level moves from 15 to 20, there is an increase in the steady-state value of the producers and a large increase in the amount of available resources. For any increase in enrichment greater than 20, only the amount of resources increase, demonstrating a food chain block (Wiegert 1979). Under these conditions all three trophic levels are space limited, with no limitation of resources.

When the resource refuge level of the carnivore's prey (herbivores) was incrementally increased from 0.5 to 4.0 (threshold saturation density is 5.0 in this case), there was no effect upon the response of the model when the maximum asymptotic density of the predator is 12 or below. Apparently, the availability threshold of the prey has little influence upon the feeding rate of the predator when the predator is itself severely limited by space; consequently, there is no effect upon other components of the food chain.

Refuge level for herbivores was important in regulating system response when the predator was not space limited (i.e., when maximum asymptotic density for the predator was 20, fig. 4). At the lowest enrichment level of 7 the predator could not survive when the refuge level was 2 or 4 (fig. 4o,p). No limit-cycle oscillations occurred until the enrichment level reached 15, even if the refuge level was reduced to 0. As enrichment proceeded, the system moved from a set of stable equilibrium points to a stable limit cycle as long as there was little or no prey refugia. When refuge levels for herbivores were increased to 1.0 or greater, the result was damped or showed no oscillations even at high enrichment levels. With enrichment the system components attain greater biomass but are less stable (as defined earlier). This result agrees with Rosenzweig's (1971) paradox of enrichment only when the model parameters include little or no refuge level for the prey.

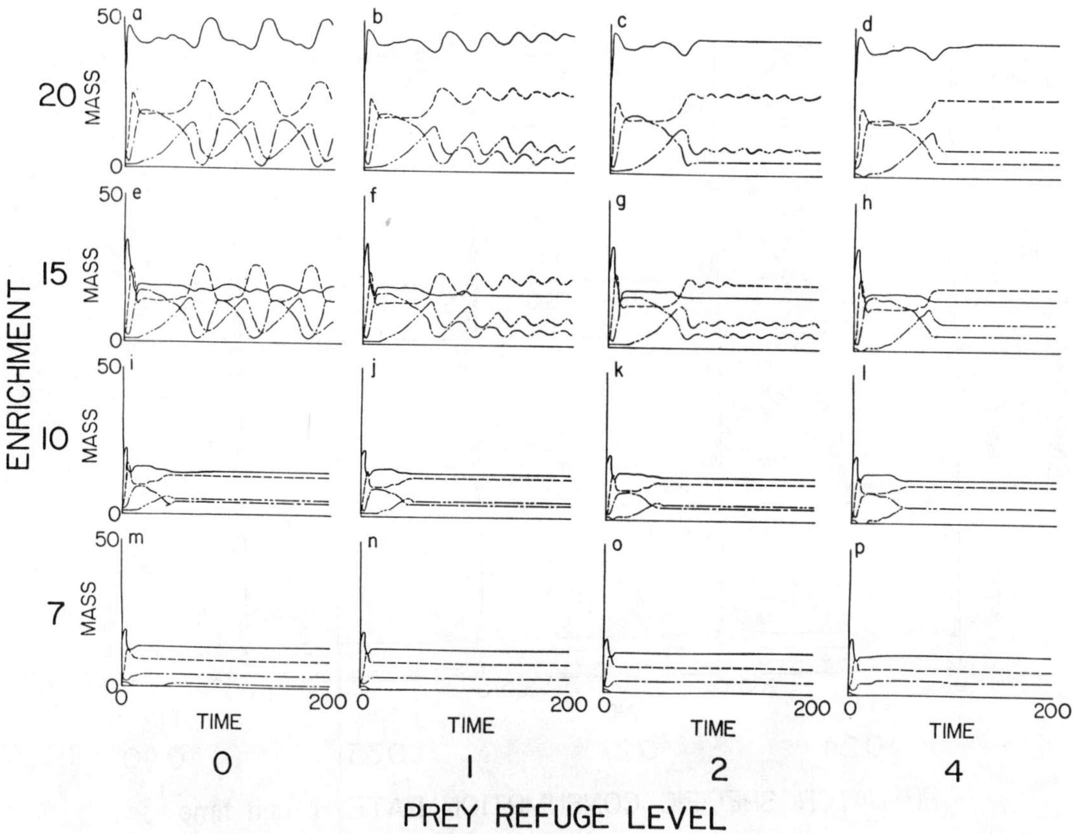

Figure 4.--Response of model to enrichment and prey (herbivore) refuge level (refuge level of X_4). Designation of components shown in figure 1.

The effect of enrichment upon model response when the predator space limitation for maximum asymptotic density was set at 12 (instead of 20.0, as in table 1) and the predator-specific consumption rate (gross specific growth potential of table 1) is varied is shown in figure 5. As discussed earlier, reducing the predator maximum asymptotic density to 12 eliminates limit cycles. Changes in the predator growth potential did not affect the final steady state. However, as predation rate increased, the amount of time it took for the system to reach equilibrium was reduced.

The influence of increasing rate of predation upon model response was identical when the maximum asymptotic density for the predator was either 12 or 20 as long as the enrichment level was low (7 or 10). When enrichment levels are increased to 15 or 20, predation rates greater than 0.27 (unit time)$^{-1}$ [i.e., 0.35 or 0.40 · (unit time)$^{-1}$] tended to damp out oscillations otherwise present when the predator space limitation was greater than 12 (fig. 6). The steady states observed in figures 6, c, d, g, and h are, in fact, damped and not continuously cyclic. These results are superficially in agreement with those of Kremer (1977) where increased predation rate reduced oscillations. As he points out, the feedback in his model is incomplete, because there is no chance for the predator biomass to respond. His results indicate that it is the algae-zooplankton interaction that is responsible for the oscillations, and the predators modify this interaction by reducing zooplankton consumption rates, resulting in reduction of the oscillations. In our model we know that the herbivore-predator interaction is the origin of the oscillations, since removal of the predator causes the model to go to a stable equilibrium. We suspect that the loss of limit-cycle oscillations in our model as the growth potential of the predator is increased is caused by the reduction in the difference between the intrinsic growth rates of the herbivores and predators.

When the enrichment level is high enough to support oscillations, they can be damped out or eliminated by decreasing the upper limit of space allowed for the predator, increasing the growth potential of the predator or increasing the amount of refuge available for prey (herbivores). When available space limits the predator in such a manner that the system is non-oscillatory, the amount of time the system takes to achieve a stable steady state decreases with an increase in growth potential of the predator. Variations in restrictive parameters can be expected

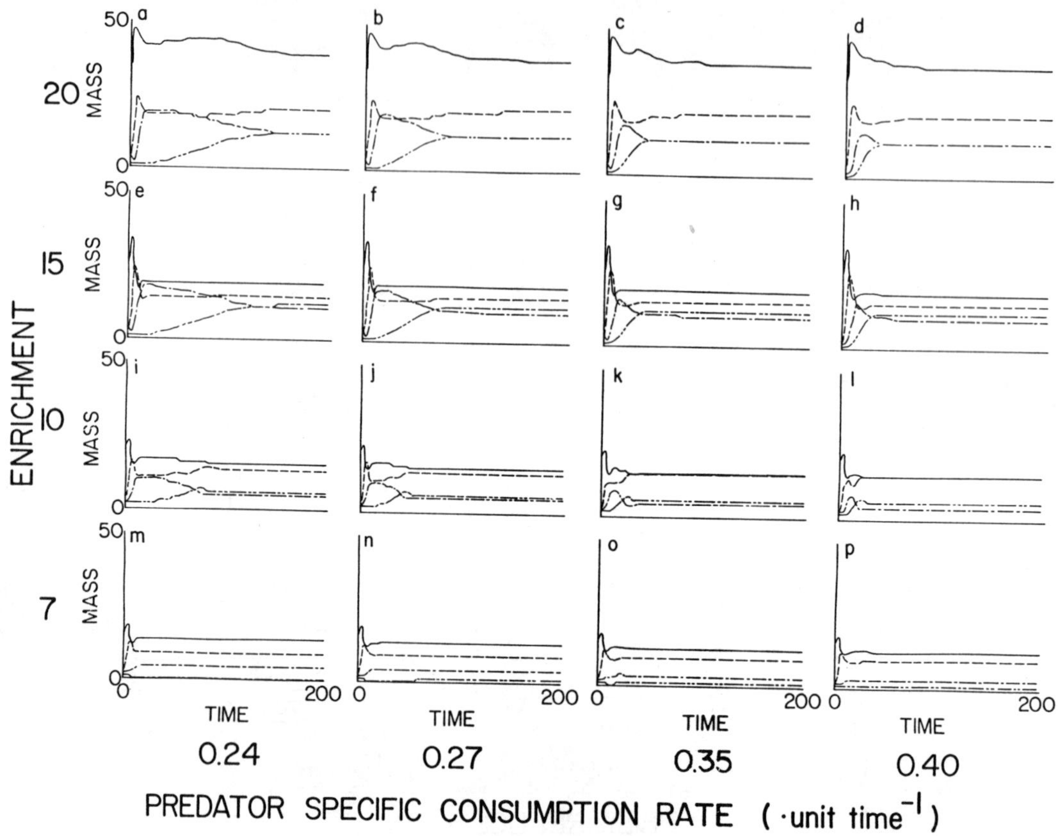

Figure 5.--Model output as influenced by enrichment and predator-specific consumption rate (gross specific growth potential of X_4 in table 1) when the predator has reduced available space (maximum asymptotic density of X_4 = 12). Designation of components shown in figure 1.

to have greater effects in models that are controlled by seasonal changes in forcing functions (i.e., light or temperature), since response time can be an important factor determining the model output. The combination of intrinsic oscillatory properties of the system under appropriate parameter values may potentially produce very complex system behavior.

ACKNOWLEDGMENTS

We would like to acknowledge National Science Foundation Grant No. DEB 8105281 to C. V. Cole for support of manuscript preparation. The Fisheries and Wildlife Department and the Department of Zoology, University of Georgia, provided monetary and physical support during the initiation of this study.

LITERATURE CITED

Dwyer, R.L., and J.N. Kremer. Frequency-domain sensitivity analyses of an estuarine ecosystem simulation model. Ecological Modeling (submitted).

Elliott, E.T., L.G. Castañares, D. Perlmutter, and K.G. Porter. 1983. Trophic-level control of production and nutrient dynamics in an experimental planktonic community. Oikos (in press).

Fretwell, S.F. 1977. Regulation of plant communities by the food chains exploiting them. Perspectives in Biology and Medicine 20:169-185.

Kremer, J.N. 1979. An analysis of the stability characteristics of an estuarine ecosystem model. In Marsh-estuarine systems simulation (R.F. Dame, editor). Belle W. Barach Library in Marine Science, No. 8. University of South Carolina Press. pp. 139-205.

Kremer, J.N., and S.W. Nixon. 1978. Coastal marine ecosystem: simulation and analysis. 217 p. Springer-Verlag, New York.

May, R.M. 1974. Stability and complexity in model ecosystems. 265 p. Princeton University Press, Princeton, NJ.

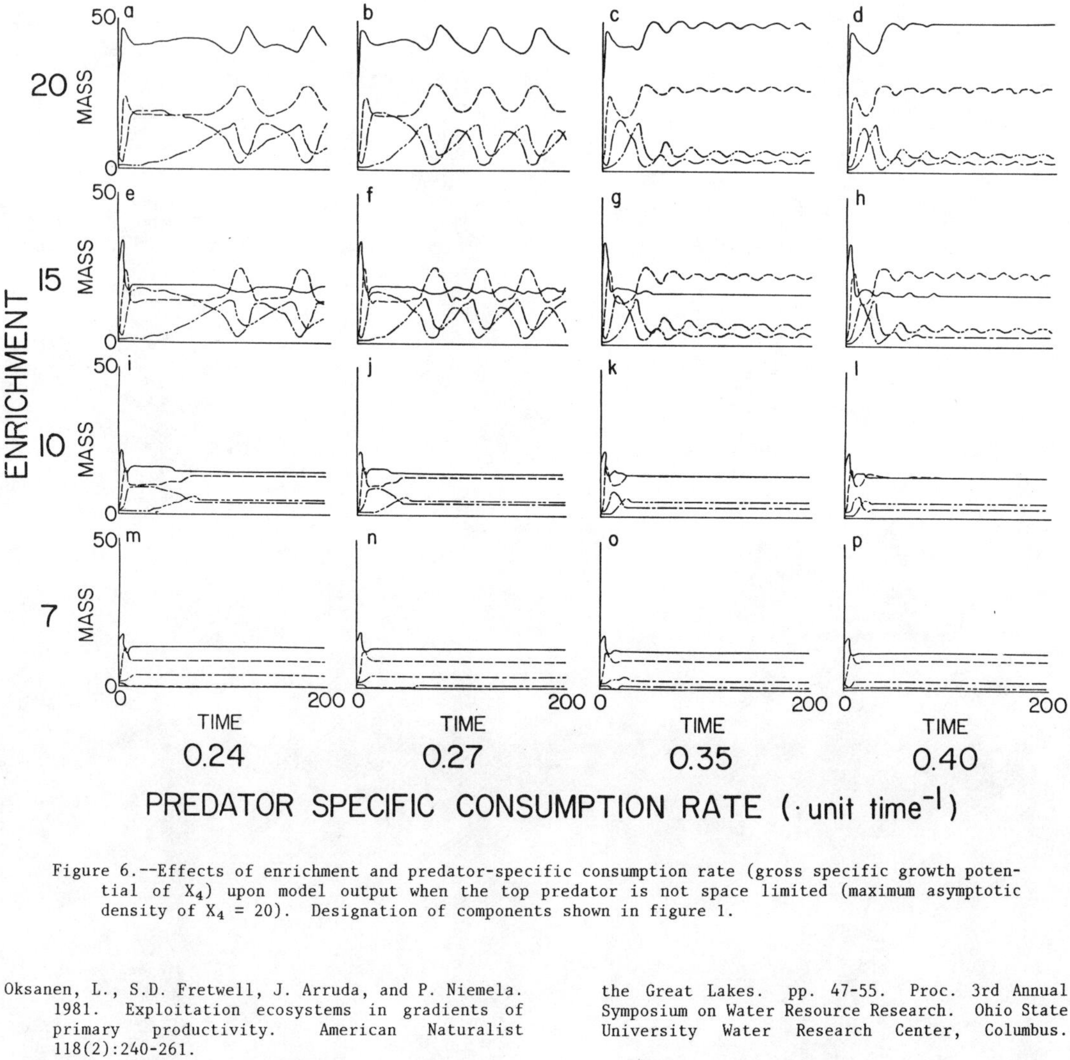

Figure 6.--Effects of enrichment and predator-specific consumption rate (gross specific growth potential of X_4) upon model output when the top predator is not space limited (maximum asymptotic density of X_4 = 20). Designation of components shown in figure 1.

Oksanen, L., S.D. Fretwell, J. Arruda, and P. Niemela. 1981. Exploitation ecosystems in gradients of primary productivity. American Naturalist 118(2):240-261.

Rosenzweig, M.L. 1971. Paradox of enrichment: Destabilization of exploitation ecosystems in ecological time. Science 171:385-387.

Rosenzweig, M.L. 1973. Exploitation in three trophic levels. American Naturalist 107:275-294.

Rosenzweig, M.L., and R. MacArthur. 1963. Graphical representation and stability conditions of predator-prey interactions. American Naturalist 97:209-223.

Smith, F.E. 1967. The role of ecosystems in water quality. In Systems approach to water quality in the Great Lakes. pp. 47-55. Proc. 3rd Annual Symposium on Water Resource Research. Ohio State University Water Research Center, Columbus.

Smith, F.E. 1969. Effects of enrichment in mathematical models. In Eutrophication: causes, consequences, correctives. 661 p. National Academy of Sciences, Washington, D.C.

Smith, F.E. 1972. Spatial heterogeneity, stability and diversity in ecosystems. Transactions Connecticut Academy of Arts and Sciences.

Wiegert, R.G. 1979. Population models: tools for analysis of ecosystems. In Analysis of ecosystems (H. Horn et al., editors). Ohio State University Press.

MODELLING GAS EXCHANGE IN CLOSED BIOLOGICAL SYSTEMS[1]

Robert K. Kaufmann[2], Berrien Moore III[3], and Maurice Averner[4]

Abstract.--Much of the fluctuation in the atmospheric concentrations of carbon dioxide and oxygen in previous closed autotroph/heterotroph systems is caused by the AQ-RQ mismatch and the partial disruption of system closure. The role that these factors play promoting atmospheric instability is evaluated using physiological models of the tomato plant _Lycopersicon esculentum_ and the lizard _Dipsosaurus dorsalis_ to simulate gas exchange in closed biological systems.

INTRODUCTION

A Controlled Ecological Life Support System (CELSS) is one option for maintaining human life during extended space flight. A CELSS uses energy to recycle matter through a variety of integrated biological and physical processes. In the primary alternative to a CELSS for space flight, all necessary materials are drawn from storage or resupply, used, and thrown away. Between these two extremes, lies a range of possible hybrid systems. Although more difficult to design, a CELSS operates with greater effectiveness than storage and resupply for long space missions (Ward et al., 1963; Mason, 1978).

In essence, a CELSS functions as a nearly closed system that uses energy to move important biological elements (e.g. C, N, O) through their respective cycles. The system requires rigorous control to assure the coordination of element residence times in various storage pools (e.g. atmosphere). Without careful synchronization, a CELSS cannot provide a habitable atmosphere, a dependable supply of potable water, and a nutritional diet.

Some researchers have suggested that a CELSS might rely on its autotrophs and heterotrophs, included for food production, to recycle respiratory gases, as well. Using biological organisms this way demands careful regulation of the rates for production and uptake of cabon dioxide and oxygen, so that their atmospheric concentrations remain within reasonable physiological limits for humans. Achieving this control is not easy because the biological cycling of the two gases is dynamically coupled. Manipulating the rate of photosynthesis and/or respiration will change the rates of oxygen and carbon dioxide exchange simultaneously. Thus, an attempt to balance the rate of oxygen exchange may disrupt the balance of carbon dioxide exchange.

Experimental research in support of a CELSS has examined the gaseous exchange of carbon dioxide and oxygen between autotrophs and heterotrophs. The concentrations of oxygen and carbon dioxide in the atmosphere of these systems fluctuated widely, regardless of the combinations and relative proportions of autotrophs and heterotrophs included. Further research must concentrate on the causes of this atmospheric instability and develop strategies to reduce fluctuations.

Models are a necessary complement to experimental analyses of gas exchange in biological systems. Simple mathematical models of gas exchange in these systems suggest that a mismatch between the assimilatory quotient (AQ) of autotrophs and the respiratory quotient (RQ) of heterotrophs may explain much of the atmospheric instability found in experimental systems. In addition, uncontrolled introduction of food or removal of algal biomass disrupts material closure in these systems, complicating the job of interpreting the dynamics of gaseous exchange. Because models can trace the flow of important biological elements throughout the system, they may help us to uncover the causes of atmospheric instability and to formulate strategies for reducing their effects.

In the first section of this paper, we briefly review the concentration of oxygen

1Paper presented at the Third International Conference on State-of-the-Art in Ecological Modelling, Colorado State University, May 24-28, 1982.

2Robert K. Kaufmann is a Research Scientist, Complex Systems Research Center, University of New Hampshire, Durham, New Hampshire 03824.

3Berrien Moore III is Director, Complex Systems Research Center, University of New Hampshire, Durham, New Hampshire 03824.

4Maurice Averner is a Research Scientist, NASA Ames Research Center, Mail Stop 239-10, Moffet Field, California 94035.

and carbon dioxide in the atmosphere of mass-closed, autotroph/heterotroph systems. Next, we examine two possible causes of the atmospheric instability displayed by experimental systems: the AQ-RQ mismatch and the disruption of material closure. Finally, we use physiological models of the tomato plant *Lycopersicon esculentum* and the lizard *Dipsosaurus dorsalis* to simulate gas exchange in a materially closed biological system and to assess the relative roles that AQ-RQ mismatch and disruption of material closure play in promoting atmospheric instability.

REVIEW OF AUTOTROPH/HETEROTROPH SYSTEMS

Modern research on autotroph-heterotroph gas exchange in mass-closed systems began in the late 1950s, when the possibility of prolonged space missions first raised the question of life-support systems. Taub (1974) has reviewed NASA-sponsored research, Gitel´zon (1975) produced a summary of the extensive Soviet program. In this section, we review the atmospheric concentrations of oxygen and carbon dioxide in closed autotroph/heterotroph systems.

Doney and Myers (1958) conducted one of the first quantitative studies on the coupling of photosynthesis and respiration in a closed system. They constructed a small (10 liters) gas-tight microcosm containing two mice, a supply of food and water, and a liquid culture of green alga *Chlorella ellipsoidea*. During the system´s month of operation, the atmospheric level of carbon dioxide ranged between 0.5% and 1.0%, and the concentration of oxygen increased from 21.0% to 25.0%.

Gafford (1959) reported autotroph-heterotroph gas exchange experiments coupling the blue-green alga *Anacystis nidulans* with mice. Testing a range of configurations, Gafford varied the ratio of autotrophic to heterotrophic biomass. In each of these experimental systems, the oxygen concentration increased, in one instance rising above 30%.

Bowman and Thomae (1960) reported another set of algae/mouse gas exchange experiments. Of the eight experiments they attempted, five were terminated when carbon dioxide reached unacceptable levels, within an average of seven days. In the most successful test, a 28 day experiment, atmospheric concentrations of CO_2 increased from 0.03% to 0.6%, and the initial 20.9% concentration of O_2 rose to 30.7% before dropping to its final level of 18.0%.

A 66-day run of a mass closed, mouse/algae gas exchange system was reported by Bowman and Thomae (1961). Atmospheric oxygen concentration fluctuated widely, ranging from an initial concentration of 21.5% to 60% after 50 days. Carbon dioxide concentration began at 0.4%, but moved as high as 5% and as low as 0.13% (final concentration).

Eley and Myers (1965) carried out a series of experiments quantitatively describing gas exchange between mice and green algae. In these experiments, researchers maintained pure cultures of the alga *Chlorellla ellipsoidea* and a dwarf variety of mouse in a closed chamber. Despite attempts to reduce the mismatch between the absolute rates of individual gas uptake and release, as well as efforts to reduce the mismatch between gas exchange ratios, oxygen concentrations always increased.

London and West (1962) attempted to create a more complex closed ecosystem in which to study gas exchange. Their system included a fungus, *Linderina pennespora*; an alga, *Chlorella pyrenoidosa*; a rat; and a sewage reactor. The system rapidly deteriorated, as indicated by drastic changes in atmospheric carbon dioxide and oxygen concentrations. Within 43 hours, CO_2 rose from 0.04% to 12.3% and O_2 concentration fell from 20.2% to 5.7%.

Zuraw (1962) reported gas exchange experiments involving a cebus monkey and *Chlorella pyrenoidosa* in a mass-closed system. During a 50-hour run, the oxygen concentration increased continuously from 21% to 25%, while carbon dioxide cycled between 0.2% and 1.0%.

Oswald et al., (1962) described an extensive study of the behavior of a closed system containing mice, algae, and bacteria. The initial series of experiments with the closed system produced rapid increases in carbon dioxide and decreases in oxygen. Subsequent modifications of the conditions for algal growth stabilized the atmospheric concentration of these gases within a more satisfactory range. During this time the oxygen concentration fluctuated in mid-20% range, and the carbon dioxide remained below 1%.

The most striking feature of all the experiments described above and summarized in Table 1 is the large fluctuation in the atmospheric concentrations of oxygen and carbon dioxide. CO_2 concentrations increased in some systems and decreased in others; O_2 behaved similarly. For both gases, these fluctuations showed no consistent pattern. In some systems both O_2 and CO_2 increased, yet in others, one gas increased while the other decreased.

The lack of any consistent pattern in the fluctuations of atmospheric concentrations of oxygen and carbon dioxide suggests that several causes of atmospheric instability were operating in these closed autotroph/heterotroph systems and that the relative importance of these causes varied between systems. The causes of these

Table 1--Summary of previous closed autotroph/heterotroph systems. A (+) indicates a net increase, (-) a net decrease, (+,-) indicates regular fluctuation.

Author	Autotroph	Hetero-troph	Oxygen	Carbon Dioxide
Doney & Myers	*Chlorella ellipsoidea*	Mouse	+	+
Gafford	*Anacystis nidulans*	Mouse	+	-
Bowman & Thomas	*Chlorella pyreniodosa*	Mouse	-	+
Bowman & Thomas	*Chlorella pyrenoidosa*	Mouse	+	-
Eley & Myers	*Chlorella ellipsoidea*	Mouse	+	+
London & West	*Chlorella pyrenoidosa*	Rat	-	+
Zuraw	*Chlorella pyrenoidosa*	Monkey	+	(+,-)
Golueke et al,	*Chlorella pyrenoidosa*	Mouse	-	+

fluctuations may be traced to one or several mechanisms: (1) An imbalance between the rate of photosynthesis and respiration; (2) the mismatch between AQ and RQ; (3) anaerobic and/or aerobic respiration by bacteria of fecal material; (4) breaches of system closure, caused either by (a) the lack of balance between inputs (e.g. food) and removals (e.g. algal harvest), or (b) physical leaks in the experimental apparatus. In the following section, we examine two of the major causes of atmospheric instability, the AQ-RQ mismatch and the partial disruption of material closure.

CAUSES OF ATMOSPHERIC INSTABILITY

In natural ecosystems, the buffering capacity of the atmosphere and the oceans maintains the concentration and relative stability of atmospheric oxygen and carbon dioxide. This buffering capacity reflects the enormous size of these bodies, as well as certain chemical equilibria and physical processes. There are, for example, 8,300 cubic meters of ocean water and 1,260 cubic meters of atmosphere for each square meter of land (MacElroy and Averner 1978). The volumes of atmosphere and water enclosed in any CELSS cannot match this ratio. In fact, no practical CELSS will be large enough to provide the dilution nor include the chemical and physical processes needed to maintain stable concentration of oxygen and carbon dioxide. This size limitation magnifies the impact of the AQ-RQ mismatch and the disruption of material closure on concentrations of gas in a CELSS.

AQ-RQ Mismatch

Ideally, in a simple autotroph/heterotroph system closed to the exchange of carbon dioxide and oxygen with the ambient atmosphere, respiration by autotrophs and heterotrophs would equal to net photosynthesis by autotrophs. This would assure an appropriate dynamic balance between atmospheric oxygen and carbon dioxide. However, an equilibrium of net photosynthesis and respiration is impossible when the RQ (moles of CO<2> produced: moles of O<2> consumed) of heterotrophs does not match the AQ (moles of CO<2> consumed: moles of O<2> produced) of autotrophs.

In the system shown in figure 1, for example, the heterotroph (RQ=.85) converts 1 volume or mole of oxygen into .85 volumes of carbon dioxide. The autotroph (AQ=.95) reconverts this 0.85 volumes of carbon dioxide to 0.89 volumes of oxygen. As a result, about 0.11 volume of oxygen is "lost" from the atmosphere during each complete cycle. The continual decline of atmospheric oxygen disrupts the stability of atmospheric gases in closed systems whose AQ exceeds RQ.

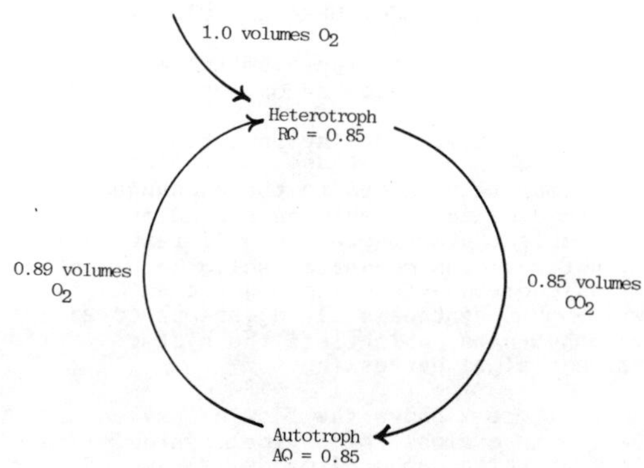

Figure 1--A simple model of the effects of the AQ-RQ mismatch on the atmosphere of closed biological systems.

Of course, oxygen is not really "lost" from a materially closed system. Despite the form of the AQ and RQ ratios, CO<2>/O<2>, oxygen is not converted directly into carbon dioxide, nor is carbon dioxide converted directly into oxygen. Instead, the metabolized carbon dioxide respired aerobically comes from the carbon skeleton of the foodstuffs, whereas the oxygen functions as the terminal electron acceptor and is respired as "metabolic" water. Similarly, the oxygen evolved from photosynthesis comes from water, and not from the carbon dioxide.

The RQ observed depends upon the balance between the quantity of oxygen needed to metabolize carbon-carbon or peptide bonds, and the number of carbon atoms "liberated" from the carbon skeleton when a bond breaks. Equivalence between the quantity of CO<2> evolved from food and the conversion of oxygen to metabolic water (RQ=1.0) occurs during carbohydrate metabolism only, because the carbon skeleton of sugar links carbon atoms with single bonds (equation 1). The skeletons of some foodstuffs, however, contain nitrogen (proteins, equation 2) or double bonds (e.g. fats, equation 3). As a result, they "require" more than one mole of oxygen, which is converted to metabolic water, for each mole of carbon dioxide evolved from the food substrate. For example, the RQ of the protein -alanyl-alanyl dipeptide is about 0.83 and the RQ for the fat triolein is about 0.71 (Lehninger 1975).

$$C_6H_{12}O_6 + 6O_2 \longrightarrow 6CO_2 + H_2O \quad (1)$$
$$2C_3H_5ON + 6O_2 \longrightarrow 5CO_2 + 3H_2O + (NH_2)_2CO_2 \quad (2)$$
$$C_{57}H_{104}O_6 + 80\ O_2 \longrightarrow 57CO_2 + 52H_2O \quad (3)$$

Thus, the overall RQ of heterotrophs depends upon the type and relative balance of food molecules respired.

Disruption of Material Closure

In addition to the mismatch of RQ and AQ, the concentration of oxygen and carbon dioxide in the atmosphere of autotroph/heterotroph systems has fluctuated partly because the systems were not completely closed to the exchange of carbon dioxide. Even when sealed against the ambient atmosphere, aerobic respiration by heterotrophs converted solid inputs of organic carbon--food--into gaseous CO<2>, while photosynthesis fixed gaseous CO<2> to carbohydrates, which left the system through algal harvesting.

Figure 2 shows the flow of oxygen and carbon in a closed biological system. The rate at which carbon dioxide is removed from the atmosphere and replaced with oxygen depends on the rate of net photosynthesis by the autotroph. Instead of an exogenous food stream, the heterotroph in the closed system receives a portion of the carbon dioxide "fixed" in this process. The rate at which this fixed carbon returns to the atmosphere depends upon the metabolic rate, the type of foodstuff metabolized (e.g. carbohydrate, fat, protein) and the relative balance between maintenance and growth in the heterotroph's energy budget. If the portion of fixed carbon dioxide assimilated exceeds the basic metabolic requirements of the heterotroph, the heterotroph is able to grow, removing carbon dioxide from the C0<2>-0<2> cycle. If the autotroph is allowed to retain some portion of the fixed carbon, it too removes C0<2> from the C0<2>-0<2> cycle. Thus, an atmospheric steady state of carbon dioxide and oxygen is possible only when the autotrophic and heterotrophic biomasses remain constant and the organic carbon sequestered in waste materials is oxidized to carbon dioxide.

These preconditions for dynamic steady state were disrupted in previous experimental systems by the introduction of fixed CO<2> in the form of food for the heterotroph. Depending on the balance between food inputs and outputs of algal biomass--and the ratio of carbohydrates, fats, and proteins in these inputs and outputs,--carbon dioxide was added or removed from the "closed system". If a smaller amount of autotrophic biomass were to be removed, a net input of CO<2> would result.

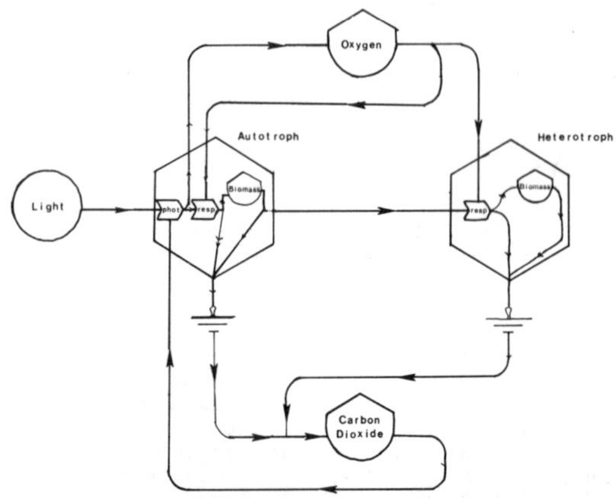

Figure 2--The flow of carbon and oxygen in a materially closed biological system.

The net flux of CO<2> to or from the system may not show up as a corresponding change in the atmospheric concentration of carbon dioxide; autotrophic or heterotrophic growth may remove any net addition of CO<2> from the atmosphere. Similarly, a net loss of biomass by either of these components will compensate for net removal of carbon dioxide. Even if the quantity of carbon input to a closed system equals the quantity removed, a net flux of carbon to or from the system is still possible, depending on the RQ of the exogenous food stream relative to the RQ of the carbon output. Without detailed records of the input of food, removal of autotrophic biomass, make-up of both streams, and net growth of autotrophs and heterotrophs, disruption of material closure can make interpreting gas exchange dynamics virtually impossible.

MODELLING GAS EXCHANGE IN CLOSED BIOLOGICAL SYSTEMS

To illustrate the effects of AQ-RQ mismatch and disruption of material closure on the cycling of oxygen and carbon dioxide in biological systems closed to the exchange of these gases with the ambient atmosphere, we constructed models of the lizard Dipsosaurus dorsalis and the tomato plant Lycopersicon esculentum.

Ectotherms have low oxygen requirements, and their rate of gas exhange can be easily manipulated by environmental changes. These factors make them an ideal type of heterotroph for closed biological systems (Moore et al., 1981). Among all ectothermic species, the physiology of Dipsosaurus dorsalis is relatively well understood, thereby making possible assembly of a fairly detailed model. Similarly, the importance of the tomato as a greenhouse crop has stimulated much research on its physiology.

Both models reflect the physiology of the respective species and incorporate many of the feedback loops between the organism and its environment that have an important effect upon the rate of gas exchange. The models are linked to simulate gas exchange between a 50-gram lizard and a 75-day-old tomato plant in a closed chamber 1.5 m3 in size. The initial concentration of carbon dioxide is set at 318ppm and oxygen at 209X10[5]ppm, reflecting the concentration of these gases in the ambient atmosphere. In accordance with their physiology, the RQ of Dipsosaurus dorsalis is set at 0.70 and the AQ of the tomato plant is set at 0.95 (Benett and Dawson 1976). Based on field observations of and laboratory work with Dipsosaurus dorsalis, the model assumes that the lizard consumes 3.5% of its weight daily (Porter et al., 1973; Kaufmann and Pough 1982).

Before describing the results of our simulations, a warning: many of the parameters relating to the system itself were chosen to facilitate the simulation. The size of the chamber was chosen not to minimize size (which may eventually be an important consideration) but to assure a reasonable lifespan for the tomato-lizard system, given the likely degree of gaseous fluctuation. The relative balance between autotrophic and heterotrophic biomass was chosen so that net photosynthesis by the tomato plant in ambient atmosphere could supply the heterotroph's oxygen requirement. As the concentration of carbon dioxide changes in the closed chamber, this requirement may or may not be satisfied. Even though simulations starting from different initial conditions would produce different gas exchange dynamics, we believe the set of conditions chosen minimized dynamics of gas exchange due solely to the initial conditions. Therefore, we feel the results of the simulations are particularly useful for examining the causes of atmospheric instability described in the previous section.

Figure 3 shows the concentration of carbon dioxide and oxygen in the tomato-lizard system for 50 days. After a rapid increase and decrease in CO_2 concentration within the first five days, the concentration of CO_2 rises slowly until day 32, after which it falls more slowly. By contrast, the atmospheric concentration of oxygen declines continuously, falling most rapidly during the initial period of the experiment. These fluctuations reflect some of the same patterns these gases exhibited in the experimental systems. Thus, some of the causes of atmospheric instability that operated on previous experimental systems may be simulated by the model.

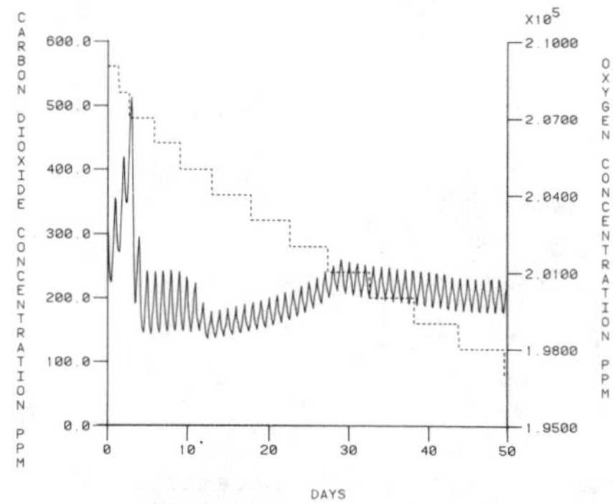

Figure 3--Atmospheric concentration of carbon dioxide (solid line) and oxygen (dotted line) in the reference scenario.

The patterns of gas exchange in figure 3 serve as a reference scenario against which we compare the behavior of modified systems. These systems incorporate changes in the RQ and AQ intended to reduce the original mismatch in these ratios. Latter scenarios increase the degree of material closure by having the lizard "eat" leaves from the tomato plant to fulfill its energy and material requirements. By comparing the behavior of carbon dioxide and oxygen in these modified systems to the behavior of these gases in the reference scenario, we can test the validity of the theoretical principles thought to cause atmospheric instability and can measure the relative importance of each one as a source of instability.

AQ-RQ Mismatch

If the AQ-RQ mismatch has been eliminated by resetting the lizard's RQ from 0.70 to 0.95, the general pattern of declining oxygen and large fluctuations in carbon dioxide concentrations remain (figure 4). From the similarity of patterns shown in figures 3 and 4, the AQ-RQ mismatch probably is not responsible for the general direction of atmospheric fluctuations.

The AQ-RQ mismatch does, however, have a significant effect on the magnitude of the fluctuations of atmospheric carbon dioxide and oxygen. The O_2 concentration in the reference scenario falls more rapidly and to lower levels than it does in the corrected system (figure 5). The decline in the oxygen concentration in the reference scenario supports the theoretical expectation that atmospheric O_2 concentrations should decline in systems where AQ exceeds RQ.

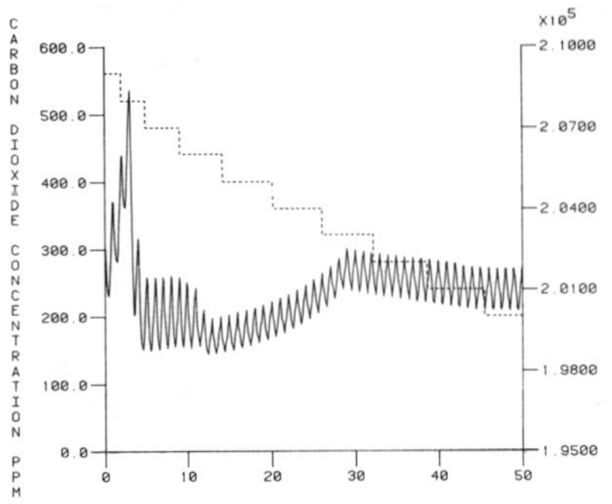

Figure 4--Atmospheric concentration of carbon dioxide (solid line) and oxygen (dotted line) in which the AQ-RQ mismatch has been eliminated by raising the lizard's RQ to 0.95.

Similarly, the atmospheric concentration of CO<2> from the reference scenario is less than its concentration in the system from which the AQ-RQ mismatch has been eliminated. However, the concentration of CO<2> increases as a result of raising the RQ from 0.70 to 0.95 and is not a direct product of the elimination of the mismatch. In the model of <u>Dipsosaurus dorsalis</u>, the rate of oxygen consumption controls the production of CO<2> so that an increase in RQ raises the CO<2> respired. Eliminating the AQ-RQ mismatch by lowering the AQ of the tomato plant to 0.70, produces no noticeable change in the concentration of atmospheric carbon dioxide relative to the reference scenario. The reduced degree of fluctuations in the oxygen concentrations shown in figure 5 remains the same. Thus, the way in which the AQ-RQ mismatch is reduced (by raising RQ, lowering AQ, or a combination of both) significantly influences the atmospheric concentration of CO<2>.

Closure of Material Loops

As described, the disruption of material closure by the exogenous food stream can have an important effect on the atmospheric concentration of carbon dioxide and oxygen. To examine the effect of this disruption of material closure, we simulated a system in which the exogenous food stream is eliminated. Instead, the lizard consumes 3.5% of it's own biomass daily from the leaves of the tomato plant, starting with the oldest, (and therefore least productive) leaves.

As shown in figure 6, the system collapses after 27 days. The lizard reduces the leaf area of the tomato plant beyond a point that it can no longer

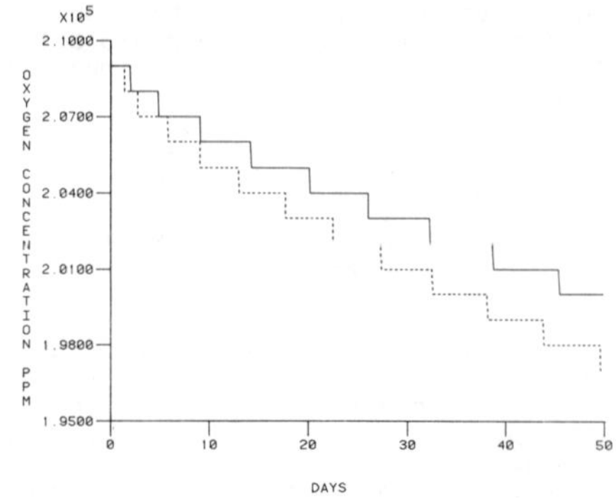

Figure 5--Atmospheric oxygen concentration in the reference scenario (dotted line) and the system illustrated in in figure 4 (solid line).

produce enough photosynthate to support itself: by day 27, the plant uses up its reserves and dies. The atmospheric concentration of CO<2> declines to levels below those of the reference scenario because the lizard's growth makes it an indirect sink for atmospheric CO<2>. Despite an identical growth rate as in the reference scenario, the loss of atmospheric carbon dioxide does not occur as rapidly in the reference scenario because the carbon that the lizard uses for growth in the partly closed system does not come from the atmosphere of the chamber.

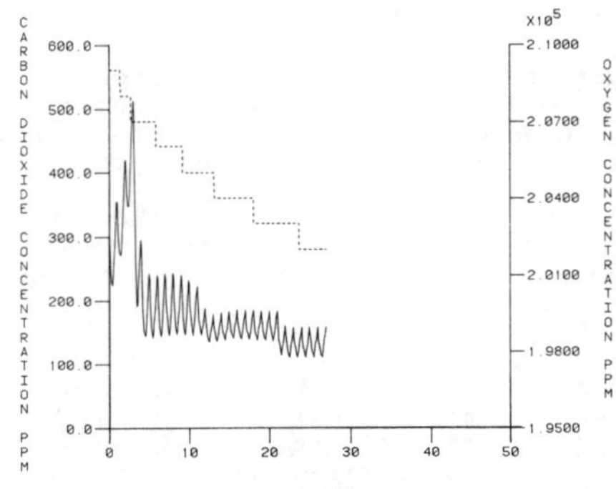

Figure 6--Atmospheric concentration of carbon dioxide (solid line) and oxygen (dotted line) without an exogenous food stream.

Disruption of material closure by exogenous food input does not destabilize the system. Instead, it subsidizes atmospheric carbon replacing what the growing lizard sequesters. When energy and carbon subsidies are eliminated, the amount of food given to the lizard and the food's distribution among maintenance and growth respiration become important factors in the cycling of carbon dioxide and oxygen.

The atmospheric concentration of the CO<2> and the longevity of system increase as the quantity of tomato fed to the lizard decreases (figure 7). The magnitude of this effect is determined by an assumption buried within the model concerning the type of material synthesized by growth. Currently, the model assumes that biosynthesis produces a carbohydrate. Glucose can be converted into carbohydrates with an 82% efficiency rate, but the rates for synthesizing protein and fat are significantly less, 66% and 33%, respectively (Penning deVries et al., 1974). Producing protein, fat, or some combination of all three materials through biosynthesis reduces the conversion efficiency of glucose to biomass decreasing the amount of atmospheric carbon dioxide fixed by the plant that can be sequestered eventually by the lizard. Therefore, changing the assumption about the products of biosynthesis could significantly reduce the effects of material closure on the concentration of CO<2>.

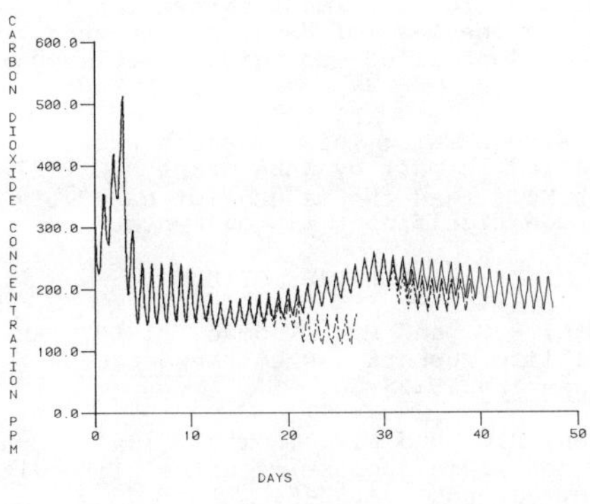

Figure 7--The effect of changing the quantity of tomato plant eaten by the lizard daily. Solid line, 0.5% of lizard biomass, dashed line 1.5%, dotted line 2.5%, and dashed-dotted line 3.5%.

Eliminating the exogenous food stream does not close the food loop completely. Unassimilated plant material in the lizard's feces remains unavailable to the simplified biological component of the system. This carbon is, in effect, "lost" from the system. To close the food loop more completely, we simulated the operation of a hypothetical "waste processor" to convert the organic matter tied up in feces to gaseous carbon dioxide. Assuming that this processor has an "RQ" of 1.0, figure 8 shows the level of atmospheric carbon dioxide relative to the partially closed system represented in figure 6. As atmospheric cabon rises with the processor in operation, the concentration of atmospheric oxygen falls, since oxygen is used to oxidize the fecal organic matter. The closure of this carbon sink enables the system to operate for an extra day.

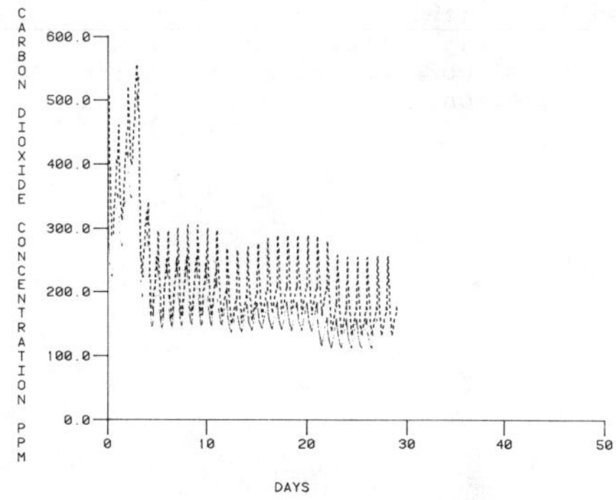

Figure 8--Atmospheric concentration of CO<2> in a system with a waste processor (dashed line) compared to CO<2> concentration from figure 6 (solid line). Spikes caused by reintroduction of CO<2> in one pulse.

Buffer-Reserviors

Most higher plants, including tomatoes, cannot photosynthesize continuously without physiological damage (Hillman 1956). As a result, the atmospheric concentration of carbon dioxide will increase during dark periods due to aerobic respiration by both autotrophs and heterotrophs. Buffer-reserviors may therefore prove useful for storing some of the excess CO<2> produced during dark periods and reintroducing it during light periods, when CO<2> removal exceeds its production.

To examine the impact that a buffer-reservior system can have on the atmospheric concentration of carbon dioxide, we simulated the operation of a buffer-reservior that removes CO<2> when atmospheric concentrations exceed 350 ppm and reintroduces it when atmospheric concentrations dip below 200 ppm. The buffer-reservior system reduces the peaks in CO<2> concentration relative to the reference scenario on days 1-3 and raises the troughs on days 4-6 (figure 9). Following day 6, the two lines merge because the buffer-reservior is empty and the atmospheric concentration of CO<2> never again reaches levels to replenish it. The operation of a buffer-reservior in this system has a slight mediating effect.

CONCLUSION

One major function of a model is to organize knowledge about a system and to evaluate our knowledge of that system based on the model's ability to simulate system operation. The value of our simulations of gas exchange in the tomato-lizard system lies not in its ability to predict specific atmospheric concentrations of oxygen and carbon dioxide, but in its use as a tool for interpreting the results of experimental systems and for pointing to research areas that may help researchers reduce atmospheric instability and increase system longevity. In conclusion, we try to link the results of our simulations to experimental work and make suggestions for future research.

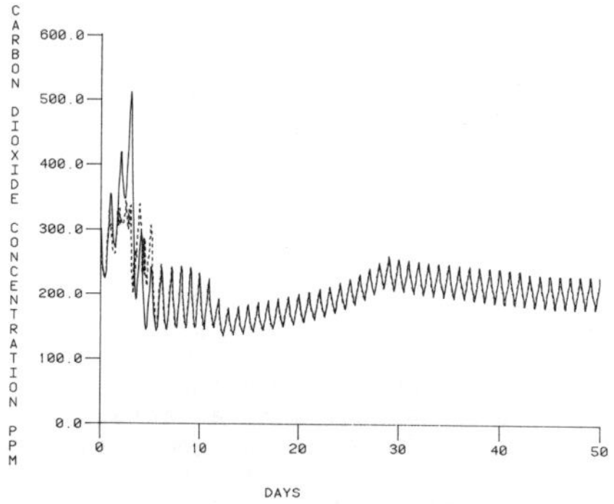

Figure 9--Carbon dioxide concentration from the reference scenario (solid line) and the system with a buffer-reservior (dotted line).

Both the theoretical discussion and the results of the simulations indicate that the mismatch between AQ and RQ has an important effect on the atmospheric concentration of oxygen and carbon dioxide. Clearly, future research with closed biological systems should attempt to reduce this mismatch. The RQ of a heterotroph may be changed by manipulating its diet and the AQ of an autotroph by varying its nitrogen source (Kaufmann et al., 1982). Experimentalists should remember that the way in which the mismatch is reduced, increasing RQ, decreasing AQ, or a combination of both, has an important effect upon the atmospheric concentration of CO_2.

Model performance can highlight areas for research in which increased understanding would help explain the operation of the system. The simulations of the tomato-lizard system show that disruption of material closure when the food stream is not balanced against the removal of autotrophic biomass may provide an energy and carbon sudsidy, thereby lengthening the longevity of previous experimental systems and reducing the magnitude of atmospheric fluctuations. In addition, the lack of a waste processor in any of these systems allowed carbon to accumulate in forms that the biological component could not reincorporate into the O_2-CO_2 cycle.

The simulations of the tomato-lizard system point to several new variables that must be monitored in order to interpret the behavior of atmospheric concentrations of oxygen and carbon dioxide. The amount and forms (e.g., fats, proteins, carbohydrates) of carbon entering the system as food and leaving it as autotrophic biomass must be carefully recorded. In addition, the amount of carbon sequestered in the autotroph or heterotroph through growth must be accounted for, along with organic carbon in waste materials. To improve the accuracy of the models, a distinction must be made for structural growth (proteins and carbohydrates) and that for storage (carbohydrates and fats).

A major shortcoming of previous interpretations of gas exchange in materially closed biological systems was the limited scope of their analysis. A major advantage of a systemic analysis of gas exchange in closed biological systems, and the use of models to represent it, stems from its ability to trace the flow of carbon and oxygen throughout the system. The results of our simulations suggest that many processes not directly related to gas exchange have an important effect on the atmospheric concentrations of CO_2 and O_2. Thus, models operate as an important complement to experimental research in amassing the base of knowledge necessary to build a controlled ecological life support system.

ACKNOWLEDGEMENTS: This research was supported in part by NASA Grant NCC 2-27 Steve Wolf read the manuscript carefully and made significant improvements.

LITERATURE CITED

Bowman, R.O. and F.W. Thomae. 1960. An algal life support system, Aerospace Engineering 19:26-30.

Bowman, R.O. and F.W. Thomae. 1961. Long term non-toxic support of animal life with algae, Science 134:55-6.

Calloway, D.H. 1975. Basic data for planning life-support systems. Foundations of space biology and medicine, eds. Melvin Calvin and Oleg G. Gazenko. Washington: NASA Scientific and Technical Information Office.

Bennett, A.F and W.R. Dawson. 1976. Metabolism. In Biology of the reptilia Vol. 5, eds. C. Gans and W.R Dawson. London: Academic Press.

Doney, R.C. and J. Myers. 1958. "Quantitative repetition of the priestly experiment. Plant Physiology 33 suppl.:xxv.

de Vries, Penning F.W.T. Brunsting and A. van Laar. 1974. Product requirements and efficiency of biosynthesis: a quantitative approach. J. Theor. Biol. 45:339-377.

Eley Jr., J.H. and J. Myers. 1965. Study of photosynthetic exchange, a quantitative repetition of the priestly experiment," Texas Journal of Science 26:296-333.

Gafford, R.D. and C.E. Craft. 1959. A photosynthetic gas exchanger capable of providing for the respiratory requirements of small animals. USAF SAM Report 58-124. U. S. Air Force, Washington, DC.

Gitel´zon et al. 1975. Problems in space biology. NASA Technical Translation F-16933. National Aeronautics and Space Administration, Washington DC.

Hillman, William S. 1956. Injury of tomato plants by continuous light and unfavorable photoperiodic cycles. Am. J. Bot. 43:884-96.

Kaufmann, R.K., B. Moore, and M. M. Averner. In press. A qualitative analysis of gas exchange in materially closed biological systems. American Society of Mechanical Engineers, New York, NY.

Kaufmann, R.K. and F.H. Pough. 1982.The effect of temperature upon the efficiency of assimilation of preformed water by the desert iguana (Dipsosaurus dorsalis) and its significance for subterranean life. Comp. Biochem. Physiol.72A:221-224.

Lehninger, R. 1975. Biochemistry. Worth Publishers, New York, New York p. 1104.

London, S.A. and A. West. 1962. Gaseous exchange in a closed ecological system." AMRL-TDR-62-139. U. S. Air Force, Wright-Patterson Air Force Base, Dayton, Ohio.

MacElroy, R.D. and M.M. Averner. 1978. Space ecosynthesis: an approach to the Design of Closed Ecosystems for Use in Space, NASA TM-78491 . National Aeronautics and Space Administration, Washington, DC.

Mason, R.M.1975. Preliminary scenario analysis model. CELSS Working Paper.WP754-1. National Aeronautics and Space Administration, Washington, DC.

Moore, B.M., R.K. Kaufmann and C. T. Reinhold. 1981. Design and control strategies for CELSS: integrating mechanistic paradigms and biological complexities. ASME #81-ENAS-43. American Society of Mechanical Engineers. (Monoprint), New York, NY.

OsWald, W.J., C.G. Golueke, J.W. Brewer and H.K. Gee. 1962. Microbiological waste conversion in control of isolated environments. ASTIA 278079. USAF Cambridge Research Laboratory, Office of Aerospace Research, Bedford, MA.

Taub, F.B. 1974. Closed ecological systems, Ann. Rev. Ecol. Systems 5:139-160.

Ward, C.H. S.S. Wilks and H.L. Craft. 1965. Use of algae and other plants in the development of life support systems." American Biological Techniques 25:512-521.

Zuraw, E.A. 1962. Algae-primate gas exchange in a closed gas system. Developments in Industrial Microbiology 3:140-9.

A SYSTEMS APPROACH TO THE ANALYSIS OF HUMAN CARRYING CAPACITY APPLIED TO THE LAKE TAHOE BASIN OF CALIFORNIA/NEVADA[1]

Martha W. Gilliland[2]

Abstract.--By combining systems methods, theoretical concepts from ecology about carrying capacity, and energy analysis, the human carrying capacity of the Lake Tahoe Basin is examined. Fundamental differences between human and ecological systems point to the need for a socially determined human carrying capacity and to a carrying capacity that is evaluated and regulated as limits on the inputs to a region.

INTRODUCTION

Interest in applying ecological concepts of carrying capacity to human systems developed in conjunction with the "limits to growth" literature about 1970.

Human populations, however, are considerably more complex than natural populations, and human carrying capacity analyses have been frought with difficulties. Using systems models and energy analysis, this paper summarizes some of the fundamental differences in the operation of natural and human systems, develops a method for analyzing human carrying capacity that recognizes those differences, and suggests an approach to managing human systems that is based in ecological carrying capacity concepts. All data and analyses are taken from and applied to the Lake Tahoe Basin, California/Nevada (Western Federal Regional Council, 1979).

THE LAKE TAHOE BASIN SYSTEM

The Lake Tahoe Basin, located on the California-Nevada border, is a unique and nationally significant scenic and recreational resource area. Its natural environment, however, is deteriorating rapidly under the stress of continuing urbanization. In fact, if current rates of urban growth and concommitant environmental deterioration continue, the very attributes that attract people to the basin will be gone and the uniqueness of Tahoe will be lost to future generations. The problems and conflicts that exist at Tahoe, while they are probably more pronounced than in most scenic recreational areas, are not unique. They arise because of competing values, limited resources, and a limited capacity to absorb the impacts of urban expansion.

The clear, blue waters of Lake Tahoe in combination with its enclosed mountain setting, fresh alpine air, snow capped peaks, coniferous forests and meadows, mountain vistas, and wildlife diversity make the Lake Tahoe Basin unique in North America. Today, that natural uniqueness is also coupled to a diverse set of year-round recreational opportunities. Visitors to the basin can camp, hike, go boating, ski, gamble, and be entertained in the resort hotels and casinos.

Existing data about Tahoe were integrated into a systems model (Fig. 1) developed at the scale of the entire Basin and with a focus on linkages between nature and people.

The model is designed to recognize people and urban environments as integral parts of the basin system; it does not presuppose that the natural environment is to be protected at the expense of the urban environment or vice verse. It does, however, imply that any concept of carrying capacity must relate to the entire basin, not just to one of its natural environments or to its urban environment.

As conceptualized in Figure 1, the Lake Tahoe Basin System is driven by two kinds of input (circles) and is comprised of four interacting subsystems (rectangles). Natural input includes solar energy, wind, and precipitation. Together these have shaped the character of Tahoe's natural environment, including the terrestrial, aquatic, and atmospheric subsystems shown in Figure 1. This natural input represents the water source for the lake and for the terrestrial ecosystems, partially determines air quality, determines rates of erosion, and determines the aquatic and terrestrial productivity of the plant and animal populations.

People-related input includes purchased goods, energy, people, dollars, and the transportation systems required to bring these into the basin. Together, these shape the character of Tahoe's urban subsystem and directly and indirectly alter the character of Tahoe's natural environment. For example, people consume energy in their houses, jobs, recreational activities, and vehicles. That energy consumption produces air pollutant emissions and heat. People also consume water which produces wastewater; they consume goods, which produce solid wastes; and their housing needs, job activities, and recreational activities "consume" natural land and convert it to urban land. That transfer of land, in turn, affects the quality of the runoff

---

[1]Paper presented at the Third International Conference on State-of-the-Art in Ecological Modelling, Colorado State University, May 24-28, 1982.

[2]Until recently, Martha W. Gilliland was Executive Director, Energy Policy Studies, Inc., Omaha, Nebraska, USA; she is now Associate Professor of Civil Engineering, University of Nebraska, Omaha, Nebraska, USA.

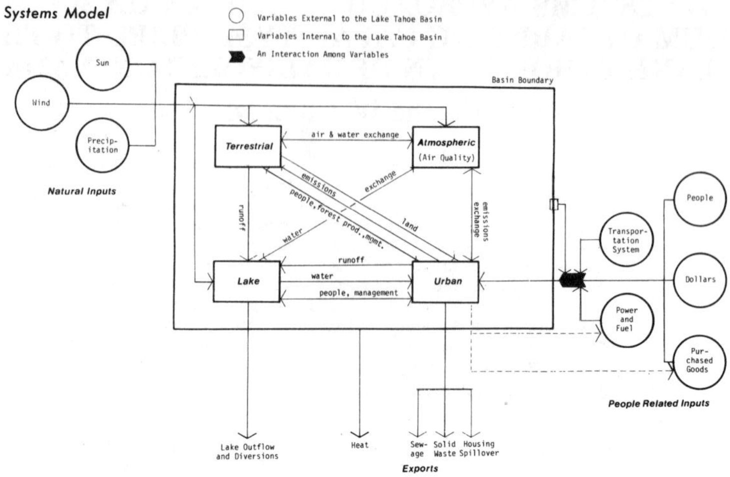

Figure 1. Systems model of the Lake Tahoe Basin. The basin occupies 132,400 hectares, of which 37% represents the lake itself; of the 83,030 hectares of land, about 12% is urbanized and 65% is National Forest. Solid lines represent material or energy transfers; dashed lines represent money transfers. From Western Federal Regional Council, 1979.

water that enters Lake Tahoe. Dollars enter the basin with visitors and as investment in the construction of casinos, commercial establishments, houses, water and sewer systems, utility systems, and recreational facilities. Dollars leave the basin with the purchase of goods and energy and as profits and taxes (Figure 1, dashed line).

Our analysis of changes that occurred in this system between 1970 and 1978 reveals exponential growth in urban activity and exponential degradation in environmental quality. For example, between 1970 and 1978, people related inputs increased an average of 130%.[3] This allowed urban activity to expand an average of 75%.[4] During the same time period, the nitrogen loading of Lake Tahoe from non-point sources increased 18%, primary productivity in the lake increased 50%, and the lake's clarity declined 6% to 13%, depending on the averaging time assumed.[5] Losses in the key wildlife habitats associated with critical stream environment zones ranged from 11 to 31%.[6]

Although all subsystems in Figure 1 are linked so that a change in one affects all others, the point to be emphasized in the context of a carrying capacity analysis is that inputs to the Basin are the primal or first-order change agents. Any workable concept of carrying capacity must deal with these inputs.

---

[3]Input increases between 1970 and 1978 were as follows: people, 101%; money, 95%; goods and services, 133%; energy, ranging from 52% to 174% depending on the form but averaging 89%; and transportation, ranging from 80% to 377% depending on the mode but averaging 228%. The average of these 5 kinds of input is 127%.

[4]Urban activity increases between 1970 and 1978 were as follows: visitation, 93%; recreational facilities ranging from 17% for outdoor facilities to 97% for gaming facilities but averaging 57%; housing 78%; employment, 107%; traffic, 90%; public utilities ranging from 22% to 51% for water, sewage, and solid waste but averaging 37%; and urban land use, 78%. The average of the 7 kinds of urban activity is 77%.

[5]Clarity was measured with a Secchi Disk; annual average values for 1970 and 1978 show a 13% decline (29.98 and 25.95 m, respectively); the 4 year average for 1968-1971 (28.71 m) compared with the 4 year average for 1975-1978 (26.95 m) shows a 6% decline (Smith and others 1973).

[6]Stream environment zones at Tahoe include riparian habitats, meadows, and marshes. Because of urban development between 1970 and 1978, the amount of each of these declined 11%, 17%, and 31% respectively. The greatest percentage decline is in marsh habitat: 421 hectares prior to urban development, 280 hectares in 1970, and 192 hectares in 1978. These zones are critical wildlife habitats and they remove nutrients from runoff water, thereby protecting the quality of Lake Tahoe.

CARRYING CAPACITY CONCEPTS

Ecological concepts of carrying capacity are based primarily in observations about population growth patterns. The most commonly observed pattern is variously known as auto-catalytic, logistic, self-regulating, or self-maintaining. It is characterized by an S-shape, where growth starts slowly, accelerates rapidly in exponential form, and finally levels off and remains more or less at a constant level. This constant level represents the system's carrying capacity. At least four characteristics of this "natural" growth pattern are salient to the Tahoe situation and to any system which includes people and urban development.

First, the S-shape develops mathematically because of feedback from the population to its growth-inducing resources. The population or the ecosystem must feed back some of its own resources to obtain more. Natural systems must pay their own way, and the expenditures required for growth originate from inside the system. Secondly, as the population or ecosystem approaches its carrying capacity, the resources required for the next increment of growth increase exponentially (the law of diminishing returns). Thirdly, growth stops when the feedback costs of obtaining the next increment of a resource is greater than the growth it allows. This equality in the costs and benefits of growth determines the ultimate limit or the carrying capacity. Finally, in natural systems, uncertainty, change, and innovation are the rule rather than the exception (Holling 1977). Such change is termed species adaptation at the population level and ecological succession at the ecosystem level. When people are involved, we call it technological innovation. Successful change is said to be due to natural selection in ecosystems and to free market economics in urban systems. In any case, innovation and new designs are allowed and can, over long time periods, change the system's carrying capacity.

Unlike natural systems, growth costs at Tahoe are only partially paid for with internal funds, but, like natural systems, these costs are rising exponentially. Both the Federal and state governments, as well as the private sector, subsidize the growth process at Tahoe with external money input (e.g., construction grants). The exponential rise in costs is evident in Forest Service expenditures for the operation and maintenance of National Forest land within the basin. These were $11.64 per hectare or $0.20 per visitor-day in 1970. By 1978, they had increased more than 200% to $36.25 per hectare or $0.70 per visitor-day (all values in 1975$). Environmental mitigation expenditures show a similar pattern. About $100 million (1975$) have been spent to treat and export municipal wastewater from the Lake Tahoe Basin. This is equivalent to $0.19 for each gram of nitrogen that is prevented from entering Lake Tahoe annually. The non-point source control program developed by the California State Water Resources Control Board (1980) is expected to cost $95 million (1979$) or $0.83 per gram of nitrogen (1975$). The second increment of nitrogen control will cost 300% more than the first.[7]

[7]Nitrogen is the primary target of mitigation because Lake Tahoe is a nitrogen-limited lake.

Finally, because Tahoe does not "pay its own way", the basin system has no way of recognizing the point at which growth benefits exceed growth costs. The system is not self-limiting or self-maintaining; it remains on the exponential portion of the growth curve. In fact, for any system in which people are involved (with the exception of the entire earth as a system), resource scarcities or environmental degradation that could and would limit growth can be provided or mitigated against by subsidies from another system. Consequently, while there may be a "natural" carrying capacity for the entire earth, at any other scale carrying capacity must be socially determined. Moreover, a socially determined carrying capacity must set limits on system inputs.

ESTABLISHING CARRYING CAPACITY BY LIMITING SYSTEM INPUTS

The numerical values for the 5 kinds of people-related inputs to the Basin are given in Table 1 for 1970 and 1978. As indicated, inputs have increased between 52 and 377% in those 8 years.

Table 1: Summary of changes in people-related inputs: 1970 - 1978

Input	1970	1978	Percent Change
People (Number)			
Residents	33,600	73,200	118
Tourists	46,600	87,900	89
Total	80,200	161,100	101
Money (million 1975 dollars)			
Federal	21	39	88
State	7	14	100
Private	55	148	169
Visitor	287	484	69
Total	370	685	85
Goods and Services (million 1975 $)	120	280	133
Energy			
Electric Power Transmission Capacity (kilovolts)	420	720	71
Peak Electric Demand (Mwe)	82.5	125.2	52
Electric Power Consumption (thousand megawatt-hours)	363.7	574.2	58
Natural Gas Consumption (million cubic meters per year)	38.6	105.7	174
External Transportation			
Traffic on Entry Roads (Average daily traffic)	25,620	46,150	80
Commercial Airline Passengers (number per year)	123,426	587,103	377

[1]All data are from Western Federal Regional Council, 1979.

An imposed limit on any of the 5 types of inputs would stop growth, would cause the input to be used more efficiently (for example, energy inputs), and/or would cause the Basin to find its own internal sources of some of these inputs. A moratorium on one input would be a starting point for establishing carrying capacity, but such a moratorium does not account for the possibility of substitutions among input types. For example, capital input of money can substitute for energy input (construction of a light rail system for gasoline or construction of a solar collector for natural gas); and people can substitute for goods and energy. A carrying capacity that allows for innovation and self-design must allow for substitutions among types of input. Providing for substitutions, in turn, requires that all types of input be measured in one common unit and limited in that unit.

In his discussion of the carrying capacity of the earth, Odum (1976) suggests that the energy unit (actually energy per unit time) is a common unit. He notes that resource inputs to any system are either energy forms themselves or require energy to manufacture and provide. In either case, it is this "embodied" energy that supports the urban activity that causes the environmental degradation. Odum and Odum (1976), Lavine and Mayburg (1978), and Costanza (1980) suggest that embodied energy is a measure of substitutability among resource types, but this point is extremely controversial. I suggest only that embodied energy consumption is an <u>indicator</u> of urban activity and therefor an <u>indicator</u> of environmental degradation. It does represent a common unit and, in theory, limiting it would limit environmental degradation while maximizing the opportunity for self design.

In order to provide an example, I have re-expressed the 1978 inputs to the Basin (from Table 1) in energy units and regrouped them by energy using activity (Table 2). The inputs of goods from Table 1 appear in Table 2 as the energy consumed to manufacture retail goods and construction materials and the energy content of food and fiber. This also accounts for money inputs to the Basin since most of the money inputs simply cycle through for the purchase of goods. Fuel and power inputs from Table 1 are evaluated directly as the energy content of each energy form consumed in buildings and for transportation at Tahoe. Finally, the energy value of people inputs is negligible insofar as their biomass is concerned; it is enormous, as expressed here, in terms of the energy costs of supporting their demands for goods and fuel and power.

Of the total 9.7×10^{12} kilocalories consumed in support of urban structures and functions in 1978 (Table 2), 56% was embodied in the retail goods and construction materials that entered the Basin; 32% was consumed in buildings, 10% was consumed in transportation activities, and 2% represented the energy content of food and fiber. Most of this energy is consumed in direct and indirect support of recreation; in fact, 93% of all people related energy that is consumed in support of urban activity is consumed in support of recreational activity (Table 2). Gaming accounts for 56% of recreational energy consumption, while outdoor recreation accounts for only 15%.

By implication a "roll back" of environmental impacts could, in theory, be brought about by a resource input limit of, for example, 7.0×10^{12} kilocalories per year. Numerous innovative alternatives for reaching that limit are available. Eliminating construction activity or restricting the availability of retail services would help greatly. An emphasis on mass transit would reduce transportation energies and could accommodate more people under the same energy limit. A switch to solar heating would increase construction energies but would decrease energy consumption in buildings.

Probably of greatest interest over the short term, however, is the effect of a change in the distribution of recreation time. From Table 2, it is clear that the number of people that could use Tahoe under the same people-related energy input constraint would change dramatically if the distribution of recreation time among gaming, outdoor, and other recreation were to change. Directly and indirectly, one visitor-day of gaming requires 490,480 kilocalories per year of embodied energy support, while one visitor-day of outdoor recreation requires only 225,690 kilocalories.

With an embodied energy limit, the total number of people that can use Tahoe remains open-ended while carrying capacity is established in terms of kilocalories of people-related embodied energy consumption. As technology alters the manner in which energy is used and the efficiency with which it is used, the number of people using Tahoe can change while the energy limit (and presumably the environmental impacts) remain the same. Similarly, as the mix of activities that people engage in change, the number of people that that embodied energy limit can accommodate also changes.

SUMMARY

By combining systems methods, theoretical concepts from ecology about carrying capacity, and energy analysis, the human carrying capacity of a region can be examined. But, because human systems differ in several fundamental ways from ecological systems, human carrying capacity must be socially determined.

In nature, the costs of growth are paid for with resources that originate from inside the natural system (as represented by the feedback in the mathematical formulation of a logistic growth curve). These growth costs increase exponentially as the carrying capacity is approached, and the limit to growth is reached when the growth costs equal the growth benefits. Additionally, in natural systems, uncertainty, change, and innovation are the rule; these innovations and new designs can change the system's carrying capacity.

In contrast, in human systems, growth costs may be subsidized from external sources. As in natural systems, the costs increase exponentially, but, since they are paid externally, the system cannot recognize its own limit. Thus, carrying capacity must be socially determined. Ideally, this carrying capacity management system would limit the external subsidies or inputs to the human system and would be structured to allow for innovation and self design.

A systems model can define the subsidies and

Table 2. People-related energy consumed in support of urban activity in the Lake Tahoe Basin in 1978 by type of recreational activity.

(10^9 kilocalories)

	Recreational Activity Gaming	Outdoor	Other[1]	Total	Total Basin
Retail Goods[2]					
Residents	978	119	167	1264	1360
Visitors	1002	608	716	2326	2326
Subtotal	1980	727	883	3590	3686
Fuels[3]					
Buildings					
Residences	705	77	583	1365	---
Commercial	618	184	257	1059	---
Casinos	341	0	0	341	---
Subtotal	1664	261	840	2765	3074
Transportation					
Automobiles	508	171	210	889	919
All other	16	13	11	40	46
Subtotal	524	184	221	929	959
Subtotal Fuels	2183	445	1061	3694	4033
Food/Fiber[4]	160	33	43	236	252
Construction	773	104	642	1519	1730
Total	5101	1309	2629	9039	9701

[1]Other recreation includes second-homes, hotel/motel, and general rest and relaxation.

[2]Calculated as the dollar value of retail sales in 1978 multiplied by the quantity of energy required to manufacture "average" retail items in the U.S. in 1978.

[3]Evaluated as the enthalpy of the fuel forms with electricity converted to its thermal equivalent at 2645.8 kcal/kwh.

[4]Evaluated as 3000 kcal per person per day for food and 1.6 million kcal/resident/year of paper.

resource inputs and energy analysis can quantify them in a common unit. Inputs to any system are either energy forms themselves or require energy to manufacture and provide. In either case, it is this "embodied" energy that supports the urban activity that causes the environmental degradation. At a minimum, embodied energy consumption is an indicator of urban activity and, therefore, an indicator of environmental degradation. It may also be a measure of substitutability among resource types. It does represent a common unit, in theory, limiting it would limit environmental degradation while maximizing the opportunity for self design.

ACKNOWLEDGEMENTS

Charles Warren, while Chairman of the Council on Environmental Quality (CEQ), provided the impetus for the Lake Tahoe Environmental Assessment (Western Federal Regional Council 1979). Ariel Lugo, also at CEQ, provided assistance in establishing the systems ecological approach to the assessment. The interagency task force included representatives from the U.S. Department of Agriculture--Forest Service, U.S. Department of Transportation, U.S. Department of Housing and Urban Development, U.S. Department of Energy, U.S. Environmental Protection Agency, U.S. Department of Labor and U.S. Department of Interior.

LITERATURE CITED

Costanza, R. 1980. Embodied Energy and Economic Valuation. Science 210: 1219-1224.

Lavine, M. J. and A. H. Mayburg, 1978. Energy Analysis and Environmental Analysis in Transportation Planning. p. 859. In Energy Use Management: III. (edited by R.A.Fazzolare and C. B. Smith, Pergamon, New York).

Odum, H. T. 1976. Energy Quality and Carrying Capacity of the Earth. Tropical Ecology: 16.

Odum, H. T. and E. C. Odum, 1976. Energy Basis for Man and Nature. 285 p. McGraw-Hill, New York, N.Y.

Smith, R. C., J. Tyler, and C. R. Goldman. 1973. Optical properties and color of Lake Tahoe and Crater Lake. Limnology and Oceanography 18:189-199.

Western Federal Regional Council. 1979. Lake Tahoe Environmental Assessment. 240 p. National Technical Information Service, Springfield, Virginia; pp 82 163437.

OPTIMAL MULTIOBSERVATION MONITORING SYSTEMS. A PRACTICAL APPROACH[1]

James F. Sisler and Donald A. Jameson[2]

Introduction

Monitoring of natural resource systems for management purposes is inherently costly and time consuming. The resource manager needs to know where, when, and how much monitoring to do. The most natural approach for model based plans is to let the model determine the monitoring program; utilization of linear filter theory provides a tractable method for such determinations. In the paper by Sisler and Jameson (in preparation) the theoretical solution is given. In this discussion a practical methodology for applying the theory will be given.

For the sake of completeness a discussion of the theory is presented first; the interested reader is referred to (Sisler and Jameson, in preparation) for a complete development. Following the theory a method of approximation is given with a numerical example.

Theory

The results from the infrequent sampling problem developed by Pimental (1975) are crucial to our formulation. Pimental's approach uses a linear state space formulation

$$\hat{x}_{k+1} = \phi x_k + q_k \ ; \ q_k \text{ is } N(0, Q)$$

with time invariate state transfer matrix ϕ; the term q_k is the model error or forecast error. The state vector x_k represents the system at time k. The vector $\hat{x}_{k+1}$ is the projected system at the next time step. With the system at time k we have the system variance P_k; the forecase error variance at time k+1 is given by $\hat{P}_{k+1}$. The two variances are related according to the equation

$$\hat{P}_{k+1} = \phi P_k \phi' + Q,$$

(ϕ' is ϕ transpose)

If at time k+1 the system is monitored for performance, the Kalman filter algorithm updates the estimate. Let y_k represent the measurement taken at time k; the measurement system is described mathematically as

$$y_k = M\hat{x}_k + v_k, \ v_k \text{ is } N(0, R)$$

The error term v_k is the error due to sampling or measurement noise. The Kalman filtered result is given by

$$G_{k+1} = \hat{P}_{k+1}M'(M\hat{P}_{k+1}M' + R)^{-1}$$

$$P_{k+1} = (I - G_{k+1}M)\hat{P}_{k+1}(I - G_{k+1}M)' + G_{k+1}RG'_{k+1}$$

$$x_{k+1} = \hat{x}_{k+1} + G_{k+1}M(y_{k+1} - M\hat{x}_{k+1}).$$

Brewer and Moore (1974) proposed that the error variance $\hat{P}_{k+1}$ be allowed to grow until it is unacceptably large. That is, do not monitor until the projected error is intolerable. This is expressed by setting a level $d_a \geq 0$. Whenever the trace of the forecast error variance $tr[\hat{P}_{k+1}] \geq d_a$ the monitor is "triggered"; and a measurement is taken to be processed by the Kalman filter. Otherwise, no measurement is taken.

For the system at hand, Pimental (1975) showed that the effect of monitoring is highly predictable. Sisler and Jameson (In Preparation) used the following results due to Pimental:

1. At each time that a measurement is needed, the required sample size will be constant as long as the accuracy requirements (d_a) remain constant.

2. The sample size required for an item at any point in time is independent of the sample size required for other items.

3. The waiting time T between sample times is a function only of sample size or measurement accuracy.

The third result represents the only way the manager can interact with the monitoring program for a given tolerance d_a. The manager can determine sampling intensity by varying n, the sample size. The larger n is, the smaller the measurement noise R (Pliska 1974 and Pimentel 1975).

The trade-off to consider is the number of samples to be taken at any one time and the waiting time $\Delta T(n)$ for the next observation. As n increases, $\Delta T(n)$ is nondecreasing. To be specific, suppose at time k trP_k is greater than d_a, then we must measure the system. By applying the Kalman filter trP_k, from the updated variance, will be less than d_a. Since trP_k is a function of the number of samples n, we have

$$0 < tr\hat{P}_k(n+1) < trP_k(n) < d_a \leq trP_k.$$

To find the waiting time, $\Delta T(n)$, associated with a sample of size n, apply the state transformation equation recursively to project the variance

---

[1]Paper presented at the Third International Conference on State-of-the-Art in Ecological Modelling, Colorado State University, May 24-28, 1982.

[2]James F. Sisler is a Graduate Research Assistant and Donald A. Jameson is a Professor in the Range Science Department at Colorado State University, Fort Collins, Colo. 80523.

$\hat{P}_{k+j}(n)$ until $tr\hat{P}_{k+j}(n) \geq d_a$. For any $j \geq 1$, $\hat{P}_{k+j}(n)$ is given by

$$P_{k+j}(n) = \phi^j P_k(n)\phi^j + \sum_{i=0}^{j-1} \phi^i Q\phi^i$$

Denote by $\Delta T(n)$ the smallest j such that $tr\hat{P}_{k+j}(n) \geq d_a$, this is the waiting time, $\Delta T(n)$, for n samples. Since $trP_k(n+1) < trP_k(n)$ we have for any j, $trP_{k+j}(n+1) < trP_{k+j}(n)$, thus $\Delta T(n+1) \leq \Delta T(n)$.

Here is the trade-off the manager must consider. The larger the number of samples becomes, The fewer observations required over the planning horizon, N. To determine the optimal decision, the cost of monitoring over the planning horizon, T_N, must be optimized. This cost is given by

$$T_N = i(n)(C + cn)$$

where:

C is the start-up cost of monitoring.
c is the cost of each sample.
$i(n) = \max_{1 \leq j \leq N} [j : j \cdot \Delta T(n) \geq N]$,

The integer i(n) represents the number of times the system will be monitored over the planning horizon for a sample of size n.

The goal of the manager is to minimize the above cost equation, T_N.

In the theoretical solution by Sisler and Jameson (in preparation) the sample size n is replaced by a continuous variable $z \geq 1$, and $\Delta T(n)$ is replaced by $f(z) \varepsilon C^\infty$ such that $f(n) = \Delta T(n)$ on $(1,\infty) \eta \{1, 2, 3, . . .\}$. This substitution will be justified later. The total cost of monitoring with these substitutions becomes

$$T_N = \frac{N}{f(z)} (C+cz).$$

Take the first derivative with respect to z. By equating to zero, the optimal sample size can be found by solving the following first order equation.

$$z = \frac{f(z)}{f'(z)} + \frac{C}{c} = g(z) + \frac{C}{c}$$

Because f is concave, $g'(z) \geq 1$ and g(z) is monotone increasing on (1,) moreover, g is unbounded. Thus, as long as $g(1) \leq 1 - C/c$, there is a solution; otherwise z = 1 represents the global optimum (Figure 1).

The Method

In application of the theory there are two distinct problems:

1. The number of samples is an integer variable.
2. The waiting time is an integer variable.

These two considerations mean that the function f(z) can only be approximated. However, if the population is very large we can consider fractional sample numbers. Suppose that each sample contains 100 individuals randomly selected from the population without replacement; then two samples contain 200 individuals and 1.5 samples contains 150 individuals. In the case of an infinitely large population, we can have any rational number greater than one for samples. In the numerical example only integer values will be entertained.

Calculation of the waiting time f(z) is not feasible for fractional values; its calculation depends on the propagation of the error variance and is valid for uniform step lengths only. Thus, for any value z, only integer values of f(z) will be used. This will be done by counting the number of times the state transfer equation must be applied until $trP_{k+j} \geq d_a$ where j is the index for counting.

Although just one algorithm could be given, for simplicity the task will be broken into three parts. The first task is the most complex; it will generate the values of f(n) and f'(n). The second task merely uses the values generated in task one to calculate the values of g(n). The last task is the easiest, the list of generated values of g(n) is examined to determine the crossing point. The crossing point occurs when g(n) becomes greater than n + C/c. The optimal number of samples is either n or n-1; this possibility arises from the fact the true optimal point is in the closed interval [n-1, n].

The approximation of f(z) will be a piecewise linear concave function (Figure 2). The corner points correspond to integer values of n, the sample size, and integral values of $\Delta T(n)$. These points are found by the following algorithm:

1. Let the initial variance $P_o < d_a$ be given.

2. Project the variance until trP_k is greater than d_a,

$$\hat{P}_k = \phi^k P_o \phi'^k + \sum_{i=0}^{k-1} \phi^i Q\phi'^i.$$

3. For each integer n find the waiting time, i.e., project $P_k(n)$ until $tr\hat{P}_{k+j}(n)$ is greater than d_a,

$$\hat{P}_{k+j} = \phi^j P_k \phi'^j + \sum_{i=0}^{j-1} \phi^i Q\phi'^i$$

Usually n = 10 is the largest n we need to consider.

4. By examining the values $\Delta T(n)$, we notice either $\Delta T(n) = \Delta T(n+1)$ or $\Delta T(n) \leq \Delta T(n+1)$. For each integral value of $\Delta T(n)$ denote by n* the smallest n that goes with that value.

5. Denote by l the number of values of $\Delta T(n)$. The points $(n_i^*, T(n_i^*))$ i = 1, 2, . . ., l, are in order from left to right, and $n_l^* = 1$.

6. To complete the approximation between any two corner points, do a linear interpolation to connect the corners (dotted lines on Figure 3).

If $n^*_{i+1} + n^*_i = 1$, there is no interpolation to do.

If $n^*_{i+1} - n^*_i = m > 1$, we do m - 1 interpolations as follows

$$\Delta T(n^*_i + i) = \Delta T(n^*_i) + \frac{[\Delta T(n^*_{i+1}) - \Delta T(n^*_i)]}{m} \cdot i$$

$$i = 1, 2, \ldots, m-1.$$

Now that the function f(z) has been approximated the next step is to approximate the function

$$g(z) = \frac{f(z)}{f'(z)}.$$

Because the approximation of f(z) is piecewise linear, the approximation of g(z) will be discontinuous. However, the points g(n), n = 1, 2, . . ., will be monotone increasing; and this is what is desired. The discontinuities arise from the fact that f'(z) is taken from the slope of the linear segments of the approximation of f(z). This approximation can be done directly from the values generated for f(z) (Figure 3).

The function g(z) is approximated as follows

1. $g(1) = \frac{f(1)}{f'(1)}$ where $f'(1) =$

$$\frac{\Delta T(n^*_2) - \Delta T(n^*_1)}{n^*_2 - n^*_1}$$

$$= \frac{\Delta T(n^*_2) - \Delta T(n^*_1)}{n^*_2 - 1}$$

$$f(1) = \Delta T(1)$$

2. Set j = 2, n = 2

3. If $n \geq n^*_j$, go to 4, otherwise set j = j+1. If $j \leq 1$, go to 4; otherwise go to 5.

4. $g(n) = \frac{f(n)}{f'(n)}$ where $f'(n) =$

$$\frac{\Delta T(n^*_{j+1}) - \Delta T(n^*_j)}{n^*_{j+1} - n^*_j}.$$

$$f(n) = \Delta T(n).$$

set n = n+1 and go to 3.

5. g(n) = 10,000, n = n+k, k = 1, 2, . . . n. The value 10,000 is a surrogate for infinity; this will be the value of g(n) for all remaining n to consider.

Now that the function g(z) has been approximated, the optimal sample size can be chosen. To do this we must know the start up cost C and the cost of one sample c. Knowing this, the choice is made as follows.

1. Set n = 1.
2. Compare n + C/c and g(n).
3. If $n + C/c \leq g(n)$, set n = n+1 and go to 2; otherwise go to 4.
4. Stop. This is the crossing point. The optimal sample size is n or n-1.
5. Choose the optimum by calculating for n and n-1 the values*,

$$T(n-1) = 2C + c(n-1)$$

$$T(n) = C + c\,n$$

The optimal sample size n goes with the optimal T.

Numerical Example

The method can easily be programmed. An interactive program named OPTI for the univariate case has been written. The listing is in Appendix 2; a sample of the output for three monitoring cycles is provided in Appendix 1.

The univariate case is easy to do by hand. For the following system the first few calculations are done; the complete results are summarized in Table 1.

State transfer equations:

$$x_{k+1} = 1.1x_k + w_k;\ w_k \text{ is } N(0, 10),$$

$$\hat{P}_{k+1} = (1.1)^2 P_k + 10 = 1.21\ P_k + 10$$

Measurement system for n samples, $n \leq 10$

$$y_k = x_k;\ v_k \text{ is } N\left(0, \frac{120}{n}\right),$$

*See Sisler and Jameson (in preparation) for a complete explanation.

Filter equations

$$G_k = \hat{P}_{k+1}(\hat{P}_{k+1} + R_n)^{-1} =$$

$$\frac{1.21\ P_{k+1} + 10}{1.21\ P_{k+1} + 10 + \frac{120}{n}}$$

$$P_{k+1} = (1 - G_k)^2\hat{P}_{k+1} + G_k^2 R_n$$

$$\left[\frac{\frac{120}{n}}{1.21\ P_{k+1} + 10 + \frac{120}{n}}\right]^2 \cdot \hat{P}_k +$$

$$\left[\frac{1.21\ P_{k+1} + 10}{1.21\ P_{k+1} + 10 + \frac{120}{n}}\right]^2 \cdot \frac{120}{n}$$

Monitoring level:

$$d_a = 120$$

Initial variance:

$$P_1 = 100$$

$$\hat{P}_2 = 1.21\ 100 + 10 = 131 > 120 = d_a$$

For n = 1

$$G_1 = \frac{131}{251} = .522$$

$$P_2 = (.478)^2(131) + (.522)^2(120)$$

$$= 29.43 + 32.70$$

$$= 62.13$$

$$\hat{P}_{2+1} = 85.141$$

$$\hat{P}_{2+2} = 113.02061$$

$$\hat{P}_{2+3} = 146.75 > 120 = d_a$$

$$\Delta T(1) = 3$$

For n = 2

$$G_1 = .68$$

$$P_2 = (.314)^2(131) + (.686)^2(60)$$

$$= 12.9 + 28.23$$

$$= 41.13$$

$$\hat{P}_{2+1} = 59.77$$

$$\hat{P}_{2+2} = 82.32$$

$$\hat{P}_{2+3} = 109.61$$

$$\hat{P}_{2+4} = 142.6 \quad 120 = d_a$$

$$\Delta T(2) = 4$$

Approximating f(n) and g(n)

$f(1) = 3 \qquad f'(1) = 1$

$f(2) = 4 \qquad g(1) = 3$

Results for all n. Assume C = 4 and c = 1

Table 1.

n	n + C/c	ΔT	g(n)
1	5.00	3.00	3.00
2	6.00	4.00	8.00
3	7.00	4.00	9.00
4	8.00	5.00	15.00
5	9.00	5.00	16.00
6	10.00	5.00	17.00
7	11.00	6.00	100000.00
8	12.00	6.00	100000.00
9	13.00	6.00	100000.00
10	14	6.00	100000.00

The crossing point occurs at n = 2
The waiting time, Delta-T, is 4.00
Start-up cost of monitoring is 4.00
The cost of each observation is 1.00

The system variance before monitoring is P(2, 1) = 131.0000
The system variance after monitoring is P(2, 2) = 41.1518
The optimal number of observations is 2
The optimal waiting time is 4.00

Conclusion

This method as presented is particularly well suited for management of natural resource systems. In the case of public lands, monitoring programs must be specified as part of a plan. Not only does this method allow managers to spell out in advance how much, where, and when monitoring will be done, it specifies the least cost monitoring scheme. The simplicity of the filter even for highly dimensional systems enables the decision maker to interact easily with various monitoring alternatives.

What makes this method useful is its tractability in the decision process. It takes a complex question and reduces it to a set of linear matrix manipulations. It is particularly well suited for systems where our only knowledge is data based as demonstrated by Dixon and Howitt (1979). The statistics for implementation of a model based monitoring system are well understood and in many cases are available.

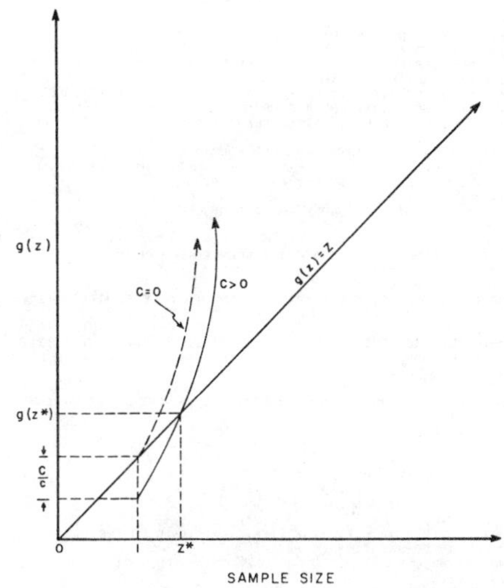

Literature Cited

Brewer, J. W., and S. F. Moore. 1974. Monitoring: An environmental state estimation problem. ASME Journal of Dynamic Systems, Measurement and Control 96:363-365.

Dixon, B. L., and R. E. Howitt. 1979. Continuous forest inventory using a linear filter. Forest Science 25:675-689.

Maybeck, P. S. 1979. Stochastic models, estimation and control. Academic Press, New York 1:129-132.

Pimental, D. D. 1975. Toward a mathematical theory of environmental monitoring: The infrequent sampling problem. Lawrence Livermore Lab Report UCRL-5187 (NTIS).

Pliska, S. R. 1974. A dynamic programming model for optimal observations of a discrete time linear stochastic process. Australian Journal of Statistics 16:156-162.

Sisler, J. F., and D. A. Jameson. (In Preparation) Optimal multiobservation monitoring systems.

Appendix 1

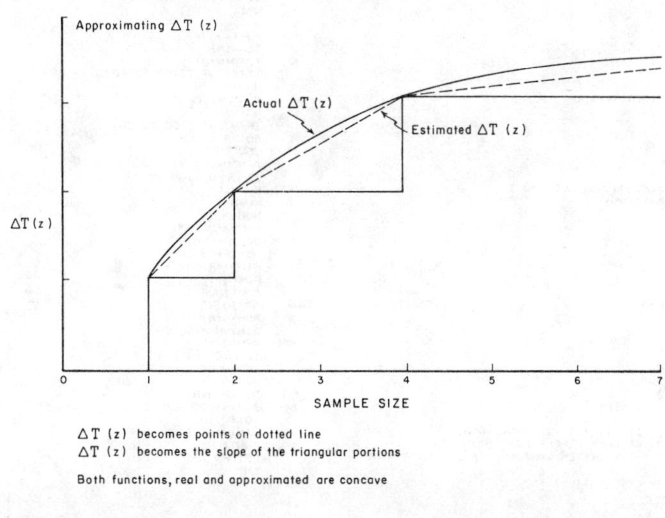

ΔT (z) becomes points on dotted line

ΔT (z) becomes the slope of the triangular portions

Both functions, real and approximated are concave

Appendix 2

```
      PROGRAM OPTI(INPUT,OUTPUT,TAPE5=INPUT)
      DIMENSION P(50,50),XK(50),XP(50),DT(50),RT(50)
      PRINT 1
1     FORMAT (30X,"PROGRAM OPTI",//,5X,"PROGRAM OPTI WAS",
     *" WRITTEN AT CSU BY:",/,20X,"JAMES F SISLER",/,
     *20X,"LATEST REVISION,MAY 25,1982")
      PRINT 2
2     FORMAT (/,5X,"PROGRAM OPTI WILL DETERMIN THE OPTIMAL",
     *" NUMBER OF SAMPLES FOR ",/,5X,"A SYSTEM DESCRIBED BY",
     *" A LINEAR STATE EQUATION IN DISCRETE TIME",/,5X,
     *"THE STATE MULTIPLIER AND PROCESS NOISE MUST BE",
     *"TIME INVARIANT.",/,5X,"THE PROCESS NOISE MUST BE SERIALLY",
     *" UNCORRELATED",/,5X,"THE MEASUREMENT NOISE MUST DEPEND",
     *" ONLY ON THE NUMBER",/,5X,"SAMPLES TAKEN")
      PRINT 3
3     FORMAT (//,5X,"THE STATE TRANSFER EQUATION IS OF THE FORM"
     *,//,20X,"X(K+1)=PHI*X(K)+U(K)",/,20X,"WHERE: U(K) IS N(0,Q)"
     *,//,5X,"THE USER MUST SUPPLY PHI AND Q.")
      PRINT 6
6     FORMAT (//," ENTER IN ORDER PHI AND Q")
      READ(5,*) PHI,Q
      PRINT 7,PHI
7     FORMAT (/," THE SYSTEM STATE TRANSFER EQUATION IS",/,
     *" X(J+1) = ",F6.4,"*X(J) + Q(J)")
      PRINT 8,PHI,PHI,Q
8     FORMAT(/," THE SYSTEM ESTIMAION ERROR VARIANCE EQUATION IS",
     */, " P(J+1,K)=",F6.4,"*P(J,K)*",F6.4," + ",F6.2)
      PRINT 9
9     FORMAT (//," NO MEASURMENTS WILL BE TAKEN UNTILL P(J,K) IS",
     */," LARGER THAN THE ALLOWABLE TOLERANCE,AD.",/," WHAT IS AD?")
      READ(5,*) AD
      PRINT 10
10    FORMAT (//," ENTER THE INITIAL ESTIMATE OF THE ERROR VARIANCE,",
     *" P(1,1)")
      READ (5,*) P(1,1)
      J=1
      K=1
 99    PRINT 111,J,K,P(J,K)
111   FORMAT(//," P(",I2,",",I2,") =",F6.2)
      CALL GROW(J,K,PHI,Q,AD,P)
      PRINT 11,J,K,P(J,K)
11    FORMAT (" P(",I2,",",I2,") = ",F6.2,/," WHAT IS THE MAXIMUM"
     *" NUMBER OF OBSERVATIONS,",/,
     *" AND THE MESUREMENT NOISE FOR ONE OBSERVATION?")
      READ (5,*) L,R
      IF(L.EQ.1) GO TO 30
      CALL MEASURE(J,K,L,R,P,XP,XK)
12    FORMAT (//," TO DETERMIN THE OPTIMAL NUMBER OF OBSERVATIONS",
     *" IT IS NECESSARY TO KNOW",/," STARTUP COSTS.BIGC, AND THE",
     *" COST OF EACH OBSERVATION, C.",/," ENTER BIGC AND C.")
      CALL DELTAT(J,K,L,PHI,Q,AD,XP,DT)
      CALL RATIO (DT,L,RT)
      PRINT 12
      READ (5,*) BIGC,C
      PRINT 13
13    FORMAT (//," DO YOU WANT THE FULL REPORT?",/,
     *" IF YES SET THE PRINT SWITCH TO 1",/," IF NO SET SWITCH TO 0")
      READ (5,*) IS
      CALL CROSS (L,RT,BIGC,C,NN)
      CALL DISPLAY(IS,L,RT,BIGC,C,NN,DT,J,K,XP)
      CALL CHOOSE(P,C,BIGC,NN,NNN,XP,J,K,DT)
      GO TO 35
30    CALL STEP(J,K,R,P)
      PRINT 111,J,K,P(J,K)
35    CONTINUE
      PRINT 15
15    FORMAT (//," DO YOU WANT TO GO AGAIN? IF YES ENTER 1 OTHERWISE 0")
      READ (5,*) III
      IF (III.EQ.0)GO TO 100
      K=J
      GO TO 99
      GO TO 99
100   CONTINUE
      END
      SUBROUTINE STEP(J,K,A,C)
      DIMENSION C(50,50)
      GAIN= C(J,K)/(C(J,K)+A)
      D=((1-GAIN)**2)*C(J,K)+(GAIN**2)*A
      K=J
      C(J,K)=D
      PRINT 10
10    FORMAT(/," THE UPDATE IS")
      RETURN
      END
      SUBROUTINE CHOOSE(C,CC,BC,NN,NNN,W,II,JJ,Z)
      DIMENSION C(50,50),W(50),Z(50)
      NNN=NN
      IF(NN.EQ.1)GO TO 15
      T1=Z(NN-1)
      T2=Z(NN)
      IF(T1.EQ.T2)GO TO 18
      FN=0.0
      DO 10 I=1,NN
10    FN=FN+1.0
      T3=BC+CC*FN
      T4=2.0*BC+CC*(FN-1.0)
      IF(T4.LT.T3)NNN=NN-1
15    C(II,II)=W(NNN)
      GO TO 19
18    CONTINUE
      NNN=NN-1
      C(II,II)=W(NNN)
19    CONTINUE
      PRINT 20,II,JJ,C(II,JJ),II,II,C(II,II),NNN,Z(NNN)
20    FORMAT(//," THE SYSTEM VARIANCE BEFORE MONITORING IS P(",
     *I2,",",I2,")=",F10.4,/," THE SYSTEM VARIANCE AFTER MONITORING
     *IS P(",I2,",",I2,")=",F10.4,/,
     *" THE OPTIMAL NUMBER OF OBSERVATIONS IS",I2,/,
     *" THE OPTIMAL WAITING TIME IS",F6.2)
      RETURN
      END
      SUBROUTINE CROSS(I,A,B,C,K)
      DIMENSION A(50)
      T1=0.
      K=0
      DO 20 J=1,I
      T1=T1+1.0
      T2=T1+B C
      T3=A(J)
      IF(K.NE.0)GO TO 20
      IF (T3.GT.T2)K=J
20    CONTINUE
      RETURN
      END
      SUBROUTINE DISPLAY(I,J,A,B,C,K,W,II,JJ,X)
      DIMENSION A(50),W(50),X(50)
      IF (I.EQ.0)GO TO 50
      PRINT 2
2     FORMAT(5X,"N",5X,"N +BIGC/C",7X,"DELTA-T",15X,"G(N)")
      FN=0.0
      DO 10 L=1,J
      FN=FN+1.0
      F=FN+B/C
      PRINT 12,L,F,W(L),A(L)
10    CONTINUE
12    FORMAT (5X,I2,4X,F6.2,11X,F6.2,11X,F10.2)
50    CONTINUE
      PRINT 51,K,W(K),B,C
51    FORMAT (//," THE CROSSING POINT OCCURES AT N =",
     *I2,/," THE WAITING TIME,DELTA-T IS",F6.2,/," START-UP COST"
     *" OF MONITORING IS",F6.2,/," THE COST OF EACH OBSERVATION IS",
     *F6.2)
      RETURN
      END
      SUBROUTINE GROW(J,K,A,B,D,C)
      DIMENSION C(50,50)
5     C(J+1,K)=A*C(J,K)*A+B
      T1=C(J+1,K)
      IF(T1.GE.D)GO TO 20
      J=J+1
      GO TO 5
20    J=J+1
      RETURN
      END
      SUBROUTINE MEASURE(J,K,L,A,C,XC,XB)
      DIMENSION C(50,50),XC(50),XB(50)
      FN=0.0
      DO 10 N=1,L
      FN=FN+1.0
      XB(N)=C(J,K)/(C(J,K)+A/FN)
10    XC(N)=((1-XB(N))**2)*C(J,K) + (XB(N)**2)*(A/FN)
      RETURN
      END
      SUBROUTINE DELTAT(J,K,L,A,B,D,XC,T)
      DIMENSION XC(50),T(50),W(50,50)
      DO 20 N=1,L
      W(J,J)=XC(N)
      I=J
      CALL GROW(I,J,A,B,D,W)
      IJ=I-J
      FIJ=0.0
      DO 15 M=1,IJ
15    FIJ=FIJ+1.0
20    T(N)=FIJ
      RETURN
      END
      SUBROUTINE RATIO(A,L,B)
      DIMENSION A(50),B(50)
      N=1
5     J=1
      FJ=1.0
      T1=A(N)
6     T2=A(N+J)
      IF(T1.NE.T2)GO TO 10
      J=J+1
      FJ=FJ+1.0
      NJ=N+J
      IF(NJ.GT.L)GO TO 50
      GO TO 6
10    CONTINUE
      DENOM=(T2-T1)/FJ
      IF(J.EQ.1)GO TO 30
      FJ=0.0
      KK=N-1
      NUM=A(N)
      R=NUM/DENOM
      DO 20 KJ=1,J
      JJ=KK+KJ
      B(JJ)=R+FJ
      FJ=FJ+1.0
20    CONTINUE
      GO TO 35
30    NUM=A(N)
      B(N)=NUM/DENOM
35    N=N+J
      IF(N.EQ.L)GO TO 70
      IF (N.GT.L)GO TO 100
      GO TO 5
50    DO 60 M=N,L
60    B(M)=100000.0
      GO TO 100
70    B(N)=100000.0
100   CONTINUE
      RETURN
      END
READY.
```

QUANTIFYING THE INTERDEPENDENCE BETWEEN MATERIAL AND ENERGY FLOWS IN ECOSYSTEMS[1]

Jean R. Fruci, Robert Costanza, and Scott G. Leibowitz[2]

Abstract-- Multicommodity analysis is critical to understanding the structure and function of ecosystems since they can be viewed as production systems involved in capturing solar energy and using it to produce a wide variety of organized structures. Three related linear programming techniques are developed and used to analyze two sets of input-output data; 1) the Silver Springs system illustrating the effect of the production of one joint product, and 2) a Louisiana cypress-tupelo swamp system illustrating the effect of the production of multiple joint products. The results of these three methods are discussed in relation to the assumptions of each and the light each sheds on the quantification of interdependence between material and energy flows in ecosystems in the presence of joint products. The Joint Products input-output model, which allocates embodied energy inputs to outputs of a process on the basis of the relative scarcities of the individual outputs, appears to be the most effective method for analyzing systems that produce multiple joint products.

INTRODUCTION

In the past, analysis of the flow of material and energy through ecosystems has been restricted to single commodities such as carbon (Richey et al. 1978), nitrogen (Finn and Leschine 1980), and calorimetric energy (Hannon 1973). The results give an indication of the relative dependencies of the processes through which the material or energy flows. By analyzing the flow of a single commodity, the problem of weighting the multiple commodities and the problem of joint products are avoided. Such analysis are useful in the comparison of single material flows in different systems as well as understanding the relative importance of ecosystem processes in the movement of a single material. This is analagous to tracing the path of a single commodity such as leather, through the economy. Through such an analysis, information regarding the connection between industries producing various leather goods (shoes, clothing, saddles, etc.) is obtained. This information is important in the determination of the allocation of leather to these industries. However, this analysis does not give any indication of the relative importance of leather to the function of the economy as a whole. Similarly, an understanding of the structure and function of the ecosystem as a whole requires an integrated analysis of the many commodities which circulate through the various processes in the system.

Expanding the analysis of material and energy flow to include multiple commodities within the input-output format would not be particularly difficult, if joint products were not involved. Unfortunately, it is almost impossible to avoid joint products in multicommodity input-output analysis. One of the assumptions of the Leontief input-output model is that each process produces only one output, ie. no joint products. All ecological processes produce joint products. Plants "produce" oxygen and waste heat in

---

(1) Paper presented at the Third International Conference on the State-of-the-Art of Ecological Modelling, sponsored by the International Society for Ecological Modelling, Colorado State University, May 24-28, 1982.

(2) Graduate Research Assistant, Assistant Professor, and Research Associate, respectively. Coastal Ecology Laboratory, Center for Wetland Resources, Louisiana State University, Baton Rouge, LA 70803.

conjunction with plant biomass.

Economists, in their analysis of production and the allocation of resources, have been able to use relative prices to allocate inputs to joint products. Ecologists, however, do not have these weighting factors with which to translate different ecological commodities into commensurable units. In fact it is precisely these relative weights which are the object of the analysis.

Costanza and Neill (1981b) used a commodity-by-process input-output model to analyze the flow of materials and energy through the biosphere, and calculate energy intensities for the individual commodities. The energy intensities are weighting factors, analagous to relative prices in the economy. Although they were able to obtain reasonable results using the standard input-output model, this was only possible when the data were in a specific form. Otherwise, the model produced negative energy intensities, which do not make sense. The model had to be sectored such that each process produced only one "major" output, minimizing the joint products problem. Some of the commodities had to be aggregated, thus destroying some of the desired detail of the analysis (eg. aggregating carbon, nitrogen, and phosphorus into nutrients).

One alternative to the use of an input-output model is to use a more general linear programming (LP) model, which does not allow negative energy intensities. In this paper, we use the LP model to analyze two sets of data. In addition we introduce another alternative, a modification of the standard input-output model, which is designed to deal with the problem of joint products. It has several advantages over the LP model as well, which encounters some of the same difficulties as the standard input-output model when handling multicommodity systems with joint products.

METHODS

The Linear Programming Model (LP)

LP models are used in economics and management to solve problems of optimal resource allocation. Evolution works to achieve optimal resource allocation in ecosystems, and LP models should therfore be useful in understanding and predicting their structure. This model is more general than the standard I-O model or the commodity-by-process input-output model (I-O) because it allows alternative technologies (different combinations of inputs which produce the same output). The I-O models are simply special, constrained cases of the LP model. The standard I-O does not allow either joint products or alternative technologies, while the commodity-by-process I-O model allows joint products but not alternative technologies. A description of these models and their applications to ecological systems can be found in Costanza and Neill (1981a and 1981b).

In the LP model and the commodity-by-process I-O there are n commodities and m processes. The model assumes a linear relationship between inputs and outputs (ie. the output is directly proportional to the inputs). We adopt this framework to calculate embodied energy intensities (which are analagous to the economist's shadow prices) for each of the commodities in the system. The embodied energy intensity is equal to the amount of direct and indirect energy required to produce a unit of a commodity. To solve for the energy intensities we set up a system of m linear, algebraic, equations setting the total input of each commodity to all processes equal to the total output of each commodity from all processes. Unlike the standard I-O model the number of commodities does not have to be equal to the number of processes. The balance equations in matrix notation are:

$$aX + Y = bX + Z \tag{1}$$

where a is an n x m matrix of gross input coefficients (ie. the amount of commodity i required as input for process j per unit activity of process j)

b is an n x m matrix of gross output coefficients (ie. the amount of commodity i produced as output from process j per unit activity of process j)

Y is an n dimensional column vector of commodity exports (which includes physical exports plus change in storage plus depreciation losses (basal metabolism))

X is an m dimensional column vector of process activity levels (ie. the output of each process)

Z is an n dimensional column vector of commodity imports

These production constraints are then used to derive a primal/dual LP pair. The primal problem is to minimize the primary resource (solar energy) cost of production for given process activity levels X:

$$P = Q\,X \tag{2}$$

subject to the production constraints

$$(b - a)X > (Y - Z) \quad \text{with } X > 0 \tag{3}$$

The dual problem solution yields the embodied energy intensities. The dual problem is to maximize the value of the net output of the system:

$$p' = e(Y - Z) \tag{4}$$

Subject to the embodied energy balance constraints, and the constraint that all the commodities have non-negative energy intensities:

$$e(b - a) < Q \quad \text{and } e < 0 \tag{5}$$

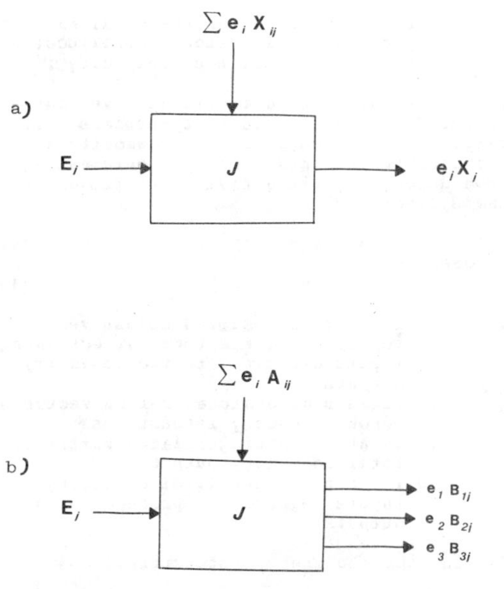

Figure 1.-- Illustration of a single process energy balance showing: a) Production of a single output, X, and b) production of joint products, B_1, B_2, and B_3.

where Q is an m dimensional column vector of net solar energy inputs to each process per unit activity level:

$$q_{ij} = \frac{E_j}{X_j} \qquad (6)$$

and E is an m dimensional column vector of direct solar energy inputs

The Commodity-by-Process Input-Output Model

The standard I-O is the result of two additional constraints on the LP model. There can be no joint products and for each output there is a unique combination of inputs (Dorfman, Samuelson, and Solow 1958). The standard I-O model (S.I-O) has been used by Hannon (1973, 1976, 1979), Richey et al. (1978), Finn and Leschine (1980), and Herendeen (1981) to study the flow of energy and materials through ecosystems, and by Bullard and Herendeen (1975) and Costanza (1980) to study these same flows through the economy. The I-O model was developed by Leontief to take account of the interdependence of the production of commodities by the various industries in the economy. The interdependence arises due to the fact that each industry uses materials produced by the other industries in the system to produce their respective outputs. The energy intensities calculated using this model are stated in terms of the solar input to the system, as in the LP model.

The commodity-by-process I-O model simply relaxes the constraint prohibiting joint products (but maintains the single technology constraint). A detailed description of this model and its application to multicommodity ecological analysis is given in Costanza and Neill (1981b). The model is briefly summarized below.

There are n commodities and m processes. The energy balance for a single process for a particular time interval is equal to the sum of direct energy input to the process and the indirect energy (ie the energy embodied in the commodity inputs to the process) (Fig. 1a.). This can be expressed as:

$$E_j + \sum e_i A_{ij} = \sum e_i B_{ij} \qquad (7)$$

where E is the direct solar energy input to process j
A is the input of commodity i to process j
B is the output of commodity i from process j
e is the weighting factor or embodied energy intensity of commodity i

In matrix notation for all m processes this becomes:

$$E + eA = eB \qquad (8)$$

where E is an m dimensional column vector of direct energy inputs
e is an n dimensional column vector of embodied energy intensities
A is an n x m matrix of commodity inputs by process
B is an n x m matrix of commodity outputs by process

If the number of commodities equal the number of processes (ie. n=m), equation 8 can be solved for the embodied energy intensities simply as

$$E = e(B - A) \qquad (9)$$
$$e = E(B - A)^{-1} \qquad (10)$$

Since this model does not contain a non-negativity constraint as the LP model, negative energy intensities can result when the matrix includes the production of joint products. Costanza and Neill (1981b) give a description of the necessary conditions for the production of all positive energy intensities, sufficient conditions are still uncertain.

Joint-Products Input-Output Model

One additional modification of the commodity-by-process I-O model was developed to deal with the negative embodied energy intensities that sometimes result when joint products are present in a multicommodity analysis. The joint products model (JP) transforms a commodity-by-process model into a more managable standard model (commodity-by-commodity) to calculate the energy intensities of the commodities in the system. To develope the commodity-by-commodity matrix from the commodity-by-process matrix the total input of each process must be allocated to the joint products according to some externally derived allocation scheme. Figure 1b. illustrates the situtation for which the allocation scheme is necessary. We based this allocation scheme on the relative percentage of the total output of each commodity that each process provides to the system. This allocation scheme assumes that the greatest percentage of process inputs should be allocated to the production of a scarce rather than a ubiquitous joint product. For example, a cypress tree has several outputs; cypress biomass , carbon dioxide, nitrogen, phosphorus, and water (Fig. 2). However, it has only one output which is unique to the cypress process--cypress biomass. Therefore, the model assumes that the greatest percentage of the inputs to the cypress process should be allocated to the production of cypress biomass.

In order to calculate the allocation matrix, the elements of the commodity-by-process output matrix are divided by the total commodity output:

$$C_{ij} = \frac{B_{ij}}{X_{ij}} \tag{11}$$

where B is the output of commodity i from process j
X is the total output of commodity i from all processes in the system
C is the fraction of commodity i produced in process j

Then the elements of the columns of matrix C are added to produce a vector of total fractional commodity outputs by process. The allocation matrix is then obtained by dividing each element of the C matrix by the total fractional commodit, outputs by process (CSUM):

$$D_{ij} = \Sigma C_{ij} \tag{12}$$

where C is the fraction of commodity i produced in process j

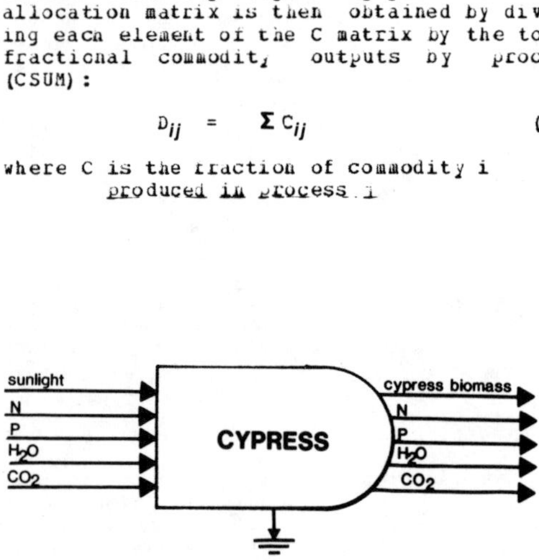

tion of five joint products.

D is the fraction of the total input to process j which should be allocated to the production of commodity i

Using the allocation matrix (D) we can now transform the commodity-by-process input matrix A into a commodity-by-commodity matrix and substitute this into the original equation describing the activity of processes in the system:

$$E = e(B - A) \tag{9}$$

becomes;

$$ED = e(X - AD) \tag{13}$$

where ED is an n dimensional column vector representing the total direct energy inputs allocated to the commodity outputs
e is an n dimensional column vector of embodied energy intensities
X is an n x n diagonalized matrix of total commodity outputs
AD is an n x n matrix of commodity inputs required to produce commodity outputs

Solving for the energy intensities:

$$e = ED(X - AD)^{-1} \tag{14}$$

The matrix resulting from the difference between X and AD is a square matrix (ie. the commodity-by-commodity matrix).

Both the LP and the JP model were used to calculate energy intensities for commodities in two ecosystems, Silver Springs, Florida and a Louisiana cypress swamp.

The data on biomass and nitrogen flows in the Silver Springs ecosystem were originally compiled by Odum (1957). This data set has since been the subject of a number of ecosystem analyses (Hannon 1973, 1976, 1979, Herendeen 1981, and Costanza and Neill (1981a).

The data on the Louisiana cypress swamp was compiled by Christopher Neill as part of a larger study of the Mississippi Deltaic Plain Region (MDPR). Since both the methods of data collection (Leibowitz and Costanza 1982) and the MDPR study (Costanza et al. 1982) are described elsewhere in this volume we will only briefly summarize the methods here. A literature search was conducted to obtain the best available data on primary and secondary production, nutrient flows, and standing stocks of various ecological commodities in a 'typical' Louisiana cypress swamp. The majority of the data were taken from studies done by various authors in the Lac Des Allemands swamp in Louisiana.

Data were placed in an I-O format such that conditions for mass balance of materials into and out of the system were met. Data which could not be obtained from studies in Louisiana were either taken from studies done in similar regions or calculated by difference.

Table 1-- Silver Springs commodity-by-process input-output table.

SILVER SPRINGS		INTERNAL PROCESSES					EXTERNAL PROCESSES		
COMMODITIES		1 PRODUCERS	2 HERBIVORES	3 CARNIVORES	4 TOP CARNIVORES	5 DECOMPOSERS	6 IMPORTS & EXPORTS	TOTAL	UNITS
PRODUCER BIOMASS	1		748.4			874.2	460.3	2082.9	g/m^2/yr
		1962.9					120.0	2082.9	g/m^2/yr
HERBIVORE BIOMASS	2			85.1		170.7	72.6	328.4	g/m^2/yr
			328.4					328.4	g/m^2/yr
CARNIVORE BIOMASS	3				4.7	6.9	3.3	14.9	g/m^2/yr
				14.9				14.9	g/m^2/yr
TOP CARNIVORE/ DECOMPOSER BIOMASS	4					80.6	22.9	103.5	g/m^2/yr
					1.3	102.2		103.5	g/m^2/yr
NITROGEN	5	121.7					4415.6	4537.3	g/m^2/yr
						70.3	4467.0	4537.3	g/m^2/yr
HEAT	6						407986.0	407986.0	kcal/m^2/yr
		401167.0	1890.0	316.0	13.0	46000.0		407986.0	kcal/m^2/yr
SUNLIGHT	7	410000.0						410000.0	kcal/m^2/yr
							410000.0	410000.0	kcal/m^2/yr

Table 2-- Louisiana cypress-tupelo swamp commodity-by-process input-output table.

HABITAT 6 CYPRESS/TUPELO		INTERNAL PROCESSES															EXTERNAL PROCESSES			
COMMODITIES		1 SURFACE WATER	2 SHALLOW SOIL/SEDIMENT	3 CYPRESS	4 TUPELO	5 OTHER WOODY PLANTS	6 HERBACEOUS PLANTS	7 DUCKWEED	8 LITTER	9 INSECTS	10 FLOATING DETRITUS	11 FLOATING HERBIVORES	12 BENTHIC DETRITUS	13 BENTHIC HERBIVORES	14 AQUATIC PREDATORS	15 PREDATORS	16 IMPORTS & EXPORTS	17 CHANGE IN STORAGE	TOTAL	UNITS
WATER	1	2.04	0.52	0.26	0.07	0.03	0.09	0.10									2.20		5.31	10^6 g H_2O/m^2/yr
		2.04	0.52	0.26	0.07	0.03	0.09	0.10									2.20		5.31	10^6 g H_2O/m^2/yr
		0.19	0.31																0.50	10^6 g H_2O/m^2
NITROGEN	2	48.57	93.84	38.8	10.4	4.1	21.3	21.6									19.31		257.92	g N/m^2/yr
		48.57	93.84	24.4	6.6	2.6	13.4	10.2		1.7	29.5	5.8	13.7	0.5	1.8	0.2	5.11		257.92	g N/m^2/yr
		0.02	347																347.02	g N/m^2
PHOSPHORUS	3	8.56	8.90	1.33	0.87	0.18	4.80	4.87									1.27	1.72	15.04	g P/m^2/yr
		8.56	8.90	0.84	0.55	0.11	3.03	2.30		0.02	4.13	0.80	1.89	0.01	0.25	0.03	1.06		15.04	g P/m^2/yr
		0.04	63.7																63.74	g P/m^2
CARBON DIOXIDE	4			4000	1078	479	1467	1490									6362.9	299	14 816.9	g CO_2/m^2/yr
				2591	679	264	923	704		41.1	706.8	138.7	328.6	12.1	42.3	5.3	8454.0		14 816.9	g CO_2/m^2/yr
TOTAL ORGANIC CARBON	5	345	322														23		989	g dry wt/m^2/yr
		345	322														322		989	g dry wt/m^2/yr
		5.9	11 475																11 480.9	g dry wt/m^2
SEDIMENT	6	266	266															266	798	g dry wt/m^2/yr
		266	266														266		798	g dry wt/m^2/yr
		20	56 025																56 045	g dry wt/m^2
CYPRESS BIOMASS	7								238	13.3							879	646	2422	g dry wt/m^2/yr
				2422															2422	g dry wt/m^2/yr
				20 919															20 919	g dry wt/m^2
TUPELO BIOMASS	8								133	50.6							353	58	653	g dry wt/m^2/yr
					653														653	g dry wt/m^2/yr
					11744														11 744	g dry wt/m^2
OTHER WOODY PLANT BIOMASS	9								46	2.6							115	45	254	g dry wt/m^2/yr
						254													254	g dry wt/m^2/yr
						4037													4037	g dry wt/m^2
HERBACEOUS BIOMASS	10								312	17.5							669		888	g dry wt/m^2/yr
							888												888	g dry wt/m^2/yr
							20												20	g dry wt/m^2
DUCKWEED BIOMASS	11										285	141					476		902	g dry wt/m^2/yr
								902											902	g dry wt/m^2/yr
								50											50	g dry wt/m^2
LITTER BIOMASS	12		227								241		161				100		729	g dry wt/m^2/yr
									729										729	g dry wt/m^2/yr
									422										422	g dry wt/m^2
INSECT BIOMASS	13										56.1					1.0	24.9		84	g dry wt/m^2/yr
										84									84	g dry wt/m^2/yr
										11.5									11.5	g dry wt/m^2
FLOATING DETRITUS	14											140					478		618	g dry wt/m^2/yr
											618								618	g dry wt/m^2/yr
											500								500	g dry wt/m^2
FLOATING HERBIVORES	15												140		51.7		89		281	g dry wt/m^2/yr
												281							281	g dry wt/m^2/yr
												5.7							5.7	g dry wt/m^2
BENTHIC DETRITUS	16		40								34			50	6		239		369	g dry wt/m^2/yr
													369						369	g dry wt/m^2/yr
													500						500	g dry wt/m^2
BENTHIC HERBIVORES	17												38		5		7.3		50.3	g dry wt/m^2/yr
														50.3					50.3	g dry wt/m^2/yr
														0.5					0.5	g dry wt/m^2
AQUATIC PREDATORS	18												29			4.5	29.6		62.6	g dry wt/m^2/yr
															62.6				62.6	g dry wt/m^2/yr
															0.9				0.9	g dry wt/m^2
PREDATORS	19												2.3				3.25		5.55	g dry wt/m^2/yr
																5.5			5.5	g dry wt/m^2/yr
																0.1			0.1	g dry wt/m^2
HEAT	20																448 158			kcal/m^2/yr
				210 699	56 806	22 096	77 253	77 829		112	1926	378	896	33	115	14			448 158	kcal/m^2/yr
SUNLIGHT	21			214 735	57 895	22 520	78 730	79 971									969 149		1 423 000	kcal/m^2/yr
																			1 423 000	kcal/m^2/yr

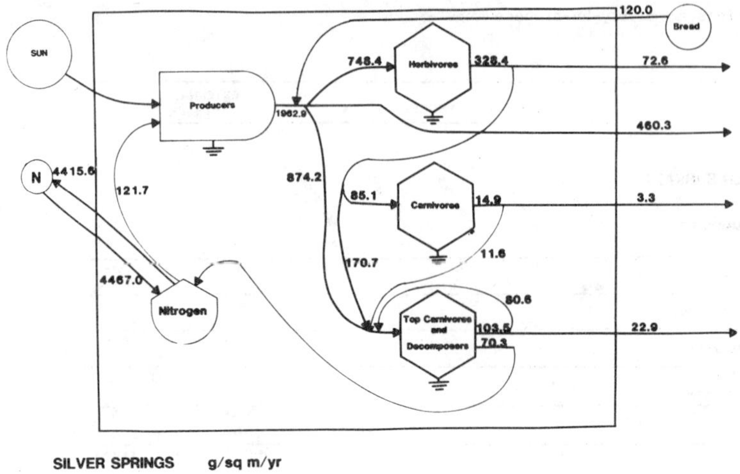

Figure 3--Illustration of material and energy flows in the Silver Springs ecosystem.

RESULTS

Data from Silver Springs and Louisiana were used to compare the LP and the JP methods of calculating energy intensities. The data from the Silver Springs and Louisiana ecosystems are presented in input-output format in Tables 1 and 2, respectively. Each commodity is represented by a row of the table, each process by a column. The inputs of the commodity to each process are listed across the top line of the row. The outputs are listed across the center line of the row. The bottom line of the row is reserved for storage estimates of the commodity within a process. The columns labeled IMPORT/EXPORT indicate materials which are brought into or exported from the system. The top line of the IMPORT/EXPORT column for each commodity indicates the input to external processes (or exports from the system). It includes physical exports, change in storage, and depreciation (basal metabolism). The center line of the IMPORT/EXPORT column indicates net imports of the commodity from outside the system.

These two data sets were selected for comparison because both systems show different degrees of the joint products problem. The Silver Springs system has only one process that produces joint products (Figure 3). The Louisiana swamp has multiple joint products produced by all of the processes in the system (Figure 4). The energy intensities for each commodity are listed in Tables 3 and 4, respectively.

The Silver Springs ecosystem has been the subject of similar energy analyses by Hannon (1973, 1976, 1979), Herendeen (1981), and Costanza and Neill (1981a). The results using the LP model on the data from Silver Springs were previously presented by Costanza and Neill (1981a), and are presented here for comparison with the results of the JP model. The energy intensities for the five commodities in the system increase along the food chain from the plant biomass to the top carnivore/decomposer biomass. Nitrogen has an intensity value between that of herbivore and carnivore biomass. In this system, nitrogen is the only joint product.

Column 2 of Table 3 lists the energy intensities obtained using the JP model. The intensities of the biomass components show the same pattern of increase along the food chain as those obtained with the LP model, although the individual values are somewhat lower. The energy intensity of the nitrogen, however, is several orders of magnitude lower than that calculated using the LP model.

Table 3. The energy intensities for the Silver Springs ecosystem calculated using the LP and JP models.

Commodity	energy intensity in kcal/g (LP)	energy intensity in kcal/g (JP)
Plant biomass	297.0464	197.0763
Herbivore biomass	676.9478	449.1228
Carnivore biomass	3866.3230	2565.1242
Top Carnivore/ Decomposer biomass	13978.2300	11372.2857
Nitrogen	1422.2010	4.0287

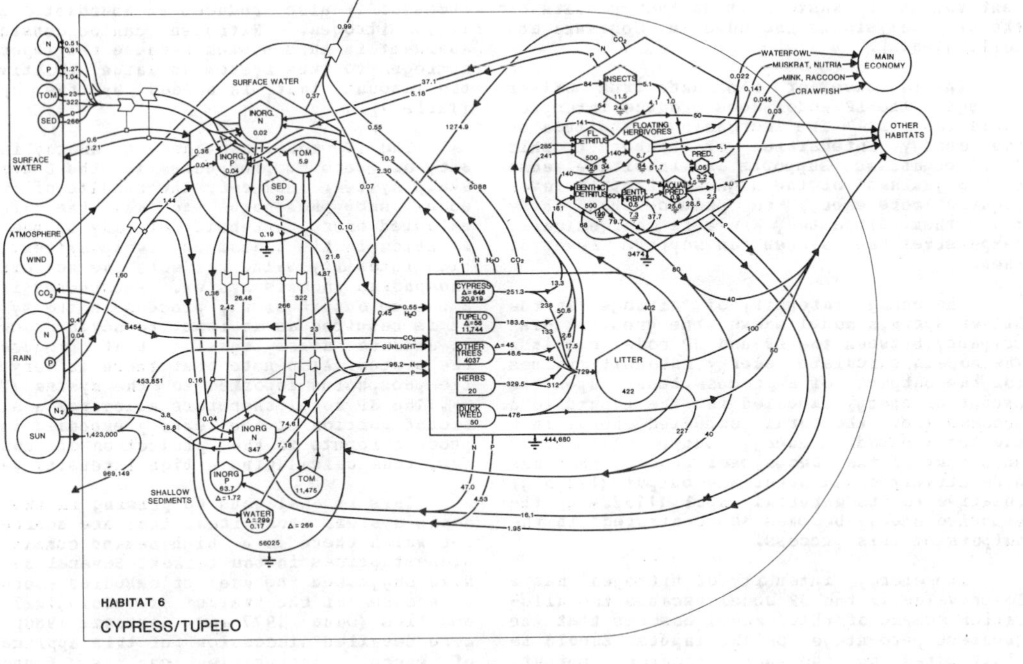

Figure 4--Illustration of material and energy flows in the Louisiana cypress-tupelo swamp ecosystem.

The LP model failed to produce sensible results in the analysis of the Louisiana swamp ecosystem (Table 4). The energy intensities of water, phosphorus, and carbon dioxide are all equal to 0. This implies that it requires no energy to cycle and transform these commodities within the swamp. The problem of zero intensities will be discussed later.

Unlike the energy intensities of biomass in the Silver Springs ecosystem, the various forms of plant and animal biomass in the swamp ecosystem have relatively similar energy intensities. The energy intensities of the nitrogen and phosphorus are much higher than that of the nitrogen in the Silver Springs system.

---

* The LP model will not obtain a solution to a matrix which has rows that are linearly dependent. Therefore, these commodities had to be eliminated from the analysis.

** The LP model does not require the number of processes and commodities to be equal. However, the model could not obtain a solution when the data were in the original format (Table 2). For this reason, certain commodities were aggregated to produce an equal number of commodities and processes. Therefore the value in Table 4 noted by a double asterisk represents the value for the two aggregated commodities.

DISCUSSION

The general LP model, although designed to handle the production of joint products and production of commodities through alternate technologies, does not always produce sensible results. For this reason, the JP model may be a more appropriate methodology for the analysis of multicommodity ecosystems with multiple joint products.

For the LP analysis of the two data sets included in this paper, the value of waste (respiratory) heat was assumed to be 0. A value of 0 was chosen because this implies an accounting scheme consistent with the second law of thermodynamics. It is probably more realistic to include some small value for waste heat. However, this would not drastically alter the results of the model, unless a first law accounting is used. In this case, waste heat is assigned a value of 1. A more complete explanation of the effect of different values of waste heat on the results of the LP analysis is included in Costanza and Neill (1981a).

In the case of the data from Silver Springs, the LP model does produce interpretable results. The increasing magnitude of the energy intensities of higher trophic level organisms supports an intuitive sense that organisms of the highest trophic level require more energy to produce than those below them. Since they are more energetically 'expensive' the system can support fewer of them.

The energy intensity of nitrogen in the Silver Springs model shows the greatest discrepancy between the LP and JP model results. The models calculate energy intensity values for the outputs of a process based upon the amount of energy embodied in the inputs to a process (ie. the total embodied energy in = the total embodied energy out). Nitrogen is an output of the Decomposer process that has a relatively small material output (172.5 g) relative to its material input (1132.4 g) the embodied energy becomes 'concentrated' in the outputs of this process.

The energy intensity of nitrogen has a lower value in the JP model because the allocation scheme of this model assumes that the greatest percentage of the inputs should be distributed to the most 'scarce' output. Scarce, in terms of the model, can be either a unique output, produced by only one process in the system, or one that is produced in quantities just large enough to meet the demand for that commodity as input to other processes in the system. In the Silver Springs system, the Decomposer process produces one unique commodity, decomposer biomass. It also produces an abundant commodity, nitrogen. Nitrogen can be considered abundant in this system because the import of nitrogen to the system is large relative to the amount that is needed by the system (Table 1).

Table 4. Energy intensities for the Louisiana Cypress Swamp ecosystem calculated using the LP and JP models.

Commodity	energy intensity in kcal/g (LP)	energy intensity in kcal/g (JP)
Water	0.0	0.2003
Nitrogen	120.5567	588.3942
Phosphorus	0.0	20936.9665
Carbon Dioxide	0.0	3.3276
Total Organic Carbon	*	136.9351
Sediment	*	168.1785
Cypress biomass	** 29.5044	63.5578
Tupelo biomass	**	107.0325
Other woody biomass	78.9764	111.7696
Herbaceous biomass	308.7383	148.4767
Duckweed	53.9236	156.0293
Litter	152.1338	110.8755
Insects	86.7696	107.4604
Floating Detritus	** 91.6396	90.4550
Floating Herbivores	43.1684	113.6403
Benthic Detritus	**	91.5253
Benthic Herbivores	89.8946	90.8314
Aquatic Predators	** 48.3556	107.2612
Predators	**	106.5166
Heat	0.0	0.2903

The LP model assigned an energy intensity of zero to phosphorus in the Louisiana swamp system. However, the results of the JP model show phosphorus to have the highest embodied energy intensity of any of the commodities in the Louisiana swamp system. The high intensity value reflects the scarcity of phosphorus in this system. Phosphorus is not a unique output of any process in the system. It is required as an input to seven processes in amounts nearly equal to that produced in the system. Also note that there is very little phosphorus imported to the system (Table 2). The JP model therefore allocates a significant portion of the energy embodied in the process inputs to the production of phosphorus, thus calculating a high intensity value.

This is analagous to pricing in the economic system. Those items that are scarce and for which there is a high demand command the highest prices in the market. Several authors have suggested the use of embodied energy as a measure of the 'value' of ecological commodities (Odum 1977 and Costanza 1980). A more detailed discussion of this application of energy intensities can be found in Costanza et al. (1982).

The LP model will sometimes assign a zero activity level to certain processes in the system to obtain an optimal solution (maximization of the energy embodied in process outputs). The assignment of zero intensities to water, carbon dioxide and phosphorus indicates that the system would achieve this optimal solution by not allocating any energy to the processes for production of these commodities. The activity levels assigned to the processes as input data to the model, did not specify the production of any of these three commodities. In addition, there are no processes which specifically produce these commodities (Table 2). While this is also true for nitrogen, the activity level assigned to the sediment process is equal to the value of the nitrogen output in this process. This in essence tells the model that nitrogen is the 'important' output of the sediment process. Carbon dioxide, water, and phosphorus are essential to the real system, and it does require energy to transform these commodities. Therefore, these results can not be considered accurate descriptions of the energy flow in this system. It is due to this type of problem that the JP model was developed.

Although the LP model produces some energy intensities which cannot be considered acceptable solutions, there are some close correlations between the energy intensities calculated by the JP and LP models for corresponding commodities. For example, the value for detritus obtained using the LP model and that obtained using the JP model are nearly equal (Table 4).

Herendeen (1981) suggested that an I-O table with numerous feedback flows should produce a reduction in the range of the energy intensity values. The implementation of the allocation assumption in the JP model should counter this tendency if it indeed exists. Intuitively, it does seem that if you increase the number of joint products (therefore the overall material output relative to the total material input) the embodied energy of the inputs would be distributed over a wider range of outputs in each process. Instead of 'concentrating' the embodied energy in one output, it would be dispersed among the joint products. More uniform values of energy intensities could then result if the number of material inputs to the processes were similar, and if the number and magnitudes of material outputs were similar. The allocation assumption of the JP model insures that the majority of the embodied energy of the inputs will be used in the production of a few primary outputs of the process, rather than being arbitrarily allocated to the outputs based on their magnitudes.

The energy intensities calculated for the biomass commodities in the Louisiana swamp system are all within a narrow range, however, the values for the nutrient show a wide range of variability. This is also true in the case of the Salt Marsh system, which was also analyzed using the JP model (Leibowitz and Costanza 1982).

The JP model has several advantages over the I-O and the LP models. One is the ability of the JP model to calculate energy intensities for a system with multiple commodities and multiple joint products. The transformation of the commodity-by-process I-O into a commodity-by-commodity or standard I-O permits the analysis of rectangular matrices as well as square. Using this format, none of the commodities or processes need to be aggregated and none of the detail of analysis is forfeited to mathematical procedure.

It is possible to calculate energy intensities using the LP and commodity by process framework, however, it is sometimes necessary to alter the data either through aggregation, omission of joint products, or change individual data points within the matrix. The JP model, however, calculates energy intensities using the unaltered input data. Although the data used in these models do have a certain degree of error associated with them, an unbiased, systemmatic approach such as a Monte Carlo variation of individual data points within a statistical error range would be more acceptable than data alterations which attempt to satisfy the specific data requirements of a particular model.

The JP model produces energy intensities via a specified allocation matrix. There is some error associated with the allocation assumption. To reduce this error an iterative improvement of the allocation matrix may be employed. We are currently pursueing this approach, which should make comparison of the LP and JP results easier.

When using a model to describe the characteristics of a system it is necessary to somehow verify the results. At this point in time there are no methods for directly verifying an embodied energy intensity value for an ecological commodity. However, the results of this model can be analyzed in conjunction with the available field data and our intuitive knowledge of the ecosystem to determine whether or not the results are reflecting the characteristics of the flow of energy and materials in the system. Also, through the use of these three linear programming methods, the consistency of the values of energy intensities can be determined.

ACKNOWLEDGEMENTS

This is contribution #LSU-CEL-82-13 to the Coastal Ecology Laboratory, Center for Wetland Resources, Louisiana State University, Baton Rouge, LA 70803. This research was made possible by a grant from the U. S. Fish and Wildlife Service, National Coastal Ecosystems Team, Contract #14-16-0009-80-073, "Mississippi Deltaic Plain Region Ecosystem Models and Narrative Report", Martha W. Young, Project Officer. We would also like to thank William Conner and Fred Sklar of the Center for Wetland Resources who provided much of the data for the Louisiana cypress swamp ecosystem, and Christopher Neill who incorporated this data into the input-output format.

REFERENCES

Bullard, C. W., and R. A. Herendeen. 1975. The energy costs of goods and services. Energy Policy. Dec. 1975: pp. 268-278.

Costanza, R. 1980. Embodied energy and economic valuation. Science 210:1219-1224.

Costanza, R. and C. Neill. 1981a. The energy embodied in the products of ecological systems: A linear programming approach. In: W. J. Mitsch, R. W. Bosseman, and J. M. Klopatek (eds.), Energy and ecological modelling: Proceedings of an international conference at Louisville, Kentucky, April, 20-23. International Society for Ecological Modelling, Copenhagen, Denmark.

Costanza, R. and C. Neill. 1981b. The energy embodied in the products of the biosphere. In: W. J. Mitsch, R. W. Bosseman, and J. M. Klopatek (eds.), Energy and ecological modelling: Proceedings of an international conference at Louisville, Kentucky, April, 20-23. International Society for Ecological Modelling, Copenhagen, Denmark.

Costanza, R., L. M. Bahr, S. G. Leibowitz, and J. R. Fruci. The Mississippi Deltaic Plain region study: An application of ecological models to the analysis and management of a complex coastal region. 1982. Paper presented at the

International Conference on Ecological Modelling, Fort Collins, Colorado, May 24-28, 1982.

Costanza, R., C. Neill, S. G. Leibowitz, J. R. Fruci, and L. M. Bahr. 1982. Mississippi Deltaic Plain Region ecological characterization: An ecological modelling study. National Coastal Ecosystems Team, U. S. Fish and Wildlife Service. In preparation.

Dorfman, R., P. A. Samuelson, and R. M. Solow. 1958. Linear programming and economic analysis. McGraw Hill, New York. 527 pp.

Finn, J. T. and T. M. Leschine. 1980. Does salt marsh fertilization enhance shellfish production? An application of flow analysis. Environmental Management 4(3): 193-203.

Hannon, B. 1973. The structure of ecosystems. Journal of Theoretical Biology. 41: 535-546.

Hannon, B. 1976. Marginal product pricing in the ecosystem. Journal of Theoretical Biology 56: 256-267.

Hannon, B. 1979. Total energy costs in ecosystems. Journal of Theoretical Biology 80: 271-293.

Herendeen, R. A. 1981. Energy intensities in ecological and economic systems. Journal of Theoretical Biology 91: 607-620.

Leibowitz, S. G., and R. Costanza. 1982. A preliminary input-output model of salt marshes in the Mississippi Deltaic Plain Region. Paper presented at the International Conference on Ecological Modelling, Fort Collins, Colorado, May 24-28, 1982.

Leontief, W. W. 1941. The structure of the American economy, 1919, 1929, an empirical application of equilibrium analysis. Harvard University Press, Cambridge, MA.

Odum, H. T. 1971. Trophic structure and productivity of Silver Springs, Florida. Ecological Monographs 27: 55-112.

Odum, H. T. 1977. Energy, value, and money. p. 174-196. In Ecosystem modeling in theory and practice. John Wiley and Sons, New York.

Richey, J. E., R. C. Wissmar, A. H. Derol, G. E. Likens, J. S. Eaton, R. G. Wetzel, W. E. Odum, H. M. Johnson, O. L. Loucks, R. T. Prentki, and P. H. Rich. 1978. Carbon flow in four lake ecosystems: a structural approach. Science 202:1183-1186.

von Neumann, J. 1945. A model of general economic equilibrium. Econ. Studies. 13:1-9.

SECTION III:

Application of Ecological Models to Animals

DEVELOPING AN ENERGETICS MODEL FOR WATERFOWL REFUGE MANAGEMENT[1]

Robert B. Frederick, William R. Clark, and Erwin E. Klaas[2]

Abstract.--A stochastic simulation model, designed to test the effects of alternative management schemes on refuging waterfowl populations, was constructed from data on fall-migrating snow geese (Anser c. caerulescens) at DeSoto National Wildlife Refuge, Iowa. Components of the energetics model include goose population, activity and energy budgets, food density and distribution, and the effects of weather, hunting pressure, and land management practices on the system. The model is useful for describing movement from the refuge core to distant feeding areas and provides insight into factors affecting emigration.

INTRODUCTION

In refuging systems, groups of individuals disperse radially from a central living place or "core" into a larger "arena" where food is aquired (Hamilton and Watt 1970). Food sources close to the core are more heavily exploited and quickly deteriorate. Animals then must move greater distances to meet energy needs. At some point, depending on food density and distribution, energy required to move back and forth between points in the arena and the core may be greater than energy gained from feeding. This concept is useful in better understanding migrating waterfowl behavior and population dynamics. Each fall, thousands of birds concentrate on midlatitude refuges on their way south, providing hunting and other recreational opportunities. Our ability to maintain refuging waterfowl at desired population levels, to influence their spacial distribution, and to control crop depredations, disease, and migration rates, depends on a full understanding of the resource utilization strategies of refuging waterfowl.

This paper documents the development and validation of a stochastic simulation model designed to test the effects of alternative management schemes on the behavior of refuging waterfowl. The model was constructed from data collected in 1976 and 1977 on fall-migrating snow geese at DeSoto National Wildlife Refuge (DNWR), Iowa (Frederick and Klaas 1982). Independent data were collected in 1981 specifically for model validation.

DNWR is a 3,166-ha multiple-use area situated on the Missouri River 32 km north of Omaha, Nebraska. It includes a 304-ha oxbow lake and 942 ha of crops. The arena, as determined by movements of geese from the refuge lake to feeding areas, includes 2,165 km^2 of intensively farmed floodplain. Approximately 50% of the cropland is corn (Frederick and Klaas 1982). Geese concentrate in large numbers on the refuge lake and feed almost exclusively on corn left in fields after harvest. They make two feeding trips into the arena daily, once in the early morning and once before dark. Although ample waste corn initially is available within a few kilometers of the refuge to meet the energy needs of the population, subsequent plowing makes much of the corn unavailable. Thus, both harvesting and plowing operations cause large seasonal fluctuations in food availability. Hunting causes a relatively small amount of mortality but causes considerable disturbance during foraging, hence increasing energy costs.

We sought a method for determining the effects of different treatments of hunting pressure and land use on the goose population. Field studies would have taken decades. We chose, instead, to incorporate refuging concepts and data already available on refuging snow geese into a simulation model. Our specific modeling objective was to simulate the effects of various agricultural practices and hunting regimes on the rectilinear distances traveled by refuging snow geese from the DNWR core and to test model validity by using independent field data.

FIELD DATA BASE

Geese usually begin to arrive in early October and increase to peak numbers in late November (fig. 1). The population declines rapidly in December as migration increases. Ten years of population data, collected by refuge personel from 1971-80, were used to determine average population levels and variance. Rate of disappearance of neckbanded geese was used to determine mean emigration rate (Frederick 1979). Average hunting mortality over the period is unknown but was estimated from information on mortality from a nearby refuge (Burgess 1980). Immigration was calculated by adding emigration and hunting mortality to the observed daily population increase.

---

[1]Paper presented at the Third International Conference on State-of-the-Art in Ecological Modelling, Colorado State University, May 24-28, 1982. This is Journal Paper No. J-10705 of the Iowa Agriculture and Home Economics Experiment Station, Ames, Iowa. Project No. 2427. Financed by the Iowa Conservation Commission, Iowa State University, U. S. Fish and Wildlife Service, and the Wildlife Management Institute through the Iowa Cooperative Wildlife Research Unit.
[2]The authors are Research Assistant, Assistant Professor, and Professor, respectively, in the Department of Animal Ecology, Iowa State University, Ames, Iowa U.S.A.

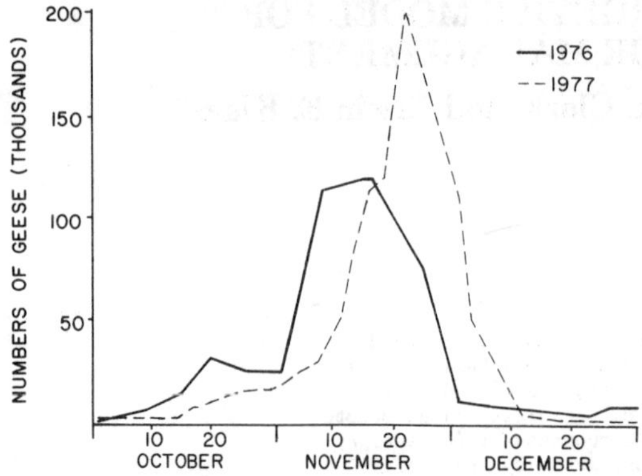

Figure 1.--Fall snow goose population at DeSoto National Wildlife Refuge, Iowa, in 1976 and 1977 (from Frederick and Klaas 1982).

Harvested and plowed cornfields were sampled to determine waste-corn density, and landsat satellite images of the study area were analyzed to determine the seasonal distribution of harvested and plowed cornfields (Klaas et al. 1978).

Weather data, including temperature and wind speed and direction, were obtained from the Ft. Calhoun Nuclear Power Plant, located 5 km upriver from DNWR.

MODEL DEVELOPMENT

The model can be summarized by briefly describing each of its sequential steps:

1. Daily weather is simulated.
2. A gridmap of the arena is created, including size, location, and potential food density for each cornfield.
3. Corn harvest and plowing are simulated at regular intervals throughout the season.
4. A hypothetical goose population is generated.
5. Flock movement and food searching are simulated.
6. Food is consumed by geese and removed from fields.
7. Daily activity and energy budgets are calculated.
8. Emigration and hunting mortality are determined for each day.

Parameter values were estimated from our field data (collected before 1981) and from literature sources. All values were input as means with associated standard deviations so that stochastic runs of the model could be used to make statistical comparisons. Flock movements, food searching, and feeding were simulated by using a 2-minute time interval. Other calculations were made on a daily basis.

Immigration was considered to be controlled by factors external to the system, but emigration was increased if geese realized an energy deficit greater than 1 day's demand and decreased if required time spent flying and feeding fell below 1 h/day. Simulated hunting mortality was zero when geese fed exclusively within refuge boundaries. When geese ventured off the refuge, hunting mortality was estimated by the input rate of 0.2% (± 0.2 SD) of the population per day (see fig. 2).

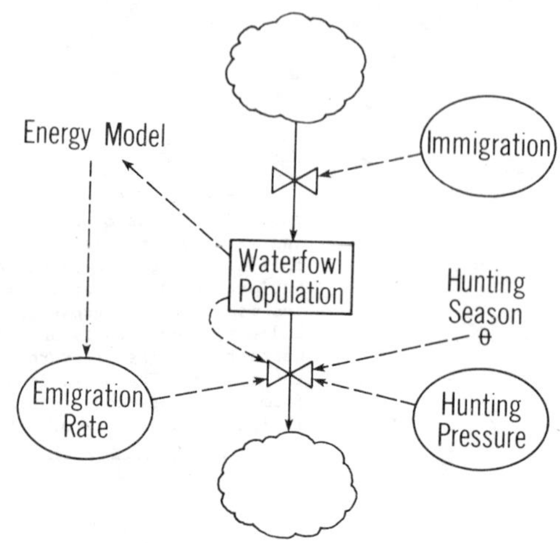

Figure 2.--Forrester diagram of the population submodel of the waterfowl refuging model.

A computerized gridmap of the arena was constructed (fig. 3). One-square-mile grids were used instead of metric units because road maps and aerial photographs clearly indicate that agricultual land use and highways follow the 1-mile section lines of original land surveys. It was more practical, therefore, to report distances in miles. Because geese avoid small fields, only the four largest fields in each section were assigned a size, crop, and agricultural treatment (unharvested, harvested, or fall-plowed), and cornfields were assigned a waste-corn density. Field sizes were based on data from aerial photograph interpretation.

Twice-daily wandering flights of geese from the core into the arena were simulated by selecting a random direction and deviating flight about this direction by using a wrapped normal distribution (Batschelet 1965). Flight speed remained constant throughout a run. If acceptable food was available within a radius of perception (Morrison 1978), the simulated flock would fly to that field and feed. The rate at which snow geese feed on various densities of waste corn is unknown. A hypothetical relative feeding rate was based on previous estimates of daily food intake and percentage of time spent feeding. If the flock was disturbed by hunters, the search would resume. When the geese met their energy demand or became satiated, they returned to the refuge core. The return trip to the core was simulated as rectilinear. On subsequent flights, geese would return directly (without searching) to the field where they had fed on the previous foraging trip. If food had been depleted or the geese had been disturbed by hunters, random search was initiated. Simulated flight paths (fig. 4) closely mimic what we observe at DNWR. Geese evidently maintain a mean bearing, but regularly deviate from it slightly.

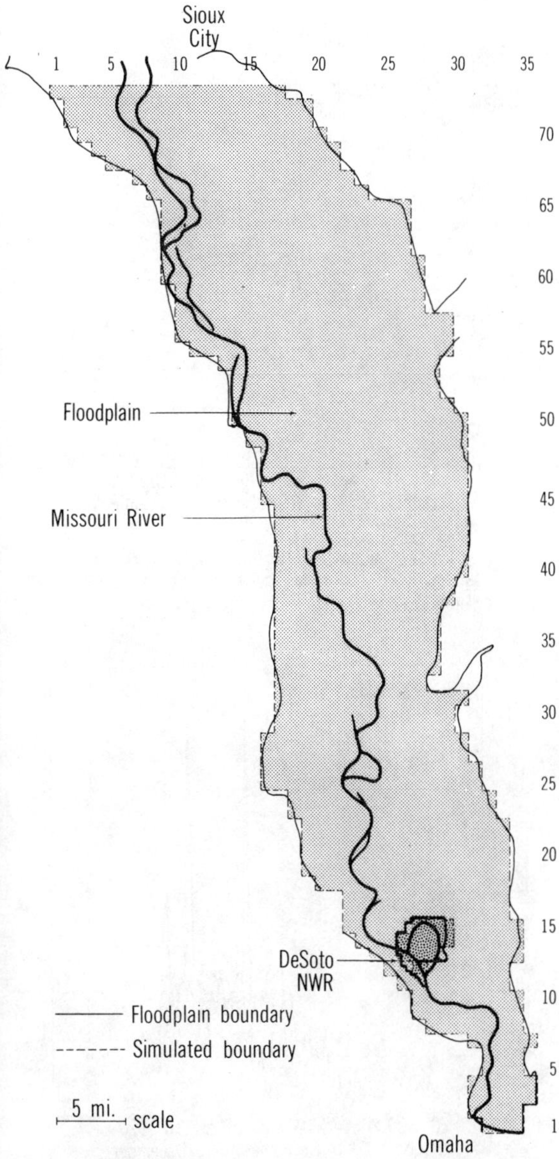

Figure 3.--Map of the refuging arena including real floodplain boundaries and the simulated gridmap boundaries.

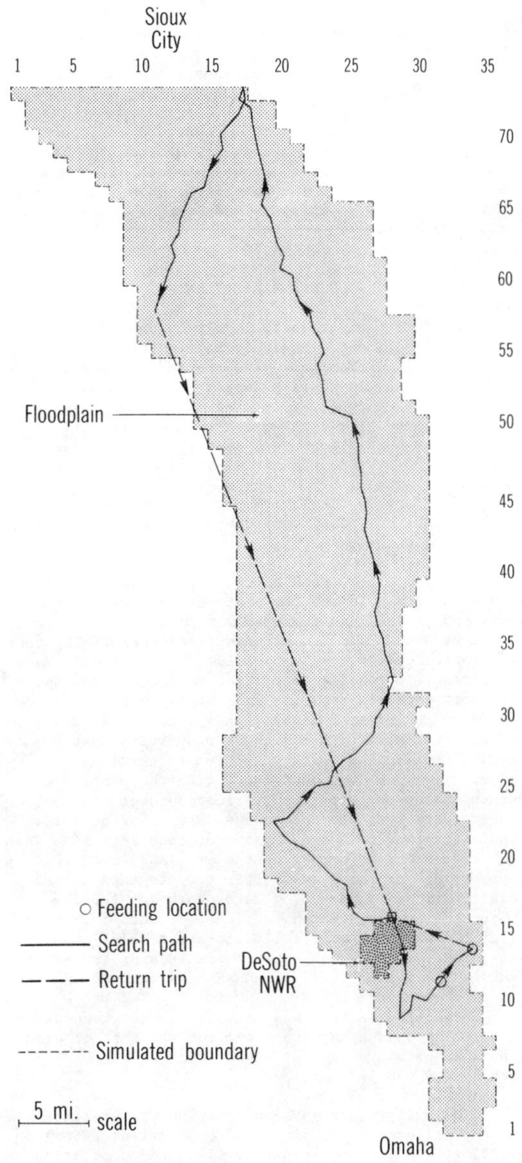

Figure 4.--Two simulated goose flight paths. The longer flight was the result of a run in which no food was available. The shorter flight, including feeding locations, is typical of normal feeding trips.

Daily energy expenditure was estimated by using Kendeigh's equations for existence energy and multiplying by a coefficient calculated from the simulated activity budget (Frederick and Klaas 1982). Activity budgets were determined primarily by simulated time spent flying and feeding. The proportions of time spent in other activities (i.e., sleeping, preening, and loafing) were estimated by considering the effects of weather on these behaviors (Frederick and Klaas 1982). Energy assimilated was calculated from simulated food consumption and the caloric equivalent of corn. An energy balance for an average goose, indicating the ability of geese to maintain desired net energy gain, was calculated by the equation:

$$\frac{\text{Energy}}{\text{balance}} = \text{Daily energy assimilated} - \text{Daily energy expended} - \text{Growth energy} - \text{Net energy demand} .$$

Growth energy depended on the proportion of juveniles in the population. The net energy demand (the daily net energy gain sought by geese) was set at zero by assuming that, except for juvenile growth, the population sought only to maintain body weight, not increase or decrease it.

MODEL VALIDATION

Validation experiments involved comparing output from 10 simulations with independent data collected at DNWR in 1981. Parameter values varied between simulations according to their standard deviations.

Food density means and confidence intervals from each of the simulations were compared with 44 real samples. Forty-four cornfields sampled in 1981 contained 229 kg/ha (183-284 kg/ha, 95% CI). The sample means from the 10 simulations ranged from 141 to 365 kg/ha, with individual confidence limits essentially identical to the observed.

Daily weather, snow goose population, and rectilinear distance to the farthest feeding areas were measured in 1981 and compared with simulation values by using a modified Van Elteren's statistic. The test statistic, called V_2^* by Reynolds et al. (1981), is a nonparametric ranking statistic, sensitive, not only to differences in location and shifts in distribution, but also to inflated simulation variance. By using a single statistic with all these properties instead of several statistics, each sensitive to particular alternatives, we were not faced with the problem of establishing an overall significance level for combined tests that are not independent. Furthermore, the assumption of normality is not required. The test is carried out by comparing the statistic with the appropriate critical value from the standard normal distribution. Time-series data, both simulated and observed, are correlated, a situation that this statistic does not specifically address. This would have the effect of inflating the statistic and, therefore, making the test for fit somewhat conservative.

Simulated weather was similar to that recorded in 1981. The validation statistic for the test of windchill temperature was not significant at the 0.05 level ($V_2^* = 0.85$).

The 1981 refuge goose population was significantly different from simulated populations ($V_2^* = 5.87$) (fig. 5). The shapes of simulated population curves, however, were similar to those observed in past years (fig. 1), suggesting some influence on population level not accounted for by the model. Some parameter values used in the model, estimated from data collected in 1976 and 1977, may have been inappropriate for making predictions about 1981. We suspect that a recent trend away from fall plowing increases goose food supply, particularly late in the season, causing geese to remain at the refuge in larger numbers later in the season.

A major objective was to simulate distance and direction traveled by geese from the core into the arena. The 1981 observed daily distances to the farthest feeding location were not significantly different from simulated distances ($V_2^* = -0.57$) (fig. 6).

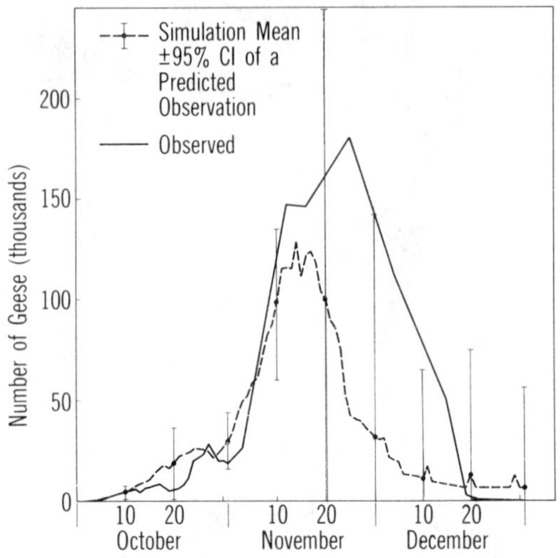

Figure 5.--Mean goose population (±95% CI for a prediction) from 10 model runs and the 1981 observed refuge population.

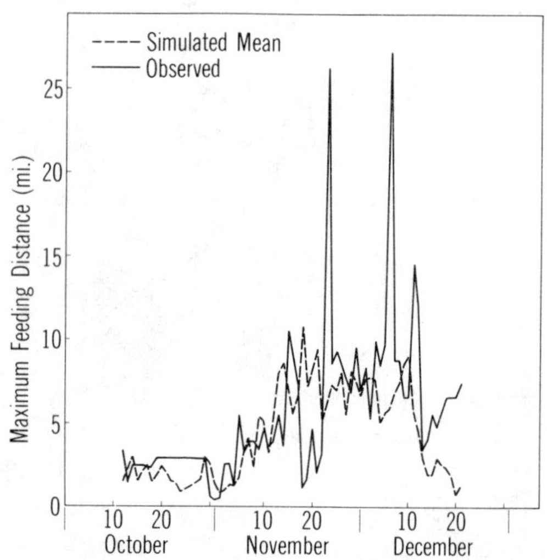

Figure 6.--The mean distance from the core to the farthest feeding location for each day from 10 simulations and the observed distances in 1981.

Simulated directions moved from the core to feeding areas were compared with observed directions by comparing the overall frequency with which feeding occurred in each quadrant. The frequency with which geese fed in each quadrant around the core in 1981 is similar to the simulated frequency (Table 1). We initially suspected that geese flew into strong winds when heading out to feed and, therefore, that wind direction affected direction of flights to feeding areas, but statistical analysis indicated no relationship (Frederick and Klaas 1982). The high percentage of time spent in the northwest quadrant is evidently due to the shape and alignment of the arena. Because geese often return to fields where they fed successfully and because most of the arena, including large cornfields, is northwest of the core, geese are more likely to feed there. That wind direction is primarily from the northwest is not necessary to explain the observed directions traveled by geese.

Table 1.--Percentage of feeding trips to each quadrant around the refuge core as observed in 1981 and as simulated by the model.

	Direction from the core			
	NE	SE	SW	NW
Observed (N=67)	28%	22%	5%	45%
Simulated (N=777)	16%	25%	15%	44%

CONCLUSIONS

We believe that this model validly simulates refuging behavior of fall-migrating snow geese and can be used to simulate the effects of various management scenarios. It would take decades to collect comparable empirical data. For the model to be useful for predicting population levels, careful consideration must be given to any shifts in agricultural practices that alter food availability. Preliminary sensitivity tests indicate that simulated populations are sensitive to these shifts. One unknown variable, which we propose to study further, is feeding rate relative to waste-corn densities. More information on sensitive and less-understood variables will increase our confidence in the model and enhance its use as a management tool.

LITERATURE CITED

Batschelet, Edward. 1965. Statistical methods for the analysis of problems in animal orientation and certain biological rhythms. American Institute of Biological Sciences Monograph, 57 p. Washington, D.C.

Burgess, Harold H. 1980. The Squaw Creek goose flocks with special attention to lesser snow geese. Administrative Report, Squaw Creek National Wildlife Refuge, 100 p. Mound City, Missouri.

Frederick, Robert B. 1979. Resource utilization and behavior of migrating snow geese at DeSoto National Wildlife Refuge. M.S. Thesis, Iowa State University, 68 p. Ames, Iowa.

Frederick, Robert B., and Erwin E. Klaas. 1982. Resource use and behavior of migrating snow geese. Journal of Wildlife Management (in press).

Hamilton, William J., and Kenneth E. F. Watt. 1970. Refuging. Annual Review of Ecology and Systematics 1:263-286.

Klaas, Erwin E., William H. Anderson, and Robert B. Frederick. 1978. Use of landsat imagery for estimating food availability to refuging lesser snow geese. Pecora IV, Proceedings of the Symposium on Application of Remote Sensing Data to Wildlife Management. National Wildlife Federation Scientific and Technical Series 3:89-94.

Morrison, Douglas W. 1978. On the optimal searching strategy for refuging predators. The American Naturalist 112:925-934.

Reynolds, Marion R., Jr., Harold E. Burkhart, and Richard F. Daniels. 1981. Procedures for statistical validation of stochastic simulation models. Forest Science 27:349-364.

A COMPUTER SIMULATION MODEL FOR THE STUDY OF DEER-ASPEN FOREST INTERACTIONS[1]

Philip J. Mello[2] and Jonathan B. Haufler[3]

Abstract. A computer simulation model was developed which represents the interaction of a white-tailed deer (Odocoileus virginianus) population with an aspen-type forest community. This model can be used by biologists as a simulation of deer responses to manipulations and as a teaching tool by instructors in wildlife and habitat management. It will aid students in the understanding of population dynamics and the effects of wildlife and habitat management practices on wildlife populations. Several simulation examples are discussed along with suggestions for future research.

INTRODUCTION

One of the major problems facing forest and wildlife professionals today is evaluating the impact of forest management policies on wildlife populations. Modern foresters and wildlife managers realize that a more holistic view of the forest ecosystem must be taken to assure proper management of both forest and wildlife resources. Unfortunately, this approach usually involves a complex set of interactions which make it increasingly difficult to forecast the outcome of a set of management decisions.

Recently, computer simulation modelling has been utilized as an aid for examining forest and wildlife relations. Models have been developed for ungulate populations which allow the manager to make decisions which will benefit both wildlife and timber production (Davis 1967, Bobek 1980). Other models have been developed which combine vegetation dynamics and population dynamics and help predict long range effects of various wildlife and habitat management strategies (Walter and Bunnell 1971, Cooperrider 1974). Models such as these, provide useful insights into the interrelations of the system. Additional models of this type dealing with different species or ecosystems are needed.

The purpose of this study was to create a computer simulation model which represents the interaction of a white-tailed deer population with an aspen type forest community. The model can be used by biologists as a simulation of deer responses to manipulations and as a teaching tool by instructors in wildlife and habitat management. The value of computer modelling as a teaching aid has been discussed (Walters and Bunnell 1971, Zarnoch and Turner 1974) and is viewed as a tool which allows the user to develop an intuitive understanding of the ecosystem dynamics and provides simulated field experience in the application of management techniques. Therefore, by using this model, the user is expected to develop an understanding of the effects of manipulating various management options on deer-aspen interactions and devise a management plan which will meet previously specified management objectives.

MODEL STRUCTURE

The sequence of calculations for a simulated year are shown in figure 1. The model was written in FORTRAN IV computer language for a CDC computer. There are 3 main components of the program: 1) vegetation dynamics; 2) deer population dynamics; and 3) deer and forest management options.

Initial Inputs

The initial deer population is created by the user by specifying values for the total number of deer, the mean age of males and females, and the proportion of males in the population. The total number of males and females is calculated by multiplying the total number of deer times the appropriate proportion. The age distribution for each sex is then computed using a negative exponential distribution with the appropriate mean age (Cooperrider 1974). Age and sex-specific weights are then assigned to each deer.

---

[1]Paper presented at the Third International Conference on State-of-the-Art in Ecological Modelling, Colorado State University, May 24-28, 1982. Michigan Agricultural Experiment Station Journal Article Number 10424.

[2]Philip J. Mello is a graduate student at Michigan State University, East Lansing, Michigan U.S.A.

[3]Jonathan B. Haufler is Assistant Professor of Wildlife Ecology, Michigan State University, East Lansing, Michigan U.S.A.

The initial aspen forest is created simply by assigning the proportion of the total area for a specific aged stand. A maximum of 7 age classes can be specified with initial ages 10 years apart and ranging from 1 to 61. Values for mean DBH and total number of trees for each aged stand given in Perala (1977) are specified within the program. Chi-square goodness of fit tests were performed on aspen DBH distributions given in Lake States (1933). The results showed that all stands with the exception of the 3 inch stand were normally distributed ($p > .05$). For the purpose of simplicity all stands greater than 11 years old were assumed normally distributed with a mean and standard deviation derived from Lake States (1933). The distribution for a 4 year old stand is shown in figure 2 (Pollard 1971).

Vegetation Dynamics

The major food for deer in this model is aspen browse. In the summer and fall herbaceous vegetation is provided to supplement the deer's diet. For three months during winter, no aspen browsing takes place and deer are provided with a specified quantity of cedar. This represents the behavioral adaptation of northern deer to gather in coniferous swamps which

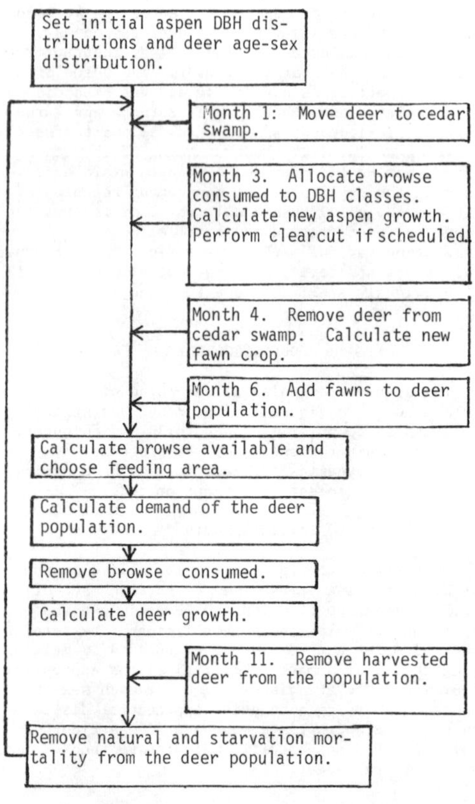

Figure 1.--Sequence of computations in 1 year of a simulation run in the aspen-deer interactions model.

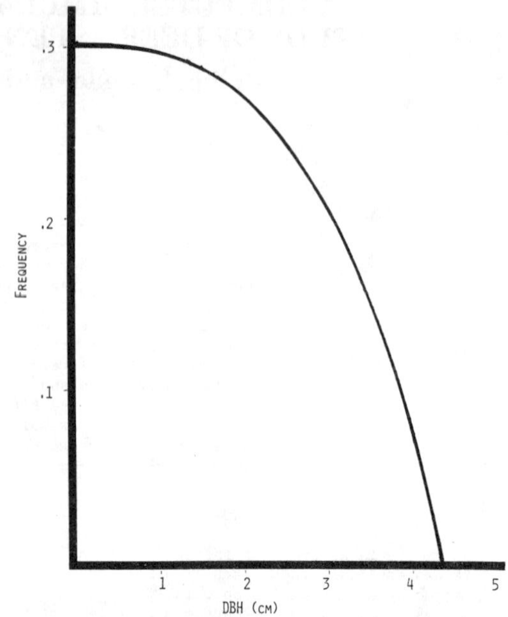

Figure 2.--DBH distribution for a 4 year old aspen stand (Pollard 1971).

provide maximum thermal protection during winter months. Winter severity is represented by the quantity of cedar supplied to deer during the winter with higher amounts supplied during mild winters and smaller amounts in severe winters.

Aspen growth is modeled as a dynamic process using a simplified version of a model developed by the North Central Forest Experimental Station (USDA 1979). Growth and mortality coefficients have been developed for various pure and mixed stand forest ecosystems of the Great Lakes region. Yearly forest growth is summarized in the following equation (Leary 1979):

$$\frac{\Delta Y}{At} = [f_1(\bar{D}, \overline{CR}, SI)][NT][f_2(Y, NT)]$$

where,

ΔY = change in the sum of tree diameters in a year
f_1 = the potential growth function
f_2 = the modifier function of potential growth actually occurring
$\bar{D}$ = mean DBH
$\overline{CR}$ = mean crown ratio
SI = site index
Y = sum of the tree diameters
NT = number of trees

Change in DBH for an individual tree is calculated as a function of its proportion to the total growing stock. This value is then modified as shown in figure 3 to represent the effect of deer browsing on aspen growth. The probability of death is then computed as a function of growth.

After yearly growth is computed, total above ground biomass of a tree is estimated using equations developed by Young and Carpenter (1967). Available browse per tree is then calculated as function of tree height (Westell 1954).

It is assumed that a tree greater than 3.66 meters in height will provide no browse to a deer. In the model, two values for browse availability are calculated. One value assumes deer will browse aspen stems and twigs to the .635 centimeter diameter as is the case under light browsing conditions, while the other assumes deer will browse to the 1.27 centimeter diameter. Heavy browsing to the 1.27 centimeter diameter usually occurs when deer populations are large and food is scarce (Westell 1954).

Deer Dynamics

The amount of food consumed by a deer is computed as a function of deer weight, the digestible energy of the food, and the time of the year. Deer show a cyclic pattern of food consumption throughout the year with the greatest amount of food consumed in the late summer and early fall and the lowest in the winter months (French et al 1955, Long et al 1965, Moen 1978). This reflects the deer's need to build up fat reserves to help survive the approaching winter and their behavioral adaptation of restricting activity and conserving energy during the winter months (Ozoga and Verme 1970).

The choice of stand in which feeding occurs is based upon the total amount of energy available to the deer in the stand. It is assumed that deer choose feeding areas and select foods which maximize their energy intake. After the feeding area has been chosen, the amount of food available to the deer is compared to the amount demanded to meet its energy requirements. The smaller value is then subtracted from the available food supply. This is repeated for every deer for each month of the year.

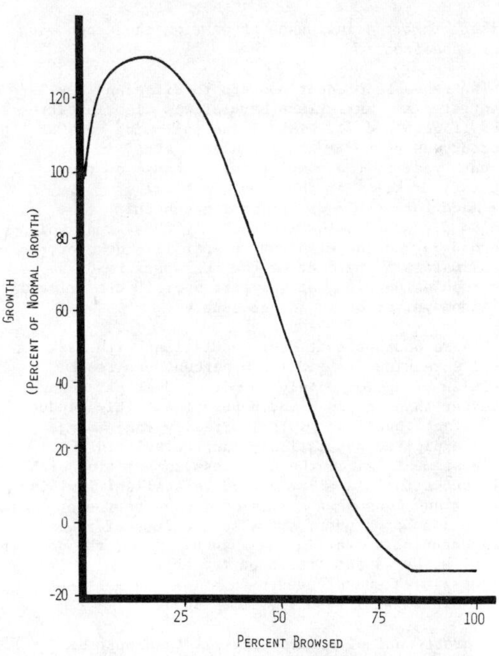

Figure 3.--The effect of deer browsing on aspen growth.

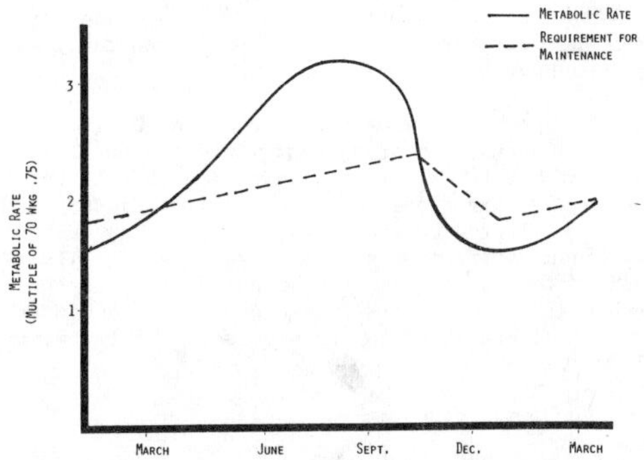

Figure 4.--Graphical estimation of deer maintenance energy requirements.

Prior to calculating yearly growth of aspen, the total amount of aspen consumed in each stand in the previous year is allocated to the tree height classes. The amount of browse consumed is proportionally allocated to each height class, so that the proportion of browse consumed in height class X to the total amount of browse consumed is equal to the proportion of browse available in height class X to the total amount of browse available.

Growth of deer in terms of weight change is computed by the following equation:

$$\Delta Y = .005357(C-M)N$$

where,

ΔY = body weight change in 1 month

C = mean digestible energy consumed in a month

M = mean energy requirements for maintenance in a month

N = the number of days in a month

The slope of this line was derived from Moen (1973). Estimates of mean monthly energy requirements for maintenance were derived graphically (fig. 4). A linear relationship was assumed between maintenance requirements and time of year with peak requirements occurring during breeding and minimum requirements occurring in mid-winter. For pregnant females peak maintenance requirements occurred during the lactation period (Moen 1973). A line was drawn which intersected the energy demand curve at points where deer fed ad libitum began to gain weight and began to lose weight (French et al 1955, Magruder et al 1957, Long et al 1965). A second line was drawn which connected peak and minimum requirements.

There are three types of mortality which occur in this model: 1) starvation; 2) natural mortality; and 3) harvest. The probability of starvation is computed as a function of weight. Natural mortality is defined as all mortality which is not starvation or harvest mortality (road kills, poaching, disease). This rate is a constant throughout the simulation for each sex and age class. This assumes density independence though in reality all three are probably density dependent. Deer harvest is controlled by the user and will be discussed further in the management options section.

The probability of a doe producing 0, 1 or 2 fawns is a function of age and winter weight loss. Does which enter the spring in poor physical condition due to low nutritional diets during the winter yarding season will produce weaker fawns with a low probability of surviving past the first few days of life (Verme 1962). The total fawn crop computed reflects the number of fawns which survive to at least one month of age and are added to the population in the month of June.

Management Options

Management options for the user include deer harvest and tree harvest. The user is able to assign a proportion of antlered and antlerless deer to be harvested in each year of the simulation. Antlered deer are defined as yearling and adult bucks while antlerless deer include all does plus buck fawns. The probability of a deer being harvested is equal to the proportion assigned by the user. This assumes that all deer in a particular category (antlered or antlerless) have an equal probability of being harvested and that hunters show no preference for particular age classes within a category. Harvest mortality is calculated and the number of deer harvested are removed in the month of November.

The only silvicultural treatment allowed in the model is clearcutting. The user is able to specify the size of the clearcut in acres and the stand age when the cut is to occur. When a stand reaches this age, the number of acres specified by the user will be removed from that stand and added to the 1 year old stand. The total number of suckers per hectare in the 1 year old stand is a function of the basal area of the cut stand (Graham et al 1963).

RESULTS

Several computer runs were performed under various biological conditions. The response of an exploited deer population (70 percent buck harvest) to varying degrees of winter severity is shown in figure 5. The importance of winter severity to deer numbers is reflected by deer densities ranging from 8 deer per square kilometer in severe conditions to 40 in mild conditions. Forty deer per square kilometer is considered extremely rare in Michigan and the model indicates that weather is a significant reason for this, although other factors such as social stress cannot be discounted. The drop in deer numbers occurring in year 10 of these runs and subsequent runs is a result of the aspen stand growing out of reach of deer.

Figure 5 also shows that a young aspen stand can support a substantial deer population for a period of 10 years. This conflicts with claims that deer densities in excess of 17.5 deer per square kilometer will cause extensive damage to an aspen stand (Graham et al 1963). Explanations for this could be that the concentration of deer in the young aspen stand examined by Graham exceeded 17.5 deer per square kilometer or that estimates of aspen browse used in this model were too high. However, since other deer foods such as grasses and forbs were not allocated to deer in this model until aspen browse had been consumed, it is

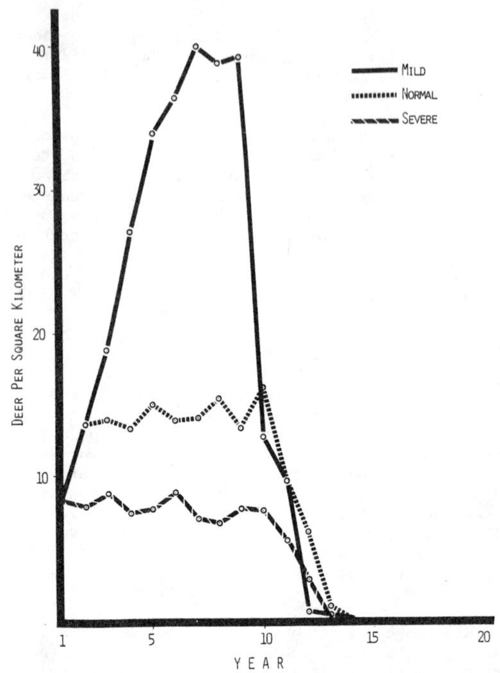

Figure 5.--Simulated response of deer densities to varying winter severity.

unlikely that the influence of deer on the aspen was underemphasized.

The response of deer numbers to different levels of antlered and antlerless harvest was also investigated (fig. 6 and 7). In all the following runs a clearcut was performed on the oldest stand in the eleventh year of the simulation. Because of the stochastic make-up of the model, a total of 5 runs were made for each deer harvesting schedule. The results discussed below include only those runs which extended beyond the clearcut in year 11. However, in approximately 50 percent of the runs performed the deer population died out in years 9 or 10 due primarily to random variation in forage supply.

Figure 6 shows that deer populations with buck harvests ranging from 50 to 70 percent can reach densities of approximately 4 more deer per square kilometer than an exploited population. This is due to the fact that less winter mortality takes place in the exploited populations, particularly in the female segment, since there is less competition with adult bucks for what little food is available. This is important because the loss of one doe represents not only the loss of 1 deer, but also the loss of fawns she would potentially bear in the future. Also, the does that do survive the winter are in better physical condition than does in an unexploited population and a higher fawn/doe ratio will result (Verme 1969).

A great deal of controversy exists about the feasibility of an antlerless harvest and its effect on deer numbers. Figure 7 shows that under initial conditions a population with a 25 percent antlerless harvest is able to reach the same population level as a population with no antlerless harvest, although it will take longer to reach that level. However, at extremely low levels this population is unable to increase. This indicates that the response potential of the population should be a major consideration when evaluating the desirability of an antlerless harvest.

Figure 6 also shows the effect of clearcutting on a deer population. All populations showed an immediate increase the year following the clearcut after the sharp decline due to the aspen stand growing out of reach of the deer. This can be compared to figure 5 where no clearcutting occurred and the populations died out in the thirteenth year. Perala (1977) prescribes aspen clearcuts on 5 to 10 year intervals when managing for white-tailed deer. The model indicated that this rate of manipulation should be followed, as longer intervals would result in an abrupt decline in deer numbers.

Several runs were made to investigate the effect of varying metabolizable energy for cedar and aspen browse on deer numbers. Reported values of metabolizable energy for cedar and aspen were used (Ullrey et al 1964, 1967, 1972) along with values derived as a function of the proximate composition of the species (Mautz et al 1974).

Figure 8 shows the relationship between metabolizable energy of cedar browse and deer numbers. At peak densities, a difference of .3 kcal/g resulted in an average difference of approximately 1 deer per square kilometer. However, in all runs using a value of 1.88 kcal/g, the population declined to 0 in the tenth year of the simulation run. This occurred in only 50 percent of the runs using higher values.

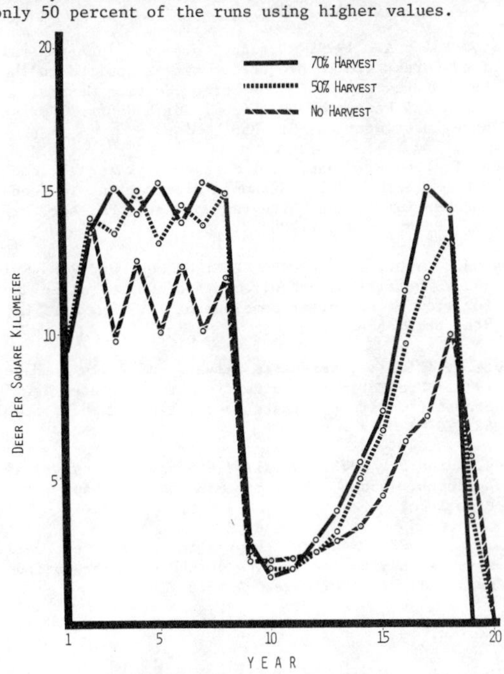

Figure 6.--Simulated effects of different intensities of antlered harvest on deer densities (no antlerless harvest on each run).

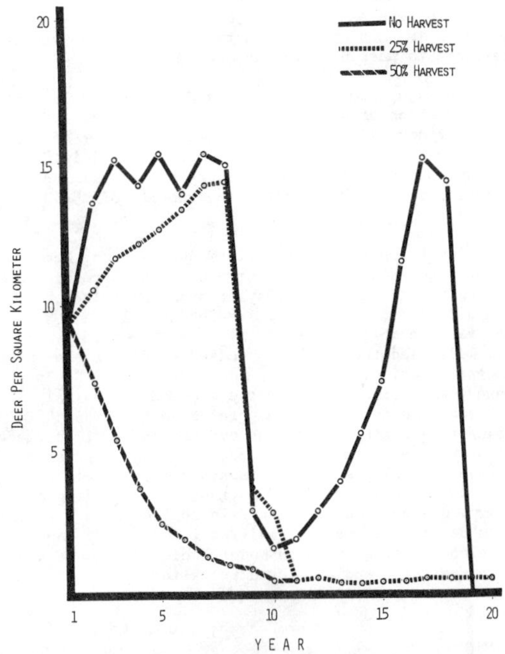

Figure 7.--Simulated effects of different intensities of antlerless harvest od deer densities (70% antlered harvest on each run).

The model showed little sensitivity to different values of aspen metabolizable energy (1.57 kcal/g and 1.47 kcal/g) for the first eight years of the simulation run. However, the population declined to 0 in year 9 in all runs in which the lower value for aspen metabolizable energy was used. Thus, as browse supply became limited, the metabolizable energy of the browse became much more influential than when browse was more abundant.

DISCUSSION

One important aspect of modelling excercises such as these is to point out areas which are ill defined and need more investigation. Once such area which needs to be more precisely defined is the question of energy requirements of wild deer. Considerable research has been carried out in this area (Ullrey et al 1969, Moen 1978) but more information is needed particularly in the area of evaluating growth to energy consumption which could lead to much better predictions of winter losses due to malnutrition.

One of the major assumptions of this model is that deer will feed in areas where there is the most energy available. However, a study using tame deer showed that deer used a mature aspen forest and a clearcut equally for foraging in the month of June (Stormer and Bauer). One could conclude from this that both stands had equal or adequate amounts of available energy or that this assumption is not valid. In either case more information on deer behavior and their choice of feeding areas could have a significant impact on models such as these.

The model also raises questions about the accuracy of deer browse estimates in young aspen stands. Further research in this area and in the area of aspen growth response to deer browsing is needed.

A major limitation in this model is its lack of consideration of spatial distributions. Other models (Cooperrider 1974) have shown that the number of acres a deer can forage in a day is quite critical in determining deer densities. Therefore, spatial considerations could also alter the results of these types of models.

What constitutes a valid model has long been a problem and a subject of controversy which remains unresolved to this date (Naylor and Finger 1967, Van Horn 1971, Goodall 1972). Shannon (1975) stated, "by far the most important test for the validity of our model and the results obtained should be the answer to the question, 'Does the model make sense?'" In this case the question is, "Are there sound biological explanations for what is happening in the model?" Using this criterion and the current state of knowledge of deer-aspen interactions, the model appears valid, although results should be regarded tentatively until more data are gathered and further understanding of deer and aspen ecology is developed. As more information becomes available, additional complexity can be developed into simulation models which will make them more realistic and add to their validity.

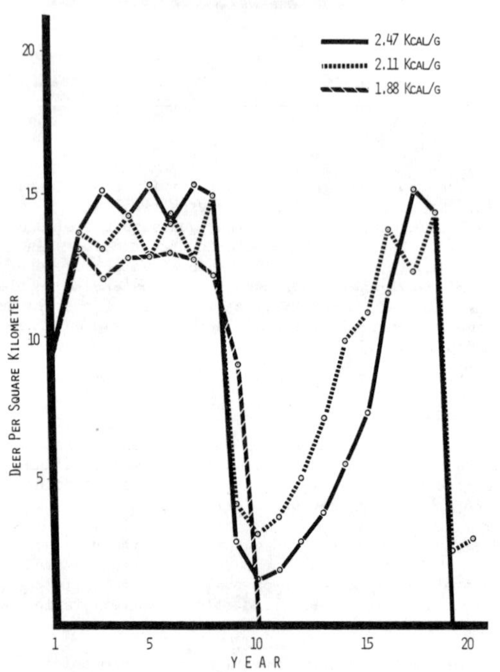

Figure 8.--Simulated effect of varying energy content of cedar browse on deer densities.

ACKNOWLEDGEMENT

This study was supported by the Michigan Agricultural Experiment Station.

LITERATURE CITED

Bobek, Boguslaw. 1980. A model for optimization of roe deer management in central Europe. J. Wildl. Manage. 44:837-848.

Cooperrider, Allen Y. 1974. Computer simulation of the interaction of deer population with northern forest vegetation. Ph.D. thesis. 220 p. SUNY Coll. Env. Sci. and For., Syracuse.

Davis, Lawrence S. 1967. Dynamic programming for deer management planning. J. Wildl. Manage. 31:667-679.

French, C.E., L.C. McEwen, N.D. Magruder, R.H. Ingram, and R.W. Swift. 1955. Nutritional requirements of white-tailed deer for growth and antler development. Pa. Agr. Exp. Sta. Bull. 600. 50 p.

Goodall, David W. 1972. Building and testing ecosystem models. p. 173-194. In J.N.R. Jeffers (Ed.). Mathematical models in ecology. Twelfth Symp. Brit. Ecol. Soc. 398 p. Blackwell Sci. Publ., Oxford, Eng.

Graham, Samuel A., Robert P. Harrison, Jr., and Casey E. Westell, Jr. 1963. Aspens, phoenix trees of the Great Lakes region. 272 pp. University of Michigan press. Ann Arbor.

Lake States Forest Experiment Station. 1933. Aspen normal stand tables. USDA Forest Service Technical Note 71. St. Paul, Minn.

Leary, Rolfe A. 1979. Design. p.5-15. In A generalized forest growth projection system applied to the Lake States region. USDA Forest Service General Technical Report NC-49, 96 p. North Central Forest Experiment Station, St. Paul, Minn.

Long, T.A., R.L. Cowan, G.D. Strawn, R.S. Wetzel, and R.C. Miller. 1965. Seasonal fluctuations in feed consumption of the white-tailed deer. Pa. Agr. Exp. Sta. Progress Report 262. 6 p.

Magruder, N.D., C.E. French, L.C. McEwen, and R.W. Swift. 1957. Nutritional requirements of white-tailed deer for growth and antler development II. Pa. Agr. Exp. Sta. Bull. 628. 21 p.

Mautz, William W., Helenette Silver, and Haven H. Hayes. 1974. Predicting the digestibility of winter deer browse from its proximate composition. Can. J. Zool. 52:1201-1205.

Moen, Aaron N. 1973. Wildlife ecology: an analytical approach. 458 p. W.H. Freeman and Co., San Francisco.

______. 1978. Seasonal changes in heart rates, activity, metabolism, and forage intake of white-tailed deer. J. Wildl. Manage. 42:715-738.

Naylor, Thomas H., and J.M. Finger. 1967. Verification of computer simulation models. Manage. Sci. 14:B92-B101.

Ozoga, John J., and Louis J. Verme. 1970. Winter feeding patterns of penned white-tailed deer. J. Wildl. Manage. 34:431-439.

Perala, Donald A. 1977. Manager's handbook for aspen in the north-central states. USDA Forest Service General Technical Report NC-36, 30 p. North Central Forest Experiment Station, St. Paul, Minn.

Pollard, D.F.W. 1971. Mortality and annual changes in distribution of above-ground biomass in an aspen sucker stand. Can. J. Forest Res. 1:262-266.

Shannon, Robert E. 1975. Systems simulation: the art and science. 387 p. Prentice-Hall Inc., Englewood Cliffs, N.J.

Stormer, Fred A., and William A. Bauer. 1980. Summer forage use by tame deer in northern Michigan. J. Wildl. Manage. 44:98-106.

Ullrey, D.E., W.G. Youatt, H.E. Johnson, L.D. Fay, P.K. Ku. 1964. Digestibility of cedar and aspen browse for the white-tailed deer. J. Wildl. Manage. 28:791-797.

_____, _____, _____, _____, and B.E. Brent. 1967. Digestibility of cedar and jack pine browse for the white-tailed deer. J. Wildl. Manage. 31:448-454.

_____, _____, _____, _____, B.L. Schoepke, and W. T. Magee. 1969. Digestible energy requirements for winter maintenance of Michigan white-tailed does. J. Wildl. Manage. 33:482-490.

_____, _____, _____, A.B. Cowan, R.L. Covert, and W.T. Magee. 1972. Digestibility and estimated metabolizability of aspen browse for white-tailed deer. 26:885-891.

U.S. Department of Agriculture, Forest Service. 1979. A generalized forest growth projection system applied to the Lake States region. USDA Forest Service General Technical Report NC-49, 96 p. North Central Forest Experiment Station, St. Paul, Minn.

Van Horn, Richard L. 1971. Validation of simulation results. Manage Sci. 17:247-258.

Verme, Louis J. 1962. Mortality of white-tailed deer fawns in relation to nutrition. p. 15-38. In Proc. 1st Nat'l. white-tailed deer disease symp., Univ. Georgia, Athens.

_____. 1969. Reproductive patterns of white-tailed deer related to nutritional plane. J. Wildl. Manage. 33:881-887.

Walters, Carl J., and Fred Bunnell. 1971. A computer management game of land use in British Columbia. J. Wildl. Manage. 41:644-657.

Westell, Casey E., Jr. 1954. Available browse following aspen logging in lower Michigan. J. Wildl. Manage. 18:266-271.

Young, Harold E., and Paul M. Carpenter. 1967. Weight, nutrient element and productivity of seedlings and saplings of eight tree species in natural ecosystems. Maine Agr. Exp. Sta. Tech. Bull 28. 39 pp.

Zarnoch, Stanley J., and Brian J. Turner. 1974. A computer simulation model for the study of wolf-moose population dynamics. J. Environ. Sys. 4:39-51.

A SPATIAL MODEL OF SAGE GROUSE HABITAT QUALITY[1]

Robert A. Nisbet, Stephen H. Berwick, and Kenneth L. Reed[2]

Computerized cartographic modeling of sage grouse habitat can predict probable lek sites. Habitat factors identified in data analysis of 41 lek sites were mapped, digitized, and processed to show areas of probable lek site location. Future invasion pathways of pinyon-juniper woodland were simulated to further screen unsuitable areas to prioritize management.

INTRODUCTION

Sage grouse inhabit upland meadows and valleys in much of the Great Basin of the western United States. Sagebrush (Artemisia spp.) is their preferred habitat, and sagebrush (along with forbs) is the primary food of adults. Sage grouse are linked to traditional and localized wintering areas and spring strutting grounds (leks) where mates are solicited. These areas are referred to as wintering - nesting complexes by Wallestad (1975). These complexes also include brood-use areas usually located within 2 miles of leks (Wallestad and Pyrah, 1974; Gill, 1975). Broods are greatly dependent on highly nutritious, succulent forbs to sustain them during their first months of life. As these forbs dry out during summer at the lower elevations, sage grouse and their broods move upward in elevation. During this time mountain meadows become very important to sage grouse survival (Oakleaf, 1971). Sage grouse are known to be negatively affected by sagebrush removal and will abandon strutting grounds, brood-use areas, and wintering grounds if adjacent habitat is disturbed (Braun et al.,1977). They apparently cannot adjust to land use patterns in which large tracts of sagebrush-grassland vegetation type have been eliminated or greatly disturbed (Patterson, 1952).

Recent population trends in Nevada/Utah indicate that sage grouse populations are either stable or increasing slightly (Molini and Barngrover, 1979). However, data for the distribution of upland game animals are generally less reliable than those for big game species for several reasons. Upland species are smaller than big game animals and, therefore, more difficult to observe. In addition, management of these species is usually less intensive than for big game, resulting in less time and effort expended in field surveys. The distribution of sage grouse in large areas of its range is unknown or uncertain.

HDR-System Sciences conducted a sage grouse lek characterization study during the summer of 1980. The objectives of this research were to describe lek sites in terms of their major environmental characteristics, to compare these leks with selected non-lek sites, and to ascertain through that comparison whether sage grouse select lek sites based upon one or a combination of environmental parameters located in obtainable data bases. The results could then be used to narrow the amount of unsurveyed sage grouse range requiring search to locate previously undiscovered leks and brood-use areas. Mapping high probability key sage grouse habitat would permit an optimal placement of any future development project features in anticipation of on-site verification. This approach is consistent with the Habitat Analysis Procedures developed by U.S. Fish & Wildlife Service (ESR 102, 1980).

METHODS AND RESULTS

Characterization of Strutting Grounds

Forty-one known strutting grounds were located during June 1980 in 18 valleys in Nevada and Utah with the help of Nevada Department of Wildlife field biologists. Since the survey was done after strutting ceased, leks were identified by the presence of dropping piles. Each lek site was characterized by vegetation type, percent shrub cover, visual obstruction, distance to escape cover, distance to nesting cover, distance to topographical change, distance to water, slope, and aspect (table 1). Vegetation type was classified according to standard BLM nomenclature. Visual obstruction readings (VOR) were recorded for each lek. This technique consisted of viewing a large wooden dowel from a distance of 4 m. The length of the dowel completely obstructed by live shrubs was recorded for views 1 m and 1 dm from the ground. The "bird's eye view" of the lek from dm may be related to the preference by sage grouse suitable lek sites. Shrub canopy cover listed in table 1 was determined by a line-intercept method. Ten sites adjacent to leks were chosen to represent non-leks, and characterized similarly to the lek sites (table 2).

Escape cover was defined as shrub vegetation at least 30 to 40 cm high. Substantial stands of big sagebrush (Artemisia tridentata) were assumed to constitute suitable nesting cover. Distance to nearest major vertical change in relief may be important as a buffer against raptors on rock outcrops and steep slopes.

Data Reduction to Determine Habitat Preference by Quadratic Assignment Analysis

Some lek characteristics were categorical (e.g., vegetation type); other variables were measured on a continuous scale (e.g., distance to water). Therefore a non-parametric quadratic assignment analysis (Hubert and Schultz, 1976) was selected to analyze data to indicate if sage grouse lek preference could be predicted by any characteristic(s). A similarity matrix is defined for each characteristic such that elements in the structure matrix (Q) are set to 1 for Q_{ij}, if both the ith and jth sites are leks or both are non-leks; otherwise $Q_{ij} = 0$. A proximity matrix (C) is defined for each characteristic, such that

---

[1]Paper presented at the Third International Conference on State-of-the-Art in Ecological Modelling, Colorado State University, May 24-28, 1982.

[2]Senior Ecologist, Chief Scientist, and Chief System Scientist, respectively, HDR-System Sciences, Santa Barbara, CA 93101.

Table 1. Sage grouse lek characteristics

Location	Vegetation Type	Environmental Variables Measured Percent Shrub Cover	VOR[1] 1 m	VOR[1] 1 dm	Distance to Escape Cover (m)	Distance to Nest Cover (m)	Distance to Type Change (m)	Distance to Topo Change (m)	Distance to Water (m)	Percent Slope	Aspect	Hydrologic Subunit (Valley)
Barley Creek	Agcr	--	--	--	1.5	792.5	792	--	1.5	5	NE	Monitor
Pott Ranch	Agcr	--	--	--	30	1,610	1,610	--	--	5	SE	Monitor
White River	Arno	31.25	0.36	6.50	--	396	3,871	--	3,220	0	--	White River
Danville Creek	Arno	9.58	0.25	5.77	--	305	305	--	305	0	--	Little Fishlake
Clover Creek	Arno Agcr	18.50	0.17	4.33	61	61	61	--	792	0	--	Little Fishlake
Jakes North	Artr Agcr	--	--	--	1.5	1.5	1.5	--	792	0	--	Jakes
Jakes Central	Arno Cela	32.25	0.14	2.80	46	46	46	--	--	0	--	Jakes
Willow Creek	Arno	20.00	0.38	2.25	640	640	640	--	640	2	SE	Little Smoky
Simpson Creek	Artr Agcr	11.50	0.33	2.34	1.5	1.5	792	--	--	2	W	Diamond
Pine Field #1	Arno	32.09	0.33	7.09	61	61	61	--	1,610	10	SE	Garden
Horse Creek	Agcr	--	--	--	1.5	1.5	1.5	--	305	3	SE	Garden
Lone Mountain	Arno	28.84	0.75	5.25	46	46	46	--	--	3	SE	Kobeh
Kobeh #1	Arno	27.83	0.75	5.67	--	396	396	--	--	10	SE	Kobeh
Kobeh #2	Arno	26.83	0.89	4.42	--	396	396	--	--	10	SE	Kobeh
Roberts Creek	Agcr	6.00	--	--	--	1,610	1,610	--	1.5	1	SW	Kobeh
Ferguson #1	Artr	16.50	0.30	6.88	1.5	1.5	2,377	--	2,377	2	S	Monitor
Ferguson #2	Artr Atco	9.84	0.19	4.81	1.5	1.5	30	--	3,170	0	--	Monitor
Ferguson #3	Artr	20.42	0.59	5.67	1.5	1.5	2,377	--	2,377	1	NW	Monitor
3-Bars	Arno	36.50	0.11	3.77	1.5	1.5	488	--	488	3	N	Monitor
Allison Creek	Atco	15.08	0.46	2.23	--	1,610	396	--	396	1	SE	Antelope
Cedar Creek	Artr	12.33	0.45	4.53	1.5	1.5	2,134	--	152	1	N	Antelope
S. Cottonwood #3	Gusa	29.58	0.74	4.49	152	152	1,610	1,610	--	3	SE	N. Steptoe
S. Cottonwood #4	Gusa	32.67	0.42	4.27	61	805	1,610	1,610	--	3	SE	N. Steptoe
Spruce Mountain	Gusa	29.08	0.89	4.36	1.5	963	963	963	--	1	SW	N. Steptoe
Butte #2	Gusa	25.08	0.35	3.00	61	61	1,610	1,610	--	2	SE	Butte
Butte #3	Gusa	20.50	0.03	1.40	21	21	1,610	1,610	--	4	NW	Butte
Black Sage #3	Arno	21.08	0.33	4.28	15	15	1,610	1,610	--	1	NW	N. Ruby
Black Sage #4	Arno	32.42	0.42	2.99	8	8	805	805	--	6	S	N. Ruby
Black Sage #5	Arno	27.92	0.20	3.09	30	30	805	805	--	4	S	N. Ruby
High Beach #1	Arno	31.02	0.30	2.22	50	30	482	482	--	1	SE	N. Ruby
Loading Chute	Arno	33.13	0.38	2.50	46	46	46	--	--	2	SE	N. Ruby
Snow Fence	Arno	26.75	1.30	5.78	1.5	1.5	1,610	61	--	3	W	Clover
Indian Creek	Atco	7.83	--	--	1.5	2,414	2,414	2,414	--	3	E	Crescent
Horse Ranch	Atco	12.25	--	--	1.5	805	2,414	--	--	5	E	Carico Lake
Cook Creek	Atco	7.75	--	--	9	9	9	152	--	2	SE	Carico Lake
Pony Express	Artr	14.50	1.37	5.87	1.5	1.5	1,610	61	--	15	NE	Big Smoky
Cape Horn	Arno	22.83	0.32	4.01	3	3,218	4,828	1,207	--	5	S	Big Smoky
Escalante Desert	Artr	16.75	0.69	5.75	1.5	1.5	1,207	1,207	--	1	NW	Escalante Desert
Hamlin Wash #1	Artr	5.00	0.03	0.91	1.5	1.5	91	91	--	0	--	Hamlin
Hamlin Wash #2	Cela	45.92	1.17	5.35	27	27	27	91	--	1	SE	Hamlin
Hamlin Wash #3	--	0	--	--	46	91	91	91	--	5	SW	Hamlin

T3359/10-2-81

[1]VOR = Visual obstruction reading.

Vegetation type symbols:

Arta - Big sage Artemisia tridentata
Arno - Black sage Artemisia nova
Gusa - Snakeweed Guterrezia sarothrae
Atco - Shadscale Atriplex confertifolia
Cela - Winterfat Cerstoides lanata
Agcr - Crested wheatgrass Agropyron cristatum

Aspect = degrees asimuth

Table 2. Characteristics of sites adjacent to sage grouse leks.

Location	Vegetation Type	Percent Shrub Cover	VOR 1 m	VOR 1 dm	Distance to Escape Cover	Distance to Nest Cover	Distance to Type Change	Distance to Topo Change	Distance to H_2O	Percent Slope	Aspect
Hamlin	Arta	24.58	2.31	7.56	0	0	40 m				220° - SW
Cottonwood Bench	Gusa-Arta	39.58	0.64	5.56	2.70 m	1300 m	860 m		4300 m	2-2	126° - SE
Clover Creek	Arno	23.33	0.25	4.12							
N. Ruby A	Arno	31.09	0.58	3.67	17 m	17 m	650 m	650 m		2	180° - S
N. Ruby B	Arno Gusa	36.08	0.34	2.77	25 m	25 m	650 m			4	31° - NE
Low Fence	Arno-Gusa	26.75	0.64	3.66	0	0	1000 m		5000 m	2	60° - NE
Danville Creek	Arno	19.25	0.01	3.22							
Pine Field	Arta	23.08	1.69	8.72							
Willow Creek	Arno	24.92	0.50	3.45							
Butte	Guso-Arta	10.84	0.06	2.28	17 m	17 m	500 m	2200 m	500 m	1-2	308° - NW

T6117/1-4-82

$C_{ij} = D_i - D_j$, where D_i and D_j are distance measurements (e.g., distance to water) of leks in the ith row and the jth column.

Comparison of the similarity matrices involves calculation of the simple observed cross-product statistic (Γ_0).

$$\Gamma_0 = \sum_{i=1}^{n} \sum_{j=1}^{n} C_{ij} Q_{ij} \tag{1}$$

The statistical significance of the observed value (Γ_0) is determined by a Z-test described in Hubert and Schultz (1976). The Z-values presented in table 3 indicate that only distance to water is significantly related to lek presence. However, consistencies in other variables listed in table 1 suggest that these variables may also be important to sage grouse in selection of lek sites. For example, no leks occurred on areas covered by pinyon-juniper or on greater than 10 percent slope.

Table 3. Significance of data in Table 1 for defining sage grouse lek sites[1].

Variable	Z
Vegetation Type	- 0.637
Aspect	1.023
Percent shrub cover	0.68
Dist. escape cover	- 0.274
Dist. nest cover	0.52
Dist. topography change	- 0.889
Dist. water	- 2.99**

[1]Percent slope and visual obstruction reading data were not analyzed.

** 99 percent probability limits are ± 2.57.

(95 percent probability limits are ± 1.96).

Perhaps the occurrence of 17 leks in black sagebrush among 41 leks characterized may indicate a proportional habitat preference for black sagebrush. Another factor which might be important (and hard to characterize) is rainfall. General observations by HDR field personnel indicate that leks are seldom, if ever, found on areas with less than 25 cm/yr rainfall. It is hypothesized that drier sites would not support forbs long enough into the summer for sage grouse brood-use. Therefore, all areas with less than 25 cm/yr rainfall might be excluded from further analysis.

Cartographic Modeling

Data maps such as vegetation types and water sources were digitized and input to the computerized Map Analysis Package--MAP (Tomlin, 1980). This facility allows data maps to be manipulated as variables in a spatial configuration. Data maps are cell-based, in which data for each map are referenced to the centroid of each cell. Cells should be assigned biologically appropriate dimensions so cell size does not mask relationships such as lek size and movements. Arithmetic (multiply, differentiate, etc.) or combinative (cross-tabulate, clump, cover, etc.) operations are performed on respective cells of one to several maps (depending on the operation chosen). The set of MAP commands can constitute a high-level spatial analysis language. Spatial models can be programmed in MAP language syntax. The command language allows multiple overlay, screening, and mathematical and topological analysis of spatial information. Any two or three dimensional topological data can be stored in one or more maps in the program data base. All operations are performed on these maps, producing output maps, which are also stored and available for future operations. The values assigned to each cell in the maps after processing can be printed out on a scale of 1 to 100 using print and overprint capabilities of a line printer. Map categories can also be output as shaded polygons on a plotter, or as three-dimensional plots.

Conventional maps are represented as 2-dimensional arrays in memory and mass storage. This allows storage of Cartesian coordinates and a category or value representation for each map cell. New maps can be created by variously combining maps or map categories, and can be referred to as output maps.

Cartographic models are similar to algebraic evaluations in which mathematical algorithms are structured to solve complex problems. Nested parentheticals represent intermediate solutions similar to the intermediate map analysis operations. Operational groups can be organized like a flowchart (fig. 1) to produce a cartographic model expressing a goal-oriented flow of spatial information. These goals may include habitat quality modeling (Tomlin et al., 1979), corridor analysis (Rasmussen et al., 1980) and impact risk assessment (Thompson, 1977, 1979), and fire gradient analysis (Kessel, 1979).

A Sage Grouse Lek Preference MAP Model

For this study, the input maps generated as a result of field data reduction were: vegetation type, elevation, rainfall, and available water. Vegetation, rainfall, and water sources were obtained from resource maps; elevation was obtained from a 1:250,000 scale USGS quad. The vegetation input map is shown in figure 2. Factors expressing lek preference in various vegetation types were calculated externally to the MAP program by equation 2.

$$PV_i = \frac{Lv/\Sigma L}{PA_v} \qquad (2)$$

where,

PV_i = the lek preference factor for vegetation

Lv = is the number of occurrences of leks in vegetation type v

ΣL = the total number of leks observed in all vegetation types

PA_v = the proportionate area of each vegetation type.

According to table 1, leks were associated with black sagebrush (Artemisia nova) at 17 of 41 sites (41 percent). Since black sagebrush constitutes only 7 percent of the total area in Pine Valley (fig. 2), the fact that 41 percent of the lek sites were located in black sagebrush suggests a preferential selection. The value of PV_i for black sagebrush was 6.2 compared to 1.7 for big sagebrush. The rainfall input map was transformed into a rainfall mask (RAINMASK) by excluding all areas with less than 25 cm rainfall per year. It was judged that these areas would not support annual forbs for brood use throughout the spring. This map was used for overlay as described below.

The vegetation map (fig. 2) consists of eleven vegetation categories, each with an assigned code value. These internal cell values were changed to PV_i values (lek preference factors) from equation 2 corresponding to each vegetation type. This intermediate map was overlaid by the RAINMASK map (see fig. 1) to produce a new map showing only those PV_i values for vegetation receiving more than 25 cm of rainfall per year. This map is illustrated in figure 3. Data in table 1 show that leks were not found in areas with greater than 10 percent slope. To consider this in the model, a slope map was generated by differentiating the elevation map. A slope mask was made to exclude all slopes greater than 10 percent and the PV_i map was overlaid by the slope mask. This step resulted in a map of PV_i values corresponding to areas with greater than 25 cm rainfall and less than 10 percent slope.

This slope exclusion was consistent with data shown in table 1, indicating that leks were not found in areas with greater than 10 percent slope. This intermediate map was then overlaid by a WATERMASK map (fig. 4) excluding all areas greater than 2,000 m from a water source. The resulting map (fig. 5) represents levels of lek preference based on the habitat variables characterized by preferred vegetation type, rainfall, slope, and accessibility to water. Cells with PV_i ranging from 0 to 2.5 were accorded a low suitability for leks, moderate suitability ranged from PV_i values of 2.6 to 5.0, and high suitability was accorded to PV_i values above 5.0. Figure 5 also includes an area bounded by Utah's Division of Wildlife Resources estimated range of sage grouse.

Such a habitat quality model permits prioritizing of costly search effort. Areas of high suitability for leks can searched first. Areas of low lek suitability can be searched last, if circumstances permit.

Selection of Areas for Lek Management

A possible mitigation to reduce impact of disturbance in sage grouse range is to create lek sites in suitable undisturbed areas. Criteria governing selection of new lek sites might include those described above as well as projected habitat change. An important change affecting sage grouse is the encroachment of pinyon-juniper woodland into suitable sage grouse areas (West et al., 1979).

A prediction of pinyon-juniper woodland encroachment was attempted using a rainfall map of Pine Valley, a radiation load prediction model (Furnival, Reifsnyder, and Siccama, unpubl.), and the estimated elevation of the cold air lake, 1750 m (Ralph Holmgren, Desert Range Experiment Station, personal communication). Radiation loads predicted from latitude, slope, and aspect data were combined with rainfall for each cell in Pine Valley. The resultant rainfall-radiation values may provide indices to heat load and moisture status. While this relationship should be tested, Reed (1982) used a similar approach to model forest growth and succession. We were unable to make a thorough evaluation of this relation, but some comparison to existing vegetation was possible. Existing vegetation was compared with the rainfall-radiation (Rain-rad) load in each cell. The distribution of big sagebrush, black sagebrush, and pinyon-juniper woodland with respect to the rain-rad value is given in figure 6. On the basis of this analysis, potential pinyon-juniper habitat was assumed to include all areas with a rain-rad value in the range of 15 to 60 and above the cold air lake. Pinyon-juniper woodland and sagebrush types overlap in the 15 to 35 rain-rad range. Figure 7 is a map of present-day sagebrush that is in the 15-35 range and above the cold air lake. These sagebrush communities are vulnerable to pinyon-juniper woodland invasion, hence figure 7 is a map of potential pinyon-juniper woodland. This cold air lake results from a sharp thermocline of about 20° which often develops at night in Great Basin closed valleys.

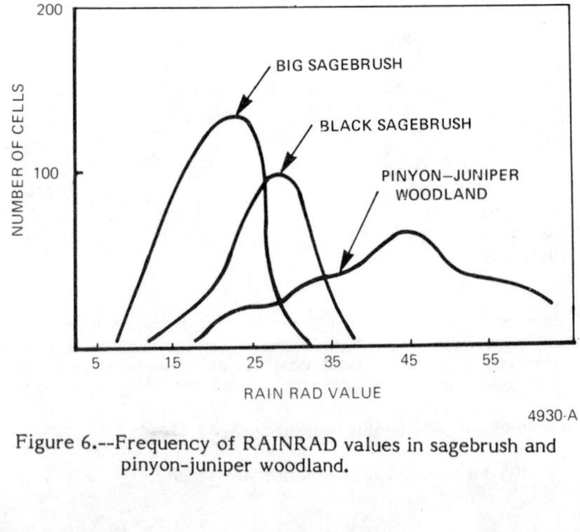

Figure 6.--Frequency of RAINRAD values in sagebrush and pinyon-juniper woodland.

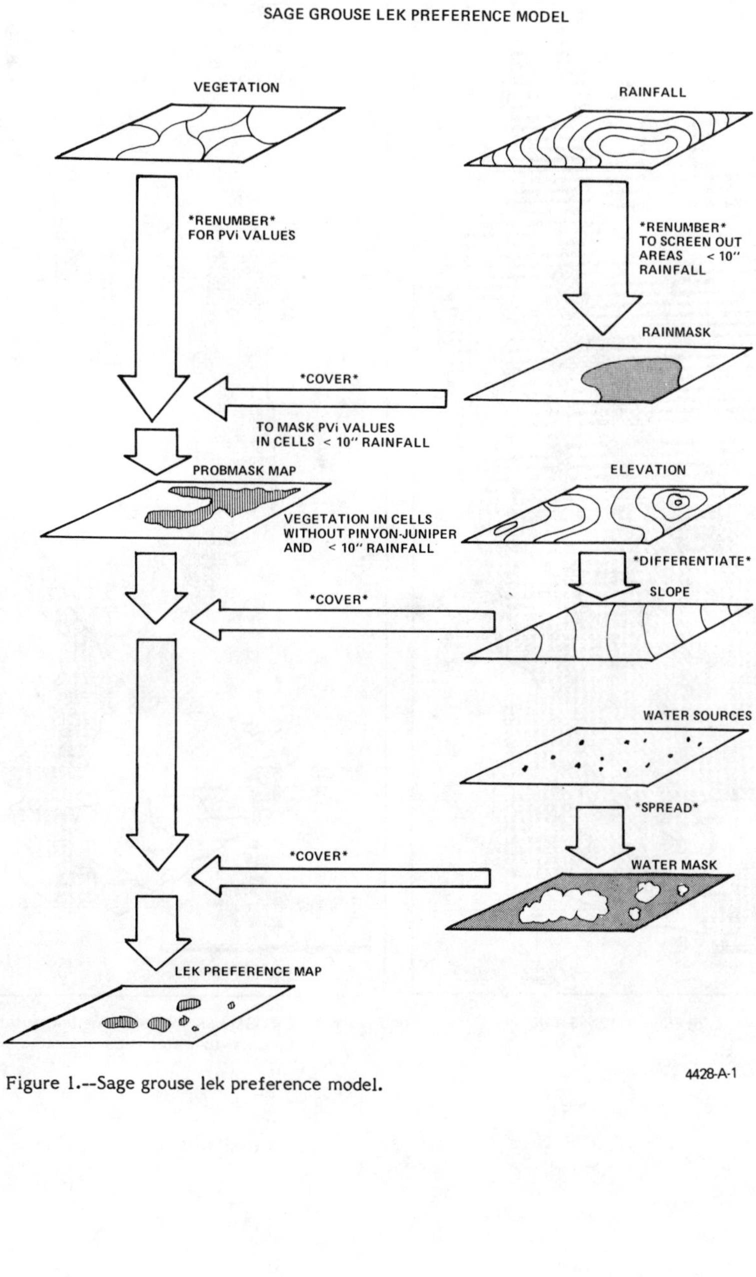

Figure 1.--Sage grouse lek preference model.

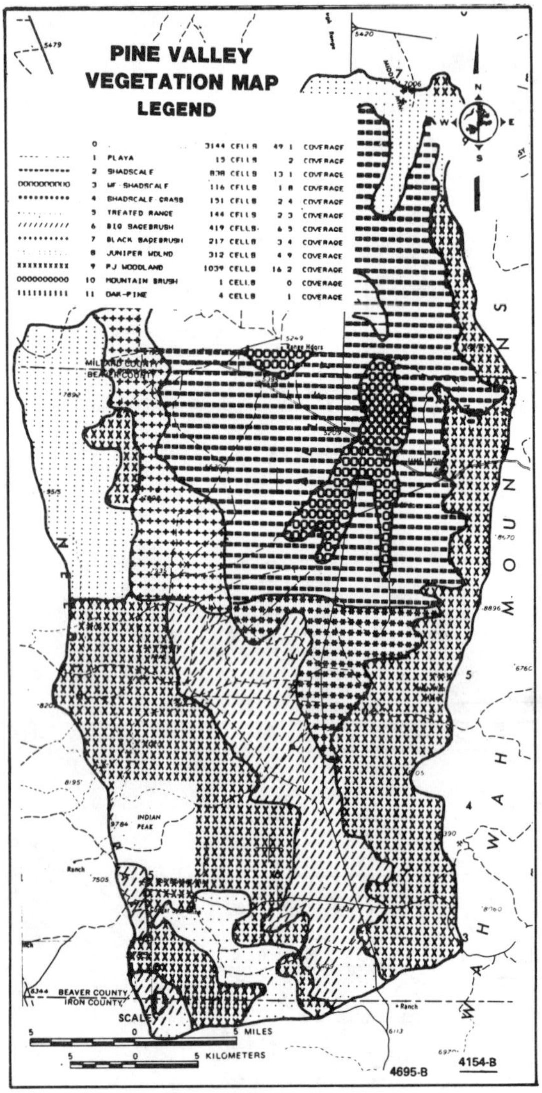

Figure 2.--Vegetation types of Pine Valley, Utah.

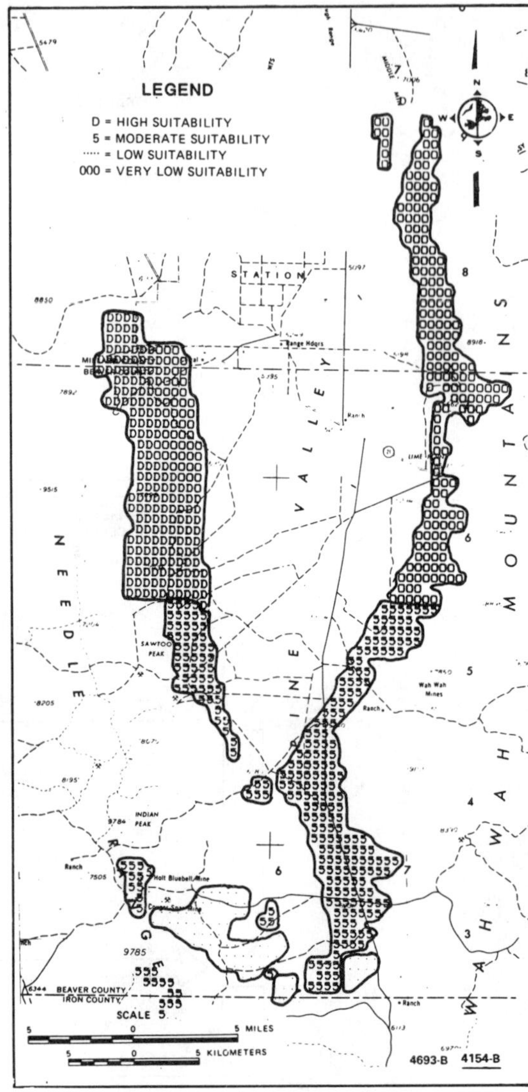

Figure 3.--PROBMASK map showing lek suitability before distance-to-water is considered.

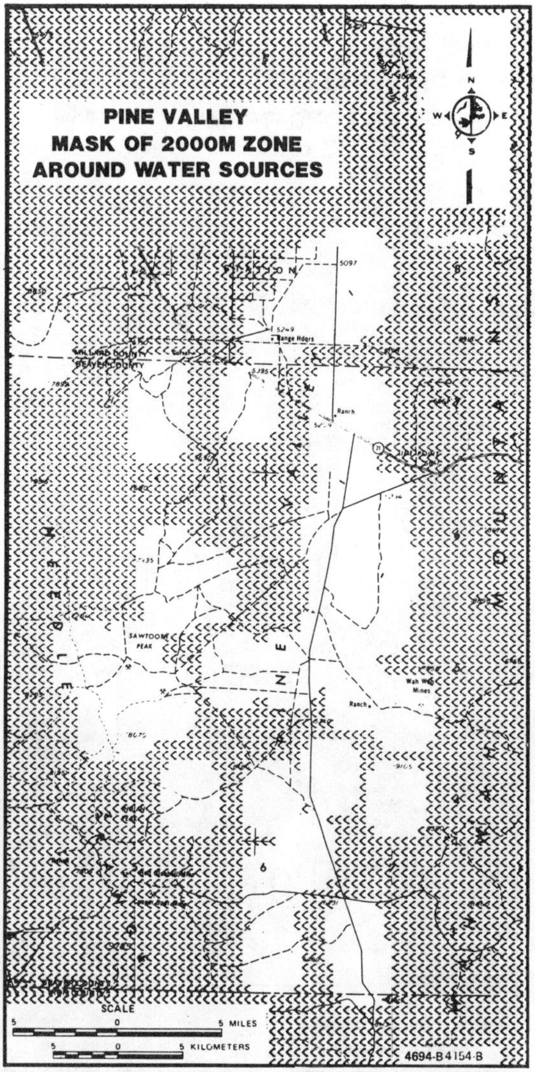

Figure 4.--WATERMASK map showing areas (blank) within 2,000 m of a water source.

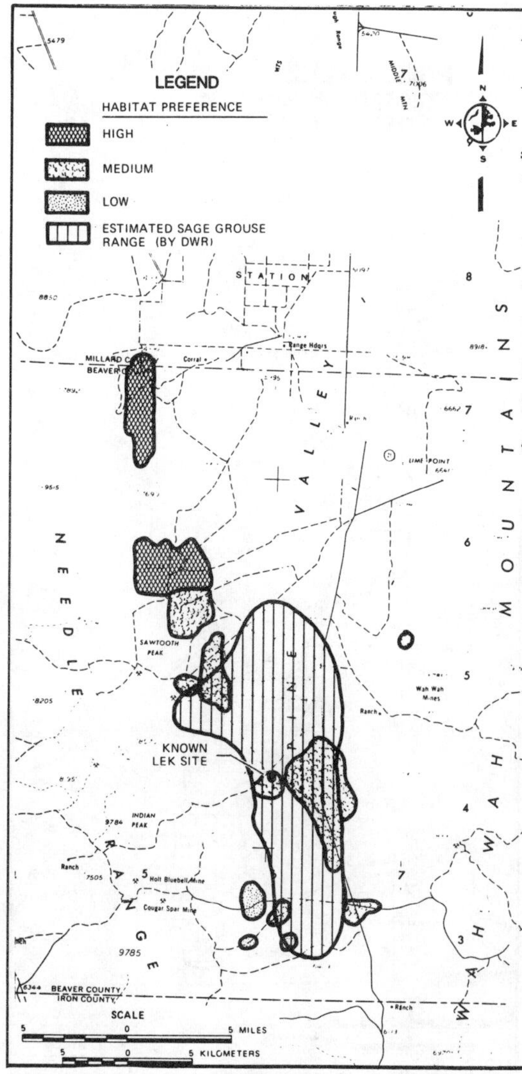

Figure 5.--Levels of lek preference, considering preferred vegetation, rainfall, slope, and accessibility to water.

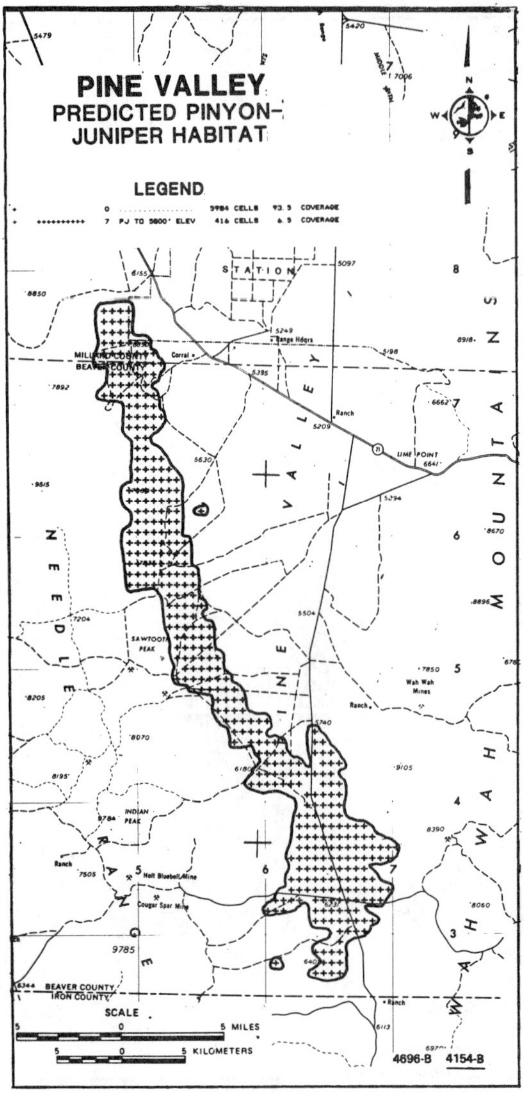

Figure 7.--Predicted pinyon-juniper woodland habitat in current sagebrush areas above 5,800 ft elevation.

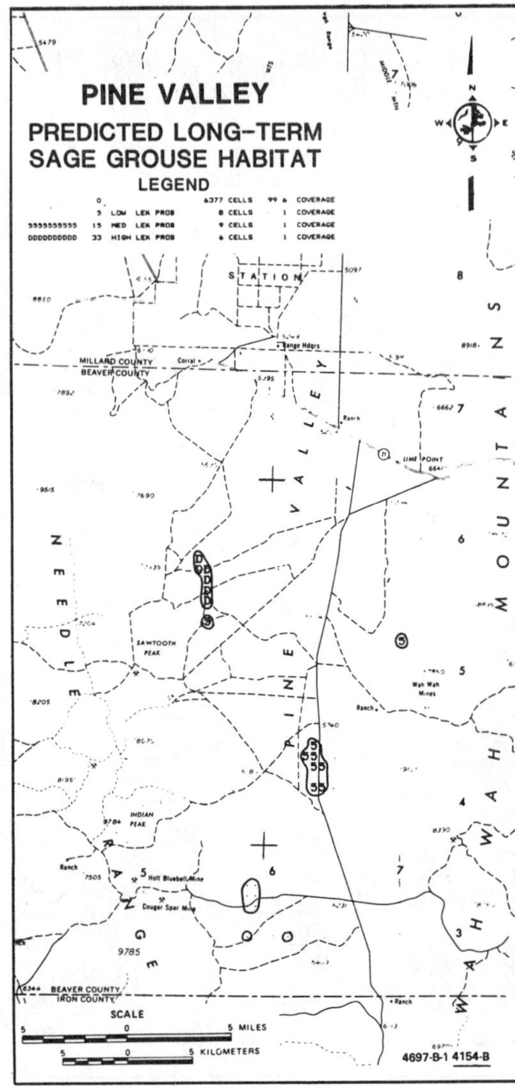

Figure 8.--Predicted long-term sage grouse habitat, excluding areas subject to invasion of pinyon-juniper woodland.

Temperatures below the inversion layer are sufficient to cause freeze damage to pinyon and juniper trees after the break of dormancy in the spring.

Cells containing predicted new pinyon-juniper woodland were excluded from subsequent analysis. Lek sites not excluded from the analysis (long-term lek sites) are displayed in fig. 8. The reduction in lek habitat can be seen by comparing figure 8 to figure 5. This difference is due to projected invasion of pinyon-juniper woodland in sagebrush communities. The obvious implications are that sage grouse management must include control of pinyon-juniper woodland invasion into sage grouse habitat, or to avoid creation of new lek sites in the invasion path.

DISCUSSION

Some variables used in the analysis were discrete and exclusionary, as in the case of elevation of the cold air lake, others were continuous and their effects are gradients of habitat suitability, viz. distance to water. In addition, future prediction is possible within the MAP context. Since this analysis is a habitat analysis, it would be amenable for use as a HEP model (ESR 102, 1980). Another application of spatial habitat analysis is identification of most probable locations for other efforts. Searches for rare species can be optimized by spatial modeling.

The distinctions of low, medium and high habitat preference in figure 6 are entirely based on vegetation preference factor values calculated by equation 1 for each cell. Better models can be constructed by incorporating other functional relationships. For example, Shurcliff (1980) found in Australian arid areas that dominance, diversity and total numbers in bird communities were primarily related to vegetation physiognomy and life form diversity. Wiens (1976) has shown that patchiness in Great Basin bird communities contributes to the distribution of species --again a demonstration of the importance of a spatial analysis. Subsequent spatial models could be improved by distinguishing various structural classes in each vegetation type. Such a vegetation structural map can be made for arid and semi-arid areas by combining vegetation type maps with classified Landsat images. Preliminary HDR Landsat classifications for portions of Utah and Nevada suggest that the major components of signature variance are vegetation height and percent cover, particularly on images with low sun angles. Subsequent classifications of sage grouse habitat should include measurements of average height and percent cover by surrounding vegetation.

The present model was developed principally as a methodological study. Identification of key habitat (high probability) areas can be useful in minimizing impacts due to development, project routing and siting. This model is currently being modified by the inclusion of more recent vegetation and precipitation data and additional field observation. Extension of similar analyses to other areas would require model validation with additional field studies. Such validation field studies are currently in the planning stage at HDR.

ACKNOWLEDGEMENTS

This analysis was performed on a remote terminal using a CDC CYBER computer (HDR-System Sciences, Omaha, Neb.). L. Neel and M. Costanzo were responsible for major portions of the field work and analysis, respectively.

LITERATURE CITED

Braun, C.E., T. Brigg, and R.O. Wallestad. 1977. Guidelines for maintenance of sage grouse habitats. Wildl. Soc. Bull. 5:99-106.

Gill, R.B. 1975. Distribution and abundance of a population of sage grouse in North Park, Colorado. M.S. Thesis, Colorado State University, Ft. Collins. 187 p.

Hubert, L. and J. Schultz. 1976. Quadratic assignment as a general data analysis strategy. British J. Math and Stat. in Psych. 29:190-241.

Kessel, S.R. 1979. Gradient Modeling. New York: Springer-Verlag. 432 p.

Molini, W.A., and L. Barngrover. 1979. Upland game, migratory game birds, fur investigations, and hunting season recommendations. Nevada Department of Wildlife. Project W-48-10, study S&II, Jobs 7-13, 15, 17.

Oakleaf, R.J. 1971. The relationship of sage grouse to upland meadows in Nevada. Job Final Rep. Fed. Aid Proj. W-48-2. Nevada Department of Fish and Game, 64 p.

Patterson, R.L. 1952. The sage grouse in Wyoming. Denver: Sage Books, Inc., 341 p.

Rasmussen, W.O., R.N. Weisz, Peter Ffolliott, and D.R. Carder. 1980. Planning for forest roads - a computer assisted procedure for selection of alternative corridors. J. Environ. Manage. 11:93-104.

Reed, K.L. 1982. Simulation of forest stands in a hypothetical watershed using models and cartographic analysis. In. (J. means, ed.) Proc. Symp. For. Growth and Succession. Oregon State University, in press.

Shurcliff, K.S. 1980. Vegetation and bird community characteristics in an Australian arid mountain range. J. Arid Environ. 3:331-348.

Thompson, L.S. 1977. Identification of critical wildlife habitat using an "impact risk" mapping technique. Proceedings of a National Symposium on Classification, Inventory, and Analysis of Fish and Wildlife Habitat. U.S. Fish and Wildlife Service.

Thompson, L.S. 1979. Probability and the assessment of impact risk in utility corridor siting. Second Symposium on Environmental Concerns in Rights-of-way Management, Oct. 16-18, Ann Arbor, Michigan.

Tomlin, C.D., 1980. Cartographic modeling techniques in environmental planning. Doctoral dissertation, Yale School of Forestry and Environmental Studies.

Tomlin, C.D., S. Berwick, and D. Tomlin. 1979. The use of computer graphics in deer habitat evaluation. International conference on computer mapping hardware, software, and data bases. Harvard University. 18 p.

U.S. Fish & Wildlife Service. 1980. Habitat Evaluation Procedure. ESR 102.

Wallestad, R. 1975. Life history and habitat requirements of sage grouse in central Montana. Game Manage. Div., Montana. Dept. of Fish and Game. 66 p.

Wallestad, R., and D.B. Pyrah. 1974. Movements and nesting requirements of sage grouse hens in central Montana. J. Wildl. Manage. 38:630-633.

West, N.E., R.J. Tausch, and Ageli A. Nabi. 1979. Patterns and rates of pinyon-juniper invasion and degree of suppression of understory vegetation in the Great Basin. Range Improvement Notes, Sept. 1979. USDA, USFS, Ogden, Utah.

Wiens, J. A. 1976. Population responses to patchy environments. Ann. Rev. Ecol. Syst. 7: 81-120.

SENSITIVITY ANALYSIS OF A SIMULATION MODEL OF BIRD-INSECT INTERACTIONS IN A DECIDUOUS FOREST ECOSYSTEM[1]

Brian A. Maurer[2]

Abstract.--A simulation model of biomass changes in a bird-insect system in a deciduous forest was constructed. The major compartments of the model were deciduous leaves, leaf-eating insects, and 15 species of insectivorous birds. Additional compartments representing alternate prey sources were also included. A sensitivity analysis was conducted to determine which parameters were important in determining the dynamics of the model. Variations of initial inputs into several compartments allowed an assessment of the importance of initial conditions. Simulation results identified 2 key parameters. The major controlling factor of insect populations was the value of a density dependent biomass removal parameter. Bird predation was insignificant in controlling the dynamics of insect populations. The major factor controlling growth of bird populations was insect densities and a parameter relating biomass intake to the difficulty of obtaining prey. Thus, though growth rates of birds were sensitive to prey population levels, insect growth rates were only slightly affected by bird predation. The model was robust to large variations in initial conditions. The dynamics of the model suggest that birds in deciduous forests may not be subject to exploitative competition.

INTRODUCTION

The role that insectivorous birds play in forest ecosystems has not been well understood. Several early studies (Otvos 1979) claimed that birds were very important in controlling insect populations. These claims were based on anecdotal accounts or reports of rather vague observations. From later studies conducted in a more scientifically rigorous fashion it became apparent that birds had little effect on insect populations during infestations. However, experimental studies by Solomon and Glen (1979) and Holmes et al. (1979) indicated that birds reduced insect population levels during the breeding and non-breeding seasons. Studies of the morphological and behavioral attributes of the lepidopteran larvae that form a great deal of the diet of birds during the breeding season demonstrated that lepidopteran larvae which are palatable to birds are cryptically colored and have behavioral traits which tend to lessen the visible evidence of their feeding activities (Heinrich 1979). Thus, evidence for the role of birds as controlling agents of insect populations though appealling, is equivocal.

An approach that has been useful in analyzing ecological systems such as the forest bird-insect system has been to construct models which simulate the flow of energy or matter through the system (Wiegert 1979). The advantage of this approach is that it allows one to manipulate the various factors which may be influencing the system and thus lead to a clearer understanding of how the system works. Previous models of forest insect systems have concentrated on dynamics of insect populations with the goal of population control (eg. Rubtsov and Shvytov 1980). In this paper I construct a model of bird-insect interactions for an eastern deciduous forest which developes in more detail the roles of the avian insectivores and report on some preliminary results

---

[1]Paper presented at the Third International Conference on State-of-the-Art in Ecological Modelling, Colorado State University, May 24-28, 1982.

[2]Brian A. Maurer is Statistician/Programmer, Statistical Support Unit, College of Agriculture, University of Arizona, Tucson, Arizona 85721, U.S.A.

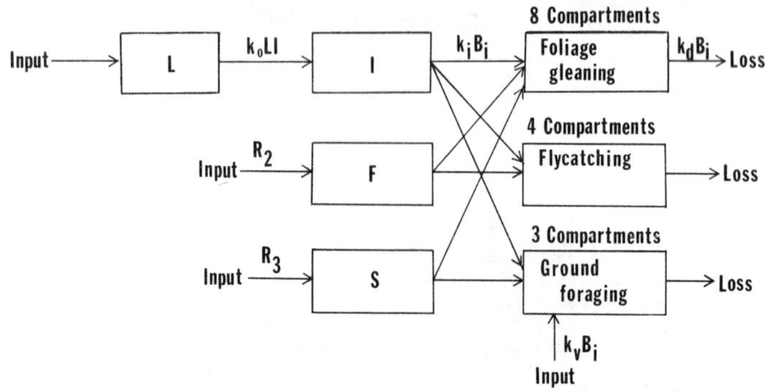

Figure 1.--Box and arrow diagram of a simulation model of forest bird-insect interactions. Symbols as in Tables 1 and 2.

which follow from the model.

THE MODEL

A box and arrow diagram of the model showing its general features is given in figure 1. All flows from one compartment to the next were converted to g dry weight $(25\times10^6 m^{-3})(day^{-1})$ to alleviate difficulties in estimating constants which would be needed to convert numbers of prey into numbers of predators. Bird densities were obtained from a 10 ha plot with an average canopy height of 25 m (McArthur 1980). This plot was a relative mature stand of eastern hardwoods. Major tree species included red oak (*Quercus rubrum*), chestnut oak (*Q. prinus*), sugar maple (*Acer saccharum*), red maple (*A. rubra*), American beech (*Fagus grandifolia*), black cherry (*Prunus serotina*), and yellow poplar (*Liriodendron tulipfera*).

The major input to the model was into the leaf compartment (fig. 1). It was assumed that leaves had a daily increment in weight proportional to the biomass present and that the rate of that increment leveled off as the season progressed. A linear relationship between the amount of leaves (L) and insect consumption (Rubtsov and Shvytov 1980) was postulated. Input into the leaf-eating insect compartment was also assumed to be influenced by weather. Input into the 15 bird compartments was a function of the amount of bird biomass present (B_i) and was related to insect biomass by the equation $k_i=f(I)$, thus the amount of biomass intake into the bird compartments was a function of both bird and insect density. In addition to leaf-eating insects, 2 other sources of prey were available to birds. Flying insects and soil arthropods were added to the model to provide the birds with an alternative food supply. Another alternative source for some birds, fruits and seeds, was assumed to be relatively abundant and always available at a constant level. The model output described changes in daily biomass accumulation in each compartment during a hypothetical breeding season lasting 100 days.

Mathematical description

Differential equations were developed to model the rates of change of biomass in each compartment. The state variables and their initial values are given in table 1. As mentioned above, these rates were given in g $(25X10^6 m^{-3})(day^{-1})$. The change in leaf biomass was given as

$$\frac{dL}{dt} = 0.25L - k_oLI \tag{1}$$

where L is leaf biomass and I is leaf eating insect biomass.

The rate of change of biomass of leaf-eating insects was

$$\frac{dI}{dt} = 0.25pk_oLI - \sum_{i=1}^{15} k_{iI}B_i - k_rI^2 \tag{2}$$

Table 1.--State variables for a model of insect bird interactions in a deciduous forest.

Variable	Symbol	units	Initial value
Leaves	L	$g(25x10^6m^{-3})$	1622611
Leaf-eating insects	I	"	variable
Flying insects	F	"	variable
Soil arthropods	S	"	variable
Red-eyed Vireo	B_1	"	660.0
Acadian Flycatcher	B_2	"	256.0
Great-crested Flycatcher	B_3	"	204.0
Black-capped Chickadee	B_4	"	64.8
Hooded Warbler	B_5	"	86.4
Rose-breasted Grosebeak	B_6	"	171.6
Solitary Vireo	B_7	"	66.8
Black-throated Blue Warbler	B_8	"	78.4
Black-throated Green Warbler	B_9	"	178.0
Scarlet Tanager	B_{10}	"	342.0
Eastern Wood Pewee	B_{11}	"	56.4
American Redstart	B_{12}	"	172.0
Wood Thrush	B_{13}	"	855.0
Ovenbird	B_{14}	"	154.4
Rufous-sided Towhee	B_{15}	"	163.6

where p is the percent maximum growth rate achieved, k_o is a constant, and B_i is the density of bird species i. The first term is multiplied by 0.25 because about 25% of the intake of an insect actually goes into production (Price 1975). This value could be adjusted to fit a specific insect population for which data on production versus intake was available. The parameter k_{iI} relates insect density to the rate of predation by birds, and k_r is the density dependent biomass loss parameter for the insects.

The rate of change of flying insects was simulated using the following equation

$$\frac{dF}{dt} = R_2 - \sum_{i=1}^{15} k_{iF}B_i - k_rF^2 \quad (3)$$

Where F is the density of flying insects, R_2 is the input into the flying insect compartment, and k_{iF} relates the density of flying insects to predation rate.

The rate of change of the biomass of soil arthropods was given by the equation

$$\frac{dS}{dt} = R_3 - \sum_{i=1}^{15} k_{iS}B_i - k_rS^2 \quad (4)$$

where S is the density of soil arthropods, R_3 is the input into soil arthropods, and k_{iS} is a parameter that relates the predation rate by birds to the density of soil arthropods.

Table 2.--Parameter values for a model of insect bird interaction in a deciduous forest.

Parameter	Symbol	Units	Value
Biomass removal for leaf-eating insects	k_r	day^{-1}	variable
Intake constant for leaf-eating insects	k_o	$(25x10^6m^3)\ (g^{-1}day^{-1})$	$1.7x10^{-6}$
Constant input of flying insects	R_2	$g(25x10^6m^{-3})\ (day^{-1})$	2679.3
Constant input of soil arthropods	R_3	$g(25x10^6m^{-3})\ (day^{-1})$	915.5
Death rate of birds	k_d	day^{-1}	0.0014
Half saturation constant	X	$g(25x10^6m^{-3})$	variable
Intake constant for bird i eating prey j	k_{ij}	day^{-1}	$k_{maxj}A_j/(X+A_j)$*
Maximal feeding constant	k_{maxj}	day^{-1}	$.25P_{ij}CS_i$*

*A_j = density of prey type j; P_{ij} = proportion of prey type j in the diet of bird species i; CS_i = clutch size of bird species i.

The rate of growth of bird species i was

$$\frac{dB_i}{dt} = 0.02B_i(k_{iI}+k_{iF}+k_{iS}) - k_dB_i \tag{5}$$

where k_d is the average daily death rate. Price (1975) estimated that birds put only about 2% of their intake into production, hence the first term of the equation (5) is multiplied by 0.02.

To estimate the density-dependent predation parameters (k_{iI}, k_{iF}, k_{iS}), it was assumed that birds exhibit a functional response to insect densities (Krebs 1973). Real (1977, 1979) noted that the behavioral phenomenon of functional response is analogous to the kinetics of enzyme reactions. The general response of a bird predator to insect prey density (k_i in Fig. 1) can be modelled as

$$k_{ij} = \frac{k_{maxj}A_j}{X + A_j} \tag{6}$$

where k_{maxj} is the maximum rate of intake of prey type j by predator i, A_j is the density of prey, and X is the density of prey at which intake is half maximal. Pulliam (1976) suggested that the functional response relation be incorporated in Lotka's (1925) predator-prey equations and used to model predator-prey systems.

Parameter estimation

Equations (1) to (5) represent a system which attempts to describe the general features of the bird-insect system in an eastern deciduous forest. An attempt was made to examine the general behavior of this model under a variety of conditions by estimating the values of the parameters in equations (1) to (6). A brief discussion of how this was done and the assumptions that were made is given below. The estimates obtained for the parameters are given in tables 2 and 3.

The parameter p is a function of temperature and was estimated from data of Andrewartha and Birch (1954)

Table 3.--Life history and parameter estimates for 15 bird species.

Species	Common name	Clutch size	Avg.adult weight (g)	Diet* %I	%F	%S	%V	k_{maxj}* I	F	S	V
Vireo olivaceus	Red-eyed Vireo	3	16.5	85	15	--	--	0.64	0.11	----	----
Empidonax virescens	Acadian Flycatcher	4	12.8	40	60	--	--	0.40	0.60	----	----
Myiarchus crinitus	Great-crested Flycatcher	5	34.0	25	75	--	--	0.31	0.94	----	----
Parus atricapillus	Black-capped Chickadee	7	10.8	90	10	--	--	1.58	0.18	----	----
Wilsonia critrina	Hooded Warbler	3	10.8	40	50	10	--	0.40	0.50	0.10	----
Pheuticus ludovicianus	Rose-breasted Grosebeak	4	42.9	50	--	--	50	0.50	----	----	0.50
Vireo solitarius	Solitary Vireo	4	16.7	90	10	--	--	0.90	0.10	----	----
Dendroica caerulescens	Black-throated Blue Warbler	4	9.8	85	15	--	--	0.85	0.15	----	----
D. virens	Black-throated Green Warbler	4	8.9	75	25	--	--	0.75	0.25	----	----
Piranga olivacea	Scarlet Tanager	4	28.5	90	10	--	--	0.90	0.10	----	----
Contopus virens	Eastern Wood Pewee	3	14.1	10	90	--	--	0.08	0.68	----	----
Setophaga ruticilla	American Redstart	4	8.6	55	45	--	--	0.55	0.45	----	----
Hylocichla mustelina	Wood Thrush	3	47.5	30	--	70	--	0.23	----	0.53	----
Seiurus aurocapillus	Ovenbird	4	19.3	35	--	65	--	0.35	----	0.65	----
Piplio erythropthalmus	Rufous-sided Towhee	4	40.9	5	--	60	35	0.05	----	0.60	0.35

*I = leaf eating insects, F = flying insects, S = soil arthropods, V = fruits and seeds.

relating the % maximum growth rate to temperature. An equation for p was obtained by regression of ln p on temperature (T) which gave

$$p = 0.002308 \exp (0.2087T)$$

The correlation coefficient for the regression was highly significant ($r = 0.97$, $df = 6$, $p < 0.0001$). In the model temperature was included as a function of time. A function that expressed temperature as a function of time was generated by trial and error. Temperatures recorded near the study area (Timber and Watershed Laboratory, Parsons, West Virginia, unpublished data) were used to give the relation

$$T = 0.63 \quad (t)^{1/2} + 24 \qquad (7)$$

where t is time measured in days since the beginning of growing season (fig. 2). For the effects a cold spell, a pre-determined value could be subtracted from equation (7).

The value of the constant k_o was estimated as follows. Waldbauer (1968) estimated that a herbivorous insect larvae needed to consume 2.72 g of vegetable material per g weight per day in order to maintain adequate growth at optimum conditions (ie., when p = 1 in equation 2). Thus, the total intake, k_oLI, (not regarding efficiency) must be equal to 2.72 x I in order for the population to grow adequately. Thus

$$k_oLI = 2.72I$$

$$k_o = \frac{2.72}{L}$$

k_o was estimated for the time at which insects first emerged and was assumed to remain constant. If leaves grew for a week before insects began feeding, $L = 1{,}622{,}611$ and $k_o = 1.7 \times 10^{-6}$.

R_2 and R_3 were assumed to be constant for the simulation runs. R_2 was calculated from data published by Holmes et al. (1978) based on captures in suction traps. They obtained an estimate of 6814 insects (18.3×10^3 m^{-3}) as a density of insects caught in their traps. If this figure is roughly proportional to the number of flying insects in a forest then

$$\frac{6814}{28.3 \times 10^3} = \frac{FI}{25 \times 10^6}$$

where FI is the number of insects. This gives FI = 6,019,434.6 insects ($25 \times 10^6 m^{-3}$). This figure estimates the number of insects in June when the sample was taken. For a rough approximation assume that growth of the population of flying insects has been linear, then dividing FI by the number of days gives an average increase, which when converted to grams is 2006.478 g ($25 \times 10^6 m^{-3}$) (day^{-1}). However, this figure is the net rate before predation. R_2 is dF/dt plus the rate at which predation occurs. Using the rate at which insects were eaten when the birds first arrived (initial values of B_i given in Table 1) R_2 was estimated as 2679.3 g ($25 \times 10^6 m^{-3}$)(day^{-1}). Data from Greenlaw (1978) was used to estimate R_3. Using estimates for 2 years the average net rate of growth of soil arthropods was calculated as 266 g ($25 \times 10^6 m^{-3}$)(day^{-1}). Correcting this for predation as was done above for flying insects gave R_3 = 915.5 g ($25 \times 10^6 m^{-3}$)(day^{-1}).

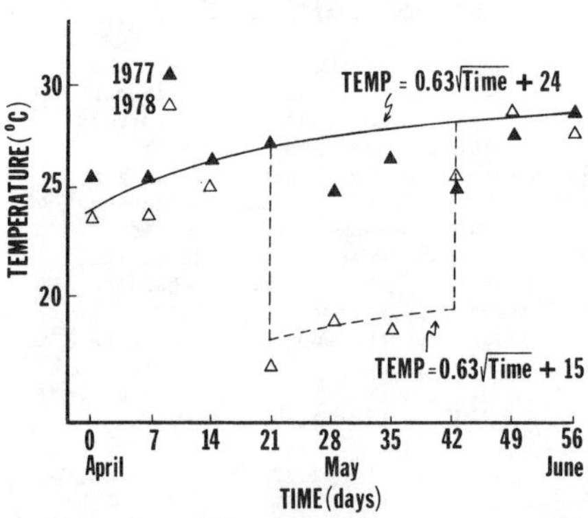

Figure 2.--Temperature as a function of time: data and functions used in simulations of bird-insect dynamics.

The daily death rate, k_d, was estimated as follows. Henny (1972) estimated yearly death rates for several avian species based on banding studies. A yearly death rate of 0.5 was about average for his data. If the death rate for each day of the year is independent of that of other days, then the average daily death rate times the number of days in a year would give the yearly death rate. Let k_d be the average daily death rate, then

$$365\ k_d = 0.5$$

$$k_d = 0.0014$$

The parameters k_{iI}, k_{iF}, k_{iS} are functions of prey density, and were included in the model using equation (6). X, the half saturation constant, was allowed to vary (table 2) and was constant for all species. By varying this parameter its effects on bird biomass accumulation could be assessed. To estimate k_{maxj}, the maximum rate of increase possible, excluding death, was calculated. Assume that when feeding is maximal the rate of biomass accumulation is also maximal. An approximation of maximal daily rate of increase (M) for a 100 day breeding season is

$$M = \frac{N_i CS_i w_i}{100}$$

where N_i is the number of breeding paris of bird species i, CS_i is the clutch size (no. of eggs) and w_i is the weight of an adult. Let P_{ij} be the proportion of prey type j in the diet of bird species i. Then the part of the maximal growth rate due to prey type j will be P_{ij} M. Thus the partial rate of biomass increase due to prey type j would be

$$P_{ij} M = .02 B_i k_{ij} \qquad (8)$$

where k_{ij} is given by equation (6). However, since feeding is maximal $k_{ij} = k_{maxj}$, so

$$k_{maxj} = \frac{P_{ij} M}{.02 B_i} = \frac{P_{ij} CS_i (N_i w_i)}{2 B_i}$$

However, $N_i w_i = B_o/2$, where B_o is the initial density of the bird species. If it is assumed that feeding is maximal when $B_i = B_o$ (ie. feeding rate is never greater) then the above relation simplifies to

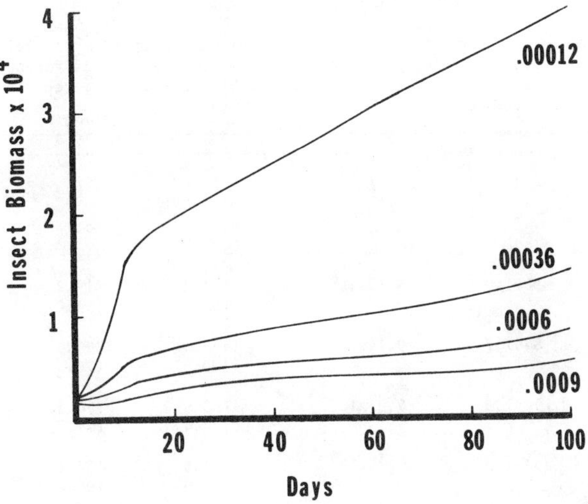

Figure 3.--Insect biomass (in $g(25x10^6m^{-3})$) trends for 100 days using various values for k_r, the density dependent biomass removal parameter excluding predation by birds.

$$k_{maxj} = .25P_{ij}CS_i \qquad (9)$$

The percentage of the diet of prey type j (P_{ij}) and the clutch size (CS_i) were estimated from the life history literature for each species bird (Bent 1942, 1946, 1949, 1950, 1952, 1953, 1958, 1968; Martin et al. 1961). Table 2 lists the values of the parameters and table 3 gives some of the life history information used in the calculations.

SENSITIVITY ANALYSIS

To assess the importance of various terms on the behavior of the model, initial inputs into several compartments and values of several parameters were varied. The sensitivity of the model to these manipulations implied that certain attributes of the bird-insect system were important in influencing model dynamics, while other attributes had little impact on the dynamics.

Simulations were run using a FORTRAN IV program written and compiled on a HP1000 computer. A simulation consisted of a numerical integration of equations (1) to (5) over a time period of 100 days (to represent 1 breeding season) using Euler's method with an interval of .025 day. The leaf compartment and all 3 insect prey compartments were allowed to accumulate biomass for 7 days before the initiation of bird predation.

Variations in k_r

The model was found to be very sensitive to variations in the density dependent biomass removal constant for leaf-eating insects (k_r). Initial input for the variable I_3 was kept constant at 2000 g $(25x10^6m^3)^{-1}$. All other state variables and constants were maintained at values given in tables 1-3. The parameter X was given the value 1000. Final densities of insects showed an almost 10 fold increase when k_r was changed from .0009 to .00012 (fig. 3). Final biomass values for the leaf compartment and for bird compartments were also affected. Leaf biomass achieved lower final biomass densities while birds achieved higher biomass densities as the value of k_r was lowered (table 4).

Variations in X

The parameter X controls the rate at which birds intake biomass from the prey compartments. As X approaches zero, biomass intake reaches its maximum rate. With k_r set to .0006, and all other initial values and parameters set as described in the previous section, X was varied from 100 to 2000. Changing the value of X had a large impact on final bird densities, but changes in insect prey densities

Table 4.--Changes in final biomass in $g(25x10^6m^{-3})$ for several state variables corresponding to different values of parameter k_r. Numbers of birds given in parentheses.

k_r	I	L	B_1 (no.)	B_2 (no.)	B_{14} (no.)
.0009	5625.9	$9.18x10^6$	1531.2 (92.8)	819.2 (64.0)	414.9 (21.5)
.0006	8401.7	$9.14x10^6$	1658.3 (100.5)	899.8 (70.3)	459.34 (23.8)
.00036	13891.5	$9.07x10^6$	1788.6 (108.4)	994.56 (77.7)	511.45 (26.5)
.00012	40381.5	$8.79x10^6$	1963.5 (119.0)	1157.1 (90.4)	617.6 (32.0)

were only slight (table 5). Raising the value of X can be viewed as increasing the difficulty of capturing a single prey item. An increase in X means that more prey must be captured before feeding reaches a maximal rate, which implies each individual prey item is less valuable in terms of its ability to contribute to the growth of the predator.

Variations in initial densities of leaf-eating insects

Initial input into the leaf-eating insect compartment was vaired using values of k_r=.0006 and k_r=.00012 for the density dependent biomass removal constant. The parameter X was set to 1000 and other state variables and parameters remained as in simulations described above. There was very little difference between final biomass densities for leaf-eating insects when low initial densities were used (I_o = 1000) when compared with high initial densities (I_o = 5000) for both values of the parameter k_r. Thus the model was insensitive to variations in initial input into the leaf-eating insect compartment.

Variations in bird densities

The parameter k_r was set to .0006 and X was set to 1000, while all other parameters retained the values as given in table 2. Initial density in the leaf-eating insect compartment was set to 2000. Under these conditions densities of birds were manipulated. In one set of simulations, birds were excluded from the model entirely. Initial densities were used as given in table 1 for a second set of simulations and then doubled for the final

Table 5.--Changes in final biomass for several state variables corresponding to different values of parameter X. Number of birds given in parentheses.

X	I	F	S	B_1	B_2	B_{14}
100	8401.5	2111.9	1234.1	2062.5 (125)	1355.5 (105.9)	785.51 (40.7)
1000	8401.7	2112.3	1234.5	1658.3 (100.5)	899.8 (70.3)	459.3 (23.8)
2000	8401.9	2112.5	1234.8	1404.2 (85.1)	689.9 (53.9)	347.4 (18.0)

Table 6.--Final biomass densities for insect prey for several densities of avian predators.

Condition	I	F	S
Birds left out of model	8403.3	2113.2	1235.2
Initial densities of birds as in Table 1	8401.7	2112.3	1234.5
Initial densities of birds twice values in Table 1	8401.7	2112.3	1234.5

set. Final densities of insects were only slightly affected by the manipulations of bird densities (table 6).

In addition, individual bird compartments had little influence on each other. Several simulations were performed where 1 or more bird compartments were removed from the model. In every case the densities of bird compartments left in the model were practically identical to densities when all bird compartments were included in the model.

In the patterns described above, the effects of a drop in temperature late in the season (fig. 2) had the greatest impact when the temperature drop caused an increase in the biomass removal constant for the prey compartments. The effects on the outcome of the model were much less when lower temperatures caused lower recruitment rates for leaf eating insects.

DISCUSSION

Studies of the importance of birds in deciduous forest ecosystems have suggested that birds play minor roles in community energetics (Holmes and Sturges 1973) and nutrient cycling (Sturges et al. 1974). The implications of the simulation model results described above are that birds may also play a relatively minor role in the population dynamics of their insect prey. Predation pressure in the bird-insect system in deciduous forests is probably important as a selection pressure in palatable prey species (Heinrich 1979, Holmes et al. 1979). Bird predation may also act as a control of individual prey species (McNamee et al. 1981), but may not control the total available prey density. The important factors controlling insect abundance seem to be those related to density dependent death rates. Though not specifically modeled, such additional factors as parasites, disease, and stress caused from high insect densities (McNamee et al. 1981) may be more important as controlling agents in preventing insect populations from reaching high levels. The parameter k_r in the simulations described above is abstracted from a larger set of density dependent factors (see Schaeffer 1981) and its importance in the simulations indicates that these combined factors need to be considered when describing the dynamics of insect population. Though bird predation may be relatively unimportant in controlling insect populations, birds may act as transmission vectors for insect parasites and diseases (Otvos 1979).

The fact that birds seem to have only a small impact on the dynamics of the simulations done above suggests that birds in natural systems may be relatively insensitive to densities of other avian species and to conspecifics when resource use is considered. Though birds could respond numerically to increases in insect densities in the simulations they could not remove enough biomass from insect populations to have any significant impact on insect population densities. This suggests that birds may not undergo exploitation competition. Notice, however, that bird biomass accumulation can be affected a great deal by making resources more difficult to obtain (see section on variations in X), hence interference competition could still play an important role in interactions between and within species.

The model suggested above appears to be amendable to validation by collection of appropriate data. Careful measurements in the field of leaf growth, insect population trends, and avian densities and reproductive success would allow estimates the parameters using the same types of calculations done above with data obtained in the literature. Though the simulations performed above have identified critical parameters, the model probably provides only a rough picture of how the bird-insect system operates. Improvements in the model currently under consideration include equations taking into account differences in the distribution of insect species, and preferences of different tree species as foraging locations by birds (Holmes and Robinson 1981). Validation of the model with appropriate data should greatly increase our insight into the processes important in determining the patterns observed in bird-insect systems in deciduous forests. The approach to avian population growth described above may be preferable to the Lotka-Volterra equations, especially in light of the controversy surrounding the estimation of the competition coefficient (eg. Emlen 1981).

ACKNOWLEDGEMENTS

This paper was generated by the author in response to a great deal of creative input by Dr. H. W. Hunt, and that input is greatfully acknowledged. L. B. McArthur was very helpful in providing unpublished data from his dissertation project. C. Maurer typed several drafts of the paper and generally provided support and encouragement. H. W. Hunt also read an earlier draft of the paper and provided many helpful comments. Barbara Warren prepared the manuscript for publication.

LITERATURE CITED

Andrewartha, H. G., and L. C. Birch. 1954. The distribution and abundance of animals. 782 p. Chicago, University of Chicago Press.

Bent, A. C. 1942. Life histories of North American flycatchers, larks, swallows, and their allies. U. S. National Mus. Bull. 179: 1-555.

______. 1946. Life histories of North American jays, crows, and titmice. U. S. National Mus. Bull. 191: 1-495.

______. 1949. Life histories of North American thrushes, kinglets, and their allies. U. S. National Mus. Bull. 196: 1-452.

______. 1950. Life histories of North American wagtails, shrikes, vireos, and their allies. U. S. National Mus. Bull. 197: 1-411.

______. 1953. Life histories of North American wood warblers. U. S. National Mus. Bull. 203: 1-734.

______. 1958. Life histories of North American blackbirds, orioles, tanagers, and allies. U. S. National Mus. Bull. 211: 1-549.

______. 1968. Life histories of North American cardinals, grosebeaks, buntings, towhees, finches, sparrows, and allies. U. S. National Mus. Bull. 237: 1-602.

Emlen, J. M. 1981. Field estimation of competition intensity. Theor. Pop. Biol. 19: 275-287.

Greenlaw, J. S. 1978. The relation of breeding schedule and clutch size to food supply in the rufous-sided towhee. Condor 80: 24-33.

Heinrich, B. 1979. Foraging strategies of caterpillars, leaf damage and possible predator avoidance strategies. Oecologia 42: 325-327.

Henny, C. J. 1972. An analysis of the population dynamics of selected avian species. USDI Fish and Wildl. Ser. Wild. Research Rep. 1. 99 p. Bureau of Sport Fisheries and Wildlife, Washington, D. C.

Holmes, R. T., and S. K. Robinson. 1981. Tree species preferences of foraging insectivorous birds in a northern hardwoods forest. Oecologia 48: 31-35.

______, J. C. Schultz, and P. Nothnagle. 1979. Bird predation on forest insects: an exclosure experiment. Sci. 206: 462-463.

______, T. W. Sherry, and S. E. Bennett. 1978. Diurnal and individual variability in the foraging behavior of American Redstarts (Setophaga ruticilla). Oecologia 36: 141-149.

______, and F. W. Sturges. 1973. Annual energy expenditure by the avifauna of a northern hardwoods ecosystem. Oikos 24: 24-29.

Krebs, J. R. 1973. Behavioral aspects of predation. Perspectives in Ethology 1: 73-111.

Lotka, A. J. 1925. Elements of physical biology. 465 p. Baltimore, Williams and Wilkins.

Martin, A. C., H. S. Zim, and A. L. Nelson. 1961. American wildlife and plants. 500 p. New York, Dover Publications.

McArthur, L. B. 1980. The impact of various forest management practices on passerine community structure. Ph.D. Dissertation. West Virginia University.

McNamee, P. J., J. M. McLeod, and C. S. Holling. 1981. The structure and behavior of defoliating insect/forest systems. Res. Pop. Ecol. 23: 280-298.

Otvos, I. S. 1979. The effects of insectivorous bird activities in forest ecosystems: an evaluation. Pp. 341-374 in The role of insectivorous birds in forest ecosystems (Dickson, J. G., et al., eds.). 381 p. New York, Academic Press.

Price, P. W. 1975. Insect ecology. 514 p. New York, John Wiley & Sons.

Pulliam, H. R. 1976. The principle of optimal behavior and the theory of communities. Perspectives in Ethology 2: 311-332.

Real, L. A. 1977. The kinetics of functional response. Amer. Natur. 111: 289-300.

. 1979. Ecological determinants of functional response. Ecology 60: 481-485.

Rubtsov, V. V., and J. A. Shvytov. 1980. Model of the dynamics of the density of forest leaf-eating insects. Ecol. Modelling 8: 39-47.

Schaffer, W. M. 1981. Ecological abstraction: the consequences of reduced dimensionality in ecological models. Ecol. Monogr. 51: 383-401.

Solomon, M. E., and D. M. Glen. 1979. Prey density and rates of predation by tits (Parus spp.) on larvae of codling moth (Cydia pomonella) under bark. J. Appl. Ecol. 16: 49-59.

Sturges, F. W., R. T. Holmes, and G. E. Likens. 1974. The role of birds in nutrient cycling in a northern hardwoods ecosystem. Ecology 55: 149-155.

Walbauer, G. P. 1968. The consumption and utilization of food by insects. Adv. Insect Physiol. 5: 229-288.

Wiegert, R. G. 1979. Population models: experimental tools for analysis of ecosystems. Pp. 233-279 in Analysis of ecological systems (Horn, D. J., R. D. Mitchell, and G. R. Stairs, eds.). 312 p. Columbus, Ohio State University Press.

IMPACTS OF COTTAGE DEVELOPMENT ON WILDLIFE IN SOUTHCENTRAL ONTARIO: THE LAKELIFE MODEL[1]

D. Euler, J. McDonnell, T. Clark, K. Clark, G. Racey[2]

Abstract.--Human recreational activity, including cottage construction, has a signficant impact on wildlife and wildlife habitat along lakeshores in central Ontario. In an effort to better understand and manage this activity a model was developed to simulate these impacts.

The model is concerned with: 19 passerine bird species, 7 small mammals, the common loon, mink, white-tailed deer, 2 hawks, and selected plant species. The model describes changes in wildlife habitat under specified conditions of cottage development and provides recommendations which are intended to minimize detrimental impacts on wildlife.

The model was tested by using an existing development, assuming it was a planning proposal, then using the model to suggest what changes would be expected. By measuring the actual impacts in the field and comparing them to the model's prediction an estimate of the error rate was made.

INTRODUCTION

The lake country of southcentral Ontario is an important recreation area for people who live in metropolitan Toronto and surrounding suburbs. Recreation activities are often centred around a cottage, usually built near the water, from which a wide variety of outdoor recreation occurs. During recent years, concern has been raised about the economic and ecological impact of this activity (Graham 1967). Areas most affected are lakeshores and waterways on which 83 percent of cottages are located (Ontario 1968).

Shoreline cottage development directly alters vegetation through road, powerline and building construction as well as through maintenance activities like cutting trees and shrubs. The extent to which these activities occur depends on many socio-economic variables and aesthetic values (Graham 1967, Sargent and Zayac 1977). Their biological impact includes the alteration of ecological sensitive areas, destruction of fish and wildlife habitat and nutrient imbalance in lakes. Aesthetic values of the landscape are also changed by the presence of man-made structures and clearings (Abelson 1979) and by alterations made to the vegetation (Klukas and Duncan 1967).

This project was initiated to study how wildlife populations are affected by cottage development. The purpose was to produce a model which would permit planners and other decision-makers to evaluate the impacts of cottage development on wildlife populations and habitat. Development of the model required 5 years of field research under the auspices of the Lakeshore Capacity Study, an inter-ministerial government task force delegated to study the problem. A complete listing of the Fortran coding, plus a user's manual, is available from the senior author.

[1]Paper presented at the Third International Conference on State-of-the-Art in Ecological Modelling, Colorado State University, May 24-28, 1982.

[2]The authors are habitat biologists at Wildlife Branch, Ontario Ministry of Natural Resources, Queens Park, Toronto, Ontario, Canada, M7A 1W3

STUDY AREA

The study area, called Region of Muskoka and Region of Haliburton (79°N 45°W) lies within the Georgian Bay-Ottawa Valley physiographic region of Ontario where the bedrock of the Precambrian Shield exerts a strong influence on the landscape (Chapman 1975) (fig. 1). Ancient igneous and metamorphic rock formations have undergone extensive folding and exposed surfaces exhibit the results of long-term differential erosion. These processes have created numerous rock basins and troughs now occupied by swamps and lakes. About one-third of the region has a very thin layer of glacial till interspersed with large areas of exposed rock. The rest of the area is covered by varying depths of sandy, stony till.

The climate is cool and moderately moist. In the southern regions of the study area, the mean annual temperature is 5.5°C, while in the northern portions it is 4.4°C. The town of Dorset, near the centre has an average of 107 frost-free days per year. Annual precipitation ranges from 104 cm on the western area facing Georgian Bay to 81 cm in the east.

The vegetation characteristic of the area is the Great Lakes - St. Lawrence Tolerent Hardwood Forest (Rowe 1972). Eastern white pine (*Pinus strobus*), red pine (P. *resinosa*), eastern hemlock (*Tsuga canadensis*), yellow birch (*Betula alleghaniensis*), sugar maple (*Acer saccarum*), red maple (A. *rubrum*),and red oak (*Quercus rubra*) are the major tree species.

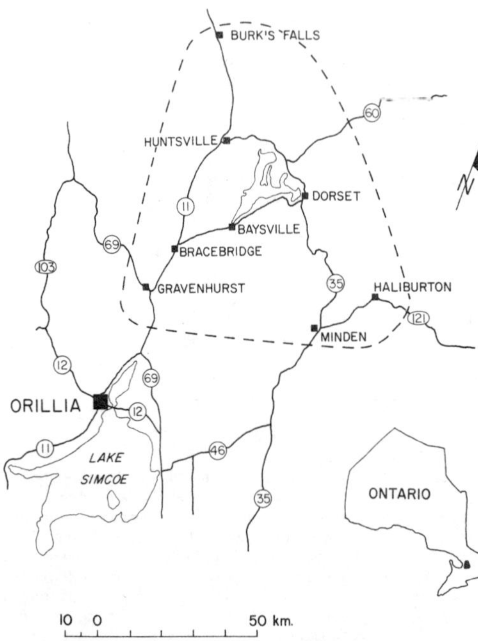

Figure 1.--The study area in Ontario.

METHODS

Wildlife Community

For the purpose of this study, the shoreline wildlife community is within 100 m of the water. Wildlife refers to all species of non-domesticated, living organisms, including mammals, birds, reptiles, amphibians, and terrestrial vegetation but excluding humans, aquatic organisms, and invertebrates. From this wildlife community certain species or groups of species were selected for study. Research concentrated on songbirds, common loons (*Gavia immer*), hawks (*Buteo* spp.), small mammals, mink (*Mustela vison*), deer (*Odocoileus virginianus*), and terrestrial vegetation. These were chosen because they were present in sufficient numbers to study; they were representative of major components of the wildlife community; and they represented the wide range of values which people place on wildlife.

Cottage Development

Cottage development is the construction of cottages and associated developments such as roads, transmission line corridors, boathouses, beaches, docks, and weirs. Cottagers usually clear portions of the shoreline for lawns, patios, and beach areas. Vegetation in the littoral zone of the lake is also altered to facilitate boating and swimming. These development-related changes were studied extensively. Clearly, other cottager activities affect wildlife populations; however, it was not possible to make detailed investigations of these problems. Research concentrated on ribbon shoreline development where a lake was ringed, partially or completely, by a single tier of cottage lots, each having its own shoreline frontage (Euler et al. 1981).

Experimental Design (Data Collection)

In order to establish which habitat conditions were important in the absence of development, wildlife habitat and specific evidence of wildlife activity were studied. For example, birds' nests, mink dens and deer winter shelter areas were identified on lakeshores where little or no cottage development existed. It was then necessary to measure how cottage development changed important habitat conditions. We accomplished this by locating study areas as similar as possible in all respects except for a range of intensities of cottage development. We measured and examined how various density development changed the habitat the animals needed and in some cases how the populations were affected. Details of this procedure are in several technical reports from the study. (Armstrong et al. 1981, Armstrong and Euler 1981, Clark et al. 1981a, 1981b, Clark and Euler 1982, Heimburger et al. 1981, Racey and Euler 1981a, 1981b, 1982).

Modelling Approach

As is often the situation, the modelling approach was dictated by management objectives. In this project the objectives were:

(1) Wildlife populations on a lake must be maintained in self-sustaining communities that are as similar as possible to undisturbed shoreline communities, on some portions of all lakes but not everywhere on all lakes.

(2) No species can be extirpated from the shoreline community of any lake.

The values implied in these statements are not the subject of this paper, but are dealt with in Euler et al. (1981). The objective of our model was to ensure that the above management objectives were met by providing local planners with a tool for predicting impacts.

Bounds

The above objectives set the lake as the fundamental planning unit for the model. Wildlife habitat was studied within 100 m of shore because cottage impact is concentrated within that band. In order to divide this unit into meaningful sub-units of wildlife habitat, the shoreline was segmented into 100 m lengths. This provided approximately 1 ha segments of shoreline which were useful for modelling wildlife habitat.

This model was intended for use by individuals new to wildlife concerns and modelling in general. For that reason, segments around lakes were sequentially numbered so that a good record of input data would be kept. This explicit system, rather than a more obtuse implicit system, also provided more flexibility in proposing hypotheses about spatial relationships between habitat types and allows the evolution of a more sophisticated model, if necessary.

In setting the temporal bounds of our model we opted for the simplest possible situation, that is a before and after type of prediction. There were two reasons for this. First, it was felt that planners and decision makers in government would be more willing to deal with a simple approach. Second, an approach using a smaller time step of development, such as a yearly progression of cottage building, would either be very time consuming to construct because of the knowledge needed to build a model, or would be too hypothetical or general for practical use (Levins 1966)

Intermediate Variables

Development is a complex process and to measure and quantify this complexity, we devised a number of indices. Although indices are disadvantageous because they often mask the cause and effect relationship present, they are an effective way of integrating numerous complex functions into a manageable tool (Pielou 1974, Holling 1978).

An Area Development Index (ADI), based on disturbance of vegetation in three layers, ground, shrub, and tree, was calculated as follows:

$$ADI = (Ag + As + At)/AREA \qquad (1)$$

where,
Ag is the area disturbed (ha) in the ground layer
As is the area disturbed (ha) in the shrub layer
At is the area disturbed (ha) in the tree layer
AREA is the size (ha) of the cottage lot (Racey and Euler 1981a)

An area was considered disturbed if a building or other man-made object occupied it, or if human activity had halted natural succession and/or returned the area to an earlier successional stage. When the vegetation of a cottage lot or a sample plot was entirely disturbed, the ADI had a value of 3.0. The ADI was used to measure the effects of cottage development on small mammals and songbirds.

One component of the ADI, the Ground Development Index (GDI), was derived from the formula:

$$GDI = Ag/AREA \qquad (2)$$

where, Ag is the area disturbed (ha) in the ground layer. there was a relationship between this index and the songbird community composition (CC) (Clark et al. 1981).

An estimate of human disturbance, the Loon Development Index (LDI), was used to quantify the relationship between human activity around a common loon nest and reproductive success. This estimate was calculated as the number of cottages within a 250 m radius of the nest (LDI 250) (Heimburger et al. 1981).

Indicator Variables

On the basis of the field research, several indicator variables were developed which index the change in wildlife habitat as development increases. These variables were used in the model to express how habitat changes initiate a change in the wildlife community. Criteria are available to enable users of the model to decide whether changes in magnitude of indicator variables are acceptable.

Songbirds.--For songbirds the unit of measurement was a 1 ha plot on which territories and nests had been studied. The model expresses the ha of songbird habitat, for 19 indicator species, both before and after development.

An index, Coefficient of Community (CC) based on Jaccard (1932), was used to quantify differences in the songbird community in areas of different intensities of development.

$$CC = (c/(a+b-c)) \times 100 \qquad (3)$$

where,

a = number of species in the most highly developed segment
b = number of species at the segment being measured
c = number of species common to both

In the model, CC is derived from its relationship to GDI (Clark et al. 1981).

Loons and Hawks.--During the course of field work it was determined that the most significant impact of development on loons and hawks was at nesting sites. Both animals have reasonably specific nest site requirements, which, if lost, mean a long-term detriment to the population. The indicator value for each (loons: Loon Nesting Disturbance Value; hawks: Hawk Nesting Disturbance Value) is the number of nest sites available which would remain after development.

Mink.--The habitat type (coniferous, mixed, deciduous) and intensity of development are used to estimate a relative measure of the potential for mink activity before and after development, called the Lake Mink Value.

The Cottager Loading Development Index (CLDI) was used to reflect both the intensity of human use of the lake surface and the distribution of cottages on its perimeter. The CLDI was given by:

$$CLDI = \log (C \times D / A \times U) \qquad (4)$$

where,

C = the number of cottages on the lake
U = the length of the undeveloped shoreline (m)
A = the surface area of the lake (ha)
D = the length of the developed shoreline (m)
(Racey and Euler 1981b)

As the CLDI increases, mink habitat quality is lowered.

Deer.--White-tailed deer require coniferous areas with overhead cover for protection from wind, deep snow, and low temperatures. Cottage developments built in these areas reduce the quality of habitat for deer in winter. The model calculates the amount of this habitat which would remain after development.

Small mammals.--The impact of development on small mammals is expressed in terms of a capture index which is the probability of capturing a given species of small mammal in a habitat type. It is presented for before and after development.

Controlling The Simulation

Development indices had to be related to development in a way meaningful to planners and decision makers. Basically there are two controls which a planner has over the model. These are cottage location and cottage density. Both of these controls relate, through the intermediate variables to the indicator variables previously described (table 1). These simulation controls provide a simple means of relating development information as provided by a subdivision plan to wildlife habitat impacts on lake shorelines.

Test Of The Model

Early in the summer of 1981, 8 lakes in southern Haliburton were selected to test the predictive ability of the model. An existing cottage development was used to compare the predictions made by LAKELIFE.

The shoreline of each lake was divided into consecutive segments of 100 m in length and 100 m in depth. A total of 93 segments, representing a range of development intensities and vegetation, was selected for study.

Songbirds were censused on these segments using the transect method of censusing breeding birds. Small mammal populations were studied by livetrapping in July and August. In order to test the relationship between ADI and cottage lot size, the amount of clearing on 59 of the 93 segments was measured and compared to the value of ADI predicted by the model for each segment (Clark et al. 1981, Fast et al. 1982, McDonnell and Euler 1982).

Table 1.--Summary of variable relationships for the simulation.

Species or Species Group	User Controls	Intermediate Variables	Wildlife Indicator Variables
Small mammals	Cottage lot size or density	ADI	Capture Index
Songbirds	Cottage lot size	GDI	Area of habitat (ha); Coefficient of Community
Loon	Cottage location	No. of nearby cottages (LDI 250)	Loon Nesting Disturbance Value
Hawk	Cottage location	Presence of cottages	Hawk Nesting Disturbance
Deer	Cottage location	Presence of cottages	Lake Deer Yard Value
Mink	Cottage location and lot size	Mink usage (scats/ segment)	Lake Mink Value
		Length of disturbed & undisturbed shoreline	Cottager loading Development Index

RESULTS

Model Development

The ADI exhibited a significantly negative correlation with the area of the cottage lot ($r = 0.875$, $P < 0.01$). The best-fitting equation for this relationship was nonlinear (figure 2)

The important finding was that the amount of disturbance was relatively constant regardless of the size of lot (Racey and Euler 1981a). Cottagers tend to clear about the same amount of land for a building and for activities near the beach on large lots as on small ones. A large number of small lots on a lakeshore has significantly more impact than a few large ones, although the same area is involved.

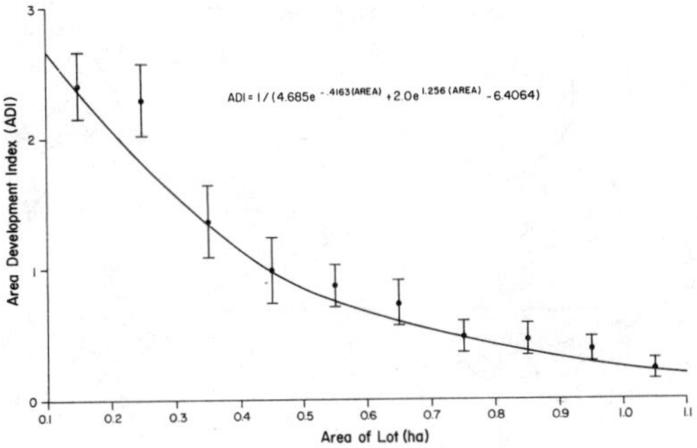

Figure 2.--Relationship between the Area Development Index (ADI) and the area of a cottage lot(Racey and Euler 1981a).

Songbirds

A definite change in the songbird community occurred as development increased (Clark et al. 1981). The ovenbird (Seiurus aurocapillus), Canada warbler (Wilsonia canadensis), winter wren (Troglodytes troglodytes), and black-throated green warbler (Dendroica virens) are examples of the group of birds associated with mature forests which are intolerant of habitat disturbance. Species found at high levels of development in this study included the robin (Turdus migratorius), eastern phoebe (Sayornis phoebe), and cowbird (Molothrus ater). These species were found in open habitats created by cottage development similar to the habitat type where they are found in other parts of their range (Howell 1942, Robertson and Flood 1980). A third group of birds relatively indifferent to development included the red-eyed vireo (Vireo olivaceus) and veery (Hylocichla fuscescens). These species are often associated with immature forest habitat (Webb et al. 1977, Possardt and Dodge 1978), and were found in undeveloped habitats where deadfalls had created openings in the canopy. These birds can also be found at the edges of cottage clearings and in cleared areas which have been allowed to revert to secondary succession.

Coefficient of Community was used to measure the degree to which the species composition on a specific plot resembled that of the most highly disturbed plot. To examine coniferous habitats, all plots were measured against a plot with a Ground Development Index (GDI) of 0.87; for deciduous habitats, a plot with a GDI of 0.51. Smaller values of CC indicated a species composition different from that of a developed habitat; larger values indicated a species composition similar to that found on the most highly developed plot. A linear regression analysis was used to express the relationship between CC and the GDI (fig. 3).

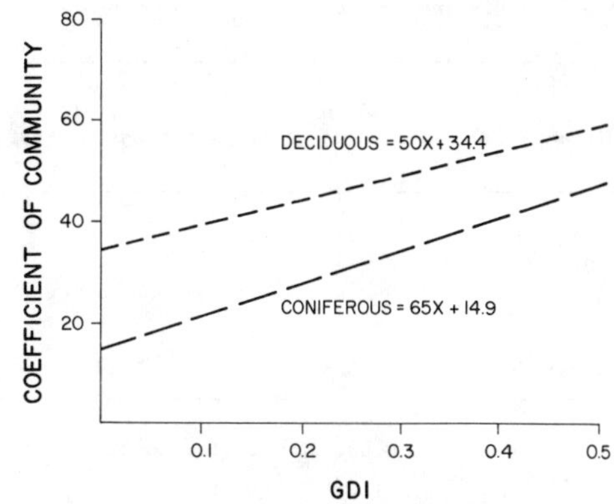

Figure 3.-- The relationship of the Ground Development Index (GDI) and Coefficient of Community for songbirds nesting near lakeshores in Ontario (Clark et al. 1981b).

Loons

Successful loon nests were located further from development than would have been caused by chance alone (table 2) and hatching success declined as cottage density increased (table 3) (Heimburger et al. 1981). This occurred because loons nest at the shore-water interface and are especially vulnerable to human activity on or near the shore. The nest is often quite visible and neonatal young are vulnerable to human activity such as boating.

Hawks

Both raptors we studied frequented deciduous, upland forests near riparian or lakeshore habitat. The broad-winged hawk (Buteo platypterus) was more commonly found in partially open forests like those along roadways; the red-shouldered (B. lineatus) was almost totally restricted to closed forest. There were no significant differences in general nesting macrohabitat. Conditions around the nest site included location on a slope near a natural clearing, close to a water source, and in a predominantly deciduous forest. Broad-winged hawk nests in the mid-west and nests of both species in the eastern U.S. show similar relationships with clearings (Keran 1978, Titus and Mosher 1981).

The results suggest that cottage development may not seriously disrupt broad-winged hawk breeding habitat. The red-shouldered hawk, however, did not respond positively to cottage development (Armstrong and Euler 1981). The nesting success rate of the red-shouldered hawk is directly proportional to the distance between adjacent sites (Heney et al. 1973) and the presence of permanent or seasonal water sources (Wiley 1975). This suggests that if available breeding habitat is lost to cottage development, the response would be a reduction in breeding population or reproductive rate.

Table 2.-- Distance of cottages from loon nests.*

Distance (m)	No. of Successful Nests	No. of Unsuccessful Nests
0 - 200	27	29
201 - 400	8	14
401 - 600	8	5
600	9	1
Total	52	49

* $X^2 = 8.76$, $p<0.05$, df = 3 (for combined 1977, 1978 and 1979 data)

Table 3.--Cottage density and loon nesting success.

No. of Cottages Within 250 m of the Nest* (LDI 250)	No. of Successful Nests	No. of Unsuccessful Nests
0	32	17
1	9	3
2	5	5
3	2	5
4	1	5
Total	49	35

* $X^2 = 10.11$, $p < 0.05$, df = 4 (for combined 1977, 1978 and 1979 data using first nesting attempts only).

Mink

Mink activity was greatest on coniferous and mixed shorelines with little or no development whereas deciduous shorelines were not used much, regardless of development levels (fig. 4).

A significant ($p < 0.05$) negative regression between mink density (number of scats/unit effort of search) and CLDI was established with data from seven lakes which were completely searched in 1978 (fig. 5). The decrease in activity was probably due to the reduction of available food, hunting locations or suitable denning sites.

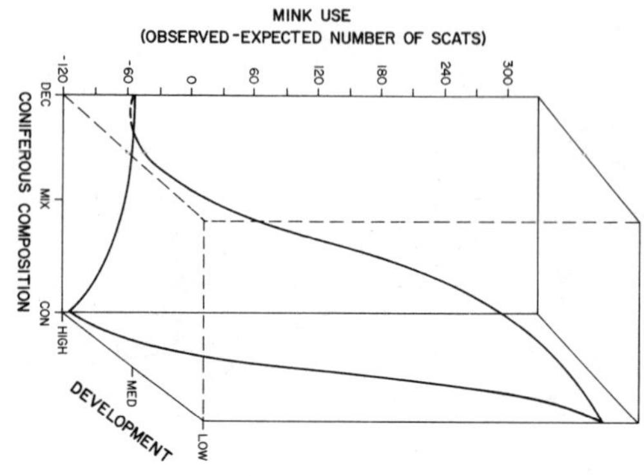

Figure 4.--Mink use of lakeshores in central Ontario as related to conifer composition and cottage development activity, (Racey and Euler 1981b).

Deer

When cottages are built in winter concentration areas of deer, the quality of deer habitat is reduced. The change occurs because cottagers tend to reduce over-head cover and food used by the animals (fig. 6) (Armstrong et al. 1981). Although careful planning of developments may alleviate this problem, the best way to preserve deer winter concentration areas is to prevent development in these areas.

Small Mammals

Best-fitting prediction equations related the capture index of seven species of small mammals to the ADI in each habitat type. The number of masked shrews (Sorex cinereus) decreased rapidly on deciduous and mixed shorelines with small increases in the ADI. Likewise the number of red-backed voles (Clethrionomys gapperi) decreased on coniferous shorelines and woodland jumping mice (Napaeozapus insignis) decreased on deciduous shorelines. Significant linear declines in number were observed for woodland jumping mice on conferous shorelines, shorelines, for red-backed voles on mixed shorelines, and for deer mice (Peromyscus maniculatus) in all three habitat types. There were no captures of woodland jumping mice on mixed shorelines with development greater than 0.0, and no prediction equation was estimated. Short-tailed shrews (Blarina brevicauda) and chipmunks (Tamias striatus) both showed a linear increase in development on deciduous shorelines (Racey and Euler 1982).

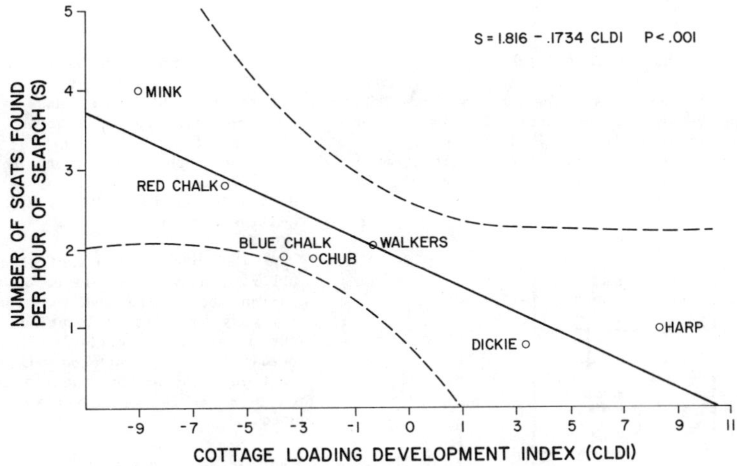

Figure 5.--The relationship between mink activity on seven lakes in Ontario and the intensity of cottage development (Racey and Euler 1981b).

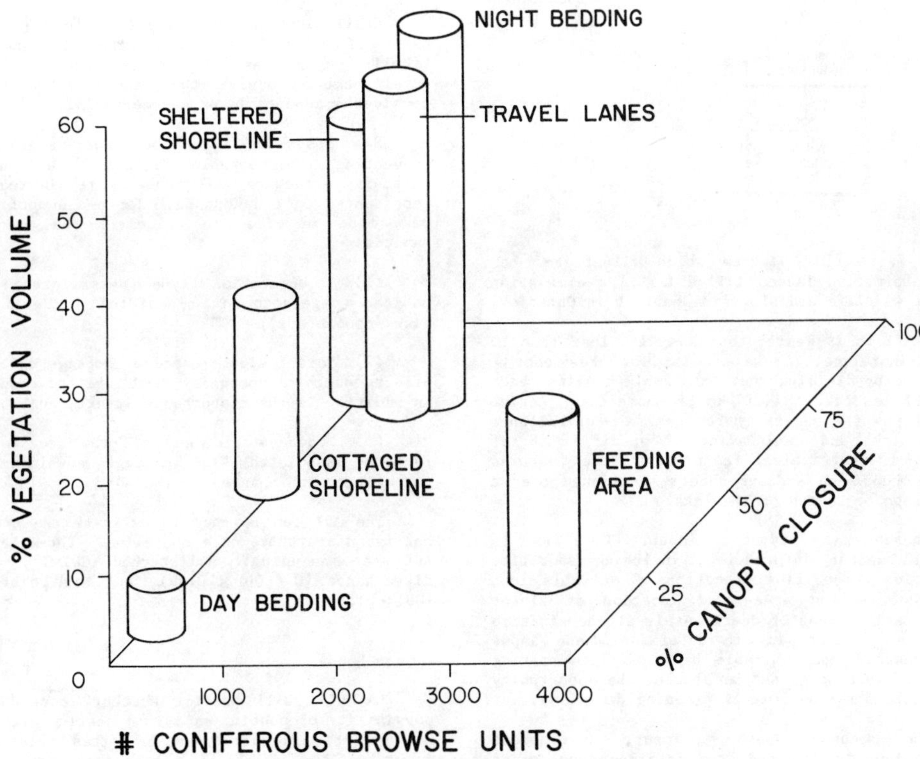

Figure 6.--Cottage development changes key components of deer winter shelter in Ontario, (Armstrong et al. 1981).

The LAKELIFE Model

Data collected from the field and the relationships between development and changes in wildlife habitat were developed into a simulation model, LAKELIFE. A flow diagram which illustrates how information is integrated in the model, is shown in figure 7.

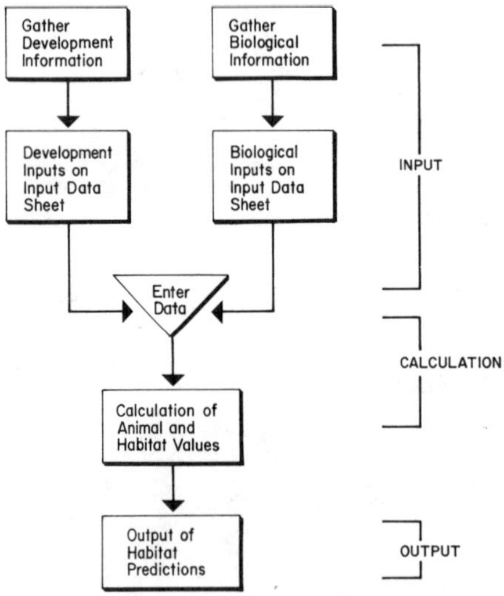

Figure 7.--A flow diagram which illustrates how information isused in the LAKELIFE simulation of wildlife and shoreline habitat in Ontario.

LAKELIFE is geared to lakes with less than 10 km of shoreline. However, larger lakes can be evaluated by dividing them into smaller units, each of which has less than 10 km of shoreline. Aerial photographs are necessary for general surveillance of the lake and delineation of specific habitat types and topographical features. These features may be marked on acetate sheets positioned over a photograph or sketch of the lake.

There are two types of input. The first is general habitat information, including shoreline vegetation composition, location of suitable loon and hawk nesting areas and location of winter concentration areas of deer. At least one visit to the site is necessary to determine these input requirments (e.g., suitable hawk and loon nesting sites). This also gives the planner the opportunity to visualize the outcome of planning decisions.

The second type of input, development information, is obtained from detailed development proposals. If the lake has been partially developed, the plans for existing cottage lots are needed. Locations and sizes of both developed and proposed lots are needed.

The intermediate variables (table 1) (ADI, GDI, LDI 250, etc.) are derived by the model from information about size and location of lots. The user may vary lot size or location, which in turn changes the intermediate variables and thus influences the prediction of impact on wildlife. The following sections discuss the model's derivation of individual species indicator variables and the assumptions used.

The intermediate variables (table 1) (ADI, GDI, LDI 250, etc.) are derived by the model from information about size and location of lots. The user may vary lot size or location, which in turn changes the intermediate variables and thus influences the prediction of impact on wildlife. The following sections discuss the model's derivation of individual species indicator variables and the assumptions used.

Songbirds

The overall change in the structure of the songbird community, rather than the fate of individual bird species, is the important concern.

Coefficient of Community (CC) is a good indicator of lakewide impact on the bird community. Lakelife estimates CC both before and after development and advises the planner if the change is considered acceptable or unacceptable.

By assessing 19 indicator species and preventing unacceptable losses, the original community structure will probably remain reasonably represented on a lakeshore. Three assumptions have been made in order to calculate the changes in habitat:

(1) vegetation type determines songbird habitat (vegetation is classified as deciduous or mixed/coniferous);

(2) forest edge created by cottage development affects songbird community structure and composition in addition to the disturbance itself, and;

(3) changes in edge and other environmental factors associated with cottage development are correlated with changes in the ADI.

The ADI can be used to estimate the extent of habitat disturbance on a shoreline. Three levels of ADI are meaningful: undisturbed (GDI=0), moderate disturbance ($0 < GDI < 0.35$), and high disturbance ($GDI > 0.35$).

Common Loons

LAKELIFE will predict whether loons have the possibility of nesting on island nesting sites or on one of the shoreline nesting sites requested as input to the model. If there are loon nesting islands, it is assumed that the loons will nest there. A loon nesting island is an island less than 0.5 ha located more than 50 m from the shoreline. An island of this size which is less than 50 ha from shore is considered a shoreline nesting site. If there is no human activity on the lake surface, it is assumed that any nesting attempts on islands would be successful (Heimburger et al. 1981, Racey et al. 1981).

The Loon Nesting Disturbance Value is equivalent to the number of potential onshore loon nesting sites (less than 50 m from shore) which would remain once development has taken place, added to the number of offshore loon nesting sites (greater than 50 m from shore).

Hawks

Generally, suitable broad-winged and red-shouldered hawk nesting sites are valleys dominated by yellow birch or maple with a small stream present. The forest should be at least 70 years old. If construction is planned for these areas, disruption of the birds is nearly inevitable. LAKELIFE provides for a Hawk Nest Disturbance Value which can range from 0 up to the number of sites available on the lake for hawks to nest in. The number of potential hawk nesting sites which remain once development has taken place is noted and available for the decision-maker to consider.

Mink

The two indicator variables for mink are based on the relationship between development, habitat composition and mink use of the shoreline as illustrated in Figure 4. The Lake Mink Value is based on actual observations of mink usage as derived from scat collection. We assumed that scats represented approximately the habitat usage by mink. Similarly the Cottager Loading Development Index is based on development variables (lengths of shoreline) which were observed to be correlated with presence of scats. Therefore the assumption is the same as for Lake Mink Value.

White-tailed Deer

Coniferous areas on a lakeshore contained within winter concentration areas provide continuous overhead and vertical cover which is needed for protection from wind, deep snow, and low temperatures. The user provides LAKELIFE with the location of these areas on a lake. LAKELIFE records the number of these segments which would remain free from development.

Small Mammals

Capture indices indicate the quality and quantity of small mammal habitat along the lake shoreline. The computer calculates these values (pre- and post-development) and the percent difference by using a two-step procedure. The first step is to estimate the changes in vegetation with development based on the ADI. In the second step, multiple regression equations are used to calculate the capture indices using the vegetation parameters from the first step, as independent variables. Different equations are used for each of the three habitat types (deciduous, coniferous, and mixed). Large index values indicate good habitat and therefore higher wildlife utilization.

Development Guidelines

The following guidelines for developments, if followed, will help the developer meet the management objectives outlined at the beginning of this paper. In most cases some cottage locations will need to be changed and lot sizes may need to be adjusted, but the entire development will seldom be prevented.

If capture index values for small mammals decrease by more than 75 percent, the habitat disturbance to this community would be considered excessive. If the Loon Nesting Disturbance Value is less than 2.0 this implies that in four years out of 10, breeding loons will have a successful nest (i.e., a nest that fledges at least one young). This is the minimum acceptable value of the Loon Nesting Disturbance Value. If the Hawk Nesting Value is not less than 2.0 there will be at least two nesting sites available for broad-winged hawks and this too is acceptable. The Lake Mink Value after development should not be permitted to fall below one-third of the Lake Mink Value before development. The Cottager Loading Development Index (CLDI) also should not exceed 1.0. At this level, mink populations will probably be sustained. At values greater than 1.0 this may not be possible. If there are any differences in the Lake Deer Yard Value before and after development, the development in that area would be unacceptable, unless the total shoreline length reserved is more than 5 percent. If more than 5 percent of the shoreline is acceptable deer winter shelter areas, then a separate decision is needed to resolve this problem. The output for songbirds is given in two parts. The first table shows the change in songbird Coefficient of Community. A development is unacceptable if more than 75% of the segments on a lake are altered. The second table shows available habitat for' breeding pairs before and after development. A decline greater than 75% in the amount of habitat available for breeding songbirds is unacceptable.

An example of how the model works is given in Appendix I.

Test of the Model

We tested the model by going to the southern extreme of the area where it was developed. The strategy was to attempt to invalidate the model in an effort to understand better how the model works (Holling 1978).

Our assessment of the test is as follows. The model accurately predicted the habitat and abundance of short-tail shrews, chipmunks, and red squirrels The data for deer mice indicate that the model was reasonably accurate in predicting the direction of change but the magnitude of the changes were inaccurate; more deer mice were caught in our study than was predicted, based on the work by Racey and Euler (1982). As a predictor of trends; however, the model is satisfactory for deer mice.

Only two masked shrews were captured in the test area where these shrews are likely to be abundant, and since the model predicted the capture of more, we must assume the model is incorrect. It appears that the model predicted the magnitude of changes accurately for redback voles in some cases, but on the whole, the direction of changes in relation to development were poorly predicted. The model's predictions for woodland jumping mice were unsatisfactory (McDonnell and Euler 1982).

In about half of the tests for songbirds (19 species tested for 6 habitat/development types) the model's prediction was significantly different from what occurred on the test site. In half of these problems the model predicted certain bird species but in the test they did not occur. The other half of the problems were caused because species were not predicted but the field test found them. Much of the error occurred with 4 species of birds: cowbird, robin, song sparrow and chipping sparrow. These species are typically adventicious and tend to exploit a wide variety of habitats.

Bird distribution in relation to climate and geography will, of course, affect the model's accuracy in different locations. For instance, the winter wren and Canada warbler are at the limit of their southern distribution in the test area, but are more abundant in the area used in the model's development. The amount of error that appears to stem from the smaller more southern test area suggests that calibration values for songbird habitat could fine-tune the model, if suitable information is available for the area in question (Fast et al. 1982).

Regressions comparing disturbance predicted by the model, based on lot size information, and disturbance actually present in the field, showed that the two were similar except at lower values of ADI. We concluded that the model did not accurately predict ADI in areas of low disturbance (Clark et al. 1982) but is satisfactory at higher levels.

DISCUSSION

LAKELIFE is a deterministic bookkeeping simulation model with application to real world management of vegetation and cottage development. It is relatively simple because it uses only a few basic indices of vegetation and development characteristics. It has a management orientation because it was constructed in response to a need by government to control development or face losing some important ecological and aesthetic values. In comparison to other models it is not complex and does not contain stochastic elements.

We tested the model's predictive ability by comparing real values with predicted values, in a subsequent step, near the limits of the models predictive range. As a result we were able to identify weak areas for further development and to understand where the model worked well. An important goal of modelling is to insure that the decision-making process is disciplined. Often decisions concerning natural resource management are made in a context lacking disciplined thought patterns. Pressing political concerns, lack of data, inherent variability of biological and ecological data, and emotional prejudices of decision-makers all mitigate against making rational decisions. Instead decisions are often made without adequate thought given to some of the more subtle implications. For example, the link between mining nickel in Sudbury, Ontario and the possible loss of red-shouldered hawks because acid rain may kill an important food source (amphibians) is obscure.

The major value of LAKELIFE is that it provides a device to record information in a disciplined way, shows how decisions are made, and allows decision-makers to assess the correctness of decisions in subsequent years. We found the model retained reasonable validity even when tested near the extreme of its developmental area.

Like most human enterprises, managing natural resources is fraught with error. LAKELIFE is not the final solution to error-free decisions. It is instead a first step in a long process of developing better management of natural resources.

LITERATURE CITED

Abelson, P.W. 1979. Property prices and the value of amenities. J. Environ. Econ. and Manage. 6: 11-28.

Armstrong, E., and D. Euler. 1981. Reproductive ecology and habitat usage of wood land buteos in central Ontario. Tech. Rep. No. 2. Lakeshore Capacity Study Wildlife Branch, Ontario Ministry of Natural Resources, Toronto, Ontario M7A 1W3.

Armstrong, E., G. Racey and D. Euler. 1981. Deer habitat quality and influences of cottage development in Central Ontario. Tech. Rep. No. 3. Lakeshore Capacity Study. Wildlife Branch. Ontario Ministry of Natural Resources, Toronto, Ontario. M7A 1W3.

Chapman, L.J. 1975. Physiography of the Georgian Bay-Ottawa Valley area of Southern Ontario. Ontario Ministry of Natural Resources., Div. Mines, Geosci. Rep. 128, Queens Park, Toronto. M7A 1W3

Clark, K., D. Euler and E. Armstrong. 1981a. Habitat associations of breeding birds in cottaged and uncottaged areas in central Ontario. Tech. Rep. No. 5. Lakeshore Capacity Study. Wildlife Branch, Ontario Ministry of Natural Resources, Toronto, Ontario. M7A 1W3.

Clark, K., D. Euler and E. Armstrong. 1981b. Predicting avian community response to lakeshore cottage development. Tech. Rep. No. 4. Lakeshore Capacity Study. Wildlife Branch, Ontario Ministry of Natural Resources, Toronto, Ontario. M7A 1W3.

Clark, T. and D. Euler. 1982. Vegetation disturbance caused by cottage development in central Ontario. Tech. Rep. No. 13. Lakeshore Capacity Study. Wildlife Branch, Ontario Ministry of Natural Resources, Toronto, Ontario. M7A 1W3.

Clark, T., D. Euler, J. McDonnell and S. Fast. 1982. Field test of a vegetation disturbance index. Tech. Rep. No. 14. Lakeshore Capacity Study. Wildlife Branch, Ontario Ministry of Natural Resources, Toronto, Ontario. M7A 1W3.

Euler, D.L., G. Racey, K. Clark, E. Armstrong, T. Clark, J. McDonnell, M. E. Stoll, 1981. Wildlife Component, Lakeshore Capacity Study, Part 1. Final Report. Wildlife Branch, Ministry Natural Resources, Toronto, Ontario. M7A 1W3.

Graham, W.W. 1967. Cottage development in rural areas. ARDA, Proj. No. 15039, Dept. of Regional Econ. Expan., Ottawa, Ontario.

Heimberger, M., D. Euler and J. Barr. 1981. The impact of cottage development on common loon reproductive success in central Ontario. Tech. Rep. No. 9. Lakeshore Capacity Study. Wildlife Branch, Ontario Ministry of Natural Resources, Toronto, Ontario. M7A 1W3.

Henney, C.J., F.C. Schmid, E.M. Martin and L.L. Hood. 1973. Territorial behaviour, pesticides and the population ecology of Red-shouldered Hawks in central Maryland, 1943-1971. Ecology 54: 545-554.

Holling, C.S., editor. 1978. Adaptive Environmental Assessment and Management. Wiley, Toronto. 377 pp.

Howell, J.C. 1942. Notes on the nesting habits of the American robin. Am. Midl. Nat. 28: 529-603.

Jaccard, P. 1932. Die statistical-floristische Methoden als Grundlage dev. Pflanzen-sociologie. Itanbuch der Biologischen Arbeitsethoden 5: 165-202.

Keran, D., 1978. Nest site selection by the broad-winged hawk in north central Minnesota and Wisconsin. Raptor Research 12: 15-20.

Klukas, R.W. and D.P. Duncan. 1967. Vegetation preferences among Ithasca Park visitors. J. For. 65: 18-21.

Levins, R. 1966. The strategy of model building in population biology. Amer. Sci. 54: 421-431.

McDonnell, J.A., and D. Euler. 1982. A field test of a Computer Model: small mammals. Tech. Rep. No. 15. Lakeshore Capacity Study, Wildlife Branch, Ontario Ministry of Natural Resources, Toronto, Ontario. M7A 1W3.

Ontario. 1968. Analysis of Ontario cottage survey. Ont. Dept. Tourism and Information, Travel Res. Rep. 55, Queens Park, Toronto, Ontario. M7A 1W3.

Pielou, E.C. 1974. Population and community ecology; princples and methods. Gordon and Breach Sci. Pub. Toronto. 424 pp.

Possardt, E.E., and W.E. Dodge. 1978. Stream channelization impacts on songbirds and small mammals in Vermont. Wildl. Soc. Bull. 6: 18-24.

Racey, G. and D. Euler. 1981b. Changes in mink habitat and food selection as influenced by cottage development in central Ontario. Tech. Rep. No. 11. Lakeshore Capacity Study. Wildlife Branch, Ontario Ministry of Natural Resources, Toronto, Ontario. M7A 1W3.

Racey, G., T. Clark, J. McDonnell and D. Euler. 1981. Wildlife Component, Lakeshore Capacity Study, Part 2, LAKELIFE User's Manual. Wildlife Branch, Ontario Ministry of Natural Resources, Toronto, Ontario. M7A 1W3.

Racey, G., and D. Euler. 1982. Small mammal and habitat response to shoreline cottage development in central Ontario. In press. Can. J. Zool.

Robertson, R.J., and N. Flood. 1980. Effects of recreational use of shorelines on breeding bird populations, Can. Field-Nat. 94: 131-138.

Rowe, J.S. 1972. Forest regions of Canada. Department of the Environment, Canadian Forest Service. Publ. No. 1300. 172 pp.

Sargent, F.O. and F.J. Zayac. 1977. Index of lake basin land use for land use planning purposes. Water Res. Bull. 13: 365-272.

Smith, K.D. 1977. Distribution of summer birds along a forest moisture gradient in an Ozark watershed. Ecology 58: 810-819.

Titus, K. and J.A. Mosher. 1981. Nest-site habitat selected by woodland hawks in the central Appalachians. Auk 98: 270-281.

Webb, W.L., D.F. Behrend, and B. Saisorn. 1977. Effect of logging on songbird populations in a northern hardwood forest. Wildl. Monogr. 55: 1-35.

Wiley, J.W. 1975. The nesting and reproductive success of red-tailed hawks and red-shouldered hawks in Orange County, California, 1973. Condor 77: 133-139.

APPENDIX I

DICKIE LAKE CASE STUDY

The Dickie Lake Case Study is an actual example of the evaluation of a development, using the LAKELIFE model. This highly developed lake is located 2 km east of the town of Baysville near the centre of the area where the model was developed. Further information about this case study is available from Racey et al. (1981).

Preparation

An enlarged copy (1:5000) of an aerial photograph, taken in a leaf-free snow-free condition, was used to obtain most of the habitat information required for the Dickie Lake case study. The development information from registered developed plan maps provided all the information required to calculate lot size. The shoreline was divided into 77 segments of 100 m each.

Output

The output for this case study illustrates the results of an execution of the model.

Original Input Parameters
Dickie Lake

Number of segments	77
Segment size (m)	100
Total length of shoreline (m)	7700
Length of shoreline developed (m)	7350
Area of the lake (ha)	89.2
No. of cottage lots	112
No. of loon nesting islands	6

	Segment No.
Loon nest site No. 1	30
Loon nest site No. 2	45
Loon nest site No. 3	70
Hawk nest valley No. 1	6
Hawk nest valley No. 2	44
Hawk nest valley No. 3	62

Output of Entire Lake Values
For Small Mammals

Animal	Before	After	% Change
MA Shrew	145.6	34.9	- 76.0
ST Shrew	198.2	339.6	71.4
Chipmunk	695.8	1773.9	154.9
R Squirl	149.0	185.3	24.3
DE Mouse	3428.9	1851.5	- 46.0
R B Vole	3089.4	1030.7	- 66.6
WJ Mouse	410.2	77.7	- 81.1

Lake Values

Loon nesting disturbance value	6.0
Hawk nesting disturbance value	1.0
Lake mink value before development	2163.
Lake mink value after development	706.
Cottager loading development index	1.4
Lake deer year values before development	0.
Lake deer year values after development	0.

Songbird Coefficient of Community

Number of segments altered	66
Number of segments on lake	77
Percent of segments altered	85

Bird Habitat Availability Before & After Development
(Calculated for the entire lake in ha)

Habitat Available Before Development (Hectares)

Bird Species	Total	Un-disturbed (GDI=0)	Some disturbance (0<GDI<0.35)	disturbed (ADI<0.35)
Ovenbird	77	77	0	0
Blthblue	77	77	0	0
Bthgreen	77	77	0	0
Cowbird	0	0	0	0
EW pewee	0	0	0	0
Canwarbl	49	49	0	0
Swthrush	49	49	0	0
Nwthrush	77	77	0	0
Revireo	28	28	0	0
Gcflycat	49	49	0	0
Mournwar	49	49	0	0
Bcreeper	77	77	0	0
Wdthrush	77	77	0	0
Wintwren	49	49	0	0
Robin	0	0	0	0
Phoebe	0	0	0	0
Songspar	0	0	0	0
Yelrumpw	49	49	0	0
Chipspar	0	0	0	0

Habitat Available After Development (Hectares)

Bird Species	Total*		Un-disturbed (GDI=0)	Some disturbance (0<ADI<0.35)	disturbed (GDI<0.35)
Ovenbird	30		10	20	0
Blthblue	10	75%loss	10	0	0
Bthgreen	10	75%loss	10	0	0
Cowbird	71		4	34	33
EW pewee	14		0	0	14
Canwarbl	30		10	20	0
Swthrush	30		10	20	0
Nwthrush	10	75%loss	10	0	0
Revireo	48		0	34	14
Gcflycat	6	75%loss	6	0	0
Mournwar	6	75%loss	6	0	0
Bcreeper	6	75%loss	6	0	0
Wdthrush	6	75%loss	6	0	0
Wintwren	10	75%loss	10	0	0
Robin	53		0	20	33
Phoebe	67		0	34	33
Songspar	47		0	14	33
Yelrumpw	44		10	34	0
Chipspar	48		0	34	14

*The statement - 75% loss - indicates that the wildlife component recommendation about maintaining bird habitat has been violated. Cottage lotsize and/or location must be modified.

Interpretation

Small Mammals

After development, the Capture Index increased in value for three species and decreased for four. Note that a decrease of 75% or more in this Index for any species after development is considered unacceptable. Therefore, the amount of habitat available for masked shrews (*Sorex cinereus*) (-76.0%) and woodland jumping mice (*Napaeozapus insignis*) (-81.1%) has decreased beyond the limits of acceptability.

Loons, Hawks, Mink, Deer

Loons.--The Loon Nesting Disturbance Value of 6.0 is greater than the acceptable value of 2.0, suggesting that adequate loon nesting sites would remain after development.

Hawks.--The Hawk Nesting Disturbance Value of 1.0 was less than the acceptable limit of 2.0. This suggests all potential hawk nesting sites previously identified were encroached upon by cottage development.

Mink.--The Lake Mink Value decreased from 2163 to 706 after development, a decrease of about 67%. According to this criteria, the level of development was marginally acceptable for mink. However, the CLDI has a value of 1.4 which is greater than the acceptable value of 1.0. The acceptability of the development plan will depend upon how the decision-maker makes trade-offs between values.

Deer.--Since no deer yards were found on Dickie Lake, the Lake Deer Yard Value was 0 before and after development.

Songbirds

LAKELIFE predicted an 85% change in the Coefficient of Community, which exceeds the acceptable limit of 75%. LAKELIFE also predicted an increase in available habitat for seven species: cowbird, eastern wood pewee, red-eyed vireo, robin, eastern phoebe, song sparrow, and chipping sparrow, and a decrease in available habitat for the other 12 species. The black-throated blue warbler, black-throated green warbler, northern waterthrush, great crested flycatcher, mourning warbler, brown creeper, wood thrush, and winter wren would experience an unacceptable loss of available habitat (greater than 75%).

RESOURCE DEVELOPMENT AND MANAGEMENT IN JACKSON HOLE, WYOMING[1]

James E. Roelle and Gregor T. Auble[2]

Abstract.--Modelers of natural resource systems are often frustrated because potential users of simulation models do not benefit from the increased understanding of system structure and behavior that accrues from the modeling process. This paper describes an attempt to extend the benefits of the modeling process by involving potential users in the initial construction and evaluation of a simple simulation model concerning resource development and management in Jackson Hole, Wyoming.

INTRODUCTION

One of the principal benefits of simulation modeling of natural resource systems is the increased understanding of system structure and behavior that the modeler gains from the modeling process. Construction of a simulation model provides a useful framework for clarifying assumptions, reflecting to the modelers the consequences of those assumptions, and organizing information bearing on their validity. This benefit usually accrues even in the absence of real predictive power in the resulting model. Modelers have often been frustrated, however, by the fact that this benefit has not extended to potential users of simulation models (Watt 1977). We believe this is largely because those users have not actively participated in the modeling process.

This paper describes an attempt to extend the utility of the modeling process by involving potential users in the construction and evaluation of a simple simulation model concerning resource development and management in Jackson Hole, Wyoming. We begin with a brief discussion of the setting and some of the potential resource management conflicts in the Jackson area. This is followed by a short description of a workshop that involved interested parties in the construction of a simulation model designed to evaluate the potential consequences of those conflicts. A more complete description of the workshop and the simulation model is given by Andrews et al. (1981). The portion of the model dealing with elk management is examined in greater detail to illustrate the general nature of the workshop model and its value in elaborating the management implications of basic assumptions. The utility of this portion of the model as a focal point for organizing information--and thus bringing that information to bear on elk management issues--is discussed. Finally, the remainder of the model is discussed in terms of its critical assumptions and the implications of those assumptions for behavior of the model and for resource management in the Jackson area.

THE SETTING

Jackson Hole is in Teton County in extreme northwestern Wyoming (fig. 1). Land ownership is a complex of public (including Grand Teton National Park, John D. Rockefeller Jr. Memorial Parkway, Yellowstone National Park, Bridger-Teton National Forest, and the National Elk Refuge) and private holdings. Mobile wildlife species such as elk, mule deer, and bald eagles utilize these lands in complex patterns to satisfy their seasonal habitat requirements. Effective management of these species, particularly elk, is critically dependent on a coordinated set of State and Federal agency objectives and policies. In the absence of a coordinated strategy, the effects of one agency's management efforts can easily be nullified or reversed by efforts of another agency as animals move between lands under different jurisdiction. Furthermore, the need for effective, coordinated wildlife management will likely be intensified in the near future as pressures for development of other resources in the Jackson area increase.

Public lands in Jackson Hole provide important opportunities for outdoor recreation activities, such as hunting, fishing, skiing, and snowmobiling, as well as resource development activities, such as oil and gas production, logging, and livestock production. Land use on private lands has historically been agricultural, with hay and cattle the major products. In recent years, however, land use has begun to change. Each year more and more people, both tourists and new residents, are attracted to the valley by its scenic beauty, diverse wildlife populations, and outdoor recreation opportunities. This increasing popularity has led to tremendous pressure for development of private lands, which comprise only about 3% of Teton County. Private land is currently so valuable for development that owners are finding it increasingly difficult to maintain their holdings in agricultural production (Hocker and Clark 1981).

Unfortunately, although the vast public holdings provide important summer range, private lands constitute a large fraction of the mule deer winter range and bald eagle nesting and wintering areas and are

[1]Paper presented at the Third International Conference on State-of-the-Art in Ecological Modelling, Colorado State University, May 24-28, 1982.

[2]James E. Roelle and Gregor T. Auble are Systems Ecologists, Adaptive Environmental Assessment Group, Western Energy and Land Use Team, U.S. Fish and Wildlife Service, Fort Collins, Colorado U.S.A.

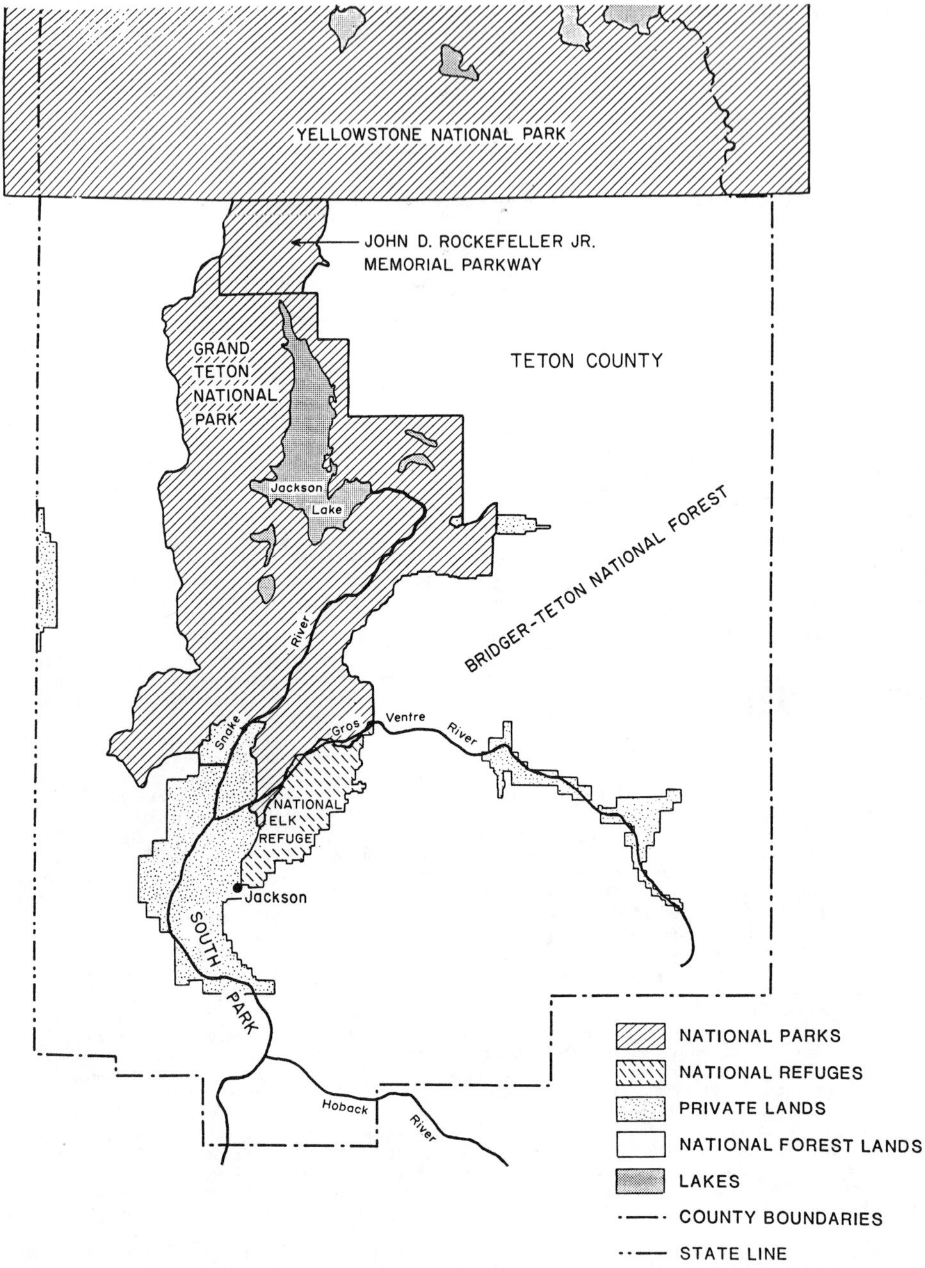

Figure 1.--Land ownership in the Jackson area.

transected by major elk migration corridors (Hocker and Clark 1981). If these habitats are significantly impacted by development, species such as deer, eagles, and elk will be more dependent on the complex of public lands to supply their life requisites. This in turn will intensify the need for coordinated management by the various State and Federal agencies. Many citizens are thus concerned that the combination of an extremely complex management situation and increasing pressure for development of private lands may have serious implications for some of the wildlife resources that contribute significantly to the attractiveness of the Jackson area.

THE APPROACH

A workshop to address these issues in the context of a simulation modeling exercise was held in Jackson in January 1981. Objectives of the workshop were to: (1) promote communication among interests and agencies concerned with resource development and management in the Jackson area; (2) identify potential conflicts within and among management agency objectives and policies, particularly with regard to elk; and (3) identify important interrelationships between wildlife management strategies and attendant issues such as population growth, recreation, logging, and energy development.

Workshop participants included representatives from the Wyoming Game and Fish Department, National Elk Refuge, Grand Teton National Park, Bridger-Teton National Forest, Teton County Planning Commission, Jackson city government, outfitting and ranching industries, Jackson Hole Chamber of Commerce, and the local environmental community. The process of constructing and evaluating a simulation model was used to provide a structured forum for discussion. A team of workshop facilitators guided participants through a series of exercises (collectively known as Adaptive Environmental Assessment) designed to elicit basic ideas concerning system structure and function, translated those ideas into a computerized simulation model, and interpreted the behavior of the resulting model in the context of the assumptions and hypotheses used in its construction (Holling 1978). Computer programming was aided by using a FORTRAN-based simulation control package (Hilborn 1973).

THE WORKSHOP MODEL

Actions and Indicators

The workshop began with the identification of the conceptual, geographic, and temporal bounds of the model. Participants first listed management actions whose effects they wished to examine using the model. Actions were defined as things that management agencies can do in their attempts to manipulate the system toward some desired end. Most of the actions in this case involved regulating the level and location of development activities and altering elk management policies concerning harvest, supplemental feeding, and range improvement (table 1).

Next, participants identified indicators of interest in the Jackson system. Indicators were defined as performance measures used to evaluate the response of the system to various management strategies. Most of the indicators pertained to levels of development, human population size, levels of recreation use, wildlife population sizes, and wildlife harvests (table 2).

Table 1.--Management actions included in the workshop model.

Regulate human population growth
Regulate housing density
Regulate access to public lands
Enlarge existing ski areas
Construct new ski areas
Regulate snowmobile use
Regulate oil and gas drilling
Regulate logging
Regulate supplemental elk food
Regulate elk harvest
Regulate natural forage production
Create nondevelopment buffer zones
Regulate recreational activity in winter habitats

Space and Time

Workshop discussions then turned to consideration of the spatial and temporal resolution necessary to represent the components and processes implied by the actions and indicators shown in tables 1 and 2.

Two aspects of spatial resolution were discussed --the geographic area to be covered by the model and the extent to which that area needed to be subdivided into smaller units in order to capture the dynamics of the processes involved. The overall geographic boundary for the model was generally identified as the total area utilized annually by elk that winter in Jackson Hole. This area was subdivided into five elk summering areas (southern Yellowstone National Park, Grand Teton National Park, the Teton Wilderness portion of Bridger-Teton National Forest, the Gros Ventre River drainage, and South Park), three elk wintering areas (the National Elk Refuge, the Gros Ventre River drainage, and South Park), and three areas of concentrated human activity (Jackson, the Gros Ventre River drainage, and South Park) (fig. 1).

Likewise, two aspects of temporal resolution were discussed--the time step, or time covered by each iteration of the model, and the time horizon, or total time represented in each model run. The time horizon was set at 50 years. The time step was annual, subdivided into three seasons defined by major elk movements and human recreational activities. The fall season (September-November) included most of the elk hunting season and the time when elk move from summer range to winter range. The winter season (December-April) represented the primary period of winter recreation (down-hill and cross-country skiing, snowmobiling) and the time when elk are on their winter ranges.

Table 2.--Performance indicators included in the workshop model.

Human population size
Levels of recreation activities
Level of energy development
Level of logging
Acres in various land use categories
Number of elk
Number of elk harvested
Cost of supplemental elk food
Number of bald eagle nests
Number of wintering bald eagles
Number of mule deer

The summer season (May-August) included the period of summer tourism and the time of elk migration to, and use of, their summer ranges.

In addition to identifying the bounds of the model, the discussions described above were useful in providing a structured framework within which all interested parties could express their concerns about the Jackson area.

Submodels and Interactions

The general structure of the simulation model was then defined in the following manner. The actions (table 1) and indicators (table 2) were subjectively grouped into three submodels: human population; elk; and other wildlife. These submodels were arrayed as row and column headings of a matrix (fig. 2). The basic content of each submodel was identified by noting the relevant actions and indicators, as well as any additional implied state variables, in the appropriate cell along the main diagonal. Thus, the human population submodel included variables such as population size, levels of recreational activities, levels of development activities, and acres of land in several ownership/use categories. Similarly, the elk submodel represented the population dynamics of elk in the Jackson area, while the other wildlife submodel dealt with population dynamics of bald eagles and mule deer.

Information exchanges between the submodels were then identified by asking the following question for each of the nondiagonal elements of the matrix: What information is needed from the submodel in this row in order to predict the behavior of the submodel in this column? For example, biologists felt that information (from the human population submodel) concerning acres of land devoted to private residences, logging, and oil and gas exploration would help them determine the impacts of development on elk biology. In addition to identifying the linkages between the submodels, the process of constructing such a matrix was extremely useful in promoting increased interdisciplinary communication and understanding.

Following this general definition of model structure, workshop participants divided into subgroups, each with a staff facilitator, to formulate the submodels. In this phase of the workshop, the task of the elk subgroup, for example, was to construct functional relationships describing basic elk reproduction and mortality rates, incorporating, insofar as possible, the information transfers identified in the interaction matrix as independent variables in those relationships. Workshop facilitators converted the relationships into a series of simple difference equations, linked the three submodels together to form an overall model of the resource system, and presented a description of the model and examples of its behavior to participants for evaluation.

THE ELK SUBMODEL

The behavior of the simulation model resulting from the exercises described above was necessarily a reflection of the assumptions used in its construction. These assumptions, in turn, have some important implications for management of resources in the Jackson area, particularly elk. We therefore present here a more detailed description of the elk submodel, both to demonstrate the general nature of the workshop simulation model and to illustrate the utility of the modeling process as a framework for clarifying assumptions, examining their consequences, and organizing information to evaluate their validity.

Submodel Structure

Elk in Jackson Hole summer in five relatively distinct areas (southern Yellowstone National Park,

FROM THIS SUBMODEL \ THIS SUBMODEL NEEDS INFORMATION CONCERNING	Population Growth	Elk	Other Wildlife
Population Growth	**Actions** Regulate human population growth Regulate housing density Regulate access to public lands Enlarge existing ski areas Regulate snowmobile use Regulate oil and gas drilling Regulate logging **Indicators** Human population size Level of recreation activities Level of energy development Level of logging Acres in various land use categories	Acres of private land developed Acres logged Acres in oil and gas development Miles of road constructed Snowmobile use	Acres of private land developed
Elk	Number of elk Number of hunters Number of elk harvested Deviation of elk populations from targets	**Actions** Regulate supplemental elk food Regulate elk harvest Regulate natural forage production **Indicators** Number of elk Number of elk harvested Cost of supplemental elk food	Number of elk wintering in South Park
Other Wildlife	Number of bald eagles Deviation of bald eagle populations from targets Number of mule deer Deviation of mule deer populations from targets		**Actions** Create nondevelopment buffer zone Regulate recreational activity in winter habitats **Indicators** Number of bald eagle nests Number of wintering bald eagles Number of mule deer

Figure 2.--Matrix of information exchanges between submodels.

the Teton Wilderness, Grand Teton National Park, the Gros Ventre River drainage, and South Park) surrounding the valley. During fall and early winter they migrate to and mix together on three principal wintering areas (the National Elk Refuge, the Gros Ventre River drainage, and South Park). Timing of the migration depends on weather conditions, being earlier in years with heavy early snowfall. Harvest is a continuous process that occurs before, during, and after migration. The relative magnitude of the harvest during various phases of the migration also depends on weather conditions. This is important because harvest that occurs early, while elk are separated on their summer ranges, is directed at a different subset of the herd than is harvest that occurs late, when elk from several summer herd segments may be mixed together on the winter ranges. Elk on the winter ranges are supported by a combination of natural forage and supplemental food. The amount of supplemental food provided also depends on weather conditions, less being necessary in years of low snowfall when more natural forage is available. Elk remain on the winter ranges until spring or early summer, when they migrate back to their summering areas.

The complex pattern of elk movement and harvest thus creates a difficult problem for management agencies interested in preserving some balance between the various herd segments while providing opportunities for both recreational harvest and nonconsumptive use. The following conceptual framework was adopted in an attempt to simplify this situation and still capture the essential elements of the processes involved.

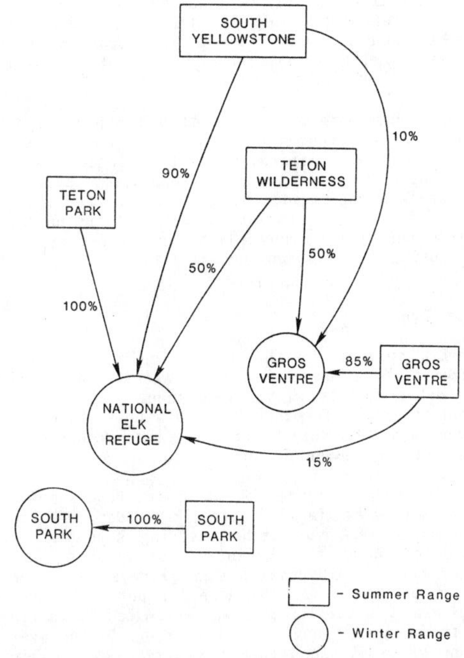

Figure 3.--Diagrammatic representation of elk herd segments and migration patterns.

Fall Dynamics

Elk were considered to summer in five discrete herd segments and migrate in fixed proportions to three wintering areas (fig. 3). Harvest was divided into two components, one occurring before the fall migration, when elk are separated into their summer segments, and one occurring after the migration, when elk are mixed on the winter ranges. In both cases the concept of a target or desired postharvest population was used to compute the number of elk available for harvest. Elk in excess of these targets, which were set for both summer and winter herds, were considered to be a harvestable surplus. It should be noted that such formal population targets do not really exist for the summer herd segments. Rather the targets were used in the model to reflect the idea that management agencies have at least general goals for the number of elk in each herd.

At the end of each winter an estimate of the preharvest population in each summer herd segment for the next fall was obtained by adding an estimated calf crop to the number of elk surviving the winter period. All elk surviving the winter were assumed to return to the summer range that they came from the previous fall. The estimated calf crop was calculated from the previous year's reproductive rate and the number of females surviving the winter. The number of elk available for the premigration harvest was then determined by subtracting the target population from the estimated preharvest population for each summer herd.

A premigration harvest was allowed on each summer herd segment except the one utilizing Grand Teton National Park. South Yellowstone elk are not actually harvested until after they leave Yellowstone National Park; however, in the conceptual framework of the model, this harvest still occurs before the migration. A harvest is conducted in Teton Park, but it is confined to the southern and eastern portions. There is considerable uncertainty as to the origin of animals taken in this harvest because the herd segments from South Yellowstone and the Teton Wilderness use the southern and eastern portions of Grand Teton National Park as a migration corridor to the National Elk Refuge. The assumption made was that the Teton Park harvest is directed at the mixed population of animals wintering on or around the National Elk Refuge and, therefore, in the conceptual framework of the model, constitutes a postmigration harvest.

Winter range populations for the next year were then estimated by subtracting the estimated premigration harvests from each of the estimated summer segments and moving the animals to the winter ranges in the proportions shown in figure 3. A desired postmigration harvest for each winter range was calculated by subtracting the target population from each of the estimated winter populations.

The number of hunters needed for each harvest was estimated by dividing the desired harvest by the hunter success ratio for the previous year. In each case the number of hunters could be limited by a maximum imposed by factors such as agency policies or the physical limitations of the area being harvested. Realized harvest for each area was then calculated on the basis of the number of hunters present and a new success ratio computed from a randomly generated snowfall pattern. The total harvest was distributed among four sex and age cohorts (calves, yearling bulls, cows, and adult bulls) according to average proportions observed in each herd segment in recent years. These proportions were not markedly different between herd segments, except in the case of the National Elk Refuge, where the harvest tends to contain a higher percentage of cows. Elk were assumed to be completely mixed on the winter ranges; therefore, postmigration harvests were assumed to occur in proportion to the number of elk contributed by each of the summer segments.

Although more formalized than the process actually used by management agencies, the above computational sequence approximates the logic used in determining harvest quotas. Population sizes are estimated, a recommended harvest is computed based on a desired or target population size, and the number of hunters required to accomplish the harvest is calculated using an estimated success ratio.

Winter Dynamics

Elk survival on the winter feeding grounds was assumed to be a function of three factors: a mortality rate determined by winter food supply; an additional mortality rate determined by food supply the previous summer; and a "natural" mortality rate occurring even with adequate summer and winter food.

The mortality rate due to winter food supply was based on food availability per elk per day. The total food requirement for each wintering herd was first calculated using the number of elk arriving on the winter range, an average number of days spent there, and an average forage requirement per elk per day. Total usable natural forage was estimated by multiplying the number of acres of winter range by the average amount of available forage per acre. The number of acres of winter range was modified by factors such as development of private lands, snowmobiling and cross-country skiing activity (fig. 2), and a randomly generated snowfall pattern. The requirement for supplemental food was then calculated by subtracting the amount of natural forage available from the total food requirement for the herd. The amount of supplemental food available each year was limited to a fixed amount (reflecting fixed agency budgets) plus any unused food carried over in storage from previous years. The total amount of food available was then converted to an amount available per animal per day, which was in turn used to calculate a winter mortality rate.

Total winter mortality was computed using this mortality rate based on winter food supply, a similar mortality rate based on food supply the previous summer (see discussion of Summer Dynamics below), and a constant natural mortality rate for each wintering herd.

Summer Dynamics

At the beginning of each summer season, elk surviving the fall and winter were assumed to move back to the summer range from which they came. Calf production per cow was computed from a pregnancy rate and a brucellosis infection rate, both of which were determined by the density of elk in the previous winter.

The mortality rate due to summer food supply was calculated by dividing the summer period into two equal parts. The total amount of food required for each summer herd segment in the first half of the season was calculated from the number of days in the period, the number of elk present, and an average amount of food required per elk per day. The amount of food available was computed from the acres of summer range--as modified by factors such as oil and gas production, road construction, and logging (fig. 2)--and the average amount of forage available per acre. If the available summer forage exceeded the requirement of the herd, animals were assumed to stay in that geographic area. If the herd forage requirement exceeded the amount available, animals not finding sufficient forage were assumed to disperse in fixed proportions to other summering areas. Following this dispersal, similar calculations were performed for the second half of the summer season, and a mortality rate was calculated based on the forage available per elk per day. As mentioned above, this mortality rate was assumed to be manifested during the subsequent winter.

Submodel Behavior

The fundamental behavior of the elk submodel can be illustrated with output from two simulations (scenarios). In both we focus on three summer herd segments (South Yellowstone, Teton Wilderness, and Teton Park) that contribute different proportions of their animals to a common wintering area (National Elk Refuge) (fig. 3).

Population Maintenance Scenario

The Population Maintenance Scenario assumes that the principal management goal is to maintain the various herd segments at approximately their present levels. Population targets for the scenario were therefore set as follows: South Yellowstone, 3,800; Teton Wilderness, 2,500; Teton Park, 4,500; and National Elk Refuge, 9,000. These targets are estimates only, and should not be construed as precise statements concerning the present status of the various populations or agency objectives.

Under the conditions of the Population Maintenance Scenario, the South Yellowstone and Teton Wilderness segments are gradually reduced for about 7 years and then are maintained near their respective targets (fig. 4). The Teton Park segment grows for about 7 years and then is maintained at a relatively constant level considerably above its target. The National Elk Refuge segment is maintained at or near its target throughout the simulation.

Simulated harvests associated with these population trends are illustrated in figure 5. Premigration harvests occur in the South Yellowstone and Teton

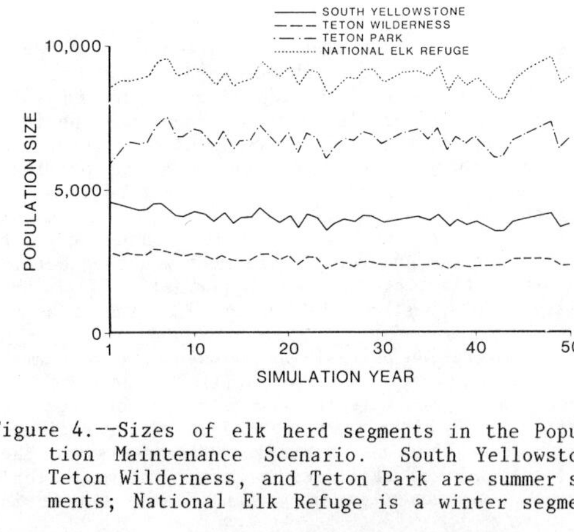

Figure 4.--Sizes of elk herd segments in the Population Maintenance Scenario. South Yellowstone, Teton Wilderness, and Teton Park are summer segments; National Elk Refuge is a winter segment.

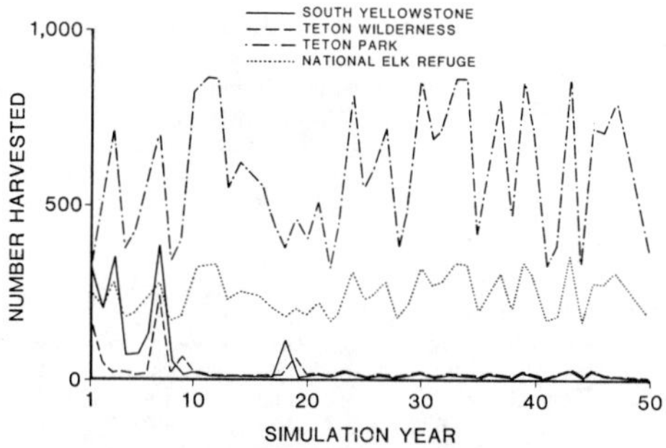

Figure 5.--Harvests from elk herd segments in the Population Maintenance Scenario. South Yellowstone and Teton Wilderness are premigration harvests; Teton Park and National Elk Refuge are postmigration harvests.

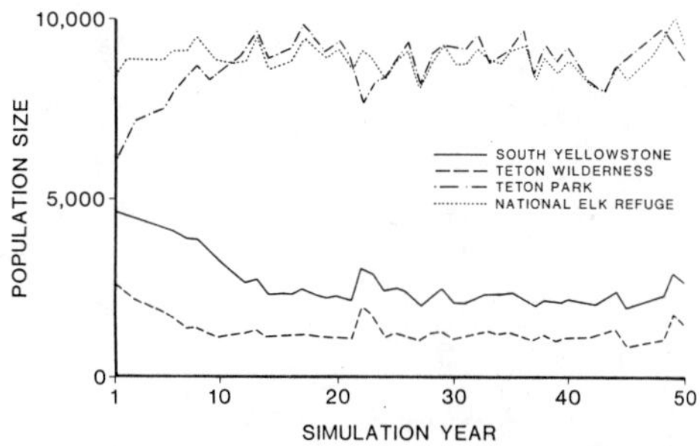

Figure 6.--Sizes of elk herd segments in the Harvest Maintenance Scenario. South Yellowstone, Teton Wilderness, and Teton Park are summer segments; National Elk Refuge is a winter segment.

Wilderness areas during the years in which those segments are being reduced toward their targets. As these segments approach their targets, however, the premigration harvests become negligible. Postmigration harvests in Teton Park and the National Elk Refuge, which are directed at the mixed population of elk wintering on or around the Refuge, vary around means that approximate current harvest levels.

The behavior of the elk submodel in this Population Maintenance Scenario is determined by several key relationships. The model assumes that agency budgets are fixed and will not be increased to provide additional supplemental food for larger populations on the National Elk Refuge. The fixed food supply results in density-dependent mortality that controls the number of elk that survive the winter period and therefore limits the total number of elk that return to the summer ranges. The distribution of this total number of elk among the summer herd segments depends on the relative net rate of increase (the sum of all production and mortality factors) of each. Herd segments having higher net rates of increase will return proportionately more elk to their summer ranges and therefore will constitute an increasing proportion of the Refuge population until something happens to equalize the net rates of increase for all three summer herd segments.

These assumptions lead to a higher net rate of increase for the Teton Park segment for about the first 7 years of the simulation due to the absence of a premigration harvest. Winter mortality rates and reproductive rates (which are determined while elk are mixed together on the winter range) are identical for all three summer segments throughout the simulation. Summer food never limits any of the segments, so differential food-related summer mortality or dispersal does not occur. Postmigration harvest rates are also identical, because the harvests in Teton Park and the National Elk Refuge are assumed to be directed equally toward animals from each of the three summer segments. The Teton Park segment therefore grows until the premigration harvest on the other two summer segments is eliminated; that is, until the numbers of animals returning to the South Yellowstone and the Teton Wilderness from the Refuge are just sufficient to reach the respective targets. At those levels there is no premigration harvest, net rates of increase for the three summer segments are equal, and the populations stabilize.

Harvest Maintenance Scenario

The Harvest Maintenance Scenario assumes that management agencies are also interested in maintaining current harvest levels and would accept lower populations in the South Yellowstone and Teton Wilderness areas if such populations resulted in higher sustained harvests. The South Yellowstone and Teton Wilderness population targets were therefore reduced to 2,000 and 1,000 elk, respectively.

Under these conditions the South Yellowstone and Teton Wilderness segments are reduced toward their new targets for 10-15 years (fig. 6). The rate of reduction is determined by the maximum number of hunters that can be accommodated and a hunter success ratio calculated from a randomly generated snowfall pattern. The Teton Park segment grows during the first 10-15 years and stabilizes at a level somewhat higher than in the Population Maintenance Scenario (fig. 4). The National Elk Refuge population again remains near its target throughout the simulation due to density-dependent winter mortality.

During the period when the South Yellowstone and Teton Wilderness segments are being reduced toward their new targets, their premigration harvests (fig. 7) are significantly higher than in the Population Maintenance Scenario (fig. 5). Furthermore, periodic

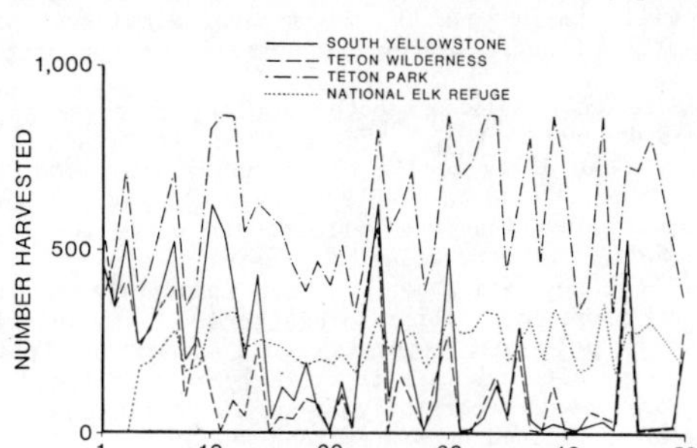

Figure 7.--Harvests from elk herd segments in the Harvest Maintenance Scenario. South Yellowstone and Teton Wilderness are premigration harvests; Teton Park and National Elk Refuge are postmigration harvests.

premigration harvests occur even after these two summer segments have stabilized near their targets. The postmigration harvest in Teton Park (fig. 7) is very similar to that generated in the Population Maintenance Scenario (fig. 5). The postmigration harvest on the National Elk Refuge (fig. 7) differs from that in the first scenario (fig. 5) only in the first 3 years. In the Harvest Maintenance Scenario, the South Yellowstone and Teton Wilderness populations are reduced faster than the Teton Park segment increases. The Refuge population is thus slightly below its target for the first 3 years and there is no harvest on the Refuge during this period.

The qualitative behavior of the elk submodel in the Harvest Maintenance Scenario is governed by the same assumptions that applied in the Population Maintenance Scenario. During the initial 10-15 years of the simulation, the Teton Park segment constitutes an increasing proportion of the National Elk Refuge herd because it has a higher net rate of increase than the South Yellowstone and Teton Wilderness segments. The higher net rate of increase for the Teton Park herd is a result of the premigration harvest directed at the other two summer segments. These premigration harvests are higher than in the Population Maintenance Scenario because of the lower population targets. Furthermore, the premigration harvests continue periodically, even after the summer segments stabilize, because a different mechanism equalizes the net rates of increase for the three summer herd segments. In the Harvest Maintenance Scenario, the population in Teton Park becomes large enough that animals are periodically forced to disperse due to summer food shortage. Dispersion effectively reduces the net rate of increase for the Teton Park herd, while increasing it for the segments to which the excess animals move. The result is a periodic influx of animals, which are available for premigration harvest, to the South Yellowstone and Teton Wilderness herd segments.

Management Implications

If model assumptions are correct, these scenarios have some interesting implications for management of elk in the Jackson area. If the Teton Park herd is being harvested at a lesser rate than the South Yellowstone and Teton Wilderness segments, and if all other production and mortality factors are equal, then the Teton Park segment will constitute an increasing proportion of the animals wintering on the Refuge. This will continue until some mechanism balances the net rates of increase for the three summer segments.

Note that although both simulations suggest an increasing population in Teton Park, this is not a necessary corollary to the above conclusion. The Teton Park herd could be stable or even decreasing; but, if model assumptions are correct, the other two segments would be decreasing at faster rates and the Teton Park population would still constitute an increasing proportion of the Refuge herd. Likewise, all three of the summer segments could theoretically be increasing, although this is unlikely because the Refuge population is currently at or above its capacity (target) as defined by winter food supply.

Because it is unlikely that all three of the summer segments contributing animals to the Refuge are increasing, there is a high probability that the South Yellowstone and Teton Wilderness segments are decreasing. If this is the case, managers will likely be faced with a choice between declining populations, declining harvests, or some combination of the two. This is exactly the situation illustrated in the two simulations described above. In the Population Maintenance Scenario, the premigration harvest is eliminated in order to maintain stable populations. In the Harvest Maintenance Scenario, the premigration harvest is maintained only by reducing the South Yellowstone and Teton Wilderness segments to lower target populations.

Finally, the last 35-40 years of the Harvest Maintenance Scenario suggest that the total number of animals ultimately available for harvest will depend on what mechanism balances the net rates of increase for the three summer herd segments. If that mechanism involves movement of Teton Park elk into other areas where they are more accessible, higher harvests will be possible than in a situation where some other mechanism balances the net rates of increase. Alternately, managers might try to balance the net rates of increase by eliminating the cow and calf harvest in the South Yellowstone and Teton Wilderness areas. Additional simulations suggested that this strategy would prevent population decline in those two areas while allowing moderate harvests to continue. Increase of the Teton Park segment would also be prevented, because the net rates of increase for the reproductive portions of the three summer herd segments would be equalized.

Implications for Data Analysis

The above implications are meaningful only insofar as the assumptions used in the model are true. Inaccurate assumptions or omission from the model of factors that significantly influence elk dynamics, either directly or indirectly, could result in different conclusions. The principal benefit of the modeling workshop probably lies not in the conclusions themselves, but rather in a clear statement of the assumptions that caused the model to behave as it did. In the case of the elk submodel, those assumptions can be stated as follows.

1. Agency budgets for supplemental winter food are largely fixed and cannot be expected to increase significantly in response to increasing elk populations.

2. Some portion of the harvest occurs while the herd segments are separated on the summer ranges and therefore does not impact the Teton Park herd.

3. The harvests in Teton Park and the National Elk Refuge take elk proportionately from all herd segments contributing to the Refuge winter herd.

4. Winter mortality rates are equal for all herd segments wintering on the National Elk Refuge.

5. At present population sizes, summer mortality rates and reproductive rates are equal for all herd segments contributing elk to the Refuge winter herd.

6. At present population sizes, net interchange of elk between summer herd segments is negligible.

We believe that such a clear statement of assumptions is extremely important for several reasons. First, many of these assumptions are being made, sometimes subconsciously, in day to day management activities. The modeling activity therefore served as a useful reminder of the underlying basis for many important decisions. Second, the above statements represent a set of clearly posed questions or hypotheses that can be used as a framework for synthesizing existing data and, if necessary, for planning additional research and data collection. Finally, by clarifying these assumptions and their implications for other community members as well as biologists, a foundation was laid for better understanding and general support of management decisions.

OTHER RELATIONSHIPS

Several additional relationships were critical in governing the general nature of the alternative development patterns portrayed by the model. They largely involve the points at which various agencies and interests must interface in their attempts to deal with the complexities of resource issues. We believe that the process of dealing with these relationships in the quantitative framework of a simulation model was very useful for participants, even though the relationships themselves may be fundamentally less tractable than the specific aspects of elk population dynamics outlined above. We summarize these relationships here in the a series of questions, the answers to which would have been valuable in formulating the simulation model and will likely be important in determining the actual future of the Jackson area.

1. How are growth rates for various sectors of the Jackson economy related? For example, to what extent does growth of the skiing industry stimulate construction of additional units for permanent or temporary lodging?

2. What are the potential feedbacks on human population growth? Would wildlife population reductions reduce the influx of people by making the Jackson area a less attractive place to live or visit? As private land is more fully developed will increasing real estate prices slow development? Does the finite amount of private land represent a real limit to human population growth or will policies and preferences simply change in favor of higher density facilities as private land becomes scarce?

3. What will be the relationship between development of private land and loss of important wildlife habitat (fig. 8)? Will a human preference for riparian areas destroy large fractions of this habitat quickly or can buffer strategies be devised to preserve those important habitats while still allowing significant development?

4. What is the relationship between development of private land and the ability of elk to migrate to and from their winter ranges?

5. How will development of private land affect hunter access to the elk for harvest management?

6. What is the relationship between the elk population size, the demand for hunting opportunity, and hunter success? What would be the impact of reduced harvest levels on the local economy?

The workshop and modeling exercise described here was not necessarily successful in providing specific answers to any of these questions. It did, however, provide both a structured forum for stating the questions clearly before a diverse audience and a mechanism for examining some of the consequences of various possible answers to these questions. Hopefully, this

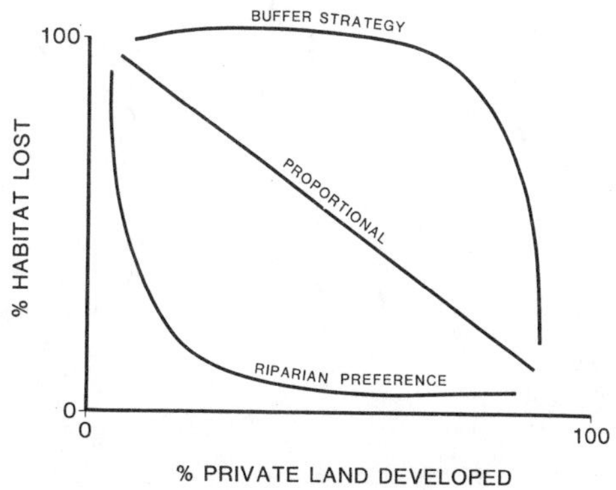

Figure 8.--Possible relationships between development of private lands and loss of wildlife habitat.

activity was useful to the participants faced with decisions concerning resource management in the Jackson area.

ACKNOWLEDGMENTS

The simulation modeling workshop described herein was attended by A. Atchison, B. Barmore, D. Bartschi, P. Bruun, S. Clark, D. Cowee, J. Hamernick, F. Herbel, J. Hocker, B. Oakleaf, C. Petersen, Jr., B. Robbins, G. Roby, J. Stark, D. Strickland, T. Toman, M. Trautman, H. Turner, J. Weaver, M. Webb, J. Wilbrecht, and B. Wood. They are to be congratulated for their hard work and sincere interest in the future of the Jackson area. Dale Strickland and Tom Toman, Wyoming Game and Fish Department, deserve special thanks for organizing the workshop. We are indebted to Kent Andrews, Dave Hamilton, and Ike Ellison, fellow members of the Adaptive Environmental Assessment Group, and Pete McNamee and Nick Sonntag, Environmental and Social Systems Analysts Ltd., Vancouver, B.C. Canada, for their assistance in facilitating the workshop. Finally, we wish to thank Cathy Short for her editorial assistance and Jenny Shoemaker for art work.

LITERATURE CITED

Andrews, A. K., R. A. Ellison, D. B. Hamilton, J. E. Roelle, and P. J McNamee. 1981. Results of a modeling workshop concerning resource development and management in Jackson Hole, Wyoming. Western Energy and Land Use Team, U.S. Fish and Wildlife Service, Fort Collins, CO. 74 p.

Hilborn, R. 1973. A control system for FORTRAN simulation programming. Simulation 20:172-175.

Hocker, J., and S. Clark. 1981. Jackson Hole: protecting public values on private lands. Izaak Walton League of America, Jackson Hole, Wyoming. 82 p.

Holling, C. S., ed. 1978. Adaptive environmental assessment and management. John Wiley and Sons, New York. 377 p.

Watt, K. E. F. 1977. Why won't anyone believe us? Simulation 28:1-3.

SIMULATED EXPERIMENTS WITH A MODEL OF DEER POPULATIONS AND NORTHERN FOREST VEGETATION[1]

Allen Y. Cooperrider and Donald F. Behrend[2]

Abstract.--The interaction of white-tailed deer (*Odocoileus virginianus*) with northern forest vegetation was investigated using a simulation model. The model, described in earlier papers, consists of a deer population model linked to a forest vegetation dynamics model. Verification analyses mimicked: (1) low browse production and low deer density in old-growth hardwoods; (2) high browse production and high deer density in cutover areas; and (3) heavy losses of deer through starvation during severe winters. Sensitivity analyses indicated the sensitivity of the simulated deer population to: (1) changes in caloric density of deer browse and (2) the length of time when herbaceous forage is available. The model was used to predict management strategies that will allow regeneration of commercial tree species in the presence of viable deer populations.

INTRODUCTION

White-tailed deer (*Odocoileus virginianus*) affect development of forests in the northeastern United States. Through selective browsing of seedlings and saplings, deer may prevent regeneration of valuable tree species and ultimately determine the species composition of mature forests. Conversely, deer densities are strongly influenced by forest management practices, making the interaction of deer with northern forest vegetation of interest to both forest and wildlife managers.

To better understand and predict these interactions, a computer simulation model that integrates dynamics of deer population and forest vegetation was developed. The model simulates the Huntington Forest, a 15,000-acre experimental forest in the central Adirondack Mountains of northern New York. However, its generality permits use, with some modification, for other northern forest areas.

The model has been described in detail elsewhere (Cooperrider 1973; Cooperrider and Behrend 1980a,b). Herein the model is used to (1) suggest explanations for observed deer-forest interactions, (2) identify critical areas for future sampling and/or research efforts, and (3) simulate possible managment strategies.

The study was supported by the SUNY College of Environmental Science and Forestry of Syracuse Universtiy. We wish to thank T. Cunia and W.C. Tierson for their assistance in all phases of the study.

Interactions of deer populations and northern forest vegetation have been studied on the Huntington Forest. Webb et al. (1956) studied effects of deer browsing on an uncut mature hardwood stand and concluded that deer in densities of 19 to 34 deer per square mile were not an important factor in the establishment of any tree, shrub or herbaceous species. Tierson et al. (1966) found that following timber harvest and stand improvement, a similar density of deer limited height growth of many hardwod species and woody shrubs. Behrend et al. (1970) demonstrated that adequate regeneration of desirable hardwoods could be attained by reducing the deer population to a fall density of 17 per square mile.

These studies were conducted on the Huntington Forest during a period of active logging on and surrounding the forest. Studies of deer movements indicated that from 2/3 to 4/5 of the fall deer population is wintering from about December 1 to April 1 in areas adjacent to the Forest (R. Sage, Pers. Comm.). These deer studies demonstrated seasonal movements of up to 10 miles, with an average of about 4 miles for deer moving off the Forest in fall (Sage 1973). Data

[1]Paper presented at the Third International Conference on State-of-the-Art in Ecolological Modeling, Colorado State University, May 24-28, 1982.
[2]Allen Y. Cooperrider is Wildlife Biologist, U.S. Bureau of Land Management, Denver Service Center, Bldg. 50, Denver Federal Center, Denver, Colo. 80225. Donald F. Behrend is Professor of Environmental and Forest Biology, SUNY College of Environmental Science and Forestry, Syracuse, New York 13210.

indicate a fall movement of deer from hardwood to conifer stands, and in some cases to areas with current logging operations.

Similar results have been obtained from studies on other areas in the Northeast. Thus forests with canopies opened up by cutting or burning may support high deer densities which, in turn, may greatly affect tree regeneration. Conversely, old growth forests, in which most growing photosynthetic parts are out of reach of deer, support few deer.

Few models of interaction of deer with northern forest vegetation have been developed, although many models of forest vegetation dynamics or deer population dynamics have been described. The simulations that follow are thus the result of using a set of simplifying assumptions to integrate data on deer and forest ecology obtained on the Huntington Forest or from literature. A brief summary of these assumptions and of the types and sources of such data follows.

THE SIMULATION MODEL

The major biological assumptions incorporated in the simulation model are described below. They are not presented as biological truths but rather as simplifications of the biological processes involved. Their utility is that when incorporated into a model, they predict reasonably well the dynamics of the major components of a deer/forest system.

Vegetation Dynamics

1. Growth of stems (trees or shrubs under 10.6 ft.) is a function of:

    (a) species specific growth rates,
    (b) the degree of competition from larger trees (over 10.6 ft.) as measured by basal area per acre, (the higher the basal area per acre, the lower the growth rate), and
    (c) the number of terminal leaders that remain unbrowsed.

2. Mortality of stems is a function of:

    (a) the number of stems/acre of all species present, and
    (b) a theoretical maximum number of stems that can be supported under a forest stand of a given basal area per acre (the higher the basal area per acre, the lower the maximum number).

3. The effect of deer browsing on the species composition of understory stems (and ultimately on overstory trees) is a function of:

    (a) deer preference for particular tree and shrub species, and
    (b) the number of stems of a given tree species that grow out of reach of deer.

Deer Dynamics

1. Increase in deer numbers is a function of:

    (a) number of adult female deer and age-specific birth rates, and
    (b) the number of deer immigrating from other areas.

2. Decrease in deer numbers is a function of:

    (a) hunting mortality rate or numbers harvested,
    (b) natural mortality rate (from accidents, disease and predators),
    (c) the number of deer emigrating out of the area, and
    (d) the number of deer starving.

3. The number of deer starving is a function of:

    (a) the nutritional state of the animals as measured by body weight. Change in body weight is determined by:

        (1) the forage intake rate,
        (2) the forages selected,
        (3) the digestible energy (DE) content of the forage, and
        (4) the relationship between DE intake and change in body weight.

    (b) a threshold starvation weight for a given sex-age class of deer, below which it is considered to have died and is removed from the population.

Weather

The primary effect of weather on deer in northern forests is a seasonal reduction in available forage caused by deep snow either covering up plants or producing an uncrusted snow pack that reduces deer mobility. Both processes are mimicked by the model.

Management

The principle management options that can be simulated by the model are removal of vegetation (timber harvest) and removal of deer (hunting). The major effect of timber harvest on the system is a decrease in the basal area per acre of overstory trees which effects growth rates of both residual trees and saplings. The major effect of hunting is direct removal of deer of given sexes and ages from the population.

Simulation Procedure

For a simulation, the user must first specify an initial state of the system (vegetation structure of the forest and size and age-sex structure of the deer population). The model will then simulate the changes in vegetation and in the deer population under a given set of weather conditions and/or management options.

Data Sources

The vegetation dynamics relationships were derived from regression analysis of data collected on the Huntington Forest, whereas deer dynamics relationships were primarily obtained from the literature.

Weather conditions are simulated through the use of actual records from the Huntington Forest.

SIMULATED EXPERIMENTS

Three categories of simulation were employed: verification analyses, sensitivity analyses, and computer experimentation.

Verification Analyses

Verification analyses are used to gauge how accurately the model mimics reality. This is done by simulating a situation in which some results are either known or can be predicted from existing knowledge. In another paper (Cooperrider and Behrend 1980b) a historical situation (low deer densities in old-growth forests) was successfully mimicked. Additional simulations showed that higher deer densities could be supported in cutover forests. Another verification analysis is described below.

In the central Adirondack region of New York, severe winters cause many deer to die of malnutrition (Severinghaus 1947; Severinghaus and Brown 1956). This is believed to result, in part, from browse being covered by snow and thus unavailable to deer, but more importantly due to deep snows reducing the ability of deer to travel.

To investigate this phenomenon, two sets of simulations were run. In both sets, three years of weather data were used for five years in a row. The winters chosen were 1956-1957, a mild winter in which snow depth never exceeded 16 inches; 1963-64, an average winter in the Adirondacks; and 1970-1971, a severe winter with snow depths of up to 54 inches and over 18 inches of snow on the ground from December 21 through April 21. In the first set of simulations, no restriction on deer travel due to snow was imposed, regardless of the type of winter. In the second set, two restrictions on travel were imposed as follows. If the snow depth exceeded 18 inches, a deer's mobility was reduced to 50% of normal; if the snow depth exceeded 22 inches, mobility was reduced to 10%. Several investigators have noted that deer travel less easily through such snow depths (Banasiak 1961; Severinghaus 1947). Hepburn (1959), in a study of the effect of snow cover on mobility, stated that in snow over 22 inches, mobility was reduced to approximately 10% of normal.

With no restriction on travel due to snow, the deer populations respond almost identically to these three dissimilar weather sequences (table 1, fig. 1). This suggests that the browse being covered by snow and thus unavailable is not a particularly important limiting factor.

However, when snow causes a reduction in mobility, the simulated results are quite different (fig. 2). Whereas the difference between consecutive mild winters and consecutive average winters is negligible, consecutive severe winters cause numerous deaths due to malnutrition (table 2). Heavy deer losses due to malnutrition were reported in the central Adirondack region during that type of winter (1970-1971) (Severinghaus 1972). Thus, the model appears to realistically mimic effects of severe winters, and restriction on mobility appears to be the factor

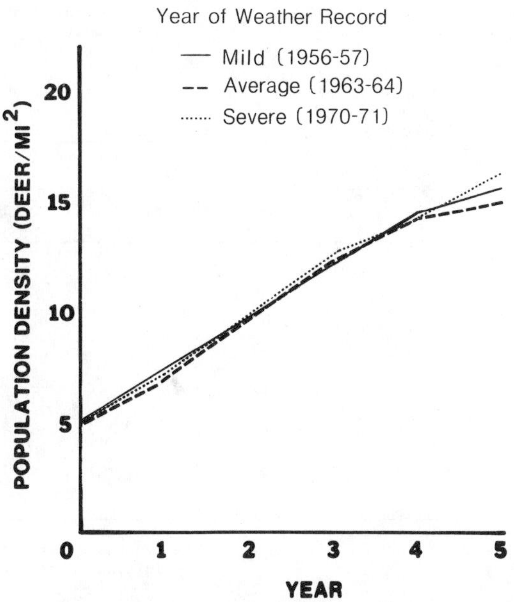

Figure 1.--Effect of continuous mild, average, and severe winters on deer population growth with no restriction on deer mobility due to snow.

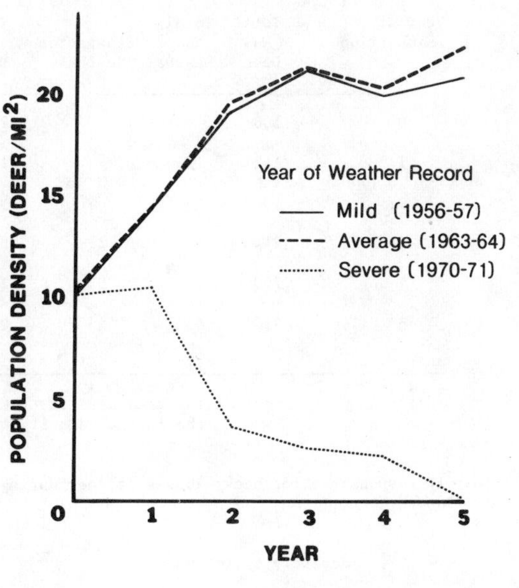

Figure 2.--Effect of continuous mild, average, and severe winters on deer population growth with restriction on deer mobility due to snow.

Table 1.--Simulated population dynamics of deer with no restrictions on mobility due to snow with different winter weather patterns.

Year of Weather Records	Mild Winters			Moderate Winters			Severe Winters		
Year of Simulation	Total No. Deer	Total No. Deaths[1]	Starvation Deaths	Total No. Deer	Total No. Deaths[1]	Starvation Deaths	Total No. Deer	Total No. Deaths[1]	Starvation Deaths
0	50			50			50		
1	72	5		69	5		70	6	
2	97	8		97	7		99	3	
3	122	16	4	123	11	3	126	13	7
4	145	21	7	142	24	11	142	29	15
5	157	34	21	150	27	15	163	21	10

[1] The area used for these simulations was 5 mi^2 so that the initial density was 10 deer/mi^2.

Table 2.--Simulated population dynamics of deer with restrictions on mobility due to snow with different winter weather patterns.

Year of Weather Records	Mild Winters			Moderate Winters			Severe Winters		
Year of Simulation	Total No. Deer	Total No. Deaths[1]	Starvation Deaths	Total No. Deer	Total No. Deaths[1]	Starvation Deaths	Total No. Deer	Total No. Deaths[1]	Starvation Deaths
0	100			100			100		
1	142	10		142	7		104	51	45
2	190	19	4	196	18	7	37	74	72
3	212	44	29	213	50	33	26	11	7
4	200	68	52	204	59	50	23	3	
5	210	65	46	224	52	33	3	20	20

[1] The area used for these simulations was 5 mi^2 so that the initial density was 10 deer/mi^2.

primarily responsible for heavy losses of deer during such winters.

Sensitivity Analyses

Sensitivity analyses are used to determine which relationships of the system are most important in determining outcomes of interest. This is done by systematically varying one or more parameters and observing outputs of interest.

An analysis of the sensitivity of the deer population to changes in the caloric density of forage has been described by Cooperrider and Behrend (1980b). It was shown that small changes in caloric density, of the order of 0.2 kilocalories of digestible energy per gram dry weight of forage, caused substantial changes in deer population density.

Further sensitivity analyses were designed to determine the sensitivity of deer population density to changes in mortality and natality rates. Many studies have dealt with the separate effects of these factors on population dynamics. It therefore seemed necessary to investigate the sensitivity of the model to natality rates and to mortality rates (from accidents, disease and predation) in order to properly interpret results of other simulations. If these parameters rather than habitat factors can have a significant effect on the simulated deer density, then habitat factors can only be interpreted in relation to a given set of mortality and natality rates.

Simulation indicated, however, that within a wide range of mortality and natality rates, the population density was relatively insensitive to such rates. With adult birth rates ranging from 1.6 to 2 fawns per doe, deer densities that can be supported are similar (fig. 3). With annual mortality rates (for all sex and age classes) of from 10 to 20%, the population trends are also quite similar (fig. 4); however, with mortality rates over 25%, populations will decline and stabilize at a lower density. Other simulations have shown that a 33% annual mortality rate is sufficient to stabilize the population. Van Etten et al. (1965), in a study of deer in a square-mile enclosure, suggested that a 33% annual harvest would have stabilized deer density. Behrend et al. (1970) reported that deer harvests of 42 to 56% were sufficient to lower the deer density on a 5,000-acre portion of the Huntington Forest. Thus, within ranges of 1.6-2 fawns per adult doe and 10-20% annual background mortality, the resulting deer density and population trends were quite similar. Therefore, by using mortality and natality rates within these ranges, one can simulate the effects of habitat factors on deer density with some assurance that the deer density obtained is not merely a function of the mortality or natality rate.

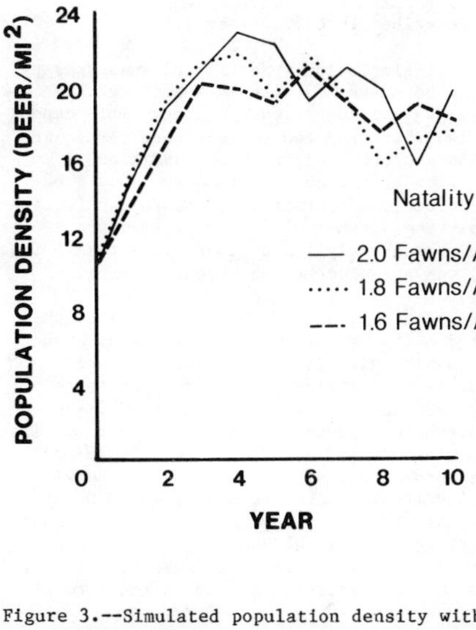

Figure 3.--Simulated population density with different natality rates.

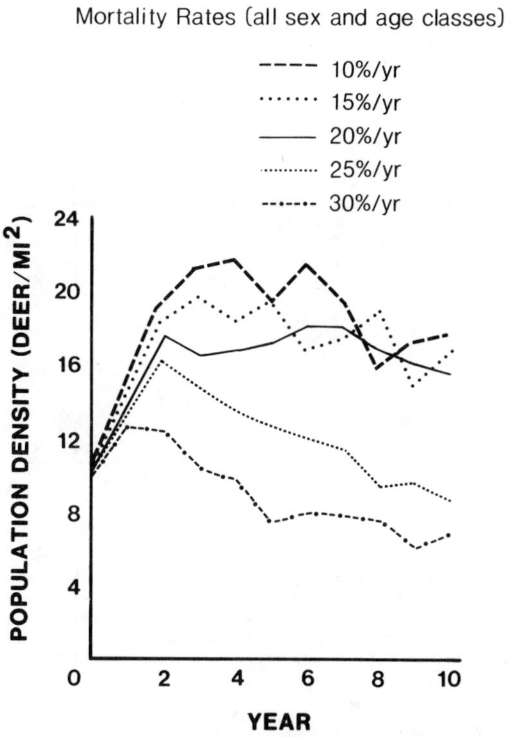

Figure 4.--Simulated population density with different mortality rates from predation, disease and accidents.

Additional sensitivity analyses investigated the importance of herbaceous forage, defined herein as all green material available to deer, including all current growth of grass and forbs as well as non-woody growth on trees and shrubs.

An initial simulation with no herbaceous forage resulted in extinction of the deer population in two years (fig. 5). This suggested that herbaceous material is crucial to deer. For standard simulations, herbaceous forage was modeled as being available in excess from June 1 to October 1. In other analyses, this period in which herbaceous forage was available was modified. An increase of eight weeks of availability of herbaceous forage (May 1 to November 1) produced a population density twice as large as the norm at the end of five years (fig. 5). Similarly, an increase of four weeks at either end of the season results in a slightly higher population density (fig. 5).

The importance of herbaceous forage may explain regional differences in deer densities that are commonly attributed to other factors. Presumably, the period in which herbaceous material is available is closely correlated with the frost-free period. The average frost-free period is 100-120 days for the Adirondacks, and averages 112 days on the Huntington Forest (Pack 1972). This compares to a mean of 150 to

180 days for the Great Lakes Plain and 120 to 150 days for the St. Lawrence Valley. Thus, the central Adirondack region has a relatively short growing season compared to the rest of the state and thus has a shorter period in which herbaceous forage is available to deer. This short growing season may explain why the central Adirondack region is a relatively poor deer range.

The importance of herbaceous forage to deer was further emphasized by a second sensitivity analysis in which the caloric density of herbaceous forage was systematically varied (fig. 6). Caloric densities of 2.8 to 3.4 kilocalories (Kcal) of digestible energy per gram (g) dry weight produced comparable deer densities after five years. However, with a caloric density of 2.6 Kcal DE/g, the resulting deer density is substantially lower.

Computer Experiments

The purpose of computer experiments is to simulate a real situation to obtain insight about the simulated system and also to make predictions about the future.

The principal system simulated was the Huntington Forest, a 15,000-acre property operated for the purpose of conducting wildlife research. This area contains about 2,000 acres of water, the rest being forest. Northern hardwood stands resembling SAF type 25 predominate. These stands have an overstory dominated by sugar maple, beech and yellow birch. Mixed-growth stands on the forest resemble SAF types 24 and 32, with overstories dominated by varying amounts of red spruce, eastern hemlock and yellow birch. During the late 19th centry most of the forest was logged lightly for conifers. Subsequent cutting, for both confiers and hardwoods has occurred principally since 1950. The areas and basal areas used for simulation of the forest are described in Cooperrider (1973).

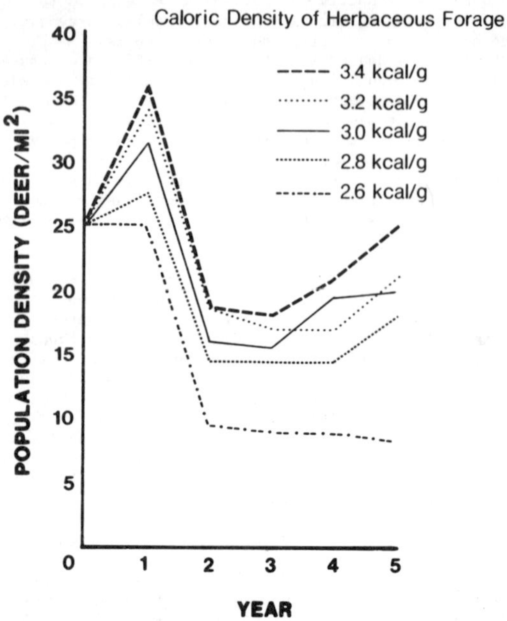

Figure 6.--Simulated deer density with varying caloric densities of herbaceous forage.

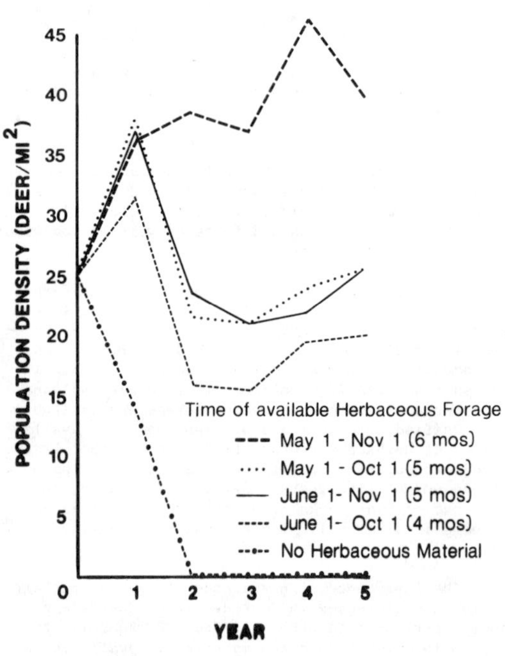

Figure 5.--Population density with herbaceous forage available for varying lengths of time.

An initial simulation investigated what density of deer could be supported with no recent cutting (within 5 years) and no current logging. This experiment indicated that from two to four deer per square mile could be supported under these conditions. In simulations with recent cutting in a few localized areas (but no current operations), the forest is capable of supporting a density of three to seven deer per square mile, though only 1,200 acres, less than 10% of the forest, was affected by the recent cutting.

With current logging, simulation indicated the supportable density is much higher, averaging around 15 deer per square mile (fig. 7). The simulated logging consisted of a reduction from approximately 150 to 50 square feet of basal area over a two-year period in northern hardwood stands of 600 acres. The higher densities are in part due to the browse production in the cutover areas, but also due to the winter forage made available in the form of slash. Other simulations have indicated that a removal of 50 square feet of basal area per acre from a mature northern hardwood stand can make up to 80 kg/acre (176 lb/acre) of hardwood browse available to deer in the form of slash.

This amount is as high as the annual browse production on a clearcut in the Adirondack region.

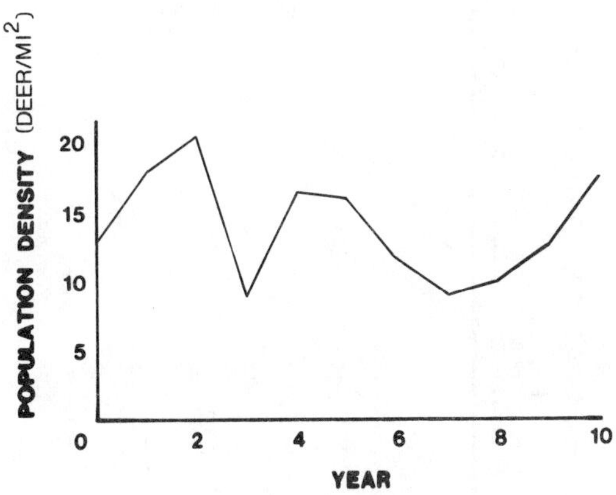

Figure 7.--Simulated year-round population density on the Huntington Forest with current logging operations.

These simulations suggest the importance of logging operations in determining deer densities. Such an inference assumes, of course, that deer can and do move into cutover areas and areas with logging operations and that browse from slash is as palatable and nutritious as other browse. Alkon (1961) reported that browse from fresh-cut or October-cut trees of sugar maple, red maple, and paper birch was accepted by penned deer during winter; however, more work on this subject needs to be done. With regard to movement, there is evidence from radio-telemetry studies on the Huntingon Forest that deer move up to 3 miles in mid-winter to current logging operations within weeks of the beginning of such operations (R. Sage, Pers. Comm.). However, the ability of deer populations to respond to new sources of food is not well documented. Behrend et al. (1970) reported that a substantial reduction of deer density on a 5,000-acre portion of the Huntington Forest did not result in an immediate influx of deer from surrounding areas.

The supportable deer densities just described are lower than densities which have been previously reported from the Huntington Forest. This may be partly explained by the evidence from radio-telemetry that a substantial proportion of the deer population winters off the Forest, whereas many of the censuses were made in the spring and fall. To investigate this, two additional subroutines were written to provide for emigration of a specified proportion of the deer population from the Forest in fall, and subsequent immigration in spring. The deer leaving in fall were chosen randomly, and the number and sex and age distribution of those returning was a function of the specified proportion and the number and sex and age distribution of those surviving the winter on the Forest.

A five-year simulation with 3/4 of the deer population wintering off the Forest is shown in figure 8. A peak fall density of 38 deer per square mile is reached, however, the results suggested that such a density cannot be maintained over a harsh winter such as Year 4 of the simulation. The year of weather data used for that year was 1969-1970. That year, the actual deer population had been reduced through hunting to a density of less than 15 deer per square mile on part of the forest, but a substantial drop in the population still occurred over the winter.

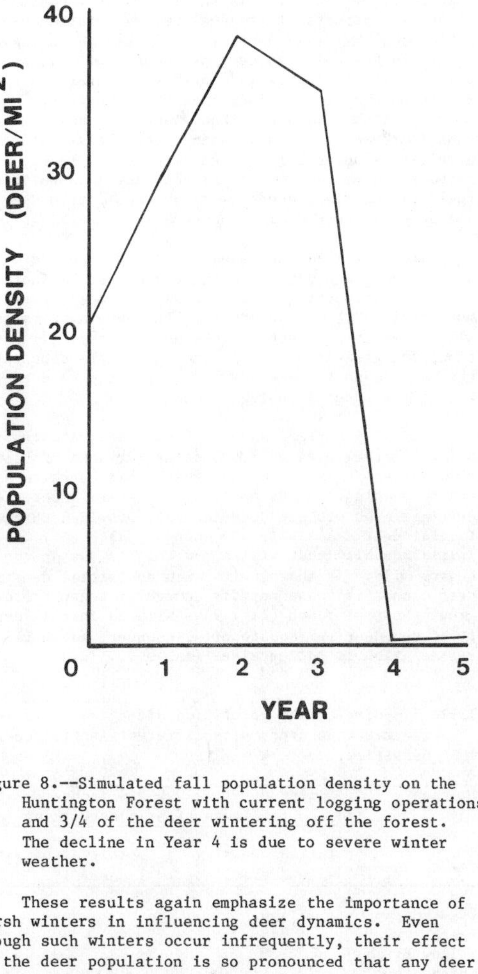

Figure 8.--Simulated fall population density on the Huntington Forest with current logging operations and 3/4 of the deer wintering off the forest. The decline in Year 4 is due to severe winter weather.

These results again emphasize the importance of harsh winters in influencing deer dynamics. Even though such winters occur infrequently, their effect on the deer population is so pronounced that any deer management scheme must take cognizance of them. Since the occurrence of such winters can be neither predicted nor controlled, a flexible and probabalistic approach to deer management is desirable.

Such simulations suggest the range of deer densities that is possible on the Huntington Forest and the conditions and types of management that will produce them. The densities range from less than 3 deer per square mile (year round) with no cutting to a maximum of 38 deer per square mile for limited periods with continuous cutting and 3/4 of the deer wintering off of the Forest. Higher densities can, of course, be attained in local areas with different managment.

The problem of how to obtain adequate regeneration of desirable hardwoods remains unanswered, however. An earlier paper (Cooperrider and Behrend 1980b) stated that a simulated year-round density of three deer per square mile was sufficient to prevent adequate regeneration after 20 years on a cutover area with a residual basal area of 60 square feet per acre. Subsequent simulations have indicated that adequate regeneration of sugar maple can be obtained after 10 to 20 years by keeping the year-round density below 0.5 deer per square mile. These densities are lower than have been reported in the literature. Behrend et al. (1970) reported that adequate regeneration can be achieved with a fall density of 17 deer per square mile. In any case 0.5 deer per square mile is an unrealistically low deer density to maintain through public hunting for periods from 10 to 20 years. Tierson (1969) has shown that electric fences can control deer use of forested areas, but their cost is prohibitive under normal circumstances. Clearly, methods are needed to shorten the period in which the regenerating stand needs to be protected from deer and/or to attract deer to other areas.

Simulation results suggest that the period until stems have grown out of reach of deer can be shortened by a heavier cutting. Since growth of most species is more rapid with less overstory, the understory stems will grow out of reach of deer more quicky. In addition, the greater number of understory stems beneath the lower density overstory can support more deer browsing without it being excessive.

Simulation results suggest that adequate regeneration, defined here as 1,000 stems per acre of valuable hardwoods (sugar maple, yellow birch, white ash and red maple), can be obtained ten years after clearcutting, even without deer control, provided that the initial deer density is low enough (table 3). The initial density that will allow this is one deer per square mile. In short, with such an initial density deer cannot increase rapidly enough to prevent browse growing out of reach (fig. 9). With an initial density of ten deer per square mile, however, adequate regeneration is not obtained (table 3).

Table 3.--Simulated regeneration after ten years in two clearcut areas with different initial deer densities.

Species	Number of stems 0.8-5.6 inches D.B.H. ten years after clearcutting	
	Initial Density = 10 deer/mi^2	Initial Density = 1 deer/mi^2
Sugar Maple	0	984
Beech	244	90
Yellow Birch	0	0
White Ash	1	4
Red Maple	263	200
Misc. Hardwood	0	3
Total Hardwood	508	1,271
Red Spruce	121	33
Hemlock	1	1
Misc. Softwood	3	1
Total Softwood	125	35
Total	633	1,305

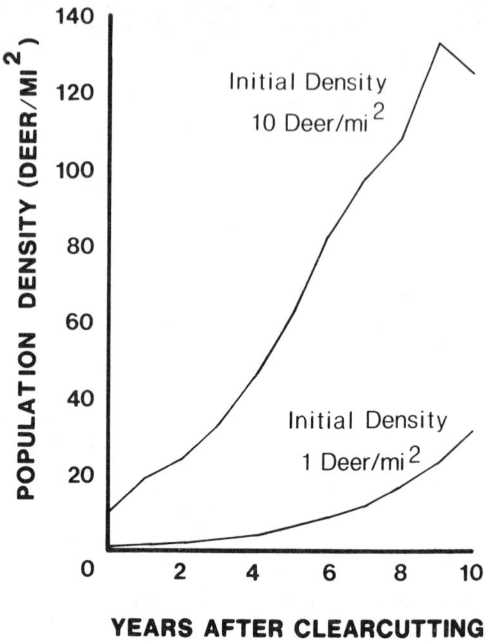

Figure 9.--Deer population growth following clearcutting with different initial densities.

SUMMARY AND CONCLUSIONS

We believe that the model used for the simulations described here can realistically mimic the effect of forest management practices on deer density as well as the effect of deer on forest tree regeneration in a realistic manner as confirmed by "on the ground" studies. A number of sensitivity analyses emphasized the importance of herbaceous forage in the diet of deer. Even though herbaceous forage is treated in this model as being available only from late spring to fall, sensitivity analyses suggest that both its quality and its duration of availability are crucial in determining the density of deer that can be supported. Since herbaceous forage in this model is available from late spring to fall, these analyses suggest that late spring and summer range is extremely important to deer.

Traditionally, deer biologists have concentrated their attention on winter range and minimized the importance of summer range as a factor limiting deer population densities. However, recent studies of deer nutrition, behavior, and physiology have suggested that deer are quite well adapted to the deep snows, cold temperatures, and lack of good quality food typical of winter deer yards in northern forests and that spring and summer range can be extremely important (Mautz 1978). Moen (1978) reviewed many of these studies and based on these and his own work concluded that the timing of the arrival of spring is an important factor in population dynamics of deer.

Simulated experiments have suggested that adequate regeneration of trees may be difficult to establish following cutting to around 60 square feet of residual basal area, even with very low deer densities. Simulated experiments to date have for the most part been unable to suggest feasible management schemes for obtaining regeneration. Simulation results have suggested that heavier cutting can both shorten the period in which deer can prevent regeneration and also lessen the vulnerability of such regeneration to deer browsing. Future modeling efforts should be able to suggest actual management schemes to allow regeneration in the presence of viable deer populations.

LITERATURE CITED

Alkon, P.U. 1961. Nutritional and acceptability values of hardwood slash as winter deer browse. J. Wildl. Manage. 25:77-81.

Banasiak, C.F. 1961. Deer in Maine. Maine Dept. Inland Fisheries and Game, Game Div. Bull. No. 6, 163. p.

Behrend, D.F., G.F. Mattfeld, W.C. Tierson and J.E. Wiley, III. 1970. Deer density control for comprehensive forest management. J. For. 68:695-700.

Cooperrider, A.Y. 1973. Computer simulation of the interaction of a deer population with northern forest vegetation [Ph.D. Thesis]: SUNY College of Environmental Science and Forestry, Syracuse. 220 p.

Cooperrider, A.Y. and D.F. Behrend. 1980a. Simulation of forest dynamics and deer browse production. J. For. 78:85-88.

______________________________. 1980b. Simulation of the interaction of deer with northern forest vegetation. New York Fish and Game J. 27:142-155.

Hepburn, R.L. 1959. Effects of snow cover on mobility and local distribution of deer in Algonquin Park [M.S.Thesis]: Univ. Toronto. 55 p.

Mautz, W.W. 1978. Sledding on a bushy hillside: the fat cycle in deer. Wildl. Soc. Bull. 6:88-90.

Moen, A.N. 1978. Seasonal changes in heartrates, activity, metabolism, and forage intake of white-tailed deer. J. Wildl. Manage. 42:715-738.

Pack, A.B. 1972. Climate of New York. Climatology of the United States No. 60-30. U.S. Dept. of Commerce, 29 p.

Sage, R.W., Jr. 1973. Movements of white-tailed deer contributing to seasonal changes in deer density. N.Y. Dept. Env. Cons. P-R Proj. Rept. W-105-R-12, 16 p.

Severinghaus, C.W. 1947. Relationship of weather to winter mortality and population levels among deer in the Adirondack region of New York. Trans. N. Am. Wildl. Nat. Resour. Conf. 12:212-223.

________________. 1972. Weather and the deer population. The Conservationist 27:28-31.

Severinghaus, C.W. and C.P. Brown. 1956. History of the white-tailed deer in New York. New York Fish and Game J. 3:129-167.

Tierson, W.C. 1969. Controlling deer use of forest vegetation with electric fences. J. Wildl. Manage. 33:922-926.

Tierson, W.C., E.F. Patric and D.F. Behrend. 1966. Influence of white-tailed deer on the logged northern hardwood forest. J. For. 64:801-805.

Van Etten, R.C., D.F. Switzenberg and L. Eberhardt. 1965. Controlled deer hunting in a square-mile enclosure. J. Wildl. Manage. 29:59-73.

Webb, W.L., R.T. King and E.F. Patric. 1956. Effect of white-tailed deer on a mature northern hardwood forest. J. For. 54:391-398.

A SIMULATION MODEL OF THE MIGRATORY GRASSHOPPER (MELANOPLUS SANGUINIPES)[1]

D. W. Hilbert and J. A. Logan[2]

Abstract.--A model is described which simulates the intraseasonal dynamics of the migratory grasshopper (Melanoplus sanguinipes) and the consumption of forage by the population. The simulated phenological progression of the population, mortality, and oviposition rates are reasonable when compared with field data in the literature. Consumption rates are also within the range expected for a moderately dense grasshopper population, about 7% of aboveground net primary production. A sensitivity analysis of the consumption submodel suggests that termperatures have a great influence on consumption rates along with mortality rates and grasshopper weight.

INTRODUCTION

The migratory grasshopper, Melanoplus sanguinipes (Fabricius), is an economically important grasshopper distributed throughout North America (Gurney and Brooks 1959). While consumption by grasshoppers is generally on the order of ten percent of primary production (Bailey and Riegert 1973, Van Hook 1971, Wiegert 1965, Smalley 1960), during outbreaks, their effect on primary production of rangelands and crops can be devastating. Most such outbreaks in North America involve melanopline species (Dempster 1963), and several major outbreaks in the past century have involved M. sanguinipes.

Because of its economic importance and the ease with which it can be raised in the laboratory, an extensive body of literature has been published that details the biology of M. sanguinipes. A truly integrated understanding of this species' population dynamics and its interaction with primary producers has, however, yet to emerge. Statistical and simulation modeling are two means of acquiring this integrated understanding, but previous modeling efforts with this species have dealt only with specific processes and do not provide a complete model of the grasshopper's life system.

Most extant models for M. sanguinipes are based on a linear degree-day approach. For example, Gage et al. (1976) developed regression equations from extensive field data, estimating the time of occurrence of the life stages of M. sanguinipes, M. bivittatus, and Camnula pellucida. Population densities were not modeled, and the three species were not separated. Mukerji et al. (1977) later developed multiple regression models to predict densities of the same three species. Randell and Mukerji (1974) developed regression models to predict hatching of M. sanguinipes, and egg development and mortality were modeled by Mukerji and Gage (1978). While these models can be valuable in management applications, they do not provide the unified structure a simulation model is capable of.

The model described here is an attempt to integrate the diverse information available on the migratory grasshopper and is primarily intended to simulate the intraseasonal dynamics of this grasshopper and its interaction with primary producers. Four specific objectives were formulated for the model; three are oriented towards prediction, and one involves the nature of the model's structure. The objectives are

1. To predict the phenological progression of a population within a season, in other words, the timing of emergence, development through the nymphal instars, adult maturation, and oviposition.
2. To simulate the density of hoppers and adults within a season.
3. To estimate the consumption by the grasshopper population, including seasonal trends.
4. To develop the model structure in a comprehensive manner, so that future improvements in model realism can be made easily.

The first three objectives are relatively straightforward. Of these, the second objective, simulation of grasshopper density, is least likely to be realized given the current state of our knowledge. A good quantitative understanding of the many sources of mortality is largely unavailable. Nevertheless, an attempt was made to simulate density as accurately as possible.

The last objective is more general and is related to the ease with which new information can be incorporated into the model. The first three objectives could possibly be met with relatively simple, nonmechanistic models. However, one of the best features

[1]Paper presented at the Third International Conference on State-of-the-Art in Ecological Modeling, Colorado State University, May 24-28, 1982. Work reported herein was supported in part by Colorado State University Agricultural Experiment Station Project 53.

[2]David W. Hilbert is a Research Associate at the Natural Resource Ecology Laboratory and Jesse A. Logan is a Population Ecologist at the Natural Resource Ecology Laboratory, Colorado State University, Fort Collins, Colorado 80523, U.S.A.

of simulation models of the type described here is their utility as research tools. The model can be thought of as a macrohypothesis incorporating knowledge of many subprocesses. Such a model can provide insight into the operation of the population system as a whole and suggest the most valuable avenues for future research, as well as clearly identify major gaps in our present knowledge. Consequently, an important objective was to create a model structure that represents the total life system of M. sanguinipes. Specific functions for many processes are not possible at this time, but the inclusion of dummy functions into the model structure at these points improves the future utility of the model as a research tool.

MODEL OVERVIEW

The model is formulated as a system of difference equations with a one-day time step. Some processes, particularly those involving temperature as a major driving variable, are calculated hourly; but these are summed for each day before the state variables are updated. The state variables are the density and developmental state (physiological age) of eggs, nymphs, and adults. Each of these is composed of a series of cohorts representing eggs laid on a particular day, nymphs hatching on a given day, or individuals that all reached the adult stage on the same day. Thus, the number and development of each cohort represents the state of the population. Total consumption and feces produced by the population are also simulated on a daily basis.

The primary driving variables for the sytem are temperature and soil water. These variables are simulated from standard weather data (Parton and Logan 1981; Parton, submitted). Temperature plays a central role in regulating most processes of grasshopper population dynamics (Dempster 1963). Soil water primarily affects plant growth and nutritional quality and thus indirectly affects grasshopper dynamics. Direct effects of soil water in this model are limited to egg mortality.

The grasshopper model was designed to interact with a primary producer model developed at the Natural Resource Ecology Laboratory (Parton et al. 1978). The model can, however, be run independently without inputs from a producer model and will simulate total daily consumption by the grasshopper population. The basic structure of the model is presented in figure 1.

All life stages of grasshoppers--eggs, nymphs, and adults--are represented in the model by daily cohorts. Thus, all nymphs hatching on a given day are placed in a cohort, and development, consumption, and mortality are calculated separately for each cohort. In order to simplify the documentation of the functions used in the model, a subscript is not used to represent cohorts. Rather, a cohort subscript is implied for all variables relating to eggs, nymphs, or adults. Thus, the number of nymphs in cohort i on day t is represented by N_t. Constants and parameters are represented by the letter P followed by a number. All of these parameters are listed in table 1, along with their definition, numerical value, and source.

DOCUMENTATION

Number of Nymphs

The number of nymphs is updated daily by

$$N_{t+1} = N_t(1 - M_t) \quad , \qquad (1)$$

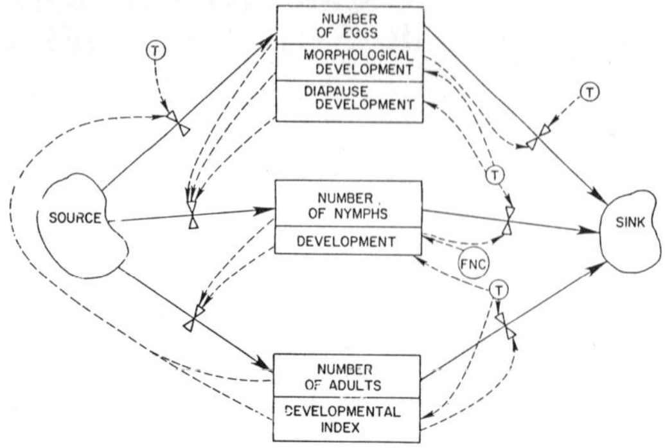

Figure 1.--Flow diagram of the population model. The state variables are divided into daily cohorts for which both numbers and developmental state are simulated. T is temperature and FNC is forage nitrogen concentration.

where,

N_t = number of nymphs in a cohort on day t

M_t = rate of mortality of the cohort on that day

Entire cohorts of nymphs are removed from the nymphal population when they reach adulthood as determined by the development submodel. New cohorts of nymphs are added when eggs hatch as determined by the egg development submodel.

Mortality of nymphs is represented by three distinct functions. A general mortality function (M_1) represents losses resulting from all the continuous and random causes of death; M_2 represents mortality caused by extremes of temperature, and M_3 is mortality caused by pesticides or other control measures.

The total mortality rate for a cohort on day t used in equation (1) is then given by

$$M_t = M_{1,t} + M_{2,t} + M_{3,t} - M_{1,t} \cdot M_{2,t} - M_{2,t} \cdot M_{3,t} - M_{1,t} \cdot M_{2,t} \cdot M_{3,t} \quad . \qquad (2)$$

Since the three causes of mortality may be thought of as the probability of death for any one nymph attributable to each cause, the simultaneous probabilities, the three products in equation (2), must be subtracted from the sum of the independent probabilities to give the total probability from all causes.

The basic mortality rate (M_1) is modified by the average nitrogen concentration (FNC) of the diet selected by nymphs. Smith and Northcott (1951) demonstrated a significant effect of dietary nitrogen on mortality rates of M. sanguinipes nymphs. The percent nitrogen content of the hydroponically grown wheat fed to their grasshoppers was extremely high. Their three treatments had 6.16, 4.29, and 3.33 percent foliar nitrogen. Very high mortality rates were found with the wheat of lowest nitrogen content. Since reasonable foliar nitrogen contents in the field would rarely exceed 3.5 percent (Allen et al. 1974), Smith and Northcott's (1951) results are difficult to interpret. Thus, their results were used in only a qualitative way in the model.

Table 1.--Parameters and constants used in the model along with the source of data upon which the parameters are based or the parameters themselves

Parameter or constant	Equation	Value	Source
P1	4	-0.00223	Smith and Northcott (1951)
P2	4	0.02195	Smith and Northcott (1951)
P3	8	0.10000	Parker (1930)
P4	9	0.10000	Parker (1930)
P5	11	1.04820	Pfadt et al. (1979), Pfadt and Smith (1972)
P6	11	71.60680	Pfadt et al. (1979), Pfadt and Smith (1972)
P7	11	7.10830	Pfadt et al. (1979), Pfadt and Smith (1972)
P8	15	0.22314	Parker (1930)
P9	16	2.23990	Parker (1930)
P10	16	0.23990	Parker (1930)
P11	16	0.00639	Parker (1930)
P12	17	0.69000	Pickford (1958), Pfadt (1979), Smith (1966, 1968)
P13	18	function	Equation (20)
P14	18	58.62000	Hilbert and Logan (in press)
P15	18	45.02000	Hilbert and Logan (in press)
P16	18	3.43300	Hilbert and Logan (in press)
P17	18	7.65700	Hilbert and Logan (in press)
P18	20	0.26760	Smith and Northcott (1951)
P19	20	2.00000	Smith and Northcott (1951)
P20	21	0.50000	Smith and Northcott (1951)
P21	22	0.67785	Hilbert and Logan (submitted)
P22	22	28.51950	Hilbert and Logan (submitted)
P23	22	144.10720	Hilbert and Logan (submitted)
P24	22	122.69980	Hilbert and Logan (submitted)
P25	22	-0.36154	Hilbert and Logan (submitted)
P26	23	0.12474	Hilbert and Logan (submitted)
P27	23	21.12270	Hilbert and Logan (submitted)
P28	23	53.99890	Hilbert and Logan (submitted)
P29	23	19.35190	Hilbert and Logan (submitted)
P30	23	-7.91103	Hilbert and Logan (submitted)
P31	31	0.47300	Van Hook (1971)
P32	31	1.24800	Van Hook (1971)
P33	31	0.06930	Van Hook (1971)
P34	31	20.00000	Van Hook (1971)
P35	32	0.05900	Van Hook (1971)
P36	32	1.27200	Van Hook (1971)
P37	33	396.66000	Pfadt et al. (1979)
P38	33	88.85000	Pfadt et al. (1979)
P39	33	-4.80600	Pfadt et al. (1979)
P40	34	628.91000	Pfadt et al. (1979)
P41	34	144.97000	Pfadt et al. (1979)
P42	34	-4.71800	Pfadt et al. (1979)
P43	35	229.70000	Pfadt et al. (1979)
P44	35	139.90000	Pfadt et al. (1979)
P45	35	10.00000	Pfadt et al. (1979)
P46	36	274.03000	Pfadt et al. (1979)
P47	36	123.16000	Pfadt et al. (1979)
P48	36	10.00000	Pfadt et al. (1979)
P49	39	5.20300	Wiegert (1965)
P50	40	0.67818	Smith (1959)
P51	40	-1.09820	Smith (1959)

Data from both the field and laboratory suggest that survival can be approximated by a negative exponential, implying that the general mortality rate is constant for a given environment. We have chosen to represent mortality rates as a constant per unit of physiological development rather than per unit of time, since physiological time is more meaningful for ectotherms than is time itself (Taylor 1981). Let S_D be the proportion of nymphs surviving to D proportion of total nymphal develompment, then

$$S_D = e^{-rD} , \qquad (3)$$

where r = the mortality rate.

Lacking better information but given results in Smith and Northcott (1951), r is represented as a linear function of percent dietary nitrogen (FNC):

$$r = P1 \cdot FNC + P2 , \qquad (4)$$

where P1 and P2 = parameters.

Combining equations (3) and (4) gives

$$S_D = e^{-(P1 \cdot FNC + P2)D} \quad . \qquad (5)$$

The number of nymphs surviving to the new developmental state $(D + \Delta D)$ is then

$$N_{D+\Delta D} = N_D e^{-(P1 \cdot FNC + P2)\Delta D} \qquad (6)$$

Thus, the mortality rate (M_1) is

$$M_1 = 1 - e^{-(P1 \cdot FNC + P2)\Delta D} \qquad (7)$$

This mortality function is plotted in figure 2 for three different values of FNC.

Mortality caused by extreme temperatures (M_2) is represented by two linear functions. Death at low temperatures is a piecewise linear function of minimum canopy temperature (T_{min})

$$M_2(\text{cold}) = \begin{cases} 0 & T_{min} > 0.0°C \\ -P3 \cdot T_{min} & -10.0°C < T_{min} < 0.0°C \\ 1 & T_{min} < -10.0°C \end{cases} \qquad (8)$$

Mortality at high temperatures, greater than 45°C, is similarly represented by

$$M_2(\text{hot}) = \begin{cases} 0 & T_{max} < 45.0°C \\ P4 \cdot T_{max} - 4.5 & 55.0°C > T_{max} > 45.0°C \\ 1 & T_{max} > 55.0°C \quad . \end{cases} \qquad (9)$$

Mortality data in the literature is insufficient to allow more precise formulation of the rates of mortality at extreme temperatures than that employed here. Equations (7) and (8) are, however, qualitatively based on information in Parker (1930).

Mortality caused by control efforts (M_3) is not modeled directly at this time, since the functional form and parameters used depend on the specific situation.

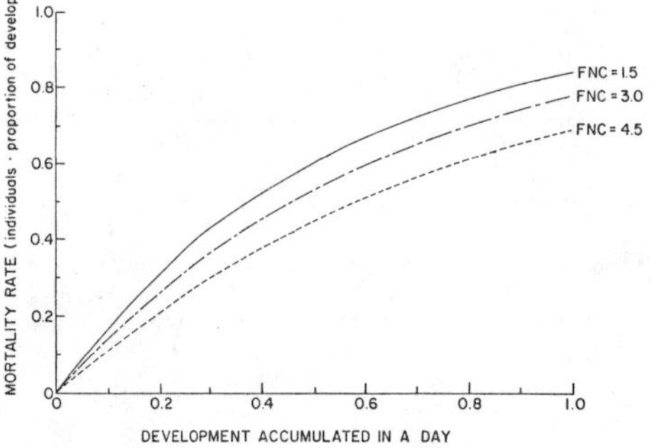

Figure. 2--Mortality rate as a function of percent forage nitrogen concentration (FNC) from equation (7).

Number of Adults

A cohort of adults (A) is added on each day that one or more cohorts of nymphs reach adulthood (i.e., D_n, the developmental state of a nymphal cohort, is greater than or equal to 1.0). Numbers of adults in each cohort are updated daily by the difference equation

$$A_{t+1} = A_t \; (1-M_t) \quad , \qquad (10)$$

where,

M_t = a composite mortality rate function as utilized for nymphs.

Once again mortality is represented by a general mortality function (M_4), mortality due to extreme temperatures (M_5), and mortality representing possible control measures (M_6). The three are combined as for nymphs (eq. 2).

A nonlinear function was found which fit survival data from three life tables in Pfadt et al. (1979) and Pfadt and Smith (1972). These life tables give the number of adults surviving as a function of time. These data were transformed to give percent survival S_A as a function of the proportion of total adult life span completed [the developmental index of adults (DI) discussed latter]. The function

$$S_A = 1 - P5/[1 + P6 \cdot e^{(-P7 \cdot DI)}] \quad , \qquad (11)$$

was fit to these data by Marquardt's non-linear least squares method (Marquardt 1963). This function is plotted in figure 3. The proportional mortality experienced in one day is

$$M_4 = \frac{A_t - A_{t+1}}{A_t} \quad . \qquad (12)$$

A_{t+1} is found from equation (11) by

$$A_{t+1} = A_o \cdot S_A \quad , \qquad (13)$$

where,

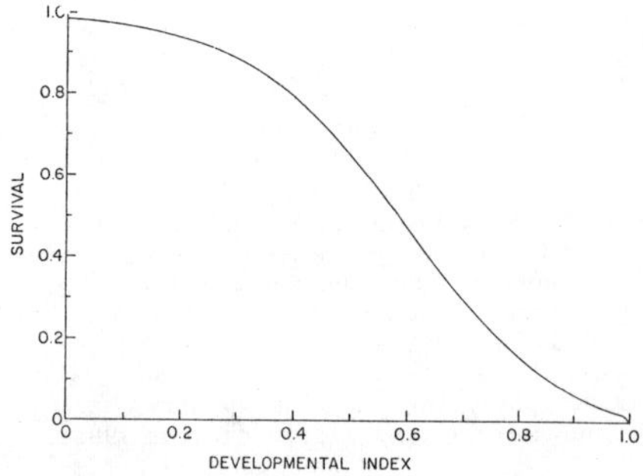

Figure 3.--Survival of adults as a function of physiological age (developmental index) from equation (11).

A_o = the initial number of adults in a cohort
S_A = the expected survival to developmental state DI.

Mortality of adults due to extreme temperatures (M_5) is identical to that for nymphs (eqs. 7 and 8). As with nymphs, M_6 is not specified at this time.

Number of Eggs

Egg mortality is modeled similarly to that for nymphs. The number of eggs at time t+1 is simply

$$E_{t+1} = E_t\ (1-M_t) \qquad (14)$$

where,
M_t = a composite function as in equation (2).

M_7 is the basic mortality rate as a function of the amount of morphological development accumulated in a day (ΔD_E). Similar to equation (7), M_7 is represented by

$$M_7 = 1 - e^{(-P8\ \cdot\ \Delta D_E)} \qquad (15)$$

where,
P8 = the constant, instantaneous rate per unit of development.

Since this mortality is a function of daily development, mortality from this source will equal zero while the egg is in diapause.

Egg mortality is also caused by extremely low soil temperatures; represented here by a quadratic function of daily minimum soil temperature (T_{SMIN}) and based on data in Parker (1930):

$$M_8 = P9 + P10\ \cdot\ T_{SMIN} + P11\ \cdot\ T^2_{SMIN}\quad . \qquad (16)$$

Egg death may also be caused by dessication (Mukerji and Gage 1978). This source of mortality (M_9) is not specified at this time since soil water is not included in the abiotic portion of the model. Once again, these three sources of mortality are combined as in equation (2). Entire cohorts of eggs are lost from the egg population when their development reaches or exceeds 100 percent. All egg cohorts hatching on a given day are thus added into a new cohort of nymphs.

A new cohort of eggs is created on any day that eggs are laid by one or more adult cohorts. Oviposition is simulated as a function of numbers of adult females, temperature, and physiological age of the adults (DI). A maximum rate of egg production per female per degree day was estimated from data in Pickford (1958), Pfadt (1979), Smith (1966), and Smith (1968). This value, P12 = 0.69 eggs per degree day greater than 20°C, is then multiplied by the average degree days greater than 20°C and the number of adults in a cohort divided by 2 (because of the assumed 1:1 sex ratio) to give the maximum egg production on that day by that cohort. This value is then adjusted by an egg production factor (EPF) that is a function of adult physiological age (DI). Thus, oviposition on day t by a cohort of adults is given by

$$O_t = EPF(DI)\ \cdot\ P12\ \cdot\ DD\ \cdot\ A_t/2 \qquad (17)$$

where,
DD = an estimate of degree days based on the average of hourly temperatures greater than 20°C on day t.

The function EPF(DI) is presented in figure 4. This piecewise linear function is based on several literature sources (Barnes 1955, Pfadt and Smith 1972, Pfadt et al. 1979, Smith 1966, Smith 1968, and Smith 1970) and adjusts oviposition rates to account for the maturation and senescense of adult females. EPF equals zero for adult cohorts with a developmental index less than 0.20 or greater than 0.93. Cohorts with developmental indices between 0.25 and 0.85 lay eggs at the maximum rate (i.e., EPF = 1).

Development of Nymphs

Rates of nymphal development are primarily a function of temperature although diet quality and quantity may also influence developmental rates (Smith and Northcott 1951, Barnes 1955). Linear degree-day models are often used to represent development (Gage et al. 1976) but this approach ignores the observed nonlinearity at high and low temperatures. Consequently, a new nonlinear model was employed, developed by the technique of matched asymptotic expansions (Lin and Segel 1974, Logan et al. 1976), and fit to data of the development of M. sanguinipes from a variety of sources (Hilbert and Logan 1981). This model is described in detail and compared to two other nonlinear models by Hilbert and Logan (in press). The equation is

$$r(T) = P13\ \frac{T^2}{T^2 + (P14)^2} - e^{-(P15-T)/P16} \qquad (18)$$

where,
r(T) = the rate of development per day
T = canopy air temperature P17
P13-P16 = shape parameters
P17 = approximately the minimum temperature at which development occurs.

A plot of this function is given in figure 5.

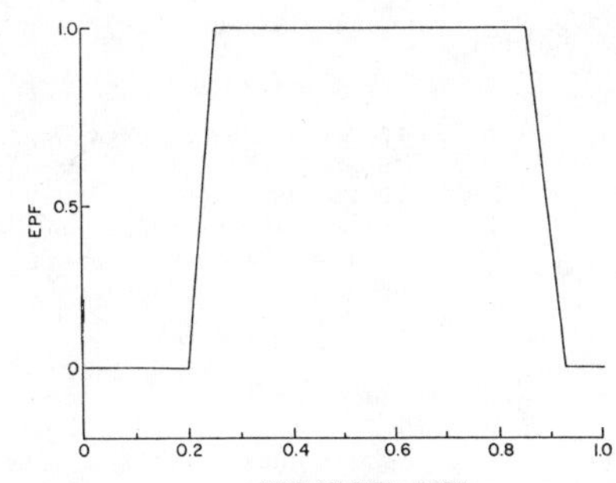

Figure 4.--The parameter (EPF) controlling egg production as a function of adult physiological age (developmental index) used in equation (17).

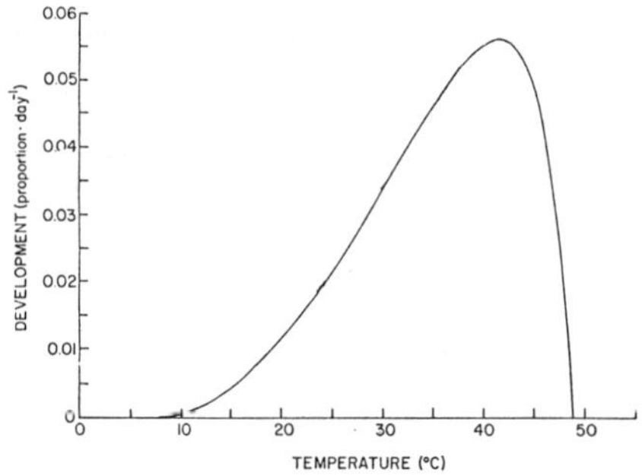

Figure 5.--Nymphal developmental rate as a function of temperature from equation (18).

Development of nymphs is updated daily by the difference equation

$$D_{N,t} = D_{N,t-1} + \sum_{h=1}^{24} r(T_h)/24 \quad , \qquad (19)$$

where,

$D_{N,t}$ = the developmental state of a cohort of nymphs on day t.

t_h = hourly canopy air temperature.

When $D_{N,t}$ is greater than or equal to 1.0 all of the individuals in the cohort are transferred into a cohort of adults.

Because diet quality may influence nymphal development strongly (Smith and Northcott 1951) an equation is included in the simulation which adjusts r(T) as a function of the percent nitrogen in the diet (FNC). As discussed in the section on nymphal mortality, Smith and Northcott's data on the effect of dietary nitrogen on nymphal development are difficult to interpret and are used here in only a qualitative way. By varying P13 as a function of FNC, the rate of development at any temperature can be adjusted up or down without changing the general shape of equation (18). The function employed is

$$P13 = P18(1 - e^{[-P19(FNC - P20)]}) \quad , \qquad (20)$$

where,

P18 = the asymptote at high FNC

P19 = controls the rate of approach to the asymptote

P20 = the minimum nitrogen concentration in the food at which development can occur.

Development of Adults

The aging of adults has not been extensively studied but the model requires some measure of the physiological age of adults to regulate oviposition and mortality rates. Thus, a developmental index (DI) was developed for adults based on maximum life span of adults held at constant temperatures in the laboratory (Pfadt and Smith 1972, Smith 1966, Smith 1968, Smith 1970). A maximum degree days (greater than 7°C) of adult life was estimated from these sources and a linear degree-day model was assumed to give

$$DI_t = DI_{t-1} + DD/P20 \qquad (21)$$

where,

DD = an estimate of degree days based on the average of hourly temperatures greater than 7°C

P20 = the maximum degree-days of adult life.

Development of Eggs

Simulating Egg development and the prediction of eclosion are complicated by the occurrence of a diapause state in which morphological development ceases. This is particularly difficult for M. sanguinipes since the occurrence and duration of diapause are quite variable in this species. Hilbert et al. (submitted) have developed a model which accounts for this variability and predicts the rates of egg development as a function of temperature. The model employed here is identical to that in Hilbert et al. (submitted) and consists of two equations; one represents the rate of morphological development (r_m) and the other models the rate of physiological processes involved in diapause "development" (r_d). The functional form of r_m and r_d is identical to that used for nymphal development (eq. 18),

$$r_m(T) = P21 \left[\frac{T^2}{T^2 + (P22)^2} - e^{-(P23 - T)/P24} \right] , \qquad (22)$$

where T = soil temperature at 5 cm -P25, and

$$r_d(T) = P24 \left[\frac{T^2}{T^2 + (P27)^2} - e^{-(P28 - T)/P29} \right] , \qquad (23)$$

where T = temperature -P30.

Plots of these equations are given in figure 6. Morphological development (D_m) and diapause development (D_d) occur simultaneously from the time the egg

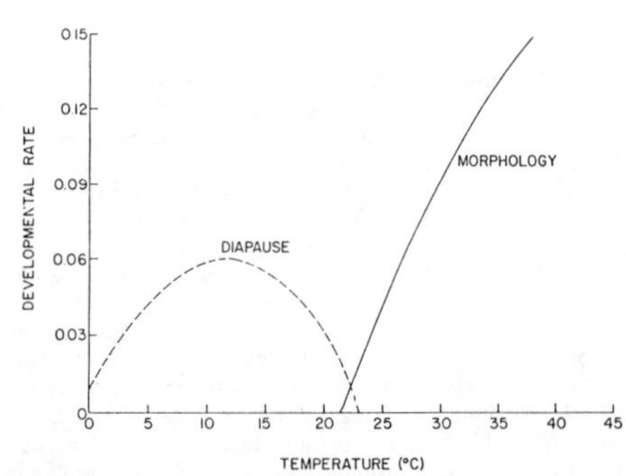

Figure 6.--Rates of morphological and diapause development as functions of temperature from equations (22) and (23).

is laid as long as temperature is above the lower threshold for both. Morphogenesis stops when D_m is 0.80 if D_d is less than 1.0 and does not resume until D_d is complete (i.e., $D_d \geq 1.0$). In this way both the occurrence and duration of diapause are simulated. A fuller explanation and test of the model are given in Hilbert et al. (submitted).

In the simulation, D_m is updated daily by

$$D_{m,t} = D_{m,t-1} + \sum_{h=1}^{24} r_m(T_h)/24 \qquad (24)$$

and D_d by

$$D_{d,t} = D_{d,t-1} + \sum_{h=1}^{24} r_d(T_h)/24 \quad , \qquad (25)$$

where r_m = zero when $D_m = 0.80$ and $D_d < 1.0$.

When D_m is greater than or equal to 1.0, that cohort of eggs is transferred to a new cohort of nymphs.

Consumption Submodel

The consumption model is based on energy flow (cal), but the objective is to estimate mass (mg) of forage consumed and wasted and mass of feces produced by the simulated grasshopper population. Although the consumption model is an integrated part of the total model, field data of population densities and developmental states could be used to drive the model and estimate consumption by the field population.

The basic structure of the consumption submodel is given in figure 7. The model is quite simple conceptually. Total consumption (cal) is given by the energy assimilated (A) divided by the efficiency of assimilation (E). This assimilation efficiency is often termed the approximate digestibility (Waldbauer 1968).

Algebraically,

$$C = \frac{A}{E} \quad , \qquad (26)$$

where,

$$E = \frac{\text{weight of food ingested} - \text{weight of feces}}{\text{weight of food ingested}} \quad . \qquad (27)$$

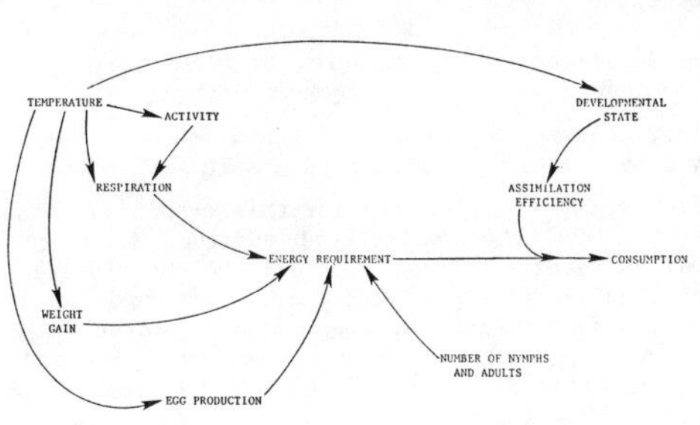

Figure 7.--Control diagram of the consumption submodel.

This efficiency could also be calculated in terms of energy ingested and egested. Assimilated energy is composed of the energy used in respiration (R) and energy used for growth (G). Thus,

$$A = R + G \qquad (28)$$

and

$$C = (R + G)/E \quad . \qquad (29)$$

Respiration is a function of grasshopper weight (W), activity (ACT), and resting metabolic rate (RMR) and is calculated hourly for each cohort of nymphs and adults:

$$R = \sum_{h=1}^{24} ACT(T_h) \cdot RMR(T_h W_t) \quad . \qquad (30)$$

ACT is a function that increases the respiration rate as a function of hourly temperature (T_h) because of increased activity of the animals. This piecewise linear function is plotted in figure 8 and is qualitatively based on information in Parker (1930) and Shotwell (1930). Resting metabolic rate is a function of hourly temperature and the average weight of a cohort. The function used is

$$RMR = P31 \cdot (DWT^{P32}) \cdot e^{[P33(T_h - P34)]} \quad , \qquad (31)$$

and the parameters are from Van Hook (1971). A plot of equation (31) is given in figure 9 with weight held constant. Dry weight (DWT) is a function of fresh weight (WT) given by

$$DWT = P35(WT^{P36}) \quad , \qquad (32)$$

also from Van Hook (1971).

Growth is represented as a deterministic function of physiological age. The fresh weights of male and female nymphs and adults are given by four functions fit to data in Pfadt et al. (1979). Fresh weight (mg) of male nymphs (WTM) is given by

$$WTM = P37/[1 + P38 \cdot e^{(P39 \cdot D_N)}] \quad , \qquad (33)$$

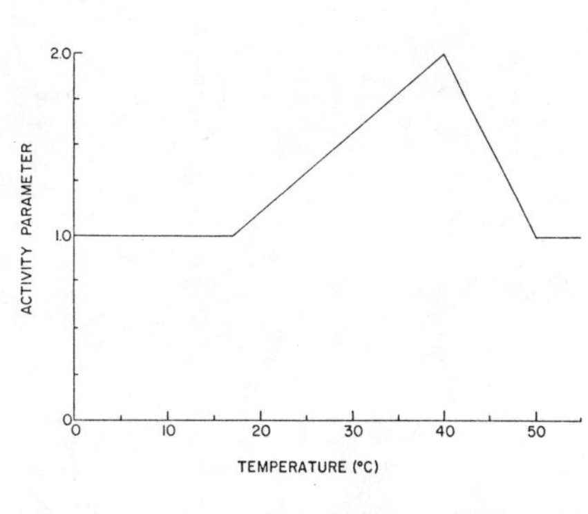

Figure 8.--Parameter (ACT) increasing respiration as a function of temperature used in equation (30).

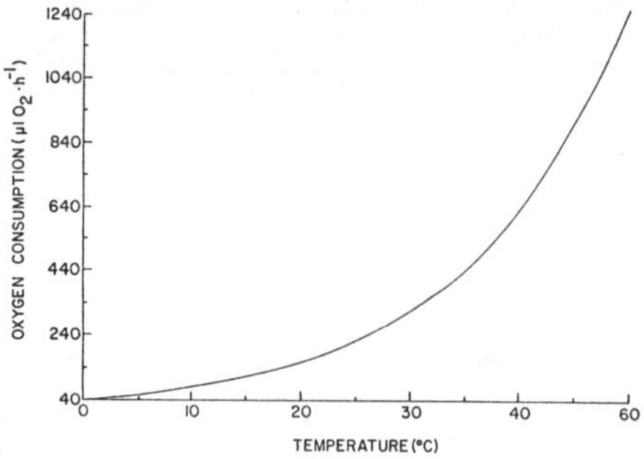

Figure 9.--Respiration rate of nymphs weighing 100 μg as a function of temperature from equation (31).

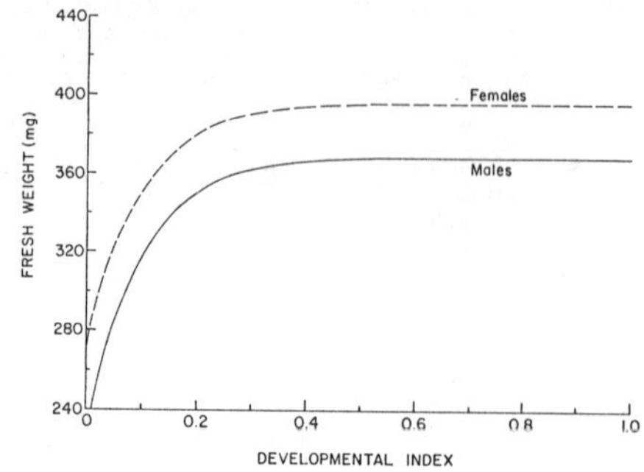

Figure 11.--Fresh weight of adult females and males as functions of physiological age from equations (35) and (36).

weight of female nymphs (WTF) is

$$WTF = P40/[1 + P41 \cdot e^{(P42 \cdot D_N)}] \quad , \tag{34}$$

weight of male adults (ADWTM) is

$$ADWTM = P43 + P44\ [1 - e^{(-P45 \cdot D_A)}] \quad , \tag{35}$$

and weight of adult females (ADWTF) is

$$ADWTF = P46 + P47[1 - e^{(-P48 \cdot D_A)}] \quad . \tag{36}$$

Plots of nymphal and adult weights are given in figures 10 and 11.

Since a 1:1 sex ratio is assumed throughout the model, the average weight of nymphs (AWN) is

$$AWN = (WTM + WTF)/2 \quad , \tag{37}$$

and the average weight of adults is

$$AWA = (ADWTM + ADWTF)/2 \quad . \tag{38}$$

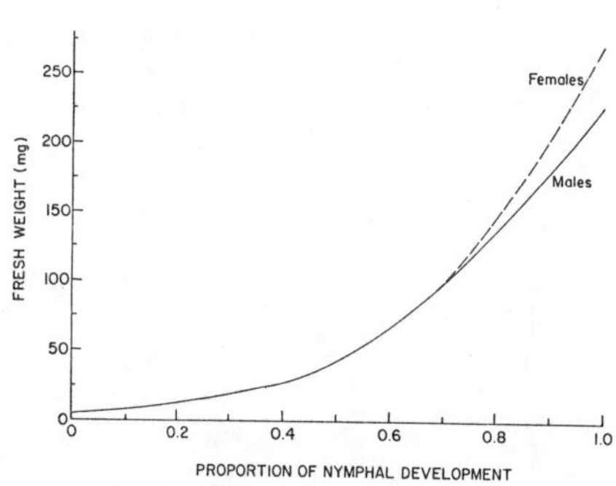

Figure 10.--Fresh weight of female and male nymphs as functions of development from equations (33) and (34).

Growth, of course, is a function of the nutritional state of an animal, but there is insufficient information available at this time to effectively model weight gain in a mechanistic way. The functions used are based on data from a laboratory population under nearly ideal conditions. Thus, weights may be considered the maximum that would be encountered in nature. Weight gain in a day is simply the change in average fresh weight from the last day, based on the change in developmental state. To calculate the energy used for growth, the fresh weight gain (WGAIN) is converted to the ash-free dry weight (AFDWG) by finding the dry weight gain (DWGAIN) from equation (32) and subtracting the proportion DWGAIN · 0.0162 of ash (Van Hook 1971). Energy used for growth is, then,

$$G = AFDWG \cdot P49 \quad , \tag{39}$$

where P49 = average caloric value of grasshopper tissue (Weigert 1965).

Assimilation efficiency is frequently a function of developmental state in grasshoppers, and the equation

$$E = P50 \cdot e^{(P51 \cdot D_N)} \tag{40}$$

was fit to data in Smith (1959). Figure 12 is a plot of this function. The reasons for the observed decline in efficiency with age are not fully understood, but several are given in Waldbauer (1968). These include changes in the portions of leaves eaten, size of bites, and ratio of gut surface area to total size of the hopper.

Assimilation efficiencies of adults are assumed to be equal to those of nymphs with $D_N = 1.0$. The energy consumed by a cohort is then found by multiplying C by the number of individuals in the cohort (N_t). Energy going to feces (F) for this cohort is consumed energy minus the assimilated energy. Consumed and fecal energies are summed for all the cohorts and then converted to milligrams of forage consumed and feces produced by dividing by the average caloric value of the forage, 4.229 cal per mg Golley 1961) and the caloric value of feces, 3.945 cal per mg (Van Hook 1971).

For adult females, assimilation also includes energy used to produce eggs. The energy for oviposition

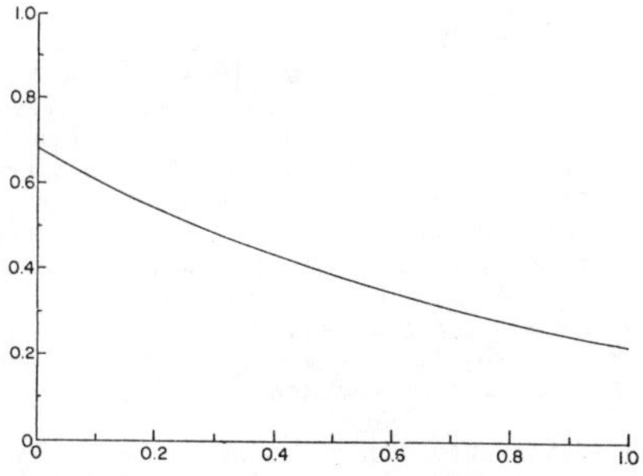

Figure 12.--Assimilation efficiency as a function of nymphal development from equation (40).

is calculated by multiplying the caloric value of the eggs (cal/mg), assumed to be equal to that of grasshopper bodies, by the number of eggs produced by that cohort on the current day, multiplied by the average dry weight of eggs (mg). This value is then divided by two, since A is the per-capita assimilation and only females produce eggs. Thus, when the total assimilation for the cohort is found by multiplying A by N_t, the correct amount of energy is calculated for oviposition. Production of sperm and energy in exuvia are not considered directly in the model.

Lacking better information, we assumed that the dry weight of eggs is equal to the dry weight of newly hatched nymphs, about 0.4 mg.

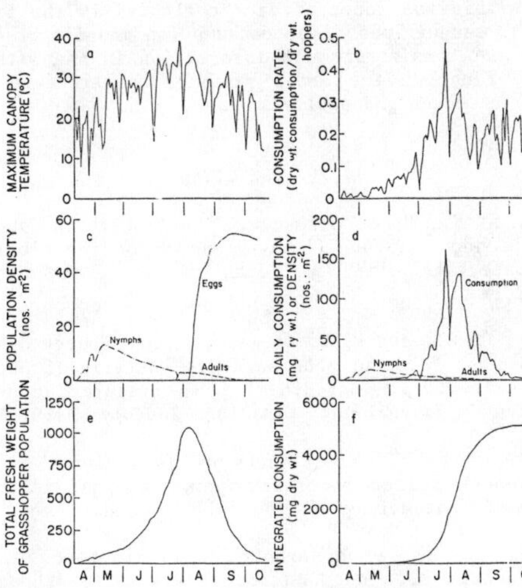

Figure 13.--Output from the simulation model (a) maximum canopy temperatures at the Central Plains Experimental Range in 1971; (b) rate of consumption per mg of hoppers through time; (c) population densities of eggs, nymphs, and adults of the simulated population; (d) daily consumption and population densities; (e) change in total fresh weight of the population through time, and (f) total integrated consumption by the population.

Plant material destroyed but not consumed is an important aspect of grasshopper feeding activity (Mitchell and Pfadt 1974). The factors regulating this process are, however, poorly understood, so the model calculates waste as a direct proportion of total consumption. Mitchell and Pfadt (1974) report that 0.69 times the amount consumed is wasted, and this value is used in the model.

DISCUSSION AND DESCRIPTION OF MODEL OUTPUT

The model was run with weather data from 1971 at the USDA Central Plains Experimental Range in northeastern Colorado. Output from the simulation is presented in figure 13. The model was initialized by calculating egg hatch as a normal distribution (number of eggs hatching as a function of time) with hatching beginning on the first day when temperature exceeded 27°C for four consecutive days (Parker 1930). The simulated density of nymphs, adults, and eggs is presented in figure 13c. The first eggs hatched on April 12 which is somewhat early for this species but is certainly possible given a warm period in early April. Two cold periods later in April caused mortality of some of these early hatching nymphs. Nymphal numbers peaked at about $14/m^2$ in the middle of May. At this time all nymphs were in the first instar. Adults first appear in the last week of July which corresponds well with field observations in Kansas (Cambell et al. 1974). At this time, densities had decreased to about 3.2 per m^2, a 77% decrease from peak density to first occurrence of adults. This mortality rate is well within the expected range (Pfadt and Smith 1972) but perhaps somewhat low compared to recent field data collected in Montana (Onsager and Hewitt 1982). Peak adult density was approximately $3/m^2$ on July 29. Thus, the simulation represents a relatively low density population. Oviposition began about one week later on August 6 and a small number of adults survived until the end of October. A total of 55 eggs per m^2 were produced by this population where the density of females midway between peak adult occurrence and disappearance of adults was approximately 0.5 per m^2. This degree of egg production is quite low compared to that for laboratory populations (Smith 1966, Pickford 1958) but within the lower range of egg production observed in the field (Shotwell 1930).

Total consumption by the simulated population was about 5 g dry-weight of forage/m^2 (fig. 13f). Based on Mitchell and Pfadt's (1974) estimate of a destroyed to consumed ratio of 0.69, 3.47 g would also be destroyed to give a total forage removal over the season of 8.47 g/m^2. Average aboveground net primary productivity (ANP) for the shortgrass prairie at the Central Plains Experimental Range is about 123 g/m^2 (Detling 1979) so simulated forage removal is 7% of ANP. _Melanoplus sanguinipes_ is primarily forbivorous on rangelands (Scharff 1954), however, and productivity of forbs and shrubs at this site is only 60 g/m^2 (Detling 1979). If grasses were not consumed then 14% of the forb and shrub ANP would be consumed. These proportions of ANP consumed by the simulated population correspond well with published values from field studies which generally report 10% or less of ANP removed by grasshoppers (Anderson 1972, Bailey and Riegert 1973, Van Hook 1971, Wiegert 1965, Smalley 1960).

Consumption by the simulated population was insignificant until midJuly, when fifth instår nymphs occur (fig. 13,f). The very low consumption rates before then are due to the small size of early instar nymphs (fig. 13e) and their high assimilation efficiencies compared to fifth instar nymphs and adults.

A sensitivity analysis of the consumption submodel was performed and results are presented in table 2. Five parameters or functions were consecutively increased by 10 percent relative to values in a control run. The consequent change in total consumption, integrated over the life of the population, was compared with the control run to give a percent change in output as a measure of model sensitivity to each of the parameters. The model is most sensitive to temperature with mortality rates and grasshopper weight being nearly as important. Sensitivity to assimilation efficiency and respiration rates is intermediate and the consumption model is least sensitive to changes in developmental rates.

CONCLUSIONS

The model presented here is the first attempt that we are aware of to integrate into one model the various processes involved in the population dynamics and consumption rates of M. sanguinipes. Rodell (1977) developed a similar model for a generalized grasshopper which is somewhat less mechanistic than the model presented here. Despite the extensive literature available concerning this species (see Hilbert and Logan 1981 for a review), many processes are poorly understood in a quantitative way. One of the benefits of simulation models of this type is the identification of such gaps in knowledge.

Temperature and moisture are generally considered to be the major driving variables affecting grasshopper population dynamics (Dempster 1963). The role of temperature in many processes has been extensively studied and drives most processes in the model. Effects of moisture are believed to be primarily mediated through direct effects on diet quality and quantity. This role of food quality in population and consumption processes was identified as an area which is quite poorly understood. With the exception of two studies, Smith (1960) which attempted to quantify the role of phosphorus concentrations in forage and Smith and Northcott (1951) which considered forage nitrogen concentrations, most studies of the role of food plants in the population dynamics of M. sanguinipes have simply indicated the relative quality of different food plants. Many authors, for instance, have shown significant differences in developmental rates (Barnes 1955, Smith 1959, Smith et al. 1952), survival rates (Barnes 1955; Pickford 1958, 1960), and fecundity (Pickford 1958) for grasshoppers reared on different diets. Unfortunately, these studies did not quantify the nutritional quality of these diets in any way so their value in modeling is limited. To our knowledge the effect of forage quality on consumption rates has not been investigated at all with this species. A quantitative understanding of the role of forage quality on grasshopper dynamics would greatly increase our ability to predict population trends and effects on primary production.

Table 2.--Results of a sensitivity analysis of the consumption submodel.

Model run	Total consumption	Change in consumption relative to control run (%)
Control	302.6	--
Temperature + 10%	352.71	16.6
Nymphal and adult mortality + 10%	255.6	-15.5
Grasshopper weight + 10%	348.95	15.3
Assimilation efficiency + 10%	273.9	-9.7
Respiration rate + 10%	327.35	8.2
Developmental rate of nymphs and adults + 10%	286.2	-5.4

A mechanistic understanding of mortality processes is also lacking and, with the exception of mortality due to extreme temperatures, is represented in the model as a constant rate. Once again diet quality may be of particular importance especially through its effect on mortality of the early instar nymphs. Mortality rates are obviously of great importance in determining rates of increase of grasshopper populations and also are shown here to have a strong influence on consumption by the population. Controlled studies of mortality factors would be of great help in modeling grasshopper population dynamics.

The model presented here provides an initial model structure for the analysis of grasshopper population dynamics and consumption which can easily be modified as more information becomes available. Despite the lack of quantitative information in some areas, the model meets the objectives for which it was designed fairly well. Developmental rates of nymphs and eggs are described by new models presented here and in Hilbert and Logan (in press) and Hilbert and Logan (submitted). The simulated phenological progression of the population through the growing season is quite reasonable and mortality and fecundity are both within the range of data collected in the field form M. sanguinipes. The consumption model's predictions are consistent with information in the literature. Finally, the model provides a structure for future research and analysis.

LITERATURE CITED

Allen, S. E., H. M. Grimshaw, J. A. Parkinson, and C. Quarmby. 1974. Chemical analysis of ecological materials. 575 p. John Wiley and Sons, New York.

Bailey, C. G., and P. W. Riegert, 1973. Energy dynamics of Encoptolophus sordidus costalis (Scudder) (Orthoptera: Acrididae) in a grassland ecosystem. Journal of Canadian Zoology 51:91-100.

Barnes, O. L. 1955. Effects of food plants on the lesser migratory grasshopper. Journal of Economic Entomology 48:119-124.

Cambell, J. B., W. H. Arnett, J. D. Lambley, O. K. Jantz, and H. Knutson. 1974. Grasshoppers (Acrididae) of the Flint Hills native tallgrass prairie in Kansas. Research paper 19, Agricultural Experiment Station, Kansas State University, Manhattan.

Dempster, J. P. 1963. The population dynamics of grasshoppers and locusts. Biological Reviews 38:490-529.

Detling, J. K. 1979. Processes controlling blue grama production on the shortgrass prairie. p.

25-42. In Perspectives in grassland ecology (N. R. French, ed.). Springer-Verlag, New York.

Gage, S. H., M. K. Mukeiji, and R. L. Randell. 1976. A predictive model for seasonal occurrence of three grasshopper species in Saskatchewan (Orthoptera: Acrididae). Canadian Entomologist 108:245-253.

Gurney, A. B., and A. R. Brooks. 1959. Grasshoppers of the mexicanus group, genus Melanoplus (Orthoptera: Acrididae). Smithsonian Institue, Proceedings U.S. National Museum 110:1-93.

Hilbert, D. W., and J. A. Logan. 1981. A review of the population biology of the migratory grasshopper, Melanoplus sanguinipes. Colorado State University Experiment Station Bulletin 577S, Fort Collins, Colo.

Hilbert, D. W., and J. A. Logan. An empirical model of nymphal development for the migratory grasshopper, Melanoplus sanguinipes. Environmental Entomology (in press).

Hilbert, D. W., J. A. Logan, and D. M. Swift. A unifying hypothesis of egg development and diapause for the migratory grasshopper, Melanoplus sanguinipes. Environmental Entomology (submitted).

Lin, C. C., and L. A. Segel. 1974. Mathematics applied to deterministic problems in natural sciences. 604 p. Macmillan Publishing Co., Inc., New York.

Logan, J. A., D. J. Wolkind, S. C. Hoyt, and L. K. Tanigoshi. 1976. An analytic model for description of temperature dependent rate phenomenon in arthropods. Environmental Entomology 5:1133-1140.

Marquardt, D. W. 1963. An algorithm for least squares estimation of nonlinear parameters. Journal of the Indiana Society of Applied Mathematics 2:431-441.

Mitchell, E., and Pfadt, R. F. 1974. A role of grasshoppers in a shortgrass prairie ecosystem. Environmental Entomology 3(2):358-359.

Mukeiji, M. K., and S. H. Gage. 1978. A model for estimating hatch and mortality of grasshopper egg populations based on soil moisture and heat. Annals of the Entomological Society of America 71:183-190.

Mukeiji. M. K., S. H. Gage, and R. L. Randell. 1977. Influence of embryonic development and heat on population trend of three grasshopper species (Orthoptera: Acrididae) in Saskatchewan. Canadian Entomolgist 109(2):229-236.

Onsager, J. A., and G. B. Hewitt. 1982. Rangeland grasshoppers: Average longevity and daily rate of mortality among six species in nature. Environmental Entomology 11:127-133.

Parker, J. R. 1930. Some effects of temperature and moisture upon Melanoplus mexicanus mexicanus Saussure and Camnula pillucida Scudder (Orthoptera). 132 p. Montana Agriculture Experiment Station Bulletin 223.

Parton, W. J. A simple soil temperature model. Soil Science (submitted).

Parton, W. J., J. E. Ellis, and D. M. Swift. 1978. The impact of strip mine reclamation practices: A simulation study. p. 584-591. In Ecology and coal resource development, volume 2 (M. Wali, ed.). Pergamon Press, Inc., New York.

Parton, W. J., and J. A. Logan. 1981. A model for diurnal variation in soil and air temperature. Agricultural Meteorology 23:205-216.

Pfadt, R. E. 1949. Food plants as factors in the ecology of the lesser migratory grasshopper Melanoplus mexicanus (Sauss.). Bulletin Wyoming Agriculture Experiment Station No. 290.

Pfadt, R. E., Y. Afzali, and J. S. Cheng. 1979. Life history and ecology of the nondiapause strain of the migratory grasshopper in the laboratory. University of Wyoming, Agricultural Experiment Station, Science Monograph No. 39, 31 p.

Pfadt, R. E., and D. S. Smith. 1972. Net reproductive rate and capacity for increase of the migratory grasshopper, Melanoplus sanguinipes sanguinipes (F.) Acrida 1:149-165.

Pickford, R. 1958. Observations on the reproductive potential of Melanoplus bilituratus (Wlk.) (Orthoptera: Acrididae) reared on different food plants in the laboratory. Canadian Entomologist 90: 483-485.

Pickford, R. 1960. Survival, fecundity and population growth of Melanoplus bilituratus (Wlk.) (Orthoptera: Acrididae) in relation to date of hatching. Canadian Entomologist 92:1-10.

Randell, R. L., and M. K. Mukerji. 1974. A technique for estimating hatching of natural egg populations of Melanoplus sanguinipes (Orthoptera: Acrididae). Canadian Entomologist 106:801-812.

Shotwell, R. L. 1930. A study of the lesser migratory grasshopper. U.S. Department of Agriculture Technical Bulletin 190, 34 p.

Smalley, A. E. 1960. Energy flow of a salt marsh grasshopper population. Ecology 41:672-677.

Smith, D. S. 1959. Utilization of food plants by the migratory grasshopper, Melanoplus bilituratus (Walker) (Orthoptera: Acrididae), with some observations on the nutritional value of the plants. Annals of the Antomological Society of America 52:674-680.

Smith, D. S. 1960. Effects of changing the phosphorus content of the food plant on the migratory grasshopper, Melanoplus bilituratus (Walker) (Orthoptera: Acrididae). Canadian Entomologist 92:103-197.

Smith, D. S. 1966. Fecundity and oviposition in the grasshoppers Melanoplus sanguinipes (F.) and Melanoplus bivattatus (Say). Canadian Entomologist 98:617-621.

Smith, D. S. 1968. Oviposition and fertility and their relation to copulation in Melanoplus sanguinipes (F.). Bulletin of Entomological Research 57:559-565.

Smith, D. S. 1970. Crowding in grasshoppers. I. Effects of crowding within one generation on *Melanoplus sanguinipes*. Annals of the Entomological Society of America 63:1775-1776.

Smith, D. S., R. H. Hanford, and W. Chefurka. 1952. Some effects of various food plants on *Melanoplus mexicanus mexicanus* (Sauss.) (Orthoptera: Acrididae). Canadian Entomologist 84:113-117.

Smith, D. S., and F. E. Northcott. 1951. The effects on the grasshopper, *Melanoplus mexicanus mexicanus* (Sauss.) (Orthoptera: Acrididae), of varying the nitrogen content in its food plant. Canadian Journal of Zoology 29:297-304.

Taylor, F. 1981. Ecology and evolution of physiological time in insects. American Naturalist 117:1-23.

Van Hook, R. I., Jr. 1971. Energy and nutrient dynamics of spider and orthopteran populations in a grassland ecosystem. Ecological Monographs 41:1-26.

Waldbauer, G. P. 1968. The consumption and utilization of food by insects. Advances in Insect Physiology 5:229-288.

Wiegert, R. G. 1965. Energy dynamics of the grasshopper populations in old field and alfalfa field ecosystems. Oikos 16:161-176.

APPLICATION OF A GRASSLAND SIMULATION MODEL TO GRASSHOPPER PEST MANAGEMENT ON THE NORTH AMERICAN SHORTGRASS PRAIRIE[1]

J. L. Capinera, W. J. Parton and J. K. Detling[2]

Abstract.--An empirical grasshopper submodel was developed and used in conjunction with existing primary producer and abiotic submodels to assess the effects of grasshoppers on blue grama rangeland in Colorado. Density dependent mortality was incorporated, and this resulted in a temporal shift in occurrence of forage consumption by grasshoppers. Under high density conditions, proportionately more forage consumption occurred in early nymphal instars, than under low density conditions.

Above-ground and below-ground plant standing crop values were estimated for various grasshopper densities, and for grasshoppers of three different size classes with three different consumption rates. Also, the influence of weather on standing crop was examined by providing recorded meteorological data from a site in Colorado for 1973-1975. Grasshopper consumption at low and moderate grasshopper densities resulted in a slight decrease in aboveground standing crop, and slight increase in below-ground standing crop. High levels of herbivory resulted in large decreases in both above-ground and below-ground standing crop. High levels of herbivory were more deleterious to below-ground standing crop during the two years with low precipitation than one year with higher precipitation.

INTRODUCTION

Grasslands in the western United States are used primarily for livestock production. On the shortgrass prairie, forage production is relatively low because of low precipitation (Lauenroth 1979); hence, large areas of grazing land are required to support the livestock industry. The amount of land required per unit livestock varies with forage availability, and estimation of forage availability is important in maximizing efficiency in ranching operations. Ranchers attempt to stock rangeland with the maximum number of livestock that the land can support without excessive depletion of rangeland forage, which would lead to long-term decrease in carrying capacity.

One of the important variables affecting availability of rangeland forage for livestock is competition from insects. Insect herbivores, and grasshoppers in particular, are a ubiquitous component of grasslands (Hewitt et al. 1974). Grasshopper abundance varies widely both spatially and temporally. While grasshoppers are known to reduce forage availability for livestock production, and they are considered to be the major invertebrate pest on western rangelands, estimation of forage lost to grasshoppers is difficult (Hewitt 1977). We used a process oriented simulation model of blue grama, *Bouteloua gracilis* (H.B.K.) Griffiths, to evaluate the potential effects of grasshopper populations on availability of blue grama forage on a Colorado shortgrass prairie. Blue grama is the dominant grass at this site, and accounts for over 40% of the total annual net primary production and over 80% of the total grass production (Detling 1979). We propose that this method of forage loss assessment may prove to be a rapid and economically feasible technique to integrate production and consumption variables into the rangeland forage protection decision-making process.

MODEL DESCRIPTION

The empirical grasshopper submodel described below was utilized in conjunction with the abiotic submodel described and validated by Parton (1978) and the primary producer submodel described and validated by Detling et al. (1978, 1979). The abiotic submodel uses daily precipitation, cloud cover, wind speed (2 m above-ground), maximum and minimum air temperatures (2 m), and humidity (2 m) data as driving variables to simulate soil and

---

[1]Paper presented at the Third International Conference on State-of-the-Art in Ecological Modelling, Colorado State University, May 24-28, 1982.

[2]John L. Capinera is Associate Professor of Entomology, Department of Zoology and Entomology, William J. Parton is Senior Systems Ecologist, Natural Resource Ecology Laboratory, and James K. Detling is Associate Director, Natural Resource Ecology Laboratory, Colorado State University, Fort Collins, Colorado U.S.A.

canopy microclimatic variables necessary to drive the primary producer submodel. The abiotic submodel is divided into a water flow section and a temperature section, and it also provides information for calculation of light attenuation through the canopy. The water flow section simulates water movement through the canopy and various soil layers, evaporation, transpiration, runoff, and deep water storage. The temperature profile section simulates daily solar radiation, canopy air temperature, and temperature throughout the soil profile.

The primary producer submodel operates on daily time steps and simulates the intraseasonal above- and below-ground biomass dynamics of blue grama as a function of soil and canopy air temperatures, soil water potential, irradiance, leaf nitrogen content, and leaf age. Simulated processes include spring regrowth, net photosynthesis, leaf dark respiration, root and crown respiration, carbohydrate translocation among leaves, crowns, and roots, plant death, and litter fall. Plant state variables include young expanding leaves, mature fully expanded leaves, crowns, and roots from three soil layers (0-4.0, 4.1-15.0, and 15.1-60.0 cm depths) which correspond to three soil water layers in Parton's (1978) abiotic submodel. Regrowth is simulated by translocating carbohydrate from crowns and upper roots to young leaves. Photosynthesis begins when young leaves are formed, and young leaves are transferred to the mature leaf category as they age. Daily photosynthetic production is subsequently allocated to the various plant parts as a function of abiotic environmental factors and physiological status of the plant. Similar mechanisms are used to control respiratory carbon losses and death. Root processes such as allocation of carbon to the various root layers were adapted from the detailed belowground model of Parton et al. (1978).

In the present modeling studies we simulated grazing by different densities and sizes of grasshoppers on pure swards of blue grama. Maximum grasshopper density varied from 0 to 360 grasshoppers/m^2 following eclosion (hatch). Grasshopper densities at eclosion were chosen to provide grasshopper numbers at the start of instar IV (i.e., about 2, 5, 10, 15/m^2, etc.) that commonly occur on Colorado rangeland, and are cause for concern by ranchers. Rates of mortality are based on average mortality rates for six common grasshopper species under field conditions in Montana for three seasons (Onsager and Hewitt 1982). No attempt was made to model the effects of weather on grasshopper mortality. Longevity parameters were assigned values that approximate grasshopper occurrence under Colorado's conditions. Grasshopper eclosion began on May 20 and was completed by June 9. Instars IV and V began on July 4 and July 20, respectively, while the adult stage began on August 14. Maximum adult longevity was 45 days. The number of days for each nymphal instar was derived from the proportions given by Pfadt et al. (1979) for *Melanoplus sanguinipes* (F.). Average longevity (Y, days) of instars IV and V, and adults was decreased as grasshopper density at the end of instar III (X, number/m^2) increased, using the relationship derived by Onsager et al. (1981):

$$\ln Y = 4.556 - 0.3965 (\ln X). \tag{1}$$

Daily death rate was assumed to be 1/Y.

Forage consumption rates reflect grasshopper size, and were determined from the average of values given by Hewitt and Onsager (1982) for three size classes of grasshoppers. Small species were assumed to consume 32 mg/day as adults, while medium and large species consumed 60 and 140 mg/day as adults, respectively. Total forage consumption by instars I-V was simulated as a proportion of the adult rate and was approximated from the values of Putnam (1962) and Hewitt and Onsager (1982) as shown in figure 1. Forage consumption was terminated with cessation of the adult stage, September 27. Grasshoppers often clip foliage from plants without ingesting it; forage consumption here refers to the total of both clipped (wasted) and ingested material. In this model, clipped foliage was assumed to be 33% of the total forage consumption (Mitchell and Pfadt 1974) and this amount was transferred to the litter compartment. Grasshoppers were assumed to consume green foliage exclusively.

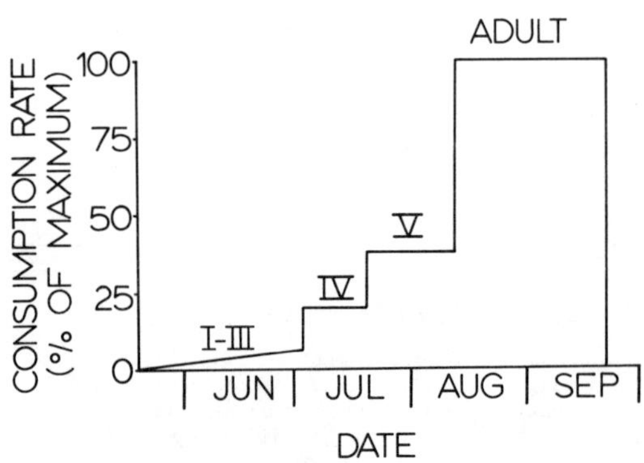

Figure 1.--Pattern of forage consumption by nymphal (I-V) and adult grasshoppers in the grasshopper submodel.

The parameters selected for grasshopper longevity were not designed to reflect occurrence of a single species. Rather, they reflect the composite occurrence of the numerous grass-feeding species known to be destructive to rangeland. Because of considerable interspecific variation in date of eclosion, our composite longevity is much longer than most individual species' longevity.

Weather, and especially precipitation, profoundly influences rangeland productivity. We simulated interseasonal variation in weather by using weather data from 1973, 1974, and 1975 at the USDA Central Plains Experimental Range in northeast Colorado. In 1973 precipitation was low, but the previous year had adequate precipitation. In 1974 precipitation also was below normal, and 1974 was the second consecutive year of low rainfall. In contrast, there was considerably more precipitation in 1975, although it followed two years of low precipitation.

RESULTS AND DISCUSSION

Intraseasonal patterns of grasshopper density occurrence are shown in figure 2, where survival curves for four maximum grasshopper densities (12, 62, 150 and 240/m^2) are plotted for comparison. Grasshopper densities at the start of instar IV were approximately 16% of maximum densities irrespective of initial population size. Thereafter, density-dependent survival rates reduced the longevity of high density grasshopper populations more than that of low density populations. At the start of the adult stage in mid-August, low density grasshopper populations were approximately 9% of the initial size, while high density populations were 2 to 3% of initial size. Thus, absolute abundance tended to converge as grasshopper populations reached maturity. The basis for reduced longevity among high density populations (Hewitt and Onsager 1982) presumably involves density-dependent mortality factors such as predators, parasitoids, pathogens, and competition for food.

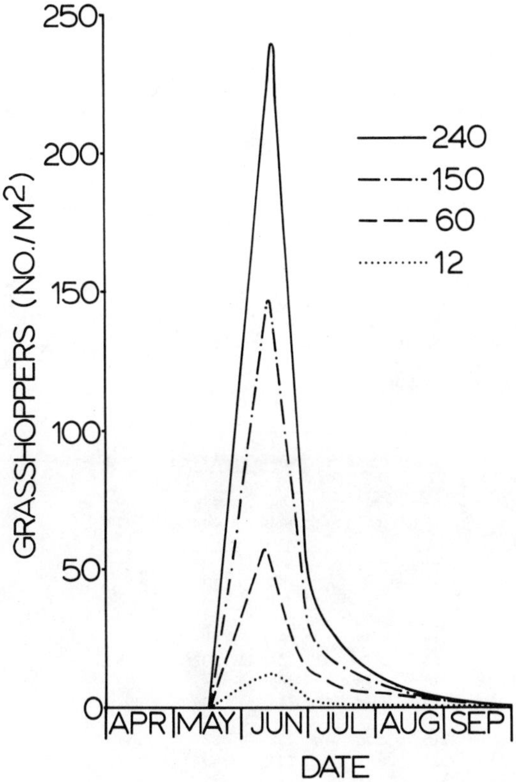

Figure 2.--Patterns of temporal occurrence of grasshoppers at four initial densities.

Differential survival rates among grasshopper populations affected the pattern of forage consumption. Forage consumption patterns for low density (12 grasshoppers/m^2 maximum or 1.1/m^2 at the beginning of the adult stage) and high density (125/m^2 maximum or 4.9/m^2 at the beginning of the adult stage) populations are shown in figure 3. In the low density population, 13% of total consumption occurred in the early nymphal instars (I to III) while 58% occurred as adults. In contrast, 26% occurred in the early instars of the high density population, while 35% occurred in the adult stage. In addition, of course, the high density population consumed a considerably greater amount of forage each day than did the low density population. An important consequence of this shift in the pattern of herbivory is that under high density conditions the decision to initiate grasshopper suppression programs must be made early in the life cycle of the grasshoppers or a significant amount of forage reduction will have occurred before grasshopper numbers can be reduced. Although some authors (e.g., Hewitt and Onsager 1982) discount the importance of early instar herbivory, apparently early instars may consume a large amount of forage if present at high densities.

Forage consumption levels associated with various population densities and low, medium, and high consumption rates are shown in table 1. Also, the resultant above- and below-ground plant standing crops at the end of the growing season (October 28) for each level of forage consumption are given for 1973, 1974 and 1975 weather patterns. For low and moderate levels of herbivory, above-ground standing crop exceeds what would be expected, on the basis of the forage consumption level. That is, except for the highest simulated grasshopper consumption rates in 1973 and 1974, the reduction in above-ground blue grama standing crop, relative to that in the nongrazed simulations, by the end of the season was less than the amount of forage ingested and wasted by grasshoppers during the year. This is more clearly visualized in figure 4, in which the solid line represents a 1:1 correlation between forage consumption and reduction in above-ground standing crop, and the shaded area represents compensatory plant growth. The compensatory growth following simulated grasshopper grazing apparently occurred as a result of greater photosynthetic rates in leaves remaining after defoliation, and allocation of a greater proportion of the photosynthetically-fixed carbon to production of new leaves. This, in turn, occurred as a result of a complex interaction among several biotic and abiotic factors in the model, including a lower average leaf age, reduced transpirational water losses and consequently higher soil potential, and reduced self-shading of leaves in the lower portion of the canopy. The somewhat different simulated plant responses in different years occurred because of the different weather patterns present during grasshopper grazing.

In contrast to the slight decrease in above-ground standings crop, below-ground standing crop (i.e., live root biomass) increased slightly in response to low and moderate levels of herbivory (table 1). However, when forage consumption

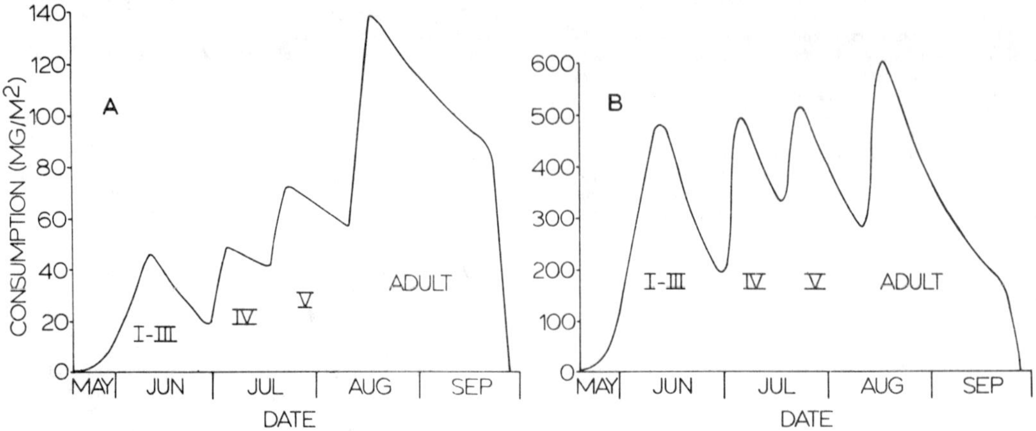

Figure 3.--Patterns of forage consumption by grasshoppers with (A) low ($12/m^2$) and (B) high ($125/m^2$) initial densities at an adult consumption rate of 140 mg/grasshopper/day.

Table 1.--Effects of several grasshopper population densities with various forage consumption rates on total forage consumption, and on above-ground and below-ground standing crop on October 28 with weather patterns observed in 1973, 1974 and 1975.

Adult consumption rate (mg/day)	No. grasshoppers at: hatch	instar IV	adult	Forage consumption (g/m^2)	Plant standing crop (g/m^2) 1973 above-ground	1973 below-ground	1974 above-ground	1974 below-ground	1975 above-ground	1975 below-ground
0	0	0	0	0	116	314	80	282	95	297
32	12	1.9	1.1	1.8	115	314	80	282	92	297
	32	5.1	2.1	3.7	114	315	79	283	91	297
	62	10.2	3.6	5.9	113	316	78	284	90	298
	125	20.4	4.9	9.4	111	318	78	285	87	299
	187	30.5	5.8	12.1	110	319	78	286	86	301
60	12	1.9	1.1	3.4	114	315	79	282	92	297
	31	5.1	2.1	6.9	112	316	77	283	89	297
	62	10.2	3.6	11.1	109	318	76	285	86	298
	125	20.4	4.9	17.6	107	321	75	287	82	299
	187	30.5	5.8	22.7	107	323	73	289	78	301
140	12	1.9	1.1	7.8	111	316	76	283	89	297
	32	5.1	2.1	16.1	106	319	73	285	82	298
	62	10.2	3.6	26.0	101	322	69	287	75	299
	95	15.5	4.3	34.4	103	321	-	-	-	-
	125	20.4	4.9	41.1	105	308	60	294	64	302
	150	24.5	5.3	46.1	98	289	57	292	60	303
	187	30.5	5.8	52.9	50	236	30	202	55	304
	240	38.9	6.4	59.4	24	190	19	166	47	309
	300	48.7	6.9	71.0	-	-	-	-	43	298
	360	58.2	7.1	75.9	-	-	-	-	27	198

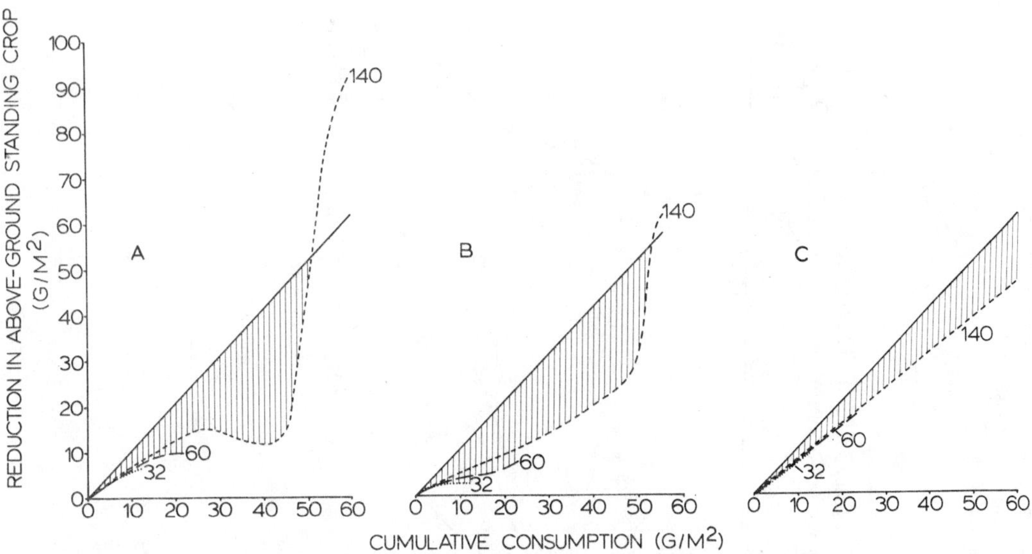

Figure 4.--Effects of various levels of consumption by grasshoppers on reduction of above-ground standing crop for (A) 1973, (B) 1974, and (C) 1975 weather. The solid, diagonal line represents a 1:1 correspondence between total forage consumed by grasshoppers during the growing season and reduction in total plant standing crop at the end of the growing season. Dashed lines represent simulated reductions in forage standing crops at three different adult consumption rates (32, 60, and 140 mg/adult/day) and a variety of grasshopper densities as specified in Table 1. The hatched area represents compensatory plant growth following grasshopper herbivory.

exceeded about 45 g/m^2 under the 1973 and 1974 precipitation patterns, both below-ground and above-ground standing crop exhibited significant declines. Under the higher precipitation conditions of 1975 a higher level of herbivory could be tolerated without deleterious effects on standing crop.

The intraseasonal below-ground and above-ground plant biomass dynamics for several grasshopper population densities at the maximum consumption rate (140 mg/adult/day) are shown in figures 5, 6, and 7 for the 1973, 1974, and 1975 weather patterns, respectively. Simulated grasshopper herbivory resulted in a delay in attainment of peak below-ground standing crop, or in the case of severe defoliation, a decline in below-ground standing crop. A noticeable difference in the pattern of response of blue grama to high grasshopper populations (5.8 and 6.9 adults/m^2) was observed over the three years. Presumably this resulted from variation in precipitation.

Peak above-ground standing crop was attained later in the season in the years (1973 and 1974) with low precipitation. Temporal response to high levels of herbivory during dry years also was markedly different than response to low and moderate levels. During the wettest year, 1975, temporal above-ground standing crop response was similar with grasshopper populations up to a population of 6.9 adults/m^2.

The temporal occurrence of grasshopper forage consumption in relation to green biomass, and the occurrence of green biomass in the absence of grasshoppers and presence of numerous grasshoppers (equivalent of 6.4 adults/m^2), also is shown in figures 5-7. In general, availability of green biomass and forage consumption by grasshoppers were synchronized. However, peak green biomass shifted slightly in the absence of grasshoppers. In the presence of high levels of herbivory during 1973 and 1974, peak green biomass occurred very early in the season.

This model is a unique demonstration of grasshopper-related forage loss assessment. It integrates variable forage consumption potential, grasshopper population dynamics, and climatically-induced variation in forage production and availability. Forage consumption estimates for small, medium and large-sized grasshoppers are provided, as are estimates of grasshopper abundance at key stages in the grasshopper life cycle. Above-ground and below-ground plant standing crop values provide estimates of available for livestock

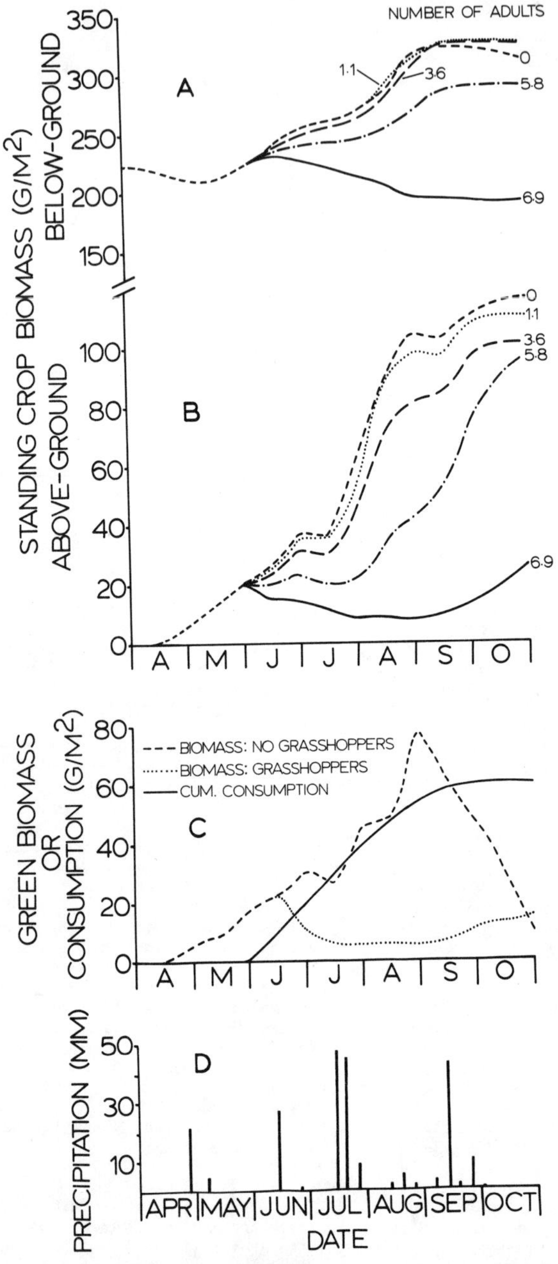

Figure 5.--Temporal sequence of (A) below-ground and (B) above-ground standing crop levels with various adult grasshopper numbers, and consumption at the maximum rate; (C) temporal occurrence of forage consumption, and green biomass in the absence of grasshoppers and presence of 6.4 adults/m^2 consuming at the maximum rate; and (D) pattern of precipitation during the summer months of 1973.

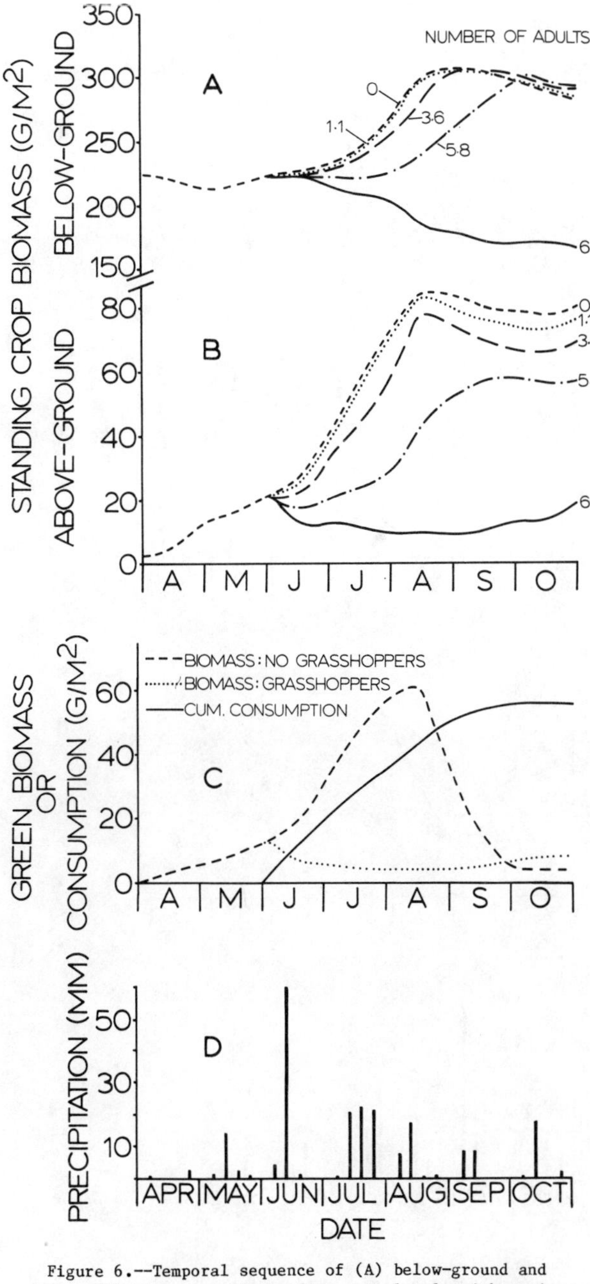

Figure 6.--Temporal sequence of (A) below-ground and (B) above-ground standing crop levels with various adult grasshopper numbers, and consumption at the maximum rate; (C) temporal occurrence of forage consumption, and green biomass in the absence of grasshoppers and presence of 6.4 adults/m^2 consuming at the maximum rate; and (D) pattern of precipitation during the summer months of 1974.

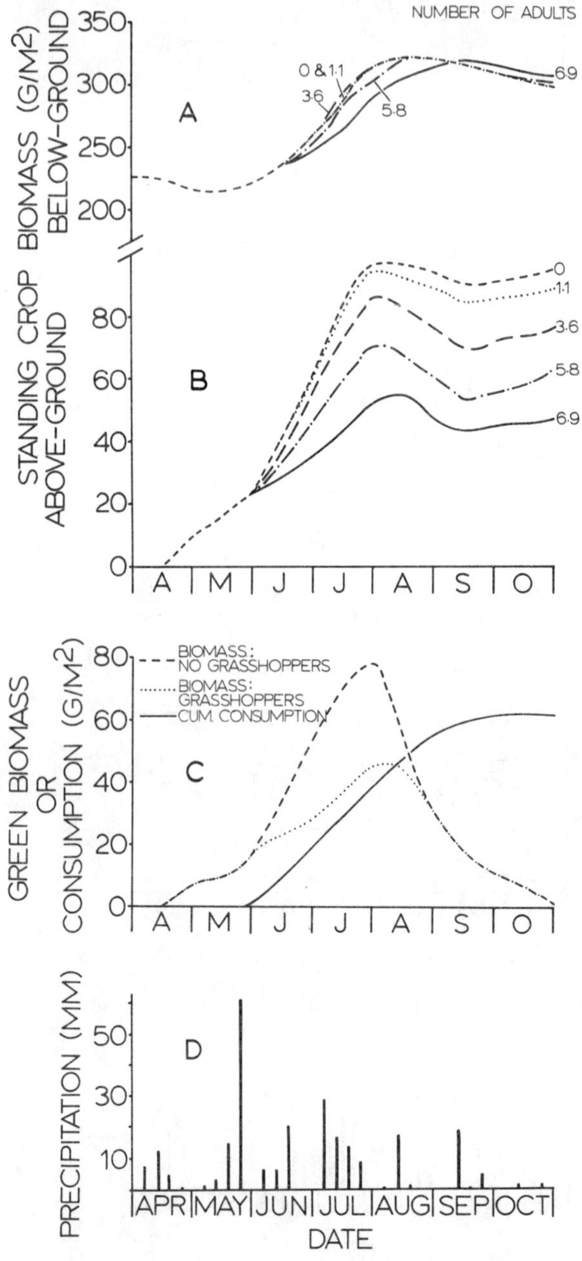

Figure 7.--Temporal sequence of (A) below-ground and (B) above-ground standing crop levels with various adult grasshopper numbers, and consumption at the maximum rate; (C) temporal occurrence of forage consumption, and green biomass in the absence of grasshoppers and presence of 6.4 adults/m^2 consuming at the maximum rate; and (D) pattern of precipitation during the summer months of 1975.

consumption, and regrowth potential in the following year, respectively.

Damaging grasshopper populations characteristically consist of a complex of species. Those species generally most damaging to rangeland in Colorado are small and medium-sized (Capinera and Sechrist 1982). Estimation of forage consumption by a species complex would require integration of forage consumption in proportion to occurrence of each grasshopper size. Estimation of forage reduction is further complicated by grasshopper host plant preference; many species feed preferentially or solely on forbs, sometimes including noxious weeds (Capinera and Sechrist 1982). Obviously this model, which does not yet consider species specific damage potential nor more than one grass species, does not purport to be definitive.

Herbivory by livestock was not included in this evaluation. Although grasshoppers reduced above-ground standing crops at all levels tested, below-ground standing crop was not affected until very high levels of grasshoppers were imposed. However, much of the rangeland supports both livestock and grasshoppers simultaneously. Our model results suggest that in a year that receives adequate precipitation, such as 1975, forage consumption levels of 71 g/m^2 could be experienced with no loss in below-ground standing crop; thus, forage production in the subsequent year may not be affected adversely. Livestock could remove 50% of the maximum standing crop (47.5 g/m^2) leaving 23.5 g/m^2 for consumption by grasshoppers, with no adverse effect on long-term production. Therefore, the equivalent of about 6 adult medium sized grasshoppers/m^2 could be tolerated. This estimate does not consider the temporal occurrence of livestock herbivory which could alter these rates. In contrast, in dry years such as 1973 and 1974, an equivalent level of forage removal by livestock could be damaging to plant growth in subsequent years. Under those conditions, no grasshopper herbivory could be tolerated. Unfortunately for ranchers, grasshopper populations characteristically are highest following periods of drought (Capinera and Sechrist 1982).

If pastures are to be reserved for winter grazing by livestock then much higher grasshopper densities can be tolerated without long-term damage. However, since grasshoppers reduced above-ground standing crop at all simulated densities, the value of replacement forage must be considered. If it is less expensive to suppress grasshoppers than to replace the amount of forage consumed by the grasshoppers, then suppression may be justified.

Current USDA recommendations suggest that "adult" grasshopper population densities of approximately 9.5/m^2 cause enough forage loss to warrant suppression (USDA, APHIS undated). The "adult" designation includes nymphal instars IV and V. Our simulations, based on data provided by Onsager and Hewitt (1982), indicate that considerable mortality occurs between instar IV and the adult stage. Adult surveys should take this mortality into consideration, and adjust adult counts accordingly. If counts are not adjusted, relatively "early" adult surveys would encounter much higher populations than relatively "late" surveys in the same area. Also, USDA suggests that counts of instars I to III be reduced to one-third to make them equivalent to adult counts. Our studies, based on data of Onsager and Hewitt (1982), suggest that they should be reduced to one-sixth to make them equivalent to counts of instar IV nymphs. Should mortality be less severe than that observed by Onsager and Hewitt (1982) in Montana (low mortality may be characteristic of "outbreak" years), however, then current USDA recommendations may be appropriate.

SUMMARY

Grasshopper pests can be managed effectively and economically only when we are able to quantify their destructive capabilities. The model utilized here integrated grasshopper abundance and consumption levels, blue grama production and abiotic factors, to provide an estimate of the effects of herbivory on above-ground and below-ground standing crop. Plant compensation reduced the impact of defoliation but above-ground standing crop decreased with increasing levels of herbivory. High levels of herbivory reduced both above-ground and below-ground standing crop. Herbivory had a less deleterious effect on below-ground standing crop when a higher level of precipitation was provided. Our simulations suggest that both livestock and moderate levels of grasshoppers could be tolerated during years of high precipitation, but that during dry years grasshoppers could not be tolerated in the presence of livestock without the potential for long-term damage.

LITERATURE CITED

Capinera, J.L. and T.S. Sechrist. 1982. Grasshoppers (Acrididae) of Colorado:. identification, biology and management. Colorado State Univ. Agric. Expt. Stn. Bull. 584S (in press).

Detling, J.K. 1979. Processes controlling blue grama production on the shortgrass prairie. Pp. 25-42 in N.R. French (ed.) Perspectives in Grassland Ecology. Ecological Studies, Vol. 32. Springer-Verlag, N.Y.

Detling, J.K., W.J. Parton and H.W. Hunt. 1979. A simulation model of Bouteloua gracilis biomass dynamics on the North American shortgrass prairie. Oecologia 38: 167-191.

Detling, J.K., W.J. Parton, and H.W. Hunt. 1978. An empirical model for estimating CO_2 exchange of Bouteloua gracilis (H.B.K.) Lag. in the shortgrass prairie. Oecologia 33: 137-147.

Hewitt, G.B. 1977. Review of forage losses caused by rangeland grasshoppers. USDA, ARS Misc. Pub. 1348. 22 pp.

Hewitt, G.B. and J.A. Onsager. 1982. A method for forecasting potential losses from grasshopper feeding on northern mixed prairie forages. J. Range. Manage. 35: 53-57.

Hewitt, G.B., E.W. Huddleston, R.J. Lavigne, D.N. Ueckert, and J.G. Watts. 1974. Rangeland Entomology. Society for Range Management, Denver. 127 pp.

Laurenroth, W.K. 1979. Grassland primary production: North American grasslands in perspective. Pp. 3-24 in N.R. French (ed.) Perspectives in Grassland Ecology. Ecological Studies, Vol. 32. Springer-Verlag, N.Y.

Mitchell, J.E. and R.E. Pfadt. 1974. A role of grasshoppers in a shortgrass prairie ecosystem. Environ. Entomol. 3: 358-360.

Onsager, J.A. and G.B. Hewitt. 1982. Rangeland grasshoppers: average longevity and daily rate of mortality among six species in nature. Environ. Entomol. 11: 127-133.

Onsager, J.A., N.E. Rees, J.E. Henry, and R.N. Foster. 1981. Integration of bait formulations of Nosema locustae and carbaryl for control of rangeland grasshoppers. J. Econ. Entomol. 74: 183-187.

Parton, W.J. 1978. Abiotic section of ELM. Pp. 31-53 in G.S. Innis (ed.) Grassland Simulation Model. Ecological Studies, Vol. 26. Springer-Verlag, Berlin.

Parton, W.J., J.S. Singh, and D.C. Coleman. 1978. A model of production and turnover of roots in shortgrass prairie. J. Appl. Ecol. 15: 515-542.

Pfadt, R.E., Y. Afzali, and J.S. Cheng. 1979. Life history and ecology of the nondiapause strain of the migratory grasshopper in the laboratory. Univ. Wyoming Agric. Expt. Stn. SM-39. 30 pp.

Putnam, L.G. 1962. The damage potential of some grasshoppers (Orthoptera: Acrididae) of the native grasslands of British Columbia. Can. J. Plant Sci. 42: 596-601.

USDA, APHIS, PPQ, undated. Grasshopper survey: a species field guide. 45 pp.

STABILITY ANALYSIS OF A FISH POPULATION UNDERGOING LIFE HISTORY CHANGES[1]

W. Peter Saunders, Jr., and John T. Finn[2]

Abstract.--The Connecticut River shad restoration program has successfully reintroduced shad to a large segment of river, but has reduced repeat-spawning by increasing mortality of emigrant adults. Stability effects of this shift towards semelparity were examined using eigenvalue analysis and simulation. Results show that iteroparity itself does not increase stability in shad.

INTRODUCTION

The American shad (Alosa sapidissima) is an anadromous fish that changes its life history characteristics both over large areas (Leggett and Carscadden 1978) and within a single river system (Carscadden and Leggett 1975a, 1975b). In southern populations, individuals spawn only once (semelparity) while in northern populations repeat-spawning individuals are common (iteroparity). However, in all populations fish mature over an age range of several years resulting in a population life history which mimics iteroparity despite semelparity of individuals. Mean age at maturity, growth, and fecundity also vary over the range of the species (Leggett and Carscadden 1978). When viewed in the context of life history theory (Schaffer 1974a, 1974b, Schaffer and Rosenzweig 1977, Goodman 1974, Pianka and Parker 1975, Stearns 1976) shad seem to provide an excellent example of how a species can solve the reproductive cost-benefit problem in a variety of ways. For populations in the southern part of the species' range, where environmental variability is low and survival of offspring is relatively certain, emphasis is placed on producing a large number of offspring early in life at the expense of adult survival. For northern populations, increased environmental variability increases the uncertainty of survival of offspring, leading to a delay in the time of first reproduction and a decrease in the number of offspring produced. This allows adult post-spawning survival through creation of an energetic reserve for downstream migration (Glebe and Leggett 1976). Differences among population life history strategies reflect mechanisms for maintaining population stability in spite of uncertainty in survival of offspring caused by environmental variability (Leggett and Carscadden 1978).

Table 1. Effects of substantial increases in fish passage at Holyoke Dam on several characteristics of the Connecticut River shad population. Data for 1974-1976 represent effects of passing less than 12% of the spawning run at Holyoke (mean=7.9%). Data for 1977-1979 represent effects of passing more than 20% of the run (mean=35.5%).

Trait	Sex	1974-1976	(N)	1977-1979	(N)
Mean Length	Females	56.0	(2,988)	54.8[1]	(512)
(cm)	Males	48.0	(6,854)	47.6	(455)
Mean Age	Females	5.30	(271)	5.27[1]	(331)
(years)	Males	4.77	(349)	4.35[1]	(291)
Repeat-	Females	10.6%	(461)	5.6%[1]	(319)
Spawners	Males	12.3%	(454)	8.4%	(298)

[1] Statistically significant difference (p=0.05)

In the Connecticut River, anadromous fish restoration programs have recently resulted in passage of a substantial portion of the spawning population over the Holyoke Dam, 137 km from the river's mouth. Data collected at Holyoke indicate that changes in shad life history parameters occurred following a substantial increase in fish passage efficiency in 1976 (table 1). The observed changes in life history appear related to increased mortality of fish passed at Holyoke dam. Mortality of fish passed at Holyoke may be attributable to entrapment and turbine passage mortality of emigrant (post-spawning) adults at the dam and to increased energetic costs of migrating further upstream.

The most notable change in life history attributes of shad since 1976 is a reduction in the percentage of repeat-spawning individuals (fig. 1). If life history traits are part of an adaptive

---

[1]Paper presented at the Third International Conference on State-of-the-Art in Ecological Modelling, Colorado State University, May24-28, 1982.

[2]W. Peter Saunders is a Graduate Research Assistant in the Massachusetts Cooperative Fishery Research Unit, and John T. Finn is Assistant Professor of Systems Ecology in the Department of Forestry and Wildlife, University of Massachusetts, Amherst, Massachusetts, USA.

strategy for dealing with environmental uncertainty (Murphy 1967, 1968, Stearns 1976, Leggett and Carscadden 1978), the effects of a loss of repeat-spawners on the population's response to short-term environmental fluctuations is cause for concern (Jones et al. 1976, Leggett 1977). This concern is not only biological but economic since the annual shad run supports commercial and sport fisheries.

In this paper we will deal with several hypotheses concerning stability of the Connecticut River shad population: 1) repeat-spawning increases population stability, and a corollary, 2) decreasing repeat spawning, through lifting fish over Holyoke dam and subjecting adults to turbine mortality, will decrease stability. These hypotheses will be addressed using stable population characteristics determined for the linear, discrete population projection matrix introduced by Leslie (1945). Finally, we will determine whether stability (as judged by our methods) relates to the effects of a fluctuating environment on the population.

METHODS

The Leslie Matrix Model

Mathematical Structure

The mathematical structure of the model used for stability analysis was developed by Leslie

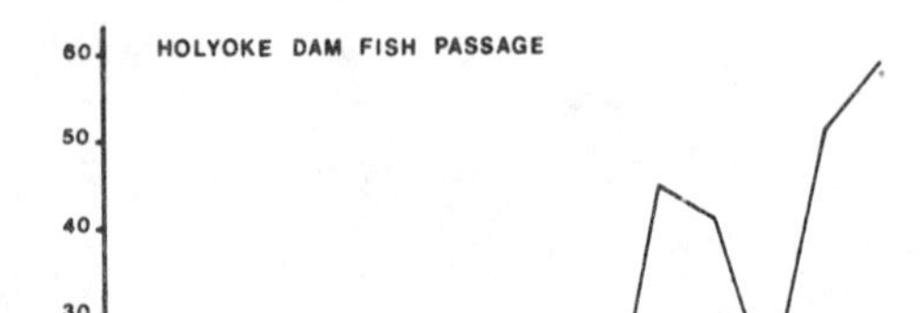

Figure 1. Trends in fish passage efficiency at Holyoke Dam and the percentage of repeat spawners entering the Connecticut River. Data are for males and females combined.

(1945). The model is:

$$N(t+1) = A \times N(t) \tag{1}$$

where,

A = a k-by-k matrix which describes the transition of the population from time t to time t+1.

N(t) = 1-by-k vector containing the number of individuals (n(i)) in each of k life stages at time t. All life stages are one time unit in duration.

The matrix A contains age-specific values for survivorship and fecundity such that:

$$A = \begin{bmatrix} F(0) & F(1) & \cdots & F(x-1) & F(x) \\ S(0) & 0 & & & 0 \\ 0 & S(1) & & & 0 \\ \cdot & & \cdot & & \cdot \\ \cdot & & & \cdot & \cdot \\ \cdot & & & \cdot & \cdot \\ 0 & \cdot & \cdots & 0 \; S(x-1) & 0 \end{bmatrix} \tag{2}$$

where,

F(i) = number of daughters born to a female of age i.

S(i) = probability of a female surviving from age i to age i+1.

x = the maximum age attained by reproductive individuals and is equal to k-1.

The population transition matrix, A, contains all the data necessary for construction of a life table for the population it describes. The computer program developed for this study produces a life table.

Eigensystem Analysis

If A is a primitive matrix, $\lambda(1)$, the maximum eigenvalue of A, is the only positive real root of A and is equal to R, the intrinsic rate of population increase (Sykes 1969). The population stable age distribution is obtained from V, the right eigenvector of A associated with the maximum eigenvalue. The distribution of reproductive value in the population, as defined by Goodman (1969), is obtained from U, the left eigenvector of A associated with the maximum eigenvalue. Caswell and Werner (1978) provide a good discussion of the derivation of reproductive value from the eigenvectors of A. Another population parameter, the time required for the population to converge on the stable age distribution following a perturbation, can be obtained from $\lambda(1)$ and $\lambda(2)$, the eigenvalue of A with the second largest modulus (Keyfitz 1977). Time to convergence is:

$$Tc = \ln(C)/(r(2) - \lambda(1)) \tag{3}$$

where,

Tc = time to convergence.

C = a convergence criterion equivalent to the maximum acceptable Euclidean distance between the age distribution of the perturbed population and the stable age distribution.

r(2) = the real part of $\lambda(2)$, if $\lambda(2)$ is complex, or $\lambda(2)$, if $\lambda(2)$ is real.

$\lambda(1)$ = the maximum eigenvalue of A.

The population parameters based on the eigenvalues and eigenvectors of A provide several criteria for comparison of the stability characteristics of different life histories of the shad population. Sensitivity analysis of models developed using different premises concerning life history (e.g. semelparity vs. iteroparity) allows an evaluation of stability consequences of changes in life history.

In our analyses we are concerned with the relative stability of populations with different life history traits. Further, since wide fluctuations in population size are characteristic of shad, we are concerned with oscillatory stability, i.e. the range and frequency of population swings. Specifically, we will use two aspects of stability, resistance and resilience. We define resistance as a measure of how fast population size changes in response to a given perturbation. Resilience is a measure of how fast a population will return to its former state following removal of a perturbation. We used the proportional change in $\lambda(1)$ resulting from a proportional change of a given magnitude in one or more of the population parameters to indicate relative resistance. We similarly used time to convergence as a measure of resilience.

Parameter Determination

Mortality Rates. Shad have a complex life cycle. Within a year class females mature and join in the spawning migration as early as age 3 and as late as age 7, with most maturing at ages 4, 5, and 6. This presents a problem when determining parameters of a Leslie matrix because there is very little data on the portion of the population which remains in the ocean during the spawning migration. Most of the available population dynamics data on Connecticut River shad are obtained either from the commercial fishery in the estuary, 0-40 km from the river's mouth, or the sport fishery at Enfield Dam, 105 km from the river's mouth. Estimates of annual mortality and the fishery exploitation rate are available for the spawning population. These estimates also allow determination of an annual natural mortality rate for spawning fish which includes both oceanic mortality and post-spawning mortality that results from exhaustion.

There is no data available on mortality rates of immature or mature shad while they are in the ocean nor is data available for the proportion of the spawning population which succumbs to exhaustion following spawning. For our model we assumed that 30% of all mature and immature shad die in the ocean each year.

Fecundity. Shad fecundity was monitored at Holyoke Dam from 1975-1979 and egg retention of downstream migrant shad was determined in 1975 and 1976 (Foote 1976, Reed and Russo 1976, 1977, Reed and Saunders 1978, Reed et al. 1979). The sex ratio of juveniles is approximately 1:1 (Scherer 1974). The mean fecundity-at-age for spawning determined for shad was corrected by the mean egg retention and divided by two to obtain the number of female offspring produced by an adult female.

Maturity Coefficients. The proportion of virgin fish in each age class which will mature and participate in the spawning migration at that age is important in determining parameters of the Leslie model for the shad population. The survival and fecundity parameters used in the Leslie model are population means and thus must consider non-reproductive individuals in their determination. Maturity coefficients provide information on this non-reproductive portion of the population.

Maturity coefficients were determined using the estimated number of virgin females at each age which entered the river to spawn. Complete data were available for the 1971-1974 year classes (cohorts). By following a cohort through time, and assuming no mortality, the vector of maximum maturity coefficients, MMAT, is:

$$\mathrm{MMAT}(k,x) = \mathrm{NV}(k,x) / \sum_{i=x}^{x(\max)} \mathrm{NV}(k,x) \qquad (4)$$

where,

MMAT(k,x) = the proportion of virgin fish from cohort k which mature at age x assuming no mortality.

NV(k,x) = the estimated number of virgin fish from cohort k which entered the river to spawn at age x.

x(max) = the maximum age of fish.

The vector of true maturity coefficients, TMAT, is determined using an estimate of the oceanic survival rate. A simple computer model of the age-specific maturation process was developed and was used to identify the TMAT which would produce MMAT for a constant survival rate. The mean TMAT for the 1971-1974 cohorts is used in developing the Leslie model parameters.

Cohort Simulation. The parameters described above are not in a form which is readily converted to a Leslie matrix. The parameters required by the Leslie matrix were developed from those above using a cohort simulation model. Four age-invariant parameters and two vectors of age specific parameters are required to simulate the population prior to operation of the Holyoke Dam fish passage facility. Four additional parameters are used to represent the effects of fish passage at Holyoke Dam. The parameters of the cohort model are:

1. Annual oceanic mortality. This rate is assumed constant for all fish following their first birthday.

2. Post-spawning natural mortality. This represents mortality following spawning due to energetic expenditures required by migration and reproduction.

3. Rate of exploitation by the commercial fishery.

4. Rate of exploitation by the sport fishery.

5. Age-specific fecundities of spawning fish.

6. True maturity coefficients for each age class.

7. Lift (or Passage) Efficiency. This is the proportion of the spawning run passed at Holyoke Dam.

8. Rate of exploitation by the fish passage facility. This includes mortality resulting from stress and mechanical injury, and removal of fish for use as "seed stock" in shad

restoration programs for other rivers.

9. Mortality rate for downstream passage of emigrant adults at Holyoke Dam. This is primarily entrapment or turbine passage mortality, but also includes mortality due to exhaustion resulting from the increased energetic cost of migration.

10. Mortality rate for emigrant age 0+ juveniles at Holyoke Dam. This represents turbine mortality of juveniles.

11. (OPTIONAL) Survival rate from the egg stage to the end of the first year of life (S(0)).

All sources of mortality are treated as discrete events occurring sequentially in time. The cohort simulation is used to determine annual mean survival rates and age-specific fecundities for the population. State variables of the cohort simulation are the numbers of individuals in each of several spatial/biological model compartments. Figure 2 presents a generalized box and arrow representation of one iteration of the cohort simulation. The cohort simulation is initiated with a fixed number of individuals in age 0 and is iterated through the oldest age (age 9 for our study). The total number of individuals of age(i) on their birthday is the sum of immature virgin spawners in the lower river (below Holyoke Dam), and spawners in the Holyoke Pool (above Holyoke Dam). Survival from age(i) to age(i+1) is the ratio of N(i+1) to N(i). Mean fecundity for individuals, F(i), of age(i) is one-half the total egg production of spawning fish divided by total number of individuals of age(i).

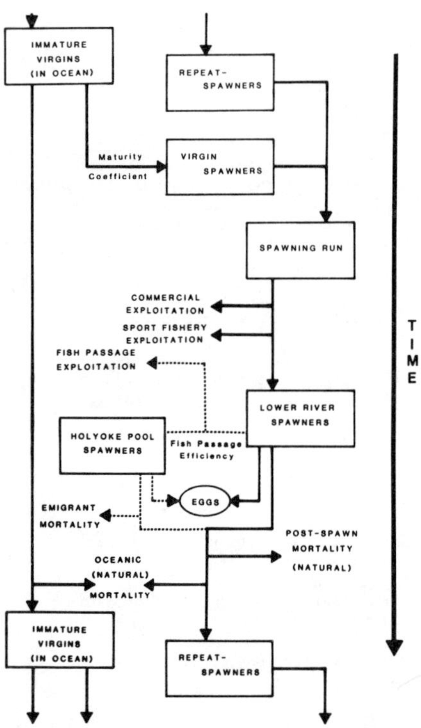

Figure 2. Schematic diagram representing one iteration of the cohort simulation. Biological/spatial partitioning of the population is indicated by boxes and mortality sources are indicated by arrows directed away from intercompartmental connections.

<u>First Year Survival</u>. The cohort simulation allows calculation of all parameters required by the Leslie model except survival from egg through the first year of life, S(0). Since population size estimates for 1967-1981 show no apparent trend, we assume that the population is at equilibrium and calculate S(0) using the method described by Vaughan and Saila (1976) for $\lambda(1) = 1$.

Stability Analyses

The stability analyses we conducted are based on comparison of sensitivity analyses of different models for the Connecticut River shad population. We constructed four models which were assumed to be in equilibrium (i.e. $\lambda(1) = 1$). represent the:

1. shad population prior to implementation of fish passage at Holyoke Dam (Base Model),

2. semelparous population, equivalent to the base model with individual iteroparity removed (i.e. post-spawning mortality of 99%),

3. shad population with 30% passage at Holyoke Dam and high mortality of emigrant adults and juveniles, representing a situation in which emigrant passage facilities do not exist or are ineffective,

4. shad population with 30% passage at Holyoke Dam and low mortality of downstream migrants, representing the operation of an effective downstream migrant passage facility.

Sensitivity analysis of each of the four models consisted of varying selected parameters by $\pm 10\%$ and $\pm 20\%$ of their initial value. The parameters selected were oceanic mortality, commercial fishery exploitation, post-spawning natural mortality, and S(0).

Population Simulation Modelling

Populations were simulated to test predictions of population stability based on the Leslie model. The aspects of stability that interest fisheries managers and fishermen are the range of year-to-year random variations in the size of the population entering the river and future trends in population size, either increases or decreases. Because changes in the resource base due to an increase in available habitat caused by the restoration program cannot be predicted, no basis is available for meaningful projection of the population size. Therefore, annual variation in population size is used to test our stability predictions.

The population simulation model is essentially a self-regenerating version of the cohort simulation discussed earlier. Environmental variability is introduced into the model as a variation in the survival rate for the first year of life. First year survival is treated as a truncated log-normal random variable with a mean equal to the survival rate required for population equilibrium and a standard deviation of 50% of the mean survival rate used for the base model. Minimum survival was set at $1x10^{-6}$, resulting in a slight truncation of the lower end of the distribution. This reflects the variability in survival observed during the shad larval and juvenile stages (Crecco 1981, Marcy 1976) which include the "critical" survival period for many fish species (Cushing 1973). All simulations were initiated with the population at the stable age distribution and a nominal spawning run size of 325,000 females. Male population dynamics were not simulated due to unreliability of data on this segment of the population. Each simulation was allowed to run for ten years prior to data "collection" to minimize effects of initial conditions on output.

For each of the four shad population models described above, seven, twenty-year simulations were averaged to produce a time series of run size estimates. Seven simulations were used because the mean for this number of simulations provides time series output from the base model which is similar in mean and range to the available time series of female population size estimates for the Connecticut River shad run. The mean run size, its standard deviation and the mean change in run size from the previous year (absolute value) were calculated for each twenty-year simulation. For the combination of seven simulations, the mean size and standard deviation of the spawning run for each year and the mean and standard deviation of the year-to-year percent change in run size (absolute value) were calculated for the twenty year period.

Table 2. Parameter sets used to construct shad population dynamics models.

Parameter	Base Model	Semelparity	Fish Passage: High Emigrant Mortality	Fish Passage: Low Emigrant Mortality
Oceanic Mortality	0.30	0.30	0.30	0.30
Commercial Fishery Exploitation	0.20	0.20	0.20	0.20
Post-Spawning Mortality	0.40	0.99	0.40	0.40
Fish Passage Efficiency	0.0	0.0	0.30	0.30
Fish Passage Exploitation	-	-	0.60	0.60
Adult Emigrant Mortality	-	-	0.90	0.30
Juvenile Emigrant Mortality	-	-	0.75	0.25
S(0)	$2.8x10^{-5}$	$4.3x10^{-5}$	$4.2x10^{-5}$	$3.3x10^{-5}$

Table 3. Comparison of spawning run parameters determined from the stable age distribution for the Leslie matrix version of the base model with spawning run parameters observed in the Connecticut River for 1965-1973.

Age	Percent Composition: base model	Percent Composition: observed 1965-1973	Percent Repeat-Spawners: base model	Percent Repeat-Spawners: observed 1965-1973
4	22.9	16.2	0	0
5	46.7	50.5	15.6	12.4
6	20.8	22.7	71.5	80.5
7	6.7	7.2	98.7	97.5
8	2.2	2.7	100.0	100.0
9	0.7	0.7	100.0	100.0
Mean	-	-	31.7	35.2
Mean Age	5.21	5.31		

RESULTS

Input Parameters and Predicted Life History

Input parameter sets developed for the four shad models differed only in the mortality parameters. Maturity coefficients and fecundity were held constant for all simulations. Table 2 compares the parameter sets used for the four models.

To determine how well our choice of parameters for the base model reflects the historical Connecticut River shad population, we estimated the age class composition and percentage of repeat-spawners in the spawning run using the stable age distribution predicted by eigenvalue analysis of the base model. These estimates were compared to averages from data for the years 1965-1973 reported by Leggett (1976). The mean age and percentage of repeat-spawners for the spawning run predicted by our model are 5.2 years and 32% while Leggett's data indicate a mean age of 5.3 years and 35% repeat spawners in the run. An age-by-age comparison of observed data with statistics predicted by the model (table 3) indicates that our age 4 maturity coefficient is probably a bit high. Overall, our Leslie model predictions agree with observed data for the population. On this basis we accepted the base model as a reasonable analog of the shad population.

The range in S(0) necessary to produce population size equilibrium is relatively narrow with a first year survivorship of $2.8x10^{-5}$ required to produce equilibrium in the base model and a survivorship of $4.3x10^{-5}$ required for equilibrium in the semelparous population model. This indicates that relatively small differences in first year survival could support seemingly radical differences in adult life history.

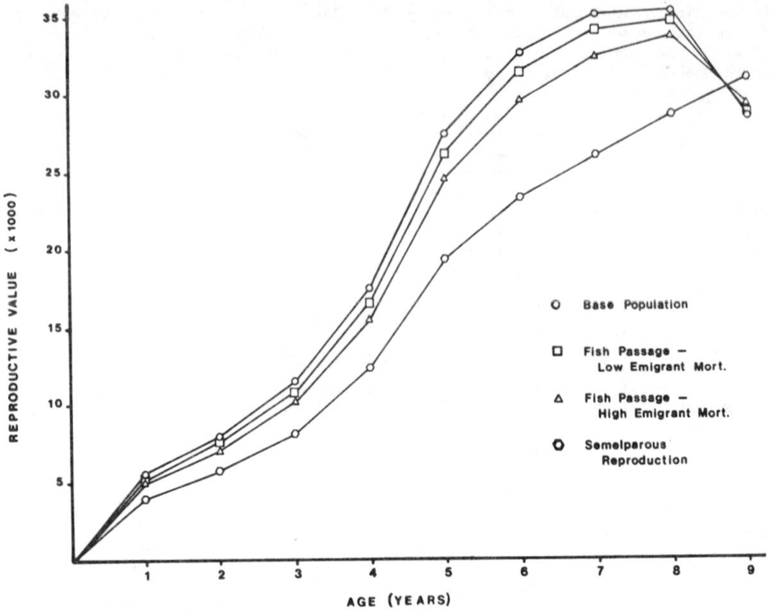

Figure 3. Reproductive value for four equilibrium shad population models.

Table 4. Comparison of predicted life history parameters determined from Leslie matrix analysis of four shad population models.

Parameter	Base Model	Semelparity	Fish Passage High Emigrant Mortality	Fish Passage Low Emigrant Mortality
		Population		
Growth Rate	1.00	1.00	1.00	1.00
Time to Convergence	13.54	12.62	13.27	13.43
Mean Age	2.37	2.24	2.32	2.34
Generation Time	6.50	4.88	6.43	6.48
		Spawning Run		
Mean Age	5.21	4.76	5.05	5.15
Percent Repeat-Spawners	31.70	0.53	23.19	28.51

Comparison of the base model with the other population models indicated relatively large differences in mean age of fish in the spawning run, percentage of repeat spawners, mean generation time, and time to convergence on the stable age distribution among models (table 4). The base model also exhibited a greater reproductive value than the other models (fig. 3). The relative ranking of the models remained the same for all parameters with the model having fish passage but low emigrant mortality most similar to the base model and the model of individual semelparity least similar to the base model.

Sensitivity Analysis

Sensitivity analysis of the four population dynamics models was restricted to four parameters: oceanic mortality, commercial fishery exploitation, post-spawning mortality, and S(0). Table 5 shows percent changes in the population growth rate and time to convergence resulting from changes of ±20% and ±10% in these four model parameters. The response of all four models to parameter changes is essentially the same. Changes in the oceanic mortality rate, which affects the survival of all age classes in the Leslie matrix except age 0, result in the greatest change in the stability indicators (the growth rate and time to convergence). Changes in S(0) have the next greatest effect followed by changes in commercial exploitation and post-spawning mortality. Despite marked differences in some life history parameters, the four models respond to parameter changes quite similarly. The magnitude of change in $\lambda(1)$ and Tc resulting from a change of a given magnitude in any of the parameters examined is nearly the same for all models.

Table 5. Percent change in Leslie model stability indicators, λ(1) (the population growth rate) and Tc (the time to convergence), resulting from changes in four key model parameters for each of four shad population dynamics models.

	Base Model		Semelparity		Fish Passage High Emigrant Mortality		Fish Passage Low Emigrant Mortality	
	λ(1)	Tc	λ(1)	Tc	λ(1)	Tc	λ(1)	Tc
Oceanic Mortality								
-20%	+5.8	-5.0	+5.5	-4.9	+5.7	-5.0	+5.8	-5.0
-10%	+2.9	-2.6	+2.8	-2.6	+2.9	-2.6	+2.9	-2.6
+10%	-2.9	+2.8	-2.8	+2.7	-2.9	+2.8	-2.9	+2.8
+20%	-5.9	+5.7	-5.7	+5.7	05.9	+5.7	-5.9	5.7
S(0)								
-20%	-3.5	+4.3	-3.8	+4.4	-3.6	+4.3	-3.5	+4.3
-10%	-1.7	+2.0	-1.8	+2.0	-1.7	+2.0	-1.7	+2.0
+10%	+1.5	-1.8	+1.7	-1.8	+1.6	-1.8	+1.5	-1.8
+20%	+2.9	-3.4	+3.2	-3.4	+3.0	-3.4	+3.0	-3.4
Commercial Exploitation								
-20%	+1.2	-1.0	+0.8	-0.9	+1.1	-1.0	+1.1	-1.0
-10%	+0.6	-0.5	+0.4	-0.5	+0.5	-0.5	+0.6	-0.5
+10%	-0.6	+0.5	-0.4	+0.5	-0.6	+0.5	-0.6	+0.5
+20%	-1.2	+1.0	-0.9	+1.0	-1.1	+1.0	-1.2	+1.0
Post-Spawning Mortality								
-20%	+1.1	-0.3	-	-	+0.8	-0.2	+1.0	-0.2
-10%	+0.6	-0.1	-	-	+0.4	-0.1	+0.5	-0.1
+10%	-0.5	+0.1	-	-	-0.4	+0.1	-0.5	+0.1
+20%	-1.1	+0.3	-	-	-0.7	+0.2	-0.9	+0.2

The response of the models to parameter changes is non-linear. Increases in mortality produce greater proportional changes in the stability indicators than do decreases in mortality. However, any change produces opposite effects on stability parameters. If λ(1) is decreased, Tc is increased or vice versa. This indicates that during transition, the state of a decreasing population will be more dependent on former states than will the state of an increasing population.

Simulation Modelling

Due to the similarities in model response indicated by the sensitivity analysis we concentrated on comparison of the base model with the semelparous population model. The responses of the two models representing fish passage at Holyoke Dam are intermediate between the base population and semelparous population models.

Outputs from the simulation models are quite variable. Four, twenty-year simulations of the base model are presented in figure 4 to illustrate the variability in predicted run size obtained by introducing variability in first year survival. Mean annual changes in spawning run size for the four models ranged from 6.7-11.9%. Mean population sizes ranged from 304,000-370,000. In the 15 years from 1967-1981 the mean annual size change observed in the Connecticut River shad population was 16.8% with a mean population size of 324,000. The lower mean annual deviations for the simulations appear to be related to the length of the time series since four, fifteen year simulations of the base model produced mean annual population size changes ranging from 9.8-14.7%. The agreement between the base model and the actual population thus appears to be reasonably good considering the limitations of the simulation model.

Table 6 compares the results of five paired simulations of the base and semelparous population models. The simulations in each pair used the same seed to initiate the pseudo-random sequence of S(0) values in the simulation. Simulation results from the two models are quite similar. The semelparous population model tends to remain nearer to initial conditions (run size = 325,000) than the base model. This is reflected in the slightly lower range in run size and mean annual size change in simulations of the semelparous population when compared with simulations of the base population. Overall differences between the two models are slight with standard deviations for simulations indicating that the observed differences are not statistically significant.

These results support those from the sensitivity analyses of the Leslie matrix models in indicating that the semelparous model is quite similar to the base model in its behavior in a fluctuating environment.

DISCUSSION

We have reached the rather surprising conclusion that iteroparity in the Connecticut River shad population does not contribute to population stability. However, our analyses deal with oscillatory stability and we have assumed that the population is able, through some undefined density dependent mechanism, to maintain growth equilibrium (λ(1)=1). This assumption is not entirely unreasonable since, to maintain equilibrium, mean first year survivorship for a semelparous population need increase only 50% from that for the base population. This is well within the range of survival rates indicated by Marcy's (1976) estimates

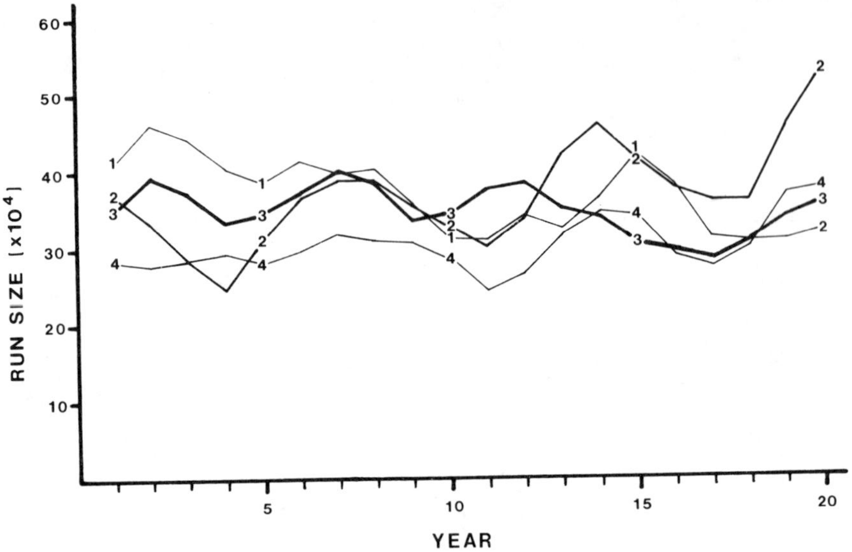

Figure 4. Results of four, twenty-year simulations of the base model. Although each simulation is an average of seven sub-simulations, the variability among simulations is evident.

of juvenile year class strength for 1967-1973 which show a range of 2-30 in the index of year class strength.

Small changes in average survival during the first year of life can support considerably different but equally stable population life histories. Although we have concentrated on comparison of the base and semelparous population models, the models which consider the impact of Holyoke Dam on the population also indicate that relatively small shifts in mean survivorship of age 0+ shad can maintain population equilibrium while the stability of the population is unaffected by loss of repeat-spawning adults. The real questions revolve around survival of young shad prior to maturity.

Unfortunately we are ill-prepared to address the question of juvenile shad survival in the Connecticut River. Opening the Holyoke Pool to spawning shad added over 50 km of potential spawning and nursery habitat to utilization by the shad population. This expansion of area available to the population will continue as fish passage facilities open access to the Turners Falls pool, where fish passage began in 1980, the Vernon pool and the Bellows Falls pool. As more area opens to the shad population, the cost of migration for adults will increase, making repeat-spawning less likely, and the available nursery habitat will increase, thereby reducing some effects of overcrowding on juveniles. However, increases in nursery area do not imply corresponding increases in the food base available

Table 6. Comparison of the results of five simulations of the base and semelparous population models. In each pair of simulations the same initial seed was used for each model in the pair to generate the random sequence of first year survivorships.

Simulation	Mean Spawning Run Size[1] (SD)		Spawning Run Size[1] Range		Mean Annual Size Change (SD)	
	base model	semelparity	base model	semelparity	base model	semelparity
1	349 (32.6)	344 (26.9)	286-400	294-384	0.067 (0.035)	0.071 (0.044)
2	304 (33.9)	312 (33.5)	245-378	254-389	0.109 (0.159)	0.097 (0.098)
3	255 (32.5)	266 (22.3)	213-309	234-304	0.100 (0.0600	0.073 (0.057)
4	370 (49.6)	355 (34.5)	308-463	310-433	0.086 (0.081)	0.084 (0.068)
5	370 (65.1)	351 (49.3)	248-532	255-447	0.119 (0.072)	0.113 (0.112)

[1] Run sizes reported in thousands of females.

to juveniles. Rosen (1981), in a study of the interaction of juvenile shad and zooplankton in the Holyoke Pool, found that shad had a strong effect on the abundance and composition of zooplankton by removing the more desirable food organisms (larger crustaceans and macroinvertebrates) from the water column within a 15 km section of the river below Turners Falls Dam, the upstream limit of migration. Smaller, less energetically suitable food items were all that remained for juveniles in the downstream reaches of the Holyoke Pool. Juvenile growth between the upstream and downstream sections of the Holyoke Pool differed significantly with fish from upstream areas larger than those further downstream. From this it seems that turnover in the zooplankton does not keep pace with grazing by juvenile shad in th river mainstem. In more lacustrine areas such as oxbow lakes and backwaters, zooplankton production is more able to keep pace with the feeding demands of juvenile shad, but these areas are limited.

Our inability to predict how spatial changes in habitat affect resources available to age 0+ shad and the effects of resources on survival of juveniles is the major problem in management of the shad restoration program.

Another aspect of the question of population stability is a population's vulnerability to disastrous events. Normally, a semelparous population would lose a cohort if no spawners were successful in any one year. However, since a given cohort spawns over several years, the semelparous population mimics iteroparity. The effects on the gene pool might not be the same for an iteroparous population and a semelparous one that mimics iteroparity, but data are not available to pursue this point.

The shad restoration program does not appear to be threatened by the trend toward reduced repeat-spawning over the last decade. Major remaining questions which must be addressed before a complete shad management model can be constructed are determination of oceanic mortality, the forces influencing maturation (maturity coefficients), and the interaction between any increases in resources available to juveniles, density dependent survival prior to emigration, and mortality of emigrant juveniles at dams.

ACKNOWLEDGEMENTS

This project was supported by the Massachusetts Cooperative Fisheries Research Unit (MCFRU) and its cooperators, the U.S. Fish and Wildlife Service, the Massachusetts Division of Fisheries and Wildlife, the Massachusetts Division of Marine Fisheries, and the University of Massachusetts. Additional support was provided by the University Computing Center and the Department of Forestry and Wildlife Management.

Data on the shad run at Holyoke dam was provided by MCFRU, which was supported by grants from Northeast Utilities Service Company and by federal Anadromous Fish Act monies. The Connecticut Department of Environmental Protection provided data which they collected at the mouth of the Connecticut River, and from the commercial and sport fisheries. Peter Minta and Victor Crecco of Connecticut DEP were particularly helpful in donating time and assistance.

Roger Reed, the leader of the MCFRU until his death in 1979, had the foresight to see the need for a model of the Connecticut River shad population and was responsible for initiating the project that led to this paper.

Finally, we wish to thank Macy Mensel for typing the manuscript, and Rick Newton for assisting with graphics.

LITERATURE CITED

Crecco, V., L. Gunn, and T. Savoy. 1981. Connecticut River shad study, 1980. 136 p. State of Connecticut, Department of Environmental Protection, Marine Fisheries, Hartford, Connecticut.

Carscadden, J.E., and W.C. Leggett. 1975a. Meristic differences in spawning populations of American shad, *Alosa sapidissima*: evidence for homing to tributaries in the St. John River, New Brunswick. Journal of the Fisheries Research Board of Canada 32(5): 653-660.

Carscadden, J.E., and W.C. Leggett. 1975b. Life history variations in populations of American shad, *Alosa sapidissima* (Wilson), spawning in tributaries of the St John River, New Brunswick. Journal of Fish Biology 7: 595-609.

Caswell, H., and P.A. Werner. 1978. Transient behavior and life history analysis of teasel. Ecology 59(1): 53-66.

Foote, P.S. 1976. Blood lactic acid levels and age structure of American shad (*Alosa sapidissima*, Wilson) utilizing the Holyoke Dam fishlift, Massachusetts, 1974 and 1975. M.S. Thesis 97 p. University of Massachusetts, Amherst.

Glebe, B.D., and W.C. Leggett. 1976. Weight loss and associated energy expenditure of Americn shad during the freshwater migration. 110 p. Final report to the U.S. National Marine Fisheries Service under Project AFC-8, Connecticut, Essex Marine Laboratory, Essex, Connecticut.

Goodman, D. 1974. Natural selection and a cost ceiling on reproductive effort. American Naturalist 108:247-268.

Goodman,L.A. 1969. The analysis of population growth when the birth and death rates depend on several factors. Biometrics 25(4): 659-681.

Jones, R.A., P. Minta, and V. Crecco. 1976. A review of American shad studies in the Connecticut River. Pages 135-162 *in* U.S. Fish and Wildlife Service and National Marine Fisheries Service, co-sponsors. Proceedings of a Workshop on American shad. U.S. Fish and Wildlife Service, Newton Corner, Massachusetts.

Keyfitz, N. 1977. Applied mathematical demography. Springer-Verlag, New York.

Leggett, W.C. 1976. The American shad (*Alosa sapidissma*) , with special reference to its migration and population dynamics in the Connecticut River. Pages 169-225 *in* D. Merriman and L.M. Thorpe, editors. Connecticut River ecological study, the impact of a nuclear power plant. Monograph No. 1. American Fisheries Society, Washington, D.C.

Leggett, W.C. 1977. Density dependence, density independence, and recruitment in the American shad (Alosa sapidissima) population of the Connecticut River. Pages 3-17 in W. Van Winkle, editor. Assessing the effects of power-plant-induced mortality on fish populations. Pergammon Press, New York.

Leggett, W.C., and J.E. Carscadden. 1978. Latitudinal variation in reproductive characteristics of American shad (Alosa sapidissima): evidence for population specific life history strategies in fish. Journal of the Fisheries Research Board of Canada 35(11): 1469-1478.

Leslie, P.H. 1945. On the use of matrices in certain population mathematics. Biometrika 33(3): 183-212.

Marcy, B.C., Jr. 1976. Early life history studies of American shad in the Lower Connecticut River and the effects of the Connecticut Yankee Plant. Pages 141-168 in D. Merriman and L.M. Thorpe, editors. The Connecticut River ecological study, the impact of a nuclear power plant. Monograph No. 1. American Fisheries Society, Washington, D.C.

Murphy, G.I. 1967. Vital statistics of the pacific sardine (Sardinops caerulea) and the population consequences. Ecology 48(5): 731-736.

Murphy, G.I. 1968. Pattern in life history and the environment. American Naturalist 102: 391-403.

Pianka, E.R., and W.S. Parker. 1975. Age-specific reproductive tactics. Americn Naturalist. 109: 453-464.

Reed, R.J., B. Kynard, and W.P. Saunders, Jr. 1980. Biology of American shad utilizing the Holyoke Dam fishlifts, Connecticut River, MA. 15 p. Unpublished report to Northeast Utilities Service Co. from Massachusetts Cooperative Fishery Research Unit, Amherst, Massachusetts.

Reed, R.J., and A.J. Russo. 1976. American shad research - Connecticut River, Massachusetts, 1976, I. Fecundity, egg retention, sex ratio and age class composition. 17 p. Unpublished report to Northeast Utilities Service Co. from Massachusetts Cooperative Fishery Research Unit, Amherst, Massachusetts.

Reed, R.J., and A.J. Russo. 1977. American shad research - Connecticut River, Massachusetts, 1977, fecundity, egg retention, sex ratio and age class composition. 19 p. Unpublished report to Norhteast Utilities Servce Co. from Massachusetts Cooperative Fishery Research Unit, Amherst, Massachusetts.

Reed, R.J., and W.P. Saunders, Jr. 1978. American shad research, Connecticut River, Massachusetts, 1978, sex ratio, age class composition and fecundity. 15 p. Unpublished report to Northeast Utilities Service Co. from Massachusetts Cooperative Fishery Research Unit, Amherst, Massachusetts.

Rosen, R.A. 1981. Seasonal cycles, distribution, and biomass of crustacean zooplankton, and feeding and growth of young American shad (Alosa sapidissima) in the Holyoke Pool, Connecticut River. Ph.D. Dissertation 301 p. University of Massachusetts, Amherst.

Schaffer, W.M. 1974a. Optimal reproductive effort in fluctuating environments. American Naturalist 108: 783-790.

Schaffer, W.M. 1974b. Selection for optimal life histories: the effects of age structure. Ecology 55: 291-303.

Schaffer, W.M., and M.L. Rosenzweig. 1977. Selection for optimal life histories. II: Multiple equilibria and the evolution of alternative reproductive srategies. Ecology 58: 60-72.

Scherer, M.D. 1974. Analysis of factors affecting passage of Americn shad (Alosa sapidissima, Wilson) at Hoyoke Dam, Massachusetts, and assessment of juvenile growth and distribution above the dam. Ph.D. Dissertation 245 p. University of Massachusetts, Amherst.

Stearns, S.C. 1976. Life-history tactics: a review of the ideas. The Quarterly Review of Biology 51(1): 3-47.

Sykes,Z.M. 1969. On discrete stable population theory. Biometrics 25(2): 285-293.

Vaughan, D.S., and S.B. Saila. 1976. A method for determining mortality rates using the Leslie matrix. Transactions of the Americn Fisheries Society 105(3): 380-383.

MODELLING FISHERIES HABITAT: PHYSICAL-HYDRAULIC SIMULATION OF STRUCTURAL MODIFICATIONS TO RIVERS[1]

Lawrence J. Schmitt[2], Edwin E. Herricks[3], and Vahid Alavian[4]

Abstract.--A physical-hydraulic model of bank modifications for the mitigation of habitat losses in rivers altered for barge navigation was investigated. Results indicate that the proposed modifications would be compatible with existing uses and could provide a significant increase in habitat diversity, thereby enhancing fisheries. Mitigation efforts such as this could prove to be valuable and cost-effective if integrated into the planning stage of water resource project development.

INTRODUCTION

In many of the streams and rivers of the U.S., essential habitat diversity is being lost at a significant rate as a result of channelization for agricultural drainage and flood control, canal construction, and navigational-use maintenance. These activities modify natural streams, providing a uniform channel with less resistance to flow. The result is a reduction in sheltered areas for fish as well as their food organisms. As Karr and Dudley (1981) note (Fig. 1), physical habitat structure is one of the four primary determinants of the "structural and functional" integrity of stream ecosystems. This implies that although flow regime, energy inputs, and water quality might not be significantly affected over the long term by channel modifications, an ecosystem may degrade as a result of the loss of habitat diversity.

Although a number of mitigation and enhancement techniques for fisheries management have been employed in the past (Schnick, et al. 1981), the success of these efforts in warmwater streams has been limited. This lack of success can be attributed to the absence of basic ecological information on warmwater fish communities and the difficulty of reconciling the fundamental incompatibilities between existing uses and mitigation requirements. Mitigation efforts have also suffered from poor planning because project specifications are often established before mitigation of aquatic habitat losses is considered. There is a need for integration of mitigation activities, such as developing specifications for physical habitat improvements, at the earliest stages of project planning.

To facilitate that early integration, a defined, quantitative analysis protocol supported by state-of-the-art modelling should be a part of preliminary water resources project design activities. An example of a methodology of this type is PHABSIM, a physical habitat simulation which is coupled with an analysis protocol for fisheries instream flow needs analysis (Stalnaker 1979). The goal of this study was the development of specifications for aquatic habitat mitigation applicable to channelization or canalization projects. The basis for these specifications is a physical-hydraulic simulation of mitigation structures integrated with known habitat requirements of a number of fish species. In developing these specifications, it is hoped that the utility of designing habitat improvements through physical-hydraulic modelling will be demonstrated.

MODELLING BACKGROUND

The development and testing of any mitigation structures for a large, complex river system requires the synthesis of some basic open channel hydraulics principles with the appropriate ecological information. In

[1]Paper presented at the Third International Conference on State-of-the-Art in Ecological Modelling, Colorado State University, May 24-28, 1982.

[2]Lawrence J. Schmitt is a Graduate Research Assistant, Department of Civil Engineering, University of Illinois, Urbana, Illinois U.S.A.

[3]Edwin E. Herricks is an Associate Professor of Environmental Biology, Department of Civil Engineering, University of Illinois, Urbana, Illinois U.S.A.

[4]Vahid Alavian is an Assistant Professor of Civil Engineering, Department of Civil Engineering, University of Illinois, Urbana, Illinois U.S.A.

attempting this synthesis, one is confronted by the problem that hydraulic engineering is highly quantifiable, while the ecological requirements of aquatic organisms tend to be more descriptive and vary widely. Nevertheless, by focusing on the limiting habitat components of a system (as a result of habitat modifications), then it should be possible to design appropriate mitigation structures.

In designing a model for habitat improvement in a river which has been modified for barge navigation, a set of constraints is developed which relates use-requirements with mitigation approaches. For example, any mitigation effort must allow for adequate depth during all seasons to meet navigation needs and an obstruction free channel. Economic efficiency in public work projects of this type often requires maintenance of navigational use at the expense of fish and wildlife habitat. The challenge is to identify methods which provide habitat, are compatible with the planned use, and are economically efficient. An analysis protocol which combines hydraulic simulation models with habitat requirements of the species of interest can provide an approach to project planning which includes mitigation.

The application of the proposed model is intended for medium to large rivers which have been altered for barge navigation. Although our analysis will be limited to this size system, the approach can be validated for other systems because the basic hydraulic relationships will be the same. In addition, community composition differences can be accounted for by including data on species typical of the size stream of interest.

STUDY AREA DESCRIPTION

The Kaskaskia River is a large, turbid stream which flows in a southwesterly direction from Champaign, Illinois to its confluence with the Mississippi River near Chester, Illinois (Fig. 2). The lower 52 miles (89 km) was altered in the early 1970's to form a 36 mile (58 km) barge canal. Alterations included the removal of meanders, increased width (necessitating the removal of riparian vegetation) and depth, and the construction of a

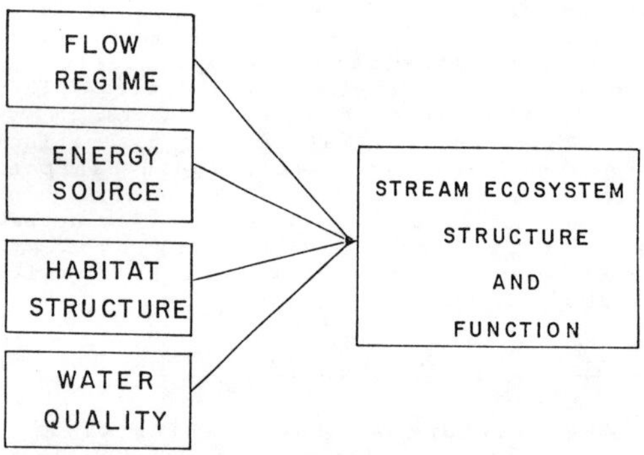

Figure 1.--Major environmental variables which determine stream ecosystem integrity (adapted from Karr and Dudley, 1981).

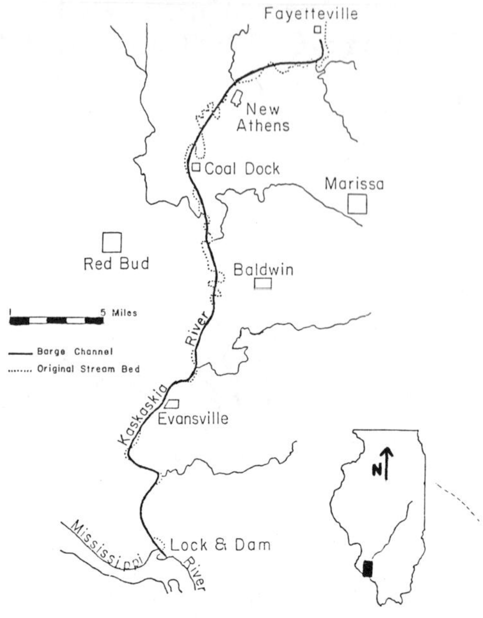

Figure 2.--Map of the Kaskaskia River Navigation Project in southern Illinois.

lock and dam just above the mouth of the river. This left a relatively straight trapezoidal channel with reduced cover, feeding, and spawning habitat for fish. For a more complete description of the former and present habitat characteristics, see Larimore (1978).

In an effort to increase fishing opportunities, the lower ends of the cutoff meanders formed during construction were left open. They were envisioned to serve as a substitute for the naturally occurring oxbow lakes and floodplain pools which were present before canal construction. Although these cutoff meanders are popular with sport fishermen, they do not provide habitat similar to backwater lakes because they are maintained at artificially high water levels and are infrequently flushed (only during the highest flows). A severe dissolved oxygen depletion has been reported which limits fisheries habitat, particularly in warm months (Mathis and Butts 1981).

PROPOSED MODIFICATIONS

As a result of physical habitat analysis and field observations, it appeared that the primary habitat limitation in the navigational reach of the Kaskaskia River was the lack of shallow quiescent areas with suitable water quality and habitat structure. Any proposed modifications needed to improve fisheries habitat without disturbing navigation activities, while being constrained by economic considerations in size and difficulty of construction. Bank modification was chosen for the development of mitigation structures. The goal was the formation of shallow pools adjacent to the main channel (Fig. 3). These pools were intended to be small enough to ensure adequate circulation yet large enough to provide protection for fish from the high velocities of the main channel. In addition, they would necessarily be lined with riprap (as are the present canal banks).

It is at this point that the suggested modelling approach became important. Although an extensive set of field trials was desirable from an ecological standpoint, this was deemed inflexible and expensive. A laboratory simulation allowed a quicker and more practical assessment of the hydraulic performance of such structures. From the results of this analysis, it was possible to evaluate the utility and practicality of the modifications for habitat improvement.

METHODS

Physical Hydraulic Model Design

A scaled physical model was chosen rather than a mathematical representation due to the complex three-dimensional nature of the flow patterns. A laboratory channel measuring 49 m in length, 1.8 m in width, and 1.2 m in depth was available for use. In order to maintain length scales of workable size, only one bank of the prototype (Kaskaskia River) was modelled. It was implicitly assumed that uniform flow would exist in the center of the channel as should exist in the prototype. The inclined bank of the model was constructed of laminated plywood 2 cm thick and designed so that 9.8 m of bank preceded the test section and 2.4 m followed (Fig. 4). This assured uniform flow conditions approaching and leaving the test area.

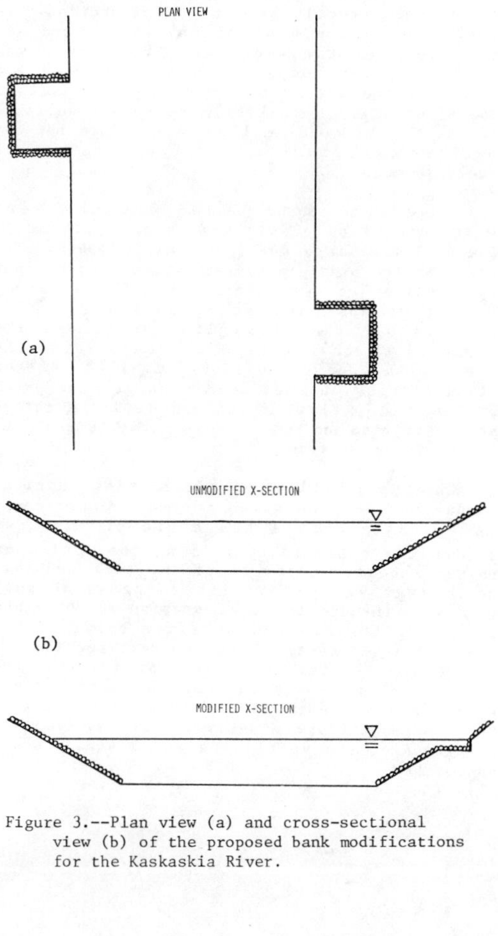

Figure 3.--Plan view (a) and cross-sectional view (b) of the proposed bank modifications for the Kaskaskia River.

In order to conserve cross-sectional area in the channel and provide adequate depth, a singly distorted model was chosen. This distortion allowed the horizontal length ratio of 1:25 (model to prototype) to be larger than the vertical ratio of 1:17. As a result, the bank slope increased from 18 degrees in the prototype to 37 degrees in the model.

The basis for similitude in physical modeling is that the significant dimensionless parameters must be equal for model and prototype (Roberson and Crowe 1980). In open channel flow hydraulics, both the Froude and Reynolds Numbers can be important. The Froude Number is the ratio of inertial to gravitational forces and the Reynolds Number is the ratio of inertial to viscous forces. Fortunately, the viscous effect of the flow becomes relatively insignificant at the high velocities and turbulence levels associated with open channel flow in large models. As a result, it is often acceptable to model solely on the basis of the Froude Number and verify the Reynolds criterion through the proper simulation of friction.

The Froude Number, F, dictates the corresponding depth-velocity combination in the model for any desired depth-velocity combination in the prototype:

$$F_m = \frac{V_m}{(gD_m)^{1/2}} = \frac{V_p}{(gD_p)^{1/2}} = F_p \cdot \quad (1)$$

which simplifies to:

$$\frac{V_m}{V_p} = (L_{r_v})^{1/2} \quad (2)$$

Where:

F_m = Model Froude No.

F_p = Prototype Froude No.

V_m = Model velocity (cm/sec)

V_p = Prototype velocity (cm/sec)

g = Acceleration of gravity (980 cm/sec)

D_m = Model depth (cm)

D_p = Prototype depth (cm)

L_{r_v} = Vertical length ratio (1:17)

In order to simulate channel roughness properly, model and prototype roughness must satisfy the following relationship:

$$\frac{n_m}{n_p} = (L_{r_v})^{1/6} \qquad (3)$$

Where:

n_m = Model Manning's Roughness Coefficient

n_p = Prototype Manning's Roughness Coefficient

This relationship is derived from the Manning's Equation commonly used in open channel flow computations. Based on data published by the U.S. Army Corps of Engineers (Franco and McKellar 1969), n_p ranges from 0.026 - 0.030. Therefore, n_m should be in the range of 0.016 - 0.019. From Chow (1959), laminated plywood has an 'n' ranging from 0.015 - 0.020 with a normal of 0.017, which incorporates the calculated range for the hydraulic simulation. For further details on model design and other aspects of this study, see Schmitt (1982).

Hydraulic Simulation

Three configurations of the test section were modelled (Table 1). This allowed the assessment of the effect of geometry on flow characteristics. Although there are a multitude of shapes which might be suitable, rectangles were chosen for flexibility and ease of construction. In addition, water levels in the Kaskaskia River do not fluctuate extensively since the prototype is maintained at a pool depth suitable for navigation. As a result, only three discharges were simulated, representing low, medium, and high flow (Table 2). Ecological considerations suggested a depth of approximately one meter in the structures at normal pool (in the prototype).

Figure 4.--Laboratory channel with the inclined bank on right side and the test section immediately below the last cross-support (upstream view).

Table 1.--Range of geometric configurations tested in the model and the corresponding range for the prototype.

Length/Width Ratio	Model Dimensions	Prototype Dimensions
1:1	41.6 cm x 41.6 cm	10 m x 10 m
2:1	41.6 cm x 20.8 cm	10 m x 5 m
3:1	41.6 cm x 13.9 cm	10 m x 3.3 m

Prototype testing involved 9 experiments with modifications of depth, velocity, and length/width (aspect) ratios. Through the adjustment of tailgate height and discharge in the laboratory channel, the desired depth and velocity in the model could be obtained. Once this was accomplished, the proper geometric configuration was attained by adjusting a movable plexiglass wall forming the back of the structure. Through the use of a dye injection system and dye granules (potassium permanganate), it was possible to assess the circulation in the test section and any effects on flow conditions in the channel for each experiment.

Data Collection

A Minolta XL-601 Super-8 movie camera with a 7.5-45 mm focal length zoom lens was mounted on a tripod directly over the test section. In order to minimize distortion, the camera was centered over the section. The camera was operated in two modes: normal film speed (18 frames/second) and timelapse (approximately 2 frames/second). Normal film speed was suitable for the high and medium flow experiments but the low flow experiments required the use of timelapse mode.

A series of bi-vaned drogues were constructed from plexiglass sheets (1.6 mm in diameter) and suspended from small floats. The length of the plexiglass piece was varied so that lighter drogues were available for use in low velocities and heavier drogues for higher velocities. The plexiglass sections were suspended so that they were centered in the water column and occupied 8.5 mm (14%) of the column for low and medium discharges and 39 mm (31%) for high flow. Each drogue had a unique pattern painted on its float so that it could be distinguished on film.

Three major flow characteristics (habitat dimensions) were measured for each experiment. The velocity across the mouth of the test section was obtained by filming the motion of drogues inserted upstream of the test section. The average velocity of the drogues as they circulated through the test section was obtained by filming their circuits after being pulled into or being inserted into the test section. A measure of the fraction of the area within the test section which is enclosed by vortices and therefore quiescent was obtained by sprinkling dye granules uniformly over the test section and photographing the vortices after a steady state was reached.

Data Analysis

Each film was analyzed using an Apple II Plus minicomputer with its Graphics Tablet. A Lafayette research projector was employed to project the film on to a mirror which reflected the image on to the tablet. The position of a drogue was tracked across a grid on the tablet as the film was advanced frame by frame. The time of travel was obtained by relating frame number to film speed. The velocity was calculated from the distance travelled per unit time. Vortex area was calculated by the computer and verified by handplotting for selected experiments.

RESULTS

General Description of Hydraulic Properties

The flow in the main channel adjacent to the test section induces secondary circulation within the section and creates an exchange zone at the channel/test section interface (Fig. 5). A series of streamlines form, all rotating in a direction generally opposite to that in the main channel. The induced pattern is directly dependent upon the flow conditions in the channel, particularly upon velocity and depth. The velocity along the streamlines is normally highest towards the perimeter of the structure and decreases towards the center. Virtually no motion occurs in the center.

Observed Results

Results are summarized in Table 3. The velocities across the mouth of the test section, which determine the induced current, are consistent for each configuration at the same discharge and increase accordingly as discharge increases. The velocity through the test section for each run was always considerably lower than the main channel velocity. The percent of the test section area which is enclosed by vortices generally decreases with both increasing discharge and increasing aspect ratio. Although the medium flow runs (4,5, and 6) deviate from this pattern, this is probably due to experimental error. As expected, the Froude Number in the test section was always lower than that in the main channel due to the lower velocities.

DISCUSSION

Ecological Implications of the Hydraulic Analysis

From a fisheries management perspective, it is important to attempt to relate these results

Table 2.--Range of depth-velocity combinations tested in the model and the corresponding range for the prototype.

	Prototype Depth (cm)	Model Depth (cm)	Prototype Velocity (cm/sec)	Model Velocity (cm/sec)
Low Flow	100	6	15	3.7
Med. Flow	100	6	30	7.5
High Flow	200	12	76	18.5

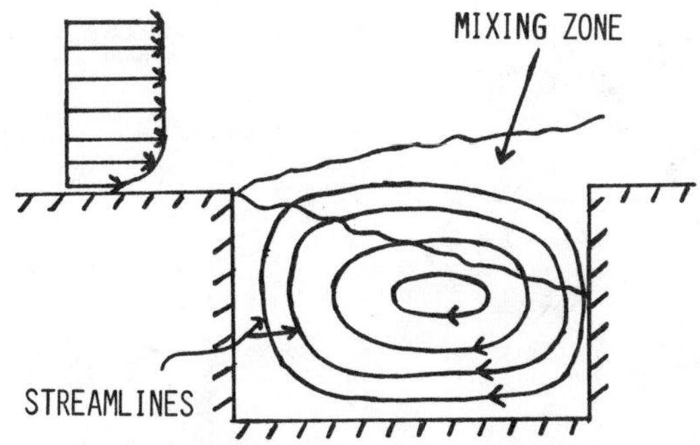

Figure 5.--Summary of the hydraulic characteristics of the test section and the adjacent area of the channel (plan view). Note that the flow direction is from left to right.

to the potential of these structures for providing improved habitat in the Kaskaskia River. It is apparent that although velocities may be quite high around the perimeter of the structure, there is normally some sort of quiescent zone somewhere in the center at all configurations and flows. This is important for fish protection during high flow periods. The results indicate that either the 1:1 or 2:1 aspect ratio would be adequate. Unless riparian property rights are a limitation, it is advisable to avoid the 3:1 configuration.

In addition, structure in the form of rubble, logs, or some similar material could be added to these quiescent areas without significantly affecting the hydraulic performance. This could alleviate problems of lack of structure and, with the riprap armoring of the entire structure, provide substantial substrate for primary and secondary production. Finally, the shallow pools should be ideal for spawning fish, particularly those which require protected areas such as the centrarchids.

Applications

Although the results of this study seem to support the use of these structures in this case, further evaluation is necessary. The flushing effect of the secondary flow at high discharge may prevent excessive siltation from occurring, but this needs to be confirmed on a site-specific basis. Assuming that siltation is not extreme, then a plan for installing the structures is necessary. Although financial constraints could be significant due to the requirement for some heavy equipment, the best option is to install the structures during normal navigational maintenance activities on the river. This would cut costs considerably and might allow the field evaluation of a smaller number of structures initially. Here the effects of a lack of foresight in planning for mitigation of habitat losses are evident, since the most economical solution would have been to install the structures during canal construction. As for placement, it seems logical to alternate the structures on the two banks with some appropriate spacing determined by economics, ease of construction, and other factors.

Table 3.--Results from the 9 experiments (all values reported are for the model).

Run No.	Aspect Ratio	Discharge (m^3/sec)	Channel Velocity (cm/sec)	Test Section Velocity (cm/sec)	Mouth Velocity (cm/sec)	Channel Depth (cm)	Test Section Depth (cm)	Channel Froude No.	Test Section Froude No.	% Area in Vortices
1	1:1	0.011	3.7	0.4	2.4	24.0	6.0	0.024	0.0052	63
2	2:1	0.011	3.7	0.5	2.4	24.0	6.0	0.024	0.0065	50
3	3:1	0.011	3.7	0.5	2.4	24.0	6.0	0.024	0.0065	43
4	1:1	0.023	7.5	1.4	6.4	24.0	6.0	0.049	0.0183	27
5	2:1	0.023	7.5	1.8	5.8	24.0	6.0	0.049	0.0235	30
6	3:1	0.023	7.5	1.8	5.7	24.0	6.0	0.049	0.0235	41
7	1:1	0.074	18.5	6.2	18.5	30.0	12.0	0.108	0.0571	31
8	2:1	0.074	18.5	6.8	20.7	30.0	12.0	0.108	0.0627	18
9	3:1	0.074	18.5	-	21.0	30.0	12.0	0.108	-	10

The use of these structures in other rivers which appear to be limited primarily by a lack of habitat diversity should be carefully evaluated. Any river which has major water level fluctuations on a continual basis would probably not be a suitable candidate due to difficulty in maintaining adequate depths and velocities in the pools as the level changes. In addition, rivers which carry extremely heavy sediment loads would probably be unsuitable due to sedimentation difficulties.

General Utility of Physical-Hydraulic Modelling

It is evident from this study that this general approach to modelling or designing habitat for improvement of fisheries may have applications in a number of situations. Although the hydraulic aspects tended to dominate the design phase, the ecological constraints determined the utility of the structures. It is these ecological requirements of fish which are not as well defined as would be desirable for many species. In this study we relied upon some broad generalizations about limiting habitat components for the fish community. In other situations it may become much more critical to know precisely what habitat link is missing in order to provide for the survival of particular species or communities of fish.

CONCLUSIONS

1.) Physical-hydraulic modelling is a valuable tool for designing aquatic habitat improvements.
2.) The structures investigated in this study may provide a realistic solution to habitat limitations in the Kaskaskia River.
3.) Results from this study should be applied to other systems with caution.
4.) Although this study focused upon a "post facto" situation, efforts such as this will be more valuable and cost-effective if incorporated into the planning stages of water resource development.

LITERATURE CITED

Chow, V. T. 1959. Open channel hydraulics. 680 p. McGraw-Hill Book Co., New York, N.Y.

Franco, J. J., and C. D. McKellar, Jr. 1969. Kaskaskia River Navigation Project, Illinois. Hydraulic model investigation. 21 p. Technical Report H-69-1. U.S. Army Engineer Waterways Experiment Station, Vicksburg, Mississippi.

Karr, J. R., and D. R. Dudley. 1981. Ecological perspectives on water quality goals. Environmental Management 5(1): 55-68.

Larimore, R. W. 1978. Profile of the habitats and biological communities of the Kaskaskia Project Area, Illinois. 29 p. U.S. Army Engineer District, St. Louis, Missouri.

Mathis, B. J., and T. A. Butts. 1981. Sediment oxygen demand and its effect on dissolved oxygen in a cut-off meander of the Kaskaskia River. 62 p. Research Report 162. Water Resources Center, Urbana, Illinois.

Roberson, J. A., and C. T. Crowe. 1980. Engineering fluid mechanics. 661 p. Houghton Mifflin Co., Boston, Mass.

Schmitt, L. J. 1982. A method of improving aquatic habitat in the navigation channel of the Kaskaskia River, Illinois. MS thesis. University of Illinois, Urbana, Illinois.

Schnick, R. A., J. M. Morton, J. C. Mochalski, and J. T. Beall. 1981. Mitigation/enhancement handbook for the Upper Mississippi River System and other large river systems. 702 p. Upper Mississippi River Basin Commission, Minneapolis, Minn.

Stalnaker, C. B. 1979. The use of habitat structure preferenda for establishing flow regulation necessary for maintenance of fish habitat. p.321-337. In Ecology of regulated streams. J. V. Ward and J. A. Stanford (Eds.). Plenum Press, New York, N.Y.

MODELING FISH SPECIES INTRODUCTION AND PREY EXTERMINATION: THE INVASION OF CICHLA OCELLARIS TO GATUN LAKE, PANAMA[1]

Gordon L. Swartzmann[2], Thomas M. Zaret[3]

Abstract. The model CICHLA simulates the colonization of Gatun Lake, Panama, by the predatory Cichla, Cichla ocellaris, and the subsequent extermination of many resident fish species. The model produces realistic Cichla diets, growth rates, migrations, timing of migrations, and extermination rates of prey fish, and can be run for several years' time. A sensitivity analysis revealed that energetics and Cichla feeding parameters were the most sensitive for controlling behavior in this energetic-based, population-dynamic model.

INTRODUCTION

Gatun Lake, a large 425 km^2) meso-oligotrophic tropical lake, is located in the Republic of Panama, Central America. Although dates for the Chagres River system indicate an existence for up to 30,000 YBP (Bartlett et al. 1969), the lake itself is relatively recent, having been formed by damming the Chagres River as part of the Panama Canal Project in 1910 (see Zaret 1980 for maps). In 1966, during the Panama rainy season, approximately 100 fingerlings of Cichla ocellaris, a predatory fish native to the large river systems of South America (Zaret 1980) escaped from rearing tanks and entered a tributary of the Chagres River system and soon afterwards entered Gatun Lake. By 1974 the fish population has advanced to the end of the approximately 50 km long lake, had in the process expanded its population to the point of providing a substantial sports fishery and had completely eliminated most populations of the formerly common native fish species from the lake. The advance of the Cichla and the subsequent demise of most resident fish populations has been documented (Zaret and Paine 1973, Zaret 1980), and reasons for this spectacular success by an introduced predator examined in several papers (Zaret 1980).

This paper reports on the development and implementation of a computer simulation model which replicates the growth and migration of Cichla ocellaris, and the concomitant prey disappearance in Gatun Lake. The model is an energetics-based, age-structured population dynamics model of Cichla with each of 3 prey fish groups represented by two size classes. The use of a computer simulation model to examine the species introduction process has allowed us to attain the following objectives: 1) the identification of those parameters of Cichla and its prey which most sensitively influence prey extinction; 2) the organization of Cichla energetics and population data, known rates of advance of Cichla in Gatun Lake and prey abundance data into a predator-prey ecosystem framework.

The fish population in Gatun Lake before the inadvertent introduction of Cichla ocellaris were dominated numerically by the atherinid Melaniris chagresi, several species of characins, the eleotrid Gobimorus dormitor, the native cichlid Cichlasoma maculicauda, and the poeciliids Gambusia nicaraguagensis and Poecilia mexicana. After passage of the Cichla wave all of these species were eliminated or severely reduced, leaving the lake dominated by cichlids, including Cichla and Cichlasoma (see Zaret and Paine 1973).

To help explain such a dramatic change in fish species composition with a single species introduction, we have identified the following factors, based on empirical information from field studies, which form the model structural basis: 1) fish grow rapidly in the warm temperature (average at 28°C) of Gatun Lake, 2) the high visibility waters year round allows feeding by these visual predators throughout the year, 3) relatively large egg sizes (about .01 gm) facilitate rapid growth of larvae, 4) prey fish distributed size range is such that they are available to the foraging schools of juvenile Cichla, 5) zooplankton and other invertebrate prey are sufficiently abundant for juvenile Cichla to reach a size sufficient to prey on resident fish species, 6) schooling habits of juvenile Cichla and subsequent migration results in more efficient foraging than by individual hunting techniques, 7) synchrony of spawning by adult Cichla ocellaris results in synchrony of school formation by juveniles, 8) guarding of larvae by adult Cichla during the first 8 weeks of life results in low juvenile mortality rates, 9) residence in lake vegetation provides a refuge, resulting in low mortality rates for post 8-week-old juveniles. These and other features of Cichla's ecology and behavior were explored using the simulation model of Cichla invasion in Gatun Lake.

MODEL DESCRIPTION

The Cichla invasion simulation model, hereafter referred to as CICHLA, divides adult Cichla ocellaris into four age classes and records changes in the weight, age, and population density of each age class. Juvenile Cichla are represented by up to four cohorts spawned at

[1]Paper presented at the Third International Conference on State-of-the-Art in Ecological Modeling, Colorado State University, May 24-28, 1982.

[2]Gordon L. Swartzman is Res. Assoc. Professor, Center for Quantitative Science, School of Fisheries, University of Washington, Seattle, WA U.S.A.

[3]Thomas M. Zaret is Res. Assoc. Professor, Dept. of Zoology, University of Washington, Seattle, U.S.A.

20 day intervals over a 60 day spawning period (Jun-Aug). The five prey types are zooplankton, non-zooplankton invertebrates, atherinids (small-sized fish), characins (medium sized fish) and cichlids (large-sized fish). The medium and large prey fish have two size classes representing juvenile and adult fish, giving a total of seven prey groups, each of which are recorded as a total biomass value.

Cichla growth depends on feeding (gains) and respiration (losses) while Cichla populations change via maturation, reproduction and natural and fishing mortality. In the model prey biomass is reduced by Cichla predation and natural mortality and is increased by reproduction (and maturation for those prey with two size classes). Migration of Cichla into previously unexploited parts of the lake is represented in CICHLA by an increase of available prey abundance and in this unique manner is a process directly controlling prey fish abundance. The model structure and important processes are diagramatically depicted in Figure 1.

The model is run by numerically solving a series of differential equations which relate changes in: 1) prey biomass, 2) Cichla weight, and 3) Cichla numbers to those processes represented in Figure 1 as controlling the changes in these variables. The solution to the differential equation is simulated using the Euler technique with a time step of one day. The spatial region considered is assumed to be homogeneous, but during the season of rapid expansion of Cichla into unexploited parts of the lake, moves as a "front", which is composed primarily of migrating sub-adults. The model is implemented in AEGIS (Aquatic Ecosystem General Impact Simulator) (Haar and Swartzman 1981), using a scheme involving "objects" and "attributes" as a way of organizing model components (see Figure 1).

Model computations in CICHLA are organized into eight main subroutines. Subroutine FEED computes Cichla feeding and prey selection and the resulting prey mortality. Subroutine FGROW computes Cichla respiration, and updates Cichla weights. Subroutine FMORT reduces Cichla abundance due to mortality. Subroutine FSPAWN calculates egg production by Cichla during the spawning season and creates new cohorts. Prey are handled by

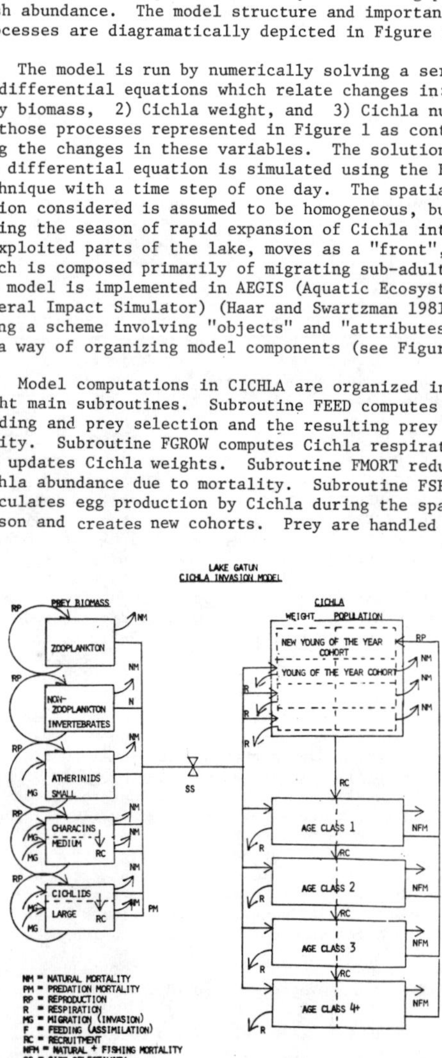

Figure 1. Compartment Diagram for CICHLA.

subroutine PRMORT, which computes total prey mortality, PRRECR which calculates prey recruitment and fecundity, PRMIGR which calculates juvenile Cichla migration by increasing of prey abundance and PRUPDT which updates prey biomass.

Notation and units for variables and parameters used in CICHLA are given in Table A1 in Appendix A. We have chosen to document CICHLA by giving a description of the equations used for each subroutine including their rationale followed by the actual equations in the notation of Table A1.

Parameter values, units and parameter data sources are given in Table B1 in Appendix B.

Although most parameters come from studies of Cichla (Zaret 1980), several were calibrated by running the model. In these cases, starting from an initial parameter value either estimated or taken from a related species, the parameter value was altered in model runs until realistic behavior resulted. Measures of realistic behavior were: 1) annual growth rates of Cichla reflecting known values, 2) realistic changes of Cichla diets with age, 3) Cichla maturing within 12 months as occurs naturally, 4) correct seasonal occurrence of schooling behavior, and 5) prey extinction when migration is absent in the model. Another major source for parameter values was data from largemouth bass, a taxonomically distant relative of _Cichla ocellaris_ but one which is an ecological analog.

A final source for parameter values is a prior model, called FISH1 (Haar and Swartzman 1981), which was developed to study fish energetics and population dynamics on Lake Keowee, a South Carolina reservoir. For example, the prey fecundity and mortality rates were obtained partly from FISH1 values for forage fish because no data were available on Gatun Lake to estimate these parameter values.

Figures 2a to 2f show output from a standard run of CICHLA. The run begins in January when pre-adult Cichla are beginning their migration as a single pre-adult age class and with other adult Cichla age classes present. Migration, which occurs when the juveniles are between 200 gm and 600 gm in weight, is reflected in the model by rapid weight gain by juvenile Cichla and by the increases in prey biomass as the model exposes non-exploited fish densities to the migrating Cichla. Following migration, prey densities drop radically except for non-zooplankton invertebrates which are no longer prey items for the ex-juvenile cohort. As the newly spawned cohorts grow they pass through states where their primary food resources are zooplankton and non-zooplankton invertebrates. Their maturation to sizes large enough to eat fishes (about day 230) results in rapid depletion of small prey fish which are again replenished, as migration for those new juveniles begins (around day 340).

EVALUATION OF MODEL BEHAVIOR

Evaluation of the CICHLA model was conducted by comparing simulated model behavior with various aspects of the natural system. Measures of comparison included:

1) weight at age for adults and juvenile Cichla

2) timing and duration of the migration period for juvenile Cichla

3) time of first maturity

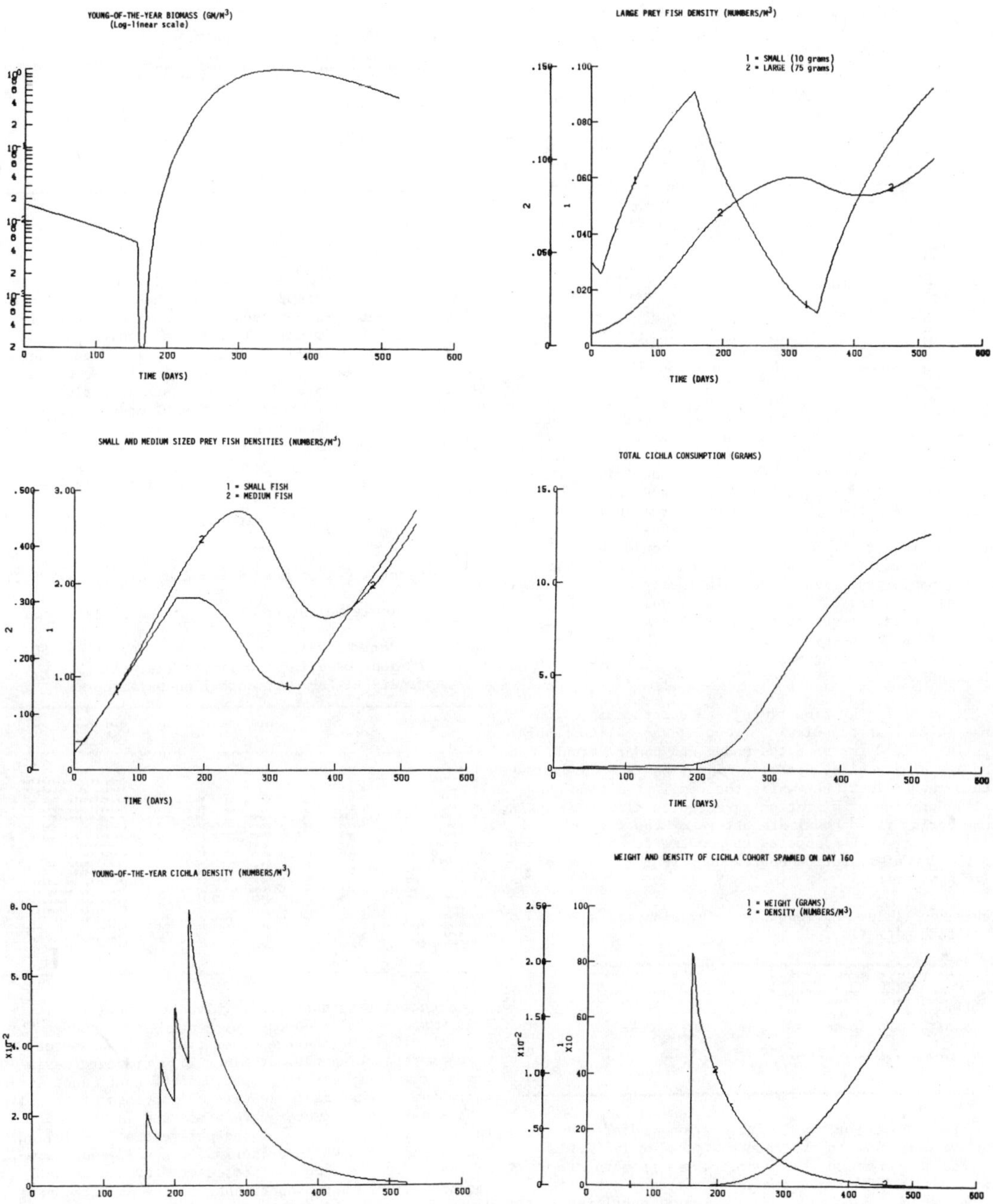

Figure 2. Plots of Cichla output versus time. (a) - biomass of young of the year Cichla (gm/m^3), (b) - densities of small and medium sized prey fish, (c) - density of young of the year Cichla, (d) - large prey fish, juveniles and adults, (e) - total Cichla consumption of prey by a single fish integrated over time, (f) - weight and density of a cohort of Cichla spawned during the first spawning season.

4) model behavior with migration deactivated (evaluates whether prey are 'decimated' in a region of Gatun Lake).

A comparison of model output to calculated weights at various ages is given in Table 1.

Table 1. Comparison of model output and Cichla weight at age data on Gatun Lake.

Time (age)	Length	Weight (grams)	
		Data	Model
4 weeks			1.7 grams
8 weeks	8-10 cm		11 grams
4 months			70 grams
7 months	18-20 cm	140-170	188-250
9 months			380-450
1 year	32 cm	600-700	540-790
2 year			1530-2090
3 year			1750-2800
4 year		largest female 1447	2930-3370
		largest male 3405	

Length at age information in Zaret (1980) is limited and weight data were collected for fish over 14 cm standard length (from the tips of the snout to the end of the vertebral column). Comparison of these data with model output is encouraging especially considering the fact that the species cannot be aged accurately and thus age values are based on life history observations and are approximate. Model predictions tend to be slightly higher than Gatun Lake data. The model shows 4-year old Cichla at 3000 gm while the largest female and male found in Gatun Lake are only 1450 and 3400 grams respectively. This difference may be due to the fact that adult Cichla do not feed for several months during the breeding and parental guarding period, a fact not included in the CICHLA model.

Juvenile migration timing and duration in Gatun Lake (time that the Cichla 'front' advances) is tabulated in Table 2 for both the model and observations. The age of juveniles starting migration is underestimated somewhat in CICHLA while the completion time and total duration of migration agree quite closely. Again, considering the approximate nature of the migration timing data, CICHLA appears to compare favorably. The conclusion of migration corresponds to the timing of Cichla maturity.

Table 2. Timing and length of migration and ages of fish migrating.

Age at start of migration (days)			Age at end of migration (days)		
Model Yr 1	Model Yr 2	Life history Data	Model Yr 1	Model Yr 2	Life history Data
198	128-188	180-210	337	310-370	1 year
Length of migration period (days)					
139	182	190			

By running the model with migration deactivated, we were able to evaluate how completely Cichla reduced a population in a year. These results are expressed in Table 3 by comparing the prey densities for each prey class with their original densities in Gatun Lake (Zaret and Paine 1973). The model shows fish weighing up to 10 grams are decimated by Cichla, having populations reduced to less than 10% of their original population in a year. Larger fish are less seriously affected with losses at around 70%. This is also reflected in the data where small poecilliids and characins were completely eliminated by Cichla, while large Cichlasome actually increased in abundance in the presence of Cichla (Zaret and Paine 1973).

Table 3. Change in prey density due to Cichla predation on five fish prey groups.

Prey group	Representative species	Ave wt (grams)	Initial density (g/m3)	Density after	Percent decrease
small	poecilliids	0.8	0.3	0.027	91
medium 1	characins	4.0	0.2	0.005	97.5
medium 2	gobies	15.0	0.03	0.01	66
large 1	atherinids	10.0	0.03	0.001	96.7
large 2	cichlids	75.0	0.006	0.0017	72

SENSITIVITY ANALYSIS

In order to evaluate how sensitive model behavior is to changes in parameters due to the lack of precision of the parameter values used we conducted a sensitivity analysis using a fractional factorial design (Box, Hunter and Hunter 1978), and an analysis of variance as outlined in Rose and Swartzman (1981). We chose 20 parameters (see Table 4) and gave each an estimated standard deviation. Four measures of model behavior were chosen to evaluate model sensitivity to changes in model parameters:

1) total annual Cichla consumption (gm/yr)--YEB

2) duration of the migration period (no. days) -- SWO2

3) total egg biomass spawned by Cichla -- YFEGTT

4) Cichla weight at the end of year one -- X$W

Table 4. Nominal values and standard deviation (SD) estimates used in sensitivity analysis (See Appendix B1 for more detail on parameters).

		Value in CICHLA	SD	SD % of tuned value	Definition
1	K$RN20	.09	.01	11	Daily ration
2	K$RNHS	.4	.1	25	Prey half saturation
3	K$03	-0.3	.03	10	Ration wt. exponent
4	K$09	2.1	1.0	48	Selectivity
5	K$08	1.25	.6	48	Selectivity var.
6	KFDWMN	.035	.0035	10	Minimum growth
7	K$MR00	.005	.0005	10	Larval mortality
8	K$MR01	.005	.0005	10	Juvenile mortality
9	K$MR02	.015	.0015	10	Pre-adult mortality
10	K$MR03	.002	.0002	10	Adult mortality
11	KFAGMT	1.0	.125	12.5	Age at maturity
12	K$WE	.01	.005	50	Egg weight
13	K$MR1, K$REC1(1)	.005	.001	20	Sm. prey mortality
14	K$MR1, K$REC1(2)	.004	.0008	20	Med. prey mortality
15	K$MR2(3)	.0004	.0001	25	Large prey mortality
16	KFWJS	200	20	10	Pre-adult weight
17	KFWMT	600	60	10	Weight at maturity
18	K$MIG2(2)	.04	.01	25	Migration med. prey
19	K$MIG2(3)	.005	.001	20	Migration lg. prey
20	K$MIG1(2)	.036	.0036	10	Migration sm. prey

A sensitivity analysis involved 64 model runs. This was done twice -- once with parameters changed by plus or minus 10% and the other time with parameter changes reflecting plus or minus 1.28 standard deviations from their model values (the 80% confidence limits for those parameters). Results of rankings of the five most sensitive parameters are presented in Table 5. The relative importance of the parameters to model output depends upon which criterion is considered, as well as whether the parameters are changed by a constant amount or an amount proportional to the expected parameter variability in Table 4.

As expected, the migration period is most sensitive to migration rates and is also sensitive to ration parameters. The most sensitive parameters overall appear to be K$03, the exponent of the effect of weight on ration, and the ration at 20 degrees centigrade, K$RN20. The prey selectivity parameters K$08 and K$09 and the prey density for half maximum ration K$RNHS also are important, especially when the variability in their estimation is considered (i.e., the + or - 1.28 standard

Table 5. Results of sensitivity analysis; ranking of top five parameters for each output criterion and sensitivity variance.

Rank	YFEGGT (tot. egg prod.)		SWO2 (duration of migration)		YEB (total consumption)		XSW (weight)	
	+/- 10%	SD	+/- 10%	SD	+/- 10%	SD	+/- 10%	SD
1	K$03	K$03	K$MIG1(2)	K$03	K$03	K$08	K$03	K$03
2	K$RN20	K$RN20	K$MIG2(3)	K$MIG1(2)	K$RN20	K$RN20	K$RN20	K$08
3	KFAGMT	K$08	K$MIG2(2)	K$RN20	KFAGMT	K$03	K$09	K$09
4	K$MIG2(3)	K$09	K$RN20	K$MIG2(3)	K$08	K$RNHS	K$08	K$RN20
5	K$MIG1(2)	K$RNHS	K$03	K$09	K$MIG1(2)	KFAGMT	K$MIG1(2)	K$RNHS

deviation runs). The important parameters for consumption, weight and egg production are more similar to each other than to those most strongly affecting the migration duration.

A "complete" sensitivity analysis involving all combinations of parameters changes would involve 2**20 runs. Having only 64 runs involving a fractional factorial design in which runs were chosen in such a way as to give information about the effect of each of the parameters and about some two way interactions -- the joint effect on model behavior of 2 parameters. In this design some two way interactions are indistinguishable from others with which they are said to be "aliased".

The design used was an R=4 design which means that main effects (single parameter effects) are indistinguishable from 3 way or greater interactions while some 2 way interactions are indistinguishable from each other. Assuming that 3 way and higher interactions are insignificant implies that the results in Table 5 really do represent the effects of the individual parameters on model behavior and not on the mutual effects of three or more interacting parameters.

We investigated the relative importance of 2 way interactions by looking at the 2 way interactions between the top 5 ranked parameters in Table 5. We assumed that a significant interaction is due primarily to the interaction between the most sensitive single effect parameters from Table 5 and not to interactions involving less significant parameters which are indistinguishable from these. Having a highly significant interaction between parameters can be said to occur when the rank of the interaction effect is higher than that for one of the main effects considered, This means that the combined effect of these parameters on the model output of concern is different than the cumulative effect of the 2 parameters changed singly.

The only case in which a significant interaction occurs is with the duration of migration where the interaction between the ration parameters K$RN20 and K$03 and the interaction of each of these parameters with migration parameters (e.g. K$MIG1(2)) are significant. This implies that growth during the migration period (which controls the duration of migration) is highly sensitive both to the fish feeding rates and to their rates of colonization of (migration into) the new territory as well as to the interrelation between these factors.

Results of the sensitivity analysis imply that the dynamics of the Cichla introduction is highly dependent on the predator's energy requirements, specifically on the changes in food demand and prey selection with increasing size, and in their ability to efficiently exploit the available resources through migration.

CONCLUSIONS

With the CICHLA model we were able to replicate the changes in fish populations resulting from introduction of Cichla ocellaris into Lake Gatun as well as to prodice a valid account of the population dynamics, energetics and growth of Cichla. This representation was accomplished using a model emphasizing Cichla energetics, life history and population parameters and representing prey dynamics in a much less detailed fashion. This produced a "Cichla-eye" view of the system.

One unique modeling feature was to represent the migrating juvenile front on Lake Gatun as an explicit increase in prey density. Thus Cichla in the model see an integrated average prey density that reflects their migration and consumption as well as the prey's population dynamics. Prey densities are not those at any one point in the lake but as the fish see it. Simulation of the situation at a point in the lake invaded by Cichla can be simulated by shutting off migration and, as expected, results in almost complete extinction of many of the resident fish in Lake Gatun.

A question we investigated further regarding extinction was why the prey were not extincted in the Chagres River where Cichla ocellaris was originally introduced (Zaret 1980). Zaret suggested this might be due to the high water turbidity in the river for 6 months of the year which reduced prey vulnerability. We simulated this phenomenon using CICHLA by reducing prey vulnerability by 80% during the 6 month turbid period. This resulted in a significant reduction in total prey consumption, slowed Cichla growth rates during the larval and juvenile periods, and allowed coexistence of Cichla and its prey.

Our sensitivity analysis revealed the energetics parameters as being the most sensitive controls on total Cichla consumption, total egg production, length of the juvenile migration period, and Cichla weight at age. The exponent for the effect of weight on ration was generally the most sensitive parameter, and maximum ration and selectivity parameters were also important. The fractional factorial design used in the sensitivity analysis (Rose and Swartzman 1981) revealed a significant interaction between migration rates and energetics parameters in their effect on the migration period. Sensitivity analyses having a single parameter changed at a time assumes no interactions and would have given different conclusions about the relative sensitivity of the parameters than in fact exist in this model.

We are presently using the CICHLA model as a base for developing scenarios for fish species introduction, to develop hypotheses on what features of introduced fish, resident species and environment lead to successful introductions, i.e., coexistence of the introduced fish and resident species. This work may prove a useful aid to managing fish introductions.

LITERATURE CITED

Andersen, K.P., and E. Ursin. 19771. A multispecies extension to the Beverton and Holt theory of fishing, with accounts of phosporus circulation and primary production. Medd. Dan. Fisk. og Havund. 7:319-435.

Bartlett, A.S., E.S, Barghoorn, and R, Berger. 1969. Fossil Maize from Panama. Science 165:389-390.

Beamish, F.W.H. 1972. Ration size and digestion in large mouth bass, Micropterus Salmoides Can. J. Zool., 50:153-164.

Box, G.E.P., W.G. Hunter, and J.S. Hunter. 1978. Statistics for experimenters. John Wiley & Sons

Carlander, K.D. 1977. Handbook of freshwater fishery biology. Volume Two. Iowa State University Press.

Haar, R.T, and G.L. Swartzman. 1980. Documentation and evaluation of data needs for FISH1: a multi-species fish feeding and population model applied to Lake Keowee, South Carolina. U.S. Nuclear Regulatory Commission Technical Report No. NUREG/CR-2436.

Hatheway, E.S. 1927. The relation of temperature to the quantity of food consumed by fishes. Ecology 814:428-434.

Heidinger, R.L. 1976. Synopsis of biological data on the large mouth bass, *Micropterus Salmoides* (Lacepede). 1802, F.A.O. Fisheries Synopsis No. 115.

Kelly, J.W. 1962. Sexual maturity and fecundity of the large mouth bass, *Micropterus salmoides*, in Maine. Trans. Amer. Fish. Soc. 91:23-28.

Kitchell, J.F., J.F. Koonce, R.V. O'Neill, H.H. Shugart, J.J. Magnuson, and R.S. Booth. 1974. Model of fish biomass dynamics. Trans. Amer. Fish. Soc. 103:786-798.

Rose, K.A., and G.L. Swartzman. 1981. A review of parameter sensitivity methods applicable to ecosystem models. U.S. Nuclear Regulatory Commission Technical Report No. NUREG/CR-2016.

Winberg, G.G. 1956. Rate of metabolism and food requirements of fishes. Nauch. Tr. Belorusskovo Gosudarstevennova Universiteta imeni V.I. Lenina, Minsk. 253 pp. (Transl. from Russian by Fish. Res. Bd. Canada Transl. Ser. No. 194, 1960).

Zaret, T.M. 1980. Life history and growth relationships of *Cichla ocellaris*, a predatory South American Cichla. Biotropica 12(2):144-157.

Zaret, T.M., and R.T. Paine. 1973. Species introduction in a tropical lake. Science 182:449-455.

APPENDIX A

Table A1. Notation and units for variables and parameters used in CICHLA.

OBJECT FISH			
	W(I) :	BODY WEIGHT OF CICHLA COHORT I	G/IND
	N(I) :	DENSITY OF CICHLA COHORT I	#/M3
	AGE :	AGE OF CICHLA COHORT I	DYS
	DWDT :	GROWTH INCREMENT	1/DY
	DNDT :	POPULATION INCREMENT	1/DY
	RN(I) :	ACTUAL RATION PER UNIT BIOMASS	1/DY
	SL(I,J):	SELECTIVITY INDEX FOR PREDATOR I ON PREY J	------
	PYTT(I):	TOTAL PREY AVAILABLE TO PREDATOR I	G/M3
	RS(I) :	RESPIRATION RATE OF CICHLA COHORT I	1/DY
	RSWT :	WT EFFECT ON RESPIRATION	------
	RNWT :	WT EFFECT ON RATION	------
OBJECT PREY			
	W1(J) :	BODY WEIGHT OF 1ST WT. CLASS OF PREY J	G
	N1(J) :	DENSITY OF 1ST WT. CLASS	G/M3
	W2(J) :	WEIGHT OF 2ND WT. CLASS	G/IND
	N2(J) :	DENSITY OF WT. CLASS 2	G/M3
	MRPR(J):	PREDATION MORTALITY FUNCTION ON PREY TYPE J	1/DY
	MRPR2(J):	PREDATION MORT. FOR 2ND WT CLASS PREY TYPE J	1/DY
	MR1(J) :	TOTAL MORTALITY RATE 1ST PREY WT CLASS	1/DY
	MR2(J) :	TOTAL ORTALITY RATE 2ND PREY WT CLASS	1/DY
	SL(I,J):	SELECTIVITY OF 1ST PREY WT. CLASS	
	SL2(I,J):	SELECTIVITY OF 2ND PREY WT. CLASS	
OBJECT SPFISH			
	K9 :	MEAN FOR LOG-NORMAL PREY SELECTIVITY	------
	K8 :	VARIANCE FOR LOG-NORMAL PREY SEL	------
	K3 :	EXP ON WT IN EFFECT OF WT ON RATION	------
	KRNHS :	HALF SAT. CONST FOR PREY DENS EFFECT	G/M3
	KRNQT :	Q10 FOR GRAZING RATE	------
	KRN20 :	20 C COEFF OF TEMP EFFECT ON RN	1/DY
	KAS :	FIXED ASSIMILATION CONSTANT	------
	KRS20 :	RESPIRATION RATE AT 20 C	1/DY
	K2 :	EXP OF WT IN EFFECT OF WT ON RESP	------
	KRSQT :	Q10 FOR RESPIRATION	------
	KMR(1) :	FIXED MORTALITY RATE FOR DEV. CLASS 0	1/DY
	KMR(2) :	CLASS 1	1/DY
	KMR(3) :	CLASS 2	1/DY
	KMR(4) :	CLASS 3	1/DY
	KWE :	EGG WEIGHT	G
	KST :	COEF FOR STARVATION RESPIRATON	------
	RNTP :	TEMP EFFECT ON RATION	------
	RSTP :	TEMP EFFECT ON RESPIRATION	------
	KFDT :	HATCHING TIME	DYS
	KFC1 :	REG. COEFF FOR FEC ON WT	------
	KFC2 :	REG. COEFF FOR FEC ON WT	------
	KFC3 :	REG. COEFF FOR FEC ON WT	------
	SW1 :	FECUNDITY SWITCH(TYPE OF FEC RELATION)	------
	KSFDT :	TIME FROM HATCHING TO FEEDING	DYS
	AGMT :	AGE AT MATURITY	YRS
	KF5 :	FEMALE FRACTION OF COHORT	------
	KWJS :	JUVENILE SCHOOLING WEIGHT	G
	KWM :	CICHLA MATURATION WEIGHT	G
	KDWM :	MIN. GROWTH RATE TO PREVENT STARVATION OF JUVENILES	1/DY
	NE :	EGGS LAID	#/M3
	KFM :	FISHING MORTALITY RATE	1/DY
	DTSP :	SPAWNING INTERVAL FOR CICHLA BETWEEN SPAWNING "EPOCHS"	DYS
	TNF :	TOTAL NUMBER OF SPAWNING "EPOCHS"	
OBJECT SPPREY			
	W1(J) :	WT. FOR 1ST WT. CLASS OF PREY TYPE J	G/IND
	W2(J) :	WT. FOR 2ND WT. CLASS OF PREY TYPE J	G/IND
	KMIG1(J):	MIGRATION RATE FOR 1ST WT. CLASS	1/DY
	KMIG2(J):	MIGRATION RATE FOR 2ND WT. CLASS	1/DY
	KREC1(J):	RECRUITMENT RATE FOR 1ST WT. CLASS	1/DY
	KREC2(J):	RECRUITMENT RATE FOR 2ND WT. CLASS	1/DY
	KMR(J) :	NATURAL MORT. FOR WEIGHT CLASS	1/DY
	KN :	NUMBER OF SIZE CLASSES	#
	KD1(J) :	DENSITY OF PREY TYPE J IN UNEXPLOITED AREA, WEIGHT 1	#/M**3
	KD2(J) :	DENSITY OF PREY TYPE J IN UNEXPLOITED AREA, WEIGHT 2	#/M**3

SUBROUTINE FEED

I = Predator type index
J = Prey index (species & weight classes)

Ration

$$RN(I) = \frac{RNTP * RNWT * KRN20 * PYTT(I)}{PYTT(I) + KRNHS} \quad (1)$$

$$RNWT = W^{K3} \quad (2)$$

$$RNTP = EXP[\ 0.1\ ln(KRNQT)(T - 20)] \quad (3)$$

$$PYTT(I) = \begin{cases} 10 & W(I) < 1.4gm \\ \sum_J W(J)\ SL(I,J)N(J) & W(I) > 1.4gm \end{cases} \quad (4)$$

$$SL(I,J) = \begin{cases} EXP(-0.5[ln((W/W(J))-K9)/K8]^2) & \text{if } ln(W(I)/W(J)) - K9 > -K8 \\ MAX(\ 0,\ \ 0.61[ln((W/W(J))-K9)/(K8 + 2)]) & \text{if } ln(W(I)/W(J)) - K9 < -K8 \\ 0 & \text{if } SL(I,J) < .0001 \end{cases} \quad (5)$$

Predation Mortality

$$MRPR(J) = \sum_I RN(I)\ W(I)*N(I)SL(I,J)/PYTT(I) \quad (6)$$

for all I such that SL(I,J)< .0001 and PYTT(I) > 0

SUBROUTINE FEED

Ration is computed as the product of a temperature effect, a predator weight effect and a prey density effect. The weight effect is an exponential function of weight after Winberg 1956. The temperature effect is an exponential function of temperature which is a constant since temperature is assumed constant year round in Lake Gatun. The prey density effect is a Michaelis Menton or Holling disk equation function (Andersen & Ursin 1977) of total prey available which is computed as the product of individual prey biomass times their selectivity by the predator in question.

Feeding selectivity is based on the ratio of weight of the predator to that of the prey with selectivity being highest at a species-specific ratio of K9 and selectivity dropping off lognormally with a 'variance' K8.

Predation Mortality

For each prey type predation mortality is the product of the ration for each predator on that type of prey times the predator biomass times selectivity for that prey type by that predator summed over all predators. Predators in this model are different age classes of Cichla.

SUBROUTINE FGROW

Respiration

$$RS = RSTP * RSWT * KRS20 \quad (7)$$

$$RSWT = W^{K2} \quad (8)$$

$$RSTP = EXP[0.1\ ln(KRSQT)(T-20)] \quad (9)$$

Growth

$$DWDT(I) = \begin{cases} RN(I) * KAS - RS & \text{if } RN * KAS > RS \\ (RN(I) * KAS - RS)KST & \text{if } RN * KAS \leqslant RS \text{ and fish are not sacfry} \\ KDWM & \text{if sacfry and } DWDT(I) < KDWM \end{cases} \quad (10)$$

$$DDDT = RN(I) * N(I) * W(I)(1 - KAS)$$

SUBROUTINE FGROW

Cichla Respiration

Respiration is the product of a constant temperature effect (exponential or G10) and a weight effect (power function of weight [Winberg 1956]).

Net Growth

If ration times assimilation is greater than respiration their difference is the net growth, unless the fish is a young juvenile and the net growth is less than some minimal acceptable growth rate KWDM in which case net growth is set equal to KDWM (the additional growth is accounted for through increased starvation mortality).

If net growth is negative and the fish are not young juveniles, the weight loss is multiplied by a starvation respiration coefficient KST.

SUBROUTINE FMORT

I = Predator type index
K = Weight class index

$$DNDT = \begin{cases} -KMR(K) - KFM & \text{if fished (adult) } (AGE > AGMT) \\ -KMR(K) & \text{if not fished or sacfry with } DWDT > KDWM \\ -KMR(K) - [DWDT(I) * (-KDWM)] & \text{if } DWDT < KDWM \text{ and sacfry} \end{cases} \quad (11)$$

SUBROUTINE FMORT

Cichla Mortality

Total Cichla mortality is size (age) class specific with additional fishing mortality for adult fish and additional mortality for juveniles whose net growth is less than the minimum acceptable growth rate in which case mortality is increased by the difference between the minimum acceptable and net growth rates.

SUBROUTINE FUPDT

Update

$$N(I) = N(I)(1 + DT * DNDT)$$
$$W(I) = W(I)(1 + DT * DWDT) \quad (12)$$

AGE = AGE + DT

Maturity

```
If  SW1 = EGG

        SW2 = SW2 + DT/KEDT
        If SW2≥1 SW1 = SACFRY                    (13)
              AGE = 0
              SW2 = 0

If SW1 = SACFRY

        SW2 = SW2 + DT/KSFDT                     (14)
        If SW2≥1 SW1 = JUVNIL

If SW1 = JUVNIL

        If AGE ≥AGMT                             (15)
            SW1 = FECUND
```

SUBROUTINE FSPAWN

$$\text{If } S3 = 0$$

$$TNF = \frac{TEND - TSTART + DTSP}{DTSP}$$

$$S3 = 1$$

$$NE(I) = N(I)\ K5/TNF\ \ KFC1 - KFC2 * [\ln(W(I)) + KFC3\ ] \qquad (16)$$

$$NE = \sum_I NE(I)$$

$$DWDT(I) = DWDT(I) - \frac{NE(I) * KWE}{W(I)} \qquad (17)$$

SUBROUTINE FSPAWN

Cichla Spawning

Spawning takes place as TNF spawning events spread over the spawning period. Each cohort produces a number of eggs based on a regression on the average weight of the cohort. The average weight of each cohort number is then reduced by the weight of the eggs produced.

SUBROUTINE PRMORT

$$MR(J) = MRPR(J) + KMR(J) \qquad (18)$$

SUBROUTINE PRRECR

If KN = 1 (1 Weight class only)

$$REC1(J) = KREC1 * N1(J)$$
$$REC2(J) = 0 \qquad (19a)$$

If KN = 2

$$REC1(J) = KREC1 * N2(J)$$
$$REC2(J) = KREC2 * N1(J) \qquad (19b)$$

SUBROUTINE PRRECR

Prey recruitment was balanced against prey natural mortality (excluding Cichla predation) so that the prey would be in equilibrium in the absence of Cichla predation.

SUBROUTINE PRMIGR

If KWJS < W(I) < KWM for any I

For all (J)

$$M1G1(J) = KM1G1(J) * KD1(J) \qquad (20)$$

else M1G1(J) = 0

If KN = 2

$$M1G2(J) = KM1G2(J) * KD2(J),\ \ M1G1(J) = 0 \qquad (21)$$

SUBROUTINE PRMIGR

Cichla migration (advancement on Lake Gatun) is simulated by increasing available prey densities at a rate concomitant with known migration rates.

"Rapid" migration occurs when the juvenile Cichla leave their protected grass beds and begin to school and search for food. In the CICHLA model we have established this by checking whether Cichla cohorts have exceeded a threshold weight but have not reached their adult weights. During this time increase of prey density for the smaller sized prey is computed as a constant fraction of pristine prey density.

There is also an additional year-round migration term for adult Cichla which is simulated by increasing prey density for larger prey size classes by a constant fraction of pristine densities.

SUBROUTINE PRUPDT

$$D1DT(J) = REC1(J) - REC2(J) - MR(J) * N1(J) + M1G1(J)$$
$$D2DT(J) = REC2(J) + M1G2(J) - MR(J) * N2(J) \qquad (22)$$

$$N1(J) = N1(J)(1 + DT * D1DT(J))$$
$$N2(J) = N2(J)(1 + DT * D2DT(J)) \qquad (23)$$

APPENDIX B

Table B1. Parameter values used in CICHLA and their data source.

PARAM		VALUE	UNITS	SOURCE
KRN20	RATION AT 20C	0.09	1/DAY	0.6 FOR LARGE MOUTH BASS (LMB) CALIBRATED IN MODEL RUN
KRNHS	HALF SATURATION FOR PREY DENSITY EFFECT ON RATION	0.4	G/M3	0.5 IN FISH1 CALIBRATED IN CICHLA
KGTRN	Q10 FOR GRAZING RATE	2.5	--	LMB VALUE FROM HATHAWAY 1927
K3	EXPONENT IN WT EFFECT ON RATION	-0.35	--	-0.45 IN FISH1 CALIBRATED IN CICHLA
K9	MEAN LOGNORMAL PREY SELECTIVITY	2.10	--	10.0 IN FISH1 FOR LMB CALIBRATED IN CICHLA ADJUSTED TO RELATIVE SIZE OF PREDATORS AND PREY
K8	VARIANCE IN LOGNORMAL SELECTIVITY	1.25	--	4.0 IN FISH1 FOR LMB CALIBRATED IN CICHLA
KAS	ASSIMILATION EFFICIENCY	0.7	--	0.9 FOR LMB FROM BEAMISH 1972 CALIBRATED IN CICHLA
KRS20	RESPIRATION AT 20C	0.024	1/DAY	0.024 FOR LMB FROM HEIDINGER 1976
K2	EXPONENT FOR WT EFFECT ON RESPIRATION	-0.2	--	-0.2 ADAPTED FROM KITCHELL 1974 YELLOW PERCH MODEL
KRSGT	Q10 FOR RESPIRATION	2.1	--	2.1 IN FISH1 ADAPTED FROM KITCHELL 1974
KST	STARVATION RESPIRATION REDUCTION	0.25	--	SAME VALUE AS FISH1 FOR LMB CALIBRATED IN FISH1
KDWM	MINIMUM GROWTH RATE FOR SACFRY	0.035	1/DAY	SAME VALUE FOR FISH1 CALCULATED FROM EXPECTED SACFRY GROWTH RATE
KREC1	PREY MATURATION FROM 1ST PREY WEIGHT CLASS	MD 0.004 LG 0.001	1/DAY	GUESSED AND CALIBRATED ADJUSTED TO APPROX. EQUILIBRIUM WITHOUT CICHLA
KREC2	PREY BIRTH RATE FOR WEIGHT CLASS PREY	SM 0.005 MD 0.003 LG 0.005 ZP 0.02 CRS 0.01	1/DAY	GUESSED AND CALIBRATED IN CICHLA, ADJUSTED TO APPROX. EQUILIBRIUM WITHOUT CICHLA PRESENT
KWJS	CICHLA JUVENILE WT. AT WHICH SCHOOLING AND MIGRATION STARTS	200	GM	ZARET 1980 & PERS. COMM.
KWM	CICHLA WEIGHT AT WHICH SCHOOLING AND MIGRATION ENDS (MATURATION WEIGHT)	600	GM	ZARET 1980 & PERS. COMM.
KMIG1 KMIG2	MIGRATION RATE FOR JUVENILES (APPARENT RATES OF INCREASE OF PREY FISH DUE TO MIGRATION)	SM 0.036 MD 0.036, MD 0.04 LG 0.036, LG 0.005	1/DAY	FIT TO REPRESENT KNOWN CICHLA 'MIGRATION' RATES IN LAKE GATUN ZARET PERS. COMM. ZARET & PAINE, 1973
KD1	PREY DENSITIES IN UNEXPLOITED AREA	SM 0.3 MD 0.2, MD 0.03 LG 0.03, LG 0.006	GM/M3	ADAPTED FROM ZARET, 1980 AND PERSONAL COMMUNICATION
KW1	WEIGHT OF PREY BY SPECIES GROUP AND WEIGHT CLASS	SM 0.8 MD 4.0, 15.0 LG 10.0, 75.0 ZP 0.0001 CRS 0.1	GM	ZARET PERSONAL COMMUNICATION AND ZARET, 1980, TABLE 1 PLUS GUESSES AND COMPUTATIONS

SM Small Prey Fish (1 size class) ZP Zooplankton

Table B1 (Continued): Parameter Values Used in CICHLA and Their Data Sources

PARAM		VALUE	UNITS	SOURCE
KMR	NAT. MORTALITY RATE BY AGE CLASS	0.005* 0.015** 0.002***	1/DAY	0.05 TO 0.001 FOR LMB CALIBRATED IN CICHLA
KFM	FISHING MORTALITY RATE	0.01-1/DAY		
KEDT	EGG HATCHING TIME	2.0	DAYS	ZARET PERSONAL COMMUNICATION
KSFDT	TIME UNTIL FEEDING BEGINS	6.0	DAYS	FOR LMB FROM CARLANDER 1977 V.2
KFAMT	AGE AT MATURITY	1.0	YRS	ZARET 1980 1 YR FOR LMB FROM CARLANDER 1977
K5	FEMALE FRACTION	0.5	--	ALSO 0.5 IN FISH1
KWE	AVERAGE EGG WEIGHT	1.0E-2	GM	7.7E-4 FOR LMB FROM KELLEY 1962 EGGS SWELL ON SPAWNING ZARET PERSONAL COMMUNICATION
KFC1 KFC2 KFC3	FECUNDITY REGRESSION COEFFICIENTS	0.4787 0.697 5.1	-- -- --	ESTIMATED BY COMBINING LENGTH ON WT. REGRESSION AND FECUNDITY ON LENGTH DATA FROM ZARET 1980 ON CICHLA
KMR	NATURAL PREY MORTALITY RATE BY SIZE CLASS	SM 0.005 MD 0.004 0.001, LG 0.0004 ZP 0.0 CRS 0.0	FISH FISH 1/DAY	GUESSED FROM FISH1 VALUES

* Eggs and Larvae
** Juveniles
*** Adults and Pre-adults
SM Small Fish Prey (1 size class)
MD Medium Fish Prey (2 size classes)
LG Large Fish Prey (2 size classes)
ZP Zooplankton
CRS Crustacea

A DISTRIBUTION AND POPULATION MODEL OF THE MUSSEL MYTILUS EDULIS IN LAKE GREVELINGEN[1]

Jan H. G. Verhagen[2]

Abstract.--It is shown that mussel beds are located there where the transport of food by wind-driven currents to the bottom is maximal. The population model, as constructed, reveals an inverse relationship between recruitment and size of the total stock, leading to long term oscillations in population density.

INTRODUCTION

Lake Grevelingen lies in the south western part of the Netherlands (fig. 1). It was created in 1971 by closing off the Grevelingen estuary. The salinity of the lake can be adjusted to a level high enough to reach the requirements for nature conservation by means of exchange sluices with the North Sea. The lake has a surface area of 108 km^2 and an average depth of 5.3 m. About 60% of the area is shallower than 5 m. Benthic filter feeders e.g. *Mytilus edulis*, *Cardium edule* and *Crepidula fornicata* are important links in the foodweb of the lake. Their average annual consumption as estimated by the Delta Institute for Hydrobiological Research is of a comparable order of magnitude as the total primary production of organic matter. The carbon cycle in Lake Grevelingen is studied by a working group of biologists and chemists of the above mentioned institute since 1972. The model presented here is one of the submodels of the carbon cycle model in Lake Grevelingen being under development by the Delft Hydraulics Laboratory under contract of the Delta Division of the Public Works Department. Main purpose of the modeling research is to enlarge the knowledge about the behaviour of salt aquatic ecosystems, especially under stress situations.

The local distribution of *Mytilus edulis* in Lake Grevelingen appeared to be controlled by the pattern of prevailing wind generated currents, the availability of substrata and the food level in those currents. The relation between the measured pattern of mussel-habitats, the bathymetric map of Lake Grevelingen, and the calculated wind driven currents will be shown.

In the second part of this paper a population model is presented which describes the growth, mortality and recruitment of a *Mytilus edulis* population expressed in biomass and numbers of individuals of nine year classes. From modelresults and available field data of the mussel standing stock in Lake Grevelingen it is postulated that the size of the standing stock is not in equilibrium with the external environmental conditions during each year. Various density

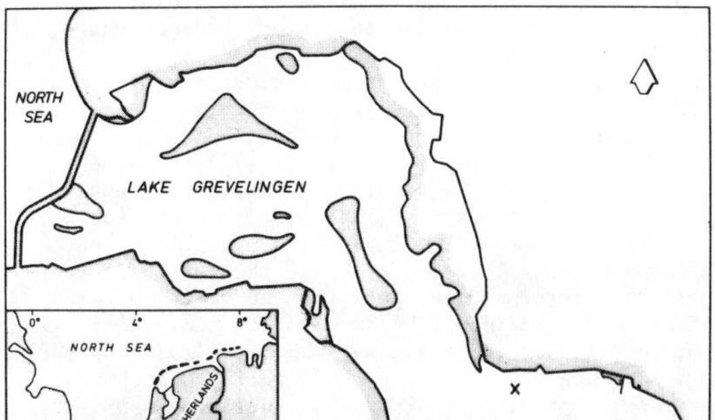

Figure 1.--Map of Lake Grevelingen in the South West Netherlands.

regulating mechanisms acting against overpopulation are effectively after a time delay which tends to have a destabilizing influence, leading to oscillatory changes in biomass of a single species even under relatively stable environmental conditions.

THE DISTRIBUTION OF *MYTILUS EDULIS* IN LAKE GREVELINGEN IN RELATION TO EXTERNAL FACTORS

As already stated in the introduction one of the outcomes of the carbon cycle study, going on for ten years in Lake Grevelingen is, that most of the net primary production (phytoplankton, eelgrass, bottom diatoms) is consumed by the benthic macrofauna mainly consisting of filter feeders. It may therefore be assumed that in general the benthic filter feeders are food limited. It will be shown in this paper that the communities of the mussel *Mytilus edulis* are located there where the vertical mass transport of food to the bottom is maximum. The mussel communities therefore appear to maximize their food supply under constraining conditions regarding the horizontal dimensions of the musselbank. These constraints might be - a limitation in availibility of suitable substrata or - a lower limit in musselbed density capable to provide a secure anchorage to one another against physical stress by action of waves and currents.

Consider now the various possible contributions in the transport of suspended particulate organic

[1]Paper presented at the Third International Conference on State-of-the-Art in Ecological Modelling, Colorado State University, May 24-28, 1982.

[2]Jan H. G. Verhagen is specialist at the Delft Hydraulics Laboratory, Wageningen Branch, Wageningen, The Netherlands.

matter (= food) to a musselbed:

. advective transport along the streamlines in a current
. dispersive transport due to a turbulent exchange of neighbouring water masses with different particle concentrations. This kind of transport is directed along the steepest descent in the concentration distribution (under the restriction of isotropic turbulence)
. transport due to a density difference of the particulates with the surrounding fluid in a gravitation field. This transport contribution is denoted by settling.

The relative importance of the vertical dispersive transport compared to the settling transport can be estimated as follows: Settling transport is v.c where c is the mean concentration of particulate matter in the water column above the musselbed and v is the settling velocity. A reasonable estimate of the order of magnitude of v is 1 m/day. So v.c is estimated as 1.c.

The vertical dispersive transport is $D_z\ \partial c/\partial z$ where $\partial c/\partial z$ is the vertical concentration gradient and D_z the vertical dispersion coefficient.

An estimate of the order of magnitude of $\partial c/\partial z$ can be c/H where H is the waterdepth. The vertical dispersion coefficient D_z under normal weather conditions is estimated as 50 m^2/day. So the vertical dispersive transport under normal weather conditions is estimated as 50 c/H. An estimate of the waterdepth above a musselbed in Lake Grevelingen is H = 5 m, leading to a transport of 10.c. In that case the vertical transport due to turbulence appears to be order 10 times higher than the transport by settling. In schematizing the food transport to the musselbed for modeling purposes we neglect therefore the settling transport. Because dispersive transport in flow direction is negligible in most mass transport problems, the advective transport along the streamlines in the current above a musselbed and the vertical dispersive transport perpendicular to the bed remain as the most important transport contributions. In that case the maximum food flow rate will be reached there where the vorticity generated by the musselbank in a current is confined to the neighbourhood of the bottom covered by the mussels. It is necessary in such cases that the fluid velocity has a component towards the bottom. This is consistent with the equation of mass conservation only, if the fluid accumulation near the bottom is removed either by draining through the bottom or by continued increase of the tangential component of velocity in the stream direction. Therefore *Mytilus*

Figure 2.--Calculated circulation pattern of vertical averaged flow in Lake Grevelingen.

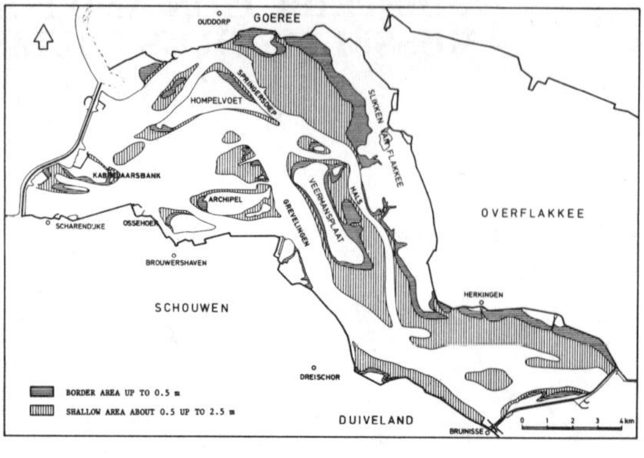

Figure 3.--Map of shallows and gullies in Lake Grevelingen.

communities are found on gently sloping bottoms in a converging flow. The hydrodynamics of such flows as a function of bottom slope and Reynolds number has been studied by various investigators. See e.g. Batchelor (1967).

The analogue between food flux and momentum flux can be useful in further clarifying a possible relation to musselbed location and places of maximum bottom erosion, in future research.

The insights developed so far has to be tested against field data.

The wind driven currents in Lake Grevelingen has been calculated by the Delft Hydraulics Laboratory (1981) using a three-dimensional semi-analytical steady state model the so called Ekman-type model. The result of such a calculation for a westerly wind is shown in figure 2. Depicted is the vector field of the horizontal velocity as integrated over the waterdepth. It appeared from these calculations that the pattern of wind driven circulation in Lake Grevelingen is dictated primarely by the pronounced bottom geometry consisting of deep gullies (10 - 35 m) and a large shallow area (0 - 4 m). See figure 3. To a lesser extent the pattern is influenced by wind direction. The prevailing direction of strong winds in the Netherlands is West-South-West.

From the calculated velocity field the pattern of wind driven circulation is obtained as shown in figure 4. It can be observed that several short closed circulation circuits exist. The exchange of water between those closed circulations cells will be limited. Combining this information with the fact already mentioned that most of the primary production is consumed by the benthic filter feeders, it will be clear that the primary produced organic material in each circulation cell will be mainly consumed by the benthic filter feeders in the same circulation cell.

Therefore optimum food utilization means that every circulation cell at least contains one musselbed. The distribution of musselbeds shown in figure 4 as measured by the Institute for Fishery Research shows a surprisingly good agreement with those calculated circulation cells. It can be seen from this figure that musselbeds are indeed located on gently sloping beaches in a converging flow according to the given theoretical explanation. The numbers along the

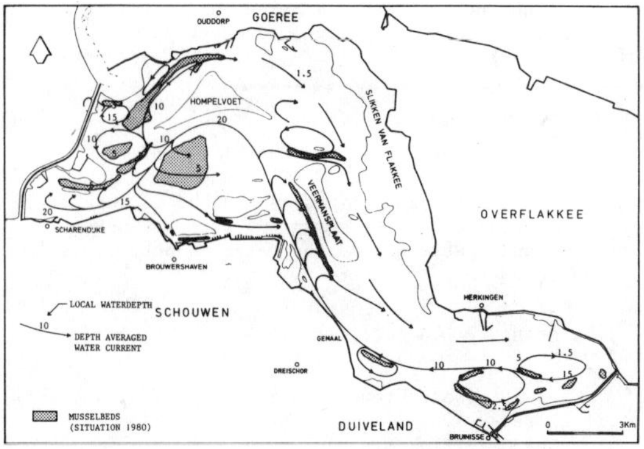

Figure 4.--Distribution of musselbeds in Lake Grevelingen and the wind driven circulation pattern.

streamlines in figure 4 refer to local waterdepth.

THE POPULATION MODEL

Introduction

A simulation model is constructed which describes the production and development of a population of *Mytilus edulis* over a sequence of years. The development of the population expressed in biomass and numbers of individuals of subsequent year classes is followed from spat settled at time zero on a bare bottom of a rectangular large scale reservoir up till a nine year class mixed ages population is reached. The dimensions of the reservoir are chosen in such a way that they are comparable to the mean area of a musselbed located at a mean depth in Lake Grevelingen. The reservoir has a constant throughflow whose magnitude is of the same order as the mean wind driven velocity.

The inlet water contains a concentration of particulate material with a given organic matter content both with an annual cycle typical of Lake Grevelingen. The concentration of particulate matter in the reservoir is assumed to be distributed horizontally homogeneous, vertically not. The vertical distribution depends on the vertical mixing and filtration activity of the mussel standing stock at the bottom.

Growth and mortality, reproduction and recruitment of *Mytilus edulis* are described as functions of external variables as temperature, salinity, season and food input to the reservoir. During the period of reproduction there is a flow of eggs and planktonic mussel-larvae through the reservoir. The simulation model keeps track of their numbers, weights and/or ages.

The growth of one individual mussel is based on the growth model of Bayne, Widdows and Thompson (1976) and Radford, Joint and Hiby (1981).

Additional information needed to build the population model is mainly derived from Bayne (1976) and Bayne (to be published).

The Model of the Growth of *Mytilus edulis* developed by Bayne and Radford

The model MYTILUS of Bayne et al. (1976) integrates the component parts of the energy budget of an individual organism.

The increase in dry flesh weight per unit time is equal to the rate of intake of assimilated dry matter minus the sum of the standard and the routine rate of oxygen consumption and the rate of egg production:

$$\frac{d}{dt}\ \mathrm{DWMYT} = \mathrm{RIADM} - \mathrm{SR} - \mathrm{RR} - \mathrm{REP}$$

where,

DWMYT	is the dry flesh weight in mg
RIADM	is the rate of intake of assimilated dry matter in mg.day^{-1}
SR	is the standard rate of oxygen consumption converted to an equivalent weight of tissue in mg. day^{-1}
RR	is the routine rate of oxygen consumption converted in the same way to mg.day^{-1}
REP	is rate of eggs produced or the fecundity in mg dry wt.day^{-1}.

The specification of each of these rates are described by Bayne et al. (1976). Some modifications have been introduced however by the present author and the description of the modified version will be presented here.

Rate of Intake of Assimilated Dry Matter

The filtration rate FR is written as:

$$\mathrm{FR} = \frac{1.25\ .\ \mathrm{DWMYT}^{0.80}}{1 + (\frac{\mathrm{DWMYT}}{5})^{0.30}}$$

The rate of filtration is optimum over a very wide range of temperatures normally encountered in natural water. Only at temperatures below say 5 ^{0}C there is a significant reduction.

The temperature reduction function FRT of the filtration rate is written as:

$\mathrm{FRT} = 0.4$ for $T < 2$ (^{0}C)
$\mathrm{FRT} = 0.1(T-2) + 0.4$ for $2 < T < 8$ (^{0}C)
$\mathrm{FRT} = 1$ for $T > 8$ (^{0}C)

The rate of filtration is specified as not being affected by the concentration of suspended particulates.

The amount of dry matter filtered per day is written as:

$$\mathrm{DMFPD} = \mathrm{FA}\ .\ \mathrm{FR}\ .\ \mathrm{FRT}$$

where

FA is the concentration of particulate material at the location of the mussel.

The proportion of filtered material rejected as pseudofaeces is specified in a way which differs from Bayne's formulation. The specification is based on the concept given by Widdows et al. (1979) and depicted in figure 5.

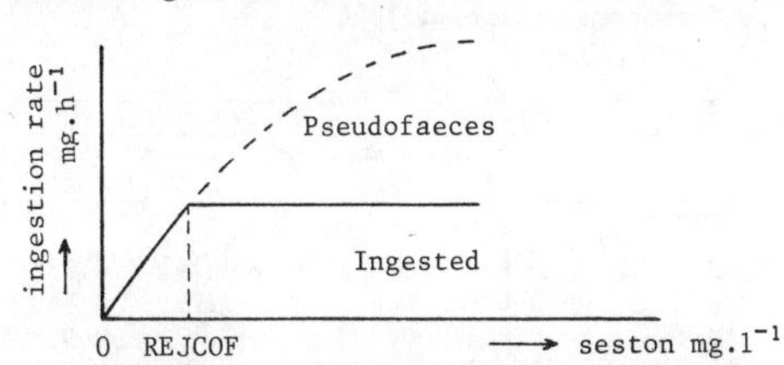

Figure 5.--Ingestion rate as a function of particle concentration.

The difference between the dry matter filtered and the dry matter rejected represents the dry matter ingested DMI.

$$DMI = DMFPD \cdot MIN(1, \frac{REJCOF}{FA})$$

where,

REJCOF is the threshold concentration above which pseudofaeces production begins, specified as

$$REJCOF = 5.4 \cdot WMG^{0.20}$$

where,

WMG is the dry flesh weight in grams.

The organic matter intake per day (OMIPD) is obtained by multiplying the dry matter ingested per day with the proportion of organic matter in the dry suspended material PPNOM. So:

$$OMIPD = PPNOM \cdot DMFPD \cdot MIN(1, \frac{REJCOF}{FA}).$$

The amount of PPNOM is a variable input to the model. A certain proportion of the organic material ingested is assimilated into the body. The assimilation efficiency OAE is written as:

$$OAE = 0.68/(1 + 0.02 \cdot OMIPD).$$

Although the specification differs from the one given by Bayne (1976), the numerical values are quite similar. The intake of assimilated organic matter per day is:

$$RIADM = OAE \cdot OMIPD.$$

Observe that the specification of OAE imposes a maximum value on RIADM reached for very large values of the organic matter intake per day which is equal to 0.68/0.02 or 34 mg dry flesh weight per day.

Rates of Oxygen Consumption

The specifications of the standard and routine rate of oxygen consumption are according to those given by Bayne et al. (1976).
In the winter the standard rate of oxygen consumption is:

$$SOC = 0.055 \cdot CSATW \cdot DWMYT^{0.70} \qquad (ml\ O_2.d^{-1})$$

In the summer:

$$SOC = 0.041 \cdot CSATS \cdot DWMYT^{0.70} \qquad (ml\ O_2.d^{-1}).$$

The effects of temperature are incorporated in CSATW and CSATS as given in table 1.

Table 1.--The ratio of oxygen consumption at 15 °C and oxygen consumption at T °C.

Temperature	Standard rate		Routine rate	
	winter	summer	winter	summer
T	CSATW	CSATS	CRATW	CRATS
----(°C)----				
0	0.617	0.466	0.791	0.575
5	0.689	0.626	0.791	0.668
10	0.803	0.803	0.791	0.953
15	1.000	1.000	1.000	1.000
20	1.284	1.162	1.265	1.048
25	1.708	1.512	1.53	1.45

In the present model version the foregoing winter and summer values are brought together as follows (Radford 1981):

$$SOC = (0.055 \cdot CSATW \cdot (1. - SEASIN) + 0.041 \cdot SEASIN \cdot CSATS) \cdot DWMYT^{0.7}$$

where SEASIN is defined as a sinusoidal function of time being 1. at the moment of water temperature maximum and 0 at the time the water temperature is lowest.

$$SEASIN = 0.5 \cdot (1. + \sin \frac{SEASON-125}{365} .2\pi)$$

(water temperature is highes at day 208)

where,
SEASON = TIME t.

The routine rate of oxygen consumption (ROC) is specified in a similar way. But ROC is also affected by ration level. Bayne and Radford specified these effects of ration in the form of a table listing ROC per unit of metabolic size ($DWMYT^{0.70}$) as a function of dry matter filtered per day (table 2).

$$ROC = (RRATW \cdot CRATW \cdot (1. - SEASIN) + RRATS \cdot CRATS \cdot SEASIN) \cdot DWMYT^{0.7}$$

$$(ml\ O_2.d^{-1})$$

Table 2.--The effect of ration on the routine rate of oxygen consumption per unit of metabolic body size.

Dry matter filtered per day	Oxygen consumption	
	in winter	in summer
DMFPD	RRATW	RRATS
-----($mg.d^{-1}$)------	---($ml\ O_2.d^{-1}.DWMYT^{-0.7}$)----	
0	0	0
2	0.001	0
3	0.015	0.008
5	0.030	0.018
7	0.039	0.025
9	0.048	0.031
11	0.054	0.035
13	0.058	0.038
15	0.060	0.039

The effects of temperature are incorporated in CRATW and CRATS as given in table 1.
The conversion from oxygen to an equivalent dry weight of tissue is 0.96. So:

$$SR = 0.96 \cdot SOC \qquad (mg.day^{-1})$$

$$RR = 0.96 \cdot ROC \qquad (mg.day^{-1}).$$

Egg Production

In the simulation model an equation for fecundity is used expressed as number of eggs produced during spawning as function of the scope for growth. The data or obtained from Bayne (personal communication).
Number of eggs produced (RNEP):

$$RNEP = 3.9 \cdot 10^6 \cdot (1. + \frac{GROW}{14}) \cdot WMG^{1.4}$$

where,
GROW = scope for growth in $mg.d^{-1}$.

Also the egg weight is a function of the scope for growth.
The data are again from Bayne (personal communication).
Egg weight:

$$EGW = 380. \;.\; 10^{-6} \;.\; (1. - \exp(- \frac{GROW + 11.1}{5.5}))$$

Dry weight loss on spawning is the product of number of eggs produced times egg weight, so:

$$REP = RNEP \;.\; EGW$$

Formulation of the Larval Development

The information used in this section is largely based on a chapter written by Bayne in the book "The biology of molluscs: reproduction and development" eds. Verdonk et al. (to be published) and on Bayne (1976) chapter 4.
Fertilisation:
Fertilisation is recorded to occur succesfully over a temperature range from 5 - 22 °C and in a salinity range from 15 - 40 o/oo (Bayne 1965).
Unlike fertilisation however, cleavage and early embryogenesis have a rather limited tolerance of environmental change. Bayne (1965) found that embryonic development of *Mytilus edulis* occurred succesfully only within a salinity of range 30 - 40 o/oo and temperatures of 8 - 18 °C. Therefore the temperature reduction effect is written as:

$$FT = 1 - (\frac{T - 13}{8})^2 \qquad T \text{ in } ^{0}C.$$

The suboptimal salinity reduction effect is specified as:

$$FS = 1 - (\frac{Cl^- - 35}{8})^2 \qquad Cl^- \text{ in o/oo}.$$

The fraction of eggs remaining unfertilized due to other reasons (i.e. no timely encounter with vital sperm) is taken as a constant. The fertilized fraction is taken as FFERT = 0.1. So the number of fertilized eggs succesful developed to embryos produced by one adult is:

$$RNLARV = FFERT \;.\; FT \;.\; FS \;.\; RNEP.$$

Larval growth in length and in weight up to the stage the larva is fully competent to metamorphose is specified as a sigmoidal curve. A generalised growth equation described by Clarke as reported by Bayne (to be published) has been used.
In terms of length:

$$L_t = L_\infty \left[1 - e^{-kt}(1 - (\frac{L_o}{L_\infty})^\delta)\right]^{1/\delta}$$

or in terms of weight:

$$W_t = W_\infty \left[1 - e^{-kt}(1 - (\frac{W_o}{W_\infty})^{\delta'})\right]^{1/\delta'}$$

where,
$W_t = const.L_t^n$ and $\delta' = \frac{\delta}{n}$.

The index t refers at time t, the indices o and ∞ refer respectively to time zero and infinite.

According to Bayne (to be published) the values of the other parameters are:

the shape parameter $\delta = 1.13$
$\delta' = 0.43$
asymptotic length $L_\infty = 293$ µm
asymptotic weight $W_\infty = 1.19$ µg.

These data give a length-weight relationship:

$$\frac{L_t}{L_\infty} = (\frac{W_t}{W_\infty})^{1/n}$$

where,

$$1/n = \delta'/\delta = 0.38.$$

The growth rate parameter k will be a function of food concentration OFA and water temperature T.

The effect of temperature on growth rate is specified as:

$$\text{sech } p(T - T_{opt})$$

The expression is used by Ursin (1963).
Bayne fitted his data for *Mytilus edulis* by this equation and obtained the parameter values $T_{opt} = 16$ °C and the temperature coefficient $p = \;\rightarrow\; 0.21\ ^{0}C^{-1}$ (Bayne to be published).
The effects of ration on growth rate is specified in the form of a Michaelis relation. Data on food concentrations which are optimal for growth in the laboratory are high relative to mean concentrations recorded in the field. Bayne (to be published) stresses this apparent anomaly between laboratory measurements and observations in nature. Therefore the parameters in the Michaelis relation are arbitrarely chosen as:

$$\frac{OFA - 0.05}{OFA + 0.1}$$

The final expression for the larval growth rate k is written as:

$$k = 0.083 \;.\; \frac{OFA - 0.05}{OFA + 0.1} \;.\; \text{sech } 0.21(T - 16)$$

The concentration of food available for the larvae is assumed to be equal to the concentration of organic particulate dry matter in the water away from the musselbed. So:

$$OFA = FAIN \;.\; PPNOM \qquad (mg.l^{-1}).$$

The Larval Mortality Rate

The mortality factors introduced are predation by adult mussels and natural mortality.
Dispersal of free swimming larvae to areas where no suitable sites for post larval survival exist are not taken into account for the Grevelingen Lake situation. It may be important however for estuarine conditions.
The ingestion rate of the free swimming larvae in a fixed water column above a musselbed is a function of the filtration rate of the stock of adult mussels on the bottom and of the concentration and size of the larvae in the water column. The filtration rate of all the adult mussels in a mussel bed is calculated in the population model and is denoted by VFILT expressed in m³/day per m² bottom surface (→ VFILT in m/day).
The concentration of free swimming larvae in a water column above a mussel bed is equal to the number of larvae produced per unit surface mussel bed multiplied by the fraction of the lake surface occupied by

(dense) mussel beds and divided by the waterdepth (DEPTH). The ingestion rate of larvae is specified as:

B . VFILT . EXP(-SIZE/SIZEM)/DEPTH

where B is taken as 0.03 based on an estimate of 6% of the Grevelingen Lake surface occupied by (dense) mussel beds and a concentration of larvae at the bottom which is about 50% of the depth averaged concentration. So B $\approx$ 0.06 . 0.5
Ingestion reduces as the size of the larvae increases. This is specified in terms of an exponential reduction where SIZEM is taken as 100 µm.

Natural mortality is specified as a function of egg weight. If the egg weight is below a given value than the succesful development of the larva will be reduced (Bayne et al. 1978). The natural mortality rate is chosen more or less arbitrarely as:

$$0.13 - 0.07 \cdot \tanh\left(\frac{egw - egwc}{egwo}\right) \qquad (day^{-1})$$

where the critical egg weight value is chosen as
$egwc = 25.10^{-6}$ mg
and $egwo = 10.10^{-6}$ mg.
It is assumed in the model that the larvae are capable to metamorphose as soon as the larval dry weight attains 85% of its maximum value. At metamorphosis which is assumed to last two days, feeding is stopped and 40% is taken as lost by mortality.

The Primary Settled Plantigrade Development

Growth of the plantigrade in size and dry weight is formulated in close analogy to the growth during the larval stage as described before. The growth rate parameter k is specified as:

$$k = 0.060 \cdot \frac{OFA - 0.05}{OFA + 0.1}$$

Temperature effects on the growth rate of plantigrades will be present. Those effects have been left out of consideration because data are missing and because it is assumed that the effects are limited if the primary settlement stage is restricted to the summer month's as is the case in this model. The shape parameters δ and δ' of the growth curves are chosen equal to those for the larval stage.
The asymptotic length of the plantigrade is 2.0 mm.
The asymptotic dry weight is 0.30 mg.
The initial length of the plantigrade is equal to the asymptotic larval length = 293 µm and the initial dry weight of the plantigrade is equal to the asymptotic larval dry weight = 1.19 µg.
The mortality rate of plantigrade is assumed constant and equal to 0.02 d^{-1}. As the plantigrade dry weight attains 85% of its maximum value they are capable to the secondary and final settlement on established mussel beds.

Recruitment

It is assumed that the recruitment of plantigrades on to established mussel beds takes place on a fixed date. This date is chosen to be day 210. The assumption is rather crude. Further research is needed to find and quantify the controlling factors in the timing of settlement. Plantigrades capable to leave their temporary substrate after day 210 are lost in the model. Maas Geesteranus (1942) showed that the plantigrades attached and detached themselves many times before finally arriving at the established bed. The number of plantigrades finally arriving at the mussel bed will be a fraction (1 - DISP) of the total number op spat. The fraction will be several times the fraction of Lake Grevelingen area which is covered by mussel beds (about 6%).
We took 1 - DISP = 0.3, which means that the trial and error procedure of the plantigrades in seeking their final place is repeated about five times.
Thorson (1957) discusses the difficulties that newly settled bivalves have in competition with adults. We specify this effect as a mortality loss, increasing with the density of the mussel bed:

$$FSS = \text{sech}\ \frac{TOTBIO}{BO}$$

where,

TOTBIO is the total biomass in g/m^2
BO is a kind of biomass carrying capacity taken as 300 g/m^2.

Not all the plantigrades arriving at the mussel bed will succeed in settling. If the number of plantigrades is very large, there will be not enough space available for attachment of everyone. A maximum possible density of recruits based on space limitation is introduced RECRM equal to 3.10^4 individuals per m^2. If the number of plantigrades arriving at the mussel bed is small compared to RECRM then all this young mussel will find an opportunity to attach. Therefore the density of recruitment is specified as:

$$RECR = RECRM\left(1 - \frac{TOTSP(1 - DISPL)}{RECRM}\right)).FSS$$

Apart from the competition for space and for food between young and older individuals of the same species, there exists the fight between predators and their prey. As Thorson (1966) points out, we usually think in terms of macropredators like sea-stars, crabs etc., but besides this we must remember that young mussels start their life on the bottom as members of the meiobenthos, which brings the various predators on the meiobenthos-level into the picture.
Lambeck, working at the Delta Institute of Hydrobiological Research, found that the prosobranch *Nassarius* (a snail commonly known as a detritus-feeder) was present in quantities up to 2 gram ashfree dry weight per m^2 on mussel beds in Lake Grevelingen (personal communication). He believes *Nassarius* is a potential predator on newly settled very young mussels. This is in agreement with results of experiments of Vinogradova reviewed by Thorson (1966).
In the simulation model as it stands now it is assumed that only 10% of the young ones settled on the bottom will grow up to the macrobenthos size (defined as the size retained by a 2 mm - mesh screen). The choice of 10% is made arbitrarely. A variety of numbers down to 1.4% (Thorson 1966) is mentioned in literature. The subject need further research.

Age Classes

The population model consists of nine age classes each with its own growth and mortality rate. Apart from the introduction of age classes in the adult population also a distinction is made between the successive development stages of the mussel larvae, as there are eggs, embryo's, planktonic larvae, metamorphosis, primary settlement stage and permanent settlement. Because the development of the mussel larvae depends not only on environmental conditions but also on the initial egg weight, a distinction has to be made in larvae classes developed from various egg

weight classes. Now the egg weight as well as the number of eggs produced depend on the physiological condition expressed in the scope for growth of the adults during spawning which are different for the various age classed. Therefore also the successive development stages of the mussel larvae are dependent on the age class of the parents they generated from. These dependency find expression in the growth rate within various stages and the rate of their development to subsequent stages.
By lumping some larval development stages the number of age class dependent state variables needed in this model could be limited to 12 x 9 = 108. Those state variables are listed below.

Dry weight adults per year class	W(i)
Dry weight eggs produced by adults of year class i	EGW(i)
Dry weight planktonic larvae developed out of egg weight class i	RLDW(i)
Dry weight plantigrades developed out of planktonic larvae class i	PLDW(i)
Number of adults per year class	F(i)
Number of eggs produced per year class	RNEP(i)
Number of planktonic larvae developed out of egg weight class i	RNLARV(i)
Number of larvae capable to metamorphose in class i	RNLACM(i)
Number of plantigrades developed out of planktonic larvae class i	RNPLA(i)
Number of spat developed out of plantigrades of class i	SPAT(i)
Age of the larvae developed out of egg weight class i	TLARV(i)
Age of the primary settled individuals of class i	PERIOD(i)

In the present model version the date of gametogenesis is chosen fixed on day 105. This crude assumption is made for reasons of simplicity. Also further research is needed to find the controlling factors in the timing of spawning.

The filtration rate of the total *Mytilus* standing stock is calculated and used as an input in the calculation of food balance of the reservoir, and as an input in the calculation of the free-swimming larval mortality due to ingestion by *Mytilus* adults.

The filtration rate of total standing stock is obtained by a summation over all year classes of the product number of individuals per year class times filtration rate of the individual per year class.

$$\text{VFILT} = \sum_{j=1}^{9} F(j) \cdot \text{FRT}\,\frac{1.25 \cdot W(j)^{0.8}}{1 + \left(\frac{W(j)}{5}\right)^{0.3}} \qquad \text{(m/day)}$$

VFILT can be interpreted as the vertical velocity of food transport to the bottom above a musselbed due to the filtration activity of the mussel standing stock. The total biomass of the standing stock is:

$$\text{TOTBIO} = \sum_{j=1}^{9} F(j) \cdot W(j)/1000 \qquad (\text{g/m}^2)$$

Mussel mortality is dependent on age.
Bayne measured annual mortality as a function of year class. He obtained high values in the first three year classes leveling of to a lower value of mortality rate for older mussels. The annual mortality parameters AM(j) introduced in the model are from Bayne (personal communication) and given in table 3.

Table 3.--Age class dependent annual mortality rate.

Year class	Percentage annual mortality
0	86.0
1	72.5
2	72.5
3	57.4
4	42.5
5	42.5
6	42.5
7	42.5
8	57.4
9?	90.0?

Model of the Food Transport to the Mussel Bed

Consider a rectangular tank, whose bottom is homogeneously covered with mussels. See figure 6.
The dimensions of the tank are chosen in such a way that they are comparable to the dimension of a water volume above a mussel bed in Lake Grevelingen:

Length	L	(L = 2000 m)
Depth	H	(H = 5 m)
Width	unit width.	

There is a uniform horizontal flow of water through the tank.
Throughflow velocity u (u = 4000 m/day).
The distribution of particulate matter FA in the tank is assumed to be horizontally homogenously, vertically not.
The vertical distribution depends on the magnitude of the vertical dispersion coefficient D_z (D_z = 50 m /day) and of the total ventilation rate of the mussels on the bottom.
The inflow contains a concentration of particulate matter FAIN which is a given function of time. FAIN is assumed independent of the vertical coordinate.
The proportion of organic matter in the dry suspended material is denoted by PPNOM. It is assumed that PPNOM remains independent of space coordinates. This means that no selection of particles in preference to others for ingestion by the mussels is assumed. Production and decomposition of suspended organic matter in the tank are neglected. Therefore the proportion of organic matter in the tank is equal to PPNOM of the suspended material FAIN in the inlet water. Also PPNOM is a given function of time.

The Conservation of Mass

The foregoing assumptions state that the suspended material FA can be considered as a conservative substance. Conservation of mass says:

$$\frac{\partial FA}{\partial t} = \frac{u}{L}\,(FAIN - FA) + D_z \frac{\partial^2}{\partial z^2} FA$$

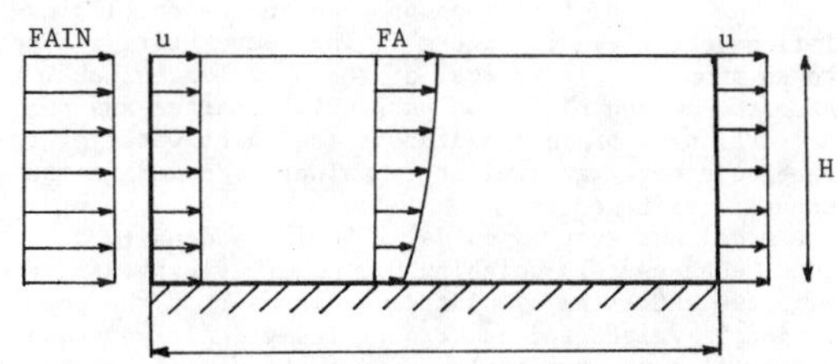

Figure 6.--Schematized food transport to a mussel bed covering a bottom area L times unit width.

A steady state assumption at each time (quasi steady state) is justified, because the time scale of adjustment to steady state is $L/u \approx 0.5$ day which is small compared to the time scale in the forcing input function FAIN(t). The latter shows only a seasonal variation, so the corresponding time scale is about 100 days. So $\partial/\partial t \approx 0$ and the solution is:

$$FA = FAIN + A\, e^{\sqrt{\frac{u}{D_z L}}\, z} + B\, e^{-\sqrt{\frac{u}{D_z L}}\, z}$$

The integration constants A and B follow from the boundary conditions at the water surface and at the bottom.
Through the water surface there is no transport of suspended material, so:

$$\frac{\partial FA}{\partial z} = 0$$

for $z = H$.
The boundary condition at the bottom $z = 0$ reads:

$$FA_{z=0} \cdot v_{filtr.} = D_z \left(\frac{\partial FA}{\partial z}\right)_{z=0}$$

where,

$v_{filtr.}$ = the ventilation rate of the mussel bed.

The formulation of this boundary condition implies that the produced faeces and pseudofaeces is assumed not to resuspend in the water column. Using these boundary conditions the distribution of suspended material over the waterdepth becomes:

$$FA = FAIN \left[1 - \frac{v_{filtr.} \cosh (H - z) \sqrt{\frac{u}{D_z L}}}{v_{filtr.} \cosh H\sqrt{\frac{u}{D_z L}} + \sqrt{\frac{u.D_z}{L}} \sinh H\sqrt{\frac{u}{DL}}} \right]$$

Mussel growth depends on the concentration of FA at the bottom which is:

$$FA_{z=0} = \frac{FAIN}{1 + v_{filtr} \cdot \sqrt{\frac{L}{uD_z}} \operatorname{cotgh} H \sqrt{\frac{u}{D_z L}}}$$

with the parametervalues given in the text we get:

$$FA_{z=0} = \frac{FAIN}{1 + 0.1313 \cdot v_{filtr.}}$$

Model Results compared with Field Data

The forcing functions as used in the model simulations are given in figure 7. The temperature was set to an annual cycle typical of the Grevelingen Lake. Also the concentration of particulate matter and the proportion of organic matter in the inlet water of the reservoir were set similar to values recorded in the Grevelingen Lake.
The model was started on day 205 with a density of 3000 young mussels weighing 0.3 mg dry flesh each. Figure 8 shows the simulated development of the population in weight and numbers of individuals per square meter. Also the available food at the mussel bed is shown.

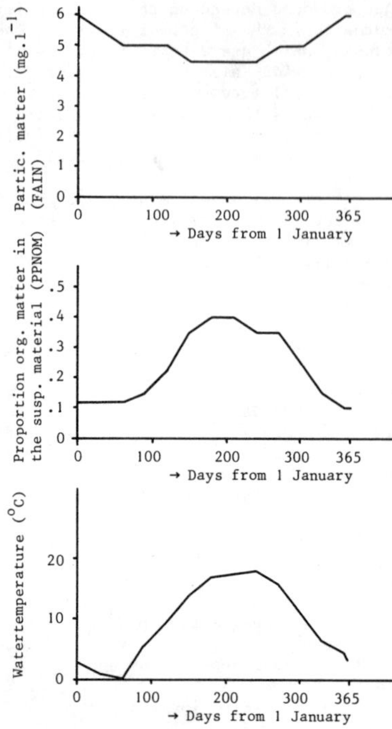

Figure 7.--The annual cycle in particulate material, the proportion of organic matter in the particulate material and the water temperature in the inlet water of the reservoir.

Field Data

During April 1977 Lambeck working at the Delta Institute for Hydrobiological Research, Yerseke, The Netherlands, gathered 65 samples from the bottom of Lake Grevelingen using a stratified sampling technique. His unpublished results (personal communication) show that the biomass of the bottom fauna consists mainly (about 80%) out of *Mytilus edulis*.
Neglecting the samples with low biomass values per m^2 (< 10 $g.m^{-2}$) the retaining samples ($N = 12$) have a mean value of 236.4 and a standard deviation of 185.12 gram dry weight per m^2. The mean value of the number of individuals per m^2 of these samples is 2404 and the standard deviation 2233 individuals.m^{-2}.
Those values are supposed to be characteristic for standing stock values in the mussel bed and are depicted also in figure 8. The agreement is reasonable.
In order to obtain an estimate of the total standing stock of *Mytilus edulis* in Lake Grevelingen, we multiplied the above mentioned mean biomass value of 236.4 in a mussel bed with the percentage of the lake bottom covered with dense mussel beds estimated as 6% based on measurements of the Institute for Fishery Research. An estimate of the biomass density averaged over the whole lake surface is therefore about 14 g ash free dry weight per m^2.
From the results of the simulation model the ratio of ingestion rate of the mussel bed per m^2 per day to the density of the mussel bed can be obtained. The yearly averaged value appeared to be 5.175%. This value is in agreement with the so called "maintenance ration" which is the ingestion rate at which the gross growth

efficiency is zero for one individual. According to Winter (1978) the maintenance ration is 3 - 5% of body weight.
Of course the value will be higher for a population which has to cope with loss rates by mortality and fishing.

The consumption of the mussel biomass will than be $14 \cdot 0.05175 \cdot 365 \approx 264\ g/m^2$ year $\approx 132\ g\ C/m^2$ year. The value is still high as compared to the primary production in Lake Grevelingen which was in 1978:

Phytoplankton	100 - 135 g $C.m^{-2}.y^{-1}$
Eelgrass	40 "
Microphytobenthos	25 - 75 "

It must be concluded that the main part of the primary production goes into the food chain.

Another set of field data on the development of biomass of macrozoobenthos in Lake Grevelingen has been obtained by Wolff et al (1975) during the first years after the closure.
They found that 81.7% of the zoobenthos biomass consisted of suspension-feeders. *Mytilus edulis* took by far the largest share of the biomass and production of all mollusc species, often over 90%. The measurements were taken bimonthly at four sampling stations. Large differences in biomass values between these sampling stations were found.
In figure 9 the monthly values for the biomass of the mollusc species *Mytilus edulis*, *Cardium edule* and *Crepidula fornicata* at the sampling station Archipel

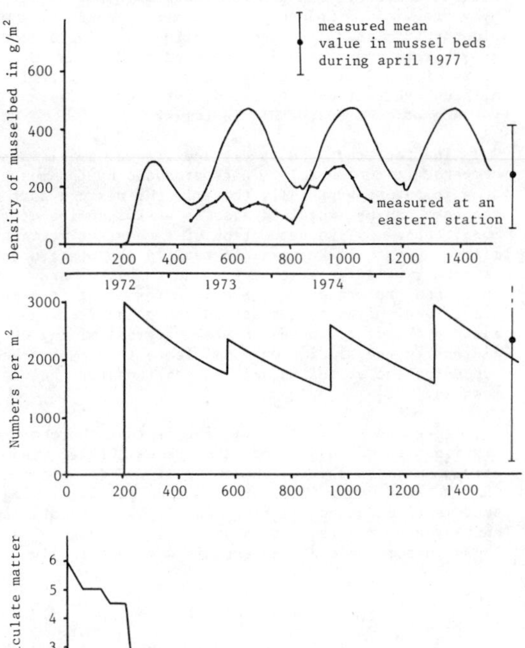

Figure 8.-- The development of the population in weight and numbers per m^2 as determined by the simulation model. Also the simulated available food at the mussel bed is shown.

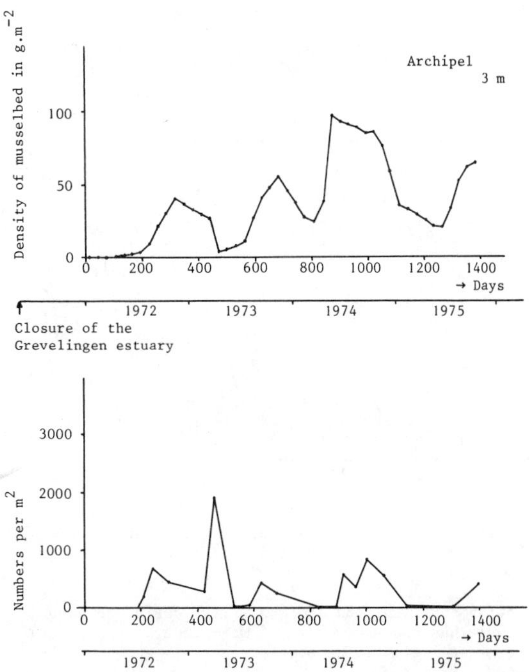

Figure 9.--The development of the population in weight and numbers per m^2 as measured.

(located in the western part of the lake at a depth of 3 m) are given for 1971 - 1975 (Wolff e.a. 1977). The bottom at this sampling station was almost free of macrozoobenthos before the closure. This barren sediment was available for colonization by benthic organisms after the closure. The situation could be compared to the initial value problem of a newly settled population as simulated with the population model. The biomass and number of individuals per m^2 appeared however much lower at this station than simulated. On the other hand the timing and the seasonal variation are more or less in agreement with model results. Monthly values for the mollusc biomass at a second station located in the eastern part also at a depth of 3 m for 1973 - 1974 are shown in figure 8 (Wolff e.a. 1975). Again the agreement with model results is only qualitatively.

The growth in weight of the mussels on the mussel bed as a function of time for various successive cohorts as calculated by the model is shown in figure 10. Field data of the Grevelingen on growth of an individual mussel from a population over several years are not available. To get some indication of the value of this calculated result a comparison is made with data obtained by Bayne and Worrall (1980). They estimated the growth of an individual mussel in two populations of *Mytilus edulis* from two sites near Plymouth (England), called Lynher and Cattewater. The method of growth estimation is by size class analysis. The assumption is made that separatable size classes represent separate age-classes. Their results are shown in figure 10 (Bayne e.a. 1980). We observe that the seasonal increase in weight was initiated in the same month (April) in the Cattewater as in Lake Grevelingen. In the Lynher it was initiated earlier in the year (February). Further it is observed that the difference

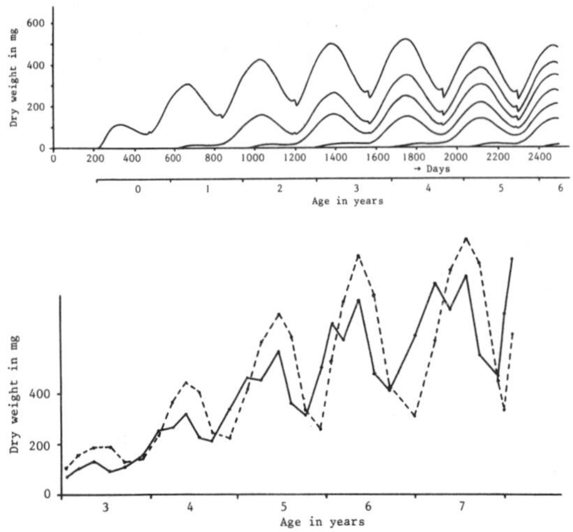

Figure 10.--Growth in weight of calculated cohorts compared with measured data from Bayne and Worrall (1980).

between maximum and minimum weight during an annual cycle has the same order of magnitude as the annual averaged weight. The same property is shown by the model results.

Summary and Conclusions

Presented here is a tentative model version. Several processes particularly those connected with spatfall and recruitment are formulated on purely hypothetical ideas and need to be studied more thoroughly. The validity of the model has to be tested against field data. It can be stated already at this moment that the prospects of the possibilities of a real validation are rather limited, because of the large number of unknown parameters and the paucity of usefull field data. However, the preliminary model can be valuable in so far as it enables the effects of a variety of theoretical processes to be investigated, giving insight into the full dynamic implication of the underlying assumptions.
The exploration of that area of research is still going on.

The most significant finding from the simulations is, that if food limitation occurs than the upper bound in animal weight is limited by the maintenance ration. The reason is that the maintenance ration is an increasing function of body weight. At low ration levels the small animal is most efficient in converting food into body tissue, although the ration is large in relation to body weight.
Most of the mussels in Lake Grevelingen remain small which corresponds with model results under food-limiting conditions. However contrary to model results, there are a number of mussel beds in the lake producing mussels with marketable value during some years. What can be the explanation for this phenomenon?
The most plausible explanation is as follows: There are indications, that the mussel population in Lake Grevelingen has not reached a steady state condition. After the closure in 1971 the mussel population recovered from the initial impact of cutting off the tidal motion. After an increase to a stock size of nearly the same order of magnitude as before the closure the size of the stock of *Mytilus edulis* decreased in 1977 - 1979 but it seems to recover again this year. So an oscillation in biomass seems to occur. By running the model under various conditions we learned that an oscillating behaviour with a period of about ten years could easily be generated even under stable environmental conditions. The reason is that various density regulating mechanisms are effectively after a time delay corresponding to the time needed for a mussel to grow up from an egg to a reproductive adult. Such a time delay tends to have a destabilizing influence leading to oscillatory changes in biomass (May 1976).
These findings are in accordance with those of Hancock (1973) who reviewed twelve years' data on the dynamics of a population of the bivalve *Cardium edule*. He showed a negative correlation to exist between recruits and the size of the total stock. The density regulating mechanisms acting against overpopulation as mentioned by Hancock are also incorporated into the described simulation model leading to the same relation between recruits and the size of the stock.
To obtain the same kind of information for a mussel population, a record of more than ten years of sufficient data are needed which are rarely available. Therefore we postulate, on the basis of model results and some field information, that the size of the total stock of *Mytilus edulis* in Lake Grevelingen is not in a steady state with the external environmental conditions during each year. Years with an overpopulation with poor recruitment and reduced growth rate alternate by years with an underpopulation merely consisting of older year classes in good conditions showing a good growth rate under non-limiting food conditions. So years of great plenty alternate with years of famine, although the periodicity may differ from the seven years mentioned in the Old Testament.

The fact that the model predicts too small mussels compared to the mussel weights produced by the mussel beds in nature especially those in the western part of the lake can be explained also as a shortcoming of the model caused by the assumption of a constant vertical mixing over the lake area. In reality the western part is more exposed to a stronger wind than the more sheltered eastern part of the lake resulting in a more homogeneous mixed food distribution going from the east to the west. Increasing the vertical mixing coefficient in the simulation model leads to a higher food concentration at the mussel bed and therefore heavier mussels.

A second significant finding is that the measured distribution of mussel beds and the calculated wind driven currents show a coherent pattern. It appeared that *Mytilus* communities are found on gently sloping bottoms in a converging flow. Each closed circulation current appeared to contain at least one mussel bed. Those phenomenon could be explained theoretically.

ACKNOWLEDGMENT

I am very grateful to Dr. B.L. Bayne of the Institute for Marine Environmental Research, Plymouth, United Kingdom, for valuable discussions and for his active collaboration by giving every information on his work, published and unpublished, needed to construct the population model.

LITERATURE CITED

Batchelor, G.K. 1967. An Introduction to Fluid Dynamics. Cambridge University Press.

Bayne, B.L. (editor) 1976. Marine mussels: their ecology and physiology. International Biological Progr. 10. Cambridge University Press, Cambridge.

Bayne, B.L. and Worrall, C.M. 1980. Growth and Production of Mussels *Mytilus edulis* from Two Populations. Marine Ecology-Progr. Ser. Vol. 3, p. 317 - 328.

Bayne, B.L. (to be published). The physiological ecology of marine molluscan larvae. Chapter 8 in: The Biology of Molluscs: Reproduction and Development. Editors: N.H. Verdonk, A. Tourpa and J.A.M. van den Biggelaar. Academic Press, London, New York.

Hancock, D.A. 1973. The relationship between stock and recruitment in exploited invertebrates. Rapport et Procès-verbaux des réunions. Conseil permanent International pour l'Exploration de la Mer, 164 p. 113 - 131.

Lindijer, G.J.H. 1981. Three-dimensional circulation models for shallow lakes and seas. Delf Hydraulics Laboratory. Report R 900-III (in Dutch).

Maas Geesteranus, R.A. 1942. On the formation of banks by *Mytilus edulis*. Archives Neerlandaises Zoologie, 6, p. 283 - 325.

Radford, P.J., Joint, I.R. and Hiby, A.R. 1981. Simulation Models of Individual Production Processes. p. 677 - 701. In "Analysis of Marine Ecosystems" edited by A.R. Longhurst, Academic Press, London, New York.

Widdows, J., Fieth, P. and Worrall, C.M. 1979. Relationships between Seston, Available Food and Feeding Activity in the Common Mussel *Mytilus edulis*. Marine Biology 50. p. 195 - 207.

Winter, J.E. 1978. A review of the knowledge of suspension feeding in lamellibranchiate bivalves, with special reference to artificial aquaculture systems. Aquaculture, 13. p. 1 - 33.

Wolff, W.J., Vegter, F., Mulder, H.G and Meijs, T. 1975. The production of benthic animals in relation to the phytoplankton production. Observations in the saline Lake Grevelingen, The Netherlands. 10th Europ. Symp. on Marine Biology, Ostend, Belgium, Vol. 2, p. 653 - 672.

Wolff, W.J., Sandee, A.J.J. and Wolf de, L. 1977. The development of a benthic ecosystem. Hydrobiologia Vol. 52, I, p. 107 - 115.

A SIMULATION OF THE PACIFIC WHITING FISHERY USING AN AGE-STRUCTURED STOCHASTIC RECRUITMENT MODEL[1]

Gordon L. Swartzman[2], Wayne M. Getz[3], Robert C. Francis[4], Robert Haar[5], Kenneth Rose[6]

Abstract. This paper applies a model developed by Getz and Swartzman (1981) to the Pacific Whiting *(Merluccius productus)* fishery off the North American Pacific continental shelf. Recruitment in this stock is lonked both to stock level and water temperature in their spawning area. This link forms the basis of a transition matrix in the model for recruitment which depends both on stock level and spawning grounds temperature. Comparison with historical data on fishery catch and stock estimates shows model output to be realistic. A number of modifications to the original Getz-Swartzman model have increased model speed, allowed a larger number of age classes, improved model stability, and eliminated errors resulting from discretization of the population in the model.

"Come all you young fishermen
listen to me
I'll sing you a song
of the fish of the sea ..."

Olde New England Ballad
Blow ye Winds Westerly

INTRODUCTION

In fisheries as with other natural resources, models have played a role as an aid to management of the resource. Classical reviews and development of quantitative methods for fisheries stock assessment and management were presented by Beverton and Holt (1957) and Ricker (1975). Most approaches to management have been deterministic and have been steady state analyses ignoring the large variability from year to year in fish spawning success and recruitment to the fishery. The most common approach used is to compute recruitment by linking it to stock size through either a linear or nonlinear function. However, the great variability in this stock-recruit relationship (Cushing 1975) does not provide substantiation for a deterministic function for most fish stocks. Walters (1975) attempted to include variability by making a parameter in a deterministic stock-function a random variable.

It appears that environmental conditions also play an important role in determining year-class strength, and in many instances, stock size appears to play a secondary role (Sissenwine 1977, Van Winkle 1979, Lett and Kohler 1976, Nelson et al. 1976). Since variability in the year-class strength appears to be an important and regular feature of most commercially important fish stocks, management policy must begin to consider the resulting year-to-year variability in stock biomass. With this in mind, Getz and Swartzman (1981) developed an age structured population model with a continuous harvest rate through the fishing season, with recruitment determined by a stock-recruitment transition matrix that reflects the probabilities with which various recruitment levels can arise from different levels of stock abundance. In this model, each element of the age distribution vector in a Leslie matrix is replaced by a probability vector giving the probability that the (biomass) level of that age class is in one of m arbitrary biomass ranges between zero and some maximum. Instead of computing deterministic yield and stock level the model computes expected values and variances for each of these variables.

In this paper, the Getz-Swartzman model is applied to the Pacific whiting (*Merluccius productus)* stock and fishery and results are compared with catch data from 1965-1980 and stock estimates from 1973-1980.

Management problems associated with the Pacific whiting offshore stock and fishery appear to lend themselves well to an application of a slightly modified version of the Getz-Swartzman model since this stock exhibits high variability in year-class strength and its recruitment appears to be loosely linked to stock level.

[1]Paper presented at the Third International Conference on State-of-the-Art in Ecological Modeling, Colorado State University, May 24-28, 1982.

[2]Gordon L. Swartzman is Research Assoc. Professor, Center for Quantitative Science, School of Fisheries, University of Washington, Seattle, WA 98195 U.S.A.

[3]Wayne M. Getz is Assistant Professor, Departments of Entomological Science and Plant Pathology, University of California, Berkeley, CA 94720 U.S.A.

[4]Robert C. Francis is Fisheries Biol. (Research), Resource Ecology and Fisheries Management Division, NMFS, 2727 Montlake Blvd. E., Seattle, WA 98112 U.S.A.

[5]Robert Haar is Statistician, Center for Quantitative Science, School of Fisheries, University of Washington, Seattle, WA 98195 U.S.A.

[6]Kenneth Rose is Pre-Doctoral Research Assistant, Center for Quantitative Science, School of Fisheries, University of Washington, Seattle, WA. 98195 U.S.A.

MODEL DESCRIPTION AND MODIFICATIONS

The model developed by Getz and Swartzman (1981) can be briefly summarized as follows:

a) The number of individuals in each of n age classes at time k is characterized by a probability vector $P_i(k)$, i = 1, ..., n, whose j^{th} element $P_{ij}(k)$, j = 1, ..., m, represents the probability that the actual size of this age class is in the interval

$$[\frac{j-1}{m}\ \bar{X}_i(k),\ \frac{j}{m}\ \bar{X}_i(k)]\ .$$

The maximum population level $\bar{X}_i(k)$ for each age class i is calculated by accruing natural and fishing mortality via a Beverton and Holt type analysis, i.e.,

$$\bar{X}_{i+1}(k) = e^{-[\alpha_i + q_i v(k)\bar{t}]}\ \bar{X}_i(k) \qquad i=1,\ \ldots,\ n-1 \qquad (1)$$

where x_i, q_i, v(k), and $\bar{t}$ are, respectively, the natural mortality rate, catchability coefficient, fishing effort, and length of fishing season as a fraction of the total season. The maximum recruitment $\bar{X}_1$ is a constant over time (k = 0, 1, 2, ...) and is determined from stock recruitment data. The probability vector $P_i(k)$ is carried forward as the age group passes into the next year class, i.e.,

$$P_{i+1}(k+1) = P_i(k) \qquad i=1,\ \ldots,\ n-2 \qquad (2)$$

with the only exception being $P_n(k+1)$ for the case where the nth age class consists of individuals not only of age n-1 but older (see Getz and Swartzman 1981 for details).

b) The fecund stock biomass level in the fishery at time k is characterized by a probability vector $P_s(k)$ whose j^{th} element $p_{sj}(k)$, j = 1, ..., m_s, represents the probability that the actual fecund stock biomass level is in the jth range

$$[\frac{j-1}{m_s}\ \bar{S},\ \frac{j}{m_s}\ \bar{S}\ ]$$

where $\bar{S}$ is the maximum fecund stock biomass level. The value of $\bar{S}$ is calculated using the expression

$$\bar{S} = \sum_{i=1}^{n} C_i W_i \bar{X}_i \qquad (3)$$

where C_i is the fraction of sexually mature fish and W_i is the average weight per individual in the ith age class, while $\bar{X}_i$ is obtained from (1) for the case v = 0.

c) The probability vector $P_1(k+1)$ [not obtainable from (2)], characterizing the probability distribution of the new recruits at time k, is obtained from the spawning stock using a stock-recruitment probability transition matrix approach. The elements t_{ij} i=1, ..., m; j=1, ..., m_s of this transition matrix T, represent the probability with which a fecund stock in the jth biomass range will result in the number of recruits falling into the range.

$$[\ \frac{i-1}{m}\ \bar{X}_1,\ \frac{i}{m}\ \bar{X}_1\ ]\ .$$

Thus, in general if newly spawned individuals are only recruited to the fishery at age r, then it follows that

$$P_1(k+1) \ = \ T\ P_s(k+1-r) \qquad (4)$$

d) The major computational problem posed by this model is the calculation of P_s from P_i i=1, ..., n based on the relationship given in (3). A method for computing P_s from P_i, based on the assumption that the vectors $P_1(k)$ are independent across k (this assumption improves as the correlation between stock and recruitment levels is reduced) is discussed in Getz and Swartzman (1981). The time required to complete this computation is a linear function of m^n, which essentially restricts the size of problems to those in which $n \le 6$ (i.e. populations with six age classes or less). A modification of this approach based on a "pairs algorithm" has reduced computational time to a linear function of $M^2 * 2^2$ (K′-1), where K′ is the integer satisfying $2^{K'-1} < n \le 2^{K'}$.

APPLICATION TO THE WHITING FISHERY

The Getz-Swartzman model, modified as discussed above, was applied to the northeast Pacific coast whiting fishery. Model parameters are given in Table 1. Specific age-dependent parameters (α_i, q_i, w_i; i=1, ..., 8) were derived from the Option 2 (modified weight-age, Laevastu-Cushing natural mortality) cohort analysis reported by Francis (in prep.). We felt that since the cohort analysis (Francis in review) was performed on 9 age classes (ages 3-11), it was important to retain this structure in the management analysis (although we combined ages 10 through 12). The maximum number of recruits and maximum parent (fecund) stock biomass ($\bar{S}$) were set to reflect levels in the fishery that would be expected to occur under optimal production conditions. These levels are well above the maximum values estimated during the period (1973-1980) from which parameter estimates were made. The age-specific catchability coefficients (q_i; i=1, ..., 8) were modified from the annual estimates, produced in the cohort analysis, so that they could be applied to the 5-month fishing season used in the management analysis. Fishing effort, therefore, is given in units of 1,000 vessel days on the fishing grounds. The fecundity coefficients were obtained from Bailey (personal communication). There

Table 1. Parameter values used in the simulation runs.

	Parameters	Value for Pacific Whiting
n	No. of age classes	8 (ages 3,4,...,9,10-12)
r	No. of years that recruitment is delayed after spawning	3
m	No. of age class subdivisions	6
m_s	No. of parent stock subdivisions	8
$\bar{x}_1$	Maximum no. of recruits	2.4×10^9
$\bar{S}$	Maximum fecund stock biomass (kg)	1.5×10^9
$\bar{t}$	Length of fishing season (yr)	0.42

Parameters	Ages 3	4	5	6	7	8	9	10-12
Natural mortality rates α_i (yr^{-1})	.278	.210	.195	.257	.357	.457	.557	.657
Catchability coefficient q_i (1000 vessel days effort)$^{-1}$	.0107	.0283	.0338	.0731	.1021	.1695	.1836	.1667
Fecundity coefficient c_i (fraction of sexually mature individuals in each age-class)	.50	.75	1.00	1.00	1.00	1.00	1.00	1.00
Weight coefficients w_i(g)	483	613	717	800	854	906	932	955

Table 2. Stock recruitment probability transition matrices. The transition matrix elements t_{ij} represent the probability that a spawning stock biomass in subdivision j ($\leq$ 8) will result in the number of recruits being in subdivision i ($\leq$ 6).

Cold + Intermediate

Average No. Recruited (10^9 ind)	Recruitment level								
2.2	6	0.00	0.00	0.00	0.00	0.00	0.00	0.00	0.00
1.8	5	0.00	0.00	0.00	0.00	0.00	0.00	0.00	0.00
1.4	4	0.00	0.00	0.00	0.00	0.00	0.00	0.00	0.00
1.0	3	0.00	0.00	0.00	0.00	0.00	0.00	0.00	0.00
0.6	2	0.00	0.45	0.60	0.65	0.68	0.68	0.68	0.68
0.2	1	1.00	0.55	0.40	0.35	0.32	0.32	0.32	0.32
Spawning stock levels		1	2	3	4	5	6	7	8
Avg Biomass (10^3 mt)		0.094	0.281	0.469	0.656	0.844	1.031	1.219	1.406

Warm

Average No. Recruited (10^9 ind)	Recruitment level								
2.2	6	0.00	0.00	0.00	0.05	0.10	0.10	0.10	0.10
1.8	5	0.00	0.00	0.05	0.07	0.15	0.15	0.15	0.15
1.4	4	0.00	0.00	0.15	0.15	0.15	0.15	0.15	0.15
1.0	3	0.00	0.20	0.20	0.23	0.20	0.20	0.20	0.20
0.6	2	0.14	0.60	0.45	0.35	0.25	0.25	0.25	0.25
0.2	1	0.86	0.20	0.15	0.15	0.15	0.15	0.15	0.15
Spawning stock levels		1	2	3	4	5	6	7	8
Avg Biomass (10^3 mt)		0.094	0.281	0.469	0.656	0.844	1.031	1.219	1.406

are 8 age classes on this model which correspond to ages 3, ..., 9, 10-12. Thus i = 1 corresponds to age 3 whiting and i = 8 corresponds to age 10-12 whiting.

The estimates of the elements of the stock-recruitment probability transition matrices are listed in Table 2. These two matrices are constructed under the hypotheses, supported by current data, that:

(1) "cold" water temperatures on the spawning grounds result in relatively high average recruitment with greater variability than in years of "cold" water temperatures; and

(2) "warm" water temperatures on the spawning grounds result in relative high average recruitment with greater variability than in years of "cold" water temperatures.

Bailey (1981) demonstrated that whiting recruitment is inversely correlated to the level of wind-driven Ekman transport on the spawning grounds at the time of spawning. We have assumed that offshore transport is positively correlated with the level of upwelling which, in turn, is negatively correlated with sea surface temperature, since colder bottom water will be brought to the surface during upwelling. Therefore, years of "cold" water temperatures on the spawning grounds are assumed to be years of high offshore transport and low larval survival and years of "warm" water temperatures are assumed to be years of low offshore transport and higher although more variable larval survival. The average sea surface temperature from January to March in the Los Angeles Bight is used as our reference temperature.

In order to estimate the elements of the stock-recruitment probability transition matrices, relationships between values of stock biomass, surface water temperature at the time of spawning, and subsequent recruitment to the fishery three years later were developed. In order to obtain as many stock-recruit data points as possible, relative and absolute measures of both stock biomass and subsequent recruitment were used. The relative measures were adjusted according to regression obtained for years when both measures were available. The relative measures were provided by Bailey, while absolute measures were obtained from Francis (in prep.). The relative stock measure was larval abundance produced by that stock on the spawning grounds (CALCOFI and Soviet cruises) during January after being adjusted for post-January spawning. The relative recruit measure is an index of year-class strength obtained from the age distribution of several year's catch surveys.

Temperature data were available from 1931 to 1979 and a frequency histogram of temperature was trimodal. On this basis, the data were divided into cold (13.4-14.7°C), intermediate (14.7-15.6°C), and warm (15.6-16.9°C) classes. Therefore, years were subjectively divided into cold (13.40-14.69°C), intermediate (14.70-15.59°C), and warm (15.60-16.90°C) so that each interval roughly contained one of the three modes found in the histogram.

A plot in Figure 1 of fecund stock biomass (S_f) against recruitment at age 3 coded by the three temperature intervals (warm, intermediate, cold) provides qualitative evidence in support of the hypothesis proposed at the beginning of this section. Recruitment from cold and intermediate years appears low and relatively invariable while for warm years, it is higher and more variable. In neither case is there any indication of a stock effect on recruitment. Two separate stock-recruit probability transition matrices were estimated from the data in Figure 1, one for cold plus intermediate years and one for warm years (see Table 2).

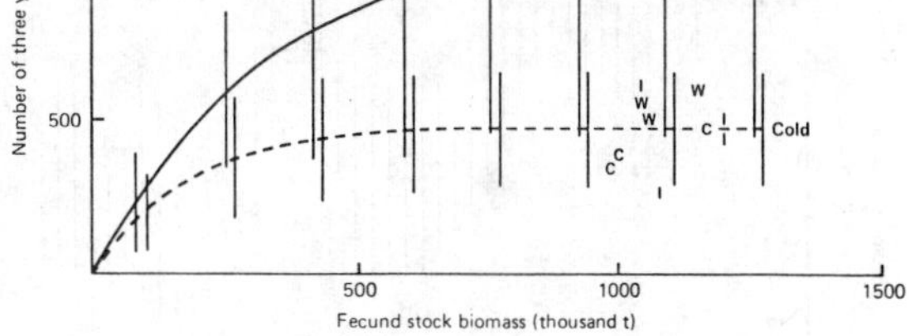

Figure 1. Stock recruit data for whiting showing recruitment for warm, intermediate and cold years and the expected recruitment plus or minus one standard deviation resulting from the stock transition matrices for warm and cold plus intermediate years.

For the runs reported in this paper, estimates were made based on fecund stock level data in the range of 0.7 to 1.2 million mt, and over this stock range, recruitment was assumed to be independent of stock and only dependent on temperature at the time of spawning (see Figure 1). In the stock range of 0 to 0.7 million mt (values below fecund stock levels actually observed), it was assumed that the expected value of the probability density function generated by the estimates of t_{kj}, the elements of the transition matrix, would roughly follow negative exponential stock-recruit curves of the form shown in Figure 1. The parameters of these curves were chosen so that the expected recruitment matched the expected recruitment obtained from the data estimated recruitment probabilities at 0.7 million metric tons.

The probabilities for the lower stock range were then chosen by trial and error such as to give the same expected recruitment as the negative exponential curve and the same variance in recruitment at the .7 million mt stock level. The resultant stock-recruitment probability transition matrices for cold plus intermediate and warm years are given in Table 2.

Validation runs of the model were made over a 47 year period (1933-1980) for which surface temperature data at the time of spawning were available. Driving variables for the population and yield model were (a) effort levels in thousands of standard vessel days from 1966 to 1980, the years of the fishery, (Bailey et al., in press) and (b) classification of environmental conditions on the whiting spawning ground into warm or cold for 1931-1977 based on mean January-March surface temperatures in the LA Bight (D. McClain, personal communication). Figure 2 is a plot of the means and standard deviations of the simulated stock biomass for 1934-80 and Figure 3, a plot of the means and standard deviations of simulated catches along with the observed catches for the history of the fishery (1966-1980). One thing that is noteworthy is that whereas the frequency of warm years was 38% between 1931 and 1979, it was 48% between 1956 and 1979, the time span within which the year classes produced could have affected the productivity of the fishery.

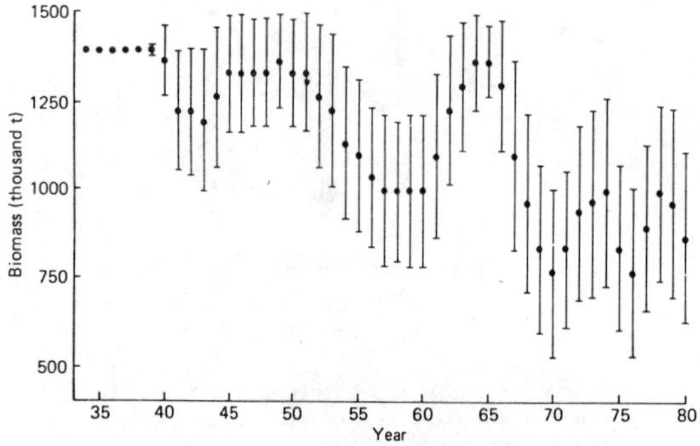

Figure 2. Stock biomass for 47 year 'historical' run.

It is interesting to observe the oscillations in stock biomass predicted by the model (Figure 2). The period between peaks and troughs is around 20 years, a reflection of the apparently periodic fluctuations in environmental conditions on the spawning grounds. Note that initial conditions for the run (1933-1938) were set arbitrarily to large stock levels to represent a virgin stock. So it appears the model reflects the kind of oscillations that the stock might be expected to undergo beginning around 1940 or 1941 when the effects of initial conditions have been damped out.

Expected fishery yields from the model agree quite well (within 1 standard deviation) with the observed yields for 1970-1980, while for 1966-1969 the yields calculated by the model are consistently larger than those reported for the fishery. This inconsistency in the early years of the fishery could be due to one or more of the following factors:

(1) Effort may have been quite inefficient in the early years of the fishery due to a lack of experence in locating the good grounds and deploying the gear to catch the fish.

(2) The gear employed by the Soviet fishery changed from predominately side trawlers at the outset to exclusively factory stern trawlers after 1972. A rough attempt was made to standardize effort (Bailey et al., in press), but the fishing power factors may be inaccurate.

(3) There may have been an underreporting of the catch in the early years.

(4) The model may have overestimated the status of the fishable stock at the time of the initiation of the fishery. Note that the simulated stock is especially high at the time (1966) the fishery started.

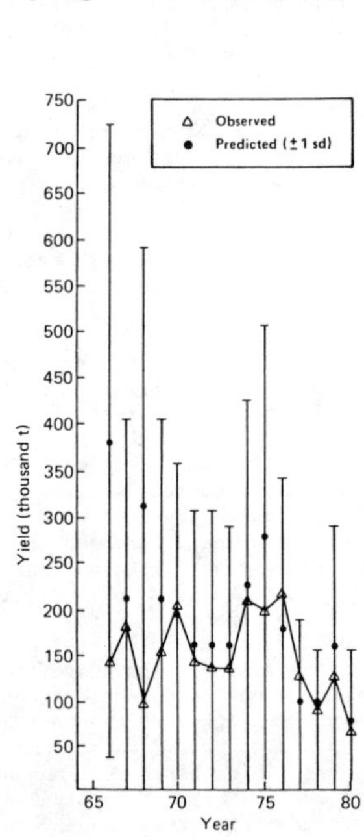

Figure 3. Yield from 'historical' run compared with yield data.

When interpreting runs of this type, it should be stressed that the 47 year trend in stock biomass in Figure 2 reflects what might have been expected to occur if there were no major changes in factors affecting stock production from what has been inferred to occur over the last 20 years; the time span over which these production parameters were estimated.

CONCLUSIONS

The major strength of the Getz-Swartzman model is that it represents realistic population and age structure features of an exploited population without requiring any underlying deterministic relationship between stock and recruitment. Furthermore, by specifying the state of the system as a discrete probability vector, the variability in the population and yield can be considered as well as the expected value.

In this application to the whiting fishery an environmental factor - Ekman transport on the spawning grounds represented by surface water temperature which it is highly correlated with - proved important in affecting recruitment year class strength and was easily included in the model structure. This extended the original applications of the model which considered only stock as affecting recruitment variability.

Comparing the model, simulated over a 47 year period for which sea surface temperatures on the whiting spawning grounds are known, with data on fishery yield and stock size estimates obtained independently showed good agreement except for the first few years of the fishery where other factors, not involved in the model, may have controlled yield. One striking feature of the model output is its extremely high variance. Variance in whiting stock and yield seems to be quite sensitive to the maximum stock level as well as to the transition matrix probabilities, especially in the region below observed stock levels. In future work we are planning to conduct a sensitivity analysis to investigate the magnitude of change resulting from changing different parameters over realistic ranges. We will especially consider varying parameters in the stock range from 0 to .7 metric tons where no observations have been made.

The whiting fishery is managed by the National Marine Fisheries Service. We are helping them by using this model to develop a management algorithm which can maintain a healthy stock level while allowing utilization by the fishery of especially strong year classes. This algorithm 'looks ahead' 5 years and uses information about the temperature over the past 2 years as well as the probability of a cold or warm year next year (cold and warm years tend to come in strings and the transition probabilities can be represented by a Markov probability) to project the future stock and yield under various harvest strategies. This algorithm and its performance relative to other management options is discussed in another paper (Swartzman et al., in review).

LITERATURE CITED

Bailey, K.M. 1981. Larval transport and recruitment of Pacific hake, Merluccius productus. Mar. Ecol. Prog. Ser., Vol. 6:1-9.

Bailey, K.M., R.C. Francis, and P.R. Stevens (in review). The life history and fishery of Pacific whiting, Merluccius productus.

Beverton, R.J.H., and S.J. Holt. 1957. On the dynamics of exploited fish populations. Fish. Invest. Minis Agric. Fish. Food (G.B.) Ser. II Salmon Freshwater Fish 19:533 p.

Clark, C.W. 1976. Mathematical bioeconomics: the optimal management of renewable resources. John Wiley & Sons, Inc., New York, NY.

Francis, R.C. (in review). On the population and trophi dynamics of Pacific whiting, Merluccius productus.

Francis, R.C., G.L. Swartzman, W.M. Getz, R. Haar, and K. Rose (in review). A management analysis of the Pacific whiting fishery.

Getz, W.M. 1980. The ultimate sustainable-yield proble in nonlinear age-structures populations. Math. Biosci. 48:279-292.

Getz, W.M., and G.L. Swartzman. 1981. A probability transition matrix model for yield estimation in fisheries with highly variable recruitment. Can. J. Fish. Aquat. Sci. 38:847-855.

Lett, P.F., and A.C. Kohler. 1976. Recruitment: a problem of multispecies interaction and environmental perturbations, with special reference to Gulf of St. Lawrence Atlantic herring (Clupea harengus harengus). J. Fish. Res. Bd. Can. 33:1353-1371.

Nelson, W.R., M.G. Ingram and W.P. Schaaf. 1976. Larva transport and year class strength of Atlantic menhaden. Brevortia tyrannis. Fish. Bull. 75:23-41.

Pacific Fisheries Management Council. 1981. Fishery management plan and supplemental environmental impact statement for the Washington, Oregon and California groundfish fishery. PMFC, Portland, Oregon.

Rose, K.A., and G.L. Swartzman. 1981. A review of parameter sensitivity methods applicable to ecosystem models. U.S. Nuclear Regulatory Comm. Technical Report NUREG/CR-2016.

Ricker, W.E. 1975. Computation and interpretation of biological strategies of fish populations. Bull. Fish. Res. Board Canada 191:382 p.

Sissenwine, M.P. 1977. A compartmentalized simulation model of the southern New England yellowtail flounder, Limanda ferruginea, fishery. Fish. Bull. 75:465-482.

Van Winkle, W., B.L. Kirk, and B.W. Rust. 1979. Periodicities in Atlantic coast striped bass (Morone saxatilis) commercial fisheries data. J. Fish. Res. Bd. Canada 36:54-62.

Walters, C.J. 1975. Optimal harvest strategies for salmon in relation to environmental variability and uncertain production parameters. F. Fish. Res. Bd. Canada 32:1777-1784.

APPLICATION OF A PHYSICAL HABITAT USABILITY MODEL TO THE FISH COMMUNITY IN A SPRING-FED DESERT STREAM[1]

Thomas B. Hardy[2], Charles G. Prewitt[3], and Kenneth A. Voos[4]

Abstract. A universal suitability function was developed for five species of fish and compared to reported habitat utilizations. The fish habitat model U = S · A was then tested for agreement between simulated and observed distributions of the fish species. The suitability functions showed agreement with the reported habitat utilization of all species using only depth and velocity attributes. The simulated versus observed distributions showed agreement in three of the five species. Universal suitability functions are supported by the data although it is recommended that other habitat variables be included in their development. This should lead to an increase in the agreement between simulated and observed distributions as reported in this study.

INTRODUCTION

It has long been recognized that lotic systems are highly variable over time and that measurements of species responses to different habitat variables is to some degree biased by attributes those variables possess at the time measurements are taken. It has also been recognized that it is desirable to eliminate these biases and to predict the response of species as habitat variables change over time. To this end a multitude of habitat models have been developed that attempt to first identify those habitat variables that may in some way either influence, or at the extreme, limit the distribution and/or abundance of a species. Second, these models in some way have attempted to eliminate the bias of the habitat from which species response patterns are measured and third, to predict the usability of a habitat as the attribute values of selected variables change over time. In most cases aquatic models have adequately dealt with the first concern but are wanting in some degree in both of the latter.

Stalnaker and Arnette (1976) concluded that a comprehensive quantitative methodology was needed to evaluate the effects of instream flow alterations on fish habitat. In order to achieve this goal, the Instream Flow and Aquatic Systems Group (IFG) of the U.S. Fish and Wildlife Service has been involved in continual improvement of fish habitat quality models.

In general, these instream fish habitat models may be represented by:

$$U = S \cdot A \tag{1}$$

where,

- U = an index of the quality (usability) of a stream segment as habitat for a target organism (species and life stage of fish)
- S = a function which models the suitability of habitat attribute values (e.g. depth, velocity, and temperature) in providing habitat for a particular fish species/life stage
- A = a function representing what habitat attribute values are available within the stream segment under study.
- · = an operation which intersects what attribute values are within a habitat with the suitability of those values relative to a particular fish species/life stage

For such a model to be useful, it is necessary that the suitability function (S) represent the preferences of the target organism for measurable habitat variables and should be transferable among streams within a region. The suitability function should not be biased by the distribution of habitat variable values which occurred in the stream system from which the function is to be developed. To develop a suitability function from observed data, it is necessary to account for: 1) the variability within the habitat which is utilized by the fish species/life stage within the studied habitat; and 2) the variability of those attribute values which are potentially available witin the habitat (A).

---

[1] Paper presented at the Third International Conference on State-of-the-Art in Ecological Modeling, Colorado State University, May 24-28, 1982.

[2] Thomas B. Hardy is a Research Scientist, University of Nevada at Las Vegas, Las Vegas, Nevada U.S.A.

[3] Charles G. Prewitt is an Aquatic Systems Analyst on IPA Assignment to USFWS, Salt Lake city, Utah from Colorado State University, Ft. Collins, Colorado U.S.A.

[4] Keneth A. Voos is a Biological Systems Engineer, OAO Corporation, consultant to the Instream Flow and Aquatic Systems Group, USFWS, Ft. Collins, Colorado U.S.A.

The purpose of this paper is, 1) to examine the hypothesis that a suitability function developed in the above manner can describe the preference pattern of particular species that is unbiased by the habitat from which it is developed, and 2) to test the hypothesis that the model U = S·A can provide a functional approach to assessing usable fish habitat.

DESCRIPTION OF STUDY AREA

Ash Spring is located 14.5 km north of Alamo, Lincoln County, Nevada and represents the southernmost of three spring systems within Pahranagat Valley still contributing perennial waters to the pluvial White River channel (Figure 1) (Miller and Hubbs 1960). Several thermal springs (33 C) originate from a carbonate aquifer and contribute a combined flow of 0.50 to 0.56 cubic meters/second which flows south through the valley. The outflow of Ash Spring presently contains the only section of relatively natural stream course that still supports the endangered Pahranagat roundtail chub, *Gila robusta jordani*, and the White River speckled dace, *Rhinichthys osculus velifer*. The White River springfish, *Crenichthys baileyi*, still occurs within the head pools of Ash Spring but is only occasionally found in the outflow. The establishment of five introduced species *Cyprinus carpio*, *Cichlasoma nigrofasciatum*, *Gambusia affinis*, *Poecilia latipinna*, and *P. mexicana* within the outflow of Ash Spring has also been documented (Deacon 1979, Deacon, et al. 1964, Hubbs and Deacon 1964, Hardy 1980a,b).

Figure 1. -- Location of Ash Springs study area near Alamo, Lincoln County, Nevada

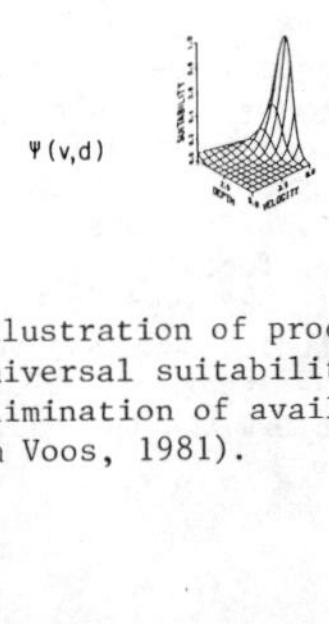

Figure 2. -- Illustration of process for development of universal suitability function through elimination of available habitat bias (from Voos, 1981).

The total amount of available linear aquatic habitat in the outflow of Ash Spring is just under 2250 meters. The riparian zone is dominated by Ash *(Fraxinus)*, willow *(Salix)*, grape *(Vitis)*, mesquite *(Prosopis)*, and cottonwood *(Populus)*. These species provide considerable overhead cover and also contribute to the aquatic cover in the form of root projections and fallen limbs. Several other species can be found along the stream margin although they contribute little in the way of cover to the aquatic system. Substrates are generally firm sand and gravel in the swiftest sections of the stream course while the slower reaches contain well developed mud and silt deposits. Runs comprise about 87% of the available habitat types while pools and riffles contribute 6% and 7% respectively. The mean width of the system is approximately 4 meters although some reaches in the lower section are over 9 meters. Most of the upper reaches have high gradients, narrow channels and a meandering stream course, while the lower reaches are low gradient, broad channeled and relatively straight. The total habitat within the outflow of Ash Spring was divided into four reaches based upon channel morphometry and flow characteristics. Although the entire system was used for development of suitability functions, only Reach IV was used in the utilization simulation analysis.

REACH IV

This section of the outflow represents the lowermost reach immediately upstream from where the stream is channelized into a cement irrigation ditch. The channel is generally broad and straight with scattered dense stands of willow and grape. Ash and cottonwood are also present and all these species contribute to aquatic cover in the form of root projections, snags and overhanging branches. Substrates are predominately firm silt and sand in the swifter sections of the stream although organic mud and silt can be found along the eastern margin of the channel where shallow slow water areas are well developed. Runs comprise about 92% of the available habitat with pools contributing the remaining 8%. There are no riffles within this section of the stream. The gradient is relatively low throughout this reach and the channel is approximately 6 meters wide. There are no well defined banks within this area, indicating a lack of active downcutting.

MATERIALS AND METHODS

A technique for deriving suitability functions has been implemented by the IFG. The mathematical development of the model is described by Voos (1981). Suitability functions are defined by:

$$\Psi(\vec{x}) = c \; \frac{f_{E|F}(\vec{x})}{f_E(\vec{x})} \tag{1}$$

where,

$(\vec{x})$ = vector of environmental attribute values which may provide fish habitat within a stream segment

$f_{E|F}(\underset{\rightarrow}{x})$ = the conditional probability density function (pdf) representing the relative utilization of environmental attribute values by the target fish species/life stage within the studied stream

$f_E(\underset{\rightarrow}{x})$ = the pdf representing the relative availability of environmental attribute values within the studied stream

c = a factor which normalizes the ratio of the two pdf s such that the maximum vaue of the suitability fuction is equal to one over the range of habitable environmental attribute values

As implemented by IFG, exponential polynomial pdf s are used for the above utilization and availability functions. Exponential polynomials were chosen for representing generalized pdf's because 1) these functions can attain shapes often encountered in habitat attribute analysis (including normal, negative exponential and multimodal forms (see Burnham et al. 1980) and 2) the ratio of two exponential polynomials is another exponential polynomial which is a convenience in expressing the suitability function (figure 2).

Collection of data for the development of $f_E(\underset{\rightarrow}{x})$ for use in calculation of the suitability was accomplished by placing transects perpendicular to the stream at 10 meter intervals or at a major change in hydraulic conditions throughout the 2250 meters of

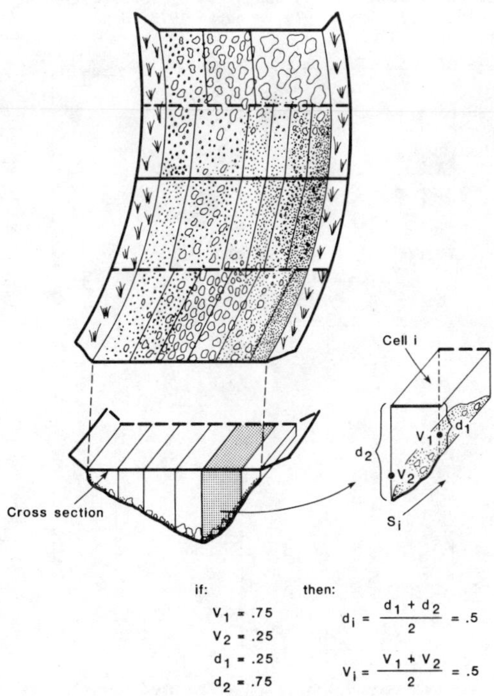

Figure 3. -- Diagrammatic representation of stream physical habitat through using cross-sectional measurement approach.

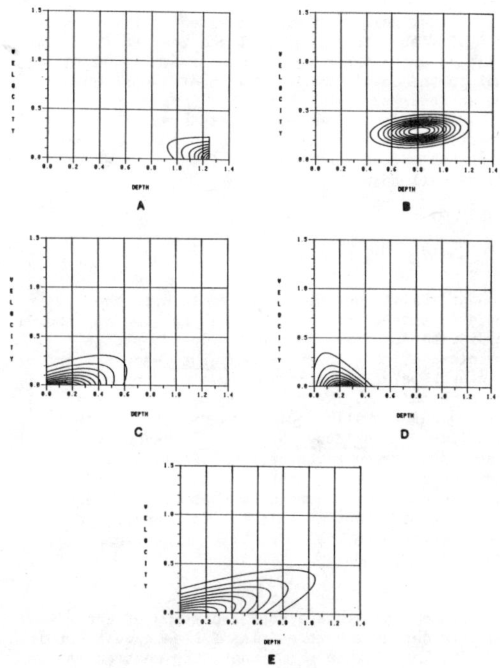

Figure 4. -- Universal (suitability contour plots for G. robusta (A), R. osculus (B), P. mexicana (c), G. affinis (D), and C. nigrofaciatum (E). (depth in f, velocity in fps)

stream course. Several readings of maximum depth and mean water column velocity were taken along each transect and paired, with an area (cell) where the length equals 1/2 the linear distance between two adjacent transects and the width equals 1/2 the distance between two adjacent readings along a given transect line (figure 3). As indicated, the total habitat was further subdivided into four reaches of which only Reach IV or the last 250 meters of habitat were utilized in the development of an availability function (A) for use in the usability simulations.

Fish data for the development of $f_{E|F}(\underset{\rightarrow}{x})$ for use in the calculation of suitability functions were obtained by random observations of adult specimens of each species throughout the entire stream and recording the associated maximum depth and mean water column velocities. In addition population counts of each species were performed within Reach IV along each transect noting their distribution on a cell by cell basis.

To test the derived suitability functions, Monte Carlo simulations were used to compare fish observations in particular cells with where, given known values of environmental attributes (i.e. depth and velocity) of given cells, the suitability function would predict utilization. A high value for the suitability of a particular cell results in a relatively high probability of it being utilized. If the Monte Carlo simulation indicates that a fish would utilize that cell, one fish is added to the running total of fish within that cell and the process is continued until the total number of simulated fish is equal to the total number of fish of a particular species observed within Reach IV. Comparisons were then made between observed and expected distributions of particular species by correlation analysis of either the presence or absence of the species on a cell by cell basis.

RESULTS

Suitability contour plots of each of the five species examined are presented in Figure 4 and show two distinct species groupings. The two native Cyprinids, G. robusta and R. osculus, show preferences for deeper water with the latter also preferring higher velocities. The remaining species show a preference for shallow quiet waters with the two Poeciliids, P. mexicana, G. affinis, having closely related preference patterns. The cichlid, C. nigrofasciatum, shows the braodest preference pattern and is most closely related to that of the Poeciliids in the habitats that it selects. This species also shows overlap with the suitability functions of G. robusta and R. osculus.

Figure 5 provides a representation of the available depths and velocities for each cell within Reach IV. This figure shows that, the eastern margin of the stream course is predominately composed of shallow, low velocity habitats while the western margin of the stream contains deeper swifter water.

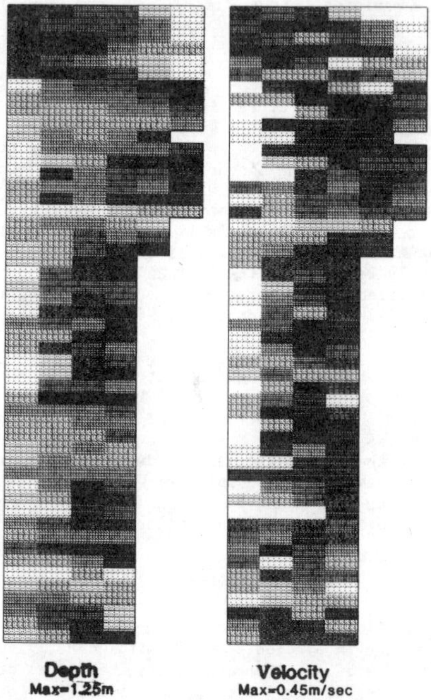

Figure 5. -- Depth and velocity distribution in Reach IV below Ash Springs outflow.

Figures 6-10 provide the observed and expected distributions of each species within individual cells as computed from utilization simulations based upon the suitability functions in figure 4. The observed distributions of G. robusta and R. osculus are more restrictive than their simulated distributions and resulted in a low overall correlation of 0.47 and 0.39 for G. robusta and R. osculus respectively. There is, however, a general agreement between the type of cell actually utilized and that simulated. The observed and simulated distributions of C. nigrofasciatum, G. affinis and P. mexicana show close agreement with correlation coefficients of 0.72, 0.62 and 0.61 respectively. Although Cichlasoma and Poecilia showed ubiquitous distributions they demonstrated a trend for selection of shallower quiet water habitats along the eastern stream margin.

DISCUSSION

Gila robusta jordani

The suitability function (figure 4) developed for this species has been truncated at 1.25 meters along the depth axis as no habitat values greater than this depth were recorded from the stream and therefore defines the upper limit of reliability for this dimension of the suitability function. However, the restricted distribution of this species in deep slow runs and pool habitats is characteristic of several related subspecies throughout the Colorado River drainage system (Barber and Minckley 1966, Deacon and Bradley 1972, Cross 1976, Schuman 1978, Hardy unpublished data).

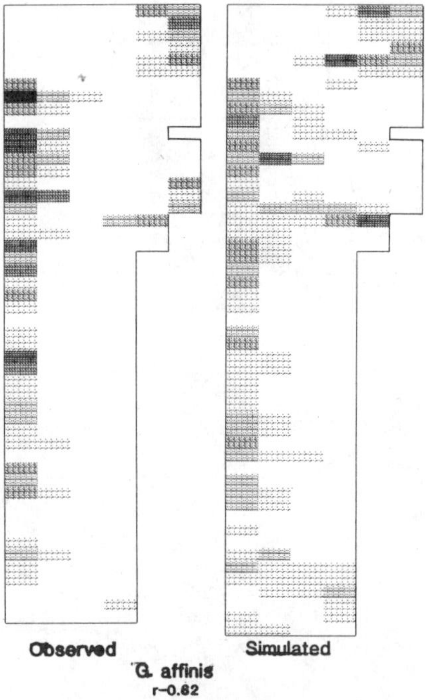

Figure 6. -- Observed vs. simulated G. robusta density at Ash Springs study site.

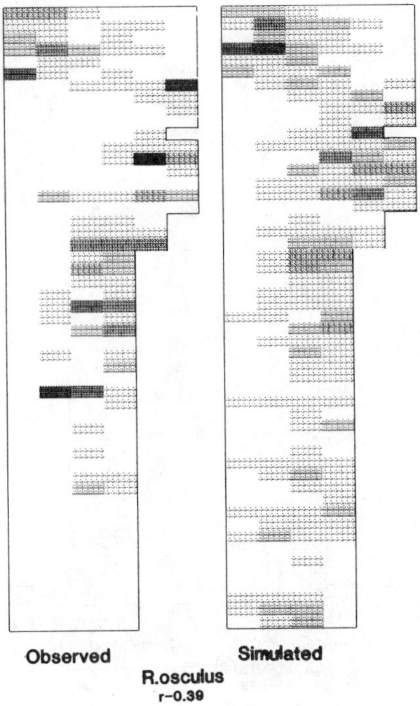

Figure 7. -- Observed vs. simulated R. osculus density at Springs study site.

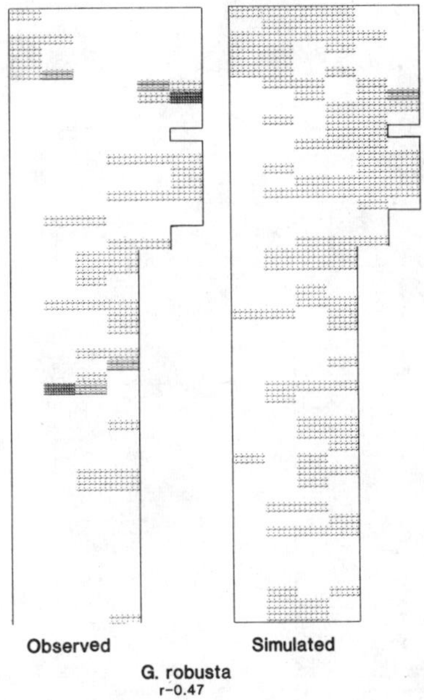

Figure 8. -- Observed vs. simulated G. affinis densities at Ash Springs study site.

The low correlation (r = 0.47) between observed and simulated distributions for this species may be related to the "artificial" truncation of the suitability function in the depth dimension or that other parameters of the environment such as cover, may influence the utilization of a particular habitat cell. This is supported by the fact that the model accurately excluded Gila from simulated cells at the stream margin containing shallow depths even though many of these cells have associated cover. Field notes also indicate that deeper waters without cover are rarely utilized by this species. These factors suggest that inclusion of cover into the suitability function should lead to a higher correlation between observed and simulated distributions and may be necessary to accurately predict usability.

Rhinichthys osculus velifer

The suitability function developed for this species demonstrates a clear preference for moderately deep swift water habitats which is consistent with collections of closely related subspecies in both spring outflows, streams and rivers throughout the Colorado River drainage system (Sigler and Miller 1963, Holden and Stalnaker 1975, Minckley 1973, Deacon and Bradley 1972, Rinne 1975, Cross 1975, Moyle 1976). Although the overall correlation of observed versus simulated distribution is low (r = 0.39) the simulation did accurately exclude Rhinichthys from the quiet water habitats at the stream margin (figure 7). As was the case for Gila, Rhinichthys apparently modifies its depth-velocity distribution dependant upon associated cover within particular cells. Examination of Figure 7 further reveals that both the observed and simulated distributions for Gila and Rhinichthys are very similar and field notes indicate that a large proportion of those simulated cells having expected fish but containing no observations are not associated with any form of cover. Inclusion of cover into dace suitabilities should lead to a significant increase in the agreement of observed versus simulated distributions.

Gambusia affinis

The suitability function developed for this species shows a strong preference for shallow, low velocity habitats. This very selective pattern of response has been shown to effectively segregate this species spatially from two co-occurring members of the same genus (Hubbs and Peden 1969) and is consistent with observed habitat utilizations in other lotic waters (Sigler and Miller 1963, Hubbs and Peden 1969, Minckley 1973, Cross 1975, Moyle 1976, Casterlin and Reynolds 1977, Hardy unpublished data). It has been shown that Gambusia orients almost exclusively near shore in lotic waters even when shallow quiet waters are available in midstream and accounts for some of the variability evident in the observed versus simulated distributions which placed Gambusia in midchannel cells (figure 8). When these cells are excluded from the analysis and only those cells along each stream margin are considered, the correlation

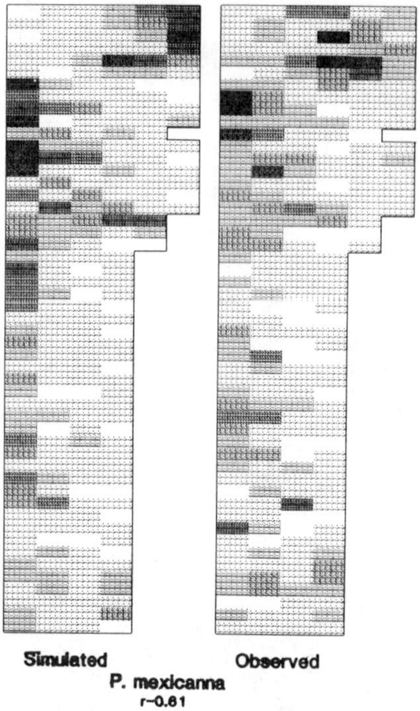

Figure 9. -- Observed vs. simulated *P. mexicana* densities at Ash Springs study site.

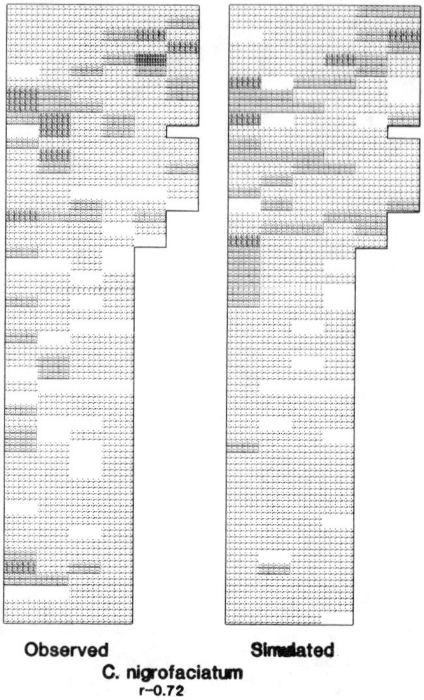

Figure 10. -- Observed vs. simulated *C. nigrofaciatum* densities at Ash Springs study site.

shows a significant increase from 0.62 to 0.76 ($p = 0.05$). Although beyond the scope of this paper, some biotic interactions between *Gambusia*, *Poecilia* and *Cichlasoma* are known to further restrict the distribution of this species (Hardy 1982 MS) and consideration of this factor should lead to a stronger relationship between observed versus simulated distributions.

Poecilia mexicana

The suitability function developed for this species is similar to that of *Gambusia* except that *Poecilia* shows a broader depth utilization (figure 4). This figure however, indicates that *Poecilia* has the potential to utilize deeper, swifter segments of the stream in contrast to *Gambusia*. The general preference patterns of *Gambusia* and *Poecilia* may be related to the similarity in body form and apparently *Poecilia's* greater size and deeper body allows it to exploit somewhat swifter velocities. The suitability function as developed within this stream is in agreement with reported habitat utilization from other lotic waters of the southwest (Deacon and Bradley 1972, Mengel and Darnell 1973, Cross 1976, Hardy unpublished data). The ubiquitous distribution of this species within the study reach is clearly demonstrated by both the observed and simulated utilizations (figure 9). The model accurately depicted the weak trend for this species to be distributed in shallow, quiet water rather than the deeper, swifter segments of the habitat. Although the agreement between observed and simulated utilization was modest ($r = 0.61$) it may have been influenced by the high mobility of this species. Field observations reveal that *Poecilia* is found throughout the habitat but is "transitory" in most of the higher velocity cells, while maintaining protracted residence in the lower velocity habitats. This may be a function of energy metabolism related to swimming energetics.

Cichlasoma nigrofasciatum

The suitability function developed for this species shows that *Cichlasoma* has a fairly broad preference range comparable to the other fish species of this system (figure 4). The comparative reliability of the suitability function is unknown as habitat utilization in lotic waters for this species has not been reported. The observed and simulated utilizations for this species show agreement ($r = 0.61$) and demonstrate the ubiquitous nature of its distribution as well as the weak tendency of increased utilization of shallow quiet waters.

CONCLUSION

The development of suitability functions using only two physical habitat variables showed agreement with the preference patterns reported by all species from a variety of regional suitability habitats and therefore supports the hypothesis of a regional function by eliminating the bias of the habitat from which the function was developed. The technique was applicable in defining suitability functions for three widely differing taxonomic families (i.e. Cyprinidae, Poeciliidae and Cichlidae) and apparently is insensitive to phylogenic group. Also, the technique adequately described small differences in the suitability functions of two closely related species groups (Gila - Rhinichthys and Gambusia - Poecilia) which are distinguished by small differences in body form and most likely behavioral responses to the environmental attributes measured.

The model demonstrated an ability to simulate utilization patterns consistent with the observed distributions of the respective species within Reach IV with only two physical environmental variables (i.e. depth and velocity) and thus supports the hypothesis that U = S·A can provide a functional method in assessing usable habitat in lotic waters. Inclusion of other environmental variables such as cover, substrate, temperature, etc. in the development of suitability functions is expected to increase the agreement between simulated and observed distributions.

LITERATURE CITED

Barber, W.E. and W.L. Minckley. 1966. Fishes of Aravaipa Creek, Graham and Pinal Counties, Arizona. Southwest. Natur. 11:313-324.

Burnham, K.P., D.R. Anderson, and J.L. Laake. 1980. Estimation of density from line transect sampling of biological populations. Wildlife Monographs. The Wildlife Society, Washington DC 202 pp.

Casterlin M.E. and W.W. Reynolds, 1977. Aspects of habitat selection in the mosquitofish Gambusia affinis Hydrobiologia 55 (1) : 125-127.

Cross, J.N. 1975. Ecological distribution of the fishes of the Virgin River (Utah, Arizona, Nevada). Unpubl. MS. Thesis, Univ. of Nevada, Las Vegas.

_____. 1976. Status of the native fish fauna of the Moapa River (Clark County, Nevada). Trans. Amer. Fish. Soc. 105 (4) : 503-508.

Deacon, J.E., C. Hubbs, and B.J. Zahuranec. 1964. Some effects of introduced fishes on the native fish fauna of southern Nevada. Copeia. 1964. (2) : 384-388.

_____, and W.G. Bradley. 1972. Ecological distribution of the fishes of the Moapa (Muddy) River in Clark County, Nevada, Trans. Amer. Fish. Soc. 191(3) : 408-419.

_____. 1979. Endangered and threatened fishes of the West. Great Basin Nat. Mem. BYU (3) : 41-64.

Hardy, T. 1980a. Interbasin report to the Desert Fishes Council-1978. Proc. Tenth Ann. Symp. Desert Fish Council. 10:68-70.

_____. 1980b. The inter-basin area report - 1979. Proc. Eleventh Ann. Symp. Desert Fish Council. 11: pp 2-21.

Harrington, R.W. Jr., and E.S. Harrington. 1961. Food selection among fishes invading a high subtropical salt marsh; from onset of flooding through the progress of a mosquito brood. Ecology 42: 646-666.

Holden, P.B. and C.B. Stalnaker. 1975. Distribution and abundance of mainstream fishes of the middle and upper Colorado River Basins. 1967-1973. Trans. Amer. Fish. Soc. 104:217-231.

Hubbs, C. and J.E. Deacon. 1964. Additional introductions of tropical fishes into southern Nevada, S.W. Nat. 9(4) : 249-251.

_____, and A.E. Peden. 1969. Gambusia georgei sp. nov. from San marcos, Texas. Copeia 1969 (2) : 357-364.

Menzel, B.W. and R.M. Darnell. 1973. Systematics of Poecilia mexicana (Pisces: Poeciliidae) in Northern Mexico. Copeia 1973. (2) 225-237.

Miller, R.R. and C.L. Hubbs. 1960. The spiry-rayed cyprinid fishes (Plagopterini) of the Colorado River System. Misc. Publ. Mus. Zool. Univ. Mich. 115: 1-34.

Minckley, W.L. 1973. Fishes of Arizona. Arizona Game and Fish Dept. 293 pp.

Moyle, P.B. 1976. Inland fishes of California. Univ. of California press. 405 pp.

Rinne, J.N. 1975. Changes in minnow populations in a small desert stream resulting from naturally and artifically induced flows. Southw. Nat. 20(2) : 185-195.

Schuman, P.B. 1978. Responses to temperature and dissolved oxygen in the roundtail chub, Gila robusta Baird and Girard. Unpubl. MS. Thesis, Univ. Nevada, Las Vegas.

Sigler, W.F., and R.R. Miller, 1963. Fishes of Utah. Utah State Dept. of Fish and Game, Salt Lake City.

Stalnaker, C.B., and J.L. Arnette, eds. 1976. Methodologies for the determination of stream resource flow requirements: An assesment. Utah State University, Logan, Utah. 199 pp.

Voos, K.A. 1981. Simulated use of the exponential polynomial/maximum likelihood technique in developing suitability of use functions for fish habitat. PhD Dissertation. Utah State University, Logan, Utah. 85 pp.

SOLVING SPECIES-HABITAT FUNCTIONS WITH GEOREFERENCED ENVIRONMENTAL DATA A NEW TOOL FOR ENVIRONMENTAL IMPACT ASSESSMENT[1]

David A. Adams[2], Margery F. Overton[3], T. C. Gopalkrishnan[4], Jing Song Wei[5], Gordon M. Cressman[6]

Abstract. ---A hydrodynamic model was linked to a species-habitat function provided by Habitat Evaluation Procedures (HEP) to evaluate the impact of either of two potential inlets upon largemouth bass habitat in Currituck Sound, NC, U.S.A. The model predicted that overall bass habitat would be reduced 39 to 52 percent if an inlet were opened. The procedure accepts georeferenced habitat data, solves species-habitat or other user functions, and provides cartographic and tabular output.

INTRODUCTION

Recent efforts to assess the impact of environmental alterations upon wildlife have emphasized the role of habitat quality. This paper describes an attempt to employ the U. S. Fish and Wildlife Service's Habitat Evaluation Procedures (HEP) to analyze the impact of possible saltwater introductions into Currituck Sound (currently an almost-fresh estuary) upon the Sound's ability to support largemouth bass (*Micropterus salmoides*). The procedure uses georeferenced environmental data and hydrodynamic modelling of saltwater distribution, and provides cartographic and tabular output.

---

[1]Paper presented at the Third International Conference on State-of-the-Art Ecological Modelling, Colorado State University, May 24-28, 1982.

[2]Associate Professor, Departments of Forestry and University Studies, North Carolina State University, Raleigh, NC, U.S.A.

[3]Assistant Professor, Department of Civil Engineering, North Carolina State University, Raleigh, NC, U.S.A.

[4]Associate Research Scientist, KISR, Hydraulic and Coastal Engineering Group, Kuwait

[5]Graduate Research Assistant, Department of Marine, Earth, and Atmospheric Sciences, North Carolina State University, Raleigh, NC, U.S.A.

[6]Graduate Research Assistant, Department of Forestry, North Carolina State University, Raleigh, NC, U.S.A.

PROJECT SETTING

Currituck Sound lies in the extreme northeastern corner of North Carolina (fig. 1). The Sound is approximately 30 miles (48 km) long (N-S), averages 4 miles (6.4 km) wide, and occupies about 97,000 acres (39,000 ha). Average depth is about 5 ft (1.5 m), with the deepest point slightly more than 10 ft (3 m) deep. Although physiographically estuarine, water levels are primarily wind, rather than lunar tide, influenced, and salinity rarely exceeds 3 ppt (parts per thousand). During historic times, however, Currituck Sound has been connected to the Atlantic Ocean by one or more narrow inlets across the barrier island (Currituck Banks).

Historically, the local economy has been dependent upon the Sound's natural resources. Most of the local population now have turned from commercial fishing toward recreational fishing and hunting and to guiding sportsmen engaged in such pursuits. At the present time, the Sound's greatest amenity is the largemouth bass fishery, which provides sport fishing for thousands of local and non-resident fishermen, serves as the focal point for several bass fishing tournaments, and attracts a major portion of the annual tourist income into the county. This fishery is also quite vulnerable to elevated salinities.

The sport fishing segment of the population seems content with the existing salinity regime. However, a portion of the local population remembers the "good old days" when the Sound was more saline and both waterfowl and commercial fish species were thought to have been more abundant.

Figure 1.--Location of Currituck Sound.

Periodically, this group requests an assessment of the salinity question by federal and/or state agencies. Thus, between 1958 and 1964, the U. S. Fish and Wildlife Service, the Virginia Commission of Game and Inland Fisheries, and the North Carolina Wildlife Resources Commission conducted an intensive study, which provided the baseline information used in this study (Back Bay - Currituck Sound Data Report 1965). During the late 1970s, concern again mounted. This latest episode led to 1) a public hearing at the Currituck County seat on May 2, 1980, 2) an excellent but brief status report prepared by state agency personnel in December, 1980 , and 3) a funding request to the federal-state Coastal Plains Regional Commission to support a more comprehensive study. Commission funds subsequently were provided to the North Carolina Department of Natural Resources and Community Development, which sponsored a multidisciplinary study through the Water Resources Research Institute of The University of North Carolina.

This paper reports a part of that study.

PROCEDURE

Environmental Data Input

The procedures used in this study were first developed as a teaching device (Adams 1980a). They were expanded under a contract from the National Coastal Ecosystems Team of the U.S. Fish and Wildlife Service (Adams 1980b), and subsequently used to compare bobcat (Lynx rufus) habitat quality as determined by HEP with the intensity of radio recoveries from animals bearing radio transmitters (Lancia et al 1982).

For this study, baseline environmental data were taken from unpublished reports of the Back Bay - Currituck Sound Study cited earlier. The original data were presented as characteristics (frequency, average) of conditions within a 1000 x 1000 yd (900 x 900 m) grid. Each datum was assigned to the center of its respective grid cell and its location identified by North Carolina Plane Coordinates (distances in feet north and east of an origination point). This step produced computer readable files containing estimates of 1) average spring salinity, 2) average summer salinity, 3) frequency of sand bottom type, and 4) frequency of aquatic vegetative cover. A separate file contained the location of each datum, expressed as coded North Carolina Plane Coordinates.

The Sound's outline was digitized from U. S. Geological Survey topographic maps at map scales of 1:24,000 and 1:62,500, using the same coding system applied to North Carolina Plane Coordinates.

These files were then entered into SYMAP.

Baseline Environmental Factor Maps

The program SYMAP (Dougenik and Sheehan 1979) produces conformant or contour maps from georeferenced data and outputs a file containing the interpolated value of each grid cell. Each input "package" contains a coordinate description of the area to be mapped, the coordinates of each input data location, values corresponding to each location, and mapping instructions. Legends, titles, and modifications of the basic program (Electives) are also available. In the Currituck example, data locations were input to the nearest 100 ft (30.5 m) and environmental factors mapped as continuous functions (contour maps) at a scale of 1 in=5,000 ft (1:60,000). The resultant map, printed at 10 characters to the inch horizontally and 8 characters to the inch vertically, thus produced a value for each 7 acre (3 ha) cell.

An elective on the SYMAP program (Elective 21 - SYMVU tape) creates a tape with an entry for each location on the output map. The original intent of this file was to provide input to SYMVU, a program which generates three dimensional line-drawing displays of data (Harvard University 1979). However, the tape also provides a dataset of georeferenced data which can be used for many kinds of analyses -- in this case habitat quality. Elective 21 was thus employed to output a file of each of the four environmental factors mapped, and these files provided input to a SAS program.

Bass-Habitat Functions

The Statistical Analysis System (SAS Institute 1979) is a computer system designed for data analysis, which runs on IBM 360/370 computers. It provides tools for information storage and retrieval, data modification and programming, report writing, statistical analysis , and graphics, and was employed primarily to process the large arrays produced by Elective 21 of SYMAP (243 x 638 in this example).

SAS merged all environmental factor files, then solved the largemouth bass-habitat function, accumulated the values computed for each cell, and outputted a file containing habitat quality values. The merged files permitted superimposition of environmental factors and solution of the habitat function for each character location. The matrix of function values outputted by SAS was processed into the form required by SAS/GRAPH (SAS Institute 1981). Row and column indices of the map matrix were transformed into North Carolina Plane Coordinates and processed by the GCONTOUR procedure of SAS/GRAPH to produce contour maps on a standard drum plotter.

In 1973, the U. S. Fish and Wildlife Service began developing uniform Habitat Evaluation Procedures (HEP) to assist in discharging its responsibilities under the Fish and Wildlife Coordination Act (116 U.S.C. 661-667), the National Environmental Policy Act (42 U.S.C. 4321-4347), the Endangered Species Act of 1973 (16 U.S.C. 1531-1543), and other federal statutes and Executive Orders.

HEP are designed to:

1. Provide a quantitative index value for existing habitat conditions.

2. Determine the difference between the index values of existing and future conditions with or without a proposed project.

3. Demonstrate, in habitat units gained or lost, the beneficial or adverse impacts anticipated as a result of alternative development strategies; and

4. Display what is required to mitigate and/or compensate for fish and wildlife resources lost as a result of a project, and demonstrate enhancement opportunities (Fish and Wildlife Service 1980).

HEP use simple mathematical functions which describe habitat quality for particular species, ascribe a quality value to each habitat type (cover type) in the project area, and sum the quantities (value x area) to obtain an index of habitat quality for a given species on the total area. In the Currituck example, basic HEP functions and procedures were employed, but the function was solved for each cell, rather than for each habitat type.

The largemouth bass function was taken from a July 1981 draft obtained from the Fish and Wildlife Service. The draft first discusses the distribution and general biology of the species, then presents specific habitat requirements for adult, juvenile, fry, and embryo bass as reported in the literature. Finally, it gives bass Habitat Suitability Index (HSI) models for riverine and lacustrine habitats, depicting habitat suitability as a function of food, cover, water quality, and reproduction. Each HSI model is composed of Suitability Index (SI) functions of aquatic vegetation cover, total dissolved solids, dissolved oxygen, pH, water temperature, salinity, substrate composition, and other measurable environmental factors. All indices are normalized to range from 0.0 to 1.0.

Although all of the factors cited (and probably more) are important to the concept of habitat quality, rarely does an investigator have access to data on all of them (particularly if he is unable to obtain original field data specifically for the HEP exercise). Furthermore, if the problem is to compare "before" and "after" conditions, data on factors which are not expected to change and are not believed to be limiting are probably unnecessary. In the Currituck example, synoptic data on temperature, pH, dissolved solids, and suspended solids were omitted; they were unavailable or were believed to be unchanged by the alterations being considered. Thus the analysis contained only the salinity, vegetative cover, and substrate elements of the HEP model.

In its abbreviated form, the HEP largemouth bass model was:

$$HSI = (CC \times CWQ \times CR)^{1/3}$$

where,

CC = a Suitability Index (SI) of the cover component; the arithmetic mean of the SIs attributable to vegetative cover for bass adults and juveniles (fig. 2) and to vegetative cover for fry (fig. 3).

CWQ = a SI of the water quality component; the arithmetic mean of the SIs attributable to the impact of maximum summer salinity upon bass adults and juveniles (fig. 4) and the impact of maximum summer salinity upon fry (fig. 5).

CR = a SI of the reproduction component; if the maximum spring salinity (fig. 6) is optimal for bass reproduction, the SI attributable to substrate (fig. 7), otherwise, the geometric mean of the SIs for spring salinity and substrate.

Further modifications of the HEP functions were necessitated by limitations in available data. The 1960-1962 baseline data contained only a single salinity observation per month per sampling location. Since HEP require "maximum" values and three observations per season were not deemed adequate to reflect a maximum, arithmetic means of monthly data were used. This concession yielded higher HSIs than otherwise would have been derived, but hopefully did not invalidate the general baseline characterization

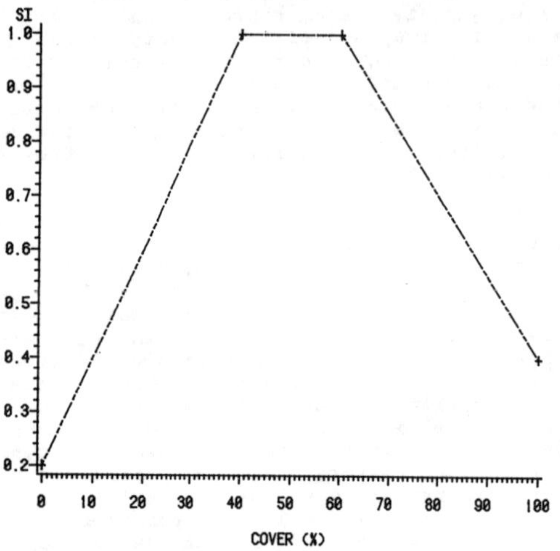

Figure 2.--SI and vegetative cover (adult, juvenile).

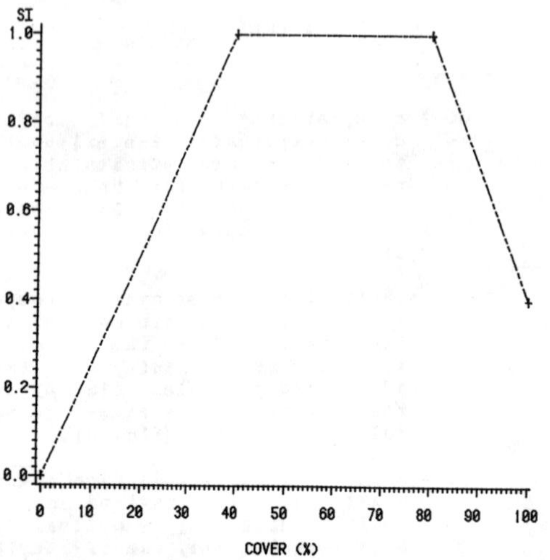

Figure 3.--SI and vegetative cover (fry).

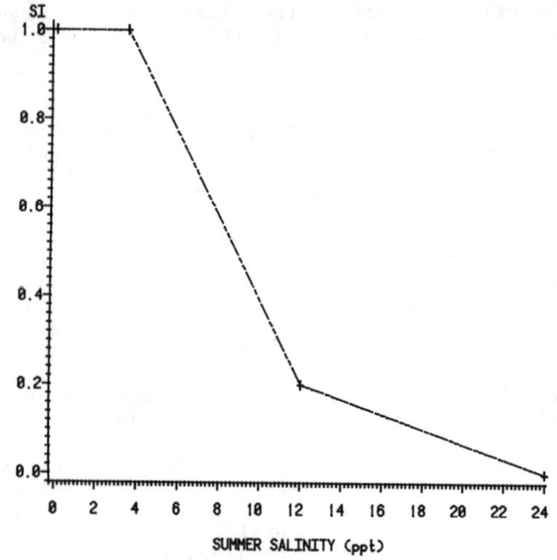

Figure 4.--SI and maximum summer salinity (adult, juvenile).

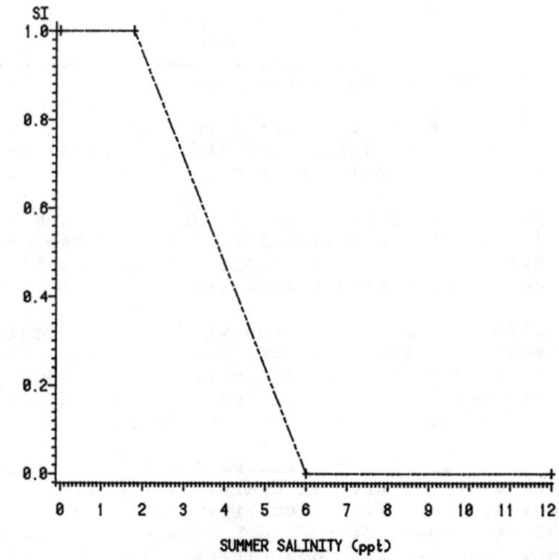

Figure 5.--SI and maximum summer salinity (fry).

nor conclusions reached.

HEP utilize "percent bottom (e.g. aquatic vegetation, logs, and debris) cover within pools, backwaters, or littoral areas during summer" as the measure of CC. The data available utilized the relative frequency with which vegetation was encountered at 10 sampling points within each 1000 x 1000 yd (900 x 900 m) grid cell. Substitution of frequency for percent cover should not have affected the ultimate results.

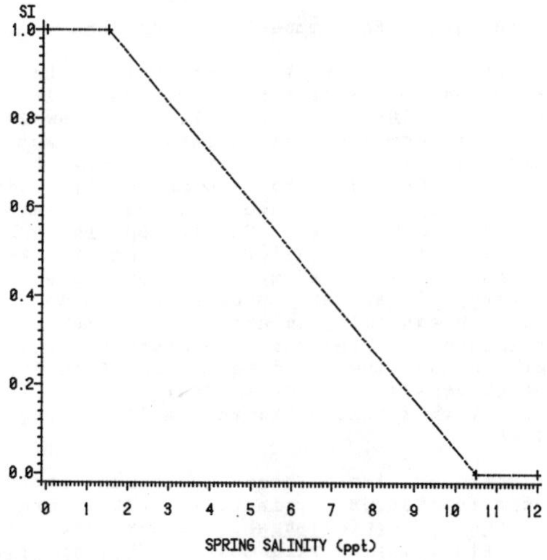

Figure 6.--SI and maximum spring salinity (embryo).

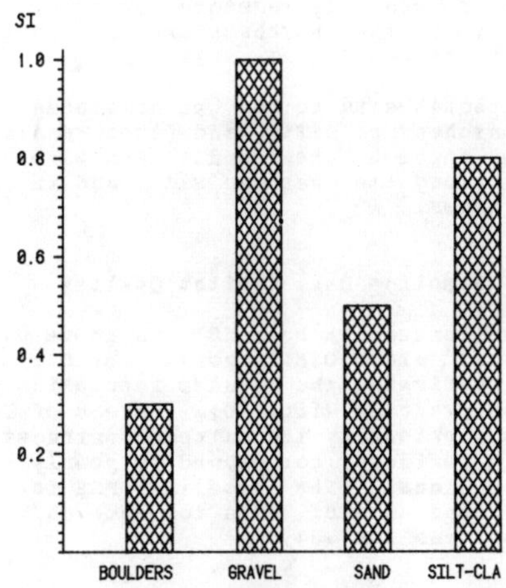

Figure 7.--SI and substrate composition (reproduction).

The SAS program computed a HSI value for each character location of input data, outputted the resultant file, and summed the values to obtain an index of largemouth bass habitat throughout the entire Sound.

Hydrodynamic Model

The numerical model is based on the Finite Element Method in which the governing partial differential equations are reduced to a system of algebraic equations using the Galerkin procedure (Taylor and Davis 1975). The governing equations are the two dimensional depth -- averaged shallow water wave equations,

$$du/dt + udu/dx + vdu/dy + gdn/dx + g(Sfx) = 0$$

$$dv/dt + udv/dx + vdv/dy + gdn/dy + g(Sfy) = 0$$

$$dn/dt + d(u(h+n)))/dx + d(v(h+n))/dy = 0$$

and the diffusion/advection equation,

$$dc/dt + udc/dx + udc/dy = k(d^2c/dx^2 + d^2c/dy^2)$$

where,

d/dx, d/dy, d/dt are partial differential operators

x,y are the axes in the horizontal plane

u,v are the corresponding velocity components

g = acceleration due to gravity

t = time

n = water level above a horizontal datum

h = depth from horizontal datum to bottom surface

Sf= friction slope

c = salinity concentration

k = diffusion coefficient.

The area of study was subdivided into two dimensional triangular elements for which discrete values were ascribed to the parameters at the nodes. Boundary conditions were applied at specific nodes as known values of water velocity or elevation and salinity concentration as a function of time.

A complete space time history of the nodal parameters (water velocity, elevation, salinity) was produced using an explicit split time integration scheme (Wang and Connor 1975). The hydrodynamic and diffusion equations were treated as sets of uncoupled equations; for each time step the water velocities and elevations were determined, the water velocities were then used as input to the diffusion/advection equation and the salinity concentrations determined.

Salinity values were stored on disc as functions of space and time for input to the bass-habitat function. Because bass would be limited by high salinity, the values derived represent the maximum concentration expected to result from the annual mean spring tide in the Atlantic Ocean at Currituck Banks.

For purposes of the overall modelling effort, salinity values should reach steady state. However, preliminary tests estimated that about one month (real time) would be required to attain equilibrium. To reduce modelling costs, computations were terminated when steady state conditions extended about 4 mi (6.4 km) from each potential inlet. All values (water velocity, elevation, and salinity concentrations) were then stored to be used as initial conditions for further runs if necessary.

Modifications

State agency personnel provided two potential inlet locations (fig. 8). For modelling purposes, each inlet was considered to be 200 ft (60 m) wide and 8 ft (2.4 m) deep at mean low tide. The impact of each

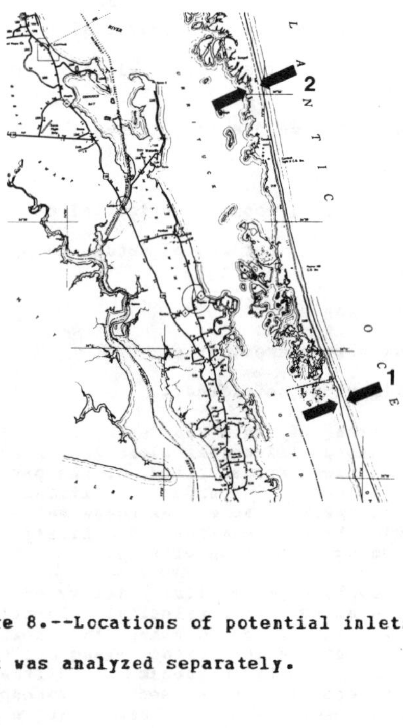

Figure 8.--Locations of potential inlets.

inlet was analyzed separately.

RESULTS

Baseline Environmental Conditions

During the late 1950s-early 1960s baseline period, Currituck Sound was essentially a freshwater body (fig. 9a). Freshwater inflow was received from Northwest and North Landing Rivers, and from overland flow. During periods of low inflow and southerly winds (particularly during summer) , saline water intruded into Currituck Sound from Albemarle Sound to the south (fig. 9b). During the baseline period, Currituck Banks presented an uninterrupted barrier to ocean water to the east. Subsequently, in early 1962, the barrier island was breached by a storm and saltwater reached the Sound for a short period. Although persisting for several years, the impact of this saline intrusion is no longer evident.

Most of the Sound contained rooted aquatic vegetation, primarily redhead grass (<u>Potomogeton perfoliatus</u>), bushy pondweed (<u>Najas flexilis</u>), wild celery (<u>Vallisneria americana</u>), and sago pondweed (<u>Potomogeton pectinatus</u>), except where bottom conditions were too unstable (mucks and flocculant organic matter) or where water depths precluded adequate light penetration to the bottom. These sparsely vegetated areas occurred primarily in the northwestern and southern parts of the Sound (fig. 9c).

Although sand bottom predominated (fig. 9d), patches of silts and finer sediments occurred through the Sound, and were more common along the western side and in protected areas.

Baseline Bass Habitat Quality

The largemouth bass HSI is above 0.8 in much, and above 0.6 in most, of Currituck Sound, confirming the Sound's reputation as a top bass producer (fig. 10). Areas of lower quality, primarily the extreme northwest and southern portions, correspond to poorly vegetated regions. The baseline HSI for the entire Sound (sum of HSIs for each character location) was 8897.34.

a. Spring Salinity b. Summer Salinity c. Vegetative Cover d. Sand Substrate

Figure 9.--Baseline environmental conditions (greater density indicates higher values).

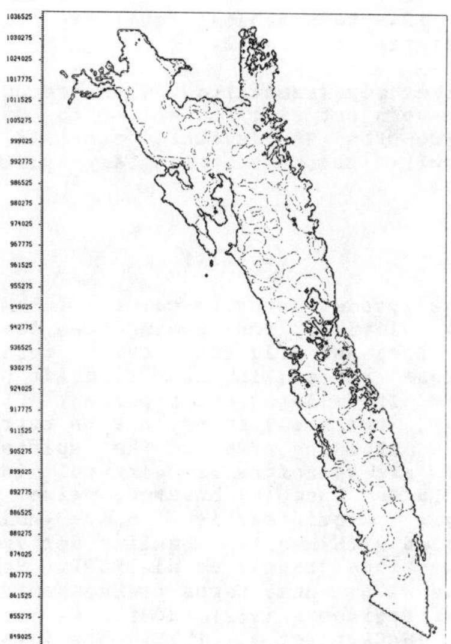

Figure 10.--Baseline bass habitat quality.

Impact of Modification

Breaching Currituck Banks would permit saline ocean water to penetrate well into the Sound, with 20 ppt salinity extending 9 mi (14 km) and 10 ppt salinity extending 12 mi (19 km) during flooding spring tides if Inlet 1 were opened (fig. 11). Inlet 2 would permit 20 ppt salinity to extend 7.5 mi (12 km) and 10 ppt to extend 10 mi (16 km) under

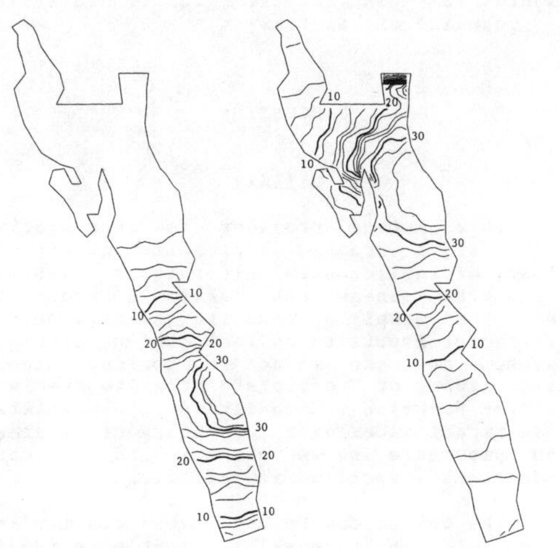

Figure 11.--Salinity intrusion following inlet opening (ppt).

comparable conditions. Under such conditions, the habitat for largemouth bass would become limited primarily by salinity, and would be greatly reduced from baseline conditions (fig. 12 and 13). Inlet 1 would reduce the HSI for the entire Sound to 5426.39 (39 percent), and Inlet 2 would reduce the HSI to

4264.13 (52 percent). Largemouth bass habitat would be virtually eliminated within 8 mi (13 km) of either inlet location.

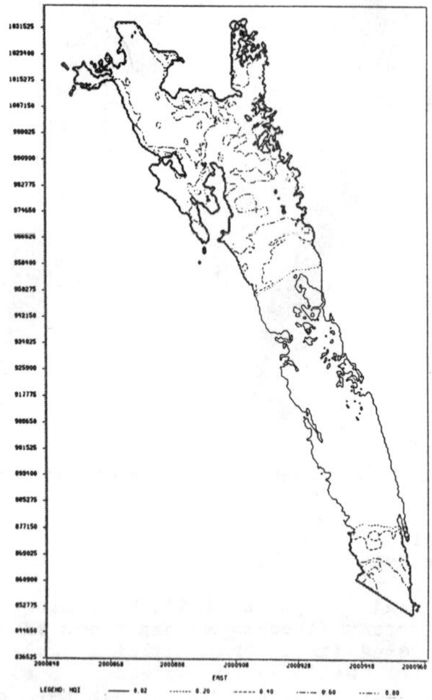

Figure 12.--Bass habitat quality following opening of Inlet 1.

DISCUSSION

Utility

This method provides a quantitative index and a graphic representation of the impact of environmental alterations. With it, a resource manager can analyze a number of "what ifs" simply by changing the environmental factor inputs to reflect project alternatives, and can produce a correspondingly large number of "so whats" from the species-habitat functions. In addition to a quantitative impact index over the entire study area, the procedure shows where, and to what degree, the functions are affected.

The method can be used as a staff planning tool, or be equally valuable in public information meetings. Maps can be used full size (in the Currituck example, about 2 x 3 ft (0.6 x 0.9 m)) as wall exhibits, or be reduced for text illustrations. In their full size form, distances and areas can easily be scaled, and scale varied to suit the situation.

In the Currituck example, only one species was considered. Before a final recommendation is made, several species, illustrating different salinity tolerances and user interests, will be modelled. In this way, the governmental agencies responsible for the project will be able to consider the trade-offs (ecological, economic, and political) before adopting any particular position.

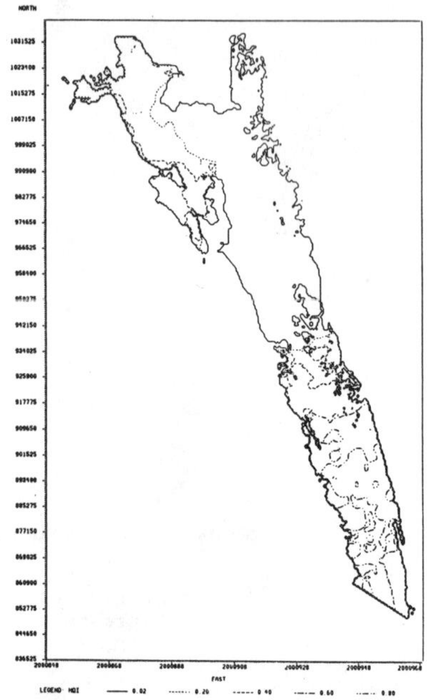

Figure 13.--Bass habitat quality following opening of Inlet 2.

Validity

The species function concept illustrated here is relatively new, having been developed by the Fish and Wildlife Service during the last eight years (Fish and Wildlife Service 1980). It alleges to represent "habitat quality", which may or may not be correlated with the standing crop of the species concerned, and therefore is difficult to validate. In one example, however, animal activity was significantly and positively correlated with habitat quality derived from such functions (Lancia et al 1982). Furthermore the U. S. Army Corps of Engineers (Army Corps of Engineers 1982) and the U. S. Forest Service (McClure et al 1979) are developing similar systems -- professional consensus, at least, that the habitat approach to impact assessment has merit.

No attempt was made to validate our model using additional field data. The hydrodynamic model was able to reproduce conditions similar to those produced by mapping baseline salinity data, however, and fisheries biologists working in the area agreed that the baseline vegetation and bass habitat quality maps generally reproduced conditions as they had observed them.

Computer Cost

Much of the project cost accrued to raw data input and computer program development and debugging. Table 1 illustrates computer costs in an operational mode exclusive of raw data input. Using the Triangle University Computing Center (TUCC), describing baseline conditions (environmental factor maps, bass-habitat computations, and a bass habitat quality map) would cost about $35 at full rates. Describing the impacts of both potential inlets would cost an additional $690. Thus, the entire exercise, in operational mode, would incur about $725 in computer costs.

Table 1.--Estimated computer costs

Element	Time CPU	I/O	Total	Cost
Environmental factor maps (4)	1:24	:40	2:04	4.80
Bass-habitat computations (1)	1:21	:43	2:04	7.10
Baseline bass hab. quality maps (1)	5:15	:54	6:09	22.98
Total base cond.	8:00	2:17	10:17	34.88
Hydro. model (7 day run x 2 inlets)	348:21	6:04	354:25	630.00
Bass-habitat computations (2)	2:42	:86	4:08	14.20
Bass habitat quality maps (2)	10:30	1:48	12:18	45.96
Total alt. cond.	361:33	9:18	370:51	690.16
TOTAL	369:33	11:35	381:08	725.04

Transferability

The basic approach illustrated here can be used with any species-habitat or other user function based upon georeferenced data. Ecologists can apply it to aquatic and terrestrial habitats and to any species for which a reasonable species-habitat function can be derived. Recreation planners can use it to identify recreational values as functions of population distribution, facilities, attractiveness, development constraints, and other geographically represented factors. Industrial developers can use it to provide an index of a site's developmental potential as a function of slope, soil conditions, utilities, zoning, etc. .

CONCLUSION

The method illustrated provides an effective means of comparing alternative environmental situations by linking physical and species-habitat functions. In addition to its utility in such ecological situations, it can be used to evaluate any user function based upon georeferenced data.

Poorly validated (or unvalidated) species-habitat functions constitute the method's greatest weakness; as these improve, so will confidence in the process.

LITERATURE CITED

Army Corps of Engineers. 1982. A habitat evaluation system for water resources planning. U.S. Army Corps of Engineers, Lower Mississippi Valley Division, Vicksburg, MS, U.S.A.

Adams, D. A. 1980a. A geobased environmental evaluation system (GEE) for use in marine and estuarine environments. The Water Resources Research Institute of the University of North Carolina, Raleigh, NC, U.S.A.

Adams, D. A. 1980b. Wildlife habitat models as aids to impact evaluation. The Environmental Professional 2:253-262.

Back Bay - Currituck Sound data report. 1968. 3 vols. U. S. Fish and Wildlife Service, Virginia Commission of Game and Inland Fisheries, North Carolina Wildlife Resources Commission.

Dougenik, James A., and David E. Sheehan. 1979. SYMAP User's Reference Manual. Laboratory for Computer Graphics and Spatial Analysis, Graduate School of Design, Harvard University, Cambridge, MA, U.S.A.

Fish and Wildife Service. 1980. Habitat Evaluation Procedures (HEP). ESM 102. U.S. Department of the Interior, Fish and Wildlife Service, Washington, DC, U.S.A.

Harvard University. 1979. SYMVU Manual. Laboratory for Computer Graphics and Spatial Analysis, Graduate School of Design, Harvard University, Cambridge, MA, U.S.A.

Lancia, R. A., and S. Douglas Miller, David A. Adams, and Dennis W. Hazel. 1982. Validating habitat quality assessment: an example. Trans. N. Amer. Wildl. and Natur. Resour. Conf. (in press).

McClure, J. P., and N. D. Cost and H. A. Knight. 1979. Multiresource inventories -- a new concept for forest survey. U.S. Dept. of Agriculture Forest Service Research Paper SE-191. Southeastern Forest Experiment Station, Asheville, NC, U.S.A.

SAS Institute. 1979. SAS user's guide, 1979 edition. SAS Institute, Cary, NC, U.S.A.

------. 1981. SAS/GRAPH user's manual, 1981 edition. SAS Institute, Cary, NC, U.S.A.

Taylor, C., and J. Davis. 1975. Tidal and long wave propagation -- a finite element approach. Computers and Fluids 3:125-148.

Wang, J. D., and J. J. Connor. 1975. Mathematical modelling of near coastal circulation. MIT Report No. 200. Ralph Parson Laboratory, Massachusetts Institute of Technology, Cambridge, MA, U.S.A.

SECTION IV:

Application of Ecological Models to Land Resources

MODELING LIVESTOCK MANAGEMENT SYSTEMS UNDER VARIABLE CLIMATIC CONDITIONS IN EAST AFRICA[1]

G. M. Sullivan[2]

Abstract.--A forage-cattle production simulation model was adapted to rangeland conditions for two ecological zones in Tanzania. Interactions of forage and cattle subsystems are examined under variable weather and stocking rates. A set of improved livestock management practices are compared to the results of the baseline herd over a ten year period of simulated weather conditions.

INTRODUCTION

Livestock management systems in tropical rangelands are important because of the need for food for family subsistence. Livestock systems in East Africa are typical of tropical conditions where growing conditions are based primarily on wet and dry seasons. Rainfall is the most limiting factor for forage production with the dry season lasting for as long as five to six months creating stress for livestock. Cattle can experience wide fluctuations in weight gains over a 12 month period (Meyn). The climatic conditions result in some of the lowest productivity for livestock systems in the world (Crotty). Other problems such as land tenure can exacerbate the productivity problem further (Sullivan and Farris).

Livestock systems in East Africa are based around the variable nature of the weather patterns. In parts of Kenya and Tanzania, herdsmen move from wet to dry season grazing areas to obtain the necessary forage for their livestock. The resource management problems become important when cattle numbers exceed the carrying capacity of the rangeland. Productivity from the herds falls with incremental increases in herd numbers by individuals. The "tragedy of the commons" leads all producers to stock heavier than if they privately owned the rangeland.

The problem of management of a tropical rangeland is examined. The effect between levels of stocking rate and rainfall over a period of time are examined to estimate the effect of stocking rate on forage production and ultimately livestock outputs. A set of improved management practices are evaluated to determine their effect on the livestock system.

[1]Paper presented at the Third International Conference on the State-of-the-Art in Ecological Modeling, Colorado State University, May 24-28, 1982.

[2]Gregory M. Sullivan is Assistant Professor of Agricultural Economics, Auburn University, Auburn, Alabama U.S.A.

METHODS

Field research was conducted in 1975-76 to document the state-of-art of livestock production in Tanzania (Texas A&M University). Two of the major livestock production areas are Sukumaland (Sub-Zone II), south of Lake Victoria, and Gogoland (Sub-Zone V), in central Tanzania. Sukumaland has an average annual rainfall of approximately 840 mm while Gogoland is drier with around 570 mm per year. From the field survey, a representative village herd was documented for each sub-zone (Sullivan and Farris, 1976). Livestock production is aggregated to the village level for each sub-zone.

To examine resource management strategies, a simulation model was adapted to the two ecological zones in Tanzania. Independently developed forage (Smith and Williams) and cattle production (Sanders; Sanders and Cartwright, 1979 a,b) models were interfaced and validated so that the models were interactive with stocking rate and grazing pressure as the feedback mechanism between cattle and forage sub-models (Sullivan, 1979; Sullivan et al. 1981). This feedback procedure acted as a control on forage production in succeeding months for various levels of stocking rate. Economic evaluation of various management options entailed estimating the value of meat, milk, and sales computed on a monthly basis.

Weather variables were the key driving factors of the forage model. The weather pattern was developed from secondary data for East Africa (East Africa Community). Daily radiation and pan evaporation levels were based on monthly averages and were held constant for the month. Daily rainfall was estimated from twenty years of data. Average number of rainy days were also used in a uniform probability distribution as a condition for whether it rained based on rain the day before or not (fig. 1).

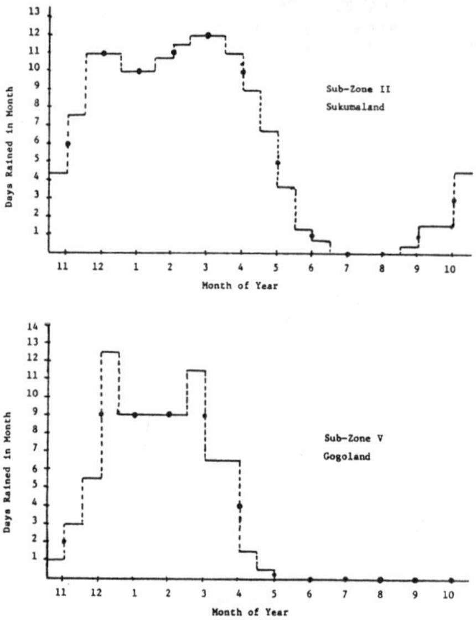

Figure 1. Distribution of rain days by 15 day periods in Sukumaland (Sub-Zone II) and Gogoland (Sub-Zone V), Tanzania

Source: East African Meteorology Service, 1975.

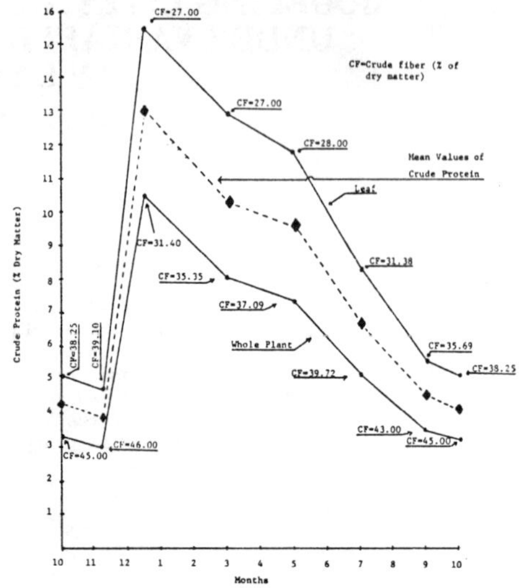

Figure 2. High and low levels of crude protein and crude fiber in leaves and whole plant for tropical forages[a]

[1] The level of crude fiber observed in the whole plant at various levels of crude protein.

Source: Bredon and Wilson (1968).

VALIDATION OF THE MODEL

The forage and cattle sub-models were validated using secondary data. Themeda triandra was the primary forage specie in Sub-Zone II (Sukumaland) and Panicum maximum for Sub-Zone V (Gogoland). From research conducted in Uganda by Bredon and Wilson, the general change in quality of the sward is illustrated in figure 2. With the onset of the rain in November in Sub-Zone II (December in Sub-Zone V), the crude protein goes from approximately 4% (% of dry matter) to an average of 13% after two months. The variance for the monthly mean quality values are indicated. Forage quality again falls to approximately 4% at the end of the dry season. A negative correlation exists between crude fiber and crude protein in the forage. The severe swing in forage quality directly effects animal productivity. The difference between the average quality in the whole plant compared to only the quality in the leaves is used to set a monthly quality value.

The actual levels of crude protein and digestibility of the forage were derived from look-up functions based on the time from first emergence and the stocking rate on the pasture. Stocking rate is hypothesized to effect both yield of forage and its quality in the following month. Cassady noted in field trials with Themeda triandra that excessive defoliation led to reduction in future yields by as much as 50 percent after three years. This effect of overgrazing is accounted for in the forage model by adjusting the beginning initial values for green weight based on estimated stocking rates at the end of the previous three dry seasons.

Forage availability was estimated and measured as green and dry grass. The quality level was different given the condition of the forage. The forage and cattle production linkages and feedback loops have been documented by Sullivan (1979, 1981). The variability in rainfall is seen as having the greatest effect on forage growth. From computer simulations based on four years of average weather conditions, the effect of stocking rates for Sub-Zones II and V are illustrated in figures 3, 4, 5 & 6. In Sub-Zone II with 12 month continuous grazing, stocking rates of .1 to .2 a.u./ha produced average daily gains (ADG) at approximately .25 kg per day. As stocking rates increased beyond .3 a.u./ha, there was a linear decline in ADG. Examining stocking rates for only the wet season, average daily gains (ADG) were higher at all stocking rates compared to twelve month grazing. In the drier region of Sub-Zone V, ADG was higher at the lower stocking rates but fell off more rapidly with increases in stocking rate. In the wet season only (a shorter season than Sub-Zone II), ADG's were the highest, but a rapid decline in ADG occured after .4 a.u./ha. The long term effect of sustained high stocking rates is range denudation with the reduction of palatable forage species and replacement with lower quality species (Heady).

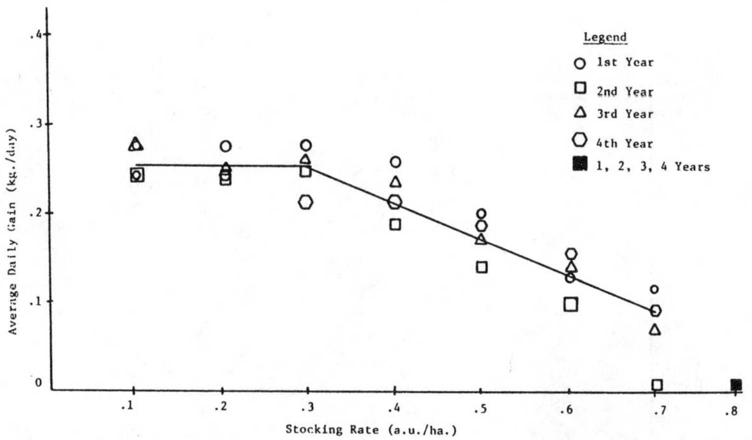

Figure 3. Effect of stock rate on average daily gain (ADG) for one year old steers for twelve months of simulation on Themeda triandra pasture with four levels of forage production in Sub-Zone II (Sukumaland), Tanzania

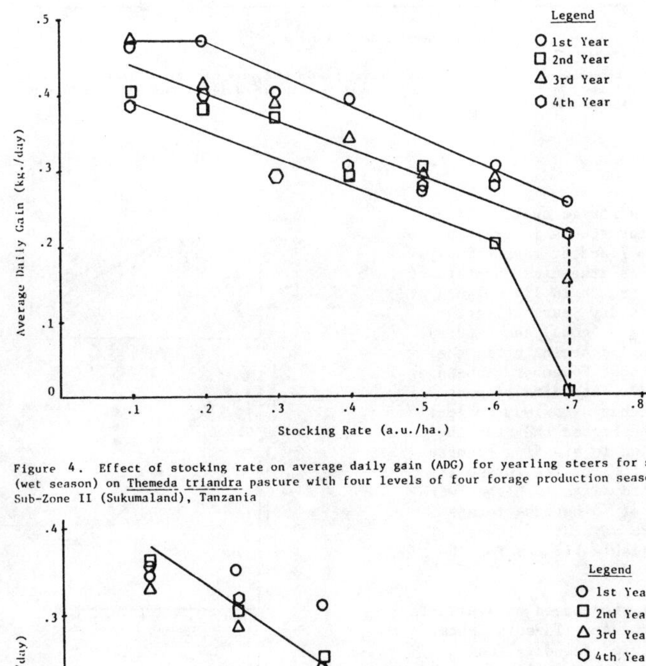

Figure 4. Effect of stocking rate on average daily gain (ADG) for yearling steers for six months (wet season) on Themeda triandra pasture with four levels of four forage production seasons, Sub-Zone II (Sukumaland), Tanzania

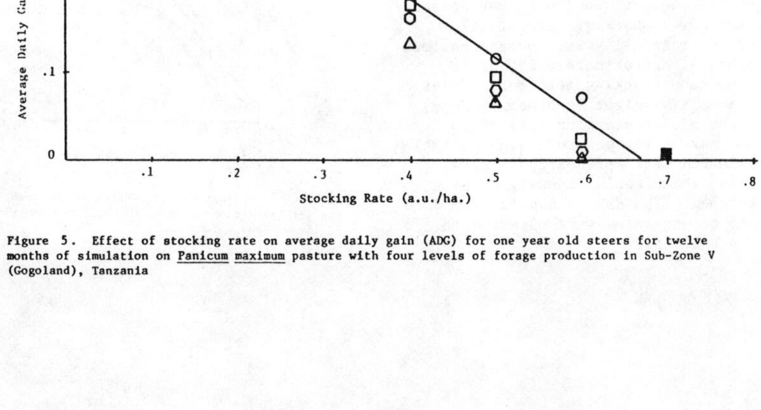

Figure 5. Effect of stocking rate on average daily gain (ADG) for one year old steers for twelve months of simulation on Panicum maximum pasture with four levels of forage production in Sub-Zone V (Gogoland), Tanzania

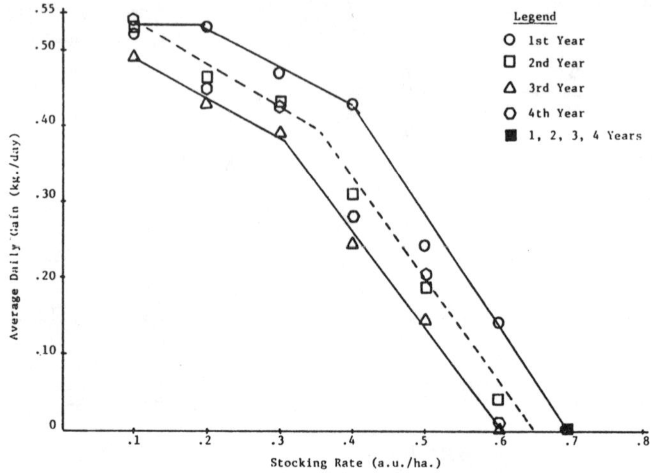

Figure 6. Effect of stocking rate on average daily gain (ADG) for one year old steers for six months (wet season) on Panicum maximum pasture with four levels of forage production in Sub-Zone V (Gogoland), Tanzania

The relationship between forage availability, quality, and rates of gain for steers is examined for Sub-Zone II and V in figures 7 and 8, respectively. The change in standing biomass, the crude protein of the forage, and the rate of growth in liveweights of steers are tracked. Good and dry years of forage production are indicated by a G (good) and D (dry). Grazing is for a one year period beginning in the first month of the rainy season, November (Sub-Zone II) and December (Sub-Zone V). Standing biomass for each stocking rate level reaches a maximum in April and May for Sub-Zone II. The higher the stocking rate the greater the reduction in standing biomass. The change in crude protein (CP) of the forage is seen in section b. Quality is slightly higher with the increased stocking rate of .5 because forage remains shorter and in a more vegetative state. The negative effect is less available biomass for the herd.

Steer weights were estimated based on available biomass and quality of the forage. These weights were compared to actual experimental data from Tanzania (Getz). Two levels of rainfall and stocking rates were examined for each sub-zone and are illustrated in section c. In Sukumaland, the beginning average steer weight was 150 kg per animal (fig. 7). Under the lower stocking rate of .17 a.u./ha and a good (G) rainfall year, average maximum liveweight is reached at approximately 260 kg in September. With the same stocking rate but a drier rainfall year, maximum liveweight (reached in July) was 249 kg. With the higher stocking rate of .5 a.u./ha and a good year of forage production, animal growth rate approximated Getz's research results. Under dry conditions, the steer's liveweight fell below 200 kg by October. The dry season is from June through October and growth rates were observed to decline from Getz's results.

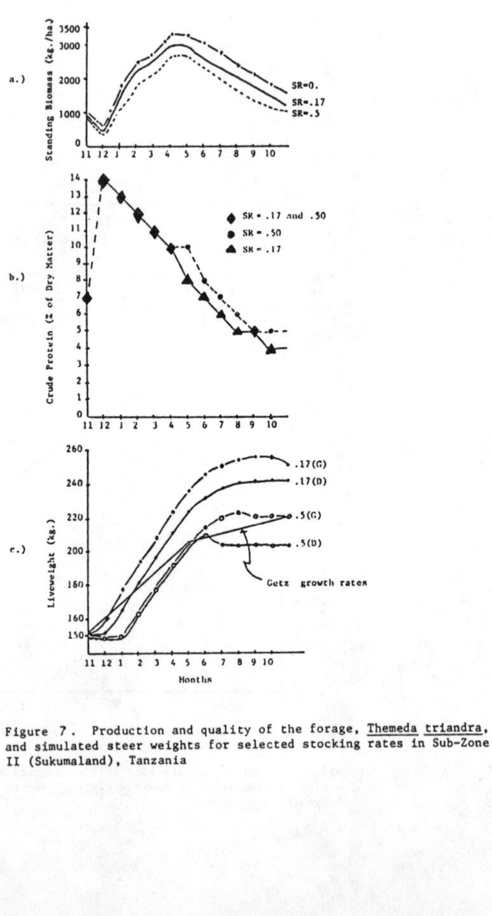

Figure 7. Production and quality of the forage, Themeda triandra, and simulated steer weights for selected stocking rates in Sub-Zone II (Sukumaland), Tanzania

In the drier zone of Sub-Zone V, standing biomass is not as great as in Sub-Zone II (fig 8). The level of quality for Panicum maximum is higher than for the Themeda triandra. As in Sub-Zone II, quality of forage increases slightly in the dry season under the heavier stocking rate of .5. Using Getz's estimated growth curve, average liveweights are tracked for the twelve months. With stocking rate of .3, simulated weights overestimated Getz's results. Getz had combination of steers and hefers and the simulation was for steers only which would add upward bias to the simulation results. With the higher stocking rates of .5 a.u./ha. final liveweights were approximately 60 kg lighter than under good weather conditions. With drier conditions, average weights were even lower. The uncertainty associated with weather is magnified when considering the level of stocking rate. Because of communal ownership of the rangeland, stocking rates will be pushed to higher levels resulting in lower productivity. The effect of stocking rates adds an important dimension for economic evaluation of livestock management pratices that could control stocking densities. Benefits would accure to all herds utilizing the rangeland.

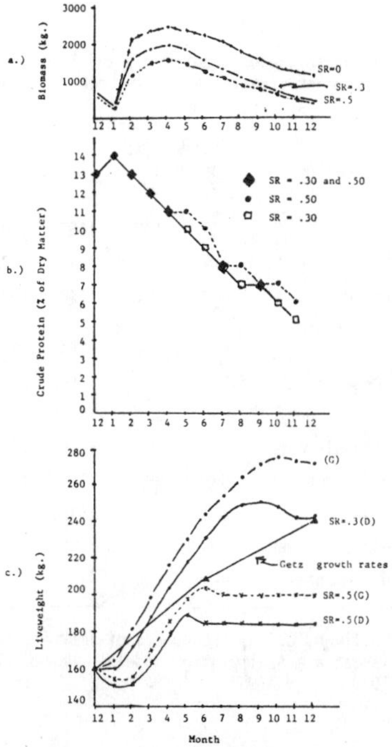

Figure 8. Production and quality of the forage, Panicum maximum, and simulated steer weights for selected stocking rates in Sub-Zone V (Gogoland), Tanzania

EVALUATION OF MANAGEMENT PRACTICES

Village cattle production was simulated over a ten year period of variable climatic conditions to measure herd performance and outputs. The representative village herd is used as the baseline herd. Incorporation of cattle management practices are tested for their feasibility and the benefits they generate for the village. The criteria for ranking the practices is weight of the herd, sale weight and milk production.

Higher Rainfall Zone (Sub-Zone II)

Simulation results are presented in Table 1 for the representative village herds for Sukumaland. Monetary and nutritional benefits are estimated in Table 2. In the higher rainfall region of Sub-Zone II (Sukumaland) with a ten year simulated average of 818 mm of rainfall per year, the weight of the baseline herd declined over the ten year period from 766 m.t. in the first year to 565 m.t. because the weather pattern included six years below the normal annual average. The only single management practice alternative which resulted in a larger herd weight than the baseline was supplemental feeding of weaner calves during the dry season, which increased the total herd weight 9 percent above the baseline. Without a supplemental hay program, total herd weight is either the same as the baseline or lower. Reducing herd numbers by selling cows twelve years or older, which are open and cycling, reduced herd weight to 84 percent of the baseline herd. Supplemental feeding allowed more forage for the rest of the herd.

More important as an evaluative criterion than herd weight was sale weight. The greatest effect on sale weight was a combination of a breeding program with supplemental feeding, a practice which provided a 9 percent increase over the baseline situation. The average weight per animal sold for the ten year period increased 10 kg per head with this combination of improved practices. Any combination using the weaning practices lowered the average weight per head. For village milk production, the cooperative management practice of supplemental feeding increased milk production by 14 percent above the baseline for the ten year period. The breeding or weaning programs resulted in either a negligible increase, or a slight decrease in milk production. Supplemental feeding plus the breeding program resulted in the second highest increase in milk production.

Another possible village policy was the sale of unproductive cows in periods when prices are higher. This practice resulted in total sale weight increasing 21 percent above the baseline for the ten years. Milk production increased ten percent above the baseline, but not to the level of supplemental feeding. The mandatory culling of cows reduced total herd weight, but made the remaining herd more productive per animal because of more available forage.

For the representative village herd in Sukumaland, the greatest economic impact was with a combination of a sales policy, supplemental feeding and a restricted breeding program. Herd size was the smallest but sale weight and milk production were the greatest for the ten year period. After the cost of hay production, the net value of meat and milk increased by 126 percent above the baseline value (Table 2). The conclusion is producers in this region would be much better off with a cooperative program than acting independently particularly given the uncertainty of weather as measured by 6 years of rainfall below the average. The productive efficiency of the herd increased with greater utilization of available forage.

Table 1. A Comparison of Traditional and Improved Management Practices for a Ten-Year Simulation of a Representative Village Livestock Herd in Sukumaland and Gogoland, Tanzania.

Management Practice	Sukumaland - 818 mm rainfall/yr. Total Weight of herd		Sale Weight		Milk Production		Gogoland - 530 mm rainfall/yr. Total Weight of herd		Sale Weight		Milk Production	
	m.t.	%	m.t.	%	m.t.	%	m.t.	%	m.t.	%	m.t.	%
Baseline	565	100	1167.	100	2618.	100	635	100	1340	100	3010	100
Breeding Season	519	92	1186.	102	2497.	95	610	96	1350	101	2900	96
Weaning Period	563	100	1170.	100	2630.	100	619	97	1360	101	2880	96
Supplemental Hay to Calves	618	109	1244.	107	2990.	114	945	149	1820	136	4940	164
Weaning + Breeding	545	96	1164.	100	2552.	97	593	93	1360	101	2780	92
Weaning + Supplement	564	100	1169.	100	2638.	101	625	98	1360	101	2870	95
Breeding + Supplement	578	102	1275.	109	2858.	109	902	142	1810	135	4710	156
Breed + Wean + Suppl.	546	97	1165.	100	2559.	98	599	94	1360	101	2770	92
Sales Policy	474	84	1410.	121	2879.	110	485	76	1670	125	3442	114
Sales + Suppl. + Breed	458	81	1550.	133	3130.	120	---		----		----	
Sales + Supplemental	---		----		----		534	84	1680	125	3443	114

SOURCE: Sullivan, G. M. "Economics of Improved Management for Transforming the Forage/Livestock System in Tanzania - A Simulation Model." Unpublished Ph.D. dissertation, Texas A&M University, August 1979.

Table 2. Average Annual Net Value of Output of Meat and Milk from Livestock Herds for Ten Year Simulation Period for Sukumaland in Tanzania.

Management Practice	Sukumaland Calories/ Cap[a]	Protein Cap[a]	Annual Net Value[c]	Percent of Baseline	Gogoland Calories/ Cap[a]	Protein Cap[a]	Annual Net Value[c]	Percent of Baseline
	'000 Cal.	'000 grms	'000 Tsh.	%	'000 Cal.	'000 grms	'000 Tsh.	%
Baseline	47.72	3.42	780	100	49.90	3.30	922	100
Breeding Season	46.29	3.32	760	97	48.90	3.23	903	98
Weaning Period	47.72	3.42	781	100	48.90	3.23	902	98
Supplemental Hay to calves	52.49	3.76	855	110	75.85	5.02	1400	152
Weaning & Breeding	46.77	3.35	764	98	47.90	3.17	883	96
Weaning & Supplemental Hay	46.29	3.32	759	97	45.91	3.04	844	92
Breeding & Supplemental Hay	51.06	3.66	835	107	74.35	4.92	1376	149
Breeding & Weaning & Suppl.	45.01	3.28	748	96	45.41	3.00	835	91
Sales Policy	56.31	4.04	918	118	65.37	4.32	1205	131
Sales & Supplemental & Breeding[b]	60.13	4.31	984	126	---	--	--	--
Sales & Supplemental Hay[b]	---	---	---	---	63.37	4.19	1172	127

[a] The calculation of the nutritional value of milk produced assumes that the milking practice of dividing milk between the calf and home consumption remains constant. The value is the equivalent in calories and protein of milk used for home consumption. The nutritional value of meat is for animals consumed which were slaughtered or found dying in the herd.

[b] The sales polcy was considered with the combination of management practices that was felt to be relevant given their respective performance in each respective region.

[c] The value of meat and milk if sold in the village; the net value has cost of hay for supplementing weaner calves subtracted where appropriate.

Drier Rainfall Zone (Sub-Zone V)

The longer dry period in Gogoland changes the effect of management practices on the village herd. The 10 year annual rainfall level used in the simulation study averaged 537 mm per year, with four years below the long term annual average for the region. Supplemental feeding of weaner calves increased the total weight of the village herd over the ten year period. Supplementing weaner calves with or without a breeding program made more forage available to other cattle in the herd, thus allowing cattle to maintain their weight during the dry season. Sale weight of cattle and milk production was greater than the baseline by 36 and 64 percent, respectively, with supplemental feeding. Implementing a weaning or breeding program only slightly increased sale weight but reduced milk production during the ten year period.

Instituting a sales policy of culling unproductive, old cows reduced total weight of the village herd but increased sale weight and milk production for the village herd by 125 and 114 percent, respectively, above the baseline. Reducing herd numbers does not have as great an impact as supplementing weaner calves. The amount of grazing was sufficient for herd size to increase, especially when supplemental feeding was available.

The sales policy, combined with supplemental feeding, slightly increased the total herd weight, but sales weight and milk production does not increase above those of a sales policy alone. Forage availability was the constraining factor in the drier zone so that reduction in herd numbers by a sales policy does not increase forage availability as much as supplemental feeding. The net benefit from supplementation was an increase in the value of meat and milk output by 152 percent above the baseline herd (Table 2). A breeding program with supplementation produced a slightly lower increase in the value of the output. In the drier zone, supplementation of weaner calves during the longer dry season improved the replacement rate of the herd allowing more available forage for the rest of the livestock herd.

CONCLUSION

Variable climatic conditions are important ecological factors which impact on mangement of a natural resource. With periods of low rainfall, management of a scarce resoure becomes crucial. Producers' decisions whether to add more cattle to their herds directly impacts on the stocking rate which is a driving variable in forage production.

Modeling the forage-cattle system, opens up research for better understanding the management of the scarce resource, forage. Supplementing weaner calves with or without other practies significantly increases output over the representative baseline herds in the two sub-zones. A cooperative village sales policy of culling old, unproductive cows increases the value of meat and milk produced above that of the baseline village herds.

Future efforts to better define relationships between stocking densities and forage qualty and its availabiity are needed. Improved management, at an effective cost, can only be incorporated if all the physical responses are known. Further research in ecological modelling of tropical rangelands can assist in raising the level of cattle productivity for the general benefit of society.

REFERENCES

Bredon, R. M. and J. Wilson. 1963. The chemical composition and nutritive value of grasses from semi-arid areas of Karamoja as related to ecology and types of soil. East African Agricultural and Forestry Journal 29: 134-142.

Cassady, J. T. 1973. The effect of rainfall, soil moisture and harvesting intensity on grass production in two rangeland sites in Kenya. East African Agricultural and Forestry Journal. 39: 26-36.

Crotty, Raymond. 1980. Cattle, economics and development. Commonwealth Agricultural Bureau. Surrey, England.

East African Community. 1975. Climatological statistics for East Africa. East African Meteorology Department, Nairobi.

Getz, W. R. D. Msura, and G. C. Kifaro, 1974. The effect of season of birth on growth and development in two breeds of beef cattle. Mimeographed. Research and Training Institute, Mpwapwa, Tanzania.

Heady, H. F. 1966. Influence of grazing on the composition of *Themeda triandra* grassland, East Africa. Journal of Ecology 54: 725-727.

Meyn, Raymond. 1980. Cattle, economics and development. Commonwealth Agricultural Bureau. Surrey, England.

Sanders, J. O. 1977. Application of a beef cattle production model to the evaluation of genetic selection criteria. (Unpublished Ph.D. dissertation). Texas A&M University.

Sanders, J. O. and T. C. Cartwright. 1979a. A general cattle production systems model I. procedures used for simulating animal performance. Agric. Systems 4. :289-309.

Sanders, J. O., and T. C. Cartwright. 1979b. A general cattle production systems model. II. Procedures Used for Simulating Animal Performance. Agric. Systems 4. :289-309.

Sullivan, G. M. and D. E. Farris. 1976. Survey of traditional sector. In: Tanzania Livestock-Meat Subsector. Vol II. International Programs Office. College Station Texas.

Sullivan, G. M. and D. E. Farris. 1978. "Economic and anthropological limitations to regional development in African livestock industries." Department of Agricultural Economics, Texas A&M University (Mimeograph). College Station,

Sullivan, G. M. 1979. Economics of improved management for transforming the forage/livestock system in Tanzania--a simulation model. (Unpublished Ph.D. dissertation). Texas A&M University.

Sullivan, G. M., T. C. Cartwright, and D. E. Farris. 1981. Simulation of production systems in East Africa by use of interfaced forage and cattle models. Agric. Systems :7 245-265.

Texas A&M University. 1976. Tanzania livestock-meat subsector. Vol. I, II, III, IV. International Programs Office, College Station, Tx.

A LINEAR PROGRAMMING/NETWORK MODEL FOR VEGETATION ALLOCATION[1]

Frederick K. Martinson[2] and Fred W. Glover[3]

Abstract.--This paper describes the linear programming model developed for and currently used by the Bureau of Land Management (BLM) for the allocation of vegetation to consumptive herbivores. The model seeks to maximize the utilization of range forage production subject to plant maintenance constraints, resource constraints and management-imposed constraints. The input data elements are: forage intake rate, seasons of use, proper use factors, dietary estimates, annual plant production, allowable use factors and grazing suitabilities. Optimization is reached at the grazing allotment level. The allotment is made up of basic or homogeneous units and the LP problem solved exhibits a "multi-divisional" structure with "divisions" corresponding to the basic units. The model is operational on the BLM Honeywell 6680 computer and it is designed to operate in a man-machine interactive mode.

Background Information

The BLM manages 173 million acres of public rangelands. Management of these rangelands is carried out on the basis of sustained yield and multiple use. A massive effort is currently underway to ascertain the present and potential impacts on the public rangelands of the current management practices in the fifty-eight grazing districts of the BLM.

To accomplish this task the BLM has developed a standardized resource data processing system. Inventory data are collected according to the Bureau's Soil and Vegetation Inventory Method (SVIM) (BLM, 1979), and are processed through the central computer site located at the Denver Service Center in Denver, Colorado. Access to these data is effected through a telecommunications computer network. Inventory data are processed together with supporting information, i.e., land surveys, stratification delineations, etc., to generate selected output reports for range, wildlife, watershed, lands and minerals and various other departmental concerns within the Bureau and to create data files that go into a vegetation allocation model.

The inventory data and vegetation allocation model are used by the BLM to allocate public land resources to various uses or user groups through the land use planning process. Conflicts invariably arise over use of the public lands. Demand for vegetation by livestock, wildlife, watershed, and other uses often exceeds the supply. The BLM manager must make an allocation among the uses, which may limit the amount of vegetation available for any one use. The vegetation allocation model described herein is intended to aid the BLM manager in the resolution of such conflicts.

---

[1]Paper presented at the Third International Conference on State-of-the-Art in Ecological Modeling, Colorado State University, May 24-28, 1982.

[2]Frederick K. Martinson is an Operations Research Analyst for the Bureau of Land Management, Denver Federal Center, Denver, Colorado.

[3]Fred W. Glover is a Professor of Management Science at the University of Colorado, Boulder, Colorado.

What follows is a description of the input data and of the conceptual structure and implementation of the model presently used within the BLM for allocation of vegetation to consumptive users.

Input Data

The following data elements are inputs to the model: forage intake rate, seasons of use, proper use factors, dietary estimates, annual plant production, allowable use factors and grazing suitability.

Forage intake rates are entered as pounds of dry matter per animal type per month. Forage intake rates are coordinated at the state level and are developed through collaboration with the state wildlife departments and the state universities and through literature search.

Seasons of use for livestock are determined by the field offices and entered into the forage allocation procedure directly off the current authorizations. Seasons of use for wildlife are determined by a wildlife inventory procedure in collaboration with state wildlife agencies. Seasons of use are entered as beginning and end dates of the grazing period.

Proper use factors (PUF) of a given plant species by a given animal type is the percent of the annual plant production that a given animal type may consume without overutilizing the surrounding plant community. Proper use factors reflect the animal's forage preference on a particular range given that all present plant species are sufficiently abundant. Historical records, scientific literature and professional judgment are used to determine PUFs for each plant and animal by season. Because PUFs reflect the animal's dietary preferences, they are used as a diet surrogate when actual dietary estimates are not available.

Dietary estimates are determined from fecal analysis and scientific literature. Unfortunately, site-specific dietary estimates are scarce. The BLM is actively engaged in supporting research in this area, and is eager to utilize this information as soon as it becomes available. The model allows the flexibility to enter actual dietary information, when existent, or to compute an estimate of the diet based on the PUFs and and composition of the plant community.

Annual plant production numbers are based on production figures, in pounds per acre (air-dried weight), estimated for homogeneous vegetative units using plots located along line transects. These estimates are adjusted for climatic conditions and extrapolated to total annual pounds per site.

Allowable use factor (AUF) of a given plant is the maximum percent of annual plant production that can be removed from a plant without reducing its vigor. Like PUFs, historical records, professional judgment, and literature searches are used to determine AUFs for each plant by season.

Grazing suitability for livestock is a function of distance to water, slope and soil surface factor. It is entered as a percent figure and it represents the fraction of productive grazing land which is suitable for grazing by livestock.

The input files to the allocation model contain all the above data elements for each of the basic or homogeneous units that constitute the grazing allotment. These basic units are referred to, in BLM SVIM terminology, as SWAs or Site Write-up Areas. An allotment is made up of several SWAs, the number of which may vary from one to as many as 1300.

Model Structure

The BLM uses a linear programming model for vegetation allocation. Linear programming (LP) is a mathematical optimization technique extensively used in allocation problems (Daellenbach and George, 1978; Cooper and Steinberg, 1974; Orchard-Hays, 1968). The BLM LP model optimizes forage allocation by allotment and seeks to maximize the amount of forage available for grazing, subject to plant maintenance constraints, resource constraints and management imposed constraints.

The LP problem to be solved exhibits what is commonly referred to as a "multidivisional" structure. The problem is almost decomposable into separate problems, where each division (SWA) is concerned only with optimizing its own operation. However, some overall coordination (at the allotment level) is required to best divide the natural resources among the divisions (SWAs). This special structure is pictorically displayed in Figure 1.

Figure 1. Structure of LP Model.

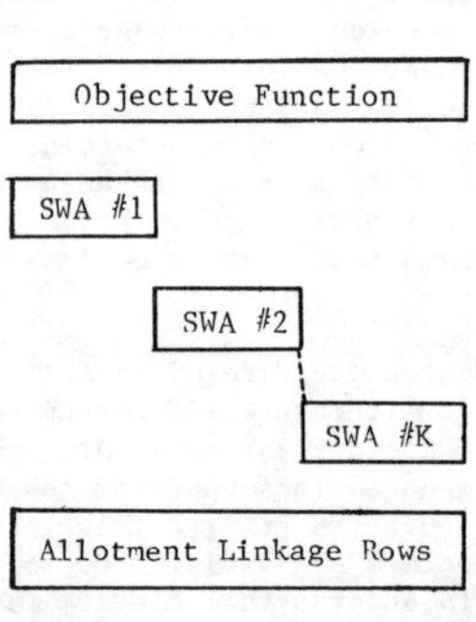

The objective function seeks to maximize the amount of forage available for grazing:

$$\text{Max} \sum_k \sum_i r_i \, m_i \, x_{ik}$$

where

r_i : Lbs. of forage intake per month by animal type i
m_i : No. of months of grazing by animal type i
x_{ik}: Number of animals of type i in SWA k

Each SWA submatrix has the following constraints (SWA subscript k is dropped to simplify notation):

$$\sum f_{ij} \le AUF_j \quad \text{for each } j \tag{1}$$

$$\sum f_{ij} \le PUF_{ij} \quad \text{for each } i \text{ and } j \tag{2}$$

$$x_i = \frac{1}{(m_i)(r_i)} \sum_j f_{ij} V_j S_i \quad \text{for each } i \tag{3}$$

$$f_{ij} V_j S_i \le (1+DRF)(RPV_{ij}) \, m_i \, r_i \, x_i \quad \text{for each } i \text{ and } j \tag{4}$$

$$f_{ij} V_j S_i \ge (1-DRF)(RPV_{ij}) \, m_i \, r_i \, x_i \quad \text{for each } i \text{ and } j \tag{5}$$

where

AUF_j : Allowable use factor of plant species j over the m_i months of grazing (allowable use factor is maximum level of use of plant growth based on plant maintenance considerations)

f_{ij}: Fraction of growth of plant species j consumed by x_i animals over the m_i months of grazing

PUF_{ij}: Proper use factor of plant j by animal type i over the m_i months of grazing (proper use factor is the percent use that is made of a plant species under proper management of the range as a whole)

x_i : Number of animals of animal type i

r_i : Lbs. of forage intake for consumption per month by animal type i

m_i : No. of months of grazing by animal type i

S_i : Fraction of productive grazing land suitable for grazing by animal type i

V_j : Lbs. of available annual production of plant species j

DRF : Dietary range factor

RPV_{ij}: Relative preference value of animal type i for plant species j

Constraint (1) ensures that the total consumption of the annual growth of plant species j by all animal types does not exceed the allowable use factor of the plant.

Constraint (2) ensures that the individual consumption of species j by animal i does not exceed the PUF for the given j plant and i animal.

Constraint (3) computes the number of animals of type i, based on the total forage intake by the given animal type and the fractional consumption of annual plant growth of each of the given plant species.

Constraints (4) and (5) enable the decision maker to set a range about the relative preference values (RPVs). The term "relative preference value" indicates the dietary composition of plant j in the diet of animal i. The RPVs are determined either from dietary sources or by normalizing the weighted PUFs, that is,

$$RPV_{ij} = \frac{f(w_j)PUF_{ij}}{\sum_j f(w_j)PUF_{ij}} \quad \text{for each i and j}$$

where $f(w_j)$ is a certain function of the percent composition of plant j. Weighting of the PUFs by a certain function of plant composition is justified on the basis that animals will chage their food habits as the forage pool changes and some plants become depleted. The $f(w_j)$ term can be selected by the decision maker to take on the value of one (if no plant composition effect is anticipated), or the log of the percent plant composition (if superabundant species of low palatability are present), or the straight percent plant composition.

To set a range about the RPVs the decision maker uses the dietary range factor (DRF). The DRF attempts to capture in a subjective way the degree of uncertainty that the decision maker attaches to the estimate of the diet. The DRF can be made to vary from 0 to p and thereby cause a fluctuation about the expected diet from 0 to p x 100 percent.

Constraints (4) and (5) are derived as follows: Let f'_{ij} be the fraction of plant j in diet of animal type i. The pounds of plant j consumed per month by each animal of type i will be expressed by $(f'_{ij})(r_i)$. The total consumption of plant j by x_i animals of type i during m_i months will be $(x_i)(f'_{ij})(m_i)(r_i)$. This can also be expressed in terms of f_{ij}, V_{ij} and S_i as the product $(f_{ij})(V_j)(S_i)$, hence we can set up the equation:

$$(x_i)(f'_{ij})(m_i)(r_i) = (f_{ij})(V_j)(S_i)$$

and solving for f'_{ij}

$$f'_{ij} = \frac{f_{ij}\ V_j\ S_i}{x_i\ r_i\ m_i}$$

The DRF enables the decision maker to set up ranges of the f'_{ij}, that is,

$$(1-DRF)(RPV_{ij}) \leq f'_{ij} \leq (1+DRF)(RPV_{ij})$$

hence

$$(1-DRF)(RPV_{ij}) \leq \frac{f_{ij}\ V_j\ S_i}{x_i\ r_i\ m_i} \leq (1+DRF)RPV_{ij})$$

which corresponds to constraints (4) and (5).

The allotment linkage rows link all the SWA submatrices in the allotment and establish upper and lower bounds at the allotment level for each animal type:

$$x_{ik} \leq U_i \text{ for each i}$$

$$x_{ik} \leq L_i \text{ for each i}$$

where

U_i: upper bound of animal type i at the allotment level

L_i: lower bound of animal type i at the allotment level

The U_i and L_i values are subject to management control and can be made to vary by the decision maker.

The AUF and PUF values that appear in the SWA submatrices are composite values that are specific to the grazing dates of each herbivore. The PUF composite values are determined by breaking down the grazing dates for each herbivore into the number of days per calendar grazing season, and weighting the standard PUF calendar season values by the number of days grazed per calendar grazing season. The AUF composite values are determined by scanning all the herbivores grazing dates, selecting the earliest starting date and the latest finishing date of the non-overlapping calendar grazing season, and weighting the standard AUF calendar season values by the number of days grazed per calendar grazing season.

Computational Aspects and Network Formulation

The proposed model results in substantial linear programming problems even for allotments of moderate size. Most of the problems encountered to date have constraints matrices that range from 500 to 2000 rows and from 200 to 1000 columns. The largest problem encountered and solved so far had a 5000 x 2500 constraint matrix.

The Honeywell Mathematical Programming System (MPS) has been used to solve the LP problem. Honeywell's MPS is a flexible package (much like APEX III or TEMPO or MPSX) that has capabilities for solving linear programming and mixed integer type problems. LP problems can be solved using primal, crashing, dual, and generalized upper bound algorithms. Mixed integer type problems are solved using a branch and bound algorithm. MPS also has extensive postoptimal capabilities.

The LP allocation problem is solved with MPS using the primal and crashing algorithms. Although MPS includes the latest advances in mathematical programming techniques, the solution times for large problems tend to be high because of the intrinsic nature of the Honeywell 6680 computer and the need to run MPS in the batch subsystem.

To speed up execution and improve man-machine interactivity the LP allocation problem has been reformulated as a generalized network with side constraints, and a special structure network code has been developed to solve the network problem (Glover et al., 1982). The gains in speed of execution are spectacular, and the larger the problem the faster the relative speed. Problems containing matrices of the order of 100 x 50 are solved 10 to 20 times faster; larger problems with matrices of the order of 4000 x 2000 are solved 100 to 200 times faster. These gains result from the simpler computational aspects of network logic, the network algorithm used (Glover, 1979), and the incorporation of the network code into the allocation program as a subroutine, as opposed to accessing MPS in the batch subsystem.

The network reformulation of the LP allocation problem is displayed in Figures 2 through 7. To help view the problem in the network-related form a

Figure 2

NETWORK REFORMULATION FOR CONSTRAINT (1)

ORIGINAL	NETWORK
$\sum_i f_{ij} \le AUF_j$	Define $y_{ij} = f_{ij} V_j$
	$U_j = AUF_j V_j$
Hence	
$\sum_i f_{ij} V_j \le AUF_j V_j$ ⟷	(1) $\sum_i y_{ij} \le U_j$

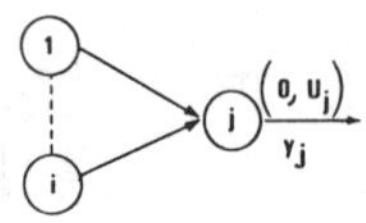

From the network picture

$$y_j = \sum_i y_{ij}$$

Hence

$$y_j \le U_j$$

Network picture says: Amount of plant j consumed by animal i (y_{ij}) summed over all animals i equals total amount of plant consumed by all animals (y_j) and this is $\le U_j$

Figure 3

NETWORK REFORMULATION FOR CONSTRAINT (2)

ORIGINAL	NETWORK
$f_{ij} \le PUF_{ij}$	Define $U_{ij} = PUF_{ij} V_j$
Hence	
$f_{ij} V_j \le PUF_{ij} V_j$ ⟷	(2) $y_{ij} \le U_{ij}$

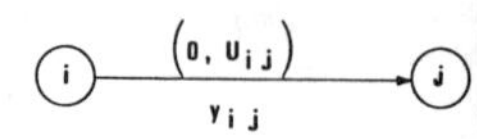

Network picture says:

y_{ij} cannot exceed U_{ij}

Figure 4

NETWORK REFORMULATION FOR CONSTRAINT (3)

ORIGINAL	NETWORK
$x_i = \frac{1}{m_i r_i} \sum_j f_{ij} V_j$	Define $M_i = m_i r_i$
Hence	
$m_i r_i x_i = \sum_j f_{ij} V_j$ ⟷	$M_i X_i = \sum_j y_{ij}$

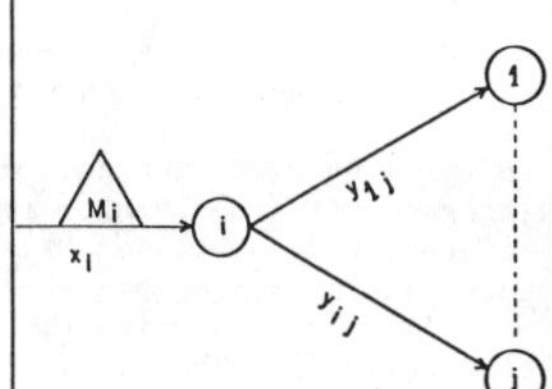

Network picture says: Total lbs consumed by all type i animals ($M_i x_i$) equals sum over every plant j of the amount animal i consumes of plant j (y_{ij})

Figure 5

NETWORK REFORMULATION FOR CONSTRAINT (4)

ORIGINAL	NETWORK
$f_{ij} V_j \le (1 + DRF)\ (RPV_{ij}) m_i r_i x_i$	Define $q_{ij} = (1 + DRF)(RPV_{ij})$
	Hence $T_i = r_i m_i x_i$
	$T_i = M_i x_i$
⟷	(4) $y_{ij} \le q_{ij} T_i$

$$\boxed{y_{ij} \le q_{ij} T_i}$$

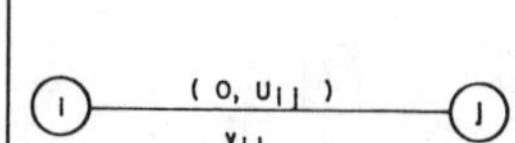

T_i = amount entering node

Network picture says: (4) is a side constraint. It can also be written

$$y_{ij} \le \bar{q}_{ij} x_i$$

for

$$\bar{q}_{ij} = M_i q_{ij}$$

The "variable" T_i is only convention, useful because the objective can also be written

$$MAX \sum_i T_i$$

Figure 6

NETWORK REFORMULATION FOR CONSTRAINT (5)

ORIGINAL	NETWORK
$f_{ij}V_j \geq (1-DRF)(RPV_{ij})\, m_i r_i x_i$ $\longleftrightarrow$ (5)	Define $p_{ij} = (1-DRF)(RPV_{ij})$ $T_i = r_i m_i x_{ij}$ $p_{ij}T_i \leq y_{ij}$ $\boxed{p_{ij}T_i \leq y_{ij}}$

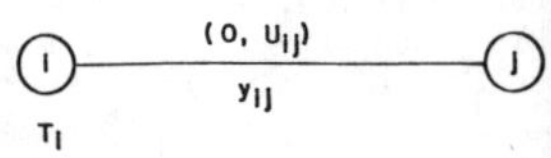

Network picture says: (5) is a side constraint. It can also be written

$\bar{p}_{ij}\ x_i \leq y_{ij}$

for $\bar{p}_{ij} = M_i\, p_{ij}$

Figure 7

NETWORK REFORMULATION FOR OBJECTIVE FUNCTION

ORIGINAL	NETWORK
$\text{MAX} \sum_j r_i m_i x_i$	Note that since y_j is the amount of plant j consumed the objective function can be expressed $\text{MAX} \sum_j y_j$ So if all y_j arcs are collected at a single sink t, we have

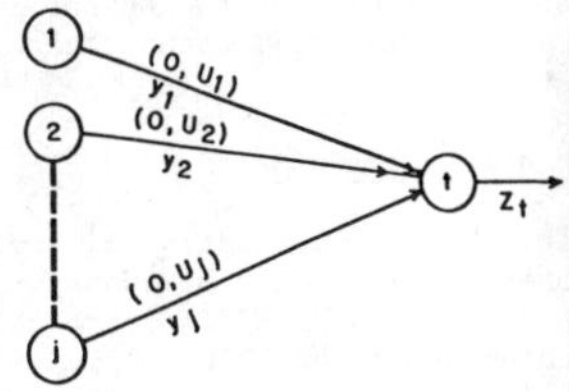

and the objective can be expresed as

$\text{MAX}\ Z_t = \sum_j y_j$

Figure 8

NETWORK DIAGRAM FOR 2 SWAs WITH 3 ANIMALS AND 2 PLANTS PER SWA

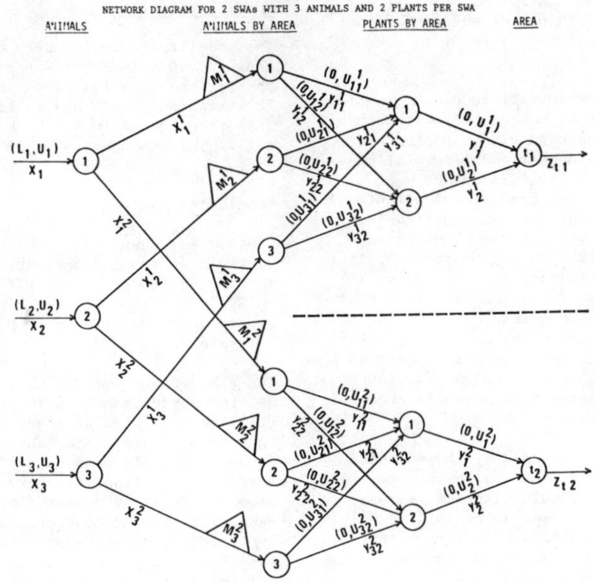

$\text{Max} \quad Z_{t1} + Z_{t2}$

Over the foregoing generalized network subject to the side constraints

$y_{ij}^k \leq \bar{q}_{ij}^k\, x_i^k$

$y_{ij}^k \geq \bar{p}_{ij}^k\, x_i^k$ for all i, j, and k.

parallel display of the original SWA submatrix constraints and the network-related alternatives is given. The suitability factor, S_i, and the k subscript have been omitted to streamline the comparison.

To help visualize the interrelationship among the network-related variables a full network diagram for a trivial case of two SWAs (designed as Areas 1 and 2) with three animal types and two plant species per SWA is illustrated in Figure 8. The areas are indicated by superscripts; X_i^k now indicates the total of all animals of type i, regardless of area, and X_i represents the number in area K (formerly represented by X_i without superscript).

Model Implementation

The LP-based vegetation allocation system is fully operational on the BLM Honeywell 6680 computer. It offers both the MPS and the network options for the solution of the LP problem. It also offers an array of options for the handling and manipulation of the SVIM inventory data. It has been designed to operate exclusively in a man-machine interactive mode.

The LP allocation system is being used extensively by the BLM to prepare grazing EISs. BLM computer records indicate that the LP allocation system has been accessed over 10,000 times in one year's time.

Allotment data are contained in an allotment data file. The decision maker needs only to enter the name of such file to begin the allocation session. The interactive steps that follow can be grouped into three categories: (a) aggregation of plant species into plant groupings, (b) change of selected data items, and (c) initialization of selective parameters.

The aggregation of plant species into plant groupings allows the decision maker to select the most abundant plants or to single out certain plant species and to agglomerate the rest into annual and perennial grasses and forbs, and shrubs. The AUFs and PUFs of these resulting plant groupings are composition-weighted to reflect their individual makeup. Aggregation (or "crunching") reduces the size of the LP matrix and shortens the execution time. More importantly, it brings down the number of plants in each SWA to a manageable level and makes it easier for the decision maker to analyze the final results. In addition to this interactive data reduction by crunching, plant species which are found to be below one percent in botanical composition are automatically aggregated into a miscellaneous category with weighted AUFs and PUFs.

Selected data items can be changed by the decision maker to examine the impact of these changes on the final results. AUFs, PUFs, suitabilities, forage intake rates and seasons of use can be individually changed on a SWA-by-SWA basis. AUFs, forage intake rates, PUFs, and seasons of use can be changed for the whole allotment at one fell swoop.

Parameter initialization gives the decision maker the capability to select the $f(w_j)$ weighting function for the RPVs and the Dietary Range Factor. The decision maker also has the choice to select strata or SWAs as the basic data units in the allotment. Other parameter initialization options include the merging of SWAs by common use and suitability, and the selection of the plant production level to modify the "normal" year production. Merging of SWAs further reduces the size of the problem and its time of execution, and gives the decision maker a quick, approximate answer which he can use as the starting point for a more elaborate run. Modification of "normal" year production can be used to study the impact of changes in production on the carrying capacities of the allotment.

Man-machine interactivity provides the decision maker with a chance to examine the results, make the necessary changes and run the model again until the answers can be regarded as satisfactory. It is easy to examine the results and determine what constraints are limiting, and whether or not the assumptions used in postulating the RPVs have led to a diet that can withstand professional scrutiny.

Example

The following interactive run illustrates the process that a field decision maker may go through when using the forage allocation model. Prior to making the run the user has decided on: which aggregation criteria to use, whether to use strata or SWAs as the basic data unit, whether or not to merge SWAs, which $f(w_j)$ weighting function to use for the RPVs, the value of the dietary range factor, and the value of the yearly production factor. An example of an interactive session in which these variables are assigned specific values is shown in Figure 9. For a more detailed account of the capabilities of the model, the reader is referred to the SVIM Forage Allocation Users Manual (BLM, 1982).

Model Specifics

The model is dimensioned to handle a maximum of 200 SWAs and 20 animal types per allotment, and 300 plant species and 16 animal types per SWA. It is written in standard FORTRAN and it is overlayed to run in less than 45k decimal, which is the current T/S ceiling on the Honeywell 6680.

Concluding Remarks

The BLM has compiled a substantial amount of PUF data in the past and is constantly attempting to refine such data. Under contract to the BLM, a recent study by Skiles, Kortopates and Van Dyne (1980) has examined the occurrence of PUFs in the scientific literature.

The BLM defines "proper use" for a particular plant as "the degree to which its current growth will be utilized by a kind of grazing animal when the range is properly used" (BLM Manual 4412.11). PUFs are derived "by determining the difference between total current production in a normal growth year and the amount left after proper use, and they indicate all removal in the process of grazing, including wastage by trampling" (BLM Manual 4412.11). PUFs are not robust data because the aggregate plant production and forage intake, and hence, they tend to mask the interaction between the two. Nevertheless, they can be regarded as an adequate surrogate for dietary preferences, until site-specific dietary data become more readily available.

Heuristic approaches have been used in the past to aid in vegetation allocation and resource planning. Single stocking rates are usually calculated, but this ignores the benefits from dietary overlap among the different herbivores, mainly because the trade-

Figure 9

Example of Interactive Session

```
/ALLOCATE

SVIM ALLOCATE VERSION 3.1.     EXPIRATION DATE 12/31/82

ENTER UNIQUE SINGLE CHARACTER IDENTIFIER
=F
ENTER DATA FORMAT OF INPUT DATA:(NEW) OR (OLD)
=NEW

DO YOU WANT TO RUN A BATCH OF ALLOTMENTS?
=NO


ENTER NAME OF ALLOTMENT DATA FILE
=TESTFILE
DO YOU WANT TO AGGREGATE PLANT SPECIES INTO PLANT GROUPS?
=YES
CRUNCH BY COMPOSITION OR SPECIES(ENTER C OR S)
=C
ENTER NO. OF MAJOR PLANT SPECIES (LESS THAN 20)
=0
DO YOU WANT TO SEE THE CRUNCH BREAKDOWN?
=YES

SWA NO. D073   % COMP
  PGRASS
     AGCR       80.18
     STC04      12.81
     ORHY        4.07
     SIHY        1.87
     BOGR2       1.06
  PFORBS
     ERIOG      59.03
     SPHAE      40.97
  T-S
     ARTR2      89.80
     CHVI8      10.20

SWA NO. D074   % COMP
  PGRASS
     SIHY      100.00
  PFORBS
     ERIOG      62.60
     ASTER      26.16
     SPHAE      11.24
  T-S
     EULA5      90.56
     ARTR2       9.09
     CHVI8       0.28
     GUSA2       0.08

SWA NO. D074   % COMP
  PGRASS
     SIHY      100.00
  PFORBS
     ERIOG      62.44
     ASTER      26.24
     SPHAE      11.31
  T-S
     EULA5      90.56
     ARTR2       9.09
     CHVI8       0.28
     GUSA2       0.08
```

```
SWA NO. D076   % COMP
  PGRASS
     ORHY      45.00
     SIHY      42.33
     BOGR2      7.09
     STC04      3.84
     AGCR       1.74
  PFORBS
     ASTER     75.72
     PHLOX     22.22
     SPHAE      1.09
     ASTRA      0.97
  AFORBS
     ACHIL    100.00
  T-S
     ARTR2     89.69
     CHVI8     10.31

IF OUTPUT FILE TO BE RESAVED, ENTER FILE NAME;
ELSE ENTER TERMINAL
=TERMINAL

DO YOU WANT TO MAKE ANY CHANGES TO THE RAW DATA?
=YES
DO YOU WANT TO ENTER A NEW ANIMAL?
=NO
DO YOU WANT TO MAKE ANY SWEEPING DATA CHANGES?
=YES
DO YOU WANT TO MAKE SWEEPING FORAGE INTAKE CHANGES?
=NO
DO YOU WANT TO MAKE SWEEPING SEASONS OF USE CHANGES?
=NO
DO YOU WANT TO MAKE SWEEPING AUF CHANGES?
=YES
ENTER PLANT NAME AND AUF VALUES
=PGRASS,.45,.45,.50,.50
    4 AUF CHANGES MADE FOR PLANT PGRASS
DO YOU WANT TO MAKE ANY MORE AUF CHANGES?
=NO
DO YOU WANT TO MAKE SWEEPING PUF CHANGES?
=NO
DO YOU WANT TO SAVE THE DATA CHANGES?
=NO
DO YOU WANT TO SEE OR CHANGE ANY OF THE RAW DATA, BY SWAS?
=NO
ENTER GROUPING DISCIPLINE FOR RAW DATA
(UNG)UNGROUPED,(STR)STRATUM,(SWA)SWA
=SWA

INPUT DATA ARE GROUPED IN    3 GROUPS

SWA NO.      NO. OF STRATA     STRATUM NO.
D073                     1
                                  0499

SWA NO.      NO. OF STRATA     STRATUM NO.
D074                     2
                                  0500
                                  0511

SWA NO.      NO. OF STRATA     STRATUM NO.
D076                     1
                                  0501

DO YOU WANT TO MERGE INPUT DATA BY COMMON USE AND SUITABILITY?
=NO


DO YOU WANT TO ENTER PARAMETERS AT ALLOTMENT LEVEL?
=YES
ENTER COMPOSITION WEIGHTING PREFERENCES FOR RPVS
(UNW)UNWEIGHTED, (LOG)LOG-WEIGHTED, (FUL)FULLY WEIGHTED
=FUL
```

```
ENTER DIETARY RANGE FACTOR, IN FRACTIONAL FORM
=.25
ENTER NORMAL YEAR PRODUCTION FACTOR,IN FRACTIONAL FORM
=1.0


GRP ENTRIES
D073
D074
D076

HAVE YOU PREVIOUSLY EXAMINED THE TARGET DIET?
=NO


TARGET DIET


SWA NUMBER D073

PLANT            CA   HO   SH

PGRASS         1.00 0.97 1.00
PFORBS         0.00 0.00 0.00
T-S            0.    0.03 0.


SWA NUMBER D074
PLANT            CA   HO   SH

T-S            1.00 1.00 1.00
MISC           0.00 0.00 0.00


SWA NUMBER D076

PLANT            AN   CA   SH

PGRASS         0.01 0.91 0.70
PFORBS         0.15 0.09 0.29
AFORBS         0.00 0.00 0.00
T-S            0.84 0.   0.


IS THE DIET OK?
=YES


ALLOTMENT ENTRIES


ENTER UPPER AND LOWER BOUNDS FOR CATTLE
=9999,0
ENTER UPPER AND LOWER BOUNDS FOR HORSES
=9999,0
ENTER UPPER AND LOWER BOUNDS FOR SHEEP
=9999,0
ENTER UPPER AND LOWER BOUNDS FOR ANTELOPE
=9999,0


PROBLEM SIZE:  NODES=    26; ARCS =   53


ALLOTMENT NUMBER 5121-01


OPTIMAL ANIMAL TYPE MIXTURE   NO. ANIMALS    MAX. NO.   MIN. NO.

CATTLE                              104.3       9999.         0.
HORSES                               13.1       9999.         0.
SHEEP                                 7.7       9999.         0.
ANTELOPE                              6.5       9999.         0.
```

FORAGE USED (LBS) 334971.

GRP NUMBER D073

	NO. ANIMALS	NO. OF MONTHS	SUITABILITY
CATTLE	26.37	2.03	1.00
HORSES	0.12	7.03	1.00
SHEEP	0.22	2.03	1.00

FORAGE USED (LBS) 43805.

FOR CATTLE

	USED PUFS	UNUSED PUFS	GIVEN PUFS	TARGET RPVS	CALC. DIET
PGRASS	0.46	0.03	0.50	1.00	1.00
PFORBS	0.06	0.02	0.08	0.00	0.00
T-S	0.	0.	0.	0.	0.

FOR HORSES

	USED PUFS	UNUSED PUFS	GIVEN PUFS	TARGET RPVS	CALC. DIET
PGRASS	0.01	0.44	0.45	0.97	0.98
PFORBS	0.	0.03	0.03	0.00	0.
T-S	0.00	0.01	0.01	0.03	0.02

FOR SHEEP

	USED PUFS	UNUSED PUFS	GIVEN PUFS	TARGET RPVS	CALC. DIET
PGRASS	0.00	0.37	0.38	1.00	1.00
PFORBS	0.00	0.15	0.15	0.00	0.00
T-S	0.	0.	0.	0.	0.

FORAGE

	USED AUFS	UNUSED AUFS	GIVEN AUFS	PLANT COMP.	PLANT PROD.
PGRASS	0.47	0.00	0.48	0.45	92453.
PFORBS	0.06	0.39	0.44	0.00	515.
T-S	0.00	0.38	0.38	0.55	111945.

FORAGE REMAINING (LBS) 42592.

DIETARY RANGE FACTOR 0.25
NORMAL YEAR PRODUCTION FACTOR 1.00

GRP NUMBER D074

	NO. ANIMALS	NO. OF MONTHS	SUITABILITY
CATTLE	74.13	2.03	1.00
HORSES	12.95	12.03	1.00
SHEEP	0.56	2.03	1.00

FORAGE USED (LBS) 276566.
FOR CATTLE

	USED PUFS	UNUSED PUFS	GIVEN PUFS	TARGET RPVS	CALC. DIET
T-S	0.18	0.	0.18	1.00	1.00
MISC	0.21	0.00	0.21	0.00	0.00

FOR HORSES

	USED PUFS	UNUSED PUFS	GIVEN PUFS	TARGET RPVS	CALC. DIET
T-S	0.23	0.11	0.34	1.00	1.00
MISC	0.08	0.07	0.16	0.00	0.00

FOR SHEEP

	USED PUFS	UNUSED PUFS	GIVEN PUFS	TARGET RPVS	CALC. DIET
T-S	0.00	0.18	0.18	1.00	1.00
MISC	0.00	0.10	0.10	0.00	0.00

FORAGE

	USED AUFS	UNUSED AUFS	GIVEN AUFS	PLANT COMP.	PLANT PROD.
T-S	0.42	0.00	0.42	1.00	663796.
MISC	0.29	0.17	0.46	0.00	2091.

FORAGE REMAINING (LBS) 429.

DIETARY RANGE FACTOR 0.25
NORMAL YEAR PRODUCTION FACTOR 1.00

GRP NUMBER D076

	NO. ANIMALS	NO. OF MONTHS	SUITABILITY
ANTELOPE	6.47	12.03	1.00
CATTLE	3.79	2.03	1.00
SHEEP	6.91	2.03	1.00

FORAGE USED (LBS) 14601.

FOR ANTELOPE

	USED PUFS	UNUSED PUFS	GIVEN PUFS	TARGET RPVS	CALC. DIET
PGRASS	0.01	-0.00	0.01	0.01	0.01
PFORBS	0.07	-0.00	0.07	0.15	0.15
AFORBS	0.02	0.00	0.02	0.00	0.00
T-S	0.07	0.	0.07	0.84	0.84

FOR CATTLE

	USED PUFS	UNUSED PUFS	GIVEN PUFS	TARGET RPVS	CALC. DIET
PGRASS	0.38	0.03	0.41	0.91	0.91
PFORBS	0.04	0.00	0.04	0.09	0.09
AFORBS	0.04	0.01	0.05	0.00	0.00
T-S	0.	0.	0.	0.	0.

FOR SHEEP

	USED PUFS	UNUSED PUFS	GIVEN PUFS	TARGET RPVS	CALC. DIET
PGRASS	0.08	0.13	0.21	0.70	0.67
PFORBS	0.04	0.05	0.09	0.29	0.32
AFORBS	0.03	0.07	0.10	0.00	0.00
T-S	0.	0.	0.	0.	0.

FORAGE

	USED AUFS	UNUSED AUFS	GIVEN AUFS	PLANT COMP.	PLANT PROD.
PGRASS	0.47	0.00	0.47	0.14	14620.
PFORBS	0.15	0.27	0.42	0.13	14076.
AFORBS	0.09	0.33	0.42	0.00	170.
T-S	0.07	0.25	0.33	0.73	77860.

```
FORAGE REMAINING (LBS)      23668.

DIETARY RANGE FACTOR      0.25
NORMAL YEAR PRODUCTION FACTOR      1.00

DO YOU WANT THE ADMINISTRATIVE REPORT?
=YES
```

ALLOTMENT NUMBER 5121-01

SWA CARRYING CAPACITIES

GRP	SUB	ACRES	PERCENT	ANIMAL	NUMBER	A-MONTHS	ACRES/AM	SUIT.
D073	0499	234	100.00	CA	26.4	53.6	4.36	1.00
				HO	0.1	0.8	276.38	1.00
				SH	0.2	0.5	517.38	1.00
D074	0500	215	70.03	CA	51.9	105.6	2.04	1.00
				HO	9.1	109.1	1.97	1.00
				SH	0.4	0.8	271.86	1.00
	0511	92	29.97	CA	22.2	45.2	2.04	1.00
				HO	3.9	46.7	1.97	1.00
				SH	0.2	0.3	271.86	1.00
D076	0501	170	100.00	AN	6.5	77.9	2.18	1.00
				CA	3.8	7.7	22.09	1.00
				SH	6.9	14.0	12.10	1.00

ALLOTMENT CARRYING CAPACITY

ACRES	ANIMAL	NUMBER	A-MONTHS	ACRES/AM	SUIT.	SEASON
711	CA	104.3	212.0	3.35	1.00	7/ 1- 8/31
	HO	13.1	156.7	4.54	1.00	11/ 1- 5/31
	SH	7.7	15.6	45.50	1.00	7/ 1- 8/31
	AN	6.5	77.9	9.13	1.00	1/ 1-12/31

ALLOTMENT NUMBER 5121-01

SWA OWNERSHIP INFORMATION

GRP	SUB	OWNER	JURIS	ADMIN	ACRES	PERCENT	ANIMAL	A-MONTHS	ACRES/AM
D073	0499	FP	BLM	BLM	234	100.00	CA	53.6	4.36
							HO	0.8	276.38
							SH	0.5	517.38
D074	0500	FP	BLM	BLM	193	62.87	CA	94.8	2.04
							HO	98.0	1.97
							SH	0.7	271.86
	0500	NP			22	7.17	CA	10.8	2.04
							HO	11.2	1.97
							SH	0.1	271.86

	0511	FP	BLM	BLM	82	26.71	CA	40.3	2.04
							HO	41.6	1.97
							SH	0.3	271.86
	0511	NP			10	3.26	CA	4.9	2.04
							HO	5.1	1.97
							SH	0.0	271.86
D076	0501	FP	BLM	BLM	95	55.88	AN	43.5	2.18
							CA	4.3	22.09
							SH	7.8	12.10
	0501	NS			75	44.12	AN	34.3	2.18
							CA	3.4	22.09
							SH	6.2	12.10

ALLOTMENT OWNERSHIP INFORMATION

OWNER	JURIS	ADMIN	ACRES	PERCENT	ANIMAL	A-MONTHS	ACRES/AM
FP	BLM	BLM	604	84.95	CA	192.9	3.13
					HO	140.4	4.30
					SH	9.3	64.86
					AN	43.5	13.88
NP			32	4.50	CA	15.7	2.04
					HO	16.2	1.97
					SH	0.1	271.86
					AN	0.	99999.99
NS			75	10.55	CA	3.4	22.09
					HO	0.	99999.99
					SH	6.2	12.10
					AN	34.3	2.18

ALLOTMENT NUMBER 5121-01

ALLOTMENT HERD UNIT INFORMATION

ANIMAL	HERD	SWA	STR	NUMBER
HO	3CC	D073		0.1
		D074		13.0
HO	3CC		TOTAL	13.1
AN	A-4	D076		6.5
AN	A-4		TOTAL	6.5

ALLOTMENT NUMBER 5121-01

BARREN SWAS INFORMATION

SWA STR ACRES

NO BARREN SWAS

```
ALLOTMENT NUMBER 5121-01



VI FILE SUITABILITY DATA

SWA  STR       SUITABLE  POT. SUIT.  LIMIT. SUIT.  UNSUITABLE

D073 0499        100.           0.            0.           0.
D074 0500        100.           0.            0.           0.
D074 0511        100.           0.            0.           0.
D076 0501        100.           0.            0.           0.

DO YOU WANT ANOTHER RUN?
=NO

*
```

End of Figure 9

off between use of each plant species cannot be computed easily. The mathematical programming approach presented above automatically takes into account dietary overlaps. It allows the decision maker to optimize forage allocation according to selected criteria. It addresses the issues of uncertainty in dietary knowledge, and provides flexibility, via man-machine interaction, to conduct sensitivity analysis and ask "what if" questions.

References

BLM. 1979. BLM Manual 4412.14. Soil-Vegetation Inventory Method. U.S. Dept. of the Interior. Bureau of Land Management.

BLM. 1982. SVIM Forage Allocation Users Manual. BLM. Denver Service Center.

Cooper, L., and D. Steinberg. 1974. Methods and Applications of Linear Programming. W. B. Saunders.

Daellenbach, H. G., and J. A. George. 1978. Introduction to Operations Research Techniques. Allyn and Bacon.

Glover, F. 1979. Enhancements of Spanning Tree Labeling Procedures for Network Optimization. INFOR, Vol. 17, No. 1, pp. 16-34.

Glover, F. W., R. Glover, and F. K. Martinson. 1982. The U.S. Bureau of Land Management's New Netform Vegetation Allocation System. University of Colorado, Division of Information Science Research.

Mathematical Programming System (MPS) User Guide. 1979. Honeywell Information System, DG10.

Orchard-Hays, W. 1968. Advanced Linear Programming Computer Techniques. McGraw-Hill.

Skiles, J. W., P. T. Kortopates, and G. M. Van Dyne. 1980. A critical review of proper use factors. Colorado State University, Dept. of Range Science.

SIMULATION ANALYSIS OF AFRICAN GRAMINOID RESPONSES TO DEFOLIATION AND POTENTIAL IMPLICATIONS FOR MANAGED GRAZING SYSTEMS[1]

Michael Coughenour[2]

Abstract.--A detailed morpho-physiological simulation model of graminoid growth was constructed and was used to dynamically integrate plant traits which determined tolerance to grazing. Responses of simulated grasses to variations in grazing height and frequency are of potential significance for design of wildlife reserves or of grazing systems in managed ranches or pastures.

INTRODUCTION

If grasses and large grazing mammals have undergone millions of years of coevolution (Stebbins 1981), potentially, the animals have learned how to utilize their grassland resource in some optimal fashion and the grasses have acquired the necessary adaptations to withstand this utilization. The Serengeti ecosystem of Tanzania is one of a few, if not the last, remaining places where large herds of grazing mammals exist in their original state with native graminoid vegetation. This paper reports on modeling work performed in conjunction with a non-management oriented ecological study concerning the effects of defoliation by large mammals on graminoid vegetation in the Serengeti ecosystem. I aim in this paper to present possibilities for extrapolation from natural to man-managed ecosystems, with the fundamental assumption that nature has solved problems which confront man the pastoralist or range manager, through evolutionary time. While nature often designs for stability maximization or low P/B ratios, man designs for maximal production or high P/B ratios (Odum 1969). Natural tropical grazing ecosystems can have especially high P/B ratios for terrestrial systems (Wiegert and Owen 1971), so they may serve as appropriate models for man-managed systems.

First, the modeling work that has been performed by myself and others will be reviewed. Simulation modeling has served as a tool to integrate the diverse plant traits which together determine tolerance to grazing. I will then examine problems confronted by range scientists, and draw analogies between the pure and applied approaches to common problems.

MODEL DEVELOPMENT AND ANALYSIS

Stage One - Biotic Processes of Graminoid Growth

Development

In this first stage of analysis, strictly biological processes were examined in the absence of grazing (Coughenour et al 1982a). A model was constructed of graminoid growth with the objective in mind that it eventually would be applied to examine grazing responses, thus ensuring that the pertinent features of grass growth would be included. Models usually cannot be applied to purposes beyond their original design objectives.

The model included a shoot number submodel that simulated numbers and types of apical meristems (fig. 1). The formation of new shoots through tillering or vegetative reproduction and the transition between shoot types during shoot maturation were simulated. Four types of shoots were distinguished; juvenile culmless vegetative shoots, mature culmless vegetative, vegetative culms, and reproductive culms.

Shoot meristem production rate was formulated as

$$SP = SPM * C_{CR} * F_T(SN_T/C_{CR}) * F_T(SN_C/C_{CR}) * F_{TG}(CHO) * F_{TG}(PNC) * F_T(LAI) \quad (1)$$

[1]Paper presented at the Third International Conference on State of the Art in Ecological Modeling, Colorado State University, May 26, 1982.

[2]Michael Coughenour is Research Associate, Department of Biology, Syracuse University, Syracuse, New York U.S.A.

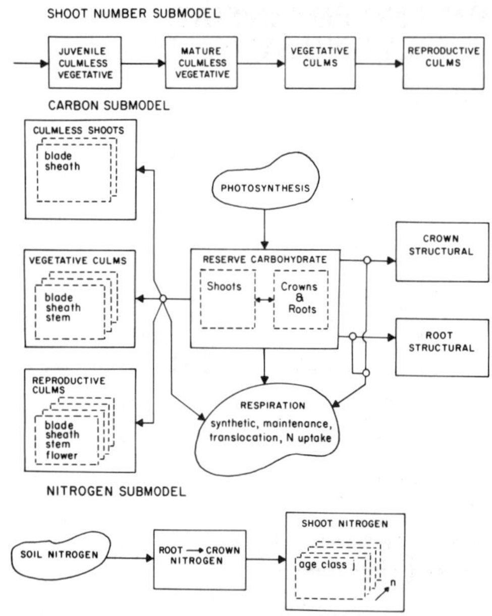

Figure 1.--Structure of the simulation model in Forrester notation. Shown are shoot number, carbon, and nitrogen submodels. (From Coughenour et al. 1982a).

where,

SP=shoot production rate, shoots/gC/m^2/d
SPM=maximum rate, shoots/gC/d
C_{CR}=crown biomass carbon, gC/m^2
F_T=any function affecting tillering
F_{TG}=any function affecting tillering and growth
SN_T=total shoot density, number/m^2
SN_C=culmed shoot density, number/m^2
CHO=plant reserve carbohydrate concentration, gC/gC
PNC=plant nitrogen/carbon ratio
LAI=leaf area index of sward, m^2leaf/m^2ground

The equation states that tillering is proportional to crown biomass (a crown being that part of the plant where roots and shoots converge). The first and second functions of shoot numbers reflect conceptions of apical dominance, where auxin production is proportional to meristem density. The second two functions correspond to observations that tillering rate is correlated with plant carbohydrate and nitrogen status. Tillering is also inhibited by canopy self shading, hence the final function.

A certain proportion of mature culmless shoots became vegetative and reproductive culms. This was a property of the species, but also varied according to plant vigor. Mortality of meristems was imposed at a constant rate, but reproductive culms were determinate, and died following floral differentiation.

Coupled with this demographic model was a biomass production and allocation submodel (fig. 1). The carbon budget of the plant was simulated and interacted with the demographic submodel via plant CHO reserves and crown biomass density. The distribution or allocation of carbon in the plant was simulated as follows. First shoot growth demands were calculated from shoot numbers and shoot growth rates. These demands were realized only if there was sufficient assimilate. Photosynthate in excess of aboveground growth demands was allocated to the reserve carbohydrate pool, and some of these carbohydrates were translocated downwards for use in belowground growth, which was dependent on reserve CHO concentration. Respiration rate was proportional to biomass, and drew from the reserve carbohydrate pool.

Shoot growth rate was calculated as,

$$SGR_k = GM_k * SN_k * F_{TG}(PNC) * F_{TG}(CHO) * F_G(SZ_k) \quad (2)$$

where,

SGR_k=shoot growth rate of the k-th shoot type, gC/m^2/d
GM_k=maximum individual shoot growth rate, gC/shoot/d
SN_k=shoot meristem density, number/m^2
SZ_k=shoot size, gC/shoot

In this formulation shoot biomass production is dependent upon the number of shoot meristems growing, on two measures of plant vigor-carbon and nitrogen concentrations, and on shoot size- where shoot growth rate declines at a maximum size characteristic for the species. The maximum size also diminishes with unfavorable nitrogen supply.

Belowground growth proceeded at the rate,

$$G_{BG} = RGR_{BG} * C_{BG} * F_{BG}(CHO) * F_{BG}(PNC) \quad (3)$$

where,

G_{BG}=belowground growth rate, gC/m^2/d

RGR_{BG}=relative growth rate, d^{-1}

C_{BG}=belowground biomass density, gC/m^2

F_{BG}=any function affecting belowground growth.

This is just a relative growth rate formulation with the introduction of substrate dependence.

The model also included a nitrogen uptake and allocation submodel. Root nitrogen uptake was simulated according to Michaelis-Menten enzyme kinetics, and nitrogen was distributed in the plant according to relative sink strengths. Younger shoot tissue presumably had a higher sink strength than old tissue, and therefore had a higher N/C ratio.

Shoot tissues were classified four ways, by tissue type (blade, sheath, stem, flower), by shoot type, by tissue age class, and by height in the canopy.

Behavior

This stage of model development indicated what parameters and processes contributed most significantly to patterns of grass growth. Sensitivity analyses were performed to ascertain model responses to variations in shoot number, photosynthetic rate, and individual shoot growth rate. Two 2-way factorial simulation experiments were performed on these parameters and the results were analysed with 2-way ANOVA. Shoot numbers had larger impacts on aboveground productivities than did individual shoot growth rates because once maximal shoot sizes were attained, growth rates became inconsequential. Increasing shoot numbers consistently decreased belowground growth. Grasses with low shoot numbers increased their belowground growth in response to increased individual shoot growth rates, but with high shoot numbers, the reverse was true. In general, greater photosynthetic rates increased belowground growth compared to aboveground growth.

In another experiment, shoot sizes and numbers were varied in a factorially designed experiment. This experiment indicated that shoot number had a greater impact on production rate than shoot size. Thus highest production could be achieved in short time periods when there were many shoots. This may be a good adaptation for grasses growing in unpredictable or short growing seasons.

Morphological differences between short and tall grasses had a greater effect on the quality than upon the total quantity of primary production, as increased shoot numbers largely compensated for decreased shoot sizes of short grasses. Morphological and physiological constraints limited the simultaneous maximization of shoot number, shoot size, and shoot growth rate. Which of these is maximized should depend on the environment.

Stage Two - Biotic Responses to Defoliation

Development

The second stage of model development was an analysis of the system to simulated defoliation (Coughenour et al 1982b). A grazing algorith was constructed which removed tissues and meristems down to a specified height at a specified frequency. The quantity of biomass removed by defoliation to a specific height depended upon the concentration of biomass per unit volume in the sward. This in turn was dependent on the simulated morphological variables shoot meristem density and shoot height.

A number of causal pathways existed within the simulated plant system which extended direct defoliation effects into indirect defoliation effects (fig. 2). Tillering rate (eqn. 1) could be directly affected by culm meristem removal or by reduction of leaf area index. Of course, defoliation removes shoot tissues, but this could have indirect consequences, through reduction of photosynthetic leaf area, reduced carbohydrate, and consequently reduced shoot and root rates (eqns. 2-3). Shoot growth could also be affected by alteration of the relative abundance of shoot types, since these had differing growth rates. Culmless shoot meristems were protected from grazing and the model assumed that these meristems would replace lost leaf tissue and grow to achieve a maximal live shoot weight. Thus grazing could maintain growth rate by keeping this shoot weight at less than the maximum. Photosynthesis could be indirectly affected by alterations in the tissue age class distribution, by changes in the types of tissues in the canopy, and by decreased self shading.

Behavior

Without water limitations, the model simulated increased above and belowground productivities under most grazing regimes (table 1). Grazing kept culmless shoots in an actively growing state, while ungrazed shoots reached maximal sizes. Belowground growth rate was also stimulated under grazing, as total photosynthesis was reduced by self shading and younger leaf tissues. Thus grazing kept the plants in an active, productive state.

Following individual grazing events, reserve carbohydrate concentrations fell, as shoot regrowth demands exceeded photosynthetic inputs (fig. 3). Reserves were replenished as leaf areas increased. Thus, sufficiently frequent and severe defoliation could prevent recovery of carbohydrate levels. Renewal of shoot growth after defoliation created a greater relative abundance of younger, high nitrogen leaf tissue (fig. 3) which also had a higher photosynthetic capacity. To some extent then, decreases in leaf area could be compensated by increases in the rate per unit leaf area.

The severity of defoliation could be expressed as a function of grazing height and frequency. The index

$$GI = 1 - g/ug \qquad (4)$$

was calculated where,

GI=grazing intensity
g=average shoot biomass of grazed plants
ug=average shoot biomass of ungrazed plants

(McNaughton 1979). From simulation results, relations of the form

$$GI = a - b*Ht - c*In - d*Ht*In \qquad (5)$$

were derived through regression analyses where,

a,b,c,d=empirical constants
H=grazing height, cm
In=grazing interval, days

Above and belowground productivities varied continuously with grazing intensity GI (fig. 4). Above or belowground growths were greatest at moderate grazing pressure, and both were stimulated above the levels of ungrazed controls. These results were corroborated by findings of stimulated productivities in the Serengeti (McNaughton 1976). This stimulation has also been termed compensatory growth (McNaughton 1979) since the plant is compensating for lost leaf tissue.

Stage Three - Addition of Abiotic Variables

Development

The third stage of analysis interrelated defoliation effects upon biotic and abiotic processes. Abiotic submodels describing water flows,

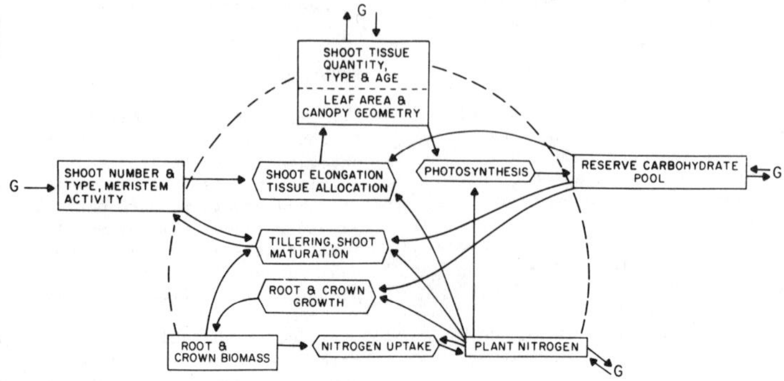

Figure 2.--Causal interrelationships among processes and state variables which determine responses to defoliation. Grazers (G) affect and are affected by aboveground plant state variables (From Coughenour et al. 1982)

Table 1. --Aboveground (AP) and belowground (BP) non water-limited productivities of simulated shortgrass and midgrass, ungrazed or grazed at specified height and frequency

Grazing Regime		Shortgrass		Midgrass	
height	interval	AP	BP	AP	BP
(cm)	(d)	(g)		(g)	
	ungrazed	482	765	617	710
2	2	840	172	565	18
	10	1172	940	590	552
	20	1332	1190	1075	777
10	10	882	1160	1135	990
	20	802	1202	1270	977
	44	700	1195	1290	942

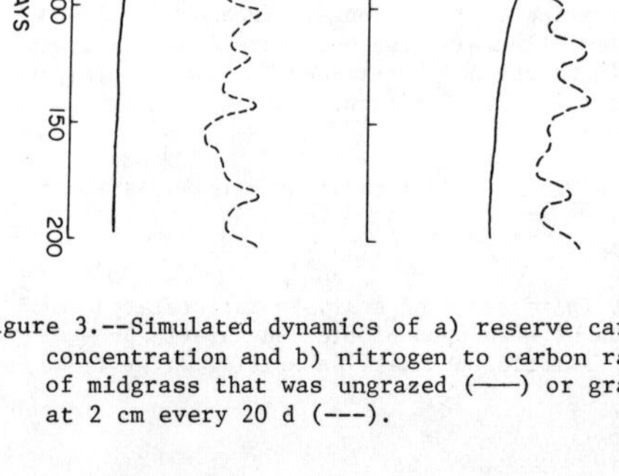

Figure 3.--Simulated dynamics of a) reserve carbon concentration and b) nitrogen to carbon ratio of midgrass that was ungrazed (——) or grazed at 2 cm every 20 d (---).

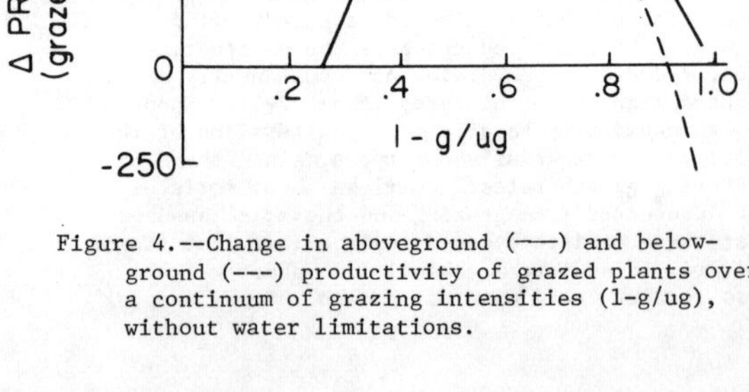

Figure 4.--Change in aboveground (——) and belowground (---) productivity of grazed plants over a continuum of grazing intensities (1-g/ug), without water limitations.

light penetration and interception, and energy balance relationships were interfaced with the graminoid growth model. A more mechanistic photosynthesis model was warranted at this stage. The following equation for photosynthesis (P_N) was introduced,

$$P_N = \frac{\Delta CO_2}{C_i^{-1} + C_s^{-1}(W) + C_a^{-1}} \qquad (6)$$

where,

ΔCO_2 = CO_2 concentration gradient across leaf surface
C_i = internal leaf conductance to CO_2
C_s = stomatal leaf conductance to CO_2
C_a = atmospheric boundary layer conductance
W = plant water content

However, transpiration rate (Tr) was proportional to the quantity,

$$Tr = \frac{\beta * \Delta H_2O}{C_s^{-1}(W) + C_a^{-1}} \qquad (7)$$

where,

β = a constant
H_2O = water vapor concentration gradient

Equations 6 and 7 reveal the central role of C_s in two mass-balance processes. Significantly, C_s was regulated by feedback as a function of plant water content.

Then, tillering and shoot growth (eqns. 1 and 2) were affected by plant water content and leaf temperature (T_1),

$$SP' = SP * F_{TG}(W) * F_{TG}(T_1) \qquad (8)$$

$$SGR_k = SGR_k * F_{TG}(W) * F_{TG}(T_1) \qquad (9)$$

where,

SP' = shoot production rate affected by W and T_1
SGR_k' = shoot growth rate affected by W and T_1, kth shoot type

Other processes affected abiotically included respiration, belowground growth, and mortality.

Imposition of abiotic processes introduced a new set of causal pathways arising from direct defoliation responses (fig. 5). A reduction in leaf area decreased total transpiration and allowed for greater penetration of light. The altered radiation flux affected the rate of bare soil evaporation and transpiration per unit leaf area, the latter through stomatal conductance. Reduced leaf area also affected interception of rainfall, and consequently, soil water. Root water uptake rate depended on the levels of soil water, plant water, and root biomass. Root biomass of course, depended on photosynthetic rate and leaf conductivity.

Behavior

As before, the model responded to increased grazing intensity with an optimal level of aboveground productivity at moderate levels of defoliation (fig. 6). However, belowground productivity was stimulated less or was depressed at comparable grazing intensities. As compensation, surviving belowground biomass decreased less at moderate grazing intensities because of reduced water stress.

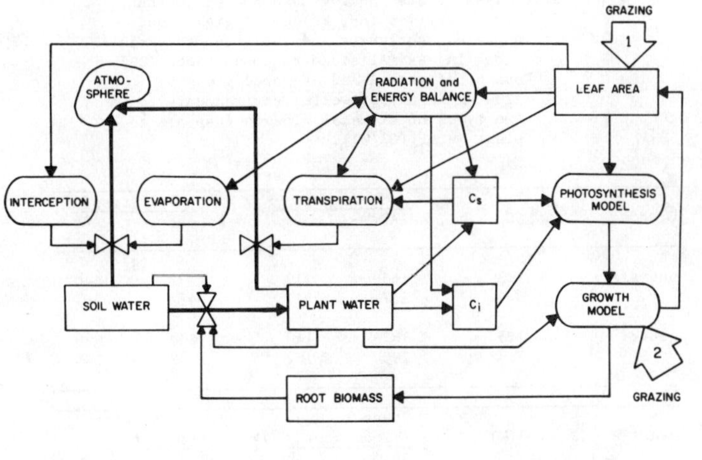

Figure 5.--Flows of water (heavy lines) and causation in the soil plant system. Processes are enclosed in ovals and states in rectangles. C_s and C_i represent stomatal and internal leaf conductances. Direct grazing effects include reduced leaf tissue (arrow 1) or loss of meristems (arrow 2). (From Coughenour 1982).

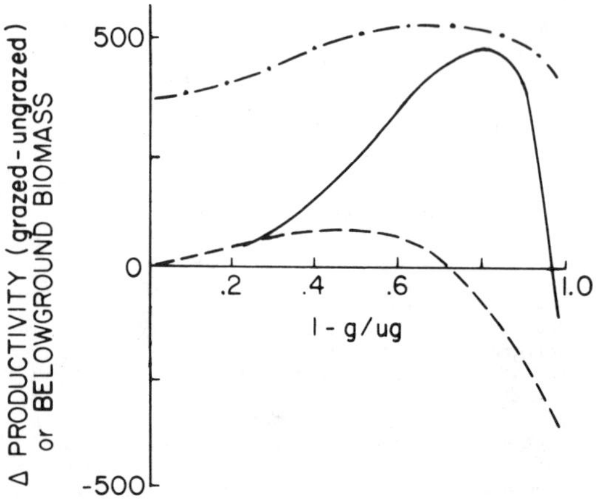

Figure 6.--Grazing induced changes in aboveground (——) and belowground (---) productivity of water-limited plants, and final belowground biomass (---) over a continuum of grazing intensities (1-g/ug).

The effect of plant water on shoot growth rate was positively related to grazing intensity, due to reduced transpiration. More favorable water levels also reduced root mortality. Water was the primary limitation over plant growth rather than self shading. As before, reductions in aboveground growth at high grazing intensities were due to low carbohydrate levels.

Improved water relations under frequent grazing provided an opportunity to increase photosynthesis by increasing stomatal conductance, with less of a negative effect on plant water. Thus, simulations showed that productivity could be increased by greater stomatal conductance in frequently grazed plants (table 2a). The way stomates close in relation to plant water is another variable plant trait. Simulations showed that rapid stomatal closure would be more successful in environments where rainfall is sporadic, but that the reverse would be true in wetter environments (table 2b).

In another simulation experiment, erect leaf angles increased light absorption, which in turn increased photosynthesis and transpiration rates (Table 3). The relative extent to which these two processes were affected determined the overall effect on productivity. Tallgrass benefited more from erect leaves than shortgrass because water was less limiting in that environment. Tallgrasses were more likely to be grazed infrequently and erect leaves increased grazing susceptibility less when grazing events were less frequent.

MANAGEMENT APPLICATIONS

Grassland productivity can be increased with four levels of capital investment. Least costly is to use the resource as it stands but adjust grazing pressure through animal density, movements, and species selection. Second through fourth levels of investment involve plant species removal, introduction, and replacement, in that order (Whyte 1974). Parallel to this, a grassland ecosystem may, over evolutionary time, change by acquiring different herbivore species, by herbivores adopting optimal grazing regimes, by plant species replacement, or by plant species evolution of traits

Table 2.--Simulated aboveground and belowground productivities (AP, BP) of a) grazed and ungrazed shortgrass and tallgrass to variations in maximal stomatal conductance (CS) and b) of grazed and ungrazed grasses growing in dry or wetter environments to two types of stomatal closure response to plant water (G'(W)).

a)

Grazing	C_S	Shortgrass AP	Shortgrass BP	Tallgrass AP	Tallgrass BP
	(cm/sec)	(g)		(g)	
ungrazed	1.3	165	312	532	440
	3.4	137	317	427	567
grazed	1.3	102	110	200	20
	3.4	295	162	1022	245

b)

Grazing	G'(W)	Grass in dry ecosystem AP	Grass in dry ecosystem BP	Grass in wetter ecosystem AP	Grass in wetter ecosystem BP
ungrazed	rapid	145	322	342	432
	gradual	132	287	365	770
grazed	rapid	287	270	822	377
	gradual	227	197	1077	655

increasing their grazing resistance. The success or failure of any of these alterations can be understood in light of the physio-morphological basis by which plants are affected by defoliation.

The Design of Wildlife Reserves

This level of human intervention operates on the assumption that native animals will distribute their own grazing pressure in an optimal way according to distribution of grass species and rainfall. Although this is the least intensive management level, it requires an understanding of native species. In the Serengeti National Park and adjoining reserves for example, there is a complex array of grassland communities superimposed upon rainfall, soil, and elevation gradients. Tallgrass communities should respond more favorably to infrequent grazing, while shortgrass communities should require frequent grazing, according to simulation experimentation. The model predicted that there should be an optimal defoliation regime for each grass type, that will maximize aboveground and belowground productivities. The Serengeti is an appropriately designed reserve because migratory herbivores are allowed to distribute their grazing pressure among these communities at will. This natural movement may be necessary for ecosystem persistence. Restriction of animal movements by inappropriate placement of reserve boundaries could enforce a suboptimal grazing system.

Manipulation of Animals - Grazing Systems

Various types of grazing systems are recognized by range scientists differing primarily in the way animals are rotated among spatial subunits of the ecosystem or ranch (Lacey and Van Poolen 1979). Fundamentally, a specific grazing management system depends upon the frequency with which a subunit should be grazed, and by the intensity of grazing in animals per unit time. This decision is made with reference to the size, number, and ecology of these subunits.

Subunits are rested or protected from grazing to encourage root growth, to build up nutrient reserves, to allow for tillering, or simply, to let herbage accumulate. The length of rest necessary is dependent upon the time required for grasses to achieve maximal production. Herbage left standing for time beyond attainment of that level of production is regarded as a loss in herbage utilization (Pratt and Gwynne 1977). Simulation modeling should be able to predict the length of time necessary to achieve maximal production, to build up reserves, to tiller, and to grow roots.

Table 3.--Effect of leaf angle on aboveground (AP) and belowground (BP) productivities of grazed and ungrazed short and tallgrasses.

Grazing	Leaf Angle	Shortgrass		Tallgrass	
		AP	BP	AP	BP
		(g)		(g)	
ungrazed	prostrate	135	282	480	310
	erect	142	212	500	657
grazed	prostrate	257	252	642	160
	erect	285	240	1070	245

Range scientists ascribe grazing on the basis of range condition, which is determined by species composition or successional state, ground cover, plant vigor, litter, and erosion (Whyte 1974). Grazing induced plant communities are dominated by defoliation resistant or unpalatable plants. Defoliation resistance may arise from morphological (Branson 1953, Hyder 1974) or physiological (McNaughton 1979, Caldwell et al. 1981) plant traits. Community composition, however, may be more affected by climatic variations than by previous grazing pressures (Hyder et al 1975). Appropriate simulation modeling should aid in prediction of community composition resulting from the interaction of plant traits with grazing pressure and climate.

Simulation analyses predicted the responses of simulated grasses to defoliation height and frequency. Response surfaces can be generated by the model to aid in the design of a grazing system. For this, it will be necessary to relate specific animal movements to height and frequency variables. Response surfaces varied with species of grass. The response surface for shortgrass indicated that these were best adapted for relatively frequent grazing compared to tallgrasses. The model performed this way because it simulated the interaction of morphological and physiological traits.

Grazing induced morphological changes in the simulated sward structure that were grazing-frequency dependent. Frequent grazing induced tillering below the grazing height and increased the relative proportion of young tissues, making herbage more nutritious and more densely concentrated in the canopy. In contrast, nutrient quantity was highest at intermediate frequencies. Since herbivores differ in their requirements for sward structure, forage quality and quantity, the model can be used to relate animal species, grazing pattern, and plant community.

The model predicted that under frequent grazing, soil water conservation counteracted the tendency of high stomatal conductance to deplete plant water. Consequently, plants with high stomatal conductance should be grazed more frequently. Another prediction was that erect leaf angles increase production more in more mesic environments. Grasses in mesic environments should be more likely to maximize shoot size rather than number, and thus be adapted to infrequent grazing. This is fortuitous because in infrequent grazing regimes, the degree of leaf erection has less effect on grazing response.

Simulation analyses can act as a link between natural and managed systems. A method of ranch management currently becoming increasingly popular in Africa, short duration grazing, simulates past game herd grazing in that it maximizes energy flow per unit of land (Savory 1979). In it, animals are rotated frequently among many small paddocks. Grazing is timed to utilize the maximum inflexion phase of the grass growth curve. Fast rotation prevents overgrazing, but maintains a constant maximum rate of energy flow into the animals.

Simulation modeling similarly suggested that grazing at specific frequencies during the growing season leads to optimal production flows. Savory (1979) suggests short bouts of grazing lasting 1-15 days and rest periods of 20-60 days, or 30-60 days for bunch grasslands. The model predicted optimal growth under similar regimes, and prescribed longer rests for tall (bunch) grasses. It is hoped that simulation modeling can help determine whether animals in the Serengeti are following optimal grazing regimes, and why.

Manipulation of Plants

Which plants are most appropriate for a specified climate and a desired grazing system? The model could act as a guide for plant community design, or for plant breeding programs, to help assess which traits are most likely to increase plant productivity or population persistence under grazing in a particular location. For example: the benefits of increasing photosynthetic rate via an increase in stomatal conductance could be outweighed by the risk of desiccation under infrequent grazing in sub-humid environments; breeding for maximal tiller density should be more successful where growing seasons are short, shoot sizes are small, and grazing is frequent; leaf angles should be erect unless frequently grazed; increased photosynthetic rate will increase yield to grazers depending on how shoot growth responds to carbohydrate concentration relative to root growth.

CONCLUSIONS

The application of simulation analyses to grassland management is recent, and the science is in a formative stage. The models and other constructed with similar objectives, can be shown to give predictions that agree with past management experience, partly because the models were constructed with the help of past experience. In the future, models will move towards increasing predictive power, provided there will be constant interaction with experimentation on the real system. As this interaction proceeds, optimization and fine tuning of the simulated and the real systems will accelerate. At the present time, simulation can be used effectively to research natural ecosystems.

ACKNOWLEDGEMENTS

This research was supported by National Science Foundation grants DEB 77-20360 and DEB 79-22991. Input from Samuel J. McNaughton was invaluable.

LITERATURE CITED

Branson, F. D. 1953. Two factors affecting resistance of grasses to grazing. Journal of Range Management 6:105-171.

Caldwell, M. M., J. H. Richards, D. A. Johnson, R.S. Nowak, and R.S. Dzurec. 1981. Coping with herbivory: Photosynthetic capacity and resource utilization in two semiarid *Agropyron* bunch grasses. Oecologia 50: 14-24.

Coughenour, M.B., 1982. A mechanistic simulation on analysis of water use, leaf angles, and grazing in East African Graminoids. submitted.

Coughenour, M. B., S. J. McNaughton, and L. L. Wallace. 1982a. A simulation model of perennial graminoid growth uniting physiological processes and morphometric traits. submitted.

Coughenour, M. B., S. J. McNaughton, and L. L. Wallace. 1982b. Simulation study of perennial graminoid responses to defoliation. submitted.

Hyder, D. N. 1974. Morphogenesis and management of perennial grasses in the United States. Pages 89-98 *in* Plant Morphogenesis as the Basis for Scientific Management of Range Resources. USDA Miscellaneous Publication 1271. Agricultural Research Service.

Hyder, D. N., R. E. Bement, E. E. Remmenga, and D. F. Hervey. 1975. Ecological Responses of Native Plants and Guidelines for Management of Shortgrass Range. USDA Technical Bulletin No. 1503, Agricultural Research Service and Colorado State University.

Lacey, J. R. and H. W. Van Poolen. 1979. Grazing system identification. Journal of Range Management. 32:38- .

McNaughton, S. J. 1979. Grazing as an optimization process: grass-ungulate relationships in the Serengeti. American Naturalist 113:691-703.

Odum, E. P. 1969. The strategy of ecosystem development. Science 164:262-270.

Pratt, D. J., and M. D. Gwynne (eds.). 1977. Rangeland Management and Ecology in East Africa. R. E. Krieger Publishing Co., Huntington, New York. 310pp.

Savory, A. 1979. A holistic approach to ranch management using short duration grazing. *In* D. N. Hyder (ed.), Proceedings of the First International Rangeland Congress, Society for Range Management 1:555-557.

Stebbins, G. L. 1981. Coevolution of grasses and herbivores. Annals of the Missouri Botanical Garden 68:75-86.

Whyte, R. O. 1974. Tropical Grazing Lands: Communities and Consistent Species. Dr. W. Junk b.v. - Publishers- The Hague. 222 pp.

Wiegert, R. G., and D. E. Owen. 1971. Trophic structure, available resources, and population density in Terrestrial vs. Aquatic Ecosystems. Journal of Theoretical Biology 30:69-81.

ESTIMATING LONG-TERM FORAGE PRODUCTION ON RANGELANDS[1]

W. H. West and W. K. Lauenroth[2]

Abstract.--We describe a method of estimating forage production on rangelands which could be of great value to rangeland managers. Input data are limited to widely available NOAA weather observations. We employed a complementary hypothesis between actual and potential evapotranspiration to estimate actual evapotranspiration. A linear relationship between water use and forage production was used to estimate forage production. Predictions were compared with data from four different rangeland types in the western U.S.

INTRODUCTION

A central issue facing agencies charged with the management of rangeland resources is setting stocking rates which reflect the average productivity of the resource. In the United States legislation has mandated environmental statements associated with agency plans for grazing management programs. While agencies such as the Bureau of Land Management (BLM), Department of the Interior, can and are collecting data relevant to questions about the current state of the range resource the problem of long-term average productivity remains. The solution revolves around a capability to either predict forage production for a sufficiently long time period into the future or to adjust current estimates of forage production based upon either averages of past weather observations or upon elements of the climate which have a very small interannual variance.

Prediction of future forage production in the absence of knowledge about weather sequences is an especially tenuous approach to the problem. Additionally the models applicable to this problem are often too complex to be practical management tools. Adjustments of current estimates of forage production based upon weather observations is widely used. Much of the work has focused upon a single water variable, usually precipitation (Rutherford 1980). While it is clear that one should not expect a linear relationship between precipitation and forage production over a broad range of precipitation (Lauenroth 1979), a linear hypothesis for the relationship is often valid for semiarid and arid regions (Noy-Meir 1973). Climate adjustment factors (CAF) based upon the linear hypothesis are in current use by the Bureau of Land Management to adjust current forage production estimates to reflect long-term production potential.

Our objective for this study was to investigate a method for deriving a CAF for use by the Bureau of Land Management which would have the following characteristics and constraints.

1. It would represent an improvement over the current method which utilizes annual precipitation.

2. It would rely exclusively upon readily available weather data collected by the National Oceanic and Atmospheric Administration (NOAA).

3. It should result in an estimate of forage production within 20% of measured production during calibration.

METHODS

Since water is clearly the most important variable controlling forage production over the semiarid and arid regions administered by BLM we decided to focus upon a hypothesized linear relationship between transpiration (Tr) and aboveground net primary production (ANPP). Transpiration was chosen to represent the interaction between weather and production because of the tight coupling between carbon dioxide (CO_2) assimilation and water vapor loss. The resistance pathways are the same for both processes (Noy-Meir 1973). Additionally, the response for the Tr-ANPP relationship has a zero-yield intercept at the origin, thus simplifying the equation (Arkley 1963, De Wit 1958):

$$ANPP = k\ Tr \quad , \qquad (1)$$

where,

ANPP = aboveground net primary production, g/m^2

k = slope of response line or average water-use efficiency, g/mm

Tr = cummulative transpiration, mm

Equation (1) can be used in a proportional equation to relate ANPP in a known year to ANPP in an unknown year. However, the slope of the response line (k) is not a constant between years for the same species or ecosystem (De Wit 1958). This would leave k in the numerator and the denominator of the proportional equation with the only way of evaluating k via field studies. Arkley (1963) suggested dividing Tr by the relative humidity deficit for the time period in which Tr was accumulated, the slope (k) then becomes a

[1]Paper presented at the Third International Conference on State-of-the-Art in Ecological Modeling, Colorado State University, May 24-28, 1982.

[2]W. H. West is a Range Conservationist with the Bureau of Land Management, USDI, Denver, Colorado; and W. K. Lauenroth is an Associate Professor, Range Science Department, Colorado State University, Fort Collins, Colorado, U.S.A.

constant between all years. Placing this adjusted equation into the proportional equation and cancelling terms yields:

$$\frac{ANPP_1}{ANPP_2} = \frac{Tr_1(100 - RH_2)}{Tr_2(100 - RH_1)} , \quad (2)$$

where,

RH = average relative humidity for the time period

Subscripts 1 and 2 indicate separate years.

Equation (2) was used to relate ANPP in a known year (subscript 2) to ANPP in the year of interest (subscript 1). The right side of the equation is the climate adjustment factor. Use of this equation assumed that no catastrophic events (i.e., hail, flooding, fire) occurred in either year to affect ANPP, that except for climatic changes all other variables remained the same and that there were no major changes in community composition between years.

The next problem was the estimation of Tr, which was accomplished by a three-step process in which the actual evapotranspiration (AET) was calculated for the site, then the bare soil evaporation (Es) was calculated and finally the Tr was calculated as the difference between AET and Es. Most equations used for estimating evapotranspiration from an ecosystem deal with potential evapotranspiration (PET). Potential evapotranspiration is an estimate of the maximum amount of water that could be lost from a system and does not consider the biological and physical processes actually experienced by the ecosystem. To convert estimates of PET to AET requires either a detailed knowledge of the local water balance or an empirical coefficient to reduce PET. Both of these corrections require field experimentation and violate the objectives of the project.

To estimate AET a complementary hypothesis between regional AET and PET from a small moist surface within the region was used (Bouchet 1963). Normally PET is considered as the driving force for AET and AET proceeds until it is equal to PET. However, PET is usually measured from a small moist surface such as a lysimeter or evaporimeter. The PET from this small moist surface is controlled by the evaporative demand of the air mass directly over the surface. The evaporative demand of the air mass is controlled by the regional moisture condition (AET) not by the PET from the small moist surface. Therefore, it is the regional AET that controls the PET from an evaporimeter, with the smaller the AET, the larger the PET (Morton 1965).

The complementary hypothesis requires three assumptions (LeDrew 1979). First, the conditions of the air mass are controlled by the regional surface only, or we are dealing with an essentially closed system. Second, the change in net radiation (Qn) for both the potential and regional surface is negligible with changes in the regional moisture condition. Third, the sensible heat flux from the regional surface to the air is equal in magnitude but opposite in sign to the heat flux from the air to the potential surface. Summing the change in the energy balance equations, with changes in regional moisture for a unit area of the regional surface and a unit area of the potential surface, applying the assumptions made above, and integrating, the equation for the complementary hypothesis results:

$$PET + AET = C \quad (3)$$

where,

C = complementary constant of integration, mm

Morton (1969) evaluated the complementary constant (C) by assuming that the regional and potential surfaces were both saturated. Under this condition AET would be equal to PET and C would be equal to 2PET'. In this case, the PET' is the potential evapotranspiration from a large moist surface and is different from the PET discussed earlier. Graphical representation of the relationship between PET and AET is shown in figure 1. Use of this complementary hypothesis allows for the estimation of AET without the need to explicitly consider the controls on evapotranspiration by biological processes or soil physics. These controls are accounted for implicitly by the complementary hypothesis (fig. 1).

Practically, both C and PET can be evaluated using semi-empirical combination equations for evapotranspiration. The Van Bavel (1966) modification of the Penman (1948) equation was used to represent PET:

$$PET = \frac{(m\ Qn + \gamma\ Ea)}{m + \gamma} , \quad (4)$$

where,

m = slope of saturation vapor pressure-temperature curve, mbar/°C

Qn = net radiation, mm/day

γ = psychrometric constant (0.66 at 1000 mbar pressure), mbar/°C

The wind function term (Ea) is evaluated by (Rosenberg 1974):

$$Ea = 1.222 \frac{u_a}{[\ln (z_a/z_o)]^2} (e_s - e_a) , \quad (5)$$

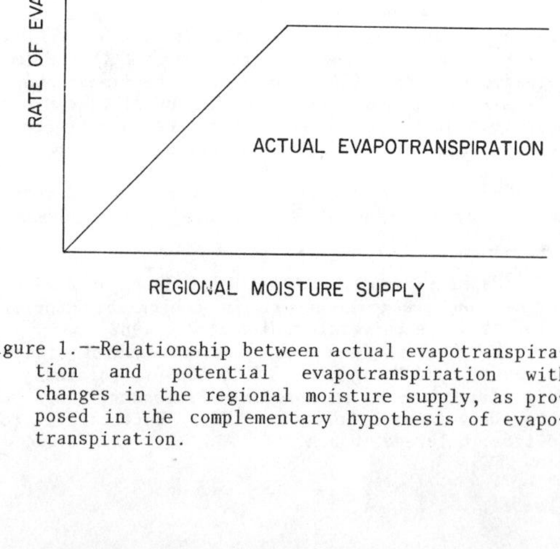

Figure 1.--Relationship between actual evapotranspiration and potential evapotranspiration with changes in the regional moisture supply, as proposed in the complementary hypothesis of evapotranspiration.

where,

u_a = daily wind travel, km/day

z_a = Height of wind measurement, cm

z_o = roughness length of surface, cm

e_s = saturation vapor pressure, mbar

e_a = actual vapor pressure, mbar

The actual vapor pressure was calculated using wet and dry bulb temperatures at two or three PM in a standard reduction of psychrometric data (List 1951). The slope term was calculated using a standard saturation vapor pressure-temperature curve at average daily temperature; slope of the curve was found by taking the first derivative of the equation of the curve (Rosenberg 1974). The roughness length (zo) was approximated by using average vegetation height in the equation (Szeicz et al. 1969):

$$\log z_o = 0.997 \log H - 0.883 \quad , \qquad (6)$$

where,

H = average vegetation height, cm.

The net radiation term used in equation (4) was developed empirically from data around the world (Linacre 1968):

$$Qn = ((1 + 2p)[Qx(0.254 + 0.016p - 0.25a) - 0.064] + 0.032)/59 \quad , \qquad (7)$$

where,

p = ratio of actual sunshine hours to possible sunshine hours

Qx = solar radiation at top of the atmosphere, ly/day

a = average surface albedo.

Finally, equation (4) results in overestimates of PET at wind speeds greater than 215 km/day, with the size of the overestimate dependent on the wind speed (Rosenberg 1969, Stone and Horton 1974). To correct for this overestimation; the PET was reduced using data from Rosenberg (1969):

$$PET = PET[1.0 - (0.0017057u_a - 0.36610496)] \quad , \qquad (8)$$

The complementary constant was calculated by doubling an equation for estimating the PET from a large moist surface (Priestley and Taylor 1972):

$$C = 2k(\frac{m\ Qn}{m + \gamma}) \quad , \qquad (9)$$

where,

k = dimensionless seasonal coefficient

The seasonal coefficient is inversely related to the Bowen ratio and ranges from 1.50 in the spring and fall to 1.21 in July, averaging 1.26 through the summer (De Bruin and Keijman 1979). The model used a table lookup function between k and the day of year to establish the coefficient for each calculation. Actual evapotranspiration was then calculated as the difference between the complementary constant and PET.

Once AET was estimated it was then partitioned into evaporation from the soil and transpiration from the plants. When the soil profile is saturated the most important factor controlling the evaporative loss is the supply of energy to the evaporating surface (Ritchie 1972). Controlling the amount or flux of energy to the surface is the density of the vegetation canopy which reflects or absorbs the incoming solar energy before it reaches the soil surface. Leaf area index (LAI) as a measure of canopy density can be used in a form of Beer's Law to derive the flux of energy at any depth in the canopy. In a study of several row crop species Ritchie (1972) found that the energy flux at depth z in the canopy could be estimated by:

$$Qz = Qo\ e^{-0.398\ LAI} \quad , \qquad (10)$$

where,

Qz = energy flux at depth z within the canopy, ly/min

Qo = energy flux at top of the canopy, ly/min

e = base of the natural logarithms

LAI = leaf area index above depth z.

By using the energy flux at the soil surface (Qz) in equation (4) and setting Ea equal to zero, the evaporation from the soil during the constant rate stage can be determined.

In contrast, the falling rate stage of soil drying is determined by the hydraulic properties of the soil and by the frequency and intensity of precipitation events. Once the cumulative evaporation has reached a threshold value for the soil type, stage two drying begins. A threshold value of 9.0 mm for a loam soil was used in the model (Ritchie 1972). After the threshold value was reached evaporative loss of water from the soil depends on the conduction coefficient and on the time since a precipitation event greater than the cumulative falling rate stage evaporative loss. For a loam soil the falling rate stage evaporative loss can be estimated by (Ritchie 1972):

$$es_2 = 4.04(t^{\frac{1}{2}}) - [4.04(t-1)^{\frac{1}{2}}] \quad , \qquad (11)$$

where,

es_2 = falling rate stage of soil drying, mm/day

t = time since start of stage two drying, day

Daily climatic data were read into an array and then converted to an average value for the time step being used. Transpiration and AET values were accumulated for the thermal growing season. The start of the growing season occurred when the 15-day mean air temperature rose above 4.4°C and ended when the temperature fell below 4.4°C (Sims et al. 1978). Calculations of Tr were made for the base year (known ANPP) and for the year of interest (unknown ANPP). Estimates of the average relative humidity for the thermal growing season were made using vapor pressure data developed in estimating PET.

SITE DESCRIPTION

Four sites were chosen for testing the model. Each site was chosen to represent a different North American grassland type: shortgrass steppe, northern mixed prairie, bunchgrass steppe, and desert grassland. The shortgrass steppe was represented by data from the Central Plains Experimental Range (CPER) located in northeastern Colorado (40°49'N, 104°46'W). Dominant species are blue grama (*Bouteloua gracilis*), buffalo grass (*Buchloe dactyloides*), fringe sagewort

(Artemisia frigida) and plains prickly pear (Opuntia polyacantha) (Sims et al. 1978). Warm-season plants accounted for 40% of the annual ANPP from estimates of ANPP for the years 1971 through 1974 (Dodd and Lauenroth 1979). Average vegetation height is 6.1 cm, average ground cover is 11% and leaf area index ranged from 0.0006 in March to 0.49 in June (Knight 1971). Soils are Mollisols from the suborder Ustolls, with a fine sandy loam texture. Average annual precipitation is 310 mm with 75% received between April and September (Sims et al. 1978). Average air temperatures range from 0°C in December and January to 22°C in July (Parton et al. 1981). A surface albedo of 0.19 was measured (Nunn 1973); however, an assumed albedo of 0.20 was used (Sellers 1965).

Cottonwood, South Dakota (43°57'N, 101°52'W) represented the northern mixed prairie. The site is located at an elevation of 744 m above sea level. Annual temperature is 8.2°C, and mean annual precipitation is 387 mm; approximately 75% received between April and September (Sims et al. 1978). The major species are Agropyron smithii, Buchloe dactyloides, and Bouteloua gracilis. Average vegetation height is 24 cm (Nichols and Johnson 1969), and LAI was approximately one (Ripley and Redmann 1976). Soils are from the order Mollisols, with a silty clay loam texture (Sims et al. 1978).

The bunchgrass steppe was represented by the Arid Lands Ecological Reserve (ALE) in Washington (46°24'N, 119°33'W). The site is at an elevation of 390 m. Mean annual temperature is 12.8°C. Average annual precipitation is 183 mm and the largest portion is received between October and March. Vegetation at the site is dominated by cool-season plants. The bunchgrasses Agropyron spicatum and Poa secunda along with the shrub Artemisia tridentata dominate the vegetation (Sims et al. 1978). Average vegetation height is approximately 36 cm (Blaisdell 1958), and LAI was assumed to be approximately 0.9. Soils are of the order Mollisols with a silt loam texture (Sims et al. 1978).

The final site was from Jornada, New Mexico (32°36'N, 106°51'W), representing a desert grassland type. The site is at an elevation of 1350 m. Mean annual temperature is 15.1°C; average annual precipitation was 225 mm (Sims et al. 1978), 52% of the precipitation comes between July and September (Buffington and Herbel 1965). Vegetation at the site is dominated by warm-season plants. Bouteloua eriopoda and Buchloe dactyloides are the most prevalent (Sims et al. 1978). Average vegetation height for Bouteloua eriopoda is 35 cm (Nelson 1934, Woodmansee and Potter 1971) and LAI was assumed to be similar to that of the CPER 0.5. Soils are Aridisols with a loamy fine sand texture (Sims et al. 1978).

Because of lack of information for specific years, it was necessary to assume a constant LAI and average vegetation height for all years at a site.

Climate data for each validation site was obtained from National Oceanic and Atmospheric Administration (NOAA) weather stations that contained the necessary data and was most representative of the site. For Cottonwood, South Dakota temperature and precipitation data came from the Cottonwood 2E station (NOAA 1970, 1971, 1972f). Wet and dry bulb temperatures, percent of possible sunshine, and daily wind travel came from Rapid City Airport (NOAA 1970, 1971, 1972b). For ALE, Washington temperature, precipitation, wet and dry bulb temperatures, and daily wind travel were from Yakima Airport (NOAA 1971, 1972d) with percent of possible sunshine from Walla Walla Airport (NOAA 1971, 1972c). For Jornada, New Mexico temperature and precipitation data were from the Jornada Experimental Range (NOAA 1970, 1971, 1972e) with wet and dry bulb temperatures, wind, and percent of possible sunshine from Albuquerque Airport (NOAA 1970, 1971, 1972a). Climate data for the CPER was collected as part of the U.S. IBP Grassland Biome Project with percent of possible sunshine from Cheyenne, Wyoming.

RESULTS

Estimates of AET were compared with evapotranspiration measurements from a weighing lysimeter at the CPER. Estimated AET values were examined for the years 1972, 1973, and 1974. Problems with the recording system for the lysimeter produced gaps in the measured data sets, reducing the number of comparisons that could be made. Some missing lysimeter values were replaced with the use of regression analysis of lysimeter values on calculated PET for that year. Each value replaced was compared with surrounding values to ensure that replaced values were reasonable. No more than four consecutive values were replaced.

Mean daily AET values were calculated for a 10-day time step for model estimates and for lysimeter measurements, for the entire thermal growing season for each of the 3 years. A student's t-test (table 1) was used to compare means, and a regression of lysimeter values on model estimates was made (fig. 2).

Estimates of ANPP were tested on single years rather than on long-term average ANPP, since the hypothesis being tested was identical, and more data points were generated using single-year estimates. The percentage error of estimation for ANPP was calculated for each site and year and the mean absolute error was calculated for each method of estimating ANPP (table 2).

To evaluate the performance of the complementary hypothesis five climate adjustment factors were compared. One of these used the most widely applied weather variable, precipitation while the remaining four used information resulting from the complementary hypothesis either Tr or AET.

CAF1 ratio of total precipitation for the 2 years

Table 1.--Statistical analysis of AET estimates from model version three compared with lysimeter measurements from the Central Plains Experimental Range for the years 1972, 1973, and 1974 combined

Time step (days)	10
Sample size	26
Lysimeter mean (mm)	1.37
Lysimeter standard deviation	0.70
Model mean (mm)	1.60
Model standard deviation	1.02
t-test[1]	-0.97
P value of t-test	0.34
Regression slope[2]	0.52
Regression intercept	0.53
Correlation coefficient	0.75

[1]Ho: lysimeter = model; ha: lysimeter ≠ model.
[2]Regression of lysimeter on model.

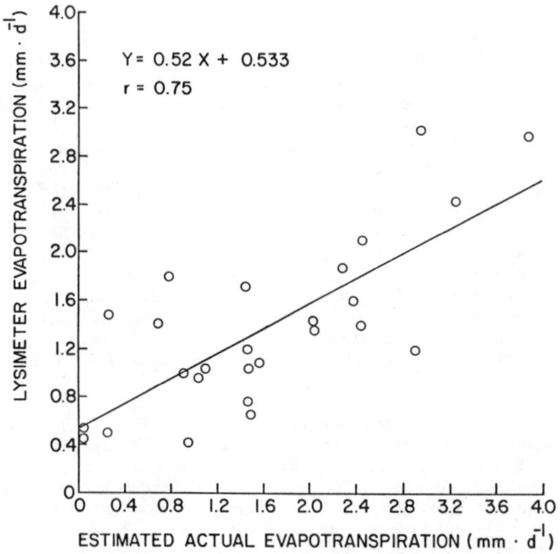

Figure 2.--Scatter diagram and regression line of lysimeter evapotranspiration on estimated actual evapotranspiration for model version three, 10-day time step.

CAF2 ratio of total transpiration estimates for the two years unadjusted for the relative humidity deficit

CAF3 ratio of total AET estimates for the 2 years unadjusted for the relative humidity deficit

CAF4 ratio of total transpiration estimates for the 2 years adjusted for the relative humidity deficit

CAF5 ratio of the total AET estimates for the 2 years adjusted for the relative humidity deficit.

Model estimates of ANPP were compared with measured estimates of ANPP (Sims and Singh 1978, Dodd and Lauenroth 1979) and no significant differences were found among any of the methods of estimating ANPP.

DISCUSSION

The model produced the best estimates of AET when run on a 10-day time step. The performance at this time step is likely related to the assumptions made in the development of the complementary hypothesis. Ben-Asher (1981) and LeDrew (1979) challenged the assumption of a negligible change in net radiation with a change in the regional moisture supply. LeDrew reported that for a daily time step the variation in net radiation with a change in moisture could account for 81% of the mean estimated AET. Changes in net radiation do occur with changes in the soil water content however, the magnitude of the radiation change depends on the presence of vegetation cover. For bare soil a change in the moisture content resulted in a change in the surface albedo and the longwave radiation flux

Table 2.--Estimated aboveground net primary production (ANPP) and percent error from measured ANPP for, ratio of annual precipitation (CAF1), ratio of unadjusted transpiration (CAF2), ratio of unadjusted actual evapotranspiration (CAF3), ratio of adjusted transpiration (CAF4), and ratio of adjusted actual evapotranspiration (CAF5)

			Method									
			CAF1		CAF2		CAF3		CAF4		CAF5	
Site	Year	Meas. ANPP gm^{-2}	Est. ANPP gm^{-2}	Error (%)	Est. ANPP gm^{-2}	Error (%)	Est. ANPP gm^{-2}	Error (%)	Est. ANPP gm^{-2}	Error (%)	Est. ANPP gm^{-2}	Error (%)
CPER	1971	107	144	35	152	42	158	48	169	58	176	64
	1972	168	137	-19	149	-11	156	-7	173	3	182	8
	1973	113[1]										
	1974	59	111	88	96	63	101	70	85	44	90	53
Cottonwood	1970	212[1]										
	1971	255	227	-11	241	-5	230	-10	252	-1	240	-6
	1972	279	204	-27	217	-22	215	-23	270	-3	268	-4
ALE	1971	82[1]										
	1972	114	77	-33	97	-15	108	-5	103	-10	115	1
Jornada	1970	134[1]										
	1971	125	177	41	128	2	133	6	125	0	130	4
	1972	186	290	56	105	-44	120	-36	108	-42	124	-33
Mean absolute error				39		26		26		20		22
Standard deviation				24		22		24		24		25

[1]Used as base year for production estimates.

consequently a change in the net radiation (Linacre 1969, Chacko et al. 1968, Idso et al. 1975). Once the surface is covered by vegetation the effect of soil moisture on net radiation becomes much less pronounced (Stanhill et al. 1968, Rijks 1968). As the vegetation becomes water stressed there may be an effect on the radiation balance as a result of changes in the color of the vegetation (Glover 1972).

The grassland types studied were expected to react in an intermediate manner between bare ground and full vegetation cover because of the small percentage cover of live vegetation and the low LAI. Changes in net radiation with changes in soil moisture at these sites will show large fluctuations over time periods as short as 1 day. By using longer time steps, the fluctuations in net radiation, because of precipitation events will be averaged out, and the assumption of minor fluctuations in net radiation will be met.

The assumption of a closed system with solar radiation the only external energy input into the system attempts to eliminate the energy input by local or regional advection. The establishment of a long fetch length, as prescribed by Morton (1965), effectively eliminates any local advection (Brakke et al. 1978), but does not reduce the regional or meso-scale advection (Brutsaert and Stricker 1979). This regional advection may account for up to 40% of the latent heat flux under certain conditions (Brakke et al. 1978). This advection is always present and may change sign, depending on the conditions present (McNaughton 1976). Use of time steps longer than 1 day will mitigate against not meeting the assumption of a closed system.

Estimates of ANPP were made from the model using a 10-day time step. Because of the small sample size and the large variability in estimates for the years 1971 and 1974 at CPER and 1972 at Jornada, no statistical difference could be detected among the methods of estimating ANPP. However, there was a trend of smaller mean absolute errors for the model estimates using CAF2-CAF5 than for the precipitation ratio (CAF1). There was a difference of 19 percent in the mean error from the best model estimate (CAF4) to the precipitation ratio. The best individual year and site estimate of 0% for CAF4, compared with 11% error for the best estimate by the precipitation ratio. The explanation for the large variation in the estimates from field data for the specific years sited resides in the difficulty of measuring biomass because of sampling error and non-normal distributions of biomass per quadrat.

Sensitivity analysis showed that the model estimates of AET were very sensitive to changes in vegetation height. A 20% increase in the vegetation height resulted in a 21% decrease in AET. The decrease in AET occurred because an increase in the vegetation height will increase the roughness length and increase the boundary layer (Oke 1978). An increase in the boundary layer will reduce the depth in the canopy to which turbulent mixing occurs and reduce the rate of evaporation from the soil surface (Rosenberg 1974). The model assumed a constant vegetation height at each site. If the vegetation height changed substantially from the base year, this could result in errors in the estimates.

No significant difference was found among ANPP estimates made with Tr (CAF2, CAF4) and estimates made with AET (CAF3, CAF5). This lack of a significant difference did not appear to be a result of the small sample size but was caused by problems in the partitioning subroutine used to separate evaporation from the soil and transpiration from the plants. The ratio of Tr to AET, for all sites and years, showed a mean ratio of 0.69, with a standard deviation of 0.047. The mean ratio was similar to that reported by Ritchie (1972); however, the small standard deviation indicated little change in the ratio from year to year and site to site. This lack of variation in the ratio was unreasonable, considering the high diversity of sites and conditions. The model assumed a constant LAI from year to year and throughout a year. An increase in the ANPP would cause an increase in the LAI and an increase in the proportion of water lost as Tr. The reverse happened in low production years (Ritchie and Burnett 1971, Al-Khafaf et al. 1978). By using a constant LAI at a site, a constant proportion of Tr to AET was assumed.

A final consideration in the lack of a significant difference in the estimates of ANPP using Tr and AET was found in the failure to consider the actual growing season. The model calculated AET for the thermal growing season and assumed that when AET was calculated, a percentage would be considered as transpiration. By accumulating Tr only when the plants were actively growing, the coupling between CO_2 assimilation and water loss would be tighter than for transpiration over the entire thermal growing season.

The adjustment of AET and Tr by the relative humidity deficit did not eppear to result in any advantage in the estimation of ANPP. No statistically significant difference were found between ANPP estimates made by CAF3 and CAF5 or CAF2 and CAF4. The adjustment for relative humidity was originally designed for the actual growing season and used relative humidity data from this period only (Arkley 1963). The model applied this adjustment to, and used data from, the thermal growing season. As in the case of estimates made by Tr, failure to consider the actual growing season may have reduced the effectiveness of the adjustment for the relative humidity deficit.

CONCLUSIONS

Research was undertaken to develop a method of adjusting aboveground net primary production data collected by the Bureau of Land Management in a single inventory process, to reflect long-term average ANPP based on long-term climatic conditions. The method developed used climatic data published by the NOAA and vegetation data collected by the inventory process or available in the literature. Estimates of ANPP should be within 20% of the actual production.

A simulation model was designed in which a linear relationship was hypothesized between ANPP and cumulative total AET or Tr. A proportional relationship was used to relate ANPP from the inventory process to the year of interest. A complementary hypothesis between regional AET and PET from a small moist surface within the region was used to estimate the AET for both years. Actual evapotranspiration was then partitioned between Tr from the plants and evaporation from the soil, using LAI and water-conducting capacity of the soil. Tests were conducted on the model to evaluate the applicability of the complementary hypothesis and the accuracy of the production estimates.

The model did a reasonable job of estimating AET from a region. A time step of approximately 10 days was needed to meet the assumptions made by the complementary hypothesis of a closed system and negligible changes in the net radiation with changes in the regional moisture condition.

Estimates of ANPP were within the 20% error limit set for the project. However, no significant difference in errors could be found among the five methods of estimating ANPP. Assumptions made about LAI and vegetation height and failure to consider the actual growing season rather than the thermal growing season contributed to the lack of a significant difference in ANPP estimation methods.

LITERATURE CITED

Al-Khafaf, S., P. J. Wierenga, and B. C. Williams. 1978. Evaporative flux from irrigated cotton as related to leaf area index, soil water, and evaporative demand. Agronomy Journal 70:912-917.

Arkley, R. J. 1963. Relationships between plant growth and transpiration. Hilgardia 34:559-584.

Ben-Asher, J. 1981. Estimating evapotranspiration from the Sonita Creek Watershed near Patagonia, Arizona. Water Resources Research 17:901-906.

Blaisdell, J. P. 1958. Seasonal development and yield of the native plants on the upper Snake River plains and their relation to certain climatic factors. U.S. Department of Agriculture Technical Bulletin 1190, 68 p. Washington, D.C.

Bouchet, R. J. 1963. Evapotranspiration réelle et potentielle, signification climatique. International Association of Science, Hydrology, General Assembly of Berkeley Publication 62, p. 134-142. De L'Association Internationale O'Hydrologie Scientifique, Gentbrugge, Belgium.

Brakke, T. W., S. B. Verma, and N. J. Rosenberg. 1978. Local and regional components of sensible heat advection. Journal of Applied Meteorology 17:955-963.

Brutsaert, W., and H. Stricker. 1979. An advection-aridity approach to estimate actual regional evapotranspiration. Water Resources Research 15:443-450.

Buffington, L C., and C. H. Herbel. 1965. Vegetation changes on a semidesert grassland range from 1858 to 1963. Ecological Mongraphs 35:139-164.

Chacko, O., C. T. Thomas, and A. Mani. 1968. Surface radiation balance measurements in India during the IQSY. Indian Journal of Meterology and Geophysics 19:93-98.

De Bruin, H. A. R., and J. Q. Keijman. 1979. The Priestly-Taylor evaporation model applied to a large, shallow lake in the Netherlands. Journal of Applied Meteorology 18:898-903.

De Wit, C. T. 1958. Trnaspiration and crop yield. Versl. Landbowk. Onderz. No. 64.6. 88 p. Institute of Biological and Chemical Research on Field Crops and Herbage, Wageningen, Netherlands.

Dodd, J. L., and W. K. Lauenroth. 1979. Analysis of the response of a grassland ecosystem to stress. p. 43-58. In Perspectives in grassland ecology, ecological studies, volume 2 (N. R. French, editor). Springer-Verlag, New York.

Glover, J. 1972. Net radiation over tall and short grasses. Agricultural Meteorology 10:455-459

Idso, S. B., Aase, J. K., and Jackson, R. D. 1975. Net radiation-soil heat flux relations as influenced by soil water content variations. Boundary-layer Meterology 9:113-122.

Knight, D. H. 1971. Some measurements of vegetation structure on the Pawnee grassland. 1970. U.S. International Biological Program, Grassland Biome, Technical Report 72. 43 p. Colorado State University, Fort Collins, Colo.

Lauenroth, W. K. 1979. Grassland primary production: North American grasslands in perspective. p. 3-24. In Perspectives in grassland ecology, ecological studies, volume 2 (N. R. French, editor). Springer-Verlag, New York.

LeDrew, E. F. 1979. Adiagnostic examination of a complementary relationship between actual and potential evapotranspiration. Journal of Applied Meteorology 18:495-501.

Linacre, E. T. 1969. Net radiation to various surfaces. Journal of Applied Ecology 6:61-75.

Linacre, E. T. 1968. Estimating the net radiation flux. Agricultural Meteorology 5:49-63.

List, Robert J., editor. 1951. Smithsonian meteorological tables. 527 p. Smithsonian Institution Publication Number 4014, Washington, D.C.

McNaughton, K. G. 1976. Evaporation and advection. I. Evaporation from extensive homogeneous surfaces. Quarterly Journal of the Royal Meteorological Society 102:181-191.

Morton, F. I. 1965. Potential evaporation and river basin evaporation. Journal of the Hydraulics Division, Proceedings of the American Society of Civil Engineers 91, HY6:67-97.

Morton, F. I. 1969. Potential evaporation as a manifestation of regional evaporation. Water Resources Research 5:1244-1255.

Nelson, E. W. 1934. The influence of precipitation and grazing upon black grama grass range. U.S. Department of Agriculture Technical Bulletin 409. 32 p.

Nichols, J. T., and Johnson, J. R. 1969. Range productivity as influenced by biennial sweetclover in western South Dakota. Journal of Range Management 22:342-347.

NOAA. 1970, 1971, 1972a. Local climatological data, monthly summaries. National Oceanic and Atmospheric Administration, Albuquerque, New Mexico.

NOAA. 1970, 1971, 1972b. Local climatological data, monthly summaries. National Oceanic and Atmospheric Administration, Rapid City, South Dakota.

NOAA. 1971, 1972c. Local climatological data, monthly summaries. National Oceanic and Atmospheric Administration, Walla Walla, Washington.

NOAA. 1971, 1972d. Local climatological data, monthly summaries. National Oceanic and Atmospheric Administration, Yakima, Washington.

NOAA. 1970, 1971, 1972e. Climatological report, New Mexico. National Oceanic and Atmospheric Administration, Albuquerque, New Mexico.

NOAA. 1970, 1971, 1972f. Climatological report, South Dakota. National Oceanic and Atmospheric Administration, Rapid City, South Dakota.

Noy-Meir, I. 1973. Desert ecosystems: Environment and producers. p. 25-51. In Annual review of ecology and systematics (Johnston, R. F., Frank, P. W, and Michener, C. D., editors). Annual Reviews, Inc., Palo Alto, California.

Nunn, J. R. 1973. Meteorological characteristics over the Grassland Biome Pawnee site. 152 p. U.S. International Biological Program, Grassland Biome, Technical Report 216.

Oke, T. R. 1978. Boundary layer climate. 372 p. Methuen and Company, Ltd., London.

Parton, W. J., Lauenroth, W. K., and Smith, F. M. 1981. Water loss from a shortgrass steppe. Agricultural Meteorology 24:97-109.

Penman, H. L. 1948. Natural evaporation from open water, bare soil and grass. Proceedings of the Royal Society of London, Series A 193:120-145.

Priestley, C. H. B., and Taylor, R. J. 1972. On the assessment of surface heat flux and evaporation using large-scale parameters. Monthly Weather Review 100:81-92.

Rijks, D. A. 1968. Water use by irrigated cotton in Sudan: II. Net radiation and soil heat flux. Journal of Applied Ecology 5:685-706.

Ripley, E. A., and Redmann, R. E. 1976. Grassland. p. 349-398. In Vegetation and the atmosphere, volume 2, case studies (Monteith, J. L., editor). Academic Press, London.

Ritchie, J. T. 1972. Model for predicting evaporation from a row crop with incomplete cover. Water Resources Research 8:1204-1213.

Ritchie, J. T., and Burnett, E. 1971. Dryland evaporative flux in a subhumid climate. II. Plant influences. Agronomy Journal 63:56-62.

Rosenberg, N. J. 1969. Seasonal patterns in evapotranspiration by irrigated alfalfa in the central Great Plains. Agronomy Journal 61:879-886.

Rosenberg, N. J. 1974. Microclimate: The biological environment. 315 p. John Wiley & Sons, New York.

Rutherford, M. C. 1980. Annual plant production-precipitation relations in arid and semi-arid regions. South African Journal of Science 76:53-56.

Sellers, W. D. 1965. Physical climatology. 272 p. University of Chicago Press, Chicago, Illinois.

Sims, P. L., and Singh, J. S. 1978. The structure and function of ten western North American grasslands. III. Net primary production, turnover and efficiencies of energy capture and water use. Journal of Ecology 66:573-597.

Sims, P. L., Singh, J. S., and Lauenroth, W. K. 1978. The structure and function of ten western North American grasslands. I. Abiotic and vegetational characteristics. Journal of Ecology 66:251-285.

Stanhill, G., Cox, J. T. H., and Moreshet, S. 1968. The effect of crop and climate factors on the radiation balance of an irrigated maize crop. Journal of Applied Ecology 5:707-720.

Stone, L. R., and Horton, M. L. 1974. Estimating evapotranspiration using canopy temperatures: Field evaluation. Agronomy Journal 66:450-454.

Szeicz, G., Endrodi, G., and Tajchman, S. 1969. Aerodynamic and surface factors in evaporation. Water Resources Research 5:380-394.

Van Bavel, C. H. M. 1966. Potential evaporation: The combination concept and its experimental verification. Water Resources Research 2:455-467.

Woodmansee, R. G., and Potter, L. D. 1971. Natural reproduction of winterfat (*Eurotia lanata*) in New Mexico. Journal of Range Management 24:24-30.

SIMULATING SUCCESSION AS A RECLAMATION ALTERNATIVE[1]

Patrick Kangas[2]

Abstract.--A simulation model was constructed to investigate the use of natural succession for reclamation of strip-mined land from phosphate mining in central Florida. The model simulates ways of increasing the rate of succession to make it a more viable alternative. Succession was subsidized in the model by adding seeds and litter. Simulations showed that succession could be accelerated from 5% to 44% faster compared to the standard run with various combinations of additions. These results suggest that succession may be a viable reclamation alternative if it is properly managed.

INTRODUCTION

The need for reclamation to restore ecologically viable landscapes after surface or "strip" mining is being recognized by state governments who are developing laws to govern the process. Surface mining is an extreme stress to natural ecosystems since aboveground biomass is stripped away and soil is removed and piled in mounds. The resulting landscape is ecologically barren and aesthetically unpleasant relative to the pre-mining system. Reclamation restores the landscape by planting and managing semi-natural ecological systems on the mined surface. A problem with reclamation, however, is that no technique is completely satisfactory to all interest groups (mining industry, government and general public). Mechanical reclamation of leveling the land and planting crops is fast but expensive and the restored system is usually highly artificial. If the landscape is abandoned, succession will restore natural ecosystems for free but at a slow rate. A reclamation alternative which involves managing natural succession is investigated in this paper through the use of a computer simulation model. If succession could be accelerated, its value for reclamation would be greatly enhanced. Here two techniques for accelerating succession are studied: addition of seeds and addition of a litter layer. The model was constructed to apply to surface mining for phosphate in central Florida but the underlying principles should apply to any successional sequence and mining landscape.

SUCCESSION IN THE SOUTHEASTERN UNITED STATES

Field studies of abandoned mine lands of various ages (Kangas, 1979) have demonstrated that succession in the phosphate regions of central Florida (fig. 1) is similar to succession on fallow agricultural fields throughout the southeastern United States. The general sequence is from "old field" vegetation (annual and perennial grasses and forbs) to pines, climaxing in mixed hardwoods. Pines are favored by periodic fires as a fire-climax, but in the absence of fire, hardwoods such as oak, hickory and magnolia invade and dominate the forest. This succession sequence is well known with many studies from the late 1930's and early 1940's (Billings, 1938; Oosting, 1942; Kurz, 1944) to the present (Christensen and Peet, 1981). One difference between Florida phosphate succession and fallow field succession is the length of time to climax. Fallow field succession takes 150-200 years while phosphate succession takes approximately half as long.

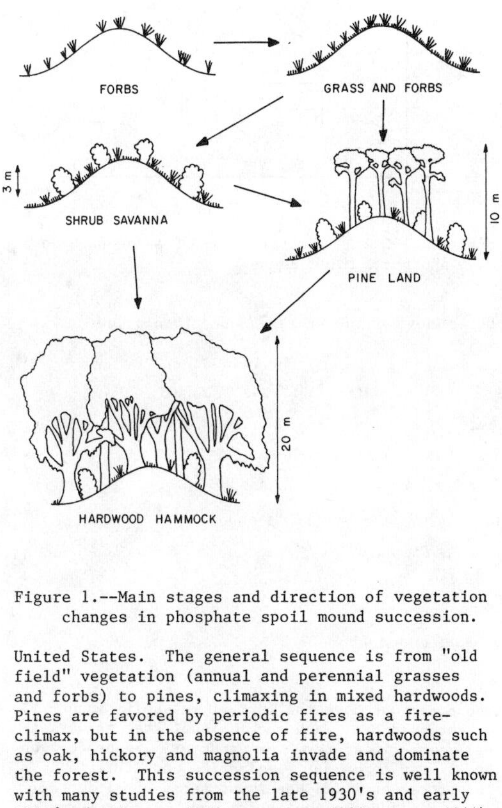

Figure 1.--Main stages and direction of vegetation changes in phosphate spoil mound succession.

[1]Paper presented at the Third International Conference on State-of-the-Art in Ecological Modelling, Colorado State University, May 24-28, 1982.

[2]Patrick Kangas is Assistant Professor of Biology, Eastern Michigan University, Ypsilanti, Michigan U.S.A.

SUCCESSIONAL MODEL

A simulation model of succession was constructed drawing from the extensive literature on southeastern U.S. ecosystems. Modelling was done using the energy circuit language. This is a symbolic language which has direct mathematical translations for each symbol and model configuration (Odum, 1972, 1982; Hall et al., 1977). Figure 2 shows the model along with the expressions for storages and pathways. Three stages of succession are shown with storages of biomass: Q3 is grasses, Q2 is pines and Q1 is hardwoods. Q4 is litter storage which receives input from all stages of succession.

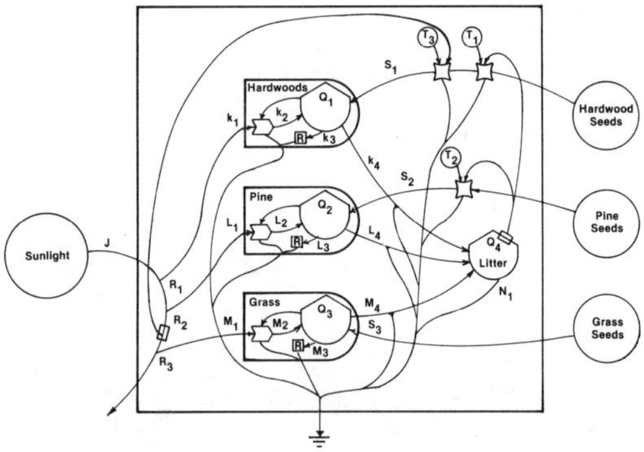

Figure 2.--Energy circuit model of succession on phosphate mined lands.

Equations for the storages are given below:

$$\dot{Q}_1 = K_2Q_1R_1 - K_3Q_1 - K_4Q_1 \ (+S_1 \text{ IF } R_2 < T_3 \text{ AND } Q_4 > T_1) \qquad (1)$$

$$\dot{Q}_2 = L_2Q_2R_2 - L_3Q_2 - L_4Q_2 \ (+S_2 \text{ IF } Q_4 > T_2) \qquad (2)$$

$$\dot{Q}_3 = M_2Q_3R_3 - M_3Q_3 - M_4Q_3 + S_3 \qquad (3)$$

$$\dot{Q}_4 = K_4Q_1 + L_4Q_2 + M_4Q_3 - N_1Q_4 \qquad (4)$$

$$R_1 = J/(1 + K_1Q_1) \qquad (5)$$

$$R_2 = R_1/(1 + L_1Q_4) \qquad (6)$$

$$R_3 = R_2/(1 + M_1Q_3) \qquad (7)$$

The equations have the form of metabolic input-output balances. Each successional stage equation has a non-linear term for gross primary production and linear terms for respiration and litterfall. The litter equation has input terms of litterfall from each successional stage and a linear term for respiration or decomposition. Values and derivations of the coefficients are given in Kangas (1982).

The main driving energy source for the model was sunlight which was apportioned according to the height of each successional stage. Hardwoods received sunlight first since they grow to be the tallest. Pines were next to use sunlight followed by the grasses which received the remainder from the pines. This way of simulating shading effects incorporates the limiting availability of sunlight as represented in the equations for R, or the remaining light.

Initially storages were started at zero and each received seed inputs from outside sources. Germination of seed inputs to pines and hardwoods was controlled by threshold switches of the litter storage. These switching controls account for changing microclimates of the forest floor which allow germination of seeds. An additional threshold of light level was also tested as control on hardwood germination. The model was run with and without this switch for comparison when light was potentially limiting to germination.

The model was calibrated with Georgia and north Florida literature data. Timing of the stages was adjusted to phosphate succession, which is faster than fallow field succession used for calibration. A one quarter year time interval was used for simulating succession though smaller time steps were investigated in runs simulating litter data for validation. The model was simulated on an Intecolor 8000 desk top computer using BASIC.

SIMULATION OF SUCCESSION

The data in table 1 provide validation of the model by comparing predicted litter biomass, run at different time integration intervals, with litter biomass from field sampling of phosphate mine ecosystems of different ages. With time steps of 0.25 years and 0.1 years only the predicted 5 and 50 year litter values fell within the standard error limits of observed litter. With a time step of 0.025 years the predicted 20 year litter also fit the data. Best fit of predicted values with those observed was found with a time interval of 0.01 year for which all predictions except the 10 year value fell within acceptable variation. Due to the length of simulation runs (70 years), however, these changes in precision did not greatly effect the form of the simulation graphs. Therefore, simulations were plotted with a time step of 0.25 years since this ran much faster on the microcomputer.

Table 1.--Comparison of litter values from simulation runs with varying time steps (DT), with litter data from field sampling for validation.

T(years)	Q4(DT=0.25)	Q4(DT=0.1)	Q4(DT=0.025)	Q4(DT=0.01)	Measured litter
1	4.0	12.4	36.6	61.8	54.0±11.8
5	309.0	317.5	340.0	375.2	377.2±61.6
10	341.0	342.3	369.7	432.8	622.8±71.5
20	524.5	557.2	604.3	668.9	688.5±96.7
50	1196.8	1223.13	1257.4	1300.5	1250.4±73.1

The standard simulation run of succession is given in Figure 3a. Grasses peaked first after about 5 years. Pines dominated next from about 10 to 40 years. Climax hardwoods supplanted pines and reached steady state at about 60 years. Litter gradually accumulated to a steady state after about 50 years. This general pattern of vegetation change fits the field observations mentioned earlier.

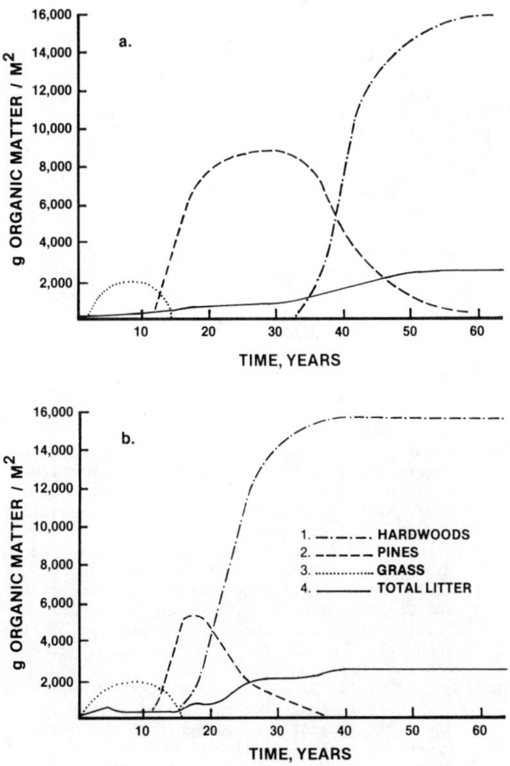

Figure 3.--Comparison of simulations of the succession model. a. standard run, b. result of increasing the seeding rate 1000 times and adding litter.

Two principal types of simulations were performed with the model: increasing seeds as a pulse after germination thresholds were reached and adding litter to improve germination success. Results of the simulations are summarized in Table 2. Performance was judged by the age at which hardwood steady state was reached. All simulations had the same general pattern of growth and transition of vegetation as the standard or control run, but they were accelerated to varying degrees.

Addition of seeds accelerated succession from 5% to 25%, with the best performance from increasing seeding rates for each successional stage 1000 times.

Table 2.--Comparison of performance of simulation experiments in accelerating succession.

Simulation description	Age of climax (years)	Percent acceleration
Standard run	59	---
Pulse grass seed by 1000	56	5
Pulse pine seed by 1000	54	9
Pulse hardwood seed by 1000	52	12
Pulse pine and hardwood seeds by 1000	47	20
Pulse all seeds by		
1000	44	25
100	51	14
10	56	5
Add litter w/light control	59	0
Add litter w/o light control	41	31
Pulse all seeds by 1000 and add litter	33	44

Performance improved with the maturity of the stage in which seeds were added and with the total amount of seeds added.

The effects of adding litter with and without the light threshold on hardwood germination were quite different from the effects of adding seeds. With the light control no addition of litter caused significant change in succession, since hardwood germination was being limited by light, not litter. Tests run on the model without the light control showed that the optimal litter addition was 300 g/m^2 on year 5 of succession. This strategy allowed the fast growing grass stage to develop its peak litter biomass to which the minimum amount of litter was added to initiate germination of hardwoods. This experiment shortened the pine stage and decreased its peak biomass, perhaps since its role in succession was taken up in part by the litter addition. Hardwood steady state was reached in year 41, which is 3 years faster than in the best seeding simulation and 18 years faster than the control simulation.

A final simulation combined the addition of seeds and litter without the light control on hardwood germination to view the maximum amount that succession may be able to be accelerated. This run is shown in Figure 3b. The general pattern of succession was similar to that found with the addition of litter. However steady state was reached by year 33 which was 44% faster than in the standard or control simulation.

SUMMARY AND DISCUSSION

The problem with natural succession as a reclamation alternative is that it is slow. Techniques for speeding up succession rates would make the alternative more attractive. Subsidizing processes of ecosystem development which regulate its rate may speed up or "abbreviate" succession. As noted by

Odum (1962) "where succession involves accumulation or consumption of quantities between starting and final climax, the successional stages may be abbreviated by artificial addition or removal of that quantity."

A simulation model of succession was constructed to test certain strategies for the use of succession as a reclamation alternative for phosphate mined lands. The model was designed as a management tool to assess the degree to which succession may be able to be accelerated as evaluated by time to climax. The role of shading and germination controls on the rate and transitions of succession were included. Other important features such as soil moisture, seed storages and mortalities and aspects of mineral cycling could be added to make the model more complete, though the additional complexity might not necessarily add to the utility of the model. Even though highly aggregated, the mini-model of succession utilized here was validated with comparison of predicted and observed litter values.

The exercise of constructing the model was itself useful in pointing out deficiencies in available data on ecosystem development, such as rates of seeding in natural communities and control functions on germination of dominant species.

Two kinds of simulations were studied, seeding and addition of litter. Perhaps the most important aspect of succession is input of seeds and animal wanderings in initiating and maintaining succession. Without seeds or animal immigration no ecosystem can develop. Furthermore, the continual "rain" of seed input may be important in development and successional transitions. Johnson et al., (1981) point out that most models of succession do not give much detail to seed input functions and, consequently, they construct a seed dispersal model with emphasis on spatial dynamics. Sell (1974) and Gunderson (1976) describe the effects of seeding deficiencies in other successional sequences. The advantage of a nearby seed source in apparently allowing succession to proceed faster than usual was observed several times in field studies of mined lands. Model results support this observation in which increased seeding did accelerate succession.

The accretion of litter layer across the soil surface of an ecosystem is another important factor in succession. As Olson (1963) states concerning succession, "some biological developments may be specifically delayed until litter and soil humus have approached near their steady state values." Perino and Risser (1972) provide an example of litter deficits in terms of biomass in early succession. Litter additions were simulated under two model structures, with and without a light threshold control on the germination of hardwoods. With light control no advantage to litter addition was found, but without this limiting action succession was accelerated by 30%. This great difference in predictions suggests the model is sensitive to germination control and thus requires further attention. One problem in the use of litter additions is finding an inexpensive analog of true forest litter. Litter is a sensitive substance which both insulates and aerates the soil, has vertical stratification of characteristics and has a long development time (in the standard simulation run litter took 50 years to reach steady state). Wood chips, hay and other kinds of mulch may be poor substitutes for natural litter.

The simulations included here suggest that natural succession can be accelerated through management and thus may be a viable reclamation alternative. An important point to be made is that with this approach natural processes do the work of reclamation with human actions only as an amplifying input at critical or sensitive times. The succession sequence is kept intact, but each stage is shortened so that climax conditions are reached faster. This is consistent with the old theory that each stage of succession is needed to "prepare" the way for the next stage in the sense of improving soil characteristics and bringing microclimate under biotic regulation. Reclamation alternatives which attempt to jump or skip stages may have limited success since site conditions may not be developed enough for growth of advanced stages.

LITERATURE CITED

Billings, W. D. 1938. The structure and development of old field shortleaf pine stands and certain associated physical properties of the soil. Ecol. Monogr. 8:437-499.

Christensen, N. L. and R. K. Peet. 1981. Secondary forest succession on the North Carolina Piedmont. p. 230-245. In Forest Succession, D. C. West, H. H. Shugart and D. B. Botkin (eds.). Springer-Verlag, New York.

Gunderson, L. 1976. Accelerating secondary succession in cypress strands. p. 738-762. In Cypress Wetlands for Water Management, Recycling and Conservation, H. T. Odum and K. C. Ewel (eds.). Third Ann. Rept. to NSF and Rockefeller Foundation, Center for Wetlands, Univ. of Florida, Gainesville.

Hall, C. A. S., J. W. Day, Jr. and H. T. Odum. 1977. A circuit language for energy and matter. p. 38-48. In Ecosystem Modeling in Theory and Practice, C. A. S. Hall and J. W. Day, Jr. (eds.). John Wiley and Sons, New York.

Johnson, W. C., D. M. Sharpe, D. L. De Angelis, D. E. Fields and R. J. Olson. 1981. Modeling seed dispersal and forest island dynamics. p. 215-139. In Forest Island Dynamics in Man-Dominated Landscapes, R. L. Burgess and D. M. Sharpe (eds.). Springer-Verlag, New York.

Kangas, P. 1979. Energy analysis of ecological succession as a reclamation alternative for phosphate mining. p. 402-429. In Energy Basis for the United States, H. T. Odum, J. Alexander, F. Wang, et al., Rept. to Dept. of Energy, Contract EY-76-S-05-4398. Systems Ecology and Energy Analysis Group, Univ. of Florida, Gainesville.

_________. 1982. Energy analysis of Landforms, Succession, and Reclamation. PhD Dissertation, Univ. of Florida, Gainesville.

Kurz, H. 1944. Secondary forest succession in the Tallahassee Red Hills. Proc. Florida Acad. Sci. 7:1-100.

Odum, H. T. 1962. Man and the ecosystem. p. 57-75. In Proc. of the Lockwood Conf. on the Suburban Forest and Ecology, P. E. Waggoner and J. D. Ovington (eds.). Conn. Agric. Exper. Station Bull. 652, New Haven, Conn.

Odum, H. T. 1972. An energy circuit language for ecological and social systems: its physical basis. p. 139-211. In Systems Analysis and Simulation in Ecology. Vol. II, B. C. Patten (ed.). Academic Press, New York.

__________. 1982. Systems, an introduction to general systems of energy, economics and environment. John Wiley and Sons, New York. in press.

Olson, J. S. 1963. Energy storage and the balance of producers and decomposers in ecological systems. Ecology 44:322-332.

Oosting, H. J. 1942. An ecological analysis of the plant communities of Piedmont, North Carolina. Amer. Midl. Natur. 28:1-126.

Perino, J. V. and P. G. Risser. 1972. Some aspects of structure and function in Oklahoma old-field succession. Bull. Torrey Bot. Club 99:233-239.

Sell, M. G., Jr. 1974. Model of managrove productivity, herbicide spraying, wood cutting and seeding availability in the Rung Sat zone of South Vietnam. p. 33-83. In Models of Herbicide, Mangroves, and War in Vietnam. National Academy of Science, National Research Council, Working Papers, Part B, Wash., D. C.

A SIMULATION MODEL OF BLUE GRAMA BIOMASS DYNAMICS, WITH SPECIAL ATTENTION TO TRANSLOCATION MECHANISMS[1]

D. Bachelet, H. W. Hunt, J. K. Detling, and D. W. Hilbert[2]

Abstract.--A simple model was created to evaluate and simulate the growth of blue grama plants in hydroponic culture. Two submodels represent the dynamics of carbon (C) and nitrogen (N) taken up from unlimited supply sources and distributed between shoot and root plus crown. Each is divided into structural material characterized by a fixed C:N ratio and metabolic and storage material. It is assumed that root and shoot growth, in a constant environment, are directly related to the C:N ratios of plant tissue, light intensity reduced by self shading being the only external factor limiting plant growth. The structure of the model agrees with the "source-sink" hypothesis in which root growth depends on available assimilates and shoot growth on N supply. The analysis of the system response to stresses such as defoliation and presence of rhizosphere bacteria leads to consideration of controversial subjects such as the relative importance of shoot control versus root control over N uptake, Humphries' (1963) "size of sink" hypothesis versus the competitive theory, and quantitative evaluation of root exudation rate in water culture. The limitations of the present formulation are considered and possible modifications to include morphological and phenological properties and environmental variables are discussed.

INTRODUCTION

What is important in a model is not that it agrees with the experiment but, on the contrary, what it says on the way processes occur, what it describes about their underlying mechanisms. For, given any empirical structure, it is always possible to exhibit a quantitative model with enough arbitrary parameters, which account for the experiment at a given approximation. If one wants a good model, one has to eliminate to a maximum those arbitrary parameters (Thom 1975, translation).

Sharing Thom's opinion about modeling, we built a simple model to simulate growth of blue grama (Bouteloua gracilis (H.B.K.) Griffiths), the dominant grass of the North American shortgrass prairie. The model was used to evaluate data from experiments on hydroponic cultures in controlled environment growth chambers. We tried to eliminate unnecessary details and to retain only factors essential to simulate growth. Since temperature and water status were constant, we focused our attention on how light intensity reduced by self-shading affects plant growth. Furthermore, the only internal factor affecting plant physiology which we considered was the nitrogen content of the various parts (Baldwin 1976). We developed a series of models representing different conceptions of plant physiology and compared their ability to account for laboratory results. The models were tested by comparing their response to shoot clipping with data from laboratory experiments.

MATERIALS AND METHODS

Laboratory Experiments

Twenty-day-old blue grama seedlings were grown in test tubes containing 1/7 strength Hoagland's solution according to the methods of Detling et al. (1980). They were set in growth chambers with 14 h photoperiod, 33°/18°C day/night temperatures, and a quantum flux density of 500 µE m^{-2} s^{-1}. Between the age of 20 and 60 days, plants were watered with the same nutrient solution, and periodically some individuals (5 to 10 plants) were analyzed. Plants were dried at 60°C for 48 hours and weighed, and their nitrogen content measured by microKjeldahl procedure (Steyermark 1961). A second experiment was conducted under the same environmental conditions, but on the 20th day plants were defoliated by clipping 0-10, 10-30, 30-60, or 60-100% of the biomass of all tillers. At regular time intervals, plant analyses were performed. In a third experiment, also with the same set of conditions, roots were isolated during growth to measure root respiration.

[1]Paper presented at the Third International Conference on State-of-the-Art in Ecological Modelling, Colorado State University, May 24-28, 1982.

[2]Dominique Bachelet, Graduate Research Assistant; H. William Hunt, Research Ecologist; James K. Detling, Associate Director, Natural Resource Ecology Laboratory; and Dave Hilbert, Graduate Research Assistant, Colorado State University, Fort Collins, CO 80523.

Optimization Method

When a model fails to mimic a system, it can be due to the lack of some key variables and processes or to a bad choice of parameter values that prevent the model from behaving realistically. To be certain we were comparing model structures and not merely sets of parameter values, an optimization routine (Powell 1965) was used. The routine calculates the difference between each data point and the corresponding model value and squares it. It then adds up all the squared differences. Given a set of parameters, it modifies their values to determine the smallest sum of squares corresponding to the best possible fit. The parameters varied were maximum photosynthetic rate, maximum N uptake rate, maximum C and N translocation rates, and C:N ratio threshold values used to define reduction factors (see below). The values assigned to those parameters by the optimization routine are in Table 1. Optimization was only used to estimate model parameters for normal growth. Thus the ability of the model to correctly predict response to clipping is a proper validation test.

GENERAL STRUCTURE OF THE MODEL

Our basic grass model BASGRAS (figure 1) divided grass into shoots and roots plus crowns (stem bases) which will be referred to as roots in the rest of the paper. Both shoots and roots were divided into structural material which cannot be translocated and metabolic and storage material considered a mobile pool inside the plant. Structural material included cell wall constituents such as cellulose, hemicelluloses, lignin, and cutin, and was assumed to have a constant C:N ratio of 150 (McGill et al. 1981). Material in

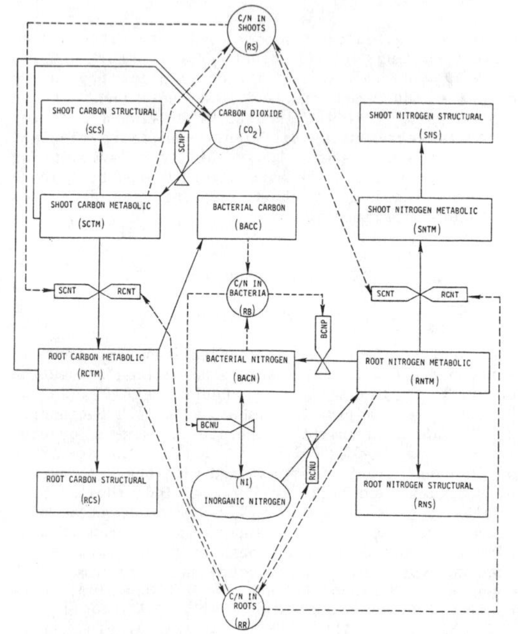

Figure 1.--Flow diagram of the plant growth model BASGRAS. (——) C and N flows; (-----) information flows.

conducting vessels and vacuoles or otherwise not actively participating in metabolism--amides, starch, fructosans, nitrogen storage compounds, etc.--were pooled with substances involved in synthesis and metabolism, such as most plant proteins, amino acids, and nucleic acid. This category will be referred to as the metabolic and storage pool or, for brevity, the metabolic pool. We assumed this second constituent had a variable C:N ratio.

The next section describes the physiological processes modeled and in so doing introduces three model variants. BASGRAS1 constitutes our basic model and starting point.

BASGRAS1--BASIC MODEL

1. Photosynthesis

The decline in irradiance in a stand of plants depends chiefly on the density of the foliage and the arrangement of the leaves (Larcher 1975). Foliage density can be expressed quantitatively by the leaf area index (LAI), ratio of total leaf area to ground area. In our data LAI ranged from 0.272 to 4.289.

When radiation passes through successive layers of leaves, its intensity decreases almost exponentially with increasing LAI in accordance with the Beer Lambert extinction law. If the layer of foliage is homogeneous, the intensity can be computed from Monsi and Saeki's (1953) modification of the extinction equation:

$$I = I_0 \, e^{-k \cdot LAI} \qquad (1)$$

where,

- I = quantum flux density at a certain distance from the top of the canopy ($\mu E\ m^{-2}\ s^{-1}$)
- I_0 = quantum flux density incident on the top of the canopy ($\mu E\ m^{-2}\ s^{-1}$)
- k = extinction coefficient for this particular plant community [between 0.3 and 0.5 for grasses according to Lemée (1978)] (unitless)

To fit our data on light response of blue grama blade and sheath, the following relation worked well:

$$P(I) = a - \frac{c}{I - b} \qquad (2)$$

where,

- a = constant (98.593 mg C $dm^{-2}\ h^{-1}$)
- b = constant (64.707 $\mu E\ m^{-2}\ s^{-1}$)
- c = constant (4734.229 mg C $\mu E\ h^{-2}\ m^{-4}$)
- I = radiation at a certain distance from the top of the canopy ($\mu E\ m^{-2}\ s^{-1}$)
- P = net photosynthetic rate (mg C $dm^{-2}\ h^{-1}$)

A similar curve was described by Patefield and Austin (1971) from an original idea by Monteith (1965). This curve was integrated over the canopy to obtain a mean value for the photosynthetic rate, later expressed per unit shoot dry weight:

$$WSP = \frac{d}{LAI} *[LAI(a - \frac{c}{b}) + \frac{c}{bk} *[\log (-b + I_0 e^{-k \cdot LAI}) - \log (-b + I_0)]] \quad (3)$$

where,

WSP = whole shoot photosynthetic rate (mg C (mg shoot C)$^{-1}$ d^{-1})

a, b, c = as defined previously

d = dm^2 (mg shoot C)$^{-1}$

LAI = Leaf area index greater or equal to the light compensation ($\frac{dm^2}{dm^2}$)

I_0 = quantum flux density of the radiation on top of the canopy (μE m^{-2} s^{-1})

k = extinction coefficient = 0.5 (unitless)

log = $\log_{10}$

Photosynthetic rates are correlated with the nitrogen concentration of the leaves (Novoa and Loomis 1981). SCNP, a function of shoot C:N ratio, reduces the rate of CO_2 assimilation when shoot nitrogen content becomes less than a given value (adjusted with the optimization routine to 2.5% of shoot dry weight).

$$P = SCNP * WSP * TOTSC \quad (4)$$

where,

P = rate of photosynthesis (mg C plant^{-1} d^{-1})

SCNP = reduction factor described in figure 2a

WSP = whole shoot photosynthetic rate [mg C (mg shoot C)$^{-1}$ d^{-1}]

TOTSC = whole shoot carbon content (mg C plant^{-1})

2. Respiration

Daytime respiration in the leaves was accounted for since we have fit the model to data on net CO_2 exchange. Nighttime respiration was assumed to be a constant fraction (15%) of daytime photosynthesis by the plant (Lambers et al. 1979, de Wit et al. 1978), since temperature remained constant and cooler at night during the experiment.

Belowground C losses as respiration were not modeled in BASGRAS1. Photosynthesis and translocation rates were lower to account for reduced C demand by roots.

3. Nitrogen Uptake

Nitrogen uptake was assumed to be proportional to the size of the metabolic and storage pool of root material. Some authors relate nitrogen uptake to shoot activity and shoot nitrogen content (Edwards and Barber 1976), while others relate it to root metabolism (Radin 1977, Thornley 1972, Raper et al. 1978). We created two reduction factors--RCNU, a function of root C:N ratio, and SCNU, a function of shoot C:N ratio--to address the hypotheses that either root nitrogen content or shoot nitrogen content controls the uptake. Thus N uptake is calculated by one of two expressions (equations assigned decimal numbers are alternative formulations for the same process):

$$U = U_{max} * RCNU * RCTM \quad (5.1)$$

or

$$U = U_{max} * SCNU * RCTM \quad (5.2)$$

where,

U = uptake rate of nitrogen (mg N plant^{-1} d^{-1})

U_{max} = maximum uptake rate of nitrogen (mg N (mg root metabolic C)$^{-1}$ d^{-1})

RCNU, SCNU = reduction factors, functions of root C:N ratio, and shoot C:N ratio, respectively, described in figure 2b and c

RCTM = metabolic C in the root (mg C plant^{-1})

Cram (1976) promoted the concept of negative feedback regulation of ion uptake. RCNU and SCNU may be interpreted as feedback parameters reducing N uptake when the nitrogen concentration of the root or shoot exceeds the maxima defined in table 1.

4. Translocation

According to Wardlaw and Moncur (1976), assimilate partitioning is regulated by both assimilate supply and demand. We extended the concept to nitrogen partitioning and defined a flow of translocated material (C and N) proportional to the source size but

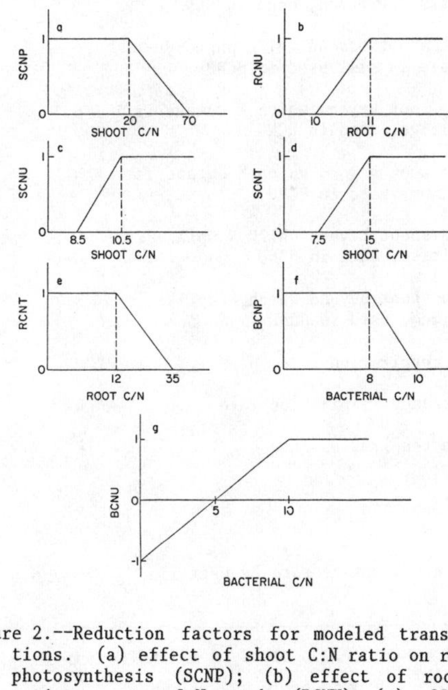

Figure 2.--Reduction factors for modeled transformations. (a) effect of shoot C:N ratio on rate of photosynthesis (SCNP); (b) effect of root C:N ratio on rate of N uptake (RCNU); (c) effect of shoot C:N ratio on rate of N uptake (SCNU); (d) effect of shoot C:N ratio on rate of translocation (SCNT); (e) effect of root C:N ratio on rate of translocation (RCNT); (f) effect of bacterial C:N ratio on exudation (BCNP); (g) effect of bacterial C:N ratio on inorganic N uptake by bacteria (BCNU).

Table 1.--Optimized model parameter values. Values in parenthesis are N percent dry weight in the plant assuming 50% C in the plants.

Definition	Equations	Units*	BASGRAS1	BASGRAS2	BASGRAS3
Maximum photosynthetic rate	3	mg C (mg shoot C)$^{-1}$ d^{-1}	0.400	0.544	0.663
Translocation rate for assimilates	6.1,6.2,6.3	mg C (mg shoot C_m)$^{-1}$ d^{-1}	0.201	0.605	0.599
Maximum uptake rate of nitrogen	5.1, 5.2	mg N (mg root C_m)$^{-1}$ d^{-1}	0.0058	0.0087	0.0098
Translocation rate for nitrogenous compounds	6.1,6.2,6.3	mg N (mg root N)$^{-1}$ d^{-1}	0.510	0.270	0.527
Fraction of shoot metabolic pool allocated to structural		mg (C or N) (mg shoot N_m)$^{-1}$ d^{-1}	0.012	0.012	0.012
Fraction of the root metabolic pool allocated to structural		mg (C or N) (mg root N_m)$^{-1}$ d^{-1}	0.012	0.018	0.012
C:N in root beyond which translocation tion decreases, used in RCNT	6.1,6.2,6.3	mg C (mg root N)$^{-1}$	22.5(2.2)	35.(1.4)	12.(4.1)
C:N in root beyond which translocation is prevented, used in RCNT	6.1,6.2,6.3	mg C (mg root N)$^{-1}$	50.(.01)	70.(.7)	35.(1.4)
C:N in shoot beyond which translocation is maximum, used in SCNT	6.1,6.2,6.3	mg C (mg shoot N)$^{-1}$	35.(1.4)	30.(1.6)	15.(3.3)
C:N in shoot beyond which translocation becomes positive, used in SCNT	6.1,6.2,6.3	mg C (mg shoot N)$^{-1}$	22.5(2.2)	15.(3.3)	7.5(6.6)
C:N in shoot beyond which photosyn- thesis decreases, used in SCNP	4	mg C (mg shoot N)$^{-1}$	36.(1.3)	35.(1.4)	20.(2.5)
C:N in shoot beyond which photosyn- thesis is nil, used in SCNP	4	mg C (mg shoot N)$^{-1}$	62.5(.8)	80.(.6)	70.(.7)
C:N in root beyond which N uptake is positive, used in RCNU	5.1	mg C (mg root N)$^{-1}$	10.(5.)	20.(2.5)	10.(5)
C:N in root beyond which N uptake is maximum, used in RCNU	5.1	mg C (mg root N)$^{-1}$	21.(2.5)	30.(1.6)	11.(4.1)
C:N in shoot beyond which N uptake is positive, used in SCNU	5.2	mg C (mg shoot N)$^{-1}$	12.5(4.1)	25.(2.)	8.5(5.8)
C:N in shoot beyond which N uptake is maximum, used in SCNU	5.2	mg C (mg shoot N)$^{-1}$	32.5(1.5)	65.(.7)	10.5(5.)
Shoot respiration rate		mg C (mg shoot C)$^{-1}$ d^{-1}	0.100	0.150	0.150
Maintenance respiration rate	8	mg C (mg root C)$^{-1}$ d^{-1}	--	0.020	0.020
Growth respiration rate	8	mg C (mg C translocated to the root)$^{-1}$ d^{-1}	--	0.200	0.200
Exudation rate	9	mg C (mg root weight)$^{-1}$ d^{-1}	--	--	0.0026
Inorganic N uptake rate by bacteria	10	mg N (mg bacterial C)$^{-1}$ d^{-1}	--	--	0.1

*C_m = metabolic C pool, N_m = metabolic N pool

controlled by root and shoot C:N ratios with a negative feedback mechanism represented by the reduction functions SCNT and RCNT (fig. 2d and e).

$$F = T_{max} * MIN (SCNT, RCNT) * POOL \qquad (6.1)$$

where,

F = translocation flow (mg C or N $plant^{-1}$ d^{-1})

Tmax = maximum translocation rate [mg C or N (mg pool C or $N)^{-1}$ d^{-1}]

MIN (SCNT, RCNT) = reduction function that chooses the lesser of its arguments. SCNT and RCNT are described in figure 2d and e (negative feedback mechanism)

POOL = source pool (mg C or N $plant^{-1}$) either metabolic shoot carbon or metabolic root nitrogen.

Results

Figures 3 and 4 show the simulated shoot and belowground C and N contents (mg C or N per plant) utilizing the basic model with equation (6.1) for nitrogen and carbohydrate translocation and equation (5.1) for nitrogen uptake. Translocation and uptake rates as well as C:N ratio threshold values determining the reduction functions were varied by the optimization routine to obtain the best possible fit (Table 1). In the latter part of the experimental growth period BASGRAS1 overestimates shoot and root nitrogen contents (table 2). When control of N uptake was assigned to shoots [eq. (5.2)] rather than roots [eq. (5.1)], the fit to data was not improved (table 2). Root carbon content was slightly reduced by that change, but the shapes of the curves (not shown) were similar. We then tried to validate this basic model by simulating the removal of 70% of shoots on day 20 without using any optimization method. The discrepancy between data for clipped plants and model output, especially shoot N (table 2), suggested that the model lacked some essential physiological properties.

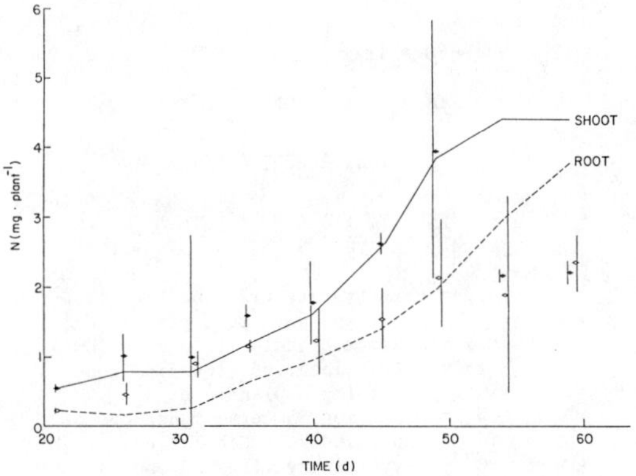

Figure 3.--Observed shoot and root nitrogen status (mg N $plant^{-1}$) and that predicted by BASGRAS1 (———) shoot N content simulated; (◆) observed (mean for 10 to 5 plants); (-----) root N content simulated; (◇) observed (mean for 10 to 5 plants) (data are plotted with their 95% CI).

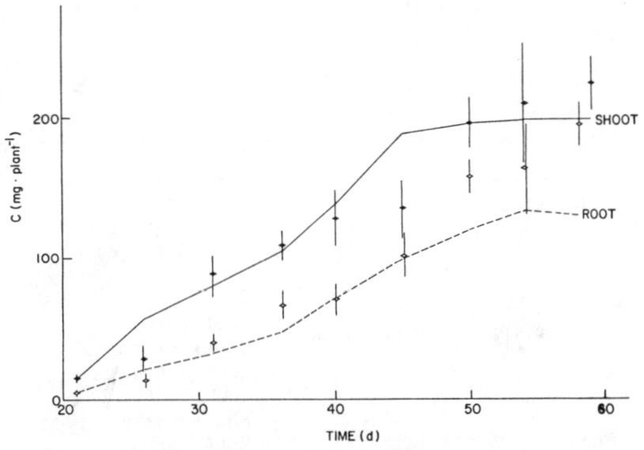

Figure 4.--Observed shoot and root C content (mg C $plant^{-1}$) and that predicted by BASGRAS1. (———) Shoot C content simulated; (◆) observed (mean for 10 to 5 plants); (-----) root C content simulated; (◇) observed (mean for 10 to 5 plants) (data are plotted with their 95% CI).

BASGRAS2--Respiration Model

This section describes the BASGRAS2 model, which is the same as BASGRAS1, except that root respiration is explicitly defined.

1. Root Respiration

McCree (1970) described respiration as the sum of a maintenance component proportional to the plant dry weight and a growth component proportional to gross photosynthesis. Penning de Vries (1972) defined maintenance respiration as production of CO_2 related to protein breakdown and synthesis, respiratory processes providing energy to maintain cell structure and gradients of ions and metabolites, and processes of physiological adaptation that maintain cells as active units. In de Wit et al.'s (1978) model, maintenance was related not directly to dry weight but to actual amounts of protein and minerals in plants, so that changes in composition were reflected in the maintenance rate. Penning de Vries described growth as biosynthesis and maintenance of cell structures and accompanying a dry weight increase. Usually, CO_2 evolution related to the cost of translocation is included in growth respiration (de Wit et al. 1978). Warembourg and Paul (1977) expressed respiratory CO_2 released by roots (Agropyron-Koeleria association) as 20 to 29% of C translocated from foliage to roots. Hatrick and Bowling (1973) and many others found that root respiration depends on the downflow of photosynthates into the root. We chose to distinguish "maintenance" respiration proportional to the size of the root and "growth" respiration proportional to the amount of carbohydrates translocated to the root:

$$FRR = CR1 * TOTRC + CR2 * F \qquad (8)$$

where,

FRR = root respiration rate (mg CO_2-C $plant^{-1}$ d^{-1})

TOTRC = root weight (mg C $plant^{-1}$)

Table 2. Goodness of fit[1] of simulated growth curves.

		BASGRAS1		BASGRAS2		BASGRAS3	
		A	R^2	A	R^2	A	R^2
Basic Runs	Shoot carbon	0.998	0.880	0.966	0.936	1.042	0.845
	Root carbon	0.823	0.953	0.974	0.907	0.769	0.930
	Shoot nitrogen	1.202	0.607	1.095	0.924	0.841	0.413
	Root nitrogen	1.138	0.725	1.026	0.556	0.873	0.934
Shoot clipping treatment	Shoot carbon	1.264	0.892	0.340	0	1.397	0.961
	Root carbon	0.820	0.830	0.198	0	0.480	0.544
	Shoot nitrogen	1.872	0.290	0.340	0.500	0.756	0.737
	Root nitrogen	1.580	0.834	0.227	0	0.598	0.756
Shoot C:N controls N uptake	Shoot carbon	1.053	0.790	1.016	0.895	1.041	0.836
	Root carbon	0.804	0.811	1.149	0.886	0.684	0.890
	Shoot nitrogen	1.348	0.781	1.250	0.669	0.736	0.431
	Root nitrogen	1.347	0.827	1.435	0.743	0.766	0.935
Source-size dependent translocation, equation (6.2)	Shoot carbon			0.975	0.924	0.817	0.997
	Root carbon			1.050	0.900	0.561	0.968
	Shoot nitrogen			1.630	0.541	0.740	0.339
	Root nitrogen			1.382	0.903	0.759	0.647
Sink-size dependent translocation equation (6.3)	Shoot carbon			1.483	0.967	1.178	0.398
	Root carbon			0.713	0	0.180	0
	Shoot nitrogen			2.533	0.717	0.870	0.499
	Root nitrogen			1.128	0.134	0.075	0

[1]R^2 is the fraction of variance in the data accounted for by the model; A is the slope of the regression line:model = A × data.

F = translocation flow (mg C $plant^{-1}$ d^{-1})
CR1 = maintenance coefficient [0.02 mg CO_2-C (mg root C)$^{-1}$ d^{-1}]
CR2 = growth coefficient [0.2 mg C respired (mg C translocated)$^{-1}$]

Equation (8) accounts for weight specific root respiration rate decreasing with age, as shown in figure 5 from actual data, since the translocation flow per unit shoot weight becomes less in larger plants when shading limits photosynthesis.

2. Translocation

Three theories, other than that embodied in BASGRAS1 above, have been proposed to explain the pattern of differential distribution of assimilates (Troughton 1977, Novoa and Loomis 1981, Neales and Incoll 1968).

(a) The competitive theory assumes that root growth is largely limited by carbohydrates from the shoot while shoot growth is limited by nitrogen from the root (Loomis 1953). Pearsall's (1923) sink hypothesis specifies that the sink nearest the site of absorption or synthesis of a metabolite has priority for its use. The excess carbohydrate hypothesis (Turner 1922, Pearsall 1923, White 1973) assumes that root growth depends upon the quantity of carbohydrates the shoot is unable to use.

(b) The size of sink theory, developed by Humphries (1963), adopts the idea that root growth typically is limited by the ability of the root system to utilize carbohydrates rather than by the ability of the shoot system to supply them.

(c) The third theory introduces hormonal regulation of translocation (Evans et al. 1964, Troughton 1977) and is not considered here. It would be difficult to model hormonal effects explicitly. However, they could be represented by introducing effects of shoot to root ratio on translocation.

Equation 6.1 (above) was modified to agree with either the competitive (eq. 6.2) or size of sink (eq. 6.3) theory, to compare their ability to account for the data.

$$FC = Tmax * SCNT * SCTM$$
$$FN = Tmax * RCNT * RNTM \qquad (6.2)$$

or

$$FC = Tmax * RCNT * SCTM$$
$$FN = Tmax * SCNT * RNTM \qquad (6.3)$$

where,

FC = carbohydrate translocation flow from shoot to root (mg C $plant^{-1}$ d^{-1})
Tmax = maximum translocation rate (d^{-1})
FN = N translocation flow from root to shoot (mg N $plant^{-1}$ d^{-1})
SCNT, RCNT = reduction factors--functions of shoot and root C:N ratios, respectively (fig. 2d and e)
RCTM = root metabolic pool (mg C $plant^{-1}$)
SCTM = shoot metabolic pool (mg C $plant^{-1}$)
RNTM = root metabolic pool (mg N $plant^{-1}$)
SNTM = shoot metabolic pool (mg N $plant^{-1}$)

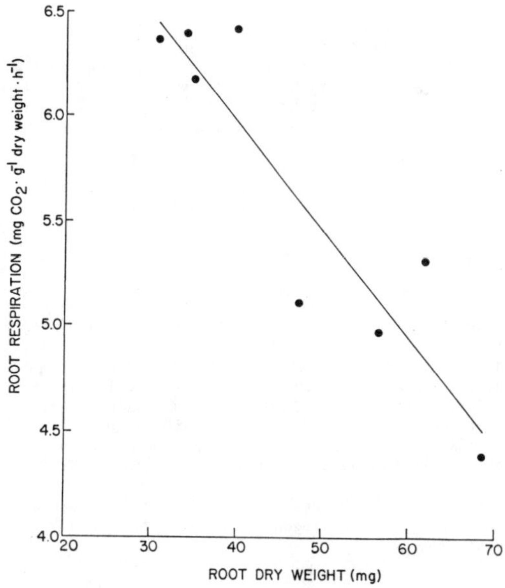

Figure 5.--Relationship between root respiration rate [mg CO_2 (g dry weight)$^{-1}$ h^{-1}] and root dry weights (mg). Each point represents 1 plant of blue grama grass grown hydroponically in a growth chamber.

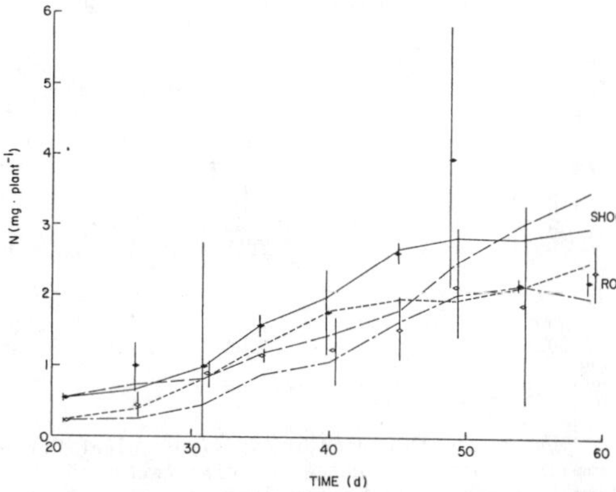

Figure 6.--Observed shoot and root nitrogen status (mg N plant^{-1}) and that predicted by BASGRAS2 and BASGRAS3. (———) shoot N content simulated by BASGRAS2, (— —) by BASGRAS3; (◆) observed (mean for 10 to 5 plants); (-----) root N content simulated by BASGRAS2, (— - —) by BASGRAS3; (◇) observed (mean for 10 to 5 plants) (data are plotted with their 95% CI).

The size of sink theory was applied to nitrogen distribution; therefore, there is a negative feedback from the sink on the translocation intensity. This was not an explicit part of Humphries' theory, although it agrees with its principle.

Results

First we used source-sink control of translocation (eq. 6.1) and root control of nitrogen uptake (eq. 5.1). Figures 6 and 7 show the result of running the respiration model. BASGRAS2 overestimated root C and N content in the middle growth stages when BASGRAS1 underestimated it and predicted shoot N content better than BASGRAS1. As a further test of the effect of including respiration in model 2, we simulated 75% shoot clipping on day 20 without using the optimization routine. Shoot clipping in the model had a strong negative influence on C and N levels.

BASGRAS2 estimated shoot and root growth accurately up to 10 days after clipping, but a discrepancy developed later on (data not shown). The data showed that root growth was slowed down just after shoot clipping, but rapidly increased after a period of 10 days, until on day 60 root biomass of clipped plants exceeded root biomass of control plants. The model as it stands does not include a mechanism that could account for such an overcompensation. Although the model underestimated biomass it more accurately simulated the shoot-root ratio (table 3).

BASGRAS2 predicted too strong an inhibition of growth after clipping but was more realistic than BASGRAS1 where neither shoot nor root growth were very different from the unclipped situation. To check whether the response to clipping could be improved by modifying translocation regulation, we tested the competitive theory using equation (6.2) and the sink size theory using equation (6.3). In the first case the model overestimated root C and N contents and shoot nitrogen content, while in the second case, overall content was overestimated and instabilities occurred (not shown). Neither source size nor sink size alone could account for the regulation of translocation. We also tried to improve the fit by using equation (5.2) for nitrogen uptake and equation (6.1) for translocation. The carbon curves exhibited some instabilities, so we rejected the hypothesis that shoot C:N ratio could directly regulate N uptake in our model.

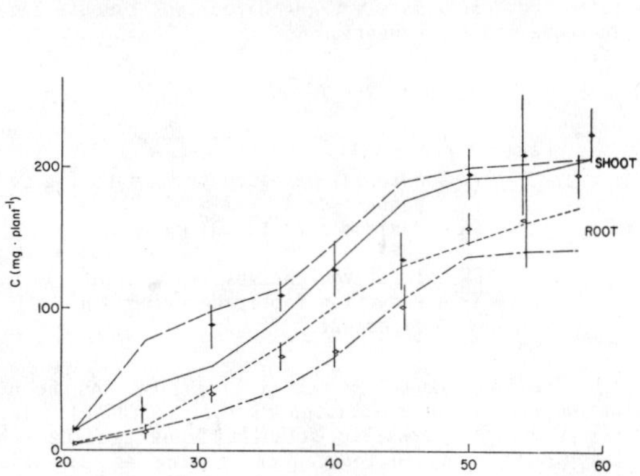

Figure 7.--Observed shoot and root carbon status (mg C plant^{-1}) and that predicted by BASGRAS2 and BASGRAS3. (———) shoot C content simulated by BASGRAS2, (— —) by BASGRAS3; (◆) observed (mean for 10 to 5 plants); (-----) root C content simulated by BASGRAS2, (— - —) by BASGRAS3; (◇) observed (mean for 10 to 5 plants) (data are plotted with their 95% CI).

Table 3.--Shoot:root ratios of unclipped plants and plants clipped when 20 days old.

Plants	Plant age (days)			
	21	31	45	59
Unclipped				
Model	2.19	1.30	1.35	1.19
Data	2.19	2.15	1.33	1.14
Clipped				
Model	0.55	1.53	1.33	1.26
Data	0.55	1.96	0.97	0.65

Evidently, the behavior of clipped plants is very complex. Additional data are being collected at this time to further test the model.

BASGRAS3--Exudation Model

The previous model could probably be improved if the parameter values could be defined using better data, but we also wanted to test if root exudation was an important process that could improve performance. This section describes a third model incorporating the hypothesis that root exudation is stimulated by associated bacteria. McDougall (1970) distinguished different mechanisms of leakage from lysing cells and exudation from still-healthy cells. Presumably, all exuded compounds could be used by associated microflora, including bacteria. To McDougall, exudates represent small excesses over reabsorption occurring by an active uptake process. Some authors argue that in nonaxenic cultures, exudation rates are much higher than generally assumed (Barber and Lynch 1977, Prikryl and Vancura 1980). Perhaps bacteria somehow impede active reabsorption. The model used a C:N ratio of exuded material of 20 which agrees with data from Bokhari et al. (1979). The rate of exudation was estimated approximately using Barber and Lynch's data. The equation for exudation is

$$FE = EX * TOTRW * BCNP \tag{9}$$

where,

FE = material released from roots (mg C $plant^{-1}$ d^{-1})
EX = maximum rate [0.005 mg C (mg root $C)^{-1}$ d^{-1}]
TOTRW = total root weight (mg C $plant^{-1}$)
BCNP = reduction factor described in figure 2f.

The root exudation rate is regulated by the nitrogen content of associated microflora (BCNP). This allows for the possible stimulation of exudation by the presence of microorganisms acting as potential additional sinks around the roots (Vancura et al. 1977, Bürström 1965). We assumed that all exuded material was absorbed by bacteria. Because the nitrogen content of exuded material was fixed, our model assumed that the bacteria could take up inorganic nitrogen to maintain a C:N ratio around 5 (McGill et al. 1981). If their C:N ratio were lower, mineralization of their own nitrogen would occur (function BCNU described in fig. 2g).

$$FU = C * BACC * BCNU \tag{10}$$

where,

FU = rate of uptake or release of inorganic nitrogen by bacteria (mg N $plant^{-1}$ d^{-1})
C = uptake rate of inorganic nitrogen by the bacteria [mg N (mg bacterial $C)^{-1}$ d^{-1}]
BACC = bacterial biomass (mg C $plant^{-1}$)
BCNU = a function of bacterial C:N ratio as described in figure 2g.

Results

Figures 6 and 7 represent the growth curves obtained using equations (5.1), (6.1), (8), (9), and (10). A discrepancy exists between shoot nitrogen data and model output. Representing nitrogen uptake by equation (5.2) does not improve the fit (table 2). Using equations (6.2) and (6.3) to modify translocation flows, BASGRAS3 underestimates C and N contents in the roots. With equation (6.3) especially, root C and N status reach unrealistic values. Therefore, both source and sink size apparently regulate translocation. BASGRAS3 responds to shoot clipping as does BASGRAS2, but nitrogen translocation to the shoot stays too low.

DISCUSSION

Table 2 compares individual runs for each of the three models and presents their fit to our data. The basic model and the respiration model simulate growth fairly accurately, but addition of a definite root respiration function improves the realism of BASGRAS. Response to clipping is linked with the presence of the respiration equation in the model whether N uptake is controlled by shoot or root nitrogen status. This idea was suggested by Winter and Mortimer (1967): "If the root of a plant exerts a controlling influence on the mechanism of translocation, then this control should be related to the respiratory activities of the root," but their experimental results did not support their idea. We can hypothesize that root respiration has an important indirect effect on translocation, which is not easily determined by experiments but certainly needs further research. The exudation model underestimates shoot N content (table 2), but we have yet to evaluate other hypotheses for exudation, such as passive loss from the root metabolic pool.

All three models were very sensitive to the light extinction coefficient (k in equation (1)) which determines the maximum shoot carbon and the root:shoot ratio. Growth patterns were not more realistic when the shoot C:N ratio was assumed to control nitrogen uptake than when root C:N ratio controlled it.

One may argue that the model, to be successful, must consider phenology. One direct justification for the lack of explicit phenological characteristics in this model is that variation of the percentages of structural and metabolic materials should account for changes in the chemical composition of the plant during its maturation. Before the model can be extended over a longer time period it may require a C:N ratio threshold value which can modify the intensity or even the direction of the flows of materials into the various compartments. This would allow for potential seed production (more nitrogen translocated to the aboveground parts), tiller production (less carbohydrate translocation to the root), and retranslocation of N from senescent shoots. The C:N ratio threshold values would be introduced in the reduction factors which regulate uptake and translocation flows. Omission of the phenology of growth confers a greater simplicity to the model. But we need more precise data about size and elasticity of structural and metabolic pools to determine parameter values accounting for chemical changes associated with aging.

Tillering mechanisms seem to be dependent upon a threshold value of shoot C:N ratio (Yu and Gounot 1981), but the hypothesis has not yet been incorporated into a model. The ratio of aboveground to belowground biomass increases when there is adequate nitrogen and decreases during nitrogen stress. In the first case there is actually a production of additional tillers which apparently divert more assimilates to the more numerous aboveground growth regions. In the second case, self-thinning probably produces an inverse effect if the root does not stop developing. Our model responds realistically to N supply and supports the idea that C and N translocation may account for tillering.

CONCLUSION

This study indicates that it is relatively easy to create a simulation model which, with a particular set of parameters, will give a good fit to a set of data. Yet it is much more difficult to find a model that also will give an adequate response to a stress situation. We believe that the exudation model should be modified, especially at the level of the exudation process, but data are still lacking and experimentation is not easy. We do not have enough information either about translocation or uptake, and their controls, and it is difficult to experimentally determine the flows. Two important findings may aid further research in this direction. One is that respiration processes are closely related to translocation and need to be carefully simulated to improve the realism of a model. The second is that both source and sink are important to control the intensity of the translocation flows.

ACKNOWLEDGEMENTS

This research was supported by National Science Foundation Grant DEB-80-22380.

We thank C. W. Ross for his helpful criticism and D. T. Winn for her help analyzing the data.

LITERATURE CITED

Baldwin, J.P. 1976. Competition for plant nutrients in soil: a theoretical approach. Journal of Agricultural Science, Cambridge 87:341-355.

Barber, D.A., and J.M. Lynch. 1977. Microbial growth in the rhizosphere. Soil Biology and Biochemistry 9:305-308.

Bokhari, U.G., D.C. Coleman, and A. Rubink. 1979. Chemistry of root exudates and rhizosphere soils of prairie plants. Canadian Journal of Botany 57:1473-1477.

Bürström, H.G. 1965. The physiology of plant roots. In K. F. Baker and W. C. Snyder (eds.) Ecology of soil-borne plant pathogens--prelude to biological control, pp. 154-163. University of California Press, Berkeley, Los Angeles.

Cram, W.J. 1976. Negative feedback regulation of transport in cells. In M. G. Pitman and U. Lüttge (eds.) Encyclopedia of plant physiology. NS Vol II. Springer-Verlag, Berlin, Heidelberg, New York.

Detling, J.K., M.I. Dyer, C. Procter-Gregg, and D.T. Winn. 1980. Plant herbivore interactions: Examination of potential effects of bison saliva on regrowth of *Bouteloua gracilis* (H.B.K.) Lag. Oecologia 45:26-31.

Edwards, J.H., and S.A. Barber. 1976. Nitrogen flux into corn roots as influenced by shoot requirement. Agronomy Journal 68:471-473.

Evans, L.T., I.F. Wardlaw, and C.N. Williams. 1964. Environmental control of growth. p. 102-125. In C. Bernard (ed.) Grasses and Grasslands. MacMillan, London.

Hatrick, A.A. and D.J.F. Bowling. 1973. A study of the relationship between root and shoot metabolism. Journal of Experimental Botany 24:607-613.

Humphries, E.C. 1963. Dependence of net assimilation rate on root growth of isolated leaves. Annals of Botany N.S. 27:175-183.

Lambers, H., R. Noord, and F. Posthumus. 1979. Response of *Senecio* roots: inhibition during photosynthesis, resistance to cyanide and relation to growth and maintenance. Physiologia Plantarum 45:351-356.

Larcher, W. 1975. Physiological Plant Ecology. Springer Verlag, Berlin.

Lemée, G. 1978. Précis d'écologie végétale. Masson, Paris.

Loomis, W.E. 1953. Growth correlation. In W.E. Loomis (ed.) Growth and differenciation in plants, a monograph of the American Society of Plant Physiologists. Iowa State College, Ames.

McCree, K.J. 1970. An equation for the rate of respiration of white clover plants grown under controlled conditions. p. 221-229. In Prediction and Measurement of Photosynthetic Productivity. Pudoc, Wageningen.

McDougall, B.M. 1970. Movement of ^{14}C photosynthate into the roots of wheat seedlings and exudation of ^{14}C from intact roots. New Phytologist 69: 37-46.

McGill, W.B., H.W. Hunt, R.G. Woodmansee, and J.O. Reuss. 1981. Phoenix, a model of the dynamics of carbon and nitrogen in grassland soils. In F.E. Clark and T. Rosswald (eds.) Terrestrial nitrogen cycles. Ecological Bulletin (Stockholm) 33:49-115.

Monsi, M., and T. Saeki. 1953. Über der Lichtfactor in den Pflanzengesellshaften und seine Bedeutung für die Stoff produktion. Japanese Journal of Botany 14:22-52.

Monteith, J.L. 1965. Light distribution and photosynthesis in field crops. Annals of Botany 29: 17-37.

Neales, T.F., and L.D. Incoll. 1968. The control of leaf photosynthesis rate by the level of assimilate concentration in the leaf: a review of the hypothesis. The Botanical Review 34:107-125.

Novoa, R., and R.S. Loomis. 1981. Nitrogen and plant production. Plant and Soil 58:177-204.

Patefield, W.M., and R.B. Austin. 1971. A model for the simulation of the growth of Beta vulgaris L. Annals of Botany 35:1227-1250.

Pearsall, W.H. 1923. Studies in growth. IV. Correlations in development. Annals of Botany 37: 261-275.

Penning de Vries, F.W.T. 1972. Respiration and growth. p. 327-347. In A.R. Rees, K.E. Cockshull, D.W. Hand, and R.J. Hurd (eds.) Crop Processes in Controlled Environments. Academic Press, London.

Powell, M.J.D. 1965. A method for minimizing the sum of squares of nonlinear functions without calculating derivatives. Computer Journal 7:303-307.

Prikryl, Z., and V. Vancura. 1980. Root exudates of plants. VI. Wheat root exudation as dependent on growth, concentration gradient of exudates and the presence of bacteria. Plant and Soil 57: 69-83.

Radin, J. 1977. Contribution of the root system to nitrate assimilation in whole cotton plants. Australian Journal of Plant Physiology 4:811-819.

Raper, C.D., D.L. Osmond, M. Wann, W.W. Weeks. 1978. Interdependence of root and shoot activities in determining N uptake rate of roots. Botanical Gazette 139:289-294.

Steyermark, A. 1961. Quantitative organic microanalysis. Academic Press, New York.

Thom, R. 1975. Modèles mathématiques de la morphogenèse. Collections 10-18. Numéro 887 (Paris).

Thornley, J.H.M. 1972. A balanced quantitative model for root:shoot ratios in vegetative plants. Annals of Botany 36:431-444.

Troughton, A. 1977. The rate of growth and partitioning of assimilates in young grass plants: a mathematical model. Annals of Botany 41:553-565.

Turner, T.W. 1922. Studies of the mechanism of the physiological effects of certain mineral salts in altering the ratio of top growth to root growth in seed plants. American Journal of Botany 9: 415-445.

Vancura, V., Z. Prikryl, L. Kalachova, and M. Wurst. 1977. Some quantitative aspects of root exudation. In Soil organisms as components of ecosystems. Ecological Bulletin (Stockholm) 25: 381-386.

Wardlaw, I.F., and L. Moncur. 1976. Source, sink and hormonal control of translocation in wheat. Planta (Berl.) 128:93-100.

Warembourg, F.R., and E.A. Paul. 1977. Seasonal transfers of assimilate ^{14}C in grassland: Plant production and turnover, translocation and respiration. In J. K. Marshall (ed.) The belowground ecosystem. Range Science Department Science Series No. 26. Colorado State University, Fort Collins, Colorado.

White, L.M. 1973. Carbohydrate reserves of grasses: a review. Journal of Range Management 26:13-18.

Winter, H., and D.C. Mortimer. 1967. Role of the root in the translocation of products of photosynthesis in sugar beet, soybean and pumpkin. Canadian Journal of Botany 45:1811-1822.

Wit, C. de et al. 1978. Simulation of assimilation, respiration and transpiration of crops. Simulation Monographs. John Wiley & Sons, New York.

Yu, O., and M. Gounot. 1981. Recherches sur le tallage chez le dactyle. II. Réflexions sur la formalisation et la modèlisation des processus de tallage. Oecologia Plantarum 17:39-47.

SIMULATION OF ABOVE- AND BELOW- GROUND CARBON AND NITROGEN DYNAMICS OF BOUTELOUA GRACILIS AND AGROPYRON SMITHII[1]

J. W. Skiles, J. D. Hanson and W. J. Parton[2]

Abstract. A difference-differential equation model has been developed which simulates phytomass and nitrogen flow of grassland primary producers. The model was developed for inclusion in a large-scale, decision oriented rangeland management model. Whether used as a submodel of the larger model or alone, the plant model should prove to be a powerful tool in making management decisions.

INTRODUCTION

Use of mathematical and simulation models with management applications by resource management agencies is a natural outcome of model development and testing over the past decade. This is due to any number of circumstances, but two deserve mention. First, resource management agencies now have within their ranks people who have participated in model building and have used models to investigate management problems in their undergraduate and graduate training (c.f. for example, Van Dyne (1977)). During the past decade, the sometimes arcane techniques of model construction were added to the standard curricula at many colleges where resource management personnel were trained. Thus, familiarity with models and modeling is now an attribute that many academically trained managers have. Secondly, the availability of in-house computers and ubiquitous programming languages have made management models portable and accessable to branch offices of management agencies. Hence, personnel already familiar with models have been able to use extant models when necessary, modify them when needed and develop new programs when model deficiencies exist or new applications arise.

Within the United States Department of Agriculture, Agricultural Research Service (USDA-ARS) two modeling programs have been initiated. The first program culminated in a large scale model for use in agricultural systems where chemical pollutants (pesticides and fertilizers), runoff and erosion were critical factors. The CREAMS (Chemical, Runoff, and Erosion from Agricultural Systems) model has proven of value in test situations and for several applications since the documentation was released (Knisel 1980). The other modeling effort began at a 1978 planning workshop when it was determined that a predictive rangeland management and production model should be built. In 1981 scientists and range specialists met to plan the construction of the range management model, assign specific tasks, adopt a time table for completion of those tasks and identify data sets for model verification and validation.

The main objective of this latter modeling effort was to formulate an operational rangeland model. The completed model would be used to provide a basis for management decisions by predicting herbage yields, runoff, erosion, water quality, etc., provide a computational basis for alternative management systems, evaluate rangeland systems, evaluate range improvement practices, optimize rangeland management systems for desired multiple-use products, forecast effect of climatic changes on rangeland, and plan and evaluate land reclamation practices. Individual components of the model were soils/hydrology, climate and weather, fauna, management, and primary producer. Subsidiary components were economic considerations and insect pests and weeds. The remainder of this paper describes the primary producer component of the model.

MODEL DESCRIPTION

The authors reviewed extant models during the construction of the primary producer model. Of primary interest was the Ecosystem Level Model (ELM) which was developed for the shortgrass prairie by the International Biome Project, U. S. Grassland Biome Study (Innis 1978) and grassland models developed by Parton et al. (1978) and Detling et al. (1979). The model described herein differs in many respects from the previous models, however, the overall structures are similar. Information presented in the three models cited above was used extensively in the development of this model.

The primary producer model flows phytomass and nitrogen. There are seven phytomass and eight nitrogen state variables in the model (Fig. 1). In the flow diagram seven of the compartments are divided into two separate components for clarity and to empahsize the concomitant existence of phytomass and nitrogen within the state variables. If nitrogen and phytomass flow between two state variables, the arrow points to the box title. However, if either phytomass or nitrogen flows independently, the arrow points to the phytomass or nitrogen subcompartment, respectively. Computationally, nitrogen and phytomass are calculated in different subroutines, thus each compartment represents two different state variables.

Amount of phytomass on each day is required output from the model. Species dependent state variables in the phytomass submodel are green shoots, live roots, propagules and standing dead. Dead roots, litter and soil prganic matter have no species identity. Conversion from phytomass to carbon is

[1]Paper presented at the Third International Conference on the State-of-the-Art in Ecological Modeling, Colorado State University, 24-28 May 1982.

[2]The authors are Research Associate, Department of Range Science, Colorado State University, Fort Collins, CO 80523; Range Scientist, USDA-ARS, Cheyenne, WY 82009; and Research Scientist, Natural Resource Ecology Laboratory, CSU, Fort Collins, CO 80523.

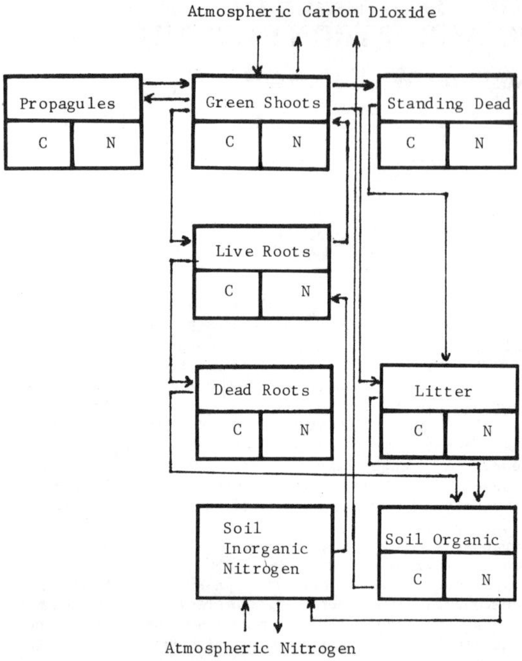

Figure 1. Flow diagram of primary production simulation model.

accomplished by multiplying phytomass by 0.4. A description of the model functional formulation will be released subsequently.

Driving Variables

Abiotic driving variables used by the primary producer model include air temperature (C), precipitation (cm/day), soil water potential (-bars), solar radiation (langley/day) and wind (km/day). For subsequent simulations, precipitation and daily maximum and minimum air temperatures from 1972 Central Plains Experimental Range weather data were used; this shortgrass prairie site is located in northeastern Colorado, USA. A 10 day running average air temperature is calculated within the model and is used in such processes as growth initiation and seed germination. Soil water potential is calculated for five soil layers (Parton unpublished). A sinusoidal function was used to simulate daily solar radiation with photoperiod and latitude as independent variables. Finally, for our purposes wind speed and soil bulk density were constant.

Phytomass Flow

Green Shoots

Change in green shoot phytomass (S) was calculated as:

dS/dt=PN-TSR-SC-DS-TSP+TRS+GERM-RESP (1).

PN is daytime net carbon dioxide assimilation rate (mg $dm^{-2} d^{-1}$) and considers ambient temperature, soil water potential, nitrogen content of green shoots, leaf area of live shoots and solar radiation (Detling et al. 1979). Translocation from shoots to roots (TSR) is calculated as a proportion of PN. The proportion is based on shoot/root phytomass ratios until senescense begins. Then a species-dependent constant fraction of the photosynthate is translocated to the roots. SC is trampling by herbivores and was a function of the number of animals. Mortality of live shoots (DS) accounts for death from frost, water stress, time of year, natural mortality and leaf area (Parton et al. 1978). Photosynthate is also transported to propagules (TSP) as a direct proportion of PN. TSP is initiated at the start of senescense. Translocation from roots to shoots (TRS) occurs when water conditions are adequate, average 10 day temperature is high enough and there are low amounts of standing green phytomass. This is the mechanism for initiating spring growth and regrowth after grazing. Biomass can also be added to standing green through germination (GERM). GERM is a function of soil water potential at the surface and average 10 day temperature. RESP is the nighttime respiration rate and utilizes soil water potential and average nighttime temperature.

Live Roots, Propagules and Standing Dead

The dynamics of live roots (R) were expressed as:

dR/dt=TSR-ROOTR-ROOTM (2).

Root respiration rate (ROOTR) and root mortality rate (ROOTM) are temperature and water stress dependent, respectively. Information concerning these processes was extracted from Parton et al. (1978).

Propagule (P) dynamics in the model were expressed as:

dP/dt=TSP-GERM-SEEDM (3)

where SEEDM is seed mortality. SEEDM was calculated as a proportion of total propagule phytomass present. The facility exists to include granivory in the model, thereby reducing seed viability.

Standing dead phytomass (D) was simply:

dD/dt=DS-TDSL (4).

Standing dead was transfered to the litter compartment (TDSL) at rates dependent on wind speed, precipitation and herbivore trampling.

Dead Roots, Litter and Soil Organic

As material moves through the system and decomposition occurs it becomes impossible to maintain species identity. Consequently, these three compartments no longer contain information concerning individual species but represent pools of phytomass. The following three equations represent the dynamics of these compartments:

dS1/dt=ROOTM(i)-TDRO (5)
dS2/dt=SC(i)+TDSL(i)+SEEDM(i)-TLO (6)
dS3/dt=0.4*TDRO+0.4*TLO-TOA (7).

Equations 5, 6 and 7 represent expressions for dead root (S1), litter (S2) and soil organic (S3) compartments. TDRO, TLO and TOA are transfers from dead roots to organic, litter to organic and organic to atmosphere, respectively. All three of these transfers are functions of water stress and temperature. The subscripted variables represent terms which reflect species identity. They have the numeric value as previously defined. It was assumed that only 40% of the phytomass was retained following decomposition (60% was released as carbon dioxide through respiration) into the organic pool (Eq. 7).

Nitrogen Flows

As can be seen from the flow diagram (Fig. 1), the nitrogen flows, for the most part, parallel the phytomass flows. Amounts of nitrogen moving between compartments were calculated by multiplying the carbon flow rate times the N/C ratio of the donor compartment. This allows the N/C ratio of the state variables to change with time. The C/N ratio for soil organic matter was assumed to be constant and the flows of N into soil organic matter were equal to the amount of N needed to form soil organic matter with a C/N ratio of 10. Decomposition of dead roots and litter form soil organic matter. Generally, the N content of litter and dead roots is fairly low and C/N ratios of phytomass flowing into soil organic matter is greater than 10. The model assumes that N from the mineral N pool is immobilized during decomposition and the amount needed to create soil organic matter with a C/N ratio of 10 is transferred to the soil organic pool. When soil organic matter decomposes, the nitrogen contained in the compartment is released into the mineral N compartment.

Several processes of the nitrogen portion of the model deserve some discussion. Nitrogen uptake by roots is calculated by the use of a Michaelis-Menton type equation with soil inorganic nitrogen as the substrate. Rate of uptake is subsequently modified depending on root biomass, soil moisture and temperature. Movement of nitrogen from roots to shoots or shoots to roots is identical to the methods used by Reuss and Innis (1977). Also, nitrogen fixation rates are calculated as a linear function of precipitation. Finally, plants attempt to retain nitrogen during the death process. Thus, instead of transfering nitrogen from green shoots to standing dead at the present C/N ratio it is transfered at a ratio of C/N + 10.0 (McGill et al. 1981).

MODEL RESPONSE

The model was developed using physiological data for Bouteloua gracilis (blue grama). This species is a warm-season or C4 plant with characteristic dicarboxylic acid photosynthesis. Plant growth begins in early summer and continues through the hotter summer months whenever moisture is available (Williams 1974). Moisture is critical for blue grama photosynthesis because carbohydrate synthesis may decrease by 75%, depending on how much water was present after the last rain, over ten days as soils dehydrate after a summer rainfall event (Detling 1979). Time series for single year and single site simulations of the plant components are subsequently discussed for single- and multiple-species cases. To this point no comparison between model results and actual plant data has been made.

Single-species

Phytomass Flow

No green shoot production was simulated until day 140, the latter part of May (Fig. 2). Peak standing phytomass of almost 100 g was reached in early June. It declined to about 80 g in mid-June and decreased to less than 2 g by day 300. As can be seen, the beginning point and the end point for standing dead were roughly the same (10g), thereby indicating an equilibrium for this variable. Peak standing dead, 53 g, occurred on day 220 (about mid-July), about 120 days after the peak for green shoots. Reduction of this biomass is through shattering and trampling. When cattle and other herbivores are linked with the plant submodel a more precipitous drop will probably occur after peak production and that peak will be lower than shown in Figure 2.

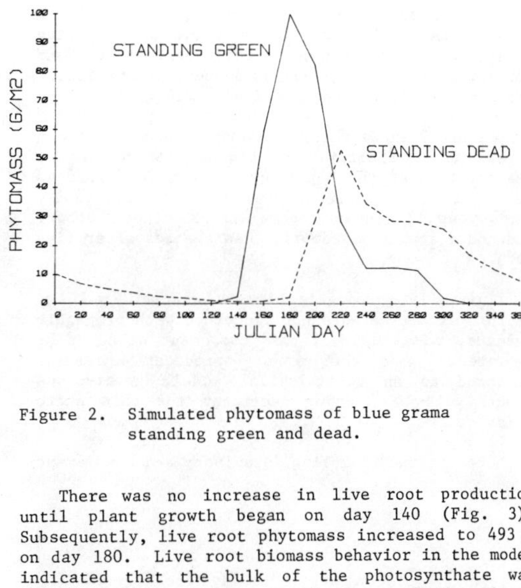

Figure 2. Simulated phytomass of blue grama standing green and dead.

There was no increase in live root production until plant growth began on day 140 (Fig. 3). Subsequently, live root phytomass increased to 493 g on day 180. Live root biomass behavior in the model indicated that the bulk of the photosynthate was translocated to below ground parts of the plant. Typically, blue grama translocates to the roots as much as 70% of carbohydrates produced (Detling 1979).

After an initial value of about 400 g on day 20, dead root biomass increased slightly during the winter and then dropped to about 350 g in mid-June (Fig. 3). An increase from that point to 445 g occurred in mid-July. This is about a 40 day time lag from maximum green shoot biomass. After peak dead shoot production, and during the remaining summer months, decomposition occurred which accounted for the loss to this variable. During cooler fall and early winter months, decomposition rates were decreased allowing dead roots to accumulate again.

On day 20, 104 g of litter was in the system (Fig. 3). Litter accumulated slightly until day 160 and dropped to a low of about 87 g within 40 days. This was the time of year when this species was

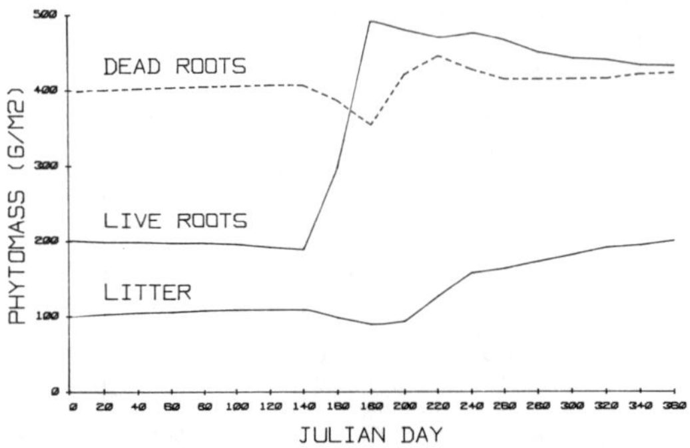

Figure 3. Simulated dead root, live root and litter phytomass in a blue grama monoculture.

producing maximum live roots and green shoots, so a decrease in litter production would be expected because of decomposition. Soil desiccation and shattering of standing dead subsequently contributed to a maximum litter accumulation of 200 g.

Figure 3 shows that live roots, dead roots and litter had more accumulated biomass at the end of one year than at the beginning. The model produced these results because of low initial conditions. Twenty-year phytomass simulations (not shown) produced a stable system for this species after five years.

Initially, propagule biomass was set to 10 g. For this effort we were not concerned with propagule dynamics, so results of this component will not be presented. When the primary producer model is connected to an entomological model, granivorous insects will be a major decreaser for this state variable.

Grazing and trampling, granivory and management influences were not reflected in these single species standard runs because the model was not linked with the animal, insect and management submodels. When these submodels are included, it is expected that stability after five years will no longer exist, and depending on the management strategies implemented, the dynamics shown in the preceding figures will not be the same.

Nitrogen Flow

Nitrogen content of green shoots showed dynamics typical of blue grama on native range (Fig. 4.). Green shoot nitrogen increased to 1.04 g by day 180 and subsequently decreased steadily to 0.1 g by day 240. The N/C ratio on day 120 was 0.067. N/C ratios decreased exponentially throughout the growing season. The other nitrogen compartments of the model were less dynamic. One general conclusion, however, was that any nitrogen in the soil inorganic state variable was immediately used by the growing plants. Nitrogen increased in live roots from 1.6 to 2.4 g and in litter from 1.0 to 1.7 g over the course of one year, indicating these state variables are systems sinks for this currency. Again, the dynamics of nitrogen flow may be expected to change with incorporation of management and animal submodels.

Multiple-species

The plant model has been dimensioned to be able to simulate up to seven species or species-groups on nine sites. The dearth of data sets necessary to test the model with the maximum number of species has precluded such simulations, though these data sets have been solicited from resource management agencies. The model was run, however, with two species; blue grama and *Agropyron smithii* (western wheatgrass), on a single site and was using actual weather data. The latter species is a cool-season or C3 grass species with and maximum growth during early spring (Williams 1974).

Green shoot biomass dynamics for two species are shown in Figure 5. The cool-season grass began to grow on about day 60 while the warm-season grass began growth on about day 140. Highest production

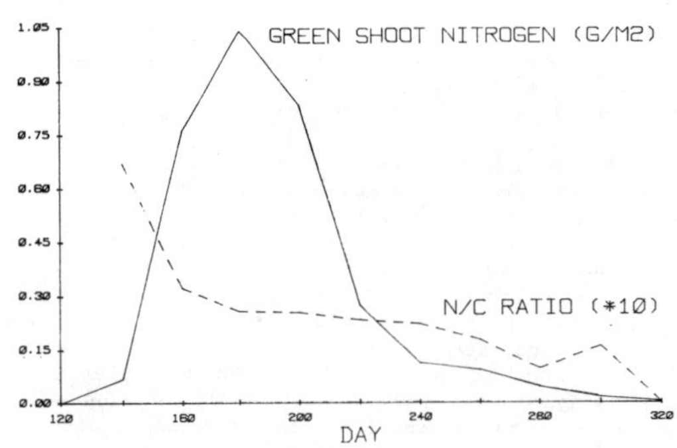

Figure 4. Green shoot nitrogen content and N/C ratios for one year simulation of blue grama.

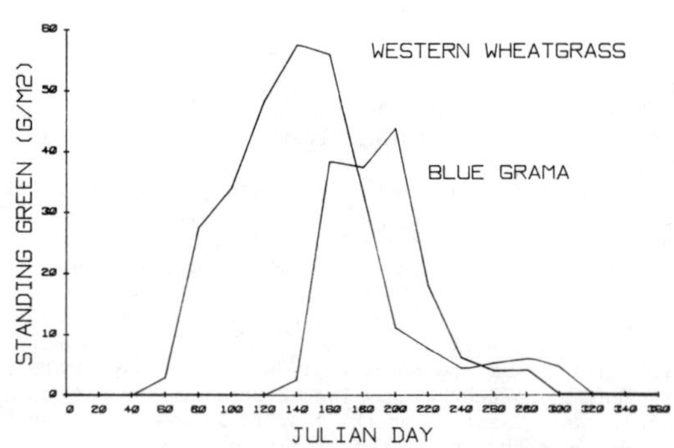

Figure 5. Simulated standing green phytomass for blue grama and western wheatgrass.

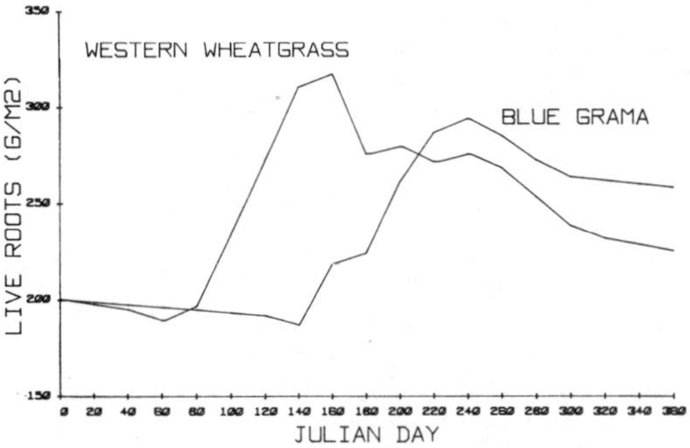

Figure 6. Simulated live root phytomass for blue grama and western wheatgrass.

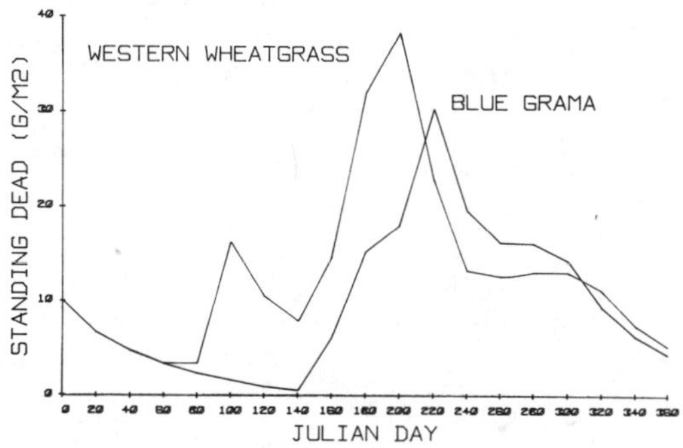

Figure 7. Simulated standing dead phytomass for blue grama and western wheatgrass.

for western wheatgrass was on about day 140 while peak production for blue grama occurred on day 160. Adding both peak standing crops yields about 100 g; this is about equal to the peak phytomass for blue grama simulations. In the two species simulation, blue grama did not produce as much above-ground green biomass as western wheatgrass because of competition for water and (minimally) nitrogen, and the earlier initiation of western wheatgrass growth.

As would be expected, western wheatgrass began production of live roots before blue grama, again owing to the earlier growth of the C3 grass (Fig. 6). Live root production of western wheatgrass peaked 20 days after the green shoot peak production while the live root production for blue grama occurred some 40 days after maximum green shoot production. Certainly this difference was due in part to competition and in part to the out-of phase growth regimes.

About 60 days after western wheatgrass reached peak green shoot phytomass, it attained maximum standing dead of about 38 g (Fig. 7). For blue grama, the difference between the two peaks was only 20 days. The interval for blue grama was shorter presumably because of a decrease in the soil moisture, which is typical for the shortgrass prairie at that time of year. Because of phytomass transfer to litter, standing dead for both species continued to decrease over the remainder of the year until it reached levels approximately equal to the initial conditions.

The production of dead root and litter phytomass represent pools which lack species identity (Fig. 8). Dead root phytomass peaked at 518 g on day 200. Forty days later, litter reached a peak for the year of about 190 g. Again, because of the growth characteristics of the two species, the greater portion of these two variables was made up of western wheatgrass in the early part of the year. Later, blue grama contributed more to dead roots and litter but never dominated western wheatgrass.

Several runs of the model using two species over an extended time were done to test the competitive interactions of the model. When a precipitation event equaling three centimeters of rainfall was added to the weather record in late February, western wheatgrass phytomass production increased greatly. Blue grama production decreased precipitously, but in no instance was it removed

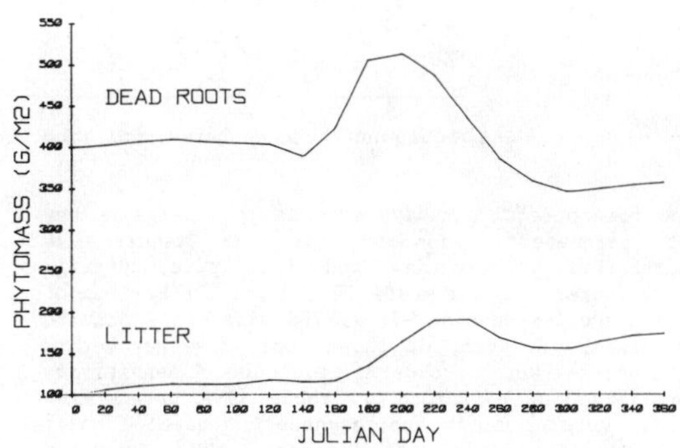

Figure 8. Simulation of dead roots and litter in a system containing blue grama and western wheatgrass.

from the system entirely. On runs over several years using the 1972 weather data each year, the model predicted a decrease in phytomass of western wheatgrass and consequent dominance of blue grama. In northeastern Colorado, from which the weather data were taken, blue grama is the dominant grass on the shortgrass prairie. Western wheatgrass does occur sympatrically with blue grama but only in small amounts, as the model predicts.

MODEL EVALUATION AND RECOMMENDATIONS

The plant model was designed to be user-friendly

Table 1. Root mean squares for selected phytomass state variables to show effect of perturbations on five parameters.

Parameter	-90	-50	-10	10	50	90
			Perturbation			
			Green Shoots			
PMAX	27.2	5.2	1.2	0.9	3.6	5.5
TOPT	*	11.8	4.3	6.2	33.1	*
W1/2	9.5	2.9	0.4	0.4	1.7	1.2
DTC	4.2	1.6	0.2	0.2	0.7	3.4
MNU	7.7	3.5	0.6	0.6	2.5	3.4
			Live Roots			
PMAX	194.1	42.3	10.6	8.5	37.1	62.8
TOPT	*	148.4	43.8	28.1	245.6	*
W1/2	60.4	29.5	4.7	4.0	17.6	27.6
DTC	24.8	10.2	0.9	0.8	4.0	8.3
MNU	48.4	24.0	4.4	18.3	18.8	26.2
			Dead Roots			
PMAX	150.3	44.2	6.9	5.9	23.4	34.5
TOPT	*	150.0	22.1	29.7	171.1	*
W1/2	86.0	15.2	2.7	2.1	10.1	15.6
DTC	30.5	6.0	0.4	0.6	1.4	4.7
MNU	108.3	50.5	8.4	10.6	35.8	47.0
			Litter			
PMAX	78.9	19.5	3.5	3.9	16.3	24.1
TOPT	*	17.1	4.1	3.4	39.0	*
W1/2	35.6	12.0	2.2	1.9	9.1	14.3
DTC	9.2	1.2	0.2	0.6	2.2	5.3
MNU	39.6	20.6	4.1	5.4	18.8	25.1

* TOPT not bounded between minimum and maximum temperatures for net photosynthesis.

and not to require managers to input a large number of parameters. However, it did require 36 parameters per species and four site specific parameters. If the model is run for the maximum of seven species on nine sites, 288 parameters must be supplied. However, of these, only five parameters per species were considered critical. A sensitivity analysis was performed on these five parameters. They were maximum photosynthetic rate (PMAX), maximum nitrogen uptake rate (MNU), optimum temperature for photosynthesis (TOPT), the soil water potential at which photosynthetic activity was one half maximum (W1/2) and the drought tolerance coefficient (DTC).

Each parameter was increased or decreased by 10%, 50% and 90%. The model was run, with these perturbations, for blue grama for one year. The results were compared to the reference run (Fig. 2 and 3) using a Root Mean Square (RMS) calculation:

$$RMS=(1/n\ \Sigma(Ri-Pi)^2)^{1/2} \qquad (8)$$

where Ri is the output on day i from the reference run, Pi is the output on day i from the perturbated run and n is the number of days output by the model. Model results were printed every twenty days, so n = 18 over 360 days, (Table 1).

Green shoots were very sensitive to perturbations in the maximum rate of photosynthesis even at plus or minus 10%. That state variable was relatively insensitive to MNU, W1/2 and DTC. Only at a decrease of 90% were green shoots sentive W1/2. Live roots showed a greater sensitivity to perturbations in all five parameters than green shoots and had RMS of 50 and above for the positive TOPT changes. Live roots also responded to decreased W1/2. Standing dead was not sensitive to W1/2 and DTC changes but somewhat responsive to changes in TOPT and decreases in PMAX. This state variable was not sensitive to MNU. Dead roots responded to changes in PMAX, MNU and TOPT, especially negative perturbations. Dead root also responded to a lesser degree changes in W1/2 and DTC. Changes in DTC did not induce large changes in litter while negative changes in PMAX and W1/2 did produce large RMS in this state variable.

SUMMARY

A primary producer model has been developed as an integral part of a range management model under the auspices of the U.S. Department of Agriculture, Agricultural Research Service. The plant model uses some techniques of previous primary producer models, but is unique in its complexity and small size (less than 40 k-bytes of storage). The model uses difference-differential equations to update fifteen state variables. Two currencies flow in the model; phytomass and nitrogen. Above-ground and below- ground plant parts are considered. The model is coded in FORTRAN IV and operates on a time step of one day.

The model produces generally good phytomass and nitrogen simulations and produces results which are typical of blue grama phytomass dynamics in northeastern Colorado. Two species runs, using blue grama and western wheatgrass, also mimic the expected dynamics of production and nitrogen flow. The model can simulate production of up to seven species or species groups on nine range sites. Plants with Crassulacean Acid Metabolism are not considered.

The model is particularly sensitive to changes in the temperature at which optimum photosynthesis takes place and in some instances, the soil water potential parameter. While 36 parameters are required per plant species, many of these are common to most plants and need not be input precisely.

Owing to the paucity of data sets however, the model has been run with only the two plant species mentioned above. The model needs to be validated against real data on a variety of range types. Experimentation is needed on desert ranges, annual grassland ranges and tallgrass prairie. When coupled with the larger management model, the plant model will help in testing grazing regimes, precipitation variation, fertilizer application and granivory. Because of its generality, however, the plant model by itself can be an important research tool for estimating the dynamics of phytomass and nitrogen on rangelands.

LITERATURE CITED

Detling, J. K., W. J. Parton and H. W. Hunt. 1979. A simulation model of Bouteloua gracilis biomass dynamics on The North American shortgrass prairie. Oecologia (Berl.) 38:167-191.

Detling, J. K. 1979. Processes controling blue grama production on the shortgrass prairie. Chapter 2. In: French, N.R.(ed.), Perspectives in grassland ecology. Springer-Verlag, New York. 204 p.

Forrester, J. W. 1961. Industrial Dynamics. Massachusetts Institute of Technology Press.

Knisel, Walter G. (ed.) 1980. CREAMS: A Field scale model for Chemicals, Runoff, and Erosion from Agriculture Management Systems, USDA-SEA Conservation Research Report No. 26. Washington D. C. 643 p.

McGill, W. B., H. W. Hunt, R. G. Woodmansee and J. O. Reuss. 1981. PHOENIX: A model of the dynamics of carbon and nitrogen in grassland soils. Ecol. Bull. (Stockholm) 33:49-115.

Parton, W. J., J. S. Singh and D. C. Coleman. 1978. A model of production and turnover rates of roots in shortgrass prairie. J. Applied Ecol. 47:515-542.

Van Dyne, G. M. 1977. Content, evolution, and educational impacts of a system ecology course sequence. pp 9-23. In Innis, G.S. (ed.). New directions in the analysis of ecological systems. Simulation Councils Proc. Ser., Soc. for Computer Simulation (Simulation Councils, Inc.), La Jolla, California.

Williams, G. J. III. 1974. Photosynthesis adaptation to temperature in C3 and C4 grasses. Plant Physiology 54:709-711.

SIMULATION OF THE HERBAGE/ANIMAL INTERFACE OF GRAZING LANDS[1]

R. W. Rice[2], M. D. MacNeil[3], T. G. Jenkins[4], L. J. Koong[4]

Abstract. A herbage/animal interface simulation model was developed as a component of a comprehensive rangelands model. The interface was designed to represent consumer influence on vegetation through dynamic simulation of animal selectivity or preference for location (site) and for plant species or plant species group. The model component includes provision for wild and domestic herbivores as potential consumers. Animal response as weight change is monitored for growing domestic cattle. Under control of user supplied data, the model can represent a wide variety of rangelands and is responsive to management options which can be compared through simulation.

INTRODUCTION

Computer simulation models are particularly useful to describe and analyze the structure and function of complex biological systems which vary with time and whose components are highly interactive. The effects of exogenous and endogenous variables on a grazing land response are manifested typically after long periods of time. Often, the consequences of an unusual event such as grazing abuse, climatological extremes or fire, set events in motion which have long lasting effects. Experimental data may not have recorded the event(s) or were not collected over a long enough time span for estimation or quantification of long term cause and effect relationships.

Application of a systems analysis technology is appropriate where data are available and where the extension of the information from discrete data sets to other situations over space and time is desirable. The extension of data for rangeland ecosystems is necessary because the infinite variety of climatological, soils, management systems and options possible make direct experimentation for each situation or real time data collection impossible.

The herbage/animal interface model component was developed as a result of the formation of a Range Model Development group by the USDA, Agricultural Research Service.[1] The group was charged to develop a model of rangeland ecosystems for the purposes of: 1) prediction of range productivity; 2) evaluation of the effects of management practices 3) transfer of research results among different range ecosystems; and 4) determination of range research needs.

The multidisciplinary Range Model Development Group identified major components for a rangeland ecosystem and formed subgroups for the development of model components to be interfaced to comprise a Range Model. The rangeland ecosystem model includes the interactive subsystems of: 1) hydrology; 2) plants; 3) animals; 4) climate; and 5) insects. The whole systems model is referred to by the acronym SPUR (Simulation of Production and Utilization of Rangelands).

A number of rangeland simulation models varying in complexity and applicability have been reported (see Van Dyne et al. 1980). Many were theoretical and not designed for application to management while others were discrete and narrow in scope and application. Each modeling effort has been important in establishing the potential of the simulation approach, resulted in improved understanding of the structure and function of the system, and revealed limitations of available data sets. Most become obsolete when developed and have found little direct use for management applications. The focus of this model development effort was to develop, test and use a rangelands ecosystem model for management analysis on rangelands.

LITERATURE REVIEW

The major factor affecting the herbage/animal interface is non-random grazing by herbivores. Animals exhibit preference for location and for plant species within an available grazing area. Selectivity by animals affects plant species survival and productivity and determines the animal production potential of rangelands. The central concepts for development of the animal component of the model were animal preference for location, plant species preference, physical limit to herbage availability, dietary digestibility and excretal return.

Location or Site

Consumers express preference for location within a grazing area. For the most part, grazing strategy is to expend the least possible effort in obtaining food and water. On most rangelands, many degrees of intensity and frequency of plant use occur. Within any rangeland, vegetation in favored locations is utilized first and often more heavily than desirable, while less favored locations receive little attention. Preferred location for grazing is governed by many factors such as herbage density, water availability, relief, slope, elevation, exposure, natural and artificial barriers, herd social interactions, climate, and prior experience.

Key concepts for location of grazers used in the model development were grazing effort, terrain, travel required for consumers to obtain food and water and forage available.

Grazing Effort

There appears to be a limit to the amount of effort an animal will expend in grazing. The time spent at grazing, the number of grazing bites and bite size affect grazing success. In situations where available forage is sparse, the grazer may not be able to consume the quantity of forage which is determined by appetite or physical capacity. Arnold (1962), from a comprehensive review, concluded that at least four factors affected

[1] USDA ARS Range Model Development Group. J. Ross Wight, Chairman, NW Watershed Res. Center, Patti Plaza, 1175 So. Orchard, Boise, ID. 83705.

[2] Univ. of Arizona, Tucson, AZ. 85721.
[3] Univ. of Nebraska, Lincoln, NE. 68523
[4] Roman L. Hruska U.S. Meat Animal Research Center, USDA,ARS, Clay Center, NE. 68933.

forage/animal interrelationships. These were: 1) green forage density, 2) animal grazing time, 3) grazing bite rate and 4) grazing bite size. When green forage availability was low (<1000 kg/ha), grazing time increased (> 8.5 hours daily), biting rate was high (765/min) and total intake was low. Allden (1962) reported similar relationships. Stobbs (1973, 1974) observed that the amount of food eaten per bite and the frequency of biting by dairy cattle was affected by forage density and maturity. The cows spent 9-12 hours grazing daily with 12 hours the maximum time. Sheep grazing strategy observed by Nicholson et al. (1970) was to repeatedly traverse a pasture, consuming preferred plants each time. If stock density and time were sufficient, uniform use of vegetation was achieved through increased consumption of less desirable vegetation or extension to less desirable locations when the more preferred vegetation or locations were depleted. The first time over, they were very selective, then tended to regraze in centers of concentration, forming a patchwork of closely grazed areas. As grazing progressed, these areas were gradually extended with repeated return to closely grazed centers for the consumption of regrowth. Over a 20 year period, cattle at light stock density on shortgrass prairie were observed to establish islands or centers where close grazing occurred, while other locations were utilized very little. At heavy stock density, the patchiness of grazing was not as clearly defined (Kipple and Costello, 1960). Similar observations of non-random grazing were made by Milne et al. (1979) and Morris (1969).

Terrain

Herbivores, especially domestic cattle, are opportunists and will utilize rangeland areas which minimize effort. Cattle prefer lowlands or bottom lands followed by benches and uplands, and use slopes or sidehills to a lesser extent (Nagle and Harris, 1966; Julander, 1958; Julander and Jeffery, 1964; Cook, 1966; Senft et al. 1980). Consumers other than cattle often exhibit different preference for location while grazing, because they are much less constrained by fences, water and other manmade features used to control grazing of domestic livestock and are less likely to be influenced by terrain.

Julander and Jeffery (1964) reported that cattle made little use of slopes greater than 30% in Utah. Stark (1973) on a rangeland with a much steeper relief observed that cattle grazed on slopes of up to 50% although they preferred level areas and south-facing slopes. Fifty percent of grazing occurred on level areas with slope use relatively uniformly distributed. A significant terrain effect on cattle use in relatively small pastures (11, 22 and 125 ha) was observed on the gently rolling shortgrass prairie of eastern Colorado (Senft et al. 1980). Cattle spent 66% of grazing time in mesic lowlands with equal amounts of time on north or south facing slopes while a modest ridge area received little attention. Low et al (1981) observed cattle grazing on a large arid rangeland (170 km^2). The topography was gently sloping with moderately hilly areas interspersed. Thirteen separate vegetation communities were identified. Cattle distribution was affected by forage availability and quality and by availability of stock water. They tended to avoid the hilly areas.

It is difficult to separate terrain from vegetation influence since the lowland, mesic areas of a rangeland are often the most productive and usually have a longer green vegetative season than drier locations. Location preference for grazing is influenced by terrain with the use dependent upon relative area, the degree of difference and vegetation.

Travel

Grazing behavior, location and success is affected by the distance and effort required for animals to obtain food and water. Other influencing factors such as shade in summer, protection from cold in winter, age, sex, condition, cover and forage density affect the distance traveled and grazing location by herbivores in food gathering. There is a paucity of data describing the distance which animals will travel for food and the basic behavioral principles governing grazing behavior. The effects of climatic stress, animal physiological condition and the character of the vegetation all influence grazing behavior and the cost of food gathering. Management systems based upon behavioral patterns will be successful in improving grazing efficiency, insuring uniform utilization of available vegetation and optimal productivity without damage to the rangeland ecosystem. Few studies report both vegetation use patterns and travel. When forage density is high and pasture size is small, variation in use which could be related to terrain, water location, fences and corrals have been reported (Senft et al. 1980; Dean and Rice, 1974).

Daily travel is highly variable for cattle. Cattle in confinement traveled about 0.6 km daily (Schake and Riggs, 1972). Under rangeland conditions, daily travel averaged 5.8 km in Utah (Malachek and Smith, 1976); cattle on Texas rangeland traveled 5 km (Anderson and Kothmann, 1980); and on arid New Mexico rangeland, travel ranged from 7-15 km (Herbel and Nelson, 1966).

Water availability is often an important influence on grazing pattern and effort, especially on arid and semi-arid rangelands. Herbivores, because of the necessity for a microbiological predigestion for forage utilization, are required to maintain conditions within the gastrointestinal tract which are conducive to fermentation. Digestive tract water content of normal ruminants is considerably greater than for monogastric animals due to the necessity to provide a good environment for microbiological activity. Domestic cattle and sheep function most efficiently in their digestive processes when water is obtained at least once daily, but can perform with less frequent watering (Squires, 1978). Lange (1969) characterized the grazing pattern in relation to water. On an arid Australian rangeland with limited water, the pattern of sheep trails tended to be radial outward from water with alterations in the pattern due to terrain, fences and natural barriers. Valentine (1947) emphasized the influence of water availability on estimation of grazing capacity of forest lands. Sneva et al. (1973) experimentally varied the distance cattle had to walk from pasture to water with a series of lanes. When water was readily accessible, cattle drank twice daily then rested and ruminated nearby. When they had to trail 1.6 km from rangeland to water, they drank less often (once daily) and their calves were lighter at weaning than those from cows trailing lesser distances to water. Intermediate distances had little effect on cow and calf performance. They concluded that distances of up to 1.5 km to feed from water was not detrimental to performance of cattle in eastern Oregon. Beck (1978) reported that vegetation utilization by cattle in New Mexico pastures was not greatly affected by distances of up to 3 km from water, however there was a trend for greater use of vegetation up to 0.8 km from water. Hodder and Lowe (1978) from aerial observations in very large Australian rangeland determined that little grazing normally occurred beyond 5 km from water. However, under drought and forage shortages, cattle were observed as much as 20 km from water.

The effort or travel required for obtaining food and water can affect food intake. Winchester and Morris (1956) reported a positive correlation between food intake and water intake. Squires (1970, 1978) reported that food intake of sheep was reduced linearly when they were required to travel from 2.5 - 6.0 km daily to obtain food and water while total daily travel of up to 2.4 km did not affect food intake. They stated that the comfortable daily travel range for cattle was up to 6 km daily. Muller et al. (1976) from aerial surveys in Australia measured an average distance of cattle from water of 2.3 km.

Plant Species Preference

Animal selection for diet is the second basic concept included in the animal/plant interface component of the model. When animals are provided a variety of potential food sources, they influence ecosystem structure, the patterns of energy flows and nutrient cycles and the structure of plant communities through the effects of selective grazing. Diet selection is viewed by theoretical ecologists as a major element of the adaptive strategies of herbivore species. Divergence of diet selection patterns is an important determinant of the species composition of a plant community and has been correlated with consumer size, food availability, dietary strategy and competition (see review by Schoener, 1971). Dietary selection tactics were summarized by Ellis et al. (1976) to include factors of food density, relative abundance of preferred and less preferred foods, consumer nutrient requirements as affected by physiological state and environment, size of consumer, previous experience, palatability and satiety. Preference was described as the desire of a consumer for a particular class of food relative to other foods when all were equally available. When foods are present in unlimited abundance, then preference should determine food choice.

Preferred foods in rangelands are rarely unlimited and realized diet is usually a function of the combination of preference and food availability. For example, when the quantity of green herbage is high, sheep and cattle will select the more digestible fractions of the green herbage (leaf vs stem, or young herbage vs old herbage) whereas, when herbage supply is low, they will consume dry herbage and exert little selectivity within the green herbage component (Hamilton et al. 1973). Some consumer types are more selective in food habits than others and habitually choose dietary items which are less abundant than other food items. A comparison of the food habits of pronghorn antelope and cattle on shortgrass prairie illustrated that the proportion of foods eaten by the antelope departed radically from the proportions available; whereas cattle diets followed herbage availability closely (Rice et al. 1971). Factors inherent in plant species chemical or nutrient composition have been reported to be important determinants of selectivity (Westoby, 1974; Van Dyne et al. 1980: Fontenot and Blaser, 1965; Heady, 1964; Cook, 1959). Sugars and soluble carbohydrates appeared to elicit a positive response, protein was generally positive and fibrous fractions, cell wall structure and cellulose content, were negative for selection. The effects of the senses of sight, smell, taste and touch were also shown to have an influence on grazing selection (Krueger, 1972; Krueger et al. 1974; Westoby, 1974).

Caswell et al (1973) advanced the hypothesis that the metabolic pathway for carbon incorporation during photosynthesis was related to grass species preference. The plants exhibiting a C_3 photosynthetic pathway were preferred to those with a C_4 pathway of carbon incorporation. Akin and Burdick (1975) determined that the chemical composition and structure of temperate (C_3) and tropical (C_4) plants was different and that the digestibility of C_4 plants was lower than C_3 plants.

There is little data which identifies the basic factors determining preference by herbivores. Many of the relationships to forage structure or composition are not necessarily casual rather they could be the effect of selection. There are numerous research reports giving values for dietary botanical composition of herbivores, however, many do not report concurrent herbage availability. Van Dyne et al. (1980) presented a summary of much of the literature on diet selection and were able to illustrate some general relationships of food preference and dietary selectivity among various wild and domestic herbivores.

Physical Limit to Forage Availability

Herbage consumption by grazing animals rarely utilizes the total aboveground biomass except under extreme conditions. It is necessary to provide input for the proportion of the total biomass which grazing animals will consume before they initiate a search for desired species in less desirable locations or extend their grazing to less desirable plants within a location.

The proportion of total biomass of a particular plant which may be grazed in one day (e.g., one grazing event) varies according to plant growth form and morphology, the total amount available, the herbivore species and the herbage allowance for the grazing animals (Jameson, 1963; Hyder 1972, 1974). Plants which have an erect growth form and an elevated apice contain a greater proportion of their total weight in the upper proportion of the plant and a larger proportion of the weight can be removed by a grazing event. Plants least affected by grazing have a low growth form and a large proportion of their weight concentrated in the basal portion; consequently a low proportion of the total weight is removed by grazing (Schmutz et al. 1963; Barnes, 1976). Browse and other woody vegetation generally have a low removal rate by grazing. The estimation of available or grazeable browse is difficult and subject to large variation (Schuster, 1965; Barnes, 1976). If possible, browse species should be divided into current annual growth biomass and old or mature biomass and the amount of "grazeable" or accessible browse would be some proportion of the current annual growth.

Dietary Digestibility

The digestibility of diets consumed by grazing animals is one of the primary factors affecting animal productivity. Digestibility is utilized to compute the productive value of the diet and cattle weight changes in a modified version of the animal growth component of the Texas A&M systems model (Sanders and Cartwright, 1979). The digestibility of forages varies with plant species and with maturity of the plant (Waldo and Jorgenson, 1981). Digestibility is related to the nutritive value of a forage or a forage diet and affects both voluntary intake and the quantity of assimilated nutrients available for support of herbivores. Crampton et al. (1960) determined that voluntary intake or maximum potential intake by ruminants varied considerably among forages. The productive value of forages was dependent to a greater extent upon intake than digestibility. They calculated that relative intake potential accounted for 70% of the variation in forage productive value while digestibility was responsible for 30% of the variation in productive value. Intake and digestibilty are related. Van Soest et al. (1978) reported a correlation of 0.61 of digestibility with intake; however, differences in intake of forages with the same digestibility are common.

The protein content of forage diets has been reported as having a significant relationship to digestibility (Bredon et al. 1963; Horton et al. 1980; Tinnimit and Thomas, 1976; Cook et al. 1977; Brown et al. 1968; Rao et al. 1973). The plant component of the rangelands model monitors the nitrogen content of herbage grown. Protein (calculated from nitrogen) in herbage is related to the maturity of the plant and to the proportions of structural carbohydrates which are known to be related to forage productive value (Van Soest, 1967). Regression equations relating protein content to digestibility of forages were reported by Horton et al. (1980), Bredon et al. (1963) and Tinnimit and Thomas (1976) and are generally similar in form. When applied within a herbage species or class of varying maturity, a high degree of prediction accuracy was achieved.

Cattle Excreta

With grazing livestock, the net removal of fertilizing constituents depends upon the intensity of grazing, the output removed (e.g., milk production, weaned calf or gain) and the livestock production level. Estimates of net removal of nitrogen from shortgrass prairie indicate that a small proportion of the nitrogen was removed by yearling heifers grazing shortgrass prairie. Dean et al. (1975) measured the quantity of dietary nitrogen retained in body tissue of steers grazing shortgrass prairie to be 20% of that consumed. When related to total biomass production and herbage nitrogen, the proportion removed from the system was 5-10%. On highly productive pastures, the excretal return may be of little importance to the nitrogen economy (Suarez et al. 1981) whereas on extensive rangelands it may have some importance (Dean et al. 1975).

The primary influence of the grazing animal upon grasslands may be related to the enhancement of nutrient cycles through the feces and urine. Ungrazed vegetation, dead material and litter is more slowly decomposed than fecal material and urinary nitrogen is readily available for plant use (Dean et al. 1975; MacDiarmid and Watkin, 1972; Suarez et al. 1981; Weeda, 1967). Fecal and urinary deposition patterns could cause a significant redistribution of nutrients in a pasture. Petersen et al (1956) showed that excreta distribution was not random and that at a stocking rate of 2.5 hectare per animal per yer, 44.5% of the pasture would receive no excreta while 28% would receive more than one excreta event. Where stock density is less, an even more non-random excreta distribution would be expected. Hilder and Mottershead (1963) reported excretal densities to be high in areas occupied when sheep were engaged in non-grazing activities (loafing, ruminating, resting). On shortgrass rangeland, Senft et al. (1980) reported that fecal pat density was related to the grazing location of cattle, however small areas used for non-grazing activities received the greatest density.

Fecal and urinary excretion by grazing animals can have a significant local effect on palatability (Omalike, 1981; MacDiarmid and Watkin, 1972) and fertility (Petersen et al. 1956; Lotero et al. 1966). The net effect upon the nitrogen economy of a productive grassland is small, but for areas with limited productivity or fertility, these events may be significant. It appears that excretal distribution is related to grazing activity and resting or other non-grazing activities and that approximately 80-90% of deposition is distributed at locations where grazing occurs (Senft et al. 1980).

The quantity of urinary nitrogen excreted is related to the dietary nitrogen intake, total food intake and animal body weight (ARC, 1965; Stillwell et al. 1982). Fecal nitrogen excretion is related to intake and digestibility of food nitrogen.

PROCEDURES

The animal/plant interface component model was developed in the same manner as that for research. The purpose was defined, the problem was stated in the form of hypotheses, the elements of the subsystem to be modelled were identified, the interrelationships of the elements were defined in the form of a subsystem flow chart which identified the state variables, flows of biomass and the interactions among the state variables.

The purpose of the development effort was to identify factors influencing the herbage/animal interface of rangelands and to formulate a dynamic subsystem simulation model of the interface. The hypotheses for development were:

1. Herbivore impact upon vegetation is a result of selective grazing. Selectivity is expressed in two ways: herbage species and location preference.

2. Vegetation determines herbivore response through the quantity and quality available for consumption and herbivore grazing success.

A systems structure was developed which identified the components and their interactions. The state variables were defined and the relationships formulated in a deterministic difference equation format with a one day time step for computation of the dynamic operation of the interface. A supply/demand concept is central to the operation of the model and forms the interface between the plant and animal subsystem components of the rangelands ecosystem model. Herbage supply and animal demand are characterized in properties which were hypothesized to influence animal selective grazing. Literature data were used to formulate the model component parameters, flows and interactions. Validation will be accomplished following integration of the animal component into the whole rangelands ecosystem model by comparison of component and whole model simulated output with research data from specific rangelands. Applicability of the model to various ecosystems will be tested by comparison of simulation results with research data from a variety of rangelands

Specific factors considered in the subsystem component development were location or site as affected by grazing effort required including travel for water and food, terrain and the growth form of the vegetation, plant species or species group preference, dietary quality and excretal return to the rangelands.

MODEL STRUCTURE AND FLOW

The structure, sequence and flow of the animal component is illustrated in figure 1. Two elements: carbon as aboveground biomass and nitrogen, enter from the plant growth component and are common elements monitored throughout the rangeland ecosystem model. Input to the consumer component is herbage available. This data is stored in the form of a matrix whose dimensions are defined in terms of the postulated animal/plant functional relationships. Data input descriptive of relative preference is required for location (site) and vegetation species or class. Insect consumers exercise their demand (offtake) followed by non-cattle consumers. The cells of the plant supply matrix (herbage values) are reduced by these consumer demands. Cattle then exert their dietary demand on the plant supply. Cattle response is monitored in detail as weight response to realized diet. The sequence of consumer demands on the supply is based upon the competitive advantage of insects and the relatively greater mobility of wild herbivores as compared to cattle.

Supply and demand matrices are compared. If supply is adequate at preferred locations and for preferred plant species, the realized intake for cattle is removed from the appropriate cells of the plant supply matrix, daily plant growth and degradation is computed and a new plant supply matrix is generated for the next daily time step.

The dietary digestibility is computed and used to drive the livestock growth portion of the model which is a modified version from the model of Sanders and Cartwright (1979).

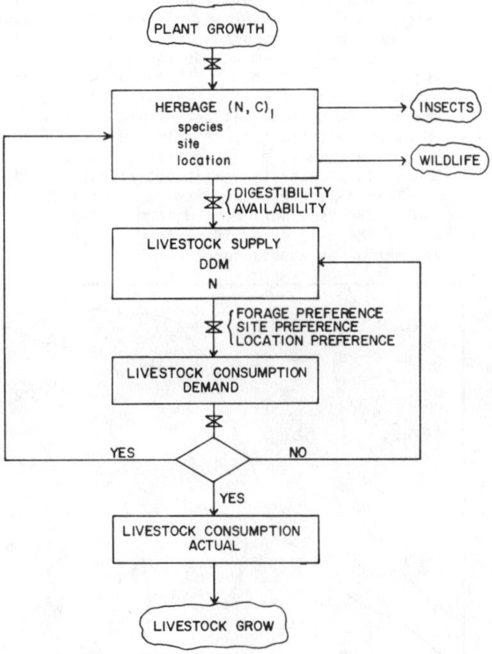

Figure 1: Structure and Flow Chart for Animal Subsystem component.

N = nitrogen (%)
C = carbon (g/m^2 forage)
DDM = digestible dry matter

MODEL DATA REQUIREMENTS AND OPERATION

The applicability of the animal model component is dependent upon data inputs which characterize the specific situation to be studied. The central concepts for model operation are grazing behavior expressed as location or site preference, plant species preference and daily physical limitations to herbage use.

Site Designation

The rangeland model, plant production component generates a matrix of forage production in the form of above ground biomass (kg dry matter per square meter) and nitrogen (percent). The matrix provides up to nine different locations or sites which are defined by the user to represent the rangeland to be studied. Within each site, there is a provision for up to seven herbage species or species groups which have two maturity states (green or dry). The species groups are defined by the user to characterize the vegetation and anticipated animal grazing response of the rangeland to be simulated.

A theoretical site overlay designation for a specific case is illustrated in figure 2. The preference for location on this rangeland was a function of terrain and distance from water. Terrain was divided into 3 general categories: lowland, sidehill or slope. Three distance zones were superimposed upon the terrain overlay. In this case, 0 - 1.5 km was designated as the distance within which water location would not be a significant factor affecting grazing use. A second distance zone was determined to be 1.5 - 3.0 km and a third greater than 3.0 km. Different areas on a rangeland may receive the same designation as is illustrated in figure 2; however no more than nine different sites are possible in the model. The user next must determine the area of each of the nine sites identified and provide this information as a data input.

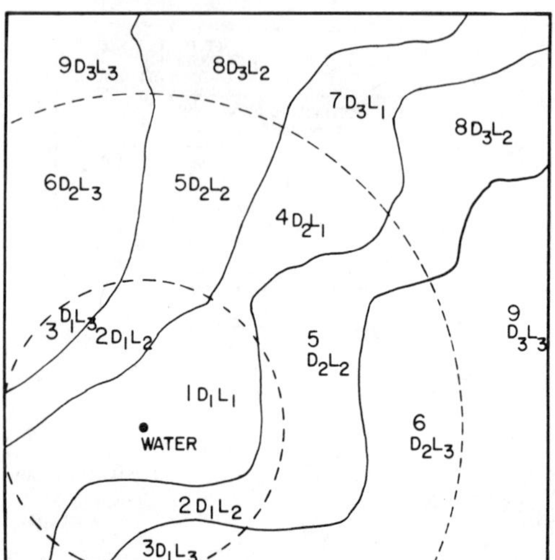

Figure 2: Site Designation of a Theoretical Rangeland, Three Terrain and Distance Classes.

Location (L)	Distance from H_2O
1 = lowland	1 = 0 - 1 km
2 = sidehill	2 = >1 - 2 km
3 = upland	3 = >2 km

The concept of location or site need not include the functional relationships used for this example. The definition of location as it affects herbivore grazing is the responsibility of the user. This insures that the model will have general applicability. Once sites are designated and area of the resulting sites determined, it is necessary to construct relative vegetation preference vectors for the consumers to be simulated.

Cattle Consumers

The model predicts the performance of growing steers. Domestic cattle are handled in detail because they are most often the consumer in a rangeland which are manipulated through stocking density and controlled grazing by management. The primary purpose for the model was to give managers a practical tool to analyze specific rangeland management alternatives and to predict the consequences of management options over a long term production horizon. The data inputs required for cattle are similar in form and concept to those for non-cattle consumers. In addition, the dry matter digestibility of the realized diet for cattle is required. It is very important that the data for cattle be as representative of the situation to be studied as possible since cattle response is the primary diagnostic information output from the consumer component of the rangeland model.

Data requirements are:

1. Number of steers or steer equivalent
2. Mature size of a typical cow representing breeding background of the steers
3. Age, in days at turnout
4. Time (days) when removed from rangeland or pasture
5. Supplements in amount (kg/hd/da), % digestibility, day to begin feeding, day to end feeding
6. Preference for site or location
7. Preference for forage species/class
8. Physical limit to forage availability

The key concepts for operation of the model are contained within the three data inputs of preference for site or location and forage species/class and the physical limit to forage availability.

Preference for site

For illustration, a preference vector for sites was constructed as shown in table 1. The concept for the location preference vector is the expected proportion of grazing use a population of cattle would be expected to expend if there were no differences or limitations in available vegetation. It is required that the vector be expressed in decimal proportions, all locations must receive a value and the vector must sum to 1.00. Preferred locations should receive a high value while those where little use is expected, a low value.

The user predicts that when vegetation is equivalent and equal distance from water, a population of cattle prefer to graze on lowlands, however, some would occasionally use slopes and uplands with uplands receiving greater attention. The estimates for terrain use are formulated into a three component vector of decimal proportions summing to 1.00 [(lowlands (1) = 0.6, slopes (2) = 0.15, uplands (3) = 0.25)]. A second three component vector is provided by the user to predict the proportion of grazing use a population of cattle would exert in the three distance zones shown in figure 2, under the concept of equivalent unlimited vegetation within each zone and no terrain differences. Under this condition, it was predicted that 0.75 of the grazing time of cattle would be in zone 1, 0.18 in zone 2 and .07 in zone 3. A nine element vector resulting from the two concepts of terrain and water is generated by multiplication of the two three element vectors for terrain and distance from water (table 1).

Table 1. Composite Preference Vector for Site

Site	Dist. Zone		Terrain[1]		Composite Vector[2]
	Zone	Pref.	Class	Pref.	D x T
1	1	.75	1	.60	.45
2	1	.75	2	.15	.11
3	1	.75	3	.25	.19
4	2	.18	1	.60	.11
5	2	.18	2	.15	.02
6	2	.18	3	.25	.05
7	3	.07	1	.60	.04
8	3	.07	2	.15	.01
9	3	.07	3	.25	.02
					1.00

[1] 1 = Lowland, 2 = Sidehill, 3 = Upland
[2] D = Distance, T = Terrain

Table 2. Composite Plant Group Preference Vector

Vector element	Species[1] group	Species pref.	G/D pref.	Composite vector
1	CSG G	.30	.90	.270
2	CSG D	.30	.10	.030
3	WSG G	.25	.90	.225
4	WSG D	.25	.10	.025
5	CSF G	.15	.90	.135
6	CSF D	.15	.10	.015
7	WSF G	.10	.90	.090
8	WSF D	.10	.10	.010
9	PB G	.15	.90	.135
10	PB D	.15	.10	.015
11	UPB G	.05	.90	.045
12	UPB D	.05	.10	.005
13	empty	.00	0.0	.00
14	empty	.00	0.0	.00
				1.00

[1] CSG = Cool season grass, WSG = Warm season grass, CSF = Cool season forbs, WSF = Warm season forbs, PB, Palatible browse, UPB = Unpalatible browse, G = green, D = dry vegetation

Preference for Plant Species Groups

The model accommodates up to seven vegetation classes at two maturities (green or dry). The bases for vegetation preference are two-fold: 1) herbivores prefer green vegetation, and 2) in a mixed vegetation rangeland, plant species are consumed selectively.

The shortgrass prairie providing this example produces six plant species groups which have been identified as determinants of animal preference. Within each group, both green and dry vegetation is normally available and has an important influence on herbivore response. The plant species classes are described in table 2. The model will accommodate up to 14 vegetation classes, however, it is not necessary to provide data for all potential elements. For cattle, it was determined that given unlimited availability, green vegetation would comprise 90% and dry material (standing dead) 10% of the diet. A two element vector containing these ratios in decimal form is multilied by preference ratios for the six plant species classes with the resulting 12 element vector generated as data input for the animal component model (table 2). The two preference vectors for location and vegetation are central to the operation and generality of the model. It is important that they are constructed carefully and that they reflect animal response under the conditions of unlimited accessibility for sites with no differences in vegetation and unlimited availability of vegetation classes.

Physical limits to forage availability

The concept for the physical limit or accessibility of the total biomass of a plant is the proportion of the total biomass readily eaten by grazing animals in a single grazing event. This value is not defined in the literature and is not biologically the same as the grazing use reported in many studies as the percent of total herbage removed during a grazing season. Season long use rates do provide an indication of relative use rankings for a species. Sensitivity tests of this parameter will be made in simulation runs when both plant and animal subsystems can be run simultaneously allowing predicted seasonal dietary intakes and patterns of vegetation to be validated with specific rangeland information. If this proportion is set too high in the model, non-biological results will be obtained as the model will allow use of preferred species and locations to be greater than the actual situation and the feedback on plant growth and survival will be strongly negative causing the disappearance of the plant from the cell. If this limit is too low, the model will yield unrealistic results on grazing distribution and use.

The vegetation classes represented include both tall, erect plants (cool season grass), for example, western wheatgrass (Agropyron smithii Rybd.), and low growing species which have a large amount of weight in the basal area and which are difficult to graze closely by cattle (for example, Blue grama [Bouteoloua gracilis (H.B.K.) Lag. ex. Steud]). We would expect the physical limitation or the proportion of the weight of the plant removed by a grazing event to be considerably greater for western wheatgrass than for blue grama. It was estimated for western wheatgrass that 60% of the weight of green material would be readily consumed; hence, a value of 0.6 was input for the physical limitation vector of cool season grass. For blue grama, a lower proportion of green material would be removed from a desired area or location before the animal begins to search farther and to consider less desirable species (input value .4). The same constraints are applied to dry cool and warm season vegetation. The forb component in this shortgrass prairie is readily accessible and a physical limit of 0.6 is input. The browse component is primarily the half shrub, fringed sagewort (Artemesia frigida Willd.) which is a readily accessible, low growing shrub. The physical limit for this shrub was set at 0.5 for green material and 0.25 for mature material. The vector data input for species class physical limitation is shown in table 3.

Table 3. Herbage Species Class Physical Limits to Utilization

Vector element	Species[1] group	Physical limitation
1	CSG G	.60
2	CSG D	.60
3	WSG G	.40
4	WSG D	.40
5	CSF G	.60
6	CSF D	.60
7	WSF G	.60
8	WSF D	.60
9	PB G	.50
10	PB D	.25
11	UPB G	.40
12	UPB D	.20
13	empty	.00
14	empty	.00

[1] CSG = cool season grass, WSG = warm season grass, CSF = cool season forbs, WSF = warm season forbs, PB = palatible browse, UPB = unpalatible browse, G = green, D = dry vegetation

This vector need not sum to 1.0 as is necessary for site and species preference vectors. From these data, a matrix containing up to 126 cells, with a relative preference value for each cell is formulated in the model. This preference matrix should contain a value for each cell, however less desirable locations and vegetation classes have low values.

Digestibility

The digestibility of the realized diet in the animal component submodel is predicted by a regression equation developed from data summarized by Cook et al. (1977). They reported protein and digestibility data for 143 typical rangeland grasses. The equation used for the calculation of dietary dry matter digestibility is shown (equation 1).

$$DMD = 37.1 + 2.4 X \qquad (1)$$

DMD is dry matter digestibility
X is % protein in the diet
$r^2 = 0.33$, N = 143

The coefficient of determination is not as high as literature values which were derived with discrete data sets; however, this equation represents the range of species digestibilities typical of rangelands and is similar in the parameters for the y intercept and slope to the prediction equations of Bredon et al. (1963) and Horton et al. (1980). The equation is applied to the nitrogen content of the realized diet, rather than to individual species groups tending to minimize some of the variability in prediction which could occur if DMD was calculated for individual species. The equation applies well to rangeland grasses and forbs, however, is inaccurate for browse species. Under most rangeland situations, browse is not a major component of cattle diets (Wilson, 1969). When small amounts are present, the equation will provide representative values for dietary digestibility, since the contribution of browse to the diet is usually small.

Excretion

The excretion component of the animal submodel distributes nitrogen excreta at locations in relation to grazing use in quantities calculated from an average nitrogen digestibility (75%) of food nitrogen (ARC, 1965; NRC, 1972) and nitrogen excretion in urine based upon 2 x endogeneous urinary nitrogen (ARC, 1965). The model accommodate the non-random distribution of excreted nitrogen by cattle in relation to grazing use, but does not account for concentrations of material in locations where animals engage in loafing, ruminating and resting.

Non-cattle Consumers

The model will accommodate up to 10 wildlife or non-cattle consumers. Wildlife exercise their demand based on dietary and location preference, population size and forage intake rate per animal. Dietary preference, body weight and average intake rates for many consumers were summarized by Van Dyne et al. (1980). The non-cattle herbivore subcomponent does not monitor performance, it accounts for the herbage demand of these animals. The sequence of calculation allocates forage to non-cattle herbivores prior to cattle demand. However, the system will handle competition for resources since a daily time step is used and non-cattle herbivores act upon the herbage supply matrix, then cattle exert their demand within the one day time-step. The herbage supply matrix is then reduced by the combined realized diets of all consumers and increased by daily incremental plant growth. Herbage supply matrix cells which are under heavy demand by the consumer moiety will be limiting on the next and subsequent time steps for both non-cattle and cattle consumers.

The applicability of the non-cattle consumers to specific rangelands to be studied is under the control of data inputs by the user. Concepts for data are the same as for cattle consumers. Data requirements for operation of this component are:

(1) Preference for site. For favored sites, a high value is input, however, for proper

Table 4. Location Preference, Non-Cattle Herbivores

	L1 Bottom	L2 Sidehill	L3 Upland
Mule deer	.4	.2	.4
Whitetail deer	.3	.3	.4
Horse	.1	.4	.5
Elk	.3	.3	.4
Antelope	.3	.1	.6
Rabbit	.7	.1	.2
Sheep	.3	.2	.5
Goat	.3	.3	.4

operation of the model, all available locations should receive a value. An example of a location or site preference vector is shown in table 4 for eight non-cattle consumers.

(2) Preference for species and/or species groups (table 5).

(3) The days or time within the scope of the model run when animals are expected to be present. Wildlife herbivores are mobile and often use certain areas during part of the year. The model will accommodate this mobility through data input control.

(4) Estimated dry matter intake (kg/da) for each wildlife species considered. This value is an average daily intake estimate for an average animal within a population. The model does not consider classes of animals within a population. If this is desirable, the animal species groupings could be subdivided by population class, however, no more than ten different animal species or species subclasses can be simulated.

(5) Population size for each non-cattle consumer. You will notice that domestic consumers other than cattle can be entered into the simulation (tables 4,5). However production of the consumers in this model subcomponent is not simulated.

The output from the non-cattle component of the model is daily herbage consumption in kg per location (site) by plant species/group. Input data for the preference matrices controls the searching strategy of non-cattle consumers. The structure of this component of the model results in a simultaneous extension of grazing to less preferred locations and to less preferred species within location when preferred forage supply is limited. There is little data available which has determined the searching priorities of herbivores, however, season long plant species intake for cattle, sheep and other consumers for a variety of pastures and rangelands is documented (Van Dyne et al. 1980).

Herbage Animal Interface

Example

A simplified example of the calculation sequence of the herbage/animal interface is presented to illustrate the procedures followed in model calculations and the influence of the key input vectors, location preference and plant species preference on animal grazing strategy. For this illustration, the plant supply is defined in the form of a 2 x 4 matrix with two locations and four plant species groups (figure 3).

Aboveground standing biomass (kg) reduced by the physical availability function (0.6 of total biomass), is entered into each cell of the supply matrix. The user estimates that given unlimited, preferred vegetation, .80 of grazing would occur in location 1 (L1) and .20 in location 2 (L2). Herbage

Table 5. Dietary Preference, Non-Cattle Herbivores

	Mule Deer	Whitetail Deer	Horse	Elk	Antelope	Rabbit	Sheep	Goat
WSG-G	.10	.10	.45	.20	.10	.10	.22	.30
WSG-D	.003	.003	.015	.025	.003	.005	.01	.003
CSG-G	.10	.10	.45	.20	.10	.10	.22	.20
CSG-D	.003	.003	.015	.025	.003	.005	.01	.003
WSF-G	.25	.25	.015	.20	.25	.15	.22	.04
WSF-D	.02	.02	.008	.02	.02	.023	.02	.005
CSF-G	.25	.25	.015	.20	.25	.15	.22	.04
CSF-D	.02	.02	.015	.02	.02	.02	.02	.005
BR-G	.25	.25	.008	.10	.25	.42	.05	.30
BR-D	.004	.004	.004	.01	.004	.02	.01	.004
DMI kg/anim	1.1	.88	11.4	4	.88	.12	1.5	.8
Av body wt,kg	54	44	380	200	44	2	70	25

	L1	L2
S1G	100	100
S1D	50	50
S2G	200	200
S2D	100	100

Figure 3. Example supply matrix (kg/cell) 2 species, 2 locations, (L = Location, S = Species, G = Green, D = Dry)

	L1 0.80	L2 0.20
S1G 0.50	.40	.10
S1D 0.05	.04	.01
S2G 0.40	.32	.08
S2D 0.05	.04	.01

Figure 4. Species and location preference matrix

	L1 SUPPLY	L1 DEMAND	L2 SUPPLY	L2 DEMAND
S1G	100	40	100	10
S1D	50	4	50	1
S2G	200	32	200	8
S2D	100	4	100	1

Figure 5. Preferred diet vs. supply demand = 100 kg/da

	L1 SUPPLY	L1 DEMAND	L2 SUPPLY	L2 DEMAND
S1G	100	160*	100	40
S1D	50	16	50	4
S2G	200	128	200	32
S2D	100	16	100	4

Figure 6. Preferred diet vs. supply demand = 400 kg/da, *Cell 1 limiting

	L1	L2
S1G	100	25
S1D	10	3
S2G	80	20
S2D	10	2

Figure 7. Partial diet, satisfied at preferred proportions. (Sum = 250 kg, deficiency = 150 kg)

	L1 SUPPLY	L1 REVISED	L2 SUPPLY	L2 REVISED
S1G	100		100	75
S1D	50	40	50	47
S2G	200	120	200	180
S2D	100	90	100	98

Figure 8. Revised supply, partial diet deducted

	L1	L2
S1G		.17
S1D	.07	.02
S2G	.53	.13
S2D	.07	.01

Figure 9. Revised preference, to satisfy balance of diet, Cell 1 depleted.

	L1	L2
S1G		26
S1D	10	3
S2G	80	20
S2D	10	1

Figure 10. Revised demand, to meet balance of diet not satisfied at first preference.

	L1	L2
S1G	100	51
S1D	20	6
S2G	160	40
S2D	20	3

Figure 11. Realized diet, demand = 400 kg, Cell 1 limiting.

		L1 Preferred 0.80	L1 Realized	L2 Preferred 0.20	L2 Realized	
S1G	0.50	.40	.25	.10	.13	0.38
S1D	0.05	.04	.05	.01	.02	0.07
S2G	0.40	.32	.40	.08	.10	0.50
S2D	0.05	.04	.05	.01	.01	0.06
			0.75		0.25	

Figure 12. Preferred vs. realized dietary proportions.

species preference for the four species classes were estimated as shown and an animal preference matrix constructed (figure 4). The concept for the herbage species preference is that with unlimited availability and accessibility, the herbivore would be expected to consume a diet consisting of 50% species 1, green (S1G), 40% species 2, green (S2G), 5% species 1 dry (S1D), and 5% species 2 dry (S2D).

Case I. If the total daily demand for the consumer is 100 kg dry matter daily based upon input data on animal weight, age, mature size and number of animals, the dietary demand would be as is shown in figure 5. The supply and demand matrices are compared. If there are no cells in the demand matrix greater than any cell in the supply matrix, dietary demand becomes the diet, is deducted from supply, and the model continues to calculate cattle response. The supply matrix is updated and computations for the next daily time step initiated.

Case II. Demand by cattle is 400 kg per day. The dietary demand matrix is computed and shown in figure 6. Herbage supply and cattle demand are compared. The first cell of the supply matrix is less than demand (Supply = 100 kg, demand = 160 kg). The proportion of the dietary demand which can be met based upon the limiting cell is computed to be 250 kg (figure 7). The deficiency of preferred supply is then 400 - 250 kg = 150 kg. A new dietary preference matrix is calculated based upon the use of all available herbage in cell 1, 1 (figure 8). The supply matrix is decreased by the partial diet consumed in preferred quantities and the reduced supply matrix compared with the new demand matrix (figure 9, figure 10). As there are no limiting cells in the revised demand matrix, the realized diet is calculated as the sum of the matrices shown in figures 7 and 9. The realized diet summing to the demand of 400 kg is computed (figure 11). The digestibility of the realized diet is computed (equation 1) and used to drive the cattle response portion of the model. A comparison of preferred dietary proportions with realized proportions is shown in figure 12. When a cell was limiting cattle consumed greater amounts of vegetation in the remaining cells. The direction and magnitude of the changed diet is under the control of the relative preferences for location and plant species. In this case, the direction of the search was greater towards Species 2, Green than towards Location 2. Species 2, Green provided 10% more food with 8 of the 10% from location 1, while location 2 provided a total of 5% more forage when there was a 37.5% deficiency in demand satisfied by the most preferred diet.

The simultaneous extension of grazing animals effort to location and species from preferred locations and species simulates a biological response. The validation and testing of the quantitative responses of the model requires season long runs as there is little data with daily resolution on food habits and grazing behavior for validation.

SUMMARY AND CONCLUSIONS

The purpose of the rangelands simulation effort is to evaluate the short and long term effects of rangeland ecosystems management. Since the impact of grazing animals is usually site specific and related to local effects, the model must simulate the effects of non-random grazing at the pasture level. Management for most rangelands is applied to domestic livestock by control of poulations, season of use and through grazing systems whose primary purpose is to override the natural tendencies to graze selectively. The animal subcomponent is structured to consider grazing behavior and is applicable to many situations through the data input requirements.

The developmental process for the animal component of the rangelands simulation model (SPUR) included three key concepts governing plant and animal response on rangeland ecosystems. These concepts; plant species preference, location preference, and daily herbage use levels coupled with stock density or animal demand, are central to the simulation of short and long term effects of grazing on plant species production and succession, herbivore competition and livestock production and grazing efficiency.

Data inputs defining these key concepts are necessary for the generalization of this model to different rangelands. Faulty data input will yield erroneous results from the simulation that may or may not be discerned. This could lead to selection of the wrong management strategy. For this reason, extensive testing and evaluation of the model will be necessary prior to general use for management analyses. Testing and validation will not be possible until the subsystem is integrated into the whole rangelands model and season long or longer term runs compared with existing data. Experimental simulation with the animal subsystem can evaluate the biological reality of the postulated concepts governing component behavior and ascertain the sensitivity and/or resiliancy of the herbage/animal interface component.

Experimental data defining daily location and plant species preference are not available, however, season long or longer term information exists and will be used for validation. The mechanism, controlled by data input for physical limits to utilization, which initiates the extension of herbivore grazing effort to less desirable locations and plant species has not been identified. Simulation runs and specific experiments will be necessary to validate this concept.

The rangelands model is intended for use by managers and researchers. Future efforts will be to evaluate and validate the whole rangelands ecosystem model in various situations. Developmental efforts have highlighted the importance of the grazing strategy of herbivores on vegetation and animal productivity.

The herbage/animal interface of the rangelands ecosystem model is unique because it provides for spatial (location or site), non random grazing use as well as for plant species preference. The subsystem component will generalize to a wide variety of rangelands because the data inputs for location and plant species preferences define the specific case to be studied. The data input requirements are of a form that experienced managers will be able to provide from practical or applied management information. The component is simple mathematically and computationally and requires no complex functions or constraints for proper operation. The evaluation of a simulation is based upon domestic cattle performance for which adequate data is available in practical situations.

LITERATURE CITED

Akin, D.E., D. Burdick. 1975. Percentage of tissue types in tropical and temperate grass leaf blades and degradation of tissues by rumen microorganisms. Crop. Sci. 15:661-668.

Anderson, D.M. and M.M. Kothmann. 1980. Relationship of distance traveled with diet and weather for Hereford heifers. J. Range Manage. 33:217-220.

ARC. 1965. The nutrient requirements of farm livestock, No. 2. Ruminants, Agricltural Research Council, Lond.

Arnold, G.W. 1962. Factors within plant associations affecting the behavior and performance of grazing animals. In: Grazing and Marine Environments (Ed. D.J. Crisp). Proc. Brit. Ecol. Soc. Bangor. p. 133-154.

Barnes, D.L. 1976. A review of plant-based methods of estimating food consumption, percentage utilization, species preferences and feeding patterns of grazing and browsing animals. Proc. Grassland Soc. Sth. Africa 11:65-71.

Beck, R.F. 1978. A grazing system for semiarid lands. Proc. 1st Int. Rangeland Cong. ed. D.N. Hyder. Soc. for Range Management, Denver, CO. p 569-572.

Bredon, R.M., K.W. Harker, B.R. Marshall. 1963. The nutritive value of grasses grown in Uganda when fed to Zebu cattle. I. The relation between the percentage of crude protein and other nutrients. J. Agric. Sci. 61:101-104.

Brown, R.H., R.E. Blaser, J.P. Fontenot. 1968. Effect of spring harvest date on nutritive value of orchard grass timothy. J. Anim. Sci. 27:562-567.

Burdick, D. 1979. Composition and microbial degradation of forage cell walls. In: Visiting Scholar Lectures, Univ. of Ark. Agr. Exp. Sta. Ed. B. Carl and F.E. King, 1979. p 15-31.

Caswell, H., F. Reed, S.N. Stephenson, P.A. Werner. 1973. Photosynthesis pathways and selective herbivory: A hypothesis. Amer. Nat. 107:465-480.

Cook, C.W. 1959. The effect of site on the palatability and nutritive content of seeded wheatgrasses. J. Range Manage. 12:289-292.

Cook, C.W. 1966. Factors affecting utilization of mountain slopes by cattle. J. Range Mgmt. 19:200-204.

Cook, C.W., R.D. Child, L.L. Larson. 1977. Digestible protein in range forages as an index to nutrient content and animal response. Colo. State Univ. Range Sci. Dept. Science Series. No. 29. 65 pp.

Crampton, E.W., E. Donefer, L.E. Lloyd. 1960. A nutritive value index for forages. J. Anim. Sci. 19:538.

Dean, R.E., R.W. Rice. 1974. Effects of fences and corrals on grazing behavior. Proc. West. Sect. Am. Soc. Anim. Sci. 26:88-89.

Dean, R., J.E. Ellis, R.W. Rice, R. E. Bement. 1975. Nutrient removed by cattle from a shortgrass prairie. J. Applied Ecology 12:25-29.

Ellis, J.E., J.A. Weins, C.F. Rodell, J.C. Anway. 1976. A conceptual model of diet selection as an ecosystem process. J. Theor. Biol. 60:93 108.

Fontenot, J.P., R.E. Blaser. 1965. Symposium on factors influencing voluntary intake of herbage by ruminants: Selection and intake of grazing animals. J. Anim. Sci. 24:1202-1208.

Hamilton, B.A., F.J. Hutchinson, P.C. Annis. 1973. Relationships between the diet selected by grazing sheep and the herbage on offer. Aust. J. Agric. Res. 24:271-277.

Heady, H.F. 1964. Palatability of herbage and animal preference. J. Range Management. 17:76-82.

Herbel, C.H., A.B. Nelson. 1966. Activities of Hereford and Santa Gertrudis cattle on a southern New Mexico range. J. Range Manage. 19:173-176.

Hilder, E.J., B.E. Mottershead. 1963. The redistribution of plant nutrients through free-grazing sheep. Aust. J. Sci. 26:88-89.

Hodder, R.M., W.A. Low. 1978. Grazing distribution of free ranging cattle at three sites in Alice Springs District, Central Australia. Australian Rangeland J. 1:95-105.

Horton, G.M.J., D.A. Christensen, G.M. Steacy. 1980. In vitro fermentation of forages with inoculum from cattle and sheep fed different diets. Agron. J. 72:601-605.

Hyder, D.N. 1972. Defoliation in relation to vegetative growth. In: Biology and Utilization of Grasses. Acad. Press. ch. 22, pp. 304-317.

Hyder, D.N. 1974. Morphogenesis and management of perennial grasses in the United States. USDA Misc. Publ. 1271. p 89-98.

Jameson, D.A. 1963. Responses of individual plants to harvesting. Bot. Rev. 29:532-594.

Julander, O. 1958. Techniques in studying competition between big game and livestock. J. Range Manage. 11:18-21.

Julander, O., D.E. Jeffery. 1964. Deer, elk and cattle range relations on summer range in Utah. Trans. N. Amer. Wildl. Conf. 29:404-414.

Kipple, G.E., D.F. Costello. 1960. Vegetation and cattle responses to different intensities of grazing on short grass ranges of the central Great Plains. USDA Tech. Bul. 1216.

Krueger, W.C. 1972. Evaluating animal forage preference. J. Range Manage. 25:471-475.

Krueger, W.D., W.A. Laycock, D.A. Price. 1974. Relationships of taste, smell, sight and touch to forage selection. J. Range Manage. 27:258-262.

Lange, R.T. 1969. The piosphere: sheep track and dung patterns. J. Range Manage. 22:396-400.

Lotero, J., W.W. Woodhouse, R.G. Peterson. 1966. Local effect on fertility of urine voided by grazing cattle. Agron. J. 58:262-265.

Low, W.A., M.L. Dudzinski, W.J. Muller. 1981. The influence of forage and climatic conditions on range community preference of Shorthorn cattle in central Australia. J. Applied Ecol. 18:11-26.

MacDiarmid, B.N., B.R. Watkin. 1972. The cattle dung patch. 3. Distribution and rate of decay of dung patches and their influence on grazing behavior. J. Br. Grassld. Soc. 27:43-54.

Malechek, J. and B.M. Smith. 1976. Behavior of range cows in a mixed prairie pasture. J. Range Manage. 29:9-12.

Milne, J.A., L. Bagley, Sheila Grant. 1979. Effects of season and level of grazing on the utilization of heather by sheep. 2. Diet selection and intake. Grass and Forage Sci. 34:45-53.

Morris, R.M. 1969. The pattern of grazing on continuously grazed swards. J. Br. Grassld. Soc. 24:65-70.

Muller, W.J., W.A. Low, C. Lendon, M.L. Dudzinski. 1976. Variation in grazing patterns of free ranging cattle in a semiarid area. Proc. Aust. Soc. Anim. Prod. 11:461-464.

Nagle, J.P., G.A. Harris. 1966. Grazing compatability of cattle and elk on mountain ranges. Proc. X. Int. Grassld. Cong. p. 994-997.

Nicholson, I.A., I.S. Patterson, A. Currie. 1970. A study of vegetation dynamics: selection by sheep and cattle in Nordus pasture. In:

Animal Populations in Relation to Their Food Resources. Ed. A Watson. p. 129-143. Blackwell Scientific Publ. Oxford.

NRC. 1972. Atlas of Nutritional Data on United States and Canadian feeds. NAS. Wash. DC. 2101 Constitution Ave., Washington, DC. 20418

Omaliko, C.P.E. 1981. Dung deposition, breakdown and grazing behavior of beef cattle at two seasons in a tropical grassland ecosystem. J. Range Manage. 34:360-362.

Petersen, R.G., W.W. Woodhouse, H.L. Lucas. 1956. The distribution of excreta by freely grazing cattle and its effect on pasture fertility. 2. Effect of returned excreta on the residual concentrations of some fertility elements. Agron. J. 48:444-449.

Rao, M.R., L.H. Harbers, E. F. Smith. 1973. Seasonal changes in the nutritive value of bluestem pastures. J. Range Manage. 23:133-136.

Rice, R.W., J.G. Nagy, D.G. Peden. 1971. Functional interaction of large herbivores on grasslands. Preliminary Analysis of Structure and Function in Grasslands. (Ed. N.R. French). p. 241-244, Range Science Dept. Sci. Series. No. 10. Colorado state Univ. Ft. Collins.

Sanders, J.O., T.C. Cartwright. 1979. A general cattle production systems model. I. Description of the model. Agri. Systems. 4:217-227.

Schake, L.M., J.K. Riggs. 1972. Behavior of cattle in confinement. Tex. Agr. Exp. Sta. Rep. No. 27. 12 pp..

Schmutz, E.M., G.A. Holt, C.C. Michaels. 1963. Grazed class method of estimating forage utilization. J. Range Manage. 16:54-60.

Schoener, T.W. 1971. Theory of feeding strategies. Ann. Rev. Ecol. Systematics. 11:364-404.

Schuster, J.L. 1965. Estimating browse from twig and stem measurements. J. Range Manage. 18:220-222.

Senft, R.L., L.R. Rittenhouse, R.F. Woodmanse. 1980. Predicting patterns of cattle behavior on shortgrass prairie. Proc. West. Sect. ASAS. 31:19-23.

Sneva, F.A., L.R. Rittenhouse, L. Foster. 1973. Stockwater restriction and trailing effects on animal gain, water drunk and mineral consumption. In: Water-animal Relations Proc. (Ed. H.F. Mayland), Water-Animal Relations Committee, Kimberly, Idaho.

Squires, V.R. 1970. Grazing behavior of sheep in relation to watering points in semiarid rangelands. Proc. 11th Int. Grassld. Cong. pp. 880-884.

Squires, V.R. 1978. Distance trailed to water and livestock response. Proc. 1st Int. Rangl. Cong. pp 431-434.

Stark, R.H. 1973. Elk-cattle interrelations on Clockum creek watershed. M.S. Thesis, Wash. State Univ. 87 pp.

Stillwell, M.A., R.L. Senft, L.R. Rittenhouse. 1982. Measurements of urinary nitrogen from yearling heifers on shortgrass range. SRM. Annual Mtg. Feb. 7-12, Calgary, Canada.

Stobbs, T.H. 1973. The effect of plant structure on the intake of tropical pastures. I. Variation in the bite size of grazing cattle. Aust. J. Agri. Res. 24:809-819.

Stobbs, T.H. 1974. Components of grazing behavior of dairy cows on some tropical and temperate pastures. Proc. Aust. Soc. Anim. Prod. 10:299-302.

Suarez, J.J., A. Senra, J.L. Galindo. 1981. Study of the contribution of nutrients to the soil by the feces of cows grazing in three rotational systems. Cuban J. Agric. Sci. 15:95-102.

Tinnimit, P., J.W. Thomas. 1976. Forage evaluation using various laboratory techniques. J. Anim. Sci. 43:1058-1065.

Valentine, K.A. 1947. Distance from water as a factor in grazing capacity of rangeland. J. For. 45:749-754.

Van Dyne, G.M., N.K. Brockington, Z. Szocs, J. Duek and C.A. Ribic. 1980. Chapt. 4. Large Herbivore Subsystem. in: Grasslands Systems Analysis and Man. ed: A.J. Breymeyer, G.M. Van Dyne. Cambridge Univ. Press, G.G. pp. 269-538.

Van Soest, P.J. 1967. Development of a comprehensive system of feed analysis and its application to forages. J. Anim. Sci. 26:119-125.

Van Soest, P.J., D.R. Mertens, B. Deinum. 1978. Preharvest factors influencing quality of conserved forage. J. Anim. Sci. 47:712-720.

Waldo, D.R., N.A. Jorgensen. 1981. Forages for high animal production: nutritional factors and effects of conservation. J. Dairy Sci. 64:1207-1229.

Weeda, W.C. 1967. The effect of cattle dung patches on pasture growth, botanical composition and pasture utilization. N.Z.J. Agric. Res. 10:150-159.

Westoby, M. 1974. An analysis of diet selection by large generalist herbivores. Amer. Nat. 108:290-304.

Wilson, A.D. 1969. A review of browse in the nutrition of grazing animals. J. Range Manage. 22:23-28.

Winchester, C.F., M.J. Morris. 1956. Water intake rates of cattle. J. Anim. Sci. 15:722-740.

DESCRIPTION AND APPLICATION OF H20TRANS A STAND LEVEL HYDROLOGIC MODEL FOR WESTERN CONIFEROUS FORESTS[1]

Steven W. Running, Dennis H. Knight, Timothy J. Fahey[2]

Abstract.--A hydrologic model has been developed to study stand level water flow and tree water stress in western coniferous forests. The model, H2OTRANS, has an hourly time resolution and is driven by air and soil temperature, humidity, incoming shortwave radiation and precipitation. Required stand parameters include leaf area index, sapwood basal area and soil water storage capacity. The model incorporates rates of snowmelt, canopy interception, litter and soil evaporation. The model predicts leaf conductance as a function of leaf water potential, absolute humidity deficit, incoming shortwave radiation and air temperature, including a special sub-freezing control. Leaf water potential is calculated from soil water content, root water uptake and soil temperature.

Primary model outputs are soil moisture depletion, subsurface outflow, transpiration and tree water stress development as measured by leaf water potential and leaf conductance. A validation exercise was done comparing model prediction to field measurements of seasonal soil moisture depletion and tree water stress development. H2OTRANS was then used to calculate hydrologic budgets for two lodgepole pine stands in Wyoming differing in leaf area index, density, basal area, and soil storage capacity. Simulation results show large differences between these stands in both magnitude and seasonal timing of transpiration, subsurface outflow and soil moisture depletion.

INTRODUCTION

Water and nutrient outflow from forest ecosystems has been the focus of considerable research. The productivity of forest land depends on water and nutrient availability, and water itself is a product of our forests. Consequently, understanding of water and nutrient movement in forest land has significant management implications.

Most research in this topic has concentrated on input-output mass and elemental balances of watersheds, where stream discharge provides convenient output measurements. However, there can be extreme diversity of topography, microclimate, soil structure, forest composition and site productivity within a watershed. Our approach has been to develop hydrologic and elemental budgets for a forest stand, whose characteristics are more homogeneous.

Given the general hydrologic mass balance that:

$$\text{Precipitation} - (\text{Outflow} + \text{Evapotranspiration}) = 0$$

watershed level research typically measures precipitation and stream outflow, obtaining evapotranspiration by difference. Because we cannot measure subsurface Outflow from a stand. our approach has turned to calculating stand evapotranspiration and predicting subsurface outflow by difference. A stand level hydrologic budget thus entails calculation of 2 of the 3 variables in the mass balance rather than only 1 variable as in the watershed approach. However, we feel the ability to calculate hydrologic budgets for stands that do not fit a closed watershed configuration and the increased capability to analyze pertubations and responses of individual stands makes our approach worthwhile.

The model presented in this paper, H2OTRANS, has been developed primarily to predict transpiration and water stress development in western coniferous forests. Earlier development of this model originated primarily from physiological research on stomatal control of conifers under stress conditions (Running et al. 1975, Waring and Running 1976). More recent efforts continued the emphasis on tree water relations but added more physical processes such as snow melt and canopy evaporation to produce a model of greater accuracy and more general application.

[1]Paper presented at the Third International Conference on State-of-the-Art in Ecological Modelling, Colorado State University, May 24-28, 1982.

[2]Steven W. Running is Assistant Professor of Forest Ecophysiology, School of Forestry, University of Montana, Missoula, Montana, U.S.A.; Dennis H. Knight is Professor of Plant Ecology, Department of Botany, University of Wyoming, Laramie, Wyoming, U.S.A.; Timothy J. Fahey is Assistant Professor of Forest Ecology, School of Natural Resources, Cornell University, Ithaca, New York, U.S.A.

The philosophy of H2OTRANS considers a stand as a collection of individual trees. This perspective is necessary because much of the physiological control of water movement in a stand occurs within the tree. A stand is a defined unit, whereas a tree is a biological unit. Consequently, while some parts of the model are stand oriented, such as leaf area index, the model can also be programmed to simulate individual trees. This is not without some difficulty, as with having to estimate lateral rooting distance of an individual tree instead of defining ground surface area as total stand area.

For simplicity the model does not layer the canopy into different compartments or consider vertical or horizontal micrometeorological profiles. A single canopy average is used for meteorological inputs and for computation of physiological responses, i.e. canopy average leaf conductance and leaf water potential. Because the model does not incorporate any terrestrial subunit stratifications, stands must be relatively homogeneous topographically for optimum model performance.

Many of the functions used in the model were developed from research at the Fraser Experimental Forest in central Colorado working on *Pinus contorta* Dougl var. *latifolia* Engelm. (lodgepole pine) and from research on *Pseudotsuga menziesii* (Mirb) Franco (Douglas-fir) from western Oregon.

Other similar models can be found in work by Sinclair et al. (1976), Tan et al. (1978), Luxmoore et al. (1978), Federer (1979), Sollins et al. (1979), Lohammar et al. (1980).

MODEL STRUCTURE

A compartment flow diagram of H2OTRANS is given in Figure 1. Precipitation enters the model at X(3), the forest floor litter compartment. Because the model is designed to run during the growing season, no attempt is made to distinguish rain from snow, which usually melts rapidly. The model is designed to begin at the initiation of snowmelt, when tree physiological activity accelerates, both events related to the onset of prolonged above-freezing air temperatures. A snowpack compartment and melt routine driven by air temperature add water from the existing snowpack to the root zone A. A soil surface infiltration coefficient determines whether incoming water is lost to surface runoff X(2) or penetrates the soil. As water enters the soil the upper compartment, root zone A is filled first to a defined field capacity. Excess water is then cascaded to the lower root zone B until it is at field capacity. Additional water then is routed to subsurface outflow. As can be seen, this part of H2OTRANS is donor controlled. The remainder of the model, evaporation and transpiration, is receiver controlled by atmospheric evaporative demand.

The general equation used to compute transpiration is:

$$E = (PCp/LV)((e_s - e_a)/(r_s + r_a)) \qquad (1)$$

where:

E = transpiration rate (g/cm^2/sec)
P = air density (g/cm^3)
Cp = specific heat of air (J/g/^{o}C)
L = latent heat of vaporization (J/g)
V = Psychrometric constant (mb/^{o}C)
e_s = saturation vapor pressure (mb)
e_a = ambient vapor pressure (mb)
r_s = stomatal resistance (sec/cm)
r_a = aerodynamic resistance (sec/cm)

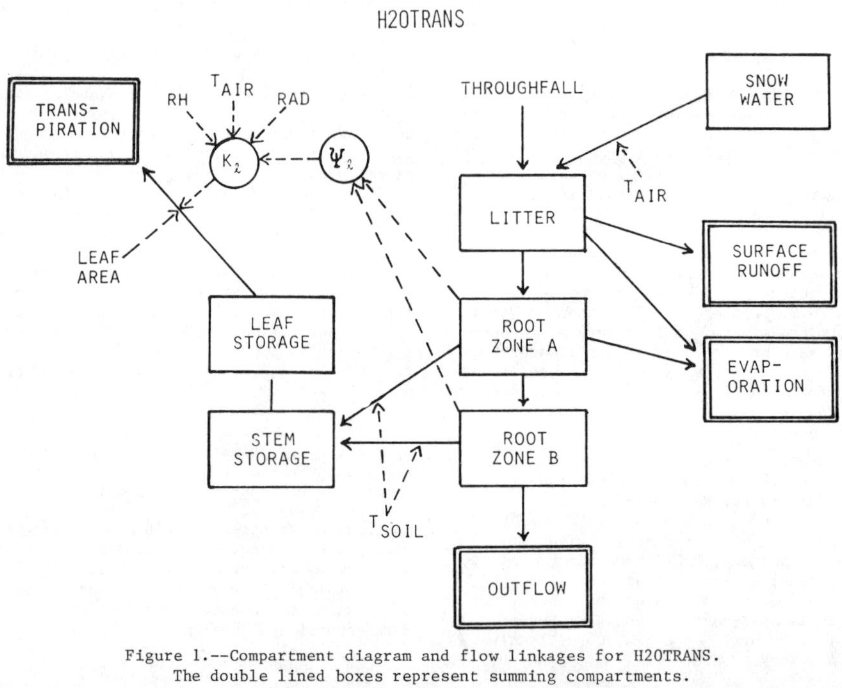

Figure 1.--Compartment diagram and flow linkages for H2OTRANS. The double lined boxes represent summing compartments.

This equation is derived from the Penman-Monteith equation but reflects the greater significance of humidity rather than radiation in driving evaporation in our arid western forests (Tan et al. 1978, Luxmoore et al. 1981).

The computer program calculates hourly stand transpiration as:

$$E = ABSHD \times k_1 \times LA \times 3600 \qquad (2)$$

where,

ABSHD = absolute humidity deficit (G/cm^3)
k_1 = leaf conductance (cm/sec)
LA = tree or stand leaf area (cm^2)
3600 = sec/hr

In this form the atmospheric constants and humidity measurements are consolidated into ABSHD, the stomatal resistance is replaced by its reciprocal, leaf conductance and aerodynamic resistance is ignored.

The calculated transpiration from equation 2 is subtracted from the Leaf Storage compartment, which in turn draws water from Stem Storage and ultimately from the Root Zone storage A and B. Feedback control of stand water loss takes place as soil water is depleted, causing tree water stress to develop which reduces k_1 in equation 2. The mechanism for this control will be considered further.

Table 1 provides an overview of the data requirements for H2OTRANS. The environmental data is input hourly. The stand and site parameters are designated initially and cannot be changed with thepresent model form. This constrains the model from being executed over extended periods of time where stand leaf area, tree height, basal area and rooting volume may increase with stand growth. Future model improvements will remove this constraint.

It is not possible in this short paper to cover all equations in the model. A complete documentation is available (Running 1982). We will present next the equations used to calculate canopy leaf conductance, as this is the section where much of our field research has focused, and is a critical control point in the water balance of our arid, snowmelt dominated stands.

Table 1.--A summary of data requirements for H2OTRANS.

Z(i) - Driving Variables	Units
Z(1) = Julian date	day
Z(2) = precipitation	cm/day
Z(3) = air temperature	^oC
Z(4) = relative humidity	%
Z(5) = soil temperature (20 cm depth)	^oC
Z(6) = incoming shortwave radiation	ly/min
Z(7) = hour of day	hr
Stand & Site Parameters	
1. ground surface area	cm^2
2. rooting zone soil water storage	cm^3
3. leaf area (tree or stand)	cm^2
4. tree height (mid-crown)	m
5. sapwood basal area	cm^2
6. maximum soil surface infiltration rate	cm/hr
Primary Output Variables	
1. Transpiration	cm^3/hr
2. Evaporation	cm^3/hr
3. Soil moisture depletion	cm^3/hr
4. Tree water stress (Leaf Ψ)	-MPa
5. Subsurface outflow	cm^3/hr

As a general rule the stomata of western conifers require about 10% of full sunlight to open completely. Figure 2 illustrates the function used to determine, for a given level of incoming radiation, how much leaf area index would be receiving 10% of full sunlight, about 100 W/m^2 or 0.13 ly/min. If stand leaf area is below the calculated value of illuminated leaf area index, then canopy leaf conductance (k_1) is unaffected. Otherwise k_1 is reduced by the proportion of the canopy not receiving sufficient radiation to open stomata.

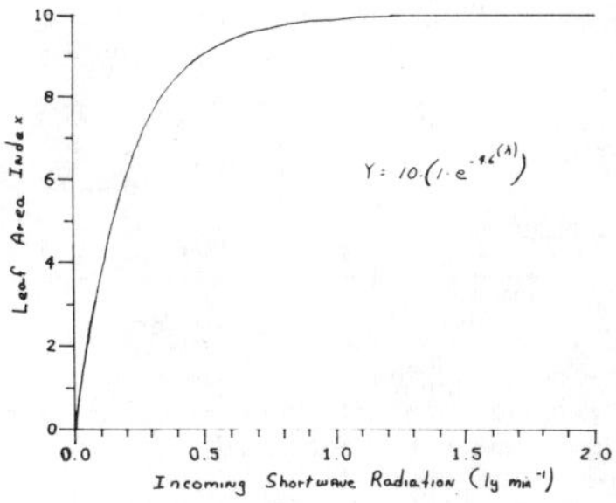

Figure 2.--Describes the leaf area index (LAI) that would receive incoming shortwave radiation in excess of 0.1 ly min^{-1} (the stomatal opening threshold) at any level of incoming radiation.

The relationship between soil water status and tree water stress is shown in figure 3. Effectively this function combines the well known curvilinear relationship between soil water content and soil water potential (Nnyamah and Black 1977) and the linear relationship between soil water potential and pre-dawn leaf water potential (Hinckley et al. 1978). As soil water is depleted, a progressively higher tree water stress is generated. Note that high water stress is technically measured as lower (more negative) leaf water potentials.

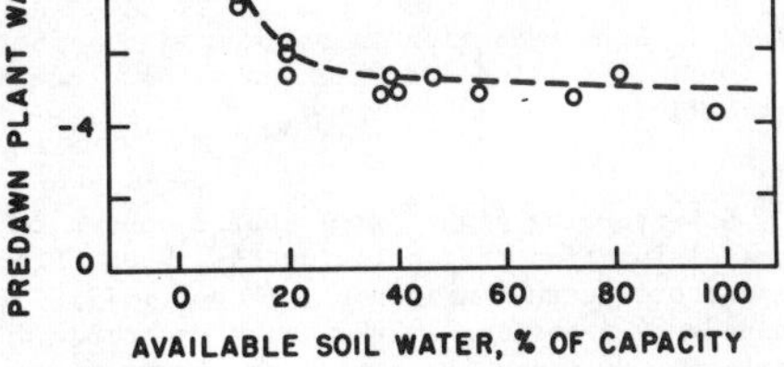

Figure 3.--The function relating soil water content (% of field capacity) to predawn leaf water potential, $B\Psi_1$ (an estimate of soil water potential.

It has been hypothesized that daily maximum canopy k_1 is controlled by predawn leaf water potential,

$B\Psi_1$ (Running 1976, 1980). Figure 4 illustrates the reduction in k_1 that occurs as tree water stress increases ($B\Psi_1$). Data from both *Pseudotsuga menziesii* and *Pinus contorta* support this hypothesis so far.

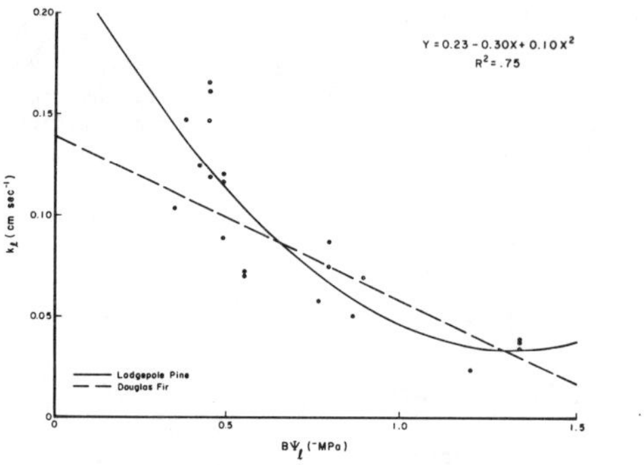

Figure 4.--The reduction in maximum leaf conductance (k_1) resulting from increasing plant water stress, (predawn leaf water potential, $B\Psi_1$). Data points are for *Pinus contorta*.

There is much evidence that canopy k_1 is reduced by low atmospheric humidity regardless of soil water status. The function in figure 5 is from *Pinus contorta* (Running 1980) although similar data is available for other western conifers (Kaufmann 1976, Tan et al. 1977). Absolute humidity deficit (ABSHD) calculated from the input driving variables is used to correct the maximum k_1 from figure 4 to lower k_1 as humidity deficit increases, figure 5.

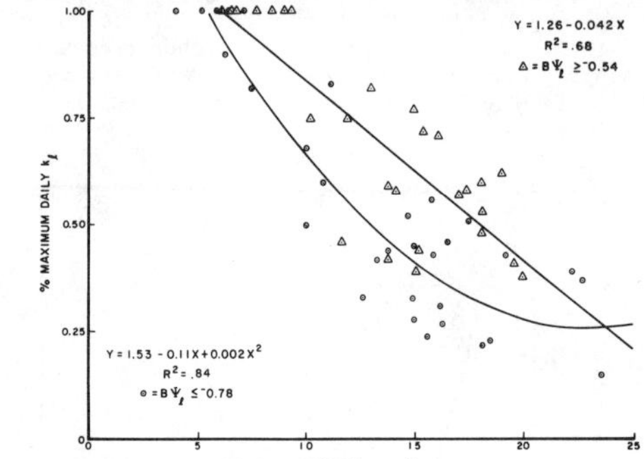

Figure 5.--The reduction in maximum k_1 of *Pinus contorta* resulting from increasing absolute humidity deficit.

Sub-freezing night temperatures appear to retard stomatal function the following day, even after air temperatures warm considerably (Fahey 1979). This can be very important in the Rocky Mountains during April and May where frequent night frost significantly lowers transpiration from levels otherwise expected. Figure 6 illustrates the effect sub-freezing night temperatures had on daily mean k_1 for *Pinus contorta* (Fahey 1979). H2OTRANS reduces canopy k_1 as a function of input air temperatures below 0°C. At air temperatures above 0°C, k_1 increases weakly with increasing air temperature.

To summarize the hierarchy of canopy k_1 control from strongest to weakest is (1) radiation, (2) predawn leaf water potential, (3) humidity and (4) air temperature (except sub-freezing nights which have (2) priority).

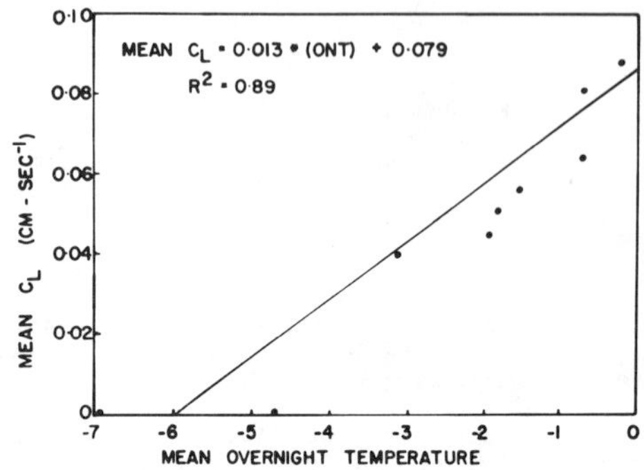

Figure 6.--The influence of overnight freezing on mean daily leaf conductance (C_1) in *Pinus contorta*.

Many other water transfer processes are covered in H2OTRANS, although not in the detail of the canopy k_1 control. Among these processes are canopy interception and evaporation, snowmelt, litter surface evaporation, soil surface runoff, leaf and sapwood water storage and exchanges, hydraulic flow resistances and diurnal leaf water potential changes. Complete documentation can be obtained in Running (1982).

RESULTS

An initial validation of H2OTRANS was attempted for a *Pinus contorta* stand in central Colorado at the U.S.F.S. Fraser Experimental Forest. The site was a *Pinus contorta* stand on level ground at 2700 m elevation. The general climate is cool and dry with annual precipitation of 58 cm, two-thirds falling as wintertime snow. Summer night temperatures are usually below 4°C and maximum air temperatures average 20-23°C during the growing season. Midday humidities are typically 15-25%. A 7.0 m tall *Pinus contorta* tree with DBH=8.2cm and a leaf area index of 6.0 was simulated for the period May 10-September 10, 1978. The calculated ground surface area occupied by this tree was 3.5m^2, with a rooting depth of 1.0m.

Validation efforts compared the field measured variables of soil moisture depletion, predawn leaf water potential and canopy leaf conductance to values simulated by H2OTRANS. Overall transpiration and subsurface outflow cannot be measured or validated.

Soil moisture depletion as predicted by H2OTRANS and as measured by neutron probe is shown in figure 7. The figure also shows the results from DAYTRANS, a sister program to H2OTRANS that computes at daily time resolution. H2OTRANS predicts the onset of soil moisture depletion and the overall extent of soil moisture drawndown fairly accurately. The trajectory of depletion in mid-summer is a little too rapid, however. The development, predicted by H2OTRANS and observed by pressure chamber measurement, of predawn leaf water potential or tree water stress is illustrated in figure 8. Three important diagnostics for interpreting the success of a seasonal tree water stress prediction are (1) the timing of the onset of water stress development, (2) the rate of stress increase and (3) the final maximum stress generated by the tree. In this simulation, H2OTRANS was very successful in diagnostics (1) and (2), while slightly overshooting the maximum stress, (3). The overall season water balance for this simulation was Transpiration = 36%, Evaporation = 15%, Subsurface Outflow = 49% from a total 42.9cm of water. Because we cannot directly validate our estimate of transpiration and subsurface outflow, our confidence in model accuracy must be gained by validation of these other, related variables.

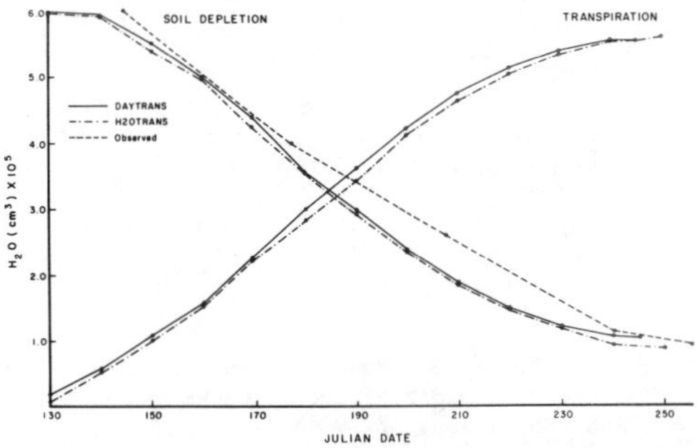

Figure 7.--A validation of soil moisture depletion, as measured by neutron probe, to predictions from H2OTRANS for a *Pinus contorta* in Colorado. Seasonal accumulation of predicted transpiration is also shown. DAYTRANS is a daily resolution model that was not presented in this paper.

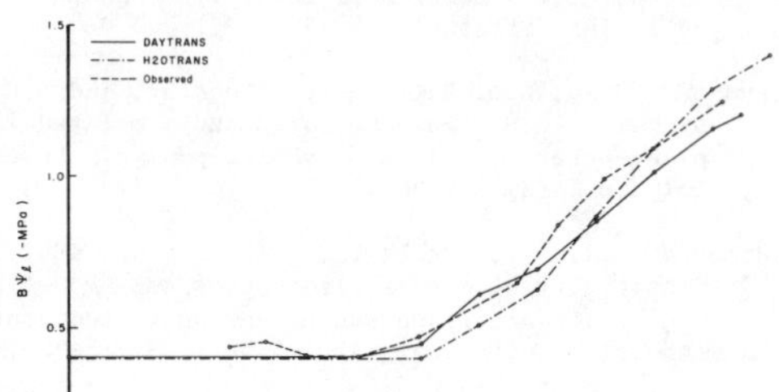

Figure 8.--An H2OTRANS prediction of seasonal water stress development in *Pinus contorta* for 1978 at the Colorado site. Measured field data was taken with a pressure chamber.

A more recent and complete validation of H2OTRANS has just been executed for six *Pinus contorta* stands of various age and structure in southern Wyoming. Complete results will be published by Knight et al. (1982). For this paper we will illustrate differences in the stand hydrologic budgets between a dry, rocky low elevation stand with a sparse (LAI = 3.9) canopy, Albany; and a moist mid-elevation site with a productive (LAI = 9.9) stand, Nash Fork. Root zone water holding capacity was only 6cm for the Albany stand and 14cm for the Nash Fork stand. Although tree density was equal on the two stands (1920 and 1840/ha), basal area at Nash Fork was $41m^2$/ha, versus $20m^2$/ha at Albany. Total above-ground biomass was 175Mg/ha at Nash Fork, 76Mg/ha at Albany.

The 1979 and 1980 seasonal simulation of soil moisture, transpiration and subsurface outflow for these two contrasting stands is shown in figures 9 and 10. Because of the low soil water holding capacity and low leaf area for transpiration in the Albany stand, predicted subsurface outflow is much higher 40cm in 1979, 32cm in 1980; than the Nash Fork stand 27cm in 1979 and 19cm in 1980. In both stands all subsurface outflow occurred at the time of snowmelt. Subsequent summer rains were not adequate to completely refill soil water storage in the rooting zone. Most summer rainfall was lost to canopy and forest floor litter interception and evaporation, although short periods of rainfall in both 1979 and 1980 resulted in some recovery of soil moisture.

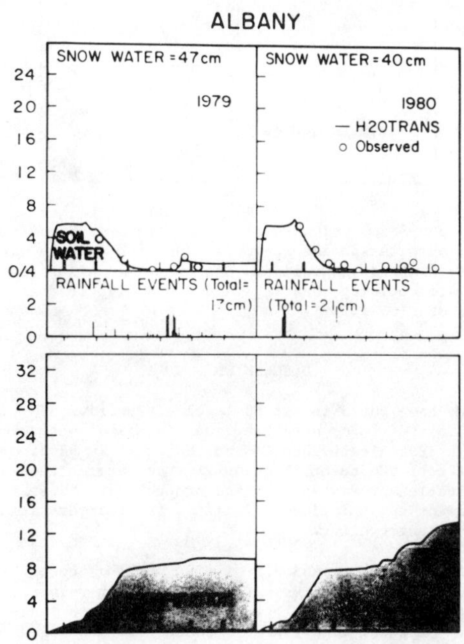

Figure 9.--A simulation of soil water dynamics, transpiration and subsurface outflow for the Albany stand had low soil water capacity (6cm) and a low leaf area index (3.9).

Most of the seasonal transpiration on the Albany site occurred before July 1, although in 1980 late summer rain provided water for some additional transpiration. After July 1 tree water stress closed stomata, reducing transpiration loss. The much greater soil water holding capacity at Nash Fork sustained transpiration longer into the summer, although the higher elevation site also began snowmelt and transpiration almost a month later. The higher leaf area at Nash Fork also generated much higher transpiration rates.

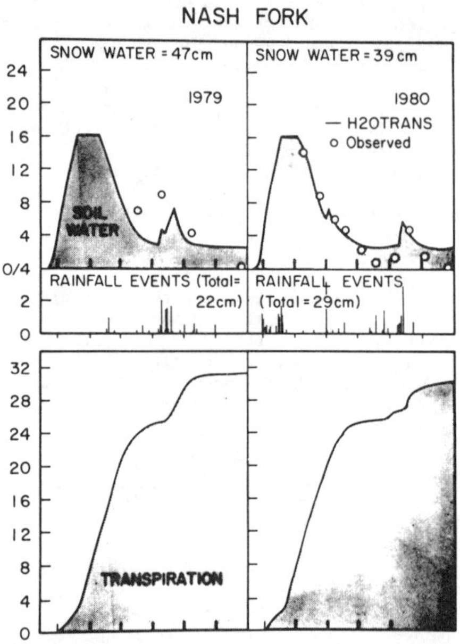

Figure 10.--Simulation results of soil water dynamics, transpiration and subsurface outflow for the Nash Fork stand in Wyoming. Nash Fork had high soil water holding capacity (14cm) and high leaf area index (9.9).

DISCUSSION

We have found the stand level perspective of H2OTRANS useful for attacking questions that watershed level hydrologic budgets cannot address. In particular we feel the quantification of tree water stress is valuable for assessing stand productivity, disease and insect epidemic susceptibility, fire danger rating and nutrient cycling.

A primary conclusion from this work has been that prediction of tree water stress development requires more than just accurate physiological response functions. Of equal importance is sound representation of hydrologic inputs and storages. The system inputs and storages control the timing of when water stress is initiated in the trees. The physiological processes primarily control the subsequent rate of tree water stress development. Interception and surface evaporation losses must be accurate to determine effective precipitation input. On wetter sites these processes could profoundly influence the success of a simulation. Of equal importance in model performance is defining the volume of the soil rooting-zone, the basic water storage reservoir for the model. This requires knowledge of tree rooting depth and density, averaged throughout the stand to be modeled. Another critical component of stand hydrology and water stress modeling is dependable measures of stand biomass, particularly leaf area, and to a lesser extent sapwood storage volume. Meteorological driving variables are critical to model performance.

On some sites we have found our treatment of soil water dynamics to be superficial. Problems of supersaturated soils, channelized flow and lateral inputs cannot be handled by the present model structure. Yet adding model complexity reduces applications as the data requirements become more demanding. Probably the single most difficult and critical parameter estimate in the model is tree rooting depth and overall rooting volume.

Our future work lies primarily with coupling nutrient budgets and cycles into the H2OTRANS structure. We are also extending the validation of physiological responses to other western conifers.

LITERATURE CITED

Fahey, T. 1979. The effect of night frost on transpiration of *Pinus contorta* spp. *latifolia*. Oecol. Plant 14:483-490.

Federer, C.A. 1979. A soil-plant-atmosphere model for transpiration and availability of soil water. Water Resources Res. 15:555-562.

Hinckley, T.M., J.P. Lassoie, and S.W. Running. 1978. Temporal and spatial variations in the water status of forest trees. 72 p. Forest Science Monograph No. 20.

Kaufmann, M.R. 1976. Stomatal response of Englemann spruce to humidity, light, and water stress. Plant Physiology 57:898-901.

Knight, D.H., T.J. Fahey and S.W. Running. 1982. Factors affecting water and nutrient outflow from lodgepole pine forests in Wyoming. Ecol. Monographs (submitted).

Lohammar, T., S. Larasson, S. Linder, and S.O. Falk. 1980. FAST-Simulation models of gaseous exchange in Scots pine. In "Structure and Function of Northern Coniferous Forests." T. Persson ed. Bulletins (Stockholm) 32:505-523.

Luxmoore, R.J., D.D. Huff, R.K. McConathy, and B.E. Dinger. 1978. Some measured and simulated plant water relations of yellow-poplar. Forest Science 24(3):327-341.

Luxmoore, R.J., J.L. Stolzy, J.T. Holdeman. 1981. Sensitivity of a soil-plant-atmosphere model to changes in air temperature, dew point temperature and solar radiation. Agric. Met. 23:115-129.

Nnyamah, J.U., and T.A. Black. 1977. Rates and patterns of water uptake in a Douglas-fir forest. Soil Science of America Journal 41:972-979.

Running, S.W. 1976. Environmental control of leaf water conductance in conifers. Canadian Journal of Forest Research 6:104-112.

Running, S.W. 1980. Environmental and physiological control of water flux through *Pinus contorta*. Canadian Journal of Forest Research. 10:82-91.

Running, S.W. 1982. Documentation and preliminary validation of H2OTRANS and DAYTRANS, two models for predicting transpiration and water stress in western coniferous forests. Rocky Mtn. Forest & Range Exp. Stn. Res. Paper (in press).

Running, S.W., R.H. Waring, and R.A. Rydell. 1975. Physiological control of water flux in conifers: A computer simulation model. Oecologia (Berl.) 18:1-16.

Sinclair, T.R., C.E. Murphy, and K.R. Knoerr. 1976. Development and evaluation of simplified models for simulating canopy photosynthesis and transpiration. Journal of Applied Ecology, 13:813-829.

Sollins, P., A.T. Brown, and G.L. Swartzman. 1979. CONIFER: A model of carbon and water flow through a coniferous forest. 152 p. US/IBP Coniferous Forest Biome Bulletin #15. AR-10, University of Washington, Seattle, WA 98195.

Tan, C.S., T.A. Black, and J.U. Nnyamah. 1977. Characteristics of stomatal diffusion resistance in a forest exposed to large soil water deficits. Canadian Journal of Forest Research 7:4, 595-604.

Tan, C.S., T.A. Black, and J.U. Nnyamah. 1978. A simple diffusion model of transpiration applied to a thinned Douglas-fir stand. Ecology 59:1221-1229.

Waring, R.H., and S.W. Running. 1976. Water uptake, storage and transpiration by conifers: A physiological model. p. 189-202. In O.L. Lange, E.D. Schulze, and L. Kappen, eds. Water and plant life, problems and modern approaches. Ecology Studies, Vol. 19, Springer-Verlag (Berl.).

MODELLING THE EFFECTS OF AIR POLLUTANTS ON FORESTED ECOSYSTEMS[1,4]

Mark A. Harwell[2], David A. Weinstein[3]

The Ecosystems Research Center is developing the CASSANDRA model to simulate the effects of various air pollutants on forested ecosystems. The approach is to build a nested model, using high resolution mechanistic submodels to create probability distributions of individual tree growth reductions in response to the air pollutants. These distributions then enter the large scale simulation model to show effects on regional ecosystem dynamics.

INTRODUCTION

Ecosystems science is currently very diverse, replete with general ideas and relationships, but lacking in a unifying conceptual framework. This status has resulted, in part, in the inadequate application of ecosystems science to the understanding and solution of environmental problems. The U.S. Environmental Protection Agency (EPA) has recognized the need for increased utilization of ecosystems science in its various roles of long-term environmental protection and management. Thus, EPA and Cornell University have cooperatively established the Ecosystems Research Center (ERC) with the mission of helping provide the ecosystems bases for environmental regulation.

A major issue associated with ERC's activities is the response of ecosystems to stress, particularly stresses of human origin. Of primary concern is the response of large-scale ecosystems to chronic, multiple stresses. Within this category, the ERC is considering the effects of air pollution stresses on forested ecosystems. A modelling effort has been initiated to deal with the variety of stresses and scales of effects associated with air pollutants. In particular, an assessment is being made of the effects of air pollutants on the productivity of forests located in the regions of greatest exposure.

A number of studies have indicated that the deciduous forests of the eastern United States are being exposed to subtle and probably significant stresses from gaseous air pollutants, acid deposition, and particulates and trace metals (cf. Loucks, 1980; Wolff and Lioy, 1980; NAS, 1981). Current information on these air pollutants and current understanding of forested ecosystems suggest that the consequences of this chronic stress may include deleterious direct and indirect effects on the recipient deciduous forests. However, empirical proof demonstrating a relationship between productivity and air pollutant stresses is quite difficult to obtain, both because of the long time periods required and the diffuse and complex nature of the anthropogenic stress inputs to the forests.

Table 1. -- Overview of existing data on the experimental response of forest trees to atmospheric pollutants (developed from Smith, 1980).

DOSE-RESPONSE OF FOREST TREES

Photosynthetic suppression by pollutants

1. Large trees
   a) Legge et al. (1977)
      1) Western pine/spruce/aspen showing reduction
      2) 131-262 $ug \cdot m^{-3}$ SO_2 events lasting several hours
   b) Kercher

2. Saplings

Ref.	Pollutant	Threshold Conc.	Time	Duration	Species
Lamoreaux and Chaney (1978)	SO_2	1.0ppm	30 m	1 treatment	Silver maple
" " " "	SO_2	.5ppm	7-11 hr.	1-2 days	Black oak
Carlson (1979)	SO_2	.5ppm	7-11 hr.	1-2 days	Sugar maple
" " "	SO_2	.5ppm	7-11 hr.	1-2 days	White ash
Botkin (1972)	O_3	.5ppm	4 hr.	1 treatment	White pine
Carlson (1979)	O_3	.5ppm	7-11 hr.	1-2 days	Black oak
" " "	O_3	.5ppm	7-11 hr.	1-2 days	Sugar maple

   a) Botkin: 3 classes of white pine responses:
      1) 0.5ppm for 10 hrs. reduced photosynthesis to zero
      2) Others reduced 50%; reversible
      3) No effect
   b) Carlson: synergisms

3. Seedlings
   Lower doses produce more drastic photosynthesis reductions.
      .1 - .2ppm yielded 59% decrease in photosynthesis

---

[1]Paper presented at the Third International Conference on State-of-the-Art in Ecological Modelling, Colorado State University, May 24-28, 1982.

[2]Assistant Director, Ecosystems Research Center, Cornell University, Ithaca, N.Y.

[3]Research Associate, Ecosystems Research Center, Cornell University, Ithaca, N.Y.

[4]Primary funding for this research was provided by the Office of Research and Development, U.S. Environmental Protection Agency, under Cooperative Agreement No. CR 807856 01. Additional funding was provided by Cornell University.

This publication is ERC Report No. 017. The ERC (Ecosystems Research Center) was established in 1980 as a unit of the Center for Environmental Research at Cornell University.

The work and conclusions published herein represent the views of the authors, and do not necessarily represent the opinions, policies, or recommendations of the Environmental Protection Agency or of Cornell University.

Because of this difficulty, the potential effects of air pollutant stresses on forests are typically examined by extrapolating from smaller scale and shorter term experiments on plant/pollutant interactions to the temporal and spatial scales of forest responses. An overview of experiments on dose/response relationships for forest species is given in Smith (1981) (table 1).

The potential reduction of forest productivity involves the alteration of many ecosystem processes, representing a mix of subtle phenomena, some acting physiologically on individuals, and others acting through longer term interspecific interactions or through linkages between the community and the abiotic environment (table 2). Thus, we are developing a simulation model to allow the temporal and spatial extrapolations, to provide a systematic framework for the integration of multiple stress events and complex effects on an ecosystem, and to aid in the understanding of how pollutants and processes interact.

Table 2. -- Sources of variation in the photosynthetic rates of forest trees.

PHOTOSYNTHETIC RATES IN FOREST TREES

Range of photosynthesis of mature intact leaves is 10-200 mg CO_2 taken up per gram dry weight per day.

Conifers	10-20 $mg \cdot g^{-1} \cdot day^{-1}$
Deciduous	40-110 $mg \cdot g^{-1} \cdot day^{-1}$

Sources of variation

1. Time of season	(Fritts, 1966)
2. Position of leaf on tree	(Pisek and Tranquillini, 1954)
3. Light intensity	(Polster, 1950)
4. Genetic variability	(Bourdeau, 1963)
5. CO_2	(Anderson, 1973)
6. Humidity	(Pisek and Tranquillini, 1954)
7. Internal accumulation of photosynthates	(Thomas and Hill, 1949)
8. Water limitation	(Slavik, 1973)
9. Temperature	(Bauer et al., 1973)
10. Nutrient limitation	(Natr, 1973; Keller, 1970)
11. Age of foliage	

AIR POLLUTION EFFECTS ON FORESTS

The eastern deciduous forests are most subject to regionally distributed air pollutant loads. The regionally transported pollutants which are known to affect plants, insects, some birds, and decomposition biota include SO_2, O_3, particulates, acid deposition, and heavy metals.

A regional pattern of annual average SO_2 concentrations has been shown, but specific SO_2 exposures are spatially highly variable (fig. 1), determined by local meteorological patterns and the existence of point sources of SO_2. In general, those areas showing elevated annual average concentrations correspond to substantial numbers of hourly occurences of SO_2 at biologically significant levels. The pattern for O_3 is rather different, in that ozone apparently tends to occur as significant pollution in urban areas, but affects exurban regions several hundred km downwind. Of particular importance, O_3 can occur as persistent regional episodes, effectively as rivers of ozone across the landscape (Wolff and Lioy, 1980). Further,

SO_2 and SO_4^{2-} concentrations in the eastern United States during a sulfate episode in 1974 (ORBES 1981).

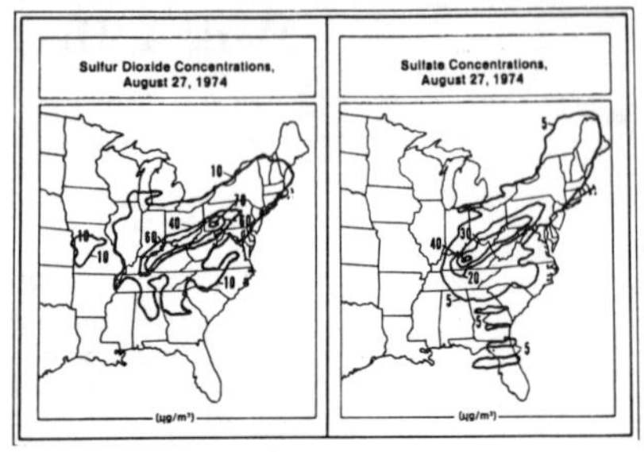

Figure 1. -- SO_2 and $SO_4^{=}$ concentrations in the eastern United States during a sulfate episode in 1974 (Loucks, 1981).

the assumption that high O_3 levels occur near urban areas has been enhanced by the fact that most O_3 monitors are located in downtown areas.

Particulates, of which $SO_4^{=}$ is a major contributor, affect acid deposition, gaseous exchange processes of many biota, and visibility and sunlight intensity. Regional patterns of visibility (fig. 2) reflect particulate loading, except that humidity and fog often are determining in the coastal areas. The markedly reduced visibility in the Ohio Valley is in large part resultant from $SO_4^{=}$.

These three pollutants thus represent the range of spatial distribution patterns which occurs for air pollutants in general. That is, SO_2 is often highly localized and intense, and regional patterns reflect frequency of occurrence of specific events, whereas particulates are less variable, more dispersed over time and space. Ozone occurs as an intermediate pattern. This disparate distribution of air pollutants provides one level of complexity which must be addressed in an air pollution model.

Visual range (V) isopleths for suburban non-urban areas, 1974-76 (Trijonis and Shapland, 1978).

Figure 2. -- Visual range (V) isopleths for suburban and non-urban areas, 1974-76 (Trijonis and Shapland, 1978).

Another layer of complexity relates to the mechanisms by which the various air pollutants can affect the forested ecosystem. Net primary productivity of the forest, an ecosystem-level measure of proper functioning, can be affected directly, such as through photosynthesis processes, indirectly, such as through reduced nutrient availability, and by effects mediated through the trophic structure, such as by increased herbivory rates. Additionally, a single pollutant may affect the ecosystem through a variety of mechanisms; these mechanisms may be quite different for another pollutant and still different when pollutants occur simultaneously. For instance, photochemical oxidants, such as O_3, are the most damaging air pollutants directly affecting tree productivity. But O_3 effects are significantly enhanced in a synergistic manner by the presence of SO_2 or NO_x. Such synergisms have also been seen to increase tree susceptibility to other stresses, such as insects and diseases, which can subsequently affect forest productivity indirectly. Indeed, it may well be that regionally the most important component of air pollution decrease in forest productivity involves indirect mechanisms such as pest outbreaks which are not readily identified in small scale experiments.

The understanding of how forests, or even individual trees, are adversely affected by air pollutants is at present incomplete. Most of the data are from short-term experiments on single species, usually crop plants rather than forest trees, exposed to single pollutants. There is currently no general agreement as to the nature, mechanisms, and consequences of plant/pollution interactions, and even less for the more indirect, pest-mediated interactions. This uncertainty, coupled with the diversity of pollutants and distribution patterns and the multiple modes of pollutant impacts on ecosystems, provides the setting for attempts at regional long-term modelling of ecosystem responses to air pollutant stresses.

ERC MODEL UNDER DEVELOPMENT

The Cornell Assessment of the Stress and Response to Air Pollution Model (CASSANDRA) is currently being developed at the ERC as a simulation model which incorporates the multiple layers of complexity associated with this problem. The key characteristics of CASSANDRA include the following:

- nested approach;
- forest-gap replacement model template;
- modular high resolution submodels;
- stochastic interface across space/time scales;
- adaptive development.

The basic structure of the ERC model is nested. This nesting allows the model to be able simultaneously to deal with ecosystem-level processes and with much finer scale mechanisms. The ecosystem-level process which is of central concern was chosen to be forest net productivity. This process provides an integrative measure of the state of the forested ecosystem; further, productivity is a measure of potential economic importance and, in general, provides an index of the health of the ecosystem which has visible meaning to humans. It is important to note, however, that the choice of a different variable of interest could lead to a substantially different modelling approach. For instance, if the primary concern for the forests' impacts from air pollution were a single economic species (e.g., a particular tree species or game animal species), then the model would be structured to emphasize all the ecosystem's processes which provide the sustenance for that population. For CASSANDRA, though, the focus is on the ecosystems level itself, rather than just on a single population being maintained by an ecosystem.

Forest productivity for deciduous forests is a long-term measure which synthesizes information from a variety of processes. The response time of productivity is appropriately measured in decades, since such a time scale is necessary to include the processes of importance. For example, a deciduous forest's productivity can be considered on a single- or few-year basis, including the direct effects of environmental conditions on annual production of each constituent tree present in the forest. However, of real importance to forest productivity is the replacement of individual trees, either by individuals from the same species and a concomitant alteration in the age distribution, or by individuals from differing species through competitive interactions. When this individual tree replacement by another individual tree is expanded across the ecosystem landscape, then shifts in the community composition and/or age structure can result in shifts in the productivity of the forested ecosystem, manifested over many years or decades. Since one aspect of chronic stress on ecosystems is that there are differential effects on the systems' biotic components, with subsequent alteration of the interactions among the components, a model of air pollution effects on forested ecosystems must explicitly include these longer term, structural alteration processes (cf. West et al., 1980, where the interplay between differential reduction in growth by each species and changes in competitive interactions resulted in non-intuitive changes in community structure).

However, a perspective at only this large spatial scale and the associated long time frames is insufficient for modelling air pollutant effects. Many of the air pollutants cause effects within very short time periods and on a very localized basis. For instance, O_3 and SO_2 depression of photosynthesis is highly dependent on the exposure and recovery period regime, measured in only hours. Further, as discussed previously, many of the air pollution episodes are of very short duration, and the large majority of empirical data are from short-term exposures. Thus, it is clear that the model must also explicitly include short-term, small scale phenomena. However, simply having both levels of processes occurring simultaneously (i.e., on hourly and up-to-decade time scales) is not practical in a computer simulation; hence, the CASSANDRA model incorporates a nested approach.

This nesting involves essentially two indepedent levels of models which are linked together. The larger scale model is a forest gap replacement type of model which uses annual time steps to simulate the forested ecosystem structure and processes over time. The finer scale level includes a number of submodels, each dealing with a single aspect of the mechanisms by which air pollution can affect biologic processes and each using an hourly or daily time step.

The large scale model is being developed around the template of an existing forested ecosystem model, FORNUT (Weinstein, 1982). This model was developed by extension of the Oak Ridge model FORET (Shugart and West, 1977) to include nutrient cycling and water availability as explicit components. FORET, in turn, incorporated many of the characteristics of JABOWA (Botkin et al., 1972), one of the earliest forest gap replacement models.

The essence of FORNUT is the individual treatment of trees, which comprise the canopy of an eastern deciduous forest. The annual diameter increment of each tree is calculated as the optimal growth, decremented by competitive interactions and by less-than-optimal environmental conditions. The model is built around a homogeneous 1/12 ha areal unit, within which the forest structure is simulated by inclusion of different tree individuals, each growing, competing, reproducing, and dying as time proceeds. The model is stochastic in that the response of each tree to a set of environmental conditions is selected randomly from a probability distribution. By simulating many of these 1/12 ha units, a pattern emerges which reflects forest-wide structure and dynamics.

The key assumption of this model is that each tree species has a maximal growth rate and a maximal size which are characteristic of that species and which would be achieved under optimal conditions. However, each tree is subject to less-than-optimal conditions through competition for nutrient, water, and sunlight resources and by existing in a stochastic physical environment. Each tree species thus has an optimal growth equation; however, even under optimal conditions, the tree would not exceed some maximal size and age, so that as the tree matures, its ideal growth rate declines and its size approaches an asymptote. Further, since different tree species have different optimal growth patterns, maximal ages, and so on, simulation over time results in a continuous, dynamic pattern of ecosystem structure and function. This process occurs through the cumulative effects of each of the various submodels (table 3) for each of the various constituent species, for multiple simulation plots, operating on the ecosystem-wide parameter of forest productivity. In practice, the FORNUT and FORET models have been demonstrated to mimic adequately real forested ecosystems dynamics (Shugart and West, 1977; Weinstein, 1982).

Table 3. -- Submodels contained in the large scale model, FORET/FORNUT, with a 1-year time step.

Submodels

- Growth
- Mortality
- Birth
- Sprout
- Decomposition/leaching
- Uptake
- Moisture
- Perturbation

Thus FORNUT provides an available template for the treatment of air pollution effects on forest productivity. The various mechanisms of the air pollutants' effects on each tree species can be categorized into the appropriate submodels within FORNUT. For example, any air pollutants which reduce the photosynthetic capability of a tree can be entered into the simulation directly by a decrease in the annual growth, as calculated by the growth submodel. Similarly, other submodels may be changed to incorporate air pollution effects, such as on sprouting, availability of nutrients, and so on. However, in many of these instances, a finer resolution is required for simulating how the air pollutant affects the process than can be instituted through a direct change in an annual submodel output. For those instances, the finer resolution level of submodels will be used. This is the major area of new model development underway at the ERC.

The finer resolution level (table 4) will include a number of processes and pollutants; in fact, this level of submodels will continue to expand as more mechanistic understanding develops. As an example of the types of submodels to be implemented, consider the process of photosynthesis. The photosynthesis high resolution model being developed will utilize an hourly or daily time step, including such phenomena as time of day, stomatal status, temperature and humidity effects, time of year, and so on. Net carbon fixed over time will be monitored during simulation, eventually to enter the larger scale model through the growth submodel. Within this photosynthesis model will be included the effects of SO_2, O_3, NO_x, and other air pollutants, acting singly or in combination to decrease the rate of photosynthesis during and subsequent to exposure. The relationship of these pollutants to photosynthesis will be modelled based on a review of experimental data which have been collected. Depending on the completeness of this data set, which currently is being analyzed, different photosynthesis models will be created for different classes of trees, reflecting differential sensitivity across species to the pollutants. Similarly, other high resolution submodels will be developed, particularly focusing on plant/herbivore and plant/disease interactions under nominal and air pollution conditions.

Table 4. -- Submodels contained in the high resolution model, with a 1-day time step.

Submodels

- Photosynthesis and allocation (including litterfall and root sloughing)
- Consumer processes
- Decomposition/mineral release
- Soil leaching

The interface of these high resolution submodels with the large scale forest model will occur through the creation of response functions (fig. 3). For instance, the specific effect on photosynthesis of an air pollution event involving SO_2 at a particular level and duration would be simulated using the high resolution model. But this would entail only a single episode among perhaps hundreds which could occur within a single year, the time period of a single time step in the large scale model. While that episode would be of the appropriate scale for the high resolution model and could directly be related to laboratory data, it is not of the appropriate time scale for the large system model. Therefore, a response function would be generated by simulating a large number of different SO_2 events. The pattern of SO_2 concentrations and episodic durations would be based upon the pattern of actual field SO_2 data at a forest site.

From such a set of simulations would come a frequency/response curve of the relative probability that a tree would experience a given reduction in growth under an assigned annual SO_2 loading. Other curves for other annual SO_2 levels and for the other pollutants would also be generated. When the larger scale model is used to simulate forest-wide phenomena, an annual pollutant load would be chosen from a probability distribution each year. The appropriate response curve generated by the high resolution model for the effective annual decrease in photosynthesis would be sampled for each tree's response to that pollutant load. This value, in turn, would enter the growth submodel and would eventually be incorporated into the net productivity of an individual tree within the plot. By repeating this for all of the trees in the plot and for all of the plots, the overall decrease in forest productivity would be simulated, and the variances resulting from difference in annual and daily SO_2 loadings, within tree species consequences, differences in effects among species, and so on, would be factored into the total forest response.

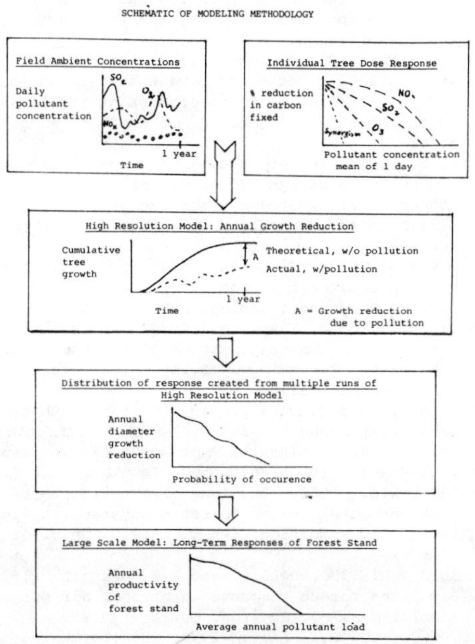

Figure 3. -- Schematic of modelling methodology used in CASSANDRA model development.

In essence, then, the high resolution model is being designed for a single purpose: to provide sets of distribution functions which provide stochastic information, useful at the scale of the ecosystem, on how pollutants affect processes. At the larger ecosystem scale, all of the small, fast events and responses need not be separately modelled; rather, at the large scale, the small scale events can be treated as part of a stochastic phenomenon. This scheme parallels the hierarchy of the real world, in that at the ecosystem level minor events and fluctuations are smoothed over, and processes which are highly important at fine scales of resolution may not even be recognized by the ecosystem. As an illustration of this point, consider that at the ecosystem level it is quite important to know the overall uptake of a nutrient during the growing season and the associated reduction in the available pool in the soil, but it is not important to the ecosystem whether or not, at a given point in time, a particular sapling is successful in attracting a nutrient molecule from an exchangeable surface in the soil.

One of the major advantages of this approach is that the ecosystem-level model requires only the distribution function for it to be able to simulate ecosystem-level responses to air pollution. This means that simulations can be conducted and the model refined currently, even though the mechanistic understanding of the fine scale processes is now incomplete. This allows substitutability of different mechanistic models very readily, as data and hypotheses are refined. A benefit here is that the overall CASSANDRA model is adaptive, and thus able to keep abreast of the state of laboratory and field air pollution research as it undergoes its current rapid expansion.

Another significant benefit of the approach here is that alternative hypotheses of mechanistic relationships can readily be tested with respect to the eventual effect on the ecosystem level. Thus two conflicting ideas of how SO_2 and O_3 synergistically affect decomposition processes, for example, can each be tested in the context of the significance to ecosystem-level characteristics, such as net productivity. Even more importantly, in those instances where there currently is such a paucity of data that no sound mechanistic relationships can be hypothesized, the approach taken here would provide a means of testing whether or not such an understanding is even necessary. That is, if a mechanism or even the nature of a fine scale relationship is unknown, an arbitrary set of substantially differing response functions could be entered into the ecosystems-level model. If the results of the model simulation are insensitive to the different response functions used, then it is clear that it is unimportant to know the causal relationship, when considered from the scale of the ecosystem. This may prove to be a powerful tool for identification of research needs and for the identification of fruitless research approaches.

SUMMARY

The Ecosystems Research Center at Cornell University is currently developing a model to simulate the effects of air pollutants on forested ecosystems of the eastern United States. The ecosystem-scale model is fairly well established, requiring relatively few major alterations. The nature of air pollutant impacts on forests is that fine scale events and effects are predominant, and that through integration over time and space, many such fine scale phenomena eventually are manifested in ecosystem-level responses. The CASSANDRA model is being structured in parallel with this real world hierarchy, so that the model ecosystem only recognizes the statistical set of short time scale events. To generate those fine scale response functions, a series of high resolution submodels are being developed. The linkage across these two distinct scales occurs through stochastic information transfer, allowing the ready replacement of the high resolution submodels. This approach allows the adaptive construction of the overall model, which will develop as the state of air pollution research develops, and it allows the direct testing of alternative hypotheses and of the sensitivity of ecosystems-level processes to reductionistically studied mechanisms.

LITERATURE CITED

Anderson, M.C. 1973. Solar radiation and carbon dioxide in plant communities - conclusions. p. 245-354. In J.P. Cooper (Ed.), Photosynthesis and productivity in different environments. Cambridge Univ. Press, New York.

Bauer, H., W. Larcher, and R.B. Walker. 1973. Influence of temperature stress on CO_2-gas exchange. p. 557-586. In J.D. Cooper (Ed.), Photosynthesis and productivity in different environments. Cambridge Univ. Press, New York.

Botkin, D.B., W.H. Smith, R.W. Carlson, and L.T. Smith. 1972. Effects of ozone on white pine saplings: Variation in inhibition and recovery of net photosynthesis. Environ. Pollut. 3:273-289.

Bourdeau, P.F. 1963. Photosynthesis and respiration of Pinus strobus seedlings in relation to provenance and treatment. Ecology 44:710-716.

Carlson, R.W. 1979. Reduction in the photosynthetic rate of Acer, Quercus, and Fraxinus species caused by sulphur dioxide and ozone. Environ. Pollut. 18:159-170.

Fritts, H.C. 1966. Growth-rings of trees: Their correlation with climate. Science 154:973-979.

Keller, T., 1970. Gaseous exchange - A good indicator of nutritional status and fertilizer response of forest trees. p. 669-676. In Proc. 6th Internat. Colloq. Plant Analysis and Fertilizer Problems. ISHS, Tel Aviv.

Lamoreaux, R.J., and W.R. Chaney. 1978. Photosynthesis and transpiration of excised silver maple leaves exposed to cadmium and sulphur dioxide. Environ. Pollut. 17:259-268.

Legge, A.H., and D.R. Jaques. 1977. Field studies of pine, spruce and aspen periodically subjected to sulfur gas emissions. Water, Soil, Air Pollut. 8:105-129.

Loucks, O.L. (Ed.). 1980. Crop and forest losses due to current projected emissions from coal-fired power plants in the Ohio River Basin. Report to the Ohio River Basin Energy Study, TIE, Indianapolis, 266 p.

Loucks, O.L. 1981. A scenario for research on the effects of stress in the eastern deciduous forest. Unpublished draft manuscript submitted to the Ecosystems Research Center, Ithaca, New York, 36 p.

National Academy of Sciences. 1981. Atmosphere-biosphere interactions: Toward a better understanding of the ecological consequences of fossil fuel combustion. Report of the U.S. National Academy of Sciences, Committee on the Atmosphere and the Biosphere, Commission on Natural Resources. Washington, D.C., 263 p.

Nátr, L. 1973. Influence of mineral nutrition on photosynthesis and the use of assimilates. p. 537-555. In J.P. Cooper (Ed.), Photosynthesis and productivity in different environments. Cambridge Univ. Press, New York.

Pisek, A., and W. Tranquillini. 1954. Assimilation und Kohlenstoffhaushalt in der Krone von Fichten (Picea excelsa Link) und Rotbuchenbäumen (Fagus silvatica L.) Flora (Jena) 141:237-270.

Polster, H. 1950. Die physiologischen Grundlagen der Stofferzeugung im Walde. Untersuchungen über Assimilation, Respiration und Transpiration unserer Hauptholzarten. Bayrischer Landwirtschaftsverlag, G.m.b.H. München, 96 p.

Shugart, H.H., and D.C. West. 1977. Development of an Appalachian deciduous forest succession model and its application to assessment of the impact of the chestnut blight. J. of Environ. Manage. 5:161-179.

Slavik, B. 1973. Water stress, photosynthesis and the use of photosynthates. p. 511-536. In J.P. Cooper (Ed.), Photosynthesis and productivity in different environments. Cambridge Univ. Press, New York.

Smith, W.H. 1981. Air pollution and forests. Interactions between air contaminants and forest ecosystems. Springer-Verlag, New York, 379 p.

Trijonis, J., and K. Shapland. 1978. Visibility in the southwest: An exploration of the historical data base. Technology Service Corporation, Santa Monica, Calif., 107 p.

Thomas, M.D., and G.R. Hill. 1949. Photosynthesis under field conditions. p. 19-52. In J. Frank and W.F. Loomis (Eds.), Photosynthesis in plants. Iowa State Univ. Press, Ames, Iowa.

Weinstein, D.A. 1982. The long term nutrient retention properties of forest ecosystems: A simulation investigation. Ph.D. thesis, Univ. of Tennessee.

West, D.C., S.B. McLaughlin, and H.H. Shugart. 1980. Simulated forest response to chronic air pollution stress. J. Environ. Qual. 9:43-49.

Wolff, G.T., and P.J. Lioy. 1980. Development of an ozone river associated with synoptic scale episodes in the eastern United States. Environ. Sci. Tech. 14(10)1257-1260.

DYNAMIC CROP MODELLING: A SYNERGISM OF COMPUTERS, EXPERIMENTAL RESEARCH, AND THE SCIENTIFIC METHOD[1]

J. M. McKinion and D. N. Baker[2]

Abstract.-- The construction of dynamic crop simulation models has become a well-defined process. This paper will describe the experimental facility needed to obtain the data base for construction of such models and the computer equipment used to build and run these models. A brief description of the crop model structure is presented.

INTRODUCTION

The methodology of dynamic crop simulation as practiced by the Crop Simulation Research Unit (C.S.R.U.) of USDA-ARS consists of three very closely linked elements: (1) The SPAR (Soil-Plant-Atmosphere-Research) research facility, (2) the computer data acquisition and control system and scientific time-sharing computer system, and (3) the crop models at the process-level of physical and physiological systems. The SPAR units enable the acquisition of the data base from which the models are constructed. The computer systems make possible the control of experiments in the SPAR units and the collection of multiple, accurate data points throughout the growth and development of the crop that is being studied. Since the time-sharing system is coupled to the control computer, the experimental status and current data can be ascertained in a timely fashion. The time-sharing computer is used not only for data reduction, analysis, and archieval but also for model development. Each scientist and technician works in an interactive mode with this system via a video display unit. The crop models constructed by the C.S.R.U. have been developed on this type of computer. This paper will describe in detail each of these three elements.

THE SPAR UNIT

In order to model a field crop such as cotton, soybeans, or wheat, a comprehensive study must be made of plant dynamics. This requires simul-taneous measurements of plant roots and tops in a controlled environment. The SPAR system provides the research tool for attaining these objectives (Phene, et al. 1978; McKinion 1980; and McKinion et al. 1982). A SPAR unit, basically, is a 2 x 0.5 x 1 m box constructed of steel plate and angle iron. The front face consists of wire-reinforced plate glass for viewing roots at the soil-glass interface. The rear face has twelve access ports approximately 10 cm in diameter cut in the steel plate and arranged in a 4 x 3 pattern in which the three rows of ports are 25 cm, 50 cm, and 75 cm from the top of the soil bin respectively. These ports provide access to the soil profile for the placement of tensiometers, gypsum blocks, oxygen probes, thermocouples, and other instrumentation for measuring soil conditions at the various depths. The bin is covered by an acrylic plastic top 1.5 m tall which is sealed to the metal bin with an air-tight gasket. The front and rear views of the SPAR unit are shown in Figures 1A and 1B. The instrumentation interface can be seen in Fig. 1B on the lower left side of the unit. In this box are barrier strips for connecting up to 12 thermocouples, 15 analog signals, 20 control lines, and up to 5 D.C. voltage sources which are connected at the other end to a patch panel for interfacing to the control computer or regulated high-capacity power supplies. The irrigation line can be seen above the interface box. The crop grown in the unit can be irrigated. The computer controlled irrigation system consists of a volumetric flow counter, a 3 - 15 psi pressure regulator, a 115 VAC solenoid valve and a 2 liter nutrient reservoir. Water and nutrients are drip irrigated at the soil surface with the drip lines spaced every 10 cm apart. At the bottom of the soil profile in the SPAR unit are a drain system and an aeration system as shown in Figure 2. The drain consists of Selas 361 FEF[3] porous porcelain tubes 0.54 cm diameter by 39.7 cm length with 0.27 micron absolute fitering capacity with these spaced at 10 cm intervals across the 2 m length. This system is used to drain excessive moisture from the soil either by gravity flow or by the use of a vacuum to simulate a soil suction. Conversely, this system can also be used to simulate a water table by pumping water into the bottom of the soil profile. The aeration system can be used to aerate the soil profile or to simulate flooding on a rapid basis by using nitrogen to flood the root zone to generate an oxygen shortage. The advantage of this technique is that the nitrogen can be rapidly flushed from the soil profile and that the level of oxygen in the profile can be controlled. This system consists of a network of 45 cm long 0.635 cm diameter copper tubing spaced every 20 cm along the 2 m length of the SPAR bin joined to a manifold of 1.27 cm diameter copper. Currently a soil temperature control system is being added. This system consists of vertical type 304 stainless steel tubing 0.95 cm diameter at a spacing of 20 cm around the periphery of the soil bin. Water cooled to the desired soil temperature is circulated through these tubes. Cooling is supplied by a heat exchanger from the SPAR air cooling system. The soil bin is surrounded by a shell of 3/4 inch styro-

---

[1] Paper presented at the Third International Conference on State-of-the-Art in Ecological modelling, Colorado State University, May 24-28, 1982.

[2] J. M. McKinion is Electronic Engineer, and D. N. Baker is Research Agronomist, USDA, Agricultural Research Service, Crop Simulation Research Unit, Mississippi State, MS U.S.A.

[3] Mention of product names is for description only, and not an endorsement by USDA-ARS.

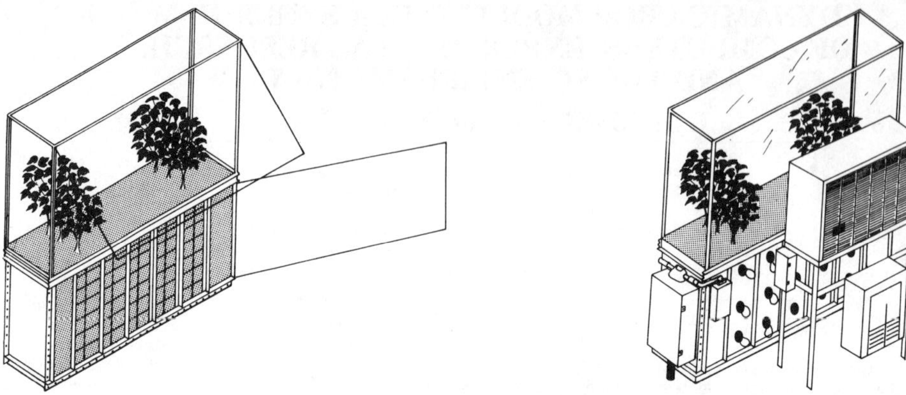

Figure 1. The Soil-Plant-Atmosphere-Research (SPAR) Unit for conducting closed-environment studies of plant growth and yield under computer control: (A) Front view; (B) Rear view. In use, screens are placed at 80 percent of plant height on the outside of the chambers to simulate between-and within-row shading.

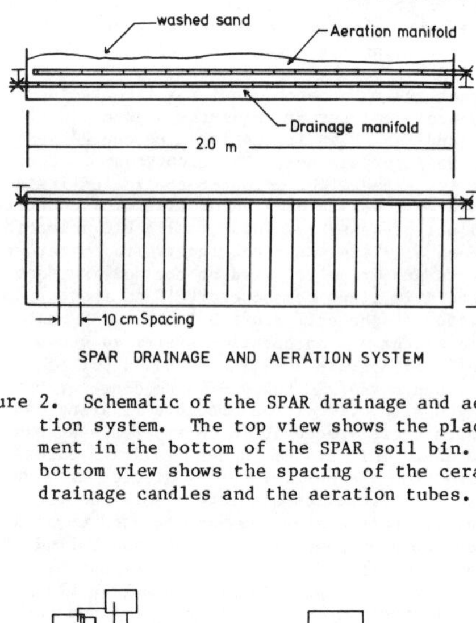

Figure 2. Schematic of the SPAR drainage and aeration system. The top view shows the placement in the bottom of the SPAR soil bin. The bottom view shows the spacing of the ceramic drainage candles and the aeration tubes.

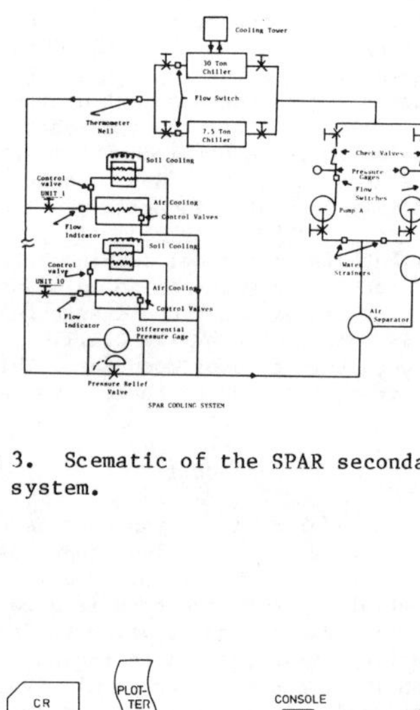

Figure 3. Scematic of the SPAR secondary cooling system.

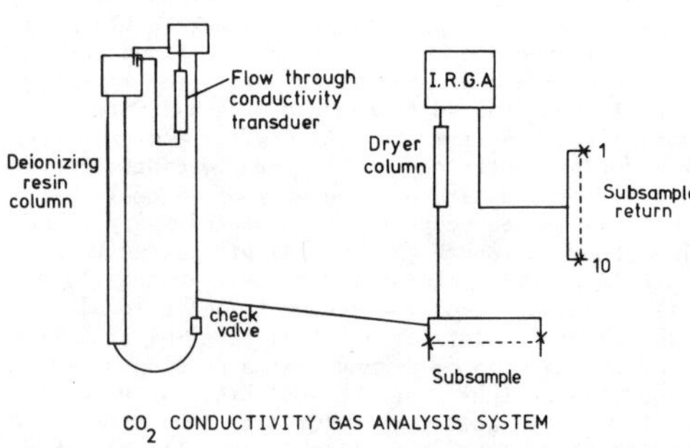

Figure 4. Schematic of the SPAR CO_2 conductivity analysis system.

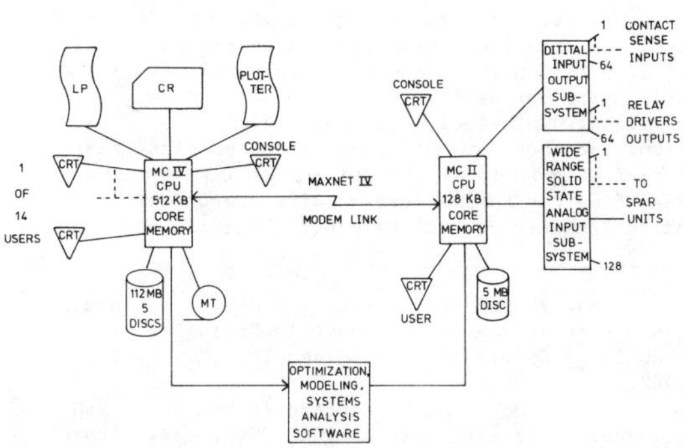

Figure 5. System diagram of information flow in the computer network for data acquisition and control of SPAR installation, information transfer and analysis application to crop models, and positive feedback to direction and scope of experimentation.

foam insulation which also provides instrumentation protection from the environment.

The top of the SPAR unit is constructed of 0.635 cm acrylic plastic which is 94% transmissive from 400 to 700 nm wavelengths of solar radiation. The front of the SPAR top consists of a 2 x 1.5 m door hinged at the top with a stainless steel piano-type hinge. The door allows access to the plant materials, soil surface, and above-soil instrumentation. The door when shut is held against a custom-made gasket by several spring-loaded quarter-turn fasteners that provide rapid access to the interior. The rear of the SPAR unit contains the ductwork for distributing controlled-temperature air from the heating and cooling system.

Air temperature control is accomplished with a secondary cooling system as shown in Figure 3. The primary refrigeration unit in this system is a 30 ton, water cooled compressor. A 7.5 ton air cooled compressor with a separate shell and tube evaporator is used in parallel as a backup unit. Dual circulation pumps are used to pump the chilled ethylene glycol-water solution which actually cools the air in the SPAR units. The second pump will automatically cut in if the first pump fails. A dual pump system is also used in the water-cooling tower of the large compressor. The air temperature in the SPAR unit is cooled by a three-circuit heat exchange coil through which the air is pulled by a blower. One circuit is on all of the time providing for continuous condensation of evapotranspiration in the unit. The two additional circuits can be switched in via solenoid valves under computer control if more cooling is needed. Fine control of air temperature is accomplished by trimming temperature with a 5.2 KW electric heater under computer control.

The carbon dioxide concentration in the SPAR unit is monitored and controlled with a conductivity system. Each SPAR unit has a set of 6 high density 0.635 cm diameter polyethylene tubes. Five of these tubes are used for the CO_2 sampling and return system. The sixth tube, run in a separate conduit from the control room, serves to furnish the system with pure CO_2 to maintain a fixed setpoint. Of the five-tube system, one tube is used to pump a sample of air from the SPAR unit to the control room located approximately 75 feet away. Most of this high flow rate sample is returned to the SPAR unit in the main sample return via the second tube. Two subsamples are taken from the sample tube and are returned to the SPAR unit in the third tube, resulting in a closed system. The first subsample is monitored continuously by the conductivity system (James 1966, Slack and Calvert 1972) in which the sample is bubbled through a CO_2-free tube of water at the end of which is a temperature-compensated conductivity probe. The second subsample is monitored periodically at minute intervals by a very stable, high accuracy infrared gas analyzer which checks the calibration of the conductivity system. The schematic of this system is shown in Figure 4. Tubes four and five are used to heat the sample and two returns to ensure that condensation does not occur.

The SPAR system is able to control atmospheric CO_2 concentration, rough control of relative humidity, air temperature, soil water, soil nutrients, soil temperature, and soil aeration. Extensive measurements can be made of crops grown in these units with the capability of measuring voltages from $\pm$ 1 microvolt to $\pm$ 100 volts DC and from $\pm$ 10 microvolts to $\pm$ 650 volts AC, resistance from 100 microohms to 10 megohms, and conductance from 0.01 to 100.0 nanosiemens. Thus any instrument or transducer, whose output that falls within the above ranges, can be measured by the data acquisition and control system.

COMPUTER FACILITIES

A diagramatic representation of the computer facility that manages the SPAR installation and modelling effort is shown in Figure 5. A real-time process control minicomputer is used to acquire data from the current ten SPAR units and to control the environments therein. (Data analysis from the SPAR units and from the literature and field experiments is performed on the mid-sized time-sharing computer.) The main requirements of the minicomputer are multiprogramming capability, response to external interrupts, and response to timed events. The computer that is being used here is a Modular Computer System Modcomp II/12 minicomputer with 128 K bytes of magnetic core memory, two 2.5 M byte cartridge disc drives, a 9 track 800 BPI tape drive, 64 bits of digital input, 128 bits of digital output, and an IEEE-488 interface bus. Data collection is performed by two programmable digital voltmeters with reed relay scanners: a Fluke 8520A DVM with a 40 channel 2204A scanner and a Hewlett-Packard 3497A 100 channel Data Acquisition/Control Unit. The DVM's under control of the computer via the IEEE-488 bus acquire data from their channels and store these data in their own internal memory until polled by the computer. The H-P unit acquires data for its 100 channels every 10 seconds. The Fluke unit is used to acquire data from slowly varying signals and a single channel is measured every 10 seconds. Most data points are scanned in the computer for a fifteen minute period and then averaged. These data are then stored on a disc file. An RS-232C port is available to transfer these data on a daily basis to the larger time-sharing system for incorporation into the data base and for archieval. A Fortran X 3.9 1966 ANSI compiler with the Purdue ISA real-time extensions is used to write all of the programs that manage the the control of and data acquisition from the SPAR units.

The mid-sized time-sharing computer consists of a central processing unit with one megabyte of memory, two disc drives with a total of 248 M bytes of storage, a 300 lpm printer/plotter, a 1600 BPI-9 track tape drive, and support for sixteen CRT terminals and remote printers. All model development, verification, and validation studies are performed on this machine. The operating system supports time-sharing and provides for memory management with a virtual-memory, demand-paged capability such that each user has a one megabyte program capability. The crop models are written in ANSI X 3.9 1966 standard Fortran. The computer supplies an ANSI 1977 Fortran compiler with a source level debug capability. This type of software tool is mandatory with process-level models because of the length of code and the complex interaction of the code.

The above described SPAR system and associated control computer and time-sharing computer provide for the collection and analysis of the data obtained from controlled environmental experiments. These data are necessary to construct process-level physiologically-based dynamic crop simulation models. The time-sharing scientific computer system makes possible the actual coding, verification, and validation of these types of models and thus helps to guide the researchers in the conduction of further SPAR experiments when gaps are found in the knowledge of the particular crop system.

PROCESS LEVEL CROP MODELS

Dynamic systems simulation as applied to field crops is the simulation of a field crop response from a causal process-oriented physiological-physical viewpoint. The required level of detail in such a simulation must be sufficient to define process level events that characterize the crop's response to environmental, cultural, and pest parameters that affect growth and yield predictions. The processes are modeled at the level of detail necessary to obtain the desired prediction accuracy. Modelling a crop at the process level provides several benefits. First, a process-level model does not have to be recalibrated each time it is moved to a different geographical location (Marani and Baker 1981). Secondly, a process-level model should simulate a crop across the ecological and economic range of that crop. Third, this type of model provides appropriate migration paths for the incorporation of new theory and science as it is developed without having to rewrite the model. An important adjunct to this capability is that any other physically or physiologically based process can easily be added to this type of model. For example, if the physiological effects are known, the result of an air pollutant on the growth and development of a crop can be predicted. Fourth, the most important aspect of this type of modelling is enhanced understanding of the crop by the researcher. In field plot experiments, the researcher can observe only one "run" per growing season. By using this type of model, the researcher can observe a hundred "runs" in a single day. Concomitant with the above capability is the ability to use the SPAR-Computer-Model interative process to determine further controlled environmental experiments to bridge the gaps in knowledge or to propose new theory or hypotheses of processes in the crop.

Currently, we are working on process-level models of three crops: cotton (Baker et al. 1982); soybean (Acock et al. 1982); and wheat (Baker et al. 1981). Of these models, the most advanced is the cotton model GOSSYM. Since all of these models are process-level models, various subsystems and algorithyms can be shared by each. The routines for handling climate, soil water and soil nutrient movement, evapotranspiration, root impedence, cultivation, and soil oxygen are essentially the same in each model.

Since the cotton model GOSSYM is the most complete model and has been successfully validated for the Mississippi delta, irrigated Arizona conditions and four geographical locations in Israel (Marani and Baker 1981), this model will be subsequently described.

The program flow of GOSSYM (Baker et al. 1976, and Whisler et al. 1982) is presented in Figure 6. GOSSYM consists of a central control program and subroutines. The subroutines are divided along lines of process levels. The solution method of GOSSYM is a semi-finite difference method for most processes which are represented by algebraic equations that calculate the rates of the individual processes. Capillary flow of water and diffusion of oxygen in the soil profile is calculated from the 2-D explicit difference solution of Darcy's law and an alternating direction solution of the 2-D diffusion equation, respectively. The parameters that drive the model are the initial input parameters such as soil type, row spacing, plant population, and inital fertilizer application. The model iterates on a daily basis and needs the following weather inputs: total daily solar radiation; total daily rainfall or irrigation; maximum daily air temperature; and minimum daily air temperature. GOSSYM starts on the first day of the emergence of the plant from the soil (an input parameter)(see Figure 7) and then proceeds to the end of the season. Many kinds of intermediate information about the plant and soil status can be printed out on a daily basis. The soil fertilizer status is shown in Figure 8. The soil is divided into a two dimensional grid of 5 x 5 cm cells. The soil profile is 1 meter wide and 1 meter deep and thus contains 20 x 20 cells. The axes of cotton plants are on the left and right corners. Since the calculations are symmetrical, only the plant on the left hand is shown. Root distribution is shown in Figure 9 and soil water status is shown in Figure 10. Soil oxygen status is shown in Figure 11. A statistical representation of the average plant simulated by the crop model is shown in Figure 12. Different types of fruiting forms are represented in this diagram. The symbol that appears is the statistical majority at that particular site. That is, if an "X" appears at a particular location on the plant, then most of the plants have a square (floral bud) at that location. Much more information can be obtained than the few examples listed above.

The MAIN program serves as an executive for the model. In this program the daily iterations are controlled here. The program starts at emergence and ends at the physiological maturity of the crop. Seven subroutines are called from MAIN.

CULVAT simulates the effect of a cultivation by prunning of roots and by changing the bulk density of the soil in the plow zone (Whisler 1982). The strategy here is to allow the soil to gradually return to its original bulk density after certain increments of rainfall (irrigation). Simulated results are shown in Figure 13.

The CLYMAT routine is responsible for reading in the weather data, converting in DATE Julian day number to day-month-year registering this information on the output listing, and calculating the soil temperatures in TMPSOL. TMPSOL uses a two-dimensional solution which assumes an initial temperature distribution and then uses air temperature and percent light interception by the crop to calculate new temperature distributions.

The SOIL subroutine and associated subprograms FRTLIZ, GRAFLO, ET, UPTAKE, CAPFLO, and NITRIF are the major improvement in GOSSYM over the previous cotton model SIMCOT II (McKinion et al. 1975). The soil section of GOSSYM solves the Darcian flow equation for flow in a porous media. This equation is solved by an iterative process where the number of iterations required for a stable solution depends on the soil type. In most cases the number of iterations (NITER) is four. The SOIL routine and its units are called daily except for UPTAKE and CAPFLO, which are called NITER and NITER*2 times, respectively.

For calculation of soil and plant water potential, the total evapotranspiration from the soil surface and plant canopy is currently estimated by

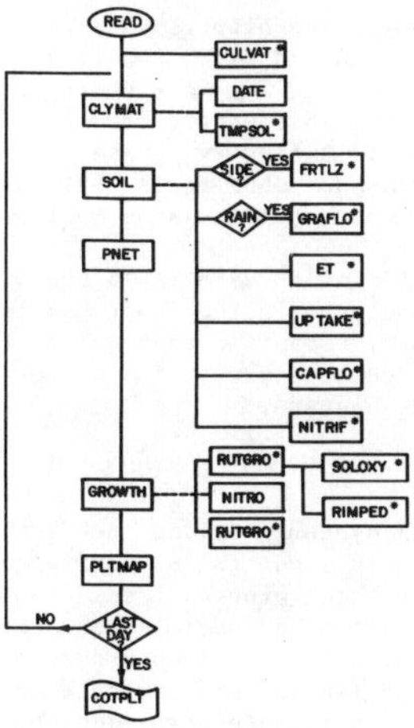

Figure 6. Block diagram of program flow in the cotton model GOSSYM.

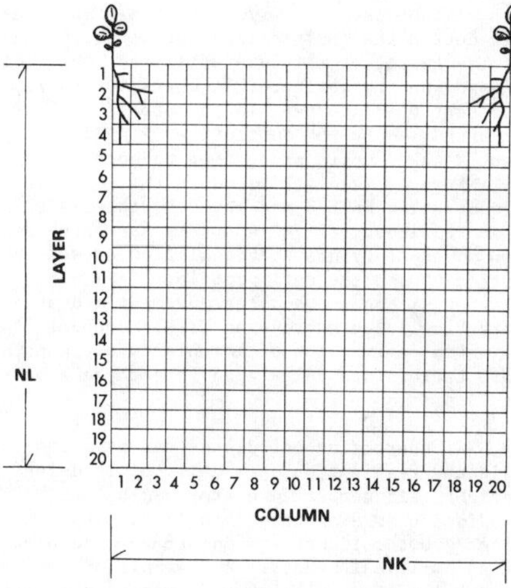

Figure 7. Schematic of the spartial configuration of the soil section of GOSSYM.

```
VOLUMETRIC NITRATE CONTENT OF SOIL                      DAY  81
AT THE END OF MAIN

UNITS  -  MG/N PER CM**3                              LEGEND

                     1 1 1 1 1 1 1 1 1 1 2
   1 2 3 4 5 6 7 8 9 0 1 2 3 4 5 6 7 8 9 0                    <=  0.0

                                                   0.0     < 0 <=  0.0100
 1 0 0 0 0 0 0                 0 0 0 0 0 0
 2                                                 0.0100 < 1 <=  0.0200
 3 0 0 0 0 0 0 0 0 0 0 0 0 0 0 0 0 0 0 0 0
 4 0 0 0 0 0 0 0 0 0 0 0 0 0 0 0 0 0 0 0 0        0.0200 < 2 <=  0.0300
 5
 6                                                 0.0300 < 3 <=  0.0400
 7 1 2 *                                 * 2 1
 8 2 5 * 1 0 0 0 0         0 0 0 0 1 * 5 2        0.0400 < 4 <=  0.0500
 9 1 4 * 3 2 2 2 1 1 1 1 1 1 2 2 2 3 * 4 1
10 0 1 * 4 4 4 4 3 3 3 3 3 3 4 4 4 4 * 1 0        0.0500 < 5 <=  0.0600
11 0 0 6 4 5 5 5 5 5 5 5 5 5 5 5 5 4 6 0 0
12 0 0 1 2 4 4 5 5 5 5 5 5 5 5 4 4 2 1 0 0        0.0600 < 6 <=  0.0700
13 0 0 0 0 2 2 3 3 4 4 4 4 3 3 2 2 0 0 0 0
14 0 0 0 0 0 1 1 2 2 2 2 2 2 1 1 0 0 0 0 0        0.0700 < 7 <=  0.0800
15 0 0 0 0 0 0 0 0 1 1 1 1 0 0 0 0 0 0 0 0
16 0 0 0 0 0 0 0 0 0 0 0 0 0 0 0 0 0 0 0 0        0.0800 < 8 <=  0.0900
17 0 0 0 0 0 0 0 0 0 0 0 0 0 0 0 0 0 0 0 0
18 0 0 0 0 0 0 0 0 0 0 0 0 0 0 0 0 0 0 0 0        0.0900 < 9 <=  0.1000
19 0 0 0 0 0 0 0 0 0 0 0 0 0 0 0 0 0 0 0 0
20 0 0 0 0 0 0 0 0 0 0 0 0 0 0 0 0 0 0 0 0        0.1000 < *

   TOTAL  =     171.7330  MG N
```

Figure 8. Soil fertilizer status of day 81 of standard simulation. The number of the soil layer is given on the extreme left (1 through 20). The number of soil columns is geven at the top (1-20). The legend is given on the right side of the figure.

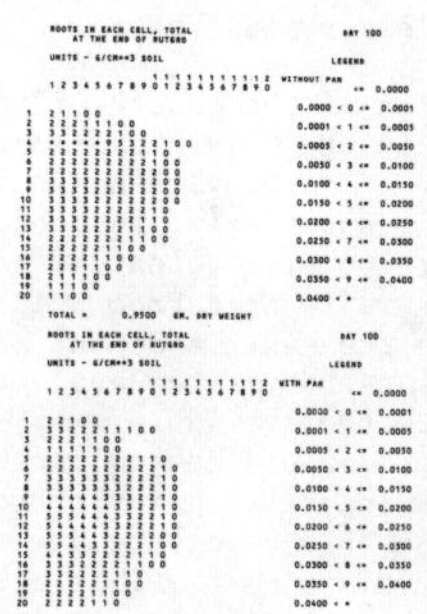

Figure 9. Soil root distribution on day 100 of simulation comparing a root zone with a hard pan at 20 cm depth and a root zone without a hard pan.

```
      VOLUMETRIC WATER CONTENT                  DAY 80

         STANDARD AAA

      UNITS - CM**3/CM**3 SOIL                 LEGEND

                         1 1 1 1 1 1 1 1 1 1 2
    1 2 3 4 5 6 7 8 9 0 1 2 3 4 5 6 7 8 9 0            <= 0.00

                                                0.00 < 0 <= 0.05
 1  5 5 5 5 5 5 5 5 5 5 5 5 5 5 5 5 5 5 5 5
 2  4 4 4 4 4 5 5 5 5 5 5 5 5 5 5 4 4 4 4 4   0.05 < 1 <= 0.10
 3  4 4 4 4 4 4 4 5 5 5 5 5 5 4 4 4 4 4 4 4
 4  5 5 5 5 5 5 5 5 5 5 5 5 5 5 5 5 5 5 5 5   0.10 < 2 <= 0.15
 5  5 5 5 5 5 5 5 5 5 5 5 5 5 5 5 5 5 5 5 5
 6  6 6 6 6 6 6 6 6 6 6 6 6 6 6 6 6 6 6 6 6   0.15 < 3 <= 0.20
 7  6 6 6 6 6 6 6 6 6 6 6 6 6 6 6 6 6 6 6 6
 8  5 5 5 6 6 6 6 6 6 6 6 6 6 6 6 6 5 5 5 5   0.20 < 4 <= 0.25
 9  4 4 4 4 5 5 6 6 6 6 6 6 6 6 5 5 4 4 4 4
10  4 4 4 4 4 5 5 6 6 6 6 6 6 5 5 4 4 4 4 4   0.25 < 5 <= 0.30
11  4 4 4 4 4 5 5 5 6 6 6 6 5 5 5 4 4 4 4 4
12  4 4 4 4 5 5 5 5 5 6 6 5 5 5 5 5 4 4 4 4   0.30 < 6 <= 0.35
13  6 6 6 6 6 6 6 6 6 6 6 6 6 6 6 6 6 6 6 6
14  6 6 6 7 7 7 7 7 7 7 7 7 7 7 7 7 7 6 6 6   0.35 < 7 <= 0.40
15  6 6 6 7 7 7 7 7 7 7 7 7 7 7 7 7 7 6 6 6
16  6 6 7 7 7 7 7 7 7 7 7 7 7 7 7 7 7 7 6 6   0.40 < 8 <= 0.45
17  7 7 7 7 7 7 7 7 7 7 7 7 7 7 7 7 7 7 7 7
18  7 7 7 7 7 7 7 7 7 7 7 7 7 7 7 7 7 7 7 7   0.45 < 9 <= 0.50
19  7 7 7 7 7 7 7 7 7 7 7 7 7 7 7 7 7 7 7 7
20  7 7 7 7 7 7 7 7 7 7 7 7 7 7 7 7 7 7 7 7   0.50 < *
```

Figure 10. Soil water status on day 80 of simulation.

```
Oxygen Concentration in Soil     Oxygen Concentration in Soil
   No Aeration Problems              Limited Aeration              Day 80
Units - ATM Partial Pressure     Unit - ATM Partial Pressure       LEGEND

                      1                            1
     1 2 3 4 5 6 7 8 9 0          1 2 3 4 5 6 7 8 9 0                     <= 0.00

                                                            0.00 < 0 <= 0.02
 1   * * * * 9 9 9 9 9 9          7 5 3 1
 2   * * 9 9 9 9 9 9 9 9          3 0                       0.02 < 1 <= 0.04
 3   9 9 9 9 9 9 9 9 9 9
 4   9 9 9 9 9 9 9 9 9 9                                    0.04 < 2 <= 0.06
 5   9 9 9 9 9 9 9 9 9 9
 6   9 9 9 9 9 9 9 9 9 9                                    0.06 < 3 <= 0.08
 7   9 9 9 9 9 9 9 9 9 9
 8   9 9 9 9 9 9 9 9 9 9                                    0.08 < 4 <= 0.10
 9   9 9 9 9 9 9 9 9 9 9
10   9 9 9 9 9 9 9 9 9 9                                    0.10 < 5 <= 0.12
11   9 9 9 9 9 9 9 9 9 9
12   9 9 9 9 9 9 9 8 8 8                                    0.12 < 6 <= 0.14
13   9 9 9 9 8 8 8 8 8 8
14   9 9 9 9 8 8 8 8 8 8                                    0.14 < 7 <= 0.16
15   9 9 9 8 8 8 8 8 8 8
16   9 9 8 8 8 8 8 8 8 8                                    0.16 < 8 <= 0.18
17   9 8 8 8 8 8 8 8 8 8
18   9 8 8 8 8 8 8 8 8 8                                    0.18 < 9 <= 0.20
19   8 8 8 8 8 8 8 8 8 8
20   8 8 8 8 8 8 8 8 8 8                                    0.20 < *
```

Figure 11. Soil oxygen status on day 80 of simulation comparing standard simulation with no aeration problem and a simulation with aeration problems due to flooding of root zone.

use of the Ritchie model (Ritchie 1972), an empirical method. In the future, a more mechanistic approach will be used. However, for most applications in cotton the empirical has proven to be adequate. The amount of moisture estimated from ET is then imposed on the soil profile where there are roots present that are capable of uptake. The water is then withdrawn by the subroutine UPTAKE. Nitrogen is taken up in proportion to soil water uptake. Nitrogen is distributed in the soil profile by the subroutine FRTLIZ and moved by the GRAFLO and CAPFLO subroutines. GRAFLO accounts for water and nitrogen movement by gravitational flow caused by percolation though the soil profile. CAPFLO accounts for nitrogen and water movement mediated by capillary flow. The subroutine NITRIF accounts for the biological conversion of ammonia to nitrates. The above factors are calculated in each cell of the soil profile.

The PNET subroutine calculates canopy photosynthesis and respiration on a daily basis using plant height, air temperature, row width, and solar radiation as exogenous data input. PNET calculates growth, light, and maintenance respiration in an empirical manner. An example of the type of data required for calculating phoytosymthesis is presented in Figure 14. Once net photosynthate is calculated on a canopy basis, then using planting density and row spacing, the amount of net photosynthate available per plant is calculated.

The simulation of the growth of the cotton crop is accounted for in the GROWTH routine. Here a materials balance of carbohydrate and nitrogen from PNET and UPTAKE, respectively, is used to form various plant parts: stems, leaves, roots, and fruit. The key here is that from SPAR experiments, one can calculate the maximum demand of various plant tissues for carbohydrates and nitrogen. One can also calculate the available supply of these materials. Using a supply/demand ratio with fruiting forms having first priority for these nutrients, one can then calculate how much growth can occur in each plant part. This strategy for calculating growth reflects the nutritional theory of plant growth. A supply/demand ratio is calculated for both nitrogen and carbohydrates. When these ratios are unity then all growth can be satified. When they are less than unity, then only a percentage of growth can occur. The amount of growth in limiting conditions is determined by an iterative process. If nitrogen is limiting and carbohydrates are abundant, then only the amount of growth allowable by the nitrogen is allowed and the remaining carbohydrates are placed in a temporary storage area. If both are limiting then an iterative process occurs in which first nitrogen is considered and then carbohydrates. In this case, if carbohydrates are limiting growth more than the nitrogen supply, then growth is further reduced and the model iterates again through the GROWTH routine until a balance is obtained. The supply/demand ratios are also used as stress indicators.

Stress factors primarily affect the morphogenetic rates and the abscission of fruiting forms. These rates are calculated in PLTMAP and are primarily temperature driven. A basic data set is shown in Figure 15. These data are used to form rate equations that calculate the time between events in the formation of internodes on the plant. As each node is formed, a floral bud and a leaf can be initiated. Stress factors primarily lengthen these times. The stress factors also are used to abscise fruit from the plant.

Root growth and nitrogen uptake are accounted for in the RUTGRO and NITRO routines which are called from GROWTH. The mass and type of roots in each cell of the soil matrix are accounted for in these routines. Basically, these are old roots that have suberized and young roots that can grow and absorb water and nutrients at a maximum rate.

At the end of the simulated season, the amount of lint accumulated in the fruit (bolls) that will open is summed, converted to a per-hectare basis, and transformed to a prediction of cotton yield in kilograms per hectare.

CONCLUSION

The SPAR system provides the research tool needed to obtain a controlled environment data base needed to construct process-level crop models. The computer systems make possible the operation of these data-intensive tools on a season basis with reliable operation and precise data acquisition. The time-sharing computer provides the research tool needed to analyze these data and to interactively construct these crop models in a timely manner.

Process-level crop models offer a tool to better understand the dynamics of crop growth and the interactions of the crop with its environment. These models also offer a method for planning of experimental research into the understanding of the mechanisms or processes within the plant. Validated crop models of this type can also be used to answer other questions such as the effect of pollutants, increased atmospheric CO_2, and future climate scenarios on crop yield.

Crop models of this type can also be used as a management tool for integrated farm management to aid the farmer in minimizing his resource inputs to a crop (fertilizer, water, pesticides) and to maximize his yield (profits).

LITERATURE CITED

Acock, B. 1982. Validation and sensitivity testing of GLYCIM: The position as of January 1982. 1981 Research Report. U. S. D.O.E./USDA Interagency Agreement No. DE AI01-81ER60001. Direct Biological Effects of Increased CO_2 on Plants. p. 1-168.

Baker, D. N., J. R. Lambert, C. J. Phene, and J. M. McKinion. 1977. GOSSYM: A simulator of cotton growth and yield. S. C. Agr. Exp. Sta. Bull. In press.

Baker, D. N., J. M. McKinion, and J. R. Lambert. 1982. GOSSYM: A simulator of cotton growth and yield. S. C. Agr. Exp. Sta. Bull. In press.

Baker, D. N., D. E. Smika, A. L. Black, W. D. Willis, and Armond Bauer. 1981. Winter wheat: A model for the simulation of growth and yield in winter wheat. AgRISTARS document YM-U2-04281 JSC-18229. NASA Johnson Space Flight Center/SK.

Marani, A., D. N. Baker. 1981. Development of a predictive dynamic simulation model of growth and yield in Acala cotton. Science Report. BSF No. of proposal 1708/78. U.S.-Israel Binational Science Foundation, Jerusalem, Israel.

James, D. B. 1966. British patent 1,018,658.

McKinion, J. M., D. N. Baker, J. D. Hesketh, and J. W. Jones. 1975. SIMCOT II: A simulation of cotton growth and yield P 27-82. In: Computer Simulation of a Cotton Production System: User's Manual. U. S. Dept. Agric., Res. Serv. ARS-5-52.

McKinion, J. M. 1980. Dynamic simulation: A positive feedback mechanism for experimental research in the biological sciences. Agricultural Systems. 5(1980). p 239-250.

McKinion, J. M.. and D. N. Baker. 1982. Modelling, experimentation, verification, and validation: Closing the feedback loop. Trans. of A.S.A.E. In press.

Phene, C. T., D. N. Baker, J. R. Lambert, J. E. Parsons, and J. M. McKinion. 1978. SPAR-a soil-plant-atmosphere research system. Trans. ASAE, 21, p 924-30.

Slack, G., A. Calvert. 1972. Control of carbon dioxide concentration in glasshouses by the use of conductimetric controllers. J. Agric. Engnr. Res., 17, p. 107-115.

Whisler, F. D., J. R. Lambert, and J. A. Landivar. 1982. Predicting tillage effects on cotton growth and yield. In: Predicting Tillage Effects on Soil Physical Properties and Processes. P. 179-198. Soc. of Agron., Soil Sci. Soc. of Am., Madison, Wis.


```
          I-X
        X-I
          I-0-X
      X-0-I
          I-0-0-X
    X-0-0-I
          I-0-0-0-X
  X-0-0-0-I
          I-0-0-0-X                 I-X
X-0-0-0-*-I                       0-I
          I-0-0-0-0-0               I-0-X
X-0-0-*-0-I                     X-0-I
          I-$-0-0-0-X               I-X-0-0-X
          I                         I
          I                         I
          I
          I
          I
```

Figure 12. A typical daily map of the fruiting in GOSSYM. "I" and "-" represent mainstem and fruiting internodes, respectively. "$", "*". "0" and "X" represent nodes with open bolls, green bolls, aborted fruit, and squares, respectively. The figure on the right represents a monopodial structure attached to the 5th mainstem node.

```
                               DRY TREATMENT

      0 Cultivations, No       5 Cultivations, No
      Wheel, No Aeration       Wheel, No Aeration
           Problems                 Problems
   Roots in Each Cell, Total  Roots in Each Cell, Total
                                                              Day 50
     Units - gm/cm**3 soil    Units - g/cm**3 soil            LEGEND

                         1 1                       1 1
     1 2 3 4 5 6 7 8 9 0 1    1 2 3 4 5 6 7 8 9 0 1                    <=  0.0000

                                                         0.0000 < 0 <=  0.0001
  1  2 2 1 0                  2 2 1
  2  3 2 2 2 1 1 0            3 2 2 2 2 1 0 0             0.0001 < 1 <=  0.0005
  3  7 6 5 3 2 2 1 0          6 6 5 3 2 2 1 0
  4  4 4 4 3 2 2 1 0          4 4 3 3 2 2 1 0             0.0005 < 2 <=  0.0050
  5  2 2 2 2 2 2 1 0          2 2 2 2 2 2 1 0
  6  2 2 2 2 2 1 0            2 2 2 2 2 1 0               0.0050 < 3 <=  0.0100
  7  2 2 2 2 2 1 0            2 2 2 2 1 1 0
  8  2 2 2 2 1 0              2 2 2 2 1 0                 0.0100 < 4 <=  0.0150
  9  2 2 2 1 1 0              2 2 2 1 0
 10  2 2 1 1 0                2 2 1 1 0                   0.0150 < 5 <=  0.0200
 11  2 1 1 0                  2 1 1 0
 12  1 1 0                    1 1 0                       0.0200 < 6 <=  0.0250
 13  0 0                      0
 14  0                                                    0.0250 < 7 <=  0.0300
 15
 16                                                       0.0300 < 8 <=  0.0350
 17
 18                                                       0.0350 < 9 <=  0.0400
 19
 20                                                       0.0400 < *

 TOTAL = 7.6419 gm dry weight  TOTAL = 7.0356 gm dry weight
```

Figure 13. Effects of a simulated cultivation on root weight and distribution comparing a cultivated treatment versus a no-cultivation treatment.

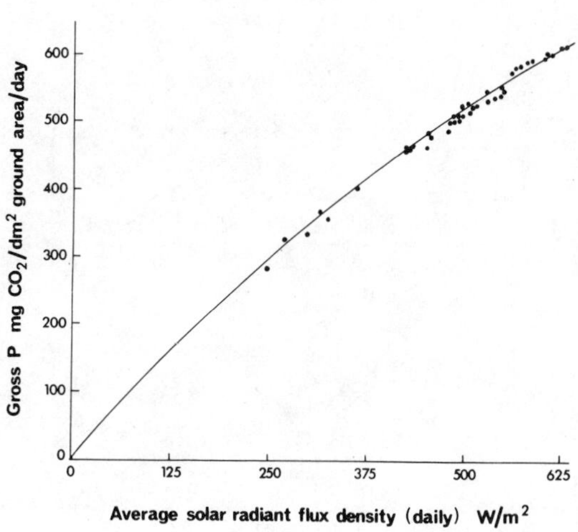

Figure 14. Total daily net photosynthesis versus total daily solar radiation.

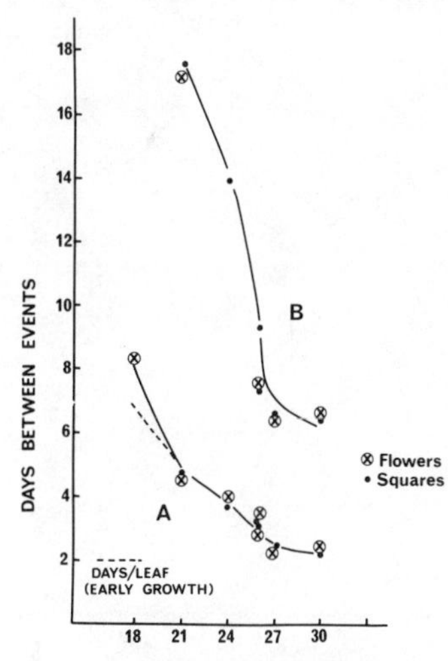

Figure 15. Time between events on the mainstem and fruiting branch versus temperature.

SIMULATION OF ORGANIC MATTER CHANGES IN SWEDISH SOILS[1]

William J. Parton, Jan Persson, and Darwin W. Anderson[2]

Abstract.--Soil carbon and nitrogen changes associated with different cultivation treatments were simulated for a long-term field experiment in Uppsala, Sweden. The model correctly represented the effects of straw incorporation and N fertilization on N mineralization, plant production, and changes in the content and composition of soil organic matter. The results suggest that the difference between treatments are primarily a result of changes in the amounts of the active soil organic fractions (3 to 30 y turnover time).

INTRODUCTION

A model developed in Parton et al. (in press) was used to simulate the effect of different cultivation practices on soil organic matter levels and crop yields. The model was designed to evaluate the impact of different agricultural practices on the transformation of carbon and nitrogen in the soil and the resulting impact upon crop yield. The model uses a one-year time step and the state variables in the model include C and N contained in the structure and metabolic plant residue and three soil organic matter fractions with turnover times of three, 30 and 1200 years.

The model was set up to simulate changes in soil organic matter levels and crop yields for a long-term (25 y) small-plot field experiment in Sweden. The experiment, located at the Swedish University of Agricultural Science, Uppsala was initiated in 1956. The experiment included 15 different organic manure, crop residues, and mineral fertilizer treatments. The crops were generally cereals and were planted annually. A detailed description of the experiments and resulting yields and soil analyses are given in Nilsson (1980) and Persson (1980). The clay content of soil is 35% and at the start of the experiments the pH was 6.5 and the soil organic carbon level was 1.5%. In this study we considered five of the treatments:

1. No crop since the start of the experiment (bare fallow),
2. Annual cropping with no aboveground C or N additions, all crop residues removed, and no fertilizer,
3. Annual cropping with nitrogen fertilization (8 g m^{-2} y^{-1}) and crop residue removed,
4. Annual cropping with incorporation of straw (183 g C m^{-2} y^{-1}), and
5. Annual cropping with straw incorporation (183 g C m^{-2} y^{-1}) and N fertilization (8 m^{-2} y^{-1}).

In this paper we will describe the results of the model simulation and compare the results with observed data. We will also present a description of the model structure and input data for the model. The results show that the model did a good job of simulating the effect of the different treatments on crop yield and soil organic matter levels.

Model Description

The model (fig. 1) was based upon the conceptual model developed by Anderson (1979). Briefly, he described a system in which dead plant and microbial residues decompose into relatively simple substrates for humus formation--e.g., carbohydrates, amino acids and proteins, and phenolics. Polymerization of these compounds results in relatively high molecular weight molecules with a high number of side chains, aliphatic in nature, and rich in nutrients such as N and S (sulfur). These largely aliphatic molecules of high molecular weight are capable of various types of chemical and physical bonding with clay minerals and amorphous mineral colloids. They may also complex among themselves with polyvalent cations such as calcium or iron. The net result is the formation of organomineral aggregates ranging in size from molecular complexes to colloids to aggregates up to hundreds of microns in size. Soil organic fractions are highly labile to microbial attack unless they are protected by adsorption to clay. Anderson (1979) describes this process as physical protection of soil organic matter. Thus, nutrient-rich organic matter is stabilized within the soil. However, even these physically stabilized fractions are not immune from microbial decomposition and mineralization to supply nutrients. The continued humification of these materials form lower- molecular weight components that are generally more aromatic, and progressively more biologically resistant.

---

[1]Paper presented at the Third International Conference on State-of-the-Art in Ecological Modeling, Colorado State University, May 24-28, 1982.

[2]William J. Parton is Senior Research Ecologist, Natural Resource Ecology Laboratory, Colorado State University, Fort Collins, Colorado U.S.A.; Jan Persson is Soil Scientist, Department of Soil Science, Swedish University of Agricultural Science, Uppsala, Sweden; and Darwin W. Anderson is Soil Scientist, Department of Soil Science, University of Saskatchewan, Saskatoon, Canada.

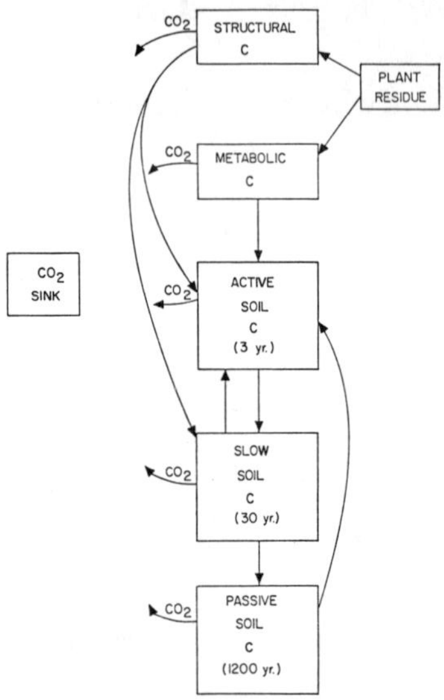

Figure 1.--Flow diagram for the carbon flow model.

From this conceptual model, a simulation model has been constructed containing three soil organic fractions. They constitute (1) an active fraction of soil C and N consisting of live and dead microbial structural and metabolic substances (active or 3 y), (2) a pool of C and N that is physically protected and/or chemical forms with more biological resistance to decomposition, with an intermediate turnover time (slow or 30 y), and (3) a fraction that is chemically recalcitrant and that may also be physically protected, with the longest turnover time (passive or 1200 y). The plant residue (shoot, leaf, and root plant biomass) was divided into structural and metabolic pools that have 1 to 5 y turnover times and a 0.1 to 1 y turnover time, respectively [based on the McGill et al. (1981)]. The model also assumes that the C:N ratio of each of these components has a fixed value that is equal to 80, 5, 8, 10, and 11, respectively, for the structural, metabolic, and active, slow, and passive soil organic fractions.

Plant residue was divided into the structural and metabolic components as a function of the C:N ratio of the plant residue. Carbon and N were allocated so that the C:N ratio of the material going into the compartments will be equal to the specific C:N ratio of each compartment (McGill et al. 1981).

Decomposition of each of the state variables was calculated using the following equation:

$$\frac{dc_i}{dt} = K_i \cdot M \cdot T \cdot c_i \quad ,$$

where, c_i = the carbon in the state variable (i = 1, 2, 3, 4, 5 for structural and metabolizing litter and active, slow, and passive soil fractions)
K_i = the decomposition rate parameter (y^{-1}) for the ith state variable (K_i = 4, 8, 2.0, 0.20, 0.004)
M = the effect of soil moisture on decomposition
T = the effect of soil temperature on decomposition.

The soil moisture term is a function of the ratio of annual precipitation to the potential evapotranspiration rate (see fig. 2a), and the temperature term was calculated as a function of growing-season average soil temperature at 10 cm depth (fig. 2b). The values of the ratio and the air temperature are 0.9 and 14°C, respectively, for Uppsala, Sweden. The soil temperature (T_s) was calculated as a function of aboveground plant biomass (B) and the average growing-season air temperature (T_A) using the following equation:

$$T_S = T_A + T_A \cdot 0.5[\ell^{(-0.0015B)} - 0.52] \quad .$$

This equation was based on soil temperature data observed under different plant canopies at a shorgrass prairie site in Colorado (Parton and Logan 1981; Parton, submitted). The main feature of this equation is that a decrease in aboveground plant biomass results in increased soil temperatures because of reduced shading of the soil. The effect of soil temperature on decomposition was based on the equations presented by Hunt (1977), and the moisture relationship was conceptually based on data presented by Hunt (1977), Wildung (1975), and Klein (1977) and quantitatively based on data presented by Weaver (unpublished).

The model assumes that 60% of the carbon loss from decomposition of structural material and soil organic fractions was lost as microbial respiration (Hunt et al. 1978, Stott et al. 1982), and the remaining carbon was transferred to different state variables and represents net microbial production (NMP) of

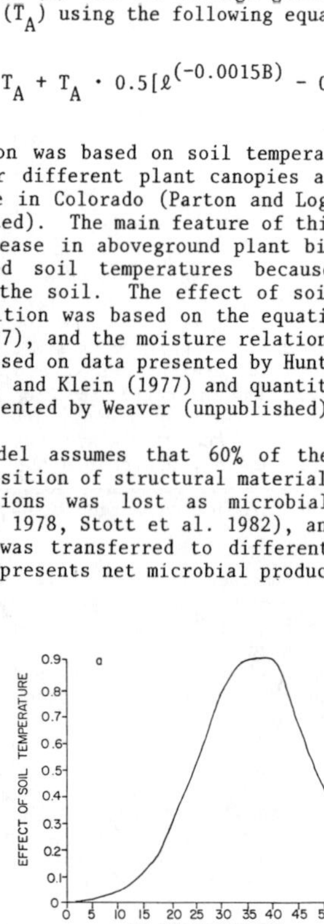

Figure 2.--Effect of soil temperature (a) and the ratio of annual precipitation to potential evapotranspiration (b) on decomposition.

carbon. Decomposition of metabolic material results in the loss of 80% of the carbon as microbial respiration and 20% to NMP (Kassim et al. 1981). For structural residue, 90% of NMP was transferred to active soil organic matter, and 10% was transferred to slow soil organic matter. Decomposition of metabolic carbon forms active soil organic matter. The decomposition of active soil organic matter results in the formation of slow organic matter as nutrients and energy are utilized and resulting simple units condense to form humic materials. Decomposition of slow organic matter partitions 92% of the material into active (microbial metabolites) and 8% forms the strongly aromatic, nutrient and energy poor passive organic matter. Passive organic matter decomposes to active organic matter. The structure of these flows was based on the conceptual model developed by Anderson (1979) and the model developed by Paul and Van Veen (1978).

Nitrogen Flows

Since N is bonded mostly to C, N flows were assumed to be stochiometrically related to C flows (fig. 3). When carbon flows from one state variable to another, the N-flow rate was equal to product of the carbon-flow rate and the fixed N:C ratio of state variables that receive the C. The N attached to the carbon lost in microbial respiration (60 or 80% of the C) was assumed to be mineralized. Analysis of the flow diagram (fig. 3) and the assumed C:N ratios for the state variable shows that decomposition of metabolic residue and active, slow, and passive soil organic matter results in a net mineralization of N; and decomposition of structural residue results in immobilization of N. The net mineralization of N was equal to the difference between the mineralization and immobilization flows. Plant production was calculated as a function of the net mineral-N pool (MIN) and the maximum plant production with N stress (MXP) such that the C:N ratios of plant matter can float between 35:1 and 30:1, depending on the net mineral-N pool. If

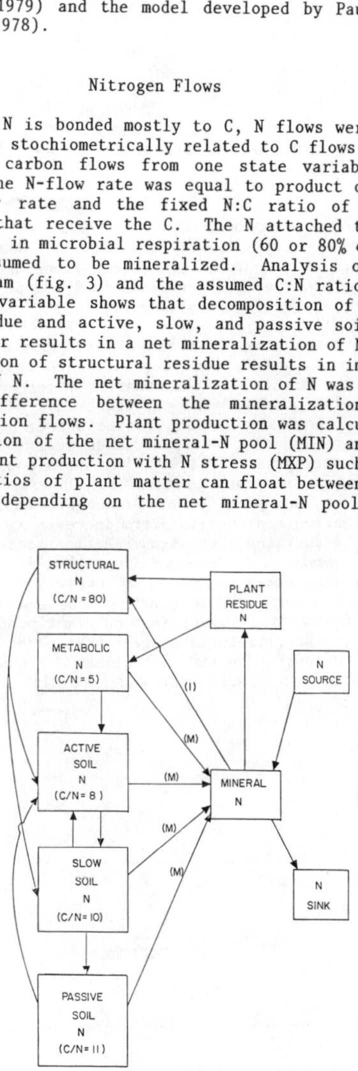

Figure 3.--Flow diagram for the nitrogen state variables.

this ratio of MXP to MIN was greater than, or equal to, 35 to 1, then actual plant production will be equal to 35 times MIN. If the ratio of MXP to MIN was between 35:1 and 30:1, then carbon production was equal to the maximum rate and N uptake was equal to MIN. When the MXP and MIN ratio was less than 30:1, then plant carbon production was at the maximum rate, and N uptake by the plants was restricted so that the C:N ratio was equal to 30:1 and the surplus N stays in the mineral-N pool.

Effect of Cultivation

The effect of the cultivation practices on N and C cycling in soil was simulated by assuming that within any one year the land was either in fallow or seeded with a crop. Soil temperatures were influenced by the aboveground crop production during the crop years, and the amount of residue on the soil surface during fallow years. The model may also be used to simulate the effect of the removal of the N and C in the grain and straw. This was calculated on the assumption that 30% of the plant carbon was in the grain, 50% in the straw and leaves, and 20% in roots; and 50% of the N was in the grain (Neghassi et al. 1975; Gregory et al. 1979).

The five treatments considered for this study include a fallow treatment (no cropping), a control (cropping with no addition of aboveground C or N), a nitrogen treatment (cropping with N fertilization), a straw treatment (cropping with incorporation of straw into the soil), and an N plus straw treatment (annual cropping with N fertilizer and straw additions). The four cropping treatments were planted annually, and 97% of the aboveground plant material was removed from the sites. The straw incorporation treatments added straw with the equivalent of 183 g m^{-2} of C and a C:N ratio of 60. The N fertilization treatments added 8 g m^{-2} of N each year.

Nitrogen inputs occur as wet and dry deposition, seed addition, and N fixation and are estimated to be equal to 3 g m^{-2} y^{-1}. Nitrogen loss occurs as a result of NH_4 volatilization (Woodmansee 1978), leaching, and volatilization of N_2O and N_2 that result from nitrification and denitrification (Mosier et al. 1981). Data from Persson (1980) were used to estimate loss of N for the different treatments, and results show that 0, 0.5, 2.0, 2.7, and 3.0 g m^{-1} y^{-1}, respectively, are lost from the straw, control, nitrogen plus straw, nitrogen, and fallow treatments.

Model Simulation

The model was set up to simulate the effect of different cultivation treatments for a field experiment in Sweden. The observed data were not used to develop the model; however, preliminary model runs for the Swedish data suggest that the model overestimates immobilization of N in the straw-addition treatments. Results from the model plus recent data from Kassim et al. (1981) prompted us to change microbial respiration rate from 60% to 80%. This change corrected the problem; however, the comparison of simulated and observed data is not a true validation of the model.

The observed data from the site includes aboveground annual crop production and total soil organic matter (0-20 cm) for the different treatments during the 25 y experiment (1956-1981). Data for the observed aboveground plant production are presented as average values for 6 years.

A comparison of the observed and simulated total soil organic matter (0-20 cm) for the different treatments (fig. 4) shows that the straw plus nitrogen treatment has the highest soil organic matter level, followed by the straw, nitrogen, control, and fallow treatments. The results suggest that the model represented the major effects of the treatments on soil organic matter (SOM). The model tended to overestimate the decline in SOM for the fallow treatment. The difference between SOM levels in the treatments was a result of differential carbon inputs and decomposition rates. The fallow treatment had the lowest SOM level as a result of minimal carbon inputs, and higher soil temperature that increase decomposition rates. Soil organic matter levels in the control are higher than in the fallow treatment, since C from dead roots was added every year (20% of total plant production). Decomposition rates were also reduced for the control because of lower soil temperature (plant biomass reduces soil temperature). The nitrogen treatment has higher SOM levels than the control because of a larger input of carbon from the roots in that plant production is more than two times larger in the nitrogen treatment, and lower decomposition rates because the greater plant production in nitrogen treatment reduces soil temperatures. The straw treatment has greater SOM levels than the nitrogen treatment because of the incorporation of straw residue, in spite of higher decomposition rates and lower root production on the straw treatment reflecting the lower plant production on the straw treatment. The highest yielding straw plus nitrogen treatment has the highest level of SOM because it has the highest input of carbon from dead roots. The results suggest that the primary factor controlling SOM level was the different carbon inputs for the treatments. The different carbon inputs are the result of straw addition, and increased plant production with nitrogen fertilization. The simulated and observed increase in soil organic matter with straw addition were similar to the increases observed by Biederbeck et al. (1980) for a 20 y straw burning experiment in Canada.

Most of the change in total SOM is a result of changes in the more active soil organic fractions, the active and slow pools (fig. 5). This is expected during a 25 y time period, since the passive soil organic pool has a slow turnover rate (1200 y). Changes in the levels of the active and slow SOM pools is very important to the nitrogen budget, since decomposition of these pools contributes more than 80% of the nitrogen mineralized in the system (fig. 9).

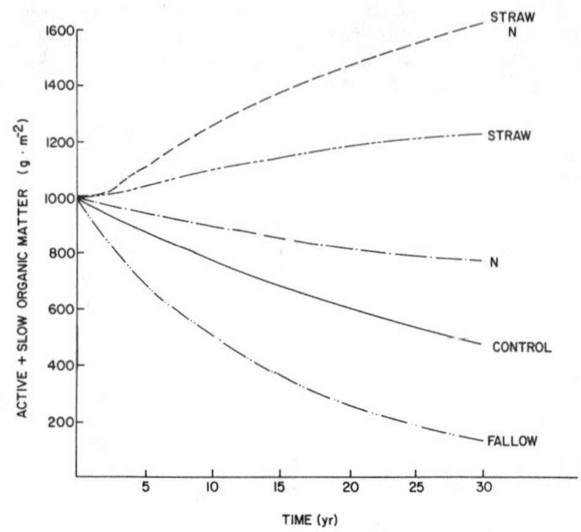

Figure 5.--Simulated active plus slow SOM for control, nitrogen, fallow, straw, and straw plus nitrogen treatments.

A comparison of observed and simulated aboveground plant production (fig. 6) shows that the model correctly simulated the major difference between treatments. Model results suggest that plant production is increased by 200% in the nitrogen fertilizer treatments and that straw addition reduced plant production during the first 8 y, but had a positive long-term effect on plant production. Observed data supports the increased plant production with nitrogen fertilization and the long-term increase in plant production associated with straw addition to the fertilization treatment. Observed data from the straw and control treatments suggest that straw addition caused a slight reduction in plant production during the first 10 y with little effect on plant production after 12 y. Model results suggest that plant production should be greater in the straw treatment than the control after ten years. The model results are similar

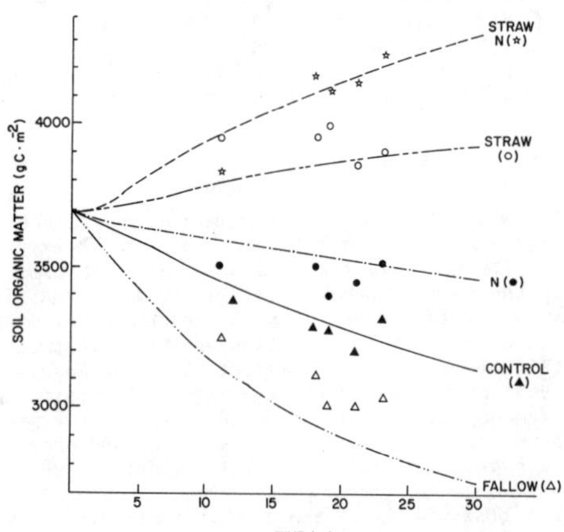

Figure 4.--Comparison of observed and simulated total SOM for the control, nitrogen, fallow, straw, and straw plus nitrogen treatments.

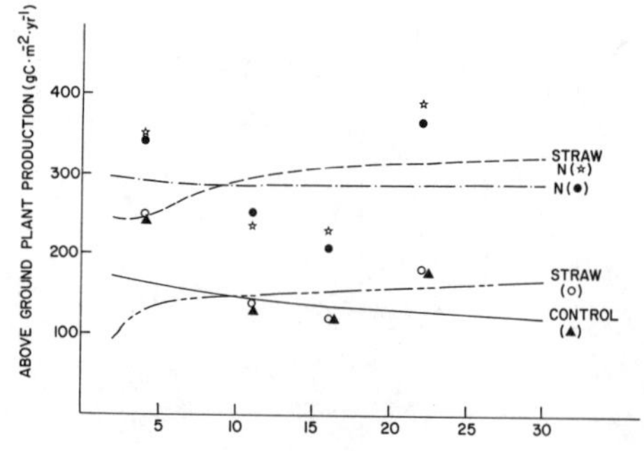

Figure 6.--Comparison of observed and simulated aboveground production for the straw, control, nitrogen, and straw plus nitrogen treatments.

to those of Biederbeck et al. (1980), which showed that straw addition reduced plant production during the first 8 y and increased plant production during the last 10 y of a 20 y experiment.

The model results suggest that the difference in aboveground plant production is primarily the result of the differece in net mineralization of nitrogen from SOM (fig. 7). Net nitrogen mineralization is equal to the sum of nitrogen mineralization flows minus the immobilization of nitrogen resulting from decomposition of structural biomass. Results show that net nitrogen mineralization is lower for the straw-incorporation treatments during the first 10 y and is greater for the last 15 y of the experiment. Lower net mineralization during the first ten years with the straw treatments is the result of nitrogen immobilization that results from straw decomposition, whereas higher mineralization rates for the last part of the experiment are the result of higher gross nitrogen mineralization rates with straw treatments (fig. 8). The similar patterns for net nitrogen mineralization and plant growth are expected, since nitrogen available for plant growth is a prime limiting factor for plant growth in this model. Plant-available nitrogen is equal to the net nitrogen mineralized plus atmospheric and fertilizer inputs minus nitrogen losses by leaching and volatilization.

Gross nitrogen mineralization is equal to the sum of nitrogen mineralization flows, which result from decomposition of metabolic, active, slow, and passive biomass. Gross nitrogen mineralization rates (see fig. 8) increase rapidly during the first 10 y for the straw treatments and then slowly approach a new equilibrium value, whereas in the control and fallow treatments nitrogen mineralization is highest initially and then decreases with time. Increases in the gross nitrogen mineralization with straw incorporation is caused by the buildup of active and slow SOM (fig. 5) in the system and the nitrogen mineralization that occurs as a result of decomposition. Similarly, the decline in gross nitrogen mineralization for the fallow and control treatments is a result of the decline in the active and slow SOM fractions associated with these treatments (see fig. 5). Gross nitrogen mineralization is fairly constant for the nitrogen-fertilized treatment, since active and slow SOM are only

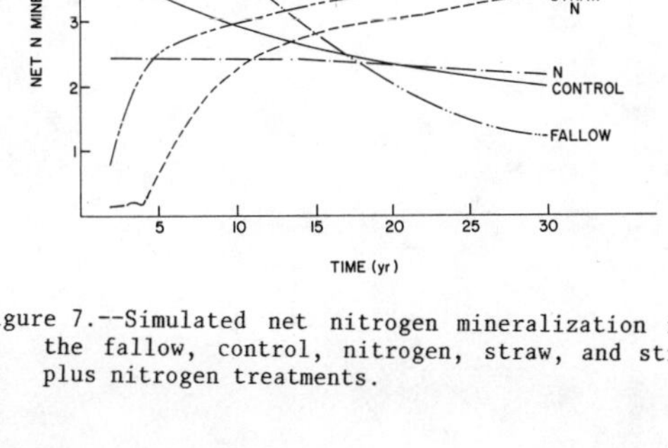

Figure 7.--Simulated net nitrogen mineralization for the fallow, control, nitrogen, straw, and straw plus nitrogen treatments.

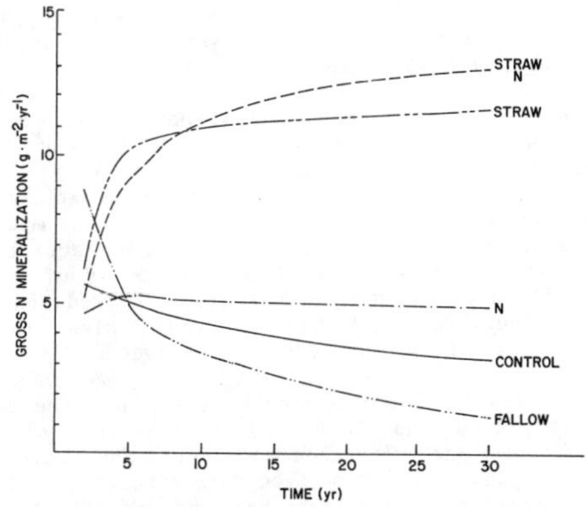

Figure 8.--Simulated gross nitrogen mineralization for the fallow, control, nitrogen, straw, and straw plus nitrogen treatments.

slightly reduced. Higher N mineralization rates with straw are consistent with the results of Biederbeck et al. (1980), who showed that N mineralization potential is higher in a long-term straw-addition experiment.

The relative contribution of nitrogen mineralization from decomposition of metabolic, active, slow, and passive SOM in the straw and control treatments (fig. 9) show that initially the active SOM contributed the most nitrogen mineralization (55%), followed by slow (29%), metabolic (15%), and passive (1%). Nitrogen mineralized by the active SOM increases rapidly during the first 10 y, whereas the nitrogen mineralization rates are fairly constant for other pools. Increases in the N mineralization from the active SOM pool are a result of a substantial increase in the

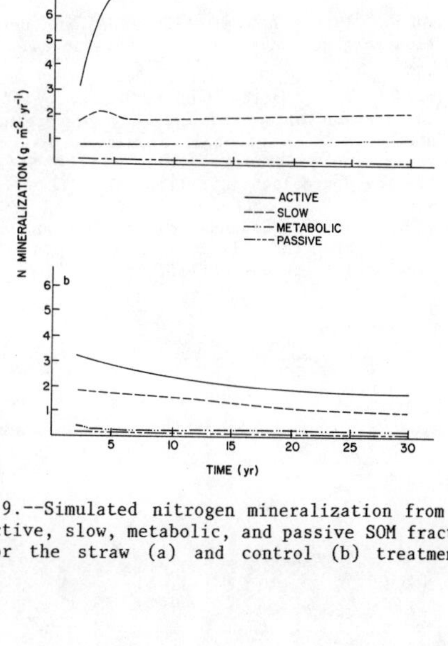

Figure 9.--Simulated nitrogen mineralization from the active, slow, metabolic, and passive SOM fraction for the straw (a) and control (b) treatments.

size of this pool. These results are supported by Biederbeck et al. (1980) and Lynch and Panting (1980), who showed that straw addition greatly increased (87 to 127%) microbial biomass.

Results for the control show that nitrogen mineralization rates decreased with time and that the fractions of nitrogen contributed by different fractions were similar with time. Decomposition of active and slow SOM contribute over 80% of the gross nitrogen mineralization of the system. Losses of active and slow SOM that result from decreased inputs of carbon into soil have the detrimental effect of reducing nitrogen mineralization and thereby reduce plant production unless nitrogen is added to the system.

In conclusion, the results show that the model correctly simulated the effect of straw incorporation and nitrogen fertilization treatments on changes in total SOM and plant production during the 25 y experiment. Model results suggest that most of the observed changes in SOM are the result of changes in active and slow organic matter pools. Changes in the level of active and slow soil organic matter have a large effect on the nitrogen mineralization rates from the soil, which, in turn, influence plant production. Model results also suggest that after 25 y the plant production and SOM levels are approaching new equilibrium values for the simulated treatments.

LITERATURE CITED

Anderson, D.W. 1979. Processes of humus formation and transformation in soils of the Canadian Great Plains. Journal of Soil Science 30:77-84.

Biederbeck, V.O., C.A. Campbell, K.E. Bowen, H. Schnitzer, and R.N. McIver. 1980. Effect of burning cereal straw on soil properties and grain yields in Saskatchewan. Soil Science Society of America Journal 44:103-111.

Gregory, P.J., D.V. Crawford, and M. McGowan. 1979. Nutrient relations of winter wheat. 1. Accumulation and distribution of Na, K, Ca, Ng, P, S, and N. Journal of Agricultural Science 93:485-494.

Hunt, H.W. 1978. A simulation model for decomposition in grasslands. Ecology 58:469-484.

Kassim, G., J. P. Martin, and K. Haider. 1981. Incorporation of a wide variety of organic substrate carbons into soil biomass as estimated by fumigation procedure. Soil Science Society of America Journal 45:1106-1112.

Klein, D. A. 1977. Seasonal carbon flow and decomposition parameter relationships in semiarid grassland soil. Ecology 58(1):184-189.

Lynch, J.M., and L.M. Panting. 1980. Variations in the size of the soil biomass. Soil Biology and Biochemistry 12:547-550.

McGill, W.B., H.W. Hunt, R.G. Woodmansee, and J.O. Reuss. 1981. PHOENIX: A model of carbon and nitrogen dynamics in grassland soils. In F.E. Clark and T. Rosswall (eds.) Terrestrial nitrogen cycles: Processes, ecosystem strategies, and management impacts, pp. 49-115. Ecological Bulletin (Stockholm).

Mosier, A.R., M. Stillwell, W. Parton, and R.G. Woodmansee. 1981. Nitrous oxide emissions from a mature shortgrass prairie. Soil Science Society of America Journal 45:614-619.

Neghassi, Habte Marian, Dale F. Harmann, and D.E. Smibree. 1975. Wheat yield models with limited soil water. Transactions of the ASAE (American Society of Agricultural Engineers) 18(3):547-557.

Nilsson, K.O. 1980. Skördeutveckling och omsättning av organisk substans vid användning av olika kvävegödselmedel och organiska material. Undersökningar i ett ramförsök under 20 år. Institutionen för markvetenskap, Avd. för växtnäringslära, nr 127.

Parton, W.J. A simplified soil temperature model. Soil Science (submitted).

Parton, W.J., and J.A. Logan. 1981. A simple model for diurnal change in soil and air temperature. Agricultural Meteorology 23:205-216.

Parton, W.J., D.W. Anderson, C.V. Cole, and J.W.B. Stewart. Soil organic matter formation model. In R.L. Todd (ed.) Agricultural Ecosystems, Athens, Georgia, September 1980. Ann Arbor Sciences Publications, Ann Arbor, Michigan (in press).

Paul, E.A., and J. Van Veen. 1978. The use of tracers to determine the dynamic nature of organic matter. 11th International Congress of Soil Science, Transactions 3:61-102.

Persson, J. 1980. Detaljstudium av den organiska substansens omsättning i ett fastliggande ramförsök. Institutionen för markvetenskap, Avd. för växtnäringslära, nr 128.

Wildung, R. E., T. R. Garland, and R. L. Buschbom. 1975. The interdependent effects of soil temperature and water content on soil respiration rate and plant root decomposition in arid grassland soils. Soil Biology and Biochemistry 7:373-378.

Woodmansee, R.G. 1978. Additions and losses of nitrogen in grassland ecosytems. BioScience 28: 448-453.

SIMULATION OF CLIMATIC AND MANAGEMENT EFFECTS ON WHEAT PRODUCTION[1,2]

Morgan, J. A., D. N. Baker, W. J. Parton, C. V. Cole, W. O. Willis, D. E. Smika, A. Bauer, and A. L. Black[3]

Abstract.--The status of an explanatory, process oriented model for the simulation of growth, development, and yield of winter wheat is presented. The objectives of this model are discussed relative to empirical and explanatory modeling strategies. Future modeling and research needs are given.

INTRODUCTION

The purpose of this presentation is to discuss the current status of the Winter Wheat growth model. Model objectives and strategies are discussed first, followed by an outline of the Winter Wheat model in which emphasis is directed towards describing new additions or corrections to the model. In the final section of the paper, future directions in modeling strategies and experimentation will be addressed.

The Winter Wheat model was first described, in abstract form, in 1978 (Smika et al. 1978). More recently, a thorough documentation of Winter Wheat was published with flow charting, Fortran code and corresponding explanatory text (Baker et al. 1982). The conceptual development of Winter Wheat is along the lines of Gossym (Baker et al. In press), which simulates cotton growth and yield. Many of the processes have been adopted or changed to accommodate differences in growth and development between cotton and wheat plants. Adjustments were made according to information found in the literature and unpublished data, although in some cases, responses understood qualitatively have had to be quantified.

Thus, while the model is currently running and simulating some reasonable output, it is still in its initial stages of development. We envision the modeling effort of Winter Wheat as an ongoing process whereby, through a systematic effort of formulating, testing and if necessary, re-formulating new hypotheses, a more successful model will evolve.

What constitutes success in a modeling effort is clearly dependent upon the objectives for the project. The objectives for Winter Wheat are "to identify and assemble in a plant growth model the factors determining winter wheat growth and yield in a format which will aid 1) system design (breeding and new cultural practices, and combinations thereof), 2) crop management decision making at the farm level, and 3) yield forecasting" (Baker et al. 1982).

It is a fair question to ask whether or not any one model can satisfy such a wide range of objectives. An explanatory, process oriented model would clearly be most appropriate for objective 1, although more empirical approaches may be taken with objective 2 and certainly with objective 3 where only yield information is required. The question is then whether or not an explanatory model would be appropriate for objectives 2 and 3 where only a small proportion of the information generated would be utilized. The answer, to a large extent, depends on the uniformity of the plant material and microclimate.

In simulating fairly simple systems with little change in the plant and climate boundaries, development of a corresponding simple empirical model may be the best approach to take since the data base may be relatively easy to acquire. Statistical or empirical models with adequate data bases generally perform better than more complex models (Norman 1982).

The appropriate model for a more complicated system is perhaps more difficult to determine. In the case of empirical approaches, data bases for constructing model algorithms may become prohibi-

[1]Paper presented at the Third International Conference on State-of-the-Art in Ecological Modelling, Colorado State University, May 24-28, 1982.

[2]Contribution from the United States Department of Agriculture, Agricultural Research Service, and Natural Resource Ecology Laboratory, Colorado State University.

[3]Research Agronomist, USDA-ARS, Fort Collins, Colorado; Research Agronomist, USDA-ARS, Mississippi State, Mississippi; Research Associate, Colorado State University, Fort Collins, Colorado; Soil Scientists, USDA-ARS, Fort Collins, Colorado; Soil Scientist, USDA-ARS, Akron, Colorado; and Soil Scientists, USDA-ARS, Mandan, North Dakota, respectively.

tively large in more complex systems because the potential plant response surface may increase dramatically. There is, thus, an impetus to develop a more explanatory, process oriented model for complicated systems because the plant response surface may be generated internally by the model instead of having to sample it. The dynamic interaction of various plant and soil processes simulated in the model at lower levels of organization actually define the potential plant response at higher levels. Much of the information for this kind of explanatory model is already in the literature (Norman 1982). There are, however, problems with this approach. The first is that many plant processes are well understood only in terms of one, two or several independent variables, and often the interactions of these variables and of the plant processes themselves are not known, or understood poorly at best. When these processes are simulated and allowed to interact dynamically in a modeling exercise, the modeler is in reality extrapolating well beyond the limits of experimentation. This kind of guesswork can lead to serious errors (Gold 1977). Secondly, as more information is obtained about the physiological responses of plants to their environment, it is becoming increasingly clear that some of the basic physiological responses of plants may vary considerably from one cultivar to another. One such example is the ability to osmoregulate, which is cultivar specific in wheat (Morgan 1977). Thus, process level models may also require rather large data bases depending upon what processes are simulated.

Another potential source of error that both empirical and more explanatory models are subject to are the simplistic assumptions implicit in the various model algorithms. Such assumptions are not necessarily bad, but actually constitute the basis for model building. However, if these assumptions cause the model response to be insensitive to environmental and genetic variability, then obviously, the model will not perform as well in atypical situations although model response may be satisfactory with "normal" conditions. An example is predicting plant growth from algorithms which estimate plant water loss and then converting that estimate to a quantity of dry matter (Hanks 1974; Russo and Seem 1980). This approach assumes a constant water use efficiency of the crop, although much experimental information would indicate that water use efficiency is not a constant (Levitt 1980; Rawson et al. 1977). Although empirical models are probably more prone to this kind of error because of the very nature of their construction, more explanatory models may also be affected because of the difficulty in estimating and understanding some of the more basic physiological plant responses required in such models. The amount of error introduced by simplifying assumptions depends to a large extent on the degree of extrapolation and interpolation from the experimental data sets, where functions or algorithms were developed, to the simulated system.

In summary, there is probably no clear cut answer to the question concerning what complexity level would be most appropriate for developing a model intended for crop management decision making and yield forecasting because the system is still poorly understood. An explanatory, physical/physiological process model of intermediate complexity (Norman 1982) depicting wheat growth and development is being developed to meet the first objective of gaining a better understanding of wheat growth and development for the purpose of altering the system to benefit man. This kind of model may or may not be most appropriate for achieving objectives 2 and 3 concerning management decision making and yield forecasting. At the very least, an explanatory model will provide direction and indicate the limitations of more empirical models. Because of the process orientation of Winter Wheat, ultimate success will depend not only on yield prediction capabilities under a wide variety of soil and climatic conditions, but also on its ability to simulate dynamic plant and soil processes. Testing and sensitivity analysis of the model is currently underway at Fort Collins.

This presentation summarizes much of the information in the publication just released by Baker et al. (1982). However, some important changes are being incorporated in the model as a result of early testing and evaluation. These changes will be clearly identified and discussed along with reasons for them. Proportionately less emphasis will be directed towards describing portions of the model which are already well documented (Baker et al. 1982; Whisler et al. 1981), although a general discussion will be given to insure continuity.

THE WINTER WHEAT MODEL

A general outline of the model is given in Figure 1. This figure is a simplified flow chart of the MAIN program of Winter Wheat. Discussion of the model will be along the lines of this outline.

Daily climatic information, consisting of radiant flux density (R), minimum and maximum ambient air temperatures, precipitation, form of precipitation (natural versus irrigation), relative humidity, and calendar day are read by the MAIN program and passed into CLYMAT. Photoperiod is calculated as a function of latitude and calendar day. The average daytime and nighttime air temperatures are calculated as functions of maximum and minimum air temperatures. This information is then used with photoperiod in calculating a daily average ambient temperature. Percentage canopy interception of R (R_{int}) is also calculated in CLYMAT as a product of a ground cover term, the maximum effective leaf length divided by row width, and a Beer-Lambert type light attenuation term based on leaf area index (LAI). The extinction coefficient, 0.4, was taken from Monteith (1965). Soil profile temperatures at 5-, 10-, 20- and 41-cm depths are calculated by the subroutine TMPSOL from regression equations of McWhorter and Brooks (1965). These equations express soil temperature as linear functions of the running average air temperature for the preceding 7 days.

The microclimatological data set generated by CLYMAT from input weather data is passed to several subroutines which require the information as driving variables. For instance, air and soil temperatures are used to calculate rates and durations of potential daily organ growth increments in the GROWTH, RUTGRO and MORPH subroutines. Plant respiration is calculated as a function of temperature in PNET. R_{int} is utilized in PNET for estimating daily canopy photosynthesis. Precipitation is added and distributed in the soil profile in GRAFLO. The relative humidity information is utilized in WATERUP for modeling plant water relations and transpiration.

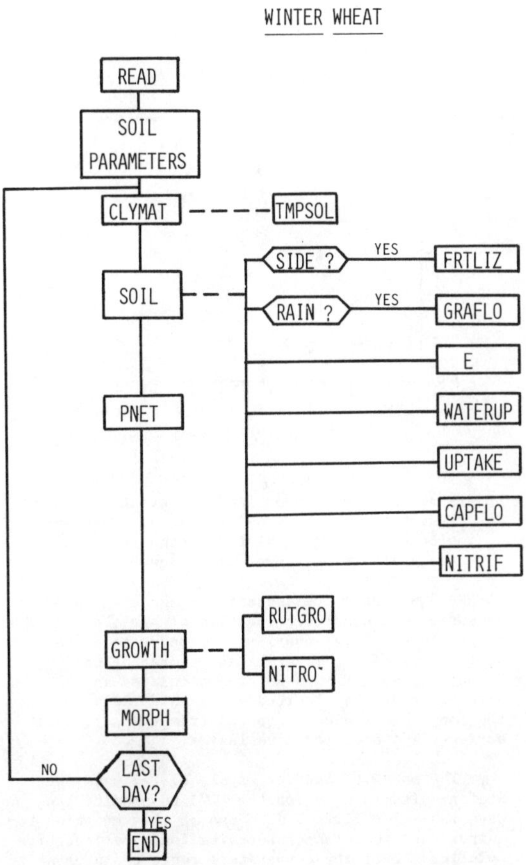

Figure 1.--General outline of MAIN program of Winter Wheat.

The subroutine SOIL and its sub-programs were adopted from Rhizos (Whisler et al. 1981), a model detailing soil and root processes in the rhizosphere. The empirical evapotranspiration subroutine (ET) (Baker et al. 1982) is being replaced by EVAP and WATERUP[4], which will be discussed in some detail. ET was adopted from Ritchie (1972). Because Ritchie's model applied to row crops with unlimited soil water, an empirical transpiration reduction factor, developed from experiments conducted in Mississippi, had been incorporated in the model for use under limiting soil water conditions. It is doubtful that an empirical transpiration factor developed in a humid climate would perform adequately in the Great Plains. The inclusion of WATERUP in Winter Wheat strengthens the water relations portion of the model considerably over the previous version. In WATERUP, transpiration is assessed by an iterative procedure that converges an energy balance and a root resistance model. The physiological process structure of WATERUP would indicate better predictive capabilities under a wide range of environmental conditions compared to the previous version. Testing is currently underway to confirm if this concept is true.

In SOIL, processes are simulated within a 6 X 20 two dimensional matrix (Figure 2). In one dimension, the width of 6 soil columns are adjusted according to the planting row width. In the other dimension, 20 soil layers of 10 cm depth each span the potential root zone depth of 2 meters. A third dimension is fixed at 1 cm thickness and is assumed to be longitudinally representative of the crop row and inter-row areas. On the assumption that soil processes occur in a symmetrical pattern in row crops, simulation occurs only in columns 1-3. Mirror images of the results are then assigned to columns 4-6, thereby saving considerable computer time.

On the first day of simulation, FRTLIZ is call to incorporate N fertilizer and crop residue into the top two soil layers. FRTLIZ may be recalled on any future day in which N fertilizer is again applied. Ammonia and nitrate N are modeled separately. Ammonium ions are assumed to be absorbed on soil colloids and to be stationary. Nitrate ions, on the other hand, are assumed to be in solution and move by mass flow with the water.

GRAFLO, EVAP, UPTAKE, WATERUP and CAPFLO simulate various phases of soil-plant-water relations. In GRAFLO, water and nitrate are distributed vertically in the profile. Evaporation from the soil is executed by EVAP in two stages according to Ritchie (1972). In stage 1, evaporation is limited only by

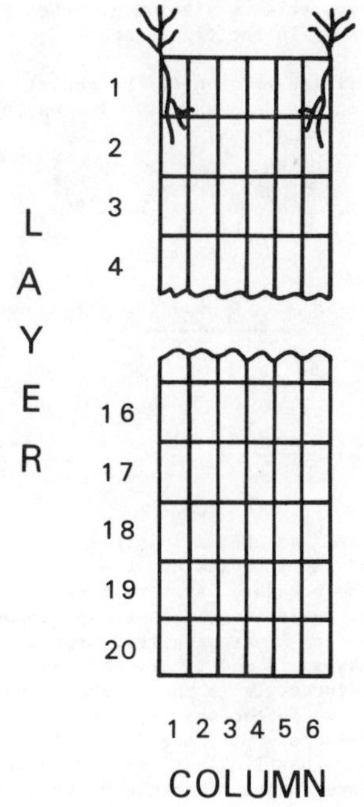

Figure 2.--Two-dimensional matrix of SOIL subroutine of Winter Wheat.

[4]WATERUP was developed by W. J. Parton, Natural Resources Ecology Laboratory, Colorado State University, Fort Collins, Colorado.

the energy supply to the surface. Hydraulic soil properties determine the slower evaporation of stage 2. The amount of water evaporated in EVAP is then imposed on UPTAKE.

UPTAKE functions to actuate and budget the water movement from the soil cells, through the plant and soil surface to the atmosphere, and to simulate N uptake by the plant. In the last version of Winter Wheat (Baker et al., 1982), N was moved into the plant passively via the transpiration stream. The results of the initial testing of Winter Wheat indicated a problem with modeling N uptake in this fashion. Throughout the simulation, average leaf N content dropped from an initialized value of 3% to less than 2% (Figure 3). Although some decline in plant N content occurs as the plant matures, mostly due to increasing proportions of structural material, the almost continual drop in leaf N from day one of simulation throughout the growing season is not characteristic of observed N dynamics. Furthermore, the importance of active mineral uptake in crop systems is well established (Russell 1977). With the assistance of Dr. Vernon Cole, an active, Michaelis-Menton type N uptake is being superimposed on the mass flow already in place. Estimates of the kinetic constants V_{max} and K_m are from Huffaker and Rains (1978). Mineralization and nitrification are simulated in NITRIF.

WATERUP, a modified version of the model suggested by Saugier et al. (1974), estimates plant water loss. A radiation balance model simulates the temperatures of dead and live leaves and transpiration water loss. The driving variables for the radiation balance model include: reference height air temperature and dewpoint, soil surface temperature, and short wave solar radiation absorbed by the dead and live leaves in the plant canopy.

A simplified version of the radiation balance equations for the dead and live leaves is shown by the following equations:

$$R_N^D = \frac{\rho C_p}{r_b} L_D (T_D - T_R) \tag{1}$$

$$R_N^L = \frac{\rho C_p L_G (e_S - e_R)}{\gamma(r_b + r_c)} + \frac{\rho C_p L_G}{r_b} (T_L - T_R) \tag{2}$$

Where R^D and R^L are net all-wave radiation for the dead and live leaves, P = density of the air, r_b = the boundary layer resistance, r_c is the stomatal resistance, γ is the psychometric constant, L_D and L_G are the leaf area of dead and live leaves, T_D and T_L are the dead and live leaf temperatures, T_R is the reference height temperature, Cp is the specific head of air at constant pressure, e_S is the saturation vapor pressure over the live leaf, and e_R is actual vapor pressure at the reference height. e_R is calculated from the reference height dewpoint temperature. Since live leaf temperature is unknown at this point, e_S is also unknown, yielding two equations with three unknowns (T_D, T_L and e_S). This problem is dealt with using Myrup's

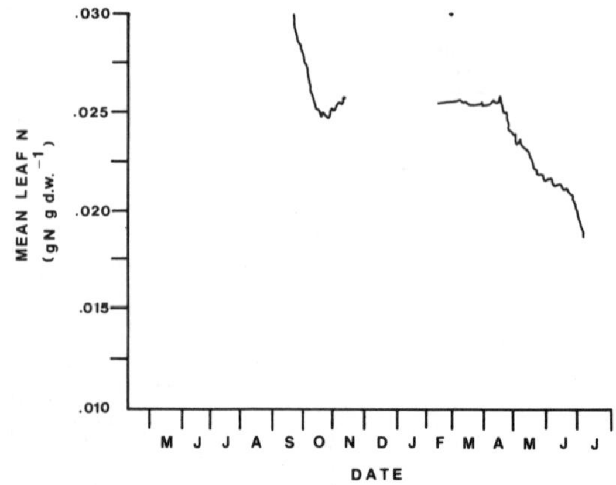

Figure 3.--Simulated mean leaf N content of canopy leaves throughout growing season. No simulation occurs during winter months in which average daily temperatures are below 0°C.

(1969) equation which describes saturation vapor pressure as a quadratic function of air temperature. This equation permits us to express E_S in terms of T_L, reducing the number of unknowns to two. R^D and R^L are calculated as a function of the absorbed short wave radiation and the long wave radiation to and from the sky, soil surface, and dead and live leaves.

The model of Saugier et al. (1974) uses the Stefan-Boltzman equation ($R = \sigma T^4$) to predict long wave radiation fluxes. The use of this equation requires that leaf temperatures be found by iterative solution. Over the temperature range of interest to us (0-50°C), the radiation flux predicted by the Stefan-Boltzman equation can be well approximated by a linear equation. Such an equation was developed by regressing the Stefan-Boltzman estimates against temperature ($R = 0.44 + .0083T$, $r^2 = .98$). Use of this equation permits a direct solution for live and dead leaf temperatures and yields a substantial savings in computation time.

The radiation balance model is connected to a root water uptake model (Belman et al. 1979). The model developed by Belman et al. (1979) calculates water uptake by layer as a direct function of the difference between the plant water potential and the soil water potential and as an inverse function of the soil and root resistance. Soil hydraulic resistance is modeled in CAPFLO, a subroutine of Rhizos (Whisler et al. 1981), and then passed into WATERUP. Root resistance is calculated in WATERUP from root length and age class estimates made in RUTGRO, another subroutine of Rhizos. The stomatal resistance term used in the radiation balance equation is a function of plant water potential. An iterative approach is used to estimate the plant water potential and corresponding stomatal resistance which results in a prediction of root water uptake that is equal to the transpiration water loss calculated by the radiation balance model. If the two initial estimates of water flux are at variance, a new estimate of plant potential is made. A second estimate of water flux is made by the root uptake model using the new plant water potential. A new stomatal resistance, corresponding to the new plant water potential is calculated and this term is used to produce a second estimate of leaf temperatures and water flux using the radiation balance model. The two estimates of water flux are compared and iteration continues if they disagree by more than 5%. When the solutions converge, estimates of transpiration water loss and leaf temperature are obtained which are consistent with each other and with both the soil water and radiant environments. As in EVAP, the potential water losses calculated in WATERUP are passed to UPTAKE.

Redistribution of water within the soil profile occurs in CAPFLO. Again, nitrate nitrogen moves with the soil water. Conductivity of the soil to water movement is also calculated in CAPFLO. These calculations require a fairly detailed description of the physical properties of the soil profile (see Whisler et al. 1981).

The synthesis of organic matter for growth and development is simulated in the subroutine PNET, which describes canopy gross photosynthesis. Daily gross photosynthesis is first estimated as a function of R_{int} from a data set collected by Baker (1982). Recall that R_{int} was estimated in CLYMAT. The maximum daily rate is adjusted if plant water or N are limiting. The effect of plant water stress on photosynthesis is realized by multiplying the maximum daily rate by a water stress reduction factor, which is a linear function of leaf water potential, taken from Figure 1 of Lawlor (1976). Similarly, a N reduction factor is used to adjust the maximum rate when average leaf N concentrations fall below 2%. The factor is arbitrarily calculated as the leaf N concentration divided by 2. The threshold value was 1% in the previous version of Winter Wheat. Although a good data base is not available for winter wheat on a canopy basis, there are considerable data in the literature showing strong correlations between leaf photosynthesis and leaf N content, with some of the information involving leaf N contents as high as 5% (Bolton and Brown 1980; Natr 1975; Osman 1977). Although it is difficult to extrapolate this information to the canopy level, a change in the threshold value from 1 to 2% seems justified at the present time. Further field experiments are being planned to more thoroughly investigate the effects of canopy plant N contents on photosynthesis.

Photosynthesis is currently modeled with no temperature effect. Given the information in the literature of photosynthetic response to temperature (Milthorpe and Moorby 1974; de Wit 1978), a temperature function needs to be developed. There are, however, considerable difficulties in extrapolating much of the temperature data to the canopy level. This topic will be discussed more thoroughly below.

After daily photosynthesis has been estimated, daily respiration by the plant is calculated and subtracted from photosynthesis to obtain daily net photosynthate production. Respiration is a function of temperature and biomass (Baker et al. 1982).

In GROWTH, photosynthate accumulated in a 24-hour period is distributed among the plant organs. Algorithms for above ground processes were developed mainly from Sofield et al. (1974) and Friend et al. (1962). The below ground growth is handled in RUTGRO, a subroutine from Rhizos.

A potential growth increment for above ground biomass is calculated for all plant organs. In the case of head parts other than the developing grain, the potential growth increments are constants which change with phenological stages of development. At the heading stage, grain growth becomes a function of daily average temperature. Stem growth potentials are also growth stage dependent constants. Potential growth of leaves is temperature dependent with maximum rates occurring at 20 C average daily temperature and no potential growth occurring below 0 or above 40 C. Leaf growth duration is a function of the average running temperature calculated for each leaf. All of the organ potential growth rates are adjusted by a water stress reduction factor calculated in RUTGRO.

Next, RUTGRO is called to calculate potential change in root weight. Potential root growth is calculated on a cell-by-cell basis, and is a function of age, daylength, layer temperature and water stress.

The potential root growth is summed for the whole soil profile and then added to the total above ground potential to obtain the total carbohydrate demand for plant growth. A plant carbohydrate pool is calculated by adding the daily total plant photosynthate production to any carbohydrate reserves. If the whole plant carbohydrate pool is greater than the whole plant demand, then all organs will receive a potential allocation of carbohydrate equal to their respective demands. Remaining carbohydrate goes into plant reserves. If the pool is less than whole plant demand, then potential allocation is to the heads until their demand is satisfied. The remainder is portioned equally (on a demand basis) among the other plant organs. Whether or not the various organs can utilize their potential carbohydrate allocations for growth depends on plant nitrogen status. Allocation of plant nitrogen to the various organs is processed in NITRO in a similar fashion as carbohydrate, with priority to heads. The reserve pool of N consists of daily root N uptake and plant N reserves which may be mobilized. Actual organ growth is then realized according to whether or not the nitrogen reserves are sufficient to meet the carbohydrate demands and C:N ratios of each organ. If nitrogen reserves are insufficient, then growth is reduced. If reserves are plentiful, then the potential allocation of carbohydrates are utilized for growth, with any excess plant nitrogen being allocated to reserves.

Maximum leaf weight is used to calculate maximum leaf length for estimating R_{int} (from CLYMAT). Finally, leaf area index is calculated by converting leaf dry weight to area.

MORPH is the last subroutine of major importance to be executed in Winter Wheat. In MORPH, the initiation and duration of various phenological and morphological events are simulated. A daily census is taken which keeps an account of all plant organs and their respective maturity status.

The initiation of tillering, head differentiation, jointing, booting, heading and anthesis for the main stem occur upon the accumulation of a specified number of degree days. The degree day accumulator has a lower limit of 0 C required for accumulation and no upper limit. The data base for these phenological events was obtained by Baker (1982). Differentiation of all the heads occurs once the main stem has accumulated the required degree days. Jointing of all stems is then spread out over a potential range of up to 15 days from the occurrence in the main stem. The spread of these jointing events is determined as an arbitrary function of plant water stress and carbohydrate and N supply demand ratios. From this point on, a range of tiller ages, with respect to morphological development, of up to 15 calendar days is possible. Booting, heading and anthesis are then modeled separately for each stem with the same degree day functions used for the main stem.

In addition to keeping track of all the plant organs, the daily census in MORPH also calculates running average temperatures which have occurred since the youngest organ in each organ class (leaves, tillers, secondary roots) was initiated. This information is then used in simulating the initiation of tillers, secondary roots, and leaves.

Briefly, tillering is modeled as a function of running average temperature, soil water potential, leaf N content, carbohydrate stress, and root/shoot ratio. The tillering period extends from its initiation to jointing. The running average temperature function was arbitrarily chosen (Baker 1982). When the plant has at least 4 primary tillers, decisions are made concerning tiller abortion. New tillers may be aborted if soil water potential is less then -1 bar in the root zone, if leaf N is less than 3 percent, or if carbohydrate demands are greater than supplies for that day. Tillers may also be aborted after spring green up if leaf water potentials are less than -20.0 bars or there are less than an average of four secondary roots per tiller.

Secondary root development is modeled as a function of running average temperature and a threshold soil water potential of -1.0 bar. Below that threshold, secondary root initiation does not occur. The period of secondary root development extends from tillering through booting.

Leaf initiation is modeled on each tiller as a function of running average temperature, with a minimum of 4 days between leaves. The maximum leaf number per tiller is nine.

Variation in grain number is realized partly through changes in potential spikelet number. Spikelet number may be reduced from a maximum of 22 as a result of shortages in carbohydrate or N. Grain number may also be affected by water stress for three days after the beginning of anthesis of each head. During this time period, up to 25% of the florets may be desiccated daily if the average leaf water potential falls below -15 bars. Floret number per head ranges from 10 to 60.

Finally, the time for grain filling is modeled separately for each head as a function of temperature. The maximum time period of grain filling is set at 50 days.

The time step for most processes is one day. The model runs from plant emergence in the full to maturity the following summer. No simulation occurs during the winter when daily average temperatures fall below 0 C.

FUTURE RESEARCH NEEDS

The changes which are currently being incorporated into Winter Wheat are the first of many to come as a result of the testing, hypothesis re-evaluation, and substitution of better process models as they become available. Baker et al. (1982) have outlined a number of improvements which need to be made in Winter Wheat, some of which have already been mentioned. Other proposed changes include incorporating N, P and carbohydrate effects on canopy photosynthesis, temperature and nutrient effects on potential dry matter accretion rates in the various organ classes, temperature effects on morphological events, head differentiation, winter survival, and nutrient remobilization within the plant. Preliminary testing of the model indicates that improvements need to be made in modeling plant senescence, phenology and canopy gas exchange. While some of the information may be extracted from the literature, there are problems with fitting that information into the structure of Winter Wheat. This indicates a need for further research efforts. The nature of these problems is shown below in relation to two plant mechanisms (water stress and temperature effects on canopy gas exchange) which need to be evaluated for eventual inclusion in the model.

Relationships between leaf water potential (ψ_1) and leaf apparent photosynthesis (AP) from two different experiments involving wheat are shown in Figure 4. Recall that the data of Lawlor (1976), indicated by the triangles and solid line, are currently used in Winter Wheat to adjust gross photosynthesis for water stress. Results from a previously unpublished experiment by J. A. Morgan were used in fitting the broken lines. Experimental procedures were essentially the same in both studies with one major exception. In Lawlor's study, plants were given only 24 hours to adjust to different root nutrient solution osmotic potentials before AP measurements were taken. The initial solution water potential before adjustment was probably greater than -0.5 bars. In contrast, Morgan gradually lowered the osmotic potential of the nutrient solution for a period of three weeks from -0.5 bars to approximately -18 bars, thereby lowering ψ_1 to -27 bars. As a result, the wheat plants in Morgan's study maintained higher AP, particularly at lower values of ψ_1. The ability to maintain AP at low ψ_1 was due, in large part, to the ability of the plant to osmoregulate. The more rapid decline in AP with ψ_1 in Lawlor's study was probably due to insufficient time for osmoregulatory adjustment to occur (Sionit et al. 1980). Consequently, there is a time element to take into account when relating AP to ψ_1. There is also evidence indicating that preliminary stress cycles may condition a plant to tolerate further stress cycles, although that conditioning may be lost if enough time has elapsed between drying cycles (Turner 1978). Because the mechanism and regulation of osmoregulatory conditioning are currently not well understood, a rather large data set may be required to properly map the response surface describing the effect of different rates of change in ψ_1 on the response of AP to ψ_1 and its components.

Another problem of modeling canopy water relations is the extrapolation of information and concepts from the organ level (the leaf in this case) to the canopy level of organization. The present assumption in the model is that the relationship between ψ_1 and photosynthesis is essentially the same for leaves and canopies. The effects of water stress are observed visually first in the older leaves, progressing from the leaf tips to the collars in each leaf (Ludlow 1975). Whether or not the progression of these events can accurately be described by a canopy model constructed from information at the organ level remains to be seen. Clearly, further experimentation in which plant water relations are examined at the two organizational levels is needed.

Another limitation of the current model is the absence of a temperature function of canopy photosynthesis. Both canopy and single leaf AP in wheat and other plants are temperature sensitive (Black 1973; Brown and Trlica 1977; Leach 1979; de Wit 1978). In the case of wheat and other C_3 plants, the optimum temperature for AP is in the range of 15 to 25 C (Black 1973; Leach 1979). This wide range is partly due to a conditioning effect of growth temperature on the optimum temperature for AP (Bjorkman 1975; Milthorpe and Moorby 1974). As in the previous example of water stress and its history of development, there is a need to collect a data base sufficient to describe temperature effects, including temperature history, on wheat photosynthesis. A complication in generating such a data set is that there is presently no adequate technique for separating CO_2 assimilation and respiration known to occur simultaneously in the light (Chollet and Ogren 1975). If temperature affects these two processes differently, as it apparently does (Chollet and Ogren 1975), then methods must be developed to measure or estimate the two fluxes to better understand their respective responses.

There are processes other than photosynthesis which will need further development in Winter Wheat, some of which were mentioned earlier. While much improvement will come from a continuing literature search, it is clear that further experimentation will be required to describe the system adequately within the framework of Winter Wheat. This is a necessary aspect of modeling a complex system. We believe that the advantages of understanding the system and the capability of predicting winter wheat yields justify the effort.

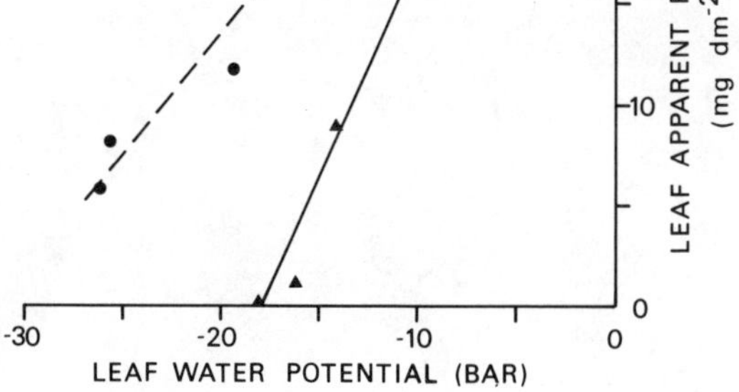

Figure 4.--Leaf apparent photosynthesis of wheat as affected by leaf water potential. Linear regression was used to fit the data of Lawlor (1976, ▲——▲) and J. A. Morgan (unpublished data, ● — ●).

REFERENCES CITED

Baker, D. N., J. R. Lambert, and J. M. McKinion. 1982. GOSSYM. Clemson Agr. Exp. Sta. Bul. In press.

Baker, D. N., D. E. Smika, A. L. Black, W. O. Willis, and A. Bauer. 1982. Winter wheat: A model for the simulation of growth and yield in winter wheat. AgRISTARS Publ. YM-U2-04281, JSC-18229.

Belman, C., J. Feyen, and D. Hillel. 1979. An attempt at experimental validation of macroscopic-scale models of soil moisture extraction by roots. Soil Sci. 127:174-186.

Bjorkman, O. 1975. Inaugural address. Environmental and biological control of photosynthesis. pp. 1-16. From R. Marcelle (ed.) Dr. W. Junk b.v. Publ., The Hague.

Black, C. C. Jr. 1973. Photosynthetic carbon fixation in relation to net CO_2 uptake. Ann. Rev. Plant Physiol. 24:253-286.

Bolton, J. K. and R. H. Brown. 1980. Photosynthesis of grass species differing in carbon dioxide fixation pathways. V. Responses of *Panicum maximum*, *Panicum miboides*, and tall fescue (*Festuca arundinacea*) to nitrogen nutrition. Plant Physiol. 66:97-100.

Brown, L. F. and M. J. Trlica. 1977. Carbon dioxide exchange of blue grama swards as influenced by several ecological variables in the field. J. Appl. Ecol. 14:205-213.

Chollet, R. and W. R. Ogren. 1975. Regulation of photorespiration in C_3 and C_4 species. Bot. Rev. 41:137-179.

Friend, D. J. C., V. A. Nelson and J. E. Fisher. 1962. Leaf growth in Marquis wheat, as regulated by temperature, light intensity, and daylength. Can. J. Bot. 40:1299-1311.

Gold, H. J. 1977. *Mathematical modeling of biological systems*. Wiley-Interscience, New York.

Hanks, R. J. 1974. Model for predicting plant yield as influenced by water use. Agy. Journ. 66:660-664.

Huffaker, R. C. and D. W. Rains. 1978. Factors influencing nitrate acquisition by plants; assimilation and fate of reduced nitrogen. In Nitrogen in the Environment. Vol. 2. (eds.) Donald R. Nielson and J. G. MacDonald. Academic Press, New York.

Lawlor, D. W. 1976. Water stress induced changes in photosynthesis, photorespiration, respiration and CO_2 compensation concentration of wheat. Photosynthetica 10:378-387.

Leach, J. E. 1979. Some effects of air temperature and humidity on crop and leaf photosynthesis, transpiration and resistance to gas transfer. Ann. Appl. Biol. 92:287-297.

Levitt, J. 1980. *Responses of plants to environmental stresses. Volume II*. Water, radiation, salt and other stresses. Academic Press, New York.

Ludlow, M. M. 1975. Effect of water stress on the decline of leaf net photosynthesis with age. In "Environmental and Biological Control of

Photosynthesis". R. Marcelle (ed.) pp. 123-134. Dr. W. Junk b.v., Publishers, The Hague.

McWhorter, J. C. and B. P. Brooks. 1965. Climatological and solar radiation relationships. Bull. 715, Miss. Agr. Exp. Sta.

Milthorpe, F. L. and J. Moorby. 1974. *An introduction to crop physiology*. Cambridge Univ. Press, New York.

Monteith, J. L. 1965. Light and crop production. Field Crop Abstracts. 18:213-219.

Morgan, J. M. 1977. Differences in osmoregulation between wheat genotypes. Nature 270:234-235.

Myrup, L. O. 1969. A numerical model of the urban heat island. J. of Appl. Met., 8:854-862.

Natr, L. 1975. Influence of mineral nutrition on photosynthesis and the use of assimilates. In *Photosynthesis and Productivity in Different Environments*, (ed.) J. P. Cooper. pp. 537-555. Cambridge University Press, Cambridge.

Norman, J. M. 1982. Data requirements for generalized physiological models. In *Computer techniques and meteorological data applied to problems of agriculture and forestry: a workshop*. (ed.) A. Weiss. Anaheim, California.

Osman, A. M., J. Goodman and J. P. Cooper. 1977. Effects of NPK on ratesof growth and photosynthesis of wheat. Photosynthetica. 11:66-75.

Rawson, H. M., J. E. Begg and R. G. Woodward. 1977. The effect of atmospheric humidity on photosynthesis, transpiration and water use efficiency of leaves of several plant species. Planta 134:5-10.

Ritchie, J. T. 1972. A model for predicting evapotranspiration from a row crop with incomplete cover. Water Resources Research 8:1024-1213.

Russell, R. S. 1977. *Plant root systems*. McGraw-Hill, New York.

Russo, J. M. and R. C. Seem. 1980. Models for integrated pest management of apple: a primer. New York State Agric. Exp. Station Publ. number 10, ISSN 0362-2754.

Saugier, B., E. A. Ripley, and P. Lueke. 1974. Modelling: VII A mechanistic model of plant growth and water use for the Matador grassland. Tech. Report #65, Canadian IBP Programme, the Univ. of Saskatchewan, Saskatoon, Saskatchewan. 96 p.

Sionit, N., I. D. Teare and P. J. Kramer. 1980. Physiol. Plant 50:11-15.

Smika, D. E., A. L. Black, D. N. Baker, W. O. Willis, and A. Bauer. 1978. Modeling the growth and grain yield of wheat. Proc. Workshop on Crop Simulation. Biological Systems Simulation Group, Clemson University.

Sofield, I., L. T. Evans, and I. F. Wardlaw. 1974. The effects of temperature and light on grain filling in wheat. In *Mechanisms of Regulation of Plant Growth* (eds.) R. L. Bielski, A. R. Ferguson, M. M. Crecswell. pp. 909-915. Bull. 12, The Royal Soc. New Zealand, Wellington.

Turner, N. C. and M. L. Tonnet. 1978. Osmotic adjustment of sorghum and sunflower crops in response to water deficits and its influence on the water potential at which stomata close. Aust. J. Plant Physiol. 5:597-608.

Whisler, F. D., J. R. Lambert, and J. A. Landivar. 1981. Predicting tillage effects in cotton growth and yield. In *Predicting the Effect of Tillage on Soil-Plant Relationships*. American Soc. of Agronomy Special Pub. Van Dorn and Unger (eds.) pp. 179-198.

Wit, C. T. de. 1978. *Simulation of assimilation, respiration and transpiration of crops*. Halsted Press Book, Wageningen.

SIMULATION OF NITROGEN, TILLAGE, AND RESIDUE MANAGEMENT EFFECTS ON SOIL FERTILITY[1]

M. J. Shaffer, S. C. Gupta, D. R. Linden, J. A. E. Molina, C. E. Clapp, and W. E. Larson[2]

A comprehensive simulation model of the soil-crop-water system has been developed as a management tool for nitrogen, tillage, and crop residues. The overall model integrates process level submodels for water flow, crop growth (including roots), nitrogen transformations, tillage effects, soil temperature, soil chemistry, and solute transport to provide an interactive process simulation. The nitrogen-tillage-residue management model is capable of seasonal and longer-term estimates of soil fertility and its effects on crop yield. Model validation and verification have been obtained for each submodel and for the overall model package. Output from the model can be used in farm and extension level management programs and models, in research and engineering studies, and in teaching programs. The NTRM1 model has been made "user friendly" by developing an extensive user operating system, computer graphics capabilities, and user documentation. The model can be directly accessed via remote terminals through Telenet or similar telephone systems, and the ANSI standard FORTRAN source code is available for adaptation to a specific computer.

INTRODUCTION

Managing the effects of nitrogen, tillage, and crop residues on soil fertility and crop yield requires consideration of a complex array of inputs and internal system interactions. Computer simulation offers a means to integrate the relevant data and processes to produce a rapid and continuous analysis of management alternatives.

Our nitrogen-tillage-residue management (NTRM) system of models is composed of submodels at several levels for water flow, crop growth (including roots), nitrogen transformations, tillage effects, soil temperature, soil chemistry, and solute transport. The submodels in various blocks are interconnected via transfer coefficients and the flow of mass and energy to produce a simulation of the soil-crop-water continuum.

The NTRM model series is composed of modeling sub-blocks ranging from research and engineering models such as NTRM1 and NTRMS3, to user models such as the soil-crop management system (SCMS) model which has direct application at the farm and extension levels. Exchange of both data and ideas takes place between the various modeling levels and efforts.

The objective of this paper is to present details of the NTRM1 research model. Some references will also be made to the NTRMS3 model and to application of both NTRM1 and NTRMS3 to research studies, and to the SCMS user model.

PROCESS SIMULATION

The major processes simulated by the NTRM1 model together with major pathways of data transfer are shown in figure 1. Submodels for these processes are designed in modular form to make it easier to isolate parts of the system for detailed study and to replace or revise sub-systems as new information becomes available. The simulation process is dynamic and step-wise over time, with data inputs on a daily or less frequent basis.

Soil physical, biological, and chemical parameters are allowed to change with depth and time. These changes respond to climate inputs such as precipitation, air temperature, pan evaporation, wind, and solar radiation, as well as management inputs including tillage, crop residue amounts and placement, fertilizer amounts, type, placement and timing, and irrigation.

The flexibility built into the model in this area is illustrated in figure 2. Note that changes in the soil physical properties due to tillage and precipitation often cause a change in the soil bulk density, which repositions the soil surface and other reference points in the affected zones. The model accomodates this feature by changing the volume or thickness of appropriate soil horizons. Other changes in soil

[1]Paper presented at the Third International Conference on State-of-the-Art in Ecological Modelling, Colorado State University, May 24-28, 1982.

[2]Soil Scientist, USDA-ARS, St. Paul, Minn.; Soil Scientist, USDA-ARS, St. Paul, Minn.; Soil Scientist, USDA-ARS, St. Paul, Minn.; Associate Professor, Dept. of Soil Science, Univ. of Minnesota, St. Paul, Minn.; Soil Chemist, USDA-ARS, St. Paul, Minn.; and Head, Dept. of Soil Science, Univ. of Minnesota, St. Paul, Minn., respectively.

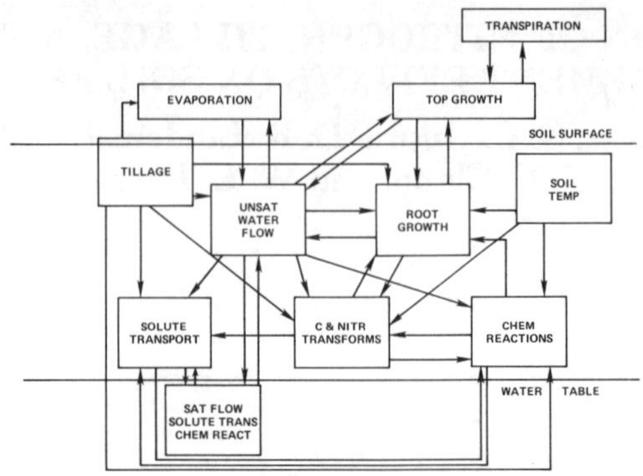

Figure 1. NTRM1 Processes Simulated and Primary Data Transfers

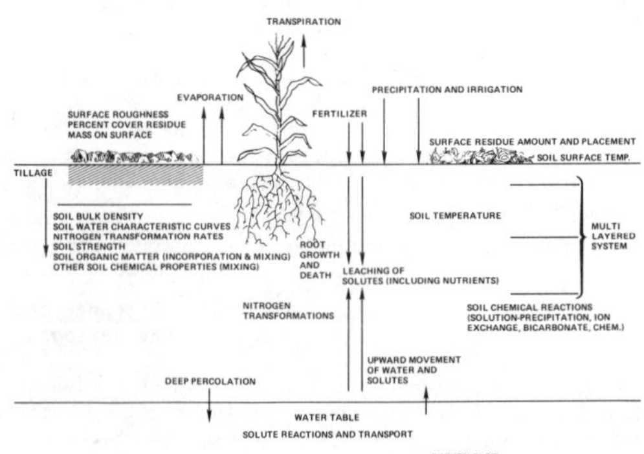

Figure 2. Changes Modeled with Depth

parameters are simulated using appropriate response equations.

Modeled changes most closely affecting soil fertility include tillage impacts on nitrogen transformation rates, mineralization-immobilization associated with crop residues, leaching of nutrients, soil temperature changes, root development and senescence, and ion exchange involving NH_4^+-N.

SUBMODEL INTERACTIONS

The major submodel blocks are identified in figure 1. With the exception of the soil temperature submodel, the routines are interdependent with information exchange in two or more directions. The soil temperature submodel provides data inputs to the remaining system, but does not receive feedback.

TOP AND ROOT GROWTH

Details of the feedback loops involving top and root growth are shown in figure 3. Note that a minimum number of parameters are used to actually operate the feedback systems. The basic parameters involved are DEFNIT for nitrogen, RATIO1 for soil water, and RATIO for root stress effects. Top growth and root growth are considered simultaneously in the model. Nitrogen deficiencies in the plant which lead to reduced growth are reflected in the DEFNIT parameter. Growth is reduced in both the tops and roots in this case. Water deficiencies in the root zone which lead to reduced growth are reflected in the RATIO1 parameter. Again both the tops and roots show reduced activity. In addition, other conditions in the root zone which limit overall root and top development such as soil temperature, soil strength, soil salinity, soil aeration, soil nitrogen, and soil suction and which are not already accounted for in the DEFNIT and RATIO1 parameters are included in the RATIO parameter. These can be thought of as indicators of situations where specific root zone conditions will not allow the development or redistribution of the overall root mass as projected.

Definitions of RATIO1, DEFNIT, and RATIO Parameters:

RATIO1 is defined as the ratio of actual to potential water extraction by the root zone over a time step. Here potential water extraction takes into account the physiological state of the crop and stage of development. Only if the crop runs out of water on this basis will RATIO1 be less than 1.0. Permanent wilt is taken as the lower limit for water extraction. Potential water extraction with depth is assumed to be proportional to current root distribution.

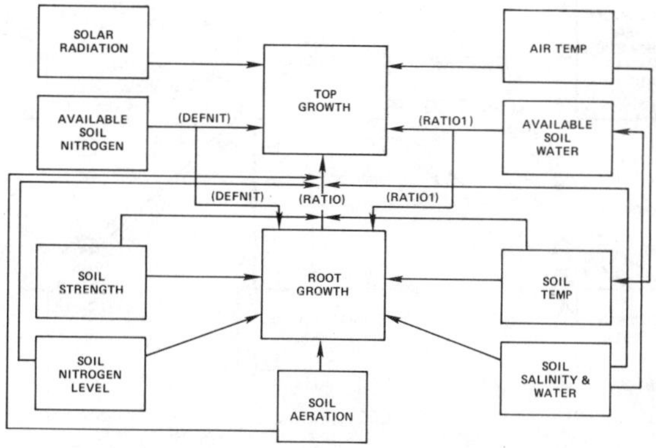

Figure 3. Interrelationships of Top and Root Growth

DEFNIT is defined as the ratio of actual to potential uptake of NO_3-N + NH_4-N by the crop for that time step. NO_3-N and soluble NH_4-N are treated as being equally accessible by the roots. Potential uptake of N is based on the current physiological state of the plant and the growth stage. Potential extraction of N with depth is assumed to be proportional to actual water uptake with depth. When a nitrogen stress condition is removed, a lag period of up to a few days is used before full nitrogen uptake resumes.

RATIO is defined as the ratio of actual to potential production of root mass for a time step. Potential root mass is defined as the mass of roots which the plant would like to produce based on its physiological state and its growth stage. This includes prior consideration of the DEFNIT and RATIO1 parameters.

TILLAGE AND SOIL PROPERTIES INTERACTIONS

Interactions between tillage and soil properties are illustrated in figure 4. Note that transient values for soil properties affect the impact tillage has on the soil properties. The model either modifies the soil properties according to input specifications or it utilizes internal submodels to estimate the changes which would occur. This applies to individual tillage events as well as more gradual changes which occur between events.

More specific data transfers between the tillage submodel and other routines are shown in figure 5. Note that tillage events (and between-tillage events) directly affect several submodels. Other submodels are affected indirectly due to the "changing" of related processes. The TILL parameter for increased biological activity and updated values for soil nitrogen concentrations and soil water content are provided for the carbon and nitrogen transformation submodel. Soil water contents and unsaturated flow properties are updated for use by the unsaturated flow submodel. Values for soil bulk density, soil water content, saturated theta (θ_s), matrix potential, osmotic potential, soil nitrogen concentrations, and root pruning are recomputed for input into the root growth submodel. Percent cover, surface reflection coefficients, and surface roughness are updated for the evaporation submodel. Values for soluble and exchangeable constituent concentrations are modified for use in the chemical equilibria submodel.

SOIL NITROGEN INTERACTIONS

Interaction of the nitrogen and carbon transformation submodel with other submodels is shown in figure 6. The COMBINE executive submodel performs the numerical integration of the overall system with time. Note that crop and root growth, root slough, nitrogen and carbon transformations, ion exchange involving NH_4^+-N, and solute transport are all considered simultaneously. Constituent masses for

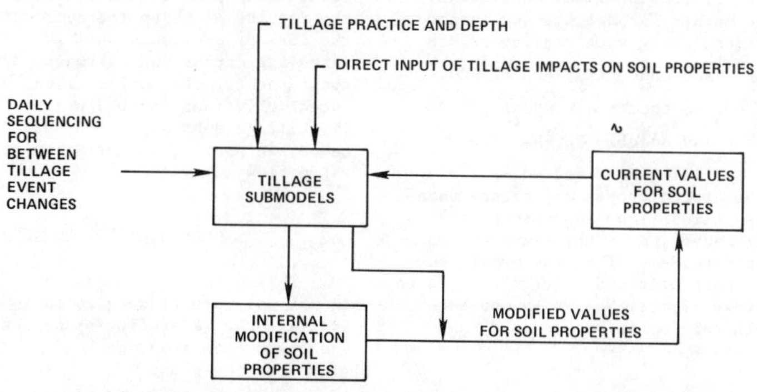

Figure 4. Interactions of Tillage with Soil Properties

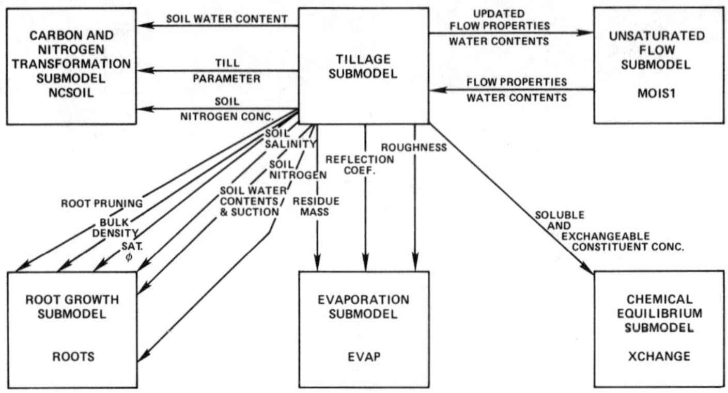

Figure 5. Exchange of Tillage Data Between Submodels

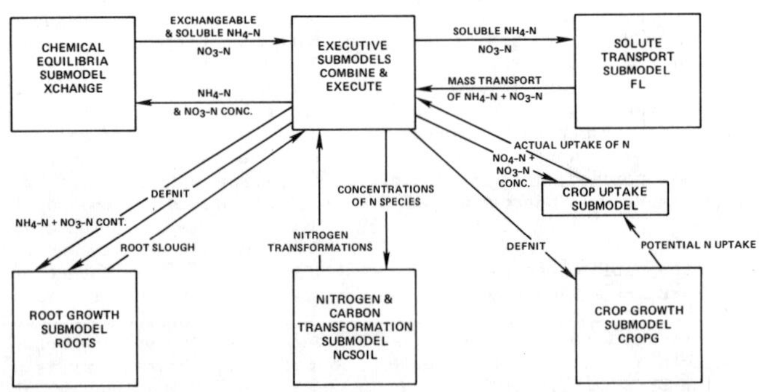

Figure 6. Exchange of Soil Nitrogen Data Between Submodels

the nitrogenous species with depth are stored in a common block area in the model. This allows access by all the submodels as required. For a given time step, each submodel uses the same nitrogen mass data. Updates are made after each submodel has completed its calculations. Any mass overruns (negative values) are redistributed back over the processes involved in order to maintain mass balance in the system.

The solute transport submodel uses soluble NH_4^+-N and NO_3^--N masses, soil water contents, and water fluxes with depth to compute transport of these species (mass/segment/time step). The chemical equilibria submodel uses current values for soluble and exchangeable NH_4^+-N concentrations along with similar values for Ca^{++}, Na^+, and Mg^{++} and concentrations of HCO_3^-, $CO_3^=$, $SO_4^=$, Cl^-, and NO_3^--N to compute updated values for exchangeable and soluble NH_4^+-N. The root growth submodel utilizes NH_4^+-N and NO_3^--N concentrations and the crop nitrogen deficiency parameter (DEFNIT) in its calculations and returns the daily root slough for recycling by the nitrogen and carbon transformation submodel. The crop growth submodel uses the DEFNIT parameter and other variables to compute potential uptake of soil N. NH_4-N and NO_3-N concentrations and other information are used to compute actual N uptake.

SOIL WATER INTERACTIONS

Submodel interactions involving soil water are illustrated in figure 7. The unsaturated flow submodel uses soil water contents and flow properties to compute updated values for soil water contents, matrix potentials, and fluxes. Soil water contents are utilized in the root growth, carbon and nitrogen transformation, chemical equilibria, solute transport, and tillage submodels. Soil suctions are used in the root growth submodel, while flow properties are input to the tillage submodel, and soil water fluxes are utilized by the solute transport submodel. Submodel feedback based on the results of using the soil water information takes the form of crop and root masses produced, transformation rates for carbon and nitrogen, movement of soluble constituents, the soil chemical equilibrium status, and potential values for evaporation and transpiration. The tillage submodel makes updates in soil properties based, in part, on information provided by the unsaturated flow submodel.

UNSATURATED FLOW SUBMODEL

This submodel uses a finite difference solution to the moisture flow equation in one dimension to simulate the water flow processes discussed previously. The basic flow equation, which includes a macroscopic sink term for plant water uptake (MS) (Dutt et al., 1972; Shaffer and Gupta, 1981), is written,

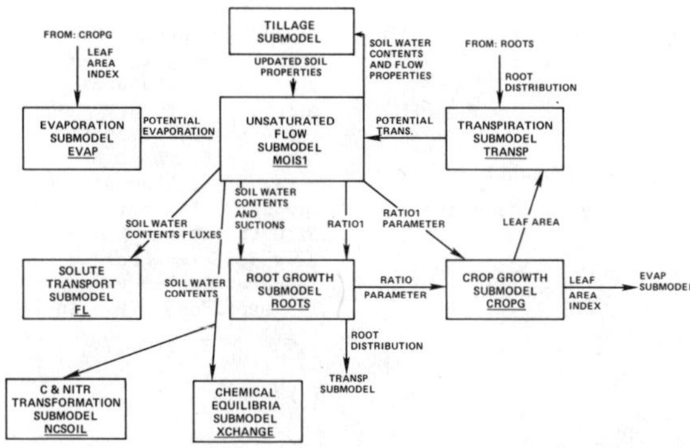

Figure 7. Exchange of Soil Water Data Between Submodels

$$\frac{\partial\theta}{\partial t} = \frac{\partial}{\partial z}\,[D(\theta)\frac{\partial\theta}{\partial z} - K(\theta)] - MS \qquad [1]$$

where: θ is volumetric water content,
t is time,
z is depth,
$D(\theta)$ is the dispersion coefficient, and
$K(\theta)$ is unsaturated hydraulic conductivity.

The finite difference solution to equation [1] can be written as,

$$\frac{\theta_j^i - \theta_j^{i-1}}{\Delta t} = [D_{j+1/2}^{i-1/2}\,(\theta_{j+1}^i + \theta_{j+1}^{i-1} - \theta_{j+1}^{i-1} - \theta_j^i - \theta_j^{i-1})$$

$$- 2\,\Delta X K_{j+1/2}^{i-1} - D_{j-1/2}^{i-1/2}\,(\theta_j^i + \theta_j^{i-1} - \theta_{j-1}^i - \theta_{j-1}^{i-1})$$

$$- 2\,\Delta X K_{j-1/2}^{i-1}]/2\,\Delta X^2 - MS_j^i \qquad [2]$$

where: i references time step i,
j references depth node j,
ΔX is the nodal spacing, and
Δt is the time step size.

The water flux between adjacent nodes is calculated for each time step, Δt, using the relationship,

$$FLUX = [K_{j+1/2}^{i-1/2} - D_{j+1/2}^{i-1/2}\,(\frac{\theta_{j+1}^i + \theta_{j+1}^{i-1} - \theta_j^i - \theta_j^{i-1}}{2\,\Delta X})]\,\Delta t \qquad [3]$$

Values for $D(\theta)$ are calculated using an estimate of the water content (θ) at the end of the next time step and averaged over two depth nodes. This approximation for anticipated water content is written,

$$\theta_{j+1/2}^{i-1/2} \approx Y\,(\theta_{j+1}^{i-1} - \theta_{j+1}^{i-2} + \theta_{j+1}^{i-1} + Y\,(\theta_j^{i-1} - \theta_j^{i-2})$$

$$+ \theta_j^{i-1} \qquad [4]$$

where Y is a weighting factor defined as,

$$Y = 0.7\,(\frac{\Delta t^i}{\Delta t^{i-1}}) \qquad [5]$$

Boundary Conditions

The lower boundary is defined in terms of a constant water content (θ_{LB}). The value used for θ_{LB} can either be the saturated theta value (θ_s) or some other value less than saturation. The value used must be determined based on the best approximation of conditions in a particular field situation.

The upper boundary (soil surface) is defined in terms of a water content ($\theta_{\mu B}$) which varies over time depending on the surface conditions. The values used for $\theta_{\mu B}$ in the finite difference equations are determined using,

$$\theta_{\mu B} = \theta_{j=1}^i = (2.0\,\Delta X/D_{j+1/2}^{i-1/2})\,(FLUX - K_{j-1/2}^{i-1/2})$$

$$+ KW\,\theta_{j=2}^{i-1} + \theta_{j=2}^{i-1} - \theta_{j=1}^{i-1} \qquad [6]$$

where: FLUX is the maximum rate of infiltration for positive water flux (water on the surface)
<u>or</u>
FLUX is the potential evaporation rate for the case where water is evaporating from the soil surface. (Note: the assumption is made that evaporation is zero when infiltration is taking place.)
<u>and</u>
KW is a weighting parameter.

The actual values computed for FLUX (+ or -) depend on the soil water contents and hydraulic properties. This allows the soil infiltration rate to vary with time during an infiltration event, and the soil evaporation rate to vary as the soil dries out.

Soil Water Characteristic Curves

The values for soil hydraulic conductivity (K) and diffusivity (D) as a function of soil water content (θ) must be provided either with a set of user supplied equations or by using the Campbell method for the soil water characteristic curves, Campbell (1974). This method assumes the moisture release curve (suction vs. theta) can be represented by the equation,

$$\psi(\theta) = \psi_r\,(\frac{\theta}{\theta_{SAT}})^{-b} \qquad [7]$$

where: $\psi(\theta)$ is soil matrix potential at water content θ,
ψ_r is soil matrix potential at the reference water content,
θ_{SAT} is saturated water content, and
b is an empirical constant;

and the hydraulic conductivity vs. theta relationship by,

$$K(\theta) = K_{SAT} \left(\frac{\theta}{\theta_{SAT}}\right)^{2b+3} \quad [8]$$

where: K_{SAT} is saturated hydraulic conductivity

Diffusivity ($D(\theta)$) is defined by the relationship,

$$D(\theta) = \frac{K(\theta)}{\partial\theta/\partial\psi} \quad [9]$$

which gives,

$$D(\theta) = \frac{b\ \psi_r\ K_{SAT}\ \theta^{b+2}}{\theta_{SAT}^{b+3}} \quad [10]$$

when equations [7], [8], and [9] are combined.

SOIL TEMPERATURE SUBMODEL

The root zone temperature model is based on the partial differential equation describing heat flow in soils (Gupta et al., 1981).

$$\frac{\partial T}{\partial t} = \frac{\partial}{\partial Z}\left(\alpha \frac{\partial T}{\partial Z}\right) \quad [11]$$

Equation [11] can be numerically solved for soil temperature, T, with depth, Z, and time, t, if the thermal diffusivity, α, along with the initial

$$T = T(Z) \quad [12]$$

and boundary conditions

$$T_{o,t} = T(t) \quad [13]$$

$$T_{L,t} = T(t) \quad [14]$$

are known.

The numerical approximation (Eq. [15]) used in this model was taken from Hanks et al. (1971).

$$\frac{T_{i,j}-T_{i,j-1}}{\Delta t} = \frac{(T_{i-1,j}-T_{i,j})\alpha_{i-1/2,j}-(T_{i,j}-T_{i+1,j})\alpha_{i+1/2,j}}{\Delta Z^2} \quad [15]$$

The procedure for solving Eq. [15] was similar to that used by Hanks et al., (1971). In our calculations ΔZ was varied from small values near the soil surface to large values near the bottom boundary. Equation [15] when written for each depth results in a tridiagonal matrix which was solved hourly for soil temperature prediction at any depth and time.

The model for estimating upper boundary temperature (UBT) is given in Eq. [16]

$$T_{o,t} = T_{a,t} + A_T \sin\left(\frac{2\pi}{P} t_x\right) \quad [16]$$

Equation [16] is the summation of the air temperature at a reference height and a sine function that describes the difference in the temperature of air and soil surface (fig. 1). However, this does not mean that the soil surface temperature is necessarily a sine function. Presence of the $T_{a,t}$ term in Eq. [16] implies that hourly temperature fluctuations at the soil surface due to cloud cover or wind are reflected in the air temperature at reference height. The reference height in our experiment was 2 m. Amplitude of the sine function (A_T) in Eq. [16] is equal to the difference between maximum temperatures at the soil surface, T_{sx}, and air, T_{ax}, or between the minimum temperatures at the soil surface, T_{sn}, and air, T_{an}, for curves above and below the reference temperature, respectively. Period, P, is twice the hourly difference between the time that the temperatures of air and soil surface are about same. In our data this corresponded to an air temperature of 21°C (reference temperature). In Eq. [16], P would have different values for predicting the upper boundary temperatures above and below the reference temperature for each day. Thus in figure 1A (see Appendix),

$$T_{o,t} = T_{a,t} + (T_{sx} - T_{ax}) \sin\left(\frac{2\pi}{P} t_x\right) \quad [17]$$

$$P = 2(t_2 - t_1) \quad [18]$$

and

$$t_x = t - t_1 \quad [19]$$

for air temperatures above the reference temperature and

$$T_{o,t} = T_{a,t} + (T_{sn} - T_{an}) \sin\left(\frac{2\pi}{P} t_x\right) \quad [20]$$

$$P = 2\ (t_3 - t_2) \quad [21]$$

and

$$t_x = t - t_2 \quad [22]$$

for air temperature below the reference temperature.

Hourly air temperatures were estimated from the daily maximum and minimum air temperatures using Eq. [23]

$$T_{a,t} = T_t^* (T_{ax} - T_{an}) + T_{an} \quad [23]$$

The procedure used to calculate the average normalized air temperature (T_t^*) is described in Gupta et al. (1982).

Lower boundary -onditions are approximated by assuming a deep profile such that temperatures at the lower boundary are nearly constant.

SOIL CHEMISTRY SUBMODEL

The soil chemistry (equilibria) submodel was originally developed by Dutt and co-workers in the 1960's, Dutt (1961), Dutt and Tanji (1962), Dutt et al. (1972), and has seen some refinement and extensive application in the 1970's.

This submodel uses a thermodynamic approach to predict concentrations of major cations and anions in neutral to alkaline soils. Acid soil reactions are not considered. Work is in progress to extend this submodel into the acid region.

Chemical processes being simulated in the model include precipitation-dissolution of $CaCO_3$ (lime) and $CaSO_4 \cdot 2H_2O$ (Gypsum); ion pair reactions involving $CaSO_4°$, $MgSO_4°$, $NaSO_4^-$, the bicarbonate buffer system; and ion exchange of Ca^{++}, Na^+, Mg^{++}, and NH_4^+.

The set of equilibrium reactions included in the model can be written as follows:

$$\text{gypsum solubility ksp} = (Ca^{2+})(SO_4^{\ 2-})\ \gamma_2^2 \quad [24]$$

$$CaSO_4°\ \text{ion pairing}\ K_D = \frac{(Ca^{2+})(SO_4^{\ 2-})}{(CaSO_4°)}\ \gamma_2^2 \quad [25]$$

$$\text{MgSO}_4{}^\circ \text{ ion pairing } K_{D1} = \frac{(Mg^{2+})(SO_4{}^{2-})}{MgSO_4{}^\circ}\gamma_2^2 \quad [26]$$

$$\text{CaCO}_3 \text{ solubility \& bicarbonate buffer system} \quad K^1\, PCO_2 = (Ca^{2+})(HCO_3{}^-)^2\,\gamma_2\,\gamma_1^2 \quad [27]$$

$$\text{NaSO}_4{}^- \text{ ion pairing } K_{D2} = \frac{(Na^+)(SO_4{}^{2-})}{(NaSO_4{}^-)}\gamma_2 \quad [28]$$

where: the ( )'s denote concentrations in solution, PCO_2 is partial pressure of carbon dioxide, and the K's are appropriate equilibrium constants.

$$\text{Ca-Na ion exchange} \quad \frac{\sqrt{a_{Ca}}}{a_{Na}} = K_{Ca\text{-}Na}\frac{E_{Ca}}{E_{Na}} \quad [29]$$

$$\text{Ca-Mg ion exchange} \quad \frac{a_{Ca}}{a_{Mg}} = K_{Ca\text{-}Mg}\frac{E_{Ca}}{E_{Mg}} \quad [30]$$

$$\text{Na-NH}_4 \text{ ion exchange} \quad \frac{a_{Na}}{a_{NH_4}} = K_{Na\text{-}NH_4}\frac{E_{Na}}{E_{NH_4}} \quad [31]$$

where: $a_{(x)}$ refers to the activity of the ion "x" in solution,
$E_{(x)}$ refers to the concentration of the ion "x" in the exchange complex,
and, $K_{(x-y)}$ is the appropriate exchange coefficient.

Activity coefficients (γ) are computed using the Debye-Hückel equation in the form

$$-\log\gamma_i = \frac{0.509\, z_i^2\sqrt{u}}{(1+\sqrt{u})} \quad [32]$$

where: $u = 1/2\sum_{i=1}^{n} C_i Z_i^2$

Z_i is the charge,
and C_i is concentration.

SOLUTE TRANSPORT SUBMODEL

Transport of soluble species in the unsaturated zone is handled in the NTRM1 model via the submodel FL1. This route uses soil water flux and soil water content data provided from the unsaturated flow submodel or elsewhere to calculate solute fluxes in and out of each soil segment for each time step. The algorithms used can be summarized as,

$$\Delta mT_i^J = C_i^{t,J-1} * F^{J-1} - C_i^{t,J} * F^J \quad [33]$$

[for positive flux]

and,

$$\Delta mT_i^J = C_i^{t,J} * F^J - C_i^{t,J+1} * F^{J+1} \quad [34]$$

[for negative flux]

where: ΔmT is the change in solute mass for segment J,
C is concentration of solute specie "i",
F is water flux,
t is the time step superscript,
J is the soil segment superscript.

Equations [33] and [34] are written for general flow in the downward or upward directions. The FL submodel also allows simultaneous upward and downward flow into any soil segment. This gives more flexibility in the solute redistribution process.

Solute transport due to dispersion or diffusion processes is not handled directly in the model. Instead, the use of finite soil segments with complete mixing assumed at each step in time and space introduces so-called "numerical dispersion" into the simulation process. This allows a first approximation of dispersion without the circular procedure of removing the numerical effects and then re-adding dispersion via a Ficks Law dispersion term.

Studies have shown that the first approximation is satisfactory for most soils. The amount of dispersion being introduced can be controlled to some extent by the number of soil segments used in the profile. If necessary, it would be possible to include a circular dispersion algorithm in the NTRM1 model.

ROOT GROWTH SUBMODEL

A one-dimensional dynamic submodel for root growth was developed to simulate root penetration, root branching, and root death. Modeled environmental effects on root growth include soil temperature, soil bulk density, soil aeration, soil strength, soil water potential, soil salinity, and soil nitrogen. The model simulates root branching, death, and penetration rates as a function of these variables and if necessary, redistributes the root growth within the rooting zone to exploit the most favorable soil conditions.

Baseline or reference rates of root growth (branching and penetration) and root death are taken as functions of crop growth stage, and dry matter production by the plant.

Root density changes are the result of root growth and root death (slough). Root slough becomes input into the NTRM1 nitrogen and carbon transformation submodel. Root death rates are a function of growth stage, soil environmental conditions, and the availability of root mass.

Physical and chemical barriers (or partial barriers) may or may not alter the total length (mass) of roots produced. In these situations, the model attempts to alter the root growth pattern, within limits, and grow the roots in a more favorable part(s) of the root zone (Taylor 1974). The model is constrained by maximum root densities, maximum upper limits on growth rates, and adverse environmental conditions elsewhere in the root zone. When the model cannot redistribute the root mass, a loss of roots occurs and a signal is sent to the top growth submodel via the RATIO coefficient to reduce dry matter production.

The rate of root extension as a function of soil temperature, soil strength, soil aeration, soil salinity, and soil nitrogen can be written,

$$R_{ex} = R_{exo} - \int_{tevo}^{tex}\frac{\partial R_{ex}}{\partial T}dt - \int_{Sexo}^{Sex}\frac{\partial R_{ex}}{\partial S}ds - \int_{Aexo}^{Aex}\frac{\partial R_{ex}}{\partial A}dA - \int_{SSexo}^{SSex}\frac{\partial R_{ex}}{\partial SS}dSS - \int_{KSexo}^{Nex}\frac{\partial R_{ex}}{\partial N}dN - \text{I A terms} \quad [35]$$

where: R_{exo} is extension rate at the reference conditions,
T is soil temperature,
S is soil strength,
A is soil aeration,
SS is soil salinity, and
N is soil nitrogen.

Each function subroutine evaluates its respective integral term in equation [35]. Note that equation [35] represents a zero-order rate process whose rate coefficient is modified by environmental variables. The extension rate given by R_{ex} is the rate under reference conditions which are taken as being ideal with respect to root growth.

Equation [35] separates the environmental effects on root growth into a series of independent and cross product IA (interaction) terms. This approach differs from the methods used by some modelers of selecting the process with the greatest negative impact on rate as being rate determining, or representing the rate as a product of a reference rate and a series of multiplicative environmental variables normalized between 0 and 1. From mathematical and theoretical standpoints, equation [35] is the most comprehensive and rigorous of the three methods. Evaluation of the interactive terms presents some difficulties which have not been thoroughly researched, but are yielding to investigative efforts. The root submodel contains equations written using all three methods. Method selection is done with a control switch.

The response to changes in temperature is assumed to be linear from 15°C up to 25°C. A plateau is then taken from 25°C to 30°C where a linear reduction is made to zero at 40°C, Nielsen (1974), Taylor et al. (1972), Allmaras et al. (1964).

The submodel is designed to generate compensatory root growth in response to temperatures which are not in the optimal range, Russel (1977). This growth is restricted only by a maximum rule of growth and an upper limit on root density both of which can be set by the user.

The function used for soil strength f (strength) contains the variables bulk density (BD) and soil matrix potential (ψ) and is written,

$$\text{func(strength)} = -5.55 + 0.379 \times 10^{-6}\,\psi^2 \; 0.152 \times 10^{-2}\,\psi + 0.102\,BD - 0.372 \times 10^{1}\,BD^2 \qquad [36]$$

The basic data were adopted from Taylor and Gardner (1963) and the working functions were derived using a second order regression approach.

The value for BD used in equation [36] is obtained from

$$BDL = (SS - 76.0) * 0.004 + BDLo, \qquad [37]$$

and

$$BD = BDo + BDLo - BDL \qquad [38]$$

or

$$BD = BDo - (SS - 76.0) * 0.004 \qquad [39]$$

where: SS is the sum of the percent sand and percent silt, and
BDLo is the soil bulk density.

These equations incorporate the work of O'Connell (1975) to extend Gardner's data for a fine sandy loam soil to other soil textures.

The role of root branching is computed using a relationship similar to equation [35], except that the rate process is taken as first order with respect to root length density (RL) in the first 3/4 of the growing season.

$$R_{br} = [R_{bro} - \int \frac{\partial R_{br}}{\partial T}\,dT - \int_{Sexo}^{Sex} \frac{\partial R_{br}}{\partial s}\,ds - \ldots]RL \qquad [40]$$

where: R_{br} is the rate of root branching
RL is the root length density (cm/cm^3)

A zero order expression is used for the remainder of the season.

COMPENSATORY ROOT GROWTH

Root growth is redistributed by the model in response to adverse conditions with respect to soil temperature, soil strength, soil water potential, soil aeration, and soil nitrogen levels. The total potential loss in root growth for each 1 cm depth element is computed for each daily time step. The ability of the active root zone to compensate for this loss is calculated for each element and summed over the root zone. This sum is compared with the total potential loss to determine the total compensatory root growth and the total actual loss. The actual compensatory growth is then proportioned over the depth elements according to their ability to grow the additional roots. Any remaining growth loss is used in the computation of the RATIO coefficient.

The compensatory growth potential is calculated as a user supplied fraction of the maximum growth rate.

CROP GROWTH SUBMODEL (TOPS)

For initial NTRM1 modeling purposes, we selected the Nebraska corn growth model (CØRNGRØ), Childs et al. (1977), Tscheschke and Gilley (1980). This submodel is somewhat specific for corn, but does contain some relationships which are being adapted to other crops. The corn crop was selected because of its agronomic importance in the midwest. The CØRNGRØ model was kept intact with respect to ideal growing conditions as a function of solar radiation. Interactions with the environment were modified or added to simulate the effects of air and soil temperature, soil water, soil aeration, soil strength, soil nitrogen, and soil salinity. The resulting submodel for corn growth was then interfaced into the overall NTRM model. Key linkages include the uptake of soil NH_4-N and NO_3-N, the DEFNIT, RATIO, and RATIO1 parameters, and other input parameters such as air temperature and solar radiation.

The corn growth submodel operates as a producer and consumer of carbohydrates. Carbohydrates are produced by photosynthesis which is simulated in the model by the relationship,

$$P = PSNK * FLEAF * FRAD * FCHO * RATIO1 \qquad [41]$$

where:

P = net photosynthesis (mg CH_2O dm^{-2} hr^{-1})
PSNK = gross photosynthesis at 1.4 ly/min = 65.0 mg CH_2 dm^{-2} hr^{-1}
FLEAF = fraction of total leaf area which is sunlit (range 0-1)

$$FRAD = \frac{2.36\,R}{1+1.65\,R}$$; R is solar radiation in ly/min

(This is an adjustment function for gross photosynthesis based on incoming solar radiation) (range 0-1)

FCHO = (84.5-A)/84.5 ; A is available carbohydrate
(This factor sets maximum available carbohydrate at 84.5 mg cm^{-2}) (range 0-1)

RATIO1 = actual transpiration/potential transpiration (range 0-1)

Carbohydrates are consumed in the model by the respiration processes for growth and maintenance. Maintenance respiration is calculated using the expression,

$$RM = C(0.044 + 1.9 \times 10^{-3}\ T + 1 \times 10^{-3}\ T^2)\ TDMP \quad [42]$$

where: RM = maintenance respiration (mg CH_2O dm^{-2} hr^{-1})
C = maintenance coefficient (0.74 mg CH_2O dm^{-2} hr^{-1} gm^{-1})
T = air temperature (°C)
TDMP = Σ total dry matter produced to date (g dry matter/plant).

Growth respiration is computed by calculating total respiration and then subtracting maintenance respiration. For total respiration,

$$RT\emptyset T = (3.857 \times 10^{-3} + 3.829 \times 10^{-3}\ T) * A * DEFNIT * RATIO \quad [43]$$

where: RTØT = total respiration (mg CH_2O dm^{-2} hr^{-1})
DEFNIT = actual N uptake/N uptake demand
RATIO = root mass produced/potential root mass production,

and for growth respiration,

$$RG = RT\emptyset T - RM \quad [44]$$

where: RG = growth respiration (mg CH_2O dm^{-2} hr^{-1}).

Crop growth is computed from growth respiration using,

$$DMP = 3.76 * RG \quad [45]$$

where: DMP = growth produced (mg CH_2O dm^{-2} hr^{-1}).

Available carbohydrate (A) is updated on an hourly (1/20 day in NTRM) basis using,

$$A = A_{t-1} + P - RT\emptyset T - DMP \quad [46]$$

where: A_{t-1} = available carbohydrate from the previous time step.

Total dry matter production in the tops is computed using the expression,

$$DRYMAT = \sum_{emergence}^{tasseling} 3.58 * RG * ALEAF * FRACT/1000 \text{ (g dry matter/plant)} \quad [47]$$

where: ALEAF = leaf area (dm^2)
FRACT = fraction of dry matter in the tops.

Total dry matter production in the ears is given by,

$$DRYEAR = \sum_{emergence}^{tasseling} 3.58 * RG * ALEAF * FRACT/1000 \text{ (g dry matter/plant)} \quad [48]$$

Leaf area from emergence to tasseling is computed on a daily basis using,

$$LA = LA\emptyset CD\ (1 + AK1 * RGDA * FRACT * FLFDRY) \quad [49]$$

where: LAØCD = previous day's leaf area (dm^2)
AK1 = AK2 - ATEMP (35-T)/25 *AK2)
RGDA = daily sum of growth respiration
FLFDRY = ratio of leaf growth to total dry matter production
(varies from 0.9 at emergence to 0.0 at tasseling)
AK2 = crop coefficient
ATEMP = temperature sensitivity coefficient = 0.72

After tasseling leaf area is computed using the equation,

$$LA = AALA\ (1 - (DDAY - (DDT + 100))/(DDM - (DDT + 100))) \quad [50]$$

where: AALA = leaf area at tasseling
DDAY = the number of accumulated degree days
DDT = number of degree days to tasseling
DDM = number of degree days to maturity

The conversion from dry matter produced (DRYM) to uptake of soil nitrogen (NH_4^+-N + NO_3^--N) is done according to the following scheme,

$$TNAPTK = NFRACT * DRYM \quad [51]$$

Days after emergence	(NFRACT) N fraction in dry matter
0-15	0.0221
16-30	0.0173
31-45	0.0150
46-maturity	0.0120

Accumulated degree days are used to estimate the dates of tasseling and maturity. The following equation is used to make the degree day calculations based on hourly (or 1/20th day) air temperatures.

$$DDAY = \Sigma\ (T-10)/20 \quad [52]$$

The summation is limited to air temperatures in the range,

$10°C < T < 35°C$

EVAPORATION SUBMODEL

Evaporation of water from the soil surface is computed as a function of tillage: the mixing and redistribution of water caused by the tillage tool, the increased radiant energy available for evaporation on roughened soil surfaces, the decreased radiant energy because of the reflectivity of residues, the increased water vapor transfer away from the soil caused by increased soil surface area, and the decreased water vapor transfer away from the soil because of a residue mulch and the water flux to the soil surface influenced by the soil hydraulic properties. Tillage-induced roughness, soil bulk density and surface residue coverage are the primary tillage inputs to the submodel.

Reflectivity of the soil is affected by the water content of the surface. Drier soils will reflect more radiation than moist soils. The effect of changing volume and therefore area (along the surface) fractions of water and air on the reflectivity of the soil surface can be written as:

$$a_w = \frac{0.06\theta + (0.5 - \theta)\ a_o}{0.5} \quad [53]$$

where a_w is the reflectivity of the soil, θ is the volumetric water content of the surface, and a_o is the reflection coefficient of dry soil. As θ increases a higher proportion of the surface is occupied by water which has a reflection coefficient of about 0.06. The reflectivity of soil a_w has limits of:

$$0.55a_o < a_w < a_o \quad [54]$$

The reflectivity of the soil a_w and the roughness index σ are used as inputs to determine the albedo of the soil surface as affected by radiation trapping. Albedo has not been reduced to a simple function of these parameters because of the effect of the altitude angle of the sun. The reflectivity of a dry smooth soil a_o is a necessary input parameter to the model.

The reflectivity of residues is affected by age and water content. The effect of age of the residues was calculated with an equation similar to one given by von Hoyningen-Huene, S. (1971). The reflectivity of the residue mulch a_R is given by:

$$a_R = a_o \left(1.06 + \left(\frac{a_{R_I}}{a_o} - 1.06\right) e^{-.0255t}\right) \quad [55]$$

where: a_o is the reflectivity of dry smooth soil, t is the residue age in days, and a_{R_I} is the reflectivity of the fresh residue mulch.

The reflectivity of the residues a_R is reduced to 75 percent of the value whenever the residues contain evaporable water (see later section on water retention and evaporation from crop residues). As time (age of residues) increases the residues approach the reflectivity of the soil material whereas fresh residues often have a reflectivity twice as high as soil material. The fraction of exposed surface B was calculated from a refitted exponential equation of data presented by Sloneker and Moldenhauer (1977) and given as:

$$B = e^{-fM} \quad [56]$$

where M is the dry weight of residues, and f is a term which depended on type of residues present. An overall reflection coefficient (a_s) was calculated from the relative proportion of the surfaces covered with residues (1-B) and uncovered (B) and the respective reflectivities. The equation is given as:

$$a_s = a_r\,(1-B) + a_w B \quad [57]$$

The total radiant energy available for evaporating water from the residue or soil surface is computed by considering the light transmitted through a crop canopy to the soil surface. A light transmission fraction F_T is computed from:

$$F_T = e^{-0.6L} \text{ for } L < L_L \quad [58]$$

or

$$F_T = e^{-0.384\,L_L} * e^{-0.216\,L} \text{ for } L > L_L \quad [59]$$

where:

$$L_L = 0.404 * \ln S + 1.49 \quad [60]$$

where S is the ground area occupied per plant and L is the leaf area index.

Equations 58, 59, and 60 are used to calculate F_T from leaf area index and plant spacing data. The light transmission fraction F_T times the total incoming solar radiation gives the radiation reaching the residue and soil surface.

An overall canopy albedo (a) is computed from weighted radiation fraction as:

$$a = a_s F_T + a_c (1-F_T) \quad [61]$$

where a_s is the surface albedo, a_c is the crop albedo, and a is the overall field albedo.

The net radiation available for evaporation at the top of the canopy would be given as:

$$R_N = (1-a)R \quad [62]$$

and the net radiation available for evaporation at the soil (or residue surface) would be given as:

$$R_N = (1 - a_s)\,F_T R \quad [63]$$

where R is the total incoming solar radiation, R_N is the net radiation, a and a_s are the weighted mean albedos, and F_T is the light transmission fraction.

Energy limited evaporation rate is described by a modified combination equation. The combination equation was presented by Van Bavel (1966) and modified by Van Doren and Allmaras (1978) and is expressed as:

$$PE = \left[\frac{\Delta}{\gamma}(R_N - H - G + R_{N_L}) + V * D_T * C\right] / \left[\frac{\Delta}{\gamma} + 1\right] \quad [64]$$

where Δ is the slope of vapor pressure vs. temperature curve, γ is the psychrometric constant, R_N is the net radiation, H is the energy used to heat air, G is the energy used to heat soil, R_{N_L} is the net long wave radiation, V is the vapor deficit, D_T is the aerodynamic transfer coefficient for water vapor, C is a combination of constants, and PE is the potential evporation rate.

The radiation-evaporation equation (Eqn. 64) is rearranged to give effective radiation R and is written as:

$$R = \frac{PE_p\,(1+\Delta/\gamma) - V*D_T*C}{\Delta/\gamma} * 0.93\,(1-a) \quad [64a]$$

where: PE_p is pan evaporation, a is the combined albedo of the field, 0.93 is a crop coefficient factor, and the remaining variables are as previously defined (Equation 64).

Soil and Air Heating

The energy consumed in heating soil and air (H and G) on a daily basis is assumed to be zero since no data was found to quantify these relationships. The assumption of zero sensible heat to air and soil is an approximation that is used extensively in estimating potential evaporation rate.

Net Long Wave Radiation

The net long wave radiation (R_{N_L}) is given by a calibration equation to daily maximum and minimum air temperatures by Ritchie (1972) and is written as:

$$R_{N_L} = (-2.61 + .00414 * T_L^{1.8})\frac{R}{R_c} \quad [65]$$

where:

$$T_L = \frac{Tmax + Tmin}{2} - 7.75 \quad [66]$$

and T_L is a calibration temperature, Tmax is the maximum daily air temperature, Tmin is the minimum daily air temperature, R is the measured incoming solar radiation and R_c is the maximum possible incoming solar radiation. Net long wave radiation is generally a small term on a daily average basis.

Delta

The slope of the vapor pressure vs. temperature curve Δ is determined from the average of the maximum and minimum temperature for the day by an equation presented by Ritchie (1972):

$$\Delta = (e^{21.255 - \frac{5304}{T_k}}) * \frac{5304}{T_k^2} \qquad [67]$$

where:

$$T_k = \frac{Tmax + Tmin}{2} + .273 \qquad [68]$$

Gamma

The psychrometric constant γ is determined from the temperature of condensation which is assumed to be at the minimum air temperature for the day by an equation taken from Ritchie (1972). The ambient air pressure is corrected for elevation before determining the psychrometric constant

$$\gamma = 6.6 \times 10^{-4} * P\ (1 + .00115 * Tmin) \qquad [69]$$

wherein P is the elevation corrected average atmospheric pressure and Tmin is the minimum daily air temperature.

Vapor deficit

The vapor deficit V was calculated from the algorithm given by Richards (1971) by assuming that the soil surface was at the assumed temperature of condensation (minimum air temperature) and the ambient air vapor pressure could be calculated from the daily mean air temperature (TMAX + TMIN)/2. Vapor pressures were obtained from the Richards (1971) algorithm. The vapor pressure deficit was calculated as the difference in the vapor pressure at these two temperatures.

Aerodynamic Transfer Coefficient

The aerodynamic transfer coefficient D_T was calculated from the daily wind travel, the logarithm ratio of the height measurement Z_A, an equivalent roughness height Z_o, and a length of path for diffusion transfer Z_D. The aerodynamic transfer coefficient is thus calculated from daily wind travel using:

$$D_T = \frac{U*C}{\left(\ln\left(\frac{Z_A+Z_c+Z_o}{Z_o+Z_D}\right)\right)^2 + \frac{.1681*U*Z_D}{D_A}} \qquad [70]$$

where:

$$C = \frac{\rho\varepsilon k^2}{P} = \frac{1.168X10^{-3} * .622*0.41^2}{1013}$$

$$= 1.2212 \times 10^{-7} \qquad [71]$$

and Z_A is the height of wind measurement, D_A is the diffusion coefficient for water vapor in still air, Z_o is a roughness parameter, U is the daily wind run, and Z_c and Z_D will be discussed more fully below.

The aerodynamic transfer of water vapor is made sensitive to tillage, residue, and crop canopy conditions by adjustments to Z_c and Z_o terms of equation 70. The height adjustment factor for the wind profile in the canopy Z_c was calculated as:

$$Z_c = 0.6H_c \qquad [72]$$

where: H_c is the height of the canopy.

Equation 72 assumes that the wind as measured at Z_A height could have been measured at $0.6*H_c + Z_A$ height over the canopy. Furthermore the wind profile is assumed to be identical so that the wind speed is zero at $0.4*H_c$ height and that vapor transfer below this height is by diffusion only so that:

$$Z_D = 0.4H_c \qquad [73]$$

The effective diffusion path length Z_D is increased to include the effects of the residue mulch when considering potential evaporation from the soil surface.

$$Z_D = 0.4*H_c + \frac{d*A*(B+A)}{1 + (A/4)} \qquad [74]$$

where:

$$A = \frac{.0127732*M}{d*\rho_R} \qquad [75]$$

and,

$$B = e^{-A} \qquad [76]$$

where: d is the diameter of residue pieces,
A is the total vertical cross section of residues,
B is the fraction of the surface not covered by residues,
M is the residue dry weight, and
ρ_R the density of residues.

TRANSPIRATION SUBMODEL

This submodel computes potential transpiration based on pan evaporation, leaf area, and root distribution. The potential values at each node are used by the unsaturated flow model in its macroscopic sink term. The actual predicted values for transpiration at each node depend on the soil hydraulic properties, the soil water contents, and the potential transpiration.

The expression used to calculate potential transpiration is written,

$$PT_i^J = RD_i^J * PAN_i * ALEAF_i * COEF \qquad [77]$$

where: PT is potential transpiration for the Jth layer.
RD is the fraction of roots in the Jth layer.
PAN is the pan evaporation for time step i.
ALEAF is the leaf area for time step i.
COEF is a crop coefficient.

Equation [77] takes into account the total energy available for evaporation, transpiration, and the size of the plant. Both the root distribution and the leaf area are dynamic properties which vary with time according to what is going on in the overall system.

The model extracts water down to the permanent wilting point (usually 15 bars suction) before water stress begins and actual predicted transpiration becomes less than the potential predicted value. The model currently makes no attempt to redistribute the water uptake pattern when a stress situation occurs in only part of the root zone. Unsaturated flow of water can occur, however, from wet to dry regions.

When water stress is occuring, a coefficient called RATIO1 becomes equal to a number in the range $1 > RATIO1 \geq 0$, otherwise RATIO1 is set equal to 1. RATIO1 is calculated by dividing the actual by the potential transpiration. RATIO1 is used in the crop growth and root growth submodels to limit dry matter production in water stress situations.

TILLAGE SUBMODEL

Tillage modifies the physical, biological, and chemical properties of the soil which, in turn, have an effect on crop yield. The manner in which the soil properties are modified in the model takes two basic forms. First, the user can specify the tillage date, depth, and resulting soil properties via appropriate input data; or secondly, the user specifies only the tillage date, depth, and type. Tillage submodels are then utilized to estimate the changes in existing soil properties. In either case, once the tillage modifications have been made, the NTRM model uses these data to compute crop yields.

Changes in the tilled system between tillage events are also considered. Again the user has the two options available to provide this information. He can enter as data input, base line conditions for the soil properties and a time period necessary to reach them after tillage, or he can allow the tillage submodel to compute these changes.

Table 1 contains a list of soil properties which the model directly changes or for which it indirectly accepts changes. The tillage submodels utilize regression and other equations to compute changes in soil bulk density, the soil water characteristic curve, material distributions, and surface properties as a function of existing values for soil texture, soil water content, soil organic matter content, and surface residues for various tillage practices. Computed soil bulk density and percent organic matter are then used to compute the updated soil water characteristic curve, soil water holding capacity, and soil strength. Carbon and nitrogen transformation rates are modified as a function of tillage depth.

Mixing of soil horizons is assumed to be either complete or partial depending on the tillage practice and soil conditions. The same holds true for incorporation of surface residues. The resulting horizons are assumed to be homogeneous within their own boundaries. This means that the tillage submodel may reconstruct the location of horizon boundaries after a tillage event. Soil segment boundaries used for internal computations are modified if the soil bulk density changes.

The effect of tillage on soil properties is illustrated in figure 8. In this example, certain residues are partially incorporated into the soil and two soil horizons are completely mixed. The resulting single horizon in the till zone is considered to be homogeneous. Some surface residues remain and the soil surface properties have been modified. The deeper soil horizons are assumed to be unaffected, although deep compaction could also be considered.

Table 1. Tillage associated soil properties modified in NTRM1 model.

Physical Properties	Biological Properties
Soil Bulk Density	Carbon & Nitrogen Transformation Rates
Soil Water Characteristic Curves	Microbial Mass & Distribution
Soil Water Holding Capacity	Root Mass & Distribution
Soil Infiltration Rate	**Chemical Properties**
	Organic Matter
Random Roughness	Residues NO_3-N & NH_4-N
% Residue Cover	Major Cations and Anions
Soil Strength	Cation Exchange Capacity
Soil Water Content (θ)	Solid Phase Salts
	Exchangeable Ions

NITROGEN AND CARBON TRANSFORMATION SUBMODEL

NCSOIL is the submodel of NTRM which describes the carbon and nitrogen transformations in soil. NCSOIL computes the changes in carbon and nitrogen organics, ammonium (exchangeable plus soluble), and nitrate concentrations which result from the processes of residue decomposition, mineralization, immobilization, nitrification, denitrification, and non-symbiotic N fixation. The rates are soil water and temperature dependent. A mechanism is included to account for the increase in the rate of mineralization following a physical disturbance (tillage) of the soil. NCSOIL,

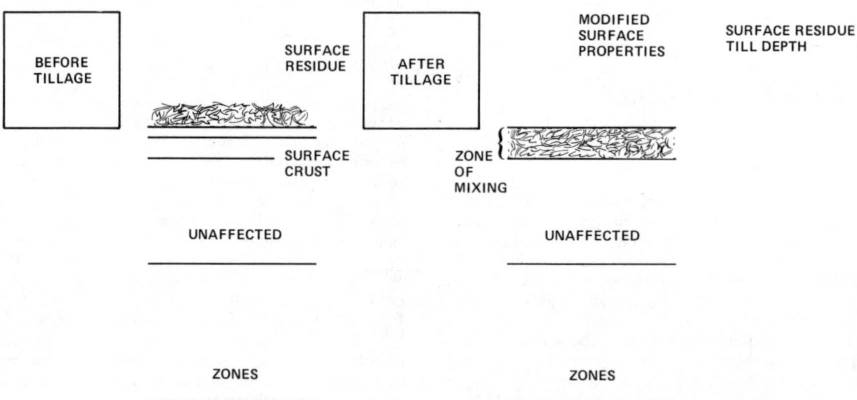

Figure 8. Modelling Tillage Effects on Soil Properties

which was devised for short term N and C dynamics, is a theoretical model (Weigert, 1975) built on the concept of catenary sequence of heterogenous substrates (Van Veen et al., 1981). The soil organic substrates are defined by their kinetic constants and their position in the model's structure. Residues are defined in reference to chemical or morphological entities.

To satisfy NTRM's specifications, input variables and parameters internal to NCSOIL were kept to a minimum, compatible with the necessity to maintain within NCSOIL a mechanistic structure to compute a variable efficiency factor and total as well as isotopic N kinetic rates. Seven C and N pools were considered: the organic residues, pool I and pool II of the soil active organic phase, exchangeable plus soluble ammonium, nitrate, dinitrogen, and carbon dioxide. The passive pool of soil organics (stable humus) was assumed to have no influence on the dynamics of the system over the short period, a growing season, considered for each run of the model. In the following, a description of NCSOIL is presented. Calibration and validation of NCSOIL can be found in other publications (Molina et al., 1982a and 1982b).

The active organic phase (Jansson, 1958) was divided into 2 substrates, pool I and pool II. Pool I serves as a sink for the organic decomposition products of the residues, and for a fraction of the decomposition products of pool I and pool II (figures 9, 10, and 11). Pool II is the recipient of that organic fraction of pool I which is not recycled. Only one pool I is considered; no distinction is made among the decomposition constants of the various pools which may arise from the decomposition of various substrates.

Pool I and pool II represent an operational simplification of the soil active organic phase. Yet analogies, if not in absolute correspondence, between the dynamic substrate pools of NCSOIL and the static, chemical, and morphological entities of the active organic phase are important to delineate for any tentative comparison of the model's decomposition constants to those found in the literature and accruing to chemical or morphological entities. Pool I and pool II, in view of their position in the model's structure, can be viewed as the soil biomass and the humads, terms which are sufficiently vague to suit the analogy and convey a qualitative comprehension of the model. The recycling of a fraction of pool I can be viewed as an expression of microbial succession.

Residues, pool I and pool II were handled as if two components, one labile and the other resistant, were decomposing exponentially and independently from each other (Hunt, 1977). Thus the potential rate of decomposition of a substrate is given by:

$$\left(\frac{dc}{dt}\right) \text{potential} = \mu\ [Sk\ C + (1-S)\ h\ C], \qquad [78]$$

where C is the carbon concentration of the substrate at the time t; S and (1-S) are the time invariant

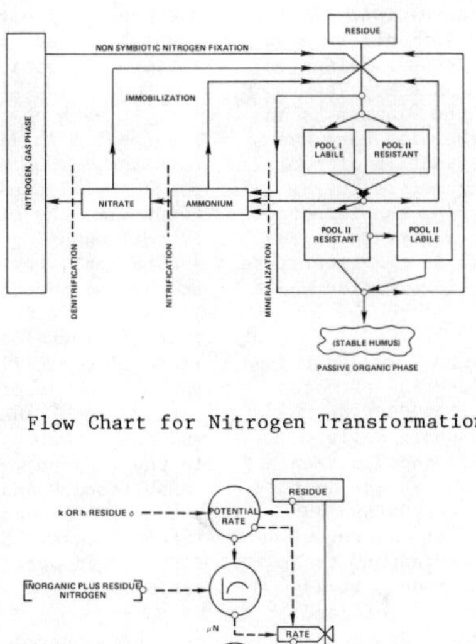

Figure 9. Flow Chart for Nitrogen Transformations

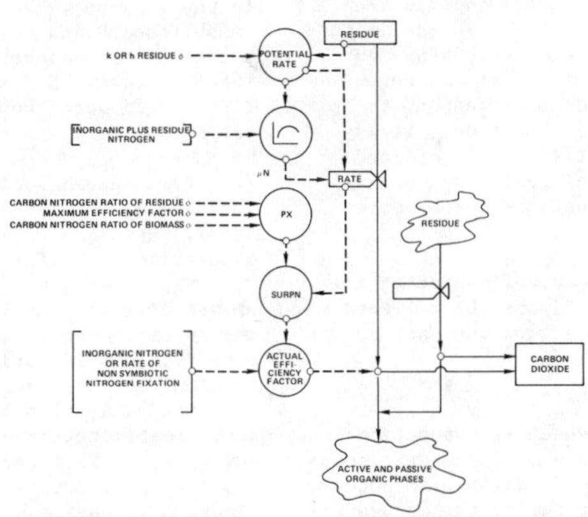

Figure 10. Relational diagram for the decomposition of a carbon residue

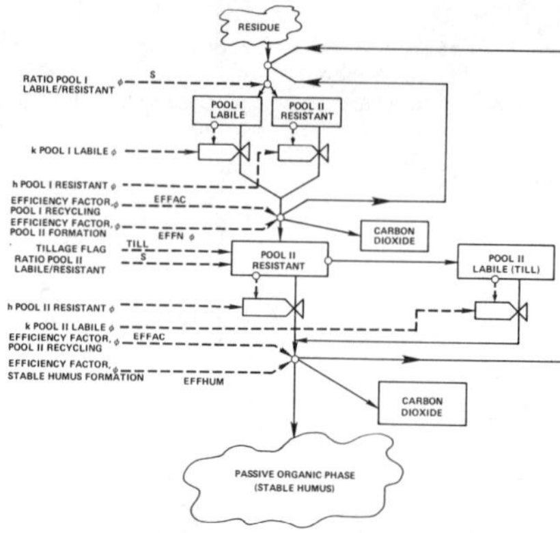

Figure 11. Relational Diagram for Carbon Pool I and Pool II Decomposition.

proportions of the labile and resistant components of the substrate; k and h are the specific rates of decomposition in each component; and μ is the lowest of the temperature or water reduction factors.

The concentrations of the labile and resistant components are defined by equation [78] and do not have any material support. Those concentrations reflect a dynamic property of the residues, pool I and pool II, characterized by kinetics of CO_2 or NH_4^+ released during decomposition experiments.

Rate equation [78] expresses a zero-order kinetics with respect to the active biomass. The influence of the degradative agent concentration is thus relegated to the decomposition constants, k, h, and S. The exclusion from the rate equation of the biomass as an independent variable (which could otherwise have been expressed by a constant or variable fraction of pool I) facilitated its integration. The parameters (k, h, and S) could then be computed by fitting the integral form against the numerous data about substrate degradation, which are presently available in the literature, and which usually do not include any information about the biomass.

Values for the constants k, h, and S were obtained for pool I by fitting the integral function of rate equation [78] against the average carbon dioxide released from a variety of C^{14}-labeled whole cells (Nelson et al., 1979), corrected (Paul and Van Veen, 1978) for a 28% C incorporation. Stanford and Smith's potentially mineralizable nitrogen N_o (Stanford and Smith, 1972) is a dynamic concept and the nitrogen content of pool I plus pool II does correspond to N_o. As a basis for the calibration of the model, it was assumed that, in the data reported by Stanford and Smith (1972), the contribution of pool I was negligible, and the decomposition constants for N_o were assigned to pool II.

The difference between the actual and potential rates of residue decomposition (figure 10) reflects a reduction in the rate of decomposition when not enough N is available. Therefore,

$$\left(\frac{dc}{dt}\right) \text{ actual} = \mu_N \left(\frac{dc}{dt}\right) \text{ potential}; \qquad [79]$$

where μ_N is a reduction factor which is a function of the ratio of the potential rate of C residue decomposition to the total amount of N available (N potentially released from the residue during the decomposition, plus the inorganic N available). The variable PX (figure 10) is identical to the one introduced by Beek and Frissel (1973) to determine whether residue decomposition will result in N mineralization or immobilization. In a situation of nitrogen immobilization (PX<0), the surplus N which has to be obtained from the inorganic pool or atmospheric N is computed as SURPN. This amount is a function of the maximum efficiency factor (EFFAC), which is reduced to an actual efficiency factor to balance SURPN with the inorganic N actually available, or the rate of non-symbiotic N fixation when the inorganic pool is depleted. In NCSOIL, the efficiency factor associated with residue decomposition is variable. Thus, through μn and the variable efficiency factor, the pool of inorganic nitrogen serves as a link for a double feedback loop in the carbon flow between each residue and pool I.

At each computational time step, the C recycled into pool I is divided into its labile and resistant portions. In contrast, the fraction (EFFNO) of decomposed pool I feeds directly into the resistant pool II. Every time the soil is physically disturbed (tillage or soil sampling and preparation for laboratory incubation), the tillage input flag (TILL) is raised, and a time invariant fraction of the resistant pool II is transferred to the labile pool II; thus, the rate of mineralization is increased. Since the C:N ratio of pool II is chosen equal to the C:N ratio of pool I, the decomposition of pool I and of pool II results in N mineralization (Beek and Frissel, 1973), and their corresponding efficiency factors are set to the maximum value EFFAC. It also follows that immobilized N can only merge with the flow of organics between the residues and pool I (figures 9, 11). The total inorganic N demand due to immobilization is distributed over the ammonium and nitrate pools in proportion to their respective concentrations.

For a given soil temperature, and water content, the rate of nitrification is constant (Sabey et al., 1969). When the water content of the soil has reached saturation, nitrification is replaced by the process of denitrification with a rate proportional to the amount of energy available, as measured by the sum of the rates of decomposition of C residues, pool I, and pool II. Nitrification and denitrification can also be simultaneously activated for water content lower than the saturation level. Boundary conditions arise when immobilization competes with nitrification for ammonium, or with denitrification for nitrate. The model assumes dominance of immobilization over nitrification for ammonium uptake. When the initial nitrate supply is not sufficient to satisfy both denitrification and immobilization, the C flow rates are recomputed with a nitrate level reduced by the nitrate demand for denitrification. Following this correction, precedence is given to immobilization over denitrification. Non-symbiotic N fixation is activated only when both immobilization calls for ammonium and nitrate and the inorganic N pool is low. The rate of fixation is proportional to the amount of energy available, as measured by the potential rate of C residues decomposition.

Rates of tagged N transformations are computed by multiplying the rates of the corresponding total N transformations by the ratio of the tagged N to total concentrations of the substrates.

ALTERNATE SUBMODEL FOR NITROGEN TRANSFORMATION

A second nitrogen submodel (TRNSFM) is contained in NTRM1. This submodel simulated nitrification immobilization-mineralization of organic-N hydrolysis of urea, and immobilization of nitrate-N. The general forms and associated coefficients for the rate equations are given in Dutt et al. (1972a) and Shaffer and Dutt (1973). The derivations of these relationships have been discussed previously (Shaffer et al., 1969; Dutt et al., 1972a; Shaffer and Dutt, 1973). The nitrification process is simulated using either the regression equation for nitrification or the transition state approach (Shaffer and Dutt, 1973). The nitrogen transformation submodel computes the mass changes for each nitrogenous species (urea-N, NO_3-N, NH_4-N, and organic-N) for a given time step and depth increment in the unsaturated zone. These changes are utilized along with changes due to convective transport and ion exchange (NH_4-N) to compute the net change in each N species with time and depth.

Denitrification is considered at the interface between the unsaturated and saturated zones. The actual denitrification equations are contained in the saturation chemistry program.

This submodel contains less detail in the areas of mineralization-immobilization relative to NCSOIL. Also, N^{15} as a tracer is not considered. However, TRNSFM has the option of using a transition state equation for nitrification and its excution time on the computer is about 0.1 that of NCSOIL.

OVERALL MODEL VALIDATION AND VERIFICATION

Development of the NTRM1 model over time is illustrated in figure 12. Note that progress on the model was rapid during 1980 and 1981 with a base having been previously established during the 1960's and 1970's. This progress included both verficiation and validation, activities on the individual submodels (see appropriate submodel chapters) and the overall integrated model.

With complex models such as NTRM1, absolute verifi- and validation cation is generally only a theoretical possibility which occurs at infinite time in the future. There is always a finite probability that the model still contains "bugs" or that the underlying concepts are inadequate or inappropriate.

The purpose of continued verification and validation activities is to minimize the risks to the extent which is practical. This usually means (based on experience with other models) that the model developers will "debug" and verify their models to the limits of their budgets or interest and that model users will identify and correct (often assisted by the developers) any remaining basic model deficiencies. Experience has shown that user feedback is one of the most effective and efficient ways to make progress with model verification and validation once a reasonable level of confidence has been reached.

The present state of the NTRM1 model is approximately that shown by the 5/82 arrow in figure 12. Some additional verification and validation work will be done by the NTRM development group. The user community will furnish the remainder. A staff will be maintained to assist the users with any model problems which may develop. The details of the model verification and validation efforts completed for the overall model and some submodels are available in Volume V of this series.

The following summary is presented to give the reader an appreciation for overall model verification and validation efforts. Model verification in terms of code checking and testing and internal mass balance checks has been done at each stage of submodel development and after incorporation of each submodel into the overall NTRM system. Members of the research team have examined in detail portions of the model developed by other members. Errors have been corrected as soon as they were identified. Model sensitivity runs have been done extensively to determine if simulated output appears reasonable over the expected

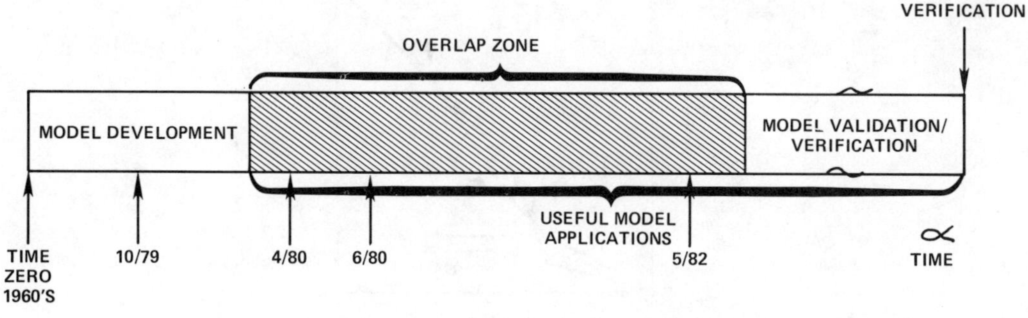

Figure 12. NTRM Model Development and Verification

range of input parameters. Graphical output has been used to a real advantage in these analyses.

Field testing of the overall integrated model has been a continuing process as additional data sets are analyzed. Corn yield results for four different locations, three in Minnesota and one in Ohio are presented in figure 13. Yield differences seen here are primarily the result of climatic differences as opposed to those induced by management.

Figures 14 and 15 contain crop dry matter production and crop nitrogen uptake curves, respectively, for predicted and observed corn data collected on a control site at Apple Valley, Minnesota. The model run indicated that nitrogen was not limiting and a relatively shallow root system would be expected due to ample nutrient availability and moist soil conditions at shallow depths. This was confirmed by root samples in the field.

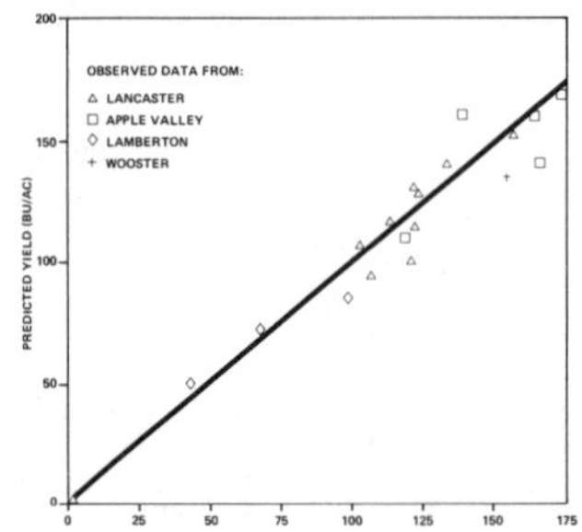

Figure 13. NTRM Model - Corn Yields

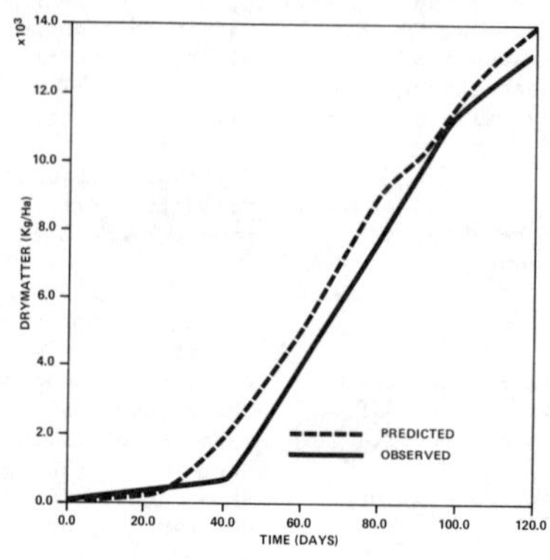

Figure 14. Dry matter production (corn) - Apple Valley 1979C.

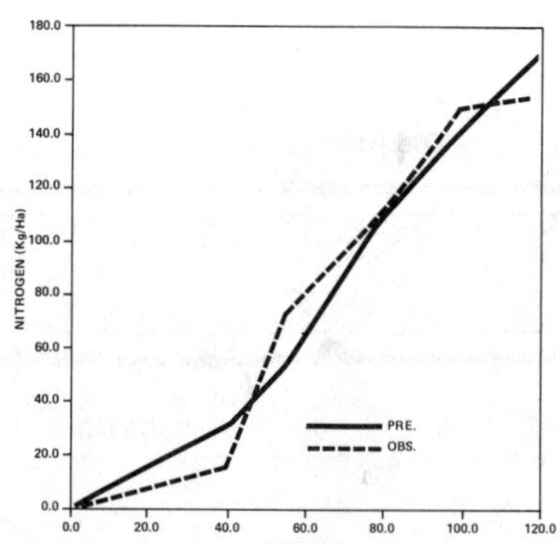

Figure 15. Nitrogen (tops) - corn - Apple Valley 1979C.

The submodels contained in NTRM1 have received individual verification and validation which are reported in the literature. Unsaturated flow submodel verification and validation are presented in Dutt et al. (1972), Shaffer et al. (1977), and Shaffer and Gupta (1981). Similar documentation for the soil temperature submodel can be found in Gupta et al. (1981) and Gupta et al. (1982). The soil chemistry and solute transport submodels are validated in Dutt et al. (1974), Shaffer et al. (1977), Wierenga et al. (1974), and Shaffer and Gupta (1981). Validation for the root growth submodel is presented in Shaffer et al. (1982). The crop growth submodel is validated in Childs et al. (1977), Tscheschke and Gilley (1980), and Shaffer (1982). Verification and validation for the evaporation submodel can be found in Linden (1979), Linden (1982), and Linden et al. (1982). Similar information for the transpiration submodel is given in Shaffer (1982). The nitrogen and carbon transformation submodel is verified in Molina et al. (1982a) and Molina et al. (1982b).

APPLICATION

Application of the NTRM1 model to simulate the effects of tillage on soil fertility is illustrated in figure 16. Moldboard plow, chisel plow, and no-till tillage practices were applied to a silt loam soil from southeastern Minnesota. In each case, corn surface residues in the amount of 2.5 metric tons/ha were contained on the soil surface prior to tillage. Before tillage NH_4SO_4 in the amount of 150 kg/ha was applied to the soil surface. Note the differences in available soil nitrogen for the modeled treatments.

The effect of using different degrees of residue incorporation with moldboard plow on soil-available nitrogen is shown in figure 17. The residue in each case was corn. Nitrogen fertilizer (150 kg/ha N) was added to all of the treatments. The residue was either partially or fully removed from the field to obtain less than complete incorporation. Note the increased soil inorganic N immobilized due to the residues incorporated.

Figure 18 contains the results for three levels of added fertilizer nitrogen. In each case, nitrogen was in the form of anhydrous ammonia. Note the response of the inorganic-N levels to the developing corn crop. With the 0 kg/ha application, N was rapidly utilized by the crop, resulting in low residual levels. The 150 and 250 kg/ha fertilizations provided an excess of nitrogen over the crop requirements.

Use of the NTRM1 model to simulate the effects of a range of tillage, crop residue, and nitrogen fertilizer treatments on soil fertility has already

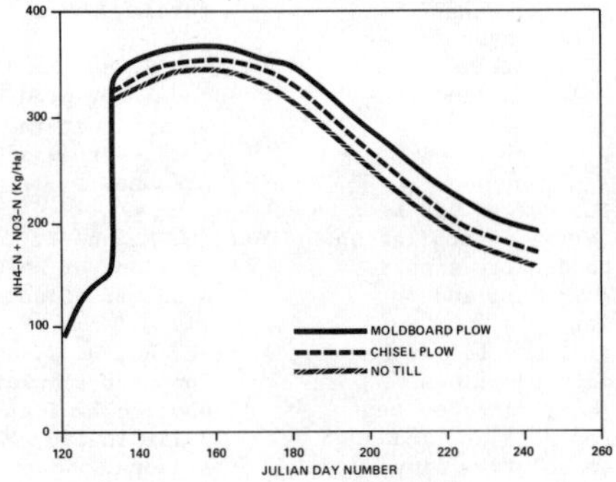

Figure 16. Simulated Effect of Three Tillage Treatments on Soil Inorganic N

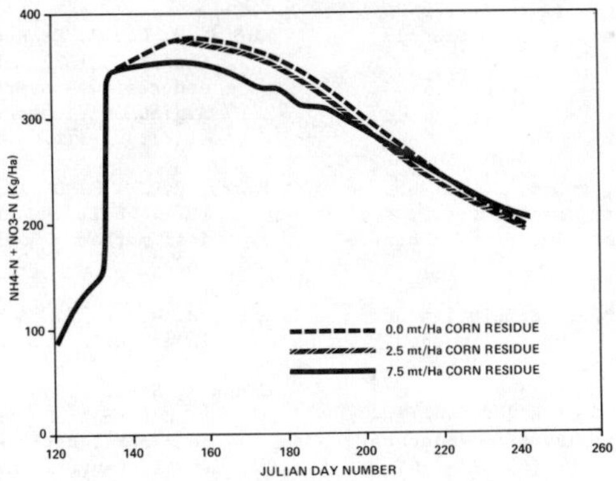

Figure 17. Simulated Effect of Three Residue Treatments on Soil Inorganic N

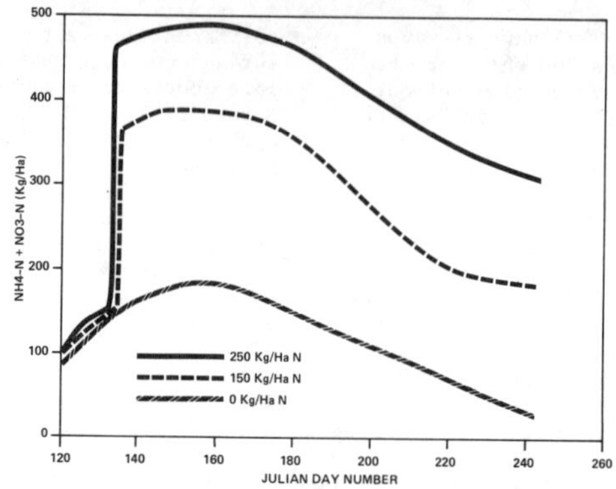

Figure 18. Simulated Effect of Three Nitrogen Fertilizer Rates on Soil Inorganic N

been demonstrated. These types of studies are generally at the research level, with scientists making use of the model's capabilities to help solve problems. Direct model use or indirect use in the development of related models such as NTRMS3 are examples of applications often made by scientists.

A model operating system has been developed which allows relatively easy operation of the program from remote time-share terminals. Access via TELENET or similar long-distance computer telephone systems, at a fixed hourly rate, is possible. This allows the user to take advantage of the more powerful and economical mainframes such as the CDC CYBER 74 and CRAY-1 computers.

Once a family of computer output curves has been generated for a particular problem area, interpolation methods and equations can be used to develop simplified relationships suitable for use in farm and extension level applications. For example, the computer results shown in figures 16, 17, and 18 can be readily developed into a set of simple equations which give the available inorganic soil nitrogen levels as a function of tillage, crop residue, and nitrogen fertilizer treatments. The solution time of these equations will be a small fraction of that used to develop the baseline curves. The only restriction is that the simplified equations should not be used to make extrapolations outside the range of the original computer results. Development of the USDA-ARS/University of Minnesota soil-crop-management system (SCMS) user model is an example of this type of application.

LITERATURE CITED

Allmaras, R. R., W. C. Burrows, and W. E. Larson. 1964. Early growth of corn as affected by soil temperature. Soil Sci. Soc. Am. Proc. 28:271-275.

Beek, J. and M. J. Frissel. 1973. Simulation of nitrogen behavior in soils. 67 p. Pudoc, Wageningen, The Netherlands.

Campbell, G. S. 1974. A simple method for determining unsaturated conductivity from moisture retention data. Soil Sci. 117(6):311-314.

Childs, S. W., J. R. Gilley, and W. E. Splinter. 1977. A simplified model of corn growth under moisture stress. Trans. ASAE 20(5):858-865.

Dutt, G. R. 1961. Quality of percolating waters, No. 1. Development of a computer program for calculating the ionic composition of percolating waters. Contribution No. 50, 35 p. Water Resources Center, Dept. of Irrigation, Univ. of California, Davis.

Dutt, G. R. and M. J. Shaffer. 1972. A computer model for predicting nitrate and other solutes of agricultural drain water. Final report to U. S. Bureau of Reclamation, 20 p. Univ. of Arizona.

Dutt, G. R. and K. Tanji. 1962. Predicting concentrations of solutes in water percolated through a column of soil. J. Geophys. Res. 64:3437-3439.

Dutt, G. R., M. J. Shaffer, and W. J. Moore. 1972. Computer simulation model of dynamic bio-physicochemical processes in soils. Technical Bulletin 196, 101 p. Agric. Exp. Stn., Univ. of Arizona.

Gupta, S. C., J. K. Radke, and W. E. Larson. 1981. Predicting temperature of bare and residue covered soils with and without a corn crop. Soil Sci. Soc. Am. J. 45(2):405-412.

Gupta, S. C., J. K. Radke, W. E. Larson, and M. J. Shaffer. 1982. Predicting temperatures of bare and residue-covered soils from daily maximum and minimum air temperatures. Soil Sci. Soc. Am. J. 46(2):372-376.

Hanks, R. J., D. D. Austin, and W. T. Ondrechen. 1971. Soil temperature estimation by a numerical method. Soil Sci. Soc. Am. Proc. 35:665-667.

Hunt, H. W. 1977. A simulation for decomposition in grasslands. Ecology. 58:469-484.

Jannson, S. L. 1958. Tracer studies on nitrogen transformations in soils with special attention to mineralization-immobilization relationships. Kungl. Lantbr. Ann. 24:101-151.

Linden, D. R. 1979. A model to predict soil water storage as affected by tillage practices. Ph.D. Dissertation. 278 p. Univ. of Minnesota.

Linden, D. R. 1982. Predicting tillage effects on evaporation from the soil. Chapter 8, pp. 117-131. In Predicting tillage effects on soil physical properties and processes. ASA, SSSA, Madison, Wisconsin.

Linden, D. R., O. M. VanDoren, J. K. Radke, and D. C. Reicosky. 1982. Tillage and surface residue sensitive evaporation submodel. 55 p. In W. E. Larson and M. J. Shaffer (eds.) Nitrogen-tillage-residue management (NTRM) model, Technical Documentation. Research Report, USDA-ARS, St. Paul, Minn.

Molina, J. A. E., C. E. Clapp, M. J. Shaffer, F. W. Chischester, and W. E. Larson. 1982a. Carbon and nitrogen transformations in soil, submodel NCSOIL. 35 p. In W. E. Larson and M. J. Shaffer (eds.) Nitrogen-tillage-residue management (NTRM) model, Technical Documentation. Research Report, USDA-ARS, St. Paul, Minn.

Molina, J. A. E., C. E. Clapp, M. J. Shaffer, F. W. Chichester, and W. E. Larson. 1982b. NCSOIL: A model of nitrogen and carbon transformations in soil. 20 p. (submitted).

Nielsen, K. F. 1974. Roots and root temperatures. pp. 293-333. In E. W. Carson (ed.) The plant root and its environment. University Press of Virginia, Charlottesville.

O'Connell, D. J. 1975. The measurement of apparent specific gravity of soils and its relationship to mechanical composition and plant root growth. pp. 298-313. In Technical Bulletin 29, Her Majesty's Stationery Office, London.

Paul, E. A. and J. A. Van Veen. 1978. The use of tracers to determine the dynamic nature of organic matter. Trans. 11th Intl. Congr. Soil Sci., Edmonton. 3:61-102.

Richards, J. M. 1971. Simple expression for the saturated vapor pressure of water in the range -50 to 140°C. Brit. J. Appl. Physics 4:L15-L18.

Ritchie, J. T. 1972. Model for predicting evaporation from a row crop with incomplete cover. Water Resour. Res. 8(5):1204-1213.

Russell, R. S. 1977. Plant root systems, their function and interaction with the soil. 298 p. McGraw-Hill Book Company (UK) Ltd., London.

Sabey, B. R., L. R. Frederick, and W. W. Bartholomew. 1969. The formation of nitrate from nitrogen in soils: IV. Use of the delay and maximum rate phases for making quantitative predictions. Soil Sci. Soc. Am. Proc. 33:276-278.

Shaffer, M. J., and G. R. Dutt. 1973. Nitrification in soil-water systems: A computerized activated complex model. pp. 284-294. Proceedings of the First World Congress on Water Resources. Chicago, Ill.

Shaffer, M. J., G. R. Dutt, and W. J. Moore. 1969. Predicting changes in nitrogenous compounds in soil-water systems. pp. 15-28. Collected Papers Regarding Nitrates in Agricultural Wastewater, Water Pollution Control Research, Series 13030 Ely 12/69.

Shaffer, M. J. and S. C. Gupta. 1981. Hydrosalinity models and field validation. Chapter 7, pp. 136-181. In I. K. Iskandar (ed.) Modeling wastewater renovation, John Wiley & Sons, New York.

Shaffer, M. J., R. W. Ribbens, and C. W. Huntley. 1977. Prediction of mineral quality of irrigation return flow. Volume V. Detailed return flow salinity and nutrient simulation model. 229 p. U. S. Environmental Protection Agency Report, EPA 600/2-77-179e, Corvallis, Oregon.

Shaffer, M. J. 1982. Transpiration submodel. 5 p. In W. E. Larson and M. J. Shaffer (eds.) Nitrogen-tillage-residue management (NTRM) model, Technical Documentation. Research Report, USDA-ARS, St. Paul, Minn.

Shaffer, M. J., C. E. Clapp, and D. C. Reicosky. 1982. Root growth submodel. 20 p. In W. E. Larson and M. J. Shaffer (eds.) Nitrogen-tillage-residue management (NTRM) model, Technical Documentation. Research Report, USDA-ARS, St. Paul, Minn.

Sloneker, L. L. and W. C. Moldenhauer. 1977. Measuring the amounts of crop residue remaining after tillage. J. Soil Water Conserv. 32:231-236.

Stanford, G. and S. J. Smith. 1972. Nitrogen mineralization potentials of soils. Soil Sci. Soc. Am. Proc. 36:465-472.

Taylor, H. M. 1974. Root behavior as affected by soil structure and strength. pp. 271-291. In E. W. Carson (ed.) The plant root and its environment. University Press of Virginia, Charlottesville.

Taylor, H. M. and H. R. Gardner. 1963. Penetration of cotton seedling taproots as influenced by bulk density, moisture content, and strength of soil. Soil Sci. 96:153-156.

Taylor, H. M., M. G. Huck, and B. Klepper. 1972. Root development in relation to soil physical conditions. pp. 57-77. In D. Hillel (ed.) Optimizing the soil physical environment toward greater crop yields. Academic Press, New York.

Tscheschke, P. D. and J. R. Gilley. 1980. Status and verification of Nebraska's corn growth model = CØRNGRØ. 32 p. Paper No. 5581, Journal Series, Nebraska Agric. Exp. Stn., Lincoln.

van Bavel, C. H. M. 1966. Potential evaporation: the combination concept and its experimental verification. Water Resour. Res. 2:455-467.

Van Doren, D. M. and R. R. Allmaras. 1978. Effect of residue management practices on the soil physical environment, micro-climate, and plant growth. pp. 49-84. Am. Soc. Agron. Spec. Publ. Crop residue management system. ASA, CSSA, SSSA, Madison, Wis.

van Hoyningen-Huene, S. 1971. Einfluss einer Strohdicke auf den Strahlungshausalt des Endbodens. (The influence of straw cover on the radiation balance of the soil.) Agr. Meteorol. 9:63-75. (English summary).

Van Veen, J. A., W. B. McGill, H. W. Hunt, M. J. Frissel, and C. V. Cole. 1981. Simulation models of the terrestrial nitrogen cycle. In F. E. Clark and T. Rosswell, (eds.) Terrestrial nitrogen cycles. Processes, ecosystem strategies and management impacts. Ecol. Bull. (Stockholm) 33:25-48.

Weigert, R. G. 1975. Simulation models of ecosystems. A. Rev. Ecol. Syst. 6:311-338.

Wierenga, P. J., M. J. Shaffer, S. P. Gomez, and G. A. O'Connor. 1975. Predicting ionic distributions in large soil columns. Soil Sci. Soc. Am. Proc. 39(6):1080-1084.

APPENDIX

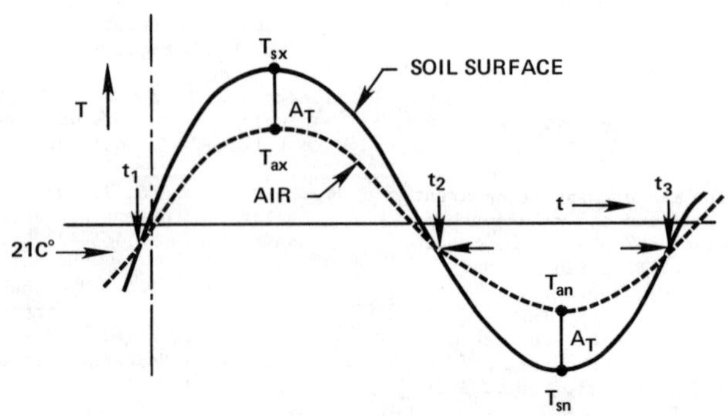

Appendix Figure 1. Upper Boundary Temperatures

MODELING THE EFFECTS OF AIR POLLUTANTS ON SOYBEAN YIELD[1]

David A. King[2], James R. Kercher[3], and Gail E. Bingham[3]

Abstract.--The National Crop Loss Assessment Network (NCLAN) is developing crop simulation models which incorporate the effects of pollutant-environment interactions for characterizing the effects of air pollutants on crop yield. The models are to be based on existing crop models and will be calibrated using the results of field chamber fumigation experiments from 6 sites located across the U.S. This paper describes the development of a model for estimating pollutant effects on soybean yield. Preliminary results demonstrate the usefulness of the model for assessing the influence of environment on the sensitivity of plants to pollutant damage.

INTRODUCTION

Recent field experiments suggest that current levels of air pollution cause significant losses to U.S. crops (Heagle *et al*. 1979). These losses are largely due to ozone (O_3) produced by photochemically induced reactions of nitrogen oxides and hydrocarbons (Mansfield 1976). A preliminary estimate of losses, based on open top field exposure chamber studies has suggested that, assuming ambient ozone levels of 0.06 ppm over the major crop-growing regions, losses might reach 3 billion in 1978 dollars (Heagle and Heck 1980).

To assess the impacts of pollutants on the U.S. economy via crop effects, EPA established the National Crop Loss Assessment Network (NCLAN). The principal goal of NCLAN is to characterize the agroeconomic effects of air pollutants as an aid to setting air pollution standards. This goal is to be accomplished by dividing the U.S. into economic regions and developing regionally averaged estimates of crop yield vs. seasonal pollutant dose for the important crops of each region. The regional dose-response estimates will then be fed into an economic model. To generate the regional dose-response estimates NCLAN is fumigating important crop plants in field chambers at six sites corresponding to the major crop growing regions of the U.S. A simple approach to regional averaging would be to use dose-response regressions from the field chamber experiments directly as regional estimates. (See for example Armentano (1980).) However, the effects of pollutants on crops are affected by both environmental conditions at the time of pollutant exposure and the previous growth environment of the plant (Dunning *et al*. 1974, Heck *et al*. 1977).

To apply the dose-response relationship generated at single sites to whole regions, the NCLAN site at Lawrence Livermore National Laboratory is developing crop growth models incorporating known environment-pollutant interactions. The models are to be based on existing crop models which already contain the environmental variables of light, temperature, etc. and their effects on plant mechanisms, e.g., stomatal opening. The models will be fit to physiological measurements and to NCLAN field experiments. Once validated models exist, regional dose response functions can be constructed by running the models repeatedly over the regional variation in environmental variables (Kercher, King, and Bingham 1982). This paper illustrates the modeling approach by presenting the present and planned development of the soybean model. We chose soybeans as the first crop for modeling because this crop is expected to show the greatest economic impact from air pollutants. Soybean is second only to corn in economic yield in the U.S. and is more sensitive to ozone than corn (Heck *et al*. 1982). The soybean model is based on a soybean growth simulator developed by Wilkerson *et al*. (1981), and the general crop simulator of de Wit *et al*. (1978). The latter will be used to simulate photosynthesis and pollutant uptake. Pollutant effects on photosynthesis will be computed in terms of cumulative dose to the leaf mesophyll and will be translated into yield changes with the soybean growth simulator.

MODELING PROCEDURE

A number of photosynthesis and plant growth simulation models have been formulated over the past two decades as an aid in relating overall plant

---

[1] Paper presented at the Third International Conference on State-of-the-Art in Ecological Modelling, Colorado State University, May 24-28, 1982. Research funded by the U.S. Environmental Protection Agency.

[2] David A. King is an ecologist employed by the Department of Agronomy and Range Science of UC Davis, working with the Environmental Sciences Division of Lawrence Livermore National Laboratory, Livermore, CA, U.S.A.

[3] James R. Kercher and Gail E. Bingham are environmental scientists of the Environmental Sciences Division, Lawrence Livermore National Laboratory, Livermore, CA, U.S.A.

growth to individual physiological processes (de Wit et al. 1978, Thornley 1976). Models have also been designed for crop management, e.g. scheduling irrigation (Swaney et al., 1981). These models divide the plant into a number of compartments and use empirically based differential or difference equations to compute plant development through time. The equations governing compartment size changes vary greatly in their physiological detail, depending on how well a particular process is understood and on the level of resolution required.

Simulation models have recently been used to assess air pollutant affects on crop growth (Kercher 1978) and on forests (West et al. 1980, Kercher and Axelrod 1981). Because gaseous pollutants enter the plant primarily through the stomata along with CO_2 (Bennett et al. 1973), the appropriate time-scale to model pollutant uptake is that which follows stomatal resistance changes (Kercher 1980). Figure 1 summarizes the general approach which we are using to model pollutant effects. The plant has an internal physiology which is subject to environmental conditions (light, temperature, etc.). Pollutants enter the plants through the stomata and ultimately affect the internal physiology. The factors which control stomatal opening (e.g., light, water stress, and even pollution history) ultimately control the internal dose which the plant experiences. It is known that the internal dose affects both short-term (Bennett and Hill 1973) and long-term (Coyne and Bingham 1981, Kercher 1978) photosynthesis as well as allocation of photosynthate between plant parts (Bennett and Oshima 1976). In the latter case, it is still unresolved whether this is due to alternation in allocation patterns which occur by virtue of reduced available photosynthate or due to the intervention of pollutants in the allocation mechanism. Our approach will be to first model the effects of pollution on photosynthesis and compare the results of these simulations with experimental data before modeling effects of pollutants on allocation.

To minimize model development time and incorporate agricultural research on soybeans we have evaluated several soybean simulation models developed by crop modeling groups in the U.S. We chose CROP, developed by Wilkerson et al. (1981), as the basic soybean simulator which we'll modify to compute pollution effects. Because CROP calculates photosynthesis in a very simplified manner and on a daily time step we

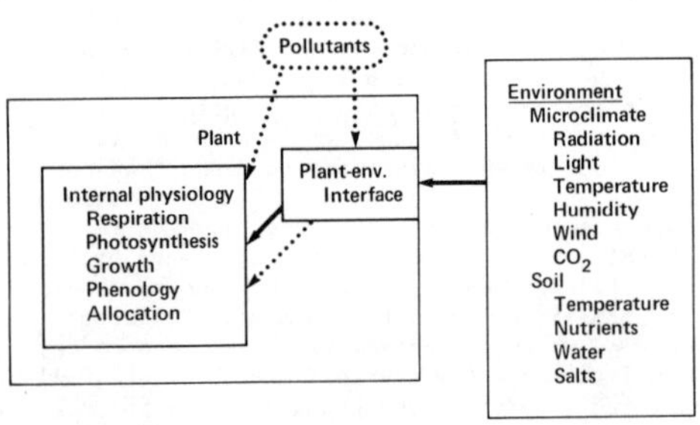

Figure 1.--Conceptualization of plant response to pollutants, as mediated by the environment. Plant-environmental interface in the figure refers to the stomata and air-mesophyll interface.

shall use BACROS of de Wit et al. (1978) to calculate photosynthesis and leaf conductance on an hourly basis. Daily photosynthesis and pollutant uptake calculated by BACROS will then be input into CROP which will in turn simulate plant growth and development. A pollution subroutine will be added to CROP which will modify the photosynthetic characteristics of leaves as a function of cumulative ozone uptake. In this fashion, pollutant uptake over the season can be translated into changes in harvest yield. The two models and the pollution subroutine will be adjusted to fit harvest yield, plant part weight, and photosynthetic data from the NCLAN field sites.

THE SOYBEAN GROWTH MODEL

An ideal model to serve as a base for assessing pollutant-caused yield reductions would have enough physiological realism to mimic crop response for stresses that in the extreme approximate pertinent attributes of pollution-stress. In addition, the model would require a minimum of varietal and environmental inputs besides daily weather. After contacting a number of soybean modelers and running two soybean models developed elsewhere, we concluded that no released model meets the above criteria for soybeans. We chose CROP of Wilkerson et al. (1981), pursuing the alternative strategy of using a relatively simple, empirically based simulator. CROP allocates photosynthate between plant parts in a conservative fashion which can be easily specified, based on field measurements. Changes in allocation or phenology caused by pollutants or other stresses must be included explicitly in the model parameters. However, the effects of pollutants on yield via different plant processes can be clearly assessed. For example, if no changes are made in the parameters controlling allocation, the proportioning of photosynthate between plant parts will vary little over a wide range of environments (fig. 2). Associated changes in yield are then due to changes in photosynthetic rate.

Wilkerson et al. (1981) describe CROP in detail and only pertinent aspects of the model will be mentioned here. Figure 3 diagrams the compartments and flows simulated by CROP, which is written in FORTRAN. Photosynthate, which is input to CROP from BACROS, is allocated between the compartments by a different algorithm during each of 6 different phenological stages. Partitioning between the vegetative compartments (leaves, stems and roots) depends only on phenological stage, i.e. on time. Allocation between reproductive parts (seeds and pod shells) depends on the sizes of these compartments, temperature, and available photosynthate. The timing of the phenological stages is given in days from planting and is a model input to the current version. The authors of CROP are now developing a subroutine which will compute phenological stages in terms of photoperiod and physiological time, for running the model for different locations and planting dates.

During the pod filling stage seed nitrogen requirements are met by the translocation of protein from pod shells, leaves and stems. Protein translocation is restricted such that no compartment is completely drained of labile nitrogen before the final phenological stage is reached. Additional nitrogen requirements are met by reduction of NO_3^-.

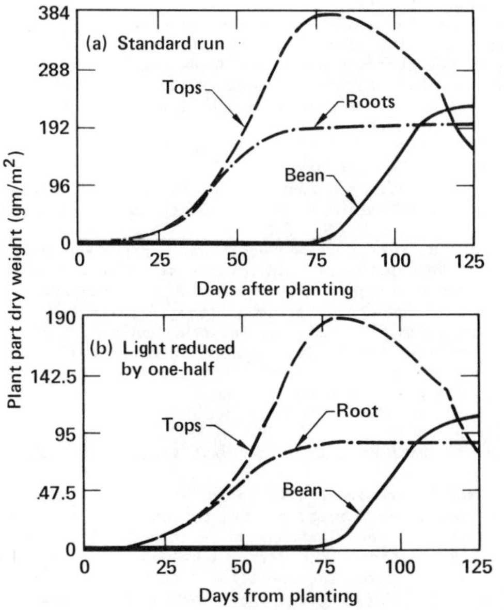

Figure 2.--Soybean simulations using CROP (1978 meteorological data from Gainesville, FL). Soybean variety is Bragg. Simulations show effect of reduced light on model response. (a) Normal light, (b) 50% normal light. Note the change in vertical scale.

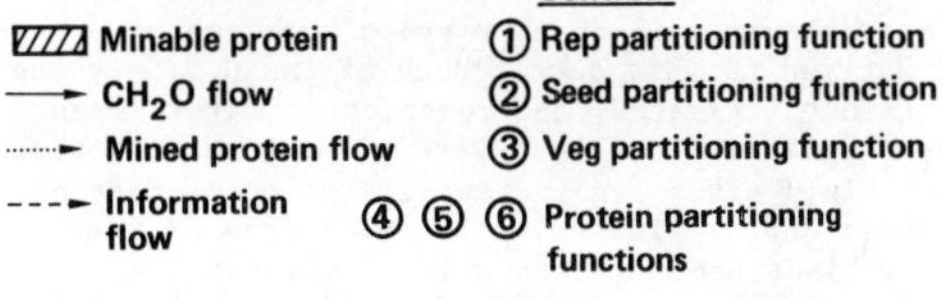

Figure 3.--Flow diagram for the soybean simulator CROP.

Model parameters which may vary with variety and location are listed in Table 1, and specifiy phenology, vegetative allocation and reproductive sink strengths.

Table 1.--CROP model parameters which may vary with location and variety.

1. Timing of phenological stages
2. Vegetative partitioning functions
3. Leaf weight per unit area
4. Ave. # seeds/shell
5. Protein and oil fractions of seed
6. Temperature dependence of potential seed and shell growth
7. Number of days a shell may grow
8. Photosynthetic dependence on temperature, etc.

THE PHOTOSYNTHESIS MODEL

Pollutant flux to the leaf mesophyll depends on stomatal conductance (Bennett *et al.* 1973). Because stomatal conductance and external pollutant concentrations can vary independently of each other, conductance must be modeled on a short term basis to compute internal pollutant dose. BACROS is appropriate for this task as it computes energy and water balances and leaf light distribution. Conductance and photosynthesis is then computed for each layer of the canopy. BACROS can be used to simulate photosynthesis for any crop by giving it the appropriate leaf angle distribution and photosynthetic response functions. De Wit *et al.* (1978) describe BACROS and list a CSMP version of the model. We are using a FORTRAN version translated by K.R. Christian (personal communication).

BACROS models stomatal conductance by assuming it to be controlled by CO_2 requirements for photosynthesis, so long as plant water content is above a certain value. Plant water content is computed as a balance between transpiration and root uptake. Root uptake is then a product of root conductance and the difference between plant and soil water potentials. Root conductance in turn depends on the suberized and nonsuberized root weight. When plant water content falls below a certain value, conductance is decreased linearly with falling water content. BACROS assumes that water content is at equilibrium between uptake and transpiration and uses an iterative routine to calculate this balance. Soil water potential and root weights are presently computed by other routines in BACROS. However, we will use CROP to simulate plant part weights and soil water potential. BACROS will be used to compute conductance and photosynthesis for a range of root weights and soil water potentials to provide input for CROP.

Simulation Procedures

Because the soybean simulator, CROP and the photosynthesis simulator, BACROS require different data inputs and run on different time steps we envision running the models separately. BACROS will be run repeatedly for each day's weather and pollutant inputs for a range of LAI's, root conductivities, soil water potentials and photosynthetic capacities. CROP will then use the tables of photosynthesis and pollutant uptake generated by BACROS to simulate crop development. The same tables can be used to run CROP for a variety of planting dates and soil types for a given weather and pollutant data set.

POLLUTANT SUBROUTINE

Ozone damages isolated enzymes, enzyme systems, cell membranes and chloroplasts with an associated reduction in photosynthesis (Heck et al. 1977). Repair of injury is associated with an increase in respiration. Coyne and Bingham (1982) found that both quantum or light harvesting efficiency and maximum photosynthesis at high light levels decrease together for sensitive ponderosa pine relative to insensitive ones in forests exposed to pollution from the Los Angeles Basin. Biochemical models of photosynthesis are often written with an overall factor multiplying the term for gross photosynthesis (Hall and Bjorkman 1975, Kercher 1978, Murphy et al., 1974). This factor is proportional to the product of viable chloroplast, enzyme, and substrate concentrations. Our initial hypothesis is that the chronic effect of ozone on gross photosynthesis can be empirically modeled as a multiplicative factor which is a product of the remaining viable chloroplast, enzyme, and substrate concentrations. This hypothesis is consistent with the observation of Coyne and Bingham cited above. The cumulative effect on photosynthesis due to each factor will depend on the turnover time for each factor. Thus for systems with rapid turnover times, such as enzymes, changes in their level would equilibrate in a short time. However for systems such as organelles with very long turnover times, the effect of small chronic doses could accumulate to be large (F.W.T. Penning de Vries, private communication).

Kercher (1978) has suggested a simple model for the change in photosynthetic capacity F, caused by a pollutant acting on that component with turnover time d^{-1}.

$$\frac{dF}{dt} = B(I,E) - (d+I)F \qquad (1)$$

where
- F = factor multiplying leaf photosynthesis
- I = internal concentration of phytoactive agent
- B = regeneration rate of F
- E = available energy for regeneration.

The function B and rate d are hard to measure. For small turnover and repair rates and chronic fumigations, a simple approximate solution to equation 1 is given by

$$F = 1 - a \int_0^t I dt \qquad (2)$$

where
- a = constant determined from photosynthesis measurements.
- I = internal concentration of phytoactive agent.

Assuming that internal loss of toxic agent is proportional to its concentration and the agent is replenished by diffusion into the leaf, then equation 2 may be rewritten

$$F = 1 - a' \int_0^t \frac{[O_3(t)]}{r(t) + b} dt \qquad (3)$$

where
- r = sum of resistance to O_3 diffusion
- $[O_3]$ = external ozone concentration (time dependent)
- a';b = constants determined from photosynthesis measurements.

It should be stressed that equation 3 was derived to estimate the cumulative effects of pollutants as opposed to the effects of severe episodes where cell death results in a nonlinear dose-response relationship (Larsen and Heck 1976). If repair mechanisms or alternative modes of loss of toxic agent are invoked, then equation 3 could be replaced by one with a threshold. Initial results of NCLAN field experiments are consistent with a model which reduces photosynthesis by a factor proportional to cummulative ozone concentration above a small threshold value.

The pollution damage equation is based on several tentative hypotheses. Environmental interactions with pollutants are only included via their influence on stomatal conductance. The ability of cells to denature damaging radicals and the sensitivity of the chloroplast itself may depend on internal cell environment and preconditioning. In such cases, a' may depend on environment.

PARAMETER ADJUSTMENT AND VALIDATION

Model parameters will be initially adjusted to fit simulation results to the soybean field chamber measurements for plants grown in charcoal filtered air. Successive plant part weight measurements from the site at Boyce Thompson Institute (BTI) will be used to fix the allocation of photosynthate between plant parts by CROP. Phenological observations from all soybean growing sites (e.g. first flowering date) will be used to adjust the phenological subroutine of CROP. Photosynthesis measurements on field chamber plants from BTI and harvested weights from all soybean sites will be used to set the parameters of BACROS.

The pollution subroutine will be adjusted to fit photosynthetic data taken at BTI from fumigated soybeans. If an adequate base of photosynthesis data is gathered, the pollution subroutine can be validated using measured yield depressions versus dose from the field sites. Otherwise, the field data will serve as an additional guide to setting the pollutant response. In either case the model can be validated by using chamber yield data from future years.

It would be very useful to determine the range of environmental conditions for which the adjusted model gives accurate results. This goal might be achieved by using data from BTI to adjust all parameters and compare simulated and measured yields for the other sites. However, all chamber grown soybeans have been well watered to this date. Much of the U.S. soybean crop is not irrigated, and reaction to water stress affects pollutant uptake. Varying soil water content at the field sites might provide a critical test of the model and is planned. The robustness of the model will determine the number of alternative experimental treatments needed to adjust the model for a regional assessment.

OZONE UPTAKE: PRELIMINARY RESULTS

As an initial analysis of pollutant-environment interactions, we used the gas exchange model (BACROS) to compute ozone uptake under a variety of weather conditions assuming typical hourly ozone concentrations. To accomplish this goal we first set the input parameters of BACROS to appropriate values for soybeans. Further adjustments were made so that the growth simulator CROP gave similar results when run with BACROS input as with its own photosynthesis subroutine for a trial simulation. The important parameters which we adjusted were the saturation rate of photosynthesis (P_{max}), internal CO_2 concentration, a function limiting stomatal conductance at low leaf water potentials, and root conductivity.

BACROS computes CO_2 assimilation as a multiplicative function of factors related to light intensity and temperature for the case of no water stress. Stomatal conductance is then calculated from assimilation by assuming that the internal CO_2 concentration at the mesophyll-air boundary is constant, independent of light and temperature. This assumption is appropriate for corn (Zea mays) and bean (Phaseolus vulgaris) according to de Wit et al., (1978). We assumed unstressed soybeans to maintain an internal CO_2 concentration of 210 ppm for the preliminary uptake calculation. The simultaneous measurements of assimilation and conductance required to estimate internal CO_2 concentration for soybeans are being taken by the Livermore NCLAN site.

To compute the effect of water stress on stomatal behavior BACROS computes stomatal conductance based on relative leaf water content (θ_1) in addition to the conductance based on assimilation, discussed above. The lower of the two conductances is then chosen as the modeled conductance. We assumed that the water limited conductance was 1.43 cm/sec for $\theta_1 \geq .90$, declining linearly to 0.01 cm/sec as θ_1 declines from .90 to .80. This relationship is consistent with soybean stomatal conductances measured by Mederski et al. (1975). BACROS also assumes leaf water potential (ψ_1) to be a function of θ_1, with $\psi_1 = -14.0$ when $\theta_1 = .80$. BACROS thus determines θ_1 and water limited conductance from the vapor pressure deficit (VPD), soil water potential (ψ_s) and plant resistance to water flow. Plant water resistance is assumed to occur entirely as a root resistance and VPD is computed by a micrometeorological subroutine. We adjusted the root resistance to initiate stomatal closing at high VPD's for well watered plants consistent with conductance measurements by Brady et al. (1975). Running CROP with BACROS we were then able to approximate the bean production of Wilkerson et al. (1981) for an input data set including some water stress during the reproductive stage.

To calculate ozone uptake we used the mean hourly concentrations given by Heagle et al. (1979) for April 9 - May 31, 1977, at Raleigh, N.C. (fig. 4). We assumed that the ozone concentration dropped to one-tenth of the external value at the mesophyll-air interface within the leaf. Ozone uptake and conductance measured on soybean by Tingey and Taylor (1981) suggest that the ozone concentration drops to a small fraction of the ambient level at the mesophyll surface, for ozone concentrations typically encountered in the field. Ozone uptake for a typical clear summer day at Raleigh, N.C. is shown in figure 5 for a soil at field capacity. The diminished uptake at lower canopy levels reflects light limitations of stomatal conductance. Ozone uptake for $\psi_s = -5$ bars shows some midday depression in the upper canopy due to the water stress limitation of conductance (fig. 6). Midday uptake depression occurs throughout the upper two-thirds of the canopy, which has a total LAI of 5, when $\psi_s = -10$ bars (fig. 7). The coalescing of the upper and mid-canopy uptake curves indicates that water stress completely determines conductance at these levels at midday. Ozone uptake is higher at 5:00 PM than 9:00 AM because the input ozone concentration is 2 ce as high at the former time (fig. 4). Ozone uptake and CO_2 assimilation are contrasted for the upper canopy layer under severe water stress in figure 8. Photosynthesis is greatly reduced at midday by a high VPD with a secondary late afternoon increase due to declining temperatures. (Temperature is assumed to peak at 2:00 PM.) Because the VPD is higher than average during the time of high ozone concentrations, water stress reduces daily ozone uptake by a greater fraction than it reduces CO_2 assimilation, as shown in figure 9.

The influence of daily insolation on ozone uptake is shown in figure 10. Uptake increases linearly with insolation with no water stress because BACROS assumes that light conditions alternate between cloudy and full sun for intermediate light levels. At a given time photosynthesis is a linear interpolation between the values for cloudy and sunny conditions. Under severe water stress, ψ_s controls uptake over a wide range of insolation and temperature. Figure 11 illustrates the temperature dependance of ozone uptake for a fixed relationship between T_{max}, T_{min} and T_{dewpt}. At high temperatures and full sunlight uptake is reduced even for well-watered plants because the root resistance was adjusted to produce this response under high VPD.

The above preliminary results show how environmental conditions can mediate pollutant uptake and presumably crop yield losses. The apparent sensitivity of crops to ozone is likely to vary between seasons and localities if there is temporal and spatial variability in the correlation between ozone concentration and VPD. The analysis suggests

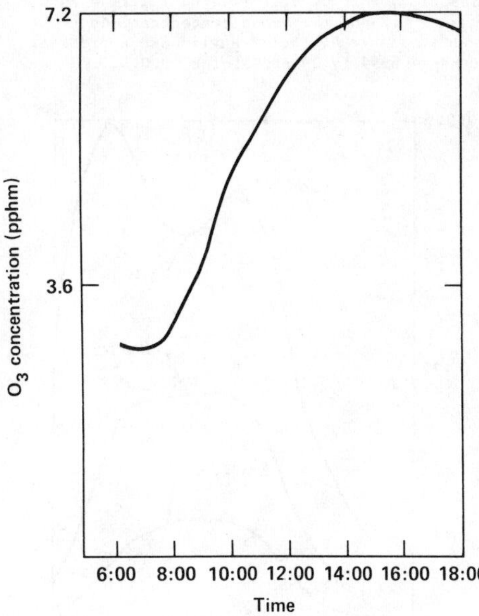

Figure 4.--Mean hourly ozone concentrations during daytime from April 9 through May 31, 1977 near Raleigh N.C. After Heagle, Riordan, and Heck (1979). These concentrations are used to compute ozone uptake in the following figures.

that important model inputs are soil water conditions root resistance and stomatal behavior. Since no data base gives soil water time series for soybeans across the U.S., we will model soil water based on a water subroutine in a recent version of CROP.

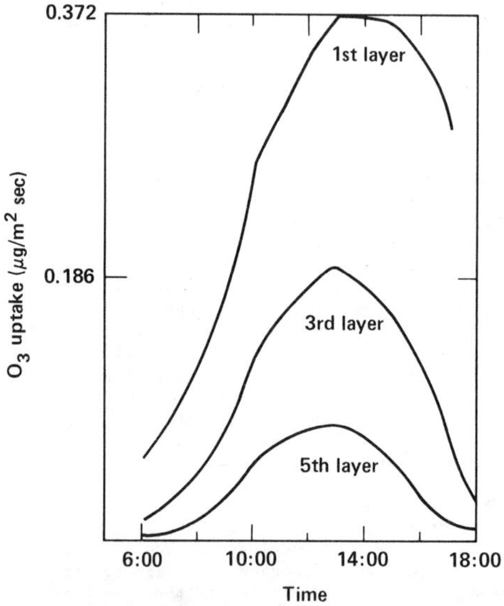

Figure 5.--Computed ozone uptake for LAI layers 0-1 (top layer), 2-3, and 4-5. Uptake is for clear skies, T_{min} = 20°C, T_{max} = 32°C, T_{dewpt} = 20°C, ψ_{soil} = -0.2 and the ozone concentrations given in figure 4. Solar angles are appropriate for May 4 or July 29 at Raleigh, N.C.

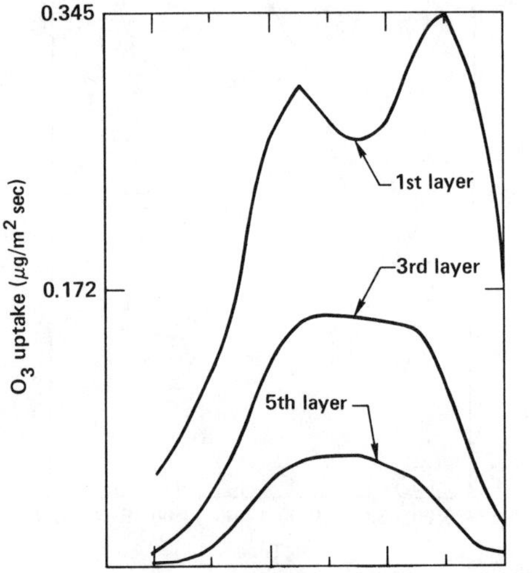

Figure 6.--Computed ozone uptake for ψ_{soil} = -5.0 and the same weather conditions as figure 5. Note vertical scale change.

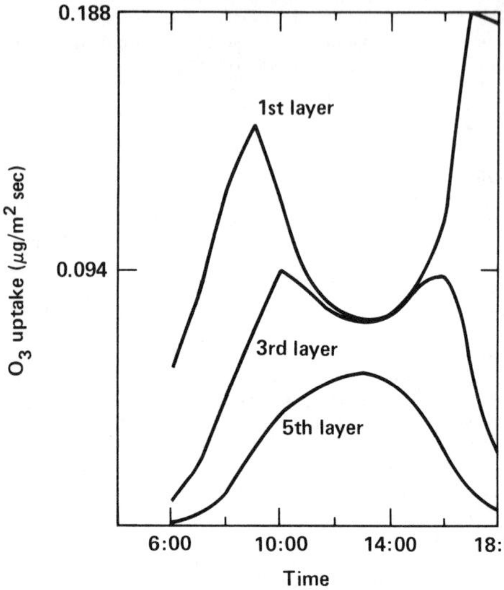

Figure 7.--Computed ozone uptake for ψ_{soil} = -10.0 and the same weather conditions as figure 5. Note vertical scale change.

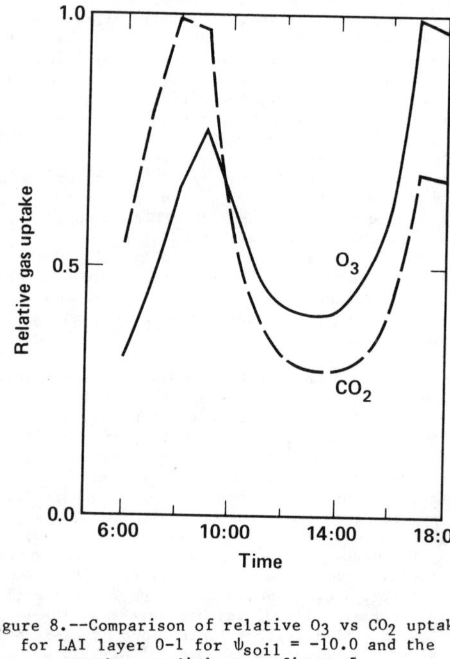

Figure 8.--Comparison of relative O_3 vs CO_2 uptake for LAI layer 0-1 for ψ_{soil} = -10.0 and the same weather conditions as figure 5.

CONCLUSIONS

We have outlined the construction of a model to simulate the effects of pollutants on soybean growth and yield. The effort is motivated by NCLAN's need to estimate regional crop losses in order to assess the

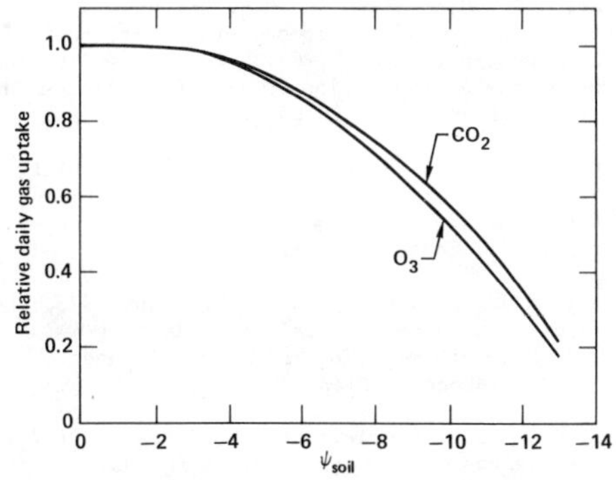

Figure 9.--Comparison of relative daily O_3 vs CO_2 uptake for the whole canopy. Weather conditions are those of figure 5.

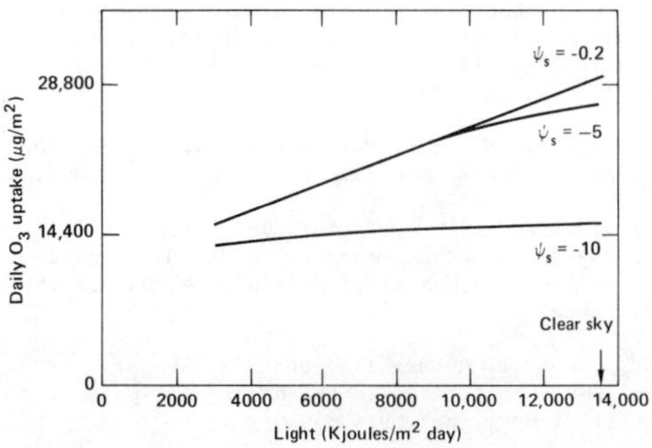

Figure 10.--Canopy O_3 uptake as a function of temperature. $T_{dewpt} = T_{min} = T_{max} - 12°C$.

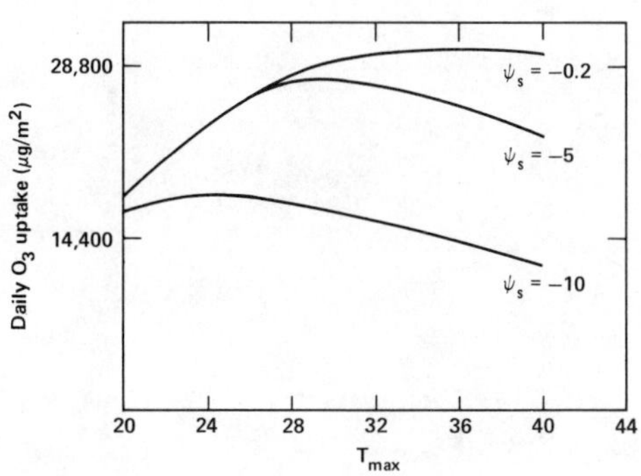

Figure 11.--Canopy O_3 uptake as a function of daily insolation. Other weather conditions are those of figure 5.

economic impact of air pollutants via agriculture. NCLAN is acquiring data bases including county crop acerages, yields, and soil properties, and weather and pollutant concentration data bases on a somewhat larger scale. Using these data bases, the model can be run repeatedly to estimate soybean yield losses for each cluster of counties with similar conditions and thus provide a regionallly averaged soybean loss estimate (Kercher, King and Bingham 1982). Once the soybean model is in place for assessment of ozone effects, we intend to produce similar models for the crops following soybeans in order of pollutant-caused economic impacts.

The modeling effort is also useful for suggesting the experimental treatments needed to better evaluate pollutant environment interactions. The preliminary results indicate the need for experiments to check the effect of water stress on pollutant caused yield depressions. Presently, all field chamber-grown soybeans are well watered, but most of the U.S. soybean crop is unirrigated. We anticipate that the model will serve as a tool to guide basic research on chronic pollutant effects, as well as playing a central role in assessment.

LITERATURE CITED

Armentano, T.V. 1980. Calculation of yield-loss coefficients for major crops in the Ohio River Basin. p. 77-120. In Crop and forest losses due to current and projected emissions from coal-field power plants in the Ohio River Basin. Office of Research and Development EPA, Washington, D.C. 266 p.

Bennett, J.H., and A.C. Hill. 1973. Inhibition of apparent photosynthesis by air pollutants. J. Env. Qual. 2:526-530.

Bennett, J.H., A.C. Hill, and D.M. Gates. 1973. A model for gaseous pollutant sorption by leaves. APCA Journal 23:957-962.

Bennett, J.P. and R.J. Oshima. 1976. Carrot injury and yield response to ozone. J. Amer. Soc. Hort. Sci. 101:638-639.

Brady, R.A., S.M. Goltz, W.L. Powers, and E.T. Kanemasu. 1975. Relation of soil water potential to stomatal resistance of soybean. Agron. Journal 67:97-99.

Coyne, P.I., and G.E. Bingham. 1981. Comparative ozone dose response of gas exchange in a ponderosa pine stand exposed to long-term fumigations. J. APCA 31:38-41.

Coyne, P.I., and G.E. Bingham. 1982. Variation in photosynthesis and stomatal conductance in an ozone-stressed ponderosa pine stand: light response. For. Sci. (In press).

de Wit, C.T., J. Goudriaan, H.H. van Laar, F.W.T. Penning de Vries, R. Rabbinge, H. van Keulen, W. Louwerse, L. Sibma, and C. deJoge. 1978. Simulation of assimilation, respiration, and transpiration of crops. Wageningen, Centre for Agricultural Publishing and Documentation. 141 p.

Dunning, J.A., W.W. Heck, and D.T. Tingey. 1974. Foliar sensitivity of pinto bean and soybean to ozone as affected by temperature, potassium nutrition and ozone dose. Water Air Soil Pollut. 3:305-313.

Hall, A.E., and O. Bjorkman. 1975. Model of leaf photosynthesis and respiration. p. 55-72. In: Perspectives of Biophysical Ecology (D. M. Gates and R. B. Schmerl, eds.) Springer-Verlag, NY 609 p.

Heagle, A.S., and W.W. Heck. 1980. Field methods to assess crop losses due to oxidant air pollutants. p. 296-305. In Crop Loss Assessment (P. S. Teng, S. V. Krupo, eds) Misc. Publ. 7-1980, Agricultural Experiment Station, University of Minnesota.

Heagle, A.S., A.J. Riordan, W.W. Heck. 1979. Field methods to assess the impact of air pollutants on crop yields. Paper 79-46.6. In: Proc. 72nd Annual Meeting of the Air Pollution Control Association, June 24-29, 1979.

Heck, W.W., J.B. Mudd, and P.R. Miller. 1977. Plants and microorganisms. p. 437-585. In Ozone and Other Photochemical Oxidants. National Academy of Sciences, Washington, D.C. 719 p.

Heck, W.W., O.C. Taylor, R. Adams, G. Bingham, J. Miller, E. Preston, and L. Weinstein. 1982. Assessment of crop loss from ozone. J. Air Pollut. Control Assoc. 32:353-361.

Kercher, J.R. 1977. GROW1: a crop growth model for asessing impacts of gaseous pollutants from goethermal technologies. UCRL-52247, Lawrence Livermore National Laboratory, Livermore, CA, 33 p.

Kercher, J.R. 1978. A model of leaf photosynthesis and the effects of simple gaseous sulfur compounds (H_2S and SO_2). UCRL-52643, Lawrence Livermore National Laboratory, Livermore, CA, 37 p.

Kercher, J.R. 1980. Developing realistic crop loss models for air pollutant stress. p. 90-97. In Crop Loss Assessment (P.S. Teng and S.V. Krupa, eds.). Misc. Publ. 7-1980, Agricultural Exp. Sta., University of Minnesota. 327 p.

Kercher, J.R. and M. C. Axelrod. 1981. SILVA: a model for forecasting the effects of SO_2 pollution on growth and succession in a western coniferous forest. UCRL-53109, Lawrence Livermore National Laboratory, Livermore, CA, 72 p.

Kercher, J.R., D.A. King, and G.E. Bingham. 1982. Approaches for modeling crop-pollutant interactions in the NCLAN program. Paper 82-69.3. In: Proc. 75th Annual Meeting of the Air Pollution Control Association New Orleans, Louisiana, June 20-25, 1982.

Larsen, R.I., and W. W. Heck. 1976. An air quality data analysis system for interrelating effects, stands, and needed source reductions: Part 3. Vegetation injury. J. Air. Pollut. Control Assoc. 26:325-333.

Mederski, H.J., R.B. Curry, and L.H. Chen. 1975. Effects of irradiance and leaf water deficit on net carbon dioxide assimilation and mesophyll and transport resistances. Plant Physiol. 55:594-597.

Murphy, C.E., T.R. Sinclair, and K.R. Knoerr. 1974. Modeling the photosynthesis of plant stands. p. 125-147. In Handbook of Vegetation Sciences Part VI; Vegetation and Environment (B.R. Strain and W.D. Billings, eds.) Dr. W. Junk B.V., The Hague, Netherlands.

Swaney, D.P., J.W. Jones, W.G. Boggess, G.G. Wilkerson, and J.W. Mishoe. 1981. A crop simulation method for evaluating within-season irrigation decisions. Paper 81-4015, Proc. 1981 Summer Meeting Am. Soc. Ag. Eng. Orlando, FL, June 21-24, 1981.

Thornley, J.H.M. 1976. Mathematical models in plant physiology. 318 p. Academic Press, New York, N.Y.

Tingey, D.T., and G.E. Taylor, Jr. Variation in plant response to ozone: A conceptual model of physiological events. For 32nd School of Agriculture, University of Nottingham. In manuscript.

West, D.C. S.B. McLaughlin, and H.H. Shugart. 1980. Simulated forest response to chronic air pollution stress. J. Env. Qual. 9:43-49.

Wilkerson, G.G., J.W. Jones, K.J. Boote, K.T. Ingram, and J.W. Mishoe. 1981. Modeling soybean growth for crop management, Paper 81-4014, Proc. 1981 Summer Meeting Am. Soc. Ag. Eng. Orlando, FL, June 21-24, 1981.

EPIC - A MODEL FOR ASSESSING THE EFFECTS OF EROSION ON SOIL PRODUCTIVITY[1]

J. R. Williams, P. T. Dyke, and C. A. Jones[2]

INTRODUCTION

Accurate estimates of future soil productivity are essential in agricultural decision making and planning from the field scale to the national level. Soil erosion depletes soil productivity, but the relationship between erosion and productivity is not well defined. Until the relationship is adequately developed, selecting management strategies to maximize long-term crop production will be impossible.

The Soil and Water Resources Conservation Act (RCA) requires a report by 1985 that establishes the current status of soil and water resources in the U.S. One important aspect of these resources is the effect of erosion on long-term soil productivity. The National Soil Erosion-Soil Productivity Research Planning Committee documented what is known about the problem, identified what additional knowledge is needed, and outlined a research approach for solving the problem (Williams et al., 1981). One of the most urgent and important needs outlined in the research approach was the development of a mathematical model for simulating erosion, crop production, and related processes. This model will be used to determine the relationship between erosion and productivity for the U.S. Thus, a national ARS erosion-productivity modeling team[3] was organized and began developing the model during 1981. The model called EPIC (Erosion-Productivity Impact Calculator) is composed of physically based components for simulating erosion, plant growth, and related processes and economic components for assessing the cost of erosion, determining optimal management strategies, etc.

EPIC simulates the physical processes involved simultaneously and realistically using readily available inputs. Commonly used EPIC input data (weather, crop, tillage, and soil parameters) are available from a computer filing system assembled especially for applying EPIC throughout the U.S. Since erosion can be a relatively slow process, EPIC is capable of simulating hundreds of years if necessary. EPIC is generally applicable, computationally efficient (operates on a daily time step), and capable of computing the effects of management changes on outputs. The model must be comprehensive to define the erosion-productivity relationship adequately.

The components of EPIC can be placed into eight major divisions for discussion--hydrology, weather, erosion, nutrients, plant growth, soil temperature, tillage, and economics. The hydrology model simulates surface runoff volume and peak discharge rate given daily rainfall amounts. Other hydrology components include evapotranspiration, percolation, lateral subsurface flow, drainage, irrigation, and snow melt. Four options are provided for inputing weather information into EPIC. The first three deal with precipitation, air temperature, and solar radiation (all three variables can be read into the program; precipitation can be read in and temperature and radiation simulated; or all three variables can be simulated). Snowfall is estimated as a function of precipitation and air temperature. The fourth option, wind (velocity and direction), is simulated if wind erosion is considered. Wind erosion is predicted using the Manhattan, Kansas (Woodruff and Siddoway, 1965) wind erosion model modified to operate with a daily time step. Water erosion is simulated with the Onstad and Foster (1975) modification to the Universal Soil Loss Equation (Wischmeier and Smith, 1978). The two plant nutrients considered in EPIC are nitrogen and phosphorus. Nitrogen processes simulated include runoff of NO_3, organic N transport by sediment, leaching, upward NO_3 movement by soil evaporation, denitrification, immobilization, mineralization, crop uptake, rainfall contribution, fertilizer, and fixation. Phosphorus processes simulated include runoff of soluble P, sediment transport of mineral and organic P, immobilization, mineralization, sorption-desorption, crop uptake, and fertilizer. A general plant growth model is used to simulate above-ground biomass, yield, and roots for corn, grain sorghum, wheat, barley, oats, peanuts, sunflowers, soybeans, alfalfa, cotton, and grasses. The plant growth model simulates energy interception; energy conversion to roots, above-ground biomass, and grain and fiber production; and water and nutrient uptake. Plant growth is constrained by water, nutrient, and air temperature stresses. Soil temperature is simulated to serve the nutrient cycling and root growth components of EPIC. Soil temperature is predicted at the center of each soil layer as a function of the previous day's soil temperature and the present day's air temperature and solar radiation. The EPIC tillage model simulates row height, surface roughness, and mixing of soil layers, nutrients, and plant residue for any tillage operation. The economics component of EPIC uses a crop budget to calculate crop production costs. Income is determined from simulated annual crop yields. Net profit (income minus cost) is subject to change as the soil erodes away.

Outputs from EPIC (crop inputs, costs, erosion rates, etc.) will be input to the CARD (Center for Agricultural and Rural Development) linear programming model (Meister and Nicol, 1975) to accomplish the national RCA analysis. The EPIC model components are described here. Also, results of limited testing are presented.

---

[1]Paper presented at the Third International Conference on State-of-the-Art in Ecological Modelling, Colorado State University, May 24-28, 1982.

[2]Hydraulic Enginer, USDA-ARS; Economist, USDA-ERS; and Plant Physiologist, USDA-ARS, Temple, TX 76503.

[3]J.R. Williams, M.J. Shaffer, K.G. Renard, G.R. Foster, M.N. Laflen, L. Lyles, C.A. Onstad, A.N. Sharpley, A.D. Nicks, C.A. Jones, C.W. Richardson, P.T. Dyke, K.R. Cooley, S.J. Smith

MODEL DESCRIPTION

The EPIC model includes eight major components: hydrology, weather, erosion, nutrients, plant growth, soil temperature, tillage, and economics. The computer program contains 49 subroutines, although there are only 2350 FORTRAN statements. Since EPIC operates on a daily time step, computer cost for overnight turn around are only about $0.15 per year of simulation on an AMDAHL 470 computer. The model can be run on a variety of computers since storage requirements are only 180 K.

The drainage area considered by EPIC is generally small (∿ 1 ha) because soils and management are assumed to be spatially homogeneous. In the vertical direction, however, the model is capable of working with any variation in soil properties--the soil profile is divided into a maximum of ten layers (the top layer thickness is set at 10 mm and all other layers may have variable thickness). When erosion occurs, the layer boundaries are moved down a distance equal to the eroded thickness and soil properties within each layer are adjusted by interpolation.

Descriptions of the EPIC components and the mathematic relationships used to simulate the processes involved follow.

Hydrology

Surface Runoff

Runoff Volume.--Surface runoff is predicted for daily rainfall using the SCS curve number (USDA, Soil Conservation Service, 1972) equation

$$Q = \frac{(R - 0.2s)^2}{R + 0.8s} \tag{1}$$

where Q is the daily runoff, R is the daily rainfall, and s is a retention parameter. The retention parameter s is related to soil water content with the equation

$$s = s_{mx}\left(\frac{UL - SW}{UL}\right) \tag{2}$$

where SW is the soil water content in the root zone minus the 15-bar water content, UL is upper limit of soil water storage in the root zone (porosity minus 15-bar water content), and s_{mx} is the maximum value of s. The maximum value of s is estimated with the curve number (CN) for the I moisture condition using the SCS equation (USDA, Soil Conservation Service, 1972)

$$s_{mx} = 254\left(\frac{100}{CN_I} - 1\right) \tag{3}$$

where CN_I is the I moisture condition (dry) CN. The constant value, 254, in Eq. 3 gives s_{mx} in mm. Thus, s, UL, SW, R, and Q are also expressed in mm. An estimate of the II moisture condition (average) CN can be obtained easily for any area using the SCS Hydrology Handbook (USDA, Soil Conservation Service, 1972). The corresponding CN_I values are also tabulated. For computing purposes CN_I was related to CN_{II} with the polynomial

$$CN_I = -16.91 + 1.348(CN_{II}) - 0.01379(CN_{II})^2 + 0.0001177(CN_{II})^3. \tag{4}$$

If soil water is distributed uniformly in the soil profile, Eq. 2 should give a good estimate of the retention parameter and thus the runoff. However, if the soil water content is greater near the surface, Eq. 2 would tend to give low runoff predictions. Conversely, runoff would be overpredicted if the soil water content was greater in the lower root zone. To account for the soil water distribution, a weighting technique was developed. The depth-weighted retention parameter is computed with the equation

$$s = s_{mx}\left(1. - \sum_{i=1}^{M} w_i \frac{SW_i}{UL_i}\right) \tag{5}$$

where w_i is the weighting factor, SW_i is the water content in mm, UL_i is the upper limit of water storage in soil layer i in mm, and M is the number of soil layers. The weighting factors decrease with depth according to the equation

$$w_i = 1.016\left(\exp\left(-4.16\frac{D_{i-1}}{RZ}\right) - \exp\left(-4.16\frac{D_i}{RZ}\right)\right) \tag{6}$$

where D_i is the depth to the bottom of layer i in mm and RZ is the root zone depth in mm. Eq. 6 assures that $\sum_{i=1}^{M} w_i = 1.$

The procedure described for simulating surface runoff is similar to the CREAMS runoff model--option one (Knisel, 1980), (Williams and Nicks, 1982). The only difference is that EPIC accommodates variable soil layers and CREAMS does not. In addition to the CREAMS surface runoff component, EPIC includes a provision for estimating runoff from frozen soil. If the temperature in the second soil layer is less than zero°C, the curve number is assigned a value of 98 regardless of the soil water content.

Peak Runoff Rate.--Peak runoff rate predictions are based on a modification of the Rational Formula.

$$q_p = (\rho)\ (r)\ (A) \tag{7}$$

where q_p is the peak runoff rate in ft^3/s, ρ is a runoff coefficient expressing the watershed infiltration characteristics, r is the rainfall intensity in in/h for the watershed's time of concentration, and A is the drainage area in acres. The runoff coefficient can be calculated for each storm if the amount of rainfall and runoff are known.

$$\rho = \frac{Q}{R} \tag{8}$$

Since R is input and Q is computed by EPIC, ρ can be calculated directly. Rainfall intensity can be expressed with the relationship

$$r = \frac{R_{t_c}}{t_c} \tag{9}$$

where r is the rainfall intensity in mm/h, R_{t_c} is the amount of rainfall in mm during the watershed's time of concentration, t_c in h. The value of R_{t_c} can be estimated by developing a relationship with total R. The Weather Service TP-40 (Hershfield, 1961) provides accumulated rainfall amounts for various durations and frequencies. Generally R_{t_c} and R_{24} (24-h duration is appropriate because EPIC uses daily rainfall) are proportional for various frequencies. Thus,

$$R_{t_c} = \alpha R_{24} \quad (10)$$

where α is a dimensionless parameter that expresses the proportion of total rainfall that occurs during t_c.

The peak runoff equation is obtained by substituting Eqs. 8, 9, and 10 into Eq. 7 and converting to metric units.

$$q_p = \frac{(\alpha)\ (Q)\ (DA)}{360\ (t_c)} \quad (11)$$

where DA is the drainage area in ha.

The time of concentration can be estimated by adding the surface and channel flow times.

$$t_c = t_{cc} + t_{cs} \quad (12)$$

where t_{cc} is the time of concentration for channel flow and t_{cs} is the time of concentration for surface flow. The t_{cc} can be computed using the equation

$$t_{cc} = \frac{L_c}{v_c} \quad (13)$$

where L_c is the average channel flow length for the watershed in km and v_c is the average channel velocity in m/s. The average channel flow length can be estimated using the equation

$$L_c = \sqrt{(L)\ (L_{ca})} \quad (14)$$

where L is the channel length from the most distant point to the watershed outlet in km and L_{ca} is the distance along the channel to the watershed centroid in km. Average velocity can be estimated using Manning's equation assuming a trapezoidal channel with 2:1 side slopes and a 10:1 bottom width-depth ratio. Proper substitution of these assumptions gives

$$t_{cc} = \frac{\sqrt{(L)\ (L_{ca})}\ (n)^{0.75}}{0.489\ (q_c)^{0.25}\ (\sigma)^{.375}} \quad (15)$$

where n is Manning's "n" value, q_c is the average flow rate in m^3/s and σ is the average channel slope in m/m. Assuming $L_{ca} = 0.5$ L and that the average flow rate is about 6.35 mm/h and is a function of the square root of drainage area yields the final equation for t_{cc}

$$t_{cc} = \frac{1.1\ (L)\ (n)^{0.75}}{(DA)^{0.125}\ (\sigma)^{0.375}} \quad (16)$$

A similar approach is used to estimate t_{cs}.

$$t_{cs} = \frac{\lambda}{v_s} \quad (17)$$

where λ is the surface slope length in m and v_s is the surface flow velocity in m/s. Considering a strip one m wide down the sloping surface and applying Manning's equation gives

$$v_s = \frac{(q_s)^{0.4}\ (S)^{0.3}}{(n)^{0.6}} \quad (18)$$

where S is the land surface slope in m/m and q_s is the average surface flow rate. Assuming that the average flow rate is about 6.35 mm/h and making substitutions into Eqs. 17 and 18 to convert from m^3/s to mm/h and from s to h, gives the equation for estimating t_{cs}.

$$t_{cs} = \frac{(\lambda \cdot n)^{0.6}}{18\ (S)^{0.3}} \quad (19)$$

Although some of the assumptions used in developing Eqs. 16 and 19 may appear liberal, Eq. 12 generally gives satisfactory results for small homogeneous watersheds. Since Eqs. 16 and 19 are based on hydraulic considerations, they are more reliable than purely empirical equations.

To properly evaluate α, variation in rainfall patterns must be considered. Some storms have short durations--most or all rain occurs during t_c causing α to approach its upper limit of 1.0. Other storms may have uniform intensities that cause α to approach a minimum value. All other patterns have higher α values than the uniform pattern, because r_{t_c} is greater than r_{24} for all patterns except the uniform. By substituting the products of intensity and time into Eq. 10 an expression for the minimum value of α is obtained.

$$\alpha_{mn} = \frac{t_c}{24} \quad (20)$$

Thus α ranges within the limits

$$\frac{t_c}{24} \leq \alpha \leq 1.0 \ .$$

Although the value of α is confined between limits, there is considerable uncertainty in assigning appropriate values given only daily rainfall and simulated runoff amounts. Thus α is generated from a gamma function with the base ranging from $t_c/24$ to 1.0.

The peak of the α distribution changes for each month because of seasonal differences in rainfall intensities. The Weather Service (U.S. Department of Commerce, 1979) provides information on monthly maximum rainfall intensities that can be used to estimate the peak α for each month.

Since the water erosion component of EPIC estimates the maximum 0.5 h amount of each daily rainfall ($\alpha_{.5}$), these estimates are used in calculating α. Besides the convenience of avoiding double calculation, it is important to assure that $\alpha_{.5}$ and α are closely related for each storm. The relationship between $\alpha_{.5}$ and α can be obtained from TP-40 (Hershfield, 1961) by fitting a log function to the 10-year frequency rainfall distribution.

$$R_t = R_6 \left(\frac{t}{6}\right)^b \quad (21)$$

where b is a parameter used to fit the TP-40 relationship at any location, R_t is the rainfall amount for any time t, and R_6 is the 6-h rainfall amount. The value of α is computed with the equation

$$\alpha = \alpha_{.5} \left(\frac{R_{t_c}}{R_{.5}}\right) \quad (22)$$

Details of the procedure for estimating $\alpha_{.5}$ are presented in the Water Erosion section.

Percolation

The percolation component of EPIC uses a storage routing technique combined with a crack-flow model to predict flow through each soil layer in the root zone. Once water percolates below the root zone it is lost from the watershed (becomes groundwater or appears as return flow in downstream basins).

The storage routing technique is based on the equation

$$SW = SW_0 \exp(-\Delta t/TT) \quad (23)$$

where SW_0 and SW are the soil water contents at the beginning and end of the day in mm, Δt is the time interval (24 h), and TT is the travel time through a particular layer in h. Thus, the percolation can be computed by subtracting SW from SW_0.

$$O = SW_0 \ (1 - \exp(-\Delta t/TT)) \quad (24)$$

where O is the percolation rate in mm/d.

The travel time, TT, is computed for each soil layer with the linear storage equation

$$TT_i = \frac{SW_i - FC_i}{H_i} \quad (25)$$

where H_i is the hydraulic conductivity in mm/h and FC is the field capacity water content for layer i in mm. The hydraulic conductivity is varied from the saturated conductivity value at saturation to near zero at field capacity.

$$H_i = SC_i \left(\frac{SW_i}{UL_i}\right)^{\beta_i} \quad (26)$$

where SC_i is the saturated conductivity for layer i in mm/h and β is a parameter that causes H_i to approach zero as SW_i approaches FC_i.

The equation for estimating β is

$$\beta_i = \frac{-2.655}{\log (FC_i/UL_i)} \quad (27)$$

the constant (-2.655) in Eq. 27 was set to assure that $H_i = 0.002 \ SC_i$ at field capacity.

Flow through a soil layer may be reduced by a saturated lower soil layer. If the layer immediately below the layer being considered is saturated, no flow can occur regardless of the results from Eq. 24. The effect of lower layer water content is expressed in the equation

$$O_{ci} = O_i \sqrt{1 - \frac{SW_{i+1}}{UL_{i+1}}} \quad (28)$$

where O_{ci} is the percolation rate for layer i in mm/d corrected for layer i+1 water content and O_i is the percolation computed with Eq. 24.

The crack-flow model allows percolation of infiltrated rainfall even though the soil water content is less than field capacity. When the soil is dry and cracked, infiltrated rainfall can flow through the cracks of a layer without becoming part of the layer's soil water. However, the portion that does become part of a layer's stored water cannot percolate until the storage exceeds field capacity.

Crack flow is simulated with relationships similar to those used to estimate percolation above field capacity. The amount of percolate caused by crack flow is estimated with the equation

$$O_i = O_{i-1} \ (1 - \exp(-\Delta t/TT_{ci})) \quad (29)$$

where O_{i-1} is the flow from the layer above in mm/d (R-Q for the top layer) and TT_{ci} is the crack flow travel time in h. Crack flow travel time is estimated with the equation

$$TT_{ci} = \frac{O_{i-1}}{(\zeta_i)\ (SC_i)} \quad (30)$$

where ζ_i is a dimensionless (0-1) soil parameter that expresses degree of cracking.

Percolation is also affected by soil temperature. If the temperature in a particular layer is zero °C or below, no percolation is allowed from that layer. Water can, however, percolate into the layer as dictated by Eqs. 24 and 29.

Since the one-day time interval is relatively long for routing flow through soils it is desirable to divide the water into several parts for routing. This is necessary because the flow rates are dependent upon soil water content which is continuously changing. For example, if the soil is extremely wet, Eqs. 24, 25, and 26 might greatly over estimate percolation if only one routing was performed using the entire amount ($SW_i - FC_i$). To overcome this problem, EPIC divides each layer's inflow into 4 mm slugs for routing. Also, by dividing the inflow into 4 mm slugs and routing each slug individually through all layers, the lower layer water content relationship (Eq. 28) is allowed to function. The lower layer water content effect cannot be calculated realistically by routing all of the input through one layer before proceeding to the next layer.

Lateral Subsurface Flow

Lateral subsurface flow is calculated simultaneously with percolation. Each 4 mm slug is given the opportunity to percolate first and then the remainder is subjected to the lateral flow function. Thus, lateral flow can occur when the storage in any layer exceeds field capacity after percolation. The lateral flow function (similar to Eq. 24) is expressed in the equation

$$QR_i = (SW_i - FC_i)\ (1 - \exp(-\Delta t/TT_{Ri})) \quad (31)$$

where QR_i is the lateral flow rate for soil layer i in mm/d and TT_{Ri} is the lateral flow travel time in d (time required for subsurface flow to travel a distance equal the land surface slope length λ).

The lateral flow travel time is input to EPIC for each soil layer. Knowledge of soils and flow conditions in a land resource area is helpful in assigning these values. Generally, the lateral flow travel time ranges from 1 to 1000 days. Initial estimates may need adjustment after the first simulation in a land resource area if the simulated subsurface flow is not satisfactory.

Drainage

Underground drainage systems are treated as a modification to the natural lateral subsurface flow of the area. Simulation of a drainage system is accomplishied by simply changing the lateral flow travel time of the soil layer that contains the drainage system. A short travel time of about one day will be appropriate for many drainage systems. Of course, this depends upon the soil properties and the drain spacing.

Evapotranspiration

The evapotranspiration component of EPIC is Ritchie's ET model (Ritchie, 1972). To compute potential evaporation, the model uses the equation

$$E_o = 1.28\ (h_o)\ (\gamma) \qquad (32)$$

where E_o is the potential evaporation rate in mm/d, h_o is the net solar radiation, and γ is a psychrometric constant. The value of h_o is calculated with the equation

$$h_o = \frac{(RA)\ (1 - AB)}{58.3} \qquad (33)$$

where RA is the daily solar radiation in ly and AB is the albedo. The albedo is evaluated by considering the soil, crop, and snow cover. If a snow cover exists with 5 mm or greater water content, the value of albedo is set to 0.8. If the snow cover is less than 5 mm and no crop is growing, the soil albedo is the appropriate value. When crops are growing, albedo is determined using the equation

$$AB = 0.23\ (1. - EA) + (AB_s)\ (EA) \qquad (34)$$

where AB_s is the soil albedo, 0.23 is the albedo for plants, and EA is a soil cover index. The value of EA ranges from 0 to 1.0 according to the equation

$$EA = \exp(-2.9 \times 10^{-5}\ (CV)) \qquad (35)$$

where CV is the sum of above ground biomass and crop residue in kg/ha.

The value of γ in Eq. 32 can be obtained from the equation

$$\gamma = \frac{\delta}{\delta + 0.68} \qquad (36)$$

where δ is the slope of the saturation vapor pressure curve at the mean air temperature. The expression for evaluating δ is

$$\delta = \frac{5304}{T_K^2}\ \exp(21.255 - \frac{5304}{T_K}) \qquad (37)$$

where T_K is the daily average air temperature in °K.

The model computes soil and plant evaporation separately. Potential soil evaporation is predicted with the equation

$$E_{so} = E_o\ \exp(-0.4\ LAI) \qquad (38)$$

where E_{so} is the potential evaporation rate at the soil surface in mm/d and LAI is the leaf area index defined as the area of plant leaves relative to the soil surface area. Actual soil evaporation is computed in two stages. In the first stage, soil evaporation is limited only by the energy available at the surface and, thus, is equal to the potential soil evaporation. When the accumulated soil evaporation exceeds the stage one upper limit (6 mm), the stage two evaporative process begins.

Stage two soil evaporation is predicted with the equation

$$E_s = 3.5\ [t^{1/2} - (t - 1)^{1/2}] \qquad (39)$$

where E_s is the soil evaporation rate for day t in mm/d and t is the number of days since stage two evaporation began.

Plant evaporation is computed with the equations

$$E_p = \frac{(E_o)(LAI)}{3}\ ,\ 0 \le LAI \le 3 \qquad (40)$$

$$E_p = E_o - E_s\ ,\ LAI > 3 \qquad (41)$$

where E_p is the predicted plant evaporation rate in mm/d. If soil water is limited, plant evaporation will be reduced as described in the Plant Growth section.

Irrigation

The EPIC user has the option to simulate dryland or irrigated agricultural areas. If irrigation is indicated, he must also specifiy a plant water stress level to start irrigation. The plant water stress factor ranges from 0 to 1.0 (1 means no stress and 0 means no growth). These stress factors are described in the plant growth section. When the user-specified stress level is reached, enough water is applied to bring the root zone up to field capacity.

Snow Melt

The EPIC snow melt component is similar to that of the CREAMS model (Knisel, 1980). If snow is present, it is melted on days when the maximum temperature exceeds 0°C using the equations

$$SML = 4.57\ T_{mx}\ , \qquad SML < SNO \qquad (42)$$

$$SML = SNO \qquad (43)$$

where SML is the snow melt rate in mm/d, T_{mx} is the daily maximum temperature in °C, and SNO is the water content of snow before melt occurs in mm. Melted snow is treated the same as rainfall for estimating runoff, percolation, etc.

Weather

The weather variables necessary for driving the EPIC model are precipitation, air temperature, solar radiation, and wind. If daily precipitation, air temperature, and solar radiation data are available, they can be input directly to EPIC. Rainfall and temperature data are available for many areas of the U.S., but solar radiation data are scarce. Even rainfall and temperature data are generally not adequate for the long-term EPIC simulations (50 years+). Thus, EPIC provides options for simulating temperature and radiation given daily rainfall or for simulating rainfall as well as temperature and radiation. If wind erosion is to be estimated, daily wind velocity and direction are simulated (There is no option for inputing wind velocity and direction). Descriptions of the models used for simulating precipitation, temperature, radiation, and wind follow.

Precipitation

The EPIC precipitation model developed by Nicks (1974) is a first-order Markov chain model. Thus the model must be provided as input monthly probabilities of receiving precipitation if the previous day was dry and monthly probabilities of receiving precipitation if the previous day was wet. Given the initial wet-dry state, the model determines if the first day of simulation had precipitation or not. A random number (0-1) is generated and compared with the appropriate wet-dry probability. If the random number is less than or equal to the wet-dry probability, precipitation occurs on that day. Random numbers greater than the wet-dry probability give no precipitation. Since the wet-dry state of the first day is established, the process can be repeated for the next day and so on throughout the simulation period.

When a precipitation event occurs, the amount is determined by generating from a skewed normal daily precipitation distribution. Inputs necessary to describe the skewed normal distribution for each month are the mean, standard deviation, and skew coefficient for daily precipitation. The amount of daily precipitation is partitioned between rainfall and snowfall using average daily air temperature. If the average daily air temperature is zero °C or below, the precipitation is snowfall--otherwise, it is rainfall.

Air Temperature and Solar Radiation

The model developed by Richardson (1981) was selected for use in EPIC because it simulates temperature and radiation that exhibit proper correlation between one another and rainfall. The residuals of daily maximum and minimum temperature and solar radiation are generated from a multivariate normal distribution. The means and coefficients of variation for each variable must be input. Since most of the variables are affected by rainfall, the means and coefficients of variation must be input separately for wet and dry days. Fortunately, a simple cosine function with two parameters adequately fits both the means and coefficients of variation. The equations are

$$XT_{IDA} = \mu_T + a_T \cos\left(\frac{2\pi}{365}(IDA - 200)\right) \qquad (44)$$

$$XR_{IDA} = \mu_R + a_R \cos\left(\frac{2\pi}{365}(IDA - 172)\right) \qquad (45)$$

where Eqs. 44 and 45 are used to estimate the means and coefficients of variation for temperature and radiation, and XT_{IDA} and XR_{IDA} are the values of the mean or coefficient of variation for temperature and radiation on julian day IDA, μ is the annual mean and a is the amplitude for the means and coefficients of variation, and subscripts T and R refer to temperature and radiation.

The multivariate generation model used implies that the residuals of maximum temperature, minimum temperature, and solar radiation are normally distributed and that the serial correlation of each variable may be described by a first-order linear autoregressive model. Details of the multivariate generation model were described by Richardson (1981). The dependence structure of daily maximum temperature, minimum temperature, and solar radiation was described by Richardson (1982).

Wind

The wind simulation model was developed by Richardson[4] for use in simulating wind erosion with EPIC. The two wind variables considered are average daily velocity and daily direction. Average daily wind velocity is generated from a two-parameter gamma distribution of the dimensionless form

$$U = \left(\frac{V}{V_p}\right)^{\eta-1} \exp\left((\eta - 1)\left(1 - \frac{V}{V_p}\right)\right) \qquad (46)$$

where U is a dimensionless variable (0-1) expressing frequency of occurrence of wind velocity V in m/s, V_p is the wind velocity at the peak frequency, and η is the gamma distribution shape parameter. The shape parameter is calculated with the equation

$$\eta = \frac{\overline{V}^2}{SDV^2} \qquad (47)$$

where $\overline{V}$ is the annual average wind velocity in m/s and SDV is the standard deviation of daily wind velocity in m/s. Values for the average annual wind velocity and the standard deviation of hourly wind are provided by the Climatic Atlas of the United States (U.S. Department of Commerce, 1968). By experimenting with standard deviations of hourly and daily wind, a correction factor of 0.7 was found to be appropriate for converting hourly standard deviations to daily. The base of the dimensionless gamma distribution (maximum $\frac{V}{V_p}$) can be determined using Newton's classical method of solving nonlinear equations. The objective function is to select the base to minimize the sum of ln(U) and 11.5. The value of V_p can be determined by differentiating the gamma function expressed in terms of V and setting the result equal to zero.

$$V_p = \frac{V_i\,(\eta - 1)}{\eta} \qquad (48)$$

where V_i is the mean daily wind velocity for month i. The rejection technique is used to generate a daily value of V/V_p. The daily wind velocity is then computed using the equation

$$V_j = (V_{pi}) \left(\frac{V}{V_p}\right) \qquad (49)$$

where V_j is the generated velocity for day j, V_{pi} is the peak velocity for month i, and $\frac{V}{V_p}$ is the generated value using the rejection technique.

Wind direction expressed as radians from north in a clockwise direction is generated from an empirical distribution specific for each location. The empirical distribution is simply the cumulative probability distribution of wind direction. The Climatic Atlas of the United States (U.S. Department of Commerce, 1968) gives monthly percentages of wind from each of 16 directions. Thus to estimate wind direction for any day, the model draws a uniformly distributed random number and locates its position on the appropriate monthly cumulative probability distribution.

Erosion

Water

The water erosion component of EPIC uses a modification of the USLE (Wischmeier and Smith, 1978) developed by Onstad and Foster (1975). The Onstad-Foster equation's energy factor is composed of both rainfall and runoff variables. In contrast, the USLE energy factor contains only rainfall variables. By including runoff variables, Onstad and Foster increased prediction accuracy, eliminated the need for delivery ratios, and produced an equation suitable for predicting individual event sediment yield instead of annual soil loss. The Onstad-Foster equation is

[4]Richardson, C.W. 1982. Personal conversation. USDA-ARS, Grassland, Soil and Water Research Laboratory, Temple, Texas.

$$Y = (0.646EI + 0.45(Q)(q_p)^{0.333})(K)(CE)(PE)(LS) , \quad Q > 0 \tag{50}$$

$$Y = 0 , \quad Q \leq 0$$

where Y is the sediment yield in t/ha, EI is the rainfall energy factor in metric units, Q is the runoff volume in mm, q_p is the peak runoff rate in mm/h, K is the soil erodibility factor, CE is the crop management factor, PE is the erosion control practice factor, and LS is the slope length and steepness factor. The value of K depends upon the soil type and is assigned before the simulation begins. Similarly, the PE value is determined initially by considering the conservation practices to be applied. The value of LS is calculated with the equation (Wischmeier and Smith, 1978)

$$LS = (\frac{\lambda}{22.1})^{\xi} \; (65.41 \; S^2 + 4.56 \; S + .065) \tag{51}$$

where S is the land surface slope in m/m, λ is the slope length in m, and ξ is a parameter dependent upon slope. To evaluate ξ, EPIC uses the equation

$$\xi = 0.6 \; (1. - \exp(-35.835 \; S)) \tag{52}$$

The crop management factor is evaluated for all days when runoff occurs using the equation

$$CE = (0.8 - CE_{mn,j}) \exp(-0.00115 \; CV) + CE_{mn,j} \tag{53}$$

where $CE_{mn,j}$ is the minimum value of the crop management factor for crop j.

The hydrology model supplies estimates of Q and q_p. To estimate the daily rainfall energy in the absence of time-distributed rainfall, it is assumed that the rainfall rate is exponentially distributed.

$$r = r_p \exp(-t/k) \tag{54}$$

where r is the rainfall rate at time t in mm/h, r_p is the peak rainfall rate in mm/h, and k is the decay constant in h. Eq. 54 contains no assumption about the sequence of rainfall rates (time distribution). The USLE energy equation in metric units is

$$E = \Delta R \; (12.1 + 8.9 \log \frac{R}{\Delta t}) \tag{55}$$

where ΔR is a rainfall amount in mm during a time interval Δt in h. The energy equation can be expressed analytically as

$$E = 12.1 \int_0^{\infty} r \, dt + 8.9 \int_0^{\infty} r \log r \, dt \tag{56}$$

Substituting Eq. 54 into Eq. 56 and integrating gives the equation for estimating daily rainfall energy

$$E = R \; (12.1 + 8.9 \; (\log r_p - 0.434)) \tag{57}$$

where R is the daily rainfall amount in mm. The rainfall energy factor, EI, is obtained by multiplying Eq. 57 by the maximum 0.5 h rainfall intensity and converting to the proper units.

$$EI = R \; (12.1 + 8.9 \; (\log r_p - 0.434))(r_{.5})/1000 \tag{58}$$

To compute values for r_p, Eq. 54 is integrated to give

$$R = (r_p) \; (k) \tag{59}$$

and

$$R_t = R \; (1 - \exp(-t/k)) \tag{60}$$

The value of $R_{.5}$ can be estimated using $\alpha_{.5}$ as mentioned in the Hydrology section

$$R_{.5} = \alpha_{.5} \; R \tag{61}$$

To determine the value of r_p, Eqs. 61 and 59 are substituted into Eq. 60 to give

$$r_p = -2 \; R \ln \; (1 - \alpha_{.5}) \tag{62}$$

Since rainfall rates vary seasonally, $\alpha_{.5}$ is evaluated for each month using Weather Service information (U.S. Department of Commerce, 1968, 1979). The frequency of the maximum 0.5 h rainfall amount is estimated using the equation

$$F = \frac{1}{2\tau} \tag{63}$$

where F is the frequency of occurrence of the largest event of a total of τ events. The total number of events for each month is the product of the number of years of record and the average number of rainfall events for the month. To estimate the mean value of $\alpha_{.5}$, it is necessary to estimate the mean value of $R_{.5}$. The value of $R_{.5}$ can be computed easily if it is assumed that the maximum 0.5 h rainfall amounts are exponentially distributed. From the exponential distribution, the expression for the mean 0.5 h rainfall amount is

$$\bar{R}_{.5,j} = \frac{R_{.5F,j}}{-\ln F_j} \tag{64}$$

where $\bar{R}_{.5,j}$ is the mean maximum 0.5 h rainfall amount, $R_{.5F,j}$ is the maximum 0.5 h rainfall amount for frequency F, and subscript j refers to the month. The mean $\alpha_{.5}$ is computed with the equation

$$\alpha_{.5j} = \frac{\bar{R}_{.5,j}}{\bar{R}_j} \tag{65}$$

where $\bar{R}$ is the mean amount of rainfall for each event (average monthly rainfall/average number of days of rainfall) and subscript j refers to the month. Daily values of $\alpha_{.5}$ are generated from a two-parameter gamma distribution in a similar manner to that described in generating wind velocity (Weather section). The base of the gamma distribution is established by examining upper and lower limits of $\alpha_{.5}$. The lower limit is determined by a uniform rainfall rate as shown in the Hydrology section for α. This gives a minimum $\alpha_{.5}$ of .5/24 or 0.0208. The upper limit of $\alpha_{.5}$ is set by considering a large rainfall event. If the event is large, it is highly unlikely that all the rainfall would occur in 0.5 h (α = 1.). The upper limit of $\alpha_{.5}$ can be estimated by substituting a high value for r_p (250 mm/h is generally near the upper limit of rainfall intensity) into Eq. 62.

$$\alpha_{.5u} = 1 - \exp(-125/R) \tag{66}$$

where $\alpha_{.5u}$ is the upper limit of $\alpha_{.5}$. The peak of the $\alpha_{.5}$ gamma distribution can be computed using Eq. 48 written in the form

$$\alpha_{.5P,j} = \frac{\alpha_{.5,j}\ (\nu - 1)}{\nu} \tag{67}$$

where $\alpha_{.5P,j}$ is the $\alpha_{.5}$ value at the peak of the gamma distribution and ν is the gamma distribution shape parameter (a value of 10 is generally satisfactory).

Wind

The Manhattan, Kansas, wind erosion equation (Woodruff and Siddoway, 1965), was modified by Cole[5] for use in the EPIC model. The original wind erosion equation is of the form

$$WE = f\ (I, WC, WK, WL, VE) \tag{68}$$

where WE is the wind erosion in t/ha, I is the soil erodibility index in t/ha, WC is the climatic factor, WK is the soil ridge roughness factor, WL is the field length along the prevailing wind direction in m, and VE is the quantity of vegetaive cover expressed as small grain equivalent in kg/ha. Eq. 68 was developed for predicting average annual wind erosion. The main modification to the model was converting from annual to daily predictions to interface with EPIC.

Two of the variables I and WC remain constant for each day of a year. The other variables, however, are subject to change from day to day. The ridge roughness is a function of a row height and row interval

$$KR = \frac{HR^2}{IR} \tag{69}$$

where KR is the ridge roughness in mm, HR is the row height in mm, and IR is the row interval in mm. The ridge roughness factor is a function of ridge roughness as expressed by the equations

$$WK = 1. \ , \quad KR < 2.27 \tag{70}$$

$$WK = 1.125 - 0.153\ \ln\ (KR)\ , \ 2.27 \leq KR < 89. \tag{71}$$

$$WK = 0.336\ \exp(0.00324\ KR)\ , \ KR \geq 89. \tag{72}$$

Field length along the prevailing wind direction is calculated by considering the field dimensions and orientation and the wind direction.

$$WL = \frac{(FL)\ (FW)}{FL\left|\cos(\frac{\pi}{2} + \theta - \phi)\right| + FW\left|\sin(\frac{\pi}{2} + \theta - \phi)\right|} \tag{73}$$

where FL is the field length in m, FW is the field width in m, θ is the wind direction clockwise from north in radians, and ϕ is the clockwise angle between field length and north in radians.

The vegetative cover equivalent factor is simulated daily as a function of standing live biomass, standing dead residue, and flat crop residue.

$$VE = 0.2533\ (g_1 B_{AG} + g_2 SR + g_3 FR)^{1.363} \tag{74}$$

where g_1, g_2, and g_3 are crop specific coefficients, B_{AG} is the above ground biomass of a growing crop in kg/ha, SR is the standing residue from the previous crop in kg/ha, and FR is the flat residue in kg/ha. Thus, all variables in Eq. 68 can be evaluated. To determine the wind erosion estimate, however, requires a special combination of the factors as follows.

$$E2 = (WK)\ (I) \tag{75}$$

$$E3 = (WK)\ (I)\ (WC) \tag{76}$$

$$WL_o = 1.56x10^6\ (E2)^{-1.26}\ \exp(-0.00156\ E2) \tag{77}$$

$$WF = E2\ (1. - 0.1218\ (WL/WL_o)^{-0.3829}\ \exp(-3.33\ WL/WL_o)) \tag{78}$$

$$E4 = (WF^{0.3484} + E3^{0.3484} - E2^{0.3484})^{2.87} \tag{79}$$

$$E5 = \psi_1\ E4^{\psi_2} \tag{80}$$

$$WE = (E5)\ (DE)\ /\ (AE) \tag{81}$$

where field lengths greater than WL_o in m give no reduction in the erosion estimate, WF is the field length factor, ψ_1 and ψ_2 are parameters, DE is the daily wind energy in KWH/m^2 and AE is the average annual wind energy in KWH/m^2. The parameters ψ_1 and ψ_2 are the functions of the vegetative cover factor described by the equations

$$\psi_2 = \frac{1.}{0.0537 + 0.9436\ \exp(-0.000112\ VE)} \tag{82}$$

$$\psi_1 = 952.5^{\ (1 - \psi_2)} \tag{83}$$

Daily wind energy is estimated with the equation

$$DE = 0.00617\ (V)^{4.35}\ \exp(-0.0620\ V) \tag{84}$$

where V is the daily average wind velocity in m/s. Average annual wind energy is estimated by integrating the monthly gamma distributions of wind velocity

$$AE = 30.4 \sum_{j=1}^{12} \frac{\int_{V_L}^{V_u} (DE)_j\ (U)_j\ dV}{\int_{V_L}^{V_u} U_j\ dV} \tag{85}$$

where U is the frequency of occurrence of wind velocity V, V_u is the upper limit of wind velocity, and V_L is the lower limit of erosive wind velocity.

Nutrients

Nitrogen

Nitrate Loss in Surface Runoff.--The amount of NO_3 in runoff is estimated by considering the top soil layer (10 mm thickness) only. The total amount of water leaving the layer is the sum of runoff, lateral subsurface flow, and percolation.

$$QT = Q + 0_1 + QR_1 \tag{86}$$

where QT is the total water lost from the first layer in mm. The amount of NO_3 lost with QT is

$$VNO3 = (QT)\ (c_{NO3}) \tag{87}$$

where VNO3 is the amount of NO_3 lost from the first layer and c_{NO3} is the concentration of NO_3 in the first layer. At the end of the day, the amount of NO_3 left in the layer is

$$WNO3 = WNO3_0 - (QT)\ (c_{NO3}) \tag{88}$$

[5]Cole, G.W. 1982. Personal correspondence. USDA-ARS, Wind Erosion Research Unit, Manhattan, Kansas.

where $WNO3_0$ and WNO3 are the weights of NO_3 contained in the layer at the beginning and ending of the day. The NO_3 concentration can be estimated by dividing the weight of NO_3 by the water storage volume (soil porosity).

$$c'_{NO3} = c_{NO3} - c_{NO3} \left(\frac{QT}{POR}\right) \qquad (89)$$

where c'_{NO3} is the concentration of NO_3 at the end of a day and POR is the porosity of the first soil layer in mm. Eq. 89 is a finite difference approximation for the exponential equation

$$c'_{NO3} = c_{NO3} \exp(-QT / POR) \qquad (90)$$

Thus VNO3 can be computed for any QT value by integrating Eq. 90

$$VNO3 = WNO3 \; (1 - \exp(-QT / POR)) \qquad (91)$$

The average concentration of QT for the day is

$$c_{NO3} = \frac{VNO3}{QT} \qquad (92)$$

Amounts of NO_3 contained in runoff, lateral flow, and percolation are estimated as the products of the volume of water and the concentration from Eq. 92.

NO_3 Leaching.--Leaching and lateral subsurface flow in lower layers are treated with the same approach used in the upper layer except surface runoff is not considered.

NO_3 Transport by Soil Evaporation.--When water is evaporated from the soil, NO_3 is moved upward into the top soil layer by mass flow. The equation for estimating this NO_3 transport is

$$ENO3 = \sum_{j=2}^{M} (E_S)_j \; (c_{NO3})_j \qquad (93)$$

where ENO3 is the amount of NO_3 moved from lower layers to the top layer by soil evaporation E_S, subscript j refers to soil layers, and M is the number of layers contributing to soil evaporation (maximum depth is 300 mm).

Organic N Transport by Sediment.--A loading function developed by McElroy et al. (1976) and modified by Williams and Hann (1978) for application to individual runoff events is used to estimate organic N loss. The loading function is

$$YON = 0.001 \; (Y) \; (c_{ON}) \; (ER) \qquad (94)$$

where YON is the organic N runoff loss in kg/ha, c_{ON} is the concentration of organic N in the top soil layer in g/t, Y is the sediment yield in t/ha, and ER is the enrichment ratio. The enrichment ratio is the concentration of organic N in the sediment divided by that of the soil. Enrichment ratios are logarithmically related to sediment concentration as described by Menzel (Knisel, 1980). An individual event enrichment ratio-sediment concentration relationship was developed for EPIC considering upper and lower bounds. The upper bound of enrichment ratio is the inverse of the sediment delivery ratio. Exceeding the inverse of the delivery ratio implies that more organic N leaves the watershed than is dislodged from the soil. The delivery ratio is estimated for each runoff event using the equation

$$DR = \left(\frac{q_p}{r_{ep}}\right)^{0.56} \qquad (95)$$

where DR is the sediment delivery ratio (sediment yield divided by gross sheet erosion), q_p is the peak runoff rate in mm/h, and r_{ep} is the peak rainfall excess rate in mm/h. Eq. 95 is based on sediment yield estimates using the Modified Universal Soil Loss Equation (MUSLE) (Williams, 1975). The rainfall excess rate cannot be evaluated directly because the hydrology model only predicts the total daily runoff volume. An estimate of the rate can be obtained, however, using the equation

$$r_{ep} = r_p - f \qquad (96)$$

where r_p is the peak rainfall rate in mm/h and f is the average infiltration rate in mm/h. The average infiltration rate can be computed from the equation

$$f = \frac{R - Q}{DUR} \qquad (97)$$

where DUR is the rainfall duration in h. The rainfall duration can be estimated by solving Eqs. 59 and 60 for t when $R_t/R = 0.01$.

$$DUR = \frac{-4.605 \; R}{r_p} \qquad (98)$$

The lower limit of enrichment ratio is 1.0--sediment particle size distribution is the same as that of the soil. Thus $1 \le ER \le 1/DR$. The logarithmic equation for estimating enrichment ratio is

$$ER = x_1 \, c_s^{\,x_2} \qquad (99)$$

where c_s is the sediment concentration g/m^3 and x_1 and x_2 are parameters set by the upper and lower limits. To approach an enrichment ratio of 1.0, the sediment concentration would be extremely high. Conversely, a very low sediment concentration would cause the enrichment ratio to approach 1/DR. The simultaneous solution of Eq. 99 at the boundaries assuming sediment concentrations range from 500 to 250,000 g/m^3 gives

$$x_2 = -\log \left(\frac{1}{DR}\right) / 2.699 \qquad (100)$$

$$x_1 = \frac{1}{(0.25)^{x_2}} \qquad (101)$$

Denitrification.--As one of the microbial processes, denitrification is a function of temperature and water content. The equation used to estimate the denitrification rate is

$$DN = WNO3(1 - \exp((CDN)(TF_N)(C)) \; , \quad SWF \ge 0.9 \qquad (102)$$

$$DN = 0. \; , \qquad SWF < 0.9$$

DN is the denitrification rate in kg/ha·d, CDN is the denitrification constant, TF_N is the nutrient temperature factor, SWF is the soil water factor, and C is the organic carbon content in percent. The temperature factor is expressed by the equation

$$TF_{Nj} = 0.1 + \frac{0.9 \; T_j}{T_j + 20551 \exp(-0.312 \; T_j)} \; , \quad T_j > 0 \qquad (103)$$

$$TFN_j = 0. \; , \qquad T_j \le 0 \; .$$

where T is soil temperature and subscript j refers to the layers. The soil water factor considers total soil water in the equation

$$SWF = \frac{SW + SW_{15}}{POR} \qquad (104)$$

where SW_{15} is the 15-bar soil water content in mm. The denitrification constant can be estimated by assuming that 10% of the NO_3 can be lost during a day if the temperature factor is 1.0 and the organic C content is 3%. Substituting these assumed values into Eq. 102 and solving gives CDN = 0.0351.

Mineralization.--The N mineralization model is a modification of the PAPRAN mineralization model (Seligman and van Keulen, 1981). The model considers two sources of mineralization: fresh organic N associated with crop residue and microbial biomass and the stable organic N associated with the soil humus pool. Mineralization from the fresh organic N pool is estimated with the equation

$$RMN = (DCR)\ (FON) \qquad (105)$$

where RMN is the N mineralization rate for fresh organic N in kg/ha•d, DCR is the decay rate constant for the fresh organic N, and FON is the amount of fresh organic N present in kg/ha. The decay rate constant is a function of C:N ratio, C:P ratio, composition of crop residue, temperature, and soil water.

$$DCR = (CNP)\ (RC)\ (SWF)\ (TF_N) \qquad (106)$$

where CNP is a C:N, C:P ratio factor and RC is a residue composition factor. The value of CNP is calculated with the equation

$$CNP = \min \begin{cases} \exp(-0.693\ (CNR - 25)\ /\ 25 \\ \exp(-0.693\ (CPR - 200)\ /\ 200 \end{cases} \qquad (107)$$

where CNR is the C:N ratio and CPR is the C:P ratio. The C:N and C:P ratios of crop residue are computed for each soil layer with the equations

$$CNR = \frac{0.4\ FR}{FON + WNO3} \qquad (108)$$

$$CPR = \frac{0.4\ FR}{FOP + AP} \qquad (109)$$

where FOP is the amount of fresh organic P in kg/ha and AP is the amount of labile P in kg/ha. The value of RC is determined by the stage of residue decomposition. Initially, when the most labile nonstructural carbohydrate and amino acid fractions of the residue are decomposing, RC = 0.8. After 20 percent of the residue has decomposed, RC drops to 0.05 (a rate appropriate for cellulose-like material). The final 10 percent of the residue is decomposed at a rate appropriate for lignin (RC = 0.0095).

Mineralization from the stable organic N pool is estimated for each soil layer with the equation

$$HMN = (CMN)\ (SWF)\ (TF_N)\ (ON) \qquad (110)$$

where HMN is the mineralization rate for the humus pool in kg/ha•d, CMN is the stable mineralization rate constant, and ON is the amount of organic N in the soil layer in kg/ha. The stable mineralization rate constant can be estimated by considering the average yield of unfertilized crops in a location. It is assumed that humus mineralization is the single N source for an unfertilized crop (mineralization of FON and immobilization cycle N but furnish no source of NO_3). Thus, the average annual amount of humus mineralization equals the average N removed with the crop yield. Setting the product of average unfertilized yield and its N concentration equal to HMN in Eq. 110 and summing the right hand side for a year leads to the equation

$$CMN = \frac{(YLD)\ (c_{NY})}{(ON)\ (SWF_a)\ \sum_{i=1}^{365} TF_i} \qquad (111)$$

where YLD is the average annual unfertilized crop yield in kg/ha, c_{NY} is the concentration of N in the crop yield in kg/kg, SWF_a is the average value of SWF, and TF_i is the temperature factor for day i. To maintain the N balance at the end of a day, the humus mineralization is subtracted from the stable organic N pool; the residue mineralization is subtracted from the FON pool; 20 percent of RMN is added to the ON pool; and 80 percent of RMN is added to the WNO3 pool.

Immobilization.--Like mineralization the immobilization model is a modification of the PAPRAN model. Immobilization is a very important process in EPIC because it determines the residue decomposition rate. Of course, residue decomposition has an important effect on erosion. The daily amount of immobilization is computed by subtracting the amount of N contained in the crop residue from the amount assimilated by the microorganisms.

$$WIM = (DCR)\ (FR)\ (0.016 - c_{NFR}) \qquad (112)$$

where WIM is the N immobilization rate in kg/ha•d, c_{NFR} is the N concentration in the crop residue in g/g, and 0.016 is the result of assuming that C = 0.4 FR, C:N of the microbial biomass and their labile products = 10, and 0.4 of C in the residue is assimilated. Immobilization may be limited by N or P availability. If the amount of N available is less than the predicted immobilization from Eq. 112, the decay rate constant is adjusted with the relationship

$$DCR' = \frac{0.95\ WNO3}{FR\ (0.016 - c_{NFR})} \qquad (113)$$

where DCR' allows 95 percent use of the available NO_3 in a soil layer. A similar adjustment is made if P is limiting. The crop residue is reduced using the equation

$$FR = FR_o - (DCR')\ (FR_o) \qquad (114)$$

where FR_o and FR are the amounts of residue in a soil layer at the start and end of a day in kg/ha. Finally the immobilized N is added to the FON pool and subtracted from the NO_3 pool.

Crop Uptake.--Crop use of N is estimated using a supply and demand approach. The daily crop N demand can be computed using the equation

$$UND_{IDA} = (c_{NB})_{IDA}(B)_{IDA} - (c_{NB})_{IDA-1}(B)_{IDA-1} \qquad (115)$$

where UND_{IDA} is the N demand of the crop in kg/ha, c_{NB} is the optimal N concentration of the crop, and B is the accumulated biomass in kg/ha. The optimal crop N concentration is computed as a function of growth stage using the equation

$$c_{NB} = bn_1 + bn_2\ \exp(-bn_3\ B_1) \qquad (116)$$

where bn_1, bn_2, and bn_3 are crop parameters expressing N concentration and B_1 is a dimensionless (0-1) expression of accumulated thermal time. The value of B_1 is estimated as a function of heat units

$$B_{1,IDA} = \sum_{i=1}^{IDA} HU_i \;/\; PHU \qquad (117)$$

where HU is daily heat units in °C above the crops base temperature and PHU is the potential heat units to mature the crop in °C.

Soil supply of N is assumed to be limited by mass flow of NO_3 to the roots

$$UNS_{IDA} = \sum_{i=1}^{M} (u_i)_{IDA} \left(\frac{WNO3_i}{SW_i}\right)_{IDA} \qquad (118)$$

where UNS is the amount of N supplied by the soil in kg/ha, u_i is the water use in mm, and subscript i refers to the soil layers. Actual N uptake on IDA is the minimum of UNS and UND.

Fixation.--Fixation of N is an important process for legumes. EPIC estimates fixation by adding N to prevent N stress that constrains plant growth. The Plant Growth section describes the determination of plant stress factors for N, P, water, and temperature. Plant growth is limited by the minimum of the four factors each day. If N is the active constraint, enough N is added to the plant to make the N stress factor equal the next most constraining factor. The amount of N added is attributed to fixation.

Rainfall.--To estimate the N contribution from rainfall, EPIC uses an average rainfall N concentration for a location for all storms. The amount of N in rainfall is estimated as the product of rainfall amount and concentration.

Fertilizer.--Both N and P fertilizer are applied according to user specified dates, rates, and depths.

Phosphorus

Soluble P Loss in Surface Runoff.--The EPIC approach is based on the concept of partitioning pesticides into the solution and sediment phases as described by Leonard and Wauchope (Knisel, 1980). Because P is mostly associated with the sediment phase, the soluble P runoff equation can be expressed in the simple form

$$YSP = 0.01\,(c_{LP1})\,(Q)\;/\;k_d \qquad (119)$$

where YSP is the soluble P in kg/ha lost in runoff volume Q in mm, c_{LP1} is the concentration of AP in soil layer 1 in g/t, and k_d is the P concentration in the sediment divided by that of the water in m^3/t. The value of k_d used in EPIC is 175.

P Transport by Sediment.--Sediment transport of P is simulated with a loading function as described in organic N transport. The P loading function is

$$YP = 0.001\,(Y)\,(c_P)\,(ER) \qquad (120)$$

where YP is the sediment phase P loss in runoff in kg/ha and c_P is the concentration of P in the top soil layer in g/t.

Mineralization.--The P mineralization model developed by Jones et al. (1982) is similar in structure to the N mineralization model. Mineralization from the fresh organic P pool is estimated for each soil layer with the equation

$$RMP = (DCR)\,(FOP) \qquad (121)$$

where RMP is the fresh organic P mineralization rate in kg/ha·d. Mineralization from the stable organic P pool associated with humus is estimated for each soil layer using the equation

$$HMP = (LF_M)\,(CMN)\,(TF_N)\,(SWF)\,(OP) \qquad (122)$$

where HMP is the humus P mineralization rate in kg/ha·d, LF_M is the labile P mineralization factor, and OP is the organic P content of the soil layer in kg/ha. The labile P mineralization factor is computed with the equations

$$LF_M = 5 - 0.16\,c_{LP}\,, \qquad c_{LP} \le 25 \qquad (123)$$

$$LF_M = 1\,, \qquad c_{LP} > 25 \qquad (124)$$

Thus, when $c_{LP} > 25$, humus P mineralization is directly proportional to humus N mineralization. However, because of increasing phosphatase activity of soil microbes, the ratio of P to N mineralization increases when labile P is inadequate.

To maintain the P balance at the end of a day, humus mineralization is subtracted from the organic P pool; residue mineralization is subtracted from the FOP pool; 20 percent of RMP is added to the OP pool and 80 percent of RMP is added to the labile P pool.

Immobilization.--The P immobilization model also developed by Jones et al. (1982) is similar in structure to the N immobilization model. The daily amount of immobilization is computed by subtracting the amount of P contained in the crop residue from the amount assimilated by the microorganisms.

$$WIP = (DCR')\,(FR)\,(0.16\,LF_I - c_{PFR}) \qquad (125)$$

where WIP is the P immobilization rate in kg/ha·d, c_{PFR} is the P concentration in the crop residue, and 0.16 is the result of assuming that c = 0.4 FR and 0.4 of the c in FR is assimilated by soil microorganisms; LF_I is the labile P immobilization factor allowing the P:C ratio of soil microorganisms to range from 0.01 to 0.02 as a function of labile P concentration. The labile P immobilization factor is computed with the equations

$$LF_I = 0.01 + 0.0004\,c_{LP}\,, \qquad c_{LP} \le 25 \qquad (126)$$

$$LF_I = 0.02\,, \qquad c_{LP} > 25 \qquad (127)$$

The immobilized P is added to the FOP pool and subtracted from the labile P pool.

Mineral P Cycling.--The mineral P model was developed by Jones et al. (1982). Mineral P is transferred among three pools: labile, active mineral, and stable mineral. When P fertilizer is applied, it is labile (available for plant use). However, it may be quickly transferred to the active mineral pool according to the equation

$$MPR = 0.1\,(AP)\,(SWF)\,\exp(0.115\,T - 2.88) \qquad (128)$$

where MPR is the rate of flow from the labile to the active mineral P pool in kg/ha·d for each soil layer. Simultaneously, P flows from the active mineral pool back to the labile pool (usually at a much slower rate) according to the equation

$$LPR = 0.1(MP_A)(SWF)\exp(0.115T - 2.88)\left(\frac{PSP}{1-PSP}\right) \qquad (129)$$

where LPR is the flow rate from the active mineral pool to the labile pool in kg/ha·d, MP_A is the amount in the active mineral P pool in kg/ha, and PSP is the P sorption coefficient defined as the fraction of fertilizer P remaining in the labile pool after an initial rapid phase of P sorption is complete. The P sorption coefficient is a function of chemical and physical soil properties as described by the equations

$$PSP = -0.172 - .00428\ CLA + 0.125\ pH - 0.00155\ CLA / C - 0.00265\ AP\ ,\ pH < 7. \qquad (130)$$

$$PSP = 0.769 - 0.0129\ CLA\ ,\ pH \geq 7.\ ,\ CaCO_3 = 0. \qquad (131)$$

$$PSP = 0.0675 - 0.00348\ CLA + 0.0447\ pH\ ,\quad pH \geq 7\ ,\ CaCO_3 > 0. \qquad (132)$$

where PSP is the P sorption coefficient for each soil layer and CLA is the percent clay content of the layer (Jones et al., 1982). Flow between the active and stable mineral P pools is governed by the equation

$$ASPR = (\omega)\ (c_{MPA} - 10) \qquad (133)$$

where ASPR is the flow rate between the active and stable mineral P pools in kg/ha·d for a particular soil layer, ω is the flow coefficient, and c_{MPA} is the concentration of active mineral P in the soil layer in g/t. Flow is from active to stable as long as the active pool concentration exceeds 10 g/t; the flow reverses when c_{MPA} becomes less than 10. The flow coefficient, ω, is a function of PSP as expressed by the equations (Jones et al., 1982)

$$\omega = \exp(-1.77\ PSP - 7.05)\ ,\quad CaCO_3 = 0. \qquad (134)$$

$$\omega = 0.00076\ ,\quad CaCO_3 > 0. \qquad (135)$$

Crop Uptake.--Crop use of P is estimated with the supply and demand approach described in the N model. The daily plant demand is computed with Eq. 115 written in the form

$$UPD_{IDA} = (c_{PB})_{IDA}(B)_{IDA} - (c_{PB})_{IDA}(B)_{IDA} \qquad (136)$$

where UPD_{IDA} is the P demand for the plant in kg/ha and c_{PB} is the optimal P concentration for the plant. The optimal plant P concentration is computed with Eq. 116 written in the form

$$c_{PB} = bp_1 + bp_2 \exp(-bp_3\ B_1) \qquad (137)$$

where bp_1, bp_2, and bp_3 are crop parameters expressing P concentration. Plant supply of P is estimated using the equation

$$UPS_{IDA} = 0.0075\ \Delta B_{IDA} \sum_{i=1}^{M} (SWF)_i (LF_u)_i \left(\frac{RW_i}{RWT_{IDA}}\right) \qquad (138)$$

where UPS is the amount of P supplied by the soil in kg/ha, ΔB is the change in biomass in kg/ha on IDA, LF_u is the labile P factor for uptake, RW is the root weight in layer i, and RWT is the total root weight on IDA in kg/ha. The labile P factor for uptake ranges from 0 to 1 according to the equations

$$LF_u = (c_{LP} - 5) / 20\ ,\qquad 5 \leq c_{LP} \leq 25 \qquad (139)$$

$$LF_u = 0\ ,\qquad c_{LP} < 5 \qquad (140)$$

$$LF_u = 1\ ,\qquad c_{LP} > 25 \qquad (141)$$

Implicit in Eq. 138 is the concept that one third of the crop root system is able to supply sufficient P to maintain a plant P concentration of 0.25% if c_{LP} is greater than or equal to 25 g/t (Jones et al., 1982). As with N, the actual P uptake is the minimum of UPD and UPS.

Soil Temperature

Daily average soil temperature is simulated at the center of each soil layer for use in nutrient cycling and hydrology. The basic soil temperature equation is

$$T\ (Z,\ t) = \bar{T} + \frac{AM}{2} \exp(-Z/DD) \cos\left(\frac{2\pi}{365}(t - 200) - Z/DD\right) \qquad (142)$$

where Z is depth from the soil surface in mm, t is time in d, $\bar{T}$ is the average annual air temperature in °C, AM is the annual amplitude in daily average temperature °C, and DD is the damping depth for the soil in mm. Eq. 142 provides estimates of air temperature (Z = 0) as well as soil temperature. Since air temperature is provided by the weather component of EPIC, the soil temperature model should be capable of using these air temperatures as drivers. Otherwise, Eq. 142 would predict the same temperatures for a given day each year. The following equation was developed to allow the EPIC air temperatures to drive the soil temperature model.

$$T\ (Z,\ t+1) = \chi(T(Z,t) + \frac{\partial T}{\partial t}(\Delta t)) + (1 - \chi)\ (T(Z-1,\ t+1)\ \frac{\partial T}{\partial Z}(\Delta Z)) \qquad (143)$$

where χ is a weighting factor (0-1) that allows a smoothing of daily air temperatures by placing more weight on the previous days soil temperature (a value of 0.7 is generally adequate). The partial derivatives of Eq. 142 are substituted into Eq. 143.

$$\frac{\partial T}{\partial t} = -\frac{\pi}{365}(AM) \exp(-Z/DD) \sin\left(\frac{2\pi}{365}(t-200) - Z/DD\right) \qquad (144)$$

$$\frac{\partial T}{\partial Z} = \frac{AM}{2DD} \exp(-Z/DD)\left(\sin\left(\frac{2\pi}{365}(t-200) - Z/DD\right) - \cos\left(\frac{2\pi}{365}(t-200) - Z/DD\right)\right) \qquad (145)$$

The damping depth is a function of soil bulk density and water content as expressed in the equations

$$DP = 1000 + \frac{2500\ BD}{BD + 686 \exp(-5.63\ BD)} \qquad (146)$$

$$\S = \frac{SW}{(0.356 - 0.144\ BD)\ (D_i - D_{i-1})} \qquad (147)$$

$$DD = DP \exp\left(\ln\left(\frac{500}{DP}\right)\left(\frac{1 - \S}{1 + \S}\right)^2\right) \qquad (148)$$

where DP is the maximum damping depth for the soil layer in mm, BD is the soil bulk density in t/m^3, and § is a scaling parameter.

Crop Growth Model

A single model is used in EPIC for simulating all the crops considered (corn, grain sorghum, wheat, barley, oats, sunflowers, soybeans, alfalfa, cotton, peanuts, and grasses). Of course, each crop has unique values for the model parameters. Energy interception is estimated with the equation

$$PAR = 0.02092\ (RA)\ (\frac{HRLT}{12})\ (1 - \exp(-0.65\ (LAI + 0.05))) \quad (149)$$

where PAR is the photosynthetic active radiation in MJ/m^2 and HRLT is the daylight time during a 24-h period in h. The daylight time is a function of the time of year and latitude as expressed in the equation

$$HRLT = 7.64 \cos^{-1}\left(\frac{-\sin(\frac{2\pi}{360}\ LAT)\ \sin(SD) - 0.044}{\cos(\frac{2\pi}{360}\ LAT)\ \cos(SD)}\right) \quad (150)$$

where LAT is the latitude of the watershed in degrees and SD is defined by the equation

$$SD = 0.4102 \sin(\frac{2\pi}{365}\ (IDA - 80.25)) \quad (151)$$

The potential increase in biomass for a day can be estimated with the equation

$$\Delta B_p = (BE)\ (PAR) \quad (152)$$

where ΔB_p is the daily potential increase in biomass in kg/ha and BE is the crop parameter for converting energy to biomass in kg/MJ. The LAI, a function of biomass, is estimated with the relationships

$$LAI = \frac{(LAI_{mx})\ (WLV)}{WLV + 5512 \exp(-0.000608\ WLV)}, \quad B_1 \leq DLAI \quad (153)$$

$$LAI = LAI_o \left(\frac{1 - B_1}{1 - DLAI}\right)^2, \quad B_1 > DLAI \quad (154)$$

where LAI_{mx} is the maximum LAI potential for the crop, WLV is the above-ground biomass minus yield in kg/ha, DLAI is the fraction of the growing season when LAI starts declining, and LAI_o is the LAI value from Eq. 153 when B_1 = DLAI. The value of WLV is computed daily using the equation

$$WLV = B_p - RWT - YLD \quad (155)$$

The daily fraction of the potential increase in biomass partitioned to yield is computed with the equation

$$\Delta YLD = (\Delta B_p)\ (B_1)\ /\ GK \quad (156)$$

where GK is the ratio of total biomass to crop yield under favorable growing conditions. The dimensionless biomass variable B_1 acts as a weighting function in Eq. 156--early in the growing season little of the biomass is partitioned into yield because B_1 is small. As the growing season progresses, however, more and more weight is given to the yield component of the biomass. For this reason, late season stresses (water, temperature, or nutrients) reduce yields more than early season stresses. The amount of root growth for a day is estimated with the equation

$$\Delta RWT = \Delta B_p\ (0.4 - 0.2\ B_1) \quad (157)$$

Daily root sloughing is estimated with the equation

$$\Delta RWS = 0.1\ (\Delta B_p)\ (B_1) \quad (158)$$

where ΔRWS is the amount of root sloughing during a day in kg/ha. The change in root weight through the root zone is simulated as a function of plant water use and accumulated root weight in each soil layer using the equations

$$RW_j = RW_{oj} + (\Delta RWL)\ (RWF)_j \quad (159)$$

where RW_o and RW are the root weights in soil layer j at the start and end of a day in kg/ha, ΔRWL is the change in live root weight during a day in kg/ha, and RWF is a root weight distribution factor. The daily change in live roots is computed by subtracting the sloughed roots from the total root growth

$$\Delta RWL = \Delta RWT - \Delta RWS \quad (160)$$

The root weight distribution factor is determined by considering the sign of ΔRWL.

$$RWF = u_i\ /\ \sum_{i=1}^{M} u_i, \quad \Delta RWL > 0 \quad (161)$$

$$RWF = RW_i\ /\ \sum_{i=1}^{M} RW_i, \quad \Delta RWL < 0 \quad (162)$$

Eqs. 161 and 162 simply distribute root growth as a function of water use if the root weight is increasing and as a function of existing root weight if root weight is decreasing. Rooting depth is simulated as a function of heat units and potential root zone depth.

$$\Delta RD = 2(RZ)\ (HU)\ /\ PHU, \quad RD \leq RZ \quad (163)$$

where ΔRD is the daily change in crop root depth in mm.

The potential biomass predicted with Eq. 152 is adjusted daily if one of the plant stress factors is less than 1.0 using the equation

$$\Delta B = (\Delta B_p)\ (REG) \quad (164)$$

Where REG is the crop growth regulating factor (the minimum stress factor). The water stress factor is computed by considering supply and demand in the equation

$$WS = \frac{E_p}{\sum_{i=1}^{M} u_i} \quad (165)$$

where WS is the water stress factor (0-1). The value of E_p is predicted in the evapotranspiration model, and u_i is a function of depth and soil water content.

$$u_{pi} = \frac{E_p}{1 - \exp(-\Lambda)}\left(1 - \exp(-\Lambda(\frac{RD_i}{RZ})\right) - \sum_{j=1}^{i-1} u_j \quad (166)$$

where u_{pi} is the potential water use rate from layer i in mm/d and Λ is a use rate-depth parameter. The details of evaluating Λ are given by Williams and Hann (1978). The value of Λ used in EPIC (3.065) was determined assuming that about 30% of the total water use comes from the top 10% of the root zone. Eq. 166 allows roots to compensate for water deficits in certain layers by using more water in layers with adequate supplies. The potential water use must be adjusted for water deficits to obtain the actual use for each layer.

$$u_i = u_{pi}, \quad SW > 0.25\ UL \quad (167)$$

$$u_i = u_{pi}\ (\frac{SW}{0.25\ UL}), \quad SW \leq 0.25\ UL \quad (168)$$

Note that the actual use for the layers above the i'th layer in Eq. 166 must be known. This presents no problems, however, since $u_j = 0$ when i = 1.

The temperature stress factor is computed with the equation

$$TS = \exp\left(\Omega\left(\frac{T_o - T}{T}\right)^2\right) \qquad (169)$$

where TS is the temperature stress factor (0-1), Ω is the temperature stress parameter for the crop, T_o is the optimal temperature for the crop in °C, and T is the daily average air temperature in °C. The stress parameter is evaluated by appropriate substituting and rearranging of Eq. 169

$$\Omega = \frac{\ln (0.9)}{\left(\frac{T_o - \frac{T_o + T_b}{2}}{\frac{T_o + T_b}{2}}\right)^2} \qquad (170)$$

where T_b is the base temperature for the crop in °C. Eq. 170 sets TS = 0.9 when the air temperature is half way between T_b and T_o.

The N and P stress factors are based on the ratio of accumulated plant N and P to the optimal values. The stress factors vary nonlinearly from 1.0 at optimal N and P levels to 0. when N or P is half the optimal level. In the case of N, the scaling equation is

$$SN_{S,IDA} = 2\left(1 - \frac{\sum_{j=1}^{M} UN_j}{(c_{NB})_{IDA}\ (B)_{IDA}}\right) \qquad (171)$$

where $SN_{S,IDA}$ is a scaling factor for the N stress factor and UN is the crop N uptake on day j in kg/ha. The N stress factor is computed with the equation

$$SN_{IDA} = 1 - \frac{SN_{S,IDA}}{SN_{S,IDA} + 29.534\ \exp(-10.93\ SN_{S,IDA})} \qquad (172)$$

where SN_{IDA} is the N stress factor for day IDA. Eqs. 171 and 172 allow SN to range from 0.0 when the ratio UN/c_{NB}·B) is equal 0.5 to 1.0 when the ratio is 1.0. The P stress factor is computed with Eqs. 171 and 172 written in P terms. Finally, the value of REG is determined as the minimum of WS, TS, NS, and PS. REG is used to adjust YLD, RWT, and RWL with equations similar to Eq. 164.

Tillage

The primary purpose of the tillage component of EPIC is to mix nutrients and crop residue within the plow depth. Each tillage operation is assigned a mixing efficiency (0-1). The tillage mixing equation is

$$X_i = (1 - EF)\ X_{oi} + \left(\frac{D_i}{PD}\right)\ (EF)\ \sum_{j=1}^{M} X_{oj} \qquad (173)$$

where X_i is the amount of the material in layer i after mixing in kg/ha, EF is the mixing efficiency of the tillage operation (0-1), X_{oj} is the amount of the material in layer j before mixing in kg/ha, and M is the number of soil layers in the plow depth, PD, in mm.

Other functions of the tillage component include simulating row height and surface roughness. Both of these variables are specified for each tillage implement. Also, the user specifies the date and depth for each tillage operation. The tillage operation is carried out on the specified date if the soil is dry enough. If not, the operation occurs on the next suitable day.

There are three means of harvest in the EPIC model--(1) traditional harvest that removes seed, fiber, etc.; (2) hay harvest (may occur on any date the user specifies); and (3) no harvest (green manure crops, etc.). The traditional harvest partitions the crop stover into 10% residue on the top soil layer and 90% standing residue. A shredder is often applied after harvest and further partitions the residue according to the height before and after cutting. Tillage operations convert standing residue to flat residue by applying the mixing efficiency coefficient

$$\Delta FR = (EF)\ (SR) \qquad (174)$$

When hay is harvested, the yield is 95% of the above ground biomass.

Economics

The crop budgets are calculated using components from the Crop Budget Generator developed at Oklahoma State University (Kletke, 1979). Budgets can be calculated for each individual year in the EPIC simulation or for the average yields and resource requirements for the period of simulation.

Inputs are divided into two catagories: fixed and variable. Fixed inputs include depreciation, interest or return on investment, insurance, and taxes on equipment, land, and capital improvements (terraces, drainage, irrigation systems, etc.). Variable inputs are defined as machinery repairs, fuel and other energy, machine lubricants, seed, fertilizer, pesticides, labor, and irrigation water.

Total variable cost is expressed with the equation

$$TVC = \sum_{i=1}^{M} (PR_i)\ (QA_i) \qquad (175)$$

where TVC is the total variable cost in dollars, PR is the price of the variable input (i) in dollars, and QA is the quantity of the input used per ha.

The price can be input at the time of the run or can be stored with the price and input file of 500 allowable inputs maintained as part of the budget generator subprogram. The quantities used (seed, fertilizer, irrigation, etc.) are consistent with the input requirements used elsewhere in EPIC.

The machinery complement file is a list of 100 pieces of equipment for use in simulating user-specified tillage operations. This file contains equipment information like purchase price, size, expected life, and repair cost. With this information, fixed costs for the machinery are allocated to each crop on an hours-of-use basis. The machinery fixed cost equations are given below.

$$DEPC = \frac{PP - SV}{(HUA)\ (YRO)} \qquad (176)$$

where DEPC is the depreciation cost in dollars/hr, PP is the purchase price in dollars, SV is the salvage value in dollars, HUA is the annual use in h, and YRO is the time the equipment is owned in yrs.

$$AI = \frac{PP + SV}{2\ HUA} \qquad (177)$$

where AI is the average investment in dollars/h.

$$IC = (AI)\ (IT) \qquad (178)$$

where IC is the interest cost in dollars/h and IT is the interest rate.

$$INSC = (AI)\ (INSR) \qquad (179)$$

where INSC is the insurance cost in dollars/h and INSR is the insurance rate.

$$MTAX = \frac{(PP)\ (TR)}{HUA} \qquad (180)$$

where MTAX is the machinery tax in dollars and TR is the tax rate.

Machine variable cost are also calculated on a per hour basis.

$$PL = \frac{(HUA)\ (YRO)}{HOL} \qquad (181)$$

where PL is the percent of machinery life used in a given year and HOL is the total machine life in h.

Total accumulated repair cost is calculated with the equation

$$TAR = (rc_1)\ (PP)\ (rc_2)\ (PL)\ (rc_3) \qquad (182)$$

where TAR is the total accumulated cost of repairs in dollars and rc_1, rc_2, and rc_3 are repair cost coefficients (American Society of Agricultural Engineers, 1971). Repair cost is placed on an hourly basis using the equation

$$rc = \frac{TAR}{(HUA)\ (YRO)} \qquad (183)$$

Fuel consumption cost is given by the equation

$$CF = (DH)\ (CFM)\ (PF) \qquad (184)$$

where CF is the fuel consumption cost in dollars/h, DH is the drawbar horsepower, CFM is the fuel consumption coefficient, and PF is the price of the fuel in dollars. Lubrication cost is estimated as 15% of the fuel cost.

$$LC = 0.15\ (CF) \qquad (185)$$

where LC is the lubrication cost in dollars/h.

Actual annual cost of operating machinery can be estimated by converting from cost per h to cost per ha using the equation

$$HPA = \frac{10}{(VT)\ (WD)\ (FE)} \qquad (186)$$

where HPA is the time required to till one ha in h, VT is the tractor velocity in km/h, WD is the implement width in m, and FE is the field efficiency of the equipment. Finally, total annual machinery is estimated by summing the cost for all the individual operations.

Fixed costs like rent, land taxes, and management are also charged on a per ha basis. Total cost is the sum of the variable and fixed cost.

Gross income from the crop is simply the market price of the crop times the yield minus any marketing or harvest cost not accounted for in the machinery cost. Net profit, of course, is the difference between gross income and total cost. Simulated annual net profits are useful in illustrating the effects of erosion.

MODEL TESTS

Although EPIC has been run on a number of trial sites, the model is still in the early stages of testing. The main purpose of testing up to this point was to determine the model's reliability under a variety of conditions. EPIC simulations have been performed on 17 test sites in the continental U.S. and 13 in Hawaii. In Table 1 we show results of the simulations for 13 of the test sites. Generally, the test results appear reasonable, although the simulated crop yields may be overpredicted in some areas. Obviously, the test sites shown in Table 1 provide a wide range in climatic conditions, soil characteristics, and management practices.

To determine the model's sensitivity to erosion, crop yields were related to accumulated erosion for the 50-year simulation periods. The resulting linear regression equations were used to compare expected yields at the start and end of the 50-year period. Generally, the regression analysis indicated a reduction in crop yield depending upon soil and climatic characteristics and fertilization rate. In some areas with high erosion rates and unfavorable subsoil, the reduction in crop yield was as much as 40%.

CONCLUSIONS

The EPIC model is operational and has produced reasonable results under a variety of climatic conditions, soil characteristics, and management practices. It has also demonstrated sensitivity to erosion in terms of reduced crop production.

More extensive testing is planned for EPIC. Although some components of the model like hydrology and erosion are based on accepted technology, other components will require rigorous testing for validation. The two components that will need testing most are crop growth and nutrients. This is true because these are newly developed and because they are extremely important to the success of the EPIC model.

EPIC has many potential uses beyond the RCA analysis including: (a) national level conservation policy studies; (b) national level program planning and evaluation; (c) project level planning and design; (d) field level studies to aid technical assistance; and (e) as a research tool.

Table 1.--Example results from long-term EPIC simulations (50 yrs each).

Location	Average Annual Values								
	Rainfall (mm)	Snowfall (mm)	Surface Runoff (mm)	Sediment Yield (t/ha)	Crop Yields (kg/ha)				
					Soybeans	Corn	Grain Sorghum	Wheat	Cotton
Bell, TX	867	0	188	23			4930	3220	2140
Caldwell, KY	1145	1	105	33	3160	8130		4070	
Canadian, OK	767	0	59	14				3280	
Fayette, KY	1087	42	55	14		9800			
Gallatin, MT	343	78	24	2				2020	
Hinds, MS	1229	0	203	112	2570				
Maricopa, AZ	181	0	2	0.2					4250*
Morton, ND	399	65	31	2				1470	
Potter, TX	517	2	80	4			10600*	3480*	
Richland, MT	347	67	27	7				1350	
Hays, KS	529	20	27	2				1270	
Fountain, IN	948	171	157	13	2730	8440			

* Irrigated crops

LITERATURE CITED

American Society of Agricultural Engineers. 1971. Agricultural machinery management data. Agricultural Engineers Yearbook, pp. 287-294.

Hershfield, D. M. 1961. Rainfall frequency atlas of the United States for durations from 30 minutes to 24 hours and return periods from 1 to 100 years. U.S. Dept. of Commerce, Technical Paper No. 40, 115 pp.

Jones, C. A., C. V. Cole, and A. N. Sharpley. 1982. A simplified soil phosphorus model, I. Documentation. Manuscript submitted for publication in Soil Science Society of America Journal.

Kletke, D. D. 1979. Operation of the enterprise budget generator. Oklahoma State University, Agricultural Experiment Station Research Report P-790.

Knisel, W. G. 1980. CREAMS, A field scale model for chemicals, runoff, and erosion from agricultural management systems. USDA Conservation Research Report No. 26, 643 pp.

McElroy, A. D., S. Y. Chiu, J. W. Nebgen, A. Aleti, and F. W. Bennett. 1976. Loading functions for assessment of water pollution from nonpoint sources. Environmental Protection Technology Series, U.S.E.P.A., EPA-600/2-76-151, 445 pp.

Meister, A. D. and K. J. Nicol. 1975. A documentation of the national water assessment model of regional agricultural production, land and water use, and environmental interaction. Center for Agricultural and Rural Development, Iowa State University Misc. Report.

Nicks, A. D. 1974. Stochastic generation of the occurrence, pattern, and location of maximum amount of daily rainfall. Proceedings Symposium on Statistical Hydrology, Tucson, AZ, Aug.-Sept. 1971. Misc. Pub. No. 1275, pp. 154-171.

Onstad, C. A. and G. R. Foster. 1975. Erosion modeling on a watershed. Trans. ASAE 18(2):288-292.

Richardson, C. W. 1981. Stochastic simulation of daily precipitation, temperature, and solar radiation. Water Resources Research 17(1):182-190.

Richardson, C. W. 1982. Dependence structure of daily temperature and solar radiation. Trans. ASAE. In press.

Ritchie, J. T. 1972. A model for predicting evaporation from a row crop with incomplete cover. Water Resources Research 8(5):1204-1213.

Seligman, N. G. and H. van Keulen. 1981. PAPRAN: A simulation model of annual pasture production limited by rainfall and nitrogen. In: Simulation of nitrogen behaviour of soil-plant systems, Proceedings of a workshop, Wageningen, Jan. 28-Feb. 1, 1980. M. J. Frissel and J. A. van Veen, Eds., pp. 192-221.

USDA, Soil Conservation Service. 1972. National Engineering Handbook, Hydrology Section 4, Chapters 4-10.

U.S. Department of Commerce. 1968. Climatic Atlas of the United States. Environmental Science Services Administration, Environmental Data Service, 80 pp.

U.S. Department of Commerce. 1979. Maximum short duration rainfall. National Summary, Climatic Data.

Williams, J. R. 1975. Sediment yield prediction with universal equation using runoff energy factor. USDA, ARS-S-40, pp. 244-252.

Williams, J. R., Chairman, National Soil Erosion-Soil Productivity Research Planning Committee, USDA-ARS. 1981. Soil erosion effects on soil productivity: A research perspective. Journal of Soil and Water Conservation 36(2):82-90.

Williams, J. R. and R. W. Hann. 1978. Optimal operation of large agricultural watersheds with water quality constraints. Texas Water Resources Institute, Texas A&M University, Technical Report No. 96, 152 pp.

Williams, J. R. and A. D. Nicks. 1982. CREAMS hydrology model--option one. In: Applied Modeling Catchment Hydrology, Proceedings of the International Symposium on Rainfall-Runoff Modeling, May 18-21, 1981, Mississippi State, Mississippi, V. P. Singh, Editor, pp. 69-86.

Wischmeier, W. H. and D. D. Smith. 1978. Predicting rainfall erosion losses, a guide to conservation planning. USDA Agricultural Handbook No. 537, 58 pp.

Woodruff, N. P. and F. H. Siddoway. 1965. A wind erosion equation. Soil Science Society of America Proceedings 29(5):602-608.

SYMBOLS

α = maximum amount of rainfall during t_c divided by the total rainfall for the day

α_{mn} = minimum value of α

$\alpha_{.5}$ = maximum amount of rainfall during 0.5 h divided by the total rainfall for the day

$\alpha_{.5P}$ = $\alpha_{.5}$ value at the peak of the gamma distribution

$\alpha_{.5U}$ = upper limit of $\alpha_{.5}$

β_i = parameter for regulating hydraulic conductivity in soil layer i

γ = psychrometric constant

δ = slope of the saturation vapor pressure curve at the mean air temperature

ζ_i = soil parameter expressing degree of cracking for layer i

η = gamma distribution shape parameter for estimating wind velocity

θ = wind direction clockwise from north (radians)

λ = land surface slope length (m)

Λ = water use rate-depth parameter (3.065)

μ_R = annual mean of solar radiation (ly) or coefficient of variation of solar radiation

μ_T = annual mean temperature (°C) or coefficient of variation of temperature

ν = gamma distribution shape parameter for estimating $\alpha_{.5}$

ξ = USLE LS factor parameter (function of slope)

ρ = rational formula runoff coefficient

σ = average channel slope (m/m)

τ = total number of maximum 0.5 h rainfall events considered

ϕ = clockwise angle between field length and north (radians)

χ = soil temperature model weighting factor (0-1)

ψ_1, ψ_2 = wind erosion parameter related to vegetative cover factor

ω = mineral P coefficient for flow between active and stable pools

Ω = temperature stress parameter for a crop

§ = soil temperature damping depth scaling parameter

a_R = amplitude of the means and coefficients of variation of daily solar radiation

a_T = amplitude of the means and coefficients of variation of daily temperature

A = drainage area (acres)

AB = albedo for solar radiation

AB_s = soil albedo

AE = average annual wind energy (KWH/m^2)

AI = average equipment investment (dollars/h)

AM = annual amplitude in daily average temperature (°C)

AP = labile P in the soil (kg/ha)

ASPR = flow rate between active and stable mineral P pools (kg/ha·d)

b = TP-40 rainfall distribution parameter

bn_1, bn_2, bn_3 = crop parameters relating c_{NB} to BD

bp_1, bp_2, bp_3 = crop parameters relating c_{PB} to BD

B = accumulated plant biomass (above ground and roots (kg/ha)

B_1 = dimensionless expression of accumulated biomass (0-1)

B_{AG} = above ground biomass of growing crop (kg/ha)

B_P = potential crop biomass (kg/ha)

BD = soil bulk density in t/m^3

BE = crop parameter--converts energy to biomass (kg/MJ)

c_{LP} = labile P concentration in the soil (g/t)

c_{MPA} = concentration of active mineral P (g/t)

c_{NB} = optimal N concentration for a plant (kg/kg)

c_{NFR} = N concentration in crop residue (kg/ha)

c_{NO3} = NO_3 concentration in water (g/m^3)

c_{NY} = N concentration in crop yield (kg/kg)

c_{ON} = concentration of organic N in the soil (g/t)

c_P = P concentration in the soil (g/t)

c_{PB} = optimal P concentration for a plant (kg/kg)

c_{PFR} = P concentration in crop residue (kg/kg)

c_s = sediment concentration in runoff (g/m^3)

C = organic carbon content in soil (%)

CDN = nitrification constant

CE = USLE crop management factor

CF = fuel consumption cost (dollars/h)

CFM = fuel consumption coefficient

CLA = percent clay in the soil

CMN = constant for N mineralization from humus

CN = SCS runoff curve number

CN_I = the SCS curve number for the I moisture (dry) condition

CN_{II} = CN for the II moisture (average) condition

CNP = C:N, C:P ratio factor

CNR = C:N ratio of crop residue

CPR = CP: ratio of crop residue

CV = soil cover (above ground biomass + residue) (kg/ha)

D_i = depth to the bottom of soil layer i (mm)

DA = drainage area (ha)

DCR = decay rate for fresh organic material (crop residue) (d^{-1})

DCR' = decay rate for fresh organic material adjusted for limited supply of N or P

DD = soil temperature damping depth (mm)

DE = daily wind energy (KWH/m^2)

DEPC = depreciation cost (dollars/h)

DH = drawbar horsepower

DP = maximum damping depth for soil temperature (mm)

DLAI = fraction of growing season when LAI starts declining

DR = sediment delivery ratio

DUR = rainfall duration (h)

EA = soil cover index (0-1)

EF = tillage mixing efficiency (0-1)

EI = USLE rainfall energy factor (metric units)

ENO3 = amount of NO_3 moved from lower layers to top layer by soil evaporation (kg/ha)

E_o = potential evaporation (mm)

E_p = potential plant evaporation rate (mm/d)

E_s = soil evaporation rate (mm/d)

E_{so} = potential evaporation at the soil surface (mm)

ER = enrichment ratio (concentration of organic N or total P in the sediment divided by the soil concentration)

f = average infiltration rate during a storm (mm/h)

F = frequency of occurrence of the largest 0.5 h rainfall recorded

FC_i = field capacity (0.3 bar-15 bar) water content of soil layer i (mm)

FE = machinery field efficiency

FL = field length (m)

FON = fresh organic N in crop residue (kg/ha)

FOP = fresh organic P in crop residue (kg/ha)

FR = flat crop residue (kg/ha)

FW = field width (m)

g_1, g_2, g_3 = crop coefficients relating crops and residue to small grain equivalent

GK = ratio of total biomass to crop yield under favorable growing conditions

h_o = net solar radiation (ly)

H_i = hydraulic conductivity (mm/h)

HMN = rate of N mineralization from humus (kg/ha·d)

HMP = rate of P mineralization from humus (kg/ha·d)

HOL = machine life (h)

HPA = time required for tillage operation (h/ha)

HR = row height (mm)

HRLT = daylight time during a 24-h period (h)

HU = daily heat units above the crops base temperature (°C)

HUA = annual equipment use (h)

I = wind erosion soil erodibility index (t/ha)

IC = interest cost (dollars/h)

IDA = julian date

INSC = insurance cost (dollars/h)

INSR = insurance rate

IR = row interval (mm)

IT = interest rate

k = decay constant in exponential rainfall distribution (h)

k_d = P concentration in sediment divided by that of the water (m^3/t)

K = USLE soil erodibility factor

KR = ridge roughness (mm)

L = channel length from the most distant point to the watershed outlet (km)

L_c = average channel flow length (km)

L_{ca} = distance along the channel to the watershed centroid (km)

LAI = leaf area index--area of plant leaves relative to the soil surface

LAI_o = LAI value when LAI starts declining

LAI_{mx} = maximum LAI potential for a crop

LAT = latitude of watershed (degrees)

LC = lubrication cost (dollars/h)

LF_I = labile P factor for P immobilization

LF_M = labile P factor for P mineralization

LF_U = labile P factor for crop uptake

LPR = flow rate from active mineral pool to labile pool (kg/ha·d)

LS = USLE slope length and steepness factor

MP_A = amount of P in active mineral pool (kg/ha)

MPR = flow rate from labile to active mineral P pool (kg/ha·d)

MTAX = machinery tax (dollars)

n = Manning's roughness coefficient

O = percolation rate (mm/d)

ON = organic N content of the soil (kg/ha)

OP = organic P content of the soil (kg/ha)

PAR = photosynthetic active radiation (MJ/m^2)

PD = plow depth (mm)

PE = USLE erosion control practice factor

PF = fuel price (dollars)

PHU = potential HU for a crop (°C)

PL = percent of machinery life

POR = soil porosity (mm)

PP = equipment purchase price (dollars

PR = price of the variable input (dollars)

PSP = P sorption coefficient

q_c = average channel flow rate (m^3/s)

q_p = peak runoff rate (mm/h)

q_s = average surface flow

Q = surface runoff volume (mm)

QA = quantity of input used per ha

QR_i = lateral subsurface flow rate from soil layer i (mm/d)

QT = total water drained from a soil layer during a day (mm)

r = rainfall intensity (mm/h)

r_{ep} = peak rate of rainfall excess (mm/h)

r_p = peak rate of rainfall (mm/h)

rc = machinery repair cost (dollars/h)

rc_1, rc_2, rc_3 = parameters relating machinery repair cost to purchase price

R = daily rainfall amount (mm)

$R_{.5F}$ = maximum 0.5 h rainfall amount for frequency F (mm)

$\overline{R}$ = mean rainfall amount per event (mm)

$\overline{R}_{.5}$ = mean maximum 0.5 h rainfall amount per event (mm)

R_t = rainfall amount at time t (mm)

RA = solar radiation (ly)

RC = residue composition factor (d^{-1})

RD = root depth (mm)

REG = crop growth regulator (minimum stress factor--water, nutrients, temperature)

RMN = mineralization rate for FON (kg/ha·d)

RMP = mineralization rate for FOP (kg/ha·d)

RW_i = root weight in soil layer i (kg/ha)

RWF = root weight distribution factor

RWL = weight of live roots (kg/ha)

RWS = weight of roots sloughed (kg/ha)

RWT = total root weight (kg/ha)

RZ = root zone depth (mm)

s = SCS curve number retention parameter (mm)

s_{mx} = maximum value of s (mm)

S = land surface slope (m/m)

SC_i = saturated conductivity of layer i (mm/h)

SD = time dependent variable used in computing HRLT (radians)

SDV = standard deviation of daily wind velocity (m/s)

SR = standing crop residue (kg/ha)

SV = equipment salvage value (dollars)

SW = soil water content minus 15-bar water content (mm)

SW_{15} = soil water content at 15-bars (mm)

SWF = soil water factor for nutrient cycling

t = time (h)

t_c = watershed time of concentration (h)

t_{cc} = time of concentration for channel flow (h)

t_{cs} = time of concentration for surface flow (h)

T = temperature (°C)

$\overline{T}$ = average annual air temperature (°C)

T_b = base temperature for a crop (plants start growing (°C)

T_K = temperature (°K)

T_o = optimal temperature for a crop (°C)

TAR = total accumulated cost for repairs (dollars)

TF_N = temperature factor for nutrient cycling

TR = tax rate

TS = temperature stress factor for crops (0-1)

TT_{ci} = crack flow travel time for layer i (h)

TT_i = travel time of flow through soil layer i (h)

TT_{Ri} = subsurface flow time to travel a distance equal the land surface slope length λ in layer i (d)

TVC = total variable cost (dollars)

u_i = plant water use in soil layer i (mm)

u_{pi} = potential plant water use in soil layer i (mm)

U = frequency of occurrence of wind velocity V

UL = upper limit of soil water storage (porosity minus 15-bar water content) (mm)

UN = accumulated N use by the crop (kg/ha)

UND = N demand by a plant (kg/ha)

UNS = N supplied by the soil (kg/ha)

UPD = P demand by a plant (kg/ha)

UPS = P supplied by the soil (kg/ha)

v_c = average channel velocity (m/s)

v_s = surface flow velocity (m/s)

V = wind velocity (m/s)

$\overline{V}$ = annual average wind velocity (m/s)

V_L = lower limit of erosive wind velocity (m/s)

V_p = wind velocity at the peak frequency of the gamma distribution (m/s)

V_u = upper limit of wind velocity (from monthly gamma distribution) (m/s)

VE = quantity of vegetative cover expressed as small grain equivalent (kg/ha)

VNO3 = amount of flow related NO_3 loss from a soil layer (kg/ha)

VT = tractor velocity (km/h)

w_i = weighing factor for soil water content in layer i

WC = wind erosion climatic factor

WD = implement width (m)

WE = daily wind erosion (t/ha)

WF = wind erosion field length factor

WIM = N immobilization rate (kg/ha·d)

WIP = P immobilization rate (kg/ha·d)

WK = wind erosion soil ridge roughness factor

WL = field length along the prevailing wind direction (m)

WL_o = maximum field length for reducing wind erosion estimate (m)

WLV = weight of crop stover (kg/ha)

WNO3 = weight of NO_3 in a soil layer (kg/ha)

x_1, x_2 = nutrient enrichment ratio parameters

X = material being mixed (nutrients or crop residue) by tillage (kg/ha)

XT_{IDA} = mean or coefficient of variation for temperature on julian day IDA

XR_{IDA} = mean or coefficient of variation for solar radiation on julian day IDA

Y = sediment yield (t/ha)

YLD = crop yield (kg/ha)

YON = loss of organic N in runoff

YP = loss of sediment phase P in runoff (kg/ha)

YRO = time equipment owned (yrs)

YSP = loss of soluble P in runoff (kg/ha)

Z = soil depth from the surface (mm)

SECTION V:

Application of Ecological Models to Water Resources

APPLICATION OF A SEASONAL PHOSPHORUS DYNAMICS MODEL TO A LAKE ECOSYSTEM: ASSESSING LAKE LOADING TOLERANCE[1]

Vicki J. Watson[2]

Abstract.--Phosphorus fluxes in a lake ecosystem were simulated with a model that assesses lake loading tolerance by determining the sensitivity of seasonal P dynamics and other indices of lake response to changes in loadings, rate parameters and lake structure. The model was applied to Lake Mendota, Wisconsin.

INTRODUCTION

In order to assess the relative importance of internal and external fluxes of phosphorus to the P dynamics of a specific lake ecosystem, a simulation model of seasonal P dynamics in such a system was constructed. The model was formulated to include an assessment of lake loading tolerance in the form of lake-specific temporal-spatial structure as well as several lake-specific processes and rates. Thus the model could be used to assess the sensitivity of seasonal P dynamics and a number of synthetic indices of lake response to manipulations of external loadings, rate parameters and year-specific lake structure. The indices include average annual standing crop of P and of algal P, annual algal uptake of P, net sedimentation and retention of P.

A SEASONAL PHOSPHORUS DYNAMICS SIMULATOR FOR MEDIUM-SIZED, TEMPERATE LAKE ECOSYSTEMS

General Framework

The exact nature and relative importance of internal and external fluxes of phosphorus are likely to be lake-specific; hence, only the basic framework of the conceptual model outlined here could be used as a general model. The desire was to construct a conceptual model that balanced generality and specificity. This was accomplished by emphasizing processes that are thought to be important in medium-sized, rich, north temperate lake ecosystems. The relative importance of these processes was then made a function of the specific lake's morphometry and thermal structure.

The pools of P simulated by this model have been reduced to those believed to be the minimum necessary to illuminate the magnitude and nature of the internal fluxes of interest. To this end, the pools simulated are: (1) dissolved P (DP), assumed to approximate P available for algal uptake and operationally defined as soluble reactive P, and (2) particulate P (PP), assumed to approximate the P tied up in biomass and detritus and operationally defined as total P minus soluble reactive P. Other forms of P were assumed to be not sufficiently important to the questions addressed here to invalidate these simplifying assumptions.

A lake may be conceptualized as well-mixed with regard to the DP and PP pools at all times except during periods of stratification (if present). During stratification the lake is treated as if it consists of two layers (epilimnion and hypolimnion) that are well-mixed internally but have restricted interchange between them.

As was stated, the relative importance of various fluxes in the lake is largely determined by the lake's constant morphometry and seasonally changing thermal structure. The latter acts on the former to produce a seasonally changing lake spatial structure. A simple conceptualization of this process and how it influences various internal fluxes is presented in figure 1. In this conceptual model, the bottom of the upper mixed zone (at depth ZE) is assumed to migrate vertically over the year. During nonstratified periods, the mixed zone extends to the bottom of the lake (ZE = mean depth of the lake) and the volume of the mixed zone (VE) is equal to the volume of the lake (V). During periods of stratification, the mixed zone extends to the sediments in shallow areas and to the thermocline in deep areas. The volume of this region divided by the lake's surface area is the mean depth of the mixed zone (ZE). Thus, a lower region, the hypolimnion, with volume VH = V-VE is defined. The depth of ZE changes over the stratified period as the thermocline migrates. The thermocline may be operationally defined as the depth interval over which temperature first begins to drop by one degree centigrade per meter; however, when a shallow secondary thermocline develops for a short period during calm sunny conditions, the lower primary thermocline is still used to define ZE.

In addition to defining the volume of the epilimnion and hypolimnion, ZE determines the area of sediments in contact with these regions and the area of contact between the two regions (AZE--see fig. 1) The latter is the area of the lake at the depth ZE. Because most lakes are very wide relative to their depth, the area of sediments in contact with the hypolimnion can often be assumed to be approximately AZE (for deeper lakes the actual area should be used). Thus, the area of sediments in contact with the epilimnion (ASE) is the area of the lake (A) minus AZE. During nonstratified periods the area of sediments in contact with the mixed zone (the entire lake) is the area A. However, it may be desirable to divide this

---

[1]Paper presented at the Third International Conference on State-of-the-Art in Ecological Modelling, Colorado State University, May 24-28, 1982.

[2]Vicki J. Watson is Research Associate of Limnology, Institute for Environmental Studies, University of Wisconsin, Madison, Wisconsin U.S.A.

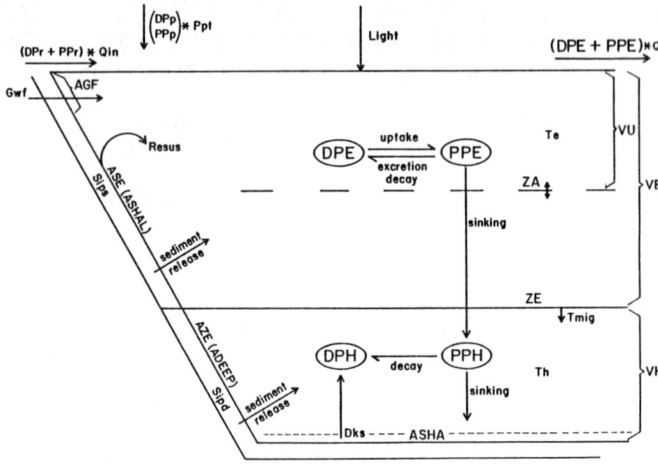

Figure 1.

Conceptual Model of Seasonal Phosphorus Dynamics Control in a Stratified Lake Ecosystem (terms defined in text)

area into a number of smaller areas if one has information on the nature of these areas that would permit more accurate estimates of sediment loading. In the present work, the sediments were divided into that area (ASHAL) shallower than the mean depth of the lake (12.4 meters) and that deeper (ADEEP). Differences in concentration of P in interstitial waters (Sips, Sipd) in these areas was evident, suggesting this division. It was also hypothesized that resuspension occurred primarily from the shallow areas and the division was necessary to show the effect of making this assumption versus assuming that resuspension occurred from all sediment areas.

Two final areas of interest are the area of groundwater flux (AGF) and the area of anoxic sediments (ASHA). The latter is determined based on lake morphometry and the depth of anoxic water (see later section on sediment processes). The area of groundwater flux AGF is not included in the area of sediments used to determine sediment release by a diffusion-like process since the yield from AGF sediments is already accounted for in the groundwater advective flux calculation.

The final structural variable that influences system processes is the depth to which algae are mixed (ZA). The buoyant algae that dominate most rich lakes during stratification often are concentrated at a depth shallower than the bottom of the epilimnion (unless this is fairly shallow). Thus, an empirically determined mean depth of algae (ZA) may be used to determine available light for algal growth and the volume (VU) where most uptake occurs. When the algae are mixed to the bottom of the mixed zone, ZA = ZE and VU = VE. Thus, lake morphometry and year-specific thermal structure define the volumes (and dilutional capacity) of certain regions of the lake as well as the magnitude of the areal fluxes that enter them.

External factors acting on the lake system include external loading, light, and the flux of water through the system (Qin, Qout). External loads include rural and urban runoff (DPr, PPr) and bulk precipitation (DPp, PPp). Groundwater loading may be considered either an external or internal load since this load is a function of groundwater flow into the system and the concentration of P in the pore water of the lake sediments in the area AGF. Loss from the system is estimated as the volumetric outflow of water (QOut) times the concentration of the P species in the mixed zone (DPE and PPE). Thus, inputs to the system enter the mixed zone and outflow losses from the system leave this region.

Internal processes have been assumed to be reducible to the following. Sediments interact with the overlying waters primarily through the release fo DP by a diffusion-like process. The physical and chemical condition of the sediments determines the magnitude of the flux, and sediment areas characterized by these conditions vary seasonally. Resuspension of particulates and the settling of same are poorly understood processes and appear to be highly stochastic (or at least not predictable from P dynamics). Consequently, these are best treated initially in an empirical manner. Exchange between the epilimnion and hypolimnion may be handled in a number of ways but the simplest approaches (bulk eddy diffusion and thermocline migration-TMIG) should be utilized if possible. Exchange between dissolved and particulate P in the water column is dominated by biological processes. These may be described by simple equations that estimate algal uptake, zooplankton excretion, and bacterial decay as functions of maximum rate parameters, modified by environmental conditions (e.g., temperature, Te and Th). Each of these internal processes will be dealt with in separate sections (hydrodynamic processes, physical-chemical processes, and biological processes).

Hydrodynamic Processes

The main hydrodynamic processes considered in this simple budget model are losses from the lake via the outlet and exchanges between the epilimnion and hypolimnion. Losses from the lake are calculated simply as the volume of water leaving the lake per unit time times the P concentration in the mixed zone over that time.

Exchanges between the hypolimnion and epilimnion may be handled in a number of ways and two methods were investigated. Typically, the exchange is estimated using a Fickian diffusion model that estimates bulk diffusion between the two layers or between a number of layers within the water column (Powell and Jassby 1974; Imboden and Gächter 1977; Lerman 1979; Berner 1980). In order to generate meaningful values for the eddy diffusion coefficient, fairly detailed and accurate temperature profiles are required (Stauffer 1974). The temperature should be recorded at the same location and the same time each day. Frequent profiles (every few days) recorded to an accuracy of 0.1°C are believed to be necessary to assess any short-term mixing phenomenon. Data points far removed in time average all the highly variable mixing periods over that time interval, which may or may not be acceptable depending on the processes being described.

Since temperature data of the above degree of accuracy and frequency are generally lacking (even for Lake Mendota), a simple approach to estimating hydrodynamic mediated exchange between the epilimnion and hypolimnion was investigated. This approach uses the migration of the thermocline to estimate exchange. As the thermocline migrates downward, the epilimnion is assumed to increase in volume while the hypolimnion decreases in volume. A volume of water equivalent to these changes in volume is assumed to leave the hypolimnion, containing an amount of P equivalent to the content of that volume of hypolimnetic water. This volume of P rich water is then mixed into the volume of the epilimnion. Since downward migration of the thermocline is often associated with the passage of fronts in Lake Mendota (Stauffer 1974), much of the exchange between the epilimnion and hypolimnion probably occurs during these discrete events.

Physical-Chemical Processes

Sediment Loading

The exchange of nutrients between lake sediments and overlying waters has received much attention in the recent literature. One finds as many approaches to estimating the impact of sediments on lake water quality as there are workers in this area. A recent Society of International Limnologists' symposium dealt with the interactions between sediments and fresh waters (Golterman 1977). Workers investigating P exchange dynamics found much evidence to support the classic work of Einsele and Mortimer; however, the generality of iron's role in controlling P release from sediments was called into question. The consensus of the symposium attendees was that the many factors that influence sediment P release (forms of iron, oxygen, pH, calcium carbonate content, etc.) operate differently in different limnological environments, so that no general quantitative model of sediment-water P exhange is likely to be developed. Instead, an investigation of each water body is necessary to determine what processes and forms of control are likely to be most important in that body. A qualitative general model may be of use as a checklist of possibly important processes and control mechanisms. Among these are bacterial decomposition of organic matter; adsorption/desorption of phosphate onto/from iron and aluminum amorphous hydroxide complexes and clay minerals as controlled by pH, Eh, and temperature; and diffusion/advection as controlled by biological and physical mixing processes.

Recently, a number of reviews of possible processes and mechanisms have appeared and together consitute such a general qualitative model (Snow and DiGiano 1976; Berner 1980; Lerman 1979). The framework within which one places the controlling processes depends on what processes are important in the lake of interest and on what sorts of questions are being asked about the lake and its sediments. Sediments may be treated as a sink or a source of nutrients and as a depletable or undepletable source. If one desires to know how long sediments will continue to act as a source once other inputs to the water column are reduced, then sediments must be treated as depletable. This involves modeling sediments as a state variable (Lorenzen 1974). If a short-term view of sediment loading (such as that associated with seasonal water column dynamics) is desired, sediments may be treated as undepletable (that is, having a fixed rate of flux to the water column or a fixed concentration).

Sediment loading models vary from simple approaches (a constant release rate per unit area of sediment or release as a function of sediment P concentration) to complex models. The latter allow many factors to control release by following a number of variable sediment pools and calculating fluxes between them and the water column as a function of changing environmental conditions (Kamp-Nielson 1977, 1978; Jorgensen et al. 1975). It should be stressed that the simple approaches are not necessarily inferior to the more complex models. The latter require the estimation of many difficult to measure parameters and have not been shown to give a more accurate prediction of sediment loading than do the simpler approaches when the latter use lake specific rates (Kamp-Nielson 1980).

In addition to the types of processes thought important and the types of questions asked, the sediment model appropriate to use in a specific case depends on the types of information available for a particular lake. The sediment chemistry of Lake Mendota has been extensively studied. Consequently, the necessary information exists to estimate sediment loading by a number of approaches. This same information is useful in determining which processes are likely to be of greatest importance and what sort of model would most simply and reliably estimate this loading. The majority of P released from sediments in laboratory experiments is in the form of phosphate and most of this is thought to originate from Al and Fe complexes in the sediment (Shukla et al. 1971, Williams et al. 1970, 1971). The remainder is thought to come from bacterial decomposition (discussed in connection with biological processes) and desorption from clay minerals (Chen et al. 1973).

The simplest approach that seems capable of describing seasonal release using the information available on Lake Mendota is a transport model analogous to a Fickian diffusion model, modified slightly to describe random transport across a two-dimensional surface between lake and sediment pools. This approach is described in a number of works (Snow and DiGiano 1976; Berner 1980; Lerman 1979). The rate of exchange is assumed to be a function of the concentration gradient between the water column P and the sediment interstitial P. This gradient is modified by a transport coefficient that is dependent on the type of flux under consideration. Rather than attempting to provide separate mechanisms for molecular diffusion and diffusion-like transport speeded by biological or physical mixing, a single bulk transport coefficient may be measured and used to provide an estimate of all these fluxes as appropriate. This is equivalent to assuming that the net effect of any advective processes (such as pumping of pore water by benthic invertebrates) is approximately random and may be described by a diffusion model.

The bulk diffusion or transport equation used to describe sediment release has the general form:

$$F = k \cdot A \cdot (SIP - DIP)/z \qquad (1)$$

where F is the flux in mg P/day, A is the area over which the flux occurs, k is the transport coefficient (in m^2/day). SIP and DIP are the concentration of dissolved inorganic P in the sediment interstitial water and in the water column, respectively (in mg P/m^3) and z is the distance over which the gradient in SIP and DIP is measured.

The approach taken in this work for the estimation of Lake Mendota sediment loading was to divide the lake's sediments into zones, each believed to be characterized by different processes and rates of release. A bulk diffusion or transport coefficient was estimated for each of these zones. Sediments were divided into areas characterized by P poor and P rich interstitial waters, areas in which bioturbation (biological mixing of sediments) and physical stirring are likely to be important and areas where they are not, and a zone characterized by high rates of groundwater flux.

The work of Holdren (1974, 1977) on Lake Mendota sediments revealed areas of P rich and P poor sediments and made possible estimating bulk transport coefficients for these sediments under various conditions of mixing.

Holdren and Armstrong (1980) measured the flux of P between intact lake sediment cores and overlying lake water under a number of conditions in the laboratory. The conditions investigated included several believed to affect the rate of P flux substantially: temperature, oxygen concentration, physical stirring of the overlying water and biological mixing of the sediments, the concentration gradient between the water and the sediment. Over the range of oxygen levels investigated (saturated to nearly anoxic), this factor was not of great significance to release. Possibly, this is a result of the high amounts of P and S relative to Fe in these sediments. Temperature had a positive effect, suggesting decomposition was of significance. Physical mixing had a positive effect until resuspension occurred and then a negative effect. Formalin treatment also had a negative effect. This was taken to mean that biological mixing (there were benthic invertebrates in most cores) and decomposition significantly increased release; however, the use of formalin may have caused other biochemical changes that confused the issue.

One may use equation 1 to estimate a bulk transport coefficient from these sediments by solving the equation for k and substituting into the equation the fluxes and gradients measured by Holdren under these experimental conditions. This approach to estimating bulk transport coefficients has been used by Snow and DiGiano (1976) and by Klump and Martens (1981). The results of this exercise yield estimates of sediment release coefficients that compare well with values estimated in other studies. The lowest value (3×10^{-5} m^2/day), obtained with formalin treatment and no stirring, is similar to molecular diffusion measured by Li and Gregory (1974). This value characterizes sediments under deep and/or undisturbed waters. Klump and Martens (1981) found that _in situ_ phosphate release from sediments in a coastal marine basin characterized by no benthic stirring and little physical stirring could best be explained using the molecular diffusion rate. McCaffrey et al. (1980) estimated a transport coefficient for phosphate in biologically disturbed bay sediments of 3×10^{-4} m^2 per day. This coefficient was said to characterize sediments undergoing slight random mixing and is similar to that estimated for cores with slight stirring from Holdren's work.

For this work the lowest transport rate coefficient was assumed to characterize deep lake sediments (12 meters or deeper) during much of the year in Lake Mendota. These sediments experience little physical mixing (except perhaps at turnover). In addition, there is little benthic activity in this region during cold winter temperatures and summer anoxia (Muttkowski 1918; Juday 1921). Shallow sediments (2m to 12m) were assumed to be characterized by one of the higher transport coefficients. Simulation results suggested that a value of 0.0003 m^2/day produced P concentrations in the water column similar to observed values. Sediments from zero to 2m were considered to be dominated by groundwater advective flux (Brock et al. 1982) and were not included in the sediment release calculations.

Sediment interstitial pools were not modeled as state variables but were input as seasonally changing driving variables, using values measured by Holdren (1974, 1977). This approach is recommended by DiToro and Connolly (1980) for short-term modeling purposes, i.e., when the emphasis is on present control of seasonal P dynamics rather than on the long-term response of the system. P released into the overlying water by diffusion is assumed not to reprecipitate with iron on encountering oxidized waters because of the very low amounts of Fe relative to P and S in Mendota sediments.

Groundwater Flux

Estimates of groundwater loading to lakes vary greatly both in magnitude and in method of estimation. Until recently, groundwater flow was estimated using the difference approach, i.e., all hydraulic inputs and outputs of the lake are measured or estimated and compared to changes in lake volume (estimated from changes in the lake level). The difference between the expected change in lake volume (based on the measured fluxes) and the observed change is assumed to be a result of groundwater flux (either in or out). Groundwater loading is then estimated by multiplying this flux (volume of water per time) by the concentration of the nutrient of interest in the local groundwater, usually measured in a nearby water well (Watson et al. 1981).

More recently, attempts have been made to measure groundwater flux more directly via the use of seepage meters (Lee 1977). However, the estimates of flow made by such devices are only as reliable as the sites selected for measurement are representative. Only those sediments covered by about two meters of water or less are believed to be involved in groundwater influx (McBride and Pffanekuch 1975). The total area of sediments from zero to two meters deep must be determined and flux measurements must be made in all the various bottom types (sand, silt, organic matter) that occur in this region. The proportion of the lake bottom in this region represented by each of these bottom types must be assessed. Some assessment of the seasonal variability in flow must be made and still the estimates may be poor because of underwater springs or other point sources missed. Thus, the integration of inputs and outputs by the lake as revealed by the difference approach estimate is a useful means of assessing the likelihood that the seepage meter estimate is reasonable.

Estimates of groundwater flux to Lake Mendota have been made using the difference approach (Lathrop 1979) and the seepage meter approach (Brock et al. 1982). The former estimates are specific to the years used in this study (1976 and 1977) while the latter averages measurements made over several years to obtain an average groundwater flux for Lake Mendota. Lathrop's estimates of groundwater flux for 1976 and 1977 were 5.9 and 8.1×10^6 m^3 per year. Sonzogni and

Lee (1978) used a value for groundwater seepage similar to that estimated by Brock et al. (29 cfs or 26 x 10^6 m^3/year). Because groundwater was not a large part of the total P load to the system, this range of values had little effect on model output.

In addition to the need to assess the groundwater flux, the concentration of the elements of interest in the inflowing groundwater must be estimated. It is unlikely that groundwater flows through nutrient rich sediments and enters the lake in the same form it had before entering the lake sediments. Since this water moves by mass flow, incoming groundwater displaces the interstitial pore water in the sediments, forcing this pore water into the lake water. Thus, it is the concentration of elements in the sediment pore water that is of interest.

The concentrations used here in the estimation of loading by groundwater were those measured by Holdren in the first 3 cm of sediment pore water at a shallow site (5 to 6 m) in Lake Mendota. These concentrations were varied monthly while the groundwater inflow rate was assumed to be constant. Brock et al. sampled pore water concentration in the area of groundwater flux in the summer and found concentrations similar to those measured by Holdren in summer at the 5-6 m depth. Thus, there is some justification for using Holdren's values throughout the rest of the year.

Settling and Resuspension

The balance between the loss of particulates from the water column by settling and the resuspension of settled particulates is of major importance to the seasonal dynamics of nonconservative elements. The time scale over which the dynamics of the lake system are followed influences the assessment of the significance of these fluxes. In an annual budget, the net flux is always to the sediments. As the time frame of observation shortens, the importance of the balance between gross sedimentation and resuspension becomes more obvious. The weekly dynamics described in the present work require a fairly accurate estimate of the magnitude of sedimentation and resuspension when these are important sinks and sources. Incorporation of these fluxes was approached in a number of ways. Biweekly gross sedimentation and resuspension rates were measured by Fallon (1978) using sedimentation traps during the years for which loading data were available (1976 and 1977). Seasonal sedimentation rates (mg P/m^2·day) were compared to water column pool sizes (mg PP/m^3) measured during the same periods to obtain sedimentation loss rate coefficients (percent of the PP pool lost per day as a result of sedimentation). These first order loss rate coefficients were applied to simulated water column pools within the model simulation to obtain sedimentation rates (mg P/m^2·day). Resuspension rates were computed from Fallon's measurements of organic matter content in the suspended matter and in the sediment trap contents, using the method described by Gasith (1975) and assuming a sediment organic matter content of 16% (Bortleson 1970). The amount of P resuspended was assumed to be proportional to the amount of organic matter resuspended. Resuspension and settling rate coefficients used in this work varied seasonally from one to 122 mg/m^2·day and from .005 to .21/day respectively (Watson 1981).

Since measured resuspension rates were thought to be inflated by large external loads of particulates that are low in organic matter, measured resuspension was reduced by an amount equivalent to the external particulate P load occurring at the same time. This had almost no effect on resuspension inputs except in the month of March when large amounts of PP entered the system. This modification improved the fit of the model simulation to observed data but the same could have been accomplished by reducing the PP load.

Gross and Net Sedimentation. Gross sedimentation is calculated both in volumetric and in areal terms. The following equations estimate gross loss of P settling from the mixed zone and from the lake as a whole (in kg):

$$GSEV = SE * PPE * VE \quad (2)$$

$$GSLV=Twit*GSEV+Swit[(GSEV(A-AZE)/A)+(SH*PPH*VH)] \quad (3)$$

where SE and SH are settling rate coefficients for the epilimnion and hypolimnion respectively, Twit = 1 during nonstratified periods and zero during stratified periods and the reverse is true for Swit.

An estimate of the gross addition of P to shallow and deep areas of the sediments (in mg/m^2) is produced by the following equations:

$$GSSA = SE * PPE * VE/A \text{ or } GSEV/A \quad (4)$$

$$GSDA = Twit * GSSA + Swit * (SH*PPH*VH/AZE) \quad (5)$$

Net sedimentation was calculated from gross sedimentation in two ways. One approach assumed that resuspension occurred from epilimnetic sediments during stratification and from all sediments during the unstratified period (i.e., resuspension comes from sediments in the mixed zone). Losses of PP from the mixed zone and the lake as a whole (in kg P) are:

$$NSEV = GSEV - RESUS \quad (6)$$

$$NSLV = GSLV - RESUS \quad (7)$$

The gain of settled particulate P to the sediments in the shallow and deep parts of the lake (in mg P/m^2) is :

$$NSSA = GSSA - (Twit*RESUS/A+Swit*RESUS/(A-AZE)) \quad (8)$$

$$NSDA = GSDA - Twit * RESUS/A \quad (9)$$

The second approach assumes that resuspension occurs only from epilimnetic sediments during stratification and only from shallow areas (those less than the mean depth of the lake) during nonstratified periods. This does not change the assessment of volumetric losses from the mixed zone and lake but does alter the estimate of areal net sedimentation. Thus, shallow and deep sediments would receive the following amount of P under this assumption:

$$NSSA = GSSA- (Twit*RESUS/ASHAL+Swit*RESUS/(A-AZE)) \quad (10)$$

$$NSDA = GSDA \quad (11)$$

This approach produces lower values of areal net sedimentation to shallow areas and higher values of net sedimentation to deeper areas than does the former approach.

Biological Processes

Algal Uptake

An attempt was made to simplify the description of biological processes so that the complexity of the description of these factors was kept in balance with that applied to the physical and chemical characterizations of the system. Thus, algal uptake was assumed to be predictable from algal growth which was in turn assumed to be a simple function of temperature, light, and external nutrients. For a number of reasons the recent approach of describing algal uptake and growth as highly nonlinear functions of internal cell nutrient contents (Droop 1973; Lehman et al. 1975; Tilman and Kilham 1976) was viewed as neither necessary nor appropriate. The "cell quota" concept requires considerable verification data concerning internal cell nutrients, the estimation of many parameters for each "functional group" of algae (e.g., blue-green algae), and the constant update of driving variables on a very short time step. The cost of running a model is largely a function of the number of times the differential equations that simulate its compartment dynamics must be modified and resolved. Thus, the desire here was to avoid formulations that required frequent updates of driving variables in order to be meaningful. The information to be gained from the cell quota approach did not appear to offset its cost; consequently, the simpler approach described below was followed.

Frequently, algal growth has been predicted using a simple function that reduces a maximum growth rate by an amount consistent with suboptimal conditions. In fact, the cell quota approach predicts algal growth μ from:

$$\mu = \mu max * TD * CD * ND \tag{12}$$

where μmax is a maximum growth rate (it may be a specific growth rate with units day^{-1}) and TD, CD, and ND are temperature, carbon, and nutrient dependence terms (Lehman et al. 1975). Ultimately, the carbon dependence term is a function of light and the nutrient term is a function of external nutrient concentration of the most limiting nutrient.

The nutrient dependence of growth has often been simulated using a Michaelis-Menten relationship (Eppley and Thomas 1969; Eppley et al. 1969; Dugdale and MacIsaac 1971; Davis et al. 1978). This function has the form:

$$\mu = \mu max * S/(S+ks) \tag{13}$$

where S is the concentration of the dissolved or available form of the limiting nutrient and ks is the half saturation constant for growth (concentration at which growth is half the maximum value). Photosynthesis (and ultimately growth) has been related to light by the Steele (1964, 1965) equation:

$$P = Pmax * (Ia/Im)\exp(1-Ia/Im) \tag{14}$$

where P is the photosynthesis rate, Pmax is the maximum rate, Ia is the amount of light available to algae (often averaged over time and/or space), Im is the light at which P = Pmax. Note that this formulation predicts photoinhibition at light levels in excess of Im. Thus, growth trends in algae may be simulated using:

$$\mu = \mu max * TD * \frac{Ia}{Im} \exp(1- \frac{Ia}{Im}) * \frac{S}{S+ks} \tag{15}$$

This formulation is similar to those used by Imboden and Gächter (1978), Riley (1965), Chen (1970), DiToro et al. (1971), Fleming (1975), and Scavia (1980). All of these workers attempted to simplify biological processes as much as possible in their nutrient dynamics models in order to consider physical and chemical processes as well.

To describe the effect of temperature (TD), many of the above workers used Q10 functions with maximum, minimum, and optimum temperature values. These formulations were found to give values similar to a simple linear formula (TD=.04*T) over the range of temperatures usually encountered in the epilimnion waters of a northern temperate lake (T is the water temperature in degrees centigrade).

The growth rate predicted by equation 15 may be used to estimate the amount of uptake that occurs under similar conditions of light, nutrients, and temperature. According to Droop (1973), if the amount of nutrients/cell (Q) is not changing drastically over the time period of interest, then,

$$u = \mu * Q \tag{16}$$

where u is the uptake per cell (as per unit biomass) and μ is a specific growth rate. Since algal cell contents are confined to a finite range and the integration step used in this model is short (0.0001 day), it was felt that the change in Q over the time step would be small enough to justify using equation 16 to approximate uptake per cell or per unit biomass. Thus, the uptake of the entire algal population could be approximated by:

$$u * N = \mu * Q * N \tag{17}$$

where N is total number of cells or the total amount of biomass in the algae population, u * N is the total uptake, Q * N is total algal nutrient pool, and μ * Q * N is the rate of growth in that pool.

These simplifying assumptions seem justified in light of the arguments of DiToro (1980) who showed that the simple Monod equation (i.e., Michaelis-Menten kinetics) predicts growth kinetics as well as the more complicated Droop kinetics when internal cell nutrients are in equilibrium. Moreover, he argues that it is reasonable to assume that cells are at equilibrium even during population growth since the internal concentration can equilibrate more rapidly than the population can change. Whenever an aggregated algal population is being considered rather than individual species, DiToro recommends using simple Monod kinetics.

The available light Ia in equation 15 was calculated from incident light Io, using Lambert-Beer's law and integrating over depth. Thus:

$$Iz = Io * \exp(-kz) \tag{18}$$

when integrated over depth becomes

$$Ia = Io\ (1-\exp(-kz))/kz \tag{19}$$

where Io is the incident light, k is the light extinction coefficient, z is depth over which light is intetrated to obtain average available light (Ia), and Iz is the amount of light at depth z.

The depth z over which light is averaged was initially set equal to the depth of the mixed zone. However, the resulting low rates of uptake suggested that the algae maintain themselves at a shallower depth despite mixing. The weighted mean depth of the chlorophyll in the water column was determined (Watson 1981), and algae were assumed to be mixed to one standard deviation below this depth. An alternative approach that gives similar results assumes that Ia is equal to Iz (equation 18) when z is the weighted mean depth of the chlorophyll.

The light extinction coefficient k was initially input to the model (using field measurements). Once the model was simulating the lake's dynamics fairly closely, the extinction coefficient was predicted using the following equation:

$$k = ECW + ECA * PP \quad (20)$$

where ECW is the extinction coefficient of filtered lake water and ECA is a proportionality constant that converts particulate P concentration into a light extinction term. Use of the predicted k gave results similar to those obtained using measured k values. Values used in this study (ECW=.325/m, ECA=.015m^2/mgP) are in close agreement with those used by other workers (James and Birge 1938, Scavia et al. 1976, DiToro et al. 1971, Imboden and Gächter 1978).

The incident light Io varies over the year as a function of the angle and distance of the sun and the day length. The total light arriving at the earth's surface each day of the year was divided by the number of daylight hours on that day to give the light energy per daylight hour (in langleys/hr). The optimum light level Im is also in these units. Thus, equation 15 gives the algal growth rate that would occur if this amount of light energy were available all day. Since it is only available for a fraction of the day (FP = daylight hours/24), equation 15 is multiplied by that fraction. Thus, equations 15 to 17 become:

$$\text{uptake} = \mu\text{max} * .04T * \frac{Ia}{Im} \exp(1- \frac{Ia}{Im}) * FP * \frac{DP}{DP+ks} * \alpha PP \quad (21)$$

where α is the percent of the particulate P (PP) that is in algae. Other terms are as previously defined. The values for the parameters used in these equations were obtained in part from the literature but whenever possible measurements made on the algal populations of Lake Mendota were used. Parameter values appear in Tables B-4 to B-6 of Watson (1981).

Some confirmation of this approach and the parameter values used is provided by Lehman (1980). If neither light nor nutrients are limiting, the uptake equation becomes

$$uN = umax * .04T * QN \quad (22)$$

At 20°C and using a value of 1.5/day for umax, this equation becomes

$$uN = 1.2/\text{day} * QN \quad (23)$$

where uN is total algal uptake and QN is total P in algae. The uptake rate coefficient 1.2/day is precisely the value determined by Lehman for nutrient sufficient Chlamydomonas cells under these conditions.

Zooplankton Excretion

The main impact of zooplankton on phosphorus dynamics is assumed to be the conversion of particulate P to dissolved P via consumption of PP and excretion of DP. Peters and Lean (1973) have shown that the predominant form of P released by Daphnia excretion is orthophosphate, the most available form of P for algal uptake. In this work the process is represented by a single highly aggregated equation that estimates zooplankton excretion from respiration. Zooplankton respiration is estimated from consumption and a basal metabolic respiration rate. The former is assumed to follow a Michaelis-Menten type relationship--i.e., a hyperbolic function describes the relationship between zooplankton consumption and algal abundance. A maximum consumption rate is modified by the ratio of prey concentration and a half saturation constant for that prey. That is:

$$ZPR = (BR + CR(A/A+K_A))) * TD * ZP \quad (24)$$

where ZPR = zooplankton respiration rate (mg C/l·day)

BR = basal respiration rate (mg C/mg C·day)

CR = constant that converts mg C consumed to mg C respired, i.e., product of maximum consumption rate and ratio of respiration to consumption

A = concentration of algae or other food (mg C/l)

K_A = half saturation constant for consumption of prey (mg C/l)

TD = temperature dependence

ZP = concentration of zooplankton C (mg C/l).

Carbon respiration may be converted to P excretion in a number of ways. The simplest approach multiplies the former by the ratio of P to C in plankton biomass (Scavia 1980). This assumes a fairly constant stoichiometry for plankton biomass over time and trophic levels. Usually the Redfield ratio of P:C is used.

Ganf and Blazka (1974) measured the ratio of O_2 uptake (which should reflect C respiration mole for mole) to P excretion by zooplankton. They found 58 moles of O_2 taken up per mole of P excreted. The ratio of P excreted to C respired would be 0.04 (by weight), suggesting that zooplankton excrete material that is richer in P than either zooplankton or their prey. Using the measured values of Ganf and Blazka as a first approximation, zooplankton phosphorus excretion may be estimated from:

$$ZPEXC = PCEXC * ZPR \quad (25)$$

where PCEXC = P:C ratio of P excretion to C respiration by weight.

Since the current work is in terms of P rather than C, it is necessary to estimate several terms in equation 24 from the analogous P term in the model. Thus, assuming fairly constant stoichiometry for zooplankton and prey:

$$A \text{ (or algal \& detrital C)} = (\alpha + \gamma) * PP * CPA \quad (26)$$

$$Z \text{ (or zooplankton C)} = \beta * PP * CPZ \quad (27)$$

where α, γ, and β are the proportions of particulate P (PP) in the algae, detritus, and zooplankton, respectively; CPA and CPZ are the ratios of C:P by weight in the algae/detritus and in the zooplankton, respectively. Thus, equations 24 to 27 become:

$$ZPEXC=PCEXC*(BR+CR(((\alpha+\gamma)*CPA*PP/((\alpha+\gamma)*CPA*PP+KA)))*TD*CPZ*\beta*PP \quad (28)$$

As with the algal uptake equation, the temperature dependence term (TD) of .04 T°C was used.

Since zooplankton do not accept all algae and

detritus as food, terms were added that make it possible to designate what percentage of the algae and detritus is being grazed (the term for algae can be varied over the season to describe the shift to non-palatable bluegreens). A lower limit or threshold concentration for feeding of 50 mg C/m^3 (or 1.2 mg P/m^3) is suggested by Scavia (1980). However, particulate P levels seldom reach this level in Mendota. Since excretion would be very low at this level, it was not considered necessary to "shut off" excretion by use of such a threshold.

By way of support of the above approach, one may look again to the work of Lehman (1980). When zooplankton grazing is saturated and the temperature is equal to 20°C, the zooplankton excretion equation reduces to 0.3/day * zooplankton P. This rate coefficient is equal to the zooplankton regeneration rate coefficient found by Lehman under these conditions. The parameters used in the zooplankton excretion equation appear in Table B-7 of Watson (1981).

Decomposition in the Water Column and
at the Sediment Surface

Decomposition in the water column is simulated by a simple first order temperature dependent rate equation that remineralizes a certain percent of the detrital phosphorus each day. Thus

$$\text{Remin} = \text{DK} * .04\text{T} * \gamma * \text{PP} \qquad (29)$$

where Remin is the amount of particulate P converted to dissolved P per day, DK is that percent of the detrital P (γ*PP) that is decomposed each day as modified by the temperature function. This formulation is similar to that used by Scavia et al. (1976), Fleming (1975), and numerous others. The decay constants and temperature constructs used by Scavia et al. yield values in the range of 2% of the detritus remineralized per day at 20°C. Fallon (1978) measured a decomposition rate for particulate P in natural Mendota algal populations of 5% per day at 20°C under aerobic conditions. Thus, a value of .05 was used for DK, yielding a decomposition rate coefficient of 4% per day at 20°C.

A significant amount of P is believed to be released by anaerobic decomposition of detritus at the sediment-water interface under natural conditions (Imboden and Lerman 1978; T.D. Brock, pers. comm.). Since models of this process are in their infancy (Kamp-Nielson 1976; Jorgenson et al. 1975) and have little predictive power at present, measured values of release per unit area of anoxic sediments were used. Phosphorus release was estimated by assuming that decomposing organic matter initially had a C:P ratio similar to that found in plankton. Thus, the amount of sulfate reduced to sulfide and the amount of methane generated from reduction of organic C should be proportional to the amount of P released by these anaerobic processes.

Recent work in Lake Mendota (Ingvorsen and Brock 1982) found six m moles of sulfate reduced and 19 m moles of C liberated as methane per square meter of anoxic sediment per day from August 8 to September 18. These values were determined by constructing a sulfate/sulfide and methane budget for the anoxic part of the lake. Since each mole of S reduced is generally assumed to indicate two moles of organic C processed, a total of 31 m moles of C per square meter per day appear to have been processed by anaerobic decomposition. Assuming that each 100 moles of C is associated with a mole of P in the plankton, then 31 m moles of C processed should result in the remineralization of .31 m moles or 9.7 mg P per square meter per day.

Fallon (1978) found the organic matter in the summer water column of Lake Mendota to be slightly richer in P than is predicted by the Redfield ratio. Using this richer ratio, a P remineralization rate of 13 mg of P per square meter per day is suggested. Release of P by this process occurs in the anoxic hypolimnion and is estimated using:

$$\text{DKSOUP} = \text{DKS} * \text{ASHA} \qquad (30)$$

where DKS = P released per m^2 of anoxic sediments per day; ASHA = area of anoxic sediments, and DKSOUP = total P released into the hypolimnion from this source each day.

The Simulation Model

The above discussion describes a conceptual model of lake P dynamics and how the various processes thought to be important are translated into mathematical equations. However, the complete sequence of translation necessary to make such a conceptual model useful is:

conceptual model→mathematical model→computer model.

A complete listing of the computer code of the model STRAT appears in Watson 1981. Figure 2 gives a highly simplified flow chart of this computer model. The differential equations that are solved to simulate lake P dynamics are stated below in simplified form (i.e., change in pool with time equals gains minus losses). The following pools are simulated: dissolved P in the epilimnion or mixed zone and hypolimnion (DPE and DPH) and particulate P in these regions (PPE and PPH).

ΔDPE = (decay+zooplankton excretion+DP inputs +groundwater inputs+sediment release) -(uptake+DP outflow)

ΔPPE = (PP inputs+uptake+resuspension)-(PP outflow+decay+sinking+zooplankton excretion)

ΔDPH = (sediment release+decay+release by anaerobic decay)

ΔPPH = (sinking from epilimnion)-(sinking from hypolimnion+decay).

APPLICATION OF MODEL TO LAKE MENDOTA, WISCONSIN

Simulation of Phosphorus Dynamics

When the previously described model was used to simulate the P dynamics of a specific lake in specific years (Mendota, 1976 and 1977), it was found to simulate reasonably well the seasonal dynamics of the dissolved and particulate P pools of the mixed lake, epilimnion, and hypolimnion. The model output is compared to observed data in Figures 3 and 4. The observed data resulted from integrating depth profiles of DP and TP (and/or PP) measured at a central lake site by T.D. Brock (University of Wisconsin, Bacteriology) and students. The standard error of the mean DP concentration in the mixed zone based on several pelagic sites was generally about 10% of the mean of the sites. The central site values differed from the mean estimates by less than this standard error (and model output generally differed from central site values by less than this amount). There was less information on PP (often it was calculated

FIGURE 2. FLOW CHART FOR STRAT--
STRATIFIED LAKE P DYNAMICS SIMULATOR

Set initial values of timestep (TS) and the dependent (system) variables (DPE and PPE in mg P/m^3; DPH and PPH = 0 mg/m^3)

Input unchanging lake morphometry and parameters (including maximum areal flux rates acted on by driving variables and lake morphometry)

Input independent (driving) variables (time varying inputs or structure, e.g., light, nutrient inputs, thermocline depth)

Calculate auxiliary equations (flux rates modified by changing environmental conditions and lake morphometry)

Call differential equation solving subroutine
Call subroutine containing system equations (coupled differential equations that describe behavior of system pools) and simulate system behavior for one week

Output final values of dependent variables after solution of equations for each time period

Calculate and output fluxes between compartments using auxiliary equations and final values of dependent variables for each time period

Update and test time step: TS = TS + 1

TS = Last	Go to end
TS < STR or > TO	New initial (i) values of dependent variables = final (f) values calculated for last time period
TS = STR	DPEi = DPEf, PPEi = PPEf DPHi = DPEf, PPHi = PPEf
TS > STR and < TO	DPEi = $(DPEf \cdot VE_{ts-1}+DPHf (VH_{ts-1}-VH_{ts}))/VE_{ts}$ PPEi = similar to DPEi DPHi = DPHf, PPHi = PPHf
TS = TO	DPEi = $(DPEf \cdot VE_{ts-1}+DPHf \cdot VH_{ts-1})/V$ PPEi = similar to DPEi DPHi = PPHi = 0

*STR = day lake stratifies; TO = day of turnover

from TP-DP). The model output agreed with central site PP values within about 10% or less. Model output was compared to data collected by Lathrop (1979) during this period also. His values were lower than Brock's in the early spring of 1976 and the model seems more in line with Lathrop's observations.

Comparison of Model Estimates with Measured Fluxes

The model estimates a number of fluxes that have been measured or estimated by other means for Lake Mendota in these or other years (Table 1). Among these are outflow losses, daily productivity, and annual sedimentation. Simulated outflow loss of P is in good agreement with "observed" outflow loss. This is no surprise since the "observed" loss is calculated by multiplying the hydraulic outflux time series (in m^3/d) times the concentration time series of DP, PP, and TP in the epilimnion. The model uses the same hydraulic flux values, and the P concentration time series has already been shown to agree well with observed values. Thus, the simulated and observed losses by outflow should agree.

Simulated annual sedimentation for 1976 (Table 1) was 1900 mg P/m^2 for shallow and epilimnetic sediment areas and 3000 mg P/m^2 for deep areas. The value was 2700 for the lake as a whole. Annual sedimentation in 1977 was estimated to be 465 mg P/m^2 for shallow areas and 2700 and 1600 for deep areas and for the lake as a whole. These much lower estimates of sedimentation are a result of the lower estimates of PP loading for this year. These estimates of net sedimentation were obtained by calculating gross sedimentation for all areas and subtracting resuspended P (divided by the area of sediments in contact with the mixed zone). Thus, resuspended P is assumed to come from all sediment areas except from hypolimnetic sediments. In this way, even deep sediment areas are assumed to lose some P during the non-stratified period.

Figure 3. Simulated and Observed Phosphorus pools in Lake Mendota, 1976

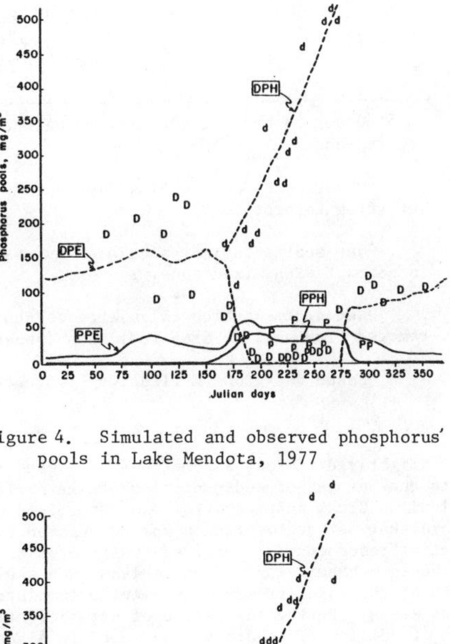

Figure 4. Simulated and observed phosphorus' pools in Lake Mendota, 1977

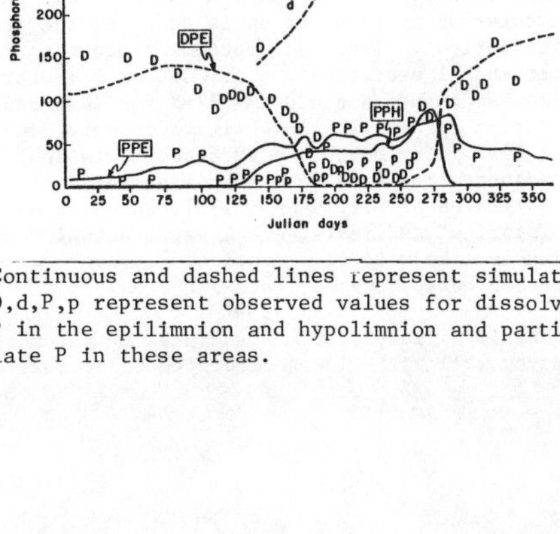

Continuous and dashed lines represent simulation; D,d,P,p represent observed values for dissolved P in the epilimnion and hypolimnion and particulate P in these areas.

TABLE 1 Comparison of some fluxes predicted by simulation of Lake Mendota P dynamics and values estimated by other methods.

Annual outflow loss (10^3 kg P/year)

Pool	Simulated 1976	Simulated 1977	Observed[a] 1976	Observed[a] 1977	Lathrop(1979)[b] 1976	Lathrop(1979)[b] 1977
DP	10.6	6.1	10.3		(4.8)	(3.8)
PP	2.1	2.1	1.4		(5.1)	(3.2)
TP	12.7	8.2	11.7		9.9	7.0

Annual net sedimentation (mg P/m^2/year)

	Simulated 1976	Simulated 1977	Bortleson (1970) 150 yr. average	Fallon & Brock (1980) June 76-77	Fallon & Brock (1980) Nov 76-77
Shallow areas	1900[c] -360[d]	465			
Deep areas	3000[c] 4200[d]	2722		4200	5000
Whole lake	2700	1600	1200-1400		

Annual production (g C/m^2/year)

Simulated 1976	Simulated 1977	Fallon & Brock (1980) June-Nov 1977	Brock (unpub.) 1979	Brock (unpub.) 1980
204	220	316	220[e]	338[e]

[a]Observed hydraulic flux time series times P concentration time series in validation data set (Appendix C).

[b]Observed hydraulic flux time series times P concentration time series of unpublished data set (from Lathrop 1979).

[c]Net sedimentation calculated by assuming that resuspended P removed from all sediment areas in contact with mixed zone.

[d]Net sedimentation calculated as above during stratification but otherwise resuspended P removed from shallow area sediments (those less than 12 m deep) only.

[e]Based on (gross C fixation) *.5, assuming half of gross C fixation is respiration.

Slightly different assumptions were used to generate another set of sedimentation estimates (also in Table 1). Gross sedimentation was determined as before, but net sedimentation was calculated by subtracting resuspended P (divided by the area of shallow sediments--those lying in waters less than the mean depth of the lake) from the sediment accumulating in these areas. During the period of stratification, the area of epilimnetic sediments was used in place of the area of shallow sediments. Consequently, shallow sediments lose more P to resuspension than in the previous estimate. Deep areas lose no P (i.e., the net sedimentation of these areas is equal to the gross sedimentation). These assumptions produce estimates of net annual sedimentation of -360 mg P/m^2 for shallow sediments and 4200 mg P/m^2 for deep sediments. The latter agrees with an estimate made by Fallon and Brock (1980) for Mendota's 1976 sedimentation, based on sediment trap data from deep areas of the lake. The estimates of Bortleson (1970), based on analysis of historical sediments, are lower. Actual sedimentation is likely to vary continuously over depth, bounded by the above estimate extremes.

Sediment release of P generated by the model compares well with fluxes predicted by another worker, using a slightly different approach. Colman (1979) estimated P flux through the layer of water immediately above the sediments, based on the concentration gradient in this water and vertical eddy diffusion coefficients estimated from radon 222 profile analysis. Colman's gradients are lower and values for k are higher since he is dealing with flux through a portion of the water column rather than flux across the sediment-water interface. However, the fluxes predicted by his approach are similar to the values predicted by the formulation used in this work. Colman's estimates were based on data from the deep hole of Lake Mendota in winter (January to early March). He estimated an average flux of 19 mg/m^2/day and predicted from this an increase in P of 950 mg/m^2. He observed an increase of only 480 mg/m^2, suggesting his flux estimate may have been twofold too high. The formulation used here predicts an average flux of 7.4 mg/m^2/day for this same period for the deep sediments of Mendota.

Annual production may be estimated from the simulated annual uptake by assuming that the net C uptake is proportional to the net P uptake by the Redfield ratio of 40:1 by weight. That is, algal biomass often exhibits a ratio of C:P of 40:1, and

P uptake estimated from algal growth should be proportional (within some limits) to net C uptake (gross C uptake minus respiration). The 1976 simulated annual uptake rate of 200,000 kg of P/year indicates a net annual production of 8 x 10^6 kg C fixed or 204 g C/m^2 (220 g in 1977). An annual mass balance analysis (see the following section) suggests that the model output underestimates the actual amount of P uptake calculated "inside" the model. An argument is made for correcting P uptake to 242,000 kg P/year and annual production to 250 g C/m^2. The same argument corrects the 1977 estimate to 265 g.

Based on C-14 uptake by Mendota plankton incubated at different light levels, Fallon and Brock (1980) estimated a higher net production for 1977 (316 g C/m^2 for the period from June to November 1977). Their estimate of respiration (based on oxygen uptake in dark bottles) was somewhat less than half the estimate of gross production throughout this part of the year. Estimates of gross annual production calculated from recent unpublished data of Brock are 440 g C/m^2/year in 1979 and 677 g in 1980. Assuming half of this is respiration, net production is about 220 and 338 g C/m^2 for these years. These estimates seem in line with those for 1976 and 1977. Obviously, there is much year-to-year variability in Lake Mendota production.

Model Assessment of Internal Fluxes

Annual Fluxes

The above comparisons suggest there is some justification for assuming that the fluxes estimated by the model can provide an approximate assessment of the importance of these fluxes to the P dynamics of the lake. The fluxes integrated over the year for the entire lake and for the mixed zone appear in figures 5 and 6. Recall that the mixed zone is assumed to be the entire lake during the nonstratified period and the epilimnion during the stratified period. Figure 7 integrates these fluxes over the stratified period and considers fluxes within and entering or leaving the epilimnion only.

The relative importance of external and internal loading and recycling is summarized in Table 2. Non-water column sources of P are summed into the assessment of internal loading (includes resuspension, sediment release, and decay at sediment surface). Recycling includes exchanges between dissolved P and particulate P, mediated by biological processes such as uptake, water column decay, and zooplankton excretion. Internal loading to the epilimnion includes inputs from the hypolimnion. The 1976 internal lake load is estimated to be five times, three times, and four times the external load for DP, PP, and TP, respectively. The ratios of internal load to external load (IL:EL) is slightly higher if the mixed zone is considered (because of the addition of hypolimnetic entrainment to this assessment). For the sake of this analysis, let us define a quantity, internal flux, as the sum of internal load and recycling. If one compares this internal flux (IF) to EL, the ratios are much larger for DP and PP (recall that recycling does not affect the "loading" of TP). The IL:EL ratios for 1977 are 9, 15, and 13 for DP, PP, and TP. Again, the assessment for the mixed zone and for internal flux increases these ratios.

A mass balance analysis of the simulated fluxes was performed to determine if the observed change in phosphorus pools over the year was in line with the

Fig. 5. Whole Lake Phosphorus Fluxes, Mendota, 1976 Simulated Annual Sums in 10^3kgP(& mgP/m^3)

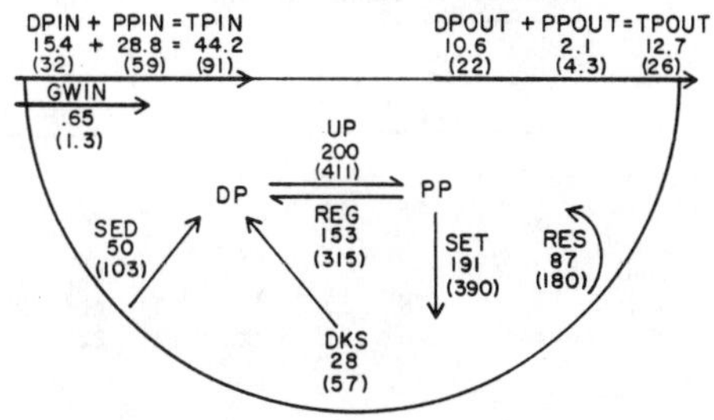

Fig. 6. Mixed Zone Phosphorus Fluxes, Mendota, 1976 Simulated Annual Sums in 10^3kgP(& mgP/m^3)

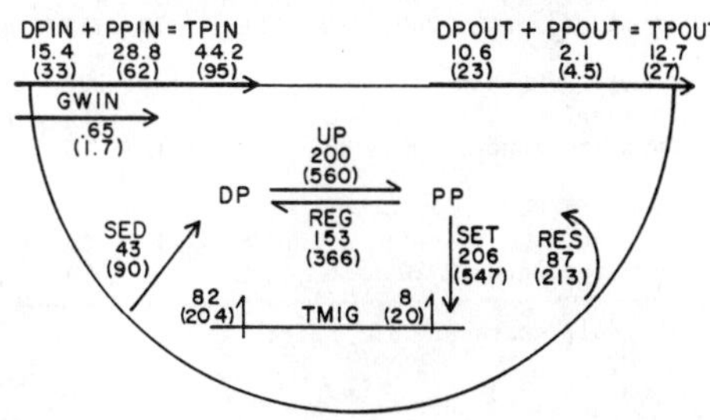

Fig. 7. Epilimnion Phosphorus Fluxes, Mendota, 1976 Simulated Sum over Stratified Period in 10^3kgP

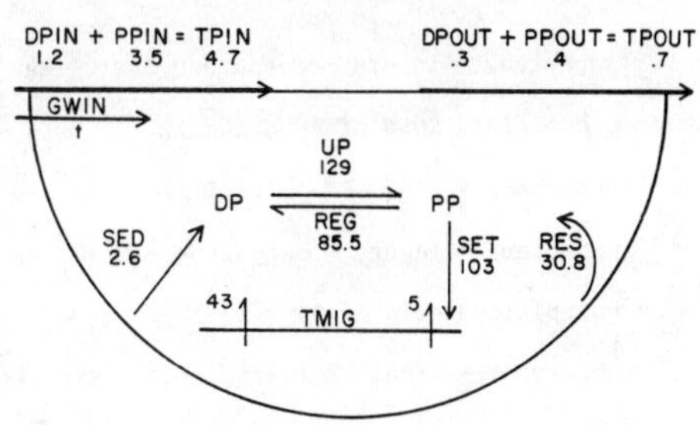

DPIN, PPIN, TPIN = inputs of P
DPOUT, PPOUT, TPOUT = outputs of P
GWIN = groundwater inputs
SED = sediment release
DKS = decay at anoxic water-sediment interface
UP = algal uptake
REG = regeneration (zooplankton excretion and decay
SET = settling of particulate P
RES = resuspension of particulate P
TMIG = thermocline migration (entrainment of hypolimnion P)

TABLE 2. Relative importance of external loading, internal loading, and recycling in Lake Mendota based on the 1976 and 1977 P dynamics simulation.[a]

	1976			1977		
	DP	PP	TP	DP	PP	TP
External load(EL) to lake	15	29	44	8	14	22
Internal load(IL)						
to lake	78	87	170	73	214	287
to mixed zone	126	95	220	121	221	340
Recycling						
in lake	153	200		175	214	
in mixed zone	138	200		161	214	
Internal flux(IF)						
in lake	230	290	170	250	430	290
in mixed zone	260	290	220	280	430	340
Ratio of IL/EL						
in lake	5	3	4	9	15	13
in mixed zone	8	3	5	15	16	15
Ratio of IF/EL						
in lake	15	10	4	32	30	13
in mixed zone	20	10	5	36	30	15

[a]All units are 10^3 kg P.

expected change (based on inputs and outputs). Using the whole lake pools and fluxes in figure 5, one finds :

$$\Delta DP(\text{observed}) = \text{final DP-initial DP} = -3.5 \times 10^3 \text{ kg}$$

$$\Delta DP(\text{expected}) = \text{inputs} - \text{outputs} = 36 \times 10^3 \text{ kg}$$

for an unexplained loss of 40 x 10^3 kg.

$$\Delta PP(\text{observed}) = \text{final PP-initial PP} = .2 \times 10^3 \text{ kg}$$

$$\Delta PP(\text{expected}) = \text{inputs} - \text{outputs} = -30 \times 10^3 \text{ kg}$$

for an unexplained gain of 30 x 10^3 kg.

$$\Delta TP(\text{observed}) = \text{final TP-initial TP} = -3 \times 10^3 \text{ kg}$$

$$\Delta TP(\text{expected}) = \text{inputs} - \text{outputs} = 6 \times 10^3 \text{ kg}$$

for an unexplained loss of 10 x 10^3 kg.

This suggests that a certain amount of inaccuracy is introduced into the model output by estimating fluxes at the end of each time step (one week long). It is likely that dynamic fluxes such as uptake and settling would be better estimated by taking shorter time steps, thus reporting out the pool sizes more often for flux calculations. Uptake, in particular, is likely to be underestimated by the current practice because it is a function of the DP concentration which has a rapid turnover rate. During the summer, a certain amount of DP is added to the epilimnion by thermocline migration at almost every time step. This input produces a higher concentration of DP just as the new initial concentrations are given to the differential equation-solver for the next week-long simulation. By the end of the time step, DP is depleted and uptake has dropped to a level sustained by the more continuous recycling processes (decay and zooplankton excretion). It is this lower level of uptake that is reported.

If one adds to uptake the amount of DP added to the epilimnion by thermocline migration over the stratified period, one finds the DP and PP equations are more nearly balanced. This addition of DP should not include the final influx of DP from thermocline migration that occurs at turnover since this final influx causes a substantial increase in the DP concentration and so is not completely absorbed by uptake.

This same bias in the model's output may explain the slight imbalance in the TP budget as well. Whenever a high settling rate coefficient is introduced at the beginning of a time step, the observed settling rate reported at the end of this interval is likely to be below the average rate for that interval as a result of the depletion of the PP pool. If one assumes the suggested correction in uptake and that another 10 x 10^3 kg of P sediments out, thus balancing the TP input-output equation, the input-output equations for DP and PP become:

$$\Delta DP = -3.6 \times 10^3 \text{ kg P} \qquad \text{(Note: total P averaged over year is } 70 \cdot 10^3 \text{ kg)}$$

$$\Delta PP = -\ .2 \times 10^3 \text{ kg P.}$$

This underestimating of fluxes only affects the model's weekly output not the rates that affect pools in the simulation.

Seasonal Fluxes

While it is instructive to look at the magnitude of the various internal fluxes summed over the year, more information can be gained from an investigation of the seasonality of these fluxes (see figures 8 to 10). The internal load of TP (figure 8) includes only those processes that provide net gains of P to the water column. These include resuspension, sediment release (by diffusion-like transport), and release by anaerobic decomposition at the sediment-water interface. Resuspension achieves greatest importance during spring and fall turnover. Sediment release is fairly uniform throughout the year. The higher concentration of P found in summer sediments which would result in greater importance from this source is balanced by the smaller transport coefficients associated with anoxic areas (where the benthos have been observed to be inactive). Thus, as the summer sediments get richer in dissolved P, mixing of the deep sediments probably decreases, resulting in the fairly uniform pattern of release predicted by the model. Anaerobic decomposition (and the P release that accompanies it) is associated with the period of summer anoxia. TPIL is greatest at fall turnover.

The internal fluxes of DP and PP (figures 9 and 10) appear to be dominated by recycling. To determine the internal flux of DP and PP exclusive of recycling (the quantity that has been referred to as the internal load or IL), note the uppermost line below the area occupied by remineralization in figure 9 and the line below the area occupied by uptake in figure 10. Thus, DPIL is characterized by a fairly constant amount of sediment loading except for the increase associated with anaerobic decomposition during summer anoxia. Remineralization (including zooplankton excretion and water column decomposition) becomes important in May and remains important until October. This is largely a function of temperature and availability of particles for grazing and decomposition.

Figure 8. Internal loading of total phosphorus in Lake Mendota, 1976

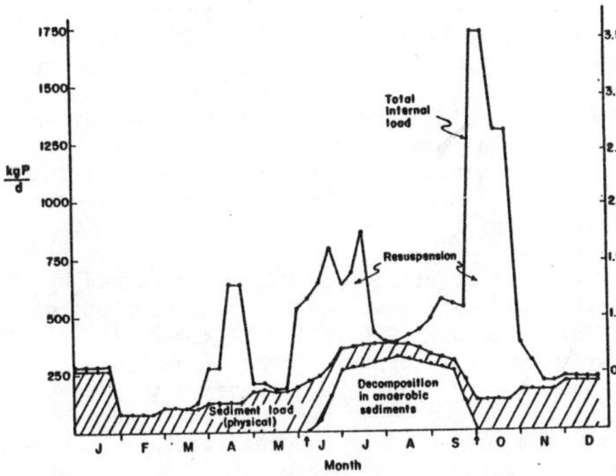

Figure 9. Internal loading and recycling of dissolved P in Lake Mendota, 1976

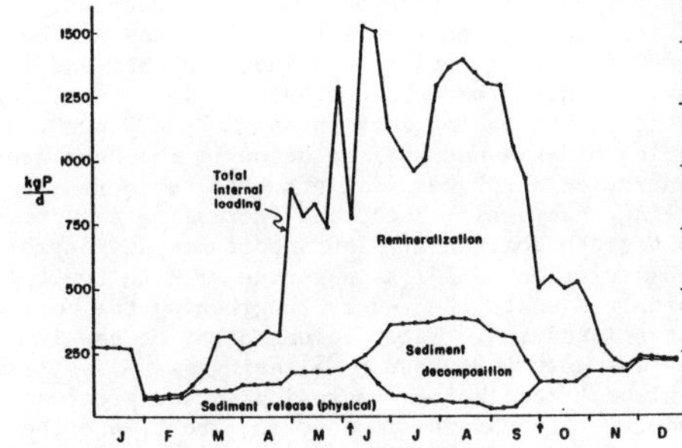

Figure 10. Internal loading and recycling of particulate P in Lake Mendota, 1976

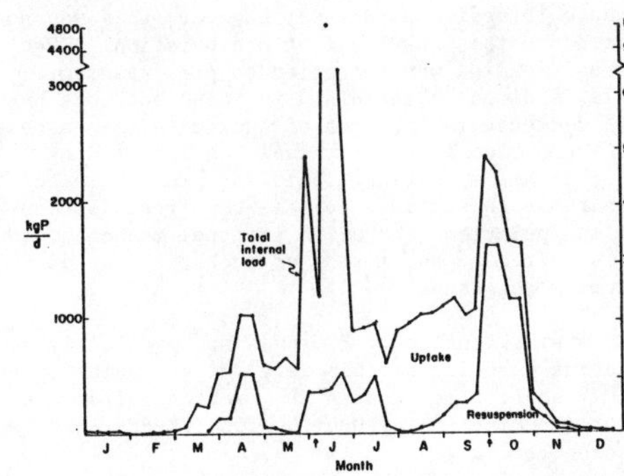

The internal flux of PP is characterized by fall and spring peaks in resuspension and an early summer peak in uptake. The latter probably occurs because dissolved P is still somewhat available and available light is high as a result of the shallow depth to which algae are mixed early in the period of stratification.

Evaluating External and Internal Control (Lake Loading Tolerance)

Using the model to simulate P dynamics and estimate internal fluxes of specific years gives some insight into the relative importance of external loading, internal loading, and recycling. The simulations made thus far suggest that internal fluxes are of much more importance in making DP available to the system for uptake. In addition, recycling is more important in this capacity during the summer stratification than either external loading or internal loading.

However, another use of this model may help to assess the relative importance of EL, IL, and recycling to the control of lake P dynamics (and, perhaps, lake trophic state and water quality as well) The model may be used to assess the effect of manipulating inputs and internal system control elements (those factors that are a part of the lake system and are internal regulated in some way by the lake). An example of an internally regulated feature of the system that affects P dynamics is lake thermal structure.

Such an analysis should be performed with great caution. Like most (if not all) models, STRAT does not have the ability to respond completely to changes in inputs and parameters that result in a radically new state of the system. Consequently, the use of a model to determine the effects of changing various inputs or internal control parameters on the system state was confined to a sensitivity analysis that involved slight changes in system inputs and internal control factors (such as rate parameters). It was hoped this would suggest how sensitive the system is to various manipulations in the "domain" near its present state. Whenever a manipulation causes a radical change in the system state, one must wonder what would be the added effect if the model could respond more fully. One should try to assess whether a more complete response would tend to push the system further from its unperturbed state or would tend to have a homeostatic effect.

To this end a simple sensitivity analysis of the model was carried out based on the 1976 nominal simulation. Here the "nominal simulation" is simply a simulation used as a baseline for comparison with simulations that result from running the model with some manipulation in an input or parameter. For the sensitivity analysis, each parameter in the system equations was varied by 10 to 20% as was external loading to the system. In addition, several structural elements of the system were varied to determine the influence of these factors on system dynamics.

The changes were analyzed both for their effects on the overall performance of the model and for their effects on certain fluxes or pools felt to be important indicators of the lake system's overall response. The former was assessed by comparing the manipulated simulation time series of P concentrations to the nominal simulation time series. The relative amounts of deviation that resulted from different manipulations were felt to be an assessment of the sensitivity of the model to changes in the various parameters and inputs. Deviation was assessed by summing the square of the differences between the P concentration time series of the nominal simulation and that of the manipulated simulation. The greatest deviations from the nominal simulation occurred with 10% changes in maximum growth rate of algae and 20% changes in: the percentage of PP in zooplankton, settling rate, the depth to which algae are mixed and the proportion of PP assumed to be algae. In fact, parameters that affected algal uptake were consistently among those to which the P dynamics appeared to be most sensitive (with the exception of the half-saturation constant).

A number of changes in the system's spatial-temporal structure were also investigated. The period of stratification was made one month longer by having stratification set up three weeks earlier and end one week later than occurred in 1976 (this period of stratification is not out of the ordinary for Mendota). In another test the period of time over which bluegreen algae dominate the system (which varies from year to year) was extended by two months (one month earlier and one month later). Only a few manipulations resulted in fits to the observed data that were better than that of the nominal simulation. Among these were: decreasing P inputs (especially PP inputs), decreasing resuspension, and increasing flushing. This suggests that loading to the lake (especially that of PP) may be overestimated.

An assessment was made of the impact of the above parameter, structure and input manipulations on certain values considered to be synthetic indices of lake system response. These include: (1) certain standing pools that are considered to be related to lake trophic state, and (2) certain internal fluxes whose response may represent the loading tolerance or buffer capacity of the lake (i.e., internal homeostatic mechanisms may respond to changes in other mechanisms in a way that decreases the net response of the lake). For this reason, all the manipulations were assessed, including those that had little effect on P dynamics since this lack of response may have been a result of compensatory mechanisms. Two points should be noted here. First, most compensatory mechanisms of lake systems probably act on time frames longer than the seasonal time frame addressed in this work. All this work can hope to assess is that small subset of mechanisms that might be observed on this shorter time frame, using a rather simple model of P dynamics. Second, one should not assume that all responses of the system parts will serve to buffer or minimize the response of the total system--some parts may respond by magnifying an effect.

Some of the indices of lake response to manipulations (see Table 3) included changes in: total lake P and algal P averaged over the year (MTLP, MAP), algal uptake summed over the year (SUPE--assumed to be reflective of productivity), net sedimentation in the deep part of the lake (SSDA), sediment release of DP, decay and zooplankton excretion, the flux of TP to the mixed zone by thermocline migration, and the amount of P retained by the system where retention = 1 - (TPOUT/TPIN).

Some of the findings were fairly predictable. When changed by 10%, most rate parameters caused a similar percentage change in the flux on which they had the most direct effect. For example, a 10% increase in PCEXC (the ratio of P excreted to C respired by zooplankton) resulted in a 9% increase in zooplankton excretion. Those parameters that had little effect on P dynamics also had little or no effect on the pools and fluxes assessed here.

The effects of some changes were not as predictable. A 10% change in the maximum uptake rate resulted in a mere 2% increase in annual uptake. This might be explained by postulating faster uptake of P earlier in the year, resulting in greater loss of P from the system by settling before the recycling mechanisms of zooplankton excretion and decay become important enough to serve to combat this loss. Some evidence for this is provided by the observation that the 10% increase in maximum uptake rate resulted in a 3% increase in net sedimentation while a 10% increase in zooplankton excretion and decay both decreased sedimentation.

The greatest effects on MTLP were caused by changes in alpha MZA, settling, and in the length of the period of stratification. Generally, MTLP was a fairly conservative quantity at this level of manipulation. A 75% reduction in loading associated with presettlement conditions (Watson 1981) produced a decrease of 10%. MAP was most affected by changes in alpha, the period of stratification, and resuspension. It was affected less than expected by changing MZA.

Flux rates tended to be more sensitive to the tested manipulations than did standing pools. Annual uptake was most reduced by increasing sinking rates and decreasing alpha and beta. Changes in resuspension, MZA, and extending the period of bluegreen dominance all caused lesser changes in annual uptake. Presettlement loads resulted in an 11% decrease. Net sedimentation was most affected by increasing the period of stratification, decreasing alpha, and increasing MZA. Other changes had little effect. Surprisingly, increasing settling by 20% only produced a 3% increase in net sedimentation in the deep area. Regeneration of DP was most affected by increasing settling, decreasing beta, and increasing the period of bluegreen dominance. The importance of thermocline migration (TMIG) as a source of P to the epilimnion was most affected by lengthening the period of stratification. Water column decay of particulates was most perturbed by altering gamma, settling, and alpha. Zooplankton excretion showed greatest sensitivity to changes in settling, beta, and the period of bluegreen dominance.

Sediment release was affected mainly by the manipulation of parameters that affect the estimation of sediment flux (KSB, SIPE, SIPH, ASHA, DS). It was thought that decreased inputs or manipulations that resulted in higher annual uptake (thus depleting DP in the water column) would result in a noticeable increase in sediment release; however, this was not observed at this low level of manipulation. When external loading was decreased to presettlement levels, sediment release did increase but only by 1%. A 20% decrease in the area of anoxic sediments causes a 19% reduction in the internal loading of P by anaerobic decomposition. This more than balances the 5% increase in sediment release that results from this manipulation. Thus, an internal mechanism that serves to reduce the impact of such a change on the system is suggested.

System retention of P proved to be a fairly conservative quantity at this level of manipulation and in this short time frame. The greatest effects were produced by increasing Qout by 20% (7% decrease) and reducing loading.

TABLE 3. Summary of sensitivity analysis of parameter and load manipulations for STRAT; impact of perturbations on indices of lake response.[a]

	% Change	MTLP	MAP	SUPE	SSDA	REG E	REG L	ZPX E	ZPX L	DK E	DK L	TMIG D	TMIG P	SED E	SED L	DKS	RET
Sediment parameters																	
KSB	10↑													9↑	9↑		
SIPE, SIPH	10↓													11↓	12↓		1↑
ASHA	20↓														5↑	19↓	
DS	.10↓													10↑	11↑		
Algal parameters																	
UPM	10↑																1↑
α	20↓	6↑	14↑	10↓	6↓					11↑	9↑	4↑	5↑				1↓
MZA	20↑	4↑	4↑	5↓	5↑												1↓
Zooplankton parameters																	
PCEXC	10↑					8↑	6↑	9↑	9↑								
CR	10↑					5↑		5↑	5↑								
BR	10↑							4↑	4↑								
CPZ	.10↑					8↑	6↑	9↑	9↑								
β	20↓			8↓		15↓	12↓	18↓	17↓				5↓				1↑
Detrital parameters																	
DK	10↑									11↑	9↑						
γ	20↓									20↓	12↓						
Other parameters																	
Qout	20↑																7↓
Resus	20↓		5↓	5↓		8↓	7↓	8↓	8↓								1↑
SET	20↑	4↓		15↓		19↓	15↓	20↓	19↓	15↓	14↓		10↓				1↑
Temporal manipulations																	
stratified period (month longer)		4↑	5↓		11↑							20↑	10↑	12↓			1↑
bluegreen period (2 months longer)				5↓		12↓	8↓	13↓	11↓	6↓	5↓		5↓				1↑
Initial conditions	10↑	5↑															1↓
Inputs																	
PPin	10↓																3↓
DPin	20↓																1↓
TPin (presettlement)	75↓	10↓	17↓	11↓	22↓	12↓	11↓	13↓	13↓	12↓	11↓	9↓	10↓				neg

[a]All values express percent change from nominal simulation value to perturbation simulation value. Changes in the following caused changes of less than 4%: DPin, PPin, KSMD, KP, KA, CPA, PACN, PACB, PDC, Io, ECW, ECA, UPM. No changes of less than 4% are shown except for changes in retention (RET). MTLP and MAP = mean annual total P and algal P; SUPE = annual sum of uptake, SSDA = net sedimentation in deep areas of lake; REGE and REGL = remineralization in epilimnion and in lake; ZPXE and ZPXL = zooplankton excretion in epilimnion and in lake; DKE and DKL = decay in these regions; TMIGD and TMIGP = entrainment of DP and PP by thermocline migration; SEDE and SEDL = sediment release in the epilimnion and lake; DKS = release from anaerobic sediments.

ASHA = area of anoxic sediments
BR = zooplankton basal respiration rate
CPA = C:P ratio in algae
CR = zooplankton respiration rate (based on consumption)
DK = decay rate coefficient
DKS = rate of release of P from anoxic sediments by anaerobic decomposition
DPin = DP loading
DS = depth into sediments used to determine concentration gradient between sediment pore water and lake water
ECA = constant that converts particles (e.g., algae) to light extinction coefficient
ECW = light extinction coefficient of lake water
Io = light optimum for algal growth
KA = algal half saturation constant for zooplankton grazing
KP = P half saturation constant for algal growth
KSB = mixed sediment release rate coefficient
KSMD = unmixed sediment release rate coefficient
MZA = depth to which algae are mixed
PACB = proportion of algae consumed by zooplankton when bluegreens dominate
PACN = proportion of algae consumed by zooplankton other times of year
PCEXC = ratio of P excreted to C respired by zooplankton
PDC = proportion of detritus consumed by zooplankton
PPin = PP loading
Qout = lake flushing rate
RES = resuspension rate
S = sinking rate
SIP = concentration of dissolved P in sediment pore water
UPM = maximum growth rate of algae
α β γ = percent of particulates assumed to be algae, zooplankton, detritus

SUMMARY OF RESULTS

A phosphorus dynamics model was applied to Lake Mendota, Wisconsin, for the years 1976 and 1977. The model's simulation followed observed P dynamics closely and was used to estimate internal fluxes (certain of these were measured and shown to agree well with the model estimates). Internal loading of P greatly exceeded external loading annually and seasonally. Biological recycling of P was the most important source of dissolved P during summer stratification; entrainment of hypolimnetic water was also significant. Most lake response indices were found to be most sensitive to the manipulation of factors considered to be greatly influenced by external driving variables (including depth of algal mixing, sinking and resuspension and the length of the period of stratification). However, short-term P dynamics were not very sensitive to changes in external loading. The community structure (reflected in the partitioning of particulate P into algae, zooplankton and detritus) was also of major significance to system behavior.

The usefulness of this approach lies in its ability to assess the effect on lake response of the simultaneous action of a number of hypothetical relationships acting within a lake specific spatial structure. A number of internal buffering mechanisms were suggested by the testing of certain hypotheses concerning the balance between conflicting internal processes. Such an approach also makes it possible to assess the amount of variability in lake response that could occur under the same loading regime as a result of variability in other controlling factors.

ACKNOWLEDGEMENTS

Funding for this research was provided by NSF grant DEB-7823826 and a Wisconsin Alumni Research Foundation Fellowship. Wendy Dickinson has my heartfelt thanks for the typing.

LITERATURE CITED

Berner, R.A. 1980. Early Diagenesis: a Theoretical Approach. 241 p. Princeton University Press, Princeton, N.J.

Bortleson, G.C. 1970. The chemical investigation of recent lake sediments from Wisconsin lakes and their interpretation. 259 p. Ph.D. Thesis, Water Chemistry, University of Wisconsin-Madison.

Brock, T.D., D.R. Lee, D. Jones and D. Winek. 1982. Groundwater seepage as a nutrient source to a drainage lake, Lake Mendota, Wisconsin. Water Res. (in press).

Chen, C.W. 1970. Concepts and utilities of ecologic models. J. Sanit. Eng. Div., Amer. Soc. Civ. Eng. 96(SA5): 1085-1097.

Chen, Y.S., J.N. Butler and W. Stumm. 1973. Kinetic study of phosphate reaction with aluminum oxide and kaolinite. Env. Sci. Tech. 7: 327-332.

Colman, J.A. 1979. Measurement of eddy diffusivity and solute transport through the near-sediment boundary layer of ice-covered lakes, using radon-222. 186 p. Ph.D. Thesis, Water Chemistry, University of Wisconsin-Madison.

Davis, C.O., N.R. Brietner and P.J. Harrison. 1978. Continuous culture of marine diatoms under silicon limitation. 3. A model of Si-limited diatom growth. Limnol. Oceanogr. 23: 41-52.

DiToro, D.M. 1980. Applicability of cellular equilibrium and Monod theory to phytoplankton growth kinetics. Ecol. Modelling 8: 201-218.

__________, and J.P. Connolly. 1980. Mathematical Models of Water Quality in Large Lakes. Part 2. Lake Erie. U.S. Environmental Protection Agency Publication Number EPA-600/3-80-065.

__________, D.J. O'Conner and R.V. Thomann. 1971. A dynamic model of the phytoplankton population in the Sacramento-San Joaquin Delta. Nonequilibrium Systems in Natural Water Chemistry. Adv. in Chem. Series 106: 131-180.

Droop, M.R. 1973. Some thoughts on nutrient limitation in algae. J. Phycol. 9: 264-272.

Dugdale, R.C. and J.J. MacIsaac. 1971. A computation model for the uptake of nitrate in the Peru upwelling region. Invest. Pesq. 35: 299-308.

Eppley, R.W., J.N. Rogers and J.J. McCarthy. 1969. Half-saturation constants for uptake of nitrate and ammonia by marine phytoplankton. Limnol. Oceanogr. 14: 912-920.

__________, and W.H. Thomas. 1969. Comparison of half-saturation constants for growth and nitrate uptake of marine phytoplankton. J. Phycol. 5: 375-379.

Fallon, R.D. 1978. The planktonic cyanobacteria: their sedimentation and decomposition in Lake Mendota, Wisconsin. 440 p. Ph.D. Thesis, Bacteriology, University of Wisconsin-Madison.

__________, and T.D. Brock. 1980. Planktonic blue-green algae: production, sedimentation and decomposition in Lake Mendota, Wisconsin. Limnol. Oceanogr. 25(1): 72-88.

Fleming, W.M. 1975. A model of the phosphorus cycle and phytoplankton growth in Skaha Lake, British Columbia. Verh. int. Ver. Limnol. 19: 229-240.

Ganf, G. and P. Blazka. 1974. Oxygen uptake, ammonia and phosphate excretion by zooplankton in a shallow equatorial lake (Lake George, Uganda). Limnol. Oceanogr. 19: 313-325.

Gasith, A. 1975. Tripton sedimentation in eutrophic lakes--a simple correction for resuspended matter. Verh. int. Ver. Limnol. 19: 116-122.

Golterman, H.L. 1977. (ed.) Interactions Between Sediments and Freshwater. Junk, The Hague.

Holdren, G.C. 1974. Measurement of sediment interstitial phosphorus and phosphorus release from intact sediment cores. 171 p. M.S. Thesis, Water Chemistry, University of Wisconsin-Madison.

__________. 1977. Factors affecting phosphorus release from lake sediments. 138 p. Ph.D. Thesis, Water Chemistry, University of Wisconsin, Madison.

__________, and D.E. Armstrong. 1980. Factors affecting phosphorus release from lake sediment cores. Env. Sci. Tech. 14(1): 79-86.

__________, D.E. Armstrong and R.F. Harris. 1977. Interstitial inorganic phosphorus concentration in lakes Mendota and Wingra. Water Res. 11: 1041-1047.

Imboden, D. and R. Gächter. 1978. A dynamic lake model for trophic state prediction. Ecological Modelling 4: 77-98.

__________, and A. Lerman. 1978. Chemical models of lakes. 363 p. In A. Lerman (ed.) Lakes: Chemistry, Geology and Physics. Springer-Verlag, New York.

Ingvorsen, K. and T.D. Brock. 1982. Electron flow via sulfate reduction and methanogenesis in the anaerobic hypolimnion of Lake Mendota. Limnol. Oceanogr. 27(3): 559-564.

James, H.R. and E.A. Birge. 1938. A laboratory study of the absorption of light by lake waters. Trans. Wisc. Acad. Sci. Arts Letts. 31: 1-154.

Jorgenson, S.E., L. Kamp-Nielson and O. Jacobsen. 1975. A submodel for anaerobic mud water exchange of phosphate. Ecological Modelling 2: 133-146.

Juday, C. 1921. Quantitative studies of the bottom fauna in the deeper waters of Lake Mendota. Trans. Wisc. Acad. Sci. Arts Letts. 20: 461-493.

Kamp-Nielson, L. 1977. Modelling the temporal variations in sedimentary phosphorus fractions. 473 p. In H.L. Golterman (ed.) Interactions Between Sediments and Freshwater. Junk, The Hague.

__________. 1978. Modeling the vertical gradients in sedimentary phosphorus fractions. Verh. int. Ver. Limnol. 20: 720-721.

__________. 1980. The influence of sediments on changed phosphorus loading to hypertrophic Lake Glumso. 29-36 p. In J. Barica and L.R. Mur (eds.) Hypertrophic Ecosystems. Junk, The Hague.

Klump, J.V. and C.S. Martens. 1981. Biogeochemical cycling in an organic rich coastal marine basin. II. Nutrient sediment-water exchange processes. Geochimica et Cosmochimica Acta 45(1): 101-121.

Lathrop, R.C. 1979. Appendix H. Lake Management. In Dane County Regional Planning Commission's (DCRPC) Dane County Water Quality Plan. 74 p. Madison, Wisconsin.

Lee, D.R. 1977. A device for measuring seepage flux in lakes and estuaries. Limnol. Oceanogr. 22: 155-163.

Lehman, J.T. 1980. Release and cycling of nutrients between planktonic algae and herbivores. Limnol. Oceanogr. 25: 620-632.

__________, D.B. Botkin and G.E. Likens. 1975. The assumptions and rationales of a computer model of phytoplankton population dynamics. Limnol. Oceanogr. 20: 343-364.

Lerman, A. 1979. Geochemical Processes--Water and Sediment Environments. 481 p. Wiley Interscience. New York.

Li, Y.H. and S. Gregory. 1974. Diffusion of ions in sea water and deep sea sediments. Geochimica et Cosmochimica Acta 38: 703-714.

Lorenzen, M.W. 1974. Predicting the effects of diversion on lake recovery. 205-210 p. In E.J. Middlebrooks, D.H. Falkenborg and T.E. Maloney (eds.) Modeling the Eutrophication Process. Ann Arbor Science Publishers, Ann Arbor, Michigan.

McBride, M.S. and H.O. Pffanekuch. 1975. The distribution of seepage within lakebeds. J. Res. U.S. Geol. Surv. 3: 505-512.

McCaffrey, R.J., A.C. Myers, E. Davey, G. Morrison, M. Bender, N. Luedtke, D. Cullen, P. Froelich and G. Klinkhammer. 1980. The relation between pore water chemistry and benthic fluxes of nutrients and manganese in Narragansett Bay, Rhode Island. Limnol. Oceanogr. 25: 31-44.

Muttkowski, R.A. 1918. The fauna of Lake Mendota; a qualitative and quantative survey with special reference to the insects. Trans. Wisc. Acad. Sci. Arts Letts. 19: 374-493.

Peters, R.H. and D. Lean. 1973. The characterization of soluble phosphorus released by limnetic zooplankton. Limnol. Oceanogr. 18: 270-279.

Powell, T. and A. Jassby. 1974. The estimation of vertical eddy diffusivites below the thermoclines in lakes. Water Resources Res. 10(2): 191-198.

Riley, G.A. 1965. A mathematical model of regional variations in plankton. Limnol. Oceanogr. 10(Suppl): R202-R215.

Scavia, D. 1980. An ecological model of Lake Ontario. Ecol. Modelling. 8: 49-78.

__________, B. Eadie and A. Robertson. 1976. An Ecological Model for Lake Ontario: Model Formulation, Calibration and Preliminary Evaluation. NOAA Tech. Report ERL 371. GLERL 12.

Shukla, S.S., J.K. Syers, J.D. Williams, D.E. Armstrong and R.H. Harris. 1971. Sorption of inorganic phosphate by lake sediments. Soil Sci. Soc. Amer. Proc. 35: 224-249.

Snow, P.D. and F.A. DiGiano. 1976. Mathematical modeling of phosphorus exchange between sediments and overlying water in shallow, eutrophic lakes. Report No. Env. E. 54-76-3 of Dept. of Civil Eng. Univ. of Massachusetts, Amherst.

Sonzogni, W.C. and G.F. Lee. 1972. Nutrient sources for Lake Mendota. Wisc. Acad. of Sci., Arts and Letters. 62: 133-164.

Stauffer, R.E. 1974. Thermocline migration--algal bloom relationships in stratified lakes. Ph.D. Thesis, 526 p. Water Chemistry, University of Wisconsin-Madison.

Steele, J.H. 1964. A study of production in the Gulf of Mexico. J. Mar. Res. 22: 211-222.

__________. 1965. Notes on some theoretical problems in production ecology. p. 383-398. In C.R. Goldman (ed.) Primary Productivity in Aquatic Environments. Mem. Ist. Ital. Idrobiol. Univ. of Calif. Press, Berkeley. 18(Suppl).

Tilman, D. and S.S. Kilham. 1976. Phosphate and silicate growth and uptake kinetics of the diatoms *Asterionella formosa* and *Cyclotella meneghiniana* in batch and semicontinuous culture. J. Phycol. 12: 375-383.

Watson, V.J. 1981. Seasonal phosphorus dynamics in a stratified lake ecosystem: an evaluation of external and internal loading and control. 279 p. Ph.D. Thesis, Botany, University of Wisconsin-Madison.

__________, V.J. and O.L. Loucks and W. Wojner. 1981. The impact of urbanization on seasonal hydrologic and nutrient budgets of a small north American watershed. Hydrobiologia. 77: 87-96.

Williams, J.D., J.K. Syers, R.F. Harris, and D.E. Armstrong. 1970. Adsorption and desorption of inorganic P by lake sediments in 0.1M NaCl system. Env. Sci. Tech. 4: 517-519.

__________, J.K. Syers, S.S. Shukla, R.F. Harris and D.E. Armstrong. 1971. Levels of inorganic and total P in lake sediments as related to other sediment parameters. Env. Sci. Tech. 5: 1113-20.

USE AND ABUSE OF EMPIRICAL PHOSPHORUS MODELS IN LAKE MANAGEMENT[1]

Dintie S. Mahamah[2], and Surinder K. Bhagat[3]

Abstract.--Empirical phosphorus models are indispensable tools in lake management. They are easy to use and provide answers to questions regarding "before" and "after" effects of lake restoration. They may be used in conjunction with error analysis to evaluate alternative control measures. They should not, however, be applied outside their constraints.

INTRODUCTION

In recent years there has been a respectable growth in the use of models in lake management. This growth is largely attributable to the equally growing demand for cleaner and less polluted lakes. The demand has in turn, placed a responsibility on engineers and others concerned with lake management for better scientific, and less vague evaluations of alternative pollution abatement schemes. In lake restoration for example, one has to justify proposed schemes in a manner more comprehensible to the taxpayer for funding and public relations reasons. The justification often depends on both long and short term prediction of effects of proposed lake restoration measures. A wise use of models represents perhaps the most available cost-effective tool for such projections.

The increased availability of computer and computer time has also played no small a part in the increased use of modelling in lake management. The cost of computations have been reduced by more than half in the two or so decades, thus making computations more affordable. The hand-held calculator is not only affordable, but very powerful for its size. Thanks to the revolution in minaturization dawned by the introduction of integrated circuitry.

Models for lake management have evolved along two (but by no means exclusive) philosophies, the "ecosystem" and "black-box" or "empirical" approaches. In the "ecosystem" approach, the lake is treated as a system of defined components and interactions. The task of the modeler is to describe these components and interactions in as realistic a manner as possible using available mathematical techniques. Often processes such as nutrients and nutrient cycling, algae, zooplankton and macrophyte successions are simulated (Parker, et al. 1979). Perhaps, the most immediate question facing the modeler who embarks on such a mission is that of extent of simulation. Obviously, every process in the system cannot be simulated for lack of knowledge of some of these processes and also for economic reasons. This has, therefore, given rise to "truncated" forms of "ecosystem" models under such names as "eutrophication models," "predator-prey models," and "food-web" models. These models dwell on processes of interest to the modeler.

Even after the question of what processes to simulate is answered, there is another important one to be answered, that of accuracy of mathematical descriptions. Mathematical descriptions of aquatic processes are oftem limited by both inadequate mathematics and incomplete understanding of process kinetics. To avoid these and other problems, some modelers have resorted to the second group of lake management models, the "black-box" or "empirical" models. As the name implies, a "black-box" model substitutes empiricism for little understood processes. This results in a greater simplification of an otherwise complex formulation. This simplicity is perhaps the most attractive quality of empirical models. It also implies that potential users do not require any formal training in modeling and a simple hand-held calculator is all that is often required for model manipulation.

In the desire to control eutrophication, phosphorus has become the most manipulated factor. The reason for this being that phosphorus is the most implicated factor of eutrophication and also the most effectively controlled using available technology (Reckhow et al. 1980). Most empirical models used in application to lake management are therefore phosphorus models. This paper deals with the application of empirical phosphorus models and its extensions to lake management.

---

[1]Paper presented at the Third International Conference of the State-of-the Art in Ecological Modelling, Colorado State University, May 24-28, 1982.

[2]Graduate Student, Department of Civil and Environmental Engineering, Washington State University, Pullman, Washington U.S.A.

[3]Professor and Chairman, Department of Civil and Environmental Engineering, Washington State University Pullman, Washington U.S.A.

REVIEW OF EMPIRICAL PHOSPHORUS MODELS

Based on earlier work, Vollenweider (1969) formulated an expression for the concentration of phosphorus in a lake as:

$$P = \frac{L}{Z/T + S(Z,T)} \quad (1)$$

where,

P = phosphorus concentration in lake (mg/liter)

L = areal phosphorus loading of lake (gm/m^2-yr)

Z = mean lake depth (m)

T = mean hydraulic detention time (yr)

$S(Z,T)$ = phosphorus removal rate by sedimentation

Details of the development of (1) are available elsewhere (Dillon 1977, Reckhow 1979a), and will not be discussed here. Suffice it to say that, the expression is a result of a phosphorus mass balance around a lake operating at a steady state with regard to contents. Some additional assumptions underlying the derivation of Equation 1 include: a) Complete mixing of lake contents, and b) Sedimentation rate being proportional to phosphorus concentration in the lake.

Actually, Vollenweider's original formulation expressed the sedimentation rate term a bit differently. Various researchers have assumed the sedimentation rate as a constant, a function of "Z", "T" or both. These variations in the formulation of the sedimentation rate term have, and continue to be the main source of differences among the several empirical phosphorus models in the literature.

Vollenweider (1975), estimated "S(Z,T)" for Swiss lakes to be equal to 10 m/yr using regression analysis. Dillon and Kirchner (1975), estimated it for 14 Canadian shield lakes as 13.2 m/yr. Chapra (1975) using data on 12 of the 14 lakes estimated "S(Z,T)" to be equal to 16 m/yr.

Jones and Bachman (1976) have estimated the sedimentation term as a function of average depth, "Z" as:

$$S(Z,T) = 0.65\ Z \quad (2)$$

The data for this regression equation was obtained from 16 Iowa lakes and from other studies reported in literature.

Other researchers have estimated "S(Z,T)" as a function of both "Z" and "T". Vollenweider (1976) estimated it as:

$$S(Z,T) = Z\sqrt{T} \quad (3)$$

The same expression (Equation 3) was obtained independently by Larsen and Mercier (1975). Walker (1977) obtained the sedimentation rate for 105 north temperate lakes as:

$$S(Z,T) = 0.824\ Z/T^{-0.546} \quad (4)$$

Reckhow et al. (1980) basing their analysis on data from 47 north temperate lakes obtained the following, using weighted least squares regression:

$$S(Z,T) = 11.6 + 0.2\ Z/T \quad (5)$$

Kirchner and Dillon (1975) re-expressed "P", the lake phosphorus concentration as a function of "R", the phosphorus retention coefficient. This expression was:

$$P = \frac{LT}{Z}(1-R) \quad (6)$$

where,

R = phosphorus retention coefficient

All the other symbols are same as described before. This expression was originally designed to get around the problem of finding an expression for the sedimentation rate term. However "R", the retention coefficient had to be estimated empirically for Equation 6. Kirchner and Dillon (1975) estimated "R", for 15 Canadian Shield lakes as:

$$R = 0.426\ \exp(-0.271) + 0.574\ \exp(0.000949\ Z/T) \quad (7)$$

Larsen and Mercier (1975) also empirically estimated R, using 36 north temperate lakes as data base, by

$$R = 0.78 - 0.121\ \log_e(Z/T) \quad (8)$$

More recently Chapra (1982) revised the basic Vollenweider model to account for a phenomenon known as "positional availability." He observed that some of the phosphorus input to a lake is associated with allochthonous particulate matter and settles quickly. This portion, he contends, has minimal effect on lake productivity. He therefore, formulated separate equations for each fraction (fast and slow settling). The total phosphorus concentration is the sum of these two fractions. Using 13 New York phosphorus concentration is the sum of these two fractions. Using 13 New York lakes as the data base, he obtained:

$$P = P_1 + P_2 = \frac{LT}{Z}\left[\frac{(1-fs)}{1 + 24.5Z/_T(1-e^{-.075z})}\right] \quad (9)$$

where,

P = total mean concentration of phosphorus in lake (mg/liter)

P_1 = mean concentration of allochthonous phosphorus fraction (mg/liter)

P_2 = mean concentration of non-allochthonous phosphorus fraction (mg/liter)

f_s = fraction of input phosphorus that does not affect midlake quality.

In his mass balance formulations for the derivation of the expression, Chapra assumed an exponential settling rate for the allochthonous portion and retained the Vollenweider (1969) formulation for the second fraction.

Osborne (1980) proposed a modified version of Vollenweider's (1969) model that included an explicit term for internal loading. In his formulation he argues that for high flushing lakes, Equation 1 reduces to:

$$P = \frac{LT}{Z} \quad (10)$$

or,

$$P = \frac{L_E T}{Z} + \frac{L_I T}{Z} \quad (11)$$

where L_E and L_I are the external and internal areal phosphorus loadings in grams per square meter per year respectively.

The introduction of error and uncertainty analysis to empirical modelling is probably one of the most useful developments since the inception of empirical phosphorus modelling. This error analysis makes possible, the evaluation of information derived from a model, thus reducing the risk of faulty decisions made using the model. The basic approach used by most modelers (Walker 1977, Reckhow 1979a 1979b, Montgomery et al. 1980) in error analysis is the "first-order error analysis" approach. This approach entails the use of a truncated form of the Taylor expansion in estimating the error propagation in a function due to contributing variables or parameters (Ditlevsen 1981). The application of error analysis in estimating uncertainty in empirical modelling is illustrated in the appendix. Some authors have shown that the use of another approach known as the "Monte Carlo method" produces comparable results (Montgomery et al. 1980).

Very often the empirical model user is not only interested in phosphorus levels, but also in the trophic state of a lake. This is because the public can relate better to the visible effect of eutrophication or its absence thereof. This has therefore, led to the use of empirical models to derive "loading criteria plots" first proposed by Vollenweider (1968). In his first work he presented a table of mean depths, and corresponding areal loading of phosphorus and nitrogen that would result in eutrophication ("dangerous loading") or oligotrophy ("permissible loading"). He later (Vollenweider 1975) revised this to include "flushing rate" (1/T) as a factor in eutrophication. His latest revision (Vollenweider 1976) also takes into account the additional effects of "Z" and "T" (Figure 1.) and is appreciated if it is recognized that a phosphorus "loading criteria plot" is actually a plot of an empirical phosphorus model. The axes of the plot are chosen such that the model equation is described by these axes. As an example, Dillon's (1975) loading criterion (Figure 2.) is actually a plot of equation (6). By settling "P" equal to 10 µg/liter and 20 µg/liter, the two lines are obtained. Ten micrograms per liter is regarded as the "permissible concentration in the lake and 20 µg/liter, the "dangerous" level. Mahamah and Bhagat (1982) have derived loading criteria plots for the Walker (1977), Kirchner and Dillon (1975) and the Jones and Bachman (1976) models. They have also evaluated the applicability of these criteria to 22 western United States lakes.

APPLICATIONS TO LAKE MANAGEMENT

The past decade can be characterized as the "clean-lakes awareness era." During the decade, lake managers have been increasingly confronted with the control of eutrophication. Many have resorted to measures such as nutrient inactivation, lake dredging, crop harvesting and watershed control. In many cases, more than one of these measures have been considered and used in an attempt to control the problem. Whatever the control measures adopted, the question of both long and short term effects of these measures must be addressed for both economic and political reasons. In the United States, most funds for lake cleaning were provided by Public Law 92-500. The public, therefore deserves the assurance that money spent to clean a lake is cost-effective. Golterman (1980) states that no European government is willing to develop a policy to control nutrient over-loading in a lake in the absence of estimates of reductions necessary to achieve a desirable amelioration. An empirical phosphorus model can provide useful information as to the effects of nutrient control measures in improving the water quality of a lake. It can provide estimates of expected phosphorus levels after the control measures are implemented. In most nutrient reduction situations, a flushing period is required before the desired results can become evident. During this period, the lake "adjusts" to the new situation. What an empirical phosphorus model simulates, are conditions approximating the end of this adjustment period. Some non-steady models can also predict the conditions during the transitional period.

If it is desired to estimate the new trophic state of a lake after the control measure, a phosphorus "loading criteria plot" like those in Figures 1 and 2 can provide it. As was observed earlier, "loading criteria plots" relate trophic states to changes in lake flushing rate, mean depth and phosphorus loading.

For the model user who wishes to estimate the effect of the change on chlorophyll levels, a "phosphorus-chlorophyll" model like that proposed by Jones and Bachman (1976) can be used. "Phosphorus-chlorophyll" models as the name implies, link ambient phosphorus changes to chlorophyll _a_ levels in lakes. Chlorophyll levels are important in these situations, because they represent the most visible effects of eutrophication in lakes.

Additionally, in conjunction with error analysis, an empirical phosphorus model may be used to evaluate alternative control measures (Reckhow 1980). This analysis is helpful in estimating the relative risks associated with a particular decision.

Error estimation in the use of an empirical phosphorus model was applied to Liberty Lake, Washington. The technique used in the analysis was modified from that proposed by Reckhow (1979a, 1979b). The results plotted in Figure 3. show the effect of volumetric phosphorus loading on the probability of exceeding a standard of 10 µg/liter. For example, at a volumetric loading of 0.04 gm/m^3/yr the probability that the standard will be exceeded is 70 percent ($\sigma = 0.25$). Reducing the volumetric loading by half reduces this probability to 53 percent. The decision concerning an acceptable risk (probability of exceeding standard) is dictated by the importance of the lake to users and their expectation of the benefits to be derived realizing the costs involved with the level of benefits desired.

Uttormark and Hutchins (1980) examined some of the techniques for estimating "before" and "after" effects of control measures. They analyzed data on 25 lakes in the United States, Europe and Canada using empirical models by Dillon and Rigler (1974) and Vollenweider (1975, 1976). They found that the empirical phosphorus models rendered accurate predictions in 70 to 80 percent of the cases considered. It is, therefore, safe to rely on predictions using empirical phosphorus models, most of the time. It is even safer when they are used in conjunction with uncertainty estimates.

Empirical phosphorus models can [illegible] .oused by applying them to the wrong condition. For example, they should never be used outside their constraints of original development without prior examinations for the new situation. The constraints include climate and geomorphology of the data base. As an illustration, a model developed on a sample of temperate lakes should not be used to predict conditions in a tropical lake without some prior knowledge of how well it will perform for the new situation. A model based on lakes with a maximum depth of 6 meters should theoretically, not be used for lakes with depths much greater than 6 meters. This does not necessarily mean that failure to observe this rule leads to wrong predictions. It does, however, mean that a model is not "accountable" for any errors that may arise under such circumstances. It constitutes the extrapolation of a curve outside the boundaries of known fit and hence could be risky.

The formation of models such as Equation 1, involves some simplifying assumptions. A violation of any of the assumptions constitutes a wrong use of the model, and can lead to wrong decisions. An important assumption involved in the derivation of Equation 1 is that of steady-state. The use of any of the models presented in this paper for non-steady state lakes constitutes a misuse of these models.

The assumption of complete-mixing is also one that can be easily over-looked. This is a simplifying assumption that is approximated in many lakes, but by no means all. For very long and oblong lakes, conditions other than complete mixing usually prevail. Phosphorus levels computed using any of the "complete-mix" phosphorus models may not be representative of whole lake conditions. Such predicted values, when obtained should be regarded as very crude estimates.

Experience has shown that conditions in lakes with high flushing rates (low detention times), and very shallow depths do not correlate well with the model predictions of phosphorus levels. This phenomenon is attributable to formulations currently employed in describing sedimentation rates in lakes. It appears most of these formulations are insensitive to changes in phosphorus levels at very high flushing rates and shallow depths. The only solution to the problem is to develop better formulations. It, however, behoves the model user to be aware of this potential weakness.

In many phosphorus mass balance formulations such as Equation 1, loss of phosphorus is assumed to occur through sedimentation and flushing at outlet only. Some lakes (closed lakes), do not possess any conventional outlets. The use of these empirical phosphorus models, without any modifications, can lead to gross errors in estimates of phosphorus levels in lakes. A technique for estimating phosphorus levels for closed lakes was proposed by Reckhow (1970a), and has been applied to Liberty Lake (see Appendix).

Finally, some empirical phosphorus models are developed on the basis of median ambient phosphorus concentrations as opposed to mean values. This by itself is not objectionable. The user, however, needs to be aware of this difference and should report computed values as such. Statistically, the median and mean are not always equal.

CONCLUSION

The preceeding discussion has examined the need for, and use of empirical phosphorus models in lake management. In some instances, the use of empirical phosphorus models without some prior testing and modifications is unwise. As correctly observed by Golterman (1980), a model must meet the three basic criteria of, precision, reality, and generality. The three are not always achieved to the same degree. It is the responsibility of the user, however, to realize that a model is only as good as the user. The use of models in decision making should, therefore, go beyond a mere "numbers game." Decisions should be based on a consideration of all available information. Used in this fashion, empirical phosphorus models do serve a very useful purpose.

ACKNOWLEDGMENTS

This work was funded in part by the Engineering Research Division, Washington State University, Pullman.

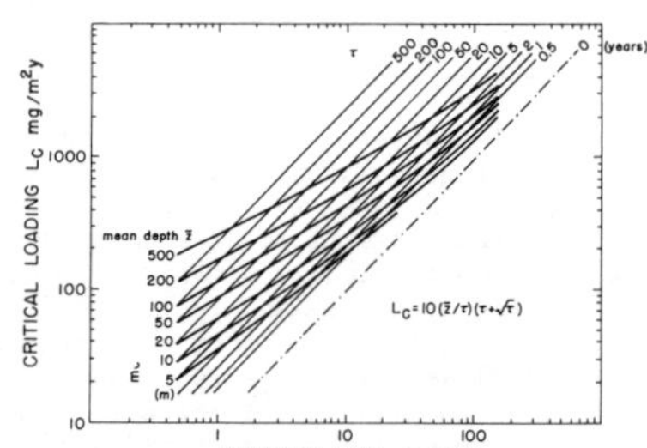

FIGURE 1. Vollenweider's phosphorus loading criteria Source: Reckhow (1979 a).

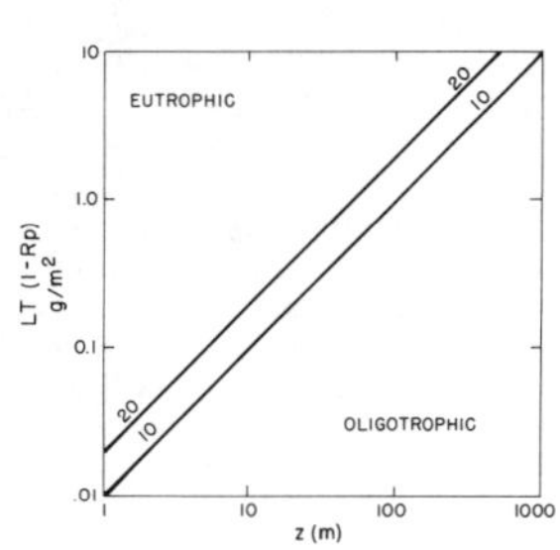

FIGURE 2. Dillon's phosphorus loading criteria. Lines of constant phosphorus concentration (μg/liter). Source: Reckhow (1979a).

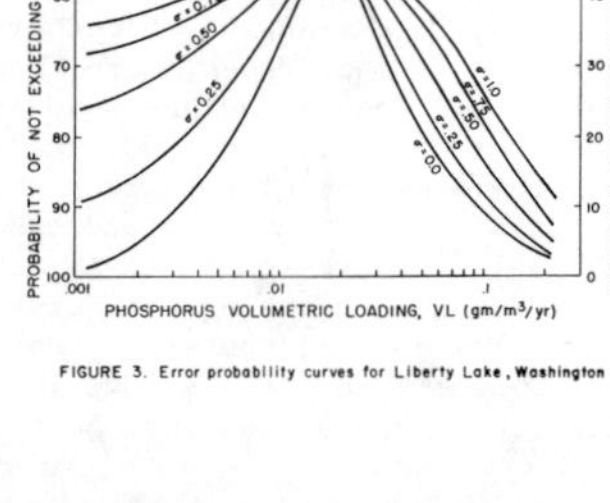

FIGURE 3. Error probability curves for Liberty Lake, Washington.

LITERATURE CITED

Chapra, S. C. 1982. A budget model accounting for the positional availability of phosphorus in lakes. p. 205-208. In Water Research (G.B.) Vol. 16.

Chapra, S. C. 1975. Comment on "An Empirical Method of Estimating the Retention of Phosphorus in Lakes." by W. B. Kirchner and P. J. Dillon. p. 1033-1034. In Water Resources Research 2 (6).

Dillon, P. J. 1977. A critical review of Vollenweider's nutrient budget model. p. 976-989. In Water Resources Bulletin.

Dillon, P. J., and W. B. Kirchner. 1975. Reply (to Chapra's comment). p. 1035-1036. In Water Resources Research 2(6).

Dillon, P. J., and F. G. Rigler. 1974. A test of a simple nutrient budget model predicting the phosphorus concentration in lake water. p. 1771-1778. In Journal Fisheries Research Board Canada. Vol. 31.

Ditlevsen, Ove. 1981. Uncertainty modeling. First edition. 412 p. McGraw-Hill International Book Company, New York, N.Y.

Funk, et al., 1982. Preliminary Assessment of multi-phase restoration efforts at Liberty Lake, Washington. Report No. 43. State of Washington Water Research Center, Pullman, Washington.

Golterman, H. L. 1980. Phosphate models, a gap to bridge. p 61-71. In Hydrobiologia. Vol. 72.

Jones, J. R., and R. W. Bachman. 1976. Prediction of phosphorus and chlorophyll levels in lakes. p. 2176-2182. In Journal of Water Pollution Control Federation. 48(9).

Larsen, D. P., and H. T. Mercier. 1975. Lake phosphorus loading graphs: an alternative. National eutrophication survey working paper No. 174, U. S. Environmental Protection Agency, Corvallis, Oregon.

Mahamah, D. S., and S. K. Bhagat. 1982. Performance of some empirical phosphorus models. Journal of Environmental Engineering Division, ASCE. (In press).

Montgomery, R. H., Lee, D. V., and Reckhow, K. H. 1980. A comparison of uncertainty analysis techniques: first order analysis vs. Monte Carlo simulation. Presented at IAGLR, Great Lakes 1980.

Orsborne, P L. 1980. Prediction of phosphorus and nitrogen concentrations in lakes from both internal and external loading rates. p. 229-232, Vol. 69.

Park, R. A., Groden, T. W., and Desormeau, C. J. 1979. Modifications to the model cleaner requiring further research. p. 87-108. In perspectives on lake ecosystem modeling. eds. Scavia, D, and A. Robertson. Ann Arbor Science. Ann Arbor Michigan.

Reckhow, K. H. 1980. An incremental approach to phosphorus lake modeling. 11 p. Prepared for the EPA/OECD Symposium on Inland Waters and Lakes Restoration.

Reckhow, K. H., Beaulac, M. N., and Simpson, J. T. 1980. Modeling phosphorus loading and lake response under uncertainty: a manual of compilation and export coefficients. Office of Water Regulations and Standards, U. S. Environmental Protection Agency, Washington, D.C. Report No. EPA 440/5-80-011.

Reckhow, K. H. 1979a. Quantitative techniques for the assessment of lake quality. 146 p. Office of Water Planning and Standards, U. S. Environmental Protection Agency, Washington, D.C. Report No. EPA-440/5-79-015.

Reckhow, K. H. 1976b. Empirical lake models for phosphorus: development, applications, limitations and uncertainty. p. 193-222. In perspectives on lake ecosystem modeling. eds. Scavia, D., and A. Robertson. Ann Arbor Science. Ann Arbor, Michigan

Uttormark, P. D., and Hutchins, M. L. 1980. Input/output models as decision aids for lake restoration. p. 494-500. In Water Resources Bulletin 3(16).

Vollenweider, R. A. 1976. Advances in defining critical loading levels for phosphorus in lake eutrophication. p. 53-83. In Mem. Institute. Ital. Idrobiol. Vol. 33.

Vollenweider, R. A. 1975. Input/output models with special reference to the phosphorus loading concept in limnology. p. 53-84. In Sch. Zeit. Hydrologic Vol. 37.

Vollenweider, R. A. 1969. Possibilities and limits of elementary models concerning the budget of substances in lakes. p. 1-36. In Arch. Hydrobiol. 66(1).

Vollenweider, R. A. 1968. The scientific basis of lake and stream eutrophication with particular reference to phosphorus and nitrogen as eutrophication factors. Technical Report DAS/DSI 68:27, OECD, Paris, France.

Walker, W. W., Jr. Some analytical methods applied to lake water quality problems. Ph.D. dissertation, Harvard University, Boston, Massachusetts,

APPENDIX

Application of Uncertainty Analysis to Liberty Lake

Liberty Lake is located in the state of Washington. A soft water lake of glacial origin, it is bordered on three sides by a small mountain range with a water shed of 3,445 hectares. The shoreline is predominantly occupied by residential areas. (Funk et al. 1982)

It is believed that algae blooms were first observed in the mid sixties which culminated in a decision in 1973 to restore the lake. The baseline studies and recommendations for restoration were undertaken by the Department of Civil and Environmental Engineering at Washington State University. Since then, two treatments with alum have been effected on the lake (1974 and 1981). Other phosphorus control measures have also been started.

The analysis presented here, does not deal with these restoration measures. The concern here is to illustrate how a model and error analysis could be used in an actual case of a lake being considered for phosphorus control

Lake Characteristics

Mean hydraulic residence time	= 3 years
Mean lake depth	= 7 meters
Maximum lake depth	= 9.1 meters
Surface area	= 299 hectares
Estimated phosphorus volumetric loading	= .02 $gm/m^3/yr$

Using non-linear regression techniques, and log-transformed data, for 79 lakes in Western United States, the following model was developed (Mahamah and Bhagat, unpublished results):

$$P = \frac{VL}{[1/T + 2.02\ VL^{.10}]} \qquad (A)$$

where,

P = mean lake phosphorus concentration (mg/liter)

VL = volumetric phosphorus loading ($gm/m^3/yr$)

T = mean hydraulic residence time (yr)

The gross probable error involved in the use of (A) is given by (Reckhow 1979a 1979b 1980, Montgomery et al. 1980):

$$S_t^2 = S_p^2 + S_{iv}^2 + S_m^2 \qquad (B)$$

where,

S_t^2 = gross probable error

S_m^2 = model derived error

S_p^2 = parameter induced error

S_{iv}^2 = independent variables induced error.

Mahamah and Bhagat (on-going research), have shown that for the model (A), the total probable error is:

$$S_t^2 = S_p^2 + S_m^2 + S_{iv}^2 = 0.0039 + 0.198 + \sigma VL \qquad (C)$$

where,

S_p^2 = 0.0039

S_m^2 = 0.198

$S_{iv}^2 = \sigma VL$*

For any given standard (to be met), the probability of violating the standard based on predictions using model (A), is given by:

$$\text{Prob [standard violation]} = \frac{P_{std} - P_m}{S_t} \qquad (D)$$

where,

P_{std} = standard to be met (mg/liter total P)

P_m = predicted total phosphorus concentration using model (mg/liter)

S_t = square root of total probable error

Figure 3 is derived from (D) for a standard of 10 micrograms per liter. For any given volumetric loading of total phosphorus, VL, the probability of not exceeding the standard may be read from the figure.

For Liberty Lake:

$$VL = 02\ gm/m^3\text{-yr}$$

Using Figure 3 and $\sigma = 0.25$, the probability that the standard will not be exceeded is 47 percent (approximately). Another way of saying that is, there is 53 percent chance that the standard of 10 μg/liter will be exceeded. Recall that 10 μg/liter is the critical phosphorus level recommended by Vollenweider (1976) and others. This result is consistent with the trophic condition of Liberty Lake. It is estimated that by control measures, the volumetric phosphorus loading of the lake can be reduced by 60 percent. Using the Figure 3 again, it is found that the new probability of exceeding the standard is only about 33 percent. This may be considered an acceptable risk.

By computations using Equation A, the phosphorus level of the lake before the reduction is estimated as 0.012 mg/liter. However, average values of as high as 0.021 mg/liter have been reported (Funk et al. 1982). This apparent discrepancy may be attributed to the fact that Liberty Lake does not possess a conventional outflow for most part of the year, as assumed in the derivation of Equation A.

---

*This is based on the assumption that the error introduced by the independent variables is proportional to the volumetric loading, VL; σ is the proportionality constant.

For lakes of that nature, lower and upper bounds of phosphorus may be estimated (Reckhow 1979a).

The lower bound is given by the value obtained using Equation A (i.e., .012 mg/liter).

The upper bound is estimated by:

$$P_{upperbound} = \frac{VL}{1/T} \qquad (E)$$

$$= \frac{0.02}{0.33} = 0.060 \text{ mg/liter}$$

Therefore, phosphorus level of Liberty Lake P, may be stated as:

$$0.011 \leq P \leq 0.060$$

A PHOSPHOROUS MANAGEMENT MODEL FOR LAKE CARNEGIE (NJ)[1]

Tzay-Rong Jenq[2], Christopher G. Uchrin[3], Marvin L. Granstrom[4], and Shing-Fu Hsueh[5]

Abstract.--A linear programming model to assess the cost-effectiveness of appropriate point and nonpoint source phosphorus control measures was constructed for Carnegie Lake, an eutrophic lake located in Mercer County, New Jersey. Appropriate in-stream total phosphorus models and a lake eutrophication model were constructed for definition of constraints. The resultant model was tested for present and future conditions and is being used as a decision making tool for water quality planning in the Lake Carnegie basin.

INTRODUCTION

The discharge of nutrient-laden wastes into receiving waters has been given considerably attention due to its influence on eutrophication processes in lakes and impoundments. The removal of nutrients required for algal growth from point source municipal wastewater discharges has historically been considered an efficient and reliable method for preventing advanced eutrophic conditions. However, due to the effect of non-point source pollution and the existence of nutrient cycles in natural water systems, the removal of nutrients contributing to algal growth is effective only when the limiting nutrient(s) can be identified, the point source discharge is the major limiting nutrient source, and the removal process is cost-effective.

Carnegie Lake, an impoundment, located in Mercer County, New Jersey, suffers from the attendent characteristics of a highly eutrophic state. Significant growths of aquatic plants in the upland tributaries has impaired in-stream water quality. Chlorophyll pigment levels, plankton densities, and nutrient concentrations are all indicative of a high degree of eutrophy. The major factors causing the eutrophication are excessive nutrient inputs and sedimentation. Phosphorus has been identified as the limiting nutrient for algal growth.

STUDY AREA CHARACTERIZATION

Lake Carnegie Basin

The Lake Carnegie Basin located in Mercer, Middlesex and Monmouth Counties, New Jersey, has a drainage area of approximately 150 square miles (388 km^2). At present, about 70,000 people populate the study area. Due to development potential for industrial, commercial and residential growth, the forcasted year 2000 population is 200,000, an increase of 300%. Land use in the entire Carnegie Basin has been estimated at 12 percent urban, 44 percent forested and 44 percent agricultural (including pasture lands).

Lake Carnegie

At present, Carnegie Lake, shown in figure 1, suffers from excessive nutrient problems. Significant growth of aquatic plants in the upland tributaries has imparied instream water quality (Hunter, 1976). As described by the Philadelphia Academy of Natural Sciences (1969), the major factors causing the eutrophication in Carnegie Lake are excessive nutrient inputs and sedimentation. Phosphorus was further identified as the limiting nutrient for algal growth. The visible symptoms of eutrophy include high turbidity, algal blooms, and dense aquatic plants during the summer months. Lake Carnegie receives waters from both the Upper Millstone River and Stony Brook. The drainage area for the lake basin itself is approximately 15 square miles (40 km^2). The lake is long and narrow with a surface area of 222 acres (0.9 km^2) and an average depth of 8 feet (2.4 m). The mean residence time of water in the lake is less than three days, based on a mean annual flow of 250 cfs (7.1 m^3/s).

---

[1]Paper presented at the Third International Conference on State-of-the-Art in Ecological Modelling, Colorado State University, May 24-28, 1982.

[2]Environmental Engineer, New Jersey Department of Environmental Protection, Trenton, NJ 08625

[3]Assistant Professor, Department of Environmental Science, Cook College, Rutgers University, New Brunswick, NJ 08903

[4]Professor, Department of Civil and Environmental Engineering, Rutgers University, Piscataway, NJ 08853

[5]Chief, Waste Load Allocation, Section, New Jersey Department of Environmental Protection, Trenton NJ 08625

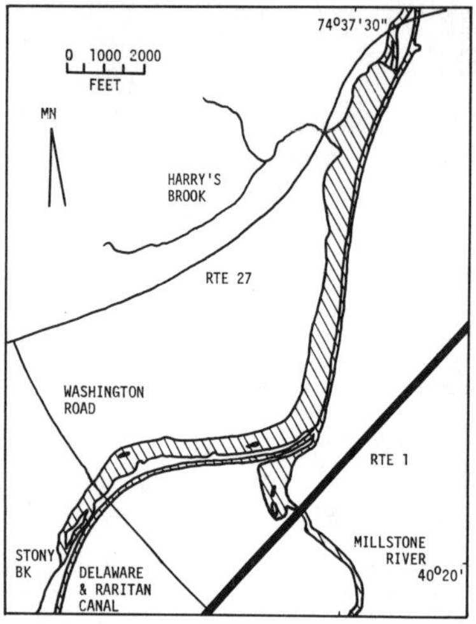

Figure 1. Carnegie Lake

Table 2.--Existing phosphorus point source discharges in the Upper Millstone River sub-watershed

Map no.	Discharger	Municipality	Existing treatment level	Flow (MGD)	Effluent T-P mg/l
1	FMC Corp.	Plainsboro Twp.	Secondary	0.012	6.0
2	RCA Corp.	West Windsor Twp.	Secondary	0.025	6.0
4	Lewis Bower Inc.	West Windsor Twp.	Secondary	0.004	2.0
6	East Windsor MUA	East Windsor Twp.	Secondary	1.00	5.5
8	Hightstown STP	Hightstown Boro	Secondary	0.7	6.5
11	NJ Turnpike Auth.	Cranbury Twp.	Advanced	0.07	2.5
12	Carter-Wallace Inc.	Cranbury Twp.	Advanced	0.288	1.0

Total Phosphorus Loads = 32,631 lb/yr (14,800 kg/yr)

PHOSPHORUS LOADINGS

Point Sources

According to the Philadelphia Academy of Natural Sciences (1969) and Hunter (1976), the most significant sources of phosphorus contributing to the eutrophication of Lake Carnegie were municipal and institutional point source dischargers. These reports also indicated that the impact of municipal sewage treatment plant effluents in the Upper Millstone River Basin appears to be largely responsible for the severe degradation of Lake Carnegie and tributary system. Listings of the present municipal and industrial point source discharges are presented in tables 1 and 2 for the Stony Brook and Millstone watersheds and are located in figures 2 and 3, respectively.

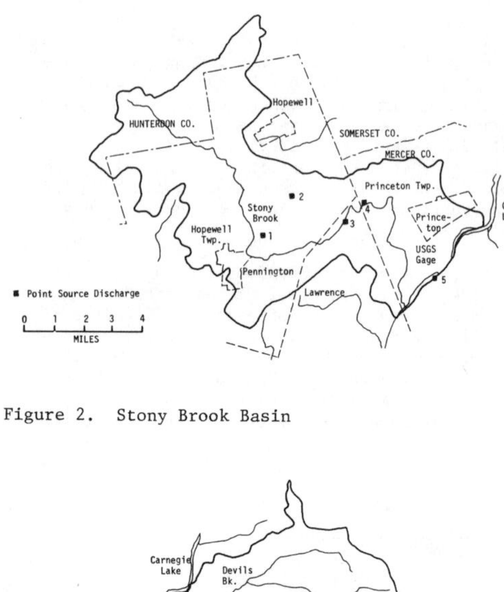

Figure 2. Stony Brook Basin

Table 1.-- Existing phosphorus point source dischargers in the Stony Brook sub-watershed

Map No.	Discharger	Municipality	Treatment Process	Flow (MGD)	Effluent T-P (mg/l)
1	Mobil Oil Co.	Hopewell Twp.	Advanced	0.004	2.0
2	Princeton Farm	Hopewell Twp.	Advanced	0.046	2.0
3	E.T.S.	Princeton Twp.	Advanced	0.05	2.0
4	Pretty Br. STP.	Princeton Boro	Secondary	0.03	6.0
5	Princeton Theo.	W. Windsor Twp.	Secondary	0.2	6.0

Total Phosphorus Loads = 4,672 lb/yr (2,119 kg/yr)

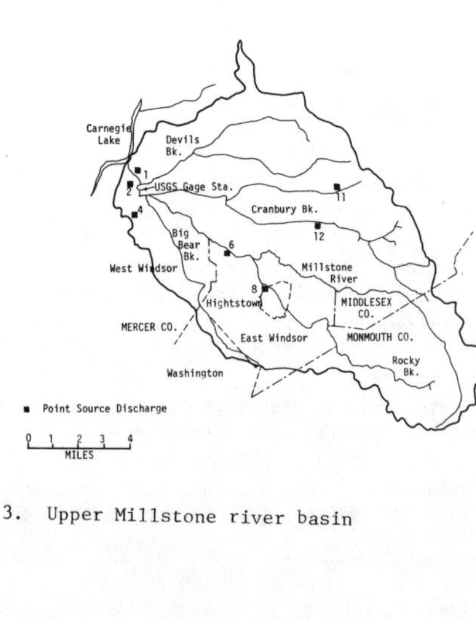

Figure 3. Upper Millstone river basin

The only intermittent-point source in this study area is the combined sewage overflow of the Princeton Sewage Treatment Plant. This overflow outfall is located at the lower part of Lake Carnegie and is considered to have insignifcant impact on the lake due to the small amount of phosphorus loading by storm overflows (Paterson, 1977). Therefore, this intermittent-point source was not addressed in this study.

Future phosphorus discharges were estimated for the year 2000 from United States Environmental Protection Agency (USEPA) facilties plans. These are presented in tables 3 and 4 for the Stony Brook and Millstone watersheds, respectively.

Table 3.--Inventory of future (year 2000) phosphorus point sources in the upper Millstone basin

Dischargers	Municipality	Flow (MGD)	T-P mg/l
Cranbury Village	Cranbury Township	0.24	6.0
East Windsor MUA	East Windsor TWP	5.32	6.0
Bergen Mills	Millstone Twp.	0.73	6.0
Union Valley Rd. STP	Monroe Township	1.40	6.0
Chate Corp. STP	Millstone Twp.	0.30	6.0
NJ Turnpike Auth.	Cranbury Township	0.25	1.0*
FMC Corp. Dis. #2	Plainsboro	0.015	6.0
RCA Corporation	West Windsor	0.042	6.0
Lewis Bower, Inc.	West Windsor	0.035	1.0*
Total Phosphorus Loads = 147,715 lb/yr (67,000 kg/yr)			

*1.0 mg/l of T-P is based on the existing advanced treatment efficiency

Table 4.--Inventory of future (year 2000) phosphorus point source dischargers in Stony Brook watershed

Dischargers	Municipality	Flow (MGD)	T-P mg/l
Mobil Oil Co.	Hopewell Township	0.015	2.0*
Princeton Farm De.	Hopewell Township	0.15	2.0*
Education Testing	Princeton Twonship	0.08	2.0*
Pretty Brook STP	Princeton Boro	0.02	6.0
Princeton Theo. Sem.	West Windsor Twp.	0.13	6.0
Pennington STP	Pennington Twp.	0.3	6.0
Total Phosphorus Loads = 9,691 lb/yr (4,400 kg/yr)			

*Based on the existing treatment plant's effluent level.

Non-Point Sources (NPS)

The effects of phosphorus non-point source pollution are felt throughout the entire watershed in rural areas as well as urban areas, both during dry weather and during storms. The phosphorus pollutant burden generated from these diffuse sources is generally attached to sediments and delivered to the stream by water movement.

Land Use Related NPS

There are approximately 12,300 acres (50 km^2) of urbanized land in the study area, and roughly 30% of this area is impervious surface which generates nutrient laden storm runoff. The approximate acreage and locations of urbanized areas in study area are listed in tables 5-8 and also displayed in figures 2 and 3. About 40,000 acres (162 km^2) of croplands are

Table 5.--Overall land use in the upper Millstone river basin

Land use	Square miles	Percent (%)
Developed	11.8	12
Open	42.5	42
Crop	45.0	46
Total	100	100

Table 6.--Land use along upper Millstone River

Stream segment	Length (mi.)	Land use (sq. mi.)		Land use (%)
I	12.25	Urban	(5.4)	20
		Agricultural	(10.8)	40
		Forest	(10.8)	40
		Total	27	100
II	8.0	Urban	(5.8)	40
		Agricultural	(4.35)	30
		Forest	(4.35)	30
		Total	14.5	100
III	12.0	Urban	(4.2)	20
		Agricultural	(8.4)	40
		Forest	(8.4)	40
		Total	21.0	100
IV	7.5	Urban	(3.1)	25
		Agricultural	(5.0)	40
		Forest	(4.3)	35
		Total	12.4	100
V	7.25	Urban	(1.16)	20
		Agricultural	(2.3)	40
		Forest	(2.3)	40
		Total	5.76	100
VI	7.75	Urban	(3.8)	20
		Agricultural	(7.77)	40
		Forest	(7.77)	40
		Total	19.34	100

Table 7.--Overall land use in the Stony Brook watershed

Land use	Square miles	Percent
Developed	8.00	16.58
Open (woods)	21.60	45.15
Crop	18.40	38.27
TOTALS	48	100

Table 8.--Land use distribution along Stony Brook

Stream segment	Length (mi.)	Land use (sq. mi.)		Land use (%)
I	4.5	Urban	(3.2)	20
		Agricultural	(6.4)	40
		Forest	(6.4)	40
		Total	16.0	100
II	6.1	Urban	(7.9)	45
		Agricultural	(5.2)	30
		Forest	(4.4)	25
		Total	17.5	100
III	5.5	Urban	(5.2)	50
		Agricultural	(2.1)	20
		Forest	(3.2)	30
		Total	10.5	100
IV	4.0	Urban	(2.2)	55
		Agricultural	(0.6)	15
		Forest	(1.2)	35
		Total	4.0	100

cultivated in the study area. A variety of agriculture is practices including livestock and poultry raising, and production of cattle feed and seed. The bulk of the dairy and beef cattle raising, as well as the concomitant production of corn grain, silage and hay, is practiced in the northern Piedmont portion of Stony Brook Watershed in Hopewell Township. The most significant crop production takes place in the Coastal Plain in the Upper Millstone Watershed, where the majority of the soybean and wheat crops are harvested in West Windsor Township. Soils of the Piedmont (mostly in the Stony Brook Basin), in the southeast as opposed to those of the Coastal Plain (mostly in the Upper Millstone Basin), are more prone to erosion. Because no silvicultural activites are presently practiced, nutrient pollution contributed from forest areas was considered not significant. The concentration of phosphorus in the runoff or groundwater directly reflects the natural equilibrium between the chemical constituents of the geology and those dissolved minerals and the pH level.

On-Site Specific NPS

In general, site specific NPS does not directly affect stream water quality. Managment programs applied to these sites are considered to improve stream background water quality. Thus on-site specific NPS were not considered in this study.

Loadings

The method used in this study (Paterson, 1977) estimates the total NPS load by first determining total in-stream loading including total NPS and point source loading at the mouth of each sub-basin. The total annual NPS loading is thus the remainder when point source load is subtracted from the total in-stream loading. Total annual in-stream loading is determined by summing the loads calculated at low, medium low, medium high and high flow conditions. In order to calculate the T-P loads under each flow condition, a relationship between T-P concentration and flow must be developed for a station selected at the mouth of each sub-basin. The evaluation of total annual NPS loads in the stream is predicated on the hypothesis that stream concentrations of phosphorus vary with flow rate and season. Paterson used the historical T-P and flow data collected at the mouth of each sub-basin in the study area to develop relationships between T-P concentrations, flow, and seasons.

The total NPS in-stream loading contributed from each sub-watershed was broken down into three categories: storm-related; steady-state (low flow); and, point source loads. Table 9 shows the total phosphorus loads and annual unit loads in each sub-basin and the Carnegie Lake Basin. Future total phosphorus NPS loads were estimated based on the type of land use and an associated unti load. Table 10 illustrates the results of these projections.

Table 9.--Average total and unit annual phosphorus (T-P) load in Carnegie basin and its sub-basins

Watershed	Drainage area (acres)	Total load (lbs/year)	Unit load (lbs/acre/yr)	% of total (yearly load)
1) Stony Brook Loadings	30,720			
Storm-Related Load		25,300	0.823	63
Steady-State Load		10,470	0.341	25
Point Source Load		4,672	0.152	12
TOTAL LOAD		40,442	1.316	100
2) Upper Millstone Loading	64,000			
Storm-Related Load		14,258	0.223	30
Steady-State Load		(- 700)	--	--
Point Source Load		32,631	0.510	70
TOTAL LOAD		46,189	0.732	100
3) Carnegie Basin Loadings	94,720			
Storm-Related Load		39,558	0.418	46
Steady-State Load		9,770	--	11
Point Source Load		37,703	0.391	43
TOTAL LOAD		86,631	0.915	100

Table 10.--Total phosphorus NPS loads in year 2000

Sources	Land use	Area (acres)	Unit load (lbs/acre/yr)		Total Load (lbs/year)	
			Storm-Related	Steady-State	Storm-Related	Steady-State
Stony Brook Loading:						
	Developed	11,520	0.72	0.45	8,294	5,184
	Crop	10,560	0.72	0.45	7,602	4,752
	Open	8,640	0.72	0.07	6,222	605
	(Sub-total)	30,720			22,118	10,541
Upper Millstone Loading:						
	Developed	12,800	0.6	0.2	7,680	2,560
	Crop	24,448	0.3	0.2	7,334	4,900
	Open	26,752	0.05	0.07	1,338	1,873
	(Sub-total)	64,000			16,352	9,333
Carnegie Basin Loading:						
	Developed	24,320	--	--	15,974	7,74
	Crop	35,008	--	--	14,936	9,652
	Open	35,392	--	--	7,560	2,478
	(Sub-total)	94,720			38,470	19,874

IN-STREAM PHOSPHORUS TRANSPORT MODEL

Model Formulation

A mathematical model describing the transport of phosphorus was calibrated and verified. The total phosphorus concentration (TP) within the ith reach of tributary k can be related to the concentration at the beginning of the reach by considering the reach analagous to a plug flow system. The resultant mass balance equation can be formulated as:

$$\frac{\delta Pki}{\delta t} = \frac{Qki}{Aki} \frac{\delta Pki}{\delta x} \pm Kki\ Pki \qquad (1)$$

where,

Pki = T-P concentration $[M/L^3]$;
t = time [T];
x = distance downstream in reach [L];
Aki = average crosss sectional area of reach $[L^2]$;
Kki = first order soure (+) or disappearance (-) of T-P [1/T]; and
Qki = average flow in reach $[L^3/T]$ (Thomann, 1972).

Assuming steady-state conditions, equation 1 can be integrated yielding.

$$Pki = Pki^a \exp\left(\pm Kki \frac{Aki}{Qki} X\right) \qquad (2)$$

Parameter Evaluation

Both the Millstone River and the Stony Brook were segmented to reflect morphologic and flow characteristics. Figures 4 and 5 display reference systems for the two streams respectively. Time of trave was computed using USGS data collected for both systems.

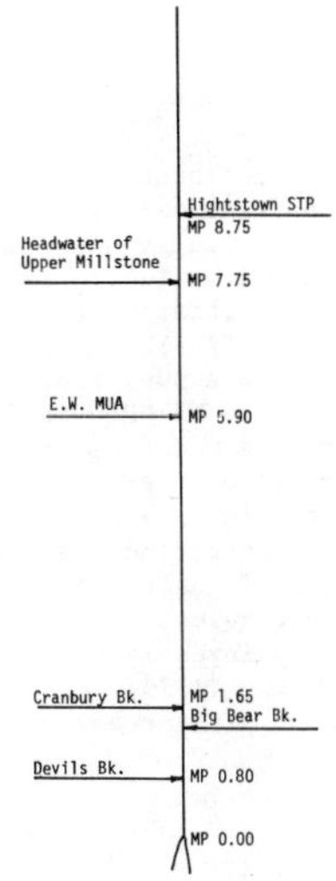

Figure 4. Millstone River schematic

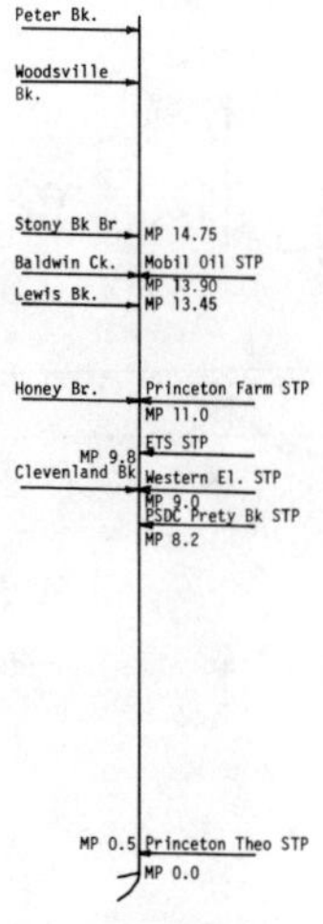

Figure 5. Stony Brook schematic

Model Calibration/Verification

The mechanisms of phosphorus decay in streams are extremely difficult to quantify due to the complicated physicochemical and biological processes occurring within stream systems. In general, the major factors causing phosphorus change (increase or decrease) can include aquatic plant uptake, settling and scouring of the deposits, and some biochemical reactions. The phosphorus decay constant (K) can be expressed as the sum of two coefficients: Kt and Ks where Kt is aquatic plant uptake (per day), and Ks is the rate of settling and scouring (per day). A single, bulk rate

K was used in this study.

Calibration

Values for K were calibrated and verified for each system. The resultant calibration traces compared to field data are displayed in figures 6-9. Figure 6 displays a low flow (18.1 cfs (0.51 m^3/s) at Plainsboro MP = 12.2) calibration and figure 7 a high flow (347 cfs [9.8 m^3/s]) calibration for the Millstone River. Average annual flow at Plainsboro is 99.2 cfs (2.8 m^3/s) and MA7CD10 is 2.8 cfs (0.08 m^3/s). The K values obtained from the low flow analysis ranged from 0.1-0.7 per day. The K values obtained from the high flow analysis ranged from -4 to 4 evidencing the occurrence of pronounced scour$^{(-)}$ and subsequent deposition$^{(+)}$. The occurrence of signifcant pollutant release due to scour has been documented by many investigators including Uchrin and Weber (1980) and Lung (1975). No attempt was made to correlate scour with stream velocity.

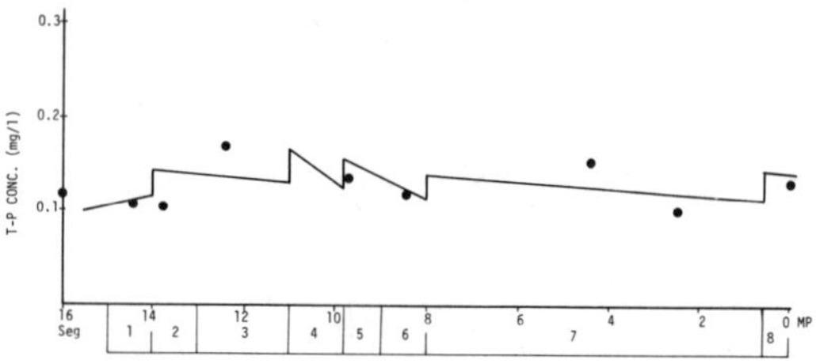

Figure 8. Stony Brook phosphorus calibration - low flow.

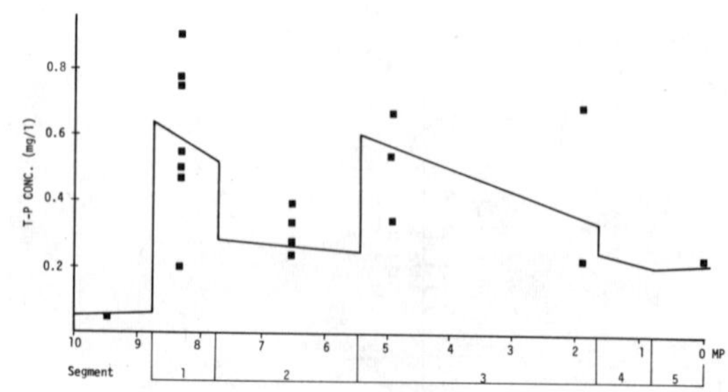

Figure 6. Millstone River phosphorus calibration - low flow

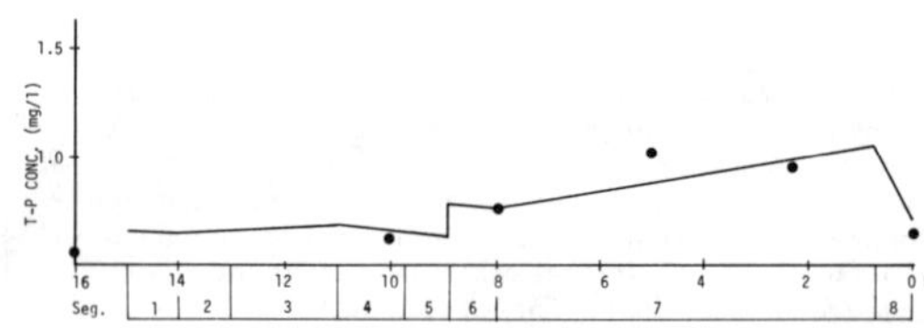

Figure 9. Stony Brook phosphorus calibration - high flow.

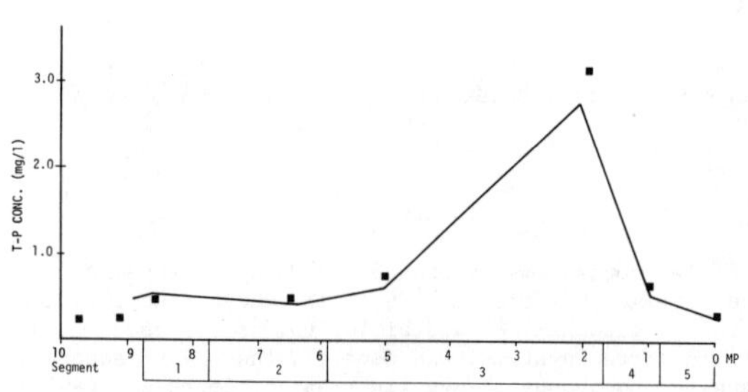

Figure 7. Millstone River phosphorus calibration - high flow

Figure 8 displays the low flow (8.6 cfs) calibration trace while figure 9 displays the high flow (155 cfs [4.4 m^3/s]) calibration for the Stony Brook (mean annual flow = 60 cfs [1.7 m^3/s] at the USGS gaging station near the Lake Carnegie confluence). The K values for the low flow analysis ranged between 0.1 and 0.2 per day while the high K values ranged from -2.0 to 2.5 per day. Once, again, scour and subsequent deposition is suggested.

Verification

The K values calibrated for the low flow case for the Millstone River were used to verify the model against medium flow (88 cfs [2.5 m^3/s]) conditions. Since the model trace fits the data reasonably well as shown in figure 10 the model can be considered well calibrated and verified. A second high flow data set was used to verify the Stony Brook Model. K values from the calibrated high flow analysis were used to generate the modeling trace shown in figure 11. This model is likewise well calibrated and verified.

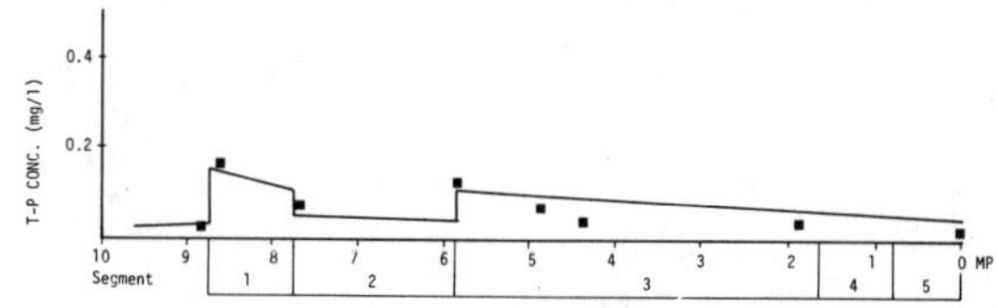

Figure 10. Millstone River phosphorus verification - medium flow.

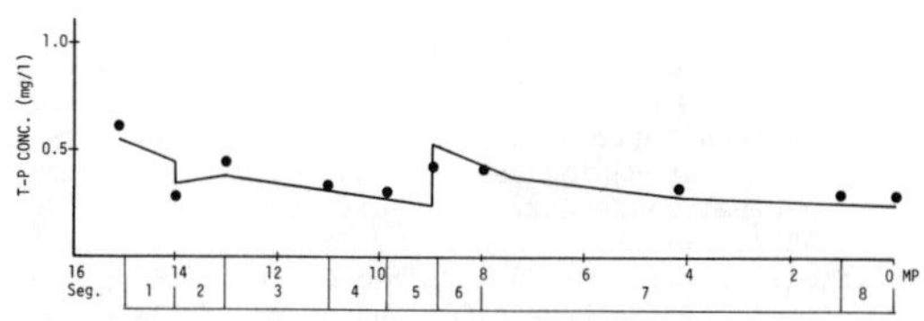

Figure 11. Stony Brook phosphorus verification - high flow.

LAKE EUTROPHICATION MODEL

Numerous empirical models have been suggested by investigators for modeling lake trophic levels and lake phosphorus concentrations as a function of phosphorus loading. Four models were used and compared in this study including Larson and Mercier (1975), Vollenweider (1976), Reckhow (1977), and Walker (1977). These models were selected on the basis of having constraints in which the study lake fit. The four models predicted phosphorus concentrations in Carnegie Lake ranging from 0.130 to 0.196 mg/l while the lake exhibits a phosphorus concentration of 0.15 ± 0.04 mg/l. Based upon a target concentration of 0.025 mg/l to achieve mesotrophic conditions in Carnegie Lake, a target loading for total phosphorus was estimated at 1,3000 lbs/yr (5900 kg/yr).

DECISION MAKING MODEL

Linear Program Formulation

A linear program to achieve the target phosphorus standard at minimum cost was formulated as:

$$\sum_{k=1}^{M} \sum_{i=1}^{N} [Cki\ (Xki) + Cki\ (Yki) + Cki\ (Zki)] \quad (3)$$

where, Xki, Yki, and Zki are respective removal efficiencies from point, urban non-point and agricultural nonpoint sources and Cki(Xki), Cki(Yki), and Cki(Zki) the respective cost functions to achieve these removals. Stream and lake water quality standards as well as the target phosphorus loading to the lake served as constriants. Treatment efficiency bounds served as additional constraints. Linear functions for municipal treatment plants were established from USEPA (1977) work. The urban non-point source linear cost function was developed from the studies of Marsalek (1978), Sullivan (1978), and the American Public Works Association (1969). Finally, the agricultural non-point source linear cost function derived from Marshall's (1977) work. The total linear programming algarithm was formulated in such a manner to utilize IBM's software package MPS/360 (Mathematical Programming System/360).

Case Selection

The linear programming model was run for five cases which attempted to bracket existing and future conditions. These are summarized as follows:

Case (1): With present loadings, under low flow condition (MA7CD10) and EPA's T-P criteria in the river - lake systems.

Case (2): With present loadings, under selected summer high flow condition and EPA's T-P criteria in river-lake systems.

Case (3): With predicted future loadings (year 2000) under low flow condition (MA7CD10) and EPA's water quality criteria in river-lake systems.

Case (4): With predicted future loadings (year 2000) under selected summer high flow condtiion and EPA's T-P criteria in river-lake systems.

Case (5): With present loadings under critical low and high flow conditions and EPA's T-P criteria in the stream, and an annual target loading the lake.

RESULTS

A feasible solution for any of the five cases was not obtained. The in-stream phosphorus water quality standard constraint of 0.05 mg/l was isolated as the critical papameter. Feasible solutions were only obtained after relaxation of this constraint. Table 11 summarizes the output of the minimum cost solution showing a revised set of in-stream standards as well as the required treatment levels to meet the standards.

Discussion

Although a solution to obtain required (albiet relaxed) levels of total phosphorus in the Carnegie Lake system was obtained, achievement of the point source treatment levels required is problematic. Indeed, it must be further realized that the eutrophication problems of Carnegie Lake may not be mitigated even if the target loadings are achieved. It does not appear that a source control solution to this particular case study is feasible.

Table 11.--Requirements of phosphorus removal and costs at various sources, considering critical low and high flow conditions

Segment No.	T-P criteria		Point source		Urban	Agriculture
	MA7CD10	H.F.	Plant STP	Eff. (%)		
1	0.12	0.10	Hightstown	98.0	70.0	55.0
2	0.10	0.12	--	--	70.0	55.0
3	0.15	0.50	E. Windsor	95.0	70.0	55.0
4	0.10	0.40	Lewis	98.0	--	10.0
4	0.10	0.40	N.J. Tpk.	98.0	--	10.0
4	0.10	0.40	Carter W.	98.0	--	10.0
5	0.10	0.10	FMC	52.0	40.0	47.8
5	0.10	0.10	RCA	52.0	40.0	47.8
Entering pt	0.08	0.08				
6	0.10	0.10	--	--	--	55.0
7	0.10	0.10	Mobil Oil	98.0	--	55.0
8	0.10	0.10	--	--	70.0	55.0
9	0.10	0.10	Princeton F.	94.5	--	55.0
10	0.10	0.10	ETS	94.7	--	55.0
11	0.10	0.10	PSOC	98.0	--	55.0
12	0.10	0.15	Western E.	96.0	70.0	55.0
13	0.10	0.15	Princ. Theo	97.5	40.0	25.0
Entering pt	0.08	0.08				
Total annual cost = $3,243,200			(PS = $1,696,000	NPS = $1,547,200)		

SUMMARY

Carnegie Lake (NJ) suffers from the attendant characteristics of a highly eutrophic state. Phosphorus has been indentified as the limiting nutrient for algal growth. A linear programming model was constructed to examine cost feasibility of point source vis-a-vis non-point source control measures. In-stream total phosphorus models were constructed for the two major tributary sources. A lake eutrophication model for Carnegie Lake was also constructed.

Feasible solutions were obtained only after significant relaxation of stream water quality standards. The high levels of point source treatment required to meet mesotrophic loading conditions suggest that the so-called feasible solution may be impractical.

ACKNOWLEDGEMENTS

This study was supported in part by the New Jersey Department of Environmental Protection and the New Jersey Agricultural Experiment Station, Publication No. J-07010-1-81.

LITERATURE CITED

American Public Works Association. 1969. "The causes and remedies of water pollution from surfaces drainage of urban areas," WPCA Contract No. WA 66-23.

Hunter, J.V. 1976. "Water quality assessment stream survey for the Upper Millstone River Basin," Rutgers University, New Brunswick, NJ.

Larson, D.P. and H.T. Mercier. 1975. "Lake Phosphorus loading graphs: An alternative," National Europhication Survey, Working Paper No. 194.

Lung, W.S. 1975. "Modeling of phosphorus sediment-water interactions in White Lake, Michigan," Ph.D. Thesis, The University of Michigan, Ann Arbor.

Marsalek, J. 1978. "Pollution due to urban runoff: Unit loads and abatement measures," Hydraulics Research Division, Canada Center for Inland Waters, Burlington, Ont.

Marshall Macklin Monaghan Limited. 1977. "Evaluation of remedial measures to control non-point sources of water pollution in the Great Lakes Basin."

Paterson, R.B. 1977. "A preliminary investigation of the phosphorus loading characteristics of Lake Carnegie, Princeton, New Jersey," Master of Science Thesis, Rutgers University, New Brunswick, NJ.

Philadephia Academy of Natural Sciences. 1969. "Ecological studies for proposed dredging of Carnegie Lake," Report to the Department of Limnology, Princeton University.

Reckhow, K.H. 1977. Phosphorus models for lake management, Ph.D. Thesis, Harvard University, Cambridge, Mass.

Sullivan, R.H., W.D. Hurst, T.M. Kipp, J.P. Heavney, W.C. Huber, and S. Nix. 1978. "Evaluation of the magnitude and signifcance of pollution loading from urban stormwater runoff - Ontario," Research Report No. 81, Canada-Ontario Agreement Ottawa, Ont.

Thomann, R.V. 1972. Systems analysis and water quality management, Environmental Research and Applications, Inc., New York, NY.

Uchrin, C.G. and W.J. Weber, Jr. 1980. "Modeling of transport processes for suspended solids and associated pollutants in river-harbor-lake systems," Chapter 19 of Contaminants and Sediments: Vol. 1 - Transport, Fate and Case Studies, ed. R.A. Baker, Ann Arbor Science, pp. 407-425.

United States Environmental Protection Agency, "Analysis of operation and maintenance costs for municipal wastewater treatment systems," EPA 430/9-77-015.

Vollenweider, R.A. 1976. "Advances in defining critical loading levels for phosphorus in lake eutrophication," Mem. Inst. Ital. Idrobiol. 33:53-83.

Walker, W.W., Jr. 1977. Some Analytical Methods Applied to Lake Water Quality Problems, Ph.D. Thesis, Harvard University, Cambridge, Mass.

SHORT TERM APPLICATION OF A MODIFIED VERSION OF THE BASIC VOLLENWEIDER NUTRIENT MODEL TO A HYPEREUTROPHIC URBAN LAKE[1]

Constantine E. Mericas[2], Ronald F. Malone[3]

Abstract.--A modified version of a basic Vollenweider nutrient model was applied to represent phosphorus kinetics in a small hypereutrophic urban lake. This application differed substantially from most investigations in the literature in that a one day time step was considered as compared to one year. A loading term was developed to represent wind induced mass exchange between the target lake and a much larger hypereutrophic lake with which it was connected. The model proved capable of simulating observed variations in lake phosphorus concentration, although radical peak events were not well represented, presumably because of sampling error resulting from incomplete mixing on a short term basis. The loading effect of interlake exchange was a significant factor in phosphorus levels.

INTRODUCTION

The use of nutrient models in lake management has become increasingly common in the past decade. Models described in the literature vary from simple equations to complex coupled structures of differential equations describing multicompartmented systems (Reckhow, 1979; Barica & Mur, 1980). From an engineering standpoint the ideal model is one that yields a satisfactory product while requiring a minimum amount of field sampling for calibration. The simpler the model, while satisfying predictive accuracy requirements, the better.

In this regard, the model originally described by Volleweider (1969) and subsequently modified by himself and others is particularly attractive. The model is essentially a mass balance equation about the lake bounds with first order reaction kinetics describing phosphorus loss to the sediments. Thus, all of the terms, with the exception of the sedimentation coefficient are readily measurable as runoff and outflow. The model has been applied to a wide variety of lakes with some success (Reckhow, 1979). These applications have emphasized use of the model in predicting steady state conditions on a long term annual basis.

It was the objective of this investigation to evaluate the usefulness of a modified Vollenweider model to predict short term phosphorus changes in a well defined hypereutrophic urban lake. The system selected is characterized by extreme fluctuations in nutrient and oxygen levels similar to those observed by others in hypereutrophic lakes (Barica, 1980). In addition to short term variations the long term (or steady state) phosphorus levels are being impacted by reclamation efforts. This application, therefore, provides a good test of the model's capabilities.

STUDY AREA

The University lakes system is composed of six urban lakes located in the southern part of Baton Rouge, La. (figure 1). The lakes are characterized by relatively shallow depths, high nutrient and algal levels, and, in most cases, some degree of

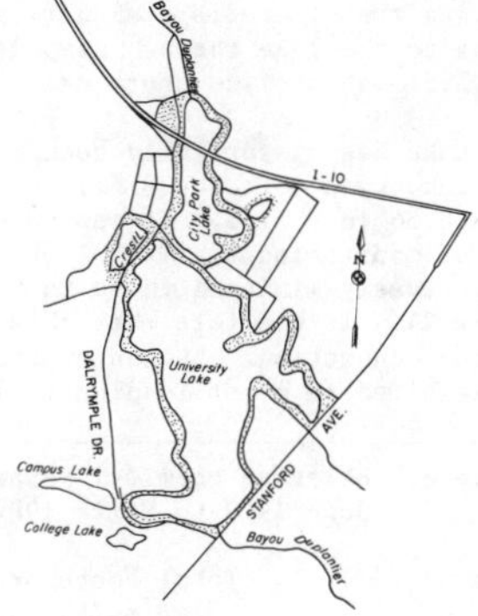

Figure 1. The six lakes of the University Lakes System.

[1]Paper presented at the Third International Conference on State-of-the-Art Ecological Modelling, Colorado State University, May 25-28, 1982.

[2]Graduate Research Assistant, Department of Civil Engineering, Louisiana State University, Baton Rouge, Louisiana, U.S.A.

[3]Assistant Professor, Department of Civil Engineering, Louisiana State University, Baton Rouge, Louisiana, U.S.A.

Table 1. Physical parameters characterizing the six lakes in the University Lakes system.

Lake	Mean Depth (meters)	Detention Time (days)	Surface Area (hectares)	Drainage Basin* (hectares)
Campus	0.46	16	2.95	41.20
College	1.10	39	2.14	28.26
City Park	0.73	25	21.00	209.00
Crest	1.52	561	3.43	2.22
Erie	0.61	14	1.21	24.63
University	0.61	39	89.20	332.70

*Excludes Lake surface area

sewage contamination. Some physical and chemical data on these lakes are presented in tables 1 and 2 respectively. Substantial fish kills have historically occurred once or twice a year during the period from June to August.

Crest lake is one of the smallest of the lakes in this system. With a surface area of 3.43 hectares and a drainage basin of 2.22 hectares the inputs to the lake are limited. Runoff is almost completely over ground from a highly developed residential district and there has been no indication of the sewage contamination common to the other lakes. University Lake is the largest lake in the system, and the second shallowest. The inputs to University are quite diverse, and include outflow from City Park Lake, direct surface runoff and discharges from some 28 storm sewer pipes. There is ample evidence that there is a significant raw sewage input to the lake through seepage, leaks and designed overflows into the storm drain system.

Crest Lake has historically been connected with University Lake by a 4 ft. by 8 ft. concrete box culvert and a 36 inch pipe. It was noted during the first year of monitoring that total phosphorus levels very closely mimicked those in University Lake (figure 2). It was felt that this was a direct result of the connection. Exchange processes between the lakes resulted in a net loading to Crest. It was hypothesized that a significant reduction in nutrient loadings could be induced by separation of the two lakes. Isolation was initiated in May of 1980 with the installation of plywood sheet piles and sand bags across the opening of the culvert. In January of 1982 the 36 inch pipe was discovered and sealed off in the same manner.

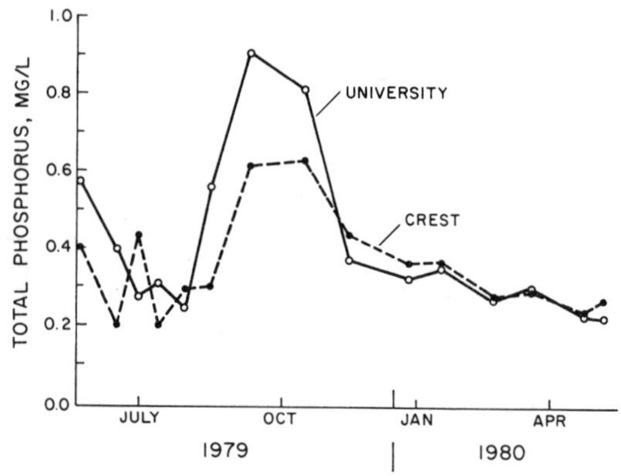

Figure 2. A comparison of observed total phosphorus in Crest and University Lakes during the period June 1979 thru June 1980.

MODEL DEVELOPMENT

The classical Vollenweider model is summarized as follows

$$V\frac{dP}{dt} = Q_i P_i - \sigma PV - Q_w P \quad (1)$$

where

V = Lake volume (m^3)

P = Inlake phosphorus concentration (g/m^3)

Q_i = Rate of water inflow (m^3/t)

P_i = Inflow phosphorus concentration (g/m^3)

σ = Sedimentation coefficient (t^{-1})

Q_w = Rate of water outflow (m^3/t)

Table 2. Selected chemical parameters for the six lakes in the University Lakes system - June 1979 to March 1982.

Lake	Total Phosphorus (mg/L)		Kjeldahl Nitrogen (mg/L)		Chlorophyll a (μg/L)	
	Mean	Range	Mean	Range	Mean	Range
Campus	0.396	0.155-0.795	2.210	0.708-6.468	143.0	25.0-343.0
College	0.428	0.136-0.849	2.464	0.833-4.970	153.4	42.0-403.0
City Park	0.332	0.149-0.868	3.240	0.874-6.380	196.6	45.0-584.0
Crest	0.270	0.144-0.624	3.769	1.802-6.916	213.2	14.0-800.0
Erie	0.469	0.189-1.340	3.149	0.805-8.880	174.2	9.5-665.2
University	0.428	0.190-0.930	5.461	2.380-12.204	313.9	37.6-1071.0

Implicit in this model are the assumptions that the lake in question is completely mixed and the loss to sedimentation is in direct proportion to inlake phosphorus concentration. Top and bottom phosphorus samples throughout the study have shown values within 10%, (Knaus, et al., 1981) indicating no significant long term stratification. Therefore the completely mixed assumption appears reasonable for modeling purposes, although it may not be strictly adhered to on a short term basis.

Application to Crest Lake required modification of this basic relationship (figure 3). Inputs are comprised of runoff and mass exchange with University Lake. Direct rainfall was assumed to have a negligible phosphorus contribution. Runoff loading was calculated by coupling a runoff flow model with a phosphorus loading model developed for a watershed in the immediate vicinity of Crest Lake. The runoff flow equation was in the form (Overton & Meadows, 1976)

$$I = \frac{(R - 0.2S)^2}{R + 0.8S} \qquad (2)$$

where

I = Runoff in inches

R = Rainfall in inches

$S = \frac{1000}{CN} - 10$

CN = Curve number based on land usage

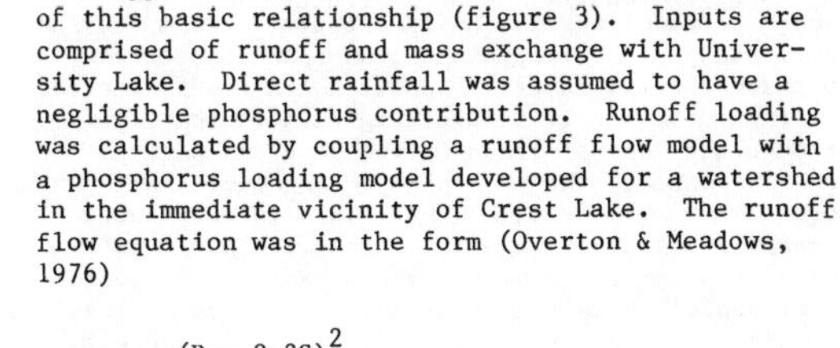

Figure 3. A schematic representation of the modified Vollenweider Model.

A phosphorus loading equation (Wegener, unpublished data) was calibrated to a nearby drainage area to permit estimation of phosphorus input to the system by runoff.

$$L = 3.026Q^{0.678} \qquad (3)$$

where

L = Mass of phosphorus in runoff (g)

Q = Runoff flow (m^3)

Equation 3 represents total phosphorus input. It has been recognized that only a portion of this total phosphorus remains in the water column following storm events. An adjustment was made to compensate for phosphorus that drops out upon entry into the lake by reducing calculated runoff mass by 25%.

The volume of water that is exchanged between lakes, Q_{ex}, is the result of wind induced head changes. As wind blows along the axis of University Lake for a period of time it induces a seiche. This results in an increase in lake level or setup on the downwind side of the lake and corresponding decrease on the upwind side. When the wind shifts the lake level returns to normal as water from the elevated side flows back towards the depressed side. Crest responds to the head differential by either receiving or releasing water through it's connections with University Lake. This exchange is on a short term basis. Q_{ex} was estimated by a function of the resultant daily wind component along the long axis of University Lake. Based upon literature equations for wind setup (Linsley & Franzini, 1979) the relationship was established as

$$Q_{ex} = (f)(w)^2 \qquad (4)$$

where

Q_{ex} = Volume of water exchanged (m^3)

f = Empirical constant (m^3/mph^2)

w = Resultant wind component along the 320° axis (mph)

The exchange of volumes of water between the lakes assumes that water enters Crest at the mean University Lake concentration and exits at the mean Crest concentration. Thus, the total mass of phosphorus introduced to Crest by this mechanism is represented by

$$M_{ex} = Q_{ex}\ (P_u - P) \qquad (5)$$

where

M_{ex} = Mass input by volume exchange (g)

Q_{ex} = Volume rate of water exchange (m^3/t)

P_u = Observed inlake phosphorus concentration in University (g/m3)

Thus, putting the model in discrete form, the modified Vollenweider model may be written

$$\frac{\Delta P}{\Delta t} V = L - \sigma PV - Q_w P + Q_{ex}(P_u - P) \qquad (6)$$

where

WP = change in water column phosphorus (g/m^3)

Wt = time step (days)

A hydraulic model was coupled to the phosphorus model to provide adjustments to the volume term. Wind and evaporation data were obtained from the Louisiana State Climatologist's office at LSU and were based on observations made at a station 3 miles from Crest Lake. Rainfall data was obtained from the LSU Forestry Department which has a recording station within one mile of the lake. Runoff flow input was calculated as described previously. Pan evaporation data were reduced by a correction factor of 0.6 to in situ conditions. The basic model may be described as

$$\frac{dV}{dt} = A(R - 0.6E) + A'Q_{RO} \qquad (7)$$

where

A = Lake surface area (m^2)

R = Rainfall Rate (m/t)

E = Pan evaporation rate (m/t)

A' = Drainage Area (m^2)

Q_{RO} = Runoff volume rate (m/t)

The differential form of the model was solved using the Euler's Technique for a one day time step. The observed data were collected over a period from June 1979 to December 1981 as part of a 13 parameter monitoring program in the University lakes system. Sampling from June 1979 to June 1981 was on a monthly basis with weekly samples taken during the months of June through August, 1980. In June of 1981 a regular weekly sampling regime began. Water samples were taken one foot below the surface and one foot above the bottom. Only top samples were considered in this study. Total phosphorus was determined by the Persulfate digestion-Ascorbic acid technique (Standard Methods, 1981). A detailed description of the monitoring program and results are presented elsewhere (Knaus et al., 1981). The exchange term was modified to reflect the three phases of isolation in terms of percent of calculated Q_{ex}. Phase I was the first 355 days of the study during which the connections between the lakes were unimpeded. During phase two, the following 475 days, effective exchange was estimated to be 15% of the open connection based on culvert and pipe sizes. Phase three assumed total isolation of the lakes with no exchange. Best fit for model calibration was judged on mean predicted total phosphorus and mean residual error.

RESULTS AND DISCUSSION

The final predictive model results are compared to observed values in figure 4. Optimum values for the sedimentation coefficient, σ and the empirical wind constant f were 0.006/day (2.17/yr) and 350 m^3/mph^2 respectively. The sedimentation term is within the range of values appearing in the literature. Jones and Bachmann (1976) reported mean annual values in the range of -25.7 to 22.4 in a survey of mid-western lakes, with a value of 0.65/yr working best overall. Considering that this lake is at the far end of the trophic scale, the value obtained seems reasonable. For this application σ and f were assumed to be constants for the three year period with no consideration of seasonal or daily variation.

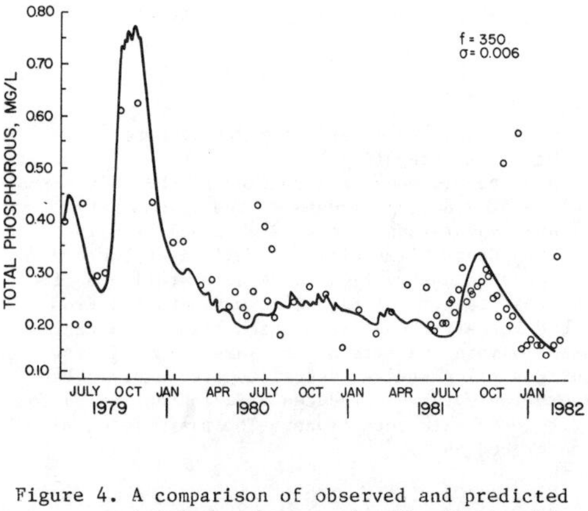

Figure 4. A comparison of observed and predicted total phosphorus using the final calibrated form of the model.

The simulated curve as shown in figure 4 compares favorably to the majority of the observed values. Peak phosphorus values observed after the first year were not well represented. For comparison, the model was run using a 100% Q_{ex} term for the full study period, simulating an open connection between the two lakes. These results are presented in figure 5. It is interesting to note that, although the general trend of the observed data is poorly modeled, the peak phosphorus concentrations are fairly closely simulated. This is attributed to inability of Crest Lake to satisfy the completely mixed assumption on a short term basis. The poorly represented peaks all occur during periods of weekly sampling. Within this timeframe it is suspected that plumes of University Lake water entering Crest Lake during periods of exchange do not completely disperse. Periodically such plumes are encountered at the sampling station. This conclusion is substantiated by the fact that peak values are not sequential but rather interspersed with observations consistent with the general trend of the data.

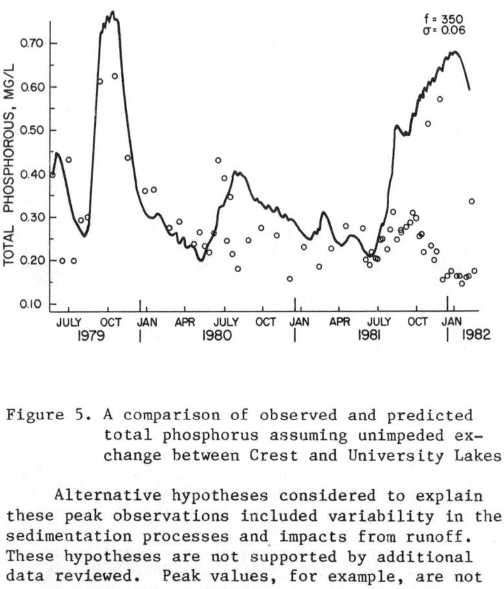

Figure 5. A comparison of observed and predicted total phosphorus assuming unimpeded exchange between Crest and University Lakes.

Alternative hypotheses considered to explain these peak observations included variability in the sedimentation processes and impacts from runoff. These hypotheses are not supported by additional data reviewed. Peak values, for example, are not correlated with rainfall events. Further, bottom samples do not indicate any vertical gradients in total phosphorus. It is believed that vertical gradients would be in evidence if the peak values were induced by sediment release during periods of weak stratification.

If the peak values can be explained by incomplete mixing as this analysis strongly suggests, then the discrepancy between the model projections and peak values (figure 4) can be attributed to sampling program, rather than the model. In this case the peak values can be considered outlyers in the statistical sense and discarded since they do not reflect the condition of the system as a whole. If, on the other hand, these peak values reflect actual variations in lake conditions the model selected to represent the system must accurately represent the peak values. The question of whether the single cell Vollenweider model is appropriate for short term application therefore rests upon the interpretation of the peak values. Based upon the data currently available, the authors must conclude that the Vollenweider model is accurately representing total phosphorus dynamics in Crest Lake and the peak values are explained by sampling error. Yet, there is a clear need for verification of this conclusion by intense spacial and temporal collection of total phosphorus data.

The applicability of the Vollenweider model to short term application implies that the complex nutrient cycling phenomena which are undoubtly occurring in this system are reasonably balanced so that only the net effect (or net sedimentation) needs be represented. This reflects, in part, the essentially completely mixed nature of Crest Lake, and the use of total phosphorus as a state variable. The authors of this paper would not expect the same results if the system displayed any significant thermal stratification. In particular, a system with weak stratification would be subject to intermittent and real variations due to nutrient accumulation in the hypolimnion and periodic overturn. Total phosphorus is a relatively stable parameter when compared to orthophosphorus and chlorophyll a which are clearly subject to high variations due to nutrient cycling. The use of a model based upon total phosphorus also implies that total phosphorus levels can be correlated to desirable or undesirable system responses. In this particular case, the undesirable system responses are fish kills believed to be caused by oxygen deficiencies induced by high algal concentrations. It should be recognized that although total phosphorus is easier to model, the modeling results are more difficult to relate to the undesirable system response. These restrictions should be considered before selecting the Vollenweider model for short term application.

SUMMARY

The modified Vollenweider model discussed proved to be reasonably capable of simulating short term phosphorus levels based on well defined external loading and a wind driven mass exchange function. There were some extreme peaks that were not well simulated under the three phase isolation assumption. It is believed that this is indicative of the coupled effect of incomplete isolation and the existence of short lived concentration gradients in the lake. An intense study of temporal and areal phosphorus concentration needs to be conducted to confirm this hypothesis.

REFERENCES

Barica, J., 1980, Why Hypertrophic Systems? In Developments in Hydrobiology, Vol. 2, pp. ix-xi, W. Junk Publishers, The Hague, Netherlands.

Barica, J. and L. R. Mur, 1980, Developments in Hydrobiology, Vol. 2, W. Junk Publishers, The Hague, Netherlands.

Jones, J. R. and R. W. Bachmann, 1976, Prediction of Phosphorus and Chlorophyll Levels in Lakes, Journal WPCF, Vol. 48, No. 9:2176-2182.

Knaus, R. M. et al., 1981, Second Annual Report, Lakes Restoration Project., Institute for Environmental Studies, 208 p. U.S. EPA Grant No. CS805836-10, Louisiana State University, Baton Rouge, La.

Overton, D. E. and M. E. Meadows, 1976, Stormwater Modeling, Academic Press, pp. 35-38.

Reckhow, K. H., 1979, Quantitative Techniques for the Assessment of Lake Quality, EPA 440/5-79-015.

Standard Methods for the Examination of Water and Wastewater., 1978, American Public Health Association, New York.

Vollenweider, R. A., 1969, Possibilities and Limits of Elementary Models Concerning the Budget of Substances in Lakes, Arch. Hydrobiol., 66(1): 1-36.

Wegener, K., Research Assistant, Department of Civil Engineering, Louisiana State University, Baton Rouge, La.

ON THE EVALUATION OF THE VERTICAL EDDY DIFFUSION COEFFICIENT IN LAKES BY MEANS OF MEASURED TEMPERATURE PROFILES[1]

G. Rossi, F. Beonio-Brocchieri, J. Sangregorio[2]

Abstract. We point out some specific calculational problems arising from the elaboration of such very detailed temperature profiles, as we have obtained by means of an automatic instrumentation installed on a limnological buoy in the Agno Bay of Lake Lugano. A particular attention is devoted to the phenomenology determined by internal gravity waves. A selection of results, i. e. vertical eddy diffusion coefficient profiles related to various time intervals during September 1978, are presented and commented.

1. INTRODUCTION

It is well known (Nihoul 1977) that, in the continuity equation for a substance dissolved in a water body, the term representing turbulent dispersion (coming from the time average of the product of fluctuations of water velocity v and concentration ρ of the considered substance) is given, in analogy with molecular diffusion, by the divergence of a flux proportional to the concentration gradient:

$$\varphi = \langle \delta\rho(t) \cdot \delta v(t) \rangle_\tau \sim \varkappa(\vec{r}, t)\ \mathrm{grad}\ \rho \qquad (1)$$

where τ is a suitable time interval, intermediate between the period of the fluctuations and the time scale of the process under study. The proportionality constant $\varkappa$ is a function of position and time, much greater than the molecular diffusion coefficient and practically independent of the substance concerned (eddy diffusion coefficient). The fact that the water temperature can be easily measured in a lake with great detail and precision suggests the use of temperature measurements in order to obtain information, via the heat transfer equation, about the diffusion properties of the water bodies, where a temperature gradient significantly $\neq 0$ can be observed.

From this point of view the mathematical description of the heat balance in a lake aims principally at the determination of the effective (i. e. molecular + eddy) heat diffusion coefficient (effective thermometric conductivity $\chi = \varkappa/\rho c$, where ρ is the density and c the specific heat of water), as a function of position and time.

If we assume that the lake water temperature depends only on the depth, the heat transfer model of the lake is simply one-dimensional, only the vertical diffusion coefficient is relevant to the results, and this is function only of z and t ($0 \leq z \leq z_{max}$). There are atmospheric factors affecting heat transfer in a lake, which have a diurnal as well as an annual evolution (like the sun radiation); other factors exhibit episodic phenomena, which cannot be foreseen on the basis of statistical information (such as wind velocity). Due to the relevance of such phenomena to the vertical transport of dissolved matter (nutrients, pollutants, etc.) the calculated $\chi(z,t)$ function should describe not only the normal seasonal evolution but also the consequences of episodic events (such as for instance the passage of a cold perturbation).

The method was carefully discussed by Jassby and Powel (1975), who applied it to Castle Lake, California. However, their complete and detailed analysis, being referred to temperature profiles recorded every 5 days, could not include the calculational problems connected with the effect of meteorological perturbations on the lake water dynamics.

2. THE EQUATIONS OF THE HEAT TRANSFER MODEL

The one-dimensionality assumption, such as for the heat transfer problem in a lake, implies the simulation of water inlet and outlet in terms of a vertical water advection velocity (Orlob 1967). It is assumed that inflow (determined by the measured inflow rate $q_{in}(t)$) concerns the water layer whose temperature T(z, t) is equal to the measured temperature of the inflowing water $T_{in}(t)$; and outflow (determined by an outflow rate $q_{out}(t)$) concerns the surface layer ($z_{max} - h$, z_{max}). Then the vertical advection velocity v(z, t) is determined by a continuity equation:

$$\frac{\partial}{\partial z}\left[A(z)v(z,t)\right] = \left[q_{in}(z,t) - q_{out}(z,t)\right]/h \qquad (2)$$

where A(z) is the area of the lake section at elevation

---

[1] Paper presented at the Third International Conference on State-of-the-Art in Ecological Modelling, Colorado State University, May 24-28, 1982.

[2] Commission of the European Communities, Joint Research Centre, Ispra Establishment, Dept. of Natural and Physical Sciences, I-21020 Ispra (Va), Italy

z, and h is the thickness of the water layer, the boundary condition at $z = z_{max}$ being given by the overall water balance of the lake.
The heat balance equation, written in the form

$$A(z)\frac{\partial T}{\partial t}(z,t) = -\frac{\partial}{\partial z}\left\{A(z)\left[v(z,t)T(z,t) - \chi(z,t)\frac{\partial T}{\partial z}(z,t)\right]\right\} - A(z)v_{out}(z,t)T(z,t) + A(z)v_{in}(z,t)T_{in}(t) - \frac{1}{\rho c}\frac{\partial}{\partial z}\left[A(z)\varphi(z,t)\right], \quad (3)$$

where $A(z)v_{in}(z,t) = q_{in}(z,t)/h$; $A(z)v_{out}(z,t) \equiv q_{out}(z,t)/h$, and $\varphi(z,t)$ is the radiation flux (directed downwards, hence < 0), is correct for a horizontally well mixed water layer, i. e. in the case of very large horizontal diffusivity, provided that heat exchange through the walls is negligible.
This condition is generally verified in the surface layers (epilimnion), but in a rather poor way in the underlying ones, where the largest vertical temperature gradients exist and horizontal temperature gradients are observed, as the consequence of internal gravity waves. As for these layers (metalimnion), the one-dimensionality assumption asks for a further discussion (see § 3).

Assuming net heat flux = 0 at the bottom surface (z = 0), eq. (3) is formally solved with respect to the diffusion coefficient:

$$\chi(z,t) = \frac{1}{A(z)\frac{\partial T}{\partial z}(z,t)}\left\{\int_0^z A(\tilde{z})\frac{\partial T}{\partial t}(\tilde{z},t)d\tilde{z} + \left[A(z)v(z,t)T(z,t) - A(0)v(0,t)T(0,t)\right] + \int_0^z A(\tilde{z})\left[v_{out}(\tilde{z},t)T(\tilde{z},t) - v_{in}(\tilde{z},t)T_{in}(t)\right]d\tilde{z} + \frac{A(z)\varphi(z,t)}{\rho c}\right\} \quad (4)$$

and the profile $\chi(z,t)$ is calculated from measured temperature profiles $T(z,t)$.

We intend to discuss in the next paragraph the case of a sub-alpine lake, with a maximal depth of ~100 m, where the measured temperature profiles concern depths down to ~30 m. At this depth the assumption of net heat flux = 0 cannot be considered realistic.

If we consider the heat balance in the $(0, z_o)$ layer, z_o being the elevation of the deepest measurement point, and we assume that the net heat flux vanishes at z = 0 and no heat sources or sinks are present between 0 and z_o, we obtain:

$$A(z_o)\chi(z_o,t)\frac{\partial T}{\partial z}(z_o,t) = \int_0^{z_o} A(z)\frac{\partial T}{\partial t}(z,t)dz - \frac{A(z_o)\,|\varphi(z_o,t)|}{\rho c} \quad (5)$$

$\varphi(z_o, t)$ being the sun radiation flux at $z = z_o$.

Referring eq. (4) to the (z_o, z) interval, and using eq. (5), we obtain:

$$\chi(z,t) = \frac{1}{A(z)\frac{\partial T}{\partial z}(z,t)}\left\{\int_0^{z_o} A(z)\frac{\partial T}{\partial t}(z,t)dz + \int_{z_o}^{z} A(\tilde{z})\frac{\partial T}{\partial t}(\tilde{z},t)d\tilde{z} - \frac{1}{\rho c}A(z)\,|\varphi(z,t)| + \text{vertical advection + inflow-outflow terms}\right\} \quad (6)$$

where $\int_0^{z_o} A(z)\frac{\partial T}{\partial t}(z,t)dz$ has to be evaluated on the basis of auxiliary coarser temperature measurements at $z < z_o$, or in want of these, of a suitable extrapolation of the measured temperatures from $T(z_o, t)$ to the bottom temperature T_o.
Disregarding the terms related to vertical advection and outflow-inflow, the positiveness condition for $\chi(z)$, which has to be satisfied by all interpretations of the experimental framework, becomes:

$$\int_0^{z_o} A(z)\frac{\partial T}{\partial t}(z,t)dz + \int_{z_o}^{z} A(\tilde{z})\frac{\partial T}{\partial t}(\tilde{z},t)d\tilde{z} > \frac{A(z)\,|\varphi(z,t)|}{\rho c} \quad (7).$$

3. APPLICATION TO TEMPERATURE PROFILES MEASURED IN THE AGNO BAY OF LAKE LUGANO DURING SEPTEMBER, 1978

In this paper we present the results of a preliminary experience in calculating effective diffusion coefficients by means of eq. (6) from temperature profiles measured in Lake Lugano.

The temperature measurements were performed at a limnological buoy in the Agno Bay (fig. 1), by means of an automatic device especially designed for the purpose and assembled at Ispra (fig. 2).

A thermometric probe (Pt resistance) is lowered and lifted by an electric drumwinch, to depths between 0 and 30 m (the depth of the lake at that point is ~87 m). Temperatures are measured at each 1.1 m after a prescribed waiting time and recorded on paper during a given time interval. The complete cycle takes 1 h 30′, a time which can be considered small with respect to the characteristic times of the phenomena investigated. The instrumentation was tested under continuous operation during the entire month of September, 1978, but the temperature profiles eventually retained for the computer elaboration only concern a 19-day period ranging from 8th to 27th September. This is a transition period between the rather complicated summer stratification, and the simpler thermal structure preceding autumn destratification. As a matter of fact, convection due to the night cooling process at the lake surface, and shear stresses induced by episodic wind perturbations, tend to create a perfectly homothermal layer, down to a thermocline, which sinks deeper and deeper.
This latter situation, which has normally been observed in the Agno Bay after mid-October, requires

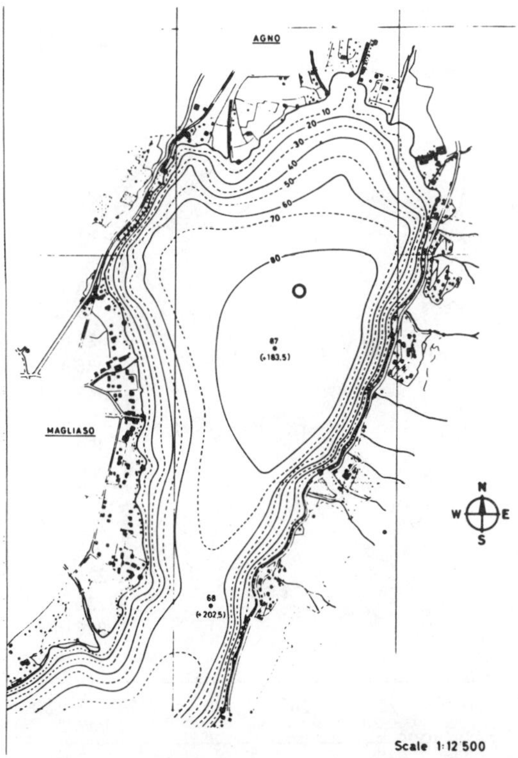

Figure 1. Bathymetric map of the Agno Bay (Lake Lugano). (O limnological buoy).

that the $\chi(z, t)$ calculation via eq. (6) is limited to $z < z_{th}$ (the thermocline elevation), because of the presence of vanishing temperature gradients for $z \geqslant z_{th}$.

The application of the model to the experimental framework considered, requires some preliminary remarks.

The measured temperatures are increasing everywhere with z, because the highest measurement point was ~ 1.5 m under the surface and this point is generally unaffected by night cooling during September. As a consequence, whenever heat loss by radiation predominates at the lake surface, the heat balance of the (z_o, z) layer, as described by eq. (6) and which does not take thermal convection into account, is compatible with the experimental temperatures, provided only that χ can assume negative values in the upper epilimnetic layers.
Furthermore, when $z \rightarrow z_{max}$, the calculation is conditioned, as for the $\varphi(z, t)$ term, by the use of semi-empirical formulae of doubtful accuracy, concerning the net long-wave radiation flux at the air-water interface.

For both these reasons numerical results obtained using eq. (6) cannot be considered as significant for the few upper metres of the epilimnion.

Figure 2. The measurement apparatus on the limnological buoy.

The temperature pattern observed at a given depth (fig. 3), is characterized by the superposition of oscillations of various frequencies to an average behaviour: this is particularly evident where the vertical temperature gradient is large enough (say $\geqslant 1^{o}C\ m^{-1}$). In general, this indicates the existence of internal gravity waves (in particular standing internal waves, or internal seiches), which are the close origin of heat's eddy diffusion at such depths.

Internal seiches are often observed in the Agno Bay during the autumn homothermy of the epilimnion (figs. 4 and 5; Beonio-Brocchieri 1979).

During summer, due to the presence of important temperature gradients, already a few metres

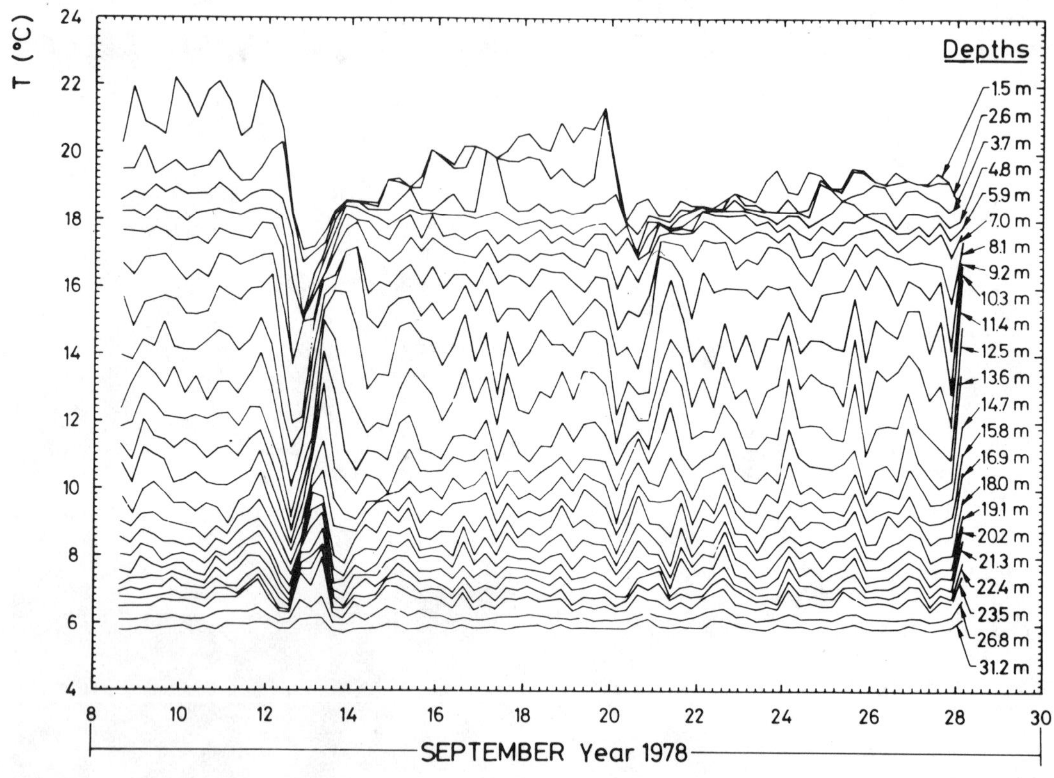

Figure 3. Temperature pattern at various depths during September 1978.

below the lake surface, a wind perturbation originates internal waves of more disordered nature, usually starting with one or two oscillations of very large amplitude (corresponding to temperature oscillations of even 5°C), which propagate down with decreasing amplitude till the deepest layers observed (see for instance the curves of figure 3 after the perturbation of September 12th, 1978).

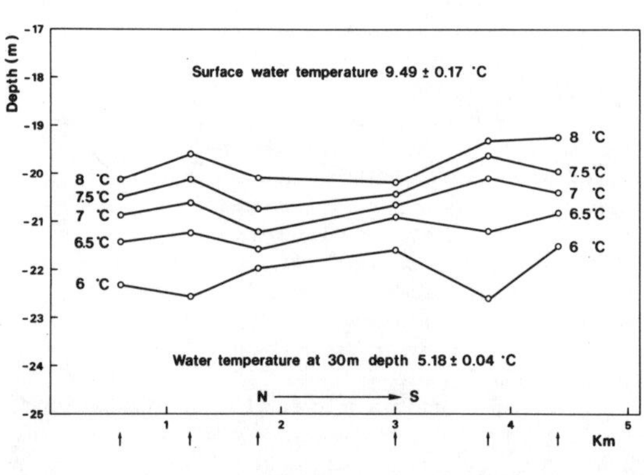

Figure 4. Isotherms in a N-S vertical section of a basin of Lake Lugano (November, 1981).

Internal gravity waves, in particular internal seiches, affect the χ-calculation in two ways. First of all, down(up)-welling of warmer (colder) water can affect the heat balance of the (z_o, z) layer, as it is accounted for by eq. (6), and possibly originate even negative χ-values at some elevations z, essentially due to the fact that such "two-dimensional" effects are not included in eq. (3).

Furthermore, internal waves of all types originate horizontal temperature gradients, and this contradicts the basic assumption of the model. The fluctuating character of various terms of eq. (6) asks for the adoption of averaging procedures, i. e. for relating the χ-profile to the mid-point t_o of a sufficiently long time interval Δt:

$$\chi(z, t_o) = \langle \chi(z, t) \rangle_{\Delta t} \tag{8}$$

In particular, if Δt is chosen as approximately a multiple of a wave-period, the effect of the oscillation concerned is removed from the average temperature profile.

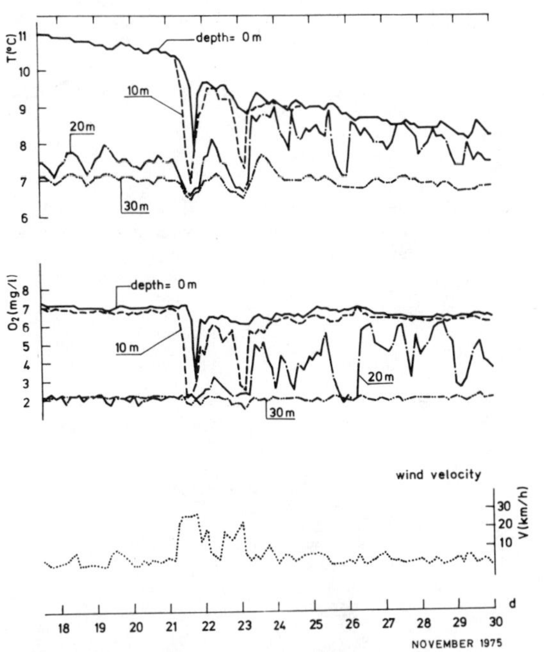

Figure 5. Oscillations of temperature and dissolved oxygen concentration at metalimnic depths in Lake Lugano after a wind perturbation (November, 1975).

As regards horizontal temperature gradients due to internal waves, they are clearly oscillating too, with periods to be considered small with respect to the scale of the time evolution of the χ-profile. Hence, we can expect that averaging a three-dimensional heat diffusion equation on the time interval Δt eliminates the horizontal diffusion term and brings us again to a one-dimensional equation such as eq. (3) with A(z) = constant (z), which is a quite acceptable approximation for the metalimnic layers. Nevertheless, further investigations on this point, using temperature profiles contemporaneously recorded at two stations, are in progress.

As for the averaging procedures on eq. (6), our experience has revealed that a particular attention has to be devoted to the "heating" term $\langle(\partial T/\partial t)(z,t)\rangle_{\Delta t}$. Indeed, the χ-value at z is essentially determined by subtracting the sun radiation contribution from the mean unit time variation of the heat content in the water layer (z_o, z), and this is heavily affected by the above analyzed temperature oscillations.

Among various possible prescriptions for averaging on Δt the time derivative $(\partial T/\partial t)(z,t)$, the following has been tentatively adopted:

$$\left\langle \frac{\partial T}{\partial t}(z,t)\right\rangle_{\Delta t} \sim \frac{1}{t_2 - t_1}\left[\langle T(z,t)\rangle_{\Delta t_2} - \langle T(z,t)\rangle_{\Delta t_1}\right] \qquad (9),$$

where Δt_1, Δt_2 (centred on times t_1, t_2) are respectively the first and second semi-interval of Δt.
It is evident that being obliged to average on sufficiently long time intervals, limits the capability of the method considerably, as for the evaluation of the effects of meteorologic perturbations on the mixing properties of the metalimnion. As a matter of fact, sufficiently long time intervals between two consecutive perturbations should elapse in order to allow drawing indications about these properties from such complicated temperature patterns.

4. CALCULATIONS, RESULTS AND DISCUSSION

Among the measurements continuously performed between September 8th and 27th, 1978, 152 temperature profiles (8 profiles per day) have been used for the computer elaboration. During the same period, wind velocities at 2 m height and radiation flux incident at the lake surface have been recorded. The two cold perturbations of September 12th and 20th (fig. 3) were characterized by two peaks of the wind velocity (respectively 30 km h^{-1} and 28.5 km h^{-1}). No precipitations occurred during the whole period in such a way that inflow rates concerning the basin have constantly been very low, and the effects of inflow and outflow, and consequently of vertical advection, on the heat balance in the basin can be considered as small.

The radiation flux at time t and elevation z ($0 < z < z_{max}$) has been calculated by means of the formula:

$$\varphi(z,t) = \varphi(z_{max}, t) \sum_{1}^{4}{}_{i}\, \epsilon_i e^{-\alpha_i(z_{max}-z)\sec\varphi(t)} \qquad (10)$$

i.e. the radiation flux penetrating the lake surface, assumed all as direct, has been subdivided into four wave-length intervals, each characterized by an energy fraction ϵ_i and an extinction coefficient α_i. The values ϵ_i, α_i are given in table 1: except for α_2 taken = 0.65 m^{-1} in the surface layer ($z_{max} > z >$ 80 m, euphotic zone) on the basis of measurements performed in the Agno Bay during previous surveys, extinction coefficients α_i refer to pure water.

Table 1

	$\Delta\lambda_i$(μm)	ϵ_i	α_i (m^{-1})
i = 1	0.3 - 0.4	0.025	0.7
i = 2	0.4 - 0.7	0.410	{0.65 0.3
i = 3	0.7 - 0.86	0.185	2.5
i = 4	0.86 - 3	0.380	100

As for the other quantities entering eq. (10):

$\Psi(t) = \arcsin\left[\frac{\sin\sigma(t)}{n}\right]$ $\quad n \equiv$ refraction index at air-water interface = 1.333,

$\sigma(t)$ = zenith angle = $\arccos\left[\sin\lambda\sin\lambda_z^{(j)} + \cos\lambda\cos\lambda_z^{(j)}\cos\omega_j(t)\right]$,

λ = latitude $\quad \lambda_z^{(j)}$ = solar declination (assumed as constant during day j),

$\omega_j(t)$ = hour angle of day j.

The calculation of $\varphi(z_{max}, t)$ has been based on the average radiation flux incident at the lake surface during day "j" R_j (kcal $m^{-2}h^{-1}$), measured by means of a pyrheliograph placed on the buoy. The ratio $|\varphi(z_{max}, t)|/R_j$ was expressed in terms of the daily average value of the cosine of the zenith angle:

$$\langle\cos\sigma(t)\rangle_j = \sin\lambda\sin\lambda_z^{(j)} + \frac{2}{\gamma_j}\cos\lambda\cos\lambda_z^{(j)}\sin\frac{\gamma_j}{2},$$

where $\gamma_j = \pi + 2\arcsin\left[\tan\lambda\tan\lambda_z^{(j)}\right]$ ($\frac{12}{\pi}\gamma_j$ = photoperiod),

and of the albedo:

$$\rho = \frac{1}{2}\left\{\frac{\tan^2[\sigma(t) - \Psi(t)]}{\tan^2[\sigma(t) + \Psi(t)]} + \frac{\sin^2[\sigma(t) - \Psi(t)]}{\sin^2[\sigma(t) + \Psi(t)]}\right\}.$$

Numerical calculations involved in eq. (6) have been performed assuming as water layer thickness h = 0.5 m for 87 m > z ⩾ 60 m, and h = 1 m for z < 60 m. Temperature values T(z, t) at the mesh-points have been drawn from the measured ones by means of linear interpolation.

As for the determination of the time interval Δt, on which the averages are performed, two aspects have to be considered: 1) the period of the temperature oscillations; 2) the time position of the meteorologic perturbations.

A Fourier-analysis of the temperature oscillations at a fixed depth can help to recognize the possible existence of one- or multi-nodal standing internal waves, but the very complicated temperature pattern characteristic of summer in an open bay seems to deny any support to the identification of a fundamental wave period. In this case the choice of Δt should be based on the time position of the meteorologic perturbations, and particularly of the strongest ones, whose effects reach the upper hypolimnic layers.

As a matter of fact the calculation should not include the time interval in which the strong heat loss towards the atmosphere affects even the upper hypolimnic layers, and successively the down-welling of the surface layers, caused by the wind action, simulates a strong heating process of the underlying ones. These two phenomena are both a possible cause of negative χ-values in the profile resulting from the model (see section 3).

Furthermore, the assumption of net heat flux = 0 at the elevation z_o of the deepest measurement point implies an additional underevaluation of χ, because

$$\int_0^{z_o}\frac{\partial T}{\partial t}(z, t)dz$$

is certainly > 0 during summer and early autumn. Hence, the term

$$\int_0^{z_o}\left\langle\frac{\partial T}{\partial t}(z, t)\right\rangle_{\Delta t}A(z)dz$$

could not be neglected in the heat balance of the (z_o, z) layer (eq. (6)), not even when z approaches the elevation of the thermocline (fig. 6).

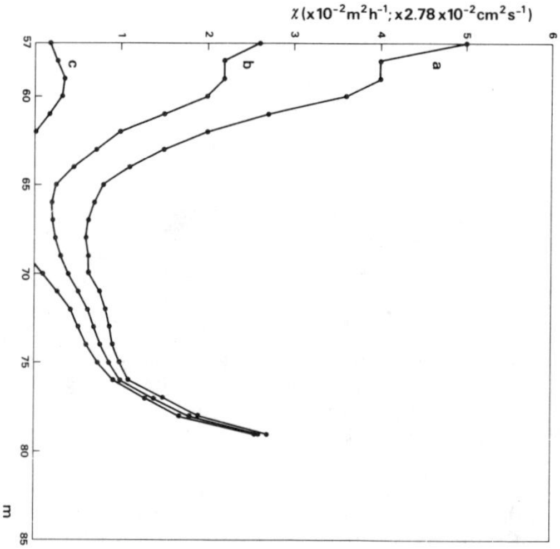

Figure 6. Eddy diffusion coefficient as function of elevation (z = 0 lake bottom; z = 87.5 lake surface; average on a 19-day period, 1978, September 8-27):

a) $\left\langle\frac{\partial T}{\partial t}(z_o, t)\right\rangle_{\Delta t} = 2. \times 10^{-4}\ ^oCh^{-1}$

b) $\left\langle\frac{\partial T}{\partial t}(z_o, t)\right\rangle_{\Delta t} = 1. \times 10^{-4}\ ^oCh^{-1}$

c) $\left\langle\frac{\partial T}{\partial t}(z_o, t)\right\rangle_{\Delta t} = 0.$

In want of temperature measurements in the deeper layers of the lake, the curves of figures 7 and 8 were calculated assuming $\left\langle\frac{\partial T}{\partial t}(z, t)\right\rangle_{\Delta t}$ as exponentially decreasing from the measured value at z_o to ~ 0. at the lake bottom z = 0. In order to avoid incorrectly evaluated additional effects of amplitude and position of Δt on the χ-profiles, we have adopted for

$$\left\langle \frac{\partial T}{\partial t}(z_o, t) \right\rangle_{\Delta t}$$

the same value $2.0 \times 10^{-4}\ ^oC.h^{-1}$, which is the mean hourly heating of the lake water at $z = z_o = 56$ m during the whole time interval 8-27 September, 1978 as it results from the direct inspection of the measured temperature profiles (in other terms, the curves are normalized to the same value of the mean hourly heat flux at $z = z_o = 56$ m).

Figure 7 shows the χ-profiles related to three time intervals of different length, all ending on September 27th.

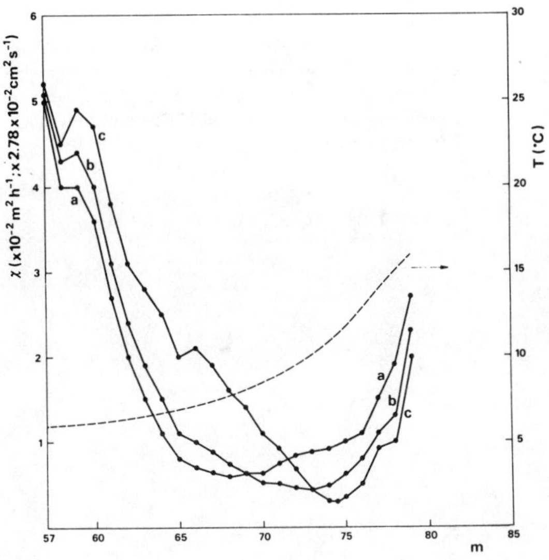

Figure 7. Eddy diffusion coefficient as function of elevation:
a) average on a 19-day period (8-27 Sept., 1978),
b) " " " 15-day " (12-27 Sept., 1978),
c) " " " 12-day " (15-27 Sept., 1978).
(------ average temperature profile for the period September 8-27).

It appears that averaging on time intervals including a meteorologic perturbation (curves a, b) could hardly reduce the numerical effects due to the large temperature oscillations in the metalimnic and upper hypolimnic layers: the curves exhibit the lowest values below the point of largest temperature gradient in the average temperature profile. The effects of internal waves seem to have been better smoothed out by averaging on a shorter time interval beginning after the main perturbation (curve c).

Figure 8 shows the χ-profiles related to three time intervals of fixed length $\Delta t = 12$ d, displaced "across" the perturbation of September 12th; curve a) exhibits the same effect of χ-underevaluation already pointed out in figure 6 ($\chi < 0$ for $65.5 < z < 72$ m); curves b) and c), in satisfactory mutual agreement, seem to confirm that, in the Agno Bay case, a 12 day interval can be considered as sufficiently long for χ- identification in the thermocline region, if strong meteorologic perturbations are absent.

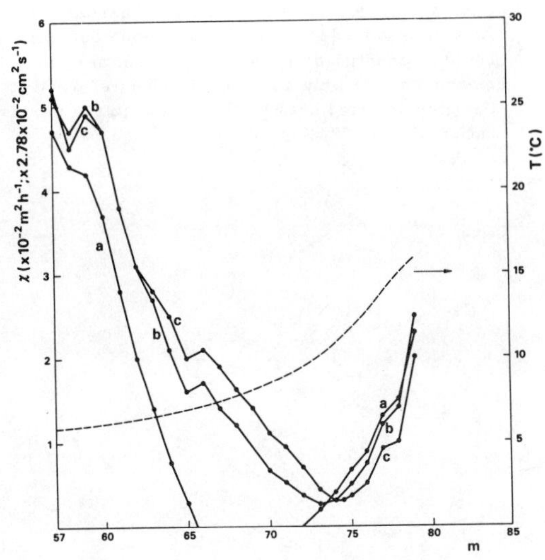

Figure 8. Eddy diffusion coefficient as function of elevation (average on a 12-day period):
a) September 12-24, 1978,
b) " 14-26, 1978,
c) " 15-27, 1978,
(------ average temperature profile for the period September 8-27).

As for the experimental errors, the accuracy of the concerned measurements (standard deviation of temperature measurements can be assumed $\leqslant 0.1^oC$; the relative standard deviation of radiation measurements $\leqslant 10\%$) appears to satisfy largely the requests of a method, which has much greater intrinsic causes of uncertainty, both in the collection and in the elaboration of experimental data.

5. CONCLUDING REMARKS

The results presented in this paper are based on measurements performed during a rather short time period, although described in great detail. Consequently, they can support only tentative conclusions. Nevertheless, the following remarks can be made, the relevance of which is due to the fact that cold wind perturbations are rather frequent in the subalpine region during summer and early autumn:

a) A temperature data collection fitted to lake hydrodynamics, i.e. performed with a sufficient time detail, is necessary in order to allow a realistic heat balance evaluation.

b) Such detailed temperature measurements put in evidence a very complex phenomenology, in particular after a strong meteorologic perturbation. As a consequence, the eddy diffusion coefficient could be identified on the basis of temperature measurements only as a mean value referred to the time interval between two consecutive perturbations.

c) A time interval between two perturbations $\Delta t \sim$ 12 days, could allow an identification of the eddy diffusion coefficient in the thermocline region accurate enough for ecological estimate purposes (transport of dissolved nutrients (Rossi 1980)).

6. LITERATURE CITED

Beonio-Brocchieri, F. et al. 1979. Preliminary experimental investigations on the thermal and oxygen stratification in Lake Lugano. External Report EUR 6557 EN.

Jassby, A. and T. Powell. 1975. Vertical patterns of eddy diffusion during stratification in Castle Lake. Limnology and Oceanography, 20, 4, p. 530-543.

Nihoul, J.C.J. 1977. Modèles mathématiques et dynamiques de l'environnement. Liège, p. 41-47.

Orlob, G.T. 1967. Prediction of thermal energy distribution in streams and reservoirs. Report WRE-TD 947 - W 325, p. 24-31.

Rossi, G. et al. 1980. Capabilities and limitations of a vertical box eutrophication model for the determination of nutrient flux from lake sediment. Proceedings of the Second ISEM Conference, Liège, p. 683-700.

MODELLING THE DISTRIBUTION OF HEAVY METALS BETWEEN SEDIMENT AND WATER IN THE UPPER NILE LAKE SYSTEM[1]

L. Kamp-Nielsen[2], L. A. Jørgensen[3], S. E. Jørgensen[4]

Abstract.--Sorption isotherms were experimentally determined for copper, lead, cadmium, and chromium with sediments from Lake Victoria, Lake Mobuto, and Lake Kioga in the Upper Nile Lake Basin. The data fitted best Langmuir equations with maximum sorption capacity at aerobic conditions for Pb of 55 µg/g dry matter and minimum sorption capacity for Cd of 2 µg/g dry matter. Under anaerobic conditions the sorption capacity increased depending on metal species. Relations between sorption capacity and pH and redox conditions were established and used in a copper submodel as an example. The submodel was coupled to an existing eutrophication model for Lake Victoria by simple partition coefficients between heavy metal and water and suspended particles generated in the eutrophication model.

INTRODUCTION

The Upper Nile Lake drainage basin covers an area of about 515.000 km^2 and includes the lakes Victoria, Mobuto Sese Seko (L. Albert), and Kyoga. As a part of the WMO/UNEP-project "Water quality Model of The Upper Nile Lakes" the distributions of heavy metal has been studied and mathematically modelled.

At present heavy metal pollution does not appear to create any serious problems in the area since only few external sources exist. However, the effluents from the Kilembe copper mines north of L. George may contribute some copper via the Kazinga Channel through L. Idd Amin (L. Edward) and further on through the Semiliki River to L. Mobuto Sese Seko (Bugeny 1979). Also copper processing plants, near Jinja north of L. Victoria, and increasing industrialization in the Kisumu area in the north-eastern and the Mwanza area in the southern part of the L. Victoria drainage basin are potential sources of heavy metal pollution.

Since the study area is very large and the distribution of heavy metals only a small part of the whole project and the data limited, a rather simple model was chosen.

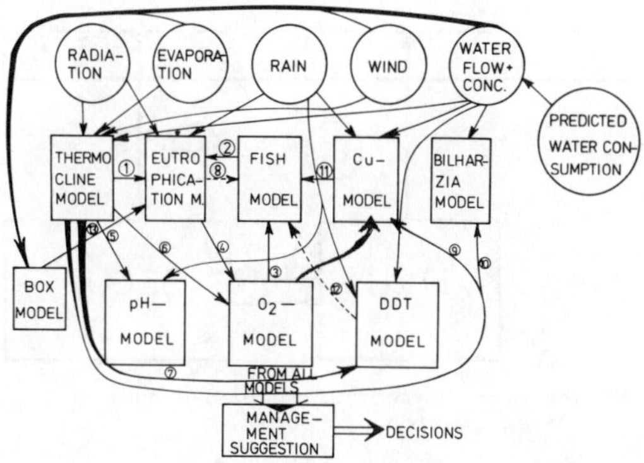

Figure 1.-- The submodels and their connections in the large scale, multi-purpose water quality model (from Jørgensen et al. 1982).

Some of the input data for the heavy metal submodels are created as output from the other submodels in the project (fig. 1) and some are experimentally determined. In the present paper only the submodel for copper distribution in L. Victoria is shown. For information on the other metals and the other lakes cf. Jørgensen et al (1982).

---

[1]Paper presented at the Third International Conference on State-of-the-Art in Ecological Modelling, Colorado State University, May 24-28, 1982.

[2]Lars Kamp-Nielsen is Associate Professor at the Freshwater Biological Laboratory, University of Copenhagen, Hillerød.

[3]Leif A. Jørgensen is Research Assistent at ISEM, Copenhagen.

[4]Svend E. Jørgensen is Associate Professor at the Royal Danish School of Pharmacy, Copenhagen.

MODEL BACKGROUND

The bulk part of the loading of heavy metals to the lakes occurs as suspended solids carried by the tributaries. Similarily, the bulk part of the heavy metals in the lakes is present in particulate forms - either as sediments or as seston. Since the toxic effects of the metals are most likely associated with the levels of the ionic and other dissolved forms, the model and the experiments are focussed on the distribution between dissolved and particulate forms of the metals.

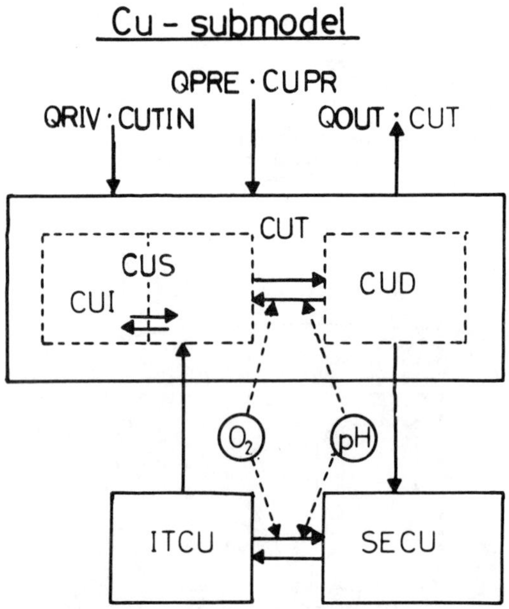

Figure 2.-- Conceptual model for the distribution of copper.

The Cu- submodel shown in figure 2 considers the loading to the lake via suspended solids from the tributaries and via the precipitation. The distribution between dissolved and the particulate copper in the water phase is described by sorption equilibria governed by pH and oxygen conditions. The dissolved copper is separated in ionic and complexed copper by an assumed partition coefficient. Particulate copper sediments out bound to inorganic particles and ton - and detritus-bound copper. In the sediment, the copper separates as dissolved, interstitial copper and particulate copper. The dissolved copper diffuses back to the water phase by an ordinary Ficks diffusion corrected for sediment porosity. The model complexity and the considerations behind are similar to those previously described by Kamp-Nielsen et al (1980) for phosphorus in the study area.

Symbols and units are presented in table 1.

MATERIALS AND METHODS

Sampling and analytical methods.

The sediment cores were taken in L. Victoria (offshore Entebbe and in the Kavirondo Gulf), in L. Mobuto Sese Seko and in L. Kyoga with a Kajak corer. The sampling stations are shown in Figure 3. For further informations on the sampling sites and sediment characteristics cf. Kamp-Nielsen et al (1980).

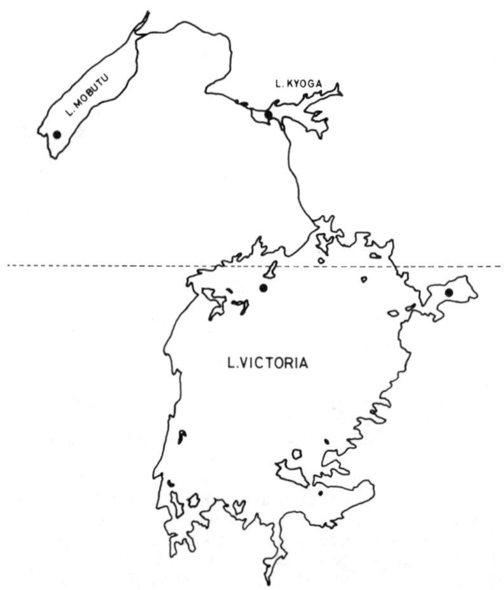

Figure 3.--The sampling stations in The Upper Nile Lake Basin.

The cores were split in sections from 0-1 cm, 1-2 cm, 2-3 cm, 3-5 cm and 5-10 cm. The subsamples were analyzed for dry matter and the dried samples were digested in half-concentrated nitric acid (Suprapur) by autoclavation. Lead, chromium, copper and cadmium were determined by atomic absorption spectrophotometry on a Perkin-Elmer model 370 equipped with a graphite furnace. All glasswares were prerinsed in 6N nitric acid.

EXPERIMENTAL PROCEDURES

From each sediment sample approximately .5 g fresh material was used and 3-4 replicate subsamples were shaken 20-40 hours at 24-26° with 50 ml of artificial lake water adjusted with respect to major ions and pH. Various concentrations of lead, chromium, copper, and cadmium were added from standard solutions prepared from Titrisol ampoules. After the incubation the suspen-

Table 1.--Symbols, descriptions and units for the copper model

Symbol	Description	Unit
ae	Subscript indicating aerobic conditions	-
an	- - anaerobic -	-
AEAN	Partition coefficient for aerobic/anaerobic sediment	-
AREA	Lake area	m^2
CUD	Sorped copper in the lake water	mg Cu/m^3
CUI	Ionic copper - - - -	mg Cu/m^3
CUPR	Total copper in precipitation	mg Cu/m^3
CUS	Dissolved copper in the lake water	mg Cu/m^3
CUSC	Copper sorption capacity	mg Cu/kg DM
CUT	Total copper in the lake water	mg Cu/m^3
CUTIN	Total copper in the tributaries	mg Cu/m^3
DEPTH	Lake depth	m
DETR	Detritus concentration in the lake water	mg DM/m^3
DIFF	Diffusion coefficient for ITCU	m^2/d
DIRF	Diffusion resistance factor	-
DM	Dry matter	
GAMMA	Partition coefficient for CUI/CUS	-
INORG	Particulate, inorganic material in lake water	mg/m^3
ITCU	Interstitial copper	mg Cu/m^3
KCU	Half-saturation constant for copper sorption	mg Cu/m^3
KORR	Ratio gross/net-sedimentation	-
MLCU	Average diffusion range for ITCU	m
NSCU	Native, sorbed copper	mg Cu/kg DM
OXYG	Oxygen concentration in bottom water	g O_2/m^3
PHYT	Phytoplankton concentration in the lake water	mg DM/m^3
QOUT	Outflow from lake	m^3/d
QPRE	Precipitation on the lake	m^3/d
QRIV	Inflow from tributaries	m^3/d
REL	Release of copper from sediment	mg $Cu/m^2/d$
SECU	Particulate copper in the sediment	mg Cu/m^3
SEDD	Sedimentation rate for detritus	m/d
SEDP	- - - phytoplankton	m/d
SEDI	- - - inorganic material	m/d
VOL	Lake volume	m^3

sions were centrifuged at 3000 rpm for 5 minutes and the supernatants decanted off and filtrated through a .45μ Millipore filter. The concentrations of lead, chromium, copper, and cadmium were determined on the filtrates. Comparisons showed the necessity of running the metals separately since they apparently were competing for the same sorption sites on the sediment particles. Equilibrium incubations were run aerobically after flushing with air,or anaerobically after flushing with nitrogen gas. Equilibrium experiments at various pH-values in the range 6.5 - 9.5 were made by adjusting the artificial lake water with HCl or NaOH.

EXPERIMENTAL RESULTS

The sorbed copper per gram dry sediment is plotted against the equilibrium concentrations of copper for the various sediments under aerobic conditions in figures 4-7 and under anaerobic conditions in figures 8-11. Similar plots have been made for lead, chromium, and cadmium, Jørgensen et al. (1982). The data were fitted to both Langmuir isotherms and Freundlich isotherms. The best fit with the lowest residuals, were obtained with Langmuir expressions of the form:

$$SM = \frac{MSC \cdot (ME}{K + (ME)} - NSM$$

where,
SM = sorbed metal
MSC = metal sorption capacity
(ME) = concentration of dissolved metal
K = half saturation constant
NSM = native, sorbed metal

The curves are shown as solid lines in figures 4-11 and the numerical values for the equations are listed in tables 2 and 3 A SAS-programme was used for the curve fitting and the missing values occur as a result of convergence criteria not met within 50 model runs. Since no significant differences could be shown between the various sediment depths,the results from all sediment depths are shown together.

The average MSC-values ranked in the order:

Pb > Cr > Cu > Cd aerobic
Pb > Cu > Cr > Cd anaerobic

Comparing these ranks with the rank of electronegativity and the affinity to $^-$SH-COOH and $^-$NH-groups a reasonable agreement is found taking into consideration that the

Table 2.--Langmuir isotherms for the sorption of various metals on various sediments under aerobic conditions.

$$SM = \frac{MSC \cdot (ME)}{K + (ME)} - NSM,$$

where,

SM = sorbed metal (mg/kg DM)
MSC = metal sorption capacity (mg/kg DM)
(ME) = concentration of dissolved metal (mg/m^3)
(K) = half saturation constant (mg/m^3)
NSM = native, sorbed metal (mg/kg DM)

Lake	AEROBIC Cd	Cr	Cu	Pb
L. Victoria (Entebbe)	$\frac{12.00\ (Cd)}{2.44 + (Cd)} - 0$	$\frac{29.6\ (Cr)}{60 + (Cr)} - 0$	$\frac{16.1\ (Cu)}{0.0001 + (Cu)} - 0$	$\frac{22.0\ (Pb}{0.0001\ (Pb)} -0$
L. Victoria (Kisumu)	$\frac{1.77\ (Cd)}{0.06 + (Cd)} + 0$	$\frac{47.3\ (Cr)}{132 + (Cr)} -0.49$	$\frac{4.50\ (Cu)}{0.0001 + (Cu)} - 0$	$\frac{24.5\ (Pb)}{0.0001 + (Pb)} -0$
L. Kioga	$\frac{5.57\ (Cd)}{0.41 + (Cd)} -0$	$\frac{10.9\ (Cr)}{29.4 + (Cr)} -0$	$\frac{15.4\ (Cu)}{0.86 + (Cu)} -0$	$\frac{55.4\ (Pb)}{2.06 + (Pb)} -0.93$
L. Mobuto	-	$\frac{3.09\ (Cr)}{47.4 + (Cr)} -0$	$\frac{6.65\ (Cu)}{2.38 + (Cu)} -0$	-

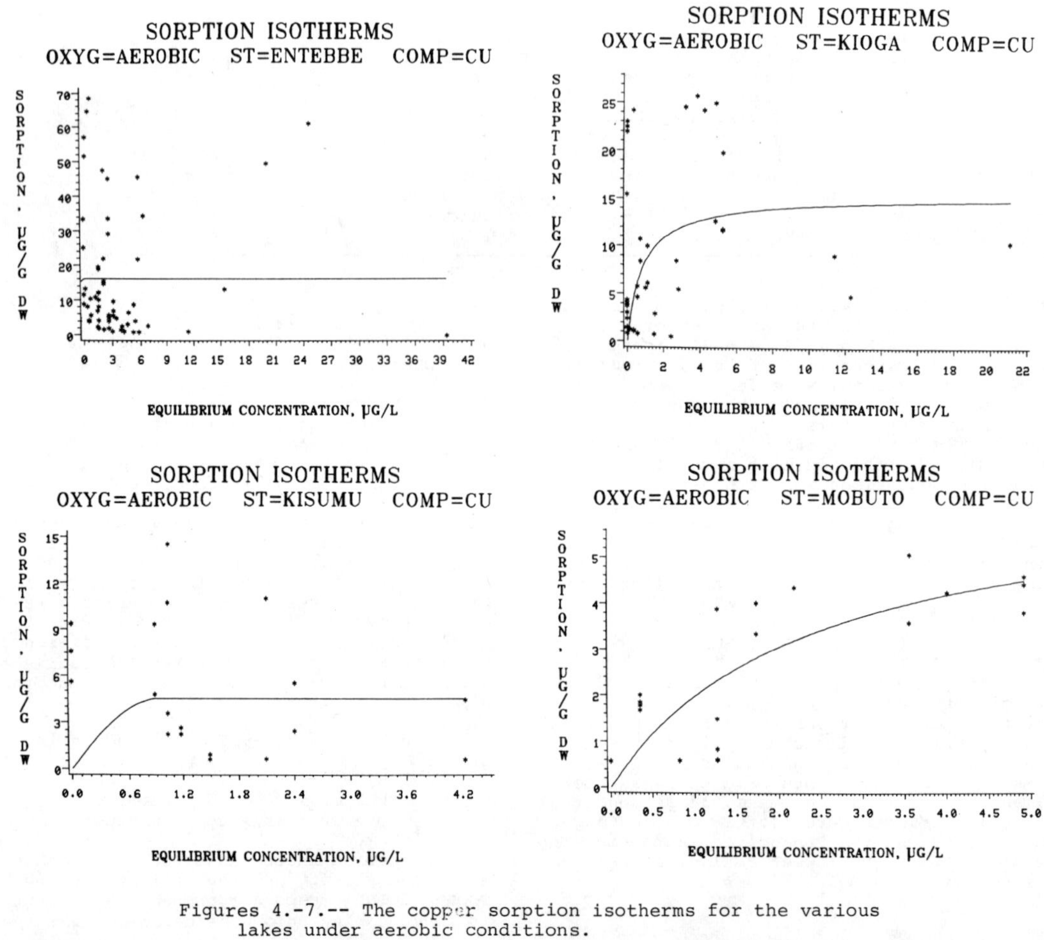

Figures 4.-7.-- The copper sorption isotherms for the various lakes under aerobic conditions.

Table 3.--Langmuir isotherms for the sorption of various metals on various sediments under anaerobic conditions.

$$SM = \frac{MSC \cdot (ME)}{K + (ME)} - NSM,$$

where,

SM = sorbed metal (mg/kg DM)
MSC = metal sorption capacity (mg/kg DM)
(ME) = concentration of dissolved metal (mg/m^3)
(K) = half saturation constant (mg/m^3)
NSM = native, sorbed metal (mg/kg DM)

	ANAEROBIC			
Lake	Cd	Cr	Cu	Pb
L. Victoria (Entebbe)	$\frac{4.61\ (Cd)}{0.02 + (Cd)} - 1.77$	$\frac{11.0\ (Cr)}{64.6 + (Cr)} - 0.18$	$\frac{21.0\ (Cu)}{2.54 + (Cu)} - 5.25$	$\frac{104\ (Pb)}{0.91 + (Pb)} - 0$
L. Victoria (Kisumu)	$\frac{1.19\ (Cd)}{0.12 + (Cd)} - 0$	$\frac{10.61\ (Cr)}{155 + (Cr)} - 0$	$\frac{3.80\ (Cu)}{0.0001\ (Cu)} - 0$	$\frac{62.5\ (Pb)}{10.7 + (Pb)} - 0$
L. Kioga	$\frac{1.23\ (Cd)}{0.30 + CD} - 0$	$\frac{9.18\ (Cr)}{85.0 + (Cr)} - 0$	$\frac{2.58\ (Cu)}{0.0001 + (Cu)} - 0$	-
L. Mobuto	$\frac{5.85\ (Cd)}{16.7 + (Cd)} - 0$	$\frac{0.87\ (Cr)}{0.0001 + (Cr)} - 0.36$	$\frac{30.5\ (Cu)}{54.6 + Cu} - 0$	$\frac{43.1\ (Pb)}{2.1 + (Pb)} - 0$

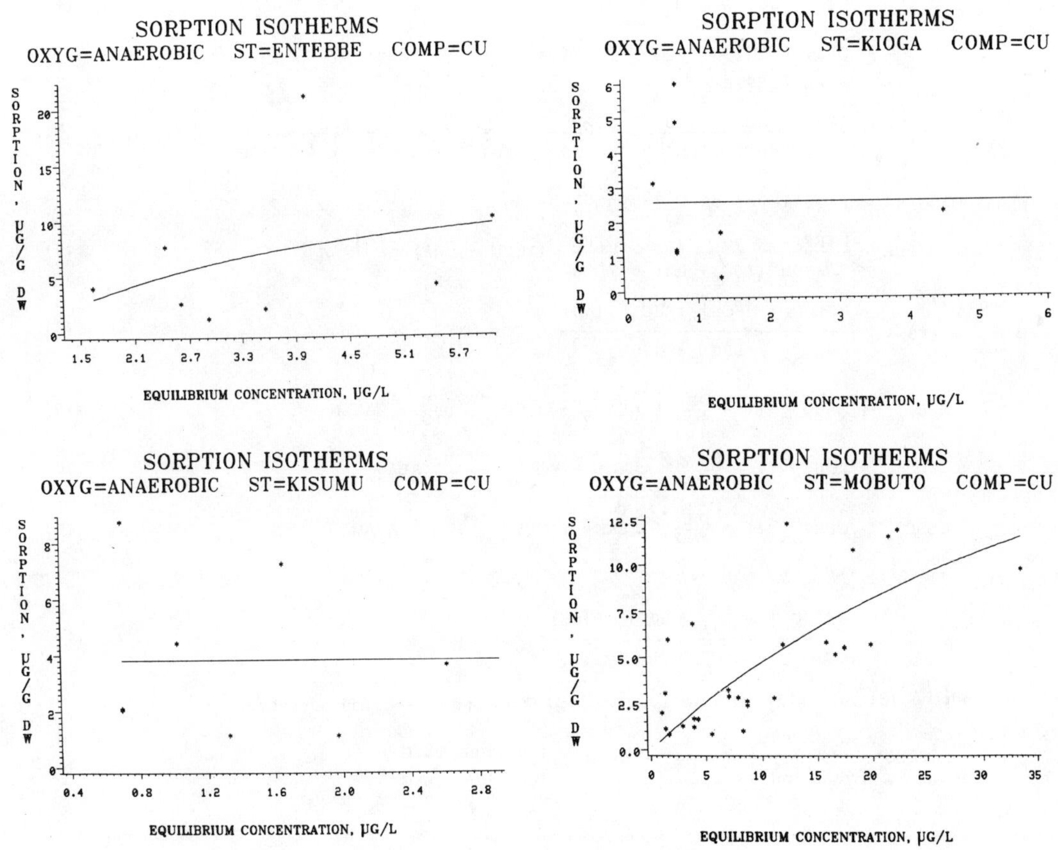

Figures 8.-11.--The copper sorption isotherms for the various lakes under anaerobic conditions.

binding to sediment particles can occur by ion-exchange as well as by physical adsorption and chemosorption to a variety of inorganic and organic compounds. However, a careful inspection of the data on chromium sorption showed that in spite of the fit to langmuir isotherms it is most likely the solubility of $Cr(OH)_3$, which controls the equilibrium concentrations as well under aerobic as under anaerobic conditions (Lu and Chen 1977).

From the conceptual model (Figure 2) a mathematical model can be formulated with 6 state variables: CUI, CUS, CUD, CUT, SECU, and ITCU, and the mass conservation equations are given by the equations 1-13. Most of the equations are simple equilibria, only CUT (eq. 5) is given by a differential equation.

$$CUI = GAMMA \cdot CUS \text{ (equilibrium)} \tag{1}$$

$$CUS = \frac{(CUD + NSCU_{ae})KCU_{ae}(pH)}{CUSC_{ae} - CUD + NSCU_{ae}} \tag{2}$$

equilibrium derived from the Langmuir isotherm with pH-dependent KCU.

$$CUD = \frac{CUSC_{ae} \cdot CUS}{KCU_{ae}\ (pH) + CUS} - NSCU_{ae} \text{ (aerobic conditions)} \tag{3}$$

$$CUD = CUT - CUS \tag{4}$$

$$\frac{d\ CUT}{dt} = QRIV \cdot CUTIN - QOUT \cdot CUT + QPRE \cdot CUPR - SEDI \cdot AREA \cdot CUD/VOL + REL/DEPTH \tag{5}$$

$$SECU = \frac{1000 \cdot SEDI \cdot CUS \cdot KORR}{SEDP \cdot PHYT \cdot SEDD \cdot DETR + SEDI \cdot INORG} - \frac{REL}{DM \cdot 10/000} \tag{6}$$

assuming an active layer of 0.1 m and where,

$$REL = \frac{DIFF}{1 + DM/(100 + DM)} \cdot \frac{DIRF}{MLCU} (ITCU - CUS) \tag{7}$$

$$ITCU = \frac{(SEDU + NSCU'_{ae})KUC'_{ae}\ (pH)}{CUSC'_{ae} - SECU + NSCU'_{ae}}, \text{ for } OXYG > 4 \tag{8}$$

where,

$$NSCU'_{ae} = NSCU_{an}\ (1 - \frac{DM}{100} \cdot AEAN) + NSCU_{ae} \cdot \frac{DM}{100} \cdot AEAN \tag{9}$$

$$KCU'_{ae} = KCU_{an}\ (1 - \frac{DM}{100} \cdot AEAN) + KCU_{ae} \frac{DM}{100} \cdot AEAN \tag{10}$$

$$CUSC'_{ae} = CUSC_{an}\ (1 - \frac{DM}{100} \cdot AEAN) + CUSC_{ae} \frac{DM}{100} \cdot AEAN \tag{11}$$

or,

$$ITCU = \frac{(SECU + NSCU_{an})\ KCU_{an}\ (pH)}{CUSC_{an} - SECU + NSCU_{an}} \text{ for } OXYG \leq 4 \tag{12}$$

and where (8) and (12) are derived as (3) from the Langmuir equation:

$$SECU = \frac{CUSC \cdot ITCU}{KCU\ (pH) + ITCU} - NSCU \quad \begin{matrix}\text{(ae when } OXYG > 4) \\ \text{(an when } OXYG \leq 4)\end{matrix} \tag{13}$$

The parameter values and the initial values for CUT and SECU are given in Table 4. Table functions for the driving variables QOUT, QRIV, QPRE and for the auxiliary variables DETR, INORG, OXYG, PHYT, and pH are derived from the eutrophication model (Jørgensen et al. 1982).

Table 4.--Parameter values for the copper model

Symbol	Value	Derived from	Reference
AEAN	0.11	Calibration	
AREA	$67.6 \cdot 10^{9}$	Literature	(Hydromet 1976)
CUPR	2.0	-	(Beavington and Cawse 1979)
$CUSC_{ae}$	16.1	Experiments	
$CUSC_{an}$	21.0	-	
InCUT	33.0	Measurements	
CUTIN	35.0	-	
DEPTH	54.0	Literature	(Hydromet 1976)
DETR	Table	Eutrophication model	(Jørgensen et al. 1982)
DIFF	0.016	Literature	(Kamp-Nielsen 1981)
DIRF	0.3	-	(Frissel & Reiniger 1974)
DM	5.09	Measurements	
GAMMA	0.2	Guess	
INORG	Table	Eutrophication model	(Jørgensen et al. 1982).
KCU_{ae}	0.0001	Experiments	
KCU_{an}	2.5	-	
KORR	1.33	Mass balance	
MLCU	0.12	Measurements	
$NSCU_{ae}$	0	Experiments	
$NSCU_{an}$	5.2	-	
OXYG	Table	Eutrophication model	(Jørgensen et al. 1982)
PHYT	Table	-	(Ibid)
QOUT	Table	-	(Ibid)
QPRE	Table	-	(Ibid)
QRIV	Table	-	(Ibid)
InSECU	0.008	Measurements	
SEDD	0.05	Eutrophication model	(Ibid)
SEDP	0.3	-	(Ibid)
SEDI	0.05	-	(Ibid)
VOL	$3.7 \cdot 10^{12}$	Literature	(Hydromet 1976)

Simulations of CUT under actual loading conditions and with 10 times increased loading are shown together in figure 12. The simulations show that the seston and the sediments buffer the copper content, when the loading is increased. A ten-fold increase in loading over a year only raises the CUT with about 5%. Similarily, figure 13 shows the effect of the actual and of a tenfold increased Cu-loading on the sediment concentration of copper. A tenfold increase in loading only raises the sediment concentration with about 2% per year.

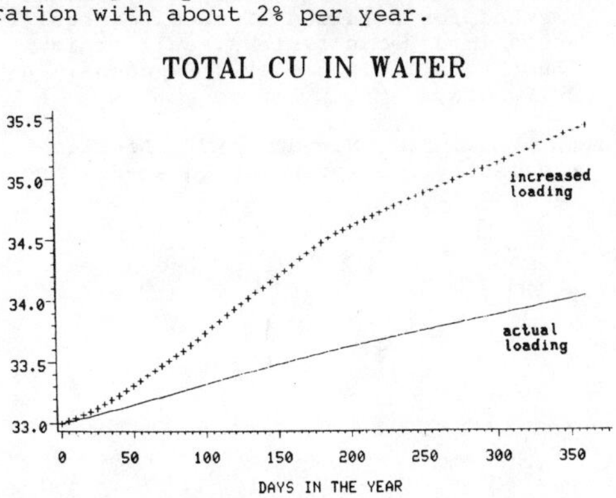

Figure 12.--A one year simulation of total copper in the water phase in L.Victoria (Entebbe) - CUT - at actual and at tenfold increased loading.

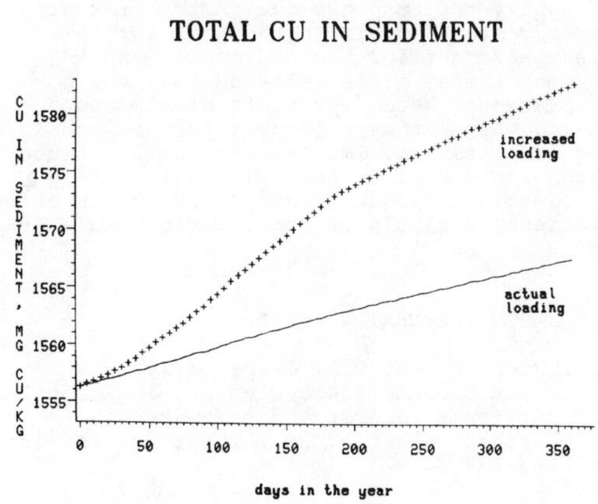

Figure 13.--A one year simulation of total copper in the sediment in L. Victoria (Entebbe) - SECU - at actual and at tenfold increased loading.

DISCUSSION

To bring the modelling efforts into a reasonable balance with the quality of the available data, and to balance the copper submodel against the total water quality model, a rather crude model with a high degree of black boxes is presented. Of course, a substantial part of the inorganic complexes could be accounted for by computing solubility diagrams, depending on equilibria constants and concentrations of major ions, as proposed by Stumm and Morgan (1970), but still our knowledge of the organic complexes is very limited. Depending on the method used - organic solvent extraction, ion selective electrodes, anodic stripping voltammetry - various values appear for the fraction of organically complexed heavy metal (Kamp-Nielsen 1971). But even if we, by using very complex models with all the important organic and inorganic ligands e.g. Morell (1968), computed the "true" ionic metal, its value would be doubtful since attempts to relate toxicity of metal to ionic concentrations have both succeeded (Davey et al 1973) and failed (Gächter et al 1978). Taking these into consideration a simple distribution coefficient for the ionic and the complexed, dissolved copper was chosen for the present model.

Besides the black box-nature of the model, there is also the uncertainty of the experimental data, the field data and the parameter values transferred from other studies, which contribute to the uncertainty of the model predictions. Since many of the experimental results and the average values of field measurements have a standard deviation of more than 100%, the model predictions also have an uncertainty not less than ± 100%.

The hydraulic residence time for L. Victoria is about 100 years (Jørgensen et al 1982) which explains the low response of a ten-fold increase in copper loading and consequently, the response does not represent a steady state condition, neither for the water phase nor for the sediment. Despite the weaknesses, it is believed that the model presented here represents a balanced approach, when large-scale multipurpose modelling is attempted. Improvements will include validation over a larger time scale and introduction of model uncertainty by use of covariance analysis or Monte Carlo simulations.

LITERATURE CITED

Beavington, F. and P.A. Cawse. 1979. The deposition of trace elements and major nutrients in dust and rainwater in Northern Nigeria, Sci. Total Environ 13: 263-274.

Bugeny, F.W.B., 1979. Copper ion distribution in the surface waters of lakes Georges and Idd Amin. Hydrobiologia 64: 9-15.

Davey, E.W., M.J. Morgan and S.J. Erickson. 1973. A biological measurement of the copper complexing capacity of seawater. Limnol. Oceanogr. 18: 993-997.

Frissel, M.J. and P. Reininger. 1974. In Simulation of accumulation and leaching in soils, 82 p. Wageningen. The Netherlands.

Gächter, R., J.S. Davis and A. Marés. 1978. Regulation of copper availability to phytoplankton by macromolecules in lake water. Environ. Sci. Tech. 12: 1416-1421.

Hydromet. 1975. Hydrometerological survey of the catchments of lakes Victoria, Kyoga and Albert. UNDP and WMO report. Vol. 1. 888 p. Nairobi. Kenya.

Jørgensen, S.E., L.A. Jørgensen, L. Kamp-Nielsen and H.F. Mejer. 1982. Water quality model for the Upper Nile Lake Basin. Final Report. 177 p. Copenhagen. Denmark.

Kamp-Nielsen, L. 1972. Some comments on the determination of copper fractions in natural waters. Deep-Sea Res. 19: 899-902.

Kamp-Nielsen, L., L.A. Jørgensen and S.E. Jørgensen. 1980. A sediment-water exchange model for lakes in the Upper Nile Basin. p. 557-581. In Progress in ecological engineering and management by mathematical modelling, 1013 p. Editions Cebedoc. Liege. Belgium.

Lu,J.C.S. and K.Y. Chen. 1977. Migration of trace metals in interface of seawater and polluted surficial sediments. Environ. Sci. Tech. 1: 174-182.

Morel, F. and J.J. Morgan. 1968. A numerical method for solution of chemical equilibria in aqueous systems. California Institute of Technology. Pasadena. California.

Stumm, W. and J.J. Morgan. 1970. Aquatic Chemistry. 583 p. Wiley-Inter-Science. New York.

A WATER QUALITY MODEL FOR THE UPPER NILE LAKE SYSTEM[1]

S. E. Jørgensen[2], L. A. Jørgensen[3], Henning F. Mejer[3] and Lars Kamp-Nielsen[4]

Abstract. During the period 1977-1981 a water quality model for the Upper Nile Lake System has been developed with the objective to give information on the changes in water quality, environmental impact and ecological changes and to show how a suitable water quality can be maintained to meet future requirements if discharge of municipal and industrial waste water to the system is increased. The model developed consists of 10 independent submodels: 1) a eutrophication model, 2) a fish model, 3) a thermocline model, 4) a box model describing the exchange of water elements and compounds between the compartments in the lakes, 5) a DDT model, 6) a model describing the interaction of nutrients between sediment and water, 7) a copper model, 8) a pH model, 9) an oxygen model and 10) a bilharzia model. The model gives only detailed information about the important variables of the system and only rough estimates about variables, which are less important or require comprehensive biological research. The model is flexible and allows addition of submodels when needed at a later stage. The interrelation between the different forcing functions and submodels can be seen in figure 1. The construction of the 10 submodels, the data on which they are based and the estimated standard deviation of the models is reviewed in the paper presented. The data available are insufficient at present to allow a comprehensive calibration. It is hoped that further data in the future will be able to improve the calibration and allow a validation.

INTRODUCTION

During the last decade, a continuous hydrological survey of the Upper Nile Basin has been carried out. The project, called HYDROMET, has 8 countries as participants: Burundi, Egypt, Kenya, Ruanda, Sudan, Tanzania, Uganda and Zaïre. The headquarters of the project is in Entebbe, Uganda, where hydrologists from the member countries are working on the project.

UNEP and WMO have agreed to expand the project with a water quality model component. This part of the project, which has the authors of this paper as consultants started in December 1977 and will be terminated during 1981. The objectives of this water quality model can be summarized as follows:

(1) to give information of the water quality of the lakes, if - due to further development of the water resource e.g. the irrigation scheme, the entire water-balance, the discharge of municipal or industrial waste water to the system is changed. The water quality should be evaluated considering the use of the water for (a) drinking purposes, (b) industrial purposes, (c) irrigation, (d) stock and wildlife watering, (e) fish and aquatic life and (f) recreational purposes.

(2) to give information about the environmental impact and ecological changes of the water bodies, if - due to further development of the water resources e.g. the irrigation schemes, the entire water-balance, the discharge of municipal or industrial waste water to the system is changed. This includes (a) the eutrophication of the lakes, (b) the use of water for stock and wildlife watering and for fish and aquatic life [see (1),(d) and (e)], (c) ecological balance of the entire ecosystem, (d) disease vector agents, (e) weed infestation intensity and (f) fishstock available for fishery.

(3) to give information as to how a suitable water quality could be achieved or maintained to meet present and future requirements. It involves maximum allowable effluent concentrations for waste water as well as first hand information about sound waste water handling methods and technology to be propagated for use by Governments. However, it will require a very comprehensive model and a lot more time-consuming biological data to meet all these objectives with a high accuracy. Furthermore, the industrial planning strategies of the countries involved are not yet known, at least not in detail, for the next one or two decades.

---

[1]Paper presented at ISEM's 3rd Internat'l Conf., Colorado State University May 24-28, 1982.

[2]Royal Danish School of Pharmacy, 2 University Park, 2100 Copenhagen Ø, Denmark

[3]Copenhagen Teknikum, Pr. Charlottegade, 2200 Copenhagen N, Denmark.

[4]University of Copenhagen, Freshwater Biological Institute, Helsingørgade, 3400 Hillerød, Denmark.

It has, therefore, been suggested to set up an ecological model, which will give more detailed and accurate information about the important variables of the system and only rough estimates about the variables which are less important, or require comprehensive biological research and/or information not available today. The model should, however, be flexible and allow addition of submodels, when needed for more detailed information at a later stage or when a more detailed development plan is available. A model essentially processes a set of input numbers to produce a set of output numbers. Whether or not a model is relevant to a problem depends on whether or not the model produces an output that corresponds to the desired objectives.

Six years ago, HYDROMET established a chemical laboratory and has included analysis of water quality parameters in the hydrological survey. In relation to the water quality project, the analytical program has been expanded during 1978 to include measurement of productivity, nutrient concentrations, oxygen profile and other relevant parameters, and the sampling program has been intensified. Heavy metal and pesticide analyses are carried out in Denmark, together with determination of phytoplankton and zooplankton species and examinations of the nutrient exchange processes for sediment-water. A comprehensive literature review on all relevant data about fishery and limnology in the 3 lakes have been carried on. A data base containing all water quality data relevant for the project has been erected.

DESCRIPTION OF THE TOTAL MODEL

The total model consists of the following independent submodels:

1. A eutrophication model
2. A fish model
3. A thermocline model
4. A box model describing the exchange of water, elements and compounds between the box-compartments in the lakes
5. A DDT model
6. A model describing the interaction of nutrients between sediment and water
7. Heavy metals (giving details for copper, included adsorption of suspended matter and interaction of heavy metals between sediment and water (for further details see L.K. Nielsen et al. 1983, in this Volume).
8. A pH model
9. An oxygen model
10. A bilharzia model

In conjunction with submodels the following two types of software are useful instruments:

a) A model for parameter estimation using intensive measurements. A description of the method can be seen from Jørgensen et al. (1981)

b) PSI - an automatic calibration method, which can be described as follows: (Paramter and System Identification).

PSI is a new general purpose simulation language with built-in facilities for parameter estimation, differential equation solving, vector handling and statistical calculations. Emphasis is put on large models and/or large amounts of input/output data. PSI is written as a CSMP-like program (Continuous System Modelling Program, IBM) using internal efficient and up-to-date software techniques. The first public release (1982) in addition to standard CSMP-facilities will provide tools for parameter estimation (several methods and options for user-supplied routines), handling of large amounts of data, vector calculus, Monte Carlo simulation, discrete simulation, statistics, weighting, smoothing, synchronizing, spline-interpolation of observations and a substantial number of functional blocks, which may be useful for ecological modelling. An initial grid search for parameter estimation can be activated and the strategy for parameter adjustment may be changed automatically during simulation. Parameters vary optionally within user specified bounds. FORTRAN statements may be mixed into the PSI-program and - since the language itself is written in FORTRAN - a high degree of machine-independence is achieved. The only software required to implement PSI is a FORTRAN compiler.
Later releases of PSI will contain improved data base and graphical communication by interfacing available software packages (SAS (Statistical Analysis System - IBM installations only) and GPGS). Also options for coding the model as flow-expressions in addition to a differential equation formulation will be provided, thereby releaving the user for keeping track of mass conservation.

The interrelation between the different forcing functions and the submodels can be seen in figure 1.

An arrow from one box (a submodel) in this figure to another indicates that the data from the first box is used in the second. The relations are further explained in table 1.

Table 1. Transfer of data from one submodel to another

Number (see fig.1)	Explanation
1,5,6,7, 9,10	The thermocline model divides the lake in epilimnion and hypolimnion. Under normal conditions used only in the eutrophication model
2	Concentration of herbivorous and carnivorous
3	The mortality of fish is oxygen dependent
4	Oxygen consumption is dependent on the mineralization rate for detritus
8	The phytoplankton concentration averaged over 6 months indicates the feed concentration for herbivorous fish species
11,12	The mortality of fish is dependent on the copper concentration
13	The box model is generally applicable to all submodels, but mainly used for the eutrophication model
14	The growth rate and mortality of phytoplankton is dependent on the copper concentration

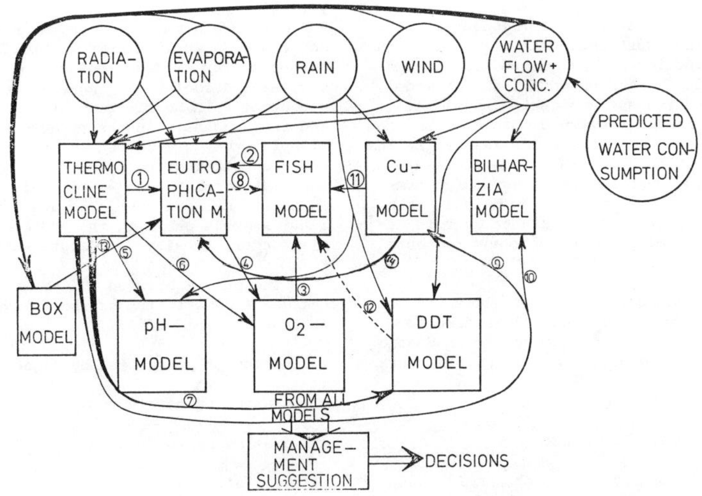

Figure 1. Interrelation between forcing functions (circles) and submodels (boxes).

Although no validation has taken place it is possible to indicate approximately the accuracy of the different submodels.

The thermocline model seems to be very accurate, giving temperature profiles where the predicted value is less than one centigrade from the measured value.

In temperate lakes it has been possible to develop eutrophication models with a standard deviation of approximately 25% relative between measured and predicted value. As this deviation is based on single values an average annual eutrophication measure will be indicated with an acceptable accuracy. When the average of 16 predicted values and 16 measured values is compared the deviation will be reduced by a factor $\sqrt{16} = 4$. However, the number of stations set up for this project is, compared with the area covered, much more scarce than used in our previous lake studies (see e.g. Jørgensen and al. 1976 and 1978). Furthermore, the quality and quantity of the data available for this project are unfortunately rather limited. The accuracy of predictions on phytoplankton concentrations at certain dates at certain stations are therefore very limited. However, an overall picture of the entire ecosystem, especially when we are considering annual average values, can be obtained from the model.

The accuracy of the fishery model is also limited due to the limited number of fishery data available. However, since it seems that the fish biomass in the lakes is, to a certain extent, in balance (the fishery and the results of trawl tests from year to year is rather constant) it will be possible based on the fishmodel to make predictions on how the fishery will be influenced if significant increase in the fishcatch will take place. It is furthermore estimated that the relation between pollution and the fishery conditions is rather well described in the model. This relation is based on quite well-known facts and also the literature review has given a great support to the parameter estimations of the fishmodel. All in all it must be estimated that the fishery model is a useful model for setting up the fishery strategy in the future, although additional data might improve the reliability of this submodel considerably.

It is not possible today to set up very accurate heavy metals submodels. However, the required accuracy is limited as it is necessary due to unknown sublethal effects to reduce heavy metals contamination to at least a magnitude less than the concentration corresponding to lethal effects.

The oxygen submodel and the pH submodel are both relatively accurate. The experience with such submodels is rather comprehensive. If additional data will be provided it will be rather easy to improve the parameter estimations of these two submodels. It might be necessary because the results of these submodels are used in other submodels (fig.1).

The DDT submodel is based on a very limited number of data. However, the processes of DDT in the environment is quite well known from other studies and this knowledge has been taken into account in the DDT submodel for the Upper Nile Lake System. It is furthermore not necessary to have a high accuracy of the DDT model to apply the model in practical environmental management.

Acceptable concentrations of DDT will be at least one order of magnitude less than what corresponds to harmful effects. Therefore it can be concluded, in spite of the scarce data used for the calibration of this submodel, that the submodel is applicable for practical management purpose in the region.

The Bilharzia model is only semi-quantitative. This submodel will indicate whether the conditions for the Bilharzia snail are changed relatively to what is known to be the present condition. However, the Bilharzia-disease is mainly dependent on the quality of water supply, the over-all health condition and other factors which, under no circumstances, can be taken into account in the model.

All in all it can be concluded that the submodels which are a part of the total model of the Upper Nile Lake System as they are presented at the end of 1981 can be used for practical management, but that the output from the model should be treated with caution. Although the model is applicable as it stands today, it is strongly recommended to improve and maintain the model in the coming years. It must especially be recommended to provide additional data to imprøve the calibration and validation of the submodels. This is of special importance for the pH model, the O_2 model, the fishery model and the eutrophication model. All four submodels give an accuracy which can only be used to give an overall picture of the water auality in the lake system. It means that the predicted value must be taken as average value over a certain period of time or combined for a number of stations to be able to be reliable, but if these precautions are taken into account the model can be of great use for environmental management in the region.

It is of course not possible to give details on all submodels. The heavy metal submodel is described by L.K. Nielsen et al. 1983, in this Volume, and some information on the eutrophication model and the thermocline model is given below.

EUTROPHICATION MODEL

Independent cycles of nitrogen, phosphorus and carbon are included in the model. The three elements are allowed to vary in phytoplankton between a minimum and maximum value and these variations are also reflected in the upper trophic levels due to the system of equations, which is based upon conservation of mass.

The state variables in the eutrophication model for P are shown in figure 2. Similar flow diagrams can be set up for nitrogen and carbon. The state variables are: dissolved P, algae P, detritus P, P in zooplankton, P in herbivorous fish (2 species are modelled independently in the fish model haplochromis and tilapia, they are lumped in the eutrophication model), and P in carnivorous fish. These 6 state variables are applied in epilimnion as well as in hypolimnion and in addition there are 2 P-state variables in the sediment: P-exchangable in the sediment and P in interstitial water, totally 14 P-state variables.

The arrows between the boxes in figure 2 are processes. A description of the following processes are included in the model:

(1) uptake of phosphorus by algae

(2) grazing by herbivorous fish

(3) grazing by zooplankton

(4)(5) predation on fish and zooplankton respectively by carnivorous fish

(6) mineralization

(7) mortality of algae

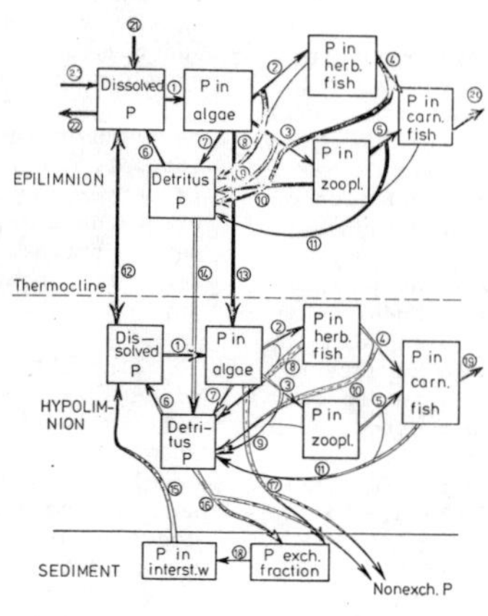

Figure 2. Eutrophication model illustrated by use of P-cycling. Arrows indicate processes. For explanation of numbers, see text.

(8)(9))10)(11) grazing and predation loss

(12) exchange of P between epilimnion and hypolimnion

(13) setttling of algae (epilimnion → hypolimnion)

(14) settling of detritus (epilimnion → hypolimnion)

(15) diffusion of P from interstitial- to lake water

(16) settling of detritus (hypolimnion → sediment) (a part goes to the non-exchangeable fraction)

(17) settling of algae (hypolimnion → sediment) (a part goes to the non-exchangeable fraction)

(18) mineralization of P in exchangeable fraction

(19)(20) fishery

(21) precipitation

(22) outflows

(23) inflows (tributaries).

It is characteristic of the model that it describes the algae growth as a 2 step process: the uptake of nitrient (a Michaelis Menten expression is used) and the growth is determined by the intracellular concentration of nutrient relative to a minimum and maximum value. It is our experience, (Jørgensen (1976) and Jørgensen et al. (1978)) that this description in some cases gives a better accord between model and observations.

The crucial parameters are determined by use of an intensive measuring program (sampling every second day during 2-3 weeks). This technique gives a considerably better parameter estimation, as many parameters (e.g. max. growth rate of algae) reflect a dynamic, requiring a high measuring frequency. The technique is described in detail in Jørgensen et al. (1981).

THE THERMOCLINE MODEL

Our termocline model is based upon Ryan and Harleman (1973). It considers the following heat fluxes:

a) incident solar radiation
b) reflected solar radiation
c) incident long wave radiation (atmospheric variation) and reflection of long wave radiation (atmospheric radiation)
d) long wave radiation from the water surface
e) evaporative heat flux
f) conduction (sensible heat flux).

Heat contents in the rain and in evaporated water are neglected; they are small and tend to balance each other.

The components in the heat fluxes are discussed below:

Incident solar radiation

The net solar radiation is emitted by the sun. It reaches the lake surface after passing through the atmosphere, where it undergoes scattering, reflection and adsorption by clouds, air and dust. The radiation reaching the furface is then a mixture of long and short wave components and of direct and diffuse components. The short wave radiation can be estimated by:

1) measurements by pyrheliometer
2) empirical equations.

Direct measurement is important when an accuracy greater than 15% relative is required.

Reflection

The reflection of the incident solar radiation is a function of the solar elevation and of the type of radiation, direct or diffuse. Typical values are for a solar elevation of 50 degrees, 2% reflection of direct sunlight and 7% of diffuse light (Golterman, 1975). In this model a reflection of 6% has been applied.

Incident long wave radiation

This radiation, E, is created by the different internal fluxes in the atmosphere. Some of these are radiation by components in the atmosphere (water vapour, ozone and carbon dioxide). No precise analytical expressions are available. The atmosphere could be considered as a black body following the Stephan-Boltzmann law; the total amount of energy emitted, integrated over all wave lengths is proportional to the fourth power of the body's absolute temperature, T, where the constant of proportionality, σ, is the Stephan-Boltzmann constant. Since the atmosphere is not a black body, a constant of emission, ε, is applied to the equation (ε is dimensionless). Then:

$$E = \varepsilon \cdot \sigma \cdot T^4$$

is the equation describing this flux.

Several formulae for ε have been proposed, containing relations to vapour pressure, e, air temperature (or different combined functions of these) and clouds. A combined formula proposed by Burtsaert (1975) is used giving the following equation for net atmospheric radiation, E_N:

$$E_N = 5.9 \cdot 10^{-3} \cdot \left(\frac{e}{T_a}\right)^{\frac{1}{7}} \cdot T^4 \cdot (1.0 + 0.17 \cdot C^2)$$

where e is the vapour pressure in m.bar, and C is the fraction of the sky covered by clouds. This formulae includes a reflectance of the waters surface of 3% to long wave radiation:

$$E_W = 0.97 \cdot E_N$$

Back radiation from the water surface is E_W

The emissivity, E_S, of a material is equal to the absorptivity (Kirschhoff's Law). This radiation is then:

$$E_S = 0.97 \cdot \sigma \cdot T_W^4$$

when T_W is the water surface temperature in ^oK.

Evaporative heat flux, E_e:

This is the latest heat flux. The evaporation from a lake is driven by the difference between the saturated vapour pressure (e_S) and the actual vapour pressure (e_Z) at height Z. The evaporation heat flux is:

$$E_e = L_V \cdot EV$$

where L_V is the latent heat of evaporation equal to:

$$L_V = (2493 - 2.26 \cdot T_S) \cdot 10^3 \text{ J/kg}$$

where T_S is the surface temperature in ^oC; EV is the mass flux of evaporation based on the law of mass transfer:

$$EV = \rho \cdot F(W_Z) \cdot (e_s - e_z)$$

where ρ is the density of water and $F(W_Z)$ is a wind speed function for mass flux including both free (wind driven) and forced (buoyance driven) convection effects. This is given as:

$$F(W_Z) = 3.08 \cdot 10^{-4} + 1.85 \cdot 10^{-4} \cdot W_Z \frac{\text{m}}{\text{24h} \cdot \text{mmHg}}$$

where W_Z is the windspeed at height Z.

A zero displacement of 6 inches has been assumed. This is the height at which the horizontal wind disappears, also known as the roughness parameter.
In total this means that:

$$E_e = (3.08 \cdot 10^{-4} + 1.85 \cdot 10^{-4} W_Z) \cdot \rho \cdot (e_s - e_z)(2493 - 2.26 \cdot T_S) 10^3$$

with the unit $J/(m^2 \cdot day)$.

Conductive heat flux

The conductive heat flux, H, can be found from the evaporative heat flux, EV. The ratio between these is the Bowen ratio:

$$B = \frac{H}{EV}$$

The Bowen ratio can be calculated as:

$$B = \frac{C_p}{L_v} \cdot \frac{(T_s - T_z)}{(e_s - e_z)}$$

where C_p is specific heat of water, L_v is the latent heat of evaporation of water.

The ratio $C_p/L_v = 0.24/590 = 269.1$ and the conductive heat flux can be found:

$$H = 269.1 \frac{(T_s - T_z)}{(e_s - e_z)} \cdot EV$$

where T_s is the water surface temperature in ^oC, and T_z is the air temperature at height, z.

Internal heat transfer

The absorption of incident radiation is not solely a surface phenomenon. The insolation at any depth is:

$$E_z = (1-K) \cdot E_s \cdot \exp(-n \cdot z)$$

where E_z is the insolation at depth z; K is a constant typical from 0.4 to 0.5; E_s is the surface insolation and n is a local parameter. It can be found from field measurements, or it can be realated to the secchi disc depth, D, with $n = 1.7/D$.

This equation is not valid close to the surface. However, for the present purpose these equations are sufficiently accurate. If the primary production of a lake was considered,a more accurate description of the absorption should be applied.

The temperature profile in a lake is also effected by advection due to through-flows. In this model the incoming water from rivers are "placed" in the lake water depth according to density calculation.

Seiching is not included in the model. Seiching occurs when a lake is tilted by the wind and an oscillatory motin is created. This would require a 3-dimensional model.

Convection caused by instability by density differencies is included together with molecular diffusion. This latter is small and included for computational convenience.

Internal mixing calculations

The basic heat transport equation for an internal element is:

$$\frac{\partial T}{\partial t} = -\frac{1}{A} \cdot \frac{\partial (Q_v \cdot T)}{\partial z} + \frac{ED}{A} \frac{\partial (A \cdot \frac{T}{z})}{\partial z} + \frac{B \cdot U_i \cdot T_i}{A} - \frac{B \cdot U_o \cdot T}{A} - \frac{1}{\rho \cdot c \cdot A} \cdot \frac{\partial (E_s \cdot A)}{\partial z}$$

where T = the temperature at z, A = area of the element considered, B = width of the element, U_i = horizontal inflow velocity of river water, T_i = temperature of inflow, U_o = horizontal flow velocity, Q_v = vertical flow velocity, t = time, E_s = internal short wave radiation, ED = vertical turbulent diffusion coefficient, c = heat capacity of water and ρ = density of water.

The quantity $\rho \cdot c \cdot T$ represents the heat per unit volume and it is assumed that $\rho \cdot c$ is constant. The equation is expressed by the explicit finite difference formulation and the left side can be computed by means of the temperature at the previous time step

$$T(z, t-1)$$

$$\frac{\partial T}{\partial z} = f(T(z, t-1), E, A, U_i, T_i B)$$

By applying the explicit scheme for the time derivate term:

$$\frac{T(z,t) - T(z, t-1)}{\Delta t} = f(T(z, t-1), ED, A, U_i, T_i B)$$

where Δt = the time step.

This gives a distribution of the temperature at time, z:

$$T(z,t) = T(z, t-1) + f(T(z, t-1), ED, A, U_i, T_i, B) \cdot \Delta t$$
$$= T(z, t-1) + \Delta T(z, t-1)$$

This equation gives the temperature distribution T(z,t) at time t by using the temperature distribution T(z, t-1) at time t-1.

Wind

There is a wind mixing routine applied. The wind effects the surface of a lake and creates a shear stress coefficient, C_z, and a shear stress at the surface element. This shear stress at the surface gives a kinetic input of energy, KE equal to:

$$KE = \tau_o \cdot U \cdot A_{surface}$$

where U = the friction velocity and

$$\tau_o = C_z \cdot \rho_{air} \cdot W_z^2$$

where ρ_{air} = the density of the air.

C_z can be related to the fetch length based on Frouds scaling:

$$\frac{1}{\sqrt{C_z}} = \frac{1}{K} \cdot \ln\left(\frac{1}{0{,}011 \cdot C_z \cdot F^2}\right)$$

where $F = W_z/\sqrt{g\, z_m}$; z_m = the wind measurement height; W_z = the wind speed at z_m and K = von Karman constant equal to 0.41. The factor 0.011 is obtained from a relation between the shear velocity and the zero plane displacement.

Potential energy

The potential energy of the ith element above the jth element is defined as:

$$PE = g \cdot \Sigma A(i) \cdot \Delta z(\rho(j,t) - \rho(i,t)) \cdot D(i,t)$$

where A(i) = the area of the ith element; $\rho(i,t)$ =

the density of the ith element at time t; $\rho(j,t)$ = the density of the element immediately below the mixed layer at time t; Δz = thickness of an element; $D(i,t)$ = the distance between the ith and the jth element at time t, and g = the gravity acceleration.

The upper layer of the lake (epilimnion) is assumed to be mixed, and then is

$$PE = g \cdot \Delta\rho \cdot V_m \cdot \Delta z \cdot \frac{m}{2} = g \cdot \Delta\rho \cdot \bar{A} \cdot H^2$$

where m = number of ith elements in the mixed layer; $m \cdot \Delta z$ = depth of the mixed layer = H; V_m = volume of mixed layer = $A \cdot H$, where A = the average cross-sectional area and $\Delta\rho$ = the density difference between the mixed layer and the elements immediately below.

A critical ratio can now be calculated. The ratio KE/PE is equivalent to the Richardson number RI, developed by Cato and Philips (1969).

The mixing algorithm calculates the Richardson number.

$$f(RI) = \frac{KE}{PE}$$

If the calculated value of the Richardson number is less than KE/PE there will be applied mixing of the two elements of the water body.

Richardsons number is calculated as:

$$RI = g \cdot H \cdot \Delta\rho/(U^2 \cdot \rho_a)$$

where $\Delta\rho$ = the density difference and ρ_a = the density

The ratio KE/PE will by the wind mixing algorithm be evaluated against a function RI.

Different functions have been used, i.e.

$$f(RI) = 1,$$

$$f1(RI) = \frac{0.57\ RI\ (29.5 - \sqrt{RI})}{14.2 + RI} \quad \text{and}$$

$$f2(RI) = \frac{RI}{14.2 + RI}$$

When KE/PE is less than the calculated RI, the two elements considered will be mixed and a new temperature will arise in the ith element. Therefore the order in calculations is important.

The following procedure has been used:

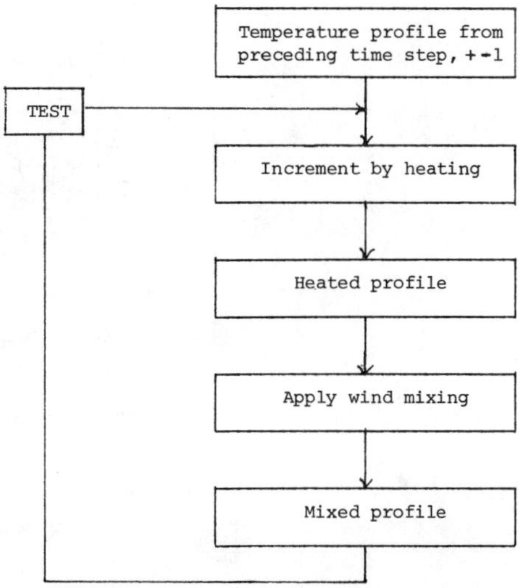

The test is done by calculation of the total outgoing heat flux and comparing the new total heat flux with the old one. Is the numeric value small, calculations of a new time step is done. Is the difference larger, the heating and mixing procedure is carried out again.

Three characteristic plots resulted from the model are shown in figures 3, 4 and 5. The first two are three dimensional plots giving the temperature as function of depth and date. The latter is an isoplet diagram.

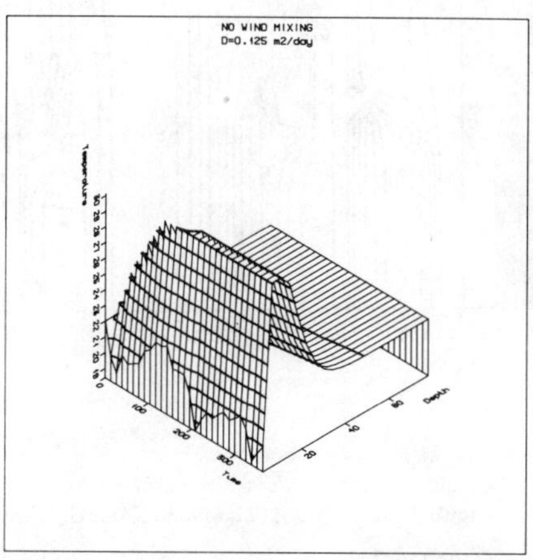

Figure 3. 3-dimensional plots. Temp = f(depth,date)

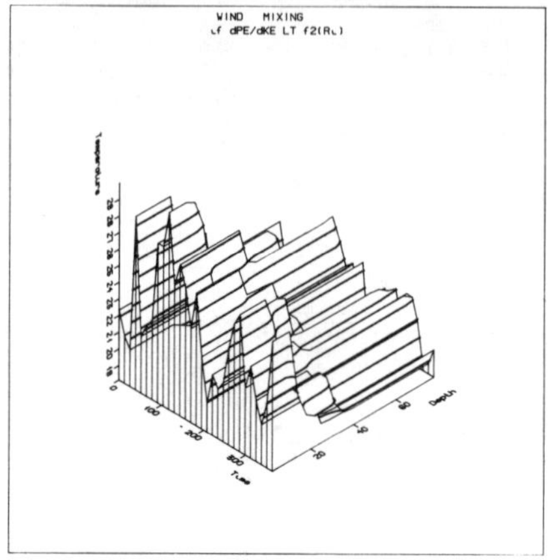

Figure 4. 3-dimensional plots. Temp = f(depth,date)

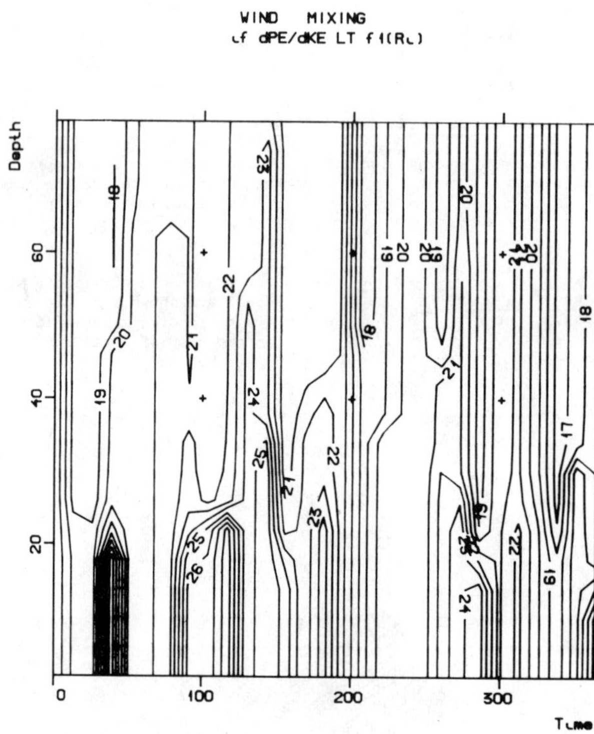

Figure 5. Isoplet diagram. (Temperature/depth/date)

CONCLUSIONS AND SUMMARY

The main features of a water quality model for the upper Nile system consisting of 10 submodels were presented.

The model seems directly applicable to solve environmental management problems, although it must be recommended to improve the reliability of the model by obtaining additional data for calibration and validation of the submodels.

The interrelation between the submodels and the main features of the eutrophication- and thermocline model were given. The first is based upon model experience with a number of lakes, the latter is based upon Ryan and Harleman (1973).

LITERATURE

Golterman, H.L. 1975. Physiological Limnology, Elsevier, Amsterdam.

Jørgensen, S.E. 1976. A eutrophication model for a lake. Ecol. Modelling 2: 147-165.

Jørgensen, S.E., Mejer, H. and Friis, M. 1978. Examination of a lake model. Ecol. Modelling 4: 253-278.

Jørgensen, S.E., Jørgensen, L.A., L. Kamp-Nielsen and H.F. Mejer 1981. Ecol. Modelling 13: 111-129.

Nielsen, L.K., Jørgensen, S.E., Jørgensen, L.A. 1983. In this Volume.

Ryan, P.J. and Harleman. D.R.F. 1973. An analytical and experimental study of transient cooling pond behavior. MIT Dept. of Civil Engl., R.M. Parsons Laboratory for Water Resources and Hydrodynamics Techn. Report, 161: 1.

USE OF MONTE CARLO WATER QUALITY SIMULATION TO EVALUATE RESERVOIR OPERATIONAL AND MANAGEMENT ALTERNATIVES[1]

Kent W. Thornton, Dennis E. Ford and Joseph L. Norton[2]

Abstract.--Natural variability or uncertainty must be incorporated in model simulations if environmentally sound, cost-effective management alternatives are to be evaluated. MANAGER, a reservoir water quality model, uses a Monte Carlo approach to incorporate variability or uncertainty in model output. This paper demonstrates how stochastic simulations permit statistical comparisons of management and operational alternatives such as selective versus bottom withdrawal and alternative downstream release rates.

INTRODUCTION

Although lakes and reservoirs are generally considered synonymous, several structural and functional differences between lakes and reservoirs make reservoirs unique ecological systems. Several of these differences include greater drainage areas, larger shoreline development ratios, shorter retention times, and pronounced longitudinal gradients in reservoirs (Baxter 1977, Baxter and Glaude 1980, Thornton et al 1981). One obvious difference is the capability to operate or regulate the withdrawal from a reservoir to meet water quality objectives both within the reservoir and in the reservoir releases. Reservoir releases can be made from different elevations (e.g. selective withdrawal) and at different rates to satisfy water quality objectives. This operational flexibility can be an effective management and/or restoration technique. In many instances, however, conflicting project purposes or objectives make reservoir operation and management difficult.

Mathematical water quality models are relatively inexpensive, highly flexible tools that can be used to evaluate eutrophication problems and different operation and management strategies to ameliorate these problems. Over the past decade, several one-dimensional reservoir water quality models have been developed that incorporate many of the ecological and operational differences between lakes and reservoirs. One model, the Water Quality for River-Reservoir Systems (WQRRS), (HEC 1978) has been extensively used by the U. S. Army Corps of Engineers. Numerous refinements and modifications have been made to WQRRS and its successor, CE-QUAL-R1 (Ford et al 1981), which have resulted in an essentially new model, MANAGER. One of the major attributes of MANAGER is its capability for performing stochastic simulations.

In the past, deterministic models have been used extensively to evaluate various management strategies. The procedure usually involved simulating each of the alternatives and comparing the results with some predetermined criteria such as downstream temperatures, release dissolved oxygen concentrations, or in-lake chlorophyll concentrations. Often overlooked in the final analysis, however, was the error or uncertainty associated with these predictions. This uncertainity is a result of experimental design, sampling variability, analytical error, and intrinsic variability in the inflow and calibration water quality data. This variability in the data must be incorporated in the predictions and the comparisons of management alternatives if reliable environmental and cost-effective decisions are to be formulated. Stochastic simulations permit the incorporation of this variability in simulation results.

Monte Carlo simulations have been the most frequently used approach in stochastic water quality simulations (Ford et al 1981, Hornberger & Spear 1980, Malone et al 1979, Scavia et al 1981, Tiwari and Hobbie 1976a). First order error analysis (Cornell 1972, Scavia et al 1981) and the Generation of Moment Equations Techniques (Malone et al 1979) have also been used to investigate prediction uncertainity. Stochastic simulations using a Monte Carlo approach permit both model coefficients and inflowing update constituents to be randomly varied according to prescribed statistical distributions. The Monte Carlo approach incorporates the variability associated with physical, chemical, and biological rate coefficients with the variability in stream water quality data in water quality simulations and provides statistical output.

This paper will discuss the use of Monte Carlo simulations in evaluating reservoir operational and management alternatives. It will consider 1) the reservoir water quality model, 2) characteristics of the random number generator, 3) convergence requirements, and 4) comparisons of operational alternatives.

---

[1]Paper presented at the Third International Conference on State-of-the-Art Ecological Modelling, Colorado State University, May 24-28, 1982.

[2]Kent W. Thornton, Dennis E. Ford, and Joseph L. Norton are Systems Ecologist, Hydrologist, and Aquatic Chemist, respectively, with Ford, Thornton, Norton & Associates, Ltd., Vicksburg, MS 39180 U.S.A.

MANAGER - A RESERVOIR WATER QUALITY MODEL

MANAGER is a comprehensive vertical one-dimensional (horizontally averaged) reservoir water quality model. It incorporates the major physical, chemical, and biological processes that occur within an impounded body of water (Ford et al 1981). MANAGER includes physical, chemical, and biological state variables such as temperature, dissolved and suspended solids, dissolved oxygen, nutrients such as nitrogen and phosphorus, phytoplankton, and higher trophic levels. The number of horizontal layers, inflowing tributaries, and selective withdrawal outlets is variable and limited only by computer size and the available simulation budget. For most reservoir simulations, however, these two factors do not present significant problems.

The model formulation consists of 22 highly coupled, nonlinear, partial differential equations. This implies that superposition does not hold and the mean trajectory of a stochastic simulation is not equivalent to the time history plot of a deterministic simulation (fig 1). Superimposing a distribution on a deterministic simulation, then, will not necessarily reproduce the output from a stochastic simulation.

In Monte Carlo simulations, coefficients and updates are selected at random from specified distributions during each computational step. Prescribed distributions incorporated in the model include four types: Gaussian, log-normal, rotated log-normal and uniform. At the end of each simulation sequence, typically one year, the initial conditions are reset and then another sequence is simulated. This process is repeated until the solution converges at some number (N) of simulations (fig 2). The output, then, consists of N trajectories, which are statistically analyzed. The results are expressed as mean and standard deviations or median and percentiles.

Monte Carlo simulations have been conducted on both the Honeywell 635 and CDC Cyber 176 computer systems. Graphics output has been obtained from the Boeing Computer System GCS plotting program and from an HP 7221 flatbed plotter driven by IBM Personal Computer software.

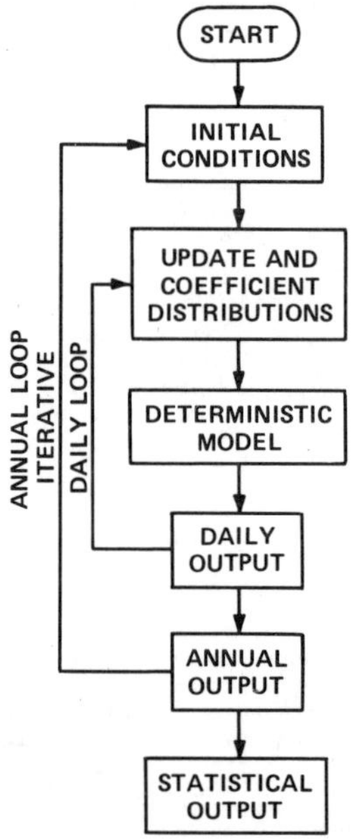

Figure 2.--Monte Carlo flow chart.

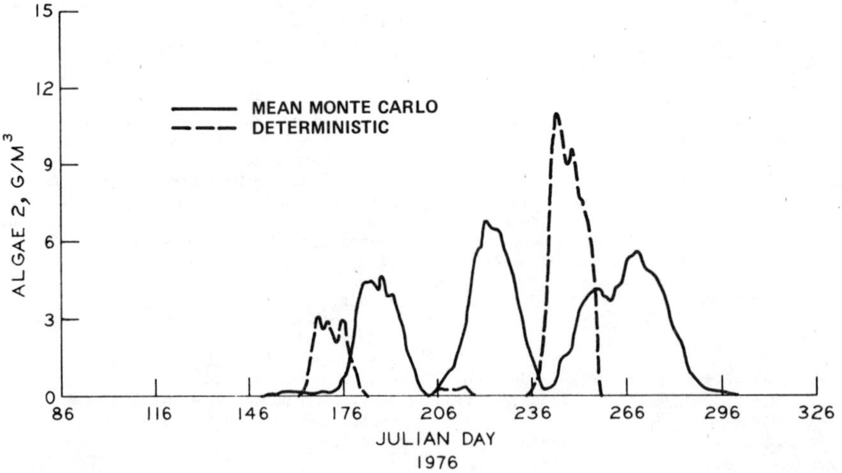

Figure 1.--Comparison of a deterministic simulation using mean coefficient values versus the mean value from a stochastic simulation.

RANDOM PROCEDURES

A critical component in Monte Carlo procedures is the random number generator used and its uniformity and independence. Random number generators are not truly random, in a theoretical sense, but rather follow a prescribed sequence of numbers once an initial seed value is specified. It is, therefore, important to assess the independence and uniformity of this random number sequence.

Several random number generators were subjected to tests of uniformity and independence. The standard test for uniformity was the Equidistance test with 10 cells. The Equidistance test was conducted 90 times with each test using a sample size of 50. The test statistic was the Chi-square with nine degrees of freedom.

The test for independence was a modified permutation test based on the distribution of pair wise, sequential permutations embedded in a given permutation of length N. Eleven tests of 30, 120, 600 and 3600 sample sizes were conducted on generated permutations of length 3, 4, 5, and 6, respectively. The test was the Chi-square with $\{((N(N-1)/2)-1)\}$ degrees of freedom.

The random number generator initially used on the Honeywell 635 computer tested uniform with the Equidistance test but failed the independence test for permutations greater than length four.

Two additional random number generators were then tested for uniformity and independence. These two random number generators were LLRANDOM and RAND, another random number generator available on the Honeywell system. Both these generators tested uniform and independent. RAND was selected for use since the model was already implemented on the Honeywell computer and would require fewer code changes.

CONVERGENCE

The random selection of values from distributions of rate coefficients and update variables produces the desired result of output variability for various state variables. For use in the decision process, however, it must be determined whether similar results would be obtained with additional iterations, i.e. if the solution converges toward a central mean and standard deviation. Running means, standard errors, medians, and interquartile ranges (QR) were computed and plotted for repetitions 1 through 10, 20, ... 100. Convergence was investigated based on biomass predictions for the phytoplankton and zooplankton state variables on four different simulation days. These days were selected to encompass both periods of rapid change and periods of slow change since convergence is a function of system dynamics.

Convergence occurred in all cases (fig 3). The median and QR converged earlier in the simulation than the mean and standard errors. Consideration should be given to reporting the median and QR since they are not as influenced by outliers as the mean and converge more rapidly.

COMPARISONS OF MANAGEMENT ALTERNATIVES

Major environmental impacts and construction and operation costs can be associated with different design, operation, and management alternatives for reservoirs. It becomes critical, therefore, that techniques used to compare these alternatives provide reliable estimates of expected reservoir water quality conditions resulting from these alternatives. The importance of stochastic simulations in evaluating these alternatives can be demonstrated with the following examples. While the examples focus on phytoplankton, other water quality variables such as temperature, dissolved oxygen, coliform bacteria, etc. were also factored into the decisions.

A comparison of selective withdrawal vs bottom withdrawal for reservoir operation was made using MANAGER in a deterministic mode (fig 4). It appeared that selective withdrawal resulted in three algae blooms compared to two blooms with bottom withdrawal. The magnitude and phasing of these blooms appeared to be different, but it was difficult to determine if one withdrawal scheme was more effective than another in reducing algae blooms. Using the Monte Carlo approach, it was clear there was no statistically significant difference between the two operational schemes since the confidence intervals completely overlapped (fig 5).

The size of the bottom flood gates was also evaluated using Monte Carlo or stochastic simulations. Two downstream release rates were proposed that required different gate sizes. The deterministic simulations indicated there might be slightly higher algae blooms in the pool with the lower release rate (fig 6). The Monte Carlo simulations indicated there would be significantly higher blooms in the pool (95% confidence level) with the lower release rate (fig 6).

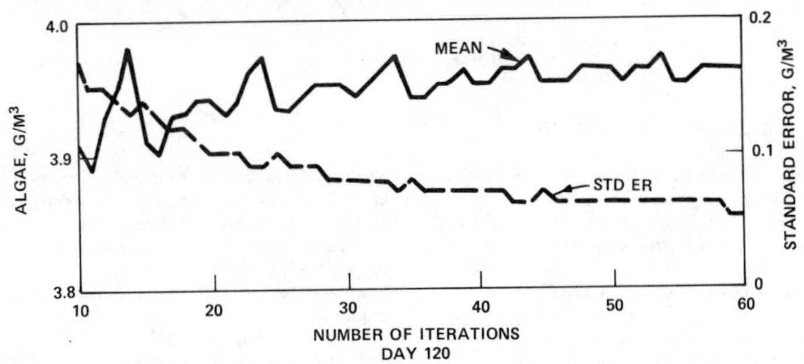

Figure 3.--Convergence of the cumulative mean and standard error of algal concentrations on Julian day 120 (from Ford et al 1981).

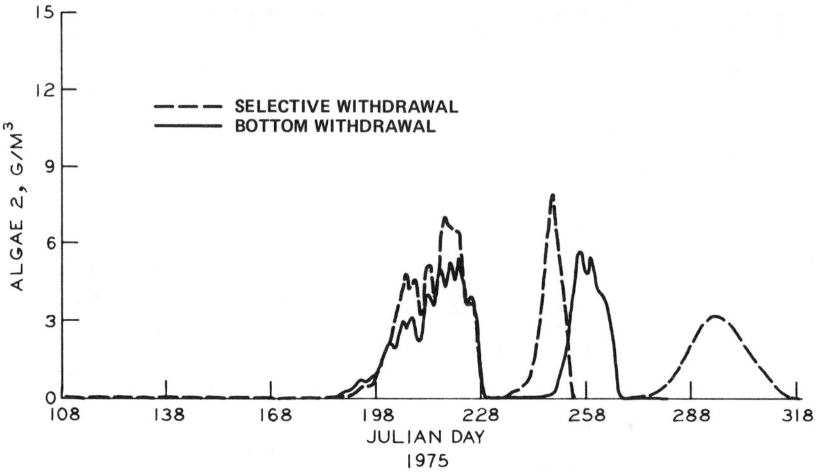

Figure 4.--Comparison of two withdrawal alternatives using a determisistic simulation.

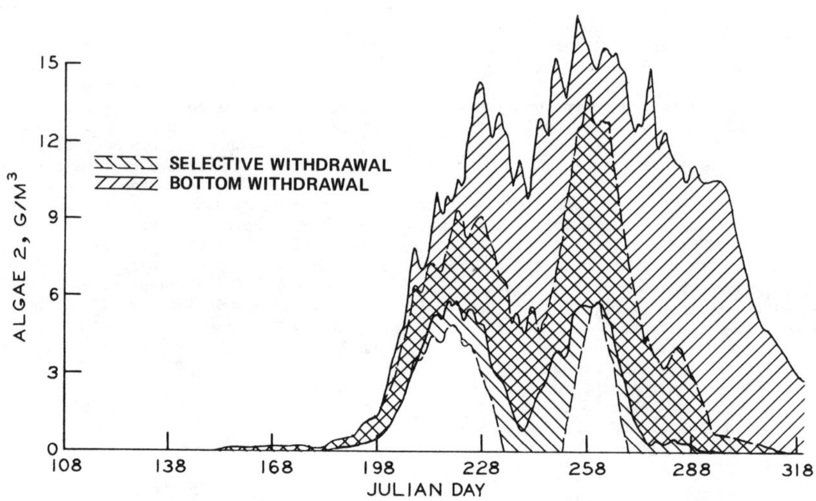

Figure 5.--Comparison of two withdrawal alternatives using a stochastic simulation.

DISCUSSION

Three major sources of error influence the output from many ecological models: 1) model formulation error, 2) parameter error, and 3) natural variability in the prototype system (O'Neill and Gardner 1979).

All models, by definition, are simplified representations of the actual prototypes. Ecological relationships and interactions, by necessity, must be simplified. Several hundred, or more, biological species exist in nearly all ecosystems and it is virtually impossible to obtain information on all of them. These species, then, must be aggregated into functional groupings or compartments for representation in the model. In nearly all cases, this aggregation is severe, e.g. two-four state variables to represent phytoplankton assemblages, a similar number of state variables to represent zooplankton assemblages. One of the benefits of this simplified representation is that the model can be manipulated for less cost and in a shorter time than experimentation on the prototype. One of the costs associated with this simplification is the number of assumptions and the errors introduced in simplifying natural systems. While it is assumed the major pathways for mass transfer and energy flow are represented, many of these pathways must be aggregated, which incorporates uncertainty about specific transfers or interactions.

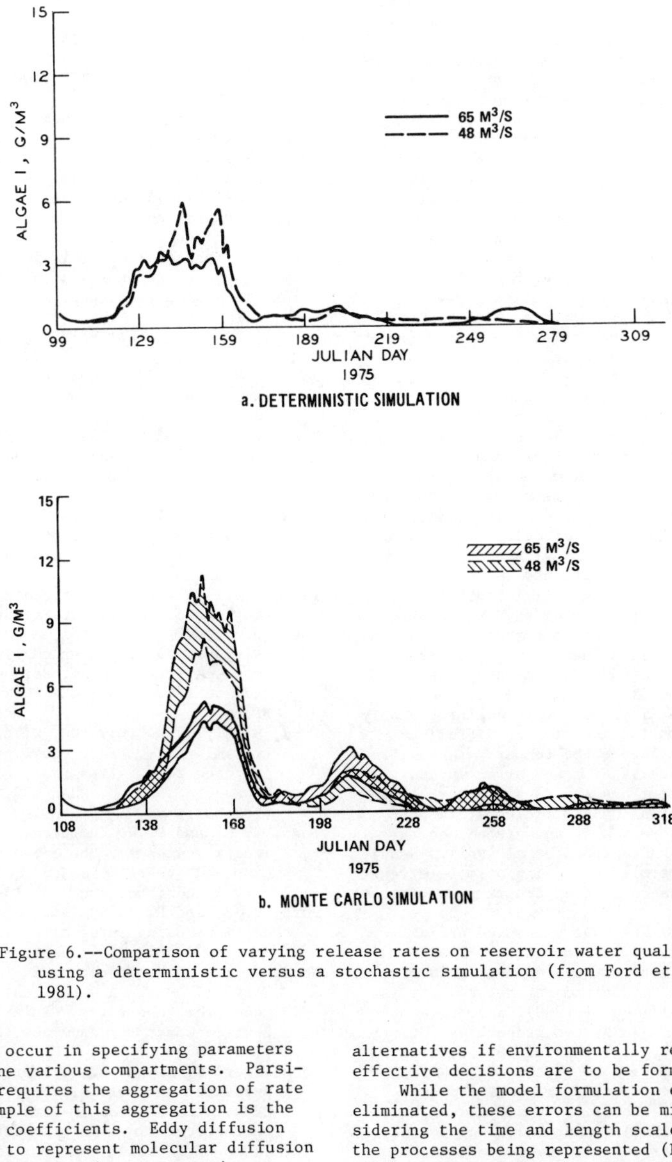

Figure 6.--Comparison of varying release rates on reservoir water quality using a deterministic versus a stochastic simulation (from Ford et al 1981).

Similar problems occur in specifying parameters or coefficients for the various compartments. Parsimony among processes requires the aggregation of rate coefficients. An example of this aggregation is the use of eddy diffusion coefficients. Eddy diffusion coefficients are used to represent molecular diffusion and turbulent mixing processes in many aquatic ecosystem models. This aggregation of processes and their representation with a single parsimonious coefficient also introduces error in the simulation.

Estimates of natural variability are functions of sampling and experimental design, sampling error, analytical error, and intrinsic variability. Additional variability or uncertainty may result from ambiguous objectives for the monitoring program, lack of knowledge about the system, or lack of funding and/or manpower. All of these sources contribute to uncertainty in the monitoring program and, therefore, in the data.

Consideration of these sources of uncertainty or variability must be incorporated both in the model predictions and in the comparison of management alternatives if environmentally reliable and cost-effective decisions are to be formulated.

While the model formulation errors can not be eliminated, these errors can be minimized by considering the time and length scales characteristic of the processes being represented (Ford and Thornton 1979). Time and length scale considerations are well established in fluid mechanics and hydrodynamics (Tennekes and Lumley 1972) but are just beginning to be recognized in the ecological disciplines. In general, an hydrodynamic equation or study that incorporates turbulence will have a length (or time) scale that characterizes its applicability. Erroneous results are possible if these equations are applied outside their characteristic range. In addition to characterize length scales, characteristic time scales are also implicit in all mathematical formulations. The characterization or conceptualization of any process, such as turbulent mixing, redox reactions, or phytoplankton production, constrains the applicability, validity, and realizations of predictions from these formulations. The use of a daily time scale, for example, would not permit the simulation of Langmuir cell formation (Faller 1978). An annual cosine relationship for solar radiation would not be sufficient to describe diurnal changes in primary productivity.

It is also important to recognize that processes conceptualized for a small time scale cannot be simply integrated over time to describe large time scale processes. The driving variables or forcing functions are usually quite different at grossly different time scales. This also has direct implications for model dimensionality. Algorithms formulated for a one-dimensional model may be inappropriate for inclusion in a multidimensional model. Consideration of appropriate time and length scales for various model formulations describing the interactions among the hydrodynamics, chemistry, and biology can reduce model error.

Parameter and coefficient error can be reduced by investigating their impact on model output through sensitivity analyses (Thornton et al 1979). Sensitive coefficients that significantly influence model output can be identified and additional effort made to accurately estimate these coefficients. Model formulations may include coefficients that cannot be measured in the field or determined easily in the laboratory. If the model output is sensitive to small changes in these coefficients, alternative formulations should be developed. Eddy diffusivities, for example, can range over three orders of magnitude depending on the sampling interval of the data and the procedure used to calculate the diffusion coefficients (Ford and Stefan 1980). Sensitivity analyses with WQRRS (HEC 1978) using a 10 percent change in the eddy diffusion coefficient resulted in less than a 10 percent change in temperature but over a 450 percent change in phytoplankton concentrations (Thornton et al 1979). In general, many water quality models are calibrated only on temperature so a small change in the diffusion coefficient may not influence the temperature calibration but could introduce significant error in the simulation of other water quality constituents. This sensitivity to eddy diffusion coefficients, in part, resulted in the incorporation of an integral energy thermal submodel in MANAGER (Ford et al 1981). Sensitivity analyses also permit the design of more effective and inclusive data collection programs complementing model applications.

Monte Carlo approaches have been used primarily as a research technique to gain additional insight into system and model functioning. Monte Carlo simulations have been used to analyze parameter error in nonlinear models (O'Neill et al 1980), investigate parameter distributions and system properties (Tiwari and Hobbie 1976a, 1976b, Tiwari et al 1978), compare variance estimates from various simulation approaches (Scavia et al 1981) and identify critical areas of uncertainty in a system that require additional research (Hornberger and Spear 1980). Monte Carlo approaches offer an excellent opportunity to evaluate management alternatives and address applied water quality problems (e.g. Ford et al 1978, Ford et al 1979).

The use of Monte Carlo simulations to compare selective withdrawal and release rate alternatives cited earlier were extremely important in arriving at environmentally sound, cost-effective decisions on the proposed reservoir. Since there was no statistically significant advantage of selective withdrawal versus bottom withdrawal in this project, an expenditure of approximately $5 million to include selective withdrawal was eliminated. Considering the environmental impact of the two release rates on algae and other water quality variables, however, it was decided the additional money for the larger gates was warranted since there would be a significant improvement in both in-lake and downstream water quality. Stochastic simulations have also been used to evaluate various management alternatives for reducing salinity in the Colorado River system (Malone et al 1979).

CONCLUSIONS

The acceptance of mathematical water quality modeling as a management tool by lake and reservoir managers and decision makers has varied from overwhelming acceptance to complete lack of trust. One reason for this is that modelers have not been able to supply managers and decision makers with an estimate of the confidence that can be placed on the model results. Monte Carlo simulations are one way of estimating the variance and reliability of model simulations when evaluating management alternatives.

LITERATURE CITED

Baxter, R. M. 1977. Environmental effects of dams and impoundments. Ann. Rev. Ecol. Syst. 8:255-283.

Baxter, R. M. and P. Glaude. 1980. Environmental effects of dams and impoundments in Canada: Experience and Prospects. Can. Bull. Fish. Aq. Sci. Bull. No. 205.

Cornell, C. A. 1972. First order analysis of model and parameter uncertainty. In Proc. Int. Symp. on Uncertainties in Hydrologic and Wat. Resour. Sys. 3:1245-1274.

Faller, A. J. 1978. Experiments with controlled Langmuir circulations. Science 201:618-620.

Ford, D. E., K. W. Thornton, A. S. Lessem, and C. Stirgus. 1978. Water quality evaluation of proposed Trexler Lake, Jordon Creek, Pennsylvania. 175 p. USAE WES TR Y-78-10. Vicksburg, MS.

Ford, D. E., K. W. Thornton, W. B. Ford III, A. S. Lessem. 1979. Water quality evaluations of proposed Twin Valley Lake, Wild Rice River, Minnesota. 200 p. USAE WES TR EL-79-5. Vicksburg, MS.

Ford, D. E. and K. W. Thornton. 1979. Time and length scales for the one-dimensional assumption and its relation to ecological models. Wat. Resour. Research 15:113-120.

Ford, D. E. and H. G. Stefan. 1980. Thermal predictions using integral energy model. ASCE Jour. Hyd. Div. 106:39-55.

Ford, D. E., K. W. Thornton, A. S. Lessem, and J. L. Norton. 1981. A water quality management model for reservoirs. In ASCE Proc. Symp. on Surface Water Impoundments, H. G. Stefan ed. 1:624-633.

Hornberger, G. M. and R. C. Spear. 1980. Eutrophication in Peel Inlet-I. The problem-defining behavior and a mathematical model for the phosphorus scenario. Wat. Research 14:29-42.

Hydrologic Engineering Center. 1978. Water quality for river-reservoir systems. Draft Manual. U.S. Corps of Engineers, Davis, CA.

Malone, R. F., D. S. Bowles, W. J. Grenney, and M. P. Windham. 1979. Stochastic analysis of water quality. 75p. Utah State University. UWRL/Q-79/01.

O'Neill, R. V. and R. H. Gardner. 1979. Sources of uncertainity in ecological models. p. 447-463. In Methods in systems modeling and simulation. B. P. Zeigler, M. S. Elzas, G. J. Klir, T. I. Oren eds. North-Holland Publishing Company.

Scavia, D., W. F. Powers, R. P. Canale and J. L. Moody. 1981. Comparison of first-order error analysis and Monte Carlo simulations in time-dependent lake eutrophication models. Wat. Resour. Research 17:1051-1059.

Tennekes, H. and J. L. Lumley. 1972. A first course in turbulence. 300 p. The MIT Press. Cambridge, MA.

Thornton, K. W., A. S. Lessem, D. E. Ford, and C. A. Stirgus. 1979. Improving ecological simulation through sensitivity analysis. Simulation 32:155-166.

Thornton, K. W., R. H. Kennedy, J. H. Carroll, W. W. Walker, R. C. Gunkel, and S. Ashby. 1981. Reservoir sedimentation and water quality - an heuristic model. In. ASCE Proc. Symp. on Surface Water Impoundments. H. G. Stefan, ed. 1:654-661.

Tiwari, J. L. and J. E. Hobbie. 1976a. Random differential equations as models of ecosystems: Monte Carlo simulation approach. Math Biosci. 28:25-44.

Tiwari, J. L. and J. E. Hobbie. 1976b. Random differential equations as models of ecosystems - II. Initial conditions and parameter specifications in terms of maximum entropy distributions. Math Biosci. 31:37-53.

Tiwari, J. L., J. E. Hobbie, J. P. Reed, D. W. Stanley, and M. C. Miller. 1978. Some stochastic differential equation models of an aquatic ecosystem. Ecol. Modeling 4:3-27.

ACKNOWLEDGEMENTS

Cpt. David Davis, US Army, U.S. Military Academy, West Point, N.Y. tested the random number generators and is gratefully acknowledged for his efforts. This work was supported, in part, by the USAE EWQOS program and Office, Chief of Engineers.

RESERVOIR DESTRATIFICATION BY WITHDRAWAL FROM THE HYPOLIMNION[1]

Craig L. Caupp[2], William J. Grenney[3]

This paper reports on the development and applications of a computer model to evaluate temperature destratification in a small reservoir by selective withdrawal. A reservoir's temperature model was developed to calculate the temperature profile and the stability index of a small reservoir. The model was developed to make the most efficient use of available meteorological and flow data. The model was applied to estimate the changes in reservoir stratification that would result from various withdrawal patterns. The model results indicate thermal stability could be reduced by 75% during mid-summer by selective withdrawal.

INTRODUCTION

Energy development and rapid population growth in the Western United States have increased the demand for a dependable supply of high quality water. In many areas of the West, reservoirs are necessary to store spring runoff for use later in the summer and fall. Water quantity is insured by reservoir storage, but the original water quality is not always maintained. For example in one study (MacKenthun 1970), 62% of water supplies obtained from reservoirs had water quality problems associated with algae. In another survey (Symons 1971) 21% had taste and odor problems with reservoir supplies. Problems with iron, maganese and sulfides were reported in reservoirs where anoxic conditions existed (Symons 1971). Thus reservoir planners need to plan not only for adequate storage but also consider steps for reducing water quality problems associated with storage.

Water quality problems associated with thermal stratification can be reduced or eliminated by artificially mixing a reservoir. Water quality has been observed to be improved by destratification in 87% of 37 cases surveyed in 1971 (Symons 1971). Not only were anoxic conditions prevented, but the troublesome blue-green algal populations were reduced. To explain the changes leading to the reduction of blue-green algae, Shapiro (1973) conducted experiments of pH and alkalinity shifts on laboratory populations of blue-green algae. Shapiro (1979) summarized these findings by explaining that the mixing of the water reintroduces oxygen into the system and reverses the chemical process to favor the more desireable green algae and diatoms over the blue-green algae. These new conditions include lower pH, higher CO concentrations, and less nutrient stress.

Various methods exist for artificially mixing reservoirs. Lorenzen and Fast (1977) discussed the design and operation of the two most common methods: mechanical mixing and diffused air mixing. These methods have both been successfully used to improve water quality (Steichen 1979 and Symons 1971). Steichen (1979) found that the pH of the water dropped, alkalinity decreased, oxygen concentrations increased and blue-green algae decreased in number when thermal stratification was eliminated. Thus thermal destratification has become an accepted means for preventing water quality deterioration during storage. However, these methods have the major disadvantage of high energy and operating costs.

The withdrawal of hypolimnion water has been observed to influence the location and shape of the thermocline in even large reservoirs (Frey 1963). For smaller reservoirs with short detention times it is reasonable to expect significant destratification if withdraws are timed properly. This paper reports on the use of a computer temperature model to examine the potential for reservoir destratification by selective hypolimnetic withdrawals. Destratification methods other than those now commonly used (mechanical mixing and air diffusion) are desirable to save energy and maintenance costs. The computer model is used to predict temperature profiles with different rates of water withdrawal from the hypolimnion. Thus it is of use for design purposes to determine the size and location of

---

[1]Paper presented at the Third International Conference on State in Ecological Modelling, Colorado State University, May 24-28, 1982.

[2]Craig L. Caupp, Graduate Student of Civil and Environmental Engineering, Utah State University, Logan, Utah, U.S.A.

[3]William J. Grenney, Professor of Civil and Environmental Engineering, Utah State University, Logan, Utah, U.S.A.

withdrawal ports. It may also be used to determine operating criteria for the reservoir to miminize storage related water quality deterioration.

DESCRIPTION OF TEMPERATURE MODEL

The temperature model in this study is similar to other state of the art temperature models now in use (HEC 1978, EPA 1978 and Grenney 1981), but this model is structured to require less extensive data input. The reservoir is divided into horizontal slices (Fig. 1) and each slice is assumed to be completely mixed.

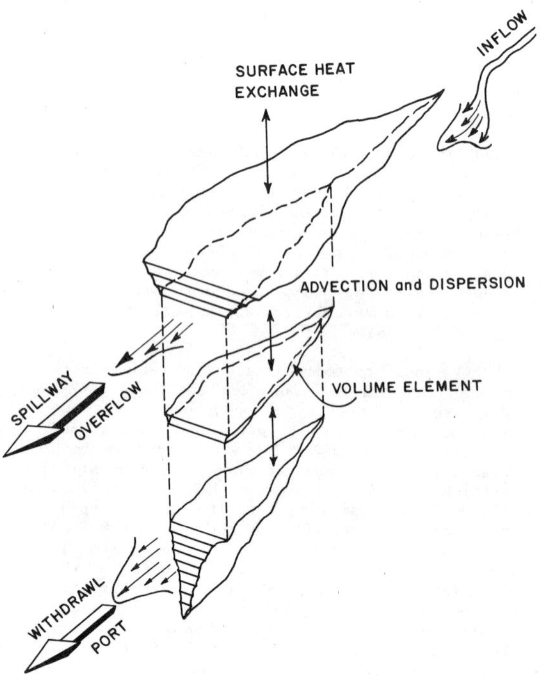

Figure 1.--Schematic diagram of reservoir configuration.

Flows entering the reservoir sink to the layers of equal density. If the density of the inflow is outside the range of densities within the reservoir, the entry is established at either the surface or bottom. Once the entry level is determined the inflow is distributed in the convectively mixed zone (Krenel 1980). Withdrawal may flow over the spillway or through as many as 10 ports located in the dam. Local velocities associated with the flows are distributed within the withdrawal zone (Krenel 1980). Heat is distributed vertically throughout the reservoir by advection and dispersion. The heat flux at the surface is represented by the heat budget equation:

$$H = H_a + H_s - H_{br} - H_e + H_c \quad (1)$$

where

H = net surface heat flux (Kcal/m^2/day),
H_s = short wave radiation entering the water (Kcal/m^2/day)
H_a = incoming longwave radiation from the atmosphere entering the water (Kcal/m^2/day),
H_{br} = back radiation emitted by the water (Kcal/m^2/day),
H_e energy utilized by evaporation (Kcal/m^2/day), and
H_c = energy conducted between the air and the water at the surface (Kcal/m^2/day).

The interelement transport of mass and conservation of heat is represented by the following partial differential equation:

$$\frac{\partial T}{\partial t} = \Delta z Q_z \frac{\partial T}{\partial z} + \Delta z A_z D_z \frac{\partial^2 T}{\partial z^2} + Q_i T_i - Q_o T + \frac{A_h H}{\rho c} - T\frac{\partial v}{\partial t} \quad (2)$$

where

T = temperature in degrees Celsius
V = volume of the fluid element in m^3
t = time in seconds
z = space coordinate in meters (vertical for the reservoir and horizontal for the stream)
Q_z = interelement flow in m^3/sec
A_z = element surface area normal to the direction of flow in m^2
D_z = effective diffusion coefficient in m^2/sec
Q_i = lateral inflow in m^3/sec
T_i = inflow water temperature in degrees Celsius
Q_o = lateral outflow in m^3/sec
A_h = element surface area in m^2
H = external heat sources and sinks in kcal/m^2/sec
ρ = water density in kg/m^3
c = specific heat of water in kcal/kg/°C

An implicit method is used to find the solution to the above partial differential equation.

One disadvantage of most temperature models is the extensive meterlogical data base required to run the model. This model avoids this difficulty by superimposing typical short term variations (3 hour) on long term trends (week, month, etc). Daily data were obtained and evaluated for March through August over a four year period. Average monthly trends and typical diurnal variations were calculated. For example, the average daily air temperature trend is assumed to either increase or decrease linearly over a month. The air temperature at the end and beginning of the period are entered and average daily values are obtained by linear interpolation. Three hour variation about the mean at the daily level are handled by using a typical temperature distribution function. This relationship was obtained by plotting the instantenous distribution of temperature over the day. A sample distribution is shown in Figure 2.

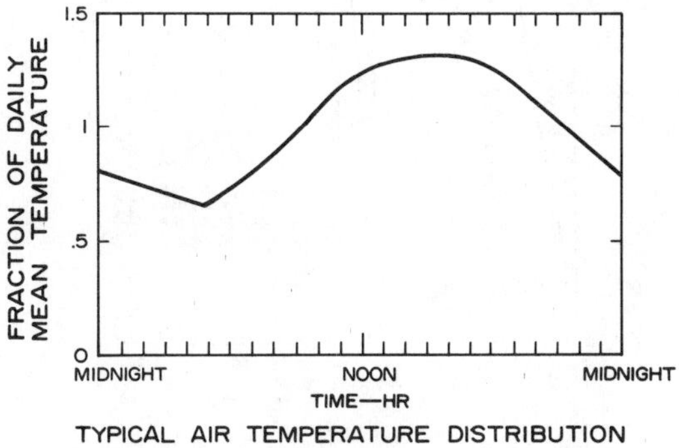

Figure 2.--Sample relationship used to calculate instanteous meterlogical value from mean daily value.

The length of time covered by each input period is dependent upon data availability and uniformity. A one month period was selected for this study. A sample of data necessary for one month is shown in Table 1.

Table 1.--Summary of Monthly Data for May

		High	Normal	Low
Inflow	cfs	2097	1510	409
Inflow temp	°C		12.9	
Withdraw	cfs	0	0	0
Solar rad.	Kcal/m^2day		6327	
Sun rise			4:30	
Sun set			19:30	
Atm. press.	in		25.1	

Time of Day							
2:00	5:00	8:00	11:00	14:00	17:00	20:00	23:00
Air temp. °C							
11.6	10.0	14.9	20.0	22.4	22.3	17.1	13.9
Cloud Cover tenth							
4	4	4	5	5	6	6	5
Wind Speed mph							
7.6	7.4	8.5	7.8	11.2	11.6	7.4	6.7
Relative Humidity %							
47	51	42	30	24	24	33	40

This method produces results that compare favorably with actual reservoir temperature data for this area. For example the model predicted a surface temperature of 20.2°C and a bottom temperature of 13.1°C in mid June; the actual surface and bottom temperatures recorded for a reservoir in the same geographic region were 21°C and 13.3°C respectively (Bureau of Land Management 1980). The flexibility of the model does not limit its usefulness to predict average temperature profiles.

STABILITY INDEX

Resistance to mixing is included in the models stability index measured in kilowatt hours (kWhr). The index represents the energy required to lift the center of gravity of a water column under stratified conditions to the height of the center of gravity under isothermal conditions. The energy input required for destratification is then:

$$E(KWhr) = 2.724 \times 10\text{-}6 \, (My1\text{-}My2) \qquad (3)$$

My1 = mass up to center of gravity for isothermal conditions (kilograms)
My2 = mass up to center of gravity for stratified conditions (kilograms).

The masses are calculated by numerical integration. This equation for calculating stability is currently being used to size equipment and determine the energy required to destratify reservoirs. The index is discussed in more detail in Krenkel (1980). This stability index has also been used in other reports (Steichen 1979) examining stability. However comparison between reports is not feasible because the stability index measures the energy required for the specific reservoir under study and is not a relative measure of stability. Volume and area differences between two reservoirs lead to different values of stability even if the temperature profiles are the same. An index value of close to zero indicates the absence of a temperature gradient. Increasing values of the stability index indicate the presence of stronger thermal stratification and thus increased energy for mixing.

DESCRIPTION OF PROPOSED RESERVOIR

A reservoir is proposed to be located on a tributary of the Colorado River in the upper Colorado River Basin. The reservoir will have a volume of about 1702 Ha-m, a surface area of 249 Ha, a maximum depth of 18 m and a typical residence time of 10 to 30 days. Flow out of the reservoir will either be over the spillway or through a single withdraw port located 16m below the spillway of the dam. Average flows into the reservoir are given in Table 2 for the period March through August.

Table 2.--Average Monthly Inflows to Reservoir

Month	Average cfs	Inflow m^3/s
March	448	13
April	515	15
May	1510	43
June	1414	40
July	498	14
August	350	10

The minimum downstream flow requirement is 2.2 m^3/s. Total phorphous input is fairly high, with an average value of 1 mg/l. Based on the above nutrient and hydraulic loading, the Vollenweider nutrient loading model (Vollenweider 1976) indicates that the reservoir would likely be eutrophic.

WITHDRAWAL PLANS FOR DESTRATIFICATION

Four operating alternatives are examined to determine if the reservoir will destratify by selective withdrawal.

Plan one: All water flow is released over the spillway except when this flow falls below the minimum downstream requirement (2.2 m^3/s). Then water is withdrawn from the port at 10 meters in order to augment the downstream flow.

Plan two: The majority of the outflow is released from the port at 10 meters. Water spills only during times of extremely high inflows to the reservoir.

Plan three: The flow is split with half over the spillway and half through the port at 10 meters.

Plan four: The majority of the flow is released through the port at 10 meters until the end of June. At this time most of the flow is then directed over the spillway.

ANALYSIS OF MODEL RESULTS

For plan one, the model predicts that the reservoir will stratify by the end of May and stay stratified for the summer months. The temperature profile for mid June (Fig. 3) is typical of profiles for stratified reservoirs.

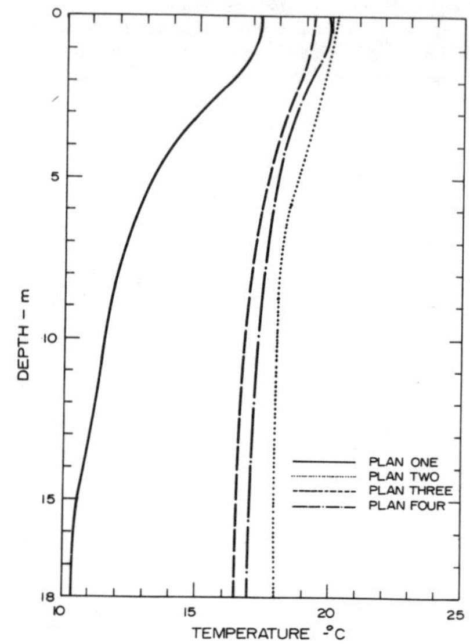

Figure 3.--Reservoir temperature profile in mid June for alternative withdrawal schemes.

The profile shows a thermocline between 2 meters and 4 meters. There is a difference of 7°C between the surface and hypolimnion temperatures. The temperature profile for mid July (Fig. 4) also indicates stratified conditions.

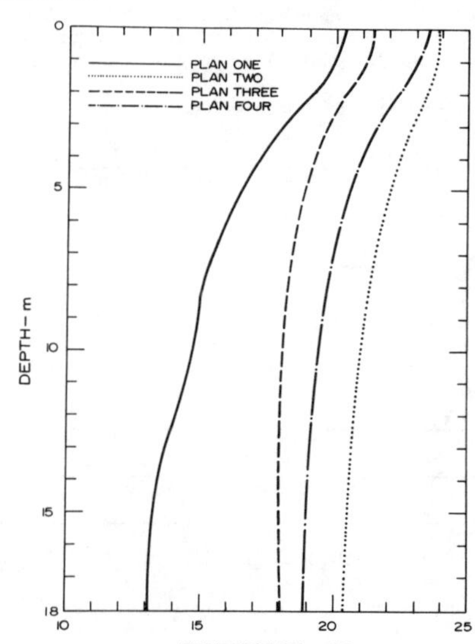

Figure 4.--Reservoir temperature profile in mid July for alternative withdrawal schemes

A thermocline and substantial temperature difference between surface and hypolimnion are present. The model predicts a surface temperature of 20.2°C and a hypolimnion temperature of 13°C. These predicted temperatures are similar to temperatures actually measured in other reservoirs in the general geographical region. The thermal stratification is also indicated by the stability index which is shown as a function of time in Figure 5.

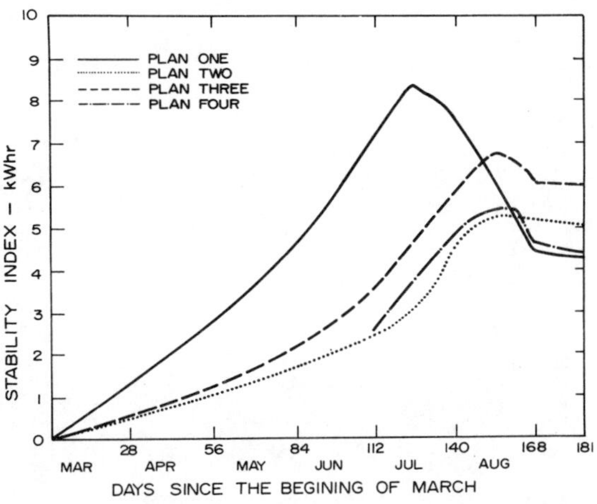

Figure 5.--Stability index values for alternative withdrawal schemes.

Stability for this alternative increases from less than 1.36 kWhr in May to 8.2 kWhr in mid July, when the thermal profile (Fig. 4) also indicates strong stratification.

In the case of plan two, the model predicts that stability is reduced by hypolimnetic withdrawal. The temperature profile for plan two for mid June (Fig. 3) shows only 2°C of difference between the surface and bottom temperatures and no thermocline. By the middle of June, stability for plan two is only 25% of the stability predicted under plan 1. During the month of July this percentage increases to 64%. By the end of August, the predicted stability for plan two exceeds the stability for plan one. This steady increase under plan two and rapid drop in stability under plan one is due to heat exchange at the surface. Solar heating during July raises the surface temperature under plan two to 23.8°C, while under plan one the maximum surface temperature is 20.2°C. Under plan one heat in the surface water layers is lost over the spillway.

The stability of plan three is between that of plan one and plan two. Stability for plan three is less than the stability of plan one until mid July when the stability for plan one drop below those of plan three. During August the stability for plan three continues to increase while decreasing for plan one. The stability of plan three is greater than that of plan two for all time periods. There is no advantage to simply splitting the flow between the spillway and withdraw port.

Plan four (a combination of plan one and plan three) is designed to take advantage of the heat loss accomplished by plan one and early destratification is accomplished by plan three. Until mid June little water is allowed over the spillway. For the rest of the summer warm surface water is taken off the top at the spillway. This plan reduces late summer stability to 4.3 kWhr. This is similar to the value obtained under plan one (4.2 kWhr) and is .5 kWhr less than the value predicted for plan two. Surface temperatures are predicted to be 20.5°C, 3.6°C less than under plan two.

CONCLUSIONS

This temperature model can be used as a design and operation tool. Results of the temperature model indicate that designing a withdraw port capable of handling more than the minimum demand flow of 2.2 m^3/s will provide the capability to decrease thermal stability. Plan 4 is the most effective opperating plan for reducing thermal stratification.

LITERATURE CITED

Bureau of Land Management. 1980. Draft Environmental Impact Statement of the White River Dam Project. Department of Interior. Richfield, Utah.

Environmental Protection Agency. 1978. Rates, constants, and kinetics formulations in surface water quality modeling. Environmental Research Laboratory, Athens, Georgia. EPA - 600/3-78-105.

Frey, D. G. ed. 1963. Limnology in North America. University of Wisconsin Press. Madison.

Grenney, W. J., A. K. Kraszewski. 1981. Description and Application of the Stream Simulation and Assessment Version IV. FWS/OBS - 81/46.

HEC. 1978. Water Quality for River-Reservoir Systems. Hydrologic Engineering Center, U.S. Army Cops of Engineers. Davis, California. 279 p.

Krenkel, P. and V. Novotny. 1980. Water Quality Management. Academic Press. New York.

Lorenzen, M. and A. Fast. 1977. A Guide to Aeration/Circulation Techniques for Lake Management. U.S. Environmental Protection Agency. Corvallis, Oregon. EPA - 600/3-77-004.

MacKenthun, K. M. and L. E. Kenp. 1970. Biological Problems Encountered in Water Supplies. Journal AWWA, 62 (8):520.

Shapiro, J. 1973. Blue-Green Algae: Why They Become Dominant. Science Volume 178, pp. 382-384.

Shapiro, Joseph. 1979. The Need for Biology in Lake Restoration. Proceedings of a National Conference. U.S. Environmental Protection Agency, Washington D.C. EPA 440/5-79-001.

Steichen, J. M., J. E. Gurton, and C. E. Rice. 1979. The Effects of Lake Destratification on Water Quality. Journal AWWA 71 (4):219.

Symons, J. M. ETAL. 1971. Committe Report: Artificial Destratification in Reservoirs. Journal AWWA 63 (9):597.

Vollenweider, R. A. 1976. Advances in Defining Critical Loading Levels for Phosphorous in Lake Eutrophication. Mem. Ist. Ital Idrobio. 33:53-83.

UNCERTAINTY ANALYSIS IN ECOLOGICAL MODELLING[1]

Efraim Halfon[2]

Abstract--A short review of recent research in uncertainty analysis applied to ecological problems is presented. Some results obtained from the analysis of a contaminant model are discussed. The analysis shows that uncertainty analysis can identify areas that should be the focus of extended field research. Out of 96 parameters associated with a fate model of Mirex, a pesticide, in Lake Ontario only a few, mainly associated with a description of sediment dynamics, were so important to influence the reliability of the predictions. The importance of the identified parameters was independently confirmed by the expert opinion of environmental chemists. Thus the methodology employed here, Monte Carlo simulations, is appropriate for the task of analyzing a model for prediction uncertainty.

INTRODUCTION

The future behaviour of most ecological systems can at best be predicted with uncertainty. The quantification of this uncertainty has been intensely investigated in the ecological modelling literature in the past several years (O'Neill, 1973; Argentesi and Olivi, 1976; Beck _et al._, 1979; Reckhow, 1979; O'Neill _et al._, 1980). Methods developed to compute uncertainty are several and diverse: Monte Carlo simulations, covariance analysis, maximum likelihood, and error analysis. Table 1 relates each method with the authors who have employed them. Other methods have been presented in the control literature even if they have not been used in ecological modelling yet. Schriber and Andrews (1981) divide these methods into five broad classes, namely: (1) the method of replications; (2) batch means methods; (3) regenerative methods; (4) time series methods, and (5) spectral analysis methods. All methods quantify the uncertainty by computing confidence limits of the simulations. Each method has its own domain of validity. For example, the method of replication, batch mean methods and regeneration methods are designed to circumvent any serial correlation structure, whereas time series and spectral analysis methods allow for a possibility of a serial correlation structure and even exploit it (Schriber and Andrews, 1981). In analyzing an ecological model, several or even all these methods could be used, however, time and budget constraints limit the method choice. Table 1 also shows the domain and the possible applications of each technique.

In reviewing the literature, I was impressed by the variety of methods used by ecological modelers to analyze these models. In this paper I shall describe the general classes of methods used to compute confidence limits, review some ecological applications and present some of my most recent results in the computation of uncertainy in a model describing the fate of a toxic chemical, Mirex, in Lake Ontario.

[1] Paper presented at the Third International Conference on State-of-the-Art In Ecological Modelling, Colorado State University, May 24 to 28, 1982.

[2] Efraim Halfon is a research scientist at the National Water Research Institute, Canada Centre for Inland Waters, Burlington, Ontario, Canada.

UNCERTAIN DYNAMIC SYSTEMS

As much as we, as ecologists, understand natural ecosystems, we still lack the comprehensive understanding to make completely reliable predictions. We know that only one future behaviour will exist, but for us this future is uncertain, i.e. indeterminate and indefinite; thus our predictions are not reliable, untrustworthy and not clearly defined, i.e. uncertain. By using systems methods we try to make determinate, reliable and clearly defined statements about uncertain ecosystem behaviour. Two sets of uncertainty models can be considered; probabilistic models and set theoretic models. Three categories of models are: _Bayesian_, _Fisher_ and _unknown-but-bounded_ (Schweppe, 1973).

In the _Bayesian_, uncertainty is modeled by random variables and/or stochastic processes with either completely specified probability distributions or completely specified first and second moments (means and variances). In the _Fisher_, uncertainty is modelled by a combination of Bayesian models and completely unknown quantities. In the _unknown-but-bounded_, uncertainty is modeled by set theoretic models; the resulting quantities are called unknown-but-bounded processes. Table 2 shows the three basic categories. Note that there are three types of Fisher models. The analysis of uncertainty in ecological models is usually done with Bayesian models. A complete analysis of the methodology is given in Schweppe (1973).

Most ecological models are in state space, a system of linear or non-linear, ordinary or partial, differential or difference, equations is used to represent the system of interest:

$$\frac{d\underset{\sim}{x}}{dt} = f\,(\underset{\sim}{x},\underset{\sim}{u},\underset{\sim}{p},t)$$

$$\underset{\sim}{y} = g\,(\underset{\sim}{x},\ \underset{\sim}{u},t)$$

Table 1.--Methods to compute uncertainty in water quality models, some of the most recent users, and domain of application of each technique

Method	Users	Domain
Monte Carlo simulations	K. Fedra *et al.*, 1981 E. Halfon, 1982a,b R.C. Spear and G.M. Hornberger, 1980 Gardner *et al.*, 1981 O'Neill, 1973 R.V. O'Neill and R.H. Gardner, 1979 Tiwari and Hobbie, 1976 Tiwari *et al.*, 1978 O'Neill *et al.*, 1980 Scavia *et al.*, 1981a	Design of models, calibration and validity. Computation of confidence limits.
Covariance Analysis	F. Argentesi and L. Olivi B. Beck and E. Halfon, 1982 Beck *et al.*, 1979 Moore, 1973 Moore *et al.*, 1976 Scavia *et al.*, 1981b	Error propagation in time. Confidence limits.
Maximum Likelihood	D.M. DiToro and G.M. van Straten, 1979	Model error, no extrapolation
Error Analysis	K.H. Reckhow, 1979	Confidence limits at steady state

Table 2.--Bayesian, Fisher and unknown-but-bounded models. Reprinted from Schweppe (1973)

	White Input Disturbance	White Observation Disturbance	Initial Condition
Bayesian	Stochastic	Stochastic	Random Variable
Fisher	1) Stochastic	Stochastic	Completely Unknown
	2) Completely Unknown	Stochastic	Random Variable
	3) Completely Unknown	Stochastic	Completely Unknown
Unknown-but-bounded	Unknown-but-bounded	Unknown-but-bounded	Unknown-but-bounded

where $\underset{\sim}{x}$ are the state variables, $\underset{\sim}{u}$ are the inputs, $\underset{\sim}{p}$ are the parameters and t is time. No state space model is unique since many different sets of values of f and g can give the same input-output behaviour. The choice of a particular f and g corresponds to the choice of a coordinate system. Halfon (1979), and Halfon *et al.* (1979) and Beghelli *et al.* (1982) have analyzed this problem of uniqness in ecological modelling for model parameters and model order, respectively. Halfon (1979) showed that a large number (250) combinations of parameter values could give an equally good fit to some phosphorus observations in Lake Ontario and Halfon *et al.* (1979) and Beghelli *et al.* (1982) have shown that the same experimental data can be described by models of different order, i.e. of different complexity and number of state variables.

When different models or the same model with a different combination of parameters is used to predict the future behaviour of an ecosystem, we enter the realm of uncertainty in ecological modelling. The computation of confidence limits associated with the simulations help us in defining and quantify this uncertainty.

CONFIDENCE INTERVAL PROCEDURES

Five broad classes were identified in the introduction. In the method of replication, a series of independent simulations is performed under the assumption that a fixed mean process is being simulated. These simulations can be performed under arbitrary rules of experimental design, as recommended by Smith and Mauro (1982) or as Monte Carlo simulations (O'Neill and Gardner, 1979; O'Neill *et al.*, 1980; Gardner *et al.*, 1981). Although the simulations are serially correlated, the averages of the simulations are by design independent and identically distributed from run to run. The main disadvantage is that a large number of simulation may have to be performed, 2^k or 3^k for a factorial experiment with the parameters at two or three levels respectively. Confounded designs can however be used (Montgomery, 1976) to reduce computer runs. With Monte Carlo methods, 500 (O'Neill and Gardner, 1979) to 10,000 (Fedra, 1979; Fedra *et al.*, 1981) runs may be needed. In the batch mean and regeneration methods, only one simulation run is performed, the output is partitioned into a series of consecutive, nonoverlapping batches of equal size (Schriber and Andrews, 1979) in the batch means methods and of unequal size in the regeneration methods (Crane and Lemoine, 1976). These latter two methods are potentially easier to use than the replication methods because each requires only one simulation run. The additional programming needed for computing the test statistics, however, make these use more demanding to the modeler (Schriber and Andrews, 1981). Ecological modellers have also been attracted by methods that use only one simulation run to compute uncertainty. These methods are based on system-theoretic approach with the assumption that the models have several intrinsic errors, errors due to the model structure, to parameters, inputs, data, natural variability of observations, time and spatial scaling. All these errors can be quantified and included in computations of uncertainty.

Covariance analysis is a method which is based on the time-update part of the Kalman filter (Gelb, 1974; Chapter 3). If P(o) is the matrix of variance covariance associated with the simulation x(o) at time = 0, then the matrix P can be propagated in time as

$$P(t+1) = \Phi(t)\ P(t)\ \Phi(t)^T$$

where $\Phi(t)$ is the transition matrix, $\Phi = e^{At}$, and A is the linearized model $\frac{d\underset{\sim}{x}}{dt} = A\underset{\sim}{x} + B\underset{\sim}{u}$ of the nonlinear model $\frac{d\underset{\sim}{x}}{dt} = f(\underset{\sim}{x},\underset{\sim}{u},\underset{\sim}{p},t)$ where $A = \frac{\partial f}{\partial \underset{\sim}{p}}$ and $B = \frac{\partial f}{\partial \underset{\sim}{u}}$. With this method only one simulation run is needed and the effects of errors in the parameters and initial conditions readily computed. The method can also be extended to include error in inputs and with more difficulty errors in the model structure. Mehra (1970) and Tsang (1980) have proposed algorithms to compute from data the elements of the matrix Q (Gelb, 1974; eq. 3.7.10) that in the Kalman filter is used to account for model error. In ecological modelling, this approach has been used by Beck et al. (1979), Scavia et al. (1981a,b) and Beck and Halfon (in preparation). Perkins and Leathrun (1971) and Scavia et al. (1981a) have compared the computational efficiences of Monte Carlo simulations and covariance analysis and have defined the conditions for optimal use of each method. In general, the larger the number of parameters, the more efficient is to use covariance analysis. Analytical pencil and paper computation may be overwhelming in some instances however.

Confidence limits have also been computed with the maximum likelihood method (DiToro and vanStraten, 1979). With this method, uncertainty can not be computed for a period when data are not available, however, an estimate of the model error can be obtained relatively simply, that is, the error due to the model error structure and to physical phenomena not included in the model. Error analysis can be used to predict confidence limits for steady state conditions. The results are proabibilities that certain limits may be exceeded due to model error, parameter error or loading error. Reckhow (1979) has applied the method to Lake Ontario using Vollenweider's model.

MIREX IN LAKE ONTARIO: UNCERTAINTY ANALYSIS

Mirex is a fully chlorinated organic compound produced by the dimerization of hexachlorocyclopentadiene. Mirex, an insecticide, has been produced by the Hooker Chemical Company in Niagara Falls and by Armstrong Chemicals in Volney, N.Y. The factories released Mirex into the Niagara and Oswego rivers, respectively. Much of the pollution occurred

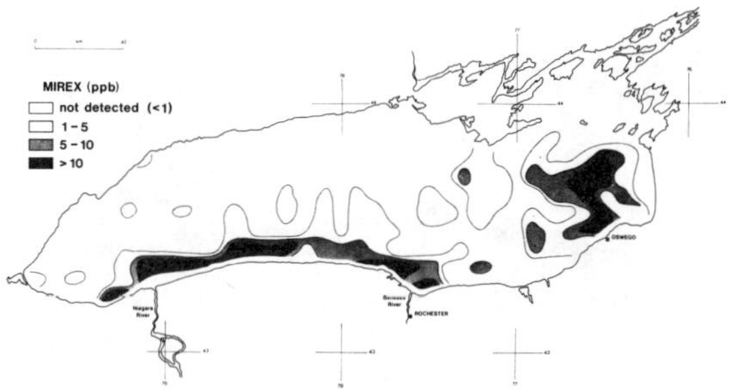

Figure 1.--Mirex distribution in the bottom sediments of Lake Ontario in 1968. Concentrations are in ppb (parts per billion or ng per gram). The two main sources were the Niagara River and the Oswego River. Computations show that in the period 1958 to 1968, 770 kilograms of Mirex entered the lake from the Niagara River and 220 kilograms from the Oswego River.

between 1958 and 1978. Kaiser (1974) found Mirex in Lake Ontario fishes and Holdrinet et al. (1978) described Mirex concentration in the bottom sediments in 1968 (fig. 1). Mirex load was reduced to 13 to 20 kg per year since 1978 (Warry and Chan, 1981) after the peaks of 70 to 140 kg per year in the previous 20 years. Mirex is very persistent and accumulates readily into living organisms and organic sediments. Mirex is not biodegraded and remains in the environment for a long time. The only possible solution to the Mirex contamination problem is burial by clean sediments after loadings stop, which is not happening yet. Figure 2 shows computed expected distribution of Mirex in the bottom sediments in 1982.

A mathematical model (Halfon, 1982a,b) was developed (fig. 3) to reproduce the observed distribution of Mirex in the bottom sediment in 1968, to make predictions about the behaviour of Mirex in Lake Ontario and to assess which processes were most important in controlling Mirex fate and which should be investigated in the future to improve predictions. An uncertainty analysis had to be performed for several reasons, the model structure, using the EXAMS framework (Lassiter et al., 1979) had several simplifications and assumptions, such as pooling the algae and fishes, and overlooking wind driven circulation and burial processes, some parameter values were poorly known, for example, the rate of turbulence at the sedimentation interface and the uptake rate of living organism and other varied widely among various locations in the lake (environmental heterogeneity). Loading rates were completely unknown and could only be estimated through simulations.

Monte Carlo simulations were chosen as the best mean of computing uncertainty. The model only took a few seconds to run, about 30 second for 4000 years, and the high number of nonlinearities and parameters made an analytical approach, such as covariance analysis, unfeasible. A model of the form

$$-\frac{ds}{dt} = V + Pd \pm Se + D - L$$

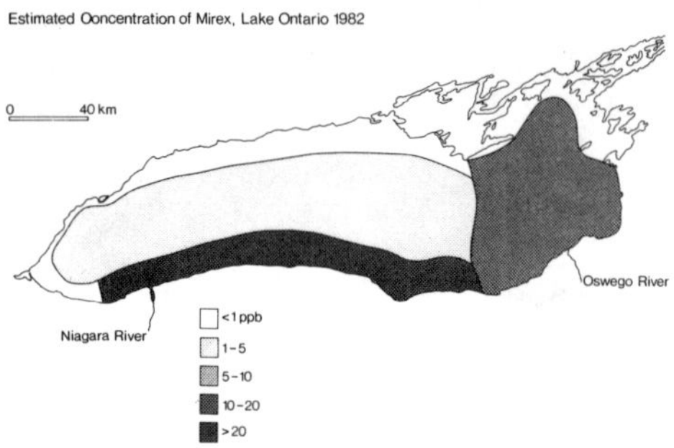

Figure 2.--Predicted Mirex distribution in the bottom sediments of Lake Ontario in 1982. Concentrations are in ppb (parts per billion or ng per gram). Since 1968 the main source of Mirex has been the Niagara River. Computations show that since 1968 an additional 1500 kilograms entered the lake. The predictions in this figure have not been verified yet.

was used, where s is the concentration of Mirex in lake water, V is the volatilization rate, Pd, is the rate of direct photolysis, Se is the rate of exchange with the sediments, D is the rate of dilution or physical transport out of the system, and L is the loading rate into the lake. This equation is fairly general and commonly accepted (Cairns, 1979). Data needed by the EXAMS (Lassiter et al., 1979) framework for a complete definition of the system are very extensive. The environmental data required cover all aspects of the lake ecology; sediment composition, plankton and benthos biomass, currents, rain, cloudiness, etc.

ERROR ANALYSIS: MIREX

Four sets of parameters account for 60 to 70% of the total variability of each state variable. The identified parameters are the bulk density of the sediments, the sediment-water exchange rate, the affinity of Mirex for organic carbon and the partition coefficient of Mirex into biomass (KPB). A nominal loading of one kilogram Mirex per year was used in the computation of table 3. Different loading scenarios produce different correlations; comparison of different loading conditions, however, show that even if numbers change, the identified important parameters do not.

Table 3 shows the correlation coefficients for five time windows. The state variables correlated with the parameters and considered in this analysis are Mirex dissolved in water and adsorbed to the suspended sediments, Mirex in pore water and adsorbed to the bottom sediments, Mirex mass (in kg) in water and in the sediments, Mirex biosorbed on living biomass in the water (fish and algae) and at the bottom (benthos and benthic fishes).

An overall analysis of Table 3 shows that the parameters important for the prediction of Mirex in the lake are not the same all the time; that is, parameters that should be measured to improve prediction ability for the first 20 years are in general different from parameters that should be measured to improve prediction over the long term. In particular we can note that biological parameters (e.g., KPB) such as uptake of Mirex by living organisms, are important only during the first 20 years. Environmental parameters later become the important parameters. Algae and fish populations change in time, and even if this aspect is not explicitly present in the model, their global importance over the fate of Mirex diminishes, and over the decades chemical and environmental properties become more important. However, over the short term, monitoring of living organisms, as it is presently done, is important.

DISCUSSION

Uncertainty analysis in ecological modelling has had a period of theoretical and methodological development over the last decade. Nowadays, most ecological modelers realize the importance of qualifying the meaning of their simulations, especially when predictions are made for management decisions. If ecological modelers realize the significance of this methodology, controversy still exists on the extension from research to routine applications. For example, McLaughlin (pers. comm.) states that uncertainty analysis may be useful to model practioners, but not much useful to users. The additional information on the validity of the simulations only confuses the users since this information increases, rather than decreases, the number of available choices or decisions to be taken in a given circumstance; whereas the user employs simulations to restrict the number of choices and indeed help him to make a decision. This opinion is important since it makes the system ecologist face a dilemma: should uncertainty analysis research in ecology be continued only for the benefit of modelers or also for the benefit of users as is the case, for example, in space sciences. My belief is that we should make an effort to let the users be aware of the possible advantages of uncertainty analysis even if at the beginning they may be confused. More discussions on this topic should be taking place in the future.

One question of general broader importance is the relation of uncertainty to model complexity, or the uncertain relations betwen distributed and lumped models . This topic is generally called the aggregation problem (Zeigler, 1976). A simple model, or a highly abstract representation of a real system, exhibits only a restricted subfraction of the interactive capabilities of the real system. In a complex model the different interactions interfere with one another in an unpredictable way. The representation of these interactions may be lost in the abstraction process that leads to the simplified model. The errors in model prediction arise because simultaneous manifestations of modes of interactions happen to interfere with one another. Therefore, a complex model would produce uncertain results sooner than a simple model: it will be less reliable for prediction. The simplified model may seem to predict a future with more reliability, but is single simplified mode does not consider all the other modes of system interaction present in the real system. As stated by Beck et al. (1979) we may predict the wrong future with a high degree of certainty, which is far less desireable than predicting the right future with low confidence. At this stage the research on the aggregation problem (Halfon et al., 1979; Beghelli et al., 1982) overlaps with uncertainty analysis. How aggregated can a model be and still maintain some behaviors of the real system, but not so large that an uncollectable amount of information is necessary for

Table 3.--Statistically significant correlation coefficients between individual parameters and state variables at five time windows. These correlations express the contribution of errors in individual parameters to uncertainty in Mirex present in the water column and sediments. The correlation coefficient when squared represents the percentage of variability in the state variable due to one parameter when the variability in the other parameters is uncontrolled. For each state variable, by summing the variability reduction due to each parameter we can compute the variability that can be obtained by better measuring those state variables.

YEAR	PARAMETERS[1]	MIREX IN WATER				MIREX IN SEDIMENTS			
		DISSOLVED	SORBED	MASS	LIVING ORGANISMS	DISSOLVED	SORBED	MASS	LIVING ORGANISMS
1968	KPB				0.212				0.223
	CHARL(H1-B2)	0.202	0.220	0.230				-0.249	
	DSP(H1-B2)	-0.336	-0.339	-0.341	-0.310			0.300	
	SDCHR(E)			0.255					
	SDCHR(B1)	-0.553	-0.580	-0.540	-0.545	-0.611	-0.677	0.583	-0.592
	SDCHR(H1)			0.244				-0.213	
	SDCHR(B2)	-0.438	-0.447	-0.419	-0.421	-0.471	-0.499	0.305	-0.446
	STRFLOW							-0.224	
	Percent Variability Explained	65	72	76	62	60	71	68	60
1978	KOC					-0.217			-0.202
	KPB								0.219
	CHARL(H1-B2)					-0.221		-0.215	-0.204
	DSP(H1-B2)	-0.239	-0.239	-0.244	-0.216			-0.290	
	DSP(H2-B3)	-0.244	-0 267	-0.270	-0.252				
	SDCHR(E)			0.234					
	SDCHR(B1)	-0.285	-0.300	-0.281	-0.289	-0.422	-0.463	-0.487	-0.441
	SDCHR(H1)				0.256			-0.239	
	SDCHR(B2)	-0.646	-0.678	-0.610	-0.628	-0.631	-0.685	-0.434	-0.559
	FROC(B2)				-0.213				
	STRFLOW							-0.313	
	Percent Variability Explained	62	68	64	70	67	68	71	64
2058	KOC					-0.213			
	KPB				0.215				0.239
	CHARL(H2-B3)	0.344	0.355	0.332	0.336				
	DSP(H2-B3)	-0.388	-0 395	-0.416	-0.394		-0.230	0.307	
	SDCHR(B1)							0.327	-0.220
	SDCHR(H1)							-0.228	
	SDCHR(B2)	-0.364	-0.373	-0.355	-0.354	-0.480	-0.475	0.487	-0.461
	SDCHR(B3)	-0.509	-0.517	-0.472	-0.466	-0.537	-0.564	0.247	-0.482
	KO2			-0.201					
	STRFLOW							-0.368	
	Percent Variability Explained	66	69	67	66	56	60	69	55
1000 years	KOC						0.223		
	CHARL(H2-B3)	-0.263	-0.237	-0.273	-0.260				
	DSP(H2-B3)	0.363	0.354	0.357	0.355	0.230	0.264		0.222
	SDCHR(H1)	-0.235	-0.236		-0.226		-0.224		
	SDCHR(B2)							0.216	
	SDCHR(H2)	-0.221				-0.305	-0.234		-0.266
	SDCHR(B3)							0.676	
	KO2	-0.349	-0.341	-0.382	-0.361	-0.345	-0.370	-0.234	-0.362
	STRFLOW	-0.525	-0.516	-0.506	-0.565	-0.539	-0.565	-0.444	-0.543
	Percent Variability Explained	70	62	62	63	58	65	76	55

Table 3. (con't.)

4000 years	KOC						0.208		
	CHARL(H2-B3)					0.221	0.206		0.216
	DSP(H2-B3)					-0.208			-0.213
	SDCHR(H2)	-0.265	-0.240		-0.246	-0.266	-0.245	-0.206	-0.246
	SDCHR(B3)	0.538	0.519	0.540	0.541	0.480	0.479	0.594	0.484
	KO2	0.294	-0.290	-0.306	-0.302	-0.281	-0.291	-0.219	-0.288
	STRFLOW	-0.495	-0.490	-0.492	-0.480	-0.487	-0.481	-0.388	-0.470
	Percent Variability Explained	69	65	63	67	71	69	59	69

[1] Parameter Definitions: KPB, Partition coefficient on biomass [(μg/g)/mg/L)]; CHARL, mixing length (metres) between compartments; DSP, eddy diffusivity between compartments (cm^2/sec); SDCHR, suspended sediment concentration (mg/L) in water column compartments and bulk density (g/cm^3) of bottom sediments; KOC, affinity of Mirex for organic carbon [((mg/kg)/(mg/L))/fraction organic carbon]; STRFLOW, stream flows entering Lake Ontario (km^3/year). Name of parameters from EXAM framework. E, epilimnion, 0 to 20 m, H1, 20 m to 40 m, H2, 40 m to bottom, see fig. 3.

its development? Zeigler (1976) and O'Neill and Gardner (1979) have recently analyzed the problem and concluded that for certain classes of ecological models minimal errors result if components with similar turnover times are aggregated, or small components are lumped with large ones.

As for Mirex, the error analysis showed that prediction errors are due only to a few parameters out of a total 96. If the parameter errors coud be reduced, we would see an improvement for Mirex prediction. A measure of turbulence at the bottom would help in understanding whether Mirex contaminated sediments are resuspended and whether this Mirex is then bioavailable. These field studies however are difficult since the mean depth of the lake is about 95 metres. Sediment characteristics are also important, for example, Scrudato (pers. comm.) has found a very high correlation between organic carbon and Mirex concentrations in the sediments. The final fate of Mirex already has had a significant impact on the development of sport and commercial fishery in Lake Ontario. Commercial fishing is banned for contaminated fishes and even European countries with standard limits for consumption higher than Canada and the United States have restricted imports of fishes and eels. Mirex is almost unsoluble in water but highly soluble in fats. Mirex accumulates and magnifies in aquatic organisms at all trophic levels (Buckler *et al.*, 1981). If the bottom sediments remains a source of biosoluble Mirex, fishery may be restricted for a long period of time. Modelling and uncertainty analysis provide some answers on the continuing research on the fate and exposure of this contaminant. Integration of field programs and set theoretic approaches may lead to the solution of this important problem.

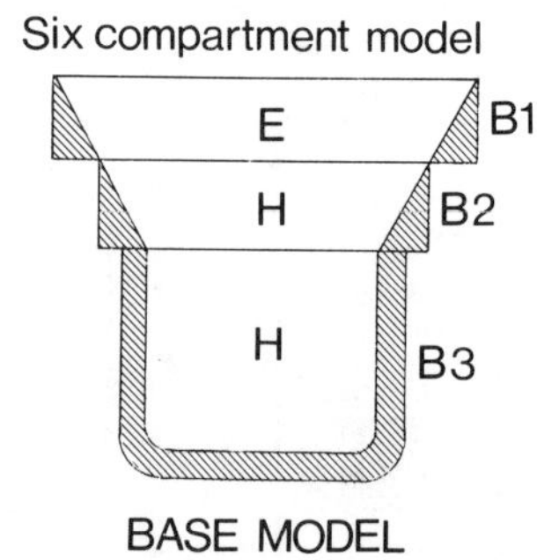

Figure 3.--Model structure. The lake is divided into three layers, considered horizontally homogeneous, 0 to 20 metres, 20 to 40 metres and 40 metres to the bottom.

ACKNOWLEDGEMENTS

Some ideas presented in this paper originated during a Task Force Meeting on uncertainty in water quality modelling at the International Institute of Applied System Analysis, in Laxemburg, Austria, November 1979.

LITERATURE CITED

Argentesi, F. and L. Olivi, 1976. Statistical sensitivity analyses of a simulation model for the biomass nutrient dynamics in aquatic ecosystem. In A.N. Other (ed.) Proceeding 4th Summer Computer Simulation Conference, Simulation Councils, Inc. La Jolla, California, 389-393.

Beck, M.B. and E. Halfon, 1982. In preparation.

Beck, M.B., E. Halfon and G. van Straten, 1979. The propagation of error and uncertainty in forecasting water quality. Part I: Method. International Institute for Applied Systems Analysis, WP-79-100, 52 pp.

Beghelli, S., R. Guidorzi, E. Halfon, and F. Terragni, 1982. Identification of phosphorus dynamics in Kootenay Lake, Canada. Ecological Modelling (in press).

Buckler, D.R., A. Witt, E.L. Mayer, and J.N. Huckins, 1981. Acute and chronic effects of kepone and Mirex on the fat head minnow. Trans. Am. Fish. Soc., 110:279-280.

Cairns, J. 1979. Estimating Hazard, Bioscience, 30:101-17.

Crane, M.A. and A.J. Lemoine, 1976. A tutorial on the regenerative approach and discrete event simulations. Proc. 1976. Winter Simulation Conf. ACM, New York, 41-47.

DiToro, D.M. and G. van Straten, 1979. Uncertainty in the parameters and predictions of phytoplankton models. 34 pp. IIASA Working Paper, WP-79-227.

Fedra, K. 1979. A stochastic approach to model uncertainty. 45 pp. IIASA Working Paper, WP-79-63.

Fedra, K., G. van Straten and M.B. Beck, 1981. Uncertainty and arbitrariness in ecosystem modelling: A lake modelling example. Ecological Modelling, 13:87-110.

Gardner, R.H., R.V. O'Neill, J.B. Mankin, and J.H. Carney, 1981. A comparison of sensitivity analysis and error analysis based on a stream ecosystem model. Ecological Modelling, 12:173-190.

Gelb, A., 1974. Applied optimal estimation. 374 pp. The M.I.T. Press, Cambridge, Massachusetts.

Halfon, E., 1979. On the parameter structure of a large scale ecological model. In G.P. Patil and M.L. Rosenzweig (Eds.) Contemporary Quantitative Ecology and Related Econometrics. International Co-operative Publishing House, Burtonsville, MD., 279-293.

Halfon, E., 1982a. Error analysis and simulation of Mirex Behaviour in Lake Ontario. 83 pp. Canada Centre for Inland Waters Technical Report, Burlington, Ontario.

Halfon, E., 1982b. Uncertainty in modeling the environmental fate of Mirex in Lake Ontario. In. L. Troncale (ed.) A general survey of systems methodology, proceeding of the 26th Annual meeting of the Society for General Systems Research, Washington, DC. January 5 to 9, 1982; Systems Science Institute, University of Louisville, KY, 553-558.

Halfon, E. H. Unbehaven and C. Schmid, 1979. Model order estimation and system identification theory and application to the modelling of ^{32}P kinetics within the trophogenic zone of a small lake. Ecological Modelling, 6:1-22.

Holdrinet, M., R. Frank, R.L. Thomas and L.J. Heiling, 1978. Mirex in the sediments of Lake Ontario. J. Great Lakes Res., 4:69-74.

Kaiser, K.L.E., 1974. Mirex: An unrecognized contaminant of fishes for Lake Ontario. Science 185:523-525.

Lassiter, R., G.L. Baughman and L.A. Burns, 1979. Fate of toxic organic substances in the aquatic environment. In. S.E. Jorgensen (ed.) State of the Art in Ecological Modelling, International Society of Ecological Modelling, Vaerlose, Copenhagen, 219-246.

Mehra, R.K., 1970. On the identification of variances and adaptive Kalman filtering. IEEE Trans. Automatic Control, Vol. AC-15:175-184.

Montgomery, D.C., 1976. Design and Analysis of Experiments, 418 pp. Wiley, NY.

Moore, S.F., 1973. Estimation theory applications to design of water quality monitoring systems. J. Hydraul. Div. Am. Soc. Civ. Eng., 99(Hy5), 815-831.

Moore, S.F., G.C. Dandy and R.J. DeLucia, 1976. Describing variance with a simple water quality model and hypothetical sampling programs, Water Resour. Res., 12:795-804.

O'Neill, R.V., 1973. Error analysis of ecological models. In. D.J. Nelson (ed.) Radionuclides in ecosystems, Conf. 710501, NTIS, Springfield, Virginia, 898-908.

O'Neill, R.V. and R.H. Gardner, 1979. Sources of uncertainty in ecological models, In B.P. Zeigler et al., (eds..) Methodology in Systems Modelling and Simulation, North-Holland Publishing Company, 447-463.

O'Neill, R.V., R.H. Gardner and J.B. Mankin, 1980. Analysis of parameter error in a nonlinear model. Ecological Modelling, 18, 297-311.

Perkins, T.R. and J.F. Leathrun, 1971. Propagation of covariance matrix for systems subjected to random noise. UNB-ETC (U) Army Material Systems Analysis Agency, Tech. Rep. No. 48 ntis Microfiche AD 725 500.

Reckhow, K.H., 1979. Empirical lake models for phosphorus: development, applications, limitations, and uncertainty. In D. Scavia and A. Robertson (eds.) Perspective on Lake Ecosystem Modelling, Ann Arbor Science Publishers, Ann Arbor, Michigan, 193-221.

Scavia, D., W.F. Powers, R.P. Canale, and J.L. Moody, 1981a. Comparison of first order error analysis and Monte Carlo simulation in time-dependent eutrophication models. Water Resources Res., 17:1051-1059.

Scavia, D., R.P. Canale, W.F. Powers, and J.L. Moody, 1981. Variance estimates for a dynamic eutrophication model of Saginaw Bay, Lake Huron. Water Resources Res., 17:1115-1124.

Schriber, T.J. and R.W. Andrews, 1979. Interactive analysis of simulation output by the method of batch means. Proc. 1979 winter simulation conference, ACM, New York, 1979, 512-524.

Schriber, T.J. and R.W. Andrews, 1981. A conceptual framework for research in the analysis of simulation output. Communications of the ACM, 24:218-232.

Schweppe, F.C., 1973. Uncertain dynamic systems. Prentice Hall, Englewood Cliffs, NJ, 563 pp.

Smith, D.E. and C.A. Mauro, 1982. Factor screening in computer simulation. Simulation, 38:49-54.

Spear, R.C. and G.M. Hornberger, 1980. Eutrophication in Peel Inlet, II. Identification of critical uncertainties via generalized sensitivity analysis.

Tiwari, J.L. and J.E. Hobbie, 1976. Random differential equations as models of ecosystems: Monte-Carlo simulation approach. Math. Biosci, 28:25-44.

Tiwari, J.L., J.E. Hobbie, J.P. Reed, D.W. Stanley, and M.C. Akler, 1978. Some stochastic differential equation models for an aquatic ecosystem. Ecological Modelling, 4:3-27.

Tsang, W.L., 1980. Identifability of unknown noise covariance matrices of a linear time invariant, discrete-time, dynamic system. Ph.D. dissertation, Northeastern University, Boston, Massachusetts.

Warry, D.W. and Chan, C.H., 1981. Organic contaminants in the suspended sediments of the Niagara River, J. Great Lakes Research, 7:394-403.

Zeigler, B.P., 1976. Theory of modelling and simulation, 435 pp., Wiley, New York.

COMPARISON OF PHYTOPLANKTON GROWTH CONSTRUCTS IN THE CONTEXT OF VERTICALLY SEGMENTED PLANKTON MODELS[1]

Michael J. McCormick[2], Bryan L. Wharram and Donald Scavia[3]

Abstract.--A previously developed and calibrated ecological model, extended to a finer vertical grid representation, was capable of simulating the vertical structure of Lake Ontario's phytoplankton, available nitrogen and silicon, but not available phosphorus. Measured phosphorus depletion extended to 40m whereas nitrogen and silicon depletion extended to only 5-10m. Analysis of models using different algal growth constructs demonstrated the ability of algal models based upon internal nutrient pools to simulate the observed divergence between phosphorus and other nutrient profiles.

INTRODUCTION

A comprehensive physical, biological, chemical and meteorological investigation of Lake Ontario in 1972 has led to a greater understanding of its structure and functioning. Several efforts to synthesize the data gathered have employed ecological models (e.g., Thomann et al. 1975, Chen et al. 1975, Scavia 1979, 1980, and Canada Centre for Inland Waters 1979). Although these works have provided useful insights into the system's behavior, they lack adequate spatial resolution to critically examine mechanisms that control detailed vertical distribution of biological and chemical properties. Herein, we investigate processes controlling these distributions by extending existing ecological model to a fine-grid vertical scheme.

Our objectives are twofold: (1) to simulate observed vertical detail of several plankton groups, detritus, phosphorus, nitrogen, and silicon and (2) to test the response characteristics of three phytoplankton growth formulations. These formulations are based on (a) external nutrient control (Michaelis-Menten, MM), (b) cellular nutrient control (kinetic, CK), and (c) cellular nutrient control (equilibrium approximation, CE). In the Michaelis-Menten approach, algal nutrient stoichiometry is fixed and growth is controlled by external nutrient concentration. The cellular nutrient methods more realistically represent cell growth controlled by intracellular nutrients.

We demonstrate herein the ability of the extended ecological model to reproduce Lake Ontario's vertical structure for all but available phosphorus. We then discuss implications of the three growth parameterizations on simulation of phosphorus dynamics.

[1]Paper presented at the Third International Conference on State-of-the-Art in Ecological Modelling, Colorado State University, May 24-28, 1982.

[2]Authors are research scientists at Great Lakes Environmental Research Laboratory, National Oceanic and Atmospheric Admin., 2300 Washtenaw Ave., Ann Arbor, Michigan 48104

[3]Also, Division of Biological Sciences, The University of Michigan, Ann Arbor, MI 48109

GLERL Contribution No. 312

DATA AND MODELS

Data.--Data used in this study were collected during the International Field Year for the Great Lakes (IFGYL) in 1972-73. Observations of temperature, nutrients, and gross biomass variables were obtained from STORET, the data management system of the U.S. Environmental Protection Agency, and are based on collections made by the Canada Centre for Inland Waters (Stadelmann and Fraser 1974). We analyzed seven of the nine cruises conducted during 1972. Methods of data collection and measurement can be found in Stadelmann and Fraser (1974), Stadelmann and Munawar (1974), and Vollenweider et al. (1974).

Model input data include solar radiation, photoperiod (percent sunlit day), eddy diffusivities, and temperature. Incident solar radiation varied between 50 and 800ly day^{-1} and was input as total daily values from Atwater and Ball (1974). Photoperiod varied between 37 and 64 percent of the day.

The lake was divided into fourteen horizontally-uniform layers partitioned from the surface down as follows: 8 layers of 5m, 2 of 10m, 2 of 20m, and 2 of 40m. Variation of area with depth was approximated by

$$A(z) = 1.11\times10^{-5}z^2 - 6.66\times10^{-3}z + 1 \qquad 0 \leq z \leq 180$$

where z is depth in m and A is dimensionless.

Ecological Model.--The ecological framework of this model was developed previously for simulation and analysis of a two-layer, lakewide averaged Lake Ontario system (Scavia 1980). The crude physical segmentation in this previous work was intentional; it reduced the complexity of physical processes and allowed investigations with emphasis on biological, chemical, and simple physical processes at lakewide averaged scale. Ecological calculations performed within the present physical context are the same as those calculated for the cruder segmentation. Detailed documentation of the system equations, demonstration of the ability of the model to simulate lakewide average conditions, and examination of biological and chemical processes on that scale can be found in Scavia (1979, 1980). This ecological framework was also used for analysis of two dimensional processes in Lake Ontario during the spring-summer transition period (Scavia and Bennett 1980). What follows are general descriptions of biological and chemical processes included in the model.

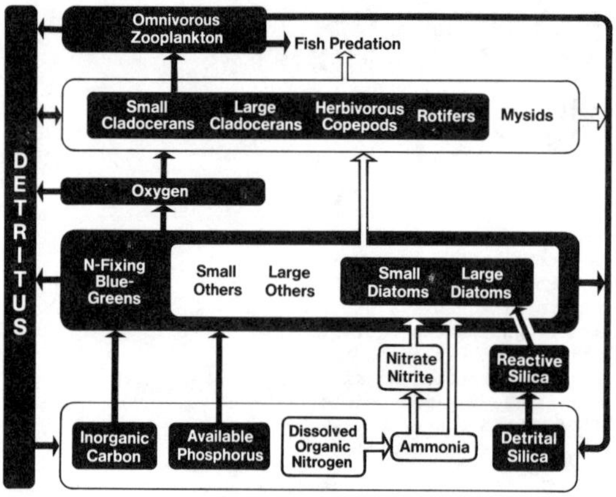

Figure 1.--Conceptualization of Lake Ontario ecological model (from Scavia 1980).

The model (fig. 1) simulates phytoplankton, zooplankton, and cycles of phosphorus, nitrogen, silicon, carbon, and oxygen. Each model compartment is represented by a differential equation for the biological and chemical processes. For example, the phytoplankton equation includes terms for gross primary production, respiration, excretion, grazing, and sinking; and the zooplankton equation includes terms for grazing, assimilation, respiration, excretion, defecation, and predation.

Zooplankton grazing is handled as a temperature-dependent saturation expression based on total food supply. The expression includes a minimum threshold for feeding as well as resource partitioning based on size selection. Feeding and assimilation efficiencies are regarded as food-specific constants. Food ingested but not assimilated is egested as detritus. Respiration is a temperature-dependent process that is the sum of a low maintenance rate and a term proportional to feeding. Nutrient excretion is proportional to respiration.

Transformation rates between detritus, dissolved organic nitrogen, ammonia, nitrite plus nitrate, available phosphorus, and available silicon pools all are temperature-dependent and first order. Sinking rates of phytoplankton and detritus are constants based on their average densities.

Physical Model.--Temperatures, calculated from the nonlinear diffusion equation using variable eddy diffusivity (McCormick and Scavia 1981), were calibrated to IFYGL observations with data requirements for the boundary conditions met by IFYGL winds and surface temperatures (Pickett 1975). Eddy diffusivities were calculated as a function of the surface wind stress divided by a modified gradient Richardson number (McCormick and Scavia 1981). During stratification a constant diffusivity, equal to the thermocline value, was used in the hypolimnion. As a computational convenience mixing was also modeled by treating the lake as a "mixed reactor" whenever the water column was isothermal.

Calculated eddy diffusivities were used to estimate turbulent flux of all the state variables except zooplankton. A crude mechanism was introduced to account for their vertical migration; migration was implicitly included by mixing the zooplankton to a uniform concentration over the top 30 meters.

Mixing and sedimentation of algae into regions of different cell quotas requires calculating new cell quotas as a weighted average. Weights are equal to the fractional amount of total algae having each specific cell quota. Afterwards, adjustments are made to available phosphorus to balance any change in the internal stores.

Growth Constructs.--Three phytoplankton growth parameterizations are used. The first represents a single-step process assuming rates of nutrient uptake and growth are in equilibrium. Growth is modeled as a temperature-dependent maximum rate times a reduction factor for light- and nutrient-limitation minus respiration. Potential light-limitation is modeled after Vollenweider (1965) and Steele (1965) and potential nutrient-limitation is expressed as a Michaelis-Menten term based on external nutrient concentration. The threshold formulation is used to determine overall limitation. Phytoplankton respiration is the sum of a low maintenance rate plus a term proportional to production and termperature. To maintain constant stoichiometry, nutrient excretion is proportional to respired carbon.

The second and third constructs represent net growth controlled by internal nutrient stores defined as the phosphorus/carbon ratio. We restrict our analyses to phosphorus because it is the major nutrient limiting production. Internal nutrient store is described kinetically (CK) by:

$$\frac{\partial Q}{\partial t} = V - \mu Q \qquad (1)$$

$$\mu = \mu_m(1 - Q_o/Q) \qquad (2)$$

$$V = V_m P/(K_p + P) \qquad (3)$$

where,

μ = growth rate (day^{-1})
μ_m = maximum growth rate
Q = cell quota (g-P/g-C)
Q_o = minimum cell quota
V = uptake rate (g-P/g-C·day)
V_m = maximum uptake rate
t = time
P = extracellular phosphorus (mg-P/L)
K_p = half-saturation constant for phosphorus uptake (mg-P/L).

When phosphorus is limiting, the cellular equilibrium approximation (CE) assumes internal and external nutrient concentrations are in equilibrium such that growth and nutrient uptake are related by

$$V_e = \mu_e Q \qquad (4)$$

where,

V_e = "cellular equilibrium" uptake rate (g-P/g-C·day)
μ_e = phosphorus-controlled growth rate (day^{-1})

and Q can be calculated by solving a nonlinear system of algebraic equations composed of equations 2, 4 and one describing the conservation of phosphorus.

DiToro (1980) presents a detailed analysis of these three constructs and demonstrates the equivalence of the latter two approaches when phosphorus is limiting. When phosphorus is not limiting CE is no longer appropriate because Q should approach its maximum value, independent of the growth rate. Under these conditions, in our CE simulations, Q is set to its maximum value ($20Q_0$) or, if constrained by an insufficient supply of available phosphorus, to the next largest Q that will conserve phosphorus.

The remaining biological and chemical calculations are performed with the same equations and coefficient values (except for growth rate) used previously (Scavia 1980). The ability of those equations to mimic dynamics of state variables has been demonstrated for the two-layer, lake-wide averaged case (Scavia 1979, 1980) and the two-dimensional transect case (Scavia and Bennett 1980). Scavia (1979) also compared simulated and measured rates of epilimnetic processes and found simulated processes to be realistic. Because process measurements are not available on finer vertical structure we cannot do similar comparisons here but rather we assume that most processes are not seriously affected by space-scale reductions from a few tens of meters to 5-10 meters.

We concentrate our analysis on the period of July-October for two reasons. First, this is the time of occurrence of subsurface peaks of algal biomass and we are concerned with the production, maintenance, and destruction of that vertical structure. Second, during this period the lake responds generally as a single water body. Large offshore gradients, caused by increasing production in surface waters and differences in vertical mixing in inshore and offshore regions (Scavia and Bennett 1980), persist during the spring-summer transition period making it difficult to treat the lake as homogenous horizontally.

RESULTS

The multi-group plankton model, extended to the finer vertical grid, reproduced both the vertical and temporal detail of the Lake Ontario observations for all state variables (e.g., figures 3, 4a) except available phosphorus.

Figure 2 shows the phasing of plankton dynamics in the top 30m. Because zooplankton are distributed uniformly in the top 30m, any line parallel to the abscissa represents the vertical distribution of phytoplankton at a given time. While detailed observations of zooplankton vertical distributions are not available, our epilimnetic estimates are similar to data from 50m vertical tows. The phytoplankton simulation is also consistent with observations (see below).

The simulation begins in the lower left corner (fig. 2), near the global minimum of the zooplankton on 10 April, 1972; and proceeds counterclockwise ending on 3 December, 1972. The trajectories do not close because of the limited time span of the simulation (240 days). In general, the plankton cycle in a smooth, continuous fashion at depths greater than 10m until fall overturn. Fall overturn, depicted as a single solid arc, represents the rapid return to a spatially uniform distribution resulting from vigorous mixing associated with surface cooling.

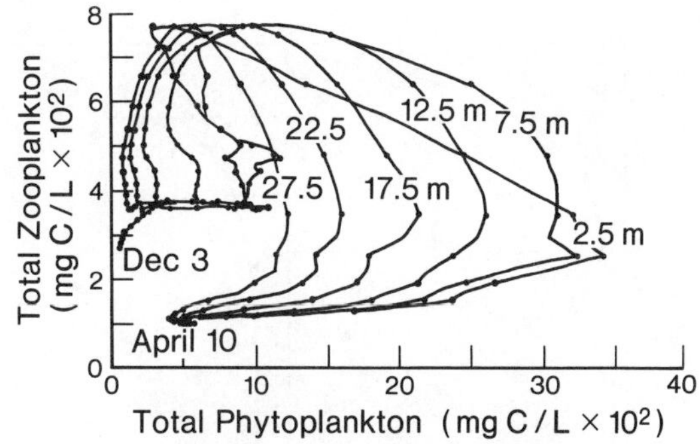

Figure 2.--Phase plane for total phytoplankton and zooplankton from multi-group Michaelis-Menten model, top 30m. The mid-depth points for each of the top six layers are indicated. Dots correspond to weekly intervals.

The top 10m were the most responsive with maximum biomass concentrations produced around 30 June in the surface layers (0-10m). Shortly after the first layer peaked, subsurface (5-10m) biomass became maximum and remained so for the next three weeks. Following this period, phytoplankton quickly converged throughout depth to a more uniform concentration and continued decreasing until late summer. In early fall, a small secondary bloom occurred, producing peak biomass again in the second layer. The bloom persisted until fall overturn.

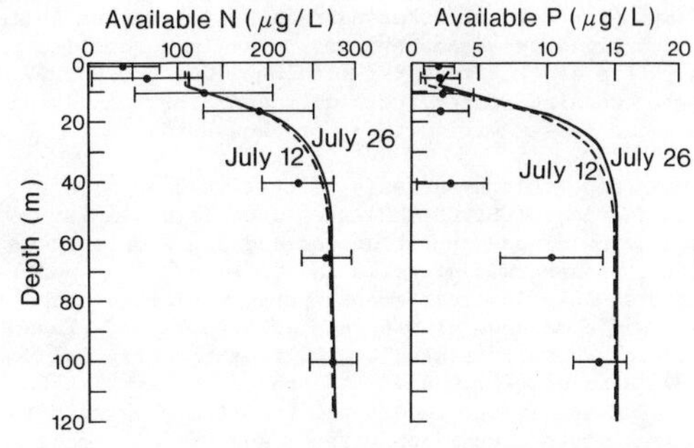

Figure 3.--Simulated and observed available nitrogen and phosphorus profiles from Michaelis-Menten (multi-group) model. Two curves are model output for July 12 and July 26. Observed data (mean ± 1SD) are for July 19.

Depth profiles (fig. 3), closest in time to the algal blooms, show inorganic nitrogen data are represented well by the model below the surface-most layer. In addition to the one data profile shown here, the model reproduced observed inorganic nitrogen quite well throughout the entire simulation. This was not the case for phosphorus. Measured nitrogen and phosphorus profiles are clearly different (fig. 3). Low concentrations of soluble reactive phosphorus (SRP) persist to depths on the order of 40m before significant increases are observed; whereas low nitrogen concentrations are limited to the top 5m. Soluble silica was also depleted only in the upper 5m. The divergence between phosphorus and other nutrient profiles persisted

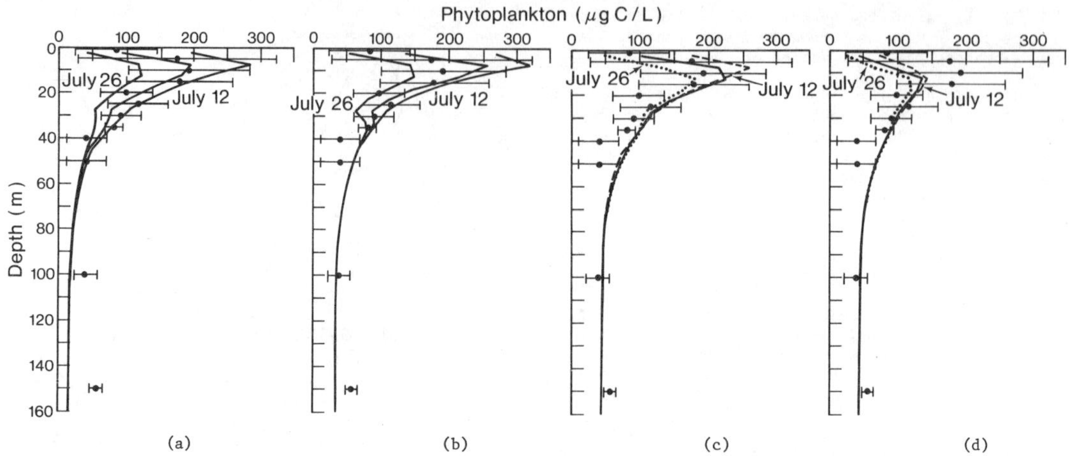

Figure 4.--Simulated and observed phytoplankton profiles. Simulated profiles for July 12, 19, 26. The center curves represent July 19. Observed data are represented as in fig. 3 for July 19. Models: a) MM, multi-group, b) MM, single group, c) CK, d) CE.

during summer and at all stations; and we were not able to mimic it. It would be impossible for a constant-stoichiometry plankton model to produce nutrient profiles that were not parallel because nutrient uptake, the dominant term of the nutrient equations in this model, is proportional for all nutrients. A logical explanation for phosphorus profiles diverging from the pattern of other nutrients lies in the ability of algae to sequester phosphorus in quantities far in excess of that required for immediate growth. While algae can also store excess nitrogen, they can do so to a far lesser extent than for phosphorus.

To test this hypothesis, different growth parameterizations (CE, CK, MM) were used in a simpler model consisting of one phytoplankton and one zooplankton group, phosphorus, nitrogen and detritus. The models differed only in their growth parameterizations and all used a maximum growth rate of 0.75 day^{-1}. The maximum growth rate was lowered from the original model's value of 1.0 day^{-1} to provide a best fit to data. Respiration was dropped from the MM simulation to facilitate comparison with CE and CK. All models were run with the same initial conditions of biomass and total phosphorus.

Each growth parameterization was capable of simulating phytoplankton depth distributions (figure 4); the poorest being the CE simulation (fig. 4d). This could have been improved by changing the maximum growth rate but it was kept the same for consistency. Effects of the growth parameterizations are manifest in simulated available phosphorus.

The MM models, using one or several algal groups, are similar. The largest differences occur at depths greater than 15m (fig. 5). As already mentioned the MM model did not simulate the SRP data adequately. The CE and CK simulations, however, show pronounced departures from the MM profiles and are in fact more representative of the data, especially for the top 30m. At depth these differences are no longer apparent which suggests that the algae below 30m maintain rather constant stoichiometry. This is particularly evident on 5 September (fig. 5) where all the single plankton group models converge to the same concentration at 100m.

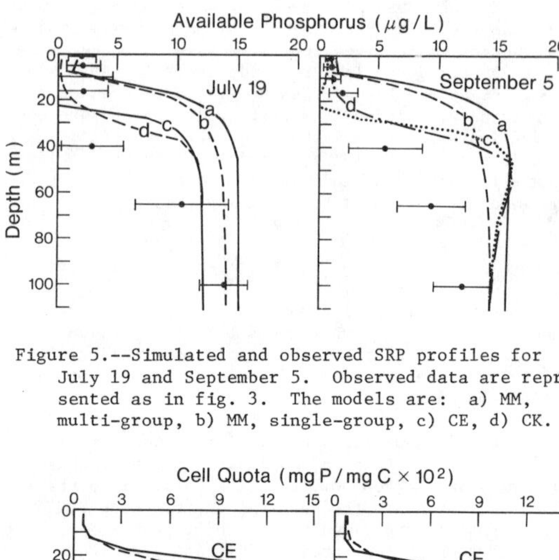

Figure 5.--Simulated and observed SRP profiles for July 19 and September 5. Observed data are represented as in fig. 3. The models are: a) MM, multi-group, b) MM, single-group, c) CE, d) CK.

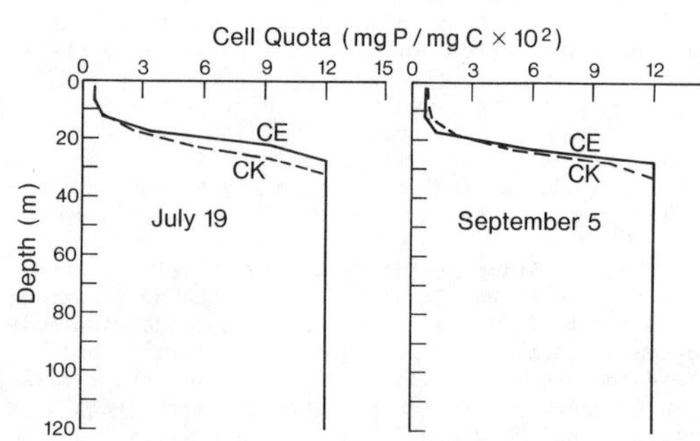

Figure 6.--Comparison of cell quota depth profiles for July 19 and September 5. Solid line represents CE and broken line represents CK.

Profiles of the cell quotas (fig. 6) provide some insight into the improved phosphorus simulations. Cell quotas are at or near their minimum values (0.006gP/gC) for the top 15m and quickly reach their maximum value (0.120gP/gC) within a depth of 25-30m. On both dates, the CE cell quota gradient for the 15-30m depth range is greater than that calculated by CK. These differences occur because equation 6 is no longer valid at these depths since light, not phosphorus, limits growth. In this context, cell quotas for CE can be viewed as the equilibrium position between instantaneously filling the algal cells to maximum capacity and the limited supply of available phosphorus to do so. This is in contrast to CK where cell quota is always limited to a finite rate of change.

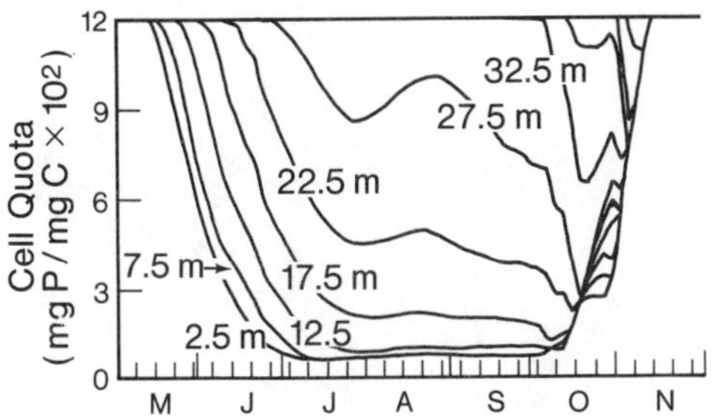

Figure 7.--Seasonal depth-variation of cell quota for CK (May 1 - Dec. 1).

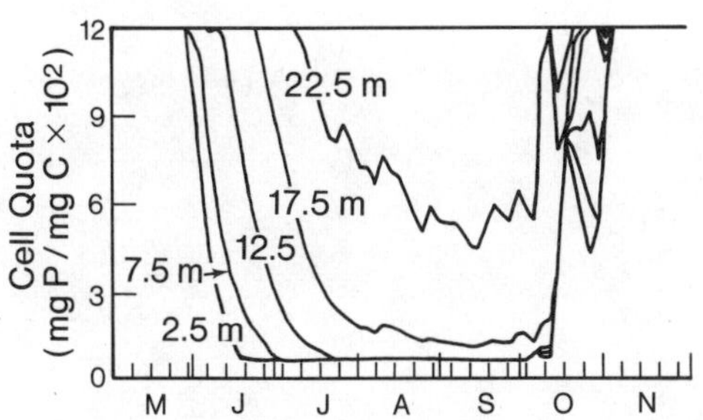

Figure 8.--Same as figure 7, but for CE.

Seasonal variation of cell quotas show CE (fig. 8) to be more dynamic than CK (fig. 7) in response to changing phosphorus concentrations. This is particularly apparent during late May to mid-July when algae are rapidly growing and again during October when vertical mixing is consistently vigorous. From late July to late September cell quotas from CE and CK are similar for the top 20m. They differ at depths of ∿25m in that only CK produces less than maximum cell quotas. This results from averaging nutrient-poor surface algae with nutrient-rich algae at depth. Since the algal flux is greater than the *in situ* nutrient uptake rate the population's internal store is maintained at less than saturation.

SUMMARY AND CONCLUSIONS

One test of model realism is whether it behaves in a consistent manner when the problem scale is changed. Successful application of an ecological model developed for a crude, 2-layer physical segmentation (Scavia 1980) to a 14-layer grid suggested that model consistency exists over a range of scales (5-20m). Some adjustments were necessary however. In order for productivity in the 14-layer model to be similar to that in the 2-layer model, growth rates had to be lowered. This was anticipated and is a consequence of the nonlinear nature of the model, where the production summed and averaged over a fine grid is greater than the average production over a coarse grid. The reduced growth rates necessitated by the fine grid are in better agreement with experimental work and emphasize the importance of selecting mechanisms that are appropriate for the scales at which variability exists (Harris 1980).

As confident as we may be with individual process constructs, success is not assured when incorporating those constructs into an ecological model where feedback is possible. Therefore, it is reassuring when problems caused by a change in scale are reduced when mechanisms more realistic for that scale are used. Such was the case with the phytoplankton growth constructs.

Examination of the multi-group plankton model output showed nitrogen and phosphorus profiles to be correlated to the degree that even the profile inflection points occurred at the same depths. This was contrary to the data and suggested the need to examine the cycling of phosphorus. It was hypothesized that the use of internal nutrient pools to describe algal growth would improve the phosphorus simulations because algae at depth should have higher cell quotas but slower growth rates because of light limitation. Depth variations in modeled cell quotas were persistent enough to make pronounced differences in the available phosphorus profiles.

From this study the following two points can be made: (1) The growth dynamics and vertical structure of phytoplankton populations can be described well by modeling their growth with respect to either internal or external phosphorus concentrations. (2) The ability to simulate the divergence of phosphorus depth profiles from those of nitrogen and silicon required allowing variable phosphorus cell stoichiometry. This suggests that the deep zone of phosphorus depletion in Lake Ontario is controlled by light-limited algae maintaining relatively high phosphorus cell quotas.

REFERENCES

Atwater, M.S., and J.T. Ball. The radiation budget and cloud cover for Lake Ontario during IFYGL. Final Report 4:130-513a, contract 2-35353; 111p. Center for the Environment and Man, Hartford, Conn.

Canada Centre for Inland Waters. 1979. Assessment of water quality simulation capability for Lake Ontario. Scientific Series No. 11; 220p. Canada Centre for Inland Waters, Burlington, Ontario.

Chen, C.W., M. Lorenzen, and D.J. Smith. 1975. A comprehensive water quality model for Lake Ontario. Report to Great Lakes Environmental Research Laboratory (NOAA). 202p. Ann Arbor, Mich.

DiToro, D.M. 1980. Applicability of cellular equilibrium and Monod theory to phytoplankton growth kinetics. *Ecol. Modelling* 8:201-218.

Harris, G.P. 1980. Temporal and spatial scales in phytoplankton ecology. Mechanisms, methods, models, and management. Can. J. Fish. Aquat. Sci. 37:877-900.

McCormick, M.J., and D. Scavia. 1981. Calculation of vertical profiles of lake-averaged temperature and diffusivity in Lakes Ontario and Washington. Water Resources Research 17:305-310.

Pickett, R.L. 1975. Distribution and variability of physical lake properties. IFYGL Bull. 14; pp. 71-73. Nat'l. Ocean. Atmos. Admin., Washington, D.C.

Scavia, D. 1979. Examination of phosphorus cycling and control of phytoplankton dynamics in Lake Ontario with an ecological model. J. Fish. Res. Board Can. 36:1336-1346.

Scavia, D. 1980. An ecological model of Lake Ontario. Ecol. Modelling 8:49-78.

Scavia, D. and J.F. Bennett. 1980. Spring transition period in Lake Ontario -- a numerical study of the causes of the large biological and chemical gradients. Can. J. Fish. Aquat. Sci. 37:823-833.

Stadelmann, P., and A. Fraser. 1974. Phosphorus and nitrogen cycle on a transect in Lake Ontario during the International Field Year 1972-1973 (IFYGL). p. 92-107. In Proc. 17th Conf. Great Lakes Res., Ann Arbor, Mich.

Stadelmann, P., and M. Munawar. 1974. Biomass parameters and primary production at a nearshore and a midlake station of Lake Ontario during IFYGL. p. 109-119. In Proc. 17th Conf. Great Lakes Res., Ann Arbor, Mich.

Steele, J.H. 1965. Notes on some theoretical problems in production ecology. p. 383-398. In C.R. Goldman (ed.), Primary production in aquatic environments. Univ. Calif. Press, Univ. Calif., Berkley, Calif.

Thomann, R.U., D.M. DiToro, R.P. Winfield, and D.J. O'Connor. 1975. Mathematical modeling of phytoplankton in Lake Ontario. Part 1: Model development and verification. Ecological Research Series, U.S. Environmental Protection Agency, EPA-660/3-75-005. 177 p. Corvallis, Oreg.

Vollenweider, R.A. 1965. Calculation models of photosynthesis -- depth curves and some implications regarding day rate estimating in primary production measurements. p. 425-457. In C.R. Goldman (ed.), Primary Production in aquatic environments. Univ. Calif. Press, Univ. Calif., Berkley, Calif.

Vollenweider, R.A., M. Munawar, and P. Stadelmann. 1974. A comparative review of phytoplankton and primary production in the Laurentian Great Lakes. J. Fish. Res. Board Can. 31:739-762.

THE MISSISSIPPI DELTAIC PLAIN REGION (MDPR) STUDY: AN APPLICATION OF ECOLOGICAL MODELS TO THE ANALYSIS AND MANAGEMENT OF A COMPLEX COASTAL REGION[1]

R. Costanza, L. M. Bahr, Jr., C. Neill, S. G. Leibowitz, and J. Fruci[2]

Abstract.--A major portion of the Mississippi Deltaic Plain Region (MDPR) in the Northern U. S. Gulf Coast was characterized for management purposes in terms of a series of quantitative models. These models were constructed at three hierarchically-nested levels of spatial resolution: a single regional model, seven hydrologic unit models, and twenty major habitat models.

The primary objective of the modeling exercise was to produce a coastal zone management/planning tool for the MDPR. An implicit goal was to develop a useful generic modeling technique for regional ecological-economic analysis. The technique developed provides a cataloguing system for the voluminous data that are available on land use, habitat types, and ecological and socioeconomic processes. It involves the incorporation of these diverse data into a series of input-output models. Average conditions of each system are represented as the flows of various ecological and economic commodities (e.g. water, nutrients, fish) between compartments (processes). The models are useful for: 1) estimating the detailed, direct and indirect impacts of small perturbations around the average conditions, 2) objectively weighting (shadow pricing) the various commodities produced (in comparable units), and 3) serving as a preliminary data base for more sophisticated dynamic simulation modeling efforts. We feel that a tool with these characteristics is an essential step toward more effective resource management.

INTRODUCTION

During the last two centuries, coastal regions in the U.S. have undergone dramatic changes. These changes include land clearing and development, impoundment and draining of wetlands, construction of flood control structures, dredging activities, fresh water diversions, pollution from a variety of sources, introduction of exotic pest organisms, and harvest pressure on coastal fish and wildlife. Most of these changes have occurred with only a rudimentary understanding of their long term impacts.

The problem of environmental management was eloquently summarized by Garret Hardin (1968) in his essay "The Tradgedy of the Commons." Many important components of the environment are "common property resources", for which individual's private cost and benefit calculations do not necessarily coincide with costs and benefits to the society as a whole. The example Hardin used had to do with the incentives to individual resource users to increase their number of sheep using a common pasture, ultimately resulting in

---

(1) Paper presented at the Third International Conference on the State-of-the-Art of Ecological Modeling, sponsored by the International Society for Ecological Modelling, Colorado State University, May 24-28, 1982).

(2) Assistant Professor, Associate Professor, Research Associate, Research Associate, and Graduate Research Assistant, respectively, Coastal Ecology Laboratory, Center for Wetland Resources, Louisiana State University, Baton Rouge, LA 70803.

overgrazing and destruction of the pasture. To avoid this fate some system of social control at a level higher than the individual must be implemented that is capable of making individuals aware of the social costs of their actions. Government performs this function by using regulations (i.e. limiting the number of sheep per acre under penalty of fines, imprisonment, etc.) or taxes and subsidies (i.e. taxing each additional sheep added according to its marginal social cost).

Environmental management in the U.S., as in most countries, has employed the regulatory approach almost exclusively. There is strong theoretical support for a tax and subsidy approach, however, since it would mesh more easily with the existing market system, allowing individuals to continue to make their own decisions while making them economically aware of the full social costs of their alternative courses of action (Page 1977). Either approach to environmental management requires accurate information on the magnitude of the social costs involved in order to work effectively.

Environmentalists have, in general, been leary of assigning economic values to environmental resources, mainly because they perceive that the standard economic methods for deriving these values consistently underestimate their worth. They have, instead, tended to take the extreme position that environmental resources are "priceless" or of infinite value. Many of our existing environmental regulations reflect this position. Recent political events indicate the danger of this position, however. Business interests have been polarized to the opposite extreme position that environmental resources are non-essential luxuries that can be sacrificed (at no cost) when things get tight.

What is needed is explicit recognition by both environmentalists and business interests that neither of these extreme positions are accurate. Effective environmental management then becomes a process of quantifying the social costs of environmental changes and implementing some system to make individuals aware of these costs. This is admittedly a very difficult task, but it is a task with a large potential benefit and it deserves the collaborative attention of everyone concerned with environmental management.

Shadow Pricing

Economists refer to the process of calculating the hidden social costs of an activity "shadow pricing." Shadow prices are those that would prevail if private and social costs coincided exactly. Since prevailing economic theory takes preferences as given, economists are most comfortable attacking this problem by asking people (directly or indirectly) to reveal their preferences for environmental goods and services. This approach does not work, however, if people are not aware of the consequences to their own welfare of the alternative allocations of environmental resources. Just as a child is unaware of the social implications of its behavior and must be instructed, most people are unaware of the larger social-ecological implications of resource use decisions and their uninformed preferences cannot be taken too seriously as measures of the true social cost.

An alternative approach to shadow pricing involves the creation of explicit mathematical models of the physical interdependence between ecological and economic systems (Costanza and Neill 1981, Odum 1978). These models can incorporate the best available information on the implications of resource use decisions. The model results can be communicated to the public in the form of taxes and subsidies based on the calculated shadow prices, or they may be used to set standards for regulations.

The Mississippi Deltaic Plain Region

The Mississippi River Deltaic Plain Region (MDPR) includes the 3.4 million ha direct result of 7,000 years of net sediment deposition onto the shallow continental shelf at the mouth of the Mississippi River (figure 1). The MDPR is perhaps as rich in resources, complex in organization, and highly altered as any coastal region in the world. It is one in a series of coastal regions that have been selected for detailed characterization by the Coastal Ecosystems Team of the U.S. Fish and Wildlife Service.

Our approach to ecological characterization was based primarily on the development of a series of nested, hierarchical, input-output multi-commodity models by which we could catalog available data and address the environmental management issues mentioned above. Three levels of spatial resolution were modeled, the entire MDPR, the hydrologic unit level, and the habitat level. Seven hydrologic units were identified within the MDPR, and twenty habitats were aggregated from quantitative aerial data on more than 100 habitats as identified by Wicker (1980). The intention was to provide quantitative information on habitat interconnections useful in implementing either direct regulation or tax-and-subsidy-based approaches to environmental management problems.

METHODS

Input-output models were first suggested by Daly (1968) as a method of integrating environmental externalities into the economic accounting system. Since then several authors have elaborated on this approach (Isard 1972, Ayres 1978, Victor 1972, Hannon 1973, 1976, 1979, Costanza 1980, Costanza and Neill 1981). The major stumbling blocks for this approach have been the availability of a suitable data base, and the working out of certain theoretical problems concerning the treatment of joint products.

Two major goals of the MDPR study were to attempt to assemble the requisite data base, and to propose solutions to the theoretical problems. This paper is a summary of the preliminary results of this effort.

The specific forms of the input-output models developed in this study and their characteristics are discussed in more detail in other contributions to this symposium (Leibowitz and Costanza 1982, Fruci et al 1982). The general strengths, weaknesses, and limitations of input-output models have been the topic of much discussion in the economics and ecological literature (Herendeen and Bullard 1978, von Neuman 1945, Ritz 1979). What follows is a general description of the method as an approach to the shadow pricing problem, and the unique modifications and problems that our particular application made necessary.

Input-Output Models for Environmental Management

Many (but certainly not all) of the questions facing managers of environmental resources concern estimating the detailed, direct and indirect impacts of proposed projects over relatively short time horizons. Usually the projects and their impacts are small relative to the overall size of the resource base. The typical environmental impact statement lists these questions but seldom does it quantify even the direct, much less the indirect, impacts.

As pointed out by Leontief (1941), Isard (1972), Victor (1972), and others, input-output models are particularily well-suited to addressing these detailed, short range questions. An input-output model can be thought of as a "snapshot" or detailed description of the structure of the average or equilibrium interconnections between components in a complex system. It is essentially an accounting system for keeping track of the production and consumption of the various "commodities" that are important in the system under study. These commodities can be economic goods or services (such as steel, retail trade services, labor services) or they can be ecological "goods and services" (such as water, plants, animals, or nutrients). Once a description of the structure of the system has beeen completed (an input-output table) various mathametical representations (models) of the system can be formulated and manipulated to yield information that is not obvious from the input-output table itself. The manipulations that are important to the environmental management problems discussed earlier are the calculation of shadow prices for ecological commodities. These shadow prices can be shown to be equivalent to embodied energy intensities under certain boundary conditions (Costanza 1980) and are therefore potentially useful as a physically-based common denominator linking ecological and economic systems. Once this common denominator has been calculated, the mixed unit original input-output table can be converted into a single unit table (either dollars or embodied energy or some other suitable unit). The single unit table calculated in this way does not ignore any of the important energy or material flows in the system (as a thermal energy or carbon flow table would). Rather, the point is to calculate the equivalence between the various commodities based on their functional interdependence, and to incorporate this information into a single unit table.

Once a single unit table has been obtained, it can be used in much the same way that standard economic input-output tables (which are in all dollar units) are used (cf. Ritz 1979). The most important of these uses is the calculation of direct and indirect impacts of proposed changes. These impacts are often summarized as "multipliers", which state, for example, the total (direct and indirect) dollar impact of a $1 change in final demand for a particular commodity. Analagously, an ecologic-economic input-output table with all flows converted to an equivalent single unit (ie. energy of sunlight quality) might be used to calculate "environmental impact multipliers" for various categories of environmental impacts. For example, we would be able to estimate the total (direct and indirect) impact (in kcal solar or dollars or whatever unit was chosen) on the overall ecological system of a 1 kcal solar change in the primary production of salt marsh resulting from canal building.

RESULTS

Figure 2 is a representative habitat-level input-output table from the MDPR study. This table shows the salt marsh habitat, which is the most thoroughly studied ecological system in the region. Details of the data collection procedures and problems, and analytical formulations and manipulations for this model are given elsewhere in this symposia volume (Leibowitz and Costanza, 1982).

Figure 2 shows the 20 commodities for which we collected data along the side and the 13 transformation processes that use or produce these commodities along the top. The intersection of each commodity and process in the table can contain two entries. The top entry is the input or use of the commodity by the process. The bottom entry is the output or production of the commodity by the process. The total input of each commodity to all internal processes plus export to external processes (which includes physical export, change in storage, and depreciation) is equal to the total output of that commodity by all internal processes plus imports from external processes. Each of the commodity rows in Figure 2 are analogous to the familiar single commodity flow tables that are popular in ecological modeling (cf. Hannon 1979, Richy et. al. 1978). Figure 2 is unique in that it incorporates several of the most important commodities into the same table. The next step is to calculate the equivalence between the various commodities, so that they can all be stated in common units and compared.

One thorny practical and theoretical issue impeding these calculations was the need to deal with joint products (Victor 1972). The joint product problem is usually circumvented by economists by using the relative prices of the joint products to allocate process inputs to outputs. In the ecological case, however, it is exactly these "prices" that are being calculated, and some other method for dealing with joint products must be formulated. Fruci et al (this volume) outlines some approaches to this problem. Additional research is needed in this area, but some preliminary results for the salt marsh model are shown in Table 1 (see Leibowitz and Costanza 1982, this volume, for details of these calculations and a discussion of some of the problems encountered). Table 1 lists the "embodied energy intensities", in kcal of solar energy per physical unit of product, for the commodities produced by the salt marsh.

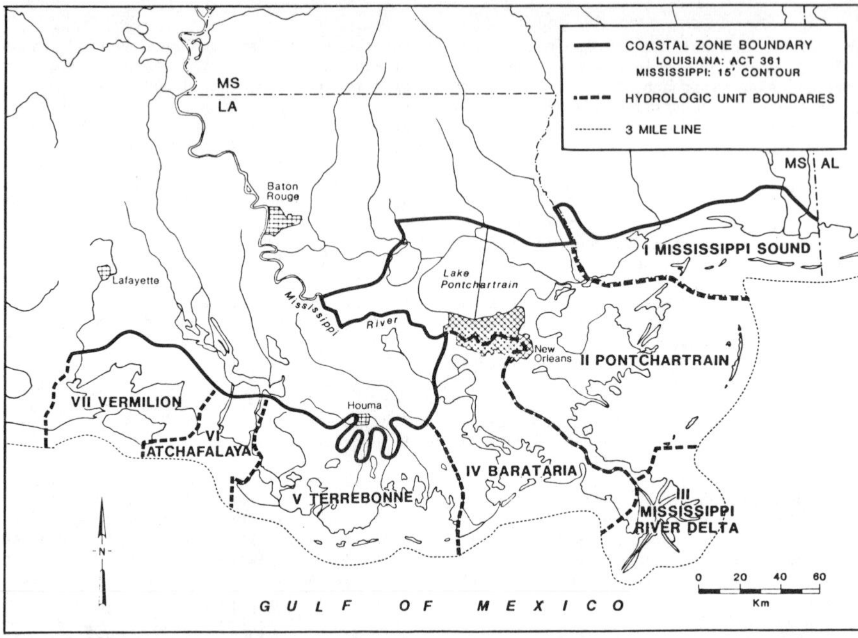

Figure 1.--The Mississippi Deltaic Plain Region (MDPR), indicating the seven hydrologic unit boundaries.

HABITAT 17
SALT MARSH

COMMODITIES		PROCESSES 1 SURFACE WATER	2 SHALLOW SOIL/SEDIMENT	3 MICROFLORA	4 *S. alterniflora*	5 OTHER MACROPHYTES	6 LITTER	7 MICROCONSUMERS	8 CRABS	9 MUSSELS	10 *Littorina*	11 INSECTS	12 BIRDS	13 MAMMALS	NET OUTPUT	TOTAL OUTPUT	UNITS
NITROGEN	1			5.1	42.2	26.3										73.6	g N/m²/yr
			3.6	1.2	15.6	9.6		26	1.0	8.2	8.2	0.02	0.01	0.07	1.0	73.6	g N/m²/yr
PHOSPHORUS	2			1.6	8.4	9.0										16.0	g P/m²/yr
			6.4	0.4	2.0	3.3		3.5	8.2 E-2	0.66	0.66	1.5 E-3	7.3 E-4	5.3 E-3		16.0	g P/m²/yr
CARBON	3			164.4	1093.5	1009.1									1936.2	4166.2	g C/m²/yr
				42.4	363.5	374.4		1004.5	16.3	23.2	74.6	1.5	6.7	12.9	2227.0	4166.2	g C/m²/yr
OTHER NUTRIENTS	4			335.7	3060.9	1757.5										5144.1	g/m²/yr
				77.1	1132.1	662.1		1177.0	17.2	19.0	61.0	1.8	8.1	16.3	1963.4	5144.1	g/m²/yr
TOTAL ORGANIC MATTER	5	1863.4	774.5					1507.6	199	307.4	44.4				2278	7174.3	g dry wt/m²/yr
		1863.4	4516.4						144.1	324.4	274.3	3.9	8.4	19.4		7174.3	g dry wt/m²/yr
SHALLOW SEDIMENTS	6	1936	2626						486		202				1936	7194	g dry wt/m²/yr
		1936	2626						486		202				1936	7194	g dry wt/m²/yr
WATER	7	1.51	0.46	0.03	0.26	0.17									1.51	3.92	10^6 g/m²/yr
		1.51	0.46	0.03	0.26	0.17									1.51	3.92	10^6 g/m²/yr
MICROFLORA BIOMASS	8						366.4				4.4	1.3	0.8	14.1	121	527.0	g dry wt/m²/yr
				527												527.0	g dry wt/m²/yr
S. alterniflora BIOMASS	9		1377				1206.8				13.3	3.9			1534	4134.0	g dry wt/m²/yr
					4134											4134.0	g dry wt/m²/yr
OTHER MACROPHYTIC BIOMASS	10		988				787.9				9.0	2.6	1.5	29.0	1040	2603.0	g dry wt/m²/yr
						2603										2603.0	g dry wt/m²/yr
LITTER BIOMASS	11		1431.9					563.6			383.6					2379.1	g dry wt/m²/yr
							2379.1									2379.1	g dry wt/m²/yr
MICROCONSUMER BIOMASS	12														2210	2210	g dry wt/m²/yr
								2210								2210	g dry wt/m²/yr
CRAB BIOMASS	13							16.86					2.29	1.34	34.04	54.6	g dry wt/m²/yr
									54.6							54.6	g dry wt/m²/yr
MUSSEL BIOMASS	14							18.12					2.39	1.39	51.1	73.0	g dry wt/m²/yr
										73.0						73.0	g dry wt/m²/yr
Littorina BIOMASS	15							21.77					2.86	1.67	164.6	190.9	g dry wt/m²/yr
											190.9					190.9	g dry wt/m²/yr
INSECT BIOMASS	16							0.51					0.06	0.03	3.3	3.9	g dry wt/m²/yr
												3.9				3.9	g dry wt/m²/yr
BIRD BIOMASS	17							6.8 E-2						0.01	14.8	14.86	g dry wt/m²/yr
													14.86			14.86	g dry wt/m²/yr
MAMMAL BIOMASS	18							1.41						0.14	28.36	29.9	g dry wt/m²/yr
														29.9		29.9	g dry wt/m²/yr
FISH BIOMASS	19												13.4	1.68		15.1	g dry wt/m²/yr
															15.1	15.1	g dry wt/m²/yr
HEAT	20														661,511	661,511	kcal/m²/yr
				44,966	364,372	240,277		10,829	74	236	404	18	86	156		661,511	kcal/m²/yr
SUNLIGHT				0.8	3.7	2.5											10^6 kcal/m²/yr

Figure 2.--MDPR salt marsh commodity-by-process input-output table.

Table 1.--Embodied energy intensity values (kcal solar/g) for ecological commodities in MDPR salt marshes.

Commodity	Value	Commodity	Value
Nitrogen	1622.4	Litter	70.3
Phosphorus	10,562.1	Microconsumers	12.2
Carbon	29.6	Crabs	170.3
Other nutrients	31.8	Mussel	74.4
TOM	19.2	Littorina	119.2
Shallow sediments	17.1	Insects	141.1
Water	0.4	Birds	66.9
Microflora	130.4	Mammals	125.2
S. alterniflora	52.4	Fish (1)	0.0
Other macrophytes	68.3	Heat (2)	0.4

(1) Fish are not produced in the salt marsh and their embodied energy intensity must be calculated using their resident habitat model.
(2) Embodied energy intensity as kcal/kcal.

USES OF THE I-O MODEL RESULTS

The MDPR study includes models and calculations analogous to those outlined above for the salt marsh, for a total of 8 of the most important ecological and economic habitats in the region. The remaining 12 habitats were treated in a more qualitative manner due to insufficient data. The habitat models were aggregated and additional information included to produce I-O tables for the 7 hydrologic units in the region, as well as for the region as a whole. This information should be useful to environmental managers as a description of the functional interdependencies of the resident systems. When the I-O models are completed and the manipulations described above performed, they are useful as estimates of the "value" of the systems and their internal components, and as tools for quantitative impact analysis (cf. Leibowitz and Costanza 1982, Fruci et. al. 1982, this volume). This is certainly not all of the information that environmental management requires, nor are the methods employed the only ones possible for producing it. It is, however, a consistent, quantitative, and systematic approach that attempts to incorporate as much of the existing ecological information as possible into a useable format.

LIMITATIONS AND SUGGESTIONS FOR FURTHER RESEARCH

The I-O modeling format is not intended to be an ultimate solution to the environmental management problem. It is, however, a potentially useful conceptual and analytical tool capable of addressing an important class of problems. There are other classes of problems for which it is ill-suited and for which other tools should be used. For example, long term or dynamic effects are better handled with simulation models. The problem of environmental management is best addressed by developing a working knowledge of all the available tools and their proper applications, just as a carpenter knows when to use a hammer and when to use a saw.

The major limitation to effective use of the I-O approach is the difficulty of assembling the requisite data base with a suitable degree of precision. Advances in automated sampling of ecological systems may one day be able to effectively solve this problem. In the meantime the approach remains an important conceptual tool for integrating ecology and economics, and produces quantitative results that are useful (although they are not extremely precise). The precision of the models can be improved as new data are generated and incorporated.

ACKNOWLEDGEMENTS

This research was funded by the U. S. Department of the Interior, Bureau of Land Management, Fish and Wildlife Service, National Coastal Ecosystems Team, Contract No. 14-16-0009-80-073, "Mississippi Deltaic Plain Region Ecosystem Models and Narrative Report, Martha W. Young, Project Officer. This is contribution #LSU-CEL-82-09 from the Coastal Ecology Laboratory, Center for Wetland Resources, Louisiana State University, Baton Rouge, LA. We thank C. Cleveland and J. W. Day Jr. for reviewing drafts and providing useful comments.

LITERATURE CITED

Ayres, R. N. 1978. Resources, environment, and economics: aplications of the materials/energy balance principle. Wiley Interscience, N. Y. 297 pp.

Costanza, R. 1980. Embodied energy and economic valuation. Science 210:1219-1224.

Costanza, R. and C. Neill. 1981. The energy embodied in the products of the biosphere. pp. 745-755 in W. J. Mitsch, R. W. Bosserman, and J. M. Klopatek (eds.) Energy and ecological modelling. Elsevier, N.Y. 839 pp.

Daly, H. E. 1968. On economics as a life science. J. Pol. Econ. 76:392-406.

Fruci, J. R., R. Costanza, and S. G. Leibowitz. 1982. Quantifying the interdependence between material and energy flows in ecosystems. Paper presented at the Third International Conference on the State-of-the-Art in Ecological Modeling, Colorado State University, May 24-28, 1982.

Hannon, B. 1973. The structure of ecosystems. J. Theor. Biol. 41:535-546.

Hannon, B. 1976. Marginal product pricing in the ecosystem. J. Theor. Biol. 56:256-267

Hannon, B. 1979. Total energy costs in ecosystems. J. Theor. Biol. 80:271-293.

Hardin, G. 1968. The tragedy of the commons. Science 162:1243-1248.

Herendeen, R. A., and C. W. Bullard. 1974 Energy costs

of goods and services, 1963 and 1967. CAC Document No. 140. Energy Research Group. University of Illinois, Champaign, IL.

Leibowitz, S. G. and R. Costanza. 1982. An input-output model of salt marshes in the Mississippi Deltaic Plain Region. Paper presented at the Third International Conference on the State-of-the-Art in Ecological Modeling, Colorado State University, May 24-28, 1982.

Leontief, W. W. 1941. The structure of American economy, 1919, 1929: an empirical application of equilibrium analysis. Harvard University Press, Cambridge, MA.

Isard, W. 1972. Ecologic-economic analysis for regional development. The Free Press, N.Y. 270 pp.

Odum, H. T. 1978. Energy analysis, energy quality, and environment. pp. 55-88 in: M. W. Gilliland (ed.), Energy analysis: a new public policy tool. AAAS Selected Symposia 9. Westview Press, New York. 110 pp.

Page, T. 1977. Conservation and economic efficiency. Johns Hopkins, Baltimore.

Richey, J. E., R. C. Wissmar, A. H. Derol, G. E. Likens, J. S. Eaton, R. G. Wetzel, W. E. Odum, H. M. Johnson, O. L. Loucks, R. T. Prentki, and P. H. Rich. 1978. Carbon flow in four lake ecosystems: a structural approach. Science 202:1183-1186.

Ritz, P. M. 1979. The input-output structure of the U. S. economy, 1972. Survey of Current Business. Feb. 1979: pp. 34-72.

Victor, P. A. 1972. Pollution, economy, and environment. University of Toronto Press, Toronto, Canada. 247 pp.

von Neumann J. 1945. A model of general economic equilibrium. Rev. Econ. Studies. 13:1-9.

Wicker, K. M. 1980. The Mississippi Deltaic Plain Region habitat mapping study. Data tapes. Office of Biological Services, U.S. Fish and Wildlife Service.

A BAYESIAN FRAMEWORK FOR INFORMATION TRANSFERABILITY IN RISK ANALYSIS[1]

Kenneth H. Reckhow[2]

Abstract.--Information and cost constraints often prevent the conduct of site-specific tests for model development, parameter estimation, or applied scientific research in water quality studies. When hazardous substances are involved, contamination due to testing becomes an additional constraint. This necessitates the transfer or extrapolation of information from other points in space or time. In the context of decision making under uncertainty, there is a need to estimate the uncertainty associated with the transfer of information. A Bayesian approach is proposed that is based on statistical measures of similarity between sites. This is developed as part of a methodology for risk analysis in the management of hazardous substances in aquatic systems.

INTRODUCTION

Many regulatory decisions are and will be made on the basis of information that is non-specific to the site of concern. The cost of information acquisition can be quite high, time constraints on the issuance of the regulation may prevent site-specific study, or in the case of hazardous substances the concern for contamination may prohibit scientific testing. As a result, information is transferred or extrapolated from other sites and additional uncertainty is incorporated into the regulatory decision. Ideally the choice of non-site-specific information should be based on an explicit analysis of the costs and risks associated with the uncertainty of information. In reality these terms are too poorly defined to permit rigorous analysis, and the pragmatic choice of readily available information from other sources often results.

In decision making under uncertainty the need to estimate the error of information transfer becomes apparent, as the alternative is an incomplete accounting of error terms. For example, risk analysis for the assessment of hazardous substance contamination of aquatic systems must often be based on model parameter estimates, dose-response relationships, etc. extrapolated from other sites and studies. However, if we are to apply a decision theoretic approach to this problem, we must attempt to estimate the additional uncertainty contributed by the information transfer. This issue is currently being confronted in studies on risk analytic approaches for the management of toxic substances in the Great Lakes watersheds, and this paper explores some methods for the estimation of information extrapolation uncertainty.

BAYESIAN DECISION ANAYSIS

Decision making under uncertainty involves an action space ($A = (a_j)$) consisting of all possible actions that may be taken in a given situation, a state space ($\Omega = (\Theta_i)$) consisting of all possible states of nature that could occur in that situation, and a loss function $L\ (a_j; \Theta_i)$ that represents the utility associated with a state-action pair. Since uncertainty is present, the actual or resultant state of nature is unknown, and therefore the appropriate action to be taken to achieve a favorable loss is also unknown. The strategy then is to identify the governing probability law from which the state of nature (Θ_i) is generated, and to select an action on the basis of a chosen optimality criterion. For example, the expected loss is:

$$\int_{\Theta} L\ (a,\Theta)\ p(\Theta)d\Theta \tag{1}$$

where,

$p(\Theta)$ is the probability model governing Θ.

An action that minimizes risk is one that results in a minimum loss in Equation 1.

The decision theoretic model in Equation 1 has two basic components, a loss function expressing the utility of action-state pairs and a probability expressing the uncertainty in knowledge concerning the resultant state of the system. Specification of the loss function is important as it can strongly affect the optimal action (see Berger 1980). The concern of this paper, however, is the encoding of uncertainty in the probability portion of Equation 1.

In a situation involving the transfer of non-site-specific information, there is often a need to pool information from several sources. Further, the information may not involve repeated measurements so that judgment concerning uncertainty may be necessary. Under those circumstances, Bayes Theorem becomes useful for the determination of the probability in Equation 1. This provides a means of expressing uncertainty as to the state of a system (Θ), given (E) and a prior probability on the system state $p(\Theta)$:

---

[1] International Symposium on the State-of-the-Art in Ecological Modeling, Colorado State University, Ft. Collins, Colorado, USA.

[2] Assistant Professor, Water Resource Systems, School of Forestry and Environmental Studies, Duke University, Durham, North Carolina, USA.

$$p(\Theta|E) = \frac{p(E|\Theta)\ p(\Theta)}{p(E)} \tag{2}$$

where,
$$p(E) = \int_\Theta p(\Theta)p(E|\Theta)d\Theta$$

Θ is a parameter characterizing the state of the system. For example, if the normal probability law governs, Θ would be defined by the mean and variance. In many situations, however, the state parameters are of less importance than is a prediction of the likely future state the system. This future state, y, may be estimated from the predictive distribution $p(y|E)$:

$$p(y|E) = \int_\Theta p(y|\Theta)\ p(\Theta|E)d\Theta \tag{3}$$

The predictive distribution is seen to be a function of the probability generated from the application of Bayes Theorem and the probability of observing a future state, y, given the parameter Θ. Predictive distributions (Aitchison and Dunsmore 1975) are a natural extension of Bayes Theorem when the objective of the analysis is prediction. They are attractive because they express the unknown parameter as a probability density function instead of as a point value. This means that uncertainty concerning the parameter, Θ, is used in the calculation of the prediction, y, whereas under the traditional methods the parameter is assumed to be known. Similar to the application of Bayes Theorem, use of the predictive distribution is facilitated by the identification of a conjugate family of distributions. This greatly simplifies the mathematics. Aitchison and Dunsmore provide several tables presenting the conjugate distributions and illustrate their use.

THE INFORMATION TRANSFER PROBLEM

The problem as presented may be generally characterterized as concerned with the transfer of a dose-response model from one system to another. In actuality, the dose-response model may relate discharge and concentration, or perhaps ambient concentration and organism concentration; in each case the former term represents dose and the latter term represents response.

At least two models are involved in the information transfer process (Larsen and Reckhow 1982). The first model presents the information to be transferred or extrapolated. As stated above, for simplicity this model is given the generic title of a dose-response model. The second model is used to determine the acceptability of information extrapolation. In essence this model is intended to estimate the similarity between two systems, one system for which the first (dose-response) model has been specified and fitted and a second system to which the information (dose-response model) may be transferred. Thus model 1 represents the information to be transferred and model 2 is used to determine the acceptability of transfer. Either or both models may be used to estimate the uncertainty of extrapolation, depending upon the characteristics of the problem.

The framework for evaluating and conducting the information transfer is Bayesian decision theory using the predictive distribution. In some situations the acceptability of transfer may be a trivial issue and all that is required is the predictive distribution to estimate the uncertainty of extrapolation. For example, the information to be transferred may be a discharge-concentration model relating the mass of a substance discharged to the concentration in a water body and to certain (e.g., hydraulic) characteristics of the water body. If this model has been confirmed ("verified") on a random sample of water bodies from a population containing the second system (to which information may be transferred), then acceptability of extrapolation is not of concern and the error of extrapolation is equivalent to the error of interpolation. In that situation, the inference space for model 1 is the entire population and extrapolation is relatively straightforward.

When it is uncertain whether or not the inference space of model 1 contains the second system, then the acceptability of extrapolation must be assessed on the basis of similarity between the two systems. This is equivalent to evaluating the inclusion of the second system in the inference space of model 1. At this point a rejection criterion must be established and statistical tests conducted to evaluate membership of the second system in the inference space of model 1. If the tests indicate that inclusion of the second system in the inference space cannot be rejected, then extrapolation is possible and the uncertainty of extrapolation is estimated using the predictive distribution for model 1 applied to the second system.

If the statistical tests result in rejection of the proposed information transfer, then it may be possible to use the test results to determine the necessary modifications to model 1 to permit extrapolation. From a substantive sense, the rejection is probably due to a mis-specification of model 1 which manifests itself in a statistical sense as additional variance or bias. In the situations addressed in this paper, that is, when data and resources are scarce preventing the inductive effort devoted to re-specification of model 1, then one must rely on the system comparison studies to suggest ad hoc modifications to model 1 to permit application on the second system. This observation must be conditional on a comparison of the risk (or net cost) or either action: attempt to re-specify model 1 for the second system or use model 1 with ad hoc modifications to permit estimation of the extrapolation uncertainty. Given the decision-theoretic context in which this problem was posed, this consideration of risks is important. More is said concerning this model selection and specification risk later in the paper.

MEASURES OF SYSTEM SIMILARITY AND EXTRAPOLATION REJECTION

As defined above, the problem of assessing system (or perhaps site-to-site) similarity concerns the specification and evaluation of model 2. This model provides a comparison of the two systems (the system studied and fitted with model 1 and the system to which the model 1 information may be extrapolated); the purpose of model 2 is to determine whether or not model 1 is applicable to both systems. Specification of model 2 is crucial. If model 2 involves variables and a functional form not appropriate for the evaluation of the use of model 1 on the second system, then it cannot provide a reliable indication of information transferability.

It is proposed here that model 2 consist of a multivariate analysis of varibles that are deemed important in the cause-effect behavior that is described in model 1 (the so-called dose-response model). Much of the model 2 specification problem is concerned with the correct identification of these variables, and their functional form, for the multivariate analysis. Model specification is an inductive, creative process based on a knowledge of theoretical principles and reinforced through statistical tests. Little can be said to improve this step save to underscore its importance.

Once a specification for model 2 is selected, statistical tests must be performed and a rejection criterion chosen. The rejection criterion is problem-specific, and it represents a trade-off between type I and type II error. The statistical tests provide a measure of multivariate similarity between the two systems. One likely measure for system similarity involves a multivariate comparison of centroids and covariance matrices for the variables from the two systems (see Green 1978). For example, Hotelling's T^2 statistic may be used to test the difference between group centroids for several variables simultaneously. The T^2 statistic is a function of the Mahalanobis distance (D^2) and the test is based on the F distribution:

$$(T^2)\ \frac{m-n-1}{n(m-2)} = (D^2)\ \frac{m_1\ m_2(m-n-1)}{n(m)\ (m-2)} \sim F_{n,m-n-1} \qquad (4)$$

where,

$m = m_1 + m_2$ = number of observations
m_1 = number of observations in group 1
m_2 = number of observations in group 2
n = number of (predictor) variables

Thus the difference between the "centers" of the two systems on the basis of pre-specified key variables could be estimated from this test and expressed as an F statistic or by the Mahalanobis distance.

Hotelling's T^2 test requires an estimate of the pooled within-groups covariance matrix, and is based on an assumption of equal variances within groups. It is generally believed that the test is relatively robust to violation of this assumption. Green presents a test (proposed by Bartlett) of the equality of covariance matrices using the chi square statistic; this may be used to test the equality assumption and to provide additional information on the similarity of the two systems.

The statistical measures of system similarity provided by the tests of equality of group centroids and covariance matrices provide the basis for acceptance or rejection of the extrapolation of model 1. If the statistical tests suggest rejection, a premise of this paper is that the constraints of the problem may require extrapolation. The issue then concerns how model 1 may be revised, given the information obtained in the statistical tests. This issue is problem-specific and is in need of research, but clearly if model 2 is well-specified, the statistics on group centroids and covariances are informative concerning expectation and uncertainty in the application of model 2 on the second system.

One possibly fruitful means of expressing both theoretial and statistical information on important system variables is through structural equations (Kenny 1979). Under this methodology, linear equations are specified on the basis of theory to describe the relationships among the variables. For each system, the parameters may be estimated using maximum likelihood. A chi square statistic can then be calculated that reflects the similarity between the two systems on the basis of the fitted linear structural equations. Again, specification of the model is critical, and again research is needed to determine how this analysis may be employed to evaluate information transferability.

A BAYESIAN FRAMEWORK

It is proposed that Bayes Theorem be used to relate the information on expectation and uncertainty in model 1, fitted on the first system, to the second system. When the statistical tests indicate that the second system is contained within the inference space of model 1, then extrapolation is relatively straightforward as shown in the example below. When rejection of extrapolation is indicated, then costs and risks must be considered with the alternatives: test and develop the dose-response model directly on the second system or use auxiliary information (theory and the statistical tests) to modify model 1 for use on the second system. For this latter alternative, Bayes Theorem may be used to construct the modification. For example, the prior (e.g., for expectation and uncertainty) is specified by the investigator on the basis of à priori judgment as to the applicability of model 1 on the second system. If he/she beleives that both systems belong to the same population and that model 1 is representative of that population, then the prior is model 1. If the investigator believes that the second system is outside the inference space of model 1, then the prior should reflect that judgment. Using the normal probability model as an example, this may result in a prior that is displaced in location and scale according to the expert judgment of the investigator. Then the likelihood, conditional on the experimental evidence, is derived from the multivariate analysis described above. The likelihood could be based on an instrumental variable; alternatively the likelihood for the location statistic could be proportional to the square root of the Mahalanobis distance while the likelihood scale factor might reflect the multivariate F ratio.

Since the statements made in the previous paragraph are just suggestions that await the examination in research on actual information transfer problems, the example presented below is primarily illustrative of straightforward determination of the predictive distribution using Bayes Theorem. To underscore the generality of the procedure (beyond the stated application of risk analysis for toxic substance management) and call upon previous experience of the author, the following problem is posed. It is desired to extrapolate a regional regression (model 1) for lake phosphorus retention to a lake not included in the original regression analysis. To test whether or not this lake is in the inference space of model 1, model 2 is defined as a multivariate comparison of the lake with those lakes in the regression data set. Variables chosen are lake phosphorus concentration, mean depth, hydraulic detention time, and monthly epilimnetic temperature. A comparison of centroids and covariance matrices (for the non-constant terms) indicates that extrapolation of model 1 to the lake is acceptable on the basis of the rejection criterion. This means that the investigator can now use the Bayesian procedure to estimate the uncertainty of extrapolation from the predictive distribution.

Although the investigator working on this problem may have experience on which to construct a prior, assume for objectivity that the prior is vague. It is further assumed that phosphorus retention is adequately described by the normal-chi-square (NCh) conjugate distribution (Aitchison and Dunsmore 1975), where the normal characterizes the mean and the chi square characterizes the variability (sum of squared deviations). According to Aitchison and Dunsmore, the vague prior for this distribution is:

$$\text{prior: vague} \quad p(\Theta) = p(\mu,\tau) = NCh(b,c,g,h) \qquad (5)$$

where,

μ,τ = mean and precision statistics

$c\to 0,\ g\to 0,\ h\to 0$

In this problem the likelihood is the regional regression, and this information may be expresed in general terms as:

$$p(E\,|\,\Theta) = N(\mu\ ,\ k\tau\ ) \qquad (6)$$

where,

$\mu,k\tau$ = expectation and precision for the regression equation

The posterior distribution is then the normal-chi-square:

$$\text{posterior:} \quad p(\Theta|E) = p(\mu,\tau\ |E) = NCh(y|x,n,df,ss) \qquad (7)$$

where,

$y|x$ = the predicted y (phosphorus retention), given x, the predictor variables

n = the number of cases (lakes) in the regresssion model development data set

df = the number of degrees of freedom for the regression equation

ss = the sum of squared deviations for y

Finally, the predictive distribution is Student (St):

$$\text{predictive:} \quad p(y|x) = St(df,\ y|x,\ s^2{}_{y|x}) \qquad (8)$$

where,

$s_{y|x}$ = the regression standard error

Thus the extrapolation of model 1 to the lake results in a prediction that is characterized by the Student distribution describing the usual regression statistics. Hardly a surprising result, but the example as a whole does illustrate how the procedure may be applied.

CONCLUSIONS

The extrapolation of information from one site to another might be considered the "third" extrapolation problem in risk analysis. Much effort has been devote to the first two extrapolation problems: the extrapolation of dose-reponse models to low doses and the extrapolation of experimental results from animal tests to man. The need for the transfer of scientific information from one site to another arises when circumstances prevent direct experimentation on the second system. A Bayesian framework is proposed that combines a loss function with a probability model for system parameters. The focus in this paper is on the probability model ultimately expressed as a preditive distribution. However, the loss function must also be considered carefully as part of the risk analysis.

Two models have been defined in this framework. This first model represents the information to be extrapolated from one site to another. The second model provides the evaluation as to whether or not extrapolation is acceptable, given a rejection criterion. In essence, model 2 determines if model 1 is mis-specified for the second site. However, model 2 may also be mis-specified, yielding misleading results on information extrapolation. Therefore, it is particularly important that a good deal of thought and theory (about all that can be called upon for this inductive problem) enter into the specification of the second model.

Finally, perhaps the most important result in calling attention to this "third" extrapolation problem may be more frequent comparison of the risk of extrapolation versus the risk of direct experimentation on the second site. In some cases, direct experimentation may be impossible (perhaps due to time constraints) or clearly of great risk (perhaps due to the introduction of an extremely toxic recalcitrant chemical into an aquatic system). However, there are likely to be situations where the extrapolation test indicates rejection of information transfer, and the resultant extrapolation uncertainty is large (and perhaps poorly defined). Depending upon the loss function, the risk of direct experimentation may favor that option. This comparison of risks should be conducted whenever the risk of information transfer is unsatisfactorily high. As experience is gained with extrapolation risk, a comparison of risks will become more sharply defined.

ACKNOWLEDGMENT

Portions of this research were supported by a Grant from the National Oceanic and Atmospheric Administration, Contract Number NA81RAC00075.

LITERATURE CITED

Aitchison, J. and I.R. Dunsmore. 1975. Statistical Prediction Analysis. Cambridge University Press. Cambridge.

Berger, J.O. 1980. Statistical Decision Theory. Springer-Verlag. New York.

Green, P.E. 1978. Analyzing Multivariate Data. Dryden Press. Hinsdale, Illinois.

Kenny, D.A. 1978. Correlation and Causality. J. Wiley and Sons. New York.

Larsen, D.P. and K.H. Reckhow. 1982. Field-to-Field Extrapolation. USEPA Workshop Proceedings. Airlie, Virginia.

ENVIRONMENTAL MODEL FOR GUAYANILLA BAY, PUERTO RICO[1]

Michael A. Chartock[2]

A two-dimensional compartment model was developed to account for water exchanges, sedimentation, and the carbon budget among trophic groups in Guayanilla Bay, an industrialized area on the Caribbean Coast of Puerto Rico. Bathymetry, wind drift, and tides were important features controlling the physical characteristics of geomorphological compartments. Exchange of 50 percent of the Bay's water occurred in about six days under average conditions. The density of biomass of benthic organism was projected to be largest in those areas bordered by mangroves. Model verification was partially accomplished through agreement with the relative spatial and seasonal distributions of animal density, abundance of particulate matter, and salinity among Bay compartments.

INTRODUCTION

The bays along the south coast of Puerto Rico provide effective case studies for both industrialized and natural ecosystems. Industrialized ecosystems (Tilley, 1979) on the south coast of Puerto Rico include the bays of Guanica, Guayanilla, Ponce, Jobos, and Yabucoa (fig. 1). The less disturbed systems include the bays of Sucia, Montalva, and Rincon, although these latter bays have been subject to some urbanization or periodic pollution events.

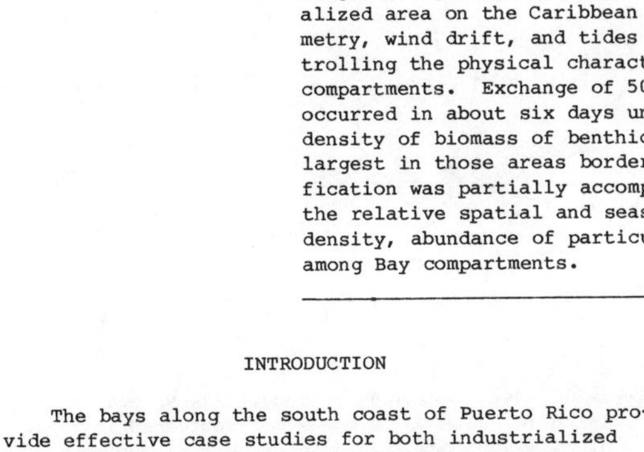

Figure 1. The major bays of Puerto Rico including Guayanilla Bay and Tallaboa Bay area.

The biological production of these south coast bays can be intensive (Odum, Burkholder, and Rivero, 1962). Contributions to productivity from mangroves, sea grass bed, and planktonic communities, and terrestrial runoff make them interesting case studies for evaluating the interactions among community types.

Guayanilla Bay and the adjacent Tallaboa Bay are perhaps the most industrialized south coast bays (fig. 2). The eastern side of Guayanilla Bay has been modified by the construction of jetties and now has three distinctive geomorphologial compartments or subbays:. an Intake Subbay that is the location of a 2,400 megawatt oil-fired power plant; Thermal Cove receiving power plant effluent; and a South East Subbay created by the construction of the Commonwealth Oil Refining Company jetty. The Central Subbay is a relatively deep harbor (Table 1). The Western Subbay is shallow and less industrialized with a border of mangroves that have been partially replaced by adjacent sugar cane. The environmental characteristics of Guayanilla Bay and its compartments are apparently distinctive in in terms of their specific patterns of community distribution, hydrography, and industrialization, but the approach described in this study may be applied to other south coast bays.

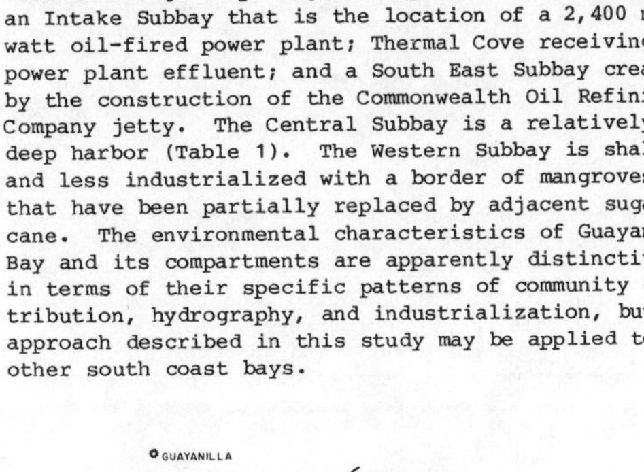

Figure 2. Guayanilla Bay compartments.

[1]Paper presented at the Third International Conference on State-of-the-Art in Ecological Modeling, Colorado State University, May 24-28, 1982.

[2]Michael A. Chartock is Associate Professor of Zoology and Research Fellow, Science and Public Policy Program, University of Oklahoma, Norman, Oklahoma; and was Visiting Scientist, Center for Energy and Environment Research, University of Puerto Rico, Mayaguez, Puerto Rico.

Table 1. Morphological characteristics of subbays.

Subbay	Compartment	Area (km^2)	Average Depth (m)	Volume (10^6 m^3)
Western (WB)	Q1	3.06	1.08	3.3
Central (CB)	Q2	4.33	8.72	37.7
Intake (I)	Q3	.38	3.5	1.4
Thermal (TC)	Q4	.24	2.19	0.5
South East (SEB)	Q5	.81	1.92	1.6
Total		8.82	5.0	44.5

The objective of this paper is to describe a two-dimensional compartment model of Guayanilla Bay that approximates the major carbon and materials exchanges among subbay compartments and the Caribbean. The purposes of the model development were to: integrate existing studies on physical, chemical, and biological characteristics; to identify major data uncertainties; and to organize information in a manner useful for identifying ecosystem characteristics that could be considered when evaluating environmental control and coastal management options. The model was intended to: (1) represent the major geomorphological, physical, and biological characteristics of the Bay; (2) to provide partial verification of properties against the relative spatial and seasonal distribution of animal density, particulate matter, and salinity among bay compartments; and (3) provide one basis for identifying possible consequences of future industrial changes to the Bay.

MODEL STRUCTURE

The Guayanilla Bay model is a multicompartment simulation model that accounts for exchanges of water among subbays and the Caribbean; sedimentation of particulate organic matter (POM) from the water column; and the flow of carbon (or other materials) through major trophic compartments. The model consists of two coupled submodels: (1) a water exchange submodel; and (2) a trophic submodel.

The model operates on either a short- or long-term time frame. As a short-term model, hourly estimates are provided for variables such as water volume, POM, and the concentration of selected conservative dissolved materials in bay compartments. This mode is suitable for projections of weeks to several months. As a long-term model, estimates of carbon biomass of trophic groups are made for periods of months to several years.

Water Exchange Submodel

The water exchange submodel simulates exchanges of water and specified conservative materials among the subbays (table 1). This submodel accounts for hourly changes in subbay volume as affected by wind-driven and tidal movement, industrial pumping and ship activity, runoff, precipitation, and evaporation. Particulate material or dissolved materials concentrations can be simulated and the deposition of settleable substances to the benthos can be estimated with specified sedimentation rates.

The flows represented in the submodel account for the identified sources of water flux on the coastal bays of Puerto Rico where the major water movements are from tides and wind drift (Kamel and Hadjitheodorou, 1970; Goldman, 1979). The water flows among subbay compartments are indicated in figure 3 and listed in table 2. The forcing functions affecting the water budget are listed in table 3. Storm tides and high winds can also be included to simulate hurricane conditions.

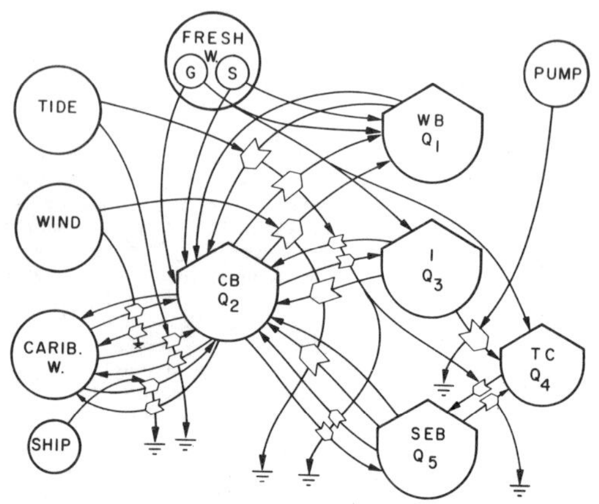

Figure 3. Water exchange submodel. Flows are identified in table 2. Symbols are indicated in table 1. G = groundwater; S = surface water; pump = industrial pumping

Table 2. System flow rates and transfer coefficients.

Flow	Transfer Coefficient	Type	Source Bay[1]	Sink Bay[1]	Average Daily Flow Rate (10^6 m^3/day)
J1	K1	Tide	CB	CAR	2.7
J2	K2	Tide	CAR	CB	2.7
J3	K3	Wind	CB	CAR	2.0
J4	K4	Wind	CAR	CB	3.0
J5	K5	Equilibrium	WB	CB	1.9
J6	K6	Wind	CB	WB	1.8
J7	K7	Tide	WB	CB	0.9
J8	K8	Tide	CB	WB	0.9
J9	K9	Stream/Ground	--	WB	0.072
J10	KA	Evaporation	WB	--	0.017
J11	KB	Rainfall	--	WB	0.007
J12	KC	Stream/Ground	--	CB	0.15
J13	KD	Evaporation	CB	--	0.024
J14	KE	Rainfall	--	CB	0.011
J15	KF	Groundwater	--	I	0.008
J16	KG	Evaporation	I	--	0.002
J17	KH	Rainfall	--	I	0.001
J18	KI	Wind	I	CB	3.3
J19	KJ	Tide	CB	I	0.11
J20	KK	Tide	I	CB	0.11
J21	KL	Pumping	I	TC	2.6
J22	KM	Evaporation	TC	--	0.0021
J23	KN	Groundwater	--	TC	0.002
J23	KN	Groundwater	--	TC	0.002
J24	KO	Rainfall	--	TC	0.0006
J25	KP	Tide	SEB	TC	0.071
J26	KQ	Tide	TC	SEB	0.071
J27	KR	Wind	TC	SEB	0.17
J28	KS	Evaporation	SEB	--	0.0005
J29	KT	Precipitation	--	SEB	0.0002
J30	KU	Tide	CB	SEB	0.24

(continued)

Table 2. (continued)

J31	KV	Tide	SEB	CB	0.24
J32	KW	Wind	SEB	CB	4.4
J33	KX	Shipping	CAR	CB	0.058
J34	KY	Shipping	CB	CAR	0.058
J35	KZ	Equilibrium	CB	I	5.9
J36	L1	Equilibrium	TC	SEB	2.4
J37	L2	Equilibrium	CB	SEB	1.7
J38	L3	Equilibrium	CB	CAR	1.7

[1]CAR = Caribbean; CB = Central Subbay; I = Intake Subbay; SEB = Southeast Subbay; TC = Thermal Cove; WB = Western Subbay.

Table 3. Factors affecting water movement.

Variable	Name	Function	Characteristics
I1	Tide	Cosine function	30 cm sea elevation change[1]
I2	Wind	Switching	Velocity average 2.9 m/sec[1]
I3	Streamflow and ground water	Annual average	87 x 10^6 m^3 per year from impounded watershed[1]
I4	Precipitation	Annual average	90 cm/year[2]
I5	Evaporation	Annual average	200 cm/year[2]
I6	Shipping	Daily average	60,000 m^3/day[3]
I7	Industrial Pumping	Daily average	2.16 x 10^6 m^3/day[4]

[1]Puerto Rico Water Resources Authority, 1972.

[2]Crooks *et al.* 1968.

[3]Puerto Rico Port Authority, 1980.

[4]Puerto Rico Electric Power Authority, 1980.

Tides

The tide at Guayanilla Bay is primarily diurnal and can be approximated as a cosine function:

$$I1 = COS\ (HR/12.5)(3.14)\ (TD)$$

where:

HR = the time on a 25 hour tidal day

TD = Tidal Range

The tidal range varies between 15 to 45 cm with a daily average range of 30 cm. Based on this range, flux for the bay is 2.7 x 10^6 m^3 daily. Tidal flux was a small proportion of Central Subbay volume (7%). However, tidal flux was a large proportion of the water volume of the Western Subbay (28%), and is significant in the other shallow-water compartments.

Wind Drift

Goldman (1979) estimates that drift in the upper (2-3 meters) mixed layer is approximately 1 percent in South Central Caribbean Bays, as supported by studies during 1977 through 1979. Water movement in Guayanilla Bay is affected by circadian and seasonal variation in wind direction. Two directions of wind predominate: northeast (60°) at night and most of the day during winter months (January through March); and southeast (120°) during the day most of the year. The northeast component is dominate approximately 50 percent of the time, and the southeast component is dominant 60 percent of the time (Puerto Rico Water Resources Authority, 1972).

The wind drift transfer coefficient is represented by:

$$I2 = (WV)(PE)(3600)(HD+HN)/24 + COS(DI-3.14/2)(WI)(WD)$$

where: WV = Wind speed (meters per second)

PE = Coefficeint of friction (water velocity/ wind velocity)(typically 0.01)

HD = Hours wind blows from 120°

HN = Hours wind blows from 60°

DI = Wind direction (in radians)

WD = Depth of channel (maximum value 3 m)

WI = Width of channel.

Thus the wind drift is proportional to wind speed and the direction acting across the channels connecting subbays. Wind drift was assumed to have an effect only on the upper 3 meters, or shallower average depth of these channels (Goldman, 1979). The average wind driven flow entering Guayanilla Bay is 2.9 x 10^6 m^3 and the average flow driven out of the Bay is 2.0 x 10^6 m^3 daily. These quantities compare with the 2.7 x 10^6 m^3 per day moving by tidal forces. However, the wind driven circulation of the bay is a critical variable for the exchange of water in Guayanilla Bay since this factor fluctuates seasonally and daily, and may control upwelling.

Industrial Pumping

Hydrologic budgets of the Intake, Thermal Cove, and Southeast embayment are affected significantly by the Costa del Sur oil-fired power plant, which operates by once-through sea water cooling. When all six units (boilers) are on-line, the cooling water flow is 38.0 m^3/sec. This plant operates with a power factor of approximately 80 percent so that estimated pumping is 2.2 x 10^6 m^3 per day. This pumping alone exchanges the volume of the Thermal Cove five times daily.

Fresh Water

The Western Subbay receives runoff from the Yauco River Watershed, and the Central Subbay receives runoff from the Guayanilla and Macana Rivers. The average annual streamflow for the Guayanilla River, measured approximately two km upstream from the bay discharge, is 30 x 10^6 m^3 per year; for the Yauco River, it is 14 x 10^6 m^3 per year; and for the Macana River, it is 9.8 x 10^6 m^3 per year (Crooks, *et al.*, 1968). Potential water entering Bay compartments from groundwater and runoff is 26 x 10^6 m^3 per year to the Western Subbay and 54 x 10^6 m^3 per year to the Central

Subbay. Actual freshwater entering may be somewhat less due to domestic freshwater use and evaporation from adjacent agricultural lands.

Direct precipitation from the annual average rainfall of 90 cm in the relatively arid south coast environment (Cintron, Lugo, Pool and Morris, 1978). This precipitation results in an addition of 79,400 m^3 of water annually, divided among bay compartments according to surface area. Much greater rainfall and runoff rates occur during tropical storms. Uncontrolled runoff from the three drainage basins can be equivalent to 1.9×10^7 m^3 of water, (10 cm of rainfall over a 24 hour period in the drainage basin). This is about 0.4 times the volume of the bay. Much of this runoff would be released within a one day period (Morris, 1980). This level of storm runoff and precipitation can be specified as an exogenous variable in the water exchange submodel.

Groundwater influx into the Bay has been estimated from average groundwater flow along the south coast of Puerto Rico (Crooks et al. 1968). A daily average discharge of 2.1 m^3 per linear foot of shoreline has been estimated (Puerto Rico Water Resources Authority, 1972).

The freshwater flow is small as an annual average, approximately 8.8 percent of the tidal flux, but seasonal variation and individual storm events can make this an important factor, with some measurements of peak runoff of 1.4×10^6 m^3 per day for the Guayanilla and Yauco Rivers (almost one-half the tidal flux during high rainfall periods). The freshwater flux is essential for maintaining the brackish water conditions of the Western Bay, and may be a critical factor in structuring the biotic community.

Evaporation

Annual pan evaporation is 200 cm at Ponce, with very similar values at Guanica and Sabana Grande (National Oceanic and Atmospheric Administration, 1979). This results in a flux of 18×10^6 m^3 annually or approximately 49,000 m^3 daily from the entire bay.

Shipping

The large container ships and tankers entering Guayanilla Bay displace approximately 100,000 metric tons, or 48,000 m^3. An average of one tanker enters and leaves the bay daily (Puerto Rico Port Authority, 1980). This exchange occurs at the interface between the Central Subbay and the coastal water. Thus, water movement due to tanker traffic is similar in magnitude to evaporation water flux, although very localized. Propeller pumping by tugboats displaces water across the Central Subbay/coastal water interface. Approximately four tugboats per day transit the mouth of the bay. This results in an estimated localized movement of 10,000 m^3 of water in each direction.

Equilibrating Flows

Equilibrating flows are the result of gravitational force to maintain a uniform (level) geopotential surface. For example, industrial pumping reduces the volume of the Intake Subbay so that water flows into the Intake Subbay from the Central Bay to reestablish equilibrium. Surface water from the Intake Subbay is also moved by wind into the Central SubbBay, which increases the equilibrating flow of bottom water from the Central Subbay. The size of equilibrating flows are mass balance estimates. They are based on the assumption that the average daily volumes represented by the Subbay compartment equations (table 4) are constant.

Table 4. Water exchange submodel equations.

1. Q1 = Western Subbbay (WB) Volume

$$\dot{Q}1 = K9\ I3 - KA\ I5\ Q1 + K8\ I1\ Q2 - K7\ I1\ Q1 + K6\ I2\ Q2 - K5\ Q1$$

2. Q2 = Central Subbay (CB) Volume

$$\dot{Q}2 = K2\ I1 - K1\ I1\ Q2 - K3\ I2\ Q2 + K4\ I2 + K5\ Q1 - K6\ I2\ Q2 + K7\ I1\ Q1 - K8\ I1\ Q2 + Kc\ I3 - KD\ I5\ Q2 + KE\ I4 + KI\ I2\ Q3 - KJ\ I1\ Q2 + KK\ I1\ Q3 - Kz\ Q2 + KW\ I2\ Q5 - KU\ I1\ Q2 + KV\ I1\ Q5 - L2\ Q2 - L3\ Q2 + KX\ I6 - KY\ I6\ Q2$$

3. Q3 = Intake Subbay (I) Volume

$$\dot{Q}3 = KF\ I3 - KG\ I5\ Q3 + KH\ I4 - KI\ I2\ Q3 + KJ\ I1\ Q2 - KK\ I1\ Q3 + KZ\ Q2 - Kl\ I7\ Q3$$

4. Q4 = Thermal Cove (TC) Volume

$$\dot{Q}4 = KL\ I7\ Q3 - KM\ I5\ Q4 + KN\ I3 + KO\ I4 + KP\ I1\ Q5 - KQ\ I1\ Q4 - KR\ I2\ Q4 - L1\ Q4$$

5. Q5 = Southeast Subbay (SEB) Volume

$$\dot{Q}5 = KQ\ I1\ Q4 - KP\ I1\ Q5 + K4\ I2\ Q4 - KS\ I5\ Q5 + KT\ I4 + KU\ I1\ Q2 - KV\ I1\ Q5 - KW\ I2\ Q5 + L2\ Q2 + L1\ Q4$$

Characteristics of the Water Exchange Submodel

The exchange of water in Guayanilla Bay is dominated by both tidal flow and wind. Wind drives about 10 percent more water into the bay (J4) than tide, but tides flush more water from the bay (J1) than does wind (J3). The largest flow is an upwelling equilibrating counter current flowing into the Intake Subbay, largely the result of westward wind drift and industral pumping. Generally, wind driven currents and equilibrating flows are the largest flows between compartments. The magnitude of the equilibrating flow out of Guayanilla Bay to the Caribbean may be directly related to wind velocity.

Figure 4 indicates the variation in volume of the subbay compartments induced from wind and tides. The patterns projected indicate that during certain periods, wind and tidal driven forces tend to cancel volume fluctuation, although significent water exchange among bay compartments would persist.

A partial verification of water movements can be based on drogue, current meter, and dye studies of current velocity. The equilibrating currents have been confirmed with drogue studies (Goldman, 1978). In addition, the general relationship between wind velocity and the magnitude of equilibrating (return) flows is supported by observations, for example, of the extension of turbid Western Subbay water into the Central Subbay during increased wind velocity. Wind and current velocity relationships are also substantiated by multidepth drogue observations (Goldman, 1978).

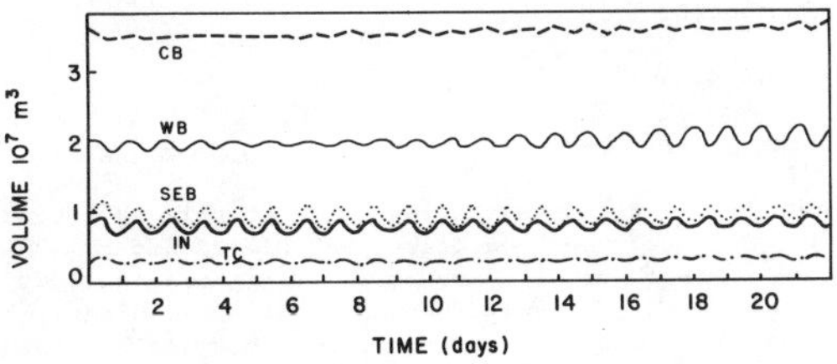

Figure 4. Volume fluctuation in five subbays within Guayanilla Bay. Scale on Central Subbay (CB) is expanded by a factor of 10.

THE TROPHIC SUBMODEL

The trophic submodel is a multicompartment simulation model that accounts for selected major factors influencing the transfer of energy (primarily as grams of carbon). The model consists of a set of feeding interactions among trophic compartments, with each set distributed in the five subbays. The movements of energy and materials are influenced by physical conditions and biological properties that can be manipulated as variables.

The major species found in Guayanilla Bay were aggregated into subgroups that represented major trophic subdivisions as indicated in table 5. The major interactions of these groups are indicated in figure 5, and the principle systems equation representing these interactions are listed in table 6. The parameters included in the differential equations are listed in table 7. Each subbay is treated as a separate set of equations, although the pool of net production (particulate organic matter) and zooplankton are transported among bay compartments as specified by water movements in the water exchange submodel.

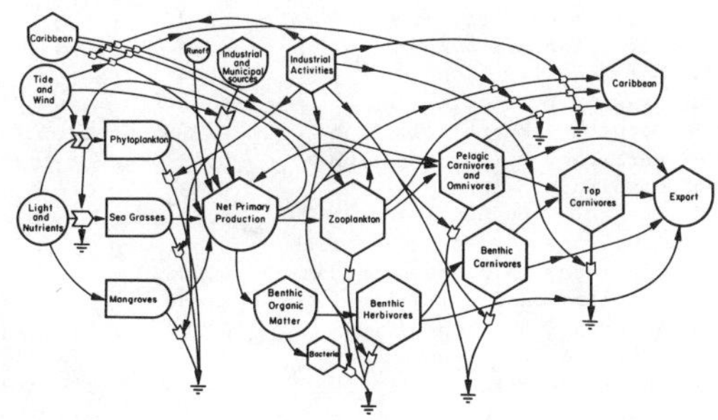

Figure 5. Trophic submodel. System equations for compartments are identified in table 6.

Table 5. Dominant representatives of trophic groups.

Common Name/Group	Representative
Net primary production	
Mangrove detritus	Rhizophora mangle
Turtle grass	Thallassia testudinium
Filamentous algae	Enteromorpha & Cladophora
Diatoms	Pyrodinium
Dinoflagellates	Microcoleus
Zooplankton	
Copepods	Acartia, Paracalanus
Chaetognaths	Sagitta
Branacle nauplii	Blanus
Decapod & fish larvae	
Pelagic carnivores and ominvores	
Anchovy	Anchoa
Slender herring	Jenkinsia
Silversides	Atherinidae
Parrot fishes	Scarus
Benthic herbivores	
Clams	Tellina, Tagelus
Polychaetes	Terebella, Prionospio,
Amphipods	Corphium
Shrimps	Paleomonetes
Mullet	Mugil
Benthic carnivores and demersal fish	
Shrimps	Penaeus, Alpheus
Crabs	Callinectes
Mantis Shrimps	Squilla
Ribbon Worms	Tubulanus
Snapper, Grunts, Crokers	
Top carnivores	
Jacks	Caranx
Snappers	Lujanus
Barricuda	

Table 6. Principle trophic submodel system equations.

M1 = Net primary production (Particulates)

$\dot{M1}$ = AM MA MM + AG TH MT IZ + AP PL MP IZ P2 - M1 G2 M2 - M1 GH M3 + D2 M2

M2 = Zooplankton

$\dot{M2}$ = M1 G2 M2 - R2 T2 M2 - D2 M2 - M2 GF M3

M3 = Pelagic carnivores and omnivores

$\dot{M3}$ = M2 GF M3 - DF M3 - R3 T2 M3 + M1 GH M3 - M3 GP M4

M4 = Top carnivores

$\dot{M4}$ = M3 GP M4 - R4 M4 T2 - DP M4 + M6 G7 M4

SD = Benthic organic matter

$\dot{SD}$ = P1/RE-SD BR T2 - SD G5 M5

M5 = Benthic herbivores

$\dot{M5}$ = SD G5 M5 - R5 M5 T2 - D5 M5 - M5 G6 M6

M6 = Benthic carnivores

$\dot{M6}$ = M5 G6 M6 - R6 T2 M6 - D6 M6 - M6 G7 M4

Table 7. Summary of major trophic submodel parameters

Symbol	Parameter	Typical Values	Units
MA	Net prod. mangroves	1-3	g C/m^2d[1]
TH	Net prod. sea grasses	0.5-1	g C/m^2d[2]
PL	Net prod. phytoplankton	0.5-1	g C/m^2d[3]
AM	Area of mangroves	-	km^2
AG	Area of sea grasses	-	km^2
AP	Area of plankton & mud bottom	-	km^2
RE	Residence time of POM	2-7	days[4]
MM	Thermal effect on mangroves	0-1	ratio[5]
MT	Thermal effect on grasses	0-1	ratio[6]
MP	Thermal effect on phytoplankton	0-1	ratio
IZ	Light intensity factor	0-1	ratio
DX	Nutrient level factor	0-1	ratio
G2	Grazing by zooplankton	0.06	g C/g C/d[7]
GF	Carnivory by pelagic fish	0.3	g C/g C/d[7]
GH	Herbivory by pelagic fish	0.01	g C/g C/d[7]
G5	Herbivory by Benthos	0.08	g C/g C/d[7]
G6	Predation by Benthos	0.01	g C/g C/d[7]
GP	Carnivory on planktonic fish	0.03	g C/g C/d[7]
G7	Carnivory on benthic predators	0.005	g C/g C/d[7]
R2	Zooplankton respiration	0.007	g C/g C/d
R3	Pelagic fish respiration	0.004	g C/g C/d[7]
BR	Bacterial respiration rate	0.002	g C/g C/d[8]
R4	Piscivore fish respiration	0.006	g C/g C/d[7]
R5	Herbivore benthos respiration	0.001	g C/g C/d[7]
R6	Carnivore benthos respiration	0.006	g C/g C/d[7]
D2	Death rate of zooplankton	0.05	Fraction dying/d[8]
DP	Death rate of top predators	0.02	"
DF	Death rate of fishes	0.02	"
D5	Death rate of benthic herb.	0.02	"
D6	Death rate of benthic pred.	0.02	"
T2	Ambient temperature factor	0-1	ratio
P1	Particulate organic matter	1-15	g C/m^2

[1]Golley, et al., 1962.

[2]Vicinte, 1980.

[3]Brackets values in Boynton, et al., 1977.

[4]Chartock, 1980a.

[5]Banus, 1978.

[6]Vicente, 1977.

[7]Kemp, et al., 1977.

[8]From Odum and de la Cruz, 1967.

Temperature

A baseline temperature is specified as an exogenous condition for each separate geographic area. This baseline temperature is then adjusted by an annual temperature cycle. This temperature cycle is simulated by the following equation:

TZ(a) = TE(a)-Cos(TA-60)/180)(3.144)(2)

where: (a) = locator for each subbay

TZ(a) = projected temperature for day TA in a subbay

TE(a) = average annual temperature

TA = day counter

The operation of the Costa del Sur power plant largely influences the baseline temperature gradient among subbays (Lopez, 1979).

Light

Light influences the level of net primary production for mangrove, sea grass, and planktonic communities. For mangrove communities, variation in light intensity is primarily the result of sun inclination and cloud cover. In addition, for seagrass beds and plankton communities, the intensity of light is reduced by attenuation from particulate matter. Average annual radiation along the south coast is estimated at 524 langleys. The variation in light reaching the surface is simulated by the following equation:

IO = 425 - Cos(TQ/180)(3.1414)(75)

where: IO = Intensity of surface radiation in langleys

TQ = day counter

Extinction of light in water is proportional to the negative exponent of the concentration of particulate matter as provided in the Lambert-Beer relationship, where a concentration of 40 mg/l dry weight attenuates 90 percent of the surface light.

Nutrients

Net primary production by phytoplankton in Guayanilla Bay is apparently limited by nutrient availability (Lair, et al. 1971). Phosphorus levels have been reported below 0.01 mg/l, and thus the model assumes that phosphorus can limit phytoplankton production. Dissolved nutrients are assumed to have negligible effect on seasonal variation of primary production from mangroves and sea grasses. Net primary production from phytoplankton is assumed proportional to phosphorus availability, which fluctuates from 0.00 mg/l to 0.21 mg/l (Puerto Rico Water Resources Authority, 1972). Nutrient levels, which originate from runoff and storm-related mixing of sediments, are specified as an exogenous seasonal variable that reaches a maximum in late summer and a minimum in winter.

Trophic Levels

The model simulates primary production from three community types: mangroves, sea grasses, and planktonic communities. The net production of these communities enters a net primary production pool that is a food source for macrophagous herbivores, phytoplankton feeders, and detritus feeders. A portion of the particulate matter from this pool settles to the bottom where it is available for benthos. The settling time for particulate organic matter from the net primary production pool is estimated from sedimentation traps and water samples from the five subbays (Chartock, 1980a).

The major flows among trophic levels are derived from parameters listed in table 7. Data on flow rates or characteristics from which flow rates can be derived are limited from Puerto Rico, and some values are based on studies in Florida bays or other subtropical and tropical locations.

MODEL PROPERTIES AND VERIFICATION

The model can be evaluated against appropriate measured criteria. As described for its hydrologic properties (Chartock, 1980), partial verification can be provided by measuring the movement of dye patches. The distribution of water exchange predicted by the model is generally consistent with the physical observations of water movement (Goldman, 1978).

Salinity measurements provide the basis to check observed measurements with model results. Exchange processes and freshwater and groundwater inflow dilute Caribbean water so that the model predicts that the bay will behave as a positive estuary. Salinity measurements in the bay range between 30 and 34 parts per thousand. Typically for most of the bay during the winter months, the values are 33 to 34 parts per thousand with the lowest salinity in the Western Subbay. When the model is run with a Caribbean salinity of 35.5 parts per thousand, salinity for the bay compartments achieves an equilibrium concentration of 33.8 for the Central Subbay, 32.9 for the Western SubbBay, 33.7 for the Southeast Subbay, 33.8 for the Intake Subbay, and 33.7 for the Thermal Cove. These values indicate that the model is generally consistent with field data on salinity distribution.

Exchange With The Caribbean

Guayanilla Bay apparently acts as a sink for some materials entering from the Caribbean. For example, there is some evidence (Lopez, 1979) that pollutants from the adjacent Tallaboa Bay enter Guayanilla Bay around Punta Guayanilla and are deposited within the Bay. Model simulation indicates that if a concentration of a conservative (soluble, nonbiodegradable) material in Caribbean water is 1 mg/l, Guayanilla Bay will receive about 10 metric tons in four days under average wind, tide, and runoff conditions. A near equilibrium concentration equivalent to 90 percent of Caribbean water will be achieved in 24 days. Ninety-five percent of Caribbean water concentration will occur in 45 days resulting in a total addition of 42.5 metric tons from the Caribbean.

The flushing of Guayanilla Bay is related to freshwater inputs, exchange processes within the bay, and transport to the Caribbean. A 50 percent dilution of the bay compartments from 10 parts per million to 5 parts per million of a conservative substance requires about six days. A dilution of 10 percent of the original concentration requires 20 days. Modification of wind, tide, runoff, or changes to bay geomorphology significantly affects simulated dilution properties.

For example, the addition (or "spill") of 100 metric tons of a conservative substance into the Central Bay is shown in figure 6. Compared with control or "average" conditions, a reduction in tides or wind speed significantly delay exchange into other bay compartments.

Trophic Groups

The distribution of particulate organic matter in the subbays provides a partial verification of the model. Generally, measurements of the Western Bay show the highest POM estimates at 3 g C/m^3 based on a 4:1 ratio of total particulates:POM (Chartock, 1980a). This compares to the model's projected 3.7 g C/m^3 (table 9), when mangrove net production is 1.5 g C/m^2, sea grass production is 1.0 g C/m^2/day and phytoplankton production is 0.5 g C/m^2/day. Estimated particulate matter of the remaining subbays, based on field

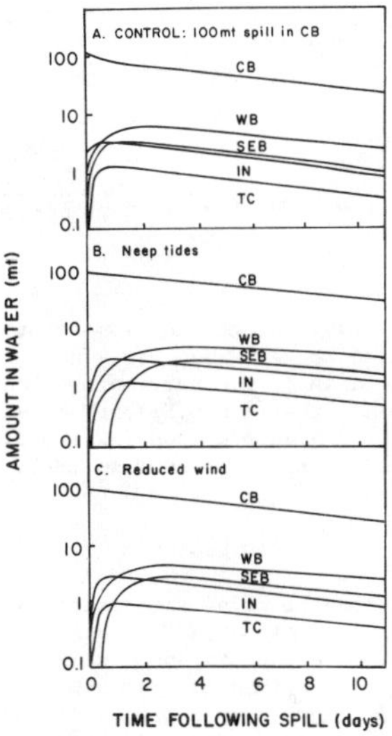

Figure 6. Dilution of a conservative matrial added to the Central Bay under control (average) condition (A), during condition of neep tides (TD=1 m)(B), and reduced wind (WV=1 m/sec)(C).

Table 8. Projected July biomass standing crop (g/m^3)

Compartment	NPP Pool (M1)	Zooplankton (M2)	Planktivorous Fish (M3)	Benthos (M5 & M6)
WB	3.7	0.22	0.18	31.1
CB	1.7	0.24	0.20	20.9
I	1.8	0.24	0.20	8.0
TC	1.9	0.23	0.20	2.3
SEB	2.0	0.22	0.20	8.3

measurements was 1.3 to 2.6 g C/m^3/day, compared to 1.7 to 2.1 g C/m^3/day projected by the model.

The seasonal change in standing crop of the net primary production pool, zooplankton, herbivirous benthos, planktivorous fish, benthic predators, and piscivorous fish (top carnivores) are indicated in figure 7. Under control conditions significant fluctuations in piscivorous fish and zooplankton occur. Data are not available to verify these fluctuations in fish populations. However, field studies indicate that zooplankton populations typically fluctuate by an order of magnitude, with low values frequently occurring during winter months (Youngbluth, 1976, 1980).

The model simulation also indicated the biomass of benthic herbivores and carnivores (M5 and M6) will be greatest in the Western Subbay (table 9). Field studies also verify the significantly higher benthic biomass in this area (Yoshioka, 1980).

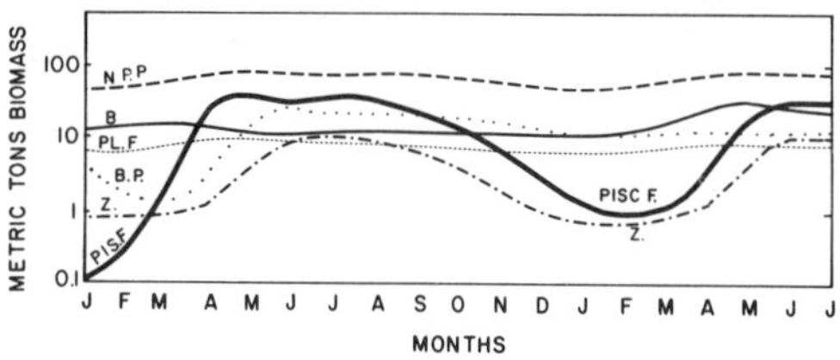

Figure 7. Seasonal changes in biomass standing crop from initial conditions. NPP = net primary production pool (M1); B = herviborous benthos (M5) PL.F planktivorous fish (M3); BP = benthic carnivores (M6); Z = zooplankton (M2); Pis.F = top carnivores and piscivorous fish (M4).

DISCUSSION

The environmental model summarized in this paper may provide an initial basis for integrating the diverse information about the south coast bays of Puerto Rico. The model provides an approximation of the flow of carbon among major tropic groups linked with water exchange among subbays. The model may be useful for indicating major system-wide factors that should be considered if changes to Bay geomorphology or industrialization are planned. For example, consideration is being given to a relocation of the Costa del Sur power plant discharge from the Thermal Cove to Tallaboa Bay in order to reduce the erosion of some valued mangroves and to protect an associated heronry in the Thermal Cove. The model projects the consequences of these changes in the water budget for Guayanilla Bay (Chartock, 1980). These changes would lead to reduction in the standing crop of particulate organic matter and reduction in the standing crop of higher trophic levels in the Western and Central subbays, primarily due to increased flushing of the Bay.

In these applications the environmental model provides an initial systematic basis to identify assumptions and data uncertainties that are important for predicting the behavior of Guayanilla Bay. However, only limited data are available for feeding and energetics for the Bays along the south coast of Puerto Rico. Best known parameters may be the total respiration and metabolism of grass beds and fringing mangrove forests. Data on standing crop and production of phytoplankton and fishes, for example, are not available, and are derived in this model from projections based on feeding and respiration rates in subtropical or temperate areas. Despite these limitations, the simulations provide agreement with some available data on abundance and distribution of benthos and zooplankton. In addition, some of the assumptions and data that need further study include:

(1) Data on vertical mixing rates between shallow and deep water, especially for the Central Bay. Data on vertical mixing would provide an informed basis for developing a three-dimensional model for Guayanilla Bay;

(2) Data on rates of biodegradation and biological uptake of representative types of compounds of management importance;

(3) Data on resuspension of sediments and associated organic matter related to patterns of currents, wind velocity, and industrial activity such as boat movement would also be useful;

Guayanilla Bay research efforts (Gonzalez, 1979) have provided extensive descriptive data on patterns of distribution of chemical, physical, and biological properties. However, research on rate phenomena described above would provide an important basis for extending and refining the model described in this report.

REFERENCES

Banus, M. D. 1978. Survival and growth of young red mangrove trees in laboratory experiments with heated seawater. Unpublished manuscript. 14 pp.

Boyton, W., D. E. Hoskins, and C. Gray. 1977. A modeling approach to regional planning in Franklin County and Apalachicola Bay, Florida. In C. A. S. Hall and J. W. Day (eds). Ecosystem modeling in theory and practice. John Wiley and Sons. pp. 478-506.

Chartock, M. A. 1980. Hydrologic model of Guayanilla Bay, Puerto Rico. CEER M-64. Mayaguez, Puerto Rico: Center for Energy and Environment Research. 42 pp.

Chartock, M.A. 1980a. Environmental transport in an industrial marine ecosystem: A simulation model of Guayanilla Bay. CEER M-76. Mayaguez, Puerto Rico: Center for Energy and Environment Research. 39 pp.

Cintron, G., A. E. Lugo, D. J. Pool, and G. Morris. 1978. Mangroves of arid environments in Puerto Rico and adjacent islands. Biotropica 10(2): 110-121.

Crooks, J. W., I. G. Grossman, and D. B. Bogart. 1968. Water resources of the Guayanilla-Yauco area, Puerto Rico. Water Resources Bulletin No. 5, U.S. Geological Survey and Commonwealth of Puerto Rico. 55 pp.

Goldman, G. C. 1979. Water circulation in Guayanilla Bay. In Jose M. Lopez (ed). Proceedings of the symposium on Energy, Industry, and the Environment in Guayanilla Bay. Center for Energy and Environment, Mayaguez, P. R. pp. 10-18.

Goldman, G. C. 1978. Physical oceanography of Guayanilla Bay. Annual report, Center for Energy and Environment Research, University of Puerto Rico. Mimeograph. 34 pp.

Golley, F., H. T. Odum, and R. F. Wilson. 1962. The structure and metabolism of a Puerto Rico red mangrove forest in May. Ecology, 43(1):9-19.

Gonzalez, J. G. 1979. Comprehensive analysis of research undertaken in Guayanilla Bay: an integrated view. In Jose M. Lopez (ed). Proceedings of the symposium on Energy, Industry, and the Environment in Guayanilla Bay. Center for Energy and Environment Research, Mayaguez, P. R. pp. 90-94.

Kamel, A. M., and C. Hadjitheodorou. 1970. Report on the current conditions in Guayanilla Bay. Water Resources Research Institute of Puerto Rico. University of Puerto Rico, Mayaguez, P.R.

Lair, M. D., R. G. Rogers, and M. R. Weldon. 1971. Environmental effects of petrochemical waste discharges on Tallaboa and Guayanilla bays, Puerto Rico. Environmental Protection Agency, Region IV, Surveillance and Analysis Division, Technical Study TS 03-71-208-02, Athens, Ga. 46 pp.

Lopez, J. M. (ed). 1979. Distribution of trace contaminants in water and sediments of Guayanilla and Tallaboa bays. In proceedings of the symposium on Energy, Industry and the Marine Environment in Guayanilla Bay. Center for Energy and Environment Research. University of Puerto Rico, Mayaguez, P.R. pp. 19-22.
Morris, G. 1980. Personal Communication, March.
National Oceanic and Atmospheric Administration. 1979. Personal communication, staff, November.
Odum, H. T., P. R. Burkholder, and J. Rivero. 1962. Measurements of productivity of turtle grass flats, reefs, and the Bahia Fosforescente of southern Puerto Rico. Publications of the Institute of Marine Science, University of Texas. 9:159-170.
Odum, E. P. and A. A. de la Cruz. 1967. Particulate organic matter in a salt marsh-estuarine ecosystem. In G. H. Lauf (ed). Estuaries. American Association for the Advancement of Science, publication 83. Washington, D.C. pp. 383-388.
Puerto Rico Electric Power Authority. 1980. Personal communication. March.
Puerto Rico Port Authority. 1980. Personal communication, staff. January.
Puerto Rico Water Resources Authority. 1972. Aquirre Power Plant Complex Environment Report. Report No. WRA E.S. -8. April. 198 pp.
Tilly, L. J. 1979. Guayanilla Bay: an industrial marine ecosystem. In proceedings of the symposium on Energy, Industry, and the Marine Environment in Guayanilla Bay. J. M. Lopez (ed). Center for Energy and Environment Research, University of Puerto Rico, Mayaguez, P.R. pp. 1-9.
Vicente, V. P. 1980. Unpublished field data, and and personal communication. May.
Vicente, V. P. 1977. The impact of heated effluents on *Thalassia* beds: a comparative study. CEER M-66. Center for Energy and Environment Research. University of Puerto Rico, Mayaguez, P.R. 50 pp.
Youngbluth, M. J. 1980. Daily, seasonal, and annual fluctuations among zooplankton populations in an unpolluted tropical embayment. Estuarine and Coastal Marine Science 10(3):265-288.
Youngbluth, M. J. 1976. Zooplanton populations in a polluted tropical embayment. Estuarine and Coastal Marine Science 4:481-496.
Yoshioka, P. 1980. Unpublished field data, and personal communication, April.

ACKNOWLEDGEMENTS

The Center for Energy and Environmental Research, University of Puerto Rico and the Research Participation Program of the Oak Ridge Associated Universities and the Research Council of the University of Oklahoma provided support for this project. This project was conducted while on sabbatical leave from the Science and Public Policy Program and Department of Zoology of the University of Oklahoma. Drs. J. A. Bonnet, K. G. Soderstrom, L. J. Tilly, J. G. Gonzalez, J. M. Lopez, and R. J. Zimmerman were helpful in implementing this work, and Drs. S. Brown, A. E. Lugo, H. T. Odum, J. J. Kimmel, and G. C. Goldman provided comments.

THE TRANSPORT AND PROCESSING OF NITROGEN IN A TIDAL, FRESHWATER MARSH AND RIVER ECOSYSTEM: MODELING THE ROLES OF WATER MOVEMENT AND BIOTIC ACTIVITY IN DETERMINING WATER QUALITY[1]

C. J. Vorosmarty[2], B. Moore[2], W. B. Bowden[3], J. E. Hobbie[3], B. J. Peterson[3], and J. Morris[4]

Abstract--A computer model was developed to integrate laboratory and field data concerning the N cycle in a tidal, freshwater marsh and river. The model links hydrodynamics to material transport and biotic processing. Preliminary results from the water transport model agree with observed tide height and salinity data collected over 24-hours. Using inorganic N as a conservative constituent, the simulation indicates that dilution processes account for most but not all of the distribution of inorganic N during low-flow conditions. For nitrate-N, the residual disparity can be accommodated by providing the constituent model with rates of biotic uptake determined by field and laboratory measurements.

INTRODUCTION

The nitrogen (N) cycle is a complex biogeochemical circuit which couples the solid, liquid and gaseous phases of a variety of organic and inorganic compounds. Inorganic N is an important plant and microbial nutrient that has been studied extensively in agricultural ecosystems. Until recently, however, few ecosystem-level studies have examined N in coastal systems. In addition, many of the specific transformations of the N cycle have eluded careful study, in part because of methodological difficulties but also because of the natural, often extreme variability of the processes considered. Notable additions to the current body of knowledge have included studies of the Delaware, Potomac, Patuxent and Chowan rivers; of Narragansett Bay; and of Massachusetts, South Carolina, Georgia and Louisiana salt marshes.[5] In general, these investigations have stressed the basic mechanisms of N transfer, not the collection and use of hydrodynamic information, even when nutrient budgets are sought. Kjerfve et al. (1981), Kjerfve and McKellar (1980), Boon (1978) and Mann (1975), however, have demonstrated the influential role hydrodynamics plays in quantifying material fluxes across estuarine boundaries and, therefore, in determining the nature and magnitude of resident biotic transformations.

Water quality engineers have recognized the importance of water transport for many years and their models have reflected this. Thus, sophisticated hydrodynamic simulations exist for numerous estuaries. Yet, most investigators within the engineering community have coupled these physical models to rather simple ecological representations, often treating isolated phenomena such as dissolved oxygen/BOD relationships or phytoplankton bloom prediction. This approach may provide useful information about narrowly-defined topics, but it has limited value in investigating ecosystem-level phenomena. This generalization admits some exceptions. Najarian and Harleman (1975), for example, summarized a N-cycle model with both a state-of-the-art hydrodynamic model and a highly detailed, mechanistic simulation to reflect biotic reality. Ironically, in applying the model to the Potomac River, they relied on literature estimates for the bulk of the biotic parameters, thereby ignoring the actual ecology of the site.

Few estuarine models from the ecological literature couple the site-specific qualities of the N cycle to a realistic representation of water flow. An important exception has been the 'eclectic' collection of models produced by Nixon and co-workers for Narragansett Bay; but, even this group's detailed,

[1]Paper presented at the Third International Conference on State-of-the Art in Ecological Modelling, Colorado State University, May 24-28, 1982.

[2]Complex Systems Research Center, UNH, Durham, NH.

[3]Ecosystems Center, Marine Biological Laboratory, Woods Hole, MA.

[4]University of South Carolina, Columbia, SC.

[5]See works of Good; MacElroy; Kemp and Boynton; Hobbie and Stanley; Nixon; Valiela and Teal; Vernberg, Dame, and McKellar; Haines; and Deluane and Patrick, respectively.

two-dimensional hydrodynamic model has been greatly simplified in order to address research questions involving long-term ecosystem behavior (Kremer 1979, Nixon and Kremer 1977). Nihoul (1981) has noted a similar lack of interaction between marine ecologists and physical oceanographers.

This paper outlines what we have learned about the processing of N in a tidal, freshwater marsh and river channel located in New England. Since autumn 1979 we have performed field, laboratory, and modeling exercises on the North River ecosystem (Massachusetts). This contribution describes our present understanding of the N cycle in this river and focuses particularly on the interaction between the study site's hydrodynamics and biology. We will report on preliminary results generated by an integrated hydrobiological model which uses data collected intensively in the field during summer, low-flow conditions.

DESCRIPTION OF THE STUDY SITE

The setting for our studies (Figure 1) lies 50 km south of Boston in a suburban area undergoing rapid development. The site's hydrodynamic regime is complex. The "seaward" boundary is the Stone Arch Bridge (SAB), located 20 km upstream from the mouth of the North River estuary which opens into Massachusetts Bay. Tidal amplitude at SAB can approach one meter during periods of low river discharge. The maximum observed salinity, about 2-3 ppt,

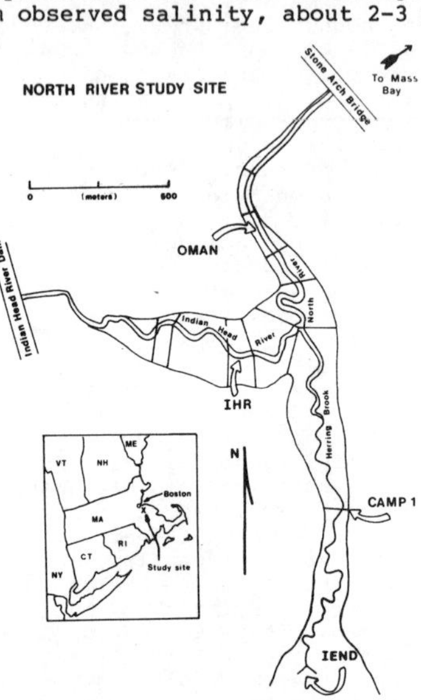

Figure 1. The North River study site, showing the river channel and flanking freshwater marshes. The sampling stations refer to points along the open-water channel where water quality has been monitored. The surrounding drainage basin is located in a suburban setting undergoing rapid development.

occurs under the same low-flow conditions, normally encountered in the summer. High runoff greatly dampens tidal amplitude and renders salinity virtually undetectable. The water column is well-mixed vertically with respect to salinity and other constituents.

Channel and marsh widths at the first upriver sampling station (OMAN) differ substantially from those at the SAB site, a variation that characterizes the study site as a whole. An important hydrologic feature of the ecosystem is the confluence of Herring Brook and the Indian Head River. Marshes flank the Indian Head River for most of its length, except in its upland reaches where it is bounded by a dam and wooded banks. In contrast, Herring Brook ends in a dendritic pattern of inlets and marshes at IEND. The elevation of the channel bottom at the DAM is approximately 2 m higher than at SAB; the elevation at IEND falls between the two. Over much of the study site, the channel bottom is composed of fine-grained sediments and organic material. However, cobbles and coarse sand cover the channel bottom directly below the DAM site and near SAB there is a sandy substrate. The wetland sediment in the marshes is highly organic and has wide variations in texture owing to the emergent vegetation and its associated litter.

The site also maintains a highly productive, diverse biota. The marsh plants yield at least 1 kg/m2 of net annual aboveground production (J.Morris, unpublished data), a typical value for freshwater wetlands (Brinson et al. 1981, Whigham and Simpson 1978). Mixed and pure stands of *Typha* (cattail), *Carax* (a sedge), and *Calimagrostis* (bent grass) dominate the vegetation. Wild rice (*Zizania*) grows in the mudflats near the marsh/river channel interface, and there are intermittent but dense beds of submerged macrophytes (primarily *Potamogeton*). The study site serves as habitat for a variety of wildfowl, muskrat, turtles, shellfish, and migrating fish such as herring and shad.

Although overall water quality is good, periodic blooms of duckweed, or *Lemna sp.*, are perceived as a nuisance by local residents. On rare occasions oxygen concentrations in the water column drop far enough to endanger fish life (unpublished data collected in August 1981). In a 1975 state survey the tidal portion of the North River received the highest attainable rating.[6] In contrast, the river above the Indian Head River Dam was assigned the lowest possible classification in 1977.

FATE OF INTRODUCED NUTRIENTS

Water entering the study site from the Indian Head River adds substantial quantities of nutrients, including inorganic N (predominately nitrate) that originates at an upriver sewage treatment

[6] Performed by the Massachusetts Divsion of Water Pollution Control.

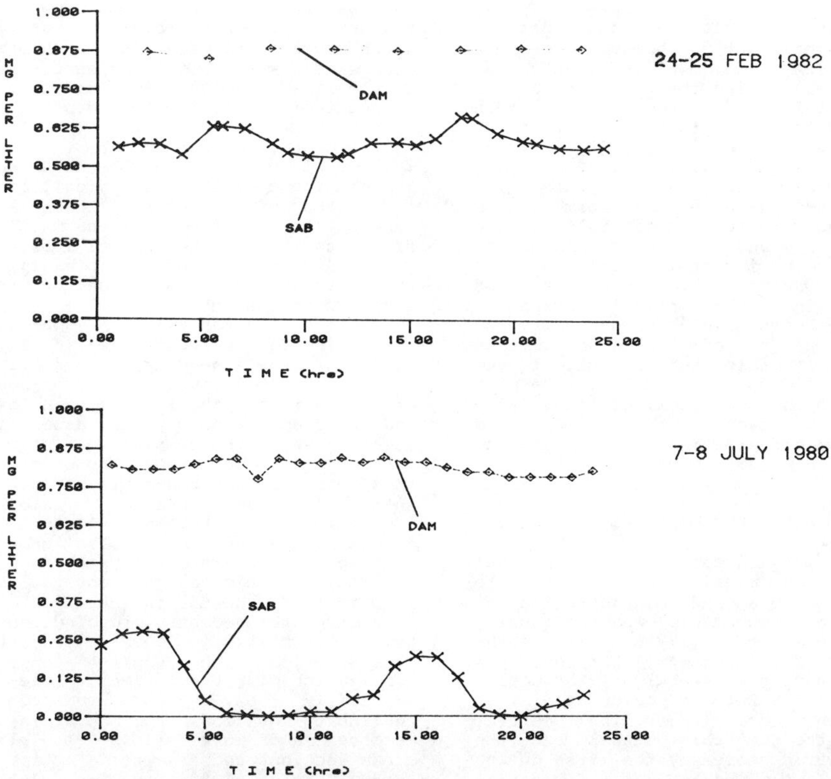

Figure 2. Observed concentration differences for (NO3+NO2)-N between upriver (DAM) and downriver (SAB) sampling sites during two 24-hour sampling runs. For the February sampling, average concentration at IEND was measured at 0.3 mg/1. Based on continued monitoring, concentrations entering the site past IEND are presumed to have approached zero for the July sampling date. Concentration loss is due either to explicit biotic or chemical processing of the nutrients or to hydrodynamic effects.

Figure 3. Idealized and empirical cross-sections sharing equal cross-sectional areas. The curvilinear transect is typical of those measured in the field. The orthogonal configuration is in the form accepted by the hydrodynamic model.

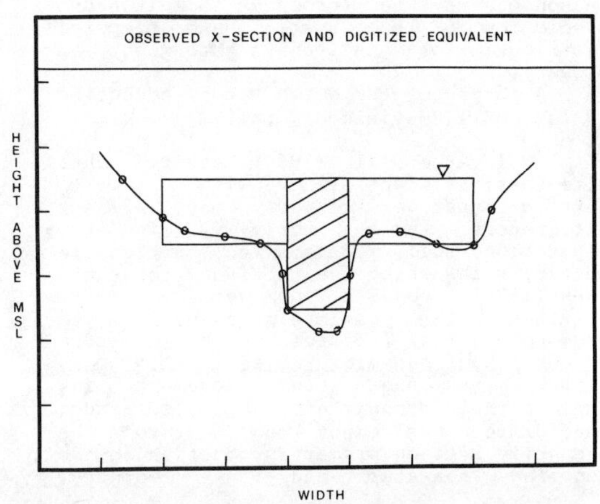

plant and at nonpoint sources. The concentration of nitrate and nitrite decreases significantly, however, between the upriver (DAM) and downriver (SAB) sites (Figure 2), suggesting that the ecosystem is successful at "processing" this artificial N loading. Although this effect exists throughout the year, it is most pronounced during the growing season.

Thus, the ecosystem appears to register a net gain of this form of N, a pattern found in estuaries generally (Nixon 1980). Ammonium and inorganic phosphorus also appear to be imported, but researchers have not observed this behavior as consistently in other tidal systems.[7] Based on the North River's productivity and on the seasonality of the concentration loss, it is plausible that an explicit and important biotic or chemical uptake of N exists within the ecosystem. It must also be recognized, however, that the observed concentration differential between sampling stations could be caused by dilution.

THE MODEL FOR WATER FLOW

We modified a detailed 1.5-dimensional tide-propagation model in order to accomodate the branched configuration of the study area. The model was originally developed by Lee (1971) at the Massachusetts Institute of Technology. The simulation enables water heights throughout the river above SAB to be calculated. It also estimates both longitudinal current velocity in the central channel and the volume of water stored by the marsh during flood stage. For calculating material budgets, such a calibrated model is an alternative to difficult in situ sampling of water flow (Correll 1981, Kjerfve et al. 1981, Boon 1978, Valiela et al. 1978). A numerical computer model also subsitutes for the hypsographic approach, which calculates water flows on the basis of basin geometry, tide height and time (Boto and Bunt 1981, Wolaver et al. 1980, Woodwell et al. 1977). Besides calculating flux across ecosystem boundaries, a detailed water flow model is the only reasonable way to produce simultaneous prediction of the spatial and temporal distribution of constituents throughout an estuary.

A survey of the North River conducted in fall of 1979 yielded detailed information on the topography of 15 transects. Figure 1 shows the positions of these cross-sections. One such transect, exaggerated vertically, appears as the curved line in Figure 3. This information showed that average water depths in the channel bottom range from approximately 1 to 2 meters, depending on the upriver inflows of water. For the marsh surface, maximal flood depths are typically lower, a few tenths of a meter. Channel width varies from 10 to 65 m, and marsh width, from 0 to 150 m. The channel measures 2700 m from SAB to DAM, and we nomially limit the distance between the channel junction and IEND to 2000 m.

The rectangular configuration overlaying the curved line in Figure 3 represents the geometry used in the simulation. Equivalent cross-sectional area establishes the link between idealized and measured morphology. By assembling all such transects the simplified system shown in Figure 4 is obtained. The model accepts this 'map' as river and marsh widths for a set of tide height calculations (corresponding elevation parameters are not displayed). Figure 4 also shows the forcing functions at the system boundaries: for SAB it is tide height, and for DAM or IEND it is freshwater inflow.

Given the geometry, forcing functions, and a bottom friction coefficient at each cross-section (to regulate longitudinal propagation of the tide), the model explicitly solves the momentum and continuity equations to determine tide heights and water flows at alternate cross-sections and at alternate time steps (see Schmidt 1982 and Lee 1971 for details). The forcing functions are applied until results from one cycle resemble those of a previous cycle within an acceptable margin of difference.

To obtain this quasi-steady-state, a timestep of 15 to 25 seconds is typically chosen, its specific size determined by the available computer space and the stability criterion of the model. For shallow estuaries, this stability relationship involves the distance between adjacent cross-sections and, for each transect, the local water velocity, tide height, hydraulic radius, and friction coefficient (Lee 1971).

We analyzed hourly information obtained during an intensive sampling of the river on 7-8 July 1980. The appropriate forcing functions were applied from the first mid-high tide to the second until the model converged upon a steady state solution. Relative tide height at SAB was transformed into the absolute height forcing function using information from a benchmarked tide gauge operating during July 1981. Gauged freshwater inflows at the DAM were 0.25 m3/sec while flows into IEND were set at 0.075 m3/sec. Figure 5 compares predictions for tide height and phase lag against field observation. The maximum and minumum flows predicted for this simulation at the SAB site were +-6 m3/sec, and the residence time of water calculated using a tidal

[7]Wolaver et al. (1980) assert that site-to-site inconsistency with respect to nutrient input or export may reflect differences in the activity of the vegetation, soil, mudflats and particulate matter, with which the nutrients interact. Mann (1975) and W.E. Odum et al (1979) attribute the various results to geomorphological features which in turn affect hydrodynamic regime. E.P. Odum (1980) agrees that water movement may explain the inconsistency, but also argues that differences might be due to the disparity between primary production in the incoming seawater and the estuary.

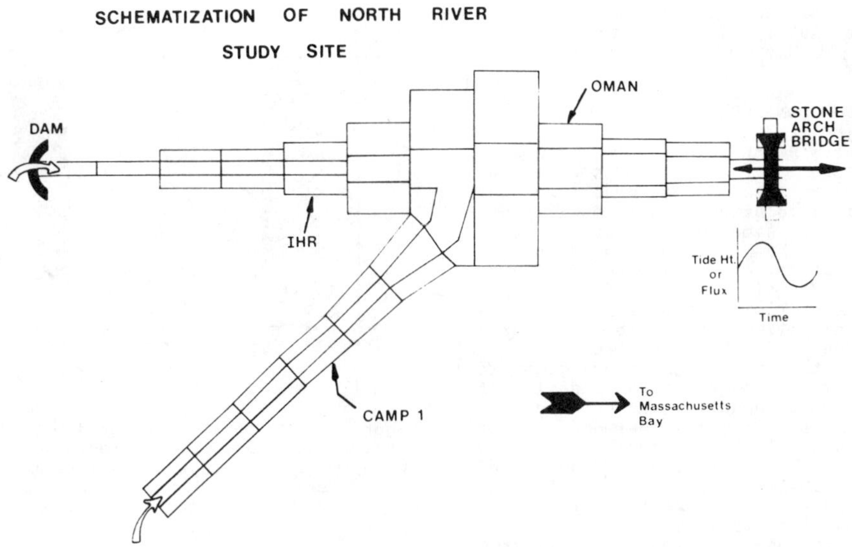

Figure 4. Integrated geometry used by the hydrodynamic and constituent models. Each rectangle represents a well-mixed reservoir. Marsh compartments represent storage volumes and flank central channels which undergo advective longitudinal water flow and material flux. Tide height at SAB and freshwater inflows drive the hydrodynamic simulation. For the conservative constituent and the linked biotic-constituent models, material fluxes into the ecosystem serve as forcing functions at the ecosystem boundaries. These fluxes are associated with freshwater inputs upriver and with flooding tidal waters at SAB.

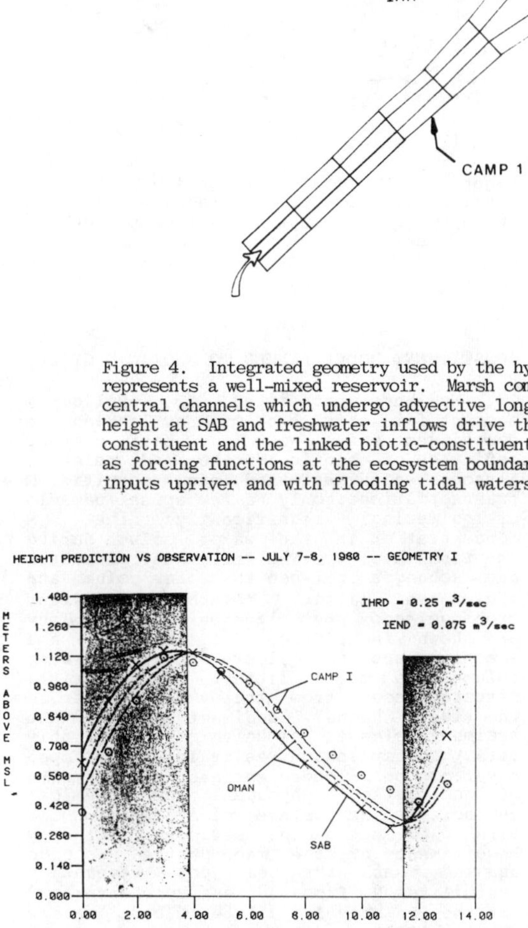

Figure 5. Height prediction vs. observation. 7-8 July sampling data shown beside model predictions for tide height. These results are for hydrodynamics calibrated using salinity. The shading represents the period of flooding tides. Ebbing tides are unshaded.

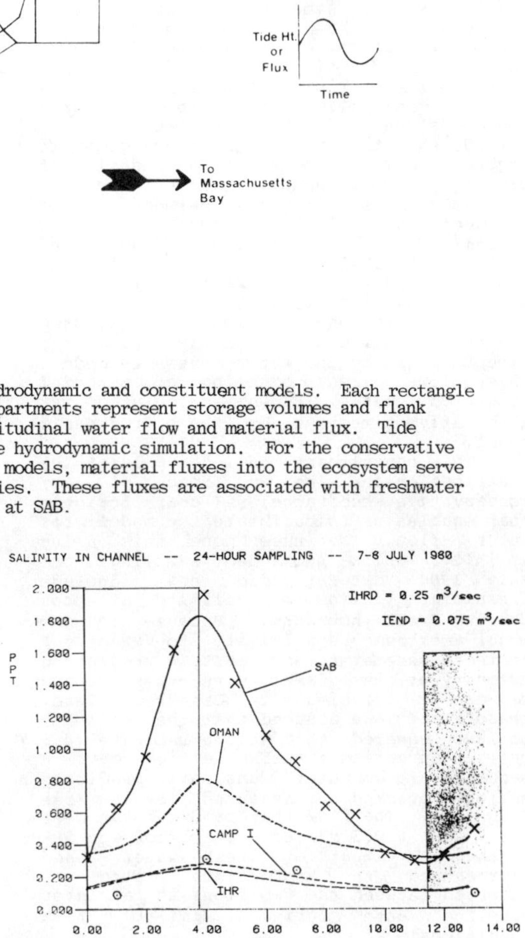

Figure 6. Plot of salinity vs. time. Measured salinity (circles) at the CAMP1 site serves as a conservative tracer in calibrating the hydrodynamic model for the July 1980 sampling. Model output appears as the solid and dashed lines.

prism approximation (maximal volume at flood stage divided by volume of tidal prism [Dyer 1973]) approached 2 tidal cycles.

MODEL FOR CONSERVATIVE CONSTITUENTS

To determine the influence of water movement alone on the cycle of N we linked a conservative constituent model to the hydrodynamic simulation. This consitituent model uses the same configuration as in Figure 4, but the upriver and downriver forcing functions are now material fluxes into the study site. Each flux simply represents the flow of water across the boundary multiplied by the measured concentration of the constituent just outside the study site. The rectangles in Figure 4 may be thought of as a series of discrete, continuously-mixed reactors. Internal mass transfers between these adjacent volumes of water are also determined by multiplying flow and concentration. As before, the forcing functions are applied repetitively until a quasi-steady state is attained, this time with regard to concentration. The timestep of the constituent model is twice the timestep used by the water movement model. This allows direct coupling of material fluxes to water flows calculated at alternative timesteps by the hydrodynamic model.

This constituent model can predict the distribution in time and space of any conservative substance, a characteristic that enables us to calibrate the model for water flow. We have done this using salinity. Unlike subsequent samplings, the July 1980 dataset did not include freshwater inflow or salinity at the Herring Brook boundary. Because of the model´s extreme sensitivity to variations in these parameters--at least in predicting salinity at the CAMP1 site--this proved especially troublesome. Based on later measurements, we assumed that the salinity at IEND equaled that at the DAM site (0.1 ppt), then varied the IEND inflow over a sequence of computer runs until predicted salinity tracked the measured levels (see Figure 6). The flow that produced the best ´fit´ was 0.075 m3/sec. This flow was 30% of the DAM input of fresh water, and represented the approximate ratio we observed between the two flows at a later sampling date during similar flow conditions.

We used the calibrated hydrodynamics with the conservative constituent model to predict how much of the concentration loss of total inorganic nitrogen (TIN) between sampling stations could be accounted for by water movement alone. Figure 7 compares model predictions for TIN with observed values at the different sampling sites and shows that much of the observed decrease can be assigned to dilution alone. The figure still suggests the presence of a net sink for TIN, since the predicted curves fall above the trend lines for measured TIN at both the OMAN and CAMP1 sites.

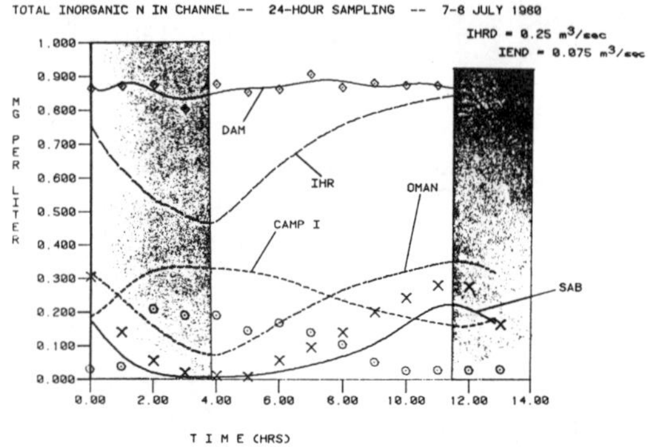

Figure 7. Total inorganic nitrogen concentration over time. Model predictions (with no biotic uptake) are compared here with observed N concentrations. The disparity represents potential biotic uptake of N.

CONSTITUENT MODEL LINKED TO BIOTIC N CYCLE

We have attempted to explicitly account for this sink of N by linking the constituent model with our best current estimates of biotic uptake and release. Although we investigated many specific N transformations, only a few appear capable of altering significantly the N concentration in the water column during one tidal cycle. The relevant processes are exchanges between the water column and the marsh litter, benthic sediment exchange, and phytoplankton uptake. They are shown in Figure 8. Nitrate and ammonium are both lost from the water column onto marsh litter. In contrast, nitrate passes from the overling water to the river channel sediment while the sediment releases ammonium at a comparable rate. Phytoplankton reside in either open river or on flooded marshes, environments of contrasting physical and chemical characteristics. Since mineralization of marsh sediment can satisfy the N requirements of the macrophytes, we have assumed that the emergent vegetation assimilates N from sediment pore water and not the relatively dilute flood waters (Bowden 1982).

To verify the linked hydrobiological model, we used the same July 1980 sampling experiment discussed above and analyzed the data for nitrate, the predominant N species at this time of the year. The exchange of nitrate and nitrite between the water column and the marsh surface litter was modeled as a zero-order process. The rate, 7000 ug (NO3+NO2)-N/m2-hr (from water to litter), was applied uniformly over all marsh within the study site. We obtained this information by tabulating differences between observed influent and effluent

N-flux in a series of flume experiments on the marsh surface (see Bowden 1982 for details). Similarly, we applied zero-order exchange between water column and river bottom sediments. This value, 3000 ug (NO3+NO2)-N/m2-hr, was derived from experiments using a modification of the circulating benthic chamber described by Hall et al. (1979).

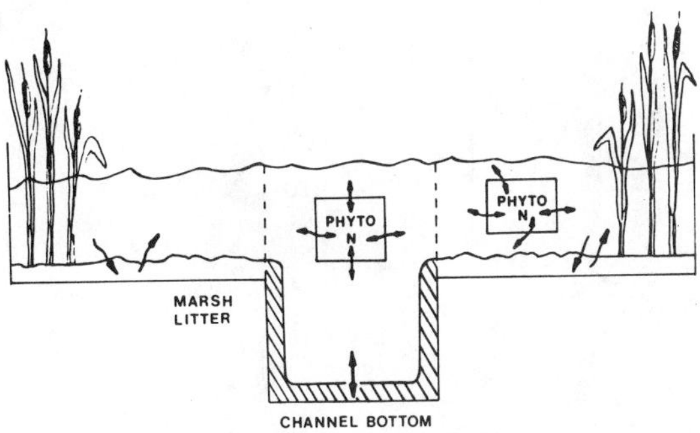

Figure 8. Relevant biotic components for short-term analysis of the N cycle.

Phytoplankton assimilation of N was treated as a first-order recipient-controlled process. The production of carbon was a multiplicative function of relative seasonal light intensity, relative temperature, and chlorophyll a in the water column as given by Ryther and Yentsch (1957). N uptake was calculated by dividing the production rate by a C:N ratio of 5.7. Assimilation of N was assumed to occur in both the light and dark (Stanley and Hobbie 1981, Kuenzler et al. 1979). Chlorophyll a was distributed throughout the channel and marshes as a conservative substance prior to its interaction with the biotic N model. The algae did not experience growth nor did it remineralize N. The preferred substrate was NH4, until the concentration in the water fell below 15 ug/l NH4-N. At that time, (NO3+NO2)-N and NH4-N were assimilated in equal proportions until exhaustion of either or both substrates (Keunzler et al. 1979).

According to USGS flow records, the study site experienced elevated freshwater inflows for several days prior to the actual 24-hour sampling run. To reflect this in our analysis, we initiated the hydrodynamic model on July 6 using gauged inflows and an elevated forcing function at the seaward boundary. We then allowed the simulation to proceed through time, approaching the extreme low-flow conditions of 7-8 July. Water transport for July 7-8 was calibrated with the available salinity data at CAMP1. Nitrate- plus nitrite-N was then applied as a conservative substance (hours 0-14 in fig. 9) until the N pools stabilized with respect to concentration. The biotic uptake rates were then applied and the resulting model proceeded to steady-state. A transient solution followed from hour 14 through 25.

The simulated and observed values for nitrate- and nitrite-N agree substantially for both OMAN and CAMP1 (Figures 10 and 11, respectively). We have been able to predict both the direction and the overall magnitude of the biotic uptake. Discrepancies between predicted and observed N curves for both sampling sites may be due in part to our hydrodynamic assumptions. It may also be attributed to the uniform application of zero-order N-transfer rates to an environment that is in fact very heterogeneous. We are currently exploring the model's sensitivity to these assumptions and modifying our scenario analysis accordingly.

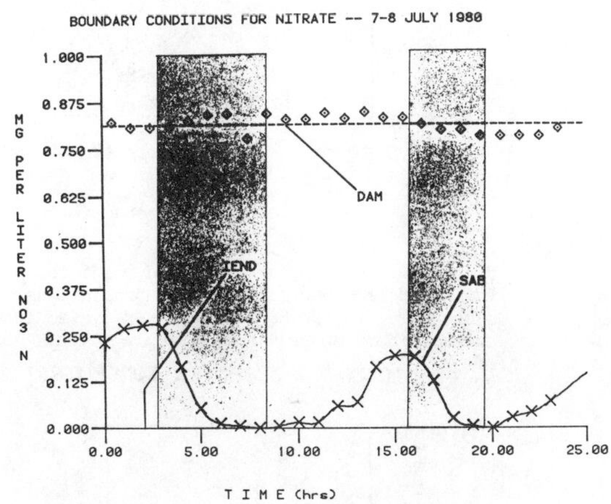

Figure 9. Forcing functions for nitrate- and nitrate-N used to analyze the coupled biotic-hydrodynamic model. The forcing functions for the first 14 hours are applied until steady state is attained with respect to concentration. After this, the model predicts a transient solution. Freshwater inflow at the DAM site was 0.27 and 0.22 m3/sec for the two cycles. IEND flows were 30% of these.

During the first complete tidal cycle, the final biotic uptake for nitrate was partitioned among benthic (66%), marsh (33%), and algal assimilation (1%). Respectively, 2.9, 1.4, and 0.04 kg of (NO3+NO2)-N were processed by the biota. The size of the apparent sink can also be estimated using the conservative constituent model which tabulates the difference between observed and predicted fluxes of N across the system boundaries. The explicit biotic uptake of N suggested by the hydrobiological model accounted for 85 percent of the 5.0 kg sink predicted by the conservative constituent model for

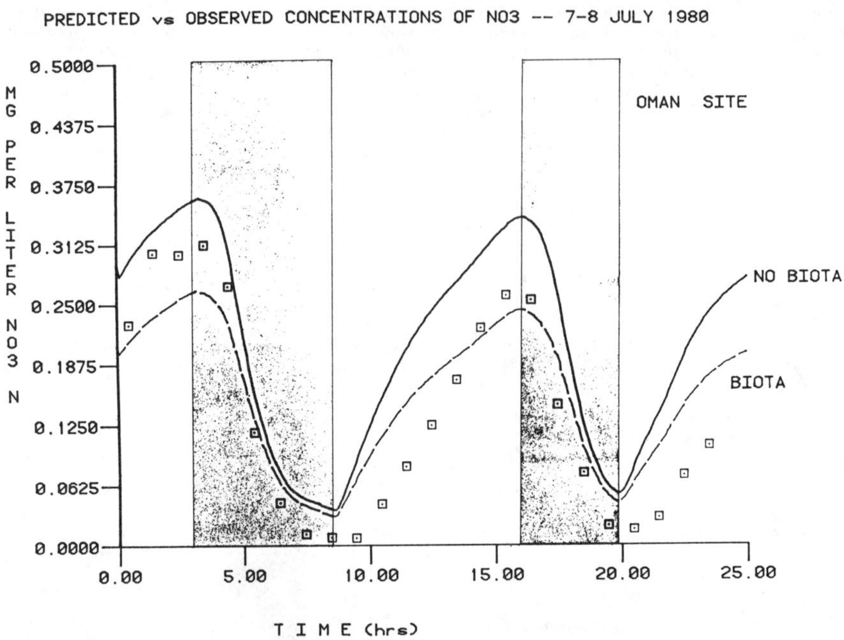

Figure 10. Plot of observed and predicted nitrate- plus nitrite-N at the OMAN site. The solid line represents model predictions for nitrate-N treated as a conservative substance. The dashed curve is predicted from the simulation incorporating biotic uptake. Zero-order kinetics apply.

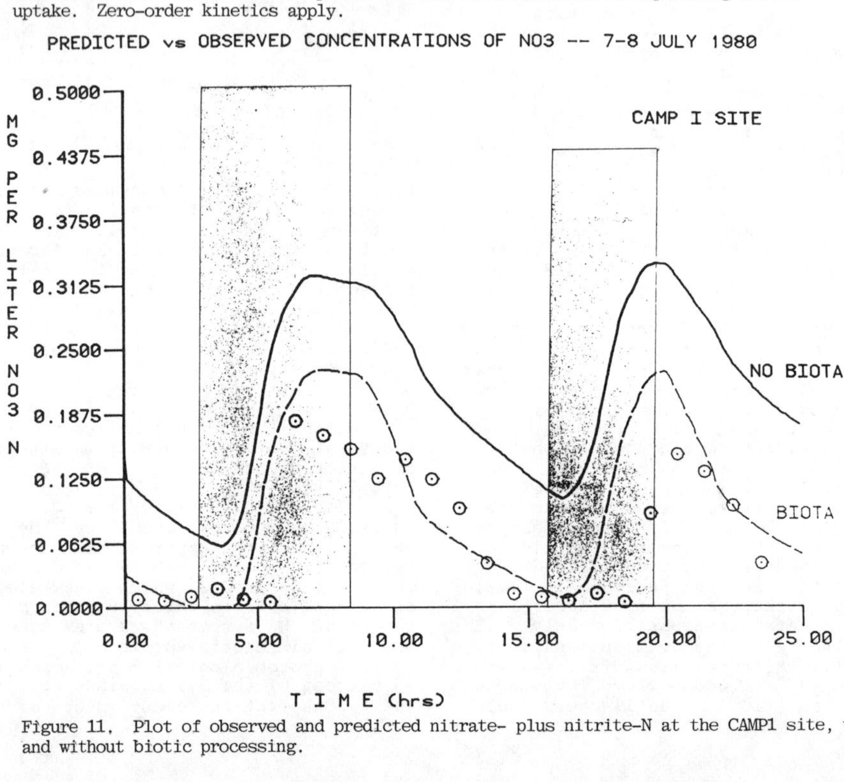

Figure 11. Plot of observed and predicted nitrate- plus nitrite-N at the CAMP1 site, with and without biotic processing.

(NO_3+NO_2)-N. Either estimate of the sink accounts for approximately 30 percent of the introduced nitrate-N.

CONCLUSIONS

During summer low-flow conditions, dilution of high-N river water by low-N tidal water is the dominant factor responsible for the distribution of inorganic N observed in the study site. Although there is a highly effective sink for N on the marsh surface, low water flows limit contact between this substrate and the water column. This renders a partial closure of the ecosystem to water column N and diminishes the capacity of the biota to influence short-term water quality.

This fact carries an important practical consequence since the ecosystem will likely undergo increased nutrient loading in response to continued urbanization (Dillon and Kirchner 1975, Kluesener and Lee 1974, Likens and Bormann 1974). Although the stationary biotic processes appear to have a finite capacity for acting on such enrichment, the open-water phytoplankton may not suffer a similar limitation. More available nutrients will support higher algal productivity and, if the loading is severe, precipitate the attendant problems of eutrophication. At present, the estuary downriver of our study area has moderately large algal populations (unpublished data), but it is difficult to assess quantitatively how the brackish-water ecosystem will respond to increased N and how this in turn will effect the coupled freshwater tidal system.

Maintaining current water quality will require careful monitoring of the quality and quantity of water passing the system boundaries. As well, additional studies of biotic uptake and release of N must continue in order to parameterize these processes in a variety of substrates and to investigate possible density-dependent responses to enhanced nutrient loading.

The hydrobiological model can provide the integration necessary to analyze such information at the ecosystem level. We are currently analyzing 24-hour datasets from other seasons of the year to refine our understanding of the roles that that dilution and biotic processing play in the nitrogen dynamics of the North River.

LITERATURE CITED

Boon, J.D. 1978. Suspended solids transport in a salt marsh creek--an analysis of errors. In Estuarine transport processes, ed. B. Kjerfve, pp. 147-59. Columbia, SC: USC Press.

Boto, K.G. and J.S. Bunt. 1981. Tidal export of particulate organic matter from a northern Australian mangrove system. Estuarine, Coastal and Shelf Science, 13: 247-55.

Bowden, W.B. 1982. Nitrogen cycling in the sediments of a tidal fresh water marsh, PhD. Dissertation. Raleigh, NC: North Carolina State University.

Brinson, M.M., A.E. Lugo and S. Brown. 1981. Primary productivity, decomposition and consumer activity in freshwater wetlands. Ann. Rev. Ecol. Sys., 12:123-61.

Correll, D.L. 1981. Nutrient mass balance for the watershed, headwaters, intertidal zone, and basin of the Rhode River estuary. Limnol. Oceanogr. 26(6):1142-49.

Dillon, P.J. and W.B. Kirchner. 1975. The effects of geology and land use on the export of phosphorus from watersheds. Water Research 9:135-48.

Dyer, K.R. 1973. Estuaries: A physical introduction. London: John Wiley and Sons.

Hall, C.A.S., N. Tempel and B.J. Peterson. 1979. A benthic chamber for intensely metabolic lotic systems. Estuaries 2:178-83.

Kjerfve, B. and H.N. McKellar. 1980. Time series measurements of estuarine material fluxes. In Estuarine perspectives, ed. V.S. Kennedy, pp. 341-57. New York: Academic Press.

Kjerfve, B., L.H. Stevenson, J.A. Proehl, T.H. Chrzanowski, and W.M. Kitchens. 1981. Estimation of material fluxes in an estuarine cross section: A critical analysis of spatial measurement density and errors. Limnol. Oceanogr. 26(2):325-35.

Kluesener, J.W., and G.F. Lee. 1974. Nutrient loading from a separate storm sewer in Madison, Wisconsin. Journal of the Water Pollution Control Federation 46(5):920-36.

Kuenzler, E.J., D.W. Stanley and J.P. Koenings. 1979. Nutrient kinetics of phytoplankton. In Pamlico River, North Carolina. Raleigh, NC: UNC Water Resources Research Institute.

Kremer, J.N. 1979. An analysis of the stability characteristics of an estuarine ecosystem model. In Marsh-estuarine systems simulation, ed. R.F. Dame, pp. 189-206. Columbia, SC: USC Press.

Lee, C.H. 1971. One-dimensional, real-time model for estuarine water quality prediction, PhD. Dissertation. Cambridge, MA: Massachusetts Institute of Technology.

Likens, G.E. and F.H. Bormann. 1974. Linkages between terrestrial and aquatic ecosystems. Bio Science 24(8):447-56.

Mann, K.H. 1975. Relationship between morphometry and biological functioning in three coastal inlets of Nova Scotia. In Estuarine Research, Vol. 1, ed. L.E. Cronin, pp. 634-44. New York: Academic Press.

Najarian, T.O. and D.R.F. Harleman. 1975. A real-time model of nitrogen- cycle dynamics in an estuarine system. Report #204, Parsons Laboratory for Water Resources and Hydrodynamics Department. Cambridge, MA: Massachusetts Institute of Technology.

Nihoul, J.C.J. 1981. Marine hydrodynamics at ecological scales. In Ecohydrodynamics, ed. J.C.J. Nihoul, pp. 1-12. Amsterdam: Elsevier.

Nixon, S. 1980. Between coastal marshes and coastal waters--A review of twenty years of speculation and research on the role of salt marshes in estuarine productivity and water chemistry. In Estuarine and wetland processes: with emphasis on modelling, eds. P. Hamilton and K.B. MacDonald, pp. 437-525. New York: Plenum Press.

Nixon, S.W. and J.N. Kremer. 1977. Naragansett Bay--the development of a composite simulaton model for a New England estuary. In Ecosystem modeling in theory and practice: An introduction with case histories, pp. 622-73. New York: John Wiley and Sons.

Odum, E.P. 1980. The status of three ecosystem hypotheses regarding salt marsh estuaries: tidal subsidy, outwelling and detritus-based food chains. In Estuarine perspectives, ed. V.S. Kennedy, pp. 485-495. New York: Academic Press.

Odum, W.E., J.S. Fisher, and J.C. Pickral. 1979. Factors controlling the flux of particulate organic carbon from estuarine wetlands. In Ecological processes in coastal and marine systems, ed. R.J. Livingston, pp. 69-80. New York: Plenum Press.

Ryther, J.H. and C.S. Yentsch. 1957. The estimation of phytoplankton production in the ocean from chlorophyll and light data. Limnol. Oceanogr. 2(13):281-86.

Schmidt, E.J. 1981. Water quality impact of non-point source contaminants in small tidal rivers, PhD Dissertation. Durham, NH: University of New Hampshire.

Stanley, D.W. and J.E. Hobbie. 1981. Nitrogen cycling in a North Carolina coastal river. Limnol. Oceanogr. 26(1):30-42.

Valiela, I., J.M. Teal, S. Volkman, D. Shafer, and E.J. Carpenter. 1978. Nutrient and particulate fluxes in a salt marsh ecosystem: Tidal exchanges and inputs by precipitation and groundwater. Limnol. Oceanogr. 23(4):798-812.

Whigham, D.F. and R.L. Simpson. 1976. The potential use of freshwater tidal marshes in the management of water quality in the Delaware River. In Biological Control of Water Pollution, eds. J. Tourbier and R.W. Pierson, Jr., pp. 173-186. Philadelphia, PA: University of Pennsylvania Press.

Wolaver, T.G. , R.L. Wetzel, J.C. Zieman and K.L. Webb. 1980. Nutrient interactions between salt marsh, mudflats, and estuarine water. In Estuarine Perspectives, ed. V.S. Kennedy, pp. 123-33. New York: Academic Press.

Woodwell, G.M., D.E. Whitney, C.A.S. Hall and R.A. Houghton. 1977. The Flax Pond ecosystem study: exchanges of carbon in water between a salt marsh Long Island Sound. Limnol. Oceanogr. 22(5):833-38.

DEVELOPMENT AND TESTING OF AN ESTUARINE PHYTOPLANKTON MODEL[1]

Joseph M. Alcamo[2]

This paper presents the development and testing of a model which describes phytoplankton-nutrient interactions in the Sacramento-San Joaquin Delta of California. The model successfully simulates spatial water quality gradients.

The paper also reviews cases for which the model (and perhaps the state-of-the-art) needs to be improved, and the data that would assist in this improvement.

INTRODUCTION

A key goal of ecological modeling is to represent exceedingly complex systems in as simplified a fashion as possible. Unfortunately, because of incomplete theoretical understanding of ecological systems, we can never be sure how detailed a model must be to accurately represent a particular ecological system unless we either (1) operate it for a range of environmental conditions or (2) test it against data.

This paper describes the results of such a modeling exercise; specifically, the rigorous testing of a state-of-the-art phytoplankton model against several sets of estuarine data. From this analysis we determined (1) the conditions for which the model adequately simulates the natural environment, (2) circumstances in which additional model detail is required, and (3) types of data which would assist in improving the detail of the model. The model describes the phytoplankton-nutrient cycle in the estuarine system of the Sacramento-San Joaquin Delta in California. The Delta consists of a network of channels and bays formed where the San Joaquin and Sacramento Rivers converge and meet brackish waters of San Francisco Bay. Figure 1 notes the portion of the Delta covered by the model.

The Delta is also where interest in California water converges since 40 percent of the state's surface runoff flows through its waterways. These interests include local municipal and industrial water supply, crop irrigation of farms on Delta islands, protection of fish migration and waterfowl, export to arid Southern California, outflow to maintain San Francisco Bay water quality, and ship navigation. Considering all these competitive uses, the ultimate goal of the Delta model is to provide insight into how human-related modifications in the Delta will impact its primary producers.

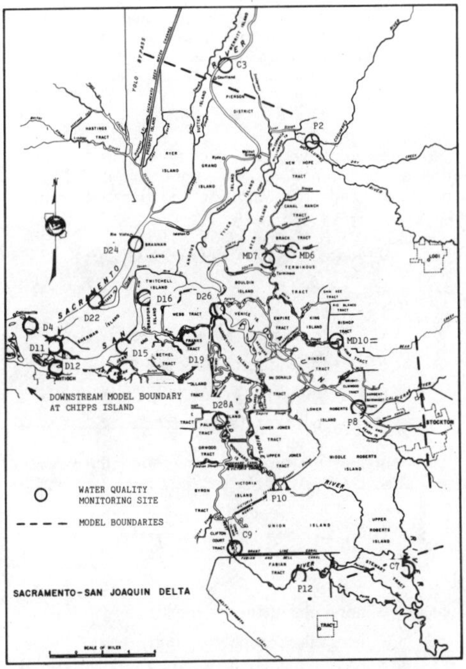

Figure 1.--Area described by model. Map from CA Department of Water Resources.

The model computes steady state concentrations of phytoplankton chlorophyll a, inorganic and organic nitrogen and phosphorus, as well as biochemical oxygen demand (BOD) and dissolved oxygen (DO). Computations are "steady state" in the sense that they are computed equilibrium conditions under specified constant inputs. These calculations in effect represent "seasonal average" conditions and are tested against observed data which are also averaged over summer or autumn periods. Specific data periods were chosen for their consistency with the steady state assumption; that is, the seasonal variability of their flow and water quality was not too great (HydroQual 1981). The five selected periods are presented in Table 1. An earlier version of the model was calibrated to two additional data periods, Summer 1966 and Summer 1970. In sum, therefore, this model has been applied to seven different data sets spanning 12 years and several different hydrologic and water quality conditions.

---

[1]Paper accepted for presentation at Third International Conference on State-of-the-Art in Ecological Modeling, Colorado State University, May 24-28, 1982.

[2]The author is currently a Doctoral Candidate in the Department of Civil Engineering, University of California at Davis. At the time of the study described by this paper, he was a Project Engineer with Hydroscience, Inc., Walnut Creek, CA.

Table 1.--Data periods

Season	Monthly Interval	Delta Outflow (cfs)
Summer, 1971	July - August	10,600
Fall, 1971	September - October	17,600
Summer, 1976	June - August	4,470
Fall, 1976	September - October	3,250
Fall, 1978	September - October	10,900

MODEL DEVELOPMENT

The theoretical basis of the model was first demonstrated by DiToro, O'Connor, and Thomann in their application of a time variable phytoplankton model to a single station on the San Joaquin River at Mossdale (DiToro, et al 1971). A steady state version of this model was applied to the Delta by Hydroscience for the California Department of Water Resources and is documented in Hydroscience (1974). This paper is based on the most recent revisions and testing of the steady state model as described by the author in HydroQual (1981).

In its most basic form, the model describes the temporal variation of phytoplankton growth dP/dt as the difference between a growth term G_p and death term D_p:

$$\frac{dP}{dt} = (G_p - D_p)\ P \qquad (1)$$

To determine G_p, the maximum growth rate k_1 is multiplied by functions of light availability, water temperature, and two limiting nutrients--nitrogen and phosphorus:

$$G_p = k_1 \cdot f(I) \cdot f(T) \cdot f(C_2, C_3) \qquad (2)$$

where,

k_1 = maximum growth rate, 1/day
$f(I)$ = light factor, fraction
$f(T)$ = temperature factor, fraction
$f(C_2, C_3)$ = nutrient factor, fraction

The nutrient limitation is conventionally expressed by the so-called "Michaelis-Menton" function:

$$f(C_2, C_3) = \frac{C_2}{C_2+K_{mn}} \cdot \frac{C_3}{C_3+K_{mp}} \qquad (3)$$

where,

C_2 = inorganic nitrogen, mg/l
C_3 = inorganic phosphorus, mg/l
K_{mn} = "half-saturation" constant for nitrogen, mg/l
K_{mp} = "half-saturation" constant for phosphorus, mg/l

The light function is taken from Steele (1965) and assumes that growth increases proportionately with light to an optimum light level after which growth decreases with increased light intensity. If we assume that light intensity diminishes exponentially with depth, we obtain the following depth-averaged expression for reduction of phytoplankton growth due to light:

$$f(I) = \frac{ef}{K_e H}\ [e^{-\alpha_1} - e^{\alpha_0}] \qquad (4)$$

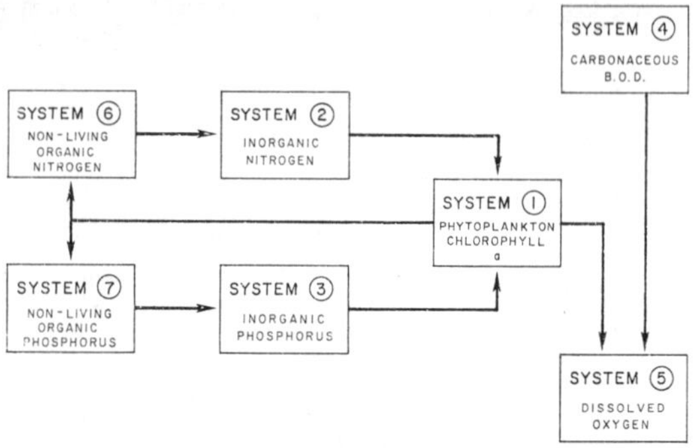

Figure 2.--Systems interactions of phytoplankton model.

where,

e = base naperian log system
f = fraction of daylight
K_e = extinction coefficient, 1/m
H = water depth, m
$\alpha_1 = \frac{I_a}{I_s} \exp(-K_e H)$
I_a = incident light intensity, langleys/day
I_s = saturated light intensity, langleys/day

and,

$\alpha_0 = \frac{I_a}{I_s}$

Growth is assumed to increase with temperature according to

$$f(T) = (1.066)^{T-20} \qquad (5)$$

where,

T = water temperature, °C

Death rate incorporates respiratory functions which decrease phytoplankton biomass. Since respiration is assumed proportional to temperature, death rate is expressed as

$$D_p = K_2 T \qquad (6)$$

where,

K_2 = respiration rate at 0°C, 1/day

The phytoplankton equation (equation 1) forms only one system of the model. The other systems include inorganic nitrogen, organic nitrogen, inorganic phosphorus, organic phosphorus, BOD and DO. These systems are linked with phytoplankton and each other in the fashion noted in Figure 2. Implicit in this figure are the following interactions:

Non-living organic nitrogen and phosphorus decompose to their inorganic forms which are in turn assimilated by phytoplankton. At the same time phytoplankton die-off and enter the pool of non-living organic nitrogen and phosphorus. In addition, phytoplankton photosynthesize and respire oxygen which links phytoplankton to the dissolved oxygen system. The carbonaceous biochemical oxygen demand (BOD) is a measure of the oxygen depleted by organic load of the estuary, and provides a link between the dissolved oxygen system and organic load of the estuary. The sum total of these interactions may be termed "phytoplankton-nutrient kinetics".

But the dynamics of phytoplankton and nutrients in the Delta do not simply depend on their kinetics. Mass transported by advective flow and mixing processes as well as waste loads, can also have an important effect on water quality levels. To account for these additional sources and sinks we can perform a mass balance around a typical elemental volume of the estuary 'j', which is bordered by one or more segments 'k'. This yields:

$$V_j \frac{dc_{ij}}{dt} = \sum_k Q_{kj} C_{ik} + \sum_k E'_{kj} (C_{ik} - C_{ij}) + W_{ij} \tag{7}$$

where,

V_j = segment volume, m^3

C_{ij} = concentration of substance i in segment j, mg/l

Q_{kj} = advective flows from segment k to j, m^3/s

C_{ik} = concentration of substance i in segment k, mg/l

W_{ij} = wasteload of i into segment j, kg/day

and,

E'_{kj} = bulk mixing coefficient between segments k and j defined by

$$E'_{kj} = \frac{E_{kj} A_{kj}}{L_{kj}} \tag{8}$$

where

E_{kj} = dispersion coefficient between segments k and j, km^2/day

A_{kj} = cross sectional area between k and j, m^2

L_{kj} = mixing length between k and j, m

Equation (7) accounts for concentration changes due mass flux of advective flow, tidal mixing, and waste loads. Combining these mass balance terms with the previously discussed kinetic relationships provides the basic model equations presented in Table 2. Additional terms in the dissolved oxygen equation in Table 2 are discussed in another publication (HydroQual 1981).

MODEL APPLICATION

To apply the model equations in Table 2 we must first divide the study area into a series of "completely mixed" segments, i.e. volumes in which we assume that phytoplankton, nutrients, and environmental parameters such as temperature and light extinction coefficients are spatially uniform. In addition we must specify: (1) boundary and initial conditions of each model system for each segment and (2) environmental parameters for each segment. Values of these parameters are presented in HydroQual (1981).

Before proceeding with the phytoplankton-nutrient model testing we must "calibrate the transport" of the model. This means (1) verifying that the specified advective flow pattern[3] is correct and (2) computing the dispersion coefficient, E_{kj} in equation (8). We accomplish this by comparing model computations with observed levels of salinity and total nitrogen which serve as conservative "tracers" of the mixing processes of the estuary. Salinity and total nitrogen are "conservative" in the sense that their concentration gradients are assumed to be independent of kinetics and due only to transport.

[3]Advective flows are taken from a hydrodynamic model of the Delta termed DYNFLO.

Table 2.--Summary of model equations. (A) mathematical equations, and (B) notation of equations.

(A)

System 1 - Phytoplankton Chlorophyll 'a'

$$V_j \frac{dp_j}{dt} = \sum_k (Q_{kj} P_k + E'_{kj} (P_k - P_j)) + (G_{pj} - D_{pj}) P_j V_j + W_{pj}$$

System 2 - Inorganic Nitrogen

$$V_j \frac{dC_{2_j}}{dt} = \sum_k (Q_{kj} C_{2_k} + E'_{kj} (C_{2_k} - C_{2_j})) - a_{21} G_p P_j V_j + k_{62} C_6 V_j + W_{2_j}$$

System 3 - Inorganic Phosphorus

$$V_j \frac{dC_{3_j}}{dt} = \sum_k (Q_{kj} C_{3_k} + E'_{kj} (C_{3_k} - C_{3_j})) - a_{31} G_p P_j V_j + k_{73} C_7 V_j + W_{3_j}$$

System 4 - Biochemical Oxygen Demand

$$V_j \frac{dC_{4_j}}{dt} = \sum_k (Q_{kj} C_{4_k} + E'_{kj} (C_{4_k} - C_{4_j})) - k_r C_4 V_j + W_{4_j}$$

System 5 Dissolved Oxygen

$$V_j \frac{dC_{5_j}}{dt} = \sum_k (Q_{kj} C_{5_k} + E'_{kj} (C_{5_k} - C_{5_j})) + [a_{51} G_p C_1 - a_{51} D_p C_1 + k_a (C_s - C_5) - k_d (\mathrm{BODU} \cdot C_4) - (B/H) - a_{56} k_{62} C_6] V_j$$

System 6 - Organic Nitrogen

$$V_j \frac{dC_{6_j}}{dt} = \sum_k (Q_{kj} C_{6_k} + E'_{kj} (C_{6_k} - C_{6_j})) - k_{62} C_6 V_j + a_{21} D_p P_j V_j + W_{6_j}$$

System 7 - Organic Phosphorus

$$V_j \frac{dC_{7_j}}{dt} = \sum_k (Q_{kj} C_{7_k} + E'_{kj} (C_{7_k} - C_{7_j})) - k_{73} C_7 V_j + a_{31} D_p P_j V_j + W_{7_j}$$

Table 2.--(continued)

(B)

Symbol	Description
a_{21}	Stochiometric ratio of nitrogen to chlorophyll 'a'
a_{31}	Stochiometric ratio of phosphorus to chlorophyll 'a'
a_{51}	Amount of oxygen produced per unit of chlorophyll 'a'
a_{56}	Amount of oxygen used per unit of oxidizable nitrogen
a_{73}	Equivalent to a_{31}
B	Benthal oxygen demand
BODU	Ultimate to 5-day BOD ratio
C_2	Inorganic nitrogen (Sum of Ammonia-N, Nitrite-N and Nitrate-N)
C_3	Inorganic phosphorus
C_4	Biochemical oxygen demand
C_5	Dissolved oxygen
C_6	Non-living organic nitrogen
C_7	Non-living organic phosphorus
D_p	Phytoplankton death rate
e	Base of naperian logarithmic system
E'	Bulk mixing coefficient
f	Fraction of daylight
G_p	Phytoplankton growth rate
H	Water column depth
I_s	"Saturated" (or optimum) light intensity for phytoplankton growth
I_{tot}	Average daily incident solar radiation
k_1	Phytoplankton maximum growth rate at 20°C
k_2	Constant in temperature—dependent phytoplankton endogenous respiration rate
k_3	Constant phytoplankton mortality rate
k_{62}	Organic to inorganic nitrogen conversion rate
k_{73}	Organic to inorganic phosphorus conversion rate
K_e	Extinction coefficient
K_{mn}	Michaelis (or "half-saturation") constant for nitrogen
K_{mp}	Michaelis constant for phosphorus
P	Phytoplankton chlorophyll 'a'
Q	Advective flow
T	Water temperature
V	Volume of model segment
W	External source of substance (from tributary or wasteload, for example).

A set of dispersion coefficients were obtained which ranged from roughly 1 km^2/d at the seaward boundary of the study area to 0.1 km^2/d in the Delta interior. This landward decrease of dispersion coefficients is consistent with observations in other estuaries[4]. It also makes physical sense because the dispersion coefficient incorporates estuarine mixing effects such as the flux of tides which, of course, decreases in intensity landward.

Once the transport of the model has been calibrated and we have confidence that we can account for the effects of transport on concentrations of phytoplankton, etc., then we can proceed to test and calibrate the kinetics of the model. This involves "fine-tuning" the coefficients of the model within justifiable limits until model computations compare favorably with observed data. Since all systems are interrelated with one another to some degree, all seven systems were tested simultaneously.

[4]See for example Thomann (1972).

MODEL RESULTS

Figure 3 compares model calculations for Summer 1976 conditions with observed data from that period. Model calculations were also compared to data from the other periods listed in Table 1. Since there is insufficient space to present results from all data periods, Figure 3 is offered as a sample of model calculations versus observed data. Model calculations were obtained by using a consistent set of kinetic coefficients for all data periods and all model segments. These coefficients are presented in Table 3. An important exception, however, was the different phytoplankton death rate used to calibrate 1971 data periods versus the later data periods. This will be discussed below.

Summarizing to this point, the model shows acceptable spatial resolution for each of the five seasonal periods. For phytoplankton chlorophyll a, for example, the model simulates high levels associated with San Joaquin River inflow from the south (Profile C, Figure 3), relatively low levels and modest growth in the Sacramento River to the north (Profile A, Figure 3) and the gradient in between due to several factors. It is valid therefore, to use the model for making water quality projections for a particular set of environmental conditions. For example, Figure 4 illustrates the projected change in phytoplankton levels due to increased turbidity under Summer, 1976 conditions. This figure indicates that a 25% increase in the extinction coefficient of each Delta segment would depress chlorophyll a concentrations of the San Joaquin River (Profile C) by 1 to 5 μg/l.

Table 3.--Values of model coefficients and constants.

Symbol	Description	Model Values
a_{21}	Stochiometric ratio of nitrogen to chlorophyll 'a'	0.007 mg N/μg chl'a'
a_{31}	Stochiometric ratio of phosphorus to chlorophyll 'a'	0.001 mg P/μg chl'a'
a_{51}	Amount of oxygen produced or respired per unit of chlorophyll 'a'	0.133 mg O_2/μg chl'a'
a_{56}	Amount of oxygen used per unit of oxidizable nitrogen	0
a_{73}	Equivalent to a_{31}	0.001 mg P/μg chl'a'
B	Benthal oxygen demand	0
BODU	Ultimate to 5-day BOD ratio	1.43
I_s	"Saturated" (or optimum) light intensity for phytoplankton growth	300 langleys/day
k_1	Phytoplankton maximum growth rate @ 20°C	2.0/day
k_2	Constant in temperature dependent phytoplankton endogenous respiration rate	0.005/day and 0.010/day
k_3	Constant phytoplankton mortality rate, temperature independent	0
k_{62}	Organic to inorganic nitrogen conversion rate	0.04/day
k_{73}	Organic to inorganic phosphorus conversion rate	0.015/day
K_{mn}	Michaelis (or "half-saturation") constant for nitrogen	0.025 mg/l
K_{mp}	Michaelis constant for phosphorus	0.005 mg/l

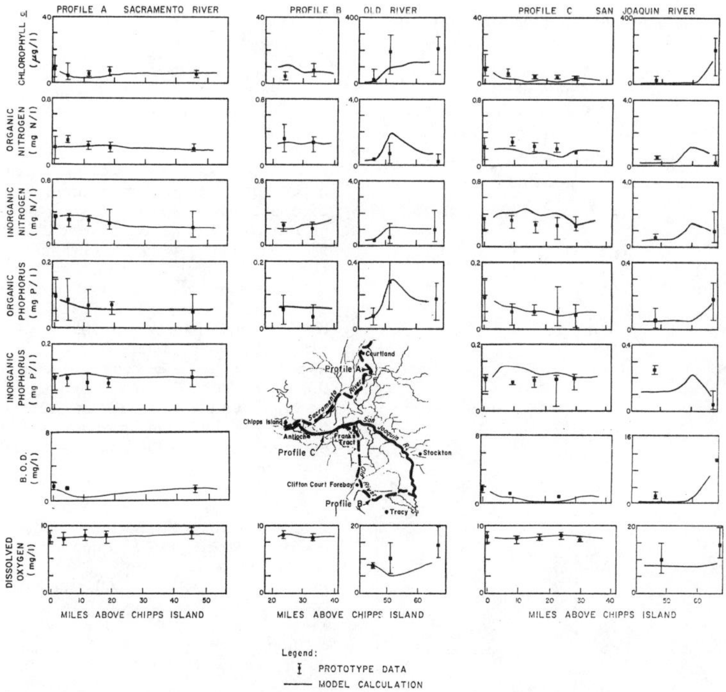

Figure 3.--Model calculations and observed data for Summer, 1976.

Although we may make projections for a specific data period, the model cannot be used to make projections between data periods. This is because, as noted, phytoplankton death rate differed significantly between the two data periods of 1971 and the three latter periods. A death rate (D_p at 20°C)[5] of 0.1/day was used for the 1971 data periods, while a death rate of 0.2/day was needed to simulate the substantially lower phytoplankton levels of 1976. As illustrated in Table 4, chlorophyll a levels in Summer and Fall, 1976 are in some reaches less than one-fourth the levels observed in 1971. It was also found that the death rate of Fall, 1978 phytoplankton was closer to the 1976 value than the 1971 value. Figure 5 notes how death rates were selected, using Profile C of the Summer 1971 data period as an example. The three graphs from top to bottom illustrate observed phytoplankton chlorophyll a, inorganic N and inorganic P along with model computations using D_p equal to 0.1/day and 0.2/day.

Not only is chlorophyll a underestimated when D_p is equal to 0.2/day, but inorganic nitrogen and phosphorus are also overestimated at this death rate because there is a smaller phytoplankton population to assimilate nutrients. At D_p=0.2/day computed levels of phytoplankton, inorganic nitrogen and phosphorus better agree with data. In a similar fashion, D_p=0.2/day agreed more closely with Summer 1976, Fall 1976 and Fall, 1978 data periods.

Note however that we cannot tell from this analysis whether death rate increased in 1976 or max growth rate decreased. In fact, we may reduce the maximum growth rate (K_1) from 2.0/day to 1.1 to 1.3/day and compute the same phytoplankton levels as we would by doubling the death rate from 0.1 to 0.2/day. Therefore, for the remainder of this paper, the apparent decrease in growth will be referred to as a change in net growth.

[5]For the remainder of this paper, cited death rates will refer to values at 20°C.

FACTORS AFFECTING PHYTOPLANKTON GROWTH

Why then didn't the model simulate reduced phytoplankton levels of 1976? Was it a problem of model formulation or a variable that the model does not account for at all? To gain insight into this question the following section briefly reviews possible factors which contributed to apparent decrease in net phytoplankton growth rate.

As a starting point, we should consider the model formulation itself. Equations (1) through (6), which describe

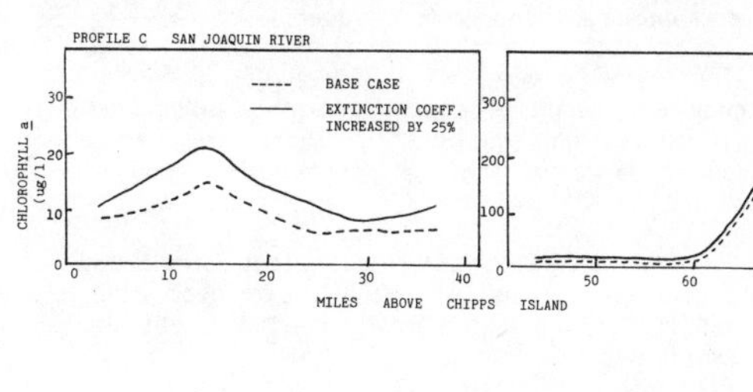

Figure 4.--Projected change in phytoplankton levels due to increased turbidity in San Joaquin River (Profile C). Consult Figure 3 for location of Profile 'C'. This projection refers to Summer, 1976 conditions.

Table 4.--Comparison of observed phytoplankton data from 1971 and 1976.

	Station[1]	Chlorophyll a			
		Summer 1971	Fall 1971	Summer 1976	Fall 1976
		(μg/l)			
D24	Sacramento River at Rio Vista	5.9	4.4	5.8	3.1
D15	San Joaquin River at Jersey Point	14.7	20.0	3.7	4.3
D19	Frank's Tract	10.7	19.6	3.6	4.1
P10	Middle River at Victoria Island	11.9	35.5	5.3	5.7

[1]Consult Figure 1 for station location.

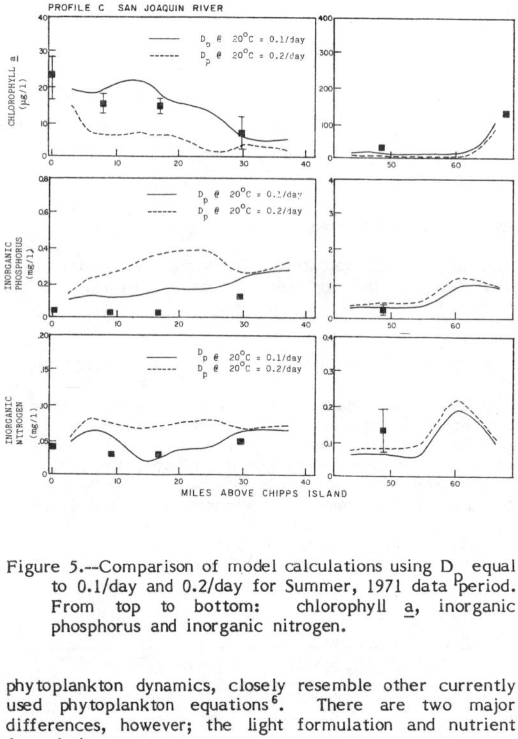

Figure 5.--Comparison of model calculations using D_p equal to 0.1/day and 0.2/day for Summer, 1971 data period. From top to bottom: chlorophyll a, inorganic phosphorus and inorganic nitrogen.

phytoplankton dynamics, closely resemble other currently used phytoplankton equations[6]. There are two major differences, however; the light formulation and nutrient formulation.

First let us consider the model's light formulation. A popular alternative to using equation (4) for expressing light dependence of phytoplankton growth is a "Michaelis-Menton" type expression:

$$f(I) = \frac{I_H}{I_H + I_2} \qquad (9)$$

where

I_H = depth average, daily average light intensity, langleys/day[7]

I_2 = "half-sautration" constant for light, i.e. intensity at which phytoplankton growth rate is one-half its maximum value, langleys/day.

This expression assumes that growth rate is a hyperbolic function of light intensity. A detailed comparison of equations (4) and (9) is unfortunately outside the scope of this paper; however, a sample calculation indicates that the two may provide similar answers for typical Delta conditions. For this calculation, let us assume water depth (H) is 6 meters, the extinction coefficient (K_e) is 3.0/meters, the fraction of daylight (f) is 0.5, the optimal light intensity (I_s) for equation (4) is 300 langleys/day and the half-saturation light intensity (I_2) in equation (9) is 250 langleys/day[8]. Furthermore, assume incident light radiation (I_a) varies from 400 to 700 langleys/day. For these values, equation (4) yields an f(I) of 0.055 to 0.068, and equation (9) 0.043 to 0.072. In other words, for this set of conditions equations (4) and (9) compute light limitation of similar magnitude.

In another analysis, Swartzman and Bentley (1979) compare f(I) versus surface light intensity using Steele's equation (Steele, 1962) (which equation (4) is based on) and Smith's equation (Smith, 1962). For a range of light intensities of 400 to 600 langleys/day, Steele's equation computes an f(I) of roughly 0.16 to 0.18 and Smith's equation about 0.15 to 0.19[9]. These results must be used with caution, however, since Swartzman and Bentley assumed conditions typical of only a few Delta channels.

One possible interpretation of the surprisingly similar answers provided by all the different light equations is that their differences are "damped out" by averaging over depth. For example, an estuary may be 5 to 10 meters deep, but only its top 2 meters or less are typically illuminated. Therefore, the differences predicted by equations (4), (9), et al for the photic zone may decrease when the light functions are integrated over the entire water column depth.

The other current debate over phytoplankton model structure concerns whether the "Michaelis" terms in equation (3) should be multiplied together or minimized[10]. By multiplying these terms together one assumes that nitrogen and phosphorus independently limit phytoplankton growth, whereas selecting the smaller of the two terms is more in accordance with "Liebig's Law of the Minimum". This law declares that "growth is limited by substance that is present in minimal quantity with respect to the needs of the organism".

But test calculations show that for the Delta the two approaches provide similar answers because phosphorus

[6]See for example, the literature review of Swartzman and Bentley (1979).

[7]I_H in this paper is a depth and daily average in order to make equation (9) comparable to equation (4). For the usual definition of I_H consult Tetratech (1975), for example. In this paper, I_H is derived from

$$I_H = f \int_H^0 I_a \exp(-K_e H) dH$$

where the symbols are as previously defined.

[8]This is a typical value from Zison (1978).

[9]These values are taken from Figure 2 of Swartzman and Bentley (1979) assuming an optimal light intensity of 350 langleys/day and an I_k of 75. langleys/day.

[10]For a more detailed discussion of this issue consult the reviews of Swartzman and Bentley (1979) or Zison, et al (1978).

levels generally exceed limiting quantities[11]. In other words, phytoplankton are probably limited by only one macronutrient, nitrogen. Silicon could also have been a limiting macronutrient if it was in short supply because the phytoplankton population consisted largely of diatoms which require silica for their cell walls. However observed silicon levels were far in excess of limiting levels (HydroQual 1981).

Based on this review of light and nutrient model formulations, we may tentatively conclude that it is unlikely that errors in the model's light and macronutrient terms can account for apparent change in phytoplankton net growth in 1976 and after.

As another alternative, the changes in water temperature and incident light radiation between data periods were examined. However these factors were discounted because no significant trends were detected between data periods.

The net growth rate of the aggregate phytoplankton population could be effected by changes in the mix of species since, for example, blue green algae have lower typical growth rates (assuming variables such as temperature and nutrients are constant) than diatoms (Canale 1974). In addition, if the zooplankton population which grazes on phytoplankton had changed in composition or size we might observe a difference in phytoplankton mortality. However, there was no evidence of major population changes to support either of these contentions.

Flow circulation merits examination as a factor because model calculations indicate that flow affects spatial gradients of phytoplankton in the Delta. However, in one respect flow circulation is not a likely factor since freshwater flows were lower in 1976 than 1971, thereby increasing residence time in the Delta's channels and bays. Increased residence time would tend to promote rather than inhibit growth in 1976. But the influence of flow on estuarine phytoplankton dynamics is not entirely understood. One conceivable way in which reduced freshwater circulation could indirectly effect phytoplankton would be to decrease channel turbulence. Some ecologists believe that many phytoplankton species would sink if turbulent eddies did not suspend them in the water column[12]. Perhaps therefore a reduction of channel flow, which in turn would lessen turbulence, may promote increased phytoplankton settling. Unfortunately, there have been few tests conductea of settling of Delta phytoplankton. It is also possible that tidal flow, which is unaffected by a reduction of freshwater flow, may promote enough turbulence to keep phytoplankton afloat.

The sharp reduction in freshwater outflows during the 1976 drought permitted greater salinity intrusion than in 1971. This raises the question of salinity impact on phytoplankton growth. However, Cloern (1978) reported that a common Delta phytoplanktor, Skeletonema costatum actually increases in growth rate up to salinity concentrations of about 20 parts per thousand. Since study area salinities did not exceed 20 parts per thousand for any of the data periods, at least this phytoplanktor probably was not inhibited by salinity intrusion. Other salinity-growth correlations support this conclusion (HydroQual 1981).

Though we may be confident that nitrogen is the only limiting macronutrient, we clearly do not understand what levels of micronutrients limit phytoplankton growth. Eyster (1964) reports ten trace elements needed in varying amounts by algae. In terms of the cellular function they serve, these micronutrients are: (1) For Photosynthesis - molybdenum, chlorine, iron, zinc and vanadium, (2) For Nitrogen Metabolism - boron, molybdenum, iron and cobalt, and (3) For other cellular functions - copper, cobalt, molybdenum, boron and silicon[13].

Not only do these metals act as micronutrients in small concentrations, but many of these same elements are toxic to phytoplankton at higher concentrations. Table 5 compares some literature data of toxic metals (and other substances) with measured levels in the Delta during 1971 and 1976. A few things may be noted from these data. First, there is no significant temporal trend. This may be attributed in part to a change in the sensitivity of the analytical methods used to measure these substances. For example, the concentration of copper at Rio Vista on September 20, 1971 was reported to be "less than 0.05 mg/l" whereas a concentration of "0.01 mg/l" was noted on May 12, 1976. Second, measured toxicant levels are close to the range of toxicants reported to have chronic effects on phytoplankton growth in laboratory tests. Even though these laboratory data do not necessarily apply to field conditions, they do raise the possibility of toxicity.

In summary, among the less easily discounted factors which influenced the apparent decrease in phytoplankton net growth are (1) toxicants (2) micronutrients and (3) increased settling rates. Of course with the limited data available we cannot confidently eliminate any other factors reviewed in the paper, nor can we ignore the possibility that several factors acted in concert.

---

[11]Inorganic phosphorus levels were usually in excess of 0.05 mg/l whereas most investigators estimate that the "half-saturation" constant of phosphorus (i.e. the level at which phytoplankton growth is one-half its maximum value) is an order of magnitude lower, 0.005 mg/l. See, for example, Zison, et al (1978).

[12]See for example Fogg (1965).

[13]Silicon is considered a macronutrient requirement of diatoms as previously noted in the text.

Table 5.--Comparison of observed heavy metals concentrations in the Delta with reported toxicity levels. (A) literature toxicity data, (B) observed levels in the Delta.

(A)

Substance and Concentration (mg/l)	Effect	Reference
Copper		
0.003-0.02	20% reduction in SF Bay phytoplankton photosynthesis	Krock and Mason (1971)
0.015-0.030	20% reduction in San Joaquin Delta phytoplankton photosynthesis	Krock and Mason (1971)
0.005-0.02	Unspecified reduction of dinoflaggelate growth	Saifullah (1978)
0.1	20% reduction in Lake Nakuru, Kenya blue-green photoplankton	Källqvist and Meadows (1978)
0.15-0.20	50% reduction in Lake Nakuru, Kenya blue-green phytoplankton	Källqvist and Meadows (1978)
Mercury		
0.003-0.01	20% reduction in SF Bay & Delta phytoplankton photosynthesis	Krock and Mason (1971)
Zinc		
0.05-0.2	20% reduction in SF Bay & Delta phytoplankton photosynthesis	Krock and Mason (1971)

Table 5.--(continued)

(B)

Substance and Concentration (mg/l)	Dates	Station[1]	
Copper			
<0.05-0.07	7/6/71 & 8/3/71	D24	Sacramento River at Rio Vista
<0.05	9/20/71		
0.01	5/12/76		
0.01	9/15/76		
<0.005-0.11	7/6/71 & 8/3/71	D15	San Joaquin River at Jersey Point
<0.05	9/20/71		
0.01	5/12/76		
0.01	9/15/76		
0.01	5/6/76	P10	Middle River at Victoria Island
0.01	9/14/76		
Mercury			
0.0	5/12/76	D24	
0.0001	9/15/76		
0.0	5/12/76	D15	
0.0002	9/15/76		
0.0	5/6/76	P10	
0.0	9/14/76		
Zinc			
<0.01-0.02	7/6/71 & 8/3/71	D24	
<0.01			
0.01	9/20/71		
0.01	9/15/76		
<0.01	7/6/71 & 8/3/71	D15	
<0.1	9/20/71		
0	5/12/76		
0.1	9/15/76		
0.0	5/6/76	P10	
0.0	9/14/76		

[1]Consult Figure 1 for station locations.

DATA FOR MODEL REFINEMENT

Based on the review in the previous section, it is apparent that the current modeling framework cannot explicitly account for factors which affected phytoplankton growth during 1976. However as noted, other popular modeling formulations cannot do better. What type of information, then would help us refine the model? One helpful type of data would be growth studies of phytoplankton under controlled laboratory conditions using field samples as media.

There are several ways in which these data could have assisted in our evaluations. First, comparative growth data from different time periods would help determine whether the apparent decrease in growth was an artifact of the model or an actual change in growth rate. Second, if growth rate between different time periods was static while death rate changed then we might further examine predation or other mechanisms grouped into the "death rate" terms of kinetics. If we then filter the medium to remove zooplankton we can examine the influence of respiration and other factors without interference by predation. Third, if growth rate changes between data periods we might look toward nutrient deficiencies or toxicity. Finally, under laboratory conditions we can vary temperature, light and investigate the relationship between growth kinetics and these factors. This, however is just a partial list of the utility of laboratory growth experiments.

What then would be ideal characteristics of such laboratory growth studies? Two of these characteristics might be:

(a) The procedure should be reproducible such that we have "benchmark" to judge temporal changes in growth kinetics.

(b) The procedure should be able to distinguish maximum growth rate from death rate rather than simply measure net growth.

The procedure commonly used to determine phytoplankton growth characteristics in the laboratory is known generically as a "Limited Volume" test, and also as Algal Growth Potential (AGP) studies, and in Standard Methods (1981) as "Biostimulation Studies". It is termed a "Limited Volume" test because it involves incubating a seed phytoplankton population in a vessel with no flow-through of media, i.e. a limited volume of media. In addition, temperature and light are kept constant at favorable levels. Under these conditions five typical phases of microbial dynamics are observed: (1) lag phase (2) exponential growth phase (3) phase of declining relative growth, (4) stationary phase and (5) death phase. In a sense these phases resemble a typical pattern of phytoplankton growth and decline in the environment during the course of the year. But despite this similarity the Limited Volume test has certain drawbacks which make it less useful for determining model coefficients.

A serious limitation of the standard AGP test concerns its measurement of death rate. Fogg (1965) notes that reduction of phytoplankton growth after the exponential phase in a Limited Volume test may be brought on by a variety of factors including nutrient exhaustion, inadequate carbon dioxide or oxygen supply/transfer, pH alteration of the medium, reduction of light through self-shading, or self inhibition through the release of autotoxins. Fogg also notes that the magnitude of the death rate (respiration in this case) can differ significantly depending on which factor or factors is acting. Though we can mitigate some factors, such as by aerating the vessel to promote CO_2 and O_2 transfer, these mitigation measures are often not taken in practice, not are they recommended or required in Standard Methods (1981).

Even if mitigation measures are taken we cannot be sure that the measures themselves will not interfere with results of the experiment in some unknown way. For example, the chemical used to maintain pH in the Limited Volume test may also directly affect phytoplankton growth.

Moreover, the death rates of the declining phase of an AGP test respresents the catastrophic decline of physiologically unhealthy cells, whereas the model coefficient K_2 reflects the combined respiration of both healthy and unhealthy cells.

One way to avoid measuring the death rate of only unhealthy cells would be to measure the respiration rate from the "healthy" part of the curve, i.e. the exponential phase. However, Figure 6 shows that the Delta AGP's were not sensitive in the exponential phase.

To illustrate this point Figure 6 presents a typical AGP test of Delta phytoplankton together with computed chlorophyll a concentrations at death rates of 0.1, 0.2 and 0.3/day. The equation used to compute these values is

$$\frac{dP}{dt} = (G_p - D_p) \cdot P \tag{10}$$

where

$$G_p = K_1 \cdot f(C_2, C_3) \cdot \frac{I_a}{I_s} \exp\left(1 - \frac{I_a}{I_s}\right)$$

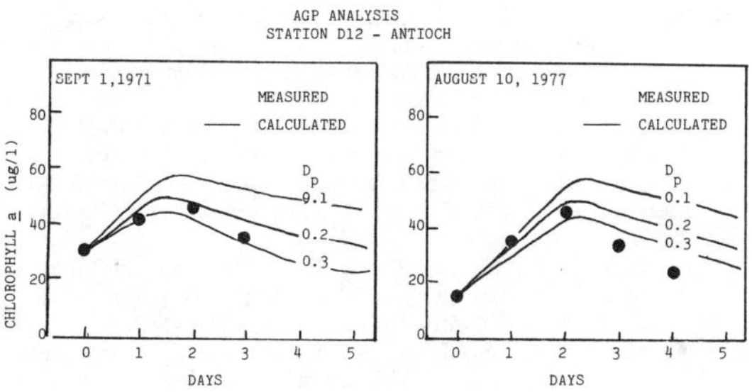

Figure. 6--Algal growth potential studies of Delta water. Courtesy, CA Department of Water Resources.

and all other terms are as previously defined.

The light term in equation (9) is from Steele (1965) who used it to evaluate laboratory phytoplankton data. As in Steele's original work, I_s is set to 2000 foot-candles.

We can see from Figure 6 that it is difficult to distinguish the best death rate on the part of the curve between zero and two days (the exponential phase). This is because the AGP test is conducted under light and temperature conditions favorable to growth. As a result, the growth term G_p (as calculated in equation 9) far exceeds the D_p term during the exponential phase. A different situation exists in the Delta where only the top meter or so of the water column is illuminated and then only for about one-half the entire day. Under these less favorable conditions, growth and death rates are more evenly balanced. Table 6 compares calculated net growth rates for field conditions with calculated net growth rates for a comparative AGP study. Equations (1) through (6) were used to compute the net growth rate of phytoplankton in the Delta, and equation (9) to calculate the net growth rate during the exponential phase of an AGP test. As Table 6 indicates, the field growth rate was an order of magnitude smaller than the laboratory growth rate. In some cases the calculated field net growth rate was negative, implying that the death rate exceeded the growth rate.

One might suspect from these observations that the phytoplankton death rate in the AGP test is more of an artifact of the test than a reflection of death rate observed in the field. For this reason, it seems that the Limited Volume approach cannot provide suitable information for estimating model growth coefficients. Considering the similarity of this model to other commonly used models in engineering studies[14], it may be fair to generalize that the Limited Volume test has limted applicability to current phytoplankton models.

Superior to the Limited Volume approach is the "Continuous Culture" represented by the turbidostat and chemostat apparatus[15]. The Continuous Culture approach involves the addition of media, and in some cases alteration of experimental conditions during the experiment, such that a microbial population is held at a particular point in its growth curve.

[14]See Zison, et al (1978), for example.

Table 6.--Computed net photoplankton growth rate at Station D15, Jersey Point on San Joaquin River. (A) computed from AGP test, (B) computed from field conditions.

	Date	Computed Net Growth[1] (day^{-1})
(A) AGP Test		
	9/1/71	0.15
	9/29/71	0.50
	7/25/77	0.62
(B) Field	**Data Period**	**Computed Net Growth[2] (day^{-1})**
	Summer 1971	0.094
	Fall 1971	0.053
	Summer 1976	-0.050
	Fall 1976	0.040
	Fall 1978	-0.020

[1]Computed with equation (9) using initial nutrient conditions of AGP test and K_1=2.0/day, D_p=0.2/day, I_a=400 ft-candles, I_s=2000 ft-candles, K_{mn}=0.025 mg/l, K_{mp}=0.005 mg/l.

[2]From HydroQual (1981)

Rather than being superior to the Limited Volume test, one may argue that the Continuous Culture approach is a less realistic representation of the environment than the Limited Volume test because a Continuous Culture does not permit the phytoplankton population to fluctuate with time. However, these fluctuations occur too quickly in a Limited Volume test to make it a useful representation of the environment. On the other hand, a Continuous Culture is more relevant to phytoplankton models because model coefficients are typically referenced to a particular set of conditions. For example, K_1 in this paper is the maximum growth of phytoplankton at 20°C under optimum nutrient and light conditions. Although a Limited Volume test can maintain constant light levels, a Continuous Culture can hold both light and nutrient levels constant at optimum levels. In addition, nutrient levels can be held constant and light varied diurnally, simulating conditions in an estuarine channel which is continuously fed nutrients by tidal and freshwater flows. Under these circumstances phytoplankton death rate may be of the same magnitude as the growth rate. Since they are of the same order, a slight increase in death rate or decrease in growth rate toxic effects (such as those noted in Table 5) would be

[15]The turbidostat apparatus operates on the principal of self-regulating turbidity. Differences in turbidity of the culture (i.e. population density) triggers a photocell system which in turn feeds new media to the culture. Meanwhile, portions of the algal population are mechanically harvested at suitable intervals. Metabolic products of the algae are also constantly diluted by addition of new media.

In a chemostat fresh media is added to the vessel at a constant rate and the population adjusts itself to this supply.

recognizable. In a Limited Volume test, small chronic effects may be either indistinguishable from mortality associated with the test itself or overshadowed by large growth rates during the exponential growth phase.

These are some of the reasons why the Continuous Culture approach is preferable to the Limited Volume test for purposes of obtaining model growth coefficients.

CONCLUSIONS

The development and testing of the phytoplankton model described in this paper provided the following insights:

1. The kinetic and transport terms of the model are adequate to simulate significant spatial gradients of phytoplankton, nutrients and water quality within a given summer or fall data period. The model, however, cannot implicitly account for apparent phytoplankton growth changes that occurred during and after a drought period.

2. Based on calculations and data analysis presented in this paper it appears that current phytoplankton modeling approaches cannot account for the factor or factors which influenced the apparent phytoplankton growth change.

3. Laboratory-type growth studies would obviously help improve the model. The Limited-Volume test however has inherent limitations which reduces its usefulness for modeling purposes. The Continuous Culture approach provides better experimental control and more sensitive data than the Limited Volume test and is therefore more useful.

ACKNOWLEDGEMENTS

Al Hunt served as Project Engineer of the Delta Study described herein and conducted many of the calculations presented in this paper. Austin Nelson, John Mueller and D.J. O'Connor provided technical guidance for the model development and applications. Much of the data for the modeling effort was provided by the California Department of Water Resources, with particular assistance by Fred Bachman.

REFERENCES

American Public Health Association, and others. 1980. Standard Methods for the Examination of Water and Wastewater. 15th Edition, 1134 p. American Public Health Association.

Canale, R.P. and A.H. Vogel. 1974. Effects of Temperature on Phytoplankton Growth. Journal Environmental Engineering Division, ASCE, EE1:231-241.

Cloern, J.E. 1978. Empirical Model of Skeletonema costatum Photosynthetic Rate with Applications in the San Francisco Bay Estuary. Advances in Water Resources 1:(5)267-274.

DiToro, D.M., O'Connor, D.J. and R.V. Thomann. 1971. A Dynamic Model of the Phytoplankton Population in the Sacramento-San Joaquin Delta. Advances in Chemistry Series. American Chemical Society Number 160, 130-180.

Eyster, C. 1964. Micronutrient Requirements for Green Plants, Especially Algae. In: Algae and Man (D.F. Jackson, ed.) Plenum Press, N.Y. pp. 86-119.

HydroQual. 1981. Analysis of Phytoplankton Growth and Death Kinetics. Interagency Ecological Study Program, Sacramento, CA.

Hydroscience. 1974. Development and Application of a Steady State Eutrophication Model of the Sacramento-San Joaquin Delta. Prepared for CA Water Resources Board.

Kållqvist, T. and B.S. Meadows. 1978. Toxic Effect of Copper and Algae and Rotifers from a Soda Lake (Lake Nakura, East Africa). Water Resources (Great Britain) 12:771.

Krock, H.J. and D.T. Mason. 1971. Bioassays of Lower Trophic Levels. Volume 6 of A Study of Toxicity and Biostimulation in S.F. Bay-Delta Water. UC Berkeley Sanitary Engineering Report 73-1.

Saifuillah, S.M. 1978. Inhibitory Effects of Copper and Marine Dinoflagellates. Marine Biology (W. Germany) 44:229.

Steele, J.H. 1965. Primary Production in Aquatic Environments. In: Notes on Some Theoretical Problems in Production Ecology, (C.R. Goldman, Ed.) U.C. Press, Berkeley, CA.

Swartzman, G.L. and R. Bentley. 1979. A Review and Comparison of Plankton Simulation Models. ISEM Journal 1:30-81.

TetraTech. 1975. A Comprehensive Water Quality Ecological Model for Lake Ontario. Prepared for NOAA.

Thomann, R.V. 1972. Systems Analysis and Water Quality Management. 286 pp. Environmental Research and Applications Inc., N.Y.

Zison, S.W., Mills, W.B., Deimer, D. and C.W. Chen. 1978. Rates Constants and Kinetics Formulations in Surface Water Quality Modeling. EPA-600/3-78-105.

MODELS AND FIELD DATA OF NEW ENGLAND TIDE POOLS[1]

Charles Puccia[2] and Judith Pederson[3]

Abstract.--Qualitative (signed digraph) or loop models were constructed for rocky shore tide pools. A discussion of the use of loop models for field research is presented. A comparison of pairwise correlation coefficients of our data with predicted correlations of the models is made. These suggest environmental changes in parameters effecting growth rates of populations is primarily through the herbivorous snail. Preliminary experimental manipulation of snails also supports this conclusion.

INTRODUCTION

Although there is a large body of literature concerning the distribution of organisms along rocky coasts, few studies have examined community dynamics (Lewis, 1964; Stephenson and Stephenson, 1972; Lubchenco, 1978; Underwood, 1980; 1981). Moreover, none of these studies develop the use of models apriori and aposteriori in determining community interactions or in understanding the processes which bring about the distribution of organisms found in high tide pools.

Tide pools are unique ecosystems in the intertidal zone. They are discrete areas which remain submerged during a tidal cycle and are relatively simple ecosystems (Lubchenco, 1978). Because of their accessibility, they provide opportunity for field manipulation essential to community studies (c.f. Paine, 1969; Connell, 1975) and are ideal for use in models.

The purpose of this paper is to present qualitative (loop) models (Levins, 1974) for high tide pools found along the rocky coast of New England. Apriori models are based on data available in the literature (Lubchenco,1978;Menge,74and aposteriori models are constructed using our own data. We illustrate steps involved in developing models, analyzing predictions and using results to guide future areas of research.

DESCRIPTION OF STUDY AREA AND GENERAL METHODS

The tide pools are located in the high (upper third) intertidal area along a rocky coast of Gloucester, Massachusetts. Eight pools are monitored monthly at 3 to 5 randomly chosen, permanently marked quadrats in each pool, permitting non-destructive sampling. Lucite sheets measuring 0.1 m^2, painted with 50 random dots are used to measure percent cover of algae and sessile invertebrates at each quadrat (Grieg-Smith, 1964). Snail densities are estimated from numbers removed within a 0.1m^2 aluminum frame, at each permanently marked quadrat.

These pools are comparable to and typical of the pools found in New England and used by other investigators. The major floral and faunal components of these pools will be briefly described.

The perennial red alga, Chondrus crispus, extends from the subtidal into the high intertidal and tide pools, under the protective cover of other algae. Its upright portions (thalli) grow from holdfasts which effectively sequester primary space and prevent settlement of other algae. Recruitment of new individuals is low; populations increase by vegetative propogation. The edible mussel, Mytilus edulis, especially in wave exposed areas, can overgrow Chondrus and other perennial algae (Paine 1969; 1974).

Numerous species of ephemeral algae occur in tide pools. These are characterized by their short growing season of 3 to 4 months and either annual recruitment or regrowth of thalli from long-lived holdfasts (Sears and Cooper, 1978; Littler and Littler, 1980).

The dominant herbivorous snail, Littorina littorea feeds preferentially on ephemeral algae and is described as avoiding Chondrus (Lubchenco, 1978) however, littorines are observed grazing Chondrus in the field (D. Cheney, pers. comm.; R. Seeley, pers. comm.). Pools with few or no Littorina snails present are found at the exposed sites. The green crab, Carcinus maenas, a likely predator of both snails and mussels, are present in all pools.

There are other inhabitants of tide pools; encrusting algae and invertebrates, such as hydroids, amphipods, crabs, starfish, tunicates, small herbivorous snails and isopods. During submersion, pools are probably visited by fish and larger crustaceans. Nevertheless, current hypotheses do not suggest these organisms are responsible for high tide pool algal patterns that are observed in New England (Lubchenco, 1978, 1980; Lubchenco and Menge, 1980). The interactions of these organisms are described in the following models.

---

[1]Paper presented at the Third International Conference on State-of-the-Art in Ecological Modelling, Colorado State University, May 24-28, 1982.

[2]Charles J. Puccia is Assistant Professor of Biology, University of Massachusetts at Boston, Boston, Mass. 02125, U. S. A.

[3]Judith B. Pederson is Assistant Director of Environmental Science Program, University of Massachusetts at Boston, Boston, Mass. 02125, U. S. A.

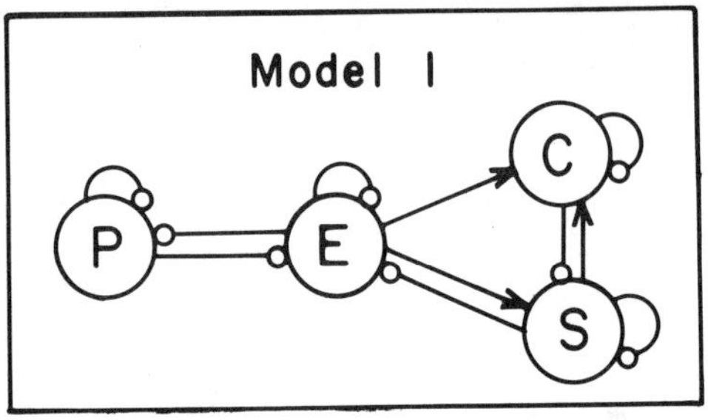

Figure 1.--Tide pool model 1 where P refers to perennials, E to ephemerals, C to crabs, S to snails, to a negative pathway, to a positive pathway, and indicates self-damping. Refer to text for further explanations.

The models we use are based on the qualitative signed digraph method, or loop analysis (Levins, 1974, 1975). The first model (fig. 1) represents a generally accepted view of the tide pool community (Lewis, 1964; Lamb and Zimmerman, 1964; Stephenson and Stephenson, 1972; Lubchenco, 1978; Underwood, 1980). Perennial algae (P), especially the red alga Chondrus crispus competes with ephemeral algae (E), such as the green alga Enteromorpha intestinalis. Herbivorous periwinkle snails (S), Littorina littorea common throughout the New England rocky coast, feed preferentially on ephemeral algae. The green crab (C), Carcinus maenas consumes periwinkles and is protected from its intertidal predator, gulls by the camouflage of ephemerals (Lubchenco, 1978). The self-damping term (links which start and end on the same variable) represent the fact that all the organisms in these models are not dependent on reproductive output of local populations, but settle from a larger, pelagic source of propagules.

Not all investigators agree with all or parts of this model (D. Cheney, pers. comm.). We include these and other reasonable objections in the following set of models.

A second model (fig. 2) does not include competitive interactions between the perennial and ephemeral algae. Instead, shading by ephemerals is considered to be harmful to the perennial, Chondrus; whereas Chondrus has no effect on ephemerals (Lubchenco, 1978).

The tide pool community represented in model 3 (fig. 3) is reduced to three components, perennial algae, ephemeral algae and snails. This model reflects the uncertainty concerning the ability of crabs to regulate snail populations (R. Hadlock, 1980; J. Carlton, pers. comm.).

Model 4 (fig. 4) separates the variable for snails into adults (S) and juveniles (S_j). In many marine ecosystems, competition and predation between juveniles and adults are important interactions (Woodin, 1976; Lee, 1972).

We introduce two types of ephemeral algae (E_1 and E_2) consumed by snails in model 5 (fig. 5). E_1 is mutualistic with P and E_2 is competitive with P. An example of E_1/P link is ephemerals shading Chondrus from intense sunlight and benefitting from a metabolite excreted by the perennial, whereas E_2/P link is Enteromorpha intestinalis competing with Chondrus (Prince and Kingsbury, 1973; Lubchenco, 1978).

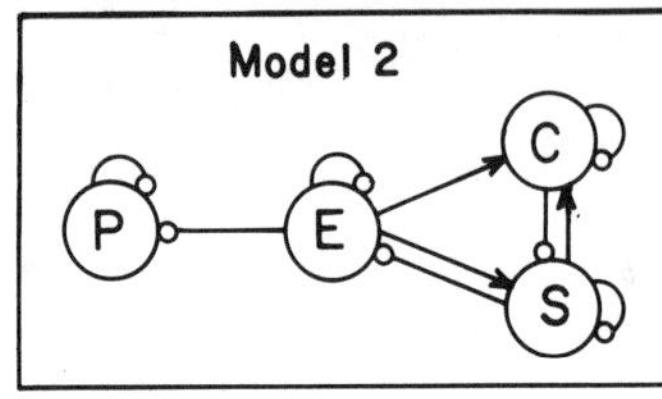

Figure 2.--Tide pool model 2 where ephemerals compete with perennials but not the converse. See text for details and Figure 1 for explanation of symbols.

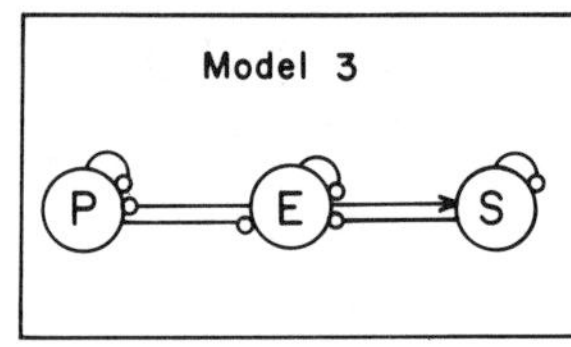

Figure 3.--Tide pool model 3 where crabs are not included. See text for details and Figure 1 for explanation of symbols.

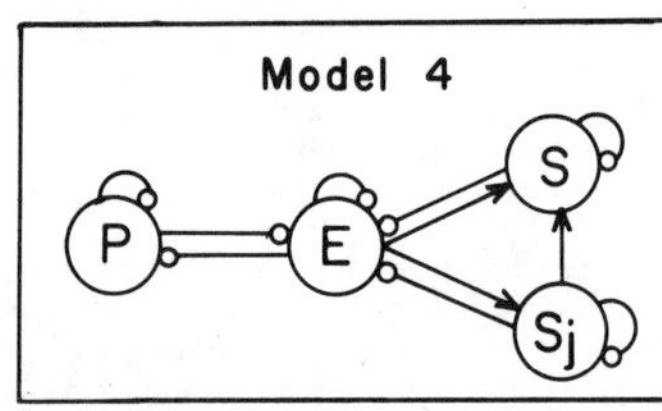

Figure 4.--Tide pool model 4 with both juvenile and adult snails. See text for details and Figure 1 for explanation of symbols.

Model 6 (fig. 6) reintroduces snails and crabs and divides them into juveniles (S_j and C_j) and adults (S and C). In the laboratory newly metamorphosed crabs feed on juvenile snails (pers. obs.). In the field crab predation may be important in some communities but we are uncertain of the crab's ability to regulate snail popualtions in our pools. In particular crabs in our high tide pools are rarely larger than 40 mm and are too small to prey on larger sized snails (Elner and Raffaeli, 1980). Because it is assumed that juveniles, alone, are recruited into pools and become the adult population, there is no self-damping of adult crabs in this model.

These six models are not an exhaustive set. The next two models explore the relationship between a sessile invertebrate, Mytilus edulis and other organisms. Mussels can outcompete algae for space in some communities (Paine, 1969; 1974). Models 7a and b (fig. 7a and b) look at just the three primary space users: perennial algae (P), ephemeral algae (E) and sessile invertebrates (I). Model 7a assumes mussels compete with both ephemeral and perennial algae for space, while the two algae provide some benefits for each other. See explanations with model 5. Model 7b (fig. 7b) has all three groups competing for space.

Model 8 (fig. 8) expands to include crabs and snails. Crabs feed both on the sessile invertebrates and on snails (Ropes, 1962). Because all of these organisms recruit from outside, they are self-damped.

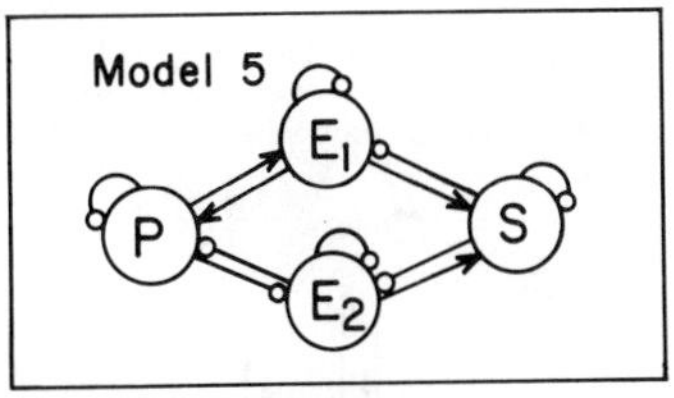

Figure 5.--Tide pool model 5 with ephemeral algae as mutualistic or competitive with perennials. See text for details and Figure 1 for explanation of symbols.

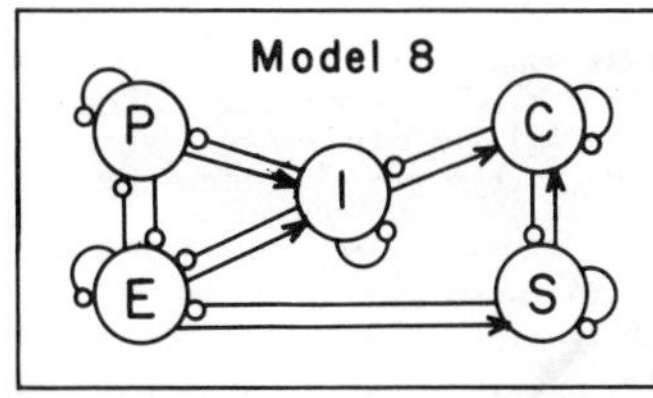

Figure 8.-- Tide pool model 8 where invertebrates are included in the basic model 1. See text for details and Figure 1 for explanation of symbols.

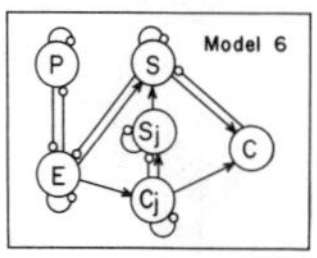

Figure 6.--Tide pool model 6 where both juvenile and adult snails and crabs are represented.

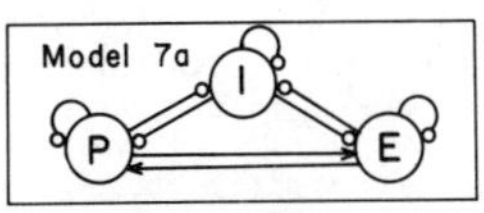

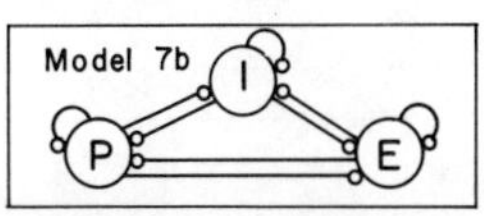

Figure 7.--Three groups using primary space. Model 7a has mutualistic relations and Model 7b competitive.

These eight models are sufficient to illustrate the use of loop models in field research. They are not meant to be the only definitive models of tide pools. This is discussed further in the following sections.

There is a sequence of steps which illustrate how theoretical predictions guide field work. First, we sort the models into groups of similar and dissimilar predictions. For models 1 through 8 we obtain a table of predictions about the direction of change in equilibrium abundances for each species if there are parameter changes affecting each. This would hold under the assumption that tide pool populations follow a moving equilibrium. This is a reasonable assumption for species with a short generation time which track environmental changes, for example phytoplankton (Lane and Levins, 1977). There is some evidence that tide pool populations are in moving equilibrium. Snail densities are higher in ephemeral algal clumps than in areas where algae are absent, i.e., mobile snails move faster than ephemeral algal growth, reproduction and dispersal. Thus, snail immigration results in the population tracking an environmental (food) change.

For model 1 the table of predictions (Table 1) indicates a change in parameters which increase the growth rate of perennials, causes an overall increase in the equilibrium abundance of the perennial algae, a decrease in the ephemeral algae, an ambiguous result for the snail population and a decline in the crab population.

Table 1. -- Predictions of the effect on equilibrium abundances when increases in growth rates occur. Qualitative changes are expressed as + and - indicating positive and negative changes; ? means direction depends on relative strength of links. In this and subsequent tables P = perennial; E= ephemeral; S = snails; C = crabs; I = invertebrates; S_j = juvenile snails; C_j = juvenile crabs. Brackets [ ] around letters refer to that subsystem. For Tables 1 through 8, vertical and horizontal headings refer to "change in parameter causing increase in growth rate of" and effect on equilibrium abundance of", respectively, as given in Table 1a. All models assume:

S---- E-----C S---- C
E-----S E-----C---- S

CHANGE IN PARAMETER CAUSING INCREASE IN GROWTH RATE OF	EFFECT ON EQUILIBRIUM ABUNDANCE OF P	E	S	C
P	a[1]	-	r[2]	-
E	-	+	-r	+
S	+	-	b[3]	b, r
C	-	+	-b	b

Table 1a. Assumes overall stability, F_4 0.

[1]a depends on stability of [ECS] subsystem.

[2]r depends on relative strengths of negative links to positive and self-damping on P.

[3]b depends on stability of [PE] subsystem.

	P	E	S	C
P	+	-	-	-
E	-	+	+	+
S	+	-	+	+
C	-	+	-	+

Table 1b. Assumes [PE] stable, [ECS] stable.

	P	E	S	C
P	+	-	-	-
E	-	+	+	+
S	+	-	-	-
C	-	+	+	-

Table 1c. Assumes [PE] unstable, [ECS] stable.

If we make assumptions either about the relative magnitudes of links or the stability of subsystems within a loop model, we can change several of the letters (a,b,r) to +'s and -'s. These are shown in the corresponding tables for each model (Tables 1b and 1c). Similarly, predictions for the remaining 7 models omit the ambiguous lettered results (Tables 2-8). Thus, for our eight models we have generated sixteen tables of predictions corresponding to sixteen models with different assumptions.

Table 2.--Predictions for model 2 of effect on equilibrium abundances when increases in growth occur. Refer to Table 1 for explanation of symbols and format.

	P	E	S	C
P	+	0	0	0
E	-	+	-	+
S	+	-	+	-
C	-	+	-	+

Table 2a.
E— S E— C— S

	P	E	S	C
P	+	0	0	0
E	-	+	+	+
S	+	-	+	+
C	-	+	-	+

Table 2b.
E— S E— C— S

Table 3.--Predicitions for model 3 of effect on equilibrium abundances when increases in growth occur. Refer to Table 1 for explanation of symbols and format.

	P	E	S
P	+	-	-
E	-	+	+
S	+	-	+

Table 3a.
[PE] stable.

	P	E	S
P	+	-	-
E	-	+	+
S	+	-	-

Table 3b.
[PE] unstable.

Table 4.--Predictions for model 4 of effect on equilibrium abundance when increases in growth occur. Refer to Table 1 for explanation of symbols and format.

	P	E	S	S_j
P	+	-	-	-
E	-	+	+	+
S	+	-	+	-
S_j	+	-	?	+

Table 4a.
[PE] stable

	P	E	S	S_j
P	+	-	-	-
E	-	+	+	+
S	+	-	+	-
S_j	+	-	-	+

Table 4b.
[PE] unstable; [PES_j] unstable; [PES] stable

Table 5.--Predictions for model 5 of effect on equilibrium abundances when increases in growth occur. Refer to Table 1 for explanation of symbols and format.

	P	E_1	E_2	S
P	+	+	-	-
E_1	+	+	-	?
E_2	-	-	+	?
S	+	?	?	+

Table 5a.
[PE_1] stable; [PE_2] stable negative links (—) positive links (—).

	P	E_1	E_2	S
P	+	+	-	-
E_1	+	+	-	-
E_2	-	-	+	-
S	+	+	?	?

Table 5b.
[PE_2] unstable; [PE_1] stable; (—) (—).

Table 6.--Predictions for model 6 of effect on equilibrium abundances when increases in growth occur. Refer to Table 1 for explanation of symbols and format. Entries with the same letter (a,b,c or d) will have the same sign, but may differ from each other. The strength of negative links are assumed to be greater than positive links.

	P	E	S	S_j	C_j	C
P	+	-	a	+	b	c
E	-	+	a	-	b	c
S	+	-	+	-	-	c
S_j	-	+	-	+	+	?
C_j	d	d	-	-	+	?
C	-	+	-	-	?	+

Table 6a. [PE] stable; [ESS_jC_jC] stable.

	P	E	S	S_j	C_j	C
P	+	-	a	+	b	c
E	-	+	a	-	b	c
S	+	-	+	-	-	c
S_j	-	+	-	+	+	?
C_j	d	-d	+	-	+	?
C	-	+	-	-	?	+

Table 6b. [PE] unstable; [ESS_jC_jC] stable.

Table 7.--Predictions for model 7 of effect on equilibrium abundances due to growth increases. Refer to Table 1 for explanation of symbols and format.

	P	E	I
P	+	+	-
E	+	+	-
I	-	-	+

Table 7a. [PI] stable; [PE] stable; [IE] stable.

	P	E	I
P	+	-a	-a
E	-a	+	-a
I	-a	-a	-

Table 7b. [PE] stable; [PE] unstable; [IE] stable.

[1]a 0 if general rule is that direct path times complement self-damping is stronger than indirect long path, i.e.,

Table 8.--Predictions for model 8 of effect on equilibrium abundances due to growth increases. See Table 1 for explanation of symbols and format.

	P	E	S	C	I
P	-	+	-	-	+
E	-	+	+	+	-
S	+	-	+[1]	+[1]	-
C	-	+	-[1]	+[1]	-
I	-	+	-	+	+

Table 8a. [PEI] stable.

	P	E	S	C	I
P	-	+	-	-	+
E	-	+	+	+	-
S	+	-	-[1]	-[1]	-
C	-	+	+[1]	-[1]	-
I	-	+	-	+	+

Table 8b. [PEI] unstable.

[1]If multiple pathways, assumes the majority of signs previals.

We do not know where or how natural environmental fluctuations effect the parameters. If we look across pools at different geographic locations, and assume parameter changes equally likely in any of the variables, then this is equivalent to looking down the columns of a table of predictions. This column examination reveals a predicted pattern of covariation betweem the variables.

A tabulation of predicted correlations across the sixteen tables generated from all eight models is provided in Table 9. Not all models provide the same resolution of prediction. For example, model 6 is highly dependent on the point of input of a parameter change. A change effecting perennials versus one effecting ephemerals produces opposite correlations of snails with the other species.

In a limited, local geographic area over a short period of time, it is improbable that parameter changes will occur with equal likelihood at all variables. We consider the case when perennial and ephemeral algae are more likely to receive parameter changes. Hence, instead of looking down the entire column of a table of predictions, we look only at the first two rows to obtain the expected correlation pattern among variables. The predicted correlation patterns (Table 10) of the perennial and ephemeral parameter changes show a slight decrease in the number of ambiguous results from those of equally likely parameter change (Table 9).

Table 9.--Predicted correlations for parameter changes being equally likely among the variables.

Model 1b

	E	S	C
P	-	?	?(-)
E		?	?(+)
S			?(+)

Model 1c

	E	S	C
P	-	-	?(-)
E		+	?(+)
S			?(+)

Model 2a

	E	S	C
P	-	+	-
E		-	+
S			-

Model 2b

	E	S	C
P	-	?(+)	?(-)
E		?(-)	?(+)
S			?(+)

Model 3a

	E	S
P	-	?(-)
E		?(+)

Model 3b

	E	S
P	-	-
E		+

Model 4a

	E	S	S_j
P	-	?	?(-)
E		?	?(+)
S			?

Model 4b

	E	S	S_j
P	-	?(-)	?(-)
E		?(+)	?(+)
S			?

Model 5a

	E_1	E_2	S
P	?(+)	?(-)	?
E_1		?(-)	?
E_2			?

Model 5b

	E_1	E_2	S
P	+	?(-)	?
E_1		?(-)	?
E_2			?

Model 6a

	E	S	S_j	C_j	C
P	?(-)	?	?	?	?
E		?	?	?	?
S			?	?	?
S_j				?	?
C					?

Table 9.--continued.

Model 6b

	E	S	S_j	C_j	C
P	-	?	?	?	?
E		?	?	?	?
S			?	?	?
S_j				?	?
C					?

Model 7a

	E	I
P	+	-
E		-

Model 7b

	E	I
P	?(-)	?(+)
E		?(+)

Model 8a

	E	S	C	I
P	-	?(+)	?	?
E		?(-)	?	?
S			?	?(-)
C				?(-)

Model 8b

	E	S	C	I
P	-	?	?	?
E		?	?	?
S			?	?(-)
C				?

Table 10.--Predicted correlations for parameter changes being most likely for the perennial or ephemeral algae.

Model 1b

	E	S	C
P	-	-	-
E		+	+
S			+

Model 1c

	E	S	C
P	-	-	-
E		+	+
S			+

Model 2a

	E	S	C
P	-	+	-
E		-	+
S			-

Model 2b

	E	S	C
P	-	-	-
E		+	+
S			+

Model 3a

	E	S
P	-	-
E		+

Model 3b

	E	S
P	-	-
E		+

Model 4a

	E	S	S_j
P	-	-	-
E		+	+
S			+

Model 4b

	E	S	S_j
P	-	-	-
E		+	+
S			+

Model 5a

	E_1	E_2	S
P	+	-	?
E_1		-	?
E_2			?

Model 5b

	E_1	E_2	S
P	+	?(-)	?
E_1		?(-)	?
E_2			?

Model 6a

	E	S	S_j	S_c	C
P	-	?	+	?	?
E		?	-	?	?
S			?	?	?
S_j				?	?
S_c					?

Model 6b

	E	S	S_j	S_c	C
P	-	?	+	?	?
E		?	-	?	?
S			?	?	?
S_j				?	?
S_c					?

Model 7a

	E	I
P	+	-
E		-

Model 7b

	E	I
P	-	?
E		?

Model 8a

	E	S	C	I
P	-	?	?	?
E		?	?	?
S			+	-
C				-

Model 8b

	E	S	C	I
P	-	?	?	?
E		?	?	?
S			+	-
C				-

RESULTS AND DISCUSSION

The predicted pairwise correlations from the models are compared with the correlation coefficients calculated for field data using linear regression analysis. The tide pools are divided into two categories: sheltered (from wave battering) and exposed. The sheltered pools have higher littorine snail densities and lower abundances of other invertebrates than the exposed pools. The data is subdivided into seasons to determine if any difference affect the ephemeral algal abundance. With the exception of combinations involving crabs, correlations are always positive (+) or negative (-) throughout all seasons; the statistical significance of the correlations varies between seasons. Crabs present a special difficulty because they migrate into the subtidal during the winter season and are sporadically observed in low numbers throughout the rest of the year (during daylight hours at low tide). Attempting a complete tide pool census of crabs would greatly damage the biota of the tide pool.

Figure 9 summarizes correlation predictions of the models with that of the data. Predicted correlations of model 2 are in agreement with the data, whereas model 4 is almost entirely in disagreement. Nearly all the models predict the negative correlation of perennial and ephemeral algae. Models 5 and 7 predict a positive correlation between ephemeral and perennial algae. In the case of model 5, ephemeral algae E_1 is hypothesized as mutualistic with the perennial and E_2 competitive with the perennial. Our data is not divided into the E_1 and E_2 classification precluding comparison with this aspect of model 5 predictions. Dividing ephemeral algae into E_1 and E_2 groups constitutes a new working hypothesis yet to be examined.

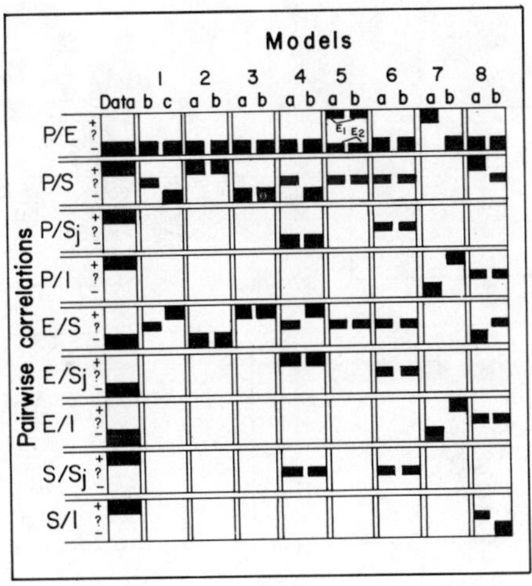

Figure 9.--Summary of pairwise correlations for data and predictions of models based on Table 9 where parameter changes occur with equal likelihood.

Model 7a results include a mutualistic relationship between the perennial and ephemerals. This differs from model 5 by assuming a direct mutualism for all ephemeral algae with the perennial and an indirect mutualism via the invertebrates. In model 7b the perennial and ephemerals are competitors, although the indirect mutualism still exists. There is some ambiguity (Table 9, model 7b) because the relative strengths of direct competitive effects versus long mutualistic ones is unknown.

A re-examination of model 1b shows the predicted correlations do not agree in two out of the three corresponding data observations. The predictions of model 1 given in Table 1 show changes entering parameters of the snails can produce a positive correlation of perennial algae and snails and a negative correlation of ephemerals and snails. Consequently, model 1b requires parameter changes at snails, not the assumption that parameter changes are equally likely at all organisms, as in Table 9, nor most likely at perennial and ephemerals, as in Table 10.

If the conclusion from model 1b that the parameter changes in our tide pools enters through the snails is carried through to model 8, then from Table 8a the majority of the predictions about correlations obtained only from the snail row disagree with the data. Model 8b is slightly better in the predictions than 8a. But even if all the ambiguous results of 8b can be resolved and agree with the data it is still not a very strong set of predictions. In our view model 8 is an unlikely candidate for use in planning experimental research of the tide pools.

Looking only at parameter inputs through snails, for model 6 and Table 6a and b, it can be seen that half the predictions agree with the observations. This is the same as 8b and suggests model 6 is probably missing important links or not all the assumptions used in either model 6 or 8 are valid. Models 2 and 7b become the prime candidates for use in further research into tide pool dynamics. In comparing the two models it should be clear that the important links are the negative effect of the ephemeral algae on the perennial and the grazing of ephemerals by snails. Moreover, the role of crabs as a predator needs further study, as the strength of the crab interactions can greatly change the results of model 2 predictions. The use of the loop models in planning research can now be amplified.

A summary of the proportion of models predicting the same correlation pattern is provided in Table 11. Not all the models contain crabs or snails, hence the total possible number of correlations varies across columns. The parameter changes of both the "equally likely among all variables case" and the "more likely for perennial and ephemeral case," show a negative correlation between perennial and ephemeral case in the majority of models (14/18 and 15/18, respectively). This is not surprising because most of the models place the two algal groups in competition with each other. Nevertheless, it also highlights this prediction as a robust result, because it is relatively unaffected by the many different assumptions manifested in the eight models.

Table 11.-- Proportion of models predicting the same correlation pattern among the variables across observations.

	P/E	P/S	P/C	E[1]/S	E[1]/C	S/C
+	3/18	3/14	0	4/16	4/8	3/8
-	15/18	4/14	4/8	3/16	0	1/8
?	0	7/14	4/8	9/16	4/8	4/8

Table 11a. Parameter changes equally likely among all variables.

	P/E[1]	P/S	P/C	E[1]/S	E[1]/C	S/C
+	3/18	0	0	7/16	2/8	4/8
-	13/18	7/14	2/8	1/16	0	
?	2/18	7/14	6/8	8/16	6/8	4/8

Table 11b. Parameter changes most likely for perennial and ephemeral algae.

[1]For purposes of this table, E_1 and E_2 are taken as separate ephemeral algal groups. Table 4 produces double sets of predictions in these categories.

The maximum information to be gained by field observations is suggested by correlations which correspond to predictions separating half of the models into positive or negative correlation only. Examination of Table 11 shows that no such cases exist because question marks are not resolved as + or -. From Table 11b two valuable correlation observations can be deduced. The perennial algae:snails (P/S) has 0 out of 14 models predicting positive correlations and 7 of 14 predicting negative ones; whereas the ephemeral:snail (E/S) shows 7 of 16 models with positive correlations and 1 of 16 with negative. If the data show ephemeral algae are negatively correlated with snails this immediately casts doubt on 1 model out of 16 and highlights 1 out of 16. Moreover, it tells us that 8 of the 16 models with ambiguous results can be reanalyzed to find the conditions which match the observations.

This approach can enhance field manipulation. With models, investigators can selectively change parameters and obtain predictions of results. By analyzing the results, decisions can be made indicating which experiments to perform in the field. We can illustrate this point by the following examples which relate to parameter inputs in the models. Clipping or removing algae is an experimental approach to decreasing population growth. Similarly adding snails to pools increases their population growth.

A manipulative experiment using two different variables such as the examples given above may or may not produce differences in expected correlations across all variables. Table 12 illustrates the differences in predicted correlation outcomes for pairwise manipulation experiments for all models. By way of illustration, model 2b shows that clipping ephemeral algae versus removing snails will produce four different predictions in correlations; clipping perennials instead of ephemerals would produce six different correlations. In model 2a the results are even more accentuated for the ephemeral:snail manipulations. Here there would be no differences in predicted outcome. Conducting the ephemeral:snail manipulation experiments would not clarify whether model 2a corresponded to the actual field conditions beyond what would be learned from doing either ephemeral or snail manipulations alone. In other words, any single prediction is a relatively weak prediction, whereas the cumulative effect of a greater number of predictions being correct leads to acceptance of a model. Those experiments which correspond to increased numbers of different predictions will yield more information about the model validity.

Table 12.--Number of predicted correlations from different inputs which differ among the tables.

Model 1a

	E	S	C
P	0	4	3
E		4	3
S			3

Model 1b

	E	S	C
P	0	0	3
E		0	3
S			3

Model 2a

	E	S	C
P	6	6	6
E		0	0
S			0

Model 2b

	E	S	C
P	6	6	6
E		4	3
S			3

Model 3a

	E	S
P	0	2
E		2

Model 3b

	E	S
P	0	0
E		0

Model 4a

	E	S	S_j
P	0	3	2+
E		3	2+
S			2+

Model 4b

	E	S	S_j
P	0	3	3
E		3	3
S			4

Model 5a

	E_1	E_2	S
P	0+	0+	1+
E_1		0+	0+
E_2			0+

Model 5b

	E_1	E_2	S
P	0	3	0+
E_1		3+	0+
E_2			0+

Table 12 (continued).

Model 6a	E	S	S_j	S_c	C
P	9+	3+	2+	0+	0+
E		4+	2+	0+	0+
S			0+	2+	3+
S_j				2+	3+
S_c					0+

Model 7a	E	I
P	0	0
E		0

Model 6b	E	S	S_j	S_c	C
P	9+	3+	2+	0+	0+
E		4+	2+	0+	0+
S			0+	2+	3+
S_j				2+	3+
S_c					1+

Model 7b	E	I
P	2	2
E		2

Model 8a	E	S	C	I
P	6	0	6	4
E		6	2	6
S			4	4
C				4

Model 8b	E	S	C	I
P	6	5	5	3
E		4	4	6
S			6	4
C				6

CONCLUSIONS

We draw these conclusions from our loop models and two years of field data. First, most models and our data indicate perennial and ephemeral algae are in competition. From a biological point of view, plausible reasons for this are that these algae are in competition for space, light, and/or nutrients.

Secondly, although known interactions are used to construct the models, these may be insufficient to explain the correlations of the field observations. Because the loop models are relatively simple to construct and inexpensive to analyze, new hypothesis can be included in new models. Hence, nontraditional hypotheses become worth exploring.

Thirdly, although parameter changes affecting ecosystems frequently enter through lower trophic levels (Lane and Levins, 1978; Sutcliffe, 1972), tide pools are being perturbed at mid and higher trophic levels. Two models suggest environmental factors enter through snails and influence the tide pool community dynamics and one (model 1b) indicates crabs are an important part of tide pool structure.

Fourthly, even with a small number of assumptions there are as many as sixteen tables of predictions -- effectively sixteen models. Observations which best sort out the stronger models use the method of comparing the number of differences in predictions. In the specific case of the tide pool models, the focus is on the perennial and ephemeral algae correlation with snails.

Fifthly, ambiguous predictions in models highlights the need for relative quantification of links. For example, the crabs predatory link on snails, other invertebrates and the link to crabs from ephemerals all need further study; these might otherwise be ignored.

Sixthly, there is a limited number of field experiments that can be carried out, but a large set of experiments possible for each model. There is, however, a subset of all possible experiments that provides the maximum number of comparisons of experimental data with model predictions. The procedure for selecting the subset is summarized in Table 10.

The use of loop models is not a claim that these models are outside of nature and outside of observation. Rather, loop models are a research tool which can guide observation, interpret data and suggest which experiments give the maximum amount of information for the questions being asked.

LITERATURE CITED

Connell, J. H. 1975. Some mechanisms producing structure in natural communities: a model and evidence from field studies. From Cody, M. L. and Diamond, J. H. (eds). *Ecology and Evolution of Communities*.

Elner, R. W. and D. G. Raffaeli. 1974. Interactions between two marine snails, *Littorina rudis* Maton and *Littorina nigrolineata* Gray, a predator, *Carcinus maenas* and a parasite *Microphallus simulus* Jagerskiold. J. Exp. Mar. Biol. Ecol. 43: 151-162.

Greig-Smith, P. 1964, 2nd Ed. *Quantitative Plant Ecology*. Butterworth, New York.

Hadlock, R. P. 1980. Alarm response on the intertidal snail *Littorina littorea* (L.) to predation by the crab *Carcinus maenas* (L.). *The Biol. Bull.* *159*: 269-279.

Lamb, McK. and M. H. Zimmerman. 1964. Marine vegetation of Cape Ann, Essex County. Rhodora 66: 217-254.

Lane, P. and R. Levins. 1977. The dynamics of aquatic systems. 2. The effects of nutrient enrichment on model plankton communities. Limnol. and Oceanogr. 22: 454-471.

Lee, W. L. 1966. Color change and the ecology of the marine isopod *Idothea* (*pentidotla*) *montereyensis*. Maloney, 1933. Ecol. 47: 930-941.

Levins, R. 1974. The qualitative analysis of partially specified systems. *Ann. N.Y. Acad. Sci.* *231*: 123-138.

Levins, R. 1975. Evolution in communities near equilibrium. pp. 16-50. In Cody, M. L. and Diamond, J. M. (eds.) *Ecology and Evolution of Communities*. Harvard University Press, Cambridge, MA.

Lewis, J. R. 1964. The Ecology of Rocky Shores. The English University Press, London, 323 pp.

Littler, M. M. and D. S. Littler. 1980. The evolution of thallus form and survival strategies in benthic marine microalgae: Field and laboratory tests of a functional form model. Amer. Natur. 116: 25-44.

Lubchenco, J. 1978. Plant species diversity in a marine intertidal community: Importance of herbivore food preferences and algal competitive abilities. Amer. Natur. 112: 23-29.

Lubchenco, J. 1980. Algal zonation in New England rocky intertidal community. An experimental analysis. Ecol. 61: 333-344.

Lubchenco, J. and B. A. Menge. 1978. Community development and persistence in a low rocky intertidal zone. Ecol. Monogr. 48: 67-94.

Menge, J. L. 1975. Effect of herbivores on community structure of the New England rocky intertidal region: distribution, abundance and diversity of algae. Ph.D. diss., Harvard University.

Paine, R. T. 1969. A note on trophic complexity and community stability. Amer. Natur. 103: 91-93.

Paine, R. T. 1974. Intertidal community structure: Experimental studies on the relationship between a dominant competitor and its principal predator. Oecologia. 15: 93-120.

Prince, J. S. and J. M. Kingsbury. 1973. The ecology of Chondrus crispus at Plymouth, Massachusetts. II. Field studies. Am. J. Bot. 60: 964-975.

Ropes, J. W. 1968. The feeding habits of the green crab, Carcinus maenas (L.) Fish. Bull. 67: 183-203.

Sears, J. R. and R. A. Cooper. 1968. Descriptive ecology of off-shore, deep-water, benthic algae in the temperate western North Atlantic Ocean. Marine Biology 44: 309-314.

Stephenson, T. A. and A. Stephenson. 1972. Life between the Tide marks on Rocky Shores. W. H. Freeman and Co., San Fransisco. 425 pp.

Underwood, A. J. 1980. The effects of grazing by gastropods and physical factors on the upper limits of distribution of intertidal macroalgae. Oecologia 46: 201-213.

Underwood, A. J. 1981. Structure of a rocky intertidal community: patterns of vertical distribution and seasonal changes. J. Exp. Mar. Biol. Ecol. 51: 57-86.

Woodin, S. A. 1976. Adult-larval interactions in dense infaunal assemblages: patterns of abundance. J. Mar. Res. 34: 35-41.

QUALITATIVE MODELS FOR THE EAST COAST BENTHOS[1]

Charles J. Puccia[2]

Abstract.--Soft-bottom estuarine environments for the East Coast are analyzed using qualitative models (loop models; signed digraphs). The benthic organisms are divided into functional groups that interact trophically or through sediment mediation. The functional groups are mobile burrowers, filter-feeders, tube-builders, and predators.

Alternative models are presented based on accepted hypotheses and some speculation. The physical substrate enters into the models through the biology of the organisms. The tube-builder/mobile burrower subsytem stability becomes an object of interest in these models. The scale of the system being modelled can alter self-damping of the organisms and change results. Finally, the trophic-amensalism hypothesis may predict the correct results but not because filter-feeders are aided directly through sediment stabilization of tube-builders but by limiting mobile burrower activity.

INTRODUCTION

Benthic communities of the soft-bottom have species distribution patterns which are predominately explained by one or more of these hypothese: (1) trophic-amensalism (Rhoads and Young 1970); (2) food limitation for suspension and deposit-feeders (Levinton 1972); and, (3) adult-larval interactions (Woodin 1976). There are, however, two other hypothesis which are getting greater attention. One is that predators can alter the species abundance patterns and diversity of the benthos (Virnstein 1979). The other hypothesis is that species which are members of a guild (sensu Root 1967) can enhance each others survival, either through increased foraging or predator avoidance (Duggins 1981).

In this paper, the consequences of the above hypotheses are examined through the use of signed-digraphs, or loop models. The models are used to explore our understanding of the benthos and not to be in one-to-one correspondence with the benthic organisms.

LOOP MODELS

The models used are those of signed digraphs, introduced by Mason (1952) but

as modified and independently developed by Levins (1974) to study complex biological systems.

For a system represented by a set of partial differential equations with the rate of change of variables written as:

$$\frac{dX_i}{dt} = f_i(X_1, X_2, \ldots, X_n) \tag{1}$$

or,

$$\underline{X}' = \underline{AX} \tag{2}$$

A correspondence can be made between a signed digraph and the $\underline{A}$ matrix. The elements of the $\underline{A}$ matrix in eq(2) are:

$$a_{ij} = \left[\frac{\partial}{\partial X_j}\frac{dX_i}{dt}\right]_{X^*} \tag{3}$$

$$= \frac{\partial}{\partial X_j}(f_i(\underline{X}))_{(\underline{X}=\underline{X}^*)}$$

evaluated at equilibrium. Each element of $\underline{A}$ is an edge or directed link in a signed digraph. Each vertex or node representes one of the X variables in the system.

In the language of loop models (Levins 1974; 1975), a simple benthic model would be represented as in Model 1 (fig.1).
The variables are the filter-feeders, FF, the tube-builders, TB, and the mobile burrowers, MB, and shown in the large circles. The lines connecting the variables represent the causal effects of one upon another. In other words, the elements of the $\underline{A}$ matrix. Lines drawn with a small circlehead indicate a negative effect. If there are positive effects then these are indicated with a line ending in an small arrowhead (see predator model, figure 3). A predator benefits (+) from consuming its prey; two competitor are harmed (-) by each other.

In model 1 the three groups, FF, TB, and MB also have negative links to themselves which indicates that even in the absence of each other their own abundance will not increase indefinitely. This is called a self-damping link and is the diagonal term of the matrix. In general the bottom level of any ecosystem is self-damped and if we choose not to represent that variable the self-damping is passed up

---

[1] Paper presented at the Third International Conference on State-of-the-Art in Ecological Modelling, Colorado State University, May 24-28, 1982.

[2] Charles J. Puccia is a Research Associate in Marine Ecology and Complex Systems at Harvard University, School of Public Health, Boston, Massachusetts, U.S.A.

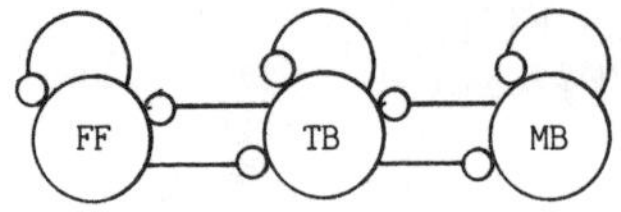

MODEL 1

FF : Filter-Feeders
TB : Tube-Builders
MB : Mobile Burrowers

→ = POSITIVE EFFECT
–O = NEGATIVE EFFECT
= SELF-DAMPING

Parameter Input to	Change in Equilibrium Abundance of FF	TB	MB
FF	–	–	+
TB	+	+	–
MB	–	–	+

Figure 1 .--Model 1 showing competitive relations between the three functional groups. See text for an explanation of the table.

to the next variable level in the model.

A path is defined as a series of links (edges) from one variable through one or more intervening variables and ending at a variable that is not the same as the initial one. A path can not cross itself, that is, it can not enter and leave a particular variable more than once. The qualitative sign of a path is calculated by multiplying the signs of the arrowheads (+) and circleheads (-) along the path.

A loop is defined as any closed path which proceeds from one variable back to itself though none to many variables. Each variable is entered and exited once. The sign of a loop is calculated in a similar manner described for paths. The length of a loop is equal to the number of variables it enters and leaves. Self-damping is a loop of length 1.

There can be many loops in a signed digraph. Some loops have variables in common and some do not. Two loops with at least one variable in common are called conjunct loops and those without any are disjunct loops. Disjunct loops enter into calculations, conjunct loops do not.

In loop models, feedback at level k is calculated as:

$$F_k = \sum_{m=1}^{n} (-1)^{m+1} L(m,k) \qquad (4)$$

where, k is the number of variables (the level) at which feedback is being calculated) out of a total of n variables ($k \leq n$). Feedback has the same value as the determinant of order k for the $\underline{A}$ matrix. F_k is the coefficient of the n-k term of the characteristic polynomial. Local stability, as determined by the Routh-Hurwitz criteria, in loop terms is:

Condition 1.
Feedback at all levels must be less than zero. $F_k < 0$, for all k.

Condition 2.
Negative feedback at the higher levels can not be too large compared with negative feedback at the lower levels. This translates into a sequence of expressions the first of which is: $F_1F_2 + F_3 > 0$.

When the first condition is violated monotonic instability ensues; when the second, oscillatory instability occurs.

A system may be at or near equilibrium, which is itself changing because the parameters of the system are changing. If the parameters are changing slowly compared to the variables, the system can be in moving equilibrium. To determine the direction of change in the equilibrium values of the variables in the system is to differentiate the systems equations with respect to the parameter or parameters that are changing. The resulting set of partial differential equations can be in principle solved for new equilibrium values of the variables. This operation in loop analysis is:

$$\frac{\delta X_i}{\delta C_h} = \frac{\sum \frac{\delta f_j}{\delta c_h} p_{ij}^{(k)} F_{(n-k)}^{(comp)}}{F_n} \qquad (5)$$

where,

c_h is the hth parameter that is changing;

$\frac{\delta f_i}{\delta c_h}$ is the effect of the c_h parameter change on the growth rate of the jth variable;

$p_{ij}^{(k)}$ is the path from the jth to the ith variable and includes k variables along the path;

$F_{(n-k)}^{(comp)}$ is the feedback of the complementary subsystem - the subsystem formed by those variables not on the path of $p_{ij}^{(k)}$ and,

F_n is the feedback of the whole system, which must be negative if the whole system is stable.

From eq (5) we can construct a table of predictions for each loop model. In the table shown in figure (1) each row consists of a increase in the growth rate of the corresponding variable due to a change in parameter and each column is the effect on the equilibrium abundance. An environmental change which causes an increase in the growth rate of filter-feeders (for short, this is called a "parameter input" to filter-feeders) results in a decline in the filter-feeder and tube-builder equilibrium abundances and an increase in mobile burrowers.

The next section will examine some benthic loop models using the above methodology and predictions as summarized in the tables.

BENTHIC MODELS

The models presented here highlight the known facts or represent plausible conjectures of interactions among benthic organisms for the middle-Atlantic coast of the U.S.A. There is no attempt to superceed or supply local information.

The habitats considered are soft-bottom sediments composed of sand, sand-mud, or mud. The spatial scale is of patches; a small portion of an estuary, bay or tidal flat. The importance of this restriction will be noted later.

Of all the working hypotheses of soft-bottom benthic communities, the most obvious is the relatin between sediment compostion of silt-clay fractions and the effect on different feeding behaviors (Sanders 1958). In general, filter-feeders have the highest diversity in sandy sediments and deposit-feeders in mud.

Organisms of the soft-bottom are divided into three functional groups: filter-feeders, tube-builders and mobile-burrowers, as represented by the clam Mya arenaria, the amphipod Ampelisca abdita, and the polychaete worm Nephtys bucera, respectively. There are some species like the polychaete worm Streblospio benedicti which is a tube-builder and uses its palps to gather suspended food particles from the water or near the sediment-water interface. There is no method for assigning every identifiable species to a single functional group, without ambiguity; the most significant attempt has been made with the Polychaeta (Fauchald and Jumars 1979).

In the first model, (fig. 1) the three primary functional groups are examined, especially with respect to their competitive interactions. Tube-builders compact the sediment and may form large areas of tube mats. The tubes inhibit foraging by the mobile burrowers (Mills 1967; Woodin, 1974). Also, the tube-builders will consume settling larvae as they enter the sediments or bury them in fecal material (Woodin 1976). The mobile burrowers interfere in the construction of the tubes or destroy those in place. This can prevent the establishment of tube-builders in an area or their ability to stay. Both tube-builders and mobile burrowers may have density-dependent growth rates which would appear as self-damping in the models. This is not the only reason for self-damping on these two functional groups. Larval settlement in any specific area or habitat will be by larvae that orginate from a larger geographic pool then the habitat. This input of organisms from outside the specific boundary produces a self-damped variable.

As with mobile burrowers, the tube-builders will either consume or bury in fecal material the settling larvae of the filter-feeders. Filter-feeders also siphon settling larvae from the water column; they will negatively effect tube-builders as well as species in their own functional group. This model assumes no interaction of filter-feeders with mobile burrowers.

Sediment characteristics are included in the model via the introduction or exclusion of links between species. If we were interested in the sediment composition itself, then we could include this as a variable of the model. In a mud habitat mobile burrowers will have a negative effect on filter-feeders, as drawn in model 2 (fig. 2). Mobile burrowers disturb the sediments, but in mud they resuspend flocculent material into the water column that interferes with the respiratory and feeding mechanism of filter-feeders.

In a sandy bottom habitat, there is less resuspended material by the mobile burrowers. On the other hand, the filter-feeders are thought to be more efficient in gathering food items (Sanders 1958). Model 3 (fig. 2) shows filter-feeders will also siphon the larvae of mobile burrowers and thus negatively effect them.

One major difference between model 2 and model 3 is that the former predicts filter-feeders and tube-builders are positively correlated while the latter predicts a negative correlation. In moving from sand to mud the filter-feeders and tube-builders become out of phase with each other. Levinton (1972) suggests that the trophic dimension of the niche explains the high specialization of deposit-feeders (we consider these tube-builders although it actually includes mobile burrowers); and, the unpredictability of food for the filter-feeders explains their patchy distribution and fluctuations. He adds that

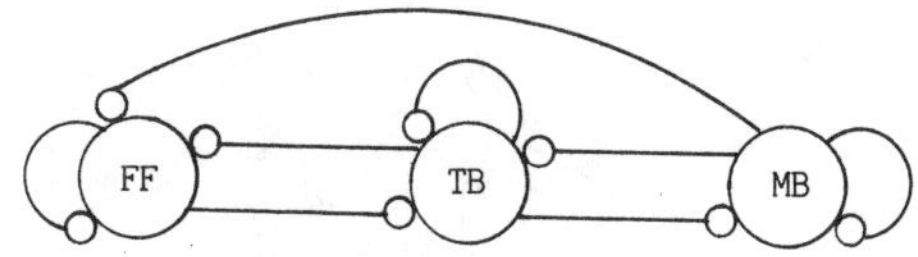

MODEL 2

Parameter Input to	Change in Equilibrium Abundance of FF	TB	MB
FF	-	-	+
TB	+	+	-
MB	-	-	+

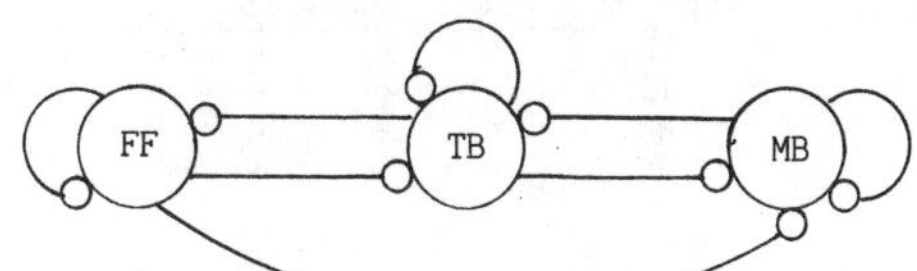

MODEL 3

Paramter Input to	Change in Equlibrium Abundance of FF	TB	MB
FF	-	+	+
TB	-	+	-
MB	+	-	+

Figure 2 .--Relationships between organisms mediated by the environment. Model 2 is for a mud habitat and Model 3 is for a sand sediment.

interference mechanisms may be important in the epifaunal filter-feeders. Models 2 and 3 indicate that this would also be true for the infaunal species. The mud habitat model indicates greater specialization when conditions are best for the filter-feeders and tube-builders. Both models agree that factors which improve the growth rate of filter-feeding populations cause an overall decline in their abundance.

Predators enter the benthic models either as generalists or specialists. In the first case, model 4 (fig. 3) is of a generalist predator in a sand environment. The most likely predators are fish, birds or small epifaunal invertebrates (Levings 1975; Reise 1978). The generalist predators that are actual members of the community being modelled are the small epifaunal invertebrates; fish and birds are too mobile and are part of a larger community. The epifaunal invertebrates, like their infaunal counterparts, will have a substantial input of newly settled larvae that originate outside the area being modelled and accounts for the self-damping. In a sand substrate the invertebrate predators will do better on the surface filter-feeders and tube-builders.

This generalist predator model coincides with model 3 sand sediment predictions that filter-feeders and tube-builders are negatively correlated. However, model 4 also indicates that environmental factors which effect the epifaunal invertebrate predators will result in a positive correlation between the filter-feeders and tube-builders. The filter-feeders are also positively correlated with mobile burrowers. Both models 1 and 2 predict a positive correlation of FF with TB, and the opposite for FF with MB. In this model, any two direct competitive

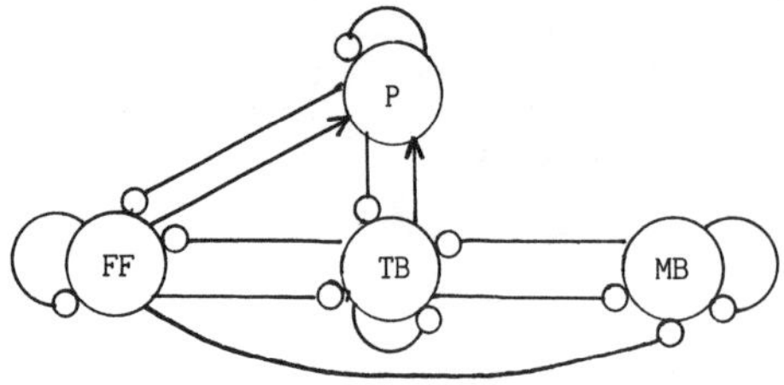

MODEL 4

Parameter Input to	Change in Equlibrium Abundance of FF	TB	MB	P
FF	+	-	+	-
TB	-	+	-	-
MB	+	-	+	+
P	+	+	+	+

Figure 3 .--A generalist predator model for a sand environment.

groups will rise and fall in the same way.

Another type of predator is the specialist as given in model 5 (fig. 4). It is rather unlikely that the predators specialize only by functional groups yet it is a reasonable first approximation. In a sand environment with specialist predators on the three functional groups we produce the predicted patterns of changes in equilibrium abundances shown (table 1). Differences with the generalist predator model occurs in the first two columns of the first row. The combined predators model reinforces the prediction that environmental change entering though the tube-builders and mobile-burrowers produces a negative correlation of filter-feeders with tube-builders, tube-builders with mobile burrowers.

A change in scale for the model can change predictions. In a specialist predator model in which filter-feeder predators become locally reproduced, because "local" is an enlarged area, then these predators are not self-damped. The filter-feeders will have no effect on any variable other than its own predator. The same would be true for the other specialist predators. If the specialist predator P_1 is not self-damped then the effect on equilibrium abundances due to parameter changes is shown in table 2.

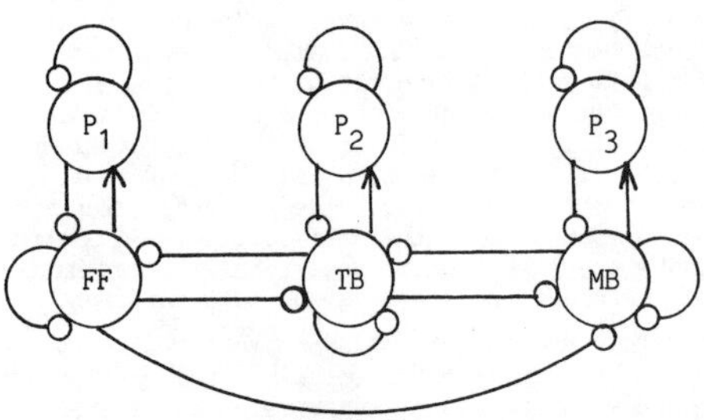

MODEL 5

Figure 4 .--A specialist predator on each functional group, in a sandy habitat.

Table 1 .--Predicted changes in equilbrium abundances for specialist predator case (Model 5) and with the generalist predator case (Model 4).

Parameter Input to	Change in Equilibrium Abundance of FF	TB	MB
SPECIALIST PREDATOR			
FF	-	+	+
TB	-	+	-
MB	+	-	+
& GENERALIST PREDATOR			
FF	?	?	+
TB	-	+	-
MB	+	-	+

Table 2 .--Predictions of changes in equilibrium abundances for model 5 without self-damping on any of the specialist predators.

Patameter Input to	Change in Equilibrium Abundance of FF	TB	MB
FF	+	-	-
TB	0	+	-
MB	0	-	+

In the table with self-damping the filter-feeders have a negative effect on their own abundance. Without self-damping on the predator P_1 the unstable tube-builder/mobile burrower subsystem has no effect on the filter-feeders. The filter-feeders are buffered from the rest of the system.

Finally, model 6 (fig. 5) considers only the consequences of trophic amensalism. In a silty sediment the filter-feeders will encounter difficulty feeding, but will take the settling larvae of the tube-builders. Tube-builders can produce dense mats in this sediment composition and retard mobile burrowers. On the other hand, the mobile-burrowers will destroy the mats and displace the tube-builders; and, their foraging resuspends sediments which clogs the filtration of the filter-feeders.

DISCUSSION

These simple benthic models can provide us with research direction and assist in answering practical questions related to environmental assessment.

First, we recognize that the physical composition of the substrate can either be an important variable in its own right, but as it relates to species abundance and distribution it is important to understand how it acts through the biology of the organisms. There may be more than one model for the same set of organisms and functional groups because of different physical substrate characteristics.

Second, the stability of the tube-builder/mobile burrower subystem is of interest. One indirect measure of the subsystem stability is the change in filter-feeder abundance. Habitat or environmental changes which laboratory experiments show have a direct positive (negative) impact on the filter-feeders and yet field observations show a decline

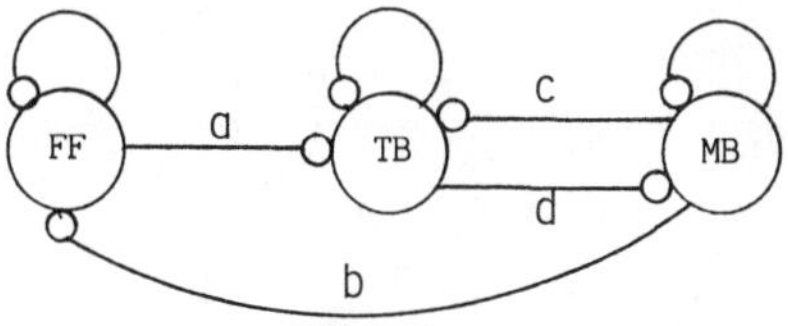

MODEL 6

Parameter Input to	Change in Equilibrium Abundance of FF	TB	MB
FF	+	-	+
TB	+	+	-
MB	-	+	+

Figure 5 .--The trophic-amensalism hypothesis depicted in a loop model. In a silty sediment , assume | ab | > | cd | .

(increase) in abundance strongly suggest the tube-builder/mobile burrower substem is unstable. It is not conclusive because other parts of the system, for example the meiofauna, have not been considered.

Third, model scale does matter if it changes self-damping terms. If any one or more of the specialist predators have no self-damping on a scale larger than patches, then in the sand sediment model mobile burrowers have no effect on filter-feeders. Furthermore, the unstable tube-builder/mobile burrower subsystem no longer effects the filter-feeders as before. The specialist predators/tube-builder/mobile burrowers subsystem will determine the effect of filter-feeders on themselves. This subsystem as shown in model 5 is stable.

Young and Rhoads (1971) report a positive correlation of filter-feeders with the percentage of silt-clay in the sediments, opposite to their own trophic-amensalism hypothesis. The explanation given by them is that the polychaete Euchone incolor builds tubes which stabilize the sediments. Sediment stabilization permits bivalve and amphipod suspension feeders to enter the area. Model 5 depicts this circumstance. Tube-builders are not effecting the filter-feeders, though mobile burrowers are. The increase in mud-silt fraction of sediments will be better for tube-builders, acting as a positive input to them. This will not only increase their own abundance but that of the filter-feeders as well. However, it is not due to the direct effect of tubes on the filter-feeders but rather acts through the mobile burrowers. If the tubes become so dense as to limit or prevent mobile burrowers from the area then it will severe the link to filter-feeder and also exclude them. But if the mud-silt sediment attracts mobile burrowers there will be a positive input to them and this will cause a decline in the filter-feeders. The community which appears will depend on whether the tube-builders arrive first and how fast they build tubes.

The loop modelling procedure, as demonstrated with these benthic models is also a strategy for scientific research. We can start with the biological knowledge to build a digraph or construct the generalized equations; or, obtain the data and see what structural aspects of a digraph would be consistent with it.

Knowledge can come from biological inferences - put the links in an see what the model predicts. Observations may require model structures with links that are previously unknown and searching for the natural phenomena consistent withthe observations and the model structure.

The actual scientific process of the loop model analysis is a back and forth between biological knowledge and data. It differs from the simulation modelling iteration in that for simulation models a discrepancy between predictions and obsevations leads to improving estimation of parameters, whereas in loop analysis the models are changed.

ACKNOWLEDGMENT

This project has been financed with Federal funds from the Environmental Protection Agency under grant number R807713 to Dr. Richard Levins. The contents do not necessarily reflect the views and policies of the Environmental Protection Agency .

LITERATURE CITED

Duggins, D. O. 1981. Interspecific facilitation in a guild of benthic marine herbivores. Oecologica 48: 157-163.

Fauchald, K. and P. A. Jumars. 1979. The diet of worms: a study of polychaete feeding guilds. Oceanogr. Mar. Biol. Annu. Rev. 17:193-384.

Levinton, J. 1972. Stability and trophic structure in deposit-feeding and suspension-feeding communities. Amer. Natural. 106: 472-486.

Levings, C. S. 1974. Seasonal changes in feeding and particle selection by winter flounder (Pseduopleuronectes americanus). Trans. Amer. Fish. Soc. 103: 828-832.

Levins, R. 1974. The qualitative analysis of partially specified systems. Ann. N.Y. Acad. Sci. 231: 123-138.

1975. Evolution in communities near equilibrium, p. 1-50. In M. L. Cody and J. M. Diamond (eds.). Ecology and Evolution of Communities. Harvard University Press, Cambridge, Mass. 545 pp.

Mason, S. J. 1953. Feedback theory --some properties of signal flow graphs. Proc. Inst. Radio Eng. 41: 1144-1156.

Mills, E. L. 1967. The biology of an ampeliscid amphipod crustacean sibling species pair. J. Fish. Res. Board Can. 24: 305-355.

Reise, K. 1978. Experiments on epibenthic predation in the Wadden Sea. Helgolander. wiss. Meeresunters. 31: 55-101.

Rhoads, D. C. and D. K. Young. 1970. The influence of deposit-feeding organisms on sediment stability and community trophic structure. J. Mar. Res. 28: 150-178.

Root, R. B. 1967. The niche exploitation pattern of the blue-gray gnatcatcher. Ecol. Monogr. 37: 317-350.

Sanders, H. L. 1958. Benthic studies in Buzzard's Bay. I. Animal-sediment relationships. Limnol. and Oceanogr. 3: 245-258.

Virnstein, R. W. 1979. Predation on estuarine infauna: response patterns of component species. Estuaries 2: 69-86.

Woodin, S. A. 1974. Polychaete abundance patterns in a marine soft-sediment environment: the importance of biological interactions. Ecol. Monogr. 44: 171-187.

1976. Adult-larval interactions in dense infaunal assemblages: patterns of abundance. J. Mar. Res. 34: 25-41.

Young, D. K. and D. C. Rhoads. 1971 . Animal-sediment relations in Cape Cod Bay, Massachusetts. I. A transect study. Marine Biology 11: 242-254.

INPUT SENSITIVITY OF AN ESTUARINE ECOSYSTEM MODEL: A FREQUENCY-DOMAIN APPROACH[1]

Robert L. Dwyer and James N. Kremer[2]

Abstract. -- Persistent oscillations were observed in the behavior of the phytoplankton and zooplankton compartments of a simulation model of Narragansett Bay, Rhode Island. The variabilities in time-series outputs from unforced and perturbed runs of a steady-state version of the model were compared in the frequency domain (using spectral analysis) to determine if the oscillations were driven by inputs.

INTRODUCTION

Oscillatory behavior is a commonly observed response in the behavior of a wide variety of ecological models. Examples include two-species predator-prey models (Rykiel and Kuenzel, 1971) and others which can be solved analytically, nonlinear models of simple food chains which switch between oscillatory and nonoscillatory stable states (Elliott et al., this volume), as well as highly complex nonlinear ecosystem models with time-varying and/or stochastic inputs, such as some of those formulated for the IBP biome studies (papers in Patten, 1975). Such common occurrences of oscillatory behavior, and the apparent complexity of the fine structure in the time-series output from models with complex trophic interactions, suggest that analyses of responses in the frequency domain may reveal their causes, and the ultimate controls on system stability.

The analyses presented in this paper were motivated by our desire to understand the causes of persistent summertime oscillations in the behavior of phytoplankton and zooplankton biomass in a simulation model of Narragansett Bay, Rhode Island, a temperate coastal ecosystem (Kremer and Nixon, 1978). We suspected that the oscillations could have been caused by exogenous forcings such as seasonal or stochastic components of inputs, or by a variety of endogenous mechanisms specified in the numerous and complex trophic interactions of the model. Thus, it was desirable to perform analyses of the sensitivity of the oscillatory output (e.g., by comparing amplitudes and frequencies in time-series model responses) to both input and parameter variations.

For simple analytical models (e.g., unforced linear models, or simple nonlinear models linearized about an equilibrium) a number of analytic methods exist for determining sensitivity (Tomovic and Vukobratovic, 1972). However, few feasible analytical methods exist to examine the sensitivity of complex nonlinear ecosystem models. Empirical methods such as Monte Carlo analyses (Scavia et al., 1981) have the disadvantage of requiring many model runs be- before the statistical properties of model responses become apparent.

In this paper, we propose an empirical alternative which examines the sensitivity of a model in the frequency domain. Specifically, we have used spectral analysis of time-series outputs from a version of the model from which seasonal and stochastic inputs have been removed. Spectra from this unforced version and from runs perturbed with random inputs show subtle differences in their variance structure which could not be revealed by more standard time-domain variance analyses such as ANOVA (Dwyer and Perez, 1983). As usually employed, spectral analysis is built around a numerical Fast Fourier Transform (FFT) algorithm which partitions the variance of time-series data among discrete bands of frequencies of oscillation. The method has recently come into widespread use to analyze field ecological data (Platt and Denman, 1975; Shugart, 1978) and to analyze the output of ecosystem simulation models (Emanuel et al., 1978a, b). Before the specifics of our method are presented, we must outline the structure of the model and how it simulates the seasonal dynamics of the plankton community.

STRUCTURE AND BEHAVIOR OF THE NARRAGANSETT BAY MODEL

Narragansett Bay is a temperate estuary on the south coast of New England which has been a focus of ecological research for several decades. In order to synthesize the large amount of data on this and other planktonic ecosystems, Kremer and Nixon (1978) developed a simulation model which reproduces the gross temporal and seasonal distributions for the major components of the pelagic ecosystem. Detailed mechanisms were developed to simulate the following state variables: phytoplankton, adult, egg, and juvenile zooplankton, nitrogen, silicon and phosphorus. Details of the phytoplankton and zooplankton compartments are shown schematically in figure 1. The effects on these state variables of additional biotic components such as benthic organisms (which may graze on phytoplankton and regenerate nutrients to the water column) and predators (which reduce both phytoplankton and zooplankton) are simulated by including their observed spatial and temporal patterns in the Bay, rather than by formulating them as additional state variables.

The model uses exact analytical integration to evaluate the growth equations, updating coefficients on a one day interval. It simulates spatial patterns by interfacing biotic components with a low-resolution tidal mixing model of the Bay. Eight spatial elements are simulated. Realistic seasonal

[1] Paper presented at the Third International Conference on State of the Art in Ecological Modeling, Colorado State University, Fort Collins, Colorado USA, May 24-28, 1982.

[2] Robert Dwyer is Research Scientist for the Environmental Center, Martin Marietta Corp., 1450 South Rolling Road, Baltimore, MD 21227. James Kremer is Assistant Professor, Dept. of Biological Sciences, University of Southern California, Los Angeles, CA 90007.

patterns of solar irradiance, temperature, and river flow are used as environmental forcing inputs to the model, and interact with model state variables via a large number of complex mechanisms (fig. 1). Phytoplankton maximum growth rate is set as a function of temperature, and the limiting effects of light and nutrients reduce maximum growth. Herbivorous zooplankton are simulated with a complex temperature-dependent age structure. Egg hatching and juvenile maturation times are functions of temperature, as are size-specific respiration rates of all life stages. Although adults feed preferentially on phytoplankton, they can also feed on their own eggs and juveniles, a mechanism which imparts considerable stability to the plant-herbivore interaction (Kremer, 1979).

Seasonal patterns of phytoplankton and zooplankton biomass in a standard run of the model are shown as the solid lines in figure 2. In winter, both components maintain stable, low levels due to temperature and/or light limitation. After an early spring bloom, both components maintain another period of relative stability, controlled primarily by nutrient limitation on the phytoplankton (Kremer, 1979). However, during the summer, both components undergo oscillations lasting until the onset of winter conditions. This transition through these different types through these different types of stability can best be seen in a phase plane plot (fig. 3a) which plots

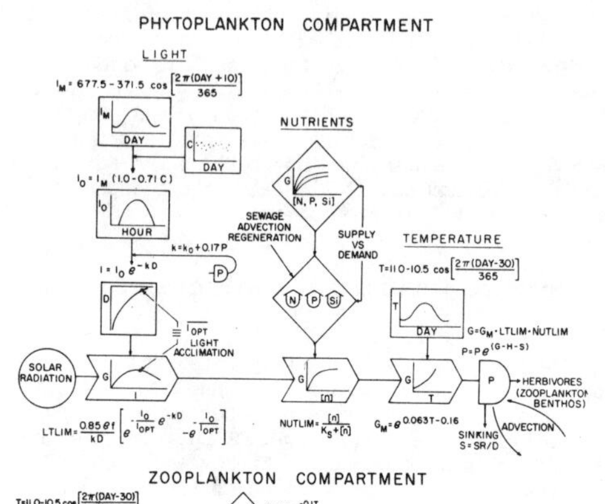

Figure 1. Mathematical and symbolic representation of the flows and controls operating on the phytoplankton and zooplankton compartments of the model. Details in text, in Kremer, 1979, and in Kremer and Nixon, 1978.

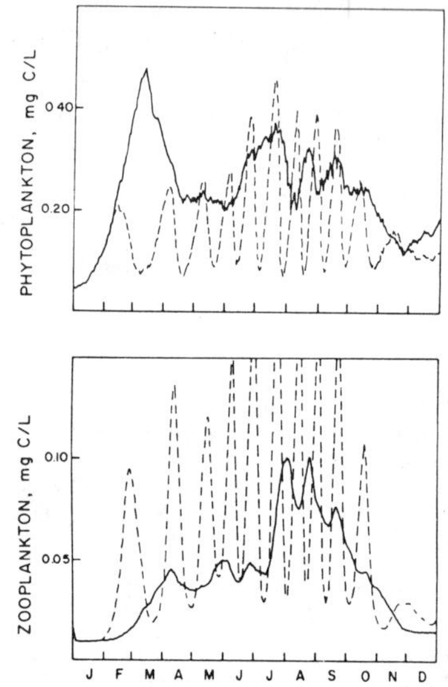

Figure 2. Simulated seasonal dynamics of phytoplankton and zooplankton biomass in a standard run of the model (solid lines), and in a run with no delay in zooplankton development (dashed lines).

zooplankton vs phytoplankton biomass. The summer oscillations suggest limit cycle behavior; however, the term applies strictly to unforced steady-state behavior of a model, which is not the case here.

The importance of the mechanistic details of the model to this dynamic pattern are illustrated by making a simple change: removing the delay in zooplankton juvenile development (fig. 2, dashed line; fig. 3b). The oscillations increase in amplitude and persist for a longer period. However, because of the overriding effect of seasonality, it is difficult to ascertain whether their frequency has changed. Other mechanisms known to affect this oscillatory behavior include cannibalism of zooplankton adults on juveniles as well as predatory pressure on the zooplankton (Kremer and Nixon, 1978) and whether or not the carnivores are modeled as full state variables able to grow in response to prey availability (Kremer, 1979).

FREQUENCY-DOMAIN METHODS FOR MODEL ANALYSES

Since seasonal and/or stochastic components of model inputs (irradiance, temperature, river flow, tide height, predation intensity) may influence the model's periodic behavior, we first converted the model to steady state by removing these environmental forcing functions. To accomplish this, all inputs were frozen at constant levels corresponding to their values at different points during the year.

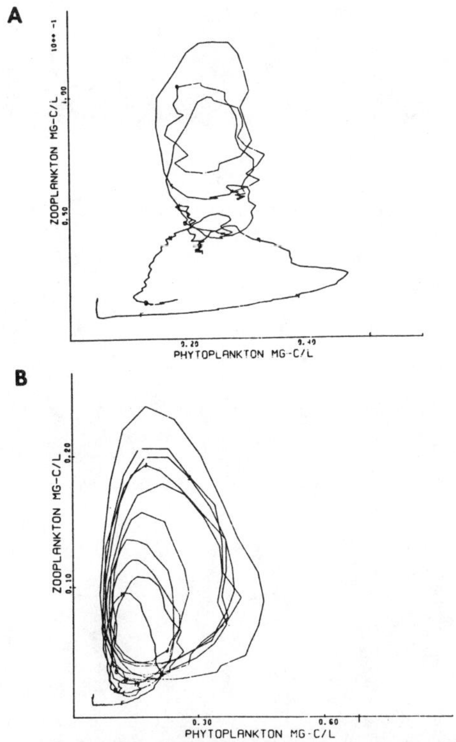

Figure 3. Phase plane plots (zooplankton vs phytoplankton biomass) of a standard model run (A), and a run with no delay in the maturation of zooplankton (B); see fig. 2.

We arbitrarily chose the conditions of January 1 and June 30 to contrast the stable vs oscillatory equilibria. Model responses to these two sets of constant inputs are free of external forcings. Thus, any periodicities in response are endogenous products of internal mechanisms alone. Conversion to steady state also facilitates application of the FFT spectral analysis, which cannot accommodate the non-stationary variance structure of the seasonal time-series model output.

With seasonality removed, the model frequently showed a transient response over the first 30 to 120 days. Consequently, the model was run for 730 days (2 years), and the first 218 data points of response time series were discarded. Analyses were done on the remaining 512 data points (2^9, since the FFT algorithm requires a time series length equal to an integral power of 2). These are steady-state segments not contaminated by transient responses and not dependent on initial conditions of the model.

Recent trends toward widespread utilization of FFT-based variance spectral analyses make a detailed exposition of the method unnecessary. Computations used in this study are completely described by Bendat and Piersol (1971), and Emanuel, et al.(1978b).

While the unforced model runs reveal how the removal of the time-varying inputs affected model response, the sensitivity of the response to perturbation of the input, or perturbation of parameters, will reveal major controlling factors. It is these sensitivity analyses which reveal the real advantage of a frequency-domain approach over others like Monte Carlo analysis. Whereas Monte Carlo analysis would require a separate model run for each new input or parameter value in order to generate a statistical distribution, input-response analyses in the frequency domain can take advantage of superposition -- the ability to add input signals (e.g., sinusoids of different frequencies) to test the model's response to all signals simultaneously.

A constant input with an added stochastic term also generates well-known responses. Using such a random time series as an input to the system is equivalent to perturbing the system with a signal composed of superimposed sinusoids at a large number of frequencies. The variance spectrum of such a signal contains equal variance contributions over a wide band of frequencies (hence the term "white" noise).

The present study takes advantage of this property to determine the input sensitivity of various model compartments to perturbations using white noice signals of varying amplitudes added to a constant input. The dependence of many mechanisms of the phytoplankton and zooplankton compartments on temperature suggested that perturbation of it from constant winter (3°C) or summer (18°C) values would be likely to have readily observable effects.

While these frequency domain analyses are obviously applicable to analyzing parameter sensitivity, such analyses are not considered here. They will be discussed in a separate paper (Kremer, in prep.).

The response spectra for unforced and forced runs of the steady-state model can reveal much about the causes of the periodicities. As discussed below, it is possible to eliminate phenomena of little ecological interest (e.g., neutrally-stable oscillations of the Lotka-Volterra type, harmonic periodicities characteristic of some nonlinearities) from the list of possible causes. Although the spectra do not of themselves suggest causal mechanisms, they are a useful diagnostic of the connectance and cycling within the system.

RESULTS

As previous described, seasonal forcing functions were removed, and the unforced spectra were calculated for some state variables subjected to constant inputs at June 30 levels. The resulting output time series of phytoplankton biomass, and the variance spectrum of the last 512 days of this time series, are shown in figure 4. After a transient response lasting approximately 100 days, phytoplankton biomass settled to a stationary variance structure with a mean near that of the summer phytoplankton response of that spatial element in the standard seasonal version of the model. Periodicities at 0.035 cycles/day (period = 28.44 days) and 0.068 cycles/day (period = 14.63 days) dominate the spectrum. The oscillation at 0.035 cycles/day can be seen in summer results of the standard seasonal version of the model (Kremer and Nixon, 1978; Kremer, 1979), but the higher frequency oscillation at 0.068 cycles/day has been hidden, probably by superimposed transient responses to seasonally changing input values in the standard version. The spectrum of total zooplankton biomass shows the same periodicities (not shown). Since all variations in environmental inputs have been removed, these periodicities can only be the result of interactions between model state variables.

The interactions which cause oscillations at these periodicities are not apparent. Kremer (1979) attributed the monthly periodicity (0.035 cycles/day) as an apparent limit cycle resulting from close coupling of phytoplankton and zooplankton. The cause of the 0.068 cycle/day oscillation is even less well understood. A strict harmonic response is characteristic of many nonlinear simulation models. However, the periodicity at 0.068 cycle/day does not appear to be the second harmonic of the 0.035 cycle/day oscillation, since the high frequency is not exactly twice the lower one. Also, the time-series data of figure 2 show a shifting phase relation between the two superimposed periodicities.

These steady state spectra showed that two major periodicities persist in the unforced model under summer conditions. These results can be contrasted

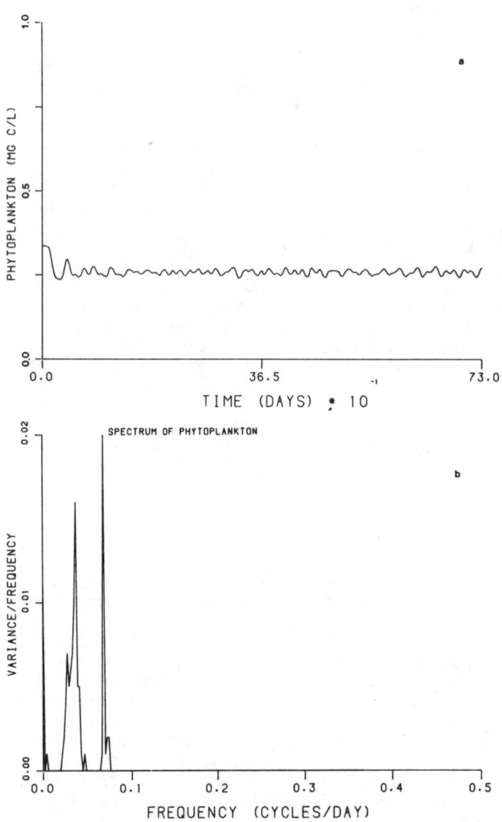

Figure 4. Model response of phytoplankton biomass to June 30 input conditions, plus constant (18°C) temperature input; a) time-domain output for 730 days, b) variance spectrum of last 512 days of time series.

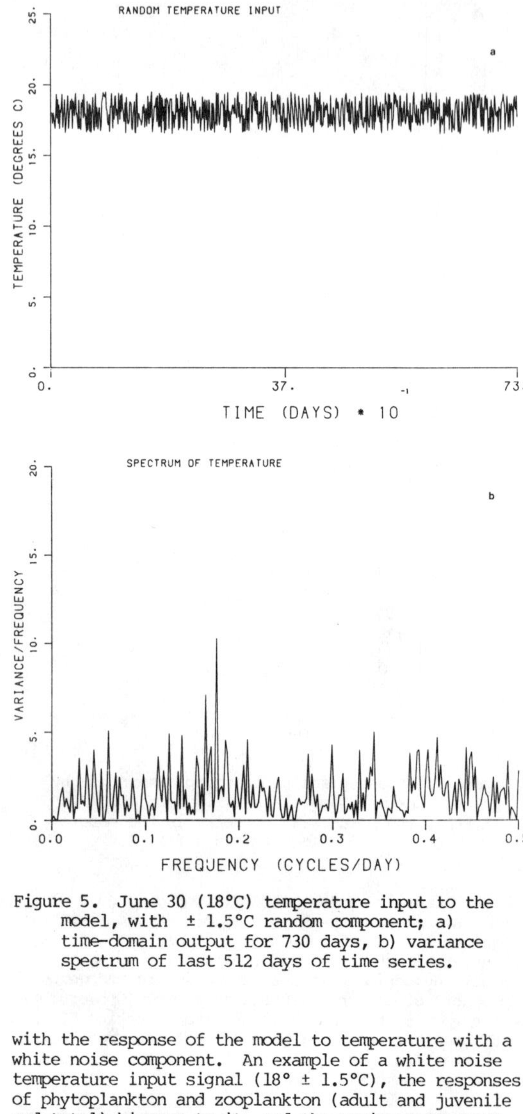

Figure 5. June 30 (18°C) temperature input to the model, with ± 1.5°C random component; a) time-domain output for 730 days, b) variance spectrum of last 512 days of time series.

with the response of the model to temperature with a white noise component. An example of a white noise temperature input signal (18° ± 1.5°C), the responses of phytoplankton and zooplankton (adult and juvenile and total) biomass to it, and the variance spectrum of each, are shown in figures 5-9. Oscillations at 0.035 and 0.068 cycles/day continue to dominate the variance spectra of these biotic compartments of the model. In fact, none of the spectra for biotic response variables resemble that of the random temperature input, which shown periodicities more or less uniformly spread over the entire frequency range of 0.0 to 0.5 cycles/day (fig. 5).

A compilation of the variances at each important frequency peak in the phytoplankton response spectra to random temperature inputs at 3 different amplitudes (18°C constant; 18° ± 1.5°C; 18° ± 3°C) is presented in table 1. It is reassuring to note that the random temperature inputs are truly near-equilibrium perturbations, since mean phytoplankton biomass does not change as input amplitude increases. However, with increasing input amplitude, several

		Summer (June 30) Conditions			Winter (January 1) Conditions
Input:		18°C (Constant)	18° ± 1.5°C	18° ± 3.0°C	3° ± 1.5°C
Mean Biomass (mg C/liter):		0.257	0.257	0.257	0.220
		Variance/Frequency (mg C/liter)^2/day)			
Periodicity:					
Frequency (cycles/day)	Period (days)				
0.0275	36.40	0.0	0.0162	0.0250	0.0
0.03516	28.44	0.0160	0.0103	0.0079	0.0120
0.06836	14.63	0.0200	0.0229	0.0190	0.0100

Table 1. Steady-state mean values, and variances at major periodicities, of model phytoplankton responses to constant and white noise temperature inputs.

changes in the variance structure of the phytoplankton response occur. First, the variance at 0.035 cycles/day decreases with increasing input amplitude, while the variance at 0.068 cycles/day remains essentially constant. In addition, variance at a new periodicity at 0.0275 cycles/day increases in the spectra with increasing input amplitude. As can be seen from the table, this periodicity dominates the phytoplankton response to the 18° ± 3°C random temperature input.

A similar set of analyses at 3°C show much weaker responses, particularly for the zooplankton, whose metabolism and reproduction are inhibited by the winter temperatures. Phytoplankton respond with periodicities at the familiar 0.035 and 0.068 cycle/day frequencies (table 1). The influence of small-amplitude random temperature inputs are thus of little consequence at winter temperatures, and will not be discussed further.

While the winter responses reveal nothing, the zooplankton responses in summer (18°C) provide additional detail about trophic interaction, because an age structure of the community is simulated. Temperature-dependent hatching and development times directly influence the proportion of juveniles in the total community (fig. 1).

Steady-state biomasses and estimates of variance at major spectral peaks are presented (table 2) for juvenile and adult zooplankton abundances which have been subjected to 18° ± 1.5°C and 18° ± 3°C white noise temperature perturbations. (Spectra for these zooplankton life stages subjected to 18° constant temperature were not calculated.) The major features of the spectra are familiar. Dominant peaks occur at 0.035 and 0.068 cycles/day, as with phytoplankton. However, oscillations at these frequencies are restricted to the juvenile component of the zooplankton community. Spectra for total zooplankton reflect the dominance of juveniles in the community, while the two periodicities are much reduced in the adult responses. The ability of development delays in populations to induce oscillations (Marchesseault, et al., 1976) might suggest that one or both of these periodicities are the result of a lag in community response due to the development period of juveniles at 18°C. Indeed, from model algorithms, hatching time plus development time is calculated to be 27 days at 18°C. This suggests that a periodicity induced by such a

		18° ± 1.5°C			18° ± 3.0°C		
		Juveniles	Adults	Total Zooplankton	Juveniles	Adults	Total Zooplankton
Mean Biomass: (mg C/liter)		0.056	0.023	0.079	0.056	0.023	0.080
		Variance/Frequency ((mg C/liter)^2/day)					
Periodicity:							
Frequency (cycles/day)	Period (days)						
0.0096	104.2						0.0015
0.0146	68.5				0.0017		
0.0156	64.1				0.0016	0.0017	
0.0195	51.3	0.0017			0.0019		
0.0215	46.5					0.0015	
0.0230	43.5				0.0010		
0.0265	37.7	0.0012	0.0030				
0.0275	36.4				0.0050	0.0019	
0.0293	34.1						0.0026
0.0350	28.4	0.0042		0.0052	0.0036		0.0050
0.0602	16.6				0.0017		
0.0680	14.6	0.0044		0.0040	0.0036		0.0042

Table 2. Steady-state mean values, and variances at major periodicities, of model zooplankton life stages to white noise temperature perturbations about June 30 values.

lag in maturation should occur close to 0.037 cycles/day. At first glance, this appears to be an explanation for the prominent 0.035 cycle/day periodicity whih is present in all biotic responses at 18°C. However, this periodicity is also present in the spectrum of phytoplankton responding to a 3°C temperature input (table 1), as is the periodicity at 0.068 cycles/day, indicating that the frequencies of these oscillations are independent of temperature. Thus, delay in development does not explain oscillations in biotic responses. No other mechanism which controls these periodicities has been identified as yet.

A number of other variance peaks occur in the zooplankton spectra (table 2). Most of these appear to be related to the amplitude of the random temperature input, since more periodicities occur in spectra responding to the higher input amplitude. However, no clear pattern is evident with respect to the frequencies at which the oscillations occur, or with respect to possible specificities of any frequency band for either life stage (juveniles or adults).

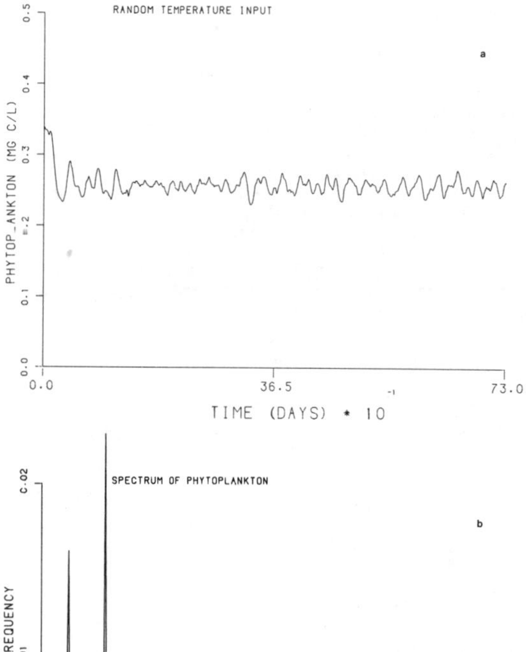

Figure 6. Model response of phytoplankton biomass to June 30 input conditions, plus 18° ± 1.5°C random component to temperature input; a) time-domain output for 730 days, b) variance spectrum of last 512 days of time series.

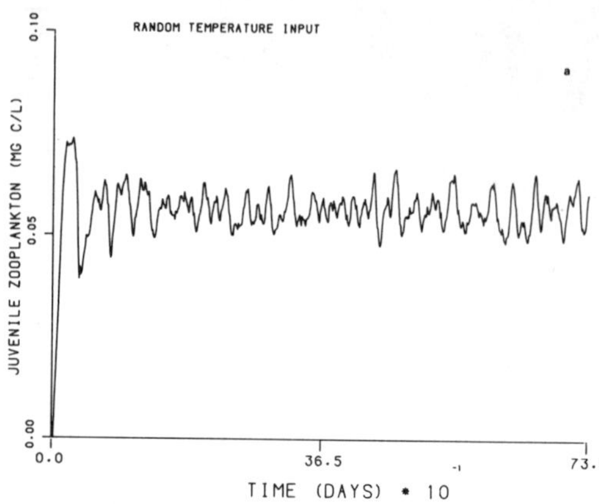

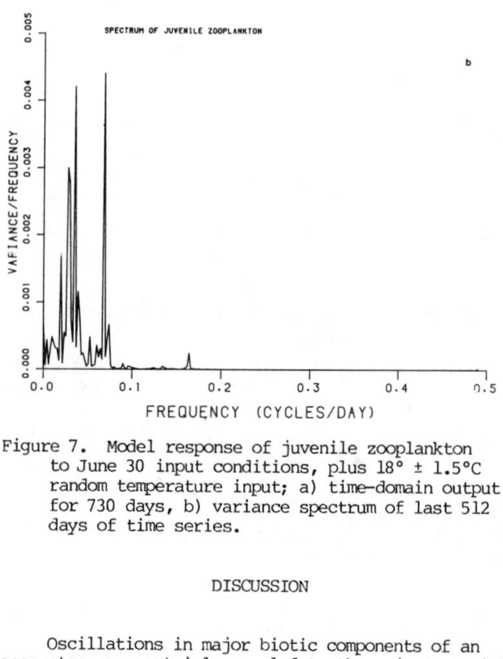

Figure 7. Model response of juvenile zooplankton to June 30 input conditions, plus 18° ± 1.5°C random temperature input; a) time-domain output for 730 days, b) variance spectrum of last 512 days of time series.

DISCUSSION

Oscillations in major biotic components of an ecosystem are certainly coupled to the major material cycles within the system. The number of such periodicities has been proposed as an index of the functional complexity of ecosystems (Van Voris et al., 1980). The characteristics of such cycles have also been suggested as reliable indicators of many important aspects of the stability of ecosystems (Webster, et al., 1975). A quantitative description of major ecosystem periodicities and their sensitivity both to environmental perturbations and to changes in internal mechanisms will yield much information about the stability characteristics of the ecosystem.

This paper has presented the results of a first step of such a description -- an analysis of the sensitivity of major periodicities in the responses of a complex ecosystem simulation model to well-defined perturbations. Although many internal mechanisms of the model are temperature-dependent, our simulations have shown that biotic responses of the model are relatively insensitive to small-scale random perturbations in temperature. Responses to white noise temperature inputs can be detected, but they are concentrated at a few discrete frequency windows. Thus, the model cannot be considered as any sort of passive filter to input signals. The effects of perturbations are small, and in general, dominant periodicities in the unforced responses of phytoplankton and zooplankton also dominate the variance spectra of the responses of these compartments to stochastic inputs. Our analyses have not revealed any specific mechanisms in the model which might control the frequencies of these dominant frequency windows, or the manner in which they might amplify or attenuate input signals.

Comparison of the spectra did permit us to eliminate some pathological causes of the periodicities which would be of little ecological interest. The oscillations do not appear to be neutrally stable because the model was run with different initial

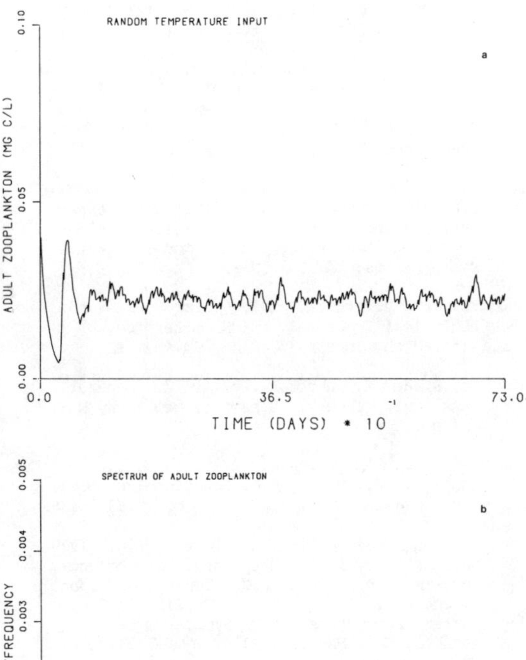

Figure 8. Model response of adult zooplankton to June 30 input conditions, plus 18° ± 1.5°C random temperature input; a) time-domain output for 730 days, b) variance spectrum of last 512 days of time series.

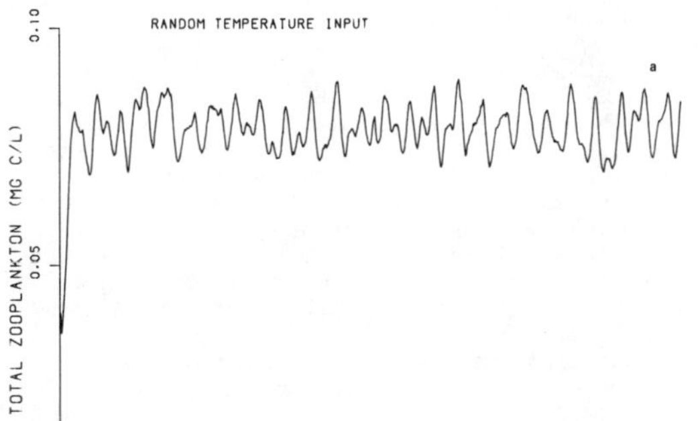

Figure 9. Model response of total zooplankton to June 30 input conditions, plus 18° ± 1.5°C random temperature input; a) time-domain output for 730 days, b) variance spectrum of last 512 days of time series.

conditions (summer vs. winter), and with a variety of input perturbations. Any of these would start the phytoplankton-zooplankton interaction off on a different trajectory (analogous to a frictionless pendulum being given different initial displacements) if the relationship were of the Lotka-Volterra type.

Similarly, there are no indications that the periodicities are harmonics of one another. (The 0.035 cycle/day periodicity is not exactly one half the frequency of the 0.068 cycle/day periodicity, indicating that there is not a harmonic relationship between them.)

Thus, it appears that the mechanistic formulations of the model play the major role in controlling the dominant periodicities of responses. An obvious next step in characterizing these periodicities is a systematic series of the displacements of rate coefficients from their nominal values in the unforced version of the Narragansett Bay model, followed by variance-spectral analyses of biotic response variables. These analyses of parameter sensitivities in the frequency domain are presently being completed (Kremer, in prep.).

Although the results of such abstract studies of a model are not amenable to validation against the responses of the natural ecosystem, they can serve as useful tools to identify critical aspects of the ecosystem which may control its behavior. These methods and results can serve as a base for designing experiments to investigate the behavior of components of the real ecosystem manipulated in the laboratory, or the behavior of the whole ecosystem examined in its natural state. We expect the integration of such model stability analyses with laboratory, microcosm, or field experimentation will soon be put into general use in ecosystem analysis.

ACKNOWLEDGEMENTS

The development of the Narragansett Bay model was supported by the Office of Sea Grant Programs, NOAA, U.S. Dept. of Commerce. The sensitivity analyses were supported by an EPA Graduate Fellowship to R.L. Dwyer. Material in this paper was first published in Ecological Modelling (Dwyer and Kremer, in press).

REFERENCES

Bendat, J.S., and Piersol, A.J., 1971. Random Data: Analysis and Measurement Procedures. Wiley-Interscience, New York. 407 pp.

Dwyer, R.L., and Kremer, J.N. In press. Frequency-domain sensitivity analyses of an estuarine ecosystem simulation model. Ecol. Modelling.

Dwyer, R.L., and Perez, K. T. In press. An experimental examination of ecosystem linearization. The American Naturalist.

Elliott, E.T., Wiegert, R., and Hunt, H.W. In press. Simulation of simple foodchains. This volume.

Emanuel, W.R., Shugart, Jr., H.H., and West D.C. 1978a. Spectral analysis and forest dynamics: the effects of perturbations on long-term dynamics. In Shugart, H.H. (Ed.). Time Series and Ecological Processes. Proceedings of a SIMS Conference, SIAM, Philadelphia. pp. 193-207.

Emanuel, W.R., West, D.C., and Shugart, H.H. 1978b. Spectral analysis of forest model time series. Ecol. Modelling 4:313-326.

Kremer, J.N., and Nixon, S.W. 1978. A Coastal Marine Ecosystem: Simulation and Analysis. Springer Verlag, New York. 217 pp.

Kremer, J.N. 1979. An analysis of the stability characteristics of an estuarine ecosystem model. In Dame, R.F. (Ed.). Marsh-Estuarine Systems Simulation. Belle W. Baruch Library in Marine Science, No. 8. University of South Carolina Press. pp. 189-205.

Marchesseault, G.D., Saila, S.B. and Palm, W.J. 1976. Delayed recruitment models and their application to the American lobster fishery. J. Fish. Res. Bd. Can. 33:1779-1787.

Patten, B.C. (Ed.). 1975. Systems Analysis and Simulation in Ecology. Vol. III. Academic Press, New York. 601 pp.

Platt T., and Denman, K.L. 1975. Spectral analysis in ecology. Ann. Rev. Ecol. Syst. 6:189-210.

Rykiel, E.J., Jr., and Kuenzel, N.T. 1971. Analog computer models of "Wolves of Isle Royale." In Patten, B.C. (Ed.). Systems Analysis and Simulation in Ecology. Vol. I. Academic Press, New York, pp. 513-541.

Scavia, D., Powers, W.F., Canale, R.P., and Moody, J.L. 1981. Comparison of first-order error analysis and Monte Carlo simulation in time-dependent lake eutrophication models. Water Resourc. Res. 17:1051-1059.

Shugart, H.H. Jr. (Ed.). 1978. Time Series and Ecological Processes. SIMS Conference, Alta, Utah, June 1977, SIAM, Phila. PA.

Tomovic, R. and M. Vukobratovic. 1972. General Sensitivity Theory. Elsevier, New York, N.Y. 258 pp.

Van Voris, P., O'Neill, R.V., Emanuel, W.R., and Shugart, Jr., H.H. 1980. Functional complexity and ecosystem stability. Ecology 6:1352-1360.

Webster, J.R., Waide, J.B., and Patten, B.C. 1975. Nutrient recycling and the stability of ecosystems. In Howell, F.G., Gentry, J.B., and Smith, M.H. (Eds.), Mineral Cycling in Southeastern Ecosystems. ERDA Symposium Series, Augusta, GA., May 1-3, 1974. CONF-740513, NTIS. pp. 1-27.

FORMULATING HYDROPERIOD EFFECTS FOR A MULTISPECIES STAND SIMULATION MODEL FOR THE OKEFENOKEE SWAMP[1]

Jane E. Glasser and M. C. Barber[2]

Abstract.--A multispecies stand simulation model of the Okefenokee Swamp is described. The model is designed to simulate the effects of hydroperiod on the spatial and temporal dynamics of vegetation in the swamp. The processes of growth, germination, and mortality are modeled as explicit or implicit functions of hydroperiod. Additional environmental forcings incorporated in the model include fire, seed sources, temperature, and competition for light. A feedback term for growth is included in the model which incorporates a tree's capacity to store net production for future use.

INTRODUCTION

Spatial and temporal patterns of vegetation in a wetland are controlled to a large extent by hydroperiod. Vegetation has been shown to vary temporally with hydroperiod in descriptions of classical bog succession (King, 1685; Wells 1928). Spatial variation of vegetation with hydroperiod has been demonstrated by gradient analysis (Frye and Quinn, 1973; Bell, 1980). In this paper we describe a multispecies forest stand model designed to simulate wetland succession under different hydrologic regimes by expressing tree growth, germination and mortality as explicit functions of hydroperiod.

The model is being developed using a data base from the Okefenokee Swamp in south Georgia and northern Florida. The Okefenokee Swamp is a natural wetland, 189,000 hectares in area, that formed approximately 8,000 years ago (Parrish and Rykiel, 1979). The vegetation of the Okefenokee is a mosaic of habitat types, ranging from aquatic macrophyte marshes to cypress, bay, and gum forests. The pattern of the mosaic varies both spatially and temporally with hydroperiod (Hamilton, 1982; Duever, 1981).

A perturbation to swamp hydroperiod occurred in 1962 when a coffer dam was built on the Suwannee River, the major river draining the swamp. As a result the total amount of water impounded has increased, and water level fluctuations have decreased (Finn and Rykiel, 1979). The long term implication of this change in hydrology to successional dynamics in the swamp is unknown. Therefore, the goal of our modelling effort has become twofold:

1) to simulate the vegetation patterns of Okefenokee habitats as a function of individual interrelationships and environmental forcings, and
2) to thereby simulate the potential long term consequences of a change in water regime to the habitat mosaic of the Okefenokee Swamp.

REVIEW OF FOREST STAND MODELS

Stand simulation models have been used previously to investigate successional relationships. Vegetation dynamics are generated by modeling germination, growth, competition, and mortality of individual trees within small "representative" quadrats of a forest. Most of these models have been recently reviewed by Shugart and West (1980). In general the models (Shugart and West 1977; Tharp 1978; Kercher and Axelrod 1981; Phipps 1979; Shugart et al. 1980) have been extensions or modifications of the forest stand model JABOWA which was developed by Botkin et al. (1972). Of special interest herein are the works by Tharp (1978) and Phipps (1979). Tharp constructed a model to simulate flood plain processes of mixed hardwood in the Mississippi

[1]Paper presented at the Third International Conference on State-of-the-Art in Ecological Modelling, Colorado State University, May 24-28, 1982.

[2]Jane E. Glasser and M.C. Barber are both Graduate Research Assistants at the Institute of Ecology, University of Georgia, Athens, Georgia, U.S.A.

Valley while Phipps formulated a simulator for an Arkansas swamp.

Ek and Monserud (1973) constructed a stand simulator independent of JABOWA. Their model FOREST simulates spatial as well as biological dynamics of individual tree within a quadrat. Consequently, FOREST unlike the JABOWA based simulators can more realistically address shading effects and competition for light. We have adopted a hybrid approach to model a small representative Okefenokee quadrat (20x20 m) that utilizes assumptions and formulations found in both JABOWA and FOREST.

MODEL DESCRIPTION

A flow chart for a preliminary Okefenokee Swamp stand model is illustrated in Figure 1. The subroutines listed are each described below, including a description of the secondary subroutines called by the primary subroutines. Table 1 lists those species being modeled.

Subroutine HYDRO

Water level is assumed to follow linear reservoir dynamics. The rate of change in water level is itself a function of evapotranspiration and runoff. Each of these processes is in turn parameterized by daily temperature and site composition.

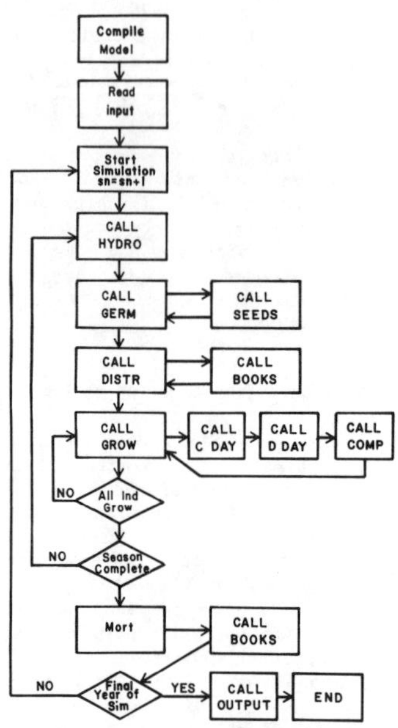

Figure 1.--Flow chart for a stand simulation model of woody communities in the Okefenokee Swamp.

Table 1.--Species simulated in Okefenokee Swamp stand model.

Species	Common Name
Acer rubrum	Red Maple
Cephalanthus occidentalis	Button Bush
Clethra alnifolia	Sweet Pepperbush
Cyrilla racemiflora	Titi
Fraxinus caroliniana	Water Ash
Gordonia lasianthus	Loblolly Bay
Ilex cassine	Dahoon
Ilex glabra	Small Gallberry
Ilex myrtifolia	--
Itea virginica	Sweet Spires
Leucothoe racemosa	Swamp Leucothoe
Lyonia ferruginea	Staggerbush
Lyonia lucida	Hurrah Bush
Magnolia grandiflora	Bull Bay
Magnolia virginiana	Sweet Bay
Myrica cerifera	Wax Myrtle
Nyssa ogeche	Ogeechee Plum
Nyssa sylvatica var. *biflora*	Black Gum
Osmanthus americana	Wild Olive
Persea palustris	Swamp Bay
Pinus elliottii	Slash pine
Pinus palustris	Longleaf pine
Pinus taeda	Loblolly pine
Quercus laurifolia	Laurel Oak
Quercus myrtifolia	Myrtle Oak
Quercus nigra	Water Oak
Quercus virginiana	Live Oak
Rhus copallina	Winged Sumac
Taxodium ascendens	Pond Cypress
Vaccinium arboreum	Sparkleberry

Subroutine GROW

The Forrester diagram presented in figure 2 is our conceptual model for tree growth. In this diagram growth is realized as incrementization of tree dbh, height, and crown diameter as well as production of leaves and reproductives. Under nominal conditions a given tree species possesses rather fixed allometric relationships between these vegetative structures. Hence, if an increase in dbh is formulated as a function of hydroperiod, the tree's remaining growth patterns can be ascertained by regression. Although we have adopted this approach herein, we realize that severe hydrologic stress can cause the formation of adventitious roots, etc., and can therefore alter these "fixed" allometric relations.

Subroutine GROW formulates a tree's seasonal growth as

$$\Delta dbh(t) = H\cdot T\cdot C\cdot(1-e^{\phi\Delta dbh(t-1)})\cdot Gmax(dbh(t))\cdot dbh(t) \qquad (1)$$

where dbh(t) represents the tree's diameter at breast height at time t, Δdbh(t) represents an

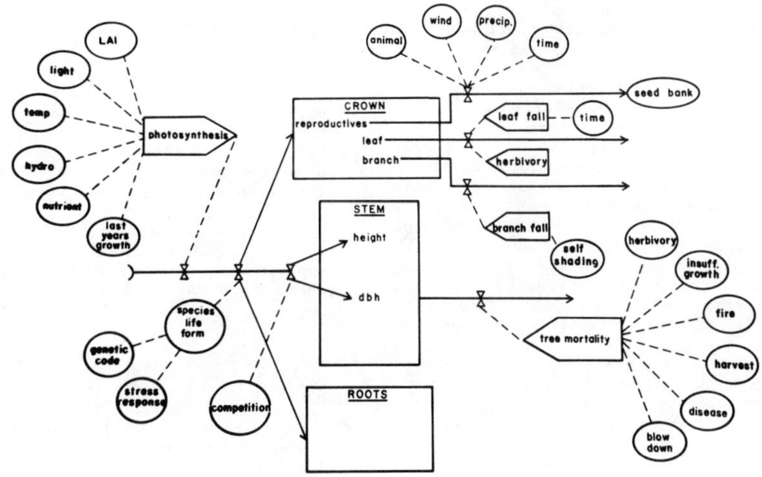

Figure 2.--Conceptual model for the allocation of net production by an individual tree using a modified Forester Diagram.

increment in that diameter, and Gmax is the species maximum size-specific growth rate. H, T, and C, are growth scaling coefficients which quantify the effect on growth of hydroperiod and temperature, and competition for light, respectively. They are formulated directly or indirectly as parabolic functions described by the equation:

$$M = \frac{4(Xmax - X) \cdot (X - Xmin)}{(Xmax - Xmin)^2} \quad (2)$$

such that:

Xmin = tree's lower tolerance limit to variable X, and

Xmax = tree's upper tolerance limit to variable X.

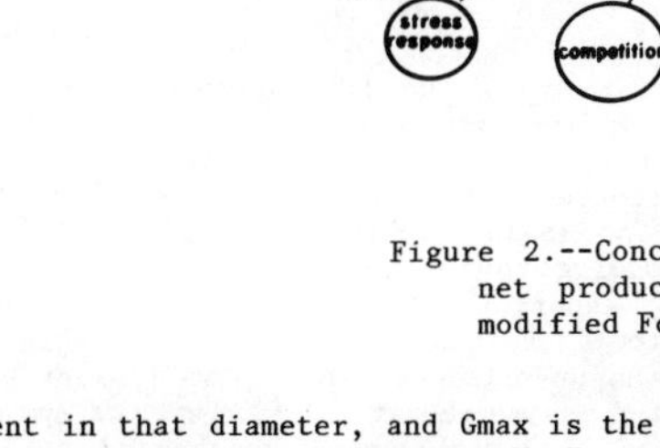

Figure 3.--Empirical parabolic relationship between degree days, DD(t), and the feedback growth multiplier T.

This formulation assumes that trees possess an optimum input for variable X function and the deviations from this optimal input slow growth symmetrically until the trees' tolerance limits are reached.

In the case of temperature the scaling factor, T≡M, is calculated in subroutine DDAY using degree-days, an integration of temperature inputs over time (Botkin et al. 1972), as the independent variable X. Consequently,

$$X \equiv DD(t) = \int_{to}^{t} T(\tau)d\tau; \quad (3)$$

and $T(\tau)$ is the temperature at time τ (fig. 3).

Similarly, to calculate the effect of hydroperiod on growth (subroutine CDAY), water level can be integrated over time to arrive at centimeter-days. Therefore, X in equation (2) becomes CD(t) i.e.,

$$X \equiv CD(t) = \int_{t_o}^{t} \delta(\tau)d\tau \quad (4)$$

where $\delta(\tau)$ is the height of the swamp's water table on day τ. During periods of recharge and inundation $\delta(\tau) > 0$, while during drawdown past the peat's surface $\delta(\tau) < 0$.

The concept of centimeter-days is implicit in the work of Phipps (1977), who has also modelled the effects of hydroperiod upon tree growth. Phipps assumed that dbh growth for Mississippi flood plain species was dependent upon the depth of an established water table from late April through late June during which 3/4 of the trees' growth is supposed to occur.

The parabolic response function in equation (2) is used to express hydroperiod effects on tree growth for common swamp species that are not restricted to bottomland/swamp habitats such as Acer rubrum (McDermott, 1954). We have adopted an alternative formulation of H for species that are endemic to wetland communities since such species probably have a much more conservative response to changing hydrologic regime. In particular figure 4 depicts one such relation. Within the range of CD_L to CD_U growth of endemic swamp species such as cypress, Taxodium ascendens, probably exhibit little or no measurable sensitivity to changing hydroperiod. However, within the ranges of CD_{min} to CD_L and CD_U to CD_{max} these species may exhibit pronounced growth suppression due to drought and flooding, respectively (Mattoon, 1916; Duever, personal communication). The relationship presented in figure 4 is basically similar to the approach taken by Tharp (1978) to model hydroperiod affects upon floodplain forest in the Mississippi Valley. Tharp, however, simply used days of inundation rather than centimeter-days as his ordinate axis.

Utilization of input and output relations of the forms given in equations (3) and (4) can be defended empirically, pragmatically, and biologically. Although trees are growing daily, in practice researchers cannot measure dbh changes on this time scale. In the Okefenokee Swamp dbh incrementization cannot be effectively recorded by dendrometers for any time interval less than six weeks (G.R. Best pers. comm.). In certain fields of mathematical analysis singleton or instantaneous value of a continuous function are of little analytical utility. For example, if ξ is a continuous random variable, then the probability that ξ equal a particular value a is zero, i.e.,

$$P[\xi = a] = 0 \tag{4}$$

The probabilistic properties of ξ are developed and investigated via its cumulative distribution function

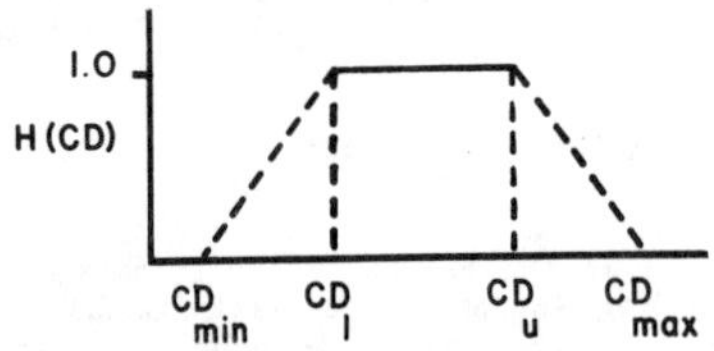

Figure 4.--Hypothetical relationship between centimeter days, CD(t), and the feedback growth multiplier H for an endemic wetland species.

$$F(a) = \int_{-\infty}^{a} P[\xi = w]\ dw. \tag{5}$$

Cumulative distribution functions are clearly analogous to eq. (3). Many organisms acclimate or condition their behavior and physiology based upon a time series of input, say DD(t) $t\varepsilon[t_o, t_1]$. That is the organism "integrates" its input.[1] Although the physiological integration is probably not isomorphic to mathematical integration, these two smoothing processors can be assumed to somehow homomorphic.

The growth scaling factor, C, is calculated in subroutine COMP, and is the product of N individual competition coefficients where N is the number of trees within the 20x20m plot. Specifically,

$$C = \prod_{\substack{j=1 \\ j\neq 1}}^{N} cij \quad , \tag{6}$$

where cij equals the light competition multiplier between individuals i and j for all j within the plot. If i and j do not compete, then $c_{ij}=c_{ji}=1$. The procedure to calculate cij is as follows:

1) Available light, λa, within the stratum corresponding to the depth of i's canopy is determined by generating a light extinction curve appropriate to j's height (fig. 5).

2) An unscaled competition multiplier, $\hat{c}ij$, is calculated using equation (2) with X≡λa and i's shade and light tolerance limits equal to λmin and λmax, respectively.

3) cij is formulated as an increasing function of the distance between i and j, dij:

$$cij = \min\{1, \Lambda \cdot dij + \hat{c}ij\} \tag{7}$$

The three growth scaling factors, H,T, and C all act to reduce the trees' maximum rate of dbh incrementization, Gmax. The multiplicative formulation of these feedback constraints implies that their influences are independent which is obviously not true. Trees can withstand cold water flooding much better than warm water inundation (Brink, 1954). Since the data which would be necessary to delineate these dependencies are generally unavailable, we accept the independence of action by these environmental factors.

The exponential function of (Δdbh_{t-1}) also reduces Gmax (fig. 6). Fritts (1976) has shown that while tree ring width may be correlated with annual climatic conditions, there is also a high degree of autocorrelation between rings of successive years. In other words in a very good year a tree might be able to store enough net production to be essentially unaffected by adverse conditions during the subsequent year.

Having calculated dbh increments for all individuals on the plot, subroutine GROW then calculates increments in tree height, basal area, and canopy

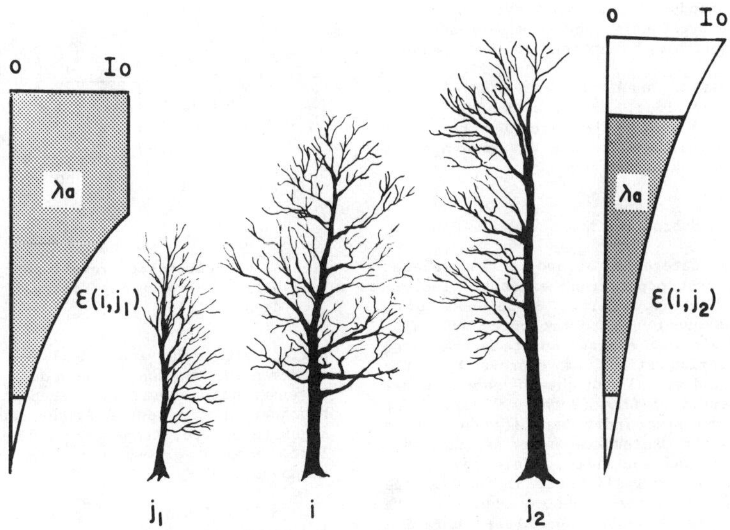

Figure 5.--Diagramatic representation of how neighboring trees j_1 and j_2 modify the maximal available light, λ_a, to individual i. $E(i,j_1)$ and $E(i,j_2)$ are light extinction curves.

diameter based on allometric relationships between these variables and dbh. These statistics as well as the spatial coordinates for each individual on the plot are incremented and stored in subroutine BOOKS.

Subroutine GERM

The subroutine GERM generates new seedlings on the plot for species m using the equation:

$$\text{Seedlings}_m = (\text{Max}_m)(\text{SB})(\text{Switch})(\text{Season}). \quad (8)$$

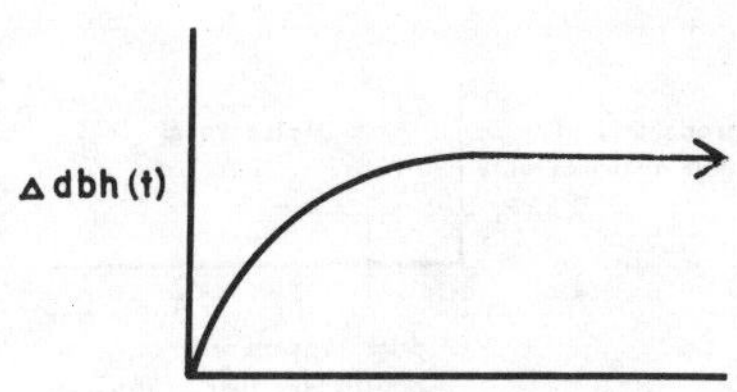

Figure 6.--Hypothetical assymtotic relationship between this year's and last year's rate of dbh growth.

Max_m equals the maximum number of seedlings of species m which could germinate under optimum conditions on the plot. SB is a scaling factor between 0 and 1 which represents the availability of seeds to the plot. In subroutine SEEDS, SB is assigned one of three values, sb_1, sb_2, or sb_3 which indicate whether a parent plant is present on the plot (sb_1), or was on the plot in the past (sb_2), or is in the vicinity of the plot (sb_3). These values are species specific and reflect the longevity and vagility of that species' seeds. In the calculation of seedlings, SB takes the highest of the three possible values.

The effect of hydroperiod on germination is incorporated in the model as the variable, Switch. Some wetland species, such as cypress and black gum, require a period of soaking and then a period of drying before they can germinate, (Mattoon, 1915; Demaree, 1932; DuBarry, 1963; DeBell and Naylor, 1972), thus Switch becomes non-zero only after both these conditions have been met. Graphically, this switching function is illustrated in figure 7 by the solid line. Due to microtopography only a portion of the seeds will be stranded initially, but microtopography effects are soon lost if water level continues to fall below the lowest point on the plot. Therefore, a more realistic representation of the effect of hydroperiod on germination might be illustrated by the dotted line in figure 7. The observation of stands of even-aged cypress in the swamp support the hypothesis that germination can be modeled as a switching function of hydroperiod.

Once germinated, seedlings are modeled as a cohort without individual coordinates until they reach a specified height. They are then recruited into the adult population and are assigned spatial coordinates by the subroutine DISTR.

Subroutine MORT

There are two categories of hydroperiod related mortality that forest populations exhibit. Individuals may be killed as a direct consequence of a single inundation event due to prolonged asphyxiation or individuals may expire due to accumulated effect of hydroperiod stress as expressed as the failure of the individual to obtain some minimal growth over a period of time (Demaree, 1932). For the most part tree size individuals (greater than one inch dbh) in the Okefenokee Swamp are not subject to direct period mortality. This particular process probably only significantly influences the mortality of germinating seeds and saplings. Growth correlated mortality, however, should be operative upon all size classes.

Mortality caused by either of these two processes can be readily modeled probabilistically. In particular having assigned or determined the probabilities for an individual's death under specified hydroperiod or growth/hydroperiod conditions, individuals can then be eliminated randomly from the stand model in accordance with these probabilities.

Fig. 8 depicts the hypothetical relationship between the probability of sapling mortality and hydroperiod. Both drought stress and inundation in excess of the criterial values CD_d and CD_f, respectively, cause 100% mortality. The function M(CD) defines the saplings' mortality probability when the swamp's water table is intermediate between these thresholds.

Figure 9 illustrates a hypothetical relation between the probability of a tree's death at time t, and the consecutive seasons, t,(t-1),...(t-n),

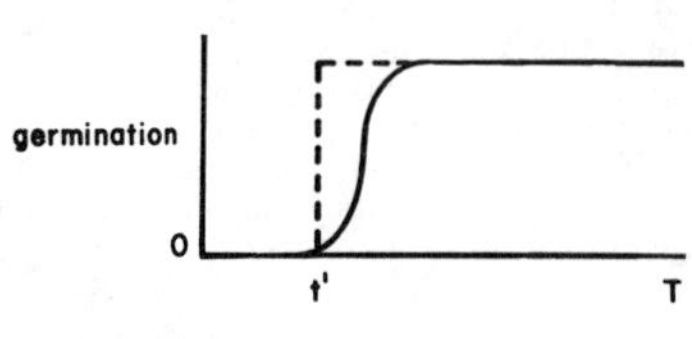

Figure 7.--Impulse response function between percent germination and centimeter days. The alternative sigmoid response function implicitly addresses plot microtopography.

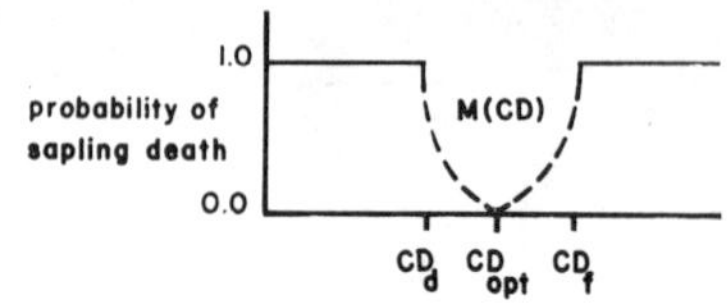

Figure 8.--Parabolic relationship between probability of sapling mortality and centimeter days, CD(t).

during which the individual has not growtn, i.e. incremented its dbh, some minimal amount necessary to ensure its survival until next season. Although hydroperiod may be the factor causing this lack of adequate growth, temperature, light competition, or last season's growth may also be modulating this response.

MORT calculates the probability of death of an individual as:

$$M = \sqrt[n]{p} \tag{9}$$

where:

p = the probability that an individual will die if Δdbh for n growing season is less than the threshold for that species,

n = the number of growing seasons during which growth has not occurred.

Phipps (1979) used an analogous formulation for mortality in his model SWAMP.

Mortality due to fire will be incorporated into the model in a subroutine FIRE, not shown in the flow chart. Fire will occur stochastically once water levels have dropped below a threshold value, with a probability inversely related to water levels. Season also influences the probability of fire. For example, frequent late summer thunderstorms would potentially increase the probability of

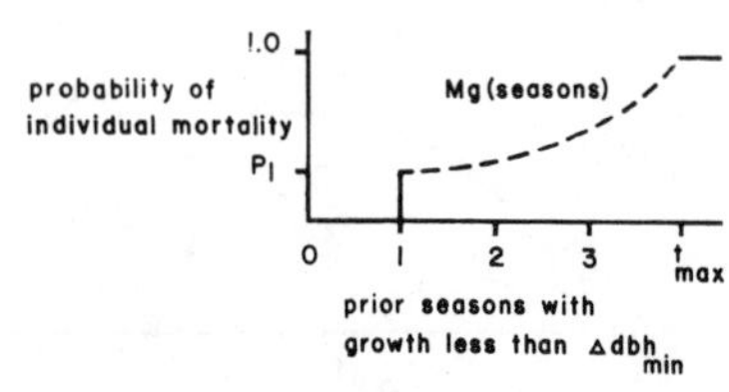

Figure 9.--Geometric relationship between the probability of tree mortality and the number of successive years of insufficient dbh growth.

lightning generated fires. Fire severity is determined by water level, season, and the length of time that has elapsed since the last fire, i.e. fuel load. The mortality of individuals caused by a fire event is determined probabilistically according to the size and species of the affected individuals, and the severity of the fire.

Model Validation

Community structure data for the Okefenokee Swamp are available from Hamilton (1982), Auble and Best (pers. comm.), and Meyers (1982), and will be used in validating the model. In addition, a time series of aerial photographs from 1944 to 1977 has been interpreted to record the actual successional pathways swamp habitats have taken in response to a major fire which occurred in 1953-4 and the 1960 construction of the sill. Model projections will be compared with these data.

SUMMARY

We have outlined the salient aspects of a stand simulation model of the Okefenokee Swamp in which the effect of hydroperiod on the processes of growth, germination, and survival of woody species is modeled. Successional dynamics are thus generated as a function of hydrology, with additional forcings on growth being temperature, competition for light, and previous years growth. Additional forcings on germination are seed source and season, and additional forcings on mortality are fire and past growth history.

The ultimate purpose of the model is dual: to determine what factors control the spatial and temporal distribution of vegetation in the Okefenokee; and to understand the long term consequences of a change in Okefenokee Swamp hydrology, a factor thought to be important in determining the distribution of vegetation in the swamp.

LITERATURE CITED

Bell, David T. 1980. Gradient trends in the streamside forest of central Illinois 107(2):172-180.

Botkin, D.B., J.F. Janak, and J.R. Wallis. 1972. Some ecological consequences of a computer model of forest growth. J. Ecol. 60:849-872.

Brink, V.C. 1954. Survival of plants under flood in the lower Fraser River valley, British Columbia, Ecology 35(1):94-95.

DeBell, D.S., and A.W. Naylor. 1972. Some factors affecting germination of swamp tupelo seeds. Ecology 53(3):504-506.

Demaree, D. 1932. Submerging experiments with Taxodium. Ecology 13:258-262.

DuBarry, A.P. 1963. Germination of bottomland tree seed while immersed in water. Journal of Forestry 61:225-226.

Duever, M.J. 1981. Hydrology-plant community relationships in the Okefenokee Swamp. Florida Scientist. In press.

Ek, A.R., and R.A.Monserud. 1974. FOREST: A computer model for simulating the growth and reproduction of mixed forest stands. Research Report A2635, College of Agricultural and Life Sciences, University of Wisconsin, Madison. Pages 1-14 and 3 Appendices.

Finn, J.T., and E.J. Rykiel, Jr. 1979. Effects of the Suwannee River sill on Okefenokee Swamp water level. Water Resources Research 15(2): 313-320.

Fritts, H.C. 1976. Tree Rings and Climate. Academic Press, New York. 567 pp.

Frye, R.J., II, and J.A. Quinn. 1979. Forest development in relation to topography and soils on a floodplain of the Raritan River, New Jersey, New Jersey. Bull. Torrey Bot. Club 106(4):334-345.

Hamilton, D.B. 1982. Plant succession and the influence of disturbance in Okefenokee Swamp. Ph.D. dissertation. Univ. of Georgia, Athens, GA.

Kercher, J.R. and M.C. Axelrod. 1981. SILVA: A model for forecasting the effects of SO_2 pollution on growth and succession in a western coniferous forest. UCRL-53109, 72 p. Lawrence Livermore National Laboratory, Livermore, California.

King, W. 1685. Of the bogs and sloughs of Ireland. R. Soc. (London) Philos. Trans. 15:948-960.

Mattoon, W.R. 1915. The southern cypress. U.S. Dept. Agriculture Bulletin 272:1-74.

Mattoon, W.R. 1916. Water requirements and growth of young cypress. Proceedings of the Society of American Foresters 11:192-197.

McDermott, R.E. 1954. Effects of saturated soil on seedling growth of some bottomland hardwood species. Ecology 35(1):36-41.

Parrish, F.K., and E.J. Rykiel, Jr. 1979. Okefenokee Swamp origin: Review and reconsideration. Journal of the Elisha Mitchell Science Society 95(1):17-31.

Phipps, R.L. 1979. Simulation of wetlands forest vegetation dynamics. Ecol. Model. 7:257-88.

Shugart, H.H. and D.C. West. 1977. Development of an Appalachian deciduous forest succession model and its application to assessment of the impact of the chestnut blight. J. Environ. Manage. 5:161-179.

Shugart, H.H. and D.C. West. 1980. Forest Succession Models. BioScience 30(5):308-313.

Shugart, H.H., A.T. Mortlock, M.S. Hopkins, and I.P. Burgess. 1980. A computer simulation model of ecological succession in Australian subtropical rainforest. ORNL/TM-7029, Oak Ridge National Laboratory, Oak Ridge, TN.

Tharp, M.L. 1978. Modeling major perturbations on a forest ecosystem. Unpublished M.S. Thesis, University of Tennessee, Knoxville.

Wells, D.W. 1928. Plant communities of the Coastal Plain of North Carolina and their successional relations. Ecology 9(2):230-243.

DYNAMIC SIMULATION OF AQUATIC MATERIAL FLOWS IN AN IMPOUNDED SWAMP HABITAT IN THE BARATARIA BASIN, LA.[1]

F. H. Sklar, R. Costanza, J. W. Day and W. H. Conner[2]

Abstract.-- Measurements of tree growth, leaf litter production, aquatic respiration, decomposition rates, benthic invertebrates and floating communities of plants and animals were taken from an impounded cypress-tupelo swamp of the Upper Barataria Basin. A dynamic computer simulation model was designed to identify the significant associations among the aquatic components as a function of environmental forcing functions (i.e. temperature, surplus precipitation, water flow, extent of dry season), and biological parameters (i.e. growth rates, predator-prey relations, assimilation efficiences). It was also developed to predict the response of an impounded swamp habitat to various management options, especially manipulation of the hydrologic regimes. Swamp impoundments, for example, were found to increase their annual energy exports by 43% when spoil banks were removed. Results indicate the sensitivity of swamp ecosystem response to changes in hydrology and the options for management of swamp ecosystems.

INTRODUCTION

The building of roads, the dredging of canals and the draining of wetlands has interfered with the laminar sheet flow of water over the backswamps of the upper Barataria Basin, La. and has produced large impoundments which accumulate water. These impoundments have large open areas where mats of Lemnaceae (duckweed) accumulate and populations of animals are very dense (Sklar,1979). The model presented here is a first attempt at describing the aquatic material flows of a cypress-tupelo system that has been inadvertedly impounded by canal dredging operations. The objectives of the model are: 1) to establish the major pathways of organic flows, 2) to discover the mechanisms operating on an ecosystem level that are controlling the dominant population densities and habitat preferences and 3) to simulate the effects of changing the height of the artifical levees on the export of materials to downstream communities.

----

(1) Paper presented at the Third International Conference on the State-of-the-Art of Ecological Modeling, sponsored by the International Society for Ecological Modelling. Colorado State University. (Fort Collins, CO, May 24-28, 1982).

(2) Graduate Research Assistant, Assistant Professor, Professor, and Research Associate, respectively, Coastal Ecology Laboratory, Center for Wetland Resources, Louisiana State University, Baton Rouge, LA 70803.

The Study Area

The des Allemands swamp forest system is located in the headwaters of the Barataria Basin Hydrologic Unit, an interdistributary basin bordered by the Mississippi River, Bayou Lafourche and the Gulf of Mexico (fig. 1). Since the institution of artifical leveeing on the Mississippi River, overbank flooding no longer occurs, and precipitation is the only source of freshwater. In general, water levels reflect the climatic forces operating over the entire basin. Differences between precipitation and evapotranspiration rates dictate the hydrology and are the basis for the water budget calculations of the model.

Within the swamp, two types of plant communities exist -- bottomland hardwood and baldcypress-water tupelo. Bottomland hardwood forests are found in swamplands of brief occasional flooding. They are composed of oak (Quercus spp.), willow (Salix nigra), elm (Ulmus americana), maple (Acer drummondii), boxelder (Acer negundo), cottonwood (Populus spp.) and a variety of other woody species. Baldcypress-water tupelo forests, found in poor drainage areas, where frequent long inundations are common, are dominated by baldcypress (Taxodium distichum), water tupelo (Nyssa aquatica), maple (Acer drummondii) and ash (Fraxinus spp.). Within the baldcypress-water tupelo forests is a floating mat of duckweed (Lemna minor, Spirodela polyrrhiza, Wolffia and Wolffiella), Riccia, and the common frog's bit (Limnobium spongia). A complete listing of the species is given in Conner et al. (1981).

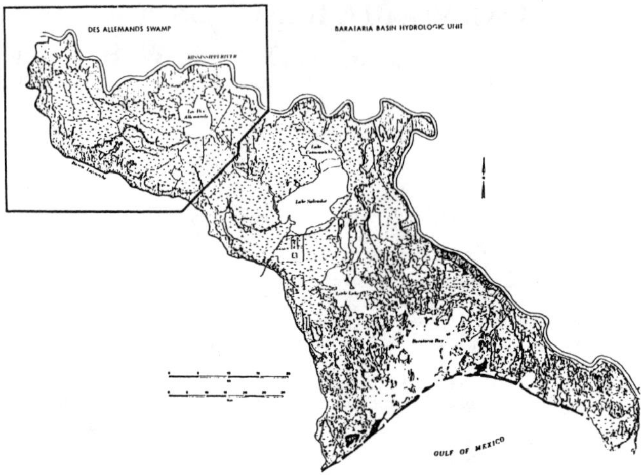

Figure 1.--Map of the Barataria Basin, Louisiana, showing uplands (dark pattern) and wetlands (light pattern). The fresh water swamp discussed in this paper is outlined.

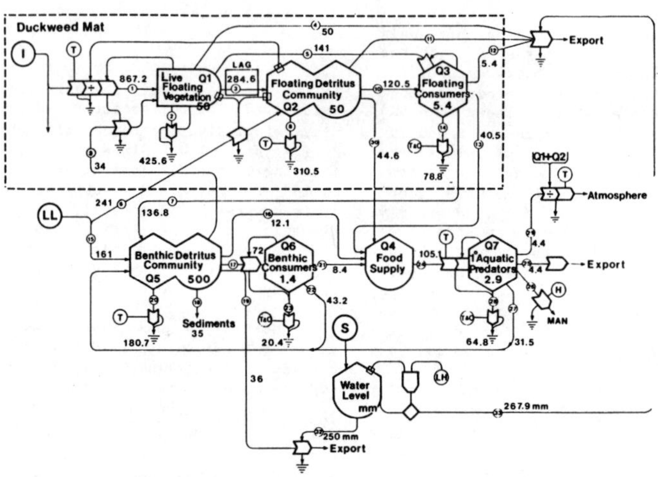

Figure 2.--Simulation model of the impounded swamp.

From previous studies, both in the Barataria swamp forest and in swamp forests throughout the southeastern United States, it is evident that much of the ecology and chemistry in these areas is determined by hydrologic conditions (Conner and Day, 1976; Conner et al., 1981; Kemp, 1978; Brown et al., 1978). Such factors as the rate, seasonality, and amount of waterflow are crucial in determining community structure, composition, and chemical cycling. Limited circulation, a result of the deposition of spoil banks during canal dredging for logging, oil extraction, navigation improvement, road construction, and drainage, can create reducing conditions in swamp surface sediments throughout most of the year (McNamara, 1978).

Benthic Ecology

There are a large variety of aquatic organisms in the backswamps of the Lac des Allemands forest and almost all of them are invertebrates, few fish species are ever found (Sklar, 1982). Only a few of the 54 species of backswamp benthic consumers dominate at any one time and they always include the ampihipoda, Hyallela azteca, and the oligochaete, Aulophorus vagus. Both animals are voracious feeders (Pennak, 1978). Hyallela is an omnivore, a general scavenger that browses on plants, animals and detritus. Aulophorus is a deposit feeder and like most aquatic oligochaetes it obtains its nutriment by ingesting bacteria-ladened substrate. In this case the substrate is Lemnacae (i.e. duckweed).

There are 4 general categories of benthic invertebrates based on feeding mechanisms (Cummins, 1973); shredders, collectors, scrapers and predators. The scrapers are strict herbivores. Their mouthparts are used for rasping plant material off rocks and sediments. They include the aquatic stage of the mayfly (Caenis) and seven species of aquatic gastropods. Scrapers usually feed by scraping off the coating of living algae which covers most submerged vegetation. However, the sparseness of submerged vegetation (Conner, 1975) and the lack of any significant algal populations (McNamara, 1978) suggests that these animals use the growing Lemnacae fronds and its' associated infuana as their principlal source of energy. This is verified by the fact that significantly greater number of gastropods are associated with the floating vegetation then are associated with the sediment (Table 1).

Shredders have mouthparts designed for chewing and mining and can be subdivided by food preferences. There are those that only eat living plant tissues, they include the Chironomidaes, the aquatic caterpillars (Neocataclysta), the weevils (Tanysphyrus spp. and Lixellus spp.), and the small Helodidae beetles (Scirtus tibialis). Those that only eat decomposing plant tissues include dipterans (Chironomus and Tipulidae) and isopods (Asellus obtusus).

Collectors are animals that gather up small particles of living or decaying organic matter by either a filtering process (suspension feeders) or by direct ingestion (deposit feeders). Collector types which use a filtering process include the springtail (Isotomurus), and the mosquito (Culicidae). The dominant suspension feeders in the upper basin swamps are juvenile crawfish (Procambarus clarki), freshwater shrimp (Paleomontes paludosus), hydra(Hydra spp.), and the fingernail clam (Sphaerium partumeium).

The shredders, scrapers and collectors in the sediment and floating vegetation support a diverse group of predators (piercers and swallowers). Swamp areas with high rates of hydrologic exchanges support as many species of aquatic invertebrate predators as there are herbivores and detritivores combined, while areas of low exchanges support only a few types at greatly reduced densities (Sklar, 1982).

Associated with the floating vegetation are populations normally associated with bottom sediments. There is a significant difference ($P<.01$) in the population size and the community structure between sediment benthos and floating vegetation "benthos" (Table 1). The mean invertebrate density in the floating mats of Lemnacae is almost 3 times that in the sediment; 16,903/m2 and 6,110/m2, respectively. In impounded swamp habitats even the deposit feeding isopod, Asellus obtusus, can be found in the Lemnacae mats. Highly reducing sediments seem to force these animals up into the water column, a remarkable feat since these animals are known to be very weak swimmers (Pennak, 1974; Barnes, 1968). When Lemnacae mats are present the oligochaete population is dominated by the rapidly reproducing, Aulophorus vagus, a Naididae which builds a dwelling from fronds of duckweed. During the cool months of the year when sediments are well oxygenated the more earthworm-like oligochaetes, Lumbricidae and Naididae, are found.

Table 1. Average density per sq. meter for each taxa. ANOV is the probability of equal values from the sediment and floating vegetation.

TAXA	Sediment Number	%	Floating Vegetation Number	%	ANOV (Pr>F)
Amphipoda	1,311	21.5	7,768	46.0	.001 *
Collembola	6	.1	7	.1	.840
Coleoptera	21	.3	91	.5	.001 *
Decapoda	3	.1	2	0	.653
Diptera	1,000	16.4	2,409	14.3	.003 *
Ephemeroptera	0	0	45	.3	.136
Gastropoda	118	1.9	854	5.1	.001 *
Bivalvia	500	8.2	8	.1	.001 *
Hemiptera	13	.2	174	1.0	.001 *
Hydroidia	2	0	278	1.6	.001 *
Isopoda	484	7.9	333	2.0	.418
Lepidoptera	15	.2	212	1.3	.06 *
Neuroptera	2	0	7	.1	.12
Odonata	11	.2	148	1.0	.001 *
Oligochata	2,606	42.7	4,062	24.0	.076
Rhychobdellia	7	.1	8	.1	.806
Tricladida	11	.2	427	2.5	.001 *
Total	6,110		16,903		.001 *

Aquatic backswamp organisms must deal with two types of stress, a hydrologic stess and a chemical stess. In a "normal" swamp the hydrology results in a seasonal drying-out period from approximately June to September. This is a time when most of the aquatic organisms must either leave the backswamp areas and migrate towards the deeper bayous and canals or enter an estivation stage if they are to survive. Gastropods, for example, like crawfish, will dig down to the water table and remain there untill the backswamps again become flooded. On the other hand, the life cycle of many aquatic insects will end at this time but not before a new generation is deposited as eggs in the backswamp sediments. In an impounded swamp the development of anoxic conditions results in the release of hydrogen sulfide--a chemical stress. Organic inputs from duckweed and leaf litter, high temperatures and continously flooded conditions combine to produce high respiration rates (McNamara, 1978) and low oxygen levels (Kemp, 1978). A diverse population of benthic macrofauna can not survive the highly reducing sediments of these areas. Numbers in the sediment drop to zero while populations in the floating Lemnacae mats are reduced by as much as 80%.

MODEL DESCRIPTION

The Water Balance Component

Nested within the larger ecosystem model, to be presented in the next section, is the hydrologic forcing function - surplus rainfall (S). Information on the water surplus, provided by the determination of the water balance, is fundamental to the understanding of hydrologic variables such as runoff rates, soil moisture changes and freshwater inflow to downstream estuaries. The water balance components developed by Thornthwaite and Mather (1955) provide quantitative estimates of these variables in a form that can be easily compared with the biological processes within the simulation model.

The water balance components are defined as follows:

PE - Potential Evapotranspiration: The maximum amount of evaporation that will take place with a continous vegetation cover and no shortage of soil moisture. Calculated by converting temperature to a heat index and correcting for latitude using the tables developed by Thornthwaite and Mather (1955).

ST - Soil Moisture Storage: A single layer of water at the surface, available for evapotranspiration and decreasing linearly as percent AE/PE decreases.

AE - Actual Evapotranspiration: The evapotranspiration rate that can be sustained by available water (ST and P) and heat.

P - Precipitation: Intensity of rainfall in inches per month.

D - Deficits: The difference between PE and AE when the water supply cannot keep up with the demand.

S - Surplus: The water available for runoff when AE equals PE and ST is fully recharged.

RO - Runoff: The continuous discharge of water from a hydrologic basin. The model assumes a 50% monthly retention rate.

Precipitation and temperature data were obtained from the National Oceanic and Atmospheric Administration's monthly publication of 'Climatological Data', and from the U.S. Weather Bureau's 'Decennial Census of U.S. Climate' for four weather stations in the Upper Barataria Basin. The capacity for soil moisture storage was taken from Gagliano et al. (1970) and Wax et al. (1978). Details on the handling of missing data, weighing factors and computer programing can be found in Sklar (1982).

The Carbon Model

The steady-state storages and flows for the backswamp aquatic community in an impoundment are shown in figure 2. The model is composed of 7 carbon components and 1 water storage component. The maximum water level is gotten according to spoil bank height (LH) and overbank runoff (flow RO) once Q9 is maximized. Runoff due to breakes in the levees is a function of their size. The local government just recently attempted to enhance the drainage from the impounded swamp by adding three new breakes. Unfortunately they only measure 1 meter across - not enough to drain a 450 ha impoundment but never-the-less capable of draining 250 mm of annual surplus precipitation as a continous daily flow (flow 40).

The carbon model divides the floating vegetation community into three storages. The primary producing duckweed component (Q1) is a function of limited light (I), temperature (T), its own accumulation rate, and consumption by herbivores (Q3) according to the following differential equation:

$$\dot{Q1}=(k1*I*T*Q1)/((k0*Q1*T)+Q2)+k9*Q1*Q5-Q1(k2*Q1)+Lag(k3)+k4*RO-k5*Q3) \quad (1)$$

The above equation also includes a lag of 30 days into the floating detrital component (Q2), an exponential respiration rate at high production rates (Q1*Q1), and an export rate as a function of runoff (RO). <3>

Lack of significant water movement allows dead floating biomass to accummulate when respiration is less then production according to differential equation 2:

$$\dot{Q2}=Lag(k3*Q1)+(k6*LL*(Q2+Q1))/(1+(k6*(Q1+Q2))-Q2(k8*T+k10+k11*RO+k30) \quad (2)$$

This component is an aggregation of detritus, leaf litter, bacteria, and meiofauna. It was assumed to be on the average 50% of the total floating biomass. As Q2 gets larger it competes with Q1 for space and donates an increasing organic flow to the floating consumers (Q3). When a large biomass of floating vegetation occurs (Q1+Q2) as much as 60% of the falling leaf litter (LL) is incorporated into Q2.

The third component of the floating community (Q3) is based on field measurements of populations of herbivores and detritivores from stations in the swamp impoundment. It is an aggregation of predominantly multivoltine species (continuous reproducers) and has inputs from both Q1 and Q2. The equation:

$$\dot{Q3}=k5*Q1*Q3+k10*Q2-Q3(k7+k12*RO+k13+K14(0.20D+1)) \quad (3)$$

includes a respiration rate of 4% of the body weight per day (Hargrave, 1971), an assimilation efficiency of 50% (Day et al., 1973), and a P/B ratio of 8.5 (Waters, 1977). Population respiration is increased by the chemical stress of a hydrogen sulfide accumulation-oxygen deficit factor (OD). The 0.20D parameter signifies a greater tolerance to an OD stress then other more sedentary organisms (Q6) and more sensitive organisms (Q7).

The truly benthic component of the model is divided into two storages, a benthic detritus component (Q5) and a macrofaunal consumer component (Q6). Inputs to Q5 include litterfall that is not processed in the Q2 component of the floating vegetation, and unassimilated organic matter from all the macrofaunal storages (Q3+Q6+Q7) according to the below equation:

<3> All equations have k's (parameters) and Q's (internal storage variables). The k number corresponds with the flow number in the model (fig. 2).

$$\dot{Q}5=LL-(k6*LL*(Q2+Q1))/(1+(k6*(Q2+Q1))+k7*Q3+ k7*Q3+k22*Q6+k27*Q7- Q5(k9*Q1+k16+k17*T*Q6+k18+ k19*(k40*Q9)+k20*T \qquad (4)$$

The benthic consumer population is a temperature sensitive (i.e. has a low temperature 'off' switch), and OD sensitive (i.e. increased respiration with high chemical stress) assemblage of detritivores. The differential equation (5),

$$\dot{Q}6=k17*Q5*Q6-Q6(k21+k22+k23(OD+1)) \qquad (5)$$

does not include an exportation parameter. The runoff velocity at the sediment-water interface is assumed to be inadequate for transporting benthic macrofauna over the top of the spoil banks. The model allows for the export of benthic juveniles if they are associated with the floating community.

The primary aquatic carnivores (Q7) are a diverse group of animals, none of which are carnivorous for their entire life cycle, and which include hemivoltine and univoltine species. The equation for Q7:

$$\dot{Q}7=k24*Q4*Q7*T-Q7(k25+k26*H+k27+ k28(T*0.2OD+1)+k29*T*1/(Q1+Q2+1)) \qquad (6)$$

estimates the general trends by allowing inputs to be a function of 1) the physiological controlling force of temperature, 2) the autocatalytic nature of predation, and 3) the availability of food (Q4) in the water column. Predators are buffered by the fact that inputs include all benthic and floating storages. Their population is controlled by harvest rate (H), respiration, death rate, emergence into the atmosphere by adults, and the natural secondary rate of production.

The seasonal changes in the temperature (T) and surplus water (S) forcing functions are the result of the water balance computations and will be discussed in the next section. Inputs of leaf litter were calculated by Conner et al. (1981). Light measurements in the canopy and subcanopy (fig. 3) were calculated

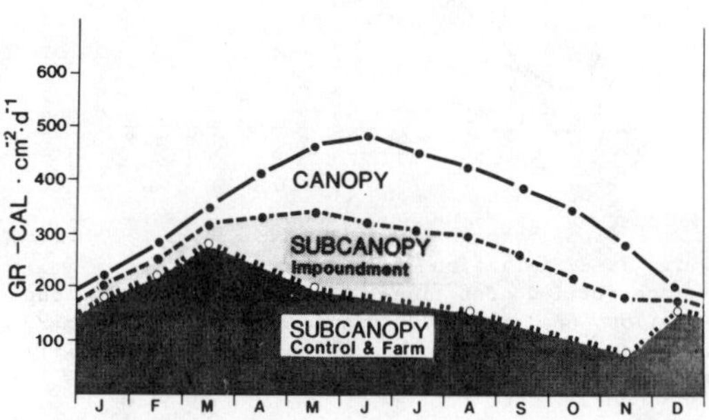

Figure 3. Average quantity of sunlight reaching the aquatic community in the impoundment.

from Bowers (1981). Our model assumes that nutrients are always present in excess. Kemp (1978) found year-round high concentrations of inorganic nitrogen and phosphorus.

RESULTS and DISCUSSION

Swamp Hydrology

Statistical parameters for the monthly water balance components were calculated based on 65 years of data (Table 2). On the average, precipitation exceeds PE by 518 mm (20.4 inches). The data are for basin uplands, with soil moisture storage of 152 mm. For basin lowlands (baldcypress-water tupelo swamps), the actual evapotranspiration will equal potential evapotranspiration (infinite soil moisture storage) and surpluses equal positive values of P-PE. Very

Table 2. Statistical parameters for the monthly water balance components.

Water Balance Component	Statistic (n = 65)	Annual Sum	Month J	F	M	A	M	J	J	A	S	O	N	D
Precipitation (inches)	Mean	63.0	5.05	4.79	5.28	4.49	5.13	5.28	7.22	6.44	5.85	3.54	4.42	5.58
	Std. Dev.	11.0	2.63	2.52	3.23	2.52	2.91	3.08	2.45	2.43	3.37	2.87	3.44	2.42
	Max.	86.3	12.61	12.18	15.22	10.92	15.58	16.06	13.09	15.72	18.04	12.58	13.12	11.62
	Min.	36.4	0.98	0.93	0.18	0.03	0.41	0.09	1.10	2.32	1.17	0	0.29	1.85
Temperature (°F)	Mean	-	55.5	57.8	62.6	69.1	74.7	80.2	81.5	81.3	78.4	70.2	61.3	56.2
	Std. Dev.	-	5.1	4.3	3.7	2.2	2.5	1.2	1.1	1.1	1.9	2.9	3.8	4.1
	Max.	-	68.7	66.8	72.0	73.7	79.3	84	84.6	84.4	83.1	79.2	70.2	63.7
	Min.	-	43.6	46.3	53.6	64.1	62.5	77.6	78.5	78.8	74.5	62.5	52.9	41.2
Actual evapotranspiration (inches)	Mean	39.2	1.00	1.15	1.99	3.15	4.65	5.41	6.17	5.86	4.57	2.71	1.50	1.02
	Std. Dev.	3.2	.49	.46	.55	.45	.57	.94	.66	.55	.84	.82	.48	.37
	Max.	45.5	2.70	2.35	3.57	4.20	5.81	6.67	6.89	6.50	5.85	4.81	2.49	1.85
	Min.	26.3	0.27	0.26	0.93	2.26	3.01	2.65	3.75	3.41	1.68	0.20	0.52	0.53
Deficit (inches)	Mean	3.4	0	0	0.02	0.10	0.31	0.85	0.55	0.52	0.58	0.45	0.07	0
	Std. Dev.	2.6	0	0	0.06	0.21	0.42	0.92	0.66	0.62	0.80	0.61	0.18	0
	Max.	13.9	0.02	0.12	.39	1.00	2.22	3.55	3.15	3.43	3.57	2.74	1.08	0
	Min.	0.3	0	0	0	0	0	0	0	0	0	0	0	0
Surplus (inches)	Mean	23.8	3.87	3.62	3.39	1.75	1.12	0.76	0.90	0.75	1.10	0.83	2.00	3.67
	Std. Dev.	9.3	2.77	2.60	3.08	2.08	2.00	1.57	1.30	1.33	1.86	1.40	2.74	2.62
	Max.	44.8	12.07	11.40	13.98	8.33	9.70	9.18	5.32	6.30	9.96	5.74	10.57	10.83
	Min.	2.9	0	0	0	0	0	0	0	0	0	0	0	0
Runoff (inches)	Mean	23.8	3.21	3.43	3.42	2.58	1.86	1.31	1.10	0.92	1.01	0.92	1.46	2.57
	Std. Dev.	8.7	1.76	1.55	1.62	1.52	1.24	1.16	0.93	0.09	1.04	0.90	1.49	1.59
	Max.	46.6	8.08	9.39	8.91	7.47	5.92	6.21	4.45	3.71	5.07	3.32	6.21	6.03
	Min.	8.6	0.26	1.12	0.69	0.46	0.23	0.16	0.08	0.04	0.02	0.01	0	0.01

high or low monthly surpluses were found to occur in any month, although winter months had the highest means and maximums. Mean AE followed PE very closely (low deficits) and had a seasonal cycle influenced more by temperature than by precipitation.

The average water budget for the Upper Barataria Basin is shown diagramatically (fig. 4). PE reflects the seasonal temperature changes and the difference between P and PE, shown as surplus, represents the 'S' forcing function used in the carbon model to recharge water storage (Q9). Large surpluses are generated during the low PE months of winter and AE equals PE every month except June.

The dry season in the backswamps is not readily apparent from the average water budget. The swamps do not dry up in June even though it is the only deficit month. The winter and spring surpluses are available as runoff during June, thus it is more likely that these swamplands would dry up during the month of August when P almost equals PE and runoff is at a yearly low.

Simulation vs. Real

Numerical integration of the differential equations and generation of the nonlinear forcing functions and results were done with a Continuous System Modeling Program (CSMP), on an IBM Model 370 digital computer. Simulations were run for three years. The first year of the simulation is the result of average forcing functions and is used to help stabilize the simulation before specific (1977-1978) forcing functions are incorporated. The average monthly biomass for backswamp components Q3, Q7, and Q1+Q2 (total

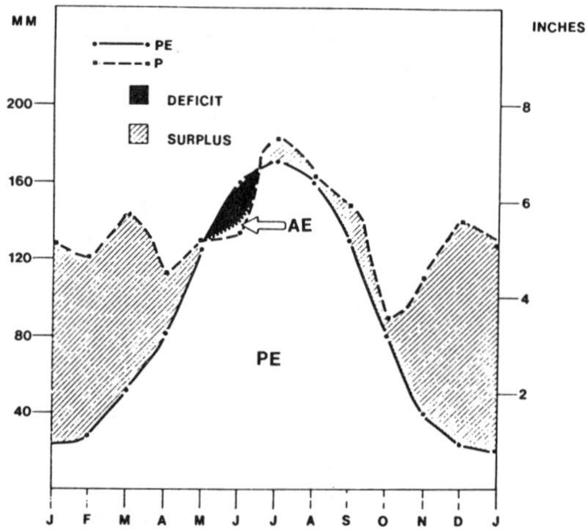

Figure 4. Average water balance budget for the upper Barataria Basin, 1914-1978.

floating vegetation) were interpolated by the CSMP integrator and graphed with a 20-35% variation line along with the results of the dynamic simulation.

The first year of the model results show seasonal variations which are reasonably consistent with the real variations (fig. 5). The simulated floating vegetation (fig. 5a) behaved 90% of the time within the bounds of the real statistics. Although not shown, the model predicts a peak in the live component of the floating vegetation (Q1) in spring which slowly decreases as the biomass of floating detritus (Q2) increases to a maximum in early winter. Low temperatures and high water levels account for the dramatic population decrease from January through Febuary (days 20-50). The second minimum in late summer (day 220) was not predicted by the model althoght a slight downward trend is noticed.

The seasonality of the simulated biomass of floating consumers (Q3) is remarkably consistent with the real world (fig. 5b). The bimodal population dynamics of these organisms is simulated as a result of the availability of duckweed in the spring, and the input of leaf litter with a decrease in the chemical stress (OD) in the fall. The same bimodal phenomenon exists for the predators (fig. 5c). What appears to be missing from the Q3 and Q7 simulation is the ability to respond quickly to changing environmental conditions. The aggregation of many species does not permit the model to respond any faster then the average animal will allow it. This does not alter the model°s usefulness as a tool. It only means that the fine details are filtered so that general ecological principles can emerge. Results for the benthic consumers (Q6), although not shown, are the same.

The simulation results for the second and third years where the forcing functions from 1977 and 1978 were used is compared to the real world data in figure 6. The floating community simulation results follow the general trends reasonably well (fig. 6a). The major difference between the simulated floating vegetation and the real data occurrs in the winter. This is due to the fact that the average "monthly"

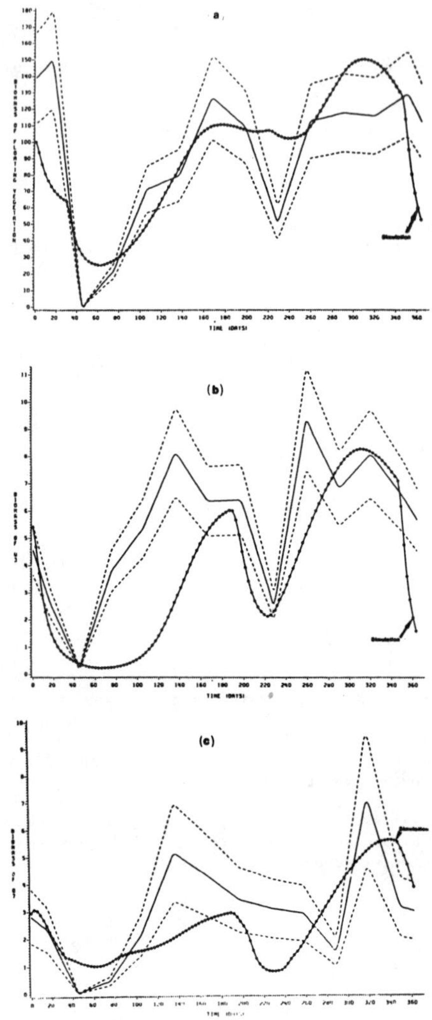

Figure 5. Simulation results for the first year (average forcing functions) for the (A), floating vegetation, (B), floating consumers, and (C), primary carnivores. Dark solid line = simulation; light solid line = average real data; dashed line = variation.

temperatures, used in the simulation, can not always reflect the duckweeds' "daily" ability to survive low temperatures. As a result the model underestimates the temperature tolerances of Q1 and Q2.

The second and third year of simulation for the floating consumers (Q3) followed the real data very closely (fig. 6b). However, the standing stock is consistently underestimated. This indicates that the ecological controls are correct but that parameters influencing respiration rates or assimilation

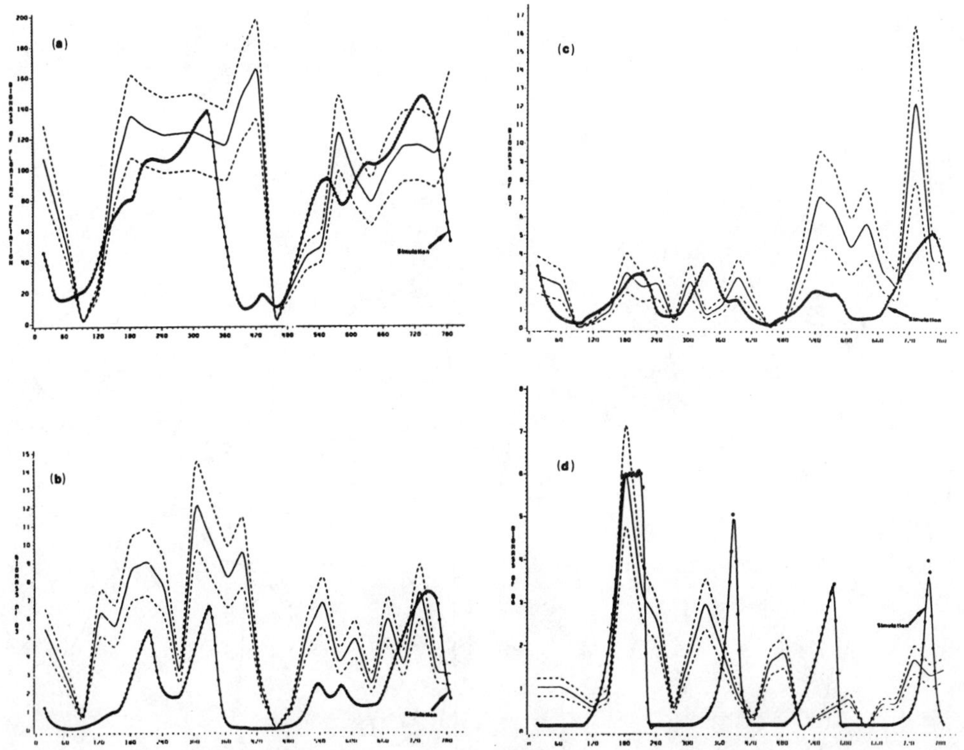

Figure 6.--Simulation results for the second and third years with 1977 and 1978 forcing functions. Dark solid line=simulation; light solid line=field data; dashed line=variation.

efficiences need slight adjustments.

The simulation results for the predators and benthos are realistic but do not follow the natural variations as well as we would like. The predator population increases in 1978 more then the model predicts (fig. 6d). The reason appears to be a lack of a feedback loop across time. Many of the species within Q7 take advantage of a good supply of prey by increasing egg production. These eggs hatch the following year thus the next year's population of predators reflect the previous years as well as current environmental conditions. It is hoped that by adding this feedback loop the simulation results for Q6 (fig. 6d) will also improve. The peaks in Q6 biomass are the proper magnitude but do not occur at the same frequency and period as was measured in the field. By adjusting the relationship between Q6 and Q7, Q6 should improve.

Removing the Levee

By now it should be obvious that, although the simulation results do not perfectly match the real data, they are in most cases a reasonably good reflection of the natural organic flows. The model can now be used as a predictive tool. Changing the levee height surrounding the simulated swamp impoundment is a relatively simple procedure. The only thing to change is the comparator switch on the water storage component. The difference between a levee height of 500 mm and 10 mm on four basic components (i.e. runoff, total animal biomass, benthic detritus, and total organic export) of the model is shown in figure 7.

As expected, the runoff rate increases. The average export of water for the 3-year simulation came to 44.0 mm per month with a 500 mm levee, and 65.6 mm per month with a 10 mm levee. This increased runoff rate has an effect on every aspect of the aquatic system living in the impoundment. For example, as shown in figure 7a, the storage of benthic detritus and its associated fauna will slowly decrease with time. This prevents the organic inputs from accumulating in the sediments, and decreases the extent of anaerobic metabolism by both exporting greater quantities downstream and by increasing the rate of comsumption by the consumers. The increase processing of organic matter by the consumers is a result of the larger biomass that exists when levee height is decreased (fig. 7d). Swamp animal populations are not as stressed hydrologically as they are chemically.

Finally, when levee height is decreased to its natural elevation, the total export of organic matter per sq. meter increases from 33.9 grams/month (LH=500

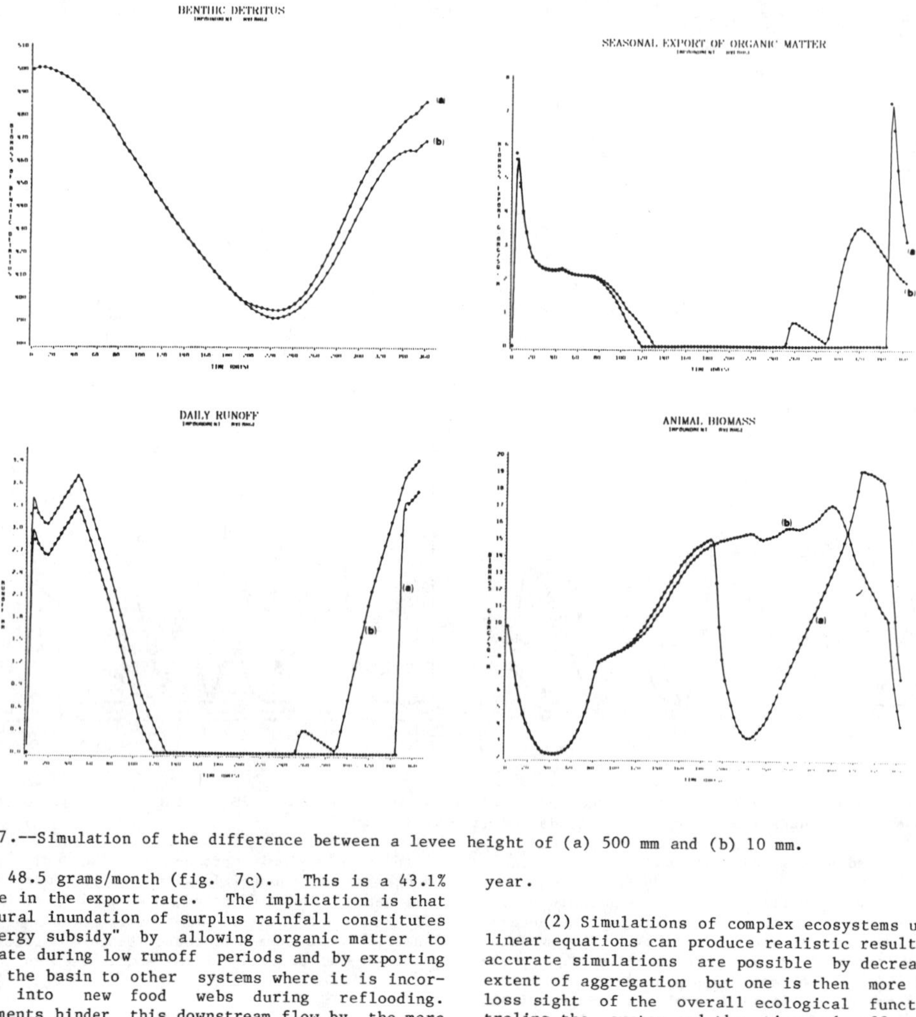

Figure 7.--Simulation of the difference between a levee height of (a) 500 mm and (b) 10 mm.

mm) to 48.5 grams/month (fig. 7c). This is a 43.1% increase in the export rate. The implication is that the natural inundation of surplus rainfall constitutes an "energy subsidy" by allowing organic matter to accumulate during low runoff periods and by exporting it down the basin to other systems where it is incorporated into new food webs during reflooding. Impoundments hinder this downstream flow by the mere presence of spoil banks and by the decreased productivity which results from the chemical stress of anoxic conditions.

SUMMARY and CONCLUSIONS

The modeling process has brought us closer to understanding how complex ecosystems, like a baldcypress-water tupelo swamp, function under natural, impacted and predicted conditions. Modeling the hydrologic regime has given us a holistic look at swamp flooding regimes. Modeling the organic flows has given us a clearer picture of ecosystem function. Besides being fun to develop, the dynamic simulation model enabled us to reach the following conclusions:

(1) The Lac des Allemands Swamp maintains high water levels because it is within a climatic region where precipitation exceeds potential evapotranspiration year.

(2) Simulations of complex ecosystems using non-linear equations can produce realistic results. More accurate simulations are possible by decreasing the extent of aggregation but one is then more likely to loss sight of the overall ecological functions controling the system and the time and effort involved are prohibitive.

(3) Impounded swamp habitats do not sufficiently drain during the months of high evapotranspiration rates. As a result the entire aquatic community is stressed by the depletion of oxygen and the accumulation of hydrogen sulfide.

(4) Removing the artifical levees surrounding a swamp impoundment will increase runoff, decrease detrital sedimentation, increase total animal biomass and most importantly, increase downstream exportation rates by almost 50%.

ACKNOWLEDGEMENTS

This research was funded by the U.S. Environmental Protection Agency and the Marine Sciences Dept., Louisiana State University. This is

State University, Baton Rouge, LA. We wish to thank N. Stevens and B. Roy for their aid in the lab. A special thanks to S. Liebowitz for assisting in the computer programing. The assistence given by F. Wang, C.S. Hopkinson, J.G. Gosselink and D. Baker was highly appreciated.

LITERATURE CITED

Barnes, R.D. 1968. Invertebrate Zoology, Second Edition. W.B. Saunders Company. Philadephia.

Bowers, L.J. 1981. Tree ring charateristics of Baldcypress growing in varying flooding regimes in the Barataria Basin, Louisiana. Ph.D. diss., Louisiana State University, Baton Rouge, LA.

Brown, S.L., M.M. Brinson, and A.E. Lugo. 1978. Structure and function of riparian wetlands. Pages 17-31 in R.R. Johnson and J. F. McCormick, tech. coord., Strategies for Protection and Management of Floodplain Wetlands and Other Riparian Ecosystems. Gen. Tech. Rep. WO-12, Forest Service, U.S. Dept. Agric., Washington, D. C.

Conner, W.H. 1975. Poductivity and composition of a freshwater swamp in Louisiana. M.S. thesis, La. State Univ., Baton Rouge.

Conner, W.H. and J.W. Day, Jr. 1976. Productivity and composition of a baldcypress-water tupelo site and a bottomland hardwood site in a Louisiana swamp. Amer. J. Bot. 63:1354-1364.

Conner, W.H., J.G. Gosselink and R.T. Parrondo. 1981. Comparison of the vegetation of three Louisiana swamp sites with different flooding regimes. Amer. J. Bot. 68:320-331.

Cummins, K.W. 1973. Trophic relations of aquatic insects. Ann. Rev. Ento. 18:183-206.

Day, J.W. Jr., W.C. Smith, P.R. Wagner and W. C. Stowe. 1973. Community structure and carbon budget of a salt marsh ans shallow bay estuarine system in Louisiana. Center for Wetland Resources Publication No. LSU-SG-72-04, Louisiana State University, Baton Rouge, LA.

Gagliano, S. M., R. Muller, P. Light, and M. Al-Awady. 1970. Hydrologic and Geologic Studies of Coastal Louisiana. Vol III. Louisiana State University, Coastal Studies Institute, Baton Rouge, LA. Contract No. DACW 29-69-C-0092.

Hargrave, B. T. 1971. The effect of a deposit-feeding amphipod on the metabolism of benthic microflora. J. Fish. Res. Bd. Can. 28:21-30.

Kemp, G.P. 1978. Agricultural runoff and nutrient dynamics of a swamp forest in Louisiana. M.S. Thesis, Louisiana State University, Baton Rouge, LA.

McNamara, S.J. 1978. Metabolism measurements of a flooded soil community in a Louisiana swamp forest. M.S. Thesis, Louisiana State University, Baton Rouge, LA.

Pennak, R.W. 1978. Fresh-Water Invertebrates of the United States. John Wiley & Sons, New York.

Sklar, F.H. 1982. The ecology of aquatic communities in the sediment and floating vegetation of a cypress-tupelo swamp in Upper Barataria Bay, Louisiana. Ph. D. dissertation. La. State Univ., Baton Rouge. (in progress)

Sklar, F.H. and W.H. Conner. 1979. Effects of altered hydrology on primary production and aquatic animal populations in a Louisiana swamp forest. Pages 191-208 in J.W. Day, Jr., D.D. Cully, Jr., R.E. Turner, and A.J. Mumphery, Jr. (eds.), Proc. Third Coastal Marsh and Estuary Management Symposium, Louisiana State University Division of Continuing Education, Baton Rouge, LA.

Thornthwaite, C.W., and J.R. Mather. 1955. The water balance. Drexel Inst. Tech., Lab of Climatology. Climatology 8(1):1-86.

Waters, T.F. 1977. Secondary production in inland waters. Adv. in Ecol. Res. 10:91-164.

Wax, C.L., M.J. Borengasser, and R.A. Muller. 1978. Barataria Basin: Synoptic weather types and environmental responces. Louisiana State University, Center for Wetland Resources, Baton Rouge, LA. Sea Grant Publ. No. LSU-T-78-001.

AQUATIC CARBON CYCLE IN OKEFENOKEE SWAMP HABITATS: ENVIRON ANALYSIS[1]

Patricia A. Flebbe[2]

Abstract.--Aquatic carbon cycles in three wetland habitats were compared by environ analysis of linear models. Identical in structure, the models differed in magnitude of flows; these differences were clarified by environ analysis. The analysis partitioned decomposition carbon flows between oxidizing and reducing pathways.

INTRODUCTION

The environ analysis technique of Matis and Patten (1981) partitions flows in steady-state, time-invariant models among the system's inputs and outputs. In the forward analysis, inputs are traced forward in time to define that portion of the system's flows, states (or storages), and outputs that are influenced by a given input: its output environment. Conversely, the reverse analysis looks backward in time from outputs to identify that portion of the system's flows, states (or storages), and inputs that produced the given output: its input environment. Environ analysis of steady state input - output models identifies flows and storages that are due to one unit of each input or that result in one unit of each output. Additional products are the expected number of times a particle visits a compartment and the mean residence time of a particle in each compartment.

Finn (1976, 1977, 1978) defined several ecosystem measures (e.g., total system throughflow [TST], straight - through TST [TST_s], mean path length [$\overline{PL}$], straight-through path length [$\overline{PL}_s$], and a cycling index [CI]) resulting from flow analysis that can be transferred directly to environ analysis. He explored the effects of changes in model structure on flow analysis and ecosystem measures. Finn (1980), compared the results of flow analysis on several Hubbard Brook nutrient models that had the same structure but differed in magnitude of flows associated with different elements. Futhermore, he evaluated the relative sensitivity of the ecosystem measures to changes in structure vs changes in magnitudes of flows.

Okefenokee Swamp is a heterogeneous wetland complex of forested and herbaceous marsh habitats that differ obviously in vegetational community structure and hydrology; consequently, biogeochemical cycles should differ among habitats. Nevertheless, the common feature of flooded, peaty soils makes these ecosystems similar. Here, environ analysis and ecosystem measures are compared among three structurally identical models of the same element, carbon, that represent different Okefenokee ecosystems. The systems differ in magnitude of flows, including carbon input in litterfall (Auble 1982) and precipitation or throughfall and output as carbon dioxide and methane (Flebbe 1982). Any differences in environ analysis results are manifestations of differences in carbon cycles in the three ecosystems.

THE MODELS

A conceptional model (fig. 1) of carbon flow in the aquatic portion of Okefenokee Swamp habitats was quantified (fig. 2) for three sites representating a spectrum of major vegetational community types: a floating-leaved aquatic bed marsh (prairie) site (LCL), dominated by *Nymphaea odorata*; a broad-leaved shrub site (SHS), dominated by *Cyrilla racemiflora* and *Lyonia lucida*; and a cypress (*Taxodium ascendens*) community (CYP) with a mixed bay (*Gordonia lasianthus*, *Persea palustris*, and *Magnolia virginiana*) subcanopy and shrub understory.

Following the terminology of Patten et al. (1976), the aquatic system is composed of components, or holons (Koestler 1967). Each of the holons, $\mathcal{H}_k$, has a state, x_k ($k = 1, \ldots, 5$ in figs. 1 and 2), here, the standing stock of carbon ($g\ C \cdot m^{-2}$). Inputs and outputs associated with $\mathcal{H}_k$ are denoted z_k and y_k, respectively.

The models were quantified from field measurements (Flebbe 1982) and literature reports; detailed documentation is beyond the scope of this paper (see Flebbe 1982). The marsh site model (LCL, fig. 2) differs from the forest site models in the additional output, y_4, of methane and carbon dioxide via macrophytes (Dacey 1981). Consequently, input of carbon dioxide from root respiration, $z_3 = 0$.

Steady-state linear donor controlled models were specified mathematically by a set of differential equations. In matrix notation

$$\dot{\underset{\sim}{x}} = A''\underset{\sim}{x} + \underset{\sim}{z}\ , \qquad (1)$$

[1]Paper presented at the Third International Conference on State-of-the-Art in Ecological Modelling, Colorado State University, May 24-28, 1982. University of Georgia, Okefenokee Ecosystem Investigations Paper No. 34 and Contributions in Systems Ecology, No. 56.

[2]Patricia A. Flebbe completed this research while a Graduate Student, Department of Zoology and Institute of Ecology, University of Georgia, Athens, Georgia, USA. She is currently an NRC Postdoctoral Research Associate, NASA Langley Research Center, Hampton, Virginia, USA.

where $\underset{\sim}{x}$ is an n-dimensional (n = 5) vector whose elements x_i are standing stocks of carbon displayed in the compartments of figure 2; $\dot{\underset{\sim}{x}}$, the first time derivative, is zero; $\underset{\sim}{z}$ is an n-dimensional vector of system inputs whose elements z_i are the sum of inputs to holon $\mathcal{H}_i$; and A" is an n x n transition matrix whose elements

$$a''_{ij} = F_{ij}/x_i,$$

where F_{ij} is the flow from $\mathcal{H}_j$ to $\mathcal{H}_i$ displayed in figure 2 and a''_{ii} is the negative sum of all intra- and intersystem outputs from $\mathcal{H}_i$. Hence, the differences between models noted above and in figure 2 appear only as differences in magnitude of elements of A" and $\underset{\sim}{z}$ and not differences in matrix structure.

METHODS

The environ analysis methodology is far too complex to present here except in condensed form. The interested reader should refer to Matis and Patten (1981). Briefly, in matrix notation, for the forward environ analysis, the systems are assumed to be in steady state:

$$\dot{\underset{\sim}{x}} \equiv \underset{\sim}{0} = A''\underset{\sim}{x} + \underset{\sim}{z}.$$

Solving for $\underset{\sim}{x}$, the steady state storages are:

$$\underset{\sim}{x} = (-A'')^{-1}\underset{\sim}{z}. \quad (2)$$

Analogously, for reverse environ analysis, models are formulated as

$$\dot{\underset{\sim}{x}} = -A'\underset{\sim}{x} - \underset{\sim}{y}, \quad (3)$$

where the elements of A' are

$$a'_{ji} = F_{ij}/x_i,$$

where F_{ij} is as above and in figure 2 and x_i is the storage (fig. 2) in the recipient compartment; a'_{ii} is the negative sum of all intra- and intersystem inputs to compartment $\mathcal{H}j$. The $y_i \varepsilon \underset{\sim}{y}$ are system outputs from compartment $\mathcal{H}i$. Assuming the systems are in steady state:

$$\dot{\underset{\sim}{x}} \equiv \underset{\sim}{0} = -A'\underset{\sim}{x} - \underset{\sim}{y}.$$

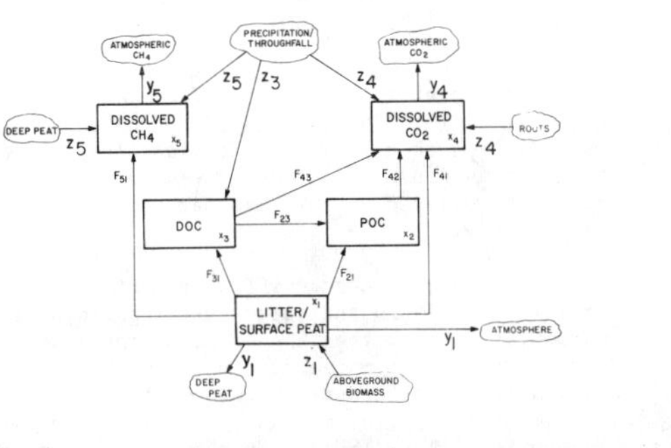

Figure 1.-- Conceptual model for carbon biogeochemistry in the aquatic subsystem of Okefenokee Swamp sites. States (x_i), inputs (z_i), and outputs (y_i) are all labeled. Within system flows, F_{ij} are: F_{21} = POC suspension, F_{31} = leaching, F_{41} = litter oxidation, F_{51} = methanogenesis, F_{42} = POC oxidation, F_{23} = POC formation from DOC, and F_{43} = DOC oxidation.

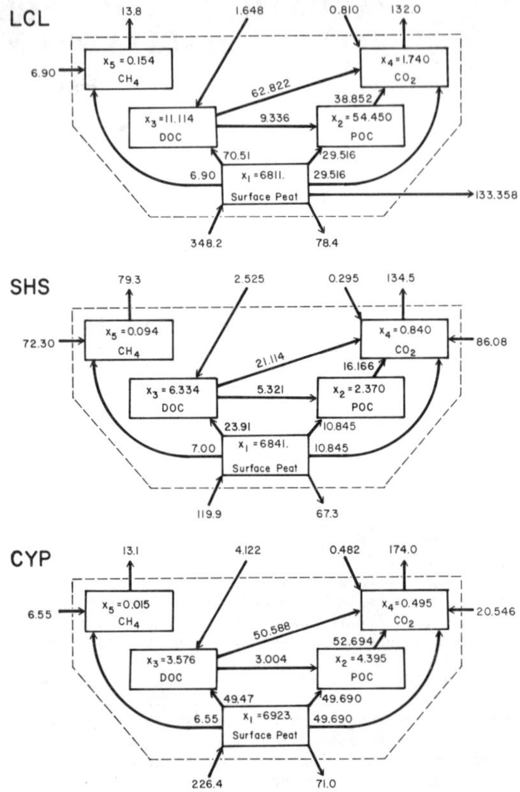

Figure 2.--Quantified carbon models for the aquatic subsystem of Okefenokee Swamp sites. LCL = Little Cooter Lake prairie site model, SHS = shrub site model, and CYP = cypress site model. Numbers within the five compartmental boxes represent standing stocks, x_i, or steady state storages (g $C \cdot m^{-2}$). Arrows are labelled with flows (g $C \cdot m^{-2} \cdot y^{-1}$) representing inputs, z_i; outputs, y_j; and intercompartmental flows, F_{ij}; $\mathcal{H}_i$ is the recipient holon, $\mathcal{H}_j$ is the donor holon, and the system boundary is denoted by a dashed line.

Solving for $\underset{\sim}{x}$, the steady state storages are:

$$\underset{\sim}{x} = (-A')^{-1}\underset{\sim}{y}. \quad (4)$$

Thus, the $(-A'')^{-1}$ and $(-A)^{-1}$ matrices partition, respectively, system inputs and outputs into steady state storages. In addition, we can determine how much of each holon's storage is attributed to each of the inputs and outputs. These matrices are the basis of environ analysis. Through a series of matrix manipulations developed by Matis and Patten (1981), flows can also be partitioned, intercompartmental transfers determined, and residence times calculated. The storages, flows, intercompartmental transfers, and residence times constitute the environ, E''_k or E'_k, for a compartment receiving input z_k or generating output y_k, respectively. In the remainder of his section, the methods developed by Matis and Patten (1981) that generate these primary products of environ analysis are presented in abbreviated form consistent with the software developed by M. Craig Barber[3].

Storage Partitions

To obtain steady state storage partitions for output environs, modify equation (2):

$$X''(Z) = (-A'')^{-1}Z, \qquad (5)$$

and for input environs, equation (4) becomes

$$X'(Y) = (-A')^{-1}Y \qquad (6)$$

where Z and Y are diagonal matrices of z and y, respectively. The kth column of X"(Z) or X'(Y) gives steady state storage partitions for inflow z_k or outflow y_k, respectively. Normalizing to one unit of each inflow or outflow, let Z = I or Y = I (I = identity) to obtain unit storage partitions.

Flow Partitions

The $(-A'')^{-1}$ and $(-A')^{-1}$ matrices also partition flows by creating F''_k, the unit internal flow matrix generated by one unit of input z_k (output environs):

$$F''_k = A''D''_k,\ k = 1, \ldots, n, \qquad (7)$$

and F'_k, the unit internal flow matrix for y_k (input environs):

$$F'_k = A'D'_k,\ k = 1, \ldots, n, \qquad (8)$$

where D''_k and D'_k are formed by diagonalizing the kth column of the unit storage partitions X"(I) = $(-A'')^{-1}$ and X'(I) = $(-A'')^{-1}$, respectively.

Off diagonal elements of F''_k, $F''_{ij,k}$ are flows from x_j to x_i due to a unit input z_k to $\mathscr{H}_k$. Diagonal elements $F''_{ii,k}$ are total outflow (always negative) from $\mathscr{H}_i$ per unit input z_k. To calculate system outflow $F''_{0j,k}$ from each x_j due to unit inflow z_k, take the negative sum of column j of F''_k.

In these models, several compartments have multiple outputs. Inputs to surface peat (x_1) in the prairie site model (Fig. 2 LCL), for example, may exit the system via two outputs from x_1, peat deposition and carbon dioxide loss via Nymphaea sp. Environ analysis provides only summed output information. Following the example of Patten and Matis (1982), the environ outflows will be allocated in proportion to relative magnitude of the original multiple outputs.

Similarly, off diagonal elements $F'_{ij,k}$ are flows from x_i to x_j resulting in a unit output y_k from $\mathscr{H}_k$. Diagonal elements $F'_{ii,k}$ are total inflows (always negative) to $\mathscr{H}_i$. To calculate system inflow $F'_{0j,k}$ to each x_j giving rise to outflow y_k, take the negative sum of the jth column of F'_k. Where a compartment has multiple inputs, the associated inflow is partitioned among input pathways in proportion to relative magnitudes of the model inputs.

Intercompartmental Transfers

Output Environs

So far, environ analysis has been deterministic. Considering flows as particles of matter (e.g., carbon) that move through the system requires a Markov chain formulation of the forward model. Matis and Patten (1981) provide development of the Markov model and subsequent analyses. The following minimal description is sufficient here. The transition matrix P" is constructed of p''_{ij}, the probability that a particle in x_j at time t will be in x_i at time t+1, and has the form:

$$P'' = \begin{bmatrix} 0 & 0 & 0 \\ I & Q'' & 0 \\ 0 & R'' & I \end{bmatrix} \qquad (9)$$

where column sums are 1.0. The central Q" matrix is of interest here and consists of intercompartmental probabilities:

$$Q'' = \begin{bmatrix} p''_{11} & \cdots & p''_{1n} \\ \cdot & \cdot & \cdot \\ \cdot & \cdot & \cdot \\ \cdot & \cdot & \cdot \\ p''_{n1} & \cdots & p''_{nn} \end{bmatrix}$$

Q" is determined directly from A":

$$Q'' = (I + hA'') \qquad (10)$$

where the constant h is chosen small enough to make all diagonal elements of Q" positive.

From Q", calculate O" by setting diagonal elements of Q" to zero and renormalize (i.e., column sums = 1.0) the resulting P" by dividing by the new column sum. Then,

$$N'' = (I - O'')^{-1} \qquad (11)$$

whose elements n''_{ij} are the expected number of times a particle in x_j will visit x_i before exiting the system. Associated variances u''_{ij} are determined as follows:

$$U'' = N''\,(2N''_D - I) - N''_2 \qquad (12)$$

where N''_D are diagonal matrices of N" and elements of N''_2 are squares of corresponding N" elements.

Input Environs

Reverse intercompartmental transfers can also be calculated from Q' in

$$P' = \begin{bmatrix} I & R' & 0 \\ 0 & Q' & I \\ 0 & 0 & 0 \end{bmatrix} \qquad (13)$$

where column sums are 1.0 and

$$Q' = \begin{bmatrix} p'_{11} & \cdots & p'_{1n} \\ \cdot & \cdot & \cdot \\ \cdot & \cdot & \cdot \\ \cdot & \cdot & \cdot \\ p'_{n1} & \cdots & p'_{nn} \end{bmatrix}$$

where p'_{ij} is the probability that a particle in x_j at time t was in x_i at time t-1. Also,

$$Q' = (I + hA') \qquad (14)$$

where h (negative) is chosen such that Q' diagonal elements are positive. O' is also calculated like O". Then,

$$N' = (I - O')^{-1} \qquad (15)$$

whose elements n'_{ij} are the expected number of times a particle in x_j was in x_i since entering the system.

[3]Institute of Ecology, University of Georgia, Athens, Georgia.

Variances u'_{ij} are determined as:

$$U' = N' (2N'_D - I) - N'_2 \quad (16)$$

where N'_D are diagonal matrices of N' and elements of N'_2 are squares of corresponsing N' elements.

Residence Times

Output Environs

The final products of the forward environ analysis are future residence times, m''_{ij}, elements of the matrix

$$M'' = h (I - Q'')^{-1}. \quad (17)$$

Each m''_{ij} is the expected time a particle in x_j will remain in x_i before exiting the system. Matis and Patten (1981) show that

$$M'' = (-A'')^{-1}. \quad (18)$$

Thus, residence times and storages are equivalent. Variances v''_{ij} associated with these means are determined as

$$V'' = M'' (2M''_D - I) - M''_2. \quad (19)$$

Input Environs

Past residence times, m'_{ij}, for the reverse analysis are found in the matrix

$$M' = h (I - Q')^{-1} \quad (20)$$

Each m'_{ij} is the expected time a particle in x_j has been in x_i since entering the system. Also,

$$M' = (-A')^{-1}. \quad (21)$$

Variances v'_{ij} are

$$V' = M' (2M'_D - I) - M'_2. \quad (22)$$

Ecosystem Measures

Finn (1976, 1977, 1978) defined a number of ecosystem measures, developed from input - output flow analysis, which may be applied here to environ analysis of the three Okefenokee carbon models. Recapitulating his development, by assuming $\dot{x}_k = 0$, throughflow (T_k) through $\#_k$ is defined:

$$T_k = \sum_{j=1}^{n} f_{kj} + z_k = \sum_{i=1}^{n} f_{ik} + y_k \quad (23)$$

where f_{ij} are as in figure 2.

Total system throughflow (TST) is the sum of all throughflow in the system:

$$TST = \sum_{k=1}^{n} T_k. \quad (24)$$

That is, TST is the mobile pool of the system. Total inflow to the mobile pool is

$$\Sigma IN = \sum_{i=1}^{n} z_i, \quad (25)$$

and total outflow from the mobile pool is

$$\Sigma OUT = \sum_{j=1}^{n} y_j. \quad (26)$$

The mean path length, $\overline{PL}$, is the average number of compartments that an inflow or outflow passes through:

$$\overline{PL} = TST/\Sigma IN = TST/\Sigma OUT. \quad (27)$$

Cycling efficiency (RE_k) of the system is defined from the N" and N' matrices defined above [equations (11) and (15)]:

$$RE_k = \frac{n''_{kk} - 1}{n''_{kk}} = \frac{n'_{kk} - 1}{n'_{kk}}. \quad (28)$$

The denominator is the number of times a particle passes through x_k, including the time it entered the system in z_k or exited in y_k, and the numerator counts only those times when cycled from or to another compartment.

Since the efficiency of cycling (RE_k) is known, the cycled portion (TST_c) and the straight-through portion (TST_s) of TST can be calculated:

$$TST_c = \sum_{k=1}^{n} RE_k \cdot T_k \quad (29)$$

and

$$TST_s = TST - TST_c. \quad (30)$$

Also, mean cycled path lengths ($\overline{PL}_c$) and mean straight-through path lengths ($\overline{PL}_s$) are

$$\overline{PL}_c = TST_c/\Sigma IN = TST_c/\Sigma OUT$$
$$\overline{PL}_s = TST_s/\Sigma IN = TST_s/\Sigma OUT. \quad (31)$$

Finn (1977) demonstrated that the latter, $\overline{PL}_s$, is fairly stable for a given model structure. For comparisons of different model structures, Finn (1980) suggested comparing

$$\overline{PL}_s/\overline{PL}_s(MAX)$$

where, for a given structure, $\overline{PL}_s(MAX)$ is maximum straight-through mean path length. $\overline{PL}_s(MAX)$, difficult to calculate, may be approximated by n, the number of compartments. Thus,

$$\overline{PL}_s/n$$

is a possible metric for comparing models.

Cycling index, CI, is

$$CI = TST_c/TST. \quad (32)$$

The cycling index ranges from zero to one. A CI = 0 means that there is no cycling, and CI = 1 means that all flow is cycled.

RESULTS

Environ analysis generates an environ for each system level input and output. For example, the first output environ, E''_1, traces the fate of fresh litter entering each of the three systems. Additional output environs trace the fates of DOC input in precipitation (E''_3), carbon dioxide input from root respiration or precipitation (E''_4), or methane input from deep peat (E''_5). Reverse environ analysis identifies three input environs for each model that reveal the sources of peat deposition output (E'_1), carbon dioxide loss (E'_4), or methane loss (E'_5). Since the original models contained no cycles, several of the seven environs generated are trivial. For example, E'_1, the source of peat deposition, consists only of a unit output that resided in $\mathcal{H}_1$, surface peat, for a period of years since entering the system in litterfall. For brevity, only three environs describing the fate of litter input (E''_1, fig. 3) and the sources of carbon dioxide (E'_4, fig. 4) and methane (E'_5, fig. 5) outputs will be discussed here. A complete treatment, including all the matrices

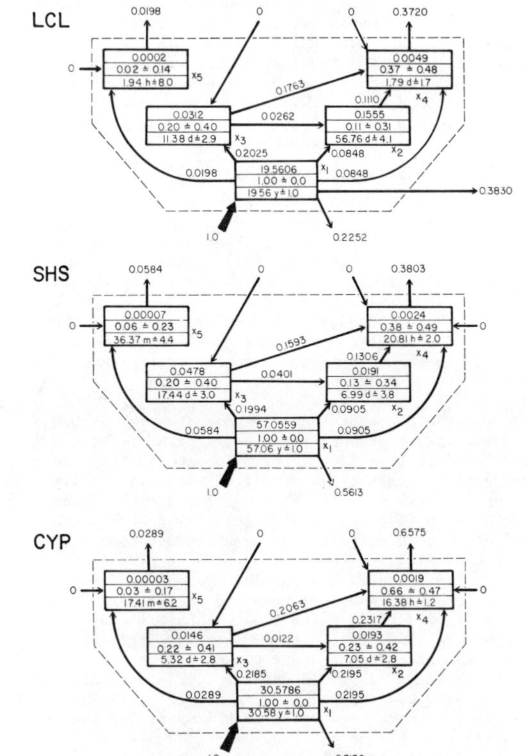

Figure 3.--Unit output environs E''_1 generated by one unit of input z_1 (bold arrow), litterfall to surface peat and litter (x_1), for the prairie site model (LCL), shrub site model (SHS), and cypress site model (CYP). Arrows are labelled with carbon flows ($g\ C{\cdot}m^{-2}{\cdot}y^{-1}$), and storages ($g\ C{\cdot}m^{-2}$) appear at the top of boxes corresponding to compartments in figures 1 and 2. Middle numbers in the boxes are mean ± standard deviation (unitless) of expected number of times carbon entering as input z_1 will cycle through the compartment. Bottom numbers are mean ± coefficient of variation of future residence time in the compartment of carbon entering as input z_1. Units for the latter means are y = years, d = days, h = hours, and m = minutes; coefficients of variation carry no units.

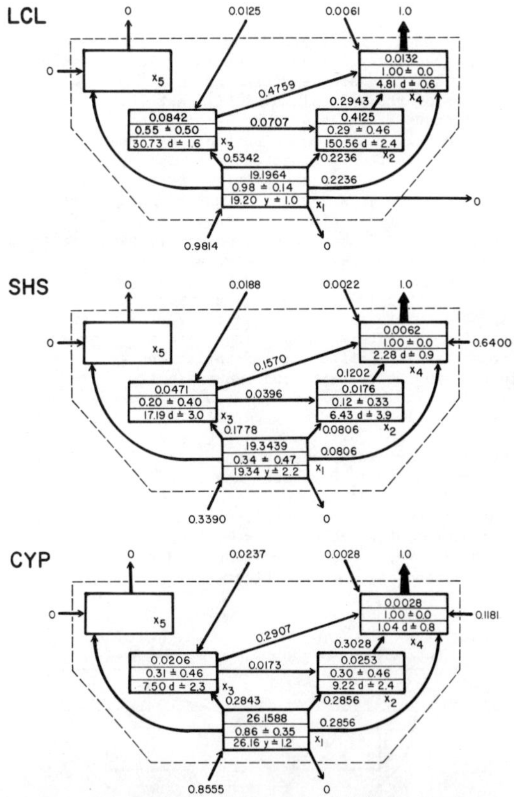

Figure 4.--Unit input environs E'_4 generating one unit of output y_4 (bold arrow), carbon dioxide flux from surface water (x_4). Details as in figure 3 except middle numbers in the boxes are expected number of times carbon exiting as output y_4 cycled through the compartment since entering the system, and bottom numbers are past residence times in the compartment of carbon exiting as output y_4.

defined by equations (1) - (22), is presented by Flebbe (1982).

The matrices produced by the analysis are summarized in figure 3-5, analogous to figures 1 and 2. All results are normalized to one unit of input (fig. 3) or output (figs. 4 and 5). Unit storage partitions x''_{ij} (j=1) and x'_{ij} (j=4, 5) are entered in the upper third of the corresponding compartment x_i in figures 3-5. Unit flow partitions, F''_k (k=1) and F'_k (k=4, 5) are incorporated in figures 3-5. For the output environ E''_1 (fig. 3), arrows are labelled with flows $F''_{ij,k}$ for intrasystem flows and $F''_{0j,k}$ for outflows; $z_k = 1$ and all other inputs are zero (k=1). For the input environs E'_4 and E'_5 (figs. 4 and 5), arrows are labeled with flows $F'_{ij,k}$ for intrasystem flows from $\mathcal{H}_i$ to $\mathcal{H}_j$ and $F'_{0j,k}$ for inflows; $y_k = 1$ and all other outputs are zero (k = 4, 5).

Elements of the kth column of N" can be interpreted as the expected number of times a particle entering x_k in input z_k will visit x_i before leaving the system. The initial entry to x_k is counted. With this interpretation, n''_{ik}'s and u''_{ik}'s (as standard deviations) are entered in the middle third of compartments in figure 3 for k = 1. Since there are no feedback flows and hence no cycles in these models, all n''_{ik} are less than or equal to 1.0. Elements of the kth column of N', past intercompartmental transfers, can also be interpreted as the expected number of times a particle exiting x_k as output y_k visited x_i since entering the system, counting the one time the particle exited from x_k. The n'_{ik} and u'_{ik} are entered in the middle third of compartments in figures 4 and 5 for k = 4, 5. Here, too, the lack of cycles makes all past visits to a compartment less than or equal to 1.0.

The kth column of M" may be interpreted as the expected future residence time in x_i of a particle entering as unit input z_k. Thus, the m''_{ik} and v''_{ik} (converted to coefficients of variation) are entered in the lower third of compartments in figure 3 (k = 1). The kth column of M' is the expected past residence times in x_i of a particle exiting as unit output y_k. Thus, m'_{ik} and v'_{ik} (converted to coefficients of variation) are entered in the lower third of compartments in figures 4 and 5 (k = 4, 5). Since there are no cycles in any of these models, residence times are purely delays (i.e, storages) rather than a combination of storages and recycling. The units of M" and M' (years) have been converted to more convenient time units in the figures.

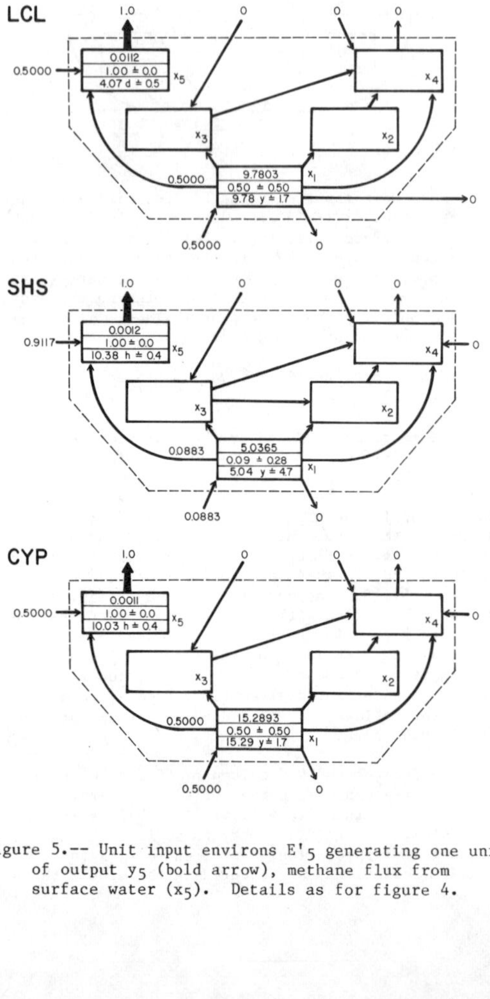

Figure 5.-- Unit input environs E'_5 generating one unit of output y_5 (bold arrow), methane flux from surface water (x_5). Details as for figure 4.

Fates of Litter Input

Storages and Residence Times

For the E''_1 environs, a unit of litter input remains longest in the surface peat compartment (x_1 in fig. 3). The unit input ($g\ C{\cdot}m^{-2}{\cdot}y^{-1}$) remains in x_1 for averages of 19.56, 57.06, and 30.58 y, accounting for the large storages, 19.56, 57.06, and 30.58 $g\ C{\cdot}m^{-2}$, generated per 1 $g\ C\ m^{-2}{\cdot}y^{-1}$ input, entered in the upper third of the compartment. Storages in the remaining compartments are relatively small. Storages are smallest and residence times shortest in prairie site model, followed by the cypress model; corresponding losses to deep peat are least in the prairie site model, followed by the cypress and shrub site models (fig. 3 LCL, CYP, SHS, respectively). Thus, in the prairie model, carbon input in litter moves relatively quickly to other compartments and out of the system. In the shrub model (fig. 3 SHS) transfers to other compartments and deposition in deep peat are very slow.

Suspended particulate carbon (x_2) remains longer in the prairie site water column compared to the forested sites (56.76 d vs 6.99 and 7.05 d; fig. 3 LCL vs SHS & CYP). Consequently, suspended POC (x_2) storage, (0.1555 $g\ C{\cdot}m^{-2}$) in that site is an order of magnitude greater than in the two remaining sites, consistent with field observations of greater POC in the prairie (Flebbe 1982). Dissolved organic carbon (x_3) resides a shorter period in the cypress site (5.32 d vs 11.38 and 17.44 d; fig. 3 CYP vs LCL & SHS), where DOC concentrations are lowest (Flebbe 1982). The longest mean residence time, 17.44 d (fig. 3 SHS), in the shrub site coincides with highest overall DOC concentrations.

Of the two dissolved gas compartments, carbon dioxide (x_4) residence times are longest, ranging from 16.38 h to 1.79 d (fig. 3). Methane (x_5) residence times are very short, 17.41 m to 1.94 h (fig. 3). Water in these sites is supersaturated with methane by three to five orders of magnitude, fostering rapid losses by both diffusion and ebullition, compared to carbon dioxide supersaturation factors of only one order of magnitude. Curiously, longest residence times for both gases are observed in the marsh site model (fig. 3 LCL); surface water there is more exposed to wind, which should increase diffusion rates compared to the other two sites. However, although total dissolved methane levels were higher (0.154 vs 0.094 and 0.015 $g\ C{\cdot}m^{-2}$; Flebbe 1982, p. 246), methane loss rates were lower (90 vs 844 and 873 y^{-1}; Flebbe 1982, p. 247). The lower measured loss rates observed in the prairie site are probably a sampling error introduced because sampling chambers shelter the water surface from wind. Thus, relatively low inputs and outputs and large standing stocks generate longer residence times. The same relations hold for carbon dioxide. These results suggest that model quantification was inadequate. Possibly, methane oxidation is more important than originally assumed, especially in the prairie site where Utricularia - periphyton systems, for example, modify their environment (Bosserman 1979).

Future intercompartmental transfers n''_{i1} for E''_1 are illustrated in figure 3. A unit input to x_1 generates one visit there and moves to one of the remaining four compartments with probabilities equal to the sum of environ flows (arrows). Since intercompartmental transfers are so closely related to flow partitions, the former will not be discussed explicitly. In models where cycles occur, the N" matrix is a valuable tool integrating total number of material visits to a holon.

Flow Partitions

One unit of input in litterfall, z_1, to surface peat (x_1) generates flows (g $C \cdot m^{-2} \cdot y^{-1}$) shown in figure 3 for the prairie, shrub, and cypress site models. The unit input (g $C \cdot m^{-2} \cdot y^{-1}$) generates storages in x_1, surface peat, of 19.5606 to 57.0559 g $C \cdot m^{-2}$. From surface peat, carbon either exits the system directly or is transferred to one of the four remaining compartments. Peat deposition is considerably more important (56 % of total x_1 outflows) in the shrub site (fig. 3 SHS) compared to the other two sites (23 and 31 %; fig. 3 LCL & CYP, respectively). Thirty-eight percent of carbon input to surface peat in the prairie model (fig. 3 LCL) is lost directly via *Nymphaea* sp. vascular fluxes of carbon dioxide and methane. Leaching losses (F''_{31}) to DOC are nearly constant across sites (20-22 %), and methanogenesis (F''_{51}) accounts for minor transfers in all sites (2-6 %; fig. 3). The transfers to POC (F''_{21}) and carbon dioxide (F''_{41}) are always equivalent and considerably more important in the cypress site (22 % vs 8-9 %; fig. 3 CYP vs LCL & SHS). Ultimately, 2-6 % of the unit litter input exits as methane (fig. 3, F''_{05} or y''_5). In the cypress site, 66 % exits as carbon dioxide from surface water (fig. 3 CYP, F''_{04} or y''_4) compared to 37-38 % in the other two sites (fig. 3 LCL & SHS, respectively). The 37 % carbon dioxide and 2 % methane losses in the prairie site (fig. 3 LCL) are deceptively low since *Nymphaea* sp. serves as a direct conduit for gases. Assuming that 2/3 of the 38 % (=0.2553) lost that way is carbon dioxide and 1/3 (=0.1277) is methane, total carbon dioxide loss constitutes 62 % (0.3720 + 0.2553) and total methane loss constitutes 15 % (0.0198 + 0.1277) of litter input (fig. 3 LCL).

The relative importance of oxidizing and reducing pathways in the three sites can be compared using these results. In the prairie model (fig. 3 LCL, F''_{01} or y''_1), only 22.5 % of a unit litter input is deposited to deep peat. Indeed, Auble (1982) found that only 15.3 % of *Nymphaea* sp. leaf weight remained after 1 year of decomposition compared to at least 30 % for other species. Certainly, rapid decomposition of this dominant plant contributes to lower peat deposition rates. This site may be characterized as one with a thick peat layer (3.8 m; Cohen 1974) that is building up slowly, promoting deep water habitat and perpetuation of the aquatic macrophyte system. The large loss of carbon via *Nymphaea* sp. (38 %; fig. 3 LCL) suggests an important subject for further research, generally unknown at the time field work was initiated. The prairie model demonstrates that aerobic decomposition is only moderately inhibited (62 % total loss calculated above compared to a maximum of 100 % in a non-peat forming terrestrial system). However, anaerobic decomposition is considerably enhanced, resulting in an additional 15 % of carbon input lost as methane via two pathways.

In the shrub site model E''_1 environ (fig. 3 SHS), aerobic decomposition is low (38 % of unit input lost as carbon dioxide in fig. 3 SHS) and methane loss (6 %) is only moderate. Thus, a large portion (56 %) of litter input is deposited as peat (fig. 3 SHS). Most of the litter input in this site appears relatively resistant to decay and the substrate is considerably built up over surrounding prairie areas. Thus, reduced aerobic and anerobic decomposition (44 % total carbon loss) contributes to rapid rates of peat buildup, even though total peat thickness is relatively low (2 m).

The cypress site model E''_1 environ (fig. 3 CYP) demonstrates the intermediate position of this site. Sixty-six percent of carbon input as litter is lost via the aerobic pathway as carbon dioxide; an additional 3 percent is lost as methane. The remaining 31 percent is deposited as deep peat (fig. 3 CYP).

Source of Carbon Dioxide Flux

Storages and Residence Times

A unit output exiting as carbon dioxide (y_4) was stored in compartments x_1 through x_4 (fig. 4). Of these past storages, those in surface peat (x_1) are most significant (19.1964 to 26.1588 g $C \cdot m^{-2}$), and those as volatile, dissolved carbon dioxide (x_4) are least important (0.0028 to 0.0132 g $C \cdot m^{-2}$).

Past residence times in E'_4 are shown in fig. 4. Residence times of carbon in surface peat (x_1) are reduced compared to those of fig. 3 because only a portion of litter input is oxidized to carbon dioxide and other inputs contributed to the stock x_4 (fig. 4). This reduction is particularly noticeable in the shrub site where root respiration is most significant (fig. 4 SHS). In the prairie site model, root respiration was not included, and the past residence time in surface peat of exiting carbon dioxide (fig. 4 LCL) nearly equals future residence time of litterfall in surface peat (fig. 3 LCL).

Flow Partitions

The most complex input environs, E'_4, are associated with a unit carbon dioxide output from x_4 (fig. 4). In the prairie site model (fig. 4 LCL), root respiration input to x_4 is absent, and inputs of carbon dioxide (0.6 %) and DOC (1.3 %) in precipitation are minimal. Thus, 98 % of the carbon dioxide output to the atmosphere originated as litter input to surface peat (x_1), and the importance of internal pathways is enhanced compared to the other two E'_4 environs (fig. 4 SHS & CYP). Forty-eight percent of the unit output came from DOC (x_3) and 53 % was once leached from surface peat and litter (fig. 4 LCL, F'_{31}), emphasizing the importance of that pathway in the marsh site.

In the shrub site, root respiration input to dissolved carbon dioxide (x_4) is a significant source (64 %) of each unit of carbon dioxide output (fig. 4 SHS). The roles of both internal flows and precipiation inputs are reduced, and only 34 % of carbon dioxide lost originated as litterfall.

The E'_4 input environ for the cypress site (fig. 4 CYP) was intermediate. Contributions from precipitation were still minor (2.7 % = 0.0237 + 0.0028). Root respiration into x_4 was only 12 % of inputs leading to carbon dioxide output, but litter inputs to x_1 were 86 %. Internal flows were moderate, although the flows from surface peat directly to carbon dioxide and to carbon dioxide via POC (both 29 % of each unit output) were more important than those in the prairie E'_4.

Table 1.--Ecosystem measures of cycling in Okefenokee aquatic subsystem carbon models compared to those calculated by Finn (1977) for several carbon models and one energy model (Finn 1980).

	TST	IN= OUT	TST_c	TST_s	$\overline{PL}$	$\overline{PL}_s$	$\overline{PL}_s/n$	CI	Original Reference
OKEFENOKEE CARBON									
Prairie	605.01	357.56	0.	605.01	1.69	1.69	0.34	0.0	This study
Shrub	376.30	281.10	0.	376.30	1.34	1.34	0.27	0.0	This study
Cypress	519.79	258.10	0.	519.79	2.01	2.01	0.40	0.0	This study
CARBON MODELS									
Marine Coprophagy					1.45	1.27	0.32	0.12	Cale & Ramsey 1970
Georgia Salt Marsh (14 Holon)					34.8	7.84	0.56	0.78	Wiegert et al., 1975
Georgia Salt Marsh (13 Holon)					2.67	2.55	0.20	0.05	Wiegert et al., 1975
Ross Sea					2.20	2.04	0.17	0.07	Green 1975, modified by Patten & Finn 1979
ENERGY									
Hubbard Brook					1.72	1.72	0.29	0.0	Gosz et al., 1978

Source of Methane Fluxes

Storages and Residence Times

Storages in dissolved methane (x_5) and surface peat (x_1) preceded methane output (y_5) in E'_5 (fig. 5). Again, storages as volatile, dissolved methane (0.0011 to 0.0112 g $C \cdot m^{-2}$) are much less significant than corresponding storages in surface peat (5.0365 to 15.2983 g $C \cdot m^{-2}$).

Past residence times in $\mathcal{H}_5$ of E'_5 (fig. 5) are turnover times. Main diagonal elements of -A" and -A' are turnover rates (T_i^{-1}), and main diagonal elements of $(-A")^{-1}$ and $(-A')^{-1}$ are their inverses, turnover times, T_i. Turnover rates (main diagonal) of A' equal those of A"; therefore, turnover times are identical for the forward and reverse analyses. The relatively low future residence times observed initially for $\mathcal{H}_5$ in $E"_1$ (fig. 3) are mean turnover times; that is, turnover times (fig. 5) weighted by the proportion ($F"_{05}$ = 0.0198, 0.0583, and 0.0289) of unit input z_1 that exits from x_5 in fig. 4 (LCL, SHS, CYP, respectively). Thus, with the exception of main diagonal elements of M" and M', future and past residence times must be referenced to donor and recipient compartments, respectively.

Flow Partitions

The input environs E'_5 for methane flux (fig. 5) emphasize the influence of model quantification decisions on environ analysis results. For the prairie and cypress site models, input of methane from deep peat was set at 50 % (= F'_{51}) of system output methane flux, and the balance came from surface peat and ultimately litterfall (see fig. 5 LCL & CYP). By contrast, total methane transferred from surface peat to surface water methane in the shrub site model was set at 7 g $C \cdot m^{-2} \cdot y^{-1}$ to resemble similar fluxes in the other two sites. That figure is 9 % (= F'_{51}) of system methane losses; the remaining 91 % (= F'_{50}) comes from deep peat (see fig. 5 SHS).

Ecosystem Measures

Ecosystem flow measures are tabulated (table 1) for the three Okefenokee carbon models. Maximum TST occurs in the prairie model, and $\Sigma IN = \Sigma OUT$ are also maximal here. The main diagonal elements of N" and N' are all 1.0; therefore RE_k = 0, TST_c = 0 and TST_s = TST. As a result cycling index (CI) is zero in these models conceptualized with no feedbacks and cycles. Mean path lengths are low, even when expressed as a fraction of the number of compartments. The highest mean path length, 2.01, was for the cypress site model.

DISCUSSION

Environ analyses of these three Okefenokee carbon models emphasize the care required in interpreting results. Environ analysis is simply a manipulation of the original model, and results mirror the original modeling decisions. Environ analysis is, however, a useful bookkeeping tool, especially for large or complex models, that provides summary information. Environ analysis presents the original model from several different viewpoints; thus, new questions may emerge.

Although model quantification is preliminary and the results of environ analysis here are compromised by the quality of the model, a few conclusions can be drawn. Output environ $E"_1$ summarizes the fate of fresh litter entering the system, one of the original questions that prompted this research. Comparing the three sites, decomposition of fresh litter is relatively slow in the shrub site and less than half is actually lost to the atmosphere as carbon dioxide or methane. The relatively greater proportion deposited in deep peat is offset by relatively low litter inputs (Auble 1982).

The importance of Nymphaea odorata in output of carbon gases from surface peat was completely unknown at the time research was initiated. Incorporating recent literature values (Dacey 1981), the prairie site model environ analysis suggests that 38 % of carbon in fresh litter may eventually exit the system via that route. Clearly, this is one important area for future research.

These models contain no feedbacks and no cycles. Although that situation is not unexpected in carbon based decomposition models, certain flows eliminated in conceptualization of figure 1 (Flebbe 1982) could be reevaluated. In particular, comparing results of environ analysis among sites for x_5, dissolved methane, suggests that methane oxidation may be more important than originally thought, especially in the prairie site. Certainly, rates of methane oxidation should be determined in field studies. In addition, the effects of wind speeds on methane and carbon dioxide flux should be evaluated.

Comparing residence turnover times (main diagonals of M" and M') among compartments in each site model, the two dissolved gas compartments are similar and shortest, remaining at most few days (x_4 residence times in fig. 4 for carbon dioxide and x_5 residence times in fig. 5 for methane). Dissolved and particulate organic carbon residence times are similar and intermediate in duration (a month or two), with the exception of POC in the prairie model. There, POC remains an average of 1.4 y. Carbon entering surface peat and litter will remain 20 to 57 years before deposition as deep peat 10 cm below the surface.

Finn (1977) found that CI's for energy models are very low, often CI=0. Carbon model CI's (table 1) were generally lowest of all nutrient models he tested, with the exception of Wiegert et al.'s (1975) 14 compartment Georgia salt marsh model (CI=0.78) where atmospheric carbon dioxide was within the system. Richey et al. (1978) compared CI's of four lake carbon models. Two of the CI's, 0.03 and 0.11, for two oligotrophic lakes were also very low, but two CI's, 0.57 and 0.66, for eutrophic and oligotrophic lakes, respectively, were relatively high. Although the structures of all four four models were similar, magnitudes of many flows were greater in the latter two models. The CI's of zero observed for these Okenenokee carbon models (table 1) are not unusually low for carbon models.

Finn (1980) suggests that ecosystem flow measures are relatively insensitive to changes in model structure and more sensitive to changes in magnitude of flows. For example, in the comparison of four lake models (Richey et al. 1978) discussed above, magnitudes of flows resulted in different CI's for models with similar structure. Of all the flow measures $\overline{PL_s}$ was most robust (Finn 1980). Indeed, these Okefenokee carbon models, identical in structure, support his statement. The ecosystem measure mean straight-through path length, normalized to the number of compartments ($\overline{PL_s}/n$), was sensitive to changes in flow magnitudes in these three models with identical structure (table 1). In fact, the range of $\overline{PL_s}/n$ observed here, 0.13, is greater than Finn (1980) observed for Hubbard Brook calcium models. Greatest path length was observed for the cypress site model; in the E''_1 environ (fig. 3), for example, 61 and 56 % of litter input is lost directly from the system via output from the first compartment in the prairie and shrub site models compared to only 31 % in the cypress site model.

Based on the results of environ analysis, the aquatic system of figure 1 can be further decomposed into two internal subsystems. The first, consisting of the surface peat (x_1), POC (x_2), DOC (x_3), and dissolved carbon dioxide (x_4) compartments plus associated inputs and outputs, corresponds functionally to oxidizing processes. The E'_4 environ (fig. 4) that generates the unit output from dissolved carbon dioxide (x_4) best defines the oxidation pathway. The second internal subsystem consists of dissolved methane (x_5), weakly linked to surface peat (x_1), plus associated inputs and outputs. The E'_5 environ of fig. 5 defines that reducing pathway.

These two pathways are combined in environ E''_1 (fig. 3), and their relative importance can be compared. For the prairie site model, only 2 % of litter input exits through the reducing pathway via output from x_5, and 37 % exits the oxidizing pathway via output from x_4. Corresponding reduced carbon losses in fig. 3 SHS and CYP for the shrub and cypress site models are 6 and 3 %, respectively, and oxidized carbon losses are 38 and 55 %. Thus, we might conclude that reduction of litter to methane is more significant in the shrub site compared to the other two sites and that oxidation of litter is most important in the cypress site. These results are tempered, however, by the finding that an additional 30 % of litter is lost directly from surface peat in the prairie site model as a mixture of carbon dioxide and methane, increasing the importance of both anaerobic and aerobic decomposition in the prairie site model.

In these peat forming systems, anaerobic and aerobic decomposition, combined, do not account for all of litter inputs. A portion of carbon in litter remains in the surface peat (x_1), covered by subsequent years of litter, until 10 cm of surface peat have covered the original carbon input, now considered deposited in deep peat. In the prairie site, it takes 19.56 years for 10 cm of surface peat to accumulate, and of the original litter input, 22.5 % remains at the end of those 19.56 years (fig. 3 LCL). In contrast, litter in the shrub site accumulates to a depth of 10 cm in 57.06 years, and 56 % of the original litter input remains. The cypress site model exhibits intermediate residence time (30.58 y) and portion of input deposited in deep peat (31 %).

The Little Cooter Lake (LCL) prairie site may be characterized, as a result of environ analysis, as the system receiving the greatest input of litter ($348.2\ g\ C{\cdot}m^{-2}{\cdot}y^1$, Flebbe 1982), of which 77.5 % is lost as methane or carbon dioxide to the atmosphere, and the remaining 22.5 % ($78.4\ g\ C{\cdot}m^{-2}$) is deposited in deep peat after a relatively short period of time. In the shrub site, although 56 % of litter input does not decompose, relatively small inputs of litter ($119.9\ g\ C{\cdot}m^{-2}{\cdot}y^{-1}$, Auble 1982 in Flebbe 1982) mean that only $67.3\ g\ C{\cdot}m^{-2}$ become peat deposits. Thus, the mass of peat remaining from litter input is slightly less than than in the prairie site and the time required to accumulate to a depth of 10 cm is greater. The Chase Prairie cypress site occupies an intermediate position in terms of litter input ($226.4\ g\ C{\cdot}m^2{\cdot}y^{-1}$, Auble 1982 in Flebbe 1982), residence time in surface peat (30.58 y) and portion deposited in deep peat (31 % = $71.9\ g\ C{\cdot}m^{-2}{\cdot}y^{-1}$). Although the original models of figure 2 are preliminary and somewhat unrealistic, environ analysis has proven to be a useful step in the iterative modeling process. Several questions suggested by the analysis are currently being investigated and will be incorporated in future versions of the model. Secondly, ecosystem measures of path length developed by Finn (1976, 1977, 1978) are sensitive to changes in magnitude of intrasystem flows in these models and provide a metric for comparison of models.

ACKNOWLEGEMENTS

This paper is based on a portion of my Ph.D. dissertation, Department of Zoology, University of Georgia. My major professor, Dr. Bernard C. Patten carefully reviewed and edited that earlier version, clarifying the environ analysis at several points. M. Craig Barber wrote the analysis software while employed by Ecology Simulations, Inc. This research was funded by NSF grant DEB 7922633 to the University of Georgia for a project titled "Integrated Studies of the Okefenokee Swamp Ecosystem. II. Biogeochemical Aspects of Hierachical Level Structures".

LITERATURE CITED

Auble, G. T. 1982. Biogeochemistry of Okefenokee Swamp: litterfall, litter decomposition and surface water dissolved cation concentrations. Ph.D. dissertation. University of Georgia, Athens, Georgia.

Bosserman, R. W. 1979. The hierarchical integrity of Ultricularia-periphyton microecosystems. Ph.D. dissertation. University of Georgia, Athens, Georgia. 266 pp.

Cale, W. G., and P. R. Ramsey. 1970. Trophic significance of coprophagy by benthic organisms. Unpublished systems ecology course report. University of Georgia, Athens, Georgia.

Cohen, A. D. 1974. Petrology and paleoecology of Holocene plants from the Okefenokee swamp - marsh complex of Georgia. Journal of Sedimentary Petrology 44:716-726.

Dacey, J. W. H. 1981. Pressurized ventilation in the yellow waterlily. Ecology 65:1137-1147.

Finn, J. T. 1976. Measures of ecosystem structure and function derived from analysis of flows. Journal of Theoretical Biology 56:363-380.

Finn, J. T. 1977. Flow analysis: a method for tracing flows through ecosystem models. Ph.D. dissertation. University of Georgia Athens, Georgia. 568 pp.

Finn, J. T. 1978. Cycling index: a general definition for cycling in compartment models. Pages 138-164 in D. C. Adriano and I. L. Brisbin, editors. Environmental Chemistry and Cycling Processes. Department of Energy Symposium Series 45 (CONF-760429). DOE Technical Information Center, Oak Ridge, Tennessee.

Finn, J. T. 1980. Flow analysis of models of the Hubbard Brook ecosystem. Ecology 61:562-571.

Flebbe, P. A. 1982. Biogeochemistry of carbon, nitrogen, and phophorus in the aquatic subsystem of selected Okefenokee Swamp sites. Ph.D. dissertation. University of Georgia, Athens, Georgia. 348 pp.

Gosz, J. R., R. T. Holmes, G. E. Likens, and F. H. Bormann. 1978. The flow of energy in a forest ecosystem. Scientific American 238(3):92-102.

Green, K. A. 1975. Simulation of the pelagic ecosystem of the Ross Sea, Antarctica: a time varying compartmental model. Ph.D. dissertation. Texas A&M University, College Station, Texas. 187 pp.

Koestler, A. 1967. The Ghost in the Machine. Macmillan, New York, New York. 384 pp.

Matis, J. H., and B. C. Patten. 1981. Environ analysis of linear, compartmental systems: the static, time invariant case. Proceedings 42nd Session, International Statistical Institute, Manilla, The Philippines, December 4-14, 1979. In press.

Patten, B. C., R. W. Bosserman, J. T. Finn, and W. G. Cale. 1976. Propagation of cause in ecosystems. Pages 457-579 in B. C. Patten, editor. Systems Analysis and Simulation in Ecology, volume IV. Academic Press, New York, New York.

Patten, B. C. and J. T. Finn. 1979. Systems approach to continental shelf ecosystems. Pages 183-212 in E. Halfon, editor. Theoretical Systems Ecology. Academic Press, New York, New York.

Richey, J. E., and 10 others. 1978. Carbon flow in four lake ecosystems: a structural approach. Science 202:1183-1186.

Wiegert, R. G., R. R. Christian, J. L. Gallagher, J. R. Hall, R. D. H. Jones, and R. L. Wetzel. 1975. A preliminary ecosystem model of coastal Georgia Spartina marsh. Pages 583-601 in L. E. Cronin, editor. Estuarine Research, volume I, Chemistry, Biology, and the Estuarine System. Academic Press, New York, New York.

CLASSIFICATION, MODELLING AND MANAGEMENT OF WETLANDS - A CASE STUDY IN WESTERN KENTUCKY[1]

William J. Mitsch, Jan R. Taylor, and Kimberly B. Benson[2]

Abstract.--A wetland classification and several models are presented for a research project in the western coal field of Kentucky. Wetlands are classified according to their hydrologic regime and the degree to which they are impacted by factors such as acid mine drainage. A wetland inventory is developed for a specific area that may have significant impact due to coal mining and drainage projects. Preliminary models, which emphasize primary productivity and nutrient cycling, are described for this region.

INTRODUCTION

The proper use of wetland classifications and wetland models can contribute to our understanding about how these ecosystems function and how to manage them in an ecologically-sound manner. Wetland classification schemes and wetland models share the common goals of 1) describing wetlands as homogeneous units, 2) providing a systematic aid to decision-makers, and 3) providing uniform concepts and terminology. In addition, classifications further aid decision-makers by providing units for inventory and mapping (Cowardin et al., 1979). Models, on the other hand, can lead to computer simulations which predict ecosystem behavior and potential impacts of future management policies. Despite the importance of both classification and modelling for wetlands, the two techniques have remained separate exercises and are often done by ecologists who have vastly different orientations. Exceptions to this statement are seen in the ecosystem models of south Florida presented by Lugo et al. (1971), Carter et al. (1973), and Odum and Brown (1975).

Wetland Classifications

Classification of wetlands has traditionally involved categories based on type of dominant vegetation, types of soils or land forms, or type and degree of flooding. Some of the earliest wetland classifications were of northern peatlands as developed by Davis (1907), Kulczynski (1949), Moore and Bellamy (1974) and others. An inventory and classification scheme of wetlands in the United States was published by the U.S. Fish and Wildlife Service in the mid 1950s (Shaw and Fredine, 1956). Wetlands were described in that classification as 20 types under four major categories of: (1) inland freshwater areas; (2) inland saline areas; (3) coastal freshwater areas; and (4) coastal saline areas. Golet and Larson (1974) and Stewart and Kantrud (1971) improved or adapted this classification to specific regions of the United States.

In 1974, the United States Fish and Wildlife Service began a new inventory of the nation's wetlands. Because this inventory was designed to facilitate management of wetlands for many uses, not only for waterfowl habitat, a new classification scheme was developed and published after several years of review in Cowardin et al. (1979) as "Classification of Wetlands and Deepwater Habitats of the United States." As described in that publication:

> This classification, to be used in a new inventory of wetlands and deepwater habitats of the United States, is intended to describe ecological taxa, arrange them in a system useful to uniformity of concepts and terms. Wetlands are defined by plants (hydrophytes), soils (hydric soils), and frequency of flooding. Ecologically related areas of deep water, traditionally not considered wetlands, are included in the classification as deepwater habitats.

This classification is based on a hierarchical approach analogous to taxonomic classifications used to identify plant and animal species.

Wetlands have also been classified according to their major forcing functions in addition to their structure. Odum et al. (1974) included several coastal wetlands in their functional classification of "Coastal Ecological Systems of the United States." This classification successfully integrated coastal ecosystems according to their forcing functions such as solar energy and tidal influence and according to stresses such as wave energy, hypersalinity and cold temperatures. Some classifications of wetlands have since been developed with a similar approach. Wharton et al. (1976) developed a classification of forested swamps in Florida which is based on hydrologic forcing functions. A similar wetland classification, which considered wetland types along a hydrodynamic gradient was developed by Gosselink and Turner (1978). The

---

[1]Paper presented at the Third International Conference on State-of-the-Art in Ecological Modelling, Colorado State University, May 24-28, 1982.

[2]William J. Mitsch is Associate Professor of Systems Science and Associate of Biology. Jan R. Taylor and Kimberly B. Benson are Graduate Research Assistants in the Department of Biology. All are at the University of Louisville, Louisville, Kentucky 40292, U.S.A.

hydrologic conditions in this scheme range from ombrotrophic bogs (raised-convex) which are fed by precipitation and capillary action to lentic (riverine) systems that receive a seasonal pulse of flood water.

Wetland Models

The diversity of wetland types and wetland classifications is matched by a wide diversity of types of models used to describe and simulate these wetlands. This is not surprising when we view wetlands as having characteristics of both aquatic and terrestrial systems. The major types of wetland models, discussed in more detail in another paper (Mitsch et al., 1982b), are given in Table 1 and are briefly described below:

Table 1.--Types of Freshwater Wetland Models and Examples in Literature. See Text for Description (from Mitsch et al., 1982b).

Model Type	Example of Model	
	Simulation	Conceptual
1. Energy/nutrient ecosystem	Mitsch (1976)	Kuenzler et al. (1980)
2. Hydrology		
a) ecosystem	Huff & Young (1980)	
b) regional	Brown (1978); Littlejohn (1977)	Rykiel (1977)
c) hydrodynamic transport	Hopkinson & Day (1980)	-
3. Spatial ecosystem	Parker & Kadlec (1974)	-
4. Tree growth	Phipps (1979)	Mitsch and Ewel (1979)
5. Process	Miller et al. (1976)	-
6. Causal	-	Bedford (1979)
7. Regional energy	-	Young & Odum (1974)

1. Energy/nutrient ecosystem models - Energy, nutrients, and/or other materials are shown to pass through or cycle among biotic and abiotic components and to exchange with the surroundings; these models are generally non-spatial aggregated models with feedbacks and interactions among components.

2. Hydrology models - There are several types of models of water flow through wetlands:

    a) Ecosystem - These models describe a water budget for a homogeneous individual wetland and do not consider uplands or other bodies of water as part of the model.

    b) Regional - These models present overall gains and losses for a large scale region or watershed that includes wetland areas. These models differ from ecosystem hydrology models in that water storages in adjacent groundwater and surface water are considered as part of the model.

    c) Hydrodynamic transport - These models, popular with engineers for streamflow and runoff calculations, can be applied to describe wetland hydrology and pollutant transport over short periods and large areas.

3. Spatial ecosystem models - These models combine the attributes of ecosystem models (1) with spatial transport models (2c). The entire ecosystem is broken into discrete blocks or nodes with each block or node having the characteristics of an ecosystem model. This type of model was applied to a peatland in Michigan by Parker and Kadlec (1974).

4. Tree growth models -Some forest models, applicable to forested wetlands, simulate the growth of individual trees. These models can be used to describe overall swamp productivity. Conceptual models (e.g. Mitsch and Ewel, 1979; Connor and Day, 1981) have described tree growth and productivity for different hydrologic conditions. Simulation models often consider a forest of individual trees and predict their growth in basal area or biomass (e.g. Phipps, 1979).

5. Process models - These models emphasize the fine-tuning of individual processes, such as photosynthesis or respiration, rather than emphasizing interactions between compartments as in ecosystem models. For example, the factors that affect the process (e.g. sunlight, CO_2, nutrients) are important input variables and the process itself (e.g. photosynthesis) may be the desired output.

6. Causal models - Flow diagrams display causal loops rather than the flow of energy and materials. The flows shown in these models generally depict interactions between components (e.g. water having positive effect on plants) without showing accompanying nutrient and energy flows. One example presents the relationship between temperature and its impact on annuals and perennials in a wetland in Wisconsin (Bedford, 1979). These models are generally conceptual in nature.

7. Regional energy models - Models display energy, material, and information exchanges between wetlands and entire economic regions in which the wetlands are found. Exchanges of money are often shown in the economic sectors. These models, such as a regional analysis developed for a major wetland basin in Louisiana (Young and Odum, 1974), are used to show relationships between natural systems and man's economic systems.

THE WESTERN KENTUCKY WETLAND STUDY

The work presented here is part of an ongoing research project on wetlands in the Western Kentucky Coal Field (Mitsch et al., 1982a). The research has as its objectives: 1) to identify and map wetlands in the Western Kentucky Coal Field, 2) to evaluate the major biotic and abiotic factors that affect those wetlands, and 3) to develop strategies for the proper management of those wetlands. Wetland classifications and wetland models have both been incorporated as tools to accomplish these research objectives.

The Western Kentucky Coal Field (Figure 1) is located in the Shawnee Hills Section of the Interior Low Plateaus Physiographic Province. Most of the region is undulating to hilly with wide, flat, silt-

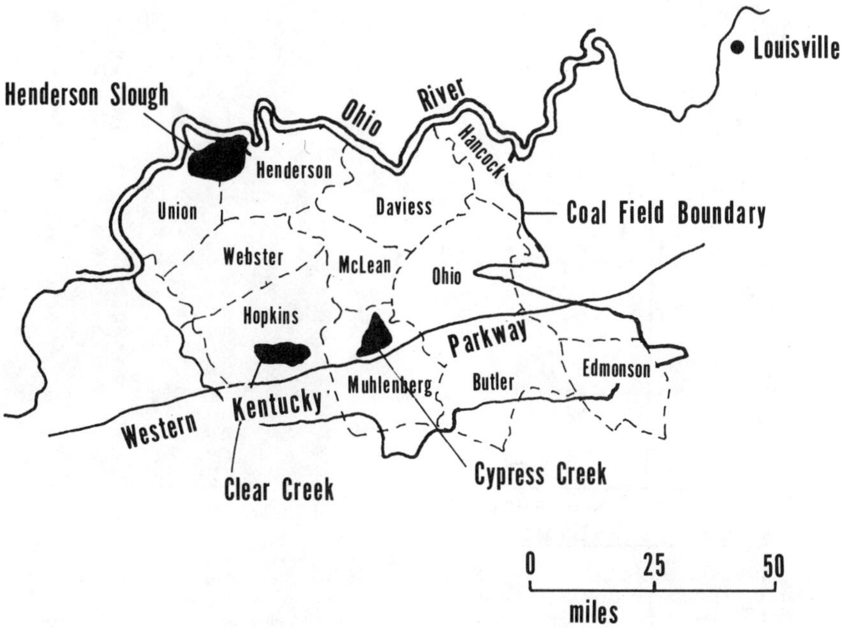

Figure 1.--Western Kentucky Coal Field showing location of intensive-study wetland sites.

filled alluvial valleys along the Green, Rough, Pond, Tradewater, Barren, and Ohio Rivers (Harker et al., 1980). The relatively flat topography supports a wide variety of wetlands. We have tentatively identified about 40,000 hectares of wetlands in the Western Kentucky Coal Field. Many of these wetlands are over 2,000 hectares in size.

Coal mining, particularly coal surface mining, has affected many of the wetlands in the region. Acid drainage and sedimentation from mining have adversely impacted the water quality, hydroperiod, and vegetation of some wetlands. Other wetlands have been formed as a direct result of mining activity and thus display characteristics of developing ecosystems. Several wetlands in the region are seemingly not influenced by mining at all. Logging, channelization, conversion to agriculture, beaver dams, and highway construction are other major influences on wetlands in the region.

Classification Scheme

A classification for wetlands in western Kentucky was developed with the following criteria in mind: 1) simplicity, 2) flexibility, and 3) inclusion of ecological stress. The classification scheme is summarized in Tables 2 and 3 and Figure 2. The major types of wetlands (palustrine systems) include seven types frequently found in our study sites (Table 2). This simple system is compatible with the classification scheme developed by Cowardin et al. (1979) but does not include as much detail. Environmental modifiers (Figure 2) describe, in the form of a two-by-two matrix, the flooding regime and environmental impacts

Table 2.--Major classification used for wetlands in Western Kentucky.

Wetland Type	Hydroperiod[1]
Persistent Emergent	Intermittently exposed
Typha Marsh	(slowly flowing water)
Phragmites Marsh	
Mixed Emergent Marsh	
Forested Wetland	
Deep Cypress Swamp	Permanently flooded
Shallow Cypress Swamp	Semi-permanently flooded
Frequently Flooded Bottomland Hardwood	Seasonally flooded
Infrequently Flooded Bottomland Hardwood	Temporarily flooded

[1]Water regime modifers of Cowardin et al. (1979)

Table 3.--Impact modifiers used for wetland classification in western Kentucky.

1. Coal Mine Drainage
2. Oil Well
3. Logging/Harvesting
4. Agricultural Runoff
5. Urban
6. Channelization or Ponding
7. Other

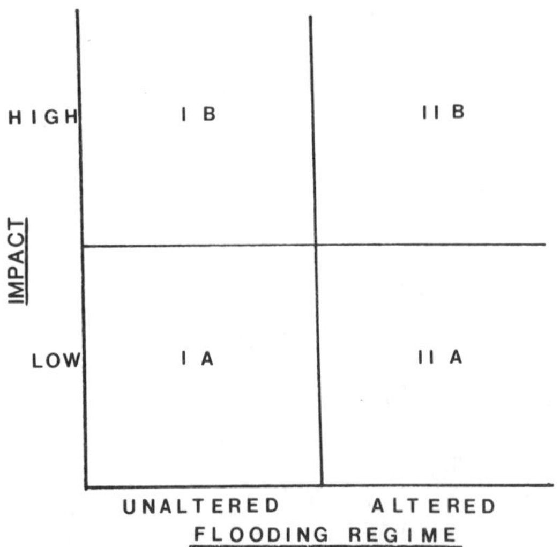

Figure 2.--Impact and flooding regime matrix used in wetland classification of western Kentucky.

Shallow Cypress Swamp IIA-1

Figure 3.--Example of wetland classification for western Kentucky. The picture is of a shallow cypress swamp with altered hydroperiod and low impact due to acid mine drainage.

on the wetlands. Along one axis, the flooding conditions are defined as unaltered (I) or altered (II). The other axis describes environmental impacts in general on the wetland with A indicating low impact and B indicating significant impact. Thus an environmental modifier of II-B indicates a wetland with an altered hydroperiod and significant environmental impact. A modifier of I-A, on the other hand, indicates a wetland in near-natural conditions. The impact can be further defined with a numbering scheme (1 through 7) to indicate coal mine drainage, oil wells, logging, agricultural runoff, urban influence, or other impacts (Table 3). An example of a typical wetland classification is shown in Figure 3.

A CASE STUDY - CYPRESS CREEK WETLANDS

An example of the use of wetland classification, mapping, and models is provided by our Cypress Creek study area in Muhlenberg County. A wetland and land-use map was developed for the upper Cypress Creek Watershed, a region of 127 square kilometers that contains both wetlands and coal mining activity (Figure 4). Information on the land-use and distribution of wetlands is included in Table 4. Wetlands are primarily riparian and are mostly found along 15 kilometers of Cypress and Little Cypress Creeks. A major feature is the presence of about 2.9 km^2 of both shallow and deep cypress (*Taxodium*) swamps. The cypress stands are among the most extensive and best developed in the interior of western Kentucky and represent some of the eastern-most cypress in the Mississippi-Ohio Basin. The dominant wetlands in the study area are bottomland hardwood forests, which comprise about 4.9 km^2. Major species in these wetlands include *Acer rubrum*, *Platanus occidentalis*, *Salix nigra*, and *Betula nigra*. About 3.1 km^2 of marshes, including one large *Typha* marsh, are also found in the study area.

Table 4.--Wetland area and land use for upper Cypress Creek Watershed shown in Figure 4.

Wetlands or Land Use	km^2
Wetlands	
Deep Cypress Swamp	0.5
Shallow Cypress Swamp	2.4
Typha Marsh	1.0
Mixed Emergent Marsh	2.1
BLH-Frequently Flooded	4.7
BLH-Infrequently Flooded	0.2
Total Wetlands	10.9
Agricultural	30.3
Residential/Commercial	8.8
Upland Forests	17.6
Active or Abandoned Surface Mines	31.4
Reclaimed Surface Mines	27.7
TOTAL WATERSHED AREA	126.7

There is considerable coal surface mining activity and signs of past mining activity in the watershed. A total of 59 km^2 or 47% of our watershed has active,

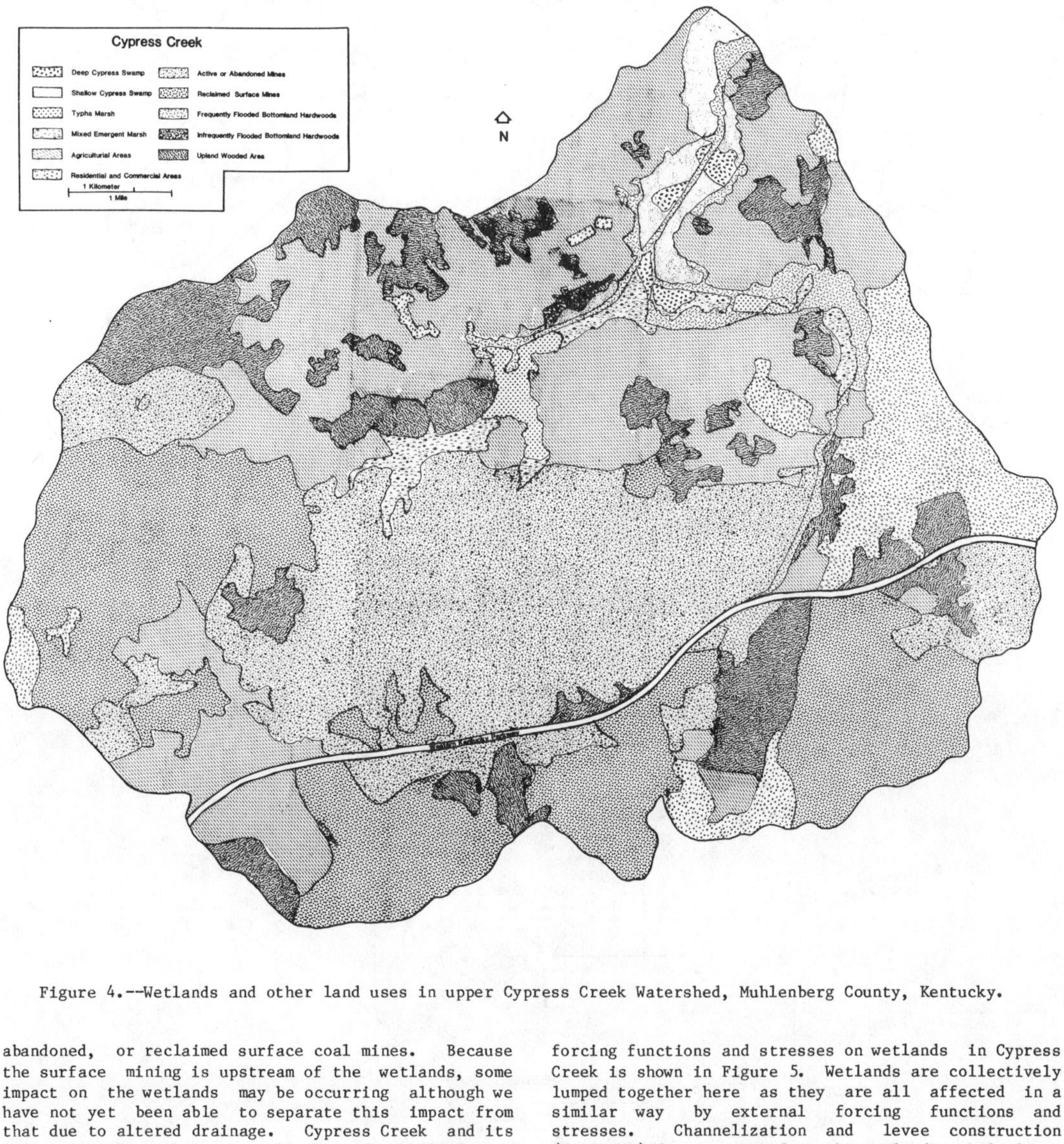

Figure 4.--Wetlands and other land uses in upper Cypress Creek Watershed, Muhlenberg County, Kentucky.

abandoned, or reclaimed surface coal mines. Because the surface mining is upstream of the wetlands, some impact on the wetlands may be occurring although we have not yet been able to separate this impact from that due to altered drainage. Cypress Creek and its tributaries have been channelized and spoil levees have been built along the streams. This has caused alteration of the hydroperiod of several of the wetlands and may be the most significant impact.

Several types of models have been constructed to describe the Cypress Creek wetlands. A previous paper (Mitsch et al., 1981) described regional, watershed and ecosystem models applicable to the western Kentucky coal field and to Cypress Creek. These models were hierarchical in nature in that one system was a sub-set of the "larger" system. In our second generation of models, we have chosen to emphasize models at the ecosystem level that are amenable to computer simulation. An overall model showing the important forcing functions and stresses on wetlands in Cypress Creek is shown in Figure 5. Wetlands are collectively lumped together here as they are all affected in a similar way by external forcing functions and stresses. Channelization and levee construction (drainage) divert water from the wetlands while levee construction often causes water to pond in wetlands, extending the duration of flooding. Surface mining can cause direct impact on wetlands due to the chemical constituents of acid mine drainage; it can indirectly affect wetlands through increased sediment loads in the streams and through hydroperiod alteration.

More detailed models of forested wetlands are shown in Figures 6 and 7. Figure 6 is a model which emphasizes primary productivity in a forested wetland. The vegetation is divided into canopy trees, emergent vegetation, and submerged vegetation. This division is somewhat operational due to the different

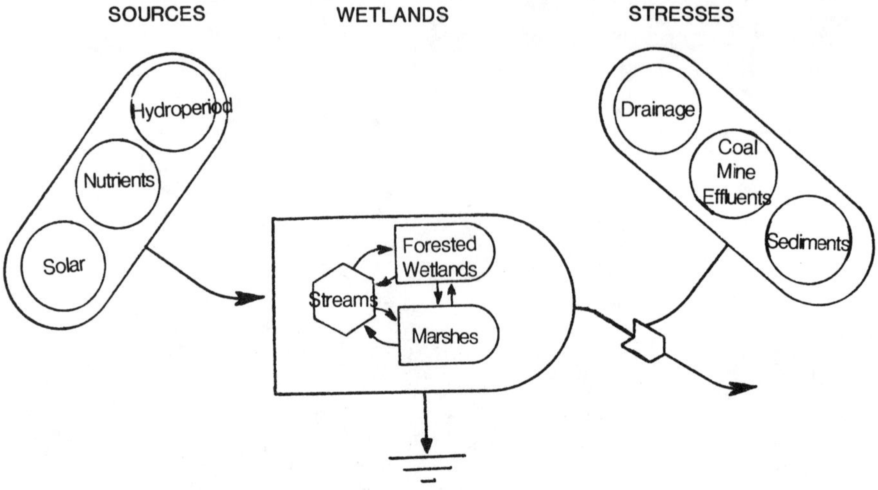

Figure 5.--Generalized model of wetland sources and stresses for Cypress Creek Watershed.

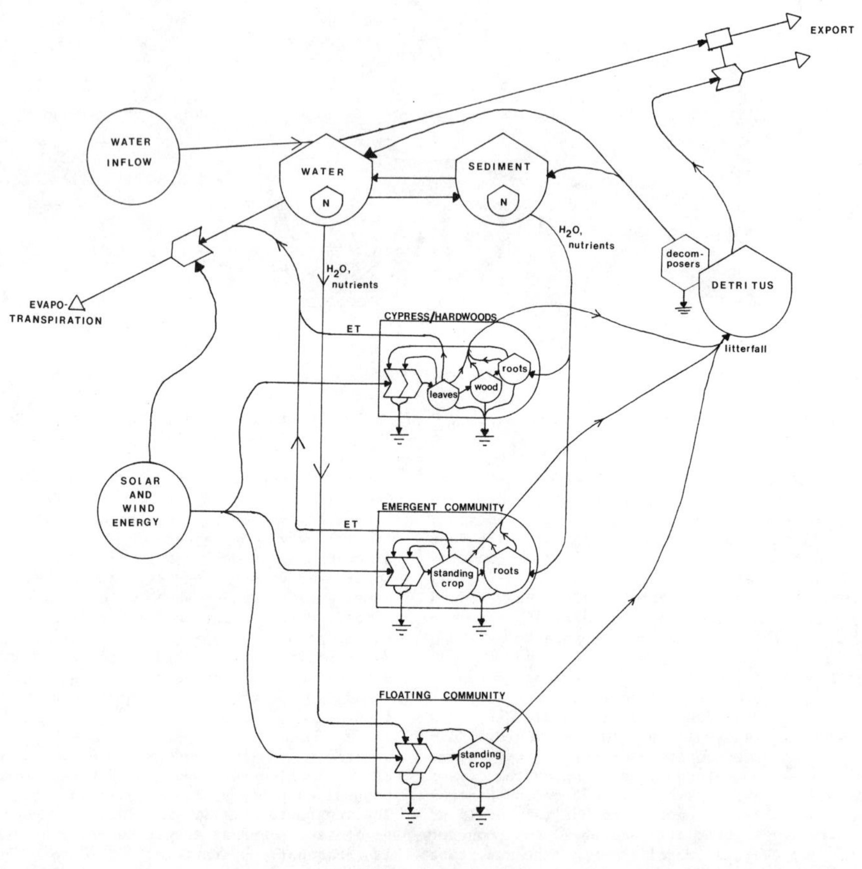

Figure 6.--General primary productivity model for forested wetlands in western Kentucky.

measurement techniques necessary for each strata but it is also functional because each component would be affected in a different way by mine drainage and altered hydrology. The dynamics of nutrients are shown in greater detail in Figure 7 which is serving as the basis of a study on the importance of phosphorus to the productivity of selected forested wetlands. In that model, phosphorus is shown to be in dissolved inorganic, particulate inorganic, or organic forms in the water column and in the sediments. The model will ultimately be quantified with measurement of fluxes into and out of the wetlands, and with transfer rates affecting intrasystem phosphorus cycling. Our hypothesis, stated in Figure 5, is that, in addition to solar energy, the hydroperiod, the availability of nutrients, and man-induced stresses are the major environmental factors that determine the structure and function of these wetlands. Wetlands classifications and models, in addition to field measurements, will assist us in testing this hypothesis.

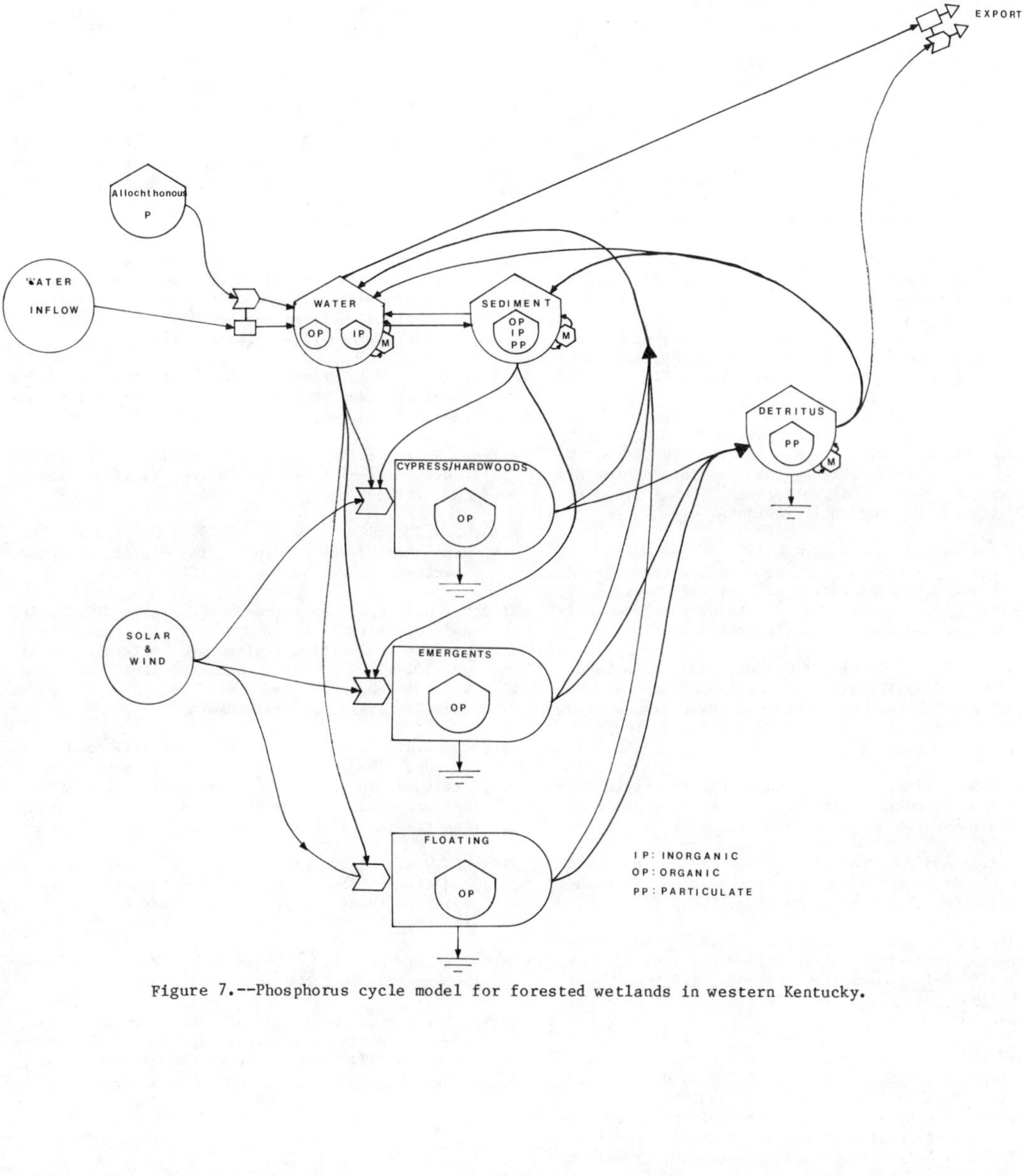

Figure 7.--Phosphorus cycle model for forested wetlands in western Kentucky.

SUMMARY

Our study of wetlands in western Kentucky has developed a classification and mapping routine that is compatible with that of the U.S. Fish and Wildlife's classification (Cowardin et al., 1979), yet is adapted to specific hydrologic and environmental conditions there. We have undertaken the task of finding, classifying, and mapping all of the wetlands in the Western Coal Field of Kentucky. Models have also been developed to guide field work at several sites. This field work has begun at several study locations, including Cypress Creek, to determine hydroperiods, primary productivity, and nutrient cycling. This information, coupled with anticipated computer simulations of specific sites within the study region will contribute to our understanding and thoughtful management of wetlands.

LITERATURE CITED

Bedford, B. 1979. Changes in the structure and function of a wetland subject to sustained perturbation: a conceptual model. pp. 44-59. In Coal-fired steam plants: Human and Environmental Exposure to Air and Water Pollutants First-year Progress Report. Water Resources Center, University of Wisconsin, Madison, Wis.

Brown, S.L. 1978. A comparison of cypress ecosystems in the landscape of Florida. Ph.D. Dissertation, University of Florida, Gainesville, Florida, 569 pp.

Carter, M.R., L.A. Burns, T.R. Cavinder, K.R. Dugger, P.L. Fore, D.B. Hicks, H.L. Revells, T.W. Schmidt. 1973. Ecosystem analysis of the big Cypress Swamp and estuaries. U.S.E.P.A. EPA 904/9-74-002, Region IV, Atlanta.

Conner, W. and J.W. Day, Jr. 1981. The ecology of forested wetlands in the southeastern United States. In Wetlands: Ecology and Management, Gopal, B. et al. (eds.). International Scientific Pub., Jaipur, India (in press).

Cowardin, L.M., V. Carter, F.C. Golet and E.T. LaRoe. 1979. Classification of wetlands and deepwater habitats of the United States. U.S. Fish & Wildlife Service Pub. FWS/OBS-79/31, Washington, D.C. 103 p.

Davis, C.A. 1907. Peat, essays on its origin uses and distribution in Michigan. Rept. State Board Geol. Surv. Mich. Fur. 1906, pp. 95-395.

Golet, F.C. and J.S. Larson. 1974. Classification of wetlands in the glaciated Northeast. U.S. Fish & Wildlife Publ. 116, Washington, D.C. 56 p.

Gosselink, J.G. and R.E. Turner. 1978. The role of hydrology in freshwater wetland ecosystems. pp. 63-78. In Freshwater wetlands, R.E. Good, D.F. Whigham and R.L. Simpson (eds.). Academic Press, New York.

Harker, D.F., R.R. Hannan, M.L. Warren, L.R. Phillippe, K.E. Camburn, R.S. Caldwell, S.M. Call, G.J. Fallo and D. Van Norman. 1980. Western Kentucky coal field: preliminary investigations of natural features and cultural resources. Kentucky Nature Preserves Commission, Frankfort, Ky.

Hopkinson, C.S., Jr. and J.W. Day, Jr. 1980b. Modeling hydrology and eutrophication in a Louisiana swamp forest ecosystem. Environmental Management 4:325-335.

Huff, D.D. and H.L. Young. 1980. The effect of a marsh on runoff I. a water-budget model. Journal of Environmental Quality 9:633-640.

Kuenzler, E.J., P.J. Molholland, L.A. Yarbro and L.A. Smock. 1980. Distributions and budgets of carbon phosphorus iron and manganese in a floodplain swamp ecosystem. Water Resources Research Report No. 157, Water Resources Research Institute of the University of North Carolina, Chapel Hill, N.C. 234 p.

Kulczynski, S. 1949. Peat bogs of Polesia. Mem. Acad. Sci. Cracovie B, pp. 1-356.

Littlejohn, C.B. 1977. An analysis of the role of natural wetlands in regional water management. pp. 451-476. In Ecosystem modeling in theory and practice, C.A.S. Hall and J.W. Day, Jr. (eds.). John Wiley and Sons, New York, N.Y.

Lugo, A.E., S.C. Snedaker, S. Bayley and H.T. Odum. 1971. Models for planning and research for the South Florida Environmental Study. Final Report to National Park Service by University of Florida, Gainseville, Florida. 123 p.

Miller, P.C., W.A. Stoner and L.L. Tieszen. 1976. A model of stand photosynthesis for the wet meadow tundra at Barrow, Alaska. Ecology 57:411-430.

Mitsch, W.J. 1976. Ecosystem modeling of waterhyacinth management in Lake Alice, Florida. Ecol. Mod. 2:69-89.

Mitsch, W.J. and K.C. Ewel. 1979. Comparative biomass and growth of cypress in Florida wetlands. American Midland Naturalist 101:417-426.

Mitsch, W.J. R,W, Bosserman, P.L. Hill, J.R. Taylor and F. Smith. 1981. Models of wetlands amid surface coal mining regions of western Kentucky. pp. 103-113. In Energy and Ecological Modelling, W.J. Mitsch, R.W. Bosserman and J.M. Klopatek (eds.). Elsevier, Amsterdam.

Mitsch, W.J., R.W. Bosserman, P.L. Hill and J.R. Taylor. 1982a. Classification and management of wetlands in the western Kentucky coal field. Kentucky Water Resources Research Institute Report No. 131, Lexington, Ky. 78 p.

Mitsch, W.J., J.W. Day, Jr., J.R. Taylor and C. Madden. 1982b. Models of North American freshwater wetlands. Int. J. Ecol. and Environ. Sci. (in press).

Moore, P.D. and D.J. Bellamy. 1974. Peatlands. Springer-Verlag, New York, N.Y. 221 p.

Odum, H.T., B.J. Copeland and E.A. McMahan (eds.). 1974. Coastal ecological systems of the United States 4 vol. Conservation Foundation, Washington, D.C.

Odum, H.T. and M. Brown (eds.). 1975. Carrying capacity for man and nature in south Florida. Final Report to U.S. Dept. of Interior and Florida Division of State Planning, Center for Wetlands, University of Florida, Gainesville, Florida. 886 p.

Parker, P.E. and R.H. Kadlec. 1974. A dynamic ecosystem simulator. A.I.Ch.E. 78th National Meeting, Paper No. 8c, Salt Lake City, Utah. 40 pp.

Phipps, R.L. 1979. Simulation of wetlands forest vegetation dynamics. Ecological Modelling 7:257-288.

Rykiel, E.J., Jr. 1977. Toward simulation and systems analysis of nutrient cycling in the Okefenokee Swamp, Georgia. Technical Completion Report. USDI/OWRT Project No. A-060-GA. University of Georgia, Athens, Ga. 139 pp.

Shaw, S.P. and C.G. Fredine. 1956. Wetlands of the United States. U.S. Fish & Wildlife Circular 39, Washington, D.C. 67 pp.

Stewart, R.E. and H.A. Kantrud. 1971. Classification of natural ponds and lakes in the glaciated prairie region. U.S. Fish & Wildlife Publ. 92, Washington, D.C. 57 p.

Wharton, C.H., H.T. Odum, K.C. Ewel _et al_. 1976. Forested wetlands of Florida - their management and use. Final Report to Florida Division of State Planning, Center for Wetlands, University of Florida, Gainesville. 421 p.

Young, D.L. and H.T. Odum. 1974. Evaluation of regional models for the alternatives in management of the Atchafalaya Basin. Report to U.S. Dept. of Int. Bur. of Sport Fisheries and Wildlife. Center for Wetlands, University of Florida, Gainesville, Fla. 34 p.

A PRELIMINARY INPUT-OUTPUT MODEL OF SALT MARSHES IN THE MISSISSIPPI DELTAIC PLAIN REGION[1]

Scott G. Leibowitz and Robert Costanza[2]

Abstract--The salt marsh habitat of the Mississippi Deltaic Plain Region (MDPR) of coastal Louisiana and Mississippi has been the object of intense study during the last two decades. A great amount of useful information has been generated concerning salt marsh structure and function, but this information has not been synthesized into a useful management model for local, state, and federal decision-makers. A study was undertaken to collect this information and present it in a comprehensible and ecologically meaningful manner. A detailed linear flow model of MDPR salt marshes is presented, making use of these data. The model includes 13 processes which produce 20 commodities, including biomass, energy, water, and nutrients. An input-output based approach which can handle multicommodity flow matrices and joint products was used to calculate the direct and indirect energy costs of the 20 ecological products included in the salt marsh model. This method allows one to quantify the interconnectedness between the 20 commodities and to estimate the direct and indirect impacts on the salt marsh system of small perturbations around the average conditions. The usefulness of this information to coastal managers is discussed, along with limitations due to incomplete and imprecise data.

INTRODUCTION

The Mississippi Deltaic Plain Region (MDPR) consists of more than 3.4 million hectares in coastal Louisiana and Mississippi (Wicker 1980). This is the portion of the Gulf coast that is or has been directly affected by the Mississippi River and its sediments. The region is an important area for commercial fishing, trapping, and agriculture, as well as oil and gas production (Larson et al. 1980). The high biological productivity of the MDPR contributes to fish and shellfish catches of over 500 million kg annually (Maruggi and Hartl 1981). The MDPR has been strongly affected by human activities, including the leveeing of the Mississippi River for flood protection as well as extensive canalization (Craig et al. 1979, Gagliano 1973). As a result of these and other factors such as subsidence, the MDPR experiences land losses of up to 39 sq mi each year (Craig et al. 1979, Gagliano et al. 1981).

The salt marsh is the third largest habitat in the MDPR, consisting of more than 180 thousand hectares in 1978 (Wicker 1980). This habitat has been the subject of intense research, including studies of primary productivity (Hopkinson et al. 1978b, Kirby and Gosselink 1976, White et al. 1978), consumers (Alexander 1976, Rodriguez-Ortega and Day 1978), sediment chemistry (DeLaune et al. 1976, DeLaune and Patrick 1980), and community dynamics (Day et al. 1973, Hopkinson et

(1) Paper presented at the Third International Conference on State-of-the-Art in Ecological Modelling, Colorado State University, May 24-28, 1982.

(2) Research Associate and Assistant Professor, respectively, Coastal Ecology Lab, Louisiana State University, Baton Rouge, LA 70808.

al. 1978a). Productivity of MDPR salt marshes is high, and contributes to estuarine productivity (Turner 1979). There is concern that loss of salt marsh will therefore have adverse effects on estuarine productivity. This is a subject of great interest to coastal managers.

As part of a U. S. Fish and Wildlife Service contract, the Center for Wetland Resources at Louisiana State University has undertaken a detailed study of MDPR salt marshes. Pertinent ecological data have been compiled to quantitatively characterize and model this habitat. Direct and indirect energy costs of marsh commodities have been calculated as an index of importance. In this paper, preliminary findings are presented, and their implications for management are discussed.

METHODS

Data Collection

A literature review was conducted to obtain data on flows and standing stocks of ecological materials within MDPR salt marshes. In this study, the salt marsh is defined to include only those areas containing vegetation; mudflats and streams are not included. Twenty ecological commodities and thirteen processes were investigated, including nutrients, sediments, biomass, water, and energy. The overall conceptual model of the marsh as well as trophic relationships were based on earlier work by Day et al. (1973). The data were synthesized from four broad categories: 1) Information collected over large enough portions of the MDPR and for long enough periods of time so as to be statistically representative of average conditions in the area. Very little data are of this category. 2) Data taken from studies of a particular site in the MDPR for a limited period of time, using statistical methods. Although this information is from sites in the MDPR, it does not necessarily represent the entire area. 3) Studies which were conducted in MDPR salt marshes but were not statistically designed, or information gathered from salt marshes in other parts of the country. Also included here are data from habitats similar to the salt marsh, such as the brackish marsh. 4) Imprecise data gained from guesstimates of other systems or by using "standard" ecological efficiencies or conversions. As a result of these qualitative differences, the salt marsh model contains data ranging from excellent to poor. The model can be upgraded as new data become available.

For each of the 13 processes, a mass balance budget was constructed. By assuming steady-state conditions, missing values could be derived by difference. A particular commodity was explicitly included in the budget only if it was consumed or produced in that form. For example, inorganic nitrogen is included as an input to marsh plants since it enters the plant as nitrogen. The nitrogen contained in plant litter was not explicitly considered as an output, however, since it is implicitly included with the litter. Once all of the process flows were derived, the data were entered into input-output form (see below). Detailed documentation of methods and flow calculations can be found in Costanza et al. (1982).

Joint Products I-C Model

The total amount of energy required both directly and indirectly to produce a commodity has been suggested as a measure of value (Costanza 1980, Odum 1977, Odum and Odum 1976). This measure is referred to as embodied energy. In this paper an input-output (I-O) technique is used to calculate the embodied energies of salt marsh commodities. Since this technique is described in detail elsewhere in this volume (Fruci et al. 1982), it will only be briefly summarized here.

In any process, the total energy input is equal to the embodied energy of the outputs. Energy inputs include direct inputs such as solar energy and fossil fuels, plus the embodied energies of any other inputs to the process (fig. 1). This is calculated as follows:

$$E_j + \Sigma e_i X_{ij} = e_j X_j \qquad (1)$$

where,

E_j = direct energy inputs to process j
X_{ij} = the commodity flow from process i to j
X_j = the total commodity output from process j
e_j = the embodied energy per unit of X

This is equivalent to the following:

$$E_j = e_j X_j - \Sigma e_i X_{ij} \qquad (2)$$

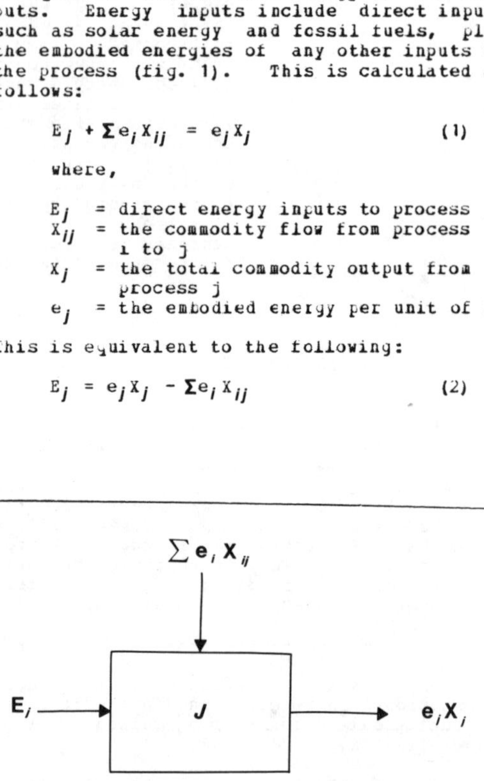

Figure 1.--Energy balance for process j with no joint products.

which can be translated into matrix notation as:

$$E = e(\hat{X} - X) \qquad (3)$$

In this notation, E is a vector of direct energy inputs, $\hat{X}$ is a diagonalized matrix of output flows, X is a commodity transactions matrix, and e is a vector of energy intensities (Costanza 1979). This equation can now be solved for e:

$$e = E(\hat{X} - X)^{-1} \qquad (4)$$

Costanza (1980), Hannon (1973), and others have used this technique to study economic and ecologic systems. In order to use this I-O procedure, it is necessary that each process produce only one type of commodity. In real ecological systems, however, a process often produces several different commodities, such as biomass, feces, respiratory by-products, etc. (fig. 2). These are referred to as joint products. For a process with a 10 kcal total energy input, 10 kcal is embodied in the output if there are no joint products. But the traditional I-O technique has no provision for distributing energy inputs to multiple commodity outputs. Joint products can be handled by using a "commodity-by-process" format instead of the more usual "commodity-by-commodity" format (Ritz 1979). A technique was developed to address this problem.

The technique used in this study distributes energy to an output according to the relative importance of that process to the total production of that commodity. For example, consider a process which is the sole source of two commodities. Since no other process produces these commodities, that process is of equal value to both. Thus the energy would be divided equally. If, however, one of the commodities was produced in great amounts elsewhere in the system, the process would not be as important to the total production of that commodity. Thus more of the energy would be allocated to the other commodity. This technique borrows from economic concepts by assuming that the energy invested in a commodity is a function of the scarcity of that commodity. A mathematical description of this technique can be found in

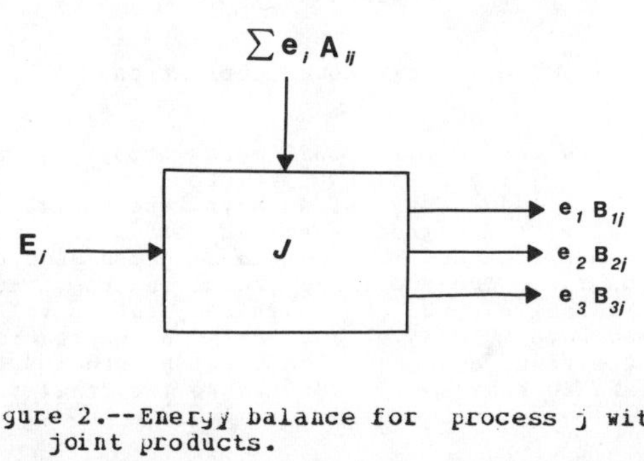

Figure 2.--Energy balance for process j with joint products.

Fruci et al. (1982).

This allocation technique allows the creation of a commodity-by-commodity transactions matrix which can then be solved by standard I-O techniques. This is then manipulated to produce a commodity-by-commodity embodied energy matrix. When this matrix is converted back to a commodity-by-process matrix, however, the total embodied energy entering each process does not always equal the total embodied energy coming out. Thus a technique is being developed to use the allocation scheme as an initial guess which is then iteratively improved by reducing the error between the embodied energy in and the embodied energy out of each process.

RESULTS

Primary production

Total aboveground net primary production of marsh macrophytes (*Spartina alterniflora*, *Spartina patens*, *Juncus roemerianus*, and *Distichlis spicata*) were estimated as 2053 g dry wt/sq m/yr (Table 1). This figure represents the average for these four plants, weighted by their percent cover. Of this value, 1223 is produced by the dominant macrophyte *S. alterniflora*. It was estimated that belowground production contributes another 2310 g, giving a total net primary production of 4363 g dry wt/sq m/yr. Belowground production was derived by considering anaerobic respiration along with subsidence rates, and is not a very reliable estimate.

A conversion factor of 0.59 was used in this study to calculate respiration from net primary production. This gives a value of 2574 g dry wt/sq m/yr. Although this is a

Table 1.--Net primary production, respiration, and gross production of salt marsh macrophytes in the MDPR (g dry wt/sq m/yr).

	Spartina alterniflora	Others (1)	Total
N.P.P.			
Aboveground	1223	830	2053
Belowground	1377	933	2310
Total	2600	1763	4363
Respiration	1534	1040	2574
G.P.P.	4134	2803	6937

(1) *Spartina patens*, *Juncus roemerianus*, and *Distichlis spicata*.

rather low value, it is based on work conducted within the MDPR (Gosselink et al. 1977), and was therefore considered more reliable than studies from marshes outside of this area.

Combining these net production and respiration values gives a gross primary production of 6937 g dry wt/sq m/yr. Another 527 g of material is produced each year by marsh microflora, such as epiphytic algae. This gives a total gross production of 7464 g dry wt/sq m/yr.

Community Carbon Budget

The flow of organic matter through the salt marsh is shown in figure 3. Of the total net primary production, only 79.9 g dry wt/sq m/yr (1.7%) is consumed directly by consumers. More important sources of food for marsh heterotrophs are litter, soil organic matter, and surface water organic matter. These contribute 390, 240, and 400 g dry wt/sq m/yr to consumers, respectively. Imports of fish from surrounding waters contribute another 15 g dry wt/sq m/yr.

Microconsumers that feed on litter consume 550 g dry wt/sq m/yr, leaving an estimated 1430 g input of litter into the soil. From there, 790 g leave the system due to sedimentation, since the lower boundary of the system only goes as far as the rooting zone.

Another loss of organic matter from the system is from harvesting of marsh animals. Harvesting of furbearers, waterfowl, and shellfish by man removes an estimated 0.5 g dry wt/sq m/yr from the marsh. The major loss of organic matter, however, is from detrital export, due to tidal action. It is estimated here that nearly 1500 g of material leave the system through this pathway. This represents 30% of the total net production of the marsh.

I-O Results

The input-output matrix used for calculating embodied energies is shown in Table 2. Energy intensities of the 20 commodities are listed in Table 3. The two largest energy intensities are for phosphorus and nitrogen, with values in the thousands. This is a reflection of the scarcity of these nutrients. Carbon and other nutrients, which are more ubiquitous, have values of less than 50 kcal/g.

Macrophytes have energy intensity values of 50-70 kcal/g. Because it is more abundant, S. alterniflora has the lower of these values. The energy intensity for microflora is double the macrophyte values.

Consumers such as crabs, Littorina, insects, and mammals all have similar energy intensities, in the range of 120-140 kcal/g. Birds and mussels have somewhat lower values. In this analysis, fish are found to have an energy intensity of zero (Table 3). This is due to the fact that fish are not produced within the system, and thus none of the energy can be allocated to them. This problem can be overcome by using a different habitat model to calculate this energy intensity.

The flows of organic matter in figure 3 can be converted to flows of embodied energy by multiplying by the appropriate energy intensity value. This is shown in figure 4. It should be noted that a commodity with a low energy intensity can have a greater flow of embodied energy than a commodity with a higher energy intensity. For example, the flow of embodied energy from S. alterniflora is 136,240 kcal, compared to 120,413 for other macrophytes. The larger number occurs because S. alterniflora makes a greater contribution to total net production. This more than compensates for the lower energy intensity.

While the contribution of organic matter from litter and surface water to consumers was roughly equal (fig. 3) litter contributes a much greater amount of embodied energy. Similarly, microconsumers receive three times more organic matter from soil than from litter, and yet litter contributes nearly 10,000 more kcal to this process than does the soil.

Harvests of salt marsh animals by man removes 80 kcal of embodied energy from the marsh. A much larger loss is due to detrital export. This is calculated here as exceeding 25,000 kcal/sq m/yr. This represents an input of very high quality materials to the surrounding estuarine areas. The marsh loses an additional 15,000 kcal to the deep sediments each year.

DISCUSSION

Ecological Characterization

In the earlier salt marsh study by Day et al. (1973), macrophyte gross productivities of 8418 g dry wt/sq m/yr were reported, along with net productivities of 1518. The large difference between this gross production value and the value found in this study is mainly due to the lower respiration value used here. Day et al. used a respiration value from a Georgia salt marsh studied by Teal and Kanwisher (1961). We feel that the

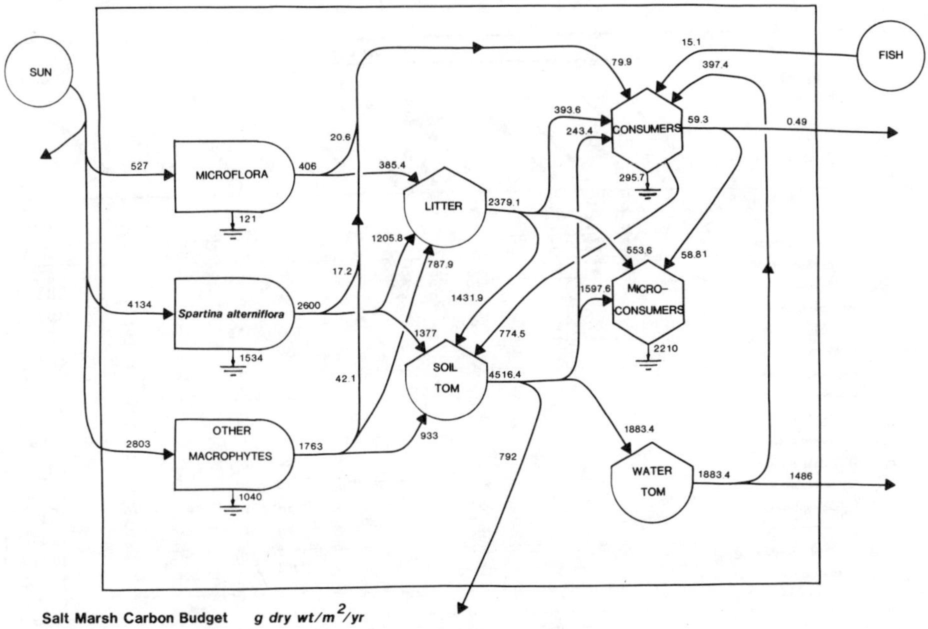

Figure 3.-- Salt marsh carbon budget. All flows in g dry wt/m^2/yr.

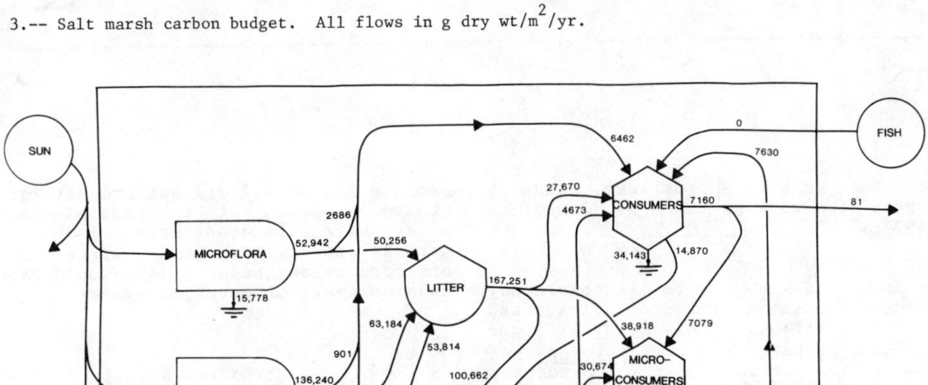

Figure 4.-- Salt marsh embodied energy in organic matter flow diagram. All flows in kcal solar/m^2/yr.

Table 2.--Salt marsh commodity-by-process input-output table.

HABITAT 17 SALT MARSH

COMMODITIES		1 SURFACE WATER	2 SHALLOW SOIL/SEDIMENT	3 MICROFLORA	4 *S. alterniflora*	5 OTHER MACROPHYTES	6 LITTER	7 MICROCONSUMERS	8 CRABS	9 MUSSELS	10 *Littorina*	11 INSECTS	12 BIRDS	13 MAMMALS	NET OUTPUT	TOTAL OUTPUT	UNITS
		PROCESSES															
NITROGEN	1			5.1	42.2	26.3										73.6	g N/m²/yr
			3.5	1.2	15.6	9.8		26	1.0	8.2	8.2	0.02	0.01	0.07	1.0	73.6	g N/m²/yr
PHOSPHORUS	2			1.6	6.4	9.0										16.0	g P/m²/yr
			6.4	0.4	2.0	3.3		3.5	8.2 E-2	0.06	0.06	1.5 E-3	7.3 E-4	6.3 E-3		16.0	g P/m²/yr
CARBON	3			164.4	1083.5	1008.1									1930.2	4166.2	g C/m²/yr
				42.4	393.5	374.4		1004.5	15.3	23.2	74.8	1.5	6.7	12.9	2227.0	4166.2	g C/m²/yr
OTHER NUTRIENTS	4			336.7	3060.9	1767.5										5144.1	g/m²/yr
				77.1	1132.1	662.1		1177.0	17.2	19.0	81.0	1.5	5.1	16.3	1963.4	5144.1	g/m²/yr
TOTAL ORGANIC MATTER	5	1063.4	774.5					1567.5	199	367.4	44.4				2278	7174.3	g dry wt/m²/yr
		1063.4	4616.4						144.1	324.4	274.3	3.9	8.4	19.4		7174.3	g dry wt/m²/yr
SHALLOW SEDIMENTS	6	1936	2626						466		202				1936	7194	g dry wt/m²/yr
		1936	2626						466		202				1936	7194	g dry wt/m²/yr
WATER	7	1.51	0.46	0.09	0.26	0.17									1.51	3.92	10⁶ g/m²/yr
		1.51	0.46	0.09	0.26	0.17									1.51	3.92	10⁶ g/m²/yr
MICROFLORA BIOMASS	8						366.4				4.4	1.3	0.8	14.1	121	527.0	g dry wt/m²/yr
				527												527.0	g dry wt/m²/yr
S. alterniflora BIOMASS	9		1377				1206.8				13.3	3.9			1534	4134.0	g dry wt/m²/yr
					4134											4134.0	g dry wt/m²/yr
OTHER MACROPHYTIC BIOMASS	10		933				757.9				9.0	2.6	1.5	29.0	1040	2903.0	g dry wt/m²/yr
						2903										2903.0	g dry wt/m²/yr
LITTER BIOMASS	11		1431.9					653.6			293.6					2379.1	g dry wt/m²/yr
							2379.1									2379.1	g dry wt/m²/yr
MICROCONSUMER BIOMASS	12														2210	2210	g dry wt/m²/yr
								2210								2210	g dry wt/m²/yr
CRAB BIOMASS	13							16.90					2.29	1.34	34.04	54.6	g dry wt/m²/yr
									54.6							54.6	g dry wt/m²/yr
MUSSEL BIOMASS	14							18.12					2.30	1.39	51.1	73.0	g dry wt/m²/yr
										73.0						73.0	g dry wt/m²/yr
Littorina BIOMASS	15							21.77					2.86	1.67	164.6	190.9	g dry wt/m²/yr
											190.9					190.9	g dry wt/m²/yr
INSECT BIOMASS	16							0.51					0.06	0.05	3.3	3.9	g dry wt/m²/yr
												3.9				3.9	g dry wt/m²/yr
BIRD BIOMASS	17							6.8 E-2						0.01	14.8	14.86	g dry wt/m²/yr
													14.86			14.86	g dry wt/m²/yr
MAMMAL BIOMASS	18							1.41						0.14	28.36	29.9	g dry wt/m²/yr
														29.9		29.9	g dry wt/m²/yr
FISH BIOMASS	19												13.4	1.66		15.1	g dry wt/m²/yr
															15.1	15.1	g dry wt/m²/yr
HEAT	20																
				44,066	364,372	240,277		10,829	74	236	494	16	66	156	661,511	661,511 / 661,511	kcal/m²/yr
SUNLIGHT				0.5	3.7	2.5											10⁶ kcal/m²/yr

value used here is more appropriate, since it was measured at a site in the MDPR (Gosselink et al. 1977).

Net production is greater in this study mainly because of the addition of belowground biomass. In addition, this study included other macrophytes besides S. alterniflora and one of these (S. patens) has a greater net production. Finally, the study by Day et al. used data from only one site, whereas this study examined several different study areas. It is hoped that this is more representative of MDPR marshes as a whole.

In the study by Day et al., 764 g dry wt/sq m/yr of detritus were exported from the marsh, accounting for 50% of net primary production. In this study, calculations show an export twice as great. In both of these studies export is calculated by difference, and thus the larger export reported here (1490 g) is due to the larger value of net production. Another possible factor contributing to this is the inclusion of belowground production. Because Day et al. did not account for belowground biomass, it is possible that they might have allocated some of their aboveground production for respiration which occurred belowground. This would lead to an underestimate of detrital export.

Embodied Energy

One of the uses of embodied energies is to point out the relative importance of commodities in the system. Commodities such as nitrogen and phosphorus were found to have high energy intensities. Thus small amounts of nutrients can have large effects on the salt marsh. This is consistent with our understanding that these scarce nutrients have important control functions on marsh dynamics.

Embodied energy calculations allow us to see that material flows which appear unimportant, such as the flow of litter to

Table 3.--Embodied energy intensity values (kcal solar/g) for ecological commodities in MDPR salt marshes.

Commodity	Value	Commodity	Value
Nitrogen	1622.4	Litter	70.3
Phosphorus	10,562.1	Microconsumers	12.2
Carbon	29.6	Crabs	170.3
Other nutrients	31.8	Mussel	74.4
TOM	19.2	*Littorina*	119.2
Shallow sediments	17.1	Insects	141.1
Water	0.4	Birds	66.9
Microflora	130.4	Mammals	125.2
S. alterniflora	52.4	Fish (1)	0.0
Other macrophytes	68.3	Heat (2)	0.4

(1) Because fish are not produced in the salt marsh, an energy intensity cannot be calculated for them by considering the salt marsh alone. Their embodied energy intensity must be calculated using the habitat model in which they are produced.
(2) Embodied energy intensity as kcal/kcal.

microconsumers, may in fact have a much greater value than can be ascertained through examining the physical flow alone. As an example, the calculations presented here suggest that altering the flow of litter would affect consumers more than a similar change in the flow of organic matter from the surface water. Although the physical flows are similar, the embodied energy of the litter is much higher. Thus a unit decrease in litter would cause a greater loss in embodied energy than a similar decrease in surface water organic matter. Similarly, microflora have a much greater effect on litter than would be expected by examining the physical flows alone.

Embodied energy can also be used to assess the effects that small perturbations on the system would have on commodities of interest to humans. For example, calculations here show that man harvests 81 kcal/sq m/yr from the salt marsh in the form of shellfish, furbearers, and waterfowl. In addition, 28,500 kcal of detritus leave the salt marsh and enter adjacent estuaries. This valuable source of organic matter contributes to the high commercial yields of these waters. Although it was not attempted here, it is possible to manipulate the marsh matrix to estimate direct and indirect effects on harvest yields, detrital export, and all other commodities of interest. Thus one could use embodied energy to assess the potential impacts of human activities on target areas. Theoretically, it would be possible to calculate the energy benefit of the manipulation and compare it with the energy cost. If the costs exceeded the benefits, the manipulation would be unwise. This could be a valuable tool to coastal managers.

CONCLUSIONS

The results of the salt marsh ecological characterization presented in this paper are not final, and will be revised in a forthcoming report (Costanza et al. 1982). In addition, there are many sources of error in the data that have been used for this study. The three most poorly understood parameters - belowground production, community respiration, and detrital export - are crucial for a proper understanding of this habitat. Our understanding of the salt marsh is fuzzy at best without further field work in these areas. In addition, the technique for calculating embodied energies in systems with joint products needs to be further developed. An iterative improvement technique is needed, along with a way of dealing with exogenous inputs (such as fish).

In light of these limitations, the question arises as to the value of undertaking a study of this kind. Can meaningful information and conclusions be gained by such an analysis, or must we wait, perhaps forever, for better data? It is our opinion that such an attempt at piecing together the salt marsh puzzle is not fruitless. By using the format described in this paper, a large amount of data have been presented in a physically consistent framework which makes the information more comprehensible and accessible. This format allows one to clearly see ecological relations, as well as to gain a better appreciation of the holes in our understanding.

While the data and methods used here will doubtless improve with time, they are an important beginning at building a framework for analyzing large, complex systems. Much work still needs to be done to test the basic hypotheses underlying the theory of embodied energy. Yet as this study demonstrates, useful information can be gained even now through such an analysis.

ACKNOWLEDGEMENTS

This research was funded by the U. S. Dept. of Interior, Bureau of Land Management, U. S. Fish and Wildlife Service, National Coastal Ecosystems Team, Contract No. 14-16-0009-80-073, "Mississippi Deltaic Plain Region Ecosystem Models and Narrative Report," Martha W. Young, Project Officer. This is contribution No. LSU-CEL-82-14 from the Coastal Ecology Laboratory, Louisiana State University, Baton Rouge, LA. Special thanks to J. W. Day, R. E. Turner, and J. G. Gosselink of CEL for their useful comments, and to K. Benkert of FWS for her reviews of the model.

LITERATURE CITED

Alexander, S. K. 1976. Relationship of macrophyte detritus to the salt marsh periwinkle, <u>Littorina irrorata</u> Say. Ph.D. Dissertation, Louisiana State University, Baton Rouge.

Costanza, R. 1979. Embodied energy basis for economic-ecologic systems. Ph.D. Dissertation, University of Florida, Gainesville.

Costanza, R. 1980. Embodied energy and economic valuation. Science 210: 1219-1224.

Costanza, R., C. Neill, S. G. Leibowitz, J. Fruci, and L. M. Bahr. 1982. Mississippi Deltaic Plain Region ecological characterization: An ecological modeling study. National Coastal Ecosystems Team, U. S. Fish and Wildlife Service. In preparation.

Craig, N. J., R. E. Turner, and J. W. Day, Jr. 1979. Land loss in coastal Louisiana. p. 227-254, in Proceedings of the third coastal marsh and estuary management symposium. Louisiana State University Division of Continuing Education, Baton Rouge.

Day, J. W., Jr., W. G. Smith, P. R. Wagner, and W. C. Stowe. 1973. Community structure and carbon budget of a salt marsh and shallow bay estuarine system in Louisiana. LSU-SG-72-04, Center for Wetland Resources, Louisiana State University, Baton Rouge.

DeLaune, R. D., and W. H. Patrick, Jr. 1980. Rate of sedimentation and its role in nutrient cycling in a Louisiana salt marsh. p. 401-412, in Estuarine and wetland processes with emphasis on modeling. Vol II. Plenum Press, N. Y.

DeLaune, R. D., W. H. Patrick, and J. M. Brannon. 1976. Nutrient transformations in Louisiana salt marsh soils. Sea Grant Publ. No. LSU-T-76-009, Center for Wetland Resources, Louisiana State University, Baton Rouge.

Fruci, J. R., R. Costanza, and S. G. Leibowitz. 1982. Quantifying the interdependence between material and energy flows in ecosystems. Paper presented at the Third International Conference on the State-of-the-Art in Ecological Modelling, Colorado State University, May 24-28, 1982.

Gagliano, S. M. 1973. Canals, dredging and land reclamation in Louisiana coastal zone. Hydrologic and geologic studies of coastal Louisiana, Report #14. Center for Wetland Resources, Lousiana State University, Baton Rouge.

Gagliano, S. M., K. J. Meyer-Arendt, and K. M. Wicker. 1981. Land loss in the Mississippi River Deltaic Plain. Transactions of the Gulf Coast Assoc. of Geol. Soc. 31: 295-300.

Gosselink, J. G., C. S. Hopkinson, Jr., and R. T. Parrondo. 1977. Common marsh plant species of the gulf coast area. Volume I: Productivity. Appendix C. Technical Report D-77-44, U. S. Army Engineer Waterways Experiment Station, Vicksburg, Miss.

Hannon, B. 1973. The structure of ecosystems. Journal of Theoretical Biology 41: 535-546.

Hopkinson, C. S., J. W. Day, Jr., and B. T. Gael. 1978a. Respiration studies in a Louisiana salt marsh. An. Centro Cienc. del Mar y Limnol. Univ. Nal. Anton. Mexico 5: 225-238.

Hopkinson, C. S., J. G. Gossilink, and R. T. Parrondo. 1978b. Aboveground production of seven marsh plant species in coastal Louisiana. Ecology 59: 760-769.

Kirby, C. J., and J. G. Gosselink. 1976. Primary production in a Louisiana Gulf Coast <u>Spartina alterniflora</u> marsh. Ecology 57: 1052-1059.

Larson, D. K., D. W. Davis, R. A. Detro, P. B. Dumond, E. B. Liebow, R. M. Motschall, D. P. Sorensen, and W. S. Guidroz. 1980. Mississippi Deltaic Plain Region ecological characterization: A socioeconomic study. Volume I. FWS/OBS-79/05, National Coastal Ecosystems Team, U. S. Fish and Wildlife Service.

Maruggi, V. and S. D. Hartl (eds.). 1981. Statistical abstract of Louisiana. University of New Orleans' Division of Business and Economic Research and the Louisiana State Planning Office. 413 pp.

Odum, H. T. 1977. Energy, value, and money. p. 174-196. In Ecosystem modeling in theory and practice. John Wiley and Sons, New York.

Odum, H. T., and E. C. Odum. 1976. Energy basis for man and nature. McGraw-Hill Book Company, New York.

Ritz, P. M. 1979. The input-output structure of the U. S. economy, 1972. Survey of Current Business, Feb. 1979, p. 34-72

Rodriguez-Ortega, D., and J. W. Day, Jr. 1978. An ecological study of <u>Modiolus demissus</u> granossissimus Sowerby in Louisiana (USA). An. Centro Cienc. del Mar y Limnol. Univ. Nal. Anton. Mexico 5: 215-224.

Teal, J. M., and J. Kanwisher. 1961. Gas exchange in a Georgia salt marsh. Limnol. Oceanogr. 6: 388-399.

Turner, R. E. 1979. Louisiana's coastal fisheries and changing environmental conditions. p. 363-370. In Proceedings of the third coastal marsh and estuary management symposium. Louisiana State University Division of Continuing Education, Baton Rouge.

White, D. A., T. E. Weiss, T. M. Trapani, and L. B. Thien. 1978. Productivity and decomposition of the dominant salt marsh plants in Louisiana. Ecology 59: 751-759.

Wicker, K. M. 1980. The Mississippi Deltaic Plain Region habitat mapping study. Data tapes. Office of Biological Services, U. S. Fish and Wildlife Service.

MODELING THE RIO GRANDE RIVER SYSTEM IN NEW MEXICO FOR ECONOMIC AND SOCIAL TRADE-OFFS: AN OVERVIEW OF A CURRENT STUDY[1]

Tim J. Ward[2], Richard A. Cole, Frank A. Ward, Paul R. Turner, George V. Sabol

Abstract.--Water in New Mexico must be applied to the most beneficial uses. Determination of the most beneficial uses, however, requires physical, social, economic, and political decisions. This paper presents the status of an interdisciplinary, multiyear study that will yield a methodology for aiding such decisions. Hypothetical, applications are discussed.

INTRODUCTION

The Rio Grande system affects three states and two nations. Beginning in the high mountains of Southern Colorado the Rio Grande flows south and east before emptying into the Gulf of Mexico. Along its course, the river becomes a life sustaining source for agricultural, domestic, and industrial users. One additional benefit of the river that is often overlooked is recreation in terms of boating and fishing.

In New Mexico, the Rio Grande drains about 20 percent of the state or 24,000 square miles. The three major cities of Santa Fe, Albuquerque, and Las Cruces are on or near it, and most New Mexicans are affected by the 500 mile long river. Although 75 to 85 percent of the water used in the Rio Grande valley of New Mexico is for agriculture, another important use of the water is for recreational pursuits.

There are six major reservoirs in the system. Heron, El Vado, and Abiquiu are located on the Rio Chama, the major tributary to the Rio Grande in the state. The other three reservoirs, all on the mainstem, are Cochiti, Elephant Butte, and Caballo (fig. 1). Each of these reservoirs has a definite recreational value as four of them are located near the northern population center, and Elephant Butte-Caballo are located near the southern population center of Las Cruces, El Paso, and Ciudad-Juarez. The reservoirs were originally built for water supply, flood control, or sediment control while recreational benefits were a fortunate by product. Reservoir operation also creates conditions on the interconnecting river reaches between reservoirs which are exploited by knowledgeable recreationists. For example, between El Vado and Abiquiu reservoirs is some of the better fishing or white water rafting, depending on flow conditions, in the state. Similarly, the reach below Elephant Butte and above Caballo has a good potential for becoming a viable instream fishery. But all these recreational benefits rely on the abundance or scarcity of water and its disposition for other uses.

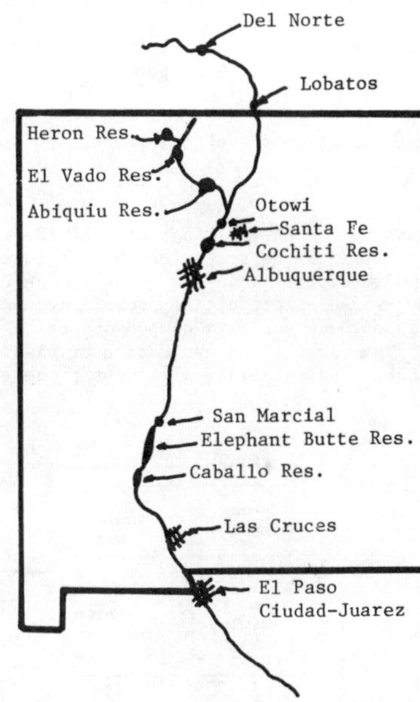

Figure 1.--Map of important locations on the Rio Grande in and near New Mexico.

The major goal of this study is develop a methodology by which the recreational benefits of water in the reservoirs and interconnecting streams can be

[1]Paper presented at the Third International Conference on State-of-the-Art in Ecological Modeling, Colorado State University, May 24-28, 1982.

[2]The authors are respectively, Associate Professor of Civil and Geological Engineering, Associate Professor of Fishery and Wildlife Science, Assistant Professor of Agricultural Economics, Assistant Professor of Fishery and Wildlife Science, and Associate Professor of Civil Engineering, New Mexico State University, Las Cruces, New Mexico, USA.

legitimately compared to benefits from other sectors of the economy. To reach this goal requires a multi-year, interdisciplinary approach.

MODELING CONCEPT AND COMPONENTS

General

The general methodology can be subdivided into three primary components (fig. 2) of hydrologic-hydraulic, biologic-ecologic, and social-economic. Each component feeds into and is fed back from other components. Because of the complexity of the system, as shown in figure 3, it was decided at the outset to base the methodology in the NMSU computer. Weekly meetings of the study team fosters an interaction that permits model develop to proceed in a parallel but separate manner for each of the components. Researchers working on each of the components know what that component must provide to the others as well as what it needs to function properly. For example, the hydrologic-hydraulic component submodel provides water level information to both the biologic-ecologic submodel and the social-economic submodel to be used for different purposes in each. All three submodels are linked by an executive program that coordinates the information flow and updating between the components.

Hydrologic-Hydraulic Submodel

General

This submodel is the driving force of the methodology in that it sets the stage for the subsequent submodels. However, it is only a means to the end, that is the desired output of the overall methodology. The modeling approach for this component has taken two directions. The first is an evaluation of historic data, with subsequent statistical and/or process models. The second is application of existing models such as HEC-3 or HEC-5Q to the system. In the first approach, all available reservoir and streamflow data for the system have been obtained and entered into a master file. To facilitate processing and analyses, the system was divided into four parts (Sabol and Gordon, 1980) of:

1) Upper river from Lobatos to Otowi
2) Rio Chama from Heron Reservoir to confluence with the Rio Grande
3) Otowi through Cochiti Reservoir, and
4) San Marcial through Caballo Reservoir

A fifth part between Cochiti Reservoir and San Marcial is being modeled by the U.S. Army Corps of Engineers for another project. The results of that study shall be incorporated into the methodology as necessary. Each part is presently being modeled separately to meet other short term objectives of the project, but all the parts will eventually be mathematically interconnected with the model.

In the second approach, the data base is used to calibrate existing reservoir models, and then use those models with proposed system changes. One drawback in applying existing models is that the time frames involved in the biologic-ecologic and social-economic submodels are on the order of one week to one month, making other models based on shorter time periods inefficient or overly exact.

Submodel performance

When completed the submodel will translate a vast amount of data into information that can be used in the subsequent components. This information will include water levels in the reservoirs at selected times, nutrient and sediment inputs, water temperature, reservoir water quality, and flows in the

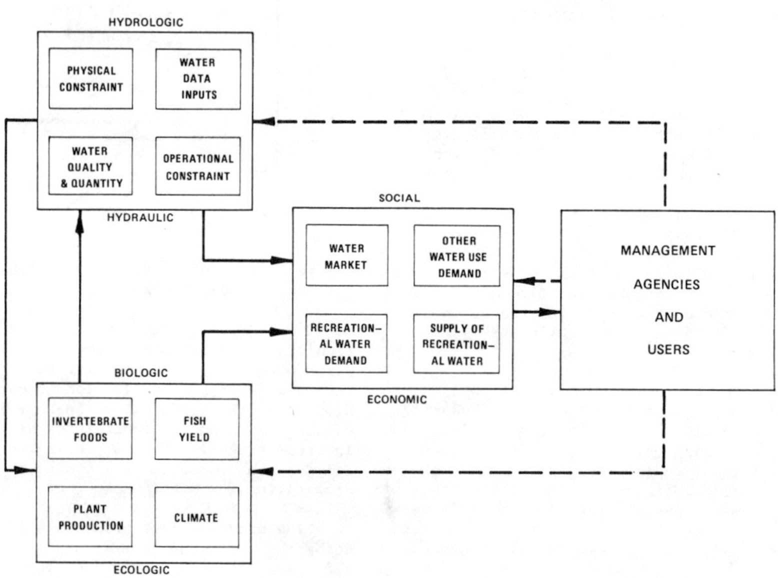

Figure 2.--Schematic of interlinked submodel components comprising the decision aiding methodology.

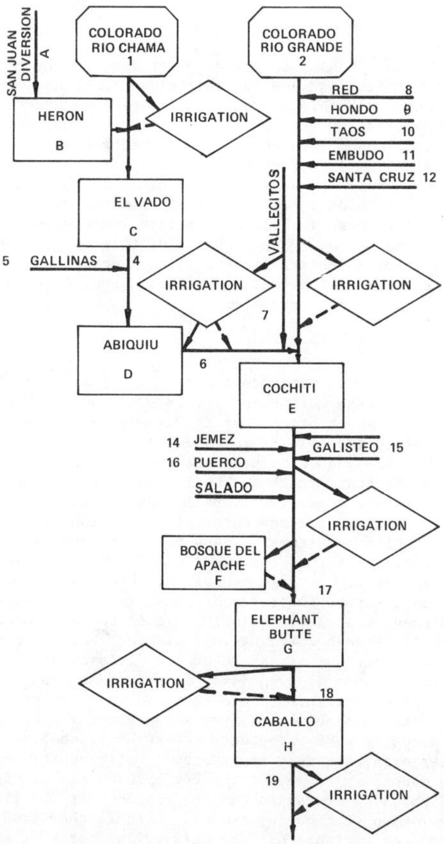

Figure 3.--Schematic of important hydrologic linkages in the Rio Grande system of New Mexico.

interconnecting river reaches. Input data will consist of current system variables such as water levels in the reservoir, historic or stochastically generated inflow records and physical constraints in the system.

Selection of a current or arbitrary starting point will permit the comparison of the same sequence of inflow records and system constraints for different initial conditions. For example, what would happen if a proposed scenario were to start with an almost full or an almost empty reservoir system? How long would the system take to recover from a low level of storage for the proposed scenario?

Different inflow sequences can also have a profound effect on system response. Continued development in the San Luis Valley of Colorado has decreased the flow in the Rio Grande near Lobatos beginning in about 1925 (fig. 4). This break has been verified by a double mass balance between an upstream station--near Del Norte, Colorado, above the San Luis Valley--and the Lobatos station (fig. 5). Water management techniques may reverse this trend, and more recent data seems to indicate such a change. Therefore, a question arises as to how should the future be depicted, as one based onlong term historic trends or based on short term upswings in the data? Fortunately, with the developed submodel structure, any and all conditions can be examined. Another inflow that will change in future years is the diversion of San Juan River water from the Colorado River Basin into the Rio Chama Drainage. Current diversions into Heron Reservoir via the Azotea tunnel are approximately 100,000 acre feet (AF) per year with an additional 125,000 AF possible. Approximately 40,000 AF per year belongs to the City of Albuquerque shich can now store this water in Abiquiu Reservoir. Increased storage could affect the recreation potential at Abiquiu and at other sites including Elephant Butte Reservoir.

Imposed on top of the initial conditions and the inflow records are physical and institutional constraints on water distribution. The Bureau of Reclamation must operate its reservoirs for water supply at the demand of its customer, the

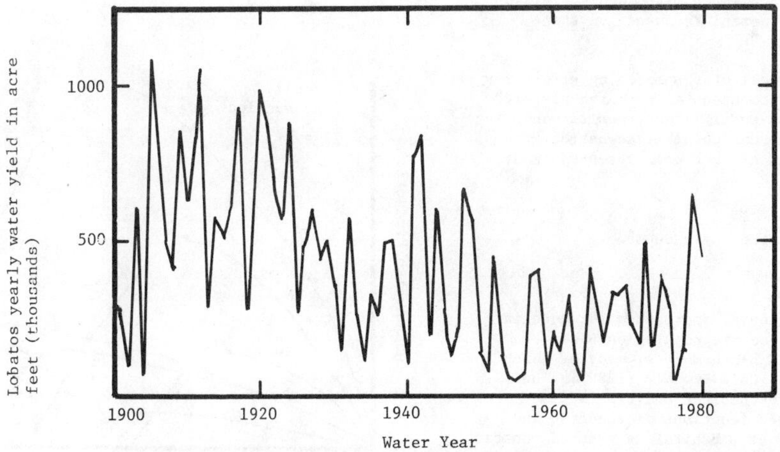

Figure 4.--Yearly water yield time series for the Rio Grande near Lobatos, Colorado (six miles north of New Mexico state line).

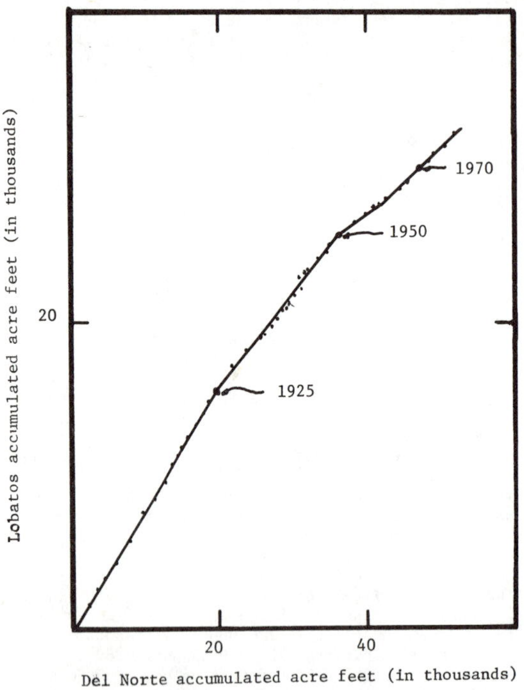

Figure 5.--Double mass curve between Lobatos and Del Norte stations showing longterm effects of development in the Colorado portion of the Rio Grande basin.

U.S. Army Corps of Engineers must satisfy flood and sediment control needs, and the State Engineer must act to meet the requirements of the Rio Grande Compact between New Mexico, Colorado, Texas, and Mexico. In addition the river is directly used by agriculture, cities, industry, and various Native American tribes, and indirectly by groundwater withdrawals. All of these groups exert powerful influence on the disposition of water.

At present the goal of the modeling effort for the hydrology-hydraulic component is to match historic conditions as best as possible and then to manipulate the system with different possible scenarios. Hypothetical examples of such are presented in a following section.

Biologic-Ecologic Submodel

General

There are two general approaches for modeling sport fisheries. These are population based and habitat and yield based models, many of which have been reviewed by Cole et. al. (1980). The first approach ignores habitat and treats fish population dynamics as functions of the fish and fishing pressures alone. Habitat is assumed constant. The second approach, and that used in this submodel, incorporates functional descriptions of habitat variables that significantly affect fish population. These are the habitat-yield models.

Morphoedaphic models

These models are the simplest form of the habitat-yield models, and are the initial approach used in the submodel. In this approach the complex interactions or processes between habitat and fish are not explicit but are lumped into coefficients which modify the different descriptions. The descriptors can include shape (morph) variables such as water depth, reservoir area, and volume, and edaphic or chemistry related factors. These include temperature, nutrients, and dissolved solids. Research during this project has an initial objective of developing a morphoedaphic model which will be refined into a higher level of model, the tropic-dynamic.

Trophic-dynamic model

Tropic-dynamic models are a highly specialized version of the morphoedaphic class of model. They have come about because of the need to partition total fish yield into yields for species or species groups. In trophic-dynamic models, the chemical and physical processes that create and modify habitat are addressed. These models trace the overall mass and energy flow in the system from raw materials into fish yields, but require an extensive data base for determining process, partition, transport and delay coefficients needed for simulating the system. Process coefficients describe the rates of mass transfer between different states of variables within the ecosystem, such as between nutrients and plants. Partitioning coefficients are a measure of the division of resources between groups of organisms, in this case fish. Transport coefficients are required when energy and material must be interlinked among diverse habitats in a system such as between the streams, shallow reservoirs, and deep reservoirs. Delay coefficients are necessary to depict the lag between the chemical and physical processes and the response of the fish population to those processes. With further study these coefficients will be derived and parts of the eventual tropic-dynamic model might be displayed, as in figure 6, for the different reservoirs. The framework and linkages are in place, what is needed are the specific coefficients plus field verification of the submodel.

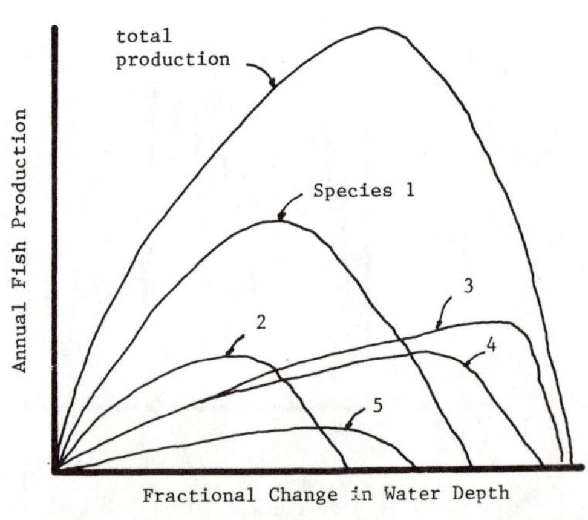

Figure 6.--One type of tropic-dynamic relationship to be developed for the biologic-ecologic submodel.

Social-Economic Submodel

The social-economic submodel is the hardest to develop as a true "process" model because it involves personal tastes, desires, and impressions that cannot be simulated with precision (fig. 7). Instead, measureable surrogate indices which effect the personal attitudes of anglers were developed. Through a shoreline census during the Spring and Summer of 1981, over 800 questionaires were completed as to number of visits, travel costs, and satisfaction with fishing conditions. This last general grouping is affected by fish yield and reservoir water levels which in turn affect fish yields. A subset of the data has been analyzed for a selected reservoir with the following model form as a result:

$$V = B_0 - B_1 W^{\frac{1}{2}} + B_2 I - B_3 C + B_4 S + B_5 H$$

where,

- V = thousands of monthly visits from Bureau of Reclamation data
- W = water level with ½ power indicating diminishing return with depth
- I = per capita income
- C = travel cost
- S = success rate as fish per hour
- H = 1 for holiday, 0 otherwise
- B_0, B_1, B_2, B_3, B_4, B_5 = regression coefficients

Monthly visits are then used to estimate gross expenditures for recreation, particularly fishing. Higher water depths reflect lower fish density and less success per unit of surface area. The initial model had an r^2 value of 0.57 (Ward et. al., 1982). Addition of other variables such as air temperature increases the predictive capability of the regression model, but air temperature is highly correlated with angling success. Improvements in this submodel are forthcoming as more surveys at more sites are incorporated into the data base. In addition, interactions between water users will be introduced with information feeding back to the hydrologic-hydraulic submodel.

HYPOTHETICAL MODEL APPLICATION

In December of 1981, a "first cut" fully linked operational model for a portion of the Rio Grande System was presented to the Project Steering Committee for this study. The committee is comprised of members from Federal, State, and local agencies as well as the project sponsors and agricultural interests. It represents a diverse set of opinions and priorities and is an important part of the study effort.

The selected site was modeled for a five year period of water years 1975 through 1979. Infows, outflows, reservoir contents and chemical data were obtained from U.S. Geological Survey, information stored in the master files. Relationships between streamflow and sediment and chemical data were developed for further use. One baseline and five modified scenarios were simulated as listed below:

Baseline - Measured inflows, outflows, and reservoir contents used to generate habitat conditions.

Scenario #1 - Suspended sediment input reduced by 75% (down to 25%) by upstream management. Water inflow remains the same.

Scenario #2 - Suspended sediment and total phosphorus reduced by 75%. Water remains the same.

Scenario #3 - Water inflow reduced by an amount proportional to evaporation losses from a 200,000 acre-foot (AF) pool at an upstream reservoir. Sediment and phosphorus loading rates returned to baseline values.

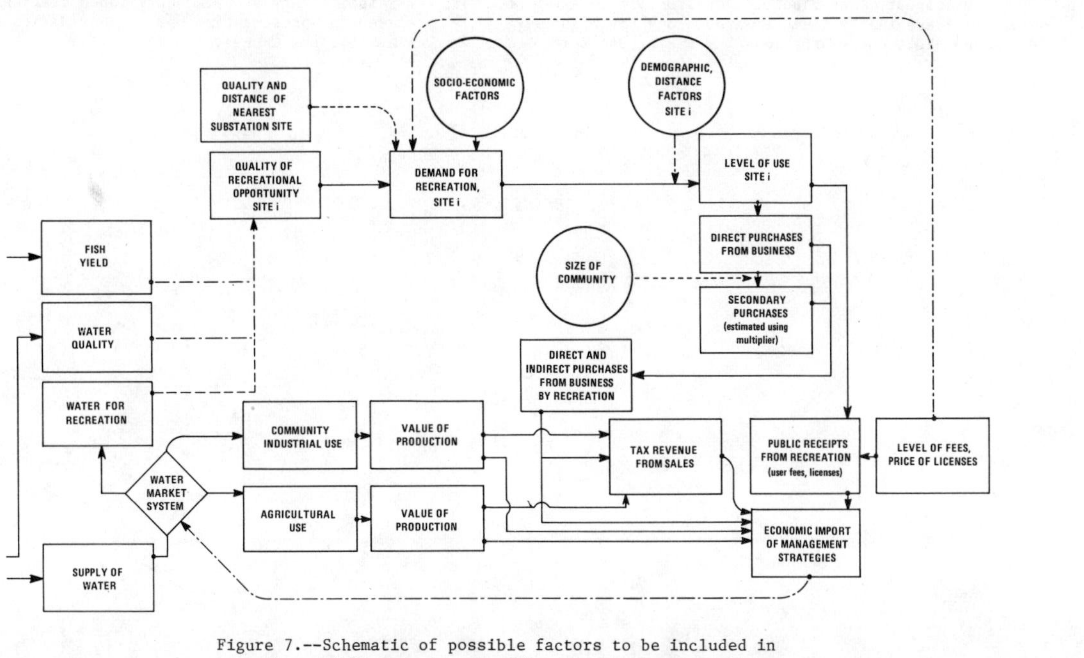

Figure 7.--Schematic of possible factors to be included in the social-economic submodel.

Scenario #4 - Reservoir held at a constant 250,000 AF level except for evaporation. Sediment and phosphorus loading rates returned to baseline values.

Scenario #5 - Compounded effects of Scenarios 2, 3, and 4 (sediment and phosporus reduced, water reduced).

The changes in water level, nutrient input, temperature, and dissolved solids were then incorporated into the trophic-dynamic submodel which generates fish production per unit area. Portions of this information were then converted into 1980 dollars using a 7-3/8% discount rate. The "bottom line" of changes in estimated benefits between scenarios is depicted below:

Table 1.--Ratio of scenario benefits to baseline benefits.

Scenario	Ratio to baseline
Baseline	1.0
#1	1.0+
#2	1.0+
#3	1.25
#4	.83
#5	.83

The above table indicated two things. First, the social-economic submodel is particularly sensitive to water level (as confirmed by surveys), and second, less sensitive to fish production. However, additional analyses of the data have begun to incorporate fish production and yield to a higher degree.

SUMMARY

A multiyear, interdisciplinary study is underway which will eventually yield a methodology for helping decision-makers choose the most beneficial use for New Mexico's limited water supply. Based on three submodels of hydrology-hydraulic, biologic-ecologic, and social-economic, the methodology will answer numerous "what if" questions about the operation of the Rio Grande system. The study is about midway in overall objectives with many pieces at or near completion. The methodology framework has been applied and initial results look promising.

ACKNOWLEDGEMENTS

The authors wish to thank the New Mexico Department of Game and Fish, the New Mexico Water Resources Research Institute, and the New Mexico Interstate Stream Commission for providing funding for this project. Special thanks go to D. Lukens, J. Duggan, and M. Libbin for their assistance in the project.

LITERATURE CITED

Cole, Richard, Diana Weigmann, and Robert Tafanelli. 1980. Model for optimization of socio-economic values on the Rio Grande - Volume II, Part B: The feasibility of biologic modeling of sport fisheries in the Rio Grande of New Mexico. Prepared for New Mexico Department of Game and Fish through the New Mexico Water Resources Research Center, Las Cruces, New Mexico, 137 p.

Sabol, George V. and Nancy Gordon. 1980. Model for optimization of socio-economic values on the Rio Grande - Volume II, Part A: Hydrology and hydraulics of the Rio Grande in New Mexico. Prepared for New Mexico Department of Game and Fish through the New Mexico Water Resources Research Center, Las Cruces, New Mexico, 93 p.

Ward, Frank A., Richard A. Cole, Tim J. Ward, and David R. Lukens. 1982. Recreation benefits from alternative water management policies on the Rio Grande of New Mexico: Overview of an interdisciplinary study. Unpublished draft paper prepared for presentation, 12 p. New Mexico State University, Las Cruces, New Mexico.

SIMULATION OF PHYTOPLANKTON GROWTH IN A SMALL, FAST-FLUSHING RESERVOIR - EUTROPHICATION MANAGEMENT IMPLICATIONS[1]

J. J. Messer, W. J. Grenney, J. Ho[2]

Abstract.--Although phosphorus based input-output (I/O) trophic state models have been applied with some success to reservoirs, factors such as rapid flushing rates, high turbidity, and selective discharge suggest caution in applying them to small, mainstem western reservoirs. The growth of competing strains of *Anabaena*, a cyanophyte, and *Fragilaria*, a diatom, in a proposed upper Colorado River Basin reservoir were simulated using continuous culture growth and nutrient uptake parameters and field and laboratory data for boundary conditions. The results of the simulation indicated that the full growth potential (standing crop chlorophyll *a*) predicted by the I/O models was reached only in low flow years, and even then only in the vicinity of the dam. The simulation also suggests a reservoir operating plan that could further reduce algal growth during late summer months.

INTRODUCTION

In the past decade water quality managers have become reasonably familiar with the value and limitations of empirical input-output (I/O) models for predicting the impacts of changes in nutrient loading and flushing rates on the trophic states of existing lakes and reservoirs (e.g., Rast and Lee 1978, Welch 1981). Often, however, these managers are called upon to predict the trophic state of a proposed reservoir, and the decision to construct such a project may depend wholly or partially on the result of the prediction. To complicate matters further, available sites for large, deep reservoirs are largely already occupied, and new reservoirs built for local or industrial water supply or low head hydropower generation are likely to be smaller and more riverine in nature than are most lakes and large reservoirs. In this paper we address what appears to be a substantial problem with applying preexisting I/O models to proposed small, fast flushing reservoirs in the semiarid upper Colorado River basin as revealed by a simulation of phytoplankton population dynamics.

[1]Paper presented at the Third International Conference on State-of-the-Art in Ecological Modelling, Colorado State University, May 24-28, 1982.

[2]J. J. Messer is Research Assistant Professor at Utah Water Research Laboratory, Utah State University, Logan, Utah U.S.A. W. J. Grenney is Professor of Civil and Environmental Engineering, Utah State University, Logan, Utah U.S.A. J. Ho is a Graduate Student in Civil and Environmental Engineering, Utah State University, Logan, Utah U.S.A.

EMPIRICAL INPUT/OUTPUT TROPHIC STATE MODELS

Scientific Basis

The purpose of all I/O trophic state models is first to predict the phosphorus concentration in the lake water as a function of phosphorus and hydraulic loadings, and subsequently to infer the resulting water quality based on the conversion efficiency of the phosphorus to algal biomass. All of the models employ a mass balance of phosphorus in the lake water column to predict a steady state phosphorus concentration. The general equation is:

$$\frac{d[P]_L}{dt} = \Sigma Q_{in}[P]_{in} - Q_{out}[P]_L - \sigma[P]_L \qquad (1)$$

Where $[P]_{in}$ and $[P]_L$ refer to mean annual phosphorus concentrations in the lake inputs and water column, respectively, Q_{in} and Q_{out}, are the mean annual hydraulic inputs and outputs (usually assumed to be equal), and σ is a sedimentation "constant". In actuality, however, the sedimentation term is not a constant, but represents a complex function of internal phosphorus sources and sinks, including types and concentrations of phosphorus in lake inputs, lake hydrology and morphometry, sediment chemistry, and oxygen content of the water in contact with the sediments. A value for σ can be calculated by difference if existing lake input-output data are available (Dillon and Rigler 1974), but it must be determined empirically as a function of lake morphometry and hydrology (Reckhow 1979) for a proposed reservoir. Equation 1 is set equal to zero, thus assuming steady state conditions, and solved for $[P]_L$ in terms of variables relating to phosphorus and hydraulic loading and mean lake depth.

Table 1 summarizes the computational forms of the more popular predictive models. The Vollenweider 1975 model employed an empirical estimation of σ based on mean depth while the 1976 version related σ to the hydraulic residence time, τ. The Larson and Mercier, Jones and Bachman, Reckhow, and Mueller models all represent attempts to parameterize the Vollenweider-type models using various data bases on lakes and reservoirs throughout the U.S. The Larson and Mercier and the Dillon and Rigler models require a retention coefficient, R, that must be estimated using the empirical Kirchner and Dillon (1976) approximation based on hydraulic loading. It is particularly important to note that although equation 1 represents a rational mass balance on phosphorus, these models have a strong empirical dependence upon the data base used to parameterize σ.

The relationship between the predicted steady state phosphorus concentrations and water quality depends upon the efficiency with which the algae comprising the phytoplankton of the lake convert phosphorus into cell biomass and/or chlorophyll. Biomass may be more the useful variable if the water quality parameters of interest are tastes and odors or filtering efficiency in water treatment plants, while chlorophyll may be more useful if the transparency of a recreational lake is the principal concern. Early models (e.g., Vollenweider 1975) employed an observation by Sawyer (1947) that average springtime phosphorus concentrations exceeding 20 mg P/m^3 resulted in algal problems in Wisconsin lakes. Subsequent refinements (e.g., Vollenweider 1976, Rast and Lee 1978) related empirical observations of mean annual (or summertime) chlorophyll a concentrations to predicted steady state phosphorus concentrations for the particular group of lakes being studied.

Table 1.--Empirical I/O models used to predict phosphorus concentrations in lakes.

Model	Equation	No.
Vollenweider (1975)	$[P] = \dfrac{1}{10 + z/\tau}$	(2)
Larson and Mercier (1975)	$[P] = \dfrac{L\tau}{Z}(1-R_p)$	(3)
	$R_p = \dfrac{1}{(1 + 1/\tau)}$	(4)
Jones and Bachmann 1976)	$[P] = \dfrac{0.84}{z(0.65 + 1/\tau}$	(5)
Dillon and Rigler (1975)	$[P] = \dfrac{L\tau}{z}(1-R)$	(6)
Kirchner and Dillon (1976)	$[P] = \dfrac{L\tau}{z}(1-R_p)$	(7)
	$R_p = 0.426 \exp(-0.271\ z/\tau) + 0.574 \exp(-0.00949\ z/\tau)$	(8)
Vollenweider (1976)	$[P] = \dfrac{L/q_s}{(1 + \tau)}$	(9)
Reckhow (1979)	$[P] = \dfrac{L}{2.77\,z + 1.05(z/\tau)\ \exp(0.0011\,z/\tau)}$	(10)
Mueller (1981)	$[P] = \dfrac{L/q_s}{(1+2.09\ \tau^{0.832})}$	(11)

[P] = Steady state total phosphorus concentration mg P/m^3)
L = Phosphorus loading rate (mg P/m^2-yr)
z = Mean depth (m)
τ = Hydraulic residence time (yr) = q_s/z
q_s = Surface hydraulic loading (m/yr)
R_p = Phosphorus retention coefficient (dimensionless)

The empirical trophic state models described here have provided the basis for most eutrophication management programs during at least the past decade, and for good reason; they have proved remarkably robust. This property is somewhat puzzling in light of the seemingly endless number of factors that affect the sedimentation coefficient, σ, and the conversion efficiency of phosphorus to chlorophyll. The empirical nature of the former has already been noted; the latter depends on light, temperature, and other nutrients, and is species specific (Kalff and Knoechel 1978). Furthermore, Smith and Shapiro (1981) have shown that linear models relating chlorophyll a to phosphorus during the summer growing season have different slopes and intercepts in different lakes, despite the fact that a good fit was found for each lake. Nonetheless, all of the models in table 1 except the Mueller model have been successfully applied to large sets of lakes different than the original data set upon which they were based (Rast and Lee 1978, Reckhow 1979).

Application of I/O Models to a Small Western Reservoir

A reconnaissance level application of the I/O models in table 1 to a proposed small upper Colorado River reservoir basin resulted in a prediction of extremely eutrophic conditions. Phosphorus and hydraulic loadings were calculated using monthly data collected at the U.S. Geological Survey gaging station near the site of the proposed dam. The resulting predicted phosphorus concentrations, along with the morphometric and loading data, are presented in table 2. Estimates were based on the arithmatic mean P and hydraulic loading values for the water years 1976-1979, and for the drought year, 1977, when low flushing rates would be expected to exacerbate the accumulation of high standing crops of algae. The model results predicted annual average phosphorus concentrations of 122-162 mg P/m^3, for 1976-1979 and 74-98 mg P/m^3 during the drought year, the lower value apparently resulting from the relatively lower phosphorus loading, due to decreased flushing of the floodplain. There was generally good agreement among all of the models. The predicted chlorophyll a concentration, based on annual phosphorus and hydraulic loading using the Vollenweider 1976 model on the OECD-North American data set (Rast and Lee 1978), is 24 mg/m^3, with 99% confidence limits of 12 to 44 mg/m^3 (fig. 1).

Several factors suggested caution in accepting the results of this analysis, however. First, although the Vollenweider, Reckhow, and Mueller models all contain lakes with predictor variables (based on some ratio of phosphorus loading and residence time) greater than those for Taylor Draw Reservoir, the particular combination of high P loading rate and short residence time in this reservoir was not duplicated in the other data sets. In the Rast and Lee study, none had a phosphorus loading rate above 10 g/m^2-yr, except for the Potomac River estuary (85 g P m^2/yr), and the shortest residence time was 0.08 yr, with all others above 0.2 yr. The Reckhow model was based on 28 lakes in the U.S. EPA National

Table 2.--Predicted phosphorus concentrations in a proposed small western reservoir. Based on several lake management models (see table 1).

Parameter	1976-1979 Mean	1977 Drought
Mean depth - z (m)	6.0	6.0
P-Loading-L (g/m^2-yr)	34.0 ± 18.9	10.4
Hydraulic loading- q_s (m/yr)	198 ± 68	105
Residence time- T_w(yr)	0.030	0.057
Phosphorus concentration $[P]_L$ (mg P/m^3)		
Vollenweider (1975)	162	98
Larson and Mercier (1976)	145	84
Jones and Bachman (1976)	140	80
Kirchner and Dillon (1975)	155	78
Vollenweider (1976)	146	80
Reckhow (1979)	122 + 200* - 122	74 + 156* - 74
Mueller (1981)	154 + 285* - 145	83 + 215* - 215

*67% confidence limits based on model error.

Eutrophication Survey (NES) having hydraulic loadings of > 50 m/yr. Residence times in these lakes were as short as 0.003 yr, with hydraulic loadings as high as 575 m/yr. However the maximum predicted phosphorus concentration was 135 mg/m^3, a value similar to that predicted for Taylor Draw Reservoir. The Mueller model was limited to western reservoirs, and included residence times as short as 0.003 years and predicted phosphorus concentrations as high as 371 mg P/m^3. However, the standard error of this model (0.325, table 2) was twice that of the Reckhow model, resulting in very poor predictive power near the extremes.

The generally poor predictive uncertainty near the upper data extremes in the I/O models may result from sampling errors in eutrophic lakes, perhaps aggravated by the low stability of phytoplankton bloom-crash sequences, and by the variation in phosphorus to chlorophyll conversion efficiency on a lake by lake basis. At any rate, there is undeniable evidence that empirical I/O models frequently overpredict phosphorus and/or chlorophyll concentrations when applied to some large reservoirs (Gloss et al. 1980, Huggens and Kim 1981, Newbry et al. 1981) or to small lakes and reservoirs characterized by high turbidity or short hydraulic residence times (Hergrader and Hammer 1973, Dillon 1975, Cieka et al. 1979, Thornton 1980, Hamm and Kucklentz 1981). The fact that Mueller (1981) was able to fit a Vollenweider-type model to the NES western reservoir with, at best, twice the model error associated with eastern lakes and reservoirs in the NES data set (Reckhow 1979), only serves to strengthen the argument that such reservoirs are characterized by different and more diverse parameter relationships than are typical of lakes. Factors that might be expected to make Taylor Draw atypical mostly relate to phosphorus conversion efficiency. They include:

1. Extremely short hydraulic residence time;
2. High inorganic turbidity;
3. Low biochemical availability of phosphorus;
4. Possible nitrogen limitation.

These complications led us to create a simulation model for the reservoir that would allow us to investigate the effects of these parameters on algal standing crop during the growth season.

RESEN, A SIMULATION MODEL FOR RIVERINE RESERVOIRS

General Description

RESEN is a mathematical model designed to simulate phytoplankton growth in small, riverine reservoirs. It is a linked box model with no dispersion, and is thus most suitable for relatively long, narrow reservoirs in which the concentrations of the constituents (nutrients, algae, etc.) vary as a function of time and downstream distance along the reservoir. At present, the model may only be applied under hydraulic conditions of either complete vertical mixing (one vertical compartment) or strong vertical stratification (two vertical compartments).

The model simulates the following state variables (water quality parameters):

1. Two or more types of phytoplankton (SNP_i)
2. Zooplankton (ZOO)
3. Available nitrogen (NA)
4. Available phosphorus (NP)
5. Biochemical oxygen demand (BOD)
6. Dissolved oxygen (DO).

We will not discuss BOD and DO here, however, except to say that the hypolimnion near the dam was predicted to become anoxic under the worst phytoplankton blooms, assuming stable stratification occurred.

The model was designed to incorporate functions to represent the most significant mechanisms felt to influence phytoplankton growth dynamics in turbid, fresh water environments. These include:

1. Phosphorus and nitrogen limitation of nutrient uptake;
2. Competition between types of phytoplankton for nitrogen and phosphorus;
3. Luxury uptake and nutrient storage in intracellular pools;
4. Phosphorus and nitrogen limitation of growth based on the intracellular nutrient pools;
5. Nitrogen fixation by cyanophyta;
6. Light limitation of growth, as influenced by suspended sediment and the self-shading effects of the phytoplankton;
7. Sensitivity of metabolic rates to temperature;
8. Selective grazing by zooplankton;
9. Sinking of phytoplankton;
10. Mineralization of nutrients from decaying phytoplankton and bottom deposits.

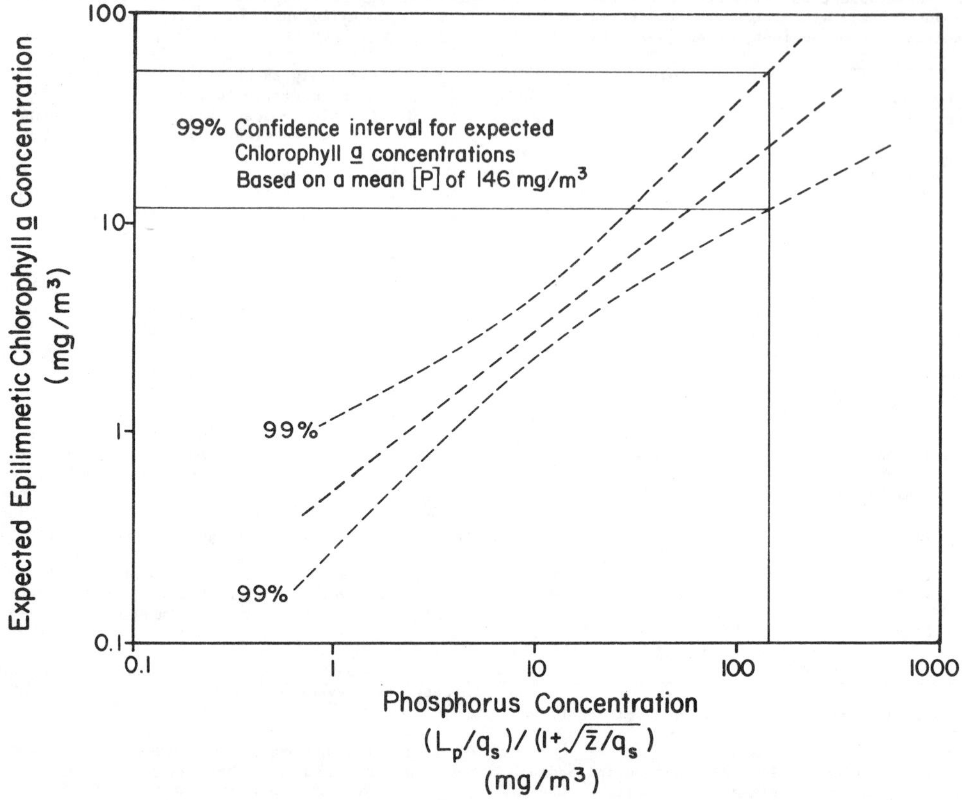

Figure 1.--Expected mean chlorophyll a concentration resulting from the average annual phosphorus and hydraulic loadings to the proposed reservoir. Diagonal lines represent mean and 99% confidence intervals for the OECD North American Lake data set.

Model Construction

Model Structure

The fundamental set of differential equations representing phytoplankton, nutrient, and oxygen dynamics is shown in table A1 in the appendix. The equations are simplified by assuming that coefficients and boundary conditions are constant over time and allowing the system to approach equilibrium. Thus, the results from the model show the conditions that would occur along the length of the reservoir if the boundary conditions were held constant for a sufficient period of time (approximately equal to the residence time of the reservoir). The model equations are shown in their reduced form in table A2 in the appendix. State variables and initial conditions are defined in table A3 and model coefficients in tables A4 and A5 of the appendix.

State Variables and Coefficients

The model simulates competition between two types (species) of phytoplankton, and subscripts are used to identify variables associated with each type. In this application of the model the subscripts "1" and "2" are assigned to a cyanophyte (*Anabaena*) and diatoms (*Fragilaria*), respectively.

Nutrient Uptake and Assimilation.--Luxury uptake of nutrients is modeled by representing each type of phytoplankton as a two compartment unit (Grenney et al. 1973). Available phosphorus and nitrogen (PA and NA) are taken up from the environment and stored in the first compartments representing intracellular nutrient pools. These compartments contain the "luxury" nutrients, PL_i and NL_i. The second compartment of the unit contains the biologically essential N and P comprising the more or less permanent structural component of the cell (SNP_i). The growth rate of the second compartment is related to the quantity of the stoichiometrically limiting nutrient (PL_i or NL_i) in the luxury pool.

The state variable SNP_i is a measure of the biomass of species in a unit volume of water. SNP_i can be related to the standing crop of cell phosphorus or nitrogen by multiplying it by the subsistence quota (QPO_i or QNO_i) and then adding PL_i or PN_i, respectively. The subsistence quota is the minimum amount of structural N or P required for survival of an algal cell, and is a characteristic of each algal species. In this model either N or P is allowed to limit algal growth, depending on the relative stoichiometric availability of the two elements, and the effect of the luxury pool size of the non-limiting nutrient on growth is assumed to be negligible (e.g., Rhee 1978).

The equations and coefficient values linking external nutrient concentrations, luxury pools, and growth rates are those of Gotham and Rhee (1981a,b). These study utilized continuous culture techniques and optimum illumination, and were performed using several algal species that cause blooms in reservoirs. The relationship between PA uptake into PL_i (VP_i) is described by Monod kinetics (equation A18), in which the first term represents an "apparent" maximum uptake rate that is affected by feedback from the luxury P pool (Gotham and Rhee 1981a). The nitrogen uptake into LN_i is treated similarly (equation A19), except that separate nitrogen uptake parameter values were not available for *Anabaena*. Therefore, the parameters were assumed equal for both species (table A4). VNF_i (equation A21) was used in place of (A19) to allow *Anabaena* to use N_2 as a nitrogen source, thus relieving its dependence on NA. Equation A10 was necessary because the half saturation constant for nitrogen uptake was found not to be constant, as in the case for phosphorus, but was found to be a function of the amount of luxury N storage, relative to the cell nitrogen quota (Gotham and Rhee 1981b).

Specific growth rates (μ_i) of each algal species were then calculated as the minimum value of P (equation A17) and N (A18) Monod equations, corrected for light and temperature. Half saturation constants were taken from Gotham and Rhee (1981a,b). The minimum function was chosen, rather than a multiplicative relationship (e.g., Zison et al. 1978), because several studies (e.g., Rhee 1980) have shown that the size of the non-limiting luxury pool has no effect on growth kinetics. A Monod-type light extinction term was chosen for simplicity (e.g., Zison et al. 1978), and half saturation coefficients were taken from Rhee and Gotham (1981) for *Fragilaria* and Chen and Orlob (1975) for *Anabaena*. Temperature corrections (equations A24-A27) followed Bierman (1976) and were applied to both growth and nutrient uptake equations.

Estimation of Chlorophyll a Standing Crop.--Managers are frequently more interested in the standing crop of chlorophyll, which most directly affects water transparency, than in standing crop biomass or organic carbon. Chlorophyll *a* concentrations per cell ($ACHL_i$) have been variously modeled as a function of cell biomass (Chen and Orlob 1975) or cell carbon (e.g., Bierman 1976). In the latter case, carbon is often estimated as a constant fraction of cell biomass, which is in turn assumed to be a constant fraction of cell volume. There are problems associated with these assumptions however. These include nonconstant ratios of cell carbon to volume or biomass across cell sizes and morphology (e.g., cell vacuoles, silica tests of diatoms, etc.), and difficulties in distinguishing between cell and detrital carbon when sampling phytoplankton populations *in situ* (Banse 1977).

Rhee (1978) found that chlorophyll/biomass ratios were determined by the stoichiometrically limiting nutrient concentration (for a given level of illumination and temperature). Therefore we opted to let the chlorophyll *a* concentration be determined by the concentration of phosphorus in the structural pool (QPO_i) of the cell. In order to determine the coefficient, it was necessary to ascertain average chlorophyll *a* to particulate phosphorus concentrations for algae under phosphorus limitation. Tett et al. (1975) provided data for a coastal phytoplankton bloom during the spring season in which a correction was made for detrital particulate phosphorus. The resulting ratio averaged over the water column during the period of virtual disappearance of orthophosphate from the water column was 1 mg chlorophyll *a*/mg particulate phosphorus. This value is consistent with the frequently asserted assumption that algal cells are 1% phosphorus (Chen and Smith 1979), and also 1% chlorophyll (Jorgensen et al. 1981), on a dry weight basis.

Phytoplankton Losses.--Phytoplankton biomass can be lost from the water column by dark respiration, sinking, and grazing by zooplankton (equations A1, A9-A11). Dark respiration was represented by a constant specific loss rate, adjusted for temperature, and sinking rates were divided by depth to give a specific loss rate (Zison et al. 1978). Losses to zooplankton are a function of zooplankton population density, which is in turn dependent on phytoplankton grazed and grazing efficiency (Zison et al. 1978) (equations A22, A23).

Initial Conditions

Water quality data for USGS monitoring station near the dam were available for water years 1975 through 1979 (e.g., USGS 1979). Table 3 shows the resulting initial conditions in the river water entering the reservoir. *Anabaena* and *Fragilaria* were chosen as the two species to be modeled because each have been found to cause summer blooms in western reservoirs (Morris et al. 1979).

Initial concentrations of *Anabaena* (SNP_1) and *Fragilaria* (SNP_2), were assumed to be 10 and 70 cells/ml based on several grab samples taken at the station during summer and winter (USGS 1977, 1979). No members of either species, which should have been observed at these concentrations, were found. The values thus may be conservative (high) during typical conditions. Benthic recruitment from resting stages may become a more important innoculum as the reservoir ages. Sensitivity analysis indicated that halving the initial populations had little effect on the maximum predicted chlorophyll *a* concentrations. Also, nutrients were shown to be taken up rapidly in the model, relative to growth, thus initial luxury nutrient concentrations were not important, and the incoming cells were assumed to be devoid of luxury nutrients.

Available phosphorus (PA) was determined by conducting both standard algal bioassays (Green et al. 1978) and the NaOH extractable phosphate procedure of Williams et al. (1980) on samples collected during the winter of 1981-82. These studies, which will be described in detail elsewhere, indicated that it would be conservative to assume that 5% of the total phosphorus (less orthophosphate) would become available to the phytoplankton. Therefore, the initial concentrations of available phosphorus were calculated as the monthly average of the orthophosphate (SRP) plus 5% of the remaining total phosphorus (non-SRP). Where no data were available for SRP, a value of 30 mg P/m^3 was used, which was thought to be conservatively high. Values for the initial concentrations of available nitrogen were calculated as the monthly average of the nitrate plus ammonium, plus 10% of the organic nitrogen. Because *Anabaena* were capable of fixing atmospheric nitrogen, nitrogen availability could affect the relative sizes of the

Table 3.--Boundary conditions for the reservoir in a normal year.

Symbol	Units	May	June	July	August	September	October
		Month					
SNP_1	cell/ml	10.	10.	10.	10.	10.	10.
PL_1	$10^{-9}\mu M$ P/cell	0.	0.	0.	0.	0.	0.
NL_1	$10^{-9}\mu M$ N/cell	0.	0.	0.	0.	0.	0.
SNP_2	cell/ml	70.	70.	70.	70.	70.	70.
PL_2	$10^{-9}\mu M$ P/cell	0.	0.	0.	0.	0.	0.
NN_2	$10^{-9}\mu M$ N/cell	0.	0.	0.	0.	0.	0.
PA	mg-P/l	.053	.028	.029	.028	.058	.025
NA	mg-N/l	0.25	0.16	0.086	0.095	0.79	0.20
BOD	mg/l	1.7	1.7	1.6	2.7	3.8	2.8
DO	mg/l	8.8	8.3	7.0	7.9	7.5	10.1
TEMP	^oC	14.1	20.0	20.6	18.4	15.1	9.9
SOL	$Kcal/m^2/s$	0.073	0.083	0.079	0.066	0.058	0.041
Q	m^3/min	3563	2842	936	671	586	690
EXTIN	1/m	21.9	15.0	16.4	7.6	6.9	9.2
ZOO	mg/l	0.005	0.005	0.005	0.005	0.005	0.005

blue green and diatom populations, but it could not limit the total primary productivity of the reservoir.

Benthic P release rates were determined using laboratory incubations of soils and sediments collected at the reservoir site. Aerobic release was <1 mg P/m^2-day for most soils, and P was absorbed from a 30 mg/m^3 P solution. Microaerobic release rates were <3 mg P/m^2-day. These rates were similar to those found by Holdren and Armstrong (1980) for intact sediment cores from both calcareous and noncalcareous lake sediments in Wisconsin. The incubations do not include remineralization from decaying phytoplankton or submerged macrovegetation (*Salix*, in this case). In the absence of more complete data, a constant release of 2 mg P/m^2-day was assumed. Nitrogen release was not measured, and set at zero.

Water temperature values (TEMP) and solar radiation (SOL) were taken from the results of a concurrent temperature modeling study. Water year 1974 was selected as an average flow year, and model runs simulating low flow conditions were based on water year 1977, the worst drought year on record.

The light extinction coefficient (EXTIN) was calculated to be the sum of the extinction due to turbidity plus that caused by shading:

$$EXTIN = [3.5 + 0.026 *SED] + [0.015 + CHLA] \quad (12)$$

The first term is based on empirical studies near the proposed reservoir (V. A. Lamarra, personal communication), and the second on a review of literature values (e.g., Zison et al. 1978). The self-shading term was found to be insignificant for this study. For example, for a bloom of approximately 21 mg/m^3 chlorophyll *a*, the self-shading light extinction coefficient was only 0.31 m^{-1}, which is small compared to that induced by sediment (>3.5 m^{-1}).

The sediment values in table 3 represent USGS data for the river entering the reservoir. A previous modeling study indicated that approximately 76% of the sediment would be removed in the reservoir. An empirical fit of the end points to an exponential model resulted in the sediment being reduced along the length of the reservoir by a factor of $e^{-0.3x}$, where x is the distance in miles from the tailwater, approximately 4.5 miles upstream from the dam. Consequently the light extinction coefficient decreased along the reservoir in accordance with equation 12.

No data were available for the initial concentration of zooplankton (ZOO). Zooplankton concentrations in reservoirs have been reported to be in the range of 0.02-0.43 mg/l (Zison et al. 1978). An initial concentration of 0.0005 mg/l seems reasonable, considering that recruitment from upstream is probably low. Reducing the initial zooplankton concentration by half in a sensitivity analysis resulted in the diatom population increasing from 17% to 42% of the total blue green-diatom community, but in less than 5% an increase in the chlorophyll *a* concentration of the total community. The higher initial zooplankton concentration results is a higher blue green population and, therefore, was used because it produces the worst case response of the model from a management standpoint.

The model was applied initially to the reservoir for the six month period, May through October. Two hydraulic regimes were considered: complete vertical mixing, and strong vertical stratification during the summer months, with the thermocline at 3 m depth. These simulations indicated that May, June, and October were not critical months and could be eliminated from the analysis. The high flows (short detention times) and cool temperatures during these periods were not conducive to phytoplankton blooms, and the simulated chlorophyll a concentrations never exceeded 2 mg/m^3. Subsequent attention was focused on July-September.

Simulation Results

Baseline Conditions

The solid line in Figure 2 displays the results of the simulation runs for the completely mixed water column during normal flow years. During July chlorophyll a concentrations would be virtually below detection limits until 0.5 miles above the dam, increasing to an average concentration of ≈3 mg/m^3 in the 0.5 mi segment nearest the dam, composed mostly of *Anabaena*. Phosphorus and nitrogen are only depleted very near the dam, in conjunction with the peak phytoplankton biomass. Similar patterns of algal growth are observed in August and September, except for higher maximum standing crops near the dam. Average values for the half-mile above the dam are ≈11 mg chlorophyll a/m^3 for August and 8 mg chlorophyll a/m^3 for September. Failure of the phytoplankton to deplete available nitrogen in September resulted in a higher proportion of diatom chlorophyll (≈33 percent), relative to total standing crop, than in other months, when N_2-fixing *Anabaena* comprises more than 80 percent of the biomass. The somewhat lower standing crop in September, despite higher initial P concentrations, results from a combination of lower water temperature and insolation.

Stratification of the reservoir, which was predicted to occur by a thermal simulation model, would further decrease the residence time in the epilimnion. The population peaks in figure 2 under stratified conditions were found to be moved completely beyond the reservoir. Simulations of flow through a three meter deep epilimnion resulted in peak chlorophyll a concentrations below 1 mg/m^3 (not shown). Consequently, if a stratified condition could be maintained, algal blooms could be minimized. However, the model predicted dissolved oxygen depletion in the hypolimnion within one week during July and August, whereas minimum dissolved oxygen levels were predicted to remain above 4 mg/l during all months if complete mixing occurred.

There is obviously a great deal of discrepancy between the chlorophyll a concentrations predicted by the simulation model and by the empirical I/O models in table 2. The volume weighted mean chlorophyll concentration for the three months of summer under the baseline conditions in figure 1 totaled 2.5 mg/m^3. If June and October were included, the average would decrease even further. Assuming for the sake of argument that the model structure and coefficient values are reasonable, the discrepancy in the chlorophyll predictions could be brought about by several factors including:

1. extremely short residence times;
2. unusually high light extinction coefficients;
3. atypically low non-SRP phosphorus availability;
4. an incorrect P:chlorophyll a conversion factor.

The adverbs in factors 1 through 3 emphazise that these values may be atypical of the lakes in the data sets used to parameterize the sedimentation coefficients in the I/O models. The results of several simulations will be used to shed some light on the importance of the first three factors listed above.

Hydraulic Residence Time

It is well known that phytoplankton populations in rivers are generally low if the current is sufficiently swift (Hynes 1970). More recently Dillon (1975) noted that in Cameron Lake, Ontario, rapid flushing led to significantly lower levels of chlorophyll than were predicted using his empirical model (table 1). The cause was apparently related to the phosphorus to chlorophyll conversion efficiency, because Dillon noted that extremely high flushing rates (short residence times) may lead to washout of sedimentary phosphorus, and thus a negative retention coefficient. Negative phosphorus retention coefficients have also been observed for some small Illinois Reservoirs (Cieka et al. 1979). Therefore, short residence times must act primarily by removing the phytoplankton from a reservoir before they can deplete the supply of the stoichiometrically limiting nutrient.

This important aspect of the short hydraulic residence time can be seen in figure 3, which compares the results of the simulations under baseline (fig. 2) conditions, and under low flow conditions encountered during the 1977 drought year. All other conditions were held at baseline levels. During July and September, much higher chlorophyll a concentrations were reached under low flow conditions. In August the maximum chlorophyll a concentration was similar in both simulations, but occurred almost 1 mile further upstream under low flow conditions. This has the result of increasing the volume weighted average chlorophyll concentration for the entire reservoir, as will be discussed below. *Anabaena* dominated *Fragilaria* in all three months, although the diatom accounted for a larger fraction of the standing crop in colder, relatively nitrogen rich July water.

Light Extinction Coefficient

Turbidity in the Colorado River Basin can be quite high during much of the year, as would be expected from the extremely dissected topography. Light limitation is usually characterized by inorganic nitrogen and phosphorus concentrations in the water column remaining at high levels, relative to the virtual disappearance of the limiting inorganic nutrient in light sufficient water columns. Light limitation in turbid reservoirs, resulting in failure of phytoplankton communities to realize their full standing crop potential relative to phosphorus supplies, has been documented for several reservoirs (Hergenrader and Hammer 1973, Oskam 1978, Brezonik et al. 1979, Gloss et al. 1980). In the extreme, light limitation may reduce algal growth to nil if the mixed depth is sufficiently deep to cause diel dark respiration to exceed the photosynthetic rate.

Figure 4 represents the results of decreasing the background (non-planktonic) extinction coefficient to

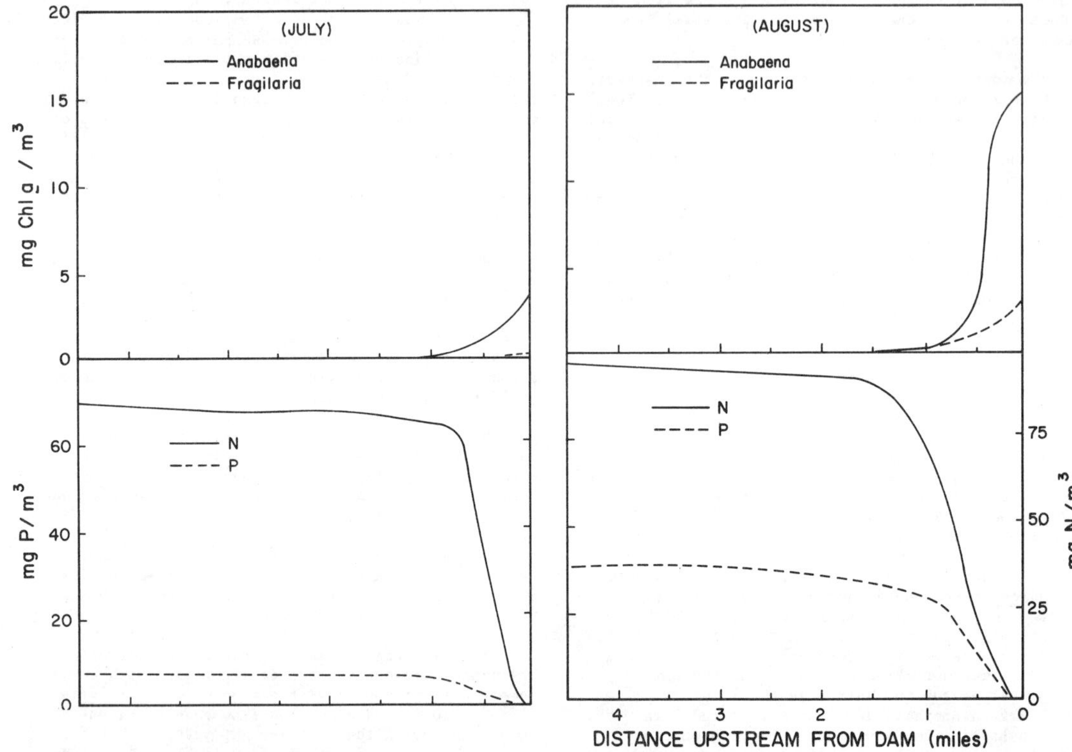

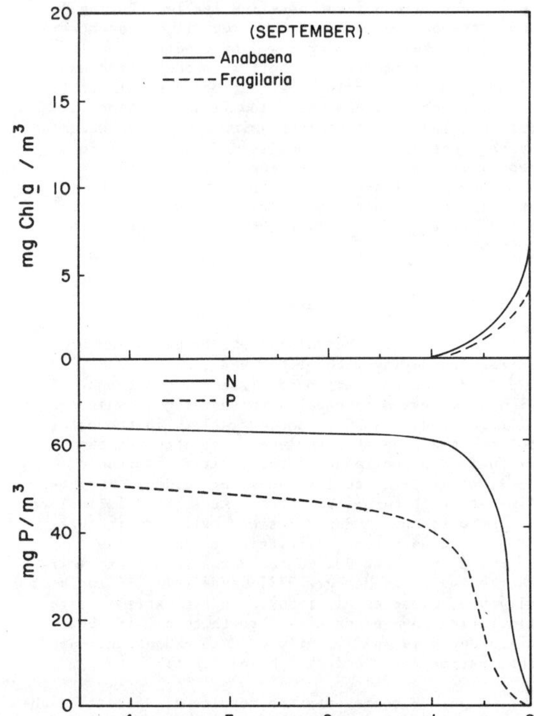

Figure 2.--Simulated chlorophyll a standing crops of *Anabaena* and *Fragilaria* and available N and P during an average year in the proposed reservoir.

0.5 m^{-1}, a value typical of a large lake without significant turbidity inputs. In each case, although significant standing crops are reached only in the reservoir reach 0.5 mi above the dam, maximum standing crops exceed baseline conditions and, to a lesser extent, low flow conditions in July and August. *Anabaena* again dominated *Fragilaria* during all three months.

Phosphorus Availability

Our experimental work suggested strongly that only about 5% of the non-SRP at the dam site would become available to reservoir phytoplankton. This value is somewhat low compared to typical values of 9-11% (maximum 19%) in the Lower Colorado Basin (Evans and Paulson 1981) and 20% in eastern U.S. lakes (Lee et al. 1980). Our lower value may be the result of wintertime sampling, when sources of available non-SRP may be absent or reduced, or merely to the absence of wastewater, urban runoff, or extensive agricultural nonpoint source inputs of P in the watershed. The importance of this point is that most I/O trophic state models use total phosphorus loadings as input values (cf Oglesby and Schaffner 1978). Therefore, the empirical chlorophyll a prediction (e.g., in the Rast and Lee model) incorporates some function of phosphorus availability. If this value is heavily weighted toward eastern lakes in more heavily populated watersheds (as are most of the I/O models), then chlorophyll concentrations will

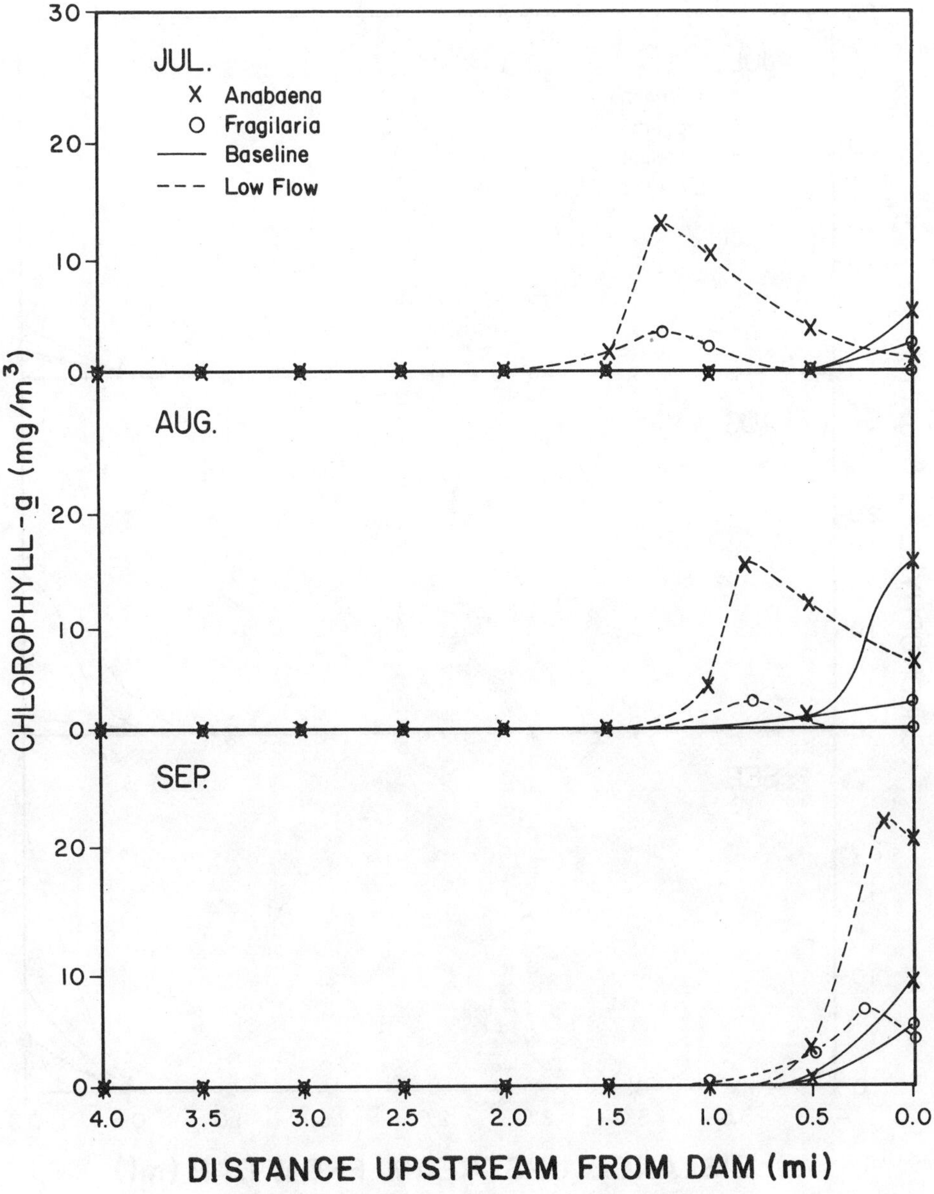

Figure 3.--Simulated chlorophylla standing crops under baseline (1974) versus low flow (1976 drought) conditions.

be overpredicted. A similar, albeit more complicated, argument may be made for water column phosphorus concentrations.

In the case of the reservoir modeled here, however, increasing non-SRP phosphorus availability had little effect on phytoplankton standing crops (fig. 5). This is probably the result of a combination of the rather high ("worst case") values of 30 mg/m^3 SRP assumed to be present when actual data were unavailable and nitrogen limitation. The resulting increases in available P loadings were 41, 11, and 90% for July, August, and September, respectively. The principal result was to improve the competitive advantage of *Fragilaria* in July and September by reducing nitrogen limitation.

Low Flow and Low Exticntion Coefficient

Perhaps the situation most analagous to "typical" lakes that characterize the data sets used to produce

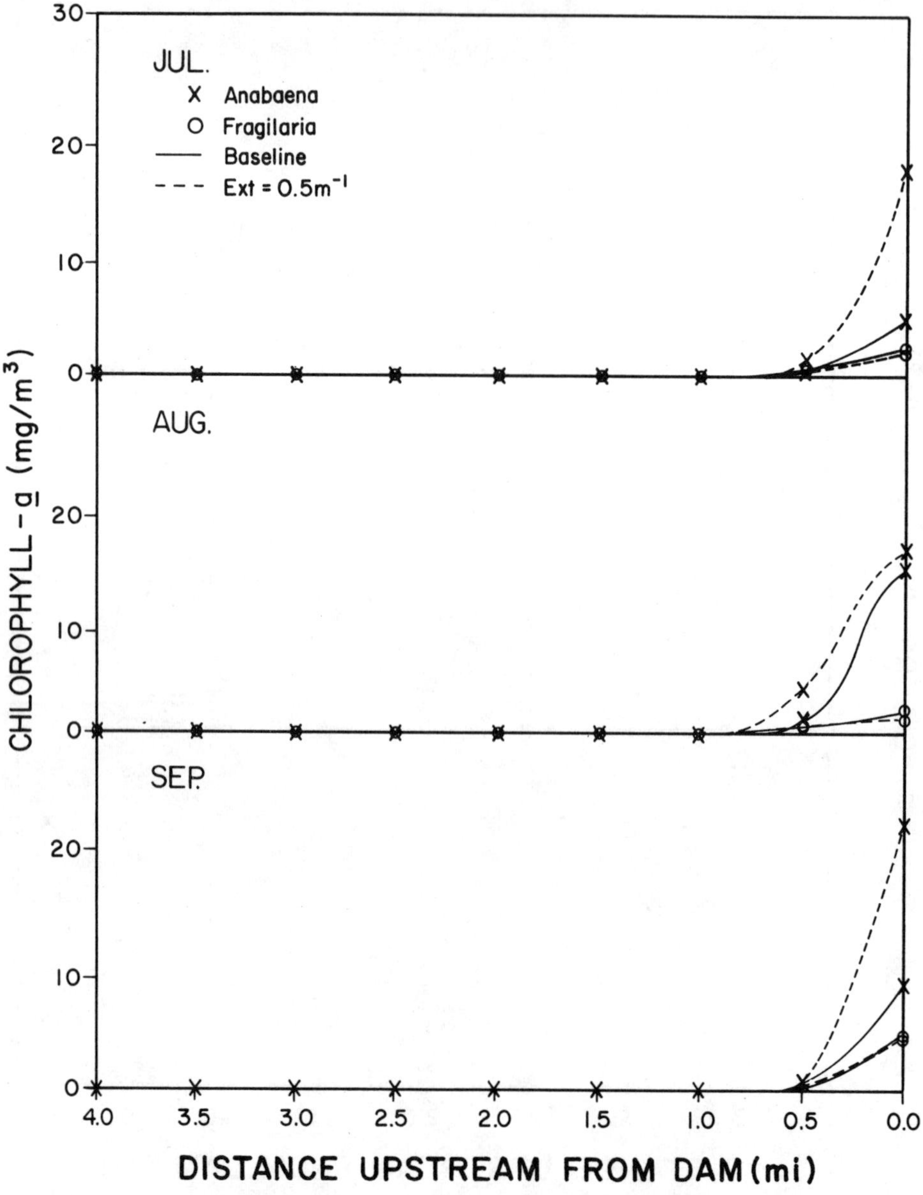

Figure 4.--Simulated chlorophyll a standing crops under baseline (EXTIN = 7-20 m^{-1}) versus low turbidity (EXTIN = 0.5 m^{-1}) conditions.

most of the I/O models in Table 1 would be low flow ("long" hydraulic residence time) and low background extinction coefficient. Figure 6 represents such a simulation using 1977 (drought) flow and EXTIN = 0.5 m^{-1}. The results indicate higher standing crops of phytoplankton than in the other simulations, occurring farther upstream from the dam. The result would, of course, mean generally much worse water quality for reservoir users. More specifically, this result suggests increasing convergence between I/O and simulation results as flow and turbidity characteristics approach the range typically found in lakes in the OECD North American Project and EPA-NES data bases.

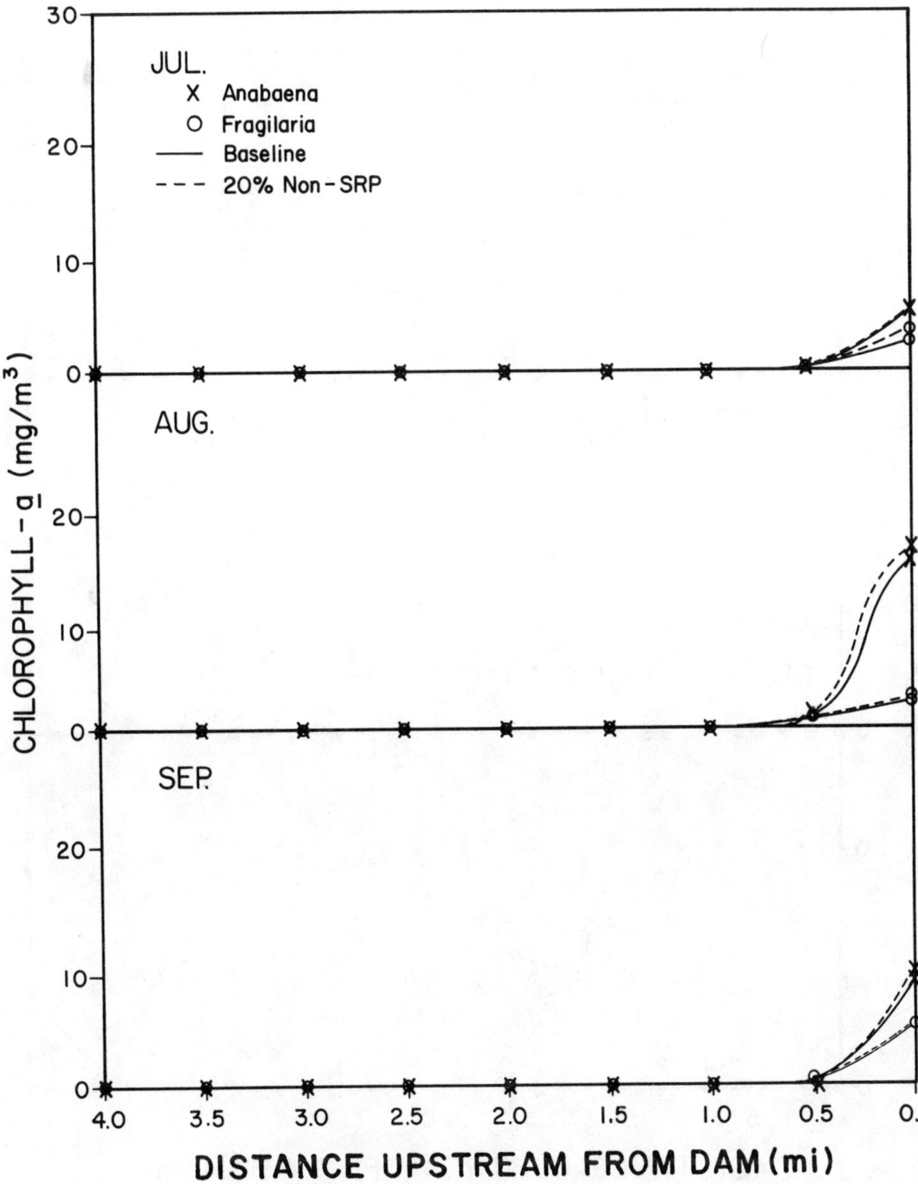

Figure 5.--Simulated chlorophyll a standing crops under baseline (5%) versus high (20%) non-SRP availability.

Comparison of I/O and Simulation Predictions

The data from the simulations in figs. 2-6 are compared to I/O model predictions in table 4. It is immediately obvious that not only seasonal, but also volume weighted monthly average chlorophyll a concentrations fall below the minimum (99% C.I.) value predicted by the I/O models. Maximum concentrations generally fall near the average predicted by the I/O models, provided the hydraulic residence time and light extinction coefficient is decreased to values more representative of North American lakes. It should be noted that the chlorophyll a concentrations for low flow conditions in Table 4 have been changed from those in Table 2 to reflect equal phosphorus loadings in order to make the treatments comparable for discussion purposes.

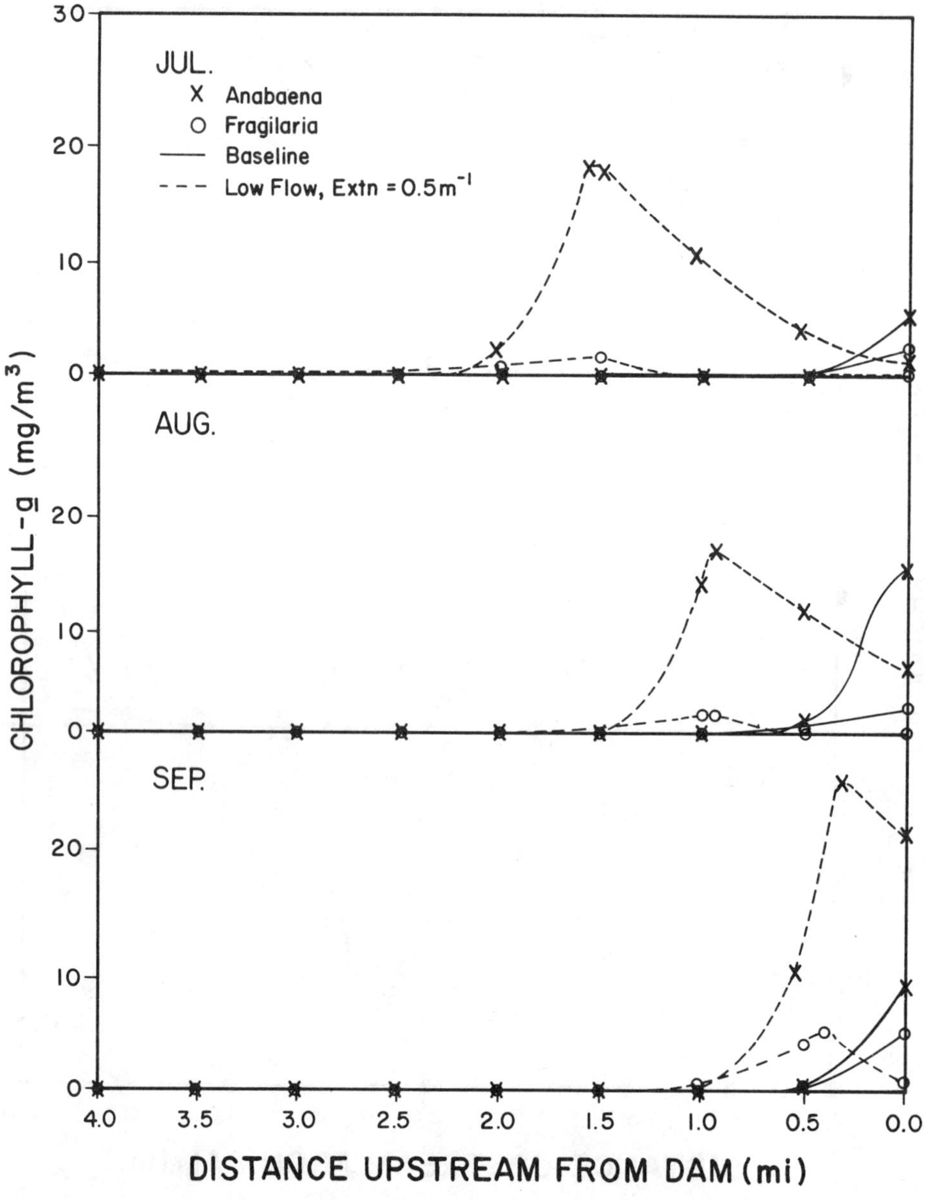

Figure 6.--Simulated chlorophyll a standing crops under baseline (1974; EXTIN = 7-20 m^{-1}) versus low flow (1977 drought) and low turbidity (EXTIN = 0.5 m^{-1}) conditions.

DISCUSSION

Thus far, the comparison of simulation results with those of I/O trophic state models lead us to two apparent management oriented conclusions:

1) traditional I/O models may greatly overpredict mean annual chlorophyll a concentrations in small, turbid reservoirs; and

2) Decreasing the hydraulic flushing rate and turbidity reduce, but do not eliminate, the discrepancy predicted by the two modeling approaches.

Factors responsible for these conclusions could result to some extent from some nitrogen limitation or incorrect model parameters, constructs or initial

Table 4.--Comparison of simulation results with I/O model predictions of standing crop chlorophyll *a* concentrations.[1]

Simulation	Simulation Results[2] July	Aug	Sept	Mean	I/O Results[3] I/O
Baseline	1.8	4.4	3.4	2.5	24 (12-44)
	(7.4)	(16.1)	(13.7)	-	-
Low flow	5.9	9.0	10.6	8.5	38 (16-100)
	(16.2)	(15.9)	(25.9)	-	-
Low EXTIN	5.3	5.7	6.8	5.9	24 (12-44)
	(19.4)	(17.2)	(27.4)	-	-
High P	1.4	4.7	4.6	3.6	24 (12-44)
	(8.6)	(17.6)	(18.9)	-	-
Low flow and EXTIN	6.5	9.1	13.7	9.8	38 (16-100)
	(19.4)	(17.1)	(30.7)	-	-

[1]All chlorophyll concentrations are in mg/m^3
[2]Mean (maximum)
[3]Mean (99% confidence limits)

conditions. A significant factor may also be the plug flow conditions imposed by the model.

Plug flow results in long upstream reaches of reservoir being relatively unproductive because of low innoculum levels of phytoplankters and generally poor growing conditions. This would appear to lower average chlorophyll concentrations, relative to a completely mixed lake with a similar residence time and phosphorus sedimentation rate, where innoculation of algae from the sediment could provide a faster start when conditions were favorable. The extent to which plug flow models are preferable to models with more complex horizontal dispersion in reservoirs is not known, but Huggens and Kim (1981) have reported better success in predicting P retention with plug flow models in Cherokee Reservoir, Tennessee.

An important corrollary to the plug flow hydrology in the simulations is the nutrient advection hypothesis of Gloss et al. (1980). This idea, developed during a limnological study of Lake Powell Reservoir on the Colorado River, holds that in stratified reservoirs, available P concentrations may be high upstream, but turbidity severaly limits algal growth. Downstream, sedimentation removes turbidity, but much of the available phosphorus, coprecipitates so that the algae switch from light to phosphorus limitation. Although RESEN presently does not reduce the available phosphorus pool (PA) by sedimentation, our bioassay results suggest that a significant amount of available P is lost from a quiescent water column after settling overnight. This may further affect the discrepancy in the two modeling approaches. It is interesting in this regard that the predicted steady state P concentration is lower in western reservoirs with short residence times (table 1, equation 1) than in lakes more typical of the earlier I/O data bases (table 1, equation 9).

In a management context, the spatial distribution of chlorophyll in a plug flow reservoir deserves special mention. Under many of the simulation conditions, maximum predicted chlorophyll *a* concentrations would reach nuisance conditions. However, such conditions may be localized in the region of the dam. Selective epilimnetic withdrawals during blooms, to the extent that they can be accomplished without hypolimnetic oxygen depletion, may represent an effective control strategy during the late summer. Location of marinas and bathing beaches some distance above the dam may also be desirable. It is noted, however, that much of the pool volume (in this case 76%) occurs within 1 mile of the dam.

Among the tasks that remain in exploring the behavior of the RESEN model are determining the importance of nitrogen limitation, the phosphorus to chlorophyll conversion efficiency, modeling of total phosphorus retention further sensitivity analyses, and validation. The dominance of *Anabaena* during the summer months, together with the virtual disappearance of nitrogen from the water column during a bloom (Fig. 2), suggest frequent nitrogen limitation. Stoichiometric N:P ratios in the influent water agree with this assessment, but the interpretation depends on the assumed available fractions of total N and P. The model is not particularly sensitive to changes in benthic P loading, which also may indicate N limitation, and suggests the importance of determining benthic N loading rates. At present, assimilation efficiency of P for the baseline conditions ranges from 41-76%, which may also indicate other limiting factors.

The relationship between structural P and chlorophyll *a* also needs validation. Although higher chlorophyll *a* ratios have been determined in the field (e.g., 1.3, Megard and Smith 1974), such studies often neglect detrital P. This problem could be avoided by modeling total P retention rather than chlorophyll concentrations. The σ value determined in the I/O models (0.9-0.19) could be compared with the retention coefficient simulated for column P alone ($[P]_L$ in equation 1). However, RESEN in its present form does not handle nonavailable P in a realistic way (cf. Gloss et al. 1980), and calculation of apparent sedimentation coefficients thus awaits further refinements.

Validation of RESEN is more difficult to deal with. Thus far, we have not been able to find a suitable data set. However, sensitivity analyses thus far have not revealed any serious instabilities, and the parameter values and constructs applied are typical of those that have been used with some success elsewhere. Indeed, at this point the only clear conclusion is that a simulation model using coefficients "typical" of those in the literature produces different results than I/O models with a relatively sound (if noisy) track record, at least near one model extreme. While the most thoroughly tested lake simulation models are not known for their simultaneous accuracy and generality (e.g., Park 1978), manipulation of initial conditions in a hypothetical reservoir near their limits for more typical lakes may nonetheless provide valuable insights into the operation of both simulations and the widely used I/O trophic state models.

REFERENCES

Bierman, V. 1976. Mathematical model of the selective enhancement of blue-green algae by nutrient enrichment. pp. 1-31 in R. Canale [ed.]. Modeling biochemical processes in aquatic ecosystems. Ann Arbor Science, New York.

Banse, K. 1977. Determining the carbon-to-chlorophyll ratio of natural phytoplankton. Marine Biol. 41:199-212.

Brezonik, P., E. Blancher, V. Meyers, C. Hilty, M. Leslie, C. Cratzer, G. Marbury, B. Snyder, T. Crisman, and J. Messer. 1979. Factors affecting primary production in Lake Okeechobee, Florida. Rept. No. 07-79-01. Dept. of Environmental Engineering Sciences, Univ. of Florida: Gainesville.

Chen, C. W. and G. T. Orlob. 1975. Ecologic simulation for aquatic environments. pp. 475-588. In B. C. Patten [ed.] Systems analysis and simulation in ecology. V. III. Academic Press.

Chen, C. W. and D. J. Smith. 1979. Preliminary insights into a three-dimensional ecological-hydrodynamic model. pp. 249-279. In D. Scavia and A. Robertson. Perspectives on lake ecosystem modeling. Ann Arbor Science.

Cieka, J., R. Fabian, and D. Merilatt. 1979. A statistical model for small lake water quality management. Water Res. Bull. 15:1318-1330.

Dillon, P. J. 1975. The phosphorus budget of Cameron Lake, Ontario: The importance of flushing rate to the degreen of eutrophy in lakes. Limnol. Oceanogr. 20:23-39.

Dillon, P. J. and F. H. Rigler. 1974. The phosphorus-chlorophyl relationship in lakes. Limnol. Oceanogr. 19:767-773.

Evans, T. D. and L. J. Paulson. 1981. The influence of Lake Powell on the suspended sediment - phosphorus dynamics of the Colorado River inflow to Lake Mead.

Gloss, S. P., Mayer, L. M., and D. E. Kidd. 1980. Advective control of nutrient dynamics in the empilimnion of a large reservoir. Limnol. Oceanogr. 25:219-228.

Gotham, Ivan J. and G.-Y. Rhee. 1981a. Comparative kinetic studies of phosphate-limited growth and phosphate uptake in phytoplankton in continuous culture. J. Phycol. 17:257-265.

Gotham, I. J. and G.-Y. Rhee. 1981b. Comparative kinetic studies of nitrate-limited growth and nitrate uptake in phytoplankton in continuous culture. J. Phycol. 17:309-314.

Green, D. B., T. J. Logan, and N. E. Smeck. 1978. Phosphate adsorption-desorption characteristics of suspended sediments in the Maumee River Basin of Ohio. J. Environ. Qualt. 7:208-212.

Grenney, William J. and Andrzej K. Kraszewski. 1982. Simplified technique for river water temperature predictions. Ecological modeling (accepted for publication).

Grenney, William J., D. A. Bella and H. C. Curl, Jr. 1973. Effects of Intracellular Nutrient Pools on Growth Dynamics of Phytoplankton. Journal Water Pollution Control Federation, 46(7):1751-1760.

Hamm, A. and V. Kucklentz. 1981. Effect of hydraulic load changes on the eutrophication of an alpine lake. Verh. Int. Ver Limnol. 21:466-472.

Hergenrader, G. L. and M. J. Hammer. 1973. Eutrophication of small reservoirs in the Great Plains pp. 560-566. In W. Ackerman, G. White, and E. Worthington [eds.] Manmade lakes: Their problems and environmental effects. Am. Geophys. Union, Washington, D. C.

Holdren, G. C., Jr. and D. Armstrong. 1980. Factors affecting phosphorus release from intact lake sediment cores. Environ. Sci. Technol. 14:79-87.

Huggins, J. and B. Kim. 1981. Phosphorus retention models for Tennessee Valley Authority reservoirs. Water Resources Res. 17:571-576.

Hynes, H. B. N. 1970. The ecology of running waters. Univ. of Toronto Press.

Jones, J. R. and R. W. Bachman. 1978. Phosphorus removal by sedimentation in some Iowa reservoirs. Verh. Internat. Verein. Limnol. 20:1576-1580.

Jorgensen, S., L. Jorgensen, L. Kamp-Nielsen, and H. Mejer. 1981. Parameter estimation in eutrophication modeling. Ecol. Modeling 13:111-129.

Kalff, J. and R. Knoechel. 1978. Phytoplankton and their dynamics in oligotrophic and eutrophic lakes. Ann. Rev. Ecol. Syst. 9:475-495.

Kirchner, W. and P. Dillon. 1975. An empirical method of estimating the retention of phosphorus in lakes. Water Resources Res. 2:182-187.

Larsen, D., H. Mercier, and K. Malueg. 1973. Modeling algal growth dynamics in Shagawa Lake, Minnesota, with comments concerning projected restoration of the lake. pp. 15-31 in E. J. Middlebrooks, D. Falkenborg, and T. Maloney. Modeling the eutrophication process. PRWG 136-1. Utah Water Research Lab., Utah State University, Logan.

Lee, G. F., W. Rast, and A. Jones. 1978. Eutrophication of waterbodies: Insights for an age-old problem. Environ. Sci. Technol. 12: 900-908.

Megard, R. and P. Smith. 1974. Mechanisms that regulate growth rates of phytoplankton in

Shagawa Lake, Minnesota. Limnol. Oceanogr. 19:279-296.

Morris, M., W. Taylor, L. Williams, S. Hern, V. Lambou, and F. Morris. 1979. Distribution of phytoplankton in Colorado lakes. EPA-600/3-79-114. U. S. Environmental Protection Agency, Las Vegas, NV.

Mueller, D. K. 1981. Mass balance model estimation of phosphorus concentration in reservoirs. Water Res. Bull.

Newbry, B., R. Jones, and G. F. Lee. 1981. Assessment and analysis of eutrophication of Tennessee River system impoundments. pp. 413-424. In. H. Stefan [ed.]. Surface water impoundments. Am. Soc. Civil Eng., New York.

Oglesby, R. and W. Schaffner. 1978. Phosphorus loadings to lakes and some of their responses. Part 2. Regression models of summer phytoplankton standing crops, winter total P, and transparency of New York lakes with known phosphorus loadings. Limnol. Oceanogr. 23: 135-145.

Oskam, G. 1978. Light and zooplankton as algae regulating factors in eutrophic biesbosch reservoirs. Verh. Internat. Verein. Limnol. 20:1612-1618.

Park, R. A. 1978. A model for simulating lake ecosystems. Rept. No. 3. CEM. Rensselaer Polytech. Inst., Troy, N.Y.

Rast, W., and G. Lee. 1978. Summary analysis of the North American (U.S. portion) OECD eutrophication project. EPA-600/3-78-008. Corvallis Environmental Research Lab., U.S. EPA., Corvallis, OR.

Reckhow, K. 1979. Quantitative techniques for the assessment of lake quality. EPA-440/5-79-015. U.S. Environ. Protection Agency, OWPS, Washington, D.C.

Rhee, G.-Y. 1978. Effects of N:P atomic ratios and nitrate limitation on algal growth. Cell composition and nitrate uptake. Limnol. Oceanogr. 23:10-25.

Rhee, G.-Y. 1980. Continuous culture in phytoplankton ecology. pp. 150-203. In M. Droop and and H. Jannasch [eds.]. Advances in aquatic microbiology. Vol. 2. Academic Press, New York.

Rhee, G.-Y. and I. J. Gotham. 1981a. The effect of environmental factors on phytoplankton growth: Temperature and the interactions of temperature with nutrient limitation. Limnol. Oceanogr. 26:635-648.

Rhee, G.-Y. and I. J. Gotham. 1981b. The effect of environmental factors on phytoplankton growth: Light and the interactions of light with nitrate limitation. Limnol. Oceanogr. 26:649-659.

Sawyer, C. N. 1947. Fertilization of lakes by agricultural and urban drainage. New Eng. Water Works Assn. 61:109-122.

Smith, V. and J. Shapiro. 1981. Chlorophyll-phosphorus relations in individual lakes. Their importance to lake restoration strategies. Env. Sci. Technol. 15:444-451.

Tett, P., J. C. Cottrell, D. O. Trew, and B. J. B. Wood. 1975. Phosphorus quota and the chlorophyll: Carbon ratio in marine phytoplankton. Limnol. Oceanogr. 20:587-603.

Thornton, J. A. 1980. A comparison of the summer phosphorus loadings to three Zimbabwean water supply reservoirs of varying trophic states. Water SA 6:163-170.

USGS. 1975-1979. Water Resources Data for Colorado. U.S. Geological Survey. Water-Data Reports. 5 Vol.

Vollenweider, R. A. 1975. Input-Output models with special reference to the phosphorus loading concept in Limnology. Schweitz. Z. Hydrol. 37:53.

Vollenweider, R. 1976. Advances in defining critical loading levels for phosphorus in lake eutrophication. Mem. Ist. Ital. Idrobiol. 33:53-83.

Welch, E. 1981. The dilution/flushing technique in lake restoration. Water Resources Bull. 17:558-564.

Williams, J. D. H., H. Shear, and R. L. Thomas. 1980. Availability to Scenedesmus Quadricaudata of different forms of phosphorus in sedimentary materials from the Great Lakes. Limnol. Oceanogr. 25:1-11.

Zison, S., W. Mills, D. Deimer, and C. Chen. 1978. Rates, constants, and kinetics formulations in surface water quality modeling. EPA-600/3-78-105. U.S. Environmental Protection Agency, ERL, Athens, GA.

Table A1

Differential equations describing phytoplankton growth dynamics.

$$\frac{\partial(SNP_i)}{\partial t} = -\frac{1}{A}*\frac{\partial Q*(SNPN_i)}{\partial x} + (\mu_i - \tau_i*KD_i - \frac{KS_i}{Z})*SNP_i - ZP_i*SEL_i*GZ*ZOO \qquad [A1]$$

$$\frac{\partial(SNP_i*PL_i)}{\partial t} = -\frac{1}{A}*\frac{\partial Q*(SNP_i*PL_i)}{\partial x} + (VP_i - \mu_i*QP\phi_i)*SNP_i - ZP_i*SEL_i*GZ*ZOO*PL_i \qquad [A2]$$

$$\frac{\partial(SNP_i*NL_i)}{\partial t} = -\frac{1}{A}*\frac{\partial Q*(SNP_i*NL_i)}{\partial x}*(VN_i - \mu_i*QN\phi_i)*SNP_i - ZP_i*SEL_i*GZ*ZOO*NL_i + (VNF_i - VN_i)*NFIX_i*SNP_i \qquad [A3]$$

$$\frac{\partial(ZOO)}{\partial t} = -\frac{1}{A}*\frac{\partial Q*(ZOO)}{\partial x} + (GZ - DZ*\tau_z)*ZOO \qquad [A4]$$

$$\frac{\partial(NA)}{\partial t} = -\frac{1}{A}*\frac{\partial Q*(NA)}{\partial x} + \frac{RN}{Z} + \sum_i(FMN*KD_i*\tau_i*QN\phi_i - VN_i)*SNP_i \qquad [A5]$$

$$\frac{\partial(PA)}{\partial t} = -\frac{1}{A}*\frac{\partial Q*(PA)}{\partial x} + \frac{RP}{Z} + \sum_i(FMP*KD_i*\tau_i*QP\phi_i - VP_i)*SNP_i - KPS*PA \qquad [A6]$$

$$\frac{\partial(BOD)}{\partial t} = -\frac{1}{A}*\frac{\partial Q*(BOD)}{\partial x} - \tau_b*KB*BOD + \frac{RB}{Z} + \sum_i KD_i*SNP_i*\tau_i*SB_i \qquad [A7]$$

$$\frac{\partial(DO)}{\partial t} = -\frac{1}{A}*\frac{\partial Q*(DO)}{\partial x} + \tau_d*KD*(DOSAT - DO) - \frac{RD}{Z} - \tau_b*KB*BOD + \sum_i (\mu_i - \tau_i*KD_i)*SNP_i*PP_i \qquad [A8]$$

See Table A2 for definitions of μ_i, VP_i, VN_i, KNA_i, VNF_i, GZ, and τ.

Table A2

Reduced differential equations for water quality in the RESEN.

$$\frac{\partial(SNP_i)}{\partial x} = [(\mu_i - \tau_i^{*}KD - \frac{KS_i}{Z})*SNP_i - ZP_i*SEL_i*GZ*ZOO]*\frac{1}{u} \quad [A9]$$

$$\frac{\partial(PL_i)}{\partial x} = [VP_i - \mu_i*QP\phi_i - (\mu_i - \tau_i*KD_i - \frac{KS_i}{Z})*PL_i]*\frac{1}{u} \quad [A10]$$

$$\frac{\partial(NL_i)}{\partial x} = [VN_i - \mu_i*QN\phi_i - (\mu_i - \tau_i*KD_i - \frac{KS_i}{Z})*NL_i + (VNF_i - VN_i)*NFIX_i]*\frac{1}{u} \quad [A11]$$

$$\frac{\partial(ZOO)}{\partial x} = [GZ - DZ*\tau_z)*ZOO]*\frac{1}{u} \quad [A12]$$

$$\frac{\partial(NA)}{\partial x} = \frac{RN}{Z} + [\sum_i(FMN*\tau_i*KD_i*QN\phi_i - VN_i)*SNP_i]*\frac{1}{u} \quad [A13]$$

$$\frac{\partial(PA)}{\partial x} = [\frac{RP}{Z} + \sum_i(FMP*\tau_i*KD_i*QP\phi_i - VP_i)*SNP_i - KPS*PA]*\frac{1}{u} \quad [A14]$$

$$\frac{\partial(BOD)}{\partial x} = [\frac{RB}{Z} - \tau_b*KB*BOD + \sum_i KD_i*SB_i*SNP_i*\tau_i]*\frac{1}{u} \quad [A15]$$

$$\frac{\partial(DO)}{\partial x} = [\tau_d*KD*(DOSAT - DO) - \frac{RD}{Z} - \tau_b*KB*BOD + \Sigma(\mu_i - \tau_i*KD\ )*SNP_i*PP_i]*\frac{1}{u} \quad [A16]$$

$$\mu_i = \min \begin{cases} \mu p_i*(\frac{PL_i}{KPL_i + PL_i})*\frac{ASOL}{KSOL_i + ASOL}*\tau_i \\ \mu n_i*(\frac{NL_i}{KNL_i + NL_i}*\frac{ASOL}{KSOL_i + ASOL_i}*\tau_i \end{cases} \quad [A17]$$

$$VP_i = \frac{VPL_i*KPI_i - VPI_i*PL_i}{KPI_i + PL_i} * \frac{PA}{KPA_i + PA}*\tau_i \quad [A18]$$

Table A2 (Continued)

$$VN_i = \frac{VNL_i*KNI_i - VNI_i*NL_i}{KNI_i + NL_i} * \frac{NA}{KNA_i + NA}*\tau_i \qquad [A19]$$

$$KNA_i = KNM_i*(1.0 - \frac{NL_i}{KNK_i}) \qquad [A20]$$

$$VNF_i = \frac{VNL_i*KNI_i - VNI_i + NL_i}{KNI_i + NL_i}*\tau_i \qquad [A21]$$

$$GZ = CG*\frac{\Sigma_j SEL_j*SNP_j}{KZO + \Sigma_j SEL_j*SNP_j}*\tau_z \qquad [A22]$$

$$SEL_i = \frac{ZP_i*SNP_i}{\Sigma_j ZP_j*SNP_j} \qquad [A23]$$

$$\tau_i = \Theta_i^{(T-20)} \qquad [A24]$$

$$\tau_b = \Theta_b^{(T-20)} \qquad [A25]$$

$$\tau_d = \Theta_d^{(T-20)} \qquad [A26]$$

$$\tau_z = \Theta_z^{(T-20)} \qquad [A27]$$

Table A3

Definitions of terms for the state variables and initial conditions.

Symbol	Units	Definitions
SNP_1	cells, ml	Concentrations of blue green phytoplankton
PL_1	$10^{-9}\mu M$ P/cell	Intracellular P pool for blue greens
NL_1	$10^{-9}\mu M$ N/cell	Intracellular N pool for blue greens
SNP_2	cells/ml	Concentrations of diatom phytoplankton
PL_2	$10^{-9}\mu M$ P/cell	Intracellular P pool for diatoms
NL_2	10-9μM N/cell	Intracellular N pool for diatoms
PA	mg/l	Available phosphorus
NA	mg/l	Available nitrogen
Zoo	mg/l	Zooplankton concentration
BOD	mg/l	Biochemical oxygen demand; 5 day, 20 C
DO	mg/l	Dissolved oxygen
TEMP	C	Temperature
SOL	$Kcal/m^2/s$	Solar radiation
Q	m^3/min	River discharge
EXTIN	m^{-1}	Light extinction coefficient

Table A4

General coefficients for RESEN.

Symbol	Descriptions	Units	Value
CG	Maximum zooplankton growth rate	min^{-1}	$1.74x10^{-4}$
KZO	Half saturation const. for zooplankton	cell/ml	$5.0x10^2$
DZ	Respiration rate for zooplankton	min^{-1}	$3.4x10^{-5}$
KPS	Physical removal rate for P	min^{-1}	$1.49x10^{-5}$
KB	BOD oxidation rate	min^{-1}	$1.39x10^{-4}$
KD	Reaeration rate	min^{-1}	$9.3x10^{-5}$
0_b	Temperature coefficient for BOD	dimensionless	1.047
0_d	Temperature coefficient for DO	dimensionless	1.029
RP	Benthic release rate for P	mg/m^2-min	$1.4x10^{-3}$
RN	Benthic release rate for N	mg/m^2-min	0
RB	Benthic release rate for BOD	mg/m^2-min	0
RD	Benthic uptake rate for DO	mg/m^2-min	1.01

Table A5

Species-Specific model coefficients for the phytoplankton simulation.

Symbol	Descriptions	Units	Coefficient Values Anabaena	Fragilaria
μp_i	P maximum growth rate	min^{-1}	$6.18x10^{-4}$	$6.25x10^{-4}$
μn_i	N Maximum growth rate	min^{-1}	$5.90x10^{-4}$	$5.90x10^{-4}$
KD_i	Respiration rate	min^{-1}	$3.40x10^{-5}$	$3.40x10^{-5}$
KPL_i	P half saturation const. for growth	$10^{-9}\mu M$ P/cell	1.13	4.64
KNL_i	N half saturation const. for growth	$10^{-9}\mu M$ N/cell	23	45.7
VPL_i	P deficient max. P uptake rate	$10^{-9}\mu M$ P/cell-min	0.403	0.0533
VPI_i	P sufficient max. P uptake rate	$10^{-9}\mu M$ P/cell-min	0.0677	$-5.0x10^{-4}$
KPI_i	P cell quota	$10^{-9}\mu M$ P/cell	4.05	2.66
KPA_i	Half saturation const. for uptake, P	μM P/l	1.15	1.06
QPO_i	P subsistance quota	$10^{-9}\mu M$ P/cell	2.63	3.34
VNL_i	N deficient max. N uptake rate	$10^{-9}\mu M$ N/cell-min	5.54	5.54
VNI_i	N sufficient max. N uptake rate	$10^{-9}\mu M$ N/cell-min	-0.49	-0.49
KNI_i	Cell N quota	$10^{-9}\mu M$ N/cell	8.04	8.04
KNM_i	True N half saturation const.	$10^{-9}\mu M$ N/cell	7.94	7.94
KNK_i	Luxury N coefficient	$10^{-9}\mu M$ N/cell	94	94
QNO_i	N subsistance quota	$10^{-9}\mu M$ N/cell	23	45.7
ZP_i	Grazing efficiency	cells/mg-zoo	0.0	$3.6x10^{7}$
O_i	Temperature coefficient	dimensionless	1.1	1.07
$KSOL_i$	Coefficient for light saturation	$Kcal/m^2$-sec	$5.0x10^{-3}$	$1.9x10^{-3}$
PP_i	Net oxygen production	mg-O_2/cell-min	0.0	0.0
FMN_i	N mineralization factor	dimensionless	0.2	0.2
FMP_i	P mineralization factor	dimensionless	0.2	0.2
KS_i	Sinking rates	m/min	0.0	$7.0x10^{-5}$
PCI_i	Ratio of Chl a to structural P	mg Chl<u>a</u>/mg P	1.0	1.0
SB_i	BOD contributions from phytoplankton	mg BOD/cell	0.0	0.0

MODELLING PHYTOPLANKTON CONCENTRATION IN A STRATIFIED LAKE[1]

Ming-pin Wang[2] and Donald R. F. Harleman[3]

Abstract. Emphasis is on linkage between hydrothermal transport and biological transformation processes. Data from a small, artificially fertilized lake is used to demonstrate importance of variable phosphorus cell quota formulation in phytoplankton modelling. Measurements of primary productivity permit determination of certain model parameters.

INTRODUCTION

The modelling of lake eutrophication in thermally stratified lakes requires a comprehensive knowledge of and a linkage of both physical mixing processes and biological processes. Unfortunately, existing eutrophication models often combine an unbalanced temporal and spatial treatment of the two processes. For example, a complex biogeochemical model may be linked to a simple two-layer, epilimnion/hypolimnion physical model. Since the latter is a crude representation of the hydrothermal processes associated with the formation and dynamics of the thermocline, it is not surprising that the transports between the two "mixed layers" are not susceptible to analysis based on hydrodynamic and thermodynamic principles. The result is the necessity of calibrating vertical "exchange coefficients" from the biogeochemical data. Inasmuch as our ability to model thermal stratification and mixing processes is much better developed than our ability to represent complex aquatic biological transformations by mathematical expressions, the above procedures only serve to compound the already difficult problem of calibrating and verifying eutrophication models.

The objective of this paper is to demonstrate the development of a one-dimensional phytoplankton model consisting of two linked sub-models: a hydrothermal model and a biological model. Figure 1 is a schematic diagram illustrating the linkages between the sub-models. The hydrothermal model (to be described later) represents the interplay of solar and atmospheric heat inputs, surface cooling and wind mixing effects. The output at a time step of one day, is the temperature profile and the depth of the uppermost fully mixed layer (FML). The biological model uses a reaction/mixing algorithm analogous to the heating/mixing algorithm of the hydrothermal model to compute the time varying vertical distribution of the biological state variables. It first computes the variation of the concentration by solving the one-dimensional mass balance equations of the following general form (one for each biological state variable):

$$\frac{\partial C}{\partial t} + \frac{1}{A}\frac{\partial}{\partial z}(Q_v C) = \frac{\alpha}{A}\frac{\partial}{\partial z}\left(A\frac{\partial C}{\partial z}\right) + \frac{r_i}{\rho} + \frac{r_e}{\rho} + \frac{BU_iC_i}{A} - \frac{BU_oC}{A} \qquad (1)$$

where C = the concentration of a biological variable (averaged over the horizontal plane)
A = the horizontal area of lake
Q_v = vertical flow rate
α = mass molecular diffusivity
r_i = the time rate of internal addition of mass of substance C per unit volume by reaction processes
r_e = the time rate of external addition of substance C per unit volume by addition across the lateral, free surface and bottom boundaries of the system
ρ = density
B = width of lake
V_i = horizontal inflow velocity
V_o = horizontal outflow velocity
C_i^o = inflow concentration

After the calculation of the one-dimensional mass balance equation, the biological model uses the mixed layer depth calculated from the temperature model to redistribute the mass in the water column. The concentration of the dissolved and suspended material affects temperature through its influence on the absorption of the solar radiation in the water column.

---

[1]Paper presented at the Third International Conference on State-of-the-Art in Ecological Modelling, Colorado State University, May 24-28, 1982.

[2]Doctoral Candidate, R.M. Parsons Laboratory, Department of Civil Engineering, Massachusetts Institute of Technology, presently Senior Engineer, Camp Dresser & McKee, Inc., Annandale, Va. U.S.A.

[3]Ford Professor of Engineering, R.M. Parsons Laboratory, Department of Civil Engineering, Massachusetts Institute of Technology, Cambridge, Mass. U.S.A.

where, S = total clear sunshine hours per day
and DL = daytime hours per day.
The surface absorption coefficient of the incident solar radiation β, is set to 0.5 and measured wind speeds are multiplied by a factor of 0.7 to account for the sheltering and possible fetch limitation effect of the small lake.

Results of Temperature Simulation

Figure 3 shows comparison between model prediction and field data. In general, the wind mixing model satisfactorily simulates the vertical temperature distribution in L227 during summer stratification.

For most of the simulation period, L227 exhibits a very strong stratification pattern. Frequently, the maximum temperature gradient was $5^{o}C/m$ and the Richardson number at the location of the thermocline was on the order of one thousand. At high Richardson numbers, dissipation of the wind generated turbulent kinetic energy in the mixed layer becomes important and the amount of energy available for vertical mixing is reduced. The thickness of the FML varies significantly on a day by day basis as shown in Figure 4. The variability is due to a combination of convective cooling (cool days following warm days) and wind induced mixing.

PHYTOPLANKTON MODELING

Analysis of the Biological Field Data

The phytoplankton simulation period is the same as the temperature simulation period. During the simulation period, biweekly measurements of total dissolved phosphorus (TDP), particulate phosphorus (PP) and chlorophyll-a are available (Schindler, unpublished data). Ortho-P concentration measured during this period by Rigler bioassay (Levine, 1975) indicated levels below 0.1 μg/ℓ. Since ortho-P concentration is so low, the reported TDP value will be referred to as soluble unreactive phosphorus (SUP) in this paper.

In 1973, L227 received 21 weekly additions of 1.14 kg PO_4-P and 14.97 kg NO_3-N during the period from May 22 to October 8 (ELA unpublished data). Nutrients were mixed with lake surface water and were placed in a boat with the drain plug of the boat in the transom removed. The boat then cruised around the lake to empty its nutrient content into the lake. The process was repeated 3 - 6 times taking 30 - 90 minutes altogether (Schindler et al., 1971). The phytoplankton standing crop was found to be governed by the amount of ortho-P added (Schindler and Fee, 1975). The epilimnion concentrations of PP and chlorophyll-a quickly responded to the addition of the nutrients while the SUP concentration remained fairly invariant.

Examples of the measurement of SUP, PP and chlorophyll-a are shown in Figure 5 with the epilimnion depths indicated by arrows. The depth of the epilimnion is defined as the depth where the temperature gradient becomes greater than $1^{o}C/m$. Profiles of PP and chlorophyll-a shown in Figure 5 have two distinctive features: high hypolimnion concentrations of both PP and chlorophyll-a in the anoxic region of the lake. The hypolimnion chlorophyll peak is located near the depth where the oxygen concentration is close to zero. This chlorophyll peak could be composed of large colonial flagellate which which do not occur in significant amounts in the water column above such as Fee et al. (1979) observed in many oligotrophic lakes in the ELA.) Another possibility is photosynthetic bacteria since the measurement method used measured substances which radiate fluorescence in addition to phytoplankton. The high hypolimnetic PP may be bound in decomposers since Levine and Schindler (1980) showed that the rapid uptake of ortho-P by hypolimnion decomposers keeps the ortho-P concentration low in this region. Our simulation is limited to fitting observed chlorophyll-a and PP concentration of the upper 4 m of the lake. 4 m represents the lower bound of the oxic layer for most of the simulation period.

Simulation with the Minimum Number of State Variables - Model I

Model Description

Of the various forms of phosphorus measured, our main interest is the amount of phosphorus retained by phytoplankton, that is, phytoplankton-P. The data shows that SUP remained fairly constant (∿ 10 μg/ℓ) for the whole period while PP in the upper layer fluctuated over the range 8 - 65 μg/ℓ. Based on this, it is assumed that SUP represents a fraction of dissolved P which is not actively involved in the ortho-P - PP dynamics and contributes very little to the PP fluctuation. This assumption further reduces the number of P compartments to two: ortho-P and phytoplankton-P.

We assume that the measurement PP is phytoplankton-P based on the rapid autolysis of phosphorus observed by scientists (Marshall and Orr, 1955; Golterman, 1960). Two measurements--chlorophyll-a and PP--represent phytoplankton. The ratio of PP to chlorophyll-a in the upper 4 m of the lake, based on the record of routine chemical measurements, is plotted in Figure 6. It may be observed that the ratio PP/chlorophyll-a is neither constant with time nor with depth. This variability of PP to chlorophyll-a ratio with time and space suggests a variable phosphorus cell quota formulation, which relates the phytoplankton growth to the intracellular content of the limiting nutrient.

Model I consisting of three state variables--ortho-P, phytoplankton-P and phytoplankton biomass--is selected as the simplest biologically plausible model with the ability to simulate the field measurements. The schematic representation of Model I and the relationship between model state variables and measurement are shown in Figure 7 and Table 1.

Table 1. Model state variables vs. field measurement Model I

$$\begin{pmatrix} 1 & 0 & 0 \\ 0 & r & 0 \\ 0 & 0 & 1 \end{pmatrix} \begin{pmatrix} S \\ X \\ XQ \end{pmatrix} = \begin{pmatrix} \text{ortho-P} \\ \text{chlo-a} \\ \text{PP} \end{pmatrix}$$

r = chlorophyll-a/biomass ratio, assumed to be 0.01

Field data plays an important role in identifying the final biological model structure and is therefore essential to any modelling effort. The phytoplankton concentrations in Lake 227 (L227) of the Experimental Lake Area (ELA) of the Freshwater Institute, Winnipeg, is used as a case study to illustrate considerations of the number of model state variables and to illustrate the importance of the physical and biological linkage.

Characteristics of Lake 227

L227 is a small circular lake (Figure 2) surrounded by tall jack pines, with a surface area of 5×10^4 m^2, a volume of 2.2×10^5 m^3 and a mean depth of 4.4 m. The deepest part of the lake is approximately 10 m. There is no stream flowing into the lake; surface runoff from the 49-hectare drainage area is the main source of inflow. At the northwest corner of the lake there is a beaver dam outlet (Johnson et al., 1971).

The outflow from the lake is highest in April and May following the spring thaw. The hydraulic retention time of L227 is estimated to be 1.4 - 2.9 years for the period 1970 to 1974 (Schindler, unpublished data). Ice covers the lake from November to April. The spring mixing is incomplete. From May to October, the lake is stratified; in late October, the lake is fully mixed; the fully mixed period is usually short (Schindler et al., 1971).

Since 1969 the lake has been artificially enriched with phosphorus (P) and nitrogen (N). L227 is generally free from nutrient input other than artificial fertilization. The artificial input of P,N is estimated to be approximately 10 times the natural input (Schindler et al., 1971). The lake responds quickly to the addition of nutrients. A few weeks after the initial addition, "the lake was transformed to a teeming, green soup" (Schindler, 1974). Since then, the lake has become eutrophic. The summer chlorophyll-a concentration reached 200 μg/ℓ. The extinction coefficient reached 2.8/m, coinciding with the occurrence of high chlorophyll-a concentrations.

TEMPERATURE SIMULATION

The MIT wind mixing model (WDMIX) (Bloss and Harleman, 1979, 1980 and Harleman, 1982) accounts for water surface heat exchange by solar and atmospheric radiation, back radiation, evaporation and convection heat losses, internal absorption of solar radiation, mixing due to convective cooling and mixing due to turbulent kinetic energy generated by wind shear at the surface.

The emphasis of this simulation is to predict the vertical temperature profile and the depth of the FML from given meteorological conditions. The basic equation for heat transport in the vertical direction is approximated by considering heat flow through an internal control volume taken as a thin horizontal layer of the water body. The amount of heat variation caused by inflow/outflow in L227 is small when compared with other heat sources. Also the vertical advection velocity is zero, therefore a simplified one-dimensional heat balance equation for an internal element applies:

$$\frac{\partial T}{\partial t} = \frac{D}{A}\frac{\partial}{\partial z}\left(A\frac{\partial T}{\partial z}\right) - \frac{1}{A\rho c}\frac{\partial(\Phi_z A)}{\partial z} \quad (2)$$

where, T = temperature
D = heat molecular diffusivity
c = heat capacity of water
Φ_z = the transmission of internal short wave solar radiation at elevation z given by Dake and Harleman (1969):

$$\Phi_z = \phi_{sn}(1-\beta)e^{-\eta}(z_s - z) \quad (3)$$

where, z_s = water surface elevation
ϕ_{sn} = net incident flux of short wave solar radiation (joules/m^2-day)
β = fraction of ϕ_{sn} absorbed at the surface (approximately 0.4 to 0.5)
η = extinction coefficient of water

The equation (2) is solved using an explicit finite difference scheme.

To satisfy the surface boundary condition, the governing equation for the surface element of thickness Δz_s includes (in addition to the terms in equation 2) the heat fluxes due to surface phenomena, i.e., back radiation ϕ_{br}, evaporation ϕ_e, convection ϕ_c, atmospheric radiation ϕ_{an}, and the portion of the solar radiation which is absorbed at or near the surface $\beta\phi_{sn}$,

$$\frac{1}{\rho c \Delta z_s}[\beta\phi_{sn} + \phi_{an} - \phi_{br} - \phi_e - \phi_c] \quad (4)$$

It is assumed that there is no heat flux to or from the earth through the bottom or sides of the lake.

Input Data to the Hydrothermal Model

L227 is divided into twenty horizontal slices of Δz = 0.5 m thickness but of variable areas which conform to the actual bathymetric data shown in Figure 2.

Meteorological data for L227 were collected at the Rawson Lake Meteorological Station which is approximately two miles from L227. The values of extinction coefficients were determined from measurements made at biweekly intervals. The simulation period is chosen from May 15 (shortly after the melting of the ice cover) to October 29, 1973 (135th to 302nd day).

Direct measurement of atmospheric longwave radiation was not available. Wunderlich's (1972) formula which modifies Swinbank's (1963) clear sky formula to include cloud effect is used for estimation:

$$\phi_a = 0.937 \times 10^{-5}\,\sigma\,T_a^6\,(1.0 + 0.17\,C^2) \quad (5)$$

where, ϕ_a = atmospheric longwave radiation
σ = Stefan-Boltzmann constant = 1.171×10^{-6} kcal/m^2-day-K^4
T_a = air temperature in K
C = fraction of the sky covered by clouds (range 0 to 1)

Cloud cover in equation (5) may be estimated from the record of total clear sunshine hours:

$$C = 1 - \frac{S}{DL} \quad (6)$$

The number of model state variables is the same as the number of field measurements. The 1-D mass balance equations for this biochemical model are shown in Table 2.

The model solves the one-dimensional mass balance equations for the biological state variables using an explicit finite difference method similar to that used in the temperature model for solving the heat balance equation. However, the computation time step for the biological model is set to 0.1 day, or one tenth of the computation time step used in the temperature simulation. The light intensity at each computing time step within the day that is needed for computing f(I) is estimated using a half sinusoidal function with a period of 24 hours:

$$I(t) = \pi I_{avg} \sin(\frac{2\pi t}{24}) \qquad (7)$$

where, I_{avg} is the daily average light intensity

The initial concentrations of ortho-P, phytoplankton-P and phytoplankton biomass are obtained from the chemical measurement record using the measured particulate-P as phytoplankton-P and 100 times the measured chlorophyll-a as the phytoplankton biomass. Temperature needed for computing h(T) is obtained from the temperature model output while the quantity and schedule of ortho-P additions are available from ELA unpublished records.

The weekly addition of ortho-P is assumed to increase the ortho-P concentration of the FML, that is, the increase of the ortho-P concentration, i_s, in the water column is computed as:

$$i_s = \frac{1140}{VOL_{FML}} \quad \text{for FML} \qquad (8a)$$

$$= 0 \quad \text{for non-FML} \qquad (8b)$$

where VOL_{FML} is the volume of the FML.

Table 2. 1-D mass balance equations - Model I

1-D Mass Balance Equations:

$$\frac{\partial S}{\partial t} = \frac{\alpha}{A}\frac{\partial}{\partial z}(A\frac{\partial S}{\partial z}) + D(S_i - S) + i_s - \rho X + (k_1 + k_2)(XQ)$$

$$\frac{\partial X}{\partial t} = \frac{\alpha}{A}\frac{\partial}{\partial z}(A\frac{\partial X}{\partial z}) + D(X_i - X) + \mu X - k_2 X - \frac{\partial}{\partial z}(gX)$$

$$\frac{\partial (XQ)}{\partial t} = \frac{\alpha}{A}\frac{\partial}{\partial z}(A\frac{\partial (XQ)}{\partial z}) + D(X_i Q_i - XQ) + \rho X - (k_1 + k_2)(XQ) - \frac{\partial}{\partial z}(gXQ)$$

Rate Expressions:

$$\mu = \mu_m f(I)\, g(Q)\, h(T)$$

$$f(I) = \frac{I}{I + k_I}$$

$$g(Q) = \frac{Q - q}{Q}$$

$$h(T) = \exp(-2.3 \left| \frac{T - T_{opt}}{T_{div}} \right|)$$

$$\rho = \rho_m \frac{S}{S + k_\rho}$$

Parameters to be Calibrated:

g, k_1, k_2, k_I, k_ρ, T_{opt}, T_{div}, ρ_m

μ_m, q

Constraints:

All variables are non-negative values
Q should have an upper bound which corresponds to the maximum phosphorus content in phytoplankton

Definitions:

- S - ortho-P
- Q - P/biomass ratio
- I - light
- μ - growth rate
- D - dilution rate
- g - settling rate
- k_1 - ortho-P excretion rate
- k_2 - mortality rate
- X - phytoplankton
- T - temperature
- ρ - uptake rate
- i_s - input of ortho-P
- α - mass molecular diffusivity
- A - lake area

*Subscript i denotes the inflow concentration

The inflow concentration of the biological variables are assumed to be negligible and the outflow is assumed to occur within the surface element.

At the end of one day, or after 10 computational time steps, the biological model uses the FML computed from the temperature model to redistribute the chemical state variables in the water column.

Simulation Results and Analysis

The satisfactory fitting between the field measurement of PP and the model prediction of phytoplankton-P with zero settling velocity (Figure 8) suggests that phosphorus added to the lake surface was effectively retained in the fully mixed layer (FML) up to the 204th day. The reduction of the PP concentration in the FML is caused by the deepening of the FML by entrainment of water from below which had a lower PP concentration.

The corresponding comparison between the predicted and measured chlorophyll-a distributions in the upper 4 m from the beginning of the simulation period (day 135) to day 204 is shown in Figure 9. The calibrated parameter values for Model I are given in Table 3.

After the 204th day, the model predicts a continuing increase of PP concentrations and of chlorophyll-a in the FML as shown in Figures 10 and 11. This is contrary to the data which shows a rather rapid decline in both state variables after day 204. Attempts were made to simulate the observed decline by introducing a non-zero settling rate for the phytoplankton after day 204. In order to reproduce the decrease in observed PP and biomass data, a time varying settling rate with a maximum value of 0.3 m/day was required. Since no convincing explanation or justification for this could be found, it was decided to increase the complexity of the phytoplankton model by introducing an additional state variable--zooplankton.

Simulation with Zooplankton as Additional Variable

Model II

Model Description

The inclusion of zooplankton in the model introduces additional model parameters. The elements which have been considered in modeling zooplankton dynamics include speciation (Thomann et al., 1975), differentiation between egg, juvenile and adult (Kremer and Nixon, 1978), and selective feeding (Park et al., 1974). The vertical migration of zooplankton compounds the difficulty (Bowers, 1979). While a detailed consideration of the zooplankton dynamics is important in its own right, the lack of information hardly justifies an elaborate approach in our present modeling attempt. Three fundamental processes of zooplankton are presently considered: growth, excretion and mortality. The description of each of these three processes are kept in the simplest possible form.

Model II is presented in Figure 12 and Tables 4 and 5.

Table 3. Model Parameters - Model I

Parameter Values

Symbol	Definition	Unit	Value	Source
ρ_m	max. PO_4^{-3} uptake rate	$(g\text{-}P)(g\text{-phytoplankton})^{-1}$ day^{-1}	0.3	Calibration
K_ρ	Michaelis constant for PO_4 uptake by phytoplankton	μg/ℓ	20.	Calibration
g	settling velocity	$m(day)^{-1}$	0.	Calibration
q	minimum P cell quota	$(g\text{-}P)(g\text{-phytoplankton})^{-1}$ day^{-1}	0.004	Calibration
T_{opt}	optimum temperature for photosynthesis	^oC	25.	Calibration
T_{div}	temperature coefficient	^oC	15.	Calibration
μ_m	maximum phytoplankton growth rate	day^{-1}	1.	Assumption
K_1	half saturation constant for light function	langley/day	15.	Calibration
K_2	phytoplankton mortality rate	day^{-1}	0	Assumption

*Initial concentrations of ortho-P, phytoplankton, phytoplankton-P are obtained from the ELA chemistry measurement notebook.

Table 4. Model State Variables vs. Field Measurement - Model II

$$\begin{bmatrix} 1 & 0 & 0 & 0 \\ 0 & r & 0 & 0 \\ 0 & 0 & 1 & 0 \\ 0 & 0 & 0 & 1 \end{bmatrix} \begin{bmatrix} S \\ X \\ XQ \\ Z \end{bmatrix} = \begin{bmatrix} \text{ortho-P} \\ \text{chlo-a} \\ \text{PP} \\ - \end{bmatrix}$$

r = chlorophyll-a/biomass ratio assumed to be 0.01

In the absence of zooplankton or phytoplankton settling, the epilimnion chlorophyll-a concentration data (Figure 11) suggest that 90% of this population was lost in two weeks time (day 218). On the other hand, the PP simulation result (Figure 10) indicates that 50% of phosphorus was lost in the same period. This suggests an increase in phosphorus cell quota (as reflected by PP/chlorophyll-a ratio) during this period of net decrease of phytoplankton population. The inclusion of zooplankton grazing followed by a subsequent excretion of ortho-phosphorus will produce this effect.

Simulation Results and Analysis

The results of the Model II simulation are presented in Figure 13 and Figure 14. The inclusion of zooplankton results in a decrease in the epilimnion chlorophyll-a and particulate P concentrations after the 204th day. Table 6 summarizes the additional parameter values and initial concentrations used to produce the results shown in Figures 13 and 14. Parameters which are common to both Models I and II are kept the same, as shown in Table 3.

In Model I, the initial concentration of each state variable is provided by the field measurement while in Model II, the initial concentration of zooplankton has to be specified without the benefit of measurement. The initial concentration of zooplankton should, therefore, be considered as a calibration parameter.

PARAMETER ESTIMATION AND FUNCTION EVALUATION FROM PRIMARY PRODUCTIVITY DATA

Primary productivity usually refers to the rate at which phytoplankton assimilates organic carbon from the dissolved inorganic carbon (DIC). Since carbon is the major cellular constituent of the phytoplankton, the primary production rate should closely reflect the increase of the phytoplankton biomass. If the assumption is made that the cellular C constituent is the least variable cellular constituent, that is the average C content among the algae is invariably 53 ± 5% (Wetzel, 1975), then the net carbon increase rate is linearly related to the growth rate and primary productivity is closely related to the phytoplankton growth rate.

Since 1973, primary productivity in ELA has been measured with an incubator technique (Shearer and Fee, 1974). The measured U_{ij} was obtained by applying liquid scintillation counting of C^{14} on the CO_2-free water sample (Schindler et al. 1972), therefore, it can be related to the carbon fluxes by:

$$U_{ij} = kX_i[(k_p)_{ij} - r_{ij} - (r_p)_{ij}] \tag{9}$$

where, U_{ij} = the measured carbon uptake rate for incubation session i under a given light intensity I_{ij}, in units of mg $C/m^3/hr$
k = the carbon to biomass ratio
X_i = represents the phytoplankton concentration during the incubation session i
r_{ij} = the respiration rate
$(r_p)_{ij}$ = the photorespiration rate

If plankton growth rate is linearly related to the net carbon increase rate, which is defined as [photosynthetic rate - respiration rate - photorespiration rate - excretion rate], then the measured primary productivity may be related to the model's definition of phytoplankton growth rate by:

$$U_{ij} - kX_i(r_e)_{ij} = kX_i\mu_{ij} \tag{10}$$

where $(r_e)_{ij}$ and μ_{ij} are the excretion and growth rates, respectively, during the incubation session i under a given light intensity I_{ij}. Phytoplankton growth rate in the model is assumed to be a function of light, temperature and cellular phosphorus content, that is μ_{ij} may be expressed as:

$$\mu_{ij} = \mu_m f(I_{ij})g(T_i)h(Q_i) \tag{11}$$

with the assumption that in each incubation period, light is the only variable. Substituting Equation (11) into Equation (10):

$$U_{ij} - kX_i(r_e)_{ij} = kX_i\mu_m f(I_{ij})g(T_i)h(Q_i) \tag{12}$$

The Light Function

Equation (12) may be rearranged into:

$$\frac{U_{ij} - kX_i(r_e)_{ij}}{kX_i\mu_m g(T_i)h(Q_i)} = f(I_{ij}) \tag{13}$$

The light function f(I) may be obtained from the ratio of the measured primary productivity, U_{ij}, to the maximum primary productivity measured during each incubation, $U_{i\,max}$, if either of the following two conditions exists:

(1) $kX_i(r_e)_{ij}$ is small compared to U_{ij}

(2) $(r_e)_{ij}$ is also a function of $f(I_{ij})$

Eighteen incubation results for the period from May 28 to October 29 were used; for each incubation there were four pairs of $(U_{ij}/U_{i\,max}, I_{ij})$. The value of $U_{i\,max}$ is estimated from the incubation result for each incubation session i. If the light dependence function is stationary, these data points may be used to estimate parameter values needed for the light dependence function. Three functions of light dependence are considered as shown in Table 7: Hyperbolic function, Smith function and Steele function. A nonlinear regression package from the Troll system developed at MIT is used to find the best fitted parameter for each function.

Table 5. 1-D Mass Balance Equations
Model II

1-D Mass Balance Equations:

$$\frac{\partial S}{\partial t} = \frac{\alpha}{A}\frac{\partial}{\partial z}\left(A\frac{\partial S}{\partial z}\right) + D(S_i - S) + i_s - \rho X + (k_1 + k_2)(XQ) + \left(\frac{1}{f} - 1\right)\mu_z QZ$$

$$\frac{\partial X}{\partial t} = \frac{\alpha}{A}\frac{\partial}{\partial z}\left(A\frac{\partial X}{\partial z}\right) + D(X_i - X) + \mu X - k_2 X - \frac{\partial}{\partial z}(gX) - \frac{1}{f}\mu_z Z$$

$$\frac{\partial (XQ)}{\partial t} = \frac{\alpha}{A}\frac{\partial}{\partial z}\left(A\frac{\partial (XQ)}{\partial z}\right) + D(X_i Q_i - XQ) + \rho X - (k_1 + k_2)(XQ) - \frac{\partial}{\partial z}(gXQ) - \frac{1}{f}\mu_z QZ$$

$$\frac{\partial Z}{\partial t} = \frac{\alpha}{A}\frac{\partial}{\partial z}\left(A\frac{\partial Z}{\partial z}\right) + D(Z_i - Z) + (\mu_z - r_z)Z$$

Rate Expressions:

$$\mu_z = \mu_{zm}\frac{X}{X+k_X}\exp\left(-2.3\left|\frac{T-T_{zopt}}{T_{zdiv}}\right|\right)$$

$$r_z = r_{zm}\exp\left(-2.3\left|\frac{T-T_{zopt}}{T_{zdiv}}\right|\right)$$

μ, ρ are the same as those defined in Table 3

Parameters to be Calibrated:

μ_{zm}, r_{zm}, k_x, f, T_{zopt}, T_{zmax}

and all the calibrated parameters listed in Table 3.

Definitions:

Z - zooplankton
r_z - zooplankton mortality rate
μ_z - zooplankton growth rate which is a constant fraction (f) of the grazing rate ρ_z
f - efficiency of zooplankton grazing
α - mass molecular diffusivity

All other symbols are the same as those defined in Table 3.

Table 6. Additional parameter values - Model II

Symbol	Definition	Unit	Value	Source
μ_{zm}	maximum specific zooplankton growth rate	day^{-1}	0.25	Calibration
r_{zm}	specific zooplankton respiration and mortality rate	day^{-1}	0.1	Calibration
T_{zopt}	optimum temperature for zooplankton growth	^oC	20	Calibration
T_{zdiv}	temperature coefficient for zooplankton growth	^oC	20	Calibration
k_X	half saturation coefficient for zooplankton grazing	μg/ℓ	100	Calibration
f	efficiency of zooplankton grazing	-	0.6	Chen & Orlob (1975)

Initial concentration for zooplankton: 100 μg/ℓ for upper 3 meters

0 for the depth below 3 meters

Table 7. Regression of specific primary productivity as a function of light

Dependent variable: U_{ij}/U_{imax}

Independent variable: I_{ij}

Parameter: K_I

	Hyperbolic	Smith	Steele
	$\frac{I}{I+K_I}$	$\frac{I}{\sqrt{K_I+I^2}}$	$\frac{I}{K_I}\exp(1-I/K_I)$
K_I	12.9 ± 1.3	628 ± 100	130 ± 10
r	80%	80%	51%

Table 7 presents the regression result for the year 1973, where K_I is the best fitted parameter value for the light function in ly/day. Both the hyperbolic and Smith functions explain 80% of the data variance. The K_I value obtained from this regression for the hyperbolic function is 13 ± 1.3, which agrees well with the value of 15 obtained by calibrating with the chlorophyll-a field measurements (see Table 3).

The Temperature and Nutrient Functions

The function $f(I_{ij})$ has a value between 0 and 1 and when it is equal to 1, the maximum primary productivity in each incubation i U_{imax} should be related to temperature and cell quota by:

$$U_{imax} - kX_i(r_e)_{imax} = kX_i\mu_m g(T_i)h(Q_i) \qquad (14)$$

U_{imax} is a function of $g(T_i)h(Q_i)$ if either of the following conditions exist:

(1) $kX_i(r_e)_{imax}$ is negligible compared to U_{imax}

(2) $(r_e)_{imax}$ is also a function of $g(T_i)h(Q_i)$

If either of the above conditions exists, then the functions g(T) and h(Q) may be obtained from the information on U_{imax}, X_i, T_i and Q_i provided that the two functions are stationary functions. Unfortunately, only T_i has been recorded for these incubation measurements of the primary production rate. However, since there remains a significant gap between growth rate and photosynthetic rate, the additional knowledge acquired from analyzing the primary production data can be of tremendous importance in gaining insight into the phytoplankton behavior. It would be a meaningful task to attempt analyzing the dependence of the specific primary activity on nutrient and temperature, even with estimated values of X_i and Q_i. The same nonlinear regression package in the Troll system is used to estimate the best fitted parameters for the following three possible functions:

Function I: $\mu = a(1-q/Q)$

Function II: $\mu = a\exp(-2.3\,\frac{T_{opt}-T}{T_{div}})$

Function III: $\mu = a(1-q/Q)\exp(-2.3\,\frac{T_{opt}-T}{T_{div}})$

The constant P cell quota formulation is not considered since it is apparent from the field measurement that no detectable amount of phosphorus was available and therefore, it will not be possible to find the correlation between primary productivity and ortho-P concentration.

The regression results are presented in Table 8. Function I can only explain 20% of the observed variance in the data while divergence occurred for Functions II and III during the process of searching for the best fitted parameters. An alternative approach to parameter estimation is simulation. The simulation results are shown for Functions II and III and again, neither of the functions can satisfactorily explain the data variation.

In view of the often delayed and variable response of the photosynthesis to the nutrient addition (Healey, 1979), care must be taken in fitting the short term photosynthesis measurement. It is assumed that a better fit for the period before the sharp decline of the phytoplankton biomass may exist for the following two reasons:

(1) The ratio of added nutrient to biomass is smaller due to the higher biomass in the earlier period, and

(2) The effect of zooplankton on ortho-P cycling is insignificant.

A regression was performed for the first 7 values which were taken from the period before the occurrence of chlorophyll-a peak. In this case, Function I is able to explain 80% of the variance; while the regression package was unable to find any realistic parameter values for both Function II and Function III. The insignificant relationship between temperature and the primary productivity may be due to the fact that there is little difference in the temperatures for the seven data sets (16 to 21°C). Although the small number of data sets prevents the making of statistically profound judgements on the estimation of two parameters, this result certainly suggests the possibility of a significant relationship between the specific primary productivity and P-cell quota.

The estimated μ_{max} from the regression is 5 µg-C/µg-(chlo-a)/hr corresponding to 60 µg-C/µg-(chlo-a)/day for 12 day-time hours. Since it was assumed that chlorophyll-a is approximately 1/100 of the phytoplankton biomass and the carbon content is estimated to be 53 ± 5% of the ash-free dry weight, the estimated maximum primary productivity then corresponds to a growth rate of μ_m = 1.0 to 1.2 per day which is close to the value 1.0/day used in the model. (See Table 3.) The reasonable agreement between the growth rate and minimum cell quota used in the model and the same parameters estimated from primary productivity data suggests that carbon excretion rate was small during the pre-peak period since the measured primary productivity represents the sum of net carbon increase rate and excretion rate. This confirms the assumption used in performing the parameter estimation for the growth function from the U_{ij} data. However, this regression gives no information on the respiration rate.

Table 8. Regression of specific primary productivity rate as a function of temperature and nutrient.

Dependent variable: μ_{max}

Independent variables: T,Q

Parameters: q, T_{opt}, T_{div}, a

	Function I		*Function II	*Function III
	all data	the first 7 data		
a	2.4	5	0.8	1.6
q	0.2	0.3	-	0.2
T_{opt}	-		50.	35.
T_{div}	-		50.	50.
r	20%	80%	3%	22%

*Divergence occurred during the search of best fitted parameters, values shown are from simulation analysis with the following specifications:

$0.1 \leq a \leq 3.0$	$\Delta a = 0.1$
$0.1 \leq q \leq 2.0$	$\Delta q = 0.1$
$10. \leq T_{opt} \leq 50.$	$\Delta T_{opt} = 5.$
$10. \leq T_{div} \leq 50.$	$\Delta T_{div} = 5.$

PARAMETER TRANSFERABILITY TO YEAR 1974

Data Analysis

In 1974, L227 received the same nutrient addition pattern as in 1973. The lake was fertilized from May 21 to October 9 with 21 weekly additions. During each fertilization, 1.14 kg and 14.97 kg of PO_4-P and NO_3-P, respectively, were dissolved and mixed in the water taken from the lake surface. These mixtures were then added to the lake surface from a boat.

The lake response to these additions can be seen from the measurement of ortho-P, TDP, PP and chlorophyll-a concentrations. Epilimnion ortho-P, measured by Rigler's bioassay method, was in the order of 0.1 μg/l which is similar to the concentration of the ortho-P measured in 1973 by the same method. Epilimnion TDP concentration, measured by regular chemical analysis, remained fairly constant (∿8 μg/ℓ) from May to October of 1974 which is also similar to the TDP concentration trend of the previous year. However, the maximum epilimnion chlorophyll-a concentration in 1974 was only 25% of the maximum chlorophyll-a concentration in the year 1973 and the maximum epilimnion concentration of PP in 1974 was 60% of the maximum PP concentration in the year 1973 (Figures 15 and 16).

Phytoplankton Simulation

As in the previous year, the phytoplankton simulation started with the temperature simulation obtained using the MIT wind mixing model. The simulation period is from May 21 to July 30 because the daily record of direct solar radiation is missing for most of the period from August 8 to November 31 due to instrument failure. In general, the wind mixing model satisfactorily simulates the measured lake temperature in 1974.

Model II, was used for phosphorus and chlorophyll-a simulation. The initial zooplankton concentration remains as a model parameter which is specified to be one fifth of the phytoplankton initial concentration as was done in the year 1973. The model predicts a continuous increase of both PP and chlorophyll-a concentration in the FML until the 210th day with peak PP and chlorophyll-a concentration one and four times higher than the measurements. Clearly, a modification of the model parameters is needed.

Information on the phytoplankton response to the nutrient addition may be gained from primary productivity data for 1974. Table 9 compares four pairs of maximum specific primary productivity under the optimum light condition and the corresponding phosphorus cell quota during the period covering the 148th to the 190th day of 1973 and 1974. Comparing 148th 1974 vs. 151st 1973 and 162nd 1974 vs. 158th 1973, it may be observed that the values of P_{max}/chl-a in 1974 are only half of the corresponding values of 1973 while the values of PP/chl-a are approximately the same in both years. For the remaining two pairs--184th 1974 vs. 166th 1973 and 190th 1974 vs. 183rd 1973--the P_{max}/chl-a values are the same for the two years while PP/chl-a in 1974 are twice the corresponding values in 1973. In other words, for the same nutrient condition, phytoplankton

Table 9. Primary productivity for L227, 1973 vs. 1974 (for the period 5/28 - 7/9)

Year 1973				Year 1974			
Date	Temp.	P_{max}/chl-a)	PP/chl-a	Date	Temp.	P_{max}/chl-a	PP/chl-a
	(C)	(μg-C/μg-(chl-a)/hr))	(μg-P/μg-(chl-a))		(C)	(μg-C/μg(chl-a)/hr))	(μg-P/μg-(chl-a))
151	17	3.4	0.8	148	12	1.8	1
158	16	3.0	0.7	162	16	1.6	0.7
166	18.5	2.2	0.5	184	20	2.4	1
183	21	1.4	0.4	190	21	1.5	1

in 1973 were able to have higher specific primary productivity than phytoplankton in 1974. If there is no difference in the responses of carbon excretion to nutrient condition, then this result suggests that the minimum cell quota in 1974 is higher than that in 1973. Phosphorus in 1974 was lost at a faster rate in the earlier period than in 1973. There are two mechanisms through which phosphorus can be lost from the water column: settling and zooplankton grazing. There is no available information on the phytoplankton settling behavior during the simulation periods of the two years but zooplankton concentration was observed to be higher in 1974 (Malley, Cheng and Schindler, 1974).

Different combinations of higher zooplankton concentration and higher minimum cell quota for the 1974 simulation were tested and a combination of zooplankton initial concentration equal to the phytoplankton concentration and a minimum cell quota value of 0.005 was found to simulate the field measurement of PP and chlorophyll-a (Figures 17 and 18). The corresponding minimum cell quota for 1973 was 0.004.

The taxonomy composition of phytoplankton of the two years has been studied by Findlay and Kling (1975), Findlay (1976). From mid-May to the end of July, Chlorophyceae dominated in both years. Other common species found in 1974 were also present in the year 1973. Therefore, the differences in phytoplankton responses cannot be explained by species variation.

SUMMARY AND CONCLUSIONS

The importance of linking a hydrothermal model which captures the diurnal variations in the vertical temperature distribution and mixing processes with a biological model which addresses the dynamics of an algal bloom has been demonstrated.

Two successive years of temperature, biological and chemical data from an artificially fertilized lake (L227) in the Canadian Experimental Lakes Area were analyzed. The hydrothermal model predicts upper, mixed-layer depths ranging from 1 to 5 meters during the summer stratification period. Daily variations in the thickness of the mixed layer are shown to affect the biochemical state variables in eutrophication models. Phosphorus cycling models of various levels of complexity are linked to the hydrothermal model. The lake data suggest that algal phosphorus to biomass ratios are not constant, hence only variable cell quota phosphorus models are considered.

The three state variables in Model I are ortho-P, phytoplankton-P and biomass (chlorophyll-a). This model cannot be calibrated to produce reasonable results for the period after the crash of the algal bloom. A fourth state variable, zooplankton, was added in Model II. Good agreement between predictions and observations of algal concentration were obtained with Model II throughout the summer period. Because of the lack of zooplankton data, the initial concentration of zooplankton was a calibrated parameter.

An effort was made to see if primary productivity data could be used to obtain an independent estimate of certain biochemical rate constants that were calibrated from the lake water column data. The maximum phytoplankton growth rate and the light limitation parameter obtained in this manner were in good agreement with the previously calibrated values.

Model II was applied, as a verification test, to data from the following year using the previously calibrated parameters. A reasonable simulation of the smaller algal bloom during the second year was achieved by increasing the initial zooplankton concentration and by a 25% increase in the cell quota.

LITERATURE CITED

Bloss, S. and D.R.F. Harleman. 1979. Effect of wind mixing on the thermocline formation in lakes and reservoirs. Ralph M. Parsons Laboratory Technical Report No. 249, Massachusetts Institute of Technology, Cambridge, Mass.

Bloss, S. and D.R.F. Harleman. 1980. Effect of wind-induced mixing on the seasonal thermocline in lakes and reservoirs. Proc. 2nd Int. Symp. on Stratified Flows, Trondheim, Norway.

Bowers, J.A. 1979. Zooplankton grazing in simulation models: the role of vertical migration. In Perspectives on Lake Ecosystem Modeling, Ann Arbor Science, p. 53-74.

Dake, J.M.K. and D.R.F. Harleman. 1969. Thermal stratification in lakes - analytical and laboratory studies. J. of WRR, Vol. 5, No. 2.

ELA. 1973. Staff of Freshwater Institute, Winnipeg, [personal communication].

Fee, E.J., J.A. Sherer, and D.R. DeClerca. 1979. In vivo chlorophyll porfiles from lakes in the experimental lakes area, northwestern Ontario, Technical Report No. 703. Fisheries and Marine Service, Canada.

Findlay, D.L. 1976. Seasonal successions of phytoplankton in seven lake basins in the experimental lakes area, northwestern Ontario, following artificial eutrophication. Data from 1974 to 1976. Can. Fish. Mar. Serv., Manuscript Report No. 1466.

Findlay, D.L., and H.J. Kling. 1975. Seasonal successions of phytoplankton in seven lake basins in the experimental lakes area, northwestern Ontario follwoing artificial eutrophication. Can. Fish. Mar. Serv. Tech. Report 513.

Golterman, H.L. 1960. Studies on the cycle of elements in freshwater. Acta Bot. Neerl., 9:1-58.

Harleman, D.R.F. 1982. Hydrothermal analysis of lakes and reservoirs. First Hunter Rouse Hydraulic Engineering Lecture, Proc. ASCE, Vol. 108, No. HY3.

Healey, F.P. 1979. Short-term responses of nutrient-deficient algae to nutrient addition. Phycol. 15:289-299.

Johnson, W.E., and J.R. Vallentyne. 1971. Rationale, background and development of experimental lake studies in northwestern Ontario. J. Fish. Res. Bd., Canada, 28:123-128.

Kremer, J.M., and S.W. Nixon. 1978. A coastal marine ecosystem simulation analysis. Springer-Verlag.

Levine, S.N., and D.W. Schindler. 1980. Radiochemical analysis of orthophosphate concentrations and seasonal changes in the flux of orthophosphate to seston in two Canadian shield lakes. Can. J. Fish. Aquatic Sci. 37:479-487.

Levine, S.N. 1975. A preliminary investigation of orthophosphate connection and the uptake of orthophosphate by seston in two Canadian shield lakes. M.S. Thesis, Univ. of Manitoba.

Malley, D.F., P.S.S. Cheng, and D.W. Schindler. 1974. Decline of zooplankton of Lake 227, Experimental Lakes Area, Ontario: 1969-1974. Manuscript (to be published).

Marshall, S.M., and A.P. Orr. 1955. On the biology of Calanus firmarchicus, VIII. Food uptake, assimilation and excretion in adult and stage V Calanus. J. Mar. Biol. Assoc., U.K., 34:495-529.

Park, R.A., R.V. O'Neil, and J.A. Bloomfield. 1974. A generalized model for simulating lake ecosystems. Simulation 23(2):33-50.

Schlindler, D.W. and K.G. Beaty. 1976. Meteorological records for Rawson Lake watershed areas, [unpublished data].

Schlindler, D.W. and E.J. Fee. 1975. The role of nutrient cycling and radiant energy in aquatic communities. In Photosynthesis and Productivity in Different Environments, J.P. Cooper (ed.), Cambridge Univ. Press.

Schlinder, D.W., D.R.S. Lean, and E.J. Fee. 1972. Productivity of world ecosystems. National Academy of Science, Proc. of Symposium.

Schlinder, D.W. 1974. Eutrophication and recovery in experimental lakes: implications for lake management. Science, Vol. 184, pp. 897-899.

Schlindler, D.W., F.A.J. Armstrong, S.K. Holmgren, and G.J. Brunskill. 1971. Eutrophication of Lake 227, experimental lake area, northwestern Ontario, by addition of phosphate and nitrate. J. Fish. Res. Bd., Canada 28:1763-1782.

Shearer, J.A., and E.J. Fee. 1974. Phytoplankton primary production in the experimental lake area using an incubator technique--1973 data. Can. Fish. Mar. Serv., Tech. Report No. 474.

Swinbank, W.C. 1963. Long wave radiation from clear skies. Quart. J. of the Royal Met. Soc. of London, Vol. 89.

Thomann, R.V., and D.M. DiToro et al. 1975. Mathematical modelling of phytoplankton in Lake Ontario. EPA-660/3-75-005.

Wetzel, R.G. 1975. Limmology. W.R. Saunders Company.

Wunderlich, W.O. 1972. Heta and mass transfer between a water surface and the atmosphere. T.V.A. Engineering Laboratory. Report No. 14, Norris, Tennessee.

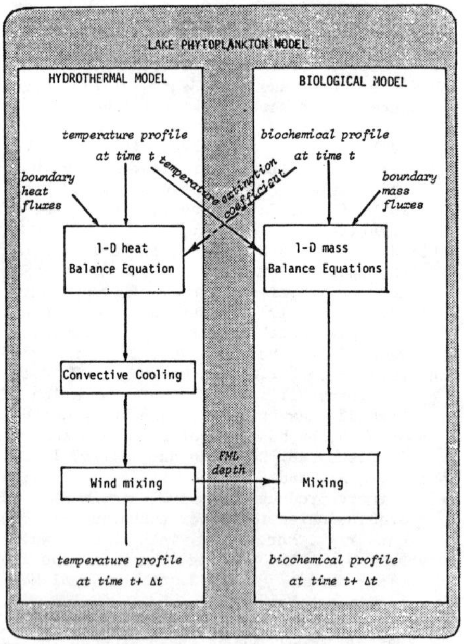

Figure 1. Schematic Diagram of Lake Phytoplankton Model

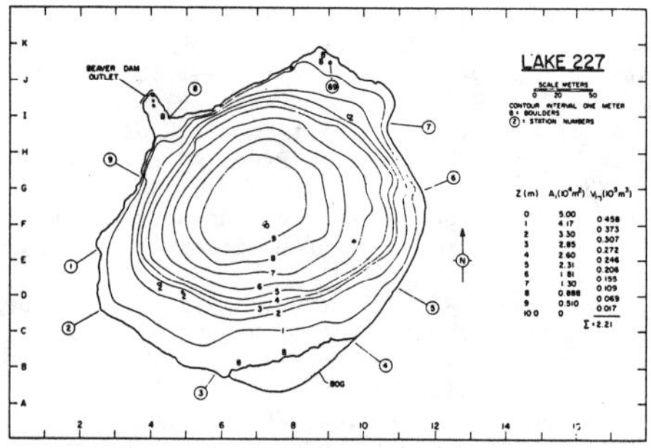

Figure 2. L227 Morphometry (from Johnson et al., 1971)

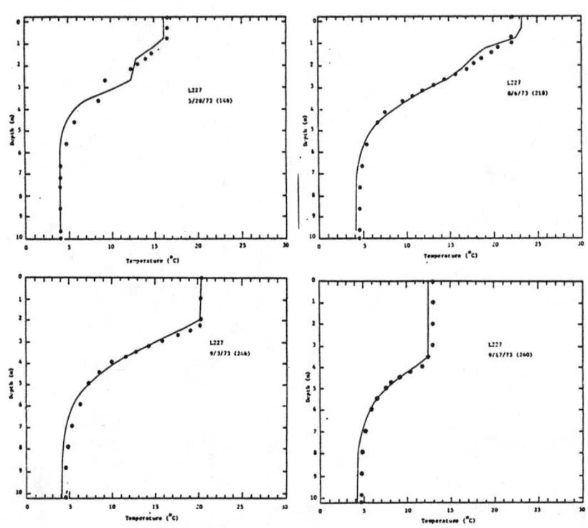

Figure 3. Temperature Simulation vs. Field Measurement

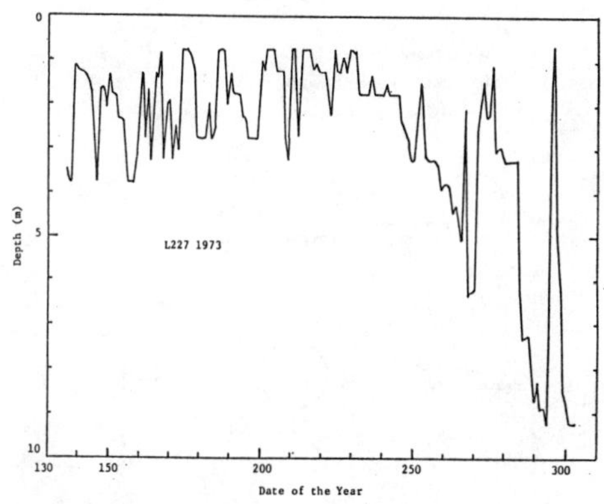

Figure 4. Variation of the Fully Mixed Layer with Time

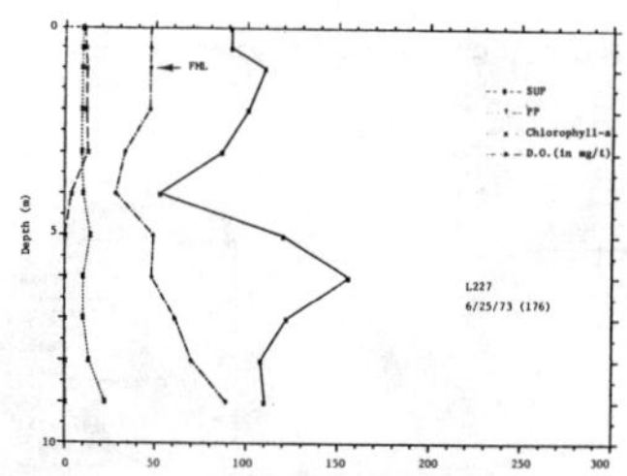

Figure 5. Examples of Chemical Measurements in 1973

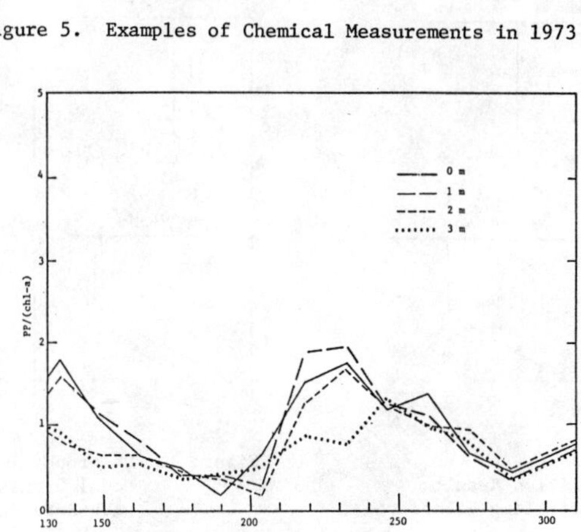

Figure 6. Measured PP/(chlorophyll-a) Ratio at Different Depths

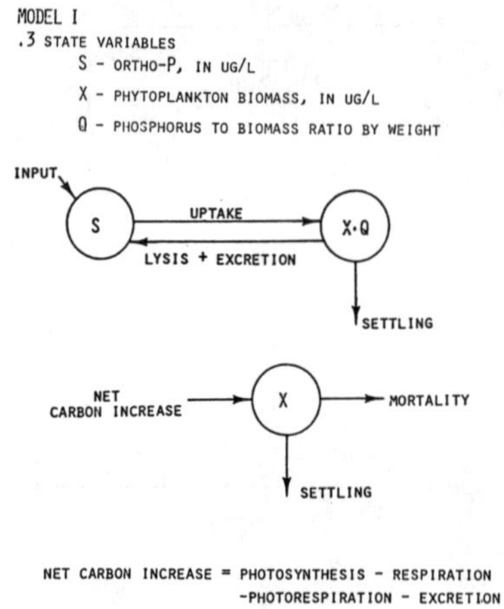

Figure 7. Schematic Diagram - Model I

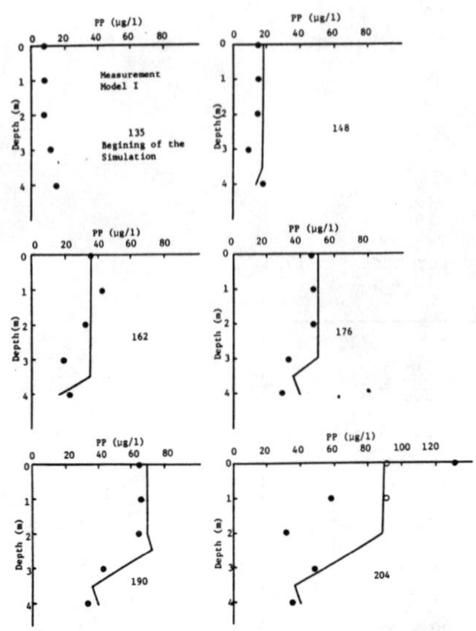

Figure 8. PP profiles - Model Simulation Results with Zero Settling Velocity

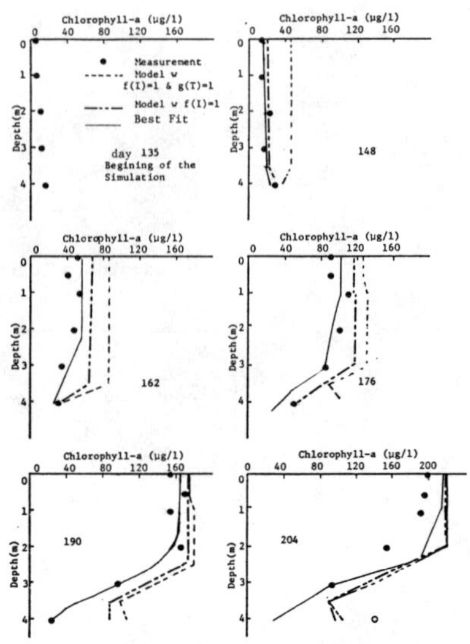

Figure 9. Chlorophyll-a profiles - Model Simulation Results with Zero Settling Velocity

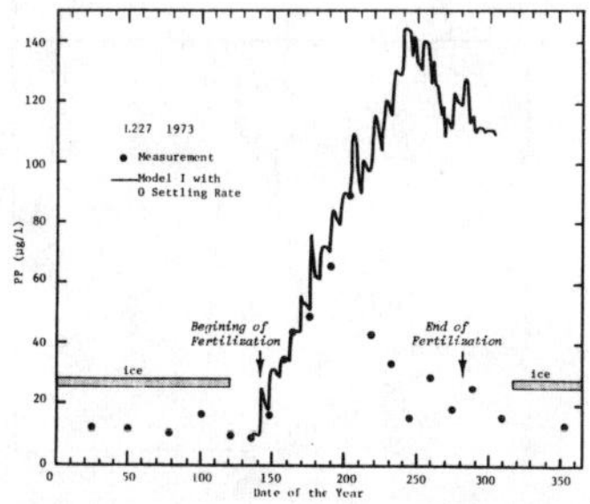

Figure 10. PP Time Series Plot - Model I Simulation Results with Zero Settling Velocity

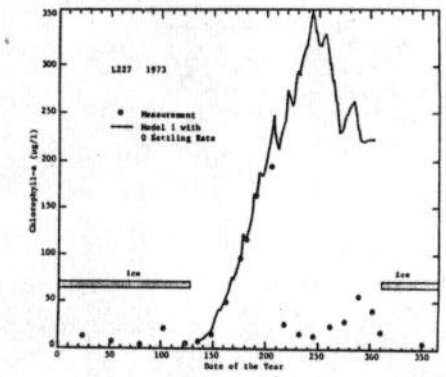

Figure 11. Chlorophyll-a Time Series Plot - Model Simulation Results with Zero Settling Velocity

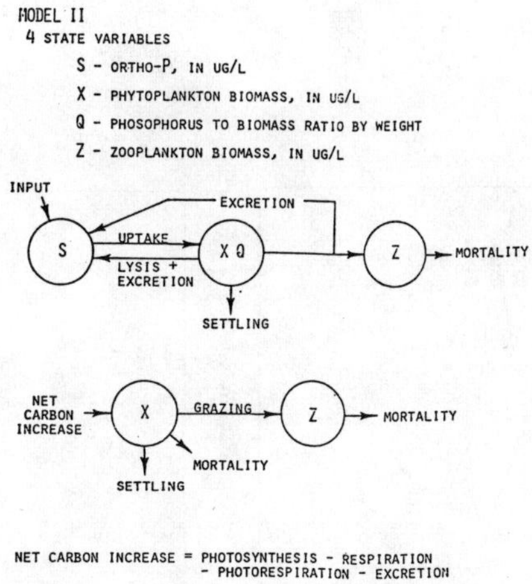

Figure 12. Schematic Diagram - Model II

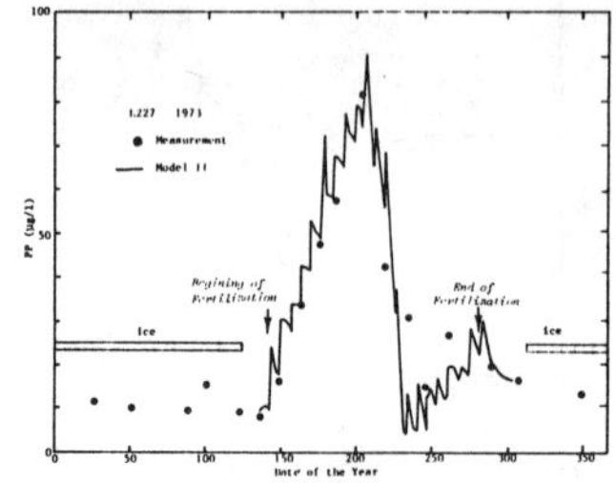

Figure 13. FML Concentration of PP - Model II Simulation Results

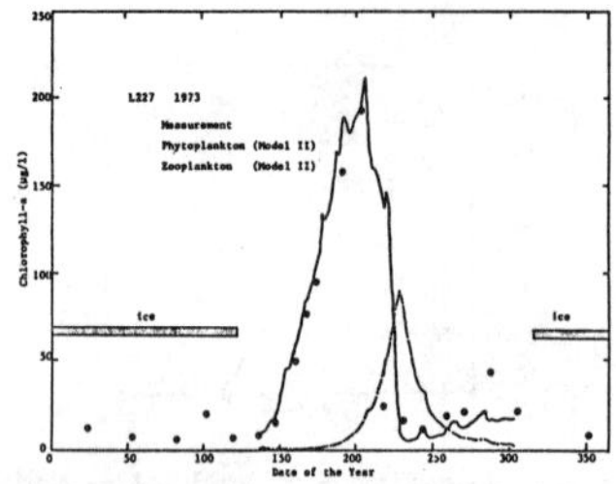

Figure 14. FML Concentration of Chlorophyll-a - Model II Simulation Results

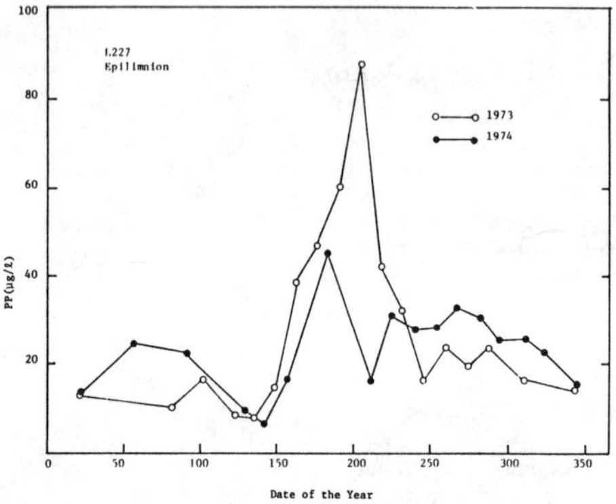

Figure 15. Epilimnion PP Concentration -1974 vs. 1973

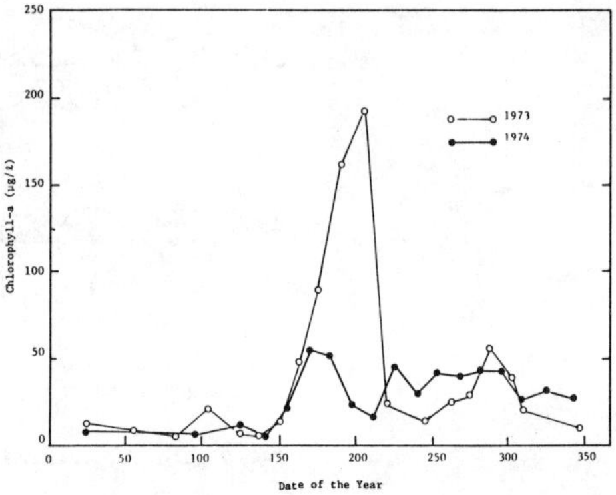

Figure 16. Epilimnion chlorophyll-a -1974 vs. 1973

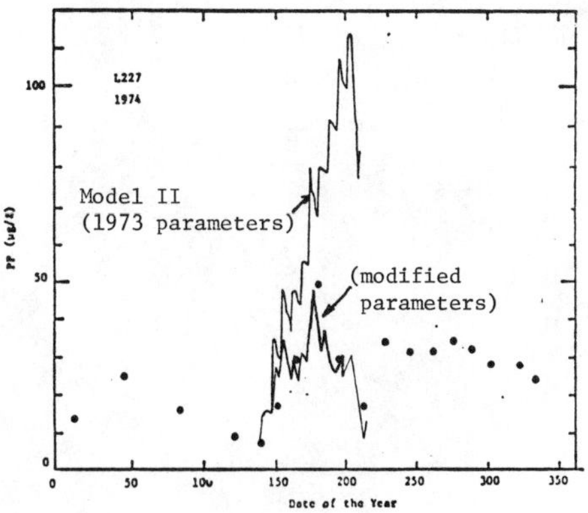

Figure 17. FML PP Concentration, 1974
- Model II Results with 1973 Parameters vs. with Modified Parameters

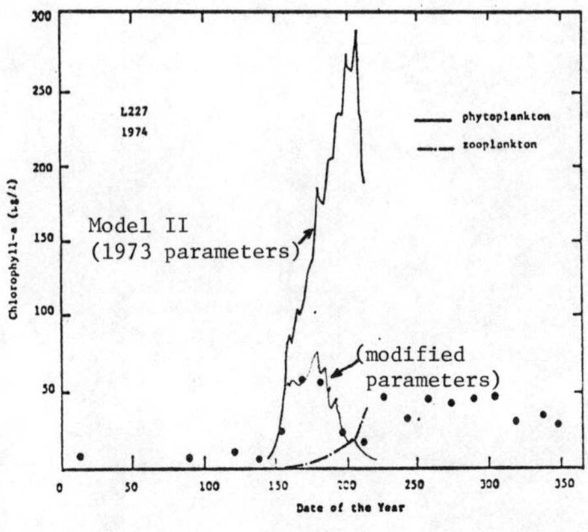

Figure 18. FML Chlorophyll-a Concentration, 1974
- Model II Results with 1973 Parameters vs. with Modified Parameters

A SYSTEMS APPROACH TO EUTROPHICATION MANAGEMENT WITH APPLICATION TO LAKE BALATON[1,2]

László Somlyódy[3]

Abstract.--The problem of the lake-region system studied is characterized by complexity, uncertainty, and by the need to incorporate different levels of analysis such as scientific understanding and policy making. The approach adopted is based on decomposition and aggregation. Detailed studies were made on subprocesses (sediment-water interaction, biochemical and hydrophysical phenomena, nutrient loads, etc.). The essential features of these processes were preserved at the level of water quality management and incorporated in an optimization model accounting for uncertainties.

INTRODUCTION

The artificial eutrophication of lakes and reservoirs is recognized worldwide as a serious problem; it encompasses both the physico-chemical changes of water quality and the economic and social processes that underlie these changes. Excessive algal blooms, provoked by augmented nutrient loads of agricultural, municipal, and industrial origin, reduce the recreational potential of the water bodies and have serious adverse effects on their suitability as sources of drinking water. Furthermore, eutrophication entails undesired changes in the ecology of the system, making it generally less stable and more vulnerable.

The major feature of man-caused eutrophication is that, although the consequences appear within the lake, the cause--the gradual increase of nutrients (various phosphorous and nitrogen compounds) reaching the lake--lies in the region. Consequently, eutrophication *management* requires analysis of the complex interactions between *the water body* vulnerable to eutrophication and its *surrounding region*. In the lake, different biological, chemical and hydrophysical processes are important, while in the region there are human activities generating nutrient residuals that are ultimately washed to the lakes and reservoirs.

To control eutrophication is not an easy task. Although in most cases a solution would include reducing nutrient loads in one form or another--in itself a difficult task--various *uncertainties* inevitably exist regarding the extent of the reductions required to achieve desirable improvements in the lake's water quality. Thus, the effectiveness of alternative solutions measured by their responses in the lake (and consequently with respect to its desired functions) calls for the explicit treatment of the problem-related uncertainties (ranging from data uncertainty to uncertainty due to a limited understanding of the processes involved).

Thus, using a *systems approach* to analyze and evaluate *alternative management strategies* offers particular advantages. In fact, this is perhaps the only way to perform an integrated study encompassing the lake, the region, and the related physical, biological, chemical, economic, and social processes. This approach structures information in a format appropriate for both the *research phase* and the subsequent phase of *policy implementation*. At the core of the approach, *models* play an important role and are effective tools. Model results can be fed back to field workers to focus their attention on areas where our knowledge of the real world is still insufficient. Model results also enable managers to visualize the effects of possible management options.

The relation of cause and effect of eutrophication is similar both for deep and shallow lakes. Still the understanding of processes in shallow lakes is much less satisfactory than for deep lakes. In contrast to deep lakes, the relatively fast and irregular dynamics influenced by the nearly complete absence of temperature stratification, and climatic and hydrologic factors being strongly stochastic in nature should be mentioned. Because the lake is shallow, wind-induced sediment/water interaction and spatial mass exchange play important roles. The light conditions in the water fluctuate a great deal more, owing to changes in the concentration of suspended solids. As a special issue, the gap between hydrophysical and ecological modeling should also be noted, the reason being that biological, chemical, and physical processes of quite different time and spatial scales are involved.

---

[1]Paper presented at the Third International Conference on State-of-the-Art in Ecological Modelling, Colorado State University, May 24-28, 1982.

[2]This study was initiated by IIASA's Resources and Environment Area and carried out in cooperation with the Hungarian Academy of Sciences and other Hungarian institutions.

[3]Dr. L. Somlyódy (on leave from the Research Center for Water Resources Development (VITUKI), Budapest, Hungary) is at present leader of the Balaton Case Study of the Resources and Environment Area at the International Institute for Applied Systems Analysis, Laxenburg, Austria.

To analyze all these issues, Lake Balaton, the largest lake in Central Europe, was selected as the subject of a case study. This lake is the most important recreational area in Hungary and has exhibited the unfavorable signs of artificial eutrophication. There is not only a scientific, but also a strong economic interest in the study (roughly 40% of the income from tourism in Hungary stems from the Balaton region). This paper gives a summary of the study, which is now nearing completion. For further details on the case study, the reader is referred to van Straten et al., 1979, van Straten and Somlyódy 1980, Somlyódy 1981a, and Somlyódy and van Straten, forthcoming.

MAJOR CHARACTERISTICS OF THE SYSTEM

The lake and its watershed are illustrated in figure 1. The length of the lake is 78 km, the average width around 8 km (surface area nearly 600 km^2) and the average depth 3.1 m. The major inflow of the lake is the River Zala at the southwestern end of the lake which drains half of the total catchment area (~ 5800 km^2). There is a single outflow at the other end of the lake, Siófok, through a control gate. The mean residence time of water is about 2 years.

The fluctuation in the water's temperature is high. There is a relatively long ice-covered period (around two months), while the temperature in summer may exceed 25° C. Concerning the chemical composition of the water, the high calcium carbonate content and pH value (8.3 to 8.7) should be mentioned. Wind action is important resulting in favorable oxygen conditions and a permanent back and forth motion (seiche) along the lake and a complicated circulation pattern. Wind strongly influences sedimentation and resuspension of the sediment (its organic material content is low) which is associated with an effective adsorption-desorption process. The yearly net desorption as internal load has the same order of magnitude as the external one (see Gelencsér et al., 1982).

In recent years, remarkable changes have been observed in the water quality due to the rapid increase in tourism, sewage discharges, fertilizer use, and other factors. The algal biomass (algae is the most important primary producer in this case) increased by a factor of 10 when compared with the past 15-20 years. The trend in primary production is similar and at the most polluted western basin, peaks of up to 13.6 gC/m^2d were observed, a hypertrophic value (Herodek and Tamás, 1975). In short, the average lake conditions moved from mesotrophic to eutrophic, thus endangering the use of the lake for recreational purposes, the prime water use in this case.

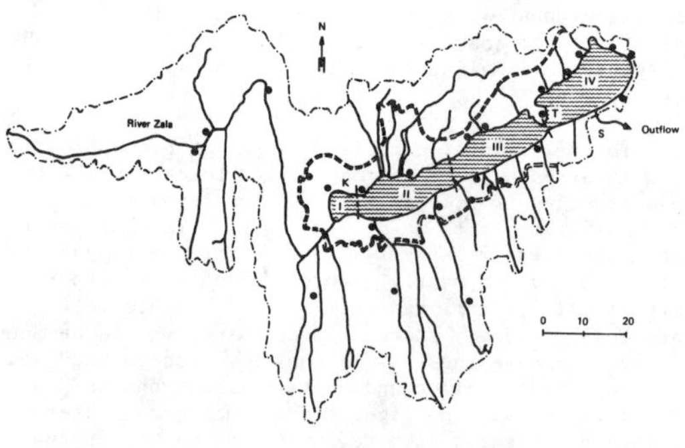

Figure 1.--Major characteristics of the system
K - Keszthely, T - Tihany, S - Siófok
I...IV typical basins of the lake
-.- boundary of the catchment
=== boundary of the recreational area
• sewage discharges in the region

Phosphorus plays a dominant role in the eutrophication of the lake. Thus, both from the point of view of understanding and managing the system, tracing the phosphorus compounds in the lake and on the watershed is of primary interest. The total phosphorus load of the lake is around 1000 kg/d (0.52 mg/m^2d in a lake-wide average) (Jolánkai and Somlyódy, 1981), half of which is estimated to be available for algal uptake. The load has many components: 33% is derived from sewage, 27% from diffuse sources, 22% is related to runoff processes in the direct vicinity of the lake, while the contribution of atmospheric pollution is 18%. The ratio of sewage discharges in the available load is higher; only the sewage released in the recreational area (figure 1) accounts for 36% of the available load. This direct load varies quite a lot in time, following the fluctuations in population due to tourism, and has a 2-4 times higher value in summer than during the off-season. The load distribution along the lake is approximately uniform (the tributary load is higher for the Western end of the lake, while the sewage load is quite the opposite); however, the volume related value is twelve times higher at the Keszthely Bay (fig. 1) than at the other end of the lake, due to differences in the volume of the four main basins. This fact is also reflected in the pronounced longitudinal gradient of various water quality parameters, e.g., for Chlorophyll-a the ratio of the maximum and minimum values ranges between 4 and 20 (van Straten et al., 1979). The gradient observed at the same time indicates that the strong wind action and the mixing associated with it are still not sufficient for leveling out the spatial nonuniformities.

From an analysis of the data it is clear that there is not only a critical state of the water quality at Keszthely Bay, but also a spreading deterioration process which extends towards other areas of the lake where the water quality is still good. Thus action is urgently required from the view of the entire lake.

Based on hydrologic and water quality considerations, the lake was divided into four basins, as indicated in figure 1. The application of the principle of segmentation proved to be a useful tool for modeling, data collection, and data handling.

Concerning data, extensive records are available on hydrology and meteorology. Regular water quality monitoring started ten years ago, in two network systems consisting of 9 and 16 spatial sampling points, respectively (10-20 measurements per year), but irregular data are also available dating back to the early sixties. Several other in situ and laboratory measurements were also taken (primary production, extinction, sediment-water interaction, velocity, etc.). A survey was done on the nutrient load between 1975-1979, which involved 20 tributaries and 27 sewage discharge points (Jolánkai and Somlyódy, 1981) (indicated in fig. 1). On the major tributary, daily observations were made during this period.

THE APPROACH

As mentioned in Section 1 the study should cover several interrelated but still quite diverse issues and processes such as scientific understanding and policy making, the lake and its region, watershed processes, biochemistry, hydrodynamics, etc. Thus the dilemma is which approach can be used for the research and within this, for modeling. Two extremes are as follows:

(i) to base it on intuition and to drastically simplify the problem a priori;
(ii) to establish one large, detailed model which accounts for all the subprocesses and the different levels (understanding and management).

As the first approach is not acceptable and the second is unrealistic, a different approach was worked out which is based on the idea of *decomposition and aggregation* (Somlyódy, 1981a).

The approach begins by decomposing the system into smaller, tractable units forming a hierarchy of issues subject to study. One can make detailed investigations of each of these issues, using in situ and laboratory observations, models and other available information. This step is followed by aggregation, the aim of which is to preserve and integrate only the issues that are essential for the higher strata of the hierarchy (at the uppermost level, ultimately for answering questions of management), ruling out the unnecessary details. From a structural viewpoint, this is an *off-line approach*, which allows application of different techniques and principles at various levels as desired, accounts for uncertainties, and leads to a realistic, yet simple, model at the highest level of the hierarchy, where evaluating management alternatives is the objective.

It is noted that in ecological modeling, there is a certain gap between "larger" models and "smaller" models (see Beck, 1982). In the first case there is nearly no hope for a precise calibration, but "smaller" models can also be just as unrealistic for complex problems because of their simplicity. The approach outlined offers a reasonable alternative for such cases.

The application of this approach for the Lake Balaton problem is explained with the help of figure 2. The first decomposition that directly comes to mind is the distinction between lake and watershed, since, as

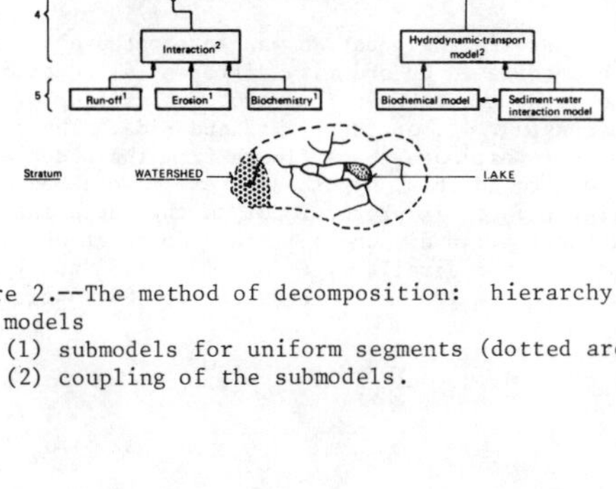

Figure 2.--The method of decomposition: hierarchy of models
(1) submodels for uniform segments (dotted areas)
(2) coupling of the submodels.

mentioned before, the water quality problem lies in the lake, but the causes, and practically all control possibilities, are to be found in the watershed.

The procedure, involving five strata, will be discussed in greater detail for the Lake Eutrophication Model (LEM), with reference to the models which have been elaborated. The parallels in the Nutrient Loading Model (NLM) can be seen in figure 2.

Stratum 5

First, those segments of the lake should be isolated which can be considered approximately uniform from the viewpoint of water quality (complete mixing inside each unit) and from the factors influencing them. The objective of the models on this stratum is to describe the algal dynamics and nutrient cycling for all the segments, involving both the water body and the sediment, since the latter is a sink and source of various materials and their interaction plays an important role in shallow lakes. These kinds of models based on the mass conservation principle and formulated through a set of nonlinear ordinary differential equations (ODEs) are well-known in the literature (Scavia and Robertson, 1979). In the frame of the present study, three submodels, BEM, BALSECT, and SIMBAL were developed (Herodek et al., 1980; Leonov, 1980; and van Straten, 1980), with respect to their comparison (van Straten and Somlyódy, 1980) which differs basically in the number of state variables (between 4 and 7) and essential parameters (10-17), as well as in the mathematical formulation of various processes and in the parameter estimation technique adopted. It is noted here that some of the parameters can be derived from further isolation up to a lower level with appropriately designed experiments. As examples, the estimation of algal growth parameters from vertical primary production measurements (van Straten and Herodek, 1981) and the study of wind induced sediment-water interaction (see the section on wind induced sediment water interaction) may be mentioned.

Stratum 4

On the next level the segment-oriented biochemical and sediment models are coupled by involving mass in- and outflows at the boundaries of the units. For this purpose, a hydrodynamic-transport model can be used. In light of the experiences gained from the study of the Great Lakes (Boyce et al., 1979), it was decided not to use a coupled multi-dimensional hydrodynamic-transport model incorporating the submodels of the lower stratum: the gain in information is not proportional to the increase in complexity. Here again, an off-line technique is applied. The basic assumption is that it is sufficient to subdivide the lake in a longitudinal direction only. This is supported by the riverine shape of the lake and the presumably extensive transversal mixing, since the prevailing wind direction is nearly perpendicular to the lake's axis (the description of the shoreline effects is not the objective here). Consequently, the parallel development of an unsteady three-, two-, and one-dimensional hydrodynamic model was decided on (Shanahan et al., 1981; Shanahan, 1981; Somlyódy, 1982; Somlyódy and Virtanen, 1982). The first two can be used to derive convection and the longitudinal dispersion coefficient (either directly or indirectly as an "empirical" function of the major wind parameters), while the 1-D model could describe convection only (its advantage lies in its simplicity and short execution time on the computer). Subsequently, the submodels of stratum 5 will be incorporated in a straightforward way into a set of longitudinal dispersion equations on stratum 4.

At this level, the 1-D model was aggregated from the 3-D version. A further aggregation can be arrived at through the use of the coupled dispersion-biochemical model (see the section on application of hydrodynamic models). Originally, in all three biochemical models (see "Major Characteristics of the System") four segments (or mixed reactors, see fig. 1) are assumed; their coupling is based on hydrologic throughflow and a wind influenced mass exchange process described globally. Since the model structure based on ODEs has many advantages, one of the objectives of the study on the 1-D coupled model is whether the four basins concept can be maintained or not.

Stratum 3

The involvement of mass exchange among segments as described before will result in the Lake Eutrophication Model (LEM) (fig. 2) which has several forcing functions, such as solar radiation, water temperature, wind, etc., (natural or uncontrollable factors) and the nutrient load. Since the latter is the only factor which can be controlled, it plays a distinguished role. A thorough data collection and the derivation of a nutrient balance for the whole lake gave a solid background (for details see Jolánkai and Somlyódy, 1981). The main conclusions have been summarized in "Major Characteristics of the System". Because of the high contribution of the sewage load and the insufficient amount of watershed data only limited effort was expended on non-point source modeling (Bogárdi and Duckstein, 1979). Rather, the analysis of historical river load data was preferred, which then satisfactorily resulted in uncertainties in the load (see the section on nutrient load under uncertainty and stochasticity) due to the stochastic character of the hydrologic regime and data scarcity. The research also allowed the derivation of the temporal and spatial pattern of the load components, both for LEM and the Water Quality Management Model (WQMM) on Stratum 2.

Stratum 2

The objective of WQMM is to generate alternative management options and strategies (the effect of these being expressed through NLM which should be used here in a planning mode) and to select from among these alternatives, on the basis of one or more objectives. Both water quality and costs can be used as objective functions or constraints, and quite often their weighting is required. Frequently the load can replace the lake's water quality in the optimization, in which case LEM is used merely to check the reaction of the lake and WQMM may have a simpler structure. Admittedly however, the inclusion of water quality is more obvious because of the nature of the problem. This formulation however leads to the dilemma: how should a complex model be incorporated into the optimization framework?

At this step aggregation is also needed. This starts with the selection of certain water quality indicators characterizing the large scale and long-term behavior of the system serving as a basis for decision making. Different parameters (yearly peak, different averages, duration of critical concentrations, frequency distributions, etc.) of typical water quality components (primary production, algal biomass, Chlorophyll-a, etc.) can be employed as indicators. Subsequently the dynamic model LEM can be used in terms of indicators established, I, under reduced loading conditions or in another way under several loading scenarios, L. Since the definition of indicators introduces temporal averaging, it is expected that the lake's response will be less complex compared to the dynamic simulation and a simple, direct I(L) type relationship can be found for the new equilibrium. If such a solution has already been attained, LEM could be replaced by I(L) in WQMM; an essential aggregation (see the section on the water quality management model).

Among the management alternatives, only the two most important options are mentioned here: (i) tertiary treatment (point source load reduction), (ii) establishing reservoirs (consisting of two segments serving for the removal of both particulate and dissolved nutrient forms, respectively (van Straten et al., 1979) at the mouth of rivers which are the recipients of point and non-point source pollutants. The optimization should then be based on the trade-off between the two basic alternatives, with respect to their locations and the spatial variation of the lake's water quality.

Stratum 1

For the sake of completeness it has to be mentioned that WQMM could be thought of as being a part of a regional development policy model forming the top of the pyramid, a field which is beyond the scope of this study.

ILLUSTRATION OF THE DIFFERENT STEPS OF THE APPROACH

Wind Induced Sediment Water Interaction (Stratum 5)

For studying the sediment-water interaction in lakes, several approaches are possible (Sheng and Lick, 1979). In this study, yet another method was chosen (Somlyódy, 1980), in recognition that when eutrophication is considered, more than just the physical processes should be examined. Daily measurements were taken for 6 months, at the mid-point of the Szemes basin (Basin 2, fig. 1). The measurements included Secchi depth, temperature, suspended solids (SS), Chlorophyll-a, and phosphorus fractions at different vertical locations. Wind velocity and direction were recorded continuously, from which hourly averages were calculated. The objective of the first part of the analysis was to describe the dynamics of the suspended solids as a function of wind. This then allowed for a characterization of the temporal changes in the light conditions, the deposition, and resuspension of particulate material and the associated sorption phenomenon. Here the behavior of SS will be discussed.

The analysis started from a simplified transport equation for describing the temporal and vertical changes of the average SS concentration for the basin, neglecting inflow and outflow (Somlyódy, 1980).

Afterward the equation was integrated along the depth leading to an ordinary differential equation incorporating the unknown *fluxes* of *deposition*, ϕ_d, and *resuspension*, ϕ_e, on the right hand side. The objective is to estimate these fluxes from the observations. In order to do this, hypotheses were made based on the literature: ϕ_d is proportional to the depth integrated SS concentration $\tilde{c}$, while ϕ_e to some power of the wind speed, W^n (for details, see Somlyódy, 1980 and 1981b). This procedure led to the equation for SS concentration

$$\frac{d\tilde{c}}{dt} = -K_1 \tilde{c} + K_2 W^n \quad (1)$$

where K_1 and K_2 comprise the unknown coefficients, derived from the hypotheses. Consequently, the structure of the model should be identified and the parameter values K_1, K_2, and n, estimated from measurements. The feasibility of Equation (1) can be appreciated from figure 3a, which clearly shows the influence of wind velocity on the concentration.

First a non-recursive deterministic estimation technique was adopted to derive the unknown coefficients which resulted in realistic values, but without proving the correctness of the hypotheses (a posteriori model structure identification, see Beck, 1982).

For this purpose, as a second step, the Extended Kalman Filter (EKF) method was applied (Beck and Somlyódy, 1982). For the power n a value near to 1 was derived which corresponded to the small Richardson number (Somlyódy, 1981b). Subsequently n was fixed to 1 since in this case the physical interpretation of the results is more obvious. The recursive estimation started from the estimates of the deterministic technique. The results are illustrated in figure 3a. As is apparent, the agreement between observations and model calculation is reasonably good, and the parameters become approximately constant after the first 40-50 days (fig. 3b), proving that the model structure is adequate and the data do not contain more information than described by the model. Some slight parameter changes can be observed at the end of the period; this may be caused, e.g., by the exclusion of inflow-outflow processes (or by other phenomena such as algal blooms). This suggests that the isolation of subprocesses is generally not complete. From the analysis, a realistic order of magnitudes follows for all the essential physical quantities; in this connection see Somlyódy, 1981b.

As can be observed in figure 3, for one month in the middle of the total period, no measurements were available, so the model was used for prediction. The appropriateness of the model is also illustrated by the fact that after getting new data, the parameter values did not change. This second period served for validation, following the identification and calibration procedure.

The advantage of the simple interaction model arrived at is that it can be easily incorporated into the modeling approach (fig. 2), for (i) characterizing transparency conditions in the water and (ii) for describing the release of the sediment layer as the internal P source. As was shown in the report by Gelencsér et al. (1982) the desorption of resuspended particles is the primary cause of the sediment phosphorus

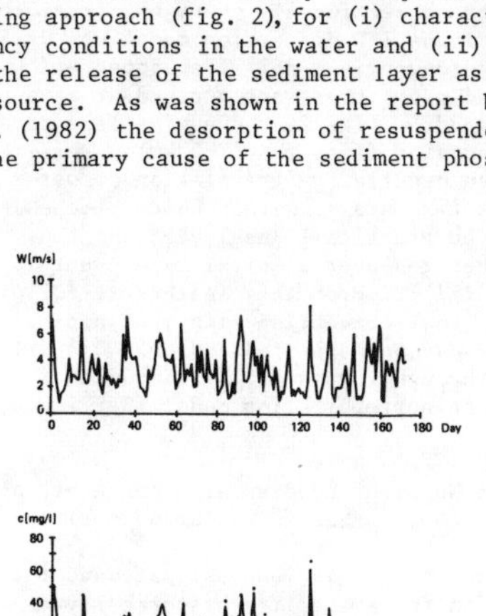

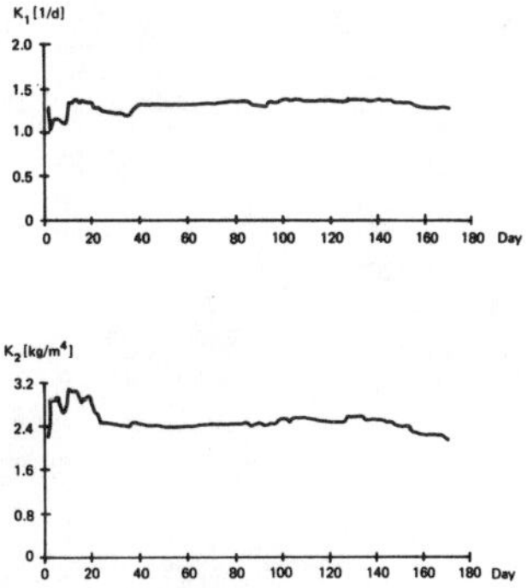

Figure 3a.--Identification and parameter estimation of a model for wind induced sediment-water interaction for Lake Balaton: recursive estimate of the suspended solids concentration; W - daily average wind speed, c - suspended solids concentration, * - observations.

Figure 3b.--Recursive parameter estimates for the sediment-water interaction model.

release (diffusion and convection of pore water contribute to a lesser extent); thus, wind-induced interaction is really of importance.

Application of Hydrodynamic Models (Stratum 4)

The results gained from the transient 3-D, 2-D (horizontal), and 1-D models (Shanahan et al., 1981; Somlyódy and Virtanen, 1982; and Somlyódy, 1982) showed that the models could be equally calibrated against the dynamic water level data. For an example of the application of the 1-D model, see figure 4. The model was already used in the validation phase. The storm was characterized by a long-lasting longitudinal wind followed by smaller shocks from different directions (fig. 4a). The agreement between measured and simulated water levels at the two ends of the lake is excellent (fig. 4b). The discharge at the Tihany peninsula shows a striking oscillation between -2000 and 3000 m^3/s (fig. 4c) associated with the seiche phenomenon. This back and forth motion is higher by two orders of magnitude than the hydrologic throughflow.

As mentioned previously, the 1-D model alone does not give satisfactory information for a water quality study as it serves the longitudinal convection term only, but not dispersion. Still this model version, the simplest, is extremely useful. Two examples discussed subsequently illustrate this statement:

(i) To find a more satisfactory agreement between model simulation and observation than that given in figure 4 is often impossible. The reason is quite simple: a small error in the wind direction may cause a drastic change in the wind force, if the direction is far from the longitudinal one. In fact, there are many kinds of uncertainties in the wind direction, such as random fluctuation (turbulence), the influence of hills on the northern side of the lake, which cause nonuniformities in the wind field, measurement errors, etc. Figure 5 illustrates the case (transversal wind conditions). A deterministic simulation did not prove acceptable. Bearing in mind the possible role of uncertainties, a random component was subsequently added to the wind direction (Gaussian distribution, zero mean,

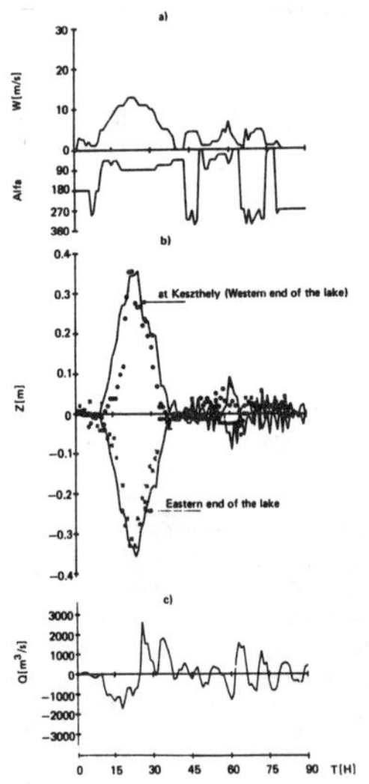

Figure 4.--Simulation of a historical event: longitudinal wind conditions. (a) wind record (Muszkalay, 1979), W speed, ALFA angle (North ≡ 0°); (b) comparison of simulated and observed water levels (T = 0 corresponds to 16/11/1966, 8 a.m.). Dots indicate measurements (Muszkalay, 1979); (c) computed streamflow rate at Tihany.

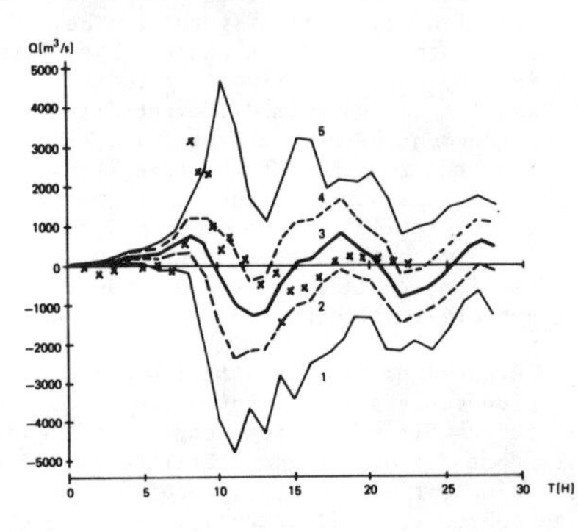

Figure 5.--The influence of wind data uncertainty on the discharge at the Tihany peninsula (T = 0 corresponds to 8/7/1963, 8 a.m.);
· discharge derived from measurements (Muszkalay, 1979).

17° standard deviation: a modest value) and a *Monte Carlo simulation* was performed (for details, see Somlyódy, 1982). Figure 5, which summarizes the results of 100 runs, does not require detailed discussion: it stresses the *extreme senstivity to input data uncertainty* (compared to this, the parameter sensitivity is negligible) and illustrates how difficult it is to validate a deterministic model (the stiuation is similar for a two-dimensional model, Somlyódy 1982).

(ii) At a given location in the lake the intensive back and forth motion causes an oscillation of various constituents within a day, which also strongly depends on the longitudinal gradient. This may result in quite a large error in the concentration determined through instantaneous sampling. For such a short time scale as a day, biological reactions can be neglected and the concentration fluctuation can be analysed through a coupled 1-D hydrodynamic-transport model assuming conservative material. Through this model an uncertainty range of historical observations can be specified. Simulations showed that for this particular lake the sampling strategy for Basin II (fig. 1) is of major importance; the error range of a single sample at a fixed location can reach ±30%, depending on the actual gradient and streamflow pattern.

The calibration of the two-dimensional hydrodynamic model (Shanahan, 1981) resulted in the same wind drag coefficient and bottom friction parameter as the 1-D model version. The simulations showed a pronounced circulation pattern observed also on some satellite photographs and a physical model (Györke, 1975). It is noted that the transient 3-D model (Shanahan et al., 1981) and also other steady state models tested, (van Straten and Somlyódy, 1980) reflected much less circulation in the lake, clearly showing that our *understanding* of the *three-dimensional water motion in shallow lakes* (and within this, of the vertical eddy viscosity) is *far from complete*.

In order to couple the horizontally 2-D hydrodynamic model to a phosphorus cycle model through a set of longitudinal dispersion equations (see "The Approach" and "The Lake Eutrophication Model (Stratum 3)"), Shanahan (1981) computed the dispersion coefficient (as a function of time and space) from the velocity field through extending the method of Fischer (1979). Dispersion is the highest near the two ends of the lake and at the vicinity of the peninsula, due to strong changes in geometry and the associated secondary currents and gyres. The large dispersion coefficient (up to 40 m^2/s) is due to strong winds. As a temporal and spatial average, 1 m^2/s was found by Shanahan to be appropriate enough for water quality simulations.

The cross sectionally averaged streamflow (or velocity) for the 1-D dispersion model was derived from the 2-D hydrodynamic model after integration. The same pattern and magnitudes were arrived at as suggested in figures 4 and 5. The fast dynamics lead to a small seiche excursion (less than 1-2 km). This means that there is an oscillatory translation of water particles of a short time scale, within which biochemical reactions can be practically neglected and thus the concentration averaged over a seiche type event is unchanged (see item (ii)). From this feature it follows (Shanahan, 1981) that convection with its uncertainties (see item (i)) can be neglected and only dispersion and hydrologic throughflow should be accounted for in the coupled lake eutrophication model (Section 4.4).

The Nutrient Load under Uncertainty and Stochasticity (Stratum 3)

The deterministic load estimate and the spatial distribution for a specific historical year (1975-79) is derived on the basis of the daily observations on the Zala River's draining 50% of the total watershed, the survey on data for other rivers and sewage treatment plants, on pilot zone studies, watershed characteristics, etc. (Jolánkai and Somlyódy, 1981). The temporal pattern is derived from the dynamics of the Zala River load, observations made in treatment plants during the off-season and summer period, respectively, and population fluctuation related to tourism. Such a load estimate is acceptable for the descriptive use of the lake model, LEM, but certainly not for planning purposes.

For *management* of the system, the *stochastic character* of the load and other existing *uncertainties* should be accounted for. In order to develop a *load scenario generator*, first the allowable integration period of the load input was tested through the dynamic lake model. The analysis showed that *monthly averages* for all the *forcing functions* can be *satisfactorily used*; an important finding, as it allows generation of the load on a monthly basis. This can be reasonably derived from the data available, while the procedure for a shorter time scale would be unrealistic.

With this conclusion, the Zala River data were a priori aggregated to monthly averages and a simple regression analysis was done between phosphorus loads (TP and PO4-P) and streamflow rate. Acceptable expressions were arrived at (of course, with error terms). Deriving the statistics of the monthly average streamflow from long-term observations (Baranyi, 1979), the load can be calculated in a stochastic fashion. Figure 6 shows the characteristics of the load pattern for 1976-79 (from observations) and the 90% confidence levels derived for the long-term load. For illustrating the influence of the hydrologic regime an event of low probability in July, 1975, is likewise indicated. The stochastic influence of the hydrologic regime for other subwatersheds was derived from the analysis outlined and available data for these catchments. The daily observations at the mouth section of the Zala River were also used to study the *implication of infrequent observations* typical for most of the tributaries (one or two samples per month). Due to scarcity of data, the contribution of floods to the load are partially unobserved. As the "accurate load" for a certain period (e.g., long-term monthly or yearly averages) for the Zala River can be gained from the original data, it allows one to study the error caused by scanty observations. The procedure is a straightforward Monte Carlo type technique which starts with a random selection on the detailed data set following the sampling strategy of the other tributaries and calculates the load of the period in question. After making a sufficient number of random selections the statistical parameters of the load can be determined. The results for the long-term monthly average load (on the basis of a four year long observation period) are illustrated in figure 7. As can be seen from figure 7, which shows the mean and extreme values, as well as the domain of ± standard deviation, the error is quite high and its fluctuation follows the change in the mean value. On the basis of this study, a random component was added to the monthly average load component (Somlyódy and Eloranta, 1982).

From figure 7, the question automatically arises: how can the uncertainty domain be reduced? In a different way, *what sampling strategy should be followed?* This issue was also studied. Besides, first order analysis (Cochran, 1963), different sampling strategies (regular, random, stratified, etc.), were realized in a Monte Carlo type fashion. Also, various kinds of estimates (simple, ratio, etc., see e.g., Dolan et al., 1981) were tested in order to reduce bias and variance of the load estimate. Without going into detail, (the reader is referred to Somlyódy and van Straten, forthcoming), it should be noted that, with proper stratified sampling (few samples when the variance is small--low-flow conditions--and frequent sampling for floods characterized by large variance (Cochran, 1965)), the total amount of samples can be reduced to one fourth or one fifth. An important conclusion of the study is that the variance of the load can be replaced by that of the discharge, Q. As Q is an easily measureable quantity, a realistic stratified sampling strategy can be worked out in practice, on this basis.

Returning to the development of the load generator, for sewage load, the same pattern is used as in the descriptive fashion, but in addition, an uncertainty component is introduced, which expresses the overload in the treatment plants due to the population increase in the main tourist season.

As a final output of the research outlined in this section, a load scenario generator was developed for the whole lake, which accounted for both uncertainty and stochasticity, discussed above. For further details see Somlyódy and Eloranta (1982).

It is noted here that using historical data, a similar analysis was made on climatic (uncontrollable) factors, which allowed the water *temperature* and *solar radiation* to be generated in harmony with each other, in a *random fashion*. Thus, future scenarios can be generated for all the essential forcing functions of the lake model--an essential tool for planning purposes (see "The Lake Eutrophication Model (Stratum 3)" and "Water Quality Management Model (Stratum 2)").

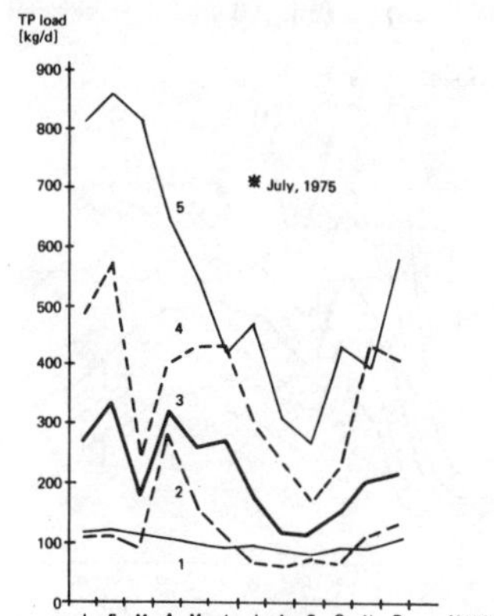

Figure 6.--Influence of the hydrologic regime on the monthly average load, Zala River: 3 - average load (1976-79), 4 and 2 - minimum and maximum values (observed), 5 and 1 - 90% confidence levels.

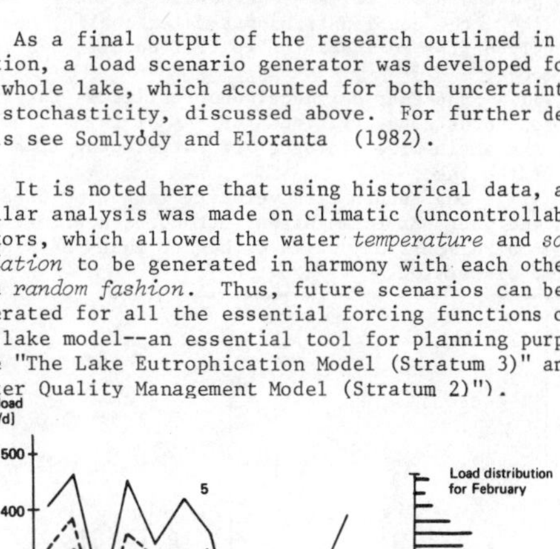

Figure 7.--Monthly average TP load: uncertainty caused by infrequent observations (Zala River, 1976-79): 3 - mean value, 4 and 2 - ± standard deviation, 5 and 1 - extreme values.

The Lake Eutrophication Model (Stratum 3)

Results gained with the simplest model, SIMBAL (van Straten, 1980), developed for Lake Balaton, are given below. The model is a phosphorus cycle model, that is, all the state variables (the essentials are two algal groups, detritus, and dissolved inorganic phosphorus) are expressed in terms of phosphorus, for the four basins indicated in Figure 1. A Monte Carlo simulation is incorporated into the model to find areas in parameter space where the model produces results fully within specified boundaries drawn around the data to account for data uncertainty and thus, is easily applicable for testing various hypotheses (van Straten, 1980; Fedra et al., 1981; Hornberger and Spear, 1980).

Among the calibration runs, results for the phytoplankton phosphorus, PPP, for the four basins, are given in figure 8 (as 1977 forcings data was used) together with the corresponding observation variable, Chlorophyll-a (basin average values). It is pointed out that Chlorophyll-a and PPP cannot be directly compared to each other; however, since a more or less linear measurement equation is expected among them, PPP should follow the pattern of Chlorophyll-a: a trend which can be generally observed. For illustration, the standard deviation around the trajectory for Basin 2 estimated through the Monte Carlo simulation is also indicated (parameter uncertainty). Further discussion on the calibration and model improvement required canbe found in van Straten (1980).

The results presented here were from the four box model (fig. 1 and "The Approach"). Whether the concept of the four box model can be preserved or not, was tested through the *coupled hydrodynamic-dispersion-P cycle model* (see "Application of Hydrodynamic Models (Stratum 4)"). In the linked model, the parameter values of the original model were maintained. From the comparison of the simulation results of the "continuous" and four box model (see Shanahan, 1981 and Shanahan and Harleman in these proceedings), we may conclude as follows:

(i) the coupled transport-water quality model obviously better reflects the spatial details and local influences;

(ii) the four box model underestimates the various phosphorus concentrations for one of the basins, while the basin wide averages are satisfactory for the rest of the lake;

(iii) the return flow velocity cannot be used since the four box formulation introduces a priori artificial dispersion, which is higher than the wind induced dispersion.

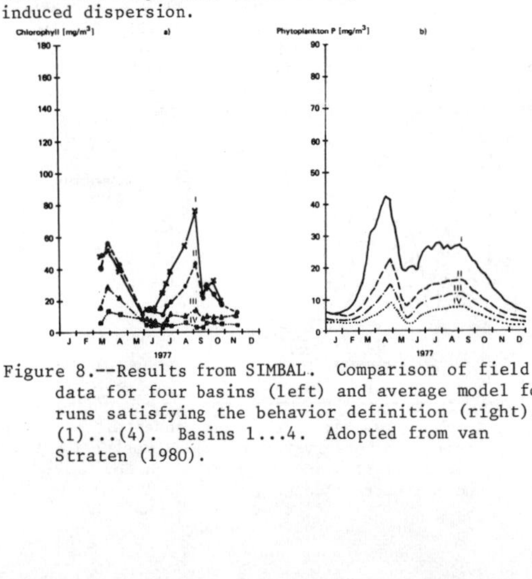

Figure 8.--Results from SIMBAL. Comparison of field data for four basins (left) and average model for runs satisfying the behavior definition (right), (1)...(4). Basins 1...4. Adopted from van Straten (1980).

(iv) *the four box model* with its ODE structure and all the advantages associated with this *can be reasonably maintained for practical purposes* and subsequent analysis.

For management purposes the simulation of historical events cannot be used. Either some critical, unfavorable environmental conditions should be introduced or the model should be considered stochastic through input data. Here the latter approach was adopted and the generators outlined in the previous section coupled to the lake model. Two essential results for Basin I are presented in figures 9 and 10.

In the first case, *uncertainties* caused by *natural factors* were considered and the 1977 load was maintained. The summary of 100 Monte Carlo runs (mean, ± standard deviation, and the extremes of PPP) suggests the relatively large sensitivity of the lake's water quality to meteorological factors and explains the essential year to year changes observed in the behavior of the lake even when the load remained unaffected. The second case (fig. 10) involved the *random generation of both natural and controllable factors*. While for the previous example the specific 1977 load was adopted, here the mean load of the input generator was derived from data for the period 1975-1979 ("The Nutrient Load under

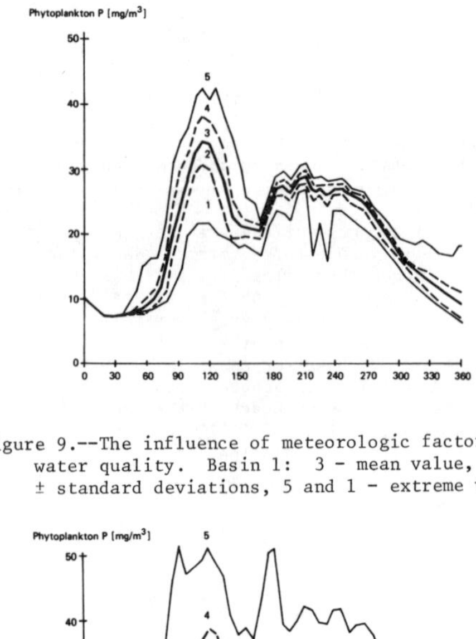

Figure 9.--The influence of meteorologic factors on the water quality. Basin 1: 3 - mean value, 4 and 2 - ± standard deviations, 5 and 1 - extreme values.

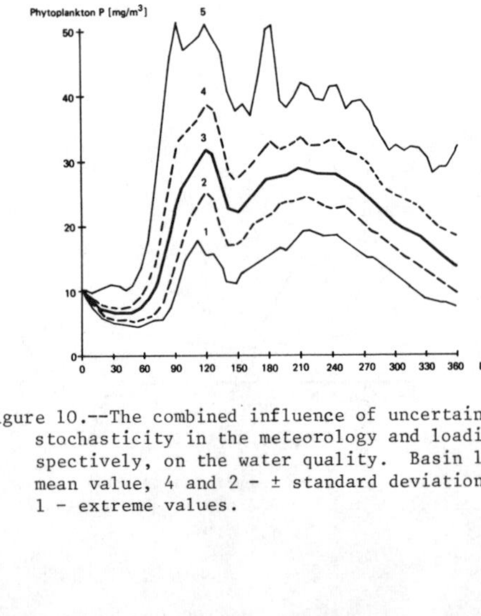

Figure 10.--The combined influence of uncertainty and stochasticity in the meteorology and loading, respectively, on the water quality. Basin 1: 3 - mean value, 4 and 2 - ± standard deviations, 5 and 1 - extreme values.

Uncertainty and Stochasticity (Stratum 3)"). This is the reason why the average trajectory shown in figure 10 differs from that shown in figure 9. The inclusion of load randomness had an obvious influence: the range of uncertainty of the water quality simulation results became much wider.

Compared to the role of parameter uncertainty (fig. 8), the meteorological factors represent the same order of magnitude, while the contribution of load to the model uncertainty is twice as high. It is stressed that in the frame of the present example, no load reduction was employed. Thus figure 10 shows in which domain the water quality may range under the present conditions, since the changes in trend are already relatively small from year to year. Control alternatives - depending on their character - can modify not only the mean load, but also the related uncertainties (see "Water Quality Management Model (Stratum 2)".

At the end of this section it is stressed that the present version of SIMBAL has no sediment compartment and thus no memory (the two other models are more complex in this respect). The internal load is represented by an equilibrium dissolved inorganic P concentration in the lake water (calibrated in the course of the Monte Carlo procedure) which controls sorption. This load is approximately half of the external one - a realistic value - but since no feedback exists it does not change from year to year. Accordingly, the model, when used under reduced load conditions gives the "immediate" response of the lake but no conclusions can be drawn for the final equilibrium and the renewal time.

In order to improve the model in both short- and long-term reactions the number of state variables was essentially increased from four to ten for each basin (Somlyódy and van Straten, forthcoming). Among the six additional variables, three are associated with the sediment (particulate inorganic-, dissolved inorganic- and detritus P), one is suspended solids (see Eq. (1)), and the rest are the particulate inorganic and dissolved organic P concentrations, respectively, in water. The establishment of such a complex model also describing resuspension, sorption, diffusion, and the changes in the active sediment layer was felt to be realistic, as recently our a priori knowledge of the behavior of the sediment has greatly improved (Gelencsér et al., 1982). Our preliminary results allow two important conclusions to be drawn:

(i) through the inclusion of *resuspension* and *sorption*, the short-term behavior became more realistic. The influence of the wind on *algal dynamics* is *apparent* and it is in harmony with observations. This phenomenon is expected to be *generally important for shallow lakes*;

(ii) the magnitude of simulation results is *quite sensitive* on the *parameters of sorption isotherms* (within the range suggested by the observations). In addition, it became clear that the long-term behavior of the lake depends on some crucial sediment parameters (initial conditions, loss from the active layer to deeper zones, etc.) for which only rough guesses are available. All these suggest that (i) the inclusion of more state variables (and parameters) still requires more quantitative knowledge in order to improve the practical use of the model, in contrast to the simpler version; and (ii) for problems where *sediment plays a role*, at present, we *can hardly say anything* about the *renewal process* of the lake.

Water Quality Management Model (Stratum 2)

First we are dealing with the issue of how to incorporate the lake model (LEM) into the management framework, a question raised in "The Approach". In the course of the analysis outlined subsequently, the different parameters of the PPP(t) distribution were selected as water quality indicators characterizing the algal behavior, and deterministic simulations were performed with the dynamic lake model, SIMBAL, under reduced loading conditions. It turned out from the study that the lake's response expressed in terms of the yearly average or peak of PPP is quite linear on the load in a wide range (Somlyódy and Eloranta, 1982)-- there was the same experience with the other two models. It is noted that a similar linearity is expressed for total phosphorus, TP, by several empirical models. However, for shallow lakes, TP may not properly characterize the process of eutrophication since this component is strongly influenced by wind induced interaction at the bottom (see "Wind Induced Sediment Water Interaction (Stratum 5)").

The recognition of the linearity leads to an important aggregation: the *dynamic lake model* can be replaced at the *level of* WQMM (see fig. 2) by a simple *linear equation* (fig. 11):

$$\underline{c} = \underline{c}_0 + \underline{\underline{A}}\ (\underline{L} - \underline{L}_0) \tag{2}$$

or

$$\Delta\underline{c} = \underline{A}\ \ \Delta\underline{L} \tag{3}$$

Here $\underline{L}_0$ and $\underline{c}_0$ are the initial volume related load (yearly mean) and concentration "vectors" respectively, defined by the number of uniform segments assumed in LEM (at present N = 4, fig. 1); $\Delta\underline{L}$ represents the reduction of volume related load achieved by various control alternatives, while $\Delta\underline{c}$ is the corresponding response of the lake.

The elements of the $\underline{\underline{A}}$ matrix maintain the essence of LEM: a transition from a "large" model to a "smaller" on a higher level of the hierarchy of figure 2. The first remarkable feature of equations (2) and (3) is that they clearly preserve the influence of subprocesses on the lower strata and show the subsequent steps of aggregation. The elements of the main diagonal are determined primarily by biochemical processes and sediment-water interaction (stratum 5), while the other elements mainly express the influence of hydrodynamics and associated mass exchange (stratum 4), showing that a management action at the region of the ith segment will

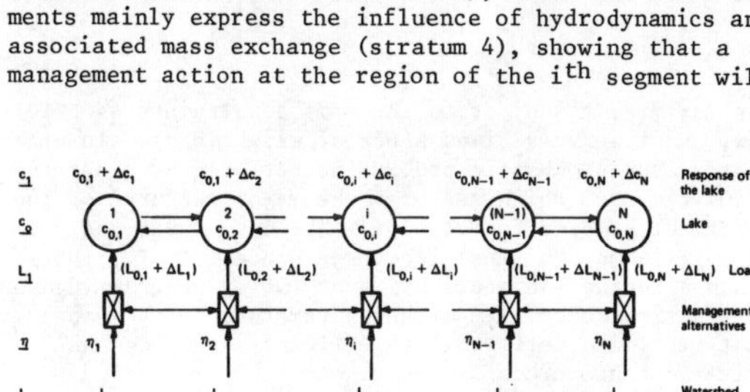

Figure 11.--Incorporation of the lake model in WQMM:
- i - uniform segments in the lake
- ⇄ - interaction among basins through mass exchange and hydrologic throughflow
- ↔ - interaction among load components through control activities.

affect the water quality of other segments as well (fig. 11). It is noted that through Monte Carlo simulations with LEM (figs. 9 and 10), it was possible *to include the uncertainties* in Equation (2). Three terms appear: two correspond to uncertainties in load and meteorology, while the third is due to their combined effect.

The second remarkable feature of Equation (2), and this is of primary importance in stratum 2, is that its *linear structure* allows its direct involvement in an *optimization framework*: a solution that we were looking for.

Optimization of this kind is not an easy task as each $L_{0,i}$ load element (fig. 11) is itself a vector, which can be controlled in many different ways. In addition, the various kinds of uncertainties discussed before should be accounted for, an issue which has not yet been explored satisfactorily for water quality problems.

Two approaches are under elaboration in this respect. Both use the same load equations with control variables for the four basins and the stochastic version of Equation (3) established for the yearly peak concentration. Control variables are the removal efficiency of treatment plants (fig. 1, $0 \leq x_k \leq 1$) and that of the reservoir projects ("The Approach", Stratum 2) assumed for the two largest rivers at the Western end of the lake ($0 \leq y_j \leq 1$). Cost functions (total annual cost) are highly nonlinear for tertiary treatment. It is stressed that x_i influences the expected value of load only, while y_i influences both expectation and variance. For those rivers which are recipients of sewage discharges and for which reservoirs can be constructed at the mouths, the product of the corresponding x_k and y_j variables will appear, leading to further nonlinearities and clearly showing the tradeoff between the two alternatives. The two approaches differ slightly in formulating the objective function and in the optimization technique adopted. For the first version, the weighted sum of the expectations of the square of the actual concentration minus the goal, for all the four basins, are minimized, subject to budget and other constraints. The technique is a stochastic quasigradient method (see e.g., Ermoliev, 1981). For the second version, the load is further aggregated from monthly to yearly average and the expectation and standard deviation of Equation (3), $E(\Delta C_i)$ and $\sigma(\Delta C_i)$ respectively, are derived analytically. Subsequently

$$\sum_{i=1}^{4} w_i \, [E\,(\Delta c_i) + t\sigma(\Delta c_i)] \qquad (4)$$

is maximized, subject to the same constraints as before (w_i and t serve various kinds of weightings). In other words, the stochastic problem is replaced by a deterministic one, which captures the major features of the original problem formulation. The cost functions and x_k, y_i terms are linearized (see Loucks et al., 1981), thus allowing the application of linear programming for the optimization. Important parameters of the model not mentioned until now are effluent standards and river detention coefficients.

It has to be noted, that the final solution of the sewage P removal will be to divert it from the watershed. Thus the present model serves as a decision making tool to find a provisional, feasible solution for the next 15-20 years. It was, however, recognized (Benedek and Szabó, 1981) that biological treatment has to be intensified in many plants before beginning on phosphorus removal (that is, a fixed cost exists, not related directly to the eutrophication problem). This feature was accounted for by (0,1) decision variables. Among the conclusions gained from the second model, three are listed below (Somlyódy and van Straten, forthcoming):

(i) in harmony with the findings of other approaches (Hughes, 1982; Dávid and Telegdi, 1982), as much P should be removed at the Western end of the lake as possible, under the given constraints;

(ii) if uncertainties are neglected tertiary treatment is chosen, but not reservoir projects (see Hughes, 1982);

(iii) however, if *uncertainties* are accounted for as given in Equation (4), (w_i = t = 1) results in a certain *combination of tertiary treatement plants* and *reservoir projects* for the Zala River catchment, clearly indicating that sewage treatment is a more effective technique for reducing the expected value of the load, but does not influence the variance in the tributary load and that of the in-lake concentrations. This shows that it is extremely *important* to include *stochastic features* at the level of *water quality management*.

REFERENCES

Baranyi, S. 1979. Hydrological Conditions of Lake Balaton. p. 128-177. In Summary of the research results related to Lake Balaton. S. Baranyi, ed. VIZDOK, Budapest, Hungary. (in Hungarian).

Beck, M.B. 1982. Identifying Models of Environmental Systems' Behaviour. J. of Mathematical Modelling (forthcoming).

Beck, M.B., and L. Somlyódy. 1982. Identification of a model for wind induced sediment-water interaction in Lake Balaton. Working Paper. Interna-Institute for Applied Systems Analysis. Laxenburg, Austria (forthcoming).

Benedek, P., and S. Szabó. 1981. Recommendation for protecting the water quality of Lake Balaton through improving sewage treatment. VITUKI Report 7783/3/22, Budapest, Hungary. (in Hungarian).

Bogárdi, I., and L. Duckstein. 1979. A Stochastic Model of Phosphorus Loading from Non-point Sources. RM-79-033. International Institute for Applied Systems Analysis. Laxenburg, Austria.

Boyce, F.M., A.S. Fraser, E. Halfon, D. Hyde, D.C.L. Lam, W.M. Schertzer, A.H. El-Shaarawi, T.J. Simons, K. Willson, D. Warry, and T.J. Simons. Assessment of Water Quality Simulation Capability for Lake Ontario. Scientific Series No. 111. 220 p. Inland Water Directorate, National Water Research Institute, Canada Center for Inland Waters, Burlington, Ontario.

Cochran, W.G. 1965. Sampling Techniques. John Wiley, New York.

Dávid, L., and L. Telegdi. 1982. Modelling the Eutrophication of Shallow Lakes on an Annual Basis: The Watershed Development Approach Applied to Lake Balaton. Collaborative Paper. International Institute for Applied Systems Analysis. Laxenburg, Austria (forthcoming).

Dolan, D.M., A.K. Yui, and R.D. Geist. 1981. Evaluation of River Load Estimation Methods for Total Phosphorus. Journal of Great Lakes Research, Michigan, USA (in press).

Ermoliev, Y. 1981. Stochastic Quasigradient Methods and their Application in Systems Optimization. WP-81-2. International Institute for Applied Systems Analysis. Laxenburg, Austria.

Fedra, K., G. van Straten, and M.B. Beck. 1981. Uncertainty and arbitrariness in ecosystems modelling: a lake modelling example. Ecological Modelling 13: 187-210.

Fischer, M.B., J. Imberger, E.J. List, R.C.Y. Koh, and N.H. Brooks. 1979. Mixing in Inland and Coastal Waters. 483 p. Academic Press, New York.

Gelencsér, P., F. Szilágyi, L. Somlyódy, and L. Lijklema. 1982. A Study on the Influence of Sediment in the Phosphorus Cycle for Lake Balaton. Collaborative Paper. International Institute for Applied Systems Analysis. Laxenburg, Austria (forthcoming).

Györke, O. 1975. Studies into the Factors Affecting the Morphological Processes in Shallow Lakes by Means of a Hydraulic Model. Proceedings of the 16th Congress of the IAHR, São Paulo, Brazil, Paper B4, Volume 2, p. 24.

Herodek, S., and G. Tamás. 1975. The Primary Production of Phytoplankton in the Keszthely basin of Lake Balaton in 1973-74. Annual Biol. Tihany 42: 175-190. (in Hungarian).

Herodek, S., P. Csáki, and T. Kutas. 1980. The BEM Modeling Approach I-III. In Proceedings of the Second Joint MTA/IIASA Task Force Meeting on Lake Balaton Modeling. G. van Straten, S. Herodek, J. Fischer, and I. Kovács, eds. MTA VEAB, Veszprém, Vol. 1, p. 81-103.

Hornberger, G.M., and R.C. Spear, Eutrophication in Peel Inlet-I. Problem-defining Behaviour and a Mathematical Model for the Phosphorus Scenario. Water Research 14: 29-42.

Hughes, T. 1982. Lake Eutrophication Management Optimization Modeling - Approaches with Application to Lake Balaton. WP-82-10. International Institute for Applied Systems Analysis. Laxenburg, Austria.

Jolánkai, G., and L. Somlyódy. 1981. Nutrient Loading Estimate for Lake Balaton. CP-81-21. International Institute for Applied Systems Analysis. Laxenburg, Austria.

Leonov, A. 1980. Simulation and Analysis of Phosphorus Transformation and Phytoplankton Dynamics in Relation to the Eutrophication of Lake Balaton. In Proceedings of the 2nd State-of-the-Art Conference in Ecological Modelling, Liège, Belgium. Also appeared as WP-80-88, International Institute for Applied Systems Analysis. Laxenburg, Austria.

Loucks, D.P., J.R. Stedinger, and D.A. Haith. 1981. Water Resource Systems Planning and Analysis. Prentice-Hall, Inc., Englewood Cliffs, New Jersey.

Muszkalay, L. 1979. Water Motions in Lake Balaton. In A Summary of Research Results for Lake Balaton. p. 34-128. S. Baranyi, ed. VIZDOK, Budapest, Hungary. (in Hungarian).

Scavia, D., and A. Robertson. 1979. Perspectives on Lake Ecosystem Modeling. 326 p. Ann Arbor Science.

Shanahan, P. 1981. Linked Hydrodynamic and Biochemical Models of Water Quality in Shallow Lakes, Submitted to the Dept. of Civil Eng. in partial fulfillment of the requirements for the degree of Doctor of Philosophy at the Massachusetts Institute of Technology.

Shanahan, P., and D.R.F. Harleman. 1982. A Linked Hydrodynamic and Biogeochemical Model of Eutrophication in Lake Balaton. In Proceedings of the Third International Conference on State-of-the-Art in Ecological Modelling, Colorado State University, May 24-28, 1982.

Shanahan, P., D.R.F. Harleman, and L. Somlyódy. 1981. Modeling Wind-Driven Circulation in Lake Balaton. WP-81-7. International Institute for Applied Systems Analysis. Laxenburg, Austria.

Sheng, T.P., and W. Lick. 1979. The Transport and Resuspension of Sediments in a Shallow Lake. Journal of Geophysical Research 84: (No. C4, April) 1809-1826.

Somlyódy, L. 1980. Preliminary Study on Wind-Induced Interaction between Water and Sediment for Lake Balaton (Szemes Basin). In Proceedings of the Second Joint MTA/IIASA Task Force Meeting on Lake Balaton Modeling. G. van Straten, S. Herodek, J. Fischer, and I. Kovacs, eds. Vol. 1, p. 26-50. MTA VEAB, Veszprém, Hungary.

Somlyódy, L. 1981a. Modeling A Complex Environmental System: The Lake Balaton Case Study. WP-81-08. International Institute for Applied Systems Analysis. Laxenburg, Austria. (also to appear in Mathematical Modelling).

Somlyódy, L. 1981b. Water Quality Modeling: A Comparison of Transport Oriented and Biochemistry Oriented Approaches. WP-81-117. International Institute for Applied Systems Analysis. Laxenburg, Austria. (also to appear in Ecological Modelling).

Somlyódy, L. 1982. Input data uncertainty and parameter sensitivity in a lake hydrodynamic model. In Uncertainty, system identification, and the prediction of water quality. M.B. Beck and G. van Straten, eds. Pergamon, Oxford (in press).

Somlyódy, L., and J. Eloranta. 1982. The Use of a Lake Eutrophication Model in the Management Framework. International Institute for Applied Systems Analysis. Laxenburg, Austria (forthcoming).

Somlyódy, L., and G. van Straten. Eutrophication Management for Lake-region Systems. With Application to Lake Balaton. IIASA-Wiley International Series on Applied Systems Analysis. International Institute for Applied Systems Analysis. Laxenburg, Austria (forthcoming).

Somlyódy, L., and M. Virtanen. 1982. Application of a One-dimensional Hydrodynamic Model to Lake Balaton. Collaborative Paper. International Institute for Applied Systems Analysis. Laxenburg, Austria (forthcoming).

van Straten, G., G. Jolánkai, and S. Herodek. 1979. Review and Evaluation of Research on the Eutrophication of Lake Balaton - A Background Report for Modeling. 94 p. WP-79-13. International Institute for Applied Systems Analysis. Laxenburg, Austria.

van Straten, G. 1980. Analysis of Model and Parameter Uncertainty in Simple Phytoplankton Models for Lake Balaton. *In* Proceedings of the 2nd State-of-the-Art Conference in Ecological Modelling, Liège, Belgium. Also appeared as WP-80-139. 28 p. International Institute for Applied Systems Analysis. Laxenburg, Austria.

van Straten, G., and S. Herodek. 1981. Estimation of Algal Growth Parameters from Vertical Primary Production Profiles. 35 p. CP-81-19. International Institute for Applied Systems Analysis. Laxenburg, Austria. (also to appear in Ecological Modelling).

van Straten, G., and L. Somlyódy. 1980. Lake Balaton Eutrophication Study: Present Status and Future Program. 89 p. WP-80-187. International Institute for Applied Systems Analysis. Laxenburg, Austria.

A LINKED HYDRODYNAMIC AND BIOGEOCHEMICAL MODEL OF EUTROPHICATION IN LAKE BALATON[1]

Peter Shanahan[2] and Donald R. F. Harleman[3]

Abstract.--A study of Lake Balaton, a large shallow freshwater lake in Hungary, critically examined fundamentally different approaches to lake water quality and ecosystem modeling. The study focused upon the importance of mass transport as the mechanism tying the effects of hydrodynamics and biogeochemical reaction in the model formulation.

INTRODUCTION

In this study, we constructed and compared two different models of the phosphorus nutrient cycle in Lake Balaton.

The linked water quality model consists of three components. The hydrodynamic component -- a two-dimensional (single-layer) model of wind-driven circulation -- is used to compute transient water motion in response to historical meteorology. The biogeochemical component is a four-compartment model of phosphorus cycling in an unstratified water column. The linking component is the phosphorus transport model, which solves the one-dimensional advective-dispersion equation using the inputs of the hydrodynamic and biogeochemical components. The linked model is a continuum model in that it employs the finite difference method to solve for phosphorus and algal distribution as essentially continuous functions along the lake.

A commonly used alternative model formulation is the multiple-box model, in which a few fully mixed boxes of relatively large areal extent form the model of the lake. These are lumped component models which do not attempt to model the lake as a continuum and which often employ simplified or averaged hydrodynamics in formulating mass transport.

The one-dimensional linked water quality model is compared with a multiple-box model in a theoretical analysis and in simulations of Lake Balaton. The theoretical analysis is summarized briefly; the analysis illustrates basic difficulties in representing dispersion in multiple-box models and in relating the transport parameters of multiple-box models to prototype observations. The comparison of the models is continued in simulations of Lake Balaton employing both model formulations. The model results from the linked water quality model are more similar to the observed character of phosphorus and algal concentrations in the lake.

DESCRIPTION OF LAKE BALATON

Lake Balaton, the largest lake of central Europe, is located in western Hungary (fig. 1). The lake is long and narrow (75 km by 8 km) but extremely shallow. On average, the lake is only 3.1 meters deep. The deepest part of the lake (11.7 m) is in Tihany Strait, where the Tihany peninsula nearly divides the lake.

The climate of the Balaton region is temperate. During the winter, temperatures drop to below freezing and the lake is ice covered for roughly two months each year. The water temperature rises to an average of 24°C in July. During the summer, pleasant weather and generally good water quality make Balaton a very popular vacation resort.

Water motion within the lake is dominated by wind-driven circulation and seiche. Wind-driven flows are magnified by the lake's shallowness, and sustained winds can cause large set-up in water surface elevation. This also results in a very prominent seiche, with major longitudinal and transverse modes as well as various secondary modes. The shallowness of the lake dramatically effects the longitudinal seiche, lengthening its period and damping its amplitude over a few cycles.

A long-term net flow through Balaton is maintained between a number of major inflows concentrated at the western end of the lake, and the single outlet at the Sió Canal near the eastern end (fig. 1). The largest inflow is the Zala River, which is also the most prominent source of nutrients to the lake. The inflow and outflow are small relative to the lake's volume; the mean hydraulic residence time is correspondingly long, approximately four years.

---

[1]Paper presented at the Third International Conference on State-of-the-Art in Ecological Modelling, Colorado State University, Fort Collins, Colorado, May 24-28, 1982.

[2]Peter Shanahan is Senior Water Resources Engineer, Environmental Research and Technology, Inc., Concord, Massachusetts U.S.A.

[3]Donald R.F. Harleman is Ford Professor of Civil Engineering and Director of the Ralph M. Parsons Laboratory at the Massachusetts Institute of Technology, Cambridge, Massachusetts U.S.A.

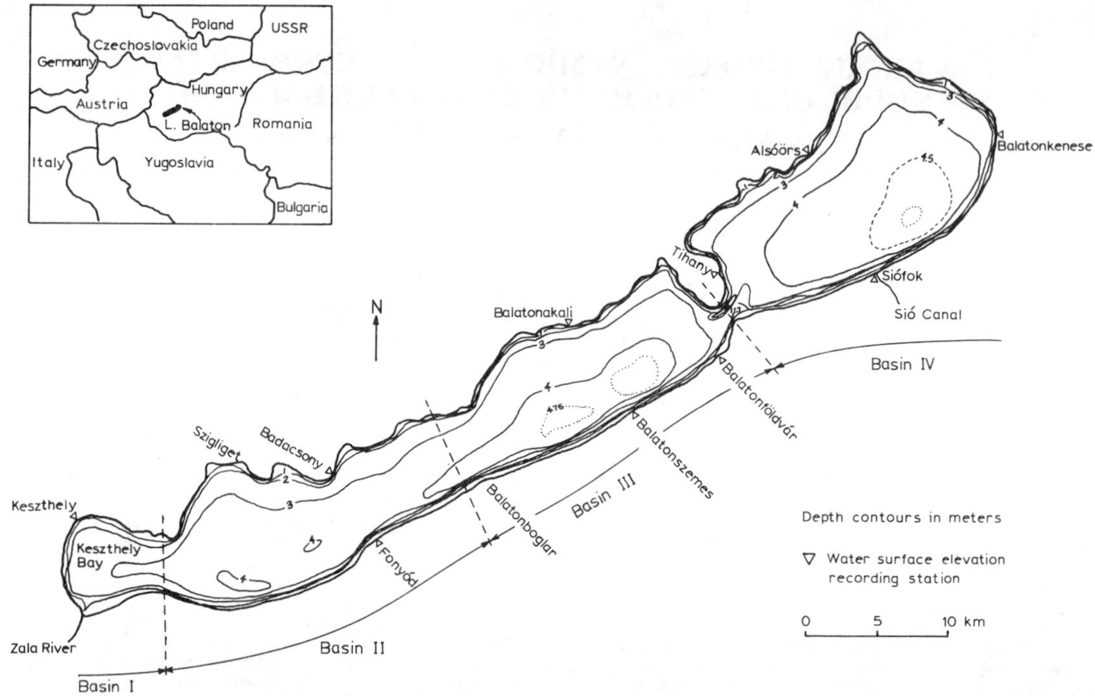

Figure 1.--Map of Lake Balaton, Hungary.

Water quality in Lake Balaton has deteriorated somewhat in recent years, and blue-green algae are a problem in Keszthely Bay, at the lake's western end. Balaton is a hardwater lake with high pH, alkalinity, calcium and magnesium. Dissolved oxygen is usually near saturation, as a consequence of wind mixing throughout the shallow depth. Phosphorus is most likely the limiting nutrient for algal growth, however phosphorus dynamics in the lake are not well understood. The dominance of calcium carbonate in the lake's chemistry should lead to nearly complete removal of orthophosphate from the lake, preventing substantial algal growth. However, algal growth causes occasional blooms, indicating that different sources and controls of orthophosphate may govern.

Although the details of Balaton's phosphorus chemistry are uncertain, it is very clear that external nutrient loading is a major factor in phosphorus availability. The concentration of nutrient inflows at the lake's western end leads to strong longitudinal gradients of phosphorus along the lake. A recent analysis by Jolánkai and Somlyódy (1981) shows that four sources each supply about one-fourth of the total phosphorus load to the lake: the Zala River, all other major streams, sewage discharges, and non-point sources.

LINKED HYDRODYNAMIC-BIOGEOCHEMICAL MODEL

A linked water quality model was developed to simulate phosphorus dynamics and eutrophication in Lake Balaton (Shanahan and Harleman, 1982). The model was constructed by linking various main components. The basic framework of the model is established by a model of water quality transport. This model component determines the changes in constituent concentrations due to the hydrodynamic influences of advection and dispersion and due to biogeochemical reactions. The water quality transport model of Lake Balaton solves the one-dimensional advection-dispersion equation for a vector of four phosphorus components. The equation is solved using a fractional step finite difference method applied to a one-dimensional finite difference grid in which the lake is divided into forty longitudinal increments. The fractional step method is advantageous since it can be constructed so as to isolate the reaction terms of the equations from the other terms. This allows the reaction step to be solved as an ordinary, rather than partial, differential equation.

The second main component of the linked water quality model is the biogeochemical model. For Lake Balaton, we use a four compartment model of the biological phosphorus cycle including phytoplankton. The model structure and reaction relations were originally developed by van Straten (1980) for application to Lake Balaton. The biogeochemical component simulates the reactions between model compartments representing dissolved inorganic phosphorus, detrital phosphorus, and two seasonal algal populations with differing temperature sensitivities. Transient meteorological conditions and phosphorus loadings are considered. The phosphorus reactions are marched through time in the reaction step of the water quality transport solution. A fourth-order Runge-Kutta technique solves the reaction equations.

The third main component of the linked model is a transient, single-layer circulation model which serves as the hydrodynamic component. This component determines the two-dimensional horizontal transport within the lake by solving the equations of motion over a two-dimensional finite difference grid. The circulation model includes the surface shear stress determined from the time-varying record of observed wind over the lake. Also considered are the forces due to non-linear bottom friction. Results from the model simulations are continuous predictions of the two-dimensional depth-averaged water motion and water surface displacement in the lake.

The hydrodynamic component is linked to the water quality transport via the dispersion coefficient. Longitudinal advection is not included in the linkage since it has relatively weak direct influence on water quality. With the exception of hydrologic through-flow, advection occurs as seiche motion of short excursion. Due to the shortness of the excursion, seiche motion has only minor influence on the net mass transport. The average dispersion coefficient over a seiche period is computed from the two-dimensional hydrodynamic model using formulae based on those of Holley, Harleman and Fischer (1970). The formulae assume lateral wind-driven advection to be the dominant source of lateral mixing, rather than the lateral diffusion considered by Holley et al. (1970). Modifications of the original equations are necessary to accommodate this change in assumptions. Otherwise, the Lake Balaton model follows Holley et al.'s procedures, including the reduction in dispersion they hypothesize for oscillatory flow. The computed dispersion coefficient is supplied as a function of time to the one-dimensional water quality transport component.

In addition to the circulation component, the linked water quality model includes a minor component which accounts for hydrologic through-flow. A water balance component budgets inflow and outflow to compute monthly average flows at each grid face along the lake. These flows are linked to the water quality transport component in the advective terms.

The various model components and linkages are shown schematically in figure 2. The components and linkages are described in detail in Shanahan and Harleman (1982). This reference includes descriptions of the component constructions, solution schemes and calibrations.

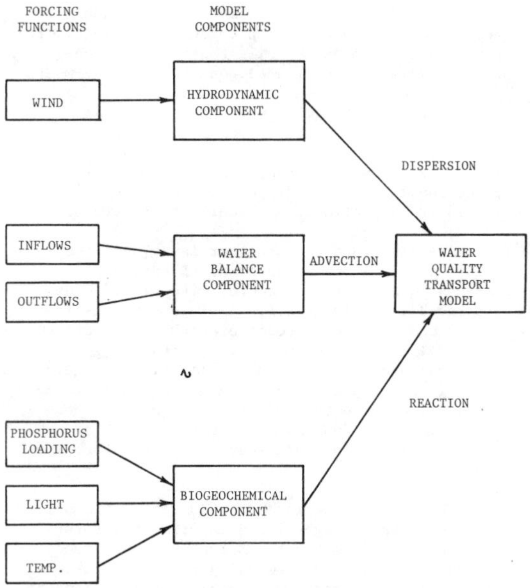

Figure 2.--Structure of linked water quality model.

MULTIPLE-BOX MODEL

Many prior lake water quality models, both of Lake Balaton (van Straten and Somlyódy, 1980) and of other lakes (DiToro et al., 1980; Thomann et al., 1979; Chen et al., 1979; and Richardson, 1976) are multiple-box types. Multiple-box models assume that the lake can be modeled as a number of interconnected fully-mixed large volume elements. The concentration is uniform within each element but differs from element to element. Changes in concentration over time are determined in each element by a simple mass balance.

The multiple-box model is conceptually very different from the finite difference mass transport component of the linked water quality model described above. The finite difference model is essentially a continuum model. If properly constructed, it serves as a close approximation of the continuous partial differential equation which describes mass transport and reaction within the lake. The multiple-box model, on the other hand, describes the lake with an essentially discrete approach. The equations solved by the multiple-box model are the integrals of the governing partial differential equation over discrete volume intervals. This conceptual difference -- continuous versus discrete -- leads to practical differences which we explore further below.

A goal of our Lake Balaton modeling investigation was to explore the effect of the model formulation on the simulation results. For this purpose, a second Lake Balaton model was programmed based upon a four-box configuration. The boxes correspond to Basins I through IV indicated in figure 1. The four-box model included the same biogeochemical model component as used by the linked water quality model. It also included the water balance component to compute advective flow. In the four-box model, however, exchange flow was specified as a model calibration parameter. This is in keeping with our findings, elaborated below, that transport processes associated with multiple-box models are not determined by hydrodynamic information. The dual model construction thus permitted the comparison of two lake model formulations with the Lake Balaton data as a reference.

ANALYTICAL MODEL ANALYSIS

In this section, we include a brief, analytical examination of one-dimensional model formulations as a preface to the results from applications to Lake Balaton. Lake Balaton behaves very nearly as a one-dimensional water quality transport system. The lake's shallowness insures that vertical concentration gradients may be ignored. The dominance of the Zala to Sió flow system, together with the lake's elongated horizontal geometry justify neglect of the lateral dimension.

A one-dimensional system is also amenable to analytical examination with conceptual models. For comparison purposes, we will consider two conceptual models roughly analogous to the lake model formulations described above. The tanks-in-series reactor with exchange flow is the conceptual equivalent of the multiple-box model. The dispersed flow reactor is a near analogue of the linked water quality model based on the one-dimensional advection-dispersion equation. These models are depicted schematically in figure 3.

The dispersed flow reactor is an analytical solution of the one-dimensional advection-dispersion equation for a reactor with closed ends separated by a distance L. A fixed through-flow rate Q and dispersion coefficient D are specified. The total reactor volume is V and its cross-sectional area is A. Mathematical flux boundary conditions are specified at each end. Thomas and McKee (1944) give the solution for the exit concentration as a function of the time after a tracer pulse is instantaneously injected into the reactor inflow.

The tanks-in-series model consists of a series of tanks with through-flow Q entering the first tank, passing from the first tank to the second, and so on, until the flow exits the last tank. The model may be made more complex by including an equal but opposite exchange flow X between the tanks. Here we will consider the simplest configuration, n equal volume tanks with total volume V. The solution for a series of tanks with no exchange flow subject to a pulse input may be determined analytically. However, the addition of exchange flow requires a numerical solution (Tuan et al., 1980).

These two reactor models may be compared in terms of their advective and dispersive character. Advection is non-dimensionally represented as the reactor hydraulic residence time, $t^* = V/Q$. The residence time will be the same for tanks-in-series and dispersed flow reactors with the same flow and total volume.

Dispersion is represented by the non-dimensional Peclet Number, Pe, the ratio of advective to dispersive transport. The Peclet Number of the dispersed flow reactor is defined as $Pe = QL/AD$. Zvirin and Shinnar (1976a, 1976b) define an apparent Peclet Number for the tanks-in-series model:

$$Pe' = \frac{2n}{1 + 2\alpha} \tag{1}$$

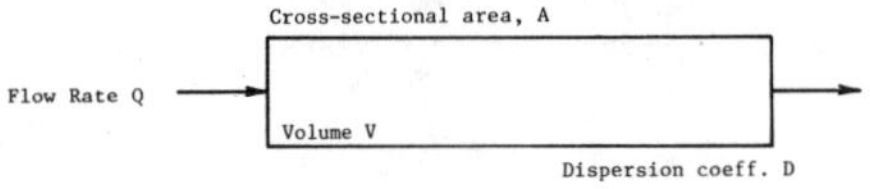

a. Dispersed flow reactor

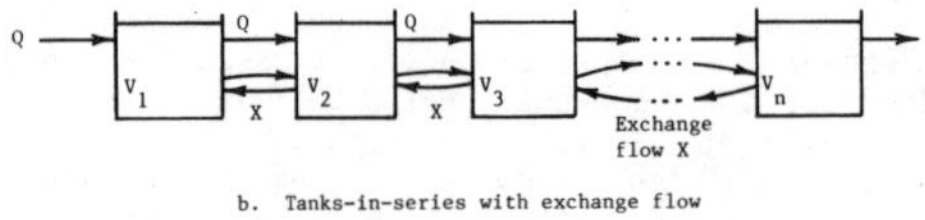

b. Tanks-in-series with exchange flow

Figure 3.--Conceptual one-dimensional reactor models.

where $\alpha = X/Q$ is the ratio of the exchange flow to the through-flow. This equation shows that even without exchange flow ($\alpha = 0$), there is implicit dispersion in the tanks-in-series model. Only in the limit $n \to \infty$, does dispersion disappear as $Pe' \to \infty$.

Ideally, the tanks-in-series model and the dispersed flow model should be able to describe any particular one-dimensional system equivalently. Equivalent advective character occurs when the hydraulic residence times of the models are equal. Equivalent dispersion requires that the reactor Peclet Numbers be equal:

$$Pe = Pe' = \frac{2n}{1 + 2\alpha} \tag{2}$$

If we take the dispersed flow reactor Peclet Number, Pe, as given, then α and possibly n must be altered to achieve a match between Pe and Pe'. This raises three possible tanks-in-series model regimes:

1. $2n < Pe$ - No match is possible. The n tanks-in-series are implicitly more dispersive than the dispersed flow reactor.
2. $2n \simeq Pe$ - The models match if $\alpha = 0$. The implicit dispersion of the n tanks is approximately equal to that of the dispersed flow reactor.
3. $2n > Pe$ - The models match if $\alpha = (2n - Pe)/2Pe$. The n tanks-in-series are implicitly less dispersive than the dispersed flow reactor, thus the additional dispersion due to an exchange flow is necessary.

These regimes clearly illustrate the importance of the implicit model dispersion to the behavior of tanks-in-series models.

We can draw important implications from the above analysis for the construction of lake water quality models. To do so, we call attention to the correspondence between the tanks-in-series model and the multiple-box model, and between the dispersed flow reactor and the finite difference model of the one-dimensional advective-dispersion equation. The latter is a close analogy if the finite difference model is a good approximation of the continuous equation. This holds if numerical dispersion is small. Numerical dispersion is similar to the implicit dispersion of the tanks-in-series model, however it differs in several respects. A critical distinction is that numerical dispersion may be greatly reduced by the choice of an appropriate finite difference technique, for example second-order accurate central differencing. However, implicit dispersion is changed only if the number of tanks-in-series is changed.

By analogy, the analysis of conceptual reactor models leads to important conclusions on constructing lake models. Most importantly, it points out the implicit dispersion contained in multiple-box lake models. The consequence of the implicit dispersion is diminished control over the mixing characteristics of the multiple-box model. The large implicit dispersion precludes the direct specification of the model exchange flow from observed hydrodynamics. And, in some cases, the implicit model dispersion will simply be too great even if the exchange flow is zero. The crucial point is that internal mixing in the multiple-box model can only be determined by calibration. The calibration parameters include not only the exchange flow, but the number of boxes as well. Consequently, the number of boxes cannot necessarily be determined solely by the lake's geometry and other characteristics.

SIMULATION RESULTS

Results of simulations of Lake Balaton reinforce the conclusions drawn above. Simulations were performed for a 250-day period beginning February 28, 1977. Data for the run came from numerous sources described by Shanahan and Harleman (1982). Included were wind data to drive the two-dimensional hydrodynamic model, phosphorus loading data, water temperature and meteorological data, and hydrologic data. The model hydrodynamic component was run only for the months of July and August -- a representative wind period. The predicted circulation was sufficiently homogeneous in time to be presumed typical, and thus it was used repeatedly for the entire 250-day water quality simulation.

Limited simulation results will be presented here. Spatial concentration profiles of total and algal phosphorus are shown in figure 4 for August 4, a period of low flow when phytoplankton are near their peak summer concentration. The figure contrasts the predictions of the linked water quality model with those of the four-box model. The four-box model includes a run with zero exchange flow and a run in which the exchange flow varies with the interfacial area between the lake boxes. The flow is determined as the interfacial area multiplied by a uniform return velocity, V_{ret}, as employed by van Straten (1980).

The effect of the box model's implicit dispersion is clear in figure 4. Local concentration gradients and peaks are seen in the finite difference model results, but are absent with the four-box model. The overall profile is flattened still more when exchange flow is added to the box model. The box models omit local concentration peaks at various points along the lake corresponding to major nutrient sources. These omissions are significant at a number of locales.

Field data collected on the lake support the character of the finite difference model results but fail to verify the concentration magnitudes predicted by either model formulation. Profiles of data collected close to August 4 are shown in figure 5. Algal population is indicated by the chlorophyll-a concentration, but this quantity is not directly convertible to the algal phosphorus concentration predicted by the model. van Straten (1980) makes the rough assumption that the algal phosphorus concentration is half that of chlorophyll-a. Similarly, the total phosphorus data determined in the field includes substantial inorganic phosphorus associated with suspended sediment. The inorganic phosphorus fraction is thus subtracted in figure 5 to permit comparison with model results. Thus, with both constituents, there is considerable uncertainty in the comparibility of field data and model results.

Despite these caveats, there remain major discrepancies between the predicted concentrations and those observed. In continuing research at MIT, Baker and Harleman are investigating the calibration of the biogeochemical model and testing alternative biogeochemical formulations in an effort to correct these discrepancies. These biogeochemical model changes do not substantially alter the spatial character of the results however. The field data show local concentration peaks and spatial trends which are in rough agreement with the finite difference model predictions. The field data thus corroborate the contention that the four-box model is overly dispersive, that it smooths out spatial structure. The results from the finite difference model, however, are much more similar in character to the field observations.

CONCLUSIONS

The analyses and simulation results presented in this paper argue that multiple-box lake water quality models must be applied with considerable care, if at all. Multiple-box models carry within their formulation a large implicit dispersion. This dispersion is controlled only if the fundamental character of the model is changed by varying the number of boxes. Further mixing may be added to multiple-box models by including equal but opposite exchange flow between the boxes. However, the character of the box models implies that the exchange flow cannot be determined directly from a knowledge of the lake's hydrodynamics. In short, neither the number of boxes nor the exchange flow can be specified a priori from the lake geometry or flow regime. Rather, they are model-dependent parameters which can only be found by calibration.

In contrast to the multiple-box model, there are the finite difference models which seek to closely approximate the continuum solution of the mass conservation equations. If properly constructed, these models reduce artificial dispersion to an insignificant level. This allows the explicit specification of the dispersion coefficient D as the determinant of the mixing properties of the model. Further, the dispersion coefficient may be found directly from the hydrodynamic behavior of the actual lake, rather than by calibration by means of water quality data.

These conclusions for Lake Balaton lead to a broader conclusion: that hydrodynamics plays an important role in lake water quality and ecosystem modeling. Thus is offered the proposition that lake modeling can be improved if the influence of lake hydrodynamics is properly accounted for in the model. This proposition challenges the notion that the biogeochemical formulation alone dominates lake model construction. It calls into question the practice of using highly complex and sophisticated ecosystem and element cycle models within the framework of large-element multiple-box structures.

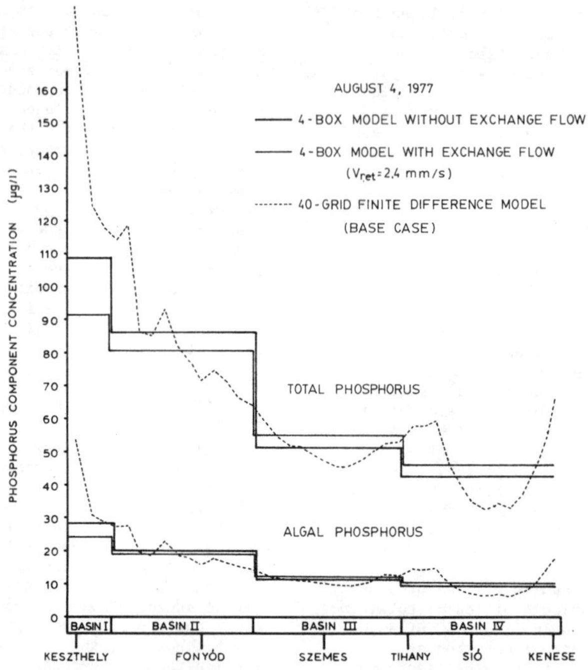

Figure 4.--Water quality profiles for August 4, 1977 predicted by the linked finite difference model and the four-box model.

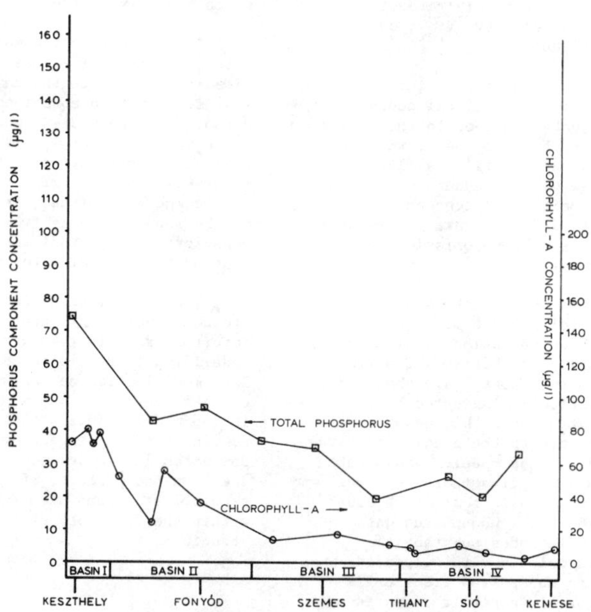

Figure 5.--Profiles of measured total phosphorus and chlorophyll-a. Total phosphorus was measured August 8, 1977; chlorophyll-a was measured August 4, 1977.

LITERATURE CITED

Chen, C.W., and D.J. Smith. 1979. Preliminary Insights into a Three-Dimensional Ecological-Hydrodynamic Model. In Perspectives on Lake Ecosystem Modeling. Edited by D. Scavia and A. Robertson. Ann Arbor Science Publishers, Inc., Ann Arbor, Michigan.

DiToro, D.M., and J.P. Connolly. 1980. Mathematical Models of Water Quality in Large Lakes, Part 2: Lake Erie. Report EPA-600/3-80-065. United States Environmental Protection Agency. July.

DiToro, D.M., and W.F. Matystick, Jr. 1980. Mathematical Models of Water Quality in Large Lakes, Part 1: Lake Huron and Saginaw Bay. Report EPA-600/3-80-056. United States Environmental Protection Agency. July.

Holley, D.R., D.R.F. Harleman, and H.B. Fischer. 1970. Dispersion in Homogeneous Estuary Flow. ASCE, Journal of the Hydraulics Division. 96:HY8, 1691. August.

Jolánkai, G., and L. Somlyódy. 1981. Nutrient Loading Estimate for Lake Balaton. Collaborative Paper CP-81-21. International Institute for Applied Systems Analysis. Laxenburg, Austria. April.

Richardson, W.L. 1976. An Evaluation of the Transport Characteristics of Saginaw Bay Using a Mathematical Model of Chloride. In Modeling Biochemical Processes in Aquatic Ecosystems. Edited by R.P. Canale. Ann Arbor Science Publishers, Inc., Ann Arbor, Michigan.

Shanahan, P., and D.R.F. Harleman. 1982. Linked Hydrodynamic and Biogeochemical Models of Water Quality in Shallow Lakes. Technical Report Number 268. Ralph M. Parsons Laboratory for Water Resources and Hydrodynamics, Department of Civil Engineering, Massachusetts Institute of Technology, Cambridge, Massachusetts. March.

Thomann, R.V., R.P. Winfield, and J.J. Segna. 1979. Verification Analysis of Lake Ontario and Rochester Embayment Three Dimensional Eutrophication Models. Report EPA-600/3-79-094. United States Environmental Protection Agency. August.

Thomas, H.A., Jr., and J.E. McKee. 1944. Longitudinal Mixing in Aeration Tanks. Sewage Works Journal. 16:1, 42. January.

Tuan, V.A., N.C. Thanh, and B.N. Lohani. 1980. Hydraulic Model for Biological Reactors: Application to Activated Sludge. Journal Water Pollution Control Federation. 52:7, 1931. July.

van Straten, G. 1980. Analysis of Model and Parameter Uncertainty in Simple Phytoplankton Models of Lake Balaton. Second ISEM Meeting. [Liege, Belgium, April 1980]. (Also published as Working Paper WP-80-139. International Institute for Applied Systems Analysis, Laxenburg, Austria. September 1980.)

van Straten, G., and L. Somlyódy. 1980. Lake Balaton Eutrophication Study: Present Status and Future Program. Working Paper WP-80-187. International Institute for Applied Systems Analysis. Laxenburg, Austria. December.

Zvirin, Y., and R. Shinnar. 1976a. A Comparison of Lumped-Parameter and Diffusional Models Describing the Effects of the Outlet Boundary Conditions on the Mixing in Flow Systems. Water Research. 10:9, 765.

Zvirin, Y., and R. Shinnar. 1976b. Interpretation of Internal Tracer Experiments and Local Sojourn Time Distributions. International Journal of Multiphase Flow. 2:5/6, 495.

EUTROPHICATION MODELLING IN RELATION TO WATER MANAGEMENT OF THE RHINE RIVER BASIN[1]

R. Klomp, F. J. Los, J. A. van Pagee, N. M. de Rooy, J. G. C. Smits, J. C. Stans[2]

Abstract. The paper deals with a modelling approach to the eutrophication in the river Rhine basin. Emphasis is laid on the mathematical models which describe the nutrient budget of the river Rhine and its main tributaries and the eutrophication process in lake IJssel in the Netherlands.

INTRODUCTION

Eutrophication can be considered as one of the most severe problems in water quality management of the fresh water lakes in the Netherlands. Approximately 65% of the annual fresh water is supplied by the river Rhine, which feeds most lakes and reservoirs. Since the nutrient concentrations in the river Rhine are high it goes without saying that the future water quality developments in its drainage basin is the key factor to the eutrophication problem in the Netherlands.

In order to predict the influence of sanitation strategies in the river Rhine drainage area on the eutrophication process of the largest lake, lake IJssel, two modelling studies were carried out:

- water quality modelling of the river Rhine in order (1) to describe the oxygen, phosphorus and nitrogen budgets and (2) to simulate the consequences of intended and possible alternative or additional sanitation measures.
- eutrophication modelling of lake IJssel in order (1) to describe the nutrient budget, phytoplankton biomass and species composition and (2) to simulate the consequences of the intended and possible local sanitation measures and changes of the pollution due to sanitation measures in the drainage area of the river Rhine.

GENERAL DESCRIPTION OF THE RIVER RHINE AND LAKE IJSSEL (FIG. 1)

The river Rhine drainage area consists of two different parts: the Swiss Alps and the remaining part of the drainage basin in Switzerland, Germany, France and the Netherlands (Anon 1978). Switzerland as well as Germany and France contribute for 50% to the mean annual discharge (2200 m^3/sec). The contribution of the Alps consists of melt water whereas the contribution of the remaining parts consists of net rainfall. The total surface of the drainage basin amounts to 160.000 km^2

The river gave among others the Ruhr area the opportunity for industrial and economic expansion, but is also the daily source of potable water for millions of people.

Therefore the water quality of the Rhine is of great concern and up till some years ago in spite of international negociations water pollution has been steadily increasing. Recently, however, some progress has been made especially with respect to BOD, DO and certain heavy metals and organic microcompounds (Dijkzeul 1980).

Figure 1.--Map of the river Rhine drainage area, including Lake IJssel.

---

[1]Paper presented at the Third International Conference on State-of-the-Art in Ecological Modelling, Colorado State University, May 24-28, 1982.

[2]Rob Klomp, Hans Los, Hans van Pagee, Nico de Rooy, Johan Smits and John Stans are resp. head and staff members of the Environmental Hydraulics Branch of the Delft Hydraulics Laboratory, Delft, the Netherlands.

Lake IJssel, the largest fresh water lake of the Deltaic area of the river Rhine, is situated in the central and northern part of the Netherlands. It is fed by some small rivers, local polders discharges and the river IJssel, which is a branch of the river Rhine. Lake IJssel has a surface area of approximately 1200 km^2 and an average depth of 4.5 meter. The average retention time of lake IJssel amounts 9 months. The bottom sediments of lake IJssel mainly consists of sandy soils. Lake IJssel has many functions; it is used for drinking, industrial and agricultural water supply but also recreation (Hosper 1981).

WATER QUALITY MODELLING OF THE RIVER RHINE

Data Collection and Analysis

Data collection and the construction of a data base are essential activities in the framework of water quality modelling of the Rhine. The data base includes data from which input for the model can be derived and data which can be used for calibration and verification.

Hydrologic data concern riverflows and hydraulic constraints such as flow dependent functions for local depths and stream velocities; they are basic inputs for the model.

The water quality data concern concentrations of dissolved and suspended solids and temperature. Except for temperature these water quality data are necessary for calibration and verification.

Data on pollutant loads to the Rhine including both point- and non-point sources are the main inputs for the model. Point sources primarily concern domestic and industrial waste loads whereas diffuse sources mainly concern agricultural run-off and small, indirect waste water loads. Since concrete reliable data with respect to the contribution of diffuse sources are lacking, the loads were calibrated within the established ranges (Van Pagee 1981).

An overview of the set-up of a data base as a part of the water quality modelling approach of the Rhine is given in figure 2.

Model Description

Based on an extensive review of available models it was decided to base the water quality modelling of the Rhine on the model QUAL-2 (Roesner et al 1977).

This model was considerably modified and extended with modules for the computation of organic nitrogen, ammonium, ortho-phosphate, particulate and organic phosphorus and a module for COD and detritus (Postma 1981). The modified QUAL-2 model (MODQUAL) was the spill in the package of programmes, which has been built for the water quality modelling the Rhine. Beside possibilities to model temperature, conservative substances and pollutants with first order kinetics, the model is especially suitable for modelling the dissolved oxygen, nitrogen and phosphorus budget, individually as well as integrated in combination with phytoplankton.

The integral approach is illustrated in figure 3.

Various processes in the model are temperature dependent, such as DO exchange with the atmosphere, decay of detritus and organic carbon, oxygen demand by the sediment, ammonification, nitrification, deni-

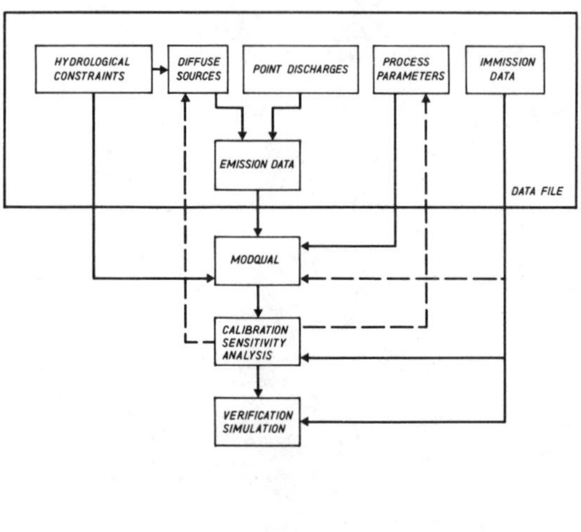

Figure 2.--The relationships between the data required for the model and the development of the model.

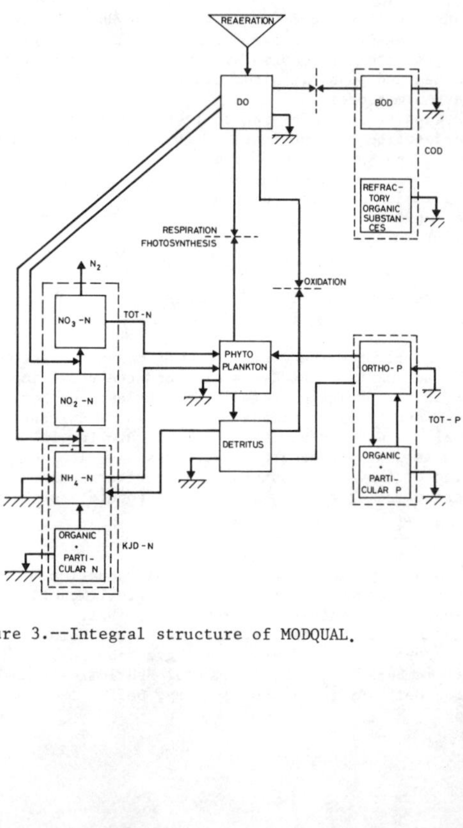

Figure 3.--Integral structure of MODQUAL.

trification and production, respiration and mortality of algae. The production of dissolved oxygen and the uptake of nutrients by algae is coupled to the algae production which is a function of irradiation, extinction, mean depth and limiting nutrient concentration.

From a numerical point of view the mass conservation equation, which is based on advective and dispersive transport, discharges and reactions is solved implicitly by the finite difference method. The river is therefore considered as a series of computational elements. In each element the water quality is assumed to be homogeneous. The length of all the computational elements is the same. Distinction has been made between the type of elements in order to include pollutant discharges and abstraction of water. Small pollutant loads are distributed equally over a reach, which consists of an even number of computational elements. Within such a reach the hydraulic properties of the river process parameters and other constraints are the same.

Application of the Model

MODQUAL has been applied to the Rhine from Stein (Switzerland) to Kampen (the entrance of lake IJssel), Vreeswijk and Gorinchem. Tributaries such as the Neckar, Main and Mosel are also modelled (fig. 4), (Klomp et al 1981). Since the main objective of this modelling study was to predict the future loads on lake IJssel and the availability of data was rather limited, no attempt has been carried out to describe the detailed dynamic water quality behaviour of the river Rhine. The yearly averages of 1973 and 1977 were the basis for respectively calibration and verification.

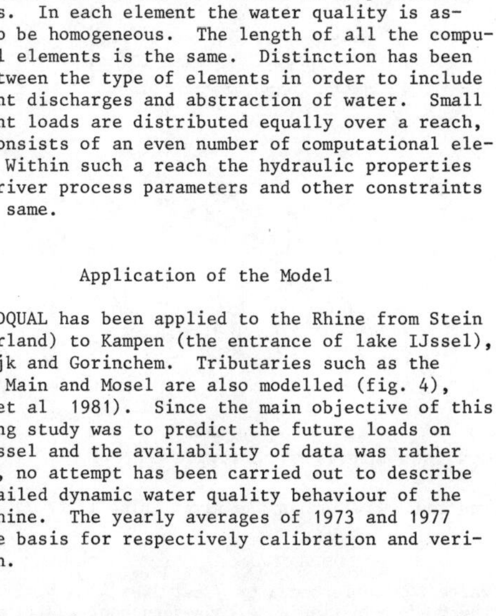

Figure 4.--The modelled part of the rivers Rhine, Neckar, Main, Moselle, Waal and IJssel.

Calibration of the model mainly consists of adjusting the process parameters within the literature ranges. Initially the set of process parameters was kept constant for the whole river. However, exceptions were necessary for the sedimentation rate in the weired part of the river, coagulation of orthophosphate and iron at the confluence with Ruhr and increased denitrification in the downstream part of the Main.

In general the results of the calibration and verification of the model were satisfactory (fig. 5 and 6). Considering the exactness and reliability of quantifying the pollutant discharges and contribution of diffuse sources, a more detailed calibration and verification would not be justified (Van Pagee 1981-2).

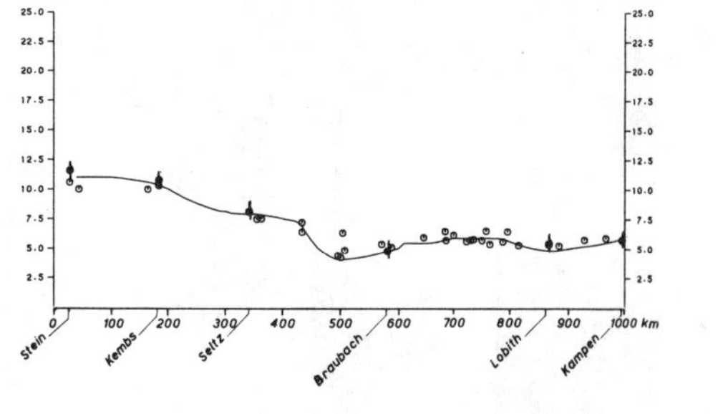

Figure 5.--Model calibration based on yearly averaged data 1973 (dissolved oxygen).

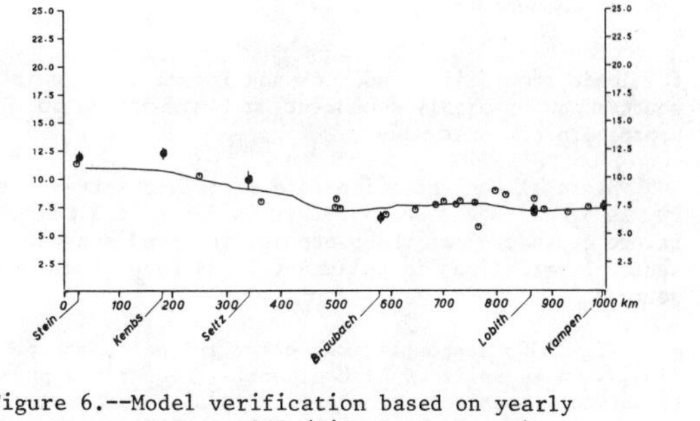

Figure 6.--Model verification based on yearly averaged data 1977 (dissolved oxygen).

For the 1972-1977 period (with exception of 1975) monthly averages of several water quality variables were computed. Despite the rather rough estimate of the pollutant loads during 1972, 1974 (based on inventory in 1973) and 1976 (based on an inventory of 1977) the model is quite able to calculate the overall changes in water quality due to changes in hydrologic and seasonal conditions. This is especially the case for dissolved substances such as ammonia and orthophosphate. The behaviour of these concentrations is primarily determined by changes in the dilution capacity of the river during the varying river flow regime (fig. 7).

Substances, which are partly subjected to sedimentation (BOD, COD, N-Kjeldahl, total-P) show a more stochastic course which makes a comparison between observations and calculated concentrations more difficult.

This character could be attributed to the interaction between sedimentation and resuspension whereas

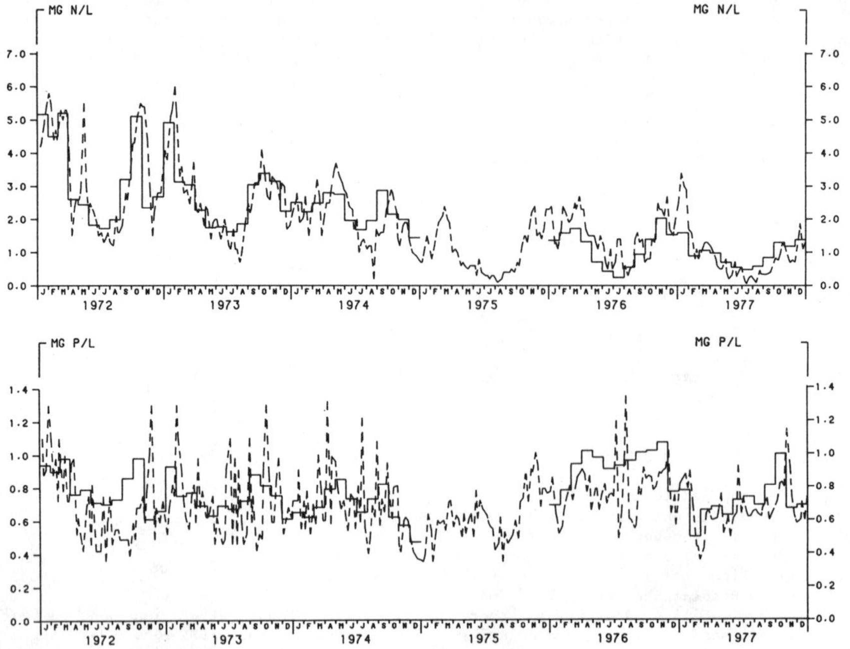

Figure 7.--Observed and computed ammonia and total phosphorus concentration in the river Rhine (IJssel) at Kampen during 1972-1977.

the loads from diffuse sources and intermittent point sources can be highly dependent on fluctuations in hydrologic circumstances.

Seasonal concentrations are clearly observed for DO, BOD5, Kjd-N and NH4-N. This is due to the temperature dependency of the processes involved and the seasonal variations in pollutant loads by diffuse sources.

High BOD5 concentrations occur in the summer period, when in spite of high mineralization rates phytoplankton is able to increase the BOD5; in other words waste BOD is replaced by algal BOD (fig. 8). Therefore the BOD5 reduction a a consequence of additional BOD removal of sewage and waste water will be limited.

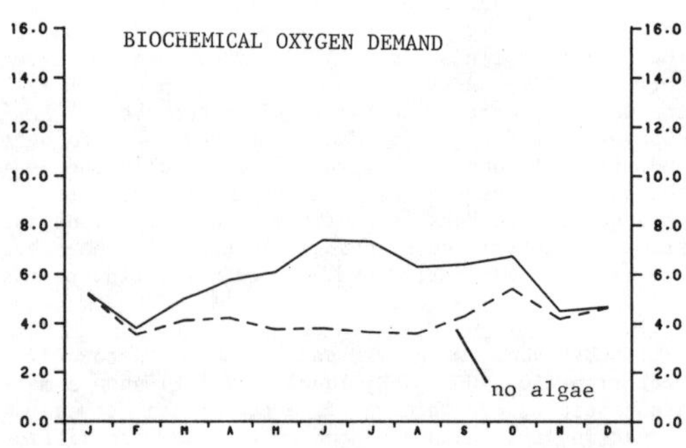

Figure 8.--The influence of phytoplankton on BOD (Lobith 1977).

It was concluded by sensitivity analysis that the contribution of waste water discharges in the concentration of Lobith is very important in the case of NH4, NO2-N and PO4-P and of minor importance in the case of NO3-N, because of the considerable influence of diffuse sources to the NO3-N load.

From the application it can be concluded that the model is able to predict the future water quality of the river Rhine under several hydrologic and seasonal conditions. Based on the results of the simulation of 1973 and 1977 it will be clear that the model can also describe the consequences of changes in the pollutant loads.

Simulations

Monthly simulations were carried out for the water quality at the entrance of lake IJssel (Kampen) and concern primarily the following scenarios:

1. Continuation of the present policy aiming at BOD reduction in the Rhine river drainage area (scenario 1)
2. As in 1, with additional replacement of polyphosphates in the detergents in the total river drainage area (scenario 4).
3. As in 1 and 2, with additional phosphate removal in the various secundary waste water treatment systems in the total river drainage area (scenario 8).

The results are shown in table 1.

Kampen	hydrology	1977	1977
	prediction	1977	1985 /1/4/8
O_2	(mg/l)	7,7	8,9
BOD_5	(mg O_2/l)	4,7	3,6
COD	(mg O_2/l)	25,9	21,9
Kjd-N	(mg N/l)	1,9	1,5
NH_4-N	(mg N/l)	0,90	0,63
NO_3-N	(mg N/l	4,2	4,7
tot-N	(mg N/l)	6,1	6,2
o-P	(mg P/l)	0,40	0,43 /0,20/0,06
tot-P	(mg P/l)	0,68	0,71 /0,42/0,21
Chl-a	(µg /l)	41,3	41,3

Table 1.--Summary of the results of the simulation for 1985 scenarios 1,4 and 8 based on the hydrological characteristics of 1977

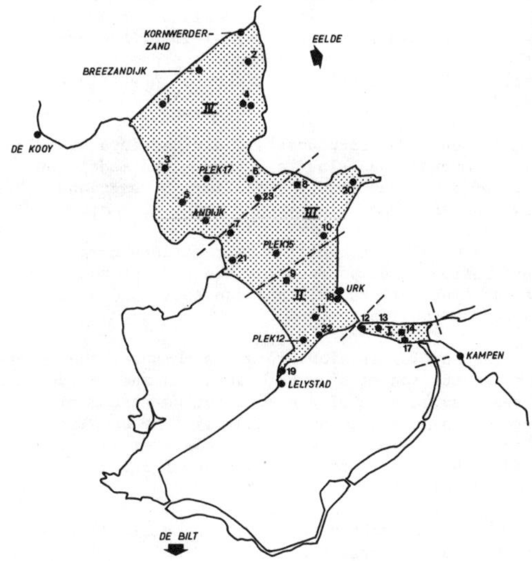

Figure 9.--Schematization of Lake IJssel.

EUTROPHICATION MODELLING OF LAKE IJSSEL

Data Collection and Analysis

Data collection and analysis showed spatial gradients of nutrient and phytoplankton concentrations. Therefore the lake was divided into 3 compartments (fig. 9).

Model Description

The eutrophication of lake IJssel is described by the (coupled) mathematical models BLOOM II (Los 1980, De Rooy et al 1982, Los 1982-1, Los 1982-2) and CHARON (De Rooy 1980). Both models were developed within the framework of the WABASIM-project (WAter BASIn Models), a joint project of the Delta Department of Rijkswaterstaat and the Delft Hydraulics Laboratory.

The phytoplanktonmodel BLOOM II uses an optimization technique (linear programming) to calculate the maximal equilibrium value of the concentrations of several phytoplankton species consistent to a given set of environmental conditions in a certain time period (generally one week). The time periods of the model are not connected but in general the input data change rather slowly from one period to the next, hence the predictions often show a smooth course during a year. If the environmental conditions do change abruptly, however, the model will respond immediately with either a sudden increase or a collapse of the bloom. The conditions are specified externally, based on measurements or in this study on the outputs of the chemical model CHARON.

The species in the model may be limited by any of the three nutrients nitrogen, phosphorus and silicon, or by the growth rate and by the light intensity, the latter which may be too high or too low. Two or more factors can be limiting simultaneously, in which case the bloom will consist of more than one phytoplankton species, because, as a consequence of the mathematical structure, the number of species in the bloom is always equal to the number of limiting factors. This property of BLOOM II is consistent to the classical formulation of Liebig's law: one species is limited by one factor only. Temperature is not included as a limiting factor explicitly, but may limit the phytoplankton bloom anyhow, because many rates in the model are temperature dependent, as for instance the rates of production and of respiration of the phytoplankton species.

Most important is that BLOOM II neglects all nutrient transports to and from the bottom, because they cannot be computed presently for the shallow Dutch waters. For calibration and validation, this is of minor importance; if there is a large change in the concentration of nutrients in the water during a week, because of an important exchange with the bottom, BLOOM II will only predict incorrectly for that particular week, because it does not use the proper nutrient concentrations. However, this error is not carried over to the next week, since the model receives a completely new set of input data, including all exchanges with the bottom from any of the foregoing weeks.

For the prediction of future situations, however, it is very important to predict the exchange of nutrients between the bottom and the water; for that purpose the chemical model CHARON has been developed and subsequently coupled to BLOOM II.

The chemical model CHARON uses two basic laws e.g. the mass conservation law and the mass action law. The model computes the chemical composition of surface waters as a result of the external load, the exchange with the bottom and the atmosphere, processes and the outflow. The chemical composition of phytoplankton is included as well.

The chemical compounds are divided into equilibrium compounds and slow reactants. The model consists of 3 modules for (1) equilibrium reactions, (2) slow processes and (3) phytoplankton.

The module for equilibrium reactions computes the distribution of components for those compounds, which are reacting very rapidly and can therefore be considered at equilibrium.

The module for slow processes computes the change in concentration of slow reactants and the influence on the mass balance of the equilibrium reactants. For each component the quantity that will be supplied to or abstracted from the equilibrium system is calculated. The kinetic formulations are first order differential equations.

The phytoplankton module is primarily designed for the coupled version of BLOOM II and CHARON (fig. 10). In case CHARON is applied individually, the model computes the net exchange of nutrients between inorganic and organic compartments.

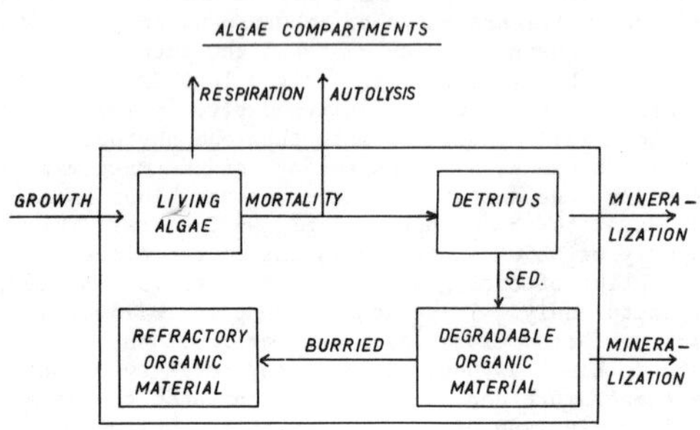

Figure 10.--Phytoplankton modelling; the magnitude of P_g-R (growth-respiration) are given by BLOOM II, the remaining processes are computed by the phytoplankton module of CHARON.

System Definition

Since BLOOM II and CHARON are based upon data from model reservoir in which the dominant circumstances differ from lake IJssel, a system definition phase was necessary.

During this phase the general structures of BLOOM II and CHARON were adjusted to lake specific circumstances with respect to parameters and processes. For BLOOM II this included some alterations in the specific extinction coefficients and of the relation between dry weight and chlorophyl of several phytoplankton species. Also the species Melosira, which is the most important diatom in lake IJssel, had not been included in the original BLOOM version, but was now added.

The system definition of CHARON comprises among others the definitions of the relevant processes and the bottom/water exchange.

Applications of the Model(s)

In the calibration phase BLOOM II as well as CHARON were considered individually.

The calibration of BLOOM II was carried out in two steps. In the first step, several coefficients were adjusted and new production-efficiency curves, which relate the production of the individual species to the light intensities were used.

During the second phase of calibration the assumptions reporting buoyancy of blue greens were adjusted and light adaptation was incorporated into the model.

As can be seen from the foregoing much attention has been given to a reconsideration of the availability of energy (light) and the way it is incorporated in the model. These activities were due to the fact that the model was mainly based on nutrient limited modelreservoirs, whereas in lake IJssel light is the only important limiting factor.

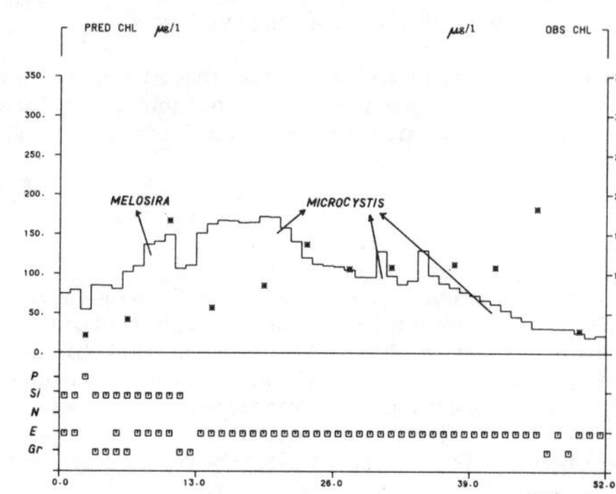

Figure 11.--Chlorophyl limiting factors and dominant species; calibration of BLOOM II (compartment 4, 1977).

The ultimate result of the calibration was acceptable (fig. 11). Although the computed chlorophyl level after the spring bloom in the second quarter is too high, there was a reasonable agreement between calculated values and observations during the remaining part of the year, except for the end of the year during which a chlorophyl peak could not be reproduced by the model.

As far as it concerns the species composition the calculated spring bloom consists of the diatom Melosira whereas the Microcystis dominates the remaining part of the year. The latter differs from the observations since after the diatom bloom, Tribonema was mainly dominant. Because of a lack of physiological information, this species could not be incorporated in the model. Phytoplankton is energy and growth limited in the model whereas silicon is limiting to the diatom bloom.

During calibration of CHARON a number of changes in the process and stochiometric constants were realized. Reaeration constant, rate of denitrification, mobilization of phosphorus and iron from bottom sediments and mineralization rate of phosphorus in suspended detritus have been increased. The sedimentation rate of calcium-phosphorus compounds has been decreased.

The ultimate result of the calibration of CHARON is satisfactory (fig. 12). However, for the northern part (compartment 3/4) differences between observed and calculated total-N and total-P remain present. These differences are presumably caused by resuspension of more or less inert material, a process, which is not incorporated in the model. Probably this process has no significant influence on the nutrient availability to phytoplankton. After calibration, the models were verified with data from another period, than used for calibration.

Verification of BLOOM II showed the same differences between observation and calculation as in the calibration. The differences are concentrated again in the second quarter and at the end of the year.

With respect to the species composition a reasonable agreement was observed between the field data and the calculation, although the model slightly overemphasizes the significance of Microcystis.

Similar to the calibration period, during the verification period energy was the main limiting factor whereas silicon was limiting the diatom bloom.

The results of the verification of CHARON were also consistent to the results of the calibration.

Verification of the coupled model does not differ significantly from the verification of the individual models (fig. 13).

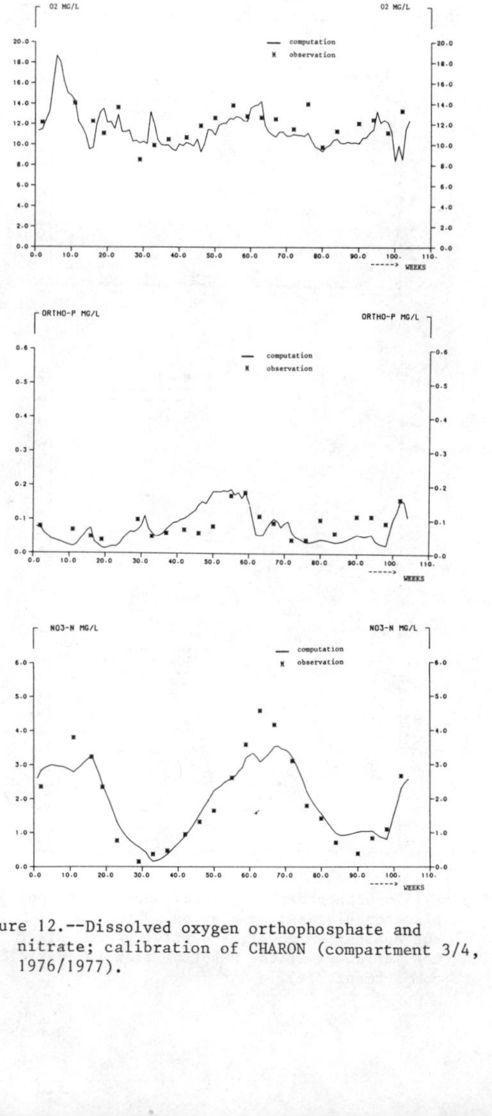

Figure 12.--Dissolved oxygen orthophosphate and nitrate; calibration of CHARON (compartment 3/4, 1976/1977).

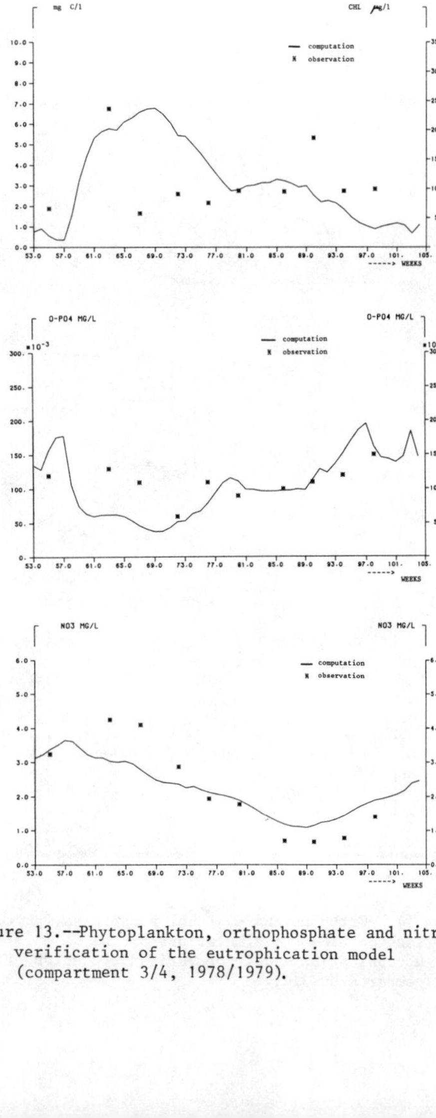

Figure 13.--Phytoplankton, orthophosphate and nitrate; verification of the eutrophication model (compartment 3/4, 1978/1979).

It is clear that the overprediction of phytoplankton biomass in the second quarter is reflected in the overprediction of pH and DO and the underprediction of nitrate and ortho-phosphate.

Simulations

With the verified coupled model and with the water quality model of the river Rhine IJssel the influence of several sanitation strategies on the eutrophication of lake IJssel is simulated.

The scenarios 1, 4 and 8, have already been mentioned. Beside these scenarios indicative calculations were carried out with respect to the influence of reduced inflow of the river IJssel in the lake and nitrogen removal in waste water treatment systems on lake IJssel.

The simulations (fig. 14) show that:

- in scenario 1 the phosphorus concentration in the lake increases due to the increased phosphorus concentration in the river Rhine (IJssel). The increased P-concentration has no impact on the phytoplankton biomass and composition, since phosphorus is not a limiting factor in the present situation.
- In spite of a considerable reduction by 43% in ortho- and total-phosphate in scenario 4 the phytoplankton biomass and species composition do not change at all
- In scenario 8 the phosphorus load to the IJssel lake is reduced by 73% in compartments 1 and 2 however energy remains the only limitation due to the short retention time. Thus the computed phytoplankton biomass and species composition do not change. In compartment 3/4, however, the phytoplankton biomass decreases by 22% (annual average). This reduction mainly occurs after the diatom bloom in the spring. With respect to the species composition the role of diatoms increases relatively. *Microcystis* remains the dominant species in the second half of the year.
- (1) Decreasing the inflow of water from the river Rhine (IJssel), (2) replacing polyphosphates in the detergents in the river Rhine drainage basin and (3) removing phosphate in waste water treatment systems in the Netherlands, a considerable reduction of phytoplankton biomass occurs (about 50%). Also a change in species composition occurs: instead of *Microcystis*, *Aphanizomenon* becomes the dominant blue green algal species.

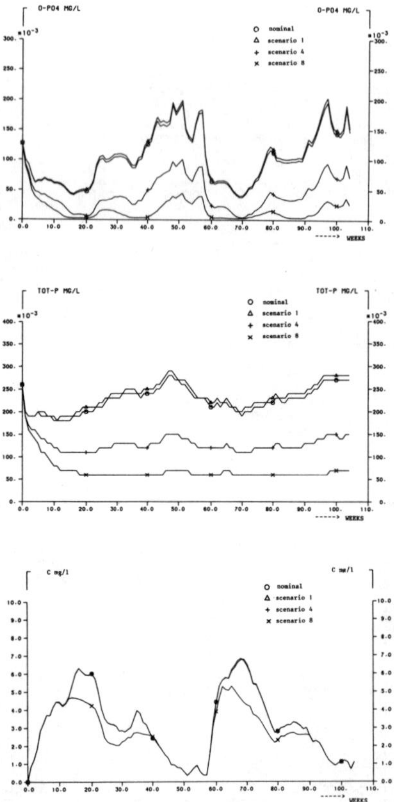

Figure 14.--Orthophosphate, total phosphorus, phytoplankton biomass; simulations by the eutrophication model for various scenarios (compartment 3/4).

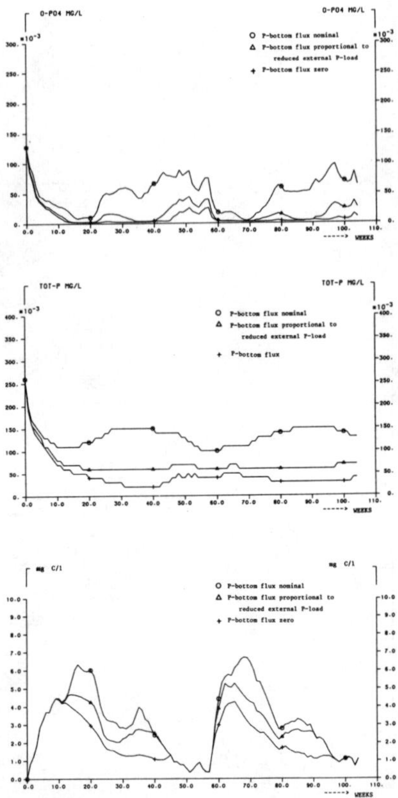

Figure 15.--Orthophosphate, total phosphorus, phytoplankton biomass; simulation by the eutrophication model, sensitivity analysis to the influence of the P-bottom flux (scenario 8 compartment 3/4).

Rather crucial assumptions had to be made about the respons of the bottom sediments to an improved water quality in the lake in the new equilibrium situation.

Since these assumptions are essential, it was the subject of additional sensitivity analysis (fig. 15).

DISCUSSION

Crucial to the results is the reliability of the predictions.

This depends on (1) the available observations, (2) the model conception and (3) the parameter values.

In order to minimize the accumulation of errors as much as possible the results of scenarios for the river Rhine (IJssel) are expressed in relative changes in relation to the calculated results for the present situation.

Based upon these relative changes the measured load from the river Rhine (IJssel) on lake IJssel is adapted for each scenario.

Uncertainties in point loads and loads from diffuse sources, however, increase the uncertainties in the prediction of effects of the various scenarios.

Especially the uncertainty in the contribution of diffuse sources becomes more and more important as the sanitation of point loads increases.

With respect to lake IJssel there is a lack of reliable information on extinction, chlorophyl measurements, relation of dry weight/chlorophyl and zooplankton data.

Sensitivity analysis shows that a considerable part of the overprediction of the phytoplankton biomass in the second quarter of the year can be attributed to zooplankton. More important is that incorporation of zooplankton in the model results in a more drastic switch to blue greens.

The same picture holds for the chemical data. There is a lack of data on alkalinity, calcium in the surface water and all relevant variables in the bottom sediment and interstitial water.

In general process-oriented information is not available, which can be considered as a severe drawback, whereas most of the state variables are measured at a low frequency and at a depth of about 0.25 meter below the surface. Because of vertical gradients this can create a non representative picture.

With respect to the model conception and parameter values main uncertainties, influencing the results of the model significantly are the bouyancy of several blue greens light adaptation of phytoplankton, the estimation of the specific (minimum) mortality rate, the production efficiency curves of several phytoplankton groups as diatoms and flagellates, the simplicity of the description of the interaction between bottom sediments and overlying water, adsorption modelling of phosphorus and the temperature dependency of equilibrium processes.

CONCLUSIONS

From the foregoing the following conclusions can be drawn:

- Against the background of quantity and quality of basic data and the present knowledge of the relevant processes, the described modelling approach results in a consistent and reasonable picture of the nutrient budget of the river Rhine and of the eutrophication process of lake IJssel.
- Uncertain and crucial assumptions in the modelling approach are related to (1) loads from diffuse sources, (2) bottom/water interaction, (3) buoyancy of blue greens, (4) light adaptation of phytoplankton, (5) the impossibility of zooplankton modelling and (6) value of the background extinction.
- The modelling approach is quite able to handle the general policy and management questions. Increasing the details of the question, means increasing the uncertainty in answering them based on the model results. In this respect more attention has to be paid to assess the uncertainties of the model results.

Acknowledgements

The authors thank Rijkswaterstaat, Delta Department, Directorate Zuiderzeewerken and Governmental Institute for Sewage and Waste Water Treatment for their sponsoring and cooperation in both projects.

REFERENCES

Anon 1978. Das Rheingebiet, Hydrologische Monographie. Internationale Kommission für die Hydrologie des Rheingebietes, Den Haag.

Dijkzeul, A. 1980. De waterkwaliteit van de Rijn in Nederland in de periode 1972-1979. Nota 80032.

Hosper, S.H. 1981. Die Eutrophierung der Gewasser im Rheineinzugsgebiet Insbesondere des IJsselmeeres: Probleme und losungsmoglichkeiten: Arbeitstagung der IAWR. Amsterdam.

Klomp, R., and Pagee, J.A. van 1981. A model approach to the problem of water quality of the river Rhine. DHL publication 256.

Los, F.J. 1980. Application of an algal bloom model (BLOOM II) to combat eutrophication. DHL publication 230.

Los, F.J. 1982. Mathematical simulation of algae blooms by the model BLOOM II. Delft Hydraulics Laboratory, R1310-7 (in press).

Los, F.J., Rooy, N.M. de, Smits, J.G.C., and Bigelow, J.H. 1982. Policy analysis of water management for the Netherlands. Vol. 11, Design of eutrophication control strategies, Rand Corp. and Delft Hydraulics Laboratory, N1500/6 - NETH.

Pagee, J.A. van 1981. Waterkwaliteitsmodel Rijn. data rapport, R1056-VI, Waterloopkundig Laboratorium.

Pagee, J.A. van 1981. Waterkwaliteitsmodel Rijn; toepassing van het waterkwaliteitsmodel MODQUAL op de Rijn, R1056-VII/S321-VI, Waterloopkundig Laboratorium.

Postma, L. 1981. Waterkwaliteitsmodel Rijn en Maas; MODQUAL, een éèndimensional steady state waterkwaliteitsmodel voor rivierstelsels. R1056-IV, R1463-I, S321-IV, Waterloopkundig Laboratorium.

Roesner, L.A., Gignere, P.R., Evenson, D.E. 1977. Computer program documentation for the stream quality model. Water Resources Engineer INC.
Rooy, N.M. de 1980. A chemical model to describe nutrient dynamics in lakes. DHL publication 245.
Rooy, N.M. de, Los, F.J., Smits, J.G.C., Stans, J.C., and Klomp, R. 1982. Waterkwaliteitsmodel IJsselmeer. Onderzoek naar de invloed van mogelijke beheersmaatregel op de eutrofiering van het IJsselmeer. R1551, Waterloopkundig Laboratorium (in press).

MODELLING THE COMMUNITY OXYGEN PRODUCTION IN LOWLAND STREAMS DOMINATED BY SUBMERGED MACROPHYTES[1]

M. Erlandsen[2] and N. Thyssen[2]

Abstract.--Eight empirical equations for the relationship between photosynthetic rate and surface light were examined in Danish streams. A hyperbolic relationship described data with good precision, whereas a linear relationship was not sufficient. Verification of the model included time series analysis of the one step ahead prediction errors between predicted and measured oxygen concentrations.

INTRODUCTION

At a study site in a stream in Denmark time series of dissolved oxygen concentration, irradiance and water temperature have been collected for the years 1979, 1980 and 1981.

In order to examine the relationship between community photosynthetic rate (P) and surface light (I) five time periods (6 - 12 days) with negligible flow variations were analysed on 8 empirical P/I functions of the form $P(t) = f(I(t))$ (t~time). The mathematical method is based on a discretization of the oxygen mass **balance equation,** which makes it possible to calculate one step ahead predictions of the oxygen concentration. The 8 P/I functions have been compared by minimizing the one step ahead prediction errors using the method of non-linear least squares and the Marquardt algorithm (1963).

Ecological modelling with non-linear parameter estimation problems and the use of the Marquardt algorithm are well-known, see e.g. Conway et al. (1970) or Benson (1979). Alternative methods could be based on the Kalman filter technique, see e.g. Beck & Young (1975) or Misra and Lee (1978).

In order to verify the model the one step ahead prediction errors have been examined by means of standard time series analysis (Box & Jenkins (1976)).

[1]Paper presented at the third international conference: State of the Art in Ecological Modelling. (Fort Collins, Colorado, May 24-28 1982).

[2]The Freshwater Laboratory, National Agency of Environmental Protection, 52 Lysbrogade, DK-8600 Silkeborg, Denmark.

STUDY SITE AND DATA ACQUISITION

Nearly continuous measurements of dissolved oxygen, water temperature, and incident light have been obtained for three years at two stations 720 meters apart in a stream situated in the north western part of Jutland, Denmark. The biological and hydraulic characteristics of the stream have been described by Thyssen (1982) and Thyssen et al. (this volume), respectively.

Details of the instrumentation and recording procedures are given in Thomsen and Thyssen (1979) and only a brief summary will be presented here.

The dissolved oxygen concentration was measured using a galvanic type oxygen transducer (phOX, model 600) without built-in temperature compensation. Corrections for variations in membrane permeability were made by computer according to Mancy et al. (1962). Precision of the oxygen measurements, based on the digital resolution of the recorder (0.24 mV), at the prevailing temperatures (10-16°C) was estimated to be better than ± 0.01 mg l^{-1}. Atmospheric pressure measurements at the study site were obtained by a pressure transducer (Penny and Giles, PP4) and were used to correct the saturation values of dissolved oxygen given by Montgomery et al. (1964). Water temperature was measured using an integrated temperature transducer (Analog Devices, AD590L) and surface light as photosynthetically active radiation (400-700 nm) was measured using a quantum sensor (Lambda Instruments, LI-190S). Units of irradiance (Langley min^{-1}) were calculated using the relationship 1 μ E cm^{-2} min^{-1} equals 5.33 · 10^{-2} Ly min^{-1} as derived from Withrow and Withrow (1956).

Transducer signals were amplified using locally designed amplifiers, digitizes and recorded on magnetic tape every fifth minute using a tape recorder (Memodyne, 821-34). After conversion of the recorded voltages to the appropriate units, the data were reduced to half hourly values by averaging six data points around each half hour.

MATHEMATICAL METHODS

Assuming that the river is sufficiently well-behaved with respect to short-term variation in flow, reaeration, respiration, light penetration and total mixing of the water, etc., we apply the well-known oxygen mass balance equation for a single station (Odum 1956)

$$\frac{dC}{dt} = K\,(C_s - C) + (P - R) \qquad (1)$$

where
$C = C\,(t)$ dissolved oxygen concentrations, mg l^{-1},
$C_s = C_s(t)$ saturation oxygen concentration, mg l^{-1},
$P = P\,(t)$ rate of community photosynthetic oxygen release, mg l^{-1} h^{-1},
$K = K\,(t)$ reaeration coefficient, h^{-1},
$R = R_1(t)$ rate of community oxygen uptake, mg l^{-1} h^{-1}.

Further let T(t) denote the water temperature at time t. It is commonly accepted that the temperature dependence of K and R can be described as

$$K = K\,(t) = K_{20}\,\theta_K^{T(t)-20}$$
$$R = R\,(t) = R_{20}\,\theta_R^{T(t)-20} \qquad (2)$$

where K_{20}, R_{20} are the reaeration coefficient and the respiration rate respectively, adjusted to 20°C. In the technical literature (e.g. Churchill et al. (1962), Benneth and Rathbun (1972)) it is well accepted that $\theta_K = 1.0241$, whereas the temperature dependency of the respiration of living organisms or communities is much more debated. The value of θ_R used in this study will be discussed below. In order to investigate the functional relationship between P(t) and I(t) (the irradiance at time t) in a macrophyte dominated river we basically selected model functions reported by Jassby and Platt (1976) and Ledermann and Tett (1981) to describe the P/I relationship for marine phytoplankton. The general form of the function is P(t) = f (I(t)) and specifically we chose the following functions

1. $f(x) = \alpha x$
2. $f(x) = \beta \dfrac{\frac{\alpha}{\beta}x}{1 + \frac{\alpha}{\beta}x}$
3. $f(x) = \beta\, x^{\frac{\alpha}{\beta}}$
4. $f(x) = \beta\,(1 - e^{-\frac{\alpha}{\beta}x})$
5. $f(x) = \beta \tanh\,(\frac{\alpha}{\beta}x)$
6. $f(x) = \beta\frac{\alpha}{\beta}x\,(1 + (\frac{\alpha}{\beta}x)^2)^{-\frac{1}{2}}$
7. $f(x) = \beta\frac{\alpha}{\beta}x(1 - \frac{\alpha}{\beta}\frac{x}{4})$
8. $f(x) = \beta\ln\,(1 + \frac{\alpha}{\beta}x)$

Except for the first function they depend on 2 unknown parameters in such a way that f'(o) = α (the initial slope), and if f(x) is bounded then β is the maximum photosynthetic rate.

We choose our time unit such that our observations are indexed $(C(t), C_s(t), I(t), T(t))$ $t = 1,\ldots\ldots,N$.

Put $\phi(t,h) = \exp\,(\int_t^{t+h} K(u)\,du)$,

$$\theta(t) = e^{-\frac{1}{2}(K(t) + K(t+1))},$$
$$S(u) = K(u)\,C_s(u) + f(I(u)) - R(u)$$

then by standard techniques (1) implies

$$C(t+1) = e^{-\phi(t,1)}\,\{\,C(t) + \int_t^{t+1} e^{\phi(t,u-t)}\,S(u)\,du\,\}. \qquad (3)$$

Approximating the integrals using the trapezoidal rule we obtain

$$C(t+1) \simeq e^{-\phi(t,1)}C(t) + \frac{1}{2}\,\{\,S(t+1) + e^{-\phi(t,1)}\,S(t)\,\}$$
$$\simeq \theta(t)\,C(t) + \frac{1}{2}\,\{\,S(t+1) + \theta\,(t)\,S(t)\}. \qquad (4)$$

By specifying values of θ_K and θ_R in (2) we can use (4) to estimate the parameters K_{20}, R_{20}, α, β. Put

$$\hat{C}(t+1|t) = \theta(t)\,C(t) + \frac{1}{2}\,\{S(t+1) + \theta(t)\,S(t)\}$$
$$Z(t+1) = C(t+1) - \hat{C}(t+1|t).$$

The estimates are obtained by minimizing the sums of squares of the prediction errors {Z(t)} for each of the 8 functions

$$SSq(f) = \sum_{t=2}^{N} Z(t)^2 \qquad (5)$$

(f = selected model function)

using standard nonlinear least squares such as the Marquardt algorithm. The sum of squares obtained from the different model functions are presented in the following section.

The analysis so far does not take account of possible measurement or system noise. Models with noise (stochastic variation) lead to stochastic differential equations. Such equations can be handled by means of the Kalman filter (Beck and Young, 1975). However, for the present purpose we will present a technique based on time series analysis. Let us make som simplifying assumptions.

Now suppose that the measurement noise on C(t) can be described by

$$\overline{C}(t) = C(t) + a(t)$$

where $\overline{C}(t)$ is the actual measurement, C(t) is the actual oxygen concentration and {a(t)} is a Gaussian white noise process with E(a(t)) = 0 and $Var(a(t)) = \sigma_a^2$. Suppose further that the system noise in equation (4) can be descreibed by another Gaussian white noise process {b(t)} with E(b(t)) = 0 and $Var(b(t)) = \sigma_b^2$. Actual values of θ(t) show that θ(t) is nearly constant, hence assume θ(t) ≃ θ. Then (4) implies that

$$Z(t) = a(t) - \theta a(t-1) + b(t) \quad (6)$$

From the theory of moving average processes it is known that $\{Z(t)\}$ can be represented by a 1st order moving average process, i.e.

$$Z(t) = u(t) - \lambda u(t-1) \quad (7)$$

where $\{u(t)\}$ is a Gaussian white noise process with mean zero. Thus the prediction errors $\{Z(t)\}$ can be analysed by standard time series technique in order to obtain some verification of the whole model.

RESULTS

The technique outlined has been applied to a total of 47 diurnal time series. These time series were selected to represent short periods (6-12 days) in early and late summer of 1979, 198o and 1981 when the river discharge did not change significantly. Each time series include 49 observations (one observation per 0.5 h). Thus the total number of observations exeeds 2000. For each day and each P/I function (1-8) the minimized sum of squares (equation (5)) was calculated, and for each of these SSq's a standard deviation (SD) is obtained as $SD^2 = SSq/DF$, where DF = 49 - (number of estimated parameters), i.e. DF = 46 for the linear P/I function and DF = 45 for the other P/I functions.

These "single-day" standard deviations are not presented here, but in table 1 the standard deviations for each of the 8 P/I functions have been pooled over days within the same season.

The magnitude of the standard deviations shows that the 2-parameter P/I functions (f = 2,.. ...., 8) are superior to the linear 1 - parameter P/I function (f = 1). This fact is well-known from the literature. The linear function is included here as a reference.

Furthermore the logarithmic expression (f = 8) gives the best over-all fit, closely followed by the hyperbola (f = 2). Also the power function (f = 3) and the exponential function (f = 4) are hardly distinguishable from the above mentioned functions (f = 2,8). The parabola (f = 7) is definitely the poorest of the selected 2-parameter P/I functions. Unfortunately, the parabola is - from the mathematical point of view - the simplest of the 2-parameter P/I functions, since it can be parametrized to be linear in (I, I^2).

A comparison of the standard deviations between the different time periods shows no strong evidence of systematic variations, although the fit in the autumn period seems to be better than in the spring. But the number of years available in the analysis is too small for more detailed studies. It should be noted that there is a considerable variation between the years.

The technique in section 3 has also been used to estimate the parameter θ_R. This analysis was carried out on time series of night time observations (in order to neglect the P-term). The results are summarized in table 2.

Table 2.--A period consists of 4-7 succeeding nights in the month. Each night consists of 12-26 observations (dependent of the time of the year). Thus a single θ_R value is based on 50-150 observations.

	period	θ_R
1979	June	1.0949
	July	1.0679
	August	1.0792
	Sept.	1.0662
	Oct.	1.0655
	Nov.	1.0466
	-----(the variation in the temperature was to small to allow estimation of θ_R in -----the winter period)	
1980	March	1.0673
	April	1.0521
	May	1.0898
primo	July	1.0866
ultimo	July	1.0681
average value =		1.0713

The analysis confirms that $\theta_R = 1.07$ is a reasonable average value.

The one step ahead prediction errors $\{Z(t)\}$ for P/I function no. 2 (the hyperbola) have been analysed for 2 diurnal time series sampled at 5 minute intervals (i.e. 289 observations in each series). The shorter sampling interval was preferred for this analysis in order to minimize the discretization error in eq (4). The autocorrelation function $\{r_k(Z)\}$ and the partial autocorrelation function $\{\phi_{kk}(Z)\}$ (see Box and Jenkins, 1976) together with the standard deviations σ_Z are shown in figure 1.

The Box-Pierce portmanteau test for lack of fit has also been calculated:

$$Q = N \sum_{k=1}^{K} r_k^2(Z) = \begin{cases} 215.5 \text{ (A)} \\ 148.7 \text{ (B)} \end{cases}$$

where N = 289, K = 48. Both values are extremely significant in a $\chi^2(48-4)$ - distribution. The same conclusion of course appears from figure 1 above. The combined behaviour of $\{r_k(Z)\}$ and $\{\phi_{kk}(Z)\}$ indicates that $\{Z(t)\}$ may be fitted to a 1st order moving average precess as the analysis in section 3 suggested. For the model

$$Z(t) = u(t) - \lambda u(t-1)$$

all parameters have been simultaneously reestimated (now minimizing $\sum_t u^2(t)$). In figure 2 are shown $\{r_k(u)\}$, $\{\phi_{kk}(u)\}$ and σ_u.

It may be noted that: 1) the decrease in the standard deviation (from σ_Z to σ_u) is significant, 2) the correlograms look more like the picture of a white noise process.

The Q-values computed as before yield Q = 119.9 (A) and Q = 78.5 (B), and now DF = 48-5. But

Table 1.--The standard deviations (mg l^{-1}) computed as pooled estimates for different time periods for each of the 8 P/I functions. n is number of days processed in the time period, i.e. the standard deviation is based on n x 49 observations.

Year	P/I function	Season May - June	Season July - September	Σ
1979	1	no observations	0.0391	
	2		0.0218	
	3		0.0223	
	4		0.0225	
	5		0.0238 n=12	
	6		0.0231	
	7		0.0256	
	8		0.0216	
1980	1	0.0888	0.0726	0.0797
	2	0.0378	0.0476	0.0438
	3	0.0365	0.0487	0.0441
	4	0.0405	0.0481	0.0451
	5	0.0453 n=7	0.0495 n=10	0.0478 n=17
	6	0.0421	0.0488	0.0462
	7	0.0508	0.0505	0.0506
	8	0.0365	0.0474	0.0432
1981	1	0.0573	0.0437	0.0486
	2	0.0366	0.0219	0.0277
	3	0.0394	0.0230	0.0295
	4	0.0363	0.0223	0.0278
	5	0.0366 n=6	0.0235 n=12	0.0285 n=18
	6	0.0365	0.0229	0.0282
	7	0.0383	0.0255	0.0304
	8	0.0374	0.0219	0.0281
Σ	1	0.0759	0.0525	0.0599
	2	0.0373	0.0317	0.0333
	3	0.0378	0.0326	0.0341
	4	0.0386	0.0322	0.0341
	5	0.0409 n=13	0.0334 n=34	0.0358 n=47
	6	0.0396	0.0328	0.0348
	7	0.0454	0.0348	0.0380
	8	0.0369	0.0315	0.0331

both values are still highly significant. A closer look at figure 2 shows a long term systematic variation in $\{r_k\}$. This probably causes the high Q-values.

DISCUSSION

The present analysis has confirmed that a 2-parameter P/I function that behaves like a hyperbola is superior to a linear 1-parameter P/I function. Among the 2-parameter P/I functions at least 4 are nearly equally good. This fact is not surprising, since the graphs of the functions are close to each other.

It does not seem likely that other "zeroorder" (i.e. $P(t) = f(I(t))$ 2-parameter P/I functions would be superior to those tested in this paper. "Better" P/I functions probably have to include other characteristics, e.g. 1) A hysteresis effect such that $\frac{dP}{dI}$ is greater for I(t) increasing than for I(t) decreasing. 2) The P/I function should be integrated over the depth thereby taking account of light attenuation. 3) The P/I function should be a sum of P/I functions, each corresponding to a single species. 4) Temperature dependence of the parameters in the P/I function. Also it seems worthwhile to analyse a model which incorporates the measurement noise of the irradiance I(t).

The analysis of the one step ahead prediction errors {Z(t)} indicates that the model is not quite satisfactory. This statement is based on the analysis of only 2 diurnal series and it may therefore not be representative. Unfortunately, we do not have 5 minute data from other periods. But we surmise that the model can be improved by the use of a P/I function that allows a hysteresis effect.

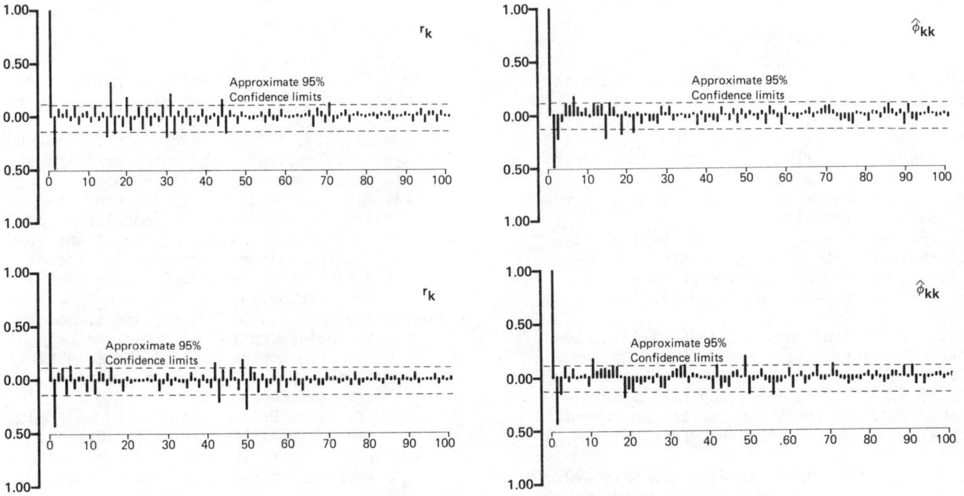

Figure 1.--Autocorrelation and partial autocorrelation function for the one step ahead prediction errors for 2 diurnal time series: (A) ~ august 15. 1979, σ_z = 0.230 mg l^{-1} and (B) ~ july 3. 1980, σ_z = 0.0422 mg l^{-1}.

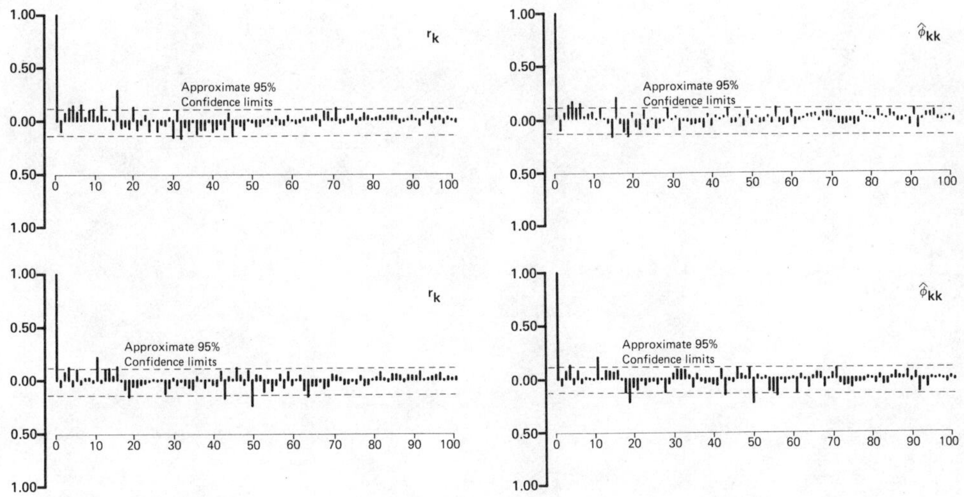

Figure 2.--Autocorrelation and partial autocorrelation function for the noise process {u(t)} for 2 diurnal time series: (C) ~ august 15. 1979, σ_u = 0.0196 mg l^{-1} and (B) ~ july 3. 1980, σ_u = 0.0378 mg l^{-1}.

LITERATURE CITED

Beck, M.B., and P.C. Young. 1975. A dynamic model for DO-BOD relationships in a nontidal stream. Wat. Res. 9: 769-776.

Benneth, J.P., and R.E. Rathbun. 1972. Reaeration in open-channel flow. U.S. Geol. Surv. Prof. Paper 737, Washington, D.C.

Benson, M. 1979. Parameter fitting in dynamic models. Ecol. Modelling 6: 97-115.

Box, G.E.P., and D.A. Pierce. 1970. Distribution of residual autocorrelations in autoregressive integrated moving average time series models. Journ. Amer. Stat. Assoc. 64: 1509-1526.

Box, G.E.P., and G.M. Jenkins. 1976. Time Series Analysis: Forecasting and Control, Holden-Day, San Francisco.

Churchill, M.A., H.L. Elmore, and R.A. Buckingham. 1962. The prediction of stream reaeration rates. Int. J. Air Wat. Poll. 6: 467-504.

Conway, G.R., N.R. Glass, and J.C. Wilcox. 1970. Fitting nonlinear models to biolocical data by Marquardts algorithm. Ecology 51: 503-507.

Jassby, A.D., and T. Platt. 1976. Mathematical formulation of the relationship between photosynthesis and light for phytoplankton. Limnol. Oceanogr. 21: 540-547.

Lederman, T.C., and P. Tett. 1981. Problems in modelling the photosynthesis-light relationship for phytoplankton. Bot. Mar. 24: 125-134.

Mancy, K.H., D.A. Okun, and C.N. Reilley. 1962. A galvanic cell oxygen analyzer, J. Electroanal. Chem. 4: 65-92.

Marquards, D.W. 1963. An algorithm for least squares estimation of nonlinear parameters. SIAM J. Numer. Anal. 11: 431-441.

Misra, P.K., and E.S. Lee. 1978. Nonlinear estimation in stream quality modelling. pp 509-580. In Chao-Lin Chiu (Ed.) Applications of Kalman filter to hydrology, hydraulics and water resources. University of Pittsburgh, Pennsylvania.

Montgomery, H.A.C., N.S. Thom, and A. Cockburn. 1964. Determination of dissolved oxygen by the Winkler method and the solubility of oxygen in pure water and seawater. J. Appl. Chem. 14: 280-296.

Odum, H.T. 1956. Primary production in flowing waters. Limnol. Oceanogr. 1: 102-117.

Thomsen, E., and N. Thyssen. 1979. Monitoring dissolved oxygen and temperature in streams. (In danish). Vand 10: 46-53.

Thyssen, N. 1982. Aspects of the oxygen balance of a macrophyte dominated lowland stream. pp 202-213. In J.J. Symoens, S.S. Hooper, and P. Compere (Eds.). Studies on Aquatic Vascular Plants. Royal Botanical Society of Belgium, Brussels.

Withrow, R.B., and A.P. Withrow. 1956. Generation, control and measurement of visible and near-visible radiant energy. pp. 125-258. In Hol-Laender (Ed.), Radiation Biology. McGraw-Hill, N.Y.

MODELLING THE REAERATION CAPACITY OF LOW-LAND STREAMS DOMINATED BY SUBMERGED MACROPHYTES[1]

Thyssen, N.[2], M. Erlandsen[2], E. Jeppesen[3], and T. F. Holm[2]

Abstract.--Eight commonly used equations for prediction of reaeration were applied to 4 small streams with discharges below 5oo l s^{-1}. Six models systematically produced erroneous predictions. Two models only gave reasonable K_2-values. Using various oxygen balance methods in 7 reaches, we obtained 44 measurements of the reaeration coefficient K_2 (base e, day^{-1}) and the primary hydraulic variables: Mean velocity U (m s^{-1}), mean depth H (m) and slope S (m m^{-1}) under different hydraulic conditions. Using multiple regression, several multiplicative models were obtained, of which

$$K_2(2o\ ^oC) = 374o6\ U^{o.846}\ H^{-o.672}\ S^{1.154}$$

best fitted the observed K_2-values. If only information on the slope is available, K_2 is obtained from

$$K_2(2o\ ^oC) = 8377\ S^{1.o63}$$

These equations are recommended for prediction of the reaeration capacity of macrophyte dominated streams where the hydraulic variation is $o.o6 < U \leq o.21$ m s^{-1}, $o.13 < H < o.6o$ m, $o.39\ 1o^{-3} < S < 2.o3\ 1o^{-3}$ m m^{-1}, respectively.

INTRODUCTION

Reaeration is a process which counteracts oxygen depletion in streams. Evaluation of stream capacity to absorb oxygen from the atmosphere is required for determination of the permissible load of oxygen depleting biodegradable organic wastes. Knowledge on reaeration is also required for calculation of rates of primary productivity and community respiration from free water measurements of dissolved oxygen.

The rate of exhange of oxygen between the atmosphere and streams is primarily considered to depend on the macroturbulence, and much effort has been devoted to relate river reaeration to various hydraulic factors which affect the turbulent motion of the water molecules. Except for a few studies (e. g. Owens et al. 1964, Tsivoglou and Neal 1976 and Thyssen and Jeppesen 198o) most investigations on reaeration have been carried out in large rivers (e.g. Churchill et al. 1962, O'Connor and Dobbins 1956), in sewers (Parkhurst and Pomeroy 1972) or in flumes (Thackston and Krenkel 1969). The reaeration capacity of small streams with mild slopes and high density of rooted submerged vegetation, common to low-land areas in Northwestern Europe, is particularly poorly known. Therefore, as part of a integrated study of the oxygen dynamics of this stream type, we have provided measurements of the reaeration coefficient and related those measurements to hydraulic characteristics of the streams.

SITE DESCRIPTION

K_2-measurements were performed in 4 different streams considered to reflect the hydraulic variation found in Danish macrophyte dominated streams (Table 1).

The Suså and Sneslev å are regulated streams

[1]Paper presented at the third international conference: State of the Art in Ecological Modelling. (Fort Collins, Colorado, May 24-28, 1982).
[2]The Freshwater Laboratory, National Agency of Environmental Protection, Lysbrogade 52, DK-86oo Silkeborg, Denmark.
[3]The Freshwater Biological Laboratory, University of Copenhagen, Helsingørsgade 51, DK-34oo Hillerød, Denmark.

Table 1.--Hydraulic and biotic variation in the study streams. Q is discharge, U mean velocity, H mean depth and S is slope. (Modified after Thyssen and Jeppesen 198o).

Stream	Q $m^3\ s^{-1}$	U $m\ s^{-1}$	H m	S $m\ m^{-1}\ 1o^{-3}$	Percentage of bottom area covered with macrophytes
Suså	o.o4-o.14	o.o6-o.17	o.13-o.23	o.39-1.3	< 5o
Sneslev å	o.o2-o.11	o.o6-o.21	o.13-o.43	o.6 -1.2	< 6o
Gryde å	o.19-o.52	o.o9-o.21	o.31-o.69	1.62-2.26	< 75
Holme å	o.2o-o.3o	o.13-o.21	o.2o-o.5o	o.7 -1.25	< 9o

located in the eastern part of the country on the island of Zealand while the other two are natural and meandering streams located on the peninsula of Jutland.

METHODS

Measurements of the reaeration coefficient were obtained using three indirect oxygen mass balance techniques and eight hydraulic prediction equations relating various turbulence indicators to reaeration.

We applied the disturbed equilibrium technique of Gameson et al. (1955), and developed two alternative methods, which we call the "impermeable surface" method and the "complex" method, respectively.

The disturbed equilibrium technique

At night the oxygen balance of a stream is affected by reaeration and respiration, and the oxygen mass balance for the study reach may be written as

$$\frac{dC}{dt} = K_2(C_s - C) - R \qquad \text{(Odum 1956)} \quad (1)$$

where,

C = dissolved oxygen concentration, mg l^{-1}

C_s= saturation concentration of dissolved oxygen, mg l^{-1}

t = time, h

K_2= reaeration coefficient (base e) at 2o °C, h^{-1}. $K_2(T\ ^oC) = K_2(2o)\ 1.o241^{T-2o}$.

R = community oxygen uptake, mg $l^{-1}\ h^{-1}$.

Substitution of measured values of C_s and C at different periods of the night into equation (1) makes it possible to obtain K_2 and R estimates from linear regression of the rate of change of oxygen versus the deficit (C_s - C). Natural streams, with moderate growths of macrophytes (biomass < 25o g dw m^{-2}) and without organic pollutional loadings, usually exhibit only minor oxygen excursions (< 2 mg $O_2\ l^{-1}$) during the dark period, leaving the K_2 estimates obtained using this method rather inaccurate. In order to improve accuracy we applied the disturbed equilibrium technique which implies an artificial saturation deficit increase at night by the continuous and controlled addition of sodium sulphite and trace amounts of a cobalteous catalyst. Constant injection of sulphite into a stream for a period of 2 - 3 hours leads to the establishment of a lower oxygen plateau, and assuming that R is not affected by the passage of deoxygenated water, K_2 can be calculated from:

$$K_2 = \frac{1}{\delta} \ln \frac{C_{n,u} - C_{p,u}}{C_{n,d} - C_{p,d}} \qquad (2)$$

where

δ = retention time, and C_n and C_p denote the oxygen concentrations in the normal and disturbed situation, respectively. The subscripts u and d refer to upstream and downstream stations which are separated by the retention time.

Prior to application of this method one must make sure that groundwater accrual is negligible, that no concentration gradients exist in the stream cross section, and that no unreacted sulphite is present when the deoxygenated water passes the upstream station.

The complex method

A more complicated oxygen balance model than equation (1) was used for the estimation of the reaeration coefficient in the organic polluted Holme river. The model was a twodimensional (time, place) finite difference model including sediment and macrophyte oxygen uptake, nitrification and denitrification dependent on the oxygen concentration and temperature, total community production, and reaeration (Jeppesen 1982). The reaeration coefficient and the oxygen uptake of macrophytes R(2o °C, 8 mg $O_2\ l^{-1}$) were calibrated on data from the dark period of a diurnal period by systematic pertubations using an optimization program CSMPOPT (Mejer et al. 1982). For a more detailed description of the model, see Jeppesen (1982).

Impermeable surface method

This study was made in the rivers Suså and Sneslev å from June 1978 to October 1978. In June the vegetation was cut and removed. In order to prevent photosynthesis the sections were kept dark by covering them with a number of wooden frames (3.6 x 5.4

m) with tarpaulins. The sections were adapted to the new conditions for three weeks before start of the measurements. The in situ measurements were made in periods of a few days. In each experimental section the tarpaulin frames were removed and a dark o.1 mm thick plastic foil of polyethylene and polyamid was rolled out on the water surface from the downstream end towards the upstream end. The foil was fastened at the banks with canes, and all air pockets were removed by pouring water over the foil. The oxygen permeability of the plastic foil is small (Schramm and Martens 1976), and the foil can therefore be considered to eliminate reaeration as well as photosynthesis.

Dissolved oxygen and temperature were then measured at the upstream and downstream end of the darkened reach during 1 - 3 diurnal periods and again one night after removal of the plastic foil. In laboratory experiments the oxygen consumption in the water was measured by the closed bottle technique at the stream temperature (Jeppesen and Iversen, in prep.).

Using the oxygen consumption data from the period when the surface was covered by the plastic foil, and the relationship between the oxygen consumption of the water and the temperature, it was possible to obtain a relation between the oxygen consumption of the sediment and oxygen concentration and temperature (Jeppesen and Iversen, in prep.). By these equations the oxygen consumption was calculated for the period when the foil was removed, whereafter the reaeration could be found by difference.

Hydraulic prediction equations

Numerous equations have been developed for the prediction of reaeration coefficients (for review see Benneth and Rathbun 1972, Wilson and Macleod 1974, and Thyssen 198o).

Theoretical models of the reaeration process are generally not suited for prediction because the model parameters have not been adequately related to bulk-flow hydraulic variables. Consequently in this study we intend only to verify 8 commonly used semiempirical and empirical types of models for the prediction of K_2 in macrophyte dominated low-land streams. The selected models are summarized in table 2, where they are grouped according to whether the data set used in their development is obtained from streams (group A), from sewers and flumes (group B) or from both (group C).

Basically all eight models include directly measurable hydraulic variables: mean velocity U (m s^{-1}), mean depth H (m), and the slope S (m m^{-1}). These primary variables are also needed to calculated the secondary variables discussed below.

Models 1 and 2 in table 2 predict K_2 from U and H whereas model 3 combines measured values of U and S into a prediction. Except for the O'Connor and Dobbins model, which in addition to U and H includes the coefficient of molecular diffusion, the remaining four models include various secondary hydraulic variables (e.g. rate of energy dissipation E, friction velocity U^*, and the Froude number N_f). The appropriate equations for calculating E, U^*, and N_f are given in table 3.

Table 3.--Secondary hydraulic variables used in reaeration modelling.

Symbol	Equation	Unit
E	SUg^1	$m^2\ s^{-3}$
U^*	$(SHg)^{o.5}$	$m\ s^{-1}$
N_f	$U(Hg)^{-o.5}$	dim. less

g^1 is the gravitational acceleration (m s^{-2}).

RESULTS AND DISCUSSION

Reaeration coefficient measurements

Using the above mentioned techniques we performed 44 experiments in four streams whose hydraulic variation during the period of measurement is presented in table 1.

19 experiments were conducted in the Gryde å using the disturbed equilibrium technique. On two occasions the measured reaeration coefficients were checked against independent and direct measurements obtained using radioactive tracers, ^{85}Kr for reaeration and ^3H for dispersion (Ursin et al. 1981). Over a two year period K_2 varied between 5.o to 15.6 day^{-1}, with the minimum value occuring in late summer to early fall, when the macrophytic biomass was highest. The maximum value was found immediately after all the submerged macrophytes had been removed to improve land drainage, causing water velocity to increase and depth to decrease.

The impermeable surface method was used to obtain 14 K_2-estimates in the Suså and Sneslev å. The reaeration coefficient showed a pronounced seasonal variation, ranging from 1.o in the summer period to 15.o in the spring when no macrophytes were present.

Finally we employed the complex oxygen mass balance method to obtain 11 K_2-estimates from the Holme å. K_2 varied seasonally from 2.o with maximum macrophyte biomass to 9.8 in periods with no macrophytes.

Predictive behavior of selected reaeration equations

In an attempt to verify the predictive equations given in table 2 we computed a K_2-value for each of the 44 data set of measured U, H, and S values and compared the predicted K_2's to the observed values.

A graphic comparison of the observed versus computed reaeration coefficients is given in figures 1 to 8. The equations by Churchill et al. 1962,

Table 2.--Reaeration models and their predictive performance

Reference	Equation	F_{slope}	$F_{slope, intercept}$
A.			
Churchill et al. (1962)	$K_2 = 5.02\ U^{0.969}\ H^{-1.673}$	38.26*	
Owens et al. (1964)	$K_2 = 5.33\ U^{0.67}\ H^{-1.85}$	53.17*	
Tsivoglou and Neal (1976)	$K_2 = 31200\ S\ U$	3.17	9.44*
B.			
Parkhurst and Pomeroy (1972)	$K_2 = 23.0\ (1+0.17\ N_F^2)\ (S\ U)^{0.375}\ H^{-1}$	0.85	104.50*
C.			
O'Connor and Dobbins (1956)	$K_2 = 294\ D_m^{0.5}\ U^{0.5}\ H^{-1.5}$	33.08*	
Dobbins (1964, 65)	$K_2(T) = 0.084\ \coth\ (B\ E^{0.125}\ C_4^{-0.5})$		
	$C_A\ A\ E^{0.375}\ C_4^{-1.5}\ H^{-1}$	0.73	3.53
	where		
	$C_A = 1.0 + N_F^2$		
	$C_4 = 0.9 + N_F$		
	$A = 9.68 + 0.054\ (T - 20)$		
	$E = 98425\ S\ U$		
	$B = 0.976 + 0.0137\ (30-T)^{1.5}$		
Thackston and Krenkel (1969)	$K_2 = 24.9\ (1+N_F^{0.5})\ U^*\ H^{-1}$	17.72*	
Cadwallader and McDonnell (1969)	$K_2 = 59.2\ E^{0.5}\ H^{-1}$	0.01	3.10

Owens et al. (1964) and O'Connor and Dobbins (1956) shown in figures 1, 2 and 5, respectively, all show similarly poor predictive behavoir. All three equations underestimate K_2 by a factor of approximately 2 in the stream with the highest slope, and they overestimate K_2 by a similar amount in the less sloped streams indicating that some causative factor(s) has been omitted from them. From figure 4 it appears that the Parkhurst and Pomeroy equation systematically underestimates K_2. In their detailed study of 16 equations for the prediction of K_2 Wilson and Macleod (1972) found that this equation gave the best general fit to a relatively large number (502) of experimental K_2 measurements. However, only 140 values were obtained from rivers and only 54 from rivers comparable to the type presently studied.

Of the remaining four equations the Thackston and Krenkel equation (figure 7) seems to be noticeably worse than the other three. It systematically predicts too high reaeration coefficients when K_2 is low and vice versa. No conspicuous difference was found when comparing the predictive performance of the Tsivoglou and Neal (1976) (figure 3), the Dobbins (1964, 65) (figure 6), and the Cadwallader and McDonnel (1969) (figure 8) equations.

In addition to the graphical comparison we included a statistical measure of the predictive performance. For each equation we calculated the slope for the best line of fit of ln-transformations of observed versus predicted K_2's and applied an F-test (α=0.05, df_1=1, df_2=42) to compare its slope to that of a regression line through the data points with a fixed slope of 1.0.

The results of this test are presented in table 2, and it appears that the actual slope for four equations (figures 1, 2, 5, and 7) deviates significantly from the theoretical slope of 1.0. The slope criterion thus falsified four equations. In order to differentiate further between the remaining equations we again applied an F-test (α=0.05, df_1= 1, df_2=43) to compare the line of best fit to that of a perfect model with a slope of 1.0 and zero ordinal intercept. The calculated F-values are given in table 2 and it appears that only the equations by Dobbins (1964, 65) and Cadwallader and McDonnell (1969) which are shown in figures 6 and 8, respectively, produce unbiased estimates of the reaeration coefficient for small macrophyte dominated low-land streams. For both of them, however, the prediction errors are relatively large.

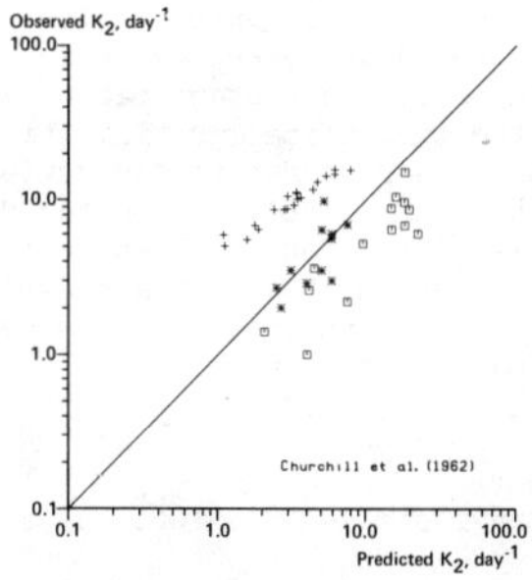

Figure 1.--Measured K_2-values versus those predicted by the equation given by Churchill et al. (1962). In figures 1 to 1o crosses indicate values from the Gryde å, stars the Holme å-values, and squares the Suså- and Sneslev å-values.

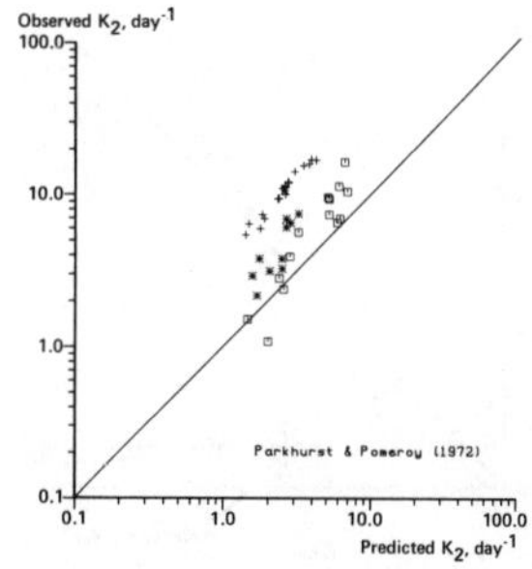

Figure 4.--Measured K_2-values versus those predicted by the equation given by Parkhurst and Pomeroy (1972).

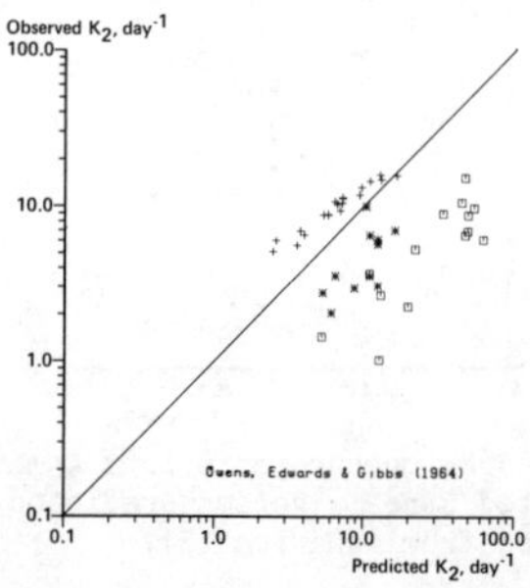

Figure 2.--Measured K_2-values versus those predicted by the equation given by Owens, Edwards and Gibbs (1964).

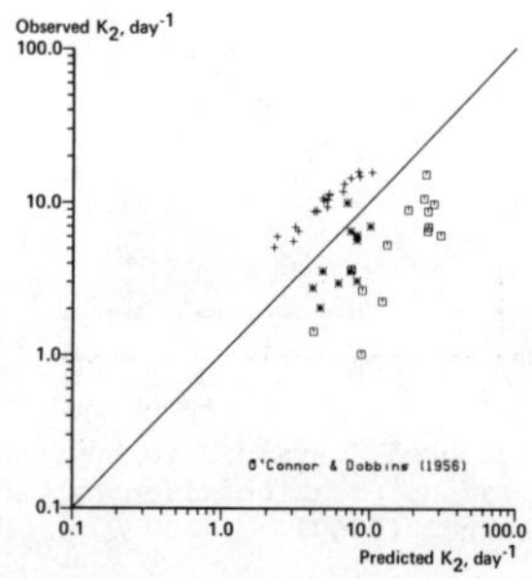

Figure 5.--Measured K_2-values versus those predicted by the equation given by O'Connor and Dobbins (1956).

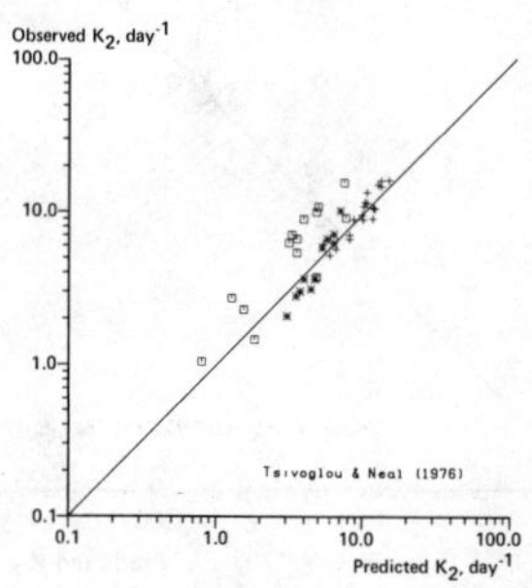

Figure 3.--Measured K_2-values versus those predicted by the equation given by Tsivoglou and Neal (1976).

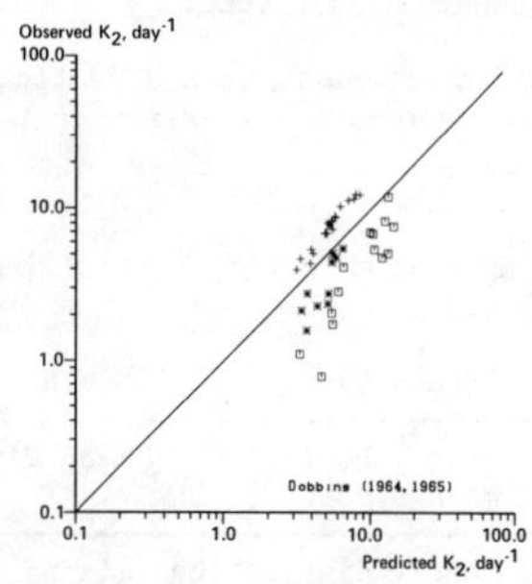

Figure 6.--Measured K_2-values versus those predicted by the equation given by Dobbins (1964, 1965).

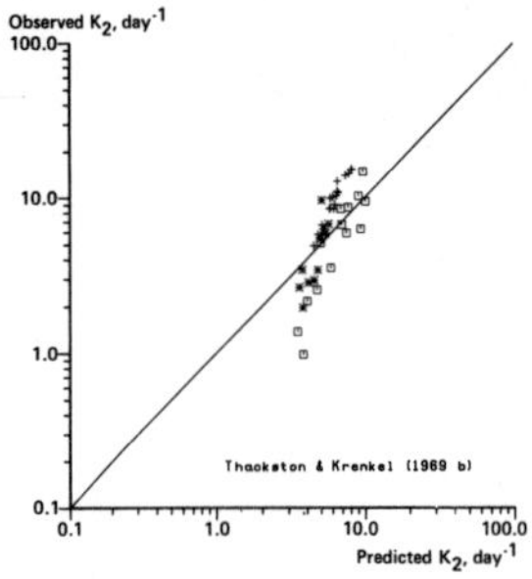

Figure 7.--Measured K_2-values versus those predicted by the equation given by Thackston and Krenkel (1969b).

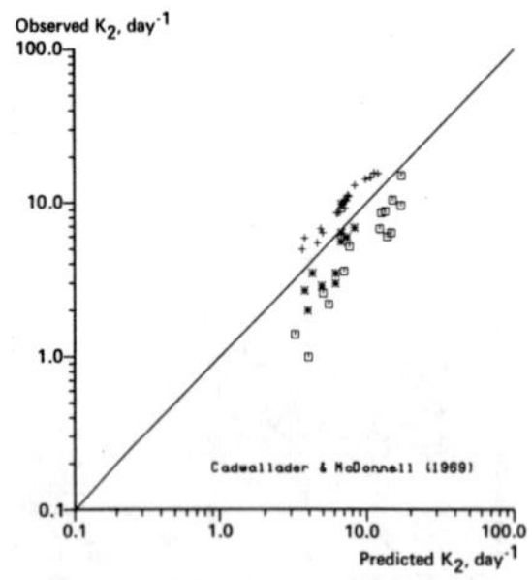

Figure 8.--Measured K_2-values versus those predicted by the equation given by Cadwallader and McDonnel (1969).

DEVELOPMENT OF PREDICTION EQUATIONS FOR SMALL MACROPHYTE DOMINATED STREAMS

Stepwise multiple linear regression analysis was applied to develope a semi-empirical prediction equation relating measured K_2-values to hydraulic variables considered to be important in controlling turbulence phenomena in rivers.

The most important factors affecting macroturbulence in rivers are summarized in table 4.

Table 4.-- Hydraulic factors controlling river turbulence.

Variable	Intensity	Scale
primary	mean velocity mean slope	mean depth
secondary	energy dissipation friction velocity Froude number	mixing length

A comprehensive discussion of the effect of each of these variables on stream reaeration is beyond the scope of the present paper. Readers interested in the theoretical justification of the above selection is referred to the review papers by Benneth and Rathbun (1972), Wilson and Macleod (1974), Thyssen (198o) and to a whole series of papers on the subject by the Tsivoglou-group. The latter papers are extensively referenced by Tsivoglou and Neal (1976).

Data analysis and model formulation

The general predictive equation can be stated as follows:

$$K_2 = f(U, H, S, E, U^*) \tag{3}$$

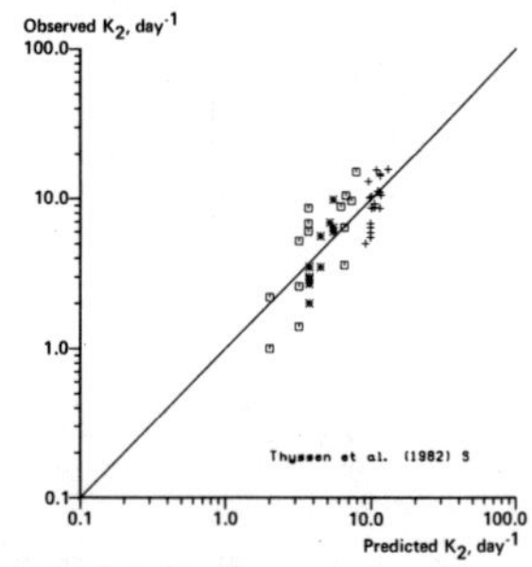

Figure 9.--44 K_2-measurements from 4 macrophyte dominated streams versus predicted values obtained from equation (5).

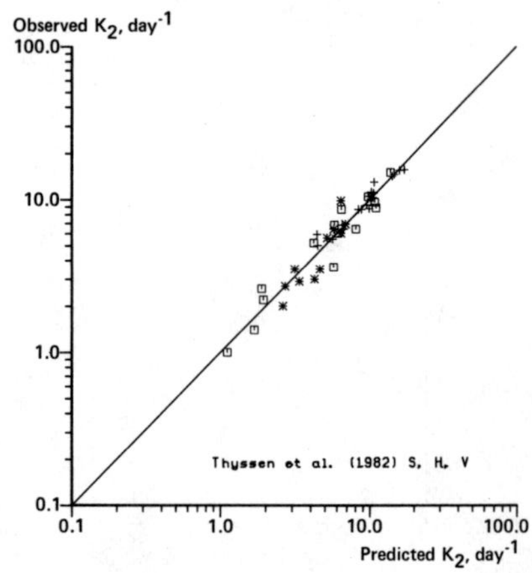

Figure 1o.--Same as 9, but predictions from equation (6).

The symbols and units of measurement in this equation are presented in tables 1 and 3.

The chosen functional relationship between the dependent variable (K_2) and the hydraulic variables was

$$K_2 = \alpha\, U^{\beta_1}\, H^{\beta_2}\, S^{\beta_3}\, E^{\beta_4}\, U^{*\beta_5} \quad (4)$$

Logarithmic transformation of this equation allowes the parameter estimation to be performed using multiple linear regression.

Stepwise inclusion of each independent variable, one at a time, and calculation of the residual variance shows which hydraulic variable explains most of the observed variation in K_2. At this step in the analysis it turned out that the rate of energy dissipation (E) explained most of the observed variation in K_2, but since E is not directly measurable we continued the analysis by selecting the primary variable (S) which explained most of the variation. In figure 9 is shown the predictive performance of the resulting slope model, equation (5)

$$K_2 = 8377\, S^{1.063} \quad (5)$$

At the second step in the analysis we combined all five remaining variables with S and found that S and V gave the smallest residual variance. Inclusion of H in the model further reduced the residual variance and the final predictive equation for small, macrophyte dominated streams became:

$$K_2 = 37406\, U^{0.846}\, H^{-0.672}\, S^{1.154}$$

The predictive behavior of this four parameter model is shown in figure 1o, and it is obvious that the reaeration capacity can be predicted quite accurately for all four streams studied. Even if their macrophytic biomass varies from practically zero up to 2oo g dw m^{-2} there does not seem to be any systematic discrepancy between observed and predicted K_2's, which means that the hydraulic effect of the submerged vegetation is already included in the primary variables U and H. Of even greater importance is this model's ability to predict low K_2's accurately.

LITERATURE CITED

Benneth, J.P., and R.E. Rathbun. 1972. Reaeration in open-channel flow. U.S. Geol. Surv. Prof. Paper 737, Washington, D.C..

Cadwallader, T.E., and A.J. McDonnell. 1969. A multivariate analysis of reaeration data. Wat. Research 3: 731-742.

Churchill, M.A., H.L. Elmore, and R.A. Buckingham. 1962. The prediction of stream reaeration rates. Int. J. Air. Wat. Poll. 6: 467-5o4.

Dobbins, W.E.. 1964. BOD and oxygen relationships in streams. J. San. Eng. Div. ASCE 9o: 53-78.

Dobbins, W.E.. 1965. Closure to BOD and oxygen relationships in streams. J. San. Eng. Div. ASCE 91: 49-55.

Gameson, A.L.H., G.A. Truesdale, and A.L. Downing. 1955. Re-aeration studies in a lakeland beck. J. Inst. Wat. Engrs. 9: 571-594.

Jeppesen, E. 1982. A comparison between four oxygen balance models for small organic polluted streams with many macrophytes. p.p. 183-195. In Dubois, D.M. (Ed.) Proceedings og the Second International Conference on the State-of-the-Art in Ecological Modelling. Liege. Belgium.

Mejer, H., S.E. Jørgensen, and L. Jørgensen. 1982. On parameter estimation in eutrofication models. pp 135-15o. In Dubois, D.M. (Ed.) Proceedings of the Second International Conference on the State-of-the-Art in Ecological Modelling. Liege. Belgium.

O'Connor, D.J., and W.E. Dobbins. 1956. Mechanism of reaeration in natural streams. J. San. Eng. Div. ASCE 82: 1-3o.

Odum, H.T. 1956. Primary production in flowing waters. Limnol. Oceanogr. 1: 1o2-117.

Owens, M., R.W., Edwards, and J.W. Gibbs. 1964. Some reaeration studies in streams. Int. J. Air Wat. Poll. 8: 469-486.

Parkhurst, J.D., and R.D. Pomeroy. 1972. Oxygen adsorption in streams. J. San. Eng. Div. ASCE 98: 1o1-124.

Schramm von, W., and V. Martens. 1976. Ein Messsystem für in situ Untersuchungen zum Stoff- und Energieumsatz in Benthosgemeinschaften. Sonderheft 3. Kieler Meeresforschungen. Inst. Für Meereskunde, Kiel Universität, West Germany.

Thackston, E.L., and P.A. Krenkel. 1969. Reaeration prediction in natural streams. J. San. Eng. Div. ASCE 95: 65-94.

Thyssen, N. 198o. The effect of physical, chemical, biological processes on the oxygen balance of a macrophyte dominated river, with special reference to the contribution of oxygen by reaeration and photosynthesis. Ph.D. thesis, Botanical Institute, University of Aarhus, Denmark.

Thyssen, N., and E. Jeppesen. 198o. Genluftning i mindre vandløb. Vatten 36: 231-248.

Tsivoglou, E.C. 1967. Tracer measurement of stream reaeration. USDI, Federal Water Pollution Control Administration, Washington, D.C..

Tsivoglou, E.C., and L.A. Neal. 1976. Tracer measurement of reaeration: III. predicting the reaeration capacity of inland streams. J. Wat. Poll. Cont. Fed. 48: 2669-2689.

Ursin, C., P.P. Madsen, and I. Drabæk. 1981. Measurement of stream reaeration by the gaseous tracer method. Vatten 37: 152-162.

Wilson, G.T., and N. Macleod. 1974. A critical appraisal of empirical equations and models for the prediction of the coefficient of reaeration of deoxygenated water. Wat. Research 8· 341-366.

SECTION VI:

Application of Ecological Models to Energy Development

M-X PROJECT IN NEVADA AND UTAH: ESTIMATION OF RECREATION DEMAND BY GRAVITY MODEL[1]

K. L. Reed and R. Villasenor[2]

Projected recreation demand by large in-migrant population in Nevada and Utah was assessed by means of a gravity model which projected the use of 983 developed and dispersed recreation sites in Nevada and Utah for 1982-1994. The model distributes recreation demand from each population center to each recreation resource in the two-state region. The model accounted for greater than 99% of the variance between the observed and predicted visitation rates and was used to compare 9 basing alternatives. Other applications are discussed.

INTRODUCTION

Proposed deployment of M-X missiles in the multiple protective shelter system in the Great Basin region of Nevada and Utah would have stimulated an influx of nearly 100,000 people into a region of relatively low current population (HDR, 1981). This influx of population would have been distributed among a number of cities, towns, and construction camps. Because of the importance of recreation in the Nevada/Utah area, this large increase in the region's population became a critical issue. Further, many of the other impacts associated with the M-X project are related to recreational activity. Off-road vehicles can damage rangelands and native vegetation; large numbers of people in an area may have an impact on animal habitat or behavior.

These problems led to the development of a model capable of predicting recreation use in the region resulting from population growth attributable to the M-X project. The requirements for the model were that the model must: (1) predict future demand for developed and undeveloped sites; (2) be sensitive to population change in time and space, and account for regional differences in recreation preferences; (3) be developed using published and otherwise available data; and finally, (4) be validated to the extent possible, in order to have a degree of acceptability.

The model that we developed predicted the use of 983 recreation sites in Nevada and Utah, over the period of 1982 to 1994. The model allocated recreation demand from each of 92 population centers in the two states. Two project alternatives involving deployment in Texas and New Mexico were also studied. This was a lesser effort. Population centers represented all of the towns and all the many construction camps that were to be located in the region. Because there were numerous siting alternatives to be considered, the model was used for comparison of the recreation demands resulting from the different project layouts.

[1]Paper presented at the Third International Conference on State-of-the-Art in Ecological Modelling, Colorado State University, May 24-28, 1982.

[2]Chief System Scientist and Scientist, respectively; HDR Sciences, 804 Anacapa Street, Santa Barbara, California.

It was not possible to consider all recreation activities so only the following outdoor recreation activities were considered: swimming, picnicking, camping, hiking, off-road vehicle (ORV) use, boating, water skiing, and fishing. These activities represent most of the popular outdoor recreation activities (Utah SCORP, 1980). Skiing and hunting were neglected because big game hunting is largely controlled by permit and skiing is a special-case commercial activity.

DERIVATION OF THE MODEL

The model is based upon the gravity model approach (Cesario, 1969; 1973), sometimes called the "travel cost" approach (Smith and Kopp, 1980). This approach is reviewed by Vickerman (1974). The formulation of the model resembles that of Klein (1979) and Sutherland (1980). The basic idea of the gravity model is that all things being equal, people tend to recreate closer to home. Since recreation sites are not equal, some are preferred over others. The patterns in figure 1 are often observed and are similar to the inverse square law (hence gravity model) or the exponential decay curve. The model can be best understood in its matrix form:

$$\mathbf{A} = \mathbf{QR} + \Omega \qquad (1)$$

where

- **A** = Vector of use of all recreation sites (visitor-days per year);
- **Q** = Matrix of probability of travel from population centers to recreation areas;
- **R** = Vector of recreation pool (visitor-days per year produced) from all population centers;
- Ω = Vector of use of recreation areas by out-of-state visitors (visitor-days per year)

The matrix **Q** is derived by multiplying each element of a matrix of the probability of travel from a population center j to recreation area i by a weighting factor that accounts for the appeal of the recreation site:

$$q_{ij} = \frac{p_{ij} w_i}{\sum_j p_{ij} w_j} \qquad (2)$$

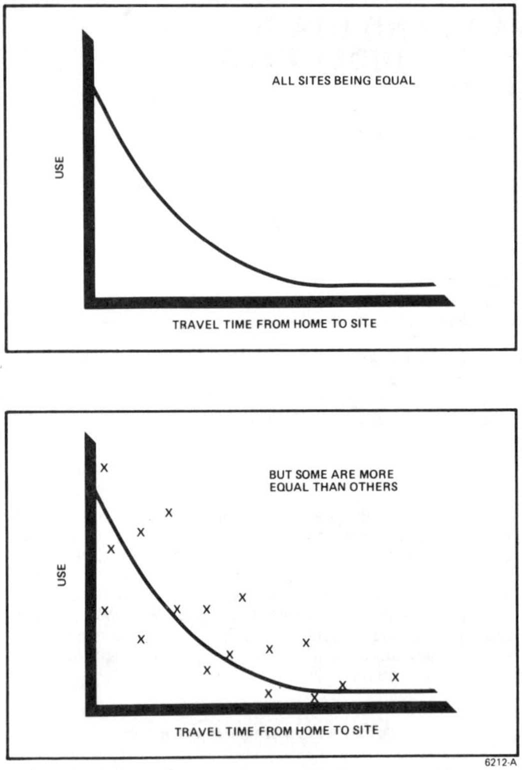

Figure 1.--Frequency of recreation area visitation as related to travel time. TOP: idealized curve with all sites being equal, BOTTOM: deviations from ideal due to different appeal among sites.

where

$$p_{ij} = \frac{\exp(-k_j x_{ij})}{\sum_j \exp(-k_j x_{ij})} \quad (3)$$

which is the probability of travel from j to i if all sites i are equally attractive. The numerator is a representation of the assumption that use of recreation areas by people from a population center j declines exponentially with travel time, x_{ij} (Klein, 1979). The constant k_j is a coefficient determining the slope of the travel-time function for various localities (e.g., people in Nevada might be willing to travel farther than people in Utah). Sutherland (1980) demonstrated that k_j also varies with activity. For example, people will travel farther to camp than to picnic. Sutherland's data show that the difference is significant and should be considered, but such a complication was impossible under the constraints of our program. Even if we had included the capacity, it would have been impossible to gather the necessary data.

The denominators in equations (2) and (3) normalize the probability matrices so that the column sums equal one. This ensures that all the visitors from each population center j are distributed among all the areas. The term w_i in equation (2) is the appeal index (weight) for recreation area i. Each area is assigned a weight by solution of the matrix equation

$$\mathbf{W} = \mathbf{S}\,\mathbf{N} \quad (4)$$

where **S** is a matrix of available activity quality scores for all recreation areas and **N** is a vector of participation rates for the activities considered in this analysis. Each recreation area was evaluated with respect to these activities. If a given activity was not available, then a score of zero was assigned; otherwise a score reflecting the quality of that recreation opportunity was assigned. Descriptions of the recreation areas, as well as 1980-use data, were used to assign scores to each area. These scores were multiplied by the fraction of time spent by an average citizen of Utah participating in that activity (the vector **N**). **N** was obtained from the Utah Statewide Comprehensive Outdoor Recreation Plan (SCORP, 1980), which listed the average number of visitor days (1 person/12 hours) spent enjoying the activities analyzed. These figures were normalized to give the vector **N** (fraction of use in table 1). Sutherland (1980) estimated appeal as a function of variables describing the resources available as each site and an accessibility function. Since accessibility is already taken into account by the gravity model itself, this step was improper for our model.

Table 1.--Activity preferences for the state of Utah.

Activity	Visitor-days	Fraction of use **N**
Swimming	0.6	0.062
Picnicking	0.7	0.072
Camping	5.2	0.540
Hiking	0.5	0.051
ORV	0.4	0.041
Boating	0.5	0.052
Water Skiing	0.2	0.021
Fishing	1.6	0.160

Source: Utah SCORP, 1980.

The appeal index expressed algebraically is

$$w_i = \sum_\sigma s_{i\sigma} n_\sigma \quad (5)$$

where σ is the index of sport (activity). If $s_{i\sigma} = 0$, then no participation in that activity is possible on site i.

The matrix **S** is, then, an inventory of available resources on a site and an index of that site's relative popularity. Further discussion of **S** is deferred to the section on model calibration.

The vector **R** in equation (1) is the set of visitor-days available from each population center. The average person in Utah spends 9.7 days per year in the activities analyzed (Utah SCORP, 1980). The visitor-day resource of each population center is given by Equation (6).

$$r_j = 9.7\,\rho_j\,(1 - t_j) \quad (6)$$

where ρ_j = population of center j, and t_j is the fraction of that population assumed to recreate in areas other than the ones considered.

We had to assume all populations in Nevada behaved similarly to those of Utah since Nevada SCORP (1977) lists activity preference by activity occasions, which are not directly convertible to visitor-days. New Mexico participation rates were used for the Texas/New Mexico analysis.

The algebraic representation of the model is equation:

$$a_i = \sum_j \left[\frac{p_{ij} \sum_\sigma s_{i\sigma} n_\sigma}{\sum_j (p_{ij} \sum_\sigma s_{i\sigma} n_\sigma)} \right] r_j + \omega_i \qquad (7)$$

where the term ω_i is out-of-state visitors to site i. Not all areas receive significant out-of-state visitation, but the so-called Golden Circle of southern Utah (i.e., Zion, Bryce Canyon, and Cedar Breaks) does. Out-of-state use was estimated as a fraction of Nevada/Utah use and added to the solution for a_i:

$$\omega_i = f_i \left\{ \sum_j \left[\frac{p_{ij} \sum_\sigma s_{i\sigma} n_\sigma}{\sum_j (p_{ij} \sum_\sigma s_{i\sigma} n_\sigma)} \right] r_j \right\} \qquad (8)$$

where f_i is out-of-state visitor-day fraction for area i. Equation (8) took the value of zero for most areas. Preliminary estimates of f_i were obtained from USFS, Park Service, and BLM district personnel, and were adjusted in the calibration procedure discussed below.

IMPLEMENTATION OF THE MODEL

The implementation of the model is diagrammed in figure 2. Project parameters are contained in the DOPAA (Description of Proposed Action and Alternatives); population data are provided by socioeconomic models or alternatively from U.S. census projections; and spatial data in the form of maps is input via special programs.

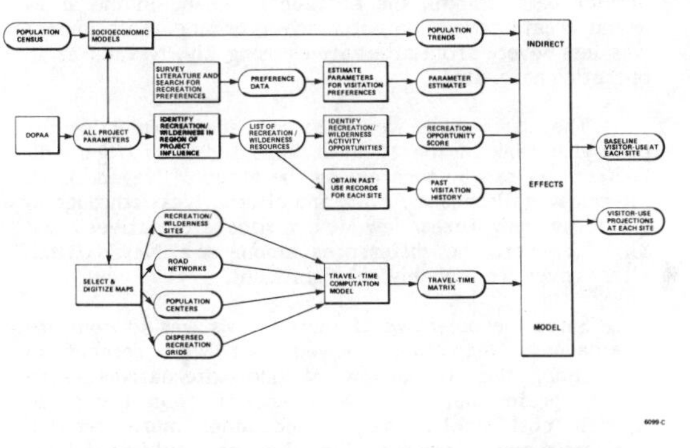

Figure 2.--Schematic of steps required for recreation analysis. The indirect effects model is the recreation model discussed here.

The states of Nevada, Utah, and New Mexico and the panhandle area of Texas were sectioned into units of approximately 290 sq mi (17 mi x 17 mi). Each grid unit was given a unique code number. Each developed recreation area, wilderness, and state and national park was located within a grid unit and dispersed recreation was referenced to each grid unit. Travel time from each population center to each grid unit was calculated by identifying the best route and dividing the mileage by average speed. Average speed was assumed to be 55 mph on paved roads, 45 mph on gravel, and 35 mph on unimproved roads. Average speed on trails was assumed to be 4 mph (hiking speed). The resulting matrix **X** of travel times was very large: 983 sites and 92 population centers were used in the Nevada/Utah analysis. Thus, the **X**, **P**, and **Q** matrices were of rank (983 x 92), or nearly 100,000 elements.

Because of the necessity to include rural and urban populations, population centers were defined by township in Nevada, Texas, and New Mexico and by division in Utah. These centers were identified by the name of a city or town in the township or division. The population centers included the proposed M-X operating base at Coyote Spring Valley and 18 Nevada/Utah construction camps (15 in Texas/New Mexico). Other operating bases were not individually identified; their populations were added to the population center to which they were closest.

U.S. Census (1980) figures were used for the calibration, and population projections for trend and high baseline with and without M-X (HDR, 1981) were used for the analysis. Because the model was to be used for comparison of nine project siting alternatives, the vector **A** was expanded to a three-dimensional matrix where $\mathbf{A}_{alt,y,l}$ is the use of recreation areas by alternative, year, and population level. Recreation use for the Proposed Action and all eight alternatives was analyzed for the years 1982 to 1994, given trend population baseline, trend baseline plus M-X, high population baseline, and high baseline plus M-X (HDR, 1981). High baseline included other large projects planned for the region. For this, **A** was defined as a matrix of rank (983 x 14 x 4). For most analyses, the difference between baseline and baseline plus project is the measure of effect. Sometimes the percent change was used.

The model was encoded in Pascal, and run on HDR's Cyber 170/730 computer, requiring 1.1 hours of CPU time for execution. Large arrays were defined as file types, and stored on disk in core image format to be read by parts. This step allowed the program to fit in approximately 45,000 octal words of central memory; over 2 million words of central memory would have been required had arrays been loaded into memory at any one time.

CALIBRATION

Use-data for 1980 in visitor-days per year were obtained for more than 500 sites in Nevada and Utah. These included USDA Forest Service developed recreation sites as well as district totals for dispersed areas. Dispersed use figures were allocated to the appropriate grid units in each state on a proportional basis. The Forest Service data were obtained from each regional office.

National Park Service data were available in the form of total visitor-days and a breakdown of backcountry use. BLM data were available for a few sites; for others, rough estimates were obtained from BLM district personnel. Data on use of non-Forest Service wilderness were especially difficult to obtain. These data were also estimates from BLM personnel.

The model was recursively fit to the data. Such parameters as the travel time coefficients, k_i; the estimates of recreationists traveling from in-state to out-of-state, t_i; and visitors entering the region to specific areas, f_i; were adjusted as necessary. For example, Forest Service personnel estimated that 60 percent of use of the portions of Dixie National Forest was by out-of-state visitors, mostly from Nevada and California. The model allocated approximately 30 percent of the users from Nevada, the rest was added by means of the out-of-state term (Equation 8).

The primary method of calibration was to adjust the recreation opportunity scores in matrix **S** by comparison of the observed use vector **A** to the predicted use $\hat{\mathbf{A}}$. The recreation opportunity scores for areas for which the predictions were poor were changed until a reasonable agreement was reached.

Decisions on change of scores were made after careful inspection of the residuals $a_i - \hat{a}_i$. Often, descriptions of two sites would be very similar but use would vary by an order of magnitude, even though the sites were equally distant from population centers. In other words, the votes of visitors were used as the final arbitrator of appeal index. The recreation opportunity scores in matrix **S** reflect the intrinsic appeal of the sites. Since we had no breakdown on activity preference on each site (the fraction preferring one activity over another), the non-zero scores for each site were adjusted together.

Recursive fit of the model to the data was surprisingly easy. The geographic distribution of the residuals provided clues to the adequacy of the out-of-state travel fractions t_i and f_i. We noticed, for example, that the model tended to underpredict in those areas where high tourism would be expected: the Flaming Gorge area and the Navajo sandstone region of southern Utah. The coefficient f_i was adjusted until the residuals for those areas were normally distributed about zero. Variance was reduced by adjusting the recreation opportunity scores, and in certain areas, the decay coefficient k_i was adjusted to account for under-or overprediction around certain large cities.

Assessment of the goodness of fit was done by examining histograms of the residuals and by analogy to ordinary regression by computing statistics based on sums of squares. The sum of squares due to the model is obtained by difference SSM = SST - SSR (Draper and Smith, 1966) where:

$$SSR = \sum_i (a_i - \hat{a}_i)^2 \text{ is the residual sum of squares,}$$

and

$$SST = \sum a_i^2 - (\sum a_i)^2/n \text{ is total sum of squares.}$$

The percent of variation explained by the model is $R^2 = 100 \cdot SSM/SST$.

The calibration procedure was repeated until R^2= 99 percent. The agreement of the model with the data is illustrated in figure 3. The residuals were normally distributed about zero. A sample comparison of the model to the data is given in table 2.

Although the model fit observed data very well, data were available for only half of the areas. Therefore, the error of prediction for the rest of the areas is unknown. However, since USFS land is the most heavily used, and since data were available for all USFS land, the general predictability of the model is probably good.

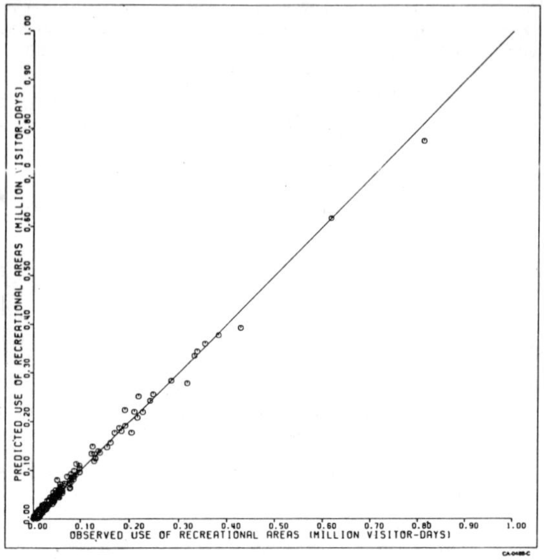

Figure 3.--Comparison of observed 1980 use of more than 500 recreation sites to model predictions using 1980 population data.

RESULTS

Recreation demand was simulated for all sites in Nevada and Utah, and selected sites in Texas and New Mexico for the period of 1982 to 1994 given four population trend projections and nine siting alternatives. The results were graphically displayed as increase over baseline projections for peak-year and long term. Peak-year varied from 1986 to 1987 but long term figures were taken from the 1994 solution. The increase in recreation demand over baseline for the Proposed Action is shown in figure 4 and long-term demand increases in figure 5. Seriously impacted areas include the Toiyabe National Forest in Nevada and the Navajo sandstone region in Utah. Since the Toiyabe areas are already over-utilized (U.S. Forest Service pers. comm), the additional demand in the areas would create severe potential overcrowding. This impact was less severe from alternatives using Ely, Nevada as an operating base site.

The alternatives were also compared by plotting the peak-year rank of the 15 most impacted sites (figure 6). Differences can be seen among the group of Nevada/Utah alternatives (Proposed Action and alternatives 1 through 6) and those using Texas/New Mexico sites (alternatives 7 and 8). However, the differences among the Nevada/Utah alternatives are probably not significant.

Since the objective of this analysis was to compare alternatives, the study showed that with respect to recreation, the Texas/New Mexico alternatives were clearly preferable. The M-X analysis was tiered, so specific cost-benefit analysis and other more detailed analyses were deferred to Tier II which would have been centered around the chosenalternative. This exercise was rendered moot by President Reagan's decision to cancel the project.

TABLE 2. SAMPLE COMPARISON OF 1980 RECREATION USE TO PREDICTED RECREATION USE IN NEVADA AND UTAH

UNIT	NAME OF AREA	OBSERVED V-DAYS	PREDICTED V-DAYS	RESIDUAL	PER CENT DIFF
N272	CHERRY CREEK	3185.0	2936.2	248.8	7.812
N273		3185.0	3890.8	-705.8	-22.160
N276	SAULSBURY WASH	17100.0	19344.6	-2244.6	-13.126
N276		14328.0	15364.8	-1036.8	-7.236
N281		10332.0	11090.4	-758.4	-7.340
N282		17220.0	14808.7	2411.3	14.003
N288		735.0	920.0	-185.0	-25.170
N356	HILLTOP	1700.0	1031.0	669.0	39.353
N356	MAHOGANY GROVE	3700.0	1933.1	1766.9	47.754
N356		64576.0	68785.8	-4209.8	-6.519
N357	DEER CREEK	3700.0	3852.5	-152.5	-4.122
N357	DOLOMITE	1500.0	880.6	619.4	41.293
N357	FOXTAIL CANYON	3800.0	963.1	2836.9	74.655
N357	MCWILLIAMS	49800.0	47880.5	1919.5	3.854
N357	OLD MILL	20000.0	18161.6	1838.4	9.192
N357		226016.0	219589.8	6426.2	2.843
N364	KYLE CANYON	23100.0	26868.0	-3768.0	-16.312
N364		60540.0	58533.8	2006.2	3.314
N365	CATHEDRAL ROCK	44200.0	46309.6	-2109.6	-4.773
N365	FLETCHER VIEW	11400.0	17033.4	-5633.4	-49.416
N365		52468.0	54294.0	-1826.0	-3.480
U003	CLEAR CREEK	2200.0	4635.1	-2435.1	-110.686
U003		86600.0	88015.6	-1415.6	-1.635
U007	SMITHFIELD CANYON	10100.0	8620.7	1479.3	14.647
U008	CHINA ROW	1100.0	5638.7	-4538.7	-412.609
U008	COTTONWOOD	300.0	704.8	-404.8	-134.933
U008	HIGH CREEK	6200.0	5638.7	561.3	9.053
U008	LEWIS M TURNER	4700.0	5638.7	-938.7	-19.972
U008	RED BANKS	6100.0	7048.4	-948.4	-15.548
U008	SUNRISE	14200.0	13438.9	761.1	5.360
U008	TONY GROVE LAKE	16400.0	17292.0	-892.0	-5.439
U008	TWIN BRIDGES	1500.0	5638.7	-4138.7	-275.913
U008	WOOD CAMP	9500.0	12405.1	-2905.1	-30.580
U008	W-MT NAOMI	7000.0	8270.1	-1270.1	-18.144
U008		128754.0	119399.2	9354.8	7.266
U016	W-WELLSVILLE MTN	5000.0	4298.3	701.7	14.034
U016		98235.0	96283.0	1952.0	1.987
U017	BEIRDNEAU	400.0	283.8	116.2	29.050
U017	BRIDGER	5800.0	4627.3	1172.7	20.219
U017	BROWNS ROLL-OFF	600.0	283.8	316.2	52.700
U017	CARD	1300.0	1986.6	-686.6	-52.815
U017	CHOKECHERRY	1300.0	1986.6	-686.6	-52.815
U017	DEWITT	1100.0	2418.5	-1318.5	-119.864
U017	FRIENDSHIP	8700.0	8329.1	370.9	4.263
U017	GUINAVAH	26000.0	24678.7	1321.3	5.082
U017	LODGE	6100.0	7403.6	-1303.6	-21.370

SOURCE: HDR SCIENCES CT 1091

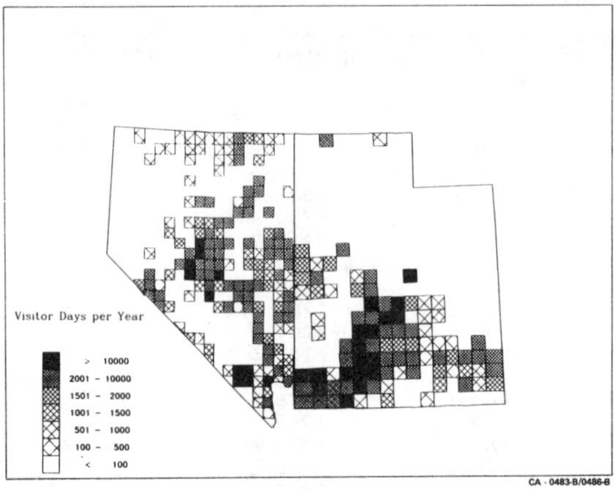

Figure 4.--Peak year increase (1987) in recreation demand, proposed action.

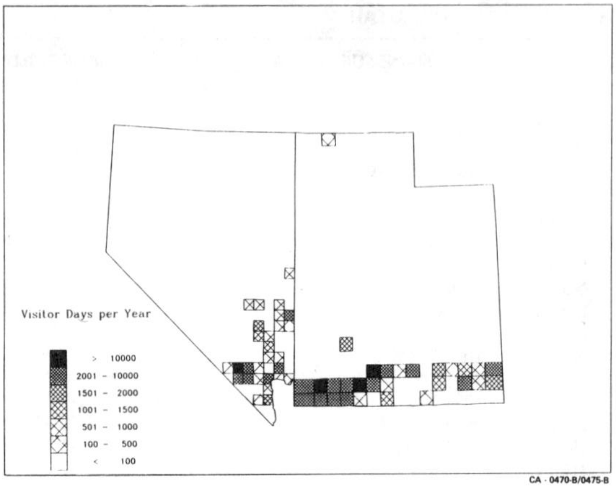

Figure 5.--Long-term increase in recreation demand, proposed action (1994).

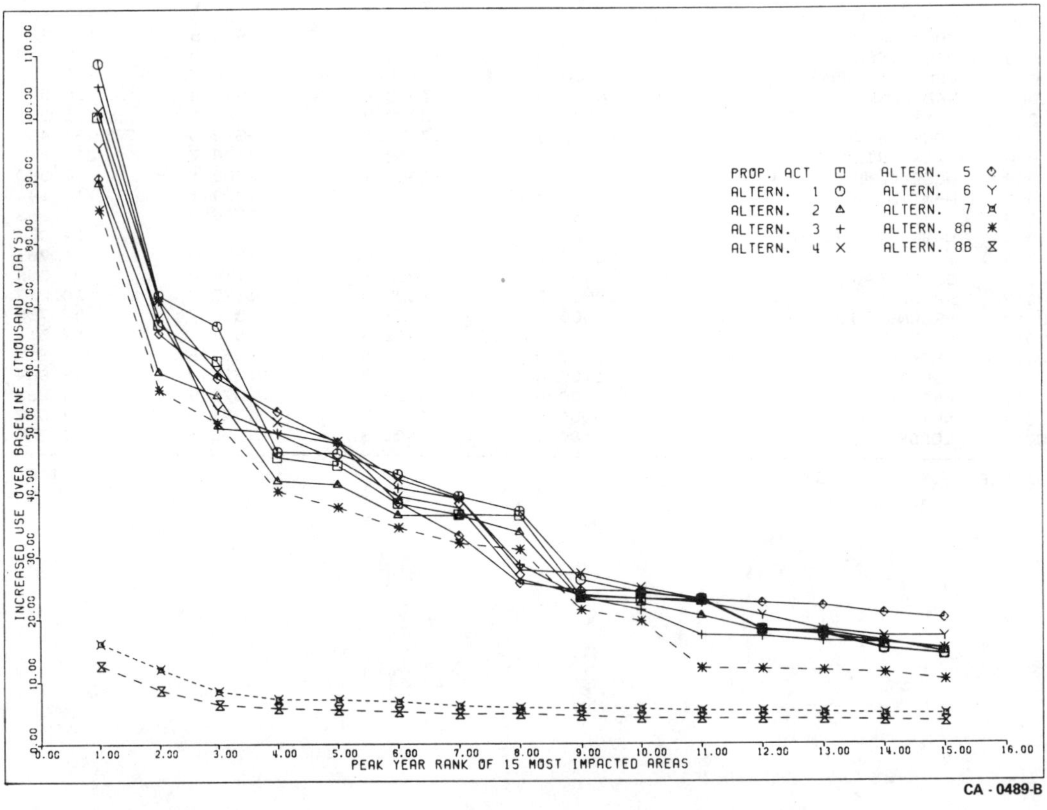

Figure 6.--Comparison of all alternatives for fifteen most impacted areas (short term-construction peak, 1986-1987).

The model does appear to provide good predictive ability and can be easily calibrated to a region or smaller area. The boundary has to be chosen, and travel across that boundary should be taken into account. The model can be expanded to accommodate more detail and survey data. For example, if recreation use at the sites can be broken down by activity, then

$$a_{i\sigma} = \sum_j \frac{p_{ij}\, s_{i\sigma} n_\sigma}{\sum_j (p_{ij} \sum_\sigma s_{i\sigma} n_\sigma)} r_j + c_\sigma\, \omega_i \qquad (9)$$

where

$a_{i\sigma}$ = visitor-days per year in activity σ at site i,

c_σ = fraction of out-of-state visits participating in activity σ

will give projected use by activity. The model can also be interfaced with economic and planning algorithms.

Ecological use of the gravity model approach might include the modeling of insect spread where the appeal index might be defined in terms of factors that could attract insects. Use of home range by animals might be another application of this approach. Certainly many questions requiring models of human activity can be addressed by this approach.

BIBLIOGRAPHY

Cesario, F. J., 1969. Operations research in outdoor recreation, J. Leisure Res. 1:33-53.

Cesario, F. J., 1973. A generalized trip distribution model, J. Regional Sci. 13:233.

Draper, N. R. and H. Smith, 1966. Applied regression analysis. New York: Wiley and Sons.

HDR Sciences, 1981. Community population allocation model. In-house document. Santa Barbara, California.

Klein, D.H., 1979. A recreation needs assessment process - an origin destination approach. Trans. Plann. and Tech. 5:169-182.

Nevada State Comprehensive Outdoor Recreation Plan (SCORP). 1977. Nevada State Park System, Carson City.

Smith, V.K., and R.J. Kopp, 1980. The spatial limits of the travel cost recreational demand model, Land. Econ. 56:64-72.

Sutherland, R.J. 1980. A regional recreation demand and benefits model. Draft Report. U.S. Environmental Protection Agency, Corvallis, Oregon. 144 p.

Utah State Comprehensive Outdoor Recreation Plan (SCORP). 1980. Utah Outdoor Recreation Agency, Salt Lake City.

Vickerman, R.W., 1974. Accessibility, attraction, and potential: a review of some concepts and their use in determining mobility. Environ. and Planning 6:675-691.

SIMULATING EFFECTS OF ACID MINE DRAINAGE[1]

Paul W. Hadley[2] and Gerald T. Orlob[3]

The model WQRRS was used to simulate water quality changes in the Allegheny River system due to acid mine drainage and to evaluate alternative strategies of flow regulation to mitigate adverse impacts.

INTRODUCTION

Acid mine drainage has long been the most serious water pollution problem throughout the Appalachian region of the United States. In the state of Pennsylvania alone it has been estimated that at one point over 2,500 miles of streams were polluted by a daily influx of one billion gallons of acid mine drainage (AMD) (Barnes and Romberger 1968). Perhaps nowhere is the potential severity of AMD more evident than in the Kiskiminetas River basin, a sub-basin of the Allegheny River basin in southwestern Pennsylvania (see fig. 1). Contaminated by AMD from the headwaters to the mouth, the Kiskiminetas River seldom exhibits a pH value above 5, while pH's below 3 are common. Under such low pH conditions aquatic life is almost precluded, limited to only acid tolerant microorganisms.

Within the Allegheny River basin the United States Army Corps of Engineers (COE) operates nine reservoirs. While the major function of these reservoirs is to provide flood control capabilities, summertime operations provide for flow augmentations to assimilate the acid polluted Kiskiminetas as it enters the lower Allegheny River.

Historically, fish kills occurred when the acid polution in the Kiskiminetas reached sufficiently high levels to cause a sharp depression in pH in the lower Allegheny. The last such fish kill happened between August 19-23, 1966, when approximately one million fish died as pH levels stayed between 4.5 and 5.0. Since the commencement of operations at Kinzua Dam in 1967, sufficient augmentation water has been available for water quality enhancement and management of the lower Allegheny, and no fish kills have occurred.

To estimate the changes in water quality that might be expected, to evaluate mitigating effects of flow regulation and to improve operating procedures, the river-reservoir system was simulated under three representative hydrological conditions:

- "Existing Conditions"; reservoir operations as documented
- "No Corps Storage"; the without-project case, utilizing no COE reservoir storage in the basin and
- "Pattern A"; reduced outflow from Kinzua Dam.

A schematic of the Allegheny River system presented in figure 2, indicates the relative locations of COE reservoirs and U.S. Geological Survey (USGS) gaging stations in the basin. Also shown is the location of the COE flow gage at Lock and Dam No. 2 on the Allegheny River, the lower boundary of the stream system.

BACKGROUND

Acid mine drainage is usually formed by the oxidation of sulfur bearing minerals in an aqueous environment, such as may exist in an abandoned coal mine. The principal source of sulfur is iron pyrite (FeS_2), commonly found in coal seams and adjacent strata.

Approximately eighty percent of the AMD pollution occurring in the Appalachian states originates in abandoned coal mines. These old mines allow natural drainage water to collect, and in the presence of oxygen facilitate the production of acid waters. The total oxidation of pyritic materials can best be described by the following sequential reactions:

$$2FeS_s + 2H_2O + 7O_2 \rightarrow 2Fe^{++} + 4H^+ + 4SO_4^= \quad (1)$$

and

$$2Fe^{++} + 5H_2O + 1/2O_2 \rightarrow 2Fe(OH)_3 + 4H^+ \quad (2)$$

The actual chemical composition of AMD is determined by the geological environment in which it has been formed. AMD from coal mines typically contains high concentrations of iron, sulfate and acidity as indicated by (1) and (2), and may also contain substantial amounts of dissolved solids, including calcium, manganese and aluminum. Such drainage usually contains a large amount of sediment. The principal adverse effects of AMD on surface waters are excessive increases in dissolved solids and sediment and the depletion of the natural alkalinity, resulting in depressed pH.

Several technologies have been utilized to prevent or control AMD pollution. Examples of these techniques include wastewater treatment to neutralize acidity and remove dissolved and suspended solids, closing off abandoned mines with hydraulic seals, and deep well injection of AMD. Where sources of AMD are many and occur over large areas, streamflow regulation for dilution of AMD is perhaps the most practical method of mitigating the effects of AMD. Such is the case in the Kiskiminetas River basin.

MODELING APPROACH

The model WQRRS (Water Quality for River-Reservoir Systems) of the Corps of Engineers Hydrologic Engineering Center (1974) was selected for the study. WQRRS consists of three separate but integrable programs:

- a stream hydraulics module (SHP)

[1]Paper presented at the Third International Conference on State-of-the-Art in Ecological Modeling, Colorado State University, May 24-28, 1982.

[2]Research Assistant, Aerojet General Corporation, Sacramento, California, 95813.

[3]Professor of Civil Engineering, University of California, Davis, California, U.S.A., 95616.

Figure 1. Allegheny River Basin

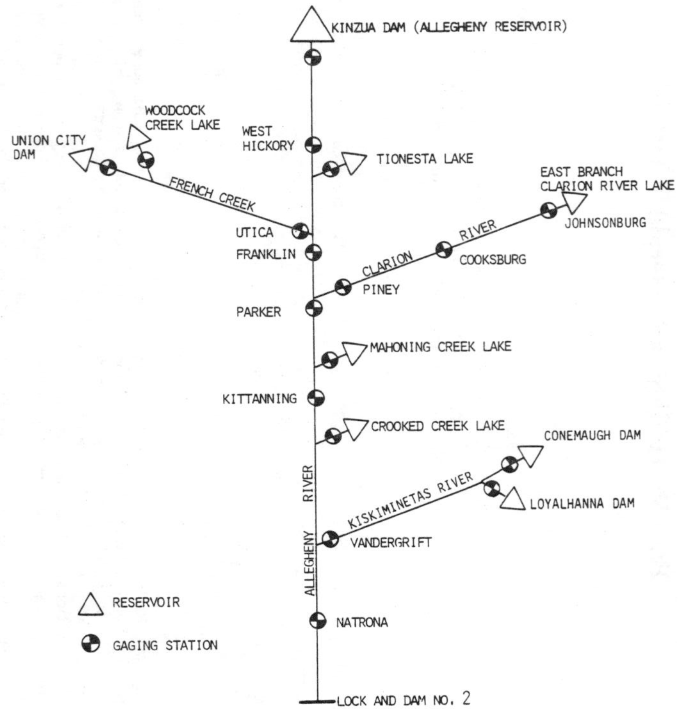

Figure 2. Schematic Representation of Allegheny River System

- a stream water quality module (STR)
- a reservoir quality module (RES) (not utilized in this study)

The Muskingum hydrologic routing option of SHP and the STR module are discussed below.

Stream Hydraulics-Muskingum Routing

The stream system in the stream hydraulics package, SHP, is represented as a linear network of stream segments with each segment characterized by length, width, cross-sectional area, volume, hydraulic radius and a special relationship between flow and depth. One option of the SHP package is the Muskingum hydrologic routing method.

The Muskingum method is derived on the assumption that the incremental storage change in a stream element is related to flow entering and leaving that segment. Formally, this assumption is represented by the following equation:

$$S = K(O) + X(K)\cdot(I-O) \qquad (3)$$

where,

S = total storage in segment (L^3)

I = inflow (L^3T^{-1})

O = outflow (L^3T^{-1})

K = empirical routing coefficient (T)

X = empirical routing coefficient (dimensionless)

Values of K for the Allegheny River system ranged from 2 to 8 hours, and values of X ranged from 0.08 to 0.33.

The Muskingum method provided the capability to route streamflow through the system dynamically, rather than assuming steady state or gradually varied flow profiles. Such a capability was well suited to the rapidly changing hydrology of the Allegheny River basin.

A first step in the application of the WQRRS model was the calibration and validation of flows made at ten stream gages for the two historic periods of interest, June through October 1975, and July through September 1977. These periods include the natural low flow period for the system when the effects of AMD have been most severe.

The modeling process required a balanced water budget for the entire system. All ungaged flow within each stream reach was distributed to tributaries based on the relative drainage area of each tributary. Subsequently, flow was routed through the stream system illustrated in figure 2, comparing simulated and observed flows at each mainstem gage located in the basin. Typical results of the flow routing performed prior to water quality simulations are presented in figure 3, for a station at Franklin, Pennsylvania during the 1977 study period. The generally excellent agreement between simulated and observed flows is confirmed by a regression analysis. The regression line of best fit is not distinguishable from a 45° line (slope=1) of perfect correlation. It is noted that the greatest scatter occurs for higher flows where differences are often accounted for by slight shifts in phase between simulated peaks and actual observations.

Water Quality Simulation

The stream water quality module, STR, simulates changes in a stream system using the same system of stream segments as in the hydraulic module. All inflows to a segment are assumed to be completely mixed within the segment. The principles of conservation of mass and energy are complied with in deriving equations to represent the dynamics of temperature and conservative and non-conservative substances. In general form, the conservation equation is stated as:

$$V\frac{dC}{dt} = \Delta z\, Q_z \frac{dC}{dz} + \Delta z\, A_z\, E_z \frac{d^2C}{dz^2} + Q_i\, C_i - Q_o\, C \pm SV \qquad (4)$$

where

C = thermal energy or concentration of a specified constitutent in a reservoir or stream in appropriate units (e.g., kcal, mg/L and MPN/100 ml)

V = volume of the fluid element, m^3

t = time coordinate, seconds

z = space coordinate, meters (vertical for reservoirs and horizontal for streams)

Q_z = advection, m^3/sec

A_z = element surface area normal to the direction of flow in m^2

E_z = coefficient of effective diffusion, m^2/sec

Q_i = lateral inflow, m^3/sec

C_i = inflow thermal energy or constitutent concentration in appropriate units

Q_o = lateral outflow, m^3/sec

S = all sources and sinks in appropriate units (e.g., kcal/sec, mg/L/sec, etc.)

Details concerning individual terms in equation 4 may be found in documentation for WQRRS (HEC 1974).

pH SIMULATION

Computer models have previously been developed to simulate the low pH conditions found in streams polluted by AMD. In general, these models simulate changes in pH based on the carbonate buffering system (Yeasted and Shane, 1976; Chadderton, 1979). This is the approach used in the WQRRS model.

Chemical Principles

The general equilibrium pathway for the solution and dissolution of atmospheric $CO_{2(g)}$ can be represented as follows:

$$CO_{2(g)} \rightleftharpoons CO_{2(aq)} + H_2O \rightleftharpoons H_2CO_3 \rightleftharpoons H^+ + HCO_3^- \qquad (5)$$

Concentrations of $CO_{2(aq)}$ are generally much greater than H_2CO_3, and the two species are difficult to measure separately. For convenience, concentrations of $CO_{2(aq)}$ and H_2CO_3 are combined toegether as:

$$CO_{2(aq)} + H_2CO_3 = H_2CO_3^* \qquad (6)$$

Substituting equation 6 into equation 5, the equilibrium pathway becomes:

$$CO_{2(g)} \rightleftharpoons H_2CO_3^* \rightleftharpoons H^+ + HCO_3^- \qquad (7)$$

When equilibrium exists, the expressions:

$$K_1 = (H^+)(HCO_3^-) \,/\, (H_2CO_3^*) \qquad (8)$$

$$K_2 = (H^+)(CO_3^=) \,/\, (HCO_3^-) \qquad (9)$$

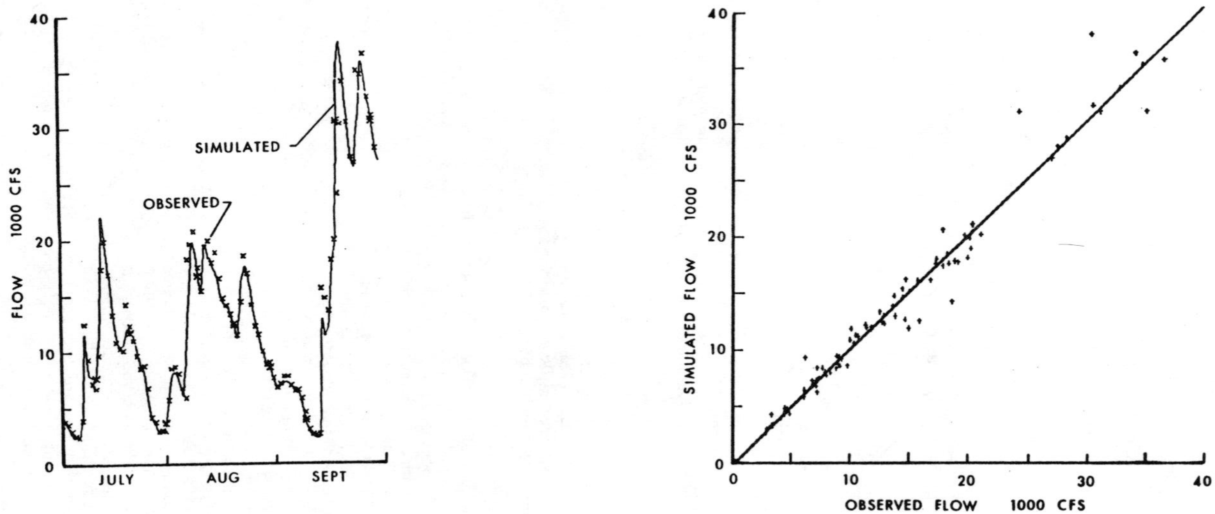

A. Simulated vs Observed Flows

B. Correlation Diagram

Figure 3. Comparison of Observed and Computed Flows, Allegheny River at Franklin, Pa., 1977 Study Period

$$K_w = (H^+)(OH^-) \qquad (10)$$

will hold. The constants K_1, K_2 and K_w are the first and second dissociation constants of carbonic acid and the ion product for water, respectively.

Total inorganic carbon concentration (C_T) is defined as:

$$(C_T) = (H_2CO_3^*) + (HCO_3^-) + (CO_3^=) \qquad (11)$$

alkalinity (Alk) is:

$$(Alk) = (HCO_3^-) + 2(CO_3^=) + (OH^-) - (H^+) \qquad (12)$$

and CO_2 acidity (CO_2-Acy) is:

$$(CO_2\text{-}Acy) = (H_2CO_3^*) + (H^+) - (OH^-) - (CO_3^=) \qquad (13)$$

These quantities are all defined in such a way that:

$$(C_T) = (Alk) + (CO_2\text{-}Acy) \qquad (14)$$

It can be shown that for a solution buffered only by the carbonate system the following fourth order equation can be solved to determine the equilibrium hydrogen ion concentration:

$$A_4(H^+)^4 + A_3(H^+)^3 + A_2(H^+)^2 + A_1(H^+) + A_0 = 0 \qquad (15)$$

where,

$$A_4 = 1.0$$
$$A_3 = (Alk) + K_1$$
$$A_2 = K_1K_2 - K_1(CO_2\text{-}Acy) - K_w$$
$$A_1 = -K_1(K_2(Alk) + 2K_2(CO_2\text{-}Acy) + K_w)$$
$$A_0 = -K_1K_2K_w$$

Equation 15 reduces to an equation of one unknown in hydrogen ion concentration when values of (Alk) and (CO_2-Acy) are known.

Mass Transfer Rates

The rate of gas transfer across the air-water interface is calculated by a subroutine of STR using the following first order equation:

$$R = k \cdot K_2(G^* - G) \qquad (16)$$

where,

R = rate of gas transfer (MT^{-1})

k = 1.0 for O_2 , 0.78 for CO_2

K_2 = reaeration coefficient (L^3T^{-1})

G^* = saturation concentration of dissolved gas (ML^{-3})

G = ambient concentration of dissolved gas (ML^{-3})

The reaeration coefficient K_2 was estimated for each stream element using the O'Connor and Dobbins method,

$$K_2 = 3.951 \frac{v^{0.5}}{h^{1.5}} \qquad (17)$$

where,

v = stream velocity (LT^{-1})

h = hydraulic depth (L)

Exchange of $CO_{2(g)}$ at the air-water interface represents an important sink for hydrogen ions. An increase in hydrogen ion concentration, as might result from a slug of AMD entering a stream, can be reduced via equilibrium pathway 7, terminating with the dissolution of $CO_{2(g)}$. Once the alkalinity of a stream has been depleted, exchange of $CO_{2(g)}$ no longer causes a net removal of hydrogen ions from solution. A solution depleted of alkalinity is in the mineral acid or H^+-acid range of the carbonate buffering system. For such a solution, where

$$(HCO_3^-) \cong (CO_3^=) \cong 0$$

it follows that

$$(C_T) \cong (H_2CO_3^*)$$

and exchange of $CO_{2(g)}$ with the atmosphere can be approximated as

$$\Delta(CO_{2(g)}) \cong \Delta(C_T).$$

WATER QUALITY DATA

Water quality responses were characterized by pH, alkalinity, D.O., BOD, temperature and total dissolved solids. To describe the general conditions observed throughout the basin for each water quality constituent and to account for the effect of rapidly changing hydrologic and meteorologic conditions on water quality, a substantial data base was necessary. Flow/quality relationships and thermographs were developed from available records, and adjusted as appropriate to describe water quality conditions where data were lacking. Personnel of the Pittsburgh district office of the COE familiar with water quality in the Allegheny basin provided valuable insight and assistance in developing the required data.

In-stream data for validation of results of water quality simulation were obtained from water treatment plants along the Allegheny River and from Ohio River Sanitation Commission (ORSANCO) robot monitors in the Allegheny and Kiskiminetas Rivers. These monitors provided data on temperature, D.O., TDS and pH on a near-continuous basis.

RESULTS

Changes in pH best characterize the response of the system to AMD pollution under alternative reservoir operation strategies. Discussion in this paper is focused on changes in pH in the lower Allegheny River basin under each study case, and are for the 1977 study period. A more extensive presentation of results is available elsewhere (Hadley & Orlob 1981).

Simulated Versus Observed Water Quality

At Freeport

Water quality simulations gave good accounts of the recorded pH histories on the Allegheny River at Freeport (river mile 31.9, above the confluence of the Kiskiminetas River). The generally steady pH conditions observed were reproduced by simulation, as shown in figure 4. Simulation results at Freeport represent the combined effects of more than three hundred miles of streamflow routing and water quality simulation. In light of the great distances and drainage areas represented by simulation, results at Freeport are considered very good when compared with observed pH histories.

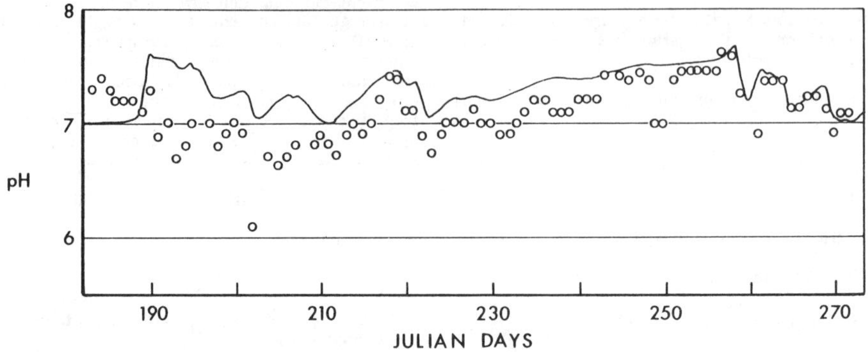

Figure 4. Observed and Simulated pH at Freeport Existing Condtions, 1977

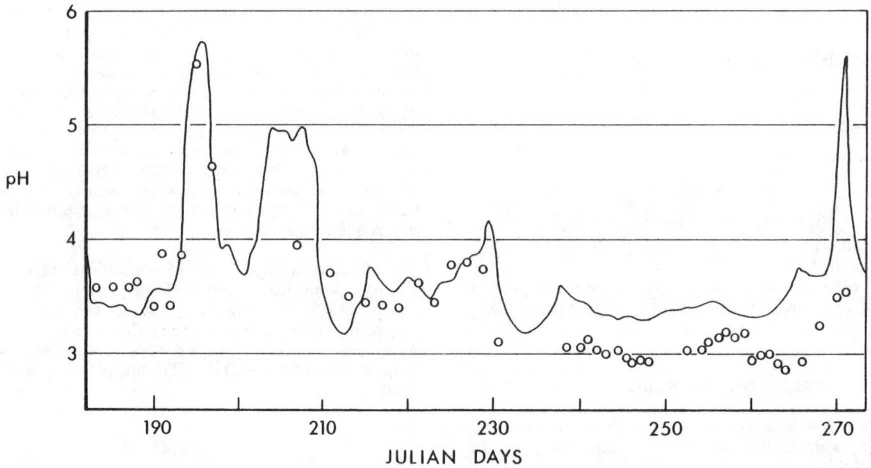

Figure 5. Observed and Simulated pH at Vandergrift Existing Conditions, 1977

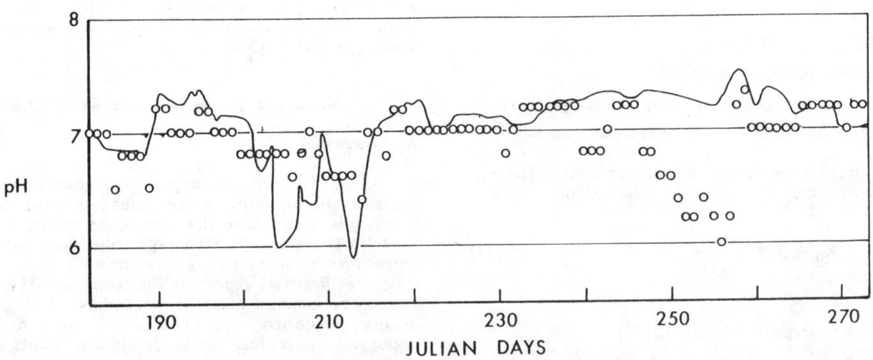

Figure 6. Observed and Simulated pH at Natrona Existing Conditions, 1977

At Vandergrift

The observed and simulated pH histories at Vandergrift in the Kiskiminetas River (river mile 10.3) are presented in figure 5. The low pH conditions observed in the Kiskiminetas were well reproduced in simulation. Sharp fluctuations in pH demonstrate the close association of flow and water quality in the basin.

Both observed and simulated pH histories include a generally lower pH condition over much of the latter half of the study period. This extended low pH period corresponds well with low flow conditions observed in the Kiskiminetas. During such low flow periods the contribution of AMD to total flow constitutes an increased fraction of the total flow, resulting in a more severe low-pH condition. The many small sources of AMD occurring along the Kiskiminetas River were not included in simulation of the basin. The complexity involved in determining the locations, magnitudes and frequencies of occurrence of such small AMD sources over the rather long study period precluded their simulation in the study. This is believed to account for the slight but consistent difference between observed and simulated pH over much of the latter half of the study period.

At Natrona

The simulated and observed pH histories at Natrona presented in figure 6 show generally good agreement. When compared with the observed and simulated pH histories at Freeport, the lower pH values resulting from the acidic Kiskiminetas inflow are evident.

Noticeable differences between the simulated and observed pH histories at this location are evident for the latter half of the study period. As noted previously these differences are attributable to the exclusion of the many small AMD sources along the Kiskiminetas.

Effects of Flow Regulation

No Corps Storage - At Vandergrift

Results of simulation of the Kiskiminetas River at Vandergrift for the "No Corps Storage" study case are presented in figure 7. Sharp peaks in the hypothetical pH history at Vandergrift characterize the response of the system to storms passing over the watershed. Increased runoff provides dilution water for the generally acid drainage occurring throughout the basin.

Comparison of results of simulation for the "No Corps Storage" case with those for the "Existing Conditions" case illustrate the effect of COE operations on water quality in the Kiskiminetas basin. Maximums in pH generally pass through the basin later in the "Existing Conditions" case than in the "No Corps Storage" case. This lag time reflects the COE storage of excessive runoff in Loyalhanna and Conemaugh Reservoirs for later release. The extended release of generally superior quality runoff water accounts for the peaks in pH being sustained for a longer time in the "Existing Conditions" case than in the "No Corps Storage" case.

No Corps Storage - At Natrona

The hypothetical pH history at Natrona under "No Corps Storage" conditions is presented in figure 8. The most notable feature of figure 8 is the sharp depression in pH to a value of 4.0, which occurred on July 19. Such a sharp depression in pH characterizes the so-called "first-flush" effect of a summer storm. During periods of no precipitation, extremely acid waters develop and collect in underground coal mines. The initial drainage from such mines when a sudden intense rain occurs is considerably more acidic than the subsequent drainage, and produces the most severe pollution of the storm event. The occurrence of the "first-flush" effect in areas contaminated by AMD supports the possibility that such a low pH event resulted from the storm event of July 19.

Pattern "A" - At Natrona

Under "Pattern A" conditions the simulated outflow at Kinzua Dam was reduced to 500 cfs over most of the study period, and the remainder of the system was simulated as in the "Existing Conditions" case. The hypothetical pH history at Natrona resulting from "Pattern A" conditions is presented in figure 9. Unlike the results of "No Corps Storage" simulations the minimum pH only reached 5.7, within the range of pH considered safe for most aquatic life.

Reduction of the outflow of Kinzua Dam to 500 cfs represented a nine-fold reduction over that corresponding to "Existing Conditions". By virtue of its large storage capacity Kinzua Dam provides most of the flow augmentation water for water quality enhancement in the lower Allegheny. That such a drastic reduction in outflow did not result in a significant deterioration in water quality suggests a potential for reserving more reservoir storage for in-project uses during the summer months.

CONCLUSIONS

Hydrologic and water quality conditions in the Allegheny River system were simulated utilizing the COE computer program WQRRSQ. Simulation for operation as historically documented ("Existing Conditions") reliably reproduced water quality conditions actually observed. Simulation of the same system in the absence of COE operations ("No Corps Storage") represented a "without project" scenario, to evaluate the effects of flow regulation. In the absence of regulation a potentially lethal low pH condition (pH=4.0) was predicted in response to an increased concentration of acid mine drainage in one of the tributaries in the lower basin, the Kiskiminetas River. The potential occurrence of such an event is supported by the commonly observed "first-flush" effect of summer storms on an acid mine drainage polluted basin.

In a third study case, the source of low flow augmentation water for water quality enhancement was restricted, with no significant deterioration in water quality resulting. The indication by the model that no major changes in water quality would result from conservation of low flow augmentation water indicates a potential for reserving more reservoir storage for in-project utilization than is currently practiced.

COE reservoirs in the Allegheny River basin are multiple use facilities for which optimum operating schedules should balance in-project benefits against those derived from mitigation of the effects of acid mine drainage. Development of system-wide schedules for such a complex system as that of the Allegheny Basin is a challenge for future research. The modeling approach utilized in this investigation should figure prominently in future efforts to achieve optimal operation for water quality control.

ACKNOWLEDGEMENTS

The authors are grateful to Mr. Jerry Willey of the Hydrologic Engineering Center, Army Corps of Engineers,

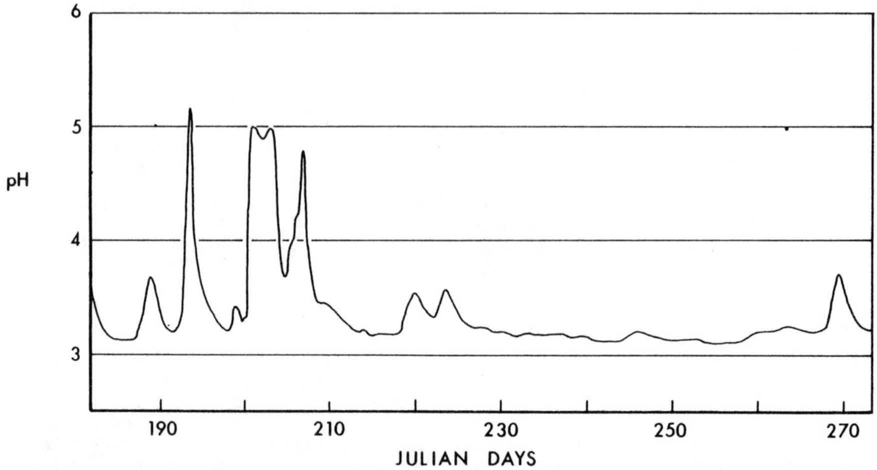

Figure 7. pH at Vandergrift, No Corps Storage, 1977

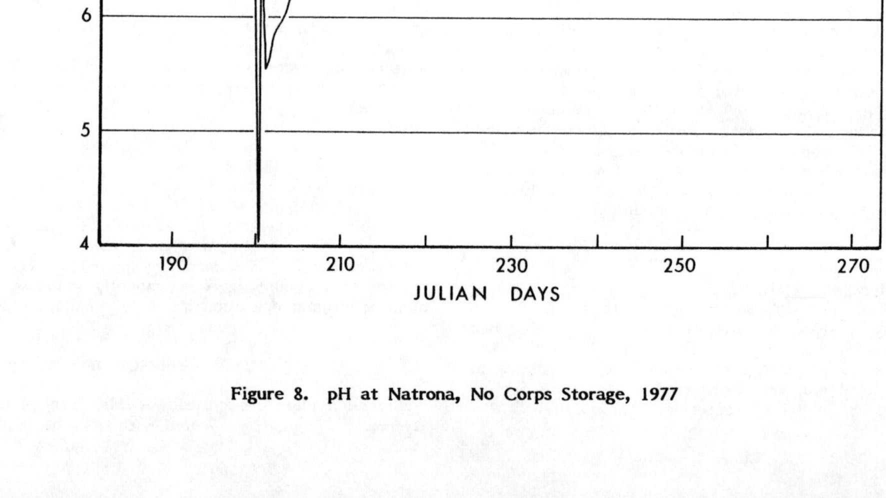

Figure 8. pH at Natrona, No Corps Storage, 1977

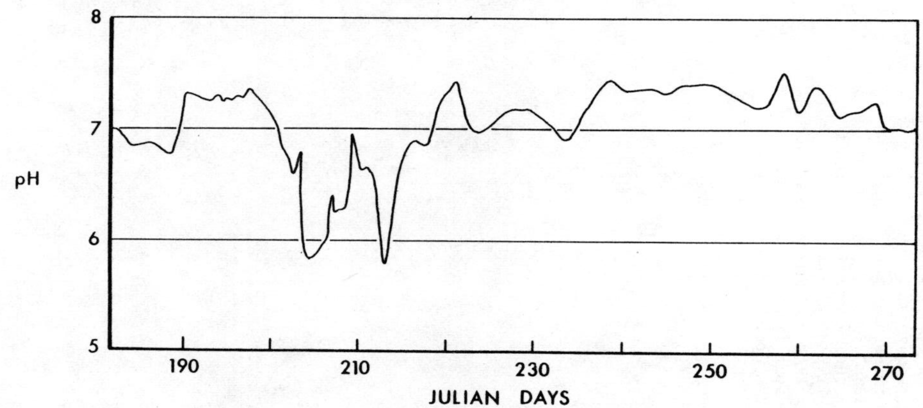

Figure 9. pH at Natrona, Pattern A, 1977

and Mr. Mike Koryak of the Pittsburgh District, Army Corps of Engineers, for their cooperation and assistance in this study.

LITERATURE CITED

Barnes, H.L. and Romberger, S.B. 1968. "Chemical Aspects of Acid Mine Drainage". Journal of the American Water Pollution Control Federation, Volume 40, #3, Part I, March 1968.

Chadderton, R.A. 1979. "A Simplified Stream Model of Acid Mine Drainage Effects". Water Resources Bulletin, American Water Resources Association, Vol. 13, No. 4, August 1979.

Hadley, P.W. and Orlob, G.T. 1981. "Allegheny River Basin Water Quality Analysis." Report to the U.S. Army Corps of Engineers, July, 1981, 42 p.

HEC. 1974. "WQRRS Generalized Computer Program for River--Reservoir Systems". U.S. Army Corps of Engineers, Hydrologic Engineering Center, Davis, California, 401-100, 100A, 210 p.

Yeasted, J.G. and Shane, R. 1976. "pH Profiles in a River System With Multiple Acid Loads". Journal of the American Water Pollution Control Federation, Volume 48, No. 1, January 1976.

NET ENERGY ANALYSIS OF GEOPRESSURED GAS RESOURCES IN THE GULF COAST REGION[1]

Cutler J. Cleveland and Robert Costanza[2]

Abstract--The Department of Energy has sponsored a test drilling program to evaluate the feasiblity of exploiting the geopressured/ geothermal energy resources in the Gulf Coast region as a means of supplementing dwindling domestic supplies of natural gas. Data from the test wells were incorporated into a model that compared potential methane production to the energy required to extract and process. Results indicate that geopressured wells are a net producer of fuel energy under the most optimistic assumptions about indirect labor and environmental costs. Both the energy return on energy investment ratio (EROI) and the total net energy production per geopressured well is well below that of conventional natural gas.

INTRODUCTION

Domestic reserves of liquid and gaseous hydrocarbons peaked in the early 1970's despite unprecedented levels of exploratory and development drilling since that time. Natural gas reserves peaked in 1967 at 293 trillion cubic feet (tcf) and consumption has exceeded new additions to reserves by an average annual rate of 10 tcf. The National Petroleum Council (1980) and the National Academy of Sciences (1976) have identified four types of "unconventional" gas resources whose exploitation in the future might supplement conventional domestic supplies: methane in "tight" sandstone formations, methane in Devonian shale formations, methane trapped in coal seams, and methane dissolved in geopressured aquifers. The last of these, the geopressured gas resources, occur almost exclusively in the northern Gulf Coast region of Texas and Louisiana (figure 1). The Department of Energy is sponsoring an on-going test drilling program at several of the more promising sites in order to gain more detailed information the characteristics and magnitude of geopressured gas resources.

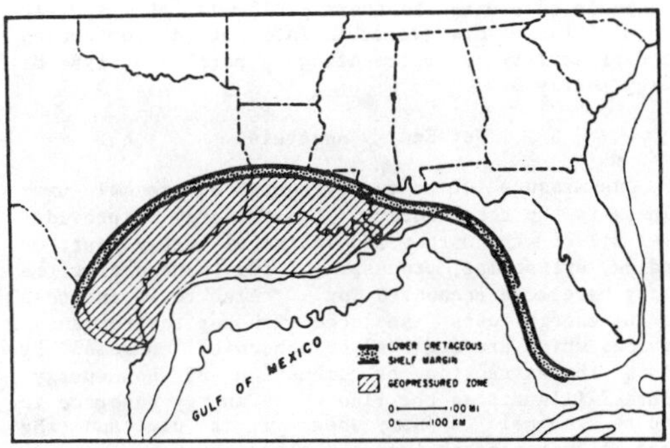

Fig. 1 Geopressured zone along the Texas and Louisiana Gulf Coast. (From Bebout and Gutierrez 1981)

The potential economic availability of gas from unconventional sources and, in general, of fuel energy from any alternative source is most often evaluated in terms of how high the price of conventional sources must go before the alternative source will be economically feasible to exploit. For instance, Kuuskraa et al (1978) estimated that if the price of natural gas were to rise to $3.00 per thousand cubic feet, 200 to 220 tcf of gas could be recovered from unconventional sources, an amount slightly larger than total proven reserves of conventional natural gas in 1982. The real price of domestic oil has increased 8 fold in the past 8 years, while natural gas prices have increased 5 fold, placing oil and gas prices at several times the supposedly "critical" level needed to stimulate development of alternative sources as predicted by many economic analyses. Investment in alternative sources by the private sector has not, however , kept pace with the predicted economic incentives despite, in some instances, substantial federal government subsidies. The question arises as to why standard economic predictions have failed in this regard, and whether an alternative measure of economic viability

----

(1) Paper presented at the Third International Conference on the State-of-the-Art of Ecological Modeling, sponsored by the International Society for Ecological Modelling. (Fort Collins, CO, May 24-28, 1982). Research funded by Mining and Minerals Research Institute, Louisiana State University, Baton Rouge, LA 70803.

(2) Graduate Research Assistant and Assistant Professor, respectively, Marine Sciences Department, Center for Wetland Resources, Louisiana State University, Baton Rouge, LA 70803.

is available that would yield more accurate predictions. We believe, for reasons pointed out below, that energy analysis is an appropriate and desirable supplemental approach to the evaluation of energy resources (and economic activity in general) that is complimentary to standard economic analysis.

Energy is valuable in an economic sense for its heat or work applications. In the U.S. and other industrialized nations, high rates of economic growth have occured simultaneously with high rates of fossil fuel energy consumption. Economic phenomona such as inflation, investment and growth have been sensitive to changes in the quality and availability of fossil fuels, where quality of the energy resource is measured by its energy return on energy investment ratio (EROI). This ratio measures the gross production of an energy supply system relative to the total direct and indirect energy costs consumed by the process. Energy analysis provides a framework for calculating trends in the EROI of our energy resources, and for predicting what impacts those changes will have on economic activity. For instance, Hall and Cleveland (1981) found that the EROI for domestically produced petroleum has decreased from about 20 to 1 in the 1940's to about 8 to 1 in the late 1970's.

Domestic resources of oil and gas will be substantially depleted by the end of this century (Hubbert 1974, Hall and Cleveland 1981). A variety of alternative energy sources have been proposed as replacements for our current sources, with large differences of opinions as to the magnitude and potential availability of these resources. Evaluating alternative fuels according to their EROI will give society and decision makers valuable information concerning the desirability of implementing a particular type of energy policy.

Net Energy Analysis

One measure of an energy source's economic usefulness is the total amount of net energy it provides after all of the direct and indirect energy costs of finding, extracting, processing, and transporting the energy have been accounted for. Also to be included are the energy costs asociated with any environmental services which are degraded or otherwise "consumed" by either the extraction or combustion of the energy. Figure 2 illustrates the flow of an energy resource in situ to the main economy where it is used and the types of energy costs incurred along the way.

Direct fuel costs are the most obvious and readily calculated energy cost components. They are the energies consumed directly in the form of fuel in the energy supply process. Examples of direct fuel costs are the diesel fuel consumed by a continous mining machine in an underground coal mine or the natural gas burned to drive a pump on an oil or gas well.

Indirect energy costs are more difficult to identify and assess, but often are more significant than direct fuel costs. There are four types of indirect energy costs to be considered. They are 1) the energy embodied in the capital structures and other industrial goods and services used in the energy suppply process, 2) the energy equivalent of the labor, 3) government, and 4) environmental services required by the process. The energy embodied in industrial goods and services is calculated by converting their dollar costs by an energy intensity factor (BTU/$) derived for the particular sector of the economy that produces that good or service. (See Herendeen 1974 for a thorough discussion of this method). The energy equivalent of labor is equal to some portion of the laborer's wage which is used to purchase food, shelter, clothing and other goods and services that keep a laborer alive and functioning in his or her economic role. The question of what portion of a laborer's wage should be included, however, is one that has yet to be resolved among energy analysts (cf. Costanza 1980).

Incorporating environmental costs into a cost/benefit framework has long been a problem in economics. When considered at all, environmental costs are usually approximated by some type of willingness-to-pay or replacement cost formula. Costanza and Neill (1981) have proposed an alternative framework by which the energy costs of environmental services can be calculated. Insufficient data on energy flows in ecological systems, however, has prevented embodied energy calculations for these services to be estimated with much precision. For these reasons, both energy and labor costs have been excluded from this analysis. We will discuss the possible impacts of this omission on the results of our model further on.

The Geopressured/Geothermal Resource Base

The term geopressured reservoir or aquifer refers to fluids trapped in sedimentary strata (usually sandstone or shale) under pressures greater than the normal vertical pressure gradient of 0.465 psi/foot. The most intensively analyzed area for the development of GP/GT energy has been the Northern Gulf of Mexico basin where, for the past 50 million years, the rivers of the Gulf coast have deposited interwoven layers of sand, clay and shale that eventually formed the reservoirs for todays geopressured zones. Concurrent with the increasing compaction of the sedimentary strata was an increase in pressure which tended to force the seawater out from between the pore spaces of the sand and clay particles. Normal fluid pressure exists where the pore fluids are free to drain in response to increasing pressure and burial, conditions which allow for the disssipation of pressure (Dickinson 1979). In some instances, however, post depositional activities such as salt tectonics and growth faults have effectively prevented fluid expulsion and thereby allowed the build up of pressure to occur (Wallace 1979). Thus, within the massive sandstone facies, pockets of geopressured fluid were formed at depths ranging from 6,000 to 22,500 feet. These pockets also tended to trap the natural heat flow from the strata below them resulting in large temperature increases.

With physical characteristics such as these, geopressured reservoirs have the potential to yield three types of energy: thermal, mechanical, and the energy embodied in the methane which is dissolved in the geopressured fluids themselves. Collectively, the three types of energy resources are termed geopressured/geothermal energies. Current information indicates that reservoir temperatures are too low to warrant the generation of electricity directly, so research and development has focused on the methane portion of the resource base.

The drilling of over 300,000 oil and gas wells in the greater Gulf coast region has provided a comprehensive data base on the geologic and geophysical setting of the region, allowing estimates of the magnitude of the GP/GT resource base to be made (table 1). Despite this experience, these estimates exhibit a rather large range due to uncertainty concerning specific characteristics of the resource like exact porosity, permeability, salinity,temperature, and reservoir volume and continuity. None of these

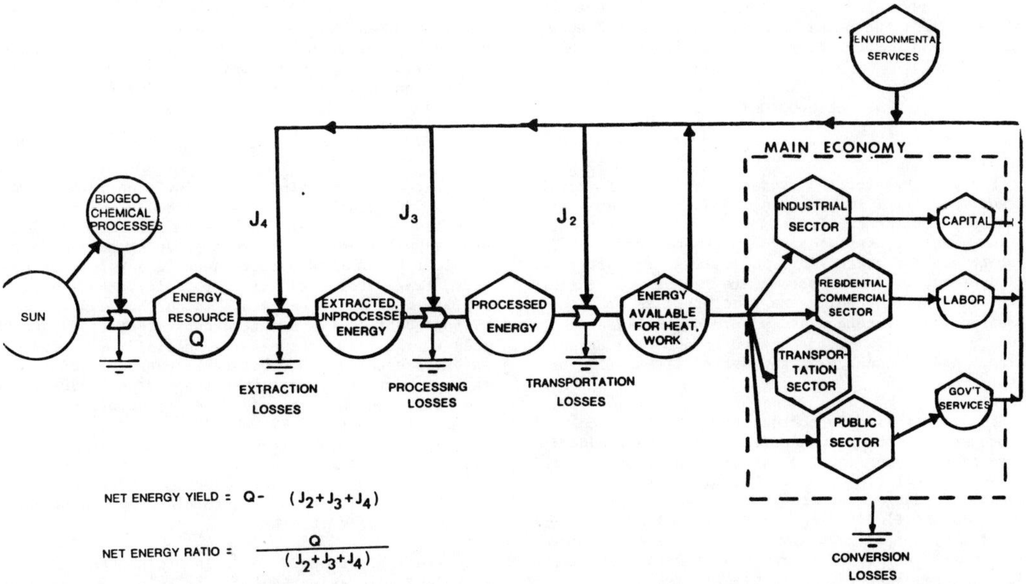

Figure 2.--A typical energy supply system that makes fuel energy available for use in the rest of the economy, and the types of energy costs incurred by that system.

Table 1.--Estimates of total in situ and recoverable geopressured and geothermal energy in the Northern Gulf Coast region (Louisiana and Texas Gulf Coast, Onshore and Offshore, in Quadrillion BTU). Modified from Johnson et al, (1980).

	GROSS ENERGY IN SITU		RECOVERABLE ENERGY		
Reference	Thermal energy	Methane energy	Thermal	Methane	electricity (MW centuries)
Brown (1976)	----	60,000-105,000	----	2,400-52,000	----
Dorfman (1974) (Texas only)	----	----	----	----	20,000
Dorfman (1977)	----	5,735	----	256	----
Elmer (1977)	----	----	529-2,058	2,000	----
Gould (1977) (onshore Texas only)	----	----	----	150	33,000
Hawkins (1977)	19.5	13.6	19.5	13.6	----
Hise (1976)	----	3,000	----	125	----
House (1975)	----	221.7	----	222	5,990
Jones (1976)					
(sandstone only)	----	100,000	----	1,146	----
(sand and shale)	----	49,000			
Karkalits (1977)	60,000	12,000-30,000	----	----	----
Myers (1974)	----	0.1-2,650	----	.0001-1,000	0.1-100,000
Papadopulos (1975) USGS (onshore only)	43,331	23,927	1,424	768	38,140
Trehan (1976)	----	----	----	----	100,000
Schnadelbach (1977)	----	111,500			
U.S. ERDA (1975)	----	3,000-40,000	----	500-2,000	33,400-110,000
U.S. ERDA (1977)	44,000-176,000	----	----	150-2,000	
Wilson (1974)	----	----	----	45.6	8,000

estimates, however, consider the energetics of trying to exploit this dilute and remote (relative to conventional oil and gas) energy source. In the next section we outline the model used to calculate the energy inputs to and output from a prototype GP/GT energy system.

METHODOLOGY

Energy Inputs

Energy costs for the project fall into five main categories: site preparation, drilling, separation and disposal plant faciities , power costs for reinjection, and operation and maintenance costs. The first three costs are incurred during the start up period of the project, assumed to be one year in this analysis, during which no methane is produced. Operation and maintenace and power costs begin in the second year of the project when production actually starts, and continue over the entire life of the project.

The largest single component of the total cost of the project is the actual drilling of the production or source well, and the 2-3 reinjection wells required per source well to reinject the brine back in to a permeable formation after the methane has been separated from the fluid at the surface. Dollar costs for the drilling of a geopressured well were calculated from a model developed by Wrighton (1981) who found that drilling costs for conventional natural gas wells in coastal Louisiana are strongly influenced by the depth of the well and the year that the well is drilled. The cost versus depth and time relationship has an exponential form and is a modified version of Fischer's (1964) original cost versus depth function. The equation for natural gas wells drilled in coastal Louisiana from 1975 to 1979 is

$$Y = K(e^{\alpha x} - 1)e^{rt}$$

where,

Y = average cost of well, in dollars
K = base drilling cost, in dollars
α = statistical parameter decribing the curvature of the function
x = depth of the well in feet
t = time in years (1975=1, 1976=2, etc.)
r = annual rate of increase in cost of well

This model correlates well with observed costs for natural gas wells in southern Louisiana (r =.988). Dollar costs calculated with this model were increased by 25% to obtain a dollar cost for a geopressured well, due mainly to the larger tubing required for a geopressured well (5.5" versus 3.5" for conventional well). The National Petroleum Coucil (1980), Quitzau (1980), and Abdulrahman (1982) also used the 25% surcharge. Disposal wells were assumed to require smaller tubing and were not cost adjusted.

The conversion of dollar costs to energy costs is illustrated in table 2. Dollar costs for geopressured wells range between 4 and 6 million dollars depending on the depth of the well which ranges from 13,000 to 17,000 feet in southern Louisiana. Dollar costs for individual components of the total drilling cost were first converted to constant 1972 dollars and then into their energy equivalents using the appropriate energy intensity factor. The total energy cost of the well divided by its total constant dollar cost produces an energy intensity factor (BTU/$) for a natural gas well as a whole. This figure represents the total energy required, directly and indirectly, to produce an "average" dollars worth of a natural gas well in southern Louisiana. The energy intensity factor can be applied to the dollar cost calculated for a particular prospect drilled to a certain depth in a given year to produce the total energy cost of a geopressured well. The energy costs for site preparation and separation and disposal plant facilities were calculated in a similiar manner (see table 2a).

Energy Outputs

Methane production was calculated using a model developed by Johnson et al (1980) for the Energy Programs Office at LSU and later modified by Abdulrahman (1982). For each prospect being analyzed, physical parameters such as depth and volume of reservoir, temperature, salinity, porosity, and permeability are read into the model which then calculates the methane content (in standard cubic feet (scf) per barrel) of the brine produced. Initial fluid production rates were set at 40,000 barrels per day, and were cut back at certain points when the bottom hole pressure dropped below a prescribed critical minimum level. The initial value of 40,000 bbl/day was based on current petroleum and reservoir engineering practices related to fluid production rates in conventional wells. Production rates above 40,000 bbl/day may be possible, but may also lead to clogging and compaction of the reservoir itself.

Four prospects, one in Texas and three in Louisiana were selected for analysis: Pleasant Bayou well, located in Brazoria County, Texas; LaFourche Crossing, located in LaFourche and Terrebonne Parishes in Louisiana; Gladys McCall in Cameron Parish, Louisiana, and Bayou Hebert in Vermillion Parish. The four prospects selected are in various stages of development as part of D.O.E's test drilling program, so the degree of accuracy and completeness of the data on their physical parameters varies.

RESULTS

The results of the simulation model for each of the four prospects are depicted in table 3. By the end of the second year of production all four prospects had recovered the large energy costs incurred during the first year of the project (see figure 3a-d). Net energy production from two of the prospects, Pleasent Bayou and Gladys McCall, fell to zero twelve years after production began, giving them only a ten year lifetime as net producers of energy. Only LaFourche Crossing and Bayou Hebert remained net producers of energy for their entire lifetime. Net energy production declined over time for all the prospects, and rather precipitously in the case of Pleasasnt Bayou and Gladys McCall. The rapid decline in gas production is a direct result of the decline in brine production after the first few years due to the relatively small volumes of these two resevoirs and the cutbacks in flow rates necessary to maintain the critical pressure level. Total net energy production ranged from 0.7 trillion BTU in the case of Pleasant Bayou to about 6.0 trillion BTU in the case of LaFourche Crossing.

Values for EROI ranged from about 1.8:1 for Pleasant Bayou to about 5.5:1 for LaFourche Crossing. Due to the precipitous decline in their production of gas in the second ten years, the net yield and EROI for Pleasant Bayou and Glayds McCall were greater after after 10 years then after 20 yeas, because energy costs for operation and maintenance and reinjection exceeded methane production during the second ten years.

Table 2.--Breakdown of dollar and energy costs of drilling a conventional natural gas well in coastal Louisiana.(1)

Item(2)	% of Total(2) Drilling Costs	SIC Code(3)	Energy Intensity Factor(4,7) (BTU/1972$)	Cost(5) (10^3 1980$)	Price Deflator(6) (1972$/1980$)	Cost (10^3 1972$)	Total Energy (10^6 BTU)
TANGIBLES:							
conductor	0.3	3701	218,072	12.42	.56	7.00	1,526
surface casing	1.3	3701	218,072	53.82	.56	30.35	6,618
intermediate "	7.7	3701	218,072	318.78	.56	179.74	39,196
production "	12.8	3701	218,072	529.92	.56	298.78	65,156
liner "	1.6	3701	218,072	66.24	.56	37.35	8,144
tubing	5.1	3701	218,072	211.14	.56	119.05	26,059
x-mas tree	3.9	4208	64,454	161.46	.56	91.04	5,868
SUBTOTAL	32.7			1,353.78		763.30	152,569
INTANGIBLES:							
location & move	5.2	6503	39,945	215.28	.56	121.38	4,849
rig-140 days @ $7200/day	26.2	4503	52,932	1,084.68	.56	611.39	32,389
mud	10.4	3619	165,954	430.56	.56	242.76	40,826
logging & perforating	6.5	7301	22,551	269.10	.56	151.72	3,421
bits	1.3	4503	52,932	53.82	.56	30.34	1,605
rental equip.	1.3	5000	53,120	53.82	.56	30.34	1,611
fuel & water	3.2	6803	47,190/639,911	132.48	.56	74.70	41,525
trucking	1.5	6503	39,945	62.10	.56	35.01	1,398
coring	0.6	7301	22,551	24.84	.56	14.00	315
geologic & engineering	3.9	7301	22,551	161.46	.56	91.04	2,053
well supplies	0.7	4503	52,932	28.98	.56	16.37	866
completion services	2.6	7301	22,551	107.64	.56	60.69	1,368
cementing	3.9	3601	378,716	161.46	.56	91.04	34,478
SUBTOTAL	67.3			2,786.22			164,579
TOTAL	100.0			4,140.00			317,148

Energy Intensity for natural gas well = $\frac{317{,}148 * 10 \text{ BTU's}}{2{,}334.2 * 10 \text{ 1972\$}}$ = 135,870 BTU/1972$

(1)Based on drilling a conventional natural gas well to 15,000 feet in coastal Louisiana.
(2)Components of drilling costs from National Petroleum Council (1980).
(3)Based on Standard Industrial Classification manual, U.S. Bureau of Census, 1972 edition.
(4)Energy intensities for capital structures from Hannon et al, (1981).
(5)Dollar costs for drilling natural gas well calculated from Wrighton (1981).
(6)Implicit price deflators from Economic Report of the President (1982).
(7)Dollar and energy costs for fuel and water consumption were estimated to be 50% water services and and 50% primary energy sector. Energy intensity factor for primary energy sector from Costanza (1978).

Table 2a--Dollar and energy costs for geopressured well site development, separation and disposal plant facilities (1).

Item	SIC Code	Energy Intensity Factor (BTU/1972$)	Current Dollar Cost(1)	Implicit Price Deflator (1972$/current)	Cost (10^3 1972$)	Total Energy (10^6 BTU)
SITE PREPARATION (2)						
site preparation						
surveying	7303	9252	9.2	.76	6.9	64.4
grading	1105	76,665	13.8	.76	10.4	800.8
fencing	4205	135,500	22.5	.76	17.4	2307.7
roads	1104	121,535	227.5	.76	172.2	20982.9
water services	6803	47,190	7.0	.76	5.3	250.4
electric power						
power service	6801	47,190	10.0	.76	7.6	357.2
flood lights	5501	46,392	5.0	.76	3.8	175.6
shops, offices	1103	64,111	75.0	.76	56.8	3639.6
contingency	1103	64,111	55.0	.76	41.6	2669.1
TOTAL SITE PREPARATION:					321.7	31,193.3
SEPARATION AND DISPOSAL FACILITIES (3)						
1st stage separator	4006	90,783	168.3	.52	86.9	7892.7
2nd stage sep.	4006	90,783				
3rd stage sep.	4006	90,783				
disposal pumps	4901	50,714	238.5	.52	123.2	6248.0
collection pipe	4208	64,454	128.7	.52	66.5	4284.9
disposal pipe	4208	64,454	344.7	.52	178.1	11,472.3
TOTAL PLANT FACILITIES:			900.0		464.9	30,831.6
OPERATION AND MAINTENANCE COSTS (4)						
FIXED COST:						
labor	1202	61,516	100.0	.61	61.4	3778.9
Transportation	6503	39,945	40.0	.61	24.6	981.4
special tests	7301	22,551	50.0	.61	30.7	692.8
laboratory	7301	22,551	25.0	.61	15.4	346.4
supervision	1202	61,516	50.0	.61	30.7	1889.8
overhead (20%)	1202	61,516	55.0	.61	33.8	2078.6
TOTAL FIXED COSTS:			320.0		196.6	9767.9
VARIABLE COSTS:						
producing well repair	1202	61,516	0.614			
disposal well repair	1202	61,516	0.203			
pump main.	1202	61,516	0.307			
chemical	7301	22,551	0.184			
well treating	7301	22,551	0.184			
overhead(20%)	1202	61,516	0.289			
TOTAL VARIABLE COST:			1.780			

Energy intensity for fixed costs = 9768 * 10^6 BTU/196.6 * 10^3 1972$ = 49,684 BTU/1972$

Variable energy costs = 0.0178 1972$/bbl. * 49,684 BTU/1972$ = 884 BTU/bbl.

(1) Current dollar costs for site preparation, plant facilities, and maintenance are in thousands of 1976, 1981 and 1979 dollars, respectively.
(2) From Carlson and Underhill (1976). Costs were scaled down for sin gle well system.
(3) From Johnson et al., (1980) and Abdulhman (1982).
(4) From National Petroleum Council (1980). Fixed costs represent annual charges, variable costs are in 1972 cents per barrel reinjected.

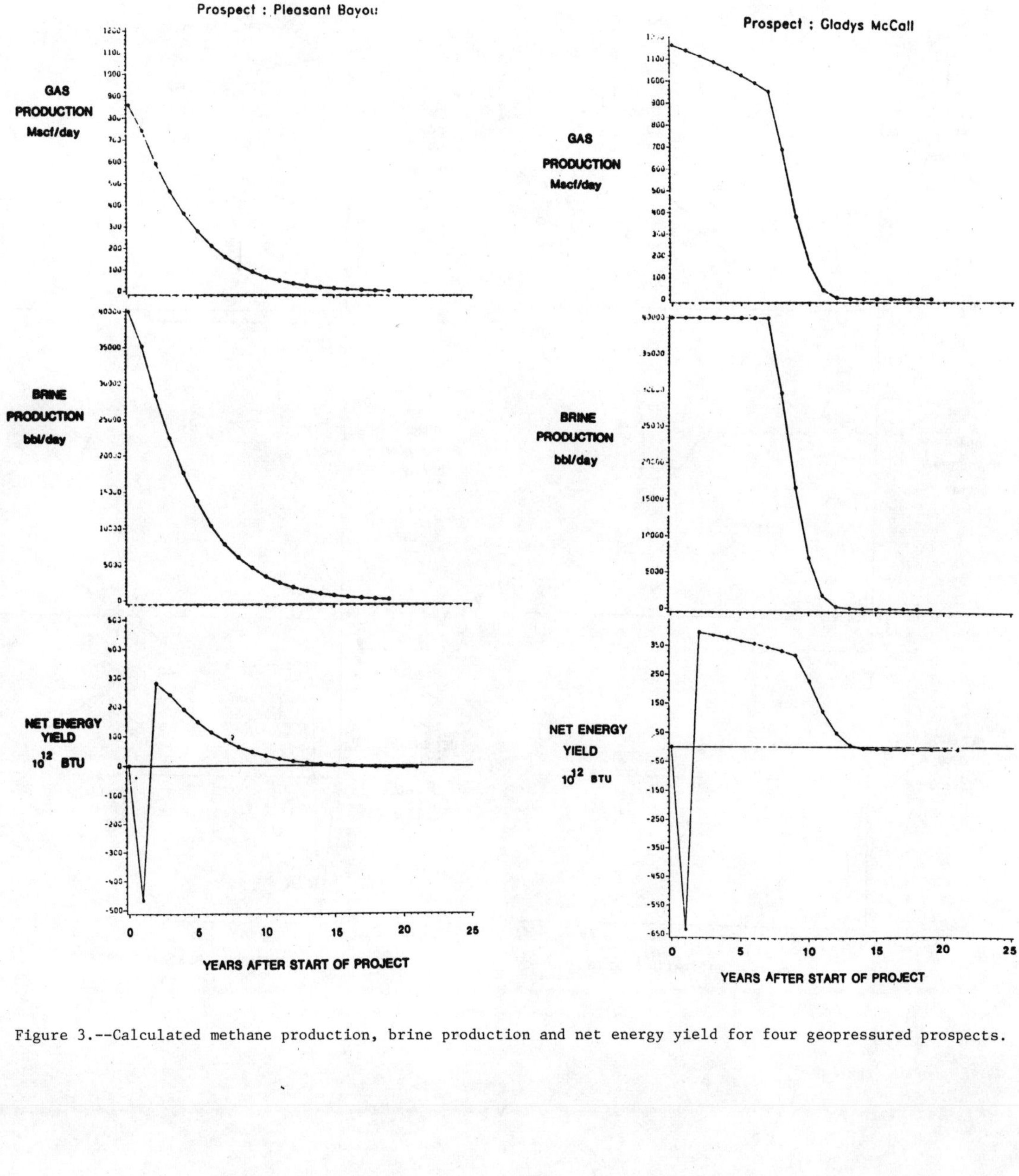

Figure 3.--Calculated methane production, brine production and net energy yield for four geopressured prospects.

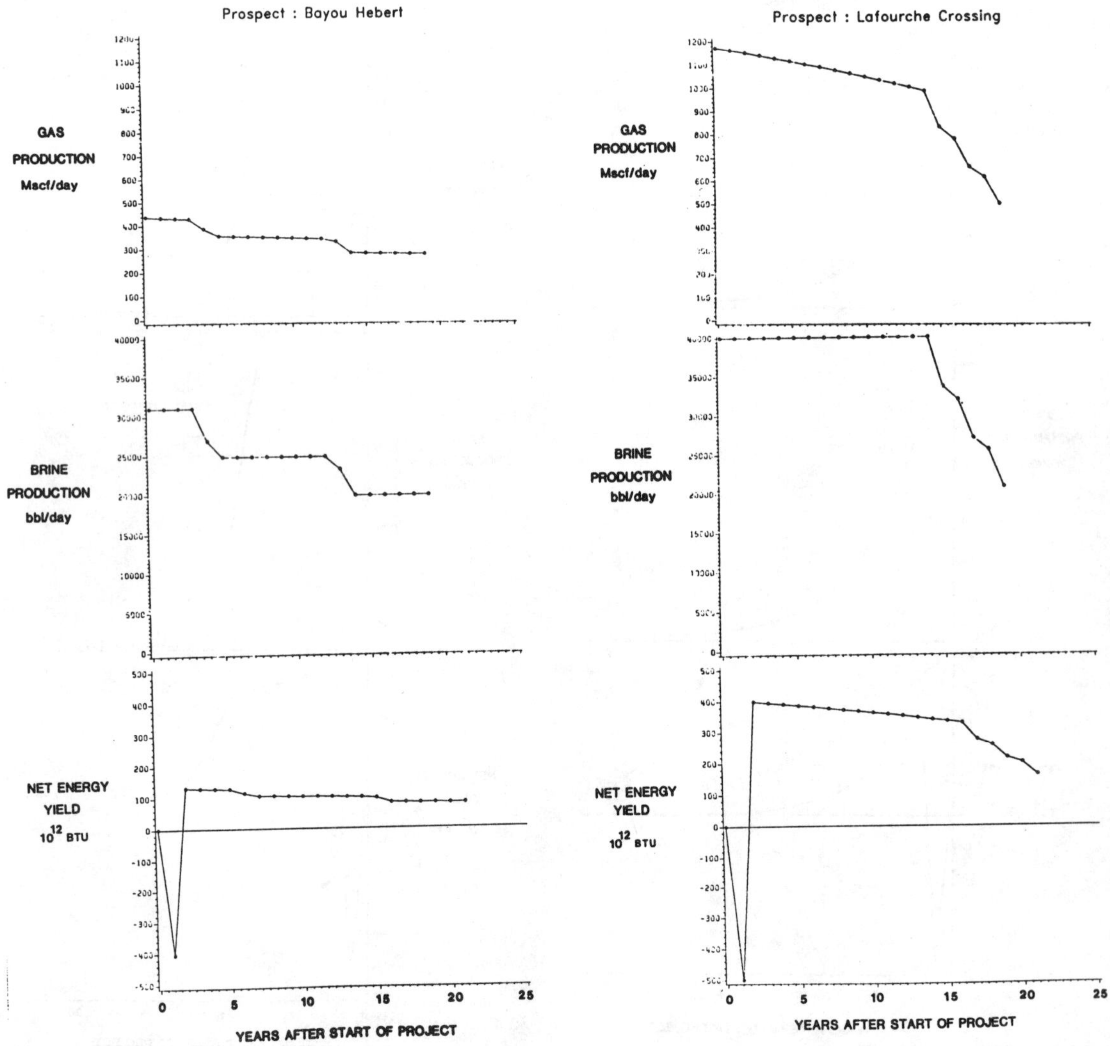

Figure 3. (continued)

Table 3.--Summary of reservoir production rates and energy inputs. All numbers are in trillions of BTU unless otherwise indicated. Totals may not add due to rounding.

Prospect	Avg. Fluid Prod. (bbl/day)	Avg. Gas Prod. (mscf/day)	Total Gas Prod.	Total Energy Cost	Net Energy Yield	EROI (ratio)
Bayou Hebert						
10 years	27478	386	1.44	0.74	0.69	1.90
20 years	24567	344	2.56	1.03	1.53	2.47
Pleasant Bayou						
10 years	18631	388	1.44	0.73	0.71	2.00
20 years	9832	206	1.54	0.84	0.70	1.84
Gladys McCall						
10 years	36627	960	3.57	1.05	2.51	3.30
20 years	18772	490	3.65	1.16	2.49	3.10
Lafourche Crossing						
10 years	40000	1050	4.56	0.95	3.61	4.80
20 years	36960	984	7.23	1.34	5.98	5.45

SUMMARY AND CONCLUSIONS

Results of this analysis indicate that wide scale investment in geopressured gas wells is not a desirable action at this time because even the most optimistic of the EROI ratios for geopressured gas at the wellhead are well below the 8:1 ratio for conventional natural gas at the wellhead reported by Hall and Cleveland (1981). It is not surprising, therefore, that there has not been widespread development of these resources by the private sector. Conventionally produced natural gas still yields a higher return in both energy and dollar terms.

The results of this analysis, which is based on preliminary test well data and pre-test well drilling information, predicts that geopressured gas does not offer the potential to significantly contribute to gas reserves. The diffuse, low quality nature of the resource in combination with the large investment of energy and other resources required to exploit it make geopressured gas a marginal, although positive net source of energy. Large numbers of geopressured wells would be necessary since the results of our model indicate that each well only produces between 0.7 and 5.5 billion scf of gas (net) over a twenty year period. This can be compared to production rates of 1.7 billion scf (net) per year for conventional deep gas wells in Louisiana as of mid 1981 (Carlson et al 1981). Wrighton (1981) cites median gross daily production rates of about 8600 mcf for 44 gas wells in the Tuscaloosa trend. The mean gross finding rate for the four projects analyzed here was about 500 mcf per day. Moreover, current information suggests that significant numbers of large, continuous fairways of geopressured reservoirs are not likely to exist.

It should be noted that the net energy yields and ratios calculated in this analysis are conservative in that certain potentially important energy costs were excluded. Only the energy costs of locating and pumping the gas to the surface were included. The energy costs of further processing and transporting the gas to its point of final consumption were not included. The energy costs of labor were excluded as well for reasons pointed out earlier. Direct labor costs are probably not a significant omission because the petroleum industry is much less labor intensive than other energy supply systems like underground coal mining. We estimated that if all the energy costs of indirect labor and government services were included, all but the highest EROI would drop below the energy break-even point.

The environmental costs associated with a wide scale geopressured gas program are not fully understood at this juncture, but potential impacts could persist for years after production has ceased. Models based on existing rock mechanism data and analogies with the effects of ground water and petroleum extraction suggest that the withdrawl of large volumes of fluids from geopressured reservoirs may cause land surface subsidence and stress release along growth faults (Groat and VanSickle 1981). Subsidence is of particular concern in southern Louisiana where loss of wetlands due to natural subsidence and human activity is already ocurring at a rapid rate.

Our analysis of geopressured gas resources illustrates how energy analysis can be useful in evaluating and ranking both existing and alternative fuel energy supply systems according to their most important economic parameter--the amount of net energy they deliver to society. Economic cost/benefit analyses of alternate energy sources are based on the calculation of an internal rate of return (IRR) for a project that can then be compared to existing or other proposed alternatives. Calculating the IRR, however, requires predictions of future economic conditions like inflation and relative price levels, a tenuous task at best even under stable economic conditions. Because energy analysis is based on physical rather than monetary flows, the results of an energy analysis do not become obsolete five or ten years from now if inflation, price levels or the discount rate do not behave as predicted. Alternative energy supply systems can be ranked according their respective EROI, and predictions can be made concerning when one type of energy will be replaced by another based on these ratios.

ACKNOWLEDGEMENTS

This is contribution # LSU-CEL-82-11 to the Coastal Ecology Laboratory, Center for Wetlands Resources, Louisiana State University, Baton Rouge, LA 70803. This research was made possible by a grant from the Mining and Minerals Research Institute, Louisiana State University. The authors would like to thank Drs. Adrian Johnson and Fred Wrighton for their assistance.

REFERENCES

Abdulrahmann, Adnan, Monte Carlo Simulation of Geopressured Energy Resource, M.S. Thesis, Louisiana State University, May, 1982.

Bebout, D.G., and D.R. Gutierrez, Geopressured geothermal resources in Texas and Louisiana--geological constraints, in D.G. Bebout and Ann L. Bachman, eds., Proceedings of fifth geopressured-geothermal energy conference (Baton Rouge, LA) 1981.

Brown, William M.,100,000 Quads of Natural Gas?, Hudson Institute Research Memorandum #31, July,

1976

Carlson, Ronald A., and Gary K. Underhill, Net Energetics Analysis, in Gary K. Underhill et al., Surface Technology and Resource Utilization, vol. 4, Second geopressured and geothermal energy conference, University of Texas at Austin, 1976.

Carlson, M.,N. Ody et al., The Natural Gas Market Under the Natural Gas Policy Act: DOE/EIA/AR-0289 (Washington:Dept. of Energy) 1981.

Costanza, Robert, Energy Costs of Goods and Services in 1967 Including Solar Energy Inputs, and Labor and Government Service Feedbacks. (ERG Document 262, Center for Advanced Computation, University of Illinois at Urbana-Champaign) 1978.

Costanza, Robert, Embodied Energy and Ecnomic Evaluation, Science 210:12 19-1224 (December, 1980).

Costanza, Robert and Christopher Neill, The Energy Embodied in the Products of the Biosphere, in, W.J. Mitsch, R.W.Bosserman, and J.M. Klopatek, eds., Energy and Ecological Modelling (Elsevier:N.Y.) 1981. International Symposium on Energy and Ecological Modeling, Louisville, KY, April, 1981.

Dickinson, G. Gelogical Aspects of Abnormal Reservoir Pressures in Gulf Coast Louisiana, Am.Assoc.Pet.Geo.Bull., 37:410-432 (1953). loc 480, Myron, and R.O. Kehle, Potential Geothermal Resources of Texas, Texas University Bureau of Economic Analysis Circular 74-4, July, 1974.

Dorfman, Myron, Natural gas from geopressured zones in the U.S. Gulf Coast Basin, Gas Supply Review, 5:6 (March 1977)

Elmer, Donald B., Geothermal energy and energy conservation: policy plannning objectives, Geothermal Energy, 5:5 (May 1977)

Elmer, Donald B. Direct heat applications of geothermal energy resources and the second law of thermodynamics, Transactions of the Geothermal Resources Council, vol. 1 (May, 1977).

Gould, S.P., and A. Spivak, Production economics of a geopressured geothermal prospect, American Institute of Chemical Engineers, paper 27E, (March 1977)

Groat, C.G., and V.R. VanSickle, An Overview of Environmental Factors in Geopressured/Geothermal Energy Development. in D.G. Bebout and Ann L. Bachman, eds., Proceedings of fifth geopressured/ geothermal energy conference (Louisiana State University, Baton Rouge) October, 1981.

Hall, Charles A.S., and Cutler J. Cleveland, Petroleum Drilling and Production in the United States: Yield Per Effort and Net Energy Analysis. Science, 211:576-579 (February 1981).

Hannon, Bruce, Robert Herendeen, and T. Blazeck, Energy and Labor Intensities for 1972. (ERG Document 307, University of Illinois, Chaimpaign-Urbans) April, 1981.

Hawkins, Murray F., Jr., Final report--investigations of the geopressured energy resource of southern Louisiana, Louisiana State University Department of Petroleum Engineering, ERDA Contract #EY-76-S-05-4889, April, 1977.

Herendeen, Robert H., Use of Input/Output Analysis to Determine the Energy Cost of Goods and Services,in Michael S. Macrakis, ed., Energy Demand, Conservation and Institutional Problems, (Cambridge:MIT Press) 1974.

Hise, Bill R., Natural gas from geopressured aquifers, in Board on Mineral Resources, ed., Natural Gas from unconventional geologic sources, National Academy of Sciences, Washington, D.C., 1976.

House, P.A., P.M. Johnson and D.F. Towse, Potential power generation andand gas production from Gulf Coast geopressured aquifers, Proceedings of First Geopressured Geothermal Conference, University of Texas at Austin, 1975.

House, P.A., P.M. Johnson and D.F. Towse, Potential power generation and gas production from Gulf Coast geopressured aquifers, Lawerence Livermore Report UCRL-51813, 1975.

Hubbert, M. King, U.S. Energy Resources: A Review as of 1972. Senate Com m. Interior Insular Affairs Ser. No. 93-40 (92-75) (1974).

Johnson, Adrian E., et al., Operations Research and Systems Analysis of geopresssured/geothermal energy in Louisana, Final Report, June 1, 1978 - August 31, 1979, DOE/ET/27085-1.

Jones, Paul H., Natural gas resources of the geopressured zones in the northern Gulf of Mexico basin, in Board on Mineral Reources, eds., Natural gas from unconventional geologic sources, National Academy of Sciences, Washington, D.C., 1976.

Karkilits, O. Carroll in Underground Louisiana, by Jim Polster, New Orleans Magazine, September 1977.

Kuuskraa, V.A., et al., Enhanced Recovery of Unconventional Gas, vols. 1 and 2, (Springfield, VA" National Technical Information Service, 1978).

Myers B., R. Nelson, J. Howard, and R. Austin, Some elements of the northern Gulf of Mexico basin geopressure energy resource, presented to the AAPG, (April 2, 1974) URCL-74807, Rev. 1, San Antonio, TX.

National Petroleum Council, Unconventional Gas Sources, vols. 1 and 480 (Washington. D.C., 1980) Papadopulos, S.S.,R.H. Wallace Jr., J.B. Wesselman, and R.E.Taylor, Assessment of Onshore geopresured geothermal energy resources in the northern Gulf of Mexico basin, in D.E. White and D.L. Williams, eds., U.S. Geologic survey 726, 1975.

Schnadelbach, T.W., Jr., Inventory and case studies of Louisiana, non-electric applications of geopressured geothermal resources, Quarterly Progress report of de Laurel Engineers, New Orleans, LA.

Trehan, Ranvir K., John G. Leigh, Felicia Williams, and Susan Pond, Analysis of geothermal energy development scenarios, MITRE Technical Report , MTR-7220, November 1976.

U.S. Energy Research and Development Administration, Division of Geothermal Energy, Definition Report,

ERDA-86, October 1975.

U.S. Energy Research and Development Administration, Market Oriented Pro gram PLanning Study (MOPPS), May 1977.

U.S. Geological Survey, Assessment of Geothermal Resources of the United States--1978, L.J.P. Muffler, ed., Circular 790, 1979, (Arlington VA.)

Wilkins, Bert, Jr., and Michael O. Waguespack, Operations research and systems analysis of geopressured geothermal reources in Louisiana (Baton Rouge, LA, Office of Advanced Studies and Research, Louisiana State University, February 1978).

Wilson, John S., B.P. Shepherd, and S. Kaufman, Economic analysis of the use of Texas geopressured geothermal energy resources for the production of electrical power, Proceedings of first geopressured geothermal energy conference, University of Texas at Austin, 1975.

Wilson, John S., B.P. Shepherd, and S. Kaufman, An analysis of the Poten tial use of geothermal energy for power generation along the Texas Gulf coast, Report by Dow Chemical USA for the Governor's Advisory Council, State of Texas, 1974.

Wrighton, Fred, Drilling Costs by Depth: The Past and the Future in South Louisiana. Unpublished manuscript, Energy Programs Office, Louisiana State University, Baton Rouge, 1981.

Wrighton, Fred, An Economic Overview of Geopressured Solution Gas, in, Don G. Bebout and Ann L. Bachman, eds., Proceedings of fifth conference on geopressured/geothermal energy (Louisiana State University, Baton Rouge) October 1981.

MEROPLANKTON ENTRAINMENT MODELING IN A COASTAL BAY[1]

D. P. Galya and T. J. Horst[2]

ABSTRACT

This study assessed the effects of entrainment due to the operation of a power plant located on a coastal bay by determining the area of the bay subject to entrainment influence. The assessment of entrainment effects consisted principally of a modeling study using CAFE-1 and DISPER-1, a pair of two dimensional, finite element models. CAFE-1, a hydrodynamic model, was used to determine water elevations and vertically averaged velocities within the bay. DISPER-1, a dispersion model, was then applied in order to determine the effects of the power plant on meroplankton concentrations. Several cases were simulated representing various spatial patterns of meroplankton concentration. The analyses indicated that, in each case, the area of power plant entrainment influence is small compared to the area of the water body.

INTRODUCTION

The present study assesses the effects of entrainment due to the operation of a power plant with a once-through cooling system, located on a coastal bay. The assessment of entrainment effects consisted principally of a modeling effort aimed at determining the area of the bay subject to entrainment influence by the plant. The area of influence is the area surrounding the water intake in which reduced concentrations of organisms are predicted. The modeling study employed CAFE-1 and DISPER-1, a pair of two-dimensional, time-dependent models.

THE MODELS

The models used in this study, CAFE-1 (Wang and Conner 1975) and DISPER-1 (Liemkuhler et al. 1975) are finite element models which have been successfully applied in several dispersion studies (Liemkuhler et al. 1975; Mangarella 1979; and Galya and Colangelo 1981).

CAFE-1 is a hydrodynamic model that uses input parameters such as study area geometry, bottom roughness, and boundary tidal fluctuations in solving the vertically integrated equations of conservation of mass and momentum. The output from CAFE-1 consists of tidal heights and vertically integrated velocities within the region.

DISPER-1 is a dispersion model, used in the present application to simulate the effects on meroplankton concentration due to entrainment by a power plant. This application required a modification to the model to properly simulate the plant as a concentration dependent point of plankton mortality. Input parameters to DISPER-1 include study area geometry, dispersion coefficient, plant intake flow, initial and boundary conditions of meroplankton, and temporally varying values of water velocity and depth. The model employs these parameters in solving the vertically integrated form of the convective-diffusion equation.

Both CAFE-1 and DISPER-1 utilize the finite element method to solve the basic equations mentioned above. The finite element method provides a high degree of solution accuracy while permitting great flexibility in the expression of geometry, boundary conditions, and input parameters. In employing the finite element method, the study area is subdivided into an array of discrete elements. The finite element array developed for a study region is determined primarily by the geometry of the area and the objectives of the study. In particular, geometric constraints dictate that smaller elements must be used in narrower regions than in wide open areas. The objective of this study is to examine the dispersion of entrained organisms toward a power plant. Therefore, resolution of the grid should be greater in the vicinity of the plant than at locations where concentrations and gradients are small. The finite element array developed for use in the present study is shown in figure 1.

[1] Paper presented at the Third International Conference State-of-the-Art in Ecological Modelling, Colorado State University, Ft. Collins, Colorado, May 24-28, 1982.

[2] Donald P. Galya is an Environmental Scientist, Stone & Webster Engineering Corporation, Boston Massachusetts U.S.A.
Thomas J. Horst is a Consultant, Stone & Webster Engineering Corporation, Boston, Massachusetts U.S.A.

MODEL APPLICATIONS

CAFE-1

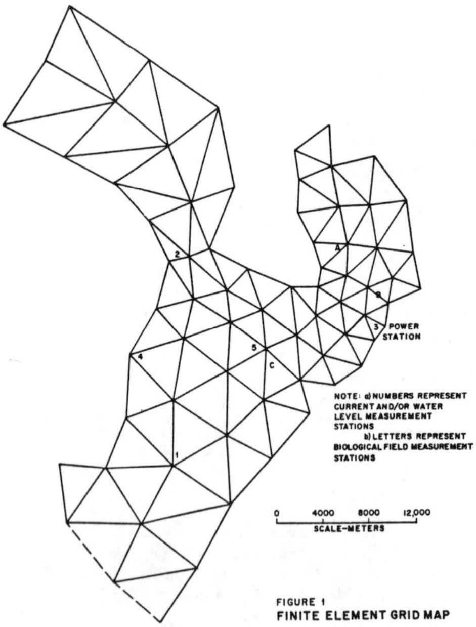

FIGURE 1
FINITE ELEMENT GRID MAP

Two types of boundary conditions, representing a fixed land boundary and an open water boundary, were employed in this application of CAFE-1. For the fixed land boundaries, an impermeable land condition was assumed. At the open water boundary, tide data were prescribed in terms of tabular values of water elevation and time.

Depths for the grid system were obtained from National Ocean Survey nautical charts. The depth at each node was determined from chart values in the vicinity of the node. The time step, eddy viscosity value, and bottom friction coefficient values for initial simulations were determined from guidelines given by Wang and Connor (1975).

Calibration of CAFE-1 was accomplished by varying the bottom friction coeffient to obtain model results which compared favorably with two sets of field data. In performing this task, tidal, wind, and river inflow conditions existing during the surveys were simulated in the model input.

The comparison between the field and model results for both surveys is presented on table 1. The comparison for the first set of data is excellent. The comparison for the second set is less satisfactory due to large freshwater inflows during and immediately preceding the second field survey. This resulted in conditions for field measurements which were not directly comparable to results from a vertically integrated model. The final values of the bottom friction coefficient vary from 0.006 in the deep channel areas to 0.025 in the shallow areas of the bay.

Subsequent to the calibration of CAFE-1, the model was used to simulate the hydrodynamics of the bay under average tidal conditions. Input parameters for this simulation are presented on table 2. The resulting current velocities predicted for selected times during the tidal cycle are shown in figures 2 through 5. The results of this model were then used as input to DISPER-1.

Table 1-Comparison of measured and predicted hydrodynamics

Survey	Station Location[1]	Max Current Vel (M/Sec) Measured	Predicted	Tidal Amplitude (M) Measured	Predicted	Tidal Phase (hr) Time Between High and Low Water Measured	Predicted
1	1	0.84	0.87	-	0.84	-	8.0
	2	0.80	0.79	-	0.99	-	8.0
	3	0.36	0.39	0.92	0.97	7.8	8.0
	4	-	0.36	0.85	0.77	7.7	8.0
2	3	0.34	0.19	0.84	0.70	8.1	8.0
	4	-	0.25	0.72	0.53	8.1	8.0
	5	0.53	0.33	-	0.50	-	8.0

[1]See figure 1

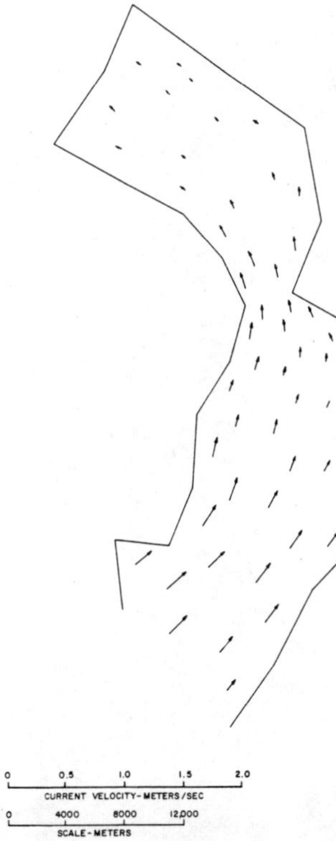

FIGURE 2
CURRENT VELOCITIES-
BEGINNING OF FLOOD

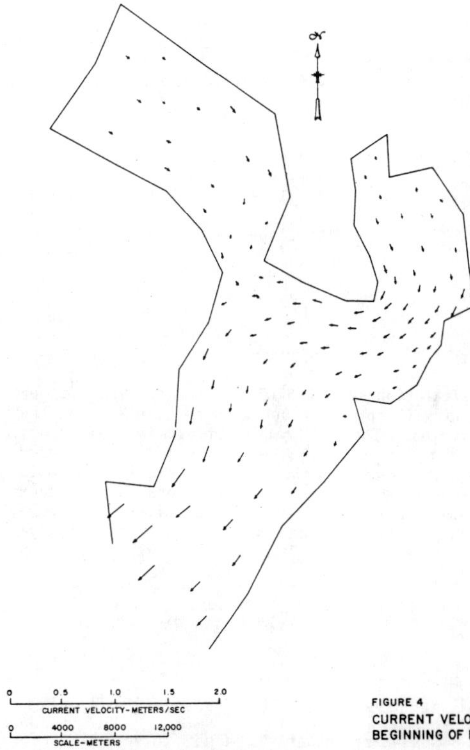

FIGURE 4
CURRENT VELOCITIES-
BEGINNING OF EBB

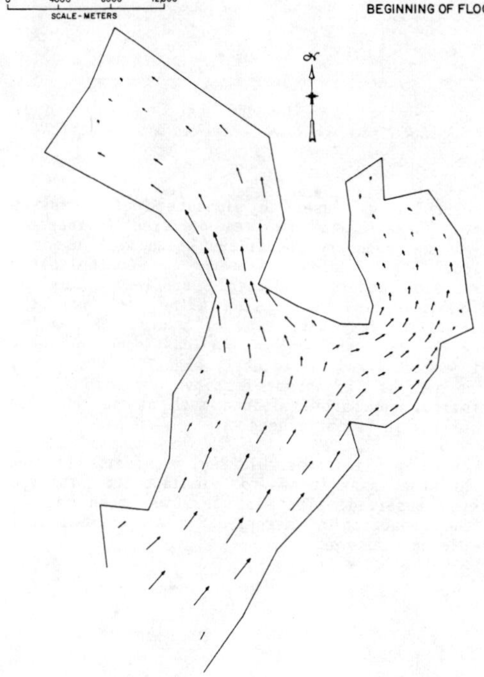

FIGURE 3
CURRENT VELOCITIES-
MID-FLOOD

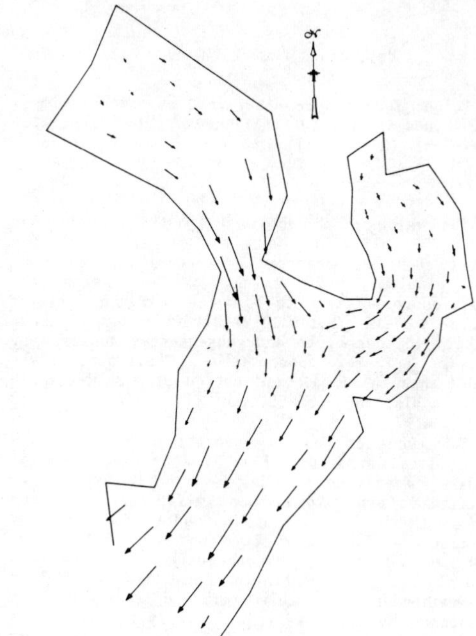

FIGURE 5
CURRENT VELOCITIES-
MID-EBB

Table 2-Input parameters for CAFE-1 applications

Parameter	Value
Eddy viscosity	200 M^2/sec
Bottom roughness	0.006-0.025
Tidal amplitude	0.64 M
Tidal period	25 hr
Time step	150 sec

DISPER-1

Calculation of DISPER-1 is accomplished by determining the proper dispersion coefficient for the study region. The calibration effort for this study involved the use of dye survey results which were analyzed statistically to obtain a single isotropic dispersion coefficient. Contour plots of dye concentration at various times during the survey were analyzed to obtain the variability (σ^2) of dye concentration (C) and plume width by using equation 1:

$$\sigma^2 = \int_{\chi_1}^{\chi_2} (\chi - \bar{\chi})^2 \frac{C}{A} d\chi \qquad (1)$$

where

χ_1, χ_2 = points on opposite sides of the plume

A = area under the C (χ) curve

$\bar{\chi}$ determined from $\int_{\chi_1}^{\bar{\chi}} C\, d\chi = A/2$

Values for σ^2 were determined at various times. These values are plotted in figure 6. The dispersion coefficient (D) was then obtained from equation 2:

$$\sigma^2 = 2\ Dt \qquad (2)$$

The value of D obtained by this method was 8 M^2/sec. Applications with the dispersion coefficient of 8 M^2/sec resulted in instability problems. These problems were solved by increasing the dispersion coefficient to 30 M^2/sec. Since a larger dispersion coefficient will result in more meroplankton being dispersed to and subsequently consumed by the sink, it is more conservative. Therefore, it was decided that the final application of DISPER-1 would utilize a dispersion coefficient of 30 M^2/sec.

Two types of boundary conditions were used in this application of DISPER-1. In the first type of boundary condition, nodal concentrations were specified in order to simulate meroplankton concentrations in the bay. In the second type of boundary condition, a dispersive flux was specified. The model was modified so that the dispersive flux term could simulate a concentration dependent sink. This modification enabled this term to represent the entrainment by the power plant simply by specifying the plant flow rate. The flow rate specified was 60.8 M^3/sec for all cases.

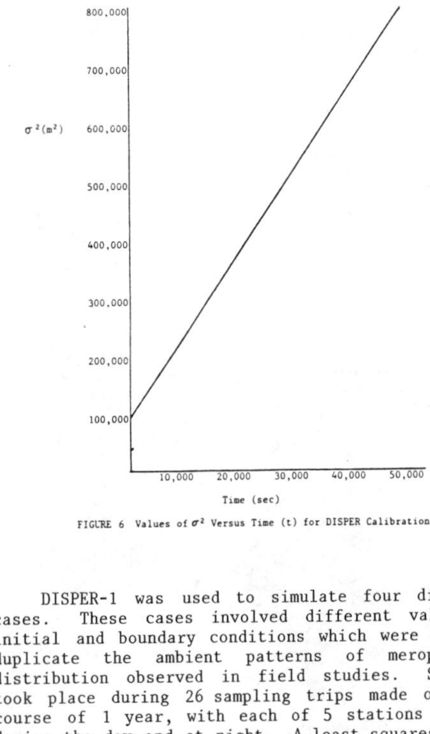

FIGURE 6 Values of σ^2 Versus Time (t) for DISPER Calibration

DISPER-1 was used to simulate four different cases. These cases involved different values of initial and boundary conditions which were used to duplicate the ambient patterns of meroplankton distribution observed in field studies. Sampling took place during 26 sampling trips made over the course of 1 year, with each of 5 stations sampled during the day and at night. A least squares linear model was utilized to identify significant ($P<0.05$) components of organism density pattern. This statistical approach explained much of the deserved variability in organism density.

In the applications, DISPER-1 was initially used without the sink term to simulate the ambient patterns observed. The sink term was then imposed and the reduction of meroplankton concentration in each element observed.

RESULTS

The output from the DISPER-1 simulations was subsequently analyzed to determine the area of entrainment influence of the plant. The extent of this area was defined to be the area which experienced at least a 5% reduction in meroplankton concentration after 12 tidal cycles with the plant in operation. Changes of less than 5% were within the error of the statistical description of the spatial pattern. The results of this analysis are presented in Figures 7 through 10. These figures also show the statistical meroplankton distributions obtained from field study results.

Figure 7 shows the results for the base case of initial uniform distribution of meroplankton in the water body. Figure 8 shows the area of entrainment influence for bay anchovy eggs which had a higher concentration in the upper end of the bay. Figures 9 and 10 show the results respectively for the egg and prolarvae stages of scaled sardines, which had higher concentrations in the lower end of the bay. The results indicate that, in each case, the area of entrainment influence of the plant is small when compared to the size of the water body.

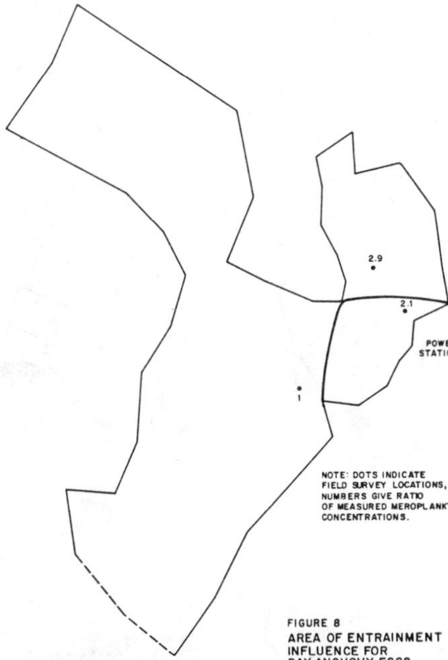

FIGURE 8
AREA OF ENTRAINMENT INFLUENCE FOR BAY ANCHOVY EGGS

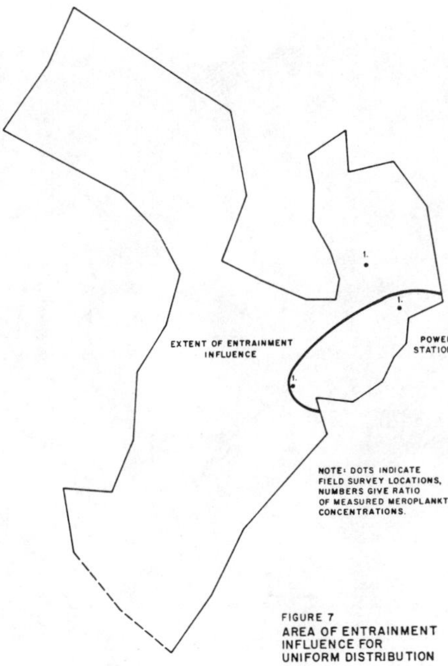

FIGURE 7
AREA OF ENTRAINMENT INFLUENCE FOR UNIFORM DISTRIBUTION

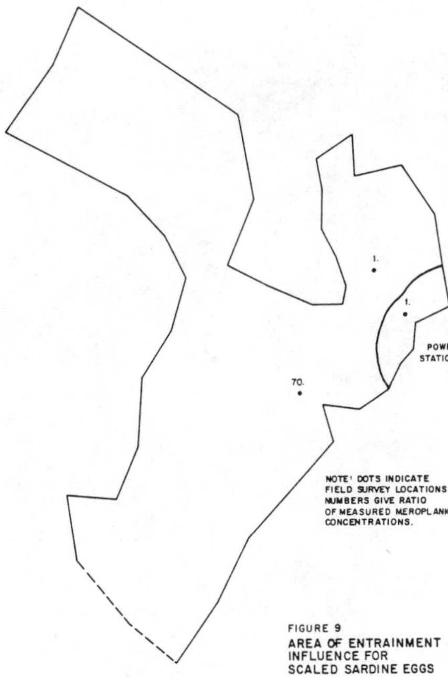

FIGURE 9
AREA OF ENTRAINMENT INFLUENCE FOR SCALED SARDINE EGGS

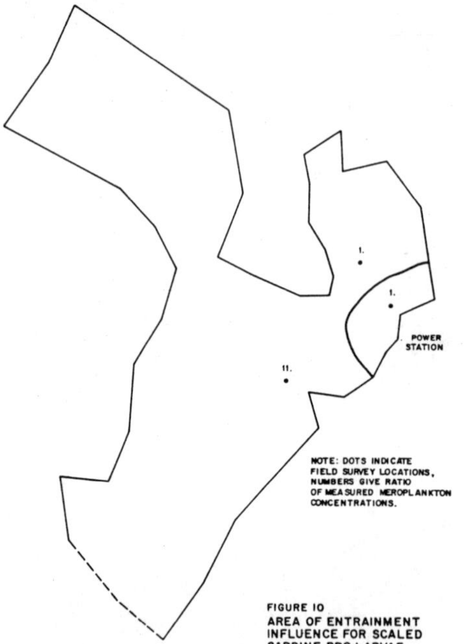

FIGURE 10
AREA OF ENTRAINMENT INFLUENCE FOR SCALED SARDINE PRO LARVAE

LITERATURE CITED

Galya, D. P. and Colangelo, P. M. 1981. Finite Elment Modeling of a Complex Embayment System, Proc. Third Waste Heat Management and Utilization Conf. Miami Beach, Florida.

Leimkuhler, W.F., et al. 1975. Two-Dimensional Finite Element Dispersion Model, Proc. Symp. on Modeling Techniques, Modeling 75 San Francisco, California.

Mangarella, P. A. 1979. Thermal Impact Modeling in San Francisco Bay, Proc. Specialty Conf. on Conservation and Utilization of Water and Energy Resources San Francisco, California.

Wang, J. D. and Connor, J. J., 1975. Mathematical Modeling of Near Coastal Circulation, Technical Report No. 200, R. M. Parsons Laboratory for Water Resources and Hydrodynamics. Massachusetts Institute of Technology, Cambridge, Massachusetts.

THE APPLICABILITY OF INSTREAM FLOW INCREMENTAL METHODOLOGY FOR IMPACT ASSESSMENT IN NEWFOUNDLAND[1]

Brian F. Bietz[2] and David J. Kiell[3]

New methods of assessing hydroelectric impact on fish stocks in Newfoundland were examined. The Instream Flow Incremental Methodology was selected and tested at Top Pond Brook. The model was used to predict Atlantic salmon habitat availability under various discharges. The results suggest that use of the model with caution will be possible in Newfoundland, and once accepted by the regulatory agencies, will likely become an important habitat management tool.

INTRODUCTION

A low head hydroelectric generating system at Burgeo, Newfoundland, has been proposed by Newfoundland and Labrado Hydro (HYDRO) to replace the diesel driven generators currently being used in the region.

The project, which has been described in detail by Shawmont (1979, 1980), would utilize the existing head produced by a series of falls on Dry Pond Brook, approximately 14 km north northwest of Burgeo. A headpond dam would be constructed to divert water to the intake structure and from there through a penstock to the powerhouse. Three generating alternatives were provided by Shawmont (1980) providing 24.7 gwh of firm energy from 4.0 m^3 plant flow and 14.2, 20.2 and 20.9 gwh of secondary energy from a maximum flow of 7.3 m^3/s, 11.0 m^3/s and 12.0 m^3/s respectively.

In order to ensure adequate flows, a second reservoir at Dry Pond (Figure 1) was also proposed. However, the relatively small size of the Dry Pond Brook catchment area reduced the economic feasibility of the project. It was, therefore, also proposed to create a third reservoir at Top Pond, and to divert water from the Top Pond drainage into the Dry Pond Basin. This would require the construction of a dam across the natural outlet of Top Pond and of a canal between Top Pond and Stephensons Pond (Figure 1). A manually operated control structure on Top Pond would regulate flows into the canal and a concrete spillway would pass excess water into Top Pond Brook.

The diversion of water from Top Pond will result in a general reduction of instream flows in Top Pond Brook. In order to assess the effect of the proposed hydroelectric project on the aquatic ecology of Top Pond Brook, HYDRO commissioned a detailed investigation of the fisheries resources of the river (deGraaf 1981). The studies included an assessment of the amount of available spawning and rearing habitat for salmonids, summer and fall surveys of fish populations and the collection of hydraulic data.

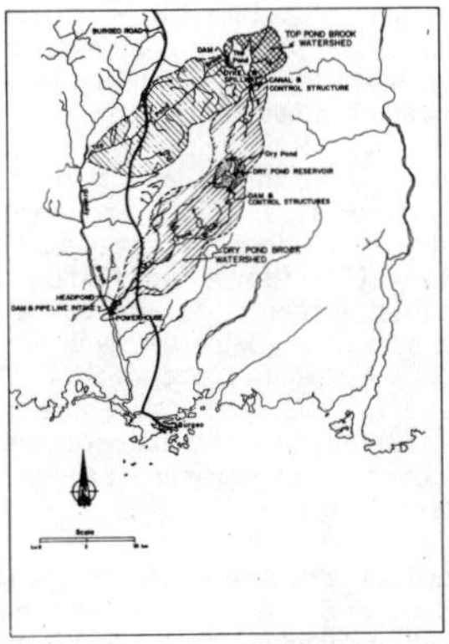

Figure 1. Dry Pond Brook hydroelectric generating facilities.

The results of these studies indicated that Top Pond Brook contained more than ninety percent of the suitable Atlantic salmon (Salmo salar) spawning habitat in the Grandy Brook watershed and at least twenty percent of the accessible nursery habitat (rearing areas for salmon fry and juveniles) (deGraaf 1981). The nursery habitat in Top Pond Brook was also felt to be superior to that in other portions of the watershed due to the more stable flow regime and less compacted substrate.

An assessment of the impact of stream flow reduction on habitat quality in Top Pond Brook was also carried out by deGraaf (1981). As part of this analysis, a technique commonly referred to as the Montana method (Tennant 1976) was used to determine minimum water release requirements from Top Pond for adequate protection of salmonid habitat. The technique is based upon calculation of mean annual discharge with instream flow requirements (minimum discharges) based on a percentage of the mean. The greater flow releases normally required during the summer months (May to

---

[1]Paper presented at the Third International Conference on State-of-the-Art in Ecological Modelling, Colorado State University, May 24-28, 1982.

[2]Brian F. Bietz is Senior Aquatic Biologist, IEC International Environmental Consultants Ltd., #345, 4400 Macleod Trail S.W., Calgary, Alberta Canada.

[3]David J. Kiell is Manager, Environmental Services, Newfoundland and Labrado Hydro, St. John's, Newfoundland Canada

September) under this method were extended into October by deGraaf (1981) in order to ensure adequate water release during the fall spawning period. The minimum flow requirements were then compared to data from Shawmont on the excess spillage of water from Top Pond reservoir which would occur assuming no flow release program into Top Pond Brook was established. These comparisons were used to determine when conflicts between instream needs and maximization of hydroelectric generation would occur in Top Pond Brook and were used to establish a flow release program.

This study, which was also initiated in 1980, was carried out in order to assess the applicability of alternative instream flow assessment techniques in Newfoundland. The major rational for carrying out this research was a recognition by HYDRO of the potential inadequacies of the Montana method. First, the system, although easily used, was felt to be only applicable for low to moderate intensity reconnaissance surveys (Stalnaker and Arnette 1976). Second, while the method is based on data from a large number of streams across the North American continent, it may not be appropriate for streams which exhibit both a strong spring and autumn flood (which is, in part, why it was necessary for deGraaf (1981) to extend the summer higher water release requirements to October). Third, the technique suggests specific minimum flows. This approach promotes the incorrect assumption that only streamflows below this "minimum" value will be detrimental to the fishery resource and does not take into account the incremental changes in habitat quality as stream flow is varied (Trihey and Wegner 1981).

The technique examined here is the Instream Flow Incremental Methodology (IFIM) developed by the U.S. Fish and Wildlife Service, Cooperative Instream Flow Service Group (CIFSG) (Bovee and Milhous 1978, Milhous 1979). The IFIM attempts to quantify the amount of potential fish habitat available for each life history stage of a species as a function of stream flow. This method is intended to be used as a decision-making tool and is specifically tailored to demonstrate the impact of incremental changes in stream flow on fish habitat (Trihey and Wegner 1981).

The core of the system is the model PHABSIM (Physical Habitat Simulation) which takes as input detailed hydraulic data (velocity, stage, slope, discharge), physical data (substrate type, cover, channel geometry) and the preferences of each life history stage of a fish species for a range of depths, velocities and substrate types (habitat suitability criteria). The hydraulic and physical data are used to simulate depths and flows under a range of discharges. The habitat suitability criteria for each life history stage is then superimposed on the hydraulic simulations and the Weighted Useable Area (WUA) for the stream reach is calculated (Bovee and Cochnauer 1977, Milhous, Wegner and Waddle 1981).

The first objective of this research was to examine the feasibility of using IFIM to assess the potential impacts of hydroelectric projects in Newfoundland. The technique, unlike the Montana method, is very labour intensive and so is potentially uneconomic. The methodology has also not as yet been recognized by the Newfoundland Department of Fisheries and Oceans, and a second function of this research was to determine its acceptibility to the regulatory agencies. A third function was to compare the water release requirements under IFIM with those of the Montana method in order to determine if water savings could be made by using a more detailed and precise technique.

A decision was made to select a single target species (Bovee 1975), rather than attempt to model the habitat requirements of all species within the drainage. Atlantic salmon were selected due to their relative economic importance and because of the effort currently being made to enhance salmon populations throughout the island.

METHODS

Study Site Selection

Top Pond Brook was initially divided into three areas of similar macro-habitat on the basis of gradients estimated from 1:50000 topographic map sheets. Segment 3 (Figure 2) extended roughly 2.5 km downstream from the outlet of Top Pond with a moderate gradient of 1.2%. Segment 2 was roughly 8 km in length with a gentle gradient of 0.6% while Segment 1 was approximately 9 km in length and relatively steep (gradient 2.0%).

Habitat mapping carried out by deGraaf in 1980 suggested that almost all available Atlantic salmon spawning habitat and a majority of high quality nursery habitat for salmon populations within the Grandy Brook watershed was contained within Segment 2 of Top Pond Brook. Since spawning habitat appeared to be potentially limiting for Grandy Brook Atlantic salmon populations, this habitat was felt to be of critical importance. Spawning habitat was also felt to be particularly vulnerable to reductions in discharge as the stream was relatively wide and shallow within these areas (deGraaf 1981). A helicopter flight was carried out in November 1980, in order to select a "representative" reach (Figure 2) in Section 2. Ironically, although the final site was selected independently of the research carried out by deGraaf, the site chosen was immediately upstream of a "representative" area of spawning habitat identified and sampled by that author earlier in the year (TPIII - deGraaf 1981).

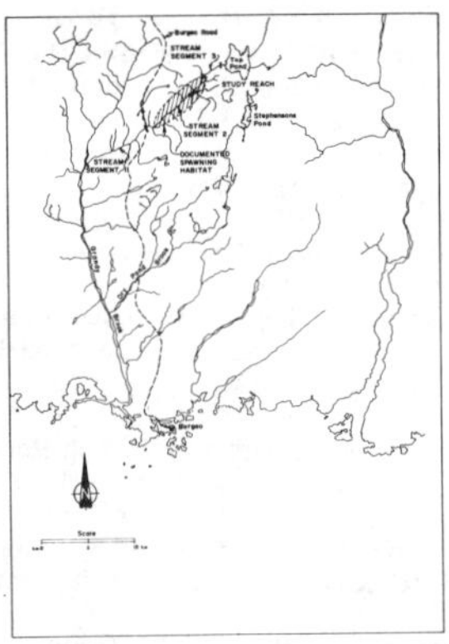

Figure 2. Study site location relative to homogeneous stream segments and documented spawning habitat.

Field Data Collection

Two multi-transect hydraulic models, WSP and IFG-4, are currently available in PHABSIM. WSP requires a single set of data collected at one discharge and uses the step backwater computational process (Milhous et al. 1981) to predict depths, velocities and widths for each cross-section. Modeling of additional discharges is carried out by inputing the appropriate stage (calculated from stage/discharge curves). The IFG-4 model requires that hydraulic data be collected at a minimum of three discharges as it uses a log-log regression analysis of stage and velocity (predicted variables) against discharge (predictor variable) to interpolate depths and velocities at unmeasured discharges. IFG-4 allows the modelling of a greater range of discharges than WSP through extrapolation. IFG-4 is also less sensitive to rapidly varied flow (waterfalls, cascades) within the simulated stream reach and to negative flows (backwaters).

Although the site selected for this study (Figure 3) was ameneable to modeling with WSP, IFG-4 was the proposed modeling technique. This was done because one major objective of the program was to test the feasibility of using the PHABSIM system in Newfoundland. Since many rivers of possible interest contain extensive areas of rapidly varied flow, IFG-4 is likely to be required in many cases. Due to another common feature of Newfoundland rivers, however, WSP was the model eventually used. Despite the fact that four field trips were made to the site in 1981 (May 26, July 12, August 30 and October 10) only two sets of useable data were obtained. This was due to the extremely unstable nature of stream discharge in Top Pond Brook. On two occasions (May and October) rainfall in the headwaters of Top Pond Brook caused significant changes in stream discharge during the time when field measurements were carried out. As neither hydraulic model is accurate under these conditions, these data were of no value. A final attempt to obtain a third data set (a relatively high water discharge was required) was made in November, but normal autumn high water levels did not occur in Top Pond Brook in 1981 and with the on-set of freeze-up in December the attempt was abandoned.

The study site selected was 320 m (1049 ft.) in length and bank full width of the river was approximately 30.5 m (100 ft.) (Figure 3). Care was taken during the final site selection to ensure the presence of a strong downstream hydraulic control and the absence of significant sources of additional stream discharge (tributaries, springs) within the study site. The substrate included bedrock, large exposed boulders, and extensive areas of smaller boulders, rubble and gravels and some sand bars.

Nine transects were placed across the stream perpendicular to the river flow. An attempt was made to locate transects at all hydraulic controls (Transects 1, 3, 7 and 9), with the final upstream transect (Transect 9) intentionally placed on a strong hydraulic control in order to ensure that the WSP hydraulic model could be used if insufficient data were gathered to make use of IFG-4 possible. Other transects were placed in order to represent stream features of biological interest. A short pool upstream of Transect 1 was not modelled, but Transect 2 did run across the riffle/pool transition point just upstream of this pool. Transect 4 was also located at a pool/riffle transition point while Transect 5 was within a very long riffle section. Transect 6 was located at a transition point between the more gentle downstream riffle cut by Transect 5 and the steeper riffle section immediately below Transect 7. Transect 8 ran through areas of sand bars on the south stream edge and a deep pool below Transect 9.

All field data were collected using the methods described by Trihey and Wegner (1981). Head pins were established at each transect and surveyed to the nearest 0.01 ft. During measurements at each cross-section, a permanent staff gauge was read before and after data collection. Left and right bank water surface elevations were recorded at each cross-section and water velocity at 60% of depth was measured at a minimum of 20 verticals located using a steel surveyors chain stretched between right and left bank head pins. Stream channel geometry was surveyed to the nearest 0.1 ft. at a minimum of fifteen verticals. Percent surface substrate composition was estimated between verticals (bedrock, boulders 250 mm, rubble 64-250 mm, gravel 2-64 mm, sand .062-2 mm, silt 0.62 mm). Substrate values were calculated using the technique described by Trihey and Wegner (1981). Water velocity was measured with a Price AA current meter. All measurements were made in Imperial units. While PHABSIM is capable of using metric data, rounding errors may occur (D. Morantz, D.F.O., Halifax) and conversions to SI units were only made after analysis was complete.

Suitability of Use Criteria (Habitat Preference)

As noted earlier, suitability of use or habitat preference criteria were only established for Atlantic salmon. The life history phases considered were:

1. Spawning requirements, adult salmon.
2. Incubation requirements, eggs and larvae.
3. Fry (under-yearlings) habitat requirements.
4. Juvenile (yearling) habitat requirements.

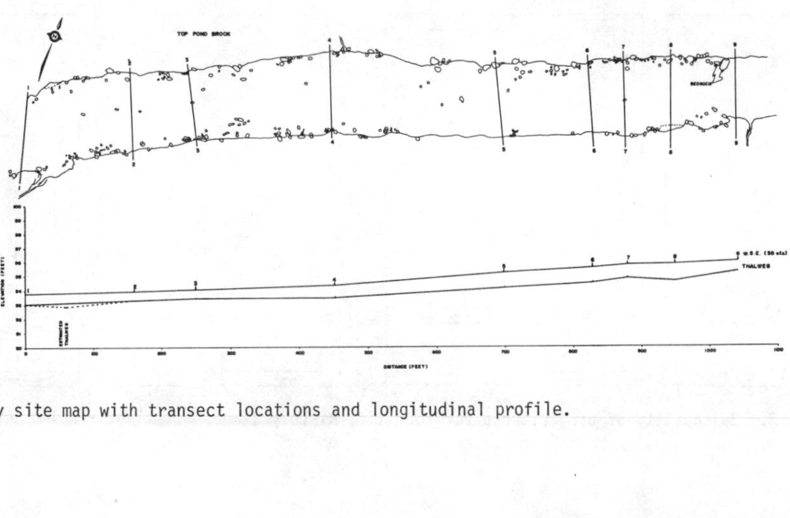

Figure 3. Study site map with transect locations and longitudinal profile.

No curves were produced for adult habitat requirements as the time spent by non-spawning adult salmon in Top Pond Brook is relatively short, and occurs primarily in habitat types not included within the study site (i.e., deep, bedrock and boulder-lined pools).

The validity of the curves is dependent on the assumption that for each physical habitat variable measured, the species in question will show a preference for an optimum value or an optimal range of values and that the frequency of occurrence of that species will decline as the physical habitat variable moves away from that optimum. The variables considered are water depth, water velocity and substrate. Weighted useable area (WUA) is a function of suitability of velocity, depth and substrate for a given life history stage where:

$$\text{WUA} = \sum_{i=1}^{n} A_i C_i \qquad (1)$$

and

A_i = surface area of a stream section element

C_i = composite suitability of that stream section

n = number of stream sections simulated

Composite suitability (C_i) is a product of the suitability index value for the stream depth, velocity and substrate within A_i. Therefore:

$$C_i = V_{si} \times D_{si} \times S_{si} \qquad (2)$$

where:

V_{si} = velocity suitability index

D_{si} = depth suitability index

S_{si} = substrate suitability index

As all suitability indices range from 0-1, C_i is always less than or equal to 1.

Data for Atlantic salmon habitat preferences during spawning, fry and juvenile life history stages were drawn primarily from Pratt (1966), Symons and Heland (1978), Wankowski and Thorpe (1979), and MacCrimmon and Gots (1979). No field data were collected. The absence of depth and surface flow data (as opposed to subsurface flow data) for salmon incubation required that a relatively simplistic approach be used to derive these curves. It was assumed that substrate preferences shown during spawning would reflect substrates most suitable for embryo survival during incubation and, therefore, the spawning substrate curve was also suitable for incubation. Since there should be very strong selective pressures for adult salmon to spawn in the gravel type which maximizes embryo survival, this is likely a reasonable assumption. The depth and velocity preferences shown during spawning were also used. However, the minimum and maximum depth ranges utilized by embryos were expanded. Flows and depths not suitable for adult spawning are still likely to maintain suitable incubation habitat.

Of major concern was the lower limit of depth and velocity suitable for incubation since the proposed hydroelectric project will always result in a reduction in natural flows. As depth and velocity diminish, movement of oxygen-enriched water through the substrate will decline, resulting in oxygen depletion and waste buildup for larvae. As well, fines will begin to drop from the water column and will fill interstitial spaces in the gravels, possibly resulting in the smothering of embryos. Since the water in Top Pond Brook is generally silt free (total suspended solids 4 mg/1 during the incubation period; (deGraaf 1981) silt deposition at low flows is not usually likely to be a problem. Greatly increased silt loads, and subsequent deposition of fines, may occur, however, during the construction of the Top Pond dam. The actual depths and water velocities needed to adequately aerate redds are also unknown, and the incubation curves may require considerable refinement.

Curves for the other three life history stages were constructed using range and optimum analysis and parameter overlap analysis as described by Bovee and Cochnauer (1977). No data appear to be currently available to permit a direct calculation of habitat preference curves (see Gore and Judy 1981), although research carried out in 1981 by the Department of Fisheries and Oceans may provide considerable additional information (Morantz pers. comm.). Once the initial curve set were generated, they were reviewed and modified by Dr. John Gibson and Dr. Michael Chadwick (Department of Fisheries and Oceans, St. John's). It must be emphasized, however, that their assistance does not indicate acceptance by Fisheries and Oceans of the final curve set or responsibility by either scientist for the results of this study. They did, however, provide an impartial review of the curves and made a number of valuable suggestions for their improvement. This is of particular importance when one considers that the general lack of juvenile predators in Newfoundland is likely to create a relatively unique set of habitat preferences (Gibson, pers. comm.).

The habitat preference curves for the four life-history stages are shown in Figures 4-7.

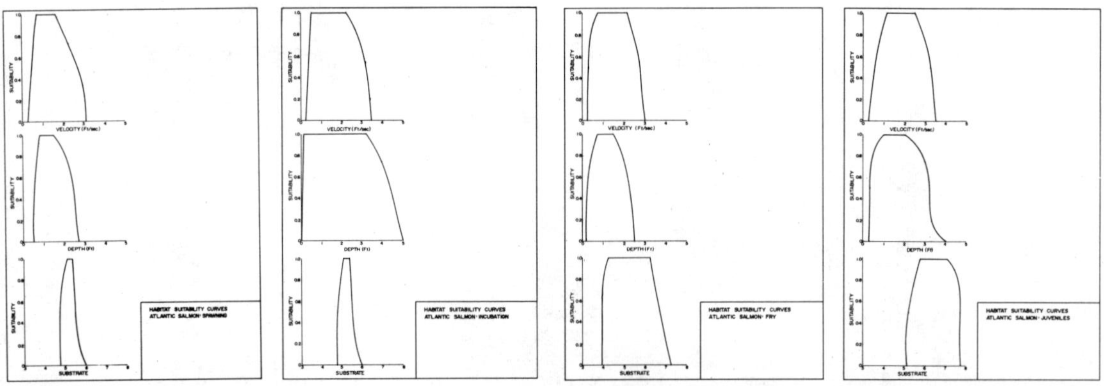

Figures 4 - 7. Suitability of use criteria for four life history stages of Atlantic salmon.

Data Analysis

The PHABSIM model was obtained from the U.S. Fish and Wildlife Service in 1981. It was adapted to and mounted on the VAX computer system of MacLaren Plansearch Inc. Use of the VAX appears to result in a considerable improvement in file modification capability over the CDC Cyber system used by the CIFSG.

The data used for calibration of the WSP hydraulic model were collected August 30, 1981. This was a low flow period, with a discharge of 0.99 m^3/sec (35 cfs) at the study site. Calibration of the WSP model was carried out using the techniques suggested by Milhous et al. (1981). All predicted water surface elevations (WSE's) were adjusted using uniform values of the roughness coefficient (Manning's N) until predicted WSE's were $\pm$ 0.1 ft. of measured WSE's. Then calibration of the averaged mean column velocities was carried out until the agreement of predicted and measured values was $\pm$ 0.2 ft./sec. Whenever a Manning's N value of less than 0.025 or greater than 0.100 was required to calibrate a column velocity, reference was made to the field notes for justification (i.e., the presence of fine, well-sorted gravels in areas of low roughness or conversely, a large boulder or series of boulders in an area of high roughness).

RESULTS

Hydraulic Simulation

An example of the results from the calibrated WSP model is shown in Figure 8. Stream channel geometry (dots) was measured at each point where a noticeable change in bottom slope or substrate characteristics occurred. Velocity verticals (solid vertical lines) were placed so that a maximum of 10 percent of the total stream volume passed through any one region. Predicted velocities and predicted water surface elevation tended to be slightly less than the values measured at this cross section. However, other patterns occurred at other transects and no difficulty was encountered balancing the model.

Following calibration of the WSP hydraulic model simulation of a range of hydraulic conditions was carried out. The recommended extrapolation range of WSP is 40 percent to 250 percent of the calibrated discharge (Milhous et al. 1981), or, in this case, 14 to 87.5 cfs (0.40 to 2.48 m^3/sec). As the model requires an estimated stage value which corresponds to the simulated discharge, a stage-discharge curve was generated (Figure 9) using data collected during other visits to the site.

Habitat Simulation

A series of hydraulic simulations for discharges between 14 and 87.5 cfs were then input into the HABTAT portion of the PHABSIM model and WUA calculated for each discharge. The WUA per thousand linear feet of stream for a range of discharges is presented in Table 1 and the percent WUA of the total available stream habitat is shown in Figure 10. Within the extrapolated range of discharges, no peak in habitat availability occurs. However, the rate of increase is declining and appears to be rapidly reaching a maximum at some discharge greater than 130 cfs (3.7 m^3/sec).

Shawmont (1980) calculated mean monthly spills and spill frequency from Top Pond reservoir for a 20 year period (Table 2). Only under Alternative 1 did any water release occur during most months, and even under this regime the frequency of spillage was generally low. Since during most months in most years, little or no water release from Top Pond could be expected, project impact on habitat availability was estimated by assuming that no flow from Top Pond occurred.

Table 3 provides an estimate of the change in monthly WUA when no discharge from Top Pond occurs. Shawmont (1980) produced a 20 year simulation of natural monthly discharge from Top Pond. Monthly means were calculated and then incremented by 42.3% to estimate the additional flow into the river from downstream sources (deGraaf 1981) in order to estimate mean monthly discharge at the study site. These flows were then reduced by the mean amount of discharge predicted to escape from Top Pond to provide regulated flows.

It was necessary in some cases to produce hydraulic simulations of discharges beyond the ranges recommended by the CIFSG. It was only necessary to model flows greater than those recommended however. Since the major impact of the project will be a reduction in flows, the reduced accuracy of the WUA estimates of high flows will not affect the periods of greatest impact. The estimation of WUA at higher discharges are likely still good indicators of the relative change in habitat availability.

The effect of curtailing discharge from Top Pond Brook appears to be most serious in September, when almost 50 percent of the available spawning habitat will

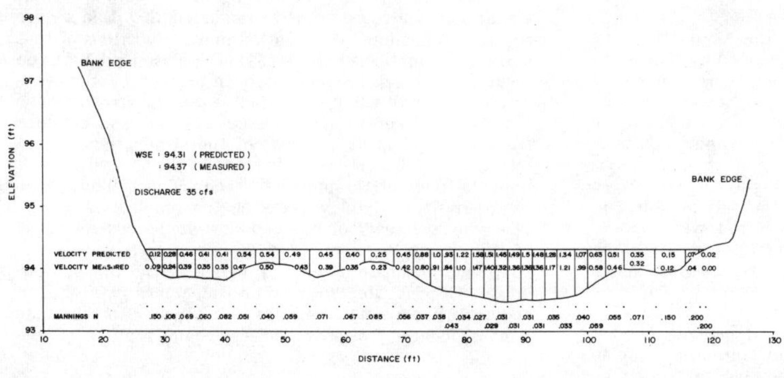

Figure 8. Stream Cross section at Transect #4.

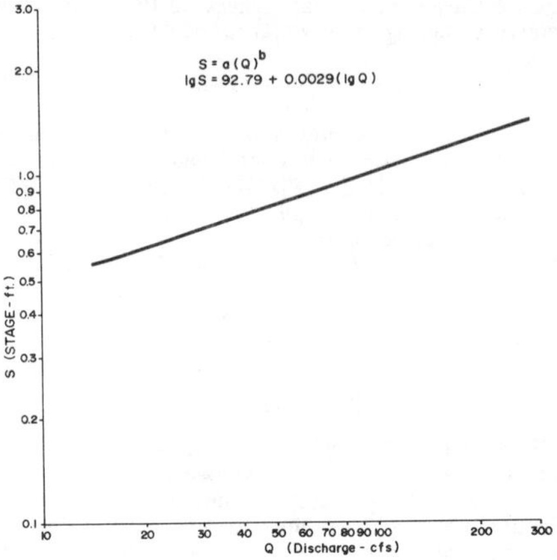

Figure 9. Stage-discharge relationship, Top Pond Brook.

Table 1. Available weighted habitat (m^2) per 1000 linear feet (305 m) of stream

Q*	Spawning	Incubation	Fry	Juvenile
0.4	76	1140	2172	1129
0.6	213	1564	3177	1687
0.8	355	1842	4107	2201
1.0	479	2083	4738	2692
1.2	692	2232	5316	3141
1.4	812	2369	5749	3532
1.6	945	2503	6037	3882
1.8	1045	2589	6272	4166
2.0	1154	2689	6511	4398
2.2	1237	2743	6684	4595
2.4	1306	2789	6805	4764
2.5	1314	2812	6860	4851

* Q = discharge (m^3/sec)

be lost. The actual impact of this loss will of course, be dependent on the total population size relative to the amount of available spawning habitat, but given the general absence of good spawning habitat within most of the Grandy Brook watershed, it is possible that spawning habitat may be a limiting resource. As spawning in Top Pond Brook is generally complete by late October (deGraaf 1981), the increase in runoff in November will likely have little effect on spawning habitat availability.

The potential impact of flow reduction on incubation habitat is somewhat confounded by impact on spawning habitat. For example, increasing flows only during the winter months would not result in any noticeable increase in embryo survival since these same areas would have been unavailable to spawning adults in the fall and, therefore, would contain no eggs. A major impact could occur, however, if higher waters were artificially maintained during only the fall months. In this case, as water levels were reduced to those proposed by Shawmont (1980), eggs laid in October would be exposed from January to March to dessication or freezing.

The greatest declines in fry and juvenile habitat occur during two periods, January to March and July to September. The loss of winter habitat may be misleading, however, as during these months, immature salmon often move into the interstitial spaces of the substrate (Gibson, pers. comm.). The loss of habitat during the late summer months may have considerable impact, however, as Atlantic salmon parr are highly territorial (Symons and Heland 1978). While some territorial compression may occur, it is more likely that a portion of the population will be forced into less suitable or unsuitable habitat (Brown 1975).

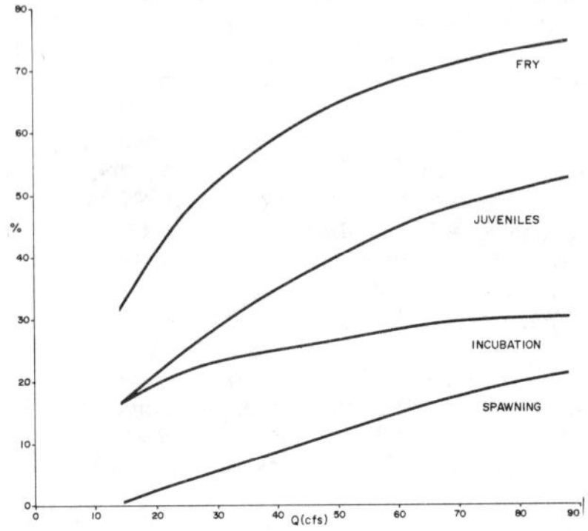

Figure 10. Discharge (Q) versus weighted useable habitat area as percentage of total area for the four salmon life history stages.

Table 2. Mean monthly spill and frequency of spills over a 20 year period from Top Pond Reservoir for the three alternative plant designs (adapted from deGraff 1981)

	Alternative 1		Alternative 2		Alternative 3	
	Mean Spill (m 3/s)	Frequency (N/20 years)	Mean Spill (m 3/s)	Frequency (N/20 years)	Mean Spill (m 3/s)	Frequency (N/20 years)
January	0.10	1	0	0	0	0
February	0	0	0	0	0	0
March	0.12	1	0	0	0	0
April	0.38	3	0.32	2	0.32	2
May	2.49	19	1.04	11	0.79	8
June	1.28	11	0.57	7	0.24	3
July	0.01	1	0	0	0	0
August	0.12	1	0	0	0	0
September	0.12	1	0	0	0	0
October	0.54	5	0	0	0	0
November	0.97	10	0.29	2	0.21	1
December	0.91	11	0.38	3	0.31	3

An assumption of the PHABSIM program is that simulation of a representative reach length within a river segment will allow extrapolation to other portions of that stream segment. deGraaf (1981) has pointed out that Top Pond Brook discharge is greatly influenced by the large number of small tributaries which enter the stream below Top Pond. Therefore, discharge will increase between the upper and lower portions of the stream segment of interest to this study. However, the magnitude of impacts found at the modeled reach are still likely to be representative of the effects on the reach as a whole. The greater impact of reduced flow at the upstream end of the study segment (due to the fewer sources of additional discharge) will be balanced by the smaller impact found at the downstream end, where greater flow augmentation from tributaries will occur. As the study site is located towards the upstream portion of the spawning area, it will provide a better estimate of the project impact in the areas with the greatest potential sensitivity.

Table 3. Changes in habitat availability as a result of flow regulation at Top Pond

	Discharge (m^3/sec)		Change in Habitat Availability							
	Unregulated	Regulated	Spawning		Incubation		Fry		Juveniles	
			%	(m^2)	%	(m^2)	%	(m^2)	%	(m^2)
January	1.27	0.74	NA		-23	(525)	-30	(1660)	-38	(1250)
February	1.27	0.74	NA		-23	(525)	-30	(1660)	-38	(1250)
March	1.19	0.68	NA		-25	(561)	-30	(1706)	-39	(1218)
April	3.11*	1.81	NA		NA		-10	(897)	-20	(1038)
May	6.51*	3.77*	NA		NA		+15	(916)	-02	(98)
June.	3.03*	1.76	NA		NA		-11	(741)	-20	(652)
July	1.36	0.79	NA		NA		-28	(1602)	-37	(1283)
August	1.84	1.08	-46	(499)	-18	(482)	-21	(1342)	-32	(1339)
September	1.81	1.05	-49	(518)	-20	(528)	-22	(1389)	-33	(1370)
October	3.03*	1.76	-24	(322)	-12	(351)	-11	(741)	-20	(652)
November	3.88*	2.24	-09	(118)	-08	(236)	-05	(346)	-15	(844)
December	2.94*	1.70	NA		-12	(358)	-11	(800)	-22	(1094)

*Discharges beyond the maximum extrapolation recommended by IFCSG. Results must be accepted with caution.

DISCUSSION

The first objective of this research program was to assess the feasibility of utilizing IFIM in Newfoundland stream systems. During this study, a number of disadvantages of IFIM became apparent. While some, such as the great care with which field data must be gathered, are well recognized by the model's designers, a few are specific to application of the system in the Newfoundland environment. The major disadvantage of the technique is directly related to three general features of Newfoundland river systems; most are remote with limited accessibility, most are subject to rapid changes in discharge and most have complex, boulder-strewn channels. The complex nature of most river channels indicates that IFG-4 will usually be the preferred hydraulic model. This requires that at least three field visits be carried out at three distinct stream discharges. However, as most rivers in Newfoundland are ungauged, or only inspected at infrequent intervals, it is often almost impossible to obtain current information on discharge. Even when these data are available, discharge may change by an order of magnitude over a few days in many river systems. Therefore,if any delay is encountered in mounting the field survey, the low flow conditions reported two days earlier may have been replaced by flood conditions (as actually occurred in this study). Therefore, it is strongly recommended that an accessible gauging system be set up before any IFIM study is initiated and that a gauge monitoring system, perhaps utilizing local inhabitants, is established.

A second difficulty found during this study was the general lack of quantitative habitat preference data for Atlantic salmon. The curves produced here must be considered only an initial attempt. Considerable modification and refinement of these curves will undoubtedly be necessary. While data currently being gathered in the Maritime provinces and the New England states will assist in improving these curves, more information specific to the Newfoundland environment will be required.

Despite the difficulties encountered in adapting IFIM to the Newfoundland environment, the potential advantages of the system are numerous. First, the system appears to provide a much more precise estimate of habitat lost or gained as a result of discharge manipulation. Since low head hydroelectric generating programs such as Dry Pond Brook are likely to become increasingly common, the value of this type of information will continue to gain importance. Second, as mentioned earlier, rather than providing a rigid set of minimum flows, the model provides a flexible basis for negotiation between project proponents and the regulatory agencies. As a result, increases in both habitat protection and economic returns may be expected. Third, the development and acceptance of a habitat preference curve set for Atlantic salmon in Newfoundland (or perhaps a curve series based on differences in response in northern and southern portions of the island) will allow the rapid estimate of the potential value of a stream system without carrying out a direct fisheries assessment. This will be of particular value in systems where other factors, such as over-exploitation of stocks, have reduced the actual population well below the carrying capacity of the environment.

A second function of the research program was to determine the acceptability of IFIM to the regulatory agencies in the Province of Newfoundland. Comments on the initial draft of this report, received from the Department of Fisheries and Oceans, the agency responsible for the protection of anadromous fish stocks, indicate that they believe it would be premature to negotiate future water release for habitat maintenance from IFIM predictions. Their major reservation appears to be with data base used to derive the suitability of use curve set for Atlantic salmon in Newfoundland. Clearly, future research must be carried out in order to provide a more refined data set, and to actually test the predictions of the model. It is somewhat ironic, however, that water release requirements in Newfoundland are currently based on techniques which appear to be based on much more tenuous assumptions than those used to produce the Atlantic salmon suitability of use curves.

A third function of the program was to compare the water release requirements of the Montana method from deGraaf (1981) with those suggested by IFIM. Unlike the Montana method, however, IFIM is not designed to provide recommended minimum flows. Rather, the model provides an estimate of habitat loss or gain under a given discharge regime. It is possible, however, to determine the relative habitat protection which would be provided by the flow release suggested by deGraaf (1981).

Table 4. Habitat availability as a function of water release from Top Pond reservoir.

Release (m^3/sec)	Protection*	Discharge (m^3/sec)	% habitat availability			
			Spawn.	Inc.	Fry	Juv.
0		1.05	-49	-18	-21	-32
0.15	Minimal	1.20	-34	-14	-16	-25
0.60	Moderate	1.65	-05	-03	-03	-05
0.90	Maximum	1.95	+08	+03	+03	+04
Unregulated		1.81				

*from Tennant (1976)

For example, deGraaf (1981) predicted that minimum instantaneous release from Top Pond of 0.9, 0.6 and 0.15 m^3/sec in summer months would provide maximum, moderate and minimum habitat protection, respectively, to downstream reaches. The effects of these release programs on habitat availability was calculated for the month of September, a period when reduced flows from Top Pond Brook appeared to produce a serious loss of habitat for all life history stages. The results are presented in Table 4. The data suggest that the moderate habitat protection flow releases predicted by the Montana method would, on average, result in almost no habitat loss. The maximum protection levels would actually increase discharge and improve habitat availability over what occurs natually. The question which remains unanswered is whether these habitat protection levels are warranted and whether the Dry Pond hydroelectric project would remain economic under these release programs.

These data were provided as an initial step in a water release negotiation program between HYDRO engineers and the regulatory agencies. Using these models, it should now be possible to determine whether a water release program into Top Pond Brook, which is acceptable to the regulatory agencies, will also provide sufficient hydro-electric generating capacity that the project will remain economically viable. From Table 3, it seems that the major periods of concern will be July to September and January to March. If flow augmentation could be maintained during those months, it is likely that major project impacts could be alleviated.

Clearly, Top Pond Brook provides a number of opportunities for the testing of the model's predictions. It would be possible to determine actual habitat utilization relative to natural discharge patterns. After construction of the reservoir, it would also be possible to manipulate discharge, and determine whether habitat utilization patterns shifted in the predicted fashion. Top Pond Brook is not likely, however, to provide these tests. HYDRO, after reviewing the available data, concluded that the potential environmental impact of the Top Pond reservoir could not be justified. Since the economics of the project were questionable without supplemental flow from the reservoir, the Dry Pond Hydroelectric project has been cancelled.

LITERATURE CITED

Bovee, K.D. 1975. Assessment and implementation of "in-stream value studies" for the Northern Great Plains, Proc. Fort Union Coal Field Symp., Vol. II. Montana Acad. Sci. pp. 112-123.

Bovee, K.D. and T. Cochnauer. 1977. Development and evaluation of weighted criteria probability-of-use curves for instream flow assessments: fisheries. Instream Flow Inform. Pap. No. 3. FWS/OBS-77/63. Cooperative Instream Flow Service Group, Fort Collins, Colorado. 38 pp.

Bovee, K.D. and R.T. Milhous. 1978. Hydraulic simulation in instream flow studies: theory and techniques. Instream Flow Inform. Pap. No. 5 FWS/OBS-78/33. U.S. Fish and Wildlife Service, Washington, D.C. 150 pp.

Brown, J.L. 1975. The evolution of behavior. W.W. Norten and Co., New York. 761 pp.

deGraaf, D.A. 1981. Fisheries investigations in the vicinity of the proposed Dry Pond Brook hydroelectric development. MS. by LGL Limited for Newfoundland and Labrado Hydro, St. John's, Newfoundland.

Gore, J.A. and R.D. Judy, Jr. 1981. Predictive models of benthic macroinvertebrate density for use in instream flow studies and regulated flow management. Can. J. Fish. Aquat. Sci. 38:1363-1370.

MacCrimmon, H.R. and B.L. Gots. 1979. World distribution of Atlantic salmon, Salmo salar. Can. J. Fish Aqat. Sci. 36:422-457.

Milhous, R.T. 1979. The PHABSIM system for instream flow studies. Proc. Summer Comp. Simulation Conf., Toronto, Ont.

Milhous, R.T., D.L. Wegner and T. Waddle. 1981. User's guide to the Physical Habitat Simulation system. Insteam Flow Inform. Pap. No. 11. FWS/OBS-81/43. U.S. Fish and Wildlife Service, Washington, D.C.

Pratt, J.D. 1968. Spawning distribution of Atlantic salmon (Salmo salar L.) in controlled flow channels. M.Sc. Thesis, Memorial University of Newfoundland.

Shawmont Ltd. 1979. Four rivers hydro study. MS by Shawmont Ltd. for Newfoundland and Labrado Hydro.

Shawmont Ltd. 1980. Four rivers - further studies on Dry Pond and Pinware River developments. MS by Shawmont Ltd. for Newfoundland and Labrador Hydro. 46 pp.

Stalnaker, C.B. and J.L. Arnette. 1976. Methodologies for determination of stream resource flow requirements; an assessment. FWS/OBS-76/03. U.S. Fish and Wildlife Service, Washington, D.C. 199 pp.

Symans, P.E.K. and M. Heland. 1978. Stream habitats and behavioural interactions of underyearling and yearling Atlantic salmon (Salmo salar). Can. J. Fish Aquat. Sci. 35:175-183.

Tennant, D.L. 1976. Instream flow regimens for fish, wildlife, recreation and related environmental resources. Fisheries 1:6-10.

Trihey, E.W. and D.L. Wegner. 1981. Field data collection procedures for use with the Physical Habitat Simulation system of the Instream Flow Group (Draft). Coopeative Instream Flow Service Group, Ft. Collins, Colorado.

Wankowski, J.W.J. and J.E. Thorpe. 1979. Spatial distributiion and feeding in Atlantic salmon, Salmo salar L. juveniles. J. Fish Biol. 14:239-248.

FACTORS CONTRIBUTING TO MARSH LAND LOSS IN LOUISIANA'S COASTAL ZONE[1]

L. A. Deegan[2], H. M. Kennedy[3], R. Costanza[4]

Abstract.--Recent studies have estimated that Louisiana is currently losing approximately 49 square miles (127 km^2) of its coastal wetland area each year. 80% of this land loss is occuring in the Mississippi River Deltaic Plain Region. Natural forces as well as manmade impacts (e.g. oil and gas access canals, and navigation channels) contribute to the loss of this valuable land and habitat. Changes in land area and habitat type were calculated from area data produced for the U.S. Fish and Wildlife Service based on aerial photographs of the Mississippi Deltaic Plain Region taken in the mid 1950's and 1970's. Data for 138 7 1/2 quadrangles for each of the two periods were used to construct a multivariate model to determine principal factors contributing to land loss. Delta age, canal density, local hydrologic regime, and total marsh area were found to affect rates of land loss. The relative contributions of these factors and their implications for Coastal Zone Management are discussed.

INTRODUCTION

Southern Louisiana comprises approximately 41% (3 million acres) of the coastal wetlands of the United States and is experiencing a startling rate of wetland loss currently predicted for the Mississippi Deltaic Plain Region alone at 24,960 acres/yr (102 km2/yr) (Gagliano et al. 1981). Of the 201 7.5 minute quadrangle maps in Louisiana's deltaic region which have wetland, 94% have lost wetlands since 1955. This high rate of land loss poses a threat to the people and their lifestyles, the natural resources, and the economy of the state. State revenues from oil and gas leases will decrease as wetlands are lost and state land converts to federal ownership. Nunez and Sour (1981) presented evidence that the state would lose approximately 20 million dollars annually, at 1981 production and price levels, if the state/federal boundary was moved inland, uniformly by 0.8km, because of shoreline erosion.

Wetland loss rate reflects the combined effects of the natural geology of the region and human induced factors. Approximately 80% of the coastal zone of Louisiana has been formed over the past several thousand years by Mississippi river deltas (Frazier 1967). The growth and decline of deltaic systems is dominated by sediment inputs, subsidence caused by sediment compaction and regional downwarping, erosion, and actual sea level rise (D. Morgan 1977, J. Morgan 1963). Regional downwarping and actual sea level rise operate over thousands of years; sediment compaction, decrease in sediment input, and erosion operate over tens of years, to contribute to land loss. Sediment compaction and erosion rates vary with the age of the delta lobe and with location within the delta lobe (J. Morgan 1963). In general, subsidence in Louisiana declines as the distance from the coast increases (Adams et al. 1976). The magnitude of the rate is determined by the age of the delta lobe (J. Morgan 1963).

The construction of levees & canals along rivers, streams, bayous, and across wetlands results in direct loss of wetlands and interrupts natural delta building processes, thereby leading to premature or accelerated landloss (Craig et al. 1979, Cleveland et al. 1981, Gagliano et al. 1981, Scaife et al. 1982). The main effect of levees and canals is believed to be a disruption of the natural hydrologic regime leading to a reduction of sediment input, salt water intrusion, and increased erosion (Stone et al. 1978, Gagliano et al. 1981).

This study investigates the relative contribution of geologic and human induced factors to marsh loss in the Deltaic Plain Region using a recently completed aerial photography study (Wicker et al. 1980). We have restricted our analysis to the effects of short term

[1]Paper presented at the Third International Conference on State-of-the-Art in Ecological Modelling, Colorado State University, May 24-28, 1982.

[2]L. A. Deegan is a doctoral student in the Marine Science Department, Louisiana State University, Baton Rouge, Louisiana U.S.A.

[3]H. M. Kennedy is a Coastal Resource Analyst, Louisiana Department of Natural Resources, Coastal Management Section, Baton Rouge, Louisiana U.S.A.

[4]R. Constanza is an Assistant Professor of Marine Science, Louisiana State University, Baton Rouge, Louisiana U.S.A.

geologic process (sediment compaction and erosion) and canaling because of the time frame of the sampling. Most of the leveeing of the Mississippi River and its distributaries had been completed prior to the first set of overflights (1955/1956), so we were not able to determine the direct effects of large scale sediment depletion. We use a multivariate approach to determine quantitatively if any of the above factors alone or interactively have a significant effect on marsh-land loss.

DESCRIPTION OF THE DATA BASE

The data base for this study was a detailed inventory of habitats prepared by Wicker et al. (1980) for the U.S. Fish and Wildlife Service. The data base consists of area data for approximately 138 habitat types by 7.5 min. quadrangle maps of the Mississippi Deltaic Plain Region (figure 1). Habitat areas, in acres, were planimetered from habitat maps constructed from interpretation of aerial photographs taken in the mid 1950's and late 1970's. A detailed description of the data base is in Wicker (1980). Analysis of canal effects was restricted to quadrangle map units that met the following criteria: (1) were underlain by sediments from a Mississippi delta lobe, (2) were not in an active delta region, (3) had a total marsh area greater than 1000 acres, (4) total map area in 1955/1956 was within 200 acres of total map area measured in 1978. Criteria 1 and 2 were used to limit the analysis to geologically similar areas; Criteria 3 and 4 to eliminate any differences due to variance in measurements. The final data set consisted of 115 7.5 min. quadrangle maps, and covers approximately 65% (2 million acres) of the Mississippi Deltaic Plain Region.

DESCRIPTION OF VARIABLES

Variables were calculated for each 7.5 min. quadrangle map unit. Marsh loss (DMARSH) was calculated as the difference in total marsh area between 1955/1956 and 1978. Total marsh area (TMARSH) in 1955/56 indicates original marsh area and indexs the loss due to natural deltaic processes. The area of natural streams, rivers and bayous in 1955/1956 (TBAYOU) was used as an index of the natural hydrologic regime of an area. The effects of canals were indicated by three parameters; (1) canal and spoil area in 1955/1956 (CANAL55) indicating pre-existing canals, (2) canal and spoilbank area in 1978 (CANAL78) indicating final canal area, and (3) change in canal and spoilbank area between 1955/1956 and 1978 (DCANAL), indicating the amount of canaling that occured over the 23 year time period. To assess the contribution of regional geology, each map was assigned an age value (AGE) corresponding to the age of the sub-delta complex associated with its progradation (Frazier 1967), and the distance from the center of the map (DIST) to the current Louisiana coastline was measured (figure 1).

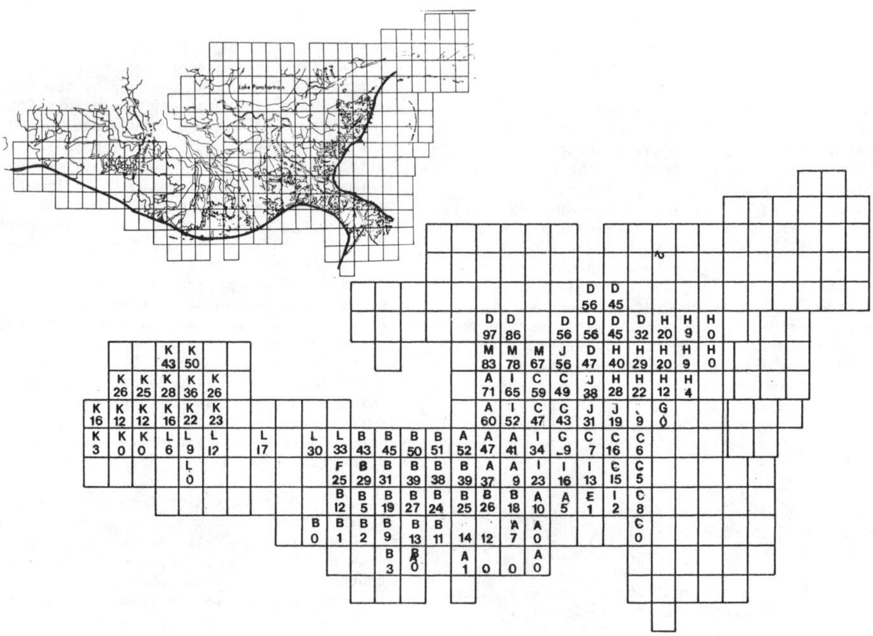

Figure 1.--Mississippi River Deltaic Plain Region study area with 7.5 min. topographic map overlay and enlargement. Letters indicate delta age (yrs); A=250, B-658, C=665, D=1,441, E=1,468, F=2,151, G=2,732, H=2,909, I=3,819, J=4,070, K=4,539, L=4,625. Numbers indicate distance from coast (Km).

A packaged statistical program (SAS) was used to construct multivariate regression equations relating marsh loss to regional geology and canaling activity.

RESULTS AND DISCUSSION

Of the two geologic factors investigated (age of the sub-delta lobe and distance from coast), only sub-delta age showed a significant relationship to marsh loss (figure 2). The form of the equation relating marsh loss to age of subdelta corresponds to the growth-deterioration curve expected from theoretical considerations first proposed by Morgan (1963) and recently confirmed by Wells et al. (1982). Immediately after abandonment of a delta, high interstitial water losses result in sediment compaction and subsidence. As dewatering continues, the increased density of the overlying sediment accelerates subsidence and land loss. Erosionalforces accelerate land loss by winnowing away the fine sediment, leaving larger size particles. The rate of land loss diminishes with time as the dewatering process is completed and erosional forces have removed most of the fine sediment. The regession equation substantiates this pattern and indicates a peak loss rate at a delta age of 600 years.

$$DMARSH = 20.37 * AGE * e^{0.00141*AGE} \qquad (1)$$

where,
DMARSH = change in marsh (acres)
AGE = age of subdelta lobe
($r^2 = 0.54$, $p < 0.0001$)

The low r^2 of this regression indicates other factors are superimposed on this basic pattern.

Adams et al. (1971) found a trend of decreased subsidence as distance from coast increased, which they attributed to differential subsidence rates within a delta lobe and erosional forces associated with proximity to the Gulf of Mexico. The data indicated no significant relationship with distance for marsh loss. Small variations in subsidence rates may not be evidenced in marsh loss rates calculated over the small time period used in this study or the use of sub-delta lobes as age indicators may incorporate this variation into the relationship with age.

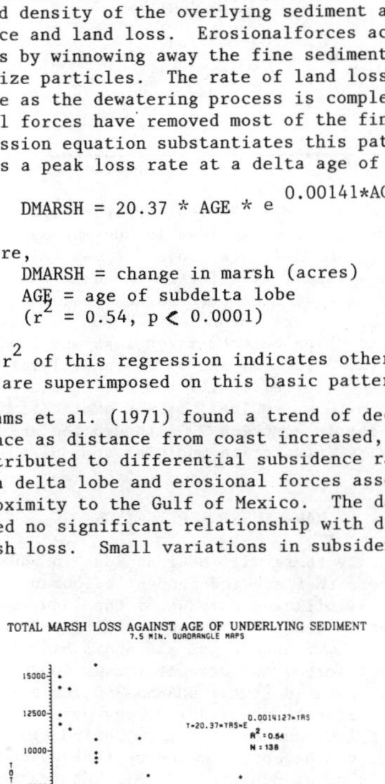

Figure 2.--Total marsh loss against age of underlying sediment. DMARSH=change in total marsh area 1955 to 1978. AGE=age of underlying sediment based on sub-delta lobe formation.

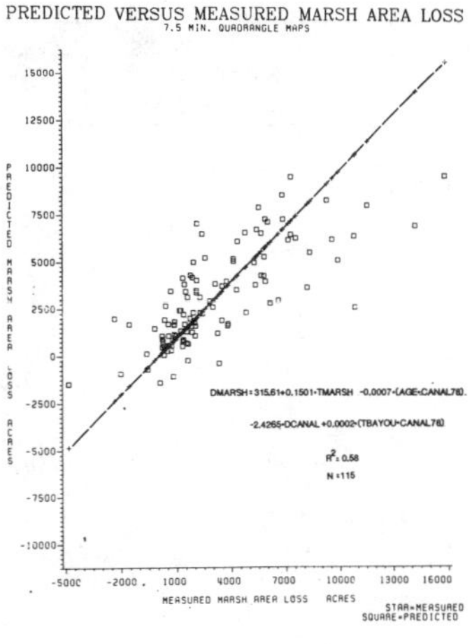

Figure 3.--Predicted versus measured marsh area loss for 7.5 min. quadrangle maps. DMARSH = change in total marsh area 1955 to 1978. TMARSH = total marsh area in 1955. AGE * CANAL78 = age of underlying sediment by canal area in 1978 interaction. DCANAL = change in canal area 1955/1956 to 1978. TBAYOU * CANAL78 = total bayou area in 1955/1956 by canal area in 1978 interaction. Regession statistics in table 1.

Regression analysis indicates canals interact with the regional geology and hydrologic regime to accelerate land loss (figure 3).

$$DMARSH = 315.61 + 0.1502 * TMARSH - 2.4265 * DCASP - 0.0007\ AGE * CANAL78 + 0.0002 * TBAYOU * CANAL78 \qquad (2)$$

where,
DMARSH = change in marsh (acres)
TMARSH = total marsh in mid 1950's (acres)
DCASP = change in canal and spoil area (acres)
AGE = age of sub-delta lobe
CANAL78 = canal and spoil area in 1978 (acres)
TBAYOU = total area of bayous in mid 1950's (acres)
($r^2 = 0.58$, $p < 0.0001$)

Although this equation does not estimate the marsh loss in a particular quadrangle map very precisely, as indicated by an r^2=0.58, the relationship of marsh loss to the predictive variables is very significant (table 1). Marsh loss rate is related to total marsh in 1955 (TMARSH). In general the larger the original marsh area the higher the rate of loss. This relationship is modified by the effects of canals. Change in canal density (DCASP) is the most canal related important factor. The regression coeffecient indicates between 2 and 2.8 acres of marsh loss can be expected from an increase of 1 acre of canal and spoil. Because this is an additive effect to natural marsh loss, areas with

Table 1.--Regression statistics for equation 2.

SOURCE	DEGREES OF FREEDOM	FVALUE (p < F)	PARAMETER ESTIMATE	STANDARD ERROR OF ESTIMATE	TVALUE (P < T)	R-SQUARE
INTERCEPT	-	-	315.61	382.81	0.82 (0.4115)	
TMARSH	1	110.86 (0.0001)	0.1502	0.0218	6.87 (0.0001)	
DCASP	1	33.05 (0.0001)	-2.4265	0.4319	-5.62 (0.0001)	
AGE * CANAL78	1	6.28 (0.0137)	-0.0007	0.0001	-6.24 (0.0001)	
TBAYOU * CANAL78	1	2.51 (0.1163)	-0.0002	0.0001	1.58 (0.1163)	
TOTAL MODEL		38.17 (0.0001)				0.58

high canal density changes that have small marsh areas will experience a higher percentage loss due to canals than would occur in large marsh areas with similar canal densities.

Age of sub-delta lobe (AGE) by final canal density (CANAL78) interaction is the second most important canal related factor. For a given final canal density the effect of the canal decreases as the age of the sub-delta lobe increases, indicating canals do not accelerate land loss in the areas of the older sediments as readily as in younger delta sediment. Sediments in the older deltas are more compacted and have a larger grain size than younger deltas making them less suceptable to erosional forces. Scaife et al. (1981) reported a similar result.

The third important canal related variable is a total bayou (TBAYOU) by final canal density (CANAL78) interaction. For a given natural bayou density, increasing canal density increases marsh loss. The model indicates the more natural bayous in an area, the more sensitive it is to disruption of the natural hydrologic regime. Stone and Mchugh (1979) demonstrated with a two dimensional hydrological model that canals draw water away from natural bayous, decreasing the amount of water and sediment that flows through bayous. Canal spoil banks also limit water exchange over wetlands, thus decreasing further the supply of sediments to marshes. Sediment inputs counteract the natural compaction of underlying sub-delta sediments (Adams et al. 1976), and are the main source of 'new' nutrients to the marshes.

Marsh loss in the coastal zone is obviously controlled by a complex interaction of factors. It is not surprising that this analysis, which treats all canals as if their only effect was due to surface area change, accounts for only 58% of the variation in marsh loss. Other studies have shown that canals have secondary effects which change with the actual dimensions of the canal (Stone and McHugh 1979), the salinity characteristics of the water bodies they connect (Craig et al. 1979), as well as their orientation relative to overland flow direction (Stone and McHugh 1979). Because the data set used for this analysis is oriented around arbitraily defined limits, not ecological or hydrological defined areas, the marsh loss in one map may actually be caused by a canal in an adjacent map. In an attempt to evaluate this factor the data was aggregated into 15 min. quadrangle maps; the results of this analysis are inconclusive, but indicate the same relationships between marsh loss and canals. In addition, the effect of age of the underlying sediment, which is known to be significant but in a nonlinear relationship to marsh loss, is not explicitly accounted for in equation 2, although the age by canal density interaction includes some of this variation.

MANAGEMENT IMPLICATION

Obviously there will be land loss in Louisiana's coastal areas that are independent of human activity. However, a significant portion of the land loss which has occured is caused by human activity. Direct conversion of marsh to canal and spoil bank (total DCASP=72,901 acres) can account for up to 20% of the marsh loss measured (total DMARSH=350,706 acres). If interactive effects are ignored, the proporation of total marsh loss due to canaling, including direct conversion, can be approximated by that caused only by change in canal density (DCASP). Estimating this loss by multipling the regression coeffecient (2.4 ± 0.8) by total change in canals indicates between 145,802 to 174,902 acres of marsh loss for a 23 year period. This represents between 33 and 67% of the actual marsh loss measured. While these numbers must be viewed with caution because the equation used to calculate them is not very precise, they are close to the 69% estimated by Craig et al. (1979). However these values are well below the 89% estimate of Scaife et al. (1981). The complexity of the relationship between canals and marsh loss is apparent from the results of the above analysis. Delta lobe age and the area of natural rivers, streams, and bayous of a given area of land have been shown here to interact with canal and spoil area to significantly effect marshloss. The data to make predictions of marsh loss due to new canals is now easily obtained from permit applications, but the results should be interpreted cautiously. The model may overpredict or underpredict marsh loss. The manager has to combine the results of the model with other information concerning the age of the delta, the relationship between the proposed canal and saltwater intrusion, the spatial relationship between the canal and the surface hydrology of the area, and hydrological and ecological factors specific to the area before making any final decisions on canal construction.

CONCLUSIONS

The major forces dominating marsh loss in coastal Louisiana are geologic factors which operate on long and short term time scales. Sediment compaction, which varies with age of the delta, operates on a time scale important to human use of coastal resources. Canaling interacts with regional geology, local hydrologic regime and physiography of the area to cause premature or accelerated marsh loss. The actual percentage of marsh loss caused by a canal varies with the characteristics of the location.

ACKNOWLEDGEMENTS

We wish to acknowledge the U.S. Fish and Wildlife Service, National Coastal Ecosystems Team in Slidell, Louisiana for providing a computer tape of the data base, Chris Neill for his advise and editorial comments, and Bill Scaife for his original thoughts concerning a relationship between marsh loss and the underlying geology of the sediment.

REFERENCES

Adams, R. D., B. B. Barrett, J. H. Blackmon, B. W. Gane, and W. G. McIntire. 1976. Barataria Basin: Geologic process and framework. Louisiana State University Center for Wetland Resources, Sea Grant Publ. No. LSU-T-76-006, Baton Rouge, La.

Cleveland, C. J., C. Neill, and J. W. Day, Jr. 1981. The impact of artificial canals on land loss in the barataria basin, Louisiana. In W. J. Mitsch, R. W. Bosserman, and J. M. Klopatek (eds), Energy and Ecological Modelling, Elsevier Publicantion Co., N.Y., N.Y. 425-434.

Craig, N. J., R. E. Turner, and J. W. Day, Jr. 1979. Land loss in coastal Louisiana (U.S.A.) and their consequences in coastal Louisiana. Environmental Management 3:133-144.

Frazier, D. E. 1967. Recent deltaic deposits of the Mississippi River, their development and chronology. Gulf Coast Assoc. Geol. Soc. Trans. 17:287-315.

Gagliano, S. M., K. J. Meyer-Arendt, and K. M. Wicker. 1981. Land loss in the Mississippi River Deltaic Plain. Trans. Gulf Coast Ass. Geol. Soc. 31:295-300.

Morgan, D. J. 1977. The Mississippi River delta legal-geomorphological evaluation of historic shoreline changes. Geosciences and Man, Vol. XII. School of Geosciences, Louisiana State University.

Morgan, J. P. 1963. Louisiana's changing shoreline. Baton Rouge, Louisiana State University, Coastal Studies, Inst., Tech. Rept. 16, Pt. D, 13 pp.

Nunez, S. B., Jr., and A. Sour. 1981. Report on special projects for coastal Louisiana to the Legislature and the Governor from the Senate Committee on Natural Resources and the House Committee on Natural Resources.

Scaife, W. W., R. E. Turner, and R. Costanza 1981. Recent land loss and canal impacts in coastal Louisiana. Paper presented at the Louisiana Marine Consortium Symposium on land loss in Louisiana.

Stone, J. H., L. M. Bahr, and J. W. Day, Jr., 1978. Effects of canals on freshwater marshes in coastal Louisiana and implications for management in Good, Whigham, and Simpson, eds. Freshwater Wetlands: Ecological Processes and Management Potential. Academic Press, N.Y., pp. 299-320.

Stone, J. H., and G. McHugh 1979. Hydrologic effects of canals in coastal Louisiana via computer simulation in proceeding of the Summer Computer Simulation Conference, Toronto, Canada. 339-346.

Wells, J. T., S. J. Chinburg, and J. M. Coleman 1982. Development of the Atchafalaya Deltas: Generic analysis. prepared for the Waterways Experiment Station, U.S. Army Corps of Engineers, Vicksburg, MS. Contract DACW 39-80-C-0082. 91 pp.

Wicker, K. M. 1980. Mississippi deltiac plain region ecological characterization: a habitat mapping study. A user's guide to the habitat maps. U.S. Fish and Wildlife Service, Office of Biological Services. FWS/OBS-79-07 and 464 maps.

Wicker, K. M., et al. 1980. The Mississippi Deltaic Plain Region Habitat Mapping Study: U.S. Fish and Wildlife Service, Office of Biological Services, FWS/OBS-79-07, 464 maps.

A REGIONAL ENERGETIC ANALYSIS OF DEVONIAN OIL SHALE DEVELOPMENT IN CENTRAL KENTUCKY[1]

Christopher G. Lind[2]

Abstract.--A hypothetical 10,000 BBL/Day Hytort oil shale retort facility was located in Nelson County, Kentucky. Estimated dollar and energy costs of the 10,000 BBL/Day oil shale operation are summarized. Net energy yield ratios ranging from 2.3:1 to 2.9:1 were obtained, depending on the final output. Finally, a regional energy model, depicting energy and economic flows through the Nelson-LaRue County study region, provided a mechanism for summarizing major environmental impacts of oil shale development.

INTRODUCTION

There has been considerable interest recently in the development of oil shale from the Eastern United States as an energy source. The U.S. Geological Survey has estimated the amount of known Devonian shale resources in the Eastern United States to be 400 billion barrels with probable extentions of known resources projected to 2600 billion barrels of oil (Janka and Dennison, 1979). Of this amount, there are approximately 190 billion barrels of oil recoverable from shales in Kentucky. According to Weil et al. (1979), the energy content in the strippable Eastern Devonian shales exceeds that of reported coal resources in Indiana, Kentucky, Ohio, and Tennessee. Estimates from a study prepared for Buffalo Trace Area Development District in north-central Kentucky (Vyas et al., 1981) show that 4.4 billion barrels of oil are recoverable from shale in Lewis and Fleming Counties, Kentucky.

No major mining operations have yet taken place in Kentucky and no pilot or full-scale retort facilities are yet operational, but much activity involving oil shale development has occurred, particularly in central Kentucky and southern Indiana. Several companies are leasing mineral rights to oil shale, retorting processes are being developed, and pilot plants are being designed. Environmental regulations have been enacted by the Kentucky Department of Natural Resources and Environmental Protection (KDNREP) specifically for oil shale development within the state.

This paper presents an analysis of oil shale development in central Kentucky. A net energy analysis is included as one means of evaluation. Finally, a regional energy diagram shows the material and energy flows through the regional system being studied and provides a mechanism for summarizing major environmental impacts of oil shale development. The gathering of a data base of information on terrestrial and aquatic ecosystems affected by oil shale development, on air and water quality impacts, on socio-economic impacts, and on reclamation techniques for specific retorting processes, will allow the Commonwealth of Kentucky to effectively regulate possible future oil shale development within the state.

Eastern Oil Shale Geology

During the middle and late Devonian and earliest Mississippian (330 to 360 million years B.P.), a majority of the eastern United States was covered by the inland Chattanooga Sea. A few major rivers flowed into the shallow sea. The prevailing stagnant conditions and prolific algal and kelp growth were most likely what led to the deposition of organic-rich (10-20 gal/ton) black shales throughought the basin (Janka and Dennison, 1979)--"Along the eastern edge of the sea, the organic-rich material was substantially diluted with river born sediments (clastics), leading to relatively lower concentrations of hydrogen carbon-rich material in very thick rock formations". Figure 1 shows the extent of Eastern oil shale deposits.

Kentucky has usable resources all along the outcrop areas (Figure 1), with the exception of an area along the far eastern thrust fault. The maximum overburden thickness was estimated at 53 m (175 ft) in the west, 61 m (200 ft) in the east, and 33 m (125 ft) in the south (Janka and Dennison 1979).

Previous Studies Of Oil Shale Development In Kentucky

Currently accepted mining methods show that strip mining will probably be used to extract Devonian oil shale deposits in Kentucky. A recent study for the Buffalo Trace Area Development District in eastern Kentucky (Vyas et al., 1981) used a ridge-top strip mining operation with an overall stripping ratio of 2:1 to analyze the engineering and economic aspects of an oil shale development in northeastern Kentucky. "Stripping ratio" is defined as the ratio of overburden material to mineable shale reserves. There are

[1]Paper presented at the Third International Conference on State-of-the-Art in Ecological Modelling, Colorado State University, May 24-28, 1982.

[2]Christopher G. Lind is Graduate Research Assistant at the Systems Science Institute, University of Louisville, Louisville, Kentucky 40292, U.S.A.

Figure 1.--Eastern oil shale resources showing semi-circular Knobs Region in Kentucky (from IGT and Dames & Moore, 1978).

two retorting processes analyzed in this study--the Paraho Process which is an internal combustion retorting method, and the Hytort Process, an external combustion retorting process.

The Institute of Gas Technology (IGT) has published several articles on Eastern oil shale development (IGT and Dames & Moore, 1978a,b; Janka and Dennison, 1979; Weil et al., 1979). IGT focused on initial process design of Hytort for hydrogen retorting of Devonian oil shales. Environmental aspects of Eastern oil shale development and environmental effects in both pilot plant and commercial plant development were discussed including water quality, land use, and socio-economic impacts.

Energy Resources Co. Inc. (ERCO) conducted a three-part study for the Kentucky DNREP. The first report was a data collection on Eastern oil shales. The second report addressed sections of the KNREP's drafted regulations for oil shale development in Kentucky that needed changing due to knowledge obtained in the first task report. The third report evaluated the specific sections of the oil shale regulations (405 KAR 30) dealing with disposal methods in shale operations and changes recommended where sections appear inadequate (ERCO, 1981a,b,c).

The University of Louisville has an ongoing research interest in Eastern oil shale development. Studies have included a preliminary net energy analysis of Eastern oil shale development based on the Buffalo Trace Study (Lind & Mitsch, 1981), a regional environmental analysis of Devonian oil shale development (Mitsch & Lind, 1981), an assesment of water quality impacts of oil shale development in Kentucky (Mitsch & Lind, in press), and a regional energy study of oil shale development in central Kentucky (Lind, 1982).

DESCRIPTION OF STUDY AREA

The Nelson-LaRue County region of Kentucky (Figure 2) was chosen for further study because of

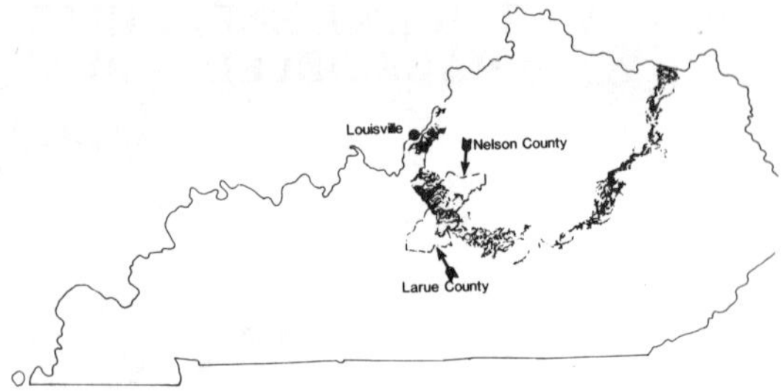

Figure 2.--Major deposits of oil shale in Kentucky Knobs Region with Nelson-LaRue County study area shown.

increased oil shale and mineral leasing activity recently in both counties, and because there have not been any site-specific studies of oil shale development in the area. There has been considerable interest shown by both local and multi-national oil companies in oil shale development in Nelson-LaRue Counties. Records at the County Clerks Office in Bardstown, Kentucky, show over 10,000 acres of land already leased for oil and mineral rights in the region. Oil companies are paying landowners for the possible eventual use of their land. Payments have been in the $5.00 to $10.00 range per-acre/year.

Climate

The climate of the oil shale region has been described as humid, temperate, continental type by Burroughs (1926). Burroughs also noted that differences in elevation or latitude did not affect climate characteristics to any noticeable degree in the Knobs region of Kentucky. Summer thunderstorms or storm systems that move from west to the east account for precipitation in the Knobs region. The greatest percentage of precipitation is accounted for by moisture-bearing, low pressure formations moving either southwest or northeast across Kentucky (Anderson, 1959). The fall season is usually the driest with the spring season the wettest. During the spring and summer months, thunderstorms associated with high intensity rainfall are common. The mean annual precipitation in the Knobs is 115.7 cm, although great variability occurs from year to year (Harker et al., 1981). Average monthly precipitation and evapotranspiration are 9.9 cm and 6.4 cm respectively (Kentucky Climate Center, 1981).

The Knobs region has a mean average temperature of 13.2° C. The summer mean maximum and minimum temperatures are 30.4° C and 16.9° C. The winter mean maximum and minimum temperatures are 7.1° C and -3.7° C. The frost-free period is usually in the 151 to 162 day range (Burroughs, 1926).

Soils and Geology

The soils of the Knobs have been characterized as naturally poor and thin (Miller, 1919), and among the least productive in Kentucky (Wharton, 1945). Soils of the Knobs are within the soil order Inceptisols, suborder Ochrepts according to Soil Conservation Service classification (Arms et al., 1971). Inceptisols are the dominant soils in large areas characterized by gentle to steep slopes with humid conditions.

Elevation in the Kentucky Knobs region varies between 120 m to 425 m. The topography varies from gently rolling hills and ridges with broad flat bottomland to hilly and mountainous terrain which rises above the foothills and eventually meets the plateau areas (Harker et al., 1981). The Devonian shales form the base upon which the Knobs lay. Overlying the Devonian shales are the New Providence shale formation and Mississippian limestone which sometimes is covered by Pennsylvanian conglomerate, sandstone, and shale (Burroughs, 1926). The upper part of the Devonian shale is the New Albany shale formation.

Aquatic Systems

Intermittent and other low order streams dominate the Knobs and hills found in Nelson-LaRue Counties. First order streams are often intermittent with second and third order streams generally displaying more frequent flow and good water quality.

Mitsch and Lind (1981) describe potential water quality problems from shale mining, shale processing, and shale disposal in a paper entitled "Potential Impact of Oil Shale Development on Water Resources in Kentucky." Water quality data availabe for the central Kentucky region was summarized.

A reading of 2 mg/l was found for total iron, a relatively high reading, indicating a potential for acid drainage problems in the region if pyrite seams are exposed. None of the other recorded water quality parameters were found to be excessive. The sparse nature of data available on water quantity in the region indicates the need for further studies quantifying the water quality of both surface and subsurface water systems to be able to accurately describe the aquatic systems.

Terrestrial Ecology

A description of the terrestrial ecology of the Kentucky Knobs oil shale regions has recently been published by the Kentucky Nature Preserves Commission (KNPC) (Harker et al., 1981). The KNPC study of the vegetation and floristics of the Knobs region was based on previous research by Wharton (1945) and Braun (1950).

The existing vegetation of the Knobs is mainly secondary forest, and little of the original forest remains. Agriculture and logging have been the primary influence on vegetation that now exist in the Knobs. Wharton (1945) found that, where excessive erosion had removed large amounts of soil, the existing forest was generally more xeric than the original. Bailey (1978) and Kuchler (1964) classified the vegetation of the oil shale region as Mixed Mesophytic Forest.

METHODS

Oil Shale Development Study

Location selection for an oil shale study was based on public records of oil shale leasing in central Kentucky. Data were gathered at the County Clerk's Office in Bardstown, Kentucky, on recent mineral and gas leasing activity. Approximately 10,000 acres of land were leased by several companies in central Kentucky for possible oil shale development. The region around New Haven, Kentucky, was initially chosen as a possible oil shale development site, primarily because there was significant leasing activity in the area, but also because of the availability of railroad lines and water supply in the Rolling Fork River.

Energy Cost Accounting (Including Net Energy)

Energy costs associated with the production process can be classified as direct and indirect energy costs. Direct energy costs include fossil fuels, electricity, solar radiation, and geothermal heat. Indirect energy costs are the summation of all energy costs that incurred in the production process and include energy costs of goods and services. There are generally two methods used in computing indirect energy costs. The first method is a vertical process analysis, which is a summation of all the energy costs incurred along the production pathway. This direct method of energy accounting was not used, because all relevant energy data was not readily available to correctly sum all contributary energy inputs.

A second approach, the input-output energy matrix model, developed by the Center of Advanced Computation Group at the University of Illinois Urbana-Champaign (Herendeen, 1973; Herendeen and Bullard, 1974), was used to compute indirect energy costs. Energy intensity coefficients have been computed for various sectors of the economy and give the average energy cost per unit of the manufacturers' final output as measured in 1967$ (Herendeen and Bullard, 1974). The energy intensity coefficients used to calculate energy costs for the oil shale development study are shown in Table 1.

Table 1.--Energy intensity coefficients of capital goods[1].

Sector	Title	Energy Intensity Coefficient (MBTU/$1967)	(BTU/BTU)
a) 10.7303	Misc. Professional Service	.0250	
b) 10.6504	Water Transport	.2236	
c) 10.4905	Power Transmission Equipment	.0596	
d) 10.4502	Mining Machinery	.0714	
e) 10.3613	Lime	.5703	
f) 10.2704	Misc. Chemical Production	.1835	
g) 10.1105	New Construction-Other	.0867	
h) 10.0900	Stone Clay Mining	.1045	
i) 10.6803	Water, Sanitation Service	.1084	
j) 10.1202	Maintenance Const-Other	.0571	
k) 10.3101	Petroleum Refined Product		1.2082

[1]Herendeen and Bullard (1974)

The input-output matrix model was an adaption of the input-output analysis model of the economy developed by Leontiff (1966). The input-output matrix model is general in scope and covers a wide spectrum of products and processes. The general assumption of the input-output matrix model is that the energy consumed in the production of a good or service is embodied in it as indirect energy.

An assumption is made that energy embodied in inputs to an economic sector is passed on as part of the sector's output. This allows one to apply the law of conservation of energy for each sector's output. In a N-sector economy, the energy balance that is worked out around the jth sector is calculated as:

$$\sum_{i=1}^{N} e_i X_{ij} + E_j = e_j X_j \qquad (1)$$

where

X_{ij} = the transaction flow from sector i to j.
X_j = is the total output of sector j.
E_j = Direct energy consumed by sector j.
e_i, e_j = energy intensity coefficient

(Herendeen and Bullard, 1974; Khan, 1977)

The above equation assumes that there is a unique energy intensity, e_j, associated with a specific sector's output. This allows one to solve the system of linear equations for the vector $\underline{e}$.

Energy costs for each sector chosen for the oil shale operation were determined by the following equation:

$$(e_i)x(\text{dollar costs-1967\$}) = \text{energy cost} \qquad (2)$$

1967$ were obtained by multiplying 4th Quarter 1980$ by a deflation factor of 0.39.

Energy costs of labor were determined by obtaining estimated dollar costs and multiplying by the GNP/energy ratio for the United States (Lind 1982). This method and the validity of including labor costs are discussed in further detail by Costanza (1980). Indirect energy costs of both Capital Investment and Annual Operating Costs were also determined using the GNP/energy ratio for comparison with the results from the input-output matrix method.

Economic data used in the energy cost calculations were obtained primarily from Vyas et al. (1981). Reservoir costs were derived from personal conversation (Jim Duck, U.S. Army Corps of Engineers). The costs for spent shale and reclamation were taken from a study published by ERCO (1981c).

Net energy yield ratios were determined by dividing the gross output measured in BTU/Day (Figure 3). The yield ratio for the production of shale oil was determined using both the CAC energy coefficient and NEC/GNP ratio. The further refining of shale oil to refined petroleum products was calculated using the CAC energy coefficient for Petroleum Refined Products (Table 1).

Regional Model

A regional and energy model was constructed using an energy systems language developed by Odum (1971) and Odum and Odum (1976). Literature dealing with each component of the model was reviewed to better understand the regional system's flows, storages, and forcing functions. The regional model was developed as an aid in describing impacts of an oil shale industry on the region's economy and natural systems. Environmental impacts were described for the oil shale study through review of published information on Devonian oil shale development. Ecological and socioeconomic data were gathered from literature and knowledgeable sources.

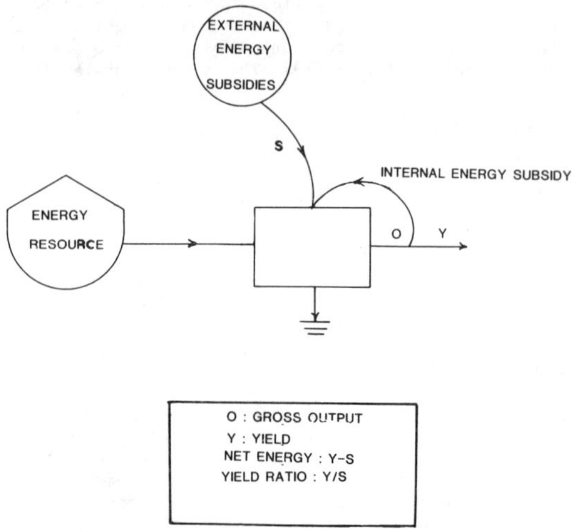

Figure 3.--General net energy model (Odum et al., 1976).

RESULTS AND DISCUSSION

Oil Shale Development Study

Location

The area chosen to site a 10,000 BBL/Day oil shale mining and retort operation is near New Haven, Kentucky (Figure 4). The study site area is located at approximately 37° 42' 35" Latitude and 85° 35' 00" Longitude. The nearest town, New Haven, is located 5 km (3 mi) south of the site.

Retorting Facilities

There have been several retorting processes mentioned for possible use in a Kentucky based oil shale operation (Feldkircher and Janka, 1979; Vyas et al., 1981). The Hytort process was chosen for this study because the process gives a higher oil yield per ton than other processes being considered for use in Kentucky. The process has also frequently been discussed for application in Kentucky (Feldkirchner and Janka, 1979; Feldkirchner et al., 1980; Weil et al., 1980; Vyas et al., 1981).

The Hytort process is based on direct, noncatalytic hydrogenation of kerogen at high pressure and controlled heat-up rates. The plant requires 4100 gpm of makeup water, approximately 30,000 tons of raw shale, 565 tons/day oxygen, 45.3 MW electric power, and 269,700 scfm of air. The processing plant produces approximately 10,217 BBL/Day of shale oil, and 104 tons/day of anhydrous ammonia. The plant itself requires approximately 65 acres (2300' x 1200').

Water Resources

Water supply in the Eastern Devonian oil shale region has often been cited as being in much greater supply than that of the Green River area in Colorado where Eocene shales are being mined. Water requirements for a 10,000 BBL/Day Hytort operation were estimated from two sources. Vyas et al. (1981) esti-

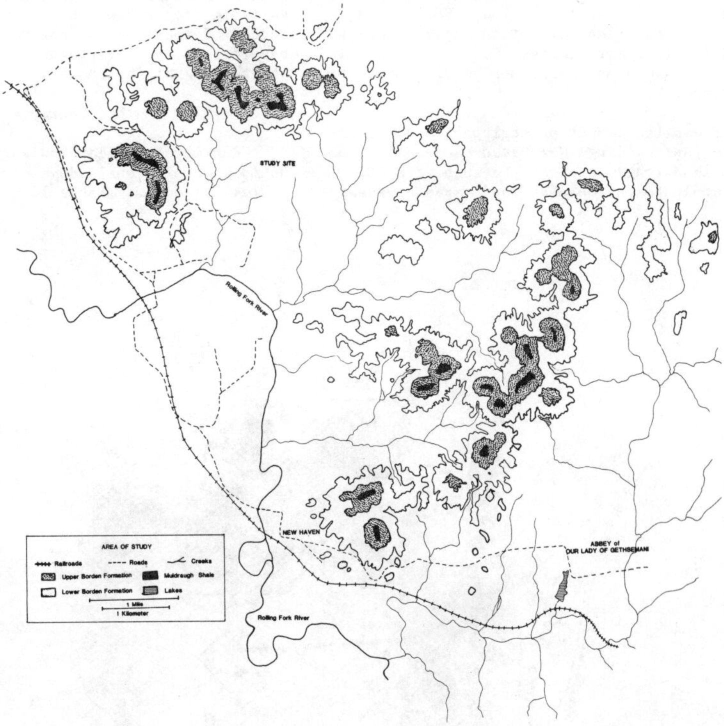

Figure 4.--Location of oil shale development study. The strippable reserves are the upper and lower formations of the Borden Formation and the Muldraugh Shale.

mated 245 liters/sec (4100 gpm) for a 10,217 BBL/Day oil shale operation, while IGT and Dames and Moore (1978) estimated 600 liters/sec (10,000 gpm) for a 46,214 BBL/Day operation. Our study assumes water supply necessary to supply the water requirements of a 10,000 BBL/Day oil shale operation near New Haven to be 245 liters/sec (4100 gpm).

It appears that the Rolling Fork River will not be able to support the water quantity needs of the 10,000 BBL/Day oil shale operation during low flow period, without a reservoir system near the plant site. The reservoir must be designed to meet the demands of the mining and processing operations in low-flow conditions of the Rolling Fork River. A mass curve of the Rolling Fork River during the 1964 water year drought was used to determine the reservoir capacity required to meet the make-up water demands. The minimum reservoir capacity needed to compensate for 4 months of low-flow conditions is estimated to be 2.7×10^6 m^3 (2.2×10^3 acre-ft) (Lind, 1982).

A reservoir was sited near the shale processing plant site, yet off the main stream, to reduce the distance of water transport. Water may be diverted from the Rolling Fork River to the reservoir during high flow periods. The water will be transported by pipeline to the reservoir from the stream and another pipeline will be constructed to carry water to the plant site. Water will also be pumped from the reservoir to meet other demands of the oil shale operation such as dust control.

Mining

The mining site-plan was developed using information from "Synthetic Fuels from Eastern Oil Shale" (Vyas et al., 1981), ERCO's report entitled "Development of Disposal Performance Standards for Oil Shale Mining and Processing" (ERCO, 1981c), and coal strip-mining practices in eastern Kentucky. Ridge-top strip-mining and a conveyor system will be the method utilized in the mining operation (Lind, 1982). Ridge-top mining is an excellent mining practice for the study area since all the resources will be mined initially. The reclaimed ore would not be disturbed by future operations.

Table 2 shows the amount of strippable shale reserve acreage surrounding New Haven and also the strippable shale acreage necessary to support a 10,000 BBL/Day oil shale operation. The land area necessary to support the mining operation would be approximately 6.8 km^2 (1660 acres) based on an average yield of 84-97 liters-per-metric ton (20-23 gal/ ton) of shale oil and a 18 m depth of seam. There are approximately 18 km^2 (4480 acres) of oil shale in the region surrounding New Haven, Kentucky. The Knobs area adjacent to the study site location contains approximately 7.3 km^2 (1800 acres) of suitable strippable acreage. Figure 5 shows those geological formations which are suitable for ridge-top mining. The upper and lower Borden Formations and the Muldraugh shale areas will be removed as overburden to strip-mine the lower strata (New Albany shale).

Net-Energy Analysis

Estimated dollar and energy costs of the 10,000 BBL/ Day oil shale development are summarized in Tables 3 and 4. Table 3 summarizes both dollar and

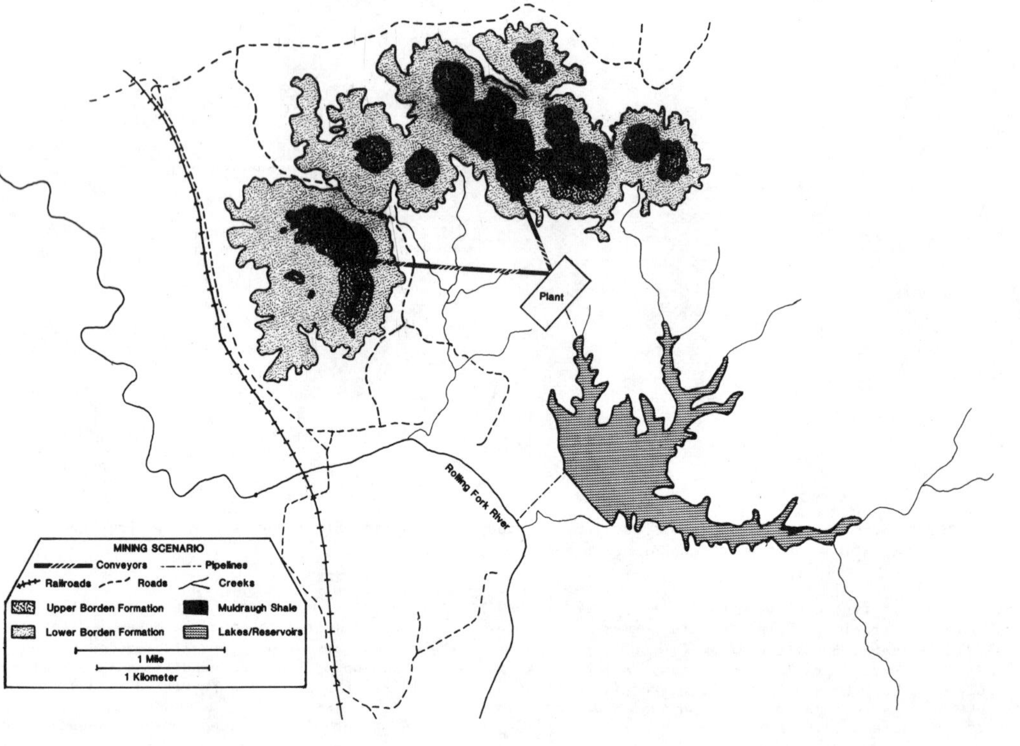

Figure 5.--Mining site-plan for oil shale development.

Table 2.--Strippable shale reserves surrounding New Haven, Kentucky

Account	Land or Mining Area Rate
10,000 BBL/Day oil shale operation (Land base required for mining)	6.8 km^2 (1660 acres)
Strippable reserves in general study area	18 km^2 (4480 acres)
Strippable reserves adjacent to study site location	7.3 km^2 (1790 acres)
Land base mined each year of 10,000 BBL/Day shale oil operation	.340 km^2/yr (84 acres/yr)

energy costs associated with total capital investment, while Table 4 outlines dollar and energy costs connected with annual operating costs.

Economic Costs

The total capital cost of the 10,000 BBL/Day operation near New Haven is $875 million (Table 3).

Table 3.--Energy costs associated with total capital investment of 10,000 BBL/Day oil shale operation.

Account	Dollar Cost ($x10^6$ 1980$)	Energy Cost $x10^9$BTU/Day)
Mining	83.0	0.46
Crushing;Screening;Storage	50.0	0.23
Reservoir	4.0	0.02
Spent Shale Disposal/ Reclamation	20.2	0.08
Retorting	144.0	1.41
Oil Recovery & Gas Cooling	15.0	.15
Ammonia Recovery & Acid Gas Cooling	37.0	0.36
Cryogenic Separation	40.0	0.39
Sulfur Recovery & Tail Gas Cleanup	15.0	0.15
Hydrogen Production	50.0	0.49
Waste Water Treatment	22.0	0.22
Shale Oil Storage	4.0	0.01
Oxygen Plant	15.0	0.15
Utilities & Offsites	33.0	0.10
Professional Services	45.0	0.06
Field Indirect Costs, Insurance & Start-up	102.0	0.36
Start-up Allowance & Contingency	33.0	0.13
Working Capital	21.8	0.08
Mine & Mobile Equipment Replacement	140.8	0.54
TOTAL	875	5.39

Retorting and mine and mobile equipment replacement represent 55% of total capital costs. Annual operating costs amounted to $86 million per year (Table 4). Capital costs amount to 33% of the total costs associated with the oil shale operation, while total annual operating costs (consumables, taxes and insurance, labor, and electricity), when summed over the 20 year operation, account for 67% of the total costs associated with the oil shale operation (Lind, 1982). Electricity represented 15% of the total annual operating costs and 10% of the total costs. Annual labor cost is estimated as $9.6 million or 29% of total annual operating costs.

Table 4.--Energy costs associated with annual operating costs of 10,000 BBL/Day oil shale operation.

Account	Dollar Cost/Year ($x10^6$ $1980/year)	Energy Cost ($x10^9$ BTU/DAY)
Consumables		
Catalysts & Chemicals	0.84	0.16
Electricity	12.57	10.90
Untreated Water	0.97	0.23
Supplies-(Oper&Main) (Oper & Main)	25.15	1.90
Labor		
Operating Labor	9.12	0.70
Maintenance Labor	5.78	0.44
General & Admin	9.67	0.74
Taxes & Insurance		
Taxes	11.50	0.88
Insurance	10.46	0.74
TOTAL	86.06	16.75

Spent shale disposal and reclamation costs were estimated from ERCO (1981c) data to be $20.2 million or 2.3% of total capital investment. The ERCO study pointed out that reclamation costs for Eastern oil shale development were derived primarily from Eastern coal reclamation costs. The spent shale material will have a greater volume than the raw shale, which will escalate the pricing for reclamation of spent shale. The general consensus is that reclamation costs for spent shale disposal and reclamation will be in the upper price range for Eastern strip-mine coal reclamation costs (ERCO, 1981c).

Energy Costs

Total direct and indirect energy costs associated with the 10,000 BBL/Day oil shale operation are $22.1x10^9$ BTU/Day (Table 3-4). Capital energy costs were estimated as $5.39x10^9$ BTU/Day (24%), while annual operating energy costs were calculated as $16.75x10^9$ BTU/Day (76%). Externally supplied energy in the form of electricity (45MW) accounted for 49% of the total energy costs. Total capital investment associated with environmental control (sulfur recovery and tail gas clean-up, waste water treatment, sediment ponds, and spent shale/disposal/reclamation) accounted for approximately 4% of the total energy costs. Energy costs associated with consumables other than electricity accounted for 10% of the total energy costs. Labor accounted for 9% of the total energy costs. Energy costs for the retort facility, $1.41x10^9$ BTU/Day, was the second costliest account in the oil operation amounting to approximately 6% of total energy costs. However, purchased direct energy costs in the form of electricity is 7.7 times greater than the retorting energy costs. It is evident from the energy calculations that the magnitude of direct energy costs of electricity is the greatest cost associated with the 10,000 BBL/Day oil shale operation.

Yield Ratio

Table 5 shows the yield ratios for shale oil produced in the 10,000 BBL/Day shale oil development study and refined petroleum. The yield ratio was calculated by dividing the gross energy output by external energy costs (Figure 6). The yield ratios for the production of unrefined oil and refined petroleum from shale were 2.9:1 and 2.3:1 respectively. The yield ratio for refined petroleum was calculated by adding the energy cost of refinery process. The total energy costs of refined petroleum was given by Herendeen and Bullard (1974) as 20.8% of the energy value of the oil.

Table 5.--Net energy yield ratio for 10,000 BBL/Day Hytort operation.

Account	Energy Costs or Output ($x10^9$ BTU/Day)	Yield ratio[1]
Gross Output	64.7	
Costs		
Annual Operating	16.75	
Total Capital	5.39	
Total Cost -Unrefined oil(shale oil)	22.14	2.9:1
Total Cost[2] -Refined petroleum	28.39	2.3:1

[1]Yield ratio = gross energy yield/external energy subsidy.

[2]Total Cost = The total cost of refined petroleum was calculated by adding the energy cost of the refinery process given by Herendeen and Bullard (1974) as 20.8% additional energy to the total energy cost of unrefined oil. The calculation was as follows: $(22.14x10^9$ BTU/Day$)(1.208) = 28.39x\ 10^9$ BTU/Day.

Using the NEC/GNP ratio for 1980, the yield ratio for the production of unrefined shale oil was calculated to be 3.20:1. This method is based on the average energy cost per dollar of investment in the U.S. economy. Specific sectors of the economy are not drawn out. The yield ratio is slightly higher using this method because the method does not account for the sectors of the economy that are more energy intensive per dollar of investment.

Although the ratio is positive for the Hytort 10,000 BBL/Day shale oil study, environmental costs could measurable reduce the yield ratio in actual operation. The reclamation costs for spent shale disposal have not been adequately researched to accurately predict the costs involved in future spent shale reclamation in Kentucky. In addition, because no pilot plant has yet been built for the Hytort process, it is difficult to determine if the capital and operating costs used to calculate the energy costs are accurate predictions of actual capital and operation costs.

The yield ratio obtained for the Hytort 10,000 BBL/Day oil shale operation is less than the yield ratios of the following energy technologies: 1) coal liquefaction (surface mining) with gasoline as final product (6.41:1), 2) coal liquefaction (underground mining) with gasoline as final product (9.61:1), 3) natural gas with High-BTU Gas as final product (24.21), 4) coal gas (surface mining) with High-BTU Gas as final product (24.75), and 5) coal gas (underground mining) with High-BTU Gas as final product (25.45:1) (CERI, 1976). The yield ratio obtained in our study was less than all the above mentioned processes due, in part, to the tremendous direct energy cost in the form of electricity that is associated with the Eastern oil shale operation. The yield ratios given for Western oil shale operations (Table III) are generally greater than the yield ratio obtained in our study, because the oil yield per ton of shale mined is greater in the Western shale deposits. This suggests that although net energy is obtained, oil shale from Kentucky is not "energetically" competitive with the above energy sources.

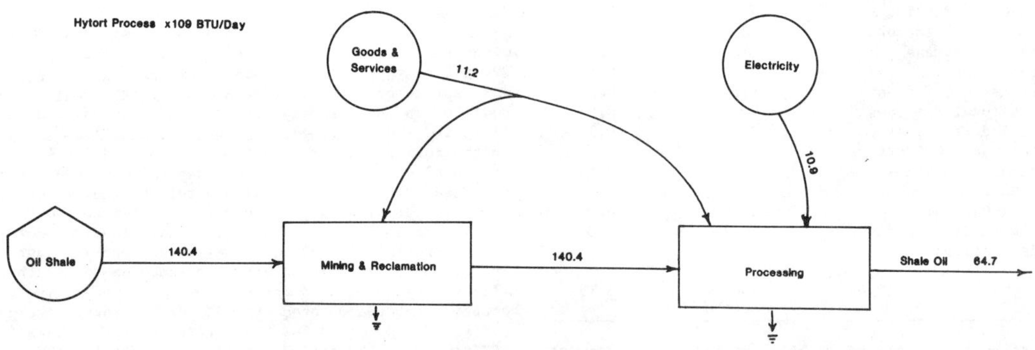

Figure 6.--Energy flow for 10,000 BBL/Day oil shale operation in central Kentucky.

Regional Impact Analysis

Regional Model

A model of the major energy and economic flows of the Nelson-LaRue study region where the shale development study takes place is shown in Figure 7. The diagram serves as a summary of environmental consequences of oil shale development discussed below. Land area is 1130 km^2 (440 mi^2) and 675 km^2 (263 mi^2) for Nelson and LaRue Counties respectively. The boundaries of the study region are specifically inexact because aquatic, terrestrial, and socio-economic impacts would be confined within watersheds, air pollution effects within the plume of any air emission, and socio-economic impacts within political and social boundaries. Major natural systems of the region include terrestrial forests of the Knobs, streams and rivers, and geologic deposits including Devonian oil shale. The local economy, centered around several small towns including Bardstown, Boston, and New Haven, is primarily supported by agriculture, forestry, tourism, alcohol production, and manufacturing. Products are exchanged with the external economy for other essential goods and services (Mitsch & Lind, 1981).

Impacts On Water Resources And Aquatic Ecosystems

The development of an oil shale industry in New Haven may present both water quantity and quality problems. The questions of water usage and water quality cannot be separated because water diversion may have an impact on in-stream water quality. Water impacts of oil shale development will be discussed separately for shale mining, shale processing, and spent shale disposal.

Mining. Major anticipated water problems of ridge-top mining of oil shale include erosion and sedimentation, hydrological modification, acid drainage, and toxic chemical pollution.

Acid drainage, typical of Kentucky coal mining, may be a major impact of shale mining in the region. Pyrite (FeS_2) is present in oil shale seams with concentrations varying between 2 and 5 percent (Mitsch & Lind, 1982). The high levels of pyrite in the shale seams coupled with the large amounts of shale that would be mined in a 10,000 BBL/Day operation suggest that acid drainage could be one of the major water quality impacts of the mining operation.

Trace metals may be found in the mine runoff. These metals may be toxic to aquatic life or detrimental to local water supplies. Vyas et al. (1981) found that cobalt (Co), chromium (Cr), copper (Cu), and molybdenum (Mo), nickel (Ni), zinc (Zn), and uranium (U) were higher in the oil shales than in surrounding geological materials. There is a possibility that some of the metals might mobilize due to acidic conditions in the mined areas, causing the metals to be transported to adjacent streams and eventually to the Rolling Fork River system. Cr, Cu, and Zn are listed

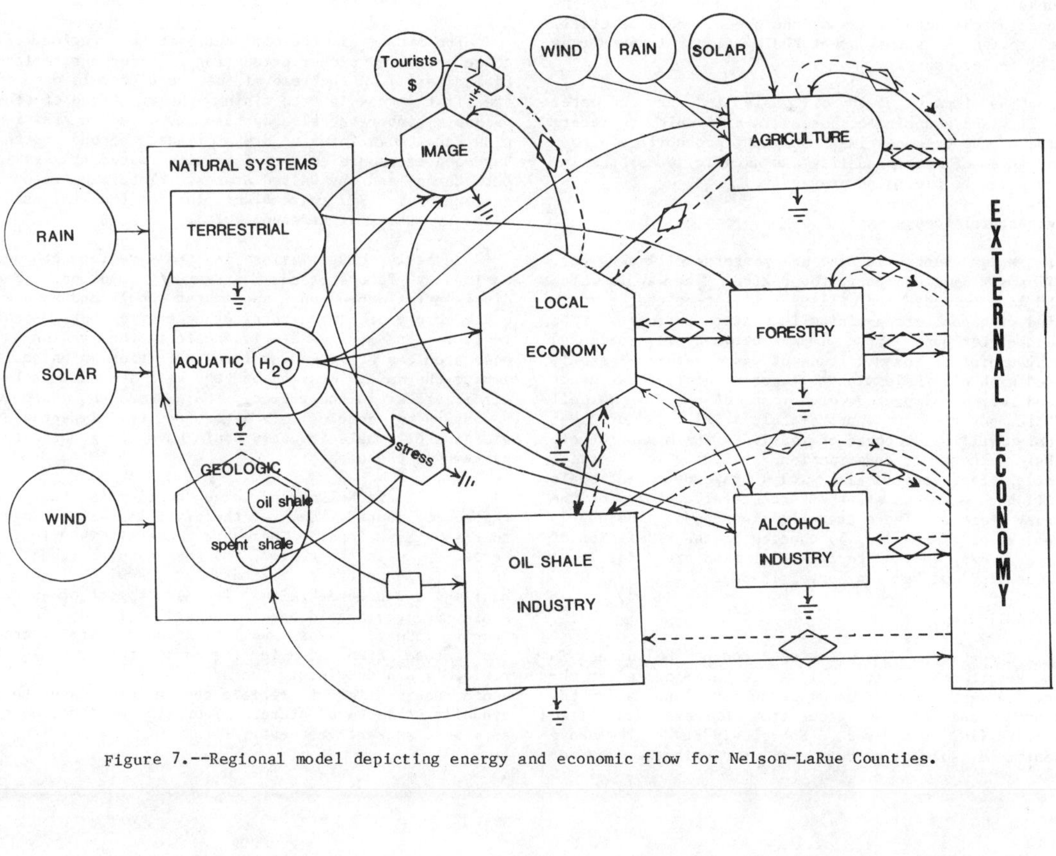

Figure 7.--Regional model depicting energy and economic flow for Nelson-LaRue Counties.

as toxic pollutants under the National Pollutant Discharge Elimination System (NPDES) program of the Federal Water Pollution Control Act (ERCO, 1981c). The oil shale industry is not regulated by the NPDES program, but eventually the operators of the industry will have to conform to NPDES permit requirements for the discharge of any toxic pollutant to navigatable water (ERCO, 1981c).

Retorting. The retorting of oil shale could possibly have several detrimental effects on the water quality of streams in the study region. The possible effects include: 1) the direct discharge of wastewaters from various unit operations into spent shale disposal sites, 2) indirect effects of water withdrawal and effects of impoundments on stream water quality, and 3) the potential of accidental oil spills.

Spent Shale Disposal. Approximately 35,000 tons per day of spent shale must be disposed of daily. Most of the material will be placed in mined areas, but the excess material will be placed in head-of-hollows causing the existing stream flow patterns to be directly effected.

Groundwater and surface water quality could be adversely affected due to leaching of raw shale and spent shale. "Preliminary tests of raw shale leachates indicate that leachate from exposed shale will be acidic and may not meet drinking water criteria for concentrations of arsenic, copper, iron, manganese, zinc, and sulfate" (Tolbert et al., in press). Other studies by McCoy and McCoy (unpublished) indicate that spent shale is not toxic and should be classified as a nonhazardous material. To determine conclusively the leaching characteristics of the spent shale material, monitoring of actual spent shale disposal operations will be necessary.

The impact of an oil shale industry on water quality and aquatic ecosystems is difficult to determine at the present time. Only the probability of occurrence of water quality and aquatic ecosystem impacts can be predicted presently.

Terrestrial Ecosystems

Major changes in land use patterns will accompany oil shale development in the region. The way in which the land is disturbed will be similiar to that associated with coal strip-mining in Eastern Kentucky. The main difference in the impact between oil shale and coal mining is the magnitude of waste material associated with oil shale mining. Approximately 84 acres of land will be impacted each year of the 20 year oil shale operation and approximately 4.3×10^7 m^3 of spent shale will be disposed of yearly. The mining of oil shale, reservoir construction, building of service roads, plant construction, and storage of spent shale will have impacts on the terrestrial ecology of the study region. These are: 1) increased erosion, 2) removal of vegetation, 3) changes in the vegetation of areas mined, and 4) depletion of land area for local fauna (IGT and Dames & Moore, 1978a).

Air Pollution

The principle sources of airborne pollutants in the Hytort process are: 1) shale handling and crushing, 2) vent gases, from Claus sulfur plant and other sources and 3) flue gases from boilers and fired heaters (IGT and Dames & Moore, 1978b). The above mentioned sources produce pollutants in the form of particulates, oxides of nitrogen and sulfur, CO, CO_2, and light hydrocarbon vapors. Airborne dust will be a result of blasting, mining, road building, and crushing and transporting of the raw and spent shale. Airborne particulates will be dispersed by winds, being eventually deposited in population centers of the region. The use of water to spray roads and crushing site will greatly reduce the amount of airborne particulates. Our oil shale study also calls for the use of water to spray the shale being transported by conveyor to the retorting site. Baghouse filters will be used in crusher units to capture the dust from the shale as it is crushed. If these steps are properly administered, the potential hazards of air pollution will be reduced significantly. One study has suggested that air pollution will not be a problem.

Socio-Economic Impacts

Agricultural production is a major industry in the two county region. The agricultural labor force was 959 and 694 in Nelson and LaRue counties respectively in 1979 (Nesbitt, 1982). Burley tobacco, beef-cows, dairy cows, hogs and pigs, hay, corn, wheat, and soybeans are the major agricultural commodities for the Nelson-LaRue study region. Total agricultural sales were estimated to be $46 million in 1978 (U.S. Dept. of Commerce, 1980). The removal of top soil from surrounding farmland may happen if top soil stockpiled from mining operations does not meet the two feet of top soil requirement for reclamation of mined areas promulgated by the Kentucky DNREP. This would cause valuable farmland to be diverted from agricultural production. There also may be detrimental effects to crops by airborn particulates from the trucking of the top soil to the mined sites.

The mining of the oil shale in the region will cause increased timber production to occur during initial clearing. The removal of the timber is one of the first steps in the mining phase. Much of the valuable timber has already been harvested in the region and most forests are at least second growth. Hardwood trees are sold to furniture manufacturers in both Europe and the United States. Total lumber product sales in 1977 were $1 million for the two counties (Ky. Dept. of Revenue, 1978).

Several distilleries in the region produce nationally famous brands of Kentucky bourbon. The distilleries depend on a good water supply and an adequate supply of agricultural products from outside the region. The mining of oil shale in the region may pose problems of groundwater contamination which could impact the water quality of the well systems of the distilleries in the region. More research would be necessary to conclusively determine the impacts an emerging oil shale industry would have on groundwater systems in the area.

Tourism in Nelson and Larue Counties is heavy in the summer months. Some of the distilleries are major tourist attractions. There are several historic interest spots in the region, and the natural landscape of the region attracts visitors for camping, fishing, hunting, and nature hiking. The development of an oil shale industry could have a negative impact on the tourist industry. Noise pollution would distract from the serene atmosphere in the Knobs for both local residents and tourists. The strip-mining of the Knobs would require time to reclaim the impacted land to a visually attractive state. Tourists may decide to seek more attractive areas.

The major town in the study region is Bardstown (pop. 7,200 in 1978), with several small towns including New Haven, Boston, Culvertown, and Howardstown. Populations for Nelson and LaRue are 26,400 and 11,400 respectively (1979 figures). The total civilian labor force was 11,091 and 5,142 in 1979 for Nelson and LaRue Counties respectively (Nesbitt, 1982). Population is estimated to rise to 49,000 by the year 2000 for the two counties (Urban Studies Center, 1979).

Vyas et al. (1981) projected an employment level of 900-1100 for a 10,000 BBL/Day oil shale mining and processing operation. Statewide and regional employment data for Kentucky were reviewed by Nesbitt (1982) to determine the primary to secondary employment ratios for the oil shale region. Primary employment included mining, manufacturing, and farming. "Since oil shale industrialization will build to some extent on existing services that will not expand proportionally, the ratio for oil shale was assumed to be two to one" (Nesbitt, 1982). Therefore, a total of approximately 1000 oil shale workers and 2000 service employees would be needed for a 10,000 BBL/Day oil shale industry in the region.

The image of the region is directly related to the maintenance of the natural ecoystems. The regional energy and economic model (Figure 7) shows the image of the region depending partially on the maintenance of the delicate balance of both the terrestrial and aquatic systems. The monitoring of environmental impacts of oil shale development in the region will be vitally important to protect the natural systems of the region. The model shows that the oil shale industry not only directly affects the economic system of the region, but also puts stress on both the terrestrial, geologic, and aquatic systems in the region. The healthy survival of both natural and manmade systems in the study region depend on continuing research to determine any detrimental effects an emerging oil shale industry will have on the local economy and natural systems of the region.

CONCLUSIONS

The proposed 10,000 BBL/Day oil shale development in central Kentucky was based on data that is preliminary in nature. The building of pilot plants or commercial operations will enable the questions on environmental and economic impacts to be quantified.

The mining of Eastern oil shales will definitely present a problem of acid drainage. Further research must be undertaken to predict quantitatively the leaching properties of spent shale material. Pilot plants will be the first stage in the development of ongoing oil shale studies in the Eastern United States. Next, commercial scale retorting operations will be developed which will allow the accumulation of data to conclusively determine the environmental impacts of mining and retorting of Devonian oil shale. Where low-flow conditions of streams in the shale region prevent the use of rivers and streams for make-up water for the retorting process, alternative cooling processes may have to be designed.

Economies of scale may necessitate the use of 50,000 BBL/Day oil shale plants in the Knobs region. If that is the case, large acreages of land will be necessary to supply shale material. Adjacent farmland surrounding the Knobs will have to be mined. Valuable farmland acreage would no longer be in production of crops. Reclamation of the farm acreage might return the land to a productive state. Reclamation and revegetation research in the Knobs region is in its infancy stage, therefore environmental impacts associated with mining of the oil shales can not be accurately concluded.

The yield ratio of 2.9:1 for unrefined oil is lower than the yield ratio for coal liquefaction with gasoline as final product and coal gas with High-BTU as final product. Kentucky has abundant coal reserves in both the eastern and western sections of the state. Investment into oil shale development with such a low yield ratio does not seem an energetically wise decision at the present. Based on the tenuous nature of the data for the energy calculations of this study, oil shale development in Kentucky needs a much more complete economic and energetic evaluation before development occurs.

REFERENCES

Anderson, O.K. 1959. The climates of Kentucky. In Climates of the States, Vol. 1, United Oceanic and Atmospheric Administration, United States Department of Commerce, National Climatic Center, Asheville, North Carolina.

Arms, F.S., J.P. Fehr, D.T. Carrol, B.L. Wilson, H.P. McDonald, and J.C. Ross. 1971. Soil survey of Nelson County, Kentucky. Soil Conservation Service.

Bailey, R.G. 1978. Description of the ecoregions of the United States. Forest Service, United States Department of Agriculture, Ogden, Utah.

Braun, E.L. 1950. Deciduous forests of eastern North America. Hafner Publishing Co., New York.

Burroughs, W.G. 1926. Geography of the Kentucky Knobs, Vol. 19, Series VI. Kentucky Geological Survey.

Costanza, R. 1980. Embodied energy and economic valuation," Science 210:1219-1224.

Energy Resources Company Inc. 1981. Development of disposal performance standards for oil shale mining and processing. Task 1 Report, prepared for Office of Special Projects (KDNREP).

Energy Resources Company Inc. 1981. Development of disposal performance standards for oil shale mining and processing. Task 2 Report, prepared for Office of Special Projects (KDNREP).

Energy Resources Company Inc. 1981. Development of disposal performance standards for oil shale mining and processing. Task 3 Report, prepared for Office of Special Projects (KDNREP).

Feldkirchner, H.L. and J.C. Janka. 1979. The Hytort Process. Institute of Gas Technology, Chicago, Ill., unpublished report.

Feldkirchner, H.L., S.A. Weil, J.C. Janka and D.V. Punwani. 1980. SNG From eastern oil shale via the Hytort Process. 1980. Paper presented at American Gas Association Transmission Conference, Salt Lake City, Utah, May 5-7.

Harker Jr., D.F., R.R. Hannan, R.R. Cicerollo, W.C. Houtcooper, L.R. Phillippe and D. VanNorman. 1981. Preliminary assessment of the ecology and ecological features of the Kentucky "Knobs" oil shale region, Vol 1. Kentucky Nature Preserves Commission, Frankfort, KY.

Herendeen, R. 1973. Affluence and energy demand. C.A.C. Doc. 102, Univ. of Ill. Urbana-Champaign, Urbana, Ill.

Institute of Gas Technology and Dames and Moore. 1978a. Environmental aspects of eastern oil shale development. IIT Center, Ill., unpublished report.

Institute of Gas Technology and Dames and Moore. 1978. Eastern oil shale and the environment. IIT Center, Ill., unpublished report.

Janka, J.C., and J.M. Dennison. 1979. Devonian oil shale - a major American energy resource. Paper presented at Synthetic Fuels from Oil Shale Symposium, Atlantic, Georgia.

Kentucky Climate Center. 1981. Annual water budget-equal availability. A computer print-out provided by Department of Geography and Geology, Western Kentucky University, Bowling Green, KY.

Kentucky Dept. of Revenue. 1978. Annual Report (1977-1978), Frankfort, Kentucky.

Khan, M.A. 1977. Energy cost analysis of advanced wastewater treatment. Masters Thesis, Department of Environmental, Engineering, Illinois Institute of Technology, Chicago, Ill.

Kuchler, A.W. 1964. Potential natural vegetation of the conterminons United States, American Geographical Society, New York.

Leontiff, W. 1966. Input-Output economics. Oxford University Press, New York.

Lind, C.G. 1982. A regional and net energy analysis of oil shale development in central Kentucky. Masters Thesis, Interdisciplinary Studies, University of Louisville, Louisville, Kentucky.

Lind, C.G., and W.J. Mitsch. 1981. A net energy analysis including environmental cost of oil shale development. In Energy and Ecological Modelling, W.J. Mitsch, R.W. Bosserman and J. Klopatek (eds.), Elsevier, Amsterdam.

McCoy and McCoy. 1981. Final comments on proposed oil shale regulations. Pyramid Minerals, Inc.

Miller, A.M. 1919. The geology of Kentucky. Kentucky Department of Geology and Forestry, Series 5, Bulletin 2.

Mitsch, W.J., and C.G. Lind. 1981. A regional environmental approach to eastern Devonian oil shale development. In Beyond The Energy Crisis, R.A. Fazzolare and C.B. Smith (eds.), Vol. IV, New York, pp. A91-A99.

Mitsch, W.J., and C.G. Lind. 1982. Potential impact of oil shale development on water resources in Kentucky. Proceedings of 1981 Eastern Oil Shale Conference, Lexington, KY., (in press).

Nesbitt, P.D., B. Lovan and J. Mays. 1982. Socio economic impacts: validity of evidence from the coal mining industry in Kentucky. Proceedings of 1981 Eastern Oil Shale Conference, Lexington, KY., (in press).

Odum, H.T. 1971. Environment, power, and society. John Wiley, New York.

Odum, H.T. 1972. An energy circuit language for ecological and social systems: its physical basis. In Systems Analysis and Simulation in Ecology, B.C. Patten (ed.), Vol. 2, Academic Press, New York.

Odum, H.T., and E.C. Odum. 1976. Energy basis for man and nature. McGraw-Hill, New York.

Odum, H.T., et al. 1976. Net energy analysis of alternatives for the United States. ERDA Contract E(40-1)-4398, Department of Environmental Engineering Sciences, University of Florida, Gainesville, Florida.

Tolbert, V.R., G.K. Eddleman, R.H. Ketelle, J.F. McBrayer, R.R. McKeown, E.P. Waits and F.W. Webb. 1982. A preliminary assessment of environmental impacts of obtaining oil from Chattanooga Shales. Proceedings of 1981 Oil Shale Conference, Lexington, Ky (in press).

Urban Studies Center. 1979. How many Kentuckians: population forecasts, 1970-2020. University of Louisville, Louisville, Kentucky.

U.S. Dept. of Commerce. 1980. Census of Agriculture. Preliminary report, Nelson County, Kentucky, Frankfort, Kentucky.

Vyas, K.C., G.D. Aho and T.L. Robl. 1981. Synthetic fuels from eastern oil shale. Report prepared for Buffalo Trace Area Development District, Maysville, Ky., Vol. 1.

Weil, S.A., D.V. Feldkirchner, F.C. Janka and D.V. Punwani. 1979. The IGT Hytort Process for hydrogen retorting of Devonian oil shales. Paper presented at 6th National Confoerence on Energy and the Environment, Pittsburgh, Pa., May 21-24.

Wharton, M.E. 1945. Floristics and vegetation of the Devonian-Mississippian black-shale region of Kentucky. Ph.D. Thesis, University of Michigan, Ann Arbor, Michigan.

SIMULATION OF RIVER FLOW AND DO DYNAMICS AFFECTED BY PEAKING DISCHARGES FROM A HYDROELECTRIC DAM[1]

Robert L. Dwyer and Martha A. Turner[2]

Abstract.--The peaking operation of the hydroelectric plant in Conowingo Dam (Susquehanna River, Maryland) causes fluctuations in water depth, velocity, temperature, and DO in the downstream nontidal reach of the river, which affect local fish populations. A hydrodynamic and dissolved oxygen model, validated against field observations of flow and DO dynamics, was used to assess the relative contributions of the discharge water and of physical and biological processes in the reach, to downstream DO distributions.

INTRODUCTION

Conowingo Dam, located 16 km upstream of the mouth of the Susquehanna River in Maryland, contains a peaking-power hydroelectric plant whose operation affects biota in the river reach below the dam. Specifically, declines in abundances of resident and anadromous fish since the 1960's and problems related to water flow and quality in the vicinity of the dam are thought to be associated with the degradation of habitat due to inadequate discharge during off-peak periods. Such periods of plant shutdown or low discharge reduce available physical habitat (in terms of adequate water depth and velocity) for fish downstream of the dam, and sometimes causes increases in water temperatures to levels stressful to fish due to solar heating of shallow, slow-flowing water. In addition, a number of studies have implicated low concentrations of dissolved oxygen (DO) in the dam discharge, or in the reach of the river from the dam downstream to the fall line, during shutdown in repeated fish kills in this river reach (Carter 1973; Tidewater Administration 1980).

Low DO values observed below the dam are a function of low DO in water discharged by the turbines (due to oxygen depletion in Conowingo Reservoir in summer and fall) and of metabolic and other processes within the downstream river reach. The analysis presented in this paper describes the relative contributions to downstream DO values of:

- The water discharged by the plant
- Biological processes in the river reach, and
- Physical processes in the river reach.

The analysis also assesses how the latter two components are affected by discharge levels.

This paper describes the formulation and validation of a hydrodynamic and dissolved oxygen simulation model for the downstream river reach that we used to quantify these phenomena. Since the DO of the discharge is controlled by processes in the reservoir rather than riverine processes, other ongoing analyses are investigating the causes of summertime DO depletion in the reservoir and the effects of modification of turbine operations on reservoir DO distributions (and thus the DO of the discharge water).

The ultimate objective of these studies is to recommend measures to correct or mitigate low DO problems in the Susquehanna River below Conowingo Dam, for example, by the use of a continuous flow release from the dam. Toward this end, the river model presented in this paper will be integrated later with the analyses of the other segments of the reservoir and river system, to provide a comprehensive assessment of the causes of the water quality problems, and to test the effects of some proposed alterations of the dam's operating schedules that are intended to improve habitat for fish and other biota.

IMPLEMENTATION OF A ONE-DIMENSIONAL HYDRODYNAMIC MODEL OF THE NONTIDAL RIVER REACH DOWNSTREAM OF CONOWINGO DAM

Description of the Problem and Approach

Summertime DO and temperature distributions downstream of the dam are affected by fluctuations in the volume of turbine discharge:

- During periods of normal discharges, DO and temperature distributions in the reach are strongly influenced by the DO and temperature of the discharge water (which are controlled by the reservoir ecosystem combined with the patterns of water withdrawal by the turbines).
- During periods of turbine shutdown or low river discharges, the oxygen metabolism of the water-column, benthic, and fish communities in the river reach dominate DO dynamics, and the intensity of solar heating controls distributions of water temperature.

---

1. Paper presented at the Third International Conference on State-of-the-Art in Ecological Modelling, Colorado State University, Fort Collins, Colo., May 24-28, 1982.
2. Robert L. Dwyer and Martha A. Turner are Research Scientists, Martin Marietta Corporation, Environmental Center, Baltimore, Md. 21227, U.S.A. Research was supported by a contract from the Maryland Power Plant Siting Program, Annapolis, Md. U.S.A.

State and federal resource agencies have suggested a continuous minimum discharge of water from Conowingo Reservoir as one possible means of alleviating these problems. In the reservoir, continuous vertical mixing (driven by the discharge) might increase the DO at the depth of the turbine intakes, whereas in the downstream reach, continuous flushing could reduce the amplitudes of daily fluctuations in DO and temperature. However, the amount of release necessary to damp out these fluctuations is not known and must be estimated from available information about system processes. As a tool to determine the optimum flows for enhancing oxygen conditions below the dam, we developed a simulation model of the hydraulics and DO dynamics of the nontidal river reach. This model involves the numerical solution of differential equations describing the physics of water flow, as well as the biochemical transformations affecting DO in the river reach. As discussed below, we calibrated and validated the model, primarily using data from field studies in the river reach in 1980 (Philipp and Klose 1981) and in 1981 (Potera et al. 1982).

Hydrodynamic Modeling Approach

The simulation of oxygen dynamics in the river below the dam first requires the development of a numerical model that reproduces the hydrodynamics in that river reach. Equations describing the conservation of momentum, plus an equation of continuity (conservation of mass), provide a complete description of water flow. The equations for one-dimensional flow in an open channel (the so-called Saint-Venant equations) are as follows:

$$\text{Continuity:}\quad \frac{\partial(AU)}{\partial x} + B\frac{\partial h}{\partial t} - q = 0 \qquad (1)$$

$$\text{Momentum:}\quad \frac{\partial h}{\partial x} + \frac{U}{g}\frac{\partial U}{\partial x} + \frac{1}{g}\frac{\partial U}{\partial t} + \frac{1}{g}\frac{q}{A}U = S_f \qquad (2)$$

where,

A = cross-sectional area of flow, m^2

U = mean velocity, m/s

x = distance along channel, m

B = water surface width, m

h = water surface elevation above MSL, m

t = time, s

q = tributary inflow per unit length along channel, $m^3/s\text{-}m$

S_f = friction slope = $\frac{n^2U|U|}{2.21\ R^{4/3}}$

(where n = Manning n-value and R = hydraulic radius).

For applications where great precision in the solution in time or space is not needed, the equations can be simplified by eliminating one or more terms. However, since river flows downstream of Conowingo Dam are characterized by rapid fluctuations due to the peaking generation cycle, the present analyses require the solution of the complete equations. We chose a one-dimensional representation of the river due to the limitations of available biological and physical data.

Application to the Susquehanna River Below Conowingo Dam

The Saint-Venant equations (1 and 2) provide an exact description of one-dimensional flow. However, practical solution of the equation for a real river reach requires the use of a method based on numerical approximation. Two general classes of numerical solution schemes--finite element and finite difference methods--have each been implemented in a number of general computer packages, any of which could be calibrated to this reach of the Susquehanna River. We selected USTFLO (a program distributed by the U.S. Army Corps of Engineers, Hydrologic Engineering Center, 1977) because it appeared that it could best reproduce the rapid discharge fluctuations characteristic of peaking-power hydroelectric dams. Once USTFLO was calibrated and validated to the hydrodynamics of the river reach below Conowingo Dam, a differential equation describing nonconservative changes in oxygen in the reach was coupled to it (see next major section).

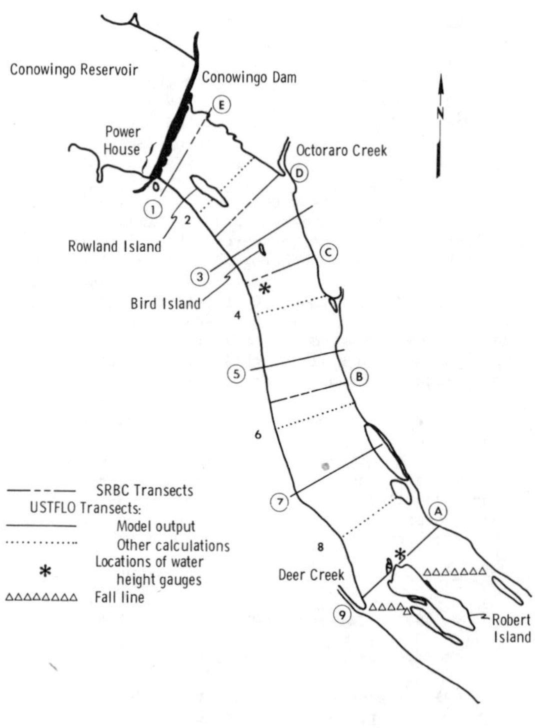

Figure 1. Susquehanna River between Conowingo Dam and the fall line, showing longitudinal segmentation: letters are boundaries of segments of SRBC hydrology studies; odd numbers are downstream boundaries of USTFLO segments at which independent output estimates are simulated; even numbers are other boundaries used in the model's finite difference calculations; asterisks are locations of water height gauges deployed in 1981.

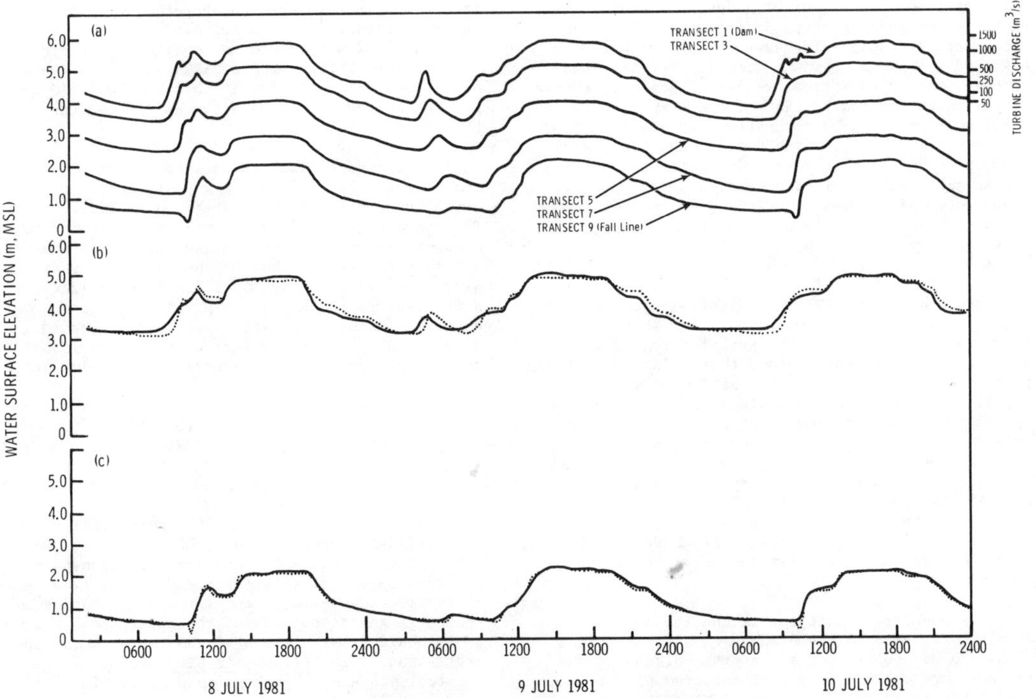

Figure 2. Validation of hydrodynamic model, 8-10 July 1981. (a) Simulated water surface elevations at five USTFLO transects. Transect 1 is water surface elevation at the dam corresponding to the turbine discharge hydrograph used as a model boundary condition (scale at right). (b) Simulated water-surface elevation at transect 3 (solid line), and observed water surface elevation at gauge at transect C (dotted line). (c) Simulated (solid line) and observed (dotted line, partially hidden) water surface elevations at transect 9 at the fall line.

Choice of River Segmentation

Our choice of river-segment length (Δx) and time step (Δt) was governed by the spatial and temporal resolution of the field data on both hydrodynamics and DO distributions, and by the Courant stability criterion needed to guarantee convergence of the explicit numerical solution (HEC 1977; Yen 1979). The upper boundary of the model was chosen as a transect 90 m downstream of the dam (to avoid having to account for nearfield effects of the turbine discharges); and the downstream model boundary was a transect 5,200 m downstream of the dam (just above the fall line; see fig. 1). Segment lengths (Δx) were set at 639 m (yielding eight segments), and Δt was chosen as 1 s, based on the constraints of data availability and the Courant criterion. Since USTFLO uses a "leapfrog" or centered-difference computation scheme in its explicit computations (HEC 1977), the model produces independent outputs for only five odd-numbered transects of the nine that bound the segments (fig. 1). Using measurements from past hydrology studies of this river reach by the Susquehanna River Basin Commission (SRBC; Jackson and Lazorchick 1980; Photo Science 1980), we supplied the model with the river-specific input data (tables of cross-sectional area, water elevation, wetted perimeter, hydraulic radius, surface width, and Manning bottom-friction coefficient for each transect) as functions of discharge volume.

Initial Conditions

Based on aerial surveys at steady-state discharges (Photo Science 1980), water surface elevations were specified for each segment (usually at 340 m^3/s steady-state discharge) as the model's initial condition at time $t = 0$.

Boundary Conditions

The specification of boundary conditions for this river reach requires the provision of the time variations of discharge, velocity, or water-surface elevation at the upper boundary (Conowingo Dam) and at the downstream boundary near the fall line. At the upstream boundary, we have a continuous time-series hydrograph [available from the US Geological

Survey's (USGS) water height gauge in Conowingo Dam] and a discharge hydrograph derived from the stage hydrograph and stage-discharge rating curve (Myron Lys, USGS, personal communication). For the downstream boundary condition, we used a stage-discharge rating curve derived by SRBC (Jackson and Lazorchick 1980) at a transect 100 m upstream of the fall line (transect A, fig. 1), which can be used for discharges of 340 m^3/s or less. In addition, data from a continuous water height recorder (Potera et al. 1982) were used to extend the SRBC stage-discharge rating to cover the full range of dam discharges (0 to 2,400 m^3/s).

Calibration Of USTFLO To Observed Flow Data

After we completed the specification of model geometry and input data needed to simulate the river reach below Conowingo Dam, our next step was to calibrate the model so that its output closely fits segments of real hydrologic data. We carried out this tuning of model parameters in two stages. First, we calibrated the model to fit detailed field elevations taken at three steady-state discharges, 25, 115, and 340 m^3/s (Photo Science 1980). Next, we fitted water surface elevations to time-series measurements made under unsteady discharge conditions, but at only two locations along the river reach.

For boundary conditions in the steady-state simulations, we used the discharge hydrograph from Conowingo Dam (upstream) and a stage-discharge rating curve derived from SRBC measurements at transect A (downstream). We adjusted Manning bottom-friction coefficients so that simulated water elevations agreed (±0.08 m) with observed elevations at SRBC transects A-D.

For calibration of the model using unsteady flows, we used data from 1981 when continuous water height gauges were deployed near SRBC transects C and A (fig. 1). The stage-discharge rating curve at transect A, which we used as the downstream boundary condition by USTFLO, was the new composite rating curve using independent height data from the gauge at transect A plus the three stage-discharge points from Photo Science's data at the same place. We adjusted the Manning bottom-friction coefficients until simulated water elevations agreed with observed elevations at the gauge at transect C under all discharges.

Validation of USTFLO

We validated the calibrated hydrodynamics of USTFLO using data from a separate period in 1981. Figure 2a shows data from the USGS stage hydrograph corresponding to the discharge-hydrograph boundary condition and for simulated water elevations at model transects 1, 3, 5, 7, and 9. The correspondences between measured and simulated water elevations at transect 3 (SRBC transect C; fig. 2b), and for measured and simulated water elevations at transect 9 (SRBC transect A; fig. 2c) show the agreement of the model with real data for fluctuating flows over the period 8-10 July 1981. The close agreement between measured and simulated water elevations (figs. 2b and 2c), and the correspondences between simulated water elevations (fig. 2a) and simulated velocities at the same transects (not shown), over a wide range of discharges, confirms the ability of the calibrated model to simulate water transport from conditions of turbine shutdown up to full turbine discharge--approximately 2,400 m^3/s. Flows above 2,400 m^3/s occur only when the dam passes water through the spillways. Under these conditions, it is difficult to make the field measurements needed to extend the calibrated range of the model.

As stated earlier, a validated hydrodynamic model is a prerequisite for any simulation of conservative transport of a substance by river flow. This is especially true for the simulation of DO, which is affected by both conservative transport and *in situ* physical and biochemical processes. Since the hydrodynamic model has been validated sufficiently, and since the physical processes embodied in the model are essentially invariant, we will not present any further hydrodynamic validation data during the discussion of DO simulations.

INCORPORATION OF DO DYNAMICS INTO THE HYDRODYNAMIC MODEL

Background

The advective transport of water described by this validated application of USTFLO can be used directly to simulate the conservative transport of any substance passively carried by water flow, after evaluating how well the model reproduces observed transport and dispersion of the substance. However, a mass-balance simulation of a material whose concentration is modified by *in situ* biological or chemical processes, such as changes in dissolved oxygen (DO), requires the quantification of a number of physical and biochemical processes affecting DO within the river reach.

Differential Equation Describing Oxygen Rate of Change

Within any segment, the rate of change in the mass of dissolved oxygen depends on the mass flux of oxygen into the segment through its upstream boundary; the flux out through the downstream boundary; diffusion of oxygen through the air-water interface; oxygen consumption due to respiration of the water-column, benthic, and nektonic communities; and photosynthetic oxygen production of the water-column and benthic communities. These rate processes are incorporated into the model by the following equation:

$$\frac{d(VC)}{dt} = Q_{in}C_{in} - Q_{out}C + K_2(C^* - C)V + V(-R_{wc} - R_n) + S(P_b - R_b + DP_{wc}) \qquad (3)$$

where,

V = volume

C = concentration of dissolved oxygen

t = time

Q = flow

K_2 = reaeration coefficient

C* = saturation DO

R = respiration

wc denotes water-column community

n denotes nekton

S = surface area

P = gross production

b denotes benthos

DP_{wc} = water-column production integrated over depth

and further:

$$V(t) = V(t-1) + (Q_{in} - Q_{out})t.$$

To simulate DO dynamics longitudinally along the river reach, we coupled eq. 3 to the hydrodynamics simulated by USTFLO (through the Q, V, and S terms) and integrated it numerically (using a fourth-order Runge-Kutta scheme) for each USTFLO segment. Numerical stability was not of concern for this solution, so we chose a Δt of 15 min. Actual DO simulations used stored hydrodynamic data from USTFLO runs, which were sampled every 15 min. As outputs, the DO model simulates DO values at transects 3, 5, 7, and 9 (fig. 1) every 15 min.

This DO model configuration is based on several assumptions. First, all oxygen passing into a given river segment is assumed to be mixed instantaneously. In addition, DO within any segment is assumed to be relatively homogeneous, so that the average DO simulated by the model is representative of actual DO concentrations anywhere in the segment. This assumption probably is not met in the field during turbine shutdowns and very low discharges when DO values in isolated pools tend to differ from one another (Philipp and Klose 1981). However, there are insufficient data both on the hydrodynamics and on DO distributions to permit refinements in the spatial resolution in the model. Actual DO concentrations at low flows will be distributed both higher and lower than the simulated average DO value, probably randomly over a range similar to that observed in the field under similar flows (Philipp and Klose 1981).

Another assumption inherent in the DO model is that longitudinal dispersion of a transported substance in the river reach is small compared to its advective transport. An explicit dispersion term is not employed in the model, but the integration scheme does generate an implicit numerical dispersion. To determine whether this numerical dispersion approximates the actual dispersion of a substance being transported in the river reach, we calculated actual dispersion rates from observations of pulses of dye released at different discharge rates, as well as numerical dispersion rates from simulations of the same dye release events.

A number of dye pulses (Rhodamine WT) were released from the dam at several discharges (Potera et al. 1982), and time-concentration curves measured downstream yielded estimates of the coefficient of dispersion (D_L; Levenspiel and Smith 1957) ranging from 1.26 to 147.0 m^2/s (table 1). Most of these values fall within the range for large rivers (D_L = 0.3 to 30 m^2/s) reported by Krenkel and Novotny (1980).

Table 1a. River dispersion coefficients (from dye releases on 26 August and 2 September 1981)

Flow (m^3/s)	Transect	D_L (m^2/s)
140	A	72.9
	A	115.0
	A	33.0
	A	32.2
	A	48.5
	A	21.9
	A	25.9
	A	17.0
	C	6.83
	C	7.14
225	A	1.26
	A	136.0
	C	8.51
	C	4.52
1,130	A	147.0
	A	102.0

Table 1b. Model dispersion coefficients (from model simulations at steady-state flows using 2 September 1981 hydrodynamic data)

Flow (m^3/s)	Transect	D_L (m^2/s)	
		Δt = 15 min	Δt = 5 min
100	3 (~ C)	13.22	
	5 (~ B)	12.06	
	9 (~ A)	12.41	
140	3 (~ C)	39.4	34.3
	5 (~ B)	39.9	35.1
	9 (~ A)	36.2	31.5
140	3 (~ C)	27.0	50.0
	5 (~ B)	22.4	48.3
	9 (~ A)	21.7	48.3
1,130	3 (~ C)	693.0	604.0
	5 (~ B)	823.0	730.0
	9 (~ A)	895.0	805.0

The implicit dispersion of the model was quantified by eliminating all nonconservative terms from eq. 3:

$$\frac{d(VC)}{dt} = Q_{in}C_{in} - Q_{out}C_{out} \quad (4)$$

and numerically integrating this equation with Δt = 5 min and again with Δt = 15 min (table 1). The results do not appear to be affected by the choice of time step. In addition, the dispersive properties implicit in the model appear very similar to those measured in the river, for all but the 1,133 m^3/s discharge. Finally, in the other simulations with all the terms of eq. 3, an artificial reduction of numerical dispersion produced no detectable differences in simulated DO values, when compared with runs having normal dispersion. Thus, we can conclude that longitudinal dispersion in this reach of the Susquehanna River is typical of other large rivers, that the implicit dispersion of the model has similar dispersion properties, and finally, that the simulated output in the model is not sensitive to this dispersion mechanism.

Based on these analyses, advective processes appear to dominate the passive transport of substances downstream. In the case of oxygen, a number of nonconservative processes, controlled by <u>in situ</u> biological and physical mechanisms, change <u>local DO</u> concentrations in the river reach. These processes are incorporated in the model by the equations described below. The effects of oxygen metabolism of biota are included in the model as rate processes, rather than as state variables, because of the short residence times of water in the reach (a few hours, at most) and because there is no evidence that the metabolism of resident biota (e.g., fish, the benthic community) changes over the summer season of prime concern. Details of parameter estimation are presented by Dwyer and Turner (1982).

Diffusion

Diffusion of oxygen through the air-water interface is incorporated into each model segment by the following equation:

$$\frac{dC}{dt} = K_2 (C^* - C) \qquad (5)$$

where the terms are defined as in eq. 3. The value for water saturated with oxygen (C*) was calculated from Kester (1975). A few estimates of the reaeration coefficient (K_2) were made in the river reach using the disturbed equilibrium technique (Philipp and Klose 1981) and the floating dome technique (Roques and Nixon in prep.; Potera et al. 1982), but the estimates showed a great deal of scatter. As an alternative, we used a general equation (Bennett and Rathbun 1972), based on data from a number of previous studies, in the model:

$$K_2 = 8.76 \frac{U^{0.607}}{h^{1.689}} \qquad (6)$$

where the symbols are defined as in eqs. 1 and 2.

Nekton Respiration

An estimate of respiration of the nekton community in the river reach (Philipp and Klose 1981) was derived from estimated biomasses of five major fish species in the river reach during 1980 (Petrimoulx and Klose 1981) (excluding Atlantic menhaden, whose presence in the area is very unpredictable). If we use estimated river volumes under shutdown conditions, average oxygen consumption of the nekton community is less than 0.01 g O_2/m^3-hr. This value is less than 25% of the rates of any other metabolic processes, so we feel justified in not including it in the DO model.

Respiration of the Water-Column Community

In 1980, a number of measurements of the respiration (and production) of the water-column community, using the oxygen light-dark bottle technique (Strickland and Parsons 1968; APHA 1975), were made in the study area and in the tidal river reach immediately downstream (Philipp and Klose 1981). The relationship between respiration and temperature probably follows the exponential Q_{10} form. However, a linear regression yielded the best fit to the data:

$$R_{wc} = 0.01 + 0.0016\ T \qquad (7)$$

where the term R_{wc} is water-column community respiration (g O_2/m^3-hr) and T is segment water temperature (°C). Under summertime conditions, the model generates spatially averaged temperatures, which generally range from 25-30°C, so the linear representation probably simulates respiration adequately.

We simulated the downstream increase in water temperature under daytime shutdown or low discharge conditions using a relationship derived from measured upstream-downstream temperature differences as functions of discharge volume and time of day (i.e., intensity of solar heating). The function is a sinusoidal temperature increment over a day, which varies downstream temperatures a maximum of 3°C about the temperature of the discharge water.

Benthic Community Respiration

Estimates of benthic community respiration rates were made in 1980 (Philipp and Klose 1981); these data indicate that benthic respiration may be a major sink for DO. A linear regression of these rates vs temperature yielded the following:

$$R_B = 0.016 + 0.0048\ T \qquad (8)$$

where the term R_B is respiration of the benthic community (g O_2/m^2-hr) and T is water temperature (°C).

The Photosynthesis-Light Relationship

The relationship between the rate of photosynthesis and available light energy is central to the estimation of primary production of oxygen in the river reach. The model uses established general relationships for the dependence of photosynthetic rate on ambient light intensity, and the attenuation of light with water depth, to estimate the oxygen production of both the water-column and benthic communities. Before we present the details of calibrating the production mechanisms to data from the Susquehanna River, it is useful to discuss the photosynthesis-light relationship in general terms.

Previous studies (reviewed by Parsons and Takahashi 1973, and Kremer and Nixon 1978) have often shown gross production vs light (hereafter P vs I) relationships in which the rate of production:

- Is light-limited at low I values,
- Reaches a (possibly temperature-dependent) maximum (P_{max}) at intermediate I values, or
- Sometimes exhibits photoinhibition at high I values.

Steele (1965) introduced a simple equation describing the relationship, which has only two parameters and remains in common use because of its versatility. The equation describes the reduction of production rate (P) when light intensity (I) is above or below the optimum light intensity (I_{opt}) at which maximum production (P_{max}) is observed:

$$P = P_{max} \frac{I}{I_{opt}} e^{(1-I/I_{opt})} . \tag{9}$$

The quantity of light energy actually available to algae for photosynthesis depends on the amount of energy reaching the water surface, as a function of time, and on the attenuation of light with water depth, due to turbidity. Light energy flux per unit area of water surface (I_o), typically measured in g-cal/cm^2-min (langleys/min), depends on solar angles due to latitude of the measurement point, time of year, and time of day, as well as atmospheric turbidity and cloud cover. Within the water column, light intensity at a depth of z meters (I_z) is related to I_o by:

$$I_z = I_o e^{(-k_{ext} z)}, \tag{10}$$

which can be substituted directly into eq. 9 to yield the estimated rate of production at depth z for surface light intensity I_o. The light extinction coefficient (k_{ext}) is commonly related to the depth of disappearance of a 20-cm Secchi disk (Wetzel 1975) by:

$$k_{ext} = \frac{2.1}{z_{Secchi}} , \tag{11}$$

which yields $I_z/I_o = 0.122$ at the Secchi-disk depth. Depending on the light absorption characteristics of the water and observer-related variations, values of I_z/I_o at the Secchi-disk depth can range from 0.05 to 0.15.

Gross Production of the Water-Column Community

During 1980, light-dark bottle pairs were incubated at several depths at stations throughout the reach. Light intensity reaching the water surface (I_o) during each incubation was estimated from solar angle (calculated from its known astronomical relationships to latitude, time of year, and time of day), which was used in a meteorological equation (Sverdrup et al. 1942) to estimate the time-dependent solar energy flux at the surface. Water transparency was measured at each station using a Secchi disk, which defined the light intensity at the depth of incubation.

Because we expected that P_{max} would depend on temperature, we separated the gross production data into three temperature groups and fitted the Steele equation (eq. 9) to each group, yielding the following three sets of parameter estimates:

$T < 20°C$: $P_{max} = 0.0017$ g O_2/m^3-min,
$I_{opt} = 0.15$ ly/min

$20° < T < 25°C$: $P_{max} = 0.0023$ g O_2/m^3-min,
$I_{opt} = 0.15$ ly/min

$T > 25°C$: $P_{max} = 0.0054$ g O_2/m^3-min,
$I_{opt} = 0.15$ ly/min.

Simulations of summer conditions, which are the focus of interest here, will generally use the last pair of coefficients (> 25°C).

The model algorithm for gross production of the water-column community numerically integrates the Steele equation (after substituting eqs. 10 and 11) between the water surface and bottom. The calculation uses an instantaneous I_o calculated from sun angle and cloud cover for that time, and a value for I_z calculated from I_o and the average Secchi-disk depth of 0.65 m. This value is used for DP_{wc} in eq. 3.

Gross Production of the Benthic Community

Production vs light relationships for benthic periphyton could not be estimated from field measurements in 1980. Given no other information, either from field data or the literature, the initial estimate for I_{opt} for the benthos was set at 0.15 ly/min. However, it was possible to fit a relationship between midday measurements of benthic gross production and temperature:

$$P_{max} = 0.1 + 0.02\ T . \tag{12}$$

The model calculates P_b (eq. 3) using eq. 12 and I_z, which is calculated as it is for the water column.

Calibration of Parameters in the DO Model

The primary purpose of the DO model is to estimate DO distributions in the river during the periods of the year when turbine shutdowns accompanied by DO depletion are likely to occur. Field measurements during 1980 and 1981 indicated that such conditions occurred primarily during summer and early fall (June to October). As discussed in the previous section, we drew a number of initial parameter estimates from field measurements of processes that were made primarily during the summer of 1980, a period of low natural river flow when periods of turbine shutdown were common. We calibrated the model using continuous DO data for a three-day interval (5-7 July 1980) that shows a typical discharge pattern for the summer. The monitor data were taken at the dam discharge and in a flowing channel area at the fall line (Philipp and Klose 1981). We used time-series DO and flow data at the dam discharge as inputs and adjusted model parameters as necessary so that simulated DO concentrations at transect A (the fall line; fig. 1) matched observed DO values at the fall-line monitor. Measured dam discharge for this period is shown in figure 3a, and the DO of the dam discharge vs DO from the fall-line monitor are shown in figure 3b.

On each of the three days during morning shutdown periods, measured DO values at the fall line (transect A) increased rapidly in early morning. On the second day (a Sunday), continuation of the shutdown throughout the day caused DO to maintain a plateau near 9 ppm. In the initial calibration runs, the simulated fall-line DO fit the observed DO quite well. However, the individual production rates (P_B and DP_{wc}) in the initial runs suggested severe photoinhibition during much of the day.

This photoinhibition is controlled by the relation between the ambient light intensity (I_z) and the photosynthetic optimum intensity (I_{opt}). Measurements of Secchi-disk depth (which partially

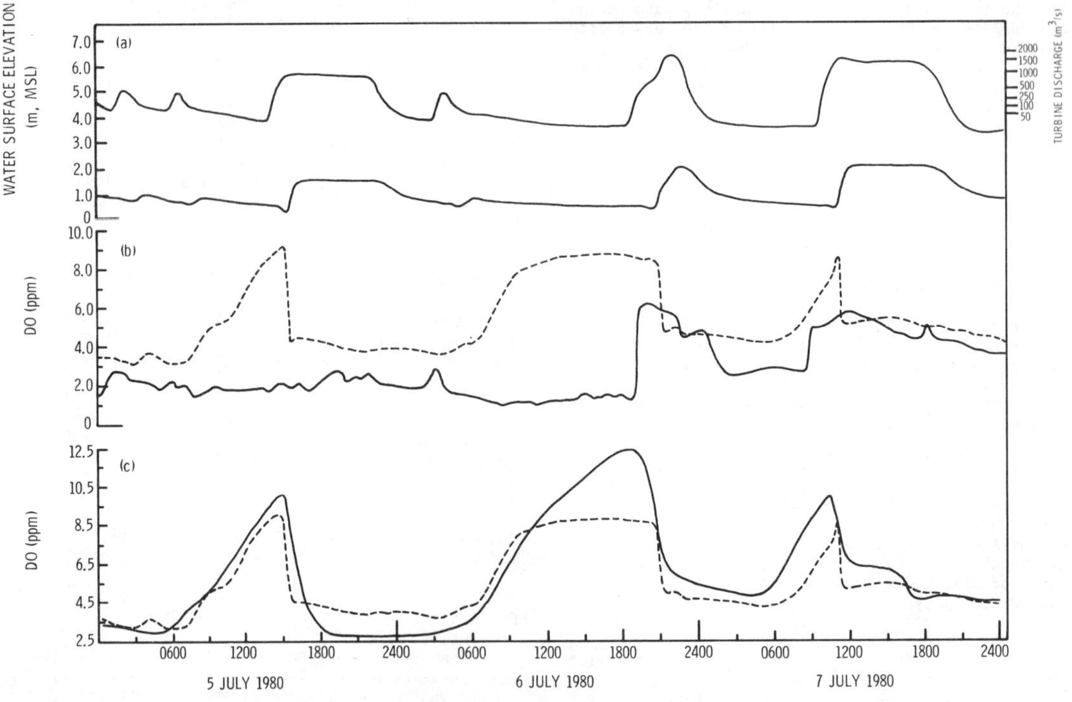

Figure 3. Calibration of the DO model, 5-7 July 1980. (a) Simulated water surface elevation and corresponding discharge at the dam (upper solid line) and simulated elevation at the fall line (lower solid line). (b) DO of the discharge from the dam (solid line) and DO measured at the fall line (dashed line). (c) DO measured at the fall line [dashed line; same as in (b) above] and DO at the fall line simulated by the model (solid line).

determines I_z) in 1980 showed considerable variation over periods as short as a few hours (Philipp and Klose 1981). Also, the initial estimate of I_{opt} = 0.15 ly/min was much lower than other values reported in the literature. To investigate the sensitivity of the model's production mechanism to variations in these light parameters, or errors in their estimation, we solved the benthic production equation (eq. 9 calculated at water depth z = 1.0 m) hourly from dawn to dusk for all combinations of Secchi-disk depth values of 0.2, 0.7, 1.2, 1.7, and 2.2 plus I_{opt} values of 0.15, 0.3, and 0.5. This yielded 15 parameterized curves of benthic production rate under a wide range of I_z values and differing sensitivities of photosynthesis to light. The curves for I_{opt} = 0.15 and 0.5 are shown in figures 4a and 4b, respectively. The curves of figure 4a show that oxygen production is very sensitive to the light intensity reaching the benthos when I_{opt} = 0.15. When Secchi-disk depth changes from 0.7 m to 1.2 m, the light level at 1 m increases by a factor of 3.5, which causes a marked depression of photosynthesis after 7 am. Since the model's water-column productivity mechanism integrates rates between the water surface and bottom, the effects of surface photoinhibition on DP_{wc} may be even more pronounced.

Actual Secchi-disk depths commonly varied by ±0.5 m during the day, and could easily have varied by more than that during the 5-7 July calibration interval. Thus, it was desirable to decrease the sensitivity of photosynthesis to such unknown variations while maintaining the general daily pattern exhibited by the combination of I_{opt} = 0.15 and Secchi-disk depth = 0.65. In figure 4b (I_{opt} = 0.5), the effects of photoinhibition are less pronounced. Secchi-disk depths of 1.2 to 2.2 all maintain a general daily production pattern similar to that observed with I_{opt} = 0.15, Secchi-disk depth = 0.65 m. Accordingly, we selected a new combination of light parameters (I_{opt} = 0.5, Secchi-disk depth = 1.7 m) to determine if it improved the fit of simulations.

Although the new parameter values differ from those derived from field measurements, there is justification for their use in the model. A review of a large body of literature (Nixon et al. 1979) indicates that an I_{opt} value of 0.5 ly/min is in

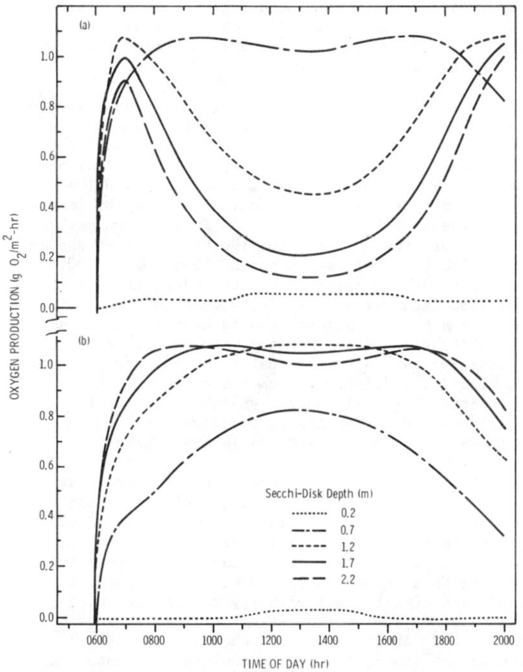

Figure 4. Sensitivity over a day (time is hours of the day) of simulated benthic oxygen production to different values of I_{opt}, and Secchi-disk depth. (a) Benthic production at depth of 1 m, I_{opt} = 0.15 ly/min. Production shows pronounced photoinhibition for all Secchi-disk depths greater than 0.7 m (i.e., for relatively nonturbid water). (b) Benthic production (1-m depth) for I_{opt} = 0.5 ly/min. Production of Secchi-disk depths of 0.7, 1.2, and 1.7 m all show similar production curves, without extreme photoinhibition.

the range of values from a number of different ecosystems. The new value for Secchi-disk depth (1.7 m) used in the model is well within the range of values measured in the river reach during the summer of 1980. During shutdown or low discharge periods, when in situ oxygen production likely has the greatest relative effect on DO values in the river, the lack of water turbulence probably permits an appreciable amount of suspended sediment to settle out. This would likely decrease the turbidity of the water, and increase Secchi-disk depths above the average value of 0.65 m.

We reran the oxygen model with the new values of I_{opt} = 0.5 (both for water-column and benthic production) and Secchi-disk depth = 1.7 m. A comparison of simulated vs observed DO values at transect A (fig. 3c) shows that simulated increases in morning DO during shutdown periods generally follow observed rates of DO increase. However, some significant discrepancies are evident. Simulated DO values on the afternoon of the second day (6 July) are significantly higher (~ 3.5 ppm) than values observed at the transect A monitor. This phenomenon can be explained by considering the differences found between model DO values, which are estimates of average DO concentrations within segments, and point DO measurements. At low discharges, such point measurements generally characterize the DO distributions of individual nonflowing pools, and the data show large pool-to-pool variations (Philipp and Klose 1981). In contrast, the DO monitor at transect A is in a continuously flowing channel, where it records the DO of water that has traveled through the river reach in minimum time. Thus, increases in DO values measured in channel areas are likely to be low relative to those in slower waters or nonflowing pool areas, since production has had a relatively short time to increase DO.

Measured discharges (fig. 3a) during these daytime DO peaks (fig. 3c) appear to represent the lowest flows observed in the record (i.e., when discharge is much lower than 30 m^3/s). Thus, it appears that the discrepancy between simulated and observed DO values may be largest when the model attempts to simulate very shallow waters or very slow flow conditions.

Another significant discrepancy between observed and simulated DO values occurs during some periods of full discharge (fig. 3c). During the first full-discharge period (1300-2100, 5 July), observed DO values are about 1 ppm higher than simulated values. Conversely, observed values are lower than simulated during the next peaking discharge (1600-2100, 6 July). Since cross-stream DO variation during periods of full discharge is negligible (Philipp and Klose 1981), the differences cannot be explained by the monitor data not being representative of average DO values in the river, as may be the case during shutdown periods.

An alternate explanation is that the DO of the discharge water is misestimated. It was not possible to measure DO in the discharge boils directly, so instead, one or more DO monitors were placed in cooling water lines that withdraw water from the turbine intake penstocks. There is evidence that these DO data are not representative of the DO of the discharge water, particularly at low discharges (Philipp and Klose 1981; McClellan et al. in prep.). We are presently attempting to obtain a better estimate of the discharge DO from analyses of reservoir and other DO data. Any errors in estimating input DO values will result in discrepancies downstream.

The incorrect estimation of input DO can explain the 2.5-ppm difference in DO between observed and simulated values for the morning shutdown on the third day (fig. 3c). The shapes of the peaks (i.e., rate of DO production) of the two curves are similar, but there is a relatively constant offset in the peak for the third day. It could well be that this difference in DO is a continuation of the difference evident during the previous night, which was probably caused by the misestimation of input DO during the nighttime peak discharge.

To summarize model calibration, we only needed to adjust parameters relating production to light to improve the fit of the model to observed data. Some discrepancies between simulated and observed DO values appear to be related to problems with the representativeness of point measurements of DO data. At low discharges, the downstream DO monitor may underestimate average DO values in the river. The amount of this misestimation is not precisely known. However, the model appears to simulate typical patterns of daytime oxygen production observed in pools during shutdowns.

The accuracy of model estimates of average river DO under shutdown conditions cannot be assessed with the data presently available. At full discharges, errors in input DO estimates have the greatest effect on simulated DO values downstream, simply because of the volume of the discharge relative to that in the river reach, and because very short residence times prevent in situ processes from appreciably affecting DO. This should not affect the usefulness of the model for assessing suitability of biotic habitat, because deleterious effects on biota are not usually associated with full-discharge conditions. With the field data presently available, it is unlikely that any further refinements of the DO model could improve the fit between simulated and observed values.

Validation of the DO Model

Validation of a simulation model consists of comparisons of simulated output with observed data that are independent of those used in model calibration. Because the only independent DO data available for this purpose are from the continuous monitor records for the summer of 1981, we selected two time intervals from that period: 27-28 June 1981, when turbine discharge was at a minimum continuous level of 140 m^3/s; and 25 July-2 August 1981, when daily intervals of turbine shutdown occurred.

However, on 6 July 1981, PECO instituted a new type of discharge regime consisting of setting the maximum duration of shutdown periods at 8 hours, and using discharge periods of a minimum of 140 m^3/s for at least four hours. A variation of this regime, used during the second validation interval (25 July-2 August 1981), did not have prolonged shutdown periods such as those used during summers prior to 1981.

Validation runs against 1981 data thus evaluate the model fit relative to high river flow (continuous discharge) conditions, or conditions under the PECO regime. To validate the model for periods of natural river flow lower than observed in 1981 and normal turbine operation (peaking followed by long shutdowns), we had to apply data from the summer of 1980. Even though data from the time segment chosen for this third validation period (8-10 July 1980) were not used in calibrating the model, the use of any 1980 data may be a strict violation of the requirement that validating data must be independent of calibration data. However, the results of this validation run (described below) do not appear to be qualitatively different from the output of the second validation interval, so the violation of the independence requirement had no practical influence on the assessment of the model's general validity.

To assess the fit of the model to DO data at transect A, we defined several quantitative indices. First, we calculated a mean value of simulated DO minus observed DO (ΔDO) using all estimates for each run. Since these values can be both positive and negative, the mean of ΔDO tends to underestimate lack of fit. Accordingly, we also calculated the mean of |ΔDO|. Then we calculated variances of ΔDO and |ΔDO|, as well as the range of ΔDO. Equivalent statistics were calculated for both temperature and water height. Calculated indices for all three variables for the three validation runs are presented in table 2. We made no attempt to calculate these statistics as a function of discharge.

Table 2. Estimates of average differences between simulated and actual values for system outputs for three validation runs

	ΔDO (ppm) (simulated minus actual)	|ΔDO|	ΔT (°C) (simulated minus actual)	|ΔT|	ΔH (m) (simulated minus actual)	|ΔH|
27-28 June 1981						
Mean	0.89	0.90	1.30	1.32	-2.46	3.69
Variance	0.41	0.40	0.71	0.65	0.86	0.61
Maximum	2.64	—	2.88	—	7.32	—
Minimum	-0.27	—	-0.40	—	27.43	—
25 July-2 August 1981						
Mean	-0.16	0.59	1.13	1.36	-0.38	4.45
Variance	0.60	0.28	1.29	0.72	1.23	0.58
Maximum	4.83	—	3.76	—	16.76	—
Minimum	-2.19	—	-1.60	—	39.93	—
8-10 July 1980						
Mean	0.53	0.68	2.26	2.06	no measurement of H in 1980	
Variance	0.61	0.43	2.64	3.49		
Maximum	4.00	—	6.30	—		
Minimum	-0.84	—	-1.10	—		

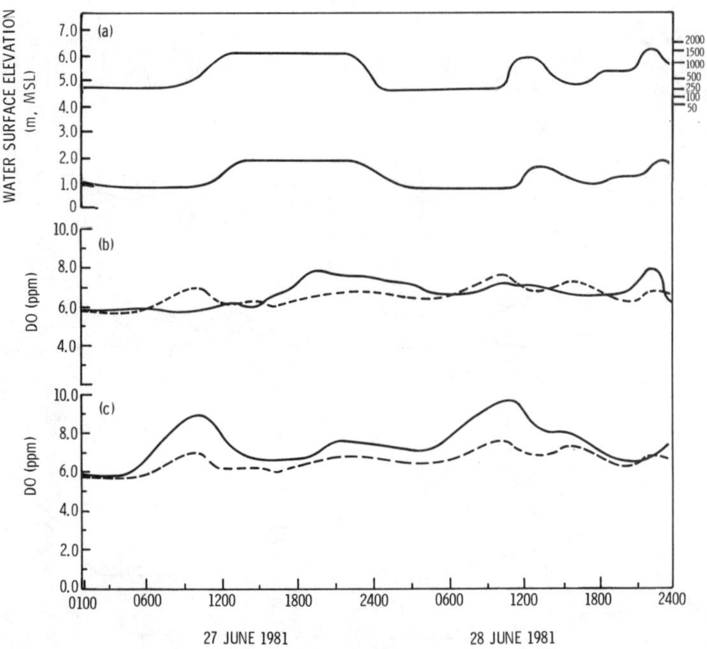

Figure 5. Validation of the DO model for the continuous discharge period, 27-28 June 1981. (a) Simulated water surface elevation at the dam (upper solid line) and simulated elevation at the fall line (lower solid line). (b) DO of the discharge from the dam (solid line) and DO measured at the fall line (dashed line). (c) DO measured at the fall line [dashed line; same as in (b) above] and DO at the fall line simulated by the model (solid line).

The 27-28 June 1981 run encompassed a Saturday-Sunday period with a minimum discharge of 180 m^3/s (fig. 5a). During this period, DO values at the input and transect A were similar, and remained in the range of 6-8 ppm (fig. 5b). Simulated DO values at transect A were always higher than observed values (fig. 5c). Estimates of mean ΔDO and $|\Delta DO|$ (table 2) were similar (~ 0.89 ppm) because the error was usually in the same direction. This bias is reflected in the range of ΔDO. The lack of fit could be due to turbidity increases during periods of high natural river flow. Continuous discharge by the turbines suggests that water in the reservoir and in the river reach is continuously mixing, thus preventing any settling of suspended solids. Any turbidity increase would depress actual oxygen production in the river reach, so the model, using a Secchi-disk depth characteristic of slower flowing conditions, would tend to overestimate DO during daylight hours. As can be seen in figure 5c, ΔDO is greatest during daylight.

Thus, under conditions of high river flow and continuous discharge, the model reproduces daily fluctuations in DO, but appears to consistently overestimate DO by slightly less than 1 ppm. Therefore, if it is necessary to use the model to simulate conditions of high river flow, it probably will be helpful to acquire more or better data (i.e., on the variation of turbidity with flow) so model mechanisms can be appropriately reformulated and recalibrated.

Although the model is not intended to simulate water temperature, a number of oxygen rate processes are controlled by temperature. Thus, it is useful to examine the fit of the algorithm for diel temperature change to data from transect A. Values for ΔT (table 2) during this period of continuous discharge appear to be within 1.3°C of actual data, and such a small difference is not likely to significantly affect metabolic rates.

We also compared simulated elevations in water surface at transect A with actual data to assure that USTFLO was generating correct hydrodynamic inputs to the oxygen model. During the period 27-28 June 1981, ΔH and $|\Delta H|$ averaged 2-4 cm from the observed data (table 2).

The second validation period (25 July to 2 August 1981) was a period of normal summer river discharges. The turbines operated on a peaking schedule, with nighttime and weekend shutdown intervals. However, during this period, PECO varied its proposed minimum discharge schedule. Shutdown periods were interrupted with 1-hour pulse discharges of 140 m^3/s and 4 hours of shutdown between pulses

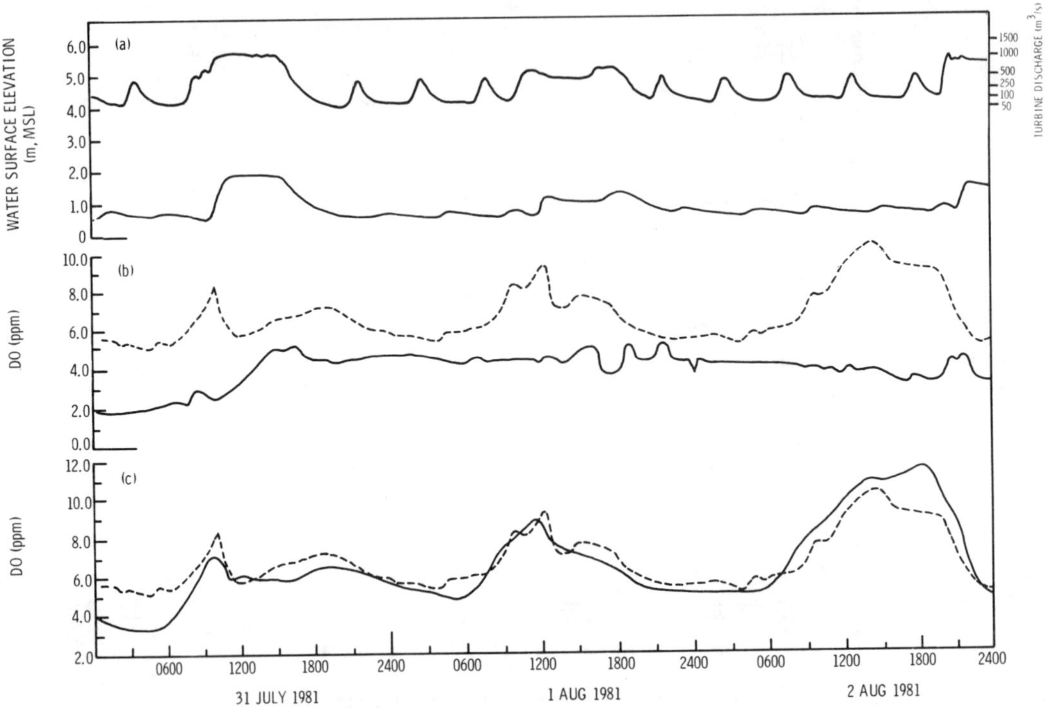

Figure 6. Validation of the DO model, 25 July-2 August 1981. (a) Simulated water surface elevation and corresponding discharge at the dam (upper solid line) and simulated elevation at the fall line (lower solid line). (b) DO of the discharge from the dam (solid line) and DO measured at the fall line (dashed line). (c) DO measured at the fall line [dashed line; same as in (b) above] and DO at the fall line simulated by the model (solid line).

(fig. 6, which shows 3 days typical of the 9-day interval). Even under this discharge regime, simulated water heights closely fit observed data (table 2).

As discussed above, the cooling-line DO data are at times unsatisfactory as the input DO values for the model. However, for the 25 July-2 August period, another DO record is available as an alternative input--the DO concentration at 10-m depth in the reservoir (near the top of the intake penstocks for the turbines). These data are likely to be representative of the DO of some portion of the water withdrawn by the turbines, and they follow the general trend shown by cooling-line DO data. In addition, the data are not subject to short-term DO depressions due to localized oxygen consumption in the penstock and cooling line during shutdowns. Accordingly, we used this reservoir DO record as the DO of input water for this simulation (fig. 6b).

Simulated DO concentrations at transect A appear to closely fit observed data (fig. 6c), averaging a 0.6-ppm difference (ΔDO) between the two data sets (table 2). As described for the calibration run, the DO record in the channel at transect A may not be a good representation of the average DO of the river (as simulated by the model) during daytime shutdown periods. The range of ΔDO for this run (up to 4.8 ppm; table 2 and fig. 6c) indicates that the model occasionally overestimates the data from transect A. However, as previously discussed, such high simulated DO values may be indicative of real DO values in the predominant pool areas of the river, but no data exist to test this hypothesis. No bias is evident in ΔDO over the entire run. Thus, the model appears to adequately fit the DO values observed over the 9-day period.

Simulated values for water temperature and height at transect A generally agree with observed data (table 2). Average values of $|\Delta T| = 1.36°C$, and this small error is not likely to significantly affect metabolic rate estimates. Average values of $|\Delta H| = 4.5$ cm, which is less than 5% of the daily fluctuations.

The two simulations of periods during 1981 assessed the fit of the model to DO under continuous high flow conditions and to DO affected by PECO's pulsing discharge regime. In contrast, the third validation period (8-10 July 1980) is characterized

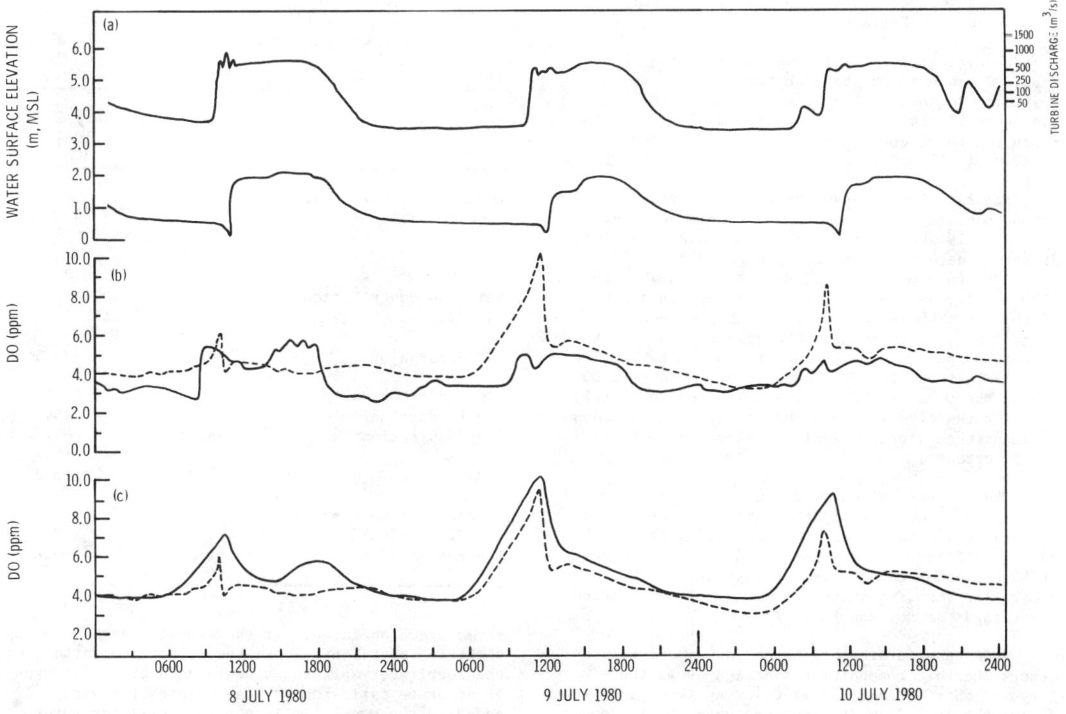

Figure 7. Validation of the DO model, 8-10 July 1980. (a) Simulated water surface elevation and corresponding discharge at the dam (upper solid line) and simulated elevation at the fall line (lower solid line). (b) DO of the discharge from the dam (solid line) and DO measured at the fall line (dashed line). (c) DO measured at the fall line [dashed line; same as in (b) above] and DO at the fall line simulated by the model (solid line).

by low natural river flows and shutdown periods not affected by PECO's pulsing discharge schedule (fig. 7a). The input data (in this case, from cooling line monitors) again show abrupt changes in DO in response to discharge changes (fig. 7b), which affect simulated DO values at transect A (fig. 7c). In spite of this, the model generally fits observed DO data at transect A (table 2), with relatively large deviations occurring only during daytime under shutdown conditions much lower than 30 m^3/s.

To summarize our discussion of validation of the DO model, it is useful to consider the conditions under which the model agrees with, or deviates from, observed DO values at transect A. First, the model appears to fit (defined arbitrarily as $|\Delta DO| < 1$ ppm) data taken during any summer intervals characterized by shutdown conditions, as shown by the validation runs of 25 July-2 August 1981 and 8-10 July 1980. These conditions are of primary importance if the model is to be used as a management tool, for instance, to estimate the relative effects of some hypothetical minimum discharge scenarios.

However, the simulated average DO values do not agree with observed data during some daytime shutdown conditions when DO conditions over a large portion of the river show much spatial variation. Also, the model overestimates DO values during periods of high river flow (e.g., 27-28 June 1981) when increased turbidity (not accounted for by the model) may decrease actual oxygen production. Another problem causing lack of fit at times is the possible unrepresentativeness of the cooling-line DO data used to estimate the DO concentration of input water from the dam discharge. We are presently investigating alternative means of calculating input DO.

In spite of minor deviations from observed data under certain conditions, the model appears to validly simulate average DO distributions in the river reach during summer periods of normal natural river flows.

Relative Contributions of Individual Rates to Simulated DO

We evaluated the relative contribution of the input DO, as well as those of the physical and biological processes in the river reach, to the DO concentration in the most downstream element of the reach (where DO would indicate the cumulative effects of all upstream sources and sinks of oxygen.

Accordingly, to quantify those individual contributions, we calculated the average DO concentration in this segment for validation runs under two different natural river flows (table 3): 27-28 June 1981 (653 m^3/s) and 25 August-2 September 1981 (370 m^3/s). As expected, the relative dominance of the input DO is larger for the higher period. In addition, we calculated the relative contributions to this output DO of each constituent process. The results indicate that the portion of downstream DO due to the DO of the discharge dominates the total, even for the time interval characterized by shutdown periods (when input DO would be expected to have little effect).

The individual contributions were calculated by running the model with a different term set to zero in each of several runs, and calculating the percentage differences from the standard run. Because of higher-order effects (e.g., feedback in the reaeration term, parameter interaction), the tabled percentages do not sum to 100%.

Among the *in situ* biological and physical processes, benthic community production shows the largest contribution, and is twice as important during the lower flow period than under high flow.

The relative contribution of reaeration is much larger for the low flow period than for high river flows. However, a large part of this apparent difference can be ascribed to larger input DO values for the first vs the second period (figs. 5b vs 6b), since reaeration is driven in part by the saturation deficit.

Benthic community respiration, and the production and respiration of the water-column community, never individually change the total DO concentration more than 10%.

The relative contributions of the terms in table 3 are long-term averages over the full range of discharges. To quantify the effects of discharge volume on their relative contributions to oxygen flux, we sorted the data from the 25 July-2 August model run into four discharge categories. The results show that the input DO concentration increases in dominance with increasing discharge--ranging from 66% at discharges less than 57 m^3/s to 85% at 850 m^3/s. The relative contributions of *in situ* processes to DO in the downstream segment decrease with increasing discharge, as was observed for the relationships of the processes to natural river flow. In general, it appears that discharge DO contributes at least 90% of the total DO at discharges above 800-1,200 m^3/s (when water residence times in the reach are 2 hours or less).

Conversely, during shutdown periods (e.g. less than 57 m^3/s), respiratory processes consume, on average, an amount of oxygen equal to 18% of the output oxygen. However, these averages also include daytime shutdown periods in which the production terms are also large. If the averages were done for nighttime shutdowns alone, the relative magnitude of the respiratory processes would undoubtedly increase. If accurate estimates of fish respiration were added to the model, it is possible that the three respiratory terms would control DO during nighttime shutdown periods.

Table 3. Relative contributions of individual rate processes (%) to the concentration of DO for two periods.

	27-28 June 1981	25 July-2 Aug. 1981
Average natural river flow (m^3/s)	653	370
Relative contribution (%) of:		
Discharge DO	90.6	75.5
Benthic Community		
--Production	7.7	15.4
--Respiration	-3.6	-9.9
Reaeration	2.3	13.3
Water-Column Community		
--Production	4.3	7.9
--Respiration	-1.9	-3.8

Sensitivity Analysis

The model uses a large number of parameters for rate processes and inputs in simulating DO distributions in the river reach. The accuracy and precision of field measurements used to derive these parameters are very variable. Thus, it is necessary to assess the relative importance of these parameters to the accuracy and precision of the results. Based on this evaluation, it is possible to determine if more or better data need to be gathered to ensure model accuracy. It is also useful to determine whether discrepancies between observed and simulated DO values occur that are caused by factors not accounted for in the model structure. In this section, we attempt to quantify the sensitivity of the DO output from the model (e.g., simulated DO values at transect A) to variations in individual rate processes affecting DO, and to some of the unpredictable physical inputs to the system that affect these processes (e.g., light intensity, input water temperature and DO, and turbidity).

To examine the effects of such variations on simulated downstream DO values, we artificially varied each individual rate process and some inputs (tables 4 and 5) in a series of simulations of the 25 July-2 August 1981 period previously used for validation (fig. 6). We multiplied each of the rate processes in eq. 4 (table 4) by 0.5 and by 1.5 in separate runs and calculated the average change in DO (ppm) relative to the standard run (fig. 6c). The rate processes of table 4 are arranged in order

Table 4. Sensitivity of simulated output DO to artificial variations of oxygen rate processes

Rate Process	Term in Equation 3	Change Factor	Mean ΔDO (ppm)	Variance ΔDO
Benthic production	P_b	x 0.5	−0.45	0.33
		x 1.5	+0.44	0.31
Atmospheric reaeration	K_2 (c* − c)	x 0.5	−0.34	0.06
		x 1.5	+0.27	0.04
Benthic respiration	R_b	x 0.5	+0.29	0.02
		x 1.5	−0.29	0.02
Water-column production	DP_{wc}	x 0.5	−0.23	0.04
		x 1.5	+0.23	0.04
Water-column respiration	R_{wc}	x 0.5	+0.11	0.002
		x 1.5	−0.11	0.002

of decreasing sensitivity to these changes. The averages cover all nine days of the runs, and thus include the full range of discharges released by the turbines. We did not attempt to describe sensitivities as functions of discharge.

Model DO output is very sensitive to benthic metabolism rates (table 4). This sensitivity may contribute to the poor fit of the model output DO to field data under certain conditions. For instance, during shutdown periods, simulated DO values are often larger than those observed at transect A (e.g., figs. 3c and 6c) for reasons relating to benthic metabolism in pools, discussed in the calibration section of this report.

Atmospheric reaeration rates also affect model output significantly (table 4). Our inability to measure the rate coefficient at high discharges forced us to use a relationship derived from data from several empirical studies in which conditions may not have been comparable to those in the Susquehanna River. We are presently unable to further assess the accuracy of the reaeration rate estimates used by the model. However, some qualitative observations indicate that the rate is reasonably accurate. Reaeration (eq. 5) is a negative feedback process, which tends to restore high or low DO values toward the saturation value (7-8 ppm for the summer temperatures of the periods simulated for calibration and validation). If the model's reaeration rate were too high, extreme supersaturated conditions (i.e., daytime DO peaks during shutdowns) or undersaturated conditions (i.e., nighttime DO minima) would be rapidly corrected by the reaeration mechanism, forcing a convergence on the saturation value. If the reaeration rate were too low, the model could not reproduce occasional nighttime increases in DO, which can be due only to reaeration. We thus feel reasonably confident in the reaeration mechanism used in the model.

Output DO levels appear to be the least sensitive to metabolic rates in the water column (table 4). Measurements of these rates were subject to some uncertainty, both because they were small in magnitude (near the detection limit of the method used) and because they were very variable in space and time (Philipp and Klose 1981). Thus, the least certain of our estimated rate processes also proved to have the smallest effect on the accuracy of our simulated results. Any errors in estimating water-column light intensities would thus have minimal effect on simulated DO levels. During shutdown periods, when DO impacts on biota are likely to be largest, the water volume in the river is near its minimum. This further decreases the effect of water-column rate processes on the total DO of the river at such times. Thus, inaccuracies in the estimation of water-column processes are not likely to significantly affect DO in the river reach.

Table 5. Sensitivity of simulated output DO to artificial variations of system inputs

Input	Parameter (model equation)	Change	Mean ΔDO (ppm)	Variance ΔDO	Maximum ΔDO	Minimum ΔDO
Input DO concentration	C_{in} (eq. 3)	−1.0 ppm	−0.73	0.03	0.0	−0.95
		+1.0 ppm	+0.73	0.03	+0.95	0.0
Light intensity at water surface	I_o (eq. 9) both benthic and water-column P)	x 0.5	−0.41	0.18	—	—
		x 1.5	+0.17	0.06	—	—
Secchi-disk depth	z_{Secchi} (eq. 11)	1.0 m	−0.38	0.11	—	—
		1.7 m	Standard run			
		2.5 m	+0.17	0.02	—	—
Input water temperature	T (eqs. 7, 8, 12)	x 0.8	+0.23	0.43	1.27	−1.50
		x 1.2	−0.49	0.38	1.13	−1.58

The magnitude of the ΔDO values of table 4 can also be compared to the overall mean DO of the 9-day run--5.8 ppm. Variations of the individual rates were ±50% about the values used in the standard run. The maximum effect on output DO from any of these variations was ±8% (due to benthic production). This indicates that, when averaged over all river conditions over many days, in situ processes, when taken individually, have a relatively small effect on average DO levels in the river. However, this first-order sensitivity analysis neglects higher-order effects, so the effects of changes in individual parameters cannot be superimposed to estimate the sensitivity of the total system. Also, this average sensitivity (over 9 days) may not be appropriate for evaluating available habitat, which is controlled by the most stringent conditions encountered over a period of time (e.g., shutdown periods).

Besides the rate processes described above, a number of system inputs affect model behavior. First, simulated output DO values are very sensitive to input DO concentrations. The addition or subtraction of 1 ppm to input DO concentrations (table 5) produces a ±0.7-ppm average change in DO at the fall line. This 9-day average hides the fact that the importance of input DO increases in proportion to discharge (i.e., input oxygen is determined by $Q_{in}C_{in}$ in eq. 3). We are currently undertaking an intensive analysis to estimate errors in measuring the DO concentration of the discharge water. The sensitivity runs also point out the importance of identifying the effects of reservoir processes on discharge DO, especially the prediction of discharge DO values under any future continuous discharge regime.

Next, factors controlling production (surface light intensity and Secchi-disk depth) also have a relatively large effect on output DO values (table 5). Most of the resulting changes in DO are probably due to the effects of ambient light intensity on benthic rather than water-column productivity.

Lastly, water temperature affects all five oxygen rate processes (table 4), so its correct estimation is important. The sensitivity runs (table 5) indicate that a ±5°C change of input water temperature (~ 25°C during summer) does change DO values by up to 0.5 ppm. Since actual temperatures of input water are relatively constant over long periods in summer, 5°C errors in its estimation are not likely to occur. Downstream, however, 5°C temperature increases are not uncommon, especially during periods with low discharges. These could increase daytime rates in pools, a process which may be reflected in large DO increases during daytime shutdown periods (figs. 3 and 6). Since the model does not include a detailed heat budget, it is unlikely that its temperature mechanism can be improved.

Discussion

The previous sections show that the model generally simulates summertime conditions over a wide range of discharges. However, fit of the model's DO output to observed data is poorest for conditions when the turbines are shut down. Under such conditions, the assumption of complete mixing may be violated, so that no single point estimate of DO may be representative of the DO distribution in a segment. Otherwise, the model can be used to assess the change in DO distributions due to biological and physical sources and sinks of oxygen within the river reach.

The validation runs indicate that the DO of the turbine discharge dominates the oxygen output of the system at the fall line, and that this relative dominance increases with increasing daily river flow and/or shorter-term discharge volume. Benthic metabolism contributed the most to the in situ oxygen change, followed by reaeration. Water-column community metabolism was of minor importance.

We also assessed the sensitivity of the model's DO output to artificial variations of rate processes or individual parameters. The sensitivity analyses showed that variations in the rates of benthic production and reaeration had the largest effects on the model's output. Despite the paucity and/or poor quality of field data on these rates as functions of space and time, the model usually reproduces observed temporal and spatial DO distributions. In contrast, the model's output appears relatively insensitive to changes in water-column metabolism, so errors in estimating these rates should have relatively little effect on model accuracy.

The model's output is also sensitive to the DO concentration of the turbine discharge water. This dependence of the model on input values, the present uncertainty about the accuracy of turbine DO data used as DO input to the model, and our present lack of knowledge about the dependence of the DO of the discharge water on reservoir processes and discharge volume, all indicate the need for further analyses to determine the source of the discharge water within the reservoir and the dependence of its DO on turbine operations. We are currently beginning such analyses. They include mechanistic simulation modeling to describe the causes of observed DO distributions in the reservoir; analyses to describe the withdrawal of water from the reservoir; and time-series analyses of the DO data from the turbine cooling line (data that are presently used as DO input to the model) relative to reservoir DO data, to generate a more precise means of estimating the DO of discharge water.

When these analyses are completed, the river DO model will be a versatile tool for estimating downstream DO distributions in summer under a variety of river flows and turbine operating schedules (e.g., hypothetical minimum discharge regimes). In the meantime, the model can still be used to make relative comparisons among the downstream DO distributions resulting from hypothetical modifications of dam operations. We are presently using the model to examine the differences in output values among a number of different discharge scenarios to provide information for an ongoing legal proceeding that will result in an interim minimum discharge requirement for Conowingo Dam.

ACKNOWLEDGEMENT

We thank T.T. Polgar and W.A. Richkus for assisting in development of the model and for their consultation during the writing of this paper.

LITERATURE CITED

APHA (American Public Health Association). 1975. Standard methods for the examination of water and wastewater. 14th edition. American Public Health Association, Washington, D.C.

Bennett, J.P., and R.E. Rathbun. 1972. Reaeration in openchannel flow. U.S. Geological Survey Professional Paper No. 737. Washington, D.C.

Carter, W.R., III. 1973. Ecological study of Susquehanna River and tributaries below the Conowingo Dam. National Marine Fisheries Service. Federal Aid Project Report No. AFSC-1. Maryland Department of Natural Resources, Fisheries Administration, Annapolis, Md.

Dwyer, R.L., and M.A. Turner. 1982. Simulation of river flow and DO dynamics affected by peaking discharges from a hydroelectric dam. Report prepared for Maryland Power Plant Siting Program by Martin Marietta Environmental Center, Baltimore, Md.

HEC (Hydrologic Engineering Center). 1977. Gradually varied unsteady flow profiles. U.S. Army Corps of Engineers, Davis, Cal.

Jackson, D.R., and G.J. Lazorchick. 1980. Instream flow needs study, Susquehanna River, vicinity of Conowingo Dam. Susquehanna River Basin Commission Technical Report No. 2. Harrisburg, Pa.

Kester, D. 1975. Dissolved gases other than carbon dioxide. In: Chemical Oceanography I, pp. 497-556. J.P. Riley and G. Skirrow, eds. London: Academic Press.

Kremer, J.N., and S.W. Nixon. 1978. A Coastal Marine Ecosystem: Simulation and Analysis. Berlin: Springer-Verlag.

Krenkel, P.A., and V. Novotny. 1980. Water Quality Management. New York: Academic Press.

Levenspiel, O., and W.K. Smith. 1957. Notes on the diffusion type model for the longitudinal mixing of fluids in flow. Chem. Eng. Sci. 6:227-233.

McClellan, E., et al. In preparation. Report of results of 1980 studies of DO and temperature in the vicinity of Conowingo Dam. Radiation Management Corp., Drumore, Pa.

Nixon, S.W., C.A. Oviatt, J.N. Kremer, and K. Perez. 1979. The use of numerical models and laboratory microcosms in estuarine ecosystem analysis --simulations of a winter phytoplankton bloom. In: Marsh-Estuarine Systems Simulation, pp. 165-188. R.F. Dame, ed. Columbia, S.C.: University of South Carolina Press.

Parsons, T.R., and M. Takahashi. 1973. Biological Oceanograpic Processes. New York: Pergamon Press.

Petrimoulx, H.J., and P.N. Klose. 1981. Resident fisheries study, lower Susquehanna River, Maryland. Maryland Power Plant Siting Program Report No. PPSP-UBLS-81-5. Annapolis, Md.

Philipp, K.R., and P.N. Klose. 1981. Lower Susquehanna River oxygen dynamics study. Maryland Power Plant Siting Program Report No. PPSP-UBLS-81-4. Annapolis, Md.

Photo Science, Inc. 1980. Operational effects of Conowingo Dam on the Susquehanna River. Report to the Susquehanna River Basin Commission, Gaithersburg, Md.

Potera, G.T., K.R. Philipp, T.D. Johnson, H.J. Petrimoulx, and P.N. Klose. 1982. Lower Susquehanna River hydrographic and water quality study, summer 1981. Maryland Power Plant Siting Program, Annapolis, Md.

Roques, P., and S. Nixon. In preparation. Direct measurement of air-water gas exchange coefficient with special reference to oxygen. University of Rhode Island, Graduate School of Oceanography, Kingston, R.I.

Steele, J.H. 1965. Notes on some theoretical problems in production ecology. In: Primary Productivity in Aquatic Environments, pp. 383-398. C.R. Goldman, ed. Mem. Ist. Ital. Idrobiol., 18 Suppl. Berkeley, Cal.: University of California Press.

Strickland, J.D.H., and T.R. Parsons. 1968. A practical handbook of seawater analysis. Fisheries Research Board of Canada Bulletin No. 167. Ottawa.

Sverdrup, H.U., M.W. Johnson, and R.H. Fleming. 1942. The Oceans. Englewood Cliffs, N.J.: Prentice-Hall.

Tidewater Administration. 1980. Susquehanna River fish kill of July 8, 1980. Maryland Department of Natural Resources, Tidewater Administration, Tidal Fisheries Division, Annapolis, Md.

Wetzel, R.G. 1975. Limnology. Philadelphia: W.B. Saunders.

Yen, B.C. 1979. Unsteady flow mathematical modeling techniques. In: Modeling of Rivers, H.W. Shen, ed. New York: Wiley-Interscience.

A PRACTICAL WATER QUALITY MODEL FOR ASSESSING LOW HEAD HYDROELECTRIC PROJECTS[1]

Barry G. Walkenshaw[2], Gerald B. Jones[3], Jeffrey R. Kerr[3]

Abstract: Water quality downstream of a proposed hydroelectric project was simulated using the SQUAL stream model. Model selection and calibration was limited by project time and budget constraints. The model projected dissolved oxygen data that was used to modify proposed project operation and monitoring.

INTRODUCTION

In November 1978, the City of Columbus, Ohio contracted with Burgess & Niple, Limited to conduct a feasibility study for hydroelectric power generation at O'Shaughnessy Dam on the Scioto River (Figure 1). Current legislation and federal regulations required that certification of the new hydroelectric facilities be acquired from the Federal Energy Regulatory Commission (FERC) - U.S. Department of Energy. FERC and the applicant share the responsibility of insuring that a proposed project is environmentally acceptable. The applicant is required to provide documents addressing the environmental issues of a proposed project in the application for license or application for license exemption.

During completion of the O'Shaughnessy study it became apparent that there was insufficient information regarding fishery resources and water quality in the Scioto River to adequately assess project effects on these resources. Consultations with Ohio Department of Natural Resources and the U.S. Fish and Wildlife Service indicated that the potential effect of the project on water quality and game fish populations were areas of major concern. The Scioto River at this location is known to support a regionally important smallmouth bass fishery.

Two possible approaches to assessing potential project effects on the Scioto River were identified: direct experimentation and water quality modeling. The direct experimentation scheme of analysis would include fitting stop logs to the water supply outlet structure to withdraw water from the reservoir at the same depth as the hydroelectric intake. The quality of this water would be monitored at downstream stations over at least a 1-year period for a variety of releases similar in both discharge and duration to those expected of the hydroelectric facility. Constraints on either approach included a budget of $60,000 and a completion schedule of 9 months.

The direct experimentation approach was screened because of limited time and high project costs related to excessive data collection requirements. Therefore, the modeling approach was recommended.

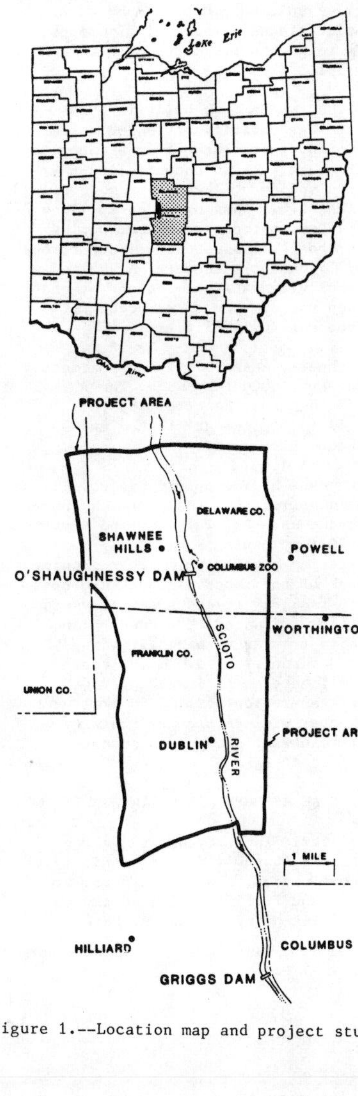

Figure 1.--Location map and project study area.

---

[1]Paper presented at Third International Conference on State-Of-The-Art in Ecological Modeling, Colorado State University, May 24-28, 1982.

[2]Barry G. Walkenshaw, Civil Engineer, Burgess & Niple, Limited Engineers and Architects, Columbus, Ohio

[3]Gerald B. Jones, Jeffrey R. Kerr, Biologists, Burgess & Niple, Limited Engineers and Architects, Columbus, Ohio

BACKGROUND

The Scioto River has a total drainage area of 6,517 square miles and is the second largest river basin in Ohio. The headwaters of the Scioto River are found in the marshy remnants of a glacial lake in Hardin County. These waters flow southeast for 60 miles, then south for another 170 miles before reaching their confluence with the Ohio River.

Two major impoundments on the Scioto River are Griggs and O'Shaughnessy Reservoirs. These reservoirs were constructed to provide water storage for Columbus and surrounding metropolitan areas. O'Shaughnessy Dam is located at River Mile 148, 16 miles north of the City of Columbus in Delaware County, Ohio. The dam is situated just north of the Delaware-Franklin county line, 9.5 miles upstream of Griggs Dam and 13.5 miles upstream of the City of Columbus, Dublin Road water treatment plant intake. The pool of Griggs Dam extends to within 3.9 miles of O'Shaughnessy Dam. (Figure 1).

O'Shaughnessy Dam was built in 1924 and has a storage capacity of 16,900 acre-feet and a head of 65 feet. The dam is a concrete gravity structure with an uncontrolled 879 foot concrete spillway. Throughout much of the year river flows pass over the spillway due to the large upstream drainage. During low flow periods, releases can be made through a water supply outlet structure. A vaulted conduit was provided at the west end of the spillway when the dam was constructed for future hydroelectric facilities.

O'Shaughnessy Reservoir is approximately 6 miles long and 1,000 feet wide. The normal depth of O'Shaughnessy Reservoir at full storage is 60 to 65 feet at the dam and 35 to 40 feet at center pool.

The proposed hydroelectric project will involve the construction of a penstock, power plant, tailrace channel, walkway, and service road. The 270 foot penstock will originate from the existing power gate chamber along the west bank of the reservoir and terminate at the hydro-facility at about the elevation of the low flow tailwater. The power plant will house two tubular turbines yielding a total rated capacity of 5.0 megawatts. The plant will have an estimated average annual power generation of 14×10^6 kilowatt hours. The power will be used by the City of Columbus to offset city power costs.

SELECTION OF AN APPROPRIATE WATER QUALITY MODEL

Hydroelectric facility water quality assessment necessitates a deterministic model for proper evaluation, since reaeration at various heads and initial dissolved oxygen concentrations is being evaluated. Deterministic models are divided into two categories time-invariant (steady-state) and time variant (dynamic) models. Steady state models are used when the system being simulated does not change with time, such as flowing streams, lakes, and reservoirs. Dynamic models are used to simulate a system which is continually changing (i.e., never reaching a steady state condition). The time-step to be used for dynamic models varies depending upon the rate of change of the system.

Data requirements are a key characteristic guiding the selection between these two types of deterministic models. Dynamic events are typically very complex and require collection of data under specific conditions (i.e., low flows and reservoir releases similar to those modeled). Accuracy demands that dynamic models be verified for every major reach and discharge modeled. Proper verification includes not only water quality data but flow and depth data as well. Steady state models require only that the stream water quality model be calibrated for a normal flow condition.

In an effort to project water quality conditions in the Scioto River below O'Shaughnessy Reservoir, a steady state stream model was applied. A steady state model was selected because of its simplicity when compared to dynamic models and the smaller data base required. Collection of the extensive data base required to run and calibrate dynamic models was clearly beyond the scope and available resources of this project.

THE SQUAL MODEL

The SQUAL stream water quality model was developed by Burgess & Niple, Limited. This model is one-dimensional (uniform mixing across the cross section), nondispersive (no mixing in the direction of flow), and advective. Nonconservative substances which were modeled are 5-day biochemical oxygen demand, dissolved oxygen, organic nitrogen, ammonia, nitrite- nitrogen, and nitrate-nitrogen. Among the assumptions made in the development of SQUAL were the following:

1. Reaeration was regarded as the only source of dissolved oxygen. Therefore, the effects of photosynthesis were not included.

2. Benthic (sediment) oxygen demand, biochemical oxygen demand, and oxygen demand resulting from the oxidation of nitrogen were regarded as the primary dissolved oxygen sinks. Other sinks of dissolved oxygen which were not included are disturbed-sediment oxygen demand and algal respiration.

The salient features of this stream water quality model are:

1. Rate Coefficients - The program can use two sources of rate coefficients: the data file and internal default values or equations. The rate coefficients are adjusted within the program, according to the mean temperature of flow.

2. Simulation of Gradually Varied Flow - The program does not simulate flow in open channels. The physical data used by the model, such as the slope of the hydraulic grade line, mean velocity, hydraulic depth, and mean top width of the water surface, were determined using the Corps of Engineers program HEC-2.

3. Reaeration at Dams and Waterfalls - The program simulates reaeration resulting from unconfined drops of the water surface utilizing dam reaeration equations.

4. Degradation of Nonconservative Substances - Degradation of nonconservative substances was recorded at both the upstream and downstream end of each reach, as well as every tenth of a river mile.

5. Settleable and Dissolved Biochemical Oxygen Demand - The 5-day biochemical oxygen demand was divided into settleable and dissolved fractions. The settleable fraction can be input as a positive number from zero to one or as a default value. When the mean velocity within the reach falls below 0.60 foot per second, the settleable 5-day biochemical oxygen demand is removed at the upstream end of the reach and creates benthic oxygen demand which is distributed throughout the reach.

MODEL LIMITATIONS

The calibration procedure was undertaken with cognizance of the model shortcomings. The model does not simulate the uptake and synthesis of various forms of nitrogen by aquatic plants and phytoplankton. No attempt was made to calibrate against measured nitrate concentrations. Additionally, very little research has been done on the development of reaeration coefficients for impoundments. It is known that the independent variables which would be used in such equations are not related to stream velocity or the slope of the hydraulic gradeline as are the stream reaeration rate equations. Constant reaeration values were selected in lieu of default reaeration rate equatious to calibrate dissolved oxygen concentrations in the upper end of Griggs Reservoir pool.

MODEL CALIBRATION

Data Requirements

Calibration may be defined as the adjustment of variables within the model to minimize the difference between measured and predicted output. The required steps for calibration of SQUAL include:

1. A water quality evaluation of tributary flows,

2. Water quality and stream discharge data collection from the main stem and principal tributaries,

3. Selection of a worst case water quality condition,

4. Collection of stream cross sections, longitudinal profile, channel roughness, and slope information,

5. Use of the HEC-2 model to develop velocity, stream width water surface elevation, slope of the hydraulic gradeline, and hydraulic depth data,

6. Adjustment of model rate constants, and,

7. Selection of an appropriate dam reaeration equation.

An initial field reconnaissance did not identify any tributary streams between O'Shaughnessy Dam and Griggs Reservoir that represented significant flow or wasteload contributions to the Scioto River.

Stream channel data were obtained using the U.S. Army Corps of Engineers HEC-2 model. This model was applied to the Scioto River between O'Shaughnessy Dam and Griggs Reservoir.

The project budget permitted collection of water quality data at weekly intervals over a 5 week period during October through November, 1979. Data was collected at one station above O'Shaughnessy Dam in the vicinity of the intake and at four river stations located downstream. Variables measured included: dissolved oxygen, temperature, biochemical oxygen demand, organic nitrogen, and ammonia-nitrogen. Samples from multiple depths were collected at the reservoir station in order to characterize water quality conditions through the entire water column in the vicinity of the intake. This information provided seasonal water quality data at specific points in the river at a given discharge. In addition, the surface dissolved oxygen and temperature above and below O'Shaughnessy Dam provided data for selection and calibration of a suitable dam reaeration equation.

The water quality data set reflecting the lowest dissolved oxygen concentrations during the study period was selected for model calibration. Ecological systems typically experience their greatest perturbations as a result of the most extreme variations in the range of environmental conditions. Therefore, in order to accurately assess the potential effect of the hydroelectric facility on water quality it was necessary to utilize that data set reflecting worst case water quality conditions during the study period.

It should be acknowledged that the conditions under which the sampling program was initiated did not reflect a typical annual worst case condition of low flow, high water temperature and heavy algal growth. Normally, the highest water temperatures in the Scioto River occur in the months of August, September, and October. Simularly, the lowest flows typically occur in September, October, and November. However, 1979 exhibited a much higher fall rainfall than normally experienced. As a result, flows were higher than normal, temperatures were lower than normal and algal blooms were flushed from the reservoir.

CALIBRATION PROCEDURE

The calibration procedure consisted of adjustment of rate coefficients. The order of coefficient calibration was organic nitrogen decay, ammonia-nitrogen decay, 5-day biochemical oxygen demand decay, and the stream reaeration. Table 1 indicates the normal range of these coefficients in deep water systems. Much higher values are typically observed in shallow waters which occur in the upstream portion of the stream segment modeled.

Sources of organic nitrogen other than that already present in the reservoir were disregarded in the model. Thus, organic nitrogen measured in O'Shaughnessy Reservoir, just upstream of the dam, was assumed to decay continuously. A default value of 0.08 (1 per day at 20° C) was used for the organic nitrogen decay rate on the Scioto River. Within Griggs Dam pool, the organic nitrogen decay rate was decreased to 0.02 (1 per day at 20° C) to account for the decreased biological decay activity in Griggs pool.

Based upon the results of research studies performed at the University of Kentucky, (Pavoni, 1977) the ammonia decay rate can be determined by the following equation:

$$k_{NH_3-N} = 0.030 + 0.019{*}S^{1.2} \qquad (1)$$

where, S is the slope of the river in feet per mile.

k_{NH_3-N} = ammonia-nitrogen decay rate coefficient

Maximum and minimum ammonia decay rates observed during data collection for the above expression were, respectively, 0.30 and 0.10 (1 per day at 20° C). This expression was used for the reach extending from O'Shaughnessy Dam to the upstream edge of Griggs pool (3.8 miles). The value 0.45 (1 per day at 20° C) was used in Griggs pool.

Table 1.--Normal ranges of rate coefficients in deep water system 1, 2

Rate Coefficient	Range (1 per day at 20° C)
Korg-N	0.02 - 0.4 (1.12)**(T-20)
KNH_3-N	0.10 - 0.50 (1.10)**(T-20)
KNO_2-N	3.0 - 10.0 (1.07)**(T-20)
K_O	0.1 - 2.5 (1.06)**(T-20)
K_A	0.0 - 100.00 (1.024)**(T-20)

[1]Bacca and Arnett, 1976

[2]U.S. Environmental Protection Agency, 1977

Nitrite-nitrogen decay rates are typically an order of magnitude larger than those for the decay of ammonia-nitrogen. The default value of 1.12 (1 per day at 20° C) was assumed for the reach extending from O'Shaughnessy Dam to the upstream end of Griggs pool.

It has been found that, in general, the carbonaceous biochemical oxygen demand deoxygenation coefficient decreases with decreasing biochemical oxygen demand concentrations. An expression for this relationship was developed at the University of Kentucky (Pavoni, 1977):

$$K_d = 0.018{*}BOD_5 \qquad (2)$$

K_d = BOD_5 decay rate coefficient

BOD_5 = 5 day biochemical oxygen demand

Maximum and minimum rates observed during data collection for the above expression were, 0.50 and 0.15 (1 per day at 20° C), respectively.

Stream reaeration equations were used in the upper 3.8 mile reach from O'Shaughnessy Dam to Griggs pool. Since wind data was not taken, constant values were used for the reaeration coefficient within Griggs pool. The stream reaeration equation which produced the best fit was that developed by Tsivoglou, (Tsivoglou, 1976):

$$K_2 = C{*}V{*}S \qquad (3)$$

where, V = mean stream velocity
S = slope of the hydraulic grade line, ft/ft
C = escape coefficient
K_2 = stream reaeration rate coefficient

The escape coefficient is very small for larger streams with relatively large, deep, and straight unobstructed channels. It is very large for smaller streams characterized by abrupt changes in slope and direction.

Table 2 relates predicted to measured concentrations of water quality variables. Predicted concentrations were within 1 milligram per liter, with the exception of organic nitrogen and nitrates. Since the model does not simulate plant synthesis of these nutrients, concentrations of organic and nitrate nitrogen were not calibrated.

The model was calibrated for a flow of 243 cubic feet per second. As flows increase, the reaeration coefficients predicted by various equations converge. Also, as flows increase, travel times and, consequently, reaction times decrease. This decrease in travel time reduces the influence of pollutant decay coefficients.

Table 2.--Results of model calibration run 1

Water Quality[3] Variable		O'Shaughnessy Reservoir	Scioto River Downstream at O'Shaughnessy	Scioto River at USGS Gaging Station
Dissolved Oxygen	(M)[2]	3.4	-	8.50
	(P)	-	8.27	8.37
BOD_5	(M)	3.0	-	2.0
	(P)	-	3.0	3.0
Org-N	(M)	1.70	-	1.10
	(P)	-	1.70	1.70
NH_3-N	(M)	0.80	-	-
	(P)	-	0.80	0.80
NO_2	(M)	0.10	-	-
	(P)	-	0.0	0.01
NO_3	(M)	3.40	-	2.50
	(P)	-	3.40	3.40
River Mile		148.39	148.19	147.57
Temp. ° C.	(M)	16.5	-	17.0
	(P)	-	16.5	16.5

Water Quality[3] Variable		Scioto River at S.R. 161	Scioto River at Hayden Run
Dissolved Oxygen	(M)[2]	8.60	8.50
	(P)	8.34	8.34
BOD_5	(M)	2.0	2.0
	(P)	2.9	2.8
Org-N	(M)	1.40	2.20
	(P)	1.68	1.64
NH_3-N	(M)	0.80	-
	(P)	0.79	0.47
NO_2	(M)	-	-
	(P)	0.02	0.02
NO_3	(M)	3.00	2.20
	(P)	1.68	3.77
River Mile		144.36	142.09
Temp. ° C.	(M)	16.0	16.5
	(P)	16.5	16.5

[1] Based on calibration data collected on October 4, 1979.

[2] M (Measured); P (Predicted Model)

[3] All values in milligrams per liter unless otherwise specified.

MODEL USE

The stream model was applied to simulate downstream conditions of dissolved oxygen, 5-day biochemical oxygen demand, organic nitrogen, ammonia-nitrogen, nitrate-nitrogen, and nitrite-nitrogen at hydroelectric facility releases of 10, 65, 360, 660, and 810 cubic feet per second. Listed below is each modeled flow and the rationale for its selection:

10 Cubic Feet Per Second. This discharge represents the flow associated with nonoperation of the hydrofacility. Ten cubic feet per second represents normal leakage through the flash-boards.

65 Cubic Feet Per Second. This flow represents the minimum flow, (55 cubic feet per second) released to deliver the daily public water supply demand, plus leakage (10 cubic feet per second).

360 Cubic Feet Per Second. A typical operating flow acceptable for efficient operation of one turbine.

660 to 810 Cubic Feet Per Second. A typical operating flow discharged through two turbines.

Reaeration results from an unconfined drop over O'Shaughnessy Dam. The dam reaeration equations evaluated included those of Foree, Jarvis, and Holler (Foree et al., 1976). Table 3 lists the dam reaeration equations and the definition of terms used in these equations. These three equations were evaluated with respect to the calibration data. The Jarvis equation proved to be most accurate, reflecting deviations of predicted dissolved oxygen concentrations of 1 to 2 percent from measured dissolved oxygen concentrations.

Table 3.--Dam reaeration equations

Foree

$$R_{25}=e^{0.16H} \tag{4}$$

where,

H = elevation difference, feet

r_{25} = dissolved oxygen deficit ratio at base temperature 25° C

$$r_T = \frac{Cs-Cu}{Cs-Cd} \tag{5}$$

where,

r_T = dissolved oxygen deficit ratio

Cu = dissolved oxygen concentration immediately upstream of the dam or falls, mg/l

Cd = dissolved oxygen concentration immediately downstream of the dam or falls, mg/l

Cs = dissolved oxygen saturation concentration at the elevation and temperature under consideration, mg/l

Cs,o = dissolved oxygen saturation concentration at the same temperature at sea level, mg/l

$$Cs = Cs,o\ (1-(0.027E)/760) \tag{6}$$

E = elevation of the site under consideration, feet above mean sea level

$$\ln r_T = (\ln r_{25})(1.022)^{(T-25)} \tag{7}$$

where,

T = temperature in degrees centigrade

Jarvis

$$r_{15} = 1.224\ H^{0.434} \tag{8}$$

Table 3.--Dam reaeration equations (continued)

$$r_T = (r_{15}-1)[1+0.027(T-15)]+1 \quad (9)$$

where,

r_{15} = dissolved oxygen deficit ratio at 15° C

H = elevation difference in meters

Holler

$$r_{20} = 1+0.275H \quad (10)$$

$$\ln r_T = (\ln r_{20})(1.1016)^{(T-20)} \quad (11)$$

where,

r_{20} = dissolved oxygen deficit ratio at 20° C

H = elevation difference, feet

MODEL RESULTS

The modeling effort quantified the difference in dissolved oxygen concentrations occuring downstream of the dam with and without hydro-facility operation. The primary finding of the stream model was the projection of dissolved oxygen recovery zones in the Scioto River downstream of the dam during hydrofacility discharges ranging from 65 to 810 cubic feet per second (Figure 2). Recovery of the stream dissolved oxygen was defined as the attainment of the Ohio Water Quality Standard of 5.0 milligrams per liter. All hydrofacility discharges modeled utilized the depth-area average dissolved oxygen concentration of 2.5 milligrams per liter.

Turbulence due to unconfined drop over the O'Shaughnessy spillway increases oxygen transfer to the water. This oxygen transfer elevates dissolved oxygen levels to Ohio Stream Water Quality Standards for all flows modeled. Flows passing through the turbine would not receive the benefit of reaeration from passing over the spillway . As flows through the turbine increase, the capacity of the stream to provide natural reaeration and maintain acceptable dissolved oxygen concentrations is exceeded. Therefore, the dissolved oxygen recovery point of the stream is shifted downstream at higher turbine flows.

At modeled turbine flows of 65 and 360 cubic feet per second, dissolved oxygen recovery was projected to run very near the upstream end of the Griggs Reservoir pool (Figure 2). At 660 to 810 cubic feet per second, dissolved oxygen recovery would not be attained until flow was mixed within Griggs Reservoir. Therefore, at all turbine discharges water quality standards would be violated over the majority of the free flowing stream segment between reservoirs.

The low dissolved oxygen concentrations predicted would have adverse effects on game fish species common to this river segment. These species, which generally have minimum dissolved oxygen requirements ranging from 3 to 5 milligrams per liter (Davis, 1975), include the smallmouth bass (*Micropterus dolomieui*), the white crappie (*Pomoxis annularis*) and bluegill (*Lepomis macrochirus*). These species would be most vulnerable during oxygen sensitive periods of the life cycle such as the late spring to early summer spawning period.

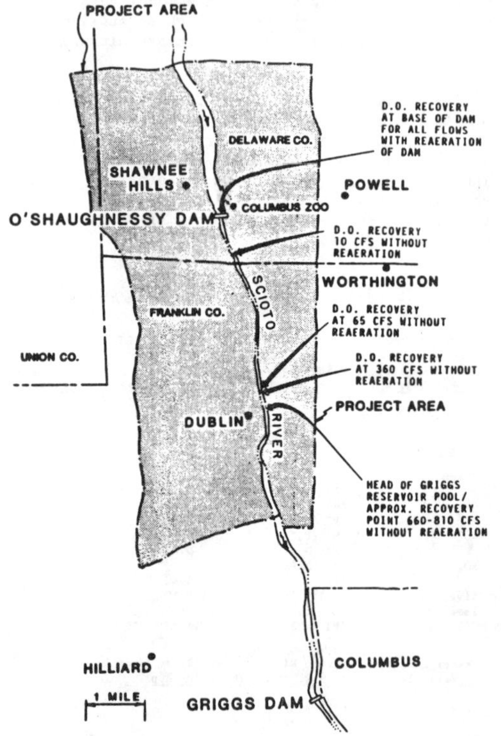

Figure 2.--Model projections of dissolved oxygen recovery points.

MITIGATION MEASURES

The following mitigation measures were recommended to preclude the adverse dissolved oxygen impacts related to facility operation or nonoperation.

1. Continuous dissolved oxygen monitors will be installed in the O'Shaughnessy pool and downstream in the Scioto River. These monitors will be telemetered to the City of Columbus Dublin Road water plant and will be used to signal when the hydroplant operation should be terminated to maintain dissolved oxygen stream standards.

2. The reaeration capacity of the existing water supply outlet structure can be increased by fully opening the outlet gate and throttling the inlet gates. This modification of operating procedures will improve downstream dissolved oxygen concentrations when the hydro-facility is not in operation.

SUMMARY

A practical means of assessment of the water quality effects of low head hydroelectric facilities has been presented. The approach incorporates stratification studies, calibration and selection of an appropriate dam reaeration equation, and calibration of a steady-state stream water quality model. The method relies on the evaluation of a worst case water quality data set rather than a statistical analysis of data taken over a long period of time.

The modeling approach was acceptable to both federal and state agencies concerned with the project. It resulted in the requirement of dissolved oxygen monitoring of the hydro-electric intake and discharge. The operation plan of the facility will include plant shutdown during low dissolved oxygen water intake conditions.

LITERATURE CITED

Baca, R. G. and R. Arnett. 1976. A limnological model for eutrophic lakes and impoundments. Battelle Pacific Northwest Laboratories.

Davis, John C. 1975. Minimal dissolved oxygen requirements of aquatic life with emphasis on Canadian species: a review. Journal Fisheries Research Board of Canada 32(12): pp. 2295-2332.

Foree, Edward G. and R. Ware. 1976. Reaeration characteristics of low-level in-channel dams in Kentucky streams. Department of Civil Engineering, University of Kentucky. pp. 19

Pavoni, Joseph L. 1977. Handbook of Water Quality Management Planning. Van Nostrand Reinhold Company, New York, N.Y. pp. 429

Tsivoglou, E. C. and L. Neal. Tracer measurement of stream reaeration, III, predicting the reaeration capacity of inland streams, J. WPCF 48(12): pp. 2669-2689.

U.S. Environmental Protection Agency, 1977. Computer program documentation for the stream quality model QUAL-II. Prepared by Water Resource Engineers, Inc. Walnut Creek, California.

PATHWAY: A SIMULATION MODEL OF RADIONUCLIDE TRANSPORT THROUGH AGRICULTURAL FOOD CHAINS[1]

T. B. Kirchner, F. W. Whicker, and M. D. Otis[2]

Abstract.--PATHWAY is a dynamic simulation model of the transport of radionuclides from fallout through an agricultural food chain. PATHWAY simulates the transport of radionuclides both by time dependent, continuous processes such as root uptake and senescence of vegetation, and by discrete events such as tillage of soil and harvest of crops.

INTRODUCTION

This paper describes PATHWAY, a model which was developed to simulate the transport of radionuclides from Nevada Test Site (NTS) fallout through agricultural food chains to man. The reconstruction of estimated radiation doses to residents of the region surrounding the NTS during the 1950s was initiated because of recent concerns about a possible relationship between nuclear fallout exposures and human cancer incidence. The internal radiation doses received by human tissues from ingested radioactivity which contaminated agricultural products can now only be estimated retrospectively through computational procedures because pertinent, direct, contemporary measurements were not made.

The goal of our project is (1) to provide estimates of the time-dependent concentrations of specified radionuclides in the major food products consumed by residents of various localities, ages, and lifestyles, and (2) to provide estimates of the rates of consumption of those foods. These data will then be used in a dosimetric analysis. We believe that the first part of our goal can only be met through the use of a simulation model which is designed specifically for dosimetric evaluation. A high degree of time dependency in the model is essential because (1) acute, time-specific fallout events are to be simulated, (2) the radionuclides that are to be simulated have a wide range of physical half lives, and (3) agricultural practices vary dramatically throughout the year. The model must also be able to accommodate either acute or chronic deposition scenarios and be amenable to validation, sensitivity, and error analyses.

Several models have been developed by other laboratories for the calculation of radionuclide transport through terrestrial food chains (Hoffman et al. 1977, Soldat et al. 1978). However, a review of the available simulation models at the outset of this project revealed none that would meet all of the requirements of the present effort. The majority of these models were developed for regulatory compliance rather than dosimetric evaluations. The development of PATHWAY was undertaken because no existing simulation model could be easily modified to meet these criteria.

PATHWAY is coded in FORTRAN and was constructed with the aid of PREMOD (Kirchner and Vevea, this volume). Thus, the model is designed so that it can be used in either an interactive or a batch mode, and can make full use of the validation, sensitivity, and error analysis routines that are available in the MODAID library associated with PREMOD. The output routines of MODAID that are used in the model provide great flexibility in the production of either tabular or graphical output.

This paper presents the model structure of PATHWAY and the basic concepts underlying the formulations of the processes. The structure and the parameters of the model reflect circumstances and conditions pertinent to selected counties in southern Nevada and Utah which received elevated NTS fallout deposition. Lengthy justification for the use of the many specific parameter values is not provided, as this will be the subject of a future communication. Ongoing efforts to validate the model and to develop error and sensitivity analyses are briefly mentioned.

GENERAL DESCRIPTION

The agro-ecosystem simulated in PATHWAY consists of several individual plots of land called land management units. Each land management unit belongs to one of three subsystems depending on whether it is managed for grazing, for growing food crops, or for growing hay. PATHWAY can simulate simultaneously one or more land management units of each of the three types.

Each land management unit is represented in PATHWAY by a minimum of six state variables, or compartments (fig. 1). These state variables represent the amount of radioactivity ($\mu Ci/m^2$) on the surface of the vegetation, in the tissues of the vegetation, in the surface layer of soil, in the rooting zone of the soil, and in the soil below the rooting zone. The radioactivity in the rooting zone of soil is partitioned into a labile and a non-labile fraction. In addition, each land management unit may have one or more compartments which represent the concentration of

[1]Paper presented at the Third International Conference on State-of-the-Art in Ecological Modeling, Colorado State University, May 24-28, 1982.

[2]T. B. Kirchner is Research Ecologist, Natural Resource Ecology Laboratory; F. W. Whicker is Professor, Department of Radiology and Radiation Biology; and M. D. Otis is Graduate Research Assistant, Department of Radiology and Radiation Biology, Colorado State University, Fort Collins, Colorado U.S.A.

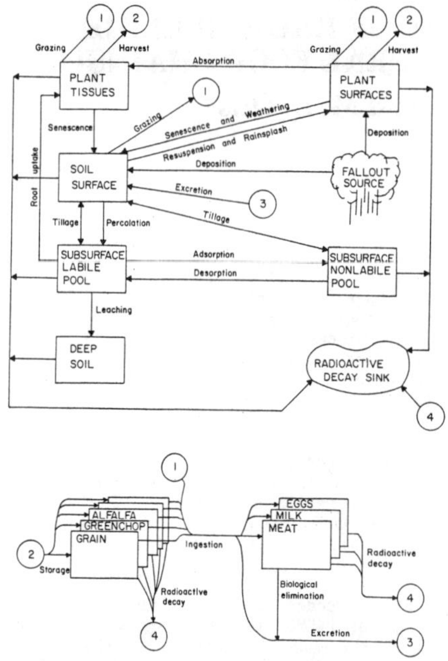

Figure 1.--A generalized flow diagram for one land management unit of the PATHWAY simulation model.

radioactivity in the agricultural products of those units, i.e., the concentration of radioactivity in fruits, vegetables, grains, hay, meat, milk, or eggs.

PATHWAY requires as input three data files: (1) a file of the initial values of the state variables and values for all parameters which are not time dependent, (2) a file of variables which change on a daily basis, and (3) a file of variables which change on a monthly basis during the simulation. The daily input file contains the amount of the radionuclide deposited during each day and for each land management unit, the aboveground biomass of the plants, rates of growth and senescence of the plants, and variables which indicate whether tillage or harvest are to occur. The monthly data file describes the dietary regimen of the livestock on a month to month basis.

Any radionuclide can be simulated in PATHWAY, providing that the appropriate radionuclide-dependent parameters are available. At present, we have assembled parameter values for 20 radionuclides, most of which are the fission products expected to contribute significantly to the radiation dose to humans in a local fallout field. Currently, data files are complete for $^{131\ 133\ 135}I$, $^{136\ 137}Ca$, $^{89\ 90\ 91}Sr$, $^{103\ 106}Ru$, $^{143\ 144}Ce$, ^{97}Zr, ^{99}Mo, ^{140}Ba, ^{132}Te, ^{105}Rh, ^{147}Nd, ^{239}Np, and ^{232}Pu. Parameters for the isotopes of iodine, cesium, strontium, and plutonium are generally based upon the greatest research foundation.

PATHWAY simulates both discrete events and continuous processes which affect the abundance and distribution of radionuclides in the system. The discrete events can occur as frequently as once per day and include the deposition of radionuclides from fallout, tillage of soil, and harvest of crops. The continuous processes include resuspension of contaminated soil, weathering from the surfaces of vegetation, percolation of radionuclides into the soil, adsorption and desorption of radionuclides within the soil, leaching of the contaminant to deep soil layers, uptake of radionuclides by roots, direct foliar absorption, and ingestion and excretion of radionuclides by animals (fig. 1). The continuous processes are represented by first order ordinary differential equations. Some of the coefficients for these equations are time dependent. The system of differential equations are solved numerically using a Runge-Kutta algorithm of order four.

DISCRETE EVENTS

The amount ($\mu Ci/m^2$) of the radionuclide which is deposited each day is read from the daily input file. The partitioning of the fallout between the surface of the soil and the surfaces of vegetation depends upon the aboveground biomass of the vegetation. The proportion of the fallout allocated to the surface of the vegetation, P_d, is given by

$$P_d = 1 - e^{-\alpha B}$$

where,

B = biomass, dry kg/m

α = a constant, m^2/kg (Chamberlain 1970).

α depends upon particle size of the fallout and has a value of about 2.8 for gases and submicron particles such as are represented in worldwide fallout (Hoffman and Baes 1979). However, direct measurements for herbaceous vegetation in areas 80 to 260 miles downwind from the Nevada Test Site indicate that 0.39 is a more realistic value for α for the 5-250 μm fallout particles depositing in Nevada and Utah from explosions on the NTS (Romney et al. 1963, Lindberg et al. 1959).

Tillage of soil redistributes radioactivity between the soil surface (0-0.1 cm) and the subsurface soil of the plow layer (0.1-25 cm). The importance of tillage in determining the concentration of radioactivity in food products is dependent upon the relative timing between the deposition and tillage events. Tillage events soon after the deposition of fallout would reduce the concentration of radioactivity in the surface layer of soil, and thus reduce the importance of resuspension and rain splash as mechanisms by which radioactivity can be transferred to food products. On the other hand, this scenario would lead to increased potential for root uptake. Tillage is simulated by redistributing the total of the inventories in the soil surface and the 0.1-25 cm soil layer in proportion to the relative masses of soil in each layer. Assuming bulk densities of 1.0 and 1.46 g/cm^3 for the surface and subsurface soil, respectively (Lyon, Buckman, and Brady 1952), the surface and subsurface soil receive 0.0027 and 0.9973 of the total radioactivity, respectively.

The effect of a harvest of a food or hay crop is simulated by removing a proportion of the total inventory of radioactivity on the surface and in the tissues of the vegetation. The proportion of radioactivity removed is equal to the proportion of the biomass which is removed during the harvest. Harvest of a crop is initiated by a variable which is read from the daily input file. The biomass of the crop on that day represents the amount of vegetation remaining after the harvest of the crop. Thus, the amount of biomass removed is equal to the difference between the standing crop on the day of the harvest and the standing crop of the previous day.

The concentration of radioactivity (μCi/g) in the harvested vegetation is calculated as the sum of the tissue and surface compartments divided by the biomass of the vegetation and this quantity is assigned to a storage compartment. Only the processes of radioactive decay or the replacement of stored vegetation by vegetation from another harvest affect the level of radioactivity in the stored products. PATHWAY simulates the storage of food crops somewhat differently than the storage of hay crops. On the one hand, there is only one storage compartment for each food crop that is simulated. Thus, this food storage compartment always contains the level of radioactivity remaining in the vegetation from the most recent harvest. On the other hand, there are twelve storage compartments for each land management unit assigned to growing hay. Up to six harvests per year can be stored and these are stored for up to two years. The individual cuttings of hay are stored separately to accomodate diets for grazers which specify both the number of the cutting of the hay and whether that hay came from the previous or the current year.

CONTINUOUS PROCESSES

Radioactive particles in the surface layer of soil can be displaced into the airstream by the action of wind or other mechanical disturbance and these resuspended particles may subsequently deposit on the surfaces of vegetation. Resuspension of radioactive particles is dependent upon such variables as wind speed, soil texture, soil moisture, plant cover, foliage characteristics, and the particle size distribution of the aerosol. Most published models for resuspension calculate the concentration of resuspended radioactivity in the air ($\mu Ci/m^3$) from the product of the density of radioactivity on the surface of the soil ($\mu Ci/m^2$) and a resuspension factor, R. R is determined empirically from the ratio of the air concentration and surface density under steady state conditions, and has units of m^{-1}. Measured values of R which have been tabulated by Stewart (1967) range from about 10^{-11} to 10^{-3} m^{-1}. Anspaugh et al. (1975) have suggested that a reasonable value of R is 10^{-4} because the few measured values which exceed 10^{-4} are associated with artificial disturbances.

Only a fraction of the resuspended radioactivity is deposited on the surface of vegetation. The rate of deposition of resuspended particles is usually estimated by the product of the concentration in air and a deposition velocity, V. The deposition velocity, which has units of m/day, depends upon most of the non-edaphic variables which affect R. V usually is determined empirically. Slade (1968) summarized the results of several experiments in which estimates of V range from 173 to 3197 m/day.

PATHWAY estimates the rate of transfer of radioactivity from the soil to the surface of vegetation directly, i.e., without explicitly calculating a concentration of radioactivity in the air. This rate of transfer is calaculated as the product of R, V, and the density of radioactivity in the surface layer of soil.

Anspaugh et al. (1975) reviewed the literature on resuspension of radioactive contaminants and noted that the concentration in the air over contaminated soil declines with time. This decline is due primarily to a reduction in the fraction of the total soil inventory that is available for resuspension rather than to a loss of radionuclide from the area. An unspecified "weathering" process has been postulated which reduces the erodibility of the contaminant with time (Olafson and Larson 1961, Stewart 1967, Larsen et al. 1966). Langham (1969, 1972) and Kathren (1968) formulated predictive models of resuspension in which the resuspension factor declined exponentially with half times of 35 and 45 days, respectively. Although these models adequately predicted the decline in resuspension for the first few weeks after a deposition event, they greatly underestimated the measured amount of resuspension of plutonium observed in an area seventeen years after the area was contaminated (Anspaugh et al. 1975). Anspaugh et al. (1975) derived another model for the rate of decline of the resuspension factor which better fit the data for long-lived radionuclides.

In contrast to Langham (1969, 1972), Kathren (1968) and Anspaugh et al. (1975), we have assumed that the resuspension factor is constant through time. We have instead modeled a decline in the amount of contaminant available for resuspension due to percolation of the radioactivity from the surface layer of soil to deeper soil layers. Percolation is modeled as a first order rate process with a rate constant of 1.98×10^{-2} day^{-1}, which corresponds to a half time of 35 days.

The impact of raindrops on the surface of soil can also elevate contaminated particles and deposit them onto foliage surfaces. Although rainsplash is in reality a discrete event which occurs only during rainfall, we have assumed that rainsplash can be represented sufficiently well as a linear, first order continuous process. We have derived a value of 8.6 $\times 10^{-4}$ day^{-1} for the rate constant representing rainsplash from the results of experiments conducted by Dreicer (1981).

Weathering consists of processes that transport radioactivity from the surfaces of vegetation to the surface of the soil. Weathering is usually described in the literature as a first order rate process. Rainfall events have little or no effect upon the rate of weathering (Chamberlain 1970). The rate of weathering seems to be most dependent upon whether the radioactivity is in the form of a particulate or a gas. Miller and Hoffman (1979) summarized from many sources the rates of weathering of radioactivity from grasses and other vegetation. These data show half times that range from 8.7 to 28 days for particulates and from 6.5 to 13 days for gaseous radioiodine.

Radionuclides which percolate from the surface layer of soil enter a layer which corresponds to the rooting zone. We assume in PATHWAY that the majority of active roots occur in the soil at depths between 0.1 and 25 cm, and that radionuclides in this subsurface layer of soil can occur as either labile or nonlabile forms. The labile fraction can be taken up by roots or transported to deeper soil layers. The nonlabile fraction represents the portion of the inventory which has been adsorbed to soil particles, particularly clays, and therefore is relatively immobile. Adsorbed radionuclides can re-enter the labile pool through the process of desorption. Adsorption and desorption are represented in PATHWAY as first order rate equations. The coefficients of these equations are highly dependent upon the chemical nature of the radionuclide.

Leaching is the process by which radionuclides of the labile pool in the rooting zone of soil are transported to deeper soil layers and is simulated in PATHWAY as a first order rate process. Although leaching occurs only at a very low rate (Schreckhise 1980), leaching can effectively isolate long lived radionuclides from transport through food chains. Thus, the soil of the deep layers acts as a sink for radionuclides.

Radionuclides can enter plant tissues via two major pathways, root uptake and absorption. Direct foliar absorption of radioactivity is probably much more important than root uptake as a pathway into tissues soon after a deposition event. However, root uptake can become more important as the deposits on the surfaces of the plants undergo weathering and as percolation occurs. Although translocation of radionuclides within the plant tissues is not explicitly simulated, we have assumed that radionuclides taken up by the roots are translocated to the aboveground tissues and that translocation in the reverse direction does not occur.

Absorption is the process by which radionuclides deposited on the aboveground surfaces of vegetation enter the plant tissues. The rate of absorption depends upon characteristics of the foliage and the chemistry of the radionuclide. Absorption is simulated using a first order rate equation. We have assumed that radionuclides which have been absorbed cannot be removed by weathering or washing and that the transfer of radionuclides from the tissues to the surfaces of plants is negligible (Olson 1965).

Only radionuclides in the labile pool of the subsurface soil layer are subject to uptake by roots. The rate of uptake is equal to the product of the concentration of the labile radionuclide, L, the rate of growth of the plants, G, and a radionuclide specific concentration factor determined from the literature, C_f, i.e.,

$$\frac{dT}{dt} = L \cdot G \cdot C_f \quad ,$$

where,

T = plant tissue inventory, $\mu Ci/m^2$
t = time.

The concentration of the radionuclide in soil is calculated by assuming that the inventory of the radionuclide in the labile pool is uniformly distributed throughout the subsurface layer, and that the density of soil in this layer is 1.46×10 g/m^3 (Lyon et al. 1952). Although the rate of root uptake is dependent upon the element, edaphic factors, and plant specific factors, we have assumed that the soils in all land management units are similar with regard to root uptake and that differences between species of plants are negligible. Thus, the value of the rate constant used in PATHWAY is dependent only upon the element.

The rate of growth of the plants in each land management unit is read from the daily data file. The daily data file is produced by PLANTS, a model which simulates the growth of plants using an algorithm similar to that for logistic growth, but which includes the effects of temperature and light on the rate of growth of new tissue and a term which represents senescence. The growth rate of plants, G, is given by

$$G = f_d \cdot f_t \cdot R_g \cdot (B_{eq} - B)/B_{eq}$$

where,

f_d and f_t = functions for the effect of light and temperature on the growth rate, respectively
R_g = the maximum instantaneous rate of growth of the plants
B = the current biomass
B_{eq} = the steady-state level of B

The terms f_d and f_t are restricted to the range of 0 to 1. The value of f_d is equal to the ratio of integrated light intensity on day d and the maximum value of that integral during the growing season (d=168). The temperature function causes the rate of growth to increase with temperature until an optimum temperature is reached, and to decline thereafter. The temperature function for alfalfa is taken from Holt et al. (1975). The temperature function for pasture grass and food crops is similar to that for alfalfa, but the optimum temperature is 30°C rather than 35°C.

As plants senesce, a portion of their aboveground biomass becomes deposited on the surface of the soil as litter. To calculate the flow of radionuclide from plants to the surface of the soil during senescence, we assume that the radionuclide is uniformly distributed in the tissues and on the surfaces of the plants. Thus, the rate of flow from plants to soil is equal to the rate of senescence of the biomass times the sum of the concentrations in the tissues and on the surfaces of the plants.

The rates of senescence of the plants in each land management unit, which are read from the daily input file, are also produced by the simulation model for plant growth. Senescence is assumed to be negligible until after July 31. The rate of senescence, S, is given by

$$S = \begin{cases} 0 & ; \text{month} \leq 7 \\ R_s \cdot (B - B_{min}) & ; \text{month} > 7 \end{cases}$$

where,

R_s = a rate constant for senescence
B_{min} = the minimum level of aboveground biomass which persists over the winter months.

The biomass, B, of the plants is simulated by integrating the difference between G and S through time.

The concentrations of radionuclides in meat, milk, and eggs depend upon the rate of ingestion of the radionuclides by the animals. The radionuclides are ingested with the forage or feed of the animals or with soil which is consumed. The animal consumers that are simulated in PATHWAY can be assigned complex diets which include both fresh forage and stored products such as hay and grain (fig. 2). The total rate of ingestion of radionuclides from forage, I_f, is calculated as the sum of the rates of ingestion of each food product, i.e.,

$$I_f = \sum_{i=1}^{n} C_i \, F_i$$

where,

n = the number of food products in the diet of the animals

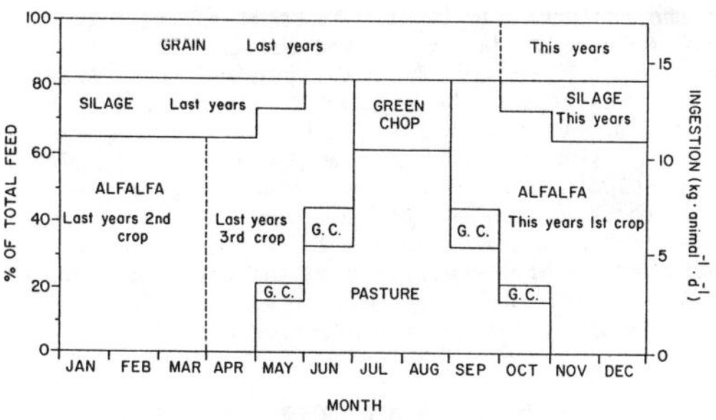

Figure 2.--The standard, or default, diet for dairy cows. The proportions of fresh forage and stored feed in the diet change on a monthly basis.

C_i = the concentration in the i-th food product, μCi/g

F_i = the rate of ingestion of food product i.

The rate of ingestion of radionuclides from soil is calculated as the product of the rate of ingestion of soil (g/day), the density of the radionuclide in the surface layer of soil (μCi/m^2), and a soil density factor (0.001 m^2/g).

Ingested radioactivity is partitioned between compartments which represent the radioactivity transferred to meat, milk or eggs, or excrement. The assimilation coefficients which determine this partitioning are dependent upon the radionuclide, the species of animal, and whether the radionculide is assimilated from soil or forage. We have assumed that milk and eggs are produced continuously at constant rates. Thus, the concentrations in milk and eggs (μCi/ℓ and μCi/g, respectively) are calculated as the quotients of the rates of transport (intake times the assimilation fraction) of the radionuclides into the animal product compartments and the rates of production of the milk and eggs (ℓ/day and g/day, respectively).

The portion of the ingested radionuclide which is not assimulated into meat, milk, or eggs is assumed to be excreted by the animal onto the surface layer of soil of the land management unit in which it resides. The animal continuously excretes some fraction of the radionuclide that is contained within its muscle tissues. This biological elimination of radionuclides has been measured in many animals and for several radionuclides, and the process usually appears to be adequately respresented by a first order rate equation. Although radionuclides can be retained in other animal tissues, we have assumed that the radionuclides contained in these tissues are negligible compared to the total inventory. Furthermore, most radionuclides reach steady state levels rapidly, so that the rate of assimilation equals the rate of biological elimination.

The radionuclide inventory in the excrement of all animals in a land management unit is assumed to be evenly distributed in the surface layer of soil of that unit. Because the animals in a land management unit can consume food products harvested from other land management units, ingestion and excretion provide the only pathways by which radionuclides can be transported between land management units.

CALIBRATION AND VALIDATION

We are in the preliminary stages of validating PATHWAY against data from studies that measured the transport of radionuclides from fallout or other sources into agricultural products. One problem encountered in the validation simulations was that some important agricultural data often was not collected during the experimental studies. For instance, data on the biomass of the vegetation in pastures, the rates of milk production, and the diets of the animals were often lacking. Therefore, we have made estimates for the values of such parameters to use when no other data are available.

The diets of livestock can greatly influence the rates of ingestion of radionuclides, and thus the concentrations of radioactivity in meat, milk, or eggs. The standard, or default, diet for dairy cattle (fig. 2) is based upon unpublished data for dairy cows in Utah dairy farms (Robert C. Pendleton, personal communication) and data from Hawthorne et al. (1976). The cattle are also assumed to ingest 500 g/day of soil (Mayland et al. 1975, Healy 1968). The default diet of poultry is based upon data from Hurd (1956), Winter and Funk (1956), Morrison (1948), Singen (1969), and Lush (1952). Hens are assumed to eat 10 g/day of soil and 95 g/day of feed, 95% of which is grain and 5% of which is live vegetation. Range cattle and sheep both are assumed to eat only native vegetation (9200 g/day and 1500 g/day, respectively; Cook 1970) throughout the year, and to ingest 500 g/day and 103 g/day of soil (Field and Purves 1964), respectively.

The parameters of the PLANTS model are calibrated to give realistic values for the biomass of vegetation in pastures, alfalfa fields, range lands, and grain crops through the year. We defined harvest dates and the biomass of alfalfa on those dates by extrapolating from data presented by Hawthorne et al. (1976). Parameters affecting the growth of pasture grass have been calibrated to fit data reported by Allred (1965) for irrigated pastures in Utah. The peak standing crop for these pastures is assumed to be 300 g/m^2. The growth and senescence rates for grain are assumed to be the same as those for grass, but the potential peak biomass is assumed to be 700 g/m^2 (Whittaker 1970). Range vegetation has parameters similar to those for pastures, but the maximum biomass is 125 g/m^2 (Cook 1970).

For our preliminary validation studies, we separated differences between model predictions and measured values into two categories, scaler errors and non-scaler errors. If model predictions differed from measured values by a more or less constant factor, we considered the model to be exhibiting scaler errors. On the other hand, if the differences between model predictions and measured values was not a constant factor, we considered the simulation to be exhibiting non-scaler errors. Scaler errors in a state variable can arise in PATHWAY from consistent errors in the driving variables (deposition data, plant biomasses, etc.) or from an error in the rate equations for that variable. Non-scaler differences are more likely to be caused by errors in parameter values which affect higher order interactions of the state variable, from a basic flaw in the way in which PATHWAY represents the system, or from a time-dependency of a parameter that is not adequately represented in the model.

In our preliminary validation exercies, we are much more interested in discovering non-scaler errors than scaler errors because the magnitude of the non-scaler errors could become quite large in simulations involving long-lived radionuclides. Furthermore, if non-scaler errors are due to a defect in model structure, then it might be fruitless to try to correct the problem by calibrating parameter values.

Examples of our validation studies are the simulations we performed based upon three different experimental studies. The source of the radionuclides in all three studies was worldwide fallout, the radionuclides were all long-lived and the experimental data were collected over intervals of from 3 to 7 years. Thus, these data are especially useful for identifying non-scaler errors. For these simulations, we selected from the range of published values those that appeared most probable (table 1). Although these parameter values were never adjusted to improve the fit between predicted and observed values, we multiplied the results of our validation simulation by a constant factor in cases where PATHWAY consistently over- or under-predicted the observed values. This re-scaling of the simulated values enables us to better compare the dynamics of the model with the dynamics observed in the field experiments. Thus, we can identify the magnitude of any scaler errors, and better assess whether PATHWAY is subject to non-scaler errors.

The first validation exercise is a simulation of the level of cesium-137 in alfalfa over a period of four years (fig. 3). The data for this experiment were collected by Hawthorne et al. (1976). This exercise is particularly valuable because it is one of the few instances in which the biomass of the vegetation, the rate of deposition of fallout, and the concentration of the radionuclide in an agricultural product were all measured. It is obvious that the model has done an excellent job in simulating the concentrations in the alfalfa.

The second validation exercise is based upon data collected by Robert C. Pendleton (personal communication) who measured the concentrations of strontium-90 and cesium-137 in milk from dairy farms near Salt Lake City, Utah (figs. 4 and 5). We used the default values for diet, milk production rate, and biomass of the pasture vegetation. Data on the amount of Sr-90 in fallout near Salt Lake City was taken from Health and Safety Laboratory (1977). The rate of deposition of cesium was estimated by multiplying the Sr-90 deposition by 1.75, the average ratio of Cs-137 to Sr-90 in worldwide fallout (Edward P. Hardy, Jr., personal communication).

The simulated values of Cs-137 in milk appear to show reasonable dynamics, but are too low by a factor of 3. Differences between simulated and actual values of certain parameters, such as milk production rates and pasture biomass, could account for this scaler error. On the other hand, the simulated values of Sr-90 in milk appear to be too low by a factor of 5, and show non-scaler errors. Re-scaling the simulated values makes it clear that PATHWAY is underestimating the concentration of Sr-90 in milk during the last 3 years of the simulation. It may be possible to eliminate the non-scaler errors by decreasing the rate of loss of Sr-90 from the surface layer of soil, or by increasing the value of the concentration factor for root uptake.

The third validation exercise is based upon data from Ward and Johnson (1963, 1964, 1965a,b) and Wilson, Ward, and Johnson (1965) for a site near Fort Collins, Colorado (figs. 6 and 7). The rate of deposition of Cs-137 was measured using collection funnels over 3 years of the experiment, and the concentrations of the radionuclide in milk and meat of the dairy cows were measured. We again used all default values for the cows' diets and the other parameters of PATHWAY. In this exercise, the simulated values appear similar to the observed values so that no re-scaling was required. Although only six measurements of the levels of Cs-137 in meat were made, PATHWAY appears to simulate these values well. However, PATHWAY seems to have underpredicted the levels of Cs-137 in milk in 1962, but to have overpredicted the levels in 1963 and 1964. We are analyzing these results in an effort to find the source of these non-scaler errors.

SENSITIVITY AND ERROR ANALYSES

We have recently initiated an analysis of the uncertainty in the predictions of PATHWAY (error analysis) and an analysis of the sensitivity of the state variables to changes in the values of the parameters (sensitivity analysis). These analyses require that distributions representing the uncertainty of each parameter be defined. We are presently developing such distributions from data reported in the literature. We are conducting these analyses using analytical routines from the MODAID (Kirchner and Vevea, this volume) library of software.

The preliminary findings of these analyses show that the lists of parameters to which the state variables are most sensitive depend upon the radionuclide and upon the date of the deposition event. Although the list of parameters is not the same for each state variable, the values of the resuspension factor, R, and the initial deposition constant, α, have significant effects on the values of all of the agricultural products. In addition, our analyses indicate that all of the distributions of the predicted values of the agricultural products can be adequately described as lognormal. These results will be discussed in greater detail in a forthcoming publication.

SUMMARY

PATHWAY simulates the transport of radionuclides from fallout through an agricultural ecosystem. The agro-ecosystem is subdivided into several land management units, each of which is used either for grazing animals, for growing hay, or for growing food crops. The model simulates the transport of radionuclides by both discrete events and continuous, time-dependent processes. The discrete events include tillage of soil, harvest and storage of crops, and deposition of fallout. The continuous processes include the transport of radionuclides due to resuspension, weathering, rain splash, percolation, leaching, adsorption and desorption of radionuclides in the soil, root uptake, foliar absorption, growth and senescence of vegetation, and the ingestion assimilation, and excretion of radionuclides by animals.

We are presently simulating an agroecosystem which produces beef, poultry, lamb, milk, eggs, alfalfa, green chop, grain, and garden vegetables. A typical simulation runs in approximately 38 seconds per year simulated in the Colorado State University Cyber 172 computer. However, we are in the process of devising techniques which will reduce the execution time of the model. Preliminary validation studies indicate that the model dynamics and simulated values of radionuclide concentrations in several agricultural products agree well with measured values when the model is driven with site specific data on deposition from worldwide fallout.

Table 1.--Values for the parameters used in the validation simulations of PATHWAY for strontium-90 and cesium-137. The animals in these simulations were dairy cows.

Parameter	Units	Definition	Value Sr-90	Value Cs-137
A1		Fraction of radionuclide in forage or feed that goes to meat	0.0099	0.5
A2		Fraction of radionuclide in ingested soil that goes to meat	0.0099	0.5
A3		Fraction of radionuclide in forage or feed that goes to milk	0.0183	0.093
A4		Fraction of radionuclide in ingested soil that goes to milk	0.0183	0.093
A5		Fraction of the total soil inventory that remains in the surface layer of soil after tillage	0.002732	0.002732
A6		Fraction of the total soil inventory that remains in the subsurface layers of soil after tillage	0.9973	0.9973
AD	no./m^2	Density of grazing animals	0.00049	0.00049
ALPHA	m^2/kg	Constant used in the function which defines the fraction of fallout that remains on vegetation of a given biomass	2.8	2.8
CF		Concentration factor for root uptake	0.2	0.01
F1	dry g/day-animal	Rate of consumption of vegetation by grazers	17000	17000
F2	dry g/day-animal	Rate of consumption of soil by grazers	500	500
GMASS	wet g/animal	Mass of meat per animal	180,000	180,000
K1	day^{-1}	Rate constant for rainsplash	0.00086	0.00086
K2	day^{-1}	Rate constant for weathering	0.0495	0.0495
K3	day^{-1}	Rate constant for percolation	0.0198	0.0198
K4	day^{-1}	Rate constant for absorption	0.001	0.0055
K5	day^{-1}	Rate constant for adsorption to soil particles	0	0.0019
K6	day^{-1}	Rate constant for desorption from soil particles	0	0.00021
K7	day^{-1}	Rate constant for leaching	0.000066	0.0000066
KB	day^{-1}	Rate constant for biological elimination	0.0932	0.232
KP	day^{-1}	Rate constant for radioactive decay	0.0000657	0.0000633
PRAP	ℓ/day-animal	Rate of production of milk	13.1	13.1
PROG		Proportion of activity remaining in stored grain after harvest	0.25	0.25
R	m^{-1}	Resuspension factor	0.00001	0.00001
RD	cm	Rooting depth of vegetation	25	25
V	m/day	Deposition velocity	173	173

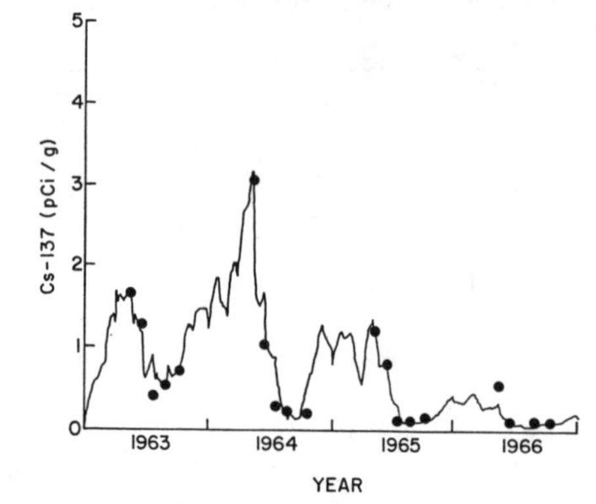

Figure 3.--Simulated (solid line) versus measured (dots) concentrations of cesium-137 in alfalfa from a dairy farm in St. George, Utah.

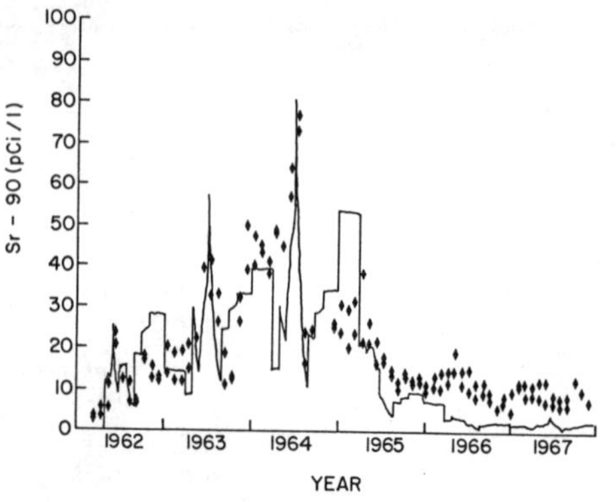

Figure 5.--Simulated (solid line) versus measured (diamonds) concentrations of strontium-90 in milk from dairy farms near Salt Lake City, Utah. After re-scaling the simulated data by multiplying it by a factor of 5, it can be seen that the model is exhibiting non-scaler errors.

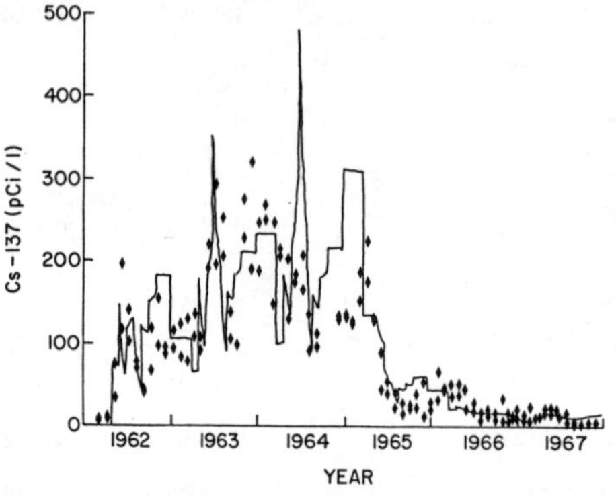

Figure 4.--Simulated (solid line) versus measured (diamonds) concentrations of cesium-137 in milk from dairy farms near Salt Lake City, Utah. The simulated values have been multiplied by a factor of 3 to compare the dynamics of the model to that of the system.

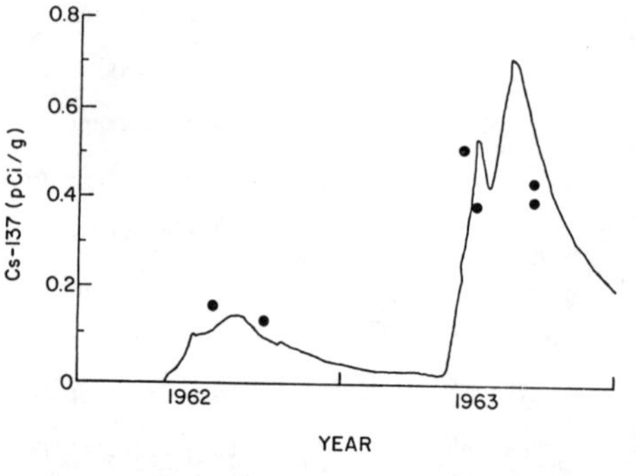

Figure 6.--Simulated (solid line) versus measured (dots) values of the concentration of cesium-137 in meat of dairy cows from Fort Collins, Colorado.

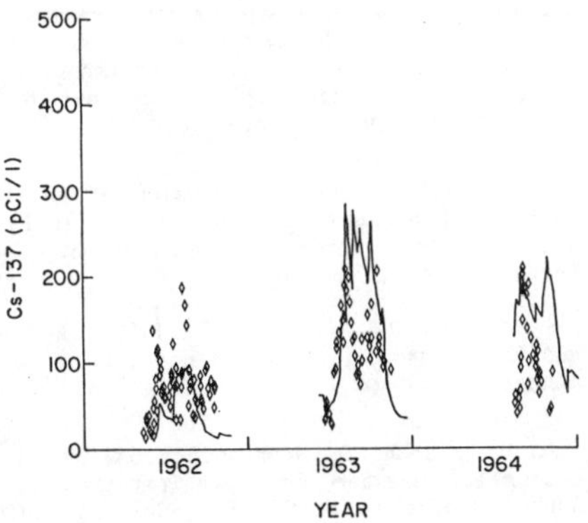

Figure 7.--Simulated (solid line) versus measured (diamonds) values for the concentration of cesium-137 in milk from dairy cows from a farm near Fort Collins, Colorado.

The PATHWAY model currently contributes to evaluations for residents of southern Nevada and Utah. Once we are satisfied with the performance of the model, we plan to expand our data base and, if necessary, make modifications to PATHWAY which will allow it to be used for similar evaluations in other geographic areas and for other sources of radionuclide contaminants.

ACKNOWLEDGMENTS

We thank Tracy Ikenberry and April Whicker for their valuable contributions to this effort. This work was supported by contract DE-AS08-79NV10057 between Colorado State University and the U.S. Department of Energy.

LITERATURE CITED

Allred, K. R. 1965. Grasses can be productive. Utah Farm and Home Science 26:78-81, 92.

Anspaugh, L. R., J. H. Shinn, P. L. Phelps, and N. C. Kennedy. 1975. Resuspension and redistribution of plutonium in soils. Health Physics 29:571-582.

Chamberlain, A. C. 1970. Interception and retention of radioactive aerosols by vegetation. Atmospheric Environment 4:57-78.

Cook, C. W. 1970. Energy budget of the range and range livestock. Bulletin TB-109. 28 p. Colorado State University Experiment Station, Fort Collins.

Dreicer, M. 1981. Rainfall as a mechanism for soil contamination of plant surfaces. M.S. Thesis. Colorado State University, Fort Collins, Colo.

Field, A. C., and D. Purves. 1964. The intake of soil by grazing sheep. Proceedings of the Nutrition Society 23:24-25.

Hawthorne, H. A., S. D. Zellmer, L. P. Eberhardt, and J. M. Thomas. 1976. ^{137}Cesium cycling in a Utah dairy farm. Health Physics 30:447-464.

Health and Safety Laboratory. 1977. Final tabulation of monthly ^{90}Sr fallout data: 1954-1976. USAEC Report HASL-329, UC-11.

Healy, W. B. 1968. Ingestion of soil by dairy cows. New Zealand Journal of Agricultural Research 11:487-499.

Hoffman, F. O., and C. F. Baes, III. A statistical analysis of selected parameters for predicting food chain transport and internal dose of radionuclides. NUREG/CR-1004:ORNL/NUREG/TM-282. 183 p. Oak Ridge National Laboratory, Oak Ridge, Tenn.

Hoffman, F. O., C. W. Miller, D. L. Shaeffer, and C. T. Garten, Jr. 1977. Computer codes for the assessment of radionuclides released to the environment. Nuclear Safety 18(3):343-354.

Holt, D. A., R. J. Bula, G. E. Miles, M. M. Schreiber, and R. M. Peart. 1975. Environmental physiology, modeling and simulation of alfalfa growth. I. Conceptual development of SIMED. Agricultural Experiment Station Research Bulletin 907, Purdue University, West Lafayette, Indiana.

Hurd, L. M. 1956. Modern poultry farming. Fourth edition. Macmillan Co., New York.

Kathren, R. L. 1968. Towards an interim acceptable surface contamination level for Pu-239 O_2 in the environment. p. 460-471. In Proceedings of the Symposium on Radiological Protection of the Public in a Nuclear Mass Disaster. [Interlaken, Swit., 26 May-1 June, 1968] EDMZ, Bern.

Kirchner, T. B., and J. M. Vevea. 1982. PREMOD and MODAID: Software tools for writing simulation models. This volume.

Langham, W. H. 1969. Biological considerations of non-nuclear incidents involving nuclear warheads. USAEC Rept. UCRL-50639. Lawrence Livermore Laboratory, Livermore, Calif.

Langham, W. H. 1972. The biological implications of the transuranium elements for man. Health Physics 22:943-952.

Larson, K. H., J. W. Neel, H. A. Hawthorne, M. H. Mork, R. H. Rowland, L. Baurmash, R. G. Lindberg, J. H. Olafson, and B. W. Kowalewsky. 1966. USAEC Report WT-1488. University of California, Los Angeles.

Lindberg, R. G., E. M. Romney, J. H. Olafson, and K. H. Larson. 1959. The factors influencing the biological fate and persistance of radioactive fallout. University of California, Los Angeles, WT-177.

Lush, R. H. 1952. Pasture production and management. Country Life Press, Garden City, New York.

Lyon, T. L., H. O. Buckman, and N. C. Brady. 1952. The nature and properties of soils. 591 p. Macmillan Co., New York.

Mayland, H. F., A. R. Florence, R. C. Rosenau, V. A. Lazar, and H. A. Turner. 1975. Soil ingestion by cattle on semiarid range as reflected by titanium analysis of feces. Journal of Range Management 28:448-452.

Miller, C. W., and F. O. Hoffman. 1979. The environmental loss constant for radionuclides deposited on the surfaces of vegetation, λ_w. In Hoffman, F. O. and C. F. Baes III. A statistical analysis of selected parameters for predicting food chain transport and internal dose of radionuclides. NUREG/CR-1004:ORNL/NUREG/TM-282. 183 p. Oak Ridge National Laboratory, Oak Ridge, Tenn.

Morrison, Frank B. 1948. Feeds and feeding. Twenty-first edition. Morrison Publishing Co., Binghamton, New York.

Olafson, J. H., and K. H. Larson. 1961. USAEC Rept. UCLA-501. UCLA, Los Angeles.

Olson, J. S. 1965. Equations for cesium transfer in a Liriodendron forest. Health Physics 11:1385-1392.

Romney, E. M., R. G. Lindberg, H. A. Hawthorne, B. G. Bystrom, and K. H. Larson. 1963. Contamination of plant foliage with radioactive fallout. Ecology 44(2):343-349.

Schreckhise, R. G. 1980. Simulation of the long-term accumulation of radiocontaminants in crop plants. PNL-2636/UC-70. Pacific Northwest Laboratory, Richland, Washington.

Singen, E. P. 1969. Forty years of change in layer feeds and feeding. Feedstuffs 41:84.

Slade, D. H. (ed.). 1968. Meteorology and atomic energy 1968. U.S. Atomic Energy Commission, Division of Technical Information, Oak Ridge, Tenn.

Soldat, J. E., et al. 1978. Terrestrial food chain transport of radionuclides. p. 85-98. In Proceedings of a workshop on the evaluation of models used for the environmental assessment of radionuclide releases. U.S. Department of Energy Report CONF-770901. 131 p. National Technical Information Service, Springfield, Va.

Stewart, K. 1967. The resuspension of particulate material from surfaces. p. 63-74. In Fish (B. R., editor). Surface contamination. 415 p. Pergamon Press, New York.

Ward, G. M. and J. E. Johnson. 1963. A study of cesium-137 passage from precipitation to milk. Annual Report to USAEC, Fort Collins, Colorado.

Ward, G. M. and J. E. Johnson. 1964. A study of cesium-137 passage from precipitation to milk. Annual Report to USAEC, Fort Collins, Colorado.

Ward, G. M. and J. E. Johnson. 1965a. A study of cesium-137 passage from precipitation to milk. Annual Report to USAEC, Fort Collins, Colorado.

Ward, G. M. and J. E. Johnson. 1965b. The cesium-137 content of beef from dairy and feedlot cattle. Health Physics 11:95-100.

Whittaker, R. H. 1970. Communities and ecosystems. 162 p. MacMillan Co., London.

Wilson, D. W., G. M. Ward, and J. E. Johnson. 1965. Deposition of fallout cesium-137 at Fort Collins, Colorado, 1962-1964. Radiological Health Data and Reports 6:675-676.

Winter, A. R., and E. M. Funk. 1956. Poultry, science and practice. Fourth edition. J. B. Lippincott Co., Chicago.

A SIMULATION MODEL OF RADIONUCLIDE MIGRATION THROUGH COASTAL MARINE ECOSYSTEMS[1]

Michael Schum and James Varnell[2]

Abstract. We present a simulation model describing the dynamic behavior of fission product radionuclides which could be input to coastal ecosystems from nuclear power reactor accidents resulting in groundwater contamination. The purpose of the model is to identify critical pathways and processes in the transfer of radioactive pollutant from the benthic sediment layer into the coastal ecosystem. The model is a highly aggregated compartmental ecosystem consisting of five abiotic and nine biotic compartments. Movement of radionuclides through the system occurs primarily by particle adsorption and hydrologic movement. However, a small fraction of the total input will be bound in sediment-benthos interactions. Our analysis indicates that in the event of a massive liquid discharge of reactor waste products to groundwater, only a few highly energetic, long-lived radioisotopes could eventually be released in the subtidal benthic boundary layer. These isotopes may persist for decades in low quantities, primarily in the sediment, and in some benthic organisms such as crabs, mollusks and kelp.

INTRODUCTION

Even though the future of nuclear power is in doubt, no alternate source of energy has been clearly shown to be a satisfactory substitute. Perhaps the most damaging criticisms leveled against nuclear power development relate to potential damage to the natural environment which may result from a large-scale accidental release of radiation in the form of ionizing, fission product radionuclides.

Pioneering research on the movement of artificial radioactivity in the environment addressed primarily the consequences of an atmospheric release, mainly because of the hazards to man, and as secondary research from atmospheric testing of nuclear weapons in the late 1950's and early 1960's. Subsequently, attention was focused on pollution of the aquatic ecosystem which might result from discharges of radioactive cooling water (Cushing, et al. 1975, NAS-NRC 1971, OECD-ENEA, 1979). One pathway that has usually been ignored is the possibility of pollution resulting from a reactor core meltdown, discharging the sumpwater or molten core into a groundwater aquifer. The contaminated groundwater might eventually reach surface waters, and those isotopes that had not decayed or become attached to soil particles could migrate across the benthic boundary layer into the aquatic foodweb.

Although there is a considerable body of data concerning the accumulation and effects of specific isotopes in isolated circumstances (Aberg and Hungate 1966, Nelson and Evans 1969, Polikarpov 1966), no general systems model has been produced which can be applied to a marine ecosystem to predict the time dependent behavior of aqueous radioisotopes in an aquatic foodweb. Our primary objective has been to develop a reasonable model which incorporates sufficient detail so that it may be used to: 1) identify critical pathways and processes; 2) determine where there are significant gaps in our knowledge concerning environmental transport of radioisotopes in marine foodwebs; and 3) guide field research studies

[1]Paper presented at the Third International Conference on State-of-the-Art in Ecological Modelling, Colorado State University, May 24-28, 1982.

[2]Michael Schum is Assistant Professor of Biology and James Varnell is a graduate student at San Diego State University, San Diego, California U.S.A.

necessary to validate the model so that it may be used to predict the fate and distribution of liquid radioactive pollutants should an accident ever occur.

STRUCTURE OF THE MODEL

There are three general approaches to ecosystem modeling - energetics, population dynamics and mass balance models. The model used in this study is based on a mass balance approach in which curies of radioisotope are traced through the environment. Fourteen biotic and abiotic compartments are defined, and a coupled system of ordinary differential equations describe the time-dependent behavior of each compartment. The model has been implemented using the GASP-IV simulation language with a Runge-Kutta-England integration scheme (Pritsker 1967). Although each compartment has its own specific input-output pathways, there are several generalized physical and biological processes that are included. The main components included are:

- SITE-SPECIFIC DATA
- BIOLOGICAL HALF LIVES AND CONCENTRATION FACTORS OF CRITICAL ISOTOPES
- SPATIAL DISTRIBUTION OF GROUNDWATER INPUT
- BIOMASS AND INGESTION RATES OF BIOTIC COMPARTMENTS
- TRANSFER COEFFICIENTS OF ABIOTIC COMPARTMENTS
- SEDIMENT-WATER INTERACTIONS
- MIXING OF DIFFERENT WATER COMPARTMENTS
- TIME OF APPEARANCE AND FLUX OF LIQUID WASTE

Many of the interactions between compartments are reversible, either because of chemical equilibria or biological phenomena such as water metabolism or trophic interactions. The amount of a toxic substance, such as a radioisotope, that can be taken in by a biotic compartment may be further limited by toxicity due to dose or dose rate. To compound the problems, many (if not most) aquatic organisms are known to selectively concentrate elemental forms of isotopes over background concentrations. As a first approximation we have assumed that concentrations in biotic compartments are directly proportional to water concentrations. The state equations are written such that the state variables cannot exceed the water concentration times some concentration factor (CF), under the assumption that CF values reported in the literature (NAS-NRC 1971, Aberg and Hungate 1966, Polikarpov 1966, Reichle, et al. 1970) represent maximum values at equilibrium. The structural forms of the state equations are shown in equations 1 and 2.

The primary input to the system is the appearance and release of a significant quantity of radioisotope at the benthic boundary layer in the near-shore coastal marine ecosystem (within 5 km). The input profile must be specified in terms of time of appearance of contaminated groundwater discharged to receiving waters, duration (i.e. acute or chronic release), and specific isotope and quantity transported by groundwater to the sea. Numerous site-specific parameters such as type and size of reactor, years of reactor operation, nature of groundwater aquifer and rates of flow, soil sorption characteristics, and geological characteristics of the coastal ocean floor interact to determine the unique pattern of source input for any given accident scenario.

Groundwater transport models developed by the U.S. Nuclear Regulatory Commission and Sandia National Laboratory (Liquid Pathways Generic Study, research in progress) have indicated that only a few, highly-energetic, long-lived isotopes (Sr-90, Ce-144, Cs-137, Ru-106, I-129, P-238, Tc-99) are likely to be produced in sufficient quantity and chemical form that they may be a potential hazard to man or to economically important marine organisms. Groundwater transport of aqueous isotopes through the clay or hardrock upon which many reactors are located appears to function as an effective filter for most fission product isotopes. Each isotope is transported at a different rate (Davis and DeWiest 1966), and it may take many years to move even a small distance when groundwater flow is slow and when cation exchange or sorption properties of the soil are high. The ecosystem simulation model presented in this report picks up where the hydrological modeling leaves off. Virtually any pattern of benthic surface release of a pollutant can be modeled with a suitable choice and scaling of parameters for the state equations.

Descriptions Of Compartments

The fourteen compartment system consists of five abiotic and nine biotic compartments. The abiotic compartments consist of groundwater, an inorganic sediment layer, an organic sediment layer, suspended particulates, and the water column above a given contaminated benthic layer. The biotic compartments were chosen on the basis of trophic interactions and relative size classes and include microbenthos, meiobenthos, macrobenthos, plankton, periphyton, larval fish, other zooplankton, adult omnivorous fish, and adult carnivorous fish. There are too many interactions to effectively present an overall compartmental diagram showing all the possible pathways. Instead, Fig. 1 shows the transfer matrix between compartments. For the mass balance model, the sign of any a_{ij} must be the reverse of any a_{ji}. No sign indicates that either there is no interaction, or that we do not consider the magnitude of the interaction relative to the other flows, to be significant.

STATE EQUATIONS

For all biotic compartments:

$$\frac{dX_j}{dt} = \sum_{i \ddagger j} \left(\frac{CXE_j - CX_j}{T_{ij}} \right) (M_j) - \sum_{j \ddagger k} R_{jk} X_j - \sum_{i \ddagger j} \left(\frac{CXE_i - CX_i}{T_{ji}} \right) (M_i) - D \pm Q \quad (1)$$

For all abiotic compartments:

$$\frac{dX_j}{dt} = \sum_{k \ddagger j} R_{kj} X_k - \sum_{j \ddagger k} R_{jk} X_j - \sum_{i \ddagger j} \left(\frac{CXE_i - CX_i}{T_{ji}} \right) (M_i) - D \pm Q \quad (2)$$

Where:
- X_j = state variable for the jth compartment (Ci)
- M_j = biomass of the jth compartment (Kg)
- CXE_j = "equilibrium" concentration of the jth compartment (Ci/Kg)
- CX_j = concentration of the jth compartment (Ci/Kg)
- T_{ij} = uptake time of compartment i by compartment j (time to consume 100% of M_j)
- R_{jk} = transfer rate of compartment j to compartment k (fraction/yr)
- D = radioactive decay (fraction/yr)
- Q = all source or sink terms

Summation i = all biotic compartments

Summation k = all abiotic compartments

The rate of input (Ci/yr/m) to the groundwater compartment underlying the marine benthic layer is specified as an input forcing function. As the groundwater comes in contact with the benthic sediment layer - seawater interface, several chemical and physical processes interact to determine the distribution of groundwater input to the system. The most soluble and non-adsorptive isotopes (Ru-106 and Sr-90) will appear first following an accident. These isotopes are likely to be quickly and easily transferred to the water compartment, subject to changes in solubility accompanying mass transport through a benthic redox layer which may alter the valance state of some isotopes of interest. The groundwater will interact with the surficial sediment layer, consisting of an inorganic, nondigestible and organic, consumable fraction as well as the receiving water above the benthic boundary layer.

Many different parameters are likely to influence the fractional distribution of pollutant to the three receiving compartments. Temperature, salinity, pH, depth (i.e. pressure) and concentration of stable isotope in seawater are all important from physico-chemical equilibria considerations. Rather than trying to model each of these complicated processes separately, our approach has been to specify the fractional distribution to the three compartments as input parameters, and to examine the sensitivity and long-term behavior of the system to minor changes in these parameters. The initial choices of these values are not arbitrary. For any aqueous isotope in contact with both water and particulate matter, an equilibrium value will be reached between the amount dissolved in the water and the amount sorbed onto particulate material. The ratio of these two

		GW	SED	WAT	OSL	SP	MIB	MEB	MAB	PLA	PER	ZOO	LF	ACF	AOF	
		ABIOTIC					BIOTIC									
ABIOTIC	1		−	−	−											GROUNDWATER
	2	+		±	±	±	−	−	−							SEDIMENT
	3	+	±		±	±	±	±	±	±	±	±	±	±	±	WATER
	4	+	±	±		±	±	±	±	+	+	±	+	±	±	ORGANIC SEDIMENT LAYER
	5		±	±	±			−	−	+	−	±	±	+	±	SUSPENDED PARTICULATES
BIOTIC	6		+	±	±			−								MICROBENTHOS
	7		+	+	±	+	+		−	+		±	±	−	−	MEIOBENTHOS
	8		+	±	±	+		+		+	+	+	+	±	±	MACROBENTHOS
	9			±	−	−		−	−			−	−		−	PLANKTON
	10			±	−	−			−						−	PERIPHYTON
	11			±	±	±		±	−	−			±	−	−	ZOOPLANKTON
	12			±	±	±		±	−	+		±		−	−	LARVAL FISH
	13			+	+	−		+	±			+	+		±	ADULT CARNIVOROUS FISH
	14			±	±	±		+	±	+	+	+	+	±		ADULT OMNIVOROUS FISH

FIGURE 1: INTERACTION MATRIX BETWEEN COMPARTMENTS

quantities is called the distribution coefficient for a specific element in a particular thermodynamic system. Published values for distribution coefficients (K_D) for most isotopes are available (Booth 1976).

Most of the isotopes of interest in this study have very large K_D values (10^3-10^6). This indicates that they will be tightly bound to particulate material, either at the sediment layer or on suspended material. The process appears to be primarily adsorption rather than absortion, and consequently the particle size distribution and resulting total particulate surface area becomes a very important consideration. The amount of isotope adsorbed onto particulate material becomes especially important in the marine environment where many different organisms must filter out suspended particulates for food resulting in direct ingestion of curie-bearing material. The whole cycle of settling and resuspension of particulates, especially organic detritus, is a complicated and poorly understood process for which only educated guesses are available.

After a radionuclide has been released into the receiving water, it will be transported by advection and diffusion. For the coastal environment, advection will probably be the dominant process. Prevailing coastal currents, tidal movements and wind driven mixing currents must all be considered. These patterns will differ markedly from the high-energy winter storm cycle, which generally resuspends, erodes and removes material, to the summer calm cycle, which settles, deposits and imports material. Until further data becomes available, we consider only long-term (yearly) averages in these parameters with no seasonal variation. We do not consider the model to be sophisticated enough at this point to attempt to model spatial differences in water concentrations by such means as finite-element methods.

Isotopes which are not removed from the system by fluid transport or burial of particulate material under sediment will become available to the aquatic foodweb by two main processes. Direct ingestion is important for the many organisms which consume sediment or suspended particulates to obtain carbon from detrital material. Because of the physical nature of the aquatic environment, direct input to biotic compartments from the water column will also become an important pathway. Most of the biomass of the biotic compartments consists chiefly of water. This cellular and interstitial or "biological" water is in a dynamic equilibrium with the surrounding water column. This biological water is constantly being exchanged with the seawater, and an average turnover time is used as a parameter in the model based on literature values for common organisms (Spector 1956).

Primary Producers

The key trophic interactions in our model of the subtidal coastal foodweb are shown in Fig 2. The arrows show the movement of radioactive material from one compartment to another. Only the primary feeding relationships are shown.

Autotrophic organisms at the base of the food chain will incorporate radioisotopes into their biomass primarily by water metabolism (i.e. at some turnover rate). As plankton grow and die, they settle out of the water column (Smayda 1970). The average settling rate of plankton is used as a model parameter to estimate the rate at which plankton may become available to filter-feeding organisms in the benthic layer.

Loss terms from biotic compartments due to predation are assumed to be recipient compartment controlled. These rates are estimated in the form of uptakes of one biotic compartment by another in the sense of "doubling time", or the average length of time (yrs) it takes a consumer to consume 100% of their biomass in prey. The reciprocal of this uptake is then the fraction of a particular prey compartment consumed per year per unit biomass of predator. The rate of uptake is then multiplied by the difference between the equilibrium concentration of the consuming compartment and the state variable value.

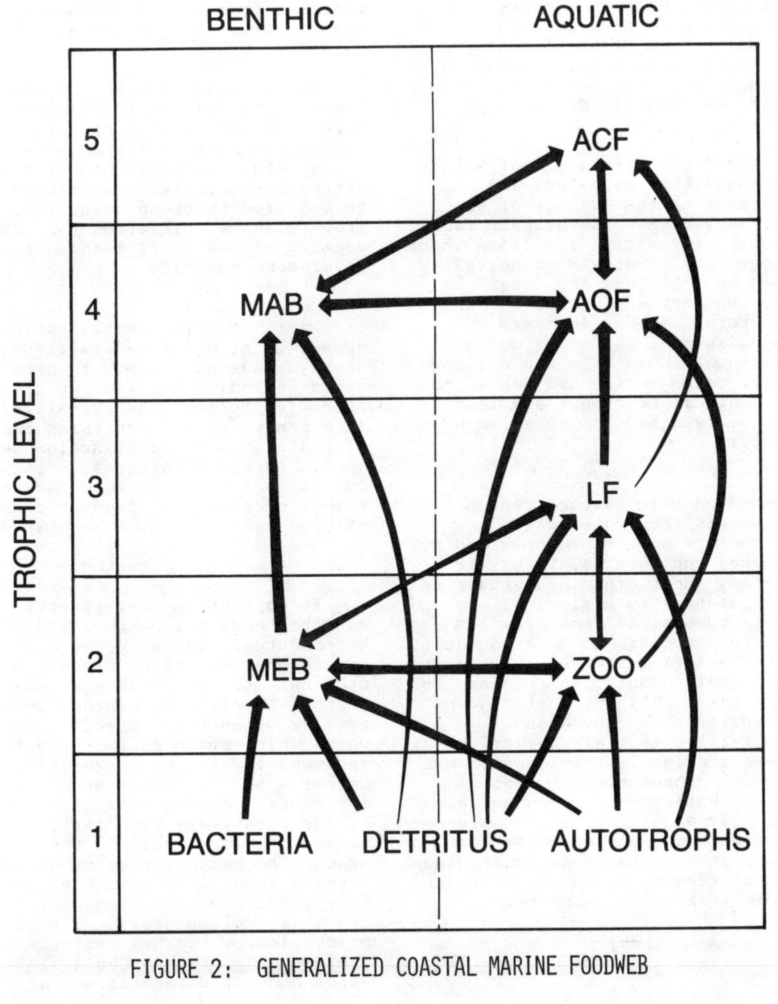

FIGURE 2: GENERALIZED COASTAL MARINE FOODWEB

Similar processes apply to periphyton and macrophytes in the coastal ecosystem. All attached primary producers are lumped into the compartment labeled "periphyton", although strictly speaking large autotrophs such as kelp and sea grasses are not usually called periphyton. We have focused our attention on kelp beds as the dominant macrophytes, which are quite important from both a biological and economical standpoint. Where kelp is present, it is usually present in large biomass. Since uptakes of radioactivity by biotic compartments are controlled by the mass of the receiving compartment, kelp may incorporate and sequester a significant amount of pollutant from the water compartment. Although kelp forests are firmly attached to the ocean's floor, the attachment is not by root systems but rather by physical holdfasts. Thus there should be no significant uptake from the benthic or sediment compartments. In addition, kelp can only remain firmly attached where the substrate is mostly rocky. This implies a relatively impervious boundary layer across which groundwater is not likely to penetrate easily. Kelp also grows extremely fast with much physical and biological deterioration of the fronds. This macrodetritus contributes significantly to both the suspended particulate and the organic sediment or "benthic detritus" compartments.

Besides these direct transfers of water-borne radioisotopes, the kelp beds are extremely important to the ecology of the coastal areas. Any damage to the kelp from chemical toxicity or ionizing radiation which might affect the rate of growth or mortality of the kelp bed could change the local composition of the marine foodweb, and thus may alter the overall ecosystem dynamic behavior of any isotope. It seems quite likely that a reduction in kelp would allow more of the isotope to be flushed out of the system by hydrological transport and less of the isotope to be retained and recycled by biological factors.

Once plankton or kelp have extracted radioactive elements from water, the pollutant may be metabolized or converted to different chemical and physical forms, or simply eliminated. Most organisms appear to remove toxic substances by biological processes. This removal is generally not linear, but instead appears to be exponential with most of the material removed quickly at first and some, small quantities retained for long periods of time. This type of process is adequately characterized by using a "biological half-life" to characterize the metabolic behavior of an isotope once it has been ingested with consumption of food or imbibed in water. Unfortunately biological half-lives for most aquatic organisms are not known. This appears to be one of the most significant gaps in our knowledge concerning the environmental transport and accumulation of radionuclides in aquatic foodwebs.

Heterotrophs

The zooplankton compartment of a marine foodweb is extremly diverse and complex (Steele 1970). Little is known about the ability of most zooplankton organisms to ingest or imbibe and concentrate radioactive elements. In the marine foodweb, the zooplankton community is dominated by filter-feeding animals, primarily copepods. A second major compartment consists of the free swimming larvae of fish and many benthic organisms. At this stage in our model, we consider the zooplankton to consist of these two compartments which we call "zooplankton" and "larval fish" respectively. The zooplankton are assumed to consume mostly plankton, and the larval fish (including the larvae of many benthos) are assumed to prey upon the plankton and zooplankton. Rates of food consumption by these two types of organisms in the marine foodweb are relatively well documented (Conover 1978). Rates of water intake are also available. Concentration factors and biological half-lives for specific isotopes have been inferred from a limited amount of data on aquatic organisms in the same general size category (0.5 - 50.0 mm).

Three separate fish compartments are used based on trophic relationships - adult carnivorous fish (top predators), adult omnivorous fish (scavengers) and larval fish of any kind. Populations of all these are extremely variable. Estimates of biomass can be inferred based on quantities and proportions of different species harvested by commercial and sport fishing reported to government agencies.

Diets of most coastal marine fishes appear to be quite generalized (Ivlev 1961). We have made no attempt to determine the diets of individual fish species, nor have we attempted to sort the overall biomass of fish into omnivores or carnivores based on species composition. The distinction between omnivores and carnivores is rather arbitrary, but it does allow us to construct a larger number of potential foodweb combinations by varying relative biomasses between the two.

Benthos

Three benthic compartments are defined: macrobenthos, meiobenthos and microbenthos. Macrobenthos consist primarily of large crustaceans (lobster, crabs), mollusks, urchins, etc. which live mainly on top of the benthic layer. Meiobenthos consist mostly of smaller organisms, especially polychaete worms which spend most of their time in the sediment layer. Microbenthos consist of bacteria and protozoa associated with marine sediments. All of these organisms, because of their intimate association with the sediment layer, would be the first organisms exposed to contaminated groundwater. Many of these are quite important economically; nearly all are food items for organisms in the water column above. The secondary production by benthos has recently been reviewed (Tenore and Coull 1979). Although a large body of data exists concerning uptake, accumulation and transport of various chemical compounds by these organisms (Aberg and Hungate 1966, McCave 1976), only a very small fraction of this data relates to the radioisotopes of interest in this study.

The macrobenthos are an important components of our model, both economically and as a possible source of pollutant transfer to man (fish and kelp are the only other compartments consumed by man). Many macrobenthos are relatively long-lived, and consequently they may be exposed to high radiation dosage at relatively low concentrations of radioisotope. Ecologically, the macrobenthos are mostly surface scavengers and predators, although some are filter-feeding organisms, gleening plankton and zooplankton from water in contact with the benthic layer.

For biological compartments in which predation occurs within the compartment in addition to between compartments, the state equations include a positive feedback term which is a function of compartment concentration. The feedback rate is a function of the degree to which a compartment consumes organisms aggregated into the same compartment compared to consumption of food items from other compartments. These fractions, for the most part, are unknown. Additional uptake rates reflecting "self-consumption" are input for those compartments which prey upon species lumped into the same compartment.

The meiofauna are an important pathway for transferring pollutant from the sediment compartment to the organic sediment compartment. As they burrow their way through the sediment, they consume organic and inorganic sediment, and effectively mix the two. This biological mixing is usually more rapid than any type of physical or chemical transfer between the sediment-organic sediment compartments (Tenore and Coull 1979, Rhoads 1974). Since many different kinds of organisms consume the organic sediment layer, this process can be a critical pathway for those isotopes that are strongly bound to subsurface sediments.

The microbenthos are an important source of food for many meiofauna. These organisms are probably in equilibrium with seawater with respect to concentration of trace elements including radionuclides. They are included mainly for completeness and biological realism in the model. However, we do not expect the microbenthos to be a major source of pollutant transfer into the foodweb.

APPLICATION OF THE MODEL

Description Of The Study Area

The results of any model are only as good as the data input to the model. Due to public scrutiny of nuclear power, there is typically a substantial body of environmental data available for any given reactor site. Unfortunately this data ranges from extremely valuable to nonsensical, and the sheer volume is usually overwhelming. Most environnmental reports seldom attempt to integrate or generalize key features of the data collected. Still they are useful for initial estimates of most the parameters, and they also serve as a check for values extrapolated from the literature.

The data base for the model presented here is for the San Onofre nuclear generating station (SONGS), located about 60 miles south of Los Angeles. Numerous environmental reports are available for this site. In addition we have a personal knowledge of the area from SCUBA diving explorations in the immediate vicinity and Southern California coastal waters in general. The data available consists mainly of the required environmental impact statements (AEC 1977) as well as a series of reports prepared by the Marine Review Committee of the California Coastal Commission (Connell, et al. 1979).

The San Onofre site is located on the Pacific Ocean about 60 miles south of Los Angeles. Should a core melt-through occur, radionuclides could enter the coastal water and be dispersed by currents. The fate of radionuclides in coastal waters is largely dependent on two factors: The hydrology of the area and the types of organisms in the area. The San Onofre site was chosen mainly due to availablity of biological data. Several institutions and agencies have been performing research and monitoring activities in the vicinity of the study area.

In the study area, a rather narrow gently sloping coastal plain extends seaward from the higher interior land. Along the beach, the plain terminates in a line of sea cliffs that have been straightened over long distances by marine erosion. Near the station site, the cliffs reach a height of 60 to 80 feet above sea level and are separated from the ocean by a narrow band of beach. The geologic structure in the vicinity of the site is dominated by sandstone which is poorly cemented but dense, and forms steep canyon walls and near-vertical cliffs along the coast. The sea floor off San Onofre is of mixed type including sand, silt, cobble and boulder.

The major ocean circulation in the North Pacific is a wind driven, clockwise gyre that is as large as the entire North Pacific Ocean. The California Current is the eastern limb of this gyre and is characterized as a meandering, diffuse eastern boundary current flowing southeast past Washington, Oregon, California and Baja California. North of Point Conception, the current has a mean speed of 0.3 knots. South of Point Conception, the current becomes more diffuse, and the inshore portion turns northward to flow through the Channel Islands and to become the nearly permanent, counter clockwise Southern California Eddy. Surface speeds in the countercurrent average from 0.01 to 0.02 knots. The flow pattern is complicated by the presence of small eddies within the Channel Islands region and fluctuates seasonally. The countercurrent is swiftest in the fall and early winter when the winds are from the southeast. In the spring and early summer, winds from the northwest strengthen the California Current and weaken the countercurrent.

Very near the California coast, local currents are influenced primarily by a combination of wind, tide and local topography. The total current is ordinarily the sum of components due to wind, tide and perhaps large-scale ocean circulation. The speed of the total current measured at San Onofre typically ranges from 0.1 to 0.75 knots, but averages 0.2 knots.

Tides in the vicinity of San Onofre are mixed semi-diurnal in nature, where two unequal high tides and two unequal low tides normally occur during a 25-hour period. The tidal wave rotates in a counterclockwise direction with the result that the tide occurs progressively later in the day proceedng northwards along the coast. As a result of this rotation, flood tide flows upcoast and ebb tide flows downcoast. Close to shore, the tidal current generally parallels the coast, periodically reversing direction with the change of tide. These complex currents and tides make it difficult to estimate the mass turnover of water in the area. It is probably on the order of days, with much seasonal variability.

The seasonal surface-water temperature ranges from 14C in the winter to about 23C in the summer. A thermocline is present in the summer, and becomes progressivley weaker in the fall and disappears in winter. This thermocline will restrict the vertical physical transport of radionuclides in the warmer months.

Our observations from SCUBA diving directly offshore of the area suggest that the concentration of suspended particulates is quite high (visibility typically 5-10 feet). The benthic interface is covered with a pronounced layer of silt, and there is considerable mass movement of sediment. The movement of sand, particulates and water at the benthic boundary layer is extremely important. Our impression is that currents are much reduced at the bottom, resuspension is quite apparent (bottom visibility nearly zero), and, except in the immediate vicinity of kelp, biomass is low. The benthic layer is a rather typical soft-benthos community.

Along the southern California coast, the ratio of soft bottom communities to hard-bottom communities is at least 9:1 and usually much higher. Given the larger distribution coefficents of most isotopes, we would expect most of the radioactivity released to nearshore benthic communities to be associated with soft-benthos communities. Production and biomass of most of the biotic compartments are less in these areas, and consequently the radiological damage to the ecosystem may be reduced.

There is a kelp bed about 1 mile from the SONGS reactor, separated by a silt-sand soft bottom substrate. Until further studies are done on groundwater flow in the area, we must assume that the kelp can receive radioactive pollutants from either groundwater or from the movement of contaminated water, sediment, particulates, and pelagic organisms, through kelp the beds.

The plankton community is typical of southern California coastal waters. There are two population biomass peaks per year, in the late spring and early fall. Average biomass is about 0.01-0.05 kg/m^3. Approximately 20-25% of the net productivity is returned to the water by excretion and secretion. Uptake of radioactivity from water is probably on the order of minutes to hours. The rapid turnover of plankton limits the ability of this compartment to accumulate a significant quantity of pollutant, unless concentration factors are extremely high.

The zooplankton and fish compartments, at least in the vicinity of the kelp beds, are very diverse in numbers of species. However on a biomass basis, zooplankton consist mostly of copepods, while sand bass, mackerel, rock fish and sheepshead make up the bulk of the fish compartment. We estimate the biomass of zooplankton to be about 0.01-0.1 kg/m^3, larval fish about the same, and fish about 0.1-1.0 kg/m^3. Zooplankton and larval fish (including benthic larvae) probably turnover every few weeks, while the fish population may take months to years. Biological half-lives of isotopes are not generally known for zooplankton but there are data available for many fish (Pentreath 1977). Fish retain most isotopes, especially Sr, Cs, and Pu for long periods of time (yrs).

Zooplankton and larval fish consume large amounts of phytoplankton, suspended organic matter, benthic detritus, and each other. Consumption rates are in the range of 5%-50% of their body weight per day (Conover 1978). Uptake of water is quite rapid - perhaps on the order of hours to a few days at most. Fish, on the other hand, probably only consume 1%-10% of their body weight per day of small organisms on the ocean floor, large zooplankton, larvae and detritus. Water turnover is considerably longer than in zooplankton, but the ability to concentrate isotopes is much greater (Polikarpov 1966).

The benthic fauna is composed mainly of polychaete worms, crabs, urchins and mollusks. Biomass for meiofauna is about 0.005-0.02 kg/m^3; macrofauna in the kelp beds is about 0.01-0.05 kg/m^3. Uptake of radioactivity comes mostly from water intake and consumption of organic sediment. Biological half-lives are generally not known; however, hard-shelled benthos would be expected to bind many types of isotopes in the mineral matrix of their shells.

Model Results And Validation

Because of the nature of this type of model, it is not appropriate to present the "results" of the model as absolute numbers. The first step in the validation process is to determine if the tightly coupled set of differential equations can produce an overall system behavior that is reasonable, based on our current knowledge of the transport of radioactivity in the marine environment. One way to do this is to attempt to balance all the parameters such that with no groundwater input to the sysyem, known background concentrations of specific isotopes (Sr,Cs,Pu) can be maintained in a quasiequilibrium over a reasonable length of time (yrs).

The GASP-IV simulation language is a combined continuous-discrete simulation program which allows for the detection of state events (Pritsker 1974). We define state events in our model to be any state variable concentration falling below the background concentration for that compartment. The occurrence of these events implies that we have either underestimated input rates or overestimated loss rates for a given compartment. As these events occur in the simulation run of the model, various parameters are automatically adjusted until background concentrations can be maintained when there is no input derived from contaminated groundwater.

If the water concentration falls below background levels, we assume that there must be a net input to the water compartment from sources external to the systems model. These sources might include advection or diffusion of water at background levels from adjacent water masses, atmospheric fallout radioactivity, or freshwater runoff from the coastal plains. Without identifying the source, the model adds a rate of input due to influx of curie-bearing material until background levels can be maintained. The initial value for background radioactivity in coastal waters is estimated to be about 80 pCi/m^3 (Bowen 1969).

The water compartment is obviously the most important part of the model. The other three abiotic compartments are tightly coupled to water by the distribution coefficients, and thus the background levels in sediment and organic particulate matter must be no greater than predicted from equilibrium concentrations between the dissolved phase and solid phase (i.e. radioisotopes adhering to particulates). Booth (1976) has modeled these processes, and obtained good agreement with predictions from field observations.

In our simulation to date, we have had no difficulty maintaining sediment background concentrations. However, suspended particulates and the organic sediment layer at the benthic boundary layer have presented some difficulties. The suspended particulate compartment has been the most difficult to balance at a background concentration. The model has predicted that 20% of the top 5 cm of sediment must be resuspended each year to maintain a background level when the distribution coefficient is set to 10^3. These values are not unreasonable, but we lack the field data to support this prediction.

The biological compartments are maintained at background levels by increasing the rate of uptake from water. There are certainly many different inputs to each biological compartment that might be adjusted but the rate of water uptake is always the most rapid input. This particular process is still one of the least understood, and there is apparently a tremendous range of variability due to individual and species differences, temperature, and water chemistry (especially stable element concentration), all superimposed on spatial and temporal variation.

It has become quite apparent from the hundreds of simulations we have run that the model output is extremely sensitive to minor changes in a few key parameters. The single most important processs is the turnover time of the water compartment, or conversely the rate of dilution of the entire system. For any given set of parameters there seems to be a minimal value of dilution above which all compartments continually increase with time and below which the water compartment and all biotic compartments rapidly approach background levels.

There are three main inputs to the water compartment: groundwater, release from sediment or suspended particulate, and net import of water at background concentrations from adjacent areas. For the equilibrium case, groundwater will not be input. There will, however, always be input to water from sediment, suspended particulates and mass transport from water constantly being flushed through the system. Release rates from sediment and particulates will be controlled by the distribution coefficent for a particular isotope. The equilibrium kinetics between a radionuclide in the dissolved phase relative to the amount adsorbed onto sediment or particulates is the second most important process; the third is the cycle of sedimentation and resuspension.

For the biological parameters, biological half-life and concentration factors for an isotope in a given chemical form are the most sensitive. Uptake of radioactivity from water is next. The model seems to be only slighty affected by rates of ingestion, and biomass of all compartments except kelp.

SUMMARY

The results which we have obtained to date have been very encouraging. We feel confident that we can reproduce the known behavior of isotopes which are currently present in coastal waters as background radioactivity. The model still has a long way to go though before it could be used for predictive purposes. Our next step in model refinement will be to couple an existing model of the physical movement of water in the vicinity of San Onofre. We will also attempt to measure the quantities and movement of suspended particulates in the area. Accurate data on these two processes will be essential to the final utility of a reasonable model describing the fate of radioactivity released to the coastal ecosystem from nuclear accidents.

Research Needs

By definition no model is perfect and no model is ever complete. The usefulness of a large scale simulation model such as the one we have presented is not so much its predictive ability as it is the identification of key processes and to guide field research where additional information is required. From our survey of the literature and the results of our simulation modeling we feel that additional work is required in the following areas:

FIELD STUDIES OF CONCENTRATION FACTORS
1. WATER CHEMISTRY
2. SPECIFIC ACTIVITY
3. BIOLOGICAL AVAILABILITY

BIOLOGICAL TURNOVER OF IMPORTANT ISOTOPES
1. SPECIES VARIABILITY
2. INGESTION RATES
3. BIOLOGICAL HALF LIFE

SEDIMENT-RADIONUCLIDE INTERACTIONS
1. K_d VALUES
2. DEPOSITION - RESUSPENSION

LITERATURE CITED

Atomic Energy Commission (AEC). 1977. San Onofre units 1 and 2. Environmental impact report, Vol. 1. Doc. 50361-151.

Aberg, B. and F. P. Hungate (eds.). 1966. Radioecological concentration processes. 1040 p. Pergamon Press, Oxford.

Booth, R. S. 1976. A systems analysis model for calculating radionuclide transport between receiving water and sediment, pp. 133-164. In: M. W. Miller and J. N. Stannard (eds.), Environmental toxicity of aquatic radionuclides: proceedings of the Rochester international conference on environmental toxicity. [Rochester, N.Y., June 2-4, 1975] Ann Arbor Science, Ann Arbor, Mich., 333p.

Bowen, V. T. et al. 1969. ^{90}Sr: concentration in surface waters of the Atlantic Ocean. Science 164: 825-827.

Connell, J. H. et al. 1979. Interim report of the Marine Review Committee to the California Coastal Commission. Part II. Appendix of technical information in the support of the general summary. MRC Doc. 79-02(II).

Conover, R. J. 1978. Transformation of organic material, pp. 221-499. In: O. Kinne, editor, Marine ecology, Vol. IV. Wiley and Sons,Inc. New York, N.Y.

Cushing, C. E., Jr., et al (eds.). 1975. Radioecology and energy resources: proceedings of the fourth national symposium on radioecology. 401 p. Dowden, Hutchinson and Ross, Inc., Stroudsburg, Penn.

Davis, S. N. and R. J. M. DeWiest. 1966. Hydrogeology. 463 p. Wiley and Sons, Inc., New York, N.Y.

Ivlev, V. S. 1961. Experimental ecology of the feeding of fishes. Yale University Press, New Haven, Conn.

McCave, I. N. (ed.). 1976. The benthic boundary layer. Plenum Press, New York, N.Y.

National Academy of Sciences - National Research Council (NAS-NRC). 1971. Radioactivity in the Marine Environment. 272 p. National Academy of Sciences, Washington, D. C.

Nelson, D.J. and F. C. Evans (eds.). 1969. Symposium on radioecology: proceedings of the second national symposium on radioecology. [Ann Arbor, Mich., May 15-17, 1967] Atomic Energy Commi-sion CONF-670503, 774 p. Washington, D. C.

Organization for Economic Co-operation and Development - European Nuclear Energy Agency (OECD-ENEA). 1979. Marine radioecology: proceedings of the third NEA Seminar. [Tokyo, Japan, October 1-5, 1979] Organization for Economic Development, 409 p. Paris, France.

Pentreath, R. J. 1977. Radionuclides in marine fishes. Oceanogr. Mar. Biol. Ann. Rev. 15: 365-460.

Polikarpov, G. G. 1966. Radioecology of marine organisms. 314 p. Reinhold Book Division, New York, N. Y.

Pritsker, A. A. 1967. The GASP-IV simulation language. Wiley and Sons, Inc., New York, N. Y.

Reichle, D. E. et al. 1970. Turnover and concentration of radionuclides in food chains. Nuclear Safety 11: 43-55.

Rhoads, D. C. 1974. Organism-sediment relations on the muddy sea floor. Oceanogr. Mar. Biol. Ann. Rev. 12: 283-308.

Smayda, T. J. 1970. The suspension and sinking of phytoplankton in the sea. Oceanogr. Mar. Biol. Ann. Rev. 8: 353-414.

Spector, W. (ed.). 1956. Handbook of biological data. W. H. Saunders, Philadelphia, Penn.

Steele, J. H. (ed.). 1970. Marine food chains. Univ. of California Press, Berkeley, California.

Tenore, K. R. and B. C. Coull (eds.). 1979. Marine Benthic Dynamics. 451 p. Univ. of South Carolina Press, Columbia, S. C.

SECTION VII:

Summary and Synthesis

ECOLOGICAL MODELLING THEORY AND METHODS: REPORT ON THE STATE OF THE ART

P. Haug, A. Blackwell, R. Fisher, and G. Swartzman

There appears to be a diversity of directions and trends emerging in ecological modelling theory and methods. Papers ranged from broad ecosystem concepts and theory to management or user orientation.

If there was a dominant trend reflected in the presentations, it appeared to be an explicit concern with making simulation models more acceptable to, and used by, managers and decisionmakers. We feel that this trend toward more responsiveness to the user is most healthy.

In the plenary session, some fundamental concerns about modelling were raised about whether models represent fact or fiction. The notion of using exergy as a unifying measure of the state of ecosystems was also discussed, and the relative importance of direct and indirect effects of perturbations on the ecosystem was formally analyzed.

Other areas of interest in theoretical ecology included qualitative analysis of ecosystem linkages and stability effects, the use of loop and compartment analysis for ecosystems, and an attempt to synthesize ecosystem theory with living systems theory.

In exploring new methods and techniques, many schools of systems ecology were represented: H.T. Odum's approach of describing ecosystems in terms of energy flows and embodiment of energy; general linear systems approaches that include direct-indirect effects analysis and input-output models; a thermodynamic principles approach; a qualitative approach using loop analysis; and a fundamental process approach in which several process models are linked into larger, nonlinear dynamic models. Both stochastic and deterministic models were discussed.

Also among the methods were various aids to modelling: sensitivity analysis, curve-fitting for input data, Monte Carlo methods, and user-friendly computer software to assist inexperienced ecologists develop simulation models. Other methods included innovative applications of parameter estimation, use of estimation theory and modelling to develop efficient and effective monitoring programs and to guide collection of field data, and ways to select equations for use in models.

There seemed to be almost a bimodal distribution between very simple models used to examine and improve techniques, on one hand, and very complicated models that attempt to deal with more global concepts of reality, on the other.

A series of papers provided a historical perspective on the use of modelling in spruce budworm management. It was reported that most of the models that had been developed for this problem seemed to be aids more for scientists than for managers and policy makers. Only two models were used at all by managers. It was observed that often managers can use something simpler, something they themselves can feel comfortable with, rather than models containing complicated formulations requiring much study and some acceptance on faith.

The spruce budworm work also demonstrated that, although models of the eastern and western species had similar components, there were different dynamics between the eastern and western systems. A variety of eastern budworm models built on quite different assumptions, <u>and</u> through quite distinct development approaches, <u>led</u> to very similar conclusions for mangement.

This theme that models must include the ultimate user, both in development and in ongoing applications, and that models should be made available to a wider audience, was reflected in several other papers. Some examples of how this could be done were: building the model so that it could be "played with" interactively by managers; making models flexible and modular; building models so that they respond quickly to interactive input and communicate results through effective output graphics. Another management-oriented modelling approach was a comparison between the relative costs of cleaning the environment of man's wastes by means of natural ecosystem processes, on one hand, and by investment of dollars and energy for man to do it technologically, on the other, the point being that saving undervalued natural systems can be shown to have real economic benefits. Several participants reiterated the importance of including the economic and social subsystems within environmental management models.

An issue that was not addressed directly, but one which we believe will have significant influence on the increased applications of ecological models to environmental management problems, is that of model transferability. Model transferability is defined as the ability to apply the model to other functionally equivalent sites by reparameterizing the model with site-specific constants. The creation of generic models has been responsible for the increased utilization of water hydrodynamic and mass transport models in water quantity and water quality management.

In summary, there are some broad observations and recommendations we would like to offer. It appears we are presently operating with a large variety of new tools, or applying old tools to qualitatively different systems. We must still experiment to find where these tools are useful and where they are being stretched beyond the limits of possible applications. In view of the uncertainties in many areas of ecological processes, some participants suggested that models would be most useful in analyzing risk, rather than in forecasting equilibrium conditions.

There seems also to be an important need for an accessible, ongoing, current data base and catalog of ecological parameters, either hard copy, computer based, or both. (See also: Sandy, J. 1982. On-line data bases vital for scientific research. Science 216:1367.) There is a healthy infusion of many people from highly varied disciplines into the field of systems ecology. These bring with them a vast pool of backgrounds, experiences, talents, perspectives, and techniques. This is good.

This varied background, however, produces a variety of symbology: block diagrams, flow diagrams, digraphs, energy-oriented and matter-oriented diagrams, conceptual diagrams, interaction matrices, etc. A more uniform system of symbology would be helpful, although prospects for this seem slim.

From among an infinite number of questions that need to be resolved, here are a few.

- How do we get good data to use in models?

- How do we get good models fast, without a data base?

- How can we be more sensitive and responsive as modellers to the needs of users?

- How can we consider the human factors involved in transferring our technology credibly?

- How can we organize and handle the burgeoning data base in ecology?

- How can we develop a set of standard units and formats for an ecological data base, and is this even desirable?

- How can we follow up and test the many theories about ecology that are emerging?

- And perhaps most importantly, how can we communicate better among ourselves to break down any barriers that may be forming among advocates of different approaches to ecological analysis?

In other words, how can we develop a more unified approach to understanding ecosystems?

STATE-OF-THE-ART AND RECOMMENDATIONS FOR FUTURE EFFORTS IN MODELING THE FUNCTIONAL ROLE OF ANIMALS IN ECOLOGICAL SYSTEMS[1]

J. A. Logan, J. E. Heasley, and C. B. Stalnaker[2]

INTRODUCTION

The task force responsible for summarizing modeling applications for animal components of ecological systems was composed of John Heasley, Jesse Logan, and Clair Stalnaker. Our backgrounds are in wildlife management, insect population ecology, and freshwater fisheries ecology, respectively. Interests of the panel, therefore, span the functional role of animals in terrestrial and aquatic habitats, for both intensively managed and natural ecosystems. In preparing this report, we each identified specific papers that we felt were most interesting considering our individual backgrounds and interests. Emphasis has been placed on the commonalities that became apparent through our individual assessments. Towards this end, the report is organized into five major topic headings: Spatial Dynamics and Spatial Modeling, Effects of Physical/Chemical Factors of the Environment, Lack of Ecosystem Perspective, Modeling Applications, and, finally, Modeling Outbreak Phenomena.

SPATIAL DYNAMICS AND SPATIAL MODELING

Modeling spatial dynamics is invariably recognized as an important and difficult problem. Analysis is difficult both conceptually and mathematically. For this reason, the modeling of spatial dynamics of animal populations is really in its infancy. Spatial modeling refers to at least three concepts: description of environmental heterogeneity; patchiness of the environment; and explicit movement of animals through a heterogeneous, patchy environment.

In recent years, analysis of environmental heterogeneity has resulted in significant advances. We believe there is emerging a relatively good understanding of how animals view a variable resource field. Significant advances are also being made in understanding how environmental heterogeneity impinges upon performance of specific animal populations (e.g., environmental assessment). Several papers in this conference, particularly those by Adams et al. and Hardy et al., indicate the relatively sophisticated state of the art in assessing and modeling effects of environmental heterogeneity on animal population performance.

The second concept of environmental patchiness describes the way in which heterogeneity is distributed. Here again many recent advances have been made. The theoretical literature is particularly rich in studies that analyze the effects of a patchy environment on persistence and stability of prey-predator systems. For example, several papers on effects of a patchy environment (Haar and Swartzman, in particular) stimulated active discussion. In modeling a patchy environment, it appears that theoretical work is well in advance of empirical studies. There is no dearth of theoretical studies dealing with habitat patchiness as it affects animal populations. What appears to be lacking are the follow-through empirical tests of hypotheses generated by these theoretical studies.

The third concept is the explicit movement of animals through a heterogeneous, patchy environment. In this area we are making far fewer advances. To our knowledge, there were no papers presented on modeling animal movement or diffusion. Dispersal and movement of animals can lead to heterogeneity as a result of differential resource utilization. Movement of animals may also influence spatial patterns of critical resources through nutrient transport, etc. The result is a complex matrix of interactions in which spatial heterogeneity influences the movement and dispersion of animal populations, and the resulting activities of animals modify habitat patchiness and heterogeneity. Trophic structure and food webs also operate within and influence the spatial matrix. The result of these complex spatial interactions is to enrich the dynamic repertoire of ecological associations. For example, the inclusion of spatial dynamics in model systems often leads to temporal persistence of point-unstable interactions. Much of the theoretical work in animal ecology has dealt with stability analysis excluding spatial considerations. Although this work has provided interesting insight, full understanding will require including explicit spatial dynamics.

In summary, we need improved methods for modeling temporal/spatial patterns and dynamics as they impinge on stability, resilience, elasticity, and persistence of animal communities. Conceptually and analytically these are difficult problems. It is becoming apparent that large simulation models are ill-suited for analysis of spatial dynamics. As R. G. Woodmansee pointed out during the Opening General Session, the large grasslands simulation model, ELM, was developed to describe one square meter. It is simply not plausible to include spatial complexity in such models. An alternate approach is to use high-resolution, complex simulation models for data generation. These data can then be used for parameter estimation in more analytically tractable models. In this respect, several papers presented dealing with sensitivity analysis of simulation models are particularly germane. We need to critically evaluate large simulation models for applications in parameter estimation and generation of meaningful analytic models.

---

[1]Paper presented at the Third International Conference on State-of-the-Art in Ecological Modeling, Colorado State University, May 24-28, 1982.

[2]J. A. Logan is a Population Ecologist, Natural Resource Ecology Laboratory, Colorado State University, Fort Collins, Colorado U.S.A; J. E. Heasley is Systems Ecologist, Department of Forest and Wood Science, Colorado State University, Fort Collins, Colorado U.S.A; and C. B. Stalnaker is Group Leader, Instream Flow and Aquatic Systems Group, Fish and Wildlife Service, USDI, Fort Collins, Colorado, U.S.A.

Further mathematical research is needed to develop efficient numerical schemes for solving complex systems of partial differential equations. The importance of mathematicians and biologist working together has been recognized throughout this conference. We should attempt to avoid not only inelegant models but also elegant solutions of uninteresting models.

EFFECTS OF PHYSICAL/CHEMICAL FACTORS OF THE ENVIRONMENT

We perceive a need for models that better relate the physical and chemical factors of the environment to population performance and community structure. For example, the modeling of physical/chemical factors of water quality is highly advanced; but coupling these models to animal community dynamics is much less advanced. The paper by Williams on fish populations in relation to instream flow was exciting because of the attempt to relate these abiotic factors to fish population performance.

There is also a need for an increased understanding of the relationship between physical/chemical factors of the environment and animal social behavior. There has been good work in modeling animal social behavior, but much less in understanding the physical environment as it influences behavior. An interesting paper by Frederick et al. discussed the effect of temperature on snow geese behavior and energetics. We also perceive a need for development of models that bridge the gap from existing physiological-ecology models to those that describe community structure.

It is apparent that we need continued development of good models of the physical environment that relate easily measured ambient variables to the biologically important microhabitat. We also recognize the importance of continuing work that couples these microhabitat models to the individual organism's physiological ecology. A current weakness is the failure to couple, in a mechanistically rigorous fashion, physiological models to the behavior and/or community interactions of involved species. Development of sound conceptual models that meld the approaches of classical population ecology to those of simulation modeling would do much to facilitate this required coupling.

LACK OF ECOSYSTEM PERSPECTIVE

One area in which plant modelers seem to be far in advance of animal modelers is the adoption of an ecosystem perspective. In general, animal models lack both an ecosystem perspective and a process orientation. This is partly because of historical reasons. The majority of work in animal population modeling has dealt with management of one or two specific, targeted species; and management criteria has most often been based on economic considerations. Usually, the currency of economics (money) does not accurately reflect the currency of ecology (energy?). For example, I am sure that if insect pests had been managed from an ecosystem point of view the face of American agriculture would be vastly different than it is today!

Several recommendations seem appropriate. First it is self-apparent that if the perspective of one species as the "center-of-the-universe" approach is continued, it is highly unlikely that animal models will advance beyond the population level of organization. At the very least we need to assess the effects of management policies on guilds and entire food webs. Work such as that presented by Elliott et al. is encouraging because it is an attempt to assess perturbances as reflected throughout a food chain. In the same sense, Eulers' paper on impacts of cottage development on wildlife was an attempt to assess perturbances on an integrated system.

Another reason for lack of an ecosystem perspective on the part of animal modelers is that most of the work in ecosystem modeling has lumped animal components into functional groups (several papers in this conference). This is not good enough! For the mainstream applications in animal modeling, we require representation of explicit species in food web representations. Work such as that presented by C. J. Puccia on qualitative analysis of East Coast benthos should help bridge the gap between the lumped functional group approach typical of ecosystem models and the population dynamics approach typical of animal modelers. The question of whether complex food webs are typically composed of many approximately equivalent links or, conversely, composed of a few strong links and many weak links is largely unanswered. Qualitative analysis of community structure provides a conceptual basis for answering such questions.

A final point is development of a common currency that reflects both economic and ecological value. Development of such a currency was the major topic of S. Jorgensen's address during the opening general session. A common currency must represent energy flows, biomass flows, information content, and feedback magnification. To date, ecosystem-level models have adequately represented both energy and biomass flows; however, representation of information content and feedback effect has been less than adequate. It is in these latter two measures of "ecological worth" that animals are most important from an ecosystem perspective. Although somewhat esoteric, work such as that described by S. Jorgensen and B. C. Patten (on feedback magnification) have obvious exoteric implications for animal modelers.

MODEL APPLICATION

There is a perceived gap between model development by researchers and model application by resource managers. Modeling as a science is ahead of and/or uncoupled with population management objectives. Several papers presented at this conference indicate significant effort is being made to close this perceived gap. In particular, papers with respect to the International Spruce Budworm Program (CANUSA) presented both a prospective plan for model application (Colbert et al.) and retrospective evaluation of past experiences (R. A. Fisher). Several interesting points were addressed in both these talks and the discussion they generated.

Many models were perceived as being so empirically based that model transfer (either in space to a new site or in time to different conditions at the same site) is difficult if not impossible. A self-apparent recommendation is to increase the mechanistic basis of animal models. Another, less obvious, recommendation is for modelers to pay more attention to parameters currently monitored by resource managers. Although these parameters (and/or variables) may not be the same as those required to drive resource management models, it is often possible to relate easily measured variables to those which drive the model.

Several recommendations were presented for facilitation of model utilization by resource managers. The concept of "friendly software" could do much to allow resource managers a more active role in model development. Kirchner and Vevea presented a discussion of two such friendly systems. Involvement of resource managers in model development has two distinct advantages. First, by direct involvement the resource manager plays an active role in deciding which variables and parameters need to be measured on a routine basis. Such involvement should help avoid model requirements for information which is difficult, if not impossible, to monitor. Second, active personal involvement in model development can do much to increase model credibility.

Several other recommendations were made by resource managers in discussions during this conference. One recurring theme was the lack of including variation in most models of animal populations. The resource manager is faced daily with the variation and indeterministic nature of the real world. It is little wonder that they distrust the deterministic nature of most models. Inclusion of variability in model structure typically increases complexity of the developed model, but it may be essential for representation of critical population events. Work, such as that presented by Swartzman et al., which included stochastic elements in a population management model, is encouraging because it indicates that such complexities do not preclude management application.

Simple infield management tools definitely have applications, particularly in remote management settings. This may seem to contradict the previous recommendation for including greater complexity. This is not necessarily the case. One of the most exciting applications of simulation modeling is development of simple models with high biological credibility. A. A. Berryman's paper dealing with outbreaks of mountain pine beetle is an example of a simulation model used to generate a simple decision criterion to judge susceptibility of a stand to beetle outbreak. It seems to be generally perceived that large simulation models are the domain of scientists, not managers. Through procedures such as those used by Berryman, it is possible to capture the essence of complex simulation in a simple decision rule capable of programming on a hand-held calculator.

A final point with respect to management applications is that we really need some spectacular successes in management of animal populations due to insights gained through modeling. Althouth models are routinely used in management decision processes, it is generally felt that spectacular successes, e.g., insights that would not have been possible without modeling applications, have been lacking. Perhaps this is more a problem of communication. As quantitative ecologists we need to better publicize model applications, particularly those which have resulted in truly unique and successful management strategies.

MODELING OUTBREAK PHENOMENA

The last area of discussion is of particular interest to those involved with pest management. Outbreaks are of major ecological importance in arthropod population systems. Outbreaks of insect populations are spectacular events in which population densities may be seven or eight orders of magnitude greater than during endemic phases. All papers on insects presented during this conference dealt with outbreak species. The reason for interest in outbreak species is twofold. First, such species are often serious competitors with man for valuable resources (e.g., major pests). Secondly, outbreaks play an important role in the structure and function of natural ecosystems as major pulsing events. These pulsing events may be retrogressive (i.e., moving the entire system to an earlier successional stage) or may, in fact, maintain the system at some particular successional state. Therefore outbreaks are interesting from both an applied and a basic science perspective.

Outbreak phenomena are difficult to study empirically for several reasons. An outbreak by its very nature is a highly transient state. Also, during non-outbreak or endemic phases, population densities may be so low that it is extremely difficult to sample populations. The combination of these two factors leads to a situation in which the most information is known about the least interesting aspect of the phenomena (e.g., populations during epidemic and collapse phase) and very little is known about the most important aspect of the phenomenon (e.g., the endemic stage and conditions which lead to the outbreak).

Since outbreaks are empirically difficult to study, they provide an ideal opportunity for model analysis to provide valuable insights. Several papers presented during this conference did, indeed, provide new insights into the general phenomena of outbreaks. A. A. Berryman, in a paper dealing with analysis of resource availability thresholds which lead to outbreaks of bark beetles, made the point that analysis of the threshold itself may be of greater interest than analysis of population dynamics at either the endemic or epidemic phase. Logan and Hilbert presented a paper in which the importance of the interaction between the physical environment and the biotic environment led to structural instability in a prey/predator community. The threshold of instability corresponded well with the observed outbreak in the field. Since outbreak species often exhibit limit cycle or bifurcation type behavior, nonlinearities are particularly important in analysis of outbreak species. For this reason, the paper presented by Puccia and Levin is valuable because it described techniques for qualitative analysis of nonlinearities in community structure which potentially lead to bifurcations.

The wide array of modeling techniques discussed at this conference for analysis of outbreaks was interesting. These techniques included use of simulation models to generate data for parameter estimation and demarcation of thresholds, structural stability analysis of systems of ordinary differential equations, loop-analysis and modifications thereof, and, topological analysis of outbreak manifolds. This variety of useful modeling techniques is once again an indication of the rich application for model analysis of outbreak phenomena in animal populations.

STATE OF THE ART IN APPLICATIONS OF ECOLOGICAL MODELS TO LAND RESOURCES

C. V. Cole, J. Hanson, and S. W. Running[1]

It is clear from the caliber and scope of papers presented at this conference that ecological modeling is making a significant contribution to our understanding of the dynamics of terrestrial ecosystems and has the potential for providing valuable tools for researchers, policy makers and resource managers. As the value of modeling as a tool for application of the scientific method to complex systems is recognized, ecological modelers will be faced with increasing challenges in efforts to protect and manage our land resources so as to sustain a stable and productive biosphere on this planet. The papers presented in sessions on applications to studies of land resources together with the extensive literature cited provide an excellent statement of the state of the art of ecological modeling in this field. As a group, these studies cover a wide range of levels of resolution in time and space consistent with the goals of the modeling efforts. Chemical and physiological processes are well represented in a number of models--the plant growth models in particular. There appears to be a trend towards simpler models, perhaps indicating that the better a system is understood, the more simply it can be represented in a model.

A general weakness in both the experimental and modeling studies of land resources is lack of adequate treatment of spatial patterns within ecosystems, as discussed in an opening paper by Woodmansee and Lauenroth.

In this conference, models applied to land resources were grouped into sessions on agroecosystems, range ecosystems, and forest ecosystems. Within each of these groups there is a range of levels of resolution and concern for providing input for policy and management decision making.

The crop growth models (McKinion et al., Morgan et al., King et al., this volume) provide excellent examples of integration of physiological processes at a high degree of resolution with time steps from hours to a day. In contrast, the century-level model (Parton et al., this volume) of management effects on soil organic matter turnover operates with annual time steps. A similar level of resolution was used in the model evaluating management effects on carbon flux for tropical forests (Skole et al., this volume).

The simulation of nitrogen, tillage, and residue management effects on soil fertility represents a team effort by the ARS-USDA group at the University of Minnesota to combine a number of mechanistic submodels into a management tool to be used in farm- and extension-level management programs, as well as for research and teaching (Shaffer et al., this volume).

Another example of a management-oriented model is EPIC (Williams et al., this volume), which is designed to assess the effects of erosion on soil productivity. This model was developed by a national ARS modeling team in response to the Soil and Water Conservation Act of 1977 (RCA) requiring an appraisal of soil and water resources of the United States. This model is of an appropriate structure and level of resolution so that it can be afforded by user agencies, and it is being welcomed by the Soil Conservation Service and other national agencies as a means of providing credible estimates for identifying erosion-susceptible soils and computing management effects on soil loss and productivity.

Mathematical modeling is becoming increasingly popular in range ecosystems. There is perhaps a hierarchical structure under which rangeland models can be discussed. These range from basic models that attempt to encapsulate the essence of a particular system to models that are of lesser resolution but aid managers in making decisions.

Four models can be discussed that reflect this hierarchy. The first model is by Parton et al. (1978). The model is very specific and attempts to describe the rate of root turnover in shortgrass prairie in a mechanistic manner. The model in its present form is useful to researchers but would have little practicality for the manager.

The Texas A&M livestock production model (Sanders and Cartwright 1978) is a theoretical model that is management oriented. The primary purpose of the model is to simulate the total realm of animal production. Even though the model is mechanistically structured, it is being used in management situations (Rice et al., this volume). Sullivan et al. (this volume) also used this model for African rangelands.

The ARS-USDA is developing a decision-oriented, range management model. Skiles et al. (this volume) and Rice et al. (this volume) discuss the primary producer and animal submodels, respectively, of this model. The purpose of the model is to aid in making management decisions.

The simplest models are the regression equations. Wight and Hanks (1981) developed a model that predicted total primary production on a site based on evapotranspiration. The model yielded relatively good estimates of production. West and Lauenroth (this volume) have a similar approach for predicting long-term range primary production.

[1]C. V. Cole is a soil scientist with the ARS-USDA and faculty affiliate at the Natural Resource Ecology Laboratory, Colorado State University, Fort Collins, Colorado; J. Hanson is a range scientist with the ARS-USDA at the High Plains Grassland Research Station, Cheyenne, Wyoming; and S. W. Running is Associate Professor of Silviculture, University of Montana, Missoula, Montana, U.S.A.

As can be seen from the dates of the publications cited, modeling of range ecosystems is a very young discipline. A hierarchy of mathematical and structural complexity is beginning to emerge, however. As our understanding of range ecosystems improves, models describing them will be developed that may be of great benefit to range managers.

A common discussion topic at this conference was "How can ecological modeling be used for management-level decision making?" For purposes of discussion concerning forestry, we will limit examples to the U.S. Forest Service and its current planning process. The National Forest Management Act of 1978 mandated that each national forest prepare a comprehensive forest plan that would direct management of all activities on the forest. This requires consideration of a wide variety of resources and issues drawing on a number of disciplines. Forest plans typically consider the following topics: Range, Recreation, Timber, Water and Soils, Wildlife, Aquatic Environment-Fisheries Habitat, Lands, Minerals, Fire, Roads, Social and Economic, and Visual Quality.

An example of the modeling used in forest management is FORPLAN. This is a linear programming optimization model that attempts to evaluate management alternatives over the complete range of resource issues. The model is developed from economic and operation research principles and analyzes resource allocation and activity scheduling.

Ecological models will rarely, if ever, be the ultimate decision-making tool in forest management. However, the data base for driving the FORPLAN-type management models could be improved significantly by sound ecological modeling. We have primary productivity, hydrology, nutrient cycling, wildlife population, entomology, microclimate, and other models that are orders of magnitude better than the back-of-envelope estimates going into the management models. Interfacing these models will be constrained by the requirement of most ecological models for intensive driving-variable data sets. In many cases the ecological models would best be run separately and final results entered into the FORPLAN model. Ecological models are one source of data that must be balanced against other resource values. Ultimately multiple-use management requires compromise and decision making; ecological models can provide data of increased accuracy for these purposes.

LITERATURE CITED

Parton, W.J., J.S. Singh, and D.C. Coleman. 1978. A model of production and turnover rates of roots in shortgrass prairie. Journal of Applied Ecology 47:515-542.

Sanders, J.O., and T.C. Cartwright. 1979. A general cattle production systems model. I. Structure of the model. Agricultural Systems 4:217-227.

Wight, J.R., and R.J. Hanks. 1981. A water-balance, climate model for range herbage production. Journal of Range Management 34(4):307-311.

STATE-OF-THE-ART IN APPLICATION OF ECOLOGICAL MODELS TO WATER RESOURCES

Reported by L. Kamp-Nielsen

In the sub-sessions on modelling of water resources 32 papers were presented. The present report is based on the impressions which my 5 co-chairmen: E. Halfon, R. Milhous, W. J. Mitsch, N. Nyholm, and N. Thyssen reported to me.

Characteristic of the conference was the diversity of models presented and the increased focusing on model uncertainty and interpretation of model results.

The models applied to water resources ranged from simple input-output models of the Vollenweider type to large-scale managing models with thousands of parameters. The systems modelled ranged from short river stretches over lakes and reservoirs to large river basins and estuaries. The basic model variables ranged from biochemical models of nutrient and carbon flows over models of toxic substances to models on energy and socio-economics. The modelling techniques presented ranged from simple steady state models over dynamic models to stochastic and structure models using various statistic and algebraic methods. Consequently, most of the water resources modelling field seemed to be covered although lotic and open marine systems were a little underrepresented.

Relevant, but not new, discussions on the choice of model and complexity took place after a number of presentations. Especially the constraints on the simple I/O-models, where high flushing rates, high background extinction, changing nutrient limitation, incomplete mixing and non-steady state conditions may strongly violate the applicability. For the dynamic eutrophication models the use of variable stochiometry was emphasized. Although these models to some extent have proven successful for managing purposes certain processes are still not adequately described e.g. water-sediment interactions and control of blue-green blooms. Concerning the state-space type of models it was agreed, that the hydrodynamic part should be well balanced against the biochemical part by considering the time scales on which the various variables are operating. Also the amount and the quality of data available for the model construction should be carefully considered when the complexity is chosen.

To account for the uncertainty associated with the variability of the data measured or with the transfer or extrapolation of information from other points in space or time, various methods were presented. The use of Monte Carlo simulations seemed useful in many cases e.g. the fate of toxic substances. To meet the uncertainty associated with the transfer and extrapolation of data in risk analysis, the use of Bayesian statistics was proposed. The advantage of the various methods was frequently discussed during the conference and it appeared, that for models with not too many parameters, Monte Carlo simulations or "squeezed" versions of it are the optimal. For larger and more complex models covariance analysis based on the time-update part of the Kalman filtering technique seemed to be the optimal. Where the implementation of uncertainty analysis associated with parameter values attracts more and more attention, the description of the uncertainty of the model structure is still in its infancy. However, a promising approach may be the use of "Environ" analysis, whereby an assumed steady state is the basis for matrix algebraic computations of a range of model solutions.

During recent years, tools have been developed to describe the uncertainty deriving from driving variables, parameter estimates and Gaussian "white noise", and model simulations supplied with these are becoming more and more common. But our predictions do still not get the full credibility, since the propagation of existing structure under changing impact is still to be validated on a more evolutionary time scale. For this purpose new concepts have to be developed and necessarily more holistic concepts have to be approached.

Years ago we had to sell the idea of mathematical modelling; consequently, we were not critical enough and "over-selling" was not uncommon. In the fascination of the new world opened for many scientists too much was promised. Now we understand our models better and through interdisciplinary efforts our understanding has made us more relaxed and more self-critical. Mathematical modelling of water resources has developed into one of the largest areas of theoretical research.

We are also being more aware of the importance of asking the right questions to our models. I have the impression that we have improved in applying the right type of model to the right question, when models are used scientifically. But concerning modelling for management purposes we still have a strong obligation to inform politicians and decisionmakers about asking the appropriate questions of our models.

TASK REPORT ON STATE OF THE ART IN THE APPLICATION OF ECOLOGICAL MODELING TO ENERGY DEVELOPMENT[1]

Marshall Flug[2], Chairman, Marie-Jose Enderle[3], Jay Messer[4]

The world is currently in a state of energy awareness. The last few years have focussed attention on developing strategic fuel reserves and new or alternative supplies for the traditional oil, gas, and coal resources. Expanded development of hydropower and solar energy, no doubt, will continue although the total impact on our growing appetite for energy will remain small. Since society is highly dependent on liquid fuel resources, much attention has been directed at converting various fossil fuels into liquid petroleum products (e.g., oil shale and tar sands into refined gasoline). A quick summary of potentially recoverable resources of fossil fuels in the world is demonstrated in Table 1:

Table 1.--Summary of potentially recoverable resources of fossil fuels:

	10^{20} joules
Oil	150.
Natural Gas	170.
Coal	1790.
Oil Shale	21100.
Tar Sands	20.

And of course, nuclear power, which potentially generates more fuel than is consumed, could provide an infinite source of energy. Quantification of energy reserves is highly questionable but the data presented above is a good relative index as to order of magnitude for each resource.

The problem of energy and the reason for our concern is that we are now looking at new and unproven technologies to produce readily usable energy on a scale which dwarfs any other industry. Much of the data or information available for converting raw fuels into usable products:

1. are based on pilot scale facilities which are still testing conversion parameters (e.g., temperature, material flow rates),

2. provide inputs and outputs which when quantified expand a larger numerical range (e.g., oil shale will produce water or require from 0.5 to 5 barrels of water per barrel of oil produced. The estimates of Exon's/Tosco shale oil facility in Colorado was projected to cost 2.5, 3.5 and then 5.5 billion dollars for a 100,000 barrel per day facility, the result was a decision to stop construction),

3. produce new and complex byproducts including hydrocarbon compounds and radionuclides which are little understood and have not been previously experienced.

4. only provide monitoring for the commonly expected discharges for which standards have been established and do not look further for other releases or synergistic effects.

Modeling efforts traditionally are used to predict either compliance or noncompliance with some preset standard and are not utilized to explore an understanding of impacts and functional interactions. Typically a worst case scenario is simulated by a quick and dirty estimation. Although parameters modeled can be classified as either physical, biological, or chemical, usually only one class and only one parameter is analyzed. Additionally, each modeling effort is developed for a site specific case (i.e., a particular energy project).

Very few, if any, attempts have been made to develop a multiparameter model that encompasses a wide range of constituents from the three parameter classes (i.e., biological, chemical and physical). Economic and energy models are large but must be based on poor and generalized input data. From an ecological modeling viewpoint, little effort is applied to include well developed, sophisticated subroutine model components from the appropriate physical, chemical and biological arenas (e.g., sediment transport models used in hydrologic analysis). Too often the ecologic modeler assumes that a poor understanding is available in these arenas, and then assumes a constant coefficient to represent the interaction of energy outputs on the transport and assimilation of a given constituent, or completely neglects this particular parameter or class of parameters. The data and methods used to calibrate these subroutines are often unfamiliar to the ecologist who is ill equipped to physically collect the needed data. Therefore, the ecological modeler must quickly become a multidisciplinary expert or assimilate a team of researchers to encompass experience in the many periphery arenas. Additionally, discharges into air, water and land must all be considered to provide a closed system; each component impacts the other when measured over a long time.

Data collection, by new energy interests, is often minimized and addresses compliance criteria only; monitoring is not formulated to gather information capable of providing an understanding of interactions of energy outputs on one or several parameters. The collection of data is expensive, difficult, and often impossible to collect (e.g., calibration of a model predicting impacts from a nuclear reactor core meltdown). Therefore, model validation is often nonexistent and calibration is severely limited. Additionally, the scale up from pilot scale energy plants is, in many cases, a poor assumption for an unproven commercial industry.

Energy developers must look beyond state and Federal standards which deal with the short-term immediate discharge of wastes. Support of long-term environmental impacts for many of the new and less understood constituents is needed. Large scale synthetic fuel development poses many new threats in that releases are unknown. The synergistic impact of such discharges is one area in which ecological models are deficient. Again, emphasis must be placed on the limited quantitative data defining energy resource inputs and outputs.

Ecological modeling is centered around efforts on single parameter models (e.g., ph, DO, radionuclides). The secondary and combined effects from several parameter discharges may prove to have a major impact when considering all pertinent parameters from the biological, chemical and physical sectors. What sense is there in developing a sophisticated model which addresses only one variable? In reality, many of the discharges not modeled may prove to be major driving forces in sensitive ecologic systems. Modeling efforts are needed which include the interaction of many parameters using subsystems for each parameter at a level of sophistication appropriate for the available data and considering the data and methodologies used for all other interacting parameters.

---

[1]Paper presented at the Third International Conference on State-of-the-Art in Ecological Modelling, Colorado State University, May 24-28, 1982.

[2]Hydrologist/Engineer, Water Resources Field Support Laboratory, National Park Service, Colorado State University, Fort Collins, Colorado.

[3]Electricite de France, Director des Etedes et Recherches, Clamart Cedex, France.

[4]Research Assistant Professor, Utah Water Research Laboratory, Utah State University, Logan Utah.